Formulas/Equations

Distance Formula

If $P_1 = (x_1, y_1)$ and $P_2 = (x_2, y_2)$, the distance from P_1 to P_2 is

$$d(P_1, P_2) = \sqrt{(x_2 - x_1)^2 + (y_2 - y_1)^2}$$

Equation of a Circle

The equation of a circle of radius r with center at (h, k) is

$$(x - h)^2 + (y - k)^2 = r^2$$

Slope Formula

The slope m of the line containing the points $P_1 = (x_1, y_1)$ and $P_2 = (x_2, y_2)$ is

$$m = \frac{y_2 - y_1}{x_2 - x_1} \qquad \text{if } x_1 \neq x_2$$

$$m \text{ is undefined} \qquad \text{if } x_1 = x_2$$

Point–Slope Equation of a Line

The equation of a line with slope m containing the point (x_1, y_1) is

$$y - y_1 = m(x - x_1)$$

Slope–Intercept Equation of a Line

The equation of a line with slope m and y-intercept b is

$$y = mx + b$$

Quadratic Formula

The solutions of the equation $ax^2 + bx + c = 0$, $a \neq 0$, are

$$x = \frac{-b \pm \sqrt{b^2 - 4ac}}{2a}$$

If $b^2 - 4ac > 0$, there are two real unequal solutions.
If $b^2 - 4ac = 0$, there is a repeated real solution.
If $b^2 - 4ac < 0$, there are two complex solutions.

Geometry Formulas

Circle

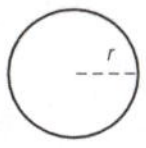

r = Radius, A = Area, C = Circumference

$A = \pi r^2 \qquad C = 2\pi r$

Triangle

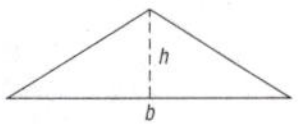

b = Base, h = Altitude (Height), A = Area

$A = \frac{1}{2}bh$

Rectangle

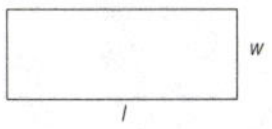

l = Length, w = Width, A = Area, P = Perimeter

$A = lw \qquad P = 2l + 2w$

Rectangular Box

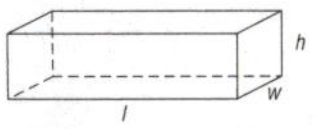

l = Length, w = Width, h = Height, V = Volume

$V = lwh$

Sphere

r = Radius, V = Volume, S = Surface area

$V = \frac{4}{3}\pi r^3 \qquad S = 4\pi r^2$

Algebra & Trigonometry

Graphing and Data Analysis

MICHAEL SULLIVAN
Chicago State University

MICHAEL SULLIVAN, III
South Suburban College

Prentice Hall, Upper Saddle River, NJ 07458

Library of Congress Cataloging-in-Publication Data
Sullivan, Michael
Algebra and trigonometry: with emphasis on graphing and data analysis/Michael Sullivan, Michael Sullivan, III.
p. cm.
Includes index.
ISBN 0-13-778481-3
1. Algebra. 2. Trigonometry. I. Sullivan, Michael. II. Title.
QA152.2.S85 1998
510—dc21 97-23428
CIP

Senior Acquisitions Editor: *Sally Denlow*
Marketing Manager: *Patrice Lumumba Jones*
Production Editor: *Robert C. Walters*
Managing Editor: *Linda Mihatov Behrens*
Executive Managing Editor: *Kathleen Schiaparelli*
Assistant Vice President of Production and Manufacturing: *David W. Riccardi*
Supplements Editor/Editorial Assistant: *April Thrower*
Marketing Assistant: *Patrick Murphy*
Creative Director: *Paula Maylahn*
Art Director: *Amy Rosen*
Art Manager: *Gus Vibal*
Interior Design: *Elm Street Publishing Services, Inc.*
Cover Design: *Jeanette Jacobs*
Photo Editor: *Lori Morris-Nantz*
Photo Research: *Beth Boyd*
Manufacturing Buyer: *Alan Fischer*
Manufacturing Manager: *Trudy Pisciotti*
Cover Photo: A PRISM. *George Dodson/Uniphoto*

Printed in the United States of America

10 9 8 7 6 5 4 3 2 1

ISBN 0-13-778481-3

Prentice-Hall International (UK) Limited, *London*
Prentice-Hall of Australia, Pty. Limited, *Sydney*
Prentice-Hall Canada Inc. *Toronto*
Prentice-Hall Hispanoamericana, S.A., *Mexico*
Prentice-Hall of India Private Limited, *New Delhi*
Prentice-Hall of Japan, Inc., *Tokyo*
Simon & Schuster Asia Pte. Ltd., *Singapore*
Editoria Prentice-Hall do Brasil, Ltda., *Rio de Janeiro*

For Michael S. (Sullivan)
and Shannon, Patrick, and Ryan (Murphy)
The Next Generation

CONTENTS

CHAPTER 4 POLYNOMIAL AND RATIONAL FUNCTIONS 241

CHAPTER 5 EXPONENTIAL AND LOGARITHMIC FUNCTIONS 349

CHAPTER 6 TRIGONOMETRIC FUNCTIONS 445

CHAPTER 7 ANALYTIC TRIGONOMETRY 527

CHAPTER 8 APPLICATIONS OF TRIGONOMETRIC FUNCTIONS 575

CHAPTER 9 POLAR COORDINATES; VECTORS 645

CHAPTER 10 ANALYTIC GEOMETRY 709

CHAPTER 11 SYSTEMS OF EQUATIONS AND INEQUALITIES 789

CHAPTER 12 SEQUENCES; INDUCTION; COUNTING; PROBABILITY 897

APPENDIX REVIEW 973

PREFACE TO THE INSTRUCTOR

Why We Wrote This Book

Times continue to change! In 1986, *Algebra and Trigonometry* was written, a traditional text that developed all the skills a student would need to succeed in college level mathematics courses. This text, now in its 4th edition, remains traditional in its approach. As technology developed, various philosophies regarding the use of incorporating graphing utilities into the curriculum emerged. These philosophies, as they relate to Algebra and Trigonometry, lead to different views regarding the use of technology in the classroom. In 1996, we wrote *Algebra and Trigonometry Enhanced with Graphing Utilities* in response to the call for a text to fully incorporate technology while maintaining strict mathematical content. Through the use of technology, the text allows students to recognize patterns, visualize, explore, and foreshadow concepts from areas such as Calculus. This book, *Algebra and Trigonometry: Graphing and Data Analysis,* was written to address the point of view that technology should not only be used to graph functions and solve equations, but also should be used to analyze data, draw scatter diagrams, and find curves of best fit. Thus, this text incorporates concepts never before discussed in a traditional Algebra and Trigonometry course while still maintaining strict mathematical content. Students are not simply "pushing buttons" on their calculators, but must answer thoughtful questions that force them to utilize mathematics.

The standards developed by MAA, AMATYC, and NCTM encourage modeling and connecting mathematics to various disciplines. This text introduces mathematical theory through the use of models developed in other disciplines. The text also expects students to develop their own models either from data or situations and proceed to analyze these models mathematically.

To this end, we begin with a discussion of univariate data, utilizing bar charts, pie charts, histograms, and frequency polygons to graphically represent the data. Then we discuss bivariate data, representing it in scatter diagrams and determining whether the relationship between the variables is linear or nonlinear. Using a graphing utility, LINear REGression techniques are used to find and interpret curves of best fit for linearly related data. As the mathematics evolves, we use a graphing utility to find curves of best fit for quadratic, power, cubic, exponential, logarithmic, logistic, and sinusoidal curves of best fit. In each instance the data is used not only to find curves of best fit, but also to analyze the functions found using concepts such as local minima and maxima. By comparing the results of their analysis to the actual data, students are able to judge the reasonableness of their answer.

The data itself is truly real, drawn from sources such as the Statistical Abstract of the United States, the U.S. Census Bureau, the U.S. Department of Justice, the Bureau of Labor Statistics, the New York and NASDAQ Stock Exchanges, and so on. Each chapter begins with a popular issue, such as domestic and global use of oil, asking students to analyze and interpret data from web site locations. This data is dynamic, changing on a regular basis. The data is widely based, drawn from disciplines such as physics, biology, chemistry, economics, finance, psychology, and statistics.

Notwithstanding the emphasis on data analysis, we fully utilize the graphers ability to promote visualization, exploration, and foreshadowing of concepts. The technology is interwoven with the mathematics. Doing mathematics by hand using traditional methods and doing mathematics using technology are often combined such as using the domain of a function to set a viewing window. Many examples are solved both graphically and algebraically. We encourage students to distinguish between solving a problem using the full power of the technology vs. recognizing when the technology is limited or the simple nature of the problem makes a solution by hand the better choice.

About This Book

Content The content of this book contains the topics found in a traditional algebra and trigonometry text. In addition, it contains topics unique to a technology approach: sections on data analysis and curves of best fit appear in chapters dealing with linear functions (LINear REGression), quadratic functions (QUADratic REGression), power functions (POWer REGression), polynomial functions (CUBic REGression), logarithmic and exponential functions (LOGarithmic REGression, EXPonential REGression, LOGISTIC REGression), and trigonometric functions (SINusoidal REGression). Also, many examples and exercises found in this text are ones that cannot be handled using traditional methods.

Organization With a fully integrated technology approach, the order of presentation is slightly different from the traditional order. In recognition of the fact that some students may not be fully prepared for this course, a specially designed Appendix has been included to provide necessary review. The Appendix is referenced at the beginning of the chapter and at the appropriate place within the chapter.

Chapter 1 begins with an analysis of univariate and bivariate data through a discussion of bar and pie charts, histograms, frequency polygons, and scatter diagrams. Graphing is done early in this chapter (Section 1.2), with emphasis on the graphs of certain key equations. A section on modeling linearly related data appears here.

Chapter 2 develops the important concept of a function from a discrete and a continuous point of view, emphasizing tables, graphs, and properties and including discussions on average rate of change, local maxima and minima, and increasing and decreasing functions. The last section focuses on the construction of functions in applications.

Chapter 3 solves equations and inequalities, utilizing both an algebraic and graphing approach. The solution of polynomial and rational inequalities is postponed to Chapter 4, so the power of graphing polynomial and rational functions can be used.

Chapter 4 discusses graphs and properties of polynomial and rational functions. Here data resulting in scatter diagrams that lead to curves of best that are quadratic, power, or cubic are analyzed. Complex numbers and quadratic equations with a negative discriminant provide motivation for the Fundamental Theorem of Algebra. Many exercises here ask students to draw scatter diagrams, find the appropriate curve of best fit, and analyze the resulting function using techniques introduced in the chapter.

Chapter 5, Exponential and Logarithmic Functions, places emphasis on graphs and properties, utilizing data with scatter diagrams that lead to curves of best fit that are exponential, logarithmic, or logistic.

Chapter 6, Trigonometric Functions, uses a right triangle approach to define the six trigonometric functions. Later the unit circle approach is used to develop certain properties of these functions. The graphs of the trigonometric functions are then discussed. The chapter concludes with the inverse trigonometric functions.

Chapter 7, Analytic Trigonometry, deals with trigonometric identities and trigonometric equations. A graphing utility is used to solve trigonometric equations for which no algebraic or trigonometric method is available.

Chapter 8, Applications of Trigonometric Functions, deals both with early applications of trigonometry (requiring solving right or oblique triangles) and with modern applications, such as simple harmonic motion, damped motion, and sinusoidal curve fitting.

Chapter 9, Polar Coordinates; Vectors, introduces polar coordinates, with an emphasis on graphing. Included is a discussion of Demoivre's Theorem. Independent of the preceding topics, sections on vectors in the plane and the dot product are also provided.

Chapter 10, Analytic Geometry, discusses conics, including rotation of axes and polar forms. Parametric equations are also discussed, including an example of motion simulation.

Chapter 11, Systems of Equations and Inequalities, discusses both graphing and algebraic approaches. Many examples and exercises here can only be worked using technology.

Chapter 12, Sequences; Induction; Counting; Probability, is mostly traditional since the use of technology usually does not lead to more efficient methods.

Instructor Supplements

Instructor Resource Manual

written by Michael Sullivan, Michael Sullivan III and Katy Murphy

ISBN: 0-13-788340-4

Contains complete step-by-step worked out solutions to all the even numbered exercises in the textbook. Also included are strategies for using the "Mission Possible" collaborative learning projects found in each chapter. Additional information on the Internet "Excursions" found at the beginning of each chapter may also be found within the IRM.

Written Test Item File

ISBN: 0-13-788357-9

Features questions that are found within the computerized test bank in a format that can be used for selecting specific problems for testing.

TestPro Computerized Test Generator

MAC ISBN: 0-13-788373-0

IBM ISBN: 0-13-788365-X

Allows instructors to generate tests from algorithms keyed to the text by chapter, section, and learning objective. Instructors select from thousands of test questions and hundreds of algorithms which generate different but equivalent questions. A user-friendly expression-building toolbar, editing and graphing capabilities are included. Customization toolbars allow for customized headers and layout options which provide instructors with the ability to add or delete workspace or add columns to conserve paper.

Tutorial Videos
ISBN: 0-13-788381-1

A new tutorial videotape series, created exclusively to accompany this text, include 15 minute segments for each section of the text. At the beginning of each section of the text, the videotape number on which the material is contained, is listed. Each segment uses both traditional and graphical ways of solving mathematical problems. These videos provide an alternative process which can add to your students' success in this course. Included with each set of tapes is a permission letter to duplicate the tapes for your mathematics lab or library.

New York Times Supplement

A free newspaper supplement from Prentice Hall and the New York Times which includes interesting and relevant articles on mathematics in the world around us. Great for getting students to talk and write about mathematics. This supplement is updated each fall. To request free copies for all of your students, contact your Prentice Hall representative.

Prentice Hall Companion Website

Located at *www.prenhall.com/sullivan* this website offers you and your students additional information, projects and exercises to enrich the learning experience using this text. The Internet "Excursions" found in the beginning of each chapter have duplicate pages on the website with the links to the appropriate data and information gathering sites described in the projects.

Acknowledgments

Textbooks are written by authors, but evolve from an idea into final form through the efforts of many people. Special thanks to Don Dellen, who first suggested this book and the other books in this series. Don's extensive contributions to publishing and mathematics are well known; we will all miss him dearly.

There are many people we would like to thank for their input, encouragement, patience and support. They have our deepest thanks and appreciation. We apologize for any omissions . . .

James Africh, *College of DuPage*
Steve Agronsky, *Cal Poly State University*
Dave Anderson, *South Suburban College*
Joby Milo Anthony, *University of Central Florida*
James E. Arnold, *University of Wisconsin-Milwaukee*
Agnes Azzolino, *Middlesex County College*
Wilson P. Banks, *Illinois State University*
Dale R. Bedgood, *East Texas State University*
Beth Beno, *South Suburban College*
William H. Beyer, *University of Akron*
Richelle Blair, *Lakeland Community College*
Trudy Bratten, *Grossmont College*
William J. Cable, *University of Wisconsin-Stevens Point*
Lois Calamia, *Brookdale Community College*
Roger Carlsen, *Moraine Valley Community College*
John Collado, *South Suburban College*
Denise Corbett, *East Carolina University*
Theodore C. Coskey, *South Seattle Community College*
John Davenport, *East Texas State University*
Duane E. Deal, *Ball State University*
Vivian Dennis, *Eastfield College*
Karen R. Dougan, *University of Florida*
Louise Dyson, *Clark College*
Paul D. East, *Lexington Community College*
Don Edmondson, *University of Texas-Austin*

Christopher Ennis, *University of Minnesota*
Garret J. Etgen, *University of Houston*
W. A. Ferguson, *University of Illinois-Urbana/Champaign*
Iris B. Fetts, *Clemson University*
Mason Flake, *student at Edison Community College*
Merle Friel, *Humboldt State University*
Richard A. Fritz, *Moraine Valley Community College*
Carolyn Funk, *South Suburban College*
Dewey Furness, *Ricke College*
Wayne Gibson, *Rancho Santiago College*
Joan Goliday, *Sante Fe Community College*
Frederic Gooding, *Goucher College*
Ken Gurganus, *University of North Carolina*
James E. Hall, *University of Wisconsin-Madison*
Judy Hall, *West Virginia University*
Edward R. Hancock, *DeVry Institute of Technology*
Brother Herron, *Brother Rice High School*
Kim Hughes, *California State College-San Bernardino*
Ron Jamison, *Brigham Young University*
Richard A. Jensen, *Manatee Community College*
Sandra G. Johnson, *St. Cloud State University*
Moana H. Karsteter, *Tallahassee Community College*
Arthur Kaufman, *College of Staten Island*
Thomas Kearns, *North Kentucky University*
Teddy Koukounas, *SUNY at Old Westbury*
Keith Kuchar, *Manatee Community College*
Tor Kwembe, *Chicago State University*
Linda J. Kyle, *Tarrant Country Jr. College*
H. E. Lacey, *Texas A & M University*
Christopher Lattin, *Oakton Community College*
Adele LeGere, *Oakton Community College*
Stanley Lukawecki, *Clemson University*
Virginia McCarthy, *Iowa State University*
James McCollow, *DeVry Institute of Technology*
Laurence Maher, *North Texas State University*
Jay A. Malmstrom, *Oklahoma City Community College*
James Maxwell, *Oklahoma State University-Stillwater*
Carolyn Meitler, *Concordia University*
Eldon Miller, *University of Mississippi*
James Miller, *West Virginia University*
Michael Miller, *Iowa State University*
Kathleen Miranda, *SUNY at Old Westbury*
Jane Murphy, *Middlesex Community College*
Bill Naegele, *South Suburban College*
James Nymann, *University of Texas-El Paso*
Sharon O'Donnell, *Chicago State University*
Seth F. Oppenheimer, *Mississippi State University*
E. James Peake, *Iowa State University*
Thomas Radin, *San Joaquin Delta College*
Ken A. Rager, *Metropolitan State College*
Elsi Reinhardt, *Truckee Meadows Community College*
Jane Ringwald, *Iowa State University*
Stephen Rodi, *Austin Community College*
Howard L. Rolf, *Baylor University*
Edward Rozema, *University of Tennessee at Chattanooga*
Dennis C. Runde, *Manatee Community College*
John Sanders, *Chicago State University*
Susan Sandmeyer, *Jamestown Community College*
A.K. Shamma, *University of West Florida*
Martin Sherry, *Lower Columbia College*
Anita Sikes, *Delgado Community College*
Timothy Sipka, *Alma College*
John Spellman, *Southwest Texas State University*
Becky Stamper, *Western Kentucky University*
Neil Stephens, *Hinsdale South High School*
Diane Tesar, *South Suburban College*
Tommy Thompson, *Brookhaven College*
Richard J. Tondra, *Iowa State University*
Marvel Townsend, *University of Florida*
Jim Trudnowski, *Carroll College*
Richard G. Vinson, *University of Southern Alabama*
Mary Voxman, *University of Idaho*
Darlene Whitkenack, *Northern Illinois University*
Chris Wilson, *West Virginia University*
Carlton Woods, *Auburn University*
George Zazi, *Chicago State University*

Recognition and thanks are due particularly to the following individuals for their valuable assistance in the preparation of this edition: Jerome Grant for his support and commitment; Sally Denlow, for her genuine interest and insightful direction; Bob Walters for his organizational skill as production supervisor; Patrice Lumumba Jones for his innovative marketing efforts; Ray Mullaney for his specific editorial comments; the entire Prentice Hall sales staff for their confidence; and to Katy Murphy for checking the answers to all the exercises.

Michael Sullivan
Michael Sullivan, III

PREFACE TO THE STUDENT

As you begin your study of Algebra and Trigonometry, you might feel overwhelmed by the number of theorems, definitions, procedures and equations that confront you. You may even wonder whether you can learn all this material in a single course. For many of you, this may be your last mathematics course, while for others, just the first in a series of many. Don't worry—either way, this text was written with you in mind.

This text was designed to help you—the student, master the terminology and basic concepts of Algebra and Trigonometry. These aims have helped to shape every aspect of the book. Many learning aids are built into the format of the text to make your study of this material easier and more rewarding. This book is meant to be a "machine for learning," one that can help you to focus your efforts and get the most from the time and energy you invest.

This book requires that you have access to a graphing utility: a graphing calculator or a computer software package that has a graphing component. Be sure you have some familiarity with the device you are using before the course begins.

Here are some hints we give our students at the beginning of the course:

1. Take advantage of the feature PREPARING FOR THIS CHAPTER. At the beginning of each chapter, we have prepared a list of topics to review. Be sure to take the time to do this. It will help you proceed quicker and more confidently through the chapter.
2. Read the material in the book before the lecture. Knowing what to expect and what is in the book, you can take fewer notes and spend more time listening and understanding the lecture.
3. After each lecture, rewrite your notes as you re-read the book, jotting down any additional facts that seem helpful. Be sure to do the Now Work Problem x as you proceed through a section. After completing a section, be sure to do the assigned problems. Answers to the Odd ones are in the back of the book.
4. If you are confused about something, visit your instructor during office hours immediately, before you fall behind. Bring your attempted solutions to problems with you to show your instructor where you are having trouble.
5. To prepare for an exam, review your notes. Then proceed through the Chapter Review. It contains a capsule summary of all the important material of the chapter. If you are uncertain of any concept, go back into the chapter and study it further. Be sure to do the Review Exercises for practice.

Remember the two "golden rules" of algebra and trigonometry:

1. DON'T GET BEHIND! The course moves too fast, and it's hard to catch up.
2. WORK LOTS OF PROBLEMS. Everyone needs to practice, and problems show where you need more work. If you can't solve the homework problems without help, you won't be able to do them on exams.

We encourage you to examine the following overview for some hints on how to use this text.

Best Wishes!

Michael Sullivan
Michael Sullivan, III

OVERVIEW

CHAPTER
2
Functions and Their Graphs

Imagine yourself as an expert for the EPA and sitting in Congressional policy meetings having to explain the importance of balancing energy resources with environmental protection. On the following page is the Internet Excursion placing you in exactly that position and asking the tough questions a member of congress might ask. Use the Sullivan website at:

www.prenhall.com/sullivan

to help link you to the Internet resources you will need to answer the questions asked.

PREPARING FOR THIS CHAPTER

Before getting started on this chapter, review the following concepts;

Topics from Algebra and Geometry *(Appendix, Section 1)*

Graphs of Certain Equations *(Example 2, p. 31)*

Example 4, p. 34, Example 5, p. 35; Example 14, p. 43

Tests for Symmetry of an Equation (p. 42)

Procedure for Finding Intercepts of an Equation (p. 37)

Linear Curve Fitting *(Chapter 1, Section 1.6)*

OUTLINE

Page 95

The list of concepts for review will help you in two major ways. ... First, it allows you to review basic concepts immeditely before using them in context. Second, it illustrates the natural building of mathematical concepts throughout the course.

Page 96

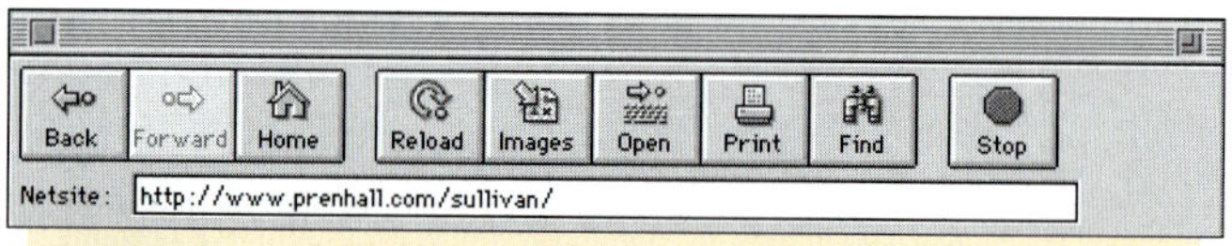

HOW LONG WILL THE OIL LAST?

The global supply of oil and other sources of energy is more than adequate to meet present needs, but most of this supply is outside of the United States. Currently, oil supplies about 40 percent of the world's energy, with the United States the biggest consumer. Since oil is a nonrenewable energy resource, plans must be made for the eventuality of diminishing supply.

Suppose that you were a consultant for the EPA and members of congress asked you to sit on a Energy Policy Steering committee as an expert analyst. You must convince the committee of the importance of environmental concerns in planning the global energy system. You are very aware of the fact that present modes of energy use and production threaten serious environmental deterioration. Your plan is to create a set of what-if functions that will allow your committee to model many possible scenarios for their policy guidelines.

1. $E(t)$, the first function that you create, will allow you to model United States oil consumption over a period of time. Make a scatterplot using the EPA Data on United States energy consumption, per capita from 1950–1990. Since oil provides 40 percent of the energy consumed, you adjust your figures to get oil consumption per capita. Use the LINear REGression tool on your graphing utility to find the linear function $E(t)$ of best fit for modeling the data. Does this seem like a good model?
2. Your second function, $P(t)$, is a population modeling tool. From the United States Census Data make a scatterplot, and then use a QUADratic REGression to model the data. Next, let $O(t) = E(t) \cdot P(t)$. What does $O(t)$ model? Graph $O(t)$. Compare to the actual figures from BP Petroleum.
3. Using our function $O(t)$, we want to estimate the area under the graph for the years 1997 until 2000. To do this we can make a trapezoid, and find its area. Now estimate the areas under $O(t)$ using 10-year intervals. For what year does the total of all the trapezoid areas equal the remaining United States oil reserve?
4. Could this sort of analysis be used to predict when the world's supply of oil will run out? Can your committee think of any methods that might improve your predictions?

Each chapter begins by placing you in an intriguing and relevant situation where you will need to collect information from the Internet, analyze data according to material available within the chapter, and answer thought-provoking questions. All the appropriate links can be found on the "live" pages at: http://www.prenhall.com/sullivan

Perhaps the most central idea in mathematics is the notion of a *function.* This important chapter deals with what a function is, how to graph functions, how to perform operations on functions, and how functions are used in applications.

The word *function* apparently was introduced by René Descartes in 1637. For him, a function simply meant any positive integral power of a variable *x*. Gottfried Wilheim von Leibniz (1646–1716), who always emphasized the geometric side of mathematics, used the word function to denote any quantity associated with a curve, such as the coordinates of a point on the curve. Leonhard Euler (1707–1783) employed the word to mean any equation or formula involving variables and constants. His idea of a function is similar to the one most often used today in courses that precede calculus. Later, the use of functions in investigating heat flow equations led to a very broad definition, due to Lejeune Dirichlet (1805–1859), which describes a function as a rule or correspondence between two sets. It is his definition that we use here.

2.1 FUNCTIONS

1. Determine Whether a Relation Represents a Function
2. Identify the Graph of a Function
3. Find the Value of a Function
4. Find the Domain of a Function
5. Obtain Information from/about the Graph of a Function

In Section 1.2, we said a **relation** is a correspondence between two variables, say x and y, and can be written as a set of ordered pairs (x, y). When relations are written as ordered pairs (x, y), we say x is related to y. Often, we are interested in specifying the type of relation (such as an equation) that might exist between the two variables. For example, when variables appear to be linearly related (based on the scatter diagram), we find a line of best fit. In Section 1.6, we found that fertilizer and crop yield appeared to be linearly related and the line of best fit was $y = 0.717x + 4.79$, where x represents the amount of fertilizer and y represents the crop yield. Thus, if we

Begin each section by reading the learning objectives. These objectives will help you organize your studies and prepare for class.

Page 97

Increase your understanding, visualize, discover, explore, and solve problems using a graphing utility.

EXAMPLE 1 Vertical Shift Down

Use the graph of $f(x) = x^2$ to obtain the graph of $h(x) = x^2 - 4$.

Solution Table 5 lists some points on the graphs of $f = Y_1$ and $h = Y_2$. The graph of h is identical to that of f, except that it is shifted down 4 units. See Figure 38.

TABLE 5

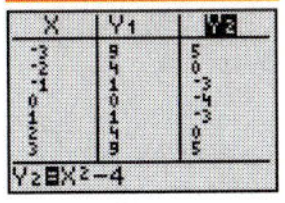

FIGURE 38

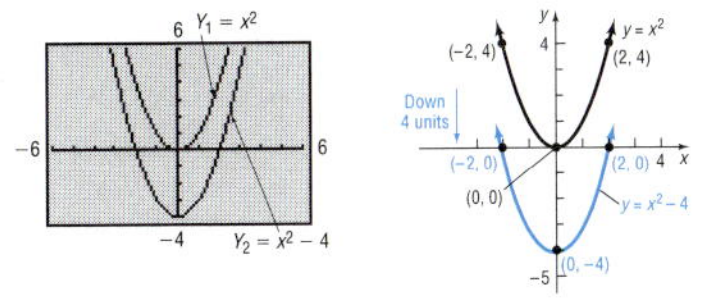

Horizontal Shifts

EXPLORATION On the same screen, graph each of the following functions:

$Y_1 = x^2$
$Y_2 = (x - 1)^2$
$Y_3 = (x - 3)^2$
$Y_4 = (x + 2)^2$

What do you observe?

Page 142

Page 141

Now work Problem 31.

Vertical and horizontal shifts are sometimes combined.

EXAMPLE 2 Combining Vertical and Horizontal Shifts

Graph the function $f(x) = (x - 1)^3 + 3$.

Solution We graph f in steps. First, we note that the rule for f is basically a cube function. Thus, we begin with the graph of $y = x^3$. Shown in (blue) in Figure 40. Next, to get the graph of $y = (x - 1)^3$, we shift the graph of $y = x^3$ horizontally 1 unit to the right. See the graph shown in (red) in Figure 40. Finally, to get the graph of $y = (x - 1)^3 + 3$, we shift the graph of $y = (x - 1)^3$ vertically up 3 units. See the graph shown in (green) in Figure 40. Note the points that have been plotted on each graph. Using key points such as these can be helpful in keeping track of just what is taking place.

FIGURE 40

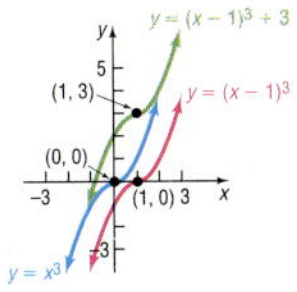

In Example 2, if the vertical shift had been done first, followed by the horizontal shift, the final graph would have been the same. (Try it for yourself.)

Now work Problem 43.

Compressions and Stretches

EXPLORATION On the same screen, graph each of the following functions:

$Y_1 = |x|$
$Y_2 = 2|x|$
$Y_3 = 3|x|$
$Y_4 = \frac{1}{2}|x|$

FIGURE 41

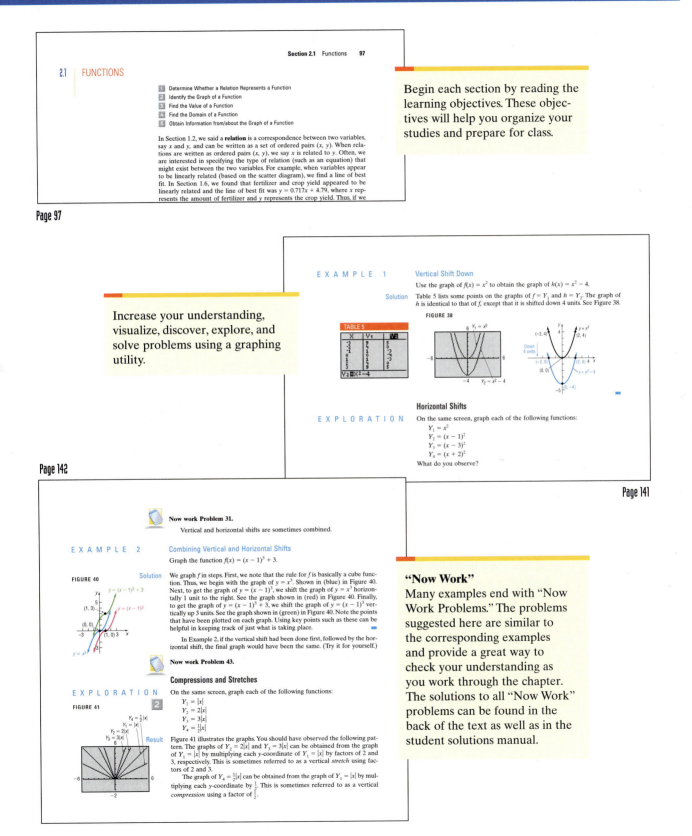

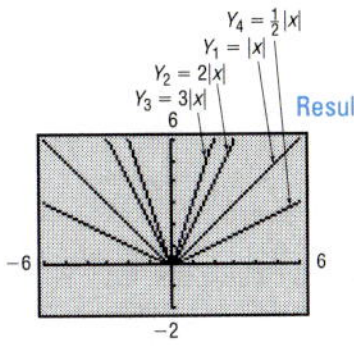

Result Figure 41 illustrates the graphs. You should have observed the following pattern. The graphs of $Y_2 = 2|x|$ and $Y_3 = 3|x|$ can be obtained from the graph of $Y_1 = |x|$ by multiplying each y-coordinate of $Y_1 = |x|$ by factors of 2 and 3, respectively. This is sometimes referred to as a vertical *stretch* using factors of 2 and 3.

The graph of $Y_4 = \frac{1}{2}|x|$ can be obtained from the graph of $Y_1 = |x|$ by multiplying each y-coordinate by $\frac{1}{2}$. This is sometimes referred to as a vertical *compression* using a factor of $\frac{1}{2}$.

"Now Work"
Many examples end with "Now Work Problems." The problems suggested here are similar to the corresponding examples and provide a great way to check your understanding as you work through the chapter. The solutions to all "Now Work" problems can be found in the back of the text as well as in the student solutions manual.

OVERVIEW

Since $b^2 - 4ac > 0$, there are two unequal real solutions.

Graphing Solution Figure 17 shows the graph of the function

$$f(x) = 3x^2 - 5x + 1$$

As expected, we see that there are two x-intercepts; one between 0 and 1, the other between 1 and 2. The solutions to the equation are 0.23 and 1.43 correct to two decimal places.

FIGURE 17

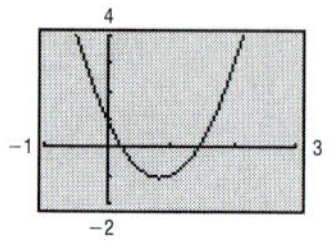

Algebraic Solution We use the quadratic formula with $a = 3$, $b = -5$, $c = 1$, and $b^2 - 4ac = 13$.

$$x = \frac{-b \pm \sqrt{b^2 - 4ac}}{2a} = \frac{5 \pm \sqrt{13}}{6}$$

The solution set is $\{(5 - \sqrt{13})/6, (5 + \sqrt{13})/6\}$. These solutions are exact. ■

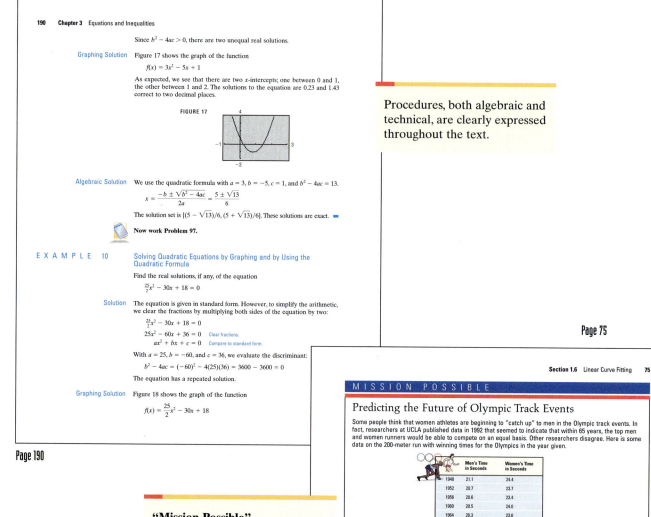

Now work Problem 97.

EXAMPLE 10 **Solving Quadratic Equations by Graphing and by Using the Quadratic Formula**

Find the real solutions, if any, of the equation

$$\frac{25}{2}x^2 - 30x + 18 = 0$$

Solution The equation is given in standard form. However, to simplify the arithmetic, we clear the fractions by multiplying both sides of the equation by two:

$$\frac{25}{2}x^2 - 30x + 18 = 0$$

$$25x^2 - 60x + 36 = 0 \quad \text{Clear fractions.}$$

$$ax^2 + bx + c = 0 \quad \text{Compare to standard form.}$$

With $a = 25$, $b = -60$, and $c = 36$, we evaluate the discriminant:

$$b^2 - 4ac = (-60)^2 - 4(25)(36) = 3600 - 3600 = 0$$

The equation has a repeated solution.

Graphing Solution Figure 18 shows the graph of the function

$$f(x) = \frac{25}{2}x^2 - 30x + 18$$

Page 190

Procedures, both algebraic and technical, are clearly expressed throughout the text.

Page 75

MISSION POSSIBLE

Predicting the Future of Olympic Track Events

Some people think that women athletes are beginning to "catch up" to men in the Olympic track events. In fact, researchers at UCLA published data in 1992 that seemed to indicate that within 65 years, the top men and women runners would be able to compete on an equal basis. Other researchers disagree. Here is some data on the 200-meter run with winning times for the Olympics in the year given.

	Men's Time in Seconds	Women's Time in Seconds
1948	21.1	24.4
1952	20.7	23.7
1956	20.6	23.4
1960	20.5	24.0
1964	20.3	23.0
1968	19.83	22.5
1972	20.00	22.40
1976	20.23	22.37
1980	20.19	22.03
1984	19.80	21.81
1988	19.75	21.34
1992	19.73	21.72

(You may notice that the introduction of better timing devices meant more accurate measures starting in 1968 for the men and 1972 for the women.)

1. To begin your investigation of the UCLA conjecture, make a graph of these data. To get the kind of accuracy you need, you should use graph paper. You will need to use only the first quadrant. On your x-axis place the Olympic years from 1948 to 2048, counting by 4's. On your y-axis place the numbers from 15 to 25 which represent the seconds; if you're using graph paper, allow about four squares per one second. Use X's to represent the men's times on the graph and O's to represent the women's times. These should form what we call a "scatter plot," not a neat line or curve.
2. Using a ruler or straightedge, draw a line that represents roughly the slope and direction indicated by the men's scores. Do the same for the women's scores. Write a sentence or two explaining why you think your line is a good representation of the scatter plot.
3. Next find the equation for each line, using your graph to estimate the slope and y-intercept for each. Try to be as accurate as possible, remembering your units and using the points where the line crosses intersections of the grid. The slope will be in seconds per year.
4. Do your two lines appear to cross? In what year do they cross? If you solve the two equations algebraically, do you get the same answer?
5. Some graphing calculators enable you do this problem in a statistics mode. They will plot the individual point that you type in and find a line that represents the data. Use your graphing calculator to find out whether its equations matches yours.
6. Make a group decision about whether or not you think that women's times will "catch up" to men's times in the future. Write out two to three sentences explaining why you believe they will or will not.

"Mission Possible"
Learn to "think beyond the box." Collaborate with fellow students to solve these extended mathematical situations based on relevant "real-world" data.

Logistic Curve Fitting

3 Logistic growth models can be used to model situations where the value of the dependent variable is limited. Many real-world situations conform to this scenario. For example, the population of the human race is limited by the availability of natural resources such as food, shelter, and so forth. When the value of the dependent variable is limited, a logistic growth model is often appropriate.

EXAMPLE 3

Fitting a Curve to a Logistic Growth Model

The following data, obtained from R. Pearl ("The Growth of Population," *Quarterly Review of Biology* 2 (1927): 532–548) represents the amount of yeast biomass after t hours in a culture.

Time (in hours)	Yeast Biomass	Time (in hours)	Yeast Biomass
0	9.6	10	513.3
1	18.3	11	559.7
2	29.0	12	594.8
3	47.2	13	629.4
4	71.1	14	640.8
5	119.1	15	651.1
6	174.6	16	655.9
7	257.3	17	659.6
8	350.7	18	661.8
9	441.0		

(a) Using a graphing utility, draw a scatter diagram of the data with time as the independent variable.

(b) Using a graphing utility, fit a logistic growth model to the data.

(c) Using a graphing utility, graph the function found in (b) on the scatter diagram.

(d) What is the predicted carrying capacity of the culture?

(e) Use the function found in (b) to predict the population of the culture at $t = 19$ hours.

Page 425

Many examples and exercises connect real-world situations to mathematical concepts.

Learning to work with models is a skill that transfers to many disciplines.

Solution (a) See Figure 54.

(b) A graphing utility fits a logistic growth model of the form $y = \dfrac{c}{1 + ae^{-bx}}$ by using the LOGISTIC regression option. See Figure 55. The logistic growth function of best fit to the data is

$$y = \frac{663.0}{1 + 71.6e^{-0.5470x}}$$

where y is the amount of yeast biomass in the culture and x is the time.

(c) See Figure 56.

(d) Based on the logistic growth function found in (b), the carrying capacity of the culture is 663.

(e) Using the logistic growth function found in (b), the predicted amount of yeast biomass at $t = 19$ is

$$y = \frac{663.0}{1 + 71.6e^{-0.5470(19)}} = 661.5$$

FIGURE 54

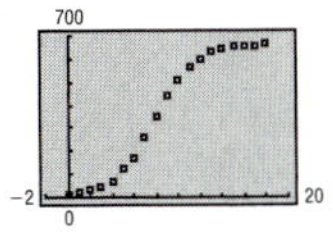

FIGURE 55

FIGURE 56

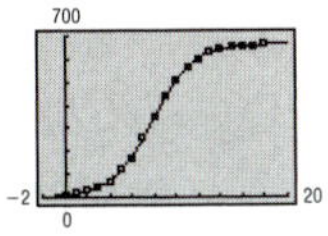

Now work Problem 13.

Choosing the Best Model

4 We have discussed seven different models thus far that can be used to explain the relation between two variables, x and y (see Table 8).

Page 426

Finding Sinusoidal Functions from Data by Hand

5 Scatter diagrams of data sometimes take the form of a sinusoidal function. Let's look at an example.

The data given in Table 1 represents the average monthly temperatures in Denver, Colorado. Since the data represents *average* monthly temperatures collected over so many years, the data will not vary much from year to year and so will essentially repeat itself each year. In other words, the data is periodic. Figure 54 shows the scatter diagram of this data repeated over two years, where $x = 1$ represents January, $x = 2$ represents February, and so on.

TABLE 1

Month, x	Average Monthly Temperature, °F
January, 1	29.7
February, 2	33.4
March, 3	39.0
April, 4	48.2
May, 5	57.2
June, 6	66.9
July, 7	73.5
August, 8	71.4
September, 9	62.3
October, 10	51.4
November, 11	39.0
December, 12	31.0

Source: U.S. National Oceanic and Atmospheric Administration

FIGURE 54

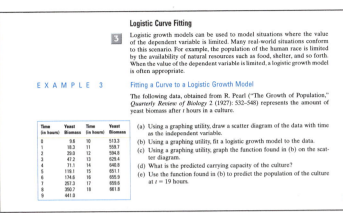

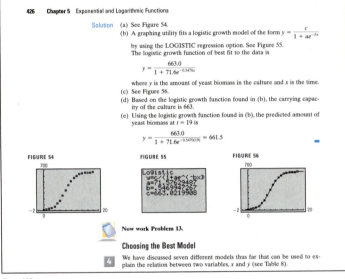

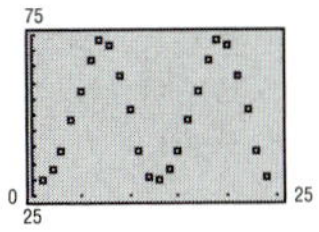

Page 618

OVERVIEW

2.2 EXERCISES

In Problems 1–8, match each graph to the function listed whose graph most resembles the one given.

A. Constant function	*B. Linear function*
C. Square function	*D. Cube function*
E. Square root function	*F. Reciprocal function*
G. Absolute value function	*H. Greatest-integer function*

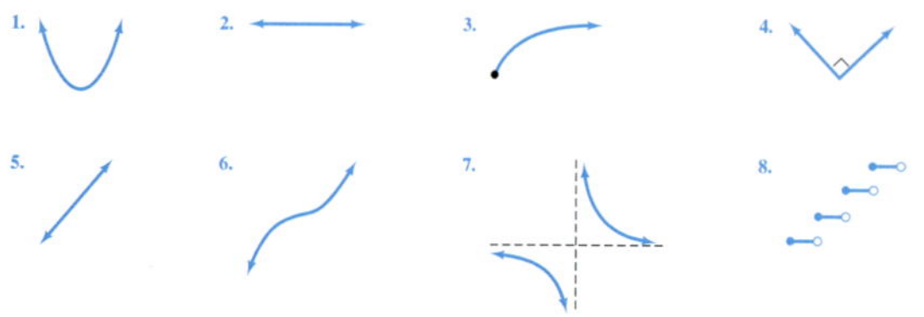

Page 134

Each end-of-section exercise set begins with visual and concept-based problems, starting you out with the basics of the section.

Challenge Yourself! Critical Thinking and Writing Questions really get you thinking.

53. How many x-intercepts can a function defined on an interval have if it is increasing on that interval? Explain.

54. How many y-intercepts can a function have? Explain.

Page 136

In Problems 85–88, the graph of a piecewise-defined function is given. Write a definition for each function.

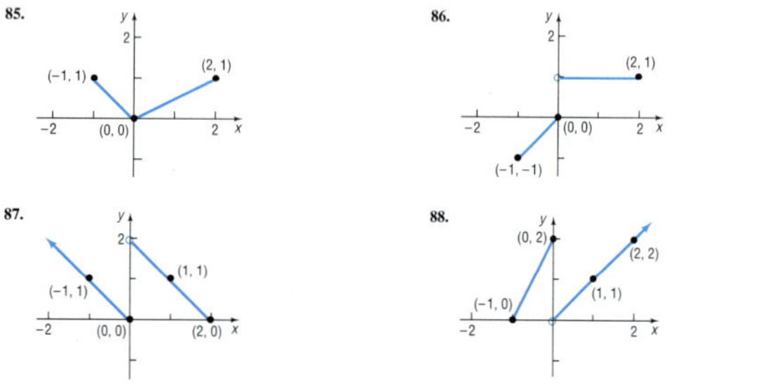

89. Growth of Bacteria The following data represents the population of an unknown bacteria.

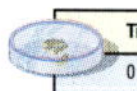

Time (Days)	Population
0	50
1	153
2	234
3	357
4	547
5	839
6	1280

(a) Draw a scatter diagram of the data by hand treating the population as the independent variable.
(b) Draw a line from the point (0, 50) to (1, 153) on the scatter diagram found in (a).
(c) Find the average rate of change of the population between 0 and 1 days.
(d) Interpret the average rate of change found in (c).
(e) Draw a line from the point (5, 839) to (6, 1280) on the scatter diagram found in (a).
(f) Find the average rate of change of the population between 5 and 6 days.
(g) Interpret the average rate of change found in (f).
(h) What is happening to average rate of change of population as time passes?

90. Cost of Manufacturing Bikes The following data represents the monthly cost of producing bicycles at Tunney's Bicycle Shop.

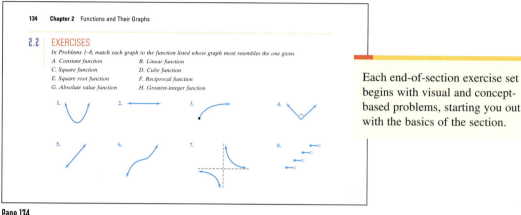

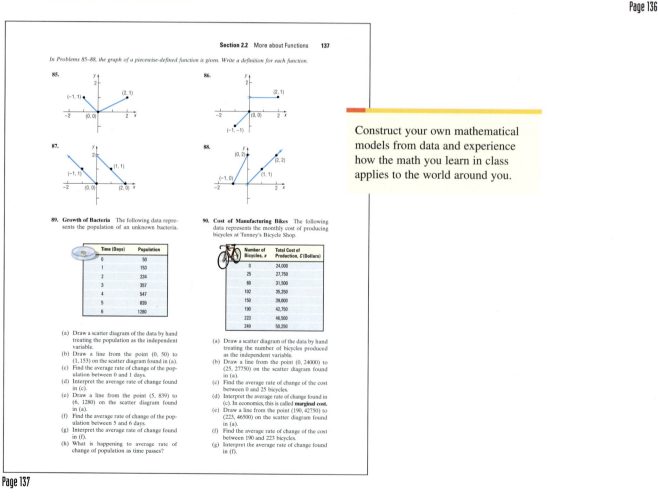

Number of Bicycles, x	Total Cost of Production, C (Dollars)
0	24,000
25	27,750
60	31,500
102	35,250
150	39,000
190	42,750
223	46,500
249	50,250

(a) Draw a scatter diagram of the data by hand treating the number of bicycles produced as the independent variable.
(b) Draw a line from the point (0, 24000) to (25, 27750) on the scatter diagram found in (a).
(c) Find the average rate of change of the cost between 0 and 25 bicycles.
(d) Interpret the average rate of change found in (c). In economics, this is called **marginal cost.**
(e) Draw a line from the point (190, 42750) to (223, 46500) on the scatter diagram found in (a).
(f) Find the average rate of change of the cost between 190 and 223 bicycles.
(g) Interpret the average rate of change found in (f).

Construct your own mathematical models from data and experience how the math you learn in class applies to the world around you.

Page 137

Chapter Review 169

CHAPTER REVIEW

THINGS TO KNOW

Function

A rule or correspondence between two sets of real numbers so that each number x in the first set, the domain, has corresponding to it exactly one number y in the second set. The range is the set of y values of the function for the x values in the domain.

x is the independent variable; y is the dependent variable.

A function f may be defined implicitly by an equation involving x and y or explicitly by writing $y = f(x)$.

A function can also be characterized as a set of ordered pairs (x, y) or $(x, f(x))$ in which no two pairs have the same first element.

Page 169

LIBRARY OF FUNCTIONS

Linear function
$f(x) = mx + b$ Graph is a straight line with slope m and y-intercept b.

Constant function
$f(x) = b$ Graph is a horizontal line with y-intercept b (see Figure 26).

Identity function
$f(x) = x$ Graph is a straight line with slope 1 and y-intercept 0 (see Figure 27).

Square function
$f(x) = x^2$ Graph is a parabola with intercept at (0, 0) (see Figure 28).

Cube function
$f(x) = x^3$ See Figure 29.

Square root function
$f(x) = \sqrt{x}$ See Figure 30.

Reciprocal function
$f(x) = 1/x$ See Figure 31.

Absolute value function
$f(x) = |x|$ See Figure 32.

Page 169

The chapter review is a great place to check your understanding of the chapter material. Start with "Things to Know." Check your understanding of the concepts listed there. Next, make sure you know "How To" solve the items within that section. "Fill in the Blanks" determines your comfort with vocabulary. "True False" is a stickler for knowing definitions. If you are uncertain of any concept, go back into the chapter and study it further. Be sure to do the Review Exercises for practice. These reviews are for your success in this course—Make good use of them.

170 **Chapter 2** Functions and Their Graphs

HOW TO

Determine whether a relation represents a function.
Find the domain and range of a function from its graph
Find the domain of a function given its equation
Determine whether a function is even or odd without graphing it
Graph certain functions by shifting, compressing, stretching, and/or reflecting (see Table 11)
Use a graphing utility to determine where the graph of a function is increasing or decreasing
Use a graphing utility to find the local maxima and local minima of a function
Find the composite of two functions
Construct functions in applications, including piece-wise-defined functions

Page 170

FILL-IN-THE-BLANK ITEMS

1. If f is a function defined by the equation $y = f(x)$, then x is called the ______________ variable and y is the ______________ variable.
2. A set of points in the xy-plane is the graph of a function if and only if no ______________ line contains more than one point of the set.
3. A(n) ______________ function f is one for which $f(-x) = f(x)$ for every x in the domain of f; a(n) ______________ function f is one for which $f(-x) = -f(x)$ for every x in the domain of f.
4. Suppose that the graph of a function f is known. Then the graph of $y = f(x - 2)$ may be obtained by a(n) ______________ shift of the graph of f to the ______________ a distance of 2 units.
5. If $f(x) = x + 1$ and $g(x) = x^3$, then ______________ $= (x + 1)^3$.

Page 170

TRUE/FALSE ITEMS

T F 1. Vertical lines intersect the graph of a function in no more than one point.
T F 2. The y-intercept of the graph of the function $y = f(x)$ whose domain is all real numbers is $f(0)$.
T F 3. Even functions have graphs that are symmetric with respect to the origin.
T F 4. The graph of $y = f(-x)$ is the reflection about the y-axis of the graph of $y = f(x)$.
T F 5. $f(g(x)) = f(x) \cdot g(x)$

Page 170

REVIEW EXERCISES

1. Given that f is a linear function, $f(4) = -5$, and $f(0) = 3$, write the equation that defines f.
2. Given that g is a linear function with slope $= -4$ and $g(-2) = 2$, write the equation that defines g.
3. A function f is defined by
$$f(x) = \frac{Ax + 5}{6x - 2}$$
If $f(1) = 4$, find A.
4. A function g is defined by
$$g(x) = \frac{A}{x} + \frac{8}{x^2}$$
If $g(-1) = 0$, find A.

Page 170

STUDENT SUPPLEMENTS

Use these text supplements as your resources. Ask your instructor how to obtain these items to complement your learning style.

Algebra Review

Written by Michael Sullivan
ISBN: 0-13-590621-0

This paperback supplement offers four chapters of Intermediate Algebra Review and the solutions to the odd exercises in the back of the supplement. Written by the author to ease the students into Precalculus courses, it offers review materials covering the basics.

Student Solutions Manual

Written by Michael Sullivan, Michael Sullivan III, and Katy Sullivan Murphy
ISBN: 0-13-788200-9

Contains complete step-by-step worked out solutions to all the odd numbered exercises in the textbook. Follows the exact procedures of the text—no magical steps.

Tutorial Videos

ISBN: 0-13-788381-1

A new tutorial videotape series which has been created to accompany this specific test, includes 15 minute segments for each section of the text. At the beginning of each section of the text, the videotape number on which the material is contained, is listed. Each segment uses both traditional and technological ways of solving the mathematical problems. Entertaining and educational, these videos provide an alternative process which can add to your success in this course. Ask your professor where the set is located for your school.

New York Times Supplement

A free newspaper from Prentice Hall and the New York Times which includes interesting and relevant articles on mathematics in the world around us. This supplement is revised each fall.

Life on the Internet—Mathematics

Guides you through the complexity of the Internet, offering navigation strategies, practice exercises and lists of mathematical resources. This supplement is updated each year to reflect the ongoing changes on the Internet.

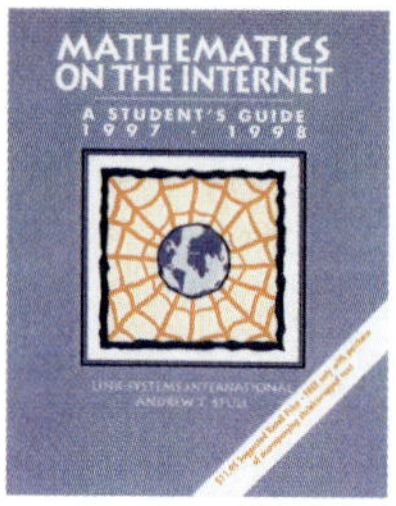

Prentice Hall Companion Website

Located at *www.prenhall.com/sullivan* this website offers you additional information, projects and exercises to enrich the learning experience using this text. The Internet "Excursions" found in the beginning of each chapter have duplicate pages on the website with the links to the appropriate data and information gathering sites described in the projects.

LIST OF APPLICATIONS

Combinatorics *(Continued)*

Construction. *See also* Engineering

Demography. *See* Population

Distance applications

Economics. *See also* Business; Finance

Education

Electricity

Electronics

Energy

Engineering

Finance. *See also* Business; Economics

PHOTO CREDITS

Chapter 1	Sir Isaac Newton	Erich Lessing/Art Resource, NY
Chapter 2	Oil well at sunset near Bakersfield, California 405 and Sunset Blvd.	Dan Suzio/Photo Researchers, Inc. Joe Sohm/Photo Researchers, Inc.
Chapter 3	Earth from space showing North and South America	NASA/GFSC/Tom Stack and Associates
Chapter 4	SETI Telescope	Jet Propulsion Laboratory/ NASA Headquarters
Chapter 5	Electron micrograph of the mite Varroae jacobeoni that occurs as an ectoparasite on the honeybee	Andrew Syred/Science Photo Library/Photo Researchers, Inc.
Chapter 6	Polyhedra	Prof. George W. Hart
Chapter 7	A Prism	George Dodson/Uniphoto
Chapter 8	Annapolis Tidal Generating Station	Nova Scotia Power
Chapter 9	Golden Gate Bridge	Brian Parker/Tom Stack & Associates
Chapter 10	Spirograph (drawing)	Kathy Enfalice
Chapter 11	Three-ring German Enigma Cypher Machine	Science Museum/Science & Society Picture Library
Chapter 12	Missey Glove—Mountain Bike Racer	Carl Yarborough/Carl Yarborough Photography

CHAPTER

1

Graphs

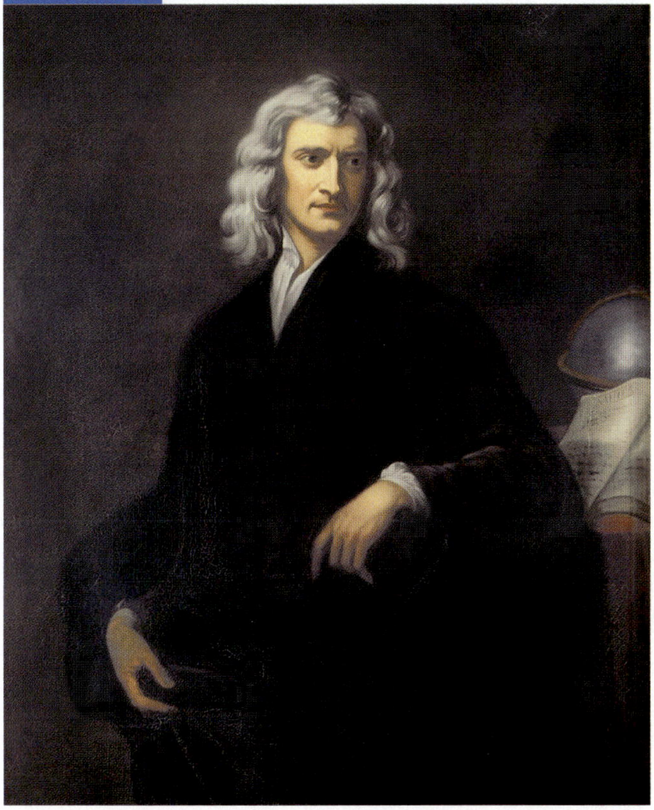

On the page following each chapter opener is an Internet Excursion. These Excursions will place you in intriguing situations where information from various Internet resources are needed in order to answer the thought provoking questions. All of the links and even more information on these Excursions can be found on your Sullivan website.
The URL address is:

www.prenhall.com/sullivan

Once at the website, find the appropriate text and explore the student's pages of the site. You will find an interactive version of the Excursion as well as many other useful and helpful materials.

Turn the page and learn all about Isaac Newton and his mathematical discoveries. . . .

PREPARING FOR THIS CHAPTER

Before getting started on this chapter, review the following concepts:

Topics from Algebra and Geometry *(Appendix, Section 1)*

Solving Equations *(Appendix, Section 4)*

Completing the Square *(Appendix, Section 5)*

OUTLINE

Back Forward Home Reload Images Open Print Find Stop

Netsite: http://www.prenhall.com/sullivan/

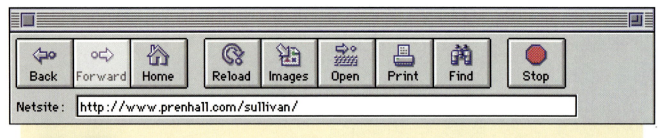

THE CATALOG OF EQUATIONS

Trinity College, 1667. You have just begun your college work and one of your teachers is a strange fellow named *Isaac Newton.* His ideas are difficult to understand and very radical. Yet, when you listen to him speak, his passion makes a profound impression on you. His curious experiments and clever devices attract you to him.

After class one day, he asks you to help him with his Catalog of Equations. He quickly explains about the coordinate geometry of Descartes, and how algebra and geometry are now joined together. He assigns to you the equation:

$$(-2x + x^2 + y^2)^2 = x^2 + y^2$$

He gives you a manuscript for *The Witch,* and asks you to do a similar work for your assigned curve. As you walk away, you feel a great sense of honor in being selected to do research for Isaac Newton.

1. Have you ever thought about such a Catalog of Equations? Take a quick tour of the famous curves at the *MacTutor Website.* Notice the cartesian formulas (x and y) for the curves. How would you organize a catalog for the classification of algebraic curves?
2. Today a student does not hear much about a Catalog of Equations. Why do you suppose that is?
3. From looking at your equation, can you use any of the symmetry principles to get some idea of its graph? Can you graph this equation on your calculator? Are you able to graph it by making a chart of x and y values?
4. Visit The Visual Dictionary of Plane Curves Website and read about catacaustic curves. Can you determine the catacaustic of your curve with a lightsource on the edge? How could you graph it?
5. Tie together your explorations into a 'manuscript.'

The idea of using a system of rectangular coordinates dates back to ancient times, when such a system was used for surveying and city planning. Apollonius of Perga, in 200 B.C., used a form of rectangular coordinates in his work on conics, although this use does not stand out as clearly as it does in modern treatments. Sporadic use of rectangular coordinates continued until the 1600s. By that time, algebra had developed sufficiently so that René Descartes (1596–1650) and Pierre de Fermat (1601–1665) could take the crucial step, which was the use of rectangular coordinates to translate geometry problems into algebra problems, and vice versa. This step was supremely important for two reasons. First, it allowed both geometers and algebraists to gain critical new insights into their subjects, which previously had been regarded as separate but now were seen to be connected in many important ways. Second, the insights gained made possible the development of calculus, which greatly enlarged the number of areas in which mathematics could be applied and made possible a much deeper understanding of these areas. With the advent of technology, in particular graphing utilities, we are now able not only to visualize the dual roles of algebra and geometry, but we are also able to solve many problems that required advanced methods before this technology.

Information, often in the form of data, is the basis for making informed decisions. In this chapter we discuss various types of data and ways data can be represented. One such way is by using rectangular coordinates.

1.1 DATA AND ITS REPRESENTATION

1 Classify Data
2 Construct and Interpret Bar Charts and Pie Charts
3 Draw and Interpret Histograms and Frequency Polygons

1 **Data** is a collection of information about a variable. Data may be a number, a word, or a letter. For example, the miles per gallon obtained by a particular make of car is data that is numeric. The colors of the cars in the South Suburban College student parking lot is data that is words. We usually classify data as either quantitative or qualitative. **Quantitative data** is data that is a numerical measure of some variable. **Qualitative data** is data that describes a nonnumeric characteristic of a variable. So, the miles per gallon for a particular car would be quantitative, while the color of a car is a qualitative variable.

EXAMPLE 1 Classifying Data as Either Quantitative or Qualitative

Determine whether the following sets of data are quantitative or qualitative.

(a) Cost of monthly phone service for households in Jacksonville, FL.
(b) Gender of students enrolled in College Algebra at South Suburban College.
(c) Method of payment for tuition bills at Chicago State University for students registering in the current semester.
(d) Number of credit card holders who have been late with their payment at least once during the past 12 months.

Solution
(a) Quantitative data because the data is numeric.
(b) Qualitative data because gender is a nonnumeric characteristic.
(c) Qualitative data because the method of payment (credit card, cash, check, etc.) is a nonnumeric characteristic.
(d) Quantitative data because the data is numeric. ■

Now work Problem 3.

When data is collected, it is generally put into table form and graphs are created from the table. Putting data into the form of a graph allows us to "see" the data. Often a graph will reveal certain characteristics of the data that may not be readily apparent from a table of data. The type of graph drawn will depend on the type of data.

Graphs of Qualitative Data

2 There are two popular methods for graphically displaying a qualitative set of data: bar charts and pie charts. A bar chart will show the number or the percent of data that are in each category, while a pie chart will show only the percent of data in each category.

EXAMPLE 2 Constructing a Bar Chart

For the fiscal year 1993 (October 1992–September 1993), the federal government spent a total of $1,444 billion. The breakdown of expenditures (in billions of dollars) is given in Table 1.

TABLE 1

Social Security, Medicare, and other retirement	$500
National defense, veterans, and foreign affairs	$344
Net interest (interest on the public debt)	$199
Physical, human, and community development	$119
Social programs	$254
Law enforcement and general government	$28

Source: Department of the Treasury

Construct a bar chart of the above data.

Solution A horizontal axis is used to indicate the category of spending and a vertical axis is used to represent the amount spent in each category. For each category of spending we draw rectangles of equal width whose height represents the amount spent in the category. The rectangles will not touch each other since the data is qualitative. See Figure 1.

FIGURE 1

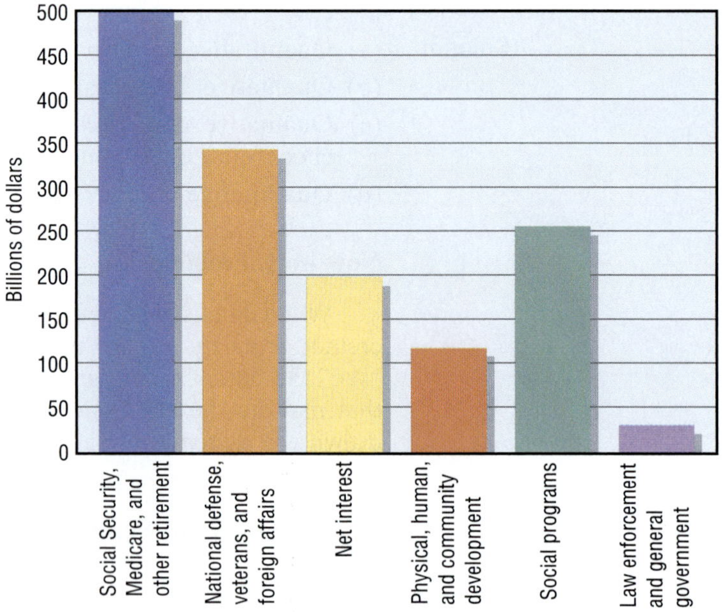

Qualitative data can also be represented graphically in a pie chart.

EXAMPLE 3 Constructing a Pie Chart

Use the data given in Example 2 to construct a pie chart.

Solution To construct a pie chart a circle is divided into sectors, one sector for each category of data. The size of each sector is proportional to the total amount spent. Since Social Security, Medicare, and other retirement is \$500 billion and total spending is \$1,444 billion, the percent of data in this category is $500/1{,}444 \approx 0.35 = 35\%$. Therefore, Social Security, Medicare, and other retirement will make up 35% of the pie chart. Since a circle has 360°, the degree measure of the sector for this category of spending is $0.35(360°) = 126°$. Following this procedure for the remaining categories of spending, we obtain Table 2.

TABLE 2

Category	Spending (in billions)	Percent of Total Spending*	Degree Measure of Sector*
Social Security, Medicare, and other retirement	\$500	0.35 = 35%	126
National defense, veterans, and foreign affairs	\$344	0.24 = 24%	86
Net interest (interest on the public debt)	\$199	0.14 = 14%	50
Physical, human, and community development	\$119	0.08 = 8%	29
Social programs	\$254	0.18 = 18%	65
Law enforcement and general government	\$ 28	0.02 = 2%	7

*The data in column three does not add up to 100% due to rounding. Similarly, the data in column four does not add up to 360° due to rounding.

To construct a pie chart by hand, we use a protractor to approximate the angles for each sector. To construct a pie chart with a computer, select a spreadsheet program that has the capability of drawing pie charts. See Figure 2.

FIGURE 2

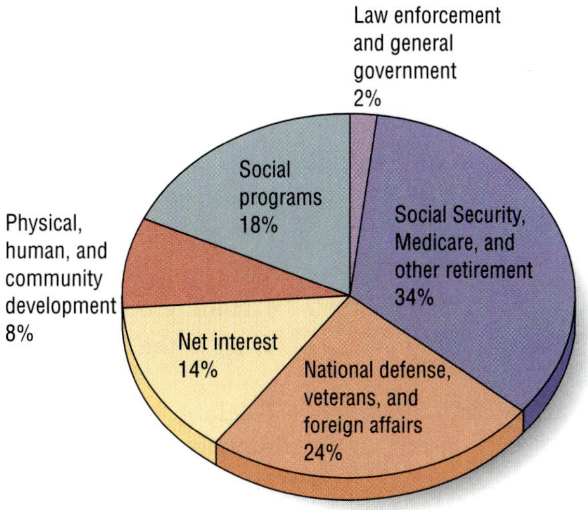

One reason for graphing data by drawing bar charts or pie charts is to quickly determine certain information about the data.

EXAMPLE 4 Analyzing a Bar Chart

The bar chart in Figure 3 represents the sources of revenue for the United States federal government in its fiscal year 1993. Answer the questions below using the bar chart.

FIGURE 3

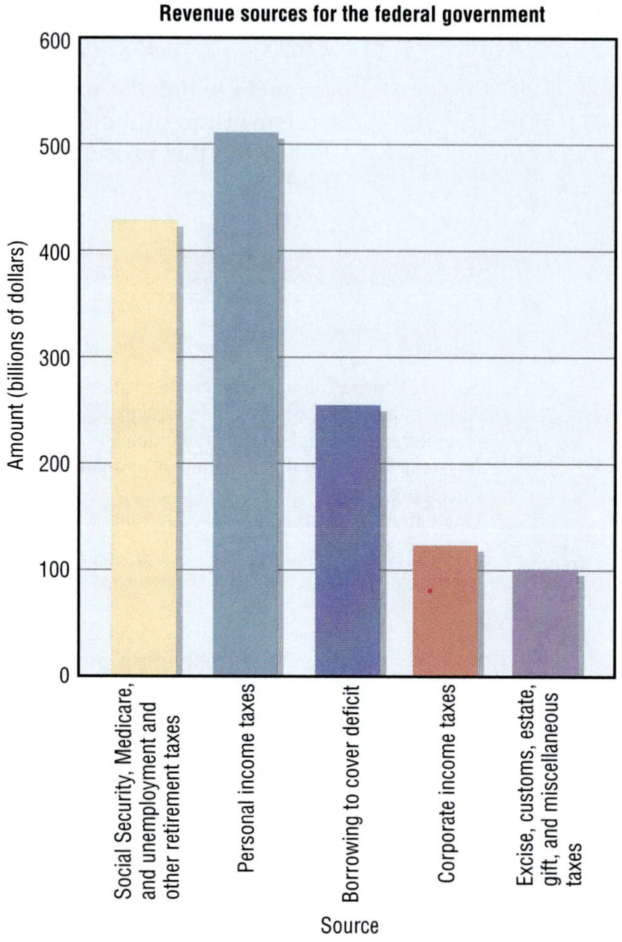

(a) What is the largest source of revenue for the federal government?

(b) What is the smallest source of revenue for the federal government?

(c) How much revenue does the government collect from all sources, excluding borrowing?

(d) What is the total revenue of the U.S. federal government?

Solution (a) The largest source of revenue for the federal government is personal income taxes. The government collects about \$510 billion from this source.

(b) The smallest source of revenue for the federal government is excise, customs, estate, gift, and miscellaneous taxes. The government collects about \$100 billion from this source.

(c) The revenue collected, excluding borrowing, totals approximately $\$430 + \$510 + \$120 + \$100 = \$1{,}160$ billion.

(d) The remaining amount represents borrowing of approximately \$250 billion. The total revenue is therefore \$1,160 + \$250 = \$1,410 billion.

Now work Problem 15.

Graphs of Quantitative Data

3

Quantitative data is often organized in a table called a **frequency distribution.** This table divides data into **class intervals** of equal width and records the number of data that fall within each class interval. The class intervals do not overlap. This prevents any data from falling into two different class intervals. The first number in any class interval is called the **lower class limit** and the second number in any class interval is called the **upper class limit.** The **class width** is the difference between consecutive lower class limits. Data organized in frequency distributions allow us to represent data using histograms and frequency polygons, discussed in Examples 5 and 7.

EXAMPLE 5 Drawing a Histogram

Table 3 represents the frequency distribution for the sale price of real estate valued between \$96,000 and \$255,000 in Bridgeview, IL in 1997.

TABLE 3

Class Interval	Frequency
96–111	9
112–127	13
128–143	10
144–159	9
160–175	4
176–191	2
192–207	1
208–223	1
224–239	2
240–255	5

(a) Determine the number of class intervals.

(b) What is the lower class limit of the first class interval? What is the upper class limit of the first class interval?

(c) Determine the class width.

(d) Draw a histogram of the frequency distribution in Table 3.

Solution

(a) The data in Table 3 has 10 class intervals.

(b) The lower class limit of the first class interval is 96. The upper class limit of the first class interval is 111.

(c) The class width is found by finding the difference between consecutive lower class limits. The lower class limit of the second class interval is 112. The lower class limit of the first class interval is 96. The class width is therefore $112 - 96 = 16$.

(d) Histograms are similar to bar charts in that we draw rectangles whose height equals the frequency of each class interval. The width of each rectangle equals the class width. The bases of the rectangles touch when drawing a histogram. This is because there are no "gaps" in the data. Figure 4 shows a histogram for the data in Table 3.

FIGURE 4

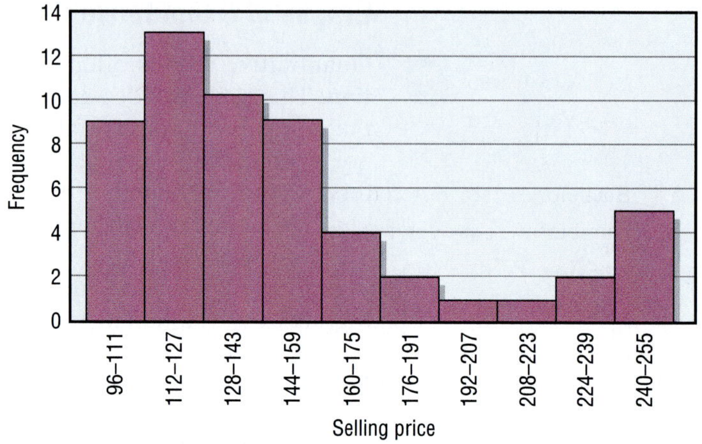

EXAMPLE 6 Analyzing a Histogram

Answer the following questions by referring to either Figure 4 or Table 3.

(a) What is the most popular price range of property sold in Bridgeview, IL?

(b) How many properties sold for prices between \$176,000 and \$191,000?

Solution (a) The most popular price range is \$112,000 to \$127,000.

(b) Two properties sold between \$176,000 and \$191,000.

EXAMPLE 7 Constructing a Frequency Polygon from a Histogram

Draw a frequency polygon for the data in Figure 4.

Solution We begin by placing a dot at the top of each rectangle in the histogram over the midpoint of the class interval. A frequency polygon is obtained by connecting the dots. See Figure 5.

FIGURE 5

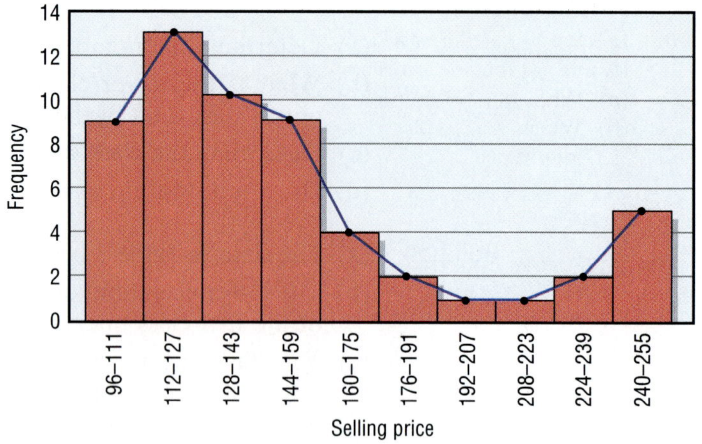

Now work Problem 21.

1.1 EXERCISES

For Problems 1–8, classify the data as either qualitative or quantitative.

1. Hair color.
2. Number of students enrolled in United States History.
3. Gas mileage in the city for a Toyota Camry.
4. Brand of stereos a local superstore sells.
5. SAT-Verbal scores of the freshman class at the University of Dayton.
6. Marital status.
7. Bachelor degrees offered by DePaul University.
8. Number of hours a bulb burns before burning out.
9. **On-Time Performance** The bar chart below represents the overall percentage of reported flight operations arriving on time for ten different airlines.
   (a) Which airline has the highest percentage of on-time flights?
   (b) Which airline has the lowest percentage of on-time flights?
   (c) What percentage of United Airlines' flights are on time?

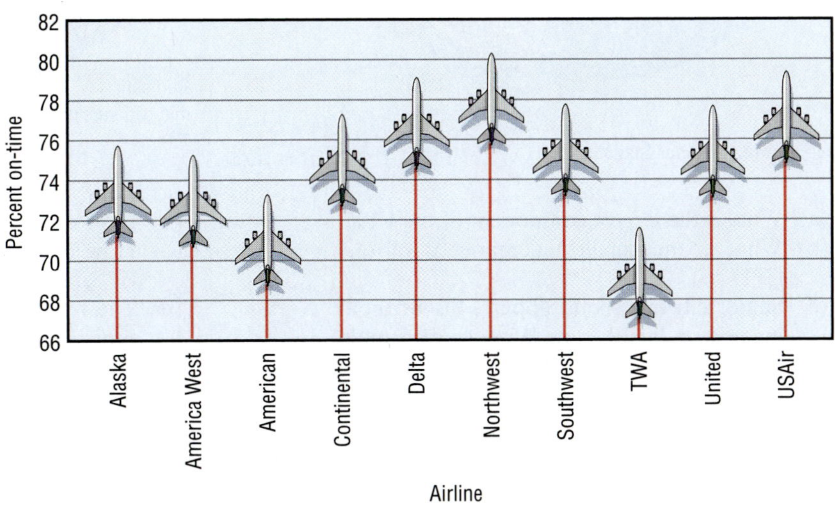

Source: United States Department of Transportation

10. **Income Required for a Loan** The bar chart below shows the minimum annual income required for a $100,000 loan using interest rates available on 12/1/96. Taxes and insurance are assumed to be $230 monthly.
   (a) What minimum annual income is needed to qualify for a 5/1 year ARM (Adjustable Rate Mortgage)?
   (b) Which loan type requires the most annual income? What minimum annual income is required for this loan type?

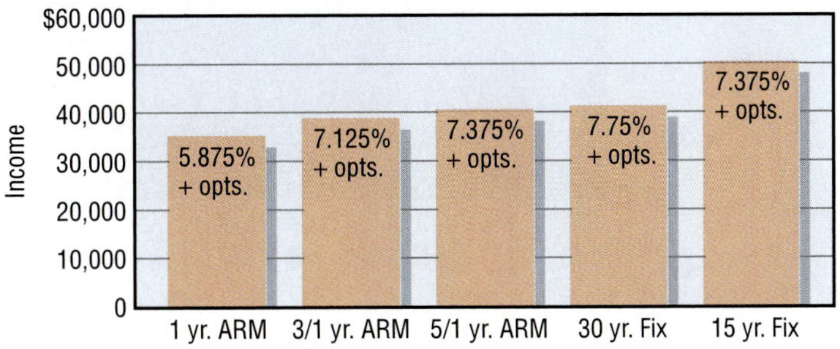

11. Consumer Price Index The Consumer Price Index (CPI) is an index that measures inflation. It is calculated by obtaining the prices of a market basket of goods each month. The market basket, along with the percentages of each product, is given in the following pie chart:

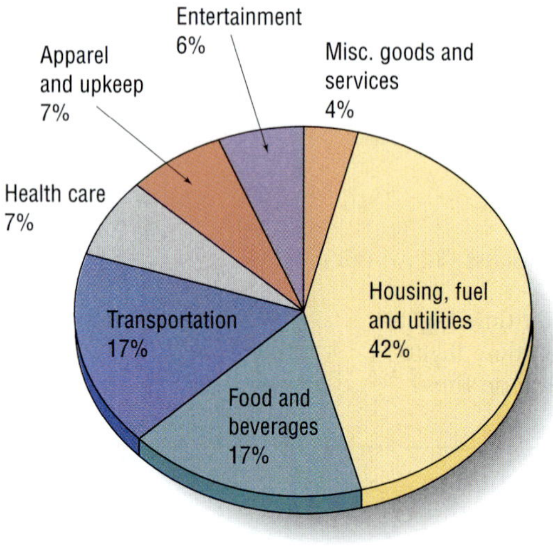

Source: Bureau of Labor Statistics

(a) What is the largest component of the CPI?
(b) What is the smallest component of the CPI?

(c) Senior citizens spend about 14% of their income on health care. Why do you think they feel the CPI weight for health care is too low?

12. Asset Allocation According to financial planners, an individual's investment mix should change over a person's lifetime. The longer an individual's time horizon, the more the individual should invest in stocks. A financial planner suggested that Jim's retirement portfolio be diversified according to the mix provided in the pie chart below, since Jim has 40 years to retirement.

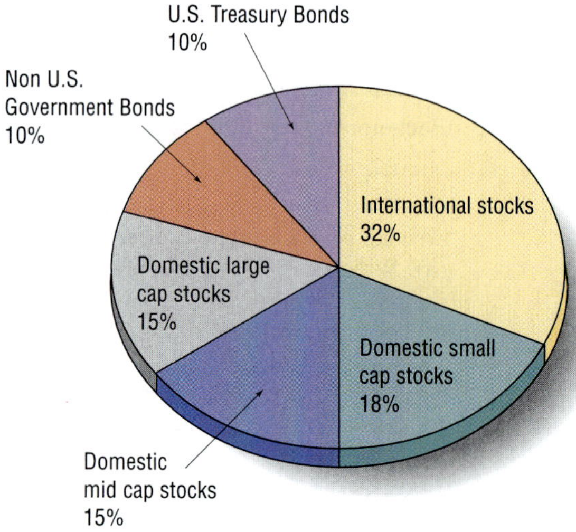

(a) How much should Jim invest in stocks?
(b) How much should Jim invest in bonds?
(c) How much should Jim invest in domestic (U.S.) stocks?
(d) The return on bonds over long periods of time is less than that of stocks. Explain why you think the financial planner recommended what she did for bonds.

13. Licensed Drivers in Florida The histogram below represents the number of licensed drivers between the ages of 20 and 84 in the state of Florida.

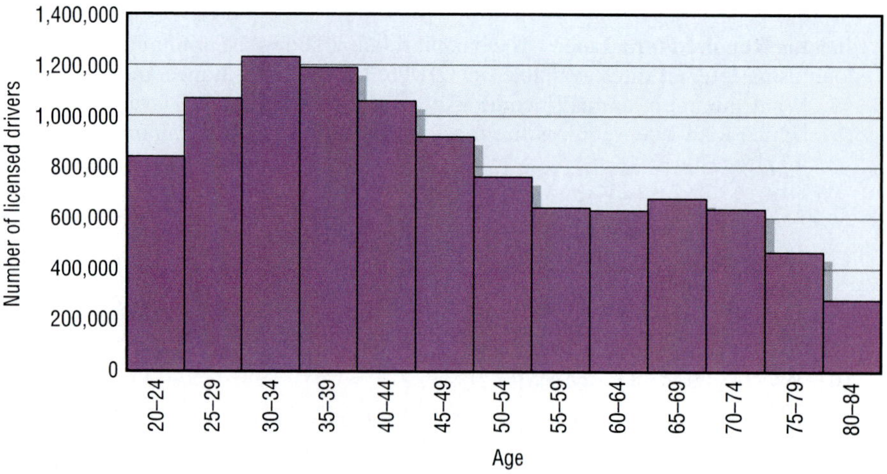

(a) Determine the number of class intervals.
(b) What is the lower class limit of the first class interval? What is the upper class limit of the first class interval?
(c) Determine the class width.
(d) How many licensed drivers are 70 to 84 years old?
(e) Which class interval has the most licensed drivers?
(f) Which class interval has the fewest licensed drivers?
(g) Draw a frequency polygon for the given data.

14. **IQ Scores** The histogram below represents the IQ scores of students enrolled in College Algebra at a local university.

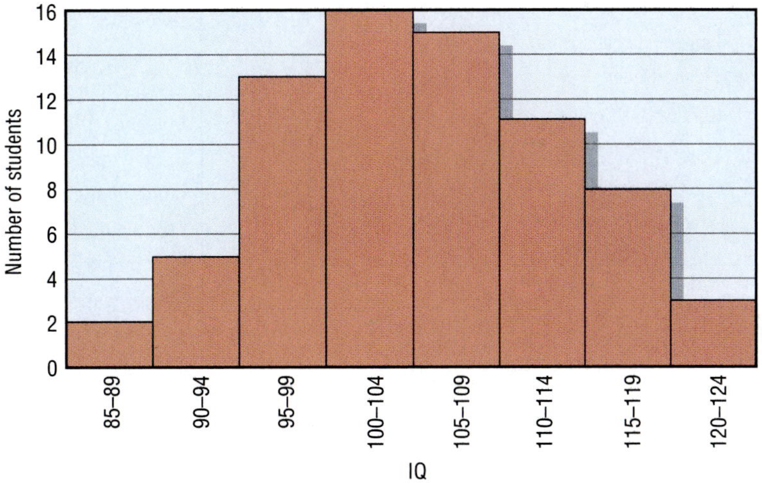

(a) Determine the number of class intervals.
(b) What is the lower class limit of the first class interval? What is the upper limit of the first class interval?
(c) Determine the class width.
(d) How many students have an IQ between 100 and 104?
(e) How many students have an IQ above 110?
(f) How many students are enrolled in College Algebra?
(g) Draw a frequency polygon for the given data.

15. **Household Income** The data below represents the median income of households (in dollars) by region of the country for 1995.

Region	Median Income
Northwest	36,111
Midwest	35,839
South	30,942
West	35,979

Source: U.S. Bureau of the Census, March 1996 Current Population Survey

(a) Draw a bar chart of the data.
(b) Draw a pie chart of the data.

(c) Which chart seems to summarize the data better?
(d) Which region has the highest median income?
(e) Which region has the lowest median income?

16. **Household Income** The data below represents the median income (in dollars) of families by type of household for 1995.

Family Household	Median Income
Married-couple families	47,129
Female householder, no husband present	21,348
Male householder, no wife present	33,534

Source: U.S. Bureau of the Census, March 1996 Current Population Survey

(a) Draw a bar chart of the data.
(b) Draw a pie chart of the data.
(c) Which chart seems to summarize the data better?
(d) Which household type has the highest median income?
(e) Which household type has the lowest median income?

17. Busing Revenue The data below represents the total operating revenues (in thousands of dollars) of regional class I motor carriers of passengers for the first quarter of 1996.

Motor Carrier	Total Revenue
Academy Lines, Inc.	5,203
Bonanza Bus Lines	3,599
Connecticut Limousine	4,769
Hudson Transit	5,763
Kerrville Bus Company	4,269
New Jersey Transit	48,524
Peter Pan Bus Lines	7,914
Texas, New Mexico & Oklahoma Coaches	4,983

Source: Bureau of Transportation Statistics

(a) Draw a bar chart of the data.
(b) Draw a pie chart of the data.
(c) Which chart seems to summarize the data better?
(d) Which motor carrier has the highest total revenue?
(e) Which motor carrier has the lowest total revenue?

18. Busing Revenue The data below represents the total operating revenues (in thousands of dollars) of regional class I motor carriers of passengers for the first quarter of 1995.

Motor Carrier	Total Revenue
Academy Lines, Inc.	4,605
Bonanza Bus Lines	3,437
Connecticut Limousine	4,846
Hudson Transit	5,793
Kerrville Bus Company	4,271
New Jersey Transit	48,113
Peter Pan Bus Lines	8,061
Texas, New Mexico & Oklahoma Coaches	4,708

Source: Bureau of Transportation Statistics

(a) Draw a bar chart of the data.
(b) Draw a pie chart of the data.
(c) Which chart seems to summarize the data better?
(d) Which motor carrier has the highest total revenue?
(e) Which motor carrier has the lowest total revenue?

19. Causes of Death The data below represents the causes of death for 15–24 year olds in 1995.

Cause of Death	Number
Accidents and adverse effects	13,532
Homicide and legal intervention	6,827
Suicide	4,789
Malignant neoplasms	1,599
Diseases of heart	964
Human immunodeficiency virus infection	643
Congenital anomalies	425
Chronic obstructive pulmonary diseases	220
Pneumonia and influenza	193
Cerebrovascular diseases	166
All other causes	4,211

Source: National Center for Health Statistics, 1996

(a) Draw a bar chart of the data.
(b) Draw a pie chart of the data.
(c) Which chart seems to summarize the data better?
(d) What was the leading cause of death for 15–24 year olds in 1995?

20. Licensed Drivers The data below represents the number of licensed drivers in the Great Lake States in 1994.

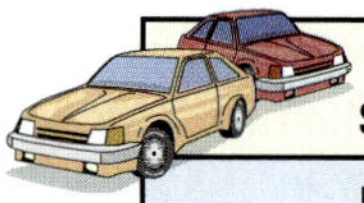

State	Number of Licensed Drivers
Illinois	7,502,201
Indiana	3,806,329
Michigan	6,601,924
Minnesota	2,705,701
Ohio	7,142,173
Wisconsin	3,554,003

Source: Each state's authorities

(a) Draw a bar chart of the data.
(b) Draw a pie chart of the data.
(c) Which chart seems to summarize the data better?
(d) Which state has the most licensed drivers?
(e) Which state has the fewest licensed drivers?

21. Licensed Drivers in Tennessee The frequency table on the following page provides the number of licensed drivers between the ages of 20 and 84 in the state of Tennessee in 1994.

Age	Number of Licensed Drivers
20–24	345,941
25–29	374,629
30–34	428,748
35–39	439,137
40–44	414,344
45–49	372,814
50–54	292,460
55–59	233,615
60–64	204,235
65–69	181,977
70–74	150,347
75–79	100,068
80–84	50,190

Source: FHWA

(a) Determine the number of class intervals.
(b) What is the lower class limit of the first class interval? What is the upper class limit of the first class interval?
(c) Determine the class width.
(d) Draw a histogram of the data.
(e) Draw a frequency polygon of the data.
(f) Which age group has the most licensed drivers?
(g) Which age group has the fewest licensed drivers?

22. Licensed Drivers in Hawaii The frequency table below provides the number of licensed drivers between the ages of 20 and 84 in the state of Hawaii in 1994.

Age	Number of Licensed Drivers
20–24	65,951
25–29	78,119
30–34	91,976
35–39	92,557
40–44	87,430
45–49	75,978
50–54	55,199
55–59	39,678
60–64	35,650
65–69	33,885
70–74	26,125
75–79	14,990
80–84	6,952

Source: FHWA

(a) Determine the number of class intervals.
(b) What is the lower class limit of the first class interval? What is the upper class limit of the first class interval?
(c) Determine the class width.
(d) Draw a histogram of the data.
(e) Draw a frequency polygon of the data.
(f) Which age group has the most licensed drivers?
(g) Which age group has the fewest licensed drivers?

23. Undergraduate Tuition The data below represents the cost of undergraduate tuition at four-year colleges for 1992–1993 having tuition amounts ranging from \$0 through \$14,999.

Tuition (Dollars)	Number of 4-Year Colleges
0–999	10
1000–1999	7
2000–2999	45
3000–3999	66
4000–4999	84
5000–5999	84
6000–6999	97
7000–7999	118
8000–8999	138
9000–9999	110
10,000–10,999	104
11,000–11,999	82
12,000–12,999	61
13,000–13,999	34
14,000–14,999	29

Source: The College Board, New York, NY, Annual Survey of Colleges 1992 and 1993.

(a) Determine the number of class intervals.
(b) What is the lower class limit of the first class interval? What is the upper class limit of the first class interval?
(c) Determine the class width.
(d) Draw a histogram of the data.
(e) Draw a frequency polygon of the data.
(f) What range of tuition occurs most frequently?

24. Undergraduate Tuition The data on the right represents the cost of undergraduate tuition at four-year colleges for 1993–1994 having tuition amounts ranging from \$0 through \$14,999.

(a) Determine the number of class intervals.

(b) What is the lower class limit of the first interval? What is the upper limit of the first class interval?

(c) Determine the class width.

(d) Draw a histogram of the data.

(e) Draw a frequency polygon of the data.

(f) What range of tuition occurs most frequently?

Tuition (Dollars)	Number of 4-Year Colleges
0–999	8
1000–1999	5
2000–2999	28
3000–3999	48
4000–4999	76
5000–5999	65
6000–6999	81
7000–7999	96
8000–8999	112
9000–9999	118
10,000–10,999	106
11,000–11,999	90
12,000–12,999	70
13,000–13,999	59
14,000–14,999	23

Source: The College Board, New York, NY, Annual Survey of Colleges 1992 and 1993.

1.2 RECTANGULAR COORDINATES; GRAPHING UTILITIES; DATA IN ORDERED PAIRS

1 Distance Formula

2 Midpoint Formula

3 Construct and Interpret Bar Charts and Line Graphs

4 Draw and Interpret Scatter Diagrams

We locate a point on the real number line by assigning it a single real number, called the *coordinate of the point.* For work in a two-dimensional plane, we locate points by using two numbers.

We begin with two real number lines located in the same plane: one horizontal and the other vertical. We call the horizontal line the ***x*-axis,** the vertical line the ***y*-axis,** and the point of intersection the **origin *O*.** We assign coordinates to every point on these number lines as shown in Figure 6, using a convenient scale. In mathematics, we usually use the same scale on each axis; in applications, a different scale is often used on each axis.

FIGURE 6

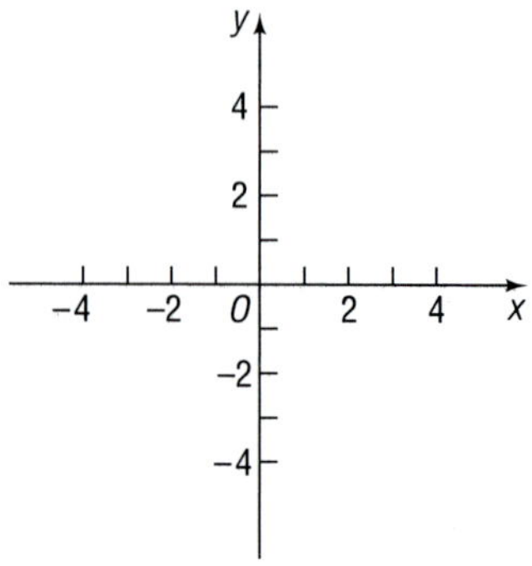

The origin O has a value of 0 on both the x-axis and the y-axis. We follow the usual convention that points on the x-axis to the right of O are associated with positive real numbers, and those to the left of O are associated with negative real numbers. Those on the y-axis above O are associated with positive real numbers, and those below O are associated with negative real numbers. In Figure 6, the x-axis and y-axis are labeled as x and y, respectively, and we have used an arrow at the end of each axis to denote the positive direction.

The coordinate system described here is called a **rectangular,** or **Cartesian* coordinate system.** The plane formed by the x-axis and y-axis is sometimes called the ***xy*-plane,** and the x-axis and y-axis are referred to as the **coordinate axes.**

FIGURE 7

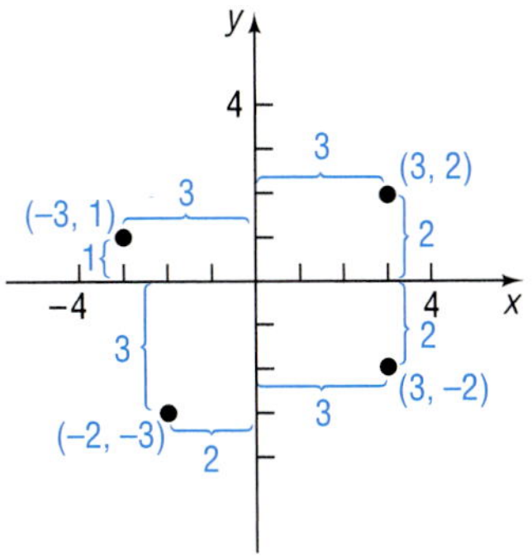

Any point P in the xy-plane can then be located by using an **ordered pair** (x, y) of real numbers. Let x denote the signed distance of P from the y-axis (*signed* in the sense that, if P is to the right of the y-axis, then $x > 0$, and if P is to the left of the y-axis, then $x < 0$); and let y denote the signed distance of P from the x-axis. The ordered pair (x, y), also called the **coordinates** of P, then gives us enough information to locate the point P in the plane.

For example, to locate the point whose coordinates are $(-3, 1)$, go 3 units along the x-axis to the left of O and then go straight up 1 unit. We **plot** this point by placing a dot at this location. See Figure 7, in which the points with coordinates $(-3, 1)$, $(-2, -3)$, $(3, -2)$, and $(3, 2)$ are plotted.

The origin has coordinates $(0, 0)$. Any point on the x-axis has coordinates of the form $(x, 0)$, and any point on the y-axis has coordinates of the form $(0, y)$.

FIGURE 8

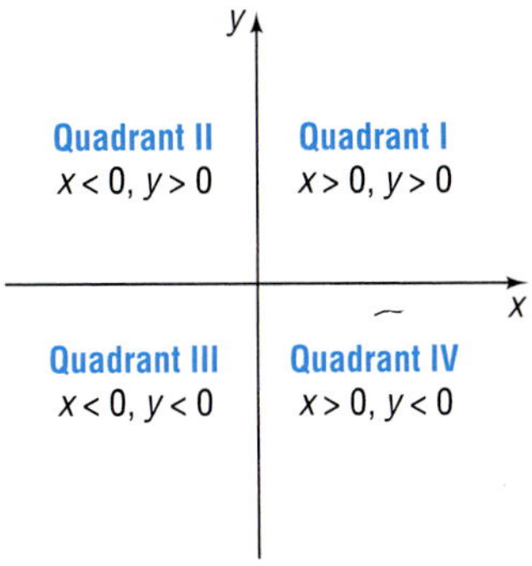

If (x, y) are the coordinates of a point P, then x is called the ***x*-coordinate,** or **abscissa,** of P and y is the ***y*-coordinate,** or **ordinate,** of P. We identify the point P by its coordinates (x, y) by writing $P = (x, y)$. Usually, we will simply say "the point (x, y)" rather than "the point whose coordinates are (x, y)."

The coordinate axes divide the xy-plane into four sections, called **quadrants,** as shown in Figure 8. In quadrant I, both the x-coordinate and the y-coordinate of all points are positive; in quadrant II, x is negative and y is positive; in quadrant III, both x and y are negative; and in quadrant IV, x is positive and y is negative. Points on the coordinate axes belong to no quadrant.

Now work Problem 1.

Graphing Utilities

All graphing utilities, that is, all graphing calculators and all computer software graphing packages, graph equations by plotting points on a screen. The screen itself actually consists of small rectangles, called **pixels.** The more pixels the screen has, the better the resolution. Most graphing calculators have 48 pixels per square inch; most computer screens have 32 to 108 pixels per square inch. When a point to be plotted lies inside a pixel, the pixel is turned on (lights up). Thus, the graph of an equation is a collection of pixels. Figure 9 shows how the graph of $y = 2x$ looks on a TI-82 graphing calculator.

FIGURE 9
$y = 2x$

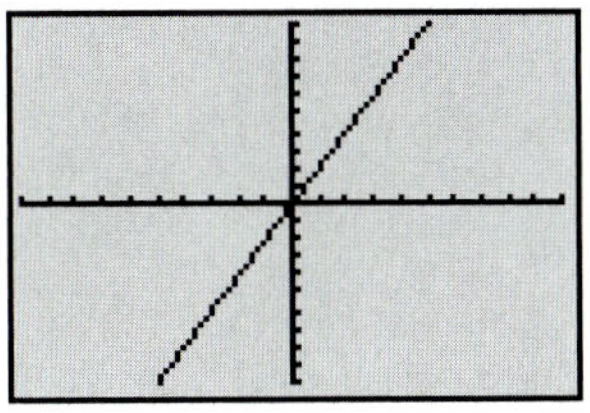

The screen of a graphing utility will display the coordinate axes of a rectangular coordinate system. However, you must set the scale on each axis. You must also include the smallest and largest values of x and y that you want included in the graph. This is called **setting the RANGE** and it gives the **viewing rectangle** or **window.**

*Named after René Descartes (1596–1650), a French mathematician, philosopher, and theologian.

FIGURE 10

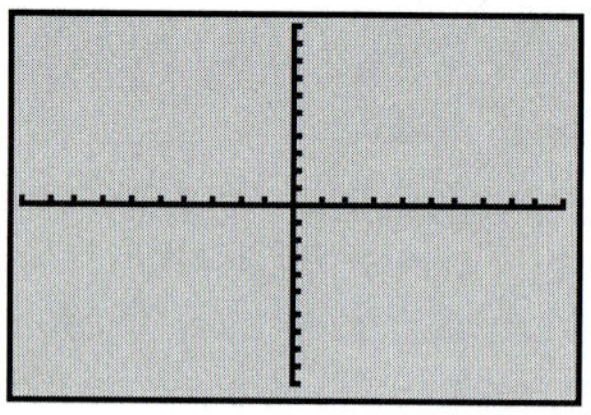

Figure 10 illustrates a typical viewing rectangle.

To select the viewing rectangle, we must give values to the following expressions:

*X*min: the smallest value of x
*X*max: the largest value of x
*X*scl: the number of units per tick mark on the x-axis
*Y*min: the smallest value of y
*Y*max: the largest value of y
*Y*scl: the number of units per tick mark on the y-axis

FIGURE 11

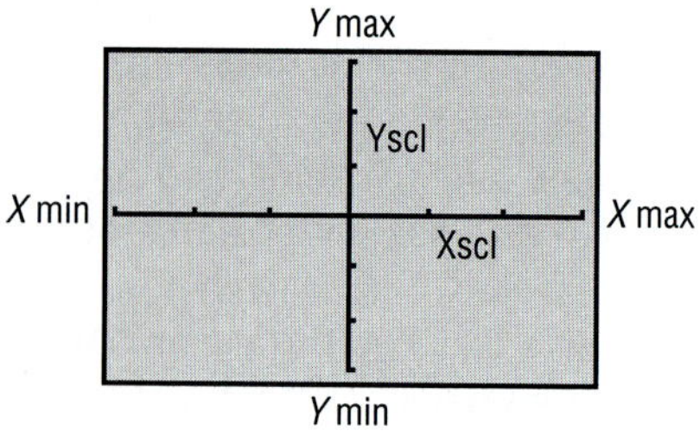

Figure 11 illustrates these settings for a typical screen.

If the scale used on each axis is known, we can determine the minimum and maximum values of x and y shown on the screen by counting the tick marks. Look again at Figure 10. For a scale of 1 on each axis, the minimum and maximum values of x are -10 and 10, respectively; the minimum and maximum values of y are also -10 and 10. If the scale is 2 on each axis, then the minimum and maximum values of x are -20 and 20, respectively; the minimum and maximum value of y are -20 and 20, respectively.

Conversely, if we know the minimum and maximum values of x and y, we can determine the scales being used by counting the tick marks displayed. We shall follow the practice of showing the minimum and maximum values of x and y in our illustrations so that you will know how the RANGE was set. See Figure 12.

FIGURE 12

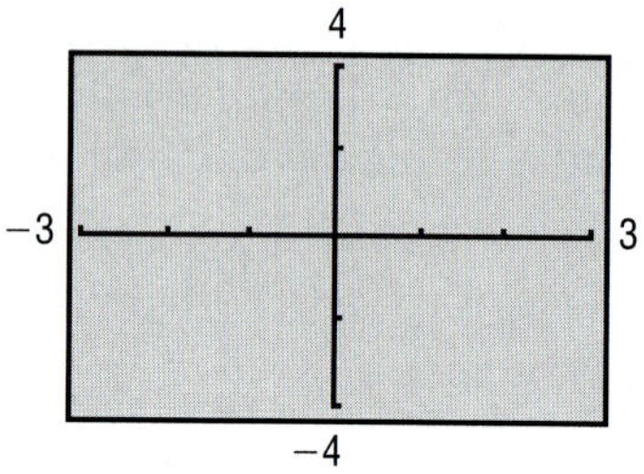

means *X*min = −3, *X*max = 3, *X*scl = 1, *Y*min = −4, *Y*max = 4, *Y*scl = 2

EXAMPLE 1 Finding the Coordinates of a Point Shown on a Graphing Utility Screen

Find the coordinates of the point shown in Figure 13.

FIGURE 13

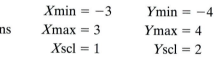

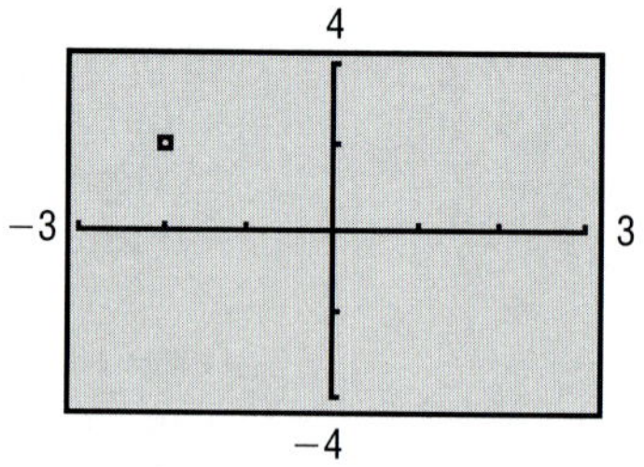

Solution First we note that the range setting used in Figure 13 is

*X*min = −3 *X*scl = 1 *Y*min = −4 *Y*scl = 2
*X*max = 3 *Y*max = 4

The point shown is 2 tic units to the left on the horizontal axis (scale = 1) and 1 tic up on the vertical scale (scale = 2). Thus, the coordinates of the point shown are $(-2, 2)$.

Now work Problems 5 and 15.

Distance between Points

1 If the same units of measurement, such as inches, centimeters, and so on, are used for both the x-axis and the y-axis, then all distances in the xy-plane can be measured using this unit of measurement.

EXAMPLE 2 Finding the Distance between Two Points

Find the distance d between the points $(1, 3)$ and $(5, 6)$.

Solution First we plot the points $(1, 3)$ and $(5, 6)$ as shown in Figure 14(a). Then we draw a horizontal line from $(1, 3)$ to $(5, 3)$ and a vertical line from $(5, 3)$ to $(5, 6)$, forming a right triangle, as in Figure 14(b). One leg of the triangle is of length 4 and the other is of length 3. By the Pythagorean Theorem (see the Appendix, Section 1), the square of the distance d we seek is

$$d^2 = 4^2 + 3^2 = 16 + 9 = 25$$
$$d = 5$$

FIGURE 14

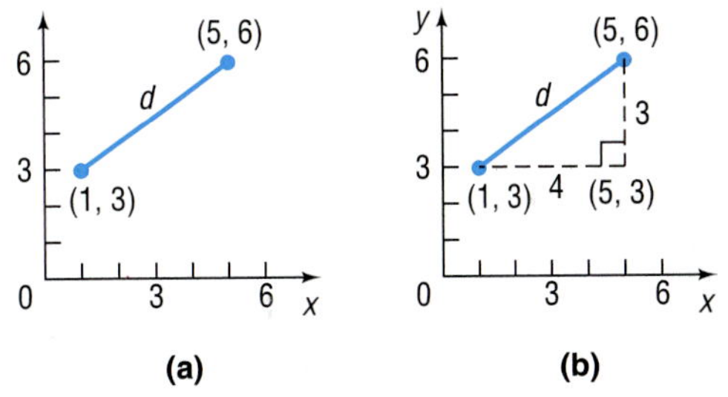

The **distance formula** provides a straightforward method for computing the distance between two points.

Theorem Distance Formula

The distance between two points $P_1 = (x_1, y_1)$ and $P_2 = (x_2, y_2)$, denoted by $d(P_1, P_2)$, is

$$d(P_1, P_2) = \sqrt{(x_2 - x_1)^2 + (y_2 - y_1)^2} \qquad (1)$$

That is, to compute the distance between two points, find the difference of the x-coordinates, square it, and add this to the square of the difference of the y-coordinates. The square root of this sum is the distance. See Figure 15.

FIGURE 15

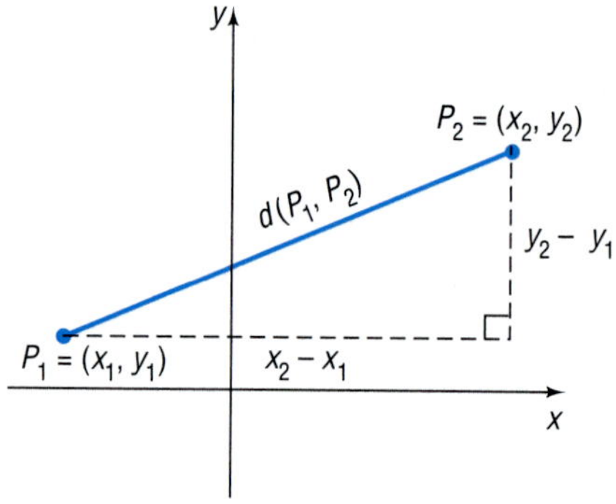

Proof of the Distance Formula Let (x_1, y_1) denote the coordinates of point P_1, and let (x_2, y_2) denote the coordinates of point P_2. Assume that the line joining P_1 and P_2 is neither horizontal nor vertical. Refer to Figure 16(a). The coordinates of P_3 are (x_2, y_1). The horizontal distance from P_1 to P_3 is the absolute value of the difference of the x-coordinates, $|x_2 - x_1|$. The vertical distance from P_3 to P_2 is the absolute value of the difference of the y-coordinates, $|y_2 - y_1|$. See Figure 16(b). The distance $d(P_1, P_2)$ that we seek is the length of the hypotenuse of the right triangle, so, by the Pythagorean Theorem, it follows that

$$\begin{aligned}[d(P_1, P_2)]^2 &= |x_2 - x_1|^2 + |y_2 - y_1|^2 \\ &= (x_2 - x_1)^2 + (y_2 - y_1)^2 \\ d(P_1, P_2) &= \sqrt{(x_2 - x_1)^2 + (y_2 - y_1)^2}\end{aligned}$$

FIGURE 16

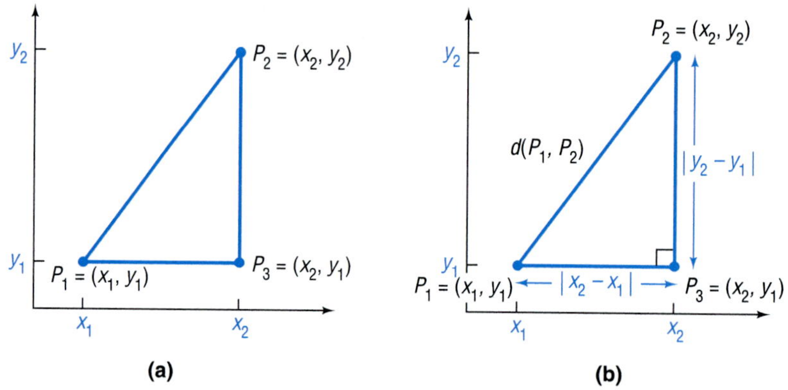

Now, if the line joining P_1 and P_2 is horizontal, then the y-coordinate of P_1 equals the y-coordinate of P_2; that is, $y_1 = y_2$. Refer to Figure 17(a). In this case, the distance formula (1) still works, because, for $y_1 = y_2$, it reduces to

$$d(P_1, P_2) = \sqrt{(x_2 - x_1)^2 + 0^2} = \sqrt{(x_2 - x_1)^2} = |x_2 - x_1|$$

A similar argument holds if the line joining P_1 and P_2 is vertical. See Figure 17(b). Thus, the distance formula is valid in all cases.

FIGURE 17

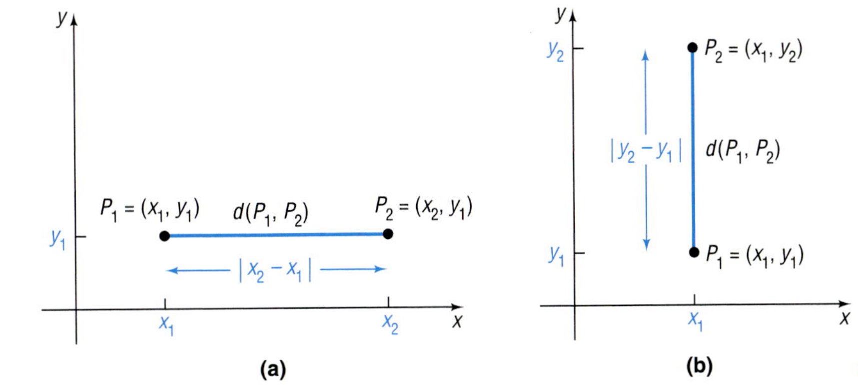

EXAMPLE 3 Finding the Length of a Line Segment

Find the length of the line segment shown in Figure 18.

FIGURE 18

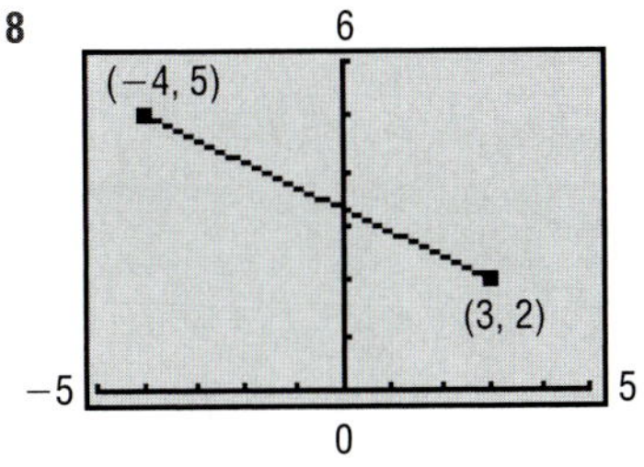

Solution The length of the line segment is the distance between the points $(-4, 5)$ and $(3, 2)$. Using the distance formula (1), the length d is

$$d = \sqrt{[3 - (-4)]^2 + (2 - 5)^2} = \sqrt{7^2 + (-3)^2}$$
$$= \sqrt{49 + 9} = \sqrt{58} \approx 7.62$$

Now work Problem 25.

The distance between two points $P_1 = (x_1, y_1)$ and $P_2 = (x_2, y_2)$ is never a negative number. Furthermore, the distance between two points is 0 only when the points are identical, that is, when $x_1 = x_2$ and $y_1 = y_2$. Also, because $(x_2 - x_1)^2 = (x_1 - x_2)^2$ and $(y_2 - y_1)^2 = (y_1 - y_2)^2$, it makes no difference whether the distance is computed from P_1 to P_2 or from P_2 to P_1; that is, $d(P_1, P_2) = d(P_2, P_1)$.

The introduction to this chapter mentioned that rectangular coordinates enable us to translate geometry problems into algebra problems, and vice versa. The next example shows how algebra (the distance formula) can be used to solve geometry problems.

EXAMPLE 4 Using Algebra to Solve Geometry Problems

Consider the three points $A = (-2, 1)$, $B = (2, 3)$ and $C = (3, 1)$.

(a) Plot each point and form the triangle ABC.
(b) Find the length of each side of the triangle.
(c) Verify that the triangle is a right triangle.
(d) Find the area of the triangle.

Solution (a) Points A, B, C, and triangle ABC are plotted in Figure 19.

FIGURE 19

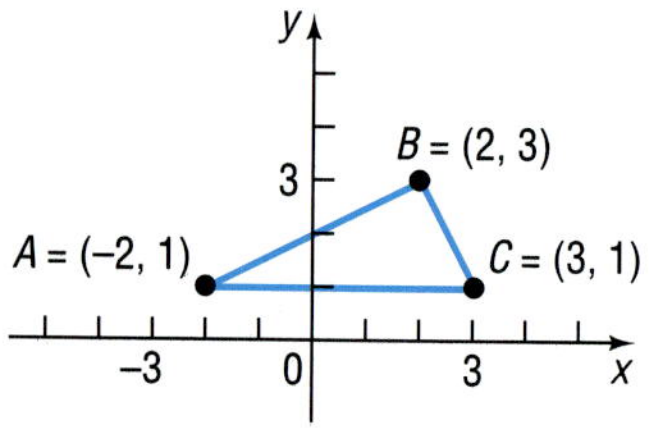

(b) $d(A, B) = \sqrt{[2 - (-2)]^2 + (3 - 1)^2} = \sqrt{16 + 4} = \sqrt{20} = 2\sqrt{5}$

$d(B, C) = \sqrt{(3 - 2)^2 + (1 - 3)^2} = \sqrt{1 + 4} = \sqrt{5}$

$d(A, C) = \sqrt{[3 - (-2)]^2 + (1 - 1)^2} = \sqrt{25 + 0^2} = 5$

(c) To show that the triangle is a right triangle, we need to show that the sum of the squares of the lengths of two of the sides equals the square of the length of the third side. (Why is this sufficient?) Looking at Figure 19, it seems reasonable to conjecture that the right angle is at vertex B. Thus, we shall check to see whether

$$[d(A, B)]^2 + [d(B, C)]^2 = [d(A, C)]^2$$

We find that

$$\begin{aligned}[d(A, B)]^2 + [d(B, C)]^2 &= (2\sqrt{5})^2 + (\sqrt{5})^2 \\ &= 20 + 5 = 25 = [d(A, C)]^2\end{aligned}$$

so it follows from the converse of the Pythagorean Theorem that triangle ABC is a right triangle.

(d) Because the right angle is at B, the sides AB and BC form the base and altitude of the triangle. Its area is therefore

$$\text{Area} = \frac{1}{2}(\text{Base})(\text{Altitude}) = \frac{1}{2}(2\sqrt{5})(\sqrt{5}) = 5 \text{ square units}$$

Now work Problem 43.

Midpoint Formula

FIGURE 20

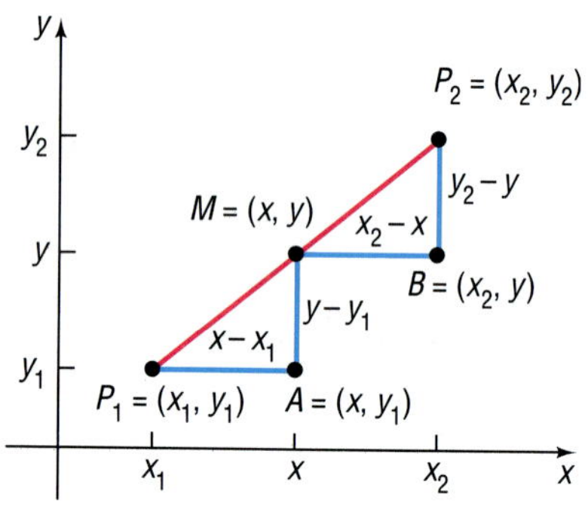

2 We now derive a formula for the coordinates of the **midpoint of a line segment.** Let $P_1 = (x_1, y_1)$ and $P_2 = (x_2, y_2)$ be the endpoints of a line segment, and let $M = (x, y)$ be the point on the line segment that is the same distance from P_1 as it is from P_2. See Figure 20. The triangles P_1AM and MBP_2 are congruent.* [Do you see why? Angle AP_1M = Angle BMP_2,† Angle P_1MA = Angle MP_2B, and $d(P_1, M) = d(M, P_2)$ is given. Thus, we have Angle–Side–Angle.] Hence, corresponding sides are equal in length. That is,

$$\begin{aligned} x - x_1 &= x_2 - x \\ 2x &= x_1 + x_2 \\ x &= \frac{x_1 + x_2}{2} \end{aligned} \qquad \text{and} \qquad \begin{aligned} y - y_1 &= y_2 - y \\ 2y &= y_1 + y_2 \\ y &= \frac{y_1 + y_2}{2} \end{aligned}$$

*The following statement is a postulate from geometry. Two triangles are congruent if their sides are the same length (SSS), or if two sides and the included angle are the same (SAS), or if two angles and the included side are the same (ASA).

†Another postulate from geometry states that the transversal $\overline{P_1P_2}$ forms equal corresponding angles with the parallel lines $\overline{P_1A}$ and $\overline{MB}$.

Theorem Midpoint Formula

The midpoint (x, y) of the line segment from $P_1 = (x_1, y_1)$ to $P_2 = (x_2, y_2)$ is

$$(x, y) = \left(\frac{x_1 + x_2}{2}, \frac{y_1 + y_2}{2}\right) \tag{2}$$

Thus, to find the midpoint of a line segment, we average the x-coordinates and the y-coordinates of the endpoints.

EXAMPLE 5

Finding the Midpoint of a Line Segment

Find the midpoint of a line segment from $P_1 = (-5, 3)$ to $P_2 = (3, 1)$. Plot the points P_1 and P_2 and their midpoint. Check your answer.

Solution We apply the midpoint formula (2) using $x_1 = -5$, $x_2 = 3$, $y_1 = 3$, and $y_2 = 1$. Then the coordinates (x, y) of the midpoint M are

$$x = \frac{x_1 + x_2}{2} = \frac{-5 + 3}{2} = -1 \quad \text{and} \quad y = \frac{y_1 + y_2}{2} = \frac{3 + 1}{2} = 2$$

That is, $M = (-1, 2)$. See Figure 21.

FIGURE 21

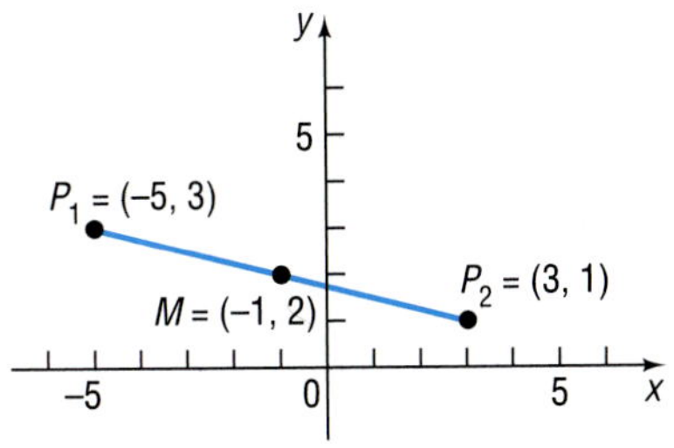

Check: Because M is the midpoint, we check the answer by verifying that $d(P_1, M) = d(M, P_2)$:

$$d(P_1, M) = \sqrt{[-1 - (-5)]^2 + (2 - 3)^2} = \sqrt{16 + 1} = \sqrt{17}$$

$$d(M, P_2) = \sqrt{[3 - (-1)]^2 + (1 - 2)^2} = \sqrt{16 + 1} = \sqrt{17}$$

Now work Problem 53.

Analysis of Data Using Ordered Pairs

3 **Time Series Data** The data analyzed in Section 1.1 was collected at a specific point in time. Often, we are interested in observing trends in data over time in order to determine any pattern that may exist. Data in which we measure a variable at different points in time is called **time series data.** For example, the price of a stock at the end of each year for the past 10 years is time series data. Time series data may be represented graphically using bar charts or line graphs.

EXAMPLE 6

Drawing a Bar Chart and Line Graph from Time Series Data

The data in Table 4 represents the closing price of Walt Disney Co. stock at the end of each month from 1/95 through 12/95.

TABLE 4

Date	Closing Price	Date	Closing Price
1/95	50 7/8	7/95	58 5/8
2/95	53 3/8	8/95	56 1/4
3/95	53 1/2	9/95	57 3/8
4/95	55 3/8	10/95	57 5/8
5/95	55 1/2	11/95	60 1/8
6/95	55 1/2	12/95	58 7/8

Source: New York Stock Exchange

(a) Draw a bar chart of the data in Table 4.
(b) Draw a line graph of the data in Table 4.
(c) Draw a line graph using a graphing utility.
(d) Using either the bar chart or line graph, at the end of which month was Disney stock highest?
(e) Using either the bar chart or line graph, at the end of which month was Disney stock lowest?

Solution (a) We draw a bar chart of the data in Table 4 using the same procedure that we followed when drawing a bar chart for qualitative data. The horizontal axis will be time and the vertical axis will be the closing price. The height of each rectangle equals the closing price at the end of each month. See Figure 22.

FIGURE 22

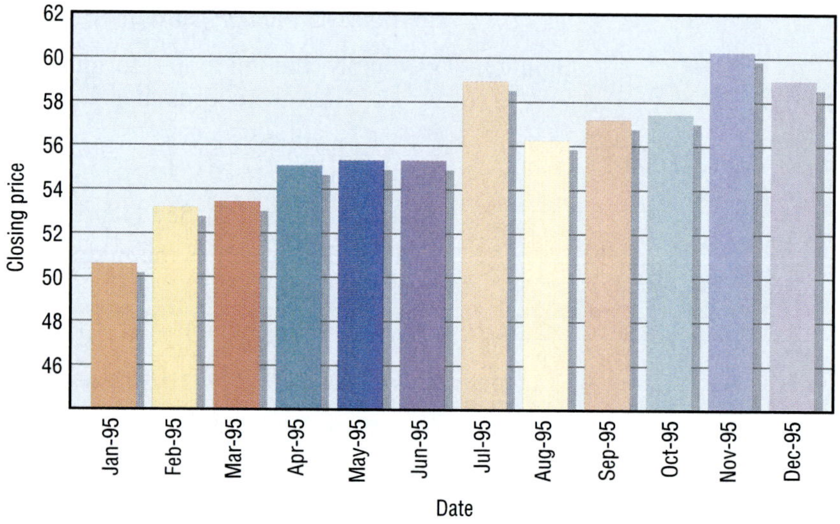

(b) A line graph of the data in Table 4 is obtained by plotting the date as the x-coordinate and the corresponding closing price as the y-coordinate. A line is then drawn connecting consecutive points. See Figure 23.

FIGURE 23

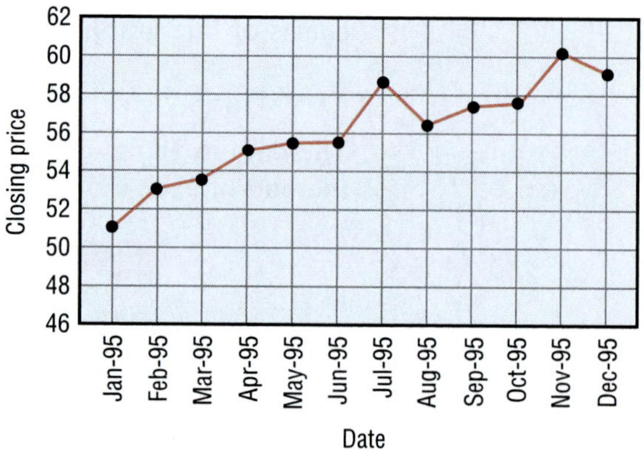

(c) To draw a line graph using a graphing utility, we enter the date numerically by letting 1 represent January, 1995, 2 represent February, 1995,

FIGURE 24

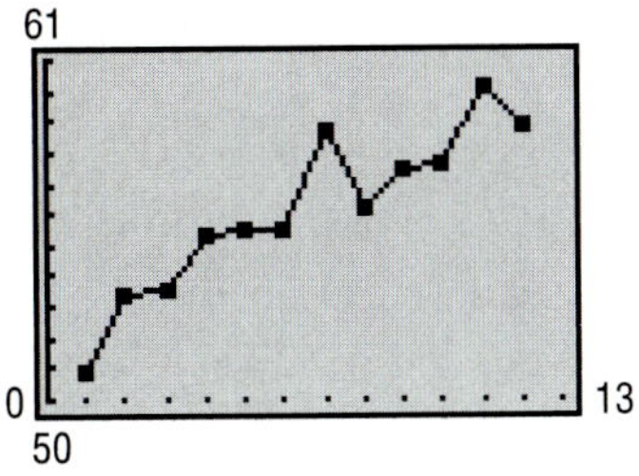

and so on. We then enter the corresponding closing price of Disney stock. Finally, we set the appropriate viewing window by letting $X\text{min} = 0$, $X\text{max} = 13$, $Y\text{min} = 50$, $Y\text{max} = 61$. The graph is shown in Figure 24.

(d) Using the bar chart, Disney stock was highest at the end of November 1995.

(e) Using the line graph, Disney stock was lowest at the end of January 1995.

Now work Problem 65.

4 **Scatter Diagrams** A **relation** is a correspondence between two variables, say, x and y, and can be written as a set of ordered pairs (x, y). When relations are written as ordered pairs, we say x is related to y. Often, we are interested in specifying the type of relation (such as an equation) that might exist between two quantitative variables. The first step in finding this relation is to plot the ordered pairs using rectangular coordinates. The resulting graph is called a **scatter diagram.**

EXAMPLE 7 Drawing a Scatter Diagram

The data in Table 5 represents the total number x of accidents for all U.S. General Aviation Flying and the corresponding total number y of fatalities as a result of the accidents for the years 1984–1993.

TABLE 5

Total Number of Accidents, x	Total Fatalities, y	(x, y)
3016	1042	(3016, 1042)
2738	955	(2738, 955)
2582	967	(2582, 967)
2494	838	(2494, 838)
2386	800	(2386, 800)
2230	768	(2230, 768)
2214	766	(2214, 766)
2170	781	(2170, 781)
2074	862	(2074, 862)
2022	715	(2022, 715)

Source: National Transportation Safety Board

(a) Draw a scatter diagram by hand.

(b) Use a graphing utility to draw a scatter diagram.

(c) Describe what happens to the total number of fatalities as the number of accidents increases.

Solution (a) To draw a scatter diagram by hand, we plot the ordered pairs listed in Table 5, with total number of accidents as the x-coordinate and total number of fatalities as the y-coordinate. See Figure 25(a). Notice the points in a scatter diagram are not connected.

(b) Figure 25(b) shows a scatter diagram of the data using a graphing utility.

FIGURE 25

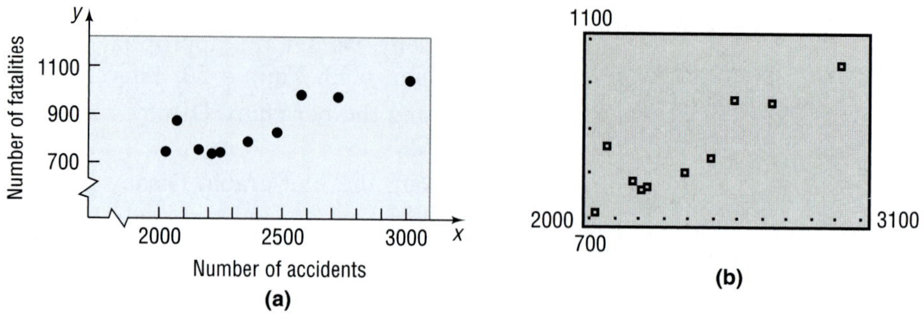

(c) From the scatter diagrams, as the number of accidents increases, we see that the number of fatalities increases. ■

Now work Problem 69.

1.2 EXERCISES

In Problems 1 and 2, plot each point in the xy-plane. Tell in which quadrant or on what coordinate axis each point lies.

1. (a) $A = (-3, 2)$ (b) $B = (6, 0)$ (c) $C = (-2, -2)$
(d) $D = (6, 5)$ (e) $E = (0, -3)$ (f) $F = (6, -3)$

2. (a) $A = (1, 4)$ (b) $B = (-3, -4)$ (c) $C = (-3, 4)$
(d) $D = (4, 1)$ (e) $E = (0, 1)$ (f) $F = (-3, 0)$

3. Plot the points $(2, 0)$, $(2, -3)$, $(2, 4)$, $(2, 1)$, and $(2, -1)$. Describe the set of all points of the form $(2, y)$, where y is a real number.

4. Plot the points $(0, 3)$, $(1, 3)$, $(-2, 3)$, $(5, 3)$ and $(-4, 3)$. Describe the set of all points of the form $(x, 3)$, where x is a real number.

In Problems 5–8, determine the coordinates of the points shown. Tell in which quadrant each point lies.

5.

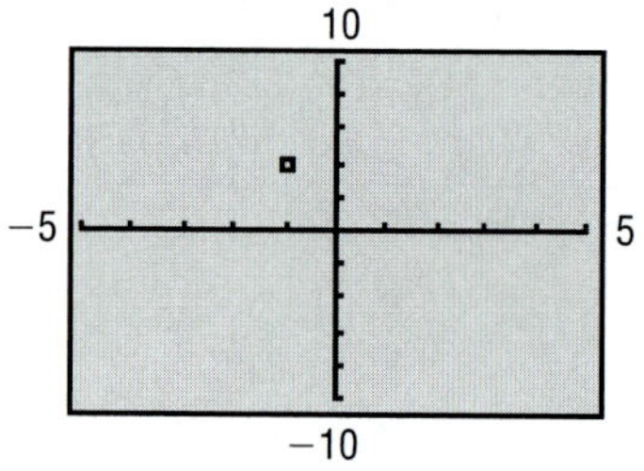

6.

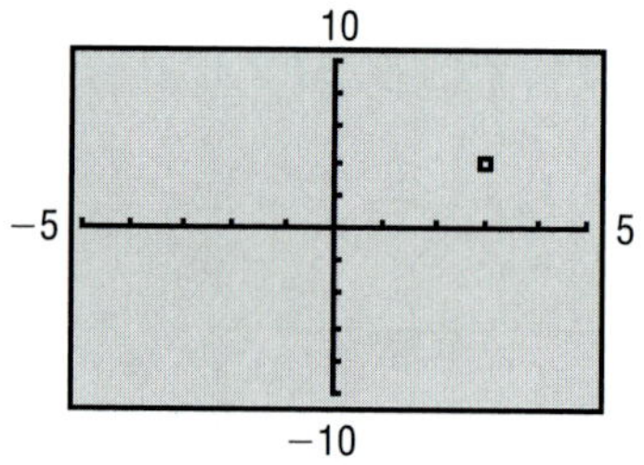

7.

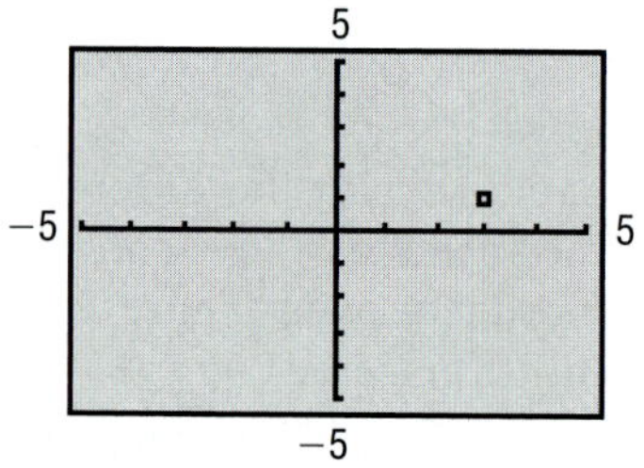

8.

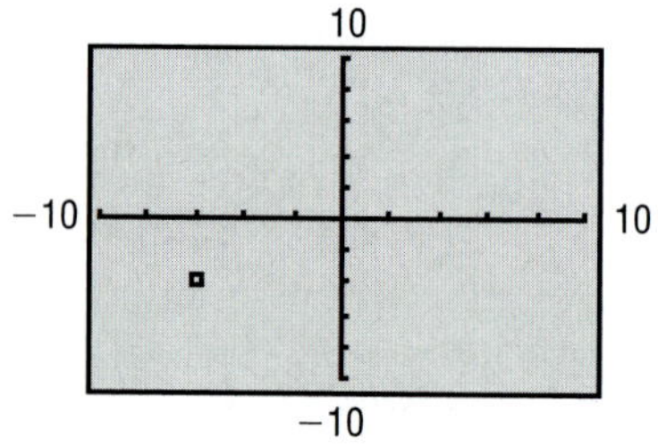

In Problems 9–14, select a RANGE setting so that each of the given points will lie within the viewing rectangle.

9. $(-10, 5), (3, -2), (4, -1)$

10. $(5, 0), (6, 8), (-2, -3)$

11. $(40, 20), (-20, -80), (10, 40)$

12. $(-80, 60), (20, -30), (-20, -40)$

13. $(0, 0), (100, 5), (5, 150)$

14. $(0, -1), (100, 50), (-10, 30)$

In Problems 15–24, determine the RANGE settings used for each viewing rectangle.

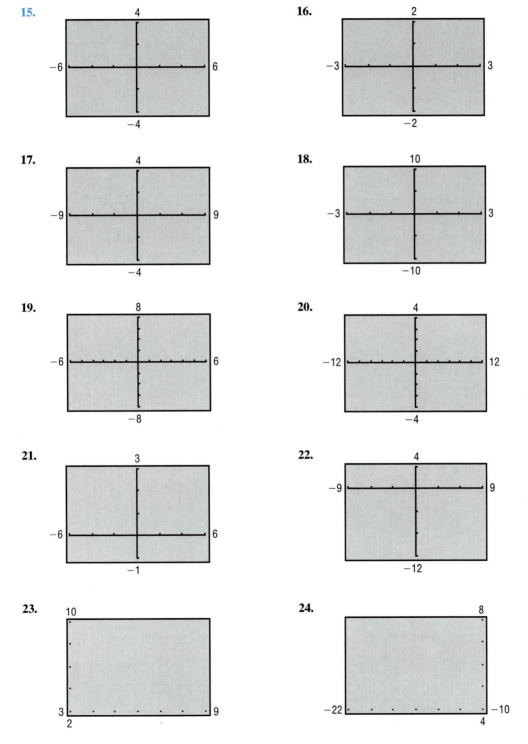

15.

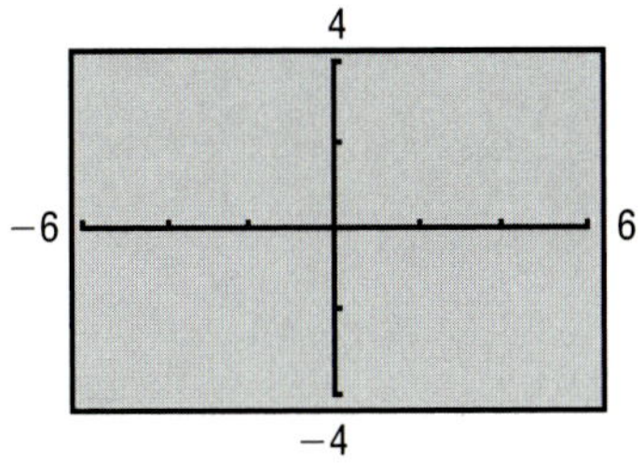

16.

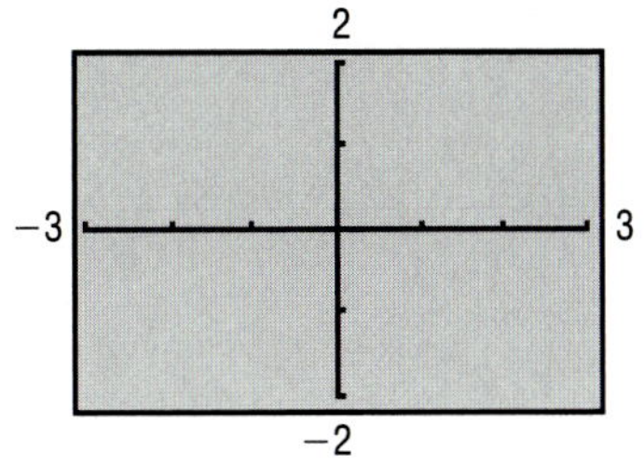

17.

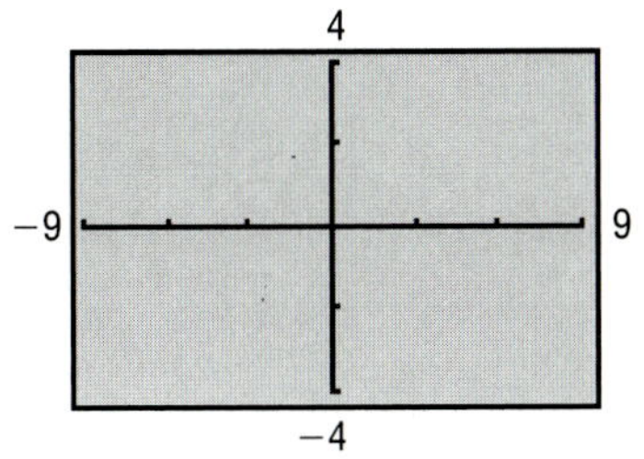

18.

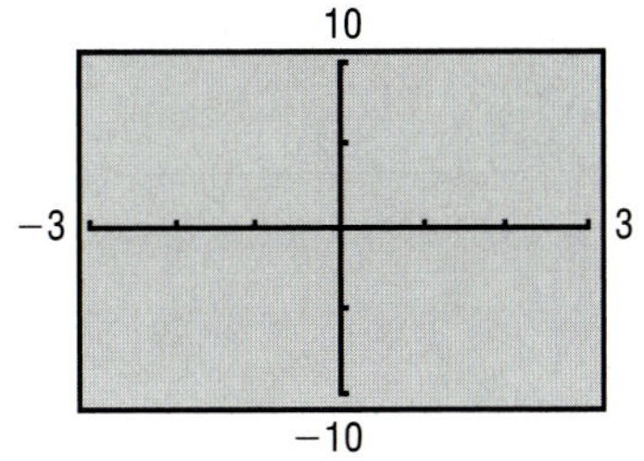

19.

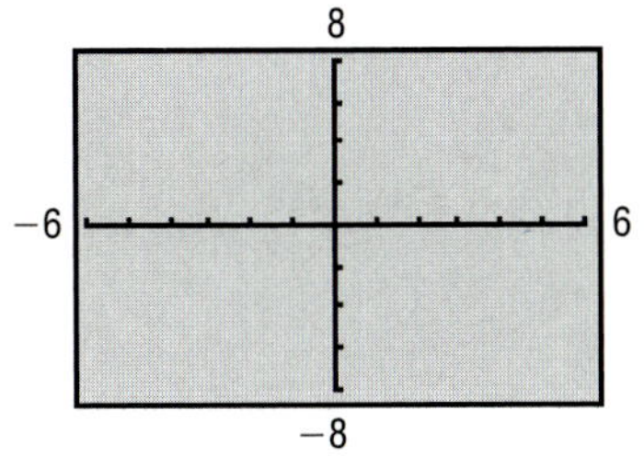

20.

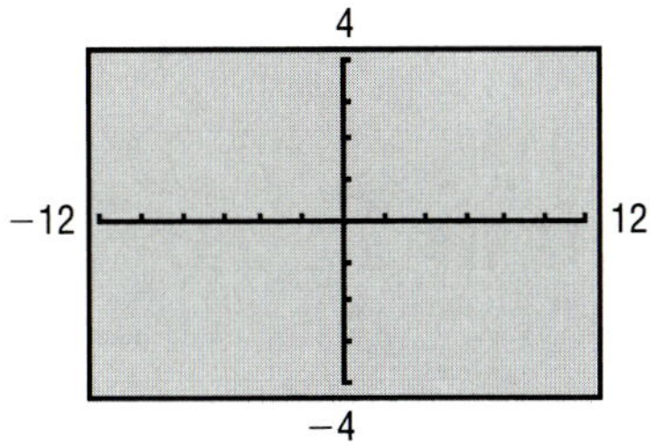

21.

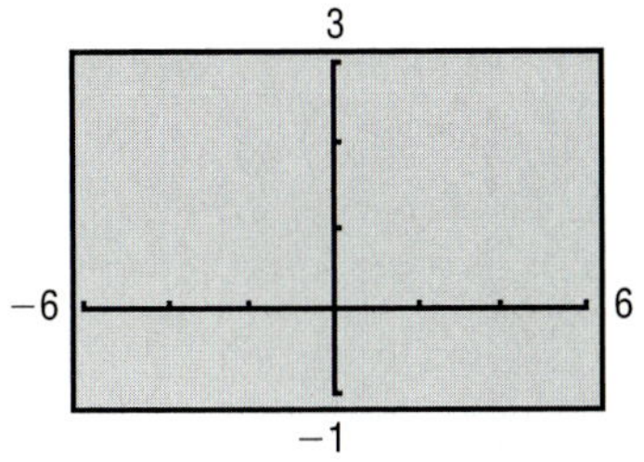

22.

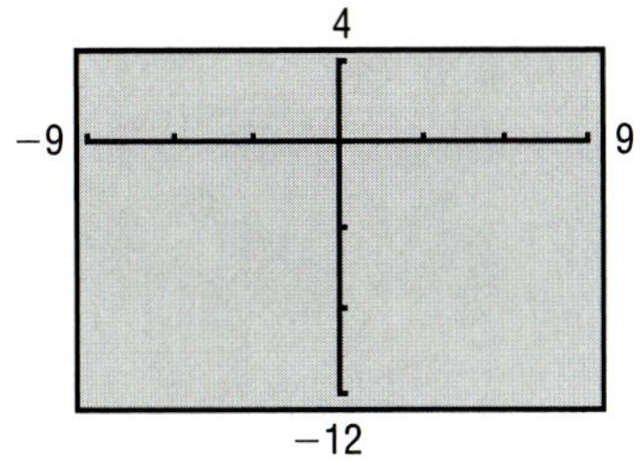

23.

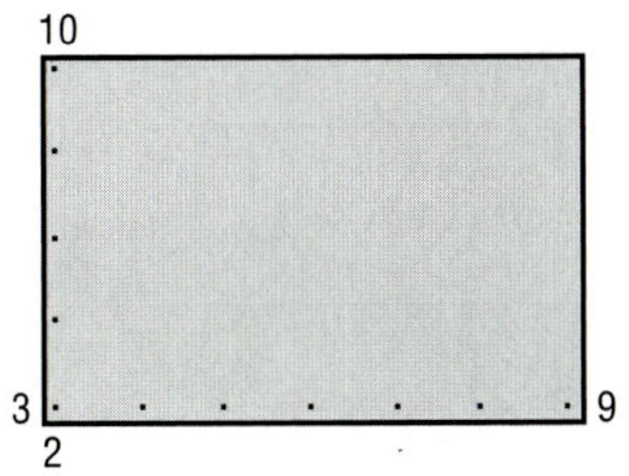

24.

In Problems 25–38, find the distance $d(P_1, P_2)$ between the points P_1 and P_2.

25.

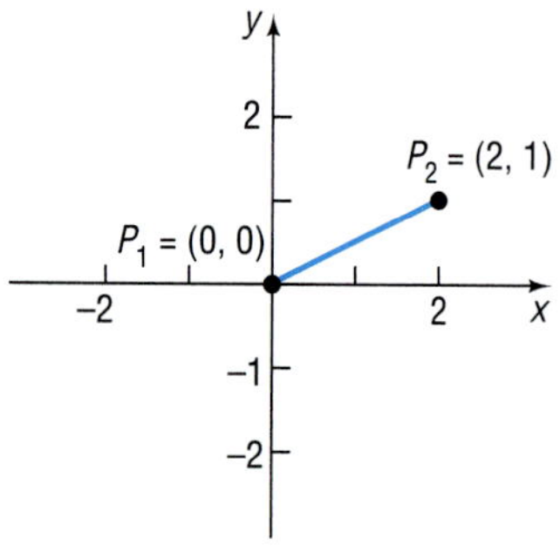

26.

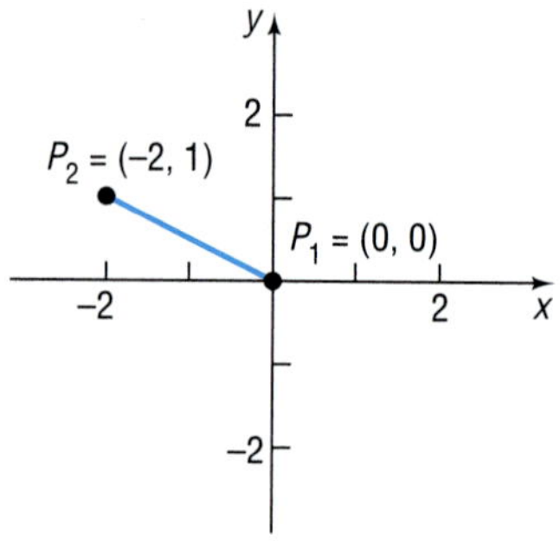

27.

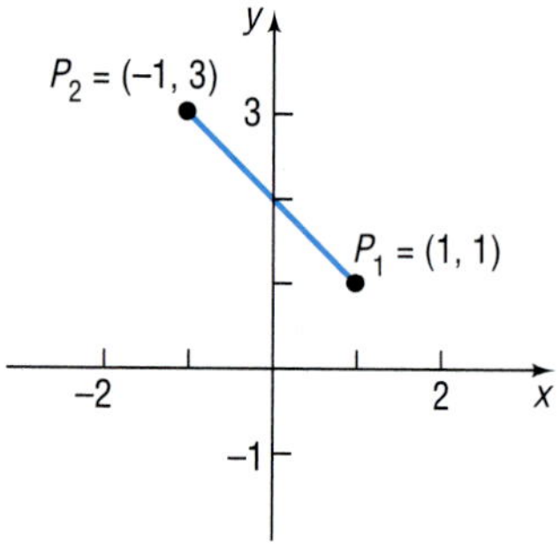

28.

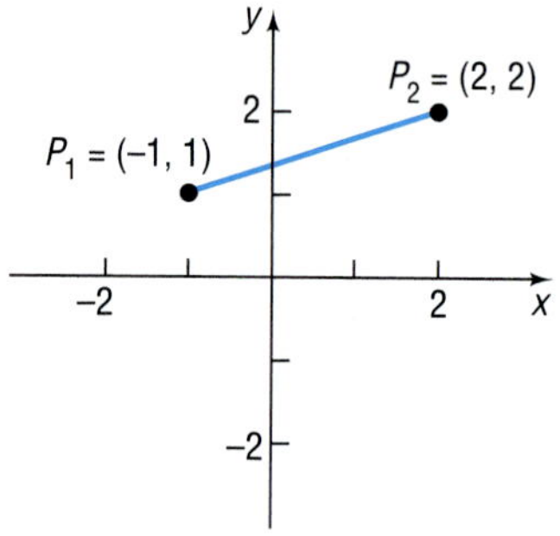

29. $P_1 = (3, -4);\quad P_2 = (5, 4)$

30. $P_1 = (-1, 0);\quad P_2 = (2, 4)$

31. $P_1 = (-3, 2);\quad P_2 = (6, 0)$

32. $P_1 = (2, -3);\quad P_2 = (4, 2)$

33. $P_1 = (4, -3);\quad P_2 = (6, 4)$

34. $P_1 = (-4, -3);\quad P_2 = (6, 2)$

35. $P_1 = (-0.2, 0.3);\quad P_2 = (2.3, 1.1)$

36. $P_1 = (1.2, 2.3);\quad P_2 = (-0.3, 1.1)$

37. $P_1 = (a, b);\quad P_2 = (0, 0)$

38. $P_1 = (a, a);\quad P_2 = (0, 0)$

In Problems 39–42, find the length of the line segment. Assume the end points of each line segment have integer coordinates.

39.

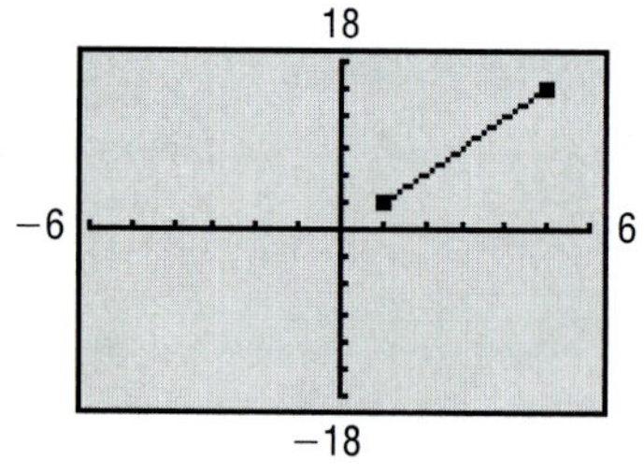

40.

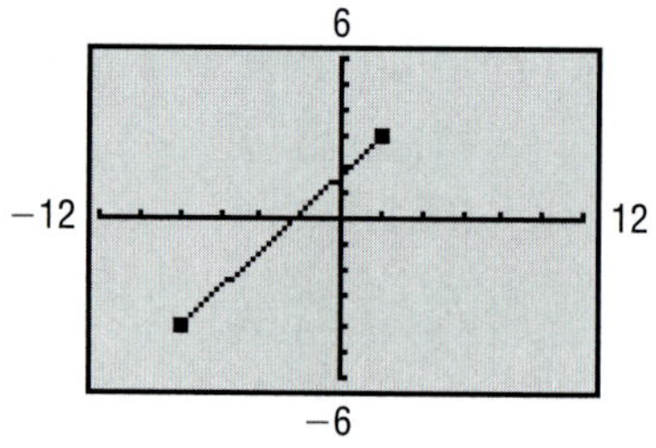

41.

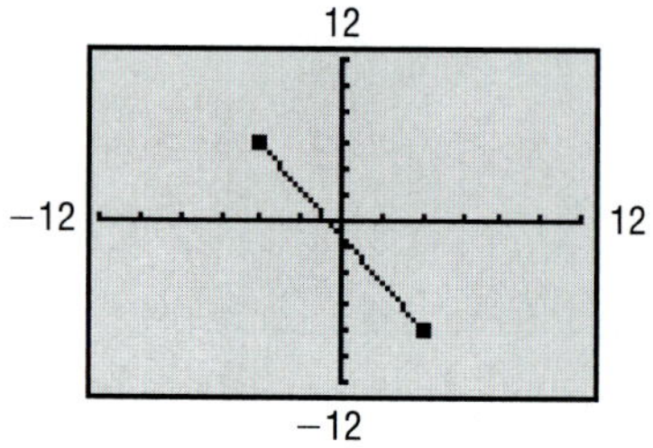

42.

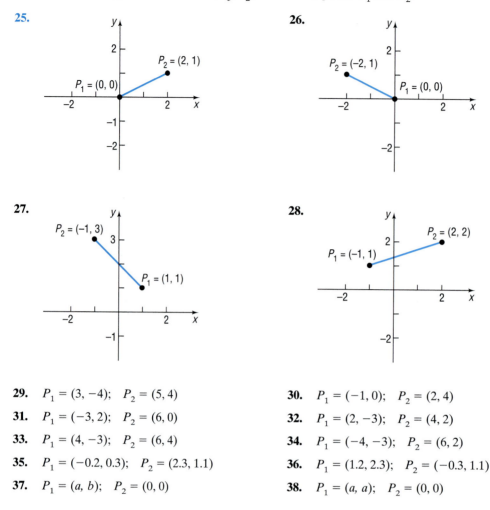

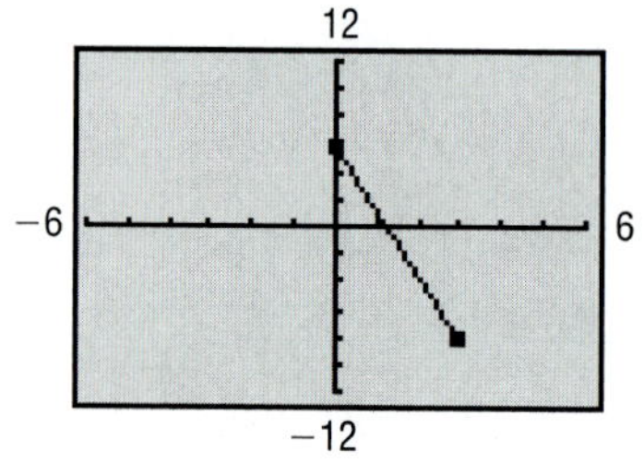

In Problems 43–48, plot each point and form the triangle ABC. Verify that the triangle is a right triangle. Find its area.

43. $A = (-2, 5)$; $B = (1, 3)$; $C = (-1, 0)$

44. $A = (-2, 5)$; $B = (12, 3)$; $C = (10, -11)$

45. $A = (-5, 3)$; $B = (6, 0)$; $C = (5, 5)$

46. $A = (-6, 3)$; $B = (3, -5)$; $C = (-1, 5)$

47. $A = (4, -3)$; $B = (0, -3)$; $C = (4, 2)$

48. $A = (4, -3)$; $B = (4, 1)$; $C = (2, 1)$

49. Find all points having an x-coordinate of 2 whose distance from the point $(-2, -1)$ is 5.

50. Find all points having a y-coordinate of -3 whose distance from the point $(1, 2)$ is 13.

51. Find all points on the x-axis that are 5 units from the point $(4, -3)$.

52. Find all points on the y-axis that are 5 units from the point $(4, 4)$.

In Problems 53–62, find the midpoint of the line segment joining the points P_1 and P_2.

53. $P_1 = (5, -4)$; $P_2 = (3, 2)$

54. $P_1 = (-1, 0)$; $P_2 = (2, 4)$

55. $P_1 = (-3, 2)$; $P_2 = (6, 0)$

56. $P_1 = (2, -3)$; $P_2 = (4, 2)$

57. $P_1 = (4, -3)$; $P_2 = (6, 1)$

58. $P_1 = (-4, -3)$; $P_2 = (2, 2)$

59. $P_1 = (-0.2, 0.3)$; $P_2 = (2.3, 1.1)$

60. $P_1 = (1.2, 2.3)$; $P_2 = (-0.3, 1.1)$

61. $P_1 = (a, b)$; $P_2 = (0, 0)$

62. $P_1 = (a, a)$; $P_2 = (0, 0)$

63. Mortgage Rates The following time series graph represents the 30-year mortgage rates in effect from September 1994 through November 1996, based on a loan origination fee of 3 points (a points equals 1% of the loan amount).

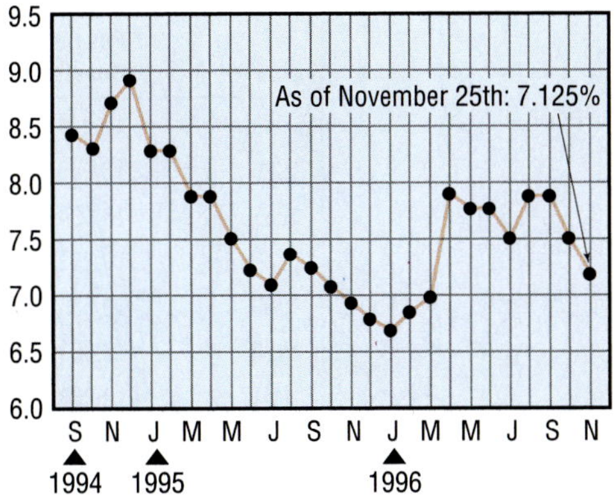

(a) When were 30-year mortgage rates at their highest?
(b) When were 30-year mortgage rates at their lowest?
(c) What were 30-year mortgages in July 1996?

64. Adjustable Rate Mortgages The following time series graph represents the 1-year Adjustable Rate Mortgage (ARM) in effect from September 1994 through November 1996, based on a loan origination fee of 3 points (a point equals 1% of the loan amount).

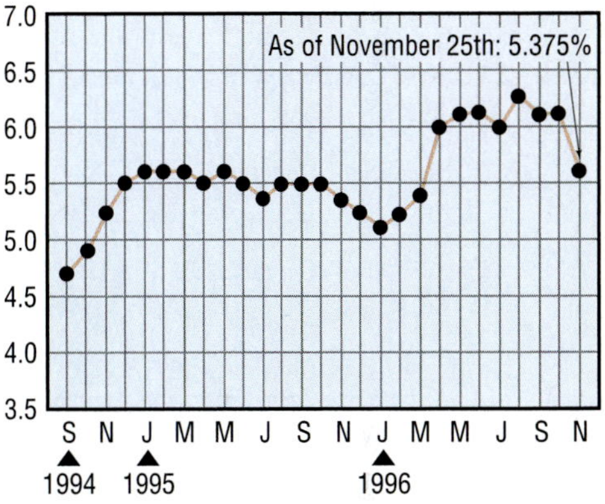

(a) When were 1-year ARMs at their highest?
(b) When were 1-year ARMs at their lowest?
(c) What were 1-year ARMs in July 1996?

65. Price of Motorola Stock The following data represents the closing price of Motorola, Inc. stock at the end of each month in 1995.

Month	Closing Price
January, 1995	$59\frac{1}{4}$
February, 1995	$57\frac{1}{2}$
March, 1995	$54\frac{5}{8}$
April, 1995	$56\frac{7}{8}$
May, 1995	$59\frac{7}{8}$
June, 1995	$67\frac{1}{8}$
July, 1995	$76\frac{1}{2}$
August, 1995	$74\frac{5}{8}$
September, 1995	$76\frac{3}{8}$
October, 1995	$65\frac{3}{4}$
November, 1995	$61\frac{1}{2}$
December, 1995	57

Source: New York Stock Exchange

(a) Draw a bar chart of the data.
(b) By hand, draw a line graph of the data.
(c) Using a graphing utility, draw a line graph of the data.
(d) At the end of which month was Motorola stock highest?
(e) At the end of which month was Motorola stock lowest?

66. Price of Microsoft Corp. Stock The following data represents the closing price of Microsoft Corporation stock at the end of each month in 1995.

Month	Closing Price
January, 1995	$29\frac{11}{16}$
February, 1995	$31\frac{1}{2}$
March, 1995	$35\frac{9}{16}$
April, 1995	$40\frac{7}{8}$
May, 1995	$42\frac{3}{8}$
June, 1995	$45\frac{3}{16}$
July, 1995	$45\frac{1}{4}$
August, 1995	$46\frac{1}{4}$
September, 1995	$45\frac{1}{4}$
October, 1995	50
November, 1995	$43\frac{9}{16}$
December, 1995	$43\frac{7}{8}$

Source: National Association of Security Dealers Automated Quotation

(a) Draw a bar chart of the data.
(b) By hand, draw a line graph of the data.
(c) Using a graphing utility, draw a line graph of the data.
(d) At the end of which month was Microsoft stock the highest?
(e) At the end of which month was Microsoft stock the lowest?

67. Fuel Consumption The following data represents average fuel consumption per automobile for the years 1984–1993.
(a) Draw a bar chart of the data.
(b) By hand, draw a line graph of the data.
(c) Using a graphing utility, draw a line graph of the data.
(d) Determine the year in which fuel consumption was highest.
(e) Determine the year in which fuel consumption was lowest.
(f) What is the trend in the data? In other words, as time passes, what is happening to average fuel consumption?

Year	Average Fuel Consumption
1984	536
1985	525
1986	526
1987	514
1988	509
1989	509
1990	502
1991	496
1992	512
1993	513

Source: U.S. Federal Highway Administration.

68. Gross Domestic Product The following data represents per capita Gross Domestic Product (GDP) of the United States for the years 1985–1994 in constant 1987 dollars (dollars adjusted for inflation). GDP is the total value of all goods and services produced in the United States.

Year	Per Capita Gross Domestic Product
1985	17,944
1986	18,299
1987	18,694
1988	19,252
1989	19,556
1990	19,593
1991	19,263
1992	19,490
1993	19,879
1994	20,476

Source: U.S. Bureau of Economic Analysis

(a) Draw a bar chart of the data.
(b) By hand, draw a line graph of the data.
(c) Using a graphing utility, draw a line graph of the data.
(d) Determine the year in which per capita GDP was highest.
(e) Determine the year in which per capita GDP was lowest.
(f) What is the trend in the data? In other words, as time passes, what is happening to per capita GDP?

69. Disposable Personal Income The data below represents per capita disposable personal income x and per capita personal consumption expenditures y for the years 1985–1994 in current dollars (dollars adjusted for inflation).

Per Capita Disposable Income, x	Per Capita Personal Consumption, y
13,258	12,013
13,552	12,336
13,545	12,568
13,890	12,903
14,005	13,025
14,101	13,063
14,003	12,860
14,279	13,110
14,341	13,361
14,696	13,711

Source: U.S. Bureau of Economic Analysis

(a) By hand, draw a scatter diagram.
(b) Using a graphing utility, draw a scatter diagram.

(c) What happens to per capita personal consumption as per capita disposable income increases?

70. Natural Gas vs. Coal The data below represents the amount of energy provided by natural gas and coal measured in trillions of Btu (British Thermal Units) for New England states. Let the amount of energy provided by natural gas be x and the amount of energy provided by coal be y.

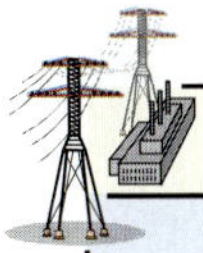

State	Natural Gas, x	Coal, y
Maine	5	22
New Hampshire	17	35
Vermont	8	1
Massachusetts	306	111
Rhode Island	79	0
Connecticut	114	22

Source: U.S. Energy Information Administration

(a) By hand, draw a scatter diagram.
(b) Using a graphing utility, draw a scatter diagram.
(c) What happens to the amount of energy provided by coal as the amount of energy provided by natural gas increases?

71. Fuel Consumption The data below represents the average fuel consumption per car (gallons) and the pollutant emissions of carbon monoxide (thousands of tons) for 1983–1993. Let x represent the average fuel consumption per car and let y represent the pollutant emissions of carbon monoxide.

Average Fuel Consumption Per Car, x	Pollutant Emissions of Carbon Monoxide, y
553	115,334
536	114,262
525	112,072
526	108,070
514	105,117
509	106,100
509	100,806
502	103,753
496	99,898
512	96,368
513	97,208

Source: Statistical Abstract of the United States, 1995.

(a) By hand, draw a scatter diagram.
(b) Using a graphing utility, draw a scatter diagram.
(c) What happens to the level of carbon monoxide as the average fuel consumption per car increases?

72. Hourly Pay and Productivity The data below represents the average hourly earnings of production workers and productivity (output per hour) for the years 1986–1995. Let x represent productivity and let y represent average hourly earnings.

Productivity, x	Average Hourly Earnings, y
94.2	8.76
94.1	8.98
94.6	9.28
95.3	9.66
96.1	10.01
96.7	10.32
100	10.57
100.2	10.83
100.7	11.12
100.8	11.44

Source: Bureau of Labor Statistics

(a) By hand, draw a scatter diagram.
(b) Using a graphing utility, draw a scatter diagram.

(c) What happens to the average hourly earnings as productivity increases?

73. Baseball A major league baseball "diamond" is actually a square, 90 feet on a side (see the figure). What is the distance directly from home plate to second base (the diagonal of the square)?

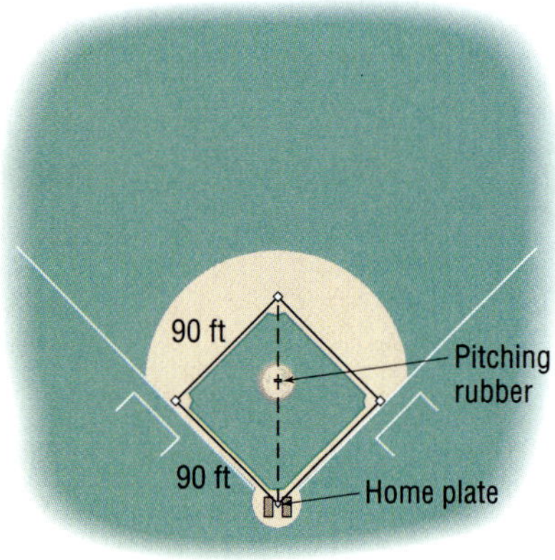

74. Little League Baseball The layout of a Little League playing field is a square, 60 feet on a side.* How far is it directly from home plate to second base (the diagonal of the square)?

75. Baseball Refer to Problem 73. Overlay a rectangular coordinate system on a major league baseball diamond so that the origin is at home plate, the positive x-axis lies in the direction from home plate to first base, and the positive y-axis lies in the direction from home plate to third base.
(a) What are the coordinates of first base, second base, and third base? Use feet as the unit of measurement.
(b) If the right fielder is located at (310, 15), how far is it from there to second base?
(c) If the center fielder is located at (300, 300), how far is it from there to third base?

Source: Little League Baseball, Official Regulations and Playing Rules, 1991.

76. Little League Baseball Refer to Problem 74. Overlay a rectangular coordinate system on a Little League baseball diamond so that the origin is at home plate, the positive x-axis lies in the direction from home plate to first base, and the positive y-axis lies in the direction from home plate to third base.
(a) What are the coordinates of first base, second base, and third base? Use feet as the unit of measurement.
(b) If the right fielder is located at (180, 20), how far is it from there to second base?
(c) If the center fielder is located at (220, 220), how far is it from there to third base?

77. A Dodge Intrepid and a Mack truck leave an intersection at the same time. The Intrepid heads east at an average speed of 30 miles per hour, while the truck heads south at an average speed of 40 miles per hour. Find an expression for their distance apart d (in miles) at the end of t hours.

78. A hot air balloon, headed due east at an average speed of 15 miles per hour and at a constant altitude of 100 feet, passes over an intersection (see the figure). Find an expression for its distance d (measured in feet) from the intersection t seconds later.

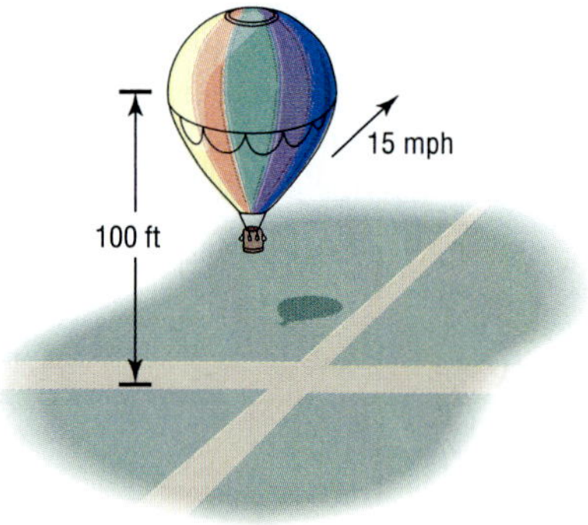

1.3 GRAPHS OF EQUATIONS

1. Graph Equations by Hand and by Using a Graphing Utility
2. Find the Intercepts on a Graph
3. Find the Intercepts from an Equation
4. Test an Equation for Symmetry with Respect to the (a) x-axis (b) y-axis (c) origin

1 Illustrations play an important role in helping us to visualize the relationships that exist between two variable quantities. For example, Figure 26 shows the variation in the price of oil from January to October 1991. Such illustrations are usually referred to as *graphs.* The **graph of an equation** in two variables x and y consists of the set of points in the xy-plane whose coordinates (x, y) satisfy the equation. Let's look at some examples.

FIGURE 26
Oil futures prices

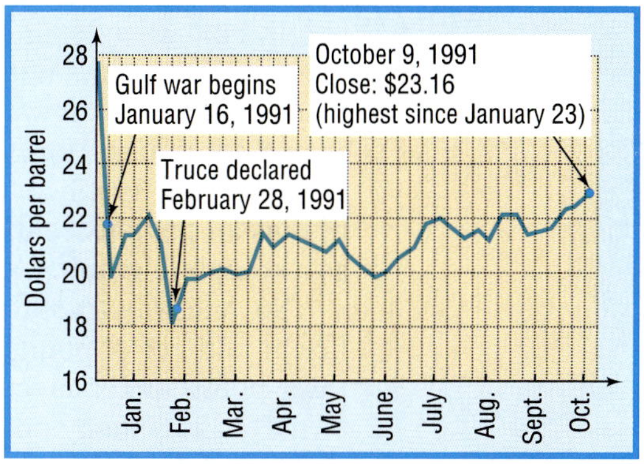

Source: Dow Jones News Retrieval

EXAMPLE 1 Graphing an Equation by Hand

Graph the equation: $y = 2x + 5$.

Solution We want to find all points (x, y) that satisfy the equation. To locate some of these points (and thus get an idea of the pattern of the graph), we assign some numbers to x and find corresponding values for y:

IF	THEN	POINT ON GRAPH
$x = 0$	$y = 2(0) + 5 = 5$	$(0, 5)$
$x = 1$	$y = 2(1) + 5 = 7$	$(1, 7)$
$x = -5$	$y = 2(-5) + 5 = -5$	$(-5, -5)$
$x = 10$	$y = 2(10) + 5 = 25$	$(10, 25)$

FIGURE 27
$y = 2x + 5$

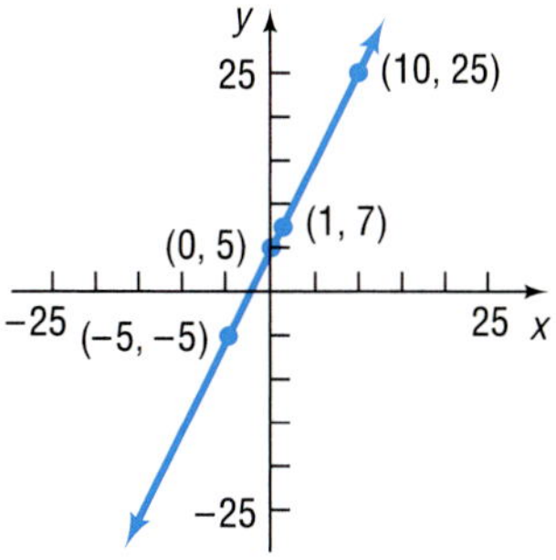

By plotting these points and then connecting them, we obtain the graph of the equation (a straight line), as shown in Figure 27.

EXAMPLE 2 Graphing an Equation by Hand

Graph the equation: $y = x^2$

Solution Table 6 provides several points on the graph. In Figure 28 we plot these points and connect them with a smooth curve to obtain the graph (a *parabola*).

TABLE 6

x	$y = x^2$	(x, y)
−4	16	(−4, 16)
−3	9	(−3, 9)
−2	4	(−2, 4)
−1	1	(−1, 1)
0	0	(0, 0)
1	1	(1, 1)
2	4	(2, 4)
3	9	(3, 9)
4	16	(4, 16)

FIGURE 28
$y = x^2$

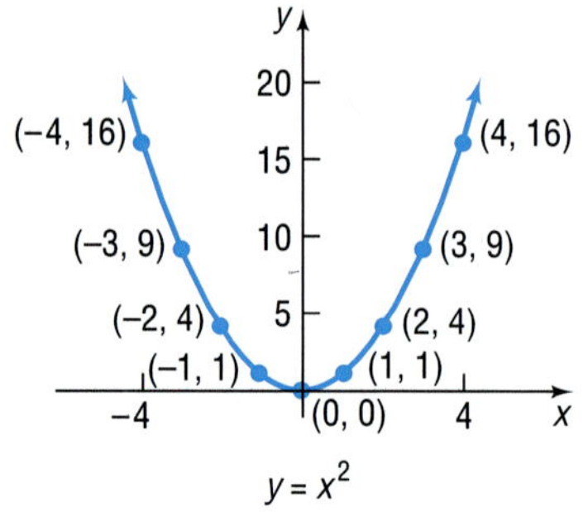

The graphs of the equations shown in Figures 27 and 28 do not show all points. For example, in Figure 27, the point (20, 45) is a part of the graph of $y = 2x + 5$, but it is not shown. Since the graph of $y = 2x + 5$ could be extended out as far as we please, we use arrows to indicate that the pattern shown continues. Thus, it is important when illustrating a graph to present enough of the graph so that any viewer of the illustration will "see" the rest of it as an obvious continuation of what is actually there. This is referred to as a **complete graph.**

So, one way to obtain a complete graph of an equation is to plot a sufficient number of points on the graph until a pattern becomes evident. Then these points are connected with a smooth curve following the suggested pattern. But how many points are sufficient? Sometimes knowledge about the equation tells us. For example, we will learn in the next section that, if an equation is of the form $y = mx + b$, then its graph is a straight line. In this case, two points would suffice to obtain the graph.

One of the purposes of this book is to investigate the properties of equations in order to decide whether a graph is complete. Sometimes we shall graph equations by hand by plotting a sufficient number of points on the graph until a pattern becomes evident; then we connect these points with a smooth curve following the suggested pattern. Shortly, we shall investigate various techniques that will enable us to graph an equation by hand without plotting so many points. Other times, we shall graph equations using a graphing utility.

Using a Graphing Utility to Graph Equations

From Examples 1 and 2, we see that a graph can be obtained by plotting points in a rectangle coordinate system and connecting them. Graphing utilities perform these same steps when graphing an equation. For example, the TI-82 determines 95 evenly spaced input values,* uses the equation to determine the output values, plots these points on the screen and finally, (if in the connected mode), draws a line between consecutive points.

*These input values depend on the values of Xmin and Xmax. For example, if $X\text{min} = -10$ and $X\text{max} = 10$, then the first input value will be -10 and the next input value will be $-10 + (10 - (-10))/94 = -9.7872$, and so on.

Most graphing utilities require the following steps in order to obtain the graph of an equation:

Steps for Graphing an Equation Using a Graphing Utility

STEP 1: Solve the equation for y in terms of x.

STEP 2: Get into the graphing mode of your graphing utility. The screen will usually display $y =$, prompting you to enter the expression involving x that you found in Step 1. (Consult your manual for the correct way to enter the expression; for example, $y = x^2$ might be entered as x^2 or as x*x or as $x\ x^Y\ 2$).

STEP 3: Select the viewing rectangle. Without prior knowledge about the behavior of the graph of the equation, it is common to select the **standard viewing rectangle** initially. The viewing rectangle is then adjusted based on the graph that appears. In this text, the standard viewing rectangle will be

$X\text{min} = -10$
$X\text{max} = 10$
$X\text{scl} = 1$
$Y\text{min} = -10$
$Y\text{max} = 10$
$Y\text{scl} = 1$

STEP 4: Execute.

STEP 5: Adjust the viewing rectangle until a complete graph is obtained.

EXAMPLE 3 Graphing an Equation on a Graphing Utility

Graph the equation $6x^2 + 3y = 36$.

Solution STEP 1: We solve for y in terms of x.

$$6x^2 + 3y = 36$$
$$3y = -6x^2 + 36 \quad \text{Subtract } 6x^2 \text{ from both sides of the equation.}$$
$$y = -2x^2 + 12 \quad \text{Divide both sides of the equation by 3.}$$

STEP 2: From the graphing mode, enter the expression $-2x^2 + 12$ after the prompt $y =$.

STEP 3: Set the viewing rectangle to the standard viewing rectangle.

STEP 4: Execute. The screen should look like Figure 29.

FIGURE 29

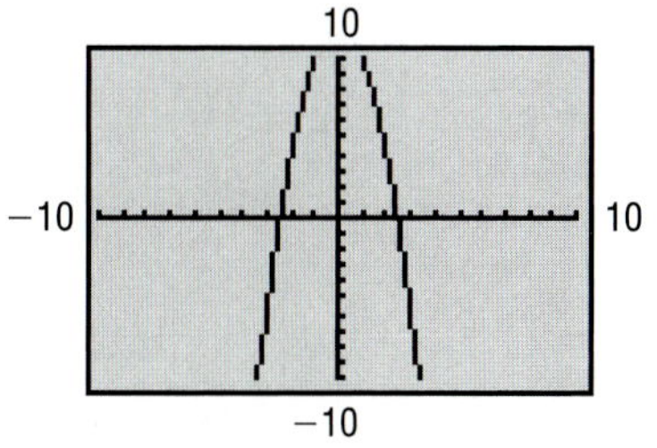

STEP 5: The graph of $y = -2x^2 + 12$ is not complete. The value of Ymax must be increased so the top portion of the graph is visible. After increasing the value of Ymax to 12 we obtain the graph in Figure 30. The graph is now complete.

FIGURE 30

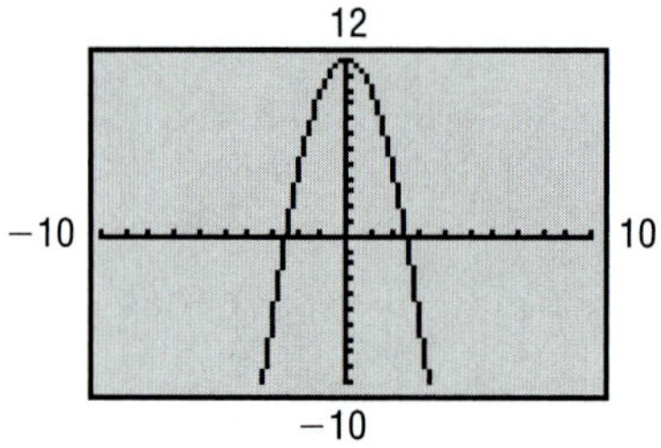

Look again at Figure 30. Although a complete graph is shown, the graph might be improved by adjusting the values of Xmin and Xmax. Figure 31 shows the graph of $y = -2x^2 + 12$ using $X\text{min} = -4$ and $X\text{max} = 4$. Do you think this is a better choice for the viewing rectangle?

FIGURE 31

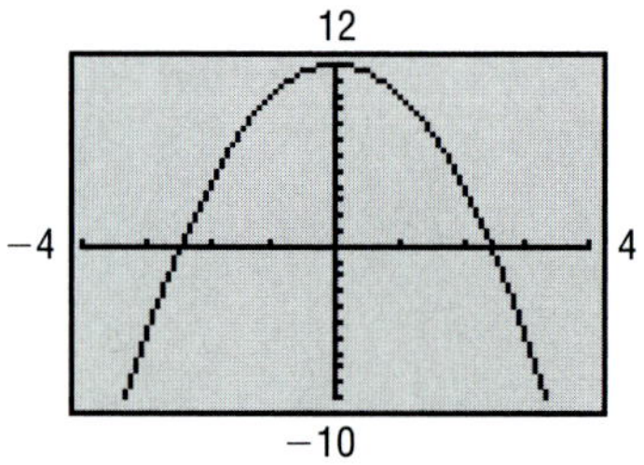

Now work Problems 11(a) and 11(b).

EXAMPLE 4

Graphing an Equation

Graph the equation $y = x^3$.

Solution Using a graphing utility, we obtain Figure 32.

FIGURE 32

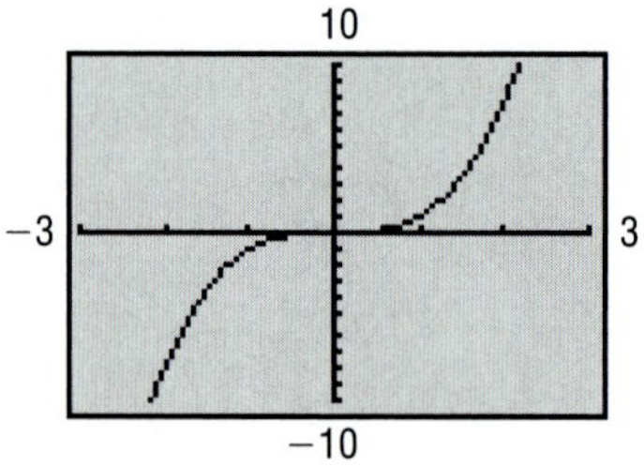

To graph by hand, we use the equation to obtain several points on the graph (Table 7) and then draw the graph on paper. See Figure 33.

TABLE 7

x	$y = x^3$	(x, y)
−3	−27	(−3, −27)
−2	−8	(−2, −8)
−1	−1	(−1, −1)
0	0	(0, 0)
1	1	(1, 1)
2	8	(2, 8)
3	27	(3, 27)

FIGURE 33

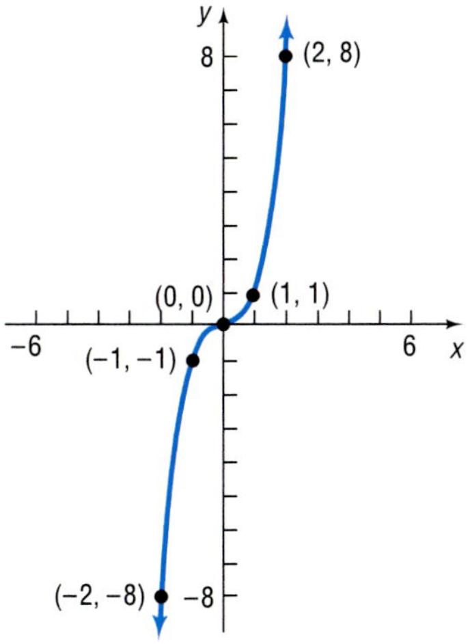

Table 7 can also be obtained using most graphing utilities. (Check your manual to see if your graphing utility has this capability.) On a TI-83, Table 7 takes the form shown in Table 8.

TABLE 8

X	Y1	
-3	-27	
-2	-8	
-1	-1	
0	0	
1	1	
2	8	
3	27	

Y1=X^3

EXAMPLE 5 Graphing an Equation

Graph the equation $x = y^2$.

Solution To graph an equation using a graphing utility, we must write the equation in the form y = expression involving x. So to graph $x = y^2$, we must solve for y:

$$x = y^2$$
$$y^2 = x$$
$$y = \pm\sqrt{x}$$

Thus, to graph $x = y^2$, we need to graph both $y = \sqrt{x}$ and $y = -\sqrt{x}$ on the same screen. Figure 34 shows the result. Table 9 shows various values of y for a given value of x when $Y_1 = \sqrt{x}$ and $Y_2 = -\sqrt{x}$. Notice when $x < 0$, we get an error. Can you explain why?

TABLE 9

X	Y1	Y2
-1	ERROR	ERROR
0	0	0
1	1	-1
2	1.4142	-1.414
3	1.7321	-1.732
4	2	-2
5	2.2361	-2.236

Y1=√(X)

FIGURE 34

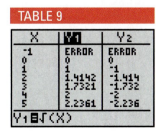

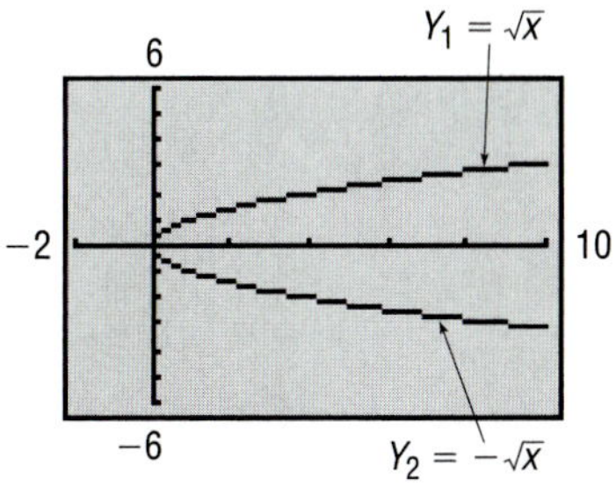

To graph by hand, we use the equation to obtain several points on the graph (Table 10) and then draw the graph on paper. See Figure 35.

TABLE 10

y	$x = y^2$	(x, y)
-3	9	$(9, -3)$
-2	4	$(4, -2)$
-1	1	$(1, -1)$
0	0	$(0, 0)$
1	1	$(1, 1)$
2	4	$(4, 2)$
3	9	$(9, 3)$
4	16	$(16, 4)$

FIGURE 35

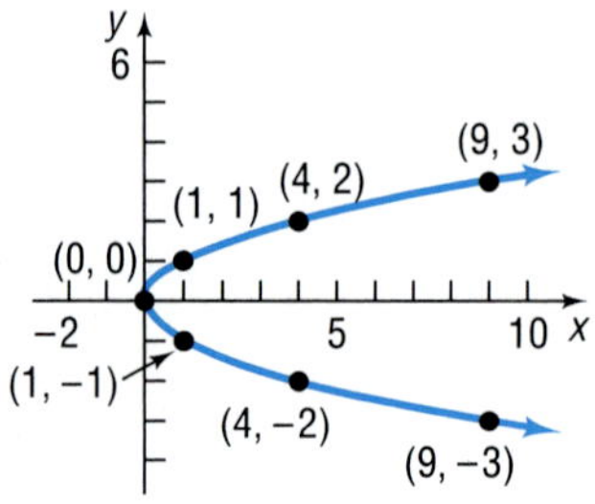

Look again at either Figure 34 or Figure 35. If we restrict y so that $y \geq 0$, the equation $x = y^2, y \geq 0$ may be written equivalently as $y = \sqrt{x}$. The portion of the graph of $x = y^2$ in quadrant I is therefore the graph of $y = \sqrt{x}$. See Figure 36 and Figure 37.

FIGURE 36

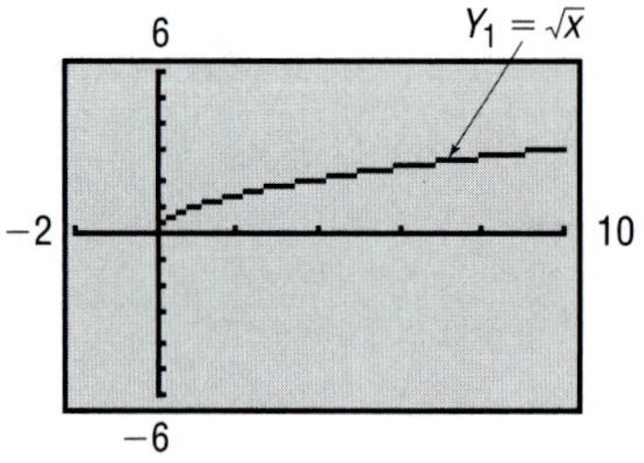

FIGURE 37

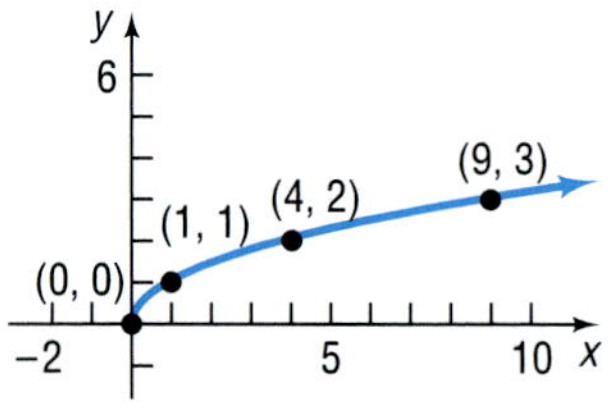

Now work Problem 47.

We said earlier that we would discuss techniques that reduce the number of points required to graph an equation by hand. Two such techniques involve finding *intercepts* and checking for *symmetry.*

Intercepts

2 The points, if any, at which a graph crosses or touches the coordinate axes are called the **intercepts.** See Figure 38. The x-coordinate of a point at which the graph crosses or touches the x-axis is an ***x*-intercept,** and the y-coordinate of a point at which the graph crosses or touches the y-axis is a ***y*-intercept.**

FIGURE 38

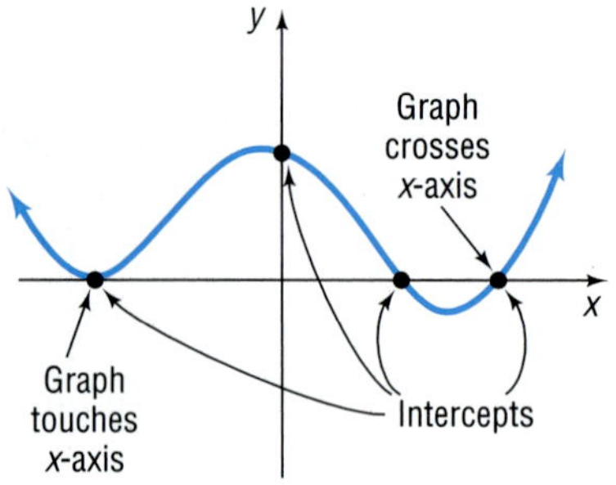

EXAMPLE 6 Finding Intercepts on a Graph

Find the intercepts of the graph in Figure 39. What are its x-intercepts? What are its y-intercepts?

FIGURE 39

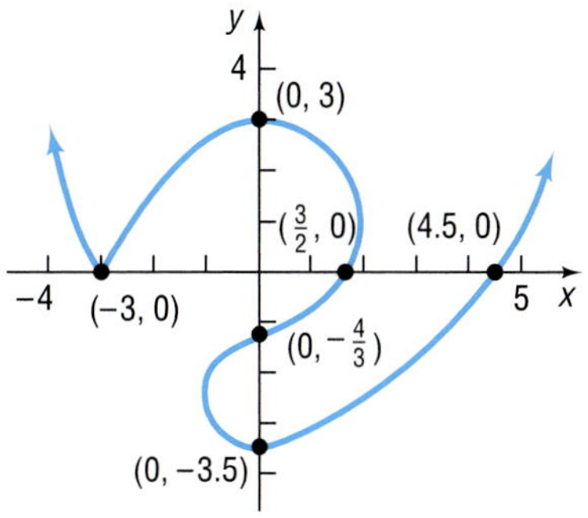

Solution The intercepts of the graph are the points

$$(-3, 0), \quad (0, 3), \quad \left(\frac{3}{2}, 0\right), \quad \left(0, -\frac{4}{3}\right), \quad (0, -3.5), \quad (4.5, 0)$$

The x-intercepts are $-3, \frac{3}{2}, 4.5$; the y-intercepts are $-3.5, -\frac{4}{3}, 3$.

3 The intercepts of the graph of an equation can be found by using the fact that points on the x-axis have y-coordinates equal to 0, and points on the y-axis have x-coordinates equal to 0.

Procedure for Finding Intercepts

1. To find the x-intercept(s), if any, of the graph of an equation, let $y = 0$ in the equation and solve for x.
2. To find the y-intercept(s), if any, of the graph of an equation, let $x = 0$ in the equation and solve for y.

EXAMPLE 7 Finding Intercepts from an Equation

Find the x-intercept(s) and the y-intercept(s) of the graph of $y = x^2 - 4$.

Solution To find the x-intercept(s), we let $y = 0$ and obtain the equation

$$x^2 - 4 = 0$$

The equation has two solutions, -2 and 2. Thus, the x-intercepts are -2 and 2.

To find the y-intercept(s), we let $x = 0$ and obtain the equation

$$y = -4$$

Thus, the y-intercept is -4.

FIGURE 40

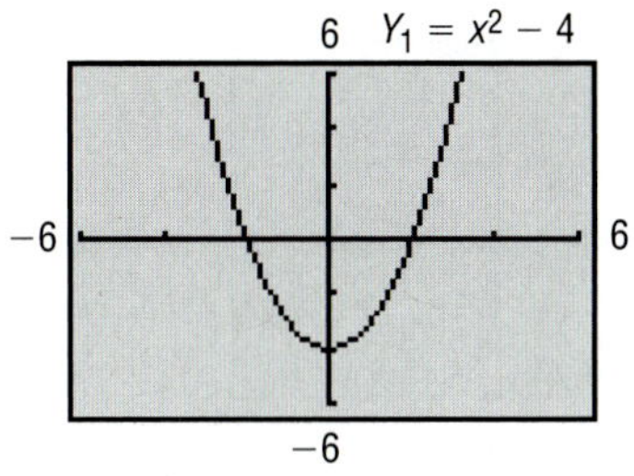

To graph $y = x^2 - 4$ using a graphing utility, we make the following observation.

Since $x^2 \geq 0$ for all x, we deduce from the equation $y = x^2 - 4$ that $y \geq -4$ for all x. This information, along with the intercepts, can now be used to set the viewing rectangle so a complete graph is obtained. See Figure 40.

To draw the graph by hand, we use the equation to obtain some additional points on the graph. See Table 11 and Figure 41.

TABLE 11

X	Y_1
-3	5
-2	0
-1	-3
0	-4
1	-3
2	0
3	5

$Y_1 = X^2 - 4$

FIGURE 41
$y = x^2 - 4$

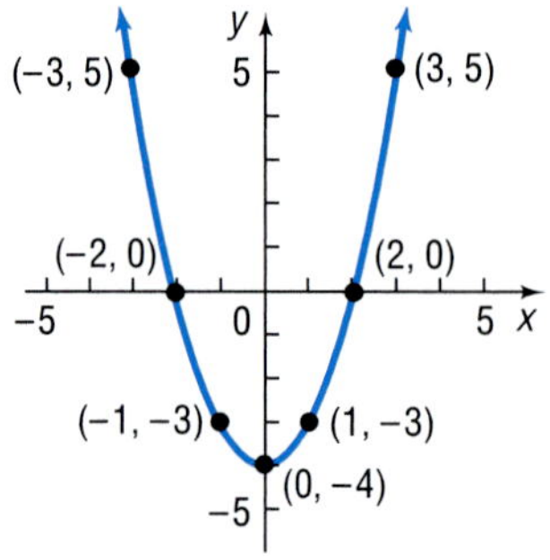

Now work Problem 31(a).

Using a Graphing Utility to Locate Intercepts

Four tools that can be used to locate intercepts using a graphing utility are TRACE, BOX, VALUE, and ROOT (or ZERO).

Most graphing utilities allow you to move from point to point along the graph, displaying on the screen the coordinates of each point. This feature is called the TRACE function.

EXAMPLE 8 Using the TRACE Function to Locate Intercepts

Graph the equation $y = x^3 - 8$. Use the TRACE function to locate the intercepts.

Solution Figure 42 shows the graph of $y = x^3 - 8$.

FIGURE 42

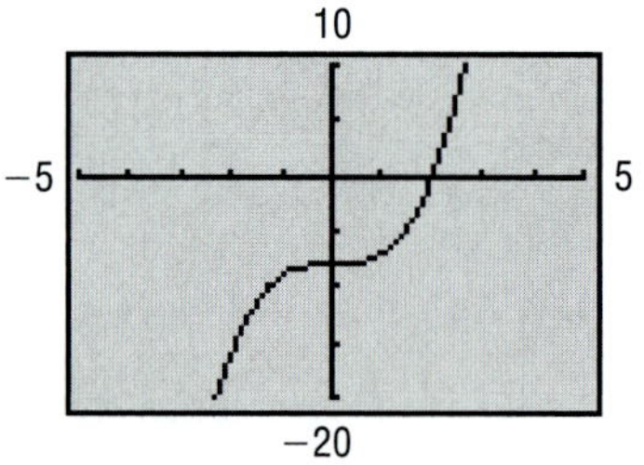

Activate the TRACE function. As you move the cursor along the graph, you will see the coordinates of each point displayed. Just before you get to the x-axis, the display will look like the one in Figure 43(a). (Due to differences in graphing utilities, your display may be slightly different than the one shown here).

FIGURE 43

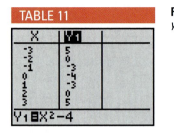

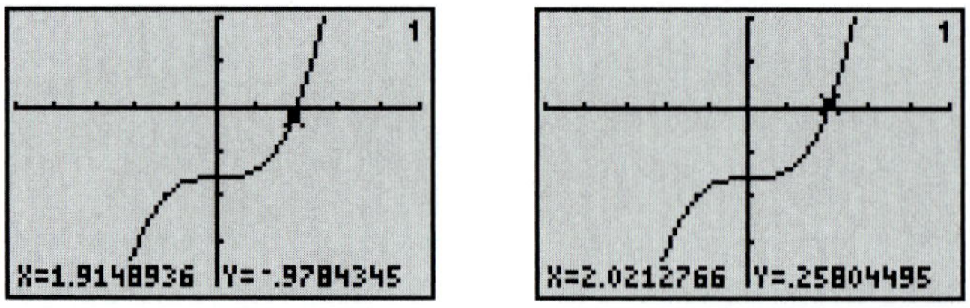

(a) (b)

In Figure 43(a), the negative value of the y-coordinate indicates we are still below the x-axis. The next position of the cursor is shown in Figure 43(b).

The positive value of the y-coordinate indicates we are now above the x-axis. This means that between these two points the x-axis was crossed. The x-intercept lies between 1.9148936 and 2.0212766.

Using the TRACE function we find the y-intercept is -8. See Figure 44.

FIGURE 44

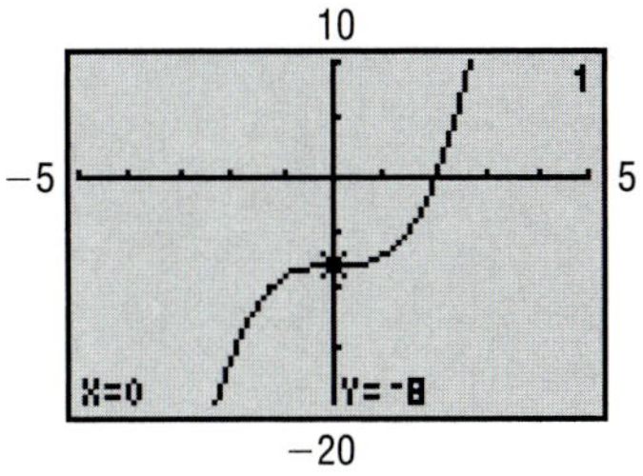

Most graphing utilities have a BOX function that allows you to box in a specific part of the graph of an equation.

EXAMPLE 9 Using the BOX Function

Graph the equation $y = x^3 - 8$ and use the BOX function to improve on the approximation found in Example 8 for the x-intercept.

Solution Using the viewing rectangle of Figure 42, graph the equation. Activate the BOX function. (With some graphing utilities, this requires positioning the cursor at one corner of the box and then tracing out the sides of the box to the diagonal corner.) See Figure 45. Once executed, the box becomes the viewing rectangle. The result is shown in Figure 46.

FIGURE 45 **FIGURE 46**

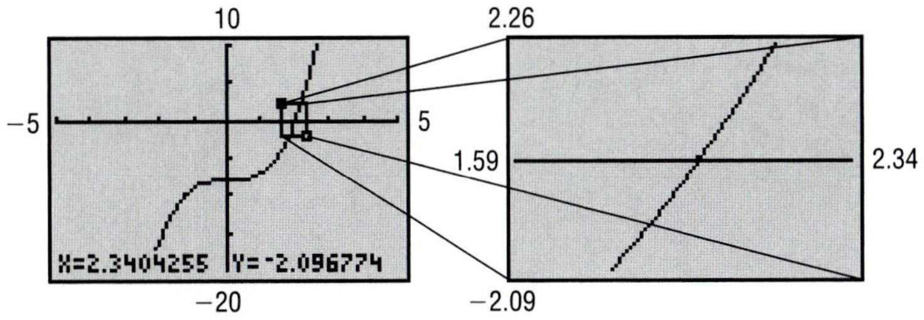

Now you can TRACE to get the approximations shown in Figure 47.

FIGURE 47

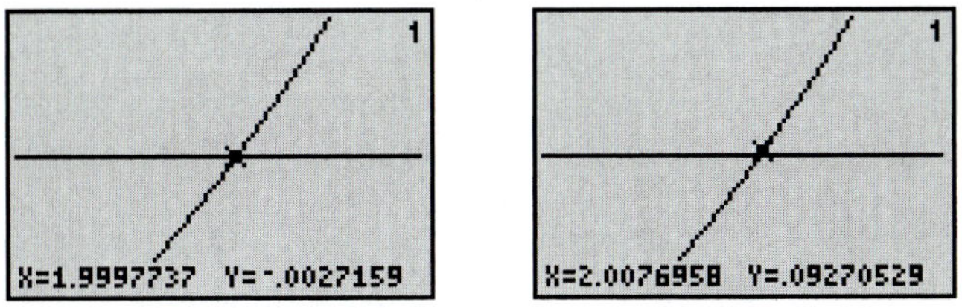

Now we know the x-intercept lies between 1.9997737 and 2.0076958.

Most graphing utilities have an eVALUEate function that, given a value of x determines the value of y for an equation. We can use this function to evaluate an equation at $x = 0$ to determine the y-intercept. Using the VALUE function, we find the y-intercept of $y = x^3 - 8$ is -8.

Most graphing utilities also have a ROOT (or ZERO) function that can be used to determine the x-intercept of an equation. Using the ROOT function on a TI-82, we find the x-intercept of $y = x^3 - 8$ is 2. Consult your owner's manual to determine the appropriate keystrokes for these functions.

Symmetry

We have just seen the role intercepts play in obtaining key points on the graph of an equation. Another helpful tool for graphing equations involves *symmetry,* particularly symmetry with respect to the x-axis, the y-axis, and the origin.

A graph is said to be **symmetric with respect to the *x*-axis** if, for every point (x, y) on the graph, the point $(x, -y)$ is also on the graph.

Figure 48 illustrates the definition. Notice that when a graph is symmetric with respect to the x-axis the part of the graph above the x-axis is a reflection or mirror image of the part below it, and vice versa.

FIGURE 48

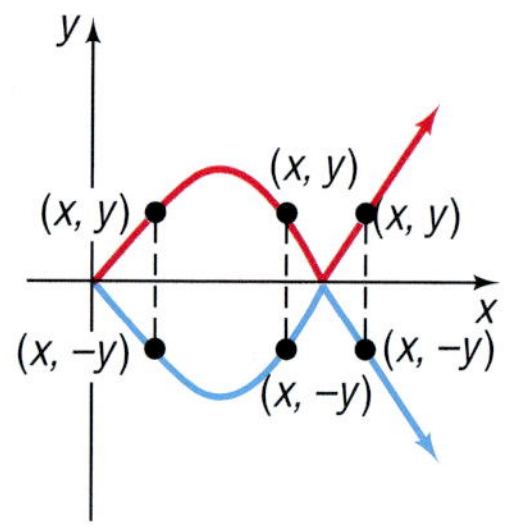

Symmetric with respect to the x-axis

EXAMPLE 10 Points Symmetric with Respect to the *x*-Axis

If a graph is symmetric with respect to the x-axis and the point (3, 2) is on the graph, then the point (3, −2) is also on the graph.

A graph is said to be **symmetric with respect to the *y*-axis** if, for every point (x, y) on the graph, the point $(-x, y)$ is also on the graph.

Figure 49 illustrates the definition. Notice that when a graph is symmetric with respect to the y-axis the part of the graph to the right of the y-axis is a reflection of the part to the left of it, and vice versa.

FIGURE 49

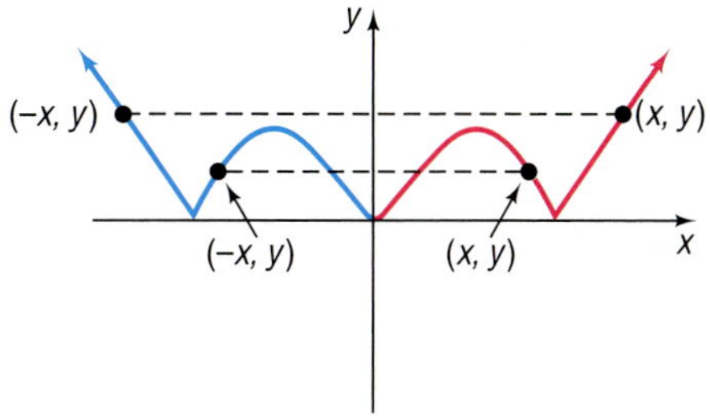

Symmetric with respect to the *y*-axis

EXAMPLE 11 Points Symmetric with Respect to the *y*-Axis

If a graph is symmetric with respect to the *y*-axis and the point (5, 8) is on the graph, then the point (−5, 8) is also on the graph.

A graph is said to be **symmetric with respect to the origin** if, for every point (*x*, *y*) on the graph, the point (−*x*, −*y*) is also on the graph.

Figure 50 illustrates the definition. Notice that symmetry with respect to the origin may be viewed in two ways:

1. As a reflection about the *y*-axis, followed by a reflection about the *x*-axis.
2. As a projection along a line through the origin so that the distances from the origin are equal.

FIGURE 50

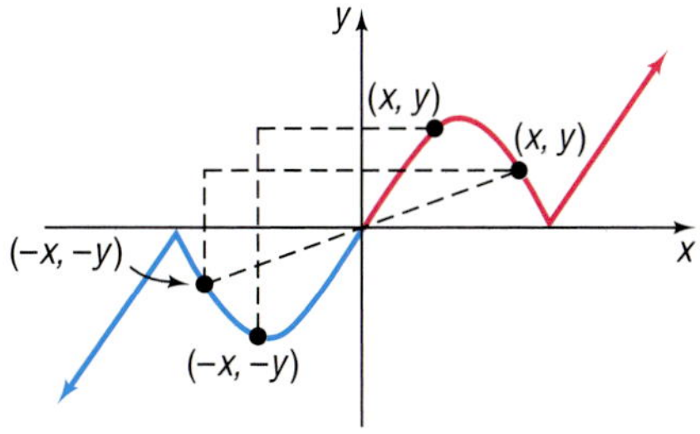

Symmetric with respect to the origin

EXAMPLE 12 Points Symmetric with Respect to the Origin

If a graph is symmetric with respect to the origin and the point (4, 2) is on the graph, then the point (−4, −2) is also on the graph.

Now work Problems 1 and 31(b).

4

When the graph of an equation is symmetric with respect to a coordinate axis or the origin, the number of points that you need to plot in order to see the pattern is reduced. For example, if the graph of an equation is symmetric with respect to the *y*-axis, then, once points to the right of the *y*-axis are plotted, an equal number of points on the graph can be obtained by reflecting them about the *y*-axis. Thus, before we graph an equation, we first want to determine whether it has any symmetry. The following tests are used for that purpose.

Tests for Symmetry

To test the graph of an equation for symmetry with respect to the

x-Axis Replace y by $-y$ in the equation. If an equivalent equation results, the graph of the equation is symmetric with respect to the x-axis.

y-Axis Replace x by $-x$ in the equation. If an equivalent equation results, the graph of the equation is symmetric with respect to the y-axis.

Origin Replace x by $-x$ and y by $-y$ in the equation. If an equivalent equation results, the graph of the equation is symmetric with respect to the origin.

Let's look at an equation we have already graphed to see how these tests are used.

EXAMPLE 13 Testing Equations for Symmetry

(a) To test the graph of the equation $x = y^2$ for symmetry with respect to the x-axis, we replace y by $-y$ in the equation, as follows:

$$x = y^2 \quad \text{Original equation.}$$
$$x = (-y)^2 \quad \text{Replace } y \text{ by } -y.$$
$$x = y^2 \quad \text{Simplify.}$$

When we replace y by $-y$, the result is the same equation. Thus, the graph is symmetric with respect to the x-axis.

(b) To test the graph of the equation $x = y^2$ for symmetry with respect to the y-axis, we replace x by $-x$ in the equation:

$$x = y^2 \quad \text{Original equation.}$$
$$-x = y^2 \quad \text{Replace } x \text{ by } -x.$$

Because we arrive at the equation $-x = y^2$, which is not equivalent to the original equation, we conclude that the graph is not symmetric with respect to the y-axis.

(c) To test for symmetry with respect to the origin, we replace x by $-x$ and y by $-y$:

$$x = y^2 \quad \text{Original equation.}$$
$$-x = (-y)^2 \quad \text{Replace } x \text{ by } -x \text{ and } y \text{ by } -y.$$
$$-x = y^2 \quad \text{Simplify.}$$

The resulting equation, $-x = y^2$, is not equivalent to the original equation. We conclude that the graph is not symmetric with respect to the origin. ▬

Figure 51(a) illustrates the graph of $x = y^2$. In forming a table of points on the graph of $x = y^2$, we can restrict ourselves to points whose y-coordinates are positive. Once these are plotted and connected, a reflection about the x-axis (because of the symmetry) provides the rest of the graph.

Figures 51(b) and (c) illustrate two other equations we graphed earlier. Notice how the existence of symmetry reduces the number of points we need to plot.

FIGURE 51

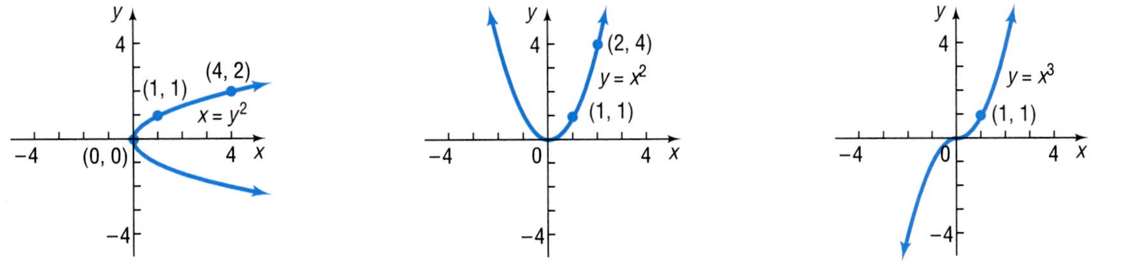

(a) Symmetry with respect to the *x*-axis **(b)** Symmetry with respect to the *y*-axis **(c)** Symmetry with respect to the origin

Now work Problem 89.

EXAMPLE 14 Graphing the Equation $y = 1/x$

Consider the equation $y = 1/x$.

(a) Graph this equation using a graphing utility. Set the viewing rectangle as

RANGE

$X\text{min} = -3$
$X\text{max} = 3$
$X\text{scl} = 1$
$Y\text{min} = -4$
$Y\text{max} = 4$
$Y\text{scl} = 1$

(b) Use algebra to find any intercepts and test for symmetry.

(c) Draw the graph by hand.

Solution

FIGURE 52

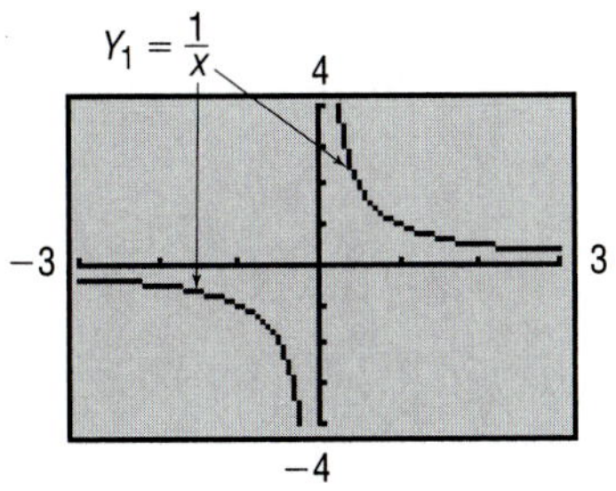

(a) Figure 52 illustrates the graph. We infer from the graph that there are no intercepts; we may also infer that symmetry with respect to the origin is a possibility. The TRACE function on a graphing utility can provide further evidence of symmetry with respect to the origin. Using TRACE we observe that for any ordered pair (x, y), the ordered pair $(-x, -y)$ is also a point on the graph. For example, the points $(0.95744681, 1.0444444)$ and $(-0.9574468, -1.044444)$ both lie on the graph.

(b) We check for intercepts first. If we let $x = 0$, we obtain a 0 denominator, which is not allowed. Hence, there is no y-intercept. If we let $y = 0$, we get the equation $1/x = 0$, which has no solution. Hence, there is no x-intercept. Thus, the graph of $y = 1/x$ does not cross the coordinate axes.

Next we check for symmetry:

x-Axis: Replacing y by $-y$ yields $-y = 1/x$, which is not equivalent to $y = 1/x$.

y-Axis: Replacing x by $-x$ yields $y = -1/x$, which is not equivalent to $y = 1/x$.

Origin: Replacing x by $-x$ and y by $-y$ yields $-y = -1/x$, which is equivalent to $y = 1/x$.

The graph is symmetric with respect to the origin. This confirms the inferences drawn in part (a) of the solution.

(c) We can use the equation to obtain some points on the graph. Because of symmetry, we need only find points (x, y) for which x is positive. Also from the equation $y = 1/x$ we infer that if x is a large and positive number, then $y = 1/x$ is a positive number close to 0. We also infer that if x is a positive number close to 0, then $y = 1/x$ is a large and positive number. Armed with this information, we can graph the equation. Table 12 lists some of these points and Figure 53 shows the graph of $y = 1/x$. Observe how the absence of intercepts and the existence of symmetry with respect to the origin were utilized.

TABLE 12

x	$y = 1/x$	(x, y)
$\frac{1}{10}$	10	$(\frac{1}{10}, 10)$
$\frac{1}{3}$	3	$(\frac{1}{3}, 3)$
$\frac{1}{2}$	2	$(\frac{1}{2}, 2)$
1	1	$(1, 1)$
2	$\frac{1}{2}$	$(2, \frac{1}{2})$
3	$\frac{1}{3}$	$(3, \frac{1}{3})$
10	$\frac{1}{10}$	$(10, \frac{1}{10})$

FIGURE 53

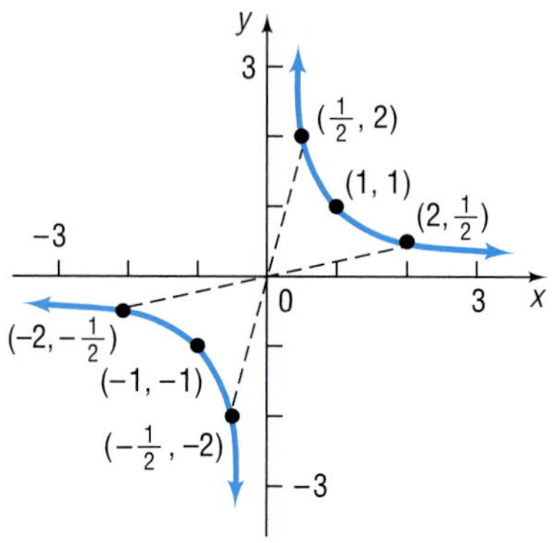

1.3 EXERCISES

In Problems 1–10, plot each point. Then plot the point that is symmetric to it with respect to (a) the x-axis; (b) the y-axis; (c) the origin.

1. (3, 4) **2.** (5, 3) **3.** (−2, 1) **4.** (4, −2) **5.** (1, 1)
6. (−1, −1) **7.** (−3, −4) **8.** (4, 0) **9.** (0, −3) **10.** (−3, 0)

In Problems 11–30, graph each equation using the following RANGE settings:

(a)	(b)	(c)	(d)
Xmin = −5	Xmin = −10	Xmin = −10	Xmin = −5
Xmax = 5	Xmax = 10	Xmax = 10	Xmax = 5
Xscl = 1	Xscl = 1	Xscl = 2	Xscl = 1
Ymin = −4	Ymin = −8	Ymin = −8	Ymin = −20
Ymax = 4	Ymax = 8	Ymax = 8	Ymax = 20
Yscl = 1	Yscl = 1	Yscl = 2	Yscl = 5

11. $y = x + 2$ **12.** $y = x - 2$ **13.** $y = -x + 2$ **14.** $y = -x - 2$
15. $y = 2x + 2$ **16.** $y = 2x - 2$ **17.** $y = -2x + 2$ **18.** $y = -2x - 2$
19. $y = x^2 + 2$ **20.** $y = x^2 - 2$ **21.** $y = -x^2 + 2$ **22.** $y = -x^2 - 2$
23. $y = 2x^2 + 2$ **24.** $y = 2x^2 - 2$ **25.** $y = -2x^2 + 2$ **26.** $y = -2x^2 - 2$
27. $3x + 2y = 6$ **28.** $3x - 2y = 6$ **29.** $-3x + 2y = 6$ **30.** $-3x - 2y = 6$

In Problems 31–46, the graph of an equation is given.

(a) List the intercepts of the graph.

(b) Based on the graph, tell whether the graph is symmetric with respect to the x-axis, y-axis, and/or origin.

31.

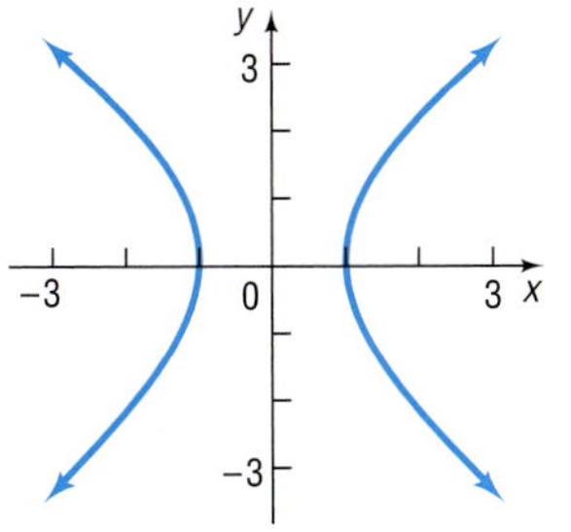

32.

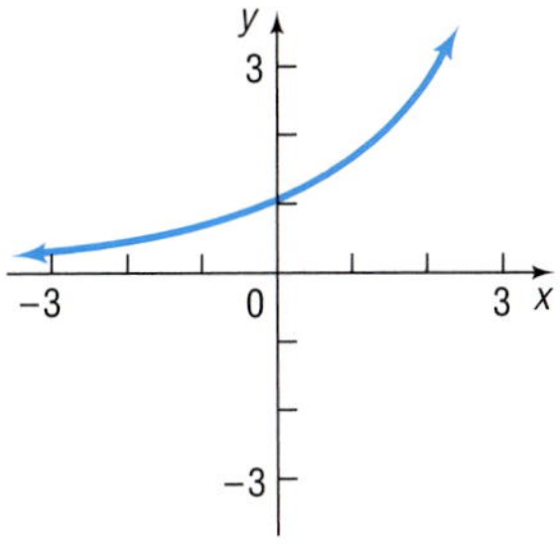

33.

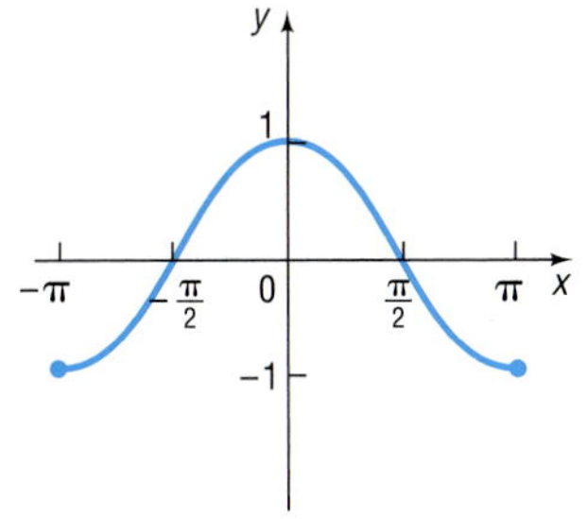

34.

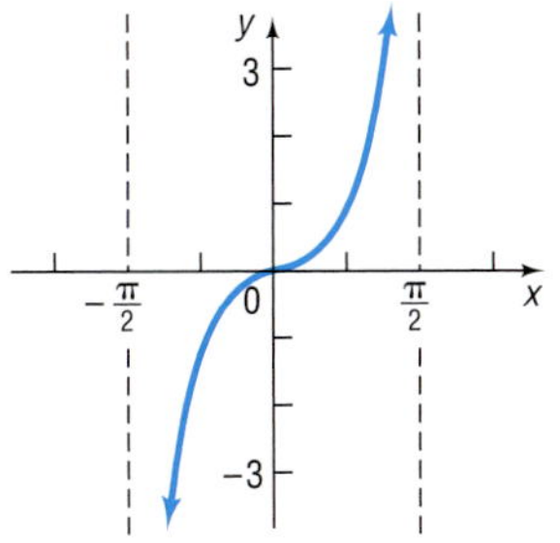

35.

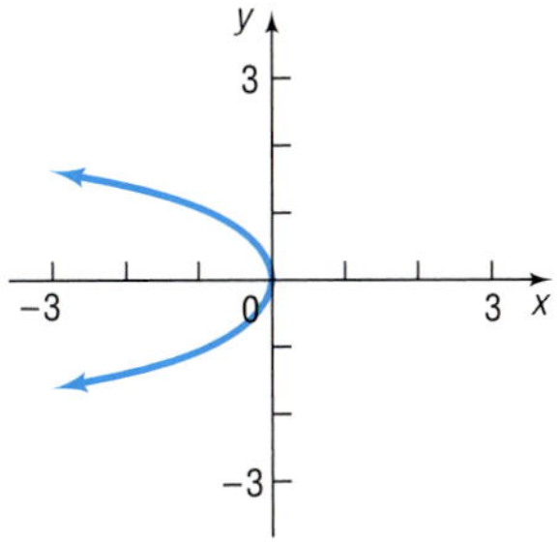

36.

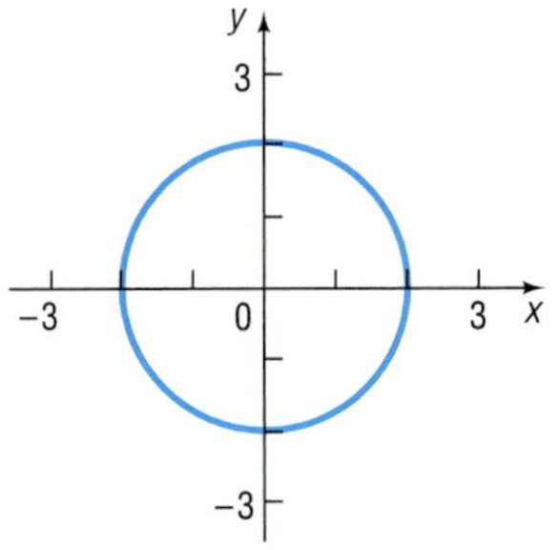

37.

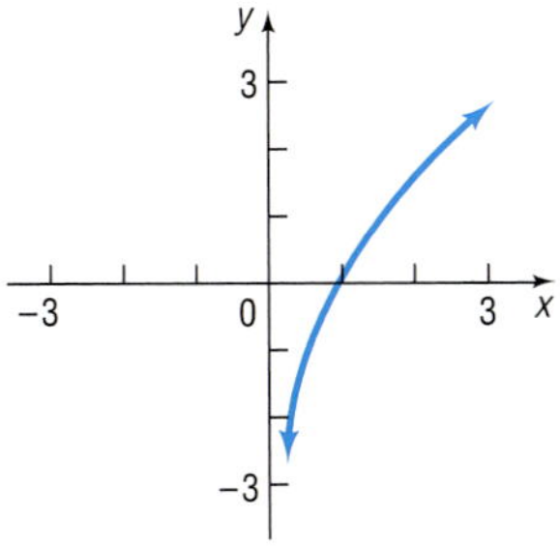

38.

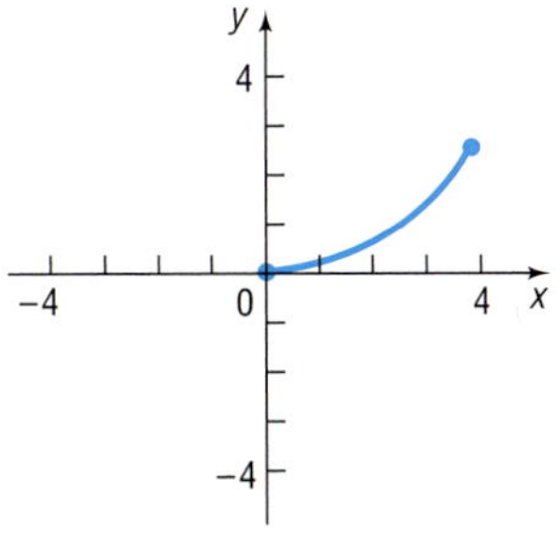

39.

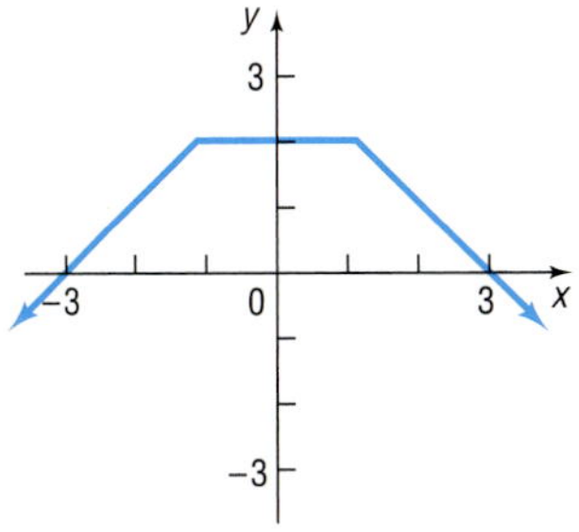

40.

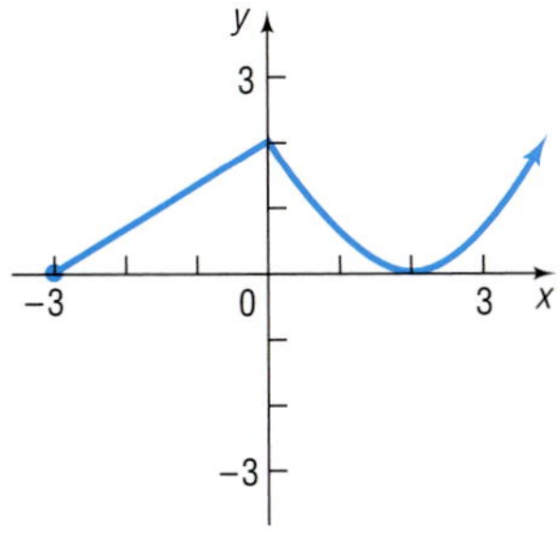

41.

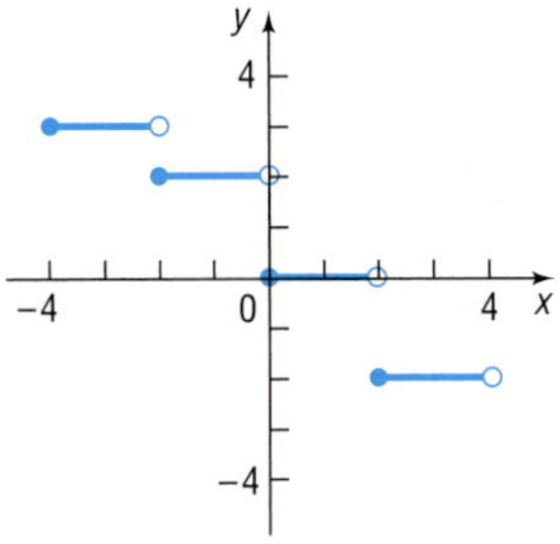

42.

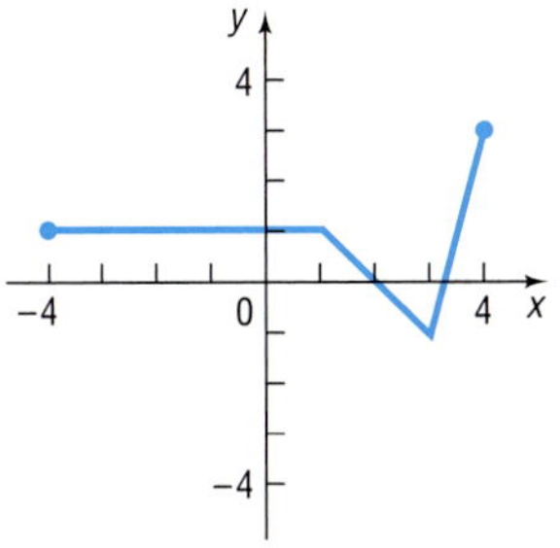

43.

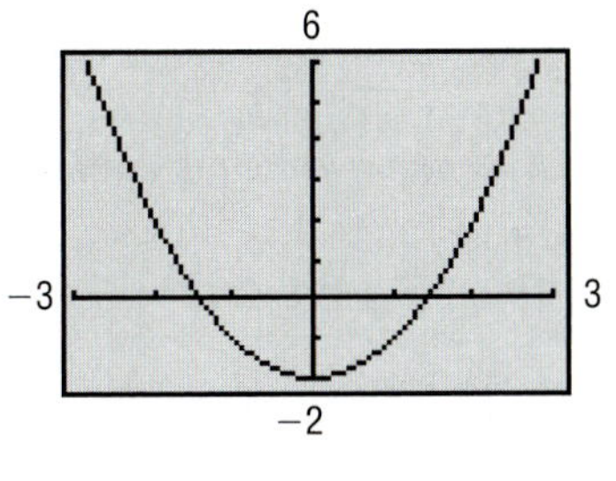

44.

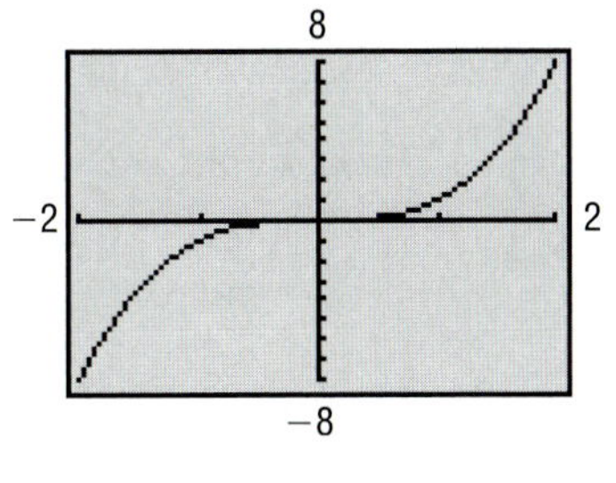

45.

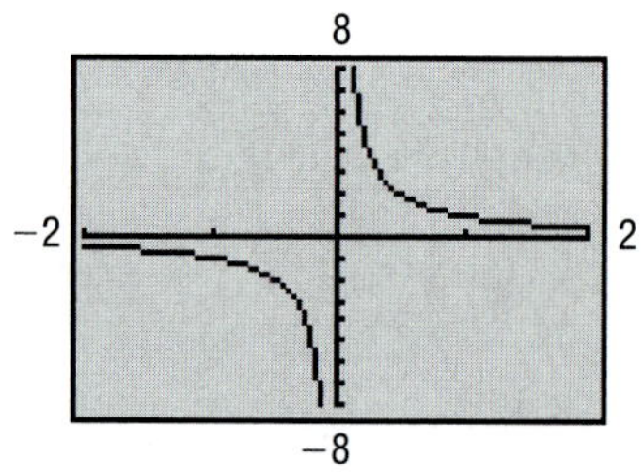

46.

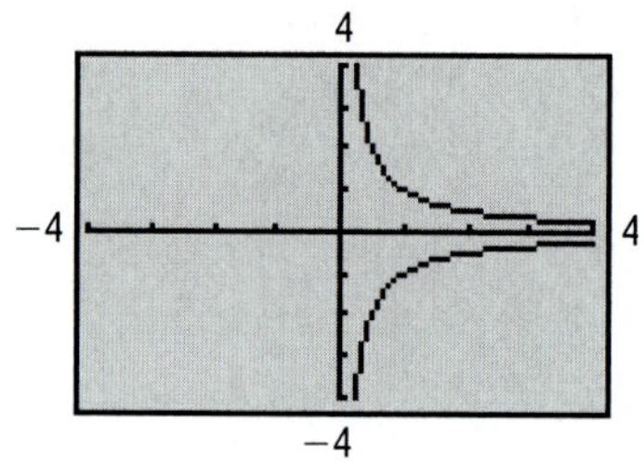

In Problems 47–52, tell whether the given points are on the graph of the equation.

47. Equation: $y = x^4 + \sqrt{x}$
Points: $(0, 0); (1, 1); (-1, 0)$

48. Equation: $y = x^3 - 2\sqrt{x}$
Points: $(0, 0); (1, 1); (1, -1)$

49. Equation: $y^2 = x^2 + 4$
Points: $(0, 2); (2, 0); (-2, 0)$

50. Equation: $y^3 = x + 1$
Points: $(1, 2); (0, 1); (-1, 0)$

51. Equation: $x^2 + y^2 = 4$
Points: $(0, 2); (-2, 2); (\sqrt{2}, \sqrt{2})$

52. Equation: $x^2 + 4y^2 = 4$
Points: $(0, 1); (2, 0); (2, \frac{1}{2})$

53. If $(a, 2)$ is a point on the graph of $y = 5x + 4$, what is a?

54. If $(2, b)$ is a point on the graph of $y = x^2 + 3x$, what is b?

55. If (a, b) is a point on the graph of $2x + 3y = 6$, write an equation that relates a to b.

56. If $(2, 0)$ and $(0, 5)$ are points on the graph of $y = mx + b$, what are m and b?

In Problems 57–76, use a graphing utility to graph each equation. State the viewing rectangle used and draw the graph by hand.

57. $3x + 5y = 75$
58. $3x - 5y = 75$
59. $3x + 5y = -75$
60. $3x - 5y = -75$
61. $y = (x - 10)^2$
62. $y = (x + 10)^2$
63. $y = x^2 - 100$
64. $y = x^2 + 100$
65. $x^2 + y^2 = 100$
66. $x^2 + y^2 = 64$
67. $3x^2 + y^2 = 900$
68. $4x^2 + y^2 = 1600$
69. $x^2 + 3y^2 = 900$
70. $x^2 + 4y^2 = 1600$
71. $y = x^2 - 10x$
72. $y = x^2 + 10x$
73. $y = x^2 - 18x$
74. $y = x^2 + 18x$
75. $y = x^2 - 36x$
76. $y = x^2 + 36x$

In Problems 77–80, use the graph on the right.

77. Draw the graph to make it symmetric with respect to the x-axis.
78. Draw the graph to make it symmetric with respect to the y-axis.
79. Draw the graph to make it symmetric with respect to the origin.
80. Draw the graph to make it symmetric with respect to the x-axis, y-axis, and origin.

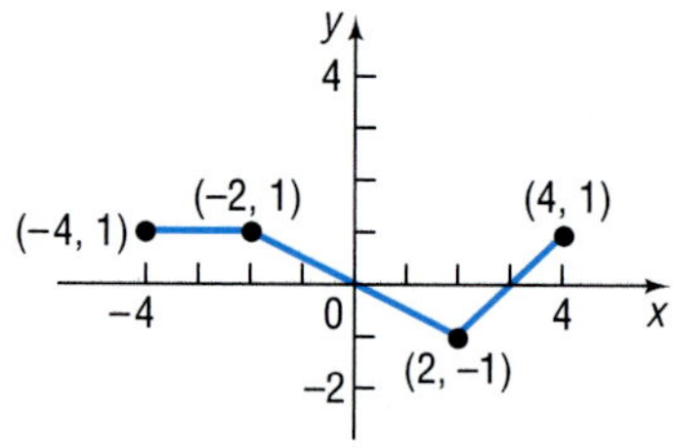

In Problems 81–84, use the graph on the right.

81. Draw the graph to make it symmetric with respect to the x-axis.
82. Draw the graph to make it symmetric with respect to the y-axis.
83. Draw the graph to make it symmetric with respect to the origin.
84. Draw the graph to make it symmetric with respect to the x-axis, y-axis, and origin.

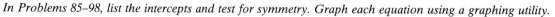

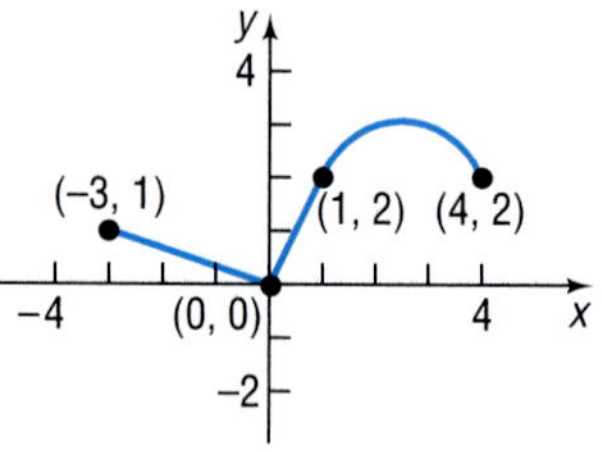

In Problems 85–98, list the intercepts and test for symmetry. Graph each equation using a graphing utility.

85. $x^2 = y$
86. $y^2 = x$
87. $y = 3x$
88. $y = -5x$
89. $x^2 + y - 9 = 0$
90. $y^2 - x - 4 = 0$
91. $9x^2 + 4y^2 = 36$
92. $4x^2 + y^2 = 4$
93. $y = x^3 - 27$
94. $y = x^4 - 1$
95. $y = x^2 - 3x - 4$
96. $y = x^2 + 4$
97. $y = \dfrac{x}{x^2 + 9}$
98. $y = \dfrac{x^2 - 4}{x}$

In Problem 99, you may use a graphing utility, but it is not required.

99. (a) Graph $y = \sqrt{x^2}$, $y = x$, $y = |x|$, and $y = (\sqrt{x})^2$ noting which graphs are the same.

(b) Explain why the graphs of $y = \sqrt{x^2}$ and $y = |x|$ are the same.

(c) Explain why the graphs of $y = x$ and $y = (\sqrt{x})^2$ are not the same.

(d) Explain why the graphs of $y = \sqrt{x^2}$ and $y = x$ are not the same.

100. Make up an equation with the intercepts $(2, 0)$, $(4, 0)$, and $(0, 1)$. Compare your equation with a friend's equation. Comment on any similarities.

101. An equation is being tested for symmetry with respect to the x-axis, the y-axis, and the origin. Explain why, if two of these symmetries are present, then the remaining one must also be present.

102. Draw a graph that contains the points $(-2, -1)$, $(0, 1)$, $(1, 3)$, and $(3, 5)$. Compare your graph with those of other students. Are most of the graphs almost straight lines? How many are "curved"? Discuss the various ways these points might be connected.

1.4 LINES

1. Calculate and Interpret the Slope of a Line
2. Graph Lines by Hand
3. Find the Equation of Vertical Lines
4. Use the Point-Slope Form of a Line; Horizontal Lines
5. Find the Equation of a Line from Two Points
6. Write the Equation of a Line in General Form
7. Write the Equation of a Line in Slope-Intercept Form
8. Identify the Slope and y-Intercept of a Line from its Equation

In this section we study a certain type of equation that contains two variables, called a *linear equation,* and its graph, a *line.*

Slope of a Line

1 Consider the staircase illustrated in Figure 54. Each step contains exactly the same horizontal **run** and the same vertical **rise.** The ratio of the rise to the run, called the *slope,* is a numerical measure of the steepness of the staircase. For example, if the run is increased and the rise remains the same, the staircase becomes less steep. If the run is kept the same, but the rise is increased, the staircase becomes more steep. This important characteristic of a line is best defined using rectangular coordinates.

FIGURE 54

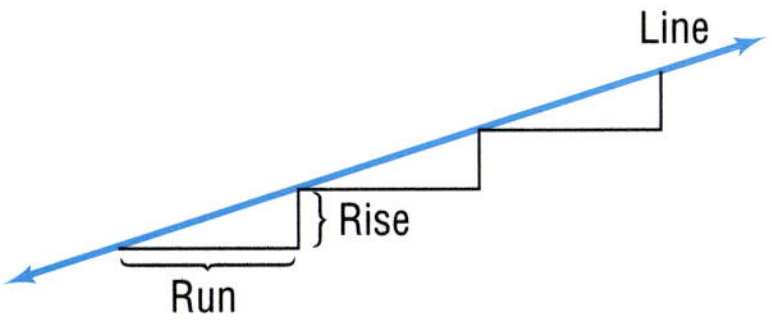

Let $P = (x_1, y_1)$ and $Q = (x_2, y_2)$ be two distinct points with $x_1 \neq x_2$. The **slope m** of the nonvertical line L containing P and Q is defined by the formula

$$m = \frac{y_2 - y_1}{x_2 - x_1} \qquad x_1 \neq x_2 \tag{1}$$

If $x_1 = x_2$, L is a **vertical line** and the slope m of L is **undefined** (since this results in division by 0).

Figure 55(a) provides an illustration of the slope of a nonvertical line; Figure 55(b) illustrates a vertical line.

FIGURE 55

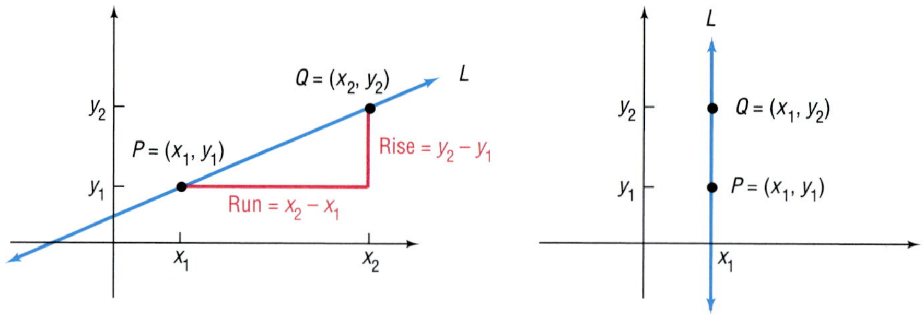

(a) Slope of L is $m = \dfrac{y_2 - y_1}{x_2 - x_1}$

(b) Slope is undefined; L is vertical

As Figure 55(a) illustrates, the slope m of a nonvertical line may be viewed as

$$m = \frac{y_2 - y_1}{x_2 - x_1} = \frac{\text{Rise}}{\text{Run}}$$

We can also express the slope m of a nonvertical line as

$$m = \frac{y_2 - y_1}{x_2 - x_1} = \frac{\text{Change in } y}{\text{Change in } x} = \frac{\Delta y}{\Delta x}$$

That is, the slope m of a nonvertical line L measures the amount y changes as x changes from x_1 to x_2. This is called the **average rate of change** of y with respect to x.

Two comments about computing the slope of a nonvertical line may prove helpful:

1. Any two distinct points on the line can be used to compute the slope of the line. (See Figure 56 for justification.)

FIGURE 56
Triangles ABC and PQR are similar (equal angles). Hence, ratios of corresponding sides are proportional. Thus:

$$\text{Slope using } P \text{ and } Q = \frac{y_2 - y_1}{x_2 - x_1}$$
$$= \text{Slope using } A \text{ and } B = \frac{d(B, C)}{d(A, C)}$$

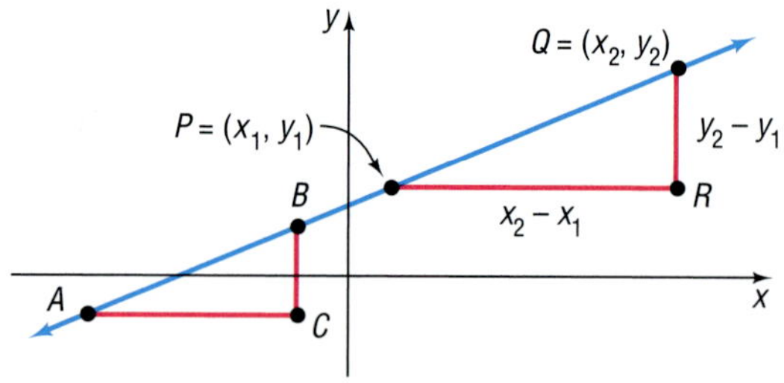

2. The slope of a line may be computed from $P = (x_1, y_1)$ to $Q = (x_2, y_2)$ or from Q to P, because

$$\frac{y_2 - y_1}{x_2 - x_1} = \frac{y_1 - y_2}{x_1 - x_2}$$

EXAMPLE 1

Finding and Interpreting the Slope of a Line Joining Two Points

The slope m of the line joining the points $(1, 2)$ and $(5, -3)$ may be computed as

$$m = \frac{-3 - 2}{5 - 1} = \frac{-5}{4} \quad \text{or as} \quad m = \frac{2 - (-3)}{1 - 5} = \frac{5}{-4} = \frac{-5}{4}$$

For every 4 unit change in x, y will change by -5 units. That is, if x increases by 4 units, then y decreases by 5 units. The average rate of change of y with respect to x is $\frac{-5}{4}$.

Now work Problem 7.

To get a better idea of the meaning of the slope m of a line L, consider the following example.

EXAMPLE 2

Finding the Slopes of Various Lines Containing the Same Point (2, 3)

Compute the slopes of the lines L_1, L_2, L_3, and L_4 containing the following pairs of points. Graph all four lines on the same set of coordinate axes.

L_1: $P = (2, 3)$ $\quad Q_1 = (-1, -2)$
L_2: $P = (2, 3)$ $\quad Q_2 = (3, -1)$
L_3: $P = (2, 3)$ $\quad Q_3 = (5, 3)$
L_4: $P = (2, 3)$ $\quad Q_4 = (2, 5)$

Solution Let m_1, m_2, m_3, and m_4 denote the slopes of the lines L_1, L_2, L_3, and L_4, respectively. Then

$$m_1 = \frac{-2 - 3}{-1 - 2} = \frac{-5}{-3} = \frac{5}{3}$$ A rise of 5 divided by a run of 3.

$$m_2 = \frac{-1 - 3}{3 - 2} = \frac{-4}{1} = -4$$

$$m_3 = \frac{3 - 3}{5 - 2} = \frac{0}{3} = 0$$

m_4 is undefined

The graphs of these lines are given in Figure 57.

FIGURE 57

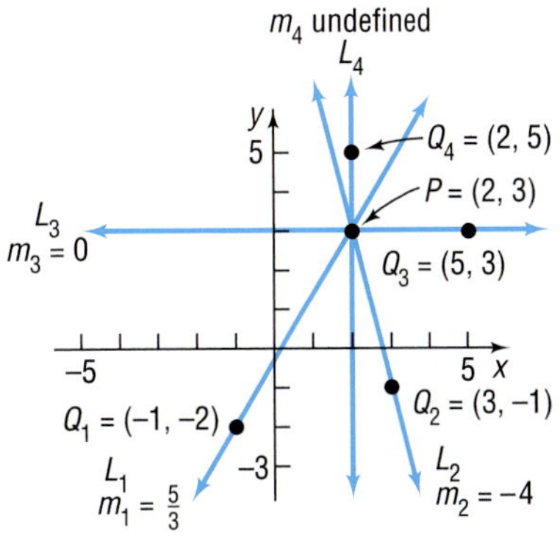

Figure 57 illustrates the following facts:

1. When the slope of a line is positive, the line slants upward from left to right (L_1).
2. When the slope of a line is negative, the line slants downward from left to right (L_2).

FIGURE 58

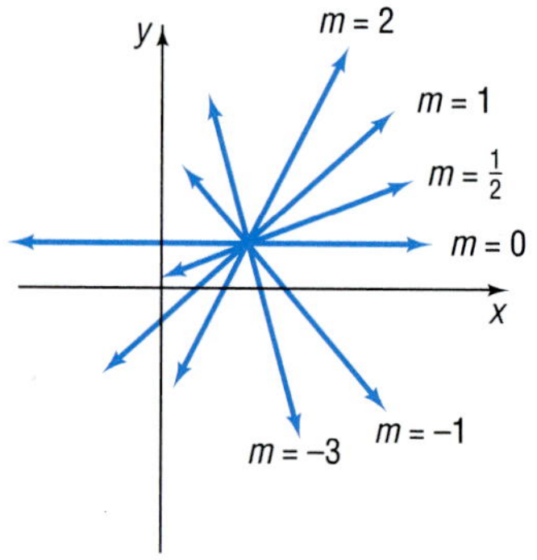

3. When the slope is 0, the line is horizontal (L_3).
4. When the slope is undefined, the line is vertical (L_4).

Figure 58 illustrates some additional facts about the slope of a line. Note that the closer the line is to the vertical position, the greater the magnitude of the slope.

Square Screens

To get an undistorted view of slope, the same scale must be used on each axis. However, most graphing utilities have a rectangular screen. Because of this, using the same RANGE for both x and y will result in a distorted view. For example, Figure 59 shows the graph of the line $y = x$ connecting the points $(-4, -4)$ and $(4, 4)$.

We expect the line to bisect the first and third quadrants, but it doesn't. We need to adjust the selections for Xmin, Xmax, Ymin, and Ymax so that a **square screen** results. On most graphing utilities, this is accomplished by setting the ratio of x to y at 3:2.* In other words,

$$2(X\text{max} - X\text{min}) = 3(Y\text{max} - Y\text{min})$$

FIGURE 59

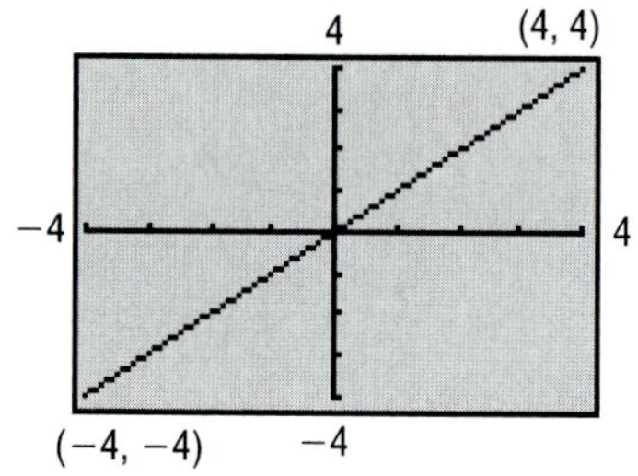

Figure 60 shows the graph of the line $y = x$ on a square screen using a TI-83. Notice that the line now bisects the first and third quadrants. Compare this illustration to Figure 59.

FIGURE 60

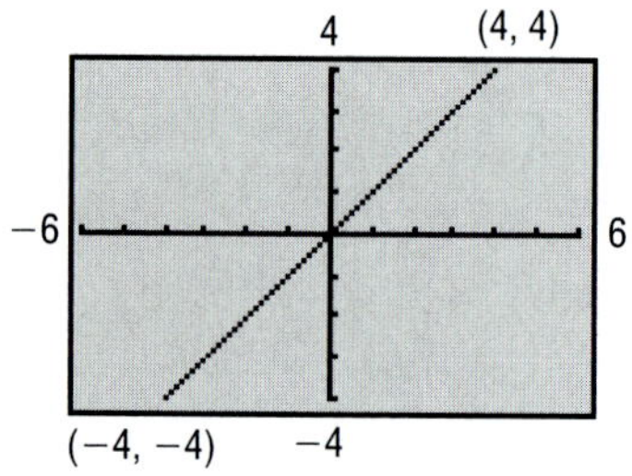

Seeing the Concept: On the same square screen, graph the following equations:

$Y_1 = 0$ Slope of line is 0.
$Y_2 = \frac{1}{4}x$ Slope of line is $\frac{1}{4}$.
$Y_3 = \frac{1}{2}x$ Slope of line is $\frac{1}{2}$.
$Y_4 = x$ Slope of line is 1.
$Y_5 = 2x$ Slope of line is 2.
$Y_6 = 6x$ Slope of line is 6.

See Figure 61.

FIGURE 61

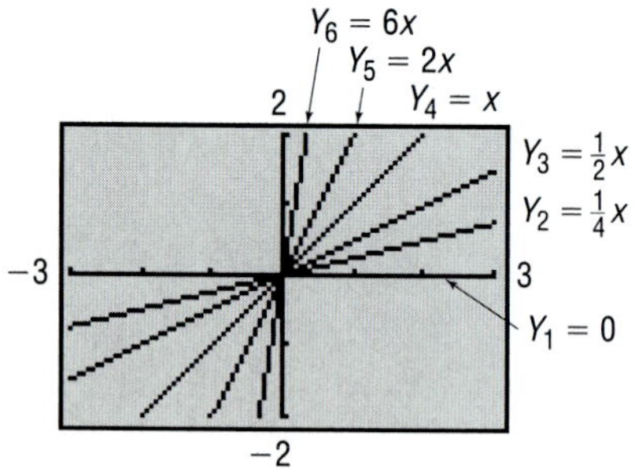

Seeing the Concept: On the same square screen, graph the following equations:

$Y_1 = 0$ Slope of line is 0.
$Y_2 = -\frac{1}{4}x$ Slope of line is $-\frac{1}{4}$.
$Y_3 = -\frac{1}{2}x$ Slope of line is $-\frac{1}{2}$.
$Y_4 = -x$ Slope of line is -1.
$Y_5 = -2x$ Slope of line is -2.
$Y_6 = -6x$ Slope of line is -6.

See Figure 62.

*Some graphing utilities have a built-in function that automatically squares the screen. For example, the TI-85 has a ZSQR function that does this. Some graphing utilities require a ratio other than 3:2 to square the screen. For example, the HP 48G requires the ratio of x to y to be 1:2 for a square screen. Consult your manual.

FIGURE 62

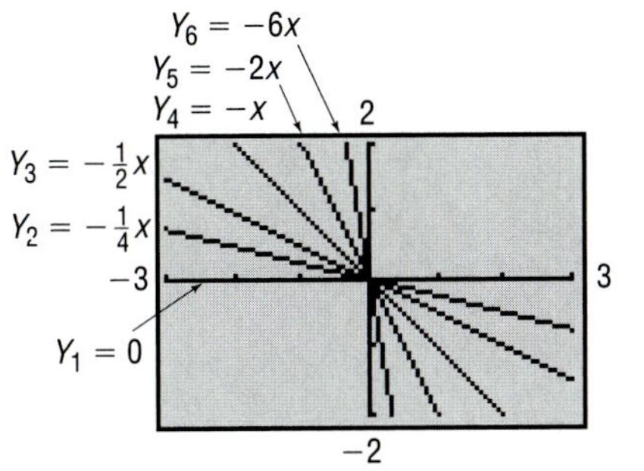

2

The next example illustrates how the slope of a line can be used to graph the line.

EXAMPLE 3

Graphing a Line Given a Point and a Slope

Draw a graph of the line that passes through the point (3, 2) and has a slope of:

(a) $\frac{3}{4}$ (b) $-\frac{4}{5}$

Solution (a) Slope = Rise/Run. The fact that the slope is $\frac{3}{4}$ means that for every horizontal movement (run) of 4 units to the right there will be a vertical movement (rise) of 3 units. If we start at the given point (3, 2) and move 4 units to the right and 3 units up, we reach the point (7, 5). By drawing the line through this point and the point (3, 2), we have the graph. See Figure 63.

FIGURE 63

Slope $= \frac{3}{4}$

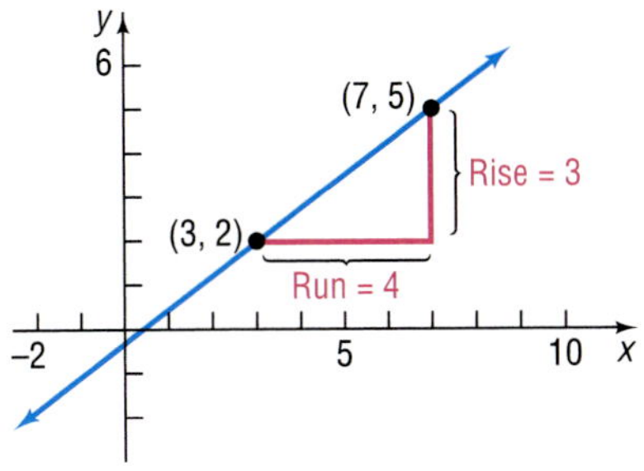

(b) The fact that the slope is

$$-\frac{4}{5} = \frac{-4}{5} = \frac{\text{Rise}}{\text{Run}}$$

means that for every horizontal movement of 5 units to the right there will be a corresponding vertical movement of −4 units (a downward movement). If we start at the given point (3, 2) and move 5 units to the right and then 4 units down, we arrive at the point (8, −2). By drawing the line through these points, we have the graph. See Figure 64.

FIGURE 64

Slope $= -\frac{4}{5}$

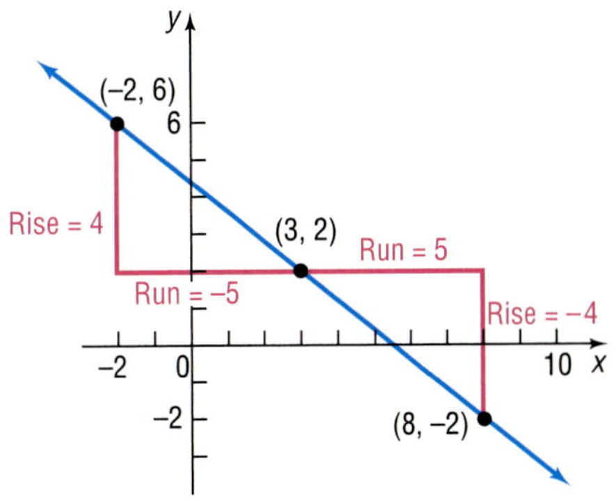

Alternatively, we can set

$$-\frac{4}{5} = \frac{4}{-5} = \frac{\text{Rise}}{\text{Run}}$$

so that for every horizontal movement of -5 units (a movement to the left) there will be a corresponding vertical movement of 4 units (upward). This approach brings us to the point $(-2, 6)$, which is also on the graph shown in Figure 64.

Now work Problem 17.

Equations of Lines

3

Now that we have discussed the slope of a line, we are ready to derive equations of lines. As we shall see, there are several forms of the equation of a line. Let's start with an example.

EXAMPLE 4

Graphing a Line

Graph the equation $x = 3$.

Solution

Using a graphing utility, we need to express the equation in the form $y =$ expression in x. But $x = 3$ cannot be put into this form. To overcome this problem, we must utilize a function in our graphing utility which allows vertical lines to be drawn. Consult your manual to determine the methodology required to draw vertical lines. Figure 65(a) shows the graph you should obtain.

To graph $x = 3$ by hand, we recall that we are looking for all points (x, y) in the plane for which $x = 3$. Thus, no matter what y-coordinate is used, the corresponding x-coordinate always equals 3. Consequently, the graph of the equation $x = 3$ is a vertical line with x-intercept 3 and undefined slope. See Figure 65(b).

FIGURE 65
$x = 3$

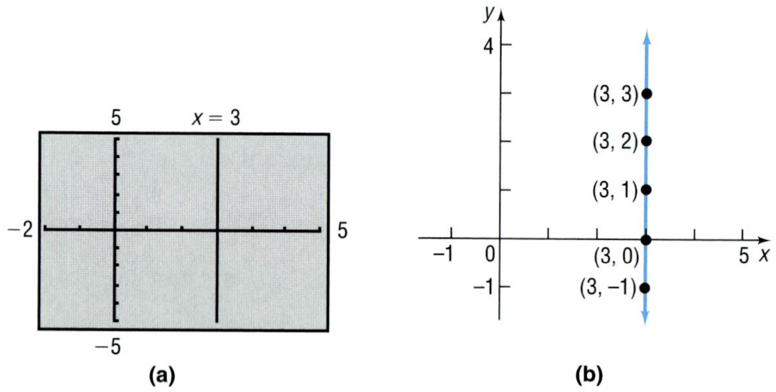

As suggested by Example 4, we have the following result:

Theorem Equation of a Vertical Line

A vertical line is given by an equation of the form

$$x = a$$

where a is the x-intercept.

4 Now let L be a nonvertical line with slope m containing the point (x_1, y_1). See Figure 66. For any other point (x, y) on L, we have

$$m = \frac{y - y_1}{x - x_1} \quad \text{or} \quad y - y_1 = m(x - x_1)$$

FIGURE 66

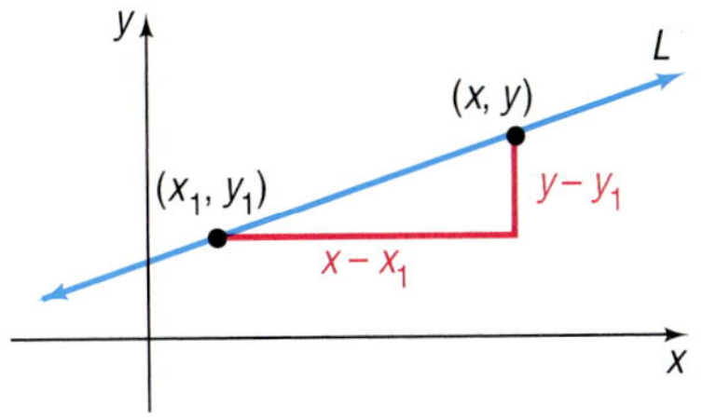

Theorem Point–Slope Form of an Equation of a Line

An equation of a nonvertical line of slope m that passes through the point (x_1, y_1) is

$$y - y_1 = m(x - x_1) \tag{2}$$

EXAMPLE 5

Using the Point–Slope Form of a Line

An equation of the line with slope 4 and passing through the point $(1, 2)$ can be found by using the point–slope form with $m = 4$, $x_1 = 1$, and $y_1 = 2$:

$$\begin{aligned} y - y_1 &= m(x - x_1) \\ y - 2 &= 4(x - 1) \\ y &= 4x - 2 \end{aligned}$$

See Figure 67.

FIGURE 67
$y = 4x - 2$

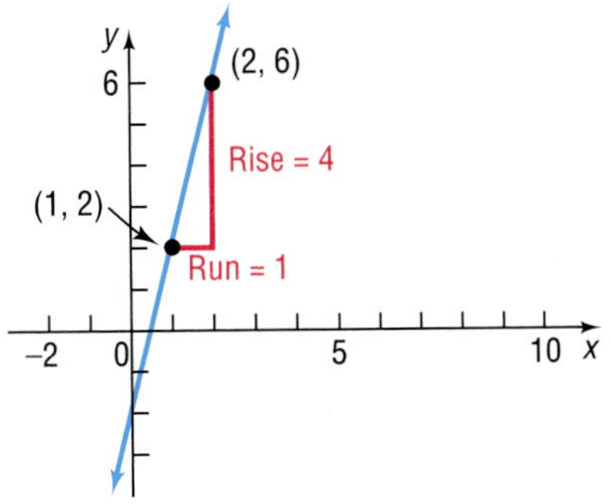

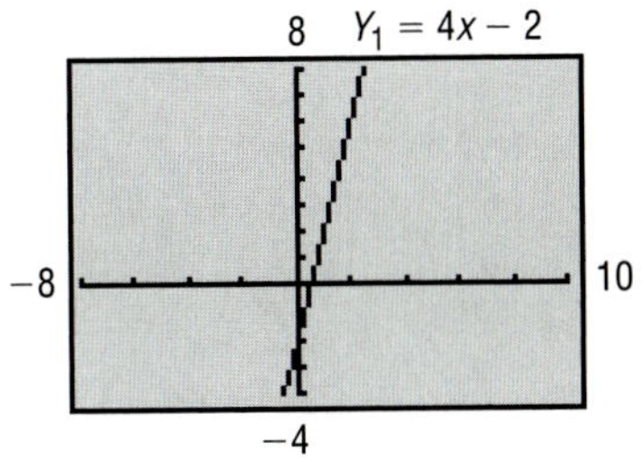

EXAMPLE 6

Finding the Equation of a Horizontal Line

Find an equation of the horizontal line passing through the point $(3, 2)$.

Solution The slope of a horizontal line is 0. To get an equation, we use the point–slope form with $m = 0$, $x_1 = 3$, and $y_1 = 2$:

$$\begin{aligned} y - y_1 &= m(x - x_1) \\ y - 2 &= 0 \cdot (x - 3) \\ y - 2 &= 0 \\ y &= 2 \end{aligned}$$

See Figure 68 for the graph.

FIGURE 68
$y = 2$

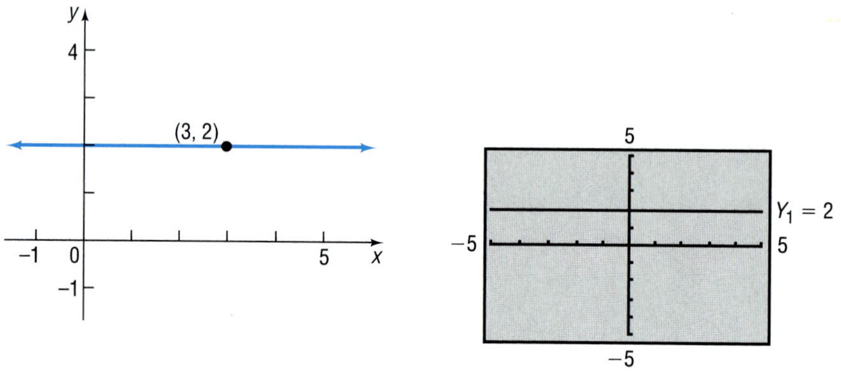

As suggested by Example 6, we have the following result:

Theorem Equation of a Horizontal Line

A horizontal line is given by an equation of the form

$$y = b$$

where b is the y-intercept.

EXAMPLE 7 Finding an Equation of a Line Given Two Points

5 Find an equation of the line L passing through the points (2, 3) and (−4, 5). Graph the line L.

Solution Since two points are given, we first compute the slope of the line:

$$m = \frac{5 - 3}{-4 - 2} = \frac{2}{-6} = \frac{-1}{3}$$

We use the point (2, 3) and the fact that the slope $m = -\frac{1}{3}$ to get the point–slope form of the equation of the line:

$$y - 3 = -\frac{1}{3}(x - 2)$$

See Figure 69 for the graph.

FIGURE 69
$y - 3 = -\frac{1}{3}(x - 2)$

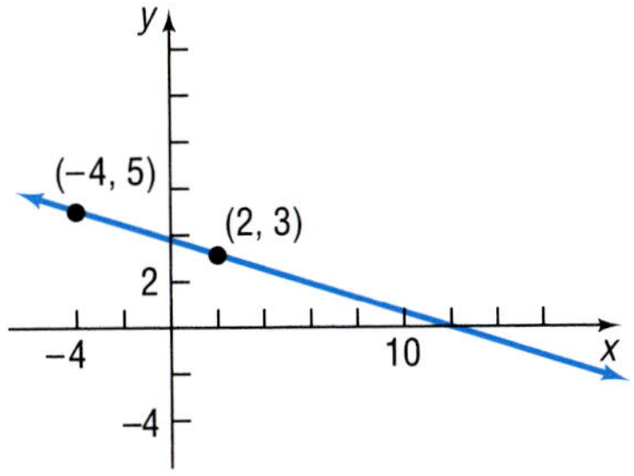

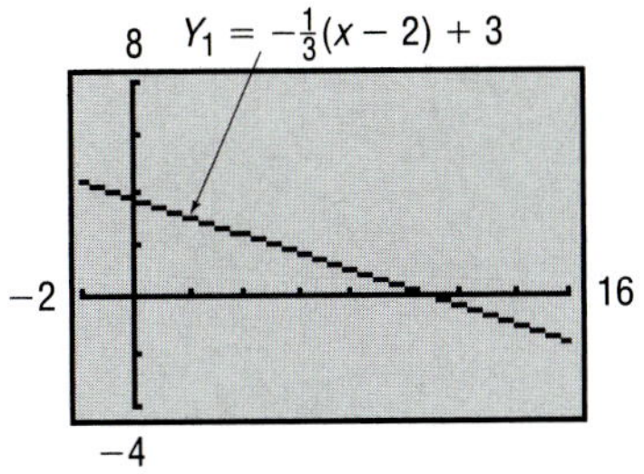

In the solution to Example 7, we could have used the other point, $(-4, 5)$, instead of the point $(2, 3)$. The equation that results, although it looks different, is equivalent to the equation we obtained in the example. (Try it for yourself.)

6 Another form of the equation of the line in Example 7 can be obtained by multiplying both sides of the point–slope equation by 3 and collecting terms:

$$y - 3 = -\frac{1}{3}(x - 2)$$

$$3(y - 3) = 3\left(-\frac{1}{3}\right)(x - 2) \quad \text{Multiply by 3.}$$

$$3y - 9 = -1(x - 2)$$

$$3y - 9 = -x + 2$$

$$x + 3y - 11 = 0$$

The equation of a line L is in **general form** when it is written as

$$Ax + By + C = 0 \tag{3}$$

where A, B, and C are three real numbers and A and B are not both 0.

Now work Problem 27.

Every line has an equation that is equivalent to an equation written in general form. For example, a vertical line whose equation is

$$x = a$$

can be written in the general form

$$1 \cdot x + 0 \cdot y - a = 0 \quad A = 1, B = 0, C = -a.$$

A horizontal line whose equation is

$$y = b$$

can be written in the general form

$$0 \cdot x + 1 \cdot y - b = 0 \quad A = 0, B = 1, C = -b.$$

Lines that are neither vertical nor horizontal have general equations of the form

$$Ax + By + C = 0 \quad A \neq 0 \text{ and } B \neq 0.$$

Because the equation of every line can be written in general form, any equation equivalent to (3) is called a **linear equation.**

The next example illustrates one way of graphing a linear equation.

EXAMPLE 8 Finding the Intercepts of a Line

Find the intercepts of the line $2x + 3y - 6 = 0$. Graph this line.

Solution To find the point at which the graph crosses the x-axis, that is, to find the x-intercept, we need to find the number x for which $y = 0$. Thus, we let $y = 0$ to get

$$\begin{aligned} 2x + 3(0) - 6 &= 0 \\ 2x - 6 &= 0 \\ x &= 3 \end{aligned}$$

FIGURE 70
$2x + 3y - 6 = 0$

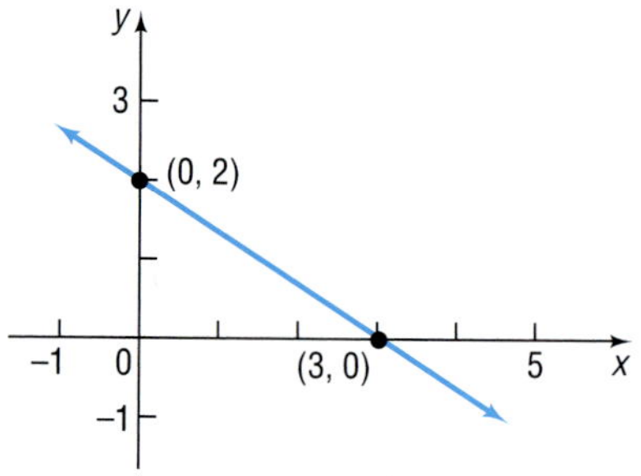

The x-intercept is 3. To find the y-intercept, we let $x = 0$ and solve for y:

$$\begin{aligned} 2(0) + 3y - 6 &= 0 \\ 3y - 6 &= 0 \\ y &= 2 \end{aligned}$$

The y-intercept is 2.

To graph the line by hand, we use the intercepts. Since the x-intercept is 3, and the y-intercept is 2, we know two points on the line: (3, 0) and (0, 2). Because two points determine a unique line, we do not need further information to graph the line. See Figure 70.

FIGURE 71

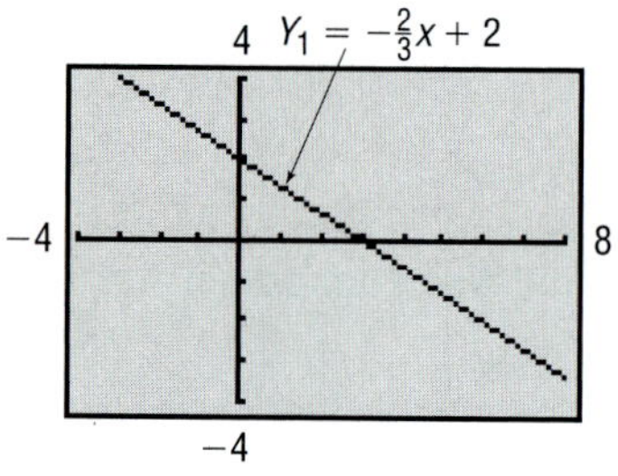

To graph the line using a graphing utility, we need to solve for y.

$$\begin{aligned} 2x + 3y - 6 &= 0 \\ 3y &= -2x + 6 \\ y &= -\tfrac{2}{3}x + 2 \end{aligned}$$

See Figure 71 for the graph.

7 Another useful equation of a line is obtained when the slope m and y-intercept b are known. In this event, we know both the slope m of the line and a point $(0, b)$ on the line; thus, we may use the point–slope form, equation (2), to obtain the following equation:

$$y - b = m(x - 0) \text{ or } y = mx + b$$

Theorem Slope–Intercept Form of an Equation of a Line

An equation of a line L with slope m and y-intercept b is

$$y = mx + b \qquad (4)$$

FIGURE 72
$y = mx + 2$

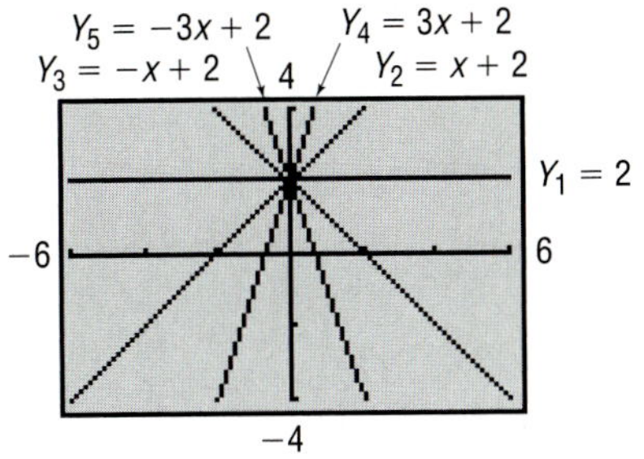

Seeing the Concept: To see the role the slope m plays, graph the following lines on the same square screen

$$Y_1 = 2$$
$$Y_2 = x + 2$$
$$Y_3 = -x + 2$$
$$Y_4 = 3x + 2$$
$$Y_5 = -3x + 2$$

See Figure 72. What do you conclude about the lines $y = mx + 2$?

Seeing the Concept: To see the role of the y-intercept b, graph the following lines on the same square screen.

$$Y_1 = 2x$$
$$Y_2 = 2x + 1$$
$$Y_3 = 2x - 1$$
$$Y_4 = 2x + 4$$
$$Y_5 = 2x - 4$$

See Figure 73. What do you conclude about the lines $y = 2x + b$?

FIGURE 73
$y = 2x + b$

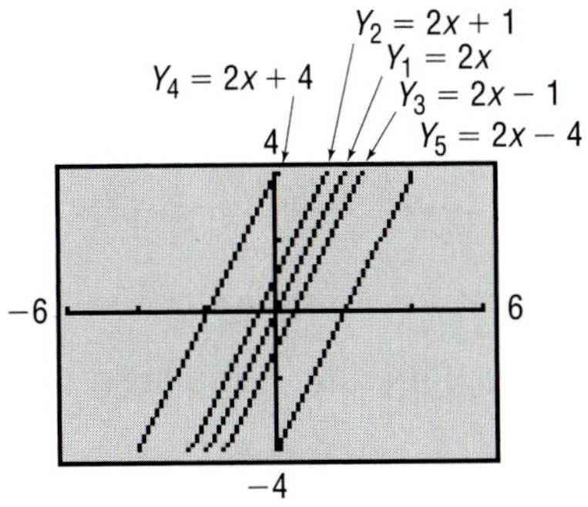

8

When the equation of a line is written in slope–intercept form, it is easy to find the slope m and y-intercept b of the line. For example, suppose the equation of a line is

$$y = -2x + 3$$

Compare it to $y = mx + b$:

$$\begin{array}{c} y = -2x + 3 \\ \qquad\;\uparrow \quad\;\; \uparrow \\ y = \;\; mx \; + b \end{array}$$

The slope of this line is −2 and its y-intercept is 3.

Let's look at another example.

EXAMPLE 9 Finding the Slope and y-Intercept of a Line

Find the slope m and y-intercept b of the line $2x + 4y - 8 = 0$. Graph the line.

Solution To obtain the slope and y-intercept, we transform the equation into its slope–intercept form. Thus, we need to solve for y:

$$\begin{aligned} 2x + 4y - 8 &= 0 \\ 4y &= -2x + 8 \\ y &= -\tfrac{1}{2}x + 2 \end{aligned}$$

FIGURE 74

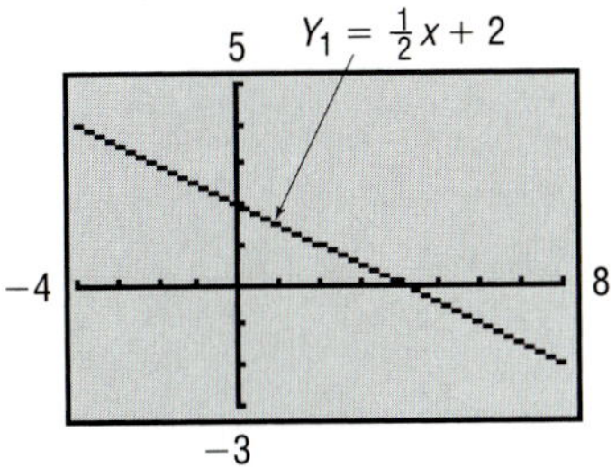

The coefficient of x, $-\frac{1}{2}$, is the slope, and the y-intercept is 2.

Figure 74 shows the graph using a graphing utility.

We can graph the line by hand in two ways:

1. Use the fact that the y-intercept is 2 and the slope is $-\frac{1}{2}$. Then, starting at the point (0, 2), go to the right 2 units and then down 1 unit to the point (2, 1). See Figure 75.

Or

2. Locate the intercepts. Because the y-intercept is 2, we know one intercept is $(0, 2)$. To obtain the x-intercept, let $y = 0$ and solve for x. When $y = 0$, we have

$$\begin{aligned} 2x + 4\cdot 0 - 8 &= 0 \\ 2x - 8 &= 0 \\ x &= 4 \end{aligned}$$

Thus, the intercepts are $(4, 0)$ and $(0, 2)$. See Figure 76.

FIGURE 75
$2x + 4y - 8 = 0$

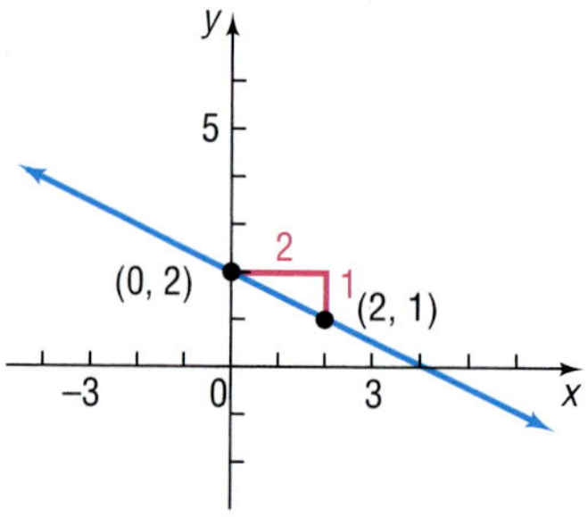

FIGURE 76
$2x + 4y - 8 = 0$

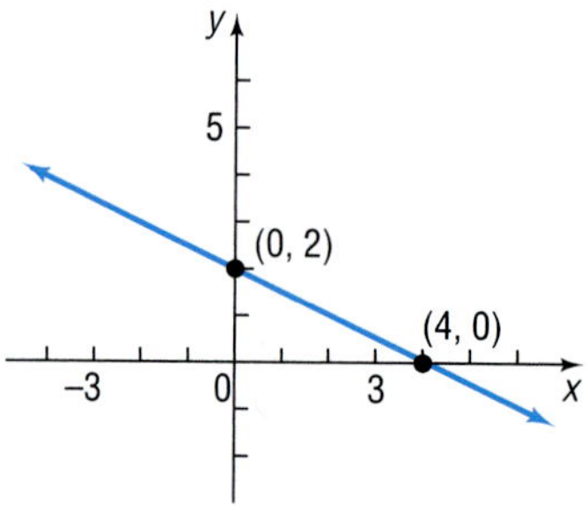

Now work Problem 45.

The next example illustrates a typical situation that requires the use of linear equations.

EXAMPLE 10 Computing the Cost of Operating a Car

The National Car Rental Company has determined that the cumulative cost of operating a vehicle is \$0.41 per mile.

(a) Write an equation that relates the cumulative cost C, in dollars, of operating a car and the number x of miles it has been driven.
(b) What is the cost of operating a car with 1,000 miles on it?
(c) What is the cost of operating a car with 2,000 miles on it?

Solution (a) If x is the number of miles the car has been driven, then the cumulative cost C, in dollars, is $0.41x$. Thus, an equation relating C and x is

$$C = 0.41x \qquad x \geq 0$$

The average cost per mile, \$0.41, is the slope of the line $C = 0.41x$. In other words, the cost increases by \$0.41 for each additional mile driven. See Figure 77.

(b) The cost of operating a car with 1,000 miles on it is

$$C = 0.41x = 0.41(1{,}000) = \$410$$

(c) The cost of operating a car with 2,000 miles on it is

$$C = 0.41x = 0.41(2{,}000) = \$820$$

FIGURE 77
$C = 0.41x$

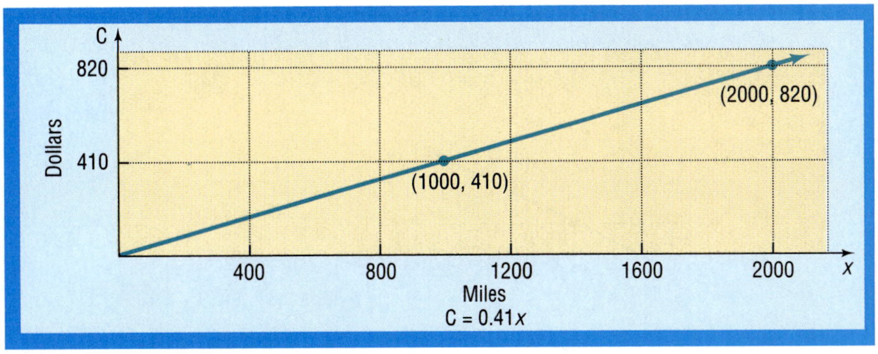

1.4 EXERCISES

In Problems 1–4 (a) find the slope of the line; (b) interpret the slope.

1.

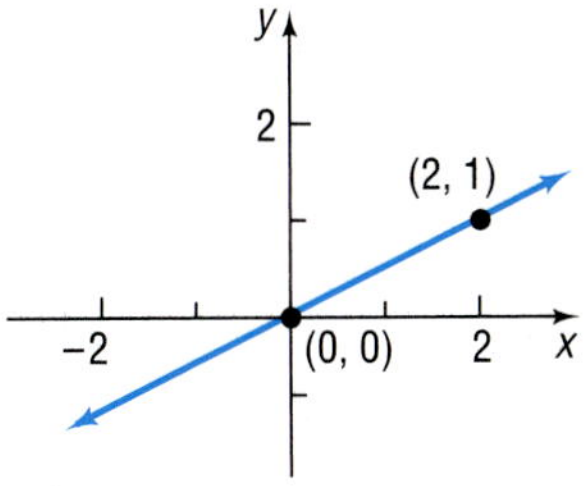

2.

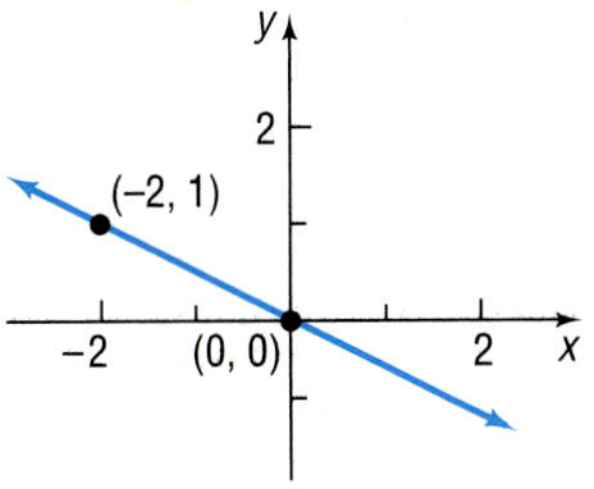

3.

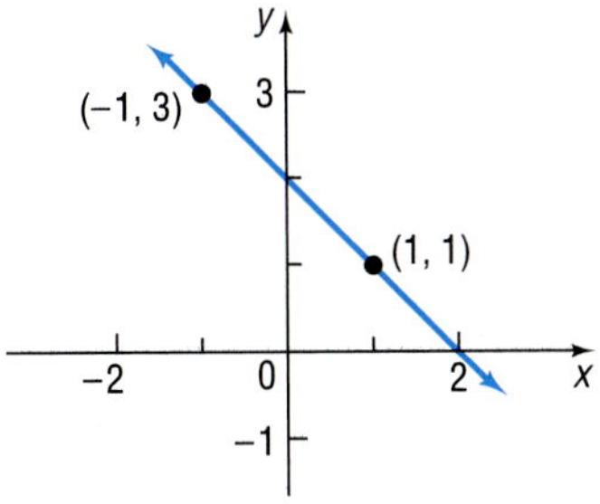

4.

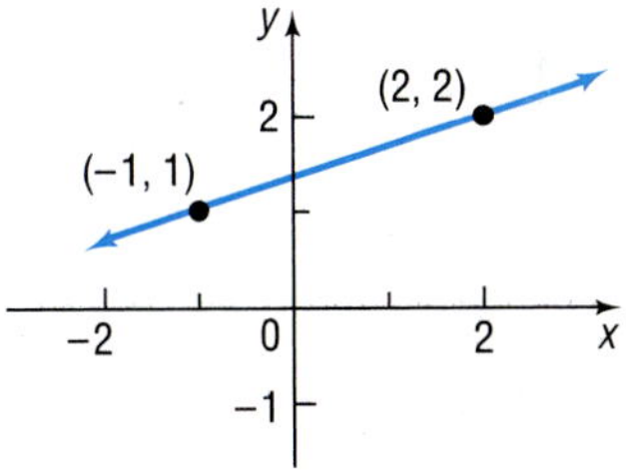

In Problems 5–14, plot each pair of points and determine the slope of the line containing them. By hand, graph the line.

5. $(2, 3); (4, 0)$
6. $(4, 2); (3, 4)$
7. $(-2, 3); (2, 1)$
8. $(-1, 1); (2, 3)$
9. $(-3, -1); (2, -1)$
10. $(4, 2); (-5, 2)$
11. $(-1, 2); (-1, -2)$
12. $(2, 0); (2, 2)$
13. $(\sqrt{2}, 3); (1, \sqrt{3})$
14. $(-2\sqrt{2}, 0); (4, \sqrt{5})$

In Problems 15–22, graph, by hand, the line passing through the point P and having slope m.

15. $P = (1, 2); m = 3$
16. $P = (2, 1); m = 4$
17. $P = (2, 4); m = \frac{-3}{4}$
18. $P = (1, 3); m = \frac{-2}{5}$
19. $P = (-1, 3); m = 0$
20. $P = (2, -4); m = 0$
21. $P = (0, 3)$; slope undefined
22. $P = (-2, 0)$; slope undefined

In Problems 23–26, find an equation of each line. Express your answer using either the general form or the slope–intercept form of the equation of a line, whichever you prefer.

23.

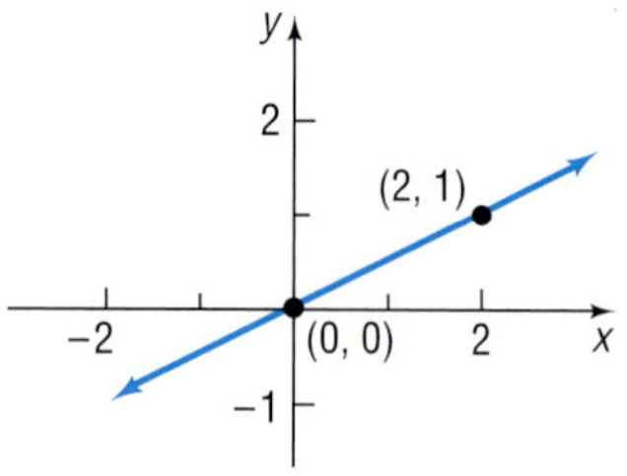

24.

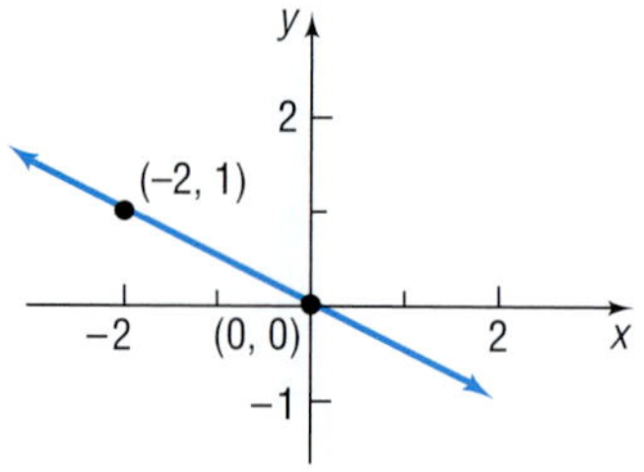

25.

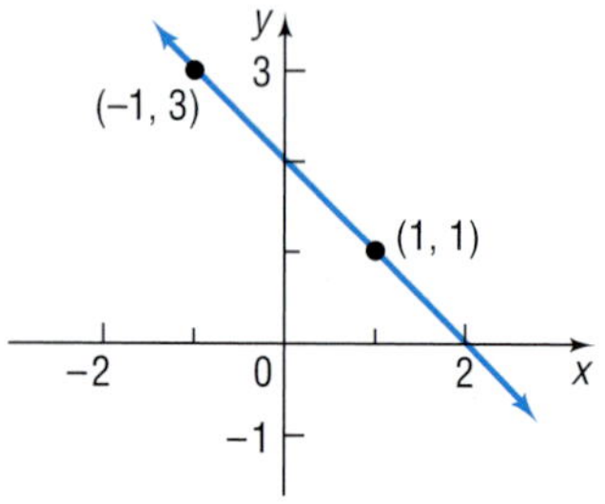

26.

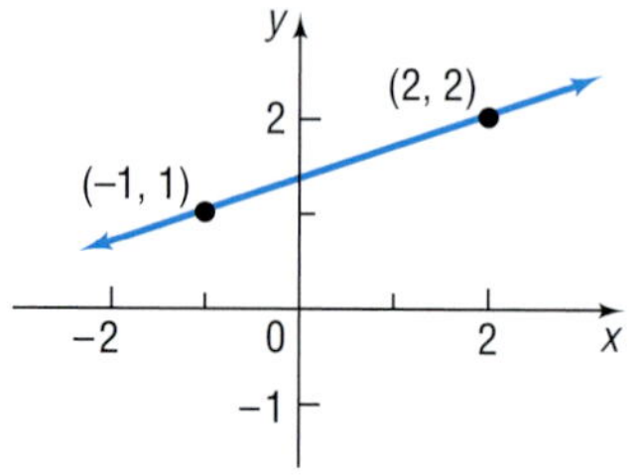

In Problems 27–38, find an equation for the line with the given properties. Express your answer using either the general form or the slope–intercept form of the equation of a line, whichever you prefer.

27. Slope = 3; passing through $(-2, 3)$

28. Slope = 2; passing through $(4, -3)$

29. Slope $= -\frac{2}{3}$; passing through $(1, -1)$

30. Slope $= \frac{1}{2}$; passing through $(3, 1)$

31. Passing through $(1, 3)$ and $(-1, 2)$

32. Passing through $(-3, 4)$ and $(2, 5)$

33. Slope $= -3$; y-intercept $= 3$

34. Slope $= -2$; y-intercept $= -2$

35. x-intercept $= 2$; y-intercept $= -1$

36. x-intercept $= -4$; y-intercept $= 4$

37. Slope undefined; passing through $(2, 4)$

38. Slope undefined; passing through $(3, 8)$

In Problems 39–58, find the slope and y-intercept of each line. By hand, graph the line. Check your graph using a graphing utility.

39. $y = 2x + 3$

40. $y = -3x + 4$

41. $\frac{1}{2}y = x - 1$

42. $\frac{1}{3}x + y = 2$

43. $y = \frac{1}{2}x + 2$

44. $y = 2x + \frac{1}{2}$

45. $x + 2y = 4$

46. $-x + 3y = 6$

47. $2x - 3y = 6$

48. $3x + 2y = 6$

49. $x + y = 1$

50. $x - y = 2$

51. $x = -4$

52. $y = -1$

53. $y = 5$

54. $x = 2$

55. $y - x = 0$

56. $x + y = 0$

57. $2y - 3x = 0$

58. $3x + 2y = 0$

59. Find an equation of the x-axis.

60. Find an equation of the y-axis.

61. Measuring Temperature The relationship between Celsius (°C) and Fahrenheit (°F) degrees for measuring temperature is linear. Find an equation relating °C and °F if 0°C corresponds to 32°F and 100°C corresponds to 212°F. Use the equation to find the Celsius measure of 70°F.

62. Measuring Temperature The Kelvin (K) scale for measuring temperature is obtained by adding 273 to the Celsius temperature.

(a) Write an equation relating K and °C.

(b) Write an equation relating K and °F (see Problem 61).

63. Business: Computing Profit Each Sunday, a newspaper agency sells x copies of a certain newspaper for \$1.00 per copy. The cost to the agency of each newspaper is \$0.50. The agency pays a fixed cost for storage, delivery, and so on, of \$100 per Sunday.

(a) Write an equation that relates the profit P, in dollars, to the number x of copies sold. Graph this equation.

(b) What is the profit to the agency if 1,000 copies are sold?

(c) What is the profit to the agency if 5,000 copies are sold?

64. Business: Computing Profit Repeat Problem 63 if the cost to the agency is \$0.45 per copy and the fixed cost is \$125 per Sunday.

65. **Cost of Electricity** In 1991, Commonwealth Edison Company supplied electricity in the summer months to residential customers for a monthly customer charge of \$9.06 plus 10.819¢ per kilowatt-hour supplied in the month.* Write an equation that relates the monthly charge C, in dollars, to the number x of kilowatt-hours in the month. Graph this equation. What is the monthly charge for using 300 kilowatt-hours? For using 900 kilowatt-hours?

66. Show that an equation for a line with nonzero x- and y-intercepts can be written as

$$\frac{x}{a} + \frac{y}{b} = 1$$

where a is the x-intercept and b is the y-intercept. This is called the **intercept form** of the equation of a line.

In Problems 67–70, match each graph with the correct equation:

(*a*) $y = x$; (*b*) $y = 2x$; (*c*) $y = x/2$; (*d*) $y = 4x$.

67.

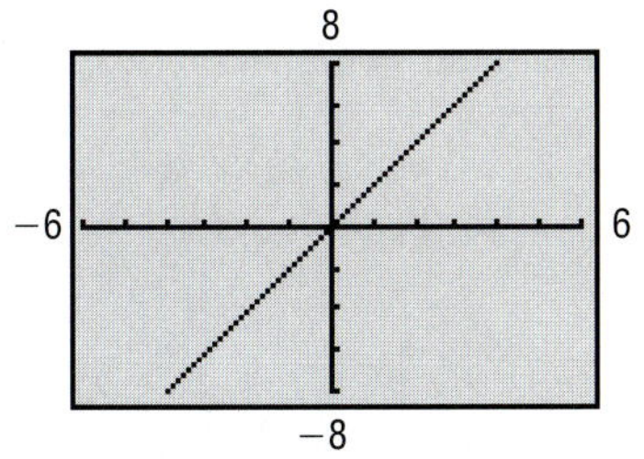

68.

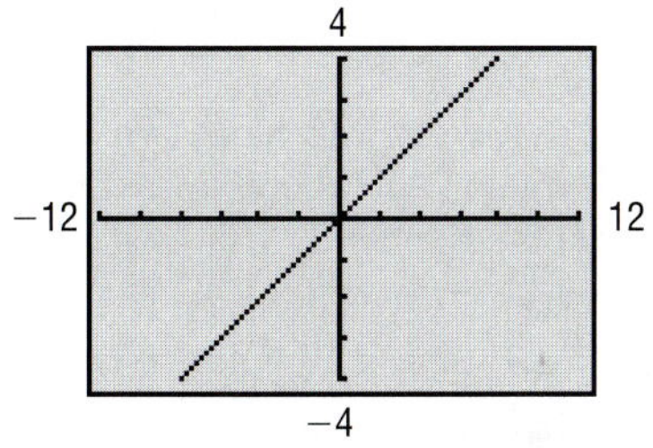

69.

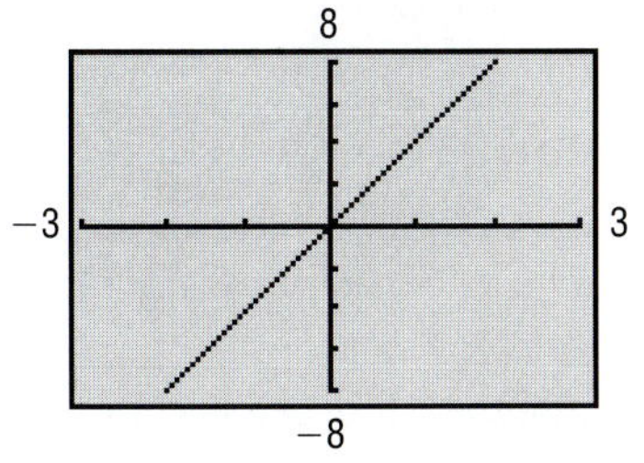

70.

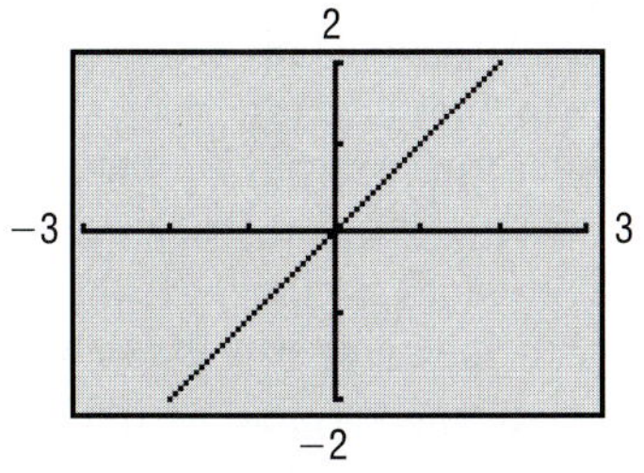

In Problems 71–76, write an equation of each line. Express your answer using either the general form or the slope–intercept form of the equation of a line, whichever you prefer.

71.

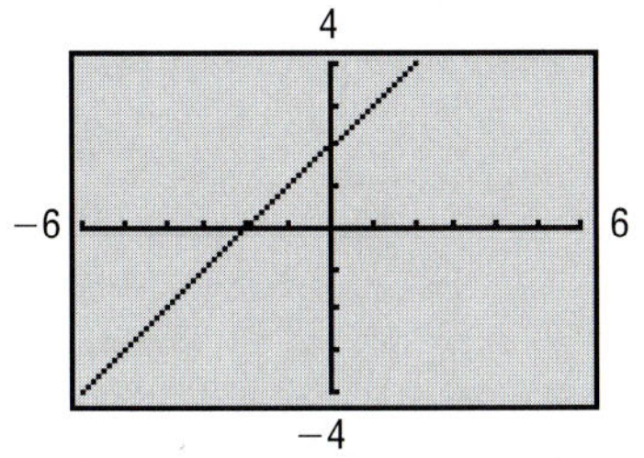

72.

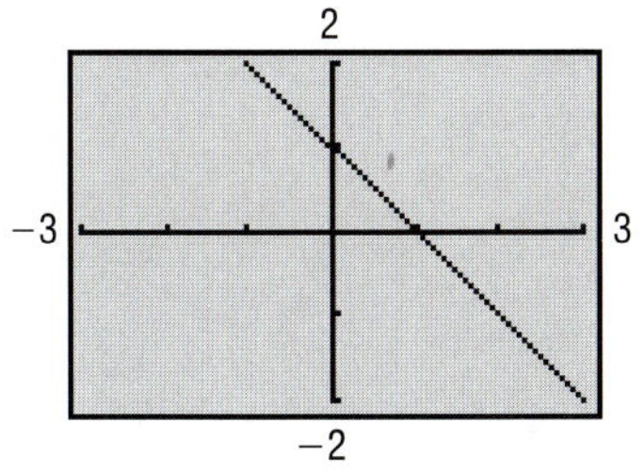

73.

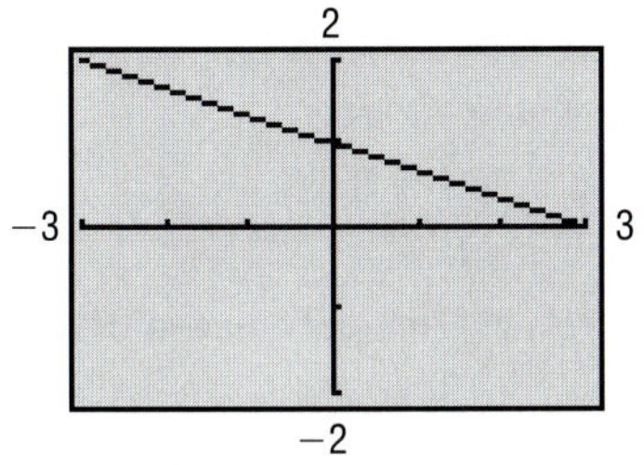

74.

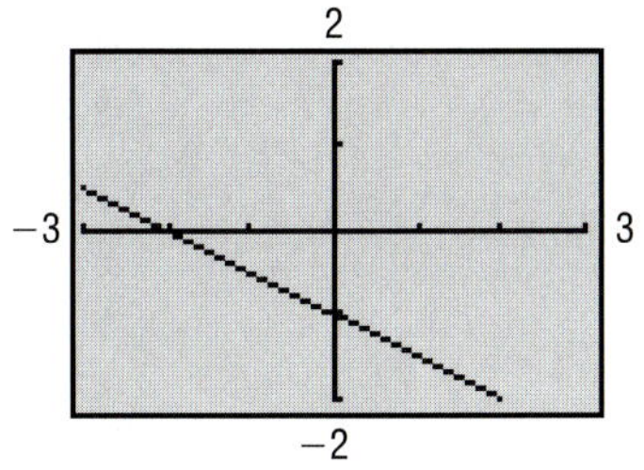

**Source:* Commonwealth Edison, Co., Chicago, Illinois, 1991.

75.

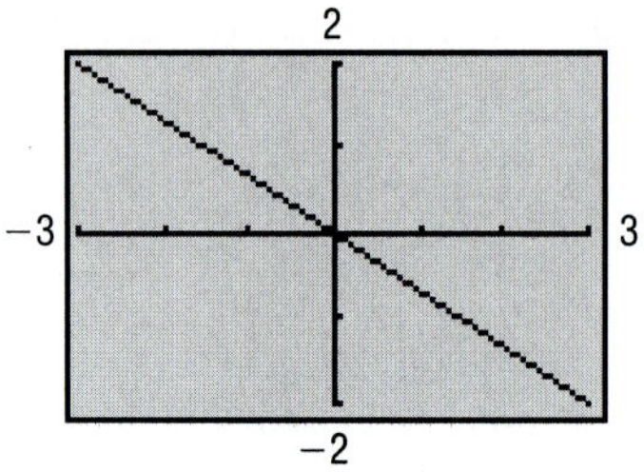

76.

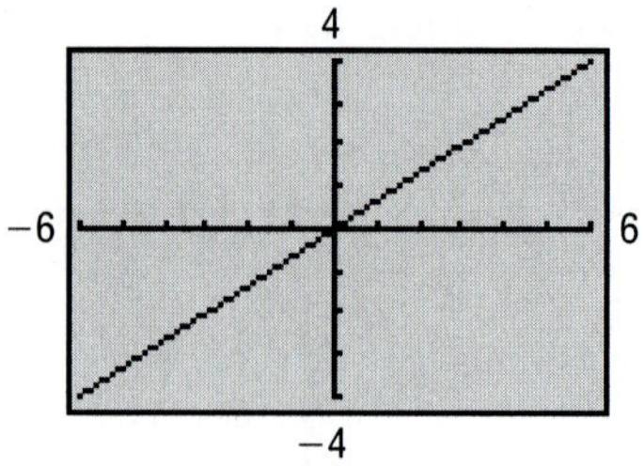

77. Which form of the equation of a line do you prefer to use? Justify your position with an example that shows that your choice is better than another. Have reasons.

78. Can every line be written in slope–intercept form? Explain.

79. Does every line have two distinct intercepts? Explain. Are there lines that have no intercepts? Explain.

80. What can you say about two lines that have equal slopes and equal y-intercepts?

81. What can you say about two lines with the same x-intercept and the same y-intercept? Assume that the x-intercept is not 0.

82. If two lines have the same slope, but different x-intercepts, can they have the same y-intercept?

83. If two lines have the same y-intercept, but different slopes, can they have the same x-intercept? What is the only way this can happen?

84. The accepted symbol used to denote the slope of a line is the letter m. Investigate the origin of this symbolism. Begin by consulting a French dictionary and looking up the French word *monter*. Write a brief essay on your findings.

85. **Grade of a Road** The term *grade* is used to describe the inclination of a road. How does this term relate to the notion of slope of a line? Is a 4% grade very steep? Investigate the grades of some mountainous roads and determine their slopes. Write a brief essay on your findings.

86. **Carpentry** Carpenters use the term pitch to describe the steepness of staircases and roofs. How does pitch relate to slope? Investigate typical pitches used for stairs and for roofs. Write a brief essay on your findings.

1.5 PARALLEL AND PERPENDICULAR LINES; CIRCLES

1 Define Parallel Lines
2 Find Equations of Parallel Lines
3 Define Perpendicular Lines
4 Find Equations of Perpendicular Lines
5 Write the Standard Form of the Equation of a Circle
6 Graph a Circle by Hand and by Using a Graphing Utility
7 Write the General Form of the Equation of a Circle
8 Identify the Center and Radius of a Circle in General Form and Graph it

Parallel and Perpendicular Lines

1 When two lines (in the plane) have no points in common, they are said to be **parallel.** Look at Figure 78. There we have drawn two lines and have constructed two right triangles by drawing sides parallel to the coordinate axes. These lines are parallel if and only if the right triangles are similar. (Do you

see why? Two angles are equal.) But the triangles are similar if and only if the ratios of corresponding sides are equal.

FIGURE 78
The lines are parallel if and only if their slopes are equal.

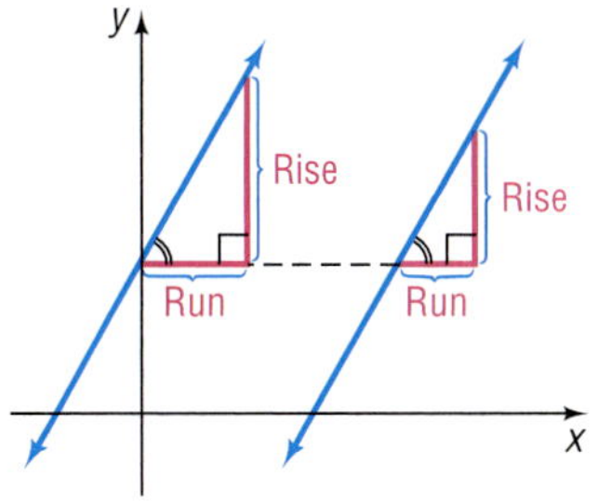

This suggests the following result:

Theorem Criterion for Parallel Lines

Two distinct nonvertical lines are parallel if and only if their slopes are equal.

The use of the words "if and only if" in the preceding theorem means that actually two statements are being made, one the converse of the other.

If two distinct nonvertical lines are parallel, then their slopes are equal.

If two distinct nonvertical lines have equal slopes, then they are parallel.

Now work Problem 1(a).

EXAMPLE 1 Showing That Two Lines Are Parallel

Show that the lines given by the following equations are parallel:

$$L:\ 2x + 3y - 6 = 0 \qquad M:\ 4x + 6y = 0$$

Solution To determine whether these lines have equal slopes, we write each equation in slope–intercept form:

$$\begin{aligned} L:\ 2x + 3y - 6 &= 0 \\ 3y &= -2x + 6 \\ y &= -\tfrac{2}{3}x + 2 \\ \text{Slope} &= -\tfrac{2}{3} \end{aligned} \qquad \begin{aligned} M:\ 4x + 6y &= 0 \\ 6y &= -4x \\ y &= -\tfrac{2}{3}x \\ \text{Slope} &= -\tfrac{2}{3} \end{aligned}$$

Because these lines have the same slope, $-\frac{2}{3}$, but different y-intercepts, the lines are parallel. See Figure 79.

FIGURE 79
Parallel lines

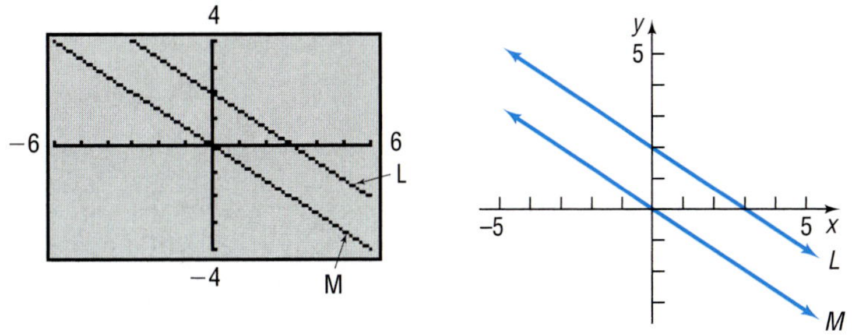

EXAMPLE 2 Finding a Line That Is Parallel to a Given Line

2 Find an equation for the line that contains the point $(2, -3)$ and is parallel to the line $2x + y - 6 = 0$.

Solution The slope of the line we seek equals the slope of the line $2x + y - 6 = 0$, since the two lines are to be parallel. Thus, we begin by writing the equation of the line $2x + y - 6 = 0$ in slope–intercept form:

$$2x + y - 6 = 0$$
$$y = -2x + 6$$

The slope is -2. Since the line we seek contains the point $(2, -3)$, we use the point–slope form to obtain

$$y + 3 = -2(x - 2)$$
$$2x + y - 1 = 0 \quad \text{General form.}$$
$$y = -2x + 1 \quad \text{Slope–intercept form.}$$

This line is parallel to the line $2x + y - 6 = 0$ and contains the point $(2, -3)$. See Figure 80.

FIGURE 80

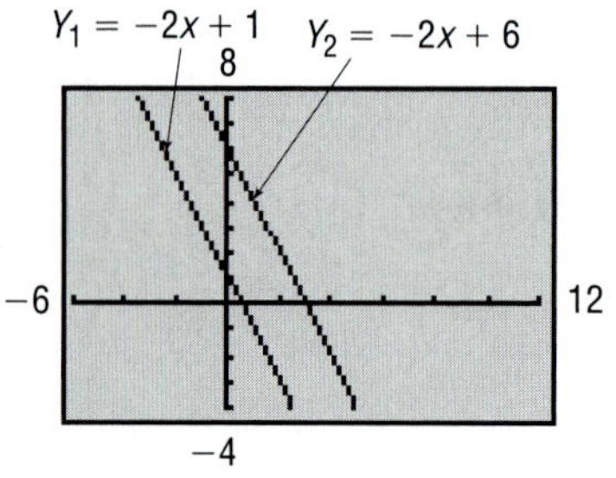

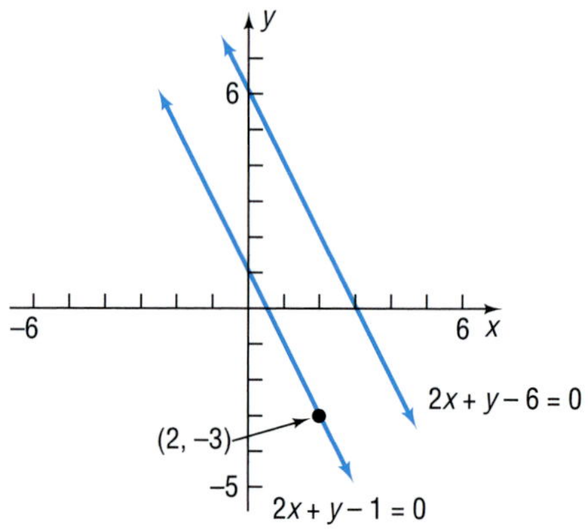

Now work Problem 15.

3 When two lines intersect at a right angle (90°), they are said to be **perpendicular.** See Figure 81.

The following result gives a condition, in terms of their slopes, for two lines to be perpendicular:

FIGURE 81
Perpendicular lines

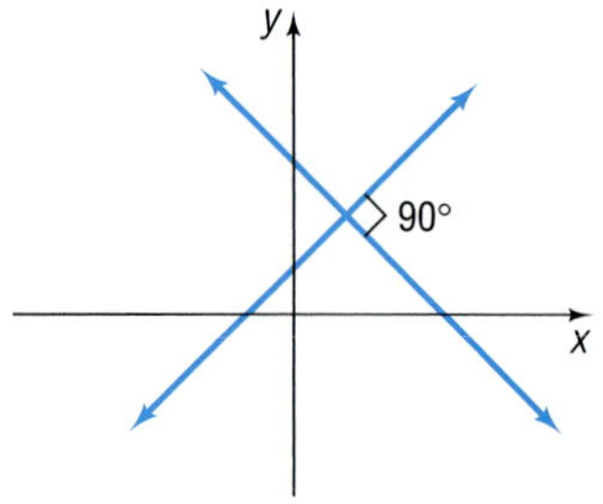

Theorem Criterion for Perpendicular Lines

Two nonvertical lines are perpendicular if and only if the product of their slopes is -1.

Here, we shall prove the "only if" part of the statement:

If two nonvertical lines are perpendicular, then the product of their slopes is -1.

FIGURE 82

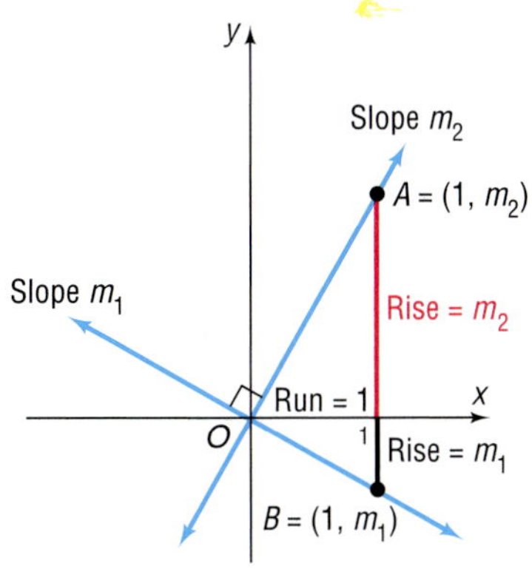

Proof Let m_1 and m_2 denote the slopes of the two lines. There is no loss in generality (that is, neither the angle nor the slopes are affected) if we situate the lines so that they meet at the origin. See Figure 82. The point $A = (1, m_2)$ is on the line having slope m_2, and the point $B = (1, m_1)$ is on the line having slope m_1. (Do you see why this must be true?)

Suppose that the lines are perpendicular. Then triangle OAB is a right triangle. As a result of the Pythagorean Theorem, it follows that

$$[d(O, A)]^2 + [d(O, B)]^2 = [d(A, B)]^2 \tag{1}$$

By the distance formula, we can write each of these distances as

$$[d(O, A)]^2 = (1 - 0)^2 + (m_2 - 0)^2 = 1 + m_2^2$$
$$[d(O, B)]^2 = (1 - 0)^2 + (m_1 - 0)^2 = 1 + m_1^2$$
$$[d(A, B)]^2 = (1 - 1)^2 + (m_2 - m_1)^2 = m_2^2 - 2m_1m_2 + m_1^2$$

Using these facts in equation (1), we get

$$(1 + m_2^2) + (1 + m_1^2) = m_2^2 - 2m_1m_2 + m_1^2$$

which, upon simplification, can be written as

$$m_1m_2 = -1$$

Thus, if the lines are perpendicular, the product of their slopes is -1. ■

You may find it easier to remember the condition for two nonvertical lines to be perpendicular by observing that the equality $m_1m_2 = -1$ means that m_1 and m_2 are negative reciprocals of each other; that is, either $m_1 = -1/m_2$ or $m_2 = -1/m_1$.

EXAMPLE 3 Finding the Slope of a Line Perpendicular to a Given Line

If a line has slope $\frac{3}{2}$, any line having slope $-\frac{2}{3}$ is perpendicular to it. ■

EXAMPLE 4 Finding the Equation of a Line Perpendicular to a Given Line

4

Find an equation of the line passing through the point $(1, -2)$ and perpendicular to the line $x + 3y - 6 = 0$. Graph the two lines.

Solution We first write the equation of the given line in slope–intercept form to find its slope:

$$x + 3y - 6 = 0$$
$$3y = -x + 6$$
$$y = -\tfrac{1}{3}x + 2$$

The given line has slope $-\frac{1}{3}$. Any line perpendicular to this line will have slope 3. Because we require the point $(1, -2)$ to be on this line with slope 3, we use the point–slope form of the equation of a line:

$$y - (-2) = 3(x - 1)$$
$$y + 2 = 3(x - 1)$$

This equation is equivalent to the forms

$$3x - y - 5 = 0 \qquad \text{General form.}$$
$$y = 3x - 5 \qquad \text{Slope–intercept form.}$$

Figure 83 shows the graphs.

FIGURE 83

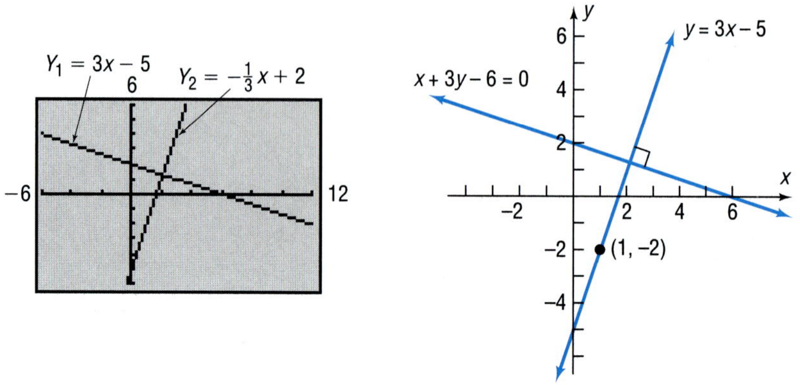

Now work Problem 23.

Warning: Be sure to use a square screen when you graph perpendicular lines. Otherwise, the angle between the two lines will appear distorted.

Circles

One advantage of a coordinate system is that it enables us to translate a geometric statement into an algebraic statement, and vice versa. Consider, for example, the following geometric statement that defines a circle.

> A **circle** is a set of points in the *xy*-plane that are a fixed distance *r* from a fixed point (h, k). The fixed distance *r* is called the **radius,** and the fixed point (h, k) is called the **center** of the circle.

FIGURE 84

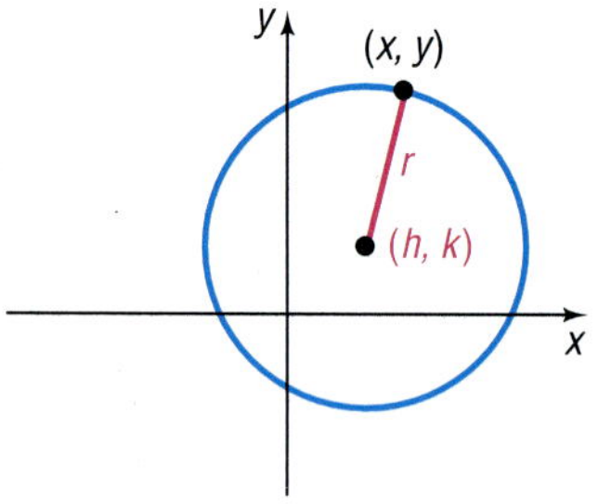

Figure 84 shows the graph of a circle. Is there an equation having this graph? If so, what is the equation? To find the equation, we let (x, y) represent the coordinates of any point on a circle with radius r and center (h, k). Then the distance between the points (x, y) and (h, k) must always equal r. That is, by the distance formula,

$$\sqrt{(x - h)^2 + (y - k)^2} = r$$

or, equivalently,

$$(x - h)^2 + (y - k)^2 = r^2$$

5 The **standard form of an equation of a circle** with radius r and center (h, k) is

$$(x - h)^2 + (y - k)^2 = r^2 \qquad (2)$$

Conversely, by reversing the steps, we conclude: The graph of any equa-

tion of the form of equation (2) is that of a circle with radius r and center (h, k).

EXAMPLE 5 Graphing a Circle

6

Graph the equation $(x + 3)^2 + (y - 2)^2 = 16$.

Solution The graph of the equation is a circle. To graph a circle on a graphing utility we must write the equation in the form $y =$ expression involving x.* Thus, we must solve for y in the equation

$$\begin{aligned}(x + 3)^2 + (y - 2)^2 &= 16\\ (y - 2)^2 &= 16 - (x + 3)^2\\ y - 2 &= \pm\sqrt{16 - (x + 3)^2}\\ y &= 2 \pm \sqrt{16 - (x + 3)^2}\end{aligned}$$

To graph the circle, we first graph the top half

$$Y_1 = 2 + \sqrt{16 - (x + 3)^2}$$

and then graph the bottom half

$$Y_2 = 2 - \sqrt{16 - (x + 3)^2}$$

Also, be sure to use a square screen. Otherwise the circle will appear distorted. Figure 85 shows the graph.

FIGURE 85

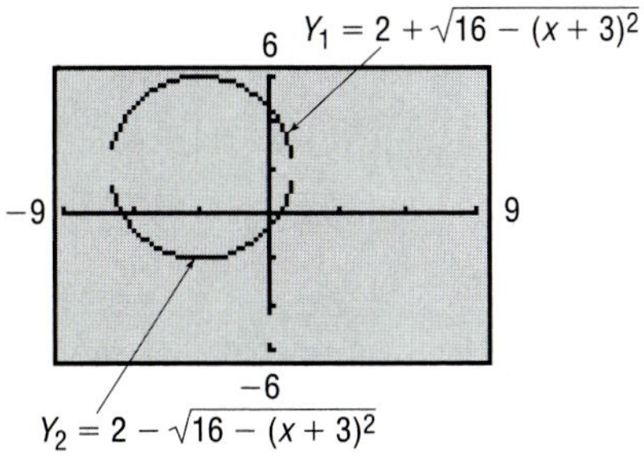

To graph the equation by hand, we first compare the given equation to the standard form of the equation of a circle. The comparison yields information about the circle:

$$\begin{aligned}(x + 3)^2 + (y - 2)^2 &= 16\\ (x - (-3))^2 + (y - 2)^2 &= 4^2\\ (x - h)^2 + (y - k)^2 &= r^2\end{aligned}$$

We see that $h = -3$, $k = 2$, and $r = 4$. Hence, the circle has center $(-3, 2)$ and a radius of 4 units. To graph this circle, we first plot the center $(-3, 2)$. Since the radius is 4, we can locate four points on the circle by going out 4 units to the left, to the right, up, and down from the center. These four points can then be used as guides to obtain the graph. See Figure 86.

FIGURE 86

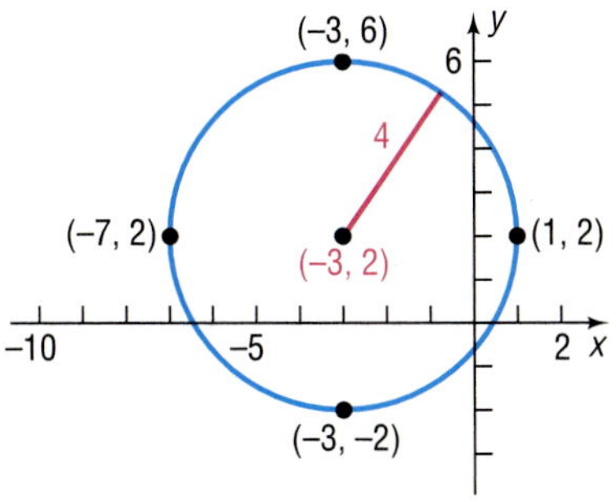

Now work Problem 43.

EXAMPLE 6 Writing the Standard Form of the Equation of a Circle

Write the standard form of the equation of the circle with radius 3 and center $(1, -2)$.

Solution Using the form of equation (2) and substituting the values $r = 3$, $h = 1$, and $k = -2$, we have

$$\begin{aligned}(x - h)^2 + (y - k)^2 &= r^2\\ (x - 1)^2 + (y + 2)^2 &= 9\end{aligned}$$

Now work Problem 27.

*Some graphing utilities (e.g., TI-82, TI-85) have a CIRCLE function which allows the user to enter only the coordinates of the center of the circle and its radius to graph the circle.

The standard form of an equation of a circle of radius r with center at the origin $(0, 0)$ is

$$x^2 + y^2 = r^2$$

If the radius $r = 1$, the circle whose center is at the origin is called the **unit circle** and has the equation

$$x^2 + y^2 = 1$$

See Figure 87.

FIGURE 87
Unit circle $x^2 + y^2 = 1$

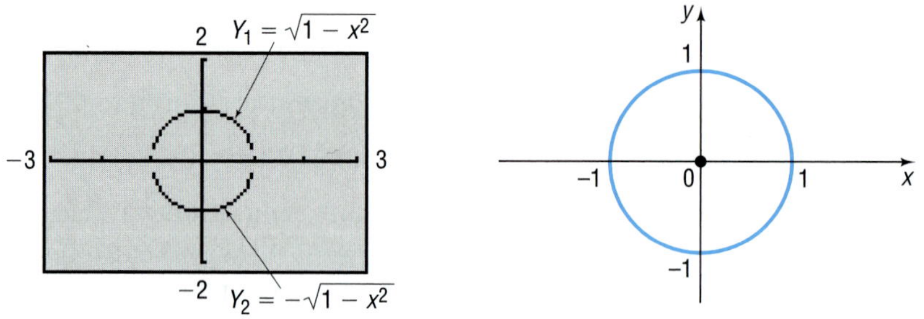

If we eliminate the parentheses from the standard form of the equation of the circle obtained in Example 6, we get

$$(x - 1)^2 + (y + 2)^2 = 9$$
$$x^2 - 2x + 1 + y^2 + 4y + 4 = 9$$

which we find, upon simplifying, is equivalent to

$$x^2 + y^2 - 2x + 4y - 4 = 0$$

By completing the squares on both the x- and y-terms, it can be shown that any equation of the form

$$x^2 + y^2 + ax + by + c = 0$$

has a graph that is a circle, or a point, or has no graph at all. For example, the graph of the equation $x^2 + y^2 = 0$ is the single point $(0, 0)$. The equation $x^2 + y^2 + 5 = 0$, or $x^2 + y^2 = -5$, has no graph, because sums of squares of real numbers are never negative. When its graph is a circle, the equation

$$x^2 + y^2 + ax + by + c = 0$$

is referred to as the **general form of the equation of a circle.**

7 The next example shows how to transform an equation in the general form to an equivalent equation in standard form. As we said earlier, the idea is to use the method of completing the square on both the x- and y-terms. See the Appendix, Section 5, for a discussion of completing the square.

EXAMPLE 7 Graphing a Circle Whose Equation Is in General Form

Graph the equation $x^2 + y^2 + 4x - 6y + 12 = 0$.

Solution We rearrange the equation as follows:

$$(x^2 + 4x) + (y^2 - 6y) = -12$$

Next, we complete the square of each expression in parentheses. Remember that any number added on the left also must be added on the right:

$$(x^2 + 4x + 4) + (y^2 - 6y + 9) = -12 + 4 + 9$$
$$(x + 2)^2 + (y - 3)^2 = 1$$

We recognize this equation as the standard form of the equation of a circle with radius 1 and center $(-2, 3)$.

To graph the equation using a graphing utility, we need to solve for y:

$$(y - 3)^2 = 1 - (x + 2)^2$$
$$y - 3 = \pm\sqrt{1 - (x + 2)^2}$$
$$y = 3 \pm \sqrt{1 - (x + 2)^2}$$

To graph the equation by hand, we use the center $(-2, 3)$ and the radius 1. Figure 88 illustrates the graph.

FIGURE 88
$x^2 + y^2 + 4x - 6y + 12 = 0$

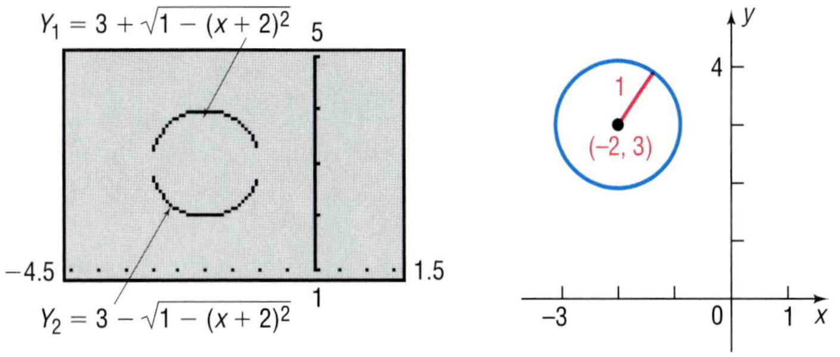

Now work Problem 45.

EXAMPLE 8 Finding the General Equation of a Circle

Find the general equation of the circle whose center is $(1, -2)$ and whose graph contains the point $(4, -2)$.

Solution To find the equation of a circle, we need to know its center and its radius. Here, we know that the center is $(1, -2)$. Since the point $(4, -2)$ is on the graph, the radius r will equal the distance from $(4, -2)$ to the center $(1, -2)$. See Figure 89. Thus,

$$r = \sqrt{(4 - 1)^2 + [-2 - (-2)]^2}$$
$$= \sqrt{9} = 3$$

The standard form of the equation of the circle is

$$(x - 1)^2 + (y + 2)^2 = 9$$

Eliminating the parentheses and rearranging terms, we get the general equation

$$x^2 + y^2 - 2x + 4y - 4 = 0$$

FIGURE 89
$x^2 + y^2 - 2x + 4y - 4 = 0$

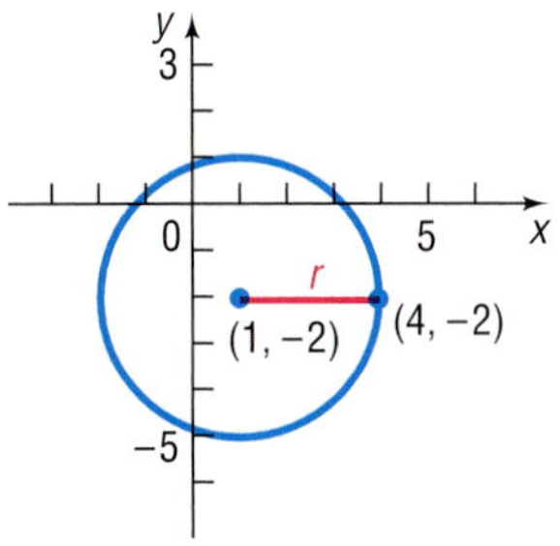

Overview

The preceding discussion about lines and circles dealt with two main types of problems that can be generalized as follows:

1. Given an equation, classify it and graph it.
2. Given a graph, or information about a graph, find its equation.

This text deals with both types of problems. We shall study various equations, classify them, and graph them. Although the second type of problem is usually more difficult to solve than the first, in many instances a graphing utility can be used to solve such problems.

1.5 EXERCISES

In Problems 1–10, the equation of a line L is given. Find the slope of a line that is (a) parallel to L; (b) perpendicular to L.

1. $y = 4x$ **2.** $y = -5x$ **3.** $y = -\frac{1}{2}x + 2$ **4.** $y = \frac{2}{3}x - 1$

5. $2x - 4y + 5 = 0$ **6.** $3x + y = 4$ **7.** $3x + 5y - 10 = 0$ **8.** $4x - 3y + 7 = 0$

9. $x = 4$ **10.** $y = 5$

In Problems 11–14, find an equation for the line L. Express your answer using either the general form or the slope–intercept form of the equation of a line, whichever you prefer.

11.

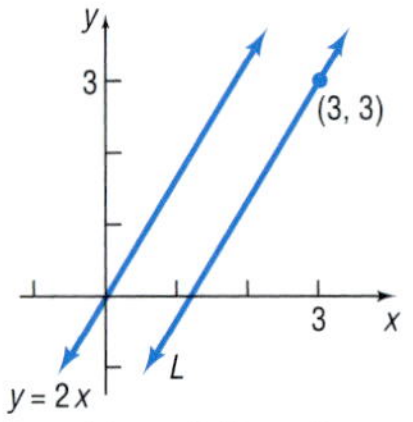

L is parallel to $y = 2x$

12.

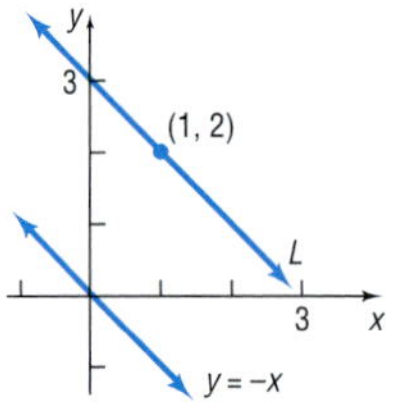

L is parallel to $y = -x$

13.

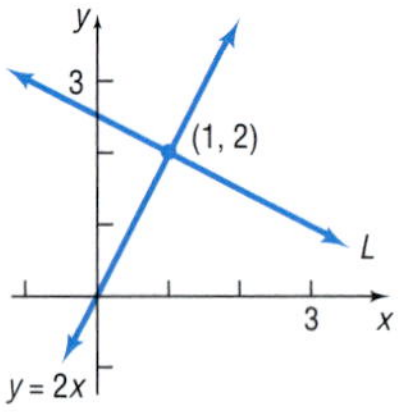

L is perpendicular to $y = 2x$

14.

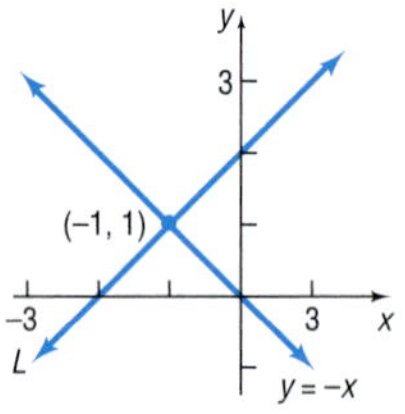

L is perpendicular to $y = -x$

In Problems 15–26, find an equation for the line with the given properties. Express your answer using either the general form or the slope–intercept form of the equation of a line, whichever you prefer.

15. Parallel to the line $y = 2x$; passing through $(-1, 2)$

16. Parallel to the line $y = -3x$; passing through $(-1, 2)$

17. Parallel to the line $2x - y + 2 = 0$; passing through $(0, 0)$

18. Parallel to the line $x - 2y + 5 = 0$; passing through $(0, 0)$

19. Parallel to the line $x = 5$; passing through $(4, 2)$

20. Parallel to the line $y = 5$; passing through $(4, 2)$

21. Perpendicular to the line $y = \frac{1}{2}x + 4$; passing through $(1, -2)$

22. Perpendicular to the line $y = 2x - 3$; passing through $(1, -2)$

23. Perpendicular to the line $2x + y - 2 = 0$; passing through $(-3, 0)$

24. Perpendicular to the line $x - 2y + 5 = 0$; passing through $(0, 4)$

25. Perpendicular to the line $x = 8$; passing through $(3, 4)$

26. Perpendicular to the line $y = 8$; passing through $(3, 4)$

In Problems 27–30, find the center and radius of each circle. Write the standard form of the equation.

27.

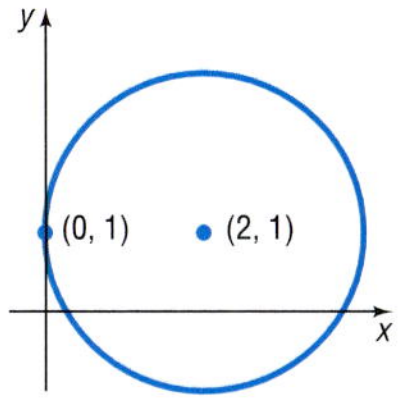

28.

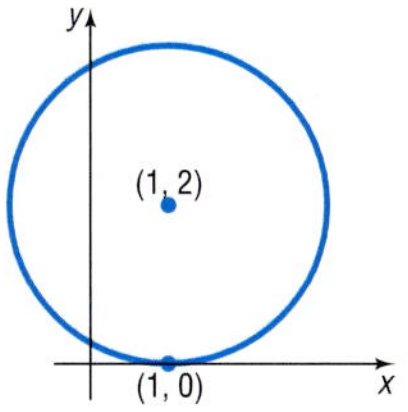

29.

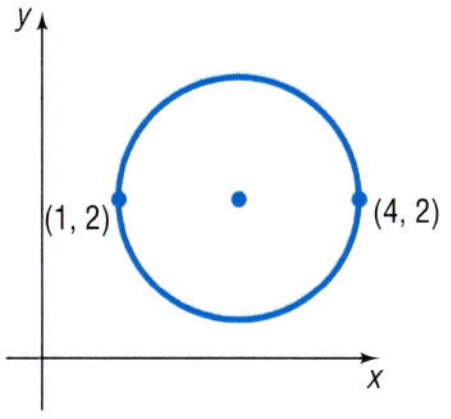

30.

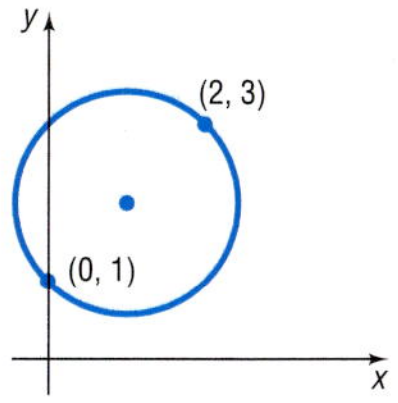

In Problems 31–40, write the standard form of the equation and the general form of the equation of each circle of radius r and center (h, k). By hand, graph each circle.

31. $r = 1$; $(h, k) = (1, -1)$

32. $r = 2$; $(h, k) = (-2, 1)$

33. $r = 2$; $(h, k) = (0, 2)$

34. $r = 3$; $(h, k) = (1, 0)$

35. $r = 5$; $(h, k) = (4, -3)$

36. $r = 4$; $(h, k) = (2, -3)$

37. $r = 2$; $(h, k) = (0, 0)$

38. $r = 3$; $(h, k) = (0, 0)$

39. $r = \frac{1}{2}$; $(h, k) = \left(\frac{1}{2}, 0\right)$

40. $r = \frac{1}{2}$; $(h, k) = \left(0, -\frac{1}{2}\right)$

In Problems 41–50, find the center (h, k) and radius r of each circle. By hand, graph each circle.

41. $x^2 + y^2 = 4$

42. $x^2 + (y - 1)^2 = 1$

43. $(x - 3)^2 + y^2 = 4$

44. $(x + 1)^2 + (y - 1)^2 = 2$

45. $x^2 + y^2 + 4x - 4y - 1 = 0$

46. $x^2 + y^2 - 6x + 2y + 9 = 0$

47. $x^2 + y^2 - x + 2y + 1 = 0$

48. $x^2 + y^2 + x + y - \frac{1}{2} = 0$

49. $2x^2 + 2y^2 - 12x + 8y - 24 = 0$

50. $2x^2 + 2y^2 + 8x + 7 = 0$

In Problems 51–56, find the general form of the equation of each circle.

51. Center at the origin and containing the point $(-2, 3)$

52. Center $(1, 0)$ and containing the point $(-3, 2)$

53. Center $(2, 3)$ and tangent to the x-axis

54. Center $(-3, 1)$ and tangent to the y-axis

55. With endpoints of a diameter at $(1, 4)$ and $(-3, 2)$

56. With endpoints of a diameter at $(4, 3)$ and $(0, 1)$

In Problems 57–60, match each graph with the correct equation.

(a) $(x - 3)^2 + (y + 3)^2 = 9$

(b) $(x + 1)^2 + (y - 2)^2 = 4$

(c) $(x - 1)^2 + (y + 2)^2 = 4$

(d) $(x + 3)^2 + (y - 3)^2 = 9$

57.

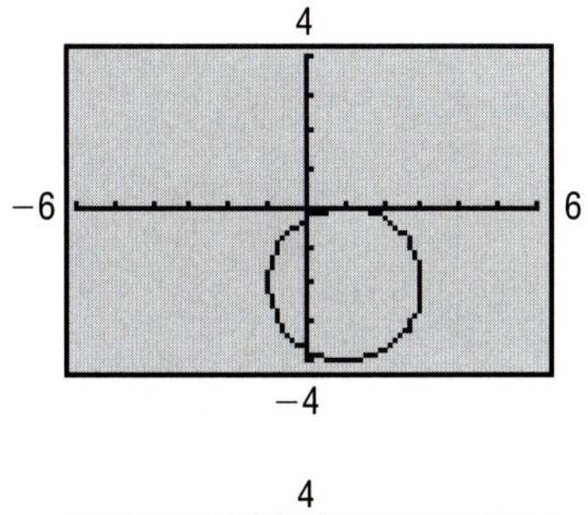

58.

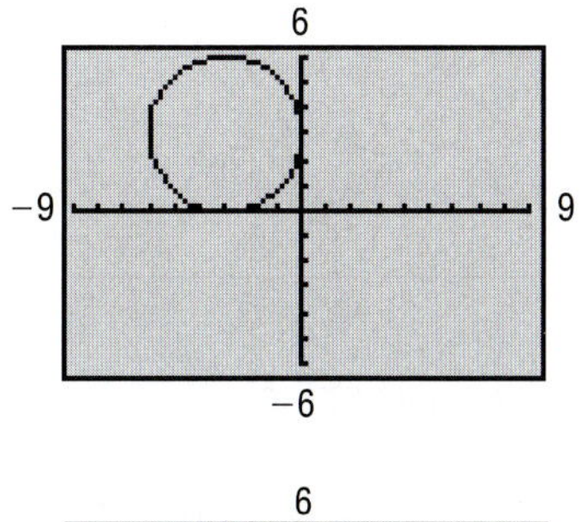

59.

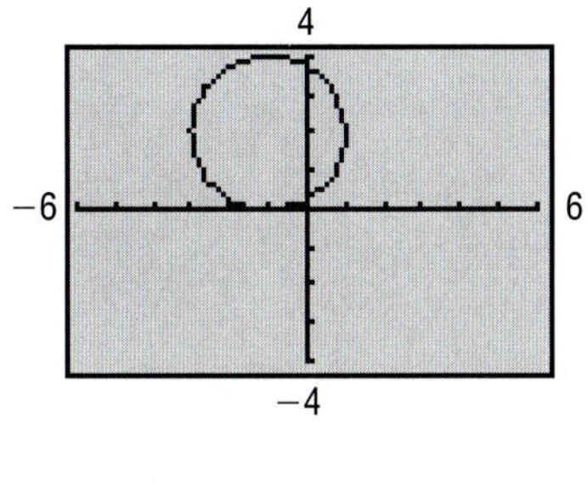

60.

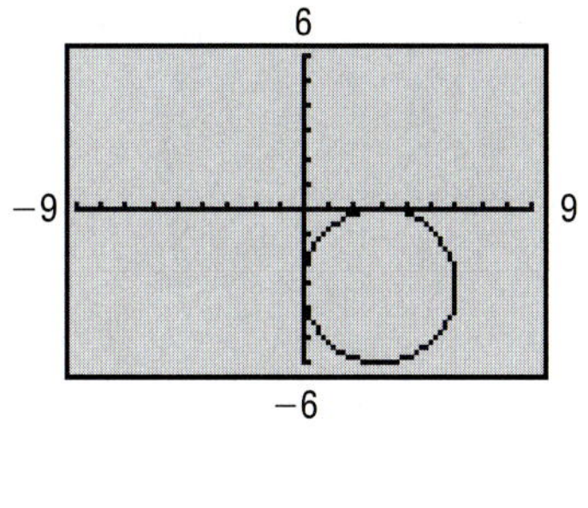

In Problems 61–64, find the standard form of the equation of each circle.

61.

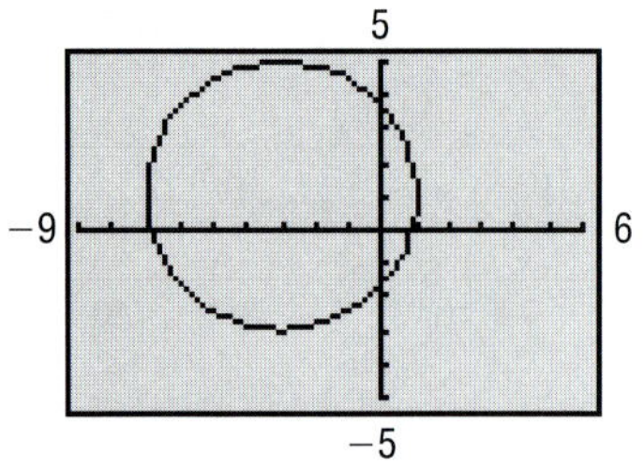

62.

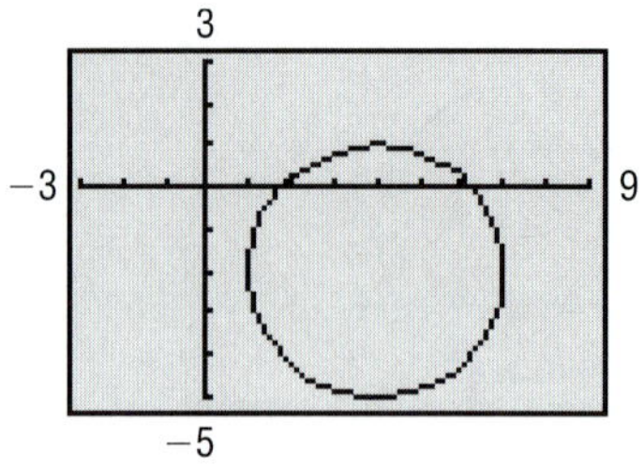

63.

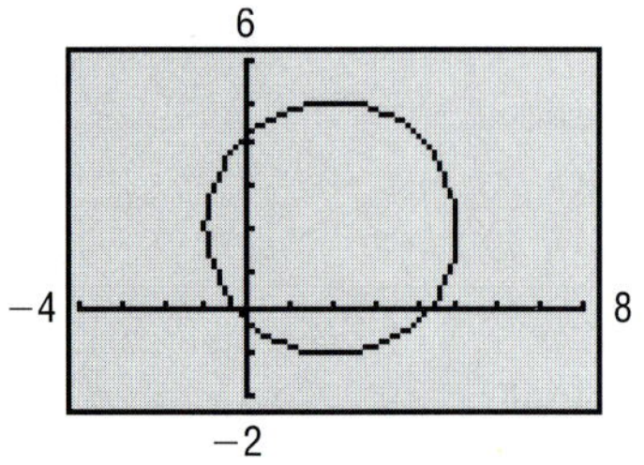

64.

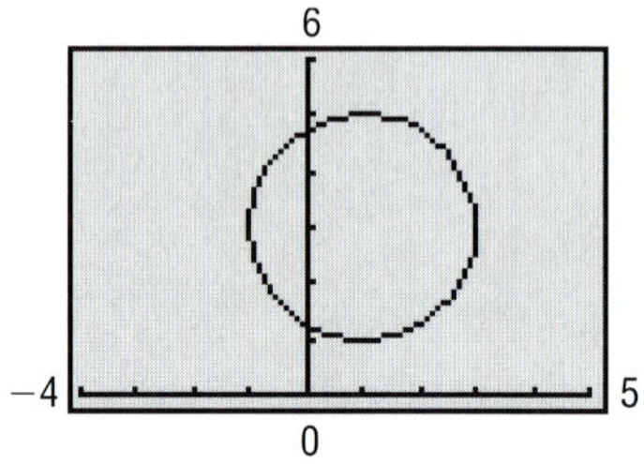

1.6 LINEAR CURVE FITTING

1. Distinguish between Linear and Nonlinear Relations
2. Use a Graphing Utility to Find the Line of Best Fit
3. Use the Line of Best Fit to Make Predictions

Curve fitting is an area of statistics in which a relation between two or more variables is explained through an equation. For example, the equation Sales = \$100,000 + 12Advertising implies that if advertising expenditures were \$0, sales would be \$100,000 + 12(0) = \$100,000 and if advertising expenditures were \$10,000, sales would be \$100,000 + 12(\$10,000) = \$220,000. In this model, the variable advertising is called the predictor (independent) variable and sales is called the response (dependent) variable because if the level of advertising is known, it can be used to predict sales. Curve fitting is used to find an equation that relates two or more variables, using observed or experimental data.

There are three steps to follow to determine whether a relation exists between two variables.

Steps for Determining a Relation Between Variables

STEP 1: Ask whether the variables are logically related to each other.
STEP 2: Obtain data and verify a relation exists. Then plot the points to obtain a **scatter diagram.**
STEP 3: Find an equation which describes this relation.

1 Scatter diagrams are used to help us see the type of relation that exists between two variables. In this text, we will concentrate on a variety of different relations that may exist between two variables. For now, we concentrate on distinguishing between linear and nonlinear relations. See Figure 90.

FIGURE 90

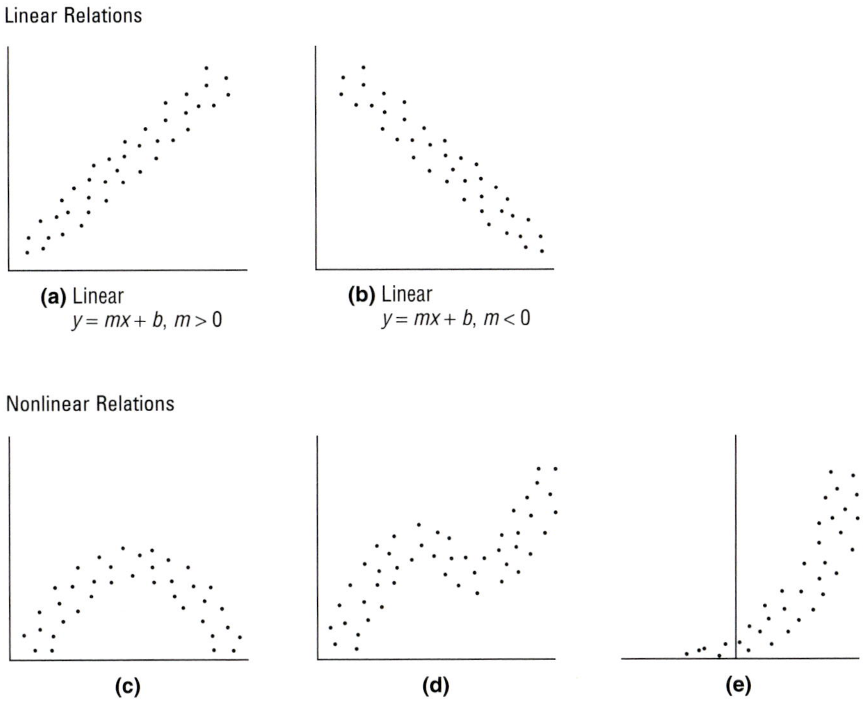

EXAMPLE 1 Distinguishing between Linear and Nonlinear Relations

Determine whether the relation between the two variables shown in Figure 91 is linear or nonlinear.

FIGURE 91

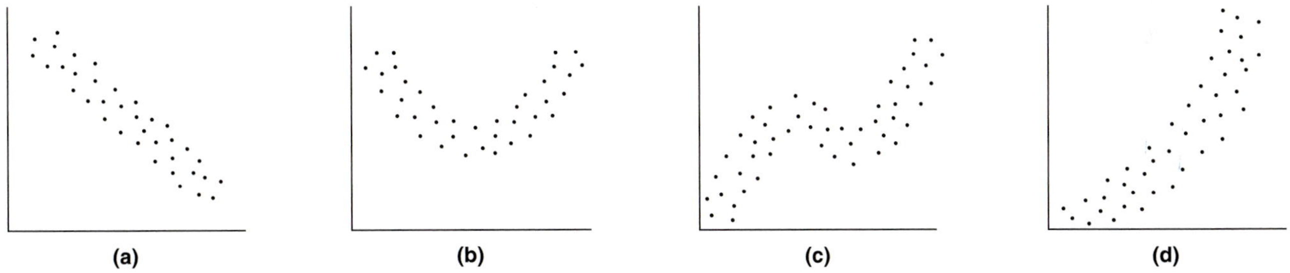

Solution (a) Linear
(b) Nonlinear
(c) Nonlinear
(d) Nonlinear

We will study linearly related data in this section. Fitting an equation to nonlinear data will be discussed in later chapters.

Suppose the scatter diagram of a set of data appears to be linearly related. One way to obtain an equation for such data is to draw a line by hand through two points on the scatter diagram and estimate the equation of the line.

EXAMPLE 2 Finding an Equation For Linearly Related Data

A farmer collected the following data, which shows crop yields for various amounts of fertilizer used.

Plot	1	2	3	4	5	6	7	8	9	10	11	12
Fertilizer, X (Pounds/100 ft2)	0	0	5	5	10	10	15	15	20	20	25	25
Yield, Y (Bushels)	4	6	10	7	12	10	15	17	18	21	23	22

(a) Using a graphing utility, draw a scatter diagram of the data.
(b) Select two points from the data and find an equation of the line containing the points.
(c) Graph the line on the scatter diagram.

Solution (a) The data collected indicates a relation exists between the amount of fertilizer used and crop yield. To draw a scatter diagram, we plot points, using fertilizer as the x-coordinate and yield as the y-coordinate. See Figure 92. From the scatter diagram, it appears a linear relation exists between the amount of fertilizer used and yield.

FIGURE 92

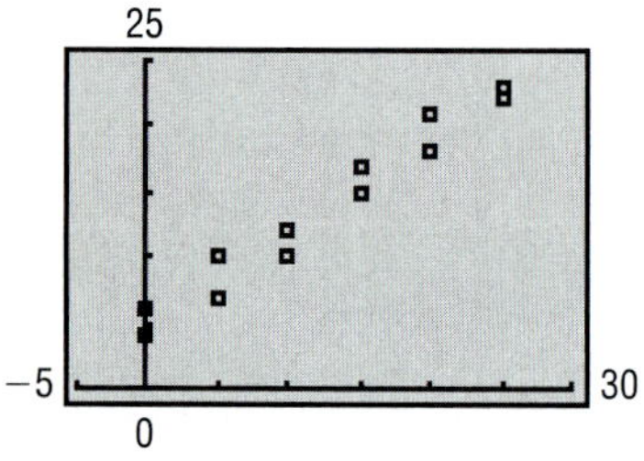

(b) Select two points, say (0, 4) and (25, 23). [You should select your own two points and complete the solution.] The slope of the line joining the points (0, 4) and (25, 23) is

$$m = \frac{23 - 4}{25 - 0} = \frac{19}{25} = 0.76$$

The equation of the line with slope 0.76 and passing through (0, 4) is found using the point–slope form with $m = 0.76$, $x_1 = 0$, $y_1 = 4$:

$$\begin{aligned} y - y_1 &= m(x - x_1) \\ y - 4 &= 0.76(x - 0) \\ y &= 0.76x + 4 \end{aligned}$$

FIGURE 93

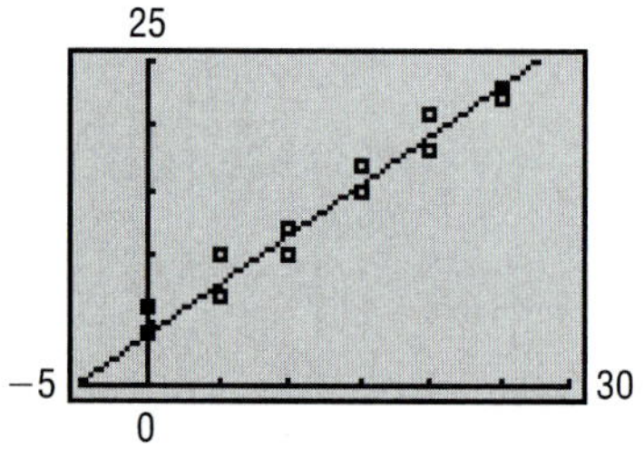

(c) Figure 93 shows the scatter diagram with the graph of the line found in (b). ■

The line obtained in Example 2 depends on the selection of points, which will vary from person to person. So the line we found might be different from the line you found. Although the line we found in Example 2 appears to "fit" the data well, there may be a line that "fits it better." Do you think your line "fits" the data better? Is there a line of "best fit"? As it turns out, there is a method for finding the line that best fits linearly related data.* This line is called the **line of best fit.**

Lines of Best Fit

EXAMPLE 3 Finding the Line of Best Fit

2 With the data from Example 2, find the line of best fit using a graphing utility.

*We shall not discuss in this book the underlying mathematics of lines of best fit. Most books in statistics and many in linear algebra discuss this topic.

MISSION POSSIBLE

Predicting the Future of Olympic Track Events

Some people think that women athletes are beginning to "catch up" to men in the Olympic track events. In fact, researchers at UCLA published data in 1992 that seemed to indicate that within 65 years, the top men and women runners would be able to compete on an equal basis. Other researchers disagree. Here is some data on the 200-meter run with winning times for the Olympics in the year given.

	Men's Time in Seconds	Women's Time in Seconds
1948	21.1	24.4
1952	20.7	23.7
1956	20.6	23.4
1960	20.5	24.0
1964	20.3	23.0
1968	19.83	22.5
1972	20.00	22.40
1976	20.23	22.37
1980	20.19	22.03
1984	19.80	21.81
1988	19.75	21.34
1992	19.73	21.72

(You may notice that the introduction of better timing devices meant more accurate measures starting in 1968 for the men and 1972 for the women.)

1. To begin your investigation of the UCLA conjecture, make a graph of these data. To get the kind of accuracy you need, you should use graph paper. You will need to use only the first quadrant. On your x-axis place the Olympic years from 1948 to 2048, counting by 4's. On your y-axis place the numbers from 15 to 25 which represent the seconds; if you're using graph paper, allow about four squares per one second. Use X's to represent the men's times on the graph and O's to represent the women's times. These should form what we call a "scatter plot," not a neat line or curve.
2. Using a ruler or straightedge, draw a line that represents roughly the slope and direction indicated by the men's scores. Do the same for the women's scores. Write a sentence or two explaining why you think your line is a good representation of the scatter plot.
3. Next find the equation for each line, using your graph to estimate the slope and y-intercept for each. Try to be as accurate as possible, remembering your units and using the points where the line crosses intersections of the grid. The slope will be in seconds per year.
4. Do your two lines appear to cross? In what year do they cross? If you solve the two equations algebraically, do you get the same answer?
5. Some graphing calculators enable you do this problem in a statistics mode. They will plot the individual point that you type in and find a line that represents the data. Use your graphing calculator to find out whether its equations matches yours.
6. Make a group decision about whether or not you think that women's times will "catch up" to men's times in the future. Write out two to three sentences explaining why you believe they will or will not.

Solution Graphing utilities contain built-in programs that find the linear equation of "best fit" for a collection of points in a scatter diagram. (Look in your owner's manual under Linear Regression or Line of Best Fit for details on how to execute the program.) Upon executing the LINear REGression program, we obtain the results shown in Figure 94. The output the utility provides shows us the equation, $y = ax + b$, where a is the slope of the line and b is the y-intercept. The line of best fit which relates fertilizer and yield is

$$\text{Yield} = 0.717(\text{fertilizer}) + 4.786$$

Figure 95 shows the graph of the line of best fit, along with the scatter diagram.

FIGURE 94

FIGURE 95

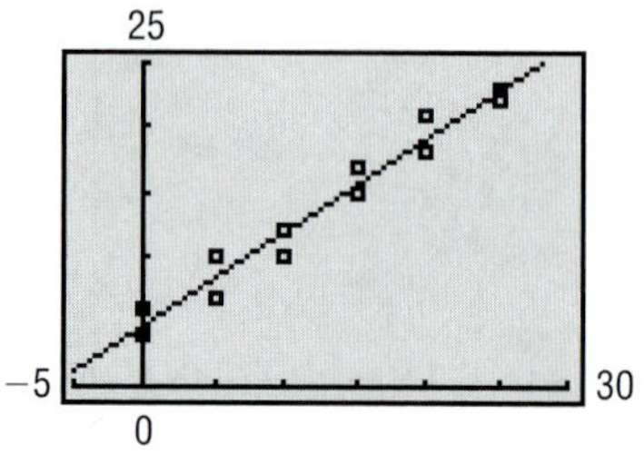

Does the line of best fit appear to be a good "fit"? In other words, does the line appear to accurately describe the relation between yield and fertilizer?

And just how "good" is this line of "best fit"? The answers are given by what is called the *correlation coefficient.*

Now work Problem 7.

Correlation Coefficients

Look again at Figure 94. The last line of output is $r = 0.98$. This number, called the **correlation coefficient, *r*,** $0 \leq |r| \leq 1$, is a measure of the strength of the *linear relation* that exists between two variables. The closer $|r|$ is to 1, the more perfect the linear relationship is. If r is close to 0, there is little or no *linear* relationship between the variables. A negative value of r, $r < 0$, indicates that as x increases, then y decreases; a positive value of r, $r > 0$, indicates that as x increases, then y does also. Thus, the data given in Example 2, having a correlation coefficient of 0.98, are strongly indicative of a linear relationship.

Figure 96 illustrates a variety of scatter diagrams and the relations they suggest.

Prediction and Accuracy

3 Once an equation has been obtained, then the equation can be used to predict values of the dependent variable for a given independent variable.

EXAMPLE 4 Prediction

Use the results of Example 3 to estimate the yield if the farmer uses 17 pounds of fertilizer/100 square feet.

FIGURE 96

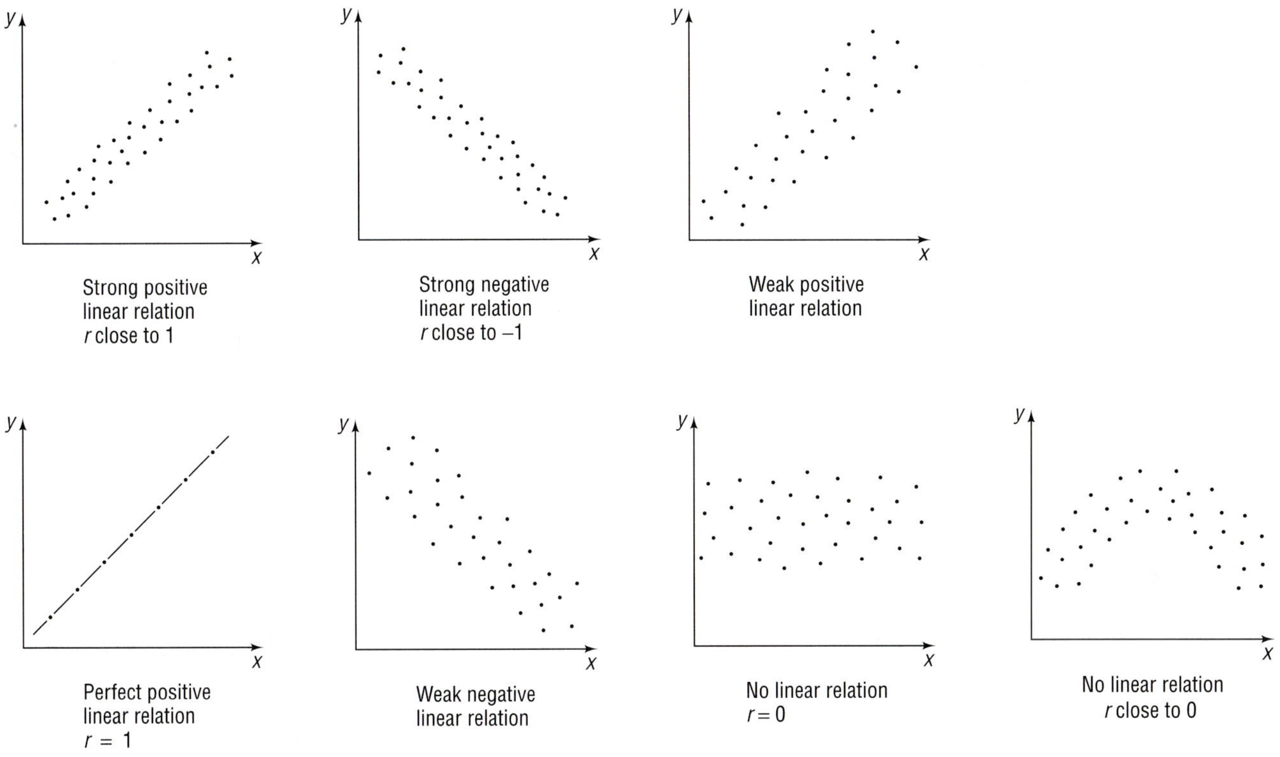

Solution To determine the yield, we substitute the value $x = 17$ pounds into the equation found in Example 3.

$$(\text{Yield}) = 0.717(17) + 4.786 \approx 17 \text{ bushels}$$

So, 17 pounds of fertilizer per 100 square feet will, on average, produce a yield of 17 bushels. Notice that we rounded our prediction to the nearest whole number. This is because our data is measured to the nearest whole number as well. Our predictions cannot be more precise than our data. ■

EXAMPLE 5 Interpreting the Slope

Use the equation of the line found in Example 3 and interpret the slope.

Solution The slope of the line is 0.717. This can be interpreted as follows: For every 1-pound per square foot increase in fertilizer, the yield is expected to increase by 0.717 bushels. ■

It is important that predictions are made within the scope of the model. That is, we can only make predictions regarding this model for $0 \leq$ fertilizer ≤ 25, since this is the range for which we have observable data. Unless we collect additional data for $x > 25$, we can't make predictions for $x > 25$. The reason is that it is not clear whether adding more fertilizer will continue to increase crop yield. In fact, it is conceivable that adding more fertilizer may actually reduce crop yield.

1.6 EXERCISES

In Problems 1–6, examine the scatter diagram and determine whether the type of relation, if any, that may exist is linear or nonlinear.

1.

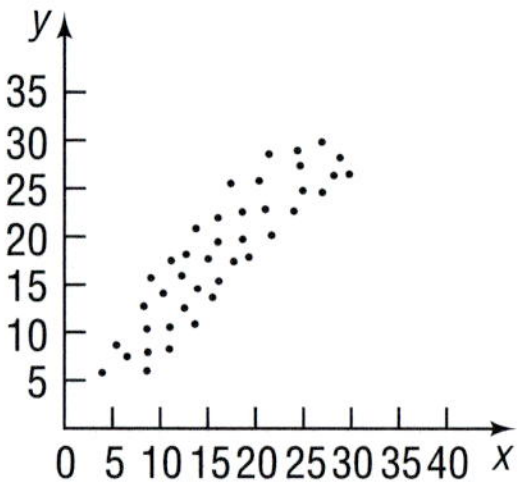

2.

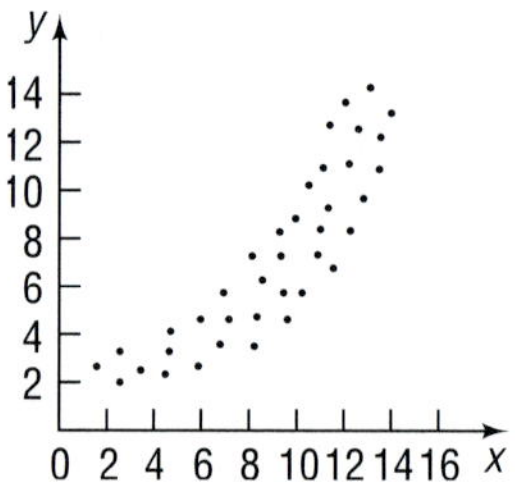

3.

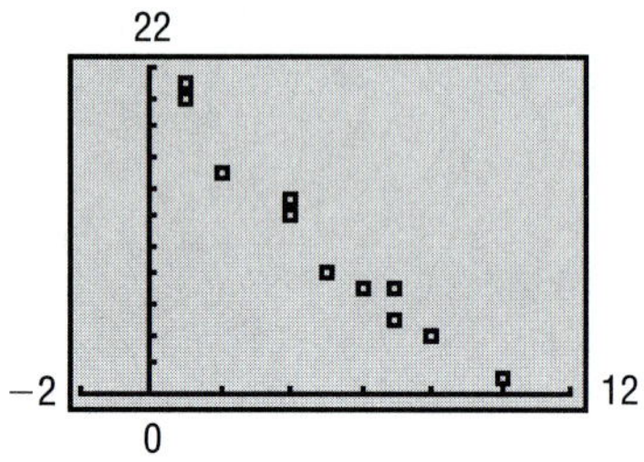

4.

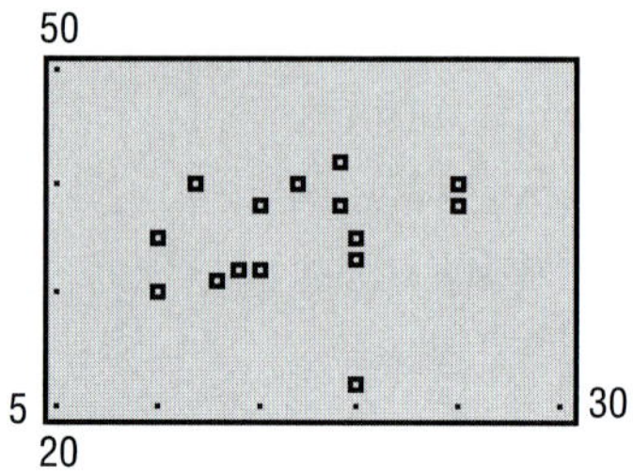

5.

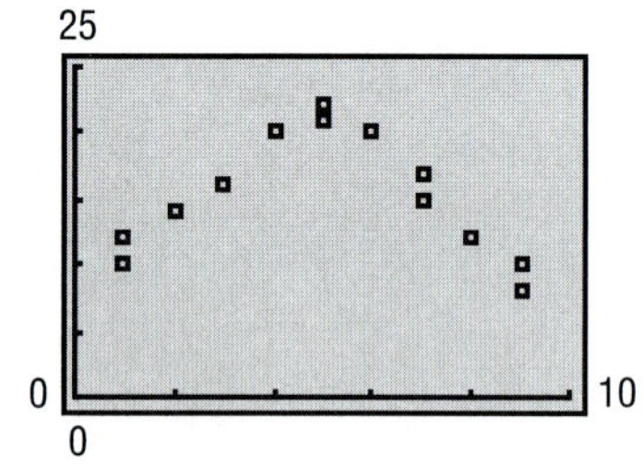

6.

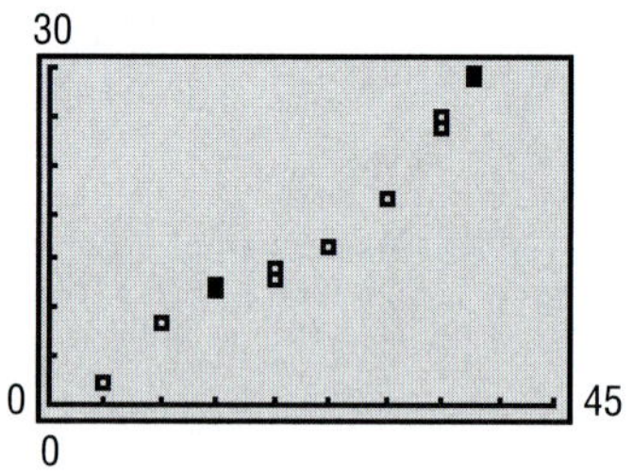

In Problems 7–14

(a) use a graphing utility to draw a scatter diagram;

(b) select two points from the scatter diagram and find the equation of the line through the points selected;

(c) use a graphing utility to graph the line found in (b) on the scatter diagram;

(d) use a graphing utility to find the line of best fit;

(e) use a graphing utility to graph the line of best fit on the scatter diagram.

7.

x	3	4	5	6	7	8	9
y	4	6	7	10	12	14	16

8.

x	3	5	7	9	11	13
y	0	2	3	6	9	11

9.

x	−2	−1	0	1	2
y	−4	0	1	4	5

10.

x	−2	−1	0	1	2
y	7	6	3	2	0

11.

x	20	30	40	50	60
y	100	95	91	83	70

12.

x	5	10	15	20	25
y	2	4	7	11	18

13.

x	−20	−17	−15	−14	−10
y	100	120	118	130	140

14.

x	−30	−27	−25	−20	−14
y	10	12	13	13	18

15. Consumption and Disposable Income An economist wishes to estimate a line which relates personal consumption expenditures (C) and disposable income (I). Both C and I are in thousands of dollars. She interviews 8 heads of households for families of size 4 and obtains the following data:

C (000)	I (000)
16	20
18	20
13	18
21	27
27	36
26	37
36	45
39	50

Let I represent the independent variable and C the dependent variable.

(a) Use a graphing utility to draw a scatter diagram.
(b) Use a graphing utility to find the line of best fit to the data.
(c) Interpret the slope. The slope of this line is called the **marginal propensity to consume.**
(d) Predict the consumption of a family whose disposable income is $42,000.

16. Marginal Propensity to Save The same economist as the one in Problem 15 wants to estimate a line which relates savings (S) and disposable income (I). Let $S = I - C$ be the dependent variable and I the independent variable. The slope of this line is called the **marginal propensity to save.**

(a) Use a graphing utility to draw a scatter diagram.
(b) Use a graphing utility to find the line of best fit to the data.
(c) Interpret the slope.
(d) Predict the savings of a family whose income is $42,000.

17. Average Speed of a Car An individual wanted to determine the relation that might exist between speed and miles per gallon of an automobile. Let X be the average speed of a car on the highway measured in miles per hour and let Y represent the miles per gallon of the automobile. The following data is collected:

X	50	55	55	60	60	62	65	65
Y	28	26	25	22	20	20	17	15

(a) Use a graphing utility to draw a scatter diagram.
(b) Use a graphing utility to find the line of best fit to the data.
(c) Interpret the slope.
(d) Predict the miles per gallon of a car traveling 61 miles per hour.

18. Height versus Weight A doctor wished to determine whether a relation exists between the height of a female and her weight. She obtained the heights and weights of 10 females aged 18–24. Let height be the independent variable, X, measured in inches, and weight be the dependent variable, Y, measured in pounds.

X	60	61	62	62	64	65	65	67	68	68
Y	105	110	115	120	120	125	130	135	135	145

(a) Use a graphing utility to draw a scatter diagram.
(b) Use a graphing utility to find the line of best fit to the data.
(c) Interpret the slope.
(d) Predict the weight of a female aged 18–24 whose height is 66 inches.

19. Sales Data versus Income The following data represent sales and net income before taxes (both are in billions of dollars) for all manufacturing firms within the United States for 1980–1989. Treat sales as the independent variable and net income before taxes as the dependent variable.

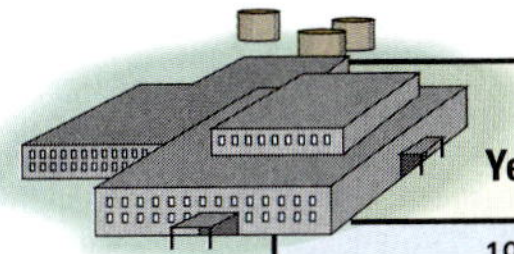

Year	Sales	Net Income Before Taxes
1980	1912.8	92.6
1981	2144.7	101.3
1982	2039.4	70.9
1983	2114.3	85.8
1984	2335.0	107.6
1985	2331.4	87.6
1986	2220.9	83.1
1987	2378.2	115.6
1988	2596.2	154.6
1989	2745.1	136.3

Source: Economic Report of the President, February 1995.

(a) Use a graphing utility to draw a scatter diagram.
(b) Use a graphing utility to find the line of best fit to the data.
(c) Interpret the slope.
(d) Predict the net income before taxes of manufacturing firms in 1990, if sales are \$2456.4 billion.

20. Employment and the Labor Force The following data represent the civilian labor force (people aged 16 years and older, excluding those serving in the military) and the number of employed people in the United States for the years 1981–1991. Treat the size of the labor force as the independent variable and the number employed as the dependent variable. Both the size of the labor force and the number employed are measured in thousands of people.

Year	Civilian Labor Force	Number Employed
1981	108,670	100,397
1982	110,204	99,526
1983	111,550	100,834
1984	113,544	105,005
1985	115,461	107,150
1986	117,834	109,597
1987	119,865	112,440
1988	121,669	114,968
1989	123,869	117,342
1990	124,787	117,914
1991	125,303	116,877

Source: Business Statistics, 1963–1991, U.S. Department of Commerce, Economics and Statistics Administration, Bureau of Economic Analysis, June 1992.

(a) Use a graphing utility to draw a scatter diagram.
(b) Use a graphing utility to find the line of best fit to the data.
(c) Interpret the slope.
(d) Predict the number of employed people if the civilian labor force is 122,340,000 people.

21. Disposable Personal Income Use the data from Problem 69, Section 1.2.
(a) With a graphing utility, find the line of best fit letting per capita disposable income represent the independent variable.
(b) Interpret the slope.
(c) Predict per capita personal consumption if per capita disposable income is \$14,989.

22. Natural Gas vs. Coal Use the data from Problem 70, Section 1.2.
(a) With a graphing utility, find the line of best fit letting the amount of natural gas represent the independent variable.
(b) Interpret the slope.
(c) Predict the amount of energy provided by coal if the amount of energy provided by natural gas is 67.

23. Fuel Consumption Use the data from Problem 71, Section 1.2.
(a) With a graphing utility, find the line of best fit letting the average fuel consumption per car represent the independent variable.
(b) Interpret the slope.
(c) Predict the pollutant emissions of carbon monoxide if the average fuel consumption per car is 505.

24. Hourly Pay and Productivity Use the data from Problem 72, Section 1.2.
(a) With a graphing utility, find the line of best fit letting productivity represent the independent variable.
(b) Interpret the slope.
(c) Predict average hourly earning if productivity is 101.3.

1.7 VARIATION

1 Construct a Model Using Direct Variation
2 Construct a Model Using Inverse Variation
3 Construct a Model Using Joint (combined) Variation

When a mathematical model is developed for a real world problem, it often involves relationships between quantities that are expressed in terms of proportionality:

Force is proportional to acceleration.

For an ideal gas held at a constant temperature, pressure and volume are inversely proportional.

The force of attraction between two heavenly bodies is inversely proportional to the square of the distance between them.

Revenue is directly proportional to sales.

Each of these statements illustrates the idea of **variation,** or how one quantity varies in relation to another quantity. Quantities may vary *directly, inversely,* or *jointly.*

Direct Variation

1 Let x and y denote two quantities. Then y **varies directly** with x, or y is **directly proportional to** x, if there is a nonzero number k such that

$$y = kx$$

The number k is called the **constant of proportionality.**

FIGURE 97
$y = kx,\ k > 0,\ x \geq 0$

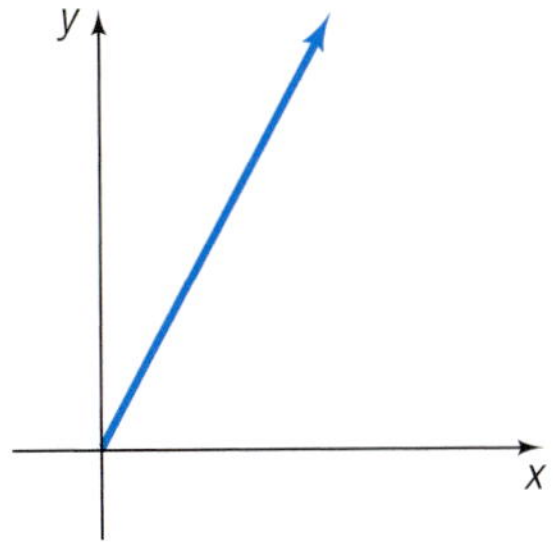

The graph in Figure 97 illustrates the relationship between y and x if y varies directly with x and $k > 0$, $x \geq 0$. Note that the constant of proportionality is, in fact, the slope of the line.

If we know that two quantities vary directly, then knowing the value of each quantity in one instance enables us to write a formula that is true in all cases.

EXAMPLE 1

Chemistry: Gas Law

For a certain gas enclosed in a container of fixed volume, the pressure P (in newtons per square meter) varies directly with temperature T (in kelvins). If the pressure is found to be 20 newtons per square meter at a temperature of 60 K, find a formula that relates pressure P to temperature T. Then find the pressure P when $T = 120$ K.

Solution Because P varies directly with T, we know that

$$P = kT$$

for some constant k. Because $P = 20$ when $T = 60$,

$$20 = k(60)$$
$$k = \frac{1}{3}$$

Thus, in all cases,

$$P = \frac{1}{3}T$$

In particular, when $T = 120\ K$, we find

$$P = \frac{1}{3}(120) = 40 \text{ newtons per square meter}$$

FIGURE 98

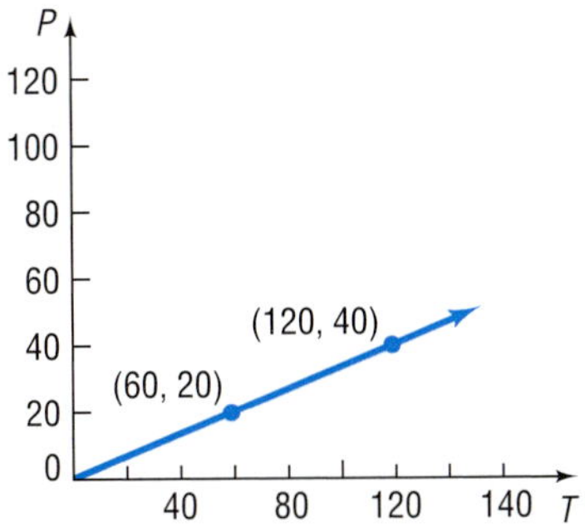

Figure 98 illustrates the relationship between the pressure P and the temperature T. ■

Now work Problem 1.

FIGURE 99
$y = \frac{k}{x}; k > 0, x > 0$

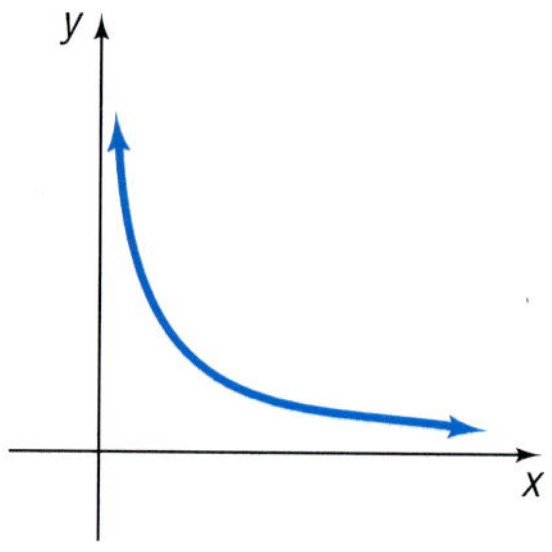

Inverse Variation

2 Let x and y denote two quantities. Then y **varies inversely** with x, or y is **inversely proportional to** x, if there is a nonzero constant k such that

$$y = \frac{k}{x}$$

The graph in Figure 99 illustrates the relationship between y and x if y varies inversely with x and $k > 0, x > 0$.

EXAMPLE 2

Safe Weight That Can Be Supported by a Piece of Pine

The weight W that can be safely supported by a 2-inch by 4-inch piece of lumber varies inversely with its length l. See Figure 100. Experiments indicate that the maximum weight a 10-foot pine 2 by 4 can support is 500 pounds. Write a general formula relating the safe weight W (in pounds) to length l (in feet). Find the maximum weight W that can be safely supported by a length of 25 feet.

FIGURE 100

Solution Because W varies inversely with l, we know that

$$W = \frac{k}{l}$$

for some constant k. Because $W = 500$ when $l = 10$, we have

$$500 = \frac{k}{10}$$
$$k = 5000$$

Thus, in all cases,

$$W = \frac{5000}{l}$$

In particular, the maximum weight W that can be safely supported by a piece of pine 25 feet in length is

$$W = \frac{5000}{25} = 200 \text{ pounds}$$

Figure 101 illustrates the relationship between the weight W and the length l. ■

FIGURE 101

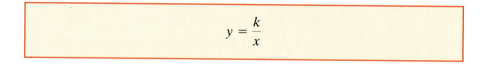

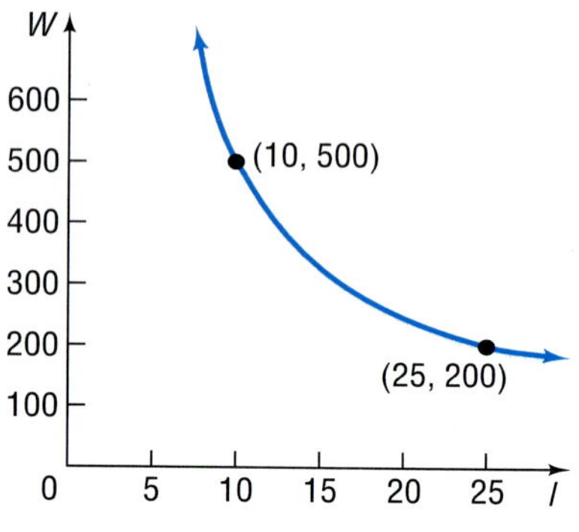

In direct or inverse variation, the quantities that vary may be raised to powers. For example, in the early seventeenth century, Johannes Kepler (1571–1630) discovered that the square of the period T of a planet varies directly with the cube of its mean distance a from the Sun. That is, $T^2 = ka^3$, where k is the constant of proportionality.

Joint Variation and Combined Variation

3 When a variable quantity Q is proportional to the product of two or more other variables, we say that Q **varies jointly** with these quantities. Finally, combinations of direct and/or inverse variation may occur. This is usually referred to as **combined variation.**

Let's look at an example.

EXAMPLE 3 Loss of Heat through a Wall

The loss of heat through a wall varies jointly with the area of the wall and the difference between the inside and outside temperatures, and varies inversely with the thickness of the wall. Write an equation that relates these quantities.

Solution We begin by assigning symbols to represent the quantities:

L = Heat loss $\qquad T$ = Temperature difference

A = Area of wall $\qquad d$ = Thickness of wall

Then

$$L = k\frac{AT}{d}$$

where k is the constant of proportionality. ■

EXAMPLE 4 Force of the Wind on a Window

The force F of the wind on a flat surface positioned at a right angle to the direction of the wind varies jointly with the area A of the surface and the square of the speed v of the wind. A wind of 30 miles per hour blowing on a window measuring 4 feet by 5 feet has a force of 150 pounds. (See Figure 102.) What is the force on a window measuring 3 feet by 4 feet caused by a wind of 50 miles per hour?

FIGURE 102

Solution Since F varies jointly with A and v^2, we have

$$F = kAv^2$$

where k is the constant of proportionality. We are told that $F = 150$ when $v = 30$ and $A = 4 \cdot 5 = 20$. Thus,

$$150 = k(20)(900)$$
$$k = \frac{1}{120}$$

The general formula is therefore

$$F = \frac{1}{120}Av^2$$

For a wind of 50 miles per hour blowing on a window whose area is $A = 3 \cdot 4 = 12$ square feet, the force F is

$$F = \frac{1}{120}(12)(2500) = 250 \text{ pounds}$$

1.7 EXERCISES

In Problems 1–12, write a general formula to describe each variation.

1. y varies directly with x; $y = 2$ when $x = 10$
2. v varies directly with t; $v = 16$ when $t = 2$
3. A varies directly with x^2; $A = 4\pi$ when $x = 2$
4. V varies directly with x^3; $V = 36\pi$ when $x = 3$
5. F varies inversely with d^2; $F = 10$ when $d = 5$
6. y varies inversely with $\sqrt{x}$; $y = 4$ when $x = 9$
7. z varies directly with the sum of the squares of x and y; $z = 5$ when $x = 3$ and $y = 4$
8. T varies jointly with the cube root of x and the square of d; $T = 18$ when $x = 8$ and $d = 3$
9. M varies directly with the square of d and inversely with the square root of x; $M = 24$ when $x = 9$ and $d = 4$
10. z varies directly with the sum of the cube of x and the square of y; $z = 1$ when $x = 2$ and $y = 3$
11. The square of T varies directly with the cube of a and inversely with the square of d; $T = 2$ when $a = 2$ and $d = 4$
12. The cube of z varies directly with the sum of the squares of x and y; $z = 2$ when $x = 9$ and $y = 4$

In Problems 13–20, write an equation that relates the quantities.

13. **Geometry** The volume V of a sphere varies directly with the cube of its radius r. The constant of proportionality is $4\pi/3$.
14. **Geometry** The square of the hypotenuse c of a right triangle varies directly with the sum of the squares of the legs a and b. The constant of proportionality is 1.
15. **Geometry** The area A of a triangle varies jointly with the lengths of the base b and the height h. The constant of proportionality is $\frac{1}{2}$.
16. **Geometry** The perimeter p of a rectangle varies directly with the sum of the lengths of its sides l and w. The constant of proportionality is 2.
17. **Geometry** The volume V of a right circular cylinder varies jointly with the square of its radius r and its height h. The constant of proportionality is π. (See the figure on the following page).

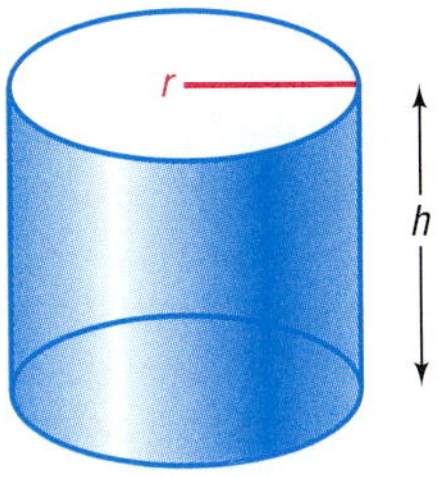

18. **Geometry** The volume V of a right circular cone varies jointly with the square of its radius r and its height h. The constant of proportionality is $\pi/3$. (See the figure).

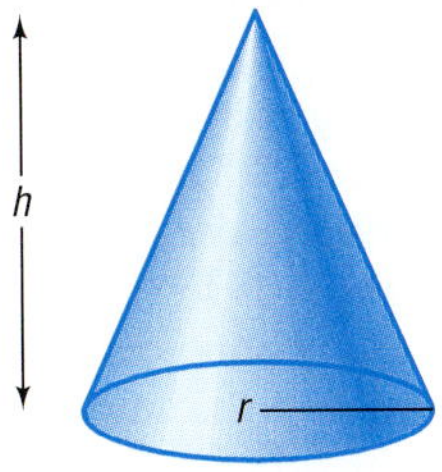

19. **Physics: Newton's Law** The force F (in newtons) of attraction between two bodies varies jointly with their masses m and M (in kilograms) and inversely with the square of the distance d (in meters) between them. The constant of proportionality is $G = 6.67 \times 10^{-11}$.

20. **Physics: Simple Pendulum** The *period* of a pendulum is the time required for one oscillation; the pendulum is usually referred to as *simple* when the angle made to the vertical is less than 5°. The period T of a simple pendulum (in seconds) varies directly with the square root of its length l (in feet). The constant of proportionality is $2\pi/\sqrt{32}$.

21. **Physics: Falling Objects** The distance s an object falls is directly proportional to the square of the time t of the fall. If an object falls 16 feet in 1 second, how far will it fall in 3 seconds? How long will it take an object to fall 64 feet?

22. **Physics: Falling Objects** The velocity v of a falling object is directly proportional to the time t of the fall. If, after 2 seconds, the velocity of the object is 64 feet per second, what will its velocity be after 3 seconds?

23. **Physics: Stretching a Spring** The elongation E of a spring balance varies directly with the applied weight W (see the figure). If $E = 3$ when $W = 20$, find E when $W = 15$.

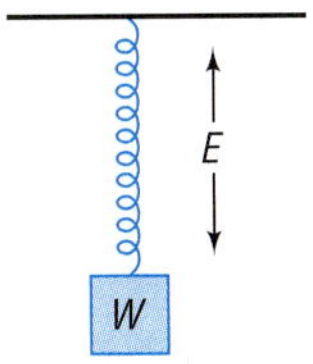

24. **Physics: Vibrating String** The rate of vibration of a string under constant tension varies inversely with the length of the string. If a string is 48 inches long and vibrates 256 times per second, what is the length of a string that vibrates 576 times per second?

25. **Weight of a Body** The weight of a body above the surface of Earth varies inversely with the square of the distance from the center of Earth. If a certain body weighs 55 pounds when it is 4×10^3 miles from the center of Earth, how much will it weigh when it is 4.4×10^3 miles from the center?

26. **Force of the Wind on a Window** The force exerted by the wind on a plane surface varies jointly with the area of the surface and the square of the velocity of the wind. If the force on an area of 20 square feet is 11 pounds when the wind velocity is 22 miles per hour, find the force on a surface area of 47.125 square feet when the wind velocity is 36.5 miles per hour.

27. **Horsepower** The horsepower that a shaft can safely transmit varies jointly with its speed (in revolutions per minute, rpm) and the cube of its diameter. If a shaft of a certain material 2 inches in diameter can transmit 36 horsepower at 75 rpm, what diameter must the shaft have in order to transmit 45 horsepower at 125 rpm?

28. **Weight of a Body** The weight of a body varies inversely with the square of its distance from the center of Earth. Assuming that the radius of Earth is 4000 miles, how much would a man weigh at an altitude of 1 mile above Earth's surface if he weighs 200 pounds on Earth's surface?

29. **Physics: Kinetic Energy** The kinetic energy K of a moving object varies jointly with its mass m and the square of its velocity v. If an object weighing 25 pounds and moving with a velocity of 100 feet per second has a kinetic energy of 400 foot-pounds, find its kinetic energy when the velocity is 150 feet per second.

30. **Electrical Resistance of a Wire** The electrical resistance of a wire varies directly with the length of the wire and inversely with the square of the diameter of the wire. If a wire 432 feet long and 4 millimeters in diameter has a resistance of 1.24 ohms, find the length of a wire of the same material whose resistance is 1.44 ohms and whose diameter is 3 millimeters.

31. **Measuring the Stress of Materials** The stress in the material of a pipe subject to internal pressure varies jointly with the internal pressure and the internal diameter of the pipe and inversely with the thickness of the pipe. The stress is 100 pounds per square inch when the diameter is 5 inches, the thickness is 0.75 inch, and the internal pressure is 25 pounds per square inch. Find the stress when the internal pressure is 40 pounds per square inch, if the diameter is 8 inches and the thickness is 0.50 inch.

32. **Safe Load for a Beam** The maximum safe load for a horizontal rectangular beam varies jointly with the width of the beam and the square of the thickness of the beam and inversely with its length. If an 8-foot beam will support up to 750 pounds when the beam is 4 inches wide and 2 inches thick, what is the maximum safe load in a similar beam 10 feet long, 6 inches wide, and 2 inches thick?

33. **Resistance due to a Conductor** The resistance (in ohms) of a circular conductor varies directly with the length of the conductor and inversely with the square of the radius of the conductor. If 50 feet of a wire with a radius of 6×10^{-3} inch has a resistance of 10 ohms, what would be the resistance of 100 feet of the same wire if the radius is increased to 7×10^{-3} inch?

34. **Chemistry: Gas Laws** The volume V of an ideal gas varies directly with the temperature T and inversely with the pressure P. Write an equation relating V, T, and P using k as the constant of proportionality. If a cylinder contains oxygen at a temperature of 300 K and a pressure of 15 atmospheres in a volume of 100 liters, what is the constant of proportionality k? If a piston is lowered into the cylinder, decreasing the volume occupied by the gas to 80 liters and raising the temperature to 310 K, what is the gas pressure?

Problems 36–40 use the result obtained in Problem 35.

35. **Satellites in Orbit** The speed v required of a satellite to maintain a near-Earth circular orbit is directly proportional to the square root of the distance r of the satellite from the center of Earth.* The constant of proportionality is $\sqrt{g}$, where g is the acceleration of gravity for Earth. Write an equation that shows the relationship between v and r. (The radius of Earth is approximately 3960 miles.)

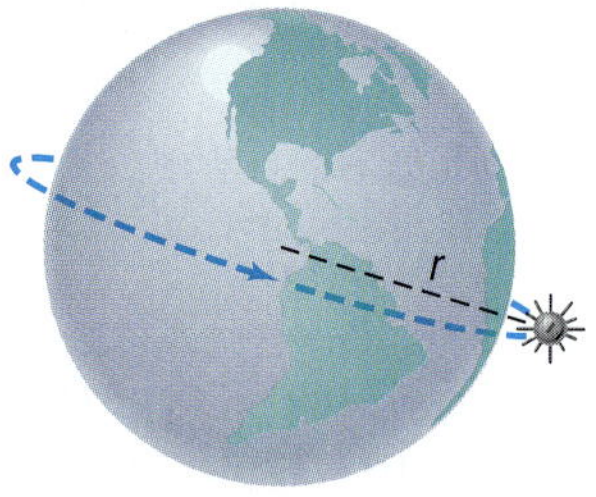

*Near-Earth orbits are at least 100 miles above Earth's surface (out of Earth's atmosphere) and up to an altitude of approximately 15,000 miles. The effect of the gravitational attraction of other bodies is ignored. Although the acceleration of gravity at such altitudes is somewhat less than $g \approx$ 32 feet per second per second $\approx$ 79,036 miles per hour per hour, we shall ignore this discrepancy in our calculations.

36. **Speed Required to Stay in Orbit** What speed is required to maintain a communications satellite in a circular orbit 500 miles above Earth's surface?

37. **Speed Required to Stay in Orbit** Find the speed of a satellite that moves in a circular orbit 140 miles above Earth's surface.

38. **Distance of a Satellite from Earth** Find the distance of a satellite from the surface of Earth as it moves around Earth in a circular orbit at a constant speed of 18,630 miles per hour.

39. **Distance of a Satellite from Earth** A weather satellite orbits Earth in a circle every 1.5 hours. How high is it above Earth?

40. **Distance and Speed of Stationary Satellites** Some communications satellites remain stationary above a fixed point on the equator of Earth's surface. How high are such satellites, and what is their common speed? (Assume that Earth turns once every 24 hours.)

Problems 42–46 use the result obtained in Problem 41.

41. **Physical Forces** The force F (in newtons) required to maintain an object in a circular path varies jointly with the mass m (in kilograms) of the object and the square of its speed v (in meters per second) and inversely with the radius r (in meters) of the circular path. The constant of proportionality is 1. Write an equation relating F, m, v, and r.

42. **Physical Forces** A motorcycle with mass 150 kilograms is driven at a constant speed of 120 kilometers per hour on a circular track with a radius of 100 meters. To keep the motorcycle from skidding, what frictional force must be exerted by the tires on the track?

43. **Physical Forces** If the speed of the motorcycle described in Problem 42 is increased by 10%, by how much is the frictional force of the tires increased?

44. **Physical Forces** If the radius of the track described in Problem 42 is cut in half, how much slower should the motorcycle be driven to maintain the same frictional force?

45. **Physical Forces** Judy is spinning a bucket of water in a horizontal plane at the end of a rope of length L (see the figure). If she triples the speed of the bucket, how many times as hard must she pull on the rope?

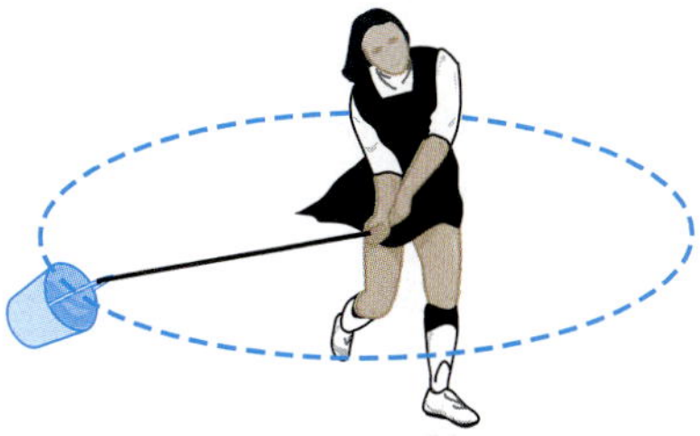

46. **Physical Forces** If Judy in Problem 45 doubles the length of the rope and maintains the same speed for the bucket, will she have to pull on the rope more or less? How much?

47. The formula on page 83 attributed to Johannes Kepler is one of the famous three Keplerian Laws of Planetary Motion. Go to the library and research these laws. Write a brief paper about these laws and Kepler's place in history.

CHAPTER REVIEW

THINGS TO KNOW

Formulas

Distance formula	$d = \sqrt{(x_2 - x_1)^2 + (y_2 - y_1)^2}$
Midpoint formula	$(x, y) = \left(\frac{x_1 + x_2}{2}, \frac{y_1 + y_2}{2}\right)$
Slope	$m = \frac{y_2 - y_1}{x_2 - x_1}$, if $x_1 \neq x_2$; undefined if $x_1 = x_2$
Parallel lines	Equal slopes ($m_1 = m_2$)
Perpendicular lines	Product of slopes is -1 ($m_1 \cdot m_2 = -1$)
Direct variation	$y = kx$
Inverse variation	$y = \frac{k}{x}$

Equations

Vertical line	$x = a$
Horizontal line	$y = b$
Point–slope form of the equation of a line	$y - y_1 = m(x - x_1)$; m is the slope of the line, (x_1, y_1) is a point on the line
General form of the equation of a line	$Ax + By + C = 0$, A, B, not both 0
Slope–intercept form of the equation of a line	$y = mx + b$; m is the slope of the line, b is the y-intercept
Standard form of the equation of a circle	$(x - h)^2 + (y - k)^2 = r^2$; r is the radius of the circle, (h, k) is the center of the circle
General form of the equation of a circle	$x^2 + y^2 + ax + by + c = 0$
Equation of the unit circle	$x^2 + y^2 = 1$

HOW TO

Graph and interpret bar charts

Graph and interpret pie charts

Draw and interpret histograms

Use the distance formula

Graph equations by plotting points

Find the intercepts of a graph

Test an equation for symmetry

Find the slope and intercepts of a line, given the equation

Graph lines by hand

Graph equations using a graphing utility

Draw and interpret frequency polygons

Draw line graphs

Draw scatter diagrams

Obtain the equation of a line

Obtain the equation of a circle

Find the center and radius of a circle, given the equation

Graph circles by hand

Solve variation problems

Find the equation of the line of best fit

FILL-IN-THE-BLANK ITEMS

1. If (x, y) are the coordinates of a point P in the xy-plane, then x is called the ______________ of P and y is the ______________ of P.

2. If three distinct points P, Q, and R all lie on a line and if $d(P, Q) = d(Q, R)$, then Q is called the ______________ of the line segment from P to R.

3. If for every point (x, y) on a graph, the point $(-x, y)$ is also on the graph, then the graph is symmetric with respect to the ______________.

4. The set of points in the xy-plane that are a fixed distance from a fixed point is called a(n) ______________. The fixed distance is called the ______________; the fixed point is called the ______________.

5. The slope of a vertical line is ______________; the slope of a horizontal line is ______________.

6. Two nonvertical lines have slopes m_1 and m_2, respectively. The lines are parallel if ______________; the lines are perpendicular if ______________.

7. If z varies jointly as x^2 and y^3 and inversely as $\sqrt{t}$, then $z =$ ______________, where k is the constant of proportionality.

TRUE/FALSE ITEMS

T F **1.** The distance between two points is sometimes a negative number.

T F **2.** The graph of the equation $y = x^4 + x^2 + 1$ is symmetric with respect to the y-axis.

T F **3.** Vertical lines have undefined slope.

T F **4.** The slope of the line $2y = 3x + 5$ is 3.

T F **5.** Perpendicular lines have slopes that are reciprocals of one another.

T F **6.** The radius of the circle $x^2 + y^2 = 9$ is 3.

T F **7.** If y varies inversely with x, then as x increases in value, y will decrease in value.

REVIEW EXERCISES

1. Male Doctorates in Mathematics The data below represents the ethnicity of male doctoral recipients in mathematics in 1995–1996.

Ethnic Group	Number of Doctoral Recipients
Asian, Pacific Islander	318
Black	15
American Indian, Eskimo, Aleut	1
Mexican American, Puerto Rican, or other Hispanic	36
White (non-Hispanic)	527
Unknown	6

Source: 1996 AMS-IMS-MAA Annual Survey

(a) Draw a bar chart of the data.
(b) Draw a pie chart of the data.
(c) Which chart seems to summarize the data better?
(d) Which ethnic group had the most doctoral recipients?

2. Female Doctorates in Mathematics The data below represents the ethnicity of female doctoral recipients in mathematics in 1995–1996.

Ethnic Group	Number of Doctoral Recipients
Asian, Pacific Islander	407
Black	19
American Indian, Eskimo, Aleut	1
Mexican American, Puerto Rican, or other Hispanic	43
White (non-Hispanic)	676
Unknown	7

Source: 1996 AMS-IMS-MAA Annual Survey

(a) Draw a bar chart of the data.
(b) Draw a pie chart of the data.
(c) Which chart seems to summarize the data better?
(d) Which ethnic group had the most doctoral recipients?
(e) Refer back to Problem 1. Which gender had more doctoral recipients?

3. Age and DUI in 1980 The data below represents the estimated number of arrests per 100,000 drivers for driving under the influence for individuals 25–64 years of age in 1980.

Age	Estimated Number of Arrests
25–29	1,347
30–34	1,076
35–39	996
40–44	944
45–49	837
50–54	686
55–59	509
60–64	335

Source: U.S. Bureau of Justice Statistics

(a) What is the class width?
(b) Draw a histogram of the data.
(c) Draw a frequency polygon of the data.
(d) Which age group has the highest number of arrests for DUI?

4. Age and DUI in 1989 The data on the following page represents the estimated number of arrests per 100,000 drivers for driving under the influence for individuals 25–64 years of age in 1989.

Age	Estimated Number of Arrests
25–29	1,869
30–34	1,486
35–39	1,123
40–44	872
45–49	725
50–54	558
55–59	400
60–64	262

Source: U.S. Bureau of Justice Statistics

(a) What is the class width?
(b) Draw a histogram of the data.
(c) Draw a frequency polygon of the data.
(d) Which age group has the highest rate of arrests for DUI?
(e) Refer back to Problem 3. What might account for the change in the number of arrests between 1980 and 1989?

5. **Concentration of Carbon Monoxide in the Air** The data below represents the average concentration of carbon monoxide in parts per million (ppm) in the air for 1987–1993.

Year	Concentration of Carbon Monoxide (ppm)
1987	6.69
1988	6.38
1989	6.34
1990	5.87
1991	5.55
1992	5.18
1993	4.88

Source: U.S. Environmental Protection Agency

(a) Draw a bar graph of the data.
(b) Draw a line graph of the data.
(c) Using either the bar graph or line graph, determine the year in which the concentration of carbon monoxide was highest.
(d) Using either the bar graph or line graph, determine the year in which the concentration of carbon monoxide was lowest.
(e) What is the trend in the data. In other words, as time passes, what is happening to the average level of carbon monoxide in the air. Why do you think this is happening?
(f) Treating the year as the x-coordinate and the average level of carbon monoxide as the y-coordinate, use a graphing utility to draw a scatter diagram of the data.
(g) What is the slope of the line joining the points (1987, 6.69) and (1990, 5.87)?
(h) Interpret this slope.
(i) What is the slope of the line joining the points (1990, 5.87) and (1993, 4.88)?
(j) Interpret this slope.
(k) Use a graphing utility to find the slope of the line of best fit for this data.
(l) Interpret this slope.
(m) How do you explain the differences among the three slopes obtained?

6. **Value of a Portfolio** The following data represents the value of the Vanguard Index Trust-500 Portfolio for 1987–1995.

Year	Value (Dollars)
1987	54.26
1988	63.06
1989	82.84
1990	80.08
1991	104.28
1992	112.03
1993	123.11
1994	124.56
1995	171.20

Source: Vanguard Index Trust, Annual Report 1995

(a) Draw a bar graph of the data.
(b) Draw a line graph of the data.
(c) Using either the bar graph or line graph, determine the year(s) in which the value of the portfolio decreased.
(d) Using either the bar graph or line graph, determine the year in which the value of the portfolio increased the most.
(e) What is the trend in the data. In other words, as time passes, what is happening to the value of the Vanguard Index Trust-500 Portfolio?
(f) Treating the year as the x-coordinate and the value of the Vanguard Index Trust-500 Portfolio as the y-coordinate, use a graphing utility to draw a scatter diagram of the data.
(g) What is the slope of the line joining the points (1987, 54.26) and (1991, 104.28)?
(h) Interpret this slope.

(i) What is the slope of the line joining the points (1991, 104.28) and (1995, 171.20)?

(j) Interpret this slope.

(k) Use a graphing utility to find the slope of the line of best fit of this data.

(l) Interpret this slope.

(m) If you were managing this trust, which of the three slopes would you use to convince someone to invest? Why?

In Problems 7–16, find an equation of the line having the given characteristics. Express your answer using either the general form or the slope–intercept form of the equation of a line, whichever you prefer.

7. Slope $= -2$; passing through $(3, -1)$

8. Slope $= 0$; passing through $(-5, 4)$

9. Slope undefined; passing through $(-3, 4)$

10. x-intercept $= 2$; passing through $(4, -5)$

11. y-intercept $= -2$; passing through $(5, -3)$

12. Passing through $(3, -4)$ and $(2, 1)$

13. Parallel to the line $2x - 3y + 4 = 0$; passing through $(-5, 3)$

14. Parallel to the line $x + y - 2 = 0$; passing through $(1, -3)$

15. Perpendicular to the line $x + y - 2 = 0$; passing through $(4, -3)$

16. Perpendicular to the line $3x - y + 4 = 0$; passing through $(-2, 4)$

In Problems 17–22, graph each line using a graphing utility. Use a graphing utility to find the x-intercept and y-intercept. Also draw each graph by hand, labeling any intercepts.

17. $4x - 5y + 20 = 0$

18. $3x + 4y - 12 = 0$

19. $\frac{1}{2}x - \frac{1}{3}y + \frac{1}{6} = 0$

20. $-\frac{3}{4}x + \frac{1}{2}y = 0$

21. $\sqrt{2}x + \sqrt{3}y = \sqrt{6}$

22. $\frac{x}{3} + \frac{y}{4} = 1$

In Problems 23–26, find the center and radius of each circle. Graph each circle using a graphing utility.

23. $x^2 + y^2 - 2x + 4y - 4 = 0$

24. $x^2 + y^2 + 4x - 4y - 1 = 0$

25. $3x^2 + 3y^2 - 6x + 12y = 0$

26. $2x^2 + 2y^2 - 4x = 0$

27. Find the slope of the line containing the points $(7, 4)$ and $(-3, 2)$. What is the distance between these points? What is their midpoint?

28. Find the slope of the line containing the points $(2, 5)$ and $(6, -3)$. What is the distance between these points? What is their midpoint?

29. The figure below shows the graph of two parallel lines. Which of the following pairs of equations might have such a graph?

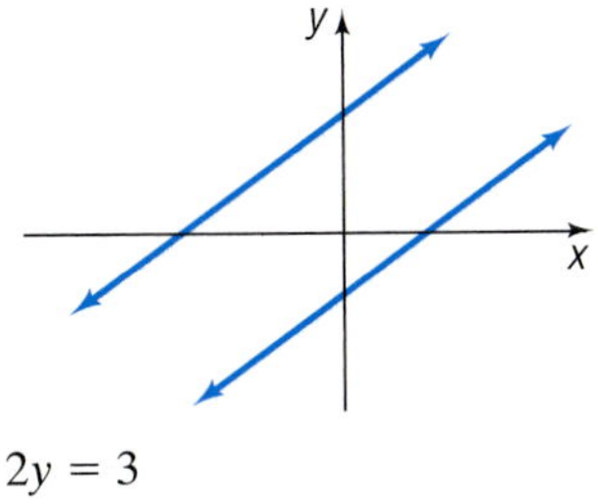

(a) $x - 2y = 3$
$x + 2y = 7$

(b) $x + y = 2$
$x + y = -1$

(c) $x - y = -2$
$x - y = 1$

(d) $x - y = -2$
$2x - 2y = -4$

(e) $x + 2y = 2$
$x + 2y = -1$

30. The figure below shows the graph of two perpendicular lines. Which of the following pairs of equations might have such a graph?

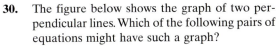

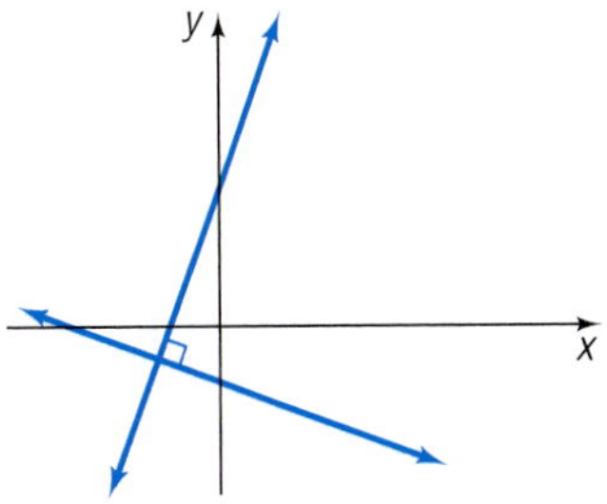

(a) $y - 2x = 2$
$y + 2x = -1$

(b) $y - 2x = 0$
$2y + x = 0$

(c) $2y - x = 2$
$2y + x = -2$

(d) $y - 2x = 2$
$x + 2y = -1$

(e) $2x + y = -2$
$2y + x = -2$

In Problems 31–38, list the intercepts and test for symmetry.

31. $2x = 3y^2$ **32.** $y = 5x$ **33.** $x^2 + 4y^2 = 16$ **34.** $9x^2 - y^2 = 9$

35. $y = x^4 + 2x^2 + 1$ **36.** $y = x^3 - x$ **37.** $x^2 + x + y^2 + 2y = 0$ **38.** $x^2 + 4x + y^2 - 2y = 0$

For Problems 39–42, (a) use a graphing utility to draw a scatter diagram; (b) use a graphing utility to find the line of best fit to the data; and (c) interpret the slope.

39.

x	3	4	5	6	7	8	9
y	3	5	6	8	10	11	13

40.

x	10	12	13	15	16	18	20
y	34	27	26	23	20	18	17

41.

x	100	110	125	130	140	145	150	160	170	175
y	300	340	365	380	400	410	425	430	450	460

42.

x	200	220	230	235	245	250	265	275	280	300
y	1000	990	975	960	955	940	935	920	910	895

43. Relating Algebra and Calculus Scores The following data represents scores in an algebra achievement test and calculus achievement test for the same student. Treat the algebra test score as the independent variable and the calculus test score as the dependent variable.

Algebra Score	Calculus Score
17	73
21	66
11	64
16	61
15	70
11	71
24	90
27	68
19	84
8	52

Source: "Factors Affecting Achievement in the First Course in Calculus," by Edge and Friedberg, *Journal of Experimental Education,* Vol. 52, No. 3.

(a) Use a graphing utility to draw a scatter diagram.
(b) Use a graphing utility to find the line of best fit to the data.
(c) Interpret the slope.
(d) Predict the score a student would receive on the calculus achievement test if she scored a 20 on the algebra achievement test.

44. Relating Emission Levels of Hydrocarbons and Carbon Monoxide The following data represents emissions levels for different vehicles. Measurements are given in grams per meter. Treat hydrocarbons (HC) as the independent variable and carbon monoxide (CO) as the dependent variable.

HC	CO
0.65	14.7
0.55	12.3
0.72	14.6
0.83	15.1
0.57	5.0
0.51	4.1
0.43	3.8
0.37	4.1

Source: "Determining Statistical Characteristics of a Vehicle Emissions Audit Procedure," by Lorenzen, *Technometrics,* Vol. 22, No. 4.

(a) Use a graphing utility to draw a scatter diagram.
(b) Use a graphing utility to find the line of best fit to the data.
(c) Interpret the slope.
(d) Predict the level of CO in a vehicle's exhaust if the level of HC is 0.67.

45. Geometry The area of an equilateral triangle varies directly with the square of the length of a side. If the area of the equilateral triangle whose sides are of length 1 centimeter is $\sqrt{3}/4$, find the length s of each side of an equilateral triangle whose area A is 16 square centimeters.

46. Vibrating Strings In a vibrating string, the pitch varies directly with the square root of the tension of the string. If a certain string vibrates 300 times per second under a tension of 9 pounds, find the tension required to cause the string to vibrate 400 times per second.

47. **Kepler's Third Law of Planetary Motion** Kepler's third law of planetary motion states that the square of the period T of revolution of a planet is proportional to the cube of its mean distance from the Sun. If the mean distance of Earth from the Sun is 93 million miles, what is the mean distance a of the planet Mercury from the Sun, given that Mercury has a "year" of 88 days?

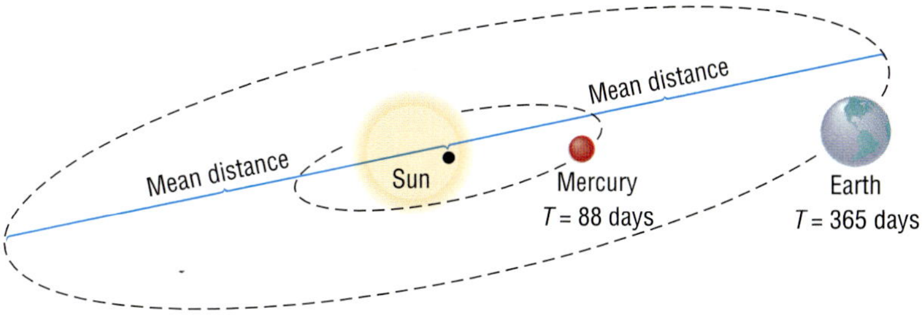

48. Use Problem 47 to find the mean distance of the planet Jupiter from the Sun, given that Jupiter circles the Sun every $5\sqrt{5}$ years.

49. Show that the points $A = (3, 4)$, $B = (1, 1)$, and $C = (-2, 3)$ are the vertices of an isosceles triangle.

50. Show that the points $A = (-2, 0)$, $B = (-4, 4)$, and $C = (8, 5)$ are the vertices of a right triangle in two ways:
   (a) By using the converse of the Pythagorean theorem
   (b) By using the slopes of the lines joining the vertices

51. Show that the points $A = (2, 5)$, $B = (6, 1)$, and $C = (8, -1)$ lie on a straight line by using slopes.

52. Show that the points $A = (1, 5)$, $B = (2, 4)$, and $C = (-3, 5)$ lie on a circle with center $(-1, 2)$. What is the radius of this circle?

53. The endpoints of the diameter of a circle are $(-3, 2)$ and $(5, -6)$. Find the center and radius of the circle. Write the general equation of this circle.

54. Find two numbers y such that the distance from $(-3, 2)$ to $(5, y)$ is 10.

55. Make up four problems you might be asked to do given the two points $(-3, 4)$ and $(6, 1)$. Each problem should involve a different concept. Be sure your directions are clearly stated.

56. Describe each of the following graphs. Give justification.
   (a) $x = 0$ (b) $y = 0$
   (c) $x + y = 0$ (d) $xy = 0$
   (e) $x^2 + y^2 = 0$

57. Suppose that you have a rectangular field that requires watering. Your watering system consists of an arm of variable length that rotates so that the watering pattern is a circle. Decide where to position the arm and what length it should be so that entire field is watered most efficiently. When does it become desirable to use more than one arm?
[**Hint:** Use a rectangular coordinate system positioned so that the axes bisect the rectangle (see the figure). Write equations for the circle(s) swept out by the watering arm(s).]

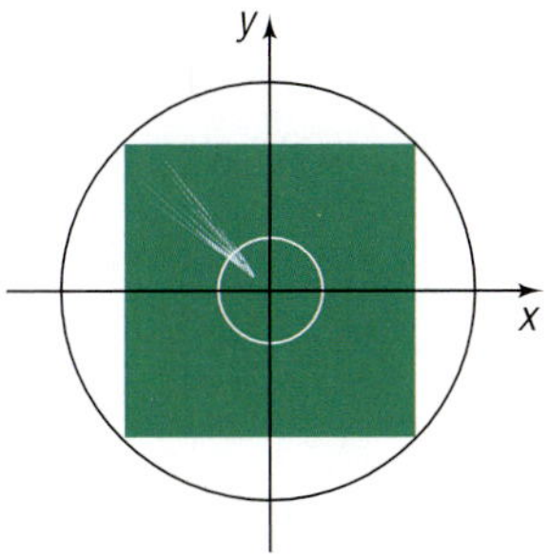

Square field

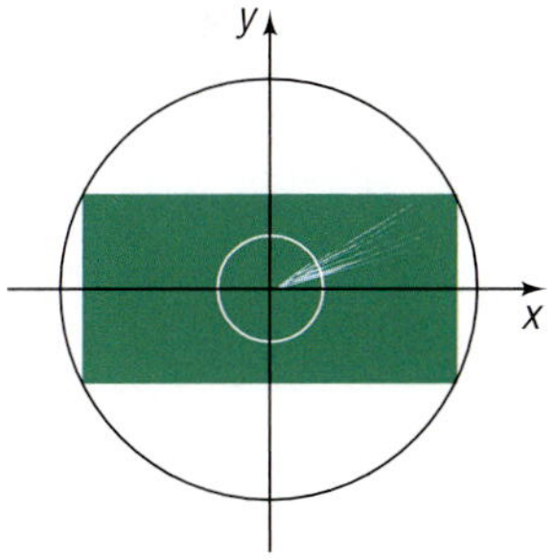

Rectangular field, one arm

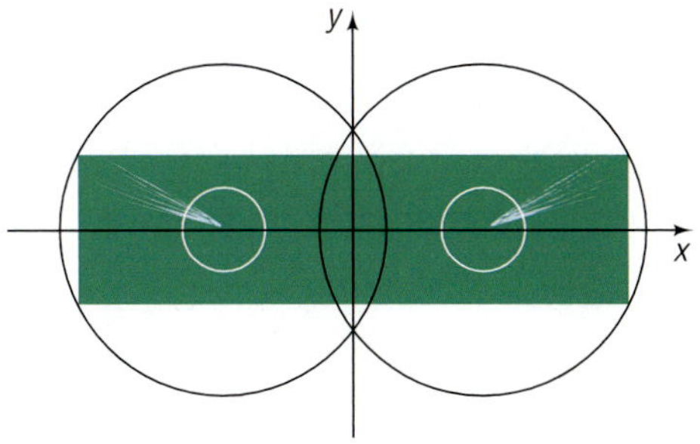

Rectangular field, two arms

58. Graph $Y_1 = x$ and $Y_2 = 0.99x + 0.01$ on the same screen. It appears the graphs are identical. Explain why this happens. Provide a way to correct this situation.

59. Why does the graph of $Y_1 = x/6$, a straight line, appear to consist of a collection of tiny horizontal line segments when drawn using a graphing utility?

CHAPTER 2

Functions and Their Graphs

Imagine yourself as an expert for the EPA and sitting in Congressional policy meetings having to explain the importance of balancing energy resources with environmental protection. On the following page is the Internet Excursion placing you in exactly that position and asking the tough questions a member of congress might ask. Use the Sullivan website at:

www.prenhall.com/sullivan

to help link you to the Internet resources you will need to answer the questions asked.

PREPARING FOR THIS CHAPTER

Before getting started on this chapter, review the following concepts;

Topics from Algebra and Geometry *(Appendix, Section 1)*

Graphs of Certain Equations *(Example 2, p. 31)*

Example 4, p. 34, Example 5, p. 35; Example 14, p. 43

Tests for Symmetry of an Equation (p. 42)

Procedure for Finding Intercepts of an Equation (p. 37)

Linear Curve Fitting *(Chapter 1, Section 1.6)*

OUTLINE

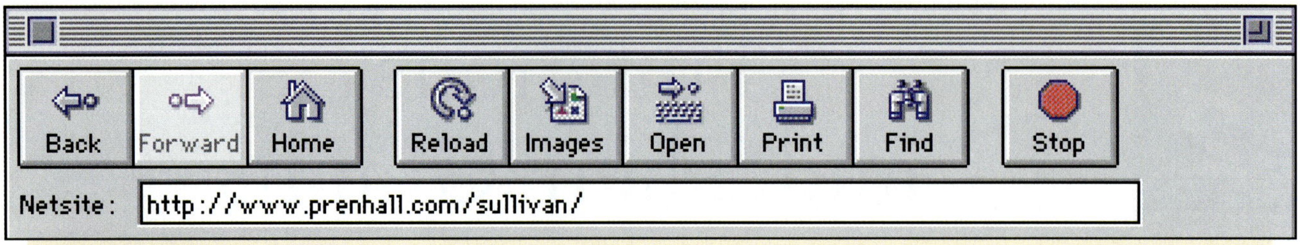

HOW LONG WILL THE OIL LAST?

The global supply of oil and other sources of energy is more than adequate to meet present needs, but most of this supply is outside of the United States. Currently, oil supplies about 40 percent of the world's energy, with the United States the biggest consumer. Since oil is a nonrenewable energy resource, plans must be made for the eventuality of diminishing supply.

Suppose that you were a consultant for the EPA and members of congress asked you to sit on a Energy Policy Steering committee as an expert analyst. You must convince the committee of the importance of environmental concerns in planning the global energy system. You are very aware of the fact that present modes of energy use and production threaten serious environmental deterioration. Your plan is to create a set of what-if functions that will allow your committee to model many possible scenarios for their policy guidelines.

1. $E(t)$, the first function that you create, will allow you to model United States oil consumption over a period of time. Make a scatterplot using the EPA Data on United States energy consumption, per capita from 1950–1990. Since oil provides 40 percent of the energy consumed, you adjust your figures to get oil consumption per capita. Use the LINear REGression tool on your graphing utility to find the linear function $E(t)$ of best fit for modeling the data. Does this seem like a good model?
2. Your second function, $P(t)$, is a population modeling tool. From the United States Census Data make a scatterplot, and then use a QUADratic REGression to model the data. Next, let $O(t) = E(t) \cdot P(t)$. What does $O(t)$ model? Graph $O(t)$. Compare to the actual figures from BP Petroleum.
3. Using our function $O(t)$, we want to estimate the area under the graph for the years 1997 until 2000. To do this we can make a trapezoid, and find its area. Now estimate the areas under $O(t)$ using 10-year intervals. For what year does the total of all the trapezoid areas equal the remaining United States oil reserve?
4. Could this sort of analysis be used to predict when the world's supply of oil will run out? Can your committee think of any methods that might improve your predictions?

Perhaps the most central idea in mathematics is the notion of a *function.* This important chapter deals with what a function is, how to graph functions, how to perform operations on functions, and how functions are used in applications.

The word *function* apparently was introduced by René Descartes in 1637. For him, a function simply meant any positive integral power of a variable *x*. Gottfried Wilheim von Leibniz (1646–1716), who always emphasized the geometric side of mathematics, used the word function to denote any quantity associated with a curve, such as the coordinates of a point on the curve. Leonhard Euler (1707–1783) employed the word to mean any equation or formula involving variables and constants. His idea of a function is similar to the one most often used today in courses that precede calculus. Later, the use of functions in investigating heat flow equations led to a very broad definition, due to Lejeune Dirichlet (1805–1859), which describes a function as a rule or correspondence between two sets. It is his definition that we use here.

2.1 FUNCTIONS

1. Determine Whether a Relation Represents a Function
2. Identify the Graph of a Function
3. Find the Value of a Function
4. Find the Domain of a Function
5. Obtain Information from/about the Graph of a Function

In Section 1.2, we said a **relation** is a correspondence between two variables, say x and y, and can be written as a set of ordered pairs (x, y). When relations are written as ordered pairs (x, y), we say x is related to y. Often, we are interested in specifying the type of relation (such as an equation) that might exist between the two variables. For example, when variables appear to be linearly related (based on the scatter diagram), we find a line of best fit. In Section 1.6, we found that fertilizer and crop yield appeared to be linearly related and the line of best fit was $y = 0.717x + 4.786$, where x represents the amount of fertilizer and y represents the crop yield. Thus, if we know how much fertilizer has been used, we can predict the yield by using the rule $y = 0.717x + 4.786$. This rule is an example of a *function.*

As another example, suppose an icicle falls off a building from a height of 64 feet above the ground. According to a law of physics, the distance s (in feet) of the icicle from the ground after t seconds is given (approximately) by the formula $s = 64 - 16t^2$. When $t = 0$ seconds, the icicle is $s = 64$ feet above the ground. After 1 second, the icicle is $s = 64 - 16(1)^2 = 48$ feet above the ground. After 2 seconds, the icicle strikes the ground. The formula $s = 64 - 16t^2$ provides a way of finding the distance s when the time t $(0 \le t \le 2)$ is prescribed. There is a correspondence between each time t in the interval $0 \le t \le 2$ and the distance s. We say that the distance s is a *function* of the time t because

1. There is a correspondence between the set of times and the set of distances.
2. There is exactly one distance s obtained for a prescribed time t in the interval $0 \le t \le 2$.

Let's now look at the definition of a function.

Definition of Function

> Let X and Y be two nonempty sets of real numbers.* A **function** from X into Y is a rule or a correspondence that associates with each element of X a unique element of Y.

*The two sets X and Y can also be sets of complex numbers, (discussed in Section 4.6), and then we have defined a complex function. In the broad definition (due to Lejeune Dirichlet), X and Y can be any two sets.

FIGURE 1

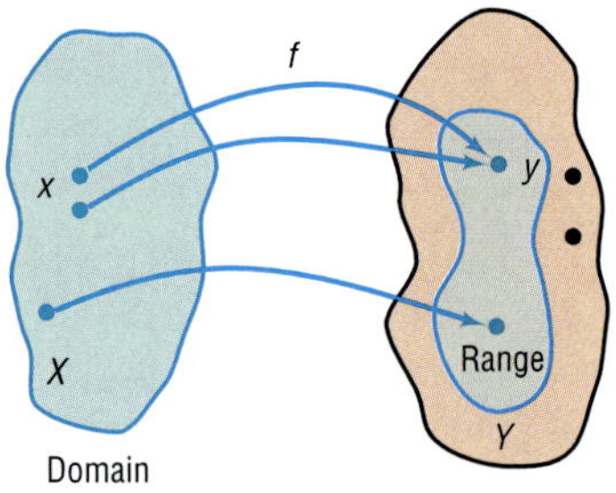

The set X is called the **domain** of the function. For each element x in X, the corresponding element y in Y is called the **value** of the function at x, or the **image** of x. The set of all images of the elements of the domain is called the **range** of the function. See Figure 1.

Since there may be some elements in Y that are not the image of some x in X, it follows that the range of a function may be a subset of Y, as shown in Figure 1.

EXAMPLE 1 Determining Whether a Relation Represents a Function

1

Determine whether the following relations represent functions.

(a) For this relation, the domain represents the employees of Sara's Pre-Owned Car Mart and the range represents their base salary.

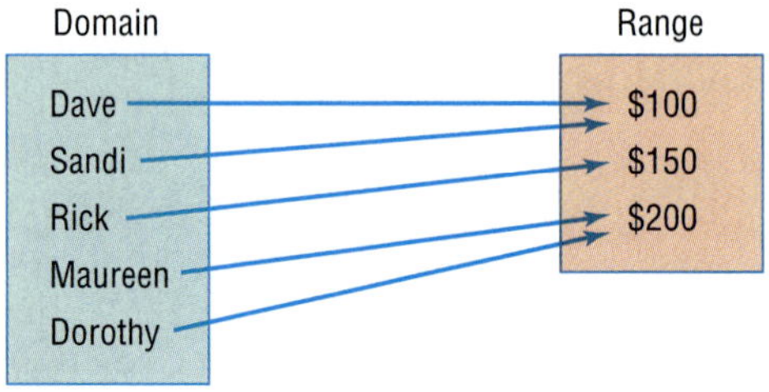

(b) For this relation, the domain represents the employees of Sara's Pre-Owned Car Mart and the range represents their phone number(*s*).

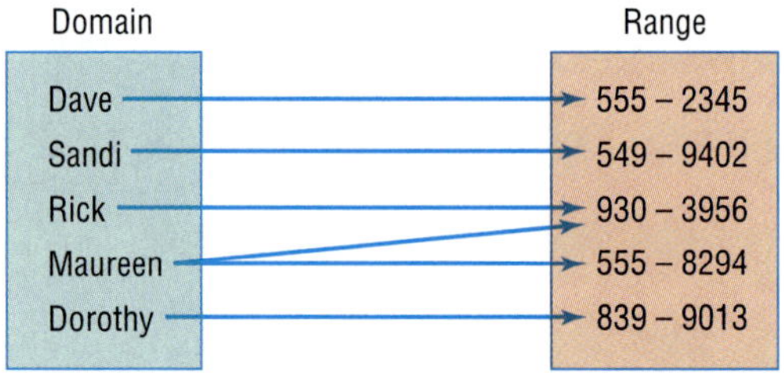

Solution (a) The relation is a function because each element in the domain corresponds to a unique element in the range. Notice that more than one element in the domain can correspond to the same element in the range.

(b) The relation is not a function because each element in the domain does not correspond to a unique element in the range. Maureen has two telephone numbers; therefore, if Maureen is chosen from the domain, a unique telephone number cannot be assigned to her. ■

If the relation is a function, then an element in the domain allows for unique identification of the corresponding element in the range. However, if the relation is not a function, then some element in the domain does not allow for unique identification of the corresponding value in the range. For example, if Maureen is chosen from the domain in Example 1(a), she is identified with a unique base salary of \$200, so base salary is a function of the person selected. However, if Maureen is chosen from the domain in Example 1(b), she is not identified with a unique telephone number (we don't know whether to call her at 555-8294 or 930-3956), so telephone number is not a function of the person selected.

We may think of a function as a set of ordered pairs (x, y), in which no two distinct pairs have the same first element. Therefore, a function is a specific type of relation. The set of all first elements x is the domain of the function, and the set of all second elements y is its range. Thus, there is associated with each element x in the domain a unique element y in the range.

EXAMPLE 2 Determining Whether a Relation Represents a Function

Determine whether each relation represents a function.

(a) $\{(1, 4), (2, 5), (3, 6), (4, 7)\}$

(b) $\{(1, 4), (2, 4), (3, 5), (6, 10)\}$

(c) $\{(-3, 9), (-2, 4), (0, 0), (1, 1), (-3, 8)\}$

Solution (a) This relation is a function because there are no distinct ordered pairs with the same first element.

(b) This relation is a function because there are no distinct ordered pairs with the same first element.

(c) This relation is not a function because there is a first element, -3, that corresponds to two different second elements, 9 and 8. ■

In Example 2(b), notice that 1 and 2 in the domain each have the same image in the range. This does not violate the definition of a function—two different first elements can have the same second element. A violation of the definition occurs when two ordered pairs have the same first element and different second elements, as in Example 2(c).

The rule (or correspondence) referred to in the definition of a function is most often given as an equation in two variables, usually denoted x and y.

EXAMPLE 3 Example of a Function

Consider the function defined by the equation

$$y = 2x - 5 \qquad 1 \le x \le 6$$

The domain $1 \le x \le 6$ specifies that the number x is restricted to the real numbers from 1 to 6, inclusive. The rule $y = 2x - 5$ specifies that the number x is to be multiplied by 2 and then 5 is to be subtracted from the result to get y. For example, the value of the function at $x = \frac{3}{2}$ (that is, the image of $x = \frac{3}{2}$) is $y = 2 \cdot \frac{3}{2} - 5 = -2$. ■

Now work Problems 1 and 5.

The Graph of a Function

In applications, a graph often demonstrates more clearly the relationship between two variables than, say, an equation or table would. For example, Table 1 shows the price per share of McDonald's Corp. stock at the end of each week from Sept. 6, 1996 to Dec. 13, 1996.

If we plot the data in Table 1, using the date as the x-coordinate and the price as the y-coordinate, and then connect the points, we obtain the graph in Figure 2.

TABLE 1

Date	Price
9/6/96	46 1/4
9/13/96	47 3/4
9/20/96	47 5/8
9/27/96	47 7/8
10/4/96	46 3/8
10/11/96	45 7/8
10/18/96	46
10/25/96	44 5/8
11/1/96	45 3/8
11/8/96	47
11/15/96	48 1/4
11/22/96	47 7/8
11/29/96	46 3/4
12/6/96	46 7/8
12/13/96	46

FIGURE 2
Weekly closing prices of McDonald's Corp. stock

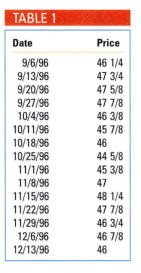

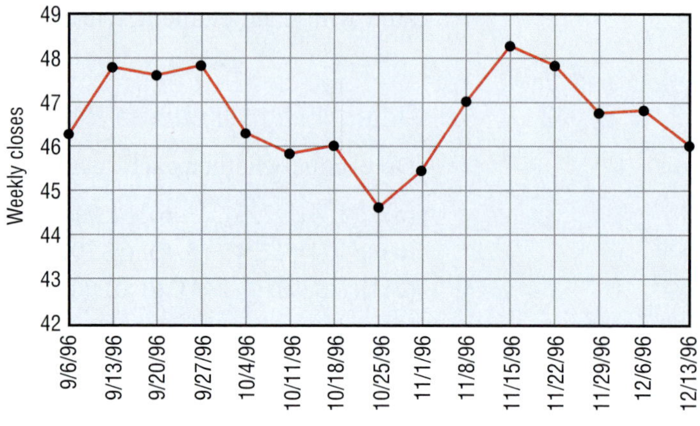

We can see from the graph that the price of the stock was falling over the few days preceding Oct. 4, 1996 and was rising over the days from Oct. 25 through Nov. 15. The graph also shows that the lowest price during this period occurred on Oct. 25 while the highest occurred on Nov. 15. Equations and tables, on the other hand, usually require some calculations and interpretation before this kind of information can be "seen."

Look again at Figure 2. The graph shows that for each time on the horizontal axis, there is only one price on the vertical axis. Thus, the graph represents a function, although the exact rule for getting from time to price is not given.

When the rule that defines a function is given by an equation in x and y, the **graph** of the function is the graph of the equation, that is, the set of points (x, y) in the xy-plane that satisfies the equation. When we use a graphing utility to graph a function we first select a viewing rectangle. The values of Xmin, Xmax give the domain we wish to view, while Ymin, Ymax give the range we wish to view. These settings usually do not represent the actual domain and range of the function.

Warning: Do not confuse the two meanings given for the word *range*. When used in connection with a function, it means the set of all images of the elements of the domain of the function. When the word RANGE is used in connection with a graphing utility, it means the settings used for the viewing rectangle.

2

Not every collection of points in the xy-plane represents the graph of a function. Remember, for a function, each number x in the domain has one and only one image y. Thus, the graph of a function cannot contain two points with the same x-coordinate and different y-coordinates. Therefore, the graph of a function must satisfy the following **vertical-line test.**

Theorem Vertical-Line Test

A set of points in the xy-plane is the graph of a function if and only if a vertical line intersects the graph in at most one point.

It follows that, if any vertical line intersects a graph at more than one point, the graph is not the graph of a function.

EXAMPLE 4 Identifying the Graph of a Function

Which of the graphs in Figure 3 are graphs of functions?

FIGURE 3

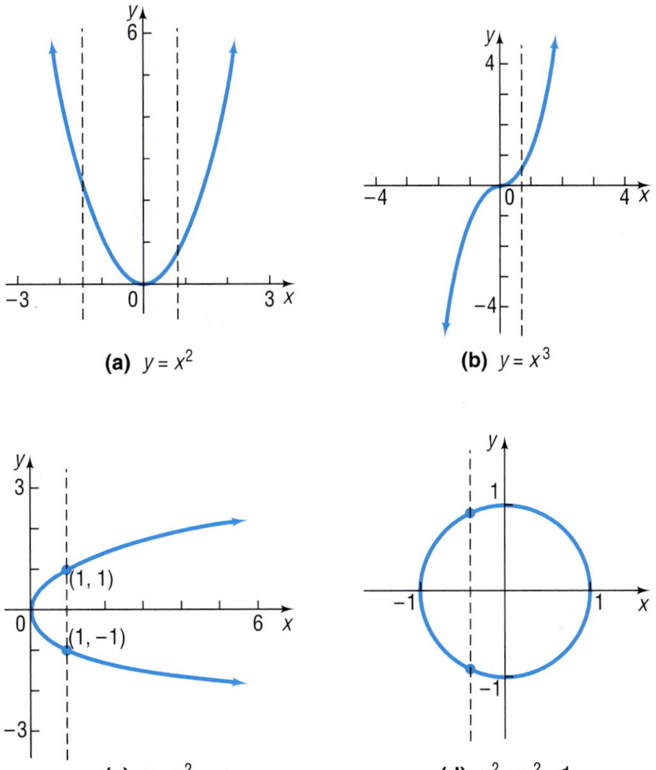

(a) $y = x^2$ **(b)** $y = x^3$ **(c)** $x = y^2$ **(d)** $x^2 + y^2 = 1$

Solution The graphs in Figures 3(a) and 3(b) are graphs of functions, because a vertical line intersects each graph in at most one point. The graphs in Figures 3(c) and 3(d) are not graphs of functions, because some vertical line intersects each graph in more than one point.

Now work Problem 41.

Function Notation

Functions are often denoted by letters such as f, F, g, G, and so on. If f is a function, then for each number x in its domain the corresponding image in the range is designated by the symbol $f(x)$, read as "f of x" or as "f at x." We refer to $f(x)$ as the **value of f at the number x.** Thus, $f(x)$ is the number that results when x is given and the rule for f is applied; $f(x)$ does *not* mean "f times x." For example, the function given in Example 3 may be written as $f(x) = 2x - 5$, $1 \le x \le 6$.

Figure 4 illustrates some other functions. Note that for each of the functions illustrated, for each x in the domain, there is one value in the range.

FIGURE 4

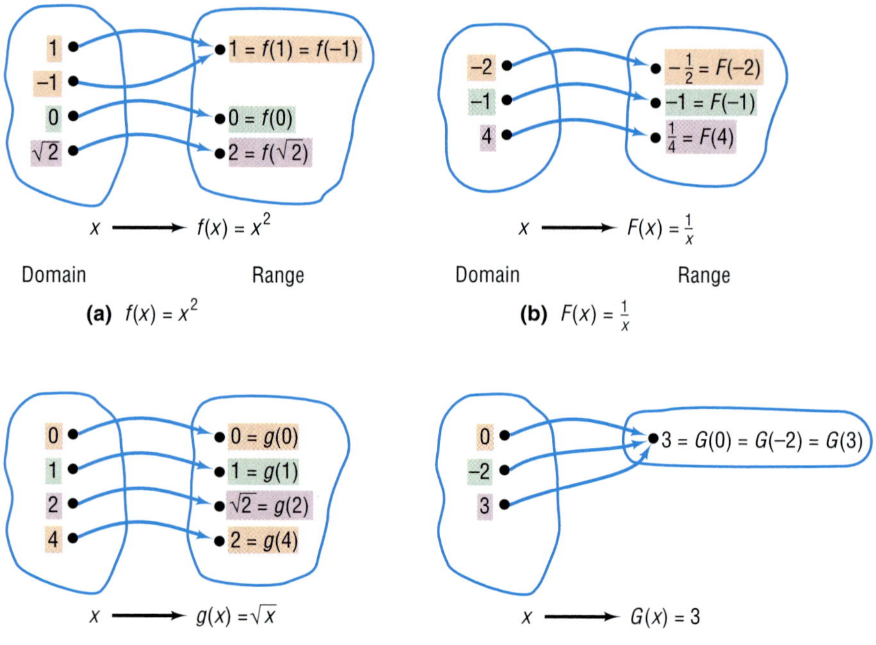

(a) $f(x) = x^2$

(b) $F(x) = \frac{1}{x}$

(c) $g(x) = \sqrt{x}$

(d) $G(x) = 3$

Sometimes it is helpful to think of a function f as a machine that receives as input a number from the domain, manipulates it, and outputs the value. See Figure 5.

FIGURE 5

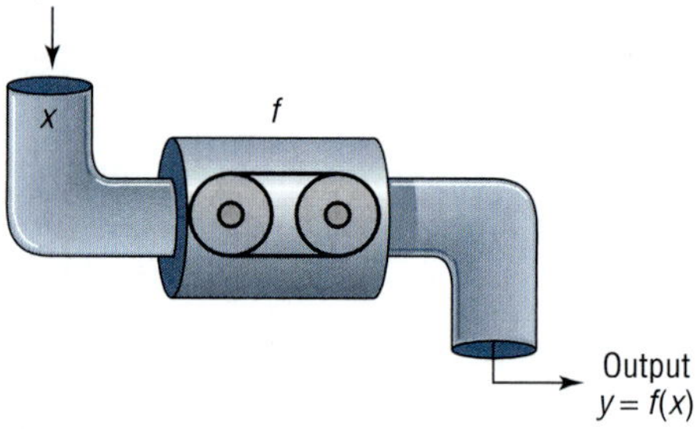

The restrictions on this input/output machine are

1. It only accepts numbers from the domain of the function.
2. For each input, there is exactly one output (which may be repeated for different inputs).

EXAMPLE 5 Finding Values of a Function

3

In physics, it is shown that the function

$$s(t) = -4.9t^2 + 100t + 5$$

relates the height s (in meters) of a bullet fired straight up to the time t (in seconds). Find the height of the bullet at

(a) $t = 0$ seconds (b) $t = 5$ seconds (c) $t = 20$ seconds

Solution (a) The height s at $t = 0$ seconds is found by replacing t by 0 in the stated rule. Thus,

$$\begin{aligned} s(0) &= -4.9(0)^2 + 100(0) + 5 \\ &= 5 \text{ meters} \end{aligned}$$

(b) The height s at $t = 5$ seconds is $s(5) = -4.9(5)^2 + 100(5) + 5 = 382.5$ meters.

(c) The height s at $t = 20$ seconds is $s(20) = -4.9(20)^2 + 100(20) + 5 = 45$ meters. ■

For a function $y = f(x)$, the variable x is sometimes called the **argument** of the function. Thinking of the independent variable as an argument sometimes can make it easier to apply the rule of the function. For example, if f is the function defined by $f(x) = x^3$, then f is the rule that tells us to cube the argument. Thus, $f(2)$ means to cube 2, $f(a)$ means to cube the number a, and $f(x + h)$ means to cube the quantity $x + h$.

EXAMPLE 6 Finding Values of a Function

For the function G defined by $G(x) = 2x^2 - 3x$, evaluate:

(a) $G(3)$ (b) $G(x) + G(3)$ (c) $G(-x)$
(d) $-G(x)$ (e) $G(x + 3)$

Solution (a) We replace x by 3 in the rule for G to get

$$G(3) = 2(3)^2 - 3(3) = 18 - 9 = 9$$

(b) $G(x) + G(3) = (2x^2 - 3x) + (9) = 2x^2 - 3x + 9$

(c) We replace x by $-x$ in the rule for G:

$$G(-x) = 2(-x)^2 - 3(-x) = 2x^2 + 3x$$

(d) $-G(x) = -(2x^2 - 3x) = -2x^2 + 3x$

(e) $$\begin{aligned} G(x + 3) &= 2(x + 3)^2 - 3(x + 3) \\ &= 2(x^2 + 6x + 9) - 3x - 9 \\ &= 2x^2 + 12x + 18 - 3x - 9 \\ &= 2x^2 + 9x + 9 \end{aligned}$$ ■

Notice the use of parentheses here.

Notice in this example that $G(x + 3) \neq G(x) + G(3)$.

Now work Problem 13.

In general, when the rule that defines a function f is given by an equation in x and y, we say that the function f is given **implicitly.** If it is possible to solve the equation for y in terms of x, then we write $y = f(x)$ and say that the function is given **explicitly.** In fact, we usually write "the function

$y = f(x)$" when we mean "the function f defined by the equation $y = f(x)$." Although this usage is not entirely correct, it is rather common and should not cause any confusion. For example,

Implicit Form	**Explicit Form**
$3x + y = 5$	$y = f(x) = -3x + 5$
$x^2 - y = 6$	$y = f(x) = x^2 - 6$
$xy = 4$	$y = f(x) = 4/x$

Not all equations in x and y define a function $y = f(x)$. If an equation is solved for y and two or more values of y can be obtained for a given x, then the equation does not define a function $y = f(x)$. For example, consider the equation $x^2 + y^2 = 1$, which defines a circle. If we solve for y; we obtain $y = \pm\sqrt{1 - x^2}$ so that two values of y will result for numbers x between -1 and 1. Thus, $x^2 + y^2 = 1$ does not define a function.

The explicit form of a function is the form required by a graphing calculator. Now do you see why it is necessary to graph a circle in two "pieces"?

We list below a summary of some important facts to remember about a function f.

SUMMARY OF IMPORTANT FACTS ABOUT FUNCTIONS

1. $f(x)$ is the image of x, or the value of f at x, when the rule f is applied to an x in the domain.
2. To each x in the domain of f, there is one and only one image $f(x)$ in the range.
3. f is the symbol we use to denote the function. It is symbolic of the domain and the rule we use to get from an x in the domain to $f(x)$ in the range.

Function Keys

Most graphing utilities have special keys that enable you to find the value of certain commonly used functions. For example, you should be able to find the square function, $f(x) = x^2$; the square root function, $f(x) = \sqrt{x}$; the reciprocal function, $f(x) = 1/x = x^{-1}$; and many others that will be discussed later in this book (such as $\ln x$, $\log x$, and so on). Verify the results of Example 7 on your graphing utility.

EXAMPLE 7 **Finding Values of a Function on a Calculator**

(a) $f(x) = x^2$; $f(1.234) = 1.522756$
(b) $F(x) = 1/x$; $F(1.234) = 0.8103727715$
(c) $g(x) = \sqrt{x}$; $g(1.234) = 1.110855526$

Domain of a Function

4 Often, the domain of a function f is not specified; instead, only a rule or equation defining the function is given. In such cases, we agree that the domain of f is the largest set of real numbers for which the rule makes sense or, more precisely, for which the value $f(x)$ is a real number. Thus, the domain of f is the same as the domain of the variable x in the expression $f(x)$.

EXAMPLE 8 Finding the Domain of a Function

Find the domain of each of the following functions:

(a) $f(x) = x^2 + 5x$ (b) $g(x) = \dfrac{3x}{x^2 - 4}$ (c) $h(x) = \sqrt{4 - 3x}$

Solution (a) The rule tells us to square a number and then add five times the number. Since these operations can be performed on any real number, we conclude that the domain of f is all real numbers.

(b) The rule g tells us to divide $3x$ by $x^2 - 4$. Since division by 0 is not allowed, the denominator $x^2 - 4$ can never be 0. Thus, x can never equal 2 or -2. The domain of the function g is $\{x \mid x \neq -2, x \neq 2\}$.

(c) The rule h tells us to take the square root of $4 - 3x$. But only nonnegative numbers have real square roots. Hence, we require that

$$\begin{aligned} 4 - 3x &\geq 0 \\ -3x &\geq -4 \\ x &\leq \tfrac{4}{3} \end{aligned}$$

The domain of h is $\{x \mid -\infty < x \leq \frac{4}{3}\}$ or the interval $(-\infty, \frac{4}{3}]$.

We can use a graphing utility to estimate the domain of a function. For example, Figure 6 shows the graph of $y = \sqrt{4 - 3x}$. We can approximate the domain by noting that the domain consists of all numbers x less than or equal to the number where the graph begins. By utilizing the TRACE and BOX features, experiment to see how close you can come to $x = \frac{4}{3}$, the largest value of x in the domain.

FIGURE 6

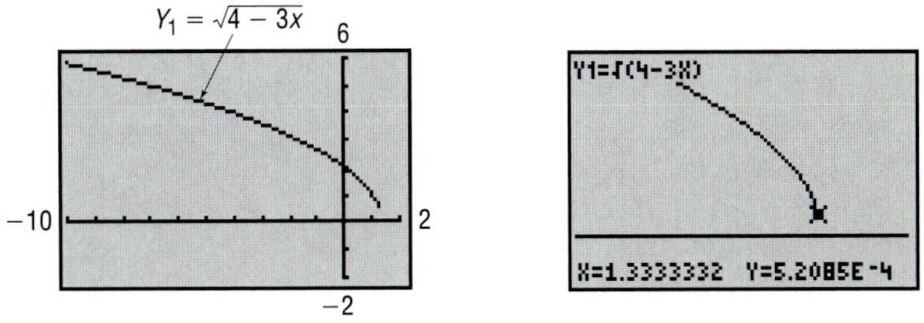

Now work Problem 53.

If x is in the domain of a function f, we shall say that ***f* is defined at *x*,** or ***f*(*x*) exists.** If x is not in the domain of f, we say that ***f* is not defined at *x*,** or ***f*(*x*) does not exist.** For example, if $f(x) = x/(x^2 - 1)$, then $f(0)$ exists, but $f(1)$ and $f(-1)$ do not exist. (Do you see why?)*

We have not said much about finding the range of a function. The reason is that when a function is defined by an equation, it is often difficult to find the range. Therefore, we shall usually be content to find just the domain of a function when only the rule for the function is given. We shall express the domain of a function using interval notation, set notation, or words, whichever is most convenient.

*On many graphing utilities, if f is not defined at x, TRACE will give no value for y. If f is not defined at x, the TABLE feature will show ERROR in the y column.

When we use functions in applications, the domain may be restricted by physical or geometric considerations. For example, the domain of the function f defined by $f(x) = x^2$ is the set of all real numbers. However, if f is used as the rule for obtaining the area of a square when the length x of a side is known, then we must restrict the domain of f to the positive real numbers, since the length of a side can never be 0 or negative.

Independent Variable; Dependent Variable

Consider a function $y = f(x)$. The variable x is called the **independent variable,** because it can be assigned any of the permissible numbers from the domain. The variable y is called the **dependent variable,** because its value depends on x.

Any symbol can be used to represent the independent and dependent variables. For example, if f is the *cube function,* then f can be defined by $f(x) = x^3$ or $f(t) = t^3$ or $f(z) = z^3$. All three rules are identical: Each tells us to cube the independent variable. In practice, the symbols used for the independent and dependent variables are based on common usage.

In applications, it is often apparent from the data which variable is the independent variable and which is the dependent variable. For example, with time-series data, the independent variable is time. Since we can only obtain one value of a variable at any point in time, time-series data always represents a function. With other applications, it is helpful to think of the independent variable as the variable that determines the value of the dependent variable.

EXAMPLE 9 Construction Cost

Sally, a builder of homes, is interested in finding a function that relates the cost C of building a house and x, the number of square feet of the house so she can easily provide estimates to clients for various sized homes. The data she obtained from constructing homes last year is listed in Table 2.

TABLE 2

Square Feet, x	Cost, C
1500	165,000
1600	176,300
1700	183,000
1700	189,500
1830	201,700
1970	217,000
2050	220,000
2100	237,400

(a) Does the relation defined by the set of ordered pairs (x, C) represent a function?

(b) Using a graphing utility, draw a scatter diagram of the data.

(c) Using a graphing utility, find the line of best fit relating square feet and cost.

(d) Interpret the slope.

(e) Express the relationship found in (c) using function notation.

(f) What is the domain of the function?

(g) What is the cost to build a 2000 square foot house?

(h) Give some reasons to explain why the cost of building a 1700 square foot house might cost \$183,000 in one case, and \$189,500 in another. Does this also explain why, perhaps, $C(2000) > C(2050)$?

Solution (a) No, because the first element 1700 has, corresponding to it, two second elements, 183,000 and 189,500.

(b) See Figure 7.

FIGURE 7

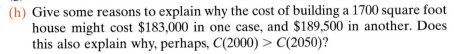

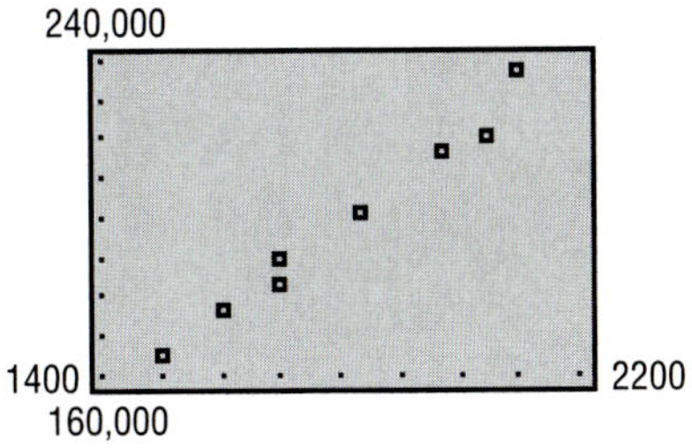

(c) The line of best fit is $y = 112.09x - 3734.02$.

(d) The cost per square foot to build a house is approximately \$112.09. Thus, for each additional square foot of house being built, the cost increases by \$112.09.

(e) Using function notation, $C(x) = 112.09x - 3734.02$.

(f) The domain is $\{x|x > 0\}$ since a house cannot have 0 or negative square feet.

(g) The cost to build a 2000 square foot house is

$$C(2000) = 112.09(2000) - 3734.02 \approx \$220,446$$

(h) One explanation is that the fixtures placed in one house are more expensive than those placed in the other house. ▬

Observe in the solution to Example 9 that we used the symbol C in two ways: It is used to name the function, and is used to symbolize the dependent variable. This double use is common in applications and should not cause any difficulty.

Refer back to Example 9. Notice the data in Table 2 does not represent a function since the same first element is paired with two different second elements ((1700, 183000), (1700, 189500)). However, when we find the line of best fit for this data, we find a function that relates the data. Thus, **curve fitting** is a process whereby we find a functional relationship between two or more variables even though the data, itself, may not represent a function.

EXAMPLE 10 Area of a Circle

Express the area of a circle as a function of its radius.

Solution We know that the formula for the area A of a circle of radius r is $A = \pi r^2$. If we use r to represent the independent variable and A to represent the dependent variable, the function expressing this relationship is

$$A(r) = \pi r^2$$

In this setting, the domain is $\{r|r > 0\}$. (Do you see why?) ▬

Now work Problems 73 and 81.

Examples 9 and 10 demonstrate that some functions are determined from data while others are based on geometric, physical, or other relations.

The next example illustrates how to determine the domain and range of a function if its graph is given.

EXAMPLE 11 Obtaining Information from the Graph of a Function

FIGURE 8

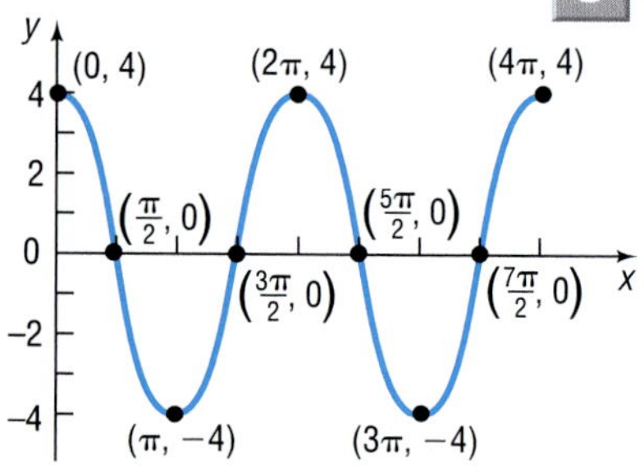

Let f be the function whose graph is given in Figure 8. The graph of f represents the distance the bob of a pendulum is from its "at rest" position. Negative values of y mean the pendulum is to the left of the "at rest" position and positive values of y mean the pendulum is to the right of the "at rest" position.

(a) What is the value of the function when $x = 0$, $x = 3\pi/2$, $x = 3\pi$?
(b) What is the domain of f?
(c) What is the range of f?
(d) List the intercepts. (Recall that these are the points, if any, where the graph crosses the coordinate axes.)

Solution

(a) Since $(0, 4)$ is on the graph of f, the y-coordinate 4 is the value of f at the x-coordinate 0; that is $f(0) = 4$. In a similar way, we find that when $x = 3\pi/2$, then $y = 0$; or $f(3\pi/2) = 0$. When $x = 3\pi$, then $y = -4$; or $f(3\pi) = -4$.
(b) To determine the domain of f, we notice that the points on the graph of f will have x-coordinates between 0 and 4π, inclusive; and, for each number x between 0 and 4π, there is a point $(x, f(x))$ on the graph. Thus, the domain of f is $\{x \mid 0 \le x \le 4\pi\}$, or the interval $[0, 4\pi]$.
(c) The points on the graph all have y-coordinates between -4 and 4, inclusive; and, for each such number y, there is at least one number x in the domain. Hence, the range of f is $\{y \mid -4 \le y \le 4\}$, or the interval $[-4, 4]$.
(d) The intercepts are $(0, 4)$, $(\pi/2, 0)$, $(3\pi/2, 0)$, $(5\pi/2, 0)$, and $(7\pi/2, 0)$. ■

When the graph of a function is given, its domain may be viewed as the shadow created by the graph on the x-axis by vertical beams of light. Its range can be viewed as the shadow created by the graph on the y-axis by horizontal beams of light. Try this technique with the graph given in Figure 8.

Now work Problems 37 and 39.

EXAMPLE 12 Obtaining Information about the Graph of a Function

Consider the function $f(x) = \dfrac{x}{x+2}$.

(a) Is the point $(1, 1/2)$ on the graph of f?
(b) If $x = -1$, what is $f(x)$? What point is on the graph of f?
(c) If $f(x) = 2$, what is x? What point is on the graph of f?

Solution

(a) When $x = 1$, then $f(x) = f(1) = 1/(1 + 2) = 1/3$. The point $(1, 1/2)$ is therefore not on the graph of f.
(b) If $x = -1$, then $f(x) = f(-1) = -1/(-1 + 2) = -1$, so the point $(-1, -1)$ is on the graph of f.

(c) If $f(x) = 2$, then

$$\begin{aligned} \frac{x}{x+2} &= 2 \\ x &= 2(x+2) \\ x &= 2x + 4 \\ x &= -4 \end{aligned}$$

The point $(-4, 2)$ is on the graph of f.

Now work Problem 33.

EXAMPLE 13 Getting from an Island to Town

An island is 2 miles from the nearest point P on a straight shoreline. A town is 12 miles down the shore from P.

(a) If a person can row a boat at an average speed of 3 miles per hour and the same person can walk 5 miles per hour, express the time T it takes to go from the island to town as a function of the distance x from P to where the person lands the boat. See Figure 9.

(b) What is the domain of T?

(c) How long will it take to travel from the island to town if the person lands the boat 4 miles from P?

(d) How long will it take if the person lands the boat 8 miles from P?

(e) Use a graphing utility to graph the function $T = T(x)$.

(f) Use TRACE to see how the time T varies as x changes from 0 to 12.

(g) What value of x results in the least time?

Solution (a) Figure 9 illustrates the situation. The distance d_1 from the island to the landing point satisfies the equation

$$d_1^2 = 4 + x^2 \text{ or } d_1 = \sqrt{4 + x^2}$$

FIGURE 9

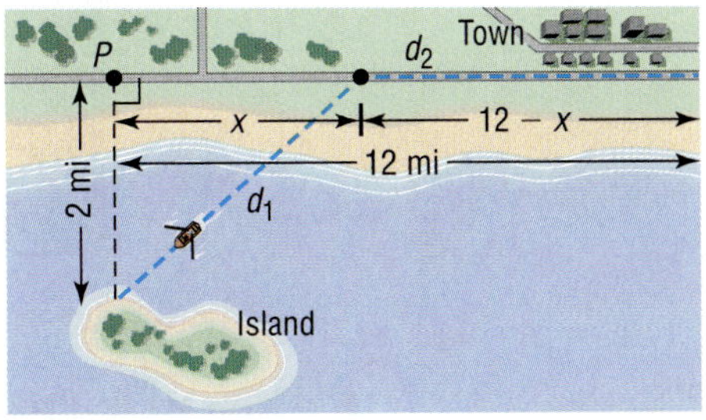

Since the average speed of the boat is 3 miles per hour, the time t_1 it takes to cover the distance d_1 satisfies

$$d_1 = 3t_1$$

Thus,

$$t_1 = \frac{d_1}{3} = \frac{\sqrt{4 + x^2}}{3}$$

The distance d_2 from the landing point to town is $12 - x$, and the time t_2 it takes to cover this distance at an average walking speed of 5 miles per hour obey the equation

$$d_2 = 5t_2$$

Thus,

$$t_2 = \frac{d_2}{5} = \frac{12 - x}{5}$$

The total time T of the trip is $t_1 + t_2$. Thus,

$$T(x) = \frac{\sqrt{4 + x^2}}{3} + \frac{12 - x}{5}$$

(b) Since x equals the distance from P to where the boat lands, it follows that the domain of T is $0 \le x \le 12$.

(c) If the boat is landed 4 miles from P, then $x = 4$. The time T the trip takes is

$$T(4) = \frac{\sqrt{20}}{3} + \frac{8}{5} \approx 3.09 \text{ hours}$$

(d) If the boat is landed 8 miles from P, then $x = 8$. The time T the trip takes is

$$T(8) = \frac{\sqrt{68}}{3} + \frac{4}{5} \approx 3.55 \text{ hours}$$

(e) See Figure 10.

(f) As x varies from 0 to 12, the time T varies from about 2.93 to about 4.05.

(g) Using the TRACE function, for x approximately 1.53 miles, the time T is least, about 2.93 hours. See Figure 11(a).

FIGURE 10

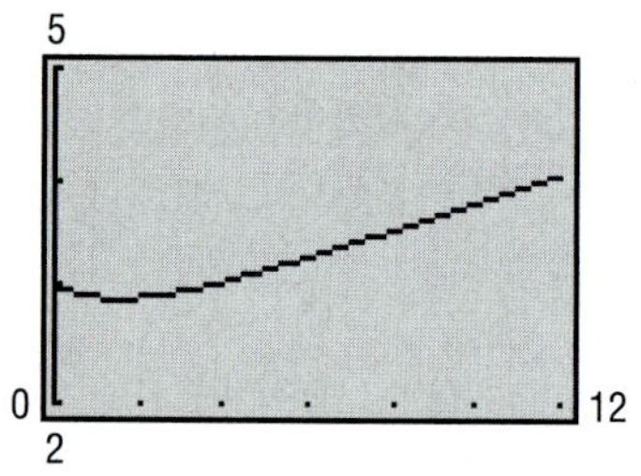

Note: Most graphing utilities have a function minimum command that determines the minimum value of a function for a specified domain. Consult your manual. Using this command, we find that x is approximately 1.50 miles with the time T about 2.93 hours. See Figure 11(b).

FIGURE 11

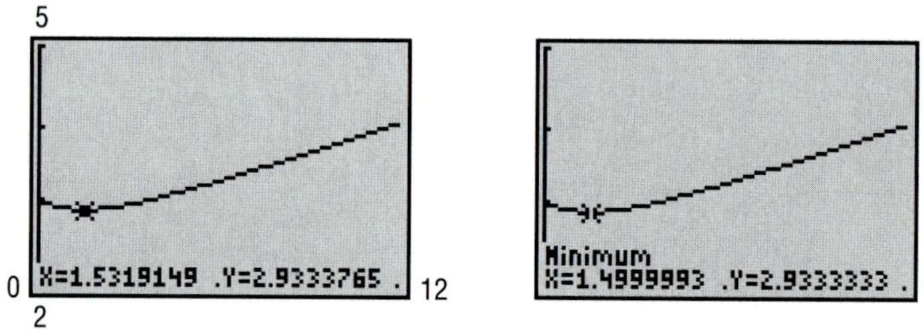

(a) Using TRACE

(b) Using minimum on a TI 83

Now work Problem 83.

SUMMARY

We list here some of the important vocabulary introduced in this section, with a brief description of each term.

Function — A rule or correspondence between two sets of real numbers so that each number x in the first set, the domain, has corresponding to it exactly one number y in the second set.

A set of ordered pairs (x, y) or $(x, f(x))$ in which no two distinct pairs have the same first element.

	The range is the set of y values of the function for the x values in the domain.
	A function f may be defined implicitly by an equation involving x and y or explicitly by writing $y = f(x)$.
Unspecified domain	If a function f is defined by an equation and no domain is specified, then the domain will be taken to be the largest set of real numbers for which the rule defines a real number.
Function notation	$y = f(x)$
	f is a symbol for the rule that defines the function.
	x is the independent variable.
	y is the dependent variable.
	$f(x)$ is the value of the function at x, or the image of x.
Graph of a function	The collection of points (x, y) that satisfies the equation $y = f(x)$.
	A collection of points is the graph of a function provided vertical lines intersect the graph in at most one point (vertical-line test).

2.1 EXERCISES

In Problems 1–12, determine whether each relation represents a function.

1.

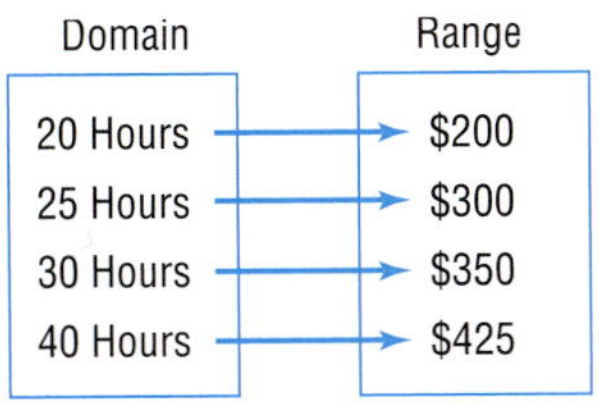

2.

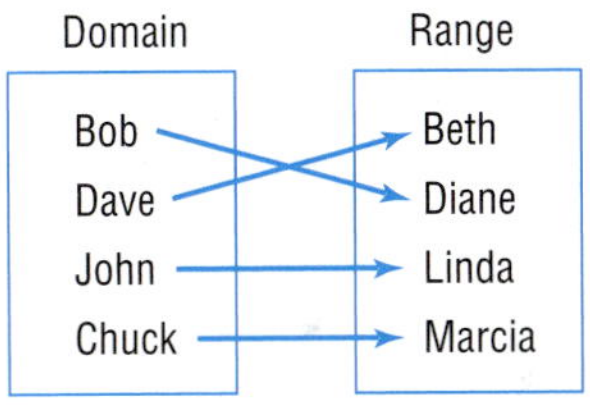

3.

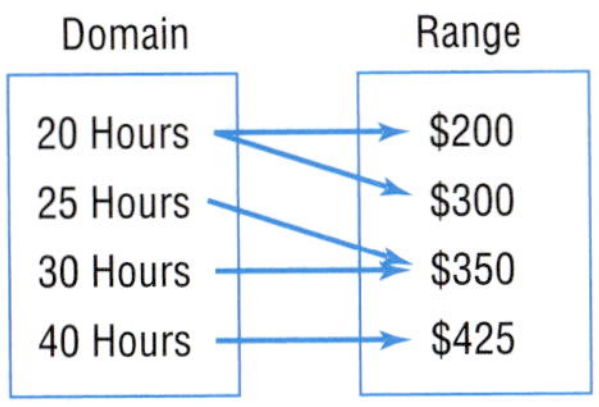

4.

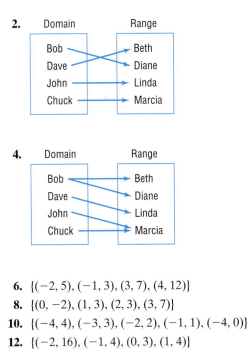

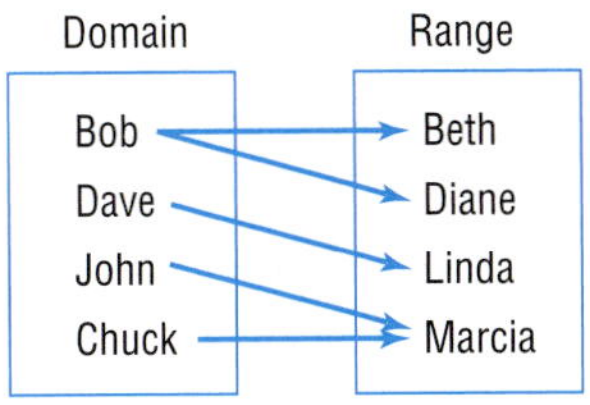

5. $\{(2, 6), (-3, 6), (4, 9), (1, 10)\}$

6. $\{(-2, 5), (-1, 3), (3, 7), (4, 12)\}$

7. $\{(1, 3), (2, 3), (3, 3), (4, 3)\}$

8. $\{(0, -2), (1, 3), (2, 3), (3, 7)\}$

9. $\{(-2, 4), (-2, 6), (0, 3), (3, 7)\}$

10. $\{(-4, 4), (-3, 3), (-2, 2), (-1, 1), (-4, 0)\}$

11. $\{(-2, 4), (-1, 1), (0, 0), (1, 1)\}$

12. $\{(-2, 16), (-1, 4), (0, 3), (1, 4)\}$

In Problems 13–20, find the following values for each function:
(a) $f(0)$; (b) $f(1)$; (c) $f(-1)$; (d) $f(-x)$; (e) $-f(x)$; (f) $f(x + 1)$

13. $f(x) = -3x^2 + 2x - 4$

14. $f(x) = 2x^2 + x - 1$

15. $f(x) = \dfrac{x}{x^2 + 1}$

16. $f(x) = \dfrac{x^2 - 1}{x + 4}$

17. $f(x) = |x| + 4$

18. $f(x) = \sqrt{x^2 + x}$

19. $f(x) = \dfrac{2x + 1}{3x - 5}$

20. $f(x) = 1 - \dfrac{1}{(x + 2)^2}$

In Problems 21–32, use the graph of the function f given in the figure.

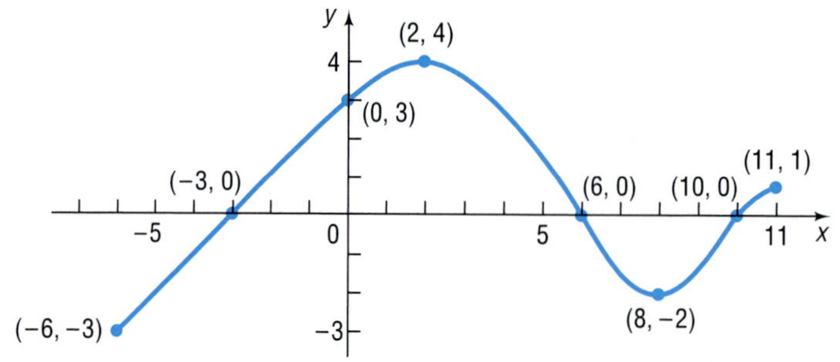

21. Find $f(0)$ and $f(-6)$.

22. Find $f(6)$ and $f(11)$.

23. Is $f(2)$ positive or negative?

24. Is $f(8)$ positive or negative?

25. For what numbers x is $f(x) = 0$?

26. For what numbers x is $f(x) > 0$?

27. What is the domain of f?

28. What is the range of f?

29. What are the x-intercepts?

30. What are the y-intercepts?

31. How often does the line $y = \frac{1}{2}$ intersect the graph?

32. How often does the line $y = 3$ intersect the graph?

In Problems 33–36, answer the questions about the given function.

33. $f(x) = \dfrac{x+2}{x-6}$

(a) Is the point $(3, 14)$ on the graph of f?
(b) If $x = 4$, what is $f(x)$? What point is on the graph of f?
(c) If $f(x) = 2$, what is x? What point is on the graph of f?
(d) What is the domain of f?

34. $f(x) = \dfrac{x^2+2}{x+4}$

(a) Is the point $(1, \frac{3}{5})$ on the graph of f?
(b) If $x = 0$, what is $f(x)$? What point is on the graph of f?
(c) If $f(x) = \frac{1}{2}$, what is x? What point is on the graph of f?
(d) What is the domain of f?

35. $f(x) = \dfrac{2x^2}{x^4+1}$

(a) Is the point $(-1, 1)$ on the graph of f?
(b) If $x = 2$, what is $f(x)$? What point is on the graph of f?
(c) If $f(x) = 1$, what is x? What point is on the graph of f?
(d) What is the domain of f?

36. $f(x) = \dfrac{2x}{x-2}$

(a) Is the point $(\frac{1}{2}, -\frac{2}{3})$ on the graph of f?
(b) If $x = 4$, what is $f(x)$? What point is on the graph of f?
(c) If $f(x) = 1$, what is x? What point is on the graph of f?
(d) What is the domain of f?

In Problems 37–48, determine whether the graph is that of a function by using the vertical-line test. If it is, use the graph to find:

(a) Its domain and range

(b) The intercepts, if any

(c) Any symmetry with respect to the x-axis, y-axis, or origin

37.

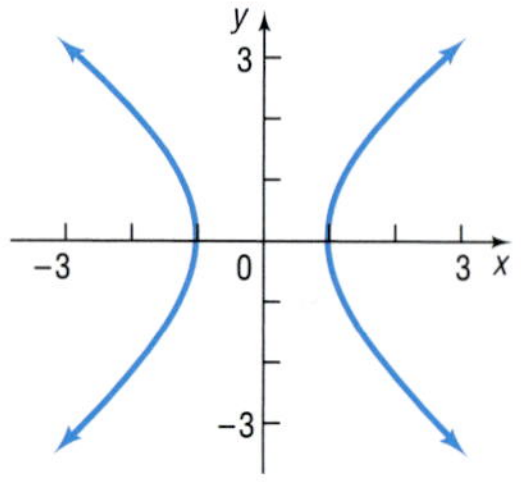

38.

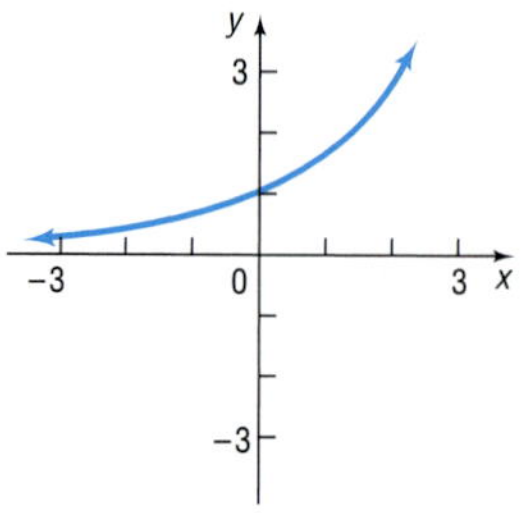

39.

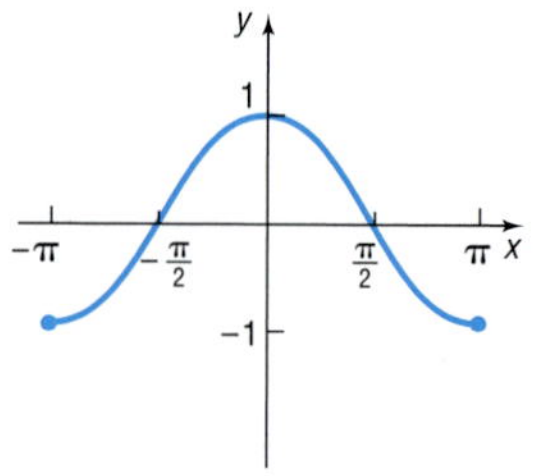

40.
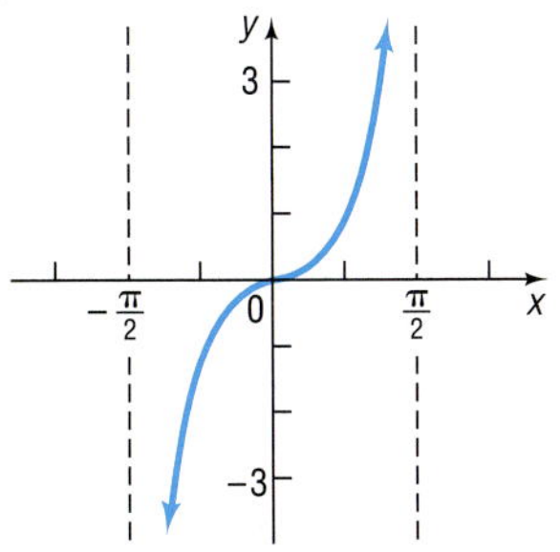

41.
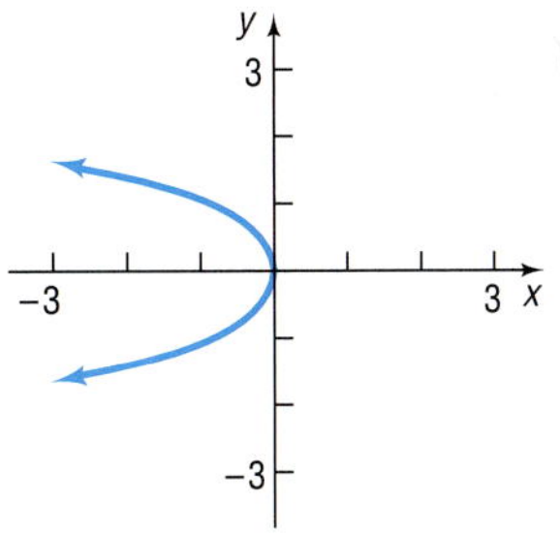

42.
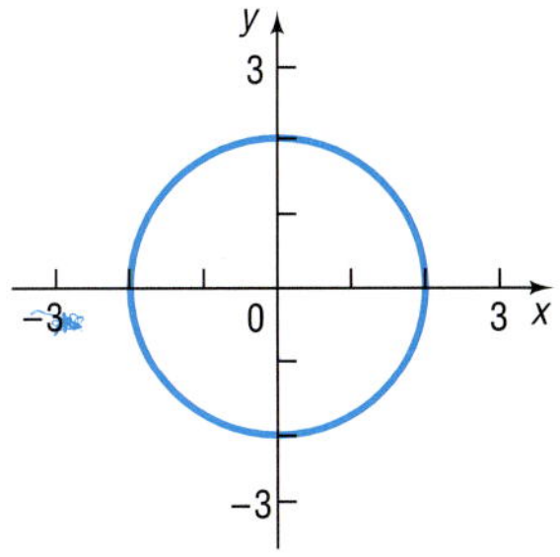

43.
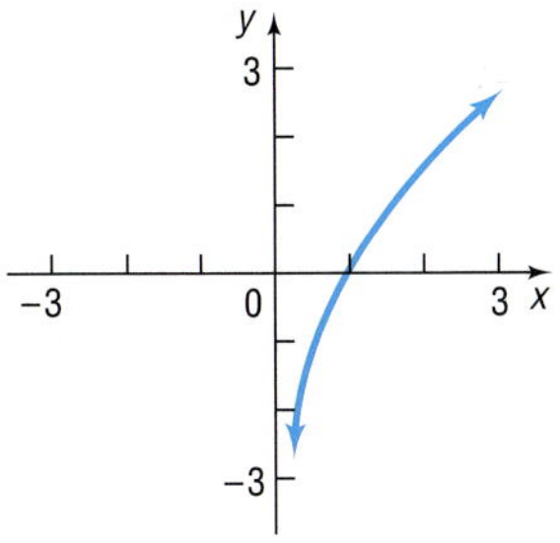

44.
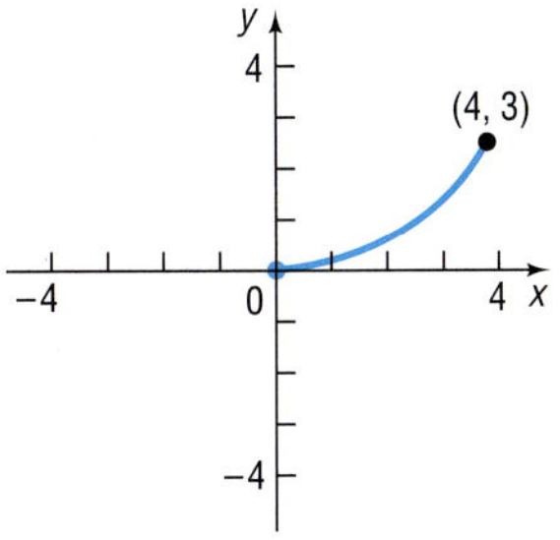

45.
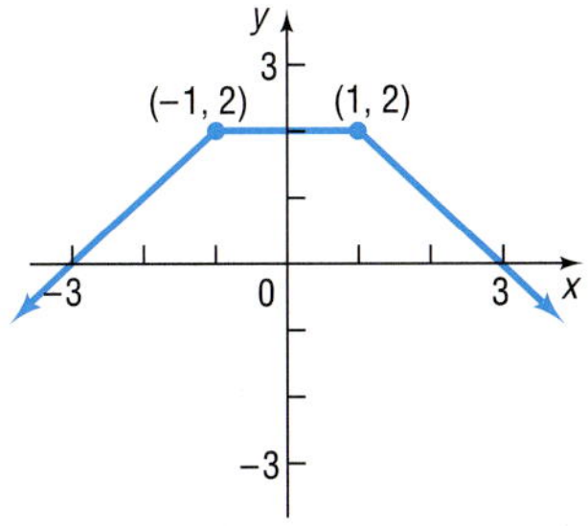

46.
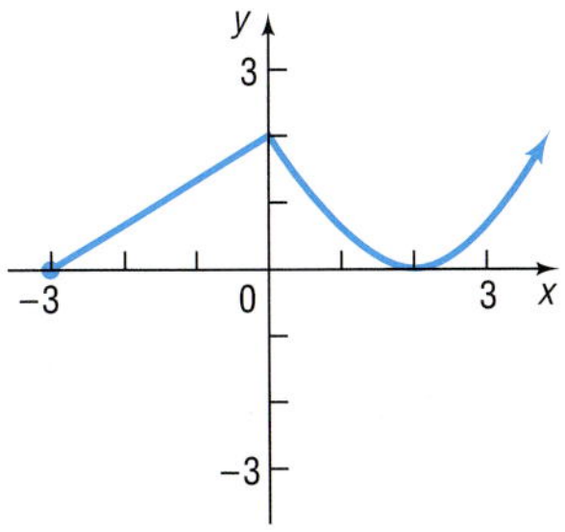

47.
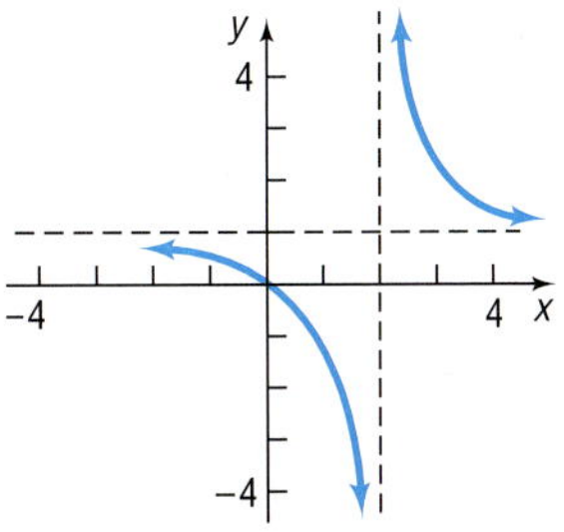

48.
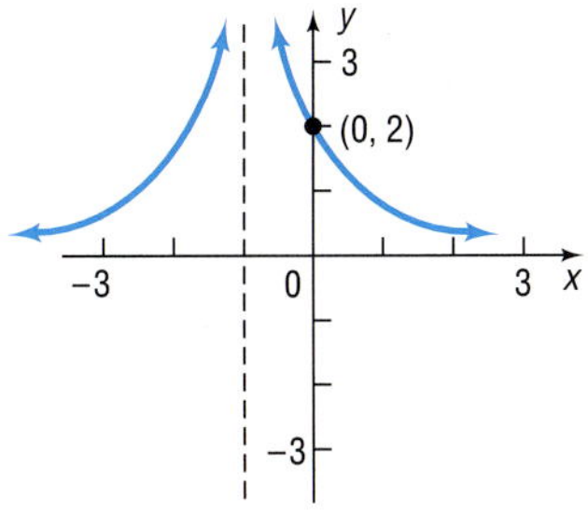

In Problems 49–62, find the domain of each function. Verify your results by graphing each function using a graphing utility.

49. $f(x) = 3x + 4$

50. $f(x) = 5x^2 + 2$

51. $f(x) = \dfrac{x}{x^2 + 1}$

52. $f(x) = \dfrac{x^2}{x^2 + 1}$

53. $g(x) = \dfrac{x}{x^2 - 1}$

54. $h(x) = \dfrac{x}{x - 1}$

55. $F(x) = \dfrac{x - 2}{x^3 + x}$

56. $G(x) = \dfrac{x + 4}{x^3 - 4x}$

57. $h(x) = \sqrt{3x - 12}$

58. $G(x) = \sqrt{1 - x}$

59. $f(x) = \dfrac{4}{\sqrt{x - 9}}$

60. $f(x) = \dfrac{x}{\sqrt{x - 4}}$

61. $p(x) = \sqrt[3]{\dfrac{x - 2}{x - 1}}$

62. $q(x) = \sqrt[3]{x^2 - x - 2}$

63. Match each of the following functions with the graphs that best describes the situation.
(a) The cost of building a house as a function of its square footage.
(b) The height of an egg dropped from a 300-foot building as a function of time.
(c) The height of a human as a function of time.
(d) The demand for Big Macs as a function of price.
(e) The height of a child on a swing as a function of time.

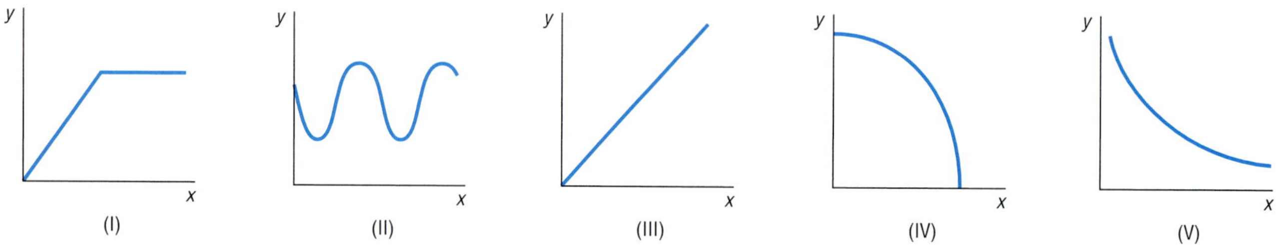

64. Match each of the following functions with the graph that best describes the situation.
(a) The temperature of a bowl of soup as a function of time.
(b) The number of hours of daylight during the year.
(c) The population of Florida as a function of time.
(d) The distance of a car traveling at a constant velocity as a function of time.
(e) The height of a golf ball hit with a 7-iron as a function of time.

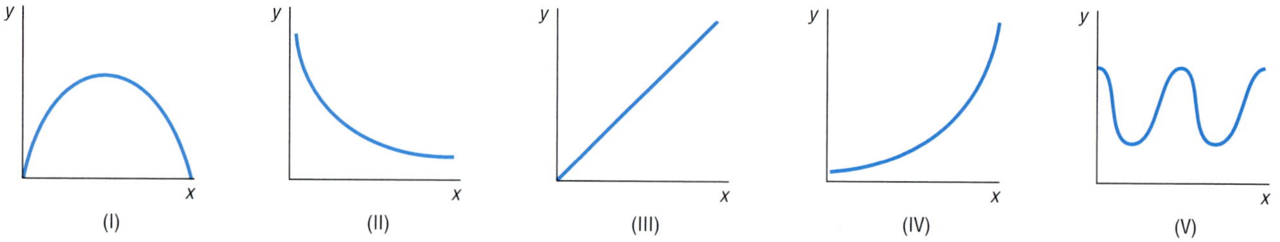

65. Consider the following scenario: Barbara decides to take a walk. She leaves home, walks 2 blocks in 5 minutes at a constant speed, and realizes she forgot to lock the door. So, Barbara runs home in 1 minute. While at her doorstep it takes her 1 minute to find her keys and lock the door. Barbara walks 5 blocks in 15 minutes, then decides to jog home. It takes her 7 minutes to get home. Draw a graph of Barbara's distance from home (in blocks) as a function of time.

66. Consider the following scenario: Jayne enjoys riding her bicycle through the woods. At the forest preserve, she gets on her bicycle and rides up a 2,000-foot hill in 10 minutes. She then travels down the hill in 3 minutes. The next 5,000 feet is level terrain and she covers the distance in 20 minutes. She rests for 15 minutes. Jayne then travels 10,000 feet in 30 minutes. Draw a graph of Jayne's distance traveled (in feet) as a function a of time.

67. If $f(x) = 2x^3 + Ax^2 + 4x - 5$ and $f(2) = 5$, what is the value of A?

68. If $f(x) = 3x^2 - Bx + 4$ and $f(-1) = 12$, what is the value of B?

69. If $f(x) = (3x + 8)/(2x - A)$ and $f(0) = 2$, what is the value of A?

70. If $f(x) = (2x - B)/(3x + 4)$ and $f(2) = \frac{1}{2}$, what is the value of B?

71. If $f(x) = (2x - A)/(x - 3)$ and $f(4) = 0$, what is the value of A? Where is f not defined?

72. If $f(x) = (x - B)/(x - A)$, $f(2) = 0$, and $f(1)$ is undefined, what are the values of A and B?

73. **Demand for Jeans** The marketing manager at Levi-Strauss wishes to find a function that relates the demand D of men's jeans and p, the price of the jeans. The following data was obtained based on a price history of the jeans.

Price ($/Pair), *p*	Demand (Pairs of Jeans Sold per Day), *D*
20	60
22	57
23	56
23	53
27	52
29	49
30	44

(a) Does the relation defined by the set of ordered pairs (p, D) represent a function?
(b) Using a graphing utility, draw a scatter diagram of the data.
(c) Using a graphing utility, find the line of best fit relating price and quantity demanded.
(d) Interpret the slope.
(e) Express the relationship found in (c) using function notation.
(f) What is the domain of the function?
(g) How many jeans will be demanded if the price is $28 a pair?

74. Advertising and Sales Revenue A marketing firm wishes to find a function that relates the sales S of a product and A, the amount spent on advertising the product. The data are obtained from past experience. Advertising and sales are measured in thousands of dollars.

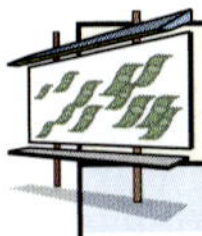

Advertising Expenditures, A	Sales, S
20	335
22	339
22.5	338
24	343
24	341
27	350
28.3	351

(a) Does the relation defined by the set of ordered pairs (A, S) represent a function?
(b) Using a graphing utility, draw a scatter diagram of the data.
(c) Using a graphing utility, find the line of best fit relating advertising expenditures and sales.
(d) Interpret the slope.
(e) Express the relationship found in (c) using function notation.
(f) What is the domain of the function?
(g) Predict sales if advertising expenditures are \$25,000.

75. Distance and Time Using the data below, find a function that relates the distance, s, in miles, driven by a Ford Taurus, and t, the time the Taurus has been driven.

(a) Does the relation defined by the set of ordered pairs (t, s) represent a function?
(b) Using a graphing utility, draw a scatter diagram of the data.

Time (Hours), t	Distance (Miles), s
0	0
1	30
2	55
3	83
4	100
5	150
6	210
7	260
8	300

(c) Using a graphing utility, find the line of best fit relating time and distance.
(d) Interpret the slope.
(e) Express the relationship found in (c) using function notation.
(f) What is the domain of the function?
(g) Predict the distance the car is driven after 11 hours.

76. High School vs. College GPA An administrator at Southern Illinois University wants to find a function that relates a student's college grade point average G to the high school grade point average, x. She randomly selects 8 students and obtains the following data:

High School GPA, x	College GPA, G
2.73	2.43
2.92	2.97
3.45	3.63
3.78	3.81
2.56	2.83
2.98	2.81
3.67	3.45
3.10	2.93

(a) Does the relation defined by the set of ordered pairs (x, G) represent a function?
(b) Using a graphing utility, draw a scatter diagram of the data.
(c) Using a graphing utility, find the line of best fit relating high school GPA and college GPA.
(d) Interpret the slope.
(e) Express the relationship found in (c) using function notation.
(f) What is the domain of the function?
(g) Predict a student's college GPA if her high school GPA is 3.23.

77. Effect of Gravity on Earth If a rock falls from a height of 20 meters on Earth, the height H (in meters) after x seconds is approximately

$$H(x) = 20 - 4.9x^2$$

(a) Use a graphing utility to graph the function H.
(b) What is the height of the rock when $x = 1$ second? $x = 1.1$ seconds? $x = 1.2$ seconds? $x = 1.3$ seconds?
(c) When is the height of the rock 15 meters? When is it 10 meters? When is it 5 meters?
(d) When does the rock strike the ground?

78. Effect of Gravity on Jupiter If a rock falls from a height of 20 meters on the planet Jupiter, its height H (in meters) after x seconds is approximately

$H(x) = 20 - 13x^2$

(a) Use a graphing utility to graph the function H.
(b) What is the height of the rock when $x = 1$ second? $x = 1.1$ seconds? $x = 1.2$ seconds?
(c) When is the height of the rock 15 meters? When is it 10 meters? When is it 5 meters?
(d) When does the rock strike the ground?

79. Geometry Express the area A of a rectangle as a function of the length x if the length is twice the width of the rectangle.

80. Geometry Express the area A of an isosceles right triangle as a function of the length x of one of the two equal sides.

81. Express the gross salary G of a person who earns \$10 per hour as a function of the number x of hours worked.

82. Tiffany, a commissioned salesperson, earns \$100 base pay plus \$10 per item sold. Express her gross salary G as a function of the number x of items sold.

83. Installing Cable TV MetroMedia Cable is asked to provide service to a customer whose house is located 2 miles from the road along which the cable is buried. The nearest connection box for the cable is located 5 miles down the road (see the figure).

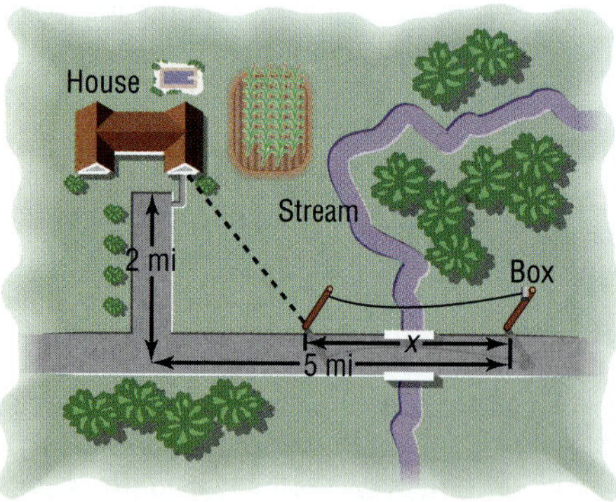

(a) If the installation cost is \$10 per mile along the road and \$14 per mile off the road, express the total cost C of installation as a function of the distance x (in miles) from the connection box to the point where the cable installation turns off the road. Give the domain.
(b) Compute the cost if $x = 1$ mile.
(c) Compute the cost if $x = 3$ miles.
(d) Graph the function $C = C(x)$. Use TRACE to see how the cost C varies as x changes from 0 to 5.
(e) What value of x results in the least cost?

84. Time Required to go from an Island to a Town An island is 3 miles from the nearest point P on a straight shoreline. A town is located 20 miles down the shore from P. (Refer to Figure 9 for a similar situation.)

(a) If a person has a boat that averages 12 miles per hour and the same person can run 5 miles per hour, express the time T it takes to go from the island to town as a function of x, where x is the distance from P to where the person lands the boat. Give the domain.
(b) How long will it take to travel from the island to town if you land the boat 8 miles from P?
(c) How long will it take if you land the boat 12 miles from P?
(d) Graph the function $T = T(x)$. Use TRACE to see how the time T varies as x changes from 0 to 20.
(e) What value of x results in the least time?

85. Page Design A page with dimensions of $8\frac{1}{2}$ inches by 11 inches has a border of uniform width x surrounding the printed matter of the page, as shown in the figure.

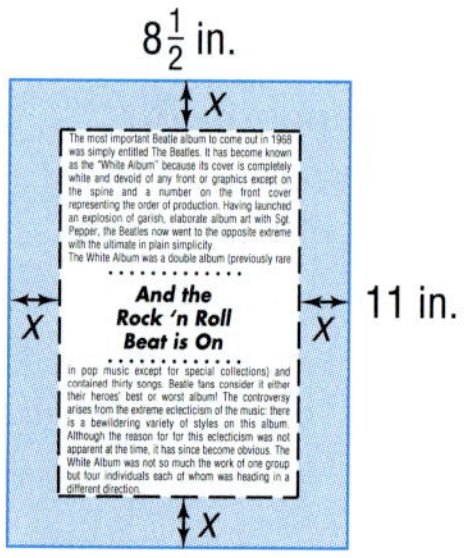

(a) Write a formula for the area A of the printed part of the page as a function of the width x of the border.
(b) Give the domain and range of A.
(c) Find the area of the printed page for borders of widths 1 inch, 1.2 inches, and 1.5 inches.
(d) Graph the function $A = A(x)$.
(e) Use TRACE to determine what margin should be used to obtain an area of 70 square inches and of 50 square inches.

86. Cost of Trans-Atlantic Travel A Boeing 747 crosses the Atlantic Ocean (3,000 miles) with an airspeed of 500 miles per hour. The cost C (in dollars) per passenger is given by

$$C(x) = 100 + \frac{x}{10} + \frac{36{,}000}{x}$$

where x is the ground speed (airspeed $\pm$ wind).

(a) What is the cost per passenger for quiescent (no wind) conditions?
(b) What is the cost per passenger with a head wind of 50 miles per hour?
(c) What is the cost per passenger with a tail wind of 100 miles per hour?
(d) What is the cost per passenger with a head wind of 100 miles per hour?
(e) Graph the function $C = C(x)$.
(f) As x varies from 400 to 600 miles per hour, how does the cost vary?

87. Period of a Pendulum The period T (in seconds) of a simple pendulum is a function of its length l (in feet) defined by the equation

$$T(l) = 2\pi\sqrt{\frac{l}{g}}$$

where $g \approx 32.2$ feet per second per second is the acceleration due to gravity.

(a) Use a graphing utility to graph the function $T = T(l)$.
(b) Use the TRACE function to see how the period T varies as l changes from 1 to 10.
(c) What length should be used if a period of 10 seconds is required?

88. Effect of Elevation on Weight If an object weighs m pounds at sea level, then its weight W (in pounds) at a height of h miles above sea level is given approximately by

$$W(h) = m\left(\frac{4000}{4000 + h}\right)^2$$

(a) If Amy weighs 120 pounds at sea level, how much will she weigh on Pike's Peak, which is 14,110 feet above sea level?
(b) Use a graphing utility to graph the function $W = W(h)$. Use $m = 120$ pounds.
(c) Use the TRACE function to see how weight W varies as h changes from 0 to 5 miles.
(d) At what height will Amy weigh 121 pounds?
(e) Does your answer to part (d) seem reasonable?

In Problems 89–96, tell whether the set of ordered pairs (x, y) defined by each equation is a function.

89. $y = x^2 + 2x$ **90.** $y = x^3 - 3x$ **91.** $y = \frac{2}{x}$ **92.** $y = \frac{3}{x} - 3$

93. $y^2 = 1 - x^2$ **94.** $y = \pm\sqrt{1 - 2x}$ **95.** $x^2 + y = 1$ **96.** $x + 2y^2 = 1$

97. Some functions f have the property that $f(a + b) = f(a) + f(b)$ for all real numbers a and b. Which of the following functions have this property?

(a) $h(x) = 2x$ (b) $g(x) = x^2$
(c) $F(x) = 5x - 2$ (d) $G(x) = 1/x$

98. Draw the graph of a function whose domain is $\{x \mid -3 \le x \le 8,\ x \ne 5\}$ and whose range is $\{y \mid -1 \le y \le 2, y \ne 0\}$. What point(s) in the rectangle $-3 \le x \le 8, -1 \le y \le 2$ cannot be on the graph? Compare your graph with those of other students. What differences do you see?

99. Are the functions $f(x) = x - 1$ and $g(x) = (x^2 - 1)/(x + 1)$ the same? Explain.

100. Describe how you would proceed to find the domain and range of a function if you were given its graph. How would your strategy change if, instead, you were given the equation defining the function?

101. How many x-intercepts can the graph of a function have? How many y-intercepts can it have?

102. Is a graph that consists of a single point the graph of a function? Can you write the equation of such a function?

103. Is there a function whose graph is symmetric with respect to the x-axis?

104. Investigate when, historically, the use of function notation $y = f(x)$ first appeared.

2.2 MORE ABOUT FUNCTIONS

1. Find the Average Rate of Change of a Function
2. Use a Graphing Utility to Determine Where a Function is Increasing and Decreasing
3. Use a Graphing Utility to Locate Local Maxima and Minima
4. Determine Even or Odd Functions from a Graph
5. Identify Even or Odd Functions from the Equation
6. Graph Certain Important Functions
7. Graph Piecewise-Defined Functions

Average Rate of Change

1 In Section 1.4 we said the slope of a straight line could be interpreted as the average rate of change. Often, we are interested in the rate at which functions change. To find the average rate of change of a function between any two points on its graph, we calculate the slope of the line containing the two points.

EXAMPLE 1 Finding the Average Rate of Change of a Function

Suppose you drop a ball from a cliff 1000 feet high. You measure the distance s the ball has fallen after time t using a motion detector and obtain the data in Table 3.

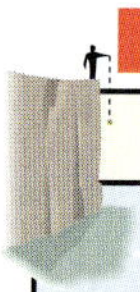

TABLE 3

Time, t (Seconds)	Distance, s (Feet)
0	0
1	16
2	64
3	144
4	256
5	400
6	576
7	784

(a) Draw a scatter diagram of the data by hand treating time as the independent variable.
(b) Draw a line from the point $(0, 0)$ to $(2, 64)$.
(c) Find the average rate of change of the ball between 0 and 2 seconds, that is, find the slope of the line in (b).
(d) Interpret the average rate of change found in (c).
(e) Draw a line from the point $(5, 400)$ to $(7, 784)$.
(f) Find the average rate of change of the ball between 5 and 7 seconds.
(g) Interpret the average rate of change found in (f).
(h) What is happening to the average rate of change as time passes?

Solution

(a) We plot the ordered pairs $(0, 0)$, $(1, 16)$, and so on, using rectangular coordinates. See Figure 12.
(b) Draw a line from $(0, 0)$ to $(2, 64)$. See Figure 13.
(c) The average rate of change is found by computing the slope of the line formed by joining the points $(0, 0)$ and $(2, 64)$.

$$\text{Average Rate of Change} = \frac{\Delta s}{\Delta t} = \frac{64 - 0}{2 - 0} = \frac{64}{2} = 32 \text{ ft/sec}$$

(d) Since the change in distance divided by the change in time represents a speed, we would interpret the slope of the line as an average speed. The average speed of the ball between 0 and 2 seconds is 32 feet per second.
(e) Draw a line from $(5, 400)$ to $(7, 784)$. See Figure 14.

FIGURE 12

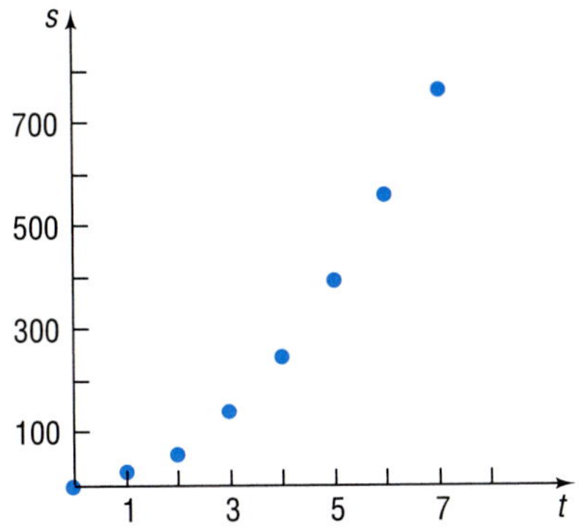

FIGURE 13

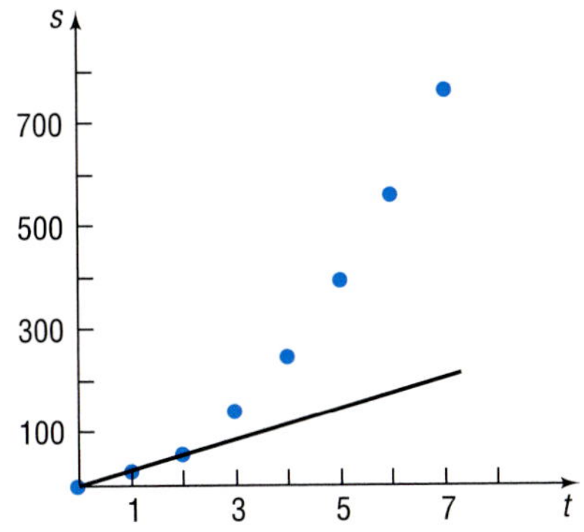

FIGURE 14

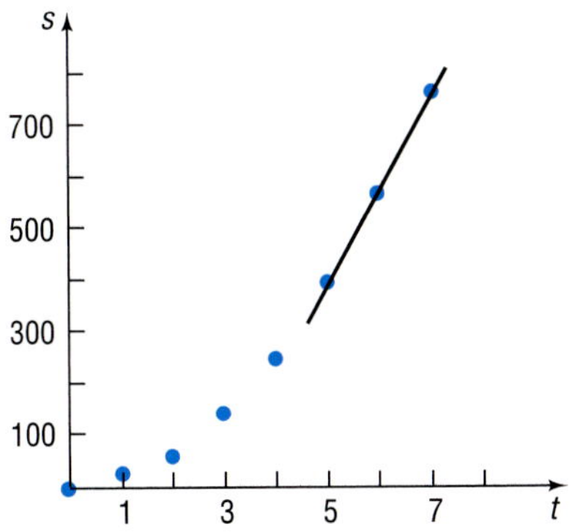

(f) The average rate of change of the ball between 5 and 7 seconds is

$$\text{Average Rate of Change} = \frac{\Delta s}{\Delta t} = \frac{784 - 400}{7 - 5} = \frac{384}{2} = 192 \text{ ft/sec}$$

(g) The average speed of the ball between 5 and 7 seconds is 192 feet per second.

(h) The average speed of the ball is increasing as time passes since the ball is accelerating due to the effect of gravity. ■

If we know the function f that relates the time t to the distance s the ball has fallen, $s = f(t)$, then the average speed of the ball between 0 and 2 seconds is

$$\frac{f(2) - f(0)}{2 - 0} = \frac{64 - 0}{2} = 32 \text{ ft/sec}$$

The average speed of the ball between 5 and 7 seconds is

$$\frac{f(7) - f(5)}{7 - 5} = \frac{784 - 400}{2} = 192 \text{ ft/sec}$$

Expressions like these occur frequently in calculus.

If c is in the domain of a function $y = f(x)$, the **average rate of change of f** between c and x is defined as

$$\text{average rate of change} = \frac{\Delta y}{\Delta x} = \frac{f(x) - f(c)}{x - c} \qquad x \neq c \qquad (1)$$

This expression is also called the **difference quotient** of f at c.

The average rate of change of a function has an important geometric interpretation. Look at the graph of $y = f(x)$ in Figure 15. We have labeled two points on the graph: $(c, f(c))$ and $(x, f(x))$. The slope of the line containing these two points is

$$\frac{f(x) - f(c)}{x - c}$$

This line is called a **secant line.** Thus, the average rate of change of a function equals the slope of a secant line containing two points on its graph.

FIGURE 15

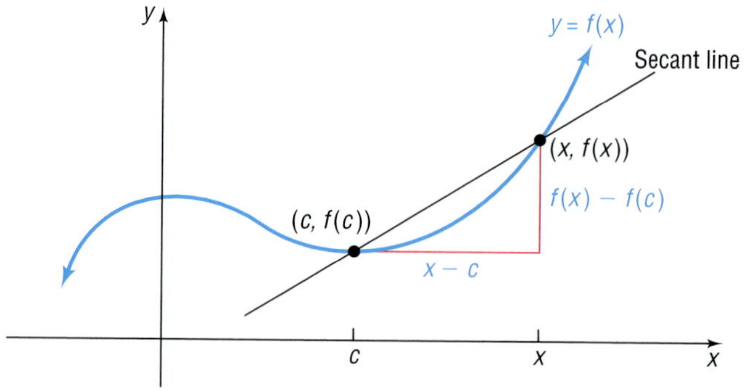

EXAMPLE 2

Finding the Average Rate of Change

The function $s(t) = -4.9t^2 + 100t + 5$ gives the height s (in meters) of a bullet fired straight up as a function of time t (in seconds).

(a) Find the average rate of change of the height of the bullet between 2 and t seconds.

(b) Using the result found in (a), find the average rate of change of the height of the bullet between 2 and 3 seconds.

Solution (a) From expression (1), we seek

$$\frac{\Delta s}{\Delta t} = \frac{s(t) - s(2)}{t - 2}, \qquad t \neq 2$$

We begin by finding $s(2)$:

$$s(2) = -4.9(2)^2 + 100(2) + 5 = -19.6 + 200 + 5 = 185.4$$

Then the average rate of change of s between 2 and t seconds is

$$\frac{s(t) - s(2)}{t - 2} = \frac{-4.9t^2 + 100t + 5 - 185.4}{t - 2} = \frac{-4.9t^2 + 100t - 180.4}{t - 2}$$
$$= \frac{(-4.9t + 90.2)(t - 2)}{t - 2} = -4.9t + 90.2$$

(b) The average rate of change between 2 and 3 seconds is found by letting $t = 3$ in the expression found in (a). Thus, the average rate of change of the bullet between 2 and 3 seconds is $-4.9(3) + 90.2 = 75.5$ meters per second.

Now work Problem 31.

Increasing and Decreasing Functions

Consider the graph given in Figure 16. If you look from left to right along the graph of the function, you will notice that parts of the graph are rising, parts are falling, and parts are horizontal. In such cases, the function is described as *increasing, decreasing,* and *constant,* respectively.

FIGURE 16

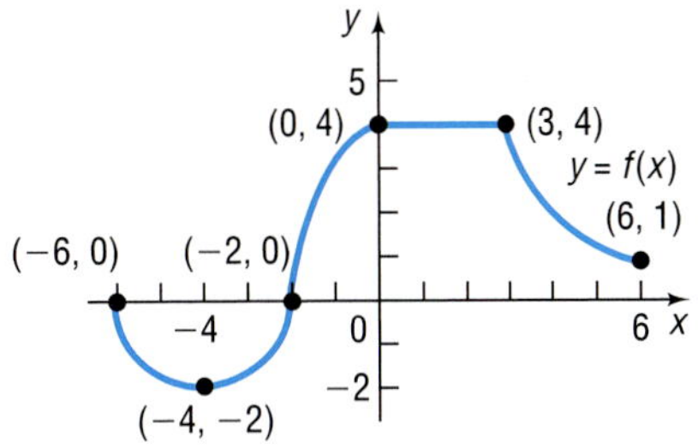

EXAMPLE 3 Determining Where a Function Is Increasing, Decreasing, or Constant

Where is the function in Figure 16 increasing? Where is it decreasing? Where is it constant?

Solution To answer the question of where a function is increasing, where it is decreasing, and where it is constant, we use inequalities involving the independent variable x or we use open intervals* of x-coordinates. The graph in Figure 16 is rising (increasing) from the point $(-4, -2)$ to the point $(0, 4)$, so we conclude that it is increasing on the open interval $(-4, 0)$ (or for $-4 < x < 0$). The graph is falling (decreasing) from the point $(-6, 0)$ to the point $(-4, -2)$ and from the point $(3, 4)$ to the point $(6, 1)$. We conclude that the graph is decreasing on the open intervals $(-6, -4)$ and $(3, 6)$ (or for $-6 < x < -4$ and $3 < x < 6$). The graph is constant on the open interval $(0, 3)$ (or for $0 < x < 3$).

More precise definitions follow.

A function f is **increasing** on an open interval I if, for any choice of x_1 and x_2 in I, with $x_1 < x_2$, we have $f(x_1) < f(x_2)$.

A function f is **decreasing** on an open interval I if, for any choice of x_1 and x_2 in I, with $x_1 < x_2$, we have $f(x_1) > f(x_2)$.

A function f is **constant** on an open interval I if, for all choices of x in I, the values $f(x)$ are equal.

Thus, the graph of an increasing function goes up from left to right, the graph of a decreasing function goes down from left to right, and the graph of a constant function remains at a fixed height. Figure 17 illustrates the definitions.

*The open interval (a, b) consists of all real numbers x for which $a < x < b$. Refer to Section 1 of the Appendix, if necessary.

FIGURE 17

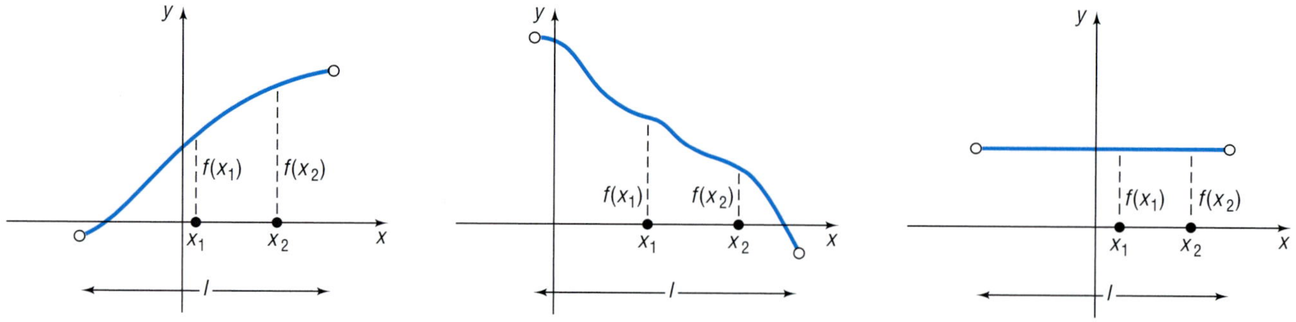

(a) For $x_1 < x_2$ in I, $f(x_1) < f(x_2)$; f is increasing

(b) For $x_1 < x_2$ in I, $f(x_1) > f(x_2)$; f is decreasing

(c) Values of f are equal; f is constant

EXAMPLE 4

Using a Graphing Utility to Determine Where a Function Is Increasing and Decreasing

2

Use a graphing utility to graph the function $f(x) = x^3 - 2x^2 + 3$ for $-1 < x < 2$. Determine where f is increasing and where it is decreasing.

FIGURE 18

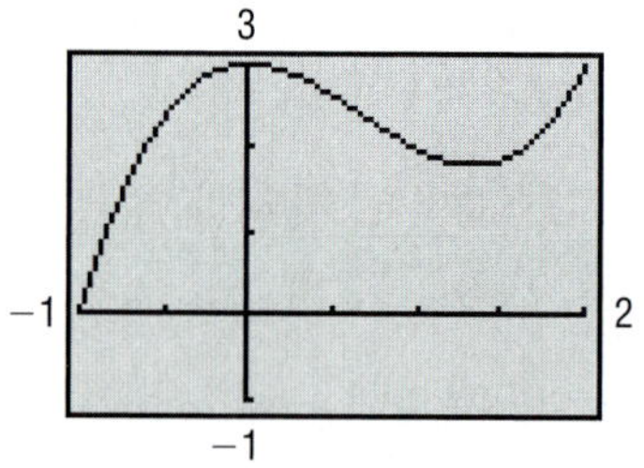

Solution Figure 18 shows the graph of f. Use TRACE, beginning at the point $(-1, 0)$. As we proceed along the graph, the values of y increase until we cross the y-axis (where $y = 3$). We infer that f is increasing on the interval $(-1, 0)$.

As we continue along the graph to the right of the y-axis, the values of f decrease until we reach $x \approx 1.33$ (where $y \approx 1.815$). We infer that f is decreasing on the interval $(0, 1.33)$.

As we continue, the values of y increase from $x = 1.33$ to $x = 2$. We infer that f is increasing on the interval $(1.33, 2)$. ■

Local Maximum; Local Minimum

When the graph of a function is increasing to the left of $x = c$ and decreasing to the right of $x = c$, then at c the value of f is largest. This value is called a *local maximum* of f.

When the graph of a function is decreasing to the left of $x = c$ and is increasing to the right of $x = c$, then at c the value of f is the smallest. This value is called a *local minimum* of f.

Thus, if f has a local maximum at c, then the value of f at c is greater than the values of f near c. If f has a local minimum at c, then the value of f at c is less than the values of f near c. The word *local* is used to suggest that it is only near c that the value $f(c)$ is largest or smallest. See Figure 19.

FIGURE 19
f has local maximum at x_1 and x_3; f has a local minimum at x_2.

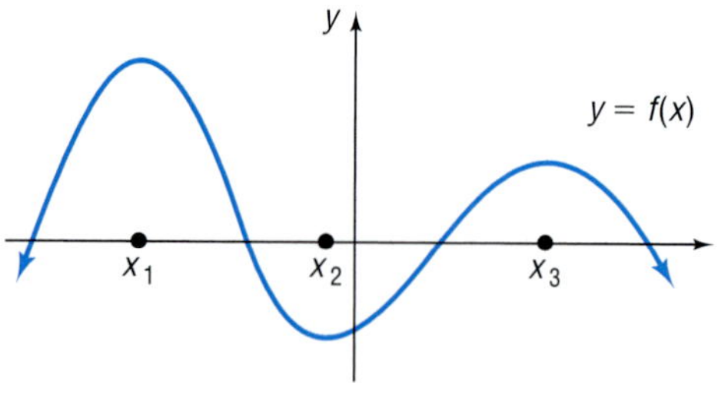

A function f has a **local maximum at c** if there is an interval I containing c so that, for all x in I, $f(x) < f(c)$. We call $f(c)$ a **local maximum of f.**

A function f has a **local minimum at c** if there is an interval I containing c so that, for all x in I, $f(x) > f(c)$. We call $f(c)$ a **local minimum of f.**

For example, in Figure 18, f has a local maximum at 0 and a local minimum at 1.33. The local maximum is $f(0) = 3$; the local minimum is $f(1.33) = 1.815$.

3 To locate the exact value at which a function f has a local maximum or a local minimum usually requires calculus. However, a graphing utility may be used to approximate these values.

EXAMPLE 5 Using a Graphing Utility to Locate Local Maxima and Minima

Use a graphing utility to graph $f(x) = 6x^3 - 12x + 5$ for $-2 < x < 2$. Determine where f has a local maximum and where f has a local minimum.

Solution Figure 20 shows the graph of f on the interval $[-2, 2]$.

Using TRACE, we find the local maximum to occur at approximately -0.80 and the local minimum to occur at 0.80, correct to two decimal places. The local maximum is 11.53 and the local minimum is -1.53, correct to two decimal places. See Figures 20(a) and (b).

FIGURE 20

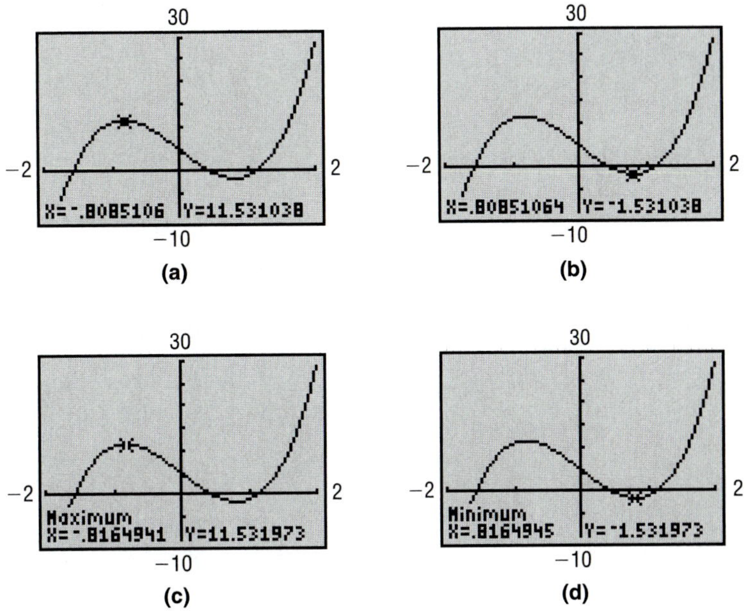

Most graphing utilities have a function that finds the maximum or minimum point of a graph within a given domain. Using this function, we find the maximum occurs at $(-0.81, 11.53)$ and the minimum occurs at $(0.81, -1.53)$, correct to two decimal places, for $-2 < x < 2$, as seen in Figures 20(c) and (d).

Now work Problem 95.

Even and Odd Functions

A function f is **even** if for every number x in its domain the number $-x$ is also in the domain and

$$f(-x) = f(x)$$

In other words, if $y = f(x)$, then f is even if and only if whenever the point (x, y) is on the graph of f, then the point $(-x, y)$ is also on the graph.

A function f is **odd** if for every number x in its domain the number $-x$ is also in the domain and

$$f(-x) = -f(x)$$

In other words, if $y = f(x)$, then f is odd if and only if whenever the point (x, y) is on the graph of f, then the point $(-x, -y)$ is also on the graph.

Refer to Section 1.3, where the tests for symmetry are listed. The following results are then evident.

Theorem

A function is even if and only if its graph is symmetric with respect to the y-axis. A function is odd if and only if its graph is symmetric with respect to the origin.

EXAMPLE 6 Determining Even and Odd Functions from the Graph

4 Determine whether each graph given in Figure 21 is the graph of an even function, an odd function, or a function that is neither even nor odd.

FIGURE 21

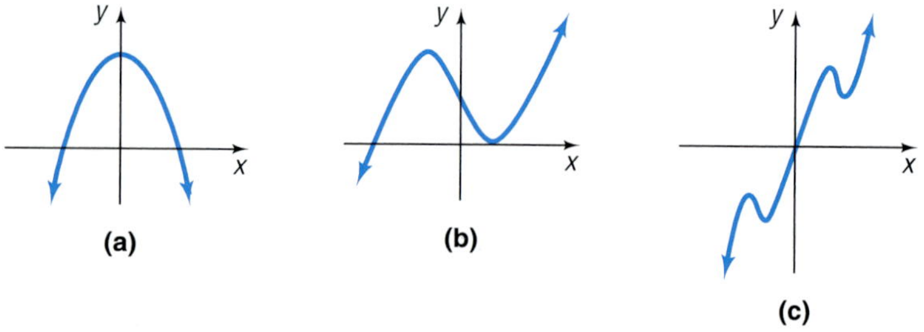

Solution The graph in Figure 21(a) is that of an even function, because the graph is symmetric with respect to the y-axis. The function whose graph is given in

Figure 21(b) is neither even nor odd, because the graph is neither symmetric with respect to the y-axis nor symmetric with respect to the origin. The function whose graph is given in Figure 21(c) is odd, because its graph is symmetric with respect to the origin.

Now work Problem 9.

A graphing utility can be used to conjecture whether a function is even, odd, or neither. As stated, a function is even if $f(-x) = f(x)$. This condition implies that when an even function contains the point (x, y) it must also contain the point $(-x, y)$. Therefore, if TRACE indicates that both the point (x, y) and the point $(-x, y)$ are on the graph for every x, then we would conjecture that the function is even.*

In addition, a function is odd if $f(-x) = -f(x)$. This condition implies that an odd function contains the points $(-x, -y)$ and (x, y). TRACE could be used in the same way to conjecture that the function is odd.*

5 In the next example, we show how to verify whether a function is even, odd, or neither.

EXAMPLE 7 Identifying Even and Odd Functions

Determine whether each of the following functions is even, odd, or neither. Then determine whether the graph is symmetric with respect to the y-axis or with respect to the origin.

(a) $f(x) = x^2 - 5$ (b) $g(x) = x^3 - 1$

(c) $h(x) = 5x^3 - x$ (d) $F(x) = |x|$

(a) Solution Graph the function. Use TRACE to determine different pairs of points (x, y) and $(-x, y)$. For example, the point $(1.957447, -1.168402)$ is on the graph. See Figure 22(a). Now move the cursor to -1.957447 and determine the corresponding y-coordinate. See Figure 22(b). Since $(1.957447, -1.168402)$ and $(-1.957447, -1.168402)$ both lie on the graph, we have evidence that the function is even. Repeating this procedure yields similar results. Therefore, we conjecture that the function is even.

FIGURE 22

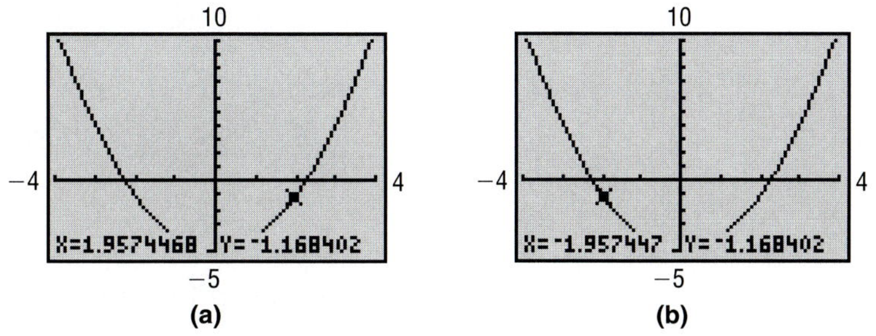

To algebraically verify the conjecture, we replace x by $-x$ in $f(x) = x^2 - 5$. Then

$$f(-x) = (-x)^2 - 5 = x^2 - 5 = f(x)$$

Since $f(-x) = f(x)$, we conclude that f is an even function, and the graph is symmetric with respect to the y-axis.

*$-X$min and Xmax must be equal for this to work.

(b) Solution Graph the function. Using TRACE, the point (1.787234, 4.7087929) is on the graph. See Figure 23(a). Now move the cursor to -1.787234 and determine the corresponding y-coordinate, -6.708793. See Figure 23(b). We conjecture that the function is neither even nor odd since (x, y) does not equal $(-x, y)$ (so it is not even) and (x, y) also does not equal $(-x, -y)$ (so it is not odd).

FIGURE 23

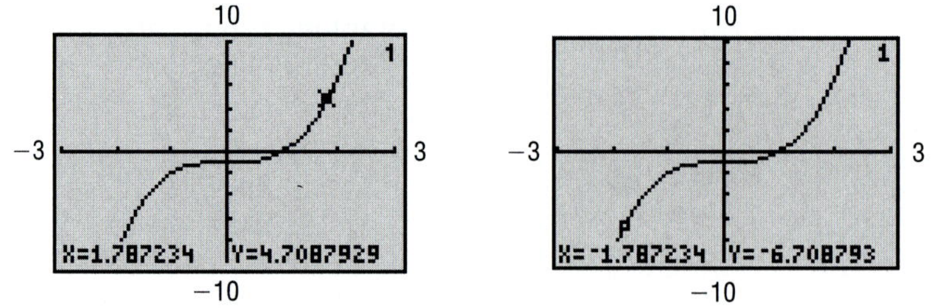

(a) (b)

To algebraically verify the conjecture, we replace x by $-x$. Then

$$g(-x) = (-x)^3 - 1 = -x^3 - 1$$

Since $g(-x) \neq g(x)$ and $g(-x) \neq -g(x)$, we conclude that g is neither even nor odd. The graph is not symmetric with respect to the y-axis nor with respect to the origin.

(c) Solution Graph the function. Using the TRACE, the point (1.0212766, 4.3047109) is on the graph. See Figure 24(a). We move the cursor to -1.021277 and determine the corresponding y-coordinate, -4.304711. See Figure 24(b). Since (x, y) equals $(-x, -y)$, correct to five decimal places, we have evidence that the function is odd. An examination of other pairs of points leads to the conjecture that the function is odd.

FIGURE 24

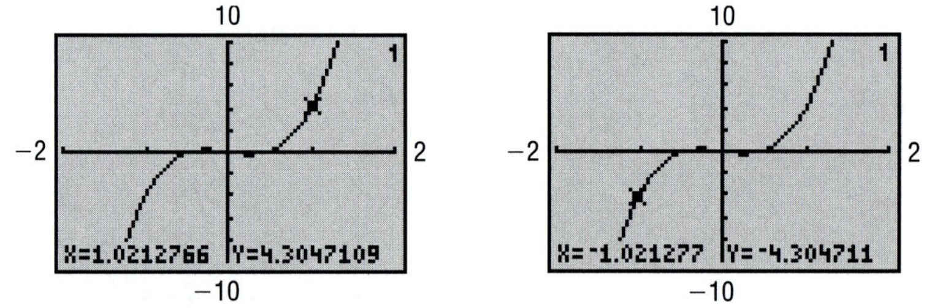

(a) (b)

To algebraically verify the conjecture, we replace x by $-x$ in $h(x) = 5x^3 - x$. Then

$$h(-x) = 5(-x)^3 - (-x) = -5x^3 + x = -(5x^3 - x) = -h(x)$$

Since $h(-x) = -h(x)$, h is an odd function, and the graph of h is symmetric with respect to the origin.

(d) Solution Graph the function. Using TRACE, the point (−5.744681, 5.7446809) is on the graph. See Figure 25(a). We move the cursor to 5.7446809 and determine the corresponding y-coordinate, 5.7446809. See Figure 25(b). Since (x, y) equals $(-x, y)$, we have evidence that the function is even. Repeating this procedure yields similar results. We conjecture that the function is even.

FIGURE 25

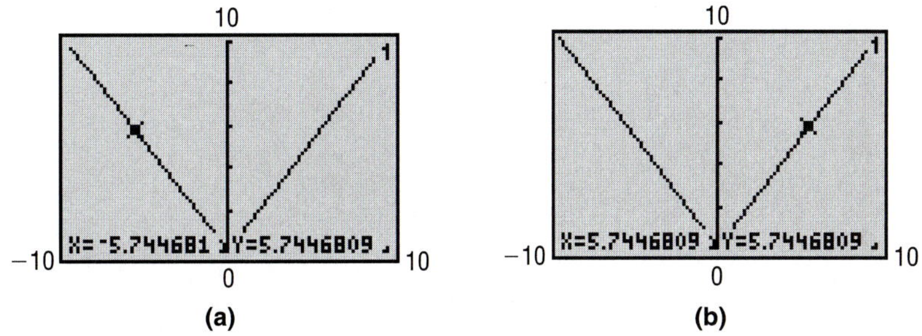

To algebraically verify the conjecture, we replace x by $-x$ in $F(x) = |x|$. Then

$$F(-x) = |-x| = |x| = F(x)$$

Since $F(-x) = F(x)$, F is an even function, and the graph of F is symmetric with respect to the y-axis.

Now work Problem 41.

Library of Functions

6

We now give names to some of the functions we have encountered. In going through this list, pay special attention to the characteristics of each function, particularly to the shape of each graph. Knowing these graphs will lay the foundation for later graphing techniques.

Linear Functions

$f(x) = mx + b$ $\quad$ m and b are real numbers

The domain of a **linear function** f consists of all real numbers. The graph of this function is a nonvertical straight line with slope m and y-intercept b. A linear function is increasing if $m > 0$, decreasing if $m < 0$, and constant if $m = 0$.

Constant Function

$f(x) = b$ $\quad$ b a real number

FIGURE 26

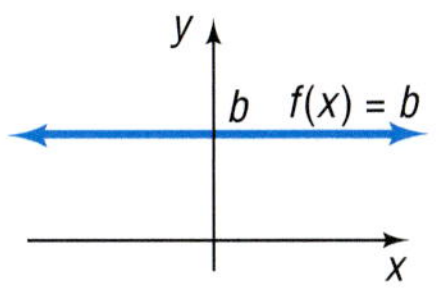

See Figure 26.

A **constant function** is a special linear function ($m = 0$). Its domain is the set of all real numbers; its range is the set consisting of a single number b. Its graph is a horizontal line whose y-intercept is b. The constant function is an even function whose graph is constant over its domain.

Identity Function

$$f(x) = x$$

See Figure 27.

FIGURE 27

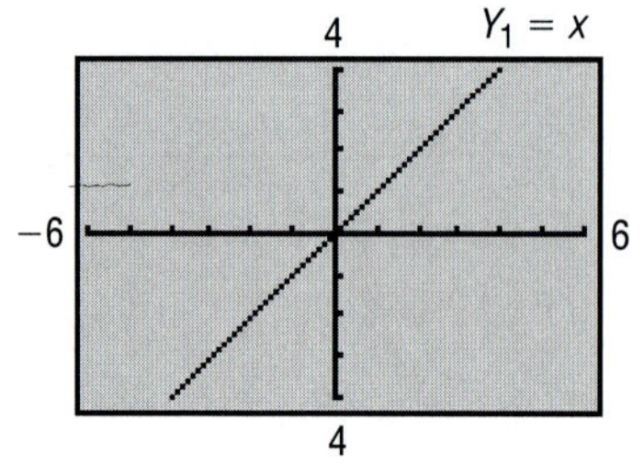

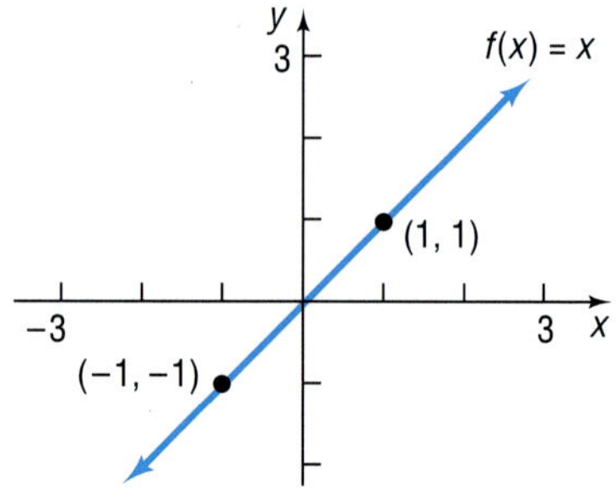

The **identity function** is also a special linear function. Its domain and its range are the set of all real numbers. Its graph is a line whose slope is $m = 1$ and whose y-intercept is 0. The line consists of all points for which the x-coordinate equals the y-coordinate. The identity function is an odd function that is increasing over its domain. Note that the graph bisects quadrants I and III.

Square Function

$$f(x) = x^2$$

See Figure 28.

FIGURE 28

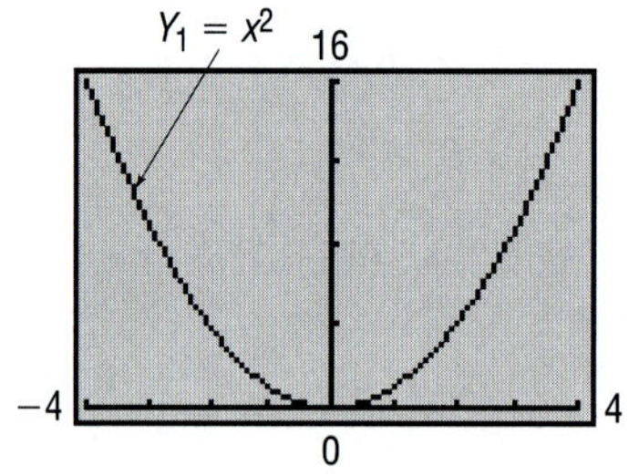

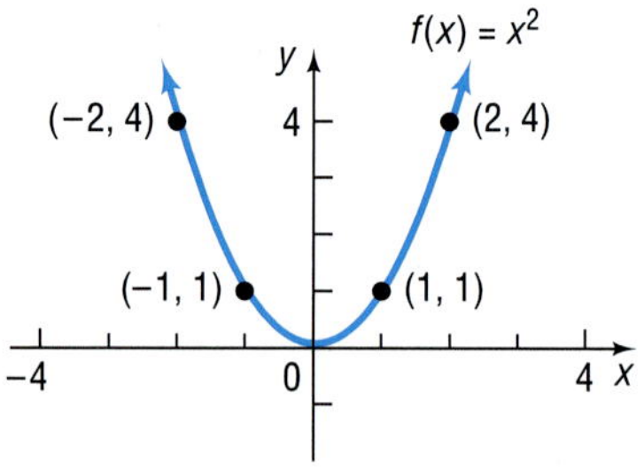

The domain of the **square function** f is the set of all real numbers; its range is the set of nonnegative real numbers. The graph of this function is a parabola, whose intercept is at (0, 0). The square function is an even function that is descreasing on the interval $(-\infty, 0)$ and increasing on the interval $(0, \infty)$.

Cube Function

$$f(x) = x^3$$

See Figure 29.

FIGURE 29

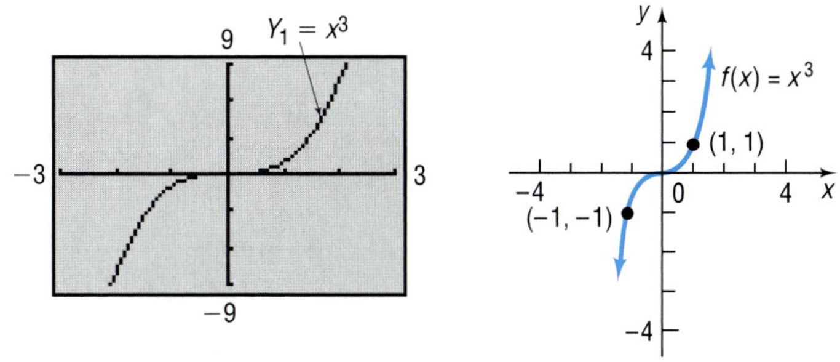

The domain and range of the **cube function** are the set of all real numbers. The intercept of the graph is at (0, 0). The cube function is odd and is increasing on the interval $(-\infty, \infty)$.

Square Root Function

$$f(x) = \sqrt{x}$$

See Figure 30.

FIGURE 30

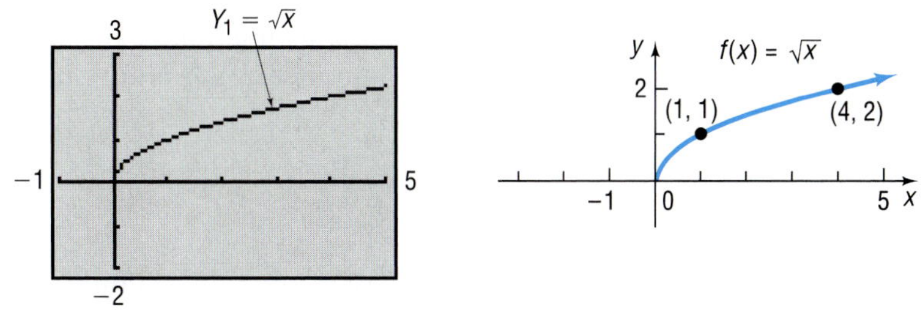

The domain and range of the **square root function** are the set of nonnegative real numbers. The intercept of the graph is at (0, 0). The square root function is neither even nor odd and is increasing on the interval $(0, \infty)$.

Reciprocal Function

$$f(x) = \frac{1}{x}$$

Refer to Example 14, p. 43, for a discussion of the equation $y = 1/x$. See Figure 31.

FIGURE 31

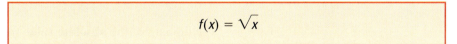

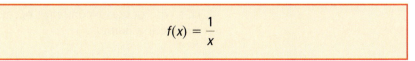

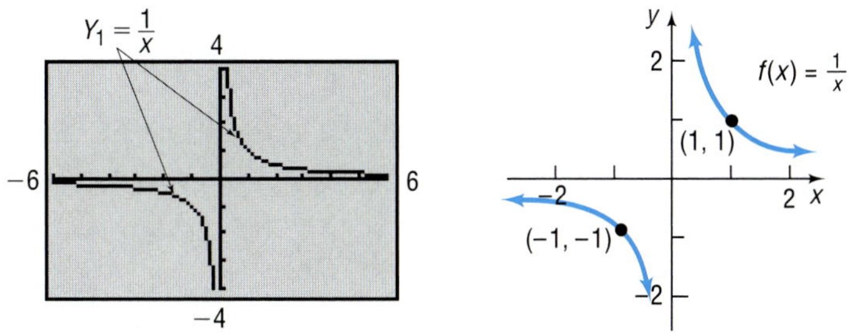

The domain and range of the **reciprocal function** are the set of all nonzero real numbers. The graph has no intercepts. The reciprocal function is decreasing on the intervals $(-\infty, 0)$ and $(0, \infty)$ and is an odd function.

Absolute Value Function

$$f(x) = |x|$$

See Figure 32.

FIGURE 32

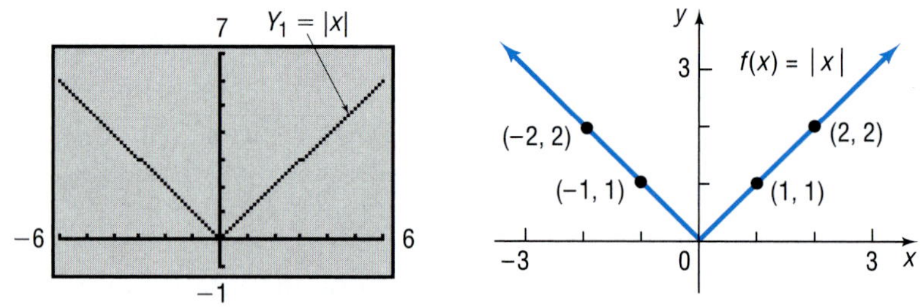

The domain of the **absolute value function** is the set of all real numbers; its range is the set of nonnegative real numbers. The intercept of the graph is at (0, 0). If $x \geq 0$, then $f(x) = x$ and the graph of f is part of the line $y = x$; if $x < 0$, then $f(x) = -x$ and the graph of f is part of the line $y = -x$. The absolute value function is an even function; it is decreasing on the interval $(-\infty, 0)$ and increasing on the interval $(0, \infty)$.

Comment: If your utility has no built-in absolute value function, you can still graph $f(x) = |x|$ by using the fact that $|x| = \sqrt{(x^2)}$.

The notation int(x) stands for the largest integer less than or equal to x. For example,

$$\text{int}(1) = 1 \qquad \text{int}(2.5) = 2 \qquad \text{int}\left(\tfrac{1}{2}\right) = 0 \qquad \text{int}\left(\tfrac{-3}{4}\right) = -1 \qquad \text{int}(\pi) = 3$$

This type of correspondence occurs frequently enough in mathematics that we give it a name.

Greatest-Integer Function

$$f(x) = \text{int}(x) = \text{Greatest integer less than or equal to } x$$

We obtain the graph of $f(x) = \text{int}(x)$ by plotting several points. See Table 4. For values of x, $-1 \leq x < 0$, the value of $f(x) = \text{int}(x)$ is -1; for values of x, $0 \leq x < 1$, the value of f is 0. See Figure 33 for the graph.

TABLE 4

X	Y1	
-1	-1	
-.75	-1	
-.5	-1	
-.25	-1	
0	0	
.25	0	
.5	0	

Y1=int(X)

FIGURE 33

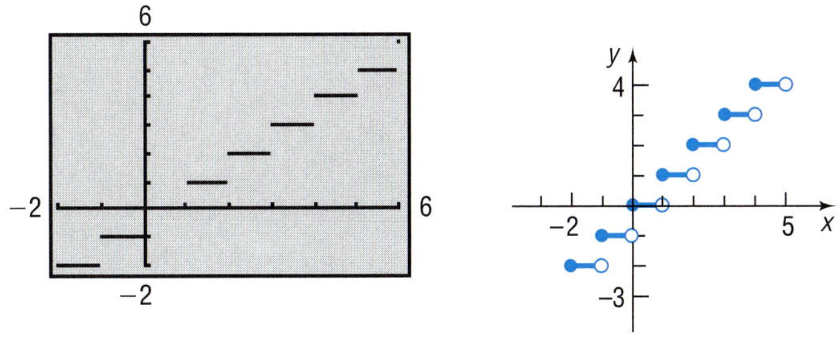

(a) **(b)**

The domain of the **greatest integer function** is the set of all real numbers; its range is the set of integers. The y-intercept of the graph is at 0. The x-intercepts lie in the interval $[0, 1)$. The greatest-integer function is neither even nor odd. It is constant on every interval of the form $[k, k + 1)$, for k an integer. In Figure 33(b), we use a solid dot to indicate, for example, that at $x = 1$ the value of f is $f(1) = 1$; we use an open circle to illustrate that the function does not assume the value of 0 at $x = 1$.

From the graph of the greatest-integer function, we can see why it is also called a **step function.** At $x = 0, x = \pm 1, x = \pm 2$, and so on, this function exhibits what is called a *discontinuity;* that is, at integer values, the graph suddenly "steps" from one value to another without taking on any of the intermediate values. For example, to the immediate left of $x = 3$, the y-coordinates are 2, and to the immediate right of $x = 3$, the y-coordinates are 3.

Comment: When graphing a function, you can choose either the **connected mode,** in which points plotted on the screen are connected, making the graph appear without any breaks, or the **dot mode,** in which only the points plotted appear. When graphing the greatest integer function with a graphing utility, it is necessary to be in the **dot mode.** This is to prevent the utility from "connecting the dots" when $f(x)$ changes from one integer value to the next. See Figure 34.

FIGURE 34

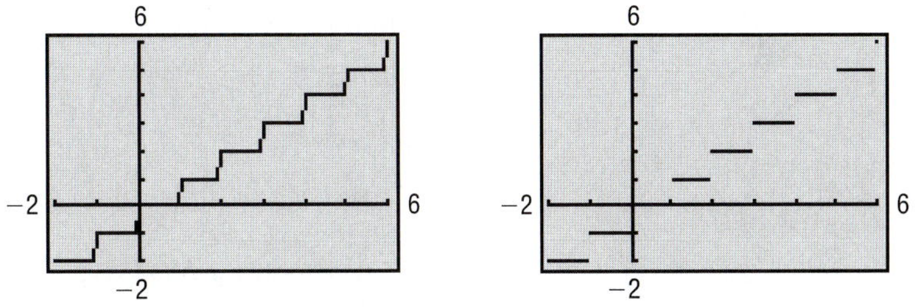

(a) Connected mode **(b)** Dot mode

The functions we have discussed so far are basic. Whenever you encounter one of them, you should see a mental picture of its graph. For example, if you encounter the function $f(x) = x^2$, you should see in your mind's eye a picture like Figure 28.

Now work Problems 1–8.

Piecewise Defined Functions

7

Sometimes, a function is defined by a rule consisting of two or more equations. The choice of which equation to use depends on the value of the independent variable x. For example, the absolute value function $f(x) = |x|$ is actually defined by two equations: $f(x) = x$ if $x \geq 0$ and $f(x) = -x$ if $x < 0$. For convenience, we generally combine these equations into one expression as

$$f(x) = |x| = \begin{cases} x & \text{if } x \geq 0 \\ -x & \text{if } x < 0 \end{cases}$$

When functions are defined by more than one equation, they are called **piecewise defined** functions.

Let's look at another example of a piecewise defined function.

EXAMPLE 8 Analyzing a Piecewise Defined Function

For the following function f,

$$f(x) = \begin{cases} -x + 1 & \text{if } -1 \leq x < 1 \\ 2 & \text{if } x = 1 \\ x^2 & \text{if } x > 1 \end{cases}$$

(a) Find $f(0)$, $f(1)$, and $f(2)$.
(b) Determine the domain of f.
(c) Graph f.
(d) Use the graph to find the range of f.

Solution (a) To find $f(0)$, we observe that when $x = 0$ the equation for f is given by $f(x) = -x + 1$. So we have

$$f(0) = -0 + 1 = 1$$

When $x = 1$, the equation for f is $f(x) = 2$. Thus

$$f(1) = 2$$

When $x = 2$, the equation for f is $f(x) = x^2$. So

$$f(2) = 2^2 = 4$$

(b) To find the domain of f, we look at its definition. We conclude that the domain of f is $\{x | x \geq -1\}$, or $[-1, \infty)$.

(c) On a graphing utility, the procedure for graphing a piecewise defined function varies depending on the particular utility. In general, you need to enter each piece as a function with a restricted domain.

To graph the middle piece [the point $(1, 2)$], use the STAT PLOT function. See Figure 35(a).

In graphing piecewise defined functions, it is usually better to be in the dot mode, since such functions may have breaks (discontinuities).

To graph f on paper, we graph "each piece." Thus, we first graph the line $y = -x + 1$ and keep only the part for which $-1 \leq x < 1$. Then we plot the point $(1, 2)$, because when $x = 1$, $f(x) = 2$. Finally, we graph the parabola $y = x^2$ and keep only the part for which $x > 1$. See Figure 35(b).

FIGURE 35

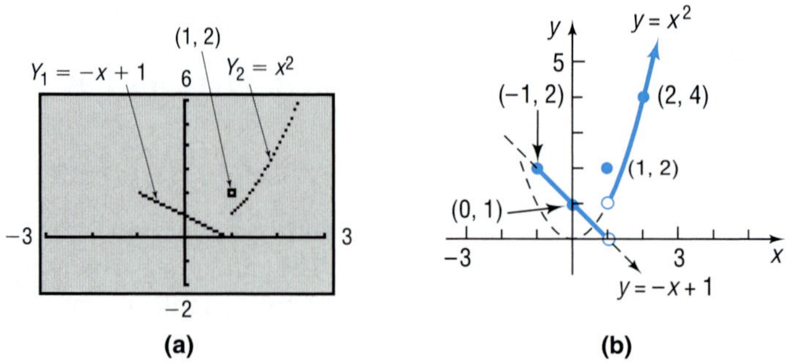

(d) From the graph, we conclude that the range of f is $\{y|y > 0\}$, or $(0, \infty)$.

Now work Problem 69.

EXAMPLE 9

Cost of Electricity

In the winter, Commonwealth Edison Company supplies electricity to residences for a monthly customer charge of \$8.91 plus 10.494¢ per kilowatt-hour (kWhr) for the first 400 kWhr supplied in the month, and 7.91¢ per kWhr for all usage over 400 kWhr in the month.*

(a) What is the charge for using 300 kWhr in a month?
(b) What is the charge for using 700 kWhr in a month?
(c) If C is the monthly charge for x kWhr, express C as a function of x.

Solution

(a) For 300 kWhr, the charge is \$8.91 plus 10.494¢ = \$0.10494 per kWhr. Thus,

$$\text{Charge} = \$8.91 + \$0.10494(300) = \$40.39$$

(b) For 700 kWhr, the charge is \$8.91 plus 10.494¢ for the first 400 kWhr plus 7.91¢ for the 300 kWhr in excess of 400. Thus,

$$\text{Charge} = \$8.91 + \$0.10494(400) + \$0.0791(300) = \$74.62$$

(c) If $0 \le x \le 400$, the monthly charge C (in dollars) can be found by multiplying x times \$0.10494 and adding the monthly customer charge of \$8.91. Thus, if $0 \le x \le 400$, then $C(x) = 0.10494x + 8.91$. For $x > 400$, the charge is $0.10494(400) + 8.91 + 0.0791(x - 400)$, since $x - 400$ equals the usage in excess of 400 kWhr, which costs \$0.0791 per kWhr. Thus, if $x > 400$, then

$$\begin{aligned} C(x) &= 0.10494(400) + 8.91 + 0.0791(x - 400) \\ &= 50.89 + 0.0791(x - 400) \\ &= 0.0791x + 19.25 \end{aligned}$$

The rule for computing C follows two equations:

$$C(x) = \begin{cases} 0.10494x + 8.91 & \text{if } 0 \le x \le 400 \\ 0.0791x + 19.25 & \text{if } x > 400 \end{cases}$$

See Figure 36 for the graph.

FIGURE 36

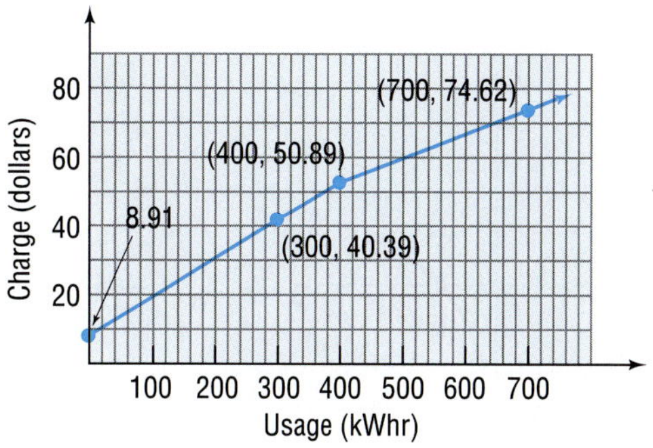

*Source: Commonwealth Edison Co., Chicago, Illinois, 1997.

2.2 EXERCISES

In Problems 1–8, match each graph to the function listed whose graph most resembles the one given.

A. Constant function *B. Linear function*

C. Square function *D. Cube function*

E. Square root function *F. Reciprocal function*

G. Absolute value function *H. Greatest-integer function*

1. 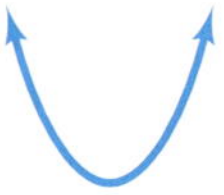

2.

3. 

4. 

5. 

6. 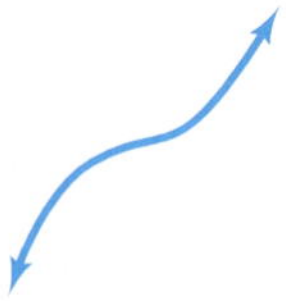

7. 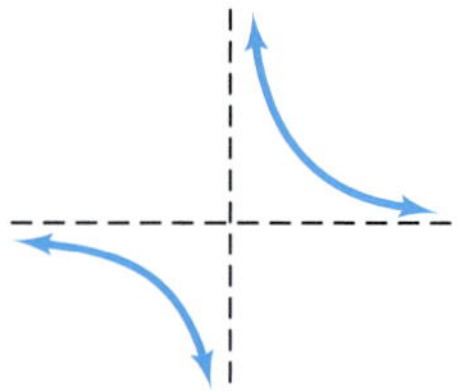

8. 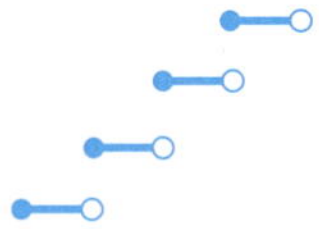

In Problems 9–24, the graph of a function is given. Use the graph to find

(a) Its domain and range

(b) The intervals on which it is increasing, decreasing, or constant

(c) Whether it is even, odd, or neither

(d) The intercepts, if any

9. 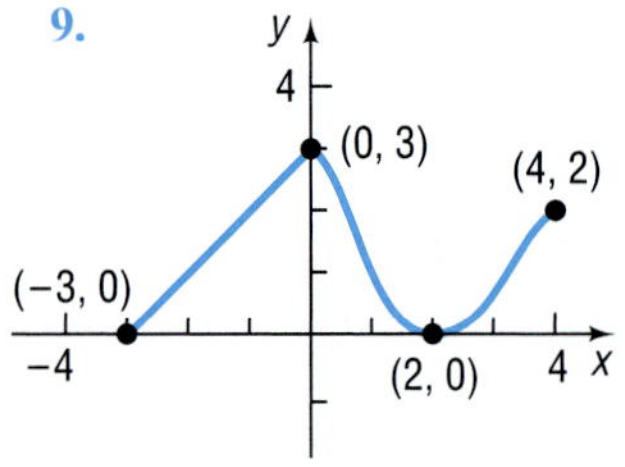

10. 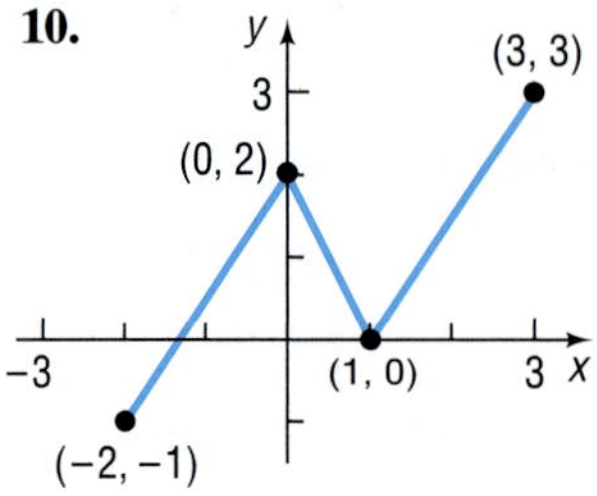

11. 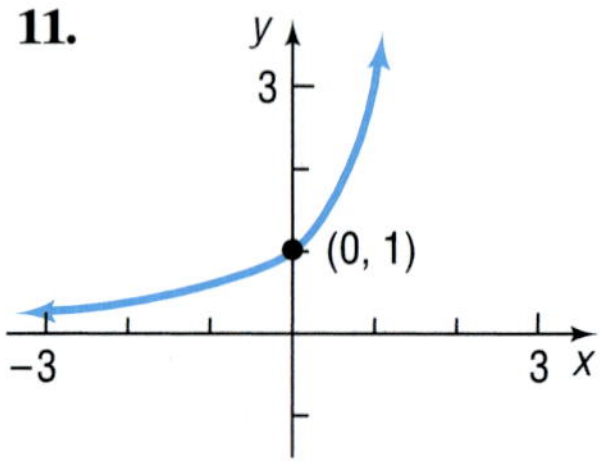

12. 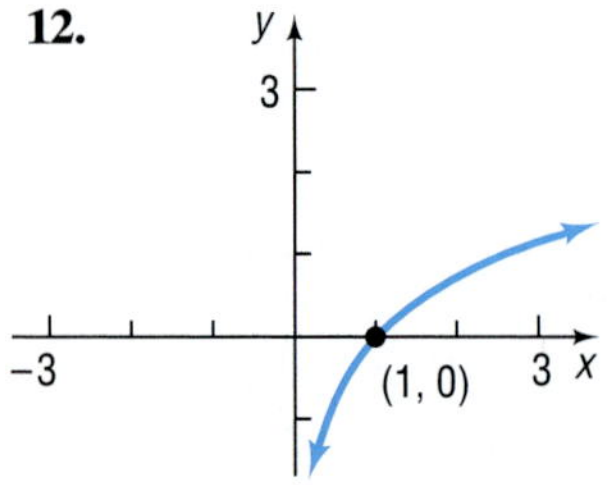

13. 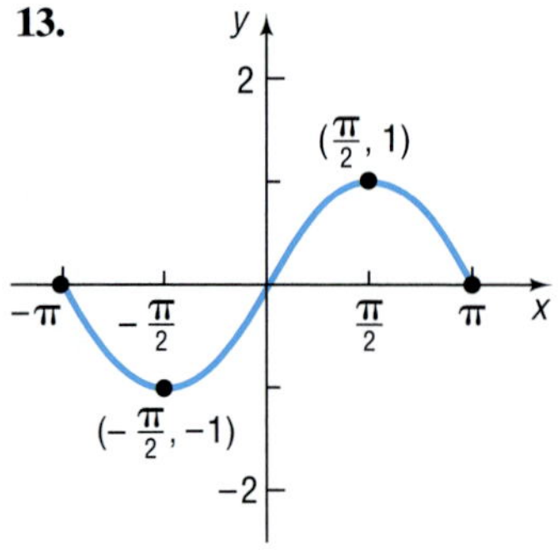

14. 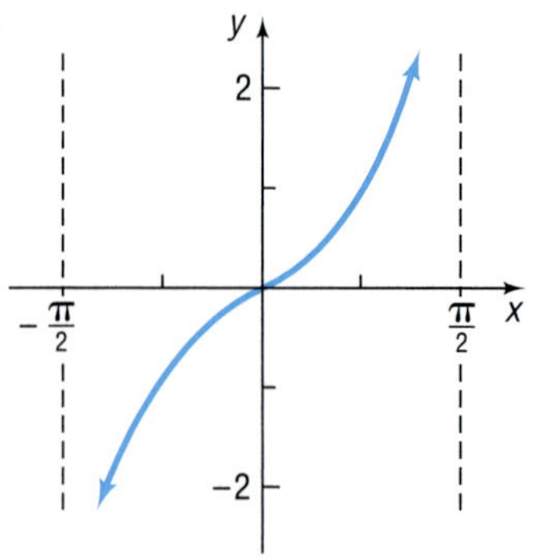

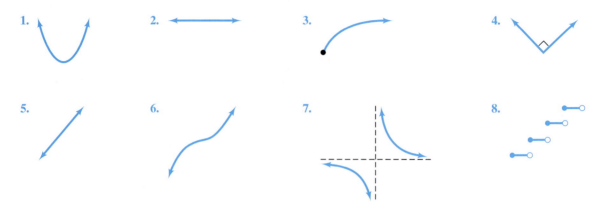

15.

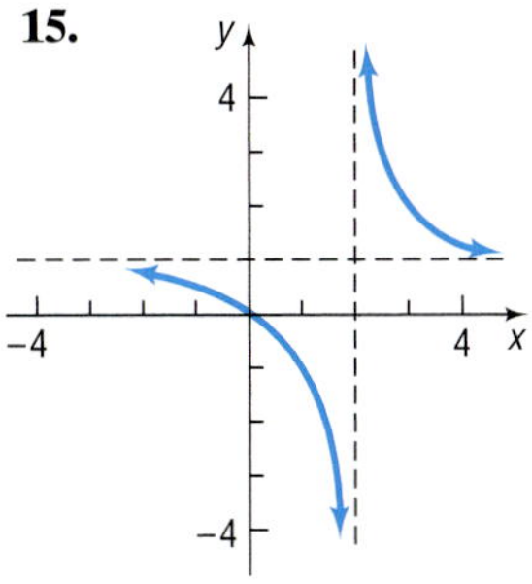

16.

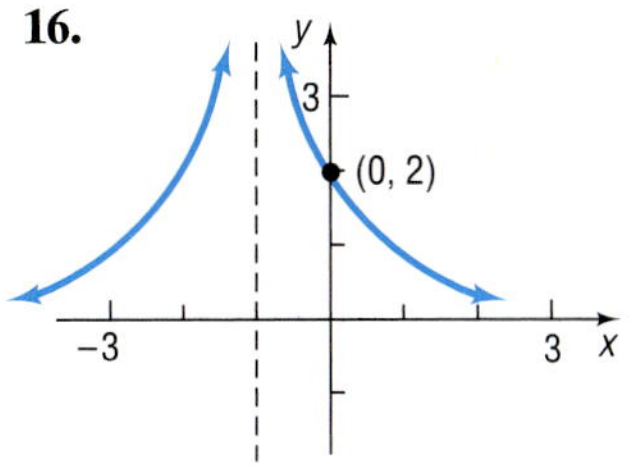

17.

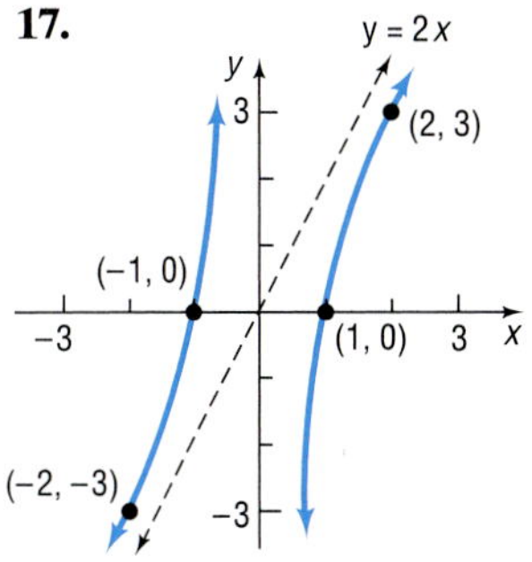

18.

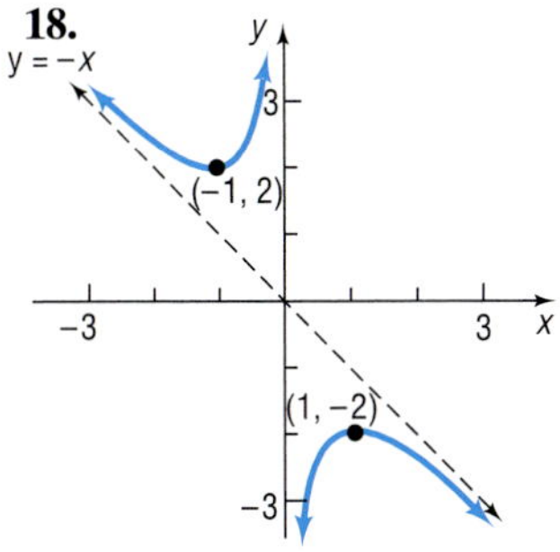

19.

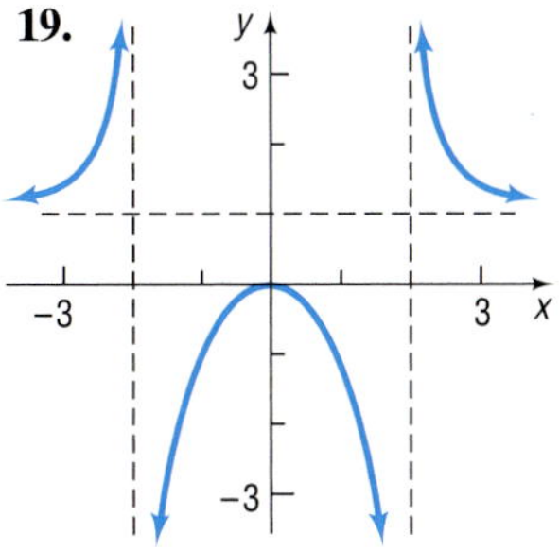

20.

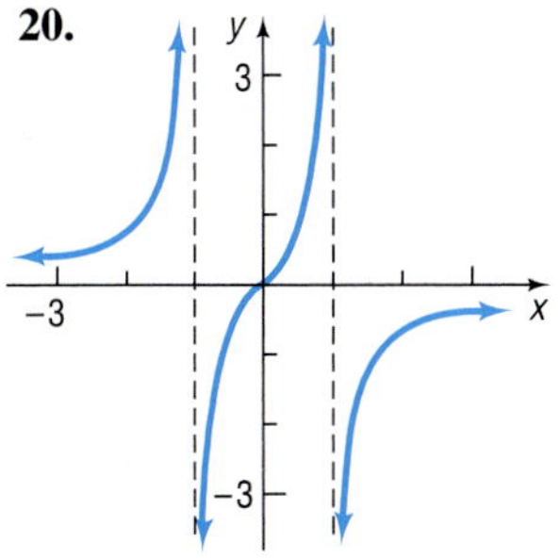

In Problems 21–24, assume that the entire graph is shown.

21.

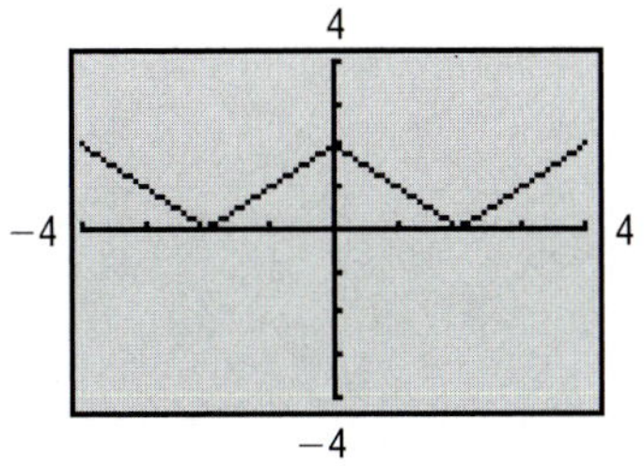

22.

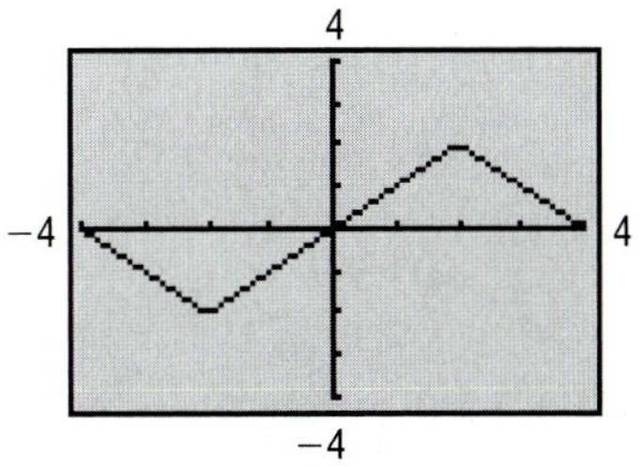

23.

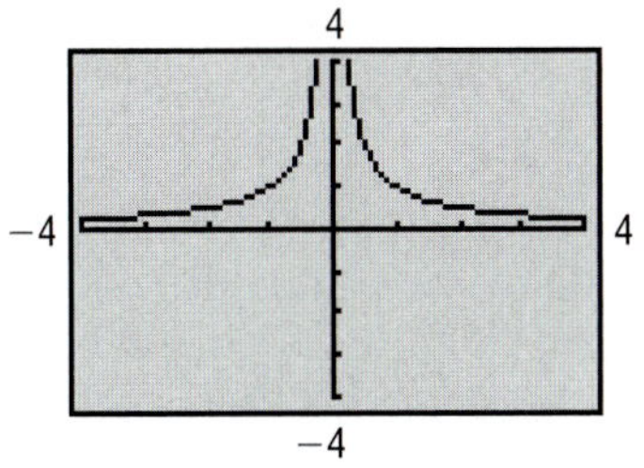

24.

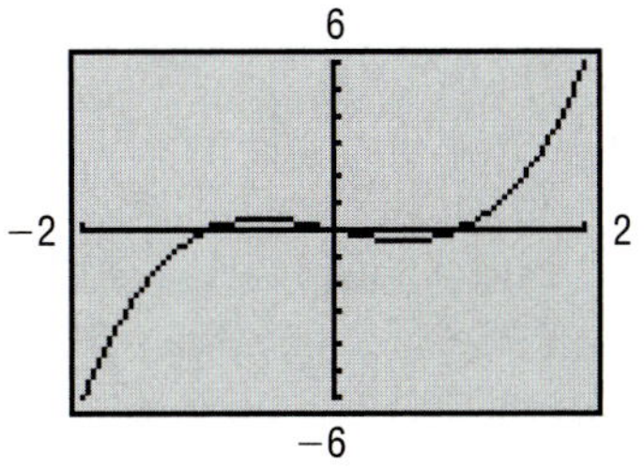

25. If

$$f(x) = \begin{cases} x^2 & \text{if } x < 0 \\ 2 & \text{if } x = 0 \\ 2x + 1 & \text{if } x > 0 \end{cases}$$

find: (a) $f(-2)$ (b) $f(0)$ (c) $f(2)$

26. If

$$f(x) = \begin{cases} x^3 & \text{if } x < 0 \\ 3x + 2 & \text{if } x \geqslant 0 \end{cases}$$

find: (a) $f(-1)$ (b) $f(0)$ (c) $f(1)$

27. If $f(x) = \text{int}(2x)$, find: (a) $f(1.2)$ (b) $f(1.6)$ (c) $f(-1.8)$

28. If $f(x) = \text{int}(x/2)$, find: (a) $f(1.2)$ (b) $f(1.6)$ (c) $f(-1.8)$

In Problems 29–40, find the average rate of change of f between 1 and x,

$$\frac{f(x) - f(1)}{x - 1} \qquad x \neq 1$$

for each function f. Be sure to simplify.

29. $f(x) = 3x$ **30.** $f(x) = -2x$ **31.** $f(x) = 1 - 3x$ **32.** $f(x) = x^2 + 1$

33. $f(x) = 3x^2 - 2x$ **34.** $f(x) = 4x - 2x^2$ **35.** $f(x) = x^3 - x$ **36.** $f(x) = x^3 + x$

37. $f(x) = \dfrac{2}{x + 1}$ **38.** $f(x) = \dfrac{1}{x^2}$ **39.** $f(x) = \sqrt{x}$ **40.** $f(x) = \sqrt{x + 3}$

In Problems 41–52, tell whether each function is even, odd, or neither without drawing a graph. Graph each function and use TRACE to verify your results.

41. $f(x) = 4x^3$ **42.** $f(x) = 2x^4 - x^2$ **43.** $g(x) = 2x^2 - 5$ **44.** $h(x) = 3x^3 + 2$

45. $F(x) = \sqrt[3]{x}$ **46.** $G(x) = \sqrt{x}$ **47.** $f(x) = x + |x|$ **48.** $f(x) = \sqrt[3]{2x^2 + 1}$

49. $g(x) = \dfrac{1}{x^2}$ **50.** $h(x) = \dfrac{x}{x^2 - 1}$ **51.** $h(x) = \dfrac{x^3}{3x^2 - 9}$ **52.** $F(x) = \dfrac{x}{|x|}$

53. How many x-intercepts can a function defined on an interval have if it is increasing on that interval? Explain.

54. How many y-intercepts can a function have? Explain.

In Problems 55–80:

(a) Find the domain of each function.

(b) Locate any intercepts.

(c) Graph each function by hand.

(d) Based on the graph, find the range.

(e) Verify your results using a graphing utility.

55. $f(x) = 3x - 3$ **56.** $f(x) = 4 - 2x$ **57.** $g(x) = x^2 - 4$ **58.** $g(x) = x^2 + 4$

59. $h(x) = -x^2$ **60.** $F(x) = 2x^2$ **61.** $f(x) = \sqrt{x - 2}$ **62.** $g(x) = \sqrt{x} + 2$

63. $h(x) = \sqrt{2 - x}$ **64.** $F(x) = -\sqrt{x}$ **65.** $f(x) = |x| + 3$ **66.** $g(x) = |x + 3|$

67. $h(x) = -|x|$ **68.** $F(x) = |3 - x|$

69. $f(x) = \begin{cases} 2x & \text{if } x \neq 0 \\ 0 & \text{if } x = 0 \end{cases}$ **70.** $f(x) = \begin{cases} 3x & \text{if } x \neq 0 \\ 4 & \text{if } x = 0 \end{cases}$

71. $f(x) = \begin{cases} 1 + x & \text{if } x < 0 \\ x^2 & \text{if } x \geq 0 \end{cases}$ **72.** $f(x) = \begin{cases} 1/x & \text{if } x < 0 \\ \sqrt{x} & \text{if } x \geq 0 \end{cases}$

73. $f(x) = \begin{cases} |x| & \text{if } -2 \leq x < 0 \\ 1 & \text{if } x = 0 \\ x^3 & \text{if } x > 0 \end{cases}$ **74.** $f(x) = \begin{cases} 3 + x & \text{if } -3 \leq x < 0 \\ 3 & \text{if } x = 0 \\ \sqrt{x} & \text{if } x > 0 \end{cases}$

75. $g(x) = \begin{cases} 1 & \text{if } x \text{ is an integer} \\ -1 & \text{if } x \text{ is not an integer} \end{cases}$ **76.** $g(x) = \begin{cases} x & \text{if } x \geq 1 \\ 1 & \text{if } x < 1 \end{cases}$

77. $h(x) = 2\text{int}(x)$ **78.** $f(x) = \text{int}(2x)$

79. $F(x) = \begin{cases} 4 - x^2 & \text{if } |x| \leq 2 \\ x^2 - 4 & \text{if } |x| > 2 \end{cases}$ **80.** $G(x) = |x^2 - 4|$

Problems 81–84 require the following definition. Secant Line: The slope of the secant line containing the two points (x, f(x)) and (x + h, f(x + h)) on the graph of a function y = f(x) may be given as

$$\frac{f(x + h) - f(x)}{h}$$

In Problems 81–84, express the slope of the secant line of each function in terms of x and h. Be sure to simplify your answer.

81. $f(x) = 2x + 5$ **82.** $f(x) = -3x + 2$ **83.** $f(x) = x^2 + 2x$ **84.** $f(x) = 1/x$

In Problems 85–88, the graph of a piecewise-defined function is given. Write a definition for each function.

85.

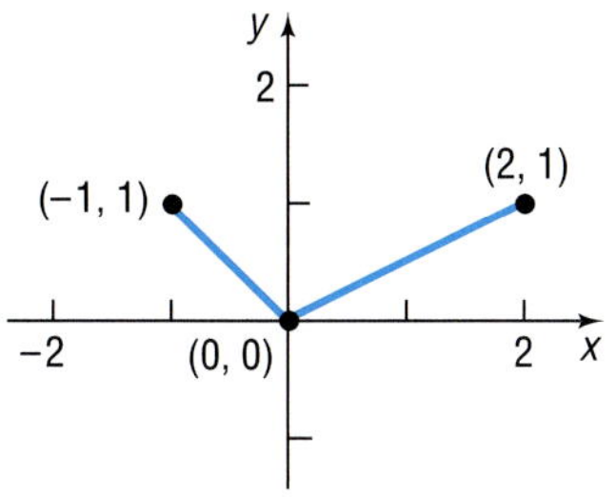

86.

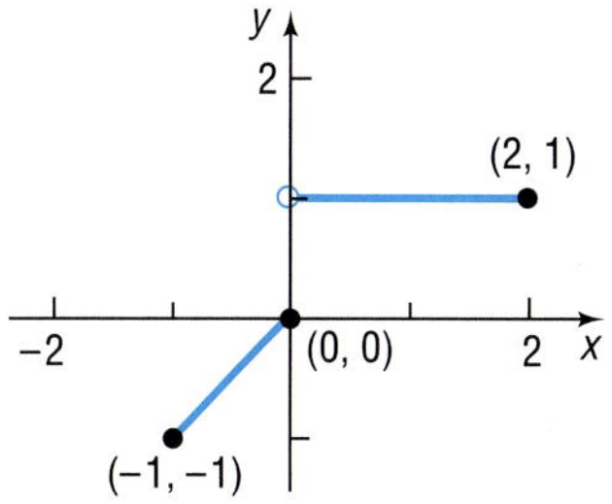

87.

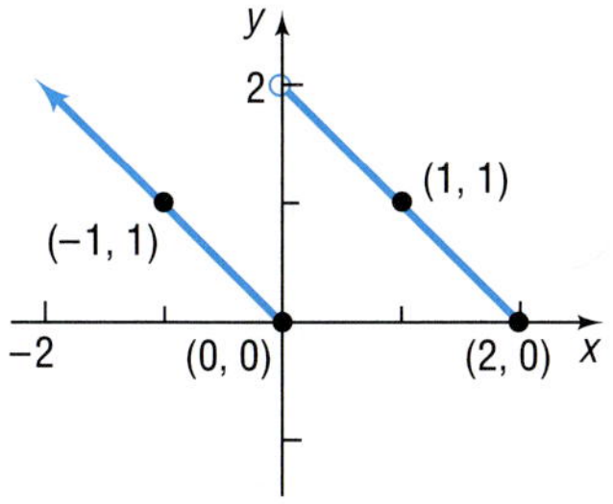

88.

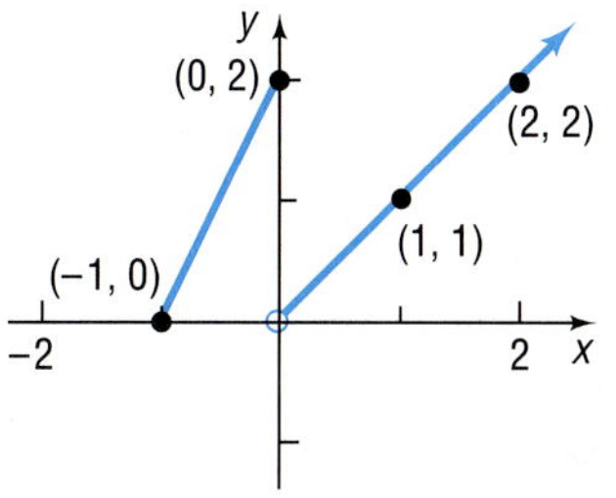

89. Growth of Bacteria The following data represents the population of an unknown bacteria.

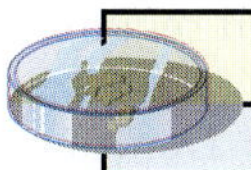

Time (Days)	Population
0	50
1	153
2	234
3	357
4	547
5	839
6	1280

(a) Draw a scatter diagram of the data by hand treating the population as the independent variable.
(b) Draw a line from the point (0, 50) to (1, 153) on the scatter diagram found in (a).
(c) Find the average rate of change of the population between 0 and 1 days.
(d) Interpret the average rate of change found in (c).
(e) Draw a line from the point (5, 839) to (6, 1280) on the scatter diagram found in (a).
(f) Find the average rate of change of the population between 5 and 6 days.
(g) Interpret the average rate of change found in (f).
(h) What is happening to average rate of change of population as time passes?

90. Cost of Manufacturing Bikes The following data represents the monthly cost of producing bicycles at Tunney's Bicycle Shop.

Number of Bicycles, x	Total Cost of Production, C (Dollars)
0	24,000
25	27,750
60	31,500
102	35,250
150	39,000
190	42,750
223	46,500
249	50,250

(a) Draw a scatter diagram of the data by hand treating the number of bicycles produced as the independent variable.
(b) Draw a line from the point (0, 24000) to (25, 27750) on the scatter diagram found in (a).
(c) Find the average rate of change of the cost between 0 and 25 bicycles.
(d) Interpret the average rate of change found in (c). In economics, this is called **marginal cost.**
(e) Draw a line from the point (190, 42750) to (223, 46500) on the scatter diagram found in (a).
(f) Find the average rate of change of the cost between 190 and 223 bicycles.
(g) Interpret the average rate of change found in (f).

91. Revenue from Selling Bikes The following data represents the total revenue that would be received from selling x bicycles at Tunney's Bicycle Shop.

Number of Bicycles, x	Total Revenue, R (Dollars)
0	0
25	28,000
60	45,000
102	53,400
150	59,160
190	62,360
223	64,835
249	66,525

(a) Draw a scatter diagram of the data by hand treating the number of bicycles produced as the independent variable.
(b) Draw a line from the point (0, 0) to (25, 28000) on the scatter diagram found in (a).
(c) Find the average rate of change of revenue between 0 and 25 bicycles.
(d) Interpret the average rate of change found in (c). In economics, this is called **marginal revenue.**
(e) Draw a line from the point (190, 62360) to (223, 64835) on the scatter diagram found in (a).
(f) Find the average rate of change of revenue between 190 and 223 bicycles.
(g) Interpret the average rate of change found in (f).
(h) What is happening to marginal revenue as the number of bicycles sold increases?
(i) Refer to Problem 90(d). A firm's profit will increase as long as marginal revenue is greater than marginal cost. At what level of output does Tunney's Bicycle Shop's profits stop increasing?

92. Cost of Tuition The following data represents the average cost of tuition and required fees (in dollars) at public four-year colleges.

Year	Average Cost of Tuition
1990	2,035
1991	2,159
1992	2,410
1993	2,604
1994	2,822

Source: U.S. National Center for Education Statistics.

(a) Draw a scatter diagram of the data by hand treating the year as the independent variable.
(b) Draw a line from the point (1990, 2035) to (1992, 2410) on the scatter diagram found in (a).
(c) Find the average rate of change of the cost of tuition and required fees between 1990 and 1992.
(d) Interpret the average rate of change found in (c).
(e) Draw a line from the point (1992, 2410) to (1994, 2822) on the scatter diagram found in (a).
(f) Find the average rate of change of the cost of tuition and required fees between 1992 and 1994.
(g) Interpret the average rate of change found in (f).
(h) What is happening to the average rate of change for the cost of tuition and required fees as time passes?

In Problems 93 and 94, decide whether each function is even. Given a reason.

93. $f(x) = \begin{cases} x^2 + 4 & \text{if } x \neq 2 \\ 6 & \text{if } x = 2 \end{cases}$

94. $f(x) = \begin{cases} x^2 + 4 & \text{if } x \neq 2 \\ 5 & \text{if } x = 2 \end{cases}$

In Problems 95–102, use a graphing utility to graph each function over the indicated interval. Determine where the function is increasing and where it is decreasing. Approximate any local maxima and local minima.

95. $f(x) = x^3 - 3x + 2 \quad (-2, 2)$

96. $f(x) = x^3 - 3x^2 + 5 \quad (-1, 3)$

97. $f(x) = x^5 - x^3 \quad (-2, 2)$

98. $f(x) = x^4 - x^2 \quad (-2, 2)$

99. $f(x) = -0.2x^3 - 0.6x^2 + 4x - 6 \quad (-6, 4)$

100. $f(x) = -0.4x^3 + 0.6x^2 + 3x - 2 \quad (-4, 5)$

101. $f(x) = 0.25x^4 + 0.3x^3 - 0.9x^2 + 3 \quad (-3, 2)$

102. $f(x) = -0.4x^4 - 0.5x^3 + 0.8x^2 - 2 \quad (-3, 2)$

103. Maximizing the Volume of a Box An open box with a square base is to be made from a square piece of cardboard 24 inches on a side by cutting out a square from each corner and turning up the sides. (See the illustration.) The volume V of the box as a function of the length x of the side of the square cut from each corner is

$$V(x) = x(24 - 2x)^2$$

Graph V and determine where V is largest.

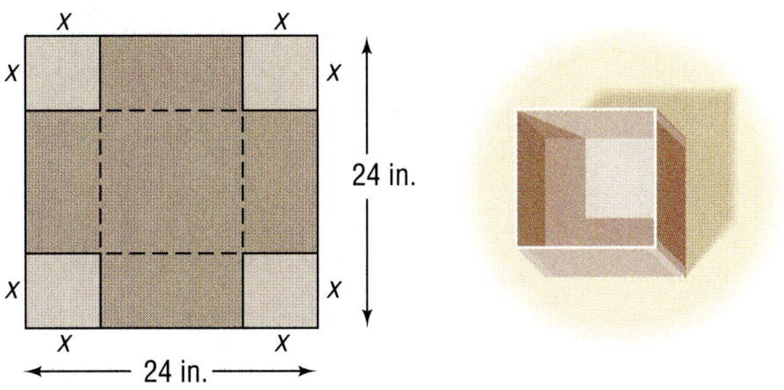

104. Minimizing the Material Needed to Make a Box An open box with a square base is required to have a volume of 10 cubic feet. The amount A of material used to make such a box as a function of the length x of a side of the square base is

$$A(x) = x^2 + 40/x$$

Graph A and determine where A is smallest.

105. Maximum Height of a Ball The height s of a ball (in feet) thrown with an initial velocity of 80 feet per second from an initial height of 6 feet is given as a function of the time t (in seconds) by

$$s(t) = -16t^2 + 80t + 6$$

(a) Graph s.
(b) Determine the time at which height is maximum.
(c) What is the maximum height?

106. Minimum Average Cost The average cost of producing x riding lawn mowers per hour is given by

$$A(x) = 0.3x^2 + 21x - 251 + \frac{2500}{x}$$

(a) Graph A.
(b) Determine the number of riding lawn mowers to produce in order to minimize average cost.
(c) What is the minimum average cost?

107. Graph $y = x^2$. Then on the same screen graph $y = x^2 + 2$, followed by $y = x^2 + 4$, followed by $y = x^2 - 2$. What pattern do you observe? Can you predict the graph of $y = x^2 - 4$? Of $y = x^2 + 5$?

108. Graph $y = x^2$. Then on the same screen graph $y = (x - 2)^2$, followed by $y = (x - 4)^2$, followed by $y = (x + 2)^2$. What pattern do you observe? Can you predict the graph of $y = (x + 4)^2$? Of $y = (x - 5)^2$?

109. Graph $y = |x|$. Then on the same screen graph $y = 2|x|$, followed by $y = 4|x|$, followed by $y = \frac{1}{2}|x|$. What pattern do you observe? Can you predict the graph of $y = \frac{1}{4}|x|$? Of $y = 5|x|$?

110. Graph $y = x^2$. Then on the same screen graph $y = -x^2$. What pattern do you observe? Now try $y = |x|$ and $y = -|x|$. What do you conclude?

111. Graph $y = \sqrt{x}$. Then on the same screen graph $y = \sqrt{-x}$. What pattern do you observe? Now try $y = 2x + 1$ and $y = 2(-x) + 1$. What do you conclude?

112. Graph $y = x^3$. Then on the same screen graph $y = (x - 1)^3 + 2$. Could you have predicted the result?

113. Graph $y = x^2$, $y = x^4$, and $y = x^6$ on the same screen. What do you notice is the same about each graph? What do you notice that is different?

114. Graph $y = x^3$, $y = x^5$, and $y = x^7$ on the same screen. What do you notice is the same about each graph? What do you notice that is different?

115. Cost of Natural Gas On December 13, 1996, the Peoples Gas Company had the following rate schedule* for natural gas usage in single-family residences:

Monthly service charge	\$9.00
Per therm service charge,	
1st 50 therms	\$0.36375/therm
Over 50 therms	\$0.11445/therm
Gas charge	\$0.3256/therm

(a) What is the charge for using 50 therms in a month?
(b) What is the charge for using 500 therms in a month?
(c) Construct a function that relates the monthly charge C for x therms of gas.
(d) Graph this function.

*Source: The Peoples Gas Company, Chicago, Illinois.

116. Cost of Natural Gas On December 19, 1996, Northern Illinois Gas Company had the following rate schedule* for natural gas usage in single-family residences:

Monthly customer charge	\$6.00
Distribution charge,	
1st 20 therms	\$0.2012/therm
Next 30 therms	\$0.1117/therm
Over 50 therms	\$0.0374/therm
Gas supply charge	\$0.3113/therm

(a) What is the charge for using 40 therms in a month?
(b) What is the charge for using 202 therms in a month?
(c) Construct a function that gives the monthly charge C for x therms of gas.
(d) Graph this function.

117. Consider the equation

$$y = \begin{cases} 1 & \text{if } x \text{ is rational} \\ 0 & \text{if } x \text{ is irrational} \end{cases}$$

Is this a function? What is its domain? What is its range? What is its y-intercept, if any? What are its x-intercepts, if any? Is it even, odd, or neither? How would you describe its graph?

118. Define some functions that pass through $(0, 0)$ and $(1, 1)$ and are increasing for $x \geq 0$. Begin your list with $y = \sqrt{x}$, $y = x$, and $y = x^2$. Can you propose a general result about such functions?

119. Can you think of a function that is both even and odd?

2.3 GRAPHING TECHNIQUES

1 Graph Functions Using Horizontal and Vertical Shifts
2 Graph Functions Using Compressions and Stretches
3 Graph Functions Using Reflections about the x-Axis or y-Axis

At this stage, if you were asked to graph any of the functions defined by $y = x$, $y = x^2$, $y = x^3$, $y = \sqrt{x}$, $y = |x|$, or $y = 1/x$, your response should be, "Yes, I recognize these functions and know the general shapes of their graphs." (If this is not your answer, review the previous section and Figures 27 through 32.)

Sometimes, we are asked to graph a function that is "almost" like one we already know how to graph. In this section, we look at some of these functions and develop techniques for graphing them. Collectively, these techniques are referred to as **transformations.**

Vertical Shifts

EXPLORATION 1

On the same screen, graph each of the following functions:

$$Y_1 = x^2$$
$$Y_2 = x^2 + 1$$
$$Y_3 = x^2 + 2$$
$$Y_4 = x^2 - 1$$
$$Y_5 = x^2 - 2$$

What do you observe?

*Source: Northern Illinois Gas Company, Naperville, Illinois.

Result Figure 37 illustrates the graphs. You should have observed a general pattern. With $Y_1 = x^2$ on the screen, the graph of $Y_2 = x^2 + 1$ is identical to that of $Y_1 = x^2$, except that it is shifted vertically up 1 unit. Similarly, $Y_3 = x^2 + 2$ is identical to that of $Y_1 = x^2$, except that it is shifted vertically up 2 units. The graph of $Y_4 = x^2 - 1$ is identical to that of $Y_1 = x^2$, except that it is shifted vertically down 1 unit.

FIGURE 37

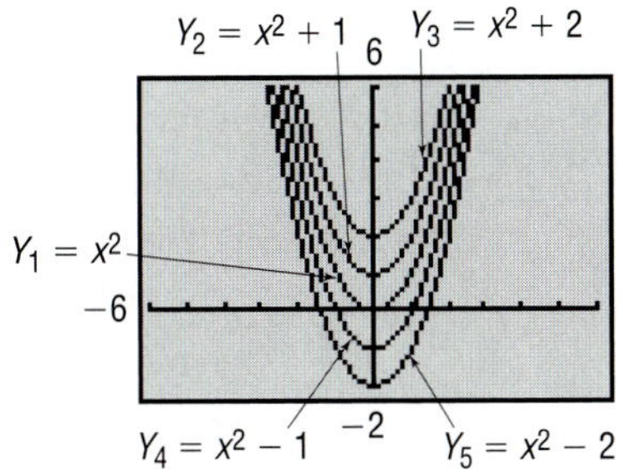

We are led to the following conclusion:

> If a real number c is added to the right side of a function $y = f(x)$, the graph of the new function $y = f(x) + c$ is the graph of f **shifted vertically** up (if $c > 0$) or down (if $c < 0$).

Let's look at an example.

EXAMPLE 1

Vertical Shift Down

Use the graph of $f(x) = x^2$ to obtain the graph of $h(x) = x^2 - 4$.

Solution Table 5 lists some points on the graphs of $f = Y_1$ and $h = Y_2$. The graph of h is identical to that of f, except that it is shifted down 4 units. See Figure 38.

TABLE 5

X	Y1	Y2
-3	9	5
-2	4	0
-1	1	-3
0	0	-4
1	1	-3
2	4	0
3	9	5

Y2=X²−4

FIGURE 38

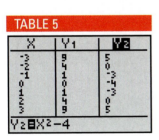

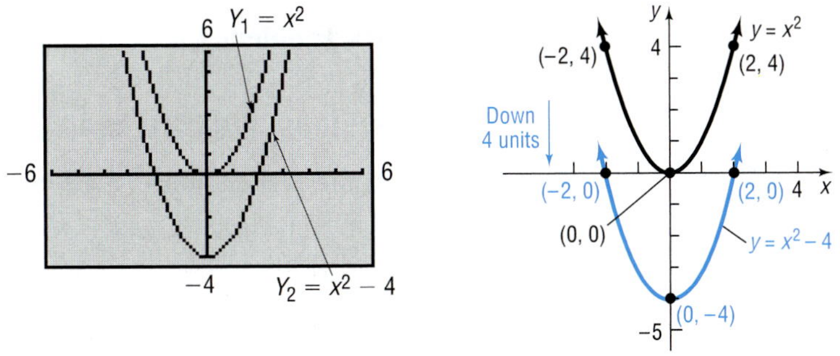

Horizontal Shifts

EXPLORATION

On the same screen, graph each of the following functions:

$Y_1 = x^2$
$Y_2 = (x - 1)^2$
$Y_3 = (x - 3)^2$
$Y_4 = (x + 2)^2$

What do you observe?

FIGURE 39

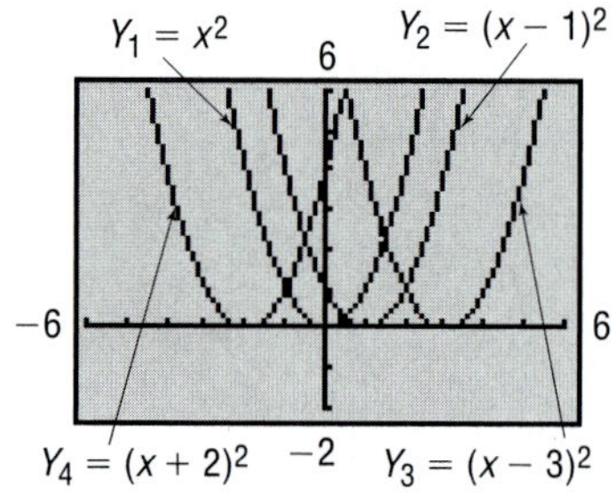

Result Figure 39 illustrates the graphs.

You should have observed the following pattern. With the graph of $Y_1 = x^2$ on the screen, the graph of $Y_2 = (x - 1)^2$ is identical to that of $Y = x^2$, except it is shifted horizontally to the right 1 unit. Similarly, the graph of $Y_3 = (x - 3)^2$ is identical to that of $Y_1 = x^2$, except it is shifted horizontally to the right 3 units. Finally, the graph of $Y_4 = (x + 2)^2$ is identical to that of $Y_1 = x^2$, except it is shifted horizontally to the left 2 units.

We are led to the following conclusion.

> If a real number c is added to the argument x of a function f, the graph of the new function $g(x) = f(x + c)$ is the graph of f **shifted horizontally** left (if $c > 0$) or right (if $c < 0$).

Now work Problem 31.

Vertical and horizontal shifts are sometimes combined.

EXAMPLE 2 Combining Vertical and Horizontal Shifts

Graph the function $f(x) = (x - 1)^3 + 3$.

FIGURE 40

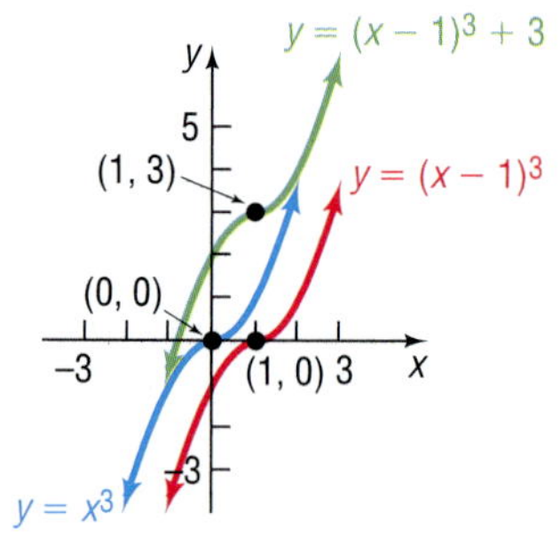

Solution We graph f in steps. First, we note that the rule for f is basically a cube function. Thus, we begin with the graph of $y = x^3$, shown in blue in Figure 40. Next, to get the graph of $y = (x - 1)^3$, we shift the graph of $y = x^3$ horizontally 1 unit to the right. See the graph shown in red in Figure 40. Finally, to get the graph of $y = (x - 1)^3 + 3$, we shift the graph of $y = (x - 1)^3$ vertically up 3 units. See the graph shown in green in Figure 40. Note the points that have been plotted on each graph. Using key points such as these can be helpful in keeping track of just what is taking place.

In Example 2, if the vertical shift had been done first, followed by the horizontal shift, the final graph would have been the same. (Try it for yourself.)

Now work Problem 43.

Compressions and Stretches

EXPLORATION 2

On the same screen, graph each of the following functions:

$$Y_1 = |x|$$
$$Y_2 = 2|x|$$
$$Y_3 = 3|x|$$
$$Y_4 = \tfrac{1}{2}|x|$$

FIGURE 41

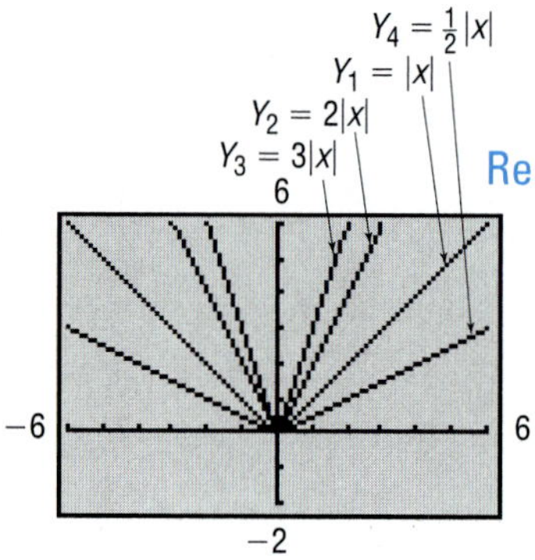

Result Figure 41 illustrates the graphs. You should have observed the following pattern. The graphs of $Y_2 = 2|x|$ and $Y_3 = 3|x|$ can be obtained from the graph of $Y_1 = |x|$ by multiplying each y-coordinate of $Y_1 = |x|$ by factors of 2 and 3, respectively. This is sometimes referred to as a vertical *stretch* using factors of 2 and 3.

The graph of $Y_4 = \frac{1}{2}|x|$ can be obtained from the graph of $Y_1 = |x|$ by multiplying each y-coordinate by $\frac{1}{2}$. This is sometimes referred to as a vertical *compression* using a factor of $\frac{1}{2}$.

Look at Tables 6 and 7, where $Y_1 = |x|$, $Y_3 = 3|x|$, and $Y_4 = \frac{1}{2}|x|$. Notice that the values for Y_3 in Table 6 are three times the values of Y_1. Therefore, the graph of Y_3 will be vertically *stretched* by a factor of 3. Likewise, the values of Y_4 in Table 7 are half the values of Y_1. Therefore, the graph of Y_4 will be vertically *compressed* by a factor of $\frac{1}{2}$.

TABLE 6

X	Y1	Y3
-2	2	6
-1	1	3
0	0	0
1	1	3
2	2	6
3	3	9
4	4	12

Y2■3abs(X)

TABLE 7

X	Y1	Y4
-2	2	1
-1	1	.5
0	0	0
1	1	.5
2	2	1
3	3	1.5
4	4	2

Y3■.5abs(X)

When the right side of a function $y = f(x)$ is multiplied by a positive number k, the graph of the new function $y = kf(x)$ is a **vertically compressed** (if $0 < k < 1$) or **stretched** (if $k > 1$) version of the graph of $y = f(x)$.

Now work Problem 33.

If the argument x of a function $y = f(x)$ is multiplied by a positive number k, the graph of the new function $y = f(kx)$ is also a compressed or stretched version of the graph of $y = f(x)$, but, in this case, it occurs horizontally rather than vertically. To see why, we look at the following Exploration.

EXPLORATION On the same screen, graph each of the following functions:

$$Y_1 = f(x) = x^2$$
$$Y_2 = f(2x) = (2x)^2$$
$$Y_3 = f\left(\frac{1}{2}x\right) = \left(\frac{1}{2}x\right)^2$$

Result You should have obtained the graphs shown in Figure 42. The graph of $Y_2 = (2x)^2$ is the graph of $Y_1 = x^2$ compressed horizontally. Look at Table 8(a). Notice (1, 1), (2, 4), (4, 16), and (16, 256) are points on the graph of $Y_1 = x^2$. Also, (0.5, 1), (1, 4), (2, 16), and (8, 256) are points on the graph of $Y_2 = (2x)^2$. Thus, the graph of $Y_2 = (2x)^2$ is obtained by multiplying the x-coordinate of each point on the graph of $Y_1 = x^2$ by 1/2. The graph of $Y_3 = \left(\frac{1}{2}x\right)^2$ is the graph of $Y_1 = x^2$ stretched horizontally. Look at Table 8(b). Notice (0.5, 0.25), (1, 1), (2, 4), (4, 16), are points on the graph of $Y_1 = x^2$. Also, (1, 0.25), (2, 1), (4, 4), (8, 16) are points on the graph of $Y_3 = \left(\frac{1}{2}x\right)^2$. Thus, the graph of $Y_3 = \left(\frac{1}{2}x\right)^2$ is obtained by multiplying the x-coordinate of each point on the graph of $Y_1 = x^2$ by a factor of 2.

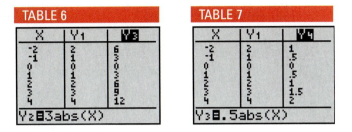

FIGURE 42

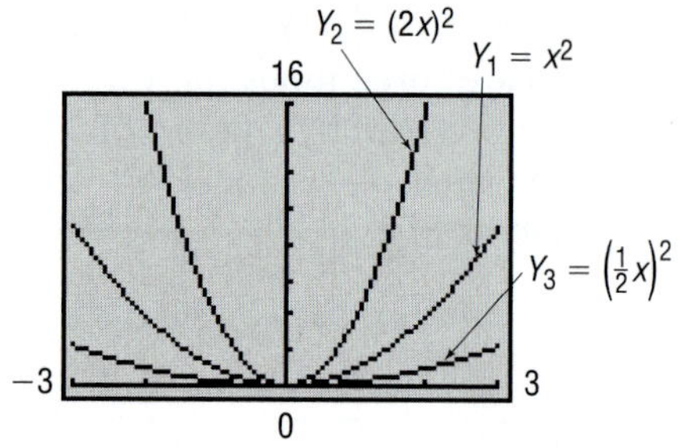

TABLE 8

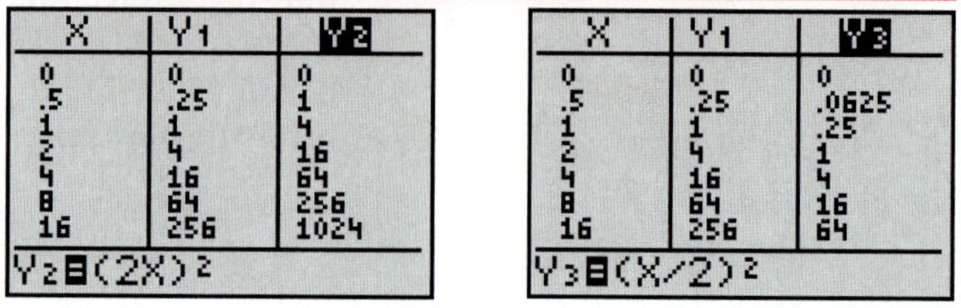

(a)

X	Y1	Y2
0	0	0
.5	.25	1
1	1	4
2	4	16
4	16	64
8	64	256
16	256	1024

Y2=(2X)²

(b)

X	Y1	Y3
0	0	0
.5	.25	.0625
1	1	.25
2	4	1
4	16	4
8	64	16
16	256	64

Y3=(X/2)²

EXAMPLE 3 Graphing Using Stretches and Compressions

The graph of $y = f(x)$ is given in Figure 43. Use this graph to find the graphs of

(a) $y = 3f(x)$ (b) $y = f(3x)$

FIGURE 43

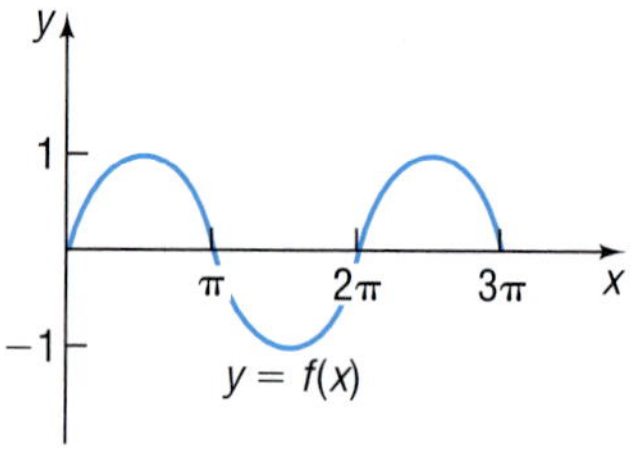

Solution (a) The graph of $y = 3f(x)$ is obtained by multiplying each y-coordinate of $y = f(x)$ by a factor of 3. See Figure 44(a).

(b) The graph of $y = f(3x)$ is obtained from the graph of $y = f(x)$ by multiplying each x-coordinate of $y = f(x)$ by a factor of $\frac{1}{3}$. See Figure 44(b).

FIGURE 44

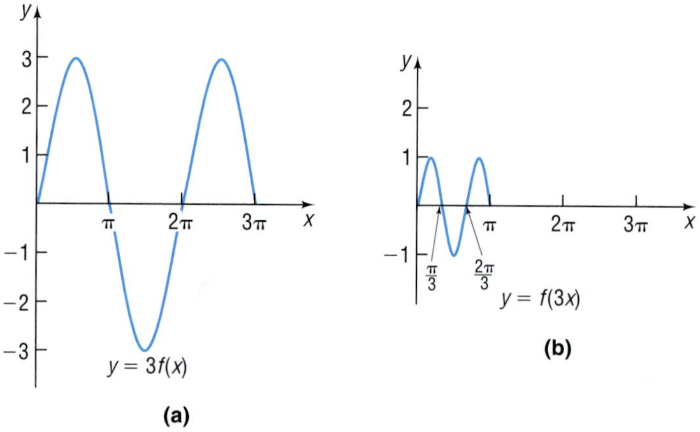

Reflections about the x-Axis and the y-Axis

EXPLORATION Reflection about the x-Axis

3

(a) Graph $Y_1 = x^2$ followed by $Y_2 = -x^2$.
(b) Graph $Y_1 = |x|$ followed by $Y_2 = -|x|$.
(c) Graph $Y_1 = x^2 - 4$ followed by $Y_2 = -(x^2 - 4) = -x^2 + 4$.

Result See Tables 9(a), (b), and (c) and Figures 45(a), (b), and (c). In each instance, the second graph is the reflection about the x-axis of the first graph.

TABLE 9

X	Y1	Y2
-3	9	-9
-2	4	-4
-1	1	-1
0	0	0
1	1	-1
2	4	-4
3	9	-9

Y2 = -X²

(a)

X	Y1	Y2
-3	3	-3
-2	2	-2
-1	1	-1
0	0	0
1	1	-1
2	2	-2
3	3	-3

Y2 = -abs(X)

(b)

X	Y1	Y2
-3	5	-5
-2	0	0
-1	-3	3
0	-4	4
1	-3	3
2	0	0
3	5	-5

Y2 = -X²+4

(c)

FIGURE 45

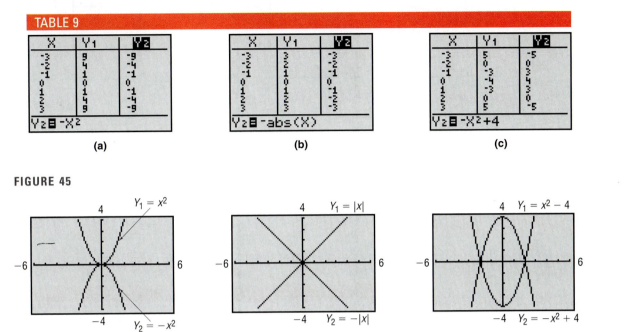

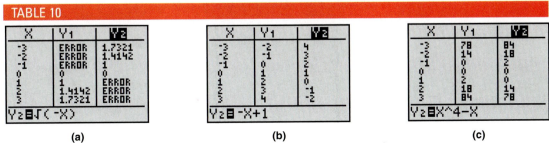

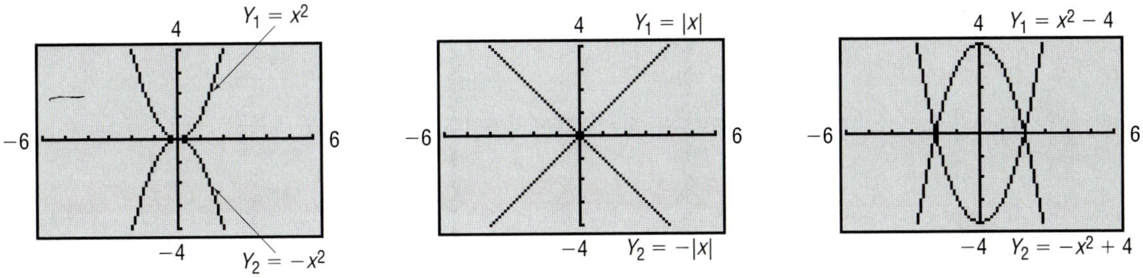

(a) (b) (c)

> When the right side of the equation $y = f(x)$ is multiplied by -1, the graph of the new function $y = -f(x)$ is the **reflection about the x-axis** of the graph of the function $y = f(x)$.

Now work Problem 39.

EXPLORATION

(a) Graph $Y_1 = \sqrt{x}$ followed by $Y_2 = \sqrt{-x}$

(b) Graph $Y_1 = x + 1$ followed by $Y_2 = -x + 1$.

(c) Graph $Y_1 = x^4 + x$ followed by $Y_2 = (-x)^4 + (-x) = x^4 - x$.

Result See Tables 10(a), (b), and (c) and Figures 46(a), (b), and (c). In each instance, the second graph is the reflection about the y-axis of the first graph.

TABLE 10

X	Y1	Y2
-3	ERROR	1.7321
-2	ERROR	1.4142
-1	ERROR	1
0	0	0
1	1	ERROR
2	1.4142	ERROR
3	1.7321	ERROR

Y2 = √(-X)

(a)

X	Y1	Y2
-3	-2	4
-2	-1	3
-1	0	2
0	1	1
1	2	0
2	3	-1
3	4	-2

Y2 = -X+1

(b)

X	Y1	Y2
-3	78	84
-2	14	18
-1	0	2
0	0	0
1	2	0
2	18	14
3	84	78

Y2 = X^4-X

(c)

FIGURE 46

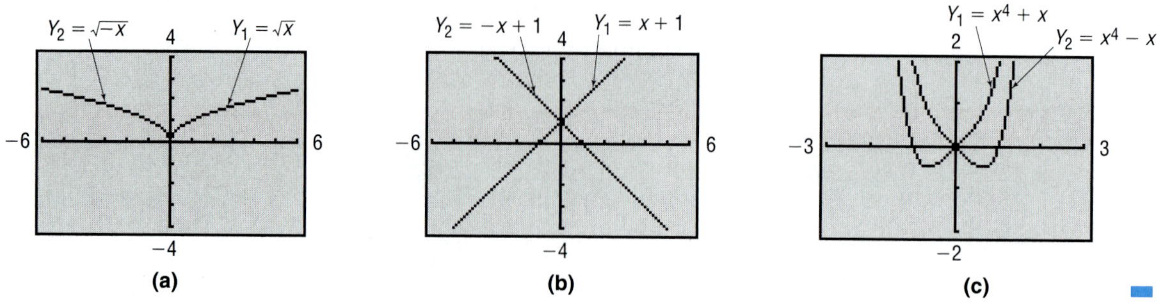

When the graph of the function $y = f(x)$ is known, the graph of the new function $y = f(-x)$ is the **reflection about the *y*-axis** of the graph of the function $y = f(x)$.

SUMMARY OF GRAPHING TECHNIQUES

Table 11 summarizes the graphing procedures we have just discussed.

TABLE 11

To Graph:	Draw the Graph of *f* and:
Vertical shifts	
$y = f(x) + c, \quad c > 0$	Raise the graph of *f* by *c* units.
$y = f(x) - c, \quad c > 0$	Lower the graph of *f* by *c* units.
Horizontal shifts	
$y = f(x + c), \quad c > 0$	Shift the graph of *f* to the left *c* units.
$y = f(x - c), \quad c > 0$	Shift the graph of *f* to the right *c* units.
Compressing or stretching	
$y = kf(x), \quad k > 0$	Multiply each *y* value of $y = f(x)$ by *k*.
$y = f(kx), \quad k > 0$	Multiply each *x* coordinate of $y = f(x)$ by $\frac{1}{k}$.
Reflection about the *x*-axis	
$y = -f(x)$	Reflect the graph of *f* about the *x*-axis.
Reflection about the *y*-axis	
$y = f(-x)$	Reflect the graph of *f* about the *y*-axis.

The examples that follow combine some of the procedures outlined in this section to get the required graph.

EXAMPLE 4 Combining Graphing Procedures

Graph the function $f(x) = \dfrac{3}{x - 2} + 1$.

Solution We use the following steps to obtain the graph of *f*:

STEP 1: $y = \dfrac{1}{x}$ Reciprocal function.

STEP 2: $y = \dfrac{3}{x}$ Vertical stretch of the graph of $y = \frac{1}{x}$ by a factor of 3; multiply by 3.

STEP 3: $y = \dfrac{3}{x-2}$ Horizontal shift to the right 2 units; replace x by $x - 2$.

STEP 4: $y = \dfrac{3}{x-2} + 1$ Vertical shift up 1 unit; add 1.

See Figure 47.

FIGURE 47

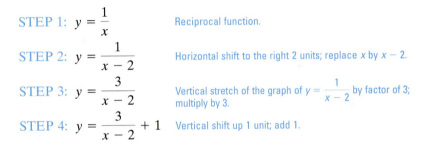

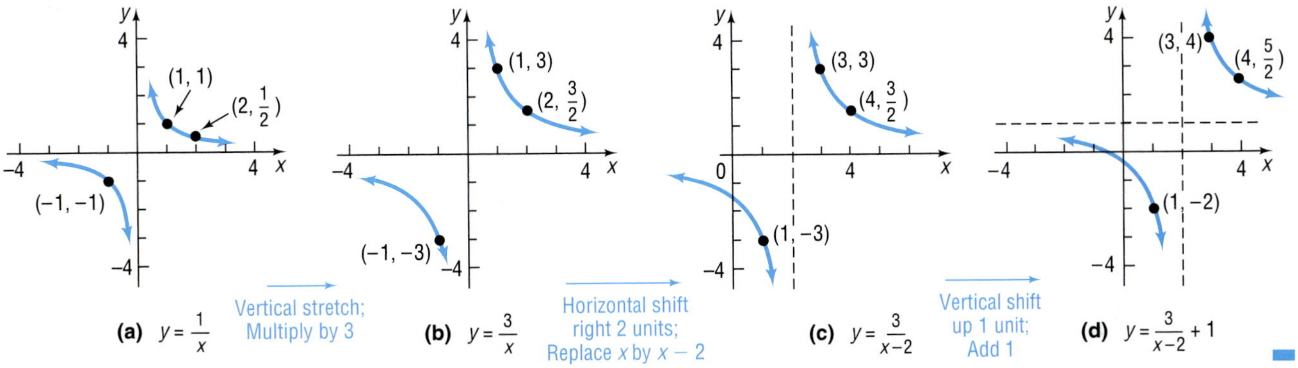

There are other orderings of the steps shown in Example 4 that would also result in the graph of f. For example, try this one:

STEP 1: $y = \dfrac{1}{x}$ Reciprocal function.

STEP 2: $y = \dfrac{1}{x-2}$ Horizontal shift to the right 2 units; replace x by $x - 2$.

STEP 3: $y = \dfrac{3}{x-2}$ Vertical stretch of the graph of $y = \dfrac{1}{x-2}$ by factor of 3; multiply by 3.

STEP 4: $y = \dfrac{3}{x-2} + 1$ Vertical shift up 1 unit; add 1.

Now work Problem 45.

EXAMPLE 5 Combining Graphing Procedures

Graph the function $f(x) = \sqrt{1-x} + 2$.

Solution We use the following steps to get the graph of $y = \sqrt{1-x} + 2$:

STEP 1: $y = \sqrt{x}$ Square root function.

STEP 2: $y = \sqrt{x+1}$ Horizontal shift left 1 unit; replace x by $x + 1$.

STEP 3: $y = \sqrt{-x+1} = \sqrt{1-x}$ Reflect about y-axis; replace x by $-x$.

STEP 4: $y = \sqrt{1-x} + 2$ Vertical shift up 2 units; add 2.

See Figure 48.

FIGURE 48

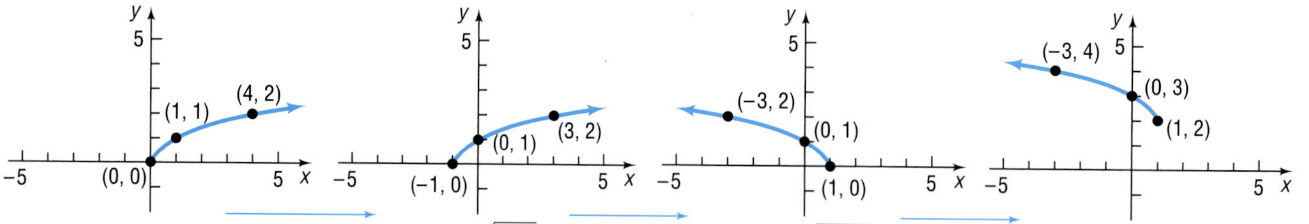

(a) $y = \sqrt{x}$ **(b)** $y = \sqrt{x+1}$ **(c)** $y = \sqrt{-x+1} = \sqrt{1-x}$ **(d)** $y = \sqrt{1-x} + 2$

2.3 EXERCISES

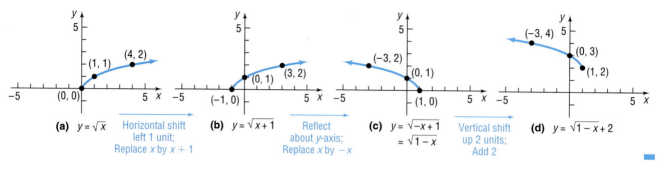

In Problems 1–12, match each graph to one of the following functions:

A. $y = x^2 + 2$	*B.* $y = -x^2 + 2$	*C.* $y = \lvert x \rvert + 2$	*D.* $y = -\lvert x \rvert + 2$
E. $y = (x-2)^2$	*F.* $y = -(x+2)^2$	*G.* $y = \lvert x - 2 \rvert$	*H.* $y = -\lvert x + 2 \rvert$
I. $y = 2x^2$	*J.* $y = -2x^2$	*K.* $y = 2\lvert x \rvert$	*L.* $y = -2\lvert x \rvert$

1.
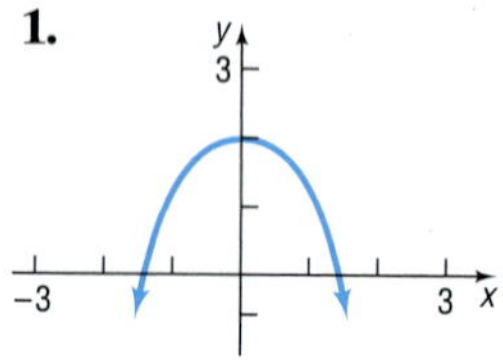

2.
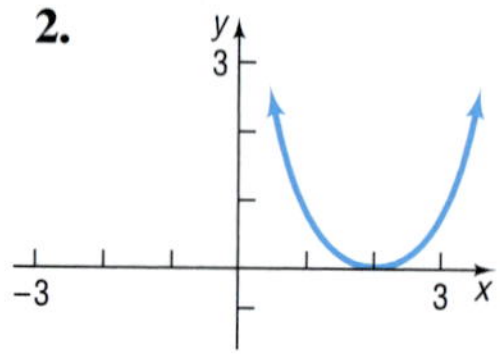

3.
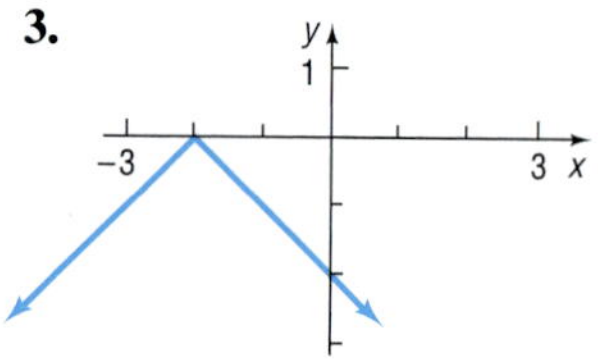

4.
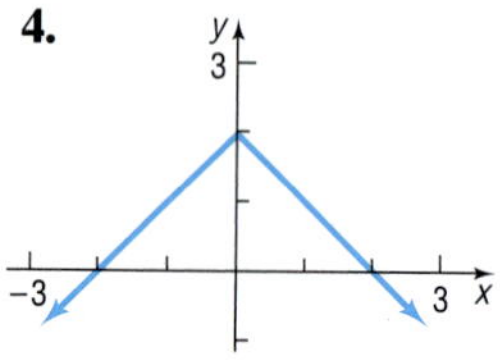

5.
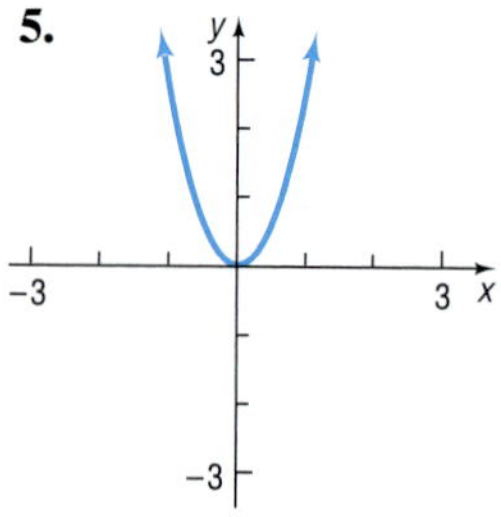

6.
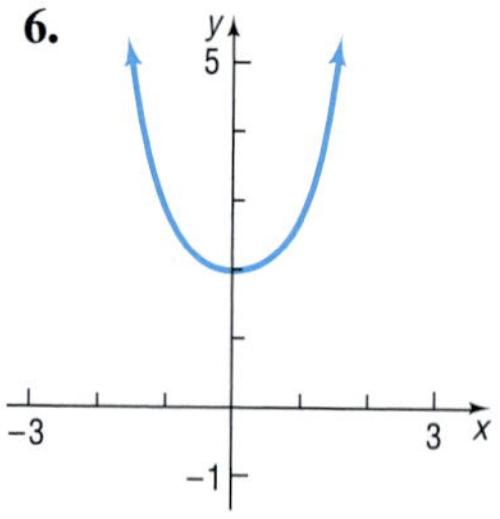

7.
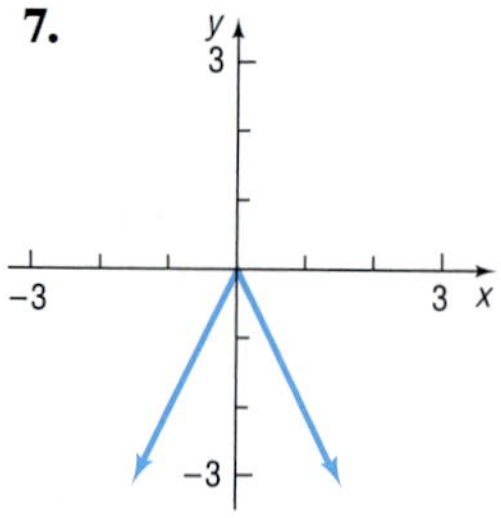

8.
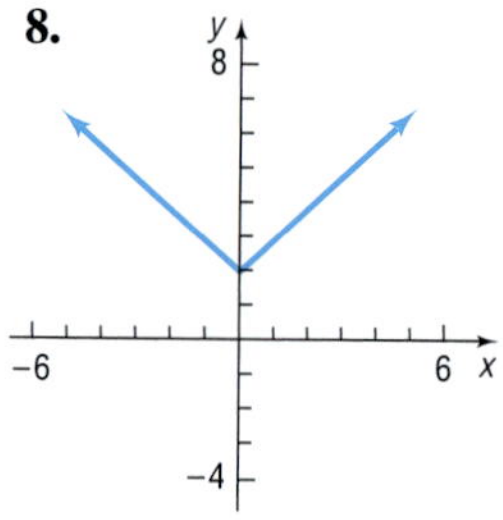

9.
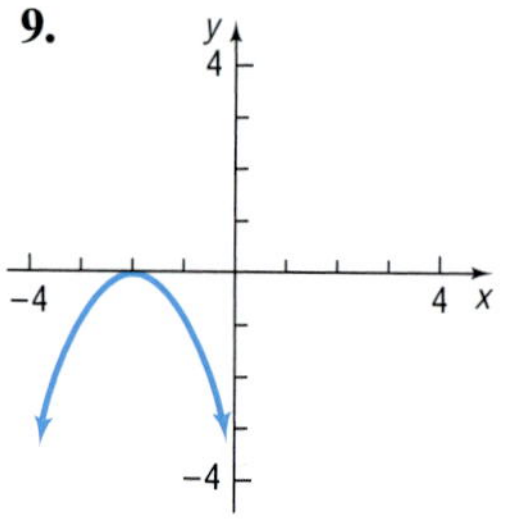

10.
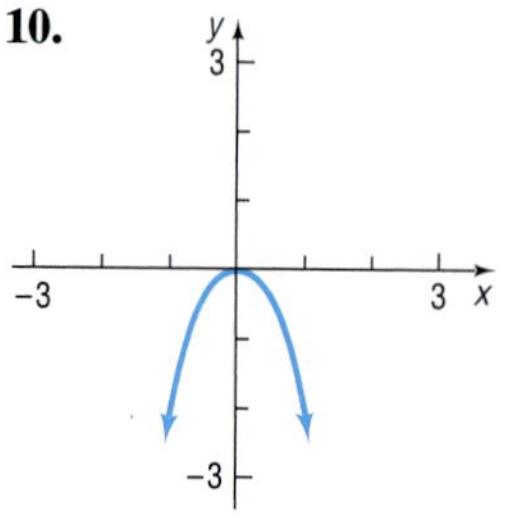

11.
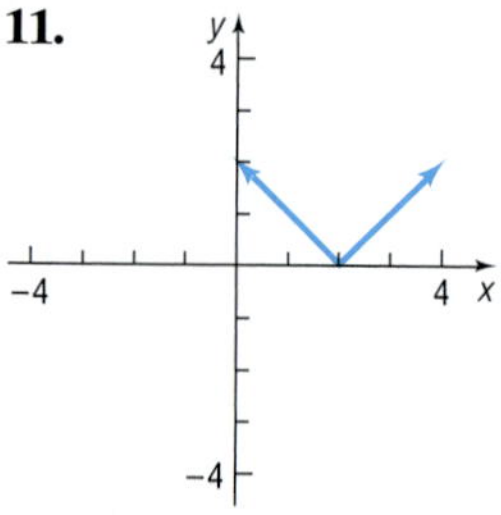

12.
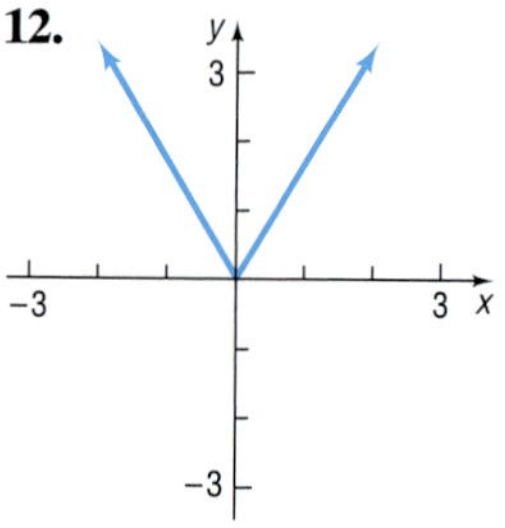

In Problems 13–16, match each graph to one of the following functions:

A. $y = x^3$ *B.* $y = (x + 2)^3$ *C.* $y = -2x^3$ *D.* $y = x^3 + 2$

13.

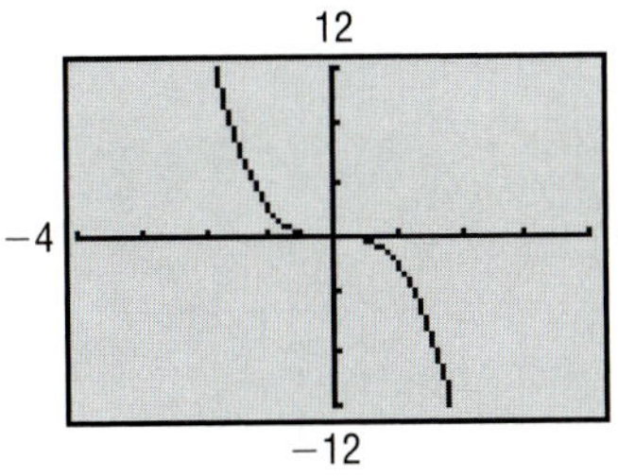

14.

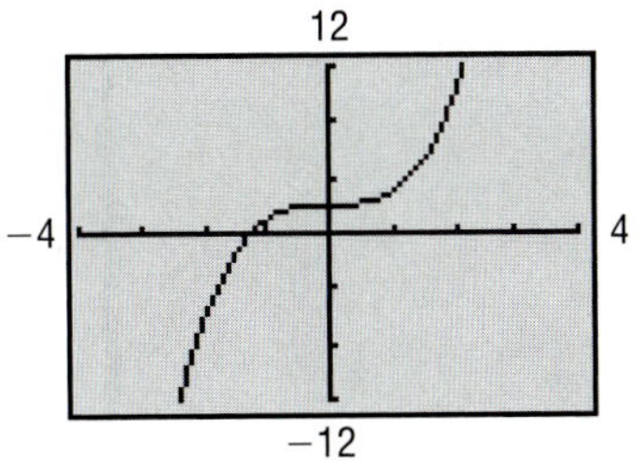

15.

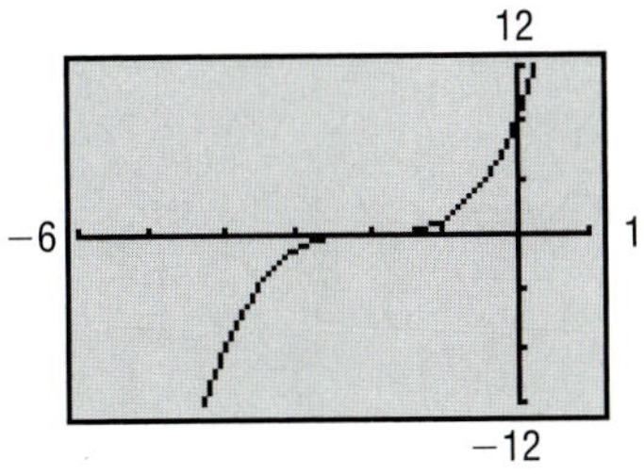

16.

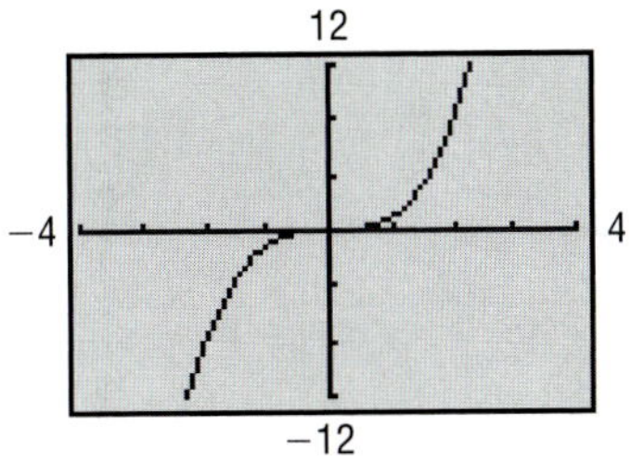

In Problems 17–24, write the function whose graph is the graph of $y = x^3$*, but is*

17. Shifted to the right 4 units

18. Shifted to the left 4 units

19. Shifted up 4 units

20. Shifted down 4 units

21. Reflected about the y-axis

22. Reflected about the x-axis

23. Vertically stretched by a factor of 4

24. Horizontally stretched by a factor of 4

In Problems 25–54, graph each function by hand using the techniques of shifting, compressing, stretching, and/or reflecting. Start with the graph of the basic function (for example, $y = x^2$*) and show all stages. Verify your answer by using a graphing utility.*

25. $f(x) = x^2 - 1$

26. $f(x) = x^2 + 4$

27. $g(x) = x^3 + 1$

28. $g(x) = x^3 - 1$

29. $h(x) = \sqrt{x - 2}$

30. $h(x) = \sqrt{x + 1}$

31. $f(x) = (x - 1)^3$

32. $f(x) = (x + 2)^3$

33. $g(x) = 4\sqrt{x}$

34. $g(x) = \frac{1}{2}\sqrt{x}$

35. $h(x) = \dfrac{1}{2x}$

36. $h(x) = \dfrac{4}{x}$

37. $f(x) = -|x|$

38. $f(x) = -\sqrt{x}$

39. $g(x) = -\dfrac{1}{x}$

40. $g(x) = -x^3$

41. $h(x) = \text{int}(-x)$

42. $h(x) = \dfrac{1}{-x}$

43. $f(x) = (x + 1)^2 - 3$

44. $f(x) = (x - 2)^2 + 1$

45. $g(x) = \sqrt{x - 2} + 1$

46. $g(x) = |x + 1| - 3$

47. $h(x) = \sqrt{-x} - 2$

48. $h(x) = \dfrac{4}{x} + 2$

49. $f(x) = (x + 1)^3 - 1$

50. $f(x) = 4\sqrt{x - 1}$

51. $g(x) = 2|1 - x|$

52. $g(x) = 4\sqrt{2 - x}$

53. $h(x) = 2\text{int}(x - 1)$

54. $h(x) = -x^3 + 2$

In Problems 55–60, the graph of a function f is illustrated. Use the graph of f as the first step toward graphing each of the following functions:

(a) $F(x) = f(x) + 3$ *(b)* $G(x) = f(x + 2)$ *(c)* $P(x) = -f(x)$

(d) $Q(x) = \frac{1}{2}f(x)$ *(e)* $g(x) = f(-x)$ *(f)* $h(x) = 3f(x)$

55.

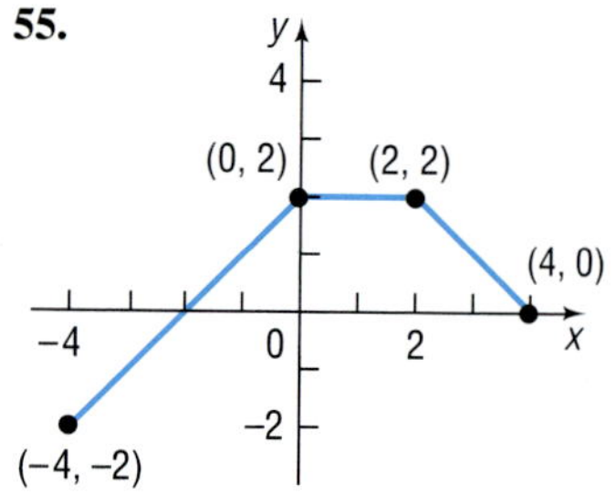

56.

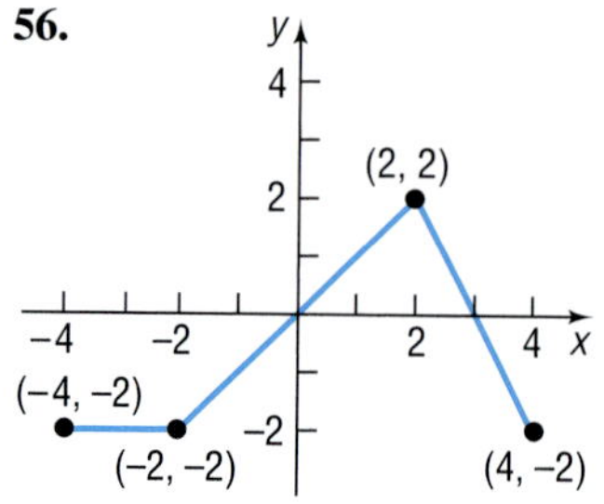

57.

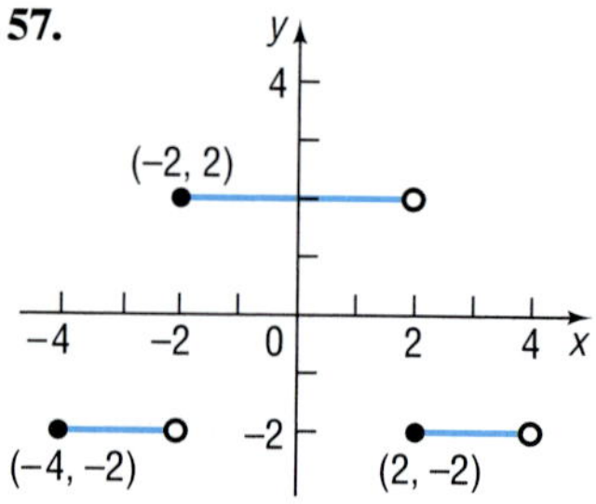

58.

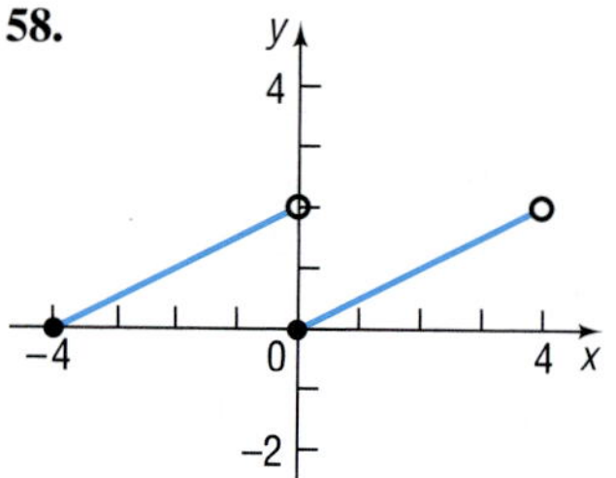

59.

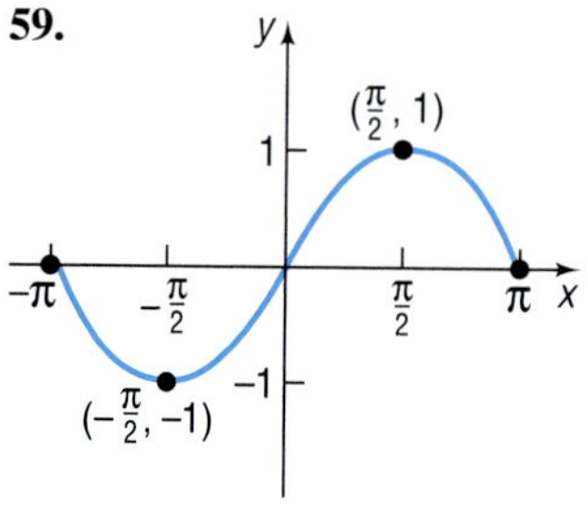

60.

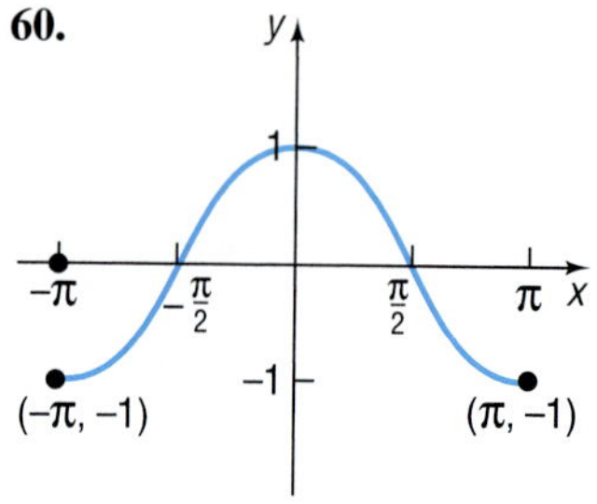

61. Exploration

(a) Use a graphing utility to graph $y = x + 1$ and $y = |x + 1|$.

(b) Graph $y = 4 - x^2$ and $y = |4 - x^2|$.

(c) Graph $y = x^3 + x$ and $y = |x^3 + x|$.

(d) What do you conclude about the relationship between the graphs of $y = f(x)$ and $y = |f(x)|$?

62. Exploration

(a) Use a graphing utility to graph $y = x + 1$ and $y = |x| + 1$.

(b) Graph $y = 4 - x^2$ and $y = 4 - |x|^2$.

(c) Graph $y = x^3 + x$ and $y = |x|^3 + |x|$.

(d) What do you conclude about the relationship between the graphs of $y = f(x)$ and $y = f(|x|)$?

63. The graph of a function f is illustrated in the figure.

(a) Draw the entire graph of $y = |f(x)|$.

(b) Draw the entire graph of $y = f(|x|)$.

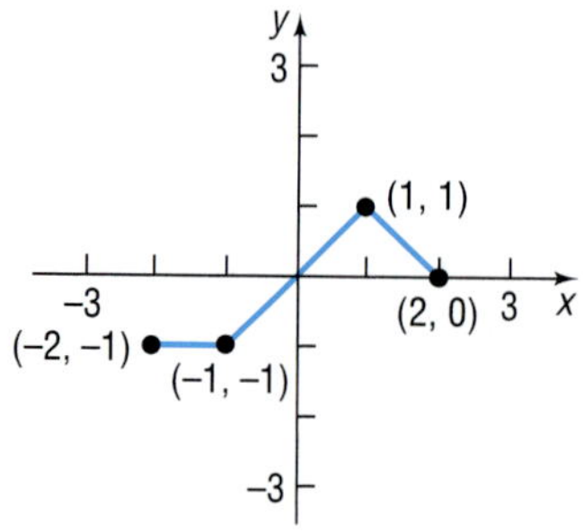

64. The graph of a function f is illustrated in the figure.

(a) Draw the entire graph of $y = |f(x)|$.

(b) Draw the entire graph of $y = f(|x|)$.

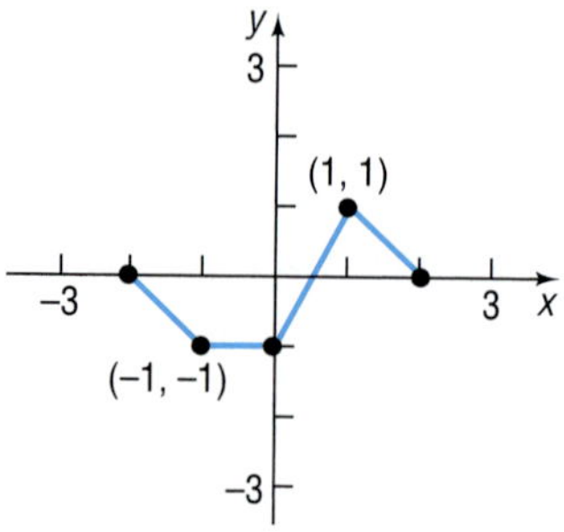

In Problems 65–70, complete the square of each quadratic expression. Then graph each function by hand using the technique of shifting. Verify your results using a graphing utility. (If necessary, refer to Appendix 5 to review Completing the Square.)

65. $f(x) = x^2 + 2x$

66. $f(x) = x^2 - 6x$

67. $f(x) = x^2 - 8x + 1$

68. $f(x) = x^2 + 4x + 2$

69. $f(x) = x^2 + x + 1$

70. $f(x) = x^2 - x + 1$

71. The equation $y = (x - c)^2$ defines a *family of parabolas,* one parabola for each value of c. On one set of coordinate axes, graph the members of the family for $c = 0$, $c = 3$, $c = -2$.

72. Repeat Problem 71 for the family of parabolas $y = x^2 + c$.

73. **Temperature Measurements** The relationship between the Celsius (°C) and Fahrenheit (°F) scales for measuring temperature is given by the equation

$$F = \frac{9}{5}C + 32$$

The relationship between the Celsius (°C) and Kelvin (K) scales is $K = C + 273$. Graph the equation $F = \frac{9}{5}C + 32$ using degrees Fahrenheit on the y-axis and degrees Celsius on the x-axis. Use the techniques introduced in this section to obtain the graph showing the relationship between Kelvin and Fahrenheit temperatures.

74. **Period of a Pendulum** The period T (in seconds) of a simple pendulum is a function of its length l (in feet) defined by the equation

$$T = 2\pi\sqrt{\frac{l}{g}}$$

where $g \approx 32.2$ feet per second per second is the acceleration of gravity.

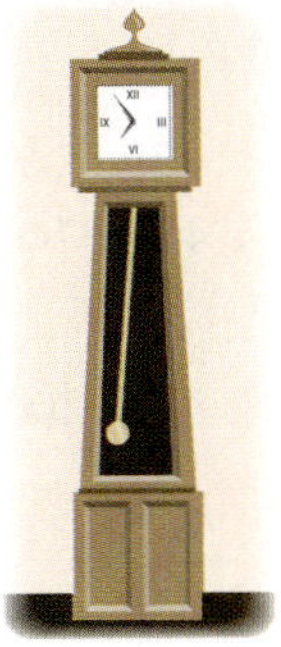

(a) Use a graphing utility to graph the function $T = T(l)$.
(b) Now graph the functions $T = T(l + 1)$, $T = T(l + 2)$, and $T = T(l + 3)$.
(c) Discuss how adding to the length l changes the period T.
(d) Now graph the functions $T = T(2l)$, $T = T(3l)$, and $T = T(4l)$.
(e) Discuss how multiplying the length l by factors of 2, 3, and 4 changes the period T.

75. **Cigarette Company Profits** The daily profits of a cigarette company from selling x packs of cigarettes are given by

$$p(x) = -0.05x^2 + 100x - 2000$$

The government wishes to impose a tax on cigarettes (sometimes called a **sin tax**) that gives the company the option of either paying a flat tax of \$10,000 per day or a tax of 10% on profits. As Chief Financial Officer (CFO) of the company, you need to decide which tax is the better option for the company.
(a) On the same viewing rectangle graph $Y_1 = p(x) - 10{,}000$ and $Y_2 = (1 - 0.10)p(x)$.
(b) Based on the graph which option would you select? Why?
(c) Using the terminology learned in this section, describe each graph in terms of the graph of $p(x)$.
(d) Suppose the government offered the option of a flat tax option of \$4,800 or a tax of 10% on profits. Which option would you select? Why?

2.4 OPERATIONS ON FUNCTIONS; COMPOSITE FUNCTIONS

1 Form the Sum, Difference, Product, and Quotient of Two Functions
2 Form the Composite Function and Find Its Domain

In this section, we introduce some operations on functions. We shall see that functions, like numbers, can be added, subtracted, multiplied, and divided. For example, if $f(x) = x^2 + 9$ and $g(x) = 3x + 5$, then

$$f(x) + g(x) = (x^2 + 9) + (3x + 5) = x^2 + 3x + 14$$

The new function $y = x^2 + 3x + 14$ is called the *sum function* $f + g$. Similarly,

$$f(x) \cdot g(x) = (x^2 + 9)(3x + 5) = 3x^3 + 5x^2 + 27x + 45$$

The new function $y = 3x^3 + 5x^2 + 27x + 45$ is called the *product function* $f \cdot g$.

The general definitions are given next.

If f and g are functions:

Their **sum** $\boldsymbol{f + g}$ is the function defined by

$$(f + g)(x) = f(x) + g(x)$$

The domain of $f + g$ consists of the numbers x that are in the domain of f and in the domain of g.

Their **difference** $\boldsymbol{f - g}$ is the function defined by

$$(f - g)(x) = f(x) - g(x)$$

The domain of $f - g$ consists of the numbers x that are in the domain of f and in the domain of g.

Their **product** $\boldsymbol{f \cdot g}$ is the function defined by

$$(f \cdot g)(x) = f(x) \cdot g(x)$$

The domain of $f \cdot g$ consists of the numbers x that are in the domain of f and in the domain of g.

Their **quotient** $\boldsymbol{f/g}$ is the function defined by

$$\left(\frac{f}{g}\right)(x) = \frac{f(x)}{g(x)}, \quad g(x) \neq 0$$

The domain of f/g consists of the numbers x for which $g(x) \neq 0$ that are

EXAMPLE 1 Operations on Functions

1 Let f and g be two functions defined as

$$f(x) = \sqrt{x + 2} \quad \text{and} \quad g(x) = \sqrt{x - 3}$$

Find the following, and determine the domain in each case:

(a) $(f + g)(x)$ (b) $(f - g)(x)$ (c) $(f \cdot g)(x)$ (d) $(f/g)(x)$

Solution

(a) $(f + g)(x) = f(x) + g(x) = \sqrt{x + 2} + \sqrt{x - 3}$

(b) $(f - g)(x) = f(x) - g(x) = \sqrt{x + 2} - \sqrt{x - 3}$

(c) $(f \cdot g)(x) = f(x) \cdot g(x) = (\sqrt{x + 2})(\sqrt{x - 3}) = \sqrt{(x + 2)(x - 3)}$

(d) $\left(\frac{f}{g}\right)(x) = \frac{f(x)}{g(x)} = \frac{\sqrt{x + 2}}{\sqrt{x - 3}} = \sqrt{\frac{x + 2}{x - 3}}$

The domain of f consists of all numbers x for which $x \geq -2$; the domain of g consists of all numbers x for which $x \geq 3$. The numbers x common to both these domains are those for which $x \geq 3$. As a result, the numbers x for which $x \geq 3$ comprise the domain of the sum function $f + g$, the difference function $f - g$, and the product function $f \cdot g$. For the quotient function f/g, we must exclude from this set the number 3, because the denominator, g, has the value 0 when $x = 3$. Thus, the domain of f/g consists of all x for which $x > 3$.

Now work Problem 1.

In calculus, it is sometimes helpful to view a complicated function as the sum, difference, product, or quotient of simpler functions. For example,

$F(x) = x^2 + \sqrt{x}$ is the sum of $f(x) = x^2$ and $g(x) = \sqrt{x}$.
$H(x) = (x^2 - 1)/(x^2 + 1)$ is the quotient of $f(x) = x^2 - 1$ and $g(x) = x^2 + 1$.

Composite Functions

2 Consider the function $y = (2x + 3)^2$. If we write $y = f(u) = u^2$ and $u = g(x) = 2x + 3$, then, by a substitution process, we can obtain the original function: $y = f(u) = f(g(x)) = (2x + 3)^2$. This process is called **composition.**

In general, suppose that f and g are two functions, and suppose that x is a number in the domain of g. By evaluating g at x, we get $g(x)$. If $g(x)$ is in the domain of f, then we may evaluate f at $g(x)$ and thereby obtain the expression $f(g(x))$. The correspondence from x to $f(g(x))$ is called a *composite function* $f \circ g$.

Look carefully at Figure 49. Only those x's in the domain of g for which $g(x)$ is in the domain of f can be in the domain of $f \circ g$. The reason is that if $g(x)$ is not in the domain of f, then $f(g(x))$ is not defined. Thus, the domain of $f \circ g$ is a subset of the domain of g; the range of $f \circ g$ is a subset of the range of f.

FIGURE 49

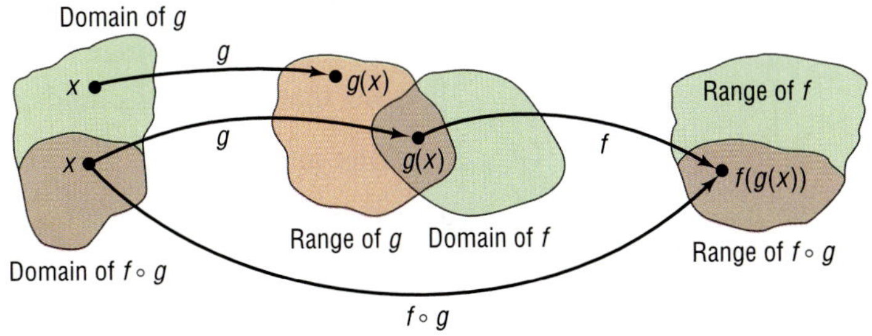

Given two functions f and g, the **composite function,** denoted by $f \circ g$ (read as "f composed with g"), is defined by

$$(f \circ g)(x) = f(g(x))$$

The domain of $f \circ g$ is the set of all numbers x in the domain of g such that $g(x)$ is in the domain of f.

Figure 50 provides a second illustration of the definition. Notice that the "inside" function g in $f(g(x))$ is done first.

FIGURE 50

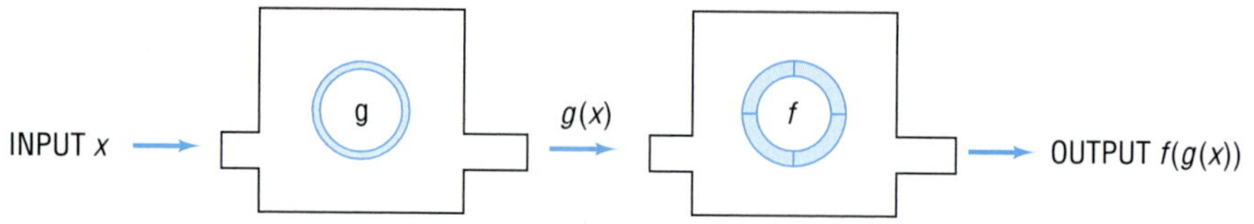

Let's look at some examples.

EXAMPLE 2 Evaluating a Composite Function

Suppose that $f(x) = 2x^2 - 3$ and $g(x) = 4x$. Find

(a) $(f \circ g)(1)$ (b) $(g \circ f)(1)$ (c) $(f \circ f)(-2)$ (d) $(g \circ g)(-1)$

Solution (a) $(f \circ g)(1) = f(g(1)) = f(4) = 2 \cdot 16 - 3 = 29$

$g(x) = 4x$, $g(1) = 4$ $\qquad$ $f(x) = 2x^2 - 3$

(b) $(g \circ f)(1) = g(f(1)) = g(-1) = 4 \cdot (-1) = -4$

$f(x) = 2x^2 - 3$, $f(1) = -1$ $\qquad$ $g(x) = 4x$

(c) $(f \circ f)(-2) = f(f(-2)) = f(5) = 2 \cdot 25 - 3 = 47$

$f(-2) = 5$

(d) $(g \circ g)(-1) = g(g(-1)) = g(-4) = 4 \cdot (-4) = -16$

$g(-1) = -4$

Now work Problem 21.

EXAMPLE 3 Finding a Composite Function

Suppose that $f(x) = \sqrt{x}$ and $g(x) = \dfrac{1}{x - 1}$. Find the following composite functions, and then find the domain of each composite function.

(a) $f \circ g$ (b) $g \circ f$ (c) $f \circ f$ (d) $g \circ g$

Solution The domain of f is $\{x|x \geq 0\}$ and the domain of g is $\{x|x \neq 1\}$.

(a) $(f \circ g)(x) = f(g(x)) = f\left(\dfrac{1}{x - 1}\right) = \sqrt{\dfrac{1}{x - 1}} = \dfrac{1}{\sqrt{x - 1}}$

The domain of $f \circ g$ consists of those x in the domain of g for which $g(x) = 1/(x - 1) \geq 0$, or equivalently, $x > 1$. Thus, the domain of $f \circ g$ is $\{x|x > 1\}$.

(b) $(g \circ f)(x) = g(f(x)) = g(\sqrt{x}) = \dfrac{1}{\sqrt{x} - 1}$

The domain of $g \circ f$ consists of those x in the domain of f for which $f(x) = \sqrt{x} \neq 1$, or equivalently, $x \neq 1$. Thus, the domain of $g \circ f$ is $\{x|x \geq 0, x \neq 1\}$.

(c) $(f \circ f)(x) = f(f(x)) = f(\sqrt{x}) = \sqrt{\sqrt{x}} = \sqrt[4]{x}$

The domain of $f \circ f$ consists of those x in the domain of f for which $f(x) = \sqrt{x} \geq 0$, or equivalently, $x \geq 0$. Thus, the domain of $f \circ f$ is $\{x|x \geq 0\}$.

(d) $(g \circ g)(x) = g(g(x)) = g\left(\frac{1}{x-1}\right) = \frac{1}{\frac{1}{x-1} - 1} = \frac{x-1}{1-(x-1)} = -\frac{x-1}{x-2}$

The domain of $g \circ g$ consists of those x in the domain of g for which $g(x) = 1/(x-1) \neq 1$, or equivalently, $x \neq 2$. Thus, the domain of $g \circ g$ is $\{x | x \neq 1, x \neq 2\}$.

Check: In Example 3(d), see for yourself! Determine what happens if $x = 2$ for the composite $(g \circ g)(x)$.

Now work Problem 35.

Examples 3(a) and 3(b) illustrate that, in general, $f \circ g \neq g \circ f$. However, sometimes $f \circ g$ does equal $g \circ f$, as shown in the next example.

EXAMPLE 4 Showing Two Composite Functions Equal

If $f(x) = 3x - 4$ and $g(x) = \frac{1}{3}(x + 4)$, show that $(f \circ g)(x) = (g \circ f)(x) = x$ for every x.

Solution

$(f \circ g)(x) = f(g(x))$

$= f\left(\frac{x+4}{3}\right)$ — $g(x) = \frac{1}{3}(x+4) = \frac{x+4}{3}$.

$= 3\left(\frac{x+4}{3}\right) - 4$ — Substitute $g(x)$ into the rule for f, $f(x) = 3x - 4$.

$= x + 4 - 4 = x$

$(g \circ f)(x) = g(f(x))$

$= g(3x - 4)$ — $f(x) = 3x - 4$.

$= \frac{1}{3}[(3x - 4) + 4]$ — Substitute $f(x)$ into the rule for g, $g(x) = \frac{1}{3}(x + 4)$.

$= \frac{1}{3}(3x) = x$

Thus, $(f \circ g)(x) = (g \circ f)(x) = x$.

In Section 5.1, we shall see that there is an important relationship between functions f and g for which $(f \circ g)(x) = (g \circ f)(x) = x$.

Now work Problem 45.

2.4 EXERCISES

In Problems 1–10, for the given functions f and g, find the following functions and state the domain of each:

(a) $f + g$ *(b)* $f - g$ *(c)* $f \cdot g$ *(d)* f/g

1. $f(x) = 3x + 4;\quad g(x) = 2x - 3$

2. $f(x) = 2x + 1;\quad g(x) = 3x - 2$

3. $f(x) = x - 1;\quad g(x) = 2x^2$

4. $f(x) = 2x^2 + 3;\quad g(x) = 4x^3 + 1$

5. $f(x) = \sqrt{x};\quad g(x) = 3x - 5$

6. $f(x) = |x|;\quad g(x) = x$

7. $f(x) = 1 + \frac{1}{x};\quad g(x) = \frac{1}{x}$

8. $f(x) = 2x^2 - x;\quad g(x) = 2x^2 + x$

9. $f(x) = \frac{2x+3}{3x-2};\quad g(x) = \frac{4x}{3x-2}$

10. $f(x) = \sqrt{x+1};\quad g(x) = \frac{2}{x}$

11. Given $f(x) = 3x + 1$ and $(f + g)(x) = 6 - \frac{1}{2}x$, find the function g.

12. Given $f(x) = 1/x$ and $(f/g)(x) = (x + 1)/(x^2 - x)$, find the function g.

In Problems 13–16, find the equation that defines the function $f + g$. Use a graphing utility to graph the functions f, g, and $f + g$ on the same screen.

13. $f(x) = x, g(x) = \dfrac{1}{x}$

14. $f(x) = x, g(x) = \dfrac{1}{x^2}$

15. $f(x) = x^2, g(x) = \dfrac{1}{x}$

16. $f(x) = x^2, g(x) = \dfrac{1}{x^2}$

In Problems 17–20, find the equation that defines the function $f \cdot g$. Use a graphing utility to graph the functions f, g, and $f \cdot g$ on the same screen.

17. $f(x) = x, g(x) = \dfrac{1}{x^2 + 1}$

18. $f(x) = x, g(x) = \dfrac{1}{x^2 - 1}$

19. $f(x) = x^2, g(x) = \dfrac{1}{x^2 + 1}$

20. $f(x) = x^2, g(x) = \dfrac{1}{x^2 - 1}$

In Problems 21–30, for the given functions f and g, find

(a) $(f \circ g)(4)$ (b) $(g \circ f)(2)$ (c) $(f \circ f)(1)$ (d) $(g \circ g)(0)$

21. $f(x) = 2x;\ \ g(x) = 3x^2 + 1$

22. $f(x) = 3x + 2;\ \ g(x) = 2x^2 - 1$

23. $f(x) = 4x^2 - 3;\ \ g(x) = 3 - \frac{1}{2}x^2$

24. $f(x) = 2x^2;\ \ g(x) = 1 - 3x^2$

25. $f(x) = \sqrt{x};\ \ g(x) = 2x$

26. $f(x) = \sqrt{x + 1};\ \ g(x) = 3x$

27. $f(x) = |x|;\ \ g(x) = \dfrac{1}{x^2 + 1}$

28. $f(x) = |x - 2|;\ \ g(x) = \dfrac{3}{x^2 + 2}$

29. $f(x) = \dfrac{3}{x^2 + 1};\ \ g(x) = \sqrt{x}$

30. $f(x) = x^3;\ \ g(x) = \dfrac{2}{x^2 + 1}$

In Problems 31–44, for the given functions f and g, find

(a) $f \circ g$ (b) $g \circ f$ (c) $f \circ f$ (d) $g \circ g$

State the domain of each composite function.

31. $f(x) = 2x + 3;\ \ g(x) = 3x$

32. $f(x) = -x;\ \ g(x) = 2x - 4$

33. $f(x) = 3x + 1;\ \ g(x) = x^2$

34. $f(x) = \sqrt{x + 1};\ \ g(x) = x + 4$

35. $f(x) = \sqrt{x};\ \ g(x) = x^2 - 1$

36. $f(x) = \sqrt{x + 1};\ \ g(x) = \dfrac{1}{x^2}$

37. $f(x) = \dfrac{x - 1}{x + 1};\ \ g(x) = \dfrac{1}{x}$

38. $f(x) = x + \dfrac{1}{x};\ \ g(x) = x^2$

39. $f(x) = x^2;\ \ g(x) = \sqrt{x}$

40. $f(x) = 2x + 4;\ \ g(x) = \frac{1}{2}x - 2$

41. $f(x) = \dfrac{1}{2x + 3};\ \ g(x) = 2x + 3$

42. $f(x) = \dfrac{x + 1}{x - 1};\ \ g(x) = \dfrac{x - 1}{x + 1}$

43. $f(x) = ax + b;\ \ g(x) = cx + d$

44. $f(x) = \dfrac{ax + b}{cx + d};\ \ g(x) = mx$

In Problems 45–52, show that $(f \circ g)(x) = (g \circ f)(x) = x$.

45. $f(x) = 2x;\ \ g(x) = \frac{1}{2}x$

46. $f(x) = 4x;\ \ g(x) = \frac{1}{4}x$

47. $f(x) = x^3;\ \ g(x) = \sqrt[3]{x}$

48. $f(x) = x + 5;\ \ g(x) = x - 5$

49. $f(x) = 2x - 6;\ \ g(x) = \frac{1}{2}(x + 6)$

50. $f(x) = 4 - 3x;\ \ g(x) = \frac{1}{3}(4 - x)$

51. $f(x) = ax + b;\ \ g(x) = \dfrac{1}{a}(x - b), a \neq 0$

52. $f(x) = \dfrac{1}{x};\ \ g(x) = \dfrac{1}{x}$

53. If $f(x) = 2x^3 - 3x^2 + 4x - 1$ and $g(x) = 2$, find $(f \circ g)(x)$ and $(g \circ f)(x)$.

54. If $f(x) = x/(x - 1)$, find $(f \circ f)(x)$.

55. If $f(x) = 2x^2 + 5$ and $g(x) = 3x + a$, find a so that the graph of $f \circ g$ crosses the y-axis at 23.

56. If $f(x) = 3x^2 - 7$ and $g(x) = 2x + a$, find a so that the graph of $f \circ g$ crosses the y-axis at 68.

57. The surface area S (in square meters) of a hot air balloon is given by

$$S(r) = 4\pi r^2$$

where r is the radius of the balloon (in meters). If the radius r is increasing with time t (in seconds) according to the formula $r(t) = \frac{2}{3}t^3$, $t \geq 0$, find the surface area S of the balloon as a function of the time t.

58. The volume V (in cubic meters) of the hot air balloon described in Problem 57 is given by $V(r) = \frac{4}{3}\pi r^3$. If the radius r is the same function of t as in Problem 57, find the volume V as a function of the time t.

59. **Automobile Production** The number N of cars produced at a certain factory in 1 day after t hours of operation is given by $N(t) = 100t - 5t^2$, $0 \leq t \leq 10$. If the cost C (in dollars) of producing N cars is $C(N) = 15{,}000 + 8{,}000N$, find the cost C as a function of the time t of operation of the factory.

60. **Environmental Concerns** The spread of oil leaking from a tanker is in the shape of a circle. If the radius r (in feet) of the spread after t hours is $r(t) = 200\sqrt{t}$, find the area A of the oil slick as a function of the time t.

61. **Production Cost** The price p of a certain product and the quantity x sold obey the demand equation

$$p = -\frac{1}{4}x + 100 \qquad 0 \leq x \leq 400$$

Suppose that the cost C of producing x units is

$$C = \frac{\sqrt{x}}{25} + 600$$

Assuming that all items produced are sold, find the cost C as a function of the price p.

[**Hint:** Solve for x in the demand equation, and then form the composite.]

62. **Cost of a Commodity** The price p of a certain commodity and the quantity x sold obey the demand equation

$$p = -\frac{1}{5}x + 200 \qquad 0 \leq x \leq 1000$$

Suppose that the cost C of producing x units is

$$C = \frac{\sqrt{x}}{10} + 400$$

Assuming that all items produced are sold, find the cost C as a function of the price p.

63. If f and g are odd functions, show that the composite function $f \circ g$ is also odd.

64. If f is an odd function and g is an even function, show that the composite functions $f \circ g$ and $g \circ f$ are also even.

2.5 MATHEMATICAL MODELS: CONSTRUCTING FUNCTIONS

1 Construct and Analyze Functions

1 Real-world problems often result in mathematical models that involve functions. These functions need to be constructed or built based on the information given. In constructing functions, we must be able to translate the verbal description into the language of mathematics. We do this by assigning symbols to represent the independent and dependent variables and then finding the function or rule that relates these variables.

EXAMPLE 1 Area of a Rectangle with Fixed Perimeter

The perimeter of a rectangle is 50 feet. Express its area A as a function of the length x of a side.

FIGURE 51

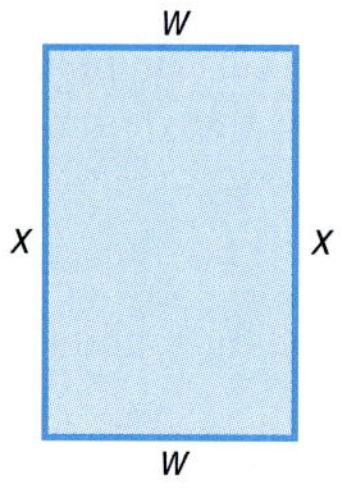

Solution Consult Figure 51. If the length of the rectangle is x and if w is its width, then the sum of the lengths of the sides is the perimeter, 50.

$$\begin{aligned} x + w + x + w &= 50 \\ 2x + 2w &= 50 \\ x + w &= 25 \\ w &= 25 - x \end{aligned}$$

The area A is length times width, so

$$A = xw = x(25 - x)$$

The area A as a function of x is

$$A(x) = x(25 - x)$$

Note that we use the symbol A as the dependent variable and also as the name of the function that relates the length x to the area A. As we mentioned earlier, this double usage is common in applications and should cause no difficulties.

EXAMPLE 2 Economics: Demand Equations

In economics, revenue R is defined as the amount of money derived from the sale of a product and is equal to the unit selling price p of the product times the number x of units actually sold. That is,

$$R = xp$$

In economics, the Law of Demand states that p and x are related: As one increases, the other decreases. Suppose that p and x are related by the following **demand equation:**

$$p = -\tfrac{1}{10}x + 20 \qquad 0 \le x \le 200$$

Express the revenue R as a function of the number x of units sold.

Solution Since $R = xp$ and $p = -\frac{1}{10}x + 20$, it follows that

$$R(x) = xp = x\left(-\tfrac{1}{10}x + 20\right) = -\tfrac{1}{10}x^2 + 20x$$

Now work Problem 3.

EXAMPLE 3 Finding the Distance from the Origin to a Point on a Graph

Let $P = (x, y)$ be a point on the graph of $y = x^2 - 1$.

(a) Express the distance d from P to the origin O as a function of x.
(b) What is d if $x = 0$?
(c) What is d if $x = 1$?
(d) What is d if $x = \sqrt{2}/2$?
(e) Use a graphing utility to graph the function $d = d(x)$, $x \ge 0$. Correct to two decimal places, find the value of x at which d has a local minimum. [This gives the point(s) on the graph of $y = x^2 - 1$ closest to the origin.]

MISSION POSSIBLE

Consulting for the Silver Satellite & Cable TV Co.

Your team works for the Silver Satellite & Cable TV Company in the Research & Development Department. You've been asked to come up with a formula to determine the cost of running cable from a connection box to a new cable household. The first example you are working with involves the Frick family who own a rural home with a driveway two miles long extending to the house from a nearby highway. The nearest connection box is along the highway, but 5 miles from the driveway.

It costs the company $10 per mile to install cable along the highway and $14 per mile to install cable off the highway. Because the Frick house is surrounded by farmland which they own, it would be possible to run the cable overland to the house directly from the connection box or from any point between the connection box and the driveway.

1. Draw a sketch of this problem situation, assuming the highway is a straight road and the driveway is also a straight road perpendicular to the highway. Include two or more possible routes for the cable.
2. Suppose x represents the distance in miles the cable runs along the highway from the connection box before turning off toward the house. Express the total cost of installation as a function of x. (You may choose to answer #3 before #2 if you would like to examine concrete instances before creating the function.)
3. Make a table of the possible integral values of x and the corresponding cost in each instance. Is there one choice which appears to cost the least?
4. If you charge the Fricks $80 for installation, would you be willing to let them choose which way the cable would go? Explain.
5. Using a graphing calculator, graph the function from #2 and determine if there is a nonintegral choice for x that would make the installation cost even cheaper. Use BOX or TRACE (or MINIMUM) to determine the least possible cost.
6. Before proceeding further with the installation, you check the local regulations for cable companies and find that there is a pending state legislation that says the cable cannot turn off the highway more than 0.5 miles from the Fricks' driveway. If this legislation passes, what will be the ultimate cost of installing the Fricks' cable?
7. If the cable company wishes to install cable in 5,000 homes in this area, and assuming the figures for the Frick installation are typical, how much will the new legislation cost the company over all if they cannot use the cheapest installation cost, but instead have to follow the new state regulations?

Solution (a) Figure 52 illustrates the graph. The distance d from P to O is

$$d = \sqrt{x^2 + y^2}$$

Since P is a point on the graph of $y = x^2 - 1$, we have

$$d(x) = \sqrt{x^2 + (x^2 - 1)^2} = \sqrt{x^4 - x^2 + 1}$$

Thus, we have expressed the distance d as a function of x.

(b) If $x = 0$, the distance d is

$$d(0) = \sqrt{1} = 1$$

(c) If $x = 1$, the distance d is

$$d(1) = \sqrt{1 - 1 + 1} = 1$$

(d) If $x = \sqrt{2}/2$, the distance d is

$$d\left(\frac{\sqrt{2}}{2}\right) = \sqrt{\left(\frac{\sqrt{2}}{2}\right)^4 - \left(\frac{\sqrt{2}}{2}\right)^2 + 1} = \sqrt{\frac{1}{4} - \frac{1}{2} + 1} = \frac{\sqrt{3}}{2}$$

(e) Figure 53 shows the graph of $Y_1 = \sqrt{x^4 - x^2 + 1}$.

Using the minimum function on a graphing utility, we find that, when $x \approx 0.70$ the value of d is smallest ($d \approx 0.86$) correct to two decimal places.

FIGURE 52

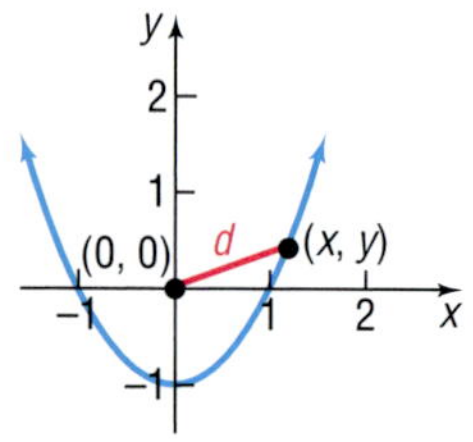

FIGURE 53

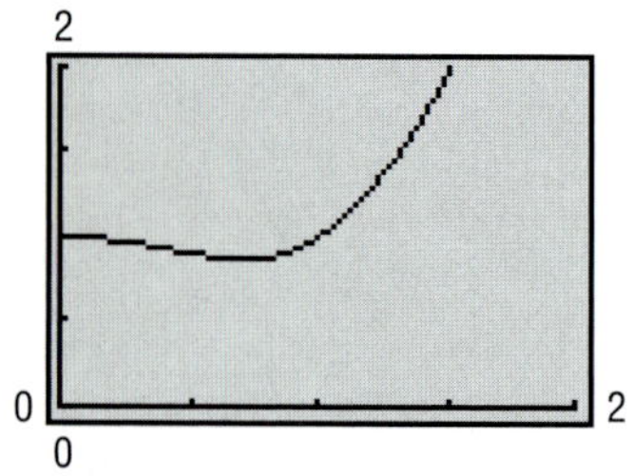

Now work Problem 13.

EXAMPLE 4 Filling a Swimming Pool

A rectangular swimming pool 20 meters long and 10 meters wide is 4 meters deep at one end and 1 meter deep at the other. Figure 54 illustrates a cross-sectional view of the pool. Water is being pumped into the pool at the deep end.

(a) Find a function that expresses the volume V of water in the pool as a function of the height x of the water at the deep end.

(b) Find the volume when the height is 1 meter.

(c) Find the volume when the height is 2 meters.

(d) Use a graphing utility to graph the function $V = V(x)$. At what height is the volume 20 cubic meters? 100 cubic meters?

FIGURE 54

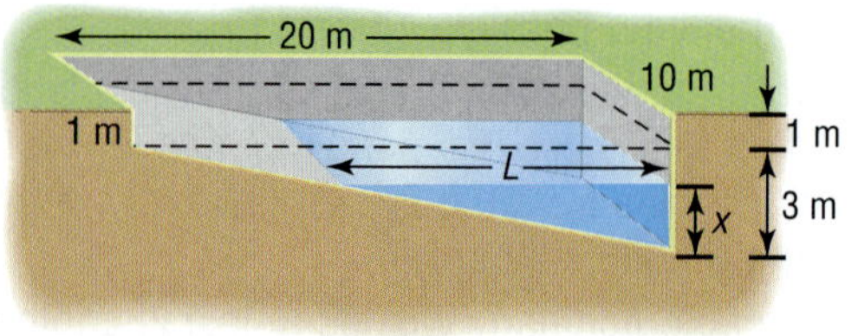

Solution (a) Let L denote the distance (in meters) measured at water level from the deep end to the short end. Notice that L and x form the sides of a triangle that is similar to the triangle whose sides are 20 meters by 3 meters. Thus, L and x are related by the equation

$$\frac{L}{x} = \frac{20}{3} \quad \text{or} \quad L = \frac{20x}{3} \qquad 0 \le x \le 3$$

The volume V of water in the pool at any time is

$$V = \begin{pmatrix}\text{Cross-sectional}\\ \text{triangular area}\end{pmatrix}(\text{Width}) = (\tfrac{1}{2}Lx)(10) \text{ cubic meters}$$

Since $L = 20x/3$, we have

$$V(x) = \left(\frac{1}{2} \cdot \frac{20x}{3} \cdot x\right)(10) = \frac{100}{3}x^2 \text{ cubic meters}$$

(b) When the height x of the water is 1 meter, the volume $V = V(x)$ is

$$V(1) = \frac{100}{3} \cdot 1^2 = 33.3 \text{ cubic meters}$$

FIGURE 55

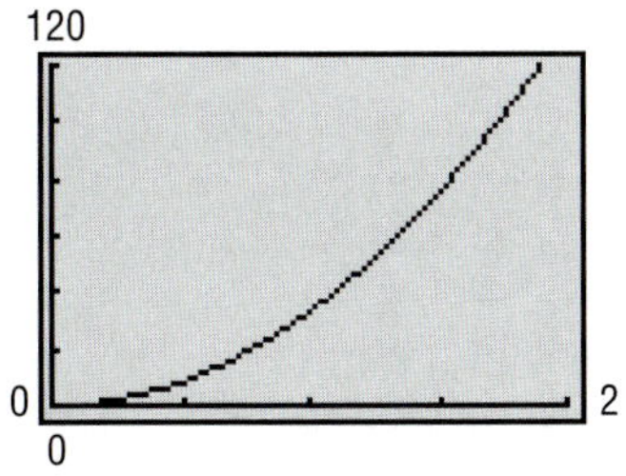

(c) When the height x of the water is 2 meters, the volume $V = V(x)$ is

$$V(2) = \frac{100}{3} \cdot 2^2 = \frac{400}{3} = 133.3 \text{ cubic meters}$$

(d) See Figure 55.

When $x \approx 0.77$ meters, the volume is 20 cubic meters. When $x \approx 1.73$ meters, the volume is 100 cubic meters.

EXAMPLE 5 Area of an Isosceles Triangle

Consider an isosceles triangle of fixed perimeter p.

(a) If x equals the length of one of the two equal sides, express the area A as a function of x.

(b) What is the domain of A?

(c) Use a graphing utility to graph $A = A(x)$ for $p = 4$.

(d) For what value of x is the area largest?

Solution (a) Look at Figure 56. Since the equal sides are of length x, the third side must be of length $p - 2x$. (Do you see why?) We know that the area A is

FIGURE 56

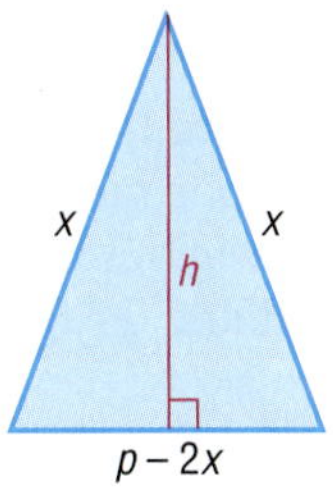

$$A = \tfrac{1}{2}(\text{Base})(\text{Height})$$

To find the height h, we drop the perpendicular to the base of length $p - 2x$ and use the fact that the perpendicular bisects the base. Then, by the Pythagorean Theorem, we have

$$\begin{aligned} h^2 &= x^2 - \left(\frac{p-2x}{2}\right)^2 = x^2 - \frac{1}{4}(p^2 - 4px + 4x^2) \\ &= px - \frac{1}{4}p^2 = \frac{4px - p^2}{4} \\ h &= \sqrt{\frac{4px - p^2}{4}} = \frac{\sqrt{p}}{2}\sqrt{4x - p} \end{aligned}$$

The area A is given by

$$A = \frac{1}{2} \cdot (p - 2x)\frac{\sqrt{p}}{2}\sqrt{4x - p} = \frac{\sqrt{p}}{4}(p - 2x)\sqrt{4x - p}$$

(b) The domain of A is found as follows. Because of the expression $\sqrt{4x - p}$, we require that

$$4x - p > 0$$
$$x > \frac{p}{4}$$

Since $p - 2x$ is a side of the triangle, we also require that

$$p - 2x > 0$$
$$-2x > -p$$
$$x < \frac{p}{2}$$

FIGURE 57

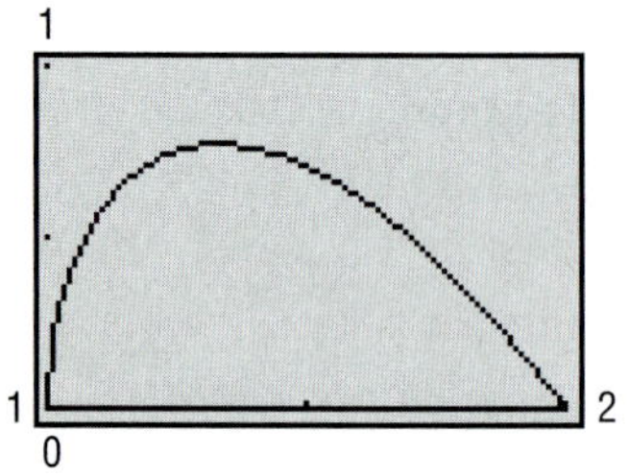

Thus, the domain of A is $p/4 < x < p/2$, or $(p/4, p/2)$, and we state the function A as

$$A(x) = \frac{\sqrt{p}}{4}(p - 2x)\sqrt{4x - p} \qquad \frac{p}{4} < x < \frac{p}{2}$$

(c) See Figure 57 for the graph of $A(x) = 2(2 - x)\sqrt{x - 1}$.

(d) The area is largest (approximately 0.77) when $x \approx 1.33$.

Now work Problem 7.

EXAMPLE 6 Minimum Payments and Interest Charged for Credit Cards

(a) Holders of credit cards issued by banks, department stores, oil companies, and so on, receive bills each month that state minimum amounts that must be paid by a certain due date. The minimum due depends on the total amount owed. One such credit card company uses the following rules: For a bill of less than \$10, the entire amount is due. For a bill of at least \$10 but less than \$500, the minimum due is \$10. There is a minimum of \$30 due on a bill of at least \$500 but less than \$1,000, a minimum of \$50 due on a bill of at least \$1,000 but less than \$1,500, and a minimum of \$70 due on bills of \$1,500 or more. Find the function f that describes the minimum payment due on a bill of x dollars. Graph f by hand.

(b) The card holder may pay any amount between the minimum due and the total owed. The organization issuing the card charges the card holder interest of 1.5% per month for the first \$1,000 owed and 1% per month on any unpaid balance over \$1,000. Find the function g that gives the amount of interest charged per month on a balance of x dollars. Graph g by hand.

Solution (a) The function f that describes the minimum payment due on a bill of x dollars is

$$f(x) = \begin{cases} x & \text{if } \quad 0 \le x < 10 \\ 10 & \text{if } \quad 10 \le x < 500 \\ 30 & \text{if } \quad 500 \le x < 1000 \\ 50 & \text{if } 1000 \le x < 1500 \\ 70 & \text{if } 1500 \le x \end{cases}$$

To graph this function f, we proceed as follows. For $0 \leq x < 10$, draw the graph of $y = x$; for $10 \leq x < 500$, draw the graph of the constant function $y = 10$; for $500 \leq x < 1000$, draw the graph of the constant function $y = 30$; and so on. The graph of f is given in Figure 58.

FIGURE 58

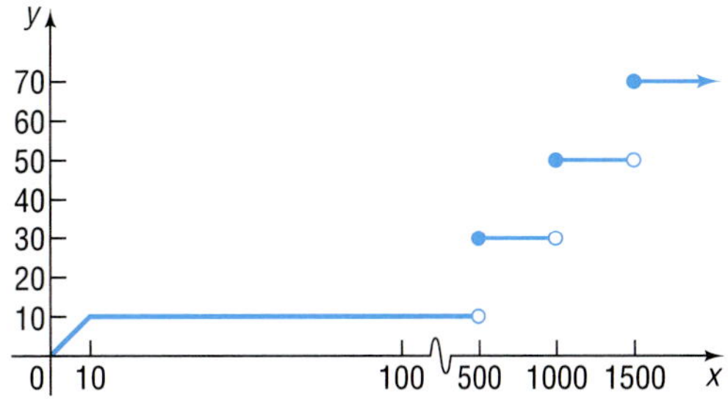

FIGURE 59

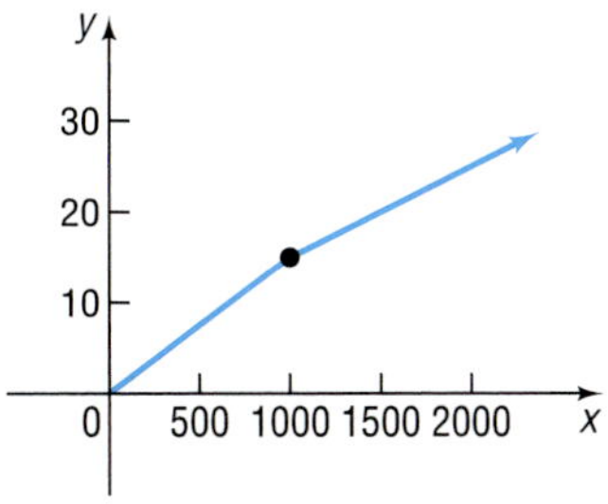

(b) If $g(x)$ is the amount of interest charged per month on a balance of x, then $g(x) = 0.015x$ for $0 \leq x \leq 1000$. The amount of the unpaid balance above \$1,000 is $x - 1000$. If the balance due is $x > 1000$, then the interest is $0.015(1000) + 0.01(x - 1000) = 15 + 0.01x - 10 = 5 + 0.01x$, so

$$g(x) = \begin{cases} 0.015x & \text{if } 0 \leq x \leq 1000 \\ 5 + 0.01x & \text{if } x > 1000 \end{cases}$$

See Figure 59.

EXAMPLE 7 Making a Playpen

FIGURE 60

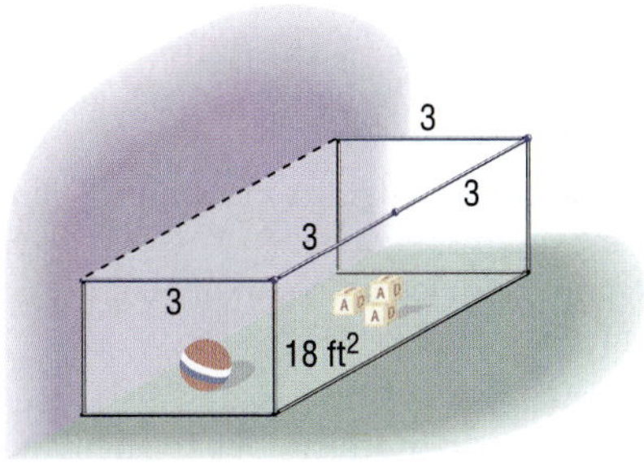

A manufacturer of children's playpens makes a square model that can be opened at one corner and attached at right angles to a wall, or, perhaps, the side of a house. If each side is 3 feet in length, the open configuration doubles the available area in which the child can play from 9 square feet to 18 square feet. See Figure 60.

Now, suppose we place hinges at the outer corners to allow for a configuration like the one shown in Figure 61.

(a) Express the area A of this configuration as a function of the distance x between the two parallel sides.

(b) Find the domain of A.

(c) Find A if $x = 5$.

(d) Graph $A = A(x)$. For what value of x is the area largest? What is the maximum area?*

FIGURE 61

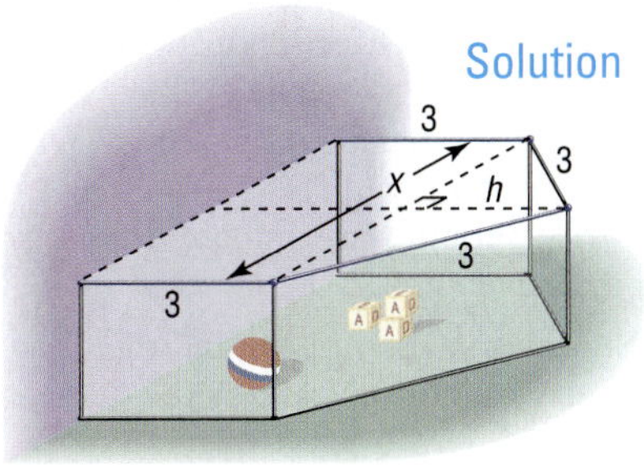

Solution (a) Refer to Figure 61. The area A that we seek consists of the area of a rectangle (with width 3 and length x) and the area of an isosceles triangle (with base x and two equal sides of length 3). The height h of the triangle may be found using the Pythagorean Theorem.

$$h^2 = 3^2 - \left(\frac{x}{2}\right)^2 = 9 - \frac{x^2}{4} = \frac{36 - x^2}{4}$$

$$h = \tfrac{1}{2}\sqrt{36 - x^2}$$

*Adapted from *Proceedings, Summer Conference for College Teachers on Applied Mathematics* (University of Missouri, Rolla), 1971.

The area A enclosed by the playpen is

$$A = \text{Area of rectangle} + \text{Area of triangle} = 3x + \tfrac{1}{2}x\left(\tfrac{1}{2}\sqrt{36 - x^2}\right)$$

$$A(x) = 3x + \frac{x\sqrt{36 - x^2}}{4}$$

Thus, we have expressed the area A as a function of x.

(b) To find the domain of A, we note first that $x > 0$, since x is a length. Also, the expression under the radical must be positive, so

$$36 - x^2 > 0$$
$$x^2 < 36$$
$$-6 < x < 6$$

Combining these restrictions, we find that the domain of A is $0 < x < 6$, or $(0, 6)$.

(c) If $x = 5$, the area is

$$A(5) = 3(5) + \frac{5}{4}\sqrt{36 - (5)^2} \approx 19.15 \text{ square feet}$$

Thus, if the width of the playpen is 5 feet, its area is 19.15 square feet.

(d) See Figure 62. The maximum area is about 19.81 square feet, obtained when x is about 5.58 feet.

FIGURE 62

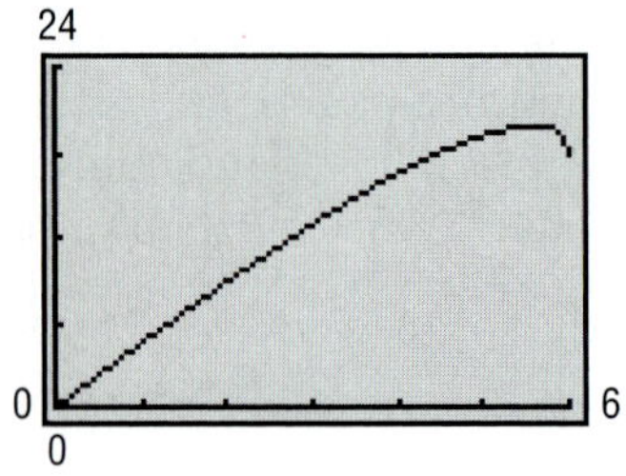

2.5 EXERCISES

1. Volume of a Cylinder The volume V of a right circular cylinder of height h and radius r is $V = \pi r^2 h$. If the height is twice the radius, express the volume V as a function of r.

2. Volume of a Cone The volume V of a right circular cone is $V = \frac{1}{3}\pi r^2 h$. If the height is twice the radius, express the volume V as a function of r.

3. Demand Equation The price p and the quantity x sold of a certain product obey the demand equation

$$p = -\tfrac{1}{6}x + 100 \qquad 0 \le x \le 600$$

(a) Express the revenue R as a function of x. (Remember, $R = xp$.)
(b) What is the revenue if 200 units are sold?
(c) Graph the revenue function using a graphing utility.
(d) What quantity x maximizes revenue? What is the maximum revenue?
(e) What price should the company charge to maximize revenue?

4. Demand Equation The price p and the quantity x sold of a certain product obey the demand equation

$$p = -\tfrac{1}{3}x + 100 \qquad 0 \le x \le 300$$

(a) Express the revenue R as a function of x.
(b) What is the revenue if 100 units are sold?
(c) Graph the revenue function using a graphing utility.
(d) What quantity x maximizes revenue? What is the maximum revenue?
(e) What price should the company charge to maximize revenue?

5. Demand Equation The price p and the quantity x sold of a certain product obey the demand equation

$$x = -5p + 100 \qquad 0 \le p \le 20$$

(a) Express the revenue R as a function of x.
(b) What is the revenue if 15 units are sold?
(c) Graph the revenue function using a graphing utility.
(d) What quantity x maximizes revenue? What is the maximum revenue?
(e) What price should the company charge to maximize revenue?

6. Demand Equation The price p and the quantity x sold of a certain product obey the demand equation

$$x = -20p + 500 \qquad 0 \le p \le 25$$

(a) Express the revenue R as a function of x.
(b) What is the revenue if 20 units are sold?

(c) Graph the revenue function using a graphing utility.
(d) What quantity x maximizes revenue? What is the maximum revenue?
(e) What price should the company charge to maximize revenue?

7. **Enclosing a Rectangular Field** David has available 400 yards of fencing and wishes to enclose a rectangular area.
(a) Express the area A of the rectangle as a function of the width x of the rectangle.
(b) What is the domain of A?
(c) Graph $A = A(x)$ using a graphing utility. For what value of x is the area largest?

8. **Enclosing a Rectangular Field along a River** Beth has 3,000 feet of fencing available to enclose a rectangular field. One side of the field lies along a river, so only three sides require fencing.
(a) Express the area A of the rectangle as a function of x, where x is the length of the side parallel to the river.
(b) Graph $A = A(x)$ using a graphing utility. For what value of x is the area largest?

9. A wire of length x is bent into the shape of a circle.
(a) Express the circumference of the circle as a function of x.
(b) Express the area of the circle as a function of x.

10. A wire of length x is bent into the shape of a square.
(a) Express the perimeter of the square as a function of x.
(b) Express the area of the square as a function of x.

11. A right triangle has one vertex on the graph of $y = x^3, x > 0$, at (x, y), another at the origin, and the third on the positive y-axis at $(0, y)$, as shown in the figure. Express the area A of the triangle as a function of x.

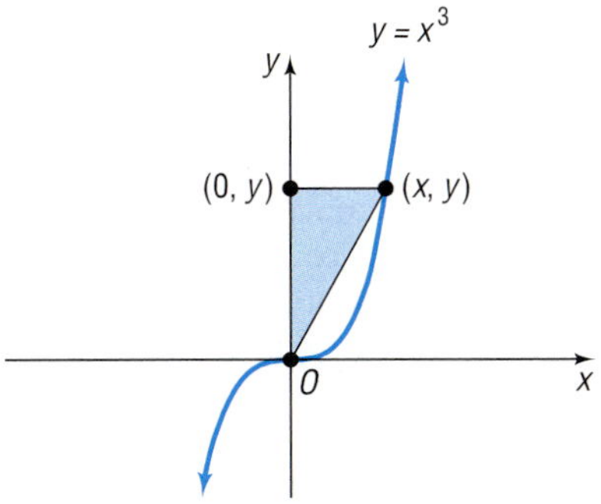

12. A right triangle has one vertex on the graph of $y = 9 - x^2, x > 0$, at (x, y), another at the origin, and the third on the positive x-axis at $(x, 0)$. Express the area A of the triangle as a function of x.

13. Let $P = (x, y)$ be a point on the graph of $y = x^2 - 8$.
(a) Express the distance d from P to the origin as a function of x.
(b) What is d if $x = 0$?
(c) What is d if $x = 1$?
(d) Use a graphing utility to graph $d = d(x)$.
(e) For what values of x is d smallest?

14. Let $P = (x, y)$ be a point on the graph of $y = x^2 - 8$.
(a) Express the distance d from P to the point $(0, -1)$ as a function of x.
(b) What is d if $x = 0$?
(c) What is d if $x = -1$?
(d) Use a graphing utility to graph $d = d(x)$.
(e) For what values of x is d smallest?

15. Let $P = (x, y)$ be a point on the graph of $y = \sqrt{x}$.
(a) Express the distance d from P to the point $(1, 0)$ as a function of x.
(b) Use a graphing utility to graph $d = d(x)$.
(c) For what values of x is d smallest?

16. Let $P = (x, y)$ be a point on the graph of $y = 1/x$.
(a) Express the distance d from P to the origin as a function of x.
(b) Use a graphing utility to graph $d = d(x)$.
(c) For what values of x is d smallest?

17. Two cars leave an intersection at the same time. One is headed south at a constant speed of 30 miles per hour; the other is headed west at a constant speed of 40 miles per hour (see the figure). Express the distance d between the cars as a function of the time t.

[**Hint:** At $t = 0$, the cars leave the intersection.]

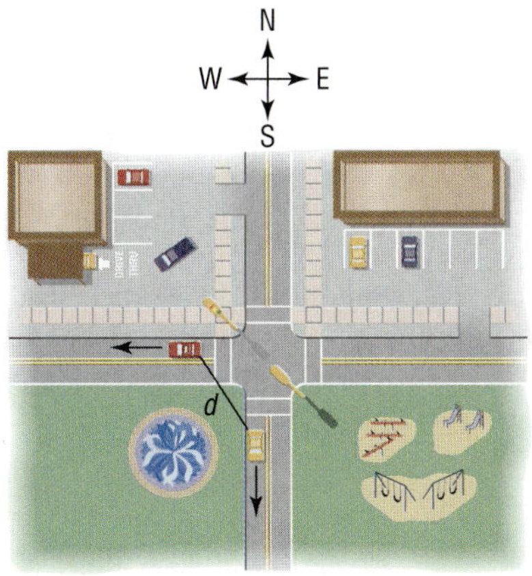

18. Two cars are approaching an intersection. One is 2 miles south of the intersection and is moving at a constant speed of 30 miles per hour. At the same time, the other car is 3 miles east of the intersection and is moving at a constant speed of 40 miles per hour.

(a) Express the distance d between the cars as a function of time T.

[**Hint:** At $t = 0$, the cars are 2 miles south and 3 miles east of the intersection, respectively.]

(b) Use a graphing utility to graph $d = d(t)$. For what value of t is d smallest?

19. Constructing an Open Box An open box with a square base is to be made from a square piece of cardboard 24 inches on a side by cutting out a square from each corner and turning up the sides (see the figure).

(a) Express the volume V of the box as a function of the length x of the side of the square cut from each corner.

(b) Graph $V = V(x)$. For what value of x is V largest?

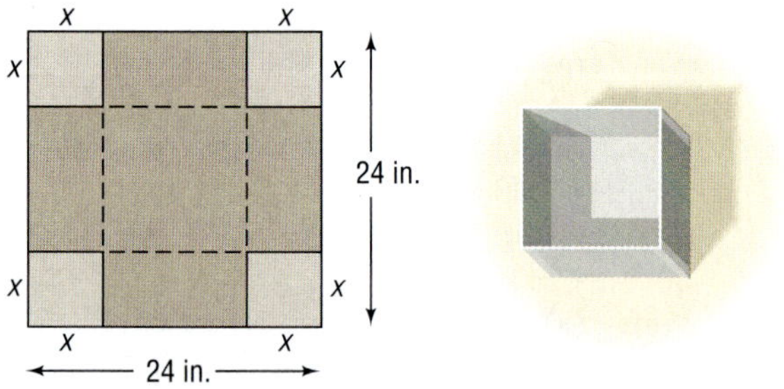

20. Constructing an Open Box An open box with a square base is required to have a volume of 10 cubic feet.

(a) Express the amount A of material used to make such a box as a function of the length x of a side of the square base.

(b) Graph $A = A(x)$. For what value of x is A smallest?

21. Constructing a Closed Box A closed box with a square base is required to have a volume of 10 cubic feet.

(a) Express the amount A of material used to make such a box as a function of the length x of a side of the square base.

(b) Graph $A = A(x)$. For what value of x is A smallest?

22. Spheres The volume V of a sphere of radius r is $V = \frac{4}{3}\pi r^3$; the surface area S of this sphere is $S = 4\pi r^2$. Express the volume V as a function of the surface area S. If the surface area doubles, how does the volume change?

23. A rectangle has one corner on the graph of $y = 16 - x^2$, another at the origin, a third on the positive y-axis, and the fourth on the positive x-axis (see the figure).

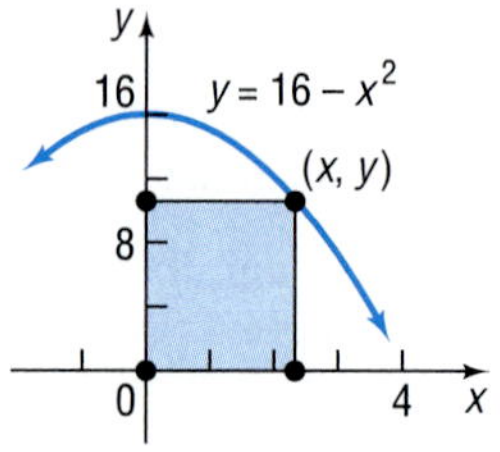

(a) Express the area A of the rectangle as a function of x.

(b) What is the domain of A?

(c) Graph $A = A(x)$. For what value of x is A largest?

24. A rectangle is inscribed in a semicircle of radius 2 (see the figure). Let $P = (x, y)$ be the point in quadrant I that is a vertex of the rectangle and is on the circle.

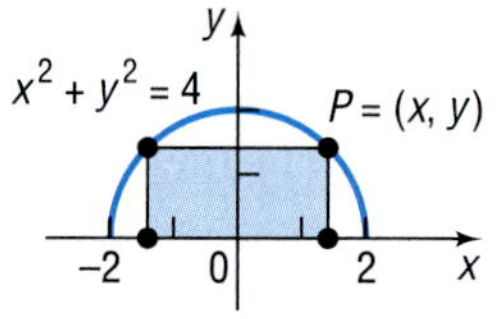

(a) Express the area A of the rectangle as a function of x.

(b) Express the perimeter p of the rectangle as a function of x.

(c) Graph $A = A(x)$. For what value of x is A largest?

(d) Graph $p = p(x)$. For what value of x is p largest?

25. A rectangle is inscribed in a circle of radius 2 (see the figure). Let $P = (x, y)$ be the point in quadrant I that is a vertex of the rectangle and is on the circle.

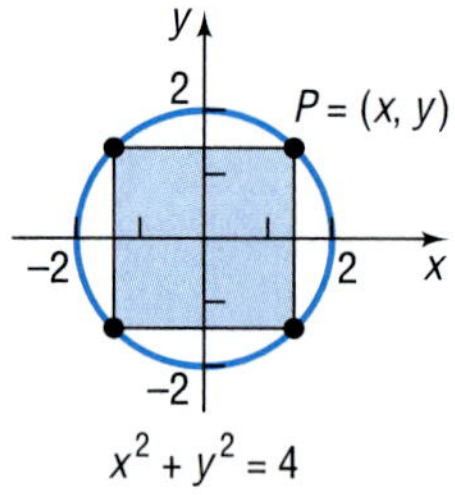

(a) Express the area A of the rectangle as a function of x.

(b) Express the perimeter p of the rectangle as a function of x.

(c) Graph $A = A(x)$. For which value of x is A largest?

(d) Graph $p = p(x)$. For what value of x is p largest?

26. A circle of radius r is inscribed in a square (see the figure).

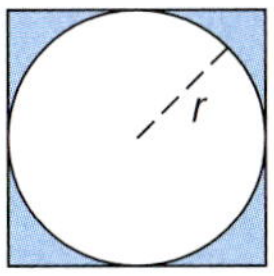

(a) Express the area A of the square as a function of the radius r of the circle.
(b) Express the perimeter p of the square as a function of r.

27. A wire 10 meters long is to be cut into two pieces. One piece will be shaped as a square, and the other piece will be shaped as a circle (see the figure).

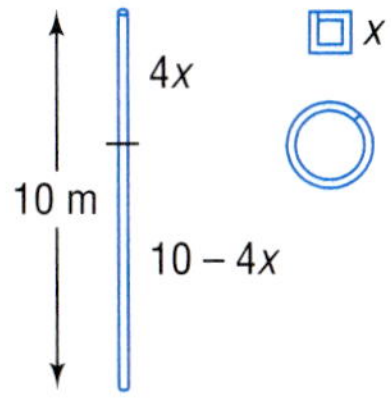

(a) Express the total area A enclosed by the pieces of wire as a function of the length x of a side of the square.
(b) What is the domain of A?
(c) Graph $A = A(x)$. For what value of x is A smallest?

28. A wire 10 meters long is to be cut into two pieces. One piece will be shaped as an equilateral triangle, and the other piece will be shaped as a circle.
(a) Express the total area A enclosed by the pieces of wire as a function of the length x of a side of the equilateral triangle.
(b) What is the domain of A?
(c) Graph $A = A(x)$. For what value of x is A smallest?

29. A semicircle of radius r is inscribed in a rectangle so that the diameter of the semicircle is the length of the rectangle (see the figure).

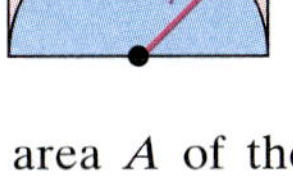

(a) Express the area A of the rectangle as a function of the radius r of the semicircle.
(b) Express the perimeter p of the rectangle as a function of r.

30. An equilateral triangle is inscribed in a circle of radius r. See the figure. Express the circumference C of the circle as a function of the length x of a side of the triangle.
[**Hint:** First show that $r^2 = x^2/3$.]

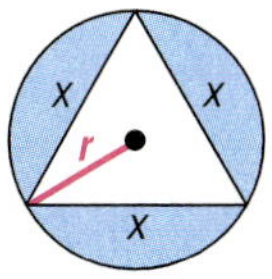

31. An equilateral triangle is inscribed in a circle of radius r. See the figure. Express the area A within the circle, but outside the triangle, as a function of the length x of a side of the triangle.

32. **Cost of Transporting Goods** A trucking company transports goods between Chicago and New York, a distance of 960 miles. The company's policy is to charge, for each pound, \$0.50 per mile for the first 100 miles, \$0.40 per mile for the next 300 miles, \$0.25 per mile for the next 400 miles, and no charge for the remaining 160 miles.
(a) Graph the relationship between the cost of transportation in dollars and mileage over the entire 960 mile route.
(b) Find the cost as a function of mileage for hauls between 100 and 400 miles from Chicago.
(c) Find the cost as a function of mileage for hauls between 400 and 800 miles from Chicago.

33. **Car Rental Costs** An economy car rented in Florida from National Car Rental® on a weekly basis costs \$95 per week.* Extra days cost \$24 per day until the day rate exceeds the weekly rate, in which case the weekly rate applies. Find the cost C of renting an economy car as a piecewise-defined function of the number x of days used, where $7 \le x \le 14$. Graph this function.
[**Note:** Any part of a day counts as a full day.]

34. Rework Problem 33 for a luxury car, which costs \$219 on a weekly basis with extra days at \$45 per day.

35. Water is poured into a container in the shape of a right circular cone with radius 4 feet and height 16 feet (see the figure). Express the volume V of water in the cone as a function of the height h of the water.
[**Hint:** The volume V of a cone of radius r and height h is $V = \frac{1}{3}\pi r^2 h$.]

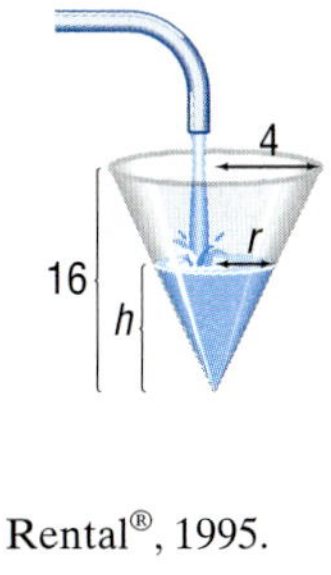

**Source:* National Car Rental®, 1995.

36. **Federal Income Tax** Two 1997 Tax Rate Schedules are given in the accompanying table. If x equals the amount on Form 1040, line 37, and y equals the tax due, construct a function f for each schedule.

1997 TAX RATE SCHEDULES

SCHEDULE X— USE IF YOUR FILING STATUS IS SINGLE

If the amount on Form 1040, line 37, is: Over—	But not over—	Enter on Form 1040, line 38			of the amount over—
$ 0	$ 24,650	$ 0	+	15%	$ 0
24,650	59,750	3,698	+	28	24,650
59,750	124,650	13,526	+	31	59,750
124,650	271,050	33,645	+	36	124,650
271,050		86,349	+	39.6	271,050

SCHEDULE Y-1—USE IF YOUR FILING STATUS IS MARRIED FILING JOINTLY OR QUALIFYING WIDOW(ER)

If the amount on Form 1040, line 37, is: Over—	But not over—	Enter on Form 1040, line 38			of the amount over—
$ 0	$ 41,200	$ 0	+	15%	$ 0
41,200	99,600	6,180	+	28	41,200
99,600	151,750	22,532	+	31	99,600
151,750	271,050	38,699	+	36	151,750
271,050		81,647	+	39.6	271,050

37. **Linear Curve Fitting** The manager of a clothing store collected the following prices and quantity demanded for shirts.

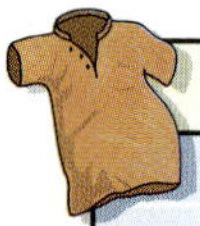

Price ($/Shirt)	Quantity Purchased
15	1000
17	920
18	890
19	870
21	800
23	720
24	670
25	620
26	580

(a) Use your graphing utility to find the line of best fit that relates price and quantity purchased. Treat price as the independent variable and quantity as the dependent variable.
(b) Determine the revenue function $R(p)$, where p is the price.
(c) Graph $R(p)$ using a graphing utility.
(d) What is the revenue maximizing price?
(e) What quantity will maximize revenue? What is the maximum revenue?

38. **Linear Curve Fitting** The manager of a clothing store collected the following prices and quantity demanded for pants.

Price ($/Pair)	Quantity Purchased
29	1000
31	985
32	960
34	920
36	875
37	845
39	790
40	765
41	740
43	695

(a) Use your graphing utility to find the line of best fit that relates price and quantity purchased. Treat price as the independent variable and quantity as the dependent variable.
(b) Determine the revenue function $R(p)$, where p is the price.
(c) Graph $R(p)$ using a graphing utility.
(d) What is the revenue maximizing price?
(e) What quantity will maximize revenue? What is the maximum revenue?

CHAPTER REVIEW

THINGS TO KNOW

Function

A rule or correspondence between two sets of real numbers so that each number x in the first set, the domain, has corresponding to it exactly one number y in the second set. The range is the set of y values of the function for the x values in the domain.

x is the independent variable; y is the dependent variable.

A function f may be defined implicitly by an equation involving x and y or explicitly by writing $y = f(x)$.

A function can also be characterized as a set of ordered pairs (x, y) or $(x, f(x))$ in which no two distinct pairs have the same first element.

Function notation

$y = f(x)$

f is a symbol for the function or rule that defines the function.

x is the argument, or independent variable.

y is the dependent variable.

$f(x)$ is the value of the function at x, or the image of x.

Domain

If unspecified, the domain of a function f is the largest set of real numbers for which the rule defines a real number.

Vertical line test

A set of points in the plane is the graph of a function if and only if every vertical line intersects the graph in at most one point.

Local maximum

A function f has a local maximum at c if there is an interval I containing c so that for all x in I, $f(x) < f(c)$.

Local minimum

A function f has a local minimum at c if there is an interval I containing c so that for all x in I, $f(x) > f(c)$.

Even function f

$f(-x) = f(x)$ for every x in the domain ($-x$ must also be in the domain).

Odd function f

$f(-x) = -f(x)$ for every x in the domain ($-x$ must also be in the domain).

LIBRARY OF FUNCTIONS

Linear function

$f(x) = mx + b$ Graph is a straight line with slope m and y-intercept b.

Constant function

$f(x) = b$ Graph is a horizontal line with y-intercept b (see Figure 26).

Identity function

$f(x) = x$ Graph is a straight line with slope 1 and y-intercept 0 (see Figure 27).

Square function

$f(x) = x^2$ Graph is a parabola with intercept at (0, 0) (see Figure 28).

Cube function

$f(x) = x^3$ See Figure 29.

Square root function

$f(x) = \sqrt{x}$ See Figure 30.

Reciprocal function

$f(x) = 1/x$ See Figure 31.

Absolute value function

$f(x) = |x|$ See Figure 32.

HOW TO

- Determine whether a relation represents a function
- Find the domain and range of a function from its graph
- Find the domain of a function given its equation
- Determine whether a function is even or odd without graphing it
- Graph certain functions by shifting, compressing, stretching, and/or reflecting (see Table 11)
- Use a graphing utility to determine where the graph of a function is increasing or decreasing
- Use a graphing utility to find the local maxima and local minima of a function
- Find the composite of two functions
- Construct functions in applications, including piecewise-defined functions

FILL-IN-THE-BLANK ITEMS

1. If f is a function defined by the equation $y = f(x)$, then x is called the ______________ variable and y is the ______________ variable.

2. A set of points in the xy-plane is the graph of a function if and only if no ______________ line contains more than one point of the set.

3. A(n) ______________ function f is one for which $f(-x) = f(x)$ for every x in the domain of f; a(n) ______________ function f is one for which $f(-x) = -f(x)$ for every x in the domain of f.

4. Suppose that the graph of a function f is known. Then the graph of $y = f(x - 2)$ may be obtained by a(n) ______________ shift of the graph of f to the ______________ a distance of 2 units.

5. If $f(x) = x + 1$ and $g(x) = x^3$, then ______________ $= (x + 1)^3$.

TRUE/FALSE ITEMS

T F **1.** Vertical lines intersect the graph of a function in no more than one point.

T F **2.** The y-intercept of the graph of the function $y = f(x)$ whose domain is all real numbers is $f(0)$.

T F **3.** Even functions have graphs that are symmetric with respect to the origin.

T F **4.** The graph of $y = f(-x)$ is the reflection about the y-axis of the graph of $y = f(x)$.

T F **5.** $f(g(x)) = f(x) \cdot g(x)$

REVIEW EXERCISES

1. Given that f is a linear function, $f(4) = -5$, and $f(0) = 3$, write the equation that defines f.

2. Given that g is a linear function with slope $= -4$ and $g(-2) = 2$, write the equation that defines g.

3. A function f is defined by

$$f(x) = \frac{Ax + 5}{6x - 2}$$

If $f(1) = 4$, find A.

4. A function g is defined by

$$g(x) = \frac{A}{x} + \frac{8}{x^2}$$

If $g(-1) = 0$, find A.

5. Tell which of the following graphs are graphs of functions.

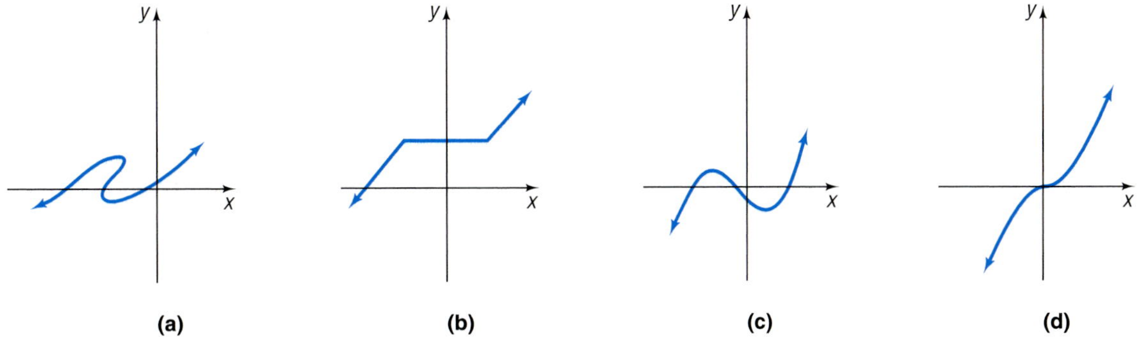

6. Use the graph of the function f shown to find
(a) The domain and range of f
(b) The intervals on which f is increasing
(c) The intervals on which f is constant
(d) The intercepts of f

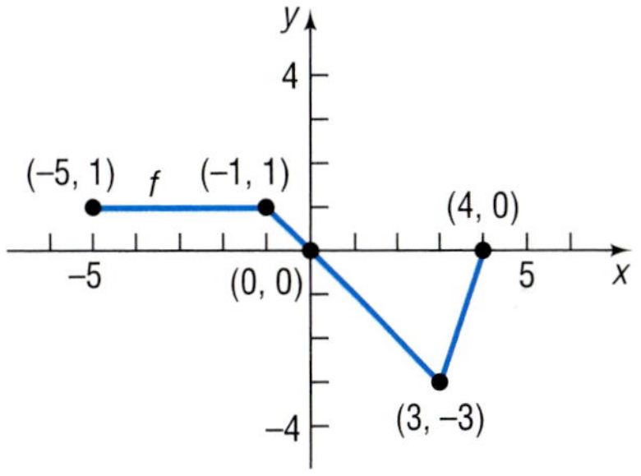

In Problems 7–12, find the following for each function:
(a) $f(-x)$ *(b)* $-f(x)$ *(c)* $f(x+2)$ *(d)* $f(x-2)$

7. $f(x) = \dfrac{3x}{x^2 - 4}$ **8.** $f(x) = \dfrac{x^2}{x + 2}$ **9.** $f(x) = \sqrt{x^2 - 4}$ **10.** $f(x) = |x^2 - 4|$

11. $f(x) = \dfrac{x^2 - 4}{x^2}$ **12.** $f(x) = \dfrac{x^3}{x^2 - 4}$

In Problems 13–18, determine whether the given function is even, odd, or neither without drawing a graph.

13. $f(x) = x^3 - 4x$ **14.** $g(x) = \dfrac{4 + x^2}{1 + x^4}$ **15.** $h(x) = \dfrac{1}{x^4} + \dfrac{1}{x^2} + 1$ **16.** $F(x) = \sqrt{1 - x^3}$

17. $G(x) = 1 - x + x^3$ **18.** $H(x) = 1 + x + x^2$

In Problems 19–30, find the domain of each function.

19. $f(x) = \dfrac{x}{x^2 - 9}$ **20.** $f(x) = \dfrac{3x^2}{x - 2}$ **21.** $f(x) = \sqrt{2 - x}$ **22.** $f(x) = \sqrt{x + 2}$

23. $h(x) = \dfrac{\sqrt{x}}{|x|}$ **24.** $g(x) = \dfrac{|x|}{x}$ **25.** $f(x) = \dfrac{x}{x^2 + 2x - 3}$ **26.** $F(x) = \dfrac{1}{x^2 - 3x - 4}$

27. $G(x) = \begin{cases} |x| & \text{if } -1 \le x \le 1 \\ 1/x & \text{if } x > 1 \end{cases}$ **28.** $H(x) = \begin{cases} 1/x & \text{if } 0 < x < 4 \\ x - 4 & \text{if } 4 \le x \le 8 \end{cases}$

29. $f(x) = \begin{cases} 1/(x-2) & \text{if } x > 2 \\ 0 & \text{if } x = 2 \\ 3x & \text{if } 0 \le x < 2 \end{cases}$ **30.** $g(x) = \begin{cases} |1 - x| & \text{if } x < 1 \\ 3 & \text{if } x = 1 \\ x + 1 & \text{if } 1 < x \le 3 \end{cases}$

In Problems 31–50:

(a) Find the domain of each function.
(b) Locate any intercepts.
(c) Graph each function.
(d) Based on the graph, find the range.
(e) Use a graphing utility to verify your answers.

31. $F(x) = |x| - 4$

32. $f(x) = |x| + 4$

33. $g(x) = -|x|$

34. $g(x) = \frac{1}{2}|x|$

35. $h(x) = \sqrt{x - 1}$

36. $h(x) = \sqrt{x} - 1$

37. $f(x) = \sqrt{1 - x}$

38. $f(x) = -\sqrt{x}$

39. $F(x) = \begin{cases} x^2 + 4 & \text{if } x < 0 \\ 4 - x^2 & \text{if } x \geq 0 \end{cases}$

40. $H(x) = \begin{cases} |1 - x| & \text{if } 0 \leq x \leq 2 \\ |x - 1| & \text{if } x > 2 \end{cases}$

41. $h(x) = (x - 1)^2 + 2$

42. $h(x) = (x + 2)^2 - 3$

43. $g(x) = (x - 1)^3 + 1$

44. $g(x) = (x + 2)^3 - 8$

45. $f(x) = \begin{cases} 2\sqrt{x} & \text{if } x \geq 4 \\ x & \text{if } 0 < x < 4 \end{cases}$

46. $f(x) = \begin{cases} 3|x| & \text{if } x < 0 \\ \sqrt{1 - x} & \text{if } 0 \leq x \leq 1 \end{cases}$

47. $g(x) = \dfrac{1}{x - 1} + 1$

48. $g(x) = \dfrac{1}{x + 2} - 2$

49. $h(x) = \text{int}(-x)$

50. $h(x) = -\text{int}(x)$

In Problems 51–56, for the given functions f and g, find:

(a) $(f \circ g)(2)$ *(b)* $(g \circ f)(-2)$ *(c)* $(f \circ f)(4)$ *(d)* $(g \circ g)(-1)$

51. $f(x) = 3x - 5;\ g(x) = 1 - 2x^2$

52. $f(x) = 4 - x;\ g(x) = 1 + x^2$

53. $f(x) = \sqrt{x + 2};\ g(x) = 2x^2 + 1$

54. $f(x) = 1 - 3x^2;\ g(x) = \sqrt{4 - x}$

55. $f(x) = \dfrac{1}{x^2 + 4};\ g(x) = 3x - 2$

56. $f(x) = \dfrac{2}{1 + 2x^2};\ g(x) = 3x$

In Problems 57–62, find $f \circ g$, $g \circ f$, $f \circ f$, *and* $g \circ g$ *for each pair of functions. State the domain of each.*

57. $f(x) = \dfrac{2 - x}{x};\ g(x) = 3x + 1$

58. $f(x) = \dfrac{2x}{x + 1};\ g(x) = \dfrac{2x}{x - 1}$

59. $f(x) = 3x^2 + x + 1;\ g(x) = |3x|$

60. $f(x) = \sqrt{3x};\ g(x) = 1 + x + x^2$

61. $f(x) = \dfrac{x + 1}{x - 1};\ g(x) = \dfrac{1}{x}$

62. $f(x) = \sqrt{x - 3};\ g(x) = \sqrt{3 - x}$

63. For the graph of the function f shown below:

(a) Draw the graph of $y = f(-x)$
(b) Draw the graph of $y = -f(x)$.
(c) Draw the graph of $y = f(x + 2)$.
(d) Draw the graph of $y = f(x) + 2$.
(e) Draw the graph of $y = f(2 - x)$.

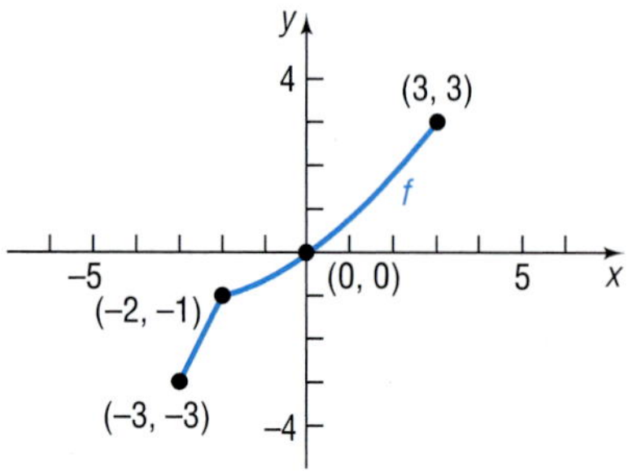

64. Repeat Problem 63 for the graph of the function g shown below.

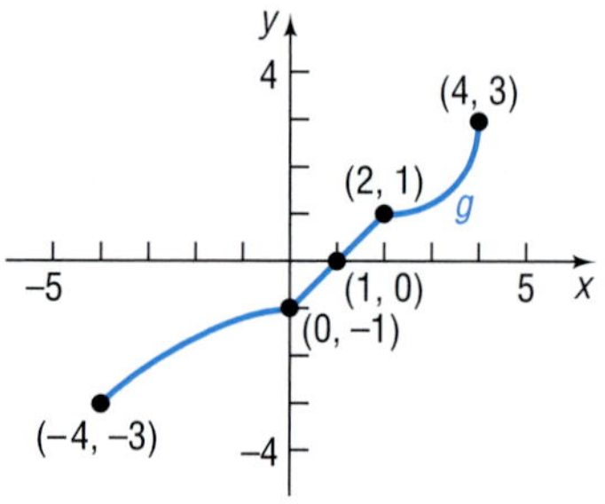

65. Temperature Conversion The temperature T of the air is approximately a linear function of the altitude h for altitudes within 10,000 meters of the surface of Earth. If the surface temperature is 30°C and the temperature at 10,000 meters is 5°C, find the function $T = T(h)$.

66. Speed as a Function of Time The speed v (in feet per second) of a car is a linear function of the time t (in seconds) for $10 \leq t \leq 30$. If after each second the speed of the car has increased by 5 feet per second, and if after 20 seconds the speed is 80 feet per second, how fast is the car going after 30 seconds? Find the function $v = v(t)$.

67. Strength of a Beam The strength of a rectangular wooden beam is proportional to the product of the width and the cube of its depth (see the figure). If the beam is to be cut from a log in the shape of a cylinder of radius 3 feet, express the strength S of the beam as a function of the width x. What is the domain of S?

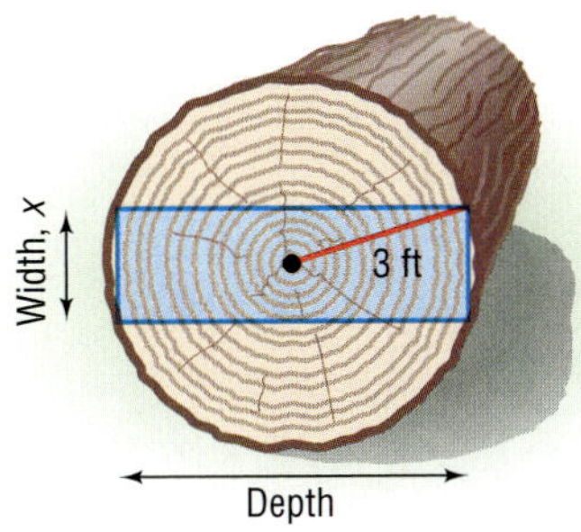

68. Productivity vs. Earnings The data below represents the average hourly earnings and productivity (output per hour) of production workers for the years 1986–1995. Let productivity x be the independent variable and average hourly earnings y be the dependent variable.

Productivity	Average Hourly Earnings
94.2	8.76
94.1	8.98
94.6	9.28
95.3	9.66
96.1	10.01
96.7	10.32
100	10.57
100.2	10.83
100.7	11.12
100.8	11.44

Source: Bureau of Labor Statistics

(a) Draw a scatter diagram of the data by hand.
(b) Draw a line from the point (94.2, 8.76) to (96.7, 10.32) on the scatter diagram found in (a).
(c) Find the average rate of change of hourly earnings for productivity between 94.2 and 96.7.
(d) Interpret the average rate of change found in (c).
(e) Draw a line from the point (96.7, 10.32) to (100.8, 11.44) on the scatter diagram found in (a).
(f) Find the average rate of change of hourly earnings for productivity between 96.7 and 100.8.
(g) Interpret the average rate of change found in (f).
(h) What is happening to the average rate of change of hourly earnings as productivity increases?

69. Speed of a Parachutist The following data represents the distance a parachutist has fallen over time, ignoring air resistance.

Time (Seconds)	Distance (Meters)
0	0
5	112.5
10	490
15	1102.5
20	1960

(a) Draw a scatter diagram of the data by hand treating time as the independent variable.
(b) Draw a line from the point (0, 0) to (5, 112.5).
(c) Find the average rate of change of distance between 0 and 5 seconds.
(d) Interpret the average rate of change found in (c).
(e) Draw a line from the points (15, 1102.5) to (20, 1960).
(f) Find the average rate of change of distance between 15 and 20 seconds.
(g) Interpret the average rate of change found in (f).
(h) What is happening to the average rate of change of distance as time passes?

CHAPTER

3 Equations and Inequalities

Environmental issues are consistently in the news these days with recycling often being the topic of discussion. On the following page you are asked to look into the impact of recycling newspaper. Use the Sullivan website at:

www.prenhall.com/sullivan

to help link you to the Internet resources you will need to answer the questions asked.

PREPARING FOR THIS CHAPTER

Before getting started on this chapter, review the following concepts:

Solving Equations *(Appendix, Section 4)*
Completing the Square *(Appendix, Section 5)*
Factoring Polynomials *(Appendix, Section 2)*
Inequalities; Intervals *(Appendix, Section 1)*
Absolute Value *(Appendix, Section 1)*

OUTLINE

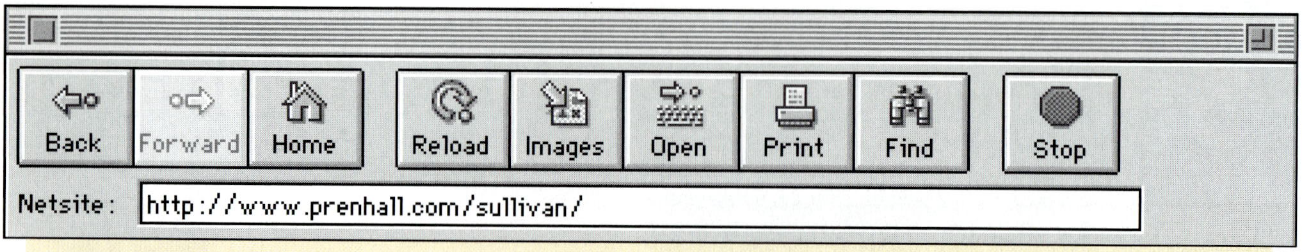

PLAN IT! EARTH RECYCLING

You work for a great company, Plan It! Earth Recycling. As a part of the company's website development team, your supervisor has just given you the assignment to design the webpage for paper recycling. You have some great ideas about how to make an interactive site that radically shows how important it is to recycle paper. You intend to create some graphics that hit home on how many trees are saved by recycling. You begin your research by browsing the Internet to find information on paper recycling. You wonder how many trees it takes to make a single edition of the Sunday *New York Times.* How many trees could be saved if it was printed on recycled paper?

1. Estimate the number of *average size trees* that are cut each day just for the making of paper. Estimate the number of trees needed to make one Sunday edition of the *New York Times.*
2. What are some of the environmental stresses created by the making of paper? We have been told that recycling is good but *does recycling really work?* Can you be objective about this question, or have you decided on the matter regardless of what evidence you might hear?
3. What do the *foresters* have to say? Is there a conflict of interest here? Look at the *Forest Service Data* on the Web.
4. Determine a line of best fit that relates the amount of paper needed to make a Sunday edition of the *New York Times* for each month for one year.
5. Determine a line of best fit that relates the total accumulated tons of paper needed to produce the *New York Times* for one year on a monthly basis.
6. Use the "fact" that 278 tons of paper is roughly equivalent to harvesting 2406 trees. Use the equation found in question 4 to determine the number of trees required for a Sunday edition of the *New York Times* in January. How many trees are required to produce the Sunday edition of the *New York Times* for one year?
7. Determine a line of best fit that relates the amount of paper needed, with recycling, to produce the Sunday edition of the *New York Times* each month for one year.

The investigation of equations and their solutions has played a central role in algebra for many hundreds of years. In fact, as early as 2000 B.C., the Babylonians had a well-developed algebra that included a solution for quadratic equations (see the Historical Feature in Section 3.2). Of late, the study of inequalities (Sections 3.5–3.6) has become equally important. For example, the problem of optimizing airline scheduling leads to linear inequalities. As you go through this chapter, you will see evidence of the importance of the topics presented here for solving problems that occur in business, engineering, and the social and natural sciences.

We shall limit our discussion in this chapter to equations and inequalities containing a single variable. Later, we will study equations and inequalities containing more than one variable.

3.1 SOLVING EQUATIONS USING A GRAPHING UTILITY

1. Use the Intermediate Value Theorem
2. Solve Equations Using a Graphing Utility
3. Approximate Solutions of Equations

You may wish to read *Solving Equations* in the Appendix, Section 4, before going on.

In solving the equation $2x = 6$, we are able to obtain an exact solution, $x = 3$. To solve the equation $x^3 = 25x$ requires some additional work:

$$\begin{aligned} x^3 &= 25x \\ x^3 - 25x &= 0 && \text{Rewrite with 0 on the right side.} \\ x(x^2 - 25) &= 0 && \text{Factor.} \\ x(x-5)(x+5) &= 0 && \text{Factor.} \\ x = 0 \quad x - 5 = 0 &\quad x + 5 = 0 && \text{Set each factor equal to 0.} \\ x = 0 \quad x = 5 &\quad x = -5 && \text{Solve.} \end{aligned}$$

Again, exact solutions are obtained.

We shall see as we proceed through this book that some equations can be solved using algebraic techniques that result in exact solutions being obtained. Whenever algebraic techniques exist, we shall discuss them and use them to obtain exact solutions.

For many equations, though, there are no algebraic techniques that lead to a solution. For such equations, a graphing utility can often be used to investigate possible solutions. When a graphing utility is used to solve an equation, usually *approximate* solutions are obtained. Unless otherwise stated, we shall follow the practice of giving approximate solutions as decimals *correct to two decimal places.*

Let's look at an example that explains this terminology.

EXAMPLE 1 Writing a Number Correct to n Decimal Places

WRITING THE NUMBER	1/7	$\sqrt{2}$	π
Correct to 1 decimal place	0.1	1.4	3.1
Correct to 2 decimal places	0.14	1.41	3.14
Correct to 3 decimal places	0.142	1.414	3.141
Correct to 4 decimal places	0.1428	1.4142	3.1415
Correct to 5 decimal places	0.14285	1.41421	3.14159
Correct to 6 decimal places	0.142857	1.414213	3.141592

As the example illustrates, **correct to *n* decimal places** means the decimal that results from truncation after the nth decimal.

Now work Problem 1.

We learned in Chapter 2 that the x-intercepts, if any, of a function $y = f(x)$ are found by letting $y = 0$ and solving the equation $f(x) = 0$. Thus, solving the equation $f(x) = 0$ is equivalent to finding the x-intercepts of the graph of the function $y = f(x)$. It is this relationship between the solutions of an equation and the graph of a function that enables us to use a graphing utility to solve equations.

We can be sure the graph of a function has an x-intercept by using the Intermediate Value Theorem.

Intermediate Value Theorem

1 The Intermediate Value Theorem requires that the function be *continuous*. Although it requires calculus to explain precisely the meaning, the *idea* of a continuous function is easy to understand. Very basically, a function f is continuous when its graph can be drawn without lifting pencil from paper, that is, when the graph contains no "holes" or "jumps" or "gaps."

Intermediate Value Theorem

Let f denote a continuous function. If $a < b$ and if $f(a)$ and $f(b)$ are of opposite sign, then the graph of f has at least one x-intercept between a and b.

Although the proof of this result requires advanced methods in calculus, it is easy to "see" why the result is true. Look at Figure 1.

FIGURE 1
If $f(a) < 0$ and $f(b) > 0$, there is an x-intercept between a and b.

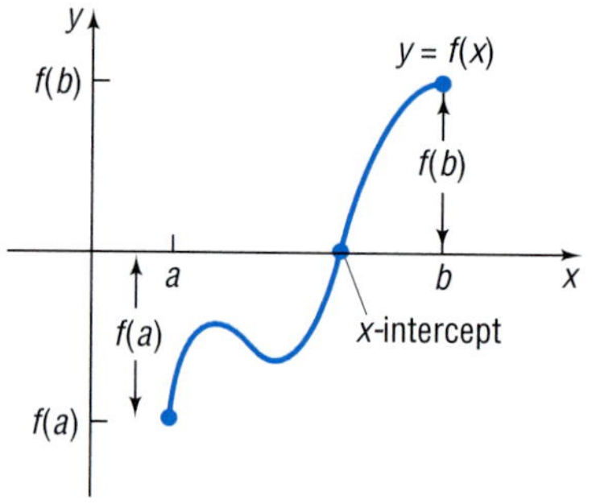

Since most of the functions we deal with are continuous, the Intermediate Value Theorem together with a graphing utility provide a basis for finding x-intercepts and solving equations.

EXAMPLE 2 Using the Intermediate Value Theorem and a Graphing Utility to Locate x-Intercepts

Show that the graph of $f(x) = x^5 - x^3 - 1$ has an x-intercept between 1 and 2.

Solution Using a graphing utility, graph f. TRACE along the function. As the cursor moves from $x = 1$ to $x = 2$, notice the value of y changes from a negative value to a positive value (Figure 2(a) and (b)). According to the Intermediate Value Theorem, there is an x-intercept between $x = 1$ and $x = 2$. In fact, there is an x-intercept between 1.212766 and 1.2553191.

FIGURE 2

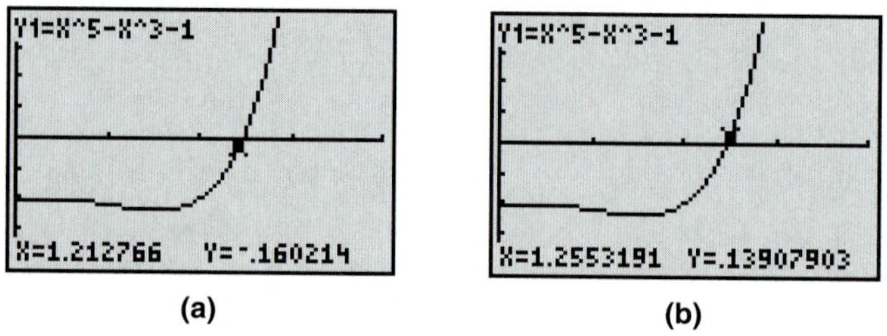

Now work Problem 7.

Solving Equations

EXAMPLE 3

Using a Graphing Utility to Approximate Solutions of an Equation

2

Find the smaller of the two solutions of the equation $x^2 - 6x + 7 = 0$. Express the answer correct to two decimal places.

Solution The solutions of the equation $x^2 - 6x + 7 = 0$ are the same as the x-intercepts of the function $f(x) = x^2 - 6x + 7$. We begin by graphing the function f using a scale of 1 on each axis.

Figure 3 shows the graph.

FIGURE 3

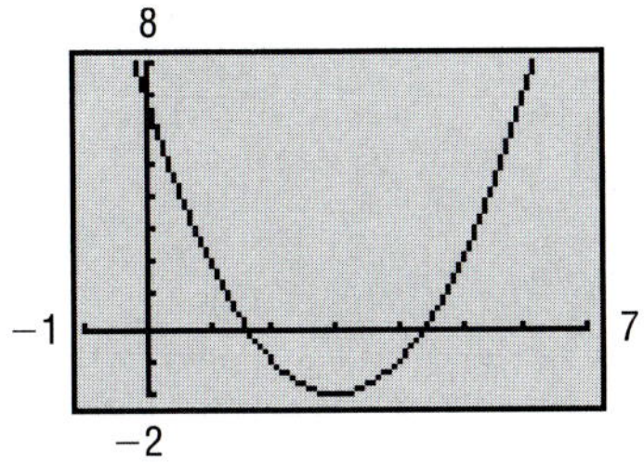

The smaller of the two x-intercepts (solutions of the equation) lies between 1 and 2. (Remember that $X\text{scl} = 1$.) Thus, correct to 0 decimal places, the smaller solution of the equation is $x = 1$.

Next, we BOX the graph from approximately $x = 1$ to $x = 2$ and from $y = -1$ to $y = 1$. See Figure 4. Adjust the RANGE setting so that $X\text{scl} = 0.1$ and $Y\text{scl} = 0.1$. Graph f again. See Figure 5.

FIGURE 4 FIGURE 5

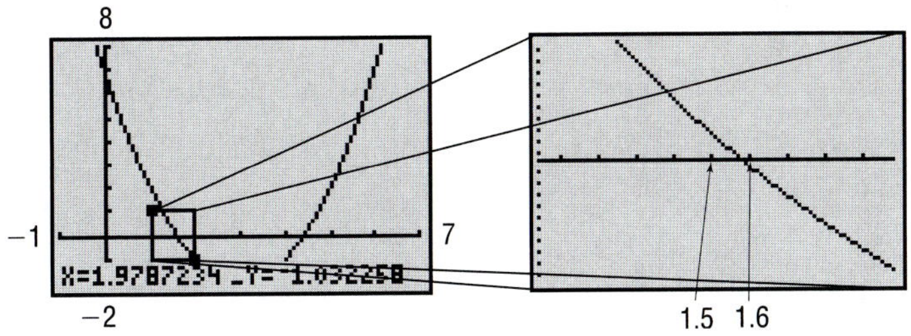

The x-intercept lies between 1.5 and 1.6. (Remember that the BOX was from $x = 1$ to $x = 2$ and $X\text{scl} = 0.1$.) Thus, correct to one decimal place, the smaller solution of the equation is $x = 1.5$.

Next, BOX again from $x = 1.5$ to $x = 1.6$ and from $y = -0.1$ to $y = 0.1$. See Figure 6. Adjust the RANGE so that $X\text{scl} = 0.01$ and $Y\text{scl} = 0.01$ and graph f. See Figure 7.

FIGURE 6 FIGURE 7

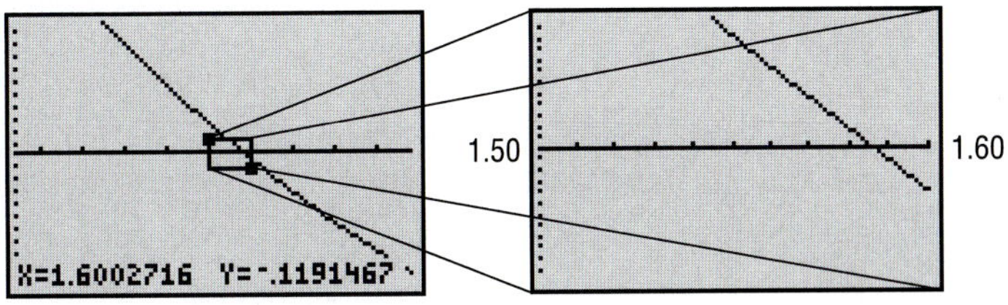

The x-intercept lies between 1.58 and 1.59. Remember that the BOX was from $x = 1.5$ to $x = 1.6$ with $X\text{scl} = 0.01$. Thus, correct to two decimal places, the smaller of the two solutions is $x = 1.58$. ■

3

The steps to follow for approximating solutions of equations correct to any desired number of decimals are given next.

Steps for Approximating Solutions of Equations

STEP 1: Write the equation in the form $f(x) = 0$.
STEP 2: Graph the function f, using $X\text{scl} = 1$.
STEP 3: BOX the graph so that the x-intercept is within the BOX. Adjust Xscl and Yscl to one-tenth of their current value and graph f again.
STEP 4: If additional accuracy is desired, repeat step 3.

In these steps, each repetition of step 3 gives one more decimal place of accuracy. Thus, when $X\text{scl} = 0.01$, the x-intercept will be known correct to two decimal places. Let's look at another example.

EXAMPLE 4 Using a Graphing Utility to Approximate the Solution(s) of the Equation

$$x^3 - 1.5x^2 + x - 1.2 = 0$$

Solution STEP 1: The equation is already in the form $f(x) = 0$.

STEP 2: Graph the function

$$f(x) = x^3 - 1.5x^2 + x - 1.2$$

FIGURE 8

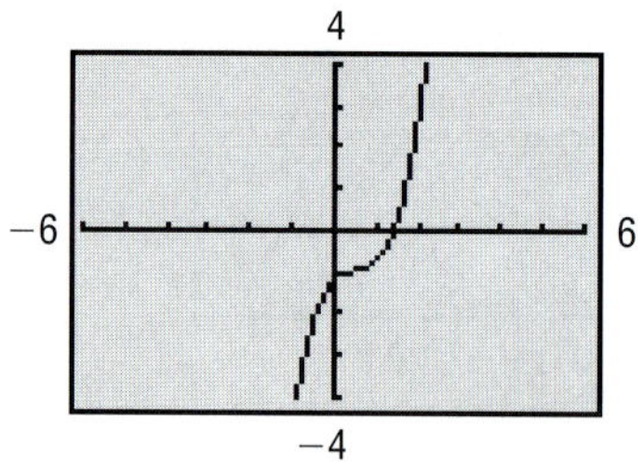

Figure 8 shows the graph of f. Notice that the scale is 1 on each axis. Also notice that the x-intercept lies between 1 and 2.

STEP 3: Figure 9 shows the box chosen. Notice that its width is about 1 unit and its height is about 2 units. Change $X\text{scl} = 0.1$, $Y\text{scl} = 0.1$ and graph.

Figure 10 shows the result. Xscl and Yscl have both been changed to 0.1. From Figure 10 we see the x-intercept is near 1.4.

FIGURE 9 FIGURE 10

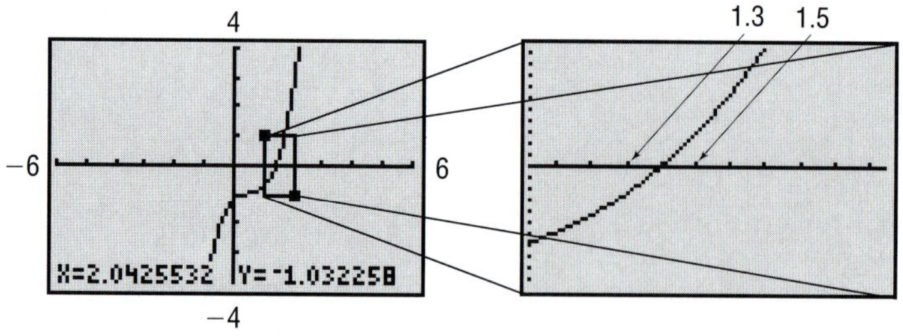

STEP 4: We repeat step 3, using a BOX from approximately $x = 1.38$ to approximately $x = 1.41$ and changing $X\text{scl} = 0.01$ and $Y\text{scl} = 0.01$. This gives us Figure 11. The x-intercept is 1.39, correct to two decimal places.

FIGURE 11

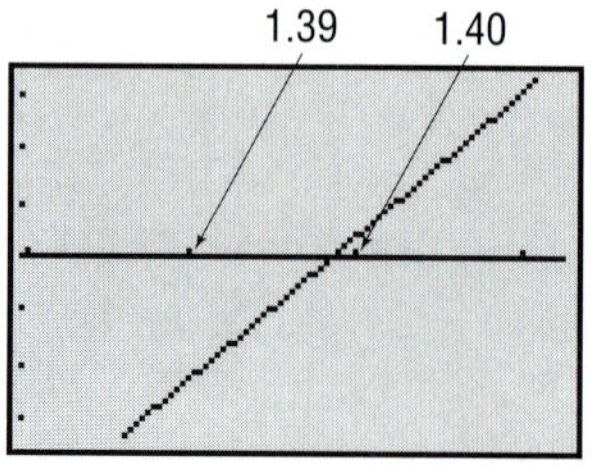

FIGURE 12

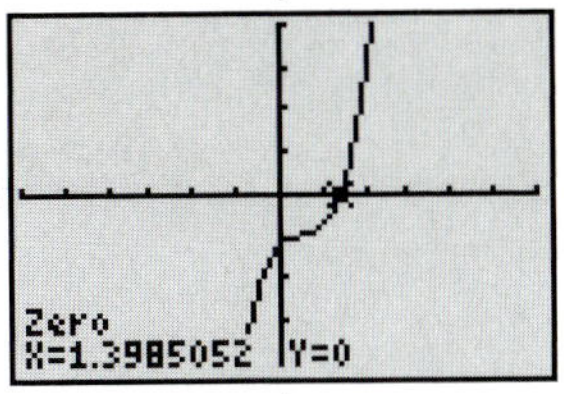

Note: If the x-intercept you are trying to approximate appears to fall on a tick mark, then you should evaluate the function at that tick mark. If the value is zero, then there is an exact solution. If the value is not zero, then you know whether the graph is above or below the x-axis at the tick mark by noting the sign at the tick mark.

If your graphing utility has a ZERO or ROOT command, you can use it to find the solution of an equation. Using the ZERO command on a TI-83, the solution of the equation $x^3 - 1.5x^2 + x - 1.2 = 0$ (Example 4) is given as $x = 1.3985052$ which is correct to six decimal places. See Figure 12.

3.1 EXERCISES

In Problems 1–6, write each expression as a decimal correct to two decimal places.

1. $\frac{3}{7}$ **2.** $\frac{2}{11}$ **3.** $\sqrt{5}$ **4.** $\sqrt{6}$ **5.** $\sqrt[3]{2}$ **6.** $\sqrt[3]{3}$

In Problems 7–12, use the Intermediate Value theorem to show that the graph of each function has an x-intercept in the given interval. Approximate the x-intercept correct to two decimal places.

7. $f(x) = 8x^4 - 2x^2 + 5x - 1;\quad [0, 1]$

8. $f(x) = x^4 + 8x^3 - x^2 + 2;\quad [-1, 0]$

9. $f(x) = 2x^3 + 6x^2 - 8x + 2;\quad [-5, -4]$

10. $f(x) = 3x^3 - 10x + 9;\quad [-3, -2]$

11. $f(x) = x^5 - x^4 + 7x^3 - 7x^2 - 18x + 18;\quad [1.4, 1.5]$

12. $f(x) = x^5 - 3x^4 - 2x^3 + 6x^2 + x + 2;\quad [1.7, 1.8]$

In Problems 13–18, use a graphing utility to approximate the smaller of the two solutions of each equation. Express the answer correct to two decimal places.

13. $x^2 + 4x + 2 = 0$ **14.** $x^2 + 4x - 3 = 0$ **15.** $2x^2 + 4x + 1 = 0$

16. $3x^2 + 5x + 1 = 0$ **17.** $2x^2 - 3x - 1 = 0$ **18.** $2x^2 - 4x - 1 = 0$

In Problems 19–26, use a graphing utility to approximate the ***positive*** *solutions for each equation. Express the answer correct to two decimal places.*

19. $x^3 + 3.2x^2 - 16.83x - 5.31 = 0$

20. $x^3 + 3.2x^2 - 7.25x - 6.3 = 0$

21. $x^4 - 1.4x^3 - 33.71x^2 + 23.94x + 292.41 = 0$

22. $x^4 + 1.2x^3 - 7.46x^2 - 4.692x + 15.2881 = 0$

23. $\pi x^3 - (8.88\pi + 1)x^2 - (42.066\pi - 8.88)x + 42.066 = 0$

24. $\pi x^3 - (5.63\pi + 2)x^2 - (108.392\pi - 11.26)x + 216.784 = 0$

25. $x^3 + 19.5x^2 - 1021x + 1000.5 = 0$

26. $x^3 + 14.2x^2 - 4.8x - 12.4 = 0$

3.2 LINEAR AND QUADRATIC EQUATIONS

1 Solve a Linear Equation Algebraically
2 Solve a Quadratic Equation By Factoring
3 Solve a Quadratic Equation Using a Graphing Utility
4 Solve a Quadratic Equation By Extracting the Root
5 Solve a Quadratic Equation Using the Quadratic Formula

In this section we discuss two types of equations that can be solved algebraically to obtain exact solutions.

Linear Equations

1 *Linear equations* are equations such as

$$3x + 12 = 0 \qquad -2x + 5 = 0 \qquad 4x - 3 = 0$$

A general definition is given next.

A **linear equation in one variable** is equivalent to an equation of the form

$$ax + b = 0$$

where a and b are real numbers and $a \neq 0$.

Sometimes, a linear equation is called a **first-degree equation,** because the left side is a polynomial in x of degree 1.

It is relatively easy to solve a linear equation:

$$\begin{aligned} ax + b &= 0 \\ ax &= -b && \text{Subtract } b \text{ from both sides.} \\ x &= \frac{-b}{a} && \text{Divide both sides by } a,\ a \neq 0. \end{aligned}$$

The linear equation $ax + b = 0$ has the single solution given by the formula $x = -b/a$.

Whenever it is possible to solve a linear equation in your head, do so. For example,

The solution of $2x = 8$ is $x = 4$.

The solution of $3x - 15 = 0$ is $x = 5$.

Often, though, some rearrangement of the terms is necessary.

EXAMPLE 1

Solving a Linear Equation

Solve the equation $\frac{1}{2}(p + 5) - 3 = \frac{1}{3}(2p - 1)$.

Solution To clear the equation of fractions, we multiply both sides by 6, the least common multiple of the denominators of the fractions $\frac{1}{2}$ and $\frac{1}{3}$.

$$\begin{aligned} \frac{1}{2}(p + 5) - 3 &= \frac{1}{3}(2p - 1) \\ 6\left[\frac{1}{2}(p + 5) - 3\right] &= 6\left[\frac{1}{3}(2p - 1)\right] && \text{Multiply both sides by 6, the LCM of 2 and 3.} \\ 3(p + 5) - 18 &= 2(2p - 1) && \text{Use the distributive property on the left and the associative property on the right.} \\ 3p + 15 - 18 &= 4p - 2 && \text{Use the distributive property.} \\ 3p - 3 &= 4p - 2 && \text{Combine like terms.} \\ 3p &= 4p + 1 && \text{Add 3 to each side.} \\ -p &= 1 && \text{Subtract } 4p \text{ from each side.} \\ p &= -1 && \text{Multiply both sides by } -1. \end{aligned}$$

Check: $\frac{1}{2}(p + 5) - 3 = \frac{1}{2}(-1 + 5) - 3 = \frac{1}{2}(4) - 3 = 2 - 3 = -1$

$\frac{1}{3}(2p - 1) = \frac{1}{3}(-2 - 1) = \frac{1}{3}(-3) = -1$

Since the two expressions are equal, the solution $p = -1$ checks.

EXAMPLE 2 Solving a Linear Equation Using a Calculator

Solve the equation $2.78x + \dfrac{2}{17.931} = 54.06$.

Express the answer correct to two decimal places.

Solution To avoid rounding errors, we solve for x before using the calculator. We proceed as in Example 1.

$$2.78x + \frac{2}{17.931} = 54.06$$

$$2.78x = 54.06 - \frac{2}{17.931} \qquad \text{Subtract 2/17.931 from each side.}$$

$$x = \frac{54.06 - (2/17.931)}{2.78} \qquad \text{Divide each side by 2.78.}$$

FIGURE 13

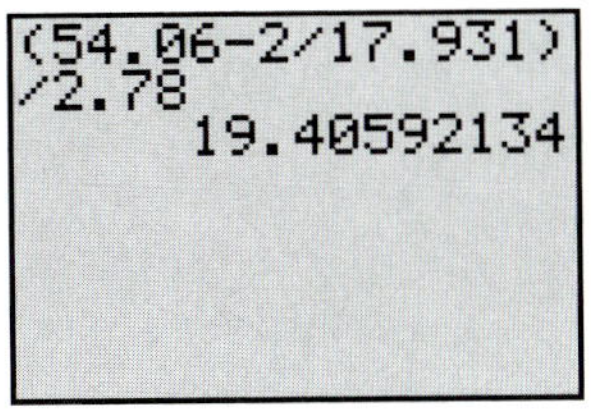

Now use your calculator. See Figure 13.

The solution, correct to two decimal places, is 19.40.

Check: We store the solution in memory and proceed to check.

$$(2.78)(19.40592134) + (2/17.931) = 54.06$$

Now work Problems 19 and 67.

Equations That Lead to Linear Equations

The next two examples illustrate the solution of equations that are not linear but lead to linear equations upon simplification.

EXAMPLE 3 Solving Equations

Solve the equation $(2y + 1)(y - 1) = (y + 5)(2y - 5)$.

Solution

$$(2y + 1)(y - 1) = (y + 5)(2y - 5)$$

$$2y^2 - y - 1 = 2y^2 + 5y - 25 \qquad \text{Multiply, and combine like terms.}$$

$$-y - 1 = 5y - 25 \qquad \text{Subtract } 2y^2 \text{ from each side.}$$

$$-y = 5y - 24 \qquad \text{Add 1 to each side.}$$

$$-6y = -24 \qquad \text{Subtract } 5y \text{ from each side.}$$

$$y = 4 \qquad \text{Divide both sides by } -6.$$

Check: $(2y + 1)(y - 1) = (8 + 1)(3) = (9)(3) = 27$

$(y + 5)(2y - 5) = (4 + 5)(8 - 5) = (9)(3) = 27$

Since the two expressions are equal, the solution $y = 4$ checks.

EXAMPLE 4 Solving Equations

Solve the equation $\dfrac{3}{x - 2} = \dfrac{1}{x - 1} + \dfrac{7}{(x - 1)(x - 2)}$.

Solution First, we note that the domain of the variable is $\{x | x \neq 1, x \neq 2\}$. As we did in Example 1, we clear the equation of fractions by multiplying both sides by the least common multiple of the denominators of the three fractions $(x - 1)(x - 2)$:

$$\frac{3}{x-2} = \frac{1}{x-1} + \frac{7}{(x-1)(x-2)}$$

$$(x-1)\cancel{(x-2)}\frac{3}{\cancel{x-2}} = (x-1)(x-2)\left[\frac{1}{x-1} + \frac{7}{(x-1)(x-2)}\right]$$

Multiply both sides by $(x-1)(x-2)$. Cancel on the left.

$$3x - 3 = \cancel{(x-1)}(x-2)\frac{1}{\cancel{x-1}} + \cancel{(x-1)}\cancel{(x-2)}\frac{7}{\cancel{(x-1)}\cancel{(x-2)}}$$

Use the distributive property on each side; cancel on the right.

$$3x - 3 = (x-2) + 7$$

$$3x - 3 = x + 5$$

Combine like terms.

$$2x = 8$$

Add 3 to each side. Subtract x from each side.

$$x = 4$$

Divide by 2.

Check: $\frac{3}{x-2} = \frac{3}{4-2} = \frac{3}{2}$

$$\frac{1}{x-1} + \frac{7}{(x-1)(x-2)} = \frac{1}{3} + \frac{7}{(3)(2)} = \frac{1}{3} + \frac{7}{6} = \frac{9}{6} = \frac{3}{2}$$

Since the two expressions are equal, the solution $x = 4$ checks. ■

Now work Problem 33.

SUMMARY

Steps for Solving a Linear Equation Algebraically

To solve a linear equation, follow these steps:

STEP 1: If necessary, clear the equation of fractions by multiplying both sides by the Least Common Multiple (LCM) of the denominators of all the fractions.

STEP 2: Remove all parentheses and simplify.

STEP 3: Collect all terms containing the variable on one side and all other terms on the other side.

STEP 4: Check your solution(s).

Quadratic Equations

Quadratic equations are equations such as

$$2x^2 + x + 8 = 0 \qquad 3x^2 - 5x + 6 = 0 \qquad x^2 - 9 = 0$$

A general definition is given next.

A **quadratic equation** is an equation equivalent to one of the form

$$ax^2 + bx + c = 0 \qquad (1)$$

where a, b, and c are real numbers and $a \neq 0$.

A quadratic equation written in the form $ax^2 + bx + c = 0$ is said to be in **standard form.**

Sometimes, a quadratic equation is called a **second-degree equation,** because the left side is a polynomial of degree 2. We shall discuss three ways of solving quadratic equations algebraically: by factoring, by extracting the root, and by use of the quadratic formula. We shall also solve quadratic equations using a graphing utility.

Factoring

2 When a quadratic equation is written in standard form, $ax^2 + bx + c = 0$, it may be possible to factor the expression on the left side as the product of two first-degree polynomials. Then, by setting each factor equal to 0 and solving the resulting linear equations, we obtain the solutions of the quadratic equation.

Let's look at an example.

EXAMPLE 5 Solving a Quadratic Equation by Graphing and by Factoring

3 Solve the equation $x^2 - 5x + 6 = 0$.

Graphing Solution Figure 14 shows the graph of $f(x) = x^2 - 5x + 6$. We see that there are two x-intercepts: one near 2, the other near 3. The equation has two solutions. We conjecture that the two x-intercepts are 2 and 3, since the graphs of the function appear to cross the x-intercept at the tick marks. To verify that 2 and 3 are x-intercepts (and hence solutions to the equation), we evaluate $f(2)$ and $f(3)$.

FIGURE 14

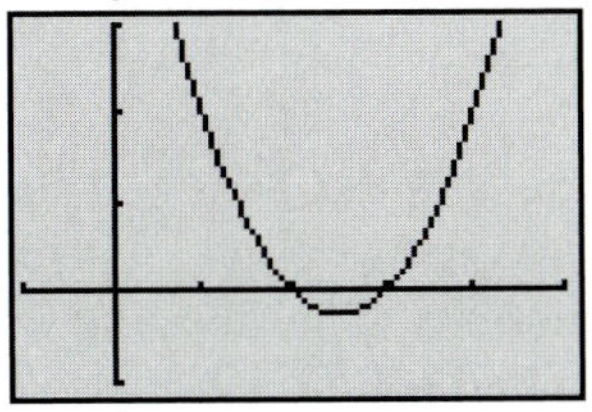

$$f(2) = 2^2 - 5(2) + 6 = 0$$
$$f(3) = 3^2 - 5(3) + 6 = 0$$

So both $x = 2$ and $x = 3$ are solutions. The solution set is $\{2, 3\}$.

Algebraic Solution The equation is in the standard form specified in equation (1). The left side may be factored as

$$x^2 - 5x + 6 = 0$$
$$(x - 2)(x - 3) = 0$$

We set each factor equal to 0 and solve the resulting first-degree equations:

$$x - 2 = 0 \quad \text{or} \quad x - 3 = 0$$
$$x = 2 \qquad\qquad x = 3$$

The solution set is $\{2, 3\}$. ■

EXAMPLE 6 Solving a Quadratic Equation by Graphing and by Factoring

Solve the equation $x^2 = 12 - x$.

Solution We put the equation in standard form by adding $x - 12$ to each side:

$$x^2 = 12 - x$$
$$x^2 + x - 12 = 0$$

Graphing Solution Figure 15 shows the graph of the function.

$$f(x) = x^2 + x - 12$$

From the graph it appears there are two x-intercepts, one near -4, the other near 3. We conjecture that the function has two x-intercepts: -4 and 3. This is verified by evaluating $f(-4)$ and $f(3)$. Since $f(-4) = f(3) = 0$, the solution set is $\{-4, 3\}$.

FIGURE 15

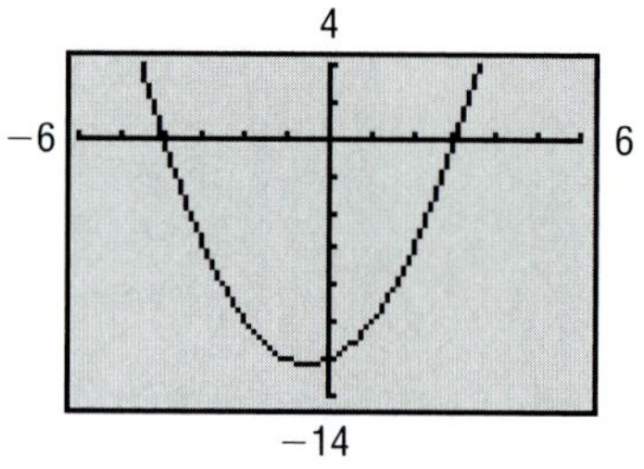

Algebraic Solution With the equation in standard form,

$$x^2 + x - 12 = 0$$

the left side may now be factored as

$$(x + 4)(x - 3) = 0$$

so that

$$x + 4 = 0 \quad \text{or} \quad x - 3 = 0$$
$$x = -4 \qquad\qquad x = 3$$

The solution set is $\{-4, 3\}$. ■

When the left side factors into two linear equations with the same solution, the quadratic equation is said to have a **repeated solution.** We also call this solution a **root of multiplicity 2,** or a **double root.**

EXAMPLE 7 Solving a Quadratic Equation by Graphing and by Factoring

Solve the equation $x^2 - 6x + 9 = 0$.

Graphing Solution Figure 16 shows the graph of the function.

$$f(x) = x^2 - 6x + 9$$

From the graph it appears there is one x-intercept, 3. Since $f(3) = 0$, the equation has only the repeated solution 3.

FIGURE 16

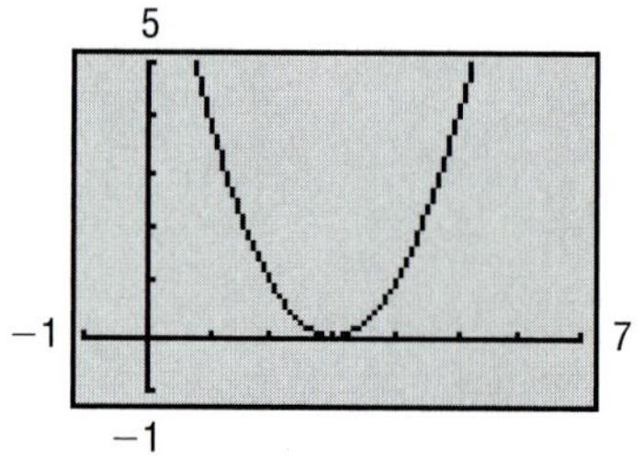

Algebraic Solution This equation is already in standard form, and the left side can be factored:

$$x^2 - 6x + 9 = 0$$
$$(x - 3)(x - 3) = 0$$

so

$$x = 3 \quad \text{or} \quad x = 3$$

This equation has only the repeated solution 3.

Now work Problems 37 and 45.

Extracting the Root

4 Suppose we wish to solve the quadratic equation

$$x^2 = p \tag{2}$$

where $p \geq 0$ is a nonnegative number. We proceed as in the earlier examples:

$$x^2 - p = 0 \qquad \text{Put in standard form.}$$
$$(x - \sqrt{p})(x + \sqrt{p}) = 0 \qquad \text{Factor (over the real numbers).}$$
$$x = \sqrt{p} \quad \text{or} \quad x = -\sqrt{p} \qquad \text{Solve.}$$

Thus, we have the following result:

> If $x^2 = p$ and $p \geq 0$, then $x = \sqrt{p}$ or $x = -\sqrt{p}$. (3)

When equation (3) is used, we refer to it as **extracting the root.** Note that if $p > 0$ the equation $x^2 = p$ has two solutions, $x = \sqrt{p}$ and $x = -\sqrt{p}$. We usually abbreviate these solutions as $x = \pm\sqrt{p}$, read as "x equals plus or minus the square root of p." For example, the two solutions of the equation

$$x^2 = 4$$

are

$$x = \pm\sqrt{4}$$

and since $\sqrt{4} = 2$, we have

$$x = \pm 2$$

The solution set is $\{-2, 2\}$.

EXAMPLE 8 **Solving Quadratic Equations By Extracting the Root**

Solve each equation.

(a) $x^2 = 5$ (b) $(x - 2)^2 = 16$

Solution (a) We use the result in equation (3) to get

$$x^2 = 5$$
$$x = \pm\sqrt{5}$$
$$x = \sqrt{5} \quad \text{or} \quad x = -\sqrt{5}$$

The solution set is $\{-\sqrt{5}, \sqrt{5}\}$.

(b) We use the result in equation (3) to get

$$(x - 2)^2 = 16$$
$$x - 2 = \pm\sqrt{16}$$
$$x - 2 = \sqrt{16} \quad \text{or} \quad x - 2 = -\sqrt{16}$$
$$x - 2 = 4 \qquad\qquad x - 2 = -4$$
$$x = 6 \qquad\qquad x = -2$$

The solution set is $\{-2, 6\}$.

The Quadratic Formula

5 We can use the method of completing the square* to obtain a general formula for solving the quadratic equation

$$ax^2 + bx + c = 0 \qquad a \neq 0$$

We begin by rearranging the terms as

$$ax^2 + bx = -c$$

Since $a \neq 0$, we can divide both sides by a to get

$$x^2 + \frac{b}{a}x = -\frac{c}{a}$$

Now the coefficient of x^2 is 1. To complete the square on the left side, add the square of $\frac{1}{2}$ the coefficient of x; that is, add

$$\left(\frac{1}{2} \cdot \frac{b}{a}\right)^2 = \frac{b^2}{4a^2}$$

to each side. Then,

$$x^2 + \frac{b}{a}x + \frac{b^2}{4a^2} = \frac{b^2}{4a^2} - \frac{c}{a}$$

$$\left(x + \frac{b}{2a}\right)^2 = \frac{b^2 - 4ac}{4a^2} \tag{4}$$

Provided $b^2 - 4ac \geq 0$, we now can apply the result in equation (3) to get

$$x + \frac{b}{2a} = \pm\sqrt{\frac{b^2 - 4ac}{4a^2}}$$

$$x = -\frac{b}{2a} \pm \frac{\sqrt{b^2 - 4ac}}{2a} = \frac{-b \pm \sqrt{b^2 - 4ac}}{2a}$$

*Refer to the Appendix, Section 5.

What if $b^2 - 4ac$ is negative? Then equation (4) states that the left expression (a real number squared) equals the right expression (a negative number). Since this occurrence is impossible for real numbers, we conclude that if $b^2 - 4ac < 0$ the quadratic equation has no *real* solution. (We discuss quadratic equations for which the quantity $b^2 - 4ac < 0$ in detail in Chapter 4).

We now state the *quadratic formula.*

Theorem Quadratic Formula

Consider the quadratic equation

$$ax^2 + bx + c = 0 \qquad a \neq 0$$

If $b^2 - 4ac < 0$, this equation has no real solution.
If $b^2 - 4ac \geq 0$, the real solution(s) of this equation is (are) given by the **quadratic formula:**

$$x = \frac{-b \pm \sqrt{b^2 - 4ac}}{2a} \tag{5}$$

The quantity $\boldsymbol{b^2 - 4ac}$ is called the **discriminant** of the quadratic equation, because its value tells us whether the equation has real solutions. In fact, it also tells us how many solutions to expect.

Discriminant of a Quadratic Equation

For a quadratic equation $ax^2 + bx + c = 0$:

1. If $b^2 - 4ac > 0$, there are two unequal real solutions.
2. If $b^2 - 4ac = 0$, there is a repeated real solution, a root of multiplicity 2.
3. If $b^2 - 4ac < 0$, there is no real solution.

Thus, when asked to find the real solutions, if any, of a quadratic equation, always evaluate the discriminant first to see how many real solutions there are.

EXAMPLE 9 Solving a Quadratic Equation by Graphing and by Using the Quadratic Formula

Find the real solutions, if any, of the equation $3x^2 - 5x + 1 = 0$.

Solution The equation is in standard form, so we compare it to $ax^2 + bx + c = 0$ to find a, b, and c:

$$3x^2 - 5x + 1 = 0$$
$$ax^2 + bx + c = 0$$

With $a = 3$, $b = -5$, and $c = 1$, we evaluate the discriminant $b^2 - 4ac$:

$$b^2 - 4ac = (-5)^2 - 4(3)(1) = 25 - 12 = 13$$

Since $b^2 - 4ac > 0$, there are two unequal real solutions.

Graphing Solution Figure 17 shows the graph of the function

$$f(x) = 3x^2 - 5x + 1$$

As expected, we see that there are two x-intercepts; one between 0 and 1, the other between 1 and 2. The solutions to the equation are 0.23 and 1.43 correct to two decimal places.

FIGURE 17

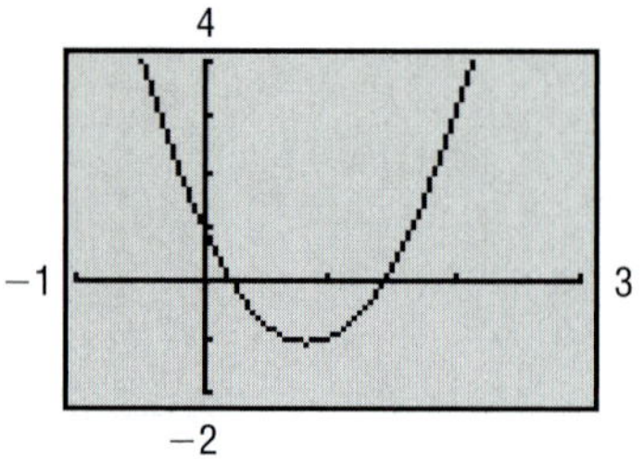

Algebraic Solution We use the quadratic formula with $a = 3$, $b = -5$, $c = 1$, and $b^2 - 4ac = 13$.

$$x = \frac{-b \pm \sqrt{b^2 - 4ac}}{2a} = \frac{5 \pm \sqrt{13}}{6}$$

The solution set is $\{(5 - \sqrt{13})/6, (5 + \sqrt{13})/6\}$. These solutions are exact. ■

Now work Problem 97.

EXAMPLE 10 Solving Quadratic Equations by Graphing and by Using the Quadratic Formula

Find the real solutions, if any, of the equation

$$\tfrac{25}{2}x^2 - 30x + 18 = 0$$

Solution The equation is given in standard form. However, to simplify the arithmetic, we clear the fractions by multiplying both sides of the equation by two:

$$\tfrac{25}{2}x^2 - 30x + 18 = 0$$

$$25x^2 - 60x + 36 = 0 \quad \text{Clear fractions.}$$

$$ax^2 + bx + c = 0 \quad \text{Compare to standard form.}$$

With $a = 25$, $b = -60$, and $c = 36$, we evaluate the discriminant:

$$b^2 - 4ac = (-60)^2 - 4(25)(36) = 3600 - 3600 = 0$$

The equation has a repeated solution.

Graphing Solution Figure 18 shows the graph of the function

$$f(x) = \frac{25}{2}x^2 - 30x + 18$$

From the graph, we see that the one x-intercept is between 1 and 2. The solution to the equation is 1.20.

FIGURE 18

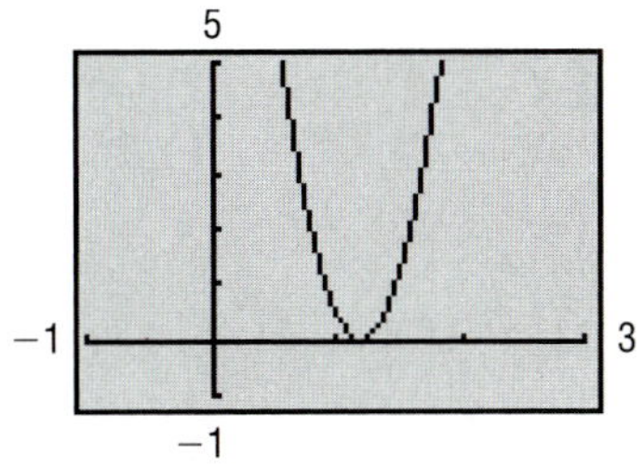

Algebraic Solution We use $a = 25$, $b = -60$, $c = 36$, and $b^2 - 4ac = 0$ to solve

$$25x^2 - 60x + 36 = 0$$

Using the quadratic formula, we find

$$x = \frac{-b \pm \sqrt{b^2 - 4ac}}{2a} = \frac{60 \pm \sqrt{0}}{50} = \frac{60}{50} = \frac{6}{5}$$

The repeated solution is $\frac{6}{5}$, which is exact. ■

EXAMPLE 11 Solving Quadratic Equations by Graphing and by Using the Quadratic Formula

Find the real solutions, if any, of the equation

$$3x^2 + 2 = 4x$$

Solution The equation, as given, is not in standard form.

$$3x^2 + 2 = 4x$$
$$3x^2 - 4x + 2 = 0 \quad \text{Put in standard form.}$$
$$ax^2 + bx + c = 0 \quad \text{Compare to standard form.}$$

With $a = 3$, $b = -4$, and $c = 2$, we find

$$b^2 - 4ac = 16 - 24 = -8$$

Since $b^2 - 4ac < 0$, the equation has no real solution.

Graphing Solution We use the standard form of the equation and graph the function

$$f(x) = 3x^2 - 4x + 2$$

See Figure 19. We see that there are no x-intercepts, so the equation has no real solution, as expected.

FIGURE 19

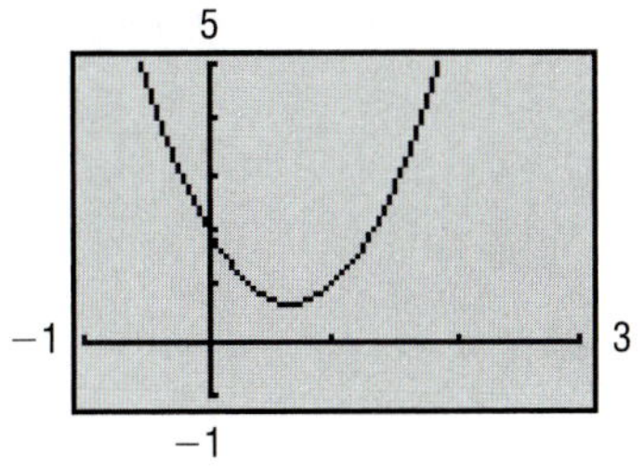

■

Now work Problem 83.

Sometimes a given equation can be transformed into a quadratic equation so that it can be solved using the quadratic formula.

EXAMPLE 12 Solving Quadratic Equations by Graphing and by Using the Quadratic Formula

Find the real solutions, if any, of the equation: $9 + \frac{3}{x} - \frac{2}{x^2} = 0, x \neq 0$

Graphing Solution Figure 20 shows the graph of the function

$$f(x) = 9 + \frac{3}{x} - \frac{2}{x^2}$$

From the graph, we conjecture that there are two x-intercepts, one between -1 and 0, the other between 0 and 1. The solutions to the equation are -0.66 and 0.33 correct to two decimal places.

FIGURE 20

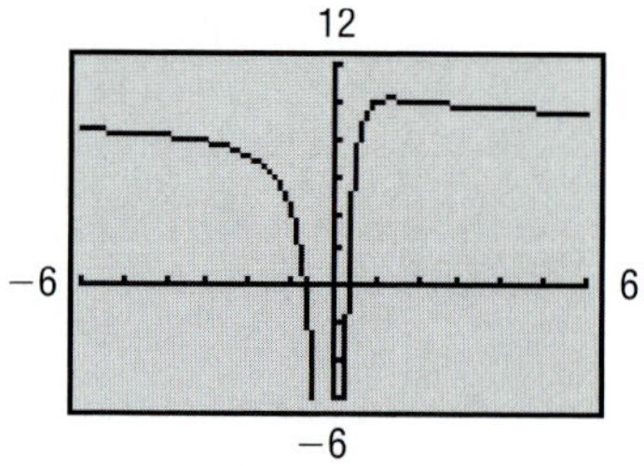

Comment: The graph of the preceding function on most graphing utilities will appear ragged near $x = 0$. This is because the function is not defined at $x = 0$. As a result, the graphing utility is unable to plot points when x gets close to zero. This problem can be easily overcome. Simply multiply both sides of the equation by x^2 and obtain $g(x) = 9x^2 + 3x - 2$. The resulting equation will have the same x-intercepts as the original equation. See Figure 21.

FIGURE 21

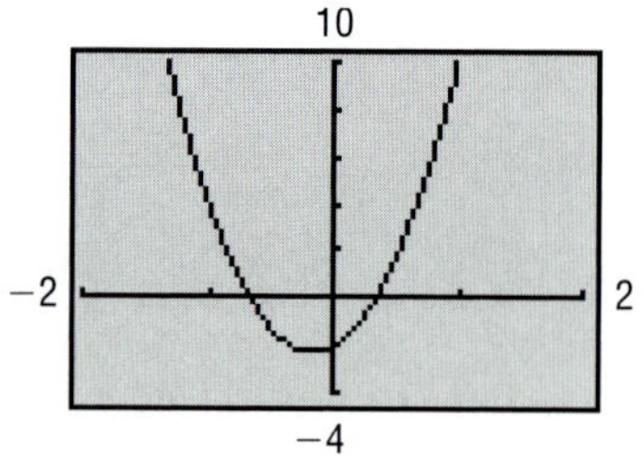

Algebraic Solution In its present form, the equation

$$9 + \frac{3}{x} - \frac{2}{x^2} = 0$$

is not a quadratic equation. However, it can be transformed into one by multiplying each side by x^2. The result is

$$9x^2 + 3x - 2 = 0$$

Although we multiplied each side by x^2, we know that $x^2 \neq 0$ (do you see why?), so this quadratic equation is equivalent to the original equation.

Using $a = 9$, $b = 3$, and $c = -2$, the discriminant is

$$b^2 - 4ac = 9 + 72 = 81$$

Since $b^2 - 4ac > 0$, the new equation has two real solutions:

$$x = \frac{-b \pm \sqrt{b^2 - 4ac}}{2a} = \frac{-3 \pm \sqrt{81}}{18} = \frac{-3 \pm 9}{18}$$

$$x = \frac{-3 + 9}{18} = \frac{6}{18} = \frac{1}{3} \quad \text{or} \quad x = \frac{-3 - 9}{18} = \frac{-12}{18} = \frac{-2}{3}$$

The solution set is $\{-\frac{2}{3}, \frac{1}{3}\}$. These are the exact solutions.

SUMMARY

Procedure for Solving a Quadratic Equation Algebraically

To solve a quadratic equation, first put it in standard form:

$$ax^2 + bx + c = 0$$

Then:

STEP 1: Identify a, b, and c.

STEP 2: Evaluate the discriminant, $b^2 - 4ac$.

STEP 3: (a) If the discriminant is negative, the equation has no real solution.

(b) If the discriminant is nonnegative, determine whether the left side can be factored. If you can easily spot factors, use the factoring method to solve the equation. Otherwise, use the quadratic formula or extracting the root.

HISTORICAL FEATURE

The solution of equations is among the oldest of mathematical activities, and efforts to systematize this activity determined much of the shape of modern mathematics.

Consider the following problem, and its solution, using only words: Solve the problem of how many apples Jim has, given that

> "Bob's five apples and Jim's apples together make twelve apples" by thinking,
>
> "Jim's apples are all twelve apples less Bob's five apples" and then concluding,
>
> "Jim has seven apples."

The mental steps translated into algebra are

$$5 + x = 12$$
$$x = 12 - 5$$
$$x = 7$$

The solution of this problem using only words is the earliest form of algebra. Such problems were solved exactly this way in Babylonia in 1800 B.C. We know almost nothing of mathematical work before this date, although most authorities believe the sophistication of the earliest known texts indicates that a long period of previous development must have occurred. The method of writing out equations in words persisted for thousands of years, and although it now seems extremely cumbersome, it was used very effectively by many generations of mathematicians. The Arabs developed a good deal of the theory of cubic equations while writing out all the equations in words. About A.D. 1500, the tendency to abbreviate words in the written equations began to lead in the direction of modern notation; for example, the Latin word *et* (meaning *and*) developed into the plus sign, +. Although the occasional use of letters to represent variables dates back to A.D. 1200, the practice did not become common until about A.D. 1600. Development thereafter was rapid, and by 1635, algebraic notation did not differ essentially from what we use now.

Problems using quadratic equations are found in the oldest known mathematical literature. Babylonians and Egyptians were solving such problems before 1800 B.C. Euclid solved quadratic equations geometrically in his *Data* (300 B.C.), and the Hindus and Arabs gave rules for solving any quadratic equation with real roots. Because negative numbers were not freely used before A.D. 1500, there were several different types of quadratic equations, each with its own rule. Thomas Harriot (1560–1621) introduced the method of factoring to obtain solutions, and François Viète (1540–1603) introduced a method that is essentially completing the square.

Until modern times it was usual to neglect the negative roots (if there were any), and equations involving square roots of negative quantities were regarded as unsolvable until the 1500s.

Historical Problems

1. *One of al-Khowârizmî's solutions* We solve $x^2 + 12x = 85$ by drawing the square shown. The area of the unshaded part is $x^2 + 12x$. We then set this expression equal to 85 to get the equation $x^2 + 12x = 85$. If we add the four shaded squares, we will have a larger square of known area. Complete the solution.

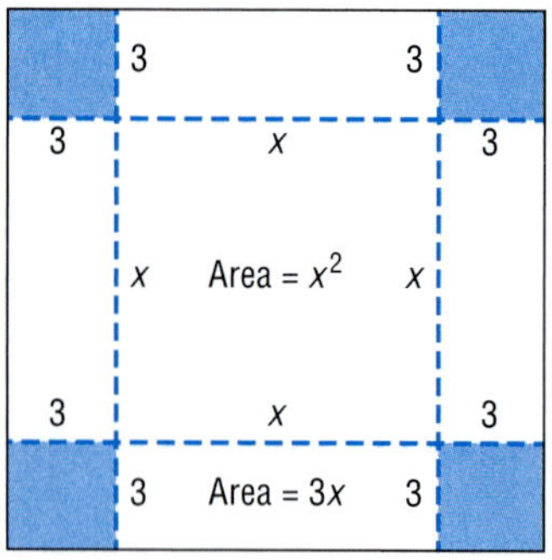

2. *Viète's method* We solve $x^2 + 12x - 85 = 0$ by letting $x = u + z$. Then

$$(u + z)^2 + 12(u + z) - 85 = 0$$
$$u^2 + (2z + 12)u + (z^2 + 12z - 85) = 0$$

Now select z so that $2z + 12 = 0$ and finish the solution.

3. *Another method to get the quadratic formula* Look at equation (4), page 188. Rewrite the right side as $(\sqrt{b^2 - 4ac}/2a)^2$ and then subtract it from each side. The right side is now 0 and the left side is a difference of two squares. If you factor this difference of two squares, you will easily be able to get the quadratic formula, and, moreover, the quadratic expression is factored, which is sometimes useful.

3.2 EXERCISES

In Problems 1–8, mentally solve each equation.

1. $3x = 21$ **2.** $3x = -24$ **3.** $5x + 15 = 0$ **4.** $3x + 18 = 0$

5. $2x - 3 = 5$ **6.** $3x + 4 = -8$ **7.** $\frac{1}{3}x = \frac{5}{12}$ **8.** $\frac{2}{3}x = \frac{9}{2}$

In Problems 9–66, solve each equation algebraically.

9. $3x + 2 = x + 6$
10. $2x + 7 = 3x + 5$
11. $2t - 6 = 3 - t$
12. $5y + 6 = -18 - y$
13. $6 - x = 2x + 9$
14. $3 - 2x = 2 - x$
15. $3 + 2n = 5n + 7$
16. $3 - 2m = 3m + 1$
17. $2(3 + 2x) = 3(x - 4)$
18. $3(2 - x) = 2x - 1$
19. $8x - (2x + 1) = 3x - 10$
20. $5 - (2x - 1) = 10$
21. $\frac{3}{2}x + 2 = \frac{1}{2} - \frac{1}{2}x$
22. $\frac{1}{3}x = 2 - \frac{2}{3}x$
23. $\frac{1}{2}x - 5 = \frac{3}{4}x$
24. $1 - \frac{1}{2}x = 6$
25. $\frac{2}{3}p = \frac{1}{2}p + \frac{1}{3}$
26. $\frac{1}{2} - \frac{1}{3}p = \frac{4}{3}$
27. $0.9t = 0.4 + 0.1t$
28. $0.9t = 1 + t$
29. $\frac{x + 1}{3} + \frac{x + 2}{7} = 5$
30. $\frac{2x + 1}{3} + 16 = 3x$
31. $\frac{2}{y} + \frac{4}{y} = 3$
32. $\frac{4}{y} - 5 = \frac{5}{2y}$
33. $\frac{1}{2} + \frac{2}{x} = \frac{3}{5}$
34. $\frac{3}{x} - \frac{1}{3} = \frac{1}{4}$
35. $x^2 = 9x$
36. $x^2 = -4x$
37. $x^2 - 25 = 0$
38. $x^2 - 9 = 0$
39. $z^2 + z - 12 = 0$
40. $v^2 + 7v + 12 = 0$
41. $2x^2 - 5x - 3 = 0$
42. $3x^2 + 5x + 2 = 0$
43. $3t^2 - 48 = 0$
44. $2y^2 - 50 = 0$
45. $x(x - 7) + 12 = 0$
46. $x(x + 1) = 12$
47. $4x^2 + 9 = 12x$
48. $25x^2 + 16 = 40x$
49. $6(p^2 - 1) = 5p$
50. $2(2u^2 - 4u) + 3 = 0$
51. $6x - 5 = \frac{6}{x}$
52. $x + \frac{12}{x} = 7$
53. $\frac{4(x - 2)}{x - 3} + \frac{3}{x} = \frac{-3}{x(x - 3)}$
54. $\frac{5}{x + 4} = 4 + \frac{3}{x - 2}$
55. $(x + 7)(x - 1) = (x + 1)^2$
56. $(x + 2)(x - 3) = (x - 3)^3$
57. $x(2x - 3) = (2x + 1)(x - 4)$
58. $x(1 + 2x) = (2x - 1)(x - 2)$
59. $z(z^2 + 1) = 3 + z^2$
60. $w(4 - w^2) = 8 - w^3$
61. $\frac{x}{x - 3} + 3 = \frac{3}{x - 3}$
62. $\frac{3x}{x + 2} = \frac{-6}{x + 2} - 2$
63. $x^2 = 4x$
64. $x^3 = x^2$
65. $t^3 - 9t^2 = 0$
66. $4z^3 - 8z^2 = 0$

In Problems 67–70, use a calculator to solve each equation. Express the solution correct to two decimal places.

67. $3.2x + \frac{21.3}{65.871} = 19.23$
68. $6.2x - \frac{19.1}{83.72} = 0.195$
69. $14.72 - 21.58x = \frac{18}{2.11}x + 2.4$
70. $18.63x - \frac{21.2}{2.6} = \frac{14x}{2.32} - 20$

In Problems 71–76, solve each equation. The letters a, b, and c are constants.

71. $ax - b = c, \quad a \neq 0$
72. $1 - ax = b, \quad a \neq 0$
73. $\frac{x}{a} + \frac{x}{b} = c, \quad a \neq 0,\ b \neq 0, \quad a \neq -b$
74. $\frac{a}{x} + \frac{b}{x} = c, \quad c \neq 0$
75. $\frac{1}{x - a} + \frac{1}{x + a} = \frac{2}{x - 1}$
76. $\frac{b + c}{x + a} = \frac{b - c}{x - a}, \quad c \neq 0, a \neq 0$

In Problems 77–96, find the real solutions, if any, of each equation. Use the quadratic formula.

77. $x^2 - 4x + 2 = 0$
78. $x^2 + 4x + 2 = 0$
79. $x^2 - 4x - 1 = 0$
80. $x^2 + 6x + 1 = 0$
81. $2x^2 - 5x + 3 = 0$
82. $2x^2 + 5x + 3 = 0$
83. $4y^2 - y + 2 = 0$
84. $4t^2 + t + 1 = 0$
85. $4x^2 = 1 - 2x$
86. $2x^2 = 1 - 2x$
87. $4x^2 = 9x$
88. $5x = 4x^2$
89. $9t^2 - 6t + 1 = 0$
90. $4u^2 - 6u + 9 = 0$
91. $3x^2 - 2x - 2 = 0$
92. $2x^2 - 3x - 1 = 0$
93. $4 - \frac{1}{x} - \frac{2}{x^2} = 0$
94. $4 + \frac{1}{x} - \frac{1}{x^2} = 0$
95. $3x = 1 - \frac{1}{x}$
96. $x = 1 - \frac{4}{x}$

In Problems 97–104, use a graphing utility to approximate the real solutions, if any, of each equation. Then use the quadratic formula to obtain exact solutions. Compare the two results. Experiment further to see how close you can make the graphing utility solution to the exact solution.

97. $x^2 - 4x + 2 = 0$
98. $x^2 + 4x + 2 = 0$
99. $x^2 + \sqrt{3}x - 3 = 0$
100. $x^2 + \sqrt{2}x - 2 = 0$
101. $\pi x^2 - x - \pi = 0$
102. $\pi x^2 + \pi x - 2 = 0$
103. $3x^2 + 8\pi x + \sqrt{29} = 0$
104. $\pi x^2 - 15\sqrt{2}x + 20 = 0$

In Problems 105–116, use a graphing utility to approximate the real solutions, if any, of each equation. Then use any algebraic method that you wish to obtain exact solutions. Compare the two results. Experiment further to see how close you can make the graphing utility solution to the exact solution.

105. $x^2 - 7 = 0$
106. $x^2 - 8 = 0$
107. $16x^2 - 8x + 1 = 0$
108. $9x^2 - 6x + 1 = 0$
109. $10x^2 - 19x - 15 = 0$
110. $6x^2 + 7x - 20 = 0$
111. $2 + z = 6z^2$
112. $2 = y + 6y^2$
113. $x^2 + \sqrt{2}x = \frac{1}{2}$
114. $\frac{1}{2}x^2 = \sqrt{2}x + 1$
115. $x^2 + x = 4$
116. $x^2 + x = 1$

In Problems 117–122, use the discriminant to determine whether each quadratic equation has two unequal real solutions, a repeated real solution, or no real solution, without solving the equation. Use a graphing utility to verify your result.

117. $2x^2 - 6x + 7 = 0$
118. $x^2 + 4x + 7 = 0$
119. $9x^2 - 30x + 25 = 0$
120. $25x^2 - 20x + 4 = 0$
121. $3x^2 + 5x - 2 = 0$
122. $2x^2 - 3x - 4 = 0$

Problems 123–128 list some formulas that occur in applications. Solve each formula for the indicated variable.

123. Electricity $\frac{1}{R} = \frac{1}{R_1} + \frac{1}{R_2}$ for R
124. Finance $A = P(1 + rt)$ for r
125. Mechanics $F = \frac{mv^2}{R}$ for R
126. Chemistry $PV = nRT$ for T
127. Mathematics $S = \frac{a}{1 - r}$ for r
128. Mechanics $v = -gt + v_0$ for t

129. Explain what is wrong in the following steps:

$$\begin{aligned} x &= 2 && (1)\\ 3x - 2x &= 2 && (2)\\ 3x &= 2x + 2 && (3)\\ x^2 + 3x &= x^2 + 2x + 2 && (4)\\ x^2 + 3x - 10 &= x^2 + 2x - 8 && (5)\\ (x - 2)(x + 5) &= (x - 2)(x + 4) && (6)\\ x + 5 &= x + 4 && (7)\\ 1 &= 0 && (8) \end{aligned}$$

130. Which of the following pairs of equations are equivalent? Explain.
(a) $x^2 = 9; x = 3$
(b) $x = \sqrt{9}; x = 3$
(c) $(x - 1)(x - 2) = (x - 1)^2; x - 2 = x - 1$

131. The equation

$$\frac{5}{x + 3} + 3 = \frac{8 + x}{x + 3}$$

has no solution, yet when we go through the process of solving it we obtain $x = -3$. Write a brief paragraph to explain what causes this to happen.

132. Make up an equation that has no solution and give it to a fellow student to solve. Ask the fellow student to write a critique of your equation.

133. Describe three ways you might solve a quadratic equation. State your preferred method; explain why you chose it.

134. Explain the benefits of evaluating the discriminant of a quadratic equation before attempting to solve it.

135. Make up three quadratic equations: one having two distinct solutions, one having no real solution, and one having exactly one real solution.

136. The word *quadratic* seems to imply four (*quad*), yet a quadratic equation is an equation that involves a polynomial of degree 2. Investigate the origin of the term *quadratic* as it is used in the expression *quadratic equation.* Write a brief essay on your findings.

137. Write a program that will solve a quadratic equation:

{Enter the coefficient of x squared} READ (a);
{Enter the coefficient of x} READ (b);
{Enter the constant term} READ (c);
IF $b^2 - 4ac < 0$
THEN {write no real solution}
ELSE IF $b^2 - 4ac = 0$
THEN {write $-b/2a$ is a double root}
ELSE {write $(-b + \text{SQRT}(b^2 - 4ac))/2a$
or $(-b - \text{SQRT}(b^2 - 4ac))/2a$
is a solution}

3.3 SETTING UP EQUATIONS: APPLICATIONS

1 Translate Verbal Descriptions into Mathematical Expressions
2 Set Up Applied Problems
3 Solve Interest Problems
4 Solve Mixture Problems
5 Solve Uniform Motion Problems
6 Solve Constant Rate Jobs Problems

The previous section provides the tools for solving equations. But, unfortunately, applied problems do not come in the form, "Solve the equation . . . ." Instead, they are narratives that supply information—hopefully, enough to answer the question that inevitably arises. Thus, to solve applied problems we must be able to translate the verbal description into the language of mathematics. We do this by using symbols (usually letters of the alphabet) to represent unknown quantities and then finding relationships (such as equations) that involve these symbols. The process of doing this is called **mathematical modeling.**

Any solution to the mathematical problem must be checked against the mathematical problem, the verbal description, and the real problem. See Figure 22 for an illustration of the modeling process.

FIGURE 22

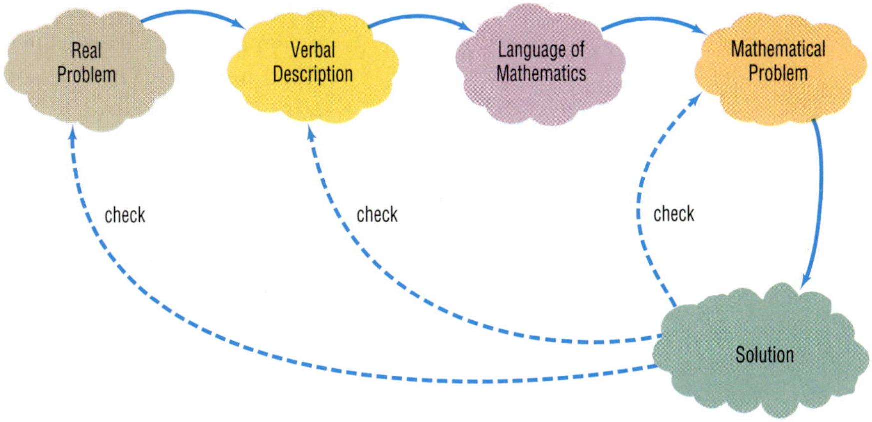

1 Let's look at a few examples that will help you translate certain words into mathematical symbols.

EXAMPLE 1 Translating Verbal Descriptions into Mathematical Expressions

(a) The area of a rectangle is the product of its length times its width.

Translation: If A is used to represent the area, l the length, and w the width, then $A = lw$.

(b) For uniform motion, the velocity of an object equals the distance traveled divided by the time required.

Translation: If v is the velocity, s the distance, and t the time, then $v = s/t$.

(c) A total of \$5000 is invested, some in stocks and some in bonds. If the amount invested in stocks is x, express the amount invested in bonds in terms of x.

Translation: If y is the amount invested in bonds, then $x + y = 5000$. Thus, if x is the amount invested in stocks, then the amount invested in bonds is $y = 5000 - x$.

(d) Let x denote a number.

The number 5 times as large as x is $5x$.

The number 3 less than x is $x - 3$.

The number that exceeds x by 4 is $x + 4$.

The number that, when added to x, gives 5 is $5 - x$. ■

Now work Problem 1.

2 Mathematical equations that represent real situations should be consistent in terms of the units used. In Example 1(a), if l is measured in feet, then w also must be expressed in feet, and A will be expressed in square feet. In Example 1(b), if v is measured in miles per hour, then the distance s must be expressed in miles and the time t must be expressed in hours. It is a good practice to check units to be sure that they are consistent and make sense.

Although each situation has its own unique features, we can provide an outline of the steps to follow in setting up applied problems.

Steps for Setting Up Applied Problems

STEP 1: Read the problem carefully, perhaps two or three times. Pay particular attention to the question being asked in order to identify what you are looking for. If you can, determine realistic possibilities for the answer.

STEP 2: Assign a letter (variable) to represent what you are looking for, and, if necessary, express any remaining unknown quantities in terms of this variable.

STEP 3: Make a list of all the known facts, and write down any relationships among them, especially any that involve the variable. These may take the form of an equation (or, later, an inequality) involving the variable. If possible, draw an appropriately labeled diagram to assist you. Sometimes, a table or chart helps.

STEP 4: Solve the equation for the variable, and then answer the question asked in the problem.

STEP 5: Check the answer with the facts in the problem. If it agrees, congratulations! If it does not agree, try again.

Let's look at an example.

EXAMPLE 2 Determining an Hourly Wage

Shannon grossed \$435 one week by working 52 hours. Her employer pays time-and-a-half for all hours worked in excess of 40 hours. With this information, can you determine Shannon's regular hourly wage?

Solution STEP 1: We are looking for an hourly wage. Our answer will be in dollars per hour.

STEP 2: Let x represent the regular hourly wage; x is measured in dollars per hour.

STEP 3: We set up a table:

	Hours Worked	Hourly Wage	Salary
Regular	40	x	$40x$
Overtime	12	$1.5x$	$12(1.5x) = 18x$

The sum of regular salary plus overtime salary will equal \$435. Thus, from the table, $40x + 18x = 435$.

STEP 4:
$$\begin{aligned} 40x + 18x &= 435 \\ 58x &= 435 \\ x &= 7.50 \end{aligned}$$

Thus, Shannon's regular hourly wage is \$7.50 per hour.

STEP 5: Forty hours yields a salary of $40(7.50) = \$300$, and 12 hours of overtime yields a salary of $12(1.5)(7.50) = \$135$, for a total of \$435. ■

Now work Problem 15.

Interest

3 The next example involves **interest.** Interest is money paid for the use of money. The total amount borrowed (whether by an individual from a bank in the form of a loan or by a bank from an individual in the form of a savings account) is called the **principal.** The **rate of interest,** expressed as a percent, is the amount charged for the use of the principal for a given period of time, usually on a yearly (that is, per annum) basis.

Simple Interest Formula

If a principal of P dollars is borrowed for a period of t years at a per annum interest rate r, expressed as a decimal, the interest I charged is

$$I = Prt \tag{1}$$

Interest charged according to formula (1) is called **simple interest.**

EXAMPLE 3 **Financial Planning**

Candy, an investor with \$70,000, decides to place part of her money in corporate bonds paying 12% per year and the rest in a certificate of deposit paying 8% per year. If she wishes to obtain an overall return of 9% per year, how much should she place in each investment?

Solution STEP 1: The question is asking for two dollar amounts: the principal to invest in the corporate bonds and the principal to invest in the certificate of deposit.

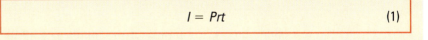

STEP 2: We let x represent the amount (in dollars) to be invested in the bonds. Then $70{,}000 - x$ is the amount that will be invested in the certificate. (Do you see why?)

STEP 3: We set up a table:

	Principal $	**Rate**	**Time yr**	**Interest $**
Bonds	x	12% = 0.12	1	$0.12x$
Certificate	$70{,}000 - x$	8% = 0.08	1	$0.08(70{,}000 - x)$
Total	70,000	9% = 0.09	1	$0.09(70{,}000) = 6300$

Since the total interest from the investments is equal to $0.09(70{,}000) = 6300$, we must have the equation

$$0.12x + 0.08(70{,}000 - x) = 6300$$

(Note that the units are consistent: the unit is dollars on each side.)

STEP 4:
$$\begin{aligned} 0.12x + 5600 - 0.08x &= 6300 \\ 0.04x &= 700 \\ x &= 17{,}500 \end{aligned}$$

Thus, Candy should place \$17,500 in the bonds and \$70,000 − \$17,500 = \$52,500 in the certificate.

STEP 5: The interest on the bonds after 1 year is 0.12(\$17,500) = \$2100; the interest on the certificate after 1 year is 0.08(\$52,500) = \$4200. The total annual interest is \$6300, the required amount. ■

Now work Problem 23.

Mixture Problems

4 The next example is a type usually referred to as a **mixture problem.**

EXAMPLE 4

Chemistry: Mixing Acids

In a chemistry laboratory the concentration of one solution is 10% hydrochloric acid (HCl) and that of a second solution is 60% HCl. How many milliliters (mL) of each should be mixed to obtain 50 mL of a 30% HCl solution?

Solution Let x represent the number of milliliters of the 10% solution. Then $50 - x$ equals the number of milliliters of the 60% solution. See Figure 23.

FIGURE 23

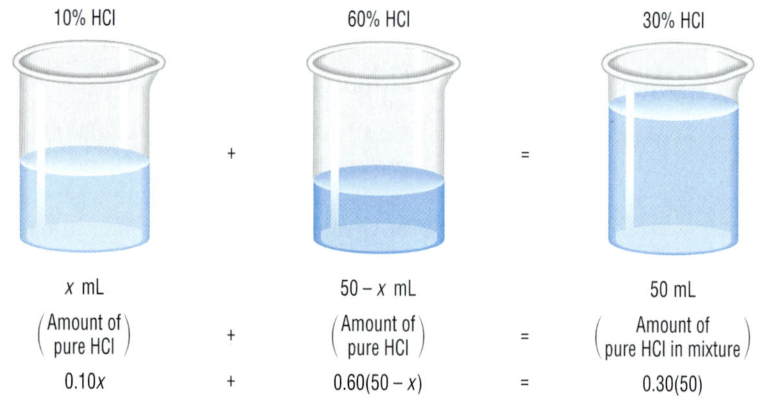

Based on the figure, we form a table:

	Amount mL	Concentration of HCl	Amount of Pure Acid mL
10% HCl	x	10% = 0.10	$0.10x$
60% HCl	$50 - x$	60% = 0.60	$0.60(50 - x)$
30% HCl	50	30% = 0.30	$0.30(50) = 15$

The amount of HCl in the 30% solution (15 milliliters) must equal the sum of the amounts of HCl found in the 10% solution and the 60% solution. Thus, we have the equation

$$\begin{aligned} 0.10x + 0.60(50 - x) &= 15 \\ 0.10x + 30 - 0.60x &= 15 \\ -0.50x &= -15 \\ x &= 30 \text{ milliliters} \end{aligned}$$

Thus, 30 milliliters of the 10% acid solution, when mixed with 20 milliliters of the 60% acid solution, yields 50 milliliters of a 30% acid solution.

Check: To check this answer, we note that there are $0.10(30) = 3$ milliliters of acid in the 10% solution and $0.60(20) = 12$ milliliters of acid in the 60% solution. The 50 milliliter mixture therefore contains 15 milliliters of acid, for an acid concentration of $\frac{15}{50} = 0.30 = 30\%$ acid solution.

Now work Problem 53.

Uniform Motion

5 The next example deals with moving objects.

Uniform Motion Formula

If an object moves at an average velocity v, the distance s covered in time t is given by the formula

$$s = vt \tag{2}$$

That is, Distance = Velocity · Time. Objects that are moving in accordance with formula (2) are said to be in **uniform motion.**

EXAMPLE 5

Physics: Uniform Motion

Tanya, who is a long-distance runner, runs at an average velocity of 8 miles per hour. Two hours after Tanya leaves your house, you leave in your Honda and follow the same route. If your average velocity is 40 miles per hour, how long will it be before you catch up to Tanya? How far will each of you be from your house?

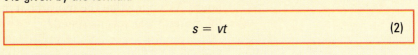

Solution Refer to Figure 24. We use t to represent the time (in hours) that it takes the Honda to catch up with Tanya. When this occurs, the total time elapsed for Tanya is $t + 2$ hours.

FIGURE 24

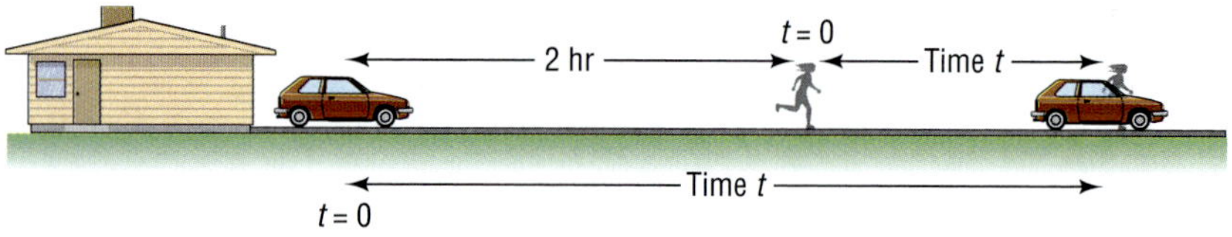

Set up the following table:

	Velocity mi/hr	Time hr	Distance mi
Tanya	8	$t + 2$	$8(t + 2)$
Honda	40	t	$40t$

Since the distance traveled is the same, we are led to the following equation:

$$8(t + 2) = 40t$$
$$8t + 16 = 40t$$
$$32t = 16$$
$$t = \frac{1}{2} \text{ hour}$$

It will take the Honda $\frac{1}{2}$ hour to catch up to Tanya. Each of you will have gone 20 miles.

Check: In 2.5 hours, Tanya travels a distance of $(2.5)(8) = 20$ miles. In $\frac{1}{2}$ hour, the Honda travels a distance of $(\frac{1}{2})(40) = 20$ miles.

Now work Problem 51.

EXAMPLE 6 Physics: Uniform Motion

A motorboat heads upstream a distance of 24 miles on the Kankakee River whose current is running at 3 miles per hour. The trip up and back takes 6 hours. Assuming that the motorboat maintained a constant speed relative to the water, what was its speed?

Solution See Figure 25. We use v to represent the constant speed of the motorboat relative to the water. Then the true speed going upstream is $v - 3$ miles per hour, and the true speed going downstream is $v + 3$ miles per hour. Since Distance = Velocity × Time, then Time = Distance/Velocity. We set up a table.

FIGURE 25

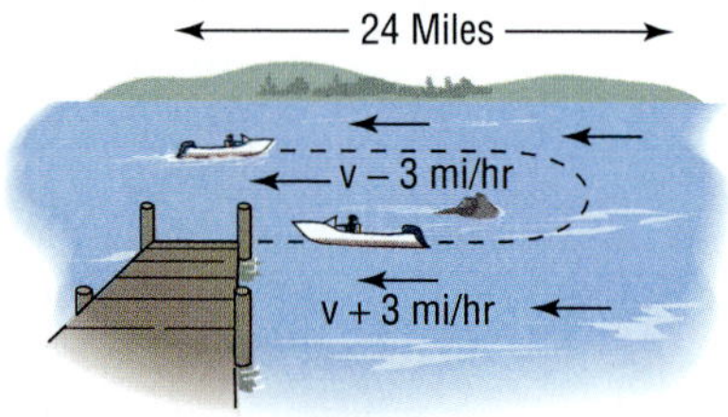

	Velocity mi/hr	Distance mi	Time = Distance/Velocity hr
Upstream	$v - 3$	24	$\frac{24}{v - 3}$
Downstream	$v + 3$	24	$\frac{24}{v + 3}$

Since the total time up and back is 6 hours, we have

$$\begin{aligned} \frac{24}{v-3} + \frac{24}{v+3} &= 6 \\ \frac{24(v+3) + 24(v-3)}{(v-3)(v+3)} &= 6 \\ \frac{48v}{v^2-9} &= 6 \\ 48v &= 6(v^2-9) \\ 6v^2 - 48v - 54 &= 0 \\ v^2 - 8v - 9 &= 0 \\ (v-9)(v+1) &= 0 \\ v = 9 \quad \text{or} \quad v &= -1 \end{aligned}$$

We discard the solution $v = -1$ mile per hour, so the speed of the motorboat relative to the water is 9 miles per hour. ■

Constant Rate Jobs

6 This section involves jobs that are performed at a **constant rate.** Our assumption is that if a job can be done in t units of time then $1/t$ of the job is done in 1 unit of time. Let's look at an example.

EXAMPLE 7 Working Together to Do a Job

At 10 A.M. Danny is asked by his father to weed the garden. From past experience Danny knows this will take him 4 hours, working alone. His older brother, Mike, when it is his turn to do this job, requires 6 hours. Since Mike wants to go golfing with Danny and has a reservation for 1 P.M., he agrees to help Danny. Assuming no gain or loss of efficiency, when will they finish if they work together? Can they make the golf date?

Solution In 1 hour, Danny does $\frac{1}{4}$ of the job, and in 1 hour, Mike does $\frac{1}{6}$ of the job. Let t be the time (in hours) it takes them to do the job together. In 1 hour, then, $1/t$ of the job is completed. We reason as follows:

$$\begin{pmatrix}\text{Part done by Danny} \\ \text{in 1 hour}\end{pmatrix} + \begin{pmatrix}\text{Part done by Mike} \\ \text{in 1 hour}\end{pmatrix} = \begin{pmatrix}\text{Part done together} \\ \text{in 1 hour}\end{pmatrix}$$

Now we set up the table in the margin. From the table,

	Hours to Do Job	Part of Job Done In 1 Hour
Danny	4	$\frac{1}{4}$
Mike	6	$\frac{1}{6}$
Together	t	$\frac{1}{t}$

$$\begin{aligned} \frac{1}{4} + \frac{1}{6} &= \frac{1}{t} \\ \frac{3+2}{12} &= \frac{1}{t} \\ t &= \frac{12}{5} \end{aligned}$$

Working together, the job can be done in $\frac{12}{5}$ hours, or 2 hours, 24 minutes. They should make the golf date, since they will finish at 12:24 P.M. ■

Now work Problem 39.

Other Applied Problems

The next two examples illustrate problems that you will probably see again in a slightly different form if you study calculus.

EXAMPLE 8 Preview of a Calculus Problem

From each corner of a square piece of sheet metal, remove a square with side 9 centimeters. Turn up the edges to form an open box. If the box is to hold 144 cubic centimeters, what should be the dimensions of the piece of sheet metal?

Solution We use Figure 26 as a guide. We have labeled by x the length of a side of the square piece of sheet metal. The box will be of height 9 centimeters and its square base will have $x - 18$ as the length of a side. The volume (Length $\times$ Width $\times$ Height) of the box is therefore

$$9(x - 18)(x - 18) = 9(x - 18)^2$$

FIGURE 26

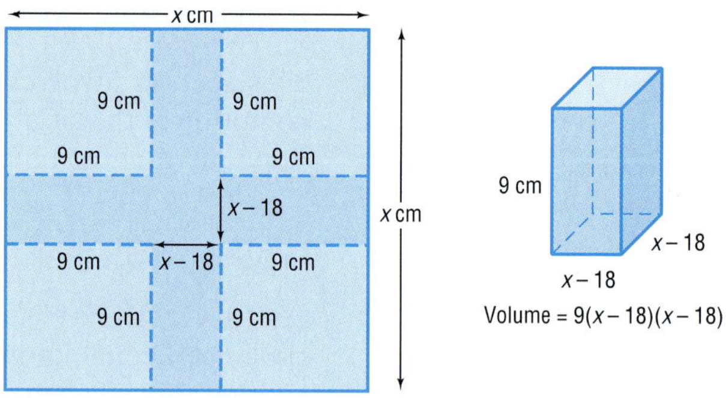

Since the volume of the box is to be 144 cubic centimeters, we have

$$\begin{aligned} 9(x - 18)^2 &= 144 \\ (x - 18)^2 &= 16 \\ x - 18 &= \pm 4 \\ x &= 18 \pm 4 \\ x = 22 \quad &\text{or} \quad x = 14 \end{aligned}$$

We discard the solution $x = 14$ (do you see why?) and conclude that the sheet metal should be 22 centimeters by 22 centimeters.

Check: If we begin with a piece of sheet metal 22 centimeters by 22 centimeters, cut out a 9 centimeter square from each corner, and fold up the edges, we get a box whose dimensions are 9 by 4 by 4, with volume $9 \times 4 \times 4 = 144$ cubic centimeters, as required. ■

Now work Problem 31.

EXAMPLE 9 Preview of a Calculus Problem

A piece of wire 8 feet in length is to be cut into two pieces. Each piece will then be bent into a square. Where should the cut in the wire be made if the sum of the areas of these squares is to be 2 square feet?

FIGURE 27

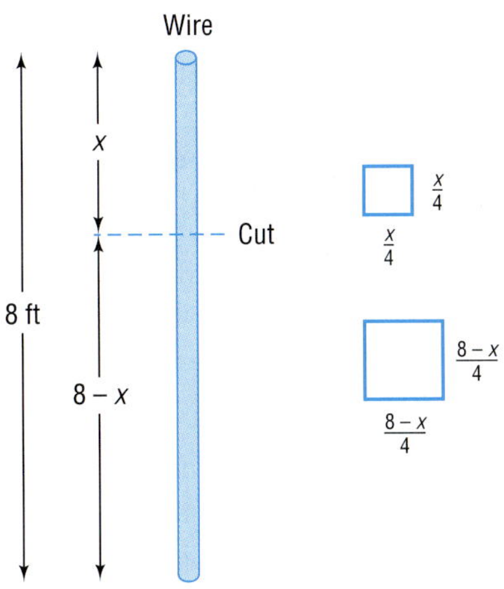

Solution We use Figure 27 as a guide. We have labeled by x the length of one of the pieces of wire after it has been cut. The remaining piece will be of length $8 - x$. If each length is bent into a square, then one of the squares has a side of length $x/4$ and the other a side of length $(8 - x)/4$. Since the sum of the areas of these two squares is 2, we have the equation

$$\begin{aligned}
\left(\frac{x}{4}\right)^2 + \left(\frac{8-x}{4}\right)^2 &= 2 \\
\frac{x^2}{16} + \frac{64 - 16x + x^2}{16} &= 2 \\
2x^2 - 16x + 64 &= 32 \\
2x^2 - 16x + 32 &= 0 && \text{Put in standard form.} \\
x^2 - 8x + 16 &= 0 && \text{Divide by 2.} \\
(x - 4)^2 &= 0 && \text{Factor} \\
x &= 4
\end{aligned}$$

Since $x = 4$, $8 - x = 4$, and the original piece of wire should be cut into two pieces, each of length 4 feet.

Check: If the length of each piece of wire is 4 feet, then each piece can be formed into a square whose side is 1 foot. The area of each square is then 1 square foot, so the sum of the areas is 2 square feet, as required. ■

3.3 EXERCISES

In Problems 1–10, translate each sentence into a mathematical equation. Be sure to identify the meaning of all symbols.

1. **Geometry** The area of a circle is the product of the number π times the square of the radius.
2. **Geometry** The circumference of a circle is the product of the number π times twice the radius.
3. **Geometry** The area of a square is the square of the length of a side.
4. **Geometry** The perimeter of a square is four times the length of a side.
5. **Physics** Force equals the product of mass times acceleration.
6. **Physics** Pressure is force per unit area.
7. **Physics** Work equals force times distance.
8. **Physics** Kinetic energy is one-half the product of the mass times the square of the velocity.
9. **Business** The total variable cost of manufacturing x dishwashers is \$150 per dishwasher times the number of dishwashers manufactured.
10. **Business** The total revenue derived from selling x dishwashers is \$250 per dishwasher times the number of dishwashers sold.
11. **Finance** A total of \$20,000 is to be invested, some in bonds and some in Certificates of Deposit (CDs). If the amount invested in bonds is to exceed that in CDs by \$2000, how much will be invested in each type of instrument?
12. **Finance** A total of \$10,000 is to be divided up between Sean and George, with George to receive \$2000 less than Sean. How much will each receive?
13. **Finance** An inheritance of \$900,000 is to be divided among Scott, Alice, and Tricia in the following manner: Alice is to receive $\frac{3}{4}$ of what Scott gets, while Tricia gets $\frac{1}{2}$ of what Scott gets. How much does each receive?
14. **Sharing the Cost of a Pizza** Canter and Carole agree to share the cost of an \$18 pizza based on how much each ate. If Carole ate $\frac{2}{3}$ the amount Canter ate, how much should each pay?

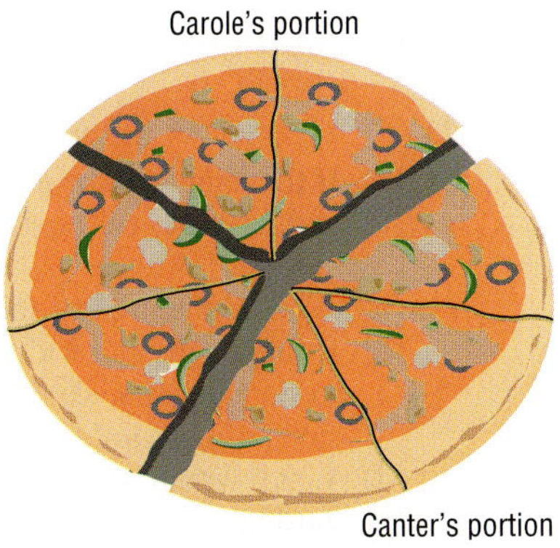

15. Computing Hourly Wages Sandra who is paid time-and-a-half for hours worked in excess of 40 hours had gross weekly wages of $442 for 48 hours worked. What is the regular hourly rate?

16. Computing Hourly Wages Leigh is paid time-and-a-half for hours worked in excess of 40 hours and double-time for hours worked on Sunday. If Leigh had gross weekly wages of $342 for working 50 hours, 4 of which were on Sunday, what is her regular hourly rate?

17. Football In an NFL football game, the BEARS scored a total of 41 points, including one safety (2 points) and two field goals (3 points each). After scoring a touchdown (6 points), a team is given the chance to score 1 or 2 extra points. The BEARS, in trying to score 1 extra point after each touchdown, missed 2 extra points. How many touchdowns did they get?

18. Basketball In a basketball game, the Bulls scored a total of 103 points and made three times as many field goals (2 points each) as free throws (1 point each). They also made eleven 3 point baskets. How many field goals did they have?

19. Geometry The perimeter of a rectangle is 60 feet. Find its length and width if the length is 8 feet longer than the width.

20. Geometry The perimeter of a rectangle is 42 meters. Find its length and width if the length is twice the width.

21. Enclosing a Garden Ruth has 46 feet of fencing to be used to enclose a rectangular garden that has a border 2 feet wide surrounding it (see the figure).

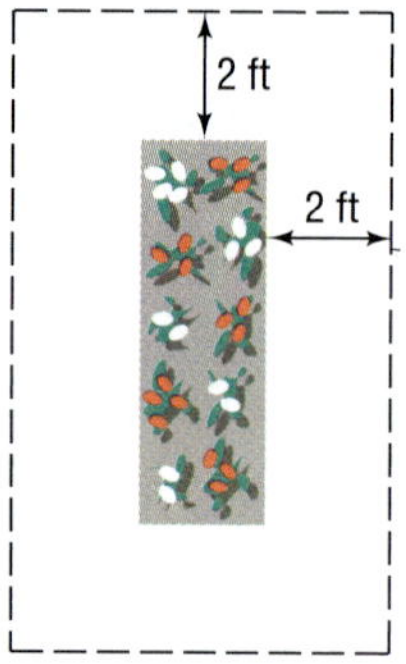

(a) If the length of the garden is to be twice its width, what will be the dimensions of the garden?
(b) What is the area of the garden?
(c) If the length and width of the garden were to be the same, what would be the dimensions of the garden?
(d) What would be the area of the square garden?

22. Chemistry: Sugar Molecules A sugar molecule has twice as many atoms of hydrogen as it does oxygen and one more atom of carbon than oxygen. If a sugar molecule has a total of 45 atoms, how many are oxygen? How many are hydrogen?

23. Financial Planning A recent retiree, Kate requires $6000 per year in extra income. She has $50,000 to invest and can invest in B-rated bonds paying 15% per year or in a Certificate of Deposit (CD) paying 7% per year. How much money should be invested in each to realize exactly $6,000 in interest per year?

24. Financial Planning After 2 years, Kate (see Problem 23) finds she now will require $7000 per year. Assuming that the remaining information is the same, how should the money be reinvested?

25. Banking A bank loaned out $12,000, part of it at the rate of 8% per year and the rest at the rate of 18% per year. If the interest received totaled $1000, how much was loaned at 8%?

26. Banking Scott, a loan officer at a bank, has $1,000,000 to lend and is required to obtain an average return of 18% per year. If he can lend at the rate of 19% or the rate of 16%, how much can he lend at the 16% rate and still meet his requirement?

27. Dimensions of a Window The area of the opening of a rectangular window is to be 143 square feet. If the length is to be 2 feet more than the width, what are the dimensions?

28. Dimensions of a Window The area of a rectangular opening is to be 306 square centimeters. If the length exceeds the width by 1 centimeter, what are the dimensions?

29. Geometry Find the dimensions of a rectangle whose perimeter is 26 meters and whose area is 40 square meters.

30. Watering a Field An adjustable water sprinkler that sprays water in a circular pattern is placed at the center of a square field whose area is 1250 square feet (see the figure). What is the shortest radius setting that can be used if the field is to be completely enclosed within the circle?

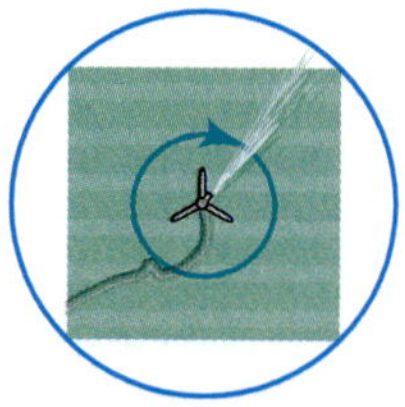

31. **Constructing a Box** An open box is to be constructed from a square piece of sheet metal by removing a square of side 1 foot from each corner and turning up the edges. If the box is to hold 4 cubic feet, what should be the dimensions of the sheet metal?

32. **Constructing a Box** Rework Problem 31 if the piece of sheet metal is a rectangle whose length is twice its width.

33. **Physics** A ball is thrown vertically upward from the top of a building 96 feet tall with an initial velocity of 80 feet per second. The distance s (in feet) of the ball from the ground after t seconds is $s = 96 + 80t - 16t^2$.
    (a) After how many seconds does the ball strike the ground?
    (b) After how many seconds will the ball pass the top of the building on its way down?

34. **Constructing a Coffee Can** A 39-ounce can of Hills Bros.® coffee requires 188.5 square inches of aluminum. If its height is 7 inches, what is its radius? (The surface area A of a right circular cylinder is $A = 2\pi r^2 + 2\pi rh$, where r is the radius and h is the height.)

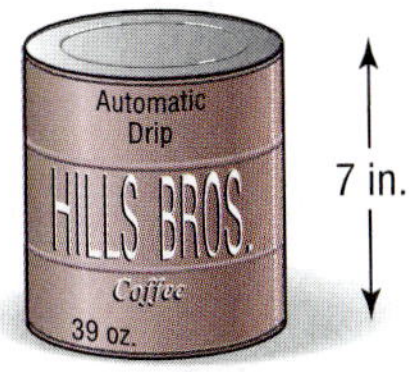

35. **Business: Discount Pricing** A builder of tract homes reduced the price of a model by 15%. If the new price is \$125,000, what was its original price? How much can be saved by purchasing the model?

36. **Business: Discount Pricing** A car dealer, at a year-end clearance, reduces the list price of last year's models by 15%. If a certain four-door model has a discounted price of \$8000, what was its list price? How much can be saved by purchasing last year's model?

37. **Business: Marking Up the Price of Books** A college book store marks up the price it pays the publisher for a book by 25%. If the selling price of a book is \$56.00, how much did the book store pay for the book?

38. **Personal Finance: Cost of a Car** The suggested list price of a new car is \$12,000. The dealer's cost is 85% of list. How much will you pay if the dealer is willing to accept \$100 over cost for the car?

39. **Working Together on a Job** Trent can deliver his newspapers in 30 minutes. It takes Lois 20 minutes to do the same route. How long would it take them to deliver the newspapers if they work together?

40. **Working Together on a Job** Alfredo by himself can paint four rooms in 10 hours. If he hires Mark to help, they can do the same job together in 6 hours. If he lets Mark work alone, how long will it take for Mark to paint four rooms?

41. **Computing Grades** Going into the final exam, which will count as two tests, Bridgette has test scores of 80, 83, 71, 61, and 95. What score does Bridgette need on the final in order to have an average score of 80?

42. **Computing Grades** Going into the final exam, which will count as two-thirds of the final grade, Roger has test scores of 86, 80, 84, and 90. What score does Roger need on the final in order to earn a B, which requires an average score of 80? What does he need to earn an A, which requires an average of 90?

43. **Football** A tight end can run the 100 yard dash in 12 seconds. A defensive back can do it in 10 seconds. The tight end catches a pass at his own 20 yard line with the defensive back at the 15 yard line. (See the figure.) If no other players are nearby, at what yard line will the defensive back catch up to the tight end?

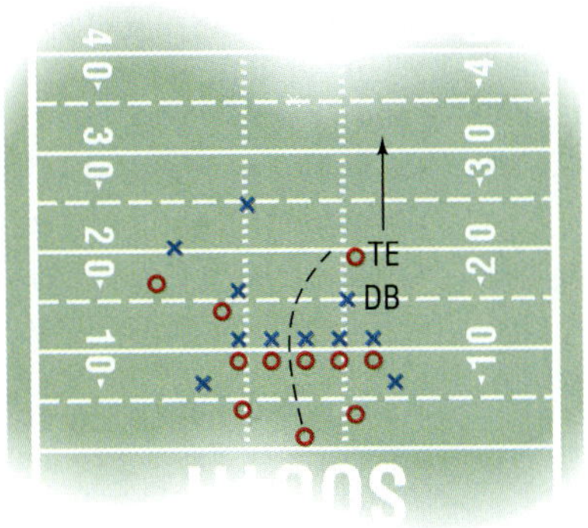

44. **Computing Business Expenses** Theresa, an outside salesperson, uses her car for both business and pleasure. Last year, she traveled 30,000 miles, using 900 gallons of gasoline. Her car gets 40 miles per gallon on the highway and 25 in the city. She can deduct all highway travel, but no city travel, on her taxes. How many miles should Theresa be allowed as a business expense?

45. **Constructing a Border Around a Garden** Wendy, a landscaper, who just completed a rectangular flower garden measuring 6 feet by 10 feet, orders 1 cubic yard of premixed cement, all of which is to be used to create a border of uniform width around the garden. If the border

is to have a depth of 3 inches, how wide will the border be? (1 cubic yard = 27 cubic feet)

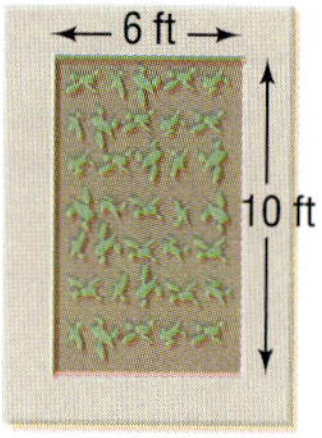

46. Physics An object is propelled vertically upward with an initial velocity of 20 meters per second. The distance s (in meters) of the object from the ground after t seconds is $s = -4.9t^2 + 20t$.

(a) When will the object be 15 meters above the ground?

(b) When will it strike the ground?

(c) Will the object reach a height of 100 meters?

(d) What is the maximum height?

47. Reducing the Size of a Candy Bar A jumbo chocolate bar with a rectangular shape measures 12 centimeters in length, 7 centimeters in width, and 3 centimeters in thickness. Due to escalating costs of cocoa, management decides to reduce the volume of the bar by 10%. To accomplish this reduction, management decides the new bar should have the same 3 centimeter thickness, but the length and width each should be reduced an equal number of centimeters. What should be the dimensions of the new candy bar?

48. Reducing the Size of a Candy Bar Rework Problem 47 if the reduction is to be 20%.

49. Constructing a Border Around a Pool A pool in the shape of a circle measures 10 feet across. One cubic yard of concrete is to be used to create a circular border of uniform width around the pool. If the border is to have a depth of 3 inches, how wide will the border be? (1 cubic yard = 27 cubic feet)

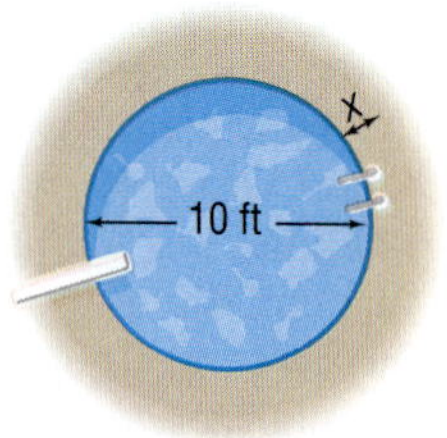

50. Constructing a Border Around a Pool Rework Problem 49 if the depth of the border is 4 inches.

51. Physics: Uniform Motion A motorboat can maintain a constant speed of 16 miles per hour relative to the water. The boat makes a trip upstream to a certain point in 20 minutes; the return trip takes 15 minutes. What is the speed of the current? (See the figure.)

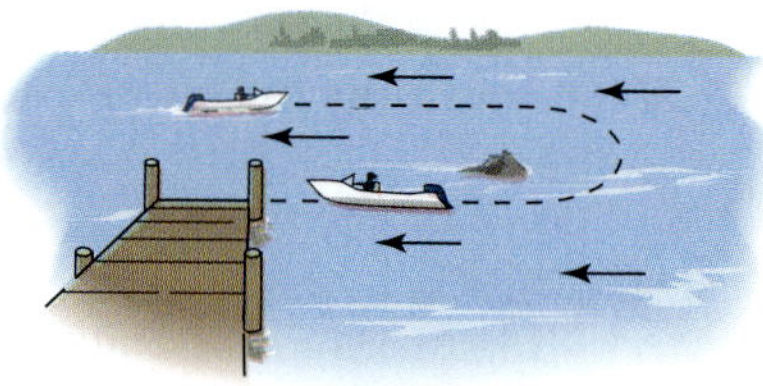

52. Purity of Gold The purity of gold is measured in karats, with pure gold being 24 karats. Other purities of gold are expressed as proportional parts of pure gold. Thus, 18 karat gold is $\frac{18}{24}$, or 75% pure gold; 12 karat gold is $\frac{12}{24}$, or 50%, pure gold; and so on. How much 12 karat gold should be mixed with pure gold to obtain 60 grams of 16 karat gold?

53. Business: Mixing Nuts A nut store normally sells cashews for \$4.00 per pound and peanuts for \$1.50 per pound. But at the end of the month, the peanuts had not sold well, so in order to sell 60 pounds of peanuts, the manager decided to mix the 60 pounds of peanuts with some cashews and sell the mixture for \$2.50 per pound. How many pounds of cashews should be mixed with the peanuts to ensure no change in the profit?

54. Business: Mixing Candy A candy store sells boxes of candy containing caramels and cremes. Each box sells for \$9.50 and holds 30 pieces of candy (all pieces are the same size). If the caramels cost \$0.25 to produce and the cremes cost \$0.45 to produce, how many of each should be in a box for no loss or gain to occur?

55. Running a Race Mike can run the mile in 6 minutes, and Dan can run the mile in 9 minutes. If Mike gives Dan a head start of 1 minute, how far from the start will Mike pass Dan? (See the figure.) How long does it take?

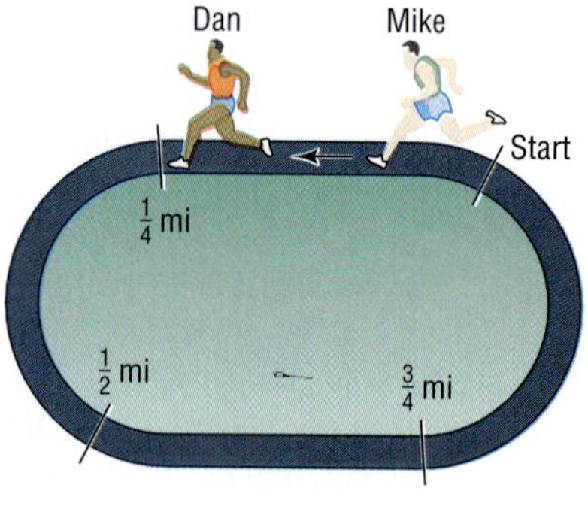

56. **Physics: Uniform Motion** A motorboat heads upstream on a river that has a current of 3 miles per hour. The trip upstream takes 5 hours, while the return trip takes 2.5 hours. What is the speed of the motorboat? (Assume that the motorboat maintains a constant speed relative to the water.)

57. **Rescue at Sea** A ship that is in danger of sinking radios the Coast Guard for assistance. When the rescue craft leaves the Coast Guard station, the ship is 60 miles away and heading directly toward the station. If the average speed of the ship is 10 miles per hour and the average speed of the rescue craft is 20 miles per hour, how long will it take for the rescue craft to reach the ship? (See the figure.)

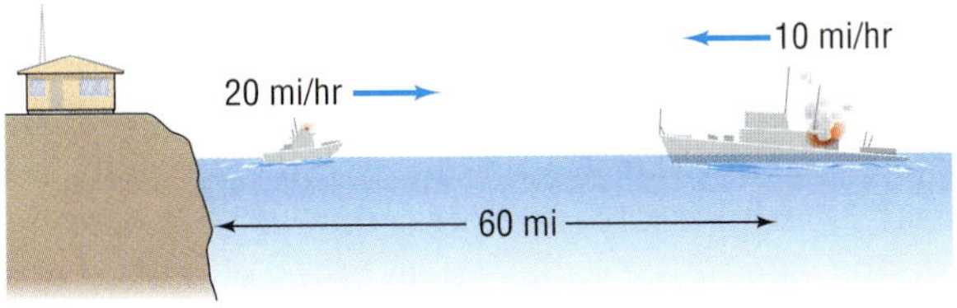

58. **Physics: Uniform Motion** Two cars enter the Florida Turnpike at Commercial Boulevard at 8:00 A.M., each heading for Wildwood. One car's average speed is 10 miles per hour more than the other's. The faster car arrives at Wildwood at 11:00 A.M., $\frac{1}{2}$ hour before the other car. What is the average speed of each car? How far did each travel?

59. **Emptying Oil Tankers** An oil tanker can be emptied by the main pump in 4 hours. An auxiliary pump can empty the tanker in 9 hours. If the main pump is started at 9 A.M., when should the auxiliary pump be started so that the tanker is emptied by noon?

60. **Cement Mix** A 20-pound bag of Economy brand cement mix contains 25% cement and 75% sand. How much pure cement must be added to produce a cement mix that is 40% cement?

61. **Emptying a Tub** A bathroom tub will fill in 15 minutes with both faucets open and the stopper in place. With both faucets closed and the stopper removed, the tub will empty in 20 minutes. How long will it take for the tub to fill if both faucets are open and the stopper is removed?

62. **Range of an Airplane** An air rescue plane averages 300 miles per hour in still air. It carries enough fuel for 5 hours of flying time. If, upon takeoff, it encounters a wind of 30 miles per hour, how far can it fly and return safely? (Assume that the wind remains constant.)

63. **Home Equity Loans** Suppose you obtain a home equity loan of $100,000 that requires only a monthly interest payment at 10% per annum, with the principal due after 5 years. You decide to invest part of the loan in a 5-year CD that pays 9% compounded and paid monthly and part in a B+ rated bond due in 5 years that pays 12% compounded and paid monthly. What is the most you can invest in the CD to ensure that the monthly home equity loan payment is made?

64. **Comparing Olympic Heroes** In the 1984 Olympics, Carl Lewis of the United States won the gold medal in the 100 meter race with a time of 9.99 seconds. In the 1896 Olympics, Thomas Burke, also of the United States, won the gold medal in the 100 meter race in 12.0 seconds. If they ran in the same race repeating their respective times, by how many meters would Lewis beat Burke?

65. **Computing Average Speed** In going from Chicago to Atlanta, a car averages 45 miles per hour, and in going from Atlanta to Miami, it averages 55 miles per hour. If Atlanta is halfway between Chicago and Miami, what is the average speed from Chicago to Miami? Discuss an intuitive solution. Write a paragraph defending your intuitive solution. Then solve the problem algebraically. Is your intuitive solution the same as the algebraic one? If not, find the flaw.

66. **Speed of a Plane** On a recent flight from Phoenix to Kansas City, a distance of 919 nautical miles, the plane arrived 20 minutes early. On leaving the aircraft, I asked the captain, "What was our tail wind?" He replied, "I don't know, but our ground speed was 550 knots." How can you determine if enough information is provided to find the tail wind? If possible, find the tail wind. (1 knot = 1 nautical mile per hour)

67. **Critical Thinking** You are the manager of a clothing store and have just purchased 100 dress shirts for $20.00 each. After one month of selling the shirts at the regular price, you plan to have a sale giving 40% off the original selling price. However, you still want to make a profit of $4 on each shirt at the sale price. What should you price the shirts at initially to ensure this?

If, instead of 40% off at the sale, you give 50% off, by how much is your profit reduced?

68. **Critical Thinking** Make up a word problem that requires solving a linear equation as part of its solution. Exchange problems with a friend. Write a critique of your friend's problem.

69. **Critical Thinking** Without solving, explain what is wrong with the following mixture problem: How many liters of 25% ethanol should be added to 20 liters of 48% ethanol to obtain a solution of 58% ethanol? Now go through an algebraic solution. What happens?

3.4 OTHER TYPES OF EQUATIONS

1 Solve Radical Equations
2 Solve Equations Quadratic in Form
3 Solve Equations That Are Factorable

In this section we look at other types of equations, most of which can be solved using variations of the techniques already discussed.

Equations Containing Radicals

1 When the variable in an equation occurs in a square root, cube root, and so on, that is, when it occurs in a radical, the equation is called a **radical equation.** Sometimes, a suitable operation will change a radical equation to one that is linear or quadratic. A commonly used procedure is to isolate the most complicated radical on one side of the equation and then eliminate it by raising each side to a power equal to the index of the radical. Care must be taken, however, because apparent solutions that are not, in fact, solutions of the original equation may result. These are called **extraneous solutions.** Thus, we need to check all answers when working with radical equations.

EXAMPLE 1 Solving a Radical Equation

Find the real solutions of the equation $\sqrt[3]{2x-4} - 2 = 0$.

Graphing Solution Figure 28(a) shows the graph of the function

$$f(x) = \sqrt[3]{2x-4} - 2$$

From the graph, we see one x-intercept near 6. We conjecture that the x-intercept is 6. Remember, if upon repeated ZOOMing it appears that the graph crosses the x-axis at a tick mark, we conjecture that the tick mark is the x-intercept. See Figure 28(b). We verify this by checking if $f(6) = 0$.

$$f(6) = \sqrt[3]{2(6)-4} - 2 = \sqrt[3]{8} - 2 = 0$$

Since $f(6) = 0$, 6 is the x-intercept. Thus, the only solution is $x = 6$.

FIGURE 28

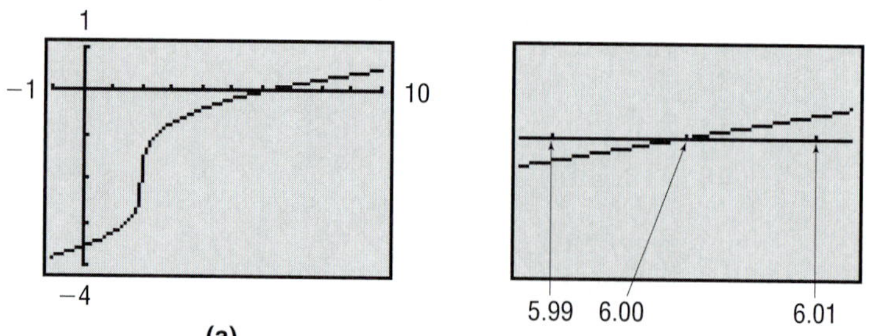

(a) (b)

MISSION POSSIBLE

Saving the Economic Future of Krispy Krunchy Candy Bar Co.

The Krispy Krunchy Candy Bar Company is facing a financial crisis because the cost of shipping cocoa beans from Ghana has increased. The CEO has decided that the way to stay afloat would be to reduce the size of their GIANT KRISPY KRUNCHY BAR by 10% but keep the price the same. He doesn't want to lose any customers, however. Therefore, he wants the change in size to be as unobtrusive as possible. The present dimensions of the GIANT KRISPY KRUNCHY BAR are 12 cm in length, 7 cm in width, and 3 cm in thickness. The CEO has asked your consulting firm to come up with the best way to shrink the candy bar. Because millions of dollars are riding on this decision, you will need to find all answers in centimeters correct to three decimal places and all percents correct to two decimal places.

1. Make a sketch of the candy bar, roughly to scale, and label it.
2. What is the present volume of the candy bar?
3. What would be the new volume after a 10% reduction?
4. What would be the new volume if each dimension were reduced by 10%. Is this the same as your answer to #3? (It shouldn't be.) Explain the difference.
5. Consider reducing only one of the dimensions. There are three possibilities. What would the new dimensions be in each case?
6. Consider reducing two (but not three) of the dimensions by the same amount. What would the new dimensions be in each case? (There are three possibilities; in each case you want the volume to be 10% less than the original volume.)
7. Consider reducing all three of the dimensions by the same amount. What would the new dimensions be?
8. Within your group make a decision about which of the seven possibilities you found would be the best one to recommend to the CEO of Krispy Krunchy. Write out two or three sentences to justify your choice.
9. Would you mind if you discovered that your favorite candy bar had been reduced in size while the price stayed the same? Do you think a candy company might actually do this to improve their financial standing? What is your protection as a consumer from being fooled?

Algebraic Solution The equation contains a radical whose index is 3. We isolate it on the left side:

$$\sqrt[3]{2x - 4} - 2 = 0$$
$$\sqrt[3]{2x - 4} = 2$$

Now raise each side to the third power (the index of the radical is 3) and solve:

$$(\sqrt[3]{2x - 4})^3 = 2^3$$
$$2x - 4 = 8$$
$$2x = 12$$
$$x = 6$$

Check: $\sqrt[3]{2(6) - 4} - 2 = \sqrt[3]{12 - 4} - 2 = \sqrt[3]{8} - 2 = 2 - 2 = 0$

The solution is $x = 6$.

Sometimes, we need to raise each side to a power more than once in order to solve a radical equation algebraically.

EXAMPLE 2 Solving a Radical Equation

Find the real solutions of the equation $\sqrt{2x + 3} - \sqrt{x + 2} = 2$.

Graphing Solution First, we put the equation in the form $f(x) = 0$. Figure 29 shows the graph of the function

$$f(x) = \sqrt{2x + 3} - \sqrt{x + 2} - 2$$

From the graph, we see one x-intercept near 23. We conjecture that the x-intercept is 23. We verify this by checking if $f(23) = 0$.

$$f(23) = \sqrt{2(23) + 3} - \sqrt{23 + 2} - 2 = \sqrt{49} - \sqrt{25} - 2 = 0$$

The solution is $x = 23$.

FIGURE 29

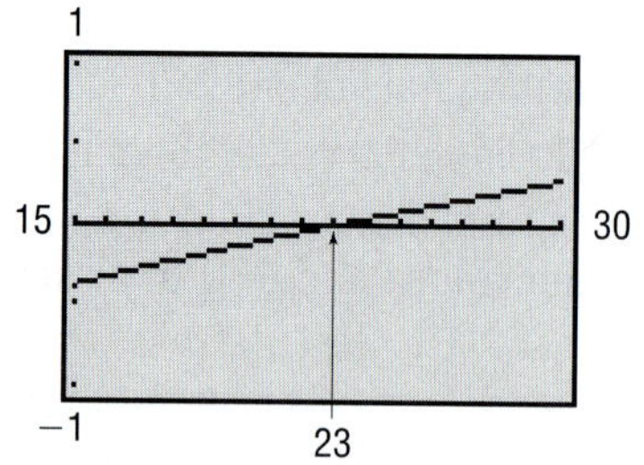

Algebraic Solution First, we choose to isolate the more complicated radical expression (in this case, $\sqrt{2x + 3}$) on the left side:

$$\sqrt{2x + 3} = \sqrt{x + 2} + 2$$

Now square both sides (the index of the radical is 2):

$$(\sqrt{2x + 3})^2 = (\sqrt{x + 2} + 2)^2$$
$$2x + 3 = (\sqrt{x + 2})^2 + 4\sqrt{x + 2} + 4$$
$$2x + 3 = x + 2 + 4\sqrt{x + 2} + 4$$

Because the equation still contains a radical, we combine like terms, isolate the remaining radical on the right side, and again square both sides:

$$\begin{aligned} x - 3 &= 4\sqrt{x + 2} \\ (x - 3)^2 &= 16(x + 2) \\ x^2 - 6x + 9 &= 16x + 32 \\ x^2 - 22x - 23 &= 0 \\ (x - 23)(x + 1) &= 0 \\ x = 23 \quad &\text{or} \quad x = -1 \end{aligned}$$

The original equation appears to have the solution set $\{-1, 23\}$. However, we have not yet checked.

Check: $\sqrt{2(23) + 3} - \sqrt{23 + 2} = \sqrt{49} - \sqrt{25} = 7 - 5 = 2$
$\sqrt{2(-1) + 3} - \sqrt{-1 + 2} = \sqrt{1} - \sqrt{1} = 1 - 1 = 0$

Thus, the equation has only one solution, 23; the solution -1 is extraneous.

Now work Problem 7.

Equations Quadratic in Form

2 The equation $x^4 + x^2 - 12 = 0$ is not quadratic in x, but it is quadratic in x^2. That is, if we let $u = x^2$, we get $u^2 + u - 12 = 0$, a quadratic equation. This equation can be solved for u and, in turn, by using $u = x^2$, we can find the solutions x of the original equation.

In general, if an appropriate substitution u transforms an equation into one of the form

$$au^2 + bu + c = 0 \qquad a \neq 0$$

then the original equation is called an **equation of the quadratic type,** or an **equation quadratic in form.**

The difficulty of solving such an equation algebraically lies in the determination that the equation is, in fact, quadratic in form. After you are told an equation is quadratic in form, it is easy enough to see it, but some practice is needed to enable you to recognize them on your own.

EXAMPLE 3

Solving Equations That Are Quadratic in Form

Find the real solutions of the equation $(x^2 - 1)^2 + (x^2 - 1) - 12 = 0$.

Graphing Solution

FIGURE 30

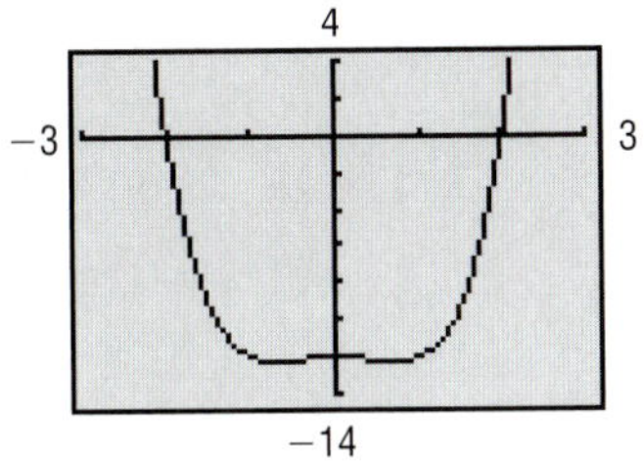

Figure 30 shows the graph of the function

$$f(x) = (x^2 - 1)^2 + (x^2 - 1) - 12$$

From the graph, we see two x-intercepts: one near -2; the other near 2. We conjecture that the x-intercepts are -2 and 2. We verify this by checking if $f(-2) = f(2) = 0$.

$$\begin{aligned} f(-2) &= [(-2)^2 - 1]^2 + [(-2)^2 - 1] - 12 = 9 + 3 - 12 = 0 \\ f(2) &= (2^2 - 1)^2 + (2^2 - 1) - 12 = 9 + 3 - 12 = 0 \end{aligned}$$

The solution set is $\{-2, 2\}$.

Algebraic Solution

For the equation $(x^2 - 1)^2 + (x^2 - 1) - 12 = 0$, we let $u = x^2 - 1$ so that $u^2 = (x^2 - 1)^2$. Then the original equation

$$(x^2 - 1)^2 + (x^2 - 1) - 12 = 0$$

becomes

$$u^2 + u - 12 = 0 \quad u = x^2 - 1.$$
$$(u + 4)(u - 3) = 0 \quad \text{Factor.}$$
$$u = -4 \quad \text{or} \quad u = 3 \quad \text{Solve.}$$

But remember that we want to solve for x. Because $u = x^2 - 1$, we have

$$x^2 - 1 = -4 \quad \text{or} \quad x^2 - 1 = 3$$
$$x^2 = -3 \qquad\qquad x^2 = 4$$

The first of these has no real solution; the second has the solution set $\{-2, 2\}$.

Check: $x = -2$: $(4 - 1)^2 + (4 - 1) - 12 = 9 + 3 - 12 = 0$
$x = 2$: $(4 - 1)^2 + (4 - 1) - 12 = 9 + 3 - 12 = 0$

Thus, $\{-2, 2\}$ is the solution set of the original equation.

The function f used in the graphing solution of Example 3 is even $[f(-x) = f(x)]$, so its graph is symmetric with respect to the y-axis. As a result, if a is an x-intercept, so is $-a$. Similarly, if a function f is odd $[f(-x) = -f(x)]$, then its graph is symmetric with respect to the origin. As a result, if a is an x-intercept, so is $-a$. When using a graphing utility to solve an equation $f(x) = 0$, check to see whether f is even or odd. This will save time in listing the solutions of the equation.

EXAMPLE 4 Solving Equations That Are Quadratic in Form

Find the real solutions of the equation $x + 2\sqrt{x} - 3 = 0$.

Graphing Solution Figure 31 shows the graph of the function

$$f(x) = x + 2\sqrt{x} - 3$$

From the graph, we see one x-intercept near 1. We conjecture that the x-intercept is 1. This is verified since $f(1) = 0$. The only solution is $x = 1$.

FIGURE 31

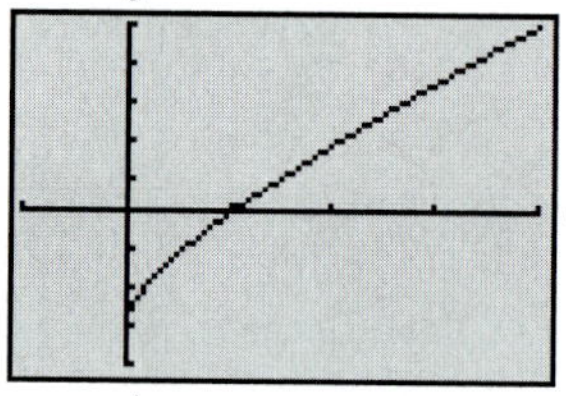

Algebraic Solution For the equation $x + 2\sqrt{x} - 3 = 0$, let $u = \sqrt{x}$. Then $u^2 = x$, and the original equation,

$$x + 2\sqrt{x} - 3 = 0$$

becomes

$$u^2 + 2u - 3 = 0 \quad u = \sqrt{x}.$$
$$(u + 3)(u - 1) = 0 \quad \text{Factor.}$$
$$u = -3 \quad \text{or} \quad u = 1 \quad \text{Solve.}$$

Since $u = \sqrt{x}$, we have $\sqrt{x} = -3$ or $\sqrt{x} = 1$. The first of these, $\sqrt{x} = -3$, has no real solution, since the square root of a real number is never negative. The second one, $\sqrt{x} = 1$, has the solution $x = 1$.

Check: $1 + 2\sqrt{1} - 3 = 1 + 2 - 3 = 0$

Thus, $x = 1$ is the only solution of the original equation.

Another algebraic method for solving Example 4 would be to treat it as a radical equation. Solve it this way for practice.

The idea should now be clear. If an equation contains an expression and that same expression squared, make a substitution for the expression. You may get a quadratic equation.

Now work Problems 31 and 35.

Factorable Equations

3 We have already used factoring as a means of algebraically solving certain quadratic equations. This method can also be used to solve *any* equation that can be factored with 0 on one side of the equation. The solutions are then found by setting each factor equal to 0.

EXAMPLE 5 Solving Equations by Factoring

Find the real solutions of the equation $x^3 - x^2 - 4x + 4 = 0$.

Graphing Solution Figure 32 shows the graph of the function

$$f(x) = x^3 - x^2 - 4x + 4$$

From the graph we see three x-intercepts, one near -2, one near 1, the other near 2. We conjecture that the x-intercepts are -2, 1, and 2. This is verified since $f(-2) = f(1) = f(2) = 0$.

FIGURE 32

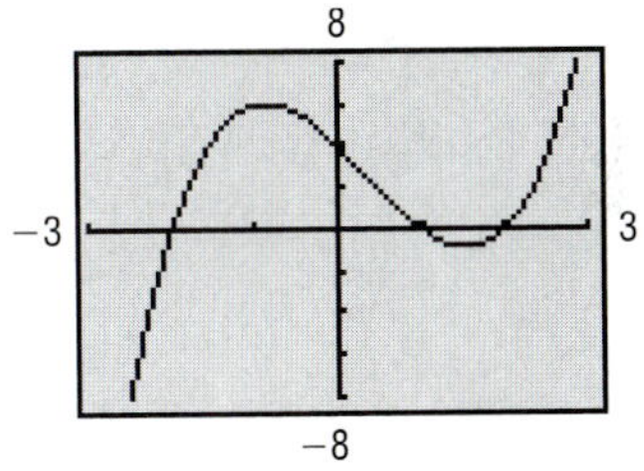

Algebraic Solution Do you recall the method of factoring by grouping? We group the terms of $x^3 - x^2 - 4x + 4 = 0$ as follows:

$$(x^3 - x^2) - (4x - 4) = 0$$

Factor out x^2 from the first grouping and 4 from the second:

$$x^2(x - 1) - 4(x - 1) = 0$$

This reveals the common factor $(x - 1)$, so we have

$$(x^2 - 4)(x - 1) = 0$$

$$(x - 2)(x + 2)(x - 1) = 0 \quad \text{Factor again.}$$

$$x - 2 = 0 \quad \text{or} \quad x + 2 = 0 \quad \text{or} \quad x - 1 = 0 \quad \text{Set each factor equal to 0.}$$

$$x = 2 \qquad x = -2 \qquad x = 1 \quad \text{Solve.}$$

The solution set is $\{-2, 1, 2\}$.

Now work Problem 67.

3.4 EXERCISES

In Problems 1–26, find the real solutions of each equation (a) graphically and (b) algebraically.

1. $\sqrt{2t-1}=1$
2. $\sqrt{3t+4}=2$
3. $\sqrt{3t+1}=-4$
4. $\sqrt{5t+4}=-3$
5. $\sqrt[3]{1-2x}-3=0$
6. $\sqrt[3]{1-2x}-1=0$
7. $\sqrt{15-2x}=x$
8. $\sqrt{12-x}=x$
9. $x=2\sqrt{x-1}$
10. $x=2\sqrt{-x-1}$
11. $\sqrt{x^2-x-4}=x+2$
12. $\sqrt{3-x+x^2}=x-2$
13. $3+\sqrt{3x+1}=x$
14. $2+\sqrt{12-2x}=x$
15. $\sqrt{2x+3}-\sqrt{x+1}=1$
16. $\sqrt{3x+7}+\sqrt{x+2}=1$
17. $\sqrt{3x+1}-\sqrt{x-1}=2$
18. $\sqrt{3x-5}-\sqrt{x+7}=2$
19. $\sqrt{3-2\sqrt{x}}=\sqrt{x}$
20. $\sqrt{10+3\sqrt{x}}=\sqrt{x}$
21. $(3x+1)^{1/2}=4$
22. $(3x-5)^{1/2}=2$
23. $(x-1)^{1/3}=2$
24. $(2x+1)^{1/3}=-2$
25. $(x^2+9)^{1/2}=5$
26. $(x^2-16)^{1/2}=9$

In Problems 27–52, find the real solutions of each equation (a) graphically and (b) algebraically.

27. $(x+1)^2+7(x+1)+12=0$
28. $(2x+3)^2-(2x+3)-6=0$
29. $(3x+4)^2-6(3x+4)+9=0$
30. $(2-x)^2+(2-x)-20=0$
31. $2(s+1)^2-5(s+1)=3$
32. $3(1-y)^2+5(1-y)+2=0$
33. $x-4\sqrt{x}=0$
34. $x+8\sqrt{x}=0$
35. $x+\sqrt{x}=20$
36. $x+\sqrt{x}-6=0$
37. $t^{1/2}-2t^{1/4}+1=0$
38. $z^{1/2}-2z^{1/4}+1=0$
39. $4x^{1/2}-9x^{1/4}+4=0$
40. $x^{1/2}-3x^{1/4}+2=0$
41. $\sqrt[4]{5x^2-6}=x$
42. $\sqrt[4]{4-5x^2}=x$
43. $x^2+3x+\sqrt{x^2+3x}=6$
44. $x^2-3x-\sqrt{x^2-3x}=2$
45. $\dfrac{1}{(x+1)^2}=\dfrac{1}{x+1}+2$
46. $\dfrac{1}{(x-1)^2}+\dfrac{1}{x-1}=12$
47. $3x^{-2}-7x^{-1}-6=0$
48. $2x^{-2}-3x^{-1}-4=0$
49. $2x^{2/3}-5x^{1/3}-3=0$
50. $3x^{4/3}+5x^{2/3}-2=0$
51. $\left(\dfrac{v}{v+1}\right)^2+\dfrac{2v}{v+1}=8$
52. $\left(\dfrac{y}{(y-1)}\right)^2=6\left(\dfrac{y}{y-1}\right)+7$

In Problems 53–72, find the real solutions of each equation by factoring. Verify any solution(s) by solving the equation graphically.

53. $x^3=x$
54. $x^4=4x^3$
55. $x^4-5x^2+4=0$
56. $x^4-10x^2+25=0$
57. $3x^4-2x^2-1=0$
58. $2x^4-5x^2-12=0$
59. $x^6+7x^3-8=0$
60. $x^6-7x^3-8=0$
61. $x=6\sqrt{x}$
62. $x=4\sqrt{x}$
63. $x^{3/2}-2x^{1/2}=0$
64. $x^{3/4}-4x^{1/4}=0$
65. $x^3+x^2-20x=0$
66. $x^3+6x^2-7x=0$
67. $x^3+x^2+x+1=0$
68. $x^3+x^2-x-1=0$
69. $x^3-3x^2-4x+12=0$
70. $x^3-3x^2-x+3=0$
71. $t^6-t^4-t^2+1=0$
72. $y^6-4y^4-y^2+4=0$

In Problems 73–78, find the real solutions of each equation. Use a calculator to express solutions correct to two decimal places.

73. $x - 4x^{1/2} + 2 = 0$

74. $x^{2/3} + 4x^{1/3} + 2 = 0$

75. $x^4 + \sqrt{3}x^2 - 3 = 0$

76. $x^4 + \sqrt{2}x^2 - 2 = 0$

77. $\pi(1 + t)^2 = \pi + 1 + t$

78. $\pi(1 + r)^2 = 2 + \pi(1 + r)$

79. If $k = \dfrac{x + 3}{x - 3}$ and $k^2 - k = 12$, find x.

80. If $k = \dfrac{x + 3}{x - 4}$ and $k^2 - 3k = 28$, find x.

81. **Physics: Using Sound to Measure Distance** The depth of a well can sometimes be found by dropping an object into the well and measuring the time elapsed until a sound is heard. If t_1 is the time (measured in seconds) it takes for the object to strike the bottom of the well, then t_1 will obey the equation $s = 16t_1^2$, where s is the distance (measured in feet). It follows that $t_1 = \sqrt{s}/4$. Suppose that t_2 is the time it takes for the sound of the impact to reach your ears. Because sound waves are known to travel at a speed of approximately 1100 feet per second, the time t_2 to travel the distance s will be $t_2 = s/1100$. Now $t_1 + t_2$ is the total time that elapses from the moment the object is dropped to the moment a sound is heard. Thus, we have the equation

$$\text{Total time elapsed} = \frac{\sqrt{s}}{4} + \frac{s}{1100}$$

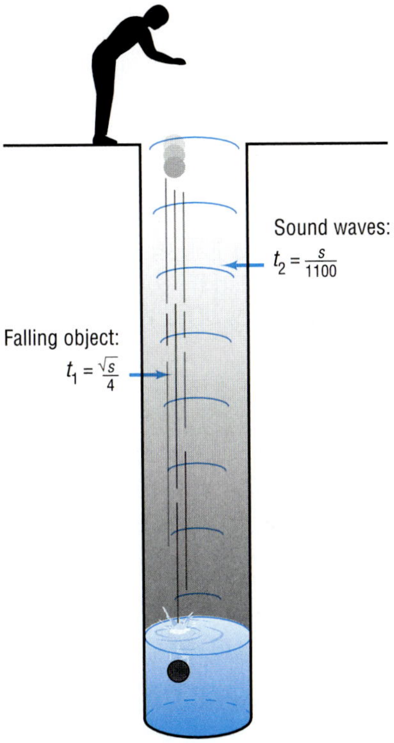

(a) Find the depth of a well if the total time elapsed from dropping a rock to hearing it hit bottom is 4 seconds.

(b) Use a graphing utility, graph the function

$$Y_1 = \frac{\sqrt{x}}{4} + \frac{x}{1100}$$

for $0 \le x \le 300$ and $0 \le Y_1 \le 5$.

(c) Set up a table with ΔTbl = 5 and TblMin = 0. From the table, we can approximate the slope of the graph found in part (b) by calculating $\dfrac{\Delta Y_1}{\Delta x}$. Approximate the slope as x increases from 0 to 5, 40 to 45, and 100 to 105. What is happening to the slope as x (the depth of the well) increases?

(d) What conclusions can you make about the relationship between the depth of a well and the time that elapses before a sound is heard based on your answers to parts (b) and (c)?

82. Make up a radical equation that has no solution.

83. Make up a radical equation that has an extraneous solution.

84. Discuss what there is in the solving process for radical equations that leads to the possibility of extraneous solutions. Why is there no such possibility for linear and quadratic equations?

3.5 INEQUALITIES

1 Use Properties of Inequalities
2 Solve Inequalities Algebraically
3 Solve Inequalities by Graphing

Properties of Inequalities

1 In working with inequalities, we will need to know certain properties that they obey.

We begin with the **trichotomy property,** which states that either two numbers are equal or one of them is less than the other.

For any pair of numbers a and b,

Trichotomy Property

$$a < b \quad \text{or} \quad a = b \quad \text{or} \quad b < a$$

If $b = 0$, the trichotomy property states that, for any real number a,

$$a < 0 \quad \text{or} \quad a = 0 \quad \text{or} \quad a > 0$$

That is, any real number is negative or 0 or positive, a fact we have already noted.

The product of two positive real numbers is positive, the product of two negative real numbers is positive, and the product of 0 and 0 is 0. Thus, for any real number a, the value of a^2 is 0 or positive; that is, a^2 is nonnegative. This is called the **nonnegative property.**

For any real number a, we have

Nonnegative Property

$$a^2 \geq 0$$

FIGURE 33

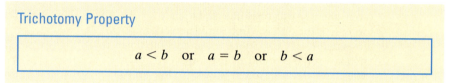

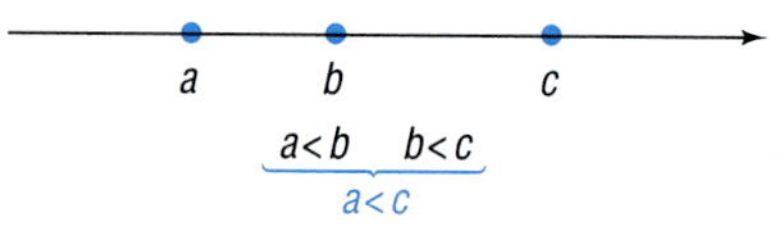

In Figure 33, we can see that, if a lies to the left of b and b lies to the left of c, then it follows that a must also lie to the left of c. This is called the **transitive property** for the inequality $<$. There is also a corresponding property for the inequality $>$.

Transitive Property of Inequalities

If $a < b$ and $b < c$, then $a < c$ (1a)
If $a > b$ and $b > c$, then $a > c$ (1b)

Draw an illustration similar to Figure 33 that depicts the transitive property (1b) for the inequality $>$.

If we add the same number to both sides of an inequality, we obtain an equivalent inequality. For example, since $3 < 5$, then $3 + 4 < 5 + 4$ or $7 < 9$. This is called the **addition property** of inequalities.

Addition Property of Inequalities

$$\text{If } a < b, \text{ then } a + c < b + c \qquad (2a)$$
$$\text{If } a > b, \text{ then } a + c > b + c \qquad (2b)$$

The addition property states that the sense, or direction, of an inequality remains unchanged if the same number is added to each side. Figure 34 illustrates the addition property (2a). In Figure 34(a), we see that a lies to the left of b. If c is positive, then $a + c$ and $b + c$ each lie c units to the right of a and b, respectively. Consequently, $a + c$ must lie to the left of $b + c$; that is, $a + c < b + c$. Figure 34(b) illustrates the situation if c is negative.

FIGURE 34

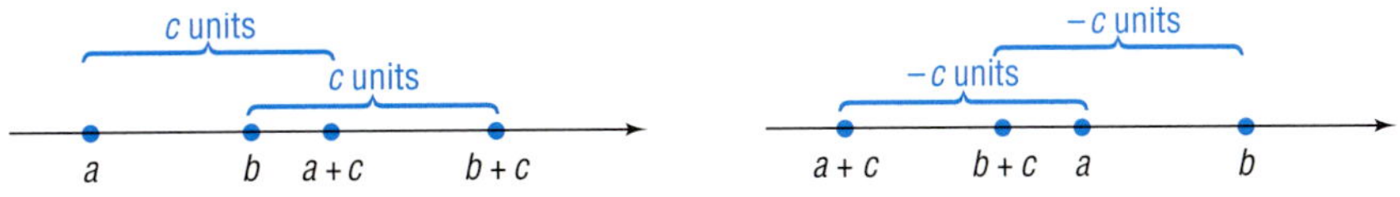

(a) If $a < b$ and $c > 0$, then $a + c < b + c$.

(b) If $a < b$ and $c < 0$, then $a + c < b + c$.

Draw an illustration similar to Figure 34 that illustrates the addition property (2b).

EXAMPLE 1 Addition Property of Inequalities

(a) If $x < -5$, then $x + 5 < -5 + 5$ or $x + 5 < 0$.

(b) If $x > 2$, then $x + (-2) > 2 + (-2)$ or $x - 2 > 0$.

Now work Problem 3.

We'll use two examples to arrive at our next property.

EXAMPLE 2 Multiplying an Inequality by a Positive Number

Express as an inequality the result of multiplying each side of the inequality $3 < 7$ by 2.

Solution We begin with

$$3 < 7$$

Multiplying each side by 2 yields the numbers 6 and 14, so we have

$$6 < 14$$

EXAMPLE 3 Multiplying an Inequality by a Negative Number

Express as an inequality the result of multiplying each side of the inequality $9 > 2$ by -4.

Solution We begin with

$$9 > 2$$

Multiplying each side by -4 yields the numbers -36 and -8, so we have

$$-36 < -8$$

Note that the effect of multiplying both sides of $9 > 2$ by the negative number -4 is that the direction of the inequality symbol is reversed.

Examples 2 and 3 illustrate the following general **multiplication properties** for inequalities:

Multiplication Properties for Inequalities

If $a < b$ and if $c > 0$, then $ac < bc$.
If $a < b$ and if $c < 0$, then $ac > bc$. (3a)

If $a > b$ and if $c > 0$, then $ac > bc$.
If $a > b$ and if $c < 0$, then $ac < bc$. (3b)

The multiplication properties state that the sense, or direction, of an inequality *remains the same* if each side is multiplied by a *positive* real number, while the direction is *reversed* if each side is multiplied by a *negative* real number.

EXAMPLE 4 Multiplication Property of Inequalities

(a) If $2x < 6$, then $\frac{1}{2}(2x) < \frac{1}{2}(6)$ or $x < 3$.

(b) If $\dfrac{x}{-3} > 12$, then $-3\left(\dfrac{x}{-3}\right) < -3(12)$ or $x < -36$.

(c) If $-4x > -8$, then $\dfrac{-4x}{-4} < \dfrac{-8}{-4}$ or $x < 2$.

(d) If $-x < 8$, then $(-1)(-x) > (-1)(8)$ or $x > -8$.

Now work Problem 7.

The **reciprocal property** states that the reciprocal of a positive real number is positive and that the reciprocal of a negative real number is negative.

Reciprocal Property for Inequalities

If $a > 0$, then $\dfrac{1}{a} > 0$. (4a)

If $a < 0$, then $\dfrac{1}{a} < 0$. (4b)

Solving Inequalities

2 An **inequality in one variable** is a statement involving two expressions, at least one containing the variable, separated by one of the inequality symbols, $<$, $\leq$, $>$, or $\geq$. To **solve an inequality** means to find all values of the variable for which the statement is true. These values are called **solutions** of the inequality.

For example, the following are all inequalities involving one variable, x:

$$x + 5 < 8 \qquad 2x - 3 \geq 4 \qquad x^2 - 1 \leq 3 \qquad \frac{x+1}{x-2} > 0$$

Two inequalities having exactly the same solution set are called **equivalent inequalities.**

As with equations, one method for solving an inequality algebraically is to replace it by a series of equivalent inequalities, until an inequality with an obvious solution, such as $x < 3$, is obtained. We obtain equivalent inequalities by applying some of the same operations as those used to find equivalent equations. The addition property and the multiplication properties form the basis for the following procedures.

Procedures That Leave the Inequality Symbol Unchanged

1. Simplify both sides of the inequality by combining like terms and eliminating parentheses:

   Replace $(x + 2) + 6 > 2x + (x + 1)$
   by $x + 8 > 3x + 1$

2. Add or subtract the same expression on both sides of the inequality:

   Replace $3x - 5 < 4$
   by $(3x - 5) + 5 < 4 + 5$

3. Multiply or divide both sides of the inequality by the same *positive* expression:

   Replace $4x > 16$ by $\frac{4x}{4} > \frac{16}{4}$

Procedures That Reverse the Sense or Direction of the Inequality Symbol

1. Interchange the two sides of the inequality:

   Replace $3 < x$ by $x > 3$

2. Multiply or divide both sides of the inequality by the same *negative* expression:

   Replace $-2x > 6$ by $\frac{-2x}{-2} < \frac{6}{-2}$

To solve an inequality using a graphing utility, we follow these steps:

Steps for Solving Inequalities Graphically

STEP 1: Write the inequality in one of the following forms:

$$f(x) < g(x) \qquad f(x) > g(x) \qquad f(x) \leq g(x) \qquad f(x) \geq g(x)$$

STEP 2: Graph: $Y_1 = f(x)$ and $Y_2 = g(x)$ on the same screen.

If the inequality is the form $f(x) < g(x)$, determine on what interval(s) Y_1 is below Y_2.

If the inequality is of the form $f(x) > g(x)$, determine on what interval(s) Y_1 is above Y_2.

If the inequality is not strict,* include the x-coordinates of the points of intersection in the solution.

*See Appendix 1 for a discussion of nonstrict inequalities.

Linear Inequalities

3

A **linear inequality in one variable** is an inequality equivalent to one of the forms

$$ax + b < 0 \qquad ax + b > 0$$
$$ax + b \le 0 \qquad ax + b \ge 0$$

where a and b are real numbers and $a \neq 0$.

The remainder of this section deals with solving linear inequalities. In the next section, we discuss the solution of inequalities involving absolute value. As the examples that follow illustrate, we solve linear inequalities algebraically using many of the same steps we would use to solve linear equations. In writing the solution of an inequality, we may use either set notation or interval notation, whichever is more convenient.

EXAMPLE 5 Solving Linear Inequalities

Solve the inequality $3 - 2x < 5$, and draw a graph to illustrate the solution.

Graphing Solution We graph $Y_1 = 3 - 2x$ and $Y_2 = 5$ on the same screen. See Figure 35. Using the INTERSECT command, we find that Y_1 and Y_2 intersect at $x = -1$. The graph of Y_1 is below that of Y_2, $Y_1 < Y_2$, to the right of the point of intersection. Since the inequality is strict, the solution set is $\{x | x > -1\}$, or, using interval notation, $(-1, \infty)$.

FIGURE 35

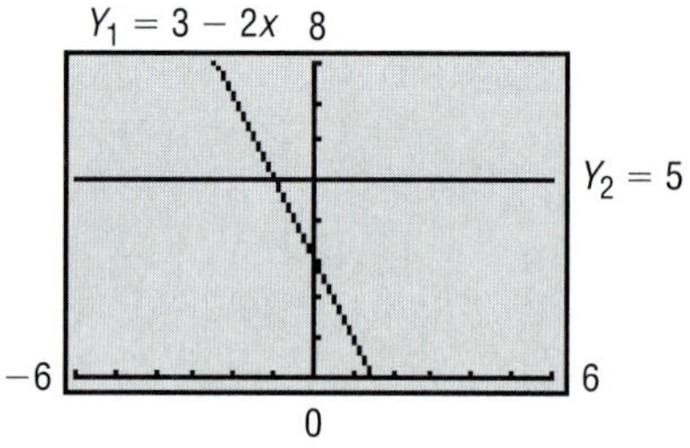

Algebraic Solution

$$3 - 2x < 5$$
$$3 - 2x - 3 < 5 - 3 \quad \text{Subtract 3 from both sides.}$$
$$-2x < 2 \quad \text{Simplify.}$$
$$\frac{-2x}{-2} > \frac{2}{-2} \quad \text{Divide both sides by } -2\text{. (The sense of the inequality symbol is reversed.)}$$
$$x > -1 \quad \text{Simplify.}$$

The solution set is $\{x | -1 < x < \infty\}$ or, using interval notation, all numbers in the interval $(-1, \infty)$. See Figure 36 for the graph of the solution set.

FIGURE 36
$-1 < x < \infty$ or $(-1, \infty)$

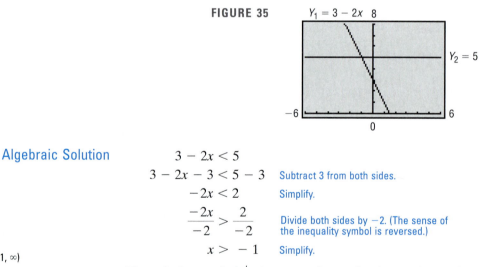

EXAMPLE 6 Solving Linear Inequalities

Solve the inequality $4x + 7 \ge 2x - 3$, and draw a graph to illustrate the solution.

Graphing Solution We graph $Y_1 = 4x + 7$ and $Y_2 = 2x - 3$ on the same screen. See Figure 37. Using the INTERSECT command, we find that Y_1 and Y_2 intersect at $x = -5$. The graph of Y_1 is above that of Y_2, $Y_1 > Y_2$, to the right of the point of intersection. Since the inequality is not strict, the solution set is $\{x|x \geq -5\}$ or, using interval notation, $[-5, \infty)$.

FIGURE 37

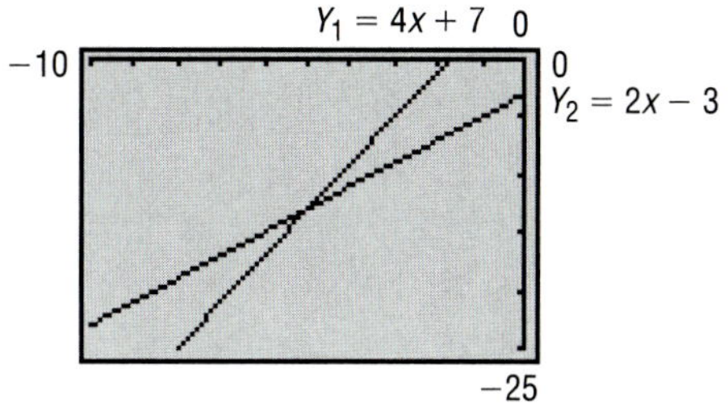

Algebraic Solution

$$4x + 7 \geq 2x - 3$$
$$4x + 7 - 7 \geq 2x - 3 - 7 \quad \text{Subtract 7 from both sides.}$$
$$4x \geq 2x - 10 \quad \text{Simplify.}$$
$$4x - 2x \geq 2x - 10 - 2x \quad \text{Subtract } 2x \text{ from both sides.}$$
$$2x \geq -10 \quad \text{Simplify.}$$
$$\frac{2x}{2} \geq \frac{-10}{2} \quad \text{Divide both sides by 2. (The sense of the inequality symbol is unchanged.)}$$
$$x \geq -5 \quad \text{Simplify.}$$

FIGURE 38
$-5 \leq x < \infty$ or $[-5, \infty)$

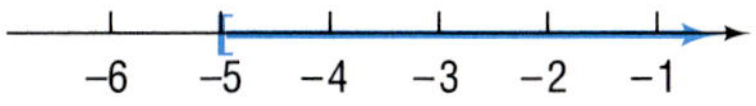

The solution set is $\{x|-5 \leq x < \infty\}$ or, using interval notation, all numbers in the interval $[-5, \infty)$.

See Figure 38 for the graph.

Now work Problem 19.

EXAMPLE 7 Solving Combined Inequalities

Solve the inequality $-5 < 3x - 2 < 1$ and draw a graph to illustrate the solution.

Graphing Solution

To solve a combined inequality, we graph each part: $Y_1 = -5$, $Y_2 = 3x - 2$, and $Y_3 = 1$. We seek the values of x for which the graph of Y_2 is between the graphs of Y_1 and Y_3. See Figure 39. The point of intersection of Y_1 and Y_2 is $(-1, -5)$, and the point of intersection of Y_2 and Y_3 is $(1, 1)$. The inequality is true for all values of x between these two intersection points. Since the inequality is strict, the solution set is $\{x| -1 < x < 1\}$ or, using interval notation, $(-1, 1)$.

FIGURE 39

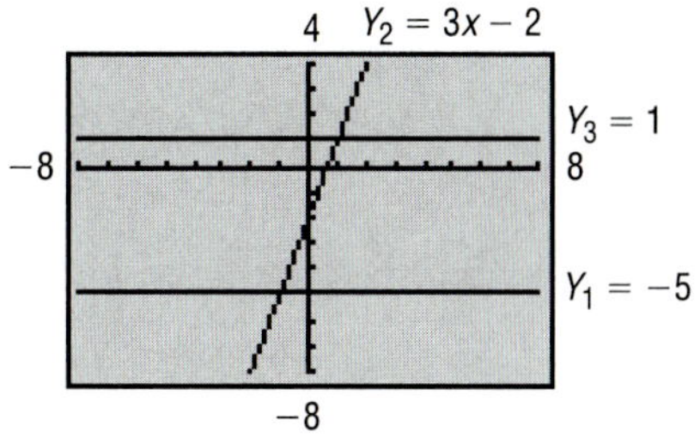

Algebraic Solution

Recall that the inequality

$$-5 < 3x - 2 < 1$$

is equivalent to the two inequalities

$$-5 < 3x - 2 \quad \text{and} \quad 3x - 2 < 1$$

We will solve each of these inequalities separately. For the first inequality,

$$\begin{aligned} -5 &< 3x - 2 \\ -5 + 2 &< 3x - 2 + 2 && \text{Add 2 to both sides.} \\ -3 &< 3x && \text{Simplify.} \\ \frac{-3}{3} &< \frac{3x}{3} && \text{Divide both sides by 3.} \\ -1 &< x && \text{Simplify.} \end{aligned}$$

The second inequality is solved as follows:

$$\begin{aligned} 3x - 2 &< 1 \\ 3x - 2 + 2 &< 1 + 2 && \text{Add 2 to both sides.} \\ 3x &< 3 && \text{Simplify.} \\ \frac{3x}{3} &< \frac{3}{3} && \text{Divide both sides by 3.} \\ x &< 1 && \text{Simplify.} \end{aligned}$$

The solution set of the original pair of inequalities consists of all x for which

$$-1 < x \quad \text{and} \quad x < 1$$

FIGURE 40
$-1 < x < 1$ or $(-1, 1)$

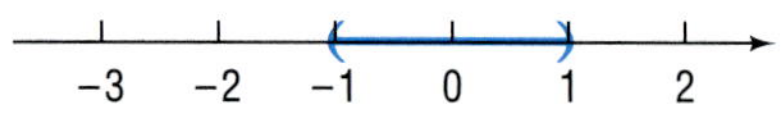

This may be written more compactly as $\{x | -1 < x < 1\}$. In interval notation, the solution is $(-1, 1)$. See Figure 40 for the graph.

We observe in the preceding process that the two inequalities we solved required exactly the same steps. A shortcut to solving the original inequality algebraically is to deal with the two inequalities at the same time, as follows:

$$\begin{aligned} -5 < \; & 3x - 2 < 1 \\ -5 + 2 < \; & 3x - 2 + 2 < 1 + 2 && \text{Add 2 to each part.} \\ -3 < \; & 3x < 3 && \text{Simplify.} \\ \frac{-3}{3} < \; & \frac{3x}{3} < \frac{3}{3} && \text{Divide each part by 3.} \\ -1 < \; & x < 1 && \text{Simplify.} \end{aligned}$$

We use this shortcut in the algebraic solution of the next example.

EXAMPLE 8 Solving Combined Inequalities

Solve the inequality:

$$-1 \le \frac{3 - 5x}{2} \le 9$$

Graph the solution set.

Graphing Solution Figure 41 shows the graphs of each part: $Y_1 = -1$, $Y_2 = \dfrac{3 - 5x}{2}$, and $Y_3 = 9$. The two points of intersection are $(-3, 9)$ and $(1, -1)$. The inequality is true for all values of x between these two intersection points. Since the inequality is not strict, the solution set is $\{x \mid -3 \le x \le 1\}$ or, using interval notation, $[-3, 1]$.

FIGURE 41

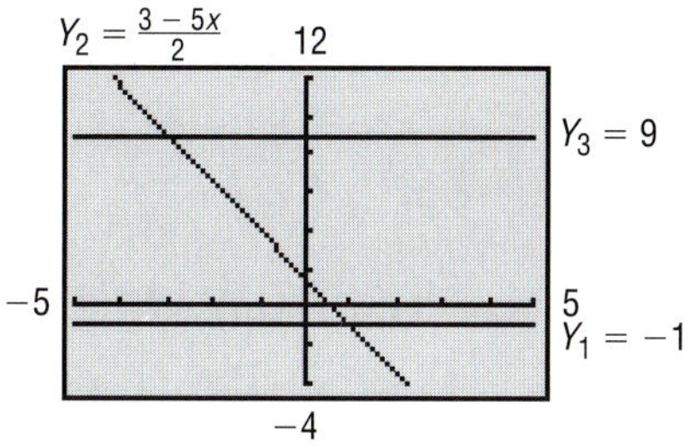

Algebraic Solution

$$-1 \le \frac{3-5x}{2} \le 9$$

$$2(-1) \le 2\left(\frac{3-5x}{2}\right) \le 2(9)$$ Multiply each part by 2 to remove the denominator.

$$-2 \le 3 - 5x \le 18$$ Simplify.

$$-2 - 3 \le 3 - 5x - 3 \le 18 - 3$$ Subtract 3 from each part to isolate the term containing x.

$$-5 \le -5x \le 15$$ Simplify.

$$\frac{-5}{-5} \ge \frac{-5x}{-5} \ge \frac{15}{-5}$$ Divide each part by −5 (change the sense of each inequality symbol).

$$1 \ge x \ge -3$$ Simplify.

$$-3 \le x \le 1$$ Reverse the order so that the numbers get larger as you read from left to right.

FIGURE 42
$-3 \le x \le 1$ or $[-3, 1]$

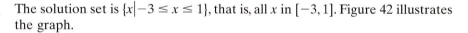

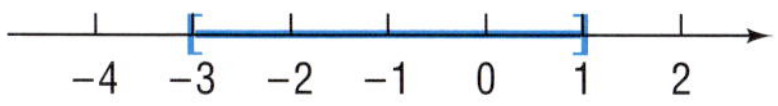

The solution set is $\{x \mid -3 \le x \le 1\}$, that is, all x in $[-3, 1]$. Figure 42 illustrates the graph.

Now work Problem 33.

EXAMPLE 9 Creating Equivalent Inequalities

If $-1 < x < 4$, find a and b so that $a < 2x + 1 < b$.

Solution

$$-1 < x < 4$$

$$-2 < 2x < 8$$ Multiply each part by 2.

$$-1 < 2x + 1 < 9$$ Add 1 to each part.

Thus, $a = -1$ and $b = 9$. ▪

Now work Problem 51.

Let's look at an applied problem involving linear inequalities.

EXAMPLE 10 Physics: Ohm's Law

In electricity, Ohm's law states that $E = IR$, where E is the voltage (in volts), I is the current (in amperes), and R is the resistance (in ohms). An air-conditioning unit is rated at a resistance of 10 ohms. If the voltage varies from 110 to 120 volts, inclusive, what corresponding range of current will the air conditioner draw?

Solution The voltage lies between 110 and 120, inclusive, so

$$\begin{aligned} 110 \le\ & E && \le 120 && \\ 110 \le\ & IR && \le 120 && \text{Ohm's law, } E = IR \\ 110 \le\ & I(10) && \le 120 && R = 10 \\ \frac{110}{10} \le\ & \frac{I(10)}{10} && \le \frac{120}{10} && \text{Divide each part by 10.} \\ 11 \le\ & I && \le 12 && \text{Simplify.} \end{aligned}$$

The air conditioner will draw between 11 and 12 amperes of current, inclusive.

HISTORICAL FEATURE

Inequalities are a relatively new component of the algebra curriculum. They have been introduced in the last 25 years for two important reasons.

First, if approximations are made in a problem, inequalities allow calculation of how serious the error is likely to be. This use is important for practical applications of mathematics and is also critical for the understanding of calculus (in its modern version, due principally to Augustin Louis Cauchy, 1789–1857, and Karl Weierstrass, 1815–1897).

Second, linear inequalities are the basis for solving a kind of problem that involves finding a maximum or minimum of a quantity, depending on variables that are subjected to certain constraints. For example, we might wish to ship several products by truck from several different factories, each factory having a limited supply of each product, and to get enough of the products to a central point within three days using the minimum amount of gas possible. Such problems, called *linear programming problems*, are discussed later in this text.

3.5 EXERCISES

In Problems 1–8, fill in the blank with the correct inequality symbol.

1. If $x < 5$, then $x - 5$ ________ 0.
2. If $x < -4$, then $x + 4$ ________ 0.
3. If $x > -4$, then $x + 4$ ________ 0.
4. If $x > 6$, then $x - 6$ ________ 0.
5. If $x > -4$, then $3x$ ________ -12.
6. If $x < 3$, then $2x$ ________ 6.
7. If $x < 6$, then $-2x$ ________ -12.
8. If $x > -2$, then $-4x$ ________ 8.

In Problems 9–12, an inequality is given. Write the equivalent inequality obtained by
(a) Adding -3 to each side of the given inequality;
(b) Subtracting 5 from each side of the given inequality;
(c) Multiplying each side of the given inequality by 3;
(d) Multiplying each side of the given inequality by -2.

9. $3 < 5$
10. $2 > 1$
11. $2x + 1 < 2$
12. $1 - 2x > 5$

In Problems 13–48, solve each inequality (a) graphically and (b) algebraically. Graph the solution set.

13. $x + 1 < 5$
14. $x - 6 < 1$
15. $1 - 2x \le 3$
16. $2 - 3x \le 5$
17. $3x - 7 > 2$
18. $2x + 5 > 1$
19. $3x - 1 \ge 3 + x$
20. $2x - 2 \ge 3 + x$
21. $-2(x + 3) < 8$
22. $-3(1 - x) < 12$
23. $4 - 3(1 - x) \le 3$
24. $8 - 4(2 - x) \le -2x$

25. $\frac{1}{2}(x - 4) > x + 8$

26. $3x + 4 > \frac{1}{3}(x - 2)$

27. $\frac{x}{2} \geq 1 - \frac{x}{4}$

28. $\frac{x}{3} \geq 2 + \frac{x}{6}$

29. $0 \leq 2x - 6 \leq 4$

30. $4 \leq 2x + 2 \leq 10$

31. $-5 \leq 4 - 3x \leq 2$

32. $-3 \leq 3 - 2x \leq 9$

33. $-3 < \frac{2x - 1}{4} < 0$

34. $0 < \frac{3x + 2}{2} < 4$

35. $1 < 1 - \frac{1}{2}x < 4$

36. $0 < 1 - \frac{1}{3}x < 1$

37. $(x + 2)(x - 3) > (x - 1)(x + 1)$

38. $(x - 1)(x + 1) > (x - 3)(x + 4)$

39. $x(4x + 3) \leq (2x + 1)^2$

40. $x(9x - 5) \leq (3x - 1)^2$

41. $\frac{1}{2} \leq \frac{x + 1}{3} < \frac{3}{4}$

42. $\frac{1}{3} < \frac{x + 1}{2} \leq \frac{2}{3}$

43. $(4x + 2)^{-1} < 0$

44. $(2x - 1)^{-1} > 0$

45. $0 < \frac{1}{x} < \frac{1}{5}$

46. $0 < \frac{1}{x} < \frac{1}{3}$

47. $0 < (2x - 4)^{-1} < \frac{1}{2}$

48. $0 < (3x + 6)^{-1} < \frac{1}{3}$

In Problems 49–58, find a and b.

49. If $-1 < x < 1$, then $a < x + 4 < b$.

50. If $-3 < x < 2$, then $a < x - 6 < b$.

51. If $2 < x < 3$, then $a < -4x < b$.

52. If $-4 < x < 0$, then $a < \frac{1}{2}x < b$.

53. If $0 < x < 4$, then $a < 2x + 3 < b$.

54. If $-3 < x < 3$, then $a < 1 - 2x < b$.

55. If $-3 < x < 0$, then $a < \frac{1}{x + 4} < b$.

56. If $2 < x < 4$, then $a < \frac{1}{x - 6} < b$.

57. If $6 < 3x < 12$, then $a < x^2 < b$.

58. If $0 < 2x < 6$, then $a < x^2 < b$.

59. If $-10 < x < 5$, solve: $\frac{2}{x + 10} > \frac{-1}{x - 5}$.

60. If $x > 3$, solve: $\frac{x}{x - 3} > 2$.

61. What is the domain of the variable in the expression $\sqrt{3x + 6}$? Verify your result using a graphing utility.

62. What is the domain of the variable in the expression $\sqrt{8 + 2x}$? Verify your result using a graphing utility.

63. If $a \leq b$ and $c > 0$, show that $ac \leq bc$.

[**Hint:** Since $a \leq b$, it follows that $a - b \leq 0$. Now multiply each side by c.]

64. If $a \leq b$ and $c < 0$, show that $ac \geq bc$.

65. **Arithmetic Mean** If $a < b$, show that $a < (a + b)/2 < b$. The number $(a + b)/2$ is called the **arithmetic mean** of a and b.

66. Refer to Problem 65. Show that the arithmetic mean of a and b is equidistant from a and b.

67. **Geometric Mean** If $0 < a < b$, show that $a < \sqrt{ab} < b$. The number $\sqrt{ab}$ is called the **geometric mean** of a and b.

68. Refer to Problems 65 and 67. Show that the geometric mean of a and b is less than the arithmetic mean of a and b.

69. **Harmonic Mean** For $0 < a < b$, let h be defined by

$$\frac{1}{h} = \frac{1}{2}\left(\frac{1}{a} + \frac{1}{b}\right)$$

Show that $a < h < b$. The number h is called the **harmonic mean** of a and b.

70. Refer to Problems 65, 67, and 69. Show that the harmonic mean of a and b equals the geometric mean squared divided by the arithmetic mean.

71. A young adult may be defined as someone older than 21, but less than 30 years of age. Express this statement using inequalities.

72. Middle-aged may be defined as being 40 or more and less than 60. Express this statement using inequalities.

73. **Life Expectancy** Metropolitan Life Insurance Co. reported that an average 25-year-old male in 1996 could expect to live at least 48.4 more years, and an average 25-year-old female in 1996 could expect to live at least 54.7 more years.

(a) To what age can an average 25-year-old male expect to live? Express your answer as an inequality.
(b) To what age can an average 25-year-old female expect to live? Express your answer as an inequality.
(c) Who can expect to live longer, a male or a female? By how many years?

74. General Chemistry For a certain ideal gas, the volume V (in cubic centimeters) equals 20 times the temperature T (in degrees Celsius). If the temperature varies from 80° to 120°C, inclusive, what is the corresponding range of the volume of the gas?

75. Real Estate Donna, a real estate agent agrees to sell a large apartment complex according to the following commission schedule: \$45,000 plus 25% of the selling price in excess of \$900,000. Assuming the complex will sell at some price between \$900,000 and \$1,100,000, inclusive, over what range does the agent's commission vary? How does the commission vary as a percent of selling price?

76. Sales Commission A used car salesperson is paid a commission of \$25 plus 40% of the selling price in excess of owner's cost. The owner claims that used cars typically sell for at least owner's cost plus \$70 and at most owner's cost plus \$300. For each sale made, over what range can the salesperson expect the commission to vary?

77. Federal Tax Withholding The percentage method of withholding for federal income tax (1995)* states that a single person whose weekly wages, after subtracting withholding allowances, are over \$476, but not over \$999, shall have \$63.90 plus 28% of the excess over \$476 withheld. Over what range does the amount withheld vary if the weekly wages vary from \$500 to \$550, inclusive?

78. Federal Tax Withholding Rework Problem 77 if the weekly wages vary from \$600 to \$700, inclusive.

79. Electricity Rates Commonwealth Edison Company's summer charge for electricity is 10.819¢ per kilowatt-hour.† In addition, each monthly bill contains a customer charge of \$9.06. If last summer's bills ranged from a low of \$82.14 to a high of \$279.63, over what range did usage vary (in kilowatt-hours)?

*Source: *Employer's Tax Guide,* Department of the Treasury, Internal Revenue Service, 1995.

†*Source*: Commonwealth Edison Co., Chicago, Illinois, 1991.

80. Water Bills The Village of Oak Lawn charges homeowners \$21.60 per quarter year plus \$1.70 per 1000 gallons for water usage in excess of 12,000 gallons.‡ In 1995, one homeowner's quarterly bill ranged from a high of \$65.75 to a low of \$28.40. Over what range did water usage vary?

81. Markup of a New Car The markup over dealer's cost of a new car ranges from 12% to 18%. If the sticker price is \$8800, over what range will the dealer's cost vary?

82. IQ Tests A standard intelligence test has an average score of 100. According to statistical theory, of the people who take the test, the 2.5% with the highest scores will have scores of more than 1.96σ above the average, where σ (sigma, a number called the *standard deviation*) depends on the nature of the test. If $\sigma = 12$ for this test and there is (in principle) no upper limit to the score possible on the test, write the interval of possible test scores of the people in the top 2.5%.

83. In your Economics 101 class, you have scores of 68, 82, 87, and 89 on the first four of five tests. To get a grade of B, the average of the five test scores must be greater than or equal to 80 and less than 90. Solve an inequality to find the range of the score you need on the last test to get a B.

84. Repeat Problem 83 if the fifth test counts double.

85. An Olds 98 averages 25 miles per gallon and has a tank that holds 20 gallons of gasoline. After a trip that covered at least 300 miles, the Olds ran out of gasoline. What is the range of the amount of gasoline (in gallons) that was in the tank at the start of the trip?

86. Repeat Problem 85 if the Olds runs out of gasoline after a trip of no more than 250 miles.

‡*Source*: Village of Oak Lawn, Illinois, 1995.

87. Make up a linear inequality that has $-3 < x < 4$ as a solution.

88. Do you prefer to use inequality notation or interval notation to express the solution to an inequality? Give your reasons. Are there particular circumstances when you prefer one to the other? Cite examples.

89. How would you explain to a fellow student the underlying reason for the multiplication property for inequalities (page 220); that is, the sense or direction of an inequality remains the same if each side is multiplied by a positive real number, while the direction is reversed if each side is multiplied by a negative real number?

3.6 EQUATIONS AND INEQUALITIES INVOLVING ABSOLUTE VALUE

1 Solve Equations Involving Absolute Value

2 Solve Inequalities Involving Absolute Value

The absolute value of a real number a has been defined as

$$|a| = \begin{cases} a & \text{if } a \geq 0 \\ -a & \text{if } a < 0 \end{cases}$$

Also, recall that, geometrically, the absolute value of a equals the distance from the origin to the point whose coordinate is a.

In this section, we shall discuss equations and inequalities involving absolute value. In solving such equations and inequalities, the following properties will prove useful.

Theorem Properties of Absolute Value

1. The absolute value of any real number a is always nonnegative; that is,

$$|a| \geq 0 \tag{1}$$

2. The absolute value of any real number a equals the principal square root of the number squared; that is,

$$|a| = \sqrt{a^2} \tag{2}$$

3. The absolute value of the product of two real numbers a and b equals the product of their absolute values; that is,

$$|ab| = |a| \cdot |b| \tag{3}$$

Proof Property (1) follows directly from the definition of absolute value. Property (2) is given in the Appendix, Section 1. Property (3) is proved by using property (2):

$$|ab| = \sqrt{(ab)^2} = \sqrt{a^2b^2} = \sqrt{a^2} \cdot \sqrt{b^2} = |a| \cdot |b|$$

1 We shall now take up the problem of solving equations and inequalities that contain absolute values.

Because there are two points whose distance from the origin is 5 units, -5 and 5, the equation $|x| = 5$ will have the solution set $\{-5, 5\}$. We are thus led to the following result:

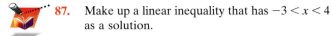

Theorem Equations Involving Absolute Value

If the absolute value of an expression equals some positive number a, then the expression itself equals either a or $-a$. Thus,

$$|u| = a \quad \text{is equivalent to} \quad u = a \quad \text{or} \quad u = -a \tag{4}$$

EXAMPLE 1 Solving an Equation Involving Absolute Value

Solve the equation $|x + 4| = 13$.

Graphing Solution For this equation, graph $Y_1 = |x + 4|$ and $Y_2 = 13$ on the same screen and find their point(s) of intersection, if any. See Figure 43. Using the INTERSECT command (twice), we find the points of intersection to be $(-17, 13)$ and $(9, 13)$. The solution set is $\{-17, 9\}$.

FIGURE 43

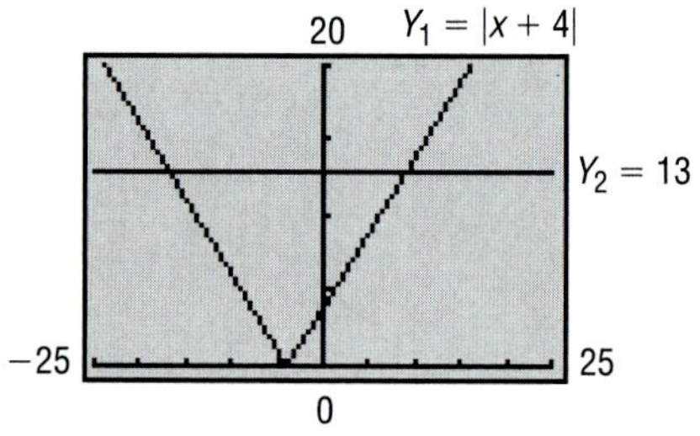

Algebraic Solution This follows the form of equation (4), where $u = x + 4$. Thus, there are two possibilities:

$$\begin{aligned} x + 4 &= 13 \quad &\text{or} \quad x + 4 &= -13 \\ x &= 9 & x &= -17 \end{aligned}$$

Thus, the solution set is $\{-17, 9\}$.

Now work Problem 3.

2

Let's look at an inequality involving absolute value.

EXAMPLE 2 Solving an Inequality Involving Absolute Value

Solve the inequality $|x| < 4$.

Graphing Solution We graph $Y_1 = |x|$ and $Y_2 = 4$ on the same screen. See Figure 44. Using the INTERSECT command, we find that Y_1 and Y_2 intersect at $x = -4$ and at $x = 4$. The graph of Y_1 is below that of Y_2, $Y_1 < Y_2$, between the points of intersection. Since the inequality is strict, the solution set is $\{x|-4 < x < 4\}$, or using interval notation, $(-4, 4)$.

FIGURE 44

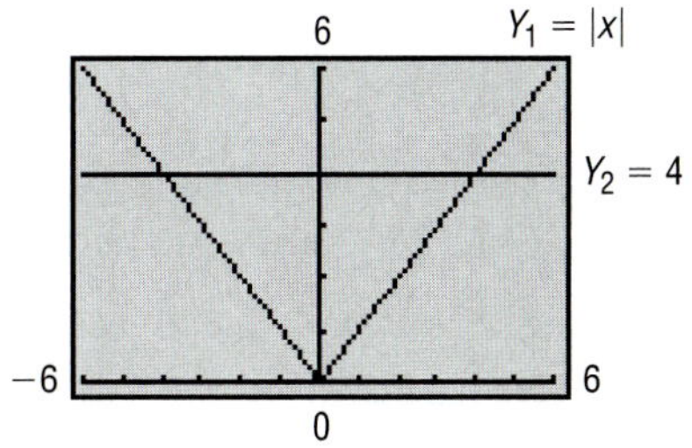

Algebraic Solution

We are looking for all points whose coordinate x is a distance less than 4 units from the origin. See Figure 45 for an illustration. Because any x between -4 and 4 satisfies the condition $|x| < 4$, the solution set consists of all numbers x for which $-4 < x < 4$; that is, all x in $(-4, 4)$.

FIGURE 45
$-4 < x < 4$ or $(-4, 4)$

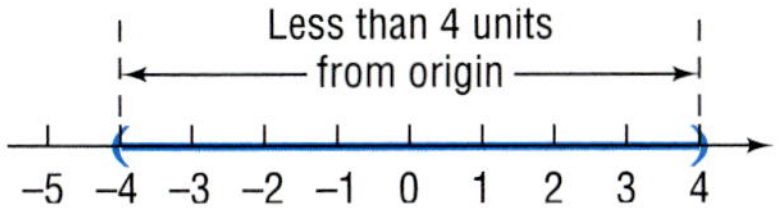

We are led to the following results:

Theorem

If a is any positive number, then

$$|u| < a \text{ is equivalent to } -a < u < a \qquad (5)$$

$$|u| \le a \text{ is equivalent to } -a \le u \le a \qquad (6)$$

In other words, $|u| < a$ is equivalent to $-a < u$ and $u < a$.

EXAMPLE 3 Solving an Inequality Involving Absolute Value

Solve the inequality $|2x + 4| \le 3$, and graph the solution set.

Graphing Solution

We graph $Y_1 = |2x + 4|$ and $Y_2 = 3$ on the same screen. See Figure 46. Using the INTERSECT command, we find that Y_1 and Y_2 intersect at $x = -3.5$ and at $x = -0.5$. The graph of Y_1 is below that of Y_2, $Y_1 < Y_2$, between the points of intersection. Since the inequality is not strict, the solution set is $\{x|-3.5 \le x \le -0.5\}$, or using interval notation, $[-3.5, -0.5]$.

FIGURE 46

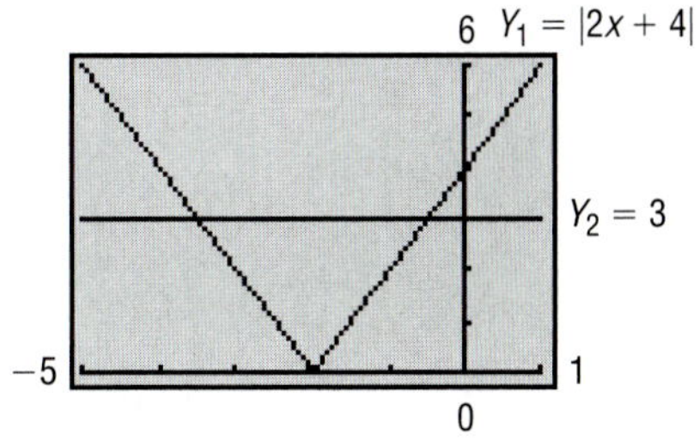

Algebraic Solution

$$\begin{aligned}
|2x+4| &\le 3 && \text{This follows the form of statement (6); the expression } u = 2x + 4 \text{ is inside the absolute value bars.}\\
-3 \le 2x+4 &\le 3 && \text{Apply statement (6).}\\
-3-4 \le 2x+4-4 &\le 3-4 && \text{Subtract 4 from each part.}\\
-7 \le 2x &\le -1 && \text{Simplify.}\\
\frac{-7}{2} \le \frac{2x}{2} &\le \frac{-1}{2} && \text{Divide each part by 2.}\\
-\frac{7}{2} \le x &\le -\frac{1}{2} && \text{Simplify.}
\end{aligned}$$

FIGURE 47

$-\frac{7}{2} \le x \le \frac{1}{2}$ or $\left[-\frac{7}{2}, -\frac{1}{2}\right]$

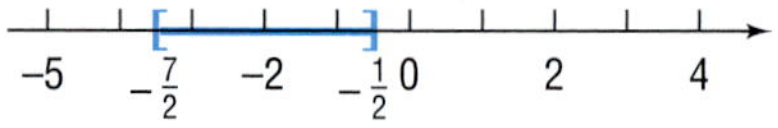

The solution set is $\{x \mid -\frac{7}{2} \le x \le -\frac{1}{2}\}$, that is, all x in $\left[-\frac{7}{2}, -\frac{1}{2}\right]$. See Figure 47.

EXAMPLE 4 Solving an Inequality Involving Absolute Value

Solve the inequality $|1 - 4x| < 5$, and graph the solution set.

Graphing Solution

We graph $Y_1 = |1 - 4x|$ and $Y_2 = 5$ on the same screen. See Figure 48. Using the INTERSECT command, we find that Y_1 and Y_2 intersect at $x = -1$ and at $x = 1.5$. The graph of Y_1 is below that of Y_2, $Y_1 < Y_2$, between the points of intersection. Since the inequality is strict, the solution set is $\{x \mid -1 < x < 1.5\}$, or using interval notation, $(-1, 1.5)$.

FIGURE 48

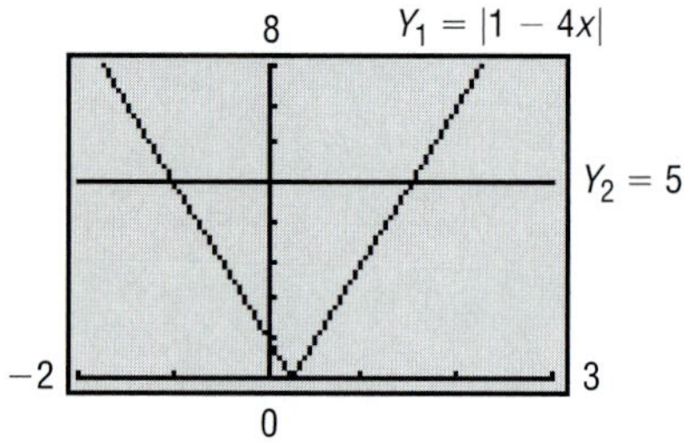

Algebraic Solution

$$\begin{aligned}
|1-4x| &< 5 && \text{This follows the form of statement (5); the expression } u = 1 - 4x \text{ is inside the absolute value bars.}\\
-5 < 1-4x &< 5 && \text{Apply statement (5).}\\
-5-1 < 1-4x-1 &< 5-1 && \text{Subtract 1 from each part.}\\
-6 < -4x &< 4 && \text{Simplify.}\\
\frac{-6}{-4} > \frac{-4x}{-4} &> \frac{4}{-4} && \text{Divide each part by } -4 \text{, which reverses the sense of the inequality symbols.}\\
\frac{3}{2} > x &> -1 && \text{Simplify.}\\
-1 < x &< \frac{3}{2} && \text{Rearrange the ordering.}
\end{aligned}$$

FIGURE 49

$-1 < x < \frac{3}{2}$ or $\left(-1, \frac{3}{2}\right)$

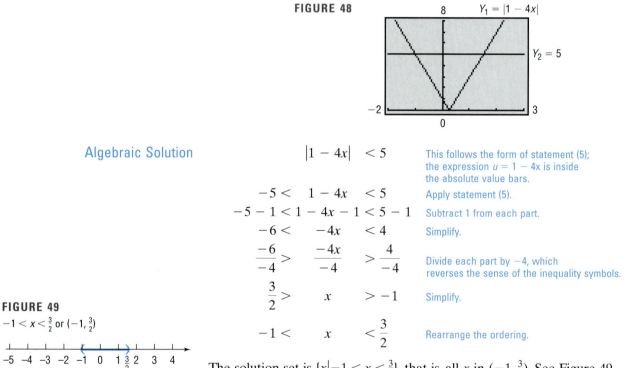

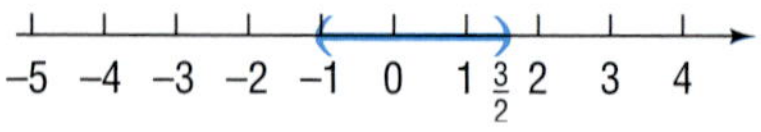

The solution set is $\{x \mid -1 < x < \frac{3}{2}\}$, that is, all x in $\left(-1, \frac{3}{2}\right)$. See Figure 49.

Now work Problem 27.

EXAMPLE 5 Solving an Inequality Involving Absolute Value

Solve the inequality $|x| > 3$, and graph the solution set.

Graphing Solution We graph $Y_1 = |x|$ and $Y_2 = 3$ on the same screen. See Figure 50. Using the INTERSECT command, we find that Y_1 and Y_2 intersect at $x = -3$ and at $x = 3$. The graph of Y_1 is above that of Y_2, $Y_1 > Y_2$, to the left of $x = -3$ and to the right of $x = 3$. Since the inequality is strict, the solution set is $\{x | -\infty < x < -3$ or $3 < x < \infty\}$. Using interval notation, the solution is $(-\infty, -3)$ or $(3, \infty)$.

FIGURE 50

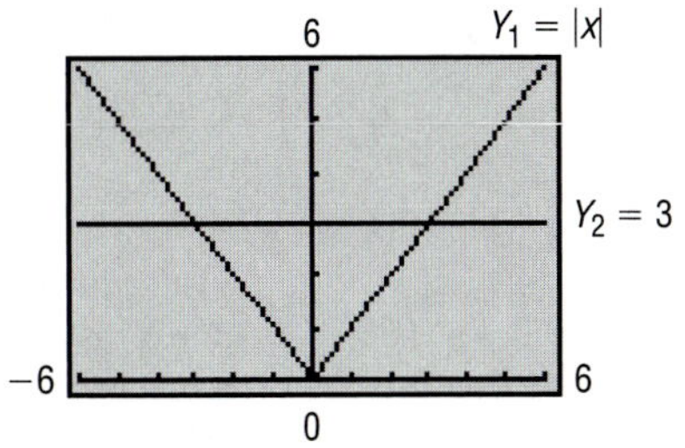

Algebraic Solution We are looking for all points whose coordinate x is a distance greater than 3 units from the origin. Figure 51 illustrates the situation. We conclude that any x less than -3 or greater than 3 satisfies the condition $|x| > 3$. Consequently, the solution set consists of all numbers x for which $-\infty < x < -3$ or $3 < x < \infty$, that is, all x in $(-\infty, -3)$ or $(3, \infty)$. ■

FIGURE 51
$-\infty < x < -3$ or $3 < x < \infty$; $(-\infty, -3)$ or $(3, \infty)$

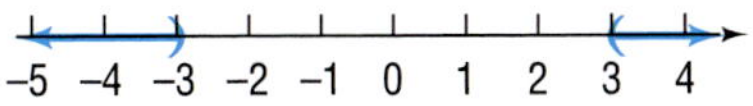

This leads to the following results:

Theorem

If a is any positive number, then

$\lvert u\rvert > a$	is equivalent to	$u < -a$	or	$u > a$	(7)
$\lvert u\rvert \ge a$	is equivalent to	$u \le -a$	or	$u \ge a$	(8)

■

EXAMPLE 6 Solving an Inequality Involving Absolute Value

Solve the inequality $|2x - 5| > 3$, and graph the solution set.

Graphing Solution We graph $Y_1 = |2x - 5|$ and $Y_2 = 3$ on the same screen. See Figure 52. Using the INTERSECT command, we find that Y_1 and Y_2 intersect at $x = 1$ and at $x = 4$. The graph of Y_1 is above that of Y_2, $Y_1 > Y_2$, to the left of $x = 1$ and to the right of $x = 4$. Since the inequality is strict, the solution set is $\{x | -\infty < x < 1$ or $4 < x < \infty\}$. Using interval notation, the solution is $(-\infty, 1)$ or $(4, \infty)$.

FIGURE 52

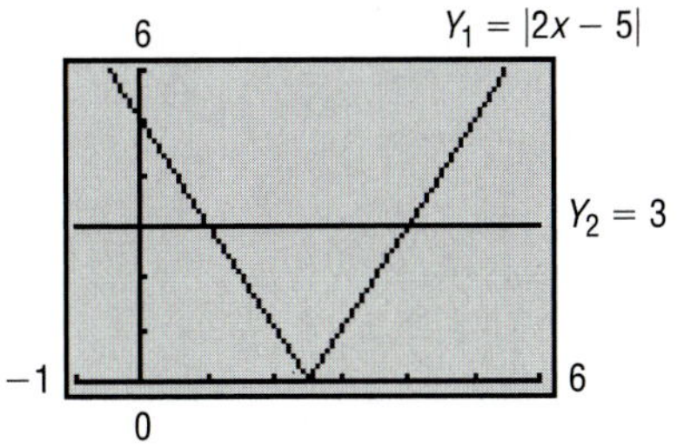

Algebraic Solution

$$|2x - 5| > 3$$ This follows the form of statement (7); the expression $u = 2x - 5$ is inside the absolute value bars.

$$2x - 5 < -3 \quad \text{or} \quad 2x - 5 > 3$$ Apply statement (7).

$$2x - 5 + 5 < -3 + 5 \quad \text{or} \quad 2x - 5 + 5 > 3 + 5$$ Add 5 to each part.

$$2x < 2 \quad \text{or} \quad 2x > 8$$ Simplify.

$$\frac{2x}{2} < \frac{2}{2} \quad \text{or} \quad \frac{2x}{2} > \frac{8}{2}$$ Divide each part by 2.

$$x < 1 \quad \text{or} \quad x > 4$$ Simplify.

The solution set is $\{x | -\infty < x < 1 \text{ or } 4 < x < \infty\}$, that is, all x in $(-\infty, 1)$ or $(4, \infty)$. See Figure 53. ■

FIGURE 53
$-\infty < x < 1$ or $4 < x < \infty$; $(-\infty, 1)$ or $(4, \infty)$

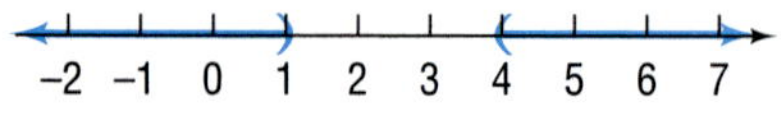

Warning: A common error to be avoided is to attempt to write the solution $x < 1$ or $x > 4$ as $1 > x > 4$, which is incorrect, since there are no numbers x for which $1 > x$ *and* $x > 4$. Another common error is to "mix" the symbols and write $1 < x > 4$, which, of course, makes no sense.

3.6 EXERCISES

In Problems 1–22, solve each equation (a) graphically and (b) algebraically.

1. $|2x| = 8$
2. $|3x| = 15$
3. $|2x + 3| = 5$
4. $|3x - 1| = 2$
5. $|1 - 4t| = 5$
6. $|1 - 2z| = 3$
7. $|-2x| = 8$
8. $|-x| = 1$
9. $|-2|x = 4$
10. $|3|x = 9$
11. $\frac{2}{3}|x| = 8$
12. $\frac{3}{4}|x| = 9$
13. $\left|\frac{x}{3} + \frac{2}{5}\right| = 2$
14. $\left|\frac{x}{2} - \frac{1}{3}\right| = 1$
15. $|u - 2| = -\frac{1}{2}$
16. $|2 - v| = -1$
17. $|x^2 - 16| = 0$
18. $|x^2 - 4| = 0$
19. $|x^2 - 2x| = 3$
20. $|x^2 + x| = 12$
21. $|x^2 + x - 1| = 1$
22. $|x^2 + 3x - 2| = 2$

In Problems 23–42, solve each inequality (a) graphically and (b) algebraically. Graph the solution set.

23. $|2x| < 8$
24. $|3x| < 15$
25. $|3x| > 12$
26. $|2x| > 6$
27. $|x - 2| < 1$
28. $|x + 4| < 2$
29. $|3t - 2| \le 4$
30. $|2u + 5| \le 7$
31. $|x - 3| \ge 2$
32. $|x + 4| \ge 2$
33. $|1 - 4x| < 5$
34. $|1 - 2x| < 3$
35. $|1 - 2x| > 3$
36. $|2 - 3x| > 1$
37. $|x + 3| > 0$
38. $|3 - x| > 0$
39. $|x + 2| > -3$
40. $|2 - x| > -2$
41. $|2x - 1| < 0.02$
42. $|3x - 2| < 0.02$

In Problems 43–48, find a and b.

43. If $|x - 1| < 3$, then $a < x + 4 < b$.

44. If $|x + 2| < 5$, then $a < x - 2 < b$.

45. If $|x + 4| \le 2$, then $a \le 2x - 3 \le b$.

46. If $|x - 3| \le 1$, then $a \le 3x + 1 \le b$.

47. If $|x - 2| \le 7$, then $a \le \frac{1}{x - 10} \le b$.

48. If $|x + 1| \le 3$, then $a \le \frac{1}{x + 5} \le b$.

49. If $b \ne 0$, prove that $\left|\frac{a}{b}\right| = \frac{|a|}{|b|}$.

50. Show that $a \le |a|$.

51. Prove the *triangle inequality* $|a + b| \le |a| + |b|$.
[**Hint:** Expand $|a + b|^2 = (a + b)^2$, and use the result of Problem 50.]

52. Prove that $|a - b| \ge |a| - |b|$.
[**Hint:** Apply the triangle inequality from Problem 51 to $|a| = |(a - b) + b|$.]

53. Express the fact that x differs from 3 by less than $\frac{1}{2}$ as an inequality involving an absolute value. Solve for x.

54. Express the fact that x differs from -4 by less than 1 as an inequality involving an absolute value. Solve for x.

55. Express the fact that x differs from -3 by more than 2 as an inequality involving an absolute value. Solve for x.

56. Express the fact that x differs from 2 by more than 3 as an inequality involving an absolute value. Solve for x.

57. Body Temperature "Normal" human body temperature is 98.6°F. If a temperature x that differs from normal by at least 1.5° is considered unhealthy, write the condition for an unhealthy temperature x as an inequality involving an absolute value, and solve for x.

58. Household Voltage In the United States, normal household voltage is 115 volts. However, it is not uncommon for actual voltage to differ from normal voltage by at most 5 volts. Express this situation as an inequality involving an absolute value. Use x as the actual voltage and solve for x.

59. If $a > 0$, show that the solution set of the inequality

$$x^2 < a$$

consists of all numbers x for which

$$-\sqrt{a} < x < \sqrt{a}$$

60. If $a > 0$, show that the solution set of the inequality

$$x^2 > a$$

consists of all numbers x for which

$$-\infty < x < -\sqrt{a} \quad \text{or} \quad \sqrt{a} < x < \infty$$

In Problems 61–68, use the results found in Problems 59 and 60 to solve each inequality.

61. $x^2 < 1$ **62.** $x^2 < 4$ **63.** $x^2 \geq 9$ **64.** $x^2 \geq 1$

65. $x^2 \leq 16$ **66.** $x^2 \leq 9$ **67.** $x^2 > 4$ **68.** $x^2 > 16$

69. Solve $|3x - |2x + 1|| = 4$.

70. Solve $|x + |3x - 2|| = 2$.

71. The equation $|x| = -2$ has no solution. Explain why.

72. The inequality $|x| > -0.5$ has all real numbers as a solution. Explain why.

CHAPTER REVIEW

THINGS TO KNOW

Quadratic equation and quadratic formula

If $ax^2 + bx + c = 0$, $a \neq 0$, and if $b^2 - 4ac \geq 0$, then $x = \dfrac{-b \pm \sqrt{b^2 - 4ac}}{2a}$.

Discriminant

If $b^2 - 4ac > 0$, there are two distinct real solutions.

If $b^2 - 4ac = 0$, there is one repeated real solution.

If $b^2 - 4ac < 0$, there are no real solutions.

Inequality properties

Trichotomy property	$a < b$ or $a = b$ or $b < a$
Transitive property	If $a < b$ and $b < c$, then $a < c$. If $a > b$ and $b > c$, then $a > c$.
Addition property	If $a < b$ then $a + c < b + c$. If $a > b$ then $a + c > b + c$.
Multiplication properties	(a) If $a < b$ and if $c > 0$, then $ac < bc$. If $a < b$ and if $c < 0$, then $ac > bc$. (b) If $a > b$ and if $c > 0$, then $ac > bc$. If $a > b$ and if $c < 0$, then $ac < bc$.
Reciprocal property	If $a > 0$, then $\dfrac{1}{a} > 0$. If $a < 0$, then $\dfrac{1}{a} < 0$.

Absolute value

If $|u| = a$, $a > 0$, then $u = -a$ or $u = a$.

If $|u| \le a$, $a > 0$, then $-a \le u \le a$.

If $|u| \ge a$, $a > 0$, then $u \le -a$ or $u \ge a$.

HOW TO

Solve linear equations in one variable

Solve quadratic equations

Solve radical equations

Solve linear inequalities in one variable

Solve equations and inequalities involving absolute value

Solve applied problems

Solve equations and inequalities using a graphing utility

FILL-IN-THE-BLANK ITEMS

1. Two equations (or inequalities) that have precisely the same solution set are called ______________.

2. To complete the square of the expression $x^2 + 5x$, you would ______________ the number ______________.

3. The quantity $b^2 - 4ac$ is called the ______________ of a quadratic equation. If it is ______________, the equation has no real solution.

4. If $a < 0$, then $|a| =$ ______________.

5. When a quadratic equation has a repeated solution, it is called a(n) ______________ root or a root of ______________ ______________.

6. When an apparent solution does not satisfy the original equation, it is called a(n) ______________ solution.

7. If each side of an inequality is multiplied by a(n) ______________ number, then the sense of the inequality symbol is reversed.

8. The equation $|x^2| = 4$ has two real solutions, ______________ and ______________.

TRUE/FALSE ITEMS

T F **1.** Equations can have no solutions, one solution, or more than one solution.

T F **2.** Quadratic equations always have two real solutions.

T F **3.** If the discriminant of a quadratic equation is positive, then the equation has two real solutions that are unequal.

T F **4.** The square of any real number is always nonnegative.

T F **5.** The expression $x^2 + x + 1$ is positive for any real number x.

6. If $a < b$ and $c < 0$, which of the following statements are true?

T F (a) $a \pm c < b \pm c$

T F (b) $a \cdot c < b \cdot c$

T F (c) $a/c > b/c$

7. If $a^2 < b^2$, $b \neq 0$, which of the following statements are true?

T F (a) $a < b$

T F (b) $a^2/b^2 < 1$

T F (c) $b^2 - a^2 > 0$

REVIEW EXERCISES

In Problems 1–38, find all real solutions, if any, of each equation (a) graphically and (b) algebraically.

1. $2 - \frac{x}{3} = 6$

2. $\frac{x}{4} - 2 = 6$

3. $-2(5 - 3x) + 8 = 4 + 5x$

4. $(6 - 3x) - 2(1 + x) = 6x$

5. $\frac{3x}{4} - \frac{x}{3} = \frac{1}{12}$

6. $\frac{4 - 2x}{3} + \frac{1}{6} = 2x$

7. $\dfrac{x}{x-1} = \dfrac{5}{6}, \quad x \neq 1$

8. $\dfrac{4x-5}{3-7x} = 4, \quad x \neq \frac{3}{7}$

9. $x(1-x) = 6$

10. $x(1+x) = 6$

11. $\dfrac{1}{2}\left(x - \dfrac{1}{3}\right) = \dfrac{3}{4} - \dfrac{x}{6}$

12. $\dfrac{1-3x}{4} = \dfrac{x+6}{3} + \dfrac{1}{2}$

13. $(x-1)(2x+3) = 3$

14. $x(2-x) = 3(x-4)$

15. $2x + 3 = 4x^2$

16. $1 + 6x = 4x^2$

17. $\sqrt[3]{x^2 - 1} = 2$

18. $\sqrt{1 + x^3} = 3$

19. $x(x+1) + 2 = 0$

20. $3x^2 - x + 1 = 0$

21. $x^4 - 5x^2 + 4 = 0$

22. $3x^4 + 4x^2 + 1 = 0$

23. $\sqrt{2x-3} + x = 3$

24. $\sqrt{2x-1} = x - 2$

25. $x^{3/2} + 5x^{1/2} = 0$

26. $x^{2/3} + x = 0$

27. $\sqrt{x+1} + \sqrt{x-1} = \sqrt{2x+1}$

28. $\sqrt{2x-1} - \sqrt{x-5} = 3$

29. $2\sqrt[3]{x^2} - \sqrt[3]{x} = 1$

30. $4\sqrt[3]{x^2} = 1$

31. $x^{-6} - 7x^{-3} - 8 = 0$

32. $6x^{-1} - 5x^{-1/2} + 1 = 0$

33. $\sqrt{x^2 + 3x + 7} - \sqrt{x^2 - 3x + 9} + 2 = 0$

34. $\sqrt{x^2 + 3x + 7} - \sqrt{x^2 + 3x + 9} = 2$

35. $|2x + 3| = 7$

36. $|3x - 1| = 5$

37. $|2 - 3x| = 7$

38. $|1 - 2x| = 3$

In Problems 39–42, find all real solutions, if any, of each equation. (a, b, m, and n are positive constants.)

39. $x^2 + m^2 = 2mx + (nx)^2$

40. $b^2x^2 + 2ax = x^2 + a^2$

41. $10a^2x^2 - 2abx - 36b^2 = 0$

42. $\dfrac{1}{x-m} + \dfrac{1}{x-n} = \dfrac{2}{x}, \quad x \neq 0, m, n$

In Problems 43–52, solve each inequality (a) graphically and (b) algebraically. Graph the solution set.

43. $\dfrac{2x-3}{5} + 2 \leq \dfrac{x}{2}$

44. $\dfrac{5-x}{3} \leq 6x - 4$

45. $-9 \leq \dfrac{2x+3}{-4} \leq 7$

46. $-4 < \dfrac{2x-2}{3} < 6$

47. $6 > \dfrac{3-3x}{12} > 2$

48. $6 > \dfrac{5-3x}{2} \geq -3$

49. $|3x + 4| < \frac{1}{2}$

50. $|1 - 2x| < \frac{1}{3}$

51. $|2x - 5| \geq 9$

52. $|3x + 1| \geq 10$

In Problems 53–56, use a graphing utility to approximate the ***positive*** *solutions for each equation. Express the answer correct to two decimal places.*

53. $x^3 + 2.9x^2 - 17.1x - 5.63 = 0$

54. $x^3 + 3.4x^2 - 7.4x - 6.2 = 0$

55. $\pi x^3 - 28.9x^2 - 123x + 42 = 0$

56. $\pi x^3 + 19.8x^2 - 328x + 217 = 0$

57. Lightning and Thunder A flash of lightning is seen, and the resulting thunderclap is heard 3 seconds later. If the speed of sound averages 1100 feet per second, how far away is the storm?

58. Physics: Intensity of Light The intensity I (in candlepower) of a certain light source obeys the equation $I = 900/x^2$, where x is the distance (in meters) from the light. Over what range of distances can an object be placed from this light source so that the range of intensity of light is from 1600 to 3600 candlepower, inclusive?

59. Extent of Search and Rescue A search plane has a cruising speed of 250 miles per hour and carries enough fuel for at most 5 hours of flying. If there is a wind that averages 30 miles per hour and the direction of search is with the wind one way and against it the other, how far can the search plane travel?

60. Extent of Search and Rescue If the search plane described in Problem 59 is able to add a supplementary fuel tank that allows for an additional 2 hours of flying, how much farther can the plane extend its search?

61. Rescue at Sea A life raft, set adrift from a sinking ship 150 miles offshore, travels directly toward a Coast Guard station at the rate of 5 miles per hour. At the time the raft is set adrift, a rescue helicopter is dispatched from the Coast Guard station. If the helicopter's average speed is 90 miles per hour, how long will it take the helicopter to reach the life raft?

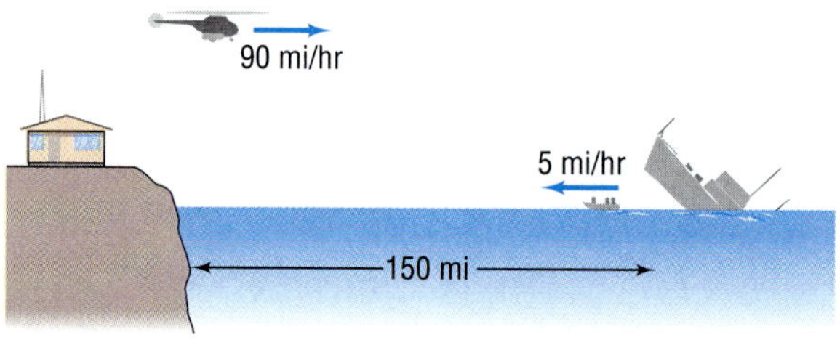

62. Physics: Uniform Motion Two bees leave two locations 150 meters apart and fly, without stopping, back and forth between these two locations at average speeds of 3 meters per second and 5 meters per second, respectively. How long is it until the bees meet for the first time? How long is it until they meet for the second time?

63. Working Together to Get a Job Done Clarissa and Shawna, working together, can paint the exterior of a house in 6 days. Clarissa by herself can complete this job in 5 days less than Shawna. How long will it take Clarissa to complete the job by herself?

64. Emptying a Tank Two pumps of different sizes, working together, can empty a fuel tank in 5 hours. The larger pump can empty this tank in 4 hours less than the smaller one. If the larger one is out of order, how long will it take the smaller one to do the job alone?

65. Chemistry: Mixing Acids For a certain experiment, a student requires 100 cubic centimeters of a solution that is 8% HCl. The storeroom has only solutions that are 15% HCl and 5% HCl. How many cubic centimeters of each available solution should be mixed to get 100 cubic centimeters of 8% HCl?

66. Business: Blending Coffee A coffee manufacturer wants to market a new blend of coffee that will cost \$3.90 per pound by mixing two coffees that sell for \$2.75 and \$5 per pound, respectively. What amounts of each coffee should be blended to obtain the desired mixture?

[**Hint:** Assume that the total weight of the desired blend is 100 pounds.]

67. Business: Theater Attendance The manager of the Coral Theater wants to know whether the majority of its patrons are adults or children. During a week in July, 5200 tickets were sold and the receipts totaled \$20,335. The adult admission is \$4.75, and the children's admission is \$2.50. How many adult patrons were there?

68. Chemistry: Salt Solutions How much water must be evaporated from 32 ounces of a 4% salt solution to make 6% salt solution?

69. Physics: Uniform Motion A man is walking at an average speed of 4 miles per hour alongside a railroad track. A freight train, going in the same direction at an average speed of 30 miles per hour, requires 5 seconds to pass the man. How long is the freight train? Give your answer in feet.

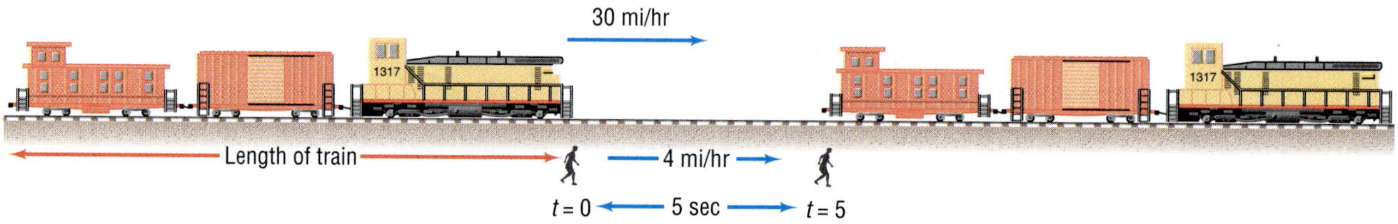

70. One formula stating the relationship between the length l and width w of a rectangle of "pleasing proportion" is $l^2 = w(l + w)$. How should a 4 foot by 8 foot sheet of plasterboard be cut so that the result is a rectangle of "pleasing proportion" with a width of 4 feet?

71. Business: Determining the Cost of a Charter A group of 20 senior citizens can charter a bus for a one-day excursion trip for \$15 per person. The charter company agrees to reduce the price of each ticket by 10¢ for each additional passenger in excess of 20 who goes on the trip, up to a maximum of 44 passengers (the capacity of the bus). If the final bill from the charter company was \$482.40, how many seniors went on the trip, and how much did each pay?

72. A new copying machine can do a certain job in 1 hour less than an older copier. Together they can do this job in 72 minutes. How long would it take the older copier by itself to do the job?

73. In a 100-meter race, Todd crosses the finish line 5 meters ahead of Scott. To even things up, Todd suggests to Scott that they race again, this time with Todd lining up 5 meters behind the start.

(a) Assuming Todd and Scott run at the same pace as before, does the second race end in a tie?

(b) If not, who wins?
(c) By how many meters does he win?
(d) How far back should Todd start so the race ends in a tie?

After running the race a second time, Scott, to even things up, suggests to Todd that he (Scott) line up 5 meters in front of the start.

(e) Assuming again that they run at the same pace as in the first race, does the third race result in a tie?
(f) If not, who wins?
(g) By how many meters?
(h) How far up should Scott start so the race ends in a tie?

74. Explain the difference between the following three problems. Are there any similarities in their solution?

(a) Write the expression as a single quotient:
$$\frac{x}{x-2}+\frac{x}{x^2-4}$$

(b) Solve: $\dfrac{x}{x-2}+\dfrac{x}{x^2-4}=0$

(c) Solve: $\dfrac{x}{x-2}+\dfrac{x}{x^2-4}<0$

CHAPTER 4

Polynomial and Rational Functions

Technology, such as radio telescopes and satellite dishes, have their foundations in the mathematical concepts of parabolas and the quadratic formula. On the following page, you have been hired as a high school physics teacher and need to introduce your class to Galileo's *Analysis of Motion.* Use the Sullivan website at:

www.prenhall.com/sullivan

to help link you to the Internet resources you will need to prepare your lesson.

PREPARING FOR THIS CHAPTER

Before getting started on this chapter, review the following concepts:

Completing the Square *(Appendix, Section 5)*

The Discriminant of a Quadratic Equation (p. 189)

Linear Curve Fitting *(Section 1.6)*

Polynomial and Rational Expressions *(Appendix, Section 2)*

Library of Functions (pp. 127–131)

Synthetic Division *(Appendix, Section 6)*

Inequalities *(Section 3.5)*

OUTLINE

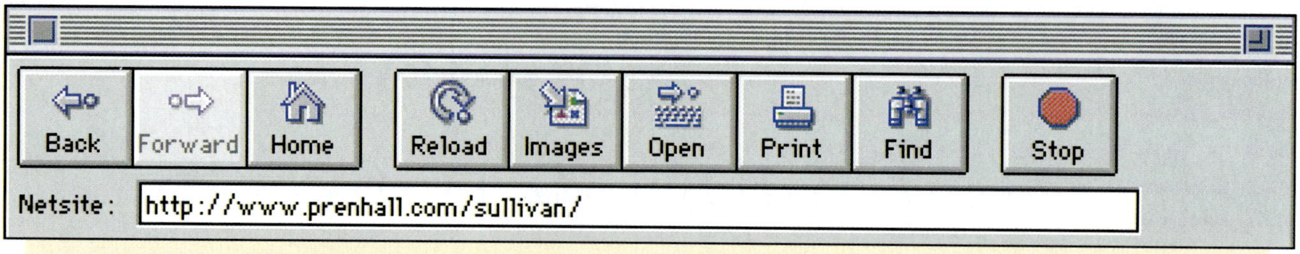

THE PARABOLA

Quadratic functions, like all polynomials, are formed using only the basic rules of arithmetic: addition, subtraction, multiplication, and division. The graph of a quadratic function is called a **parabola.** The parabola has properties that make it useful for satellite dishes, car headlights, radio telescopes, and reflecting telescopes, including *the liquid mirror telescope.* Although the Greeks knew about the parabola as a conic section, it was *Galileo* who first observed that a quadratic function can be used to describe the motion of a falling body.

Suppose that you are a high school physics teacher and you want to put together an Internet lesson for your class. You would like to begin with a short lecture on Galileo's *Analysis of Motion.* His analysis was a major advance in science. The quadratic function was like his sword for fighting against the bondage of the dark ages. You plan to duplicate *his experiments* with in-class hands-on activity.

1. As you start to prepare your Internet lesson you search the net for "parabolas." You are overwhelmed with the number of links. How might you handle this information overload?
2. How many different kinds of applications of the parabola did you find on the web? Did you find any links that refer to suspension bridges?
3. For the math side you want to make a list of important definitions, and find some good illustrations of the essential features of parabolas. You want to try to find interactive sites. Did you find any?
4. Given the physical description of the *largest radio telescope on the planet* at Arecibo, Puerto Rico, approximate the dish with a quadratic function. Then approximate it with a circle. Graph both functions with your calculator. Can you determine the amount of difference between the two graphs? What advantages are there in making the radio telescope circular?
5. Suppose you perform Galileo's *parabola experiment.* Model the data with a quadratic function. How well does the graph of your quadratic function fit the data? How well does your *data* fit the ideal law?

In Chapters 1 and 2, we graphed linear functions $f(x) = ax + b$, $a \neq 0$; the square function $f(x) = x^2$; and the cube function $f(x) = x^3$. Each of these functions belongs to the class of functions called *polynomial functions,* which we discuss further in this chapter. We will also discuss *rational functions,* which are ratios of polynomial functions. In this chapter, we place special emphasis on the graphs of polynomial and rational functions. This emphasis will demonstrate the importance of evaluating polynomials and solving polynomial equations (Section 4.5). Section 4.6 introduces complex numbers and the role they play in solving quadratic equations with a negative discriminant. This paves the way for the Fundamental Theorem of Algebra (Section 4.7). Finally, polynomial and rational inequalities are discussed in Section 4.8.

4.1 QUADRATIC FUNCTIONS; CURVE FITTING

1. Graph a Quadratic Function by Hand and by Using a Graphing Utility
2. Identify the Vertex and Axis of Symmetry of a Quadratic Function
3. Determine the Maximum or Minimum Value of a Quadratic Function
4. Use a Graphing Utility to Find the Quadratic Function of Best Fit

A **quadratic function** is a function of the form

$$f(x) = ax^2 + bx + c \qquad (1)$$

where a, b, and c are real numbers and $a \neq 0$. The domain of a quadratic function consists of all real numbers.

Many applications require a knowledge of quadratic functions. For example, suppose Texas Instruments collects the data shown in Table 1 that relates the number c of calculators sold at the price p per calculator. Since the price of a product determines the quantity that will be purchased, we treat price as the independent variable.

TABLE 1

Price per Calculator, p (Dollars)	Number of Calculators, c
60	11,100
65	10,115
70	9,652
75	8,731
80	8,087
85	7,205
90	6,439

Using the LINear REGression option on a graphing utility, we find the relation between the number c of calculators and the price p per calculator is

$$c = 20208.43 - 152.63p$$

Then the revenue R derived from selling c calculators at the price p per calculator is

$$\begin{aligned} R &= cp \\ &= (20208.43 - 152.63p)p \\ &= -152.63p^2 + 20208.43p \end{aligned}$$

FIGURE 1

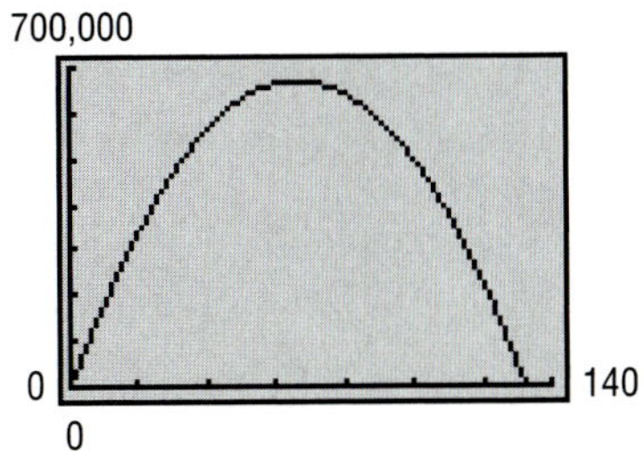

So, the revenue R is expressed as a quadratic function of the price p. Figure 1 illustrates the graph of this revenue function. Since both c and p must be nonnegative, we only show the graph of R in the first quadrant.

A second situation in which a quadratic function appears involves the motion of a projectile. Based on Newton's second law of motion (force equals mass times acceleration, $F = ma$), it can be shown that, ignoring air resistance, the path of a projectile propelled upward at an inclination to the horizontal is the graph of a quadratic function. See Figure 2 for an illustration.

FIGURE 2
Path of a cannonball

Graphing Quadratic Functions

We know how to graph quadratic functions of the form $f(x) = ax^2$, $a \neq 0$, based on prior discussions. Figure 3 shows the graph of three functions of the form $f(x) = ax^2$, $a > 0$, for $a = 1$, $a = \frac{1}{2}$, and $a = 3$. Notice that the larger the value of a, the "narrower" the graph is and the smaller the value of a the "wider" the graph is.

FIGURE 3

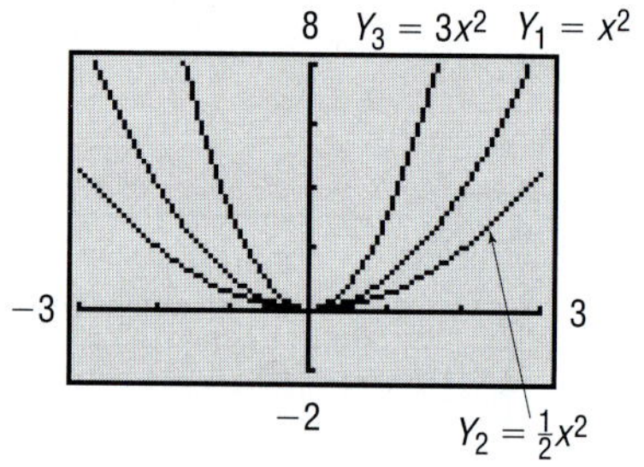

FIGURE 4

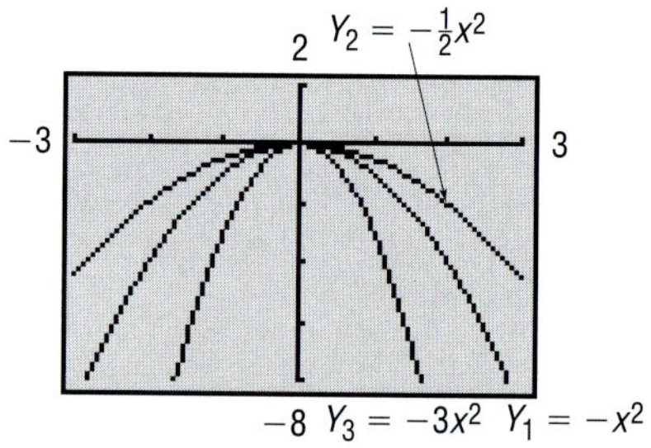

Figure 4 shows the graphs of $f(x) = ax^2$ for $a < 0$. Notice these graphs are reflections about the x-axis of the graphs in Figure 3. Based on the results of these two figures, we can draw some general conclusions about the graph of $f(x) = ax^2$. First, as $|a|$ increases, the graph becomes "narrower" and as $|a|$ gets closer to zero, the graph gets "wider." Second, if a is positive, then the graph opens "up" and if a is negative, then the graph opens "down."

The graphs in Figures 3 and 4 are typical of the graphs of all quadratic functions, which we call **parabolas.*** Refer to Figure 5, where two parabolas are pictured. The one on the left **opens up** and has a lowest point; the one on the right **opens down** and has a highest point. The lowest or highest point of a parabola is called the **vertex.** The vertical line passing through the vertex in each parabola in Figure 5 is called the **axis of symmetry** (usually abbreviated to **axis**) of the parabola. Because the parabola is symmetric about its axis, the axis of symmetry of a parabola can be used to advantage in graphing the parabola by hand.

FIGURE 5
Graphs of a quadratic function, $f(x) = ax^2 + bx + c$, $a \neq 0$

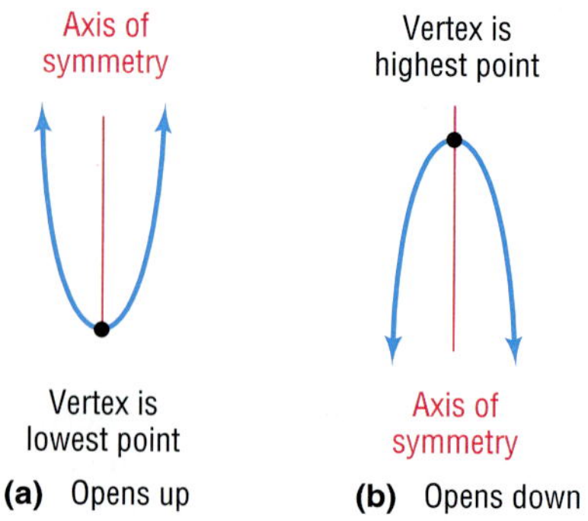

The parabolas shown in Figure 5 are the graphs of a quadratic function $f(x) = ax^2 + bx + c$, $a \neq 0$. Notice that the coordinate axes are not included in the figure. Depending on the values of a, b, and c, the axes could be placed anywhere. The important fact is that, except possibly for compression or

*We shall study parabolas using a geometric definition later in this book.

stretching, the shape of the graph of a quadratic function will look like one of the parabolas in Figure 5.

EXAMPLE 1 Graphing a Quadratic Function

1

Graph the function $f(x) = 2x^2 + 8x + 5$. Find the vertex and axis of symmetry.

Graphing Solution Before graphing, notice the leading coefficient, 2, is positive and, therefore, the graph will open up and the vertex will be the lowest point. Now graph f. See Figure 6. We observe the graph does in fact open up. To estimate the vertex of the parabola, we use the MINIMUM command. Correct to two decimal places, the vertex is $(-2, -3)$. Therefore, the axis of symmetry is the line $x = -2$.

FIGURE 6

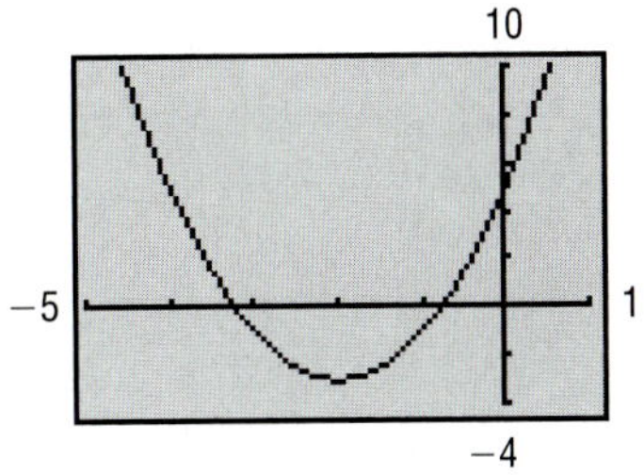

Algebraic Solution We begin by completing the square* on the right side:

$$\begin{aligned} f(x) &= 2x^2 + 8x + 5 \\ &= 2(x^2 + 4x) + 5 && \text{Factor out the 2 from } 2x^2 + 8x. \\ &= 2(x^2 + 4x + 4) + 5 - 8 && \text{Complete the square of } 2(x^2 + 4x)\text{. Notice that the factor of 2 requires that 8 be added and subtracted.} \\ &= 2(x + 2)^2 - 3 && (2) \end{aligned}$$

The graph of f can be obtained in three stages, as shown in Figure 7. Now compare this graph to the graph in Figure 5(a). The graph of $f(x) = 2x^2 + 8x + 5$ is a parabola that opens up and has its vertex (lowest point) at $(-2, -3)$. Its axis of symmetry is the line $x = -2$.

FIGURE 7

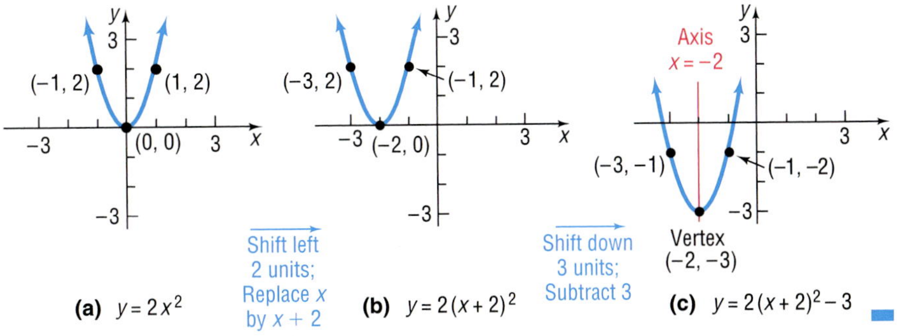

(a) $y = 2x^2$ **(b)** $y = 2(x+2)^2$ **(c)** $y = 2(x+2)^2 - 3$

*Refer to the Appendix, Section 5, for a review.

The method used in the Algebraic Solution to Example 1 can be used to graph any quadratic function $f(x) = ax^2 + bx + c$, $a \neq 0$, as follows:

$$\begin{aligned} f(x) &= ax^2 + bx + c \\ &= a\left(x^2 + \frac{b}{a}x\right) + c && \text{Factor out } a \text{ from } ax^2 + bx. \\ &= a\left(x^2 + \frac{b}{a}x + \frac{b^2}{4a^2}\right) + c - a\left(\frac{b^2}{4a^2}\right) && \text{Complete the square by adding and subtracting } a(b^2/4a^2)\text{. Look closely at this step!} \\ &= a\left(x + \frac{b}{2a}\right)^2 + c - \frac{b^2}{4a} \\ &= a\left(x + \frac{b}{2a}\right)^2 + \frac{4ac - b^2}{4a} \end{aligned}$$

Based on the above results, we conclude:

> If $h = -b/2a$ and $k = (4ac - b^2)/4a$, then
>
> $$f(x) = ax^2 + bx + c = a(x - h)^2 + k \qquad (3)$$

The graph of f is the parabola $y = ax^2$ shifted horizontally h units and vertically k units. As a result, the vertex is at (h, k), and the graph opens up if $a > 0$ and down if $a < 0$. The axis is the vertical line $x = h$.

For example, compare equation (3) with equation (2) of Example 1.

$$f(x) = 2(x + 2)^2 - 3$$
$$f(x) = a(x - h)^2 + k$$

We conclude that $a = 2$, so the graph opens up. Also, we find that $h = -2$ and $k = -3$, so its vertex is at $(-2, -3)$.

2 It is not required to complete the square to obtain the vertex. In almost every case, it is easier to obtain the vertex of a quadratic function f by remembering that its x-coordinate is $h = -b/2a$. The y-coordinate can then be found by evaluating f at $-b/2a$.

These results are summarized next:

Characteristics of the Graph of a Quadratic Function

> $$f(x) = ax^2 + bx + c$$
>
> $$\text{Vertex} = \left(\frac{-b}{2a}, f\left(\frac{-b}{2a}\right)\right) \quad \text{Axis of Symmetry: The line } x = \frac{-b}{2a} \qquad (4)$$
>
> Parabola opens up if $a > 0$. Parabola opens down if $a < 0$.

EXAMPLE 2 Locating the Vertex without Graphing

Without graphing, locate the vertex and axis of the parabola defined by $f(x) = -3x^2 + 6x + 1$. Does it open up or down?

Solution For this quadratic function, $a = -3$, $b = 6$, and $c = 1$. The x-coordinate of the vertex is

$$\frac{-b}{2a} = \frac{-6}{-6} = 1$$

The y-coordinate of the vertex is therefore

$$f\left(\frac{-b}{2a}\right) = f(1) = -3 + 6 + 1 = 4$$

The vertex is located at the point $(1, 4)$. The axis of symmetry is the line $x = 1$. Finally, because $a = -3 < 0$, the parabola opens down.

The information we gathered in Example 2, together with the location of the intercepts, usually provides enough information to graph $f(x) = ax^2 + bx + c$, $a \neq 0$, by hand. The y-intercept is the value of f at $x = 0$, that is, $f(0) = c$. The x-intercepts, if there are any, are found by solving the equation

$$f(x) = ax^2 + bx + c = 0$$

This equation has two, one, or no real solutions, depending on whether the discriminant $b^2 - 4ac$ is positive, 0, or negative. Thus, it has corresponding x-intercepts, as follows:

The x-intercepts of a Quadratic Function

1. If the discriminant $b^2 - 4ac > 0$, the graph of $f(x) = ax^2 + bx + c$ has two distinct x-intercepts and so will cross the x-axis in two places.
2. If the discriminant $b^2 - 4ac = 0$, the graph of $f(x) = ax^2 + bx + c$ has one x-intercept and touches the x-axis at its vertex.
3. If the discriminant $b^2 - 4ac < 0$, the graph of $f(x) = ax^2 + bx + c$ has no x-intercept and so will not cross or touch the x-axis.

Figure 8 illustrates these possibilities for parabolas that open up.

FIGURE 8
$f(x) = ax^2 + bx + c, a > 0$

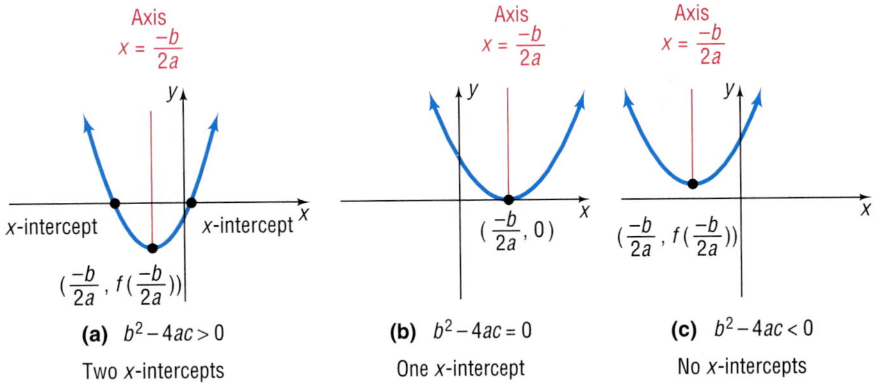

EXAMPLE 3 Graphing a Quadratic Function By Hand Using Its Vertex, Axis, and Intercepts

Use the information from Example 2 and the locations of the intercepts to graph $f(x) = -3x^2 + 6x + 1$.

Solution In Example 2, we found the vertex to be at $(1, 4)$ and the axis of symmetry to be $x = 1$. The y-intercept is found by letting $x = 0$. Thus, the y-intercept is $f(0) = 1$. The x-intercepts are found by letting $f(x) = 0$. This results in the equation

$$-3x^2 + 6x + 1 = 0$$

The discriminant $b^2 - 4ac = (6)^2 - 4(-3)(1) = 36 + 12 = 48 > 0$, so the equation has two real solutions and the graph has two x-intercepts. Using the quadratic formula, we find

FIGURE 9
$f(x) = -3x^2 + 6x + 1$

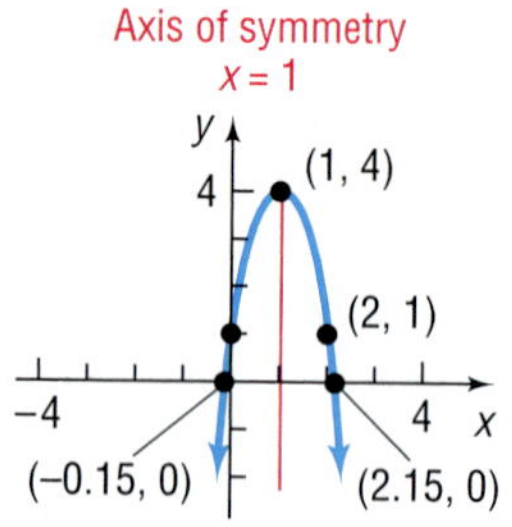

$$x = \frac{-b + \sqrt{b^2 - 4ac}}{2a} = \frac{-6 + \sqrt{48}}{-6} = \frac{-6 + 4\sqrt{3}}{-6} \approx -0.15$$

and

$$x = \frac{-b - \sqrt{b^2 - 4ac}}{2a} = \frac{-6 - \sqrt{48}}{-6} = \frac{-6 - 4\sqrt{3}}{-6} \approx 2.15$$

The x-intercepts are approximately -0.15 and 2.15.

The graph is illustrated in Figure 9. Notice how we used the y-intercept and the axis of symmetry, $x = 1$, to obtain the additional point $(2, 1)$ on the graph.

EXAMPLE 4 Analyzing the Graph of a Quadratic Function

Graph $f(x) = x^2 - 6x + 9$. Determine whether the graph opens up or down. Find its vertex, axis of symmetry, y-intercept, and x-intercepts, if any.

FIGURE 10

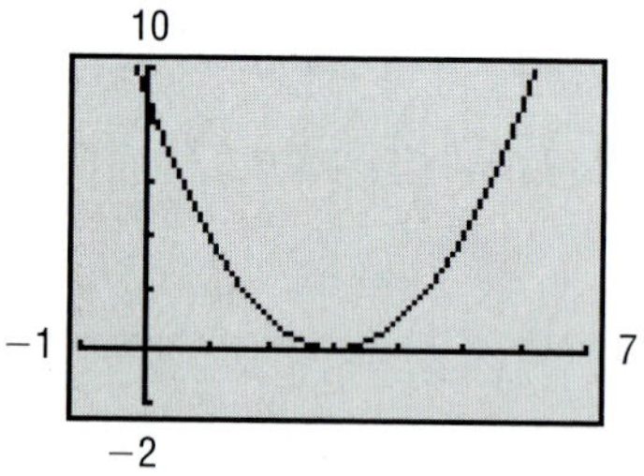

Solution Figure 10 shows the graph of $f(x) = x^2 - 6x + 9$.

From the graph, we conjecture there is one y-intercept near 9 and one x-intercept near 3. The parabola opens up.

Now we confirm these conclusions using algebra. The parabola opens up since $a = 1 > 0$. Since $a = 1$, $b = -6$, and $c = 9$, the x-coordinate of the vertex is

$$\frac{-b}{2a} = \frac{-(-6)}{2(1)} = 3$$

The y-coordinate of the vertex is

$$f(3) = (3)^2 - 6(3) + 9 = 0$$

So the vertex is at $(3, 0)$. The axis of symmetry is the line $x = 3$. The y-intercept is $f(0) = 9$. Since the vertex $(3, 0)$ lies on the x-axis, the graph touches the x-axis at the x-intercept.

Now work Problem 25.

EXAMPLE 5 Analyzing the Graph of a Quadratic Function

Graph $f(x) = 2x^2 + x + 1$. Determine whether the graph opens up or down. Find its vertex, axis of symmetry, y-intercept, and x-intercepts, if any.

Solution Graph the function. See Figure 11. From the graph we can see the graph opens up. We also observe the graph has no x-intercepts and one y-intercept 1.

Now we confirm these conclusions. Since $a = 2$, $b = 1$, and $c = 1$, the x-coordinate of the vertex is

$$\frac{-b}{2a} = -\frac{1}{4}$$

FIGURE 11

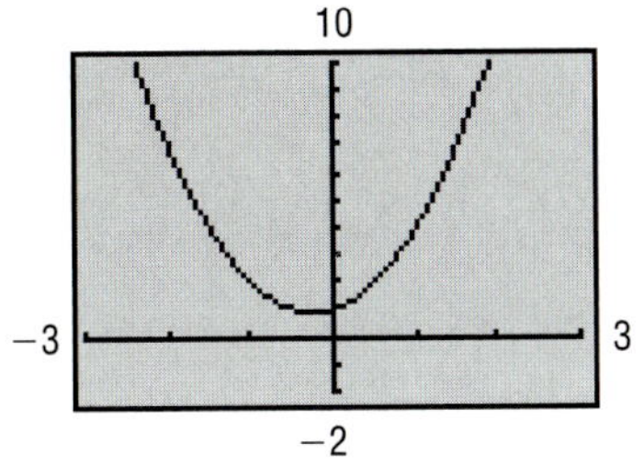

The y-coordinate of the vertex is

$$f\left(-\frac{1}{4}\right) = 2\left(\frac{1}{16}\right) + \left(-\frac{1}{4}\right) + 1 = \frac{7}{8}$$

So the vertex is at $\left(-\frac{1}{4}, \frac{7}{8}\right)$. The axis of symmetry is the line $x = -\frac{1}{4}$. The y-intercept is $f(0) = 1$. Since the discriminant $b^2 - 4ac = (1)^2 - 4(2)(1) = -7 < 0$, the equation $2x^2 + x + 1 = 0$ has no real solutions and therefore the graph has no x-intercepts, confirming the conclusions reached earlier.

Now work Problem 31.

Applications

3

We have already seen that the graph of a quadratic function $f(x) = ax^2 + bx + c$ is a parabola with vertex at $(-b/2a, f(-b/2a))$. This vertex is the highest point on the graph if $a < 0$ and the lowest point on the graph if $a > 0$. If the vertex is the highest point ($a < 0$), then $f(-b/2a)$ is the **maximum value** of f. If the vertex is the lowest point ($a > 0$), then $f(-b/2a)$ is the **minimum value** of f. These ideas give rise to many applications.

EXAMPLE 6

Maximizing Revenue

The marketing department at Texas Instruments has found that, when certain calculators are sold at a price of p dollars per unit, the revenue R (in dollars) as a function of the price p is

$$R(p) = -152.63p^2 + 20208.43p$$

What unit price should be established in order to maximize revenue? If this price is charged, what is the maximum revenue?

Solution

The revenue R is

$$R(p) = -152.63p^2 + 20208.43p = ap^2 + bp + c$$

The function R is a quadratic function with $a = -152.63$, $b = 20{,}208.43$, and $c = 0$. Because $a < 0$, the vertex is the highest point of the parabola. The revenue R is therefore a maximum when the price p is

$$p = \frac{-b}{2a} = \frac{-20208.43}{2(-152.63)} = \frac{-20208.43}{-305.26} = \$66.20$$

The maximum revenue R is

$$R(66.20) = -152.63(66.20)^2 + 20208.43(66.20) \approx \$668{,}906.25$$

See Figure 12 for an illustration.

FIGURE 12
$R(p) = -152.63p^2 + 20208.43p$

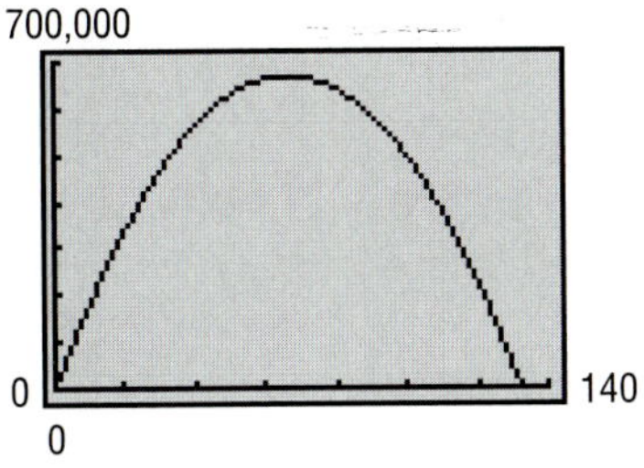

EXAMPLE 7

Analyzing the Motion of a Projectile

A projectile is fired from a cliff 500 feet above the water at an inclination of 45° to the horizontal, with a muzzle velocity of 400 feet per second. In physics, it is established that the height h of the projectile above the water is given by

$$h(x) = \frac{-32x^2}{(400)^2} + x + 500$$

where x is the horizontal distance of the projectile from the base of the cliff. See Figure 13.

FIGURE 13

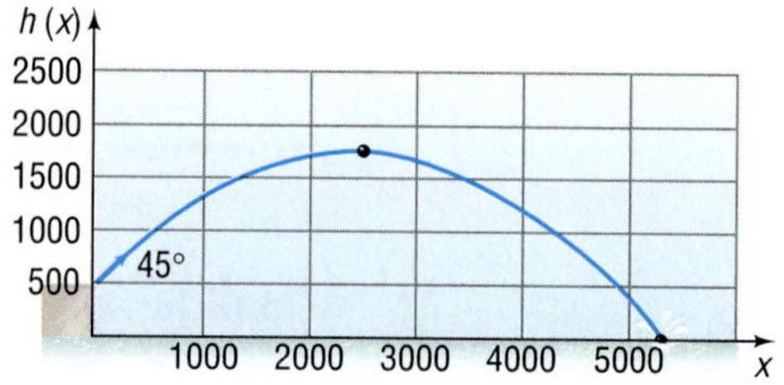

(a) Using a graphing utility, graph the function h, $0 \le x \le 6000$.
(b) Find the maximum height of the projectile.
(c) How far from the base of the cliff will the projectile strike the water?
(d) Use a graphing utility to verify the solutions found in (b) and (c).

FIGURE 14

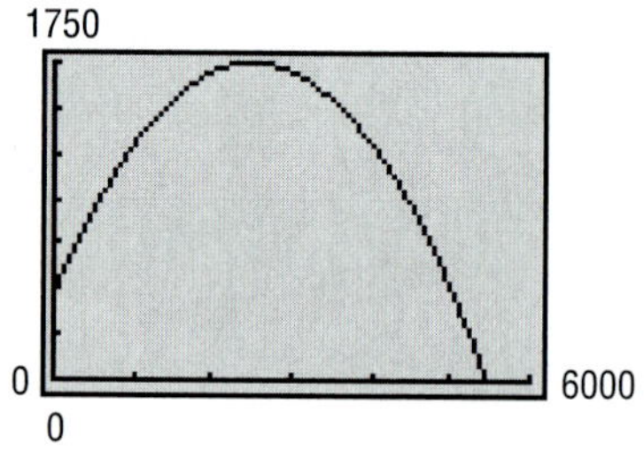

Solution (a) The graph of h is shown in Figure 14.

(b) The height of the projectile is given by a quadratic function:

$$h(x) = \frac{-32x^2}{(400)^2} + x + 500 = \frac{-1}{5000}x^2 + x + 500$$

We are looking for the maximum value of h. Since the maximum value is obtained at the vertex, we compute

$$x = \frac{-b}{2a} = \frac{-1}{2(-1/5000)} = \frac{5000}{2} = 2500$$

The maximum height of the projectile is

$$h(2500) = \frac{-1}{5000}(2500)^2 + 2500 + 500 = -1250 + 2500 + 500 = 1750 \text{ ft}$$

(c) The projectile will strike the water when its height is zero. To find the distance x traveled, we need to solve the equation

$$h(x) = \frac{-1}{5000}x^2 + x + 500 = 0$$

We use the quadratic formula with

$$b^2 - 4ac = 1 - 4\left(\frac{-1}{5000}\right)(500) = 1.4$$

$$x = \frac{-1 \pm \sqrt{1.4}}{2(-1/5000)} = \begin{cases} -458 \\ 5458 \end{cases}$$

We discard the negative solution and find that the projectile will strike the water a distance of 5458 feet from the base of the cliff.

(d) Using the MAXIMUM command on a graphing utility, we find the projectile is at its maximum height of $y = 1750$ feet at $x = 2500$ feet. See Figure 15(a). Using the ZERO or ROOT command, we find the projectile strikes the water 5458 feet from the base of the cliff. See Figure 15(b).

FIGURE 15

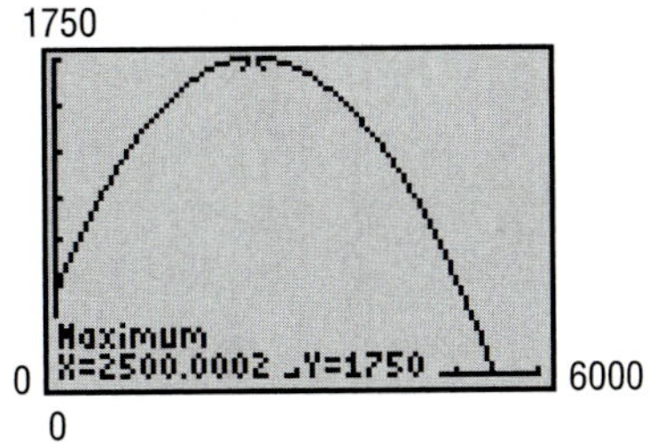

(a)

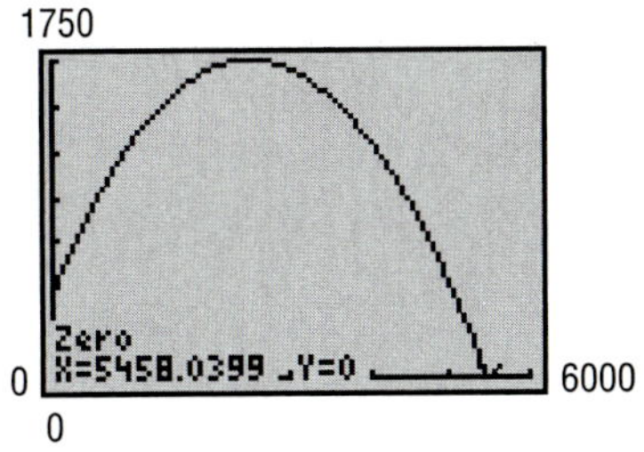

(b)

In a suspension bridge, the main cables are of parabolic shape because, if the total weight of a bridge is uniformly distributed along its length, the only cable shape that will bear the load evenly is that of a parabola. The Golden Gate Bridge in San Francisco is an example of a suspension bridge.

EXAMPLE 8 The Golden Gate Bridge

The Golden Gate Bridge, a suspension bridge, spans the entrance to San Francisco Bay. Its 746-foot-tall towers are 4200 feet apart. The bridge is suspended from two huge cables more than 3 feet in diameter; the 90-foot-wide roadway is 220 feet above the water. The cables are parabolic in shape and touch the road surface at the center of the bridge. Find the height of the cable at a distance of 1000 feet from the center.

Solution We begin by choosing the placement of the coordinate axes so that the x-axis coincides with the road surface and the origin coincides with the center of the bridge. As a result, the twin towers will be vertical (height $746 - 220 = 526$ feet above the road) and located 2100 feet from the center. Also, the cable, which has the shape of a parabola, will extend from the towers, open up, and have its vertex at $(0, 0)$. As illustrated in Figure 16, the choice of placement of the axes enables us to identify the equation of the parabola as $y = ax^2$, $a > 0$. We also can see that the points $(-2100, 526)$ and $(2100, 526)$ are on the graph.

FIGURE 16

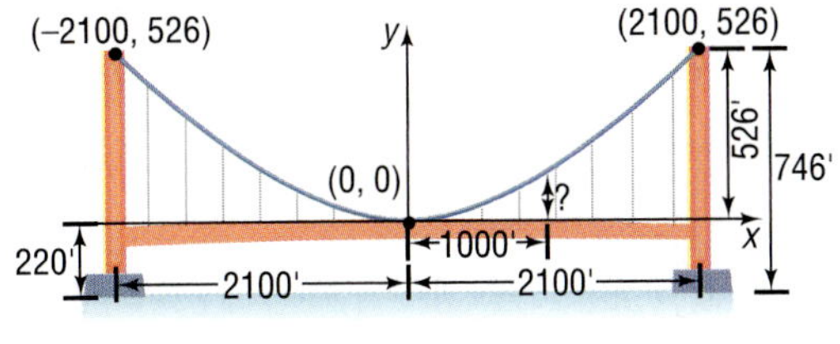

Based on these facts, we can find the value of a in $y = ax^2$:

$$y = ax^2$$
$$526 = a(2100)^2$$
$$a = \frac{526}{(2100)^2}$$

The equation of the parabola is therefore

$$y = \frac{526}{(2100)^2}x^2$$

The height of the cable when $x = 1000$ is

$$y = \frac{526}{(2100)^2}(1000)^2 \approx 119.3 \text{ feet}$$

Thus, the cable is 119.3 feet high at a distance of 1000 feet from the center of the bridge.

Curve Fitting

4 In Section 1.6, we found the line of best fit for data that appeared to be linearly related. It was noted that data may also follow a non-linear relation. Figures 17(a) and (b) show scatter diagrams of data that follow a quadratic relation.

FIGURE 17

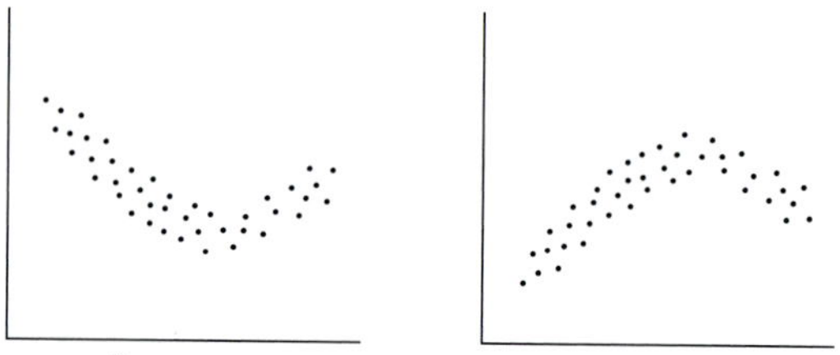

$y = ax^2 + bx + c,\ a > 0$
(a)

$y = ax^2 + bx + c,\ a < 0$
(b)

EXAMPLE 9

Fitting a Quadratic Function to Data

Refer back to Example 2 in Section 1.6. Suppose the farmer collected additional data which shows crop yields Y for various amounts of fertilizer used, x.

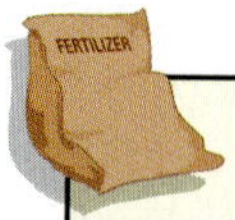

Plot	Fertilizer, x (Pounds/100 ft^2)	Yield (Bushels)
1	0	4
2	0	6
3	5	10
4	5	7
5	10	12
6	10	10
7	15	15
8	15	17
9	20	18
10	20	21
11	25	20
12	25	21
13	30	21
14	30	22
15	35	21
16	35	20
17	40	19
18	40	19

(a) Draw a scatter diagram of the data. Comment on the type of relation that may exist between the two variables.

(b) Find the quadratic function of best fit.

(c) Draw the quadratic function of best fit on the scatter diagram.

(d) Use the function found in (b) to find the optimal amount of fertilizer to use.

(e) Use the function found in (b) to predict the crop yield if the optimal amount of fertilizer is used.

(f) Compare the results found in (d) and (e) with the data.

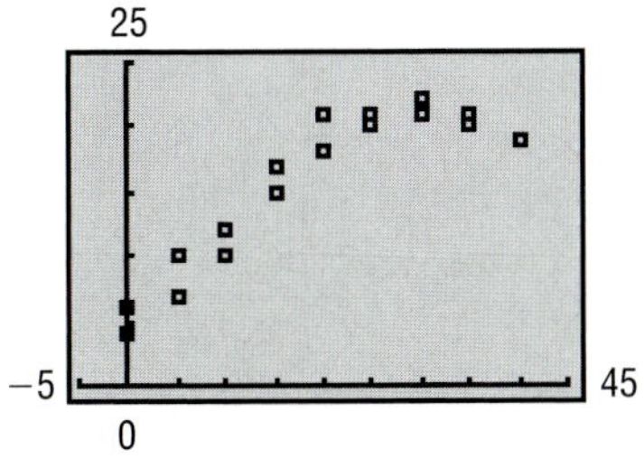

FIGURE 18

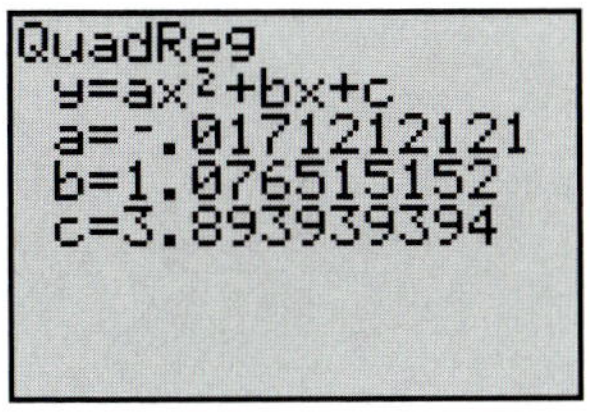

FIGURE 19

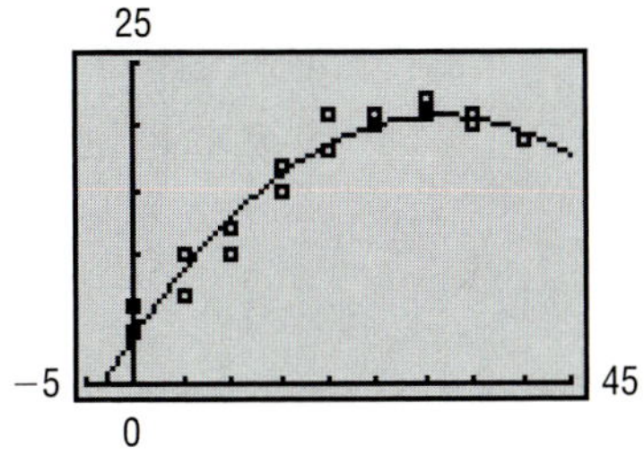

FIGURE 20

Solution (a) Figure 18 shows the scatter diagram. From the scatter diagram, it appears the data follow a quadratic relation, with $a < 0$.

(b) Upon executing the QUADratic REGression program, we obtain the results shown in Figure 19. The output the utility provides shows us the equation, $y = ax^2 + bx + c$. The quadratic function of best fit is $Y(x) = -0.0171x^2 + 1.0765x + 3.8939$, where x represents the amount of fertilizer used and Y represents crop yield.

(c) Figure 20 shows the graph of the quadratic function found in (b) drawn on the scatter diagram.

(d) Based on the quadratic function of best fit, the optimal amount of fertilizer to use is

$$\frac{-b}{2a} = \frac{-1.0765}{2(-0.0171)} \approx 31.5 \text{ pounds of fertilizer per 100 square feet.}$$

(e) We evaluate the function, $Y(x)$ for $x = 31.5$:

$$Y(31.5) = -0.0171(31.5)^2 + 1.0765(31.5) + 3.8939 \approx 20.8 \text{ bushels}$$

If we use 31.5 pounds of fertilizer per 100 square feet, the crop yield will be 20.8 bushels according to the quadratic function of best fit.

(f) From the data, 30 pounds of fertilizer per 100 square feet results in the most crop 22 bushels. The result found in (d) is close to the level of fertilizer that yields the most crop from the data. However, the prediction found in (e) appears to underestimate the crop yield based on the data.

Look again at Figure 19. Notice the output given by the graphing calculator does not include r the correlation coefficient. Recall, the correlation coefficient is a measure of the strength of a **linear** relation that exists between two variables. The graphing calculator does not provide an indication of how well the function "fits" the data in terms of r since the function cannot be expressed as a linear function.

It is not recommended that we use the functions of best fit to make predictions outside the scope of the model since we cannot be certain the data will continue to follow the pattern it has for the observed data. Therefore, we will limit our predictions to values of the independent variable that are close to the observed data.

4.1 EXERCISES

In Problems 1–8, match each graph to one of the following functions:

A. $y = x^2 - 1$ B. $y = -x^2 - 1$ C. $y = x^2 - 2x + 1$ D. $y = x^2 + 2x + 1$

E. $y = x^2 - 2x + 2$ F. $y = x^2 + 2x$ G. $y = x^2 - 2x$ H. $y = x^2 + 2x + 2$

1.

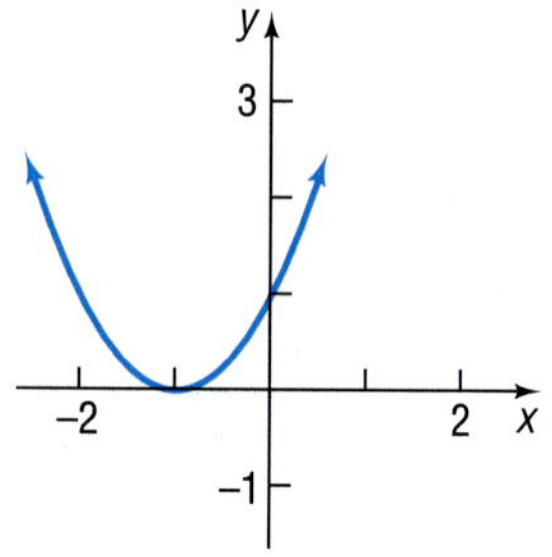

2.

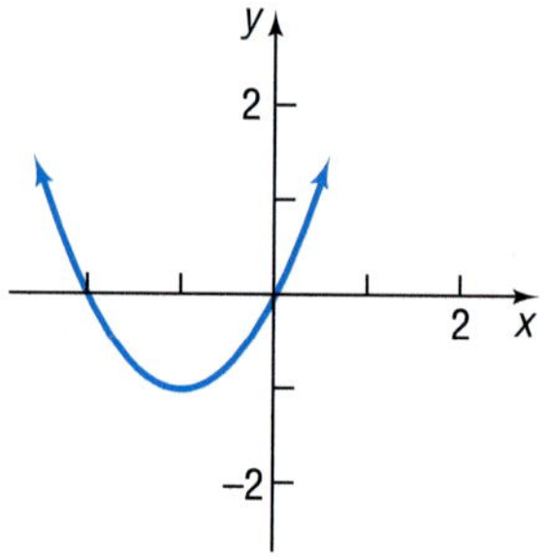

3.

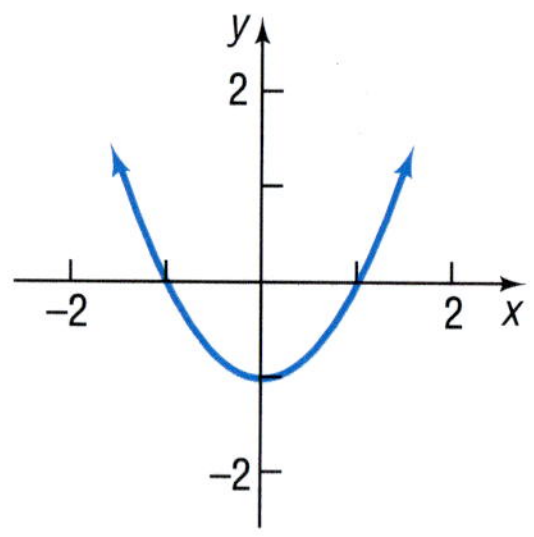

4.

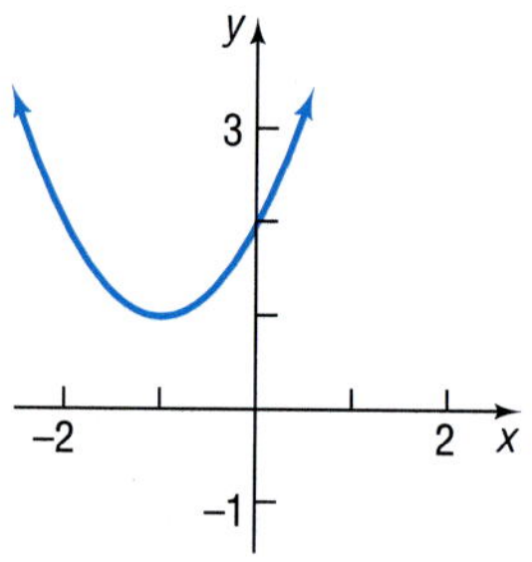

5.

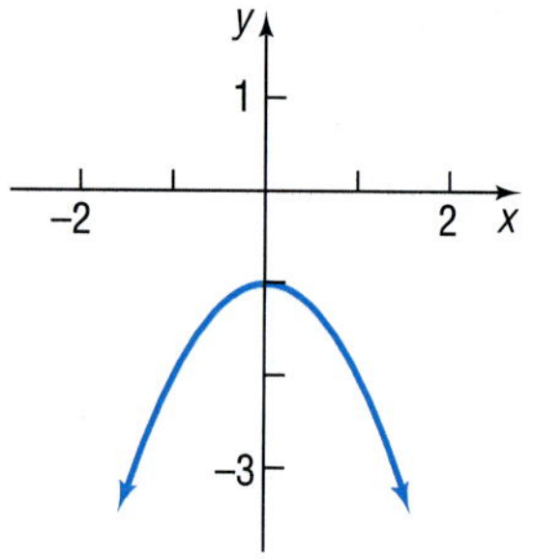

6.

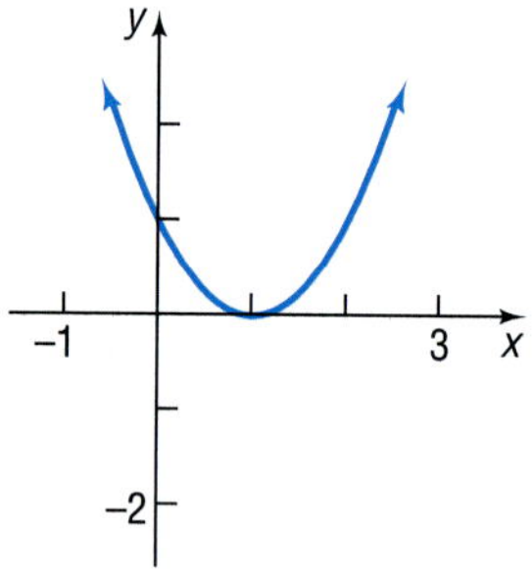

7.

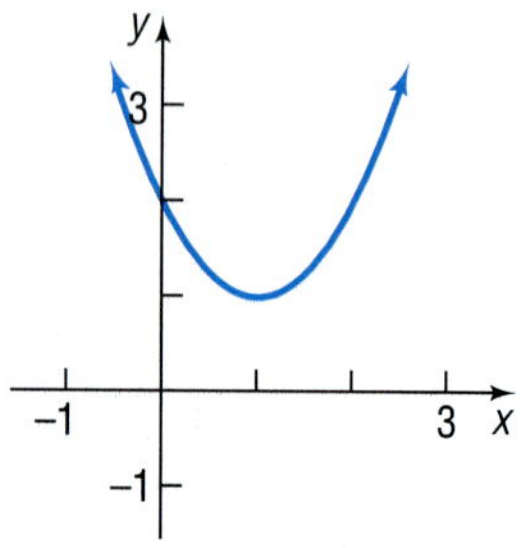

8.

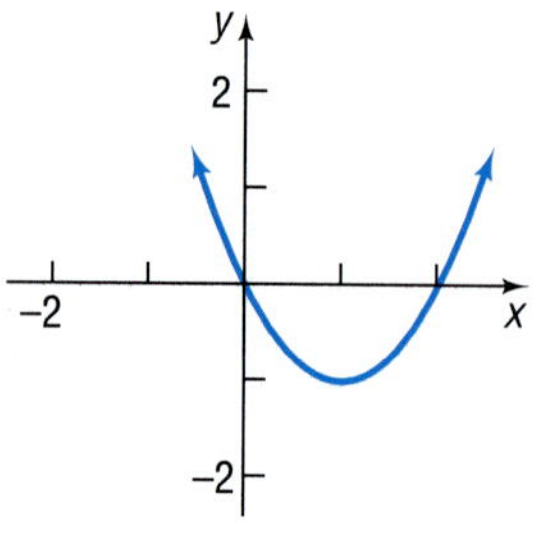

In Problems 9–12, match each graph to one of the following functions.

A. $y = (1/3)x^2 + 2$ *B.* $y = 3x^2 + 2$

C. $y = x^2 + 5x + 1$ *D.* $y = -x^2 + 5x + 1$

9.

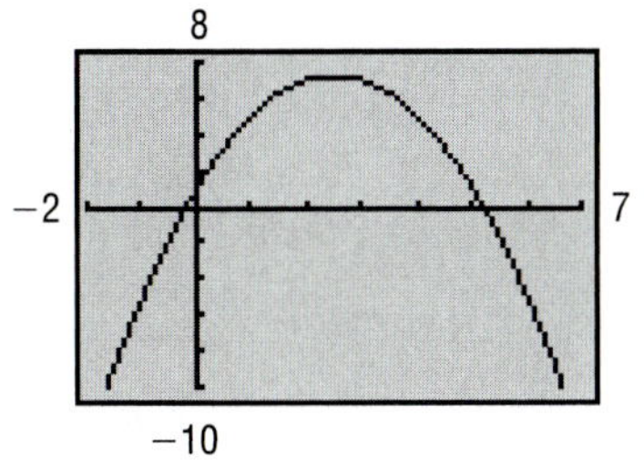

10.

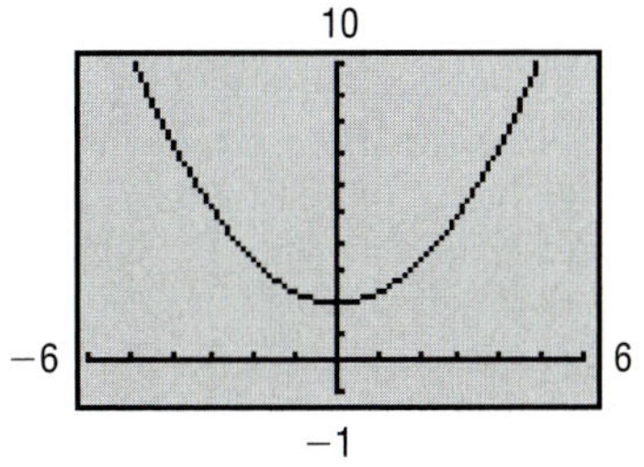

11.

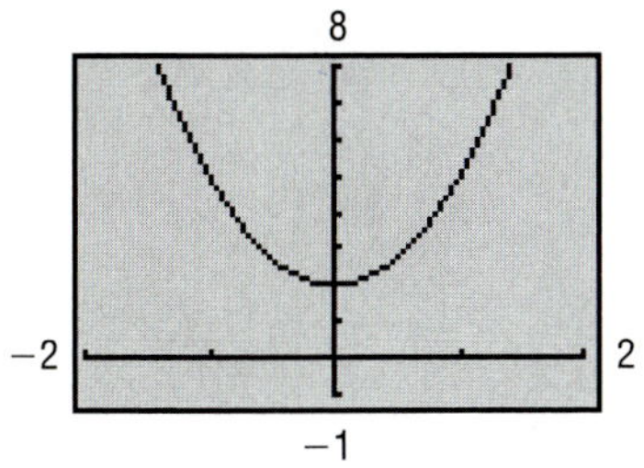

12.

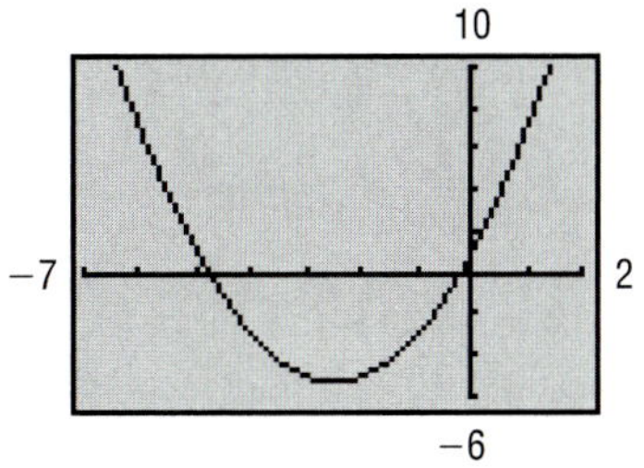

In Problems 13–42, graph each quadratic function by hand by determining whether its graph opens up or down and by finding its vertex, axis of symmetry, y-intercept, and x-intercepts, if any. Verify your results using a graphing utility.

13. $f(x) = \frac{1}{4}x^2$ **14.** $f(x) = 2x^2$ **15.** $f(x) = \frac{1}{4}x^2 - 2$

16. $f(x) = 2x^2 - 3$ **17.** $f(x) = \frac{1}{4}x^2 + 2$ **18.** $f(x) = 2x^2 + 4$

19. $f(x) = -\frac{1}{4}x^2 + 1$ **20.** $f(x) = -2x^2 - 2$ **21.** $f(x) = x^2 + 4x + 2$

22. $f(x) = x^2 - 6x - 1$ **23.** $f(x) = 2x^2 - 4x + 1$ **24.** $f(x) = 3x^2 + 6x$

25. $f(x) = -x^2 - 2x$ **26.** $f(x) = -2x^2 + 6x + 2$ **27.** $f(x) = \frac{1}{2}x^2 + x - 1$

28. $f(x) = \frac{2}{3}x^2 + \frac{4}{3}x - 1$ **29.** $f(x) = x^2 + 2x - 8$ **30.** $f(x) = x^2 - 2x - 3$

31. $f(x) = -x^2 - 3x + 4$ **32.** $f(x) = -x^2 + x + 2$ **33.** $f(x) = x^2 + 2x + 1$

34. $f(x) = -x^2 + 4x - 4$ **35.** $f(x) = 2x^2 - x + 2$ **36.** $f(x) = 4x^2 - 2x + 1$

37. $f(x) = -2x^2 + 2x - 3$ **38.** $f(x) = -3x^2 + 3x - 2$ **39.** $f(x) = 3x^2 + 6x + 2$

40. $f(x) = 2x^2 + 5x + 3$ **41.** $f(x) = -4x^2 - 6x + 2$ **42.** $f(x) = 3x^2 - 8x + 2$

In Problems 43–48, determine without graphing, whether the given quadratic function has a maximum value or a minimum value and then find the value. Verify your results using a graphing utility.

43. $f(x) = 2x^2 + 12x - 3$ **44.** $f(x) = 4x^2 - 8x + 3$ **45.** $f(x) = -x^2 + 10x - 4$

46. $f(x) = -2x^2 + 8x + 3$ **47.** $f(x) = -3x^2 + 12x + 1$ **48.** $f(x) = 4x^2 - 4x$

49. On one set of coordinate axes, graph the family of parabolas $f(x) = x^2 + 2x + c$ for $c = -3$, $c = 0$ and $c = 1$. Describe the characteristics of a member of this family.

50. On one set of coordinate axes, graph the family of parabolas $f(x) = x^2 + cx + 1$ for $c = -4$, $c = 0$, and $c = 4$. Describe the general characteristics of this family.

51. Graph $y = x^2 + 1$. Then graph $y = x^2 + x + 1$, followed by $y = x^2 + 2x + 1$, followed by $y = x^2 + 3x + 1$. What is happening? Do you see any pattern?

52. Graph $y = x^2 + x + 1$. Then graph $y = 2x^2 + x + 1$, followed by $y = 3x^2 + x + 1$, followed by $y = 4x^2 + x + 1$. What is happening? Do you see any pattern?

53. **Maximizing Revenue** Suppose that the manufacturer of a gas clothes dryer has found that, when the unit price is p dollars, the revenue R (in dollars) is

$$R = -4p^2 + 4000p$$

What unit price should be established for the dryer to maximize revenue? What is the maximum revenue?

54. **Maximizing Revenue** The John Deere company has found that the revenue from sales of heavy-duty tractors is a function of the unit price p it charges. If the revenue R is

$$R = -\frac{1}{2}p^2 + 1900p$$

what unit price p should be charged to maximize revenue? What is the maximum revenue?

55. **Rectangles with Fixed Perimeter** What is the largest rectangular area that can be enclosed with 400 feet of fencing? What are the dimensions of the rectangle?

56. **Rectangles with Fixed Perimeter** What are the dimensions of a rectangle of a fixed perimeter P that result in the largest area?

57. **Enclosing the Most Area with a Fence** A farmer with 4000 meters of fencing wants to enclose a rectangular plot that borders on a river. If the farmer does not fence the side along the river, what is the largest area that can be enclosed? (See the figure.)

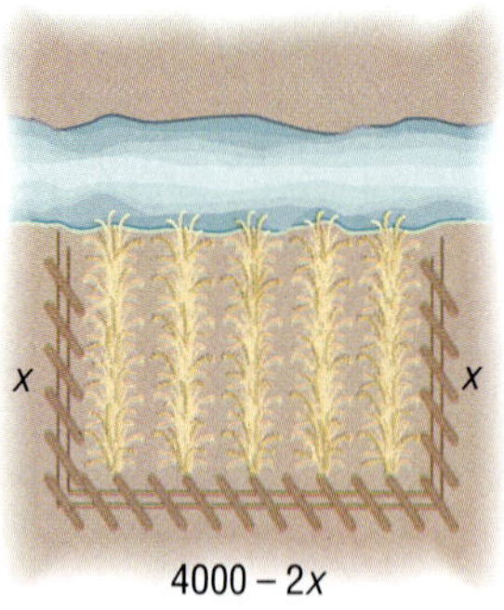

58. **Enclosing the Most Area with a Fence** A farmer with 2000 meters of fencing wants to enclose a rectangular plot that borders on a straight highway. If the farmer does not fence the side along the highway, what is the largest area that can be enclosed?

59. **Enclosing the Most Area with a Fence** A farmer with 10,000 meters of fencing wants to enclose a rectangular field and then divide it into two plots with a fence parallel to one of the sides (see the figure). What is the largest area that can be enclosed?

60. **Enclosing the Most Area with a Fence** A farmer with 10,000 meters of fencing wants to enclose a rectangular field and then divide it into three plots with two fences parallel to one of the sides. What is the largest area that can be enclosed?

61. **Analyzing the Motion of a Projectile** A projectile is fired from a cliff 200 feet above the water at an inclination of 45° to the horizontal, with a muzzle velocity of 50 feet per second. The height h of the projectile above the water is given by

$$h(x) = \frac{-32x^2}{(50)^2} + x + 200$$

where x is the horizontal distance of the projectile from the base of the cliff.

(a) Using a graphing utility, graph the function h, $0 \le x \le 200$.
(b) Find the maximum height of the projectile.
(c) How far from the base of the cliff will the projectile strike the water?
(d) Use a graphing utility to verify the solutions found in (b) and (c).
(e) When the height of the projectile is 100 feet above the water, how far is it from the cliff?

62. **Analyzing the Motion of a Projectile** A projectile is fired at an inclination of 45° to the hor-

izontal, with a muzzle velocity of 100 feet per second. The height h of the projectile is given by

$$h(x) = \frac{-32x^2}{(100)^2} + x$$

where x is the horizontal distance of the projectile from the firing point.

(a) Using a graphing utility, graph the function h, $0 \le x \le 350$.
(b) Find the maximum height of the projectile.
(c) How far from the firing point will the projectile strike the ground?
(d) Use a graphing utility to verify the results obtained in parts (b) and (c).
(e) When the height of the projectile is 50 feet above the ground, how far has it traveled horizontally?

63. Navigation The *U.S.S. Independence* maintains a constant speed of 10 knots heading due north. At 4:00 P.M. the ship's radar detects a destroyer 100 nautical miles due east of the carrier. If the destroyer is heading due west at 20 knots, when will the two ships be the closest? (1 knot = 1 nautical mile per hour)

64. Air Traffic Control An air traffic controller sees two aircraft flying at the same altitude on his screen. One, a Piper Cub, is headed due west at 150 miles per hour. The other, a Lear jet, is 15 miles due north of the Piper and is headed due south at 400 miles per hour. How close will the two aircraft come to each other?

65. Suspension Bridge A suspension bridge with weight uniformly distributed along its length has twin towers that extend 75 meters above the road surface and are 400 meters apart. The cables are parabolic in shape and are suspended from the tops of the towers. The cables touch the road surface at the center of the bridge. Find the height of the cables at a point 100 meters from the center. (Assume that the road is level.)

66. Architecture A parabolic arch has a span of 120 feet and a maximum height of 25 feet. Choose suitable rectangular coordinate axes and find the equation of the parabola. Then calculate the height of the arch at points 10 feet, 20 feet, and 40 feet from the center.

67. Constructing Rain Gutters A rain gutter is to be made of aluminum sheets that are 12 inches wide by turning up the edges 90°. What depth will provide maximum cross-sectional area and hence allow the most water to flow?

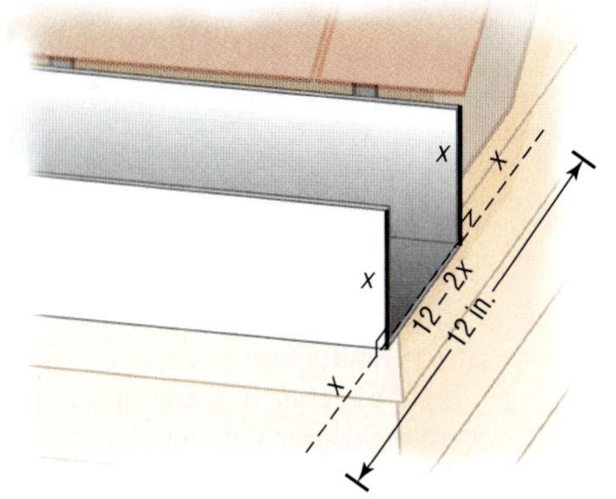

68. Navigation At 4 P.M. the *Ecstasy* leaves the Port of Miami heading due east at a constant speed of 15 knots. At the same time, a pleasure boat located 100 nautical miles northeast of the Port of Miami is headed due south at a constant speed of 12 knots. When are the two ships closest? How close do they get to each other? (Express your answer in nautical miles; 1 knot = 1 nautical mile per hour.)

69. Norman Windows A Norman window has the shape of a rectangle surmounted by a semicircle of diameter equal to the width of the rectangle (see the figure). If the perimeter of the window is 20 feet, what dimensions will admit the most light (maximize the area)?

[**Hint:** Circumference of circle = $2\pi r$; Area of circle = πr^2, where r is the radius of the circle.]

70. Constructing a Stadium A track and field playing area is in the shape of a rectangle with semicircles at each end (see the figure). The inside perimeter of the track is to be 1500 meters.

What should the dimensions of the rectangle be so that the area of the rectangle is a maximum?

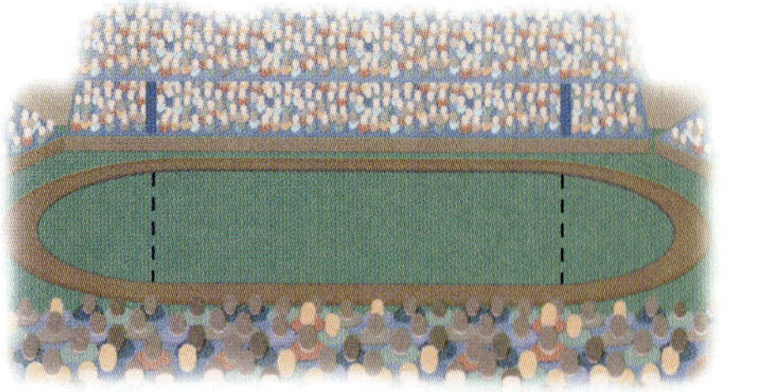

71. Architecture A special window has the shape of a rectangle surmounted by an equilateral triangle (see the figure). If the perimeter of the window is 16 feet, what dimensions will admit the most light?

[**Hint:** Area of an equilateral triangle = $(\sqrt{3}/4)x^2$, where x is the length of a side of the triangle.]

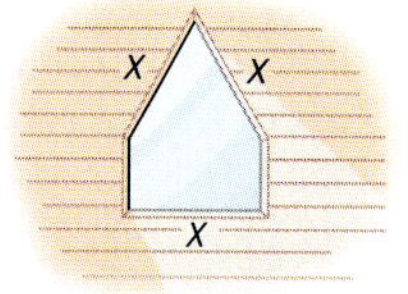

72. Analyzing the Motion of a Projectile A projectile is fired at an inclination of 45° to the horizontal with an initial velocity of v_0 feet per second. If the starting point is the origin, the x-axis is horizontal, and the y-axis is vertical, then the height y (in feet) after a horizontal distance x has been traversed is approximately

$$y = \frac{-32x^2}{v_0^2} + x$$

(a) Find the maximum height in terms of the initial velocity v_0.
(b) If the initial velocity is doubled, what happens to the maximum height?
(c) Assuming the ground is flat, how far from the starting point will the projectile land if the initial velocity is 64 feet per second?

73. Charter Club Memberships A charter flight club charges its members $400 per year. But, for each new member in excess of 60, the charge for every member is reduced by $5. What number of members leads to a maximum revenue?

74. Car Rentals A car rental agency has 24 identical cars. The owner of the agency finds that all the cars can be rented at a price of $10 per day. However, for each $2 increase in rental, one of the cars is not rented. What should be charged to maximize income?

75. Life Cycle Hypothesis An individual's income varies with his/her age. Below is the median income I of individuals of different age groups within the United States for 1995. For each class, let the class midpoint represent the independent variable, x. For the class "65 years and older," we will assume the class midpoint is 69.5.

Age	Class Midpoint, x	Median Income, I
15–24 years	19.5	$20,979
25–34 years	29.5	$34,701
35–44 years	39.5	$43,465
45–54 years	49.5	$48,058
55–64 years	59.5	$38,077
65 years and older	69.5	$19,096

Source: U.S. Census Bureau.

(a) Using a graphing utility, draw a scatter diagram of the data. Comment on the type of relation that may exist between the two variables.
(b) Find the quadratic function of best fit.
(c) Using a graphing utility, draw the quadratic function of best fit on your scatter diagram.
(d) Using the function found in (b), determine the age at which an individual can expect to earn the most income.
(e) Predict the peak income earned.
(f) Compare your result in (d) and (e) to the data and comment.

76. Advertising A small manufacturing firm collected the data below on advertising expenditures (in thousands of dollars) and total revenue (in thousands of dollars).

Advertising	Total Revenue
20	$6101
22	$6222
25	$6350
25	$6378
27	$6453
28	$6423
29	$6360
31	$6231

(a) Using a graphing utility, draw a scatter diagram of the data. Comment on the type of relation that may exist between the two variables.

(b) Find the quadratic function of best fit.
(c) Using a graphing utility, draw the quadratic function of best fit on your scatter diagram.
(d) Using the function found in (b), determine the optimal level of advertising for this firm.
(e) Determine the revenue the firm can expect if it uses the optimal level of advertising.

(f) Compare your result in (d) and (e) to the data and comment.

77. Fuel Consumption The data below represents the average fuel consumption C by cars (in billions of gallons) for the years 1980–1993.

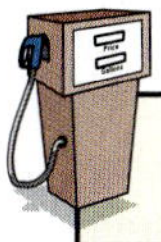

Year, t	Average Fuel Consumption, C
1980	71.9
1981	71.0
1982	70.1
1983	69.9
1984	68.7
1985	69.3
1986	71.4
1987	70.6
1988	71.9
1989	72.7
1990	72.0
1991	70.7
1992	73.9
1993	75.1

Source: U.S. Federal Highway Administration.

(a) Using a graphing utility, draw a scatter diagram of the data. Comment on the type of relation that may exist between the two variables.
(b) Find the quadratic function of best fit.
(c) Using a graphing utility, draw the quadratic function of best fit on your scatter diagram.
(d) Using the function found in (b), determine the year in which average fuel consumption was lowest.
(e) Compare your result in (d) to the data and comment.
(f) Use the function found in (b) to predict the average fuel consumption for 1994.

78. Miles Per Gallon An engineer collects data showing the speed s of a Ford Taurus and its average miles per gallon, M. See the table.

Speed, s	Miles per Gallon, M
30	18
35	20
40	23
40	25
45	25
50	28
55	30
60	29
65	26
65	25
70	25

(a) Using a graphing utility, draw a scatter diagram of the data. Comment on the type of relation that may exist between the two variables.
(b) Find the quadratic function of best fit.
(c) Using a graphing utility, draw the quadratic function of best fit on your scatter diagram.
(d) Using the function found in (b), determine the speed that maximizes miles per gallon.
(e) Compare your result in (d) to the data.
(f) Using the function found in (b), predict the miles per gallon of the car if you travel an average of 63 miles per hour.

79. Height of a Ball A physicist throws a ball at an inclination of 45° to the horizontal. The data below represents the height of the ball h at the instant it has traveled x feet horizontally.

Distance, x	Height, h
20	25
40	40
60	55
80	65
100	71
120	77
140	77
160	75
180	71
200	64

(a) Using a graphing utility, draw a scatter diagram of the data. Comment on the type of relation that may exist between the two variables.

(b) Find the quadratic function of best fit.
(c) Using a graphing utility, draw the quadratic function of best fit on your scatter diagram.
(d) Using the function found in (b), how far will the ball travel before it reaches its maximum height?
(e) Using the function found in (b), determine the maximum height of the ball.
(f) Compare your result in (d) and (e) to the data.
(g) Determine the horizontal distance the ball will travel based on the function found in (b).

80. Enrollment in Public Schools The data below represents the enrollment E in all public schools (both elementary and high school) for the academic years 1980–1981 to 1988–1989. Let 1 represent the academic year 1980–1981, 2 the academic year 1981–1982, and so on.

Year, t	Enrollment, E
1	41.5
2	40.8
3	40.1
4	39.6
5	39.1
6	39.1
7	39.6
8	39.8
9	40.1

(a) Using a graphing utility, draw a scatter diagram of the data. Comment on the type of relation that may exist between the two variables.
(b) Find the quadratic function of best fit.
(c) Using a graphing utility, draw the quadratic function of best fit on your scatter diagram.
(d) Using the function found in (b), determine when enrollment was lowest.
(e) Compare your result in (d) to the data.
(f) Using the function found in (b), predict the enrollment for 1991–1992.

81. Chemical Reactions A self-catalytic chemical reaction results in the formation of a compound that causes the formation ratio to increase. If the reaction rate V is given by

$$V(x) = kx(a - x) \qquad 0 \le x \le a$$

where k is a positive constant, a is the initial amount of the compound, and x is the variable amount of the compound, for what value of x is the reaction rate a maximum?

82. Make up a quadratic function that opens down and has only one x-intercept. Compare yours with others in the class. What are the similarities? What are the differences?

83. CBL Experiment As a ball bounces up and down, the maximum height the ball attains continually decreases from one bounce to the next. For a given bounce, plotting the height of the ball against time results in a parabola. The motion of the ball will be analyzed. (Activity 9, Real-World Math with the CBL System, 1994.)

84. CBL Experiment A cart is pushed up a ramp and allowed to return down the ramp. Plotting the distance the cart is from a motion detector against time results in the graph of a parabola. The parabola will be analyzed to determine the characteristics of the cart's motion. (Activity 8, Real-World Math with the CBL System, 1994.)

4.2 POWER FUNCTIONS; CURVE FITTING

1 Determine the Coefficient and Degree of a Monomial
2 Graph Power Functions in Stages
3 Find the Power Function of Best Fit from Data

1 A **monomial in one variable** is the product of a constant times a variable raised to a nonnegative integer power. Thus, a monomial is of the form ax^n where a is a constant, x is a variable, and $n \ge 0$ is an integer. The constant a is called the coefficient of the monomial. If $a \neq 0$, then n is called the degree of the monomial.

EXAMPLE 1

Identifying the Coefficient and Degree of a Monomial

Determine the coefficient and degree of the following monomials.

(a) $-4x^3$ (b) $10x$ (c) 7

Solution (a) The coefficient is -4 and the degree is 3.

(b) The coefficient is 10 and the degree is 1.

(c) The coefficient is 7 and the degree is 0 since $7 = 7x^0$.

We now discuss *power functions,* functions whose rule is given by a monomial.

Power Functions

A **power function of degree *n*** is a function of the form

$$f(x) = ax^n \tag{1}$$

where a is a real number, $a \neq 0$, and $n > 0$ is an integer.

The graph of a power function of degree 1, $f(x) = ax$, is a straight line, with slope a, that passes through the origin. The graph of a power function of degree 2, $f(x) = ax^2$, is a parabola, with vertex at the origin, that opens up if $a > 0$ and down if $a < 0$.

If we know how to graph a power function of the form $f(x) = x^n$, then a compression or stretch and, perhaps, a reflection about the x-axis will enable us to obtain the graph of $g(x) = ax^n$. Consequently, we shall concentrate on graphing power functions of the form $f(x) = x^n$.

We begin with power functions of even degree of the form $f(x) = x^n, n \geq 2$ and n *even*. Using your graphing utility and a RANGE of $-2 \leq x \leq 2$, $-4 \leq y \leq 16$, graph the function $f(x) = x^4$. Can you conjecture the domain of this function from the graph? The range? On the same screen, graph $f(x) = x^8$. Now, also on the same screen, graph $f(x) = x^{12}$. What do you notice about the graphs as the magnitude of the exponent increases? Repeat the procedure given above for a RANGE of $-1 \leq x \leq 1; 0 \leq y \leq 1$. What do you notice? See Figures 21(a) and (b).

FIGURE 21

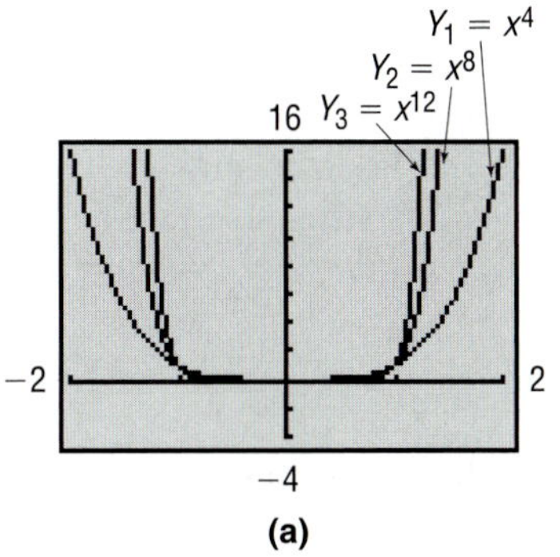

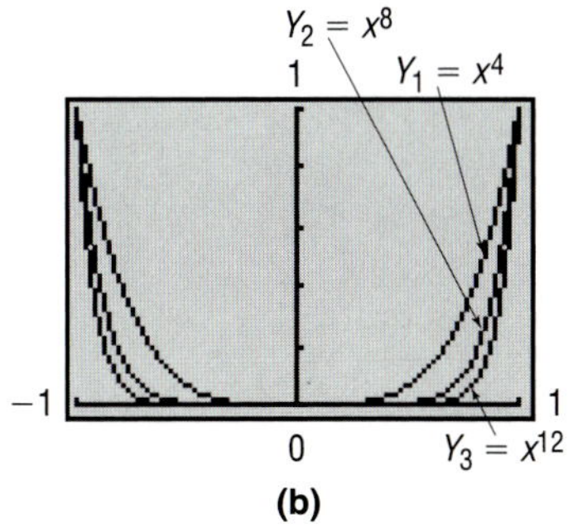

From the graphs, we can conjecture a few characteristics of even power functions. First, even power functions are symmetric with respect to the y-axis. Second, the domain is the set of all real numbers and the range is the set of nonnegative real numbers. Third, the graph always contains the points $(0, 0)$, $(1, 1)$, and $(-1, 1)$. Fourth, as the exponent increases in magnitude, the graph increases very rapidly as x increases, but for x near the origin the graph tends to flatten out and lie closer to the origin.

Note that for large n, it appears the graph coincides with the x-axis near the origin, but it does not; the graph actually touches the x-axis only at the origin. See Table 2, where $Y_1 = x^8$ and $Y_2 = x^{12}$. In Table 2, the expression 1E-8 means $1 \cdot 10^{-8}$. Also, for large n, it may appear that for $x < -1$ or for $x > 1$, the graph is vertical, but it is not; it is only increasing very rapidly. If you use TRACE along one of the graphs these distinctions would be clear.

TABLE 2

X	Y1	Y2
-1	1	1
1	1	1
.5	.00391	2.4E-4
.1	1E-8	1E-12
.01	1E-16	1E-24
.001	1E-24	1E-36
0	0	0

Y1■X^8

Now we consider power functions of odd degree of the form $f(x) = x^n$, *n odd*. Using your graphing utility and a RANGE of $-2 \leq x \leq 2$, $-16 \leq y \leq 16$, graph the function $f(x) = x^3$. You have seen this graph several times before. Based on the graph, can you conjecture the domain and range of this function? On the same screen, graph $f(x) = x^7$ and $f(x) = x^{11}$. What do you notice about the graphs as the magnitude of the exponent increases? Repeat the procedure given above for a RANGE of $-1 \leq x \leq 1$; $-1 \leq y \leq 1$. What do you notice? The graphs on your screen should look like Figures 22(a) and (b).

FIGURE 22

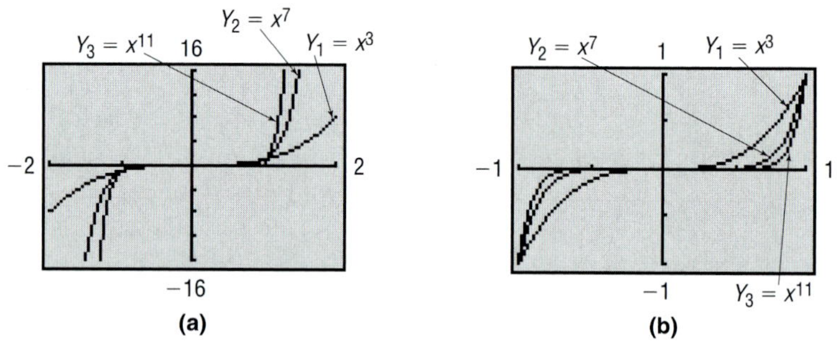

From the graphs in Figure 22(a) and (b), we can conjecture a few characteristics of odd power functions. First, odd power functions are symmetric with respect to the origin. Second, the domain and range of odd power functions are the set of all real numbers. Third, the graph of an odd power function always contains the points $(0, 0)$, $(1, 1)$, and $(-1, -1)$. Fourth, as the exponent increases in magnitude, the graphs become more vertical when $x > 1$ or $x < -1$ and the graphs tend to flatten out and lie closer to the x-axis when x is near the origin. Note that it appears the graph coincides with the x-axis near the origin, but it does not; the graph actually touches the x-axis only at the origin. Also it appears that as x increases, the graph is vertical, but it is not, it is increasing very rapidly. TRACE along the graphs to verify these distinctions.

2 The methods of shifting, compression, stretching, and reflection studied in Section 2.3, when used with the facts just presented, will enable us to graph and analyze a variety of polynomials.

EXAMPLE 2 Graphing In Stages

Using a graphing utility, show each stage to obtain the graph of $f(x) = 1 - x^5$.

Solution STAGE 1: Graph $y = x^5$. See Figure 23 and Table 3.

FIGURE 23

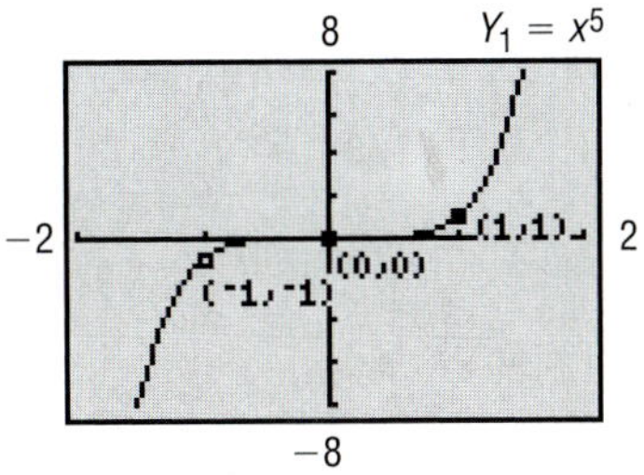

TABLE 3

X	Y1
-3	-243
-2	-32
-1	-1
0	0
1	1
2	32
3	243

Y1=X^5

STAGE 2: Graph $y = -x^5$ (Reflection about y-axis). See Figure 24 and Table 4.

FIGURE 24

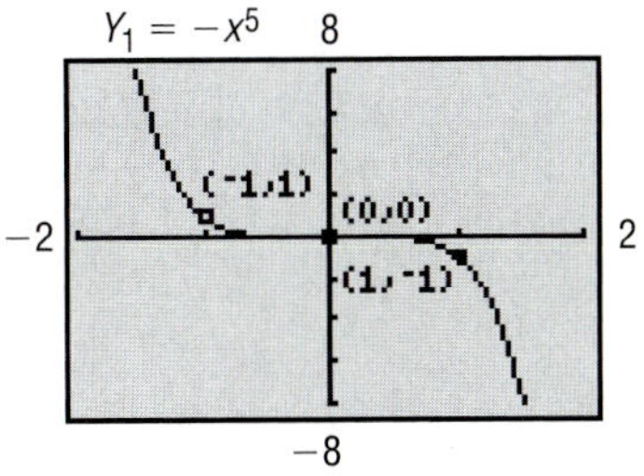

TABLE 4

X	Y1
-3	243
-2	32
-1	1
0	0
1	-1
2	-32
3	-243

Y1=-X^5

STAGE 3: Graph $y = 1 - x^5$ (Shift up 1 unit). See Figure 25 and Table 5.

FIGURE 25

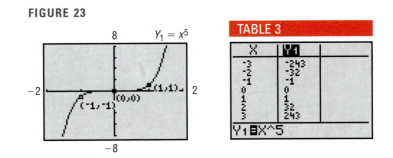

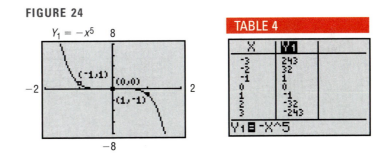

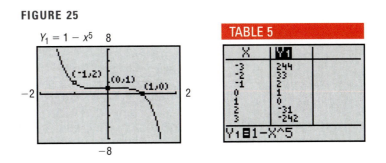

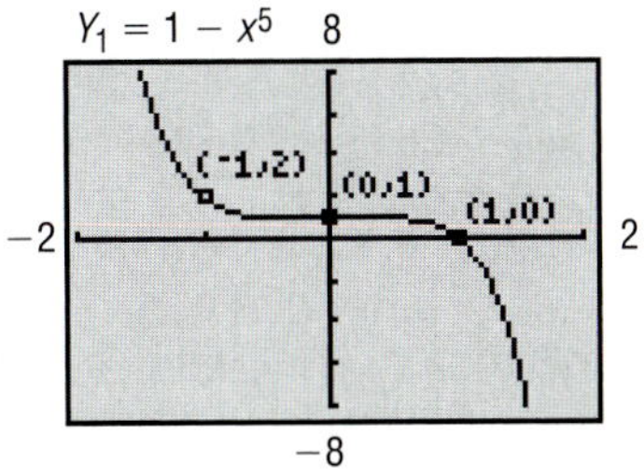

TABLE 5

X	Y1
-3	244
-2	33
-1	2
0	1
1	0
2	-31
3	-242

Y1=1-X^5

By following key points as we shift, reflect, and so forth, we determine the intercepts of f are (0, 1) and (1, 0).

EXAMPLE 3 Graphing In Stages

Using a graphing utility, show each stage to obtain the graph of $f(x) = \frac{1}{2}(x - 1)^4$.

Solution STAGE 1: Graph $y = x^4$. See Figure 26 and Table 6.

FIGURE 26

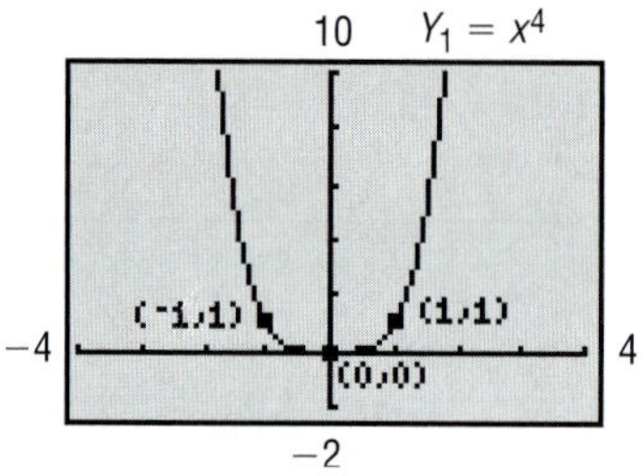

TABLE 6

X	Y1
-3	81
-2	16
-1	1
0	0
1	1
2	16
3	81

Y1=X^4

STAGE 2: Graph $y = \frac{1}{2}x^4$ (compression). See Figure 27 and Table 7.

FIGURE 27

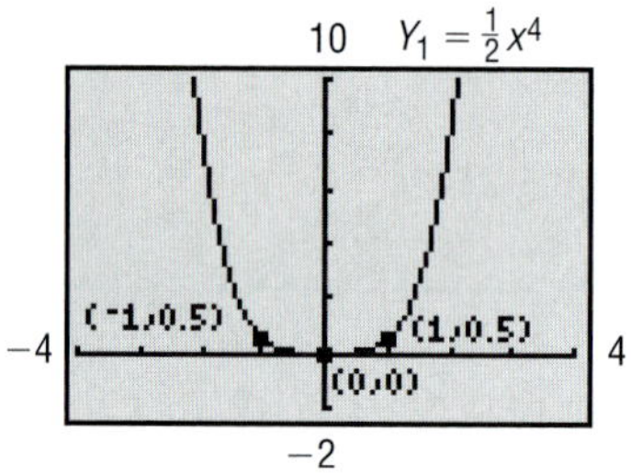

TABLE 7

X	Y1
-3	40.5
-2	8
-1	.5
0	0
1	.5
2	8
3	40.5

Y1=(1/2)X^4

STAGE 3: Graph $y = \frac{1}{2}(x - 1)^4$ (Shift right 1 unit). See Figure 28 and Table 8.

FIGURE 28

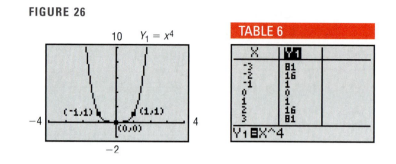

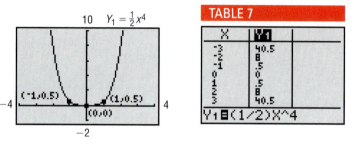

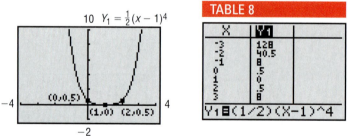

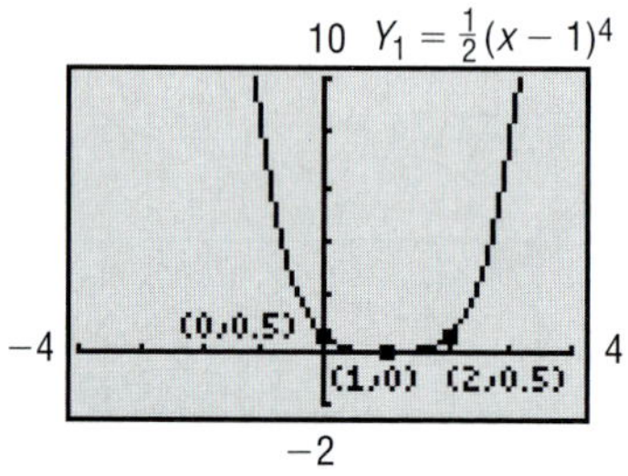

TABLE 8

X	Y1
-3	128
-2	40.5
-1	8
0	.5
1	0
2	.5
3	8

Y1=(1/2)(X-1)^4

The intercepts of f are $(0, \frac{1}{2})$ and $(1, 0)$.

Now work Problem 7.

Curve Fitting

3 In many applications, the relation that exists between two variables x and y follows a function of the form

$$y = ax^b$$

where $a > 0$ and $b > 0$ are two real numbers. This function is called a **generalized power function.** Figure 29 shows scatter diagrams that typically follow a generalized power function. Below each scatter diagram are restrictions on the values of the parameters, a and b, and the independent and dependent variables x and y.

FIGURE 29

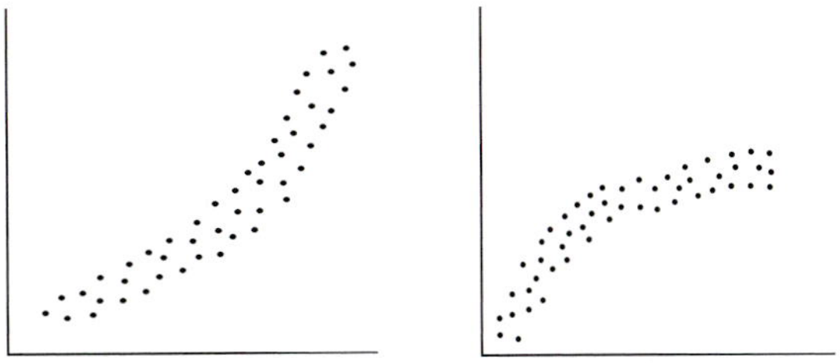

$y = ax^b$, $a > 0$, $b > 1$
$x, y > 0$

$y = ax^b$, $a > 0$, $0 < b < 1$
$x, y > 0$

EXAMPLE 4 Fitting a Curve to a Generalized Power Function

The *period* of a pendulum is the time required for one oscillation; the pendulum is usually referred to as *simple* when the angle made to the vertical is less than 5°. An experiment is conducted in which simple pendulums are constructed with different lengths, l, and the corresponding period T is recorded. The data below is collected:

Length l (in feet)	1	2	3	4	5	6	7
Period T (in seconds)	1.10	1.55	1.89	2.24	2.51	2.76	2.91

(a) Using a graphing utility, draw a scatter diagram of the data with length as the independent variable and the period as the dependent variable.

(b) Using a graphing utility, fit a generalized power function to the data.

(c) Graph the generalized power function found in (b) on the scatter diagram.

(d) If the length of a simple pendulum is known to be 2.3 feet, what is the period of the pendulum?

Solution (a) After entering the data into the graphing utility, we obtain the scatter diagram shown in Figure 30.

FIGURE 30

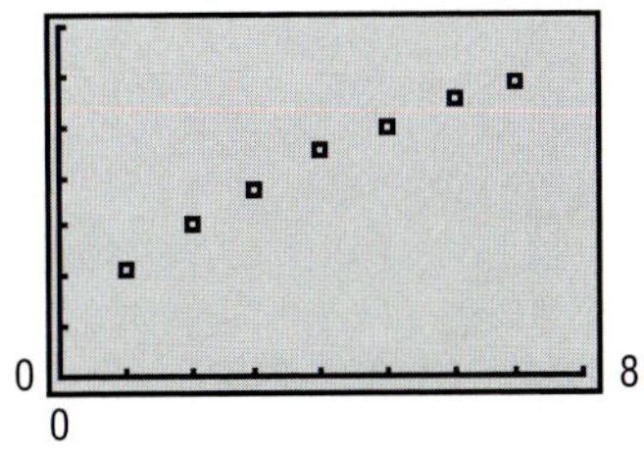

(b) A graphing utility fits the data in Figure 30 to a generalized power function of the form $y = ax^b$ by using the PoWeR REGression option. See Figure 31.

FIGURE 31

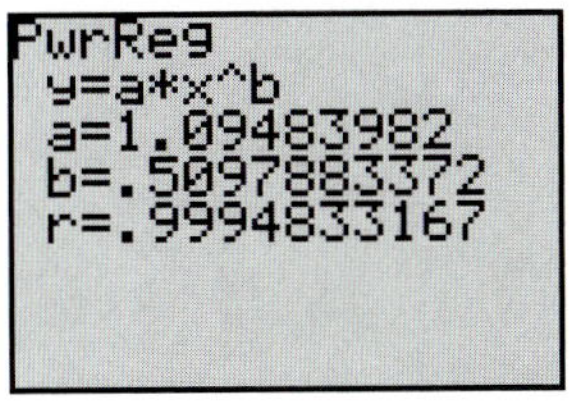

The period T of a simple pendulum is related to its length l by the power function

$$T = 1.0948l^{0.5098}$$

For generalized power functions, a value for r is given, since such functions can be expressed as linear functions using logarithms. (See Chapter 5).

Notice that the value of r is close to 1, indicating a good fit.*

(c) Figure 32 shows the graph of the power function on the scatter diagram.

FIGURE 32

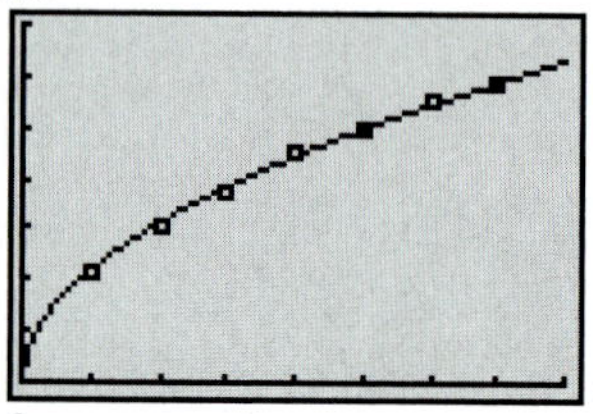

(d) The period T of a pendulum whose length l is 2.3 feet is

$$T = (1.0948)2.3^{0.5098} = 1.674 \text{ seconds}$$

4.2 EXERCISES

For Problems 1–6, determine the coefficient and degree of each monomial.

1. $8x^5$ **2.** $-12x^3$ **3.** 4 **4.** $10x^5$ **5.** $-7x$ **6.** 5

In Problems 7–14, using a graphing utility, show the steps required to graph each function.

7. $f(x) = (x + 1)^4$ **8.** $f(x) = x^4 + 2$ **9.** $f(x) = \frac{1}{2}x^4$

10. $f(x) = -x^4$ **11.** $f(x) = 2(x + 1)^4 + 1$ **12.** $f(x) = 3 - (x + 2)^4$

13. $f(x) = -\frac{1}{2}(x - 2)^4 - 1$ **14.** $f(x) = 1 - 2(x + 1)^4$

15. CBL Experiment Cathy is conducting an experiment to measure the relation between a light bulb's intensity and the distance from the light source. She measures a 100-watt light bulb's intensity 1 meter from the bulb and at 0.1-meter intervals up to 2 meters from the bulb and obtains the following data.

*In physics, it is proven that $T = \frac{2\pi}{\sqrt{32}}\sqrt{l}$, provided friction is ignored. Since $\sqrt{l} = l^{1/2} = l^{0.5}$ and $\frac{2\pi}{\sqrt{32}} \approx 1.1107$, the power function obtained is reasonably close to what is expected.

Distance (Meters)	Intensity
1.0	0.29645
1.1	0.25215
1.2	0.20547
1.3	0.17462
1.4	0.15342
1.5	0.13521
1.6	0.11450
1.7	0.10243
1.8	0.09231
1.9	0.08321
2.0	0.07342

(a) Using your graphing utility, draw a scatter diagram with distance as the independent variable and intensity as the dependent variable.

(b) Using your graphing utility, fit a generalized power function to the data.

(c) Graph the function found in (b) on the scatter diagram.

(d) What will the intensity of a 100-watt light bulb be if you stand 2.3 meters away?

(e) It is known that intensity I is inversely proportional to the square of the distance x, so that $I = \dfrac{a}{x^2}$. How close is the estimate of the exponent?

16. **CBL Experiment** Cathy repeats the experiment from Problem 15, but this time uses a 40-watt light bulb and obtains the following data.

Distance (Meters)	Intensity
1.0	0.0972
1.1	0.0804
1.2	0.0674
1.3	0.0572
1.4	0.0495
1.5	0.0433
1.6	0.0384
1.7	0.0339
1.8	0.0294
1.9	0.0268
2.0	0.0224

(a) Using your graphing utility, draw a scatter diagram with distance as the independent variable and intensity as the dependent variable.

(b) Using your graphing utility, fit a generalized power function to the data.

(c) Graph the function found in (b) on the scatter diagram.

(d) What will the intensity of a 40-watt light bulb be if you stand 2.3 meters away?

17. **CBL Experiment** David is conducting an experiment in order to estimate the acceleration of an object due to gravity. David takes a ball and drops it from different heights and records the time it takes for the ball to hit the ground. Using an optic laser connected to a stop watch in order to determine the time, he collects the following data:

Time (Seconds)	Distance (Feet)
1.003	16
1.365	30
1.769	50
2.093	70
2.238	80

(a) Draw a scatter diagram using time as the independent variable and distance as the dependent variable.

(b) Fit a generalized power function to the data.

(c) Graph the function found in (b) on the scatter diagram.

(d) Predict how long it will take an object to fall 100 feet.

(e) Physics theory states the distance an object falls is directly proportional to the time squared. Rewrite the model found in (b) so it is of the form $s = \frac{1}{2}gt^2$, where s is distance, g is the acceleration due to gravity, and t is time. What is David's estimate of g? (It is known that the acceleration due to gravity is approximately 32 feet/sec^2.)

18. **CBL Experiment** Paul, David's friend, doesn't believe David's estimate of acceleration due to gravity is correct. Paul repeats the experiment described in Problem 17 and obtains the following data:

Time (Seconds)	Distance (Feet)
0.7907	10
1.1160	20
1.4760	35
1.6780	45
1.9380	60

(a) Draw a scatter diagram using time as the dependent variable and distance as the independent variable.
(b) Fit a generalized power function to the data.
(c) Graph the function found in (b) on the scatter diagram.
(d) Predict how long it will take an object to fall 100 feet. Why do you think this estimate is different than the one from Problem 17?
(e) Physics theory states the distance an object falls is directly proportional to the time squared. Rewrite the model found in (b) so it is of the form $s = \frac{1}{2}gt^2$, where s is distance, g is the acceleration due to gravity, and t is time. What is Paul's estimate of g? [It is known that the acceleration due to gravity is approximately 32 feet/sec^2.]

4.3 POLYNOMIAL FUNCTIONS; CURVE FITTING

1 Identify Polynomials and Their Degree
2 Identify the Zeros of a Polynomial and Their Multiplicity
3 Analyze the Graph of a Polynomial
4 Find the Cubic Function of Best Fit from Data

A **polynomial function** is a function of the form

$$f(x) = a_n x^n + a_{n-1}x^{n-1} + \cdots + a_1 x + a_0 \qquad (1)$$

where $a_n, a_{n-1}, \ldots, a_1, a_0$ are real numbers and n is a nonnegative integer. The domain consists of all real numbers.

1 Thus, a polynomial function is a function whose rule is given by a polynomial in one variable.*

EXAMPLE 1 Identifying Polynomial Functions

Determine which of the following are polynomial functions. For those that are, state the degree; for those that are not, tell why not.

(a) $f(x) = 2 - 3x^4$ (b) $g(x) = \sqrt{x}$ (c) $h(x) = \dfrac{x^2 - 2}{x^3 - 1}$

(d) $F(x) = 0$ (e) $G(x) = 8$

Solution

(a) f is a polynomial function of degree 4.
(b) g is not a polynomial function. The variable x is raised to the $\frac{1}{2}$ power, which is not a nonnegative integer.
(c) h is not a polynomial function. It is the ratio of two polynomials, and the polynomial in the denominator is of positive degree.
(d) F is the zero polynomial function; it is not assigned a degree.
(e) G is a nonzero constant function, a polynomial function of degree 0.

Now work Problems 1 and 5.

*A review of polynomials may be found in the Appendix, Section 2.

We have already discussed in detail polynomial functions of degrees 0, 1, and 2. See Table 9 for a summary of the characteristics of the graphs of these polynomial functions.

TABLE 9

Degree	Form	Name	Graph
No degree	$f(x) = 0$	Zero function	The x-axis
0	$f(x) = a_0, \quad a_0 \neq 1$	Constant function	Horizontal line with y-intercept a_0
1	$f(x) = a_1x + a_0, \quad a_1 \neq 0$	Linear function	Nonvertical, nonhorizontal line with slope a_1 and y-intercept a_0
2	$f(x) = a_2x^2 + a_1x + a_0, \quad a_2 \neq 0$	Quadratic function	Parabola: Graph opens up if $a_2 > 0$; graph opens down if $a_2 < 0$

Graphing Polynomials

FIGURE 33

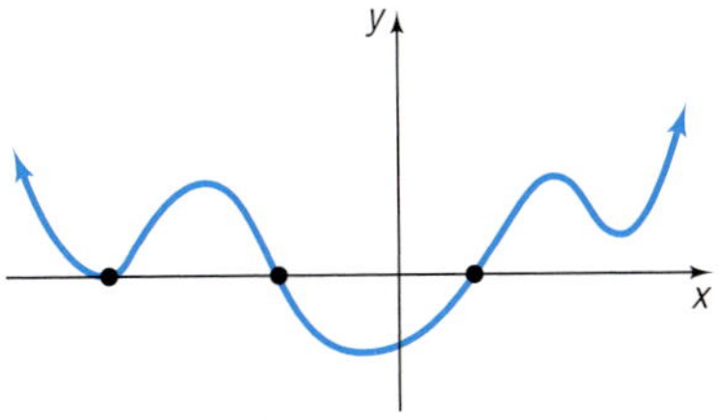

(a) Graph of a polynomial function: smooth, continuous

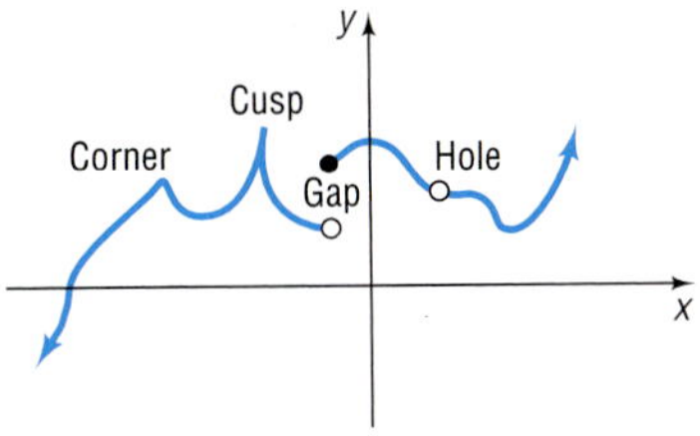

(b) Cannot be the graph of a polynomial function

To graph most polynomial functions of degree 3 or higher by hand requires techniques beyond the scope of this text. If you take a course in calculus you will learn that the graph of every polynomial function is both smooth and continuous. By **smooth,** we mean that the graph contains no sharp corners or cusps; by **continuous,** we mean that the graph has no gaps or holes and can be drawn without lifting pencil from paper. See Figures 33(a) and 33(b).

Figure 34 shows the graph of a polynomial function with four x-intercepts. Notice that at the x-intercepts the graph must either cross the x-axis or touch the x-axis. Consequently, between consecutive x-intercepts the graph is either above the x-axis or below the x-axis. We will make use of this characteristic of the graph of a polynomial shortly.

FIGURE 34

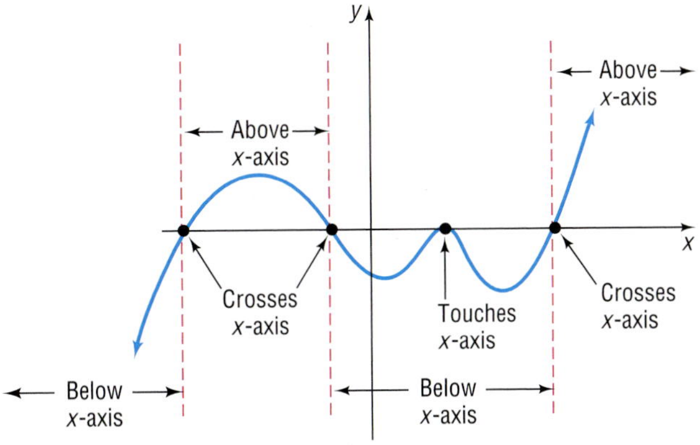

If a polynomial function f is factored completely, it is easy to solve the equation $f(x) = 0$ and locate the x-intercepts of the graph. For example, if $f(x) = (x - 1)^2(x + 3)$, then the solutions of the equation

$$f(x) = (x - 1)^2(x + 3) = 0$$

are easily identified as 1 and -3. Based on this result, we make the following observations:

If f is a polynomial function and r is a real number for which $f(r) = 0$, then r is called a (real) **zero of f**, or **root of f**.

If r is a (real) zero of f, then

(a) r is an x-intercept of the graph of f.

(b) $(x - r)$ is a factor of f.

2 If the same factor $x - r$ occurs more than once, then r is called a **repeated,** or **multiple, zero of f.** More precisely, we have the following definition.

If $(x - r)^m$ is a factor of a polynomial f and $(x - r)^{m+1}$ is not a factor of f, then r is called a **zero of multiplicity m of f.**

EXAMPLE 2 Identifying Zeros and Their Multiplicities

For the polynomial

$$f(x) = 5(x - 2)(x + 3)^2\left(x - \frac{1}{2}\right)^4$$

2 is a zero of multiplicity 1

-3 is a zero of multiplicity 2

$\frac{1}{2}$ is a zero of multiplicity 4

Suppose it is possible to factor completely a polynomial function and, as a result, locate all the x-intercepts of its graph (the real zeros of the function). The following example illustrates the role the multiplicity of the x-intercept plays.

EXAMPLE 3 Investigating the Role of Multiplicity

For the polynomial $f(x) = x^2(x - 2)$.

(a) Find the x- and y-intercepts of the graph.

(b) Using a graphing utility, graph the polynomial.

(c) For each x-intercept, determine whether it is of odd or even multiplicity.

Solution (a) The y-intercept is $f(0) = 0^2(0 - 2) = 0$. The x-intercepts satisfy the equation

$$f(x) = x^2(x - 2) = 0$$

from which we find

$$x^2 = 0 \quad \text{or} \quad x - 2 = 0$$
$$x = 0 \qquad\qquad x = 2$$

The x-intercepts are 0 and 2.

(b) See Figure 35.

FIGURE 35

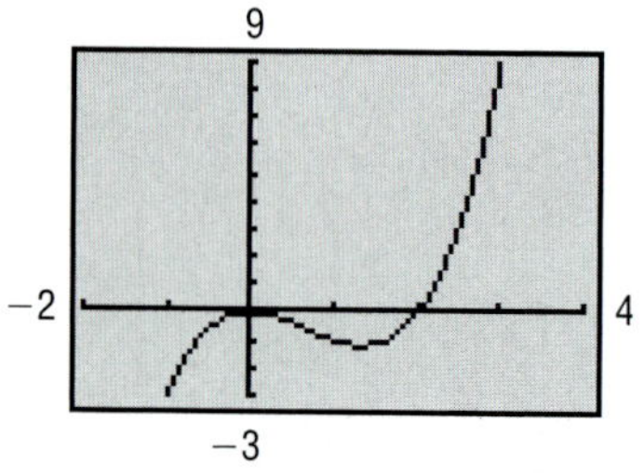

(c) We can see from the factored form of f that 0 is a zero or root of multiplicity 2 and 2 is a zero or root of multiplicity 1, so 0 is of even multiplicity and 2 is of odd multiplicity.

Notice from the graph in (b) that the graph of the polynomial at the root 0 just touches the x-axis so that the sign of $f(x)$ is the same on each side of the root. At the root 2, the graph of the polynomial crosses the x-axis so that the sign of $f(x)$ changes from one side of the root to the other. This observation leads us to the following result:

If r is a Zero of Even Multiplicity

Sign of $f(x)$ does not change from one side to the other side of r.	Graph **touches** x-axis at r.

If r is a Zero of Odd Multiplicity

Sign of $f(x)$ changes from one side to the other side of r.	Graph **crosses** x-axis at r.

Now work Problem 11.

Look again at Figure 35. We can use a graphing utility to determine the graph's minimum point in the interval $0 < x < 2$. After utilizing BOX (or MINIMUM if your utility has this feature), we find the graph's minimum point in the interval $0 < x < 2$ is $(1.33, -1.18)$ correct to two decimal places. In calculus, a method is presented for locating minimum and maximum points. There, such points are called *local minima* or *local maxima. Local minima* and *local maxima* are points where the graph changes direction (i.e., changes from an increasing function to a decreasing function or vice versa). We call these points **turning points.**

Look again at Figure 35. The graph of $f(x) = x^2(x - 2) = x^3 - 2x^2$, a polynomial of degree 3, has two turning points.

Exploration Graph $Y_1 = x^3$, $Y_2 = x^3 - x$, and $Y_3 = x^3 + 3x^2 + 4$. How many turning points do you see? How does the number of turning points relate to the degree? Graph $Y_1 = x^4$, $Y_2 = x^4 - \frac{4}{3}x^3$, and $Y_3 = x^4 - 2x^2$. How many turning points do you see? How does the number of turning points compare to the degree?

The following theorem from calculus supplies the answer.

Theorem

If f is a polynomial function of degree n, then f has at most $n - 1$ turning points.

One last remark about Figure 35. Notice that the graph of $f(x) = x^2(x - 2)$ looks somewhat like the graph of $y = x^3$. In fact, for very large values of x, either positive or negative, there is little difference.

Exploration Consider the functions Y_1 and Y_2 given on page 272 in (a), (b), (c). Graph Y_1 and Y_2 on the same viewing window. TRACE for large positive and large negative values of x. What do you notice about the graphs of Y_1 and Y_2 as x becomes very large and positive or very large and negative?

(a) $Y_1 = x^2(x - 2); Y_2 = x^3$
(b) $Y_1 = x^4 - 3x^3 + 7x - 3; Y_2 = x^4$
(c) $Y_1 = -2x^3 + 4x^2 - 8x + 10; Y_2 = -2x^3$

Theorem

For large values of x, either positive or negative, the graph of the polynomial

$$f(x) = a_n x^n + a_{n-1}x^{n-1} + \cdots + a_1 x + a_0$$

resembles the graph of the power function

$$y = a_n x^n$$

The following display summarizes some features of the graph of a polynomial function.

SUMMARY: GRAPH OF A POLYNOMIAL FUNCTION

$f(x) = a_n x^n + a_{n-1}x^{n-1} + \cdots + a_1 x + a_0,\ a_n \neq 0$

Degree of the polynomial f: n
Maximum number of turning points: $n - 1$
At zero of even multiplicity: graph of f touches x-axis
At zero of odd multiplicity: graph of f crosses x-axis
Between zeros, graph of f is either above the x-axis or below it.
For large x, graph of f behaves like graph of $y = a_n x^n$.

EXAMPLE 4 Analyzing the Graph of a Polynomial Function

For the polynomial $f(x) = x^4 - 3x^3 - 4x^2$,

(a) Using a graphing utility, graph f.
(b) Find the x- and y-intercepts.
(c) Determine whether each x-intercept is of odd or even multiplicity.
(d) Find the power function that the graph of f resembles for large values of x.
(e) Determine the number of turning points on the graph of f.
(f) Determine the *local maxima* and *local minima*, if any exist, correct to two decimal places.

FIGURE 36

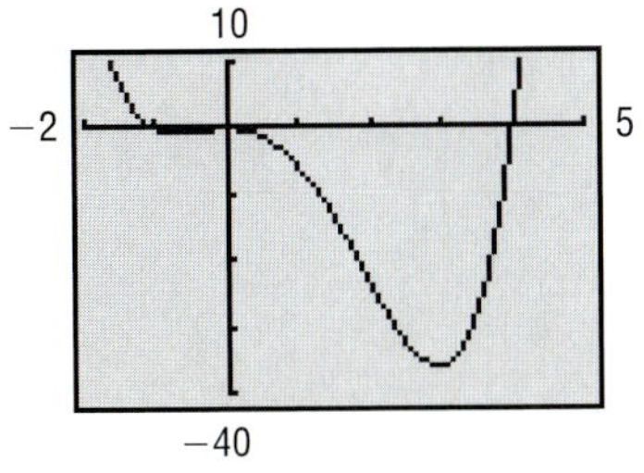

Solution

(a) See Figure 36.

(b) The y-intercept is $f(0) = 0$. Factoring, we find $f(x) = x^4 - 3x^3 - 4x^2 = x^2(x - 4)(x + 1)$. We find the x-intercepts by solving the equation

$$f(x) = x^2(x - 4)(x + 1) = 0$$

So

$$x^2 = 0 \quad \text{or} \quad x - 4 = 0 \quad \text{or} \quad x + 1 = 0$$
$$x = 0 \qquad\qquad x = 4 \qquad\qquad x = -1$$

The x-intercepts are -1, 0, and 4.

(c) The intercept 0 is a zero of even multiplicity, 2, so the graph of f will touch the x-axis at 0; 4 and -1 are zeros of odd multiplicity, 1, so the graph of f will cross the x-axis at 4 and -1. This means the graph will cross the x-axis at -1 and 4, and touch the x-axis at 0. Look again at Figure 36. The graph is below the x-axis for $-1 < x < 0$!

(d) The graph of f behaves like $f(x) = x^4$ for large x.

(e) Since f is of degree 4, the graph can have at most three turning points. From the graph, we see the graph has three turning points: one between -1 and 0, one at $(0, 0)$, and one between 2 and 4.

(f) Correct to two decimal places, the *local maximum* is $(0, 0)$ and the *local minima* are $(-0.68, -0.69)$ and $(2.93, -36.10)$.

EXAMPLE 5 Analyzing the Graph of a Polynomial Function

Follow the instructions of Example 4 for the polynomial below:

$$f(x) = x^3 + 2.48x^2 - 4.3155x + 1.484406$$

FIGURE 37

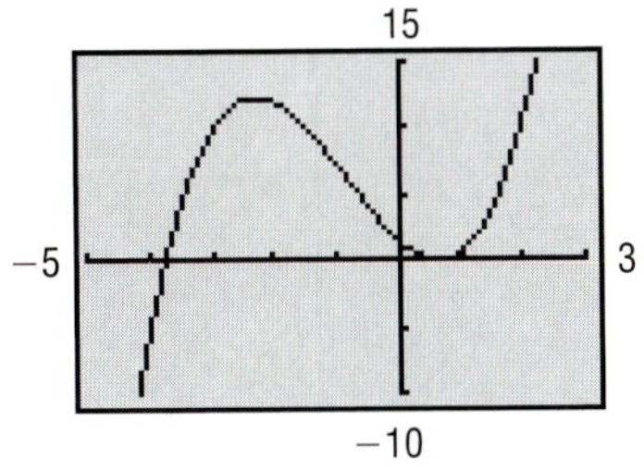

Solution

(a) See Figure 37.

(b) The y-intercept is $f(0) = 1.484406$. In Example 4 we could easily factor $f(x)$ to find the x-intercepts. However, it is not readily apparent how $f(x)$ factors in this example. Therefore, we use a graphing utility and find the x-intercepts to be -3.74 and 0.63.

(c) The x-intercept -3.74 is of odd multiplicity since the graph of f crosses the x-axis at -3.74; the x-intercept 0.63 is of even multiplicity since the graph of f touches the x-axis at 0.63.

(d) The graph behaves like $f(x) = x^3$ for large x.

(e) Since f is of degree 3, the graph can have at most two turning points. From the graph we see the graph has two turning points: one between -3 and -2; the other at $(0.63, 0)$.

(f) Correct to two decimal places, the *local maximum* is $(-2.28, 12.36)$ and the *local minimum* is $(0.63, 0)$.

Now work Problem 23.

Curve Fitting

4 In Section 1.6, we discussed finding a line of best fit from data. In Section 4.1, we discussed finding a quadratic function of best fit. It is possible to find

functions of best fit for any polynomial of degree n. However, most statisticians recommend finding polynomials of best fit of degree no higher than 3.*

Data that follows a cubic relation should look like Figure 38(a) or (b).

FIGURE 38

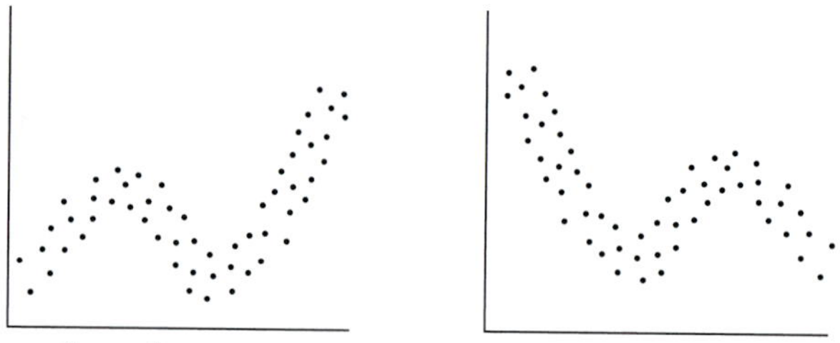

$y = ax^3 + bx^2 + cx + d,\ a > 0$

(a)

$y = ax^3 + bx^2 + cx + d,\ a < 0$

(b)

EXAMPLE 6 Finding a Cubic Function of Best Fit

The data in Table 10 represents the number of barrels of refined oil products imported into the United States for 1977–1992, where 1 represents 1977, 2 represents 1978, and so on.

TABLE 10

Year, x	Refined Oil Products (1000 Barrels per Day), R
1977, 1	2193
1978, 2	2008
1979, 3	1937
1980, 4	1646
1981, 5	1599
1982, 6	1625
1983, 7	1722
1984, 8	2011
1985, 9	1866
1986, 10	2045
1987, 11	2004
1988, 12	2295
1989, 13	2217
1990, 14	2123
1991, 15	1844
1992, 16	1805

Source: U.S. Energy Information Administration, *Monthly Energy Review*, February, 1995.

*Two points determine a unique line or function. Three noncollinear points determine a unique quadratic function. Four points determine a unique cubic function, and n points will determine a unique polynomial of degree $n - 1$. Therefore, higher degree polynomials will always "fit" data at least as well as lower degree polynomials. However, higher degree polynomials yield highly erratic predictions. Since the ultimate goal of curve fitting is not necessarily to find the model that best fits the data, but instead to explain relationships between two or more variables, polynomial models of degree three or less are usually used.

(a) Draw a scatter diagram of the data. Comment on the type of relation that may exist between the two variables.

(b) Find the cubic function of best fit.

(c) Draw the cubic function of best fit on your scatter diagram.

(d) Use the function found in (b) to predict the number of barrels of refined oil products imported in 1993 ($x = 17$).

(e) Do you think the function found in (b) will be useful in predicting the number of barrels of refined oil products for the year 1999? Why?

Solution (a) Figure 39 shows the scatter diagram.

FIGURE 39

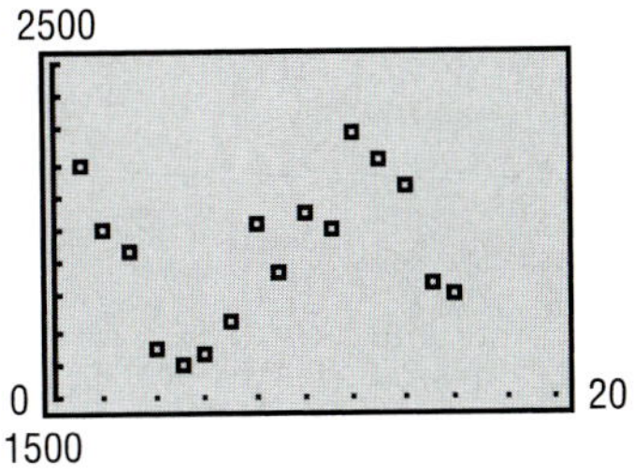

A cubic relation with $a < 0$ may exist between the two variables.

(b) Upon executing the CUBIC REGression program, we obtain the results shown in Figure 40. The output the utility provides shows us the equation $y = ax^3 + bx^2 + cx + d$. The cubic function of best fit is

$$R(x) = -2.4117x^3 + 63.4297x^2 - 453.7614x + 2648.0027,$$

where x represents the year and R represents the number of barrels of refined oil products imported.

FIGURE 40

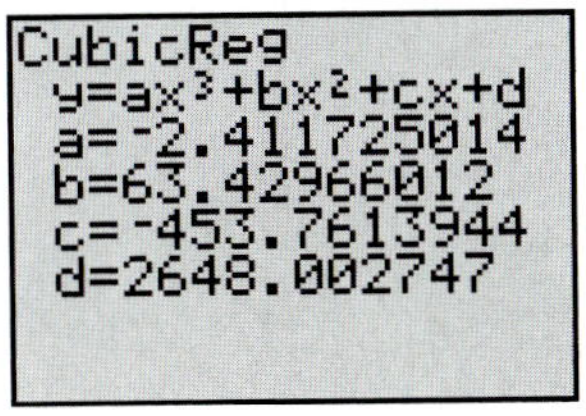

(c) Figure 41 shows the graph of the cubic function of best fit on the scatter diagram. The function seems to fit the data well.

FIGURE 41

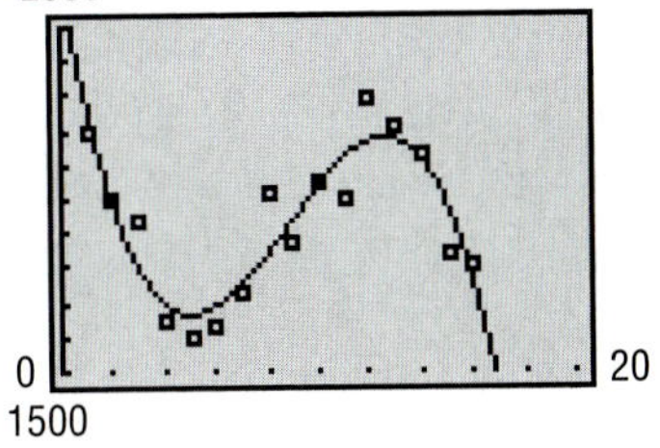

(d) We evaluate the function $R(x)$ for $x = 17$:

$$R(17) = -2.4117(17)^3 + 63.4297(17)^2 - 453.7614(17) + 2648.0027 \approx 1417$$

So, we predict 1417 barrels of refined oil products will be imported into the United States in 1993.

(e) The graph of the function will continually decrease for years beyond 1993, indicating the number of barrels of refined oil will continue to decline. This is not a likely scenario. So, the function probably will not be useful in predicting the number of barrels of refined oil products imported into the United States in 1999.

4.3 EXERCISES

In Problems 1–10, determine which functions are polynomial functions. For those that are, state the degree. For those that are not, tell why not.

1. $f(x) = 4x + x^3$
2. $f(x) = 5x^2 + 4x^4$
3. $g(x) = \dfrac{1 - x^2}{2}$
4. $h(x) = 3 - \dfrac{1}{2}x$
5. $f(x) = 1 - \dfrac{1}{x}$
6. $f(x) = x(x - 1)$
7. $g(x) = x^{3/2} - x^2 + 2$
8. $h(x) = \sqrt{x}(\sqrt{x} - 1)$
9. $F(x) = 5x^4 - \pi x^3 + \dfrac{1}{2}$
10. $F(x) = \dfrac{x^2 - 5}{x^3}$

In Problems 11–20, for each polynomial function, list each real zero and its multiplicity. Determine whether the graph crosses or touches the x-axis at each x-intercept.

11. $f(x) = 3(x - 7)(x + 3)^2$
12. $f(x) = 4(x + 4)(x + 3)^3$
13. $f(x) = 4(x^2 + 1)(x - 2)^3$
14. $f(x) = 2(x - 3)(x + 4)^3$
15. $f(x) = -2\left(x + \dfrac{1}{2}\right)^2(x^2 + 4)^2$
16. $f(x) = \left(x - \dfrac{1}{3}\right)^2(x - 1)^3$
17. $f(x) = (x - 5)^3(x + 4)^2$
18. $f(x) = (x + \sqrt{3})^2(x - 2)^4$
19. $f(x) = 3(x^2 + 8)(x^2 + 9)^2$
20. $f(x) = -2(x^2 + 3)^3$

In Problems 21–56, for each polynomial function f:

(a) Using a graphing utility, graph f.

(b) Find the x- and y-intercepts.

(c) Determine whether each x-intercept is of odd or even multiplicity.

(d) Find the power function that the graph of f resembles for large values of x.

(e) Determine the number of turning points on the graph of f.

(f) Determine the local maxima and local minima, if any exist, correct to two decimal places.

21. $f(x) = (x - 1)^2$
22. $f(x) = (x - 2)^3$
23. $f(x) = x^2(x - 3)$
24. $f(x) = x(x + 2)^2$
25. $f(x) = 6x^3(x + 4)$
26. $f(x) = 5x(x - 1)^3$

27. $f(x) = -4x^2(x + 2)$

28. $f(x) = -\frac{1}{2}x^3(x + 4)$

29. $f(x) = x(x - 2)(x + 4)$

30. $f(x) = x(x + 4)(x - 3)$

31. $f(x) = 4x - x^3$

32. $f(x) = x - x^3$

33. $f(x) = x^2(x - 2)(x + 2)$

34. $f(x) = x^2(x - 3)(x + 4)$

35. $f(x) = x^2(x - 2)^2$

36. $f(x) = x^3(x - 3)$

37. $f(x) = x^2(x - 3)(x + 1)$

38. $f(x) = x^2(x - 3)(x - 1)$

39. $f(x) = x(x + 2)(x - 4)(x - 6)$

40. $f(x) = x(x - 2)(x + 2)(x + 4)$

41. $f(x) = x^2(x - 2)(x^2 + 3)$

42. $f(x) = x^2(x^2 + 1)(x + 4)$

43. $f(x) = x^3 + 0.2x^2 - 1.5876x - 0.31752$

44. $f(x) = x^3 - 0.8x^2 - 4.6656x + 3.73248$

45. $f(x) = x^3 + 2.56x^2 - 3.31x + 0.89$

46. $f(x) = x^3 - 2.91x^2 - 7.668x - 3.8151$

47. $f(x) = x^4 - 2.5x^2 + 0.5625$

48. $f(x) = x^4 - 18.5x^2 + 50.2619$

49. $f(x) = x^4 + 0.65x^3 - 16.6319x^2 + 14.209335x - 3.1264785$

50. $f(x) = x^4 + 3.45x^3 - 11.6639x^2 - 5.864241x - 0.69257738$

51. $y = \pi x^3 + \sqrt{2}x^2 - x - 2$

52. $y = -2x^3 + \pi x^2 + \sqrt{3}x + 1$

53. $y = 2x^4 - \pi x^3 + \sqrt{5}x - 4$

54. $y = -1.2x^4 + 0.5x^2 - \sqrt{3}x + 2$

55. $y = -2x^5 - \sqrt{2}x^2 - x - \sqrt{2}$

56. $y = \pi x^5 + \pi x^4 + \sqrt{3}x + 1$

57. Consult the illustration. Which of the following polynomial functions might have this graph? (More than one answer may be possible.)

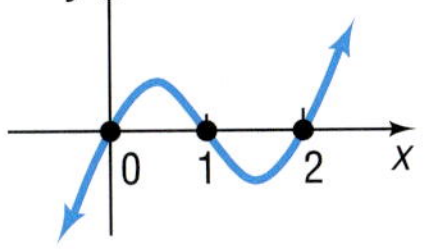

(a) $y = -4x(x - 1)(x - 2)$
(b) $y = x^2(x - 1)^2(x - 2)$
(c) $y = 3x(x - 1)(x - 2)$
(d) $y = x(x - 1)^2(x - 2)^2$
(e) $y = x^3(x - 1)(x - 2)$
(f) $y = -x(1 - x)(x - 2)$

58. Consult the illustration. Which of the following polynomial functions might have this graph? (More than one answer may be possible.)

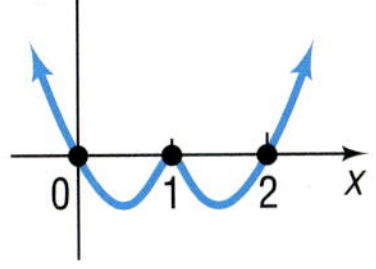

(a) $y = 2x^3(x - 1)(x - 2)^2$
(b) $y = x^2(x - 1)(x - 2)$
(c) $y = x^3(x - 1)^2(x - 2)$
(d) $y = x^2(x - 1)^2(x - 2)^2$
(e) $y = 5x(x - 1)^2(x - 2)$
(f) $y = -2x(x - 1)^2(2 - x)$

59. Consult the illustration. Which of the following polynomial functions might have this graph? (More than one answer may be possible.)

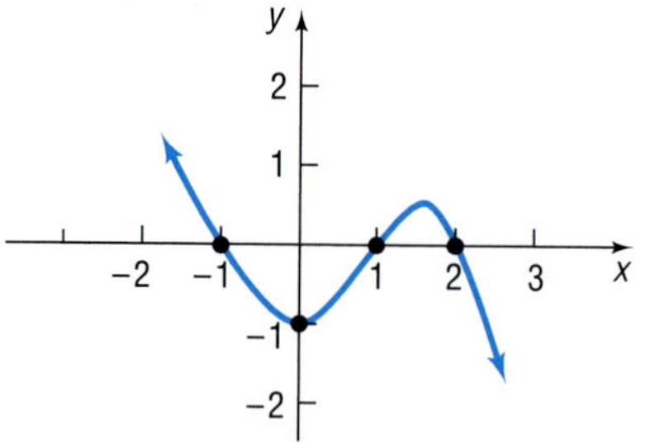

(a) $y = \frac{1}{2}(x^2 - 1)(x - 2)$

(b) $y = -\frac{1}{2}(x^2 + 1)(x - 2)$

(c) $y = (x^2 - 1)\left(1 - \frac{x}{2}\right)$

(d) $y = -\frac{1}{2}(x^2 - 1)^2(x - 2)$

(e) $y = \left(x^2 + \frac{1}{2}\right)(x^2 - 1)(2 - x)$

(f) $y = -(x - 1)(x - 2)(x + 1)$

60. Consult the illustration. Which of the following polynomial functions might have this graph? (More than one answer may be possible.)

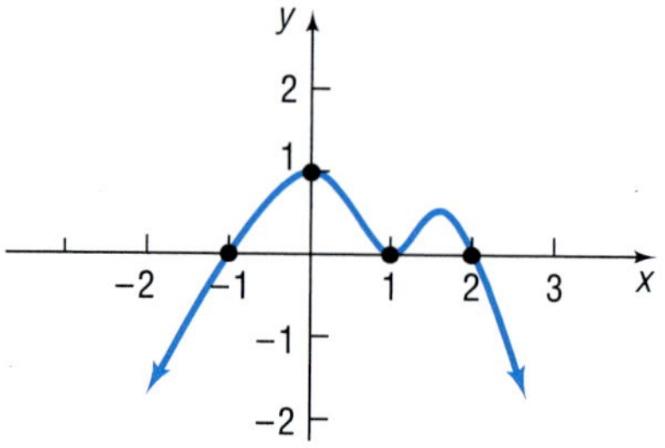

(a) $y = -\frac{1}{2}(x^2 - 1)(x - 2)(x + 1)$

(b) $y = -\frac{1}{2}(x^2 + 1)(x - 2)(x + 1)$

(c) $y = -\frac{1}{2}(x + 1)^2(x - 1)(x - 2)$

(d) $y = (x - 1)^2(x + 1)\left(1 - \frac{x}{2}\right)$

(e) $y = -(x - 1)^2(x - 2)(x + 1)$

(f) $y = -\left(x^2 + \frac{1}{2}\right)(x - 1)^2(x + 1)(x - 2)$

61. Motor Vehicle Thefts The following data represents the number of motor vehicle thefts (in thousands) in the United States for the years 1983–1993, where 1 represents 1983, 2 represents 1984, and so on.

Year, x	Motor Vehicle Thefts, M
1983, 1	1,008
1984, 2	1,032
1985, 3	1,103
1986, 4	1,224
1987, 5	1,289
1988, 6	1,433
1989, 7	1,565
1990, 8	1,636
1991, 9	1,662
1992, 10	1,611
1993, 11	1,561

Source: U.S. Federal Bureau of Investigation.

(a) Using a graphing utility, draw a scatter diagram of the data. Comment on the type of relation that may exist between the two variables.
(b) Find the cubic function of best fit.
(c) Using a graphing utility, draw the cubic function of best fit on your scatter diagram.
(d) Use the function found in (b) to predict the number of motor vehicle thefts in 1994.
(e) Do you think the function found in (b) will be useful in predicting the number of motor vehicle thefts in 1999?

62. Larceny Thefts The following data represents the number of larceny thefts (in thousands) in the United States for the years 1983–1993, where 1 represents 1983, 2 represents 1984, and so on.

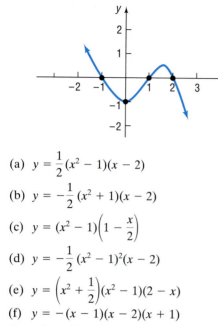

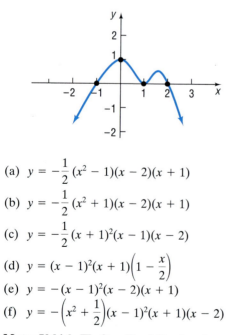

Year, x	Larceny Thefts, L
1983, 1	6713
1984, 2	6592
1985, 3	6926
1986, 4	7257
1987, 5	7500
1988, 6	7706
1989, 7	7872
1990, 8	7946
1991, 9	8142
1992, 10	7915
1993, 11	7821

Source: U.S. Federal Bureau of Investigation.

(a) Using a graphing utility, draw a scatter diagram of the data. Comment on the type of relation that may exist between the two variables.

(b) Find the cubic function of best fit.
(c) Using a graphing utility, draw the cubic function of best fit on your scatter diagram.
(d) Use the function found in (b) to predict the number of larceny thefts in 1994.
(e) Do you think the function found in (b) will be useful in predicting the number of larceny thefts in 1999?

63. Cost of Manufacturing The following data represents the cost C of manufacturing Chevy Cavaliers (in thousands of dollars) and the number x of Cavaliers produced.

Number of Cavaliers Produced, x	Cost, C
0	10
1	23
2	31
3	38
4	43
5	50
6	59
7	70
8	85
9	105
10	135

(a) Using a graphing utility, draw a scatter diagram of the data. Comment on the type of relation that may exist between the two variables.
(b) Find the average rate of change of cost between four and five Cavaliers. Economists call this the **marginal cost.**
(c) What is the marginal cost of producing between eight and nine Cavaliers?
(d) Find the cubic function of best fit.
(e) Using a graphing utility, draw the cubic function of best fit on your scatter diagram.
(f) Use the function found in (d) to predict the cost of manufacturing eleven Cavaliers.
(g) Interpret the y-intercept.

64. Cost of Printing The following data represents the weekly cost of printing textbooks C (in thousands) and the number of texts printed (in thousands), x.

(a) Using a graphing utility, draw a scatter diagram of the data. Comment on the type of relation that may exist between the two variables.

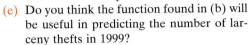

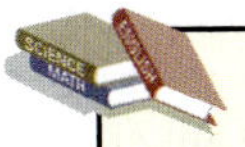

Number of Text Books, x	Cost, C
0	100
5	128.1
10	144
13	153.5
17	161.2
18	162.6
20	166.3
23	178.9
25	190.2
27	221.8

(b) Find the cubic function of best fit.
(c) Find the average rate of change of cost between 10 and 13 textbooks. Economists call this the **marginal cost.**
(d) What is the marginal cost of producing between 18 and 20 textbooks?
(e) Using a graphing utility, draw the cubic function of best fit on your scatter diagram.
(f) Use the function found in (b) to predict the cost of printing 22 thousand texts per week.
(g) Interpret the y-intercept.

65. Can the graph of a polynomial function have no y-intercept? Can it have no x-intercepts? Explain.

66. Write a few paragraphs that provide a general strategy for graphing a polynomial function. Be sure to mention the following: degree, intercepts, and turning points.

67. Make up a polynomial that has the following characteristics: crosses the x-axis at -1 and 4, touches the x-axis at 0 and 2, and is above the x-axis between 0 and 2. Give your polynomial to a fellow classmate and ask for a written critique of your polynomial.

68. Make up two polynomials, not of the same degree, with the following characteristics: crosses the x-axis at -2, touches the x-axis at 1, and is above the x-axis between -2 and 1. Give your polynomials to a fellow classmate and ask for a written critique of your polynomials.

69. The graph of a polynomial function is always smooth and continuous. Name a function studied earlier that is smooth and not continuous. Name one that is continuous, but not smooth.

70. Which of the following statements are true regarding the graph of the cubic polynomial $f(x) = x^3 + bx^2 + cx + d$? (Give reasons for your conclusions.)

(a) It intersects the y-axis in one and only one point.
(b) It intersects the x-axis in at most three points.
(c) It intersects the x-axis at least once.
(d) For x very large, it behaves like the graph of $y = x^3$.
(e) It is symmetric with respect to the origin.
(f) It contains the origin.

4.4 RATIONAL FUNCTIONS

1. Find the Domain of a Rational Function
2. Determine the Vertical Asymptotes of a Rational Function
3. Determine the Horizontal or Oblique Asymptotes of a Rational Function
4. Analyze the Graph of a Rational Function

Ratios of integers are called *rational numbers*. Similarly, ratios of polynomial functions are called *rational functions*.

A **rational function** is a function of the form

$$R(x) = \frac{p(x)}{q(x)}$$

where p and q are polynomial functions and q is not the zero polynomial. The domain consists of all real numbers except those for which the denominator q is 0.

EXAMPLE 1 Finding the Domain of a Rational Function

1

(a) The domain of $R(x) = \dfrac{2x^2 - 4}{x + 5}$ consists of all real numbers x except -5.

(b) The domain of $R(x) = \dfrac{1}{x^2 - 4}$ consists of all real numbers x except -2 and 2.

(c) The domain of $R(x) = \dfrac{x^3}{x^2 + 1}$ consists of all real numbers.

(d) The domain of $R(x) = \dfrac{-x^2 + 2}{3}$ consists of all real numbers.

(e) The domain of $R(x) = \dfrac{x^2 - 1}{x - 1}$ consists of all real numbers x except 1.

It is important to observe that the functions

$$R(x) = \frac{x^2 - 1}{x - 1} \quad \text{and} \quad f(x) = x + 1$$

are not equal, since the domain of R is $\{x | x \neq 1\}$ and the domain of f is all real numbers.

Now work Problem 3.

If $R(x) = p(x)/q(x)$ is a rational function and if p and q have no common factors, then the rational function R is said to be in **lowest terms.** For a rational function $R(x) = p(x)/q(x)$ in lowest terms, the zeros, if any, of the numerator are the x-intercepts of the graph of R and so will play a major role in the graph of R. The zeros of the denominator of R [that is, the numbers x, if any, for which $q(x) = 0$], although not in the domain of R, also play a major role in the graph of R. We will discuss this role shortly.

We have already discussed the characteristics of the rational function $f(x) = 1/x$. (Refer to Example 14, page 43). The next rational function we take up is $H(x) = 1/x^2$.

EXAMPLE 2 Graphing $y = \frac{1}{x^2}$

Analyze the graph $H(x) = \frac{1}{x^2}$.

Solution See Figure 42.

FIGURE 42

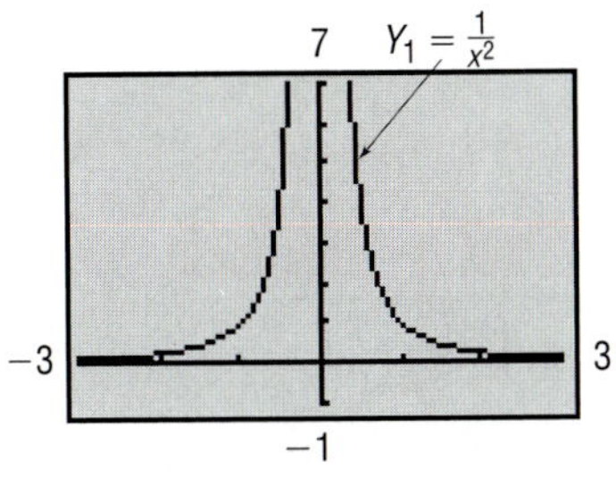

The domain of $H(x) = 1/x^2$ consists of all real numbers x except 0. Thus, the graph has no y-intercept, because x can never equal 0. The graph has no x-intercept because the equation $H(x) = 0$ has no solution. Therefore, the graph of H will not cross either of the coordinate axes.

Because

$$H(-x) = \frac{1}{(-x)^2} = \frac{1}{x^2} = H(x)$$

H is an even function, so its graph is symmetric with respect to the y-axis.

Look again at Figure 42. What happens to the function as the values of x get closer and closer to 0? We use a TABLE to answer the question. See Table 11. There, we see the values of $H(x)$ become larger and larger positive numbers. When this happens we say that H is **unbounded in the positive direction.** We symbolize this by writing $H \to \infty$ (read as "H **approaches infinity**"). In calculus, this idea is symbolized by writing $\lim_{x \to 0} H = \infty$, read as "the limit of $H(x)$ as x approaches 0 is infinity.

TABLE 11

X	Y1
-.01	10000
.1	100
0	ERROR
1	1
2	.25
10	.01
100	1E-4

Y1 = 1/X²

Look again at Table 11. As $x \to \infty$, the values of $H(x)$ approach 0 (symbolized in calculus by writing $\lim_{x \to \infty} H(x) = 0$).

EXAMPLE 3 Graphing in Stages

Using a graphing utility, show each step to graph $R(x) = \dfrac{1}{(x-2)^2} + 1$.

Solution STAGE 1: Graph $y = \dfrac{1}{x^2}$. See Figure 43 and Table 12.

FIGURE 43

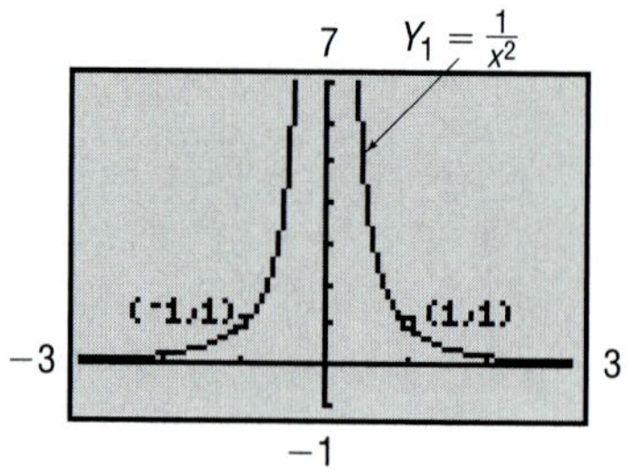

TABLE 12

X	Y1
-3	.11111
-2	.25
-1	1
0	ERROR
1	1
2	.25
3	.11111

Y1=1/X²

STAGE 2: Graph $y = \dfrac{1}{(x-2)^2}$. (Replace x by $x-2$; shift to the right 2 units.) See Figure 44 and Table 13.

FIGURE 44

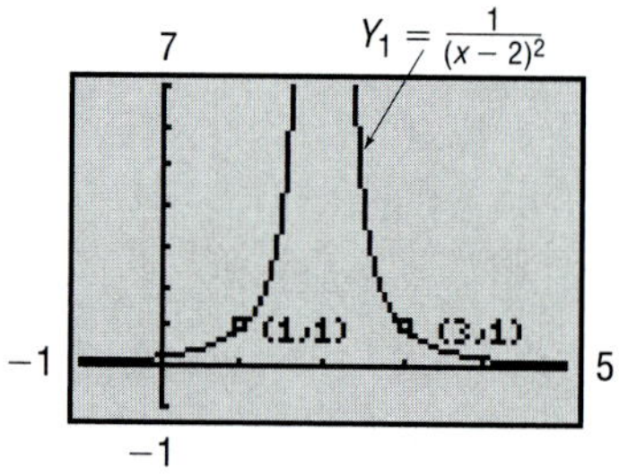

TABLE 13

X	Y1
-1	.11111
0	.25
1	1
2	ERROR
3	1
4	.25
5	.11111

Y1=1/(X-2)²

STAGE 3: Graph $y = \dfrac{1}{(x-2)^2} + 1$. (Add 1; shift up 1 unit.) See Figure 45 and Table 14.

FIGURE 45

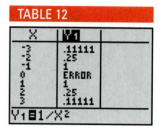

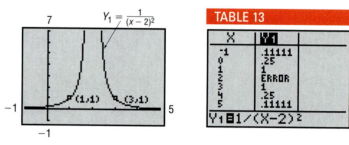

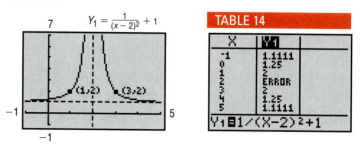

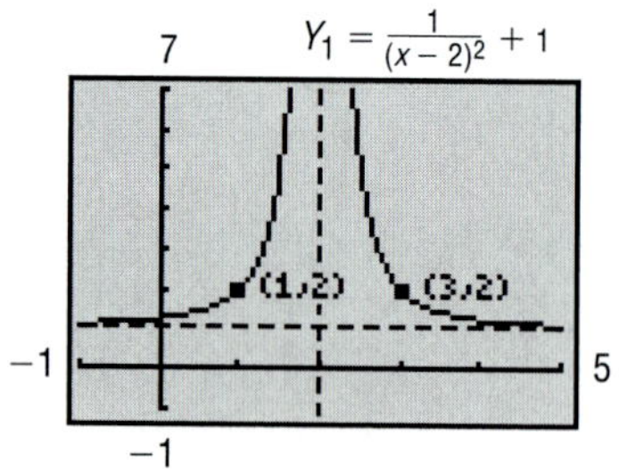

TABLE 14

X	Y1
-1	1.1111
0	1.25
1	2
2	ERROR
3	2
4	1.25
5	1.1111

Y1=1/(X-2)²+1

Notice that the y-axis in Figure 43 is represented by the vertical line $x = 2$ in Figure 45 and the x-axis in Figure 43 is represented by the horizontal line $y = 1$ in Figure 45.

Now work Problem 21.

Asymptotes

2 In Figure 45, notice that as the values of x become more negative, that is, as x becomes **unbounded in the negative direction** ($x \to -\infty$, read as "x **approaches negative infinity**"), the values $R(x)$ approach 1. In fact, we can conclude the following from Figure 45:

1. As $x \to -\infty$, the values $R(x)$ approach 1. $[\lim_{x\to-\infty} R(x) = 1]$.
2. As x approaches 2, the values $R(x) \to \infty$. $[\lim_{x\to 2} R(x) = \infty]$.
3. As $x \to \infty$, the values $R(x)$ approach 1. $[\lim_{x\to\infty} R(x) = 1]$.

Using a graphing utility and the TRACE function, verify the results discussed above for the graph shown in Figure 45. The behavior of the graph is depicted by the vertical line $x = 2$ and the horizontal line $y = 1$. These lines are called *asymptotes* of the graph, which we define as follows.

Let R denote a function:

If, as $x \to -\infty$ or as $x \to \infty$, the values of $R(x)$ approach some fixed number L, then the line $y = L$ is a **horizontal asymptote** of the graph of R.

If, as x approaches some number c, the values $|R(x)| \to \infty$, then the line $x = c$ is a **vertical asymptote** of the graph of R.

Even though asymptotes of a function are not part of the graph of the function, they provide information about the way the graph looks. Figure 46 illustrates some of the possibilities.

FIGURE 46

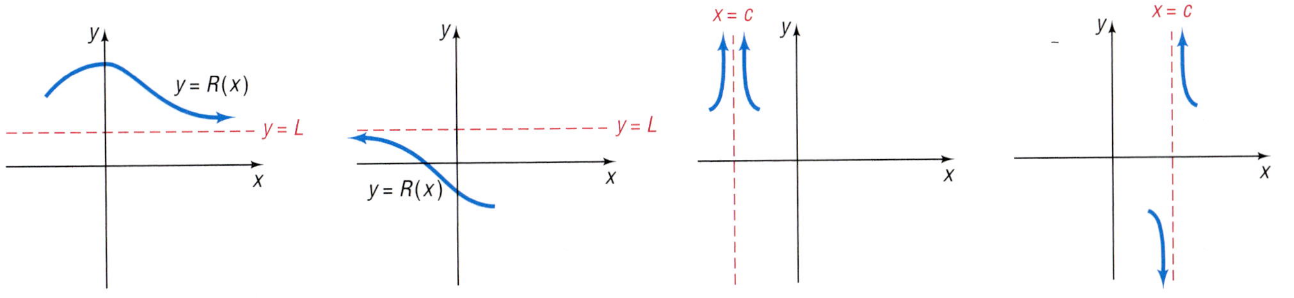

(a) As $x \to \infty$, the values of $R(x)$ approach L. That is, the points on the graph of R are getting closer to the line $y = L$; $y = L$ is a horizontal asymptote.

(b) As $x \to -\infty$, the values of $R(x)$ approach L. That is, the points on the graph of R are getting closer to the line $y = L$; $y = L$ is a horizontal asymptote.

(c) As x approaches c, the values of $|R(x)| \to \infty$. That is, the points on the graph of R are getting closer to the line $x = c$; $x = c$ is a vertical asymptote.

(d) As x approaches c, the values of $|R(x)| \to \infty$. That is, the points on the graph of R are getting closer to the line $x = c$; $x = c$ is a vertical asymptote.

FIGURE 47
Oblique asymptote

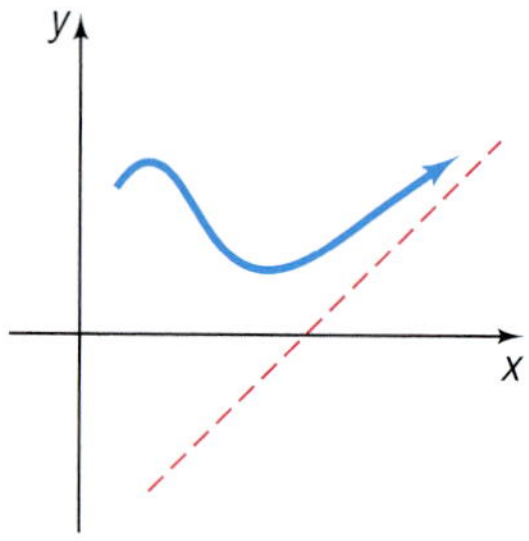

Thus, an asymptote is a line that a certain part of the graph of a function gets closer and closer to but never touches. However, other parts of the graph of the function may intersect a nonvertical asymptote. The graph of the function will never intersect a vertical asymptote. Notice that a horizontal asymptote, when it occurs, describes a certain behavior of the graph as $x \to \infty$ or as $x \to -\infty$, while a vertical asymptote, when it occurs, describes a certain behavior of the graph when x is close to some number c.

If an asymptote is neither horizontal nor vertical, it is called **oblique.** Figure 47 shows an oblique asymptote.

Finding Asymptotes

The vertical asymptotes, if any, of a rational function $R(x) = p(x)/q(x)$, in lowest terms, are found by factoring the denominator of $q(x)$. Suppose that $x - r$ is a factor of the denominator. Now, as x approaches r, symbolized as $x \to r$, the values of $x - r$ approach 0, causing the ratio to become unbounded, that is, causing $|R(x)| \to \infty$. Based on the definition, we conclude that the line $x = r$ is a vertical asymptote.

Theorem Locating Vertical Asymptotes

A rational function $R(x) = p(x)/q(x)$, *in lowest terms,* will have a vertical asymptote $x = r$, if $x - r$ is a factor of the denominator q.

Thus, if r is a zero of the denominator of a rational function $R(x) = p(x)/q(x)$, in lowest terms, then R will have the vertical asymptote $x = r$.

Warning: If a rational function is not in lowest terms, an application of this theorem may result in an incorrect listing of vertical asymptotes.

EXAMPLE 4 Finding Vertical Asymptotes

Find the vertical asymptotes, if any, of the graph of each rational function.

(a) $R(x) = \dfrac{x}{x^2 - 4}$ (b) $F(x) = \dfrac{x + 3}{x - 1}$ (c) $H(x) = \dfrac{x^2}{x^2 + 1}$

(d) $G(x) = \dfrac{x^2 - 9}{x^2 + 4x - 21}$

Solution

(a) The zeros of the denominator $x^2 - 4$ are -2 and 2. Hence, the lines $x = -2$ and $x = 2$ are the vertical asymptotes of the graph of R.

(b) The only zero of the denominator is 1. Hence, the line $x = 1$ is the only vertical asymptote of the graph of F.

(c) The denominator has no zeros. Hence, the graph of H has no vertical asymptotes.

(d) Factor $G(x)$ to determine if it is in lowest terms

$$G(x) = \frac{x^2 - 9}{x^2 + 4x - 21} = \frac{(x + 3)(x - 3)}{(x + 7)(x - 3)} = \frac{x + 3}{x + 7}.$$

The only zero of the denominator of $G(x)$ in lowest terms is -7. Hence, the line $x = -7$ is the only vertical asymptote of the graph of G. ■

As Example 4 points out, rational functions can have no vertical asymptotes, one vertical asymptote, or more than one vertical asymptote. However, the graph of a rational function will never intersect any of its vertical asymptotes. (Do you know why?)

Exploration Graph each of the following rational functions:

$$y = \frac{1}{x - 1} \qquad y = \frac{x}{x - 1} \qquad y = \frac{x^2}{x - 1} \qquad y = \frac{x^3}{x - 1}$$

Each has the vertical asymptote $x = 1$. Use TRACE (on a TABLE) to see what happens as x approaches 1. Be sure to look at both sides of $x = 1$.

3

The procedure for finding horizontal and oblique asymptotes is somewhat more involved. To find such asymptotes, we need to know how the values of a function behave as $x \to -\infty$ or as $x \to \infty$.

If a rational function $R(x)$ is **proper,** that is, if the degree of the numerator is less than the degree of the denominator, then as $x \to -\infty$ or as $x \to \infty$, the values of $R(x)$ approach 0. Consequently, the line $y = 0$ (the x-axis) is a horizontal asymptote of the graph.

Theorem

If a rational function is proper, the line $y = 0$ is a horizontal asymptote of its graph.

EXAMPLE 5 Finding Horizontal Asymptotes

Find the horizontal asymptotes, if any, of the graph of

$$R(x) = \frac{x - 12}{4x^2 + x + 1}$$

Solution The rational function R is proper, since the degree of the numerator, 1, is less than the degree of the denominator, 2. We conclude that the line $y = 0$ is a horizontal asymptote of the graph of R.

To see why $y = 0$ is a horizontal asymptote of the function R in Example 5, we need to investigate the behavior of R for x unbounded. When x is unbounded, the numerator of R, which is $x - 12$, can be approximated by the power function $y = x$, while the denominator of R, which is $4x^2 + x + 1$, can be approximated by the power function $y = 4x^2$.

Thus,

$$R(x) = \frac{x - 12}{4x^2 + x + 1} \underset{\substack{\uparrow \\ \text{For } x \text{ unbounded}}}{\approx} \frac{x}{4x^2} = \frac{1}{4x} \underset{\substack{\uparrow \\ \text{As } x \to -\infty \text{ or } x \to \infty}}{\to} 0$$

This shows that the line $y = 0$ is a horizontal asymptote of the graph of R.

If a rational function $R(x)$ is **improper,** that is, if the degree of the numerator is greater than or equal to the degree of the denominator, we must use long division to write the rational function as the sum of a polynomial $f(x)$ plus a proper rational function $r(x)$. That is, we write

$$R(x) = \frac{p(x)}{q(x)} = f(x) + r(x)$$

where $f(x)$ is a polynomial and $r(x)$ is a proper rational function. Since $r(x)$ is proper, then $r(x) \to 0$ as $x \to -\infty$ or as $x \to \infty$. Thus,

$$R(x) = \frac{p(x)}{q(x)} \to f(x) \qquad \text{as} \qquad x \to -\infty \text{ or as } x \to \infty$$

The possibilities are listed next:

1. If $f(x) = b$, a constant, then the line $y = b$ is a horizontal asymptote of the graph of R.
2. If $f(x) = ax + b$, $a \neq 0$, then the line $y = ax + b$ is an oblique asymptote of the graph of R.
3. In all other cases, the graph of R approaches the graph of f, and there are no horizontal or oblique asymptotes.

The following examples demonstrate these conclusions.

EXAMPLE 6 Finding Horizontal or Oblique Asymptotes

Find the horizontal or oblique asymptotes, if any, of the graph of

$$H(x) = \frac{3x^4 - x^2}{x^3 - x^2 + 1}$$

Solution The rational function H is improper, since the degree of the numerator, 4, is larger than the degree of the denominator, 3. To find any horizontal or oblique asymptotes, we use long division:

$$\begin{array}{r|rrrrr}
 & 3x & +3 & & & \\
x^3 - x^2 + 1 & 3x^4 & & -x^2 & & \\
 & 3x^4 & -3x^3 & & +3x & \\
\hline
 & & 3x^3 & -x^2 & -3x & \\
 & & 3x^3 & -3x^2 & & +3 \\
\hline
 & & & 2x^2 & -3x & -3
\end{array}$$

Thus,

$$H(x) = \frac{3x^4 - x^2}{x^3 - x^2 + 1} = 3x + 3 + \frac{2x^2 - 3x - 3}{x^3 - x^2 + 1}$$

Then, as $x \to -\infty$ or as $x \to \infty$, the remainder will behave as follows:

$$\frac{2x^2 - 3x - 3}{x^3 - x^2 + 1} \approx \frac{2x^2}{x^3} = \frac{2}{x} \to 0$$

Thus, as $x \to -\infty$ or as $x \to \infty$, we have $H(x) \to 3x + 3$. We conclude that the graph of the rational function H has an oblique asymptote $y = 3x + 3$. ■

EXAMPLE 7 Finding Horizontal or Oblique Asymptotes

Find the horizontal or oblique asymptotes, if any, of the graph of

$$R(x) = \frac{8x^2 - x + 2}{4x^2 - 1}$$

Solution The rational function R is improper, since the degree of the numerator, 2, equals the degree of the denominator, 2. To find any horizontal or oblique asymptotes, we use long division:

$$\begin{array}{r@{}l}
 & 2 \\
4x^2 - 1 & \overline{)8x^2 - x + 2} \\
 & \underline{8x^2 - 2} \\
 & -x + 4
\end{array}$$

Thus,

$$R(x) = \frac{8x^2 - x + 2}{4x^2 - 1} = 2 + \frac{-x + 4}{4x^2 - 1}$$

Then, as $x \to -\infty$ or as $x \to \infty$, the remainder will behave as follows:

$$\frac{-x + 4}{4x^2 - 1} \approx \frac{-x}{4x^2} = \frac{-1}{4x} \to 0$$

Thus, as $x \to -\infty$ or as $x \to \infty$, we have $R(x) \to 2$. We conclude that $y = 2$ is a horizontal asymptote of the graph. ■

In Example 7, note that the quotient 2 obtained by long division is the quotient of the leading coefficients of the numerator polynomial and the denominator polynomial $(\frac{8}{4})$. This means we can avoid the long division process for rational functions whose numerator and denominator *are of the same degree* and conclude that the quotient of the leading coefficients will give us the horizontal asymptote.

Now work Problem 35.

EXAMPLE 8 Finding Horizontal or Oblique Asymptotes

Find the horizontal or oblique asymptotes, if any, of the graph of

$$G(x) = \frac{2x^5 - x^3 + 2}{x^3 - 1}$$

Solution The rational function G is improper, since the degree of the numerator, 5, is larger than the degree of the denominator, 3. To find any horizontal or oblique asymptotes, we use long division:

$$\begin{array}{r@{}l}
 & 2x^2 - 1 \\
x^3 - 1 & \overline{)2x^5 - x^3 + 2} \\
 & \underline{2x^5 - 2x^2 } \\
 & -x^3 + 2x^2 + 2 \\
 & \underline{-x^3 + 1} \\
 & 2x^2 + 1
\end{array}$$

Thus,

$$G(x) = \frac{2x^5 - x^3 + 2}{x^3 - 1} = 2x^2 - 1 + \frac{2x^2 + 1}{x^3 - 1}$$

Then, as $x \to -\infty$ or as $x \to \infty$, the remainder will behave as follows:

$$\frac{2x^2 + 1}{x^3 - 1} \approx \frac{2x^2}{x^3} = \frac{2}{x} \to 0$$

Thus, as $x \to -\infty$ or as $x \to \infty$, we have $G(x) \to 2x^2 - 1$. We conclude that, for large values of x, the graph of G approaches the graph of $y = 2x^2 - 1$. ■

We summarize next the procedure for finding horizontal and oblique asymptotes.

Finding Horizontal and Oblique Asymptotes of a Rational Function *R*

Consider the rational function

$$R(x) = \frac{p(x)}{q(x)} = \frac{a_n x^n + a_{n-1}x^{n-1} + \cdots + a_1 x + a_0}{b_m x^m + b_{m-1}x^{m-1} + \cdots + b_1 x + b_0}$$

in which the degree of the numerator is n and the degree of the denominator is m.

1. If R is a proper rational function ($n < m$), the graph will have the horizontal asymptote $y = 0$ (the x-axis).
2. If R is improper ($n \geq m$), use long division.
   (a) If $n = m$, the quotient obtained will be a number L ($= a_n/b_m$) and the line $y = L$ ($= a_n/b_m$) is a horizontal asymptote.
   (b) If $n = m + 1$, the quotient obtained is of the form $ax + b$ (a polynomial of degree 1) and the line $y = ax + b$ is an oblique asymptote.
   (c) If $n > m + 1$, the quotient obtained is a polynomial of degree 2 or higher, and R has neither a horizontal nor an oblique asymptote. In this case, for x unbounded, the graph of R will behave like the graph of the quotient.

Note: The graph of a rational function either has one horizontal or one oblique asymptote or else has no horizontal and no oblique asymptote.

Now we are ready to graph rational functions.

Graphing Rational Functions

4

With a graphing utility, the task of graphing a rational function is as simple as entering the function into the utility and pressing graph. However, without appropriate knowledge of the behavior of rational functions, it is dangerous to draw conclusions from the graph. The analysis of a rational function $R(x) = p(x)/q(x)$ requires the following steps:

Analyzing the Graph of a Rational Function

STEP 1: Locate the intercepts, if any, of the graph. The x-intercepts, if any, of $R(x) = p(x)/q(x)$ satisfy the equation $p(x) = 0$. The y-intercept, if there is one, is $R(0)$.

STEP 2: Test for symmetry. Replace x by $-x$ in $R(x)$. If $R(-x) = R(x)$, there is symmetry with respect to the y-axis; if $R(-x) = -R(x)$, there is symmetry with respect to the origin.

STEP 3: Locate the vertical asymptotes, if any, by factoring the denominator $q(x)$ of $R(x)$ and identifying its zeros. If R is in lowest terms, each zero will give rise to a vertical asymptote.
STEP 4: Locate the horizontal or oblique asymptotes, if any, using the procedure given earlier. Determine points, if any, at which the graph of R intersects these asymptotes.
STEP 5: Graph R.

After graphing, it is important to confirm that the graph is consistent with the conclusions found in steps 1–4. Let's look at an example.

EXAMPLE 9 Analyzing the Graph of a Rational Function

Analyze the graph of the rational function $R(x) = \dfrac{x - 1}{x^2 - 4}$.

Solution First, we factor both the numerator and the denominator of R:

$$R(x) = \frac{x - 1}{(x + 2)(x - 2)}$$

R is in lowest terms.

STEP 1: We locate the x-intercepts by finding the zeros of the numerator. By inspection, 1 is the only x-intercept. The y-intercept is $R(0) = \frac{1}{4}$.
STEP 2: Because

$$R(-x) = \frac{-x - 1}{x^2 - 4}$$

we conclude that R is neither even nor odd. Thus, there is no symmetry with respect to the y-axis or the origin.
STEP 3: We locate the vertical asymptotes by factoring the denominator: $x^2 - 4 = (x + 2)(x - 2)$. The graph of R thus has two vertical asymptotes: the lines $x = -2$ and $x = 2$.
STEP 4: The degree of the numerator is less than the degree of the denominator, so R is proper and the line $y = 0$ (the x-axis) is a horizontal asymptote of the graph. To determine if the graph of R intersects the horizontal asymptote, we solve the equation $R(x) = 0$. The only solution is $x = 1$, so the graph of R intersects the horizontal asymptote at $(1, 0)$.
STEP 5: The analysis just completed helps us set the viewing rectangle to obtain a complete graph. Figure 48(a) shows the graph of $R(x) = \dfrac{x - 1}{x^2 - 4}$ in connected mode and Figure 48(b) shows it in dot mode. Notice in Figure 48(a), the graph has vertical lines at $x = -2$ and $x = 2$. This is due to the fact that when the graphing utility is in connected mode, it will "connect the dots" between consecutive pixels. We know the graph of R does not cross the lines $x = -2$ and $x = 2$, since R is not defined at $x = -2$ or $x = 2$. Thus, when graphing rational functions, dot mode should be used if repeated experimenting with the viewing rectangle dimensions does not remove

the extraneous vertical lines. You should confirm that all of the algebraic conclusions we arrived at in steps 1–4 are part of the graph. For example, the graph has vertical asymptotes at $x = -2$ and $x = 2$ and the graph has a horizontal asymptote at $y = 0$. The y-intercept is $\frac{1}{4}$, the x-intercept is 1.

FIGURE 48

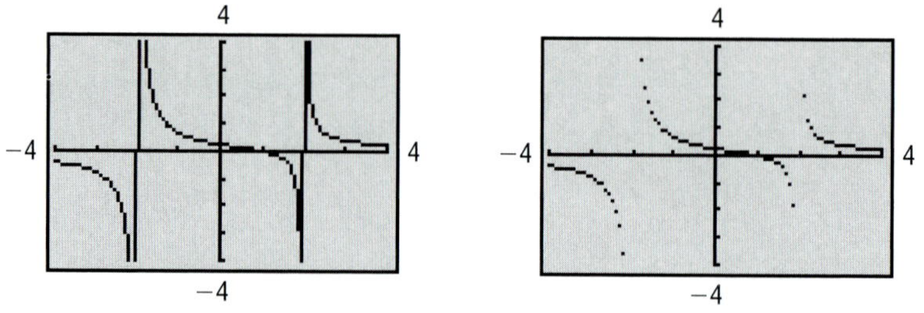

Connected mode
(a)

Dot mode
(b)

Now work Problem 41.

EXAMPLE 10 Analyzing the Graph of a Rational Function

Analyze the graph of the rational function $R(x) = \dfrac{x^2 - 1}{x}$.

Solution

STEP 1: The graph has two x-intercepts: -1 and 1. There is no y-intercept.

STEP 2: Since $R(-x) = -R(x)$, the function is odd and the graph is symmetric with respect to the origin.

STEP 3: Since R is in lowest terms, the graph of $R(x)$ has the line $x = 0$ (the y-axis) as a vertical asymptote.

STEP 4: The rational function R is improper since the degree of the numerator, 2, is larger than the degree of the denominator, 1. To find any horizontal or oblique asymptotes, we use long division:

$$\begin{array}{r} x \\ x\overline{)x^2 - 1} \\ \underline{x^2 } \\ -1 \end{array}$$

The quotient is x, so the line $y = x$ is an oblique asymptote of the graph. To determine whether the graph of R intersects the asymptote $y = x$, we solve the equation $R(x) = x$:

$$R(x) = \frac{x^2 - 1}{x} = x$$
$$x^2 - 1 = x^2$$
$$-1 = 0 \quad \text{Impossible.}$$

We conclude that the equation $(x^2 - 1)/x = x$ has no solution, so the graph of $R(x)$ does not intersect the line $y = x$.

FIGURE 49

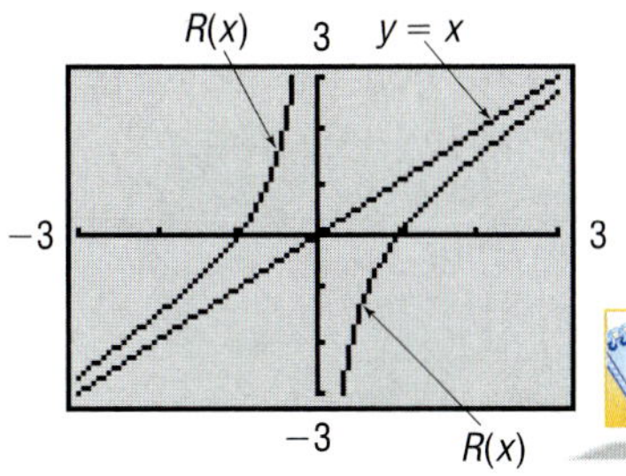

STEP 5: See Figure 49. We see from the graph that there is no y-intercept and two x-intercepts, -1 and 1. The symmetry with respect to the origin is also evident. We can also see that there is a vertical asymptote at $x = 0$ and an oblique asymptote, $y = x$. Finally, it is not necessary to graph this function in dot mode since there are no extraneous vertical lines present. ■

Now work Problem 49.

EXAMPLE 11 Analyzing the Graph of a Rational Function

Analyze the graph of the rational function $R(x) = \dfrac{3x^2 - 3x}{x^2 + x - 12}$.

Solution We factor R to get

$$R(x) = \frac{3x(x - 1)}{(x + 4)(x - 3)}$$

R is in lowest terms.

STEP 1: The graph has two x-intercepts: 0 and 1. The y-intercept is $R(0) = 0$.

STEP 2: Because

$$R(-x) = \frac{-3x(-x - 1)}{(-x + 4)(-x - 3)} = \frac{3x(x + 1)}{(x - 4)(x + 3)}$$

we conclude that R is neither even nor odd. Thus, there is no symmetry with respect to the y-axis or the origin.

STEP 3: The graph of R has two vertical asymptotes: $x = -4$ and $x = 3$.

STEP 4: Since the degree of the numerator equals the degree of the denominator, the graph has a horizontal asymptote. To find it, we form the quotient of the leading coefficient of the numerator, 3, and the leading coefficient of the denominator, 1. Thus, the graph of R has the horizontal asymptote $y = 3$. To find out whether the graph of R intersects the asymptote, we solve the equation $R(x) = 3$.

$$\begin{aligned} R(x) = \frac{3x^2 - 3x}{x^2 + x - 12} &= 3 \\ 3x^2 - 3x &= 3x^2 + 3x - 36 \\ -6x &= -36 \\ x &= 6 \end{aligned}$$

Thus, the graph intersects the line $y = 3$ only at $x = 6$, and $(6, 3)$ is a point on the graph of R.

STEP 5: Figure 50(a) shows the graph of R in connected mode. Notice the extraneous vertical lines at $x = -4$ and $x = 3$ (the vertical asymptotes). As a result, we will also graph R in dot mode. See Figure 50(b). We use TRACE to verify $y = 3$ is a horizontal asymptote by TRACEing for large values of x and observing y gets closer to 3.

We can also use TRACE to verify that the graph of R crosses the horizontal asymptote, $y = 3$, at $(6, 3)$.

FIGURE 50

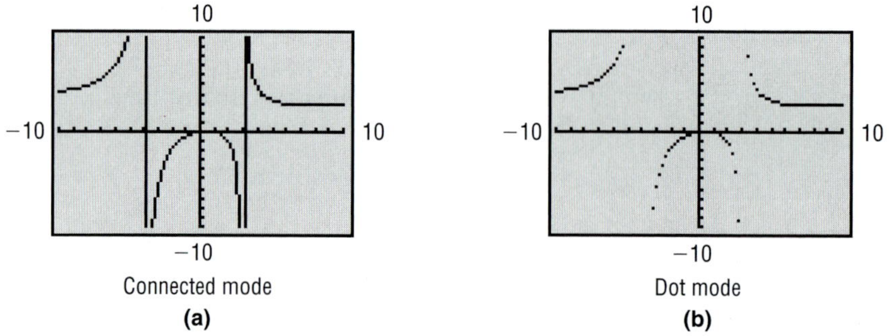

Figure 50 does not display the graph between the two x-intercepts, 0 and 1. Nor does it show the graph crossing the horizontal asymptote at $(6, 3)$. To see these parts better, we graph R for $-1 \le x \le 2$ (Figure 51(a)), and for $4 \le x \le 60$ (Figure 51(b)).

FIGURE 51

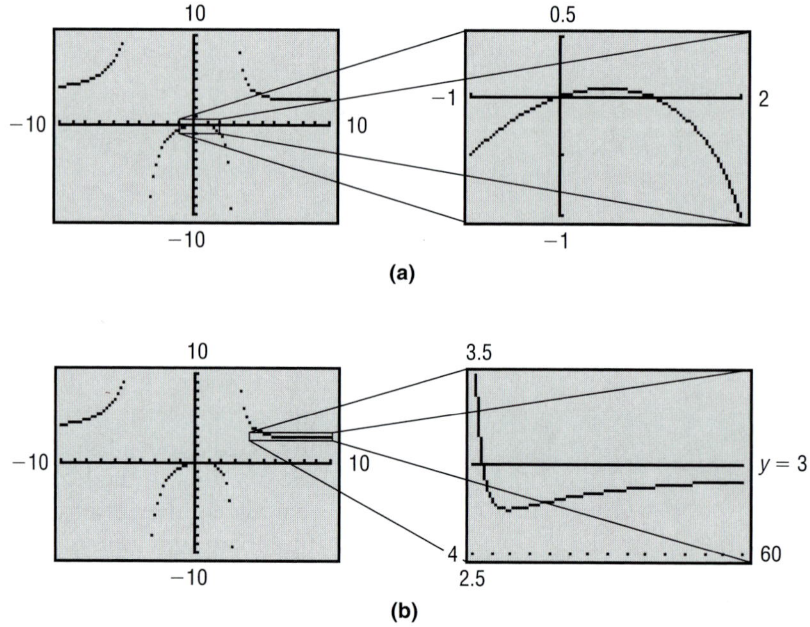

These graphs now reflect the behavior produced by the analysis. Further, we observe two turning points, one between 0 and 1; the other to the right of 4. Correct to two decimal places, these turning points are $(0.52, 0.06)$ and $(11.47, 2.74)$.

EXAMPLE 12 Analyzing the Graph of a Rational Function

Analyze the graph of the rational function

$$R(x) = \frac{0.5x^3 - 2x^2 + 3}{0.1x^3 + x - 3}$$

Solution STEP 1: The x-intercepts obey the equation

$$0.5x^3 - 2x^2 + 3 = 0$$

The x-intercepts, correct to two decimal places, are $-1.08, 1.57$ and 3.51. The y-intercept is $R(0) = -1$.

STEP 2: Because

$$R(-x) = \frac{-0.5x^3 - 2x^2 + 3}{-0.1x^3 - x - 3}$$

we conclude that R is neither even nor odd. Thus, there is no symmetry with respect to the y-axis or the origin.

STEP 3: We solve the equation

$$0.1x^3 + x - 3 = 0$$

The only solution, correct to two decimal places, is 2.08. Comparing this result with that of step 1, we conclude R is in lowest terms. Thus, the only vertical asymptote is the line $x = 2.08$.

STEP 4: The graph of R has $y = 5$ as a horizontal asymptote. To see if the graph of R intersects $y = 5$, we solve the equation

$$\frac{0.5x^3 - 2x^2 + 3}{0.1x^3 + x - 3} = 5$$

$$0.5x^3 - 2x^2 + 3 = 0.5x^3 + 5x - 15$$

$$-2x^2 - 5x + 18 = 0$$

The solutions are $x = -4.5$ and $x = 2$.

STEP 5: The graph of R is shown in Figure 52.

FIGURE 52

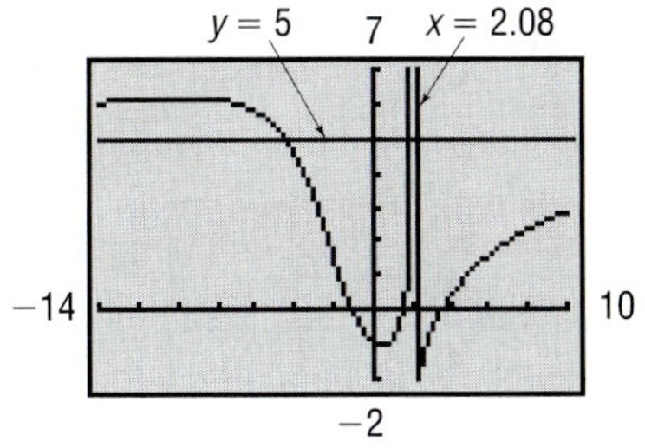

The two turning points, correct to two decimal places, are at $(-9.54, 6.17)$ and $(0.30, -1.05)$.

EXAMPLE 13 Analyzing the Graph of a Rational Function With a Hole

Analyze the graph of the rational function $R(x) = \dfrac{2x^2 - 5x + 2}{x^2 - 4}$.

Solution We factor R and obtain

$$R(x) = \frac{(2x - 1)(x - 2)}{(x + 2)(x - 2)}$$

In lowest terms,

$$R(x) = \frac{2x - 1}{x + 2}$$

STEP 1: The graph has one x-intercept: 0.5. The y-intercept is $R(0) = -0.5$.

STEP 2: Because

$$R(-x) = \frac{2x^2 + 5x + 2}{x^2 - 4}$$

We conclude that R is neither even nor odd. Thus, there is no symmetry with respect to the y-axis or the origin.

STEP 3: The graph has one vertical asymptote: $x = -2$, since $x + 2$ is the only factor of the denominator of $R(x)$ *in lowest terms.* However, the rational function is undefined at both $x = 2$ and $x = -2$.

STEP 4: Since the degree of the numerator equals the degree of the denominator, the graph has a horizontal asymptote. To find it, we form the quotient of the leading coefficient of the numerator, 2, and the leading coefficient of the denominator, 1. Thus, the graph of R has the horizontal asymptote $y = 2$. To find out whether the graph of R intersects the asymptote, we solve the equation $R(x) = 2$.

$$\begin{aligned} R(x) = \frac{2x - 1}{x + 2} &= 2 \\ 2x - 1 &= 2(x + 2) \\ 2x - 1 &= 2x + 4 \\ -1 &= 4 \qquad \text{Impossible.} \end{aligned}$$

Thus, the graph does not intersect the line $y = 2$.

STEP 5: Figure 53 shows the graph of $R(x)$. Notice the graph has one vertical asymptote at $x = -2$ and the function appears to be continuous at $x = 2$. However, looking at Table 15, we see R is undefined at $x = 2$. Thus, the graph leads us to the incorrect conclusion that the graph of R is continuous at $x = 2$. In fact, there is a **hole** in the graph where $x = 2$. So, we must be careful when using a graphing utility to analyze rational functions.

FIGURE 53

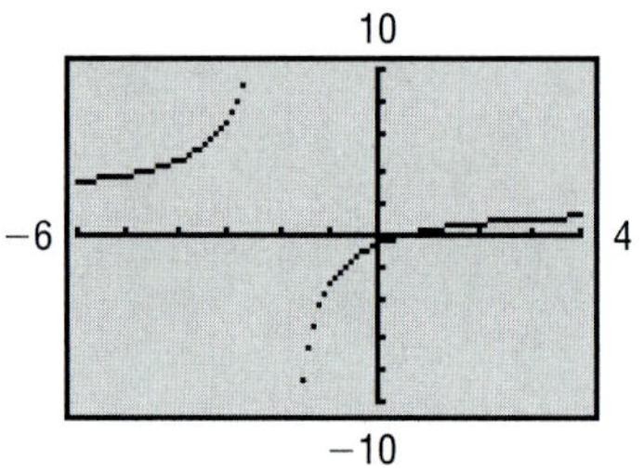

TABLE 15

X	Y1
-3	7
-2	ERROR
-1	-3
0	-.5
1	.33333
2	ERROR
3	1

Y1■(2X²−5X+2)/(...

EXAMPLE 14 Finding the Least Cost of a Can

Reynolds Metal Co. manufactures aluminum cans in the shape of a cylinder with a capacity of 500 cubic centimeters ($\frac{1}{2}$ liter). The top and bottom of the can will be made of a special aluminum alloy that costs 0.05¢ per square centimeter. The sides of the can are to be made of material that costs 0.02¢ per square centimeter.

(a) Express the cost of material for the can as a function of the radius r of the can.

(b) Use a graphing utility to graph the function $C = C(r)$.

(c) What value of r will result in the least cost?

(d) What is this least cost?

FIGURE 54

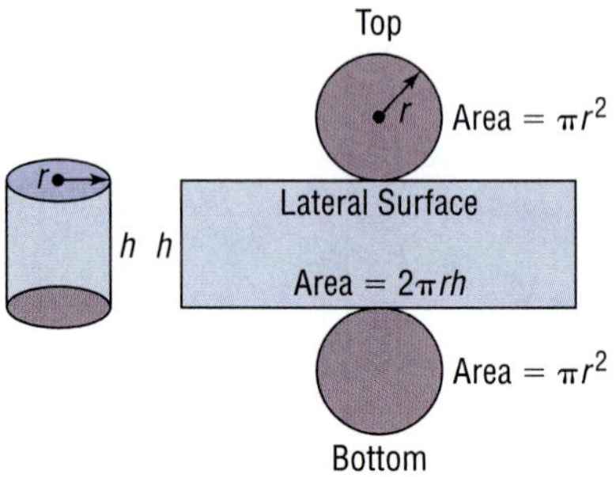

Solution

(a) Figure 54 illustrates the situation. Notice that the material required to produce a cylindrical can of height h and radius r consists of a rectangle of area $2\pi rh$ and two circles, each of area πr^2. The total cost C (in cents) of manufacturing the can is therefore

$$\begin{aligned} C &= \text{Cost of top and bottom} + \text{Cost of side} \\ &= \underbrace{2(\pi r^2)}_{\text{Total area of top and bottom}} \underbrace{(0.05)}_{\text{Cost/unit area}} + \underbrace{(2\pi rh)}_{\text{Total area of side}} \underbrace{(0.02)}_{\text{Cost/unit area}} \\ &= 0.10\pi r^2 + 0.04\pi rh \end{aligned}$$

But we have the additional restriction that the height h and radius r must be chosen so that the volume V of the can is 500 cubic centimeters. Since $V = \pi r^2 h$, we have

$$500 = \pi r^2 h \quad \text{or} \quad h = \frac{500}{\pi r^2}$$

Thus, the cost C, in cents, as a function of the radius r is

$$C(r) = 0.10\pi r^2 + 0.04\pi r\left(\frac{500}{\pi r^2}\right) = 0.10\pi r^2 + \frac{20}{r} = \frac{0.10\pi r^3 + 20}{r}$$

FIGURE 55

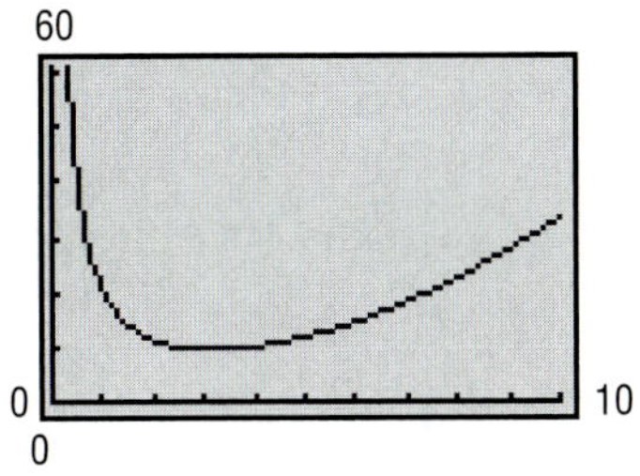

(b) See Figure 55.

(c) Using the MINIMUM command, the cost is least for a radius of about 3.16 centimeters.

(d) The least cost is 9.46¢.

4.4 EXERCISES

In Problems 1–10, find the domain of each rational function.

1. $R(x) = \dfrac{4x}{x - 3}$

2. $R(x) = \dfrac{5x^2}{3 + x}$

3. $H(x) = \dfrac{-4x^2}{(x - 2)(x + 4)}$

4. $G(x) = \dfrac{6}{(x + 3)(4 - x)}$

5. $F(x) = \dfrac{3x(x - 1)}{2x^2 - 5x - 3}$

6. $Q(x) = \dfrac{-x(1 - x)}{3x^2 + 5x - 2}$

7. $R(x) = \dfrac{x}{x^3 - 8}$

8. $R(x) = \dfrac{x}{x^4 - 1}$

9. $H(x) = \dfrac{3x^2 + x}{x^2 + 4}$

10. $G(x) = \dfrac{x - 3}{x^4 + 1}$

In Problems 11–20, use the graph shown to find:

(a) The domain and range of each function

(b) The intercepts, if any

(c) Horizontal asymptotes, if any

(d) Vertical asymptotes, if any

(e) Oblique asymptotes, if any

11.

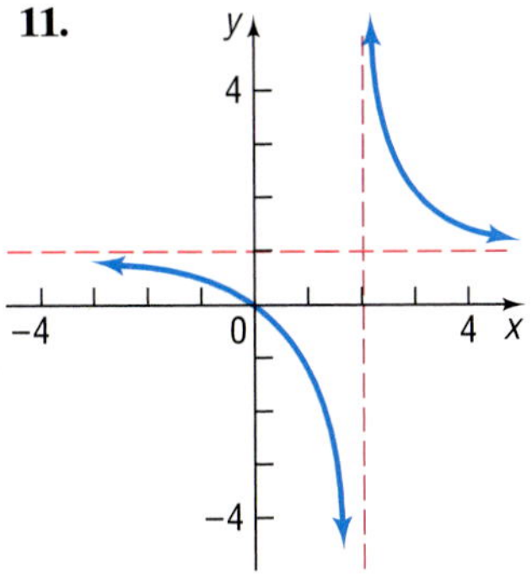

12.

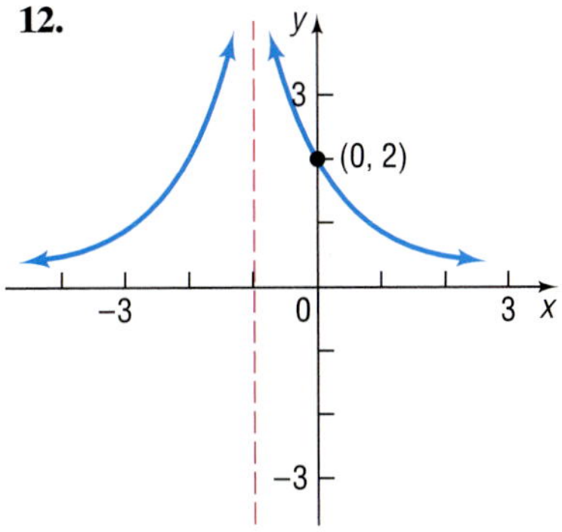

13.

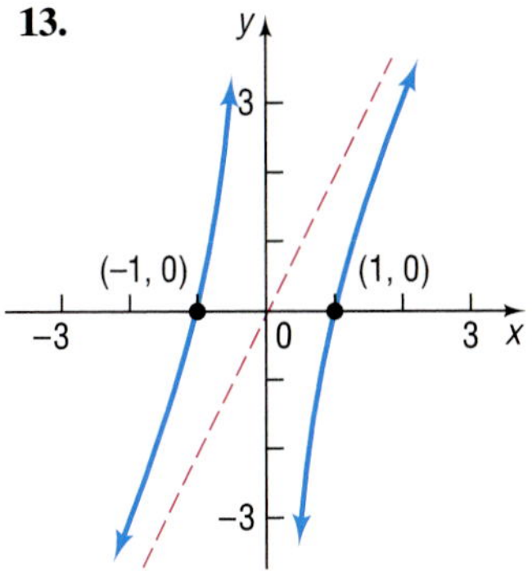

14.

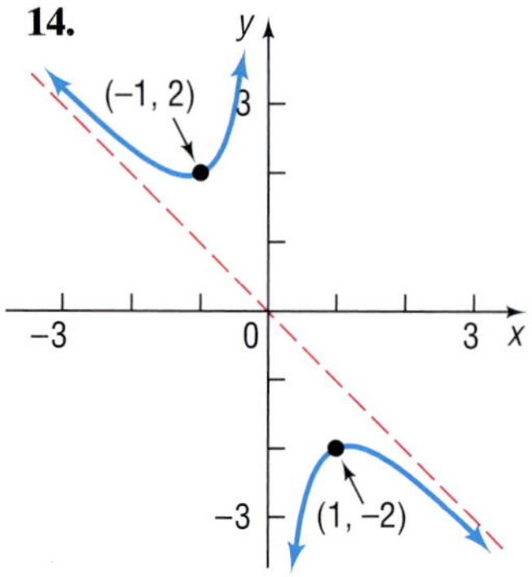

15.

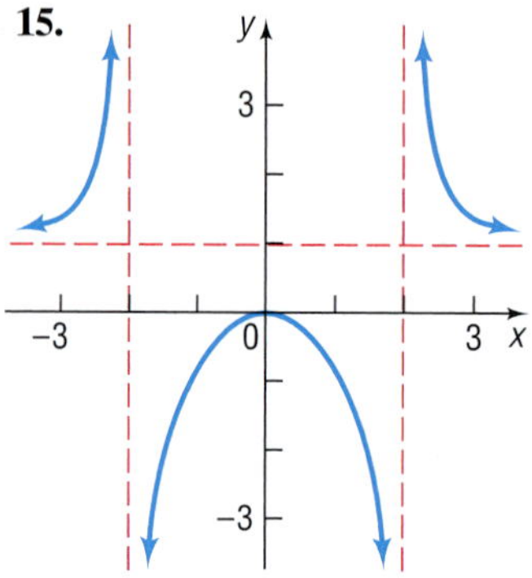

16.

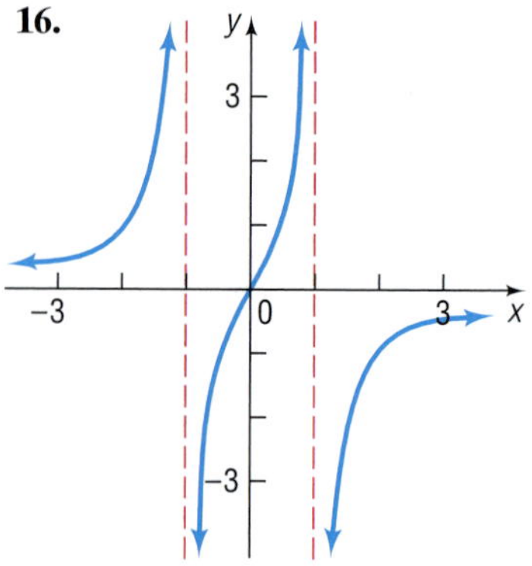

17.

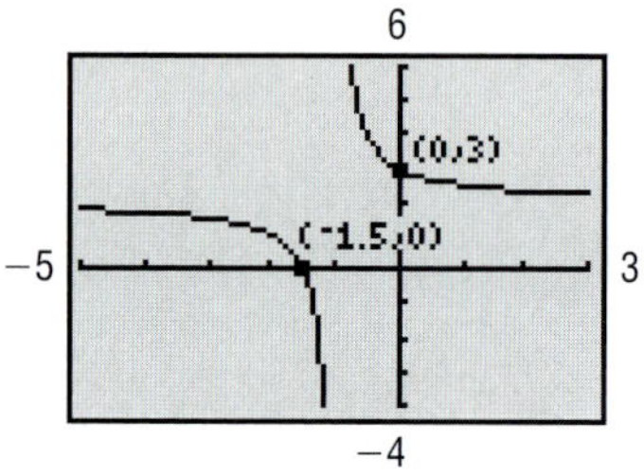

18.

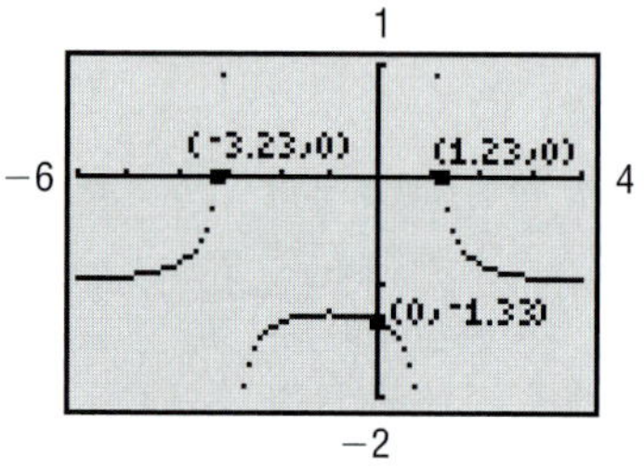

19.

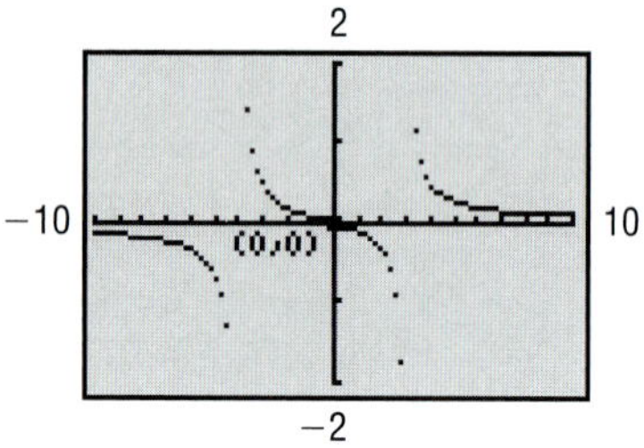

20.

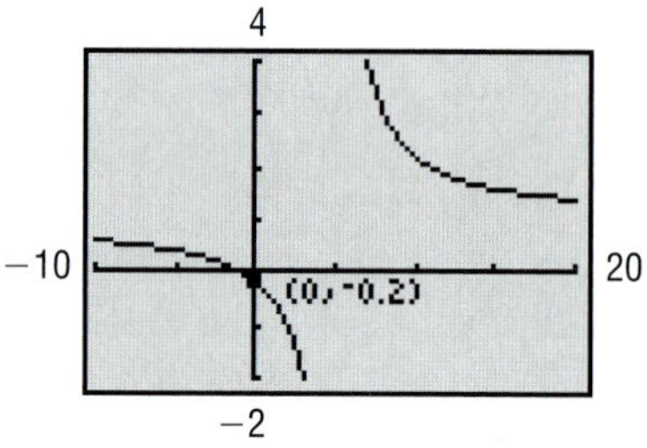

In Problems 21–30, using a graphing utility, show the steps required to graph each rational function.

21. $R(x) = \dfrac{1}{(x-1)^2}$

22. $R(x) = \dfrac{3}{x}$

23. $H(x) = \dfrac{-2}{x+1}$

24. $G(x) = \dfrac{2}{(x+2)^2}$

25. $R(x) = \dfrac{1}{x^2 + 4x + 4}$

26. $R(x) = \dfrac{1}{x-1} + 1$

27. $F(x) = 1 - \dfrac{1}{x}$

28. $Q(x) = 1 + \dfrac{1}{x}$

29. $R(x) = \dfrac{x^2 - 4}{x^2}$

30. $R(x) = \dfrac{x-4}{x}$

In Problems 31–40, find the vertical, horizontal, and oblique asymptotes, if any, of each rational function, without graphing.

31. $R(x) = \dfrac{3x}{x + 4}$

32. $R(x) = \dfrac{3x + 5}{x - 6}$

33. $H(x) = \dfrac{x^4 + 2x^2 + 1}{x^2 - x + 1}$

34. $G(x) = \dfrac{-x^2 + 1}{x + 5}$

35. $T(x) = \dfrac{x^3}{x^4 - 1}$

36. $P(x) = \dfrac{4x^5}{x^3 - 1}$

37. $Q(x) = \dfrac{5 - x^2}{3x^4}$

38. $F(x) = \dfrac{-2x^2 + 1}{2x^3 + 4x^2}$

39. $R(x) = \dfrac{3x^4 + 4}{x^3 + 3x}$

40. $R(x) = \dfrac{6x^2 + x + 12}{3x^2 - 5x - 2}$

In Problems 41–84, follow steps 1 through 5 on pages 288–289 to analyze the graph of each function.

41. $R(x) = \dfrac{x + 1}{x(x + 4)}$

42. $R(x) = \dfrac{x}{(x - 1)(x + 2)}$

43. $R(x) = \dfrac{3x + 3}{2x + 4}$

44. $R(x) = \dfrac{2x + 4}{x - 1}$

45. $R(x) = \dfrac{3}{x^2 - 4}$

46. $R(x) = \dfrac{6}{x^2 - x - 6}$

47. $P(x) = \dfrac{x^4 + x^2 + 1}{x^2 - 1}$

48. $Q(x) = \dfrac{x^4 - 1}{x^2 - 4}$

49. $H(x) = \dfrac{x^3 - 1}{x^2 - 9}$

50. $G(x) = \dfrac{x^3 + 1}{x^2 + 2x}$

51. $R(x) = \dfrac{x^2}{x^2 + x - 6}$

52. $R(x) = \dfrac{x^2 + x - 12}{x^2 - 4}$

53. $G(x) = \dfrac{x}{x^2 - 4}$

54. $G(x) = \dfrac{3x}{x^2 - 1}$

55. $R(x) = \dfrac{3}{(x - 1)(x^2 - 4)}$

56. $R(x) = \dfrac{-4}{(x + 1)(x^2 - 9)}$

57. $H(x) = 4\dfrac{x^2 - 1}{x^4 - 16}$

58. $H(x) = \dfrac{x^2 + 4}{x^4 - 1}$

59. $F(x) = \dfrac{x^2 - 3x - 4}{x + 2}$

60. $F(x) = \dfrac{x^2 + 3x + 2}{x - 1}$

61. $R(x) = \dfrac{x^2 + x - 12}{x - 4}$

62. $R(x) = \dfrac{x^2 - x - 12}{x + 5}$

63. $F(x) = \dfrac{x^2 + x - 12}{x + 2}$

64. $G(x) = \dfrac{x^2 - x - 12}{x + 1}$

65. $R(x) = \dfrac{x(x - 1)^2}{(x + 3)^3}$

66. $R(x) = \dfrac{(x - 1)(x + 2)(x - 3)}{x(x - 4)^2}$

67. $R(x) = \dfrac{4x^3 - 0.5x + 2}{2x^3 + 0.3x^2 - 1}$

68. $R(x) = \dfrac{3x^4 - 4x^2 + 8}{x^4 - x^3 + 2}$

69. $R(x) = \dfrac{3x^3 + 5x^2 - 3}{0.1x^4 - 2x^2 + 1}$

70. $R(x) = \dfrac{x^2 - 0.9x + 2}{x^3 + 0.5x^2 + 1}$

71. $R(x) = \dfrac{0.5x^4 - x^3 + 1}{0.1x^3 - \pi x + 1}$

72. $R(x) = \dfrac{2x^3 + \pi x + 3}{-x^2 + 0.9x + 1}$

73. $R(x) = \dfrac{x^2 + x - 12}{x^2 - x - 6}$

74. $R(x) = \dfrac{x^2 + 3x - 10}{x^2 + 8x + 15}$

75. $R(x) = \dfrac{6x^2 - 7x - 3}{2x^2 - 7x + 6}$

76. $R(x) = \dfrac{8x^2 + 26x + 15}{2x^2 - x - 15}$

77. $R(x) = \dfrac{x^3 + 2x^2 - 5x - 6}{x^3 + 7x^2 + 7x - 15}$

78. $R(x) = \dfrac{x^3 - 6x^2 - x + 30}{x^3 - 4x^2 + x + 6}$

79. $f(x) = x + \dfrac{1}{x}$

80. $f(x) = 2x + \dfrac{9}{x}$

81. $f(x) = x^2 + \dfrac{1}{x}$

82. $f(x) = 2x^2 + \dfrac{9}{x}$

83. $f(x) = x + \dfrac{1}{x^3}$

84. $f(x) = 2x + \dfrac{9}{x^3}$

85. If the graph of a rational function R has the vertical asymptote $x = 4$, then the factor $x - 4$ must be present in the denominator of R. Explain why.

86. If the graph of a rational function R has the horizontal asymptote $y = 2$, then the degree of the numerator of R equals the degree of the denominator of R. Explain why.

87. Consult the illustration. Which of the following rational functions might have this graph? (More than one answer might be possible.)

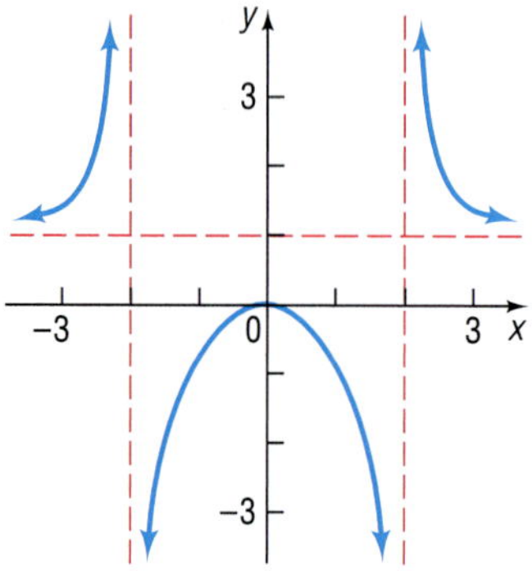

(a) $y = \dfrac{4x^2}{x^2 - 4}$

(b) $y = \dfrac{x}{x^2 - 4}$

(c) $y = \dfrac{x^2}{x^2 - 4}$

(d) $y = \dfrac{x^2(x^2 + 1)}{(x^2 + 4)(x^2 - 4)}$

(e) $y = \dfrac{x^3}{x^2 - 4}$

(f) $y = \dfrac{x^2 - 4}{x^2}$

88. Consult the illustration. Which of the following rational functions might have this graph? (More than one answer may be possible.)

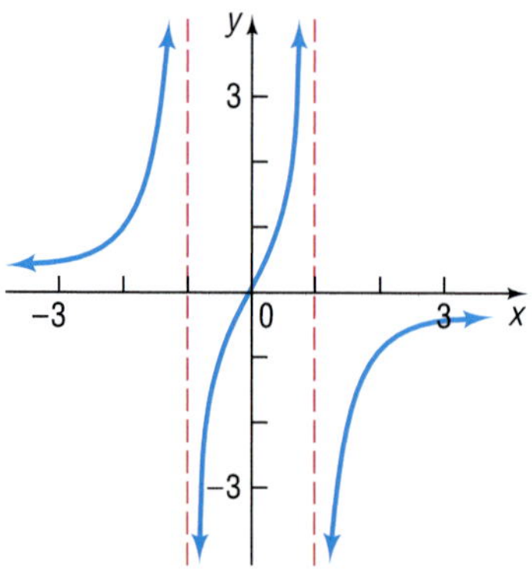

(a) $y = \dfrac{2x}{x^2 - 1}$

(b) $y = \dfrac{-3x}{x^2 - 1}$

(c) $y = \dfrac{x^3}{x^2 - 1}$

(d) $y = \dfrac{x^2 - 1}{-3x}$

(e) $y = \dfrac{-x^3}{(x^2 + 1)(x^2 - 1)}$

(f) $y = \dfrac{-x^2}{(x^2 + 1)(x^2 - 1)}$

89. **Gravity** In physics, it is established that the acceleration due to gravity, g, at a height h meters above sea level is given by

$$g(h) = \frac{3.99 \times 10^{14}}{(6.374 \times 10^6 + h)^2}$$

where 6.374×10^6 is the radius of Earth in meters.

(a) What is the acceleration due to gravity at sea level?
(b) The Sears Tower in Chicago, IL is 443 meters tall. What is the acceleration due to gravity at the top of the Sears Tower?
(c) The peak of Mount Everest is 8848 meters above sea level. What is the acceleration due to gravity on the peak of Mount Everest?
(d) Find the horizontal asymptote of $g(h)$.
(e) Using your graphing utility, graph $g(h)$.
(f) Solve $g(h) = 0$. How do you interpret your answer?

90. **Population Model** A rare species of insect was discovered in the Amazon Rain Forest. In order to protect the species, environmentalists declare the insect endangered and transplant the insects into a protected area. The population of the insect t months after being transplanted is given by P

$$P(t) = \frac{50(1 + 0.5t)}{(2 + 0.01t)}$$

(a) How many insects were discovered? In other words, what was the population when $t = 0$?
(b) What will the population be after 5 years?
(c) Using your graphing utility, graph $P(t)$.
(d) Determine the horizontal asymptote of $P(t)$. What is the largest population the protected area can sustain?
(e) TRACE $P(t)$ for large values of t to verify your answer to (d).

91. **Drug Concentration** The concentration C of a certain drug in a patient's bloodstream t hours after injection is given by

$$C(t) = \frac{t}{2t^2 + 1}$$

(a) Using your graphing utility, graph $C(t)$.
(b) Determine the time at which the concentration is highest.
(c) Find the horizontal asymptote of $C(t)$. What happens to the concentration of the drug as t increases?

92. Drug Concentration The concentration C of a certain drug in a patient's bloodstream t minutes after injection is given by

$$C(t) = \frac{50t}{(t^2 + 25)}$$

(a) Using your graphing utility, graph $C(t)$.
(b) Determine the time at which the concentration is highest.
(c) Find the horizontal asymptote of $C(t)$. What happens to the concentration of the drug as t increases?

93. Average Cost In Problem 63, Exercise 4.3, the cost function for manufacturing Chevy Cavaliers was found to be

$$C(x) = 0.2156x^3 - 2.3473x^2 + 14.3275x + 10.2238$$

Economists define the **average cost function** as

$$\overline{C}(x) = \frac{C(x)}{x}$$

(a) Find the average cost function.
(b) What is the average cost of producing six Cavaliers per hour?
(c) What is the average cost of producing nine Cavaliers per hour?
(d) Using your graphing utility, graph the average cost function.
(e) Using your graphing utility, find the number of Cavaliers that should be produced per hour to minimize average cost.
(f) What is the minimum average cost?

94. Average Cost In Problem 64, Exercise 4.3, the cost function for printing textbooks was found to be

$$C(x) = 0.0155x^3 - 0.5951x^2 + 9.1502x + 98.4327$$

(a) Find the average cost function (refer to Problem 93).
(b) What is the average cost of printing 13 thousand textbooks per week?
(c) What is the average cost of printing 25 thousand textbooks per week?
(d) Using your graphing utility, graph the average cost function.
(e) Using your graphing utility, find the number of textbooks that should be printed to minimize average cost.
(f) What is the minimum average cost?

95. Minimizing Surface Area United Parcel Service has contracted you to design a closed box with a square base that has a volume of 10,000 cubic inches. See the illustration.

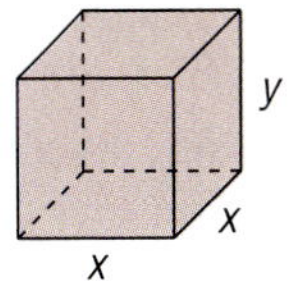

(a) Find a function for the surface area of the box.
(b) Using a graphing utility, graph the function found in (a).
(c) What is the minimum amount of cardboard that can be used to construct the box?
(d) What are the dimensions of the box that minimizes surface area?
(e) Why might UPS be interested in designing a box that minimizes surface area?

96. Minimizing Surface Area United Parcel Service has contracted you to design a closed box with a square base that has a volume of 5000 cubic inches. See the illustration.

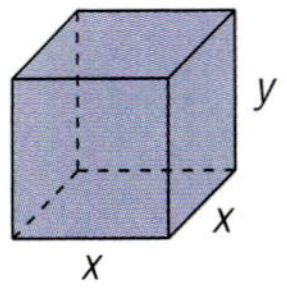

(a) Find a function for the surface area of the box.
(b) Using a graphing utility, graph the function found in (a).
(c) What is the minimum amount of cardboard that can be used to construct the box?
(d) What are the dimensions of the box that minimizes surface area?
(e) Why might UPS be interested in designing a box that minimizes surface area?

97. Cost of a Can A can in the shape of a right circular cylinder is required to have a volume of 500 cubic centimeters. The top and bottom are made of material that costs 6¢ per square centimeter, while the sides are made of material that costs 4¢ per square centimeter.
(a) Express the total cost C of the material as a function of the radius r of the cylinder. (Refer to Figure 54.)
(b) Graph $C = C(r)$. For what value of r is the cost C least?

98. Material Needed to Make a Drum A steel drum in the shape of a right circular cylinder is required to have a volume of 100 cubic feet.
(a) Express the amount A of material required to make the drum as a function of the radius r of the cylinder.
(b) How much material is required if the drum is of radius 3 feet?

(c) Of radius 4 feet?
(d) Of radius 5 feet?
(e) Graph $A = A(r)$. For what value of r is A smallest?

99. Consult the illustration. Make up a rational function that might have this graph. Compare yours with a friend's. What similarities do you see? What differences?

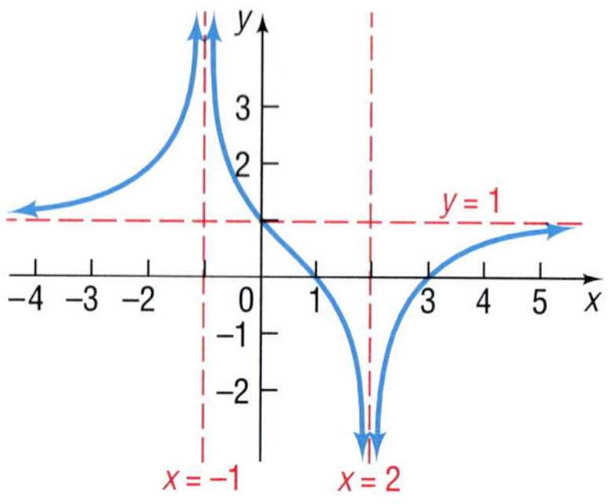

100. Can the graph of a rational function have both a horizontal and an oblique asymptote? Explain.

101. Write a few paragraphs that provide a general strategy for graphing a rational function. Be sure to mention the following: proper, improper, intercepts, and asymptotes.

102. Make up a rational function that has the following characteristics: crosses the x-axis at 3; touches the x-axis at -2; one vertical asymptote, $x = 1$; and one horizontal asymptote, $y = 2$. Give your rational function to a fellow classmate and ask for a written critique of your rational function.

103. Make up a rational function that has $y = 2x + 1$ as an oblique asymptote. Explain the methodology you used.

4.5 THE REAL ZEROS OF A POLYNOMIAL FUNCTION

1 Utilize the Remainder and Factor Theorems
2 Use Descartes' Rule of Signs to Determine the Number of Positive and the Number of Negative Real Zeros of a Polynomial Function
3 Use the Rational Zeros Theorem to List the Potential Rational Zeros of a Polynomial Function
4 Find the Real Zeros of a Polynomial Function
5 Solve Polynomial Equations
6 Utilize the Upper and Lower Bounds Test

Remainder and Factor Theorems

1 Recall that when we divide one polynomial (the dividend) by another (the divisor) we obtain a quotient polynomial and a remainder, the remainder being either the zero polynomial or a polynomial whose degree is less than the degree of the divisor. To check our work, we verify that

$$\text{(Divisor)(Quotient)} + \text{Remainder} = \text{Dividend}$$

This checking routine is the basis for a famous theorem called the **division algorithm* for polynomials,** which we now state without proof.

*A systematic process in which certain steps are repeated a finite number of times is called an **algorithm.** Thus, long division is an algorithm.

MISSION POSSIBLE

Responding to a Challenge by the Cardassians

While playing with virtual reality in your school's computer lab, you are suddenly faced with some very real and very nasty-looking Cardassians who have burst into your school through your computer network. As usual, they are very scornful of the intelligence of Earthlings, and when you try to protest, they throw this challenge at you. "Find a rational function defined by this graph," they say, "and we promise to support your reputation for intellectual achievement in every intergalactic community this side of the Macklin Nebula. Otherwise, consider yourselves the laughing stock of the universe." And they departed back through the computer screen, laughing at what they perceived to be their extreme cleverness.

As you look at the computer screen now, all you can see is this graph:

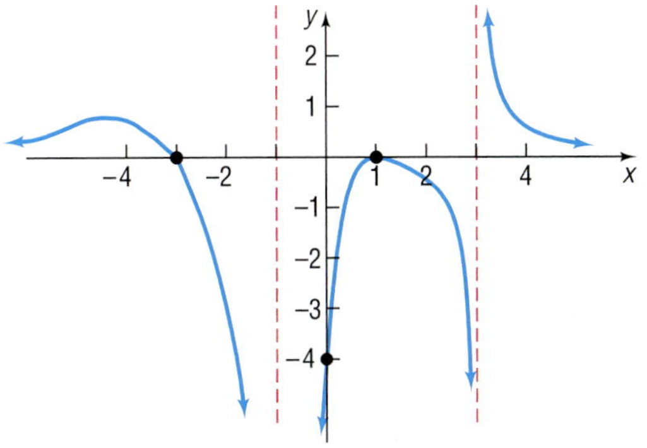

You decide to tackle this as a team project to save the reputation of all the humans on this planet! The Cardassians mentioned "rational function," so you know the solution will have the form

$$R(x) = \frac{a_n x^n + a_{n-1}x^{n-1} + \cdots + a_0}{b_m x^m + b_{m-1}x^{m-1} + \cdots + b_0}$$

Sometimes it is easier to think of it in factored form.

1. You notice some things right away. There are two vertical asymptotes. How do they fit into the solution? Put them in.
2. There are two x-intercepts. Where do they show up in the solution? Put them in.
3. Use a graphing utility to check what you have so far. Does it look like the original graph? Notice that 1 and -3 appear to be the only real zeros. What can you tell about the multiplicity of each one? Do you need to change the multiplicity of either zero in your solution?
4. How can you account for the fact that the graph of the function goes to $-\infty$ on both sides of one vertical asymptote and at the other vertical asymptote it goes to $-\infty$ on one side and to ∞ on the other? Can you change the denominator in some way to account for this?
5. Is the information you have so far consistent with the horizontal asymptote $y = 0$? If not, how should you adjust your function so that it is consistent?
6. Is the information you have so far consistent with the y-intercept? If not, how should you adjust your function so that it is consistent?
7. Are you finished? Have you checked your solution on a graphing utility? Are you satisfied you have it right?
8. Are there other functions whose graph might look like the one given? If so, what common characteristics do they have? Will the Cardassians be more impressed if more than one correct answer is sent?
9. Compare your group's solution to those of other groups before typing the solution into the computer. Then send all the correct solutions off to cardass@slime.uni.

Theorem Division Algorithm for Polynomials

If $f(x)$ and $g(x)$ denote polynomial functions and if $g(x)$ is not the zero polynomial, then there are unique polynomial functions $q(x)$ and $r(x)$ such that

$$\frac{f(x)}{g(x)} = q(x) + \frac{r(x)}{g(x)} \quad \text{or} \quad \underset{\text{dividend}}{f(x)} = \underset{\text{divisor}}{g(x)}\underset{\text{quotient}}{q(x)} + \underset{\text{remainder}}{r(x)} \tag{1}$$

where $r(x)$ is either the zero polynomial or a polynomial of degree less than that of $g(x)$.

In equation (1), $f(x)$ is the **dividend,** $g(x)$ is the **divisor,** $q(x)$ is the **quotient,** and $r(x)$ is the **remainder.**

If the divisor $g(x)$ is a first-degree polynomial of the form

$$g(x) = x - c \qquad c \text{ a real number}$$

then the remainder $r(x)$ is either the zero polynomial or a polynomial of degree 0. Thus, for such divisors, the remainder is some number, say, R, and we may write

$$f(x) = (x - c)q(x) + R \tag{2}$$

This equation is an identity in x and is true for all real numbers x. In particular, it is true when $x = c$. Thus, if $x = c$, then equation (2) becomes

$$f(c) = (c - c)q(c) + R$$
$$f(c) = R$$

and equation (2) takes the form

$$f(x) = (x - c)q(x) + f(c) \tag{3}$$

We have now proved the following result, called the **Remainder Theorem.**

Remainder Theorem

Let f be a polynomial function. If $f(x)$ is divided by $x - c$, then the remainder is $f(c)$.

EXAMPLE 1

Using the Remainder Theorem

Find the remainder if $f(x) = x^3 - 4x^2 + 2x - 5$ is divided by

(a) $x - 3$ (b) $x + 2$

Solution (a) We could use long division. However, it is much easier here to use the Remainder Theorem, which says the remainder is

$$f(3) = (3)^3 - 4(3)^2 + 2(3) - 5 = 27 - 36 + 6 - 5 = -8$$

(b) To find the remainder when $f(x)$ is divided by $x + 2 = x - (-2)$, we evaluate

$$f(-2) = (-2)^3 - 4(-2)^2 + 2(-2) - 5 = -8 - 16 - 4 - 5 = -33$$

Thus, the remainder is -33.

An important and useful consequence of the Remainder Theorem is the **Factor Theorem.**

Factor Theorem

Let f be a polynomial function. Then $x - c$ is a factor of $f(x)$ if and only if $f(c) = 0$.

The Factor Theorem actually consists of two separate statements:

1. If $f(c) = 0$, then $x - c$ is a factor of $f(x)$.
2. If $x - c$ is a factor of $f(x)$, then $f(c) = 0$.

Thus, the proof requires two parts.

Proof

1. Suppose that $f(c) = 0$. Then, by equation (3), we have

$$f(x) = (x - c)q(x)$$

for some polynomial $q(x)$. That is, $x - c$ is a factor of $f(x)$.

2. Suppose that $x - c$ is a factor of $f(x)$. Then there is a polynomial function q such that

$$f(x) = (x - c)q(x)$$

Replacing x by c, we find that

$$f(c) = (c - c)q(c) = 0 \cdot q(c) = 0$$

This completes the proof.

One use of the Factor Theorem is to determine whether a polynomial has a particular factor.

EXAMPLE 2

Using the Factor Theorem

Use the Factor Theorem to determine whether the function $f(x) = 2x^3 - x^2 + 2x - 3$ has the factor

(a) $x - 1$ (b) $x + 3$

Solution (a) Because $x - 1$ is of the form $x - c$ with $c = 1$, we find the value of $f(1)$:

$$f(1) = 2(1)^3 - (1)^2 + 2(1) - 3 = 2 - 1 + 2 - 3 = 0$$

By the Factor Theorem, $x - 1$ is a factor of $f(x)$.

(b) To test the factor $x + 3$, we first need to write it in the form $x - c$. Since $x + 3 = x - (-3)$, we find the value of $f(-3)$:

$$f(-3) = 2(-3)^3 - (-3)^2 + 2(-3) - 3 = -54 - 9 - 6 - 3 = -72$$

Because $f(-3) \neq 0$, we conclude from the factor theorem that $x - (-3) = x + 3$ is not a factor of $f(x)$.

Now work Problem 1.

Descartes' Rule of Signs

2 The real zeros of a polynomial function f are the real solutions, if any, of the equation $f(x) = 0$. They are also the x-intercepts of the graph of f. For polynomial and rational functions, we have seen the importance of the zeros for graphing. In most cases, however, the zeros of a polynomial function are difficult to find using algebraic methods. There are no nice formulas like the quadratic formula available to help us for polynomials of degree higher than 2. Although formulas do exist for solving any third- or fourth-degree polynomial equation, they are somewhat complicated. It has been proved that no general formulas exist for polynomial equations of degree 5 or higher. In this section, we shall learn some ways of detecting information about the character of the zeros, which, in turn, will help us use our graphing utility more effectively.

Our first theorem concerns the number of zeros a polynomial function may have. In counting the zeros of a polynomial, we count each zero as many times as its multiplicity.

Theorem Number of Zeros

A polynomial function cannot have more zeros than its degree.

Proof The proof is based on the Factor Theorem. If r is a zero of a polynomial function f, then $f(r) = 0$ and, hence, $x - r$ is a factor of $f(x)$. Thus, each zero corresponds to a factor of degree 1. Because f cannot have more first-degree factors than its degree, the result follows.

The next theorem, called **Descartes' Rule of Signs,** provides information about the number and location of the real zeros of a polynomial function, so we know where to look for zeros. Descartes' Rule of Signs assumes that the polynomial is written in descending powers of x and requires that we count the number of variations in sign of the coefficients of $f(x)$ and $f(-x)$.

For example, the following polynomial function has two variations in the signs of coefficients:

$$\begin{aligned} f(x) &= -3x^7 + 4x^4 + 3x^2 - 2x - 1 \\ &= -3x^7 + 0x^6 + 0x^5 + 4x^4 + 0x^3 + 3x^2 - 2x - 1 \end{aligned}$$

$-$ to $+$ $\qquad$ $+$ to $-$

Notice that we ignored the zero coefficients in $0x^6$, $0x^5$ and $0x^3$ in counting the number of variations in sign of $f(x)$. Replacing x by $-x$, we get

$$f(-x) = 3x^7 + 4x^4 + 3x^2 + 2x - 1$$

$+$ to $-$

which has one variation in sign.

Theorem Descartes' Rule of Signs

Let f denote a polynomial function.

The number of positive real zeros of f either equals the number of variations in sign of the nonzero coefficients of $f(x)$ or else equals that number less an even integer.

The number of negative real zeros of f either equals the number of variations in sign of the nonzero coefficients of $f(-x)$ or else equals that number less an even integer.

We shall not prove Descartes' Rule of Signs. Let's see how it is used.

EXAMPLE 3 Using Descartes' Rule of Signs to Locate Zeros

Discuss the real zeros of $f(x) = 3x^6 - 4x^4 + 3x^3 + 2x^2 - x - 3$.

Solution There are at most six zeros, because the polynomial is of degree 6. Since there are three variations in sign of the nonzero coefficients of $f(x)$, by Descartes' Rule of Signs we expect either three, or one, positive real zero(s). To continue, we look at $f(-x)$:

$$f(-x) = 3x^6 - 4x^4 - 3x^3 + 2x^2 + x - 3$$

There are three variations in sign, so we expect either three, or one, negative real zero(s). Equivalently, we now know that the graph of f has either three or one positive x-intercept(s) and three or one negative x-intercept(s).

FIGURE 56

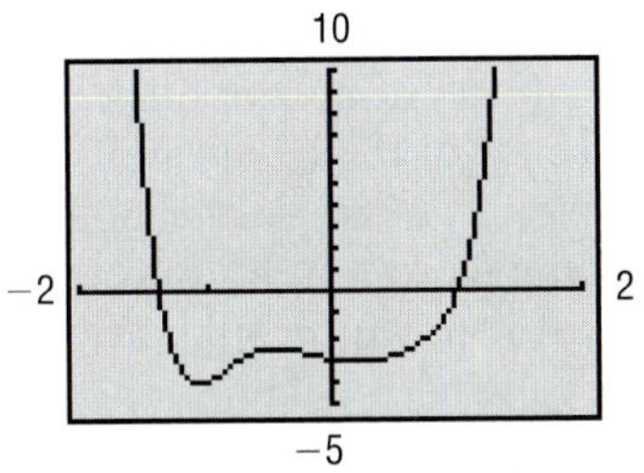

We see in Figure 56 that f has one negative zero near -1.3 and one positive zero near 1. Thus, the conclusions of Descartes' Rule of Signs are confirmed by the graph.

Now work Problem 11.

Rational Zeros Theorem

3 The next result, called the **Rational Zeros Theorem,** provides information about the rational zeros of a polynomial *with integer coefficients.*

Theorem Rational Zeros Theorem

Let f be a polynomial function of degree 1 or higher of the form

$$f(x) = a_n x^n + a_{n-1} x^{n-1} + \cdots + a_1 x + a_0 \qquad a_n \neq 0, a_0 \neq 0$$

where each coefficient is an integer. If p/q, in lowest terms, is a rational zero of f, then p must be a factor of a_0 and q must be a factor of a_n.

EXAMPLE 4 Listing Potential Rational Zeros

List the potential rational zeros of

$$f(x) = 2x^3 + 11x^2 - 7x - 6$$

Solution Because f has integer coefficients, we may use the Rational Zeros Theorem. First, we list all the integers p that are factors of $a_0 = -6$ and all the integers q that are factors of $a_3 = 2$:

$$p: \quad \pm 1, \pm 2, \pm 3, \pm 6$$
$$q: \quad \pm 1, \pm 2$$

Now we form all possible ratios p/q:

$$\frac{p}{q}: \quad \pm 1, \pm 2, \pm 3, \pm 6, \pm \frac{1}{2}, \pm \frac{3}{2}$$

If f has a rational zero, it will be found in this list, which contains 12 possibilities.

Now work Problem 23.

Be sure you understand what the Rational Zeros Theorem says: For a polynomial with integer coefficients, *if* there is a rational zero, it is one of those listed. There may not be any rational zeros.

The Rational Zeros Theorem provides a list of potential rational zeros of a function f. If we graph f, we can get a better sense of the location of the x-intercepts and test to see if they are rational. We can also use the potential rational zeros to select our initial viewing rectangle to graph f and then adjust the rectangle based on the results. The graphs shown throughout the text will be those obtained after setting the final viewing rectangle.

EXAMPLE 5 Finding the Rational Zeros of a Polynomial Function

Continue working with Example 4 to find the rational zeros of

$$f(x) = 2x^3 + 11x^2 - 7x - 6$$

Solution We gather all the information we can about the zeros:

STEP 1: There are at most three zeros.

STEP 2: By Descartes' Rule of Signs, there is one positive real zero. Also, because

$$f(-x) = -2x^3 + 11x^2 + 7x - 6$$

there are two, or no, negative real zeros.

STEP 3: Now we use the list of potential rational zeros obtained in Example 4: $\pm 1, \pm 2, \pm 3, \pm 6, \pm\frac{1}{2}, \pm\frac{3}{2}$.

We could, of course, test each potential rational zero to see if the value of f there is zero. This is not very efficient. The graph of f will tell us approximately where the real zeros are. So we only need to test those rational zeros that are nearby. Figure 57 shows the graph of f. We see f has three zeros: one near -6, one between -1 and 0, and one near 1. From our original list of potential rational zeros, we will test:

FIGURE 57

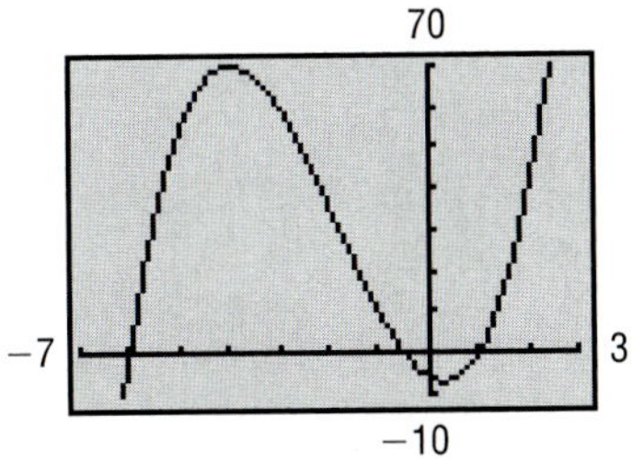

near -6: test -6; between -1 and 0: test $-\frac{1}{2}$; near 1: test 1

$$f(-6) = 0 \qquad f\left(-\frac{1}{2}\right) = 0 \qquad f(1) = 0$$

The three zeros of f are -6, $-\frac{1}{2}$, and 1; each one is a rational zero. ■

We can also use synthetic division* to determine whether a potential rational zero is, in fact, a zero. For example, to test -6, divide f by $x - (-6) = x + 6$

$$\begin{array}{r|rrrr} -6 & 2 & 11 & -7 & -6 \\ & & -12 & 6 & 6 \\ \hline & 2 & -1 & -1 & 0 \end{array}$$

The fact that the remainder is zero tells us -6 is a zero and $x - (-6) = x + 6$ is a factor. Similarly, we have $x + \frac{1}{2}$ and $x - 1$ as factors of f. As a result, we can write f in factored form as

$$f(x) = 2x^3 + 11x^2 - 7x - 6 = 2\left(x^3 + \frac{11}{2}x^2 - \frac{7}{2}x - 3\right)$$

$$= 2(x + 6)\left(x + \frac{1}{2}\right)(x - 1)$$

The next two examples show an advantage of using synthetic division.

*Synthetic division is discussed in the Appendix, Section 6.

EXAMPLE 6 Finding the Real Zeros of a Polynomial Function

4 Find the real zeros of $f(x) = 3x^5 - 2x^4 - 15x^3 + 10x^2 + 12x - 8$. Then factor f over the real numbers.

Solution We gather all the information we can about the zeros.

STEP 1: There are at most five zeros.

STEP 2: By Descartes' Rule of Signs, there are three, or one, positive real zero(s). Also, because

$$f(-x) = -3x^5 - 2x^4 + 15x^3 + 10x^2 - 12x - 8$$

there are two, or no, negative real zeros.

STEP 3: To obtain the list of potential rational zeros, we write the factors p of $a_0 = -8$ and the factors q of $a_5 = 3$:

$$p: \quad \pm 1, \pm 2, \pm 4, \pm 8$$
$$q: \quad \pm 1, \pm 3$$

The potential rational zeros consist of all possible quotients p/q:

$$\frac{p}{q}: \quad \pm 1, \pm 2, \pm 4, \pm 8, \pm\frac{1}{3}, \pm\frac{2}{3}, \pm\frac{4}{3}, \pm\frac{8}{3}$$

FIGURE 58

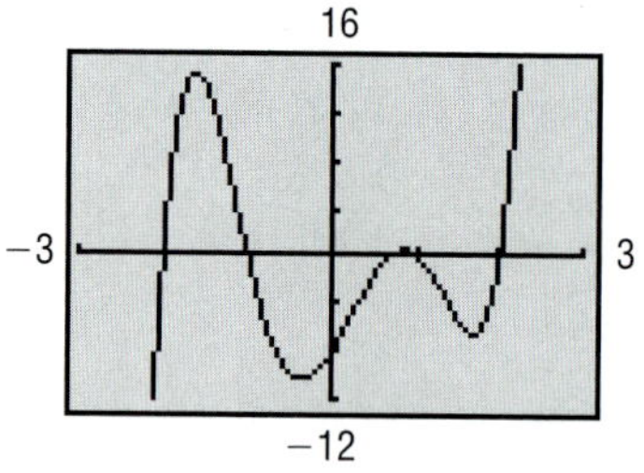

STEP 4: Figure 58 shows the graph of f. The graph has the characteristics we expect of the given polynomial of degree 5: four turning points, y-intercept -8, and behaves like $y = 3x^5$ for large x. The information found in steps 1 and 2 is also confirmed by the graph. The polynomial has 2 negative real zeros and 3 or 1 positive real zeros. The graph has five or three x-intercepts: one near -2, one near -1, two (possibly repeated), or none near 1, and one near 2.

Since -2 appears to be a zero and -2 is a potential rational zero, we use synthetic division to determine if $x + 2$ is a factor of f.

$$\begin{array}{r|rrrrrr} -2 & 3 & -2 & -15 & 10 & 12 & -8 \\ & & -6 & 16 & -2 & -16 & 8 \\ \hline & 3 & -8 & 1 & 8 & -4 & 0 \end{array}$$

Since the remainder is 0, we know -2 is a zero and $x + 2$ is a factor of f. We use the entries in the bottom row of the synthetic division to factor f:

$$\begin{aligned} f(x) &= 3x^5 - 2x^4 - 15x^3 + 10x^2 + 12x - 8 \\ &= (x + 2)(3x^4 - 8x^3 + x^2 + 8x - 4) \end{aligned}$$

Now any solution of the equation $3x^4 - 8x^3 + x^2 + 8x - 4 = 0$ is a zero of f. Because of this we call the equation $3x^4 - 8x^3 + x^2 + 8x - 4 = 0$ a **depressed equation** of f. Since the degree of the depressed equation of f is less than that of the original equation, we work with the depressed equation to find the zeros of f.

We check the potential rational zero -1 next using synthetic division.

$$\begin{array}{r|rrrrr} -1 & 3 & -8 & 1 & 8 & -4 \\ & & -3 & 11 & -12 & 4 \\ \hline & 3 & -11 & 12 & -4 & 0 \end{array}$$

Since the remainder is 0, we know -1 is a zero and $x + 1$ is a factor of f. Thus, we can write f as

$$f(x) = (x + 2)(x + 1)(3x^3 - 11x^2 + 12x - 4)$$

We work with the second depressed equation and check the potential rational zero 1 using synthetic division,

$$\begin{array}{r|rrrr} 1) & 3 & -11 & 12 & -4 \\ & & 3 & -8 & 4 \\ \hline & 3 & -8 & 4 & 0 \end{array}$$

Since the remainder is 0, we conclude that 1 is a zero and $x - 1$ is a factor of f. Thus, we have

$$f(x) = (x + 2)(x + 1)(x - 1)(3x^2 - 8x + 4)$$

The new depressed equation of f, $3x^2 - 8x + 4 = 0$, is a quadratic equation with a discriminant of $b^2 - 4ac = (-8)^2 - 4(3)(4) = 16$. Therefore, this equation has two real solutions, and, in this case, we can find them by factoring:

$$\begin{aligned} 3x^2 - 8x + 4 &= 0 \\ (3x - 2)(x - 2) &= 0 \\ 3x - 2 = 0 \quad &or \quad x - 2 = 0 \\ x = \frac{2}{3} \quad & \quad\quad x = 2 \end{aligned}$$

FIGURE 59

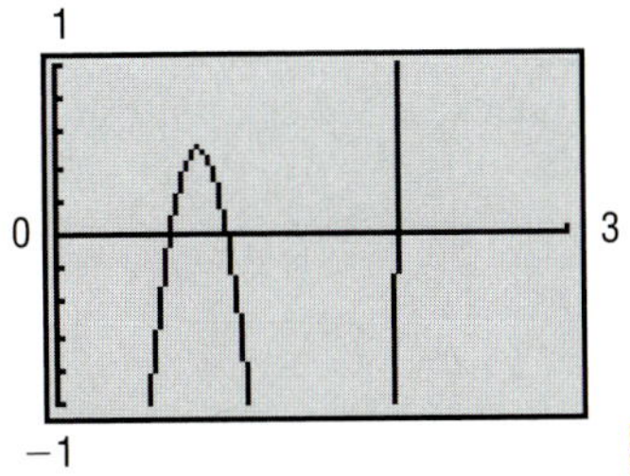

The zeros of f are $-2, -1, \frac{2}{3}, 1$, and 2. The factored form of f is

$$f(x) = 3(x + 2)(x + 1)\left(x - \frac{2}{3}\right)(x - 1)(x - 2)$$

Now we know that f has two distinct roots near 1, 1 and $\frac{2}{3}$.

Based on the information gathered, we change the viewing rectangle to obtain the graph of f for $0 \le x \le 3$. See Figure 59. Now we can see the three distinct positive zeros between 0 and 3.

Now work Problem 41.

5 The procedure outlined in Example 6 for finding the zeros of a polynomial can also be used to solve polynomial equations.

EXAMPLE 7 Solving a Polynomial Equation

Solve the equation $x^5 - 5x^4 + 12x^3 - 24x^2 + 32x - 16 = 0$.

Solution The solutions of this equation are the zeros of the polynomial function

$$f(x) = x^5 - 5x^4 + 12x^3 - 24x^2 + 32x - 16$$

STEP 1: There are at most five real solutions.

STEP 2: By Descartes' Rule of Signs, there are five, three, or one positive real solution(s). Because

$$f(-x) = -x^5 - 5x^4 - 12x^3 - 24x^2 - 32x - 16$$

there are no negative real solutions.

STEP 3: Because $a_5 = 1$ and there are no negative real solutions, the potential rational solutions are the positive integers 1, 2, 4, 8, and 16.

STEP 4: Figure 60 shows the graph of f. The graph has the characteristics we expect of this polynomial of degree 5: it behaves like $y = x^5$ for large x, has y-intercept -16, and has two turning points. The information obtained in steps 1 and 2 is confirmed by the graph: no

FIGURE 60

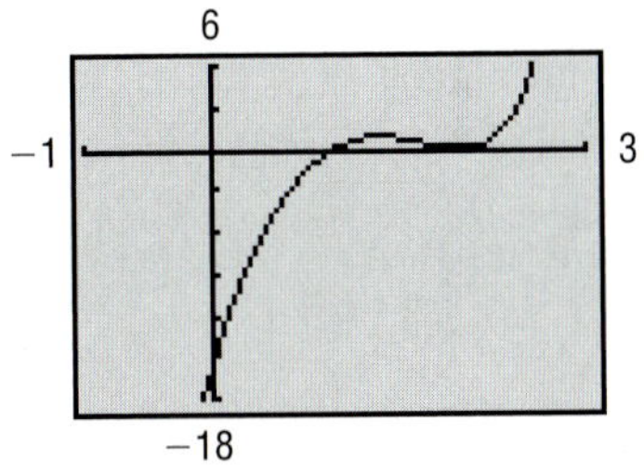

negative zeros. There is an x-intercept near 1 (where the graph crosses the x-axis) and another near 2 (which may be a root of even multiplicity, since the graph appears to touch the x-axis near 2).

Since 1 appears to be a zero and 1 is a potential rational zero, we use synthetic division to determine if $x - 1$ is a factor of f:

$$\begin{array}{r|rrrrrr} 1 & 1 & -5 & 12 & -24 & 32 & -16 \\ & & 1 & -4 & 8 & -16 & 16 \\ \hline & 1 & -4 & 8 & -16 & 16 & 0 \end{array}$$

Since the remainder is 0, we know 1 is a zero and $x - 1$ is a factor of f. Thus, 1 is a solution to the equation.

We now work with the first depressed equation of f:

$$x^4 - 4x^3 + 8x^2 - 16x + 16 = 0$$

We check the potential rational zero 2 next using synthetic division.

$$\begin{array}{r|rrrrr} 2 & 1 & -4 & 8 & -16 & 16 \\ & & 2 & -4 & 8 & -16 \\ \hline & 1 & -2 & 4 & -8 & 0 \end{array}$$

Since the remainder is 0, we conclude 2 is a zero and $x - 2$ is a factor of f. Thus, 2 is a solution to the equation.

We work with the second depressed equation of f: $x^3 - 2x^2 + 4x - 8 = 0$. Earlier, we conjectured 2 may be a root of even multiplicity, so we check 2 again.

$$\begin{array}{r|rrrr} 2 & 1 & -2 & 4 & -8 \\ & & 2 & 0 & 8 \\ \hline & 1 & 0 & 4 & 0 \end{array}$$

Since the remainder is 0, 2 is a zero and $x - 2$ is a factor of f. Thus, $x = 2$ is a zero of multiplicity 2.

The third depressed equation is $x^2 + 4 = 0$. This equation has no real solutions.

Thus, the real solutions are 1 and 2 (the latter being a repeated solution). ■

Based on the solution to Example 7, the polynomial function f can be factored as follows:

$$f(x) = x^5 - 5x^4 + 12x^3 - 24x^2 + 32x - 16 = (x - 1)(x - 2)^2(x^2 + 4)$$

Now work Problem 61.

EXAMPLE 8

Finding the Zeros of a Polynomial

Use Descartes' Rule of Signs and the Rational Zeros Theorem to find the real zeros of the polynomial function

$$g(x) = x^5 - 7x^4 + 8x^3 - 20x^2 + 20x - 2$$

Express any irrational zeros correct to two decimal places.

Solution STEP 1: There are at most five zeros.

STEP 2: There are five, three, or one positive real zeros. Also, because

$$g(-x) = -x^5 - 7x^4 - 8x^3 - 20x^2 - 20x - 2$$

there are no negative real zeros.

STEP 3: The potential rational zeros of g are 1 and 2.

STEP 4: Because there are no negative real zeros, we graph f only for $x \geq -1$. See Figure 61. The graph has the characteristics we expect: it behaves like $y = x^5$ for large x and has y-intercept -2. The information obtained in steps 1 and 2 is confirmed. ZOOMing-IN near $(\frac{1}{2}, 1)$, we see the graph has three x-intercepts: one between 0 and 1, one near 1, and one near 6. We test the potential rational zero 1 using synthetic division.

FIGURE 61

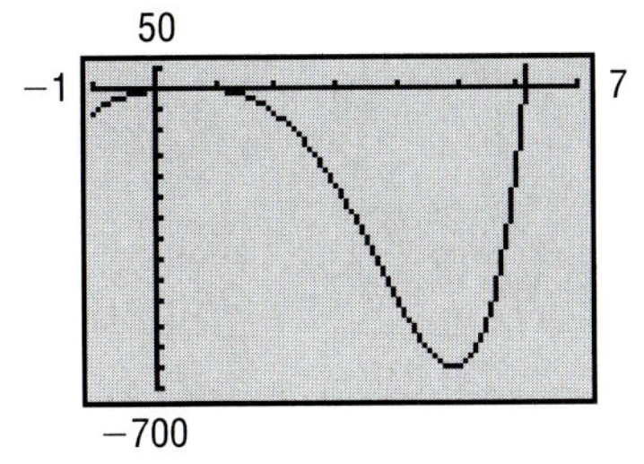

$$\begin{array}{r|rrrrrr} 1 & 1 & -7 & 8 & -20 & 20 & -2 \\ & & 1 & -6 & 2 & -18 & 2 \\ \hline & 1 & -6 & 2 & -18 & 2 & 0 \end{array}$$

The remainder is 0 so 1 is a zero and $x - 1$ is a factor of g.

STEP 5: The first depressed equation of g

$$x^4 - 6x^3 + 2x^2 - 18x + 2 = 0$$

has only 1 and 2 as potential rational zeros. Neither of these is, in fact, a zero. So the remaining zeros of g are not rational. Correct to two decimal places, these irrational zeros are: 0.10 and 6.14. ▪

In Example 7, the quadratic factor $x^2 + 4$ that appears in the factored form of $f(x)$ is called *irreducible*, because the polynomial $x^2 + 4$ cannot be factored over the real numbers. In general, we say that a quadratic factor $ax^2 + bx + c$ is **irreducible** if it cannot be factored over the real numbers, that is, if it is prime over the real numbers.

Refer back to the polynomial function f of Example 6. We found that f has five real zeros, so, by the Factor Theorem, its factored form will contain five linear factors. The polynomial function g of Example 8 has three real zeros, so its factored form will contain three linear factors and one irreducible quadratic factor. The next result tells us what to expect when we factor a polynomial.

Theorem

Every polynomial function (with real coefficients) can be uniquely factored into a product of linear factors and/or irreducible quadratic factors.

▪

We shall prove this result in Section 4.7, and, in fact, we shall draw several additional conclusions about the zeros of a polynomial function. One conclusion is worth noting now. If a polynomial (with real coefficients) is of odd degree, then it must contain at least one linear factor. (Do you see why?) Therefore, it will have at least one real zero.

Corollary

A polynomial function (with real coefficients) of odd degree has at least one real zero.

▪

One of the challenges in using a graphing utility is to set the viewing window so that a complete graph is obtained. The next theorem is a tool which can be used to find bounds on the zeros. This will assure the function does not have any zeros above or below these bounds.

Upper and Lower Bounds

6 The search for the zeros of a polynomial function can be reduced somewhat if upper and lower bounds to the zeros can be found. A number M is an **upper bound** to the zeros of a polynomial f if no zero of f exceeds M. The number m is a **lower bound** if no zero of f is less than m.

Thus, if m is a lower bound and M is an upper bound to the zeros of a polynomial f, then

$$m \leq \text{Any zero of } f \leq M$$

One immediate advantage of knowing the values of a lower bound m and an upper bound M is that, for polynomials with integer coefficients, it may allow you to eliminate some potential rational zeros, that is, any that lie outside the interval $[m, M]$. The next result tells us how to locate lower and upper bounds.

Theorem Bounds on Zeros

Let f denote a polynomial function whose leading coefficient is positive.

If $M > 0$ is a real number and if the third row in the process of synthetic division of f by $x - M$ contains only numbers that are positive or 0, then M is an upper bound to the zeros of f.

If $m \leq 0$ is a real number and if the third row in the process of synthetic division of f by $x - m$ contains numbers that are alternately positive (or 0) and negative (or 0), then m is a lower bound to the zeros of f.

Proof (Outline) We shall give only an outline of the proof of the first part of the theorem. Suppose that M is a positive real number, and the third row in the process of synthetic division of the polynomial f by $x - M$ contains only numbers that are positive or 0. Then there is a quotient q and a remainder R so that

$$f(x) = (x - M)q(x) + R$$

where the coefficients of $q(x)$ are positive or 0 and the remainder $R \geq 0$. Then, for any $x > M$, we must have $x - M > 0$, $q(x) > 0$, and $R \geq 0$, so that $f(x) > 0$. That is, there is no zero of f larger than M.

We can use the Upper and Lower Bounds Test to determine if the RANGE we have selected for our graph is large enough to show all the real zeros of the function, ensuring no zeros lie off the viewing rectangle. Consider Example 9.

EXAMPLE 9 Using the Upper and Lower Bounds Test

Use the Upper and Lower Bounds Test on the polynomial function $f(x) = x^5 - 5x^4 + 12x^3 - 24x^2 + 32x - 16$ to verify there are no zeros larger than 2 and no zeros smaller than 1.

Solution To get an upper bound to the zeros, the usual practice is to start with the value of the next highest integer larger than the largest zero found and continue with consecutive integers until the third row of the process of synthetic division yields only numbers that are positive or 0.

We begin with the graph of f. See Figure 62.

FIGURE 62

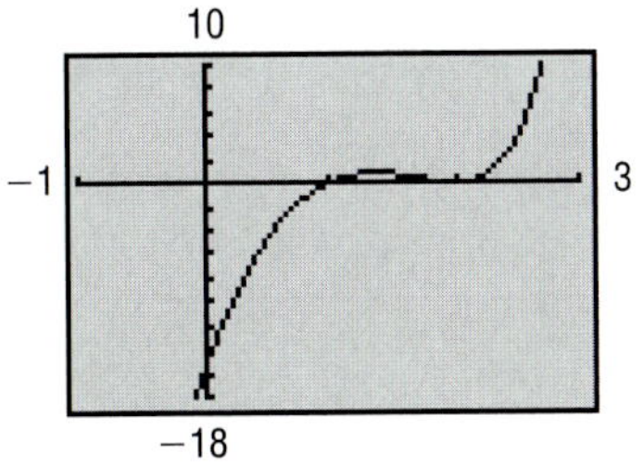

The largest zero we see is near 2. The next highest integer, 3, is where we begin. We use synthetic division to divide f by $x - 3$. If the third row yields only numbers that are positive or zero, then 3 is an upper bound.

3)	1	−5	12	−24	32	−16
		3	−6	18	−18	42
	1	−2	6	−6	14	26

Since the third row yields some negative numbers, 3 is not an upper bound. We will try 4, the next consecutive integer.

4)	1	−5	12	−24	32	−16
		4	−4	32	32	256
	1	−1	8	8	64	240

Since the third row yields some negative numbers, 4 is not an upper bound. We will try 5.

5)	1	−5	12	−24	32	−16
		5	0	60	180	1060
	1	0	12	36	212	1044

The last row yields only numbers that are positive or 0; therefore, 5 is an upper bound. We are certain there are no zeros larger than 5. This means 5 is a good choice for Xmax when we next graph f.

To get a lower bound to the zeros, we start with the next integer smaller than the smallest zero we see.

Look again at Figure 62. The smallest zero we see is near 1, so we begin by testing 0.

0)	1	−5	12	−24	32	−16
		0	0	0	0	0
	1	−5	12	−24	32	−16

Since the third row of the process of synthetic division yields numbers that alternate in sign, 0 is a lower bound; we are certain there are no zeros less than 0.*

This means 0 is a good choice for Xmin. ▬

Now work Problem 35.

*__Note:__ In determining lower bounds, a 0 in the bottom row following a nonzero entry may be counted as positive or negative, as needed. If the next entry is also a 0, it must be counted opposite to the way the preceding 0 was counted (i.e., if the third row yields two consecutive zeros and the first 0 was counted as positive, the next zero would count as negative).

Steps for Finding the Zeros of a Polynomial

STEP 1: Use the degree of the polynomial to determine the maximum number of zeros.

STEP 2: Use Descartes' Rule of Signs to determine the possible number of positive real zeros and negative real zeros.

STEP 3: If the polynomial has integer coefficients, use the Rational Zeros Theorem to identify those rational numbers that are potential zeros.

STEP 4: (a) Using a graphing utility, graph the polynomial.
(b) Conjecture possible zeros based on the graph and the list of potential rational zeros.
(c) Test each potential rational zero.
(d) Approximate any irrational zeros correct to two decimal places.

STEP 5: Use the Upper and Lower Bounds test to be sure no zeros exist off the viewing window.

Let's work one more example that shows the STEPS listed above.

EXAMPLE 10 Finding the Zeros of a Polynomial

Find all the real zeros of the polynomial function

$$f(x) = x^5 - 1.8x^4 - 17.79x^3 + 31.672x^2 + 37.95x - 8.712$$

Solution STEP 1: There are at most 5 zeros.

STEP 2: By Descartes' Rule of Signs there are three or one positive real zero(s). Also because

$$f(-x) = -x^5 - 1.8x^4 + 17.79x^3 + 31.672x^2 - 37.95x - 8.712$$

there are two or no real negative zeros.

STEP 3: Since there are noninteger coefficients, the Rational Zeros Theorem does not apply.

FIGURE 63

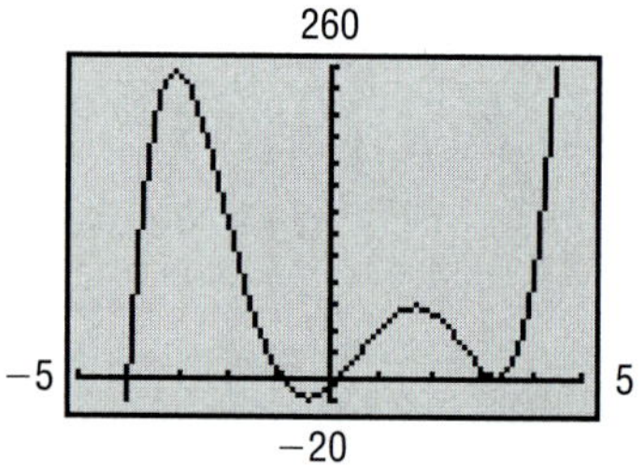

STEP 4: See Figure 63 for the graph of f. We see f appears to have four x-intercepts: one near -4, one near -1, one between 0 and 1, and one near 3. The x-intercept near 3 might be a zero of even multiplicity since the graph seems to touch the x-axis at the point. From step 2, we learned the function would have two or no negative real zeros and three or one positive real zeros. Thus, we know there are no other negative real zeros off the viewing window. However, we may only be seeing two positive real zeros and Descartes' Rule of Signs says there are three.

We use the factor theorem to determine if -4, and -1 are zeros. Since $f(-4) = f(-1) = 0$, we know -4 and -1 are the two negative zeros. Correct to two decimal places, the remaining zeros are 0.20 and 3.30.

STEP 5: We will check to see if 4 is an upper bound to the zeros.

4)	1	−1.8	−17.79	31.672	37.95	−8.712
		4	8.8			
	1	2.2	−8.99			

We stop the process since the upper bound theorem fails. We try 5 next.

5)1	−1.8	−17.79	31.672	37.95	−8.712
	5	16			
1	3.2	−1.79			

Again we stop the process and try 6.

6)1	−1.8	−17.79	31.672	37.95	−8.712
	6	25.2	44.46	456.792	2968.452
1	4.2	7.41	76.132	494.742	2959.74

So, 6 is an upper bound to the zeros. Therefore, either 3.30 is a root of multiplicity 2 or there are two distinct zeros, both of which equal 3.30 correct to two decimal places.

Note: Notice f has 4 turning points. Thus, we know there are no zeros off the viewing rectangle. (Do you know why?)

Historical Feature

Formulas for the solution of third- and fourth-degree polynomial equations exist, and, while not very practical, they do have an interesting history.

In the 1500s in Italy, mathematical contests were a popular pastime, and persons possessing methods for solving problems kept them secret. (Solutions that were published were already common knowledge.) Niccolo of Brescia (1499–1557), commonly referred to as Tartaglia ("the stammerer"), had the secret for solving cubic (third-degree) equations, which gave him a decided advantage in the contests. Girolamo Cardano (1501–1576) found out that Tartaglia had the secret, and, being interested in cubics, he requested it from Tartaglia. The reluctant Tartaglia hesitated for some time, but finally, swearing Cardano to secrecy with midnight oaths by candlelight, told him the secret. Cardano then published the solution in his book *Ars Magna* (1545), giving Tartaglia the credit but rather compromising the secrecy. Tartaglia exploded into bitter recriminations, and each wrote pamphlets that reflected on the other's mathematics, moral character, and ancestry.

The quartic (fourth-degree) equation was solved by Cardano's student Lodovico Ferrari, and this solution also was included, with credit and this time with permission, in the *Ars Magna*.

Attempts were made to solve the fifth-degree equation in similar ways, all of which failed. In the early 1800s, P. Ruffini, Niels Abel, and Evariste Galois all found ways to show that it is not possible to solve fifth-degree equations by formula, but the proofs required the introduction of new methods. Galois' methods eventually developed into a large part of modern algebra.

4.5 EXERCISES

In Problems 1–10, use the Factor Theorem to determine whether $x - c$ is a factor of $f(x)$.

1. $f(x) = 4x^3 - 3x^2 - 8x + 4; \quad c = 2$

2. $f(x) = -4x^3 + 5x^2 + 8; \quad c = -3$

3. $f(x) = 3x^4 - 6x^3 - 5x + 10; \quad c = 2$

4. $f(x) = 4x^4 - 15x^2 - 4; \quad c = 2$

5. $f(x) = 3x^6 + 82x^3 + 27; \quad c = -3$

6. $f(x) = 2x^6 - 18x^4 + x^2 - 9; \quad c = -3$

7. $f(x) = 4x^6 - 64x^4 + x^2 - 15; \quad c = -4$

8. $f(x) = x^6 - 16x^4 + x^2 - 16; \quad c = -4$

9. $f(x) = 2x^4 - x^3 + 2x - 1; \quad c = \frac{1}{2}$

10. $f(x) = 3x^4 + x^3 - 3x + 1; \quad c = -\frac{1}{3}$

In Problems 11–22, tell the maximum number of zeros each polynomial function may have. Then, use Descartes' Rule of Signs to determine how many positive and how many negative zeros each polynomial function may have. Do not attempt to find the zeros.

11. $f(x) = -4x^7 + x^3 - x^2 + 2$
12. $f(x) = 5x^4 + 2x^2 - 6x - 5$
13. $f(x) = 2x^6 - 3x^2 - x + 1$
14. $f(x) = -3x^5 + 4x^4 + 2$
15. $f(x) = 3x^3 - 2x^2 + x + 2$
16. $f(x) = -x^3 - x^2 + x + 1$
17. $f(x) = -x^4 + x^2 - 1$
18. $f(x) = x^4 + 5x^3 - 2$
19. $f(x) = x^5 + x^4 + x^2 + x + 1$
20. $f(x) = x^5 - x^4 + x^3 - x^2 + x - 1$
21. $f(x) = x^6 - 1$
22. $f(x) = x^6 + 1$

In Problems 23–34, list the potential rational zeros of each polynomial function. Do not attempt to find the zeros.

23. $f(x) = 3x^4 - 3x^3 + x^2 - x + 1$
24. $f(x) = x^5 - x^4 + 2x^2 + 3$
25. $f(x) = x^5 - 6x^2 + 9x - 3$
26. $f(x) = 2x^5 - x^4 - x^2 + 1$
27. $f(x) = -4x^3 - x^2 + x + 2$
28. $f(x) = 6x^4 - x^2 + 2$
29. $f(x) = 3x^4 - x^2 + 2$
30. $f(x) = -4x^3 + x^2 + x + 2$
31. $f(x) = 2x^5 - x^3 + 2x^2 + 4$
32. $f(x) = 3x^5 - x^2 + 2x + 3$
33. $f(x) = 6x^4 + 2x^3 - x^2 + 2$
34. $f(x) = -6x^3 - x^2 + x + 3$

In Problems 35–40, find integer-valued upper and lower bounds to the zeros of each polynomial function.

35. $f(x) = 2x^3 + x^2 - 1$
36. $f(x) = 3x^3 - 2x^2 + x + 4$
37. $f(x) = x^3 - 5x^2 - 11x + 11$
38. $f(x) = 2x^3 - x^2 - 11x - 6$
39. $f(x) = x^4 + 3x^3 - 5x^2 + 9$
40. $f(x) = 4x^4 - 12x^3 + 27x^2 - 54x + 81$

In Problems 41–60, find the rational zeros of f. List any irrational zeros correct to two decimal places.

41. $f(x) = x^3 + 2x^2 - 5x - 6$
42. $f(x) = x^3 + 8x^2 + 11x - 20$
43. $f(x) = 2x^3 - x^2 + 2x - 1$
44. $f(x) = 2x^3 + x^2 + 2x + 1$
45. $f(x) = x^4 + x^2 - 2$
46. $f(x) = x^4 - 3x^2 - 4$
47. $f(x) = 4x^4 + 7x^2 - 2$
48. $f(x) = 4x^4 + 15x^2 - 4$
49. $f(x) = x^4 + x^3 - 3x^2 - x + 2$
50. $f(x) = x^4 - x^3 - 6x^2 + 4x + 8$
51. $f(x) = 4x^5 - 8x^4 - x + 2$
52. $f(x) = 4x^5 + 12x^4 - x - 3$
53. $f(x) = x^3 + 3.2x^2 - 16.83x - 5.31$
54. $f(x) = x^3 + 3.2x^2 - 7.25x - 6.3$
55. $f(x) = x^4 - 1.4x^3 - 33.71x^2 + 23.94x + 292.41$
56. $f(x) = x^4 + 1.2x^3 - 7.46x^2 - 4.692x + 15.2881$
57. $f(x) = \pi x^3 - (8.88\pi + 1)x^2 - (42.066\pi - 8.88)x + 42.066$
58. $f(x) = \pi x^3 - (5.62\pi + 2)x^2 - (108.392\pi - 11.26)x + 216.784$
59. $f(x) = x^3 + 19.5x^2 - 1021x + 1000.5$
60. $f(x) = x^3 + 42.2x^2 - 664.8x + 1490.4$

In Problems 61–70, solve each equation in the real number system.

61. $x^4 - x^3 + 2x^2 - 4x - 8 = 0$
62. $2x^3 + 3x^2 + 2x + 3 = 0$
63. $3x^3 + 4x^2 - 7x + 2 = 0$
64. $2x^3 - 3x^2 - 3x - 5 = 0$
65. $3x^3 - x^2 - 15x + 5 = 0$
66. $2x^3 - 11x^2 + 10x + 8 = 0$
67. $x^4 + 4x^3 + 2x^2 - x + 6 = 0$
68. $x^4 - 2x^3 + 10x^2 - 18x + 9 = 0$
69. $x^3 - \frac{2}{3}x^2 + \frac{8}{3}x + 1 = 0$
70. $x^3 - \frac{2}{3}x^2 + 3x - 2 = 0$

71. Average Cost of Manufacturing In Problem 93 of Section 4.4, you found the average cost function for manufacturing Chevy Cavaliers. Use the methods learned in this section to determine how many Cavaliers can be manufactured if the average cost is \$10,500. In other words, solve the equation $\overline{C}(x) = 10.5$.

72. Average Cost of Printing In Problem 94 of Section 4.4, you found the average cost function for printing textbooks. Use the methods learned in this section to determine how many textbooks can be printed if the average cost is \$8.90. In other words, solve the equation $\overline{C}(x) = 8.9$.

73. Find k such that $f(x) = x^3 - kx^2 + kx + 2$ has the factor $x - 2$.

74. Find k such that $f(x) = x^4 - kx^3 + kx^2 + 1$ has the factor $x + 2$.

75. What is the remainder when $f(x) = 2x^{20} - 8x^{10} + x - 2$ is divided by $x - 1$?

76. What is the remainder when $f(x) = -3x^{17} + x^9 - x^5 + 2x$ is divided by $x + 1$?

77. Is $\frac{1}{3}$ a zero of $f(x) = 2x^3 + 3x^2 - 6x + 7$? Explain.

78. Is $\frac{1}{3}$ a zero of $f(x) = 4x^3 - 5x^2 - 3x + 1$? Explain.

79. Is $\frac{3}{5}$ a zero of $f(x) = 2x^6 - 5x^4 + x^3 - x + 1$? Explain.

80. Is $\frac{2}{3}$ a zero of $f(x) = x^7 + 6x^5 - x^4 + x + 2$? Explain.

81. What is the length of the edge of a cube if, after a slice 1 inch thick is cut from one side, the volume remaining is 294 cubic inches?

82. What is the length of the edge of a cube if its volume could be doubled by an increase of 6 centimeters in one edge, an increase of 12 centimeters in a second edge, and a decrease of 4 centimeters in the third edge?

4.6 COMPLEX NUMBERS; QUADRATIC EQUATIONS WITH A NEGATIVE DISCRIMINANT

1 Add, Subtract, Multiply, and Divide Complex Numbers

2 Solve Quadratic Equations with a Negative Discriminant

One property of a real number is that its square is nonnegative. For example, there is no real number x for which

$$x^2 = -1$$

To remedy this situation, we introduce a number called the **imaginary unit,** which we denote by $\boldsymbol{i}$ and whose square is -1. Thus,

$$i^2 = -1$$

This should not surprise you. If our universe were to consist only of integers, there would be no number x for which $2x = 1$. This unfortunate circumstance was remedied by introducing numbers such as $\frac{1}{2}$ and $\frac{2}{3}$, the *rational numbers.* If our universe were to consist only of rational numbers, there would be no number x whose square equals 2. That is, there would be no number x for which $x^2 = 2$. To remedy this, we introduced numbers such as $\sqrt{2}$ and $\sqrt[3]{5}$, the *irrational numbers.* The *real numbers,* you will recall, consist of the rational numbers and the irrational numbers. Now, if our universe were to consist only of real numbers, then there would be no number x whose square is -1. To remedy this, we introduce a number i, whose square is -1.

In the progression outlined, each time we encountered a situation that was unsuitable, we introduced a new number system to remedy this situation. And each new number system contained the earlier number system as a subset. The number system that results from introducing the number i is called the **complex number system.**

Complex numbers are numbers of the form $\boldsymbol{a + bi}$, where a and b are real numbers. The real number a is called the **real part** of the number $a + bi$; the real number b is called the **imaginary part** of $a + bi$.

For example, the complex number $-5 + 6i$ has the real part -5 and the imaginary part 6.

When a complex number is written in the form $a + bi$, where a and b are real numbers, we say it is in **standard form.** However, if the imaginary part of a complex number is negative, such as in the complex number $3 + (-2)i$, we agree to write it instead in the form $3 - 2i$.

Also, the complex number $a + 0i$ is usually written merely as a. This serves to remind us that the real numbers are a subset of the complex numbers. The complex number $0 + bi$ is usually written as bi. Sometimes the complex number bi is called a **pure imaginary number.**

1 Equality, addition, subtraction, and multiplication of complex numbers are defined so as to preserve the familiar rules of algebra for real numbers. Thus, two complex numbers are equal if and only if their real parts are equal and their imaginary parts are equal. That is,

Equality of Complex Numbers

$$a + bi = c + di \quad \text{if and only if} \quad a = c \text{ and } b = d \tag{1}$$

Two complex numbers are added by forming the complex number whose real part is the sum of the real parts and whose imaginary part is the sum of the imaginary parts. That is,

Sum of Complex Numbers

$$(a + bi) + (c + di) = (a + c) + (b + d)i \tag{2}$$

To subtract two complex numbers, we follow the rule

Difference of Complex Numbers

$$(a + bi) - (c + di) = (a - c) + (b - d)i \tag{3}$$

EXAMPLE 1 Adding and Subtracting Complex Numbers

(a) $(3 + 5i) + (-2 + 3i) = [3 + (-2)] + (5 + 3)i = 1 + 8i$

(b) $(6 + 4i) - (3 + 6i) = (6 - 3) + (4 - 6)i = 3 + (-2)i = 3 - 2i$

Now work Problem 5.

Products of complex numbers are calculated as illustrated in Example 2.

EXAMPLE 2 Multiplying Complex Numbers

$$(5 + 3i) \cdot (2 + 7i) \underset{\text{Distributive property}}{=} 5 \cdot (2 + 7i) + 3i(2 + 7i) \underset{\text{Distributive property}}{=} 10 + 35i + 6i + 21i^2$$

$$\underset{i^2 = -1}{=} 10 + 41i + 21(-1)$$

$$= -11 + 41i$$

Based on the procedure of Example 2, we define the **product** of two complex numbers by the formula

Product of Complex Numbers

$$(a + bi) \cdot (c + di) = (ac - bd) + (ad + bc)i \tag{4}$$

Do not bother to memorize formula (4). Instead, whenever it is necessary to multiply two complex numbers, follow the usual rules for multiplying two binomials, as in Example 2, remembering that $i^2 = -1$. For example,

$$(2i)(2i) = 4i^2 = -4$$
$$(2 + i)(1 - i) = 2 - 2i + i - i^2 = 3 - i$$

Now work Problem 11.

Algebraic properties for addition and multiplication, such as the commutative, associative, and distributive properties, hold for complex numbers. Of these, the property that every nonzero complex number has a multiplicative inverse, or reciprocal, requires a closer look.

Conjugates

If $z = a + bi$ is a complex number, then its **conjugate,** denoted by $\bar{z}$, is defined as

$$\bar{z} = \overline{a + bi} = a - bi$$

For example, $\overline{2 + 3i} = 2 - 3i$ and $\overline{-6 - 2i} = -6 + 2i$.

EXAMPLE 3 Multiplying a Complex Number by Its Conjugate

Find the product of the complex number $z = 3 + 4i$ and its conjugate $\bar{z}$.

Solution Since $\bar{z} = 3 - 4i$, we have

$$z\bar{z} = (3 + 4i)(3 - 4i) = 9 + 12i - 12i - 16i^2 = 9 + 16 = 25$$

The result obtained in Example 3 has an important generalization:

Theorem

The product of a complex number and its conjugate is a nonnegative real number. Thus, if $z = a + bi$, then

$$z\bar{z} = a^2 + b^2 \qquad (5)$$

Proof If $z = a + bi$, then

$$z\bar{z} = (a + bi)(a - bi) = a^2 - (bi)^2 = a^2 - b^2i^2 = a^2 + b^2$$

To express the reciprocal of a nonzero complex number z in standard form, multiply the numerator and denominator by its conjugate $\bar{z}$. Thus, if $z = a + bi$ is a nonzero complex number, then

$$\frac{1}{a+bi} = \frac{1}{z} = \frac{1}{z}\cdot\frac{\bar{z}}{\bar{z}} = \frac{\bar{z}}{z\bar{z}} = \frac{a-bi}{(a+bi)(a-bi)}$$

$$= \frac{a-bi}{a^2+b^2}$$

↑ Use (5).

$$= \frac{a}{a^2+b^2} - \frac{b}{a^2+b^2}i$$

EXAMPLE 4 Writing the Reciprocal of a Complex Number in Standard Form

Write $\dfrac{1}{3+4i}$ in standard form $a + bi$; that is, find the reciprocal of $3 + 4i$.

Solution The idea is to multiply the numerator and denominator by the conjugate of $3 + 4i$, that is, the complex number $3 - 4i$. The result is

$$\frac{1}{3+4i} = \frac{1}{3+4i}\cdot\frac{3-4i}{3-4i} = \frac{3-4i}{9+16} = \frac{3}{25} - \frac{4}{25}i$$

To express the quotient of two complex numbers in standard form, we multiply the numerator and denominator of the quotient by the conjugate of the denominator.

EXAMPLE 5 Writing the Quotient of Complex Numbers in Standard Form

Write each of the following in standard form:

(a) $\dfrac{1+4i}{5-12i}$ (b) $\dfrac{2-3i}{4-3i}$

Solution (a) $$\frac{1+4i}{5-12i} = \frac{1+4i}{5-12i}\cdot\frac{5+12i}{5+12i} = \frac{5+20i+12i+48i^2}{25+144}$$

$$= \frac{-43+32i}{169} = \frac{-43}{169} + \frac{32}{169}i$$

(b) $$\frac{2-3i}{4-3i} = \frac{2-3i}{4-3i}\cdot\frac{4+3i}{4+3i} = \frac{8-12i+6i-9i^2}{16+9} = \frac{17-6i}{25} = \frac{17}{25} - \frac{6}{25}i$$

Now work Problem 19.

EXAMPLE 6 Writing Other Expressions in Standard Form

If $z = 2 - 3i$ and $w = 5 + 2i$, write each of the following expressions in standard form:

(a) $\dfrac{z}{w}$ (b) $\overline{z+w}$ (c) $z + \bar{z}$

Solution (a) $\dfrac{z}{w} = \dfrac{z \cdot \overline{w}}{w \cdot \overline{w}} = \dfrac{(2 - 3i)(5 - 2i)}{(5 + 2i)(5 - 2i)} = \dfrac{10 - 15i - 4i + 6i^2}{25 + 4}$

$= \dfrac{4 - 19i}{29} = \dfrac{4}{29} - \dfrac{19}{29}i$

(b) $\overline{z + w} = \overline{(2 - 3i) + (5 + 2i)} = \overline{7 - i} = 7 + i$

(c) $z + \overline{z} = (2 - 3i) + (2 + 3i) = 4$

The conjugate of a complex number has certain general properties that we shall find useful later.

For a real number $a = a + 0i$, the conjugate is $\overline{a} = \overline{a + 0i} = a - 0i = a$. That is,

Theorem

The conjugate of a real number is the real number itself.

Other properties of the conjugate that are direct consequences of the definition are given next. In each statement, z and w represent complex numbers.

Theorem

The conjugate of the conjugate of a complex number is the complex number itself:

$$\overline{(\overline{z})} = z \tag{6}$$

The conjugate of the sum of two complex numbers equals the sum of their conjugates:

$$\overline{z + w} = \overline{z} + \overline{w} \tag{7}$$

The conjugate of the product of two complex numbers equals the product of their conjugates:

$$\overline{z \cdot w} = \overline{z} \cdot \overline{w} \tag{8}$$

We leave the proofs of equations (6), (7), and (8) as exercises.

Powers of *i*

The **powers of *i*** follow a pattern that is useful to know:

$i^1 = i$	$i^5 = i^4 \cdot i = 1 \cdot i = i$
$i^2 = -1$	$i^6 = i^4 \cdot i^2 = -1$
$i^3 = i^2 \cdot i = -i$	$i^7 = i^4 \cdot i^3 = -i$
$i^4 = i^2 \cdot i^2 = (-1)(-1) = 1$	$i^8 = i^4 \cdot i^4 = 1$

And so on. Thus, the powers of i repeat with every fourth power.

EXAMPLE 7 Evaluating Powers of i

(a) $i^{27} = i^{24} \cdot i^3 = (i^4)^6 \cdot i^3 = 1^6 \cdot i^3 = -i$
(b) $i^{101} = i^{100} \cdot i^1 = (i^4)^{25} \cdot i = 1^{25} \cdot i = i$

EXAMPLE 8 Writing the Power of a Complex Number in Standard Form

Write $(2 + i)^3$ in standard form.

Solution We use the special product formula for $(x + a)^3$:

$$(x + a)^3 = x^3 + 3ax^2 + 3a^2x + a^3$$

Thus,

$$\begin{aligned}(2 + i)^3 &= 2^3 + 3 \cdot i \cdot 2^2 + 3 \cdot i^2 \cdot 2 + i^3 \\ &= 8 + 12i + 6(-1) + (-i) \\ &= 2 + 11i\end{aligned}$$

Now work Problem 33.

Quadratic Equations with a Negative Discriminant

2 Quadratic equations with a negative discriminant have no real number solution. However, if we extend our number system to allow complex numbers, quadratic equations will always have a solution. Since the solution to a quadratic equation involves the square root of the discriminant, we begin with a discussion of square roots of negative numbers.

If N is a positive real number, we define the **principal square root of $-N$,** denoted by $\sqrt{-N}$, as

$$\sqrt{-N} = \sqrt{N}\, i$$

where i is the imaginary unit and $i^2 = -1$.

EXAMPLE 9 Evaluating the Square Root of a Negative Number

(a) $\sqrt{-1} = \sqrt{1}i = i$ (b) $\sqrt{-4} = \sqrt{4}i = 2i$
(c) $\sqrt{-8} = \sqrt{8}i = 2\sqrt{2}i$

EXAMPLE 10 Solving Equations

Solve each equation in the complex number system.

(a) $x^2 = 4$ (b) $x^2 = -9$

Solution (a) $x^2 = 4$

$$x = \pm\sqrt{4} = \pm 2$$

The equation has two solutions, -2 and 2.

(b) $x^2 = -9$

$$x = \pm\sqrt{-9} = \pm\sqrt{9}i = \pm 3i$$

The equation has two solutions, $-3i$ and $3i$.

Now work Problem 45.

Warning: When working with square roots of negative numbers, do not set the square root of a product equal to the product of the square roots (which can be done with positive numbers). To see why, look at this calculation: We know that $\sqrt{100} = 10$. However, it is also true that $100 = (-25)(-4)$, so

$$10 = \sqrt{100} = \sqrt{(-25)(-4)} \underset{\uparrow}{=} \sqrt{-25}\sqrt{-4} = (\sqrt{25}i)(\sqrt{4}i) = (5i)(2i) = 10i^2 = -10$$

Here is the error.

Because we have defined the square root of a negative number, we now can restate the quadratic formula without restriction.

Theorem

In the complex number system, the solutions of the quadratic equation $ax^2 + bx + c = 0$, where a, b, and c are real numbers and $a \neq 0$, are given by the formula

$$x = \frac{-b \pm \sqrt{b^2 - 4ac}}{2a} \qquad (9)$$

EXAMPLE 11 Solving Quadratic Equations in the Complex Number System

Solve the equation $x^2 - 4x + 8 = 0$ in the complex number system.

Solution Here $a = 1$, $b = -4$, $c = 8$, and $b^2 - 4ac = 16 - 4(8) = -16$. Using equation (9), we find

$$x = \frac{4 \pm \sqrt{-16}}{2} = \frac{4 \pm \sqrt{16}i}{2} = \frac{4 \pm 4i}{2} = 2 \pm 2i$$

The equation has the solution set $\{2 - 2i, 2 + 2i\}$.

Check:

$$2 + 2i: \quad (2 + 2i)^2 - 4(2 + 2i) + 8 = 4 + 8i + 4i^2 - 8 - 8i + 8$$
$$= 4 - 4 = 0$$
$$2 - 2i: \quad (2 - 2i)^2 - 4(2 - 2i) + 8 = 4 - 8i + 4i^2 - 8 + 8i + 8$$
$$= 4 - 4 = 0$$

Now work Problem 51.

The discriminant, $b^2 - 4ac$, of a quadratic equation still serves as a way to determine the character of the solutions.

Discriminant of a Quadratic Equation

In the complex number system, consider a quadratic equation $ax^2 + bx + c = 0$ with real coefficients.

1. If $b^2 - 4ac > 0$, the equation has two unequal real solutions.
2. If $b^2 - 4ac = 0$, the equation has a repeated real solution—a double root.
3. If $b^2 - 4ac < 0$, the equation has two complex solutions that are not real. The solutions are conjugates of each other.

The third conclusion in the display is a consequence of the fact that if $b^2 - 4ac = -N < 0$ then, by the quadratic formula, the solutions are

$$x = \frac{-b + \sqrt{b^2 - 4ac}}{2a} = \frac{-b + \sqrt{-N}}{2a} = \frac{-b + \sqrt{N}i}{2a} = \frac{-b}{2a} + \frac{\sqrt{N}}{2a}i$$

and

$$x = \frac{-b - \sqrt{b^2 - 4ac}}{2a} = \frac{-b - \sqrt{-N}}{2a} = \frac{-b - \sqrt{N}i}{2a} = \frac{-b}{2a} - \frac{\sqrt{N}}{2a}i$$

which are conjugates of each other.

EXAMPLE 12 Determining the Character of the Solution of a Quadratic Equation

Without solving, determine the character of the solution of each equation in the complex number system.

(a) $3x^2 + 4x + 5 = 0$
(b) $2x^2 + 4x + 1 = 0$
(c) $9x^2 - 6x + 1 = 0$

Solution

(a) Here, $a = 3, b = 4$, and $c = 5$, so $b^2 - 4ac = 16 - 4(3)(5) = -44$. The solutions are complex numbers that are not real and are conjugates of each other.

(b) Here, $a = 2$, $b = 4$, and $c = 1$, so $b^2 - 4ac = 16 - 8 = 8$. The solutions are two unequal real numbers.

(c) Here, $a = 9, b = -6$, and $c = 1$, so $b^2 - 4ac = 36 - 4(9)(1) = 0$. The solution is a repeated real number, that is, a double root.

4.6 EXERCISES

In Problems 1–38, write each expression in the standard form $a + bi$.

1. $(2 - 3i) + (6 + 8i)$ **2.** $(4 + 5i) + (-8 + 2i)$ **3.** $(-3 + 2i) - (4 - 4i)$ **4.** $(3 - 4i) - (-3 - 4i)$

5. $(2 - 5i) - (8 + 6i)$ **6.** $(-8 + 4i) - (2 - 2i)$ **7.** $3(2 - 6i)$ **8.** $-4(2 + 8i)$

9. $2i(2 - 3i)$ **10.** $3i(-3 + 4i)$ **11.** $(3 - 4i)(2 + i)$ **12.** $(5 + 3i)(2 - i)$

13. $(-6 + i)(-6 - i)$ **14.** $(-3 + i)(3 + i)$ **15.** $\dfrac{10}{3 - 4i}$ **16.** $\dfrac{13}{5 - 12i}$

17. $\dfrac{2 + i}{i}$ **18.** $\dfrac{2 - i}{-2i}$ **19.** $\dfrac{6 - i}{1 + i}$ **20.** $\dfrac{2 + 3i}{1 - i}$

21. $\left(\dfrac{1}{2} + \dfrac{\sqrt{3}}{2}i\right)^2$ **22.** $\left(\dfrac{\sqrt{3}}{2} - \dfrac{1}{2}i\right)^2$ **23.** $(1 + i)^2$ **24.** $(1 - i)^2$

25. i^{23} **26.** i^{14} **27.** i^{-15} **28.** i^{-23}

29. $i^6 - 5$ **30.** $4 + i^3$ **31.** $6i^3 - 4i^5$ **32.** $4i^3 - 2i^2 + 1$

33. $(1 + i)^3$ **34.** $(3i)^4 + 1$ **35.** $i^7(1 + i^2)$ **36.** $2i^4(1 + i^2)$

37. $i^6 + i^4 + i^2 + 1$ **38.** $i^7 + i^5 + i^3 + i$

In Problems 39–44, perform the indicated operations and express your answer in the form $a + bi$.

39. $\sqrt{-4}$ **40.** $\sqrt{-9}$ **41.** $\sqrt{-25}$

42. $\sqrt{-64}$ **43.** $\sqrt{(3 + 4i)(4i - 3)}$ **44.** $\sqrt{(4 + 3i)(3i - 4)}$

In Problems 45–64, solve each equation in the complex number system.

45. $x^2 + 4 = 0$ **46.** $x^2 - 4 = 0$ **47.** $x^2 - 16 = 0$ **48.** $x^2 + 25 = 0$

49. $x^2 - 6x + 13 = 0$ **50.** $x^2 + 4x + 8 = 0$ **51.** $x^2 - 6x + 10 = 0$ **52.** $x^2 - 2x + 5 = 0$

53. $8x^2 - 4x + 1 = 0$ **54.** $10x^2 + 6x + 1 = 0$ **55.** $5x^2 + 2x + 1 = 0$ **56.** $13x^2 + 6x + 1 = 0$

57. $x^2 + x + 1 = 0$ **58.** $x^2 - x + 1 = 0$ **59.** $x^3 - 8 = 0$ **60.** $x^3 + 27 = 0$

61. $x^4 - 16 = 0$ **62.** $x^4 - 1 = 0$ **63.** $x^4 + 13x^2 + 36 = 0$ **64.** $x^4 + 3x^2 - 4 = 0$

In Problems 65–70, without solving, determine the character of the solutions of each equation in the complex number system. Verify your answer using a graphing utility.

65. $3x^2 - 3x + 4 = 0$ **66.** $2x^2 - 4x + 1 = 0$ **67.** $2x^2 + 3x - 4 = 0$

68. $x^2 + 2x + 6 = 0$ **69.** $9x^2 - 12x + 4 = 0$ **70.** $4x^2 + 12x + 9 = 0$

71. $2 + 3i$ is a solution of a quadratic equation with real coefficients. Find the other solution.

72. $4 - i$ is a solution of a quadratic equation with real coefficients. Find the other solution.

In Problems 73–76, $z = 3 - 4i$ and $w = 8 + 3i$. Write each expression in the standard form $a + bi$.

73. $z + \bar{z}$ **74.** $w - \bar{w}$ **75.** $z\bar{z}$ **76.** $\overline{z - w}$

77. Use $z = a + bi$ to show that $z + \bar{z} = 2a$ and that $z - \bar{z} = 2bi$.

78. Use $z = a + bi$ to show that $\overline{(\bar{z})} = z$.

79. Use $z = a + bi$ and $w = c + di$ to show that $\overline{z + w} = \bar{z} + \bar{w}$.

80. Use $z = a + bi$ and $w = c + di$ to show that $\overline{z \cdot w} = \bar{z} \cdot \bar{w}$.

81. Explain to a friend how you would add two complex numbers and how you would multiply two complex numbers. Explain any differences in the two explanations.

82. Write a brief paragraph that compares the method used to rationalize the denominator of a rational expression and the method used to write a complex number in standard form.

4.7 COMPLEX ZEROS; FUNDAMENTAL THEOREM OF ALGEBRA

1 Utilize the Conjugate Pairs Theorem to find the Complex Zeros of a Polynomial
2 Find a Polynomial Function with Specified Zeros
3 Find the Complex Zeros of a Polynomial

In Section 4.5 we found the **real** zeros of a polynomial function. In this section, we will find the **complex** zeros of a polynomial function. Since the set of real numbers is a subset of the set of complex numbers, finding the complex zeros of a function requires finding all zeros of the form $a + bi$. These zeros will be real if $b = 0$.

A variable in the complex number system is referred to as a **complex variable.** A **complex polynomial function** f of degree n is a complex function of the form

$$f(x) = a_n x^n + a_{n-1} x^{n-1} + \cdots + a_1 x + a_0 \qquad (1)$$

where $a_n, a_{n-1}, \ldots, a_1, a_0$ are complex numbers, $a_n \neq 0$, n is a nonnegative integer, and x is a complex variable. Here, a_n is called the **leading coefficient** of f. A complex number r is called a (complex) **zero** of a complex function f if $f(r) = 0$.

We have learned that some quadratic equations have no real solutions, but that in the complex number system every quadratic equation has a solution, either real or complex. The next result, proved by Karl Friedrich Gauss (1777–1855) when he was 22 years of age,* gives an extension to complex polynomials. In fact, this result is so important and useful it has become known as the **Fundamental Theorem of Algebra.**

Fundamental Theorem of Algebra

Every complex polynomial function $f(x)$ of degree $n \geq 1$ has at least one complex zero.

We shall not prove this result, as the proof is beyond the scope of this book. However, using the Fundamental Theorem of Algebra and the Factor Theorem, we can prove the following result:

Theorem

Every complex polynomial function $f(x)$ of degree $n \geq 1$ can be factored into n linear factors (not necessarily distinct) of the form

$$f(x) = a_n(x - r_1)(x - r_2) \cdot \cdots \cdot (x - r_n) \qquad (2)$$

where $a_n, r_1, r_2, \ldots, r_n$ are complex numbers.

*In all, Gauss gave four different proofs of this theorem, the first one in 1799 being the subject of his doctoral dissertation.

Proof Let

$$f(x) = a_n x^n + a_{n-1}x^{n-1} + \cdots + a_1 x + a_0$$

By the Fundamental Theorem of Algebra, f has at least one zero, say, r_1. Then, by the Factor Theorem, $x - r_1$ is a factor, and

$$f(x) = (x - r_1)q_1(x)$$

where $q_1(x)$ is a complex polynomial of degree $n - 1$ whose leading coefficient is a_n. Again, by the Fundamental Theorem of Algebra, the complex polynomial $q_1(x)$ has at least one zero, say, r_2. By the Factor Theorem, $q_1(x)$ has the factor $x - r_2$, so

$$q_1(x) = (x - r_2)q_2(x)$$

where $q_2(x)$ is a complex polynomial of degree $n - 2$ whose leading coefficient is a_n. Consequently,

$$f(x) = (x - r_1)(x - r_2)q_2(x)$$

Repeating this argument n times, we finally arrive at

$$f(x) = (x - r_1)(x - r_2)\cdot \cdots \cdot (x - r_n)q_n(x)$$

where $q_n(x)$ is a complex polynomial of degree $n - n = 0$ whose leading coefficient is a_n. Thus, $q_n(x) = a_n x^0 = a_n$, and so

$$f(x) = a_n(x - r_1)(x - r_2)\cdot \cdots \cdot (x - r_n)$$

Complex Polynomials with Real Coefficients

1 We can use the Fundamental Theorem of Algebra to obtain valuable information about the zeros of complex polynomials whose coefficients are real numbers.

Conjugate Pairs Theorem

Let $f(x)$ be a complex polynomial whose coefficients are real numbers. If $r = a + bi$ is a zero of f, then the complex conjugate $\bar{r} = a - bi$ is also a zero of f.

In other words, for complex polynomials whose coefficients are real numbers, the zeros occur in conjugate pairs.

Proof Let

$$f(x) = a_n x^n + a_{n-1}x^{n-1} + \cdots + a_1 x + a_0$$

where $a_n, a_{n-1}, \ldots, a_1, a_0$ are real numbers and $a_n \neq 0$. If r is a zero of f, then $f(r) = 0$, so

$$a_n r^n + a_{n-1}r^{n-1} + \cdots + a_1 r + a_0 = 0$$

We take the conjugate of both sides to get

$$\overline{a_n r^n + a_{n-1} r^{n-1} + \cdots + a_1 r + a_0} = \overline{0}$$

$$\overline{a_n r^n} + \overline{a_{n-1} r^{n-1}} + \cdots + \overline{a_1 r} + \overline{a_0} = \overline{0}$$

The conjugate of a sum equals the sum of the conjugates (see Section 4.6).

$$\overline{a_n}(\overline{r})^n + \overline{a_{n-1}}(\overline{r})^{n-1} + \cdots + \overline{a_1}\,\overline{r} + \overline{a_0} = \overline{0}$$

The conjugate of a product equals the product of the conjugates.

$$a_n(\overline{r})^n + a_{n-1}(\overline{r})^{n-1} + \cdots + a_1\overline{r} + a_0 = 0$$

The conjugate of a real number equals the real number.

This last equation states that $f(\overline{r}) = 0$; that is, $\overline{r}$ is a zero of f. ■

The value of this result should be clear. Once we know that, say, $3 + 4i$ is a zero of a polynomial with real coefficients, then we know that $3 - 4i$ is also a zero. This result has an important corollary.

Corollary

A complex polynomial f of odd degree with real coefficients has at least one real zero.

■

Proof Because complex zeros occur as conjugate pairs in a complex polynomial with real coefficients, there will always be an even number of zeros that are not real numbers. Consequently, since f is of odd degree, one of its zeros has to be a real number. ■

For example, the polynomial $f(x) = x^5 - 3x^4 + 4x^3 - 5$ has at least one zero that is a real number, since f is of degree 5 (odd) and has real coefficients.

EXAMPLE 1 Using the Conjugate Pairs Theorem

A polynomial f of degree 5 whose coefficients are real numbers has the zeros 1, $5i$, and $1 + i$. Find the remaining two zeros.

Solution Since complex zeros appear as conjugate pairs, it follows that $-5i$, the conjugate of $5i$, and $1 - i$, the conjugate of $1 + i$, are the two remaining zeros. ■

Now work Problem 1.

EXAMPLE 2 Finding a Polynomial Function Whose Zeros Are Given

(a) Find a polynomial f of degree 4 whose coefficients are real numbers and has the zeros 1, 1, and $-4 + i$.

(b) Graph the polynomial found in (a) to verify your result.

Solution (a) Since $-4 + i$ is a zero, by the Conjugate Pairs Theorem, $-4 - i$ must also be a zero of f. Because of the Factor Theorem, if $f(c) = 0$, then $x - c$ is a factor of $f(x)$. So, we can now write f as

$$f(x) = a(x - 1)(x - 1)[x - (-4 + i)][x - (-4 - i)]$$

where a is any real number. If we let $a = 1$, we obtain

$$\begin{aligned} f(x) &= (x - 1)(x - 1)[x - (-4 + i)][x - (-4 - i)] \\ &= (x^2 - 2x + 1)[x^2 - (-4 + i)x - (-4 - i)x + (-4 + i)(-4 - i)] \\ &= (x^2 - 2x + 1)(x^2 + 4x - ix + 4x + ix + 16 + 4i - 4i - i^2) \\ &= (x^2 - 2x + 1)(x^2 + 8x + 17) \\ &= (x^4 + 8x^3 + 17x^2 - 2x^3 - 16x^2 - 34x + x^2 + 8x + 17) \\ &= x^4 + 6x^3 + 2x^2 - 26x + 17 \end{aligned}$$

FIGURE 64

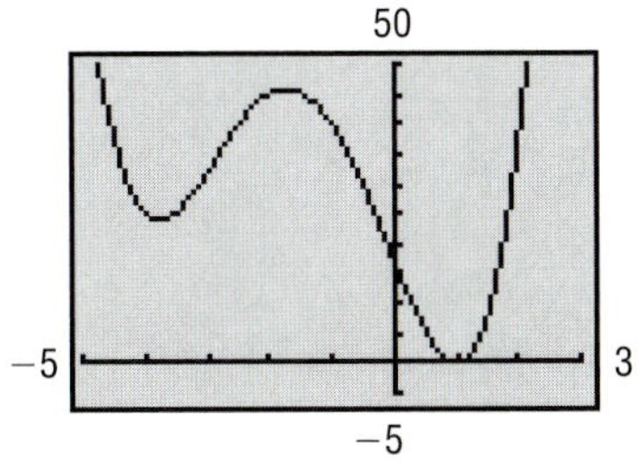

(b) A quick analysis of the polynomial f tells us what to expect:

At most 3 turning points.
For large x, the graph will behave like $y = x^4$.
A repeated real zero at 1 so that the graph will touch the x-axis at 1.
The only x-intercept is at 1.

Figure 64 shows the complete graph. (Do you see why? The graph has exactly 3 turning points.)

Exploration Graph the function found in Example 2 for $a = 2$ and $a = -1$. Does the value of a affect the zeros of f? How does the value of a affect the graph of f?

Now we can prove the theorem we conjectured earlier in Section 4.5.

Theorem

Every polynomial function with real coefficients can be uniquely factored over the real numbers into a product of linear factors and/or irreducible quadratic factors.

Proof Every complex polynomial f of degree n has exactly n zeros and can be factored into a product of n linear factors. If its coefficients are real, then those zeros that are complex numbers will always occur as conjugate pairs. As a result, if $r = a + bi$ is a complex zero, then so is $\bar{r} = a - bi$. Consequently, when the linear factors $x - r$ and $x - \bar{r}$ of f are multiplied, we have

$$(x - r)(x - \bar{r}) = x^2 - (r + \bar{r})x + r\bar{r} = x^2 - 2ax + a^2 + b^2$$

This second-degree polynomial has real coefficients and is irreducible (over the real numbers). Thus, the factors of f are either linear or irreducible quadratic factors.

EXAMPLE 3 Finding the Complex Zeros of a Polynomial

3

(a) It is known that $2 + i$ is a zero of $f(x) = x^4 - 8x^3 + 64x - 105$. Find the remaining zeros.

(b) Graph f to verify your results.

Solution (a) Since $2 + i$ is a zero of f and f has real coefficients, then $2 - i$ must also be a zero of f. (Complex zeros appear as conjugate pairs in a polynomial with real coefficients.) So, $x - (2 + i)$ and $x - (2 - i)$ are factors of f. Therefore, their product,

$$[x - (2 + i)][x - (2 - i)]$$
$$= x^2 - (2 - i)x - x(2 + i) + (2 + i)(2 - i) = x^2 - 4x + 5$$

is also a factor of f. We use long division to obtain the other factor:

$$\begin{array}{r} x^2 - 4x - 21 \\ x^2 - 4x + 5 \overline{\smash{)}\, x^4 - 8x^3 \qquad\quad + 64x - 105} \\ \underline{x^4 - 4x^3 + 5x^2} \qquad\qquad\quad \\ -4x^3 - 5x^2 + 64x \qquad\; \\ \underline{-4x^3 + 16x^2 - 20x} \qquad\; \\ -21x^2 + 84x - 105 \\ \underline{-21x^2 + 84x - 105} \\ 0 \end{array}$$

So, $f(x) = (x^2 - 4x + 5)(x^2 - 4x - 21) = (x^2 - 4x + 5)(x - 7)(x + 3)$. By the Factor Theorem, the zeros of f are $-3, 7, 2 + i$, and $2 - i$.

(b) A quick analysis of the polynomial f tells us what to expect:

At most 3 turning points.

Two x-intercepts: at -3 and 7.

For large x, the graph will behave like $y = x^4$.

FIGURE 65

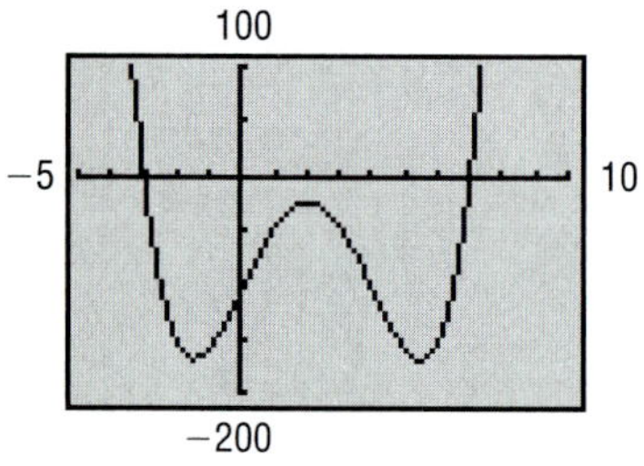

Figure 65 shows the complete graph. (Do you see why? The graph has exactly 3 turning points.)

Now work Problem 17.

The steps on page 314 in Section 4.5 used to find the real zeros of a polynomial function can also be used to find the complex zeros of a polynomial.

EXAMPLE 4

Finding the Complex Zeros of a Polynomial

Find the complex zeros of the polynomial function

$$f(x) = 3x^4 + 5x^3 + 25x^2 + 45x - 18.$$

STEP 1: The degree of f is 4. So, f will have 4 complex zeros.

STEP 2: Descartes' Rule of Signs provides information about the real zeros. For this polynomial, there is one positive real zero. Because

$$f(-x) = 3x^4 - 5x^3 + 25x^2 - 45x - 18$$

there are three, or, one negative real zero(s).

STEP 3: The Rational Zeros Theorem provides information about the potential rational zeros of polynomials with integer coefficients. For this polynomial (which has integer coefficients), the potential rational zeros are

$$\pm\frac{1}{3}, \pm\frac{2}{3}, \pm 1, \pm 2, \pm 3, \pm 6, \pm 9, \pm 18$$

FIGURE 66

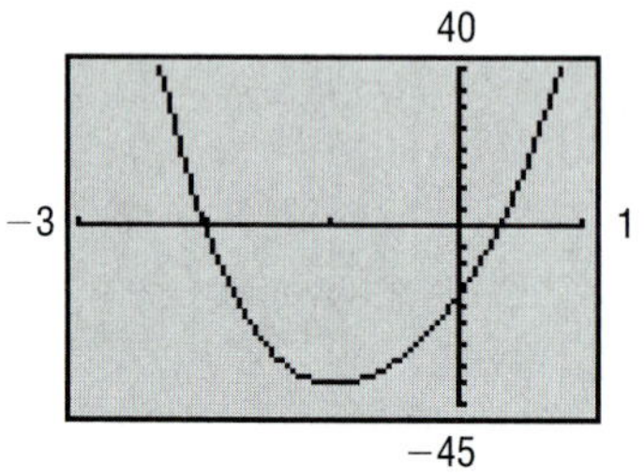

STEP 4: Figure 66 shows the graph of f. The graph has the characteristics we expect of this polynomial of degree 4: It behaves like $y = 3x^4$

for large x and has y-intercept -18. There are x-intercepts near -2 and between 0 and 1. Since $f(-2) = 0$, we know -2 is a zero of f. We use synthetic division to factor f.

$$\begin{array}{r|rrrrr} -2 & 3 & 5 & 25 & 45 & -18 \\ & & -6 & 2 & -54 & 18 \\ \hline & 3 & -1 & 27 & -9 & 0 \end{array}$$

So, $f(x) = (x + 2)(3x^3 - x^2 + 27x - 9)$. From the graph of f and the list of potential rational zeros, it appears 1/3 may also be a zero of f. We use synthetic division on the depressed equation of f to confirm this.

$$\begin{array}{r|rrrr} 1/3 & 3 & -1 & 27 & -9 \\ & & 1 & 0 & 9 \\ \hline & 3 & 0 & 27 & 0 \end{array}$$

Thus, 1/3 is a zero of f and $x - 1/3$ is therefore a factor. Using the bottom row of the synthetic division, we find

$$f(x) = (x + 2)\left(x - \frac{1}{3}\right)(3x^2 + 27) = 3(x + 2)\left(x - \frac{1}{3}\right)(x^2 + 9).$$

The factor $x^2 + 9$ does not have any real zeros; its complex zeros are $\pm 3i$. Thus, the complex zeros of $f(x) = 3x^4 + 5x^3 + 25x^2 + 45x - 18$ are $-2, 1/3, 3i, -3i$. ■

4.7 EXERCISES

In Problems 1–10, information is given about a complex polynomial $f(x)$ whose coefficients are real numbers. Find the remaining zeros of f.

1. Degree 3; zeros: $3, 4 - i$

2. Degree 3; zeros: $4, 3 + i$

3. Degree 4; zeros: $i, 1 + i$

4. Degree 4; zeros: $1, 2, 2 + i$

5. Degree 5; zeros: $1, i, 2i$

6. Degree 5; zeros: $0, 1, 2, i$

7. Degree 4; zeros: $i, 2, -2$

8. Degree 4; zeros: $2 - i, -i$

9. Degree 6; zeros: $2, 2 + i, -3 - i, 0$

10. Degree 6; zeros: $i, 3 - 2i, -2 + i$

In Problems 11–16, form a polynomial $f(x)$ with real coefficients having the given degree and zeros.

11. Degree 4; zeros: $3 + 2i$; 4, multiplicity 2

12. Degree 4; zeros: $i, 1 + 2i$

13. Degree 5; zeros: 2, multiplicity 1; $1 - i$; $1 + i$

14. Degree 6; zeros: $i, 4 - i$; $2 + i$

15. Degree 4; zeros: 3, multiplicity 2; $-i$

16. Degree 5: zeros: 1, multiplicity 3; $1 + i$

In Problems 17–24, use the given zero to find the remaining zeros of each function. Graph the function to verify your results.

17. $f(x) = x^3 - 4x^2 + 4x - 16$; zero: $2i$

18. $g(x) = x^3 + 3x^2 + 25x + 75$; zero: $-5i$

19. $f(x) = 2x^4 + 5x^3 + 5x^2 + 20x - 12$; zero: $-2i$

20. $h(x) = 3x^4 + 5x^3 + 25x^2 + 45x - 18$; zero: $3i$

21. $h(x) = x^4 - 9x^3 + 21x^2 + 21x - 130$; zero: $3 - 2i$

22. $f(x) = x^4 - 7x^3 + 14x^2 - 38x - 60$; zero: $1 + 3i$

23. $h(x) = 3x^5 + 2x^4 + 15x^3 + 10x^2 - 528x - 352$; zero: $-4i$

24. $g(x) = 2x^5 - 3x^4 - 5x^3 - 15x^2 - 207x + 108$; zero: $3i$

In Problems 25–34, find the complex zeros of each polynomial function.

25. $f(x) = x^3 - 1$

26. $f(x) = x^4 - 1$

27. $f(x) = x^3 - 8x^2 + 25x - 26$

28. $f(x) = x^3 + 13x^2 + 57x + 85$

29. $f(x) = x^4 + 5x^2 + 4$

30. $f(x) = x^4 + 13x^2 + 36$

31. $f(x) = x^4 + 2x^3 + 22x^2 + 50x - 75$

32. $f(x) = x^4 + 3x^3 - 19x^2 + 27x - 252$

33. $f(x) = 3x^4 - x^3 - 9x^2 + 159x - 52$

34. $f(x) = 2x^4 + x^3 - 35x^2 - 113x + 65$

In Problems 35 and 36, tell why the facts given are contradictory.

35. $f(x)$ is a complex polynomial of degree 3 whose coefficients are real numbers; its zeros are $4 + i$, $4 - i$, and $2 + i$.

36. $f(x)$ is a complex polynomial of degree 3 whose coefficients are real numbers; its zeros are 2, i, and $3 + i$.

37. $f(x)$ is a complex polynomial of degree 4 whose coefficients are real numbers; three of its zeros are 2, $1 + 2i$, and $1 - 2i$. Explain why the remaining zero must be a real number.

38. $f(x)$ is a complex polynomial of degree 4 whose coefficients are real numbers; two of its zeros are -3 and $4 - i$. Explain why one of the remaining zeros must be a real number. Write down one of the missing zeros.

4.8 POLYNOMIAL AND RATIONAL INEQUALITIES

1 Solve Polynomial Inequalities Graphically and Algebraically

2 Solve Rational Inequalities Graphically and Algebraically

1 In this section, we consider inequalities that involve polynomials of degree 2 and higher, as well as some that involve rational expressions. We will solve these inequalities using both a graphing approach and an algebraic approach.

To solve polynomial and rational inequalities algebraically, we follow these steps:

Steps for Solving Polynomial and Rational Inequalities Algebraically

STEP 1: Write the inequality so that a polynomial or rational expression f is on the left side and zero is on the right side in one of the forms shown below:

$$f(x) > 0 \qquad f(x) \geq 0 \qquad f(x) < 0 \qquad f(x) \leq 0$$

For rational expressions, be sure the left side is written as a single quotient.

STEP 2: Determine the numbers at which the expression f on the left side equals zero and the numbers at which the expression f on the left side is undefined. These are called **critical numbers** of the expression f.

STEP 3: Use the critical numbers to separate the real number line into intervals.

STEP 4: Select a **test number** in each interval and evaluate f at the test number.

(a) If the value of f is positive, then $f(x) > 0$ for all numbers x in the interval.

(b) If the value of f is negative, then $f(x) < 0$ for all numbers x in the interval.

If the inequality is not strict, include the solutions of $f(x) = 0$ in the solution set.

EXAMPLE 1 Solving Quadratic Inequalities

Solve the inequality $x^2 \le 4x + 12$, and graph the solution set.

Graphing Solution

We graph $Y_1 = x^2$ and $Y_2 = 4x + 12$ on the same screen. See Figure 67. Using the INTERSECT command, we find that Y_1 and Y_2 intersect at $x = -2$ and at $x = 6$. The graph of Y_1 is below that of Y_2, $Y_1 < Y_2$, between the points of intersection. Since the inequality is not strict, the solution set is $\{x | -2 \le x \le 6\}$ or, using interval notation, $[-2, 6]$.

FIGURE 67

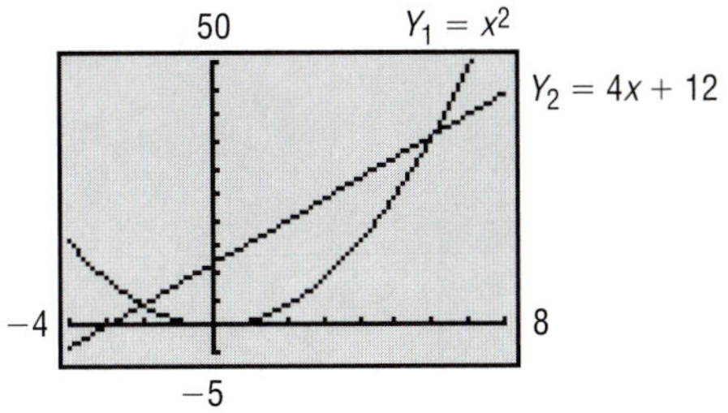

Algebraic Solution

STEP 1: Rearrange the inequality so that 0 is on the right side.

$$x^2 \le 4x + 12$$
$$x^2 - 4x - 12 \le 0 \quad \text{Subtract } 4x + 12 \text{ from both sides of the inequality.}$$

STEP 2: Find the critical numbers of $f(x) = x^2 - 4x - 12$ by solving the equation $x^2 - 4x - 12 = 0$.

$$x^2 - 4x - 12 = 0$$
$$(x + 2)(x - 6) = 0 \quad \text{Factor.}$$
$$x = -2 \quad \text{or} \quad x = 6$$

STEP 3: We use the critical numbers to separate the real number line into three intervals:

$$-\infty < x < -2 \qquad -2 < x < 6 \qquad 6 < x < \infty$$

STEP 4: We select a test number in each interval found in step 3 and evaluate $f(x) = x^2 - 4x - 12$ at each test number to determine if $f(x)$ is positive or negative. See Table 16.

TABLE 16

Interval	Test Number	$f(x)$	Positive/Negative
$-\infty < x < -2$	-3	$f(-3) = (-3)^2 - 4(-3) - 12 = 9$	Positive
$-2 < x < 6$	0	$f(0) = -12$	Negative
$6 < x < \infty$	7	$f(7) = 9$	Positive

Since we want to know where $f(x)$ is negative, we conclude that the solutions are all x such that $-2 < x < 6$. However, because the original inequality is not strict, numbers x that satisfy the equation $x^2 = 4x + 12$ are also solutions of the inequality $x^2 \le 4x + 12$. Thus, we include -2 and 6. The solution set of the given inequality is $\{x | -2 \le x \le 6\}$ or, using interval notation, $[2, 6]$.

Figure 68 shows the graph of the solution set. ■

FIGURE 68
$-2 \le x \le 6$

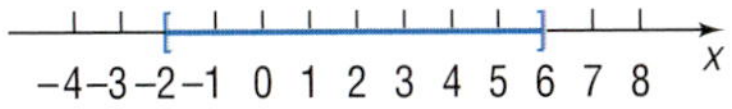

Now work Problems 3 and 7.

EXAMPLE 2 Solving a Polynomial Inequality

Solve the inequality $x^4 > x$, and graph the solution set.

Graphing Solution We graph $Y_1 = x^4$ and $Y_2 = x$ on the same screen. See Figure 69. Using the INTERSECT command, we find that Y_1 and Y_2 intersect at $x = 0$ and at $x = 1$. The graph of Y_1 is above that of Y_2, $Y_1 > Y_2$, to the left of $x = 0$ and to the right of $x = 1$. Since the inequality is strict, the solution set is $\{x \mid x < 0$ or $x > 1\}$ or, using interval notation, $(-\infty, 0)$ or $(1, \infty)$.

FIGURE 69

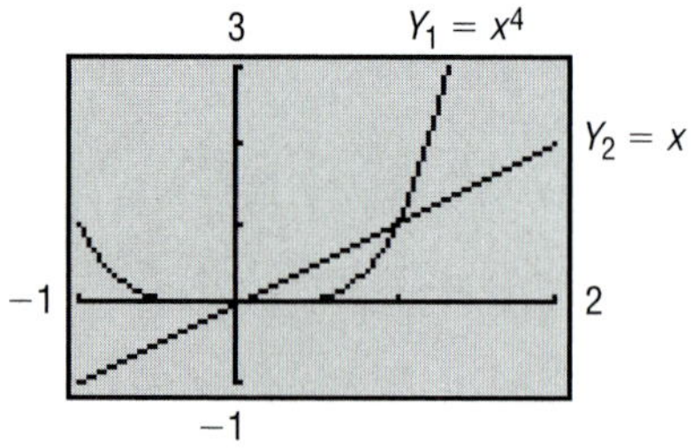

Algebraic Solution STEP 1: Rearrange the inequality so that 0 is on the right side.

$$x^4 > x$$
$$x^4 - x > 0 \quad \text{Subtract } x \text{ from both sides of the inequality.}$$

STEP 2: Find the critical numbers of $f(x) = x^4 - x$ by solving $x^4 - x = 0$.

$$x^4 - x = 0$$
$$x(x^3 - 1) = 0 \quad \text{Factor out } x.$$
$$x(x - 1)(x^2 + x + 1) = 0 \quad \text{Factor difference of two cubes.}$$
$$x = 0 \quad \text{or} \quad x - 1 = 0 \quad \text{or} \quad x^2 + x + 1 = 0 \quad \text{Set each factor equal to zero and solve.}$$
$$x = 0 \quad \text{or} \quad x = 1$$

The equation $x^2 + x + 1 = 0$ has no real solutions. (Do you see why?)

STEP 3: We use the critical numbers to separate the real number line into three intervals

$$-\infty < x < 0 \qquad 0 < x < 1 \qquad 1 < x < \infty$$

STEP 4: We select a test number in each interval found in step 3 and evaluate $f(x) = x^4 - x$ at each test number to determine if $f(x)$ is positive or negative. See Table 17.

TABLE 17

Interval	Test Number	$f(x)$	Positive/Negative
$-\infty < x < 0$	-1	$f(-1) = (-1)^4 - (-1) = 2$	Positive
$0 < x < 1$	$1/2$	$f(1/2) = -7/16$	Negative
$1 < x < \infty$	2	$f(2) = 14$	Positive

Since we want to know where $f(x)$ is positive, we conclude that the solutions are numbers x for which $-\infty < x < 0$ or $1 < x < \infty$. Because the original inequality is strict, numbers x that satisfy the equation $x^4 = x$ are not solutions. Thus, the solution set of the given inequality is $\{x \mid -\infty < x < 0$ or $1 < x < \infty\}$ or, using interval notation, $(-\infty, 0)$ or $(1, \infty)$.

Figure 70 shows the graph of the solution set.

FIGURE 70 $x < 0$ or $x > 1$

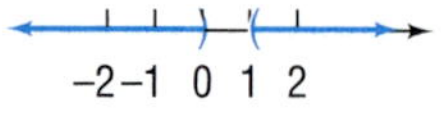

Now work Problem 17.

2 Let's solve a rational inequality.

EXAMPLE 3 Solving a Rational Inequality

Solve the inequality $\frac{4x + 5}{x + 2} \geq 3$, and graph the solution set.

Graphing Solution We first note that the domain of the variable consists of all real numbers except -2. We graph $Y_1 = (4x + 5)/(x + 2)$ and $Y_2 = 3$ on the same screen. See Figure 71. Using the INTERSECT command, we find that Y_1 and Y_2 intersect at $x = 1$. The graph of Y_1 is above that of Y_2, $Y_1 > Y_2$, to the left of $x = -2$ and to the right of $x = 1$. Since the inequality is not strict, the solution set is $\{x \mid -\infty < x < -2 \text{ or } 1 \leq x < \infty\}$. Using interval notation, the solution set is $(-\infty, -2)$ or $[1, \infty)$.

FIGURE 71

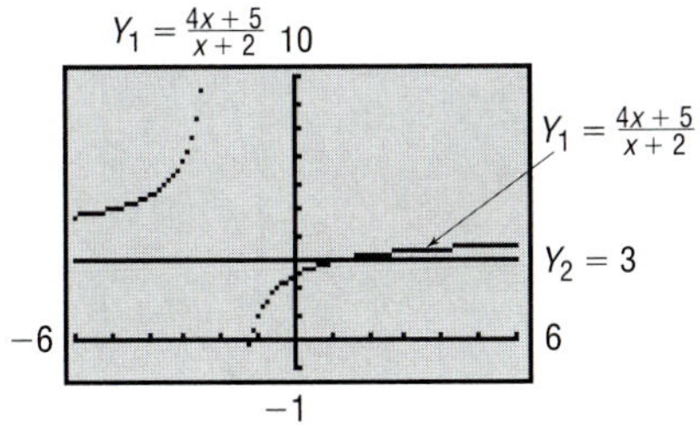

Algebraic Solution STEP 1: We first note that the domain of the variable consists of all real numbers except -2. We rearrange terms so that 0 is on the right side:

$$\frac{4x + 5}{x + 2} - 3 \geq 0$$ Subtract 3 from both sides of the inequality.

STEP 2: Find the critical numbers of $f(x) = \frac{4x + 5}{x + 2} - 3$. To find the critical numbers we must express f as a quotient.

$$f(x) = \frac{4x + 5}{x + 2} - 3$$ Least Common Denominator: $x + 2$.

$$f(x) = \frac{4x + 5}{x + 2} - 3\,\frac{x + 2}{x + 2}$$ Multiply -3 by $\frac{x + 2}{x + 2}$.

$$f(x) = \frac{4x + 5 - 3x - 6}{x + 2}$$ Distribute -3.

$$f(x) = \frac{x - 1}{x + 2}$$ Collect like terms.

The critical numbers are -2 and 1.

STEP 3: We use the critical numbers to separate the real number line into three intervals:

$$-\infty < x < -2 \qquad -2 < x < 1 \qquad 1 < x < \infty$$

STEP 4: We select a test number in each interval found in step 3 and evaluate $f(x) = \frac{4x + 5}{x + 2} - 3$ at each test number to determine if $f(x)$ is positive or negative. See Table 18.

TABLE 18

Interval	Test Number	$f(x)$	Positive/Negative
$-\infty < x < -2$	-3	$f(-3) = 4$	Positive
$-2 < x < 1$	0	$f(0) = -1/2$	Negative
$1 < x < \infty$	2	$f(2) = 1/4$	Positive

We want to know where $f(x)$ is positive. We conclude that the solutions are all x such that $-\infty < x < -2$ or $1 < x < \infty$. Because the original inequality is not strict, numbers x that satisfy the equation $\frac{x-1}{x+2} = 0$ are also solutions of the inequality. Since $\frac{x-1}{x+2} = 0$ only if $x = 1$, we conclude that the solution set is $\{x | -\infty < x < -2 \text{ or } 1 \leq x < \infty\}$ or using interval notation $(-\infty, -2)$ or $[1, \infty)$.

Figure 72 shows the graph of the solution set.

FIGURE 72
$-\infty < x < -2$ or $1 \leq x < \infty$
$(-\infty, -2)$ or $[1, \infty)$

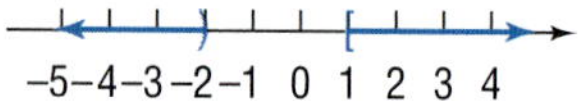

Now work Problem 37.

EXAMPLE 4 Minimum Sales Requirements

Tami is considering leaving her \$30,000 a year job and buying a cookie company. According to the financial records of the firm, the relationship between pounds of cookies sold and profit is exhibited by Table 19.

TABLE 19

Pounds of Cookies (in Hundreds), x	Profit, P
0	−20,000
50	−5990
75	412
120	10,932
200	26,583
270	36,948
340	44,381
420	49,638
525	49,225
610	44,381
700	34,220

(a) Draw a scatter diagram of the data in Table 19 with the pounds of cookies sold as the independent variable.

(b) Find the quadratic function of best fit using a graphing utility.

(c) Using the function found in (b), determine the number of pounds of cookies Tami must sell in order for the profits to exceed \$30,000 a year and therefore make it worthwhile for her to quit her job.

(d) Using the function found in (b), determine the number of pounds of cookies Tami should sell in order to maximize profits.

(e) Using the function found in (b), determine the maximum profit Tami can expect to earn.

Solution (a) Figure 73 shows the scatter diagram.

FIGURE 73

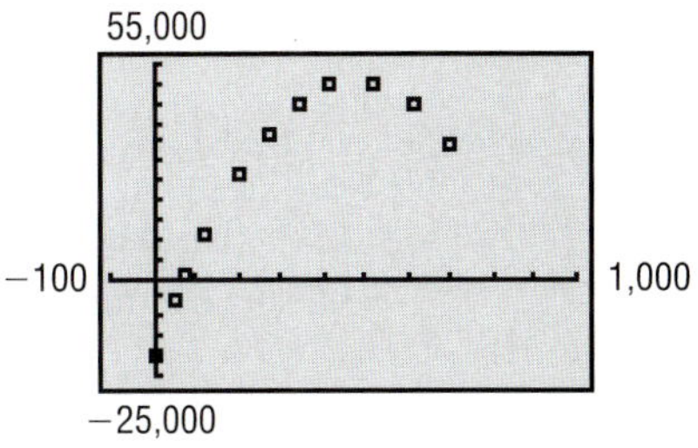

(b) Using a graphing utility, the quadratic function of best fit is

$$p(x) = -0.31x^2 + 295.86x - 20042.52$$

See Figure 74.

FIGURE 74

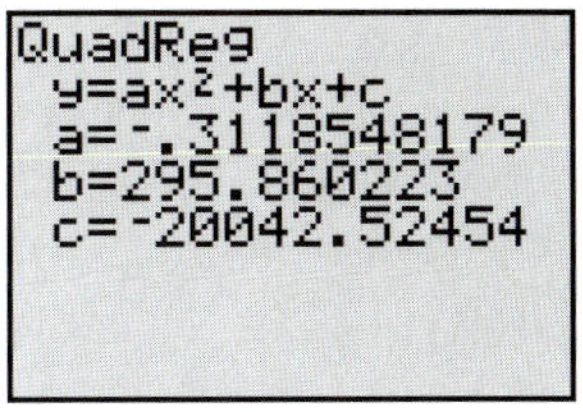

(c) We want profit to exceed \$30,000; therefore, we want to solve the inequality

$$-0.31x^2 + 295.86x - 20042.52 > 30000$$

We graph

$$Y_1 = p(x) = -0.31x^2 + 295.86x - 20042.52 \quad \text{and} \quad Y_2 = 30000$$

on the same screen. See Figure 75.

FIGURE 75

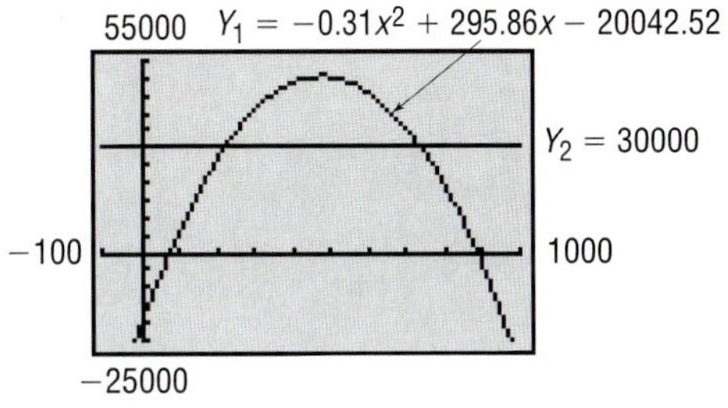

Using the INTERSECT command, we find that Y_1 and Y_2 intersect at $x = 220$ and at $x = 735$. The graph of Y_1 is above that of Y_2, $Y_1 > Y_2$, between the points of intersection. Since the inequality is strict, the solution set is $\{x | 220 < x < 735\}$.

(d) The function $p(x)$ found in (b) is a quadratic function, whose graph opens down. The vertex is therefore the highest point. The x-coordinate of the vertex is

$$x = \frac{-b}{2a} = \frac{-295.96}{2(-0.31)} = 477$$

To maximize profit, 47,700 pounds of cookies must be sold.

(e) The maximum profit is

$$p(477) = -0.31(477)^2 + 295.86(477) - 20042.52 = \$50,549$$

4.8 EXERCISES

In Problems 1–58, solve each inequality (a) graphically and (b) algebraically.

1. $(x-5)(x+2) < 0$

2. $(x-5)(x+2) > 0$

3. $x^2 - 4x > 0$

4. $x^2 + 8x > 0$

5. $x^2 - 9 < 0$

6. $x^2 - 1 < 0$

7. $x^2 + x > 12$

8. $x^2 + 7x < -12$

9. $2x^2 < 5x + 3$

10. $6x^2 < 6 + 5x$

11. $x(x-7) > 8$

12. $x(x+1) > 20$

13. $4x^2 + 9 < 6x$

14. $25x^2 + 16 < 40x$

15. $6(x^2 - 1) > 5x$

16. $2(2x^2 - 3x) > -9$

17. $(x-1)(x^2+x+1) > 0$

18. $(x+2)(x^2-x+1) > 0$

19. $(x-1)(x-2)(x-3) < 0$

20. $(x+1)(x+2)(x+3) < 0$

21. $x^3 - 2x^2 - 3x > 0$

22. $x^3 + 2x^2 - 3x > 0$

23. $x^4 > x^2$

24. $x^4 < 4x^2$

25. $x^3 > x^2$

26. $x^3 < 3x^2$

27. $x^4 > 1$

28. $x^3 > 1$

29. $x^2 - 7x - 8 < 0$

30. $x^2 + 12x + 32 \ge 0$

31. $x^3 + x - 12 \ge 0$

32. $x^3 - 3x + 1 \le 0$

33. $x^4 - 3x^2 - 4 > 0$

34. $x^4 - 5x^2 + 6 < 0$

35. $x^3 - 4 \ge 3x^2 + 5x - 3$

36. $x^4 - 4x \le -x^2 + 2x + 1$

37. $\dfrac{x+1}{x-1} > 0$

38. $\dfrac{x-3}{x+1} > 0$

39. $\dfrac{(x-1)(x+1)}{x} < 0$

40. $\dfrac{(x-3)(x+2)}{x-1} < 0$

41. $\dfrac{(x-2)^2}{x^2-1} \ge 0$

42. $\dfrac{(x+5)^2}{x^2-4} \ge 0$

43. $6x - 5 < \dfrac{6}{x}$

44. $x + \dfrac{12}{x} < 7$

45. $\dfrac{x+4}{x-2} \le 1$

46. $\dfrac{x+2}{x-4} \ge 1$

47. $\dfrac{3x-5}{x+2} \le 2$

48. $\dfrac{x-4}{2x+4} \ge 1$

49. $\dfrac{1}{x-2} < \dfrac{2}{3x-9}$

50. $\dfrac{5}{x-3} > \dfrac{3}{x+1}$

51. $\dfrac{2x+5}{x+1} > \dfrac{x+1}{x-1}$

52. $\dfrac{1}{x+2} > \dfrac{3}{x+1}$

53. $\dfrac{x^2(3+x)(x+4)}{(x+5)(x-1)} > 0$

54. $\dfrac{x(x^2+1)(x-2)}{(x-1)(x+1)} > 0$

55. $\dfrac{2x^2-x-1}{x-4} \le 0$

56. $\dfrac{3x^2+2x-1}{x+2} > 0$

57. $\dfrac{x^2+3x-1}{x+3} > 0$

58. $\dfrac{x^2-5x+3}{x-5} < 0$

59. For what positive numbers will the cube of a number exceed 4 times its square?

60. For what positive numbers will the square of a number exceed twice the number?

61. What is the domain of the variable in the expression $\sqrt{x^2 - 16}$?

62. What is the domain of the variable in the expression $\sqrt{x^3 - 3x^2}$?

63. What is the domain of the variable in the expression $\sqrt{\dfrac{x-2}{x+4}}$?

64. What is the domain of the variable in the expression $\sqrt{\dfrac{x-1}{x+4}}$?

65. **Physics** A ball is thrown vertically upward with an initial velocity of 80 feet per second. The distance s (in feet) of the ball from the ground after t seconds is $s = 80t - 16t^2$.

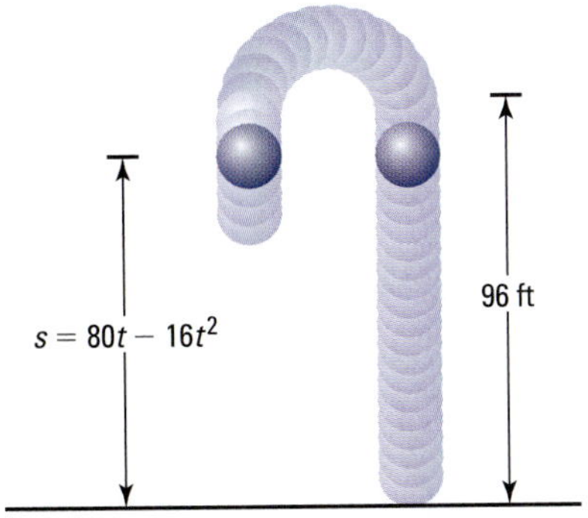

(a) For what time interval is the ball more than 96 feet above the ground? (See the figure.)
(b) Using a graphing utility, graph the relation between s and t.
(c) What is the maximum height of the ball?
(d) After how many seconds does the ball reach the maximum height?

66. **Physics** A ball is thrown vertically upward with an initial velocity of 96 feet per second. The distance s (in feet) of the ball from the ground after t seconds is $s = 96t - 16t^2$.
(a) For what time interval is the ball more than 112 feet above the ground?
(b) Using a graphing utility, graph the relation between s and t.
(c) What is the maximum height of the ball?
(d) After how many seconds does the ball reach the maximum height?

67. **Business** The monthly revenue achieved by selling x wristwatches is figured to be $x(40 - 0.2x)$ dollars. The wholesale cost of each watch is \$32.
(a) How many watches must be sold each month to achieve a profit (revenue − cost) of at least \$50?
(b) Using a graphing utility, graph the revenue function.
(c) What is the maximum revenue this firm could earn?
(d) How many wristwatches should the firm sell to maximize revenue?
(e) Using a graphing utility, graph the profit function.
(f) What is the maximum profit this firm can earn?
(g) How many watches should the firm sell to maximize profit?
(h) Provide a reasonable explanation as to why the answers found in parts (d) and (g) differ. Is the shape of the revenue function reasonable in your opinion? Why?

68. **Business** The monthly revenue achieved by selling x boxes of candy is figured to be $x(5 - 0.05x)$ dollars. The wholesale cost of each box of candy is \$1.50.
(a) How many boxes must be sold each month to achieve a profit of at least \$60?
(b) Using a graphing utility, graph the revenue function.
(c) What is the maximum revenue this firm could earn?
(d) How many boxes of candy should the firm sell to maximize revenue?
(e) Using a graphing utility, graph the profit function.
(f) What is the maximum profit this firm can earn?
(g) How many boxes of candy should the firm sell to maximize profit?
(h) Provide a reasonable explanation as to why the answers found in parts (d) and (g) differ. Is the shape of the revenue function reasonable in your opinion? Why?

69. **Cost of Manufacturing** In Problem 63 of Section 4.3, a cubic function of best fit relating the cost C of manufacturing x Chevy Cavaliers in a day was found. Budget constraints will not allow Chevy to spend more than \$97,000 per day. Determine the number of Cavaliers that should be produced in a day.

70. Cost of Printing In Problem 64 of Section 4.3, a cubic function of best fit relating the cost C of printing x textbooks in a week was found. Budget constraints will not allow the printer to spend more than \$170,000 per week. Determine the number of textbooks that should be printed in a week.

71. Minimum Sales Requirements Marissa is thinking of leaving her \$1000 a week job and buying a computer resale shop. According to the financial records of the firm, the profits (in dollars) of the company for different amounts of computers sold and the corresponding profits are:

Number of Computers Sold, x	Profit, p
0	−1500
4	−522
7	54
12	775
18	1184
23	1132
29	653

(a) Using a graphing utility, draw a scatter diagram of the data with the number of computers sold as the independent variable.
(b) Find the quadratic function of best fit using a graphing utility.
(c) Using the function found in (b), determine the number of computers Marissa must sell in order for the profits to exceed \$1000 a week and therefore make it worthwhile for her to quit her job.
(d) Using the function found in (b), determine the number of computers Marissa should sell in order to maximize profits.
(e) Using the function found in (b), determine the maximum profit Marissa can expect to earn.

72. Minimum Sales Requirements Barry is considering the purchase of a gas station. According to the financial records of the gas station, the monthly sales of the gas station (in thousands of gallons of gasoline) and the corresponding profits are:

Gallons of Gasoline (000s), x	Profit, p
50	3947
54	4214
74	4942
92	4838
82	5003
75	4965
100	4521
88	4933
63	4665

(a) Using a graphing utility, draw a scatter diagram of the data with the number of gallons of gasoline sold as the independent variable.
(b) Find the quadratic function of best fit using a graphing utility.
(c) Using the function found in (b), determine the number of gallons of gasoline Barry must sell in order for the profits to exceed \$4000 a month and therefore make it worthwhile for him to quit his job.
(d) Using the function found in (b), determine the number of gallons of gasoline Barry should sell in order to maximize profits.
(e) Using the function found in (b), determine the maximum profit John can expect to earn.

73. Prove that if a, b are real numbers and $a \geq 0$, $b \geq 0$ then

$$a \leq b \quad \text{is equivalent to} \quad \sqrt{a} \leq \sqrt{b}$$

[**Hint:** $b - a = (\sqrt{b} - \sqrt{a})(\sqrt{b} + \sqrt{a})$].

74. Make up an inequality that has no solution. Make up one that has exactly one solution.

75. The inequality $x^2 + 1 < -5$ has no solution. Explain why.

CHAPTER REVIEW

THINGS TO KNOW

Quadratic function	$f(x) = ax^2 + bx + c, a \neq 0$	Vertex: $(-b/2a, f(-b/2a))$ Axis: The line $x = -b/2a$ Parabola opens up if $a > 0$. Parabola opens down if $a < 0$.

Quadratic equation and quadratic formula

If $ax^2 + bx + c = 0, a \neq 0$, then $x = \dfrac{-b \pm \sqrt{b^2 - 4ac}}{2a}$.

Discriminant

If $b^2 - 4ac > 0$, there are two distinct real solutions.
It $b^2 - 4ac = 0$, there is one repeated real solution.
If $b^2 - 4ac < 0$, there are two distinct complex solutions that are not real; the solutions are conjugates of each other.

Power function	$f(x) = x^n, n \geq 2$ even	Even function: Passes through $(-1, 1), (0, 0), (1, 1)$ Opens up
	$f(x) = x^n, n \geq 3$ odd	Odd function: Passes through $(-1, -1), (0, 0), (1, 1)$ Increasing
Polynomial function	$f(x) = a_n x^n + a_{n-1}x^{n-1} + \cdots + a_1 x + a_0, a_n \neq 0$	At most $n - 1$ turning points; behaves like $y = a_n x^n$ for large x.
Rational function	$R(x) = \dfrac{p(x)}{q(x)}$, p, q are polynomial functions in lowest terms	See Steps 1 through 5 on pages 288–289.

Zeros of a polynomial f	Numbers for which $f(x) = 0$; these are the x-intercepts of the graph of f.
Remainder Theorem	If a polynomial $f(x)$ is divided by $x - c$, then the remainder is $f(c)$.
Factor Theorem	$x - c$ is a factor of a polynomial $f(x)$ if and only if $f(c) = 0$.
Descartes' Rule of Signs	Let f denote a polynomial function. The number of positive zeros of f either equals the number of variations in sign of the nonzero coefficients of $f(x)$ or else equals that number less some even integer. The number of negative zeros of f either equals the number of variations in sign of the nonzero coefficients of $f(-x)$ or else equals that number less some even integer.
Rational Zeros Theorem	Let f be a polynomial function of degree 1 or higher of the form $f(x) = a_n x^n + a_{n-1}x^{n-1} + \cdots + a_1 x + a_0, \quad a_n \neq 0, a_0 \neq 0$ where each coefficient is an integer. If p/q, in lowest terms, is a rational zero of f, then p must be a factor of a_0 and q must be a factor of a_n.
Fundamental Theorem of Algebra	Every complex polynomial function $f(x)$ of degree $n \geq 1$ has at least one complex zero.
Conjugate Pairs Theorem	Let $f(x)$ be a complex polynomial whose coefficients are real numbers. If $r = a + bi$ is a zero of f, then the complex conjugate $\bar{r} = a - bi$ is also a zero of f.

HOW TO

Graph quadratic functions

Graph power functions

Graph polynomial functions

Graph rational functions (see steps 1 through 5, pages 288–289)

Use synthetic division to divide a polynomial by $x - c$

Find the real zeros of a polynomial by using Descartes' Rule of Signs, the Rational Zeros Theorem, and depressed equations

Solve polynomial equations using Descartes' Rule of Signs, the Rational Zeros Theorem, and depressed equations

Use the Upper and Lower Bounds Test

Find the complex zeros of a polynomial

Solve polynomial and rational inequalities

FILL-IN-THE-BLANK ITEMS

1. The graph of a quadratic function is called a(n) ______________. Its lowest or highest point is called the ______________.

2. In the process of long division,

(Divisor)(Quotient) + ______________ = ______________

3. When a polynomial function f is divided by $x - c$, the remainder is ______________.

4. A polynomial function f has the factor $x - c$ if and only if ______________.

5. A number r for which $f(r) = 0$ is called a(n) ______________ of the function f.

6. The polynomial function $f(x) = x^5 - 2x^3 + x^2 + x - 1$ has either ______________ or ______________ positive real zeros; it has ______________ or ______________ negative real zeros.

7. The possible rational zeros of $f(x) = 2x^5 - x^3 + x^2 - x + 1$ are ______________.

8. The line ______________ is a horizontal asymptote of $R(x) = \dfrac{x^3 - 1}{x^3 + 1}$.

9. The line ______________ is a vertical asymptote of $R(x) = \dfrac{x^3 - 1}{x^3 + 1}$.

10. If $3 + 4i$ is a zero of a polynomial of degree 5 with real coefficients, then so is ______________.

TRUE/FALSE ITEMS

T F **1.** Every polynomial of degree 3 with real coefficients has exactly three real zeros.

T F **2.** If $2 - 3i$ is a zero of a polynomial with real coefficients, then so is $-2 + 3i$.

T F **3.** The graph of $R(x) = \dfrac{x^2}{x - 1}$ has exactly one vertical asymptote.

T F **4.** The graph of $f(x) = x^2(x - 3)(x + 4)$ has exactly three x-intercepts.

T F **5.** If f is a polynomial function of degree 4 and if $f(2) = 5$, then

$$\frac{f(x)}{x - 2} = p(x) + \frac{5}{x - 2}$$

where $p(x)$ is a polynomial of degree 3.

REVIEW EXERCISES

In Problems 1–10, (a) graph each quadratic function by hand by determining whether its graph opens up or down and by finding its vertex, axis of symmetry, y-intercept, and x-intercepts, if any; (b) verify your results using a graphing utility.

1. $f(x) = (x - 2)^2 + 2$
2. $f(x) = (x + 1)^2 - 4$
3. $f(x) = \frac{1}{4}x^2 - 16$
4. $f(x) = -\frac{1}{2}x^2 + 2$
5. $f(x) = -4x^2 + 4x$
6. $f(x) = 9x^2 - 6x + 3$
7. $f(x) = \frac{9}{2}x^2 + 3x + 1$
8. $f(x) = -x^2 + x + \frac{1}{2}$
9. $f(x) = 3x^2 + 4x - 1$
10. $f(x) = -2x^2 - x + 4$

In Problems 11–16, using a graphing utility, show the stages required to graph each function.

11. $f(x) = (x + 2)^3$
12. $f(x) = -x^3 + 3$
13. $f(x) = -(x - 1)^4$
14. $f(x) = (x - 1)^4 - 2$
15. $f(x) = (x - 1)^4 + 2$
16. $f(x) = (1 - x)^3$

In Problems 17–22, determine whether the given quadratic function has a maximum value or a minimum value, and then find the value. Verify your results using a graphing utility.

17. $f(x) = 3x^2 - 6x + 4$
18. $f(x) = 2x^2 + 8x + 5$
19. $f(x) = -x^2 + 8x - 4$
20. $f(x) = -x^2 - 10x - 3$
21. $f(x) = -3x^2 + 12x + 4$
22. $f(x) = -2x^2 + 4$

In Problems 23–30:

(a) Using a graphing utility graph f.

(b) Find the x- and y-intercepts.

(c) Determine whether each x-intercept is of odd or even multiplicity.

(d) Find the power function that the graph of f resembles for large values of x.

(e) Determine the number of turning points on the graph of f.

(f) Determine the local maxima and local minima, if any exist, correct to two decimal places.

23. $f(x) = x(x + 2)(x + 4)$
24. $f(x) = x(x - 2)(x - 4)$
25. $f(x) = (x - 2)^2(x + 4)$
26. $f(x) = (x - 2)(x + 4)^2$
27. $f(x) = x^3 - 4x^2$
28. $f(x) = x^3 + 4x$
29. $f(x) = (x - 1)^2(x + 3)(x + 1)$
30. $f(x) = (x - 4)(x + 2)^2(x - 2)$

In Problems 31–40, follow steps 1–5 on pages 288–289 to analyze each rational function.

31. $R(x) = \dfrac{2x - 6}{x}$
32. $R(x) = \dfrac{4 - x}{x}$
33. $H(x) = \dfrac{x + 2}{x(x - 2)}$
34. $H(x) = \dfrac{x}{x^2 - 1}$
35. $R(x) = \dfrac{x^2 + x - 6}{x^2 - x - 6}$
36. $R(x) = \dfrac{x^2 - 6x + 9}{x^2}$
37. $F(x) = \dfrac{x^3}{x^2 - 4}$
38. $F(x) = \dfrac{3x^3}{(x - 1)^2}$
39. $R(x) = \dfrac{2x^4}{(x - 1)^2}$
40. $R(x) = \dfrac{x^4}{x^2 - 9}$

In Problems 41 and 42 use Descartes' Rule of Signs to determine how many positive and negative zeros each polynomial function may have. Do not attempt to find the zeros.

41. $f(x) = 12x^8 - x^7 + 8x^4 - 2x^3 + x + 3$
42. $f(x) = -6x^5 + x^4 + 5x^3 + x + 1$

43. List all the potential rational zeros of $f(x) = 12x^8 - x^7 + 6x^4 - x^3 + x - 3$.

44. List all the potential rational zeros of $f(x) = -6x^5 + x^4 + 2x^3 - x + 1$.

In Problems 45–56, use the steps listed on page 314 to find the real zeros of f(x). Approximate any irrational zeros correct to two decimal places.

45. $f(x) = x^3 - 3x^2 - 6x + 8$
46. $f(x) = x^3 - x^2 - 10x - 8$
47. $f(x) = 4x^3 + 4x^2 - 7x + 2$
48. $f(x) = 4x^3 - 4x^2 - 7x - 2$
49. $f(x) = x^4 - 4x^3 + 9x^2 - 20x + 20$
50. $f(x) = x^4 + 6x^3 + 11x^2 + 12x + 18$

51. $f(x) = 2x^3 - 11.84x^2 - 9.116x + 82.46$

52. $f(x) = 12x^3 + 39.8x^2 - 4.4x - 3.4$

53. $g(x) = 15x^4 - 21.5x^3 - 1718.3x^2 + 5308x + 3796.8$

54. $g(x) = 3x^4 + 67.93x^3 + 486.265x^2 + 1121.32x + 412.195$

55. $f(x) = 3x^3 + 18.02x^2 + 11.0467x - 53.8756$

56. $f(x) = x^3 - 3.16x^2 - 39.4611x + 151.638$

In Problems 57–60, solve each equation in the real number system.

57. $2x^4 + 2x^3 - 11x^2 + x - 6 = 0$

58. $3x^4 + 3x^3 - 17x^2 + x - 6 = 0$

59. $2x^4 + 7x^3 + x^2 - 7x - 3 = 0$

60. $2x^4 + 7x^3 - 5x^2 - 28x - 12 = 0$

In Problems 61–64, find integer-valued upper and lower bounds to the zeros of each polynomial function.

61. $f(x) = 2x^3 - x^2 - 4x + 2$

62. $f(x) = 2x^3 + x^2 - 10x - 5$

63. $f(x) = 2x^3 - 7x^2 - 10x + 35$

64. $f(x) = 3x^3 - 7x^2 - 6x + 14$

In Problems 65–68, each polynomial has exactly one positive zero. Approximate the zero correct to two decimal places.

65. $f(x) = x^3 - x - 2$

66. $f(x) = 2x^3 - x^2 - 3$

67. $f(x) = 8x^4 - 4x^3 - 2x - 1$

68. $f(x) = 3x^4 + 4x^3 - 8x - 2$

In Problems 69–72, information is given about a complex polynomial $f(z)$ whose coefficients are real numbers. Find the remaining zeros of f.

69. Degree 3; zeros: $4 + i, 6$

70. Degree 3; zeros: $3 + 4i, 5$

71. Degree 4; zeros: $i, 1 + i$

72. Degree 4; zeros: $1, 2, 1 + i$

73. Find the quotient and remainder if $x^4 + 2x^3 - 7x^2 - 8x + 12$ is divided by $(x - 2)(x - 1)$.

74. Find the quotient and remainder if $x^4 + 2x^3 - 4x^2 - 5x - 6$ is divided by $(x - 2)(x + 3)$.

In Problems 75–88, solve each equation in the complex number system.

75. $x^2 + x + 1 = 0$

76. $x^2 - x + 1 = 0$

77. $2x^2 + x - 2 = 0$

78. $3x^2 - 2x - 1 = 0$

79. $x^2 + 3 = x$

80. $2x^2 + 1 = 2x$

81. $x(1 - x) = 6$

82. $x(1 + x) = 2$

83. $x^4 + 2x^2 - 8 = 0$

84. $x^4 + 8x^2 - 9 = 0$

85. $x^3 - x^2 - 8x + 12 = 0$

86. $x^3 - 3x^2 - 4x + 12 = 0$

87. $3x^4 - 4x^3 + 4x^2 - 4x + 1 = 0$

88. $x^4 + 4x^3 + 2x^2 - 8x - 8 = 0$

In Problems 89–92, find the complex zeros of each polynomial function.

89. $f(x) = x^3 + 5x^2 + 36x + 180$

90. $f(x) = x^3 - 2x^2 + 16x - 32$

91. $f(x) = x^4 - 3x^3 - 52x^2 + 262x - 208$

92. $f(x) = x^4 + 5x^3 + 55x^2 - 435x - 1746$

93. Find integer-valued upper and lower bounds to the zeros of

$$f(x) = 4x^5 - 3x^4 + 8x^2 + x + 2$$

94. Find integer-valued upper and lower bounds to the zeros of

$$f(x) = 8x^6 - x^4 + 6x^2 + 24x + 15$$

In Problems 95–104, solve each inequality (a) graphically and (b) algebraically.

95. $2x^2 + 5x - 12 < 0$

96. $3x^2 - 2x - 1 \geq 0$

97. $\dfrac{6}{x + 3} \geq 1$

98. $\dfrac{-2}{1 - 3x} < 1$

99. $\dfrac{2x - 6}{1 - x} < 2$

100. $\dfrac{3 - 2x}{2x + 5} \geq 2$

101. $\dfrac{(x - 2)(x - 1)}{x - 3} > 0$

102. $\dfrac{x + 1}{x(x - 5)} \leq 0$

103. $\dfrac{x^2 - 8x + 12}{x^2 - 16} > 0$

104. $\dfrac{x(x^2 + x - 2)}{x^2 + 9x + 20} \leq 0$

105. Relating the Length and Period of a Pendulum Tom constructs simple pendulums with different lengths, l, and uses a light probe to record the corresponding period, T. The following data is collected:

Length, l (in feet)	0.5	1.2	1.8	2.3	3.7	4.9
Period, T (in seconds)	0.79	1.20	1.51	1.65	2.14	2.43

(a) Using a graphing utility, draw a scatter diagram of the data with length as the independent variable and the period as the dependent variable.

(b) Using a graphing utility, fit a power model to the data.

(c) Graph the power model found in (b) on the scatter diagram.

(d) If the period of a simple pendulum is known to be 1.35 seconds, what is the length of the pendulum?

106. Relating Time and Distance a Ball Falls Scott drops a ball from various heights and records the time it takes for the ball to hit the ground. Using a motion detector connected to his graphing calculator, he collects the following data:

Time (seconds)	1.528	2.015	3.852	4.154	4.625
Distance (meters)	11.46	19.99	72.41	84.45	104.23

(a) Using a graphing utility, draw a scatter diagram using time as the independent variable and distance as the dependent variable.

(b) Using a graphing utility, fit a power model to the data.

(c) Graph the power model found in (b) on the scatter diagram.

(d) Physics theory states the distance an object falls is directly proportional to the time squared. Rewrite the model found in (b) so it is of the form $s = \frac{1}{2}gt^2$, where s is distance, g is acceleration due to gravity, and t is time. What is Scott's estimate of g? (It is known that the acceleration due to gravity is approximately 9.8 meters/sec^2.)

(e) Predict how long it will take the ball to fall 100 meters.

107. Gravity on the Moon Neil Armstrong wishes to estimate the acceleration due to gravity on the Moon. He throws a ball (with a special chip that measures the distance to the ground in two second intervals) straight down into a crater that is known to be 1000 feet deep. The ball records the following data:

Time (seconds)	0	2	4	6	8	10	12	14	16
Distance (feet)	1000	969	917	844	749	633	496	337	157

(a) Using a graphing utility, draw a scatter diagram using time as the independent variable and distance as the dependent variable.

(b) Using a graphing utility, find the quadratic function of best fit.

(c) Graph the quadratic function found in (b) on the scatter diagram.

(d) According to physics theory,

$$s(t) = -\frac{1}{2}gt^2 + v_0t + s_0,$$

where $s(t)$ is the height of the object at time t, g is acceleration due to gravity v_0 is the initial velocity of the object (this number will be negative if the object is thrown down), and s_0 is the initial height of the object. What is the acceleration due to gravity on the moon? (Compare your answer with the fact that the acceleration due to gravity on the moon is approximately -5.33 feet/sec^2.)

(e) What was the initial velocity of the ball?

(f) Predict how long it will take for the ball to hit the ground.

108. Average Cost of Baking Bread The average cost of production is found by taking the total cost of producing a good and dividing it by the number of units produced. Mary owns a bakery and collects the following data relating the average cost per day for baking loaves of bread:

Average Cost, $\bar{C}$ ($/loaf)	14.50	12.20	8.92	3.68	2.04	1.09	0.57	2.26
Loaves of Bread, x	4	10	20	40	50	60	80	100

(a) Using a graphing utility, draw a scatter diagram using loaves of bread as the independent variable and average cost as the dependent variable.
(b) Using a graphing utility, find the quadratic function of best fit.
(c) Graph the quadratic function found in (b) on the scatter diagram.
(d) Using the function found in (b), determine the number of loaves of bread Mary should bake if she wishes to minimize average cost.
(e) What is the minimum average cost?
(f) If Mary charges $1.59 per loaf of bread, what will her daily profit be if she is able to sell all the loaves of bread she bakes?

109. Aids Cases in the US The data below represents the cumulative number of reported AIDS cases in the United States from 1983–1994.

Year, t	Number of AIDS Cases, A
1983	4,589
1984	10,750
1985	22,399
1986	41,256
1987	69,592
1988	104,644
1989	146,574
1990	193,878
1991	251,638
1992	326,648
1993	399,613
1994	457,280

Source: Center for Disease Control.

(a) Using a graphing utility, draw a scatter diagram of the data. Comment on the type of relation that appears to exist between the two variables.
(b) Using your graphing utility, find the cubic function of best fit.
(c) Draw the cubic function of best fit found in (b) on your scatter diagram. Comment on the fit.
(d) Use the function found in (b) to predict the cumulative number of AIDS cases reported in the United States in 1995.
(e) Do you think the function found in (b) will be useful in predicting the number of AIDS cases in 1999?

110. Cost of Manufacturing Bicycles The following data represents the monthly cost of manufacturing bicycles C and the number of bicycles produced, x.

Number of Bicycles Produced, x	Cost, C
0	10,000
100	30,000
150	39,000
180	43,950
205	47,825
220	50,075
225	50,850
240	53,325
265	57,750
280	60,675
300	65,075

(a) Using a graphing utility, draw a scatter diagram of the data. Comment on the type of relation that appears to exist between the two variables.
(b) Using your graphing utility, find the cubic function of best fit.
(c) Draw the cubic function of best fit found in (b) on your scatter diagram. Comment on the fit.
(d) Use the function found in (b) to predict the cost of manufacturing 230 bicycles.
(e) Interpret the y-intercept.
(f) How many bicycles can be produced if costs are equal to $55,000?

111. Find the point on the line $y = x$ that is closest to the point (3, 1).

[**Hint:** Find the minimum value of the function $f(x) = d^2$, where d is the distance from (3, 1) to a point on the line.]

112. Find the point on the line $y = x + 1$ that is closest to the point (4, 1).

113. A horizontal bridge is in the shape of a parabolic arch. Given the information shown in the figure, what is the height h of the arch 2 feet from shore?

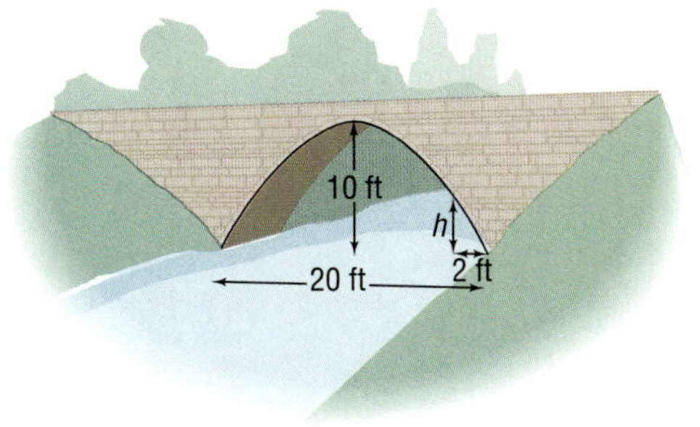

114. Find the length and width of a rectangle whose perimeter is 20 feet and whose area is 16 square feet.

115. Design a polynomial function with the following characteristics: degree 6; four real zeros, one of multiplicity 3; y-intercept 3; behaves like $y = -5x^6$ for large values of x. Is this polynomial unique? Compare your polynomial with those of other students. What terms will be the same as everyone else's? Add some more characteristics such as symmetry or naming the real zeros. How does this modify the polynomial?

116. Design a rational function with the following characteristics: three real zeros, one of multiplicity 2; y-intercept 1; vertical asymptotes $x = -2$ and $x = 3$; oblique asymptote $y = 2x + 1$. Is this rational function unique? Compare yours with those of other students. What will be the same as everyone else's? Add some more characteristics such as symmetry or naming the real zeros. How does this modify the rational function?

117. The illustration shows the graph of a polynomial function.

(a) Is the degree of the polynomial even or odd?
(b) Is the leading coefficient positive or negative?
(c) Is the function even, odd, or neither?
(d) Why is x^2 necessarily a factor of the polynomial?
(e) What is the minimum degree of the polynomial?
(f) Formulate five different polynomials whose graphs could look like the one shown. Compare yours to those of other students. What similarities do you see? What differences?

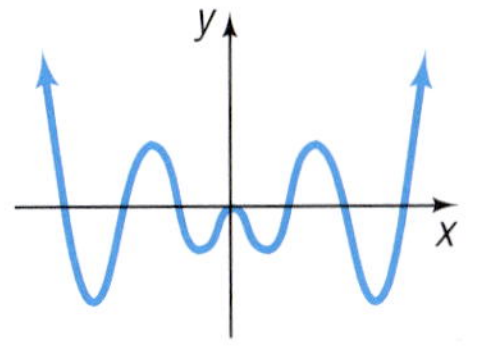

CHAPTER 5 Exponential and Logarithmic Functions

Most people view bees as pests, but farmers and scientists view bees rather differently. On the following page you will find your project as a research entomologist and will need to use the Sullivan website at:

www.prenhall.com/sullivan

to help link you to the Internet resources you will need to prepare your report.

PREPARING FOR THIS CHAPTER

Before getting started on this chapter, review the following concepts:

Radicals; Rational Exponents *(Appendix, Section 3)*

Functions *(Sections 2.1, 2.2, and 2.3)*

Simple Interest *(p. 199)*

Linear Curve Fitting *(Section 1.6)*

Quadratic Curve Fitting *(Section 4.1)*

Power Curve Fitting *(Section 4.2)*

Cubic Curve Fitting *(Section 4.3)*

OUTLINE

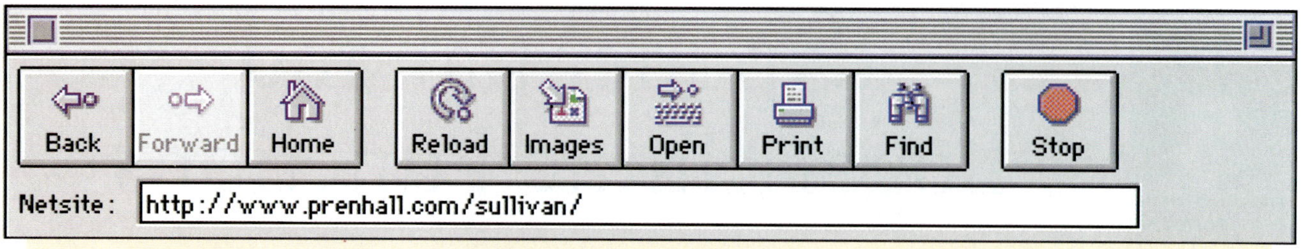

BEE PARASITIC MITE SYNDROME (BPMS)

The Bee Crisis. If you've noticed a lack of bees buzzing around your home, you're not alone. In recent years, there has been a *serious loss* of bee colonies in the United States. The cause is from two parasitic mites: Acarapis woodi and Varroa jacobsoni. Varroa first showed up in 1987. Both mites now infest bees in every state except Hawaii. There is no official wild bee census, but estimates go as high as 90 percent of the wild bee colonies have been wiped out! Scientists expect that all the bee colonies in the United States will be exposed to the Varroa mite.

Varroa mites are a serious threat to the United States honeybee industry. In 1986, more than 211,000 beekeepers produced about 200 million pounds of honey valued at $103.1 million, according to the National Honey Board. To agriculture, in general, the mite poses an even bigger threat. Crops valued at $20 billion, ranging from blueberries in Maine to almonds in California, depend on bees for pollination each year. Serious economic losses could result if Varroa mites were to devastate the United States bee population.

As a research entomologist, you want to find alternative methods of control other than the use of pesticides. Your research into Varroa resistant hybrids may offer some hope. You need to create some mathematical models that reflect possible outcomes of your research. These models will help you to determine a focus for your genetic work.

1. What are some of the *difficulties* in fighting the Varroa mite? What are likely results unless some control measures are used?
2. Pesticides are recommended to curb the Varroa epidemic. This is the main defense currently in use. How effective are the pesticides? Does the short-term gain result in any long-term costs?
3. If your research in bee hybridization allows you to breed a mite-resistant honey bee such that only one-third of your bees die from Varroa per year, would this be considered a success?
4. If you study the graphs of bee-mite interaction, you will notice some fairly well defined *logistic curves.* You search the Internet for references to logistic growth models. Using a graphing utility, determine a logistic growth function of best fit that models the situation.
5. Collect and assemble the data on *honey production* for the 1990's. Can the Varroa infestation be seen in this data?
6. Given the results of questions 4 and 5, could you justify removing the use of pesticides for the treatment of a colony?

Until now, our study of functions has concentrated primarily on polynomial and rational functions. These functions belong to the class of **algebraic functions,** that is, functions that can be expressed in terms of sums, differences, products, quotients, powers, or roots of polynomials. Functions that are not algebraic are termed **transcendental** (they transcend, or go beyond, algebraic functions).

In this chapter, we study two transcendental functions: the *exponential* and *logarithmic functions.* These functions occur frequently in a wide variety of applications, such as biology, chemistry, economics, and psychology.

The chapter begins with a discussion of inverse functions.

5.1 ONE-TO-ONE FUNCTIONS; INVERSE FUNCTIONS

1 Determine Whether a Function Is One-to-One

2 Obtain the Graph of the Inverse Function from the Graph of the Function

3 Find an Inverse Function

In Section 2.1, we said a function f can be thought of as a machine that receives as input a number, say x, from the domain, manipulates it, and outputs the value $f(x)$. The **inverse** of f receives as input a number $f(x)$, manipulates it, and outputs the value x. The inverse of f reverses the process of f. If the function f is the set of ordered pairs (x, y), then the inverse of f is the set of ordered pairs (y, x).

EXAMPLE 1 Finding the Inverse of a Function

Find the inverse of the following functions.

(a) Let the domain of the function represent the employees of Yolanda's Preowned Car Mart and let the range represent their base salaries.

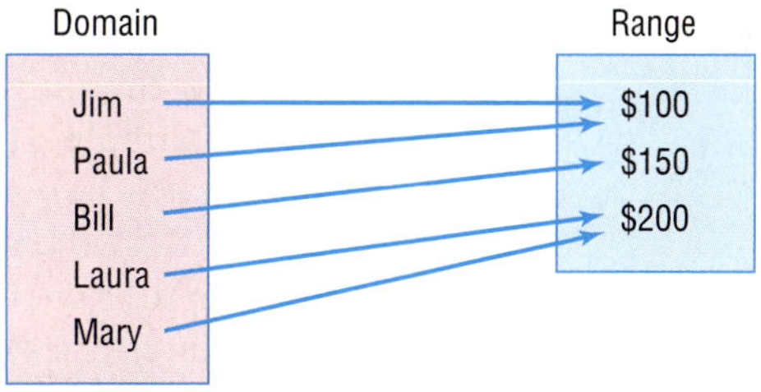

(b) Let the domain of the function represent the employees of Yolanda's Preowned Car Mart and let the range represent their spouse's names.

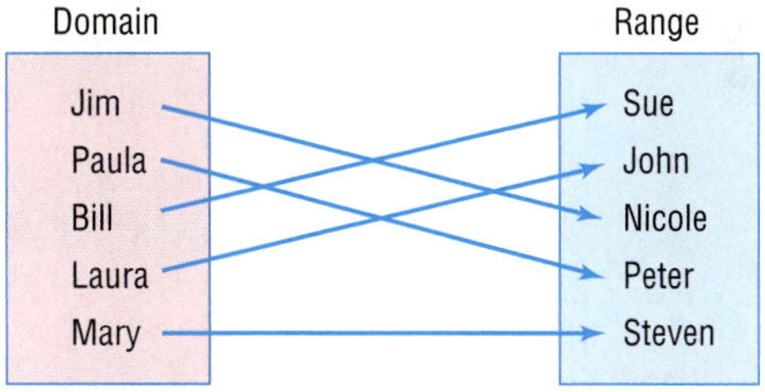

(c) $\{(-3, -27), (-2, -8), (-1, -1), (0, 0), (1, 1), (2, 8), (3, 27)\}$

(d) $\{(-3, 9), (-2, 4), (-1, 1), (0, 0), (1, 1), (2, 4), (3, 9)\}$

Solution (a) The elements in the domain represent inputs to the function and the elements in the range represent the outputs. To find the inverse, interchange the elements in the domain with the elements in the range. For example, the function receives as input Bill and outputs $150. So, the inverse receives as input $150 and outputs Bill.

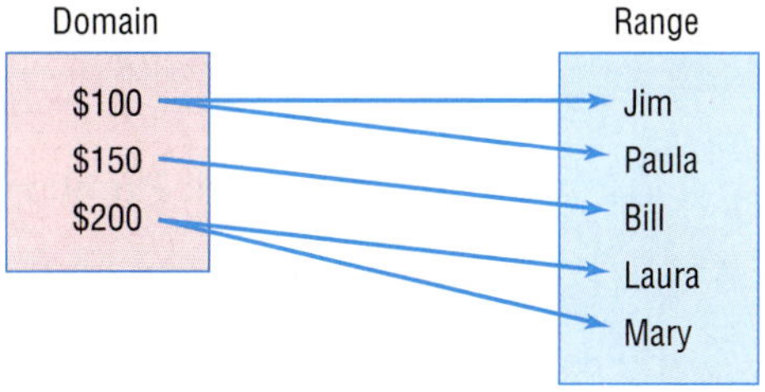

(b)

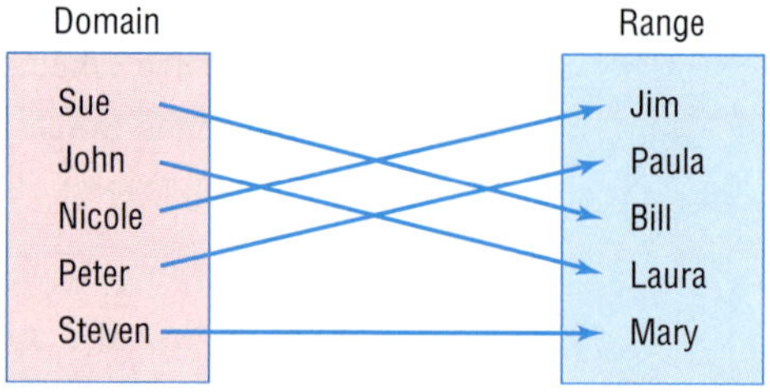

(c) The inverse of the function is found by interchanging the entries in each ordered pair. So the inverse of the given function is $\{(-27, -3), (-8, -2), (-1, -1), (0, 0), (1, 1), (8, 2), (27, 3)\}$.

(d) The inverse of the given function is $\{(9, -3), (4, -2), (1, -1), (0, 0), (1, 1), (4, 2), (9, 3)\}$.

1 We notice the inverses found in Examples 1(b) and (c) represent functions since each element in the domain corresponds to a unique element in the range. The inverses found in Examples 1(a) and (d) do not represent functions since each element in the domain does not correspond to a unique element in the range. Compare the function in Example 1(c) with the function in Example 1(d). For the function in Example 1(c), to every unique x-coordinate there corresponds a unique y-coordinate; for the function in Example 1(d), every unique x-coordinate does not correspond to a unique y-coordinate. Functions where unique x-coordinates correspond to unique y-coordinates are called *one-to-one* functions. So, in order for the inverse of a function, f, to be a function itself, f must be one-to-one.

A function f is said to be **one-to-one** if, for any choice of numbers x_1 and x_2, $x_1 \neq x_2$, in the domain of f, then $f(x_1) \neq f(x_2)$.

In other words, if f is a one-to-one function, then for each x in the domain of f, there is exactly one y in the range, and no y in the range is the image of more than one x in the domain. See Figure 1.

FIGURE 1

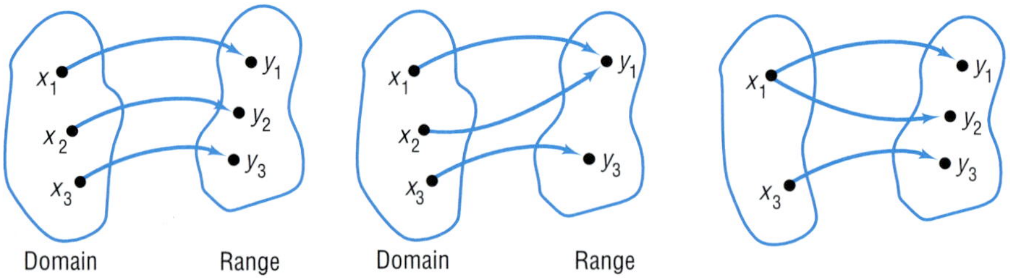

(a) One-to-one function: Each x in the domain has one and only one image in the range

(b) Not a one-to-one function: y_1 is the image of both x_1 and x_2

(c) Not a function: x_1 has two images, y_1 and y_2

Now work Problem 1.

If the graph of a function f is known, there is a simple test, called the **horizontal line test,** to determine whether f is one-to-one.

FIGURE 2
$f(x_1) = f(x_2) = h$, but $x_1 \neq x_2$; f is not a one-to-one function.

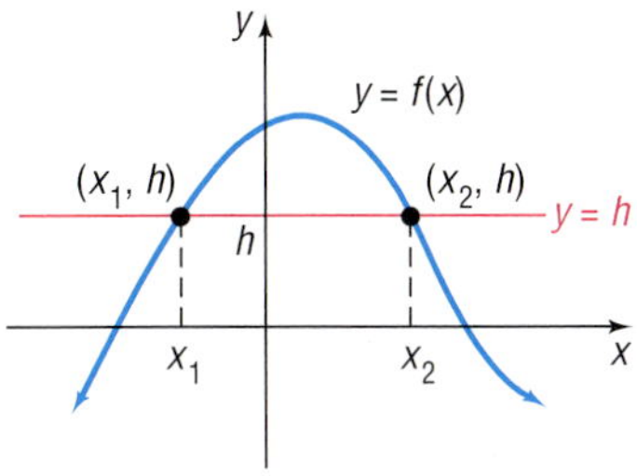

Theorem Horizontal Line Test

If horizontal lines intersect the graph of a function f in at most one point, then f is one-to-one.

The reason this test works can be seen in Figure 2, where the horizontal line $y = h$ intersects the graph at two distinct points, (x_1, h) and (x_2, h), with the same second element. Thus, f is not one-to-one.

EXAMPLE 2 Using the Horizontal Line Test

For each given function, use the graph to determine whether the function is one-to-one.

(a) $f(x) = x^2$ (b) $g(x) = x^3$

Solution (a) Figure 3(a) illustrates the horizontal line test for $f(x) = x^2$. The horizontal line $y = 1$ meets the graph of f twice, at $(1, 1)$ and at $(-1, 1)$, so f is not one-to-one.

(b) Figure 3(b) illustrates the horizontal line test for $g(x) = x^3$. Because each horizontal line will intersect the graph of g exactly once, it follows that g is one-to-one.

FIGURE 3

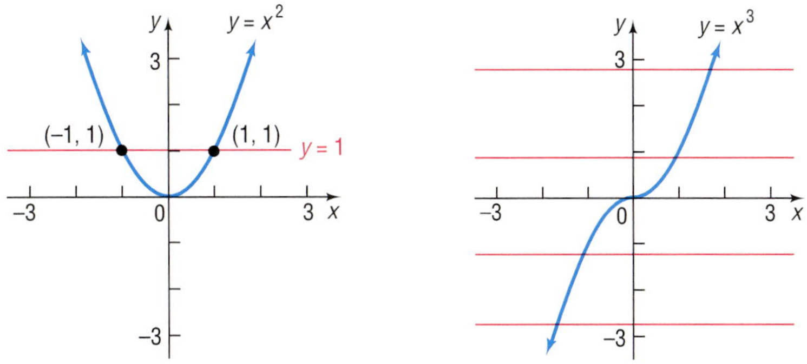

(a) A horizontal line intersects the graph twice; thus, f is not one-to-one

(b) Horizontal lines intersect the graph exactly once; thus, g is one-to-one

Now work Problem 9.

Let's look more closely at the one-to-one function $g(x) = x^3$. This function is an increasing function. Because an increasing (or decreasing) function will always have different y values for unequal x values, it follows that a function that is increasing (or decreasing) on its domain is also a one-to-one function.

> **Theorem**
>
> An increasing (decreasing) function is a one-to-one function.

Inverse of a Function $y = f(x)$

We mentioned in Chapter 2 that a function $y = f(x)$ can be thought of as a rule that tells us to do something to the argument x. For example, the function $f(x) = 2x$ multiplies the argument by 2. An *inverse function* of f undoes whatever f does. For example, the function $g(x) = \frac{1}{2}x$, which divides the argument by 2, is an inverse of $f(x) = 2x$. See Figure 4.

FIGURE 4
$f(x) = 2x$; $g(x) = \frac{1}{2}x$

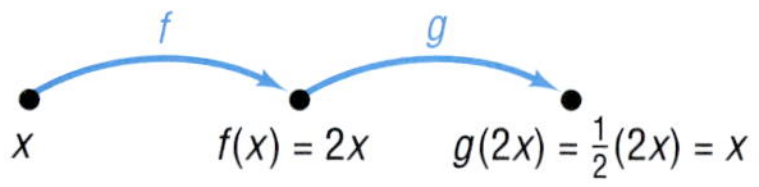

To put it another way, if we think of f as an input/output machine that processes an input x into $f(x)$, then the inverse function reverses this process, taking $f(x)$ back to x.

For a function $y = f(x)$ to have an inverse *function,* f must be one-to-one. Then for each x in its domain there is exactly one y in its range; furthermore, to each y in the range, there corresponds exactly one x in the domain. The correspondence from the range of f onto the domain of f is, therefore, also a function. It is this function that is the *inverse of f.* A definition is given next.

> Let f denote a one-to-one function $y = f(x)$. The **inverse of *f*,** denoted by f^{-1}, is a function such that $f^{-1}(f(x)) = x$ for every x in the domain of f and $f(f^{-1}(x)) = x$ for every x in the domain of f^{-1}.

Warning: Be careful! The -1 used in f^{-1} is not an exponent. Thus, f^{-1} does *not* mean the reciprocal of f; it means the inverse of f.

FIGURE 5

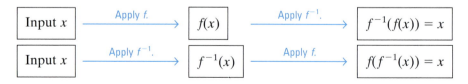

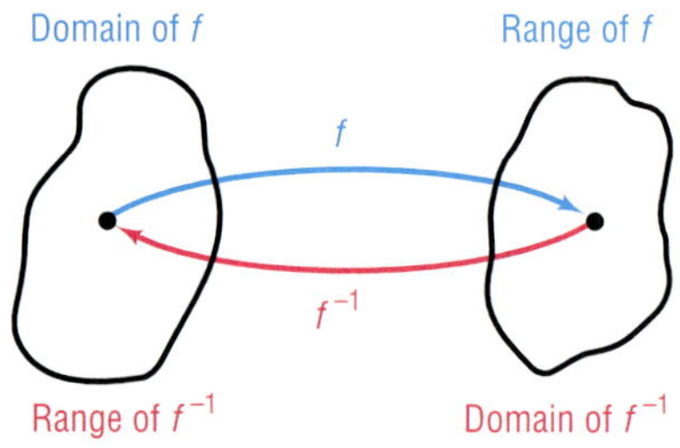

Figure 5 illustrates the definition.

Two facts are now apparent about a function f and its inverse f^{-1}.

> Domain of f = Range of f^{-1} Range of f = Domain of f^{-1}

Look again at Figure 5 to visualize the relationship. If we start with x, apply f, and then apply f^{-1}, we get x back again. If we start with x, apply f^{-1}, and then apply f, we get the number x back again. To put it simply, what f does, f^{-1} undoes, and vice versa:

Input x —Apply f.→ $f(x)$ —Apply f^{-1}.→ $f^{-1}(f(x)) = x$

Input x —Apply f^{-1}.→ $f^{-1}(x)$ —Apply f.→ $f(f^{-1}(x)) = x$

In other words,

$$f^{-1}(f(x)) = x \quad \text{and} \quad f(f^{-1}(x)) = x$$

The preceding conditions can be used to verify that a function is, in fact, the inverse of f, as Example 3 demonstrates.

EXAMPLE 3 Verifying Inverse Functions

(a) We verify that the inverse of $g(x) = x^3$ is $g^{-1}(x) = \sqrt[3]{x}$ by showing that

$$g^{-1}(g(x)) = g^{-1}(x^3) = \sqrt[3]{x^3} = x$$

and

$$g(g^{-1}(x)) = g(\sqrt[3]{x}) = (\sqrt[3]{x})^3 = x$$

(b) We verify that the inverse of $h(x) = 3x$ is $h^{-1}(x) = \frac{1}{3}x$ by showing that

$$h^{-1}(h(x)) = h^{-1}(3x) = \tfrac{1}{3}(3x) = x$$

and

$$h(h^{-1}(x)) = h(\tfrac{1}{3}x) = 3(\tfrac{1}{3}x) = x$$

(c) We verify that the inverse of $f(x) = 2x + 3$ is $f^{-1}(x) = \frac{1}{2}(x - 3)$ by showing that

$$f^{-1}(f(x)) = f^{-1}(2x + 3) = \tfrac{1}{2}[(2x + 3) - 3] = \tfrac{1}{2}(2x) = x$$

and

$$f(f^{-1}(x)) = f(\tfrac{1}{2}(x - 3)) = 2[\tfrac{1}{2}(x - 3)] + 3 = (x - 3) + 3 = x$$

Now work Problem 21.

Exploration Simultaneously graph $Y_1 = x$, $Y_2 = x^3$, and $Y_3 = \sqrt[3]{x}$ on a square screen, using the viewing rectangle $-3 \le x \le 3$, $-2 \le y \le 2$. What do you observe about the graphs of $Y_2 = x^3$, its inverse $Y_3 = \sqrt[3]{x}$, and the line $Y_1 = x$?

Do you see the symmetry of the graph of f and its inverse with respect to the line $Y_1 = x$?

For the functions in Example 3(c), we list points on the graph of $f = Y_1$ and on the graph of $f^{-1} = Y_2$ in Table 1.

TABLE 1

X	Y1
-5	-7
-4	-5
-3	-3
-2	-1
-1	1
0	3
1	5

Y1=2X+3

X	Y2
-7	-5
-5	-4
-3	-3
-1	-2
1	-1
3	0
5	1

Y2=(1/2)(X-3)

FIGURE 6

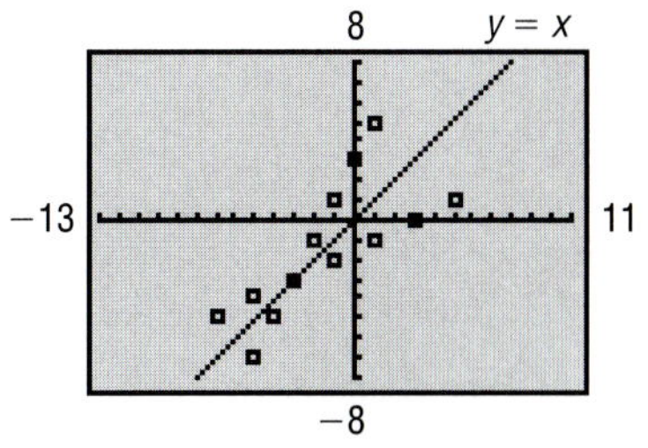

We notice that whenever (a, b) is on the graph of f then (b, a) is on the graph of f^{-1}. Figure 6 shows these points plotted. Also shown there is the graph of $y = x$, which you should observe is a line of symmetry of the points.

FIGURE 7

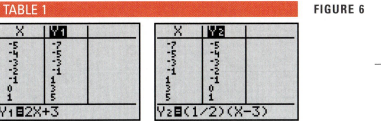

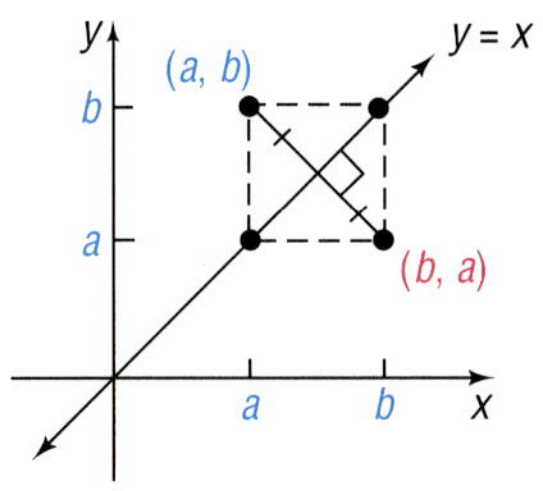

Geometric Interpretation

2 Suppose that (a, b) is a point on the graph of the one-to-one function f defined by $y = f(x)$. Then $b = f(a)$. This means that $a = f^{-1}(b)$, so (b, a) is a point on the graph of the inverse function f^{-1}. The relationship between the point (a, b) on f and the point (b, a) on f^{-1} is shown in Figure 7. The line

joining (a, b) and (b, a) is perpendicular to the line $y = x$ and is bisected by the line $y = x$. (Do you see why?) It follows that the point (b, a) on f^{-1} is the reflection about the line $y = x$ of the point (a, b) on f.

Theorem

The graph of a function f and the graph of its inverse f^{-1} are symmetric with respect to the line $y = x$.

Figure 8 illustrates this result. Notice that, once the graph of f is known, the graph of f^{-1} may be obtained by folding the paper along the line $y = x$.

FIGURE 8

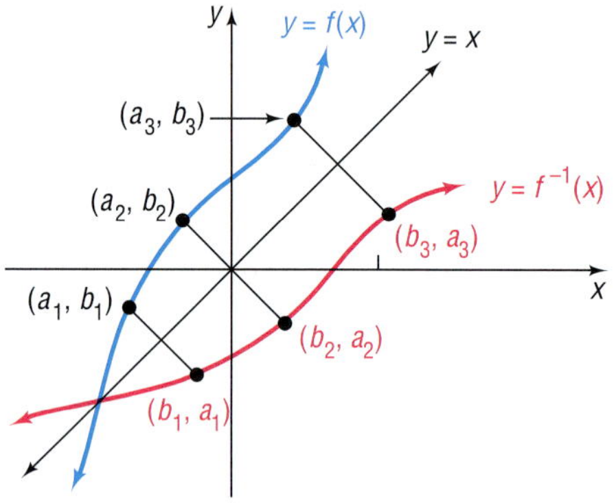

EXAMPLE 4 Graphing the Inverse Function

The graph in Figure 9(a) is that of a one-to-one function $y = f(x)$. Draw the graph of its inverse.

Solution We begin by adding the graph of $y = x$ to Figure 9(a). Since the points $(-2, -1)$, $(-1, 0)$, and $(2, 1)$ are on the graph of f, we know that the points $(-1, -2)$, $(0, -1)$, and $(1, 2)$ must be on the graph of f^{-1}. Keeping in mind that the graph of f^{-1} is the reflection about the line $y = x$ of the graph of f, we can draw f^{-1}. See Figure 9(b).

FIGURE 9

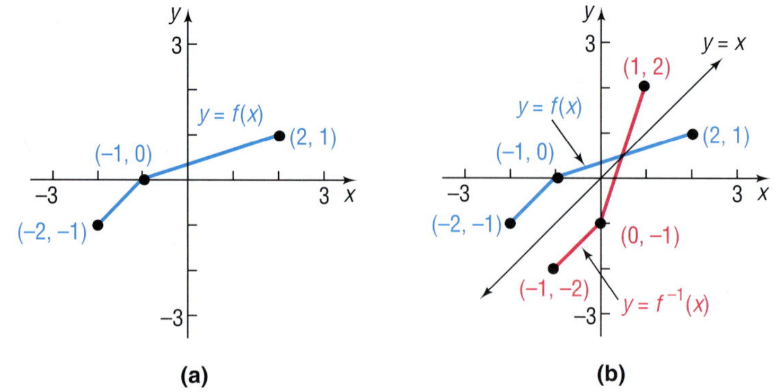

Now work Problem 15.

Finding the Inverse Function

3 The fact that the graph of a one-to-one function f and its inverse are symmetric with respect to the line $y = x$ tells us more. It says that we can obtain

f^{-1} by interchanging the roles of x and y in f. Look again at Figure 8. If f is defined by the equation

$$y = f(x)$$

then f^{-1} is defined by the equation

$$x = f(y)$$

The equation $x = f(y)$ defines f^{-1} *implicitly*. If we can solve this equation for y, we will have the *explicit* form of f^{-1}, that is,

$$y = f^{-1}(x)$$

Let's use this procedure to find the inverse of $f(x) = 2x + 3$. (Since f is a linear function and is increasing, we know that f is one-to-one.)

EXAMPLE 5 Finding the Inverse Function

Find the inverse of $f(x) = 2x + 3$. Also find the domain and range of f and f^{-1}. Graph f and f^{-1} on the same coordinate axes.

Solution In the equation $y = 2x + 3$, interchange the variables x and y. The result,

$$x = 2y + 3$$

is an equation that defines the inverse f^{-1} implicitly. Solving for y, we obtain

$$\begin{aligned} 2y + 3 &= x \\ 2y &= x - 3 \\ y &= \tfrac{1}{2}(x - 3) \end{aligned}$$

The explicit form of the inverse f^{-1} is therefore

$$f^{-1}(x) = \tfrac{1}{2}(x - 3)$$

which we verified in Example 3(c).

Then we find

$$\begin{aligned} \text{Domain } f = \text{Range } f^{-1} &= (-\infty, \infty) \\ \text{Range } f = \text{Domain } f^{-1} &= (-\infty, \infty) \end{aligned}$$

The graphs of $Y_1 = f(x) = 2x + 3$ and its inverse $Y_2 = f^{-1}(x) = \frac{1}{2}(x - 3)$ are shown in Figure 10. Note the symmetry of the graphs with respect to the line $Y_3 = x$.

FIGURE 10

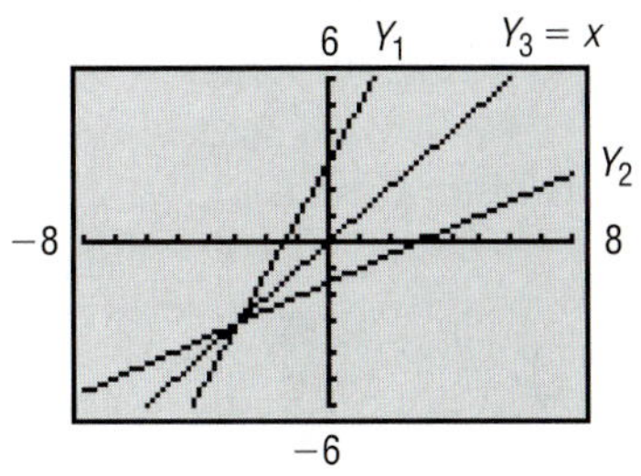

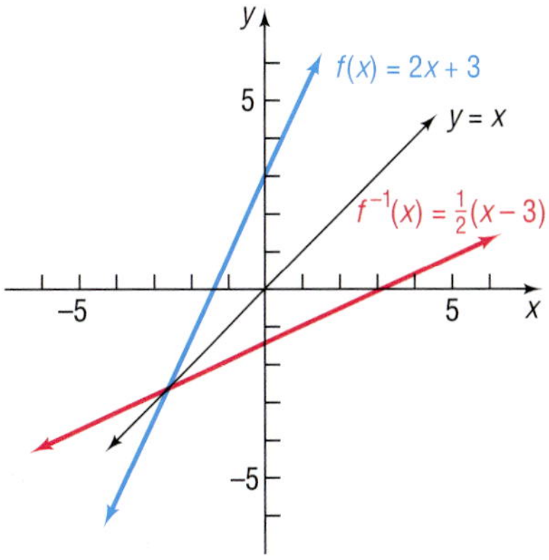

We outline next the steps to follow for finding the inverse of a one-to-one function.

Procedure for Finding the Inverse of a One-to-One Function

STEP 1: In $y = f(x)$, interchange the variables x and y to obtain

$$x = f(y)$$

This equation defines the inverse function f^{-1} implicitly.

STEP 2: If possible, solve the implicit equation for y in terms of x to obtain the explicit form of f^{-1}:

$$y = f^{-1}(x)$$

STEP 3: Check the result by showing that

$$f^{-1}(f(x)) = x \quad \text{and} \quad f(f^{-1}(x)) = x$$

EXAMPLE 6 Finding the Inverse Function

The function

$$f(x) = \frac{2x+1}{x-1}, \quad x \neq 1$$

is one-to-one. Find its inverse and check the result.

Solution STEP 1: Interchange the variables x and y in

$$y = \frac{2x+1}{x-1}$$

to obtain

$$x = \frac{2y+1}{y-1}$$

STEP 2: Solve for y:

$$\begin{aligned} x &= \frac{2y+1}{y-1} \\ x(y-1) &= 2y+1 \\ xy - x &= 2y+1 \\ xy - 2y &= x+1 \\ (x-2)y &= x+1 \\ y &= \frac{x+1}{x-2} \end{aligned}$$

The inverse is

$$f^{-1}(x) = \frac{x+1}{x-2}, \quad x \neq 2$$

STEP 3: *Check:*

$$f^{-1}(f(x)) = f^{-1}\left(\frac{2x+1}{x-1}\right) = \frac{\dfrac{2x+1}{x-1}+1}{\dfrac{2x+1}{x-1}-2} = \frac{2x+1+x-1}{2x+1-2(x-1)} = \frac{3x}{3} = x$$

$$f(f^{-1}(x)) = f\left(\frac{x+1}{x-2}\right) = \frac{2\left(\dfrac{x+1}{x-2}\right)+1}{\dfrac{x+1}{x-2}-1} = \frac{2(x+1)+x-2}{x+1-(x-2)} = \frac{3x}{3} = x$$

Exploration We found that, if $f(x) = (2x + 1)/(x - 1)$, then $f^{-1}(x) = (x + 1)/(x - 2)$. Graph $y = f(f^{-1}(x))$ on a square screen. What do you see? Are you surprised?

Now work Problem 33.

If a function is not one-to-one, then it will have no inverse function. Sometimes, though, an appropriate restriction on the domain of such a function will yield a new function that is one-to-one. Let's look at an example of this common practice.

EXAMPLE 7 Finding the Inverse Function

Find the inverse of $y = f(x) = x^2$ if $x \geq 0$.

Solution The function $f(x) = x^2$ is not one-to-one. [Refer to Example 2(a).] However, if we restrict f to only that part of its domain for which $x \geq 0$, as indicated, we have a new function that is increasing and therefore is one-to-one. As a result, the function defined by $y = x^2$, $x \geq 0$, has an inverse function, f^{-1}.

We follow the steps given previously to find f^{-1}:

STEP 1: In the equation $y = x^2$, $x \geq 0$, interchange the variables x and y. The result is

$$x = y^2 \qquad y \geq 0$$

This equation defines (implicitly) the inverse function.

STEP 2: We solve for y to get the explicit form of the inverse. Since $y \geq 0$, only one solution for y is obtained:

$$y = \sqrt{x}$$

so that $f^{-1}(x) = \sqrt{x}$.

STEP 3: *Check:* $f^{-1}(f(x)) = f^{-1}(x^2) = \sqrt{x^2} = |x| = x, \quad$ since $x \geq 0$

$$f(f^{-1}(x)) = f(\sqrt{x}) = (\sqrt{x})^2 = x$$

Figure 11 illustrates the graphs of $Y_1 = f(x) = x^2$, $x \geq 0$, and $Y_2 = f^{-1}(x) = \sqrt{x}$.

FIGURE 11

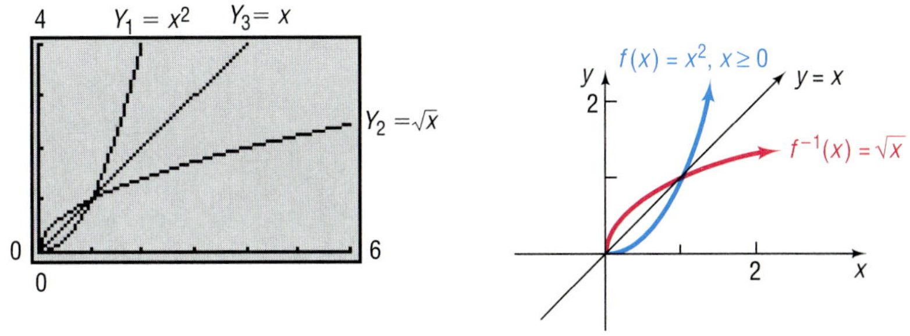

SUMMARY

1. If a function f is one-to-one, then it has an inverse function f^{-1}.
2. Domain f = Range f^{-1}; Range f = Domain f^{-1}.
3. To verify that f^{-1} is the inverse of f, show that $f^{-1}(f(x)) = x$ and $f(f^{-1}(x)) = x$.
4. The graphs of f and f^{-1} are symmetric with respect to the line $y = x$.

5.1 EXERCISES

In Problems 1–8, (a) find the inverse and (b) determine whether the inverse represents a function.

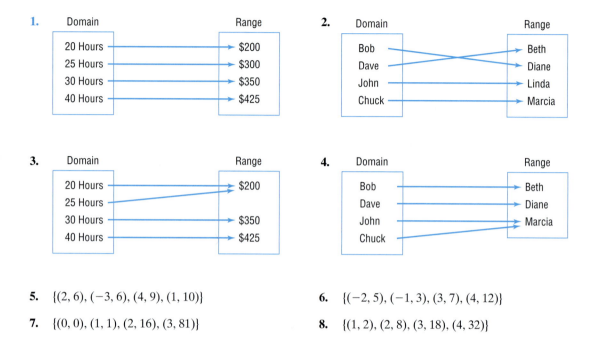

1.

Domain
Range
20 Hours
25 Hours
30 Hours
40 Hours
$200
$300
$350
$425

2.

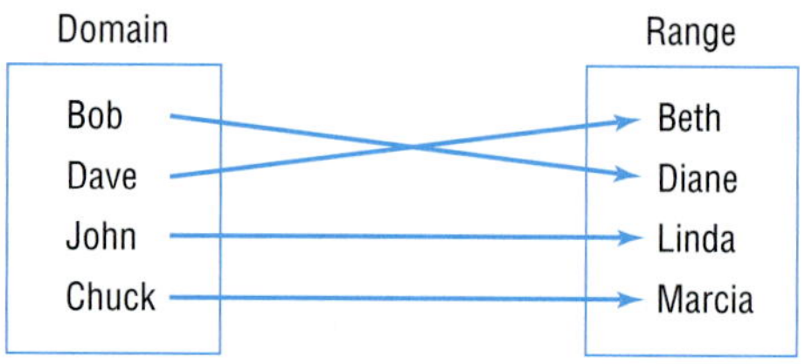
Domain
Range
Bob
Dave
John
Chuck
Beth
Diane
Linda
Marcia

3.

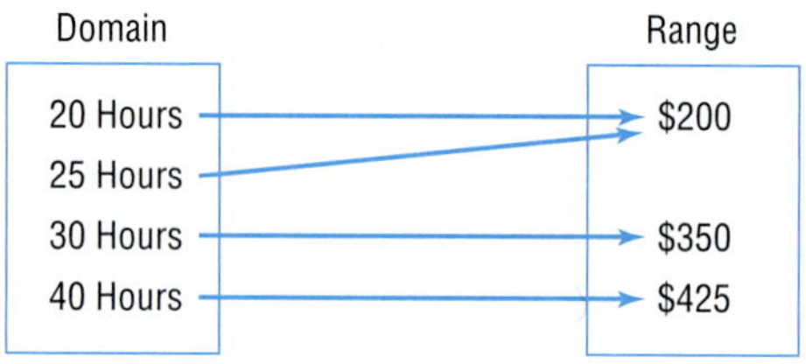
Domain
Range
20 Hours
25 Hours
30 Hours
40 Hours
$200
$350
$425

4.

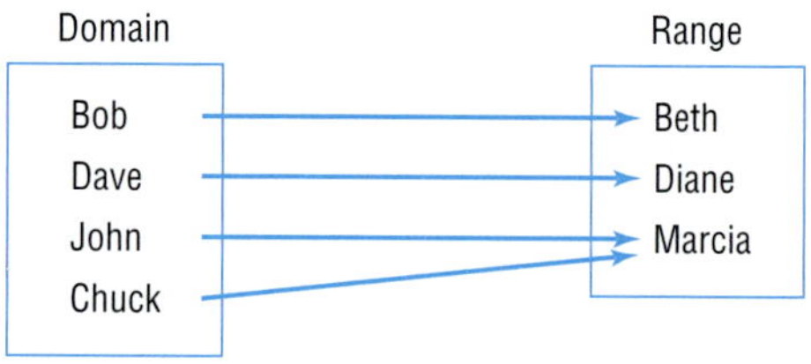
Domain
Range
Bob
Dave
John
Chuck
Beth
Diane
Marcia

5. $\{(2, 6), (-3, 6), (4, 9), (1, 10)\}$

6. $\{(-2, 5), (-1, 3), (3, 7), (4, 12)\}$

7. $\{(0, 0), (1, 1), (2, 16), (3, 81)\}$

8. $\{(1, 2), (2, 8), (3, 18), (4, 32)\}$

In Problems 9–14, the graph of a function f is given. Use the horizontal line test to determine whether f is one-to-one.

9.

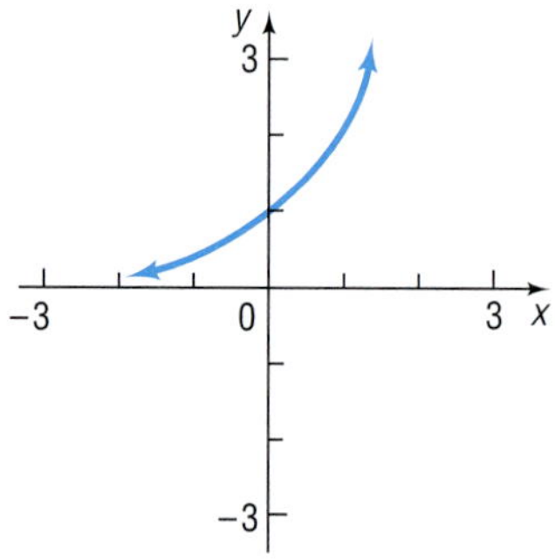

10.

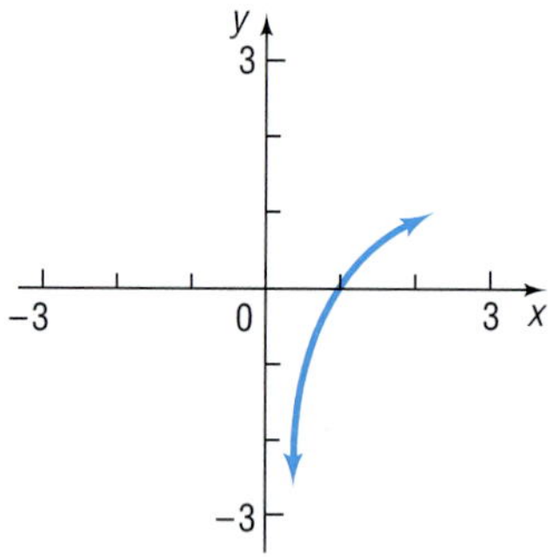

11.

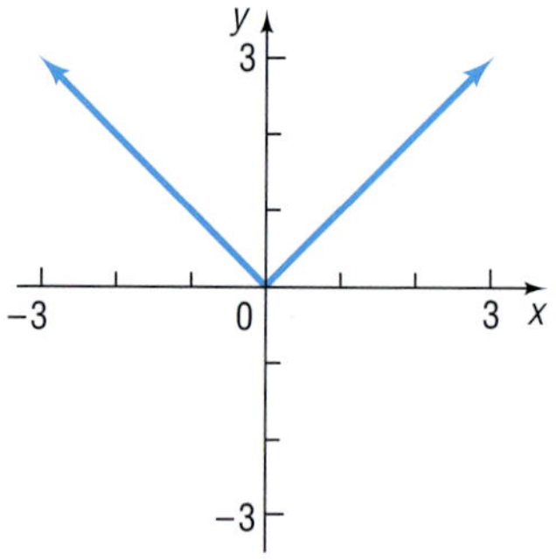

12.

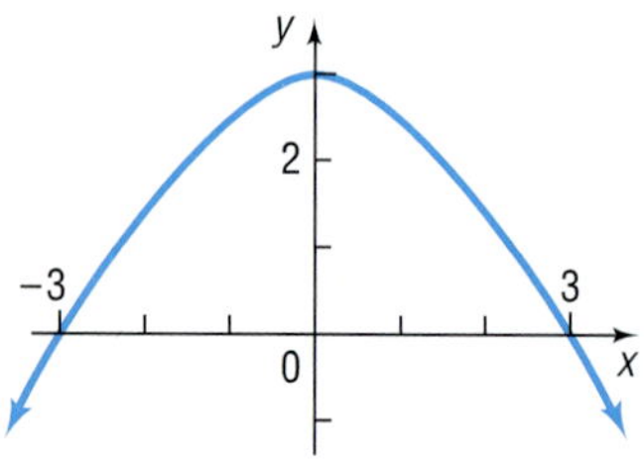

13.

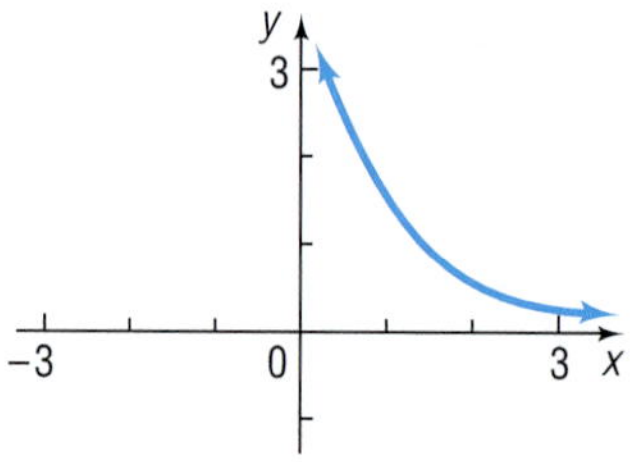

14.

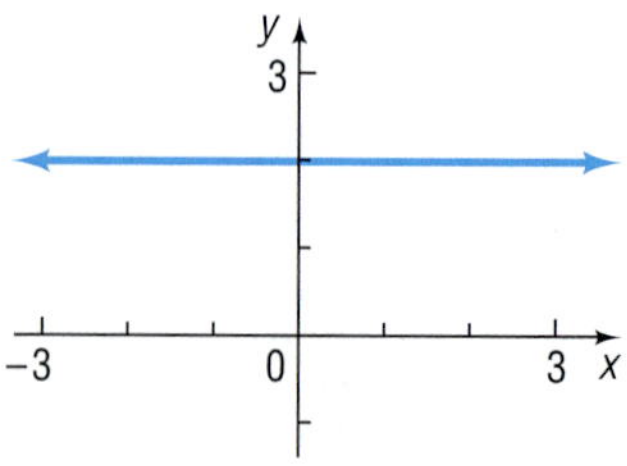

In Problems 15–20, the graph of a one-to-one function f is given. Find the graph of the inverse function f^{-1}. For convenience (and as a hint), the graph of $y = x$ is also given.

15.

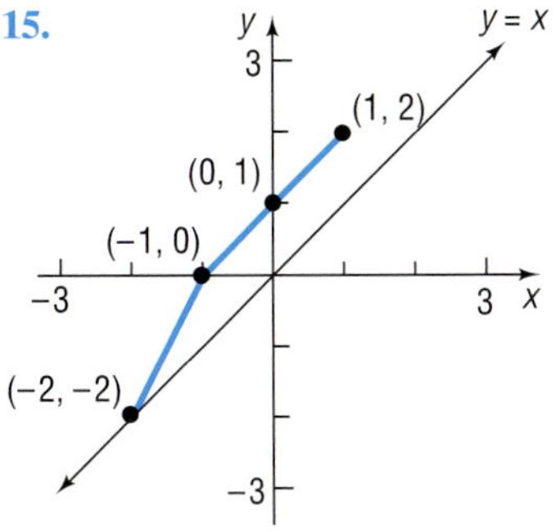

16.

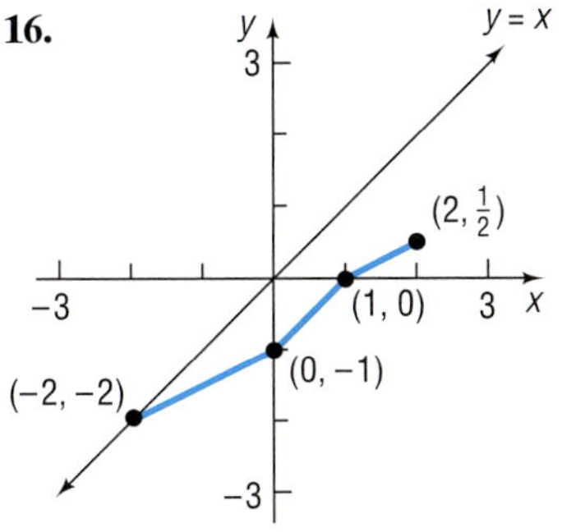

17.

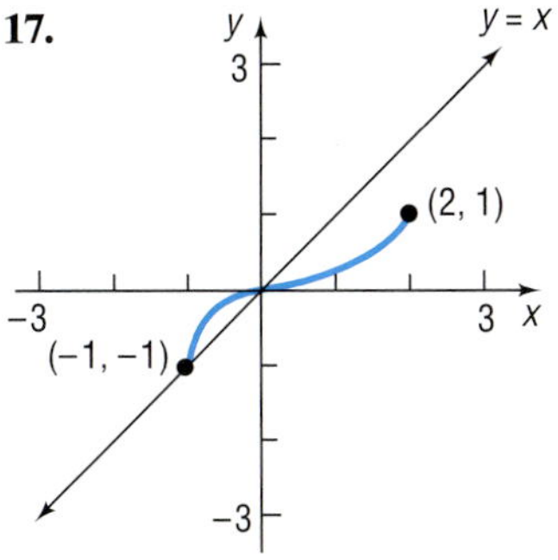

18.

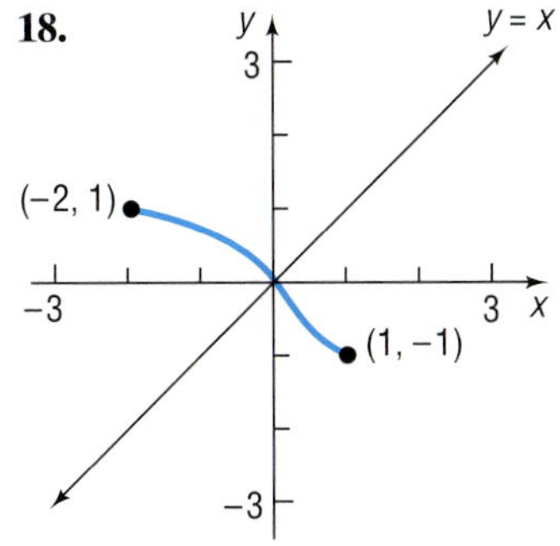

19.

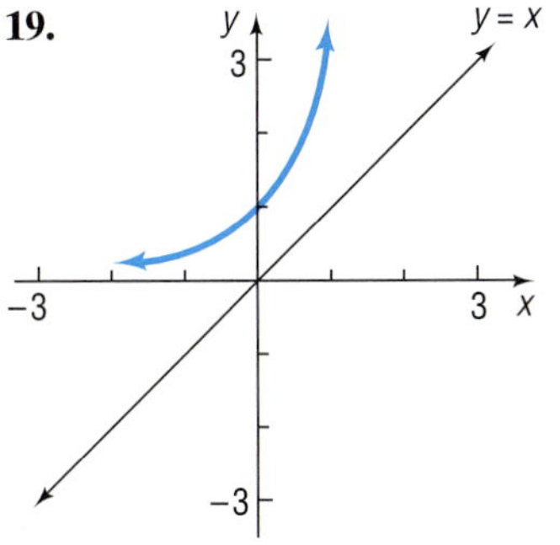

20.

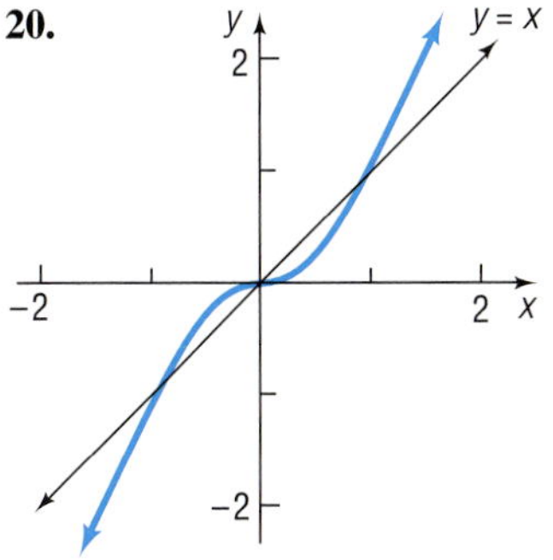

In Problems 21–30, verify that the functions f and g are inverses of each other by showing that $f(g(x)) = x$ and $g(f(x)) = x$. Using a graphing utility, simultaneously graph f, g, and $y = x$ on the same square screen.

21. $f(x) = 3x + 4;\quad g(x) = \frac{1}{3}(x - 4)$

22. $f(x) = 3 - 2x;\quad g(x) = -\frac{1}{2}(x - 3)$

23. $f(x) = 4x - 8;\quad g(x) = \frac{x}{4} + 2$

24. $f(x) = 2x + 6;\quad g(x) = \frac{1}{2}x - 3$

25. $f(x) = x^3 - 8;\quad g(x) = \sqrt[3]{x + 8}$

26. $f(x) = (x - 2)^2;\ \ x \geq 2;\quad g(x) = \sqrt{x} + 2,\ x \geq 0$

27. $f(x) = \frac{1}{x};\quad g(x) = \frac{1}{x}$

28. $f(x) = x;\quad g(x) = x$

29. $f(x) = \frac{2x + 3}{x + 4};\quad g(x) = \frac{4x - 3}{2 - x}$

30. $f(x) = \frac{x - 5}{2x + 3};\quad g(x) = \frac{3x + 5}{1 - 2x}$

In Problems 31–42, the function f is one-to-one. Find its inverse and check your answer. State the domain and range of f and f^{-1}. By hand, graph f, f^{-1}, and $y = x$ on the same coordinate axes. Check your results using a graphing utility.

31. $f(x) = 3x$

32. $f(x) = -4x$

33. $f(x) = 4x + 2$

34. $f(x) = 1 - 3x$

35. $f(x) = x^3 - 1$

36. $f(x) = x^3 + 1$

37. $f(x) = x^2 + 4;\quad x \geq 0$

38. $f(x) = x^2 + 9,\quad x \geq 0$

39. $f(x) = \frac{4}{x}$

40. $f(x) = -\frac{3}{x}$

41. $f(x) = \frac{1}{x - 2}$

42. $f(x) = \frac{4}{x + 2}$

In Problems 43–54, the function f is one-to-one. Find its inverse and check your answer. State the domain and range of f and f^{-1}. Using a graphing utility, simultaneously graph f, f^{-1}, and $y = x$ on the same square screen.

43. $f(x) = \frac{2}{3 + x}$

44. $f(x) = \frac{4}{2 - x}$

45. $f(x) = (x + 2)^2,\quad x \geq -2$

46. $f(x) = (x - 1)^2, \quad x \geq 1$

47. $f(x) = \dfrac{2x}{x - 1}$

48. $f(x) = \dfrac{3x + 1}{x}$

49. $f(x) = \dfrac{3x + 4}{2x - 3}$

50. $f(x) = \dfrac{2x - 3}{x + 4}$

51. $f(x) = \dfrac{2x + 3}{x + 2}$

52. $f(x) = \dfrac{-3x - 4}{x - 2}$

53. $f(x) = 2\sqrt[3]{x}$

54. $f(x) = \dfrac{4}{\sqrt{x}}$

55. Find the inverse of the linear function $f(x) = mx + b, m \neq 0$.

56. Find the inverse of the function $f(x) = \sqrt{r^2 - x^2}, 0 \leq x \leq r$.

57. Can an even function be one-to-one? Explain.

58. Is every odd function one-to-one? Explain.

59. A function f has an inverse. If the graph of f lies in quadrant I, in which quadrant does the graph of f^{-1} lie?

60. A function f has an inverse. If the graph of f lies in quadrant II, in which quadrant does the graph of f^{-1} lie?

61. The function $f(x) = |x|$ is not one-to-one. Find a suitable restriction on the domain of f so that the new function that results is one-to-one. Then find the inverse of f.

62. The function $f(x) = x^4$ is not one-to-one. Find a suitable restriction on the domain of f so that the new function that results is one-to-one. Then find the inverse of f.

63. **Temperature Conversion** To convert from x degrees Celsius to y degrees Fahrenheit, we use the formula $y = f(x) = \frac{9}{5}x + 32$. To convert from x degrees Fahrenheit to y degrees Celsius, we use the formula $y = g(x) = \frac{5}{9}(x - 32)$. Show that f and g are inverse functions.

64. **Demand for Corn** The demand for corn obeys the equation $p(x) = 300 - 50x$, where p is the price per bushel (in dollars) and x is the number of bushels produced, in millions. Express the production amount x as a function of the price p.

65. **Period of a Pendulum** The period T (in seconds) of a simple pendulum is a function of its length l (in feet), given by $T(l) = 2\pi\sqrt{l/g}$, where $g \approx 32.2$ feet per second per second is the acceleration of gravity. Express the length l as a function of the period T.

66. Give an example of a function whose domain is the set of real numbers and that is neither increasing nor decreasing on its domain, but is one-to-one.

[**Hint:** Use a piecewise defined function.]

67. Given

$$f(x) = \frac{ax + b}{cx + d}$$

find $f^{-1}(x)$. If $c \neq 0$, under what conditions on a, b, c, and d is $f = f^{-1}$?

68. We said earlier that finding the range of a function f is not easy. However, if f is one-to-one, we can find its range by finding the domain of the inverse function f^{-1}. Use this technique to find the range of each of the following one-to-one functions:

(a) $f(x) = \dfrac{2x + 5}{x - 3}$

(b) $g(x) = 4 - \dfrac{2}{x}$

(c) $F(x) = \dfrac{3}{4 - x}$

For Problems 69–74, write a program that will graph the inverse of a function $y = f(x)$. Then graph the function f and its inverse on the same screen. Compare your answers with those of Problems 31–36.

69. $f(x) = 3x$

70. $f(x) = -4x$

71. $f(x) = 4x + 2$

72. $f(x) = 1 - 3x$

73. $f(x) = x^3 - 1$

74. $f(x) = x^3 + 1$

75. If the graph of a function and its inverse intersect, where must this necessarily occur? Can they intersect anywhere else? Must they intersect?

76. Can a one-to-one function and its inverse be equal? What must be true about the graph of f for this to happen? Give some examples to support your conclusion.

77. Draw the graph of a one-to-one function that contains the points $(-2, -3)$, $(0, 0)$, and $(1, 5)$. Now draw the graph of its inverse. Compare your graph to those of other students. Discuss any similarities. What differences do you see?

5.2 EXPONENTIAL FUNCTIONS

1 Evaluate Exponential Functions
2 Graph Exponential Functions
3 Define the Number *e*

1 In the Appendix, Section 3, we gave a definition for raising a real number a to a rational power. Based on that discussion, we gave meaning to expressions of the form

$$a^r$$

where the base a is a positive real number and the exponent r is a rational number.

But what is the meaning of a^x, where the base a is a positive real number and the exponent x is an irrational number? Although a rigorous definition requires methods discussed in calculus, the basis for the definition is easy to follow: Select a rational number r that is formed by truncating (removing) all but a finite number of digits from the irrational number x. Then it is reasonable to expect that

$$a^x \approx a^r$$

For example, take the irrational number $\pi = 3.14159 \ldots$. Then, an approximation to a^π is

$$a^\pi \approx a^{3.14}$$

where the digits after the hundredths position have been removed from the value for π. A better approximation would be

$$a^\pi \approx a^{3.14159}$$

where the digits after the hundred-thousandths position have been removed. Continuing in this way, we can obtain approximations to a^π to any desired degree of accuracy.

Graphing calculators can easily evaluate expressions of the form a^x as follows. Enter the base a, then press the caret key (^), enter the exponent x, and press enter.

EXAMPLE 1 Using a Calculator to Evaluate Powers of 2

Using a calculator, evaluate:

(a) $2^{1.4}$ (b) $2^{1.41}$ (c) $2^{1.414}$ (d) $2^{1.4142}$ (e) $2^{\sqrt{2}}$

Solution Figure 12 shows the solution to (a) using a T1-83 graphing calculator.

(a) $2^{1.4} \approx 2.639015822$ (b) $2^{1.41} \approx 2.657371628$

FIGURE 12

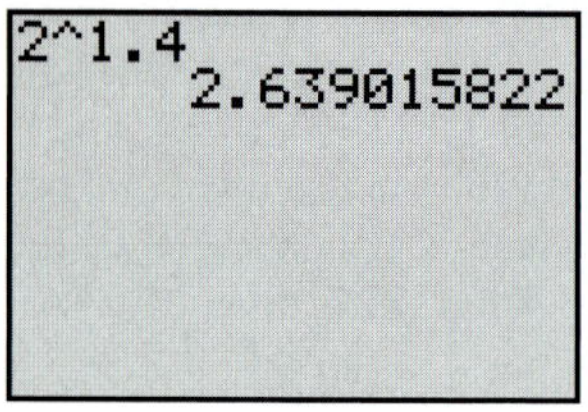

(c) $2^{1.414} \approx 2.66474965$ (d) $2^{1.4142} \approx 2.665119089$

(e) $2^{\sqrt{2}} \approx 2.665144143$

Now work Problem 1.

It can be shown that the familiar laws of rational exponents hold for real exponents.

Theorem Laws of Exponents

If s, t, a, and b are real numbers with $a > 0$ and $b > 0$, then

$$a^s \cdot a^t = a^{s+t} \qquad (a^s)^t = a^{st} \qquad (ab)^s = a^s \cdot b^s$$
$$1^s = 1 \qquad a^{-s} = \frac{1}{a^s} = \left(\frac{1}{a}\right)^s \qquad a^0 = 1 \tag{1}$$

We are now ready for the following definition.

An **exponential function** is a function of the form

$$f(x) = a^x$$

where a is a positive real number ($a > 0$) and $a \neq 1$. The domain of f is the set of all real numbers.

We exclude the base $a = 1$, because this function is simply the constant function $f(x) = 1^x = 1$. We also need to exclude bases that are negative, because, otherwise, we would have to exclude many values of x from the domain, such as $x = \frac{1}{2}$, $x = \frac{3}{4}$, and so on. [Recall that $(-2)^{1/2}$, $(-3)^{3/4}$, and so on, are not defined in the system of real numbers.]

Graphs of Exponential Functions

2 First, we graph the exponential function $f(x) = 2^x$.

EXAMPLE 2 Graphing an Exponential Function

Graph the exponential function $f(x) = 2^x$.

Solution The domain of $f(x) = 2^x$ consists of all real numbers. We begin by locating some points on the graph of $f(x) = 2^x$, as listed in Table 2.

Since $2^x > 0$ for all x, the range of f is $(0, \infty)$. From this, we conclude that the graph has no x-intercepts, and, in fact, the graph will lie above the x-axis. As Table 2 indicates, the y-intercept is 1. Table 2 also indicates that as $x \to -\infty$ the value of $f(x) = 2^x$ gets closer and closer to 0. Thus, the x-axis is a horizontal asymptote to the graph as $x \to -\infty$. Look again at Table 2. As $x \to \infty$, $f(x) = 2^x$ grows very quickly, causing the graph of $f(x) = 2^x$ to rise very rapidly. Thus, it is apparent that f is an increasing function and, hence, is one-to-one.

Figure 13 shows the graph of $f(x) = 2^x$. Notice that all the conclusions given earlier are confirmed by the graph.

TABLE 2

X	Y1
-10	9.8E-4
-3	.125
-1	.5
0	1
1	2
3	8
10	1024

Y1=2^X

FIGURE 13

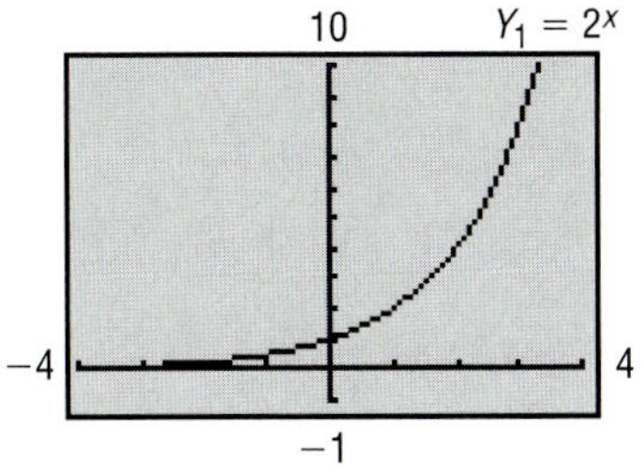

As we shall see, graphs that look like the one in Figure 13 occur very frequently in a variety of situations. For example, look at the graph in Figure 14, which illustrates the population in Ethiopia. Researchers might conclude from this graph that the population in Ethiopia is "behaving exponentially"; that is, the graph exhibits "rapid, or exponential, growth." We shall have more to say about situations that lead to exponential growth later in this chapter. For now, we continue to seek properties of the exponential functions.

FIGURE 14
Ethiopia's population is growing at one of the fastest paces in the world, exacerbating continuing shortfalls in food needed to feed its millions of people. The nation, which is only about three-quarters the size of Alaska, grows with more than 2 million births each year.

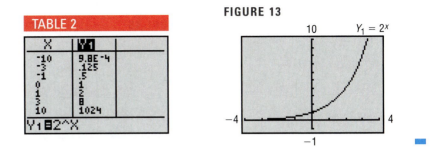

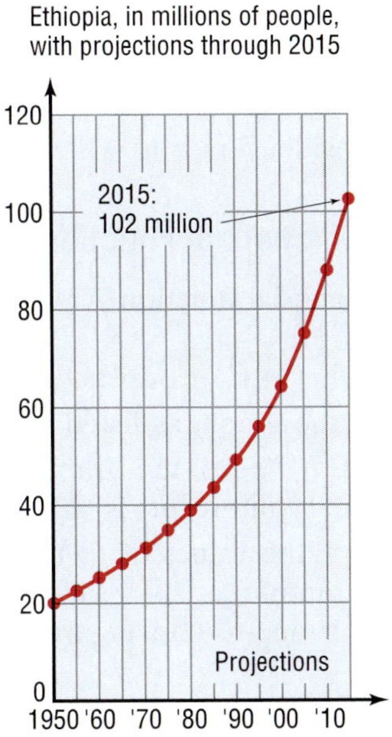

Sources: News reports, World Bank, U.S. Agency for International Development, UN Food and Agriculture Organization, Population Reference Bureau, UNICEF, U.S. Department of Agriculture, Human Nutrition Information Service.

The graph of $f(x) = 2^x$ in Figure 13 is typical of all exponential functions that have a base larger than 1. Such functions are increasing functions and hence are one-to-one. Their graphs lie above the x-axis, pass through the point $(0, 1)$, and thereafter rise rapidly as $x \to \infty$. As $x \to -\infty$, the x-axis is a horizontal asymptote. There are no vertical asymptotes. Finally, the graphs are smooth and continuous, with no corners or gaps.

Figure 15 illustrates the graphs of two more exponential functions whose bases are larger than 1. Notice that for the larger base the graph is steeper when $x > 0$. Figure 16 shows that when $x < 0$, the graph of the equation with the larger base is closer to the x-axis.

FIGURE 15

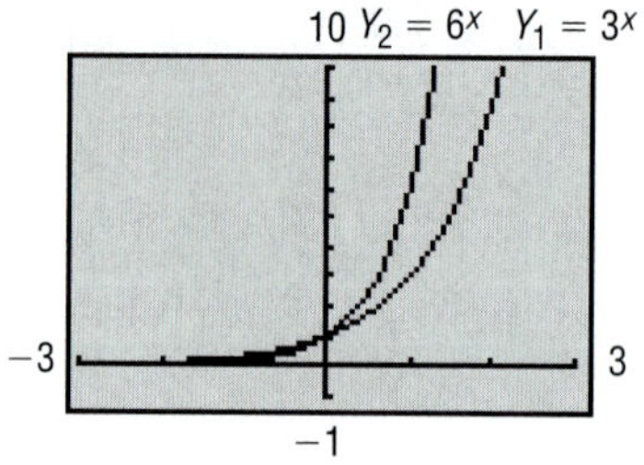

FIGURE 16

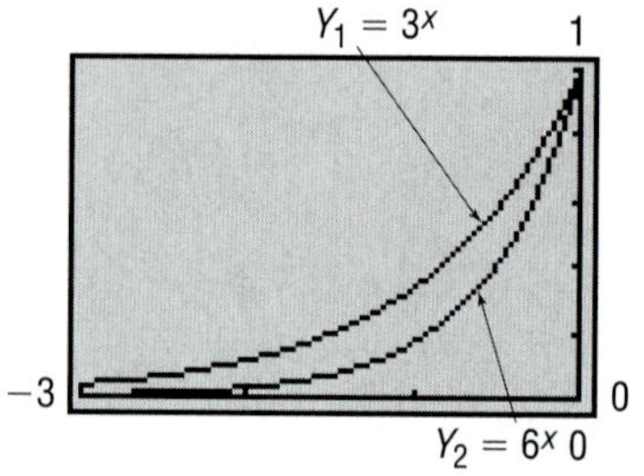

The display summarizes the information we have about $f(x) = a^x$, $a > 1$:

> $f(x) = a^x$ $a > 1$
> Domain: $(-\infty, \infty)$ Range: $(0, \infty)$
> x-intercepts: None y-intercept: 1
> Horizontal asymptote: x-axis, as $x \to -\infty$
> f is an increasing function
> f is one-to-one
> The graph of f contains the points $(0, 1)$ and $(1, a)$

Now we consider $f(x) = a^x$ when $0 < a < 1$.

EXAMPLE 3 Graphing an Exponential Function

Graph the exponential function $f(x) = \left(\frac{1}{2}\right)^x$.

Solution The domain of $f(x) = \left(\frac{1}{2}\right)^x$ consists of all real numbers. As before, we locate some points on the graph, as listed in Table 3. Since $\left(\frac{1}{2}\right)^x > 0$ for all x, the range of f is $(0, \infty)$. Thus, the graph lies above the x-axis and so has no x-intercepts. The y-intercept is 1. As $x \to -\infty$, $f(x) = \left(\frac{1}{2}\right)^x$ grows very quickly. As $x \to \infty$, the values of $f(x)$ approach 0. Thus, the x-axis ($y = 0$) is a horizontal asymptote as $x \to \infty$. It is apparent that f is a decreasing function and, hence, is one-to-one. Figure 17 illustrates the graph.

TABLE 3

X	Y1
-10	1024
-3	8
-1	2
0	1
1	.5
3	.125
10	9.8E-4

Y1■(1/2)^X

FIGURE 17

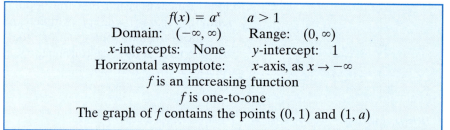

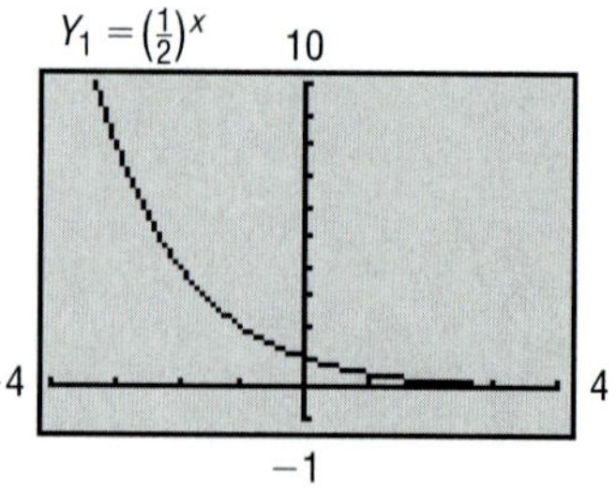

Note that we could have obtained the graph of $y = \left(\frac{1}{2}\right)^x$ from the graph of $y = 2^x$. If $f(x) = 2^x$, then $f(-x) = 2^{-x} = \frac{1}{2^x} = \left(\frac{1}{2}\right)^x$. Thus, the graph of $y = \left(\frac{1}{2}\right)^x = 2^{-x}$ is a reflection about the y-axis of the graph of $y = 2^x$. Compare Figures 13 and 17.

Seeing the Concept Using a graphing utility, simultaneously graph:

(a) $Y_1 = 3^x, Y_2 = \left(\dfrac{1}{3}\right)^x$ (b) $Y_1 = 6^x, Y_2 = \left(\dfrac{1}{6}\right)^x$

Conclude the graph of $Y_2 = \left(\dfrac{1}{a}\right)^x$ for $a > 0$ is the reflection about the y-axis of the graph of $Y_1 = a^x$.

The graph of $f(x) = \left(\frac{1}{2}\right)^x$ in Figure 17 is typical of all exponential functions that have a base between 0 and 1. Such functions are decreasing, one-to-one functions. Their graphs lie above the x-axis and pass through the point $(0, 1)$. The graphs rise rapidly as $x \to -\infty$. As $x \to \infty$, the x-axis is a horizontal asymptote. There are no vertical asymptotes. Finally, the graphs are smooth and continuous, with no corners or gaps.

Figure 18 illustrates the graphs of two more exponential functions whose bases are between 0 and 1. Notice that the choice of a base closer to 0 results in a graph that is steeper when $x < 0$. Figure 19 shows that when $x > 0$, the graph of the equation with the smaller base is closer to the x-axis.

FIGURE 18

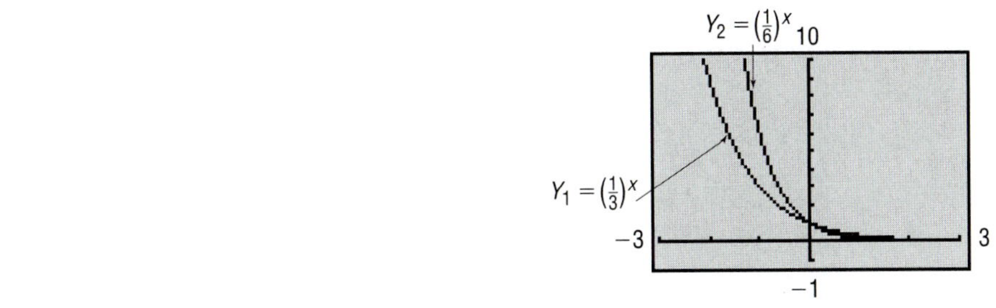

FIGURE 19

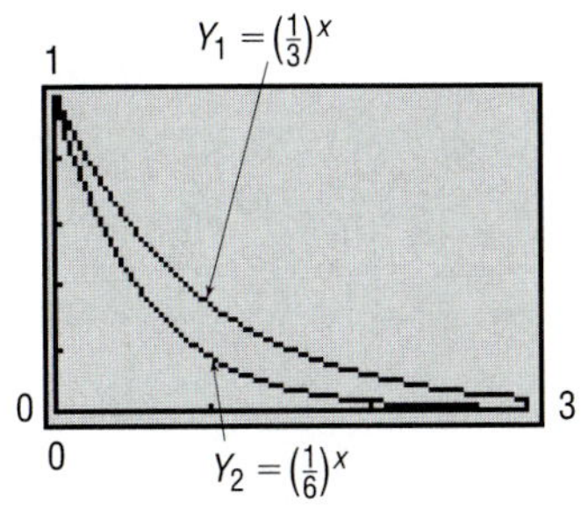

The display summarizes the information we have about the function $f(x) = a^x$, $0 < a < 1$:

> $f(x) = a^x \quad 0 < a < 1$
> Domain: $(-\infty, \infty)$ Range: $(0, \infty)$
> x-intercepts: None y-intercept: 1
> Horizontal asymptote: x-axis, as $x \to \infty$
> f is a decreasing function
> f is one-to-one
> The graph of f contains the points $(0, 1)$ and $(1, a)$

The techniques of shifting, compression, stretching, and reflection may be used to graph many functions that are basically exponential functions.

EXAMPLE 4 Graphing Exponential Functions

Graph $f(x) = 2^{-x} - 3$ and determine the domain, range and horizontal asymptote of f.

Solution Figure 20 shows the various steps.

FIGURE 20

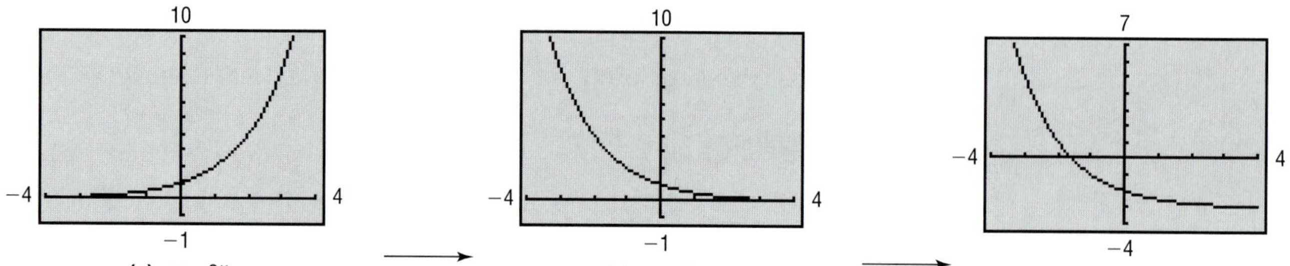

(a) $y = 2^x$ → Reflection about y-axis; replace x by $-x$ → **(b)** $y = 2^{-x}$ → Shift down 3 units; subtract 3 → **(c)** $y = 2^{-x} - 3$

As Figure 20(c) illustrates, the domain of $f(x) = 2^{-x} - 3$ is $(-\infty, \infty)$ and the range is $(-3, \infty)$. The horizontal asymptote of f is the line $y = -3$.

EXAMPLE 5 Graphing Exponential Functions

Graph $f(x) = -(2^{x-3})$ and determine the domain, range, and horizontal asymptote of f.

Solution Figure 21 shows the various steps.

FIGURE 21

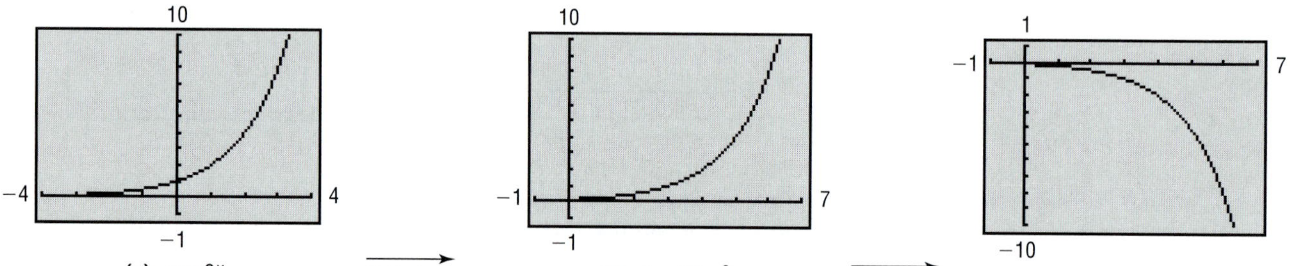

(a) $y = 2^x$ → Shift right 3 units; replace x by $x - 3$ → **(b)** $y = 2^{x-3}$ → Reflection about x-axis; multiply by -1 → **(c)** $y = -(2^{x-3})$

The domain of $f(x) = -(2^{x-3})$ is $(-\infty, \infty)$ and the range is $(-\infty, 0)$. The horizontal asymptote of f is $y = 0$.

Now work Problems 11 and 13.

The Base e

3 As we shall see shortly, many problems that occur in nature require the use of an exponential function whose base is a certain irrational number, symbolized by the letter e.

Let's look now at one way of arriving at this important number e.

The **number e** is defined as the number that the expression

$$\left(1 + \frac{1}{n}\right)^n \qquad (2)$$

approaches as $n \to \infty$. In calculus, this is expressed using limit notation as

$$e = \lim_{n \to \infty} \left(1 + \frac{1}{n}\right)^n$$

Table 4 illustrates what happens to the defining expression (2) as n takes on increasingly large values. The last number in the last column in the table is correct to nine decimal places and is the same as the entry given for e on your calculator (if expressed correct to nine decimal places).

TABLE 4

n	$\frac{1}{n}$	$1 + \frac{1}{n}$	$\left(1 + \frac{1}{n}\right)^n$
1	1	2	2
2	0.5	1.5	2.25
5	0.2	1.2	2.48832
10	0.1	1.1	2.59374246
100	0.01	1.01	2.704813829
1,000	0.001	1.001	2.716923932
10,000	0.0001	1.0001	2.718145927
100,000	0.00001	1.00001	2.718268237
1,000,000	0.000001	1.000001	2.718280469
1,000,000,000	10^{-9}	$1 + 10^{-9}$	2.718281827

The exponential function $f(x) = e^x$, whose base is the number e, occurs with such frequency in applications that it is usually referred to as *the* exponential function. Indeed, graphing calculators have the key $\boxed{e^x}$ or $\boxed{exp(x)}$, which may be used to evaluate the exponential function for a given value of x. Now use your calculator to find e^x for $x = -2$, $x = -1$, $x = 0$, $x = 1$, and $x = 2$, as we have done to create Table 5. The graph of the exponential function $f(x) = e^x$ is given in Figure 22. Since $2 < e < 3$, the graph of $y = e^x$ lies between the graphs of $y = 2^x$ and $y = 3^x$. (Refer to Figures 13 and 15.)

TABLE 5

X	Y1
-2	.13534
-1	.36788
0	1
1	2.7183
2	7.3891

Y1 = e^(X)

FIGURE 22

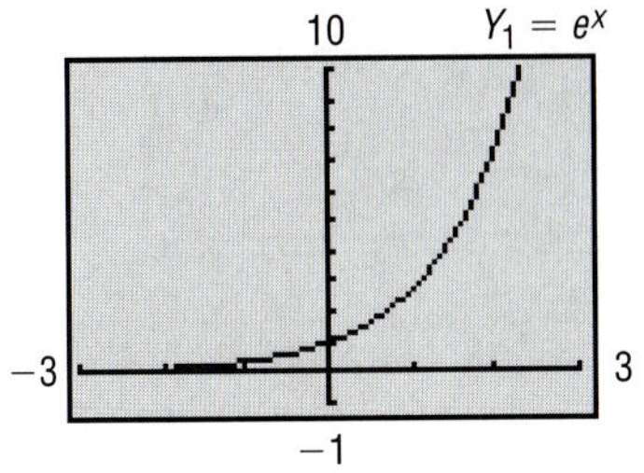

Now work Problem 25.

There are many applications involving the exponential function. Let's look at one.

EXAMPLE 6 Exponential Probability

Between 9:00 P.M. and 10:00 P.M. cars arrive at Burger King's drive-thru at the rate of 12 cars per hour (0.2 cars per minute). The following formula from statistics can be used to determine the probability a car will arrive within t minutes of 9:00 P.M.:

$$F(t) = 1 - e^{-0.2t}$$

(a) Determine the probability a car will arrive within 5 minutes of 9 P.M. (that is, before 9:05 P.M.).

(b) Determine the probability a car will arrive within 30 minutes of 9 P.M. (before 9:30 P.M.).

(c) Graph F using your graphing utility.

(d) Using TRACE, determine how many minutes are needed for the probability to reach 75%.

(e) What value does F approach as t becomes unbounded in the positive direction?

Solution (a) We have $t = 5$. The probability a car will arrive within 5 minutes is

$$F(5) = 1 - e^{-0.2(5)}$$

FIGURE 23

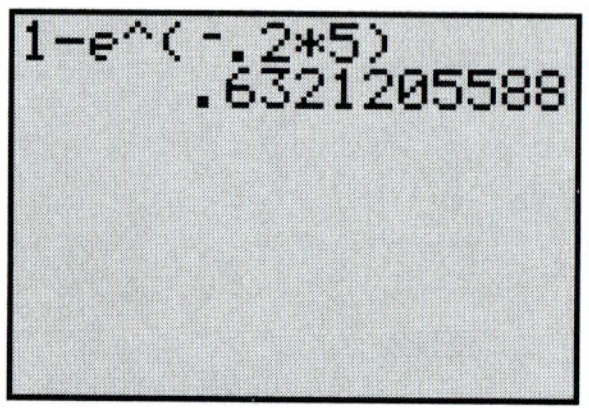

We evaluate this expression in Figure 23. We conclude there is a 63% probability a car will arrive within 5 minutes.

(b) We have $t = 30$. The probability a car will arrive within 30 minutes is

$$F(30) = 1 - e^{-0.2(30)} \approx 0.9975$$

FIGURE 24

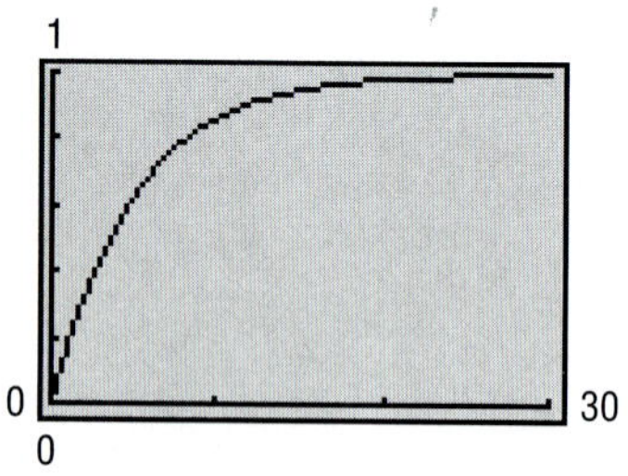

There is a 99.75% probability a car will arrive within 30 minutes.

(c) See Figure 24.

(d) It takes about 7 minutes for the probability to equal 75%. Thus, there is a 75% probability a car will arrive before 9:07 P.M.

(e) As time passes, the probability a car will arrive increases. The value F approaches can be found by letting $t \to \infty$. Since $e^{-0.2t} = 1/e^{0.2t}$, it follows that $e^{-0.2t} \to 0$ as $t \to \infty$. Thus, F approaches 1 as t gets large.

Now work Problem 43.

SUMMARY

Properties of the Exponential Function

$f(x) = a^x, a > 1$ Domain: $(-\infty, \infty)$; Range: $(0, \infty)$; x-intercepts: none; y-intercept: 1; horizontal asymptote: x-axis as $x \to -\infty$; increasing; one-to-one
See Figure 13 for a typical graph.

$f(x) = a^x, 0 < a < 1$ Domain: $(-\infty, \infty)$; Range: $(0, \infty)$; x-intercepts: none, y-intercept: 1; horizontal asymptote: x-axis as $x \to \infty$; decreasing; one-to-one
See Figure 17 for a typical graph.

5.2 EXERCISES

In Problems 1–10, approximate each number using a calculator. Express your answer rounded to three decimal places.

1. (a) $3^{2.2}$ (b) $3^{2.23}$ (c) $3^{2.236}$ (d) $3^{\sqrt{5}}$

2. (a) $5^{1.7}$ (b) $5^{1.73}$ (c) $5^{1.732}$ (d) $5^{\sqrt{3}}$

3. (a) $2^{3.14}$ (b) $2^{3.141}$ (c) $2^{3.1415}$ (d) 2^{π}

4. (a) $2^{2.7}$ (b) $2^{2.71}$ (c) $2^{2.718}$ (d) 2^{e}

5. (a) $3.1^{2.7}$ (b) $3.14^{2.71}$ (c) $3.141^{2.718}$ (d) π^{e}

6. (a) $2.7^{3.1}$ (b) $2.71^{3.14}$ (c) $2.718^{3.141}$ (d) e^{π}

7. $e^{1.2}$ **8.** $e^{-1.3}$ **9.** $e^{-0.85}$ **10.** $e^{2.1}$

In Problems 11–18, the graph of an exponential function is given. Match each graph to one of the following functions:

A. $y = 3^x$ *B.* $y = 3^{-x}$ *C.* $y = -3^x$ *D.* $y = -3^{-x}$

E. $y = 3^x - 1$ *F.* $y = 3^{x-1}$ *G.* $y = 3^{1-x}$ *H.* $y = 1 - 3^x$

11.

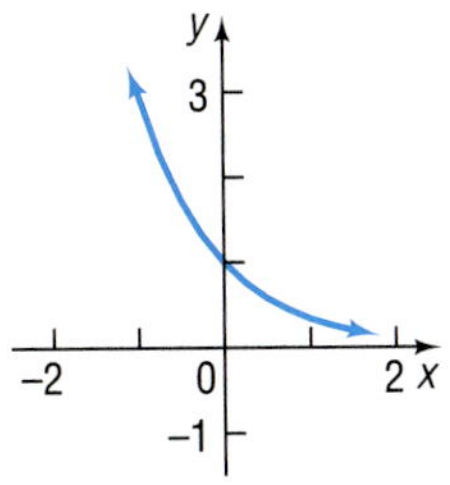

12.

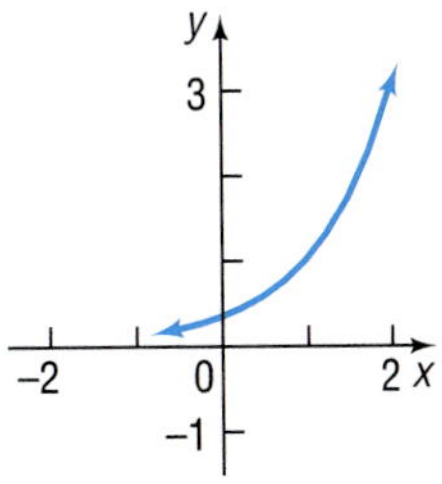

13.

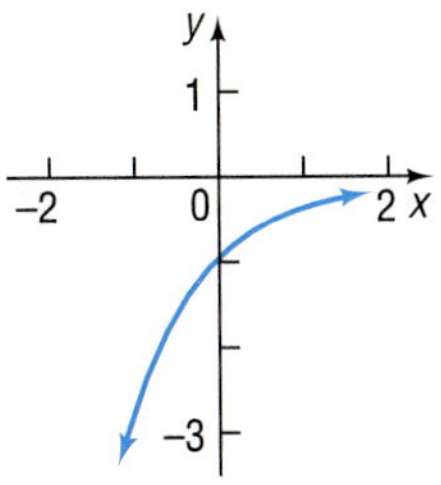

14.

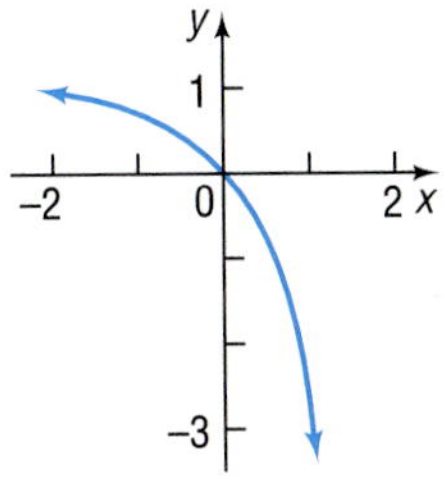

15.

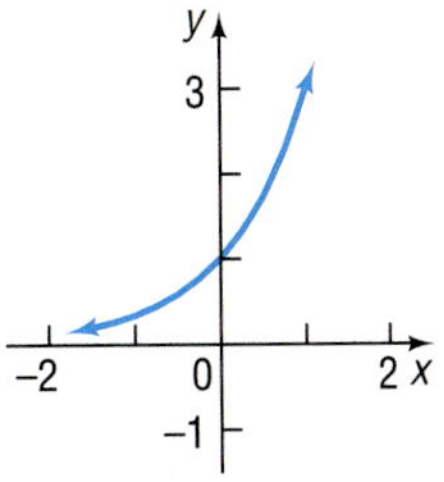

16.

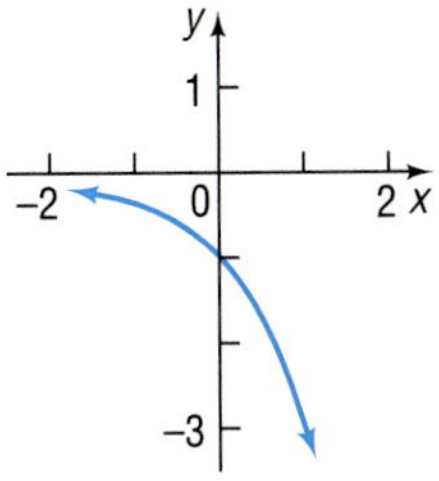

17.

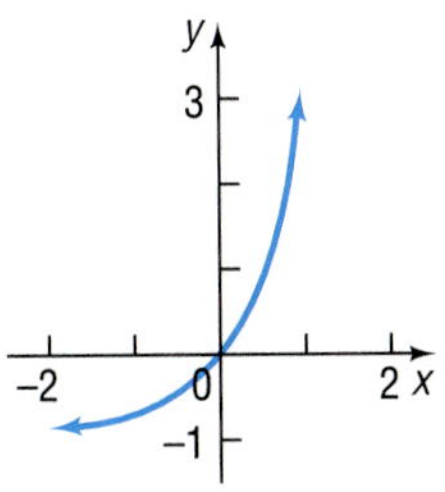

18.

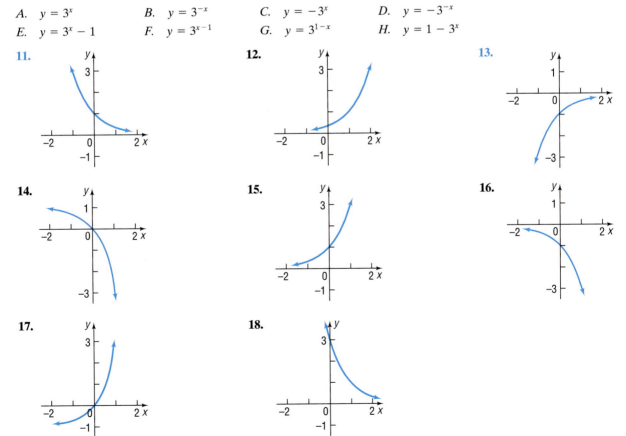

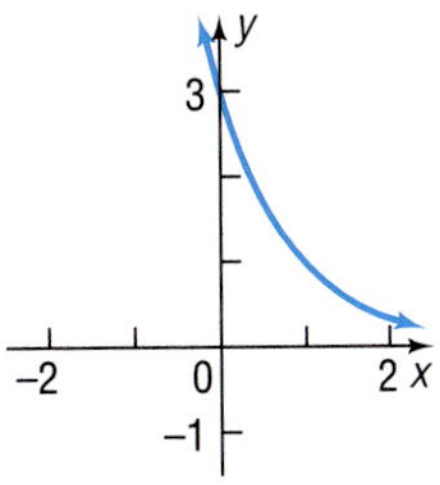

In Problems 19–24, the graph of an exponential function is given. Match each graph to one of the following functions:

A. $y = 4^x$ *B.* $y = 4^{-x}$ *C.* $y = 4^{x-1}$ *D.* $y = 4^x - 1$ *E.* $y = -4^x$ *F.* $y = 1 - 4^x$

19.

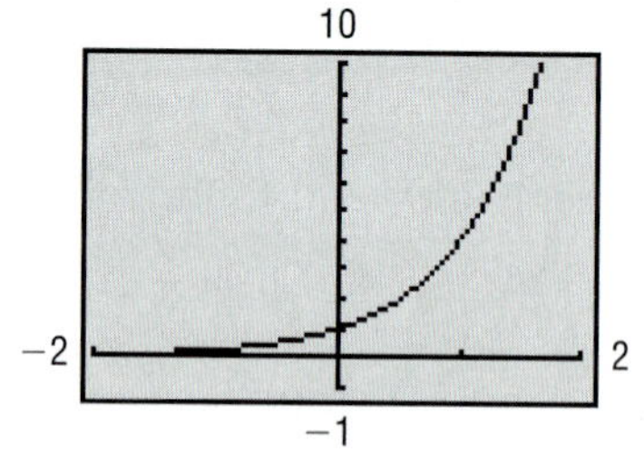

20.

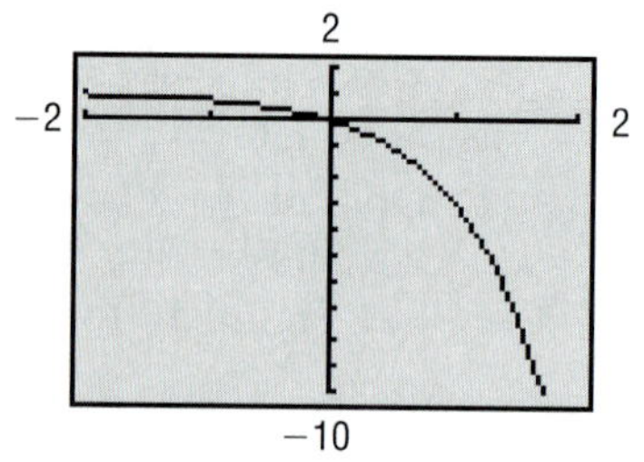

21.

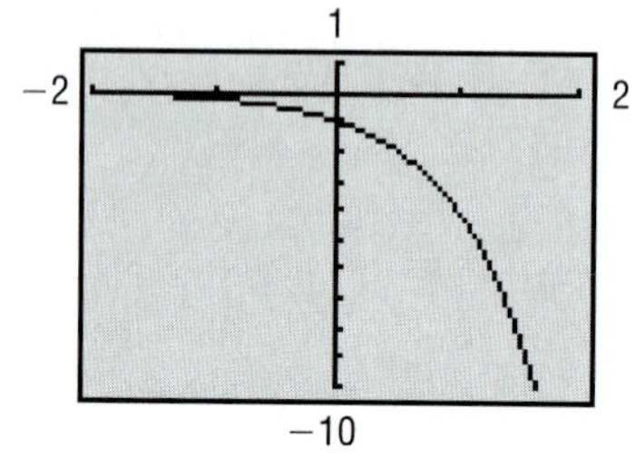

22.

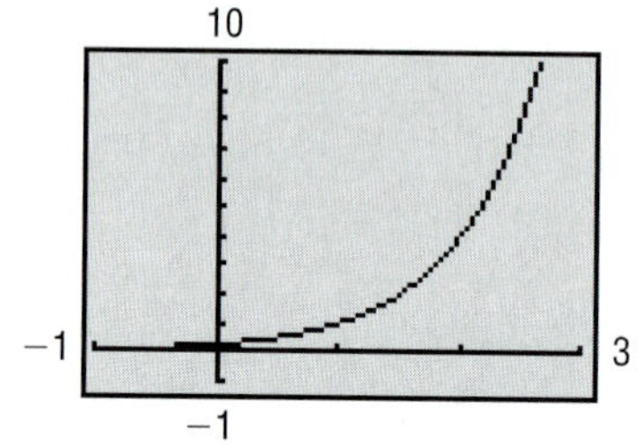

23.

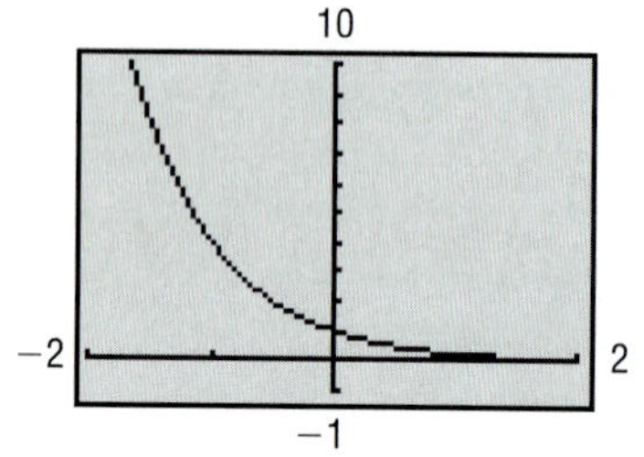

24.

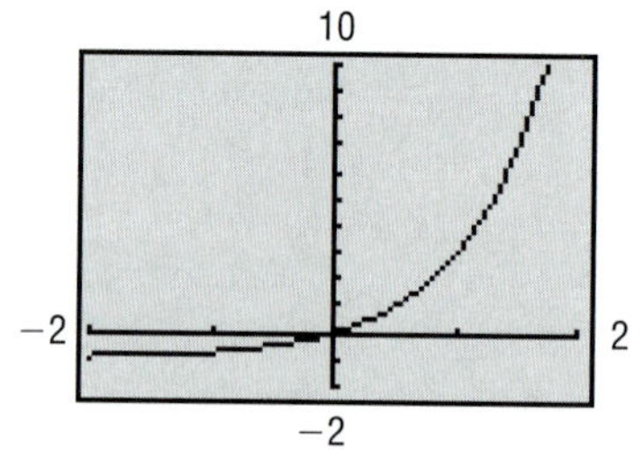

In Problems 25–32, using a graphing utility, show the stages required to graph each function. Determine the domain, range and horizontal asymptote of each function.

25. $f(x) = e^{-x}$
26. $f(x) = -e^x$
27. $f(x) = e^{x+2}$
28. $f(x) = e^x - 1$
29. $f(x) = 5 - e^{-x}$
30. $f(x) = 9 - 3e^{-x}$
31. $f(x) = 2 - e^{-x/2}$
32. $f(x) = 7 - 3e^{-2x}$

33. If $4^x = 7$, what does 4^{-2x} equal?

34. If $2^x = 3$, what does 4^{-x} equal?

35. If $3^{-x} = 2$, what does 3^{2x} equal?

36. If $5^{-x} = 3$, what does 5^{3x} equal?

37. Optics If a single pane of glass obliterates 3% of the light passing through it, then the percent p of light that passes through n successive panes is given approximately by the equation

$$p = 100e^{-0.03n}$$

(a) What percent of light will pass through 10 panes?
(b) What percent of light will pass through 25 panes?

38. Atmospheric Pressure The atmospheric pressure p on a balloon or plane decreases with increasing height. This pressure, measured in millimeters of mercury, is related to the number of kilometers h above sea level by the formula

$$p = 760e^{-0.145h}$$

(a) Find the atmospheric pressure at a height of 2 kilometers (over a mile).
(b) What is it at a height of 10 kilometers (over 30,000 feet)?

39. Space Satellites The number of watts w provided by a space satellite's power supply over a period of d days is given by the formula

$$w = 50e^{-0.004d}$$

(a) How much power will be available after 30 days?

(b) How much power will be available after 1 year (365 days)?

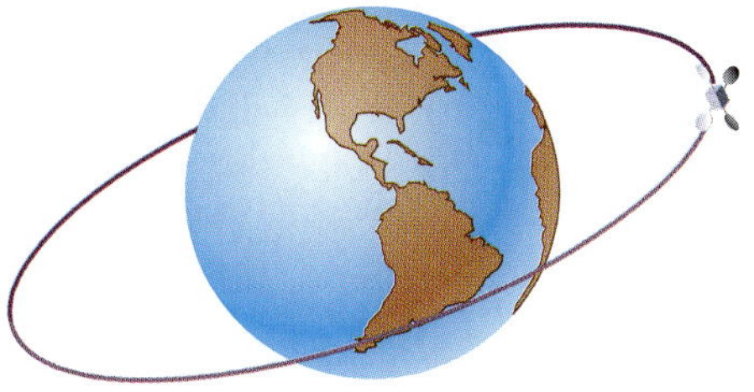

40. Healing of Wounds The normal healing of wounds can be modeled by an exponential function. If A_0 represents the original area of the wound and if A equals the area of the wound after n days, then the formula

$$A = A_0 e^{-0.35n}$$

describes the area of a wound on the nth day following an injury when no infection is present to retard the healing. Suppose a wound initially had an area of 100 square centimeters.

(a) If healing is taking place, how large should the area of the wound be after 3 days?

(b) How large should it be after 10 days?

41. Drug Medication The formula

$$D = 5e^{-0.4h}$$

can be used to find the number of milligrams D of a certain drug that is in a patient's bloodstream h hours after the drug has been administered. How many milligrams will be present after 1 hour? After 6 hours?

42. Spreading of Rumors A model for the number of people N in a college community who have heard a certain rumor is

$$N = P(1 - e^{-0.15d})$$

where P is the total population of the community and d is the number of days that have elapsed since the rumor began. In a community of 1000 students, how many students will have heard the rumor after 3 days?

43. Exponential Probability Between 12:00 P.M. and 1:00 P.M. cars arrive at Citibank's drive-thru at the rate of 6 cars per hour (0.1 cars per minute). The following formula from statistics can be used to determine the probability a car will arrive within t minutes of 12:00 P.M.

$$F(t) = 1 - e^{-0.1t}$$

(a) Determine the probability a car will arrive within 10 minutes of 12:00 P.M. (That is, before 12:10 P.M.).

(b) Determine the probability a car will arrive within 40 minutes of 12:00 P.M. (before 12:40 P.M.).

(c) Graph F using your graphing utility.

(d) Using TRACE, determine how many minutes are needed for the probability to reach 50%?

(e) What value does F approach as t becomes unbounded in the positive direction?

44. Exponential Probability Between 5:00 P.M. and 6:00 P.M. cars arrive at Jiffy Lube at the rate of 9 cars per hour (0.15 cars per minute). The following formula from statistics can be used to determine the probability a car will arrive within t minutes of 5:00 P.M.:

$$F(t) = 1 - e^{-0.15t}$$

(a) Determine the probability a car will arrive within 15 minutes of 5:00 P.M. (That is, before 5:15 P.M.).

(b) Determine the probability a car will arrive within 30 minutes of 5:00 P.M. (before 5:30 P.M.).

(c) Graph F using your graphing utility.

(d) Using TRACE, determine how many minutes are needed for the probability to reach 60%?

(e) What value does F approach as t becomes unbounded in the positive direction?

45. Response to TV Advertising The percent R of viewers who respond to a television commercial for a new product after t days is found by using the formula

$$R = 70 - 100e^{-0.2t}$$

(a) What percent is expected to respond after 10 days?

(b) What percent has responded after 20 days?

(c) What is the highest percent of people expected to respond?

(d) Graph $R = 70 - 100e^{-0.2t}$, $t > 0$. TRACE and compare the values of R for $t = 10$ and $t = 20$ to the ones obtained in parts (a) and (b). How many days are required for R to exceed 40%?

46. Profit The annual profit P of a company due to the sales of a particular item after it has been on the market x years is determined to be

$$P = \$100{,}000 - \$60{,}000\left(\tfrac{1}{2}\right)^x$$

(a) What is the profit after 5 years?

(b) What is the profit after 10 years?

(c) What is the most profit the company can expect from this product?

(d) Graph the profit function. TRACE and compare the values of P for $x = 5$ and $x = 10$ to the one obtained in parts (a) and (b). How many years does it take before a profit of $65,000 is obtained?

47. Alternating Current in a *RL* Circuit The equation governing the amount of current I (in amperes) after time t (in seconds) in a single RL circuit consisting of a resistance R (in ohms), an inductance L (in henrys), and an electromotive force E (in volts) is

$$I = \frac{E}{R}[1 - e^{-(R/L)t}]$$

(a) If $E = 120$ volts, $R = 10$ ohms, and $L = 5$ henrys, how much current I_1 is available after 0.3 second? After 0.5 second? After 1 second?
(b) What is the maximum current?
(c) Graph this function $I = I_1(t)$, measuring I along the y-axis and t along the x-axis.
(d) If $E = 120$ volts, $R = 5$ ohms, and $L = 10$ henrys, how much current I_2 is available after 0.3 second? After 0.5 second? After 1 second?
(e) What is the maximum current?
(f) Graph this function $I = I_2(t)$ on the same viewing window as $I_1(t)$.

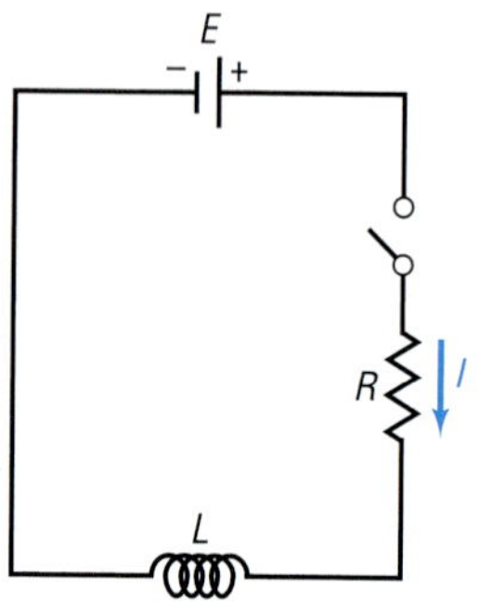

48. Alternating Current in a *RC* Circuit The equation governing the amount of current I (in milliamperes) after time t (in milliseconds) in a single RC circuit consisting of a resistance R (in ohms), a capacitance C (in microfarads), and an electromotive force E (in volts) is

$$I = \frac{E}{R}e^{-t/(RC)}$$

(a) If $E = 120$ volts, $R = 2000$ ohms, and $C = 1.0$ microfarad, how much current I_1 is available initially ($t = 0$)? After 1000 milliseconds? After 3000 milliseconds?
(b) What is the maximum current?
(c) Graph this function $I = I_1(t)$, measuring I along the y-axis and t along the x-axis.
(d) If $E = 120$ volts, $R = 1000$ ohms, and $C = 2.0$ microfarads, how much current I_2 is available initially? After 1000 milliseconds? After 3000 milliseconds?
(e) What is the maximum current?
(f) Graph this function $I = I_2(t)$ on the same viewing window as $I_1(t)$.

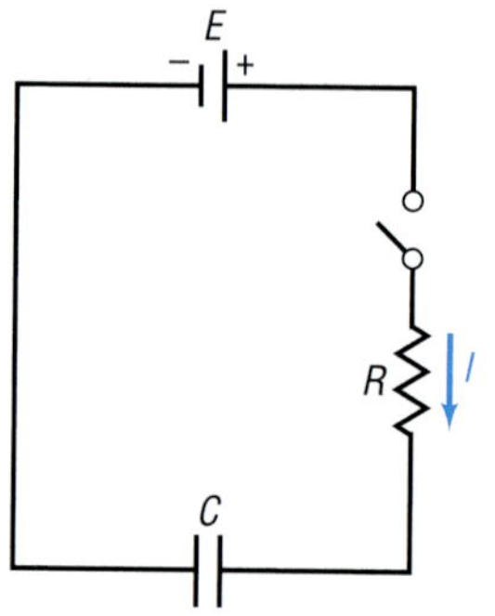

49. The Challenger Disaster* After the *Challenger* disaster in 1986, a study of the 23 launches that preceded the fatal flight was made. A mathematical model was developed involving the relationship between the Fahrenheit temperature x around the O-rings and the number y of eroded or leaky primary O-rings. The model stated that

$$y = \frac{6}{1 + e^{-(5.085 - 0.1156x)}}$$

where the number 6 indicates the 6 primary O-rings on the spacecraft.
(a) What is the predicted number of eroded or leaky primary O-rings at a temperature of 100°F?
(b) What is the predicted number of eroded or leaky primary O-rings at a temperature of 60°F?
(c) What is the predicted number of eroded or leaky primary O-rings at a temperature of 30°F?

*Linda Tappin, "Analyzing Data Relating to the *Challenger* Disaster," *Mathematics Teacher,* Vol. 87, No. 6, September 1994, pp. 423–426.

(d) Graph the equation and TRACE. At what temperature is the predicted number of eroded or leaky O-rings 1? 3? 5?

50. **Postage Stamps*** The cumulative number y of different postage stamps (regular and commemorative only) issued by the U.S. Post Office can be approximated (modeled) by the exponential function

$$y = 78e^{0.025x}$$

where x is the number of years since 1848.

(a) What is the predicted cumulative number of stamps that will have been issued by the year 1995? Check with the Postal Service and comment on the accuracy of using the function.

(b) What is the predicted cumulative number of stamps that will have been issued by the year 1998?

(c) The cumulative number of stamps actually issued by the United States was 2 in 1848, 88 in 1868, 218 in 1888, and 341 in 1908. What conclusion can you draw about using the given function as a model over the first few decades in which stamps were issued?

51. **Historical Problem** Pierre de Fermat (1601–1665) conjectured that the function

$$f(x) = 2^{(2^x)} + 1$$

for $x = 1, 2, 3, \ldots$, would always have a value equal to a prime number. But Leonhard Euler (1707–1783) showed that this formula fails for $x = 5$. Use a calculator to determine the prime numbers produced by f for $x = 1, 2, 3, 4$. Then show that $f(5) = 641 \times 6{,}700{,}417$, which is not prime.

52. The bacteria in a 4-liter container double every minute. After 60 minutes the container is full. How long did it take to fill half the container?

53. Explain in your own words what the number e is. Provide at least two applications that require the use of this number.

54. Do you think there is a power function that increases more rapidly than an exponential function whose base is greater than 1? Explain.

5.3 LOGARITHMIC FUNCTIONS

1 Change Exponential Expressions to Logarithmic Expressions
2 Change Logarithmic Expressions to Exponential Expressions
3 Evaluate Logarithmic Functions
4 Determine the Domain of a Logarithmic Function
5 Graph Logarithmic Functions

Recall that a one-to-one function $y = f(x)$ has an inverse that is defined (implicitly) by the equation $x = f(y)$. In particular, the exponential function $y = f(x) = a^x$, $a > 0$, $a \neq 1$, is one-to-one and, hence, has an inverse that is defined implicitly by the equation

$$x = a^y \qquad a > 0, a \neq 1$$

This inverse is so important that it is given a name, the *logarithmic function.*

*David Kullman, "Patterns of Postage-stamp Production," *Mathematics Teacher,* Vol. 85, No. 3, March 1992, pp. 188–189.

The **logarithmic function to the base *a*,** where $a > 0$ and $a \neq 1$, is denoted by $y = \log_a x$ (read as "y is the logarithm to the base a of x") and is defined by

$$y = \log_a x \quad \text{if and only if} \quad x = a^y$$

EXAMPLE 1 Relating Logarithms to Exponents

(a) If $y = \log_3 x$, then $x = 3^y$. Thus, if $x = 9$, then $y = 2$, so $9 = 3^2$ is equivalent to $2 = \log_3 9$.

(b) If $y = \log_5 x$, then $x = 5^y$. Thus, if $x = \frac{1}{5} = 5^{-1}$, then $y = -1$, so $\frac{1}{5} = 5^{-1}$ is equivalent to $-1 = \log_5 \left(\frac{1}{5}\right)$.

EXAMPLE 2 Changing Exponential Expressions to Logarithmic Expressions

1 Change each exponential expression to an equivalent expression involving a logarithm.

(a) $1.2^3 = m$ (b) $e^b = 9$ (c) $a^4 = 24$

Solution We use the fact that $y = \log_a x$ and $x = a^y$, $a > 0$, $a \neq 1$, are equivalent.

(a) If $1.2^3 = m$, then $3 = \log_{1.2} m$.
(b) If $e^b = 9$, then $b = \log_e 9$.
(c) If $a^4 = 24$, then $4 = \log_a 24$.

Now work Problem 1.

EXAMPLE 3 Changing Logarithmic Expressions to Exponential Expressions

2 Change each logarithmic expression to an equivalent expression involving an exponent.

(a) $\log_a 4 = 5$ (b) $\log_e b = -3$ (c) $\log_3 5 = c$

Solution (a) If $\log_a 4 = 5$, then $a^5 = 4$.
(b) If $\log_e b = -3$, then $e^{-3} = b$.
(c) If $\log_3 5 = c$, then $3^c = 5$.

Now work Problem 13.

3 To find the exact value of a logarithm, we write the logarithm in exponential notation and use the following fact:

$$\text{If } a^u = a^v, \quad \text{then} \quad u = v. \tag{1}$$

The result (1) is a consequence of the fact that exponential functions are one-to-one.

EXAMPLE 4

Finding the Exact Value of a Logarithmic Function

Find the exact value of

(a) $\log_2 8$ (b) $\log_3 \frac{1}{3}$ (c) $\log_5 25$

Solution

(a) For $y = \log_2 8$, we have the equivalent exponential equation $2^y = 8 = 2^3$, so, by (1), $y = 3$. Thus, $\log_2 8 = 3$.

(b) For $y = \log_3 \frac{1}{3}$, we have $3^y = \frac{1}{3} = 3^{-1}$, so $y = -1$. Thus, $\log_3 \frac{1}{3} = -1$.

(c) For $y = \log_5 25$, we have $5^y = 25 = 5^2$, so $y = 2$. Thus, $\log_5 25 = 2$.

Now work Problem 25.

Domain of a Logarithmic Function

4 The logarithmic function $y = \log_a x$ has been defined as the inverse of the exponential function $y = a^x$. That is, if $f(x) = a^x$, then $f^{-1}(x) = \log_a x$. Based on the discussion given in Section 5.1 on inverse functions, we know that for a function f and its inverse f^{-1}

$$\text{Domain } f^{-1} = \text{Range } f \quad \text{and} \quad \text{Range } f^{-1} = \text{Domain } f$$

Consequently, it follows that

> Domain of logarithmic function = Range of exponential function = $(0, \infty)$
> Range of logarithmic function = Domain of exponential function = $(-\infty, \infty)$

In the next box, we summarize some properties of the logarithmic function:

> $y = \log_a x$ (defining equation: $x = a^y$)
> Domain: $0 < x < \infty$ Range: $-\infty < y < \infty$

Notice that the domain of a logarithmic function consists of the *positive* real numbers.

EXAMPLE 5

Finding the Domain of a Logarithmic Function

Find the domain of each logarithmic function:

(a) $F(x) = \log_2 (1 - x)$ (b) $g(x) = \log_5 \left(\dfrac{1 + x}{1 - x}\right)$ (c) $h(x) = \log_{1/2} |x|$

Solution (a) The domain of F consists of all x for which $(1 - x) > 0$; that is, all $x < 1$, or $(-\infty, 1)$.

(b) The domain of g is restricted to

$$\frac{1 + x}{1 - x} > 0$$

Solving this inequality, we find that the domain of g consists of all x between -1 and 1; that is, $-1 < x < 1$, or $(-1, 1)$.

(c) Since $|x| > 0$ provided $x \neq 0$, the domain of h consists of all nonzero real numbers.

Now work Problem 39.

Graphs of Logarithmic Functions

5 Since exponential functions and logarithmic functions are inverses of each other, the graph of a logarithmic function $y = \log_a x$ is the reflection about the line $y = x$ of the graph of the exponential function $y = a^x$, as shown in Figure 25.

FIGURE 25

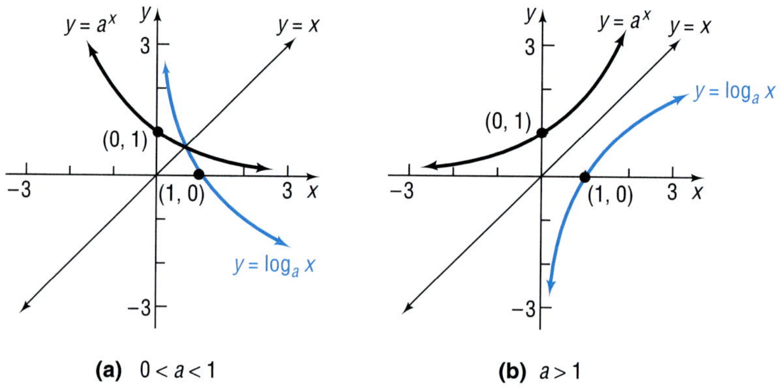

Facts about the Graph of a Logarithmic Function $f(x) = \log_a x$

1. The x-intercept of the graph is 1. There is no y-intercept.
2. The y-axis is a vertical asymptote of the graph.
3. A logarithmic function is decreasing if $0 < a < 1$ and increasing if $a > 1$.
4. The graph is smooth and continuous, with no corners or gaps.

If the base of a logarithmic function is the number e, then we have the **natural logarithm function.** This function occurs so frequently in applications that it is given a special symbol, **ln** (from the Latin, *logarithmus naturalis*). Thus,

$$y = \ln x \quad \text{if and only if} \quad x = e^y$$

Since $y = \ln x$ and the exponential function $y = e^x$ are inverse functions, we can obtain the graph of $y = \ln x$ by reflecting the graph of $y = e^x$ about the line $y = x$. See Figure 26.

Table 6 displays other points on the graph of $f(x) = \ln x$. Notice for $x < 0$ we obtain an error message. Do you recall why?

FIGURE 26

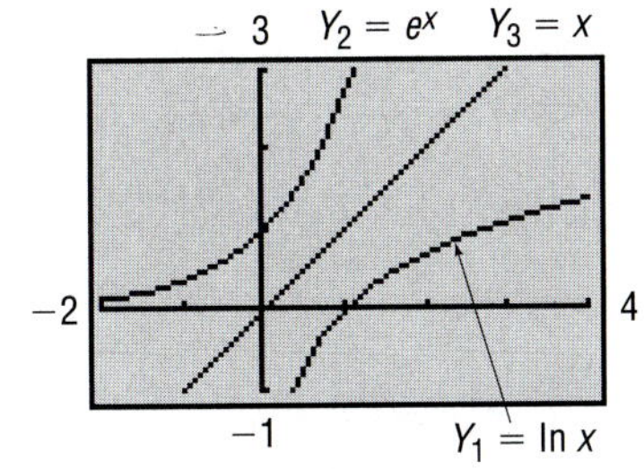

TABLE 6

X	Y1
-1	ERROR
.5	-.6931
1	0
2	.69315
2.7183	1
3	1.0986

Y1=ln(X)

EXAMPLE 6 Graphing Logarithmic Functions

Graph $f(x) = -\ln x$ by starting with the graph of $y = \ln x$. Determine the domain, range, and vertical asymptote of $f(x) = -\ln x$.

Solution The graph of $f(x) = -\ln x$ is obtained by a reflection about the x-axis of the graph of $f(x) = \ln x$. See Figure 27.

FIGURE 27

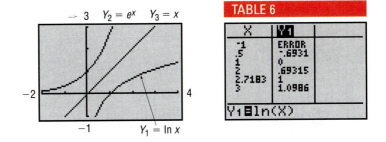

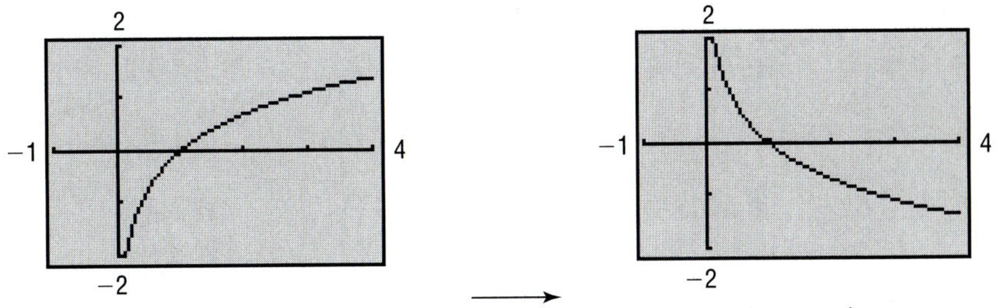

(a) $y = \ln x$ → Reflection about x-axis; multiply by −1 **(b)** $y = -\ln x$

The domain of $f(x) = -\ln x$ is $(0, \infty)$, the range is $(-\infty, \infty)$, and the vertical asymptote is $x = 0$.

EXAMPLE 7 Graphing Logarithmic Functions

Determine the domain, range, and vertical asymptote of $f(x) = \ln(x + 2)$. Graph $f(x) = \ln(x+ 2)$.

Solution The graph is obtained by applying a horizontal shift to the left 2 units, as shown in Figure 28.

FIGURE 28

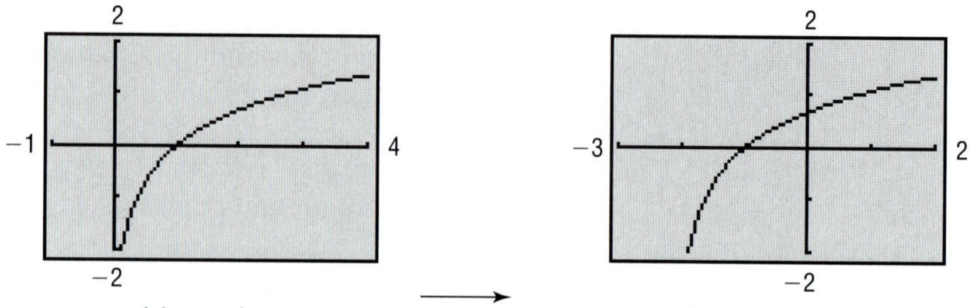

(a) $y = \ln x$

Shift left 2 units; replace x by $x + 2$

(b) $y = \ln(x + 2)$

The domain of $f(x) = \ln(x + 2)$ is $(-2, \infty)$, the range is $(-\infty, \infty)$, and the vertical asymptote is $x = -2$.

EXAMPLE 8 Graphing Logarithmic Functions

Graph $f(x) = \ln(1 - x)$. Determine the domain, range, and vertical asymptote of $f(x) = \ln(1 - x)$.

Solution To obtain the graph of $y = \ln(1 - x)$, we use the steps illustrated in Figure 29.

FIGURE 29

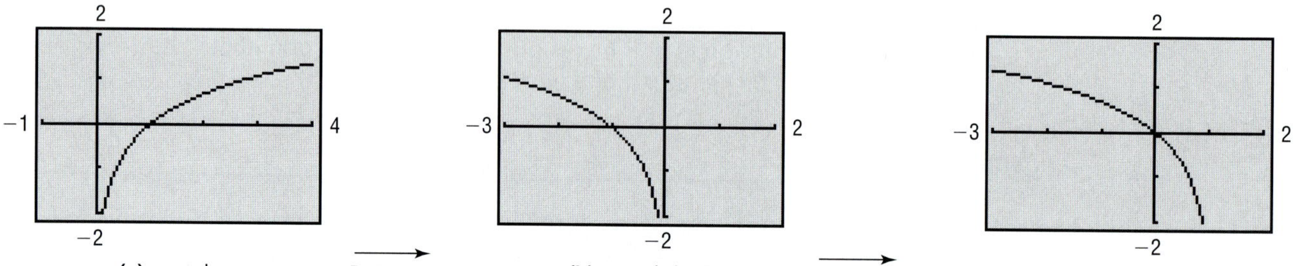

(a) $y = \ln x$

Reflection about y-axis; replace x by $-x$

(b) $y = \ln(-x)$

Shift right 1 unit; replace x by $x - 1$

(c) $y = \ln[-(x - 1)]$
$= \ln(1 - x)$

The domain of $f(x) = \ln(1 - x)$ is $(-\infty, 1)$, the range is $(-\infty, \infty)$, and the vertical asymptote is $x = 1$.

Now work Problem 67.

EXAMPLE 9 Alcohol and Driving

The concentration of alcohol in a person's blood is measurable. Recent medical research suggests that the risk R (given as a percent) of having an accident while driving a car can be modeled by the equation

$$R = 6e^{kx}$$

where x is the variable concentration of alcohol in the blood and k is a constant.

(a) Suppose that a concentration of alcohol in the blood of 0.04 results in a 10% risk ($R = 10$) of an accident. Find the constant k in the equation. Graph $R = 6e^{kx}$, using this value of k.

(b) Using this value of k, what is the risk if the concentration is 0.17?

(c) Using the same value of k, what concentration of alcohol corresponds to a risk of 100%?

(d) If the law asserts that anyone with a risk of having an accident of 20% or more should not have driving privileges, at what concentration of alcohol in the blood should a driver be arrested and charged with a DUI—Driving Under the Influence?

Solution (a) For a concentration of alcohol in the blood of 0.04 and a risk of 10%, we let $x = 0.04$ and $R = 10$ in the equation and solve for k.

$$R = 6e^{kx}$$

$$10 = 6e^{k(0.04)}$$

$$\frac{10}{6} = e^{0.04k} \quad \text{Change to a logarithmic expression.}$$

$$0.04k = \ln\frac{10}{6} = 0.5108256$$

$$k = 12.77$$

See Figure 30 for the graph of $R = 6e^{12.77x}$.

FIGURE 30

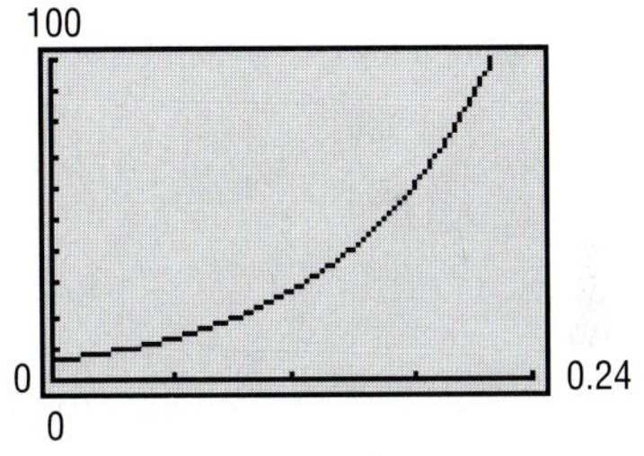

(b) Using $k = 12.77$ and $x = 0.17$ in the equation, we find the risk R to be

$$R = 6e^{kx} = 6e^{(12.77)(0.17)} = 52.6$$

For a concentration of alcohol in the blood of 0.17, the risk of an accident is about 52.6%. Verify this answer by TRACEing the graph.

(c) Using $k = 12.77$ and $R = 100$ in the equation, we find the concentration x of alcohol in the blood to be

$$R = 6e^{kx}$$

$$100 = 6e^{12.77x}$$

$$\frac{100}{6} = e^{12.77x} \quad \text{Change to a logarithmic expression.}$$

$$12.77x = \ln\frac{100}{6} = 2.8134$$

$$x = 0.22$$

For a concentration of alcohol in the blood of 0.22, the risk of an accident is 100%. Verify this answer by TRACEing the graph.

(d) Using $k = 12.77$ and $R = 20$ in the equation, we find the concentration x of alcohol in the blood to be

$$\begin{aligned} R &= 6e^{kx} \\ 20 &= 6e^{12.77x} \\ \frac{20}{6} &= e^{12.77x} \\ 12.77x &= \ln\frac{20}{6} = 1.204 \\ x &= 0.094 \end{aligned}$$

A driver with a concentration of alcohol in the blood of 0.094 or more should be arrested and charged with DUI. Verify by TRACEing the graph.

Note: Most states use 0.10 as the blood alcohol content at which a DUI citation is given. A few states use 0.08.

SUMMARY

Properties of the Logarithmic Function

$f(x) = \log_a x, \quad a > 1$ ($y = \log_a x$ means $x = a^y$) Domain: $(0, \infty)$; Range: $(-\infty, \infty)$; x-intercept: 1; y-intercept: none; vertical asymptote: y-axis; increasing; one-to-one
See Figure 25(b) for a typical graph.

$f(x) = \log_a x, \quad 0 < a < 1$ ($y = \log_a x$ means $x = a^y$) Domain: $(0, \infty)$; Range: $(-\infty, \infty)$; x-intercept: 1; y-intercept: none; vertical asymptote: y-axis; decreasing; one-to-one
See Figure 25(a) for a typical graph.

5.3 EXERCISES

In Problems 1–12, change each exponential expression to an equivalent expression involving a logarithm.

1. $9 = 3^2$ **2.** $16 = 4^2$ **3.** $a^2 = 1.6$ **4.** $a^3 = 2.1$ **5.** $1.1^2 = M$ **6.** $2.2^3 = N$
7. $2^x = 7.2$ **8.** $3^x = 4.6$ **9.** $x^{\sqrt{2}} = \pi$ **10.** $x^\pi = e$ **11.** $e^x = 8$ **12.** $e^{2.2} = M$

In Problems 13–24, change each logarithmic expression to an equivalent expression involving an exponent.

13. $\log_2 8 = 3$ **14.** $\log_3(\frac{1}{9}) = -2$ **15.** $\log_a 3 = 6$ **16.** $\log_b 4 = 2$
17. $\log_3 2 = x$ **18.** $\log_2 6 = x$ **19.** $\log_2 M = 1.3$ **20.** $\log_3 N = 2.1$
21. $\log_{\sqrt{2}} \pi = x$ **22.** $\log_\pi x = \frac{1}{2}$ **23.** $\ln 4 = x$ **24.** $\ln x = 4$

In Problems 25–36, find the exact value of each logarithm without using a calculator.

25. $\log_2 1$ **26.** $\log_8 8$ **27.** $\log_5 25$ **28.** $\log_3(\frac{1}{9})$ **29.** $\log_{1/2} 16$ **30.** $\log_{1/3} 9$
31. $\log_{10}\sqrt{10}$ **32.** $\log_5 \sqrt[3]{25}$ **33.** $\log_{\sqrt{2}} 4$ **34.** $\log_{\sqrt{3}} 9$ **35.** $\ln\sqrt{e}$ **36.** $\ln e^3$

In Problems 37–46, find the domain of each function.

37. $f(x) = \ln(3 - x)$ **38.** $g(x) = \ln(x^2 - 1)$ **39.** $F(x) = \log_2 x^2$ **40.** $H(x) = \log_5 x^3$

41. $h(x) = \log_{1/2}(x^2 - x - 6)$ **42.** $G(x) = \log_{1/2}\left(\frac{1}{x}\right)$ **43.** $f(x) = \frac{1}{\ln x}$ **44.** $g(x) = \ln(x - 5)$

45. $g(x) = \log_5\left(\frac{x + 1}{x}\right)$ **46.** $h(x) = \log_3\left(\frac{x^2}{x - 1}\right)$

In Problems 47–50, use a calculator to evaluate each expression. Round your answer to three decimal places.

47. $\ln \dfrac{5}{3}$ **48.** $\dfrac{\ln 5}{3}$ **49.** $\dfrac{\ln (10/3)}{0.04}$ **50.** $\dfrac{\ln (2/3)}{-0.1}$

51. Find a such that the graph of $f(x) = \log_a x$ contains the point $(2, 2)$.

52. Find a such that the graph of $f(x) = \log_a x$ contains the point $(\frac{1}{2}, -4)$.

In Problems 53–60, the graph of a logarithmic function is given. Match each graph to one of the following functions:

A. $y = \log_3 x$ *B.* $y = \log_3(-x)$ *C.* $y = -\log_3 x$ *D.* $y = -\log_3(-x)$

E. $y = \log_3 x - 1$ *F.* $y = \log_3(x - 1)$ *G.* $y = \log_3(1 - x)$ *H.* $y = 1 - \log_3 x$

53.

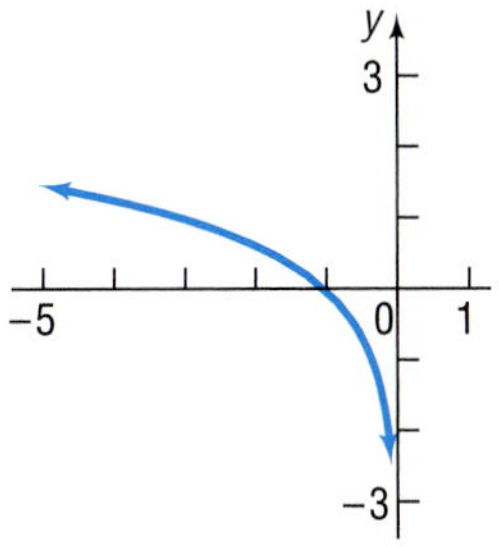

54.

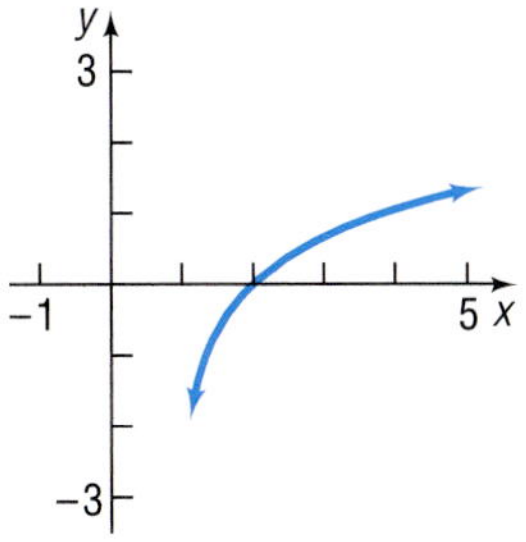

55.

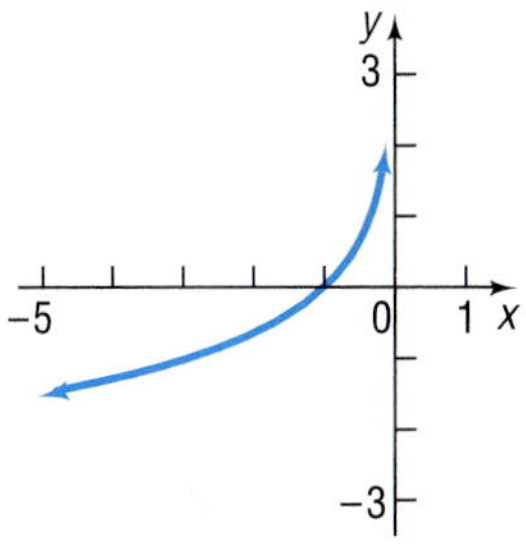

56.

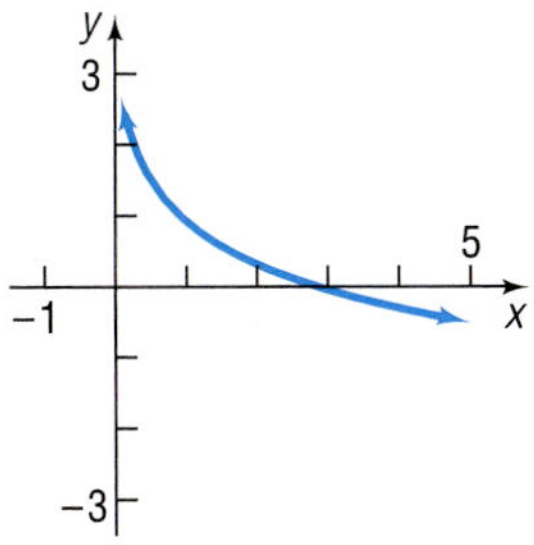

57.

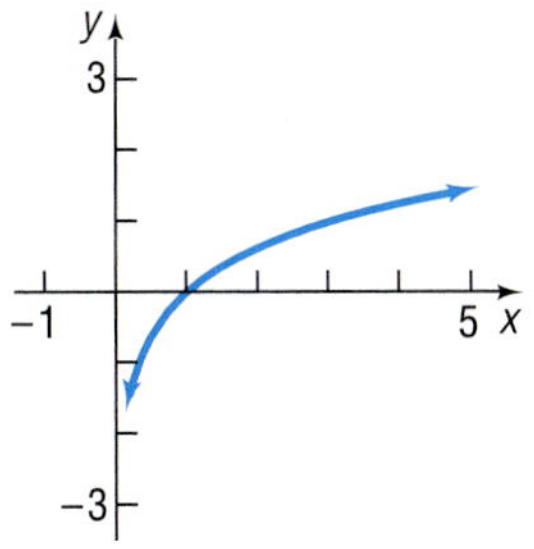

58.

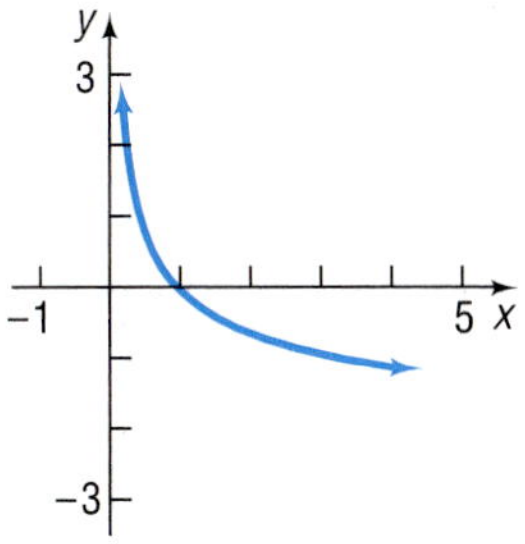

59.

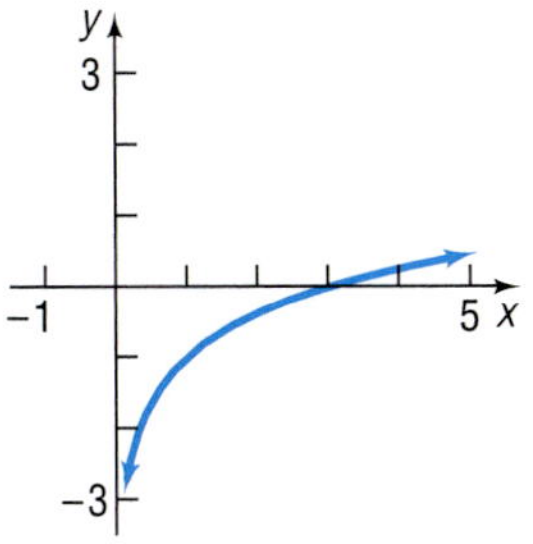

60.

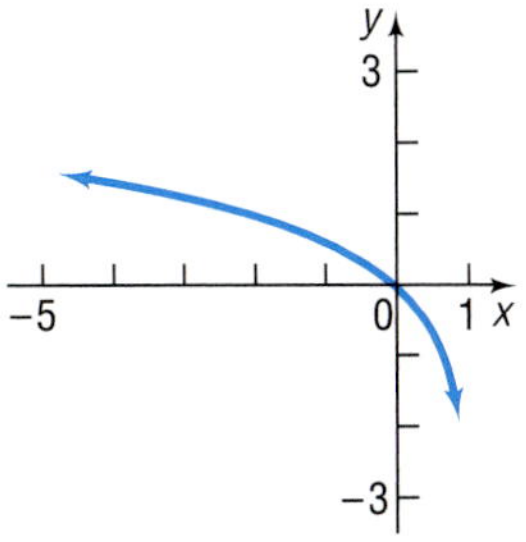

In Problems 61–66, the graph of a logarithmic function is given. Match each graph to one of the following functions:

A. $y = \log_4 x$ *B.* $y = \log_4(-x)$ *C.* $y = \log_4(x - 1)$

D. $y = -\log_4 x$ *E.* $y = 1 - \log_4 x$ *F.* $y = -\log_4(-x)$

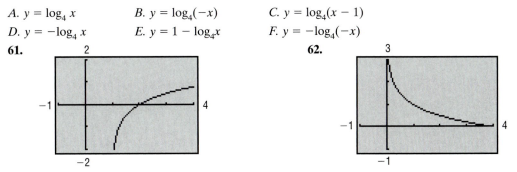

61.

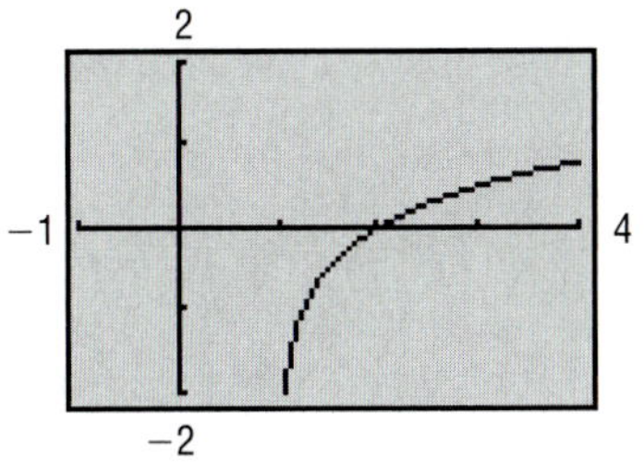

62.

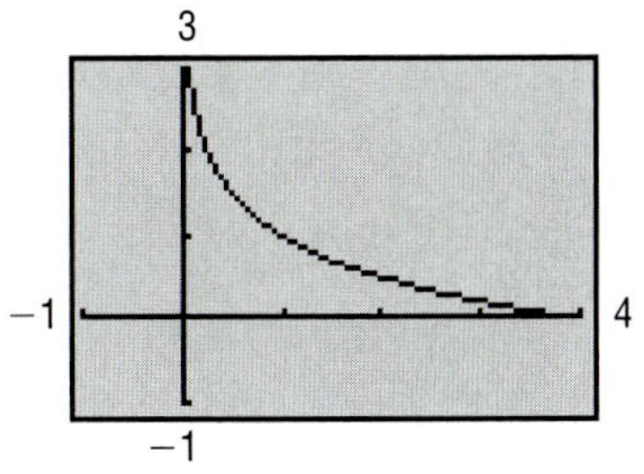

63.

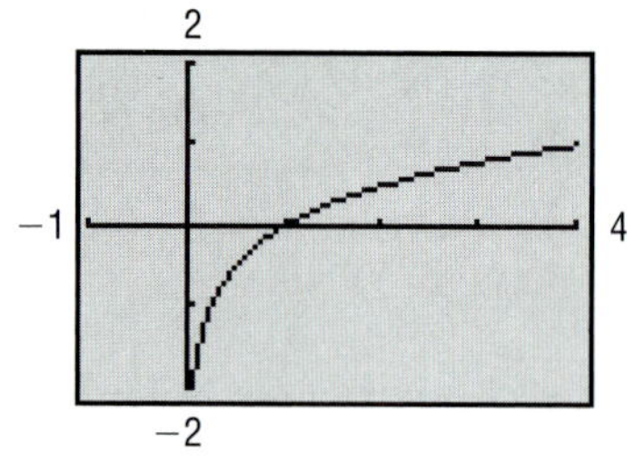

64.

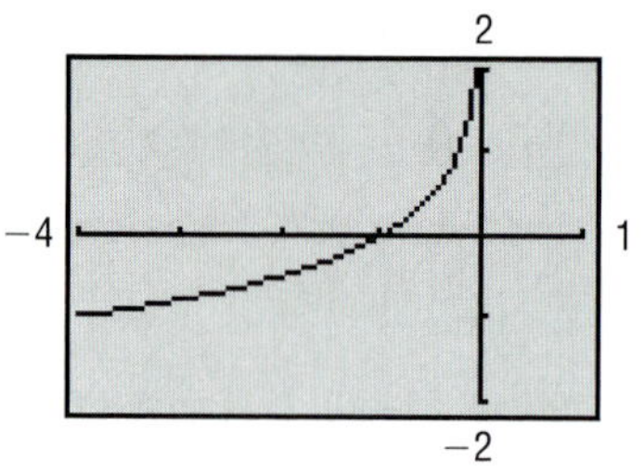

65.

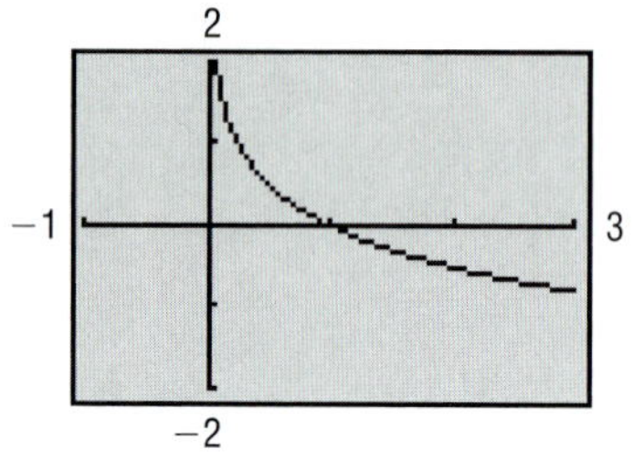

66.

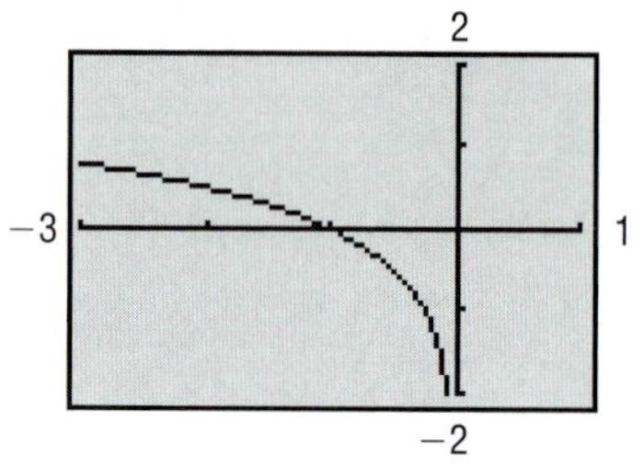

In Problems 67–76, using a graphing utility, show the stages required to graph each function. Determine the domain, range, and vertical asymptote of each function.

67. $f(x) = \ln(x + 4)$

68. $f(x) = \ln(x - 3)$

69. $f(x) = \ln(-x)$

70. $f(x) = -\ln(-x)$

71. $g(x) = \ln 2x$

72. $h(x) = \ln \frac{1}{2}x$

73. $f(x) = 3 \ln x$

74. $f(x) = -2 \ln x$

75. $g(x) = \ln(3 - x)$

76. $h(x) = \ln(4 - x)$

77. Optics If a single pane of glass obliterates 10% of the light passing through it, then the percent P of light that passes through n successive panes is given approximately by the equation

$$P = 100e^{-0.1n}$$

(a) How many panes are necessary to block at least 50% of the light?

(b) How many panes are necessary to block at least 75% of the light?

78. Chemistry The pH of a chemical solution is given by the formula

$$\text{pH} = -\log_{10} [H^+]$$

where $[H^+]$ is the concentration of hydrogen ions in moles per liter. Values of pH range from 0 (acidic) to 14 (alkaline).

(a) Find the pH of a 1 liter container of water with 0.0000001 mole of hydrogen ion.

(b) Find the hydrogen ion concentration of a mildly acidic solution with a pH of 4.2.

79. Space Satellites The number of watts w provided by a space satellite's power supply after d days is given by the formula

$$w = 50e^{-0.004d}$$

(a) How long will it take for the available power to drop to 30 watts?

(b) How long will it take for the available power to drop to only 5 watts?

80. Healing of Wounds The normal healing of wounds can be modeled by an exponential function. If A_0 represents the original area of the wound and if A equals the area of the wound after n days, then the formula

$$A = A_0e^{-0.35n}$$

describes the area of a wound on the nth day following an injury when no infection is present to retard the healing. Suppose a wound initially had an area of 100 square centimeters.

(a) If healing is taking place, how many days should pass before the wound is one-half its original size?

(b) How long before the wound is 10% of its original size?

81. **Drug Medication** The formula

$$D = 5e^{-0.4h}$$

can be used to find the number of milligrams D of a certain drug that is in a patient's bloodstream h hours after the drug has been administered. When the number of milligrams reaches 2, the drug is to be administered again. What is the time between injections?

82. **Spreading of Rumors** A model for the number of people N in a college community who have heard a certain rumor is

$$N = P(1 - e^{-0.15d})$$

where P is the total population of the community and d is the number of days that have elapsed since the rumor began. In a community of 1000 students, how many days will elapse before 450 students have heard the rumor?

83. **Current in a *RL* Circuit** The equation governing the amount of current I (in amperes) after time t (in seconds) in a simple RL circuit consisting of a resistance R (in ohms), an inductance L (in henrys), and an electromotive force E (in volts) is

$$I = \frac{E}{R}[1 - e^{-(R/L)t}]$$

If $E = 12$ volts, $R = 10$ ohms, and $L = 5$ henrys, how long does it take to obtain a current of 0.5 ampere? Of 1.0 ampere? Graph the equation.

84. **Learning Curve** Psychologists sometimes use the function

$$L(t) = A(1 - e^{-kt})$$

to measure the amount L learned at time t. The number A represents the amount to be learned, and the number k measures the rate of learning. Suppose that a student has an amount A of 200 vocabulary words to learn. A psychologist determines that the student learned 20 vocabulary words after 5 minutes.

(a) Determine the rate of learning k.
(b) Approximately how many words will the student have learned after 10 minutes?
(c) After 15 minutes?
(d) How long does it take for the student to learn 180 words?

85. **Alcohol and Driving** The concentration of alcohol in a person's blood is measurable. Suppose the risk R (given as a percent) of having an accident while driving a car can be modeled by the equation

$$R = 3e^{kx}$$

where x is the variable concentration of alcohol in the blood and k is a constant.

(a) Suppose a concentration of alcohol in the blood of 0.06 results in a 10% risk ($R = 10$) of an accident. Find the constant k in the equation.
(b) Using this value of k, what is the risk if the concentration is 0.17?
(c) Using the same value of k, what concentration of alcohol corresponds to a risk of 100%?
(d) If the law asserts that anyone with a risk of having an accident of 15% or more should not have driving privileges, at what concentration of alcohol in the blood should a driver be arrested and charged with a DUI?
(e) Compare this situation with that of Example 9. If you were a lawmaker, which situation would you support? Give your reasons.

86. Is there any function of the form $y = x^{\alpha}$, $0 < \alpha < 1$, that increases more slowly than a logarithmic function whose base is greater than 1? Explain.

87. **Constructing a Function** Look back at Figure 14 in Section 5.2. Assuming that the points (1950, 20) and (1990, 50) are on the graph, find an exponential equation $y = Ae^{bt}$, that fits the data.

[**Hint:** Let $t = 0$ correspond to the year 1950. Then show that $A = 20$. Now find b.]

Is the projection of 102 million in 2015 confirmed by your model? Try to obtain similar data about the United States birthrate and construct a function to fit those data.

88. **Critical Thinking** In buying a new car, one consideration might be how well the price of the car holds up over time. Different makes of cars have different depreciation rates. One way to compute a depreciation rate for a car is given here. Suppose the current prices of a certain Mercedes automobile are as follows:

New	1 Year Old	2 Years Old	3 Years Old	4 Years Old	5 Years Old
\$38,000	\$36,600	\$32,400	\$28,750	\$25,400	\$21,200

Use the formula New = Old(e^{Rt}) to find R, the annual depreciation rate, for a specific time t. When might be the best time to trade the car in? Consult the NADA ("blue") book and compare two like models that you are interested in. Which has the better depreciation rate?

5.4 PROPERTIES OF LOGARITHMS

1. Write a Logarithmic Expression as a Sum/Difference of Logarithms
2. Write a Logarithmic Expression as a Single Logarithm
3. Evaluate Logarithms Whose Base Is Neither 10 nor *e*
4. Graph Logarithmic Functions Whose Base Is Neither 10 nor *e*

Logarithms have some very useful properties that can be derived directly from the definition and the laws of exponents.

EXAMPLE 1 Establishing Properties of Logarithms

(a) Show that $\log_a 1 = 0$. (b) Show that $\log_a a = 1$.

Solution (a) This fact was established when we graphed $y = \log_a x$ (see Figure 25). Algebraically, for $y = \log_a 1$, we have $a^y = 1 = a^0$, so $y = 0$.

(b) For $y = \log_a a$, we have $a^y = a = a^1$, so $y = 1$. ■

To summarize

$$\log_a 1 = 0 \qquad \log_a a = 1$$

Theorem Properties of Logarithms

In the properties given next, M and a are positive real numbers, with $a \neq 1$, and r is any real number.

The number $\log_a M$ is the exponent to which a must be raised to obtain M. That is,

$$a^{\log_a M} = M \qquad (1)$$

The logarithm to the base a of a raised to a power equals that power. That is,

$$\log_a a^r = r \qquad (2)$$

■

Proof of Property (1) Let $x = \log_a M$. Change this logarithmic expression to the equivalent exponential expression:

$$a^x = M$$

But $x = \log_a M$, so

$$a^{\log_a M} = M$$

Proof of Property (2) Let $x = a^r$. Change this exponential expression to the equivalent logarithmic expression

$$\log_a x = r$$

But $x = a^r$, so

$$\log_a a^r = r$$

EXAMPLE 2 Using Properties (1) and (2)

(a) $2^{\log_2 \pi} = \pi$ (b) $\log_{0.2} 0.2^{-\sqrt{2}} = -\sqrt{2}$ (c) $\ln e^{kt} = kt$

Other useful properties of logarithms are given now.

Theorem Properties of Logarithms

In the following properties, M, N, and a are positive real numbers, with $a \neq 1$, and r is any real number.

The Log of a Product Equals the Sum of the Logs

$$\log_a (MN) = \log_a M + \log_a N \quad (3)$$

The Log of a Quotient Equals the Difference of the Logs

$$\log_a\left(\frac{M}{N}\right) = \log_a M - \log_a N \quad (4)$$

$$\log_a\left(\frac{1}{N}\right) = -\log_a N \quad (5)$$

$$\log_a M^r = r \log_a M \quad (6)$$

We shall derive properties (3) and (6) and leave the derivations of properties (4) and (5) as exercises (see Problems 63 and 64).

Proof of Property (3) Let $A = \log_a M$ and let $B = \log_a N$. These expressions are equivalent to the exponential expressions

$$a^A = M \quad \text{and} \quad a^B = N$$

Now

$$\begin{aligned}\log_a(MN) = \log_a(a^A a^B) &= \log_a a^{A+B} && \text{Law of exponents.}\\ &= A + B && \text{Property (2) of logarithms.}\\ &= \log_a M + \log_a N\end{aligned}$$

Proof of Property (6) Let $A = \log_a M$. This expression is equivalent to

$$a^A = M$$

Now

$$\begin{aligned}\log_a M^r = \log_a(a^A)^r &= \log_a a^{rA} && \text{Law of exponents.}\\ &= rA && \text{Property (2) of logarithms.}\\ &= r \log_a M\end{aligned}$$

1 Logarithms can be used to transform products into sums, quotients into differences, and powers into factors. Such transformations prove useful in certain types of calculus problems.

EXAMPLE 3

Writing a Logarithmic Expression as a Sum of Logarithms

Write $\log_a(x\sqrt{x^2+1})$ as a sum of logarithms. Express all powers as factors.

Solution

$$\begin{aligned}\log_a(x\sqrt{x^2+1}) &= \log_a x + \log_a \sqrt{x^2+1} && \text{Property (3).}\\ &= \log_a x + \log_a(x^2+1)^{1/2}\\ &= \log_a x + \tfrac{1}{2}\log_a(x^2+1) && \text{Property (6).}\end{aligned}$$

EXAMPLE 4

Writing a Logarithmic Expression as a Difference of Logarithms

Write

$$\log_a \frac{x^2}{(x-1)^3}$$

as a difference of logarithms. Express all powers as factors.

Solution

$$\log_a \frac{x^2}{(x-1)^3} \underset{\text{Property (4).}}{\underset{\uparrow}{=}} \log_a x^2 - \log_a(x-1)^3 \underset{\text{Property (6).}}{\underset{\uparrow}{=}} 2\log_a x - 3\log_a(x-1)$$

Now work Problem 13.

EXAMPLE 5

Writing a Logarithmic Expression as a Sum and Difference of Logarithms

Write

$$\log_a \frac{x^3\sqrt{x^2+1}}{(x+1)^4}$$

as a sum and difference of logarithms. Express all powers as factors.

Solution

$$\log_a \frac{x^3\sqrt{x^2+1}}{(x+1)^4} = \log_a(x^3\sqrt{x^2+1}) - \log_a(x+1)^4$$
$$= \log_a x^3 + \log_a \sqrt{x^2+1} - \log_a(x+1)^4$$
$$= \log_a x^3 + \log_a(x^2+1)^{1/2} - \log_a(x+1)^4$$
$$= 3\log_a x + \tfrac{1}{2}\log_a(x^2+1) - 4\log_a(x+1)$$

2 Another use of properties (3) through (6) is to write sums and/or differences of logarithms with the same base as a single logarithm.

EXAMPLE 6 Writing Expressions as a Single Logarithm

Write each of the following as a single logarithm:

(a) $\log_a 7 + 4\log_a 3$
(b) $\frac{2}{3}\log_a 8 - \log_a(3^4 - 8)$
(c) $\log_a x + \log_a 9 + \log_a(x^2+1) - \log_a 5$

Solution

(a) $\log_a 7 + 4\log_a 3 = \log_a 7 + \log_a 3^4$ Property (6).
$= \log_a 7 + \log_a 81$
$= \log_a(7 \cdot 81)$ Property (3).
$= \log_a 567$

(b) $\frac{2}{3}\log_a 8 - \log_a(3^4 - 8) = \log_a 8^{2/3} - \log_a(81 - 8)$ Property (6).
$= \log_a 4 - \log_a 73$
$= \log_a(\frac{4}{73})$ Property (4).

(c) $\log_a x + \log_a 9 + \log_a(x^2+1) - \log_a 5 = \log_a 9x + \log_a(x^2+1) - \log_a 5$
$= \log_a[9x(x^2+1)] - \log_a 5$
$= \log_a\left[\frac{9x(x^2+1)}{5}\right]$

Warning: A common error made by some students is to express the logarithm of a sum as the sum of logarithms:

$\log_a(M+N)$ is not equal to $\log_a M + \log_a N$

Correct statement. $\log_a(MN) = \log_a M + \log_a N$ Property (3).

Another common error is to express the difference of logarithms as the quotient of logarithms:

$\log_a M - \log_a N$ is not equal to $\dfrac{\log_a M}{\log_a N}$

Correct statement. $\log_a M - \log_a N = \log_a\left(\dfrac{M}{N}\right)$ Property (4).

Now work Problem 23.

EXAMPLE 7 Writing a Power Equation as a Linear Equation

Write the equation $y = ax^b$ as a linear equation.

Solution The equation $y = ax^b$ is the model for the generalized power function. Earlier, we obtained a value for the correlation coefficient, r, of this model even though it is not linear. We said this is due to the fact the model can be written in linear form using logarithms. Let's see how.

$$y = ax^b$$
$$\ln y = \ln(ax^b) \quad \text{Take the natural log of both sides.}$$
$$\ln y = \ln a + \ln x^b \quad \text{Property (3).}$$
$$\ln y = \ln a + b \ln x \quad \text{Property (6).}$$
$$Y = \ln a + bX$$

This is an equation in linear form with y-intercept $\ln a$ and slope b. ■

Seeing The Concept Redo Example 4 from Section 4.2 by letting the independent variable be $x = \ln$ (length) and the dependent variable be $y = \ln$ (period). Find the line of best fit. What is the slope of your line of best fit? Compare it to the value of b found in Example 4. Why is the value of a found in Example 4 different from the y-intercept of the line of best fit? Compare the values of the correlation coefficient. Are they the same? ■

There remain two other properties of logarithms we need to know. They are a consequence of the fact that the logarithmic function $y = \log_a x$ is one-to-one.

Theorem

In the following properties, M, N, and a are positive real numbers, with $a \neq 1$:

$$\text{If} \quad M = N, \quad \text{then} \quad \log_a M = \log_a N. \qquad (7)$$
$$\text{If} \quad \log_a M = \log_a N, \quad \text{then} \quad M = N. \qquad (8)$$

■

Properties (7) and (8) are useful for solving *logarithmic equations*, a topic discussed in the next section.

Using a Calculator to Evaluate and Graph Logarithms with Bases Other Than *e* or 10

3 Logarithms to the base 10, called **common logarithms,** were used to facilitate arithmetic computations before the widespread use of calculators. (See the Historical Feature at the end of this section.) Natural logarithms, that is, logarithms whose base is the number e, remain very important because they arise frequently in the study of natural phenomena.

Common logarithms are usually abbreviated by writing **log,** with the base understood to be 10, just as natural logarithms are abbreviated by **ln,** with the base understood to be e.

Graphing calculators have both [log] and [ln] keys to calculate the common logarithm and natural logarithm of a number. Let's look at an example to see how to calculate logarithms having a base other than 10 or e.

EXAMPLE 8 Evaluating Logarithms Whose Base Is Neither 10 Nor *e*

Evaluate $\log_2 7$.

Solution Let $y = \log_2 7$. Then $2^y = 7$, so

$$\begin{aligned} 2^y &= 7 \\ \ln 2^y &= \ln 7 && \text{Property (7).} \\ y \ln 2 &= \ln 7 && \text{Property (6).} \\ y &= \frac{\ln 7}{\ln 2} && \text{Solve for } y. \\ &= 2.8074 && \text{Use calculator (} \boxed{\ln} \text{ key).} \end{aligned}$$

Example 8 shows how to change the base from 2 to *e*. In general, to change from the base *b* to the base *a*, we use the **Change-of-Base Formula.**

Theorem Change-of-Base Formula

If $a \neq 1$, $b \neq 1$, and M are positive real numbers, then

$$\log_a M = \frac{\log_b M}{\log_b a} \qquad (9)$$

Proof We derive this formula as follows: Let $y = \log_a M$. Then $a^y = M$, so

$$\begin{aligned} \log_b a^y &= \log_b M && \text{Property (7).} \\ y \log_b a &= \log_b M && \text{Property (6).} \\ y &= \frac{\log_b M}{\log_b a} && \text{Solve for } y. \\ \log_a M &= \frac{\log_b M}{\log_b a} && y = \log_a M. \end{aligned}$$

Since calculators have only keys for $\boxed{\log}$ and $\boxed{\ln}$, in practice, the Change-of-Base Formula uses either $b = 10$ or $b = e$. Thus,

$$\log_a M = \frac{\log M}{\log a} \quad \text{and} \quad \log_a M = \frac{\ln M}{\ln a} \qquad (10)$$

EXAMPLE 9 Using the Change-of-Base Formula

Calculate:

(a) $\log_5 89$ (b) $\log_{\sqrt{2}} \sqrt{5}$

Solution (a) $\log_5 89 = \dfrac{\log 89}{\log 5} \approx \dfrac{1.94939}{0.69897} = 2.7889$

or

$$\log_5 89 = \frac{\ln 89}{\ln 5} \approx \frac{4.4886}{1.6094} = 2.7889$$

(b) $\log_{\sqrt{2}} \sqrt{5} = \dfrac{\log \sqrt{5}}{\log \sqrt{2}} = \dfrac{\frac{1}{2}\log 5}{\frac{1}{2}\log 2} \approx \dfrac{0.69897}{0.30103} = 2.3219$

or

$$\log_{\sqrt{2}} \sqrt{5} = \frac{\ln \sqrt{5}}{\ln \sqrt{2}} = \frac{\frac{1}{2}\ln 5}{\frac{1}{2}\ln 2} \approx \frac{1.6094}{0.6931} = 2.3219$$

Now work Problem 35.

4 We also use the Change-of-Base Formula to graph logarithmic functions whose base is neither 10 nor *e*.

EXAMPLE 10 Graphing a Logarithmic Function Whose Base Is Neither 10 Nor *e*

Use a graphing utility to graph $y = \log_2 x$.

FIGURE 31

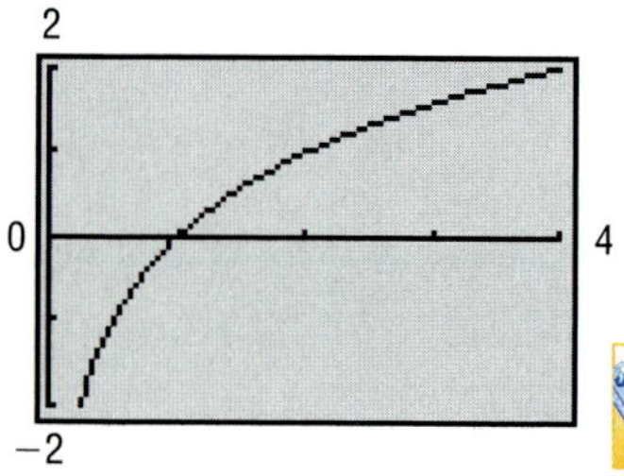

Solution Since graphing utilities only have logarithms with the base 10 or the base *e*, we need to use the Change-of-Base Formula to express $y = \log_2 x$ in terms of logarithms with base 10 or base *e*. We can graph either $y = \ln x/\ln 2$ or $y = \log x/\log 2$ to obtain the graph of $y = \log_2 x$. See Figure 31.

Check: Verify that $y = \ln x/\ln 2$ and $y = \log x/\log 2$ result in the same graph by graphing each one on the same viewing rectangle.

Now work Problem 43.

SUMMARY OF PROPERTIES OF LOGARITHMS

In the list that follows, $a > 0$, $a \neq 1$, and $b > 0$, $b \neq 1$; also, $M > 0$ and $N > 0$.

Definition	$y = \log_a x$ means $x = a^y$
Properties of logarithms	$\log_a 1 = 0; \quad \log_a a = 1$
	$a^{\log_a M} = M; \quad \log_a a^r = r$
	$\log_a (MN) = \log_a M + \log_a N$
	$\log_a\left(\frac{M}{N}\right) = \log_a M - \log_a N$
	$\log_a\left(\frac{1}{N}\right) = -\log_a N$
	$\log_a M^r = r \log_a M$
Change-of-Base Formula	$\log_a M = \frac{\log_b M}{\log_b a}$

HISTORICAL FEATURE Logarithms were invented about 1590 by John Napier (1550–1617) and Jobst Bürgi (1552–1632), working independently. Napier, whose work had the greater influence, was a Scottish lord, a secretive man whose neighbors were inclined to believe him to be in league with the devil. His approach to logarithms was quite different from ours; it was based on the relationship between arithmetic and geometric sequences (see the chapter on induction and sequences), and not on the inverse function relationship of logarithms to exponential functions (described in Section 5.3). Napier's tables, published in 1614, listed what would now be called *natural logarithms* of sines and were rather difficult to use. A London professor, Henry Briggs, became interested in the tables and visited Napier. In their conversations, they developed the idea of common logarithms, and Briggs then converted Napier's tables into tables of common logarithms, which were published in 1617. Their importance for calculation was immediately recognized, and by 1650 they were being printed as far away as China. They remained an important calculation tool until the advent of the inexpensive handheld calculator about 1972, which has decreased their calculational, but not their theoretical, importance.

A side effect of the invention of logarithms was the popularization of the decimal system of notation for real numbers. ■

5.4 EXERCISES

In Problems 1–12, suppose $\ln 2 = a$ *and* $\ln 3 = b$*. Use properties of logarithms to write each logarithm in terms of a and b.*

1. $\ln 6$

2. $\ln \frac{2}{3}$

3. $\ln 1.5$

4. $\ln 0.5$

5. $\ln 2e$

6. $\ln\left(\frac{3}{e}\right)$

7. $\ln 12$

8. $\ln 24$

9. $\ln \sqrt[5]{18}$

10. $\ln \sqrt[4]{48}$

11. $\log_2 3$

12. $\log_3 2$

In Problems 13–22, write each expression as a sum and/or difference of logarithms. Express powers as factors.

13. $\ln(x^2\sqrt{1 - x})$

14. $\ln(x\sqrt{1 + x^2})$

15. $\log_2\left(\frac{x^3}{x - 3}\right)$

16. $\log_5\left(\frac{\sqrt[3]{x^2 + 1}}{x^2 - 1}\right)$

17. $\log\left[\frac{x(x + 2)}{(x + 3)^2}\right]$

18. $\log\frac{x^3\sqrt{x + 1}}{(x - 2)^2}$

19. $\ln\left[\frac{x^2 - x - 2}{(x + 4)^2}\right]^{1/3}$

20. $\ln\left[\frac{(x - 4)^2}{x^2 - 1}\right]^{2/3}$

21. $\ln\frac{5x\sqrt{1 - 3x}}{(x - 4)^3}$

22. $\ln\left[\frac{5x^2\sqrt[3]{1 - x}}{4(x + 1)^2}\right]$

In Problems 23–32, write each expression as a single logarithm.

23. $3 \log_5 u + 4 \log_5 v$

24. $\log_3 u^2 - \log_3 v$

25. $\log_{1/2} \sqrt{x} - \log_{1/2} x^3$

26. $\log_2\left(\frac{1}{x}\right) + \log_2\left(\frac{1}{x^2}\right)$

27. $\ln\left(\frac{x}{x - 1}\right) + \ln\left(\frac{x + 1}{x}\right) - \ln(x^2 - 1)$

28. $\log\left(\frac{x^2 + 2x - 3}{x^2 - 4}\right) - \log\left(\frac{x^2 + 7x + 6}{x + 2}\right)$

29. $8 \log_2 \sqrt{3x - 2} - \log_2\left(\frac{4}{x}\right) + \log_2 4$

30. $21 \log_3 \sqrt[3]{x} + \log_3 9x^2 - \log_5 25$

31. $2 \log_a 5x^3 - \frac{1}{2} \log_a(2x + 3)$

32. $\frac{1}{3} \log(x^3 + 1) + \frac{1}{2} \log(x^2 + 1)$

33. Write the exponential model, $y = ab^x$, as a linear model.

34. Is the logarithmic model, $y = a + b \ln x$, linear? If so, what is the slope? What is the y-intercept?

In Problems 35–42, use the Change-of-Base Formula and a calculator to evaluate each logarithm. Round your answer to three decimal places.

35. $\log_3 21$
36. $\log_5 18$
37. $\log_{1/3} 71$
38. $\log_{1/2} 15$
39. $\log_{\sqrt{2}} 7$
40. $\log_{\sqrt{5}} 8$
41. $\log_{\pi} e$
42. $\log_{\pi} \sqrt{2}$

For Problems 43–48, graph each function using a graphing utility and the Change-of-Base Formula.

43. $y = \log_4 x$
44. $y = \log_5 x$
45. $y = \log_2(x + 2)$
46. $y = \log_4(x - 3)$
47. $y = \log_{x-1}(x + 1)$
48. $y = \log_{x+2}(x - 2)$

In Problems 49–58, express y as a function of x. The constant C is a positive number.

49. $\ln y = \ln x + \ln C$
50. $\ln y = \ln(x + C)$
51. $\ln y = \ln x + \ln(x + 1) + \ln C$
52. $\ln y = 2 \ln x - \ln(x + 1) + \ln C$
53. $\ln y = 3x + \ln C$
54. $\ln y = -2x + \ln C$
55. $\ln(y - 3) = -4x + \ln C$
56. $\ln(y + 4) = 5x + \ln C$
57. $3 \ln y = \frac{1}{2} \ln(2x + 1) - \frac{1}{3} \ln(x + 4) + \ln C$
58. $2 \ln y = -\frac{1}{2} \ln x + \frac{1}{3} \ln(x^2 + 1) + \ln C$
59. Find the value of $\log_2 3 \cdot \log_3 4 \cdot \log_4 5 \cdot \log_5 6 \cdot \log_6 7 \cdot \log_7 8$.
60. Find the value of $\log_2 4 \cdot \log_4 6 \cdot \log_6 8$.
61. Find the value of $\log_2 3 \cdot \log_3 4 \cdot \ldots \cdot \log_n(n + 1) \cdot \log_{n+1} 2$.
62. Find the value of $\log_2 2 \cdot \log_2 4 \cdot \ldots \cdot \log_2 2^n$.
63. Show that $\log_a(M/N) = \log_a M - \log_a N$, where a, M, and N are positive real numbers, with $a \neq 1$.
64. Show that $\log_a (1/N) = -\log_a N$, where a and N are positive real numbers, with $a \neq 1$.
65. Find the domain of $f(x) = \log_a x^2$ and the domain of $g(x) = 2 \log_a x$. Since $\log_a x^2 = 2 \log_a x$, how do you reconcile the fact that the domains are not equal? Write a brief explanation.

5.5 LOGARITHMIC AND EXPONENTIAL EQUATIONS

1 Solve Logarithmic Equations
2 Solve Exponential Equations
3 Solve Logarithmic and Exponential Equations Using a Graphing Utility

Logarithmic Equations

1 Equations that contain terms of the form $\log_a x$, where a is a positive real number, with $a \neq 1$, are often called **logarithmic equations.**

As before, our practice will be to solve equations, whenever possible, by finding exact solutions using algebraic methods. In such cases, we will also verify the solution obtained by using a graphing utility. When algebraic methods cannot be used, approximate solutions will be obtained using a graphing utility. The reader is encouraged to pay particular attention to the form of equations for which exact solutions are possible.

MISSION POSSIBLE

McNewton's Coffee

Your team has been called in to solve a problem encountered by a fast food restaurant. They believe that their coffee should be brewed at 170° Fahrenheit. However, at that temperature it is too hot to drink, and a customer who accidentally spills the coffee on himself might receive third-degree burns.

What they need is a special container that will heat the water to 170°, brew the coffee at that temperature, then cool it quickly to a drinkable temperature, say 140°F, and hold it there, or at least keep it at or above 120°F for a reasonable period of time without further cooking. To cool down the coffee, three companies have submitted proposals with these specifications:

(a) The CentiKeeper Company has a container that will reduce the temperature of a liquid from 200°F to 100°F in 90 minutes by maintaining a constant temperature of 70°F.
(b) The TempControl Company has a container that will reduce the temperature of a liquid from 200°F to 110°F in 60 minutes by maintaining a constant temperature of 60°F.
(c) The Hot'n'Cold, Inc., has a container that will reduce the temperature of a liquid from 210°F to 90°F in 30 minutes by maintaining a constant temperature of 50°F.

Your job is to make a recommendation as to which container to purchase. For this you will need Newton's law of cooling which follows:

$$u(t) = T + (u_0 - T)e^{kt}, \quad k < 0$$

In this formula, T represents the temperature of the surrounding medium, u_0 is the initial temperature of the heated object, t is the length of time in minutes, k is a negative constant, and u represents the temperature at time t.

1. Use Newton's Law of Cooling to find the constant k of the formula for each container.
2. Use a graphing utility to graph each relation.
3. How long does it take each container to lower the coffee temperature from 170°F to 140°F?
4. How long will the coffee temperature remain between 120°F and 140°F?
5. On the basis of this information, which company should get the contract with McNewton's? What are your reasons?
6. Define "capital cost" and "operating cost." How might they affect your choice?

EXAMPLE 1 Solving a Logarithmic Equation

Solve $\log_3(4x - 7) = 2$.

Solution We can obtain an exact solution by changing the logarithm to exponential form.

$$\begin{aligned} \log_3(4x - 7) &= 2 \\ 4x - 7 &= 3^2 \\ 4x - 7 &= 9 \\ 4x &= 16 \\ x &= 4 \end{aligned}$$

We verify the solution by graphing $Y_1 = \log_3(4x - 7)$ and $Y_2 = 2$ to determine where the graphs intersect. See Figure 32. The graphs intersect at $x = 4$.

FIGURE 32

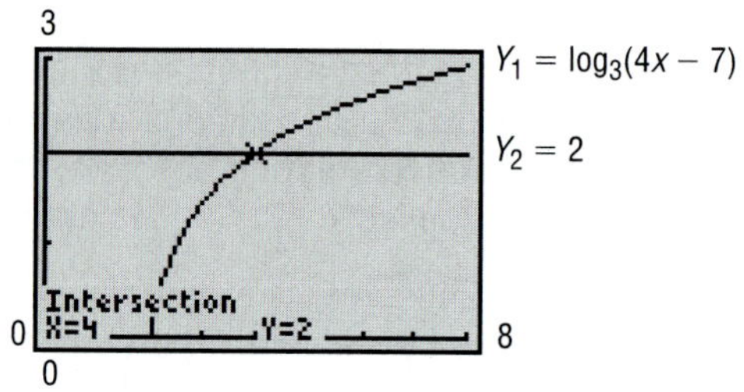

EXAMPLE 2 Solving a Logarithmic Equation

Solve $2 \log_5 x = \log_5 9$.

Solution Because each logarithm is to the same base, 5, we can obtain an exact solution as follows:

$$\begin{aligned} 2 \log_5 x &= \log_5 9 \\ \log_5 x^2 &= \log_5 9 && \text{Property (6), Section 5.4.} \\ x^2 &= 9 && \text{Property (8), Section 5.4.} \\ x = 3 \quad &\text{or} \quad \cancel{x = -3} \end{aligned}$$

Recall that logarithms of negative numbers are not defined, so, in the expression $2 \log_5 x$, x must be positive. Therefore, -3 is extraneous and we discard it.

The equation has only one solution, 3.

You should verify for yourself that 3 is the only solution using a graphing utility.

Now work Problem 1.

EXAMPLE 3

Solving a Logarithmic Equation

Solve $\log_4(x + 3) + \log_4(2 - x) = 1$.

Solution To obtain an exact solution, we need to express the left side as a single logarithm. Then we will change the expression to exponential form.

$$\log_4(x + 3) + \log_4(2 - x) = 1$$
$$\log_4[(x + 3)(2 - x)] = 1 \qquad \text{Property (3), Section 5.4.}$$
$$(x + 3)(2 - x) = 4^1 = 4$$
$$-x^2 - x + 6 = 4$$
$$x^2 + x - 2 = 0$$
$$(x + 2)(x - 1) = 0$$
$$x = -2 \quad \text{or} \quad x = 1$$

You should verify that both of these are solutions using a graphing utility.

Now work Problem 11.

Care must be taken when solving logarithmic equations algebraically. Be sure to check each apparent solution in the original equation and discard any that are extraneous. In the expression $\log_a M$, remember that a and M are positive and $a \neq 1$.

Exponential Equations

2 Equations that involve terms of the form a^x, $a > 0$, $a \neq 1$, are often referred to as **exponential equations.** Such equations sometimes can be solved by appropriately applying the laws of exponents and equation (1), namely,

$$\text{If } a^u = a^v, \quad \text{then } u = v. \qquad (1)$$

To use equation (1), each side of the equality must be written with the same base.

EXAMPLE 4

Solving an Exponential Equation

Solve the equation $3^{x+1} = 81$.

Solution Since $81 = 3^4$, we can write the equation as

$$3^{x+1} = 81 = 3^4$$

Now we have the same base, 3, on each side, so we can apply (1) to obtain

$$x + 1 = 4$$
$$x = 3$$

We verify the solution by graphing $Y_1 = 3^{x+1}$ and $Y_2 = 81$ to determine where the graphs intersect. See Figure 33. The graphs intersect at $x = 3$.

FIGURE 33

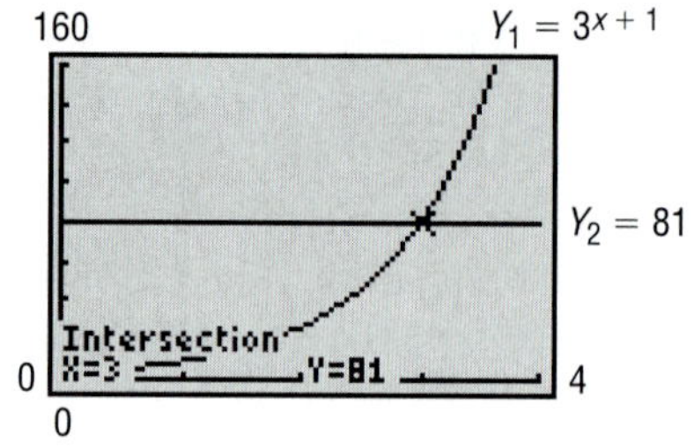

Now work Problem 19.

EXAMPLE 5 Solving an Exponential Equation

Solve the equation $e^{-x^2} = (e^x)^2 \cdot \dfrac{1}{e^3}$.

Solution We use some laws of exponents first to get the same base e on each side:

$$e^{-x^2} = (e^x)^2 \cdot \frac{1}{e^3} = e^{2x} \cdot e^{-3} = e^{2x-3}$$

Now apply (1) to get

$$\begin{aligned} -x^2 &= 2x - 3 \\ x^2 + 2x - 3 &= 0 \\ (x+3)(x-1) &= 0 \\ x = -3 \quad \text{or} \quad x &= 1 \end{aligned}$$

You should verify these solutions using a graphing utility.

EXAMPLE 6 Solving an Exponential Equation

Solve the equation $4^x - 2^x - 12 = 0$.

Solution We note that $4^x = (2^2)^x = 2^{2x} = (2^x)^2$, so the equation is actually quadratic in form, and we can rewrite it as

$$(2^x)^2 - 2^x - 12 = 0$$

Now we can factor as usual:

$$\begin{aligned} (2^x - 4)(2^x + 3) &= 0 \\ 2^x - 4 = 0 \quad &\text{or} \quad 2^x + 3 = 0 \\ 2^x = 4 \quad &\quad\quad\; 2^x = -3 \end{aligned}$$

The equation on the left has the solution $x = 2$, since $2^x = 4 = 2^2$; the equation on the right has no solution, since $2^x > 0$ for all x. ■

In each of the preceding three examples, we were able to write each exponential expression using the same base, obtaining exact solutions to the equation. When this is not possible, logarithms can sometimes be used to obtain the solution.

EXAMPLE 7 Solving an Exponential Equation

Solve for x: $2^x = 5$.

Solution We write the exponential equation as the equivalent logarithmic equation:

$$2^x = 5$$

$$x = \log_2 5 \underset{\uparrow}{=} \frac{\ln 5}{\ln 2}$$

Change-of-Base Formula (10).

Alternatively, we can solve the equation $2^x = 5$ by taking the natural logarithm (or common logarithm) of each side. Taking the natural logarithm,

$$\begin{aligned} 2^x &= 5 \\ \ln 2^x &= \ln 5 \\ x \ln 2 &= \ln 5 \\ x &= \frac{\ln 5}{\ln 2} \end{aligned}$$

Using a calculator, the solution, rounded to three decimal places, is

$$x = \frac{\ln 5}{\ln 2} \approx 2.322$$ ■

Now work Problem 33.

EXAMPLE 8 Solving an Exponential Equation

Solve for x: $8 \cdot 3^x = 5$.

Solution

$$\begin{aligned} 8 \cdot 3^x &= 5 && \text{Isolate } 3^x \text{ on the left side.} \\ 3^x &= \tfrac{5}{8} && \text{Proceed as in Example 7.} \\ x &= \log_3\left(\tfrac{5}{8}\right) = \frac{\ln \frac{5}{8}}{\ln 3} \end{aligned}$$

Using a calculator, the solution, rounded to three decimal places, is

$$x = \frac{\ln\left(\frac{5}{8}\right)}{\ln 3} \approx -0.428$$ ■

EXAMPLE 9 Solving an Exponential Equation

Solve for x: $5^{x-2} = 3^{3x+2}$.

Solution Because the bases are different, we take the natural logarithm of each side and apply appropriate properties of logarithms. The result is an equation in x that we can solve.

$$5^{x-2} = 3^{3x+2}$$

$$\ln 5^{x-2} = \ln 3^{3x+2} \quad \text{Property (7).}$$

$$(x-2)\ln 5 = (3x+2)\ln 3 \quad \text{Property (6).}$$

$$(\ln 5)x - 2\ln 5 = (3\ln 3)x + 2\ln 3$$

$$(\ln 5 - 3\ln 3)x = 2\ln 3 + 2\ln 5$$

$$x = \frac{2(\ln 3 + \ln 5)}{\ln 5 - 3\ln 3} \approx -3.212$$

Now work Problem 41.

Graphing Utility Solutions

3 The techniques introduced in this section only apply to certain types of logarithmic and exponential equations. Solutions for other types are usually studied in calculus, using numerical methods. However, we can use a graphing utility to approximate the solution.

EXAMPLE 10 Solving Equations Using a Graphing Utility

Solve $\log_3 x + \log_4 x = 4$.
Express the solution(s) correct to two decimal places.

Solution The solution is found by graphing

$$Y_1 = \log_3 x + \log_4 x = \frac{\log x}{\log 3} + \frac{\log x}{\log 4} \text{ and } Y_2 = 4.$$

(Remember you must use the Change-of-Base Formula to graph Y_1). Y_1 is an increasing function (do you know why?) and so there is only one point of intersection for Y_1 and Y_2. Figure 34 shows the graphs of Y_1 and Y_2. The solution is 11.60 correct to two decimal places.

FIGURE 34

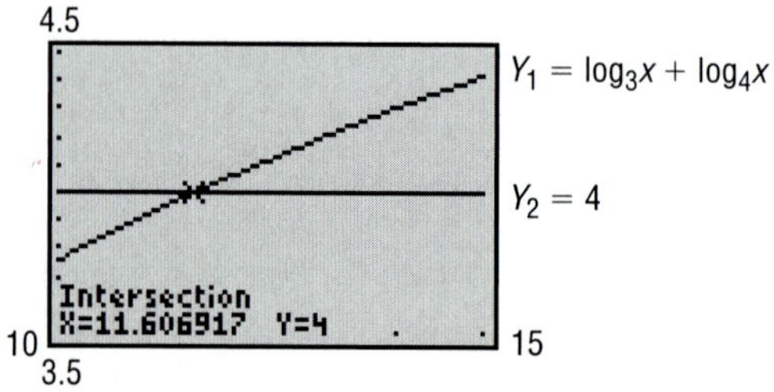

Can you discover an algebraic solution to Example 10?

[**Hint:** Factor $\log x$ from Y_1.]

EXAMPLE 11 Solving Equations Using a Graphing Utility

Solve $x + e^x = 2$.
Express the solution(s) correct to two decimal places.

Solution The solution is found by graphing $Y_1 = x + e^x$ and $Y_2 = 2$. Y_1 is an increasing function (do you know why?) and so there is only one point of intersection for Y_1 and Y_2. Figure 35 shows the graphs of Y_1 and Y_2. The solution is 0.44 correct to two decimal places.

FIGURE 35

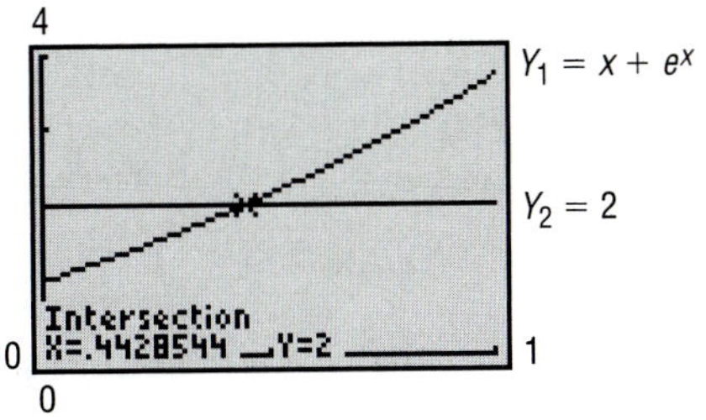

5.5 EXERCISES

In Problems 1–58, solve each equation. Verify your solution using a graphing utility.

1. $\log_2(2x + 1) = 3$
2. $\log_3(3x - 2) = 2$
3. $\log_3(x^2 + 1) = 2$
4. $\log_5(x^2 + x + 4) = 2$
5. $\frac{1}{2}\log_3 x = 2\log_3 2$
6. $-2\log_4 x = \log_4 9$
7. $2\log_5 x = 3\log_5 4$
8. $3\log_2 x = -\log_2 27$
9. $3\log_2(x - 1) + \log_2 4 = 5$
10. $2\log_3(x + 4) - \log_3 9 = 2$
11. $\log x + \log(x + 15) = 2$
12. $\log_4 x + \log_4(x - 3) = 1$
13. $\log_x 4 = 2$
14. $\log_x(\frac{1}{8}) = 3$
15. $\log_3(x - 1)^2 = 2$
16. $\log_2(x + 4)^3 = 6$
17. $\log_{1/2}(3x + 1)^{1/3} = -2$
18. $\log_{1/3}(1 - 2x)^{1/2} = -1$
19. $2^{2x+1} = 4$
20. $5^{1-2x} = \frac{1}{5}$
21. $3^{x^3} = 9^x$
22. $4^{x^2} = 2^x$
23. $8^{x^2-2x} = \frac{1}{2}$
24. $9^{-x} = \frac{1}{3}$
25. $2^x \cdot 8^{-x} = 4^x$
26. $(\frac{1}{2})^{1-x} = 4$
27. $2^{2x} + 2^x - 12 = 0$
28. $3^{2x} + 3^x - 2 = 0$
29. $3^{2x} + 3^{x+1} - 4 = 0$
30. $4^x - 2^x = 0$
31. $4^x = 8$
32. $9^{2x} = 27$
33. $2^x = 10$
34. $3^x = 14$
35. $8^{-x} = 1.2$
36. $2^{-x} = 1.5$
37. $3^{1-2x} = 4^x$
38. $2^{x+1} = 5^{1-2x}$
39. $(\frac{3}{5})^x = 7^{1-x}$
40. $(\frac{4}{3})^{1-x} = 5^x$
41. $1.2^x = (0.5)^{-x}$
42. $(0.3)^{1+x} = 1.7^{2x-1}$
43. $\pi^{1-x} = e^x$
44. $e^{x+3} = \pi^x$
45. $5(2^{3x}) = 8$
46. $0.3(4^{0.2x}) = 0.2$
47. $400e^{0.2x} = 600$
48. $500e^{0.3x} = 600$
49. $\log_a(x - 1) - \log_a(x + 6) = \log_a(x - 2) - \log_a(x + 3)$
50. $\log_a x + \log_a(x - 2) = \log_a(x + 4)$
51. $\log_{1/3}(x^2 + x) - \log_{1/3}(x^2 - x) = -1$
52. $\log_4(x^2 - 9) - \log_4(x + 3) = 3$
53. $\log_2 8^x = -3$
54. $\log_3 3^x = -1$
55. $\log_2(x^2 + 1) - \log_4 x^2 = 1$
[**Hint:** Change $\log_4 x^2$ to base 2.]
56. $\log_2(3x + 2) - \log_4 x = 3$
57. $\log_{16} x + \log_4 x + \log_2 x = 7$
58. $\log_9 x + 3\log_3 x = 14$

In Problems 59–74, use a graphing utility to solve each equation. Express your answer correct to two decimal places.

59. $\log_5 x + \log_3 x = 1$

60. $\log_2 x + \log_6 x = 3$

61. $\log_5(x + 1) - \log_4(x - 2) = 1$

62. $\log_2(x - 1) - \log_6(x + 2) = 2$

63. $e^x = -x$

64. $e^{2x} = x + 2$

65. $e^x = x^2$

66. $e^x = x^3$

67. $\ln x = -x$

68. $\ln 2x = -x + 2$

69. $\ln x = x^3 - 1$

70. $\ln x = -x^2$

71. $e^x + \ln x = 4$

72. $e^x - \ln x = 4$

73. $e^{-x} = \ln x$

74. $e^{-x} = -\ln x$

5.6 COMPOUND INTEREST

1 Determine the Future Value of a Lump Sum of Money

2 Calculate Effective Rates of Return

3 Determine the Present Value of a Lump Sum of Money

4 Determine the Time Required to Double or Triple a Lump Sum of Money

1 Interest is money paid for the use of money. The total amount borrowed (whether by an individual from a bank in the form of a loan or by a bank from an individual in the form of a savings account) is called the **principal.** The **rate of interest,** expressed as a percent, is the amount charged for the use of the principal for a given period of time, usually on a yearly (that is, per annum) basis.

If a principal of P dollars is borrowed for a period of t years at a per annum interest rate r, expressed as a decimal, the interest I charged is

Simple Interest Formula

$$I = Prt \qquad (1)$$

Interest charged according to formula (1) is called **simple interest.**

In working with problems involving interest we use the term **payment period** as follows:

Annually	Once per year	Monthly	12 times per year
Semiannually	Twice per year	Daily	365 times per year*
Quarterly	4 times per year		

When the interest due at the end of a payment period is added to the principal so that the interest computed at the end of the next payment period is based on this new principal amount (old principal + interest), the interest is said to have been **compounded.** Thus, **compound interest** is interest paid on previously earned interest.

EXAMPLE 1 Computing Compound Interest

A credit union pays interest of 8% per annum compounded quarterly on a certain savings plan. If \$1000 is deposited in such a plan and the interest is left to accumulate, how much is in the account after 1 year?

*Most banks use a 360 day "year." Why do you think they do?

Solution We use the simple interest formula, $I = Prt$. The principal P is \$1000 and the rate of interest is $8\% = 0.08$. After the first quarter of a year, the time t is $\frac{1}{4}$ year, so the interest earned is

$$I = Prt = (\$1000)(0.08)(\tfrac{1}{4}) = \$20$$

The new principal is $P + I = \$1000 + \$20 = \$1020$. At the end of the second quarter, the interest on this principal is

$$I = (\$1020)(0.08)(\tfrac{1}{4}) = \$20.40$$

At the end of the third quarter, the interest on the new principal of $\$1020 + \$20.40 = \$1040.40$ is

$$I = (\$1040.40)(0.08)(\tfrac{1}{4}) = \$20.81$$

Finally, after the fourth quarter, the interest is

$$I = (\$1061.21)(0.08)(\tfrac{1}{4}) = \$21.22$$

Thus, after 1 year the account contains \$1082.43.

The pattern of the calculations performed in Example 1 leads to a general formula for compound interest. To fix our ideas, let P represent the principal to be invested at a per annum interest rate r, which is compounded n times per year. (For computing purposes, r is expressed as a decimal.) The interest earned after each compounding period is the principal times r/n. Thus, the amount A after one compounding period is

$$A = P + P\left(\frac{r}{n}\right) = P\left(1 + \frac{r}{n}\right)$$

After two compounding periods, the amount A, based on the new principal $P(1 + r/n)$, is

$$A = \underbrace{P\left(1 + \frac{r}{n}\right)}_{\text{New principal.}} + \underbrace{P\left(1 + \frac{r}{n}\right)\left(\frac{r}{n}\right)}_{\text{Interest on new principal.}} = P\left(1 + \frac{r}{n}\right)\left(1 + \frac{r}{n}\right) = P\left(1 + \frac{r}{n}\right)^2$$

After three compounding periods,

$$A = P\left(1 + \frac{r}{n}\right)^2 + P\left(1 + \frac{r}{n}\right)^2\left(\frac{r}{n}\right) = P\left(1 + \frac{r}{n}\right)^2\left(1 + \frac{r}{n}\right) = P\left(1 + \frac{r}{n}\right)^3$$

Continuing this way, after n compounding periods (1 year),

$$A = P\left(1 + \frac{r}{n}\right)^n$$

Because t years will contain $n \cdot t$ compounding periods, after t years we have

$$A = P\left(1 + \frac{r}{n}\right)^{nt}$$

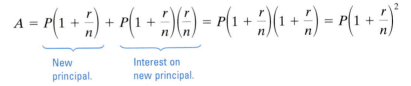

Theorem Compound Interest Formula

The amount A after t years due to a principal P invested at an annual interest rate r compounded n times per year is

$$A = P\left(1 + \frac{r}{n}\right)^{nt} \tag{2}$$

EXAMPLE 2 Comparing Investments Using Different Compounding Periods

Investing \$1,000 at an annual rate of 10% compounded annually, quarterly, monthly, and daily will yield the following amounts after 1 year:

Annual compounding: $A = P(1 + r)$
$= (\$1000)(1 + 0.10) = \$1100.00$

Quarterly compounding: $A = P\left(1 + \frac{r}{4}\right)^4$
$= (\$1000)(1 + 0.025)^4 = \$1103.81$

Monthly compounding: $A = P\left(1 + \frac{r}{12}\right)^{12}$
$= (\$1000)(1 + 0.00833)^{12} = \$1104.71$

Daily compounding: $A = P\left(1 + \frac{r}{365}\right)^{365}$
$= (\$1000)(1 + 0.000274)^{365} = \$1105.16$

Now work Problem 1.

From Example 2, we can see that the effect of compounding more frequently is that the amount after 1 year is higher: \$1000 compounded 4 times a year at 10% results in \$1103.81; \$1000 compounded 12 times a year at 10% results in \$1104.71; and \$1000 compounded 365 times a year at 10% results in \$1105.16. This leads to the following question: What would happen to the amount after 1 year if the number of times the interest is compounded were increased without bound?

Let's find the answer. Suppose P is the principal, r is the per annum interest rate, and n is the number of times the interest is compounded each year. The amount after 1 year is

$$A = P\left(1 + \frac{r}{n}\right)^n$$

Now suppose that the number n of times the interest is compounded per year gets larger and larger; that is, suppose that $n \to \infty$. Then,

$$A = P\left(1 + \frac{r}{n}\right)^n = P\left[1 + \frac{1}{n/r}\right]^n = P\left(\left[1 + \frac{1}{n/r}\right]^{n/r}\right)^r \underset{h = \frac{n}{r}}{\uparrow}= P\left[\left(1 + \frac{1}{h}\right)^h\right]^r \tag{3}$$

In (3), as $n \to \infty$, then $h = n/r \to \infty$, and the expression in brackets equals e. [Refer to (2) on p. 369]. Thus, $A \to Pe^r$. Table 7 compares $(1 + r/n)^n$, for large values of n, to e^r for $r = 0.05$, $r = 0.10$, $r = 0.15$, and $r = 1$. The larger n gets, the closer $(1 + r/n)^n$ gets to e^r. Thus, no matter how frequent the compounding, the amount after 1 year has the definite ceiling Pe^r.

TABLE 7

	$\left(1+\frac{r}{n}\right)^n$			
	$n = 100$	$n = 1000$	$n = 10{,}000$	e^r
$r = 0.05$	1.0512579	1.05127	1.051271	1.0512711
$r = 0.10$	1.1051157	1.1051654	1.1051703	1.1051709
$r = 0.15$	1.1617037	1.1618212	1.1618329	1.1618342
$r = 1$	2.7048138	2.7169239	2.7181459	2.7182818

When interest is compounded so that the amount after 1 year is Pe^r, we say the interest is **compounded continuously.**

Theorem Continuous Compounding

The amount A after t years due to a principal P invested at an annual interest rate r compounded continuously is

$$A = Pe^{rt} \tag{4}$$

EXAMPLE 3 Using Continuous Compounding

The amount A that results from investing a principal P of \$1000 at an annual rate r of 10% compounded continuously for a time t of 1 year is

$$A = \$1000e^{0.10} = (\$1000)(1.10517) = \$1105.17$$

Now work Problem 9.

2 The **effective rate of interest** is the equivalent annual simple rate of interest that would yield the same amount as compounding after 1 year. For example, based on Example 3, a principal of \$1000 will result in \$1105.17 at a rate of 10% compounded continuously. To get this same amount using a simple rate of interest would require that interest of \$1105.17 − \$1000.00 = \$105.17 be earned on the principal. Since \$105.17 is 10.517% of \$1000, a simple rate of interest of 10.517% is needed to equal 10% compounded continuously. Thus, the effective rate of interest of 10% compounded continuously is 10.517%.

Based on the results of Examples 2 and 3, we find the following comparisons:

	Annual Rate	Effective Rate
Annual compounding	10%	10%
Quarterly compounding	10%	10.381%
Monthly compounding	10%	10.471%
Daily compounding	10%	10.516%
Continuous compounding	10%	10.517%

Now work Problem 21.

EXAMPLE 4 Computing the Value of an IRA

On January 2, 1996, \$2000 is placed in an Individual Retirement Account (IRA) that will pay interest of 10% per annum compounded continuously. What will the IRA be worth on January 1, 2016?

Solution The amount A after 20 years is

$$A = Pe^{rt} = \$2000e^{(0.10)(20)} = \$14,778.11$$

Exploration How long will it be until $A = \$40,000$?
[**Hint:** Graph $Y_1 = 2000e^{0.1x}$ and $Y_2 = 40,000$. Use INTERSECT to find x.]

FIGURE 36

Time is money

3 When people engaged in finance speak of the "time value of money," they are usually referring to the **present value** of money. The present value of A dollars to be received at a future date is the principal you would need to invest now so that it would grow to A dollars in the specified time period. Thus, the present value of money to be received at a future date is always less than the amount to be received, since the amount to be received will equal the present value (money invested now) *plus* the interest accrued over the time period.

We use the compound interest formula (2) to get a formula for present value. If P is the present value of A dollars to be received after t years at a per annum interest rate r compounded n times per year, then, by formula (2),

$$A = P\left(1 + \frac{r}{n}\right)^{nt}$$

To solve for P, we divide both sides by $(1 + r/n)^{nt}$, and the result is

$$\frac{A}{(1 + \frac{r}{n})^{nt}} = P \quad \text{or} \quad P = A\left(1 + \frac{r}{n}\right)^{-nt}$$

Theorem Present Value Formulas

The present value P of A dollars to be received after t years, assuming a per annum interest rate r compounded n times per year, is

$$P = A\left(1 + \frac{r}{n}\right)^{-nt} \qquad (5)$$

If the interest is compounded continuously, then

$$P = Ae^{-rt} \qquad (6)$$

To prove(6), solve formula (4) for P.

EXAMPLE 5 Computing the Value of a Zero-Coupon Bond

A zero-coupon (noninterest-bearing) bond can be redeemed in 10 years for \$1000. How much should you be willing to pay for it now if you want a return of

(a) 8% compounded monthly?
(b) 7% compounded continuously?

Solution (a) We are seeking the present value of \$1000. Thus, we use formula (5) with $A = \$1000$, $n = 12$, $r = 0.08$, and $t = 10$:

$$\begin{aligned} P &= A\left(1 + \frac{r}{n}\right)^{-nt} \\ &= \$1000\left(1 + \frac{0.08}{12}\right)^{-12(10)} \\ &= \$450.52 \end{aligned}$$

For a return of 8% compounded monthly, you should pay \$450.52 for the bond.

(b) Here, we use formula (6) with $A = \$1000$, $r = 0.07$, and $t = 10$:

$$\begin{aligned} P &= Ae^{-rt} \\ &= \$1000e^{-(0.07)(10)} \\ &= \$496.59 \end{aligned}$$

For a return of 7% compounded continuously, you should pay \$496.59 for the bond.

Now work Problem 11.

EXAMPLE 6 Rate of Interest Required to Double an Investment

4

What annual rate of interest compounded annually should you seek if you want to double your investment in 5 years?

Algebraic Solution If P is the principal and we want P to double, the amount A will be $2P$. We use the compound interest formula with $n = 1$ and $t = 5$ to find r:

$$\begin{aligned} 2P &= P(1 + r)^5 \\ 2 &= (1 + r)^5 \\ 1 + r &= \sqrt[5]{2} \\ r &= \sqrt[5]{2} - 1 = 1.148698 - 1 = 0.148698 \end{aligned}$$

The annual rate of interest needed to double the principal in 5 years is 14.87%.

Graphing Solution We solve the equation

$$2 = (1 + r)^5$$

for r by graphing the two functions $Y_1 = 2$ and $Y_2 = (1 + x)^5$. The x-coordinate of their point of intersection is the rate r we seek. See Figure 37. Using the INTERSECT command,* we find the point of intersection of Y_1 and Y_2 is (0.14869835, 2).

FIGURE 37

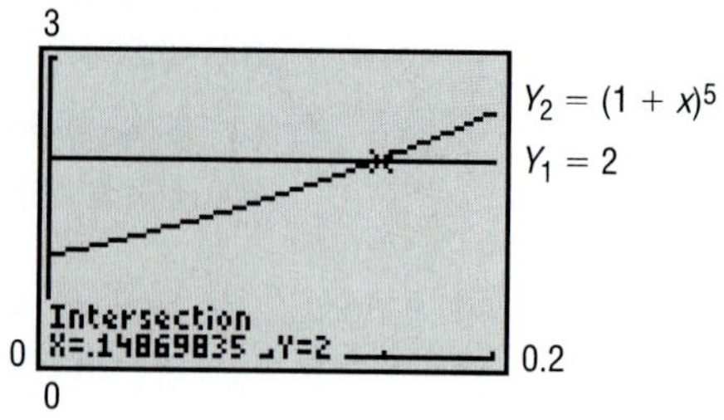

Now work Problem 23.

EXAMPLE 7 Doubling and Tripling Time for an Investment

(a) How long will it take for an investment to double in value if it earns 5% compounded continuously?

(b) How long will it take to triple at this rate?

Algebraic Solution

(a) If P is the initial investment and we want P to double, the amount A will be $2P$. We use formula (4) for continuously compounded interest with $r = 0.05$. Then

$$\begin{aligned} A &= Pe^{rt} \\ 2P &= Pe^{0.05t} \\ 2 &= e^{0.05t} \\ 0.05t &= \ln 2 \\ t &= \frac{\ln 2}{0.05} = 13.86 \end{aligned}$$

It will take about 14 years to double the investment.

Graphing Solution

We solve the equation

$$2 = e^{0.05t}$$

for t by graphing the two functions $Y_1 = 2$ and $Y_2 = e^{0.05x}$. Their point of intersection is (13.86, 2). See Figure 38.

FIGURE 38

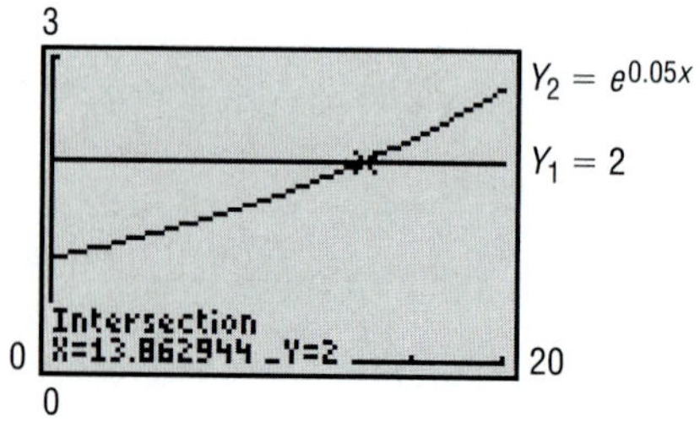

*Note: If your graphing utility does not have an INTERSECT command, you will have to use BOX or ZOOM-IN with TRACE to find the point of intersection.

Algebraic Solution (b) To triple the investment, we set $A = 3P$ in formula (4).

$$A = Pe^{rt}$$
$$3P = Pe^{0.05t}$$
$$3 = e^{0.05t}$$
$$0.05t = \ln 3$$
$$t = \frac{\ln 3}{0.05} = 21.97$$

It will take about 22 years to triple the investment.

Graphing Solution We solve the equation

$$3 = e^{0.05t}$$

for t by graphing the two functions $Y_1 = 3$ and $Y_2 = e^{0.05x}$. Their point of intersection is (21.97, 3). See Figure 39.

FIGURE 39

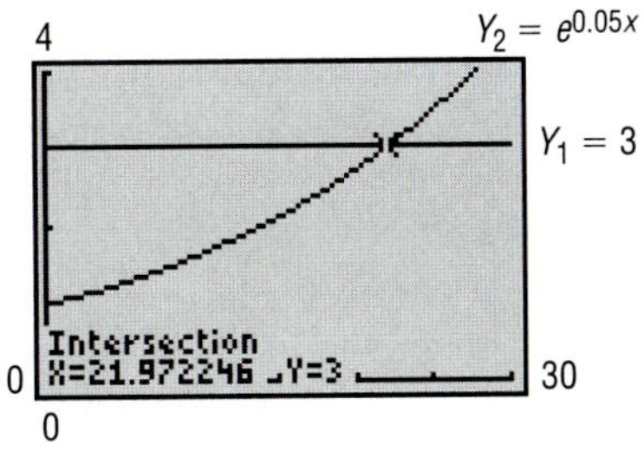

Now work Problem 29.

5.6 EXERCISES

In Problems 1–10, find the amount that results from each investment.

1. \$100 invested at 4% compounded quarterly after a period of 2 years
2. \$50 invested at 6% compounded monthly after a period of 3 years
3. \$500 invested at 8% compounded quarterly after a period of $2\frac{1}{2}$ years
4. \$300 invested at 12% compounded monthly after a period of $1\frac{1}{2}$ years
5. \$600 invested at 5% compounded daily after a period of 3 years
6. \$700 invested at 6% compounded daily after a period of 2 years
7. \$10 invested at 11% compounded continuously after a period of 2 years
8. \$40 invested at 7% compounded continuously after a period of 3 years
9. \$100 invested at 10% compounded continuously after a period of $2\frac{1}{4}$ years
10. \$100 invested at 12% compounded continuously after a period of $3\frac{3}{4}$ years

In Problems 11–20, find the principal needed now to get each amount; that is, find the present value.

11. To get \$100 after 2 years at 6% compounded monthly
12. To get \$75 after 3 years at 8% compounded quarterly
13. To get \$1,000 after $2\frac{1}{2}$ years at 6% compounded daily
14. To get \$800 after $3\frac{1}{2}$ years at 7% compounded monthly
15. To get \$600 after 2 years at 4% compounded quarterly
16. To get \$300 after 4 years at 3% compounded daily

17. To get $80 after $3\frac{1}{4}$ years at 9% compounded continuously

18. To get $800 after $2\frac{1}{2}$ years at 8% compounded continuously

19. To get $400 after 1 year at 10% compounded continuously

20. To get $1,000 after 1 year at 12% compounded continuously

21. Find the effective rate of interest for $5\frac{1}{4}$% compounded quarterly.

22. What interest rate compounded quarterly will give an effective interest rate of 7%?

23. What annual rate of interest is required to double an investment in 3 years? Verify your answer using a graphing utility.

24. What annual rate of interest is required to double an investment in 10 years? Verify your answer using a graphing utility.

In Problems 25–28, which of the two rates would yield the larger amount in 1 year?

[**Hint:** Start with a principal of $10,000 in each instance.]

25. 6% compounded quarterly or $6\frac{1}{4}$% compounded annually?

26. 9% compounded quarterly or $9\frac{1}{4}$% compounded annually?

27. 9% compounded monthly or 8.8% compounded daily?

28. 8% compounded semiannually or 7.9% compounded daily?

29. How long does it take for an investment to double in value if it is invested at 8% per annum compounded monthly? Compounded continuously? Verify your answer using a graphing utility.

30. How long does it take for an investment to double in value if it is invested at 10% per annum compounded monthly? Compounded continuously? Verify your answer using a graphing utility.

31. If Tanisha has $100 to invest at 8% per annum compounded monthly, how long will it be before she has $150? If the compounding is continuous, how long will it be?

32. If Angela has $100 to invest at 10% per annum compounded monthly, how long will it be before she has $175? If the compounding is continuous, how long will it be?

33. How many years will it take for an initial investment of $10,000 to grow to $25,000? Assume a rate of interest of 6% compounded continuously.

34. How many years will it take for an initial investment of $25,000 to grow to $80,000? Assume a rate of interest of 7% compounded continuously.

35. What will a $90,000 house cost 5 years from now if the inflation rate over that period averages 3% compounded annually?

36. Sears charges 1.25% per month on the unpaid balance for customers with charge accounts (interest is compounded monthly). A customer charges $200 and does not pay her bill for 6 months. What is the bill at that time?

37. Jerome will be buying a new car for $15,000 in 3 years. How much money should he ask his parents for now so that, if he invests it at 5% compounded continuously, he will have enough to buy the car?

38. John will require $3000 in 6 months to pay off a loan that has no prepayment privileges. If he has the $3000 now, how much of it should he save in an account paying 3% compounded monthly so that in 6 months he will have exactly $3000?

39. George is contemplating the purchase of 100 shares of a stock selling for $15 per share. The stock pays no dividends. The history of the stock indicates that is should grow at an annual rate of 15% per year. How much will the 100 shares of stock be worth in 5 years?

40. Tracy is contemplating the purchase of 100 shares of a stock selling for $15 per share. The stock pays no dividends. Her broker says the stock will be worth $20 per share in 2 years. What is the annual rate of return on this investment?

41. A business purchased for $650,000 in 1994 is sold in 1997 for $850,000. What is the annual rate of return for this investment?

42. Tanya has just inherited a diamond ring appraised at $5000. If diamonds have appreciated in value at an annual rate of 8%, what was the value of the ring 10 years ago when the ring was purchased?

43. Jim places $1000 in a bank account that pays 5.6% compounded continuously. After 1 year, will he have enough money to buy a computer system that costs $1060? If another bank will pay Jim 5.9% compounded monthly, is this a better deal?

44. On January 1, Kim places \$1000 in a certificate of deposit that pays 6.8% compounded continuously and matures in 3 months. Then Kim places the \$1000 and the interest in a passbook account that pays 5.25% compounded monthly. How much does Kim have in the passbook account on May 1?

45. Will invests \$2000 in a bond trust that pays 9% interest compounded semiannually. His friend Henry invests \$2000 in a Certificate of Deposit (CD) that pays $8\frac{1}{2}\%$ compounded continuously. Who has more money after 20 years, Will or Henry?

46. Suppose that April has access to an investment that will pay 10% interest compounded continuously. Which is better: To be given \$1000 now so that she can take advantage of this investment opportunity or to be given \$1325 after 3 years?

47. Colleen and Bill have just purchased a house for \$150,000, with the seller holding a second mortgage of \$50,000. They promise to pay the seller \$50,000 plus all accrued interest 5 years from now. The seller offers them three interest options on the second mortgage:
(a) Simple interest at 12% per annum
(b) $11\frac{1}{2}\%$ interest compounded monthly
(c) $11\frac{1}{4}\%$ interest compounded continuously
Which option is best; that is, which results in the least interest on the loan?

48. The First National Bank advertises that it pays interest on saving accounts at the rate of 4.25% compounded daily. Find the effective rate if the bank uses (a) 360 days or (b) 365 days in determining the daily rate.

Problems 49–52 involve zero-coupon bonds. A zero-coupon bond is a bond that is sold now at a discount and will pay its face value at some time in the future when it matures; no interest payments are made.

49. A zero-coupon bond can be redeemed in 20 years for \$10,000. How much should you be willing to pay for it now if you want a return of:
(a) 10% compounded monthly?
(b) 10% compounded continuously?

50. A child's grandparents are considering buying a \$40,000 face value zero-coupon bond at birth so that she will have enough money for her college education 17 years later. If they want a rate of return of 8% compounded annually, what should they pay for the bond?

51. How much should a \$10,000 face value zero-coupon bond, maturing in 10 years, be sold for now if its rate of return is to be 8% compounded annually?

52. If Pat pays \$12,485.52 for a \$25,000 face value zero-coupon bond that matures in 8 years, what is his annual rate of return?

53. Explain in your own words what the term *compound interest* means. What does *continuous compounding* mean?

54. Explain in your own words the meaning of present value.

55. Write a program that will calculate the amount after n years if a principal P is invested at $r\%$ per annum compounded quarterly. Use it to verify your answers to Problems 1 and 3.

56. Write a program that will calculate the principal needed now to get the amount A in n years at $r\%$ per annum compounded daily. Use it to verify your answer to Problem 13.

57. Write a program that will calculate the annual rate of interest required to double an investment in n years. Use it to verify your answers to Problems 23 and 24.

58. Write a program that will calculate the number of months required for an initial investment of x dollars to grow to y dollars at $r\%$ per annum compounded continuously. Use it to verify your answer to Problems 33 and 34.

59. Time to Double or Triple an Investment The formula

$$y = \frac{\ln m}{n \ln\left(1 + \frac{r}{n}\right)}$$

can be used to find the number of years y required to multiply an investment m times when r is the per annum interest rate compounded n times a year.
(a) How many years will it take to double the value of an IRA that compounds annually at the rate of 12%?
(b) How many years will it take to triple the value of a savings account that compounds quarterly at an annual rate of 6%?
(c) Give a derivation of this formula.

60. Time to Reach an Investment Goal The formula

$$y = \frac{\ln A - \ln P}{r}$$

can be used to find the number of years y required for an investment P to grow to a value

A when compounded continuously at an annual rate r.

(a) How long will it take to increase an initial investment of \$1000 to \$8000 at an annual rate of 10%?

(b) What annual rate is required to increase the value of a \$2000 IRA to \$30,000 in 35 years?

(c) Give a derivation of this formula.

61. Critical Thinking You have just contracted to buy a house and will seek financing in the amount of \$100,000. You go to several banks. Bank 1 will lend you \$100,000 at the rate of 8.75% amortized over 30 years with a loan origination fee of 1.75%. Bank 2 will lend you \$100,000 at the rate of 8.375% amortized over 15 years with a loan origination fee of 1.5%. Bank 3 will lend you \$100,000 at the rate of 9.125% amortized over 30 years with no loan origination fee. Bank 4 will lend you \$100,000 at the rate of 8.625% amortized over 15 years with no loan origination fee. Which loan would you take? Why? Be sure to have sound reasons for your choice. If the amount of the monthly payment does not matter to you, which loan would you take? Again, have sound reasons for your choice. Use the information in the table to assist you. Compare your final decision with others in the class. Discuss.

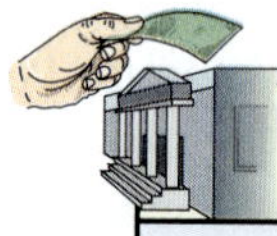

	Monthly Payment	Loan Origination Fee
Bank 1	\$786.70	\$1,750.00
Bank 2	\$977.42	\$1,500.00
Bank 3	\$813.63	\$0.00
Bank 4	\$990.68	\$0.00

5.7 GROWTH AND DECAY

1 Find Equations of Populations That Obey the Law of Uninhibited Growth
2 Find Equations of Populations That Obey the Law of Decay
3 Use Newton's Law of Cooling
4 Use Logistic Growth Models

1 Many natural phenomena have been found to follow the law that an amount A varies with time t according to

$$A = A_0 e^{kt} \quad (1)$$

where A_0 is the original amount ($t = 0$) and $k \neq 0$ is a constant.

If $k > 0$, then equation (1) states that the amount A is increasing over time; if $k < 0$, the amount A is decreasing over time. In either case, when an amount A varies over time according to equation (1), it is said to follow the **exponential law** or the **law of uninhibited growth** ($k > 0$) **or decay** ($k < 0$). See Figure 40.

FIGURE 40

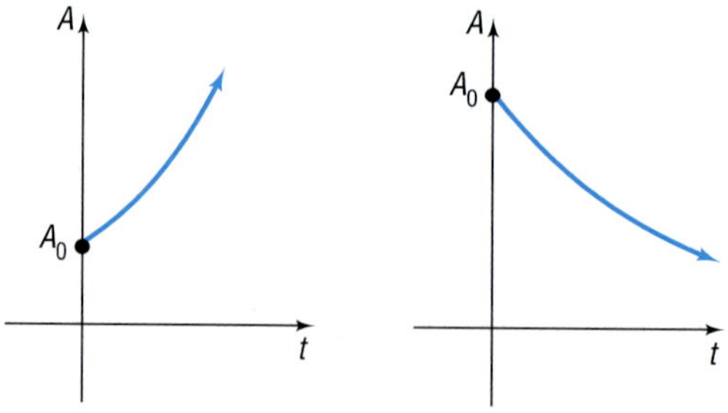

(a) $A = A_0 e^{kt}, k > 0$ **(b)** $A = A_0 e^{kt}, k < 0$

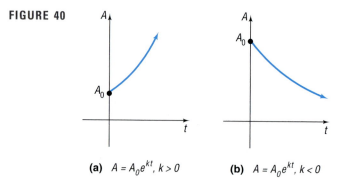

For example, we saw in Section 5.6 that continuously compounded interest follows the law of uninhibited growth. In this section we shall look at three additional phenomena that follow the exponential law.

Biology

Cell division is a universal process in the growth of living organisms such as amoebas, plants, human skin cells, and many others. Based on an ideal situation in which no cells die and no byproducts are produced, the number of cells present at a given time follows the law of uninhibited growth. Actually, however, after enough time has passed, growth at an exponential rate will cease due to the influence of factors such as lack of living space, dwindling food supply, and so on. The law of uninhibited growth accurately reflects only the early stages of the cell division process.

The cell division process begins with a culture containing N_0 cells. Each cell in the culture grows for a certain period of time and then divides into two identical cells. We assume that the time needed for each cell to divide in two is constant and does not change as the number of cells increases. These cells then grow, and eventually each divides in two, and so on.

A model that gives the number N of cells in the culture after a time t has passed (in the early stages of growth) is

Uninhibited Growth of Cells

$$N(t) = N_0 e^{kt} \qquad k > 0 \tag{2}$$

where k is a positive constant.

In using equation (2) to model the growth of cells, we are using a function that yields positive real numbers, even though we are counting the number of cells, which must be an integer. This is a common practice in many applications.

EXAMPLE 1 Bacterial Growth

A colony of bacteria increases according to the law of uninhibited growth.

(a) If the number of bacteria doubles in 3 hours, how long will it take for the size of the colony to triple?

(b) Using a graphing utility, verify that the population doubles in 3 hours.

(c) Using a graphing utility, approximate the time it takes for the population to double a second time (i.e., increase four times). Does this answer seem reasonable?

Solution (a) Using formula (2), the number N of cells at a time t is

$$N(t) = N_0 e^{kt}$$

where N_0 is the initial number of bacteria present and k is a positive number. We first seek the number k. The number of cells doubles in 3 hours; thus, we have

$$N(3) = 2N_0$$

But $N(3) = N_0 e^{k(3)}$, so

$$N_0 e^{k(3)} = 2N_0$$
$$e^{3k} = 2$$
$$3k = \ln 2 \quad \text{Write the exponential equation as a logarithm.}$$
$$k = \tfrac{1}{3}\ln 2 \approx \tfrac{1}{3}(0.6931) = 0.2310$$

Formula (2) for this growth process is therefore

$$N(t) = N_0 e^{0.2310t}$$

The time t needed for the size of the colony to triple requires that $N = 3N_0$. Thus, we substitute $3N_0$ for N to get

$$3N_0 = N_0 e^{0.2310t}$$
$$3 = e^{0.2310t}$$
$$0.2310t = \ln 3$$
$$t = \frac{1}{0.2310}\ln 3 \approx \frac{1.0986}{0.2310} = 4.756 \text{ hours}$$

It will take about 4.756 hours for the size of the colony to triple.

(b) Figure 41 shows the graph of $Y_1 = 2$ and $Y_2 = e^{0.2310x}$, where x represents the time in hours. Using INTERSECT, it takes 3 hours for the population to double.

(c) Figure 42 shows the graph of $Y_1 = 4$ and $Y_2 = e^{0.2310x}$, where x represents the time in hours. Using INTERSECT, it takes 6 hours for the population to double a second time.

FIGURE 41

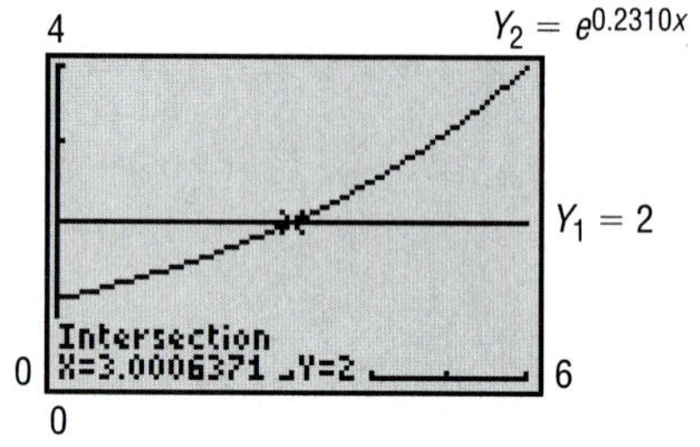

FIGURE 42

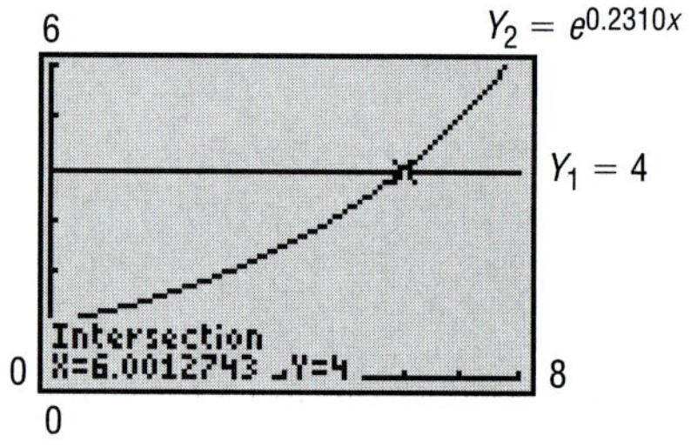

Now work Problem 1.

Radioactive Decay

2 Radioactive materials follow the law of uninhibited decay. Thus, the amount A of a radioactive material present at time t is given by the model

Uninhibited Radioactive Decay

$$A = A_0 e^{kt} \qquad k < 0 \tag{3}$$

where A_0 is the original amount of radioactive material and k is a negative number.

All radioactive substances have a specific **half-life,** which is the time required for half of the radioactive substance to decay. In **carbon dating,** we use the fact that all living organisms contain two kinds of carbon, carbon-12 (a stable carbon) and carbon-14 (a radioactive carbon, with a half-life of 5600 years). While an organism is living, the ratio of carbon-12 to carbon-14 is constant. But when an organism dies, the original amount of carbon-12 present remains unchanged, whereas the amount of carbon-14 begins to decrease. This change in the amount of carbon-14 present relative to the amount of carbon-12 present makes it possible to calculate when an organism died.

EXAMPLE 2

Estimating the Age of Ancient Tools

Traces of burned wood found along with ancient stone tools in an archaeological dig in Chile were found to contain approximately 1.67% of the original amount of carbon-14.

(a) If the half-life of carbon-14 is 5600 years, approximately when was the tree cut and burned?

(b) Using a graphing utility, graph the relation between the percentage of carbon-14 remaining and time.

(c) Determine the time that elapses until half of the carbon-14 remains. This answer should equal the half-life of carbon-14.

(d) Verify the answer found in (a).

Solution

(a) Using equation (3), the amount A of carbon-14 present at time t is

$$A = A_0 e^{kt}$$

where A_0 is the original amount of carbon-14 present and k is a negative number. We first seek the number k. To find it, we use the fact that after 5600 years half of the original amount of carbon-14 remains. Thus,

$$\begin{aligned} \frac{1}{2}A_0 &= A_0 e^{k(5600)} \\ \frac{1}{2} &= e^{5600k} \\ 5600k &= \ln\frac{1}{2} \\ k &= \frac{1}{5600}\ln\frac{1}{2} \approx -0.000124 \end{aligned}$$

Formula (3) therefore becomes

$$A = A_0 e^{-0.000124t}$$

If the amount A of carbon-14 now present is 1.67% of the original amount, it follows that

$$\begin{aligned} 0.0167A_0 &= A_0 e^{-0.000124t} \\ 0.0167 &= e^{-0.000124t} \\ -0.000124t &= \ln 0.0167 \\ t &= \frac{1}{-0.000124}\ln 0.0167 \approx 33{,}000 \text{ years} \end{aligned}$$

The tree was cut and burned about 33,000 years ago. Some archaeologists use this conclusion to argue that humans lived in the Americas 33,000 years ago, much earlier than is generally accepted.

(b) Figure 43 shows the graph of $y = e^{-0.000124x}$, where y is the fraction of carbon-14 present and x is the time.

FIGURE 43

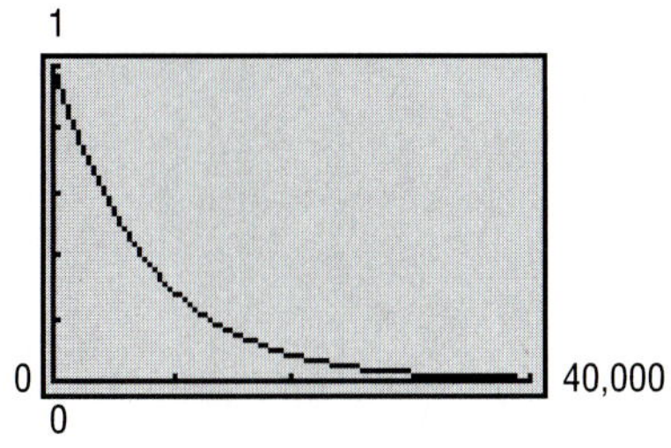

(c) By graphing $Y_1 = 0.5$ and $Y_2 = e^{-0.000124x}$, where x is time, and using INTERSECT, we find it takes 5,590 years until half the carbon-14 remains.

(d) By graphing $Y_1 = 0.0167$ and $Y_2 = e^{-0.000124x}$, where x is time, and using INTERSECT, we find it takes 33,003 years until 1.67% of the carbon-14 remains.

Now work Problem 3.

Newton's Law of Cooling

3 **Newton's Law of Cooling*** states that the temperature of a heated object decreases exponentially over time toward the temperature of the surrounding medium. That is, the temperature u of a heated object at a given time t can be modeled by the function

Newton's Law of Cooling

$$u(t) = T + (u_0 - T)e^{kt} \qquad k < 0 \tag{4}$$

where T is the constant temperature of the surrounding medium, u_0 is the initial temperature of the heated object, and k is a negative constant.

EXAMPLE 3 Using Newton's Law of Cooling

An object is heated to 100°C and is then allowed to cool in a room whose air temperature is 30°C.

(a) If the temperature of the object is 80°C after 5 minutes, when will its temperature be 50°C?

(b) Using a graphing utility, graph the relation found between the temperature y and time x.

(c) Using a graphing utility, verify that when $x = 18.6$, then $y = 50$.

(d) Using a graphing utility, determine the elapsed time before the object is 35°C.

(e) What do you notice about y, the temperature, as x, time, increases?

Solution (a) Using equation (4) with $T = 30$ and $u_0 = 100$, the temperature (in degrees Celsius) of the object at time t (in minutes) is

$$u(t) = 30 + (100 - 30)e^{kt} = 30 + 70e^{kt} \tag{5}$$

*Named after Sir Isaac Newton (1642–1727), one of the cofounders of calculus.

where k is a negative constant. To find k, we use the fact that $u = 80$ when $t = 5$. Then

$$\begin{aligned} 80 &= 30 + 70e^{k(5)} \\ 50 &= 70e^{5k} \\ e^{5k} &= \frac{50}{70} \\ 5k &= \ln\frac{5}{7} \\ k &= \frac{1}{5}\ln\frac{5}{7} \approx -0.0673 \end{aligned}$$

Formula (5) therefore becomes

$$u = 30 + 70e^{-0.0673t}$$

Now, we want to find t when $u = 50°\text{C}$, so

$$\begin{aligned} 50 &= 30 + 70e^{-0.0673t} \\ 20 &= 70e^{-0.0673t} \\ e^{-0.0673t} &= \frac{20}{70} \\ -0.0673t &= \ln\frac{2}{7} \\ t &= \frac{1}{-0.0673}\ln\frac{2}{7} \approx 18.6 \text{ minutes} \end{aligned}$$

Thus, the temperature of the object will be 50°C after about 18.6 minutes.

FIGURE 44

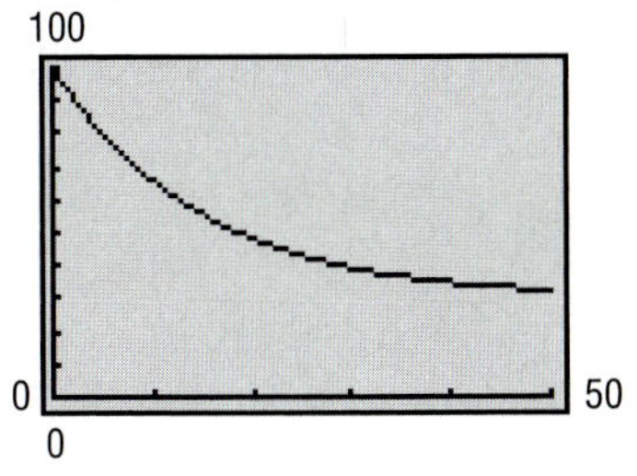

(b) Figure 44 shows the graph of $y = 30 + 70e^{-0.0673x}$, where y is the temperature and x is the time.

(c) By graphing $Y_1 = 50$ and $Y_2 = 30 + 70e^{-0.0673x}$, where x is time, and using INTERSECT, we find it takes $x = 18.6$ minutes for the temperature to cool to 50°C.

(d) By graphing $Y_1 = 35$ and $Y_2 = 30 + 70e^{-0.0673x}$, where x is time, and using INTERSECT, we find it takes $x = 39.21$ minutes for the temperature to cool to 35°C.

(e) As x increases, the value of $e^{-0.0673x}$ approaches zero, so the value of y approaches 30°C.

Logistic Models

4 The exponential growth model $A = A_0e^{kt}, k > 0$ assumes uninhibited growth, meaning the value of the function grows without limit. Recall we stated that cell division could be modeled using this function assuming no cells die and no byproducts are produced. However, cell division would eventually be limited by factors such as living space, food supply, and so forth. The **logistic growth model** is an exponential function that can model situations where the growth of the dependent variable is limited. The logistic growth model is given by the function

Logistic Growth Model

$$P(t) = \frac{c}{1 + ae^{-bt}}$$

where a, b, and c are constants with $c > 0$ and $b > 0$.

The number c is called the **carrying capacity** because the value $P(t)$ approaches c as t approaches infinity; that is $\lim_{t\to\infty} P(t) = c$. Thus, c represents the maximum value the function can attain.

EXAMPLE 4 Fruit Fly Population

Fruit flies are placed in a half-pint milk bottle with a banana (for food) and yeast plants (for food and to provide a stimulus to lay eggs). Suppose the fruit fly population after t days is given by

$$P(t) = \frac{230}{1 + 56.5e^{-0.37t}}$$

(a) Using a graphing utility, graph $P(t)$.

(b) What is the carrying capacity of the half-pint bottle? That is, what is $P(t)$ as $t \to \infty$?

(c) How many fruit flies were initially placed in the half-pint bottle?

(d) When will the population of fruit flies be 180?

FIGURE 45

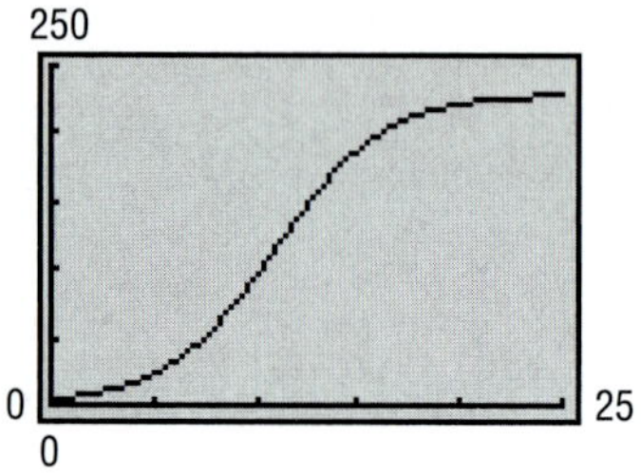

Solution (a) See Figure 45.

(b) As $t \to \infty$, $e^{-0.37t} \to 0$, and $P(t) \to 230/1$. The carrying capacity of the half-pint bottle is 230 fruit flies.

(c) To find the initial number of fruit flies in the half-pint bottle, we evaluate $P(0)$:

$$P(0) = \frac{230}{1 + 56.5e^{-0.37(0)}} = \frac{230}{1 + 56.5} = 4$$

So, initially there were four fruit flies in the half-pint bottle.

(d) To determine when the population of fruit flies will be 180, we solve the equation

$$\begin{aligned} \frac{230}{1 + 56.5e^{-0.37t}} &= 180 \\ 230 &= 180(1 + 56.5e^{-0.37t}) \\ 1.2778 &= 1 + 56.5e^{-0.37t} \\ 0.2778 &= 56.5e^{-0.37t} \\ 0.0049 &= e^{-0.37t} \\ \ln(0.0049) &= -0.37t \\ t &\approx 14.4 \text{ days} \end{aligned}$$

It will take approximately 14.4 days for the population to reach 180 fruit flies.

We could have also solved this problem by graphing $Y_1 = \dfrac{230}{1 + 56.5e^{-0.37t}}$ and $Y_2 = 180$, using INTERSECT to find the solution shown in Figure 46.

FIGURE 46

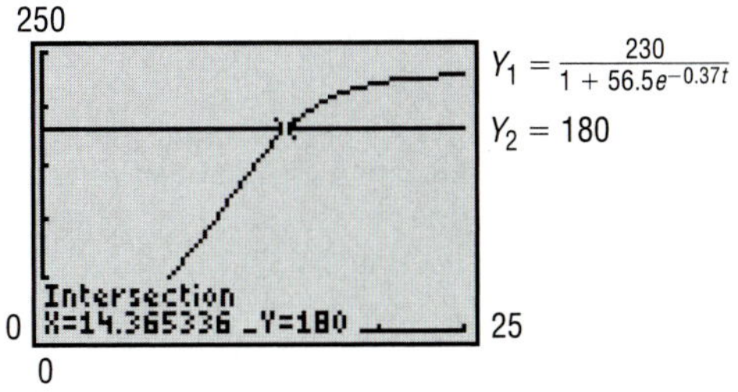

Exploration On the same viewing rectangle, graph $Y_1 = \dfrac{500}{1 + 24e^{-0.03t}}$ and $Y_2 = \dfrac{500}{1 + 24e^{-0.08t}}$. What effect does b have on the logistic growth function?

5.7 EXERCISES

1. **Growth of an Insect Population** The size P of a certain insect population at time t (in days) obeys the equation $P = 500e^{0.02t}$.
   (a) After how many days will the population reach 1000? 2000?
   (b) Using a graphing utility, graph the relation between P and t. Verify your answers in (a) using INTERSECT.

2. **Growth of Bacteria** The number N of bacteria present in a culture at time t (in hours) obeys the equation $N = 1000e^{0.01t}$.
   (a) After how many hours will the population equal 1500? 2000?
   (b) Using a graphing utility, graph the relation between N and t. Verify your answers in (a) using INTERSECT.

3. **Radioactive Decay** Strontium-90 is a radioactive material that decays according to the equation $A = A_0e^{-0.0244t}$, where A_0 is the initial amount present and A is the amount present at time t (in years). What is the half-life of strontium-90?

4. **Radioactive Decay** Iodine-131 is a radioactive material that decays according to the equation $A = A_0e^{-0.087t}$, where A_0 is the initial amount present and A is the amount present at time t (in days). What is the half-life of iodine-131?

5. (a) Use the information in Problem 3 to determine how long it takes for 100 grams of strontium-90 to decay to 10 grams.
   (b) Graph the relation $A = 100e^{-0.0244t}$ and verify your answer found in (a).

6. (a) Use the information in Problem 4 to determine how long it takes for 100 grams of iodine-131 to decay to 10 grams.
   (b) Graph the relation $A = 100e^{-0.087t}$ and verify your answer found in (a).

7. **Growth of a Colony of Mosquitoes** The population of a colony of mosquitoes obeys the law of uninhibited growth. If there are 1000 mosquitoes initially, and there are 1800 after 1 day, what is the size of the colony after 3 days? How long is it until there are 10,000 mosquitoes?

8. **Bacterial Growth** A culture of bacteria obeys the law of uninhibited growth. If 500 bacteria are present initially, and there are 800 after 1 hour, how many will be present in the culture after 5 hours? How long is it until there are 20,000 bacteria?

9. **Population Growth** The population of a southern city follows the exponential law. If the population doubled in size over an 18-month period and the current population is 10,000, what will the population be 2 years from now?

10. **Population Growth** The population of a midwestern city follows the exponential law. If the population decreased from 900,000 to 800,000 from 1993 to 1995, what will the population be in 1997?

11. **Radioactive Decay** The half-life of radium is 1690 years. If 10 grams are present now, how much will be present in 50 years?

12. **Radioactive Decay** The half-life of radioactive potassium is 1.3 billion years. If 10 grams are present now, how much will be present in 100 years? In 1000 years?

13. **Estimating the Age of a Tree** A piece of charcoal is found to contain 30% of the carbon-14 it originally had.
    (a) When did the tree from which the charcoal came die? Use 5600 years as the half-life of carbon-14.
    (b) Using a graphing utility, graph the relation between the percentage of carbon-14 remaining and time.
    (c) Using INTERSECT, determine the time that elapses until half of the carbon-14 remains.
    (d) Verify the answer found in (a).

14. **Estimating the Age of a Fossil** A fossilized leaf contains 70% of its normal amount of carbon-14.
    (a) How old is the fossil?
    (b) Using a graphing utility, graph the relation between the percentage of carbon-14 remaining and time.
    (c) Using INTERSECT, determine the time that elapses until half of the carbon-14 remains.
    (d) Verify the answer found in (a).

15. **Cooling Time of a Pizza** A pizza baked at 450°F is removed from the oven at 5:00 P.M. into a room that is a constant 70°F. After 5 minutes, the pizza is at 300°F.
    (a) At what time can you begin eating the pizza if you want its temperature to be 135°F?
    (b) Using a graphing utility, graph the relation between temperature and time.
    (c) Using INTERSECT, determine the time that needs to elapse before the pizza is 160°F?
    (d) TRACE the function for large values of time. What do you notice about y, the temperature?

16. **Newton's Law of Cooling** A thermometer reading 72°F is placed in a refrigerator where the temperature is a constant 38°F. If the thermometer reads 60°F after 2 minutes, what will it read after 7 minutes?
    (a) How long will it take before the thermometer reads 39°F?
    (b) Using a graphing utility, graph the relation between temperature and time.
    (c) Using INTERSECT, determine the time needed to elapse before the thermometer reads 45°F.
    (d) TRACE the function for large values of time. What do you notice about y, the temperature?

17. **Newton's Law of Cooling** A thermometer reading 8°C is brought into a room with a constant temperature of 35°C. (a) If the thermometer reads 15°C after 3 minutes, what will it read after being in the room for 5 minutes? For 10 minutes? (b) Graph the relation between temperature and time. TRACE to verify your answers are correct.

    [**Hint:** You need to construct a formula similar to equation (4).]

18. **Thawing Time of a Steak** A frozen steak has a temperature of 28°F. It is placed in a room with a constant temperature of 70°F. After 10 minutes, the temperature of the steak has risen to 35°F. What will the temperature of the steak be after 30 minutes? How long will it take the steak to thaw to a temperature of 45°F? [See the hint given for Problem 17.] Graph the relation between temperature and time. TRACE to verify your answer is correct.

19. **Decomposition of Salt in Water** Salt (NaCl) decomposes in water into sodium (NA^+) and chloride (Cl^-) ions according to the law of uninhibited decay. If the initial amount of salt is 25 kilograms and, after 10 hours, 15 kilograms of salt are left, how much salt is left after 1 day? How long does it take until $\frac{1}{2}$ kilogram of salt is left?

20. **Voltage of a Conductor** The voltage of a certain condenser decreases over time according to the law of uninhibited decay. If the initial voltage is 40 volts, and 2 seconds later it is 10 volts, what is the voltage after 5 seconds?

21. **Radioactivity from Chernobyl** After the release of radioactive material into the atmosphere from a nuclear power plant at Chernobyl (Ukraine) in 1986, the hay in Austria was contaminated by iodine-131 (half life 8 years). If it is all right to feed the hay to cows when 10% of the iodine-131 remains, how long do the farmers need to wait to use this hay?

22. **Pig Roasts** The hotel Bora-Bora is having a pig roast. At noon, the chef put the pig in a large earthen oven. The pig's original temperature was 75°F. At 2:00 P.M., the chef checked the pig's temperature and was upset because it had reached only 100°F. If the oven's temperature remains a constant 325°F, at what time may the hotel serve its guests, assuming that pork is done when it reaches 175°F?

23. **Proportion of the Population That Owns a VCR** The logistic growth model

$$P(t) = \frac{0.9}{1 + 6e^{-0.32t}}$$

relates the proportion of U.S. households that own a VCR to the year. Let $t = 0$ represent 1984, $t = 1$ represent 1985, and so on.

(a) What proportion of U.S. households owned a VCR in 1984?
(b) Determine the maximum proportion of households that will own a VCR.
(c) Using a graphing utility, graph $P(t)$.
(d) When will 0.8 (80%) of U.S. households own a VCR?

24. **Market Penetration of Intel's Coprocessor** The logistic growth model

$$P(t) = \frac{0.90}{1 + 3.5e^{-0.339t}}$$

relates the proportion of new personal computers sold at Best Buy that have Intel's latest coprocessor t months after it has been introduced.

(a) What proportion of new personal computers sold at Best Buy will have Intel's latest coprocessor when it is first introduced (i.e., at $t = 0$)?
(b) Determine the maximum proportion of new personal computers sold at Best Buy that will have Intel's latest coprocessor.
(c) Using a graphing utility, graph $P(t)$.
(d) When will 0.75 (75%) of new personal computers sold at Best Buy will have Intel's latest coprocessor?

25. **Population of a Bacteria Culture** The logistic growth model

$$P(t) = \frac{1000}{1 + 32.33e^{-0.439t}}$$

represents the population of a bacteria after t hours.

(a) Using a graphing utility, graph $P(t)$.
(b) What is the carrying capacity of the environment?
(c) What was the initial amount of bacteria in the population?
(d) When will the amount of bacteria be 800?

26. **Population of a Endangered Species** Often environmentalists will capture an endangered species and transport the species to a controlled environment where the species can produce offspring and regenerate its population. Suppose 6 American Bald Eagles are captured and transported to Montana and set free. Based on experience, the environmentalists expect the population to grow according to the model

$$P(t) = \frac{500}{1 + 83.33e^{-0.162t}}$$

(a) Using a graphing utility, graph $P(t)$.
(b) What is the carrying capacity of the environment?
(c) What is the predicted population of the American Bald Eagle in 20 years?
(d) When will the population be 300?

5.8 EXPONENTIAL, LOGARITHMIC, AND LOGISTIC CURVE FITTING

1 Use a Graphing Utility to Obtain the Exponential Function of Best Fit
2 Use a Graphing Utility to Obtain the Logarithmic Function of Best Fit
3 Use a Graphing Utility to Obtain the Logistic Function of Best Fit
4 Determine the Model of Best Fit Based on "r"

In Section 1.6 we discussed how to find the linear equation of best fit ($y = ax + b$). In Chapter 4, we discussed how to find power equations

($y = ax^b$), quadratic equations ($y = ax^2 + bx + c$), and cubic equations ($y = ax^3 + bx^2 + cx + d$) of best fit.

In this section we will discuss how to use a graphing utility to find equations of best fit that describe the relation between two variables when the relation is thought to be exponential ($y = ab^x$), logarithmic ($y = a + b \ln x$), or logistic $\left(y = \dfrac{c}{1 + ae^{-bx}}\right)$. As before, we draw a scatter diagram of the data to help determine the appropriate model to use.

Figure 47 shows scatter diagrams that will typically be observed for the three models. Below each scatter diagram are any restrictions on the values of the parameters.

FIGURE 47

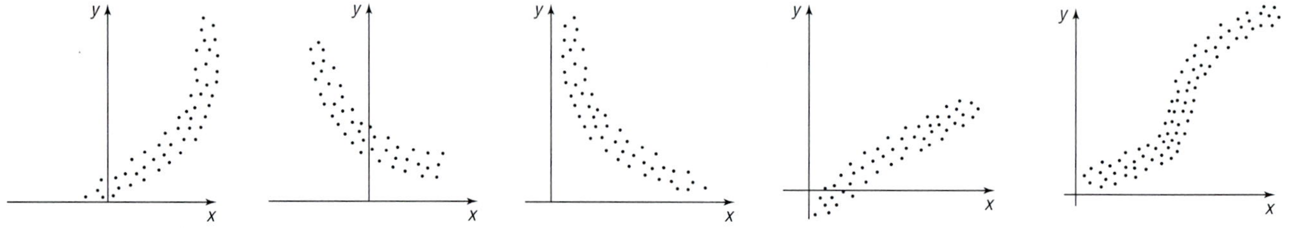

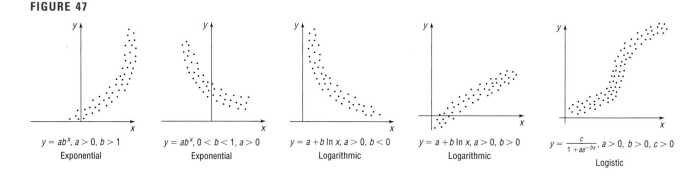

$y = ab^x$, $a > 0$, $b > 1$
Exponential

$y = ab^x$, $0 < b < 1$, $a > 0$
Exponential

$y = a + b \ln x$, $a > 0$, $b < 0$
Logarithmic

$y = a + b \ln x$, $a > 0$, $b > 0$
Logarithmic

$y = \dfrac{c}{1 + ae^{-bx}}$, $a > 0$, $b > 0$, $c > 0$
Logistic

Determining the appropriate model to use requires statistical techniques that are beyond the scope of this text. However, there are some rudimentary techniques we can use that will allow us to determine which model is most appropriate. Before we do this, however, we need to discuss how graphing utilities are used to fit curves to data.

Most graphing utilities have REGression options that fit data to a specific type of curve. Once the data has been entered and a scatter diagram obtained, the type of curve you want to fit to the data is selected. Then that REGression option is used to obtain the curve of "best fit" of the type selected.

The correlation coefficient, r, will appear only if the model can be written as a linear expression. Thus, r will appear for the linear, power, exponential, and logarithmic models since these models can be written as a linear expression (see Section 5.4). Remember, the closer $|r|$ is to 1, the better the fit.

Let's look at some examples.

Exponential Curve Fitting

1

As we saw in Section 5.7, many growth and decay models are exponential.

EXAMPLE 1

Fitting a Curve to an Exponential Model

Beth is interested in finding a function that explains the relation between the amount of an unknown radioactive material and time. She obtains 100 grams of the radioactive material and every day for 9 days measures the amount of radioactive material in the sample and obtains the following data:

Day	Weight (in Grams)
0	100
1	91.6
2	84.1
3	74.2
4	68.3
5	63.7
6	58.5
7	53.8
8	50.2
9	47.3

(a) Using a graphing utility, draw a scatter diagram.

(b) Using a graphing utility, fit an exponential model to the data.

(c) Express the curve in the form $A = A_0 e^{kt}$.

(d) Graph the exponential function found in (b) or (c) on the scatter diagram.

(e) Using the solution to (c), estimate the half-life of the unknown radioactive material.

(f) Predict how much of the material will remain after 30 days.

Solution

(a) After entering the data into the graphing utility, we obtain the scatter diagram shown in Figure 48.

FIGURE 48

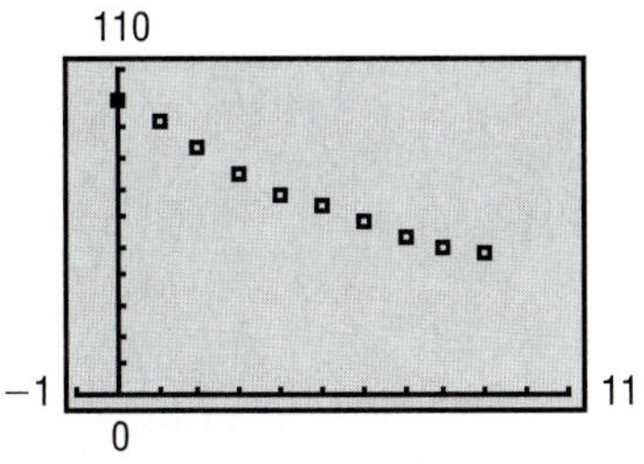

(b) A graphing utility fits the data in Figure 48 to an exponential model of the form $y = ab^x$ by using the EXPonential REGression option.* See Figure 49.

The exponential function of best fit to the data is

$$y = 98.2463(0.9188)^x$$

Notice that $r = -0.997178162$, so $|r|$ is close to 1, indicating a good fit.

(c) To express $y = ab^x$ in the form $A = A_0 e^{kt}$, we proceed as follows.

FIGURE 49

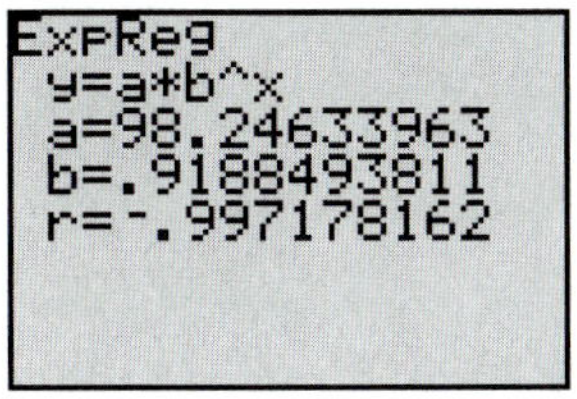

*If your utility does not have such an option but does have a LINear REGression option, you can transform the exponential model to a linear model by the following technique (see Section 5.4):

$$\begin{aligned} y &= ab^x && \text{Exponential form.} \\ \ln y &= \ln(ab^x) && \text{Take natural logs of each side.} \\ \ln y &= \ln a + \ln b^x && \text{Property of logarithms.} \\ \ln y &= \ln a + (\ln b)x && \text{Property of logarithms.} \\ Y &= \ln a + (\ln b)x && \end{aligned}$$

Now apply the LINear REGression techniques discussed in Chapter 1. Be careful though. The dependent variable is now $Y = \ln y$, while the independent variable remains x. Once the line of best fit is obtained, you can find a and b and obtain the exponential curve $y = ab^x$.

$$ab^x = A_0e^{kt}$$
$$a = A_0 \qquad b^x = e^{kt} = (e^k)^t$$
$$b = e^k$$

Thus,

$$A_0 = 98.2463 \text{ and } 0.9188 = e^k$$
$$k = -0.0846$$

As a result, $A = A_0e^{kt} = 98.2463e^{-0.0846}$.

(d) See Figure 50.

FIGURE 50

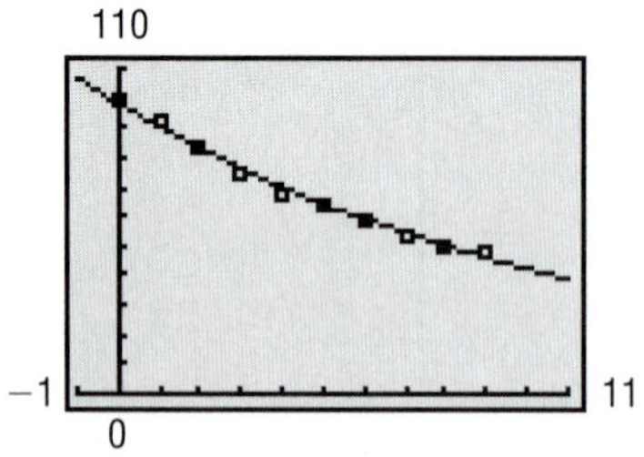

(e) We set $A = \frac{1}{2}A_0$ and solve the equation $\frac{1}{2}A_0 = A_0e^{-0.08463t}$ for t. The result is that the half-life is $t = 8.2$ days.

(f) After $t = 30$ days, the amount A of the radioactive material left is

$$A = 98.2463e^{-0.0846(30)} = 7.8 \text{ grams}$$

Now work Problem 1.

Logarithmic Curve Fitting

2 Many models do not follow uninhibited exponential growth but instead, as the independent variable increases, the growth of the dependent variable diminishes. Under these circumstances, a logarithmic model may be appropriate.

EXAMPLE 2 Fitting a Curve to a Logarithmic Model

Jodi, a meteorologist, is interested in finding a function that explains the relation between the height of a weather balloon (in kilometers) and the atmospheric pressure (measured in millimeters of mercury) on the balloon. She collects the data in the table to the left:

Atmospheric Pressure, p	Height, h
760	0
740	0.184
725	0.328
700	0.565
650	1.079
630	1.291
600	1.634
580	1.862
550	2.235

(a) Using a graphing utility, draw a scatter diagram of the data with atmospheric pressure as the independent variable.

(b) Using a graphing utility, fit a logarithmic model to the data.

(c) Draw the logarithmic function found in (b) on the scatter diagram.

(d) Use the function found in (b) to predict the height of the weather balloon if the atmospheric pressure is 560 millimeters of mercury.

Solution (a) After entering the data into the graphing utility, we obtain the scatter diagram shown in Figure 51.

(b) A graphing utility fits the data in Figure 51 to a logarithmic model of the form $y = a + b \ln x$ by using the Logarithm REGression option. See Figure 52. The logarithmic function of best fit to the data is

$$h = 45.7863 - 6.9025 \ln p$$

where h is the height of the weather balloon and p is the atmospheric pressure. Notice that $|r|$ is close to 1, indicating a good fit.

(c) Figure 53 shows the graph of $h = 45.7863 - 6.9025 \ln p$ on the scatter diagram.

(d) Using the function found in (b), Jodi predicts the height of the weather balloon when the atmospheric pressure is 560 to be

$$h = 45.7863 - 6.9025 \ln 560$$
$$\approx 2.108 \text{ kilometers}$$

FIGURE 51

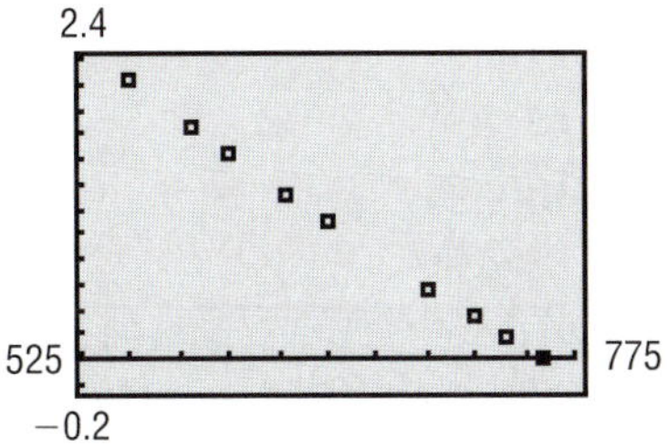

FIGURE 52

FIGURE 53

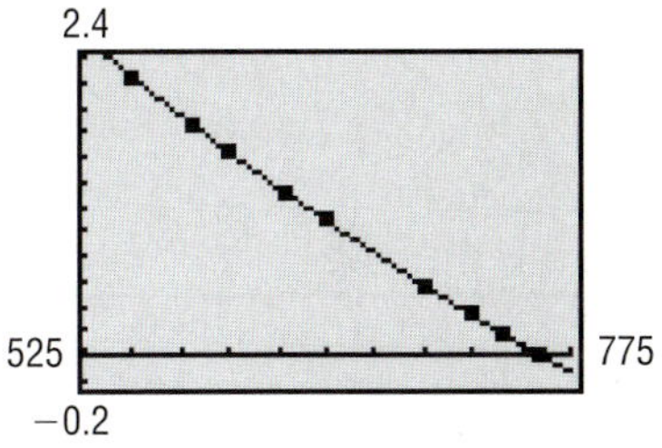

Now work Problem 11.

Logistic Curve Fitting

3 Logistic growth models can be used to model situations where the value of the dependent variable is limited. Many real-world situations conform to this scenario. For example, the population of the human race is limited by the availability of natural resources such as food, shelter, and so forth. When the value of the dependent variable is limited, a logistic growth model is often appropriate.

EXAMPLE 3

Fitting a Curve to a Logistic Growth Model

The data on the left, obtained from R. Pearl ("The Growth of Population," *Quarterly Review of Biology* 2 (1927): 532–548) represents the amount of yeast biomass after t hours in a culture.

Time (in hours)	Yeast Biomass	Time (in hours)	Yeast Biomass
0	9.6	10	513.3
1	18.3	11	559.7
2	29.0	12	594.8
3	47.2	13	629.4
4	71.1	14	640.8
5	119.1	15	651.1
6	174.6	16	655.9
7	257.3	17	659.6
8	350.7	18	661.8
9	441.0		

(a) Using a graphing utility, draw a scatter diagram of the data with time as the independent variable.

(b) Using a graphing utility, fit a logistic growth model to the data.

(c) Using a graphing utility, graph the function found in (b) on the scatter diagram.

(d) What is the predicted carrying capacity of the culture?

(e) Use the function found in (b) to predict the population of the culture at $t = 19$ hours.

Solution

(a) See Figure 54.

(b) A graphing utility fits a logistic growth model of the form $y = \dfrac{c}{1 + ae^{-bx}}$ by using the LOGISTIC regression option. See Figure 55. The logistic growth function of best fit to the data is

$$y = \frac{663.0}{1 + 71.6e^{-0.5470x}}$$

where y is the amount of yeast biomass in the culture and x is the time.

(c) See Figure 56.

(d) Based on the logistic growth function found in (b), the carrying capacity of the culture is 663.

(e) Using the logistic growth function found in (b), the predicted amount of yeast biomass at $t = 19$ is

$$y = \frac{663.0}{1 + 71.6e^{-0.5470(19)}} = 661.5$$

FIGURE 54

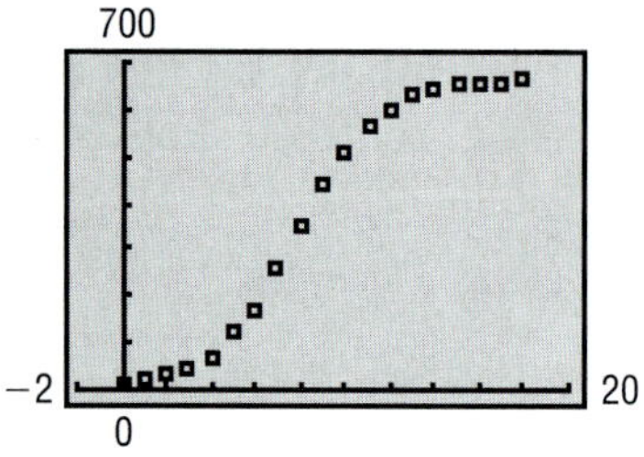

FIGURE 55

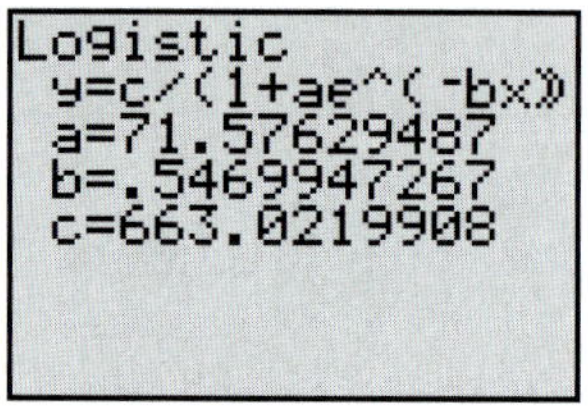

FIGURE 56

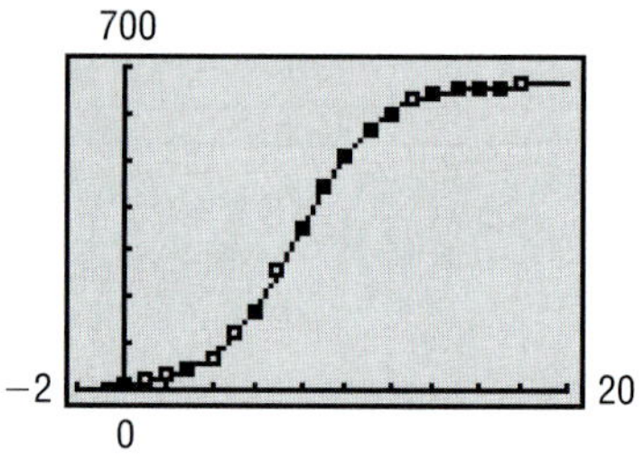

Now work Problem 13.

Choosing the Best Model

4 We have discussed seven different models thus far that can be used to explain the relation between two variables, x and y (see Table 8).

TABLE 8

Model	Functional Form	Section
1. Linear	$y = ax + b$	1.6
2. Quadratic	$y = ax^2 + bx + c$	4.1
3. Power	$y = ax^b$	4.2
4. Cubic	$y = ax^3 + bx^2 + cx + d$	4.3
5. Exponential	$y = ab^x$	5.8
6. Logarithmic	$y = a + b \ln x$	5.8
7. Logistic	$y = \dfrac{c}{1 + ae^{-bx}}$	5.8

How can we be certain the *model* we used is the "best" *model* to explain this relation? For example, why did we use an exponential model in Example 1 of this section?

Unfortunately, there is no such thing as *the* correct model. Modeling is not only a science but also an art form. Selecting an appropriate model requires experience and skill in the field in which you are modeling. For example, knowledge of economics is imperative when trying to determine a model to predict unemployment. The main reason for this is that there are theories in the field that can help the modeler select appropriate relations and variables.

Examining the methodologies used to determine the most appropriate model are beyond the scope of this text. However, we can use scatter diagrams and the correlation coefficient, r, to decide the "best" model among linear, power, exponential, and logarithmic models.* For our purposes, we will say the "best" model is the one which yields the value of r for which $|r|$ is closest to 1.

When there is no theoretical basis for choosing a particular model, the best approach to follow is to first draw a scatter diagram and obtain a general idea of the shape of the data. Then fit several models to the data whose graph exhibits the same behavior as the scatter diagram and determine which seems to explain the relation between the two variables best. Again, we will define "best" among linear, power, exponential, and logarithmic models as the one that yields the value of $|r|$ closest to 1.

EXAMPLE 4 Selecting the Best Model

Suppose an astronomer claims to have discovered a tenth planet that is 4.2 billion miles from the sun. How long would it take for this planet to completely orbit the sun?

Solution In order to solve this problem, we need to develop a model that relates the distance a planet is from the sun and the time it takes to make a complete orbit. This time is called the *sidereal year*. The data in the following table shows the distances the planets are from the sun and their sidereal years.

Planet	Distance from Sun (millions of miles)	Sidereal Year
Mercury	36	0.24
Venus	67	0.62
Earth	93	1.00
Mars	142	1.88
Jupiter	483	11.9
Saturn	887	29.5
Uranus	1,785	84.0
Neptune	2,797	165.0
Pluto	3,675	248.0

FIGURE 57

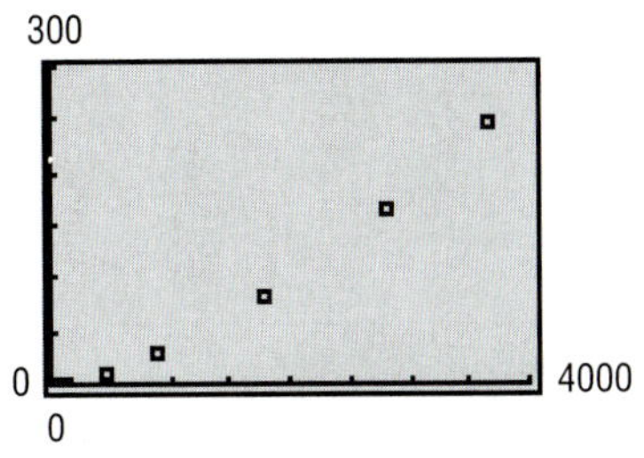

The first step in determining the best model is to draw a scatter diagram. Figure 57 shows a scatter diagram for the planet data. From the scatter diagram in Figure 57 it appears the data follows either a linear model, an exponential model ($k > 0$), or a power model ($b > 1$). Therefore, we shall find functions for all three models and use the value of r to determine which is best. Figure 58(a) shows the results obtained from the graphing utility for an

*Graphing utilities do not provide an r value for the remaining three models.

exponential model, Figure 58(b) shows the results for a power model, and Figure 58(c) shows the results for a linear model.

FIGURE 58

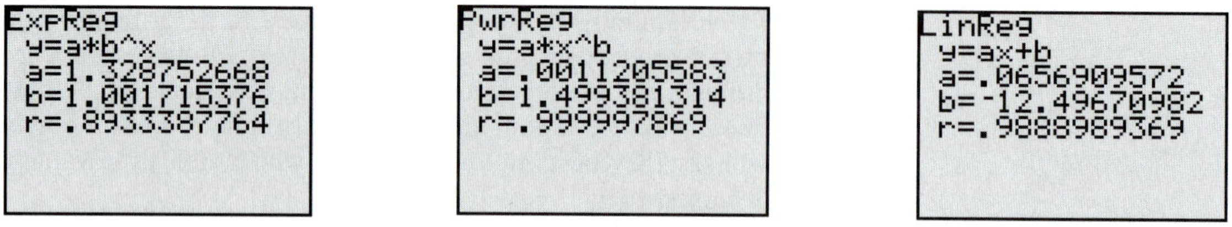

(a) (b) (c)

By looking at the value of r, we conclude that the power model is superior to both the exponential model and the linear model. Thus, we conclude the relation between the distance a planet is from the sun (x) and its sidereal year (y) is $y = 0.00112x^{1.49938}$.*

We can predict the sidereal year of the newly discovered planet by substituting $x = 4200$ (4.2 billion = 4200 million) into the function just found:

$$y = 0.00112(4200)^{1.49938}$$
$$\approx 303.3 \text{ sidereal years}$$

Now work Problem 17.

5.8 EXERCISES

1. **Biology** A certain bacteria initially increases according to the law of uninhibited growth. A biologist collects the following data for this bacteria:

Time (Hours)	Population
0	1000
1	1415
2	2000
3	2828
4	4000
5	5656
6	8000

(a) Using a graphing utility, draw a scatter diagram.
(b) Using a graphing utility, fit an exponential curve to the data.
(c) Express the curve in the form $N = N_0e^{kt}$.
(d) Graph the exponential function found in (b) or (c) on the scatter diagram.
(e) Using the solution to (b), predict the population at $t = 7$ hours.

2. **Biology** A colony of bacteria initially increases according to the law of uninhibited growth. A biologist collects the following data for this bacteria:

Time (Days)	Population
0	50
1	153
2	234
3	357
4	547
5	839
6	1280

*If we square both sides of this equation we obtain

$$y^2 \approx 0.0000013x^3$$

which implies the square of the sidereal year is proportional to the cube of the distance from the sun to the planet. This is Kepler's Third Law of Planetary Motion, named after Johann Kepler. Kepler discovered this relation in 1618 through experimentation without the aid of data analysis.

(a) Using a graphing utility, draw a scatter diagram.
(b) Using a graphing utility, fit an exponential curve to the data.
(c) Express the curve in the form $N = N_0e^{kt}$.
(d) Graph the exponential function found in (b) or (c) on the scatter diagram.
(e) Using the solution to (b), predict the population at $t = 7$ days.

3. **Economics/Marketing** A store manager collected the following data regarding price and quantity demanded of shoes:

Price ($/Unit)	Quantity Demanded
79	10
67	20
54	30
46	40
38	50
31	60

(a) Using a graphing utility, draw a scatter diagram with quantity demanded on the x-axis and price on the y-axis.
(b) Using a graphing utility, fit an exponential curve to the data.
(c) Graph the exponential function found in (b) on the scatter diagram.
(d) Predict the quantity demanded if the price is $60.

4. **Economics/Marketing** A store manager collected the following data regarding price and quantity supplied of dresses:

Price ($/Unit)	Quantity Supplied
25	10
32	20
40	30
46	40
60	50
74	60

(a) Using a graphing utility, draw a scatter diagram with quantity supplied on the x-axis and price on the y-axis.
(b) Using a graphing utility, fit an exponential curve to the data.
(c) Graph the exponential function found in (b) on the scatter diagram.
(d) Predict the quantity supplied if the price is $45.

5. **Chemistry** A chemist has a 100-gram sample of a radioactive material. He records the amount of radioactive material every week for 6 weeks and obtains the following data:

Week	Weight (in Grams)
0	100.0
1	88.3
2	75.9
3	69.4
4	59.1
5	51.8
6	45.5

(a) Using a graphing utility, draw a scatter diagram with week as the independent variable.
(b) Using a graphing utility, fit an exponential curve to the data.
(c) Express the curve in the form $A = A_0e^{kt}$.
(d) Graph the exponential function found in (b) or (c) on the scatter diagram.
(e) From the result found in (b), determine the half-life of the radioactive material.
(f) How much radioactive material will be left after 50 weeks?

6. **Chemistry** A chemist has a 1000-gram sample of a radioactive material. She records the amount of radioactive material remaining in the sample every day for a week and obtains the following data:

Day	Weight (in Grams)
0	1000.0
1	897.1
2	802.5
3	719.8
4	651.1
5	583.4
6	521.7
7	468.3

(a) Using a graphing utility, draw a scatter diagram with day as the independent variable.
(b) Using a graphing utility, fit an exponential curve to the data.
(c) Express the curve in the form $A = A_0 e^{kt}$.
(d) Graph the exponential function found in (b) or (c) on the scatter diagram.
(e) From the result found in (b), find the half-life of the radioactive material.
(f) How much radioactive material will be left after 20 days?

7. Finance The following data represents the amount of money an investor has in an investment account each year for 10 years. She wishes to determine the effective rate of return on her investment.

Year	Value of Account
1985	$10,000
1986	$10,573
1987	$11,260
1988	$11,733
1989	$12,424
1990	$13,269
1991	$13,968
1992	$14,823
1993	$15,297
1994	$16,539

(a) Using a graphing utility, draw a scatter diagram with time as the independent variable and the value of the account as the dependent variable.
(b) Using a graphing utility, fit an exponential curve to the data.
(c) Based on the answer to (b), what was the effective rate of return from this account over the past 10 years?
(d) If the investor plans on retiring in 2020, what will the predicted value of this account be?

8. Finance The following data shows the amount of money an investor has in an investment account each year for 7 years. He wishes to determine the effective rate of return on his investment.

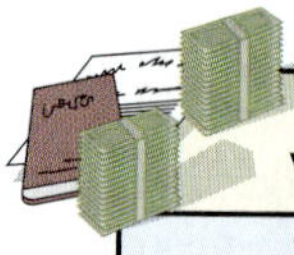

Year	Value of Account
1988	$20,000
1989	$21,516
1990	$23,355
1991	$24,885
1992	$27,434
1993	$30,053
1994	$32,622

(a) Using a graphing utility, draw a scatter diagram with time as the independent variable and the value of the account as the dependent variable.
(b) Using a graphing utility, fit an exponential curve to the data.
(c) Based on the answer to (b), what was the effective rate of return from this account over the past 7 years?
(d) If the investor plans on retiring in 2020, what will the predicted value of this account be?

9. CBL Experiment The following data was collected by placing a temperature probe in a cup of hot water, removing the probe, and then recording temperature over time.

According to Newton's Law of Cooling, this data should follow an exponential model.

(a) Using a graphing utility, draw a scatter diagram for the data.
(b) Using a graphing utility, fit an exponential model to the data.
(c) Graph the exponential function found in (b) on the scatter diagram.
(d) Predict how long it will take for the water to reach a temperature of 110°F.

Time	Temperature (F°)
0	175.69
1	173.52
2	171.21
3	169.07
4	166.59
5	164.21
6	161.89
7	159.66
8	157.86
9	155.75
10	153.70
11	151.93
12	150.08
13	148.50
14	146.84
15	145.33
16	143.83
17	142.38
18	141.22
19	140.09
20	138.69
21	137.59
22	136.78
23	135.70
24	134.91
25	133.86

10. CBL Experiment The following data was collected by placing a temperature probe in a portable heater, removing the probe, and then recording temperature over time.

According to Newton's Law of Cooling, this data should follow an exponential model.

(a) Using a graphing utility, draw a scatter diagram for the data.
(b) Using a graphing utility, fit an exponential model to the data.
(c) Graph the exponential function found in (b) on the scatter diagram.
(d) Predict how long it will take for the air blown out of the heater to reach a temperature of 110°F.

Time	Temperature (F°)
0	165.07
1	164.77
2	163.99
3	163.22
4	162.82
5	161.96
6	161.20
7	160.45
8	159.35
9	158.61
10	157.89
11	156.83
12	156.11
13	155.08
14	154.40
15	153.72

11. Economics/Marketing The following data represents the price and quantity demanded in 1997 for IBM personal computers at Best Buy.

(a) Using a graphing utility, draw a scatter diagram of the data with price as the dependent variable.
(b) Using a graphing utility, fit a logarithmic model to the data.
(c) Using a graphing utility, draw the logarithmic function found in (b) on the scatter diagram.
(d) Use the function found in (b) to predict the number of IBM personal computers that would be demanded if the price were $1650.

Price ($/Computer)	Quantity Demanded
2300	152
2000	159
1700	164
1500	171
1300	176
1200	180
1000	189

12. Economics/Marketing The following data represents the price and quantity supplied in 1997 for IBM personal computers.

Price ($/Computer)	Quantity Supplied
2300	180
2000	173
1700	160
1500	150
1300	137
1200	130
1000	113

(a) Using a graphing utility, draw a scatter diagram of the data with price as the dependent variable.
(b) Using a graphing utility, fit a logarithmic model to the data.
(c) Using a graphing utility, draw the logarithmic function found in (b) on the scatter diagram.
(d) Use the function found in (b) to predict the number of IBM personal computers that would be supplied if the price were $1650.

13. Population Model The following data obtained from the U.S. Census Bureau represents the population of the United States. An ecologist is interested in finding a function that describes the population of the United States.

Year	Population
1900	76,212,168
1910	92,228,496
1920	106,021,537
1930	123,202,624
1940	132,164,569
1950	151,325,798
1960	179,323,175
1970	203,302,031
1980	226,542,203
1990	248,709,873

(a) Using a graphing utility, draw a scatter diagram of the data using the year as the independent variable and population as the dependent variable.
(b) Using a graphing utility, fit a logistic model to the data.
(c) Using a graphing utility, draw the function found in (b) on the scatter diagram.
(d) Based on the function found in (b), what is the carrying capacity of the United States?
(e) Use the function found in (b) to predict the population of the United States in 2000.

14. Population Model The following data obtained from the U.S. Census Bureau represents the world population. An ecologist is interested in finding a function that describes the world population.

Year	Population (in Billions)
1981	4.533
1982	4.614
1983	4.695
1984	4.775
1985	4.856
1986	4.941
1987	5.029
1988	5.117
1989	5.205
1990	5.295
1991	5.381
1992	5.469
1993	5.556
1994	5.644
1995	5.732

(a) Using a graphing utility, draw a scatter diagram of the data using year as the independent variable and population as the dependent variable.
(b) Using a graphing utility, fit a logistic model to the data.
(c) Using a graphing utility, draw the function found in (b) on the scatter diagram.
(d) Based on the function found in (b), what is the carrying capacity of the world?
(e) Use the function found in (b) to predict the population of the world in 2000.

15. Population Model The following data obtained from the U.S. Census Bureau represents the population of Illinois. An urban economist is interested in finding a model that describes the population of Illinois.

Year	Population
1900	4,821,550
1910	5,638,591
1920	6,485,280
1930	7,630,654
1940	7,897,241
1950	8,712,176
1960	10,081,158
1970	11,110,285
1980	11,427,409
1990	11,430,602

(a) Using a graphing utility, draw a scatter diagram of the data using year as the independent variable and population as the dependent variable.
(b) Using a graphing utility, fit a logistic model to the data.
(c) Using a graphing utility, draw the function found in (b) on the scatter diagram.
(d) Based on the function found in (b), what is the carrying capacity of Illinois?
(e) Use the function found in (b) to predict the population of Illinois in 2000.

16. Population Model The following data obtained from the U.S. Census Bureau represents the population of Pennsylvania. An urban economist is interested in finding a model that describes the population of Pennsylvania.

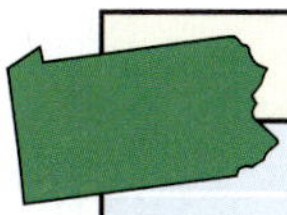

Year	Population
1900	6,302,115
1910	7,665,111
1920	8,720,017
1930	9,631,350
1940	9,900,180
1950	10,498,012
1960	11,319,366
1970	11,800,766
1980	11,864,720
1990	11,881,643

(a) Using a graphing utility, draw a scatter diagram of the data using year as the independent variable and population as the dependent variable.
(b) Using a graphing utility, fit a logistic model to the data.
(c) Using a graphing utility, draw the function found in (b) on the scatter diagram.
(d) Based on the function found in (b), what is the carrying capacity of Pennsylvania?
(e) Use the function found in (b) to predict the population of Pennsylvania in 2000.

For Problems 17–22 the independent variable is x and the dependent variable is y. (a) For each set of data draw a scatter diagram of the data. (b) Use an exponential model, logarithmic model, power model, and linear model to find the "best" model that explains the relationship between the variables.

17.

x	1	2	3	4	5
y	164	269	450	740	1220

18.

x	1	2	3	4	5
y	110	245	553	1228	2728

19.

x	10	20	30	40	50
y	6.94	13.78	20.45	21.23	22.95

20.

x	10	20	30	40	50
y	4.95	6.46	7.23	7.78	8.23

21.

x	1	1.3	1.7	2.1	2.4	2.7
y	9.92	4.56	1.42	0.37	0.12	0.032

22.

x	1	1.3	1.7	2.1	2.4	2.7
y	1.72	1.94	2.48	3.28	4.01	5.07

23. Economics The data that follows on p. 434 are levels of the Consumer Price Index for food items
(a) Using a graphing utility, draw a scatter diagram of the data using time as the independent variable and the CPI as the dependent variable.
(b) Use an exponential model, logarithmic model, power model, and linear model to find the "best" model to describe the relation between time and the CPI.
(c) Use this model to predict the CPI for food items for 1994.

Year	CPI
1984	103.2
1985	105.6
1986	109.0
1987	113.5
1988	118.2
1989	125.1
1990	132.4
1991	136.3
1992	137.9
1993	140.9

Source: Bureau of Labor Statistics.

24. Economics The following data is the level of the Consumer Price Index for electricity.

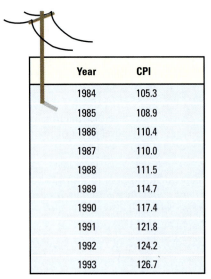

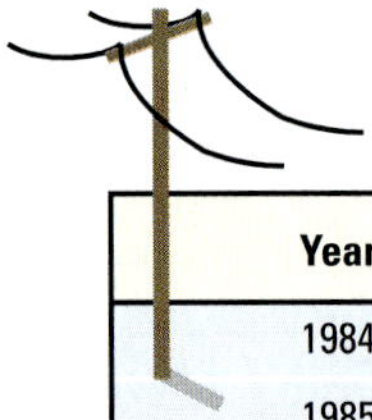

Year	CPI
1984	105.3
1985	108.9
1986	110.4
1987	110.0
1988	111.5
1989	114.7
1990	117.4
1991	121.8
1992	124.2
1993	126.7

Source: Bureau of Labor Statistics.

(a) Using a graphing utility, draw a scatter diagram of the data using time as the independent variable and the CPI as the dependent variable.
(b) Use an exponential model, logarithmic model, power model, and linear model to find the "best" model to describe the relation between time and the CPI.
(c) Use this model to predict the CPI for electricity for 1994.

25. Economics The following data represents the per capita Gross Domestic Product (GDP) in constant 1987 dollars (this removes the effects of inflation).

GDP

Year	Per Capita Gross Domestic Product (1987 Dollars)
1988	19,252
1989	19,556
1990	19,593
1991	19,238
1992	19,518
1993	19,888

Source: U.S. Bureau of Economic Analysis.

(a) Using a graphing utility, draw a scatter diagram of the data using time as the independent variable and the per capita GDP as the dependent variable.
(b) Use an exponential model, logarithmic model, power model, and linear model to find the "best" model to describe the relation between time and the per capita GDP.
(c) Use this model to predict per capita GDP for 1994.

26. Economics The following data represents the per capita Gross National Product (GNP) in constant 1987 dollars (this removes the effects of inflation).

GNP

	Per Capita Gross National Product (1987 Dollars)
1988	19,284
1989	19,615
1990	19,670
1991	19,290
1992	19,548
1993	19,897

Source: U.S. Bureau of Economic Analysis.

(a) Using a graphing utility, draw a scatter diagram of the data using time as the independent variable and the GNP as the dependent variable.
(b) Use an exponential model, logarithmic model, power model, and linear model to find the "best" model to describe the relation between time and the per capita GNP.
(c) Use this model to predict the per capita GNP for 1994.

5.9 LOGARITHMIC SCALES

1 Decibels (Loudness of Sound)
2 Richter Scale (Earthquake Magnitude)

Common logarithms often appear in the measurement of quantities, because they provide a way to scale down positive numbers that vary from very small to very large. For example, if a certain quantity can take on values from $0.0000000001 = 10^{-10}$ to $10{,}000{,}000{,}000 = 10^{10}$, the common logarithms of such numbers would be between -10 and 10, respectively.

Loudness of Sound

1 Our first application utilizes a logarithmic scale to measure the loudness of a sound. Physicists define the **intensity of a sound wave** as the amount of energy the wave transmits through a given area. For example, the least intense sound that a human ear can detect at a frequency of 100 Hertz is about 10^{-12} watt per square meter. The **loudness** $L(x)$, measured in **decibels** (named in honor of Alexander Graham Bell), of a sound of intensity x (measured in watts per square meter) is defined as

$$L(x) = 10 \log \frac{x}{I_0} \qquad (1)$$

where $I_0 = 10^{-12}$ watt per square meter is the least intense sound that a human ear can detect. If we let $x = I_0$ in equation (1), we get

$$L(I_0) = 10 \log \frac{I_0}{I_0} = 10 \log 1 = 0$$

Thus, at the threshold of human hearing, the loudness is 0 decibels. Figure 59 on page 436 gives the loudness of some common sounds.

Note that a decibel is not a linear unit like the meter. For example, a noise level of 10 decibels is 10 times as great as a noise level of 0 decibels. [If $L(x) = 10$, then $x = 10I_0$.] A noise level of 20 decibels is 100 times as great as a noise level of 0 decibels. [If $L(x) = 20$, then $x = 100I_0$.] A noise level of 30 decibels is 1000 times as great as a noise level of 0 decibels, and so on.

EXAMPLE 1 Finding the Intensity of a Sound

Use Figure 59 to find the intensity of the sound of a dripping faucet.

Solution From Figure 59, we see that the loudness of the sound of a dripping faucet is 30 decibels. Thus, by equation (1), its intensity x may be found as follows:

$$30 = 10 \log\left(\frac{x}{I_0}\right)$$

$$3 = \log\left(\frac{x}{I_0}\right) \qquad \text{Divide by 10.}$$

$$\frac{x}{I_0} = 10^3 \qquad \text{Write in exponential form.}$$

$$x = 1000I_0$$

FIGURE 59
Loudness of common sounds (in decibels)

Decibels		
140	Shotgun blast, jet 100 feet away at takeoff	Pain
130	Motor test chamber	Human ear pain threshold
120	Firecrackers, severe thunder, pneumatic jackhammer, hockey crowd	Uncomfortably loud
110	Amplified rock music	
100	Textile loom, subway train, elevated train, farm tractor, power lawn mower, newspaper press	Loud
90	Heavy city traffic, noisy factory	
80	Diesel truck going 40 mi/hr 50 feet away, crowded restaurant, garbage disposal, average factory, vacuum cleaner	Moderately loud
70	Passenger car going 50 mi/hr 50 feet away	
60	Quiet typewriter, singing birds, window air conditioner, quiet automobile	Quiet
50	Normal conversation, average office	
40	Household refrigerator, quiet office	Very quiet
30	Average home, dripping faucet, whisper 5 feet away	
20	Light rainfall, rustle of leaves	Average person's threshold of hearing
10	Whisper across room	Just audible
0		Threshold for acute hearing

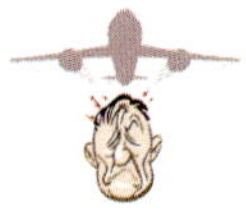

where $I_0 = 10^{-12}$ watt per square meter. Thus, the intensity of the sound of a dripping faucet is 1000 times as great as a noise level of 0 decibels; that is, such a sound has an intensity of $1000 \cdot 10^{-12} = 10^{-9}$ watt per square meter.

Now work Problem 5.

EXAMPLE 2 Finding the Loudness of a Sound

Use Figure 59 to determine the loudness of a subway train if it is known that this sound is 10 times as intense as the sound due to heavy city traffic.

Solution The sound due to heavy city traffic has a loudness of 90 decibels. Its intensity, therefore, is the value of x in the equation

$$90 = 10 \log\left(\frac{x}{I_0}\right)$$

A sound 10 times as intense as x has loudness $L(10x)$. Thus, the loudness of the subway train is

$$\begin{aligned}
L(10x) &= 10\log\left(\frac{10x}{I_0}\right) && \text{Replace } x \text{ by } 10x.\\
&= 10\log\left(10\cdot\frac{x}{I_0}\right)\\
&= 10\left[\log 10 + \log\left(\frac{x}{I_0}\right)\right] && \text{Log of product = Sum of logs.}\\
&= 10\log 10 + 10\log\left(\frac{x}{I_0}\right) && \log 10 = 1.\\
&= 10 + 90 = 100 \text{ decibels}
\end{aligned}$$

Magnitude of an Earthquake

2 Our second application uses a logarithmic scale to measure the magnitude of an earthquake.

The **Richter scale*** is one way of converting seismographic readings into numbers that provide an easy reference for measuring the magnitude M of an earthquake. All earthquakes are compared to a **zero-level earthquake** whose seismographic reading measures 0.001 millimeter at a distance of 100 kilometers from the epicenter. An earthquake whose seismographic reading measures x millimeters has **magnitude** $M(x)$ given by

$$M(x) = \log\left(\frac{x}{x_0}\right) \tag{2}$$

where $x_0 = 10^{-3}$ is the reading of a zero-level earthquake the same distance from its epicenter.

EXAMPLE 3

Finding the Magnitude of an Earthquake

What is the magnitude of an earthquake whose seismographic reading is 0.1 millimeter at a distance of 100 kilometers from its epicenter?

Solution If $x = 0.1$, the magnitude $M(x)$ of this earthquake is

$$M(0.1) = \log\left(\frac{x}{x_0}\right) = \log\left(\frac{0.1}{0.001}\right) = \log\left(\frac{10^{-1}}{10^{-3}}\right) = \log 10^2 = 2$$

This earthquake thus measures 2.0 on the Richter scale.

Now work Problem 7.

Based on formula (2), we define the **intensity of an earthquake** as the ratio of x to x_0. For example, the intensity of the earthquake described in Example 3 is $\frac{0.1}{0.001} = 10^2 = 100$. That is, it is 100 times as intense as a zero-level earthquake.

EXAMPLE 4

Comparing the Intensity of Two Earthquakes

The devastating San Francisco earthquake of 1906 measured 6.9 on the Richter scale. How did the intensity of that earthquake compare to the Papua New Guinea earthquake of 1988, which measured 6.7 on the Richter scale?

*Named after the American scientist, C.F. Richter, who devised it in 1935.

FIGURE 60

Solution Let x_1 and x_2 denote the seismographic readings, respectively, of the 1906 San Francisco earthquake and the Papua New Guinea earthquake. Then, based on formula (2),

$$6.9 = \log\left(\frac{x_1}{x_0}\right) \qquad 6.7 = \log\left(\frac{x_2}{x_0}\right)$$

Consequently,

$$\frac{x_1}{x_0} = 10^{6.9} \qquad \frac{x_2}{x_0} = 10^{6.7}$$

The 1906 San Francisco earthquake was $10^{6.9}$ times as intense as a zero-level earthquake. The Papua New Guinea earthquake was $10^{6.7}$ times as intense as a zero-level earthquake. Thus,

$$\frac{x_1}{x_2} = \frac{10^{6.9}x_0}{10^{6.7}x_0} = 10^{0.2} \approx 1.58$$

$$x_1 \approx 1.58x_2$$

Hence, the San Francisco earthquake was 1.58 times as intense as the Papua New Guinea earthquake.

Example 4 demonstrates that the relative intensity of two earthquakes can be found by raising 10 to a power equal to the difference of their readings on the Richter scale.

5.9 EXERCISES

1. **Loudness of a Dishwasher** Find the loudness of a dishwasher that operates at an intensity of 10^{-5} watt per square meter. Express your answer in decibels.
2. **Loudness of a Diesel Engine** Find the loudness of a diesel engine that operates at an intensity of 10^{-3} watt per square meter. Express your answer in decibels.
3. **Loudness of a Jet Engine** With engines at full throttle, a Boeing 727 jetliner produces noise at an intensity of 0.15 watt per square meter. Find the loudness of the engines in decibels.
4. **Loudness of a Whisper** A whisper produces noise at an intensity of $10^{-9.8}$ watt per square meter. What is the loudness of a whisper in decibels?
5. **Intensity of a Sound at Threshold of Pain** For humans, the threshold of pain due to sound averages 130 decibels. What is the intensity of such a sound in watts per square meter?

6. If one sound is 50 times as intense as another, what is the difference in the loudness of the two sounds? Express your answer in decibels.

7. **Magnitude of an Earthquake** Find the magnitude of an earthquake whose seismographic reading is 10.0 millimeters at a distance of 100 kilometers from its epicenter.

8. **Magnitude of an Earthquake** Find the magnitude of an earthquake whose seismographic reading is 1,210 millimeters at a distance of 100 kilometers from its epicenter.

9. **Comparing Earthquakes** The Mexico City earthquake of 1985 registered 8.1 on the Richter scale. What would a seismograph 100 kilometers from the epicenter have measured for this earthquake? How does this earthquake compare in intensity to the 1906 San Francisco earthquake, which registered 6.9 on the Richter scale?

10. **Comparing Earthquakes** Two earthquakes differ by 1.0 when measured on the Richter scale. How would the seismographic readings differ at a distance of 100 kilometers from the epicenter? How do their intensities compare?

CHAPTER REVIEW

THINGS TO KNOW

One-to-one function f

If $x_1 \neq x_2$, then $f(x_1) \neq f(x_2)$ for any choice of x_1 and x_2 in the domain.

Horizontal line test

If horizontal lines intersect the graph of a function f in at most one point, then f is one-to-one.

Inverse function f^{-1} of f

Domain of f = Range of f^{-1}; Range of f = Domain of f^{-1}

$f^{-1}(f(x)) = x$ and $f(f^{-1}(x)) = x$.

Graphs of f and f^{-1} are symmetric with respect to the line $y = x$.

Properties of the exponential function

$f(x) = a^x$, $a > 1$	Domain: $(-\infty, \infty)$; Range: $(0, \infty)$; x-intercepts: none; y-intercept: 1; horizontal asymptote: x-axis as $x \to -\infty$; increasing; one-to-one See Figure 13 for a typical graph.
$f(x) = a^x$, $0 < a < 1$	Domain: $(-\infty, \infty)$; Range: $(0, \infty)$; x-intercepts: none, y-intercept: 1; horizontal asymptote: x-axis as $x \to \infty$; decreasing; one-to-one See Figure 17 for a typical graph.

Properties of the logarithmic function

$f(x) = \log_a x$, $a > 1$ ($y = \log_a x$ means $x = a^y$)	Domain: $(0, \infty)$; Range: $(-\infty, \infty)$; x-intercept: 1; y-intercept: none; vertical asymptote: y-axis; increasing; one-to-one See Figure 25(b) for a typical graph.
$f(x) = \log_a x$, $0 < a < 1$ ($y = \log_a x$ means $x = a^y$)	Domain: $(0, \infty)$; Range: $(-\infty, \infty)$; x-intercept: 1; y-intercept: none; vertical asymptote: y-axis; decreasing; one-to-one See Figure 25(a) for a typical graph.

Number e

Value approached by the expression $\left(1 + \dfrac{1}{n}\right)^n$ as $n \to \infty$; that is, $\lim_{n \to \infty} \left(1 + \dfrac{1}{n}\right)^n = e$

Natural logarithm

$y = \ln x$ means $x = e^y$

Properties of logarithms

$\log_a 1 = 0 \qquad \log_a a = 1 \qquad a^{\log_a M} = M \qquad \log_a a^r = r$

$\log_a (MN) = \log_a M + \log_a N \qquad \log_a\left(\dfrac{M}{N}\right) = \log_a M - \log_a N \qquad \log_a\left(\dfrac{1}{N}\right) = -\log_a N \qquad \log_a M^r = r \log_a M$

FORMULAS

Change-of-Base Formula	$\log_a M = \dfrac{\log_b M}{\log_b a}$
Compound interest	$A = P\left(1 + \dfrac{r}{n}\right)^{nt}$
Continuous compounding	$A = Pe^{rt}$
Present value	$P = A\left(1 + \dfrac{r}{n}\right)^{-nt}$ or $P = Ae^{-rt}$
Growth and decay	$A = A_0 e^{kt}$

HOW TO

Find the inverse of certain one-to-one functions (see the procedure given on page 358).
Graph f^{-1} given the graph of f.
Graph exponential and logarithmic functions.
Solve certain exponential equations.
Solve certain logarithmic equations.
Solve problems involving compound interest.
Solve problems involving growth and decay.
Obtain exponential, logarithmic, and logistic functions of best fit.
Choose models of best fit.
Solve problems involving intensity of sound and intensity of earthquakes.

FILL-IN-THE-BLANK ITEMS

1. If every horizontal line intersects the graph of a function f at no more than one point, then f is a(n) ______________ function.
2. If f^{-1} denotes the inverse of a function f, then the graphs of f and f^{-1} are symmetric with respect to the line ______________.
3. The graph of every exponential function $f(x) = a^x, a > 0, a \neq 1$, passes through the two points ______________.
4. If the graph of an exponential function $f(x) = a^x$, $a > 0$, $a \neq 1$, is decreasing, then its base must be less than ______________.
5. If $3^x = 3^4$, then $x =$ ______________.
6. The logarithm of a product equals the ______________ of the logarithms.
7. For every base, the logarithm of ______________ equals 0.
8. If $\log_8 M = \log_5 7/\log_5 8$, then $M =$ ______________.
9. The domain of the logarithmic function $f(x) = \log_a x$ consists of ______________.
10. The graph of every logarithmic function $f(x) = \log_a x, a > 0, a \neq 1$, passes through the two points ______________.
11. If the graph of a logarithmic function $f(x) = \log_a x$, $a > 0$, $a \neq 1$, is increasing, then its base must be larger than ______________.
12. If $\log_3 x = \log_3 7$, then $x =$ ______________.

TRUE/FALSE ITEMS

T F 1. If f and g are inverse functions, then the domain of f is the same as the domain of g.

T F 2. If f and g are inverse functions, then their graphs are symmetric with respect to the line $y = x$.

T F 3. The graph of every exponential function $f(x) = a^x$, $a > 0$, $a \neq 1$, will contain the points $(0, 1)$ and $(1, a)$.

T F 4. The graphs of $y = 3^{-x}$ and $y = \left(\frac{1}{3}\right)^x$ are identical.

T F 5. The present value of \$1000 to be received after 2 years at 10% per annum compounded continuously is approximately \$1205.

T F 6. If $y = \log_a x$, then $y = a^x$.

T F 7. The graph of every logarithmic function $f(x) = \log_a x$, $a > 0$, $a \neq 1$, will contain the points $(1, 0)$ and $(a, 1)$.

T F 8. $a^{\log_M a} = M$, where $a > 0$, $a \neq 1$, $M > 0$.

T F 9. $\log_a(M + N) = \log_a M + \log_a N$, where $a > 0$, $a \neq 1$, $M > 0$, $N > 0$.

T F 10. $\log_a M - \log_a N = \log_a(M/N)$, where $a > 0$, $a \neq 1$, $M > 0$, $N > 0$.

REVIEW EXERCISES

In Problems 1–6, the function f is one-to-one. Find the inverse of each function and check your answer. Find the domain and range of f and f^{-1}. Use a graphing utility to simultaneously graph f, f^{-1}, and $y = x$ on the same square screen.

1. $f(x) = \dfrac{2x + 3}{5x - 2}$
2. $f(x) = \dfrac{2 - x}{3 + x}$
3. $f(x) = \dfrac{1}{x - 1}$
4. $f(x) = \sqrt{x - 2}$
5. $f(x) = \dfrac{3}{x^{1/3}}$
6. $f(x) = x^{1/3} + 1$

In Problems 7–12, evaluate each expression.

7. $\log_2\left(\frac{1}{8}\right)$
8. $\log_3 81$
9. $\ln e^{\sqrt{2}}$
10. $e^{\ln 0.1}$
11. $2^{\log_2 0.4}$
12. $\log_2 2^{\sqrt{3}}$

In Problems 13–18, write each expression as a single logarithm.

13. $3 \log_4 x^2 + \dfrac{1}{2} \log_4 \sqrt{x}$
14. $-2 \log_3\left(\dfrac{1}{x}\right) + \dfrac{1}{3} \log_3 \sqrt{x}$
15. $\ln\left(\dfrac{x - 1}{x}\right) + \ln\left(\dfrac{x}{x + 1}\right) - \ln(x^2 - 1)$
16. $\log(x^2 - 9) - \log(x^2 + 7x + 12)$
17. $2 \log 2 + 3 \log x - \dfrac{1}{2}[\log(x + 3) + \log(x - 2)]$
18. $\dfrac{1}{2} \ln(x^2 + 1) - 4 \ln \dfrac{1}{2} - \dfrac{1}{2}[\ln(x - 4) + \ln x]$

In Problems 19–26, find y as a function of x. The constant C is a positive number.

19. $\ln y = 2x^2 + \ln C$
20. $\ln(y - 3) = \ln 2x^2 + \ln C$
21. $\frac{1}{2} \ln y = 3x^2 + \ln C$
22. $\ln 2y = \ln(x + 1) + \ln(x + 2) + \ln C$
23. $\ln(y - 3) + \ln(y + 3) = x + C$
24. $\ln(y - 1) + \ln(y + 1) = -x + C$
25. $e^{y+C} = x^2 + 4$
26. $e^{3y-C} = (x + 4)^2$

In Problems 27–36, using a graphing utility, show the stages required to graph each function. Determine the domain, range, and any asymptotes.

27. $f(x) = e^{-x}$
28. $f(x) = \ln(-x)$
29. $f(x) = 1 - e^x$
30. $f(x) = 3 + \ln x$
31. $f(x) = 3e^x$
32. $f(x) = \frac{1}{2} \ln x$
33. $f(x) = e^{|x|}$
34. $f(x) = \ln|x|$
35. $f(x) = 3 - e^{-x}$
36. $f(x) = 4 - \ln(-x)$

In Problems 37–56, solve each equation. Verify your result using a graphing utility.

37. $4^{1-2x} = 2$
38. $8^{6+3x} = 4$
39. $3^{x^2+x} = \sqrt{3}$
40. $4^{x-x^2} = \frac{1}{2}$
41. $\log_x 64 = -3$
42. $\log_{\sqrt{2}} x = -6$
43. $5^x = 3^{x+2}$
44. $5^{x+2} = 7^{x-2}$
45. $9^{2x} = 27^{3x-4}$

46. $25^{2x} = 5^{x^2-12}$

47. $\log_3 \sqrt{x-2} = 2$

48. $2^{x+1} \cdot 8^{-x} = 4$

49. $8 = 4^{x^2} \cdot 2^{5x}$

50. $2^x \cdot 5 = 10^x$

51. $\log_6(x+3) + \log_6(x+4) = 1$

52. $\log_{10}(7x - 12) = 2\log_{10} x$

53. $e^{1-x} = 5$

54. $e^{1-2x} = 4$

55. $2^{3x} = 3^{2x+1}$

56. $2^{x^3} = 3^{x^2}$

In Problems 57–60, use the following result: If x is the atmospheric pressure (measured in millimeters of mercury), then the formula for the altitude h(x) (measured in meters above sea level) is

$$h(x) = (30T + 8000)\log\left(\frac{P_0}{x}\right)$$

where T is the temperature (in degrees Celsius) and P_0 is the atmospheric pressure at sea level, which is approximately 760 millimeters of mercury.

57. Finding the Altitude of an Airplane At what height is a Piper Cub whose instruments record an outside temperature of 0°C and a barometric pressure of 300 millimeters of mercury?

58. Finding the Height of a Mountain How high is a mountain if instruments placed on its peak record a temperature of 5°C and a barometric pressure of 500 millimeters of mercury?

59. Atmospheric Pressure Outside an Airplane What is the atmospheric pressure outside a Boeing 737 flying at an altitude of 10,000 meters if the outside air temperature is −100°C?

60. Atmospheric Pressure at High Altitudes What is the atmospheric pressure (in millimeters of mercury) on Mt. Everest, which has an altitude of approximately 8900 meters, if the air temperature is 5°C?

61. Amplifying Sound An amplifier's power output P (in watts) is related to its decibel voltage gain d by the formula $P = 25e^{0.1d}$.

(a) Find the power output for a decibel voltage gain of 4 decibels.

(b) For a power output of 50 watts, what is the decibel voltage gain?

62. Limiting Magnitude of a Telescope A telescope is limited in its usefulness by the brightness of the star it is aimed at and by the diameter of its lens. One measure of a star's brightness is its *magnitude:* the dimmer the star, the larger its magnitude. A formula for the limiting magnitude L of a telescope, that is, the magnitude of the dimmest star it can be used to view, is given by

$$L = 9 + 5.1 \log d$$

where d is the diameter (in inches) of the lens.

(a) What is the limiting magnitude of a 3.5 inch telescope?

(b) What diameter is required to view a star of magnitude 14?

63. Product Demand The demand for a new product increases rapidly at first and then levels off. The percent P of actual purchases of this product after it has been on the market t months is

$$P = 90 - 80\left(\frac{3}{4}\right)^t$$

(a) What is the percent of purchases of the product after 5 months?

(b) What is the percent of purchases of the product after 10 months?

(c) What is the maximum percent of purchases of the product?

(d) How many months does it take before 40% of purchases occur?

(e) How many months before 70% of purchases occur?

64. Disseminating Information A survey of a certain community of 10,000 residents shows that the number of residents N who have heard a piece of information after m months is given by the formula

$$m = 55.3 - 6\ln(10{,}000 - N)$$

How many months will it take for half of the citizens to learn about a community program of free blood pressure readings?

65. Salvage Value The number of years n for a piece of machinery to depreciate to a known salvage value can be found using the formula

$$n = \frac{\log s - \log i}{\log(1 - d)}$$

where s is the salvage value of the machinery, i

is its initial value, and d is the annual rate of depreciation.

(a) How many years will it take for a piece of machinery to decline in value from \$90,000 to \$10,000 if the annual rate of depreciation is 0.20 (20%)?

(b) How many years will it take for a piece of machinery to lose half of its value if the annual rate of depreciation is 15%?

66. Funding a College Education A child's grandparents purchase a \$10,000 bond fund that matures in 18 years to be used for her college education. The bond fund pays 4% interest compounded semiannually. How much will the bond fund be worth at maturity?

67. Funding a College Education A child's grandparents wish to purchase a bond fund that matures in 18 years to be used for her college education. The bond fund pays 4% interest compounded semiannually. How much should they purchase so that the bond fund will be worth \$85,000 at maturity?

68. Funding an IRA First Colonial Bankshares Corporation advertised the following IRA investment plans.

(a) Assuming continuous compounding, what was the annual rate of interest they offered?

(b) First Colonial Bankshares claims that \$4000 invested today will have a value of over \$32,000 in 20 years. Use the answer found in part (a) to find the actual value of \$4000 in 20 years. Assume continuous compounding.

Target IRA Plans

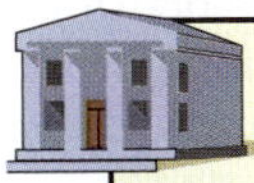

For each \$5000 Maturity Value Desired

Deposit:	At a Term of:
\$620.17	20 Years
\$1045.02	15 Years
\$1760.92	10 Years
\$2967.26	5 Years

69. Loudness of a Garbage Disposal Find the loudness of a garbage disposal unit that operates at an intensity of 10^{-4} watts per square meter. Express your answer in decibels.

70. Comparing Earthquakes On September 9, 1985, the western suburbs of Chicago experienced a mild earthquake that registered 3.0 on the Richter scale. How did this earthquake compare in intensity to the great San Francisco earthquake of 1906, which registered 6.9 on the Richter scale?

71. Estimating the Date a Prehistoric Man Died The bones of a prehistoric man found in the desert of New Mexico contain approximately 5% of the original amount of carbon-14. If the half-life of carbon-14 is 5600 years, approximately how long ago did the man die?

72. Temperature of a Skillet A skillet is removed from an oven whose temperature is 450°F and placed in a room whose temperature is 70°F. After 5 minutes, the temperature of the skillet is 400°F. How long will it be until its temperature is 150°F?

In Problems 73–74, the independent variable is x and the dependent variable is y. For each set of data (a) draw a scatter diagram of the data; (b) use an exponential model, logarithmic model, power model, and linear model to find the "best" model that explains the relationship between the variable.

73.

x	1	4	5	7	10	12
y	1.3084	0.8932	0.8123	0.7737	0.6954	0.6659

74.

x	0.8	1.2	1.7	2.4	2.9	3.4
y	1.1843	2.8634	4.3104	5.6459	6.4683	7.2081

75. AIDS Infections The following data represent the cumulative number of HIV/AIDS cases reported worldwide for the years 1980–1995. Let $x = 1$ represent 1980, $x = 2$ represent 1981, and so on.

(a) Using a graphing utility, draw a scatter diagram of the data using the year as the independent variable and the number of HIV infections as the dependent variable.

(b) Using a graphing utility, determine the logistic function of best fit.

(c) Graph the function found in (b) on the scatter diagram.

(d) Based on the function found in (b), what is the maximum number of HIV infections predicted to be?

(e) Using a graphing utility, determine the exponential function of best fit.

(f) Graph the function found in (e) on the scatter diagram.

(g) Based on the graphs drawn in (c) and (f), which model appears to describe the relation between the year and the number of HIV infections reported?

Year	HIV Infections (Millions)
1980, 1	0.2
1981, 2	0.6
1982, 3	1.1
1983, 4	1.8
1984, 5	2.7
1985, 6	3.9
1986, 7	5.3
1987, 8	6.9
1988, 9	8.7
1989, 10	10.7
1990, 11	13.0
1991, 12	15.5
1992, 13	18.5
1993, 14	21.9
1994, 15	25.9
1995, 16	30.6

Source: Global AIDS Policy Coalition, Harvard School of Public Health, Cambridge, Mass., private communication, January 18, 1996.

76. Economics The following data represents personal consumption expenditures and per capita gross national product figures in the United States from 1984 to 1993.

Year	Personal Per Capita Consumption	Gross National Per Capita Product
1984	11,617	17,659
1985	12,015	18,007
1986	12,336	18,337
1987	12,568	18,712
1988	12,903	19,284
1989	13,029	19,615
1990	13,093	19,670
1991	12,895	19,290
1992	13,081	19,548
1993	13,372	19,897

Source: U.S. Bureau of Economic Analysis

(a) Using a graphing utility, draw a scatter diagram of the data using gross national product as the independent variable and personal consumption as the dependent variable.

(b) Use an exponential model, logarithmic model, power model, and linear model to find the best model to describe the relationship between the variables.

(c) Using the model found in (b), predict personal consumption expenditures if per capita gross national product is $20,000.

77. In a room whose constant temperature is 70°F, will a pizza heated to 450°F ever reach exactly 70°F? Newton's law of cooling seems to say no! What really happens? What assumptions are made in using Newton's law? Write a brief paragraph explaining your conclusions.

CHAPTER 6

Trigonometric Functions

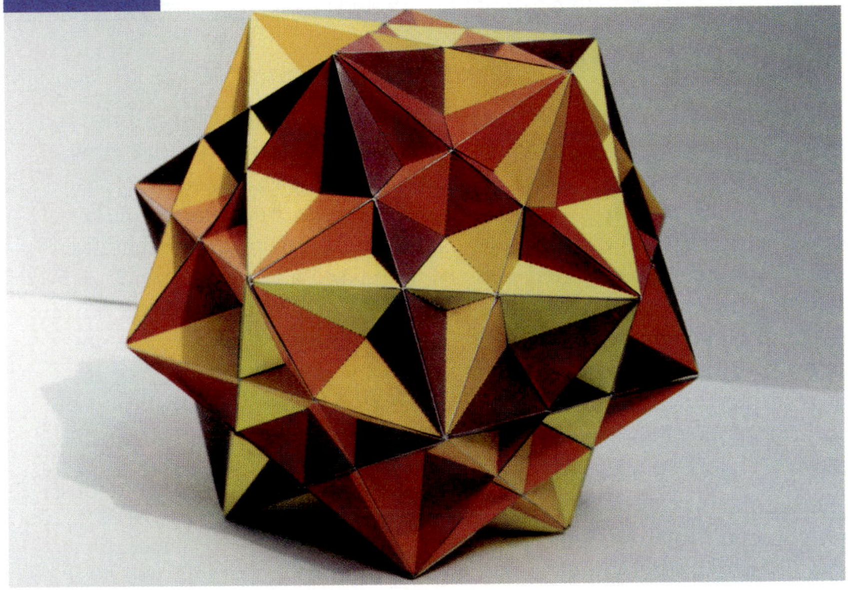

Many of the everyday items around us are based upon mathematics. On the following page, you will take on a commission to create a stained-glass lamp with a mathematical shape. The information you will need to build this lamp can be found on the Sullivan website at:

www.prenhall.com/sullivan

PREPARING FOR THIS CHAPTER

Before getting started on this chapter, review the following concepts:

Pythagorean Theorem *(Appendix, Section 1)*

Unit Circle *(p. 68)*

Functions (Sections 2.1, 2.2, and 2.3)

Inverse Functions (Section 5.1)

OUTLINE

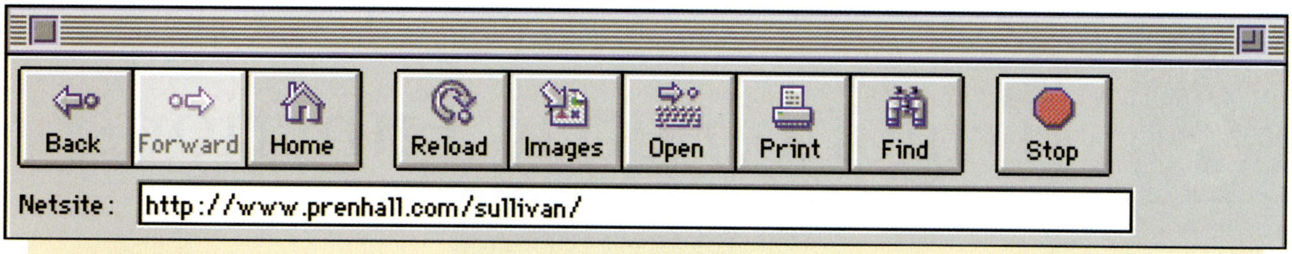

POLYHEDRA STAINED-GLASS LAMPS

Polyhedra, or many sided 3-D objects, have been known and studied for a long time. The Greek philosopher Plato studied the five regular sided polyhedra, now known as *platonic solids*. Many objects found in nature, such as crystals, some types of virus, and buckyballs, have the shape of polyhedra. Ever since *Buckminster Fuller* popularized the geodesic dome there has been a resurgence of interest in polyhedra.

Suppose that you are the owner and operator of a stained-glass business and a customer who lives in a geodesic dome wants to commission you to make her a stained-glass lamp in the shape of a polyhedron. As a master craftsman, building such a lamp will not pose much of a problem, but you have no ready-made designs drawn up for the customer to choose from, so you ask the customer if she would like to choose a pattern from the *World Wide Web Library of Polyhedra*. The virtual reality displays make a big impression on your client. She selects a *5cube* as her choice for a lamp.

1. The stained-glass lamp that you are about to build at first appears to be very complicated, but as you continue to look at the 5cube you begin to see an underlying structure. Can you see the underlying structure? Do you think you could build a 5cube?
2. Since the cost of a stained-glass lamp runs about \$5 per piece of glass, (and each triangle will be a piece), what do you estimate is the cost of the lamp?
3. You will be able to build the lamp by making six square sections. Each of these sections is made up only of triangles. How many different types of triangles are there? How many pieces will you need to cut for each of the different triangles?
4. Next you have the challenge of figuring out how to make the patterns for cutting the glass. You plan to make a *paper model*. You decide to make the lamp with an overall diameter of 20 inches. Can you use your calculator and a formula to determine the size of the cubes?
5. Once you have the side lengths of the cubes that compose the 5cube, what do you need to know in order to calculate the triangles? Can you determine these lengths by computation?
6. If you spend 20 minutes per piece making the lamp, how much would you make per hour if you charge the price that you estimated in question 2?

Trigonometry was developed by Greek astronomers who regarded the sky as the inside of a sphere, so it was natural that triangles on a sphere were investigated early (by Menelaus of Alexandria about A.D. 100) and that triangles in the plane were studied much later. The first book containing a systematic treatment of plane and spherical trigonometry was written by the Persian astronomer Nasîr ed-dîn (about A.D. 1250).

Regiomontanus (1436–1476) is the man most responsible for moving trigonometry from astronomy into mathematics. His work was improved by Copernicus (1473–1543) and Copernicus's student Rhaeticus (1514–1576). Rhaeticus's book was the first to define the six trigonometric functions as ratios of sides of triangles, although he did not give the functions their present names. Credit for this is due to Thomas Fincke (1583), but Fincke's notation was by no means universally accepted at the time. The notation was finally stabilized by the textbooks of Leonhard Euler (1707–1783).

Trigonometry has since evolved from its use by surveyors, navigators, and engineers to present applications involving ocean tides, the rise and fall of food supplies in certain ecologies, brain wave patterns, and many other phenomena.

There are two widely accepted approaches to the development of the trigonometric functions: One uses right triangles; the other uses circles, especially the unit circle. In this book, we develop the trigonometric functions using right triangles. In Section 6.5, we introduce trigonometric functions using the unit circle and show that this approach leads to the definition using right triangles.

6.1 ANGLES AND THEIR MEASURE

1. Convert between Degrees, Minutes, Seconds, and Decimal Forms for Angles
2. Find the Arc Length of a Circle
3. Convert from Degrees to Radians
4. Convert from Radians to Degrees
5. Find the Linear Speed of an Object Traveling in Circular Motion

A **ray,** or **half-line,** is that portion of a line that starts at a point V on the line and extends indefinitely in one direction. The starting point V of a ray is called its **vertex.** See Figure 1.

FIGURE 1

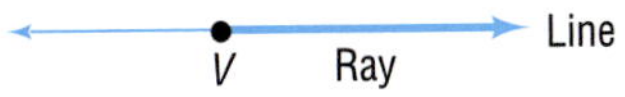

If two rays are drawn with a common vertex, they form an **angle.** We call one of the rays of an angle the **initial side** and the other the **terminal side.** The angle that is formed is identified by showing the direction and amount of rotation from the initial side to the terminal side. If the rotation is in the counterclockwise direction, the angle is **positive;** if the rotation is clockwise, the angle is **negative.** See Figure 2. Lowercase Greek letters, such as α (alpha), β (beta), γ (gamma), θ (theta), and so on, will be used to denote angles. Notice in Figure 2(a) that the angle α is positive because the direction of the rotation from the initial side to the terminal side is counterclockwise. The angle β in Figure 2(b) is negative because the rotation is clockwise. The angle γ in Figure 2(c) is positive. Notice that the angle α in Figure 2(a) and the angle γ have the same initial side and the same terminal side. However, α and γ are unequal because the amount of rotation required to go from the initial side to the terminal side is greater for angle γ than for angle α.

FIGURE 2

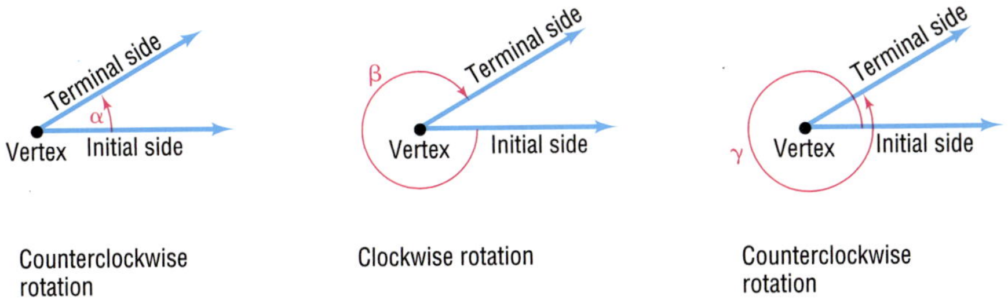

(a) Positive angle **(b)** Negative angle **(c)** Positive angle

An angle θ is said to be in **standard position** if its vertex is at the origin of a rectangular coordinate system and its initial side coincides with the positive x-axis. See Figure 3.

When an angle θ is in standard position, the terminal side either will lie in a quadrant, in which case we say θ **lies in that quadrant,** or it will lie

FIGURE 3

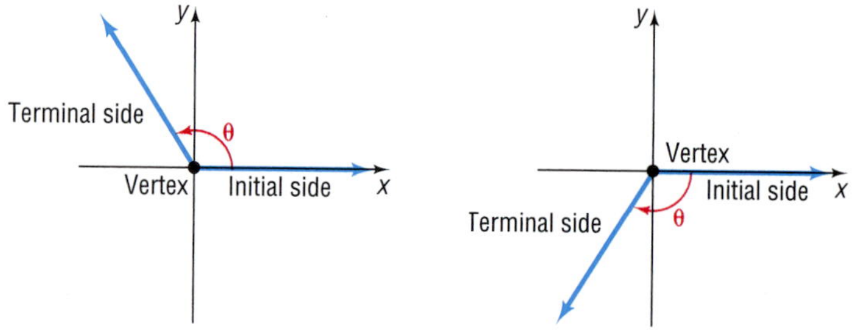

(a) θ in standard position; θ positive

(b) θ in standard position; θ negative

on the x-axis or the y-axis, in which case we say θ is a **quadrantal angle.** For example, the angle θ in Figure 4(a) lies in quadrant II, the angle θ in Figure 4(b) lies in quadrant IV, and the angle θ in Figure 4(c) is a quadrantal angle.

FIGURE 4

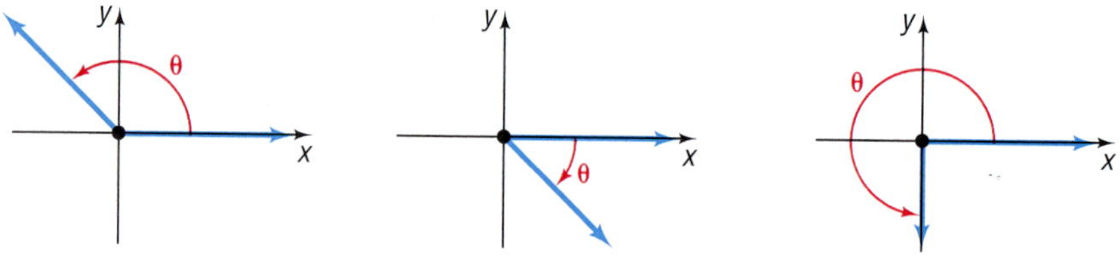

(a) θ lies in quadrant II

(b) θ lies in quadrant IV

(c) θ is a quadrantal angle

We measure angles by determining the amount of rotation needed for the initial side to become coincident with the terminal side. There are two commonly used measures for angles: *degrees* and *radians.*

Degrees

The angle formed by rotating the initial side exactly once in the counterclockwise direction until it coincides with itself (1 revolution) is said to measure 360 degrees, abbreviated 360°. Thus **one degree, 1°,** is $\frac{1}{360}$ revolution. A **right angle** is an angle of 90°, or $\frac{1}{4}$ revolution; a **straight angle** is an angle of 180°, or $\frac{1}{2}$ revolution. See Figure 5. As Figure 5(b) shows, it is customary to indicate a right angle by using the symbol ⦜.

FIGURE 5

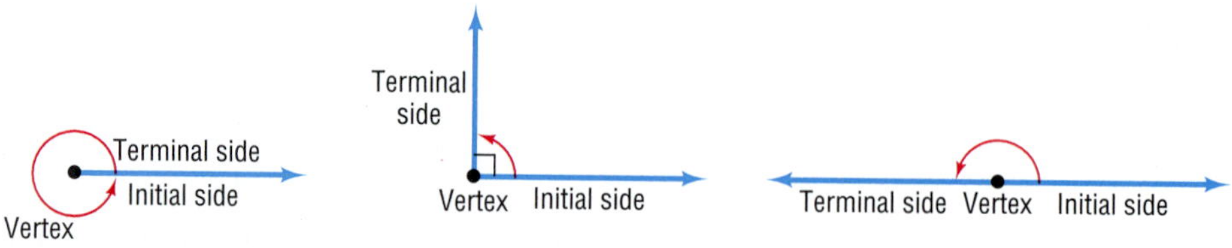

(a) 1 revolution counterclockwise, 360°

(b) $\frac{1}{4}$ revolution counterclockwise, 90°

(c) $\frac{1}{2}$ revolution counterclockwise, 180°

EXAMPLE 1 Drawing an Angle

Draw each angle:

(a) $45°$ (b) $-90°$ (c) $225°$ (d) $405°$

Solution

(a) An angle of $45°$ is $\frac{1}{2}$ of a right angle. See Figure 6.

FIGURE 6

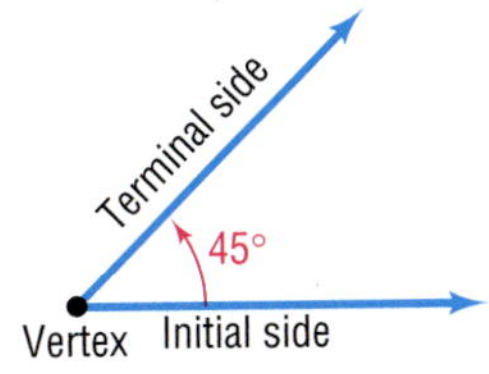

(b) An angle of $-90°$ is $\frac{1}{4}$ revolution in the clockwise direction. See Figure 7.

FIGURE 7

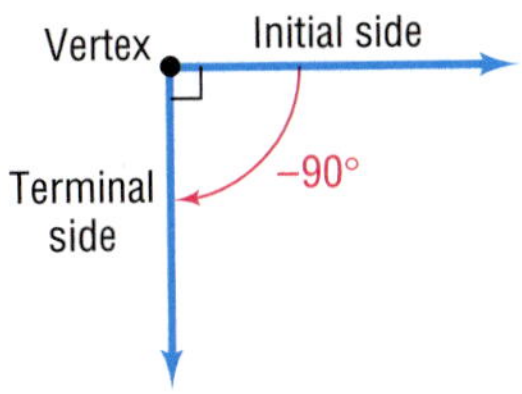

(c) An angle of $225°$ consists of a rotation through $180°$ followed by a rotation through $45°$. See Figure 8.

FIGURE 8

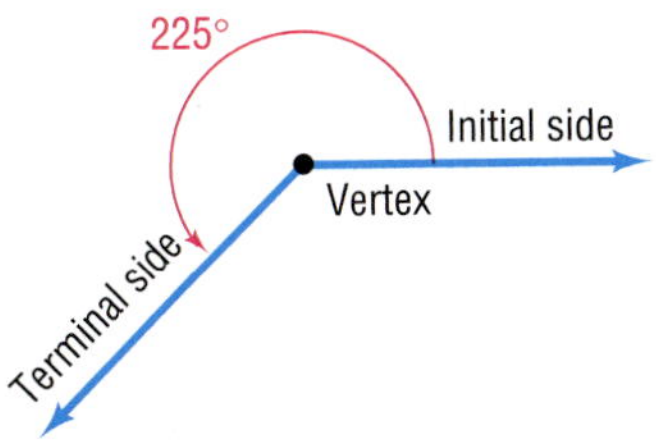

(d) An angle of $405°$ consists of 1 revolution ($360°$) followed by a rotation through $45°$. See Figure 9.

FIGURE 9

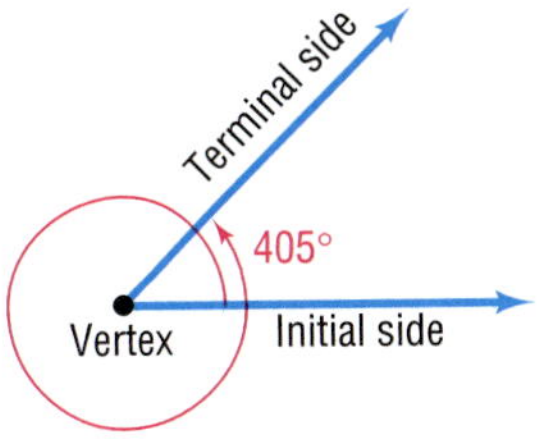

Now work Problem 1.

1 Although subdivisions of a degree may be obtained by using decimals, we also may use the notion of *minutes* and *seconds*. **One minute,** denoted by **1′,** is defined as $\frac{1}{60}$ degree. **One second,** denoted by **1″,** is defined as $\frac{1}{60}$ minute, or equivalently, $\frac{1}{3600}$ degree. An angle of, say, 30 degrees, 40 minutes, 10 seconds is written compactly as $30°40'10''$. To summarize:

$$1 \text{ counterclockwise revolution} = 360° \qquad (1)$$
$$60' = 1° \qquad 60'' = 1'$$

Because calculators use decimals, it is sometimes necessary to convert from the degree, minute, second notation (D°M′S″) to a decimal form, and vice versa. Check your calculator; it should be capable of doing the conversion for you.

Before getting started, though, you must set the mode to degrees, because there are two common ways to measure angles: degree mode and radian mode. (We will define radians shortly.) Usually, a menu is used to change from one mode to another. Check your instruction manual to find out how your particular calculator works.

EXAMPLE 2 Converting between Degrees, Minutes, Seconds, and Decimal Forms

(a) Convert $50°6'21''$ to a decimal in degrees.
(b) Convert $21.256°$ to the $D°M'S''$ form.

Algebraic Solution (a) Because $1' = \frac{1}{60}°$ and $1'' = \frac{1}{60}' = \left(\frac{1}{60} \cdot \frac{1}{60}\right)°$, we convert as follows:

$$\begin{aligned} 50°6'21'' &= \left(50 + 6 \cdot \tfrac{1}{60} + 21 \cdot \tfrac{1}{60} \cdot \tfrac{1}{60}\right)° \\ &\approx (50 + 0.1 + 0.005833)° \\ &= 50.105833° \end{aligned}$$

(b) We start with the decimal part of $21.256°$, that is, $0.256°$:

$$0.256° = (0.256)(1°) \underset{\substack{\uparrow \\ 1° = 60'}}{=} (0.256)(60') = 15.36'$$

Now we work with the decimal part of $15.36'$, that is, $0.36'$:

$$0.36' = (0.36)(1') = (0.36)(60'') = 21.6'' \approx 22''$$

Thus,

$$\begin{aligned} 21.256° &= 21° + 0.256° = 21° + 15.36' = 21° + 15' + 0.36' \\ &= 21° + 15' + 21.6'' \approx 21°15'22'' \end{aligned}$$

Graphing Solution (a) Figure 10 shows the solution using a TI-83 graphing calculator.

FIGURE 10

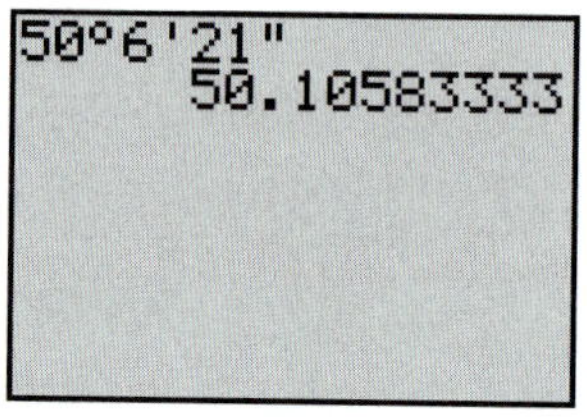

(b) Figure 11 shows the solution using a TI-83 graphing calculator.

FIGURE 11

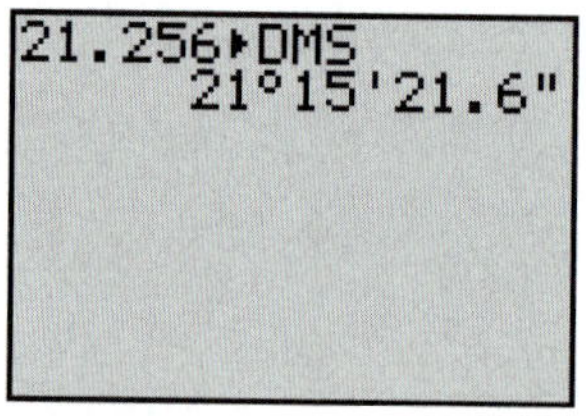

Now work Problems 57 and 63.

In many applications, such as describing the exact location of a star or the precise position of a boat at sea, angles measured in degrees, minutes, and even seconds are used. For calculation purposes, these are transformed to decimal form. In many other applications, especially those in calculus, angles are measured using *radians*.

Radians

Consider a circle of radius r. Construct an angle whose vertex is at the center of this circle, called a **central angle,** and whose rays subtend an arc on the circle whose length equals r. See Figure 12(a). The measure of such an angle is **1 radian.** Thus, for a circle of radius 1, the rays of a central angle with measure 1 radian would subtend an arc of length 1. For a circle of radius 3 the rays of a central angle with measure 1 radian would subtend an arc of length 3. See Figure 12(b).

FIGURE 12

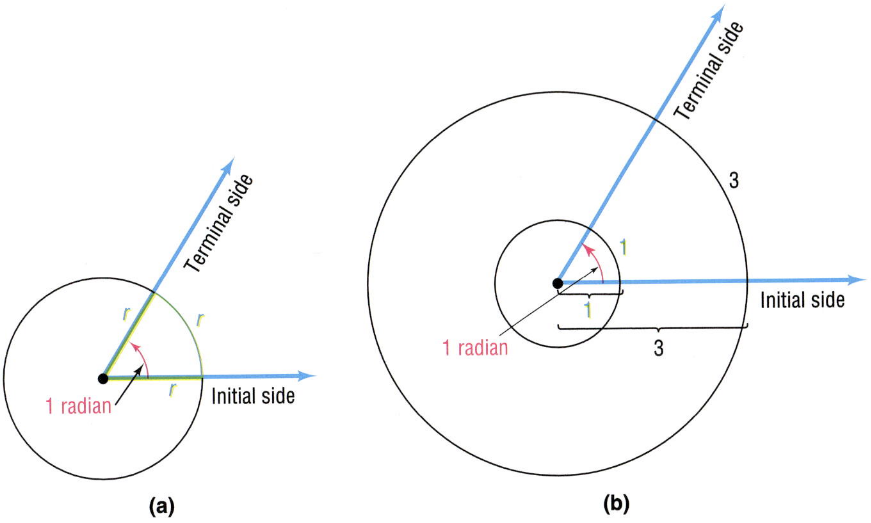

2

Now consider a circle and two central angles, θ and θ_1. Suppose that these central angles subtend arcs of lengths s and s_1, respectively, as shown in Figure 13. From geometry, we know that the ratio of the measures of the angles equals the ratio of the corresponding lengths of the arcs subtended by those angles; that is,

FIGURE 13 $\frac{\theta}{\theta_1} = \frac{s}{s_1}$

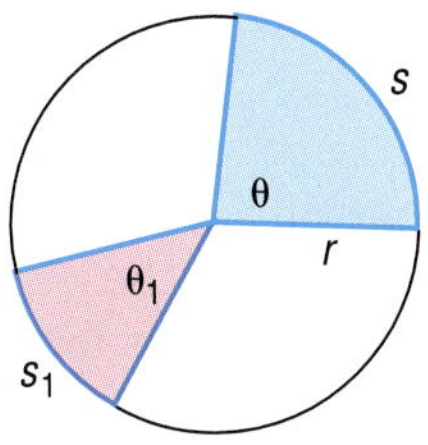

$$\frac{\theta}{\theta_1} = \frac{s}{s_1} \tag{2}$$

Suppose that θ and θ_1 are measured in radians, and suppose that $\theta_1 = 1$ radian. Refer again to Figure 13. Then the amount of arc s_1 subtended by the central angle θ_1 equals the radius r of the circle. Thus, $s_1 = r$, so formula (2) reduces to

$$\frac{\theta}{1} = \frac{s}{r} \quad \text{or} \quad s = r\theta \tag{3}$$

Theorem Arc Length

For a circle of radius r, a central angle of θ radians subtends an arc whose length s is

$$s = r\theta \tag{4}$$

Note: Formulas must be consistent with regard to the units used. In equation (4), we write

$$s = r\theta$$

To see the units, however, we must go back to equation (3) and write

$$\frac{\theta \text{ radians}}{1 \text{ radian}} = \frac{s \text{ length units}}{r \text{ length units}}$$

$$s \text{ length units} = (r \text{ length units})\frac{\theta \text{ radians}}{1 \text{ radian}}$$

Since the radians cancel, we are left with

$$s \text{ length units} = (r \text{ length units})\theta \qquad s = r\theta$$

where θ appears to be "dimensionless" but, in fact, is measured in radians. Thus, in using the formula $s = r\theta$, the dimension of radians for θ is usually omitted, and any convenient unit of length (such as inches or meters) may be used for s and r.

EXAMPLE 3

Finding the Length of Arc of a Circle

Find the length of the arc of a circle of radius 2 meters subtended by a central angle of 0.25 radian.

Solution We use equation (4) with $r = 2$ meters and $\theta = 0.25$. The length s of the arc is

$$s = r\theta = 2(0.25) = 0.5 \text{ meter}$$

Now work Problem 33.

Relationship between Degrees and Radians

Consider a circle of radius r. A central angle of 1 revolution will subtend an arc equal to the circumference of the circle (Figure 14). Because the circumference of a circle equals $2\pi r$, we use $s = 2\pi r$ in equation (4) to find that, for an angle θ of 1 revolution,

FIGURE 14
1 revolution $= 2\pi$ radians

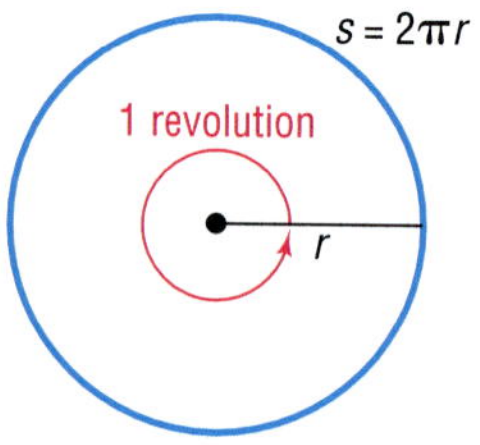

$$s = r\theta$$
$$2\pi r = r\theta$$
$$\theta = 2\pi \text{ radians}$$

Thus,

$$1 \text{ revolution} = 2\pi \text{ radians} \qquad (5)$$

so that

$$360° = 2\pi \text{ radians}$$

or

$$180° = \pi \text{ radians} \qquad (6)$$

Divide both sides of equation (6) by 180. Then

$$1 \text{ degree} = \frac{\pi}{180} \text{ radian}$$

Divide both sides of (6) by π. Then

$$\frac{180}{\pi} \text{ degrees} = 1 \text{ radian}$$

Thus, we have the following two conversion formulas:

$$1 \text{ degree} = \frac{\pi}{180} \text{ radian} \qquad 1 \text{ radian} = \frac{180}{\pi} \text{ degrees} \qquad (7)$$

EXAMPLE 4 Converting from Degrees to Radians

3 Convert each angle in degrees to radians:

(a) $60°$ (b) $150°$ (c) $-45°$ (d) $90°$

Solution

(a) $60° = 60 \cdot 1 \text{ degree} = 60 \cdot \frac{\pi}{180} \text{ radian} = \frac{\pi}{3} \text{ radians}$

(b) $150° = 150 \cdot \frac{\pi}{180} \text{ radian} = \frac{5\pi}{6} \text{ radians}$

(c) $-45° = -45 \cdot \frac{\pi}{180} \text{ radian} = -\frac{\pi}{4} \text{ radian}$

(d) $90° = 90 \cdot \frac{\pi}{180} \text{ radian} = \frac{\pi}{2} \text{ radians}$

Now work Problem 13.

Example 4 illustrates that angles that are fractions of a revolution are expressed in radian measure as fractional multiples of π, rather than as decimals. Thus, a right angle, as in Example 4(d), is left in the form $\pi/2$ radians, which is exact, rather than using the approximation $\pi/2 \approx 3.1416/2 = 1.5708$ radians.

EXAMPLE 5 Converting Radians to Degrees

4 Convert each angle in radians to degrees.

(a) $\frac{\pi}{6}$ radian (b) $\frac{3\pi}{2}$ radians (c) $-\frac{3\pi}{4}$ radians (d) $\frac{7\pi}{3}$ radians

Solution (a) $\frac{\pi}{6}\text{ radian} = \frac{\pi}{6} \cdot 1 \text{ radian} = \frac{\pi}{6} \cdot \frac{180}{\pi} \text{ degrees} = 30°$

(b) $\frac{3\pi}{2}\text{ radians} = \frac{3\pi}{2} \cdot \frac{180}{\pi} \text{ degrees} = 270°$

(c) $-\frac{3\pi}{4}\text{ radians} = -\frac{3\pi}{4} \cdot \frac{180}{\pi} \text{ degrees} = -135°$

(d) $\frac{7\pi}{3}\text{ radians} = \frac{7\pi}{3} \cdot \frac{180}{\pi} \text{ degrees} = 420°$

Now work Problem 23.

Table 1 lists the degree and radian measures of some commonly encountered angles. You should learn to feel equally comfortable using degree or radian measure for these angles.

TABLE 1

Degrees	0°	30°	45°	60°	90°	120°	135°	150°	180°
Radians	0	$\frac{\pi}{6}$	$\frac{\pi}{4}$	$\frac{\pi}{3}$	$\frac{\pi}{2}$	$\frac{2\pi}{3}$	$\frac{3\pi}{4}$	$\frac{5\pi}{6}$	π
Degrees	210°	225°	240°	270°	300°	315°	330°	360°	
Radians	$\frac{7\pi}{6}$	$\frac{5\pi}{4}$	$\frac{4\pi}{3}$	$\frac{3\pi}{2}$	$\frac{5\pi}{3}$	$\frac{7\pi}{4}$	$\frac{11\pi}{6}$	2π	

EXAMPLE 6

Finding the Length of Arc of a Circle

Find the length of the arc of a circle of radius $r = 3$ feet subtended by a central angle of 30°.

Solution We use equation (4), but first we must convert the central angle of 30° to radians. Since $30° = \pi/6$ radian, we use $\theta = \pi/6$ and $r = 3$ feet in equation (4). The length of the arc is

$$s = r\theta = 3 \cdot \frac{\pi}{6} = \frac{\pi}{2} \approx \frac{3.14}{2} = 1.57 \text{ feet}$$

When an angle is measured in degrees, the degree symbol always will be shown. However, when an angle is measured in radians, we will follow the usual practice and omit the word *radians*. Thus, if the measure of an angle is given as $\pi/6$, it is understood to mean $\pi/6$ radian.

Circular Motion

5 We have already defined the average speed of an object as the distance traveled divided by the elapsed time. Suppose that an object moves along a circle of radius r at a constant speed. If s is the distance traveled in time t along this circle, then the **linear speed** v of the object is defined as

$$v = \frac{s}{t} \tag{8}$$

FIGURE 15

$v = \frac{s}{t}, \omega = \frac{\theta}{t}$

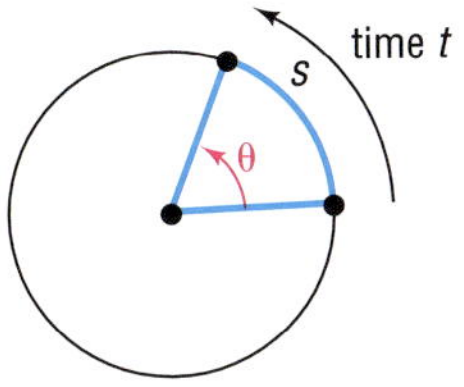

As this object travels along the circle, suppose that θ (measured in radians) is the central angle swept out in time t (see Figure 15). Then the **angular speed** ω (the Greek letter omega) of this object is the angle (measured in radians) swept out divided by the elapsed time; that is,

$$\omega = \frac{\theta}{t} \tag{9}$$

Angular speed is the way the speed of a phonograph record is described. For example, a 45 rpm (revolutions per minute) record is one that rotates at an angular speed of

$$\frac{45 \text{ revolutions}}{\text{Minute}} = \frac{45 \text{ revolutions}}{\text{Minute}} \cdot \frac{2\pi \text{ radians}}{\text{Revolution}} = \frac{90\pi \text{ radians}}{\text{Minute}}$$

There is an important relationship between linear speed and angular speed. In the formula $s = r\theta$, divide each side by t:

$$\frac{s}{t} = r\frac{\theta}{t}$$

Then, using equations (8) and (9), we obtain

$$v = r\omega \tag{10}$$

When using equation (10), remember that $v = s/t$ (the linear speed) has the dimensions of length per unit of time (such as feet per second or miles per hour), r (the radius of the circular motion) has the same length dimension as s, and ω (the angular speed) has the dimensions of radians per unit of time. As noted earlier, we leave the radian dimension off the numerical value of the angular speed ω so that both sides of the equation will be dimensionally consistent (with "length per unit of time"). If the angular speed is given in terms of *revolutions* per unit of time (as is often the case), be sure to convert it to *radians* per unit of time before attempting to use equation (10).

EXAMPLE 7 Finding Linear Speed

Find the linear speed of a $33\frac{1}{3}$ rpm record at the point where the needle is 3 inches from the spindle (center of the record).

FIGURE 16

Solution Look at Figure 16. The point P is traveling along a circle of radius $r = 3$ inches. The angular speed ω of the record is

$$\omega = \frac{33\frac{1}{3} \text{ revolutions}}{\text{Minute}} = \frac{100 \text{ revolutions}}{3 \text{ minutes}} \cdot \frac{2\pi \text{ radians}}{\text{Revolution}}$$
$$= \frac{200\pi \text{ radians}}{3 \text{ minutes}}$$

From equation (10), the linear speed v of the point P is

$$v = r\omega = 3 \text{ inches} \cdot \frac{200\pi \text{ radians}}{3 \text{ minutes}} = \frac{200\pi \text{ inches}}{\text{Minute}} \approx \frac{628 \text{ inches}}{\text{Minute}}$$

Now work Problem 71.

6.1 EXERCISES

In Problems 1–12, draw each angle.

1. $30°$ **2.** $60°$ **3.** $135°$ **4.** $-120°$ **5.** $450°$ **6.** $540°$

7. $3\pi/4$ **8.** $4\pi/3$ **9.** $-\pi/6$ **10.** $-2\pi/3$ **11.** $16\pi/3$ **12.** $21\pi/4$

In Problems 13–22, convert each angle in degrees to radians. Express your answer as a multiple of π.

13. $30°$ **14.** $120°$ **15.** $240°$ **16.** $330°$ **17.** $-60°$

18. $-30°$ **19.** $180°$ **20.** $270°$ **21.** $135°$ **22.** $-225°$

In Problems 23–32, convert each angle in radians to degrees.

23. $\pi/3$ **24.** $5\pi/6$ **25.** $-5\pi/4$ **26.** $-2\pi/3$ **27.** $\pi/2$

28. 4π **29.** $\pi/12$ **30.** $5\pi/12$ **31.** $2\pi/3$ **32.** $5\pi/4$

In Problems 33–40, s denotes the length of arc of a circle of radius r subtended by the central angle θ. Find the missing quantity.

33. $r = 10$ meters, $\theta = \frac{1}{2}$ radian, $s = ?$

34. $r = 6$ feet, $\theta = 2$ radians, $s = ?$

35. $\theta = \frac{1}{3}$ radian, $s = 2$ feet, $r = ?$

36. $\theta = \frac{1}{4}$ radian, $s = 6$ centimeters, $r = ?$

37. $r = 5$ miles, $s = 3$ miles, $\theta = ?$

38. $r = 6$ meters, $s = 8$ meters, $\theta = ?$

39. $r = 2$ inches, $\theta = 30°$, $s = ?$

40. $r = 3$ meters, $\theta = 120°$, $s = ?$

In Problems 41–48, convert each angle in degrees to radians. Express your answer in decimal form, rounded to two decimal places.

41. $17°$ **42.** $73°$ **43.** $-40°$ **44.** $-51°$ **45.** $125°$ **46.** $200°$ **47.** $340°$ **48.** $350°$

In Problems 49–56, convert each angle in radians to degrees. Express your answer in decimal form, rounded to two decimal places.

49. 3.14 **50.** π **51.** 10.25 **52.** 0.75 **53.** 2 **54.** 3 **55.** 6.32 **56.** $\sqrt{2}$

In Problems 57–62, convert each angle to a decimal in degrees. Round off your answer to two decimal places.

57. $40°10'25''$ **58.** $61°42'21''$ **59.** $1°2'3''$ **60.** $73°40'40''$ **61.** $9°9'9''$ **62.** $98°22'45''$

In Problems 63–68, convert each angle to D°M′S″ form. Round off your answer to the nearest second.

63. $40.32°$ **64.** $61.24°$ **65.** $18.255°$ **66.** $29.411°$ **67.** $19.99°$ **68.** $44.01°$

69. Minute Hand of a Clock The minute hand of a clock is 6 inches long. How far does the tip of the minute hand move in 15 minutes? How far does it move in 25 minutes?

70. Movement of a Pendulum A pendulum swings through an angle of $20°$ each second. If the pendulum is 40 inches long, how far does its tip move each second?

71. An object is traveling around a circle with a radius of 5 centimeters. If in 20 seconds a central angle of $\frac{1}{3}$ radian is swept out, what is the angular speed of the object? What is its linear speed?

72. An object is traveling around a circle with a radius of 2 meters. If in 20 seconds the object travels 5 meters, what is its angular speed? What is its linear speed?

73. Bicycle Wheels The diameter of each wheel of a bicycle is 26 inches. If you are traveling at a speed of 35 miles per hour on this bicycle, through how many revolutions per minute are the wheels turning?

74. Car Wheels The radius of each wheel of a car is 15″. If the wheels are turning at the rate of 3 revolutions per second, how fast is the car moving? Express your answer in inches per second and in miles per hour.

75. Windshield Wipers The windshield wiper of a car is 18 inches long. How many inches will the tip of the wiper trace out in $\frac{1}{3}$ revolution?

76. Windshield Wipers The windshield wiper of a car is 18 inches long. If it takes 1 second to trace out $\frac{1}{3}$ revolution, how fast is the tip of the wiper moving?

77. Speed of the Moon The mean distance of the Moon from Earth is 2.39×10^5 miles. Assuming that the orbit of the Moon around Earth is circular and that 1 revolution takes 27.3 days, find the linear speed of the Moon. Express your answer in miles per hour.

78. Speed of the Earth The mean distance of Earth from the Sun is 9.29×10^7 miles. Assuming that the orbit of Earth around the Sun is circular and that 1 revolution takes 365 days, find the linear speed of Earth. Express your answer in miles per hour.

79. Pulleys Two pulleys, one with radius 2 inches and the other with radius 8 inches, are connected by a belt. (See the figure.) If the 2-inch pulley is caused to rotate at 3 revolutions per minute, determine the revolutions per minute of the 8-inch pulley.

[**Hint:** The linear speeds of the pulley, that is, the speed of the belt, are the same.]

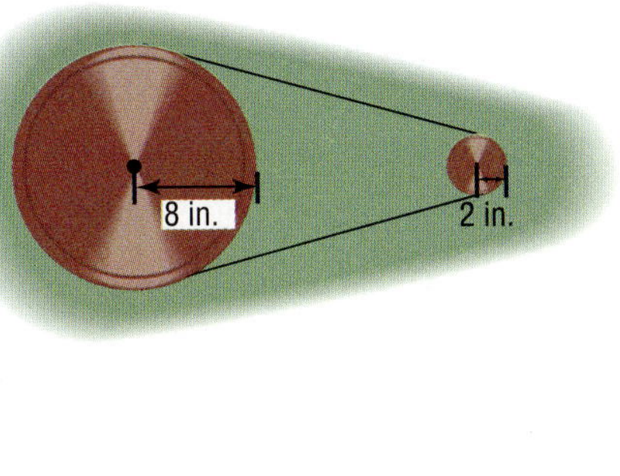

80. Pulleys Two pulleys, one with radius r_1 and the other with radius r_2, are connected by a belt. The pulley with radius r_1 rotates at ω_1 revolutions per minute, whereas the pulley with radius r_2 rotates at ω_2 revolutions per minute. Show that $r_1/r_2 = \omega_2/\omega_1$.

81. Computing the Speed of River Current To approximate the speed of the current of a river, a circular paddle wheel with radius 4 feet is lowered into the water. If the current causes the wheel to rotate at a speed of 10 revolutions per minute, what is the speed of the current? Express your answer in miles per hour.

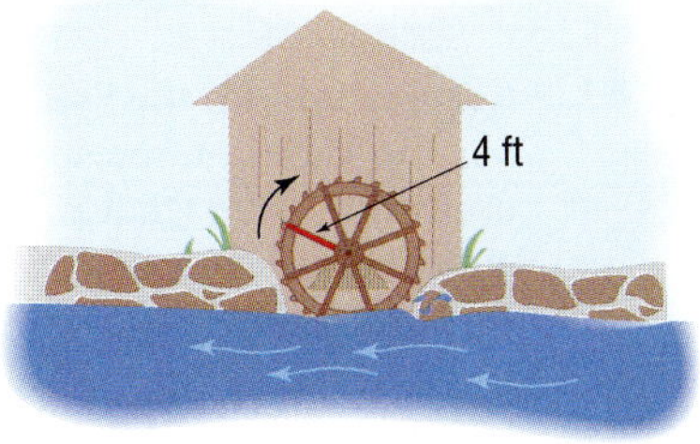

82. Spin Balancing Tires A spin balancer rotates the wheel of a car at 480 revolutions per minute. If the diameter of the wheel is 26 inches, what road speed is being tested? Express your answer in miles per hour. At how many revolutions per minute should the balancer be set to test a road speed of 80 miles per hour?

83. The Cable Cars of San Francisco At the Cable Car Museum you can see the four cable lines that are used to pull cable cars up and down the hills of San Francisco. Each cable travels at a speed of 9.55 mi/hr, caused by a rotating wheel whose diameter is 8.5 feet. How fast is the wheel rotating? Express your answer in revolutions per minute.

84. Difference in Time of Sun Rise Naples, Florida, is approximately 90 miles due west of Ft. Lauderdale. How much sooner would a person in Ft. Lauderdale first see the rising Sun than a person in Naples?

[**Hint:** Consult the figure (p. 458). When a person at Q sees the first rays of the Sun, a person at P is still in the dark. The person at P sees the first rays after Earth has rotated so that P is at the location Q.]

Now use the fact that in 24 hours a length of arc of $2\pi(3960)$ miles is subtended.]

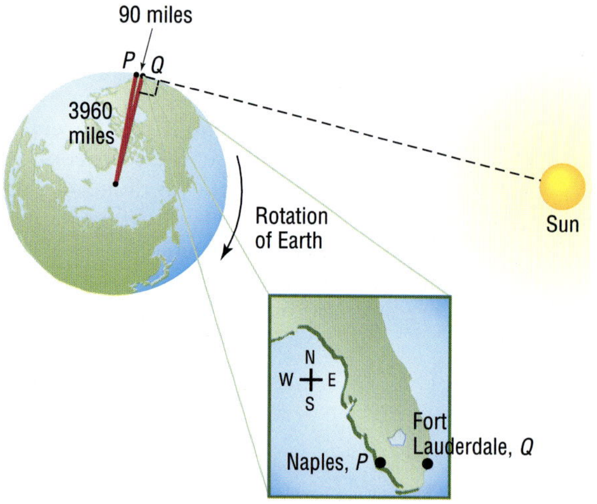

85. Keeping up with the Sun How fast would you have to travel on the surface of Earth to keep up with the Sun (that is, so that the Sun would appear to remain in the same position in the sky)?

86. Nautical Miles A **nautical mile** equals the length of arc subtended by a central angle of 1 minute on a great circle* on the surface of Earth. (See the figure.) If the radius of Earth is taken as 3960 miles, express 1 nautical mile in terms of ordinary, or **statute,** miles (5280 feet).

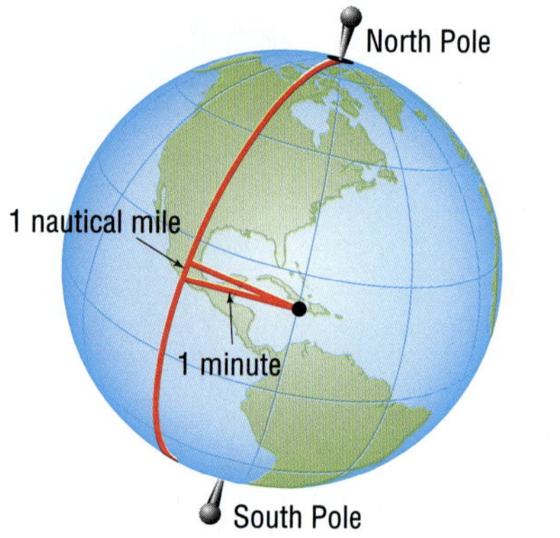

87. Do you prefer to measure angles using degrees or radians? Provide justification and a rationale for your choice.

88. Discuss why ships and airplanes use nautical miles to measure distance. Explain the difference between a nautical mile and a statute mile.

89. Investigate the way speed bicycles work. In particular, explain the differences and similarities between 10-speed and 18-speed derailleurs. Be sure to include a discussion of linear speed and angular speed.

6.2 RIGHT TRIANGLE TRIGONOMETRY

1. Find the Value of Trigonometric Functions of Acute Angles
2. Find the Value of Trigonometric Functions Utilizing Fundamental Identities
3. Use the Complementary Angle Theorem

A triangle in which one angle is a right angle (90°) is called a **right triangle.** Recall that the side opposite the right angle is called the **hypotenuse,** and the remaining two sides are called the **legs** of the triangle. In Figure 17 we have labeled the hypotenuse as c, to indicate its length is c units, and, in a like manner, we have labeled the legs as a and b. Because the triangle is a right triangle, the Pythagorean Theorem tells us that

FIGURE 17

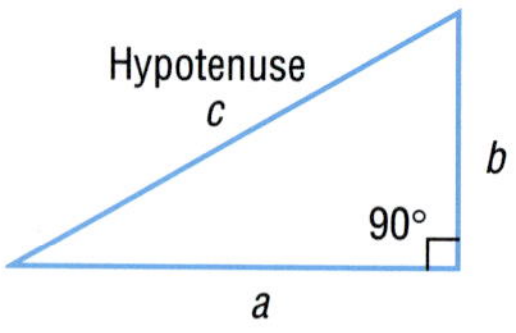

$$c^2 = a^2 + b^2$$

Now, suppose that θ is an **acute angle;** that is, $0° < \theta < 90°$ (if θ is measured in degrees) and $0 < \theta < \pi/2$ (if θ is measured in radians). See Figure 18(a). Using this acute angle θ, we can form a right triangle, like the one il-

*Any circle drawn on the surface of Earth that divides Earth into two equal hemispheres.

lustrated in Figure 18(b), with hypotenuse of length c and legs of lengths a and b. Using the three sides of this triangle, we can form exactly six ratios:

$$\frac{b}{c}, \frac{a}{c}, \frac{b}{a}, \frac{c}{b}, \frac{c}{a}, \frac{a}{b}$$

FIGURE 18

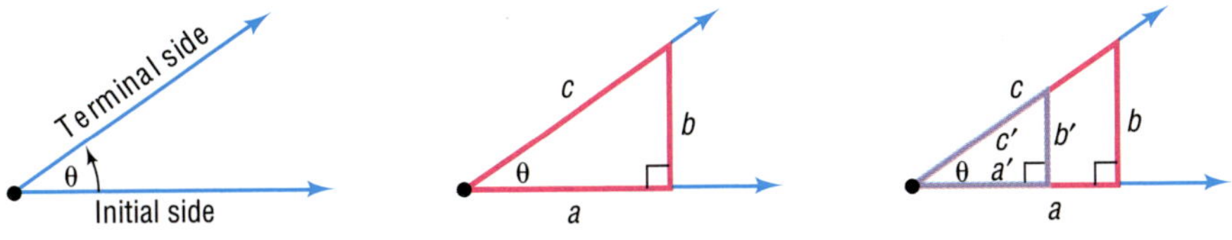

(a) Acute angle **(b)** Right triangle **(c)** Similar triangles

In fact, these ratios depend only on the size of the angle θ and not on the triangle formed. To see why, look at Figure 18(c). Any two right triangles formed using the angle θ will be similar and, hence, corresponding ratios will be equal. As a result,

$$\frac{b}{c} = \frac{b'}{c'} \quad \frac{a}{c} = \frac{a'}{c'} \quad \frac{b}{a} = \frac{b'}{a'} \quad \frac{c}{b} = \frac{c'}{b'} \quad \frac{c}{a} = \frac{c'}{a'} \quad \frac{a}{b} = \frac{a'}{b'}$$

Because the ratios depend only on the angle θ and not on the triangle itself, we give each ratio a name that involves θ: sine of θ, cosine of θ, tangent of θ, cosecant of θ, secant of θ, and cotangent of θ.

The six ratios of a right triangle are called **trigonometric functions of acute angles** and are defined as follows:

Function Name	Abbreviation	Value
sine of θ	$\sin\theta$	$\frac{b}{c}$
cosine of θ	$\cos\theta$	$\frac{a}{c}$
tangent of θ	$\tan\theta$	$\frac{b}{a}$
cosecant of θ	$\csc\theta$	$\frac{c}{b}$
secant of θ	$\sec\theta$	$\frac{c}{a}$
cotangent of θ	$\cot\theta$	$\frac{a}{b}$

FIGURE 19

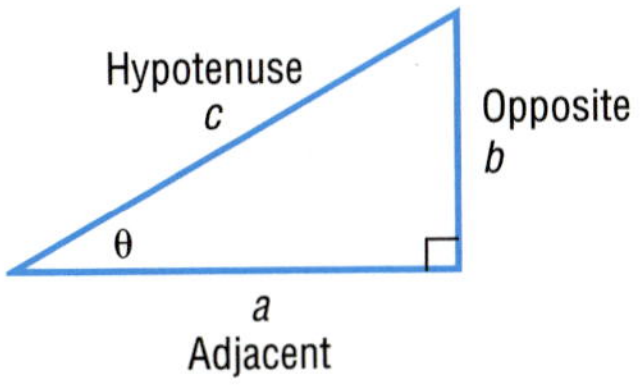

As an aid to remembering these definitions, it may be helpful to refer to the lengths of the sides of the triangle by the names *hypotenuse (c)*, *opposite (b)*, and *adjacent (a)*. See Figure 19. In terms of these names, we have the following ratios:

$$\sin\theta = \frac{\text{Opposite}}{\text{Hypotenuse}} = \frac{b}{c} \qquad \cos\theta = \frac{\text{Adjacent}}{\text{Hypotenuse}} = \frac{a}{c}$$

$$\tan\theta = \frac{\text{Opposite}}{\text{Adjacent}} = \frac{b}{a} \qquad \csc\theta = \frac{\text{Hypotenuse}}{\text{Opposite}} = \frac{c}{b} \qquad (1)$$

$$\sec\theta = \frac{\text{Hypotenuse}}{\text{Adjacent}} = \frac{c}{a} \qquad \cot\theta = \frac{\text{Adjacent}}{\text{Opposite}} = \frac{a}{b}$$

Notice that each of the trigonometric functions of the acute angle θ is positive.

EXAMPLE 1

Finding the Value of Trigonometric Functions

FIGURE 20

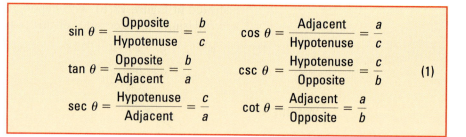

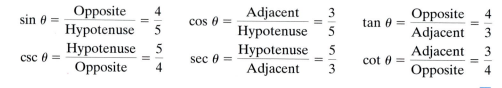

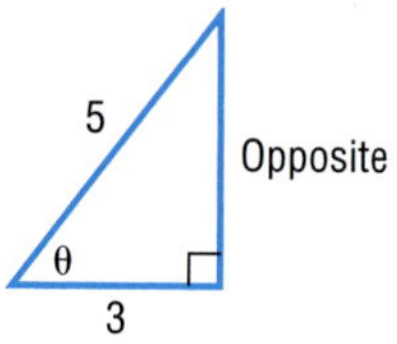

1 Find the value of each of the six trigonometric functions of the angle θ in Figure 20.

Solution We see in Figure 20 that the two given sides of the triangle are

$$c = \text{Hypotenuse} = 5 \qquad a = \text{Adjacent} = 3$$

To find the length of the opposite side, we use the Pythagorean Theorem:

$$\begin{aligned}(\text{Adjacent})^2 + (\text{Opposite})^2 &= (\text{Hypotenuse})^2\\ 3^2 + (\text{Opposite})^2 &= 5^2\\ (\text{Opposite})^2 &= 25 - 9 = 16\\ \text{Opposite} &= 4\end{aligned}$$

Now that we know the lengths of the three sides, we use the ratios in (1) to find the value of each of the six trigonometric functions:

$$\sin\theta = \frac{\text{Opposite}}{\text{Hypotenuse}} = \frac{4}{5} \qquad \cos\theta = \frac{\text{Adjacent}}{\text{Hypotenuse}} = \frac{3}{5} \qquad \tan\theta = \frac{\text{Opposite}}{\text{Adjacent}} = \frac{4}{3}$$

$$\csc\theta = \frac{\text{Hypotenuse}}{\text{Opposite}} = \frac{5}{4} \qquad \sec\theta = \frac{\text{Hypotenuse}}{\text{Adjacent}} = \frac{5}{3} \qquad \cot\theta = \frac{\text{Adjacent}}{\text{Opposite}} = \frac{3}{4}$$

Now work Problem 1.

Fundamental Identities

You should have observed some relationships that exist among the six trigonometric functions of acute angles. For example, the **reciprocal identities** are

Reciprocal Identities

$$\csc\theta = \frac{1}{\sin\theta} \qquad \sec\theta = \frac{1}{\cos\theta} \qquad \cot\theta = \frac{1}{\tan\theta} \qquad (2)$$

Two other fundamental identities that are easy to see are the **quotient identities:**

Quotient Identities

$$\tan\theta = \frac{\sin\theta}{\cos\theta} \qquad \cot\theta = \frac{\cos\theta}{\sin\theta} \tag{3}$$

Seeing the Concept To see the identity $\tan\theta = (\sin\theta)/(\cos\theta)$, graph $Y_1 = \tan x$ and $Y_2 = (\sin x)/(\cos x)$ on the same screen.

2 If $\sin\theta$ and $\cos\theta$ are known, formulas (2) and (3) make it easy to find the values of the remaining trigonometric functions.

EXAMPLE 2 Finding the Value of the Remaining Trigonometric Functions, Given $\sin\theta$ and $\cos\theta$

Given $\sin\theta = 1/\sqrt{5}$ and $\cos\theta = 2/\sqrt{5}$, find the value of each of the four remaining trigonometric functions of θ.

Solution Based on formula (3), we have

$$\tan\theta = \frac{\sin\theta}{\cos\theta} = \frac{1/\sqrt{5}}{2/\sqrt{5}} = \frac{1}{2}$$

Then we use the reciprocal identities from formula (2) to get

$$\csc\theta = \frac{1}{\sin\theta} = \frac{1}{1/\sqrt{5}} = \sqrt{5} \qquad \sec\theta = \frac{1}{\cos\theta} = \frac{1}{2/\sqrt{5}} = \frac{\sqrt{5}}{2} \qquad \cot\theta = \frac{1}{\tan\theta} = \frac{1}{\frac{1}{2}} = 2$$

FIGURE 21
$a^2 + b^2 = c^2$

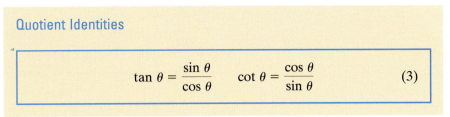

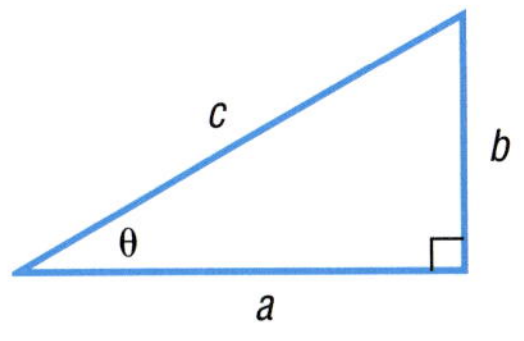

Refer now to the triangle in Figure 21. The Pythagorean Theorem states that

$$b^2 + a^2 = c^2$$

Dividing each side by c^2, we get

$$\frac{b^2}{c^2} + \frac{a^2}{c^2} = 1 \quad \text{or} \quad \left(\frac{b}{c}\right)^2 + \left(\frac{a}{c}\right)^2 = 1$$

In terms of trigonometric functions of the angle θ, this equation states that

$$(\sin\theta)^2 + (\cos\theta)^2 = 1 \tag{4}$$

Equation (4) is, in fact, an identity, since the equation is true for any acute angle θ.

It is customary to write $\sin^2\theta$ instead of $(\sin\theta)^2$, $\cos^2\theta$ instead of $(\cos\theta)^2$, and so on. With this notation, we can rewrite equation (4) as

$$\sin^2\theta + \cos^2\theta = 1 \tag{5}$$

Another identity can be obtained from equation (5) by dividing each side by $\cos^2 \theta$:

$$\frac{\sin^2 \theta}{\cos^2 \theta} + 1 = \frac{1}{\cos^2 \theta}$$

Now use formulas (2) and (3) to get

$$\tan^2 \theta + 1 = \sec^2 \theta \qquad (6)$$

Similarly, by dividing each side of equation (5) by $\sin^2 \theta$, we get

$$1 + \cot^2 \theta = \csc^2 \theta \qquad (7)$$

Collectively, the identities in equations (5), (6), and (7) are referred to as the **Pythagorean identities.**

Let's pause here to summarize the fundamental identities.

Fundamental Identities

$$\tan \theta = \frac{\sin \theta}{\cos \theta} \qquad \cot \theta = \frac{\cos \theta}{\sin \theta}$$

$$\cot \theta = \frac{1}{\tan \theta} \qquad \sec \theta = \frac{1}{\cos \theta} \qquad \csc \theta = \frac{1}{\sin \theta}$$

$$\sin^2 \theta + \cos^2 \theta = 1 \qquad \tan^2 \theta + 1 = \sec^2 \theta \qquad 1 + \cot^2 \theta = \csc^2 \theta$$

Once the value of one trigonometric function is known, it is possible to find the value of each of the remaining five trigonometric functions.

EXAMPLE 3

Finding the Value of the Remaining Trigonometric Functions: Given $\sin \theta$, θ Acute

Given that $\sin \theta = \frac{1}{3}$ and θ is an acute angle, find the exact value of each of the remaining five trigonometric functions of θ.

Solution We solve this problem in two ways: The first way uses the definition of the trigonometric functions; the second method uses the fundamental identities.

Solution 1 We draw a right triangle with acute angle θ, opposite side of length 1, and hypotenuse of length 3 (because $\sin \theta = \frac{1}{3} = \frac{b}{c}$). See Figure 22. The adjacent side a can be found by using the Pythagorean Theorem:

FIGURE 22

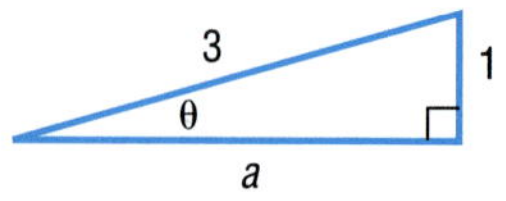

$$a^2 + 1 = 9$$
$$a^2 = 8$$
$$a = 2\sqrt{2}$$

Now the definitions given in equation (1) can be used to find the value of each of the remaining five trigonometric functions. (Refer back to the method used in Example 1.)

$$\cos\theta = \frac{2\sqrt{2}}{3} \qquad \tan\theta = \frac{1}{2\sqrt{2}} = \frac{\sqrt{2}}{4}$$

$$\csc\theta = \frac{3}{1} = 3 \qquad \sec\theta = \frac{3}{2\sqrt{2}} = \frac{3\sqrt{2}}{4} \qquad \cot\theta = \frac{2\sqrt{2}}{1} = 2\sqrt{2}$$

Solution 2 We begin by seeking $\cos\theta$, which can be found by using equation (5):

$$\begin{aligned} \sin^2\theta + \cos^2\theta &= 1 \\ \tfrac{1}{9} + \cos^2\theta &= 1 \\ \cos^2\theta &= 1 - \tfrac{1}{9} = \tfrac{8}{9} \end{aligned}$$

Because $\cos\theta > 0$ for an acute angle θ, we have

$$\cos\theta = \sqrt{\frac{8}{9}} = \frac{2\sqrt{2}}{3}$$

Now we know that $\sin\theta = \frac{1}{3}$ and $\cos\theta = 2\sqrt{2}/3$, so we can proceed as we did in Example 2:

$$\tan\theta = \frac{\sin\theta}{\cos\theta} = \frac{\frac{1}{3}}{2\sqrt{2}/3} = \frac{1}{2\sqrt{2}} = \frac{\sqrt{2}}{4} \qquad \cot\theta = \frac{1}{\tan\theta} = \frac{1}{1/2\sqrt{2}} = 2\sqrt{2}$$

$$\sec\theta = \frac{1}{\cos\theta} = \frac{1}{2\sqrt{2}/3} = \frac{3}{2\sqrt{2}} = \frac{3\sqrt{2}}{4} \qquad \csc\theta = \frac{1}{\sin\theta} = \frac{1}{\frac{1}{3}} = 3$$

Finding the Values of the Trigonometric Functions When One is Known

Given the value of one trigonometric function, the exact value of each of the remaining five trigonometric functions can be found in either of two ways:

Method 1

STEP 1: Draw a right triangle showing the angle θ.

STEP 2: Two of the sides can then be assigned values based on the given trigonometric function.

STEP 3: Find the length of the third side by using the Pythagorean Theorem.

STEP 4: Use the definitions in equation (1) to find the value of each of the remaining trigonometric functions.

Method 2

Use appropriately selected identities to find the value of each of the remaining trigonometric functions.

Now work Problem 15.

Complementary Angles; Cofunctions

FIGURE 23

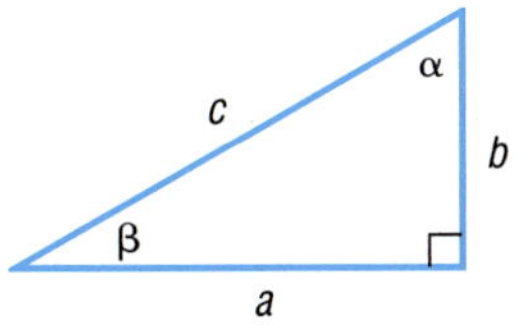

3 Two acute angles are called **complementary** if their sum is a right angle. Because the sum of the angles of any triangle is 180°, it follows that, for a right triangle, the two acute angles are complementary.

Refer now to Figure 23; we have labeled the angle opposite side b as β and the angle opposite side a as α. Notice that side a is adjacent to angle β

and is opposite angle α. Similarly, side b is opposite angle β and is adjacent to angle α. As a result,

$$\begin{aligned} &\sin\beta = \frac{b}{c} = \cos\alpha \qquad \cos\beta = \frac{a}{c} = \sin\alpha \qquad \tan\beta = \frac{b}{a} = \cot\alpha \\ &\csc\beta = \frac{c}{b} = \sec\alpha \qquad \sec\beta = \frac{c}{a} = \csc\alpha \qquad \cot\beta = \frac{a}{b} = \tan\alpha \end{aligned} \tag{8}$$

Because of these relationships, the functions sine and cosine, tangent and cotangent, and secant and cosecant are called **cofunctions** of each other. The identities (8) may be expressed in words as follows:

Theorem Complementary Angle Theorem

Cofunctions of complementary angles are equal.

Here are examples of this theorem.

Complementary angles: $\sin 30° = \cos 60°$ — Cofunctions

Complementary angles: $\tan 40° = \cot 50°$ — Cofunctions

Complementary angles: $\sec 80° = \csc 10°$ — Cofunctions

If an angle θ is measured in degrees, we will use the degree symbol when writing a trigonometric function of θ, as, for example, in $\sin 30°$ and $\tan 45°$. If an angle θ is measured in radians, then no symbol is used when writing a trigonometric function of θ, as, for example, in $\cos \pi$ and $\sec \pi/3$.

If θ is an acute angle measured in degrees, the angle $90° - \theta$ (or $\pi/2 - \theta$, if θ is in radians) is the angle complementary to θ. Table 2 restates the preceding theorem on cofunctions.

TABLE 2

θ (Degrees)	θ (Radians)
$\sin\theta = \cos(90° - \theta)$	$\sin\theta = \cos(\pi/2 - \theta)$
$\cos\theta = \sin(90° - \theta)$	$\cos\theta = \sin(\pi/2 - \theta)$
$\tan\theta = \cot(90° - \theta)$	$\tan\theta = \cot(\pi/2 - \theta)$
$\csc\theta = \sec(90° - \theta)$	$\csc\theta = \sec(\pi/2 - \theta)$
$\sec\theta = \csc(90° - \theta)$	$\sec\theta = \csc(\pi/2 - \theta)$
$\cot\theta = \tan(90° - \theta)$	$\cot\theta = \tan(\pi/2 - \theta)$

Although the angle θ in Table 2 is acute, we will see later (Section 7.2) that these results are valid for any angle θ.

Seeing the Concept Graph $Y_1 = \sin x$ and $Y_2 = \cos(90° - x)$ on the same screen. Be sure the mode is set to degrees.

EXAMPLE 4 Using the Complementary Angle Theorem

(a) $\sin 62° = \cos(90° - 62°) = \cos 28°$

(b) $\tan\dfrac{\pi}{12} = \cot\left(\dfrac{\pi}{2} - \dfrac{\pi}{12}\right) = \cot\dfrac{5\pi}{12}$

(c) $\cos\dfrac{\pi}{4} = \sin\left(\dfrac{\pi}{2} - \dfrac{\pi}{4}\right) = \sin\dfrac{\pi}{4}$

(d) $\csc\dfrac{\pi}{6} = \sec\left(\dfrac{\pi}{2} - \dfrac{\pi}{6}\right) = \sec\dfrac{\pi}{3}$

Now work Problem 31.

Historical Feature

The name *sine* for the sine function is due to a medieval confusion. The name comes from the Sanskrit word *jīva* (meaning chord), first used in India by Aryabhata the Elder (A.D. 510). He really meant half-chord, but abbreviated it. This was brought into Arabic as *jība,* which was meaningless. Because the proper Arabic word *jaib* would be written the same way (short vowels are not written out in Arabic), *jība* was pronounced as *jaib,* which meant bosom or hollow, and *jaib* remains as the Arabic word for sine to this day. Scholars translating the Arabic works into Latin found that the word *sinus* also meant bosom or hollow, and from *sinus* we get the word *sine.*

The name *tangent,* due to Thomas Fincke (1583), can be understood by looking at Figure 24. The line segment $\overline{DC}$ is tangent to the circle at C. If $d(O, B) = d(O, C) = 1$, then the length of the line segment $\overline{DC}$ is

$$d(D, C) = \frac{d(D, C)}{1} = \frac{d(D, C)}{d(O, C)} = \tan\alpha$$

FIGURE 24

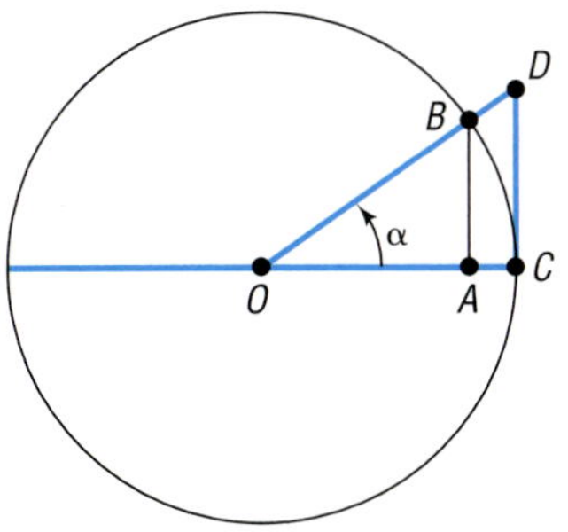

The old name for the tangent is *umbra versa* (meaning turned shadow), referring to the use of the tangent in solving height problems with shadows.

The names of the cofunctions came about as follows. If α and β are complementary angles, then $\cos\alpha = \sin\beta$. Because β is the complement of α, it was natural to write the cosine of α as *sin co* α. Probably for reasons involving ease of pronunciation, the *co* migrated to the front, and then cosine received a three-letter abbreviation to match sin, sec, and tan. The two other cofunctions were similarly treated, except that the long forms *cotan* and *cosec* survive to this day in some countries.

6.2 EXERCISES

In Problems 1–10, find the value of the six trigonometric functions of the angle θ in each figure.In Problems 11–14,

1.

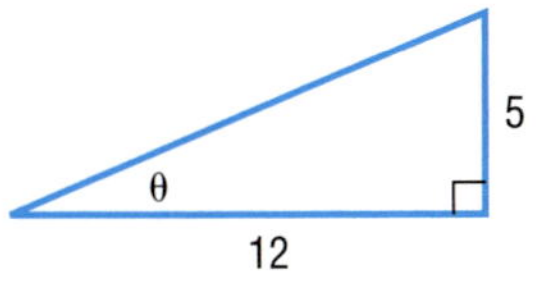

2.

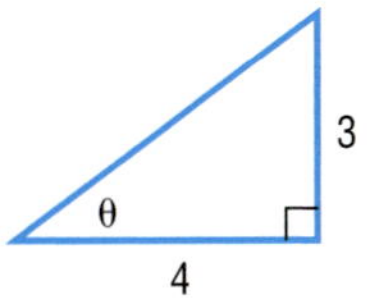

3.

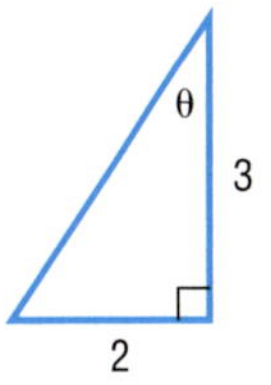

4.

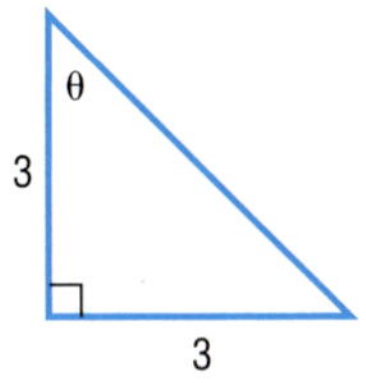

5.

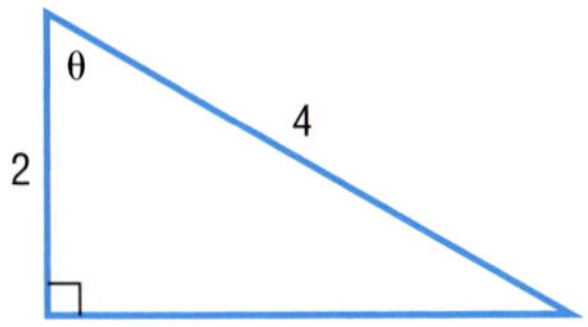

6.

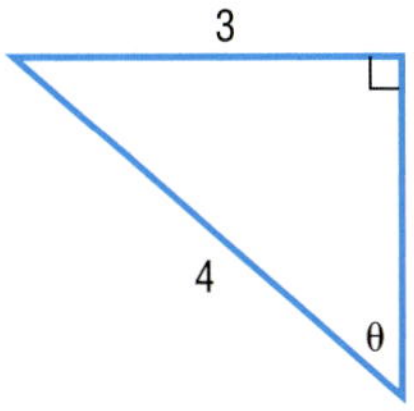

7.

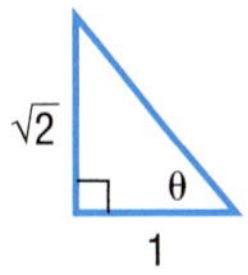

8.

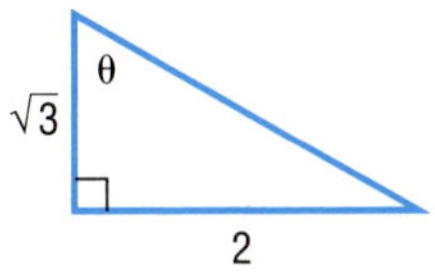

9.

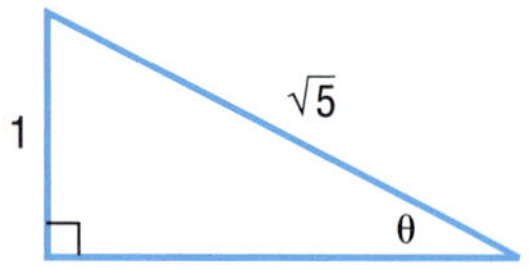

10.

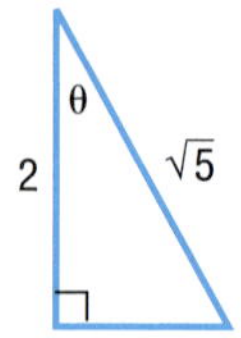

In Problems 11–14, find the exact value of each of the four remaining trigonometric functions of the acute angle θ.

11. $\sin\theta = \frac{1}{2}$, $\cos\theta = \sqrt{3}/2$

12. $\sin\theta = \sqrt{3}/2$, $\cos\theta = \frac{1}{2}$

13. $\sin\theta = \frac{2}{3}$, $\cos\theta = \sqrt{5}/3$

14. $\sin\theta = \frac{1}{3}$, $\cos\theta = 2\sqrt{2}/3$

In Problems 15–24, find the exact value of each of the remaining five trigonometric functions of the acute angle θ.

15. $\sin\theta = \sqrt{2}/2$

16. $\cos\theta = \sqrt{2}/2$

17. $\cos\theta = \frac{1}{3}$

18. $\sin\theta = \sqrt{3}/4$

19. $\tan\theta = \frac{1}{2}$

20. $\cot\theta = \frac{1}{2}$

21. $\sec\theta = 3$

22. $\csc\theta = 5$

23. $\tan\theta = \sqrt{2}$

24. $\sec\theta = \frac{5}{3}$

In Problems 25–40, find the exact value of each expression. Do not use a calculator.

25. $\sin^2 20° + \cos^2 20°$

26. $\sec^2 28° - \tan^2 28°$

27. $\sin 80° \csc 80°$

28. $\tan 10° \cot 10°$

29. $\tan 50° - \dfrac{\sin 50°}{\cos 50°}$

30. $\cot 25° - \dfrac{\cos 25°}{\sin 25°}$

31. $\sin 38° - \cos 52°$

32. $\tan 12° - \cot 78°$

33. $\dfrac{\cos 10°}{\sin 80°}$

34. $\dfrac{\cos 40°}{\sin 50°}$

35. $1 - \cos^2 20° - \cos^2 70°$

36. $1 + \tan^2 5° - \csc^2 85°$

37. $\tan 20° - \dfrac{\cos 70°}{\cos 20°}$

38. $\cot 40° - \dfrac{\sin 50°}{\sin 40°}$

39. $\cos 35° \sin 55° + \cos 55° \sin 35°$

40. $\sec 35° \csc 55° - \tan 35° \cot 55°$

41. Given $\sin 30° = \frac{1}{2}$, use trigonometric identities to find the exact value of
(a) $\cos 60°$ (b) $\cos^2 30°$
(c) $\csc\dfrac{\pi}{6}$ (d) $\sec\dfrac{\pi}{3}$

42. Given $\sin 60° = \sqrt{3}/2$, use trigonometric identities to find the exact value of
(a) $\cos 30°$ (b) $\cos^2 60°$
(c) $\sec\dfrac{\pi}{6}$ (d) $\csc\dfrac{\pi}{3}$

43. Given $\tan\theta = 4$, use trigonometric identities to find the exact value of
(a) $\sec^2\theta$ (b) $\cot\theta$
(c) $\cot\left(\dfrac{\pi}{2} - \theta\right)$ (d) $\csc^2\theta$

44. Given $\sec\theta = 3$, use trigonometric identities to find the exact value of
(a) $\cos\theta$ (b) $\tan^2\theta$
(c) $\csc(90° - \theta)$ (d) $\sin^2\theta$

45. Given $\csc\theta = 4$, use trigonometric identities to find the exact value of
(a) $\sin\theta$ (b) $\cot^2\theta$
(c) $\sec(90° - \theta)$ (d) $\sec^2\theta$

46. Given $\cot\theta = 2$, use trigonometric identities to find the exact value of
(a) $\tan\theta$ (b) $\csc^2\theta$
(c) $\tan\left(\dfrac{\pi}{2} - \theta\right)$ (d) $\sec^2\theta$

47. Given the approximation $\sin 38° \approx 0.62$, use trigonometric identities to find the approximate value of
(a) $\cos 38°$ (b) $\tan 38°$
(c) $\cot 38°$ (d) $\sec 38°$
(e) $\csc 38°$ (f) $\sin 52°$
(g) $\cos 52°$ (h) $\tan 52°$

48. Given the approximation $\cos 21° \approx 0.93$, use trigonometric identities to find the approximate value of
(a) $\sin 21°$ (b) $\tan 21°$
(c) $\cot 21°$ (d) $\sec 21°$
(e) $\csc 21°$ (f) $\sin 69°$
(g) $\cos 69°$ (h) $\tan 69°$

49. If $\sin\theta = 0.3$, find the exact value of $\sin\theta + \cos\left(\dfrac{\pi}{2} - \theta\right)$.

50. If $\tan\theta = 4$, find the exact value of $\tan\theta + \tan\left(\dfrac{\pi}{2} - \theta\right)$.

51. Find an acute angle θ that satisfies the equation $\sin\theta = \cos(2\theta + 30°)$.

52. Find an acute angle θ that satisfies the equation $\tan\theta = \cot(\theta + 45°)$.

53. **Calculating the Time of a Trip** From a parking lot you want to walk to a house on the ocean. The house is located 1500 feet down a paved path that parallels the beach, which is 500 feet wide. Along the path you can walk 300 ft/min, but in the sand on the beach you can only walk 100 ft/min. See the illustration.

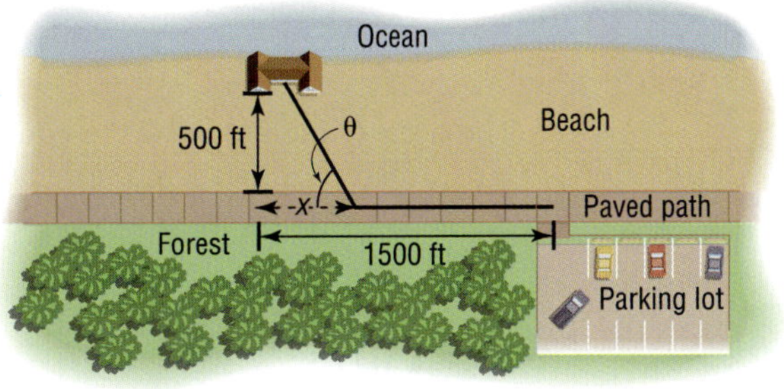

(a) Calculate the time T if you walk 1500 feet along the paved path and then 500 feet in the sand to the house.

(b) Calculate the time T if you walk in the sand directly toward the ocean for 500 feet and then turn left and walk along the beach for 1500 feet to the house.

(c) Express the time T to get from the parking lot to the beachhouse as a function of the angle θ shown in the illustration.

(d) Calculate the time T if you walk directly from the parking lot to the house.
[**Hint:** $\tan\theta = 500/1500$.]

(e) Calculate the time T if you walk 1000 feet along the paved path and then walk directly to the house.

(f) Graph $T = T(\theta)$. For what angle θ is T least? What is x for this angle? What is the minimum time?

54. **Carrying a Ladder Around a Corner** A ladder of length L is carried horizontally around a corner from a hall 3 feet wide into a hall 4 feet wide. See the illustration. Find the length L of the ladder as a function of the angle θ shown in the illustration.

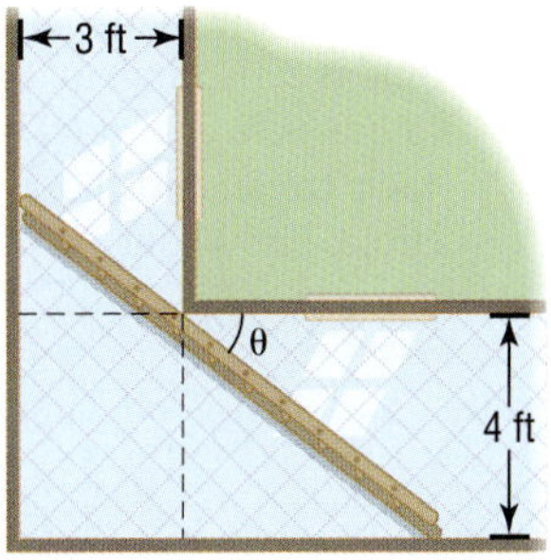

55. Suppose that the angle θ is a central angle of a circle of radius 1 (see the figure). Show that

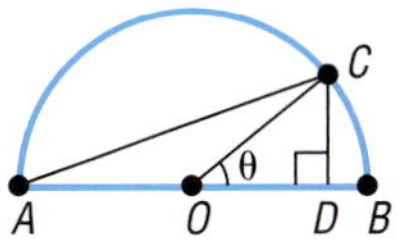

(a) Angle $OAC = \theta/2$

(b) $|CD| = \sin\theta$ and $|OD| = \cos\theta$

(c) $\tan\dfrac{\theta}{2} = \dfrac{\sin\theta}{1 + \cos\theta}$

56. Show that the area of an isosceles triangle is $A = a^2\sin\theta\cos\theta$, where a is the length of one of the two equal sides and θ is the measure of one of the two equal angles (see the figure).

57. Let $n \ge 1$ be any real number and let θ be any angle for which $0 < n\theta < \pi/2$. Then we can draw a triangle with the angles θ and $n\theta$ and included side of length 1 (do you see why?) and place it on the unit circle as illustrated. Now, drop the perpendicular from C to $D = (x, 0)$, and show that

$$x = \frac{\tan n\theta}{\tan\theta + \tan n\theta}$$

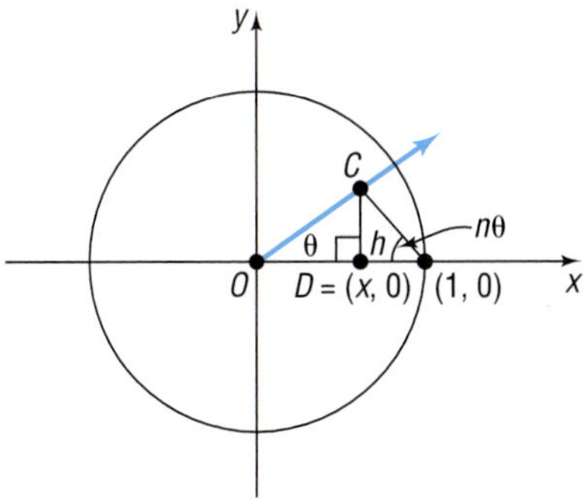

58. Refer to the figure. The smaller circle, whose radius is a, is tangent to the larger circle, whose radius is b. The ray OA contains a diameter of each circle, and the ray OB is tangent to each circle. Show that

$$\cos\theta = \frac{\sqrt{ab}}{\dfrac{a+b}{2}}$$

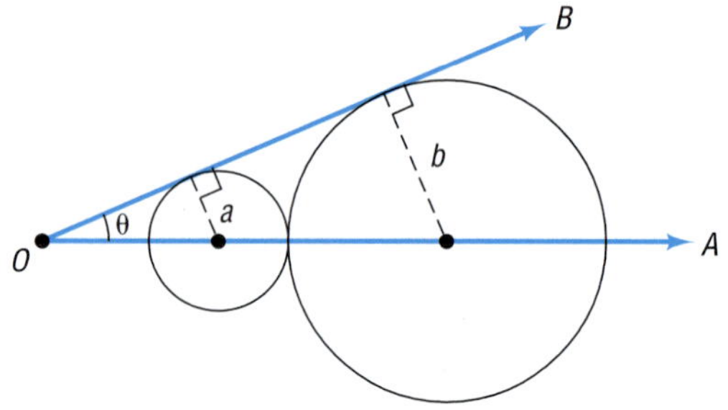

(that is, $\cos\theta$ equals the ratio of the geometric mean of a and b to the arithmetic mean of a and b).

[**Hint:** First show that $\sin\theta = (b-a)/(b+a)$.]

59. Refer to the figure. If $|OA| = 1$, show that:

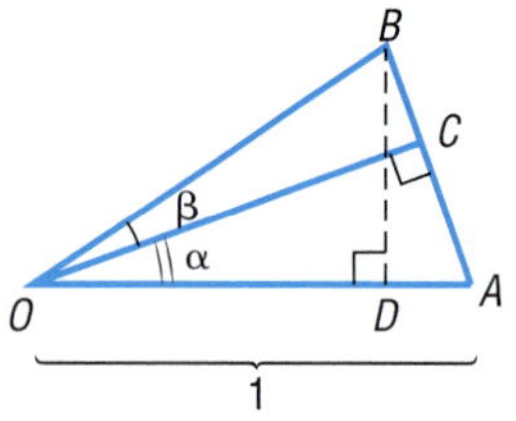

(a) Area $\triangle OAC = \frac{1}{2}\sin\alpha\cos\alpha$

(b) Area $\triangle OCB = \frac{1}{2}|OB|^2\sin\beta\cos\beta$

(c) Area $\triangle OAB = \frac{1}{2}|OB|\sin(\alpha+\beta)$

(d) $|OB| = \dfrac{\cos\alpha}{\cos\beta}$

(e) $\sin(\alpha+\beta) = \sin\alpha\cos\beta + \cos\alpha\sin\beta$

[**Hint:** Area $\triangle OAB$ = Area $\triangle OAC$ + Area $\triangle OCB$.]

60. Refer to the figure, where a unit circle is drawn. The line DB is tangent to the circle.

(a) Express the area of $\triangle OBC$ in terms of $\sin\theta$ and $\cos\theta$.

(b) Express the area of $\triangle OBD$ in terms of $\sin\theta$ and $\cos\theta$.

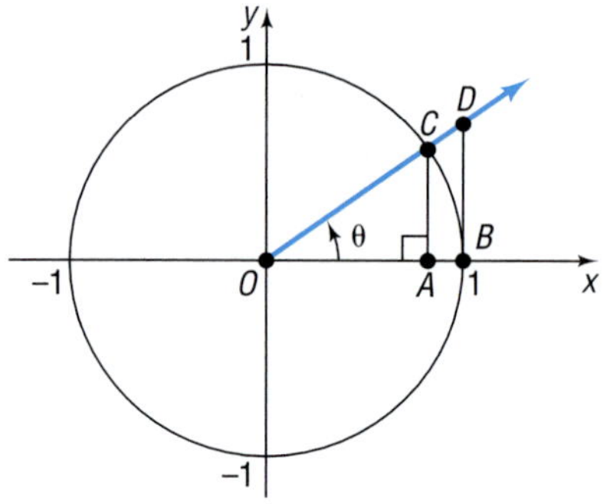

(c) The area of the sector of the circle OBC, is $\frac{1}{2}\theta$, where θ is measured in radians. Use the results of parts (a) and (b) and the fact that

$$\text{Area } \triangle OBC < \text{Area } \overset{\frown}{OBC} < \text{Area } \triangle OBD$$

to show that

$$1 < \frac{\theta}{\sin\theta} < \frac{1}{\cos\theta}$$

61. If $\cos\alpha = \tan\beta$ and $\cos\beta = \tan\alpha$, where α and β are acute angles, show that

$$\sin\alpha = \sin\beta = \sqrt{\frac{3-\sqrt{5}}{2}}$$

62. Explain why, if θ is an acute angle, then $\sec\theta > 1$.

63. Explain why, if θ is an acute angle, then $0 < \sin\theta < 1$.

64. How would you explain the meaning of the sine function to a fellow student who has just completed college algebra?

6.3 COMPUTING THE VALUES OF TRIGONOMETRIC FUNCTIONS OF GIVEN ANGLES

1 Find the Exact Value of the Trigonometric Functions for 30°, 45°, 60° Angles

2 Use a Graphing Calculator to Approximate the Value of the Trigonometric Functions of Acute Angles

In the previous section, we developed ways to find the value of each of the trigonometric functions of an acute angle when one of the functions is known.

FIGURE 25

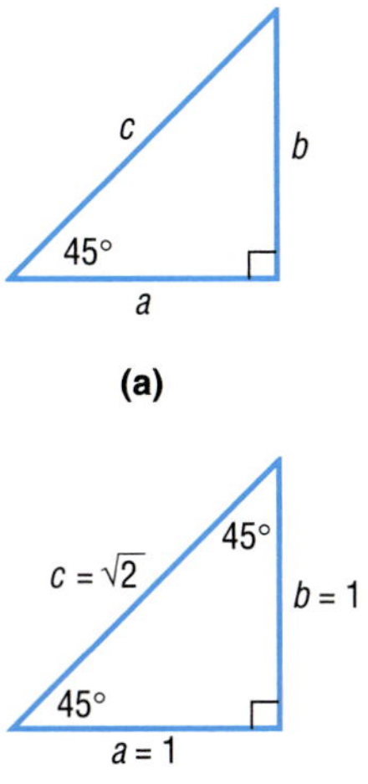

(a) (b)

In this section, we discuss the problem of finding the value of each of the trigonometric functions of an acute angle when the angle is given.

For three special acute angles, we can use some results from plane geometry to find the exact value of each of the six trigonometric functions.

45°, 30°, and 60° Angles

1 Consider the right triangle in Figure 25(a), in which one of the angles is 45°. It follows that the other acute angle is also 45°, and hence the triangle is isosceles. As a result, side a and side b are equal in length. Because the values of the trigonometric functions of an angle depend only on the angle and not on the size of the triangle, we may choose to use the triangle for which

$$a = b = 1$$

Then, by the Pythagorean Theorem,

$$c^2 = a^2 + b^2 = 1 + 1 = 2$$
$$c = \sqrt{2}$$

As a result, we have the triangle in Figure 25(b), from which we find

$$\sin 45° = \frac{b}{c} = \frac{1}{\sqrt{2}} = \frac{\sqrt{2}}{2} \qquad \cos 45° = \frac{a}{c} = \frac{1}{\sqrt{2}} = \frac{\sqrt{2}}{2}$$

Thus,

$$\tan 45° = \frac{\sin 45°}{\cos 45°} = \frac{\sqrt{2}/2}{\sqrt{2}/2} = 1 \qquad \cot 45° = \frac{1}{\tan 45°} = \frac{1}{1} = 1$$

$$\sec 45° = \frac{1}{\cos 45°} = \frac{1}{1/\sqrt{2}} = \sqrt{2} \qquad \csc 45° = \frac{1}{\sin 45°} = \frac{1}{1/\sqrt{2}} = \sqrt{2}$$

FIGURE 26

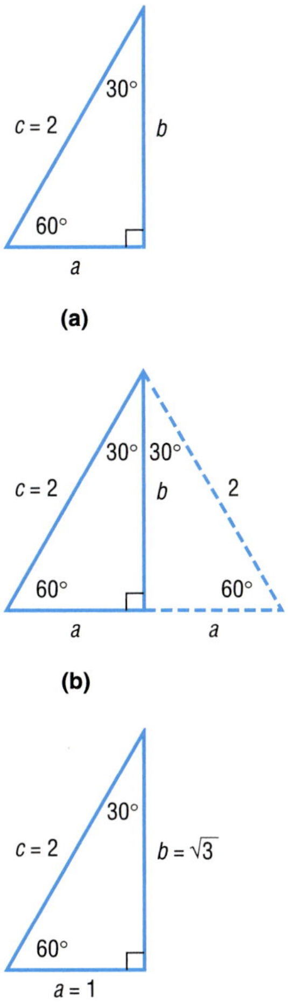

(a) (b) (c)

Now, consider a right triangle in which one of the angles is 30°. It then follows that the other angle is 60°. Figure 26(a) illustrates such a triangle with hypotenuse of length 2. Our problem is to determine a and b.

We begin by placing next to the triangle in Figure 26(a) another triangle congruent to the first, as shown in Figure 26(b). Notice that we now have a triangle whose angles are each 60°. This triangle is therefore equilateral, so each side is of length 2. In particular, the base is $2a = 2$, and so $a = 1$. By the Pythagorean Theorem, b satisfies the equation $a^2 + b^2 = c^2$, so we have

$$a^2 + b^2 = c^2$$
$$1^2 + b^2 = 2^2$$
$$b^2 = 4 - 1 = 3$$
$$b = \sqrt{3}$$

Using the triangle in Figure 26(c) and the fact that 30° and 60° are complementary angles, we find

$$\sin 30° = \frac{\text{Opposite}}{\text{Hypotenuse}} = \frac{1}{2} \qquad \cos 60° = \frac{1}{2}$$

$$\cos 30° = \frac{\text{Adjacent}}{\text{Hypotenuse}} = \frac{\sqrt{3}}{2} \qquad \sin 60° = \frac{\sqrt{3}}{2}$$

$$\tan 30^\circ = \frac{\sin 30^\circ}{\cos 30^\circ} = \frac{\frac{1}{2}}{\sqrt{3}/2} = \frac{1}{\sqrt{3}} = \frac{\sqrt{3}}{3} \qquad \cot 60^\circ = \frac{\sqrt{3}}{3}$$

$$\csc 30^\circ = \frac{1}{\sin 30^\circ} = \frac{1}{\frac{1}{2}} = 2 \qquad \sec 60^\circ = 2$$

$$\sec 30^\circ = \frac{1}{\cos 30^\circ} = \frac{1}{\sqrt{3}/2} = \frac{2}{\sqrt{3}} = \frac{2\sqrt{3}}{3} \qquad \csc 60^\circ = \frac{2\sqrt{3}}{3}$$

$$\cot 30^\circ = \frac{1}{\tan 30^\circ} = \frac{1}{\sqrt{3}/3} = \frac{3}{\sqrt{3}} = \sqrt{3} \qquad \tan 60^\circ = \sqrt{3}$$

Table 3 summarizes the information just derived for the angles 30°, 45°, and 60°. Until you memorize the entries in Table 3, you should draw the appropriate triangle to determine the values given in the table.

TABLE 3

θ (Radians)	θ (Degrees)	$\sin\theta$	$\cos\theta$	$\tan\theta$	$\csc\theta$	$\sec\theta$	$\cot\theta$
$\pi/6$	30°	$\frac{1}{2}$	$\sqrt{3}/2$	$\sqrt{3}/3$	2	$2\sqrt{3}/3$	$\sqrt{3}$
$\pi/4$	45°	$\sqrt{2}/2$	$\sqrt{2}/2$	1	$\sqrt{2}$	$\sqrt{2}$	1
$\pi/3$	60°	$\sqrt{3}/2$	$\frac{1}{2}$	$\sqrt{3}$	$2\sqrt{3}/3$	2	$\sqrt{3}/3$

EXAMPLE 1

Finding the Exact Value of a Trigonometric Expression

Find the exact value of each expression:

(a) $\sin 45^\circ \cos 30^\circ$ (b) $\tan\frac{\pi}{4} - \sin\frac{\pi}{3}$

Solution

(a) $\sin 45^\circ \cos 30^\circ = \frac{\sqrt{2}}{2}\cdot\frac{\sqrt{3}}{2} = \frac{\sqrt{6}}{4}$

(b) $\tan\frac{\pi}{4} - \sin\frac{\pi}{3} = 1 - \frac{\sqrt{3}}{2} = \frac{2-\sqrt{3}}{2}$

Now work Problems 5 and 17.

The exact values of the trigonometric functions for the angles 30°, 45°, and 60° are relatively easy to calculate, because the triangles that contain such angles have "nice" geometric features. For most other angles, we can only approximate the value of each of the trigonometric functions. To do this, we will need a calculator.

Using a Calculator to Find Values of Trigonometric Functions

2 Before getting started, you must first decide whether to enter the angle in the calculator using radians or degrees and then set the calculator to the correct MODE. Your calculator probably has only the keys marked sin, cos, and tan. To find the values of the remaining three trigonometric functions, secant, cosecant, and cotangent, we use the reciprocal identities:

$$\sec\theta = \frac{1}{\cos\theta} \qquad \csc\theta = \frac{1}{\sin\theta} \qquad \cot\theta = \frac{1}{\tan\theta}$$

EXAMPLE 2 Using a Calculator to Approximate the Value of Trigonometric Functions

Use a calculator to find the approximate value of

(a) $\cos 48^\circ$ (b) $\csc 21^\circ$ (c) $\tan \dfrac{\pi}{12}$

Express your answer rounded to two decimal places.

Solution (a) First, we set the MODE to receive degrees. See Figure 27(a). Figure 27(b) shows the solution using a TI-83 graphing calculator.

FIGURE 27

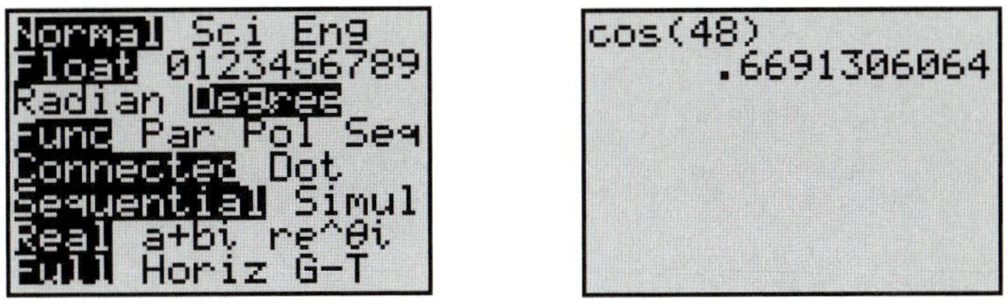

(a) (b)

Thus,

$$\cos 48^\circ = 0.67$$

rounded to two decimal places.

FIGURE 28

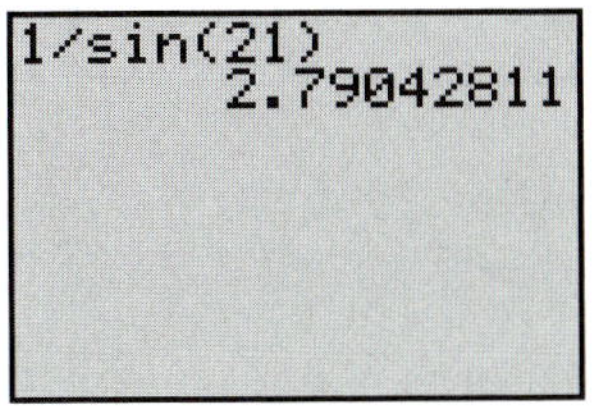

(b) Most calculators do not have a csc key. The manufacturers assume the user knows some trigonometry. Thus, to find the value of $\csc 21^\circ$, we use the fact that $\csc 21^\circ = 1/(\sin 21^\circ)$. Figure 28 shows the solution using a TI-83 graphing calculator.
Thus,

$$\csc 21^\circ = 2.79$$

rounded to two decimal places.

(c) Set the MODE to receive radians. Figure 29 shows the solution using a TI-83 graphing calculator.

FIGURE 29

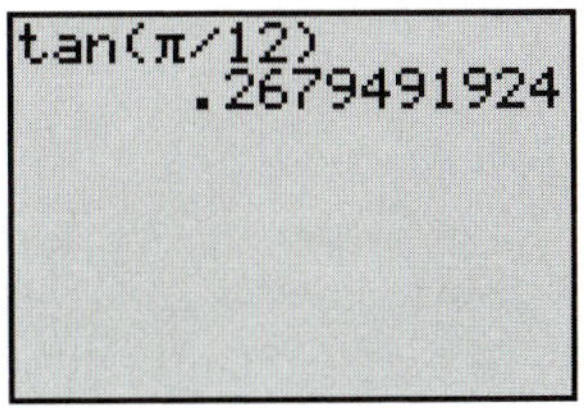

Thus,

$$\tan \frac{\pi}{12} = 0.27$$

rounded to two decimal places.

Note: In Figure 29, the parentheses are necessary. Without them, the calculator will evaluate $\tan \pi$ first, then divide by 12.

Now work Problem 25.

EXAMPLE 3 Constructing a Rain Gutter

A rain gutter is to be constructed of aluminum sheets 12 inches wide. After marking off a length of 4 inches from each edge, this length is bent up at an angle θ. See Figure 30.

FIGURE 30

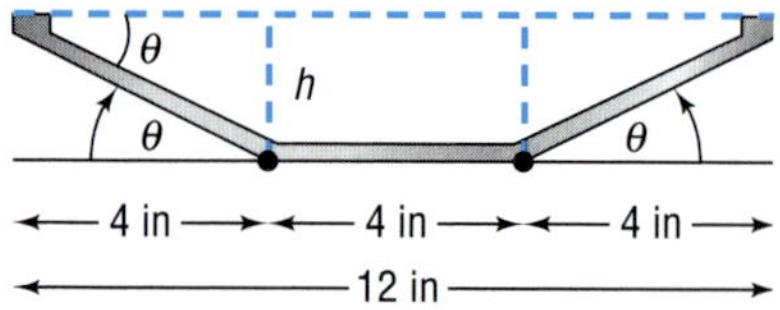

(a) Express the area A of the opening as a function of θ.

[**Hint:** Let h denote the vertical height of the bend].

(b) Find the area A of the opening for $\theta = 30°$, $\theta = 45°$, $\theta = 60°$, and $\theta = 75°$.

(c) Graph $A = A(\theta)$.

(d) Find the angle θ that makes A largest. [This bend will allow the most water to flow through the gutter.]

Solution (a) The area A of the opening is the sum of the areas of two right triangles and one rectangle. Each triangle has legs h and $\sqrt{16 - h^2}$ and hypotenuse 4. The rectangle has length 4 and height h. Thus,

$$A = 2\left(\tfrac{1}{2}\right)h\sqrt{16 - h^2} + 4h \underset{\substack{\uparrow \\ \sin\theta = \frac{h}{4},\ \cos\theta = \frac{\sqrt{16-h^2}}{4}}}{=} (4\sin\theta)(4\cos\theta) + 4(4\sin\theta)$$

$$A(\theta) = 16\sin\theta(\cos\theta + 1)$$

(b) For $\theta = 30°$:
$$A(30°) = 16\sin 30°\,(\cos 30° + 1) = 16\left(\tfrac{1}{2}\right)\left(\sqrt{3}/2 + 1\right) = 4\sqrt{3} + 8$$

The area of the opening for $\theta = 30°$ is about 14.9 square inches.

For $\theta = 45°$:
$$A(45°) = 16\sin 45°\,(\cos 45° + 1) = 16\left(\sqrt{2}/2\right)\left(\sqrt{2}/2 + 1\right) = 8 + 8\sqrt{2}$$

The area of the opening for $\theta = 45°$ is about 19.3 square inches.

For $\theta = 60°$:
$$A(60°) = 16\sin 60°(\cos 60° + 1) = 16\left(\sqrt{3}/2\right)\left(\tfrac{1}{2} + 1\right) = 12\sqrt{3}$$

The area of the opening for $\theta = 60°$ is about 20.8 square inches.

For $\theta = 75°$:
$$A(75°) = 16\sin 75°(\cos 75° + 1) \approx 19.5$$

The area of the opening for $\theta = 75°$ is about 19.5 square inches.

FIGURE 31

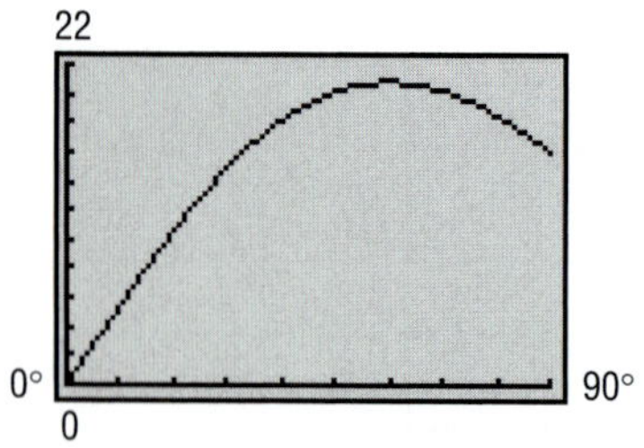

(c) Figure 31 shows the graph $A = A(\theta)$.

(d) Using MAXIMUM, the angle θ that makes A largest is 60°. ■

6.3 EXERCISES

1. Write down the exact value of each of the six trigonometric functions of 45°.
2. Write down the exact value of each of the six trigonometric functions of 30°.
3. Write down the exact value of each of the six trigonometric functions of 60°.

In Problems 4–14, find the exact value of each expression. Do not use a calculator.

4. $2 \sin 45° + 4 \sin 30°$

5. $6 \tan 45° - 8 \cos 60°$

6. $\sin 30° \cdot \tan 60°$

7. $\sec \frac{\pi}{4} + 2 \csc \frac{\pi}{3}$

8. $\tan \frac{\pi}{4} + \cot \frac{\pi}{4}$

9. $\sec^2 \frac{\pi}{6} - 4$

10. $4 + \tan^2 \frac{\pi}{3}$

11. $\sin^2 40° + \cos^2 40°$

12. $\sec^2 18° - \tan^2 18°$

13. $1 - \cos^2 20° - \cos^2 70°$

14. $1 + \tan^2 5° - \csc^2 85°$

In Problems 15–24, find the exact value of each expression if $\theta = 60°$. Do not use a calculator.

15. $\sin \theta$

16. $\cos \theta$

17. $\sin \frac{\theta}{2}$

18. $\cos \frac{\theta}{2}$

19. $(\sin \theta)^2$

20. $(\cos \theta)^2$

21. $2 \sin \theta$

22. $2 \cos \theta$

23. $\frac{\sin \theta}{2}$

24. $\frac{\cos \theta}{2}$

In Problems 25–48, use a graphing utility to find the approximate value of each expression rounded to two decimal places.

25. $\sin 28°$

26. $\cos 14°$

27. $\tan 21°$

28. $\sin 15°$

29. $\sec 41°$

30. $\csc 55°$

31. $\cot 70°$

32. $\tan 80°$

33. $\sin \frac{\pi}{10}$

34. $\cos \frac{\pi}{8}$

35. $\tan \frac{5\pi}{12}$

36. $\sin \frac{3\pi}{10}$

37. $\sec \frac{\pi}{12}$

38. $\csc \frac{5\pi}{13}$

39. $\cot \frac{\pi}{18}$

40. $\sin \frac{\pi}{18}$

41. $\sin 1$

42. $\tan 1$

43. $\sin 1°$

44. $\tan 1°$

45. $\cos 21.5°$

46. $\cos 35.2°$

47. $\tan 0.3$

48. $\tan 0.1$

Projectile Motion The path of a projectile fired at an inclination θ to the horizontal with initial speed v_0 is a parabola (see the figure). The range R of the projectile, that is, the horizontal distance the projectile travels, is found by using the formula

$$R = \frac{2v_0^2 \sin \theta \cos \theta}{g}$$

where $g \approx 32.2$ feet per second per second ≈ 9.8 meters per second per second is the acceleration due to gravity. The maximum height H of the projectile is

$$H = \frac{v_0^2 \sin^2 \theta}{2g}$$

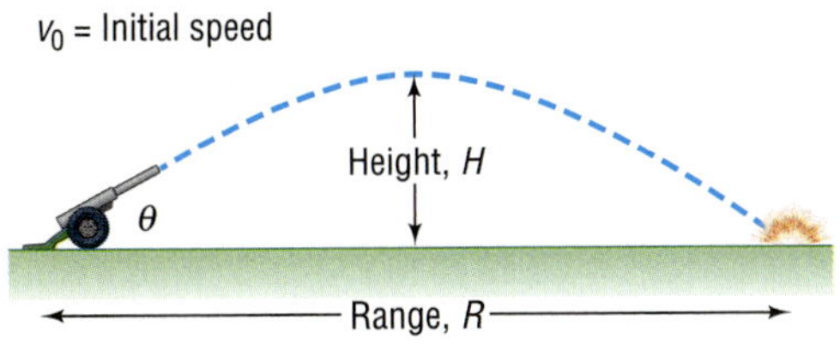

In Problems 49–52, find the range R and maximum height H.

49. The projectile is fired at an angle of 45° to the horizontal with initial speed of 100 feet per second.

50. The projectile is fired at an angle of 30° to the horizontal with initial speed of 150 meters per second.

51. The projectile is fired at an angle of 25° to the horizontal with initial speed of 500 meters per second.

52. The projectile is fired at an angle of 50° to the horizontal with initial speed of 200 feet per second.

53. **Inclined Plane** If friction is ignored, the time t (in seconds) required for a block to slide down an inclined plane (see the figure on page 474) is given by the formula

$$t = \sqrt{\frac{2a}{g \sin \theta \cos \theta}}$$

where a is the length (in feet) of the base and $g \approx 32$ feet per second per second is the acceleration of gravity. How long does it take a block to slide down an inclined plane with base $a = 10$ feet when

(a) $\theta = 30°$?
(b) $\theta = 45°$?
(c) $\theta = 60°$?

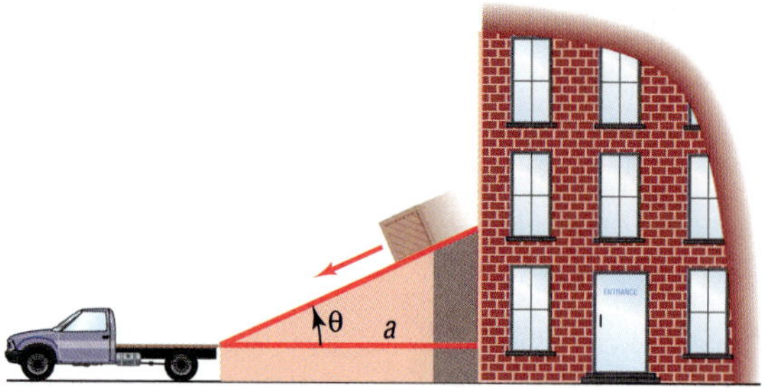

54. Piston Engines In a certain piston engine, the distance x (in meters) from the center of the drive shaft to the head of the piston is given by

$$x = \cos\theta + \sqrt{16 + 0.5(2\cos^2\theta - 1)}$$

where θ is the angle between the crank and the path of the piston head (see the figure). Find x when $\theta = 30°$ and when $\theta = 45°$.

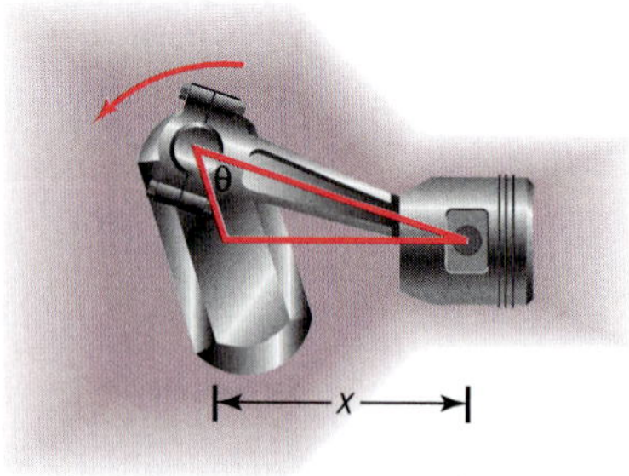

55. Calculating the Time of a Trip Two ocean front homes are located 8 miles apart on a straight stretch of beach, each a distance of 1 mile from a paved road that parallels the ocean. Sally can jog 8 mi/hr along the paved road, but only 3 mi/hr in the sand that separates the road from the beach. Because of a river directly between the two houses, it is necessary to jog on the sand to the road, continue on the road, and then jog directly back on the sand to get from one house to the other. See the illustration.

(a) Express the time T to get from one house to the other as a function of the angle θ shown in the illustration.

(b) Calculate the time T for $\theta = 30°$. How long is Sally on the paved road?

(c) Calculate the time T for $\theta = 45°$. How long is Sally on the paved road?

(d) Calculate the time T for $\theta = 60°$. How long is Sally on the paved road?

(e) Calculate the time T for $\theta = 90°$. Describe the path taken.

(f) Calculate the time T for $\tan\theta = 1/4$. Describe the path taken. Explain why θ must be larger than 14°.

(g) Graph $T = T(\theta)$. What angle θ results in the least time. What is the least time? How long is Sally on the paved road?

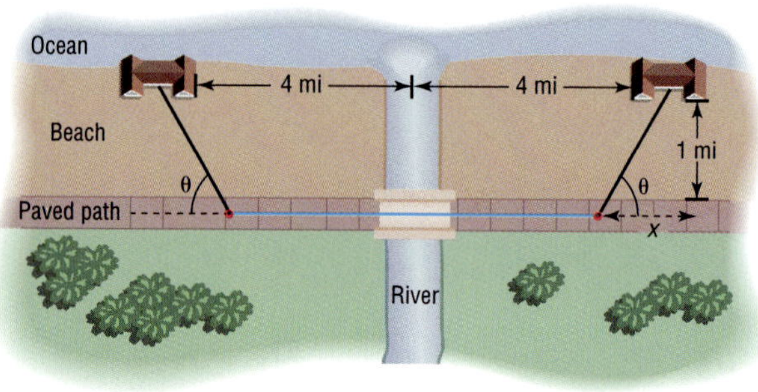

56. Designing Fine Decorative Pieces A designer of decorative art plans to market solid gold spheres encased in clear crystal cones. Each sphere is of fixed radius R and will be enclosed in a cone of height h and radius r. See the illustration. Many cones can be used to enclose the sphere, each having a different slant angle θ.

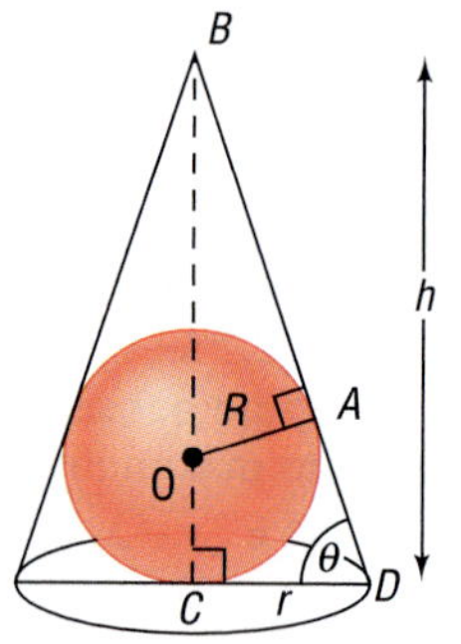

(a) Express the volume V of the cone as a function of the slant angle θ of the cone.

[**Hint:** The volume V of a cone of height h and radius r is $V = 1/3\ \pi r^2 h$.]

(b) What volume V is required to enclose a sphere of radius 2 cm in a cone whose slant angle θ is 30°? 45°? 60°?

(c) What slant angle θ should be used for the volume V of the cone to be a minimum? [This choice minimizes the amount of crystal required and gives maximum emphasis to the gold sphere.]

57. Use a graphing utility set in radian mode to complete the table below. What can you conclude about the ratio $(\sin \theta)/\theta$ as θ approaches 0?

θ	0.5	0.4	0.2	0.1	0.01	0.001	0.0001	0.00001
$\sin \theta$								
$\frac{\sin \theta}{\theta}$								

58. Use a graphing utility set in radian mode to complete the table below. What can you conclude about the ratio $(\cos \theta - 1)/\theta$ as θ approaches 0?

θ	0.5	0.4	0.2	0.1	0.01	0.001	0.0001	0.00001
$\cos \theta - 1$								
$\frac{\cos \theta - 1}{\theta}$								

59. Find the exact value of $\tan 1° \cdot \tan 2° \cdot \tan 3° \cdot \ldots \cdot \tan 89°$.

60. Find the exact value of $\cot 1° \cdot \cot 2° \cdot \cot 3° \cdot \ldots \cdot \cot 89°$.

61. Find the exact value of $\cos 1° \cdot \cos 2° \cdot \ldots \cdot \cos 45° \cdot \csc 46° \cdot \ldots \cdot \csc 89°$.

62. Find the exact value of $\sin 1° \cdot \sin 2° \cdot \ldots \cdot \sin 45° \cdot \sec 46° \cdot \ldots \cdot \sec 89°$.

63. Write a brief paragraph that explains how to quickly compute the trigonometric functions of 30°, 45°, and 60°.

6.4 TRIGONOMETRIC FUNCTIONS OF GENERAL ANGLES

1 Find the Exact Value of the Trigonometric Functions for General Angles

2 Determine the Sign of the Trigonometric Functions

3 Find the Reference Angle

In order to extend the definitions of the trigonometric functions of acute angles to include angles that are not acute, we employ a rectangular coordinate system.

Let θ be any angle in standard position, and let (a, b) denote the coordinates of any point, except the origin $(0, 0)$, on the terminal side of θ. If $r = \sqrt{a^2 + b^2}$ denotes the distance from $(0, 0)$ to (a, b), then the **six trigonometric functions of θ** are defined as the ratios

$$\sin \theta = \frac{b}{r} \qquad \cos \theta = \frac{a}{r} \qquad \tan \theta = \frac{b}{a}$$

$$\csc \theta = \frac{r}{b} \qquad \sec \theta = \frac{r}{a} \qquad \cot \theta = \frac{a}{b}$$

provided no denominator equals 0. If a denominator equals 0, that trigonometric function of the angle θ is not defined.

FIGURE 32

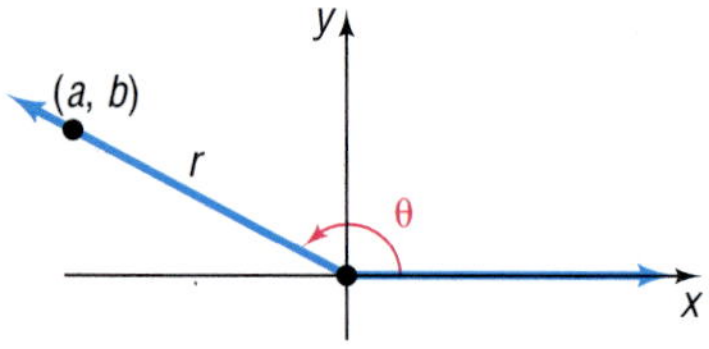

FIGURE 33

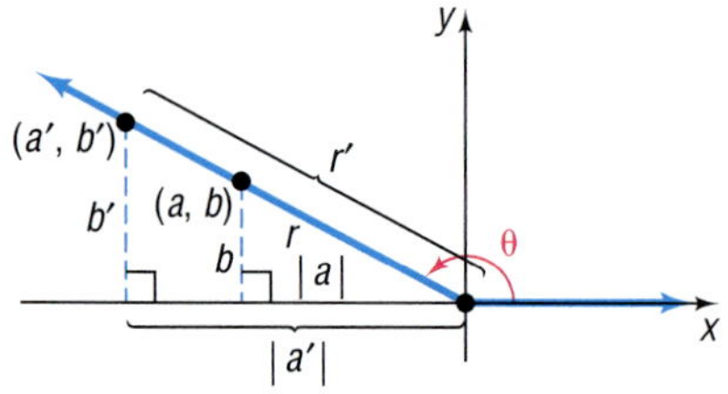

Figure 32 illustrates the definition.

Notice in the preceding definitions that if $a = 0$, that is, if the point $(0, b)$ is on the y-axis, then the tangent function and the secant function are undefined. Also, if $b = 0$, that is, if the point $(a, 0)$ is on the x-axis, then the cosecant function and the cotangent function are undefined.

By constructing similar triangles, you should be convinced that the values of the six trigonometric functions of an angle θ do not depend on the selection of the point (a, b) on the terminal side of θ, but rather depend only on the angle θ itself. See Figure 33 for an illustration when θ lies in quadrant II. Since the triangles are similar, the ratio b/r equals the ratio b'/r', the common value being $\sin\theta$. Also, the ratio $|a|/r$ equals the ratio $|a'|/r'$, so $a/r = a'/r'$, the common value being $\cos\theta$. And so on.

Also, observe that if θ is acute these definitions reduce to the right triangle definitions given in Section 6.2, as illustrated in Figure 34.

FIGURE 34

$$\sin\theta = \frac{b}{r} = \frac{\text{Opposite}}{\text{Hypotenuse}}$$

$$\cos\theta = \frac{a}{r} = \frac{\text{Adjacent}}{\text{Hypotenuse}}$$

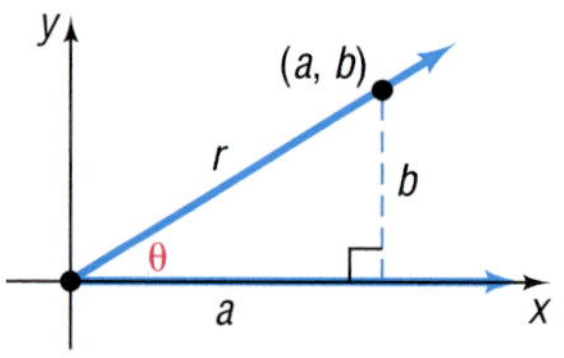

EXAMPLE 1 Finding the Exact Value of the Six Trigonometric Functions

1

Find the exact value of each of the six trigonometric functions of a positive angle θ if $(4, -3)$ is a point on its terminal side.

Solution Figure 35 illustrates the situation. For the point $(a, b) = (4, -3)$, we have $a = 4$ and $b = -3$. Then $r = \sqrt{a^2 + b^2} = \sqrt{16 + 9} = 5$. Thus,

$$\sin\theta = \frac{b}{r} = -\frac{3}{5} \qquad \cos\theta = \frac{a}{r} = \frac{4}{5} \qquad \tan\theta = \frac{b}{a} = -\frac{3}{4}$$

$$\csc\theta = \frac{r}{b} = -\frac{5}{3} \qquad \sec\theta = \frac{r}{a} = \frac{5}{4} \qquad \cot\theta = \frac{a}{b} = -\frac{4}{3}$$

FIGURE 35

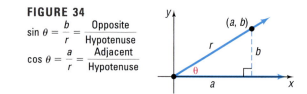

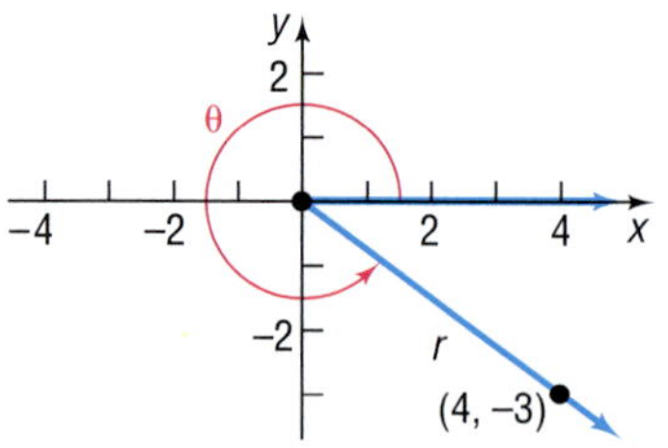

Now work Problem 1.

In the next example, we find the exact value of each of the six trigonometric functions at the quadrantal angles 0, $\pi/2$, π, and $3\pi/2$.

EXAMPLE 2 Finding the Exact Value of the Trigonometric Functions of Quadrantal Angles

Find the exact value of each of the trigonometric functions at

(a) $\theta = 0 = 0°$ (b) $\theta = \pi/2 = 90°$ (c) $\theta = \pi = 180°$
(d) $\theta = 3\pi/2 = 270°$

Solution (a) The point $P = (1, 0)$ is on the terminal side of $\theta = 0 = 0°$ and is a distance of 1 unit from the origin. See Figure 36. Thus,

FIGURE 36
$\theta = 0 = 0°$

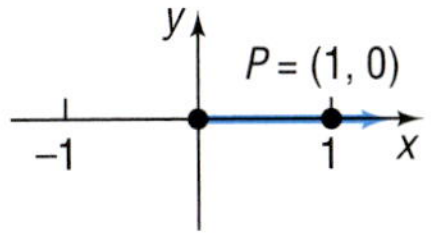

$$\sin 0 = \sin 0° = \frac{0}{1} = 0 \qquad \cos 0 = \cos 0° = \frac{1}{1} = 1$$

$$\tan 0 = \tan 0° = \frac{0}{1} = 0 \qquad \sec 0 = \sec 0° = \frac{1}{1} = 1$$

Since the y-coordinate of P is 0, csc 0 and cot 0 are not defined.

(b) The point $P = (0, 1)$ is on the terminal side of $\theta = \pi/2 = 90°$ and is a distance of 1 unit from the origin. See Figure 37. Thus,

FIGURE 37
$\theta = \pi/2 = 90°$

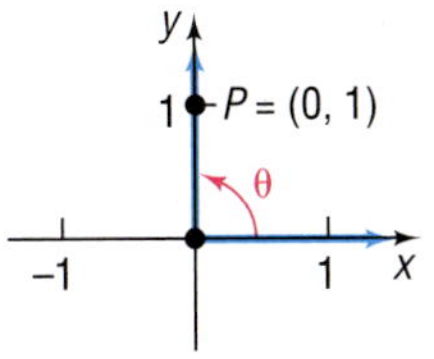

$$\sin \frac{\pi}{2} = \sin 90° = \frac{1}{1} = 1 \qquad \cos \frac{\pi}{2} = \cos 90° = \frac{0}{1} = 0$$

$$\csc \frac{\pi}{2} = \csc 90° = \frac{1}{1} = 1 \qquad \cot \frac{\pi}{2} = \cot 90° = \frac{0}{1} = 0$$

Since the x-coordinate of P is 0, tan $(\pi/2)$ and sec $(\pi/2)$ are not defined.

(c) The point $P = (-1, 0)$ is on the terminal side of $\theta = \pi = 180°$ and is a distance of 1 unit from the origin. See Figure 38. Thus,

FIGURE 38
$\theta = \pi = 180°$

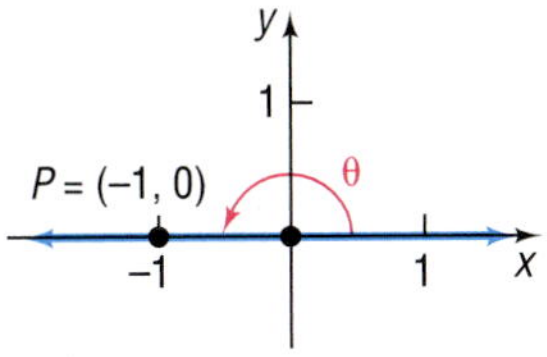

$$\sin \pi = \sin 180° = \frac{0}{1} = 0 \qquad \cos \pi = \cos 180° = \frac{-1}{1} = -1$$

$$\tan \pi = \tan 180° = \frac{0}{-1} = 0 \qquad \sec \pi = \sec 180° = \frac{1}{-1} = -1$$

Since the y-coordinate of P is 0, csc π and cot π are not defined.

(d) The point $P = (0, -1)$ is on the terminal side of $\theta = 3\pi/2 = 270°$ and is a distance of 1 unit from the origin. See Figure 39. Thus,

FIGURE 39
$\theta = 3\pi/2 = 270°$

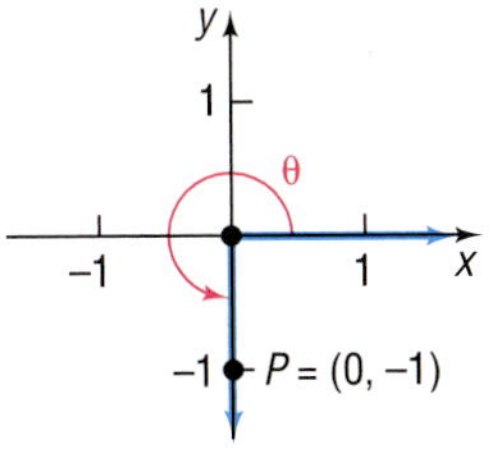

$$\sin \frac{3\pi}{2} = \sin 270° = \frac{-1}{1} = -1 \qquad \cos \frac{3\pi}{2} = \cos 270° = \frac{0}{1} = 0$$

$$\csc \frac{3\pi}{2} = \csc 270° = \frac{1}{-1} = -1 \qquad \cot \frac{3\pi}{2} = \cot 270° = \frac{0}{-1} = 0$$

Since the x-coordinate of P is 0, tan $(3\pi/2)$ and sec $(3\pi/2)$ are not defined. ■

Table 4 summarizes the values of the trigonometric functions found in Example 2.

TABLE 4

θ (Radians)	θ (Degrees)	sin θ	cos θ	tan θ	csc θ	sec θ	cot θ
0	0°	0	1	0	Not defined	1	Not defined
$\pi/2$	90°	1	0	Not defined	1	Not defined	0
π	180°	0	−1	0	Not defined	−1	Not defined
$3\pi/2$	270°	−1	0	Not defined	−1	Not defined	0

The Signs of the Trigonometric Functions

2 If θ is not a quadrantal angle, then it will lie in a particular quadrant. In such a case, the signs of the x-coordinate and the y-coordinate of a point (a, b) on the terminal side of θ are known. Because $r = \sqrt{a^2 + b^2} > 0$, it follows that the signs of the trigonometric functions of an angle θ can be found if we know in which quadrant θ lies.

FIGURE 40
θ in quadrant II
$a < 0$, $b > 0$, $r > 0$

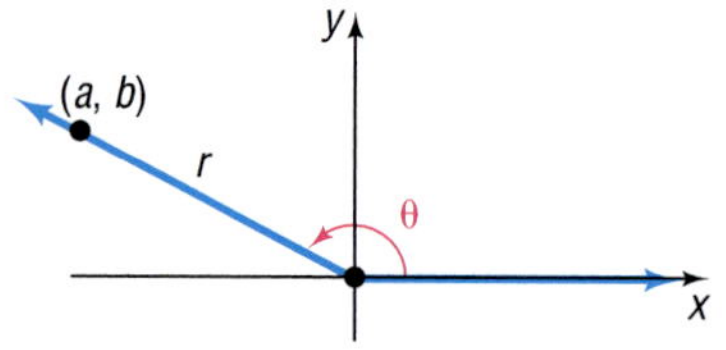

For example, if θ lies in quadrant II, as shown in Figure 40, then a point (a, b) on the terminal side of θ has a negative x-coordinate and a positive y-coordinate. Thus,

$$\sin\theta = \frac{b}{r} > 0 \qquad \cos\theta = \frac{a}{r} < 0 \qquad \tan\theta = \frac{b}{a} < 0$$

$$\csc\theta = \frac{r}{b} > 0 \qquad \sec\theta = \frac{r}{a} < 0 \qquad \cot\theta = \frac{a}{b} < 0$$

Table 5 lists the signs of the six trigonometric functions for each quadrant. See also Figure 41.

TABLE 5

Quadrant of θ	$\sin\theta$, $\csc\theta$	$\cos\theta$, $\sec\theta$	$\tan\theta$, $\cot\theta$
I	Positive	Positive	Positive
II	Positive	Negative	Negative
III	Negative	Negative	Positive
IV	Negative	Positive	Negative

FIGURE 41

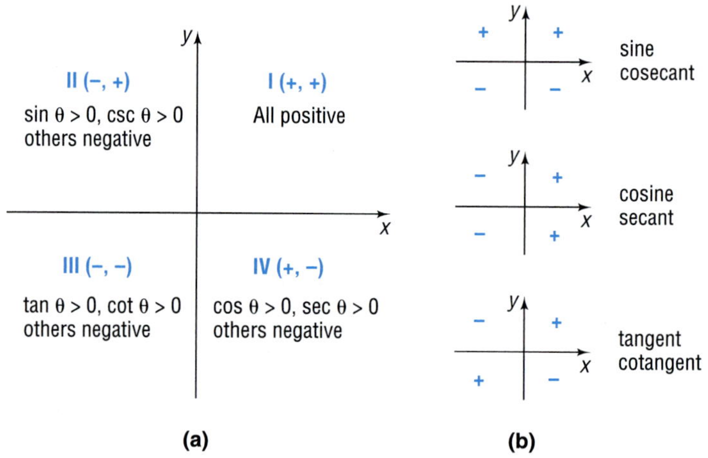

EXAMPLE 3 Finding the Quadrant in Which an Angle Lies

If $\sin\theta < 0$ and $\cos\theta < 0$, name the quadrant in which the angle θ lies.

Solution Let $P = (a, b)$ be a point on the terminal side of the angle θ a distance r from the origin. Then $\sin\theta = b/r < 0$ and $\cos\theta = a/r < 0$, so $b < 0$ and $a < 0$. Thus $P = (a, b)$ is in quadrant III and so θ lies in quadrant III. ■

Now work Problem 9.

Coterminal Angles

Two angles in standard position are said to be **coterminal** if they have the same terminal side.

See Figure 42.

FIGURE 42

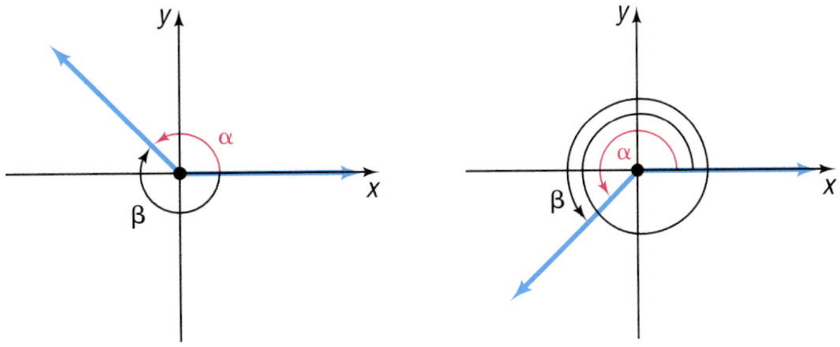

(a) α and β are coterminal **(b)** α and β are coterminal

In general, if θ is an angle measured in degrees, then $\theta \pm 360°k$, where k is any integer, is an angle coterminal with θ. If θ is measured in radians, then $\theta \pm 2\pi k$, where k is any integer, is an angle coterminal with θ.

Because coterminal angles have the same terminal side, it follows that the values of the trigonometric functions of coterminal angles are equal. We use this fact in the next example.

EXAMPLE 4 Using the Coterminal Angle to Find the Exact Value of a Trigonometric Function

Find the exact value of each of the following trigonometric functions:

(a) $\sin 390°$ (b) $\cos 420°$ (c) $\tan \dfrac{9\pi}{4}$ (d) $\sec\left(-\dfrac{7\pi}{4}\right)$ (e) $\csc(-270°)$

Solution (a) It is best to sketch the angle first. See Figure 43. The angle 390° is coterminal with 30°. Thus,

$$\sin 390° = \sin(360° + 30°) = \sin 30° = \tfrac{1}{2}$$

(b) See Figure 44. The angle 420° is coterminal with 60°. Thus,

$$\cos 420° = \cos(360° + 60°) = \cos 60° = \tfrac{1}{2}$$

FIGURE 43

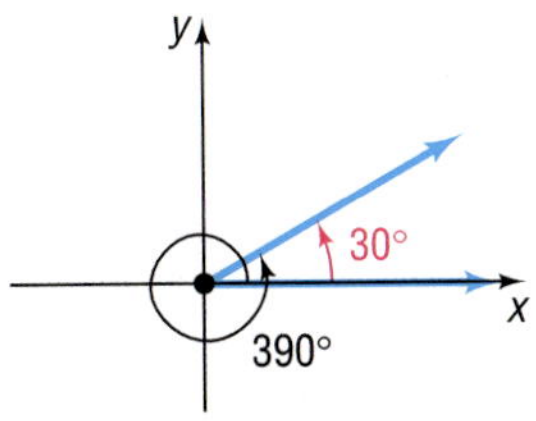

FIGURE 44

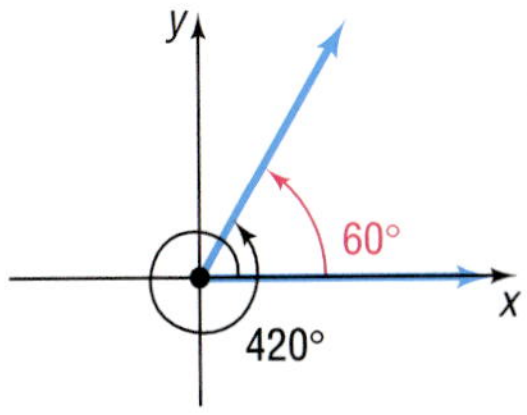

FIGURE 45

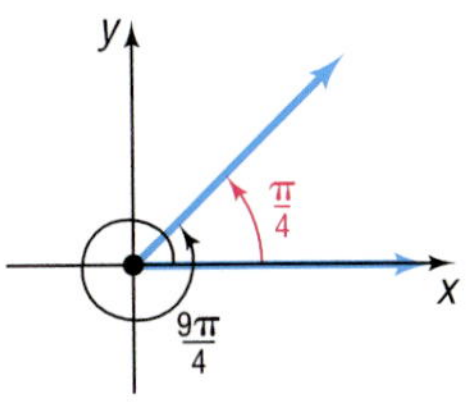

(c) See Figure 45. The angle $9\pi/4$ is coterminal with $\pi/4$. Thus,

$$\tan \frac{9\pi}{4} = \tan\left(2\pi + \frac{\pi}{4}\right)$$

$$= \tan \frac{\pi}{4} = 1$$

FIGURE 46

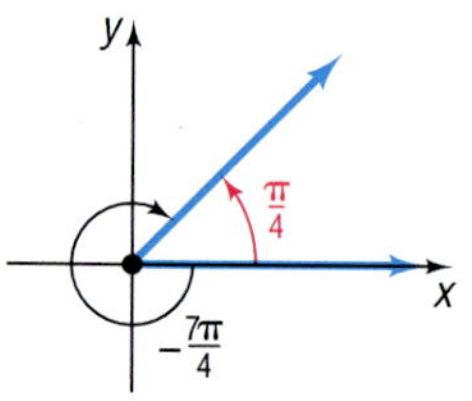

FIGURE 47

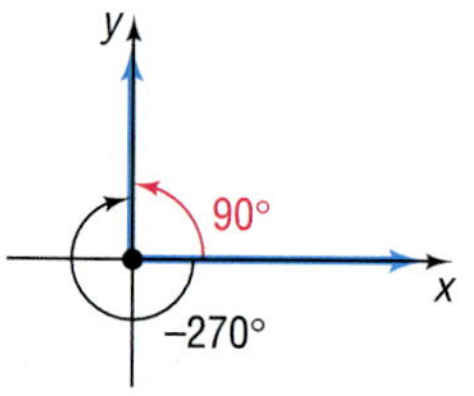

(d) See Figure 46. The angle $-7\pi/4$ is coterminal with $\pi/4$. Thus,

$$\sec\left(-\frac{7\pi}{4}\right) = \sec\left(-2\pi + \frac{\pi}{4}\right) = \sec\frac{\pi}{4} = \sqrt{2}$$

(e) See Figure 47. The angle $-270°$ is coterminal with $90°$. Thus,

$$\begin{aligned}\csc(-270°) &= \csc(-360° + 90°)\\ &= \csc 90° = 1\end{aligned}$$

As Example 4 illustrates, the value of a trigonometric function of any angle is equal to the value of the same trigonometric function of an angle θ coterminal to it where $0° \le \theta < 360°$ (or $0 \le \theta < 2\pi$). Moreover, because the angles θ and $\theta \pm 360°k$ (or $\theta \pm 2\pi k$), where k is any integer, are coterminal, and because the values of the trigonometric functions are equal for coterminal angles, it follows that

$$\begin{aligned}
\sin(\theta \pm 360°k) &= \sin\theta & \sin(\theta \pm 2\pi k) &= \sin\theta\\
\cos(\theta \pm 360°k) &= \cos\theta & \cos(\theta \pm 2\pi k) &= \cos\theta\\
\tan(\theta \pm 360°k) &= \tan\theta & \tan(\theta \pm 2\pi k) &= \tan\theta\\
\csc(\theta \pm 360°k) &= \csc\theta & \csc(\theta \pm 2\pi k) &= \csc\theta\\
\sec(\theta \pm 360°k) &= \sec\theta & \sec(\theta \pm 2\pi k) &= \sec\theta\\
\cot(\theta \pm 360°k) &= \cot\theta & \cot(\theta \pm 2\pi k) &= \cot\theta
\end{aligned} \qquad (1)$$

These formulas show that the values of the trigonometric functions repeat themselves every 360° (or 2π radians).

Now work Problem 33.

Reference Angle

3 Once we know in which quadrant an angle lies, we know the sign of each of the trigonometric functions of that angle. The use of a certain reference angle may help us to evaluate the trigonometric functions of such an angle.

> Let θ denote a nonacute angle that lies in a quadrant. The acute angle formed by the terminal side of θ and either the positive *x*-axis or the negative *x*-axis is called the **reference angle** for θ.

Figure 48 illustrates the reference angle for some general angles θ. Note that a reference angle is always an acute angle, that is, an angle whose measure is between 0° and 90°.

FIGURE 48

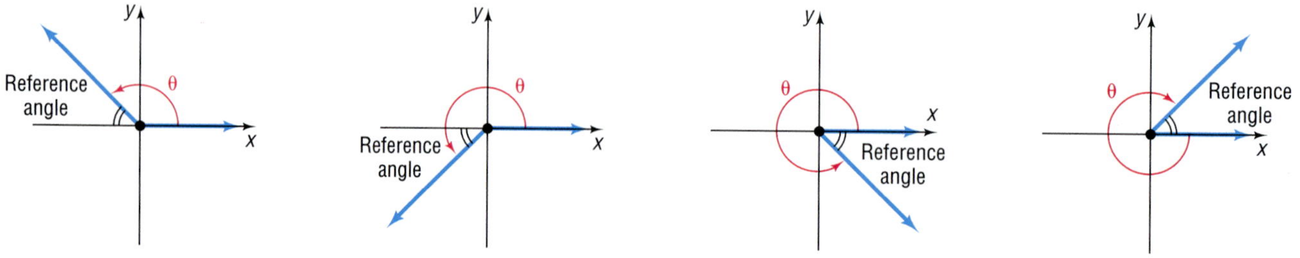

(a) (b) (c) (d)

Although formulas can be given for calculating reference angles, usually it is easier to find the reference angle for a given angle by making a quick sketch of the angle.

EXAMPLE 5 Finding Reference Angles

Find the reference angle for each of the following angles:

(a) $150°$ (b) $-45°$ (c) $9\pi/4$ (d) $-5\pi/6$

Solution (a) Refer to Figure 49. The reference angle for $150°$ is $30°$.

FIGURE 49

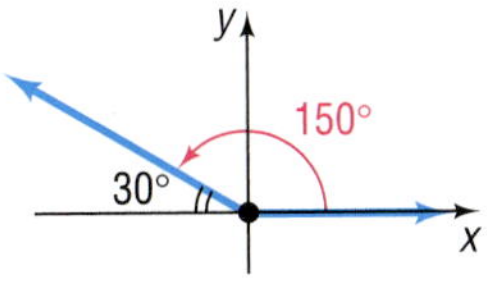

(b) Refer to Figure 50. The reference angle for $-45°$ is $45°$.

FIGURE 50

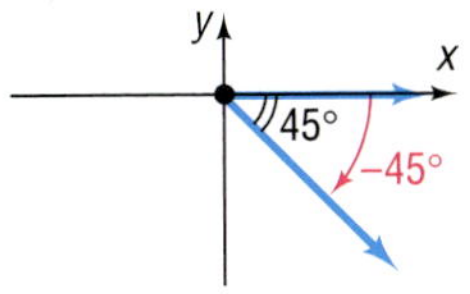

(c) Refer to Figure 51. The reference angle for $9\pi/4$ is $\pi/4$.

FIGURE 51

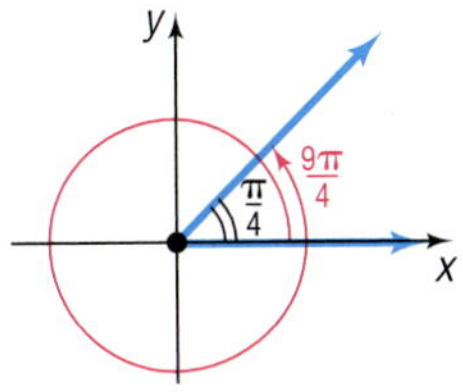

(d) Refer to Figure 52. The reference angle for $-5\pi/6$ is $\pi/6$.

FIGURE 52

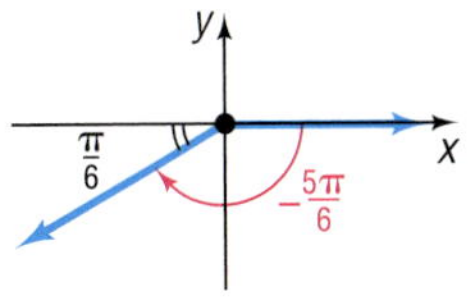

Now work Problem 17.

The advantage of using reference angles is that, except for the correct sign, the values of the trigonometric functions of a general angle θ equal the values of the trigonometric functions of its reference angle.

Theorem Reference Angles

If θ is an angle that lies in a quadrant and if α is its reference angle, then

$$\sin\theta = \pm\sin\alpha \qquad \cos\theta = \pm\cos\alpha \qquad \tan\theta = \pm\tan\alpha$$
$$\csc\theta = \pm\csc\alpha \qquad \sec\theta = \pm\sec\alpha \qquad \cot\theta = \pm\cot\alpha \tag{2}$$

where the + or − sign depends on the quadrant in which θ lies.

For example, suppose that θ lies in quadrant II and α is its reference angle. See Figure 53. If (a, b) is a point on the terminal side of θ and if $r = \sqrt{a^2 + b^2}$, we have

$$\sin\theta = \frac{b}{r} = \sin\alpha \qquad \cos\theta = \frac{a}{r} = \frac{-|a|}{r} = -\cos\alpha$$

and so on.

FIGURE 53 $\sin\theta = b/r$, $\sin\alpha = b/r$, $\cos\theta = a/r$, $\cos\alpha = |a|/r$

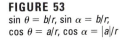

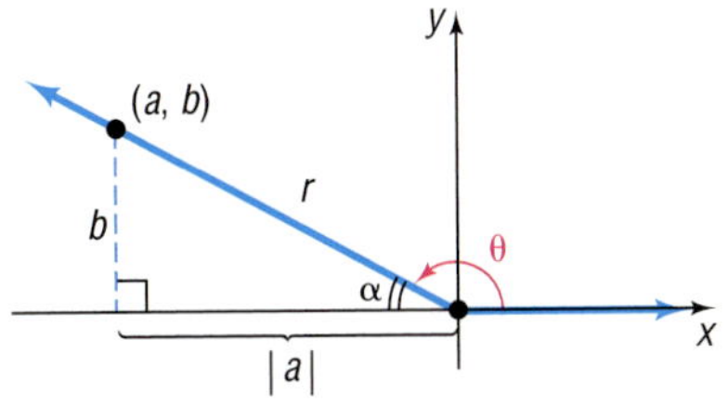

The next example illustrates how the theorem on reference angles is used.

EXAMPLE 6 Using the Reference Angle to Find the Exact Value of a Trigonometric Function

FIGURE 54

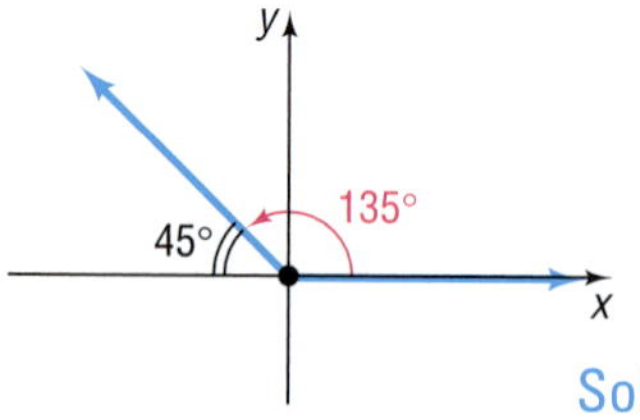

Find the exact value of each of the following trigonometric functions using reference angles:

(a) $\sin 135°$ (b) $\cos 240°$ (c) $\cos \dfrac{5\pi}{6}$ (d) $\tan\left(-\dfrac{\pi}{3}\right)$

Solution (a) Refer to Figure 54. The angle 135° is in quadrant II, where the sine function is positive. The reference angle for 135° is 45°. Thus,

$$\sin 135° = \sin 45° = \frac{\sqrt{2}}{2}$$

FIGURE 55

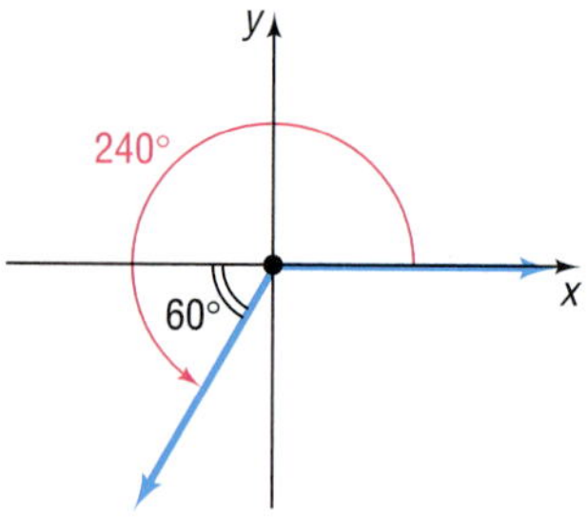

(b) Refer to Figure 55. The angle 240° is in quadrant III, where the cosine function is negative. The reference angle for 240° is 60°. Thus,

$$\cos 240° = -\cos 60° = -\tfrac{1}{2}$$

(c) Refer to Figure 56. The angle $5\pi/6$ is in quadrant II, where the cosine function is negative. The reference angle for $5\pi/6$ is $\pi/6$. Thus,

FIGURE 56

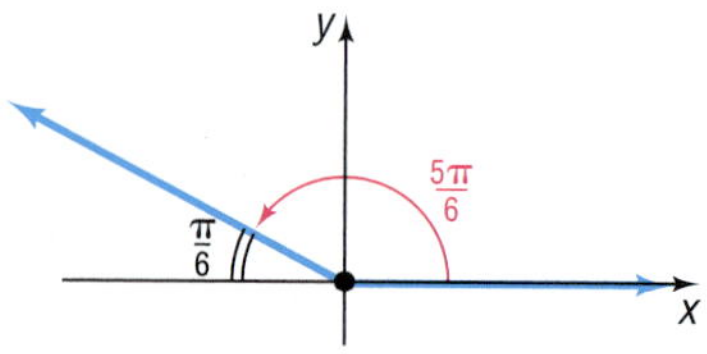

$$\cos\frac{5\pi}{6} = -\cos\frac{\pi}{6} = -\frac{\sqrt{3}}{2}$$

(d) Refer to Figure 57. The angle $-\pi/3$ is in quadrant IV, where the tangent function is negative. The reference angle for $-\pi/3$ is $\pi/3$. Thus,

FIGURE 57

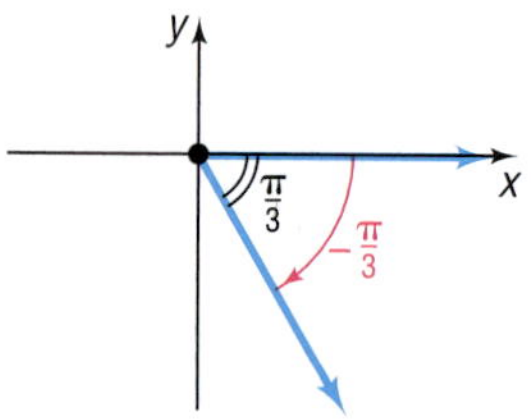

$$\tan\left(-\frac{\pi}{3}\right) = -\tan\frac{\pi}{3} = -\sqrt{3}$$

Now work Problem 49.

EXAMPLE 7 Finding the Exact Value of Trigonometric Functions

Given that $\cos\theta = -\frac{2}{3}$, $\pi/2 < \theta < \pi$, find the exact value of each of the remaining trigonometric functions.

Solution The angle θ lies in quadrant II, so we know that $\sin\theta$ and $\csc\theta$ are positive, whereas the other trigonometric functions are negative. If α is the reference angle for θ, then $\cos\alpha = \frac{2}{3}$. The values of the remaining trigonometric func-

FIGURE 58

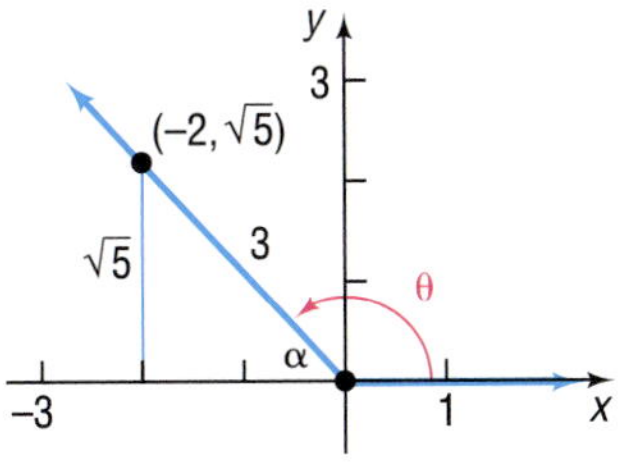

tions of the angle α can be found by drawing the appropriate triangle. We use Figure 58 to obtain

$$\sin\alpha = \frac{\sqrt{5}}{3} \qquad \cos\alpha = \frac{2}{3} \qquad \tan\alpha = \frac{\sqrt{5}}{2}$$

$$\csc\alpha = \frac{3}{\sqrt{5}} = \frac{3\sqrt{5}}{5} \qquad \sec\alpha = \frac{3}{2} \qquad \cot\alpha = \frac{2}{\sqrt{5}} = \frac{2\sqrt{5}}{5}$$

Now we assign the appropriate signs to each of these values to find the values of the trigonometric functions of θ:

$$\sin\theta = \frac{\sqrt{5}}{3} \qquad \cos\theta = -\frac{2}{3} \qquad \tan\theta = -\frac{\sqrt{5}}{2}$$

$$\csc\theta = \frac{3\sqrt{5}}{5} \qquad \sec\theta = -\frac{3}{2} \qquad \cot\theta = -\frac{2\sqrt{5}}{5}$$

Now work Problem 69.

EXAMPLE 8

Finding the Exact Value of Trigonometric Functions

If $\tan\theta = -4$ and $\sin\theta < 0$, find the exact value of each of the remaining trigonometric functions of θ.

Solution

Since $\tan\theta = -4 < 0$ and $\sin\theta < 0$, it follows that θ lies in quadrant IV. If α is the reference angle for θ, then $\tan\alpha = 4$. See Figure 59. Thus,

FIGURE 59

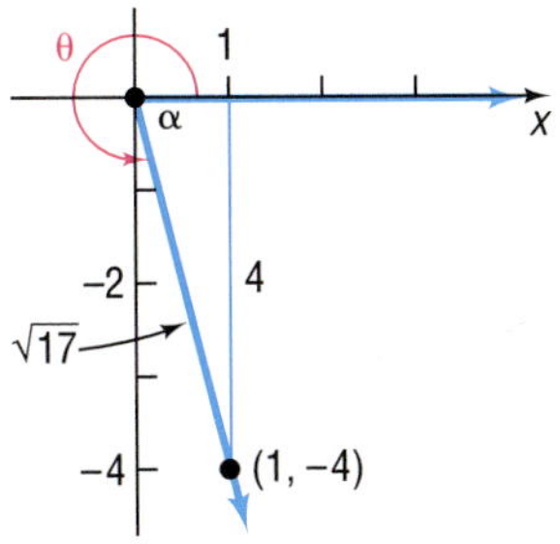

$$\sin\alpha = \frac{4}{\sqrt{17}} = \frac{4\sqrt{17}}{17} \qquad \cos\alpha = \frac{1}{\sqrt{17}} = \frac{\sqrt{17}}{17} \qquad \tan\alpha = \frac{4}{1} = 4$$

$$\csc\alpha = \frac{\sqrt{17}}{4} \qquad \sec\alpha = \frac{\sqrt{17}}{1} = \sqrt{17} \qquad \cot\alpha = \frac{1}{4}$$

We assign the appropriate sign to each of these to obtain the values of the trigonometric functions of θ:

$$\sin\theta = \frac{-4\sqrt{17}}{17} \qquad \cos\theta = \frac{\sqrt{17}}{17} \qquad \tan\theta = -4$$

$$\csc\theta = \frac{-\sqrt{17}}{4} \qquad \sec\theta = \sqrt{17} \qquad \cot\theta = -\frac{1}{4}$$

SUMMARY

To find the values of the trigonometric functions of a general angle:

STEP 1: If the angle is not between 0° and 360° (0 and 2π radians), replace it by an angle θ, $0° \le \theta < 360°$ $(0 \le \theta < 2\pi)$, that is coterminal with it.

STEP 2: If the angle θ is a quadrantal angle, pick a point on its terminal side and apply the definitions of the trigonometric functions.

STEP 3: If the angle θ lies in a quadrant, determine the correct signs of the trigonometric functions in that quadrant and determine the reference angle α for θ. Now express each trigonometric function of θ in terms of the same value (except for the sign) of the trigonometric function of α, an acute angle.

6.4 EXERCISES

In Problems 1–8, a point on the terminal side of an angle θ is given. Find the exact value of each of the six trigonometric functions of θ.

1. $(-3, 4)$ **2.** $(5, -12)$ **3.** $(2, -3)$ **4.** $(-1, -2)$

5. $(-3, -3)$ **6.** $(2, -2)$ **7.** $(-3, -2)$ **8.** $(2, 2)$

In Problems 9–16, name the quadrant in which the angle θ lies.

9. $\sin\theta > 0, \cos\theta < 0$ **10.** $\sin\theta < 0, \cos\theta > 0$ **11.** $\sin\theta < 0, \tan\theta < 0$

12. $\cos\theta > 0, \tan\theta > 0$ **13.** $\cos\theta > 0, \cot\theta < 0$ **14.** $\sin\theta < 0, \cot\theta > 0$

15. $\sec\theta < 0, \tan\theta > 0$ **16.** $\csc\theta > 0, \cot\theta < 0$

In Problems 17–32, find the reference angle of each angle.

17. $-30°$ **18.** $60°$ **19.** $120°$ **20.** $300°$ **21.** $210°$ **22.** $330°$

23. $5\pi/4$ **24.** $5\pi/6$ **25.** $8\pi/3$ **26.** $7\pi/4$ **27.** $-135°$ **28.** $-240°$

29. $-2\pi/3$ **30.** $-7\pi/6$ **31.** $440°$ **32.** $490°$

In Problems 33–68, find the exact value of each expression. Do not use a calculator.

33. $\sin 405°$ **34.** $\cos 420°$ **35.** $\tan 405°$ **36.** $\sin 390°$ **37.** $\csc 450°$ **38.** $\sec 540°$

39. $\cot 390°$ **40.** $\sec 420°$ **41.** $\cos\dfrac{33\pi}{4}$ **42.** $\sin\dfrac{9\pi}{4}$ **43.** $\tan(21\pi)$ **44.** $\csc\dfrac{9\pi}{2}$

45. $\sec\dfrac{17\pi}{4}$ **46.** $\cot\dfrac{17\pi}{4}$ **47.** $\tan\dfrac{19\pi}{6}$ **48.** $\sec\dfrac{25\pi}{6}$ **49.** $\sin 150°$ **50.** $\cos 210°$

51. $\cos 315°$ **52.** $\sin 120°$ **53.** $\sec 240°$ **54.** $\csc 300°$ **55.** $\cot 330°$ **56.** $\tan 225°$

57. $\sin\dfrac{3\pi}{4}$ **58.** $\cos\dfrac{2\pi}{3}$ **59.** $\cot\dfrac{7\pi}{6}$ **60.** $\csc\dfrac{7\pi}{4}$ **61.** $\cos(-60°)$ **62.** $\tan(-120°)$

63. $\sin\left(-\dfrac{2\pi}{3}\right)$ **64.** $\cot\left(-\dfrac{\pi}{6}\right)$ **65.** $\tan\dfrac{14\pi}{3}$ **66.** $\sec\dfrac{11\pi}{4}$ **67.** $\csc(-315°)$ **68.** $\sec(-225°)$

In Problems 69–84, find the exact value of each of the remaining trigonometric functions of θ.

69. $\sin\theta = \frac{12}{13}$, $90° < \theta < 180°$ **70.** $\cos\theta = \frac{3}{5}$, $270° < \theta < 360°$

71. $\cos\theta = -\frac{4}{5}$, $\pi < \theta < 3\pi/2$ **72.** $\sin\theta = -\frac{5}{13}$, $\pi < \theta < 3\pi/2$

73. $\sin\theta = \frac{5}{13}$, $\cos\theta < 0$ **74.** $\cos\theta = \frac{4}{5}$, $\sin\theta < 0$

75. $\cos\theta = -\frac{1}{3}$, $\csc\theta > 0$ **76.** $\sin\theta = -\frac{2}{3}$, $\sec\theta > 0$

77. $\sin\theta = \frac{2}{3}$, $\tan\theta < 0$ **78.** $\cos\theta = -\frac{1}{4}$, $\tan\theta > 0$

79. $\sec\theta = 2$, $\sin\theta < 0$ **80.** $\csc\theta = 3$, $\cot\theta < 0$

81. $\tan\theta = \frac{3}{4}$, $\sin\theta < 0$ **82.** $\cot\theta = \frac{4}{3}$, $\cos\theta < 0$

83. $\tan\theta = -\frac{1}{3}$, $\sin\theta > 0$ **84.** $\sec\theta = -2$, $\tan\theta > 0$

85. Find the exact value of
$\sin 45° + \sin 135° + \sin 225° + \sin 315°$.

86. Find the exact value of $\tan 60° + \tan 150°$.

87. If $\sin \theta = 0.2$, find $\sin(\theta + \pi)$.

88. If $\cos \theta = 0.4$, find $\cos(\theta + \pi)$.

89. If $\tan \theta = 3$, find $\tan(\theta + \pi)$.

90. If $\cot \theta = -2$, find $\cot(\theta + \pi)$.

91. If $\sin \theta = \frac{1}{5}$, find $\csc \theta$.

92. If $\cos \theta = \frac{2}{3}$, find $\sec \theta$.

93. Find the exact value of $\sin 1° + \sin 2° + \sin 3° + \cdots + \sin 358° + \sin 359°$.

94. Find the exact value of $\cos 1° + \cos 2° + \cos 3° + \cdots + \cos 358° + \cos 359°$.

95. Projectile Motion An object is propelled upward at an angle θ, $45° < \theta < 90°$, to the horizontal with an initial velocity of v_0 feet per second from the base of a plane that makes an angle of 45° with the horizontal. See the illustration. If air resistance is ignored, the distance R it travels up the inclined plane is given by

$$R = \frac{v_0^2\sqrt{2}}{32}(\sin 2\theta - \cos 2\theta - 1)$$

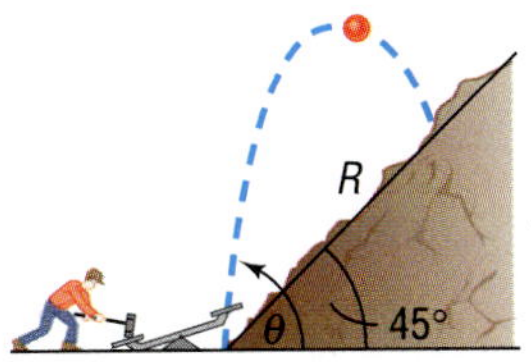

(a) Find the distance R the object travels along the inclined plane if the initial velocity is 32 ft/sec and $\theta = 60°$.
(b) Graph $R = R(\theta)$ if the initial velocity is 32 ft/sec.
(c) What value of θ makes R largest?

96. Give three examples that demonstrate how to use the theorem on reference angles.

97. Write a brief paragraph that explains how to quickly compute the value of the trigonometric functions of 0°, 90°, 180°, and 270°.

6.5 PROPERTIES OF THE TRIGONOMETRIC FUNCTIONS

1 Find the Exact Value of the Trigonometric Functions Using the Unit Circle
2 Determine the Domain and Range of the Trigonometric Functions
3 Determine the Period of the Trigonometric Functions
4 Use Even–Odd Properties to Find the Exact Value of the Trigonometric Functions

In this section, we develop important properties or characteristics of the trigonometric functions. We begin by introducing the trigonometric functions using the unit circle. We show that this approach leads to the definition given earlier using right triangles.

The Unit Circle

Recall that the **unit circle** is a circle whose radius is 1 and whose center is at the origin of a rectangular coordinate system. Because the radius r of the unit circle is 1, we see from the formula $s = r\theta$ that, on the unit circle, a central angle of θ radians subtends an arc whose length s is

$$s = \theta$$

FIGURE 60

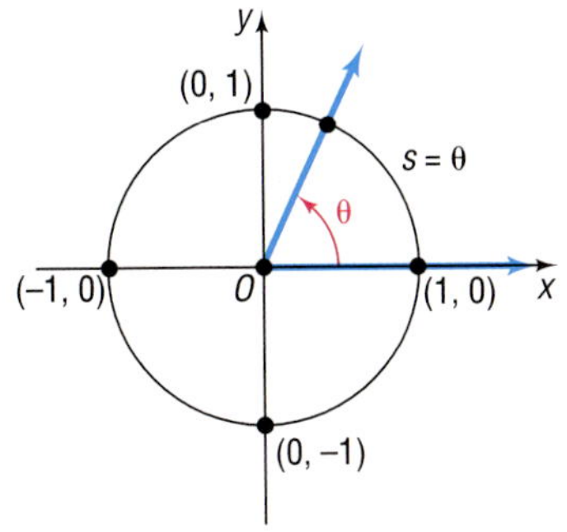

See Figure 60. Thus, on the unit circle, the length measure of the arc s equals the radian measure of the central angle θ. In other words, on the unit circle, the real number used to measure an angle θ in radians corresponds exactly with the real number used to measure the length of the arc subtended by that angle.

For example, suppose that $r = 1$ foot. Then, if $\theta = 3$ radians, $s = 3$ feet; if $\theta = 8.2$ radians, then $s = 8.2$ feet; and so on.

Now let t be any real number and let θ be the angle, in standard position, equal to t radians. Let P be the point on the unit circle that is also on the terminal side of θ. If $t \geq 0$, this point P is reached by moving *counterclockwise* along the unit circle, starting at $(1, 0)$, for a length of arc equal to t units. See Figure 61(a). If $t < 0$, this point P is reached by moving *clockwise* along the unit circle, starting at $(1, 0)$, for a length of arc equal to $|t|$ units. See Figure 61(b).

FIGURE 61

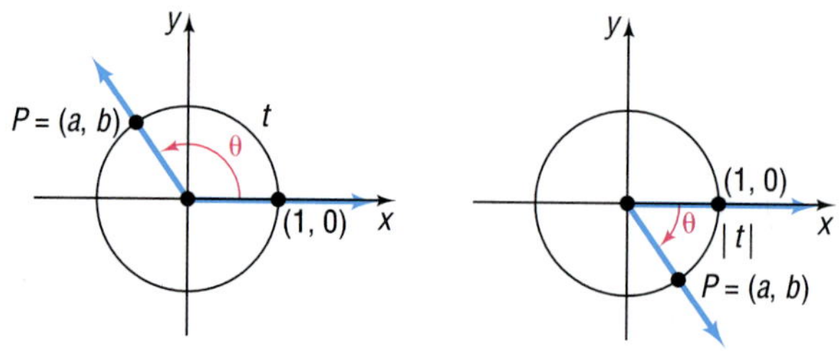

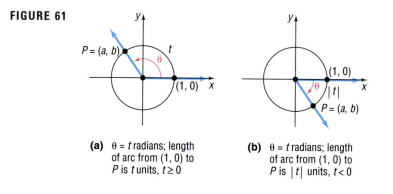

(a) $\theta = t$ radians; length of arc from $(1, 0)$ to P is t units, $t \geq 0$

(b) $\theta = t$ radians; length of arc from $(1, 0)$ to P is $|t|$ units, $t < 0$

Thus, to each real number t, there corresponds a unique point $P = (a, b)$ on the unit circle. This is the important idea here. No matter what real number t is chosen, there corresponds a unique point P on the unit circle. We use the coordinates of the point $P = (a, b)$ on the unit circle corresponding to the real number t to define the six trigonometric functions of the real number t.

Let t be a real number and let $P = (a, b)$ be the point on the unit circle that corresponds to t.

The **sine function** associates with t the y-coordinate of P and is denoted by

$$\sin t = b$$

The **cosine function** associates with t the x-coordinate of P and is denoted by

$$\cos t = a$$

If $a \neq 0$, the **tangent function** is defined as

$$\tan t = \frac{b}{a}$$

If $b \neq 0$, the **cosecant function** is defined as

$$\csc t = \frac{1}{b}$$

If $a \neq 0$, the **secant function** is defined as

$$\sec t = \frac{1}{a}$$

If $b \neq 0$, the **cotangent function** is defined as

$$\cot t = \frac{a}{b}$$

Once again, notice in these definitions that if $a = 0$, that is, if the point $P = (0, b)$ is on the y-axis, then the tangent function and the secant function are undefined. Also, if $b = 0$, that is, if the point $P = (a, 0)$ is on the x-axis, then the cosecant function and the cotangent function are undefined.

Because we use the unit circle in these definitions of the trigonometric functions, they are also sometimes referred to as **circular functions.**

EXAMPLE 1

Finding the Value of the Six Trigonometric Functions

1 Let t be a real number and let $P = \left(-\frac{1}{2}, \sqrt{3}/2\right)$ be the point on the unit circle that corresponds to t. See Figure 62. Then

FIGURE 62
$\theta = t$ radians

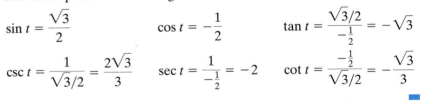

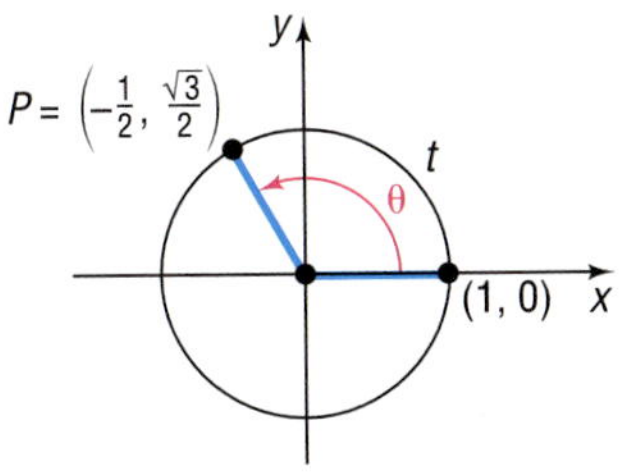

$$\sin t = \frac{\sqrt{3}}{2} \qquad \cos t = -\frac{1}{2} \qquad \tan t = \frac{\sqrt{3}/2}{-\frac{1}{2}} = -\sqrt{3}$$

$$\csc t = \frac{1}{\sqrt{3}/2} = \frac{2\sqrt{3}}{3} \qquad \sec t = \frac{1}{-\frac{1}{2}} = -2 \qquad \cot t = \frac{-\frac{1}{2}}{\sqrt{3}/2} = -\frac{\sqrt{3}}{3}$$

Now work Problem 1.

Trigonometric Functions of Angles

Let P be the point on the unit circle corresponding to the real number t. Then the angle θ, in standard position and measured in radians, whose terminal side is the ray from the origin through P is

$$\theta = t \text{ radians}$$

FIGURE 63

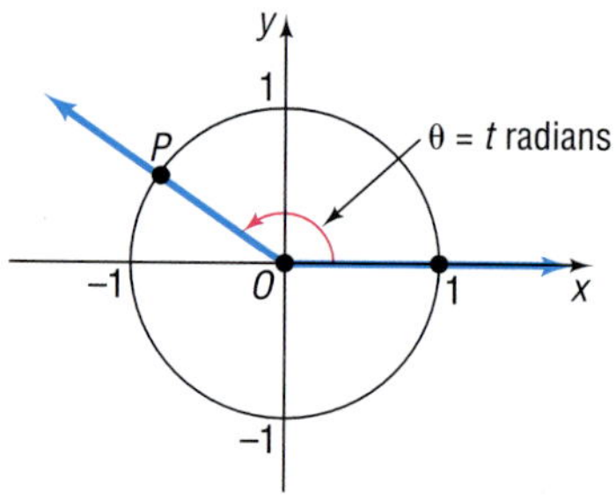

See Figure 63. Thus, on the unit circle, the measure of the angle θ in radians equals the value of the real number t. As a result, we can say that

$$\sin t = \sin \theta$$

↑ Real number ↑ $\theta = t$ radians

and so on. We now can define the trigonometric functions of the angle θ.

If $\theta = t$ radians, the six **trigonometric functions of the angle θ** are defined as

$\sin \theta = \sin t$	$\cos \theta = \cos t$	$\tan \theta = \tan t$
$\csc \theta = \csc t$	$\sec \theta = \sec t$	$\cot \theta = \cot t$

Even though the distinction between trigonometric functions of real numbers and trigonometric functions of angles is important, it is customary to refer to trigonometric functions of real numbers and trigonometric functions of angles collectively as *the trigonometric functions.* We will follow this practice from now on.

Since the values of the trigonometric functions of an angle θ are determined by the coordinates of the point $P = (a, b)$ on the unit circle corresponding to θ, the units used to measure the angle θ are irrelevant. For example, it does not matter whether we write $\theta = \pi/2$ radians or $\theta = 90°$. The point on the unit circle corresponding to this angle is $P = (0, 1)$. Hence,

$$\sin \frac{\pi}{2} = \sin 90° = 1 \quad \text{and} \quad \cos \frac{\pi}{2} = \cos 90° = 0$$

To find the exact value of a trigonometric function of an angle θ requires that we locate the corresponding point P on the unit circle. In fact, though, any circle whose center is at the origin can be used.

Let θ be any nonquadrantal angle placed in standard position. Let $P^* = (a^*, b^*)$ be the point where the terminal side of θ intersects the unit circle. See Figure 64.

FIGURE 64

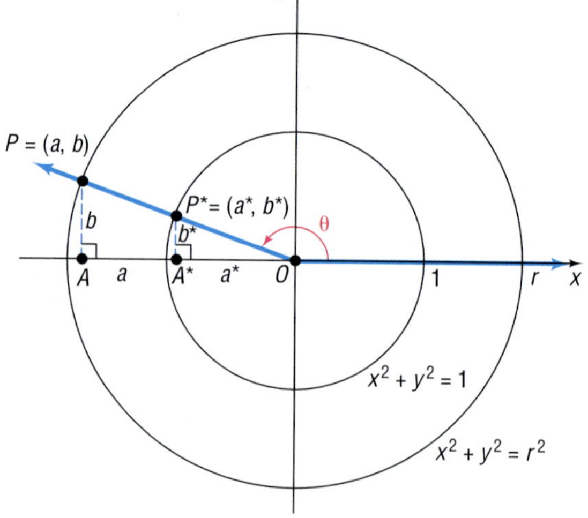

Let $P = (a, b)$ be any point on the terminal side of θ. Suppose that r is the distance from the origin to P. Then P is on the circle $x^2 + y^2 = r^2$. Refer again to Figure 64. Notice that the triangles OA^*P^* and OAP are similar; thus, ratios of corresponding sides are equal:

$$\frac{b^*}{1} = \frac{b}{r} \qquad \frac{a^*}{1} = \frac{a}{r} \qquad \frac{b^*}{a^*} = \frac{b}{a}$$

$$\frac{1}{b^*} = \frac{r}{b} \qquad \frac{1}{a^*} = \frac{r}{a} \qquad \frac{a^*}{b^*} = \frac{a}{b}$$

These results lead us to formulate the following theorem:

Theorem

For an angle θ in standard position, let $P = (a, b)$ be any point on the terminal side of θ. Let r equal the distance from the origin to P. Then

$$\sin\theta = \frac{b}{r} \qquad \cos\theta = \frac{a}{r} \qquad \tan\theta = \frac{b}{a},\quad a \neq 0$$

$$\csc\theta = \frac{r}{b},\quad b \neq 0 \qquad \sec\theta = \frac{r}{a},\quad a \neq 0 \qquad \cot\theta = \frac{a}{b},\quad b \neq 0$$

This result coincides with the definition given in Section 6.4 for the six trigonometric functions of a general angle θ.

EXAMPLE 2 Finding the Value of Trigonometric Functions, Given $\sin\theta$

Given that $\sin\theta = \frac{1}{3}$ and $\cos\theta < 0$, find the exact value of the remaining five trigonometric functions.

FIGURE 65

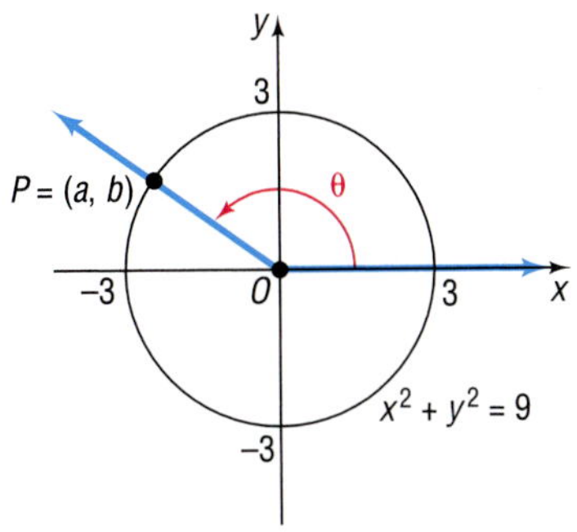

Solution Suppose that $P = (a, b)$ is a point on the terminal side of θ that lies a distance of $r = 3$ units from the origin. See Figure 65. (Do you see why we chose 3 units? Notice that $\sin\theta = \frac{1}{3} = \frac{b}{r}$. The choice $r = 3$ will make our calculations easy.) With this choice, $b = 1$ and $r = 3$. Since $\cos\theta = a/r < 0$, it follows that $a < 0$. Thus, we have

$$a^2 + b^2 = r^2 \qquad b = 1,\ r = 3,\ a < 0.$$

$$a^2 + 1^2 = 3^2$$

$$a^2 = 8$$

$$a = -2\sqrt{2}$$

Thus,

$$\cos\theta = \frac{a}{r} = \frac{-2\sqrt{2}}{3} \qquad \tan\theta = \frac{b}{a} = \frac{1}{-2\sqrt{2}} = \frac{-\sqrt{2}}{4}$$

$$\csc\theta = \frac{r}{b} = \frac{3}{1} = 3 \qquad \sec\theta = \frac{r}{a} = \frac{3}{-2\sqrt{2}} = \frac{-3\sqrt{2}}{4} \qquad \cot\theta = \frac{a}{b} = \frac{-2\sqrt{2}}{1} = -2\sqrt{2}$$

You should compare this method with the two methods used earlier to solve Example 3, Section 6.2.

Now work Problem 9.

Domain and Range of the Trigonometric Functions

FIGURE 66

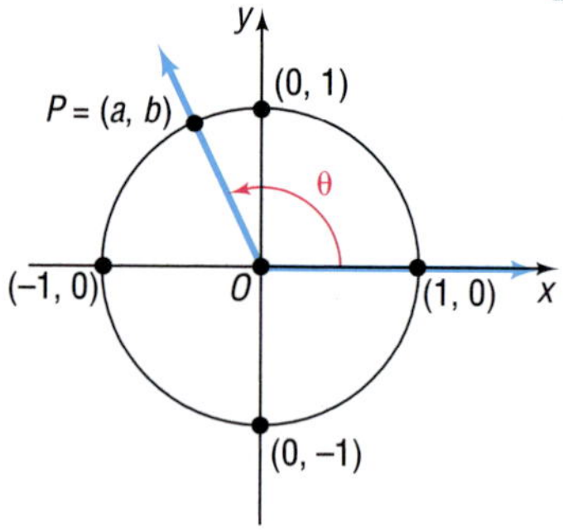

2 Let θ be an angle in standard position, and let $P = (a, b)$ be a point on the terminal side of θ. Suppose, for convenience, that P also lies on the unit circle. See Figure 66. Then, by definition

$$\sin\theta = b \qquad \cos\theta = a \qquad \tan\theta = \frac{b}{a}, \quad a \neq 0$$

$$\csc\theta = \frac{1}{b}, \quad b \neq 0 \qquad \sec\theta = \frac{1}{a}, \quad a \neq 0 \qquad \cot\theta = \frac{a}{b}, \quad b \neq 0$$

For $\sin\theta$ and $\cos\theta$, θ can be any angle, so it follows that the domain of the sine function and cosine function is the set of all real numbers.

The domain of the sine function is the set of all real numbers.

The domain of the cosine function is the set of all real numbers.

If $a = 0$, then the tangent function and the secant function are not defined. Thus, for the tangent function and secant function, the x-coordinate of $P = (a, b)$ cannot be 0. On the unit circle, there are two such points, $(0, 1)$ and $(0, -1)$. These two points correspond to the angles $\pi/2$ (90°) and $3\pi/2$ (270°) or, more generally, to any angle that is an odd multiple of $\pi/2$ (90°), such as $\pi/2$ (90°), $3\pi/2$ (270°), $5\pi/2$ (450°), $-\pi/2$ (−90°), and $-3\pi/2$ (−270°). Such angles must therefore be excluded from the domain of the tangent function and secant function.

The domain of the tangent function is the set of all real numbers, except odd multiples of $\pi/2$ (90°).

The domain of the secant function is the set of all real numbers, except odd multiples of $\pi/2$ (90°).

If $b = 0$, then the cotangent function and the cosecant function are not defined. Thus, for the cotangent function and cosecant function the y-coordinate of $P = (a, b)$ cannot be 0. On the unit circle, there are two such points, $(1, 0)$ and $(-1, 0)$. These two points correspond to the angles 0 (0°) and π (180°) or, more generally, to any angle that is an integral multiple of π (180°), such as 0 (0°), π (180°), 2π (360°), 3π (540°), and $-\pi$ (−180°). Such angles must therefore be excluded from the domain of the cotangent function and cosecant function.

The domain of the cotangent function is the set of all real numbers, except integral multiples of π (180°).

The domain of the cosecant function is the set of all real numbers, except integral multiples of π (180°).

Next, we determine the range of each of the six trigonometric functions. Refer again to Figure 66. Let $P = (a, b)$ be the point on the unit circle that corresponds to the angle θ. It follows that $-1 \le a \le 1$ and $-1 \le b \le 1$. Consequently, since $\sin\,\theta = b$ and $\cos\,\theta = a$, we have

$$-1 \le \sin\,\theta \le 1 \quad \text{and} \quad -1 \le \cos\,\theta \le 1$$

Thus, the range of both the sine function and the cosine function consists of all real numbers between -1 and 1, inclusive. In terms of absolute value notation, we have $|\sin\,\theta| \le 1$ and $|\cos\,\theta| \le 1$.

Similarly, if θ is not a multiple of π (180°), then $\csc\,\theta = 1/b$. Since $b = \sin\,\theta$ and $|b| = |\sin\,\theta| \le 1$, it follows that $|\csc\,\theta| = 1/|\sin\,\theta| = 1/|b| \ge 1$. Thus, the range of the cosecant function consists of all real numbers less than or equal to -1 or greater than or equal to 1. That is,

$$\csc\,\theta \le -1 \quad \text{or} \quad \csc\,\theta \ge 1$$

If θ is not an odd multiple of $\pi/2$ (90°), then, by definition, $\sec\,\theta = 1/a$. Since $a = \cos\,\theta$ and $|a| = |\cos\,\theta| \le 1$, it follows that $|\sec\,\theta| = 1/|\cos\,\theta| = 1/|a| \ge 1$. Thus, the range of the secant function consists of all real numbers less than or equal to -1 or greater than or equal to 1.

$$\sec\,\theta \le -1 \quad \text{or} \quad \sec\,\theta \ge 1$$

The range of both the tangent function and the cotangent function consists of all real numbers. That is,

$$-\infty < \tan\,\theta < \infty \quad \text{and} \quad -\infty < \cot\,\theta < \infty$$

You are asked to prove this in Problems 35 and 36.

Table 6 summarizes these results.

TABLE 6

Function	Symbol	Domain	Range
sine	$f(\theta) = \sin\,\theta$	All real numbers	All real numbers from -1 to 1, inclusive
cosine	$f(\theta) = \cos\,\theta$	All real numbers	All real numbers from -1 to 1, inclusive
tangent	$f(\theta) = \tan\,\theta$	All real numbers, except odd multiples of $\pi/2$ (90°)	All real numbers
cosecant	$f(\theta) = \csc\,\theta$	All real numbers, except integral multiples of π (180°)	All real numbers greater than or equal to 1 or less than or equal to -1
secant	$f(\theta) = \sec\,\theta$	All real numbers, except odd multiples of $\pi/2$ (90°)	All real numbers greater than or equal to 1 or less than or equal to -1
cotangent	$f(\theta) = \cot\,\theta$	All real numbers, except integral multiples of π (180°)	All real numbers

Now work Problem 17.

Period of the Trigonometric Functions

FIGURE 67

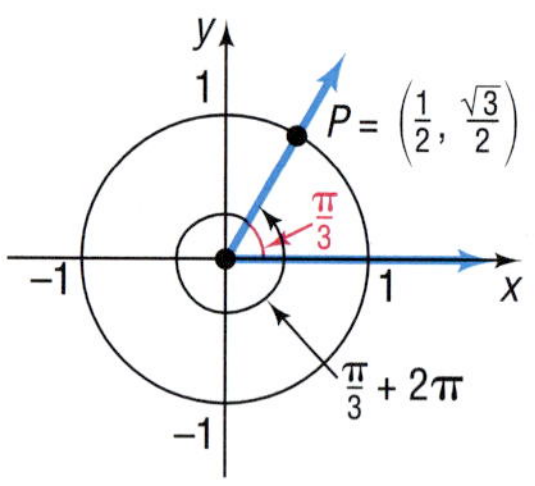

3 Look at Figure 67. This figure shows that for an angle of $\pi/3$ radians the corresponding point P on the unit circle is $(1/2, \sqrt{3}/2)$. Notice that for an angle of $\pi/3 + 2\pi$ radians the corresponding point P on the unit circle is also $(1/2, \sqrt{3}/2)$. Thus,

$$\sin\frac{\pi}{3} = \frac{\sqrt{3}}{2} \quad \text{and} \quad \sin\left(\frac{\pi}{3} + 2\pi\right) = \frac{\sqrt{3}}{2}$$

$$\cos\frac{\pi}{3} = \frac{1}{2} \quad \text{and} \quad \cos\left(\frac{\pi}{3} + 2\pi\right) = \frac{1}{2}$$

This example illustrates a more general situation. For a given angle θ, measured in radians, suppose that we know the corresponding point $P = (a, b)$ on the unit circle. Now add 2π to θ. The point on the unit circle corresponding to $\theta + 2\pi$ is identical to the point P on the unit circle corresponding to θ. See Figure 68. Thus, the values of the trigonometric functions of $\theta + 2\pi$ are equal to the values of the corresponding trigonometric functions of θ.

FIGURE 68

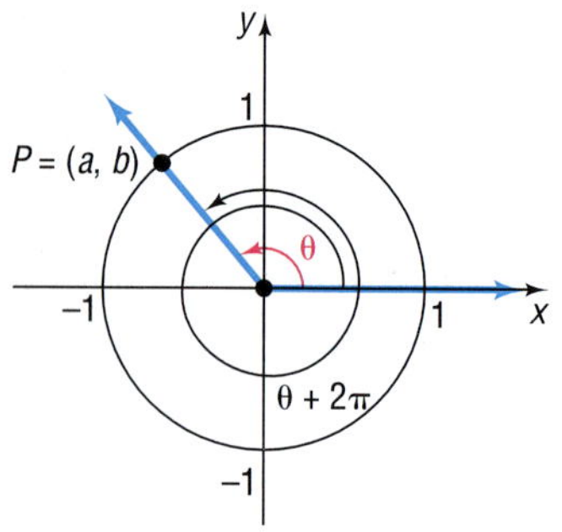

If we add (or subtract) integral multiples of 2π to θ, the trigonometric values remain unchanged. That is, for all θ,

$$\sin(\theta + 2\pi k) = \sin\theta \qquad \cos(\theta + 2\pi k) = \cos\theta \tag{1}$$

where k is any integer.

Functions that exhibit this kind of behavior are called *periodic functions.*

Seeing the Concept To see the periodic behavior of the sine function, graph $Y_1 = \sin x$ and $Y_2 = \sin(x + 2\pi)$, $Y_3 = \sin(x - 2\pi)$, and $Y_4 = \sin(x + 4\pi)$ on the same screen.

A function f is called **periodic** if there is a positive number p such that whenever θ is in the domain of f, so is $\theta + p$, and

$$f(\theta + p) = f(\theta)$$

If there is a smallest such number p, this smallest value is called the **(fundamental) period** of f.

Thus, based on equation (1), the sine and cosine functions are periodic. In fact, the sine and cosine functions have period 2π. You are asked to prove this fact in Problems 37 and 38. The secant and cosecant functions are also periodic with period 2π; the tangent and cotangent functions are periodic with period π. You are asked to prove these statements in Problems 39 to 42.

These facts are summarized as follows:

Periodic Properties

$\sin(\theta + 2\pi) = \sin\theta$	$\cos(\theta + 2\pi) = \cos\theta$	$\tan(\theta + \pi) = \tan\theta$
$\csc(\theta + 2\pi) = \csc\theta$	$\sec(\theta + 2\pi) = \sec\theta$	$\cot(\theta + \pi) = \cot\theta$

EXAMPLE 3 **Using Periodic Properties to Find Exact Values**

Find the exact value of (a) $\sin 420°$; (b) $\tan 5\pi/4$.

Solution (a) $\sin 420° = \sin(360° + 60°) = \sin 60° = \sqrt{3}/2$

(b) $\tan 5\pi/4 = \tan(\pi + \pi/4) = \tan \pi/4 = 1$

Because the sine, cosine, secant, and cosecant functions have period 2π, once we know their values for $0 \le \theta < 2\pi$, we know all their values; similarly, since the tangent and cotangent functions have period π, once we know their values for $0 \le \theta < \pi$, we know all their values.

The periodic properties of the trigonometric functions will be very helpful to us when we study their graphs in the next section.

Even–Odd Properties

4 Recall that a function f is even if $f(-\theta) = f(\theta)$ for all θ in the domain of f; a function f is odd if $f(-\theta) = -f(\theta)$ for all θ in the domain of f. We will now show that the trigonometric functions sine, tangent, cotangent, and cosecant are odd functions, whereas the functions cosine and secant are even functions.

Theorem Even–Odd Properties

$\sin(-\theta) = -\sin\theta$	$\cos(-\theta) = \cos\theta$	$\tan(-\theta) = -\tan\theta$
$\csc(-\theta) = -\csc\theta$	$\sec(-\theta) = \sec\theta$	$\cot(-\theta) = -\cot\theta$

FIGURE 69

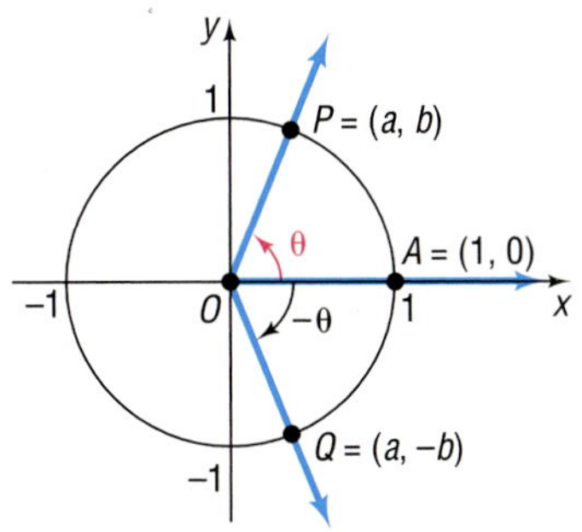

Proof Let $P = (a, b)$ be the point on the terminal side of the angle θ that is on the unit circle. (See Figure 69.) The point Q on the terminal side of the angle $-\theta$ that is on the unit circle will have coordinates $(a, -b)$. Using the definition for the trigonometric functions, we have

$$\sin\theta = b \qquad \cos\theta = a \qquad \sin(-\theta) = -b \qquad \cos(-\theta) = a$$

so

$$\sin(-\theta) = -\sin\theta \qquad \cos(-\theta) = \cos\theta$$

Now, using these results and some of the fundamental identities, we have

$$\tan(-\theta) = \frac{\sin(-\theta)}{\cos(-\theta)} = \frac{-\sin\theta}{\cos\theta} = -\tan\theta$$

$$\cot(-\theta) = \frac{1}{\tan(-\theta)} = \frac{1}{-\tan\theta} = -\cot\theta$$

$$\sec(-\theta) = \frac{1}{\cos(-\theta)} = \frac{1}{\cos\theta} = \sec\theta$$

$$\csc(-\theta) = \frac{1}{\sin(-\theta)} = \frac{1}{-\sin\theta} = -\csc\theta$$

EXAMPLE 4 Using Even–Odd Properties to Find Exact Values

Find the exact value of (a) $\cos(-\pi/3)$; (b) $\tan(-60°)$.

Solution (a) $\cos(-\pi/3) = \cos\frac{\pi}{3} = \frac{1}{2}$

(b) $\tan(-60°) = -\tan 60° = -\sqrt{3}$

Seeing the Concept To see that the cosine function is even, graph $Y_1 = \cos x$. Do you see the symmetry? Why would you conclude the cosine function is even? Clear the screen. Now graph $Y_1 = \sin x$. Do you see the symmetry? Do you conclude the sine function is odd?

6.5 EXERCISES

In Problems 1–8, t is a real number and P is the point on the unit circle that corresponds to t. Find the exact value of each of the six trigonometric functions of t.

1. $P = (\sqrt{3}/2, \frac{1}{2})$
2. $P = (\sqrt{2}/2, \sqrt{2}/2)$
3. $P = (-\frac{1}{2}, -\sqrt{3}/2)$
4. $P = (\sqrt{2}/2, -\sqrt{2}/2)$
5. $P = (-\frac{3}{5}, \frac{4}{5})$
6. $P = (-\frac{12}{13}, \frac{5}{13})$
7. $P = (\sqrt{5}/5, -2\sqrt{5}/5)$
8. $P = (-2\sqrt{5}/5, -\sqrt{5}/5)$

In Problems 9–16, find the exact value of the remaining five trigonometric functions of θ. Use the method shown in Example 2.

9. $\sin\theta = \frac{2}{3}, \quad \pi/2 < \theta < \pi$
10. $\cos\theta = \frac{1}{3}, \quad 3\pi/2 < \theta < 2\pi$
11. $\tan\theta = \frac{1}{2}, \quad \pi < \theta < 3\pi/2$
12. $\sec\theta = 3, \quad 0 < \theta < \pi/2$
13. $\sin\theta = -\frac{1}{4}, \quad \cos\theta > 0$
14. $\cos\theta = -\frac{3}{4}, \quad \sin\theta < 0$
15. $\sec\theta = 3, \quad \tan\theta < 0$
16. $\csc\theta = -4, \quad \tan\theta > 0$
17. What is the domain of the sine function?
18. What is the domain of the cosine function?
19. For what numbers θ is $f(\theta) = \tan\theta$ not defined?
20. For what numbers θ is $f(\theta) = \cot\theta$ not defined?
21. For what numbers θ is $f(\theta) = \sec\theta$ not defined?
22. For what numbers θ is $f(\theta) = \csc\theta$ not defined?
23. What is the range of the sine function?
24. What is the range of the cosine function?
25. What is the range of the tangent function?
26. What is the range of the cotangent function?
27. What is the range of the secant function?
28. What is the range of the cosecant function?
29. Is the sine function even, odd, or neither? Is its graph symmetric? With respect to what?
30. Is the cosine function even, odd, or neither? Is its graph symmetric? With respect to what?
31. Is the tangent function even, odd, or neither? Is its graph symmetric? With respect to what?
32. Is the cotangent function even, odd, or neither? Is its graph symmetric? With respect to what?
33. Is the secant function even, odd, or neither? Is its graph symmetric? With respect to what?
34. Is the cosecant function even, odd, or neither? Is its graph symmetric? With respect to what?
35. Show that the range of the tangent function is the set of all real numbers.
36. Show that the range of the cotangent function is the set of all real numbers.
37. Show that the period of $f(\theta) = \sin\theta$ is 2π.

    [**Hint:** Assume that $0 < p < 2\pi$ exists so that $\sin(\theta + p) = \sin\theta$ for all θ. Let $\theta = 0$ to find p. Then let $\theta = \pi/2$ to obtain a contradiction.]

38. Show that the period of $f(\theta) = \cos\theta$ is 2π.

39. Show that the period of $f(\theta) = \sec\theta$ is 2π.

40. Show that the period of $f(\theta) = \csc\theta$ is 2π.

41. Show that the period of $f(\theta) = \tan\theta$ is π.

42. Show that the period of $f(\theta) = \cot\theta$ is π.

43. If θ $(0 < \theta < \pi)$ is the angle between a horizontal ray directed to the right (say, the positive x-axis) and a nonhorizontal, nonvertical line L, show that the slope m of L equals $\tan\theta$. The angle θ is called the **inclination** of L.

[**Hint:** See the illustration, where we have drawn the line L^* parallel to L and passing through the origin. Use the fact that L^* intersects the unit circle at the point $(\cos\theta, \sin\theta)$.]

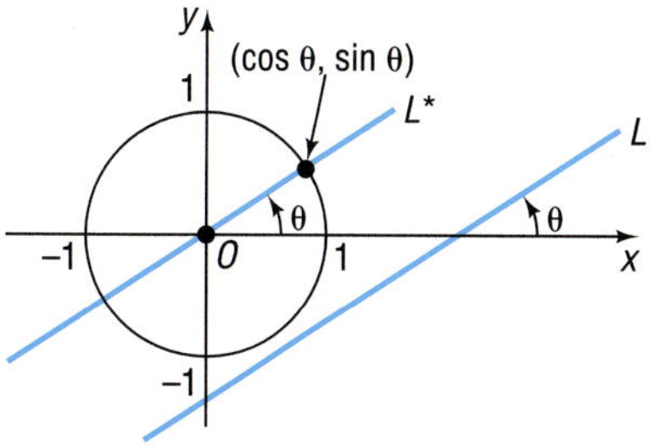

44. Write down five characteristics of the tangent function. Explain the meaning of each.

45. Describe your understanding of the meaning of a periodic function.

6.6 GRAPHS OF THE TRIGONOMETRIC FUNCTIONS

1. Graph Transformations of the Sine Function
2. Graph Transformations of the Cosine Function
3. Graph Transformations of the Tangent Function
4. Graph Transformations of the Cosecant, Secant, and Cotangent Functions

We have discussed properties of the trigonometric functions $f(\theta) = \sin\theta$, $f(\theta) = \cos\theta$, and so on. In this section, we shall use the traditional symbols x to represent the independent variable (or argument) and y for the dependent variable (or value at x) for each function. Thus, we write the six trigonometric functions as

$$y = \sin x \qquad y = \cos x \qquad y = \tan x$$
$$y = \csc x \qquad y = \sec x \qquad y = \cot x$$

Our purpose in this section is to graph each of these functions. Unless indicated otherwise, we shall use radian measure throughout for the independent variable x.

The Graph of $y = \sin x$

TABLE 7

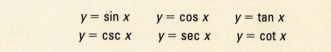

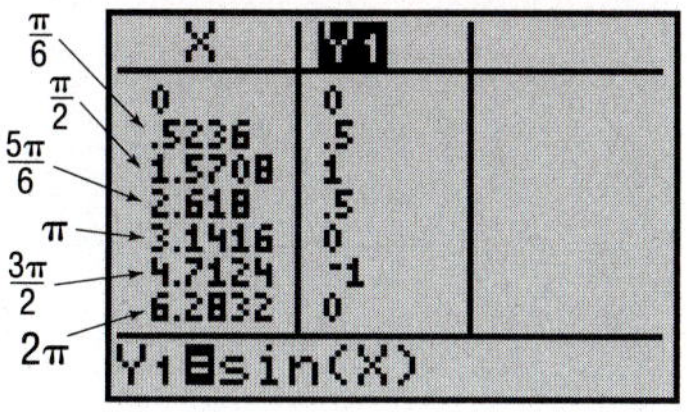

	X	Y1
0	0	0
$\frac{\pi}{6}$	.5236	.5
$\frac{\pi}{2}$	1.5708	1
$\frac{5\pi}{6}$	2.618	.5
π	3.1416	0
$\frac{3\pi}{2}$	4.7124	-1
2π	6.2832	0

Y1=sin(X)

Since the sine function has period 2π, we need to graph $y = \sin x$ only on the interval $[0, 2\pi]$. The remainder of the graph will consist of repetitions of this portion of the graph.

We begin by constructing Table 7, which lists some points on the graph of $y = \sin x$, $0 \le x \le 2\pi$. As the table shows, the graph of $y = \sin x$, $0 \le x \le 2\pi$, begins at the origin. As x increases from 0 to $\pi/2$, the value of $y = \sin x$

increases from 0 to 1; as x increases from $\pi/2$ to π to $3\pi/2$, the value of y decreases from 1 to 0 to -1; as x increases from $3\pi/2$ to 2π, the value of y increases from -1 to 0. Based on this, we set the viewing rectangle as shown in Figure 70(a) and graph $y = \sin x$, $0 \le x \le 2\pi$. See Figure 70(b). Figure 70(c) shows the graph drawn by hand.

FIGURE 70

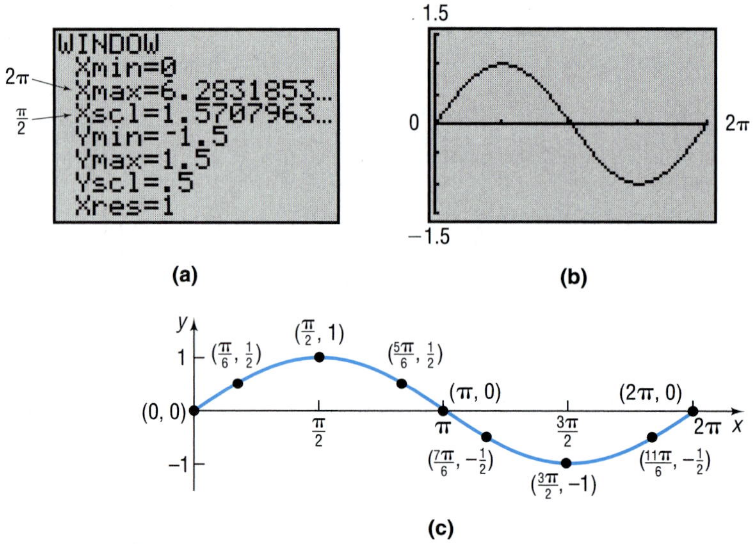

The graph in Figure 70 is one period of the graph of $y = \sin x$. To obtain a more complete graph of $y = \sin x$, we repeat this period in each direction, as shown in Figure 71.

FIGURE 71
$y = \sin x$,
$-\infty < x < \infty$

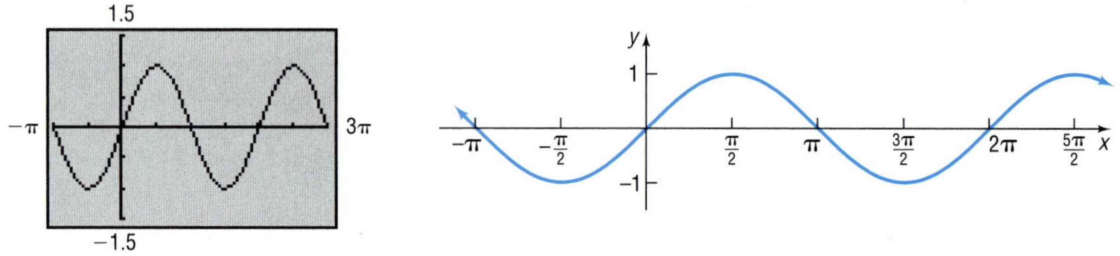

The graph of $y = \sin x$ illustrates some of the facts we already know about the sine function:

Characteristics of the Sine Function

1. The domain is the set of all real numbers.
2. The range consists of all real numbers from -1 to 1, inclusive.
3. The sine function is an odd function, as the symmetry of the graph with respect to the origin indicates.
4. The sine function is periodic, with period 2π.
5. The x-intercepts are $\ldots, -2\pi, -\pi, 0, \pi, 2\pi, 3\pi, \ldots$; the y-intercept is 0.
6. The maximum value is 1 and occurs at $x = \ldots, -3\pi/2, \pi/2, 5\pi/2, 9\pi/2, \ldots$; the minimum value is -1 and occurs at $x = \ldots, -\pi/2, 3\pi/2, 7\pi/2, 11\pi/2, \ldots$.

Now work Problems 1, 3, and 5.

The graphing techniques introduced in Chapter 2 may be used to graph functions that are transformations of the sine function (refer to Chapter 2, Section 2.3).

EXAMPLE 1 Graphing Transformations of $y = \sin x$ Using Shifts, Reflections, and the Like

Use the graph of $y = \sin x$ to graph $y = -\sin x + 2$.

Solution Figure 72 illustrates the steps.

FIGURE 72

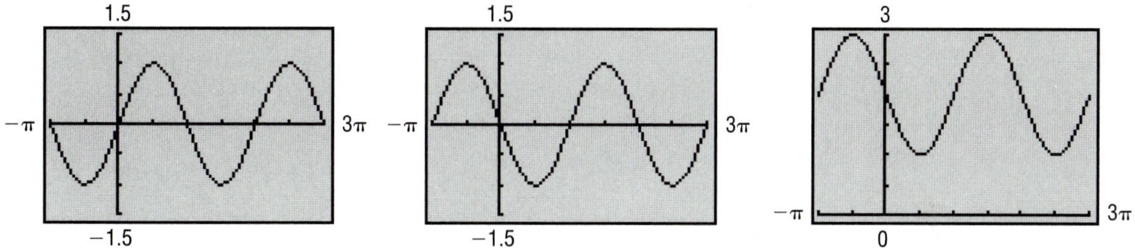

(a) $y = \sin x$ → Reflection about x-axis; Multiply by −1 → (b) $y = -\sin x$ → Vertical shift; Add 2 → (c) $y = -\sin x + 2$

EXAMPLE 2 Graphing Transformations of $y = \sin x$ Using Shifts, Reflections, and the Like

Use the graph of $y = \sin x$ to graph $y = \sin\left(x - \frac{\pi}{4}\right)$.

Solution Figure 73 illustrates the steps.

FIGURE 73

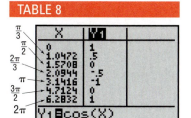

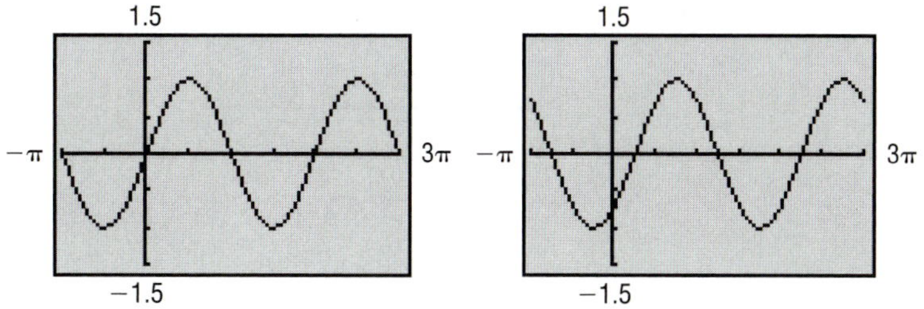

(a) $y = \sin x$ → Horizontal shift to right $\frac{\pi}{4}$ units; Replace x by $x - \frac{\pi}{4}$ → (b) $y = \sin\left(x - \frac{\pi}{4}\right)$

Now work Problem 35.

The Graph of $y = \cos x$

TABLE 8

	X	Y1
	0	1
$\frac{\pi}{3}$	1.0472	.5
$\frac{\pi}{2}$	1.5708	0
$\frac{2\pi}{3}$	2.0944	-.5
π	3.1416	-1
$\frac{3\pi}{2}$	4.7124	0
2π	6.2832	1

Y1=cos(X)

The cosine function also has period 2π. Thus, we proceed as we did with the sine function by constructing Table 8, which lists some points on the graph of $y = \cos x$, $0 \le x \le 2\pi$. As the table shows, the graph of $y = \cos x$, $0 \le x \le 2\pi$, begins at the point $(0, 1)$. As x increases from 0 to $\pi/2$ to π, the value of y

decreases from 1 to 0 to -1; as x increases from π to $3\pi/2$ to 2π, the value of y increases from -1 to 0 to 1. As before, we set the viewing rectangle as shown in Figure 74(a) and graph $y = \cos x$, $0 \leq x \leq 2\pi$. See Figure 74(b). Figure 74(c) shows the graph drawn by hand.

FIGURE 74
$y = \cos x, 0 \leq x \leq 2\pi$

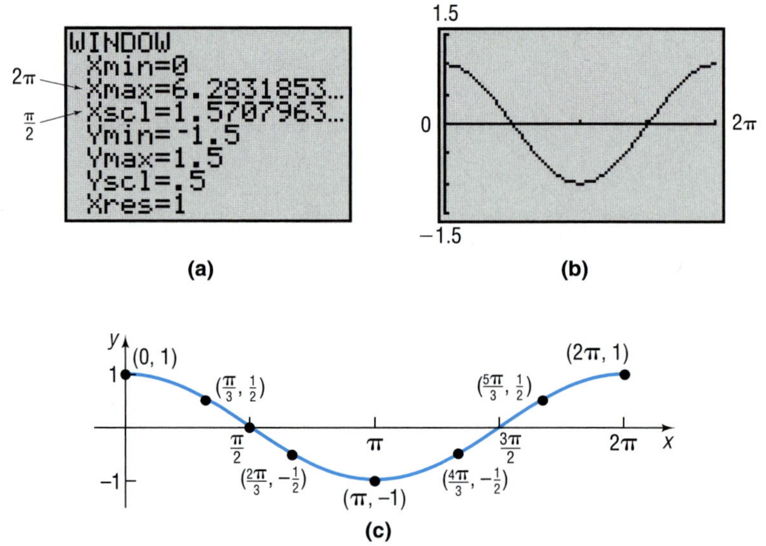

A more complete graph of $y = \cos x$ is obtained by repeating this period in each direction, as shown in Figure 75.

FIGURE 75
$y = \cos x, -\infty < x < \infty$

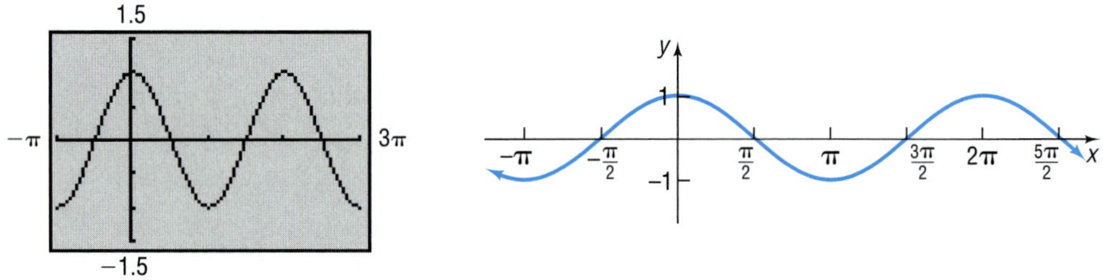

The graph of $y = \cos x$ illustrates some of the facts we already know about the cosine function:

Characteristics of the Cosine Function

1. The domain is the set of all real numbers.
2. The range consists of all real numbers from -1 to 1, inclusive.
3. The cosine function is an even function, as the symmetry of the graph with respect to the y-axis indicates.
4. The cosine function is periodic, with period 2π.
5. The x-intercepts are $\ldots, -3\pi/2, -\pi/2, \pi/2, 3\pi/2, 5\pi/2, \ldots$; the y-intercept is 1.
6. The maximum value is 1 and occurs at $x = \ldots, -2\pi, 0\ 2\pi, 4\pi, 6\pi, \ldots$; the minimum value is -1 and occurs at $x = \ldots, -\pi, \pi, 3\pi, 5\pi, \ldots$.

2 Again, the graphing techniques from Chapter 2 may be used to graph transformations of the cosine function.

EXAMPLE 3 Graphing Transformations of $y = \cos x$ Using Shifts, Reflections, and the Like

Use the graph of $y = \cos x$ to graph $y = 2 \cos x$.

Solution Figure 76 illustrates the graph, which is a vertical stretch of the graph of $y = \cos x$. (Multiply each y-coordinate by 2).

FIGURE 76
$y = 2 \cos x$

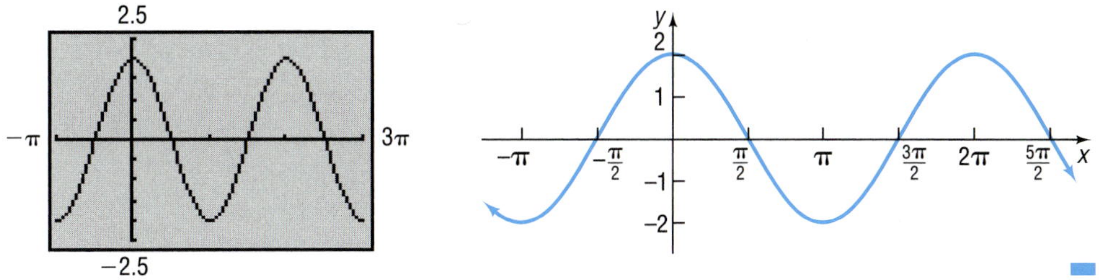

EXAMPLE 4 Graphing Transformations of $y = \cos x$ Using Shifts, Reflections, and the Like

Use the graph of $y = \cos x$ to graph $y = \cos 3x$.

Solution Figure 77 illustrates the graph, which is a horizontal compression of the graph of $y = \cos x$. (Multiply each x-coordinate by $\frac{1}{3}$). Notice that, due to this compression, the period of $y = \cos 3x$ is $2\pi/3$, whereas the period of $y = \cos x$ is 2π. We will comment more about this in Section 8.5.

FIGURE 77
$y = \cos 3x$

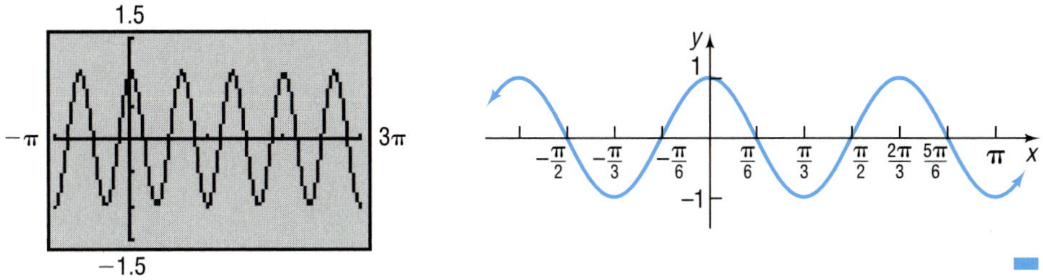

Now work Problem 43.

The Graph of $y = \tan x$

Because the tangent function has period π, we only need to determine the graph over some interval of length π. The rest of the graph will consist of repetitions of that graph. Because the tangent function is not defined at $\ldots$, $-3\pi/2$, $-\pi/2$, $\pi/2$, $3\pi/2$, $\ldots$, we will concentrate on the interval $(-\pi/2, \pi/2)$, of length π, and construct Table 9, which lists some points on the graph of $y = \tan x$, $-\pi/2 < x < \pi/2$. We plot the points in the table and connect them with a smooth curve. See Figure 78 for a partial graph of $y = \tan x$, where $-\pi/3 \le x \le \pi/3$.

TABLE 9

x	$y = \tan x$	(x, y)
$-\pi/3$	$-\sqrt{3} \approx -1.73$	$(-\pi/3, -\sqrt{3})$
$-\pi/4$	-1	$(-\pi/4, -1)$
$-\pi/6$	$-\sqrt{3}/3 \approx -0.58$	$(-\pi/6, -\sqrt{3}/3)$
0	0	$(0, 0)$
$\pi/6$	$\sqrt{3}/3 \approx 0.58$	$(\pi/6, \sqrt{3}/3)$
$\pi/4$	1	$(\pi/4, 1)$
$\pi/3$	$\sqrt{3} \approx 1.73$	$(\pi/3, \sqrt{3})$

FIGURE 78
$y = \tan x$,
$-\pi/3 \le x \le \pi/3$

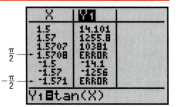

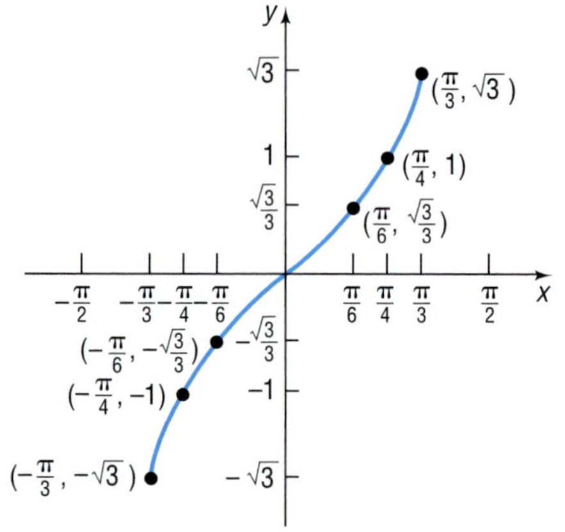

To complete the graph of $y = \tan x$, we need to investigate the behavior of the function as x approaches $-\pi/2$ and $\pi/2$. We must be careful, though, because $y = \tan x$ is not defined at these numbers. To determine this behavior, we use the identity

$$\tan x = \frac{\sin x}{\cos x}$$

If x is close to $\pi/2$, but remains less than $\pi/2$, then $\sin x$ will be close to 1 and $\cos x$ will be positive and close to 0. (Refer back to the graphs of the sine function and the cosine function.) Hence, the ratio $(\sin x)/(\cos x)$ will be positive and large. In fact, the closer x gets to $\pi/2$, the closer $\sin x$ gets to 1 and $\cos x$ gets to 0, so $\tan x$ approaches ∞. In other words, the vertical line $x = \pi/2$ is a vertical asymptote to the graph of $y = \tan x$.

If x is close to $-\pi/2$, but remains greater than $-\pi/2$, then $\sin x$ will be close to -1 and $\cos x$ will be positive and close to 0. Hence, the ratio $(\sin x)/(\cos x)$ approaches $-\infty$. In other words, the vertical line $x = -\pi/2$ is also a vertical asymptote to the graph.

Table 10 confirms these conclusions.

TABLE 10

	X	Y1
	1.5	14.101
	1.57	1255.8
	1.5707	10381
$\frac{\pi}{2}$ →	1.5708	ERROR
	-1.5	-14.1
	-1.57	-1256
$-\frac{\pi}{2}$ →	-1.571	ERROR

Y1 = tan(X)

Figure 79 shows the graph of $y = \tan x$, $-\infty < x < \infty$. Notice we used dot mode when graphing $y = \tan x$ using a graphing utility. Do you see why?

FIGURE 79
$y = \tan x$, $-\infty < x < \infty$, x not equal to odd multiples of $\pi/2$

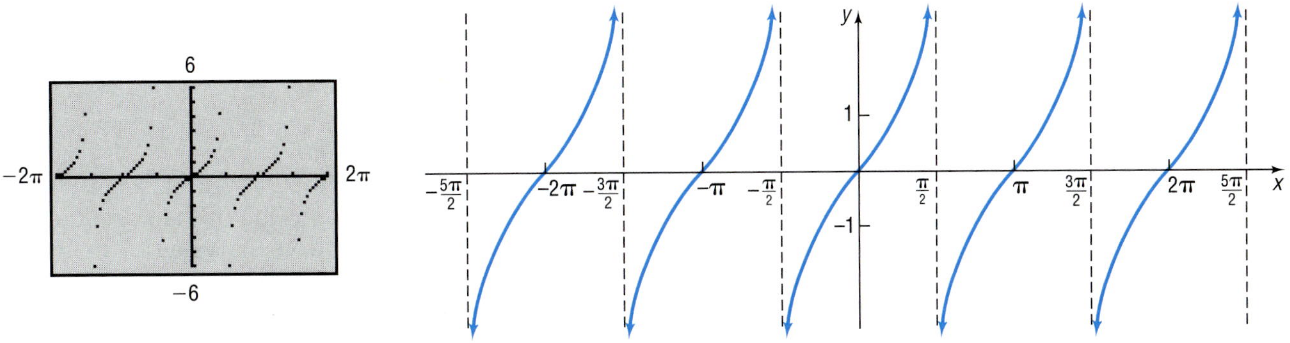

The graph of $y = \tan x$ illustrates some of the facts we already know about the tangent function:

Characteristics of the Tangent Function

1. The domain is the set of all real numbers, except odd multiples of $\pi/2$.
2. The range consists of all real numbers.
3. The tangent function is an odd function, as the symmetry of the graph with respect to the origin indicates.
4. The tangent function is periodic, with period π.
5. The x-intercepts are $\ldots, -2\pi, -\pi, 0, \pi, 2\pi, 3\pi, \ldots$; the y-intercept is 0.
6. Vertical asymptotes occur at $x = \ldots, -3\pi/2, -\pi/2, \pi/2, 3\pi/2, \ldots$.

Now work Problems 11 and 19.

EXAMPLE 5 Graphing Transformations of $y = \tan x$ Using Shifts, Reflections, and the Like

3

Graph $y = \tan\left(x + \dfrac{\pi}{4}\right)$.

Solution We start with the graph of $y = \tan x$ and shift it horizontally to the left $\pi/4$ unit. See Figure 80.

FIGURE 80
$y = \tan\left(x + \dfrac{\pi}{4}\right)$

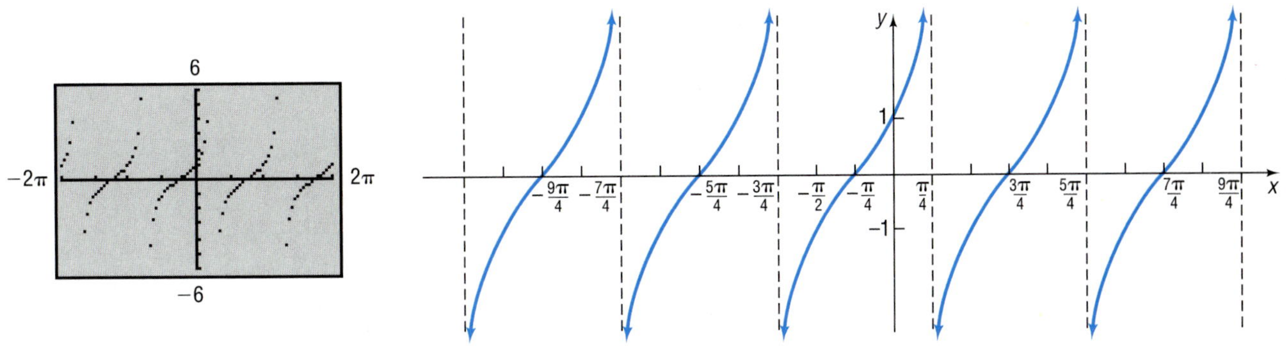

Now work Problem 51.

The Graphs of $y = \csc x$, $y = \sec x$, and $y = \cot x$

4 The cosecant and secant functions, sometimes referred to as **reciprocal functions,** are graphed by making use of the reciprocal identities

$$\csc x = \frac{1}{\sin x} \quad \text{and} \quad \sec x = \frac{1}{\cos x}$$

For example, the value of the cosecant function $y = \csc x$ at a given number x equals the reciprocal of the corresponding value of the sine function, provided the value of the sine function is not 0. If the value of $\sin x$ is 0, then, at such numbers x, the cosecant function is not defined. In fact, the graph of the cosecant function has vertical asymptotes at integral multiples of π. Figure 81 shows the graph drawn by hand and Figure 82 shows the graph using a graphing utility.

FIGURE 81
$y = \csc x$, $-\infty < x < \infty$, x not equal to integral multiples of π, $|y| \geq 1$

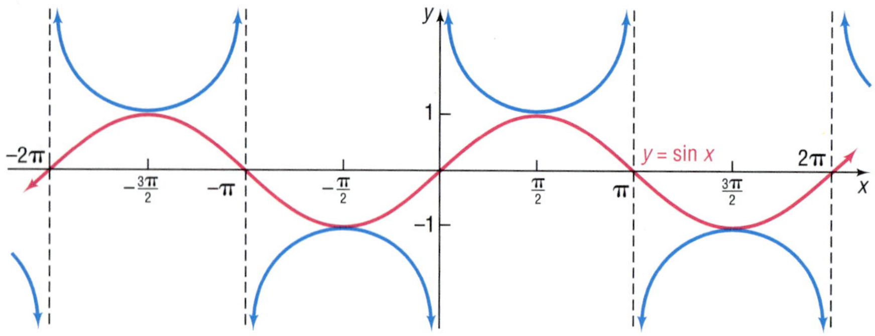

FIGURE 82

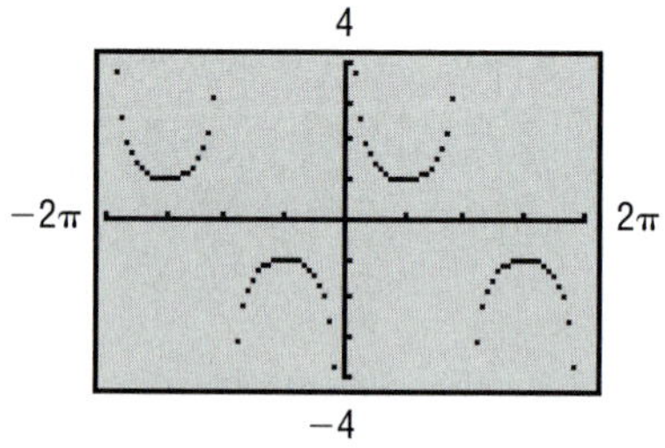

EXAMPLE 6 Graphing Transformations of $y = \csc x$ Using Shifts, Reflections, and the Like

Graph $y = 2\csc(x - \pi/2)$.

Solution Figure 83 shows the required steps.

FIGURE 83

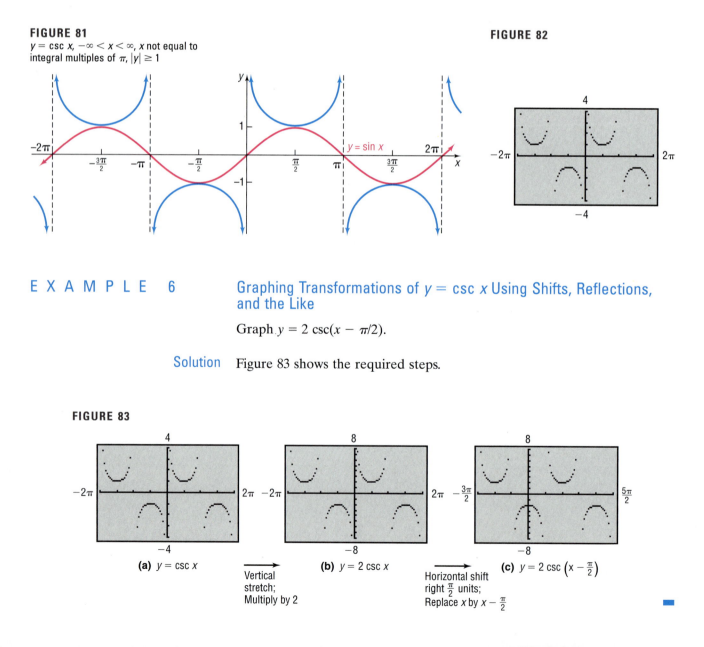

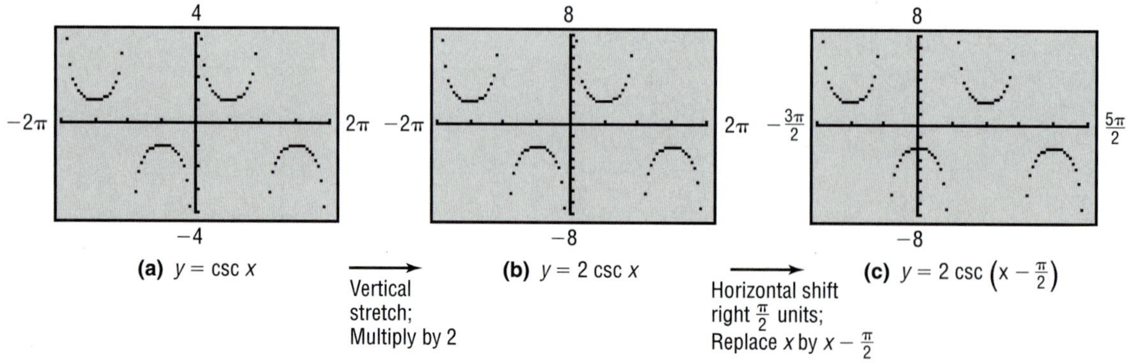

(a) $y = \csc x$ **(b)** $y = 2\csc x$ **(c)** $y = 2\csc\left(x - \frac{\pi}{2}\right)$

Using the idea of reciprocals, we can similarly obtain the graph of $y = \sec x$. See Figure 84 and Figure 85.

FIGURE 84
$y = \sec x$, $-\infty < x < \infty$, x not equal to odd multiples of $\pi/2$, $|y| \geq 1$

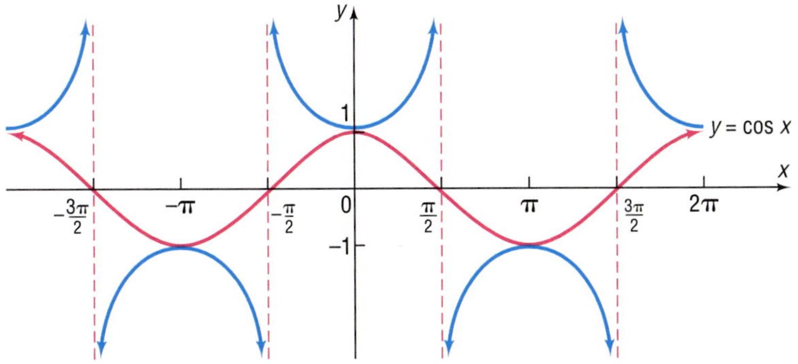

FIGURE 85

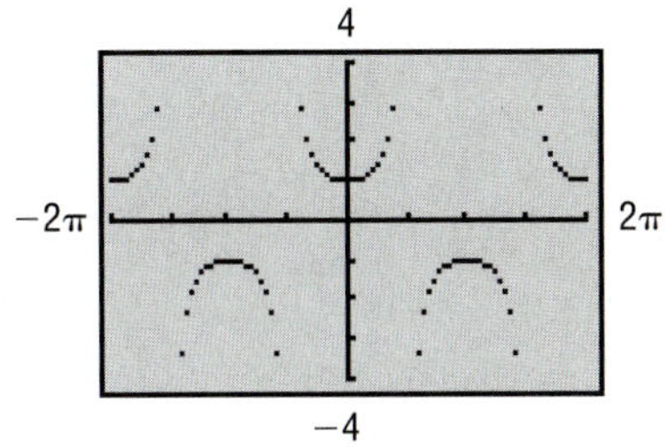

TABLE 11

x	$y = \cot x$	(x, y)
$\pi/6$	$\sqrt{3}$	$(\pi/6, \sqrt{3})$
$\pi/4$	1	$(\pi/4, 1)$
$\pi/3$	$\sqrt{3}/3$	$(\pi/3, \sqrt{3}/3)$
$\pi/2$	0	$(\pi/2, 0)$
$2\pi/3$	$-\sqrt{3}/3$	$(2\pi/3, -\sqrt{3}/3)$
$3\pi/4$	-1	$(3\pi/4, -1)$
$5\pi/6$	$-\sqrt{3}$	$(5\pi/6, -\sqrt{3})$

We obtain the graph of $y = \cot x$ as we did the graph of $y = \tan x$. The period of $y = \cot x$ is π. Because the cotangent function is not defined for integral multiples of π, we will concentrate on the interval $(0, \pi)$. Table 11 lists some points on the graph of $y = \cot x$, $0 < x < \pi$. As x approaches 0, but remains greater than 0, the value of $\cos x$ will be close to 1 and the value of $\sin x$ will be positive and close to 0. Hence, the ratio $(\cos x)/(\sin x) = \cot x$ will be positive and large, so as x approaches 0, $\cot x$ approaches ∞. Similarly, as x approaches π, but remains less than π, the value of $\cos x$ will be close to -1 and the value of $\sin x$ will be positive and close to 0. Hence, the ratio $(\cos x)/(\sin x) = \cot x$ will be negative and will approach $-\infty$ as x approaches π. Figure 86 shows the graph.

FIGURE 86
$y = \cot x$, $-\infty < x < \infty$, x not equal to integral multiples of π, $-\infty < y < \infty$

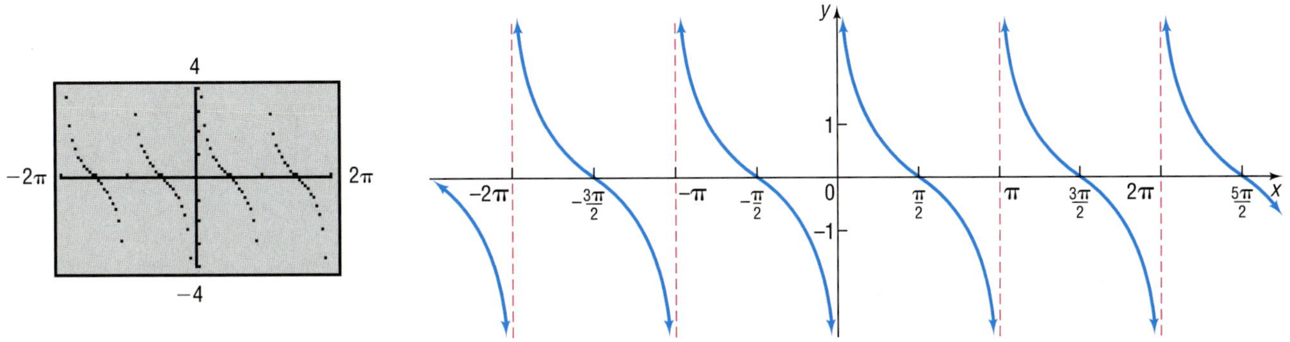

6.6 EXERCISES

In Problems 1–20, refer to the graphs to answer each question, if necessary.

1. What is the y-intercept of $y = \sin x$?
2. What is the y-intercept of $y = \cos x$?
3. For what numbers x, $-\pi \leq x \leq \pi$ is the graph of $y = \sin x$ increasing?
4. For what numbers x, $-\pi \leq x \leq \pi$ is the graph of $y = \cos x$ decreasing?
5. What is the largest value of $y = \sin x$?
6. What is the smallest value of $y = \cos x$?

7. For what numbers x, $0 \le x \le 2\pi$, does $\sin x = 0$?

8. For what numbers x, $0 \le x \le 2\pi$, does $\cos x = 0$?

9. For what numbers x, $-2\pi \le x \le 2\pi$, does $\sin x = 1$? What about $\sin x = -1$?

10. For what numbers x, $-2\pi \le x \le 2\pi$, does $\cos x = 1$? What about $\cos x = -1$?

11. What is the y-intercept of $y = \tan x$?

12. What is the y-intercept of $y = \cot x$?

13. What is the y-intercept of $y = \sec x$?

14. What is the y-intercept of $y = \csc x$?

15. For what numbers x, $-2\pi \le x \le 2\pi$, does $\sec x = 1$? What about $\sec x = -1$?

16. For what numbers x, $-2\pi \le x \le 2\pi$, does $\csc x = 1$? What about $\csc x = -1$?

17. For what numbers x, $-2\pi \le x \le 2\pi$, does the graph of $y = \sec x$ have vertical asymptotes?

18. For what numbers x, $-2\pi \le x \le 2\pi$, does the graph of $y = \csc x$ have vertical asymptotes?

19. For what numbers x, $-2\pi \le x \le 2\pi$, does the graph of $y = \tan x$ have vertical asymptotes?

20. For what numbers x, $-2\pi \le x \le 2\pi$, does the graph of $y = \cot x$ have vertical asymptotes?

In Problems 21–22, match the graph to a function. Three answers are possible.

A. $y = -\sin x$ *B.* $y = -\cos x$ *C.* $y = \sin\left(x - \frac{\pi}{2}\right)$

D. $y = -\cos\left(x - \frac{\pi}{2}\right)$ *E.* $y = \sin(x + \pi)$ *F.* $y = \cos(x + \pi)$

21.

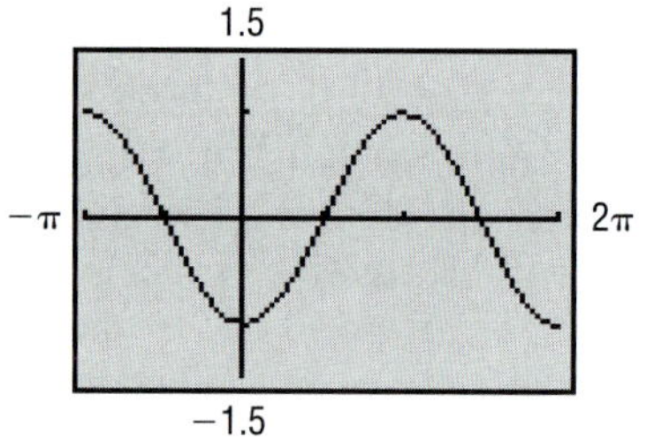

22.

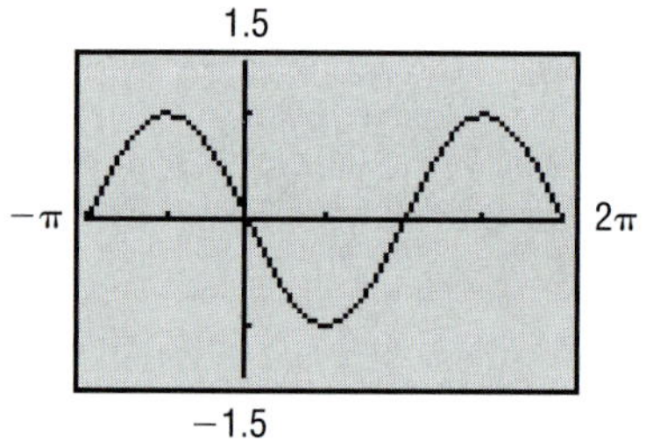

In Problems 23–26, match each function to its graph.

A. $y = \sin 2x$ *B.* $y = \sin 4x$ *C.* $y = \cos 2x$ *D.* $y = \cos 4x$

23.

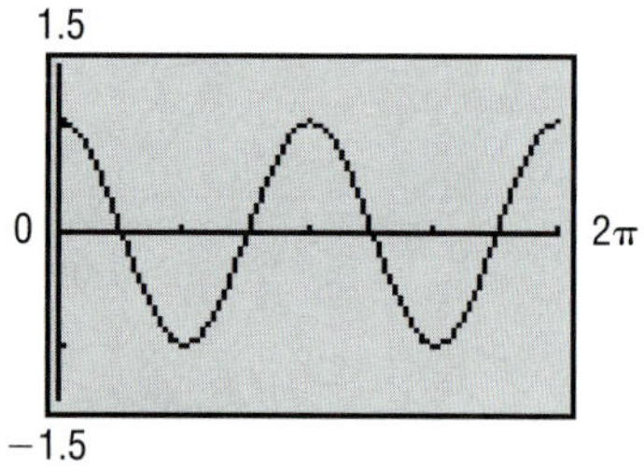

24.

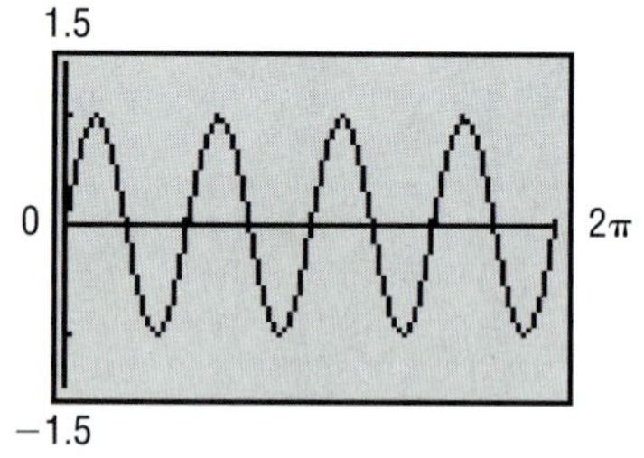

25.

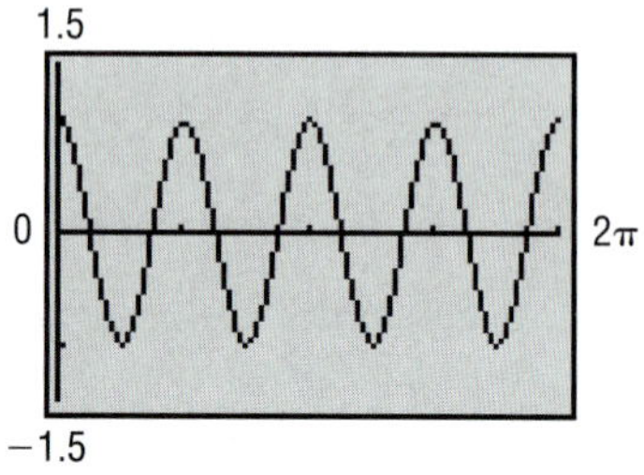

26.

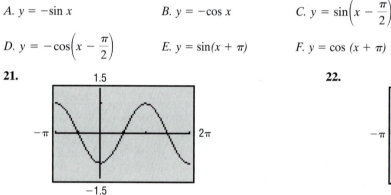

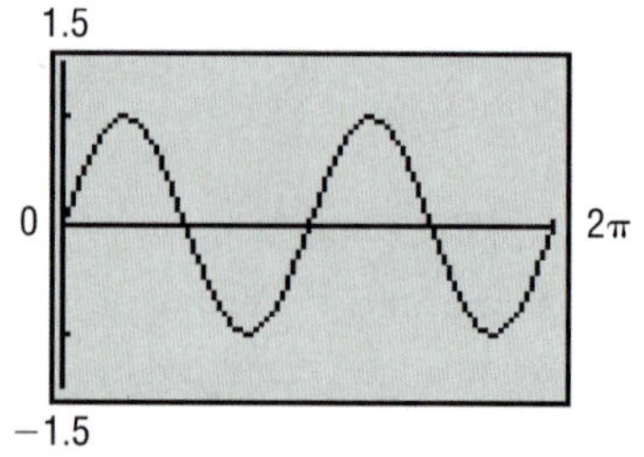

In Problems 27–30, match each function to its graph.

A. $y = -\tan x$ *B.* $y = \tan\left(x + \dfrac{\pi}{2}\right)$ *C.* $y = \tan\,(x + \pi)$ *D.* $y = -\tan\left(x - \dfrac{\pi}{2}\right)$

27.

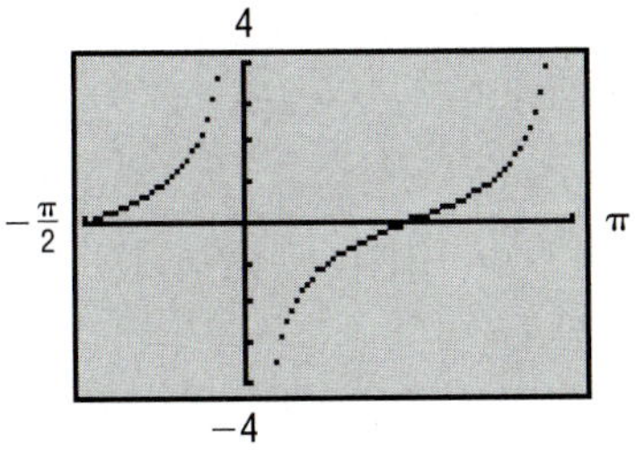

28.

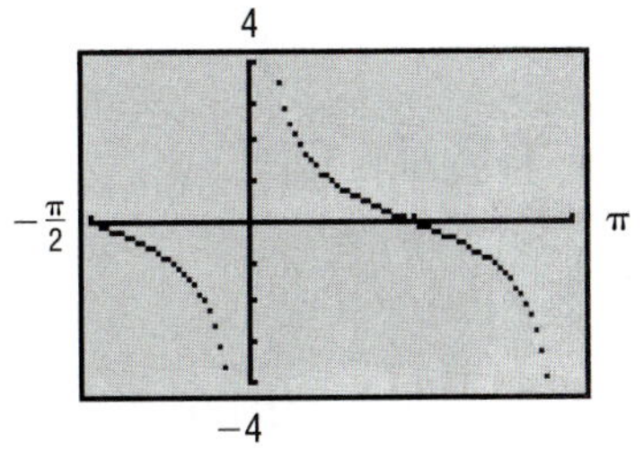

29.

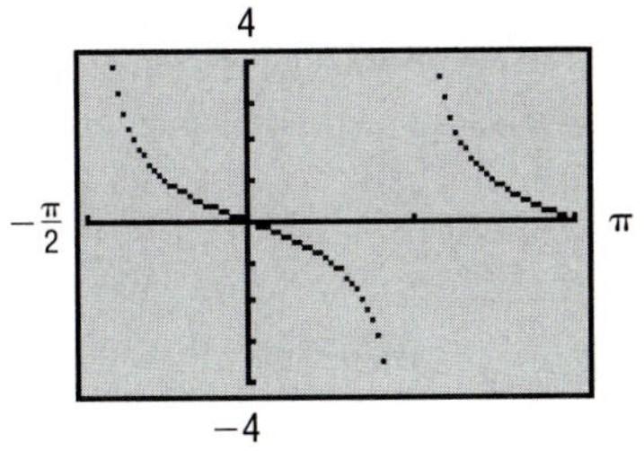

30.

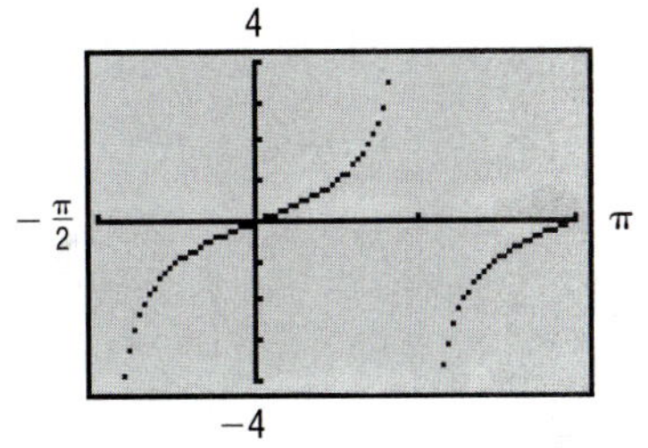

In Problems 31–62, show the stages required to graph each function.

31. $y = 3 \sin x$
32. $y = 4 \cos x$
33. $y = \cos\left(x + \dfrac{\pi}{4}\right)$
34. $y = \sin(x - \pi)$
35. $y = \sin x - 1$
36. $y = \cos x + 1$
37. $y = -2 \sin x$
38. $y = -3 \cos x$
39. $y = \sin \pi x$
40. $y = \cos \dfrac{\pi}{2} x$
41. $y = 2 \sin x + 2$
42. $y = 3 \cos x + 3$
43. $y = -2 \cos\left(x - \dfrac{\pi}{2}\right)$
44. $y = -3 \sin\left(x + \dfrac{\pi}{2}\right)$
45. $y = 3 \sin\,(\pi - x)$
46. $y = 2 \cos\,(\pi - x)$
47. $y = -\sec x$
48. $y = -\cot x$
49. $y = \sec\left(x - \dfrac{\pi}{2}\right)$
50. $y = \csc(x - \pi)$
51. $y = \tan(x - \pi)$
52. $y = \cot(x - \pi)$
53. $y = 3 \tan 2x$
54. $y = 4 \tan \frac{1}{2} x$
55. $y = \sec 2x$
56. $y = \csc \frac{1}{2} x$
57. $y = \cot \pi x$
58. $y = \cot 2x$
59. $y = -3 \tan 4x$
60. $y = -3 \tan 2x$
61. $y = 2 \sec \frac{1}{2} x$
62. $y = 2 \sec 3x$

63. **Carrying a Ladder Around a Corner** A ladder of length L is carried horizontally around a corner from a hall 3 feet wide into a hall 4 feet wide. See the illustration.

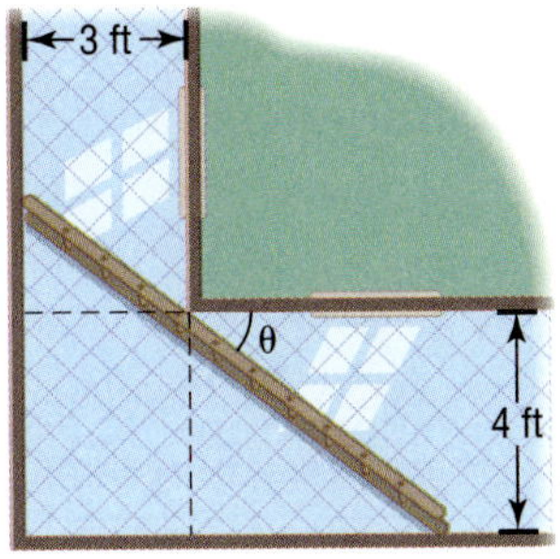

(a) Show that the length L of the ladder as a function of the angle θ is

$L = 3 \sec \theta + 4 \csc \theta$

(b) Graph L, $0 < \theta < \dfrac{\pi}{2}$.

(c) Where is L the least?

(d) What is the length of the largest ladder that can be carried around the corner? Why is this also the least value of L?

64. Graph $y = \sin x$, $y = 2 \sin x$, $y = \frac{1}{2} \sin x$, and $y = 8 \sin x$. What do you conclude about the graph of $y = A \sin x$, $A > 0$?

65. Graph $y = \sin x$, $y = \sin 2x$, $y = \sin 4x$, and $y = \sin \frac{1}{2} x$. What do you conclude about the graph of $y = \sin \omega x$?

66. Graph $y = \sin x$, $y = \sin[x - (\pi/3)]$, $y = \sin[x - (\pi/4)]$, and $y = \sin[x - (\pi/6)]$. What do you conclude about the graph of $y = \sin(x - \phi)$, $\phi > 0$?

67. Graph

$$y = \sin x \quad \text{and} \quad y = \cos\left(x - \frac{\pi}{2}\right)$$

Do you think that $\sin x = \cos\left(x - \dfrac{\pi}{2}\right)$?

68. Graph

$$y = \tan x \quad \text{and} \quad y = -\cot\left(x + \frac{\pi}{2}\right)$$

Do you think that $\tan x = -\cot\left(x + \dfrac{\pi}{2}\right)$?

6.7 THE INVERSE TRIGONOMETRIC FUNCTIONS

1 Find the Exact Value of an Inverse Trigonometric Function

2 Find the Approximate Value of an Inverse Trigonometric Function

In Section 5.1 we discussed inverse functions, and we noted that if a function is one-to-one it will have an inverse function. We also observed that if a function is not one-to-one, it may be possible to restrict its domain in some suitable manner such that the restricted function is one-to-one. In this section, we use these ideas to define inverse trigonometric functions. (You may wish to review Section 5.1 at this time.) We begin with the inverse of the sine function.

The Inverse Sine Function

In Figure 87, we reproduce the graph of $y = \sin x$. Because every horizontal line $y = b$, where b is between -1 and 1, intersects the graph of $y = \sin x$ infinitely many times, it follows from the horizontal-line test that the function $y = \sin x$ is not one-to-one.

FIGURE 87
$y = \sin x$, $-\infty < x < \infty$, $-1 \le y \le 1$

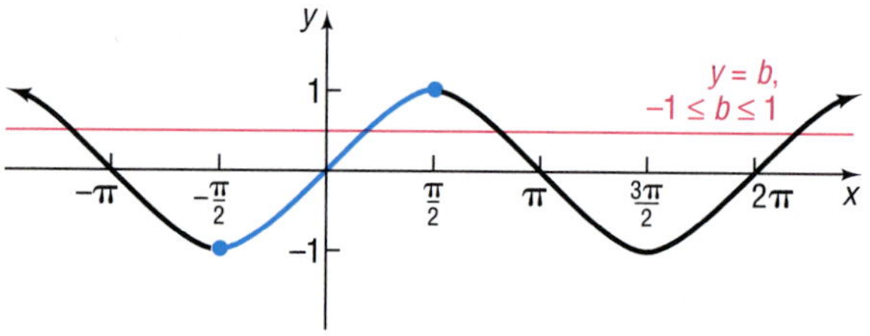

However, if we restrict the domain of $y = \sin x$ to the interval $[-\pi/2, \pi/2]$, the restricted function

$$y = \sin x, \qquad -\frac{\pi}{2} \le x \le \frac{\pi}{2}$$

is one-to-one and, hence, will have an inverse function.* See Figure 88 on page 508.

*Although there are many other ways to restrict the domain and obtain a one-to-one function, mathematicians have agreed on a consistent use of the interval $[-\pi/2, \pi/2]$ in order to define the inverse of $y = \sin x$.

MISSION POSSIBLE

Identifying the Mountains of the Hawaiian Islands Seen from Oahu

A wealthy tourist with a strong desire for the perfect scrapbook has called in your consulting firm for help in labeling a photograph he took on a clear day from the southeast shore of Oahu. He shows you the photograph in which you see mostly sea and sky. But on the horizon are three mountain peaks, equally spaced and apparently all of the same height. The tourist tells you that the T-shirt vendor at the beach informed him that the three mountain peaks are the volcanoes of Lanai, Maui, and the Big Island (Hawaii). The vendor even added, "You almost never see the Big Island from here."

The tourist thinks his photograph may have some special value. He has some misgivings, however. He is wondering if, in fact, it is ever really possible to see the Big Island from Oahu. He wants accurate labels on his photos, so he asks you for help.

You realize that some trigonometry is needed as well as a good atlas and encyclopedia. After doing some research you discover the following facts:

The heights of the peaks vary. Lanaihale, the tallest peak on Lanai, is only about 3370 feet above sea level. Maui's Haleakala is 10,023 feet above sea level. Mauna Kea on the Big Island is the highest of all, 13,796 feet above sea level. (If measured from its base on the ocean floor, it would qualify as the tallest mountain on Earth.)

Measuring from the southeast corner of Oahu, the distance to Lanai is about 65 miles, to Maui it's about 110 miles, and to the Big Island it is about 190 miles. These distances, measured along the surface of the Earth, represent the length of arc that originates at sea level on Oahu and terminates in an imaginary location directly below the peak of each mountain at what would be sea level.

1. These volcanic peaks are all different heights. If your wealthy client took a photo of three mountain peaks that all look the same height, how could they possible represent these three volcanoes?
2. The radius of the earth is approximately 3960 miles. Based on that figure, what is the circumference of the earth? How is the distance between islands related to the circumference of the earth?
3. To determine which of these mountain peaks would actually be visible from Oahu, you need to consider that the tourist standing on the shore and looking "straight out" would have a line of sight tangent to the surface of the earth at that point. Make a rough sketch of the right triangle formed by the tourist's line of sight, the radius from the center of the earth to the tourist, and a line from the center of the earth that passes through Mauna Kea. Can you determine the angle formed at the center of the earth?
4. What would the length of the hypotenuse of the triangle be? Can you tell from that whether or not Mauna Kea would be visible from Oahu?
5. Repeat this procedure with the information about Maui and Lanai. Will they be visible from Oahu?
6. There is another island off Oahu that the T-shirt vendor did not mention. Molokai is about 40 miles from Oahu, and its highest peak, "Kamakou," is 4961 feet above sea level. Would Kamakou be visible from Oahu?
7. Your team should now be prepared to name the three volcanic peaks in the photograph. How do you decide which one is which?

FIGURE 88
$y = \sin x$,
$-\pi/2 \le x \le \pi/2,\ -1 \le y \le 1$

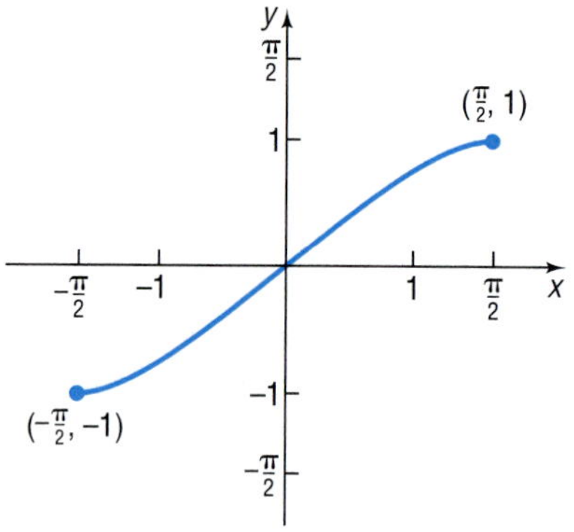

The inverse function is called the **inverse sine** of x and is symbolized by $y = \sin^{-1} x$. Thus,

$$y = \sin^{-1} x \quad \text{means} \quad x = \sin y \qquad (1)$$
$$\text{where } -\frac{\pi}{2} \le y \le \frac{\pi}{2} \quad \text{and} \quad -1 \le x \le 1$$

Because $y = \sin^{-1} x$ means $x = \sin y$, we read $y = \sin^{-1} x$ as "y is the angle whose sine equals x." Alternatively, we can say that "y is the inverse sine of x." Be careful about the notation used. The superscript -1 that appears in $y = \sin^{-1} x$ is not an exponent, but is reminiscent of the symbolism f^{-1} used to denote the inverse of a function f. (To avoid this notation, some books use the notation $y = \arcsin x$ instead of $y = \sin^{-1} x$.)

Based on the general discussion of functions and their inverses (Section 5.1), we have the following results:

$$\sin^{-1}(\sin u) = u \quad \text{where} \quad -\frac{\pi}{2} \le u \le \frac{\pi}{2}$$
$$\sin(\sin^{-1} v) = v \quad \text{where} \quad -1 \le v \le 1$$

Let's examine the function $y = \sin^{-1} x$. Its domain is $-1 \le x \le 1$, and its range is $-\pi/2 \le y \le \pi/2$. Its graph can be obtained by reflecting the restricted portion of the graph of $y = \sin x$ about the line $y = x$, as shown in Figure 89(a). Figure 89(b) shows the graph using a graphing utility.

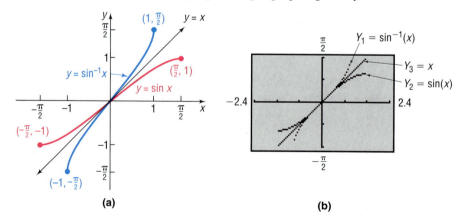

FIGURE 89
$y = \sin^{-1} x,\ -1 \le x \le 1,$
$-\pi/2 \le y \le \pi/2$

1 For some numbers x it is possible to find the exact value of $y = \sin^{-1} x$.

EXAMPLE 1 Finding the Exact Value of an Inverse Sine Function

Find the exact value of $\sin^{-1} 1$.

Solution Let $\theta = \sin^{-1} 1$. We seek the angle θ, $-\pi/2 \le \theta \le \pi/2$, whose sine equals 1:

$$\theta = \sin^{-1} 1, \qquad -\frac{\pi}{2} \le \theta \le \frac{\pi}{2}$$

$$\sin \theta = 1, \qquad -\frac{\pi}{2} \le \theta \le \frac{\pi}{2} \qquad \text{By definition.}$$

From Figure 90 we see that the only angle θ within the interval $[-\pi/2, \pi/2]$ whose sine is 1 is $\pi/2$. (Note that $\sin(5\pi/2)$ also equals 1, but $5\pi/2$ lies outside the interval $[-\pi/2, \pi/2]$ and hence is not admissible.)

FIGURE 90

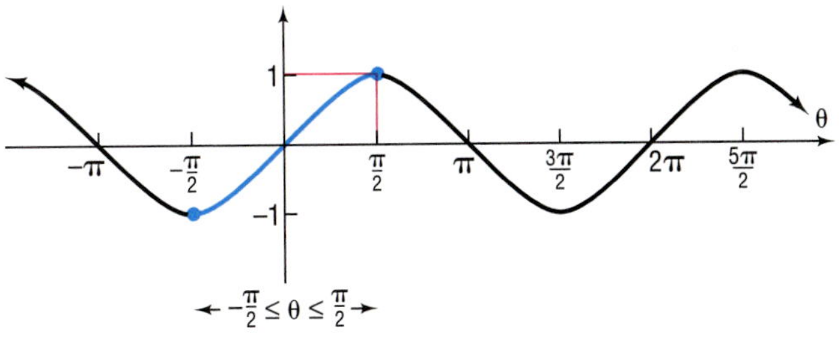

We conclude that

$$\theta = \frac{\pi}{2}$$

So

$$\sin^{-1} 1 = \frac{\pi}{2}$$

Now work Problem 1.

EXAMPLE 2 Finding the Exact Value of an Inverse Sine Function

Find the exact value of $\sin^{-1}(\sqrt{3}/2)$.

Solution Let $\theta = \sin^{-1}(\sqrt{3}/2)$. We seek the angle θ, $-\pi/2 \le \theta \le \pi/2$, whose sine equals $\sqrt{3}/2$:

$$\theta = \sin^{-1} \frac{\sqrt{3}}{2}, \qquad -\frac{\pi}{2} \le \theta \le \frac{\pi}{2}$$

$$\sin \theta = \frac{\sqrt{3}}{2}, \qquad -\frac{\pi}{2} \le \theta \le \frac{\pi}{2} \qquad \text{By definition.}$$

From Figure 91, we see that the only angle θ within the interval $[-\pi/2, \pi/2]$ whose sine is $\sqrt{3}/2$ is $\pi/3$. (Note that $\sin(2\pi/3)$ also equals $\sqrt{3}/2$, but $2\pi/3$ lies outside the interval $[-\pi/2, \pi/2]$ and hence is not admissible.)

FIGURE 91

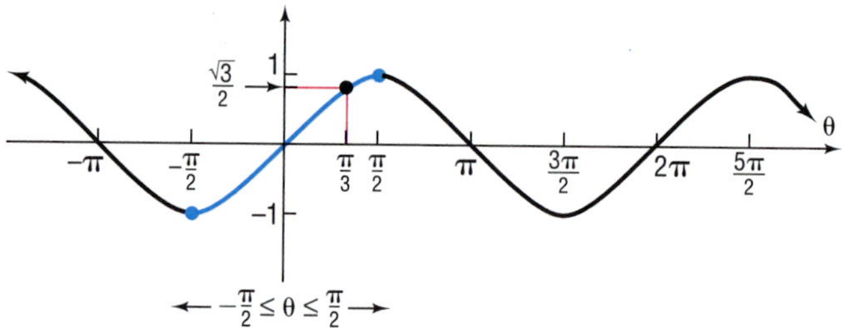

We conclude that

$$\theta = \frac{\pi}{3}$$

So

$$\sin^{-1}\frac{\sqrt{3}}{2} = \frac{\pi}{3}$$

EXAMPLE 3 Finding the Exact Value of an Inverse Sine Function

Find the exact value of $\sin^{-1}(-\frac{1}{2})$.

Solution Let $\theta = \sin^{-1}(-\frac{1}{2})$. We seek the angle θ, $-\pi/2 \le \theta \le \pi/2$, whose sine equals $-\frac{1}{2}$:

$$\theta = \sin^{-1}\left(-\frac{1}{2}\right), \qquad -\frac{\pi}{2} \le \theta \le \frac{\pi}{2}$$

$$\sin\theta = -\frac{1}{2}, \qquad -\frac{\pi}{2} \le \theta \le \frac{\pi}{2}$$

(Refer to Figure 91, if necessary.) The only angle within the interval $[-\pi/2, \pi/2]$ whose sine is $-\frac{1}{2}$ is $-\pi/6$. Thus,

$$\theta = -\frac{\pi}{6}$$

So

$$\sin^{-1}\left(-\frac{1}{2}\right) = -\frac{\pi}{6}$$

Now work Problem 3.

For most numbers x, the value $y = \sin^{-1} x$ must be approximated.

EXAMPLE 4 Finding an Approximate Value of an Inverse Sine Function

2

Find the approximate value of

(a) $\sin^{-1}\frac{1}{3}$ (b) $\sin^{-1}\left(-\frac{1}{4}\right)$

Express the answer in radians rounded to two decimal places.

Solution Because we want the angle measured in radians, we first set the mode to radians.

(a) Figure 92 shows the solution using a TI-83 graphing calculator.

FIGURE 92

Thus, $\sin^{-1}\frac{1}{3} = 0.34$ rounded to two decimal places.

(b) Figure 93 shows the solution using a TI-83 graphing calculator.

FIGURE 93

Thus, $\sin^{-1}\left(-\frac{1}{4}\right) = -0.25$ rounded to two decimal places.

Now work Problem 13.

The Inverse Cosine Function

In Figure 94 we reproduce the graph of $y = \cos x$. Because every horizontal line $y = b$, where b is between -1 and 1, intersects the graph of $y = \cos x$ infinitely many times, it follows that the cosine function is not one-to-one.

FIGURE 94
$y = \cos x,$
$-\infty < x < \infty, -1 \le y \le 1$

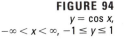

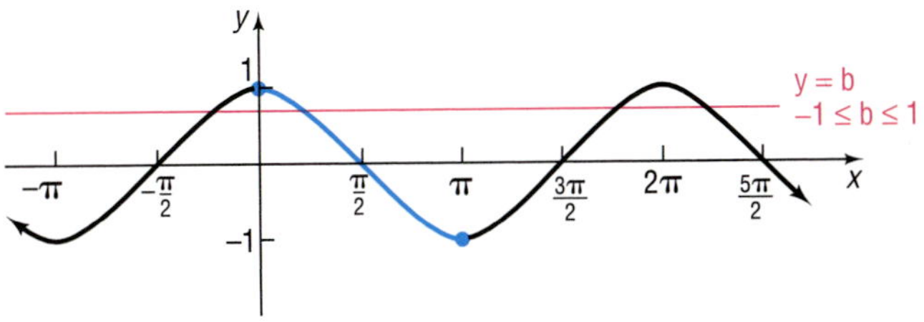

However, if we restrict the domain of $y = \cos x$ to the interval $[0, \pi]$, the restricted function

$$y = \cos x, \qquad 0 \le x \le \pi$$

is one-to-one and hence will have an inverse function.* See Figure 95.

FIGURE 95
$y = \cos x$,
$0 \le x \le \pi$, $-1 \le y \le 1$

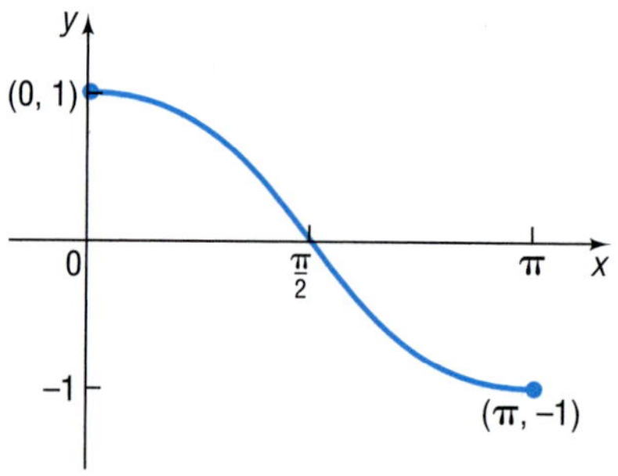

The inverse function is called the **inverse cosine** of x and is symbolized by $y = \cos^{-1} x$ (or by $y = \arccos x$). Thus,

$$y = \cos^{-1} x \quad \text{means} \quad x = \cos y \qquad (2)$$
$$\text{where} \quad 0 \le y \le \pi \quad \text{and} \quad -1 \le x \le 1$$

Here, y is the angle whose cosine is x. The domain of the function $y = \cos^{-1} x$ is $-1 \le x \le 1$, and its range is $0 \le y \le \pi$. The graph of $y = \cos^{-1} x$ can be obtained by reflecting the restricted portion of the graph of $y = \cos x$ about the line $y = x$, as shown in Figure 96(a). Figure 96(b) shows the graph using a graphing utility.

FIGURE 96
$y = \cos^{-1} x$,
$-1 \le x \le 1$, $0 \le y \le \pi$

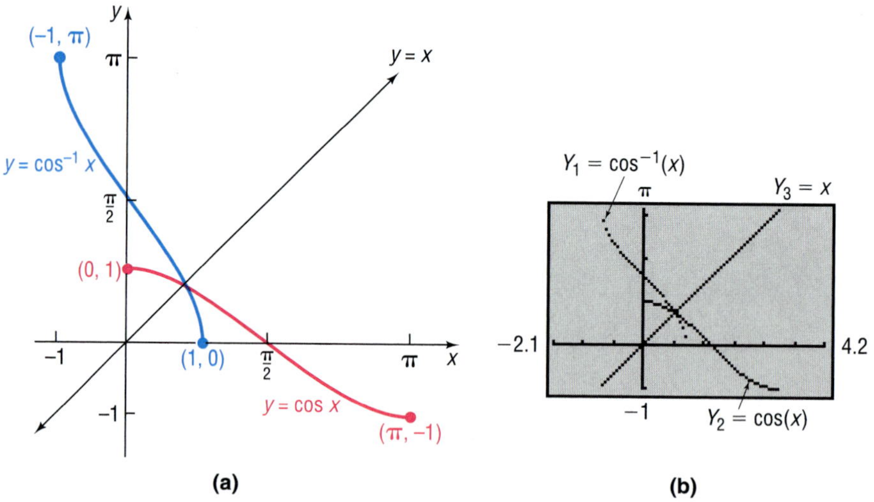

In general,

$$\cos^{-1}(\cos u) = u, \qquad \text{where} \quad 0 \le u \le \pi$$
$$\cos(\cos^{-1} v) = v, \qquad \text{where} \quad -1 \le v \le 1$$

*This is the generally accepted restriction to define the inverse.

EXAMPLE 5 Finding the Exact Value of an Inverse Cosine Function

Find the exact value of $\cos^{-1} 0$.

Solution Let $\theta = \cos^{-1} 0$. We seek the angle θ, $0 \le \theta \le \pi$, whose cosine equals 0:

$$\theta = \cos^{-1} 0, \qquad 0 \le \theta \le \pi$$
$$\cos \theta = 0, \qquad 0 \le \theta \le \pi$$

From Figure 97, we see that the only angle θ within the interval $[0, \pi]$ whose cosine is 0 is $\pi/2$. (Note that $\cos 3\pi/2$ also equals 0, but $3\pi/2$ lies outside the interval $[0, \pi]$ and hence is not admissible.)

FIGURE 97

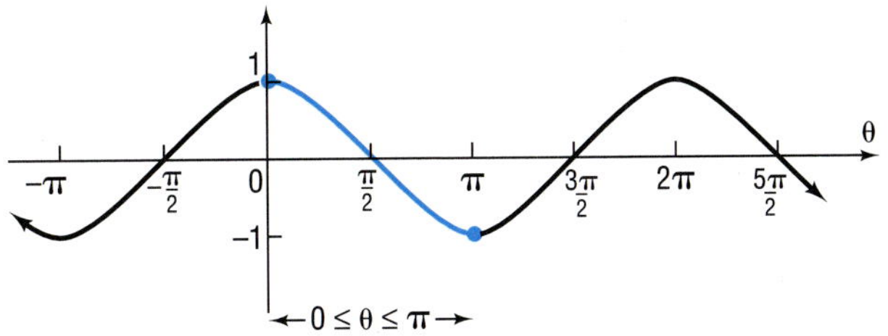

We conclude that

$$\theta = \frac{\pi}{2}$$

So

$$\cos^{-1} 0 = \frac{\pi}{2}$$

EXAMPLE 6 Finding the Exact Value of an Inverse Cosine Function

Find the exact value of $\cos^{-1}(\sqrt{2}/2)$.

Solution Let $\theta = \cos^{-1}(\sqrt{2}/2)$. We seek the angle θ, $0 \le \theta \le \pi$, whose cosine equals $\sqrt{2}/2$:

$$\theta = \cos^{-1} \frac{\sqrt{2}}{2}, \qquad 0 \le \theta \le \pi$$
$$\cos \theta = \frac{\sqrt{2}}{2}, \qquad 0 \le \theta \le \pi$$

From Figure 98, we see that the only angle θ within the interval $[0, \pi]$ whose cosine is $\sqrt{2}/2$ is $\pi/4$.

FIGURE 98

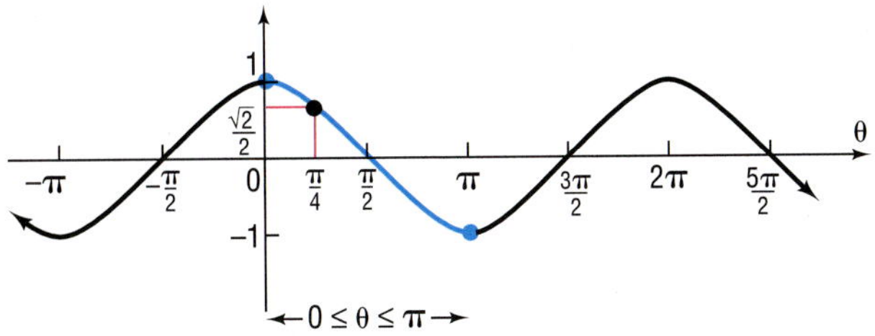

We conclude that

$$\theta = \frac{\pi}{4}$$

So

$$\cos^{-1}\frac{\sqrt{2}}{2} = \frac{\pi}{4}$$

EXAMPLE 7 Finding the Exact Value of an Inverse Cosine Function

Find the exact value of $\cos^{-1}(-\frac{1}{2})$.

Solution Let $\theta = \cos^{-1}(-\frac{1}{2})$. We seek the angle θ, $0 \le \theta \le \pi$, whose cosine equals $-\frac{1}{2}$:

$$\theta = \cos^{-1}\left(-\tfrac{1}{2}\right), \qquad 0 \le \theta \le \pi$$
$$\cos\theta = -\tfrac{1}{2}, \qquad 0 \le \theta \le \pi$$

(Refer to Figure 98, if necessary.) The only angle within the interval $[0, \pi]$ whose cosine is $-\frac{1}{2}$ is $2\pi/3$. Thus,

$$\theta = \frac{2\pi}{3}$$

So

$$\cos^{-1}\left(-\frac{1}{2}\right) = \frac{2\pi}{3}$$

The Inverse Tangent Function

In Figure 99 we reproduce the graph of $y = \tan x$. Because every horizontal line intersects the graph infinitely many times, it follows that the tangent function is not one-to-one.

FIGURE 99
$y = \tan x$, $-\infty < x < \infty$, x not equal to odd multiples of $\pi/2$, $-\infty < y < \infty$

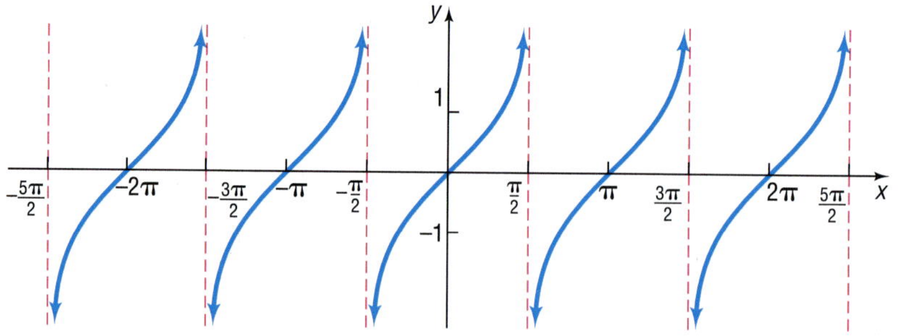

However, if we restrict the domain of $y = \tan x$ to the interval $(-\pi/2, \pi/2)$,* the restricted function

$$y = \tan x, \qquad -\frac{\pi}{2} < x < \frac{\pi}{2}$$

*This is the generally accepted restriction.

is one-to-one and hence has an inverse function. See Figure 100.

FIGURE 100
$y = \tan x$,
$-\pi/2 < x < \pi/2$, $-\infty < y < \infty$

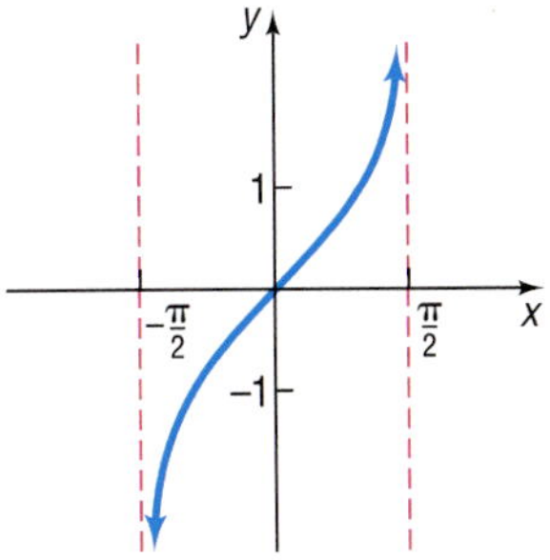

The inverse function is called the **inverse tangent** of x and is symbolized by $y = \tan^{-1} x$ (or by $y = \arctan x$). Thus,

$$y = \tan^{-1} x \quad \text{means} \quad x = \tan y \qquad (3)$$
$$\text{where} \quad -\frac{\pi}{2} < y < \frac{\pi}{2} \quad \text{and} \quad -\infty < x < \infty$$

Here, y is the angle whose tangent is x. The domain of the function $y = \tan^{-1} x$ is $-\infty < x < \infty$, and its range is $-\pi/2 < y < \pi/2$. The graph of $y = \tan^{-1} x$ can be obtained by reflecting the restricted portion of the graph of $y = \tan x$ about the line $y = x$, as shown in Figure 101(a). Figure 101(b) shows the graph using a graphing utility.

FIGURE 101
$y = \tan^{-1} x$, $-\infty < x < \infty$,
$-\pi/2 < y < \pi/2$

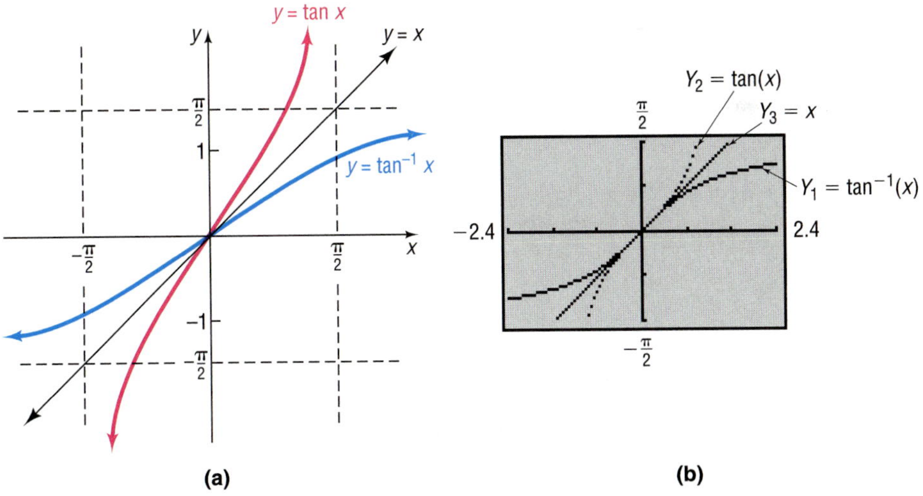

In general,

$$\tan^{-1}(\tan u) = u, \quad \text{where} \quad -\frac{\pi}{2} < u < \frac{\pi}{2}$$
$$\tan(\tan^{-1} v) = v, \quad \text{where} \quad -\infty < v < \infty$$

EXAMPLE 8 Finding the Exact Value of an Inverse Tangent Function

Find the exact value of $\tan^{-1} 1$.

Solution Let $\theta = \tan^{-1} 1$. We seek the angle θ, $-\pi/2 < \theta < \pi/2$, whose tangent equals 1:

$$\theta = \tan^{-1} 1, \qquad -\frac{\pi}{2} < \theta < \frac{\pi}{2}$$

$$\tan\theta = 1, \qquad -\frac{\pi}{2} < \theta < \frac{\pi}{2}$$

Refer to Figure 100. The only angle θ within the interval $(-\pi/2, \pi/2)$ whose tangent is 1 is $\pi/4$.

We conclude that

$$\theta = \frac{\pi}{4}$$

So

$$\tan^{-1} 1 = \frac{\pi}{4}$$

EXAMPLE 9 Finding the Exact Value of an Inverse Tangent Function

Find the exact value of $\tan^{-1}(-\sqrt{3})$.

Solution Let $\theta = \tan^{-1}(-\sqrt{3})$. We seek the angle θ, $-\pi/2 < \theta < \pi/2$, whose tangent equals $-\sqrt{3}$:

$$\theta = \tan^{-1}(-\sqrt{3}), \qquad -\frac{\pi}{2} < \theta < \frac{\pi}{2}$$

$$\tan\theta = -\sqrt{3} \qquad -\frac{\pi}{2} < \theta < \frac{\pi}{2}$$

Refer again to Figure 100. The only angle θ within the interval $(-\pi/2, \pi/2)$ whose tangent is $-\sqrt{3}$ is $-\pi/3$. Thus,

$$\theta = -\frac{\pi}{3}$$

So

$$\tan^{-1}(-\sqrt{3}) = -\frac{\pi}{3}$$

Now work Problem 5.

EXAMPLE 10 Finding the Exact Value of Expressions Involving Inverse Trigonometric Functions

Find the exact value of $\sin[\cos^{-1}(\sqrt{3}/2)]$.

Solution Let $\theta = \cos^{-1}(\sqrt{3}/2)$. We seek the angle θ, $0 \le \theta \le \pi$, whose cosine equals $\sqrt{3}/2$:

$$\cos\theta = \frac{\sqrt{3}}{2}, \qquad 0 \le \theta \le \pi$$

$$\theta = \frac{\pi}{6}$$

Now

$$\sin\left(\cos^{-1}\frac{\sqrt{3}}{2}\right) = \sin\theta = \sin\frac{\pi}{6} = \frac{1}{2}$$

EXAMPLE 11 Finding the Exact Value of Expressions Involving Inverse Trigonometric Functions

Find the exact value of $\cos[\tan^{-1}(-1)]$.

Solution Let $\theta = \tan^{-1}(-1)$. We seek the angle θ, $-\pi/2 < \theta < \pi/2$, whose tangent equals -1:

$$\tan\theta = -1, \qquad -\frac{\pi}{2} < \theta < \frac{\pi}{2}$$

$$\theta = -\frac{\pi}{4},$$

Now

$$\cos[\tan^{-1}(-1)] = \cos\theta = \cos\left(-\frac{\pi}{4}\right) = \frac{\sqrt{2}}{2}$$

EXAMPLE 12 Finding the Exact Value of Expressions Involving Inverse Trigonometric Functions

Find the exact value of $\sec(\sin^{-1}\frac{1}{2})$.

Solution Let $\theta = \sin^{-1}\frac{1}{2}$. We seek the angle θ, $-\pi/2 \le \theta \le \pi/2$, whose sine equals $\frac{1}{2}$:

$$\sin\theta = \frac{1}{2}, \qquad -\frac{\pi}{2} \le \theta \le \frac{\pi}{2}$$

$$\theta = \frac{\pi}{6}$$

Now

$$\sec\left(\sin^{-1}\frac{1}{2}\right) = \sec\theta = \sec\frac{\pi}{6} = \frac{2}{\sqrt{3}} = \frac{2\sqrt{3}}{3}$$

Now work Problem 25.

It is not necessary to be able to find the angle in order to solve problems like those given in Examples 10–12.

EXAMPLE 13 Finding the Exact Value of Expressions Involving Inverse Trigonometric Functions

Find the exact value of $\sin(\tan^{-1}\frac{1}{2})$.

Solution Let $\theta = \tan^{-1}\frac{1}{2}$. Then $\tan\theta = \frac{1}{2}$, where $-\pi/2 < \theta < \pi/2$. Because $\tan\theta > 0$, it follows that $0 < \theta < \pi/2$. Now, in Figure 102, we draw a triangle in the appropriate quadrant depicting $\tan\theta = \frac{1}{2}$. The hypotenuse of this triangle is easily found to be of length $\sqrt{5}$. Hence, the sine of θ is $1/\sqrt{5}$, so

$$\sin\left(\tan^{-1}\frac{1}{2}\right) = \sin\theta = \frac{1}{\sqrt{5}} = \frac{\sqrt{5}}{5}$$

FIGURE 102
$\tan\theta = \frac{1}{2}$

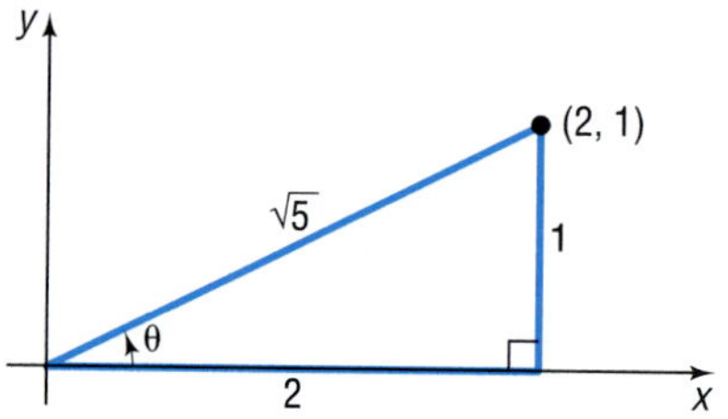

EXAMPLE 14 Finding the Exact Value of Expressions Involving Inverse Trigonometric Functions

Find the exact value of $\tan(\cos^{-1}\frac{1}{3})$

Solution Let $\theta = \cos^{-1}\frac{1}{3}$. Then $\cos\theta = \frac{1}{3}$, where $0 \le \theta \le \pi$. Because $\cos\theta > 0$, it follows that $0 \le \theta < \pi/2$. Look at the triangle in Figure 103. The side opposite angle θ is $\sqrt{8} = 2\sqrt{2}$. Thus,

$$\tan(\cos^{-1}\tfrac{1}{3}) = \tan\theta = 2\sqrt{2}$$

FIGURE 103
$\cos\theta = \frac{1}{3}$

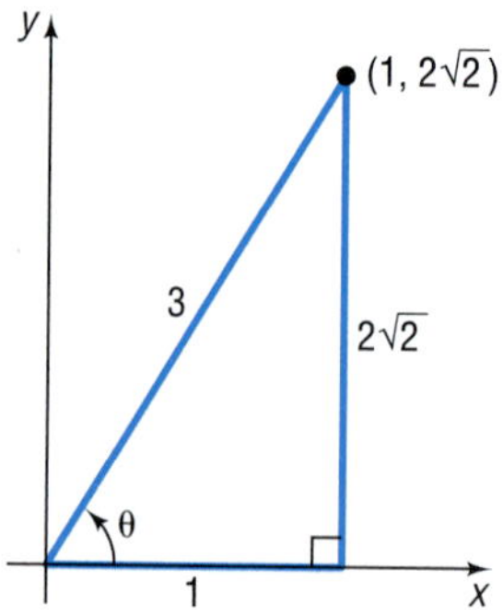

EXAMPLE 15 Finding the Exact Value of Expressions Involving Inverse Trigonometric Functions

Find the exact value of $\cos[\sin^{-1}(-\frac{1}{3})]$.

Solution Let $\theta = \sin^{-1}(-\frac{1}{3})$. Then $\sin\theta = -\frac{1}{3}$ and $-\pi/2 \le \theta \le \pi/2$. Because $\sin\theta < 0$, it follows that $-\pi/2 \le \theta < 0$. Based on Figure 104, we conclude that

$$\cos\left[\sin^{-1}\left(-\frac{1}{3}\right)\right] = \cos\theta = \frac{2\sqrt{2}}{3}$$

FIGURE 104
$\sin\theta = -\frac{1}{3}$

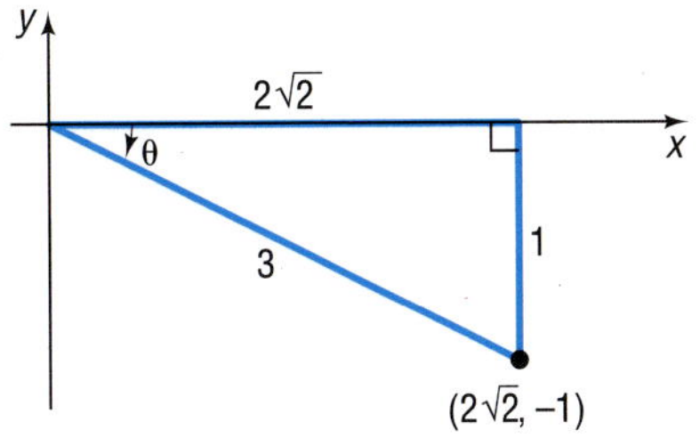

EXAMPLE 16 Finding the Exact Value of Expressions Involving Inverse Trigonometric Functions

Find the exact value of $\tan[\cos^{-1}(-\frac{1}{3})]$.

FIGURE 105
$\cos\theta = -\frac{1}{3}$

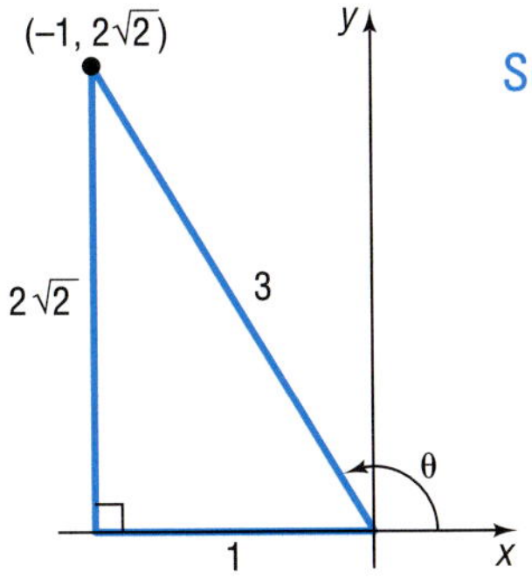

Solution Let $\theta = \cos^{-1}(-\frac{1}{3})$. Then $\cos\theta = -\frac{1}{3}$ and $0 \le \theta \le \pi$. Because $\cos\theta < 0$, it follows that $\pi/2 < \theta \le \pi$. Based on Figure 105, we conclude that

$$\tan\left[\cos^{-1}\left(-\frac{1}{3}\right)\right] = \tan\theta = \frac{2\sqrt{2}}{-1} = -2\sqrt{2}$$

Now work Problem 39.

EXAMPLE 17 Establishing an Identity Involving Inverse Trigonometric Functions

Show that $\sin(\tan^{-1} v) = \dfrac{v}{\sqrt{1 + v^2}}$.

Solution Let $\theta = \tan^{-1} v$ so that $\tan\theta = v$, $-\pi/2 < \theta < \pi/2$. There are two possibilities: either $-\pi/2 < \theta < 0$ or $0 \le \theta < \pi/2$. If $0 \le \theta < \pi/2$, then $\tan\theta = v \ge 0$. Based on Figure 106, we conclude that

$$\sin(\tan^{-1} v) = \sin\theta = \frac{v}{\sqrt{1 + v^2}}$$

If $-\pi/2 < \theta < 0$, then $\tan\theta = v < 0$. Based on Figure 107, we conclude that

$$\sin(\tan^{-1} v) = \sin\theta = \frac{v}{\sqrt{1 + v^2}}$$

FIGURE 106
$v > 0$

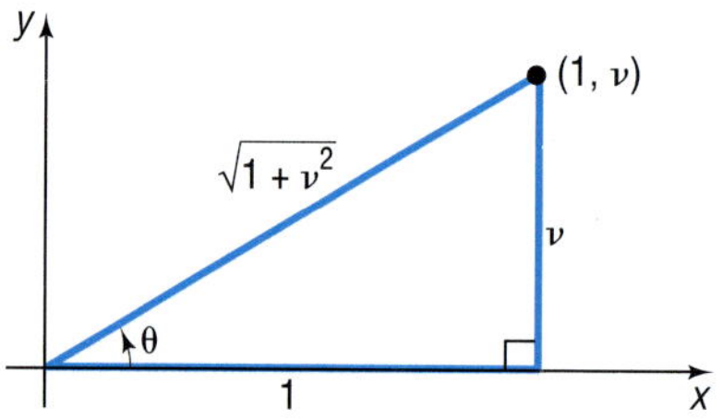

FIGURE 107
$v < 0$

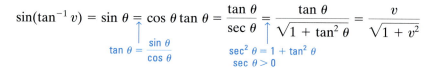

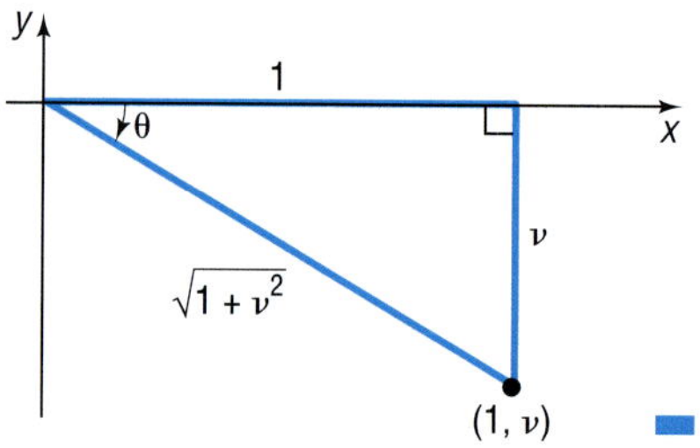

An alternative solution to Example 17 uses the fundamental identities. If $\theta = \tan^{-1} v$, then $\tan \theta = v$ and $-\pi/2 < \theta < \pi/2$, as before. As a result, we know that $\sec \theta > 0$. Thus,

$$\sin(\tan^{-1} v) = \sin\theta \underset{\tan\theta = \frac{\sin\theta}{\cos\theta}}{=} \cos\theta\tan\theta = \frac{\tan\theta}{\sec\theta} \underset{\substack{\sec^2\theta = 1 + \tan^2\theta \\ \sec\theta > 0}}{=} \frac{\tan\theta}{\sqrt{1 + \tan^2\theta}} = \frac{v}{\sqrt{1 + v^2}}$$

Now work Problem 57.

The Remaining Inverse Trigonometric Functions

The inverse cotangent, inverse secant, and inverse cosecant functions are defined as follows:

$$y = \cot^{-1} x \quad \text{means} \quad x = \cot y \tag{4}$$
$$\text{where} \quad -\infty < x < \infty \quad \text{and} \quad 0 < y < \pi$$

$$y = \sec^{-1} x \quad \text{means} \quad x = \sec y \tag{5}$$
$$\text{where} \quad |x| \geq 1 \quad \text{and} \quad 0 \leq y \leq \pi, y \neq \frac{\pi}{2}$$

$$y = \csc^{-1} x \quad \text{means} \quad x = \csc y \tag{6}$$
$$\text{where} \quad |x| \geq 1 \quad \text{and} \quad -\frac{\pi}{2} \leq y \leq \frac{\pi}{2}, y \neq 0$$

Most calculators do not have keys for evaluating these inverse trigonometric functions. The easiest way to evaluate them is to convert to an inverse trigonometric function whose range is the same as the one to be evaluated. In this regard, notice that $y = \cot^{-1} x$ and $y = \sec^{-1} x$ (except where undefined) each have the same range as $y = \cos^{-1} x$, while $y = \csc^{-1} x$ (except where undefined) has the same range as $y = \sin^{-1} x$.

EXAMPLE 18

Approximating the Value of Inverse Trigonometric Functions

Use a calculator to approximate each expression in radians rounded to two decimal places:

(a) $\sec^{-1} 3$ (b) $\csc^{-1}(-4)$ (c) $\cot^{-1}\frac{1}{2}$ (d) $\cot^{-1}(-2)$

Solution First, set your calculator to radian mode.

(a) Let $\theta = \sec^{-1} 3$. Then $\sec\theta = 3$ and $0 \le \theta \le \pi$, $\theta \ne \pi/2$. Thus, $\cos\theta = \frac{1}{3}$ and

$$\sec^{-1} 3 = \theta = \cos^{-1}\tfrac{1}{3} \approx 1.23$$

Use a calculator.

(b) Let $\theta = \csc^{-1}(-4)$. Then $\csc\theta = -4$, $-\pi/2 \le \theta \le \pi/2$, $\theta \ne 0$. Thus, $\sin\theta = -\frac{1}{4}$ and

$$\csc^{-1}(-4) = \theta = \sin^{-1}(-\tfrac{1}{4}) \approx -0.25$$

FIGURE 108
$\cot\theta = \frac{1}{2}$, $0 < \theta < \pi$

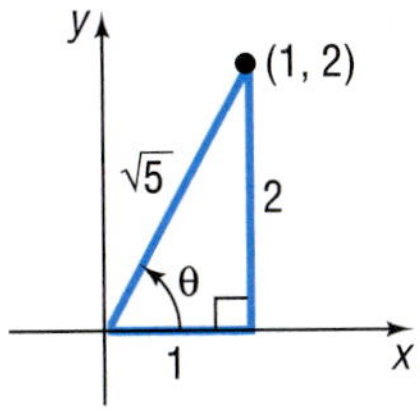

(c) Let $\theta = \cot^{-1}\frac{1}{2}$. Then $\cot\theta = \frac{1}{2}$, $0 < \theta < \pi$. From these facts we know that θ is in quadrant I. We draw Figure 108 to help us to find $\cos\theta$. Thus, $\cos\theta = 1/\sqrt{5}$, $0 < \theta < \pi/2$, and

$$\cot^{-1}\frac{1}{2} = \theta = \cos^{-1}\frac{1}{\sqrt{5}} \approx 1.11$$

FIGURE 109
$\cot\theta = -2$, $0 < \theta < \pi$

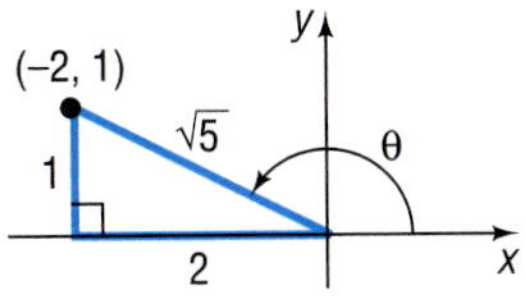

(d) Let $\theta = \cot^{-1}(-2)$. Then $\cot\theta = -2$, $0 < \theta < \pi$. From these facts we know that θ lies in quadrant II. We draw Figure 109 to help us to find $\cos\theta$. Thus, $\cos\theta = -2/\sqrt{5}$, $\pi/2 < \theta < \pi$, and

$$\cot^{-1}(-2) = \theta = \cos^{-1}\left(-\frac{2}{\sqrt{5}}\right) \approx 2.68$$

Now work Problem 67.

6.7 EXERCISES

In Problems 1–12, find the exact value of each expression.

1. $\sin^{-1} 0$ **2.** $\cos^{-1} 1$ **3.** $\sin^{-1}(-1)$ **4.** $\cos^{-1}(-1)$

5. $\tan^{-1} 0$ **6.** $\tan^{-1}(-1)$ **7.** $\sin^{-1}\frac{\sqrt{2}}{2}$ **8.** $\tan^{-1}\frac{\sqrt{3}}{3}$

9. $\tan^{-1}\sqrt{3}$ **10.** $\sin^{-1}\left(-\frac{\sqrt{3}}{2}\right)$ **11.** $\cos^{-1}\left(-\frac{\sqrt{3}}{2}\right)$ **12.** $\sin^{-1}\left(-\frac{\sqrt{2}}{2}\right)$

In Problems 13–24, use a calculator to find the value of each expression rounded to two decimal places.

13. $\sin^{-1} 0.1$ **14.** $\cos^{-1} 0.6$ **15.** $\tan^{-1} 5$ **16.** $\tan^{-1} 0.2$

17. $\cos^{-1}\frac{7}{8}$ **18.** $\sin^{-1}\frac{1}{8}$ **19.** $\tan^{-1}(-0.4)$ **20.** $\tan^{-1}(-3)$

21. $\sin^{-1}(-0.12)$ **22.** $\cos^{-1}(-0.44)$ **23.** $\cos^{-1}\frac{\sqrt{2}}{3}$ **24.** $\sin^{-1}\frac{\sqrt{3}}{5}$

In Problems 25–46, find the exact value of each expression.

25. $\cos\left(\sin^{-1}\frac{\sqrt{2}}{2}\right)$ **26.** $\sin\left(\cos^{-1}\frac{1}{2}\right)$ **27.** $\tan\left[\cos^{-1}\left(-\frac{\sqrt{3}}{2}\right)\right]$ **28.** $\tan\left[\sin^{-1}\left(-\frac{1}{2}\right)\right]$

29. $\sec\left(\cos^{-1}\frac{1}{2}\right)$ **30.** $\cot\left[\sin^{-1}\left(-\frac{1}{2}\right)\right]$ **31.** $\csc(\tan^{-1} 1)$ **32.** $\sec(\tan^{-1}\sqrt{3})$

33. $\sin[\tan^{-1}(-1)]$ **34.** $\cos\left[\sin^{-1}\left(-\frac{\sqrt{3}}{2}\right)\right]$ **35.** $\sec\left[\sin^{-1}\left(-\frac{1}{2}\right)\right]$ **36.** $\csc\left[\cos^{-1}\left(-\frac{\sqrt{3}}{2}\right)\right]$

37. $\tan\left(\sin^{-1}\dfrac{1}{3}\right)$ **38.** $\tan\left(\cos^{-1}\dfrac{1}{3}\right)$ **39.** $\sec\left(\tan^{-1}\dfrac{1}{2}\right)$ **40.** $\cos\left(\sin^{-1}\dfrac{\sqrt{2}}{3}\right)$

41. $\cot\left[\sin^{-1}\left(-\dfrac{\sqrt{2}}{3}\right)\right]$ **42.** $\csc[\tan^{-1}(-2)]$ **43.** $\sin[\tan^{-1}(-3)]$ **44.** $\cot\left[\cos^{-1}\left(-\dfrac{\sqrt{3}}{3}\right)\right]$

45. $\sec\left(\sin^{-1}\dfrac{2\sqrt{5}}{5}\right)$ **46.** $\csc\left(\tan^{-1}\dfrac{1}{2}\right)$

In Problems 47–56, use a calculator to find the value of each expression rounded to two decimal places.

47. $\sin^{-1}(\tan 0.5)$ **48.** $\cos^{-1}(\tan 0.4)$ **49.** $\tan^{-1}(\sin 0.1)$ **50.** $\tan^{-1}(\cos 0.2)$

51. $\cos^{-1}(\sin 1)$ **52.** $\tan^{-1}(\cos 1)$ **53.** $\sin^{-1}\left(\tan\dfrac{\pi}{8}\right)$ **54.** $\cos^{-1}\left(\sin\dfrac{\pi}{8}\right)$

55. $\tan^{-1}\left(\sin\dfrac{\pi}{8}\right)$ **56.** $\tan^{-1}\left(\cos\dfrac{\pi}{8}\right)$

57. Show that $\sec(\tan^{-1} v) = \sqrt{1 + v^2}$.

58. Show that $\tan(\sin^{-1} v) = v/\sqrt{1 - v^2}$.

59. Show that $\tan(\cos^{-1} v) = \sqrt{1 - v^2}/v$.

60. Show that $\sin(\cos^{-1} v) = \sqrt{1 - v^2}$.

61. Show that $\cos(\sin^{-1} v) = \sqrt{1 - v^2}$.

62. Show that $\cos(\tan^{-1} v) = 1/\sqrt{1 + v^2}$.

63. Show that $\sin^{-1} v + \cos^{-1} v = \pi/2$.

64. Show that $\tan^{-1} v + \cot^{-1} v = \pi/2$.

65. Show that $\tan^{-1}(1/v) = \pi/2 - \tan^{-1} v$.

66. Show that $\cot^{-1} e^v = \tan^{-1} e^{-v}$.

In Problems 67–78, use a calculator to find the value of each expression rounded to two decimal places.

67. $\sec^{-1} 4$ **68.** $\csc^{-1} 5$ **69.** $\cot^{-1} 2$ **70.** $\sec^{-1}(-3)$

71. $\csc^{-1}(-3)$ **72.** $\cot^{-1}\left(-\frac{1}{2}\right)$ **73.** $\cot^{-1}(-\sqrt{5})$ **74.** $\cot^{-1}(-8.1)$

75. $\csc^{-1}\left(-\frac{3}{2}\right)$ **76.** $\sec^{-1}\left(-\frac{4}{3}\right)$ **77.** $\cot^{-1}\left(-\frac{3}{2}\right)$ **78.** $\cot^{-1}(-\sqrt{10})$

79. Being the First To See the Rising Sun Cadillac Mountain, elevation 1530 feet, is located in Acadia National Park, Maine, and is the highest peak on the east coast of the United States. It is said that a person standing on the summit will be the first person in the United States to see the rays of the rising Sun. How much sooner would a person atop Cadillac Mountain see the first rays than a person standing below, at sea level?

[**Hint:** Consult the figure. When the person at D sees the first rays of the Sun, the person at P does not. The person at P sees the first rays of the Sun only after Earth has rotated so that P is at location Q. Compute the length of arc s subtended by the central angle θ. Then use the fact that in 24 hours a length of $2\pi(3960)$ miles is subtended, and find the time it takes to subtend the length s.]

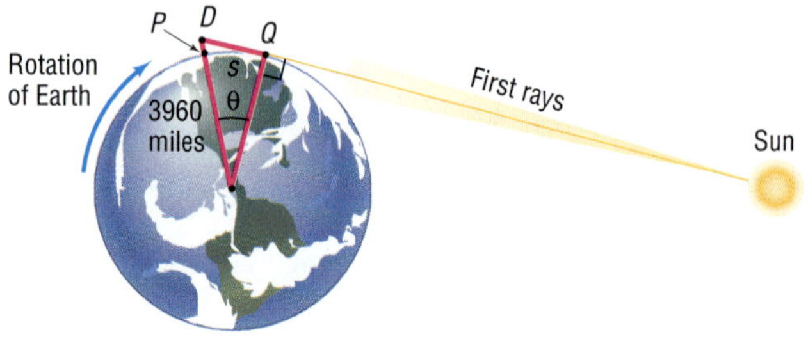

80. For what numbers x does $\sin(\sin^{-1} x) = x$?

81. For what numbers x does $\cos(\cos^{-1} x) = x$?

82. For what numbers x does $\sin^{-1}(\sin x) = x$?

83. For what numbers x does $\cos^{-1}(\cos x) = x$?

84. Graph $y = \cot^{-1} x$.

85. Graph $y = \sec^{-1} x$.

86. Graph $y = \csc^{-1} x$.

87. Explain in your own words how you would use your calculator to find the value of $\cot^{-1} 10$.

88. Consult three books on calculus and write down the definition in each of $y = \sec^{-1} x$ and $y = \csc^{-1} x$. Compare these with the definitions given in this book.

CHAPTER REVIEW

THINGS TO KNOW

Definitions

Angle in standard position	Vertex is at the origin; initial side is along the positive x-axis
Degree ($1°$)	$1° = \frac{1}{360}$ revolution
Radian (1)	The measure of a central angle whose rays subtend an arc whose length is the radius of the circle
Acute angle	An angle θ whose measure is $0° < \theta < 90°$ (or $0 < \theta < \pi/2$)
Trigonometric functions	$P = (a, b)$ is the point on the terminal side of θ a distance r from the origin;

$$\sin\theta = \frac{b}{r} \qquad \cos\theta = \frac{a}{r}$$

$$\tan\theta = \frac{b}{a},\quad a \neq 0; \qquad \cot\theta = \frac{a}{b},\quad b \neq 0$$

$$\csc\theta = \frac{r}{b},\quad b \neq 0; \qquad \sec\theta = \frac{r}{a},\quad a \neq 0$$

Complementary angles	Two acute angles whose sum is $90°$ ($\pi/2$)
Cofunction	The following pairs of functions are cofunctions of each other: sine and cosine; tangent and cotangent; secant and cosecant
Reference angle of θ	The acute angle formed by the terminal side of θ and either the positive or negative x-axis
Unit circle	Center at origin; radius = 1 Equation: $x^2 + y^2 = 1$
Periodic function	$f(\theta + p) = f(\theta)$, for all θ, $p > 0$, where the smallest such p is the fundamental period

Formulas

1 revolution $= 360°$

$= 2\pi$ radians

$s = r\theta$	θ is measured in radians; s is the length of arc subtended by the central angle θ of the circle of radius r
$v = r\omega$	v is the linear speed along the circle of radius r; ω is the angular speed (measured in radians per unit time)

Table of Values

θ (Radians)	θ (Degrees)	$\sin\theta$	$\cos\theta$	$\tan\theta$	$\csc\theta$	$\sec\theta$	$\cot\theta$
0	$0°$	0	1	0	Not defined	1	Not defined
$\pi/6$	$30°$	$\frac{1}{2}$	$\sqrt{3}/2$	$\sqrt{3}/3$	2	$2\sqrt{3}/3$	$\sqrt{3}$
$\pi/4$	$45°$	$\sqrt{2}/2$	$\sqrt{2}/2$	1	$\sqrt{2}$	$\sqrt{2}$	1
$\pi/3$	$60°$	$\sqrt{3}/2$	$\frac{1}{2}$	$\sqrt{3}$	$2\sqrt{3}/3$	2	$\sqrt{3}/3$
$\pi/2$	$90°$	1	0	Not defined	1	Not defined	0
π	$180°$	0	-1	0	Not defined	-1	Not defined
$3\pi/2$	$270°$	-1	0	Not defined	-1	Not defined	0

Identities

$$\tan\theta = \frac{\sin\theta}{\cos\theta}, \quad \cot\theta = \frac{\cos\theta}{\sin\theta}$$

$$\cot\theta = \frac{1}{\tan\theta}, \quad \sec\theta = \frac{1}{\cos\theta}, \quad \csc\theta = \frac{1}{\sin\theta}$$

$$\sin^2\theta + \cos^2\theta = 1, \quad \tan^2\theta + 1 = \sec^2\theta, \quad 1 + \cot^2\theta = \csc^2\theta$$

Properties of the Trigonometric Functions

$y = \sin x$ Domain: $-\infty < x < \infty$
Range: $-1 \le y \le 1$
Periodic: period $= 2\pi$ (360°)
Odd function

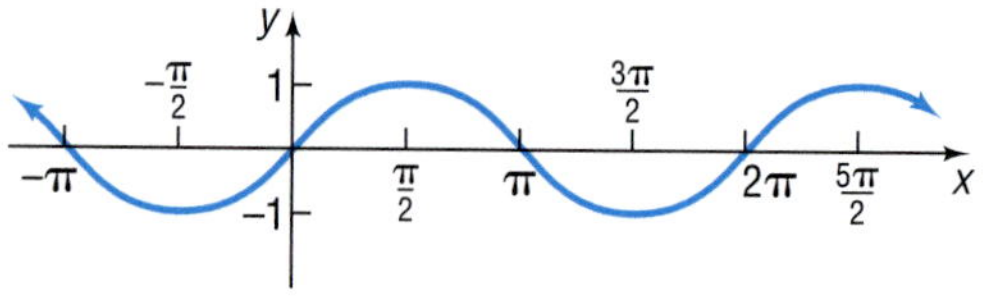

$y = \cos x$ Domain: $-\infty < x < \infty$
Range: $-1 \le y \le 1$
Periodic: period $= 2\pi$ (360°)
Even function

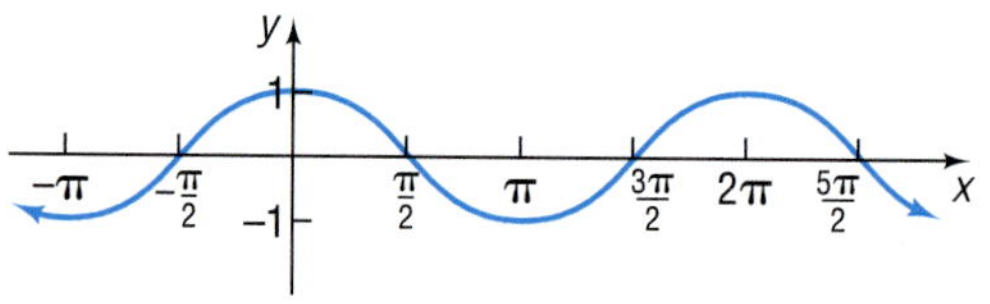

$y = \tan x$ Domain: $-\infty < x < \infty$, except odd multiples of $\pi/2$ (90°)
Range: $-\infty < y < \infty$
Periodic: period $= \pi$ (180°)
Odd function

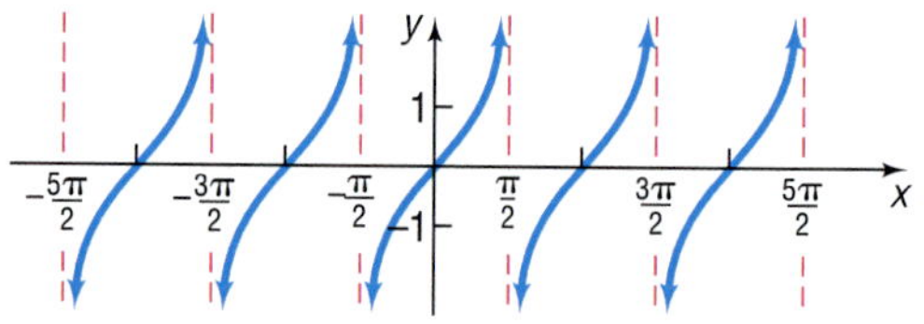

$y = \csc x$ Domain: $-\infty < x < \infty$, except integral multiples of π (180°)
Range: $|y| \ge 1$
Periodic: period $= 2\pi$ (360°)
Odd function

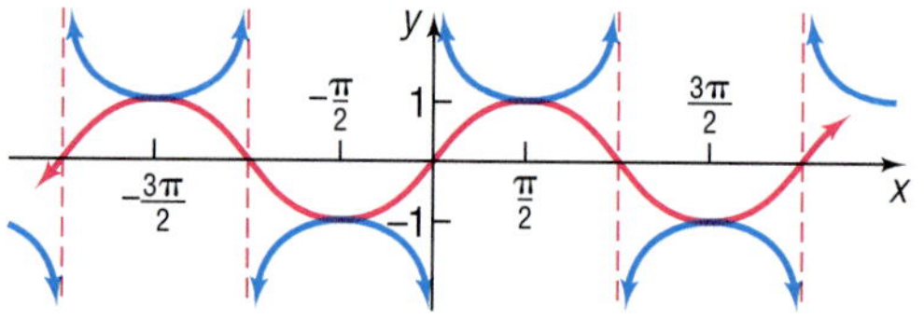

$y = \sec x$ Domain: $-\infty < x < \infty$, except odd multiples of $\pi/2$ (90°)
Range: $|y| \ge 1$
Periodic: period $= 2\pi$ (360°)
Even function

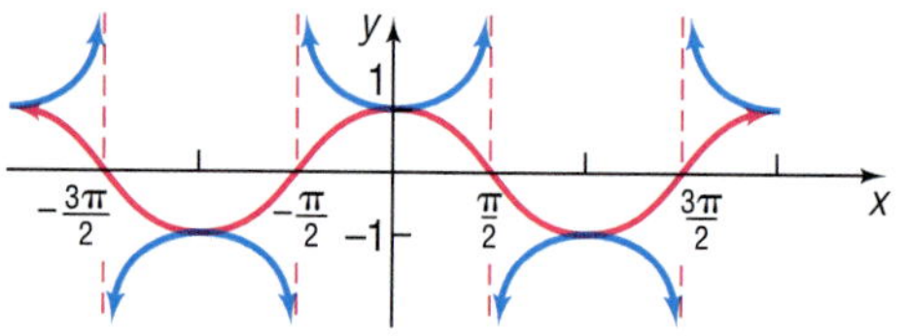

$y = \cot x$ Domain: $-\infty < x < \infty$, except integral multiples of π (180°)
Range: $-\infty < y < \infty$
Periodic: period $= \pi$ (180°)
Odd function

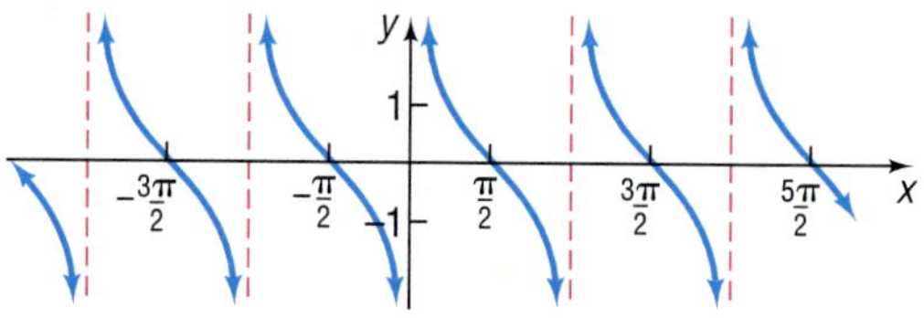

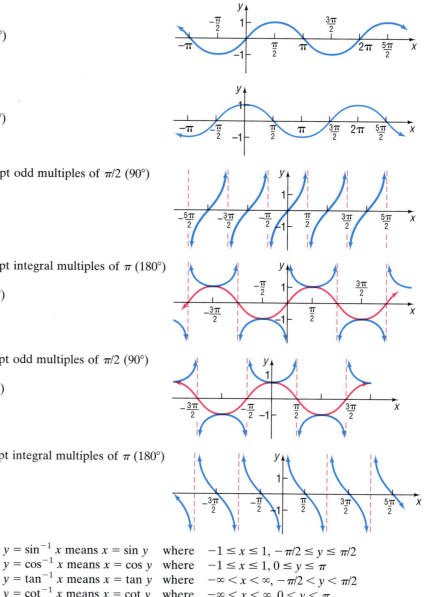

Definitions of the six inverse trigonometric functions

$y = \sin^{-1} x$ means $x = \sin y$ where $-1 \le x \le 1, -\pi/2 \le y \le \pi/2$
$y = \cos^{-1} x$ means $x = \cos y$ where $-1 \le x \le 1, 0 \le y \le \pi$
$y = \tan^{-1} x$ means $x = \tan y$ where $-\infty < x < \infty, -\pi/2 < y < \pi/2$
$y = \cot^{-1} x$ means $x = \cot y$ where $-\infty < x < \infty, 0 < y < \pi$
$y = \sec^{-1} x$ means $x = \sec y$ where $|x| \ge 1, 0 \le y \le \pi, y \ne \pi/2$
$y = \csc^{-1} x$ means $x = \csc y$ where $|x| \ge 1, -\pi/2 \le y \le \pi/2, y \ne 0$

HOW TO

Convert an angle from radian measure to degree measure

Convert an angle from degree measure to radian measure

Find the value of each of the remaining trigonometric functions if the value of one function and the quadrant of the angle are given

Use the theorem on cofunctions of complementary angles

Graph the trigonometric functions, including transformations.

Use reference angles to find the value of a trigonometric function

Use a calculator to find the value of a trigonometric function

Find the exact value of certain inverse trigonometric functions

Use a calculator to find the approximate values of inverse trigonometric functions

FILL-IN-THE-BLANK ITEMS

1. Two rays drawn with a common vertex form a(n) ______________. One of the rays is called the ______________ ______________; the other is called the ______________ ______________.
2. In the formula $s = r\theta$ for measuring the length s of arc along a circle of radius r, the angle θ must be measured in ______________.
3. 180 degrees = ______________ radians.
4. Two acute angles whose sum is a right angle are called ______________.
5. The sine and ______________ functions are cofunctions.
6. An angle is in ______________ ______________ if its vertex is at the origin and its initial side coincides with the positive x-axis.
7. The reference angle of 135° is ______________.
8. The sine, cosine, cosecant, and secant functions have period ______________; the tangent and cotangent functions have period ______________.
9. Which of the trigonometric functions have graphs that are symmetric with respect to the y-axis?
10. Which of the trigonometric functions have graphs that are symmetric with respect to the origin?
11. The function $y = \sin^{-1} x$ has domain ______________ and range ______________.
12. The value of $\sin^{-1}[\cos(\pi/2)]$ is ______________.

TRUE/FALSE ITEMS

T F 1. In the formula $s = r\theta$, r is the radius of a circle and s is the arc subtended by a central angle θ, where θ is measured in degrees.

T F 2. $|\sin \theta| \leq 1$

T F 3. $1 + \tan^2 \theta = \csc^2 \theta$

T F 4. The only even trigonometric functions are the cosine and secant functions.

T F 5. $\tan 62° = \cot 38°$

T F 6. $\sin 182° = \cos 2°$

T F 7. The graphs of $y = \tan x$, $y = \cot x$, $y = \sec x$, and $y = \csc x$ have infinitely many vertical asymptotes.

T F 8. The domain of $y = \sin^{-1} x$ is $-\pi/2 \leq x \leq \pi/2$.

T F 9. $\cos(\sin^{-1} 0) = 1$ and $\sin(\cos^{-1} 0) = 1$.

REVIEW EXERCISES

In Problems 1–4, convert each angle in degrees to radians. Express your answer as a multiple of π.

1. 135° **2.** 210° **3.** 18° **4.** 15°

In Problems 5–8, convert each angle in radians to degrees.

5. $3\pi/4$ **6.** $2\pi/3$ **7.** $-5\pi/2$ **8.** $-3\pi/2$

In Problems 9–30, find the exact value of each expression. Do not use a calculator.

9. $\tan \frac{\pi}{4} - \sin \frac{\pi}{6}$

10. $\cos \frac{\pi}{3} + \sin \frac{\pi}{2}$

11. $3 \sin 45° - 4 \tan \frac{\pi}{6}$

12. $4 \cos 60° + 3 \tan \frac{\pi}{3}$

13. $6 \cos \frac{3\pi}{4} + 2 \tan\left(-\frac{\pi}{3}\right)$

14. $3 \sin \frac{2\pi}{3} - 4 \cos \frac{5\pi}{2}$

15. $\sec\left(-\frac{\pi}{3}\right) - \cot\left(-\frac{5\pi}{4}\right)$

16. $4 \csc \frac{3\pi}{4} - \cot\left(-\frac{\pi}{4}\right)$

17. $\tan \pi + \sin \pi$

18. $\cos \frac{\pi}{2} - \csc\left(-\frac{\pi}{2}\right)$

19. $\cos 180° - \tan(-45°)$

20. $\sin 270° + \cos(-180°)$

21. $\sin^2 20° + \dfrac{1}{\sec^2 20°}$ **22.** $\dfrac{1}{\cos^2 40°} - \dfrac{1}{\cot^2 40°}$ **23.** $\sec 50° \cos 50°$

24. $\tan 10° \cot 10°$ **25.** $\dfrac{\sin 50°}{\cos 40°}$ **26.** $\dfrac{\tan 20°}{\cot 70°}$

27. $\dfrac{\sin(-40°)}{\cos 50°}$ **28.** $\tan(-20°) \cot 20°$ **29.** $\sin 400° \sec(-50°)$

30. $\cot 200° \cot(-70°)$

In Problems 31–46, find the exact value of each of the remaining trigonometric functions.

31. $\sin\theta = -\frac{4}{5}$, $\cos\theta > 0$ **32.** $\cos\theta = -\frac{3}{5}$, $\sin\theta < 0$ **33.** $\tan\theta = \frac{12}{5}$, $\sin\theta < 0$

34. $\cot\theta = \frac{12}{5}$, $\cos\theta < 0$ **35.** $\sec\theta = -\frac{5}{4}$, $\tan\theta < 0$ **36.** $\csc\theta = -\frac{5}{3}$, $\cot\theta < 0$

37. $\sin\theta = \frac{12}{13}$, θ in quadrant II **38.** $\cos\theta = -\frac{3}{5}$, θ in quadrant III **39.** $\sin\theta = -\frac{5}{13}$, $3\pi/2 < \theta < 2\pi$

40. $\cos\theta = \frac{12}{13}$, $3\pi/2 < \theta < 2\pi$ **41.** $\tan\theta = \frac{1}{3}$, $180° < \theta < 270°$ **42.** $\tan\theta = -\frac{2}{3}$, $90° < \theta < 180°$

43. $\sec\theta = 3$, $3\pi/2 < \theta < 2\pi$ **44.** $\csc\theta = -4$, $\pi < \theta < 3\pi/2$ **45.** $\cot\theta = -2$, $\pi/2 < \theta < \pi$

46. $\tan\theta = -2$, $3\pi/2 < \theta < 2\pi$

In Problems 47–54, graph each function. Each graph should contain at least one period.

47. $y = 2 \sin 4x$ **48.** $y = -3 \cos 2x$ **49.** $y = -2\cos\left(x + \dfrac{\pi}{2}\right)$ **50.** $y = 3 \sin(x - \pi)$

51. $y = \tan(x + \pi)$ **52.** $y = -\tan\left(x - \dfrac{\pi}{2}\right)$ **53.** $y = -2 \tan 3x$ **54.** $y = 4 \tan 2x$

In Problems 55–70, find the exact value of each expression.

55. $\sin^{-1} 1$ **56.** $\cos^{-1} 0$ **57.** $\tan^{-1} 1$ **58.** $\sin^{-1}\left(-\dfrac{1}{2}\right)$

59. $\cos^{-1}\left(-\dfrac{\sqrt{3}}{2}\right)$ **60.** $\tan^{-1}(-\sqrt{3})$ **61.** $\sin\left(\cos^{-1}\dfrac{\sqrt{2}}{2}\right)$ **62.** $\cos(\sin^{-1} 0)$

63. $\tan\left[\sin^{-1}\left(-\dfrac{\sqrt{3}}{2}\right)\right]$ **64.** $\tan\left[\cos^{-1}\left(-\dfrac{1}{2}\right)\right]$ **65.** $\sec\left(\tan^{-1}\dfrac{\sqrt{3}}{3}\right)$ **66.** $\csc\left(\sin^{-1}\dfrac{\sqrt{3}}{2}\right)$

67. $\sin\left(\tan^{-1}\dfrac{3}{4}\right)$ **68.** $\cos\left(\sin^{-1}\dfrac{3}{5}\right)$ **69.** $\tan\left[\sin^{-1}\left(-\dfrac{4}{5}\right)\right]$ **70.** $\tan\left[\cos^{-1}\left(-\dfrac{3}{5}\right)\right]$

71. Find the length of arc subtended by a central angle of 30° on a circle of radius 2 feet.

72. The minute hand of a clock is 8 inches long. How far does the tip of the minute hand move in 30 minutes? How far does it move in 20 minutes?

73. **Angular Speed of a Race Car** A race car is driven around a circular track at a constant speed of 180 miles per hour. If the diameter of the track is $\frac{1}{2}$ mile, what is the angular speed of the car? Express your answer in revolutions per hour (which is equivalent to laps per hour).

74. **Angular Speed of a Race Car** Repeat Problem 73 if the car goes only 150 miles per hour.

75. **Lighthouse Beacons** The Montauck Point Lighthouse on Long Island has dual beams (two light sources opposite each other). Ships at sea observe a blinking light every 5 seconds. What rotation speed is required to do this?

76. **Spin Balancing Tires** The radius of each wheel of a car is 16 inches. At how many revolutions per minute should a spin balancer be set to balance the tires at a speed of 90 mi/hr? Is the setting different for a wheel of radius 14 inches? What is this setting?

CHAPTER 7

Analytic Trigonometry

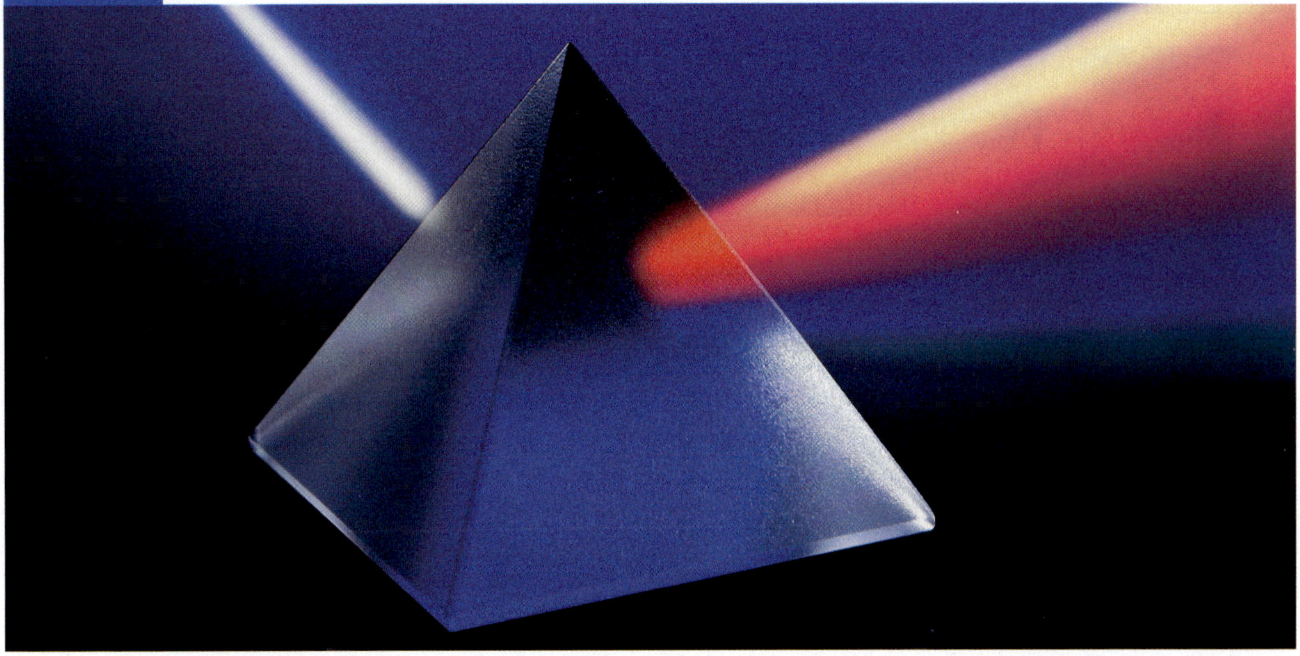

Extra credit is always welcome in school and this time your physics professor has given you a golden opportunity to shine! All you need to do is to create a display on rainbows for the college open house. You will want to use the Sullivan website at:

www.prenhall.com/sullivan

to find the information needed to complete your display.

PREPARING FOR THIS CHAPTER

Before getting started on this chapter, review the following concepts:

Fundamental Identities *(p. 462)*

Values of the Trigonometric Functions of Certain Angles *(p. 470 and p. 477)*

Solving Equations *(Appendix, Section 4)*

OUTLINE

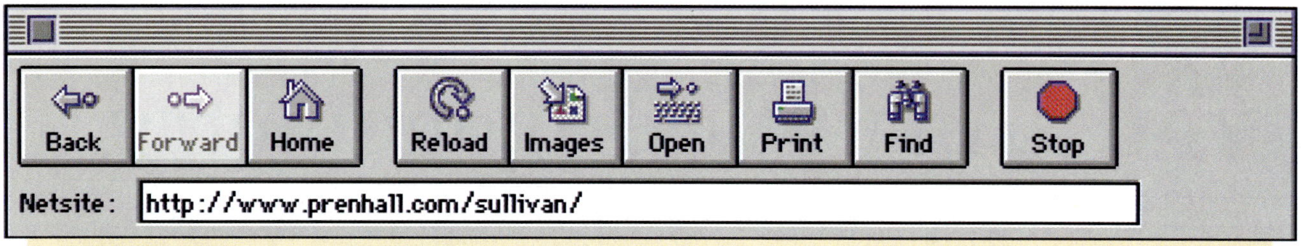

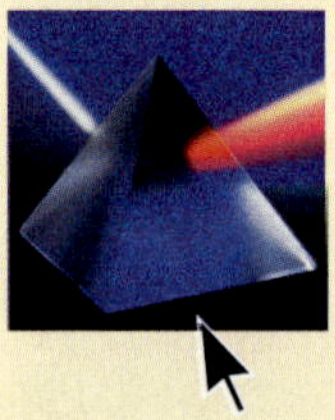

THE MATHEMATICS OF RAINBOWS

Rainbows, those incredible displays of light and water, will always be a source of wonder and inspiration. From pots-o-gold to bridges for the gods, countless myths and legends have been told to explain rainbows.

In 1621, the basic law of refraction, was proposed by the Dutch scientist Willebrord Snell. The law of refraction and the law of reflection provide the keys to the modern explanation for the formation of rainbows. Refraction is the bending of light's path as it passes from one medium into another. When this happens the different wavelengths of light separate. The algebraic expression for Snell's Law looks similar to the Law of Sines, except that the constants no longer refer to the lengths of the sides of a triangle, but instead refer to a "refractive index".

You have been asked by your physics professor to make a display on rainbows for the college open house.

1. You begin working on the project by brainstorming. Since this is the age of the World Wide Web, you decide to create an interactive display! This gets the adrenaline going. You make a web search to collect some ideas. Your web search generates thousands of sites! You select some of your favorites. At this point you want to make a list of questions that your display might answer. For example, did you ever notice that you generally will see rainbows only in the morning or afternoon? If you were to make a display of a rainbow, how would you do it?
2. Since your display is for the physics department, you want to be sure to find some examples that reveal the physical properties of rainbows. What would you add? For some examples, access *keys*.
3. To get the visitors involved in your booth, you decide to perform the following experiment: Draw a circle on a piece of paper. This will represent a water droplet. Now, with a ruler draw a line, a light ray, from the edge of your paper to your "water drop". Make a sketch of your guess of the path of the light ray as it passes through the water drop. You try this out on some of your friends to see if it will work. Do you want to use it? Why or why not?
4. Next imagine a curtain of water droplets and a light source. Try to make an illustration as to where the light source, the water curtain, the rainbow, and an observer will be located with respect to one another. For a solution check out: *Descartes and Rainbows.*
5. For the graphic of your display you decide to make a large scale illustration of the experiment of question 3. (For this exercise you will want to review *Snell's Law*.) You use colored felt tip pens, a compass, a ruler, a calculator, and a protractor to make an *exact drawing*. You want this to be right since visitors will ask you to explain your drawing to them. Make the drawing and then try to explain it to a friend. How did you do?

In this chapter we continue the derivation of identities involving the trigonometric functions. These identities play an important role in calculus, the physical and life sciences, and economics, where they are used to simplify complicated expressions.

The last section of this chapter provides techniques for solving equations that contain trigonometric functions.

7.1 TRIGONOMETRIC IDENTITIES

1 Establish Identities

First, we review a fundamental definition:

Two functions f and g are said to be **identically equal** if

$$f(x) = g(x)$$

for every value of x for which both functions are defined. Such an equation is referred to as an **identity.** An equation that is not an identity is called a **conditional equation.**

For example, the following expressions are identities:

$$(x + 1)^2 = x^2 + 2x + 1 \qquad \sin^2 x + \cos^2 x = 1 \qquad \csc x = \frac{1}{\sin x}$$

The following expressions are conditional equations:

$2x + 5 = 0$ True only if $x = -\frac{5}{2}$.

$\sin x = 0$ True only if $x = k\pi$, k an integer.

$\sin x = \cos x$ True only if $x = \frac{\pi}{4} + 2k\pi$ or $x = \frac{5\pi}{4} + 2k\pi$, k an integer.

The following boxes summarize the trigonometric identities we have established thus far:

Quotient Identities

$$\tan\theta = \frac{\sin\theta}{\cos\theta} \qquad \cot\theta = \frac{\cos\theta}{\sin\theta}$$

Reciprocal Identities

$$\csc\theta = \frac{1}{\sin\theta} \qquad \sec\theta = \frac{1}{\cos\theta} \qquad \cot\theta = \frac{1}{\tan\theta}$$

Pythagorean Identities

$$\sin^2\theta + \cos^2\theta = 1 \qquad \tan^2\theta + 1 = \sec^2\theta$$
$$1 + \cot^2\theta = \csc^2\theta$$

Even-Odd Identities

$\sin(-\theta) = -\sin\theta$	$\cos(-\theta) = \cos\theta$	$\tan(-\theta) = -\tan\theta$
$\csc(-\theta) = -\csc\theta$	$\sec(-\theta) = \sec\theta$	$\cot(-\theta) = -\cot\theta$

This list of identities comprises what we shall refer to as the **basic trigonometric identities.** These identities should not merely be memorized, but should be *known* (just as you know your name rather than have it memorized). In fact, the use made of a basic identity is often a minor variation of the form listed here. For example, we might want to use $\sin^2\theta = 1 - \cos^2\theta$ instead of $\sin^2\theta + \cos^2\theta = 1$. For this reason, among others, you need to know these relationships and be quite comfortable with variations of them.

1 In the examples that follow, the directions will read "Establish the identity . . . ." As you will see, this is accomplished by starting with one side of the given equation (usually the one containing the more complicated expression) and, using appropriate basic identities and algebraic manipulations, arriving at the other side. The selection of an appropriate basic identity to obtain the desired result is learned only through experience and lots of practice.

EXAMPLE 1 Establishing an Identity

Establish the identity: $\sec\theta \cdot \sin\theta = \tan\theta$

Solution We start with the left side, because it contains the more complicated expression, and apply a reciprocal identity:

$$\sec\theta \cdot \sin\theta = \frac{1}{\cos\theta} \cdot \sin\theta = \frac{\sin\theta}{\cos\theta} = \tan\theta$$

Having arrived at the right side, the identity is established. ■

Comment: A graphing utility can be used to provide evidence of an identity. For example, if we graph $Y_1 = \sec\theta \cdot \sin\theta$ and $Y_2 = \tan\theta$, the graphs appear to be the same. This provides evidence that $Y_1 = Y_2$. However, it does not prove their equality. A graphing utility *cannot be used to establish an identity*—identities must be established algebraically.

Now work Problem 1.

EXAMPLE 2 Establishing an Identity

Establish the identity: $\sin^2(-\theta) + \cos^2(-\theta) = 1$

Solution We begin with the left side and apply even–odd identities:

$$\begin{aligned}\sin^2(-\theta) + \cos^2(-\theta) &= [\sin(-\theta)]^2 + [\cos(-\theta)]^2\\ &= (-\sin\theta)^2 + (\cos\theta)^2\\ &= (\sin\theta)^2 + (\cos\theta)^2\\ &= 1\end{aligned}$$

■

EXAMPLE 3 Establishing an Identity

Establish the identity: $\dfrac{\sin^2(-\theta) - \cos^2(-\theta)}{\sin(-\theta) - \cos(-\theta)} = \cos\theta - \sin\theta$

Solution We begin with two observations: The left side appears to contain the more complicated expression. Also, the left side contains expressions with the argument $-\theta$, whereas the right side contains expressions with the argument θ. We decide, therefore, to start with the left side and apply even–odd identities:

$$\begin{aligned}
\frac{\sin^2(-\theta) - \cos^2(-\theta)}{\sin(-\theta) - \cos(-\theta)} &= \frac{[\sin(-\theta)]^2 - [\cos(-\theta)]^2}{\sin(-\theta) - \cos(-\theta)} \\
&= \frac{(-\sin\theta)^2 - (\cos\theta)^2}{-\sin\theta - \cos\theta} && \text{Even–odd identities.} \\
&= \frac{(\sin\theta)^2 - (\cos\theta)^2}{-\sin\theta - \cos\theta} && \text{Simplify.} \\
&= \frac{(\sin\theta - \cos\theta)\cancel{(\sin\theta + \cos\theta)}}{-\cancel{(\sin\theta + \cos\theta)}} && \text{Factor.} \\
&= \cos\theta - \sin\theta && \text{Cancel and simplify.}
\end{aligned}$$

EXAMPLE 4 Establishing an Identity

Establish the identity: $\dfrac{1 + \tan\theta}{1 + \cot\theta} = \tan\theta$

Solution

$$\frac{1 + \tan\theta}{1 + \cot\theta} = \frac{1 + \tan\theta}{1 + \dfrac{1}{\tan\theta}} = \frac{1 + \tan\theta}{\dfrac{\tan\theta + 1}{\tan\theta}} = \frac{\tan\theta\cancel{(1 + \tan\theta)}}{\cancel{\tan\theta + 1}} = \tan\theta$$

Now work Problem 9.

When sums or differences of quotients appear, it is usually best to rewrite them as a single quotient, especially if the other side of the identity consists of only one term.

EXAMPLE 5 Establishing an Identity

Establish the identity: $\dfrac{\sin\theta}{1 + \cos\theta} + \dfrac{1 + \cos\theta}{\sin\theta} = 2\csc\theta$

Solution The left side is more complicated, so we start with it and proceed to add:

$$\begin{aligned}
\frac{\sin\theta}{1 + \cos\theta} + \frac{1 + \cos\theta}{\sin\theta} &= \frac{\sin^2\theta + (1 + \cos\theta)^2}{(1 + \cos\theta)(\sin\theta)} \\
&= \frac{\sin^2\theta + 1 + 2\cos\theta + \cos^2\theta}{(1 + \cos\theta)(\sin\theta)} \\
&= \frac{(\sin^2\theta + \cos^2\theta) + 1 + 2\cos\theta}{(1 + \cos\theta)(\sin\theta)} \\
&= \frac{2 + 2\cos\theta}{(1 + \cos\theta)(\sin\theta)} \\
&= \frac{2\cancel{(1 + \cos\theta)}}{\cancel{(1 + \cos\theta)}(\sin\theta)} \\
&= \frac{2}{\sin\theta} \\
&= 2\csc\theta
\end{aligned}$$

EXAMPLE 6 Establishing an Identity

Establish the identity: $\dfrac{1}{\cos\theta} - \dfrac{\cos\theta}{1+\sin\theta} = \tan\theta$

Solution

$$\begin{aligned}\frac{1}{\cos\theta} - \frac{\cos\theta}{1+\sin\theta} &= \frac{1+\sin\theta-\cos^2\theta}{\cos\theta(1+\sin\theta)}\\ &= \frac{\sin\theta+(1-\cos^2\theta)}{\cos\theta(1+\sin\theta)}\\ &= \frac{\sin\theta+\sin^2\theta}{\cos\theta(1+\sin\theta)} \qquad 1-\cos^2\theta=\sin^2\theta\\ &= \frac{\sin\theta\cancel{(1+\sin\theta)}}{\cos\theta\cancel{(1+\sin\theta)}}\\ &= \tan\theta\end{aligned}$$

Now work Problem 23.

Sometimes, multiplying the numerator and denominator by an appropriate factor will result in a simplification.

EXAMPLE 7 Establishing an Identity

Establish the identity: $\dfrac{1-\sin\theta}{\cos\theta} = \dfrac{\cos\theta}{1+\sin\theta}$

Solution We start with the left side and multiply the numerator and the denominator by $1+\sin\theta$. (Alternatively, we could multiply the numerator and denominator of the right side by $1-\sin\theta$.)

$$\begin{aligned}\frac{1-\sin\theta}{\cos\theta} &= \frac{1-\sin\theta}{\cos\theta}\cdot\frac{1+\sin\theta}{1+\sin\theta}\\ &= \frac{1-\sin^2\theta}{\cos\theta(1+\sin\theta)}\\ &= \frac{\cos^2\theta}{\cos\theta(1+\sin\theta)}\\ &= \frac{\cos\theta}{1+\sin\theta}\end{aligned}$$

Now work Problem 35.

Although a lot of practice is the only real way to learn how to establish identities, the following guidelines should prove helpful:

Guidelines for Establishing Identities

1. It is almost always preferable to start with the side containing the more complicated expression.
2. Rewrite sums or differences of quotients as a single quotient.
3. Sometimes rewriting one side in terms of sines and cosines only will help.
4. Always keep your goal in mind. As you manipulate one side of the expression, you must keep in mind the form of the expression on the other side.

Warning: Be careful not to handle identities to be established as if they were equations. You *cannot* establish an identity by such methods as adding the same expression to each side and obtaining a true statement. This practice is not allowed, because the original statement is precisely the one that you are trying to establish. You do not know until it has been established that it is, in fact, true.

7.1 EXERCISES

In Problems 1–80, establish each identity.

1. $\csc\theta\cdot\cos\theta=\cot\theta$

2. $\csc\theta\cdot\tan\theta=\sec\theta$

3. $1+\tan^2(-\theta)=\sec^2\theta$

4. $1+\cot^2(-\theta)=\csc^2\theta$

5. $\cos\theta(\tan\theta+\cot\theta)=\csc\theta$

6. $\sin\theta(\cot\theta+\tan\theta)=\sec\theta$

7. $\tan\theta\cot\theta-\cos^2\theta=\sin^2\theta$

8. $\sin\theta\csc\theta-\cos^2\theta=\sin^2\theta$

9. $(\sec\theta-1)(\sec\theta+1)=\tan^2\theta$

10. $(\csc\theta-1)(\csc\theta+1)=\cot^2\theta$

11. $(\sec\theta+\tan\theta)(\sec\theta-\tan\theta)=1$

12. $(\csc\theta+\cot\theta)(\csc\theta-\cot\theta)=1$

13. $\sin^2\theta(1+\cot^2\theta)=1$

14. $(1-\sin^2\theta)(1+\tan^2\theta)=1$

15. $(\sin\theta+\cos\theta)^2+(\sin\theta-\cos\theta)^2=2$

16. $\tan^2\theta\cos^2\theta+\cot^2\theta\sin^2\theta=1$

17. $\sec^4\theta-\sec^2\theta=\tan^4\theta+\tan^2\theta$

18. $\csc^4\theta-\csc^2\theta=\cot^4\theta+\cot^2\theta$

19. $\sec\theta-\tan\theta=\dfrac{\cos\theta}{1+\sin\theta}$

20. $\csc\theta-\cot\theta=\dfrac{\sin\theta}{1+\cos\theta}$

21. $3\sin^2\theta+4\cos^2\theta=3+\cos^2\theta$

22. $9\sec^2\theta-5\tan^2\theta=5+4\sec^2\theta$

23. $1-\dfrac{\cos^2\theta}{1+\sin\theta}=\sin\theta$

24. $1-\dfrac{\sin^2\theta}{1-\cos\theta}=-\cos\theta$

25. $\dfrac{1+\tan\theta}{1-\tan\theta}=\dfrac{\cot\theta+1}{\cot\theta-1}$

26. $\dfrac{\csc\theta-1}{\csc\theta+1}=\dfrac{1-\sin\theta}{1+\sin\theta}$

27. $\dfrac{\sec\theta}{\csc\theta}+\dfrac{\sin\theta}{\cos\theta}=2\tan\theta$

28. $\dfrac{\csc\theta-1}{\cot\theta}=\dfrac{\cot\theta}{\csc\theta+1}$

29. $\dfrac{1+\sin\theta}{1-\sin\theta}=\dfrac{\csc\theta+1}{\csc\theta-1}$

30. $\dfrac{\cos\theta+1}{\cos\theta-1}=\dfrac{1+\sec\theta}{1-\sec\theta}$

31. $\dfrac{1-\sin\theta}{\cos\theta}+\dfrac{\cos\theta}{1-\sin\theta}=2\sec\theta$

32. $\dfrac{\cos\theta}{1+\sin\theta}+\dfrac{1+\sin\theta}{\cos\theta}=2\sec\theta$

33. $\dfrac{\sin\theta}{\sin\theta-\cos\theta}=\dfrac{1}{1-\cot\theta}$

34. $1-\dfrac{\sin^2\theta}{1+\cos\theta}=\cos\theta$

35. $\dfrac{1-\sin\theta}{1+\sin\theta}=(\sec\theta-\tan\theta)^2$

36. $\dfrac{1-\cos\theta}{1+\cos\theta}=(\csc\theta-\cot\theta)^2$

37. $\dfrac{\cos\theta}{1-\tan\theta}+\dfrac{\sin\theta}{1-\cot\theta}=\sin\theta+\cos\theta$

38. $\dfrac{\cot\theta}{1-\tan\theta}+\dfrac{\tan\theta}{1-\cot\theta}=1+\tan\theta+\cot\theta$

39. $\tan\theta+\dfrac{\cos\theta}{1+\sin\theta}=\sec\theta$

40. $\dfrac{\sin\theta\cos\theta}{\cos^2\theta-\sin^2\theta}=\dfrac{\tan\theta}{1-\tan^2\theta}$

41. $\dfrac{\tan\theta+\sec\theta-1}{\tan\theta-\sec\theta+1}=\tan\theta+\sec\theta$

42. $\dfrac{\sin\theta-\cos\theta+1}{\sin\theta+\cos\theta-1}=\dfrac{\sin\theta+1}{\cos\theta}$

43. $\dfrac{\tan\theta-\cot\theta}{\tan\theta+\cot\theta}=\sin^2\theta-\cos^2\theta$

44. $\dfrac{\sec\theta-\cos\theta}{\sec\theta+\cos\theta}=\dfrac{\sin^2\theta}{1+\cos^2\theta}$

45. $\dfrac{\tan\theta-\cot\theta}{\tan\theta+\cot\theta}=2\sin^2\theta-1$

46. $\dfrac{\tan\theta-\cot\theta}{\tan\theta+\cot\theta}=1-2\cos^2\theta$

47. $\dfrac{\sec\theta+\tan\theta}{\cot\theta+\cos\theta}=\tan\theta\sec\theta$

48. $\dfrac{\sec\theta}{1+\sec\theta}=\dfrac{1-\cos\theta}{\sin^2\theta}$

49. $\dfrac{1-\tan^2\theta}{1+\tan^2\theta}=2\cos^2\theta-1$

50. $\dfrac{1-\cot^2\theta}{1+\cot^2\theta}=1-2\cos^2\theta$

51. $\dfrac{\sec\theta-\csc\theta}{\sec\theta\csc\theta}=\sin\theta-\cos\theta$

52. $\dfrac{\sin^2\theta-\tan\theta}{\cos^2\theta-\cot\theta}=\tan^2\theta$

53. $\sec\theta-\cos\theta=\sin\theta\tan\theta$

54. $\tan\theta+\cot\theta=\sec\theta\csc\theta$

55. $\dfrac{1}{1-\sin\theta}+\dfrac{1}{1+\sin\theta}=2\sec^2\theta$

56. $\dfrac{1+\sin\theta}{1-\sin\theta}-\dfrac{1-\sin\theta}{1+\sin\theta}=4\tan\theta\sec\theta$

57. $\dfrac{\sec\theta}{1-\sin\theta}=\dfrac{1+\sin\theta}{\cos^3\theta}$

58. $\dfrac{1-\sin\theta}{1+\sin\theta}=(\sec\theta-\tan\theta)^2$

59. $\dfrac{(\sec\theta-\tan\theta)^2+1}{\csc\theta(\sec\theta-\tan\theta)}=2\tan\theta$

60. $\dfrac{\sec^2\theta-\tan^2\theta+\tan\theta}{\sec\theta}=\sin\theta+\cos\theta$

61. $\dfrac{\sin\theta+\cos\theta}{\cos\theta}-\dfrac{\sin\theta-\cos\theta}{\sin\theta}=\sec\theta\csc\theta$

62. $\dfrac{\sin\theta+\cos\theta}{\sin\theta}-\dfrac{\cos\theta-\sin\theta}{\cos\theta}=\sec\theta\csc\theta$

63. $\dfrac{\sin^3\theta+\cos^3\theta}{\sin\theta+\cos\theta}=1-\sin\theta\cos\theta$

64. $\dfrac{\sin^3\theta+\cos^3\theta}{1-2\cos^2\theta}=\dfrac{\sec\theta-\sin\theta}{\tan\theta-1}$

65. $\dfrac{\cos^2\theta-\sin^2\theta}{1-\tan^2\theta}=\cos^2\theta$

66. $\dfrac{\cos\theta+\sin\theta-\sin^3\theta}{\sin\theta}=\cot\theta+\cos^2\theta$

67. $\dfrac{(2\cos^2\theta-1)^2}{\cos^4\theta-\sin^4\theta}=1-2\sin^2\theta$

68. $\dfrac{1-2\cos^2\theta}{\sin\theta\cos\theta}=\tan\theta-\cot\theta$

69. $\dfrac{1+\sin\theta+\cos\theta}{1+\sin\theta-\cos\theta}=\dfrac{1+\cos\theta}{\sin\theta}$

70. $\dfrac{1+\cos\theta+\sin\theta}{1+\cos\theta-\sin\theta}=\sec\theta+\tan\theta$

71. $(a\sin\theta+b\cos\theta)^2+(a\cos\theta-b\sin\theta)^2=a^2+b^2$

72. $(2a\sin\theta\cos\theta)^2+a^2(\cos^2\theta-\sin^2\theta)^2=a^2$

73. $\dfrac{\tan\alpha+\tan\beta}{\cot\alpha+\cot\beta}=\tan\alpha\tan\beta$

74. $(\tan\alpha+\tan\beta)(1-\cot\alpha\cot\beta)+(\cot\alpha+\cot\beta)(1-\tan\alpha\tan\beta)=0$

75. $(\sin\alpha+\cos\beta)^2+(\cos\beta+\sin\alpha)(\cos\beta-\sin\alpha)=2\cos\beta(\sin\alpha+\cos\beta)$

76. $(\sin\alpha-\cos\beta)^2+(\cos\beta+\sin\alpha)(\cos\beta-\sin\alpha)=-2\cos\beta(\sin\alpha-\cos\beta)$

77. $\ln|\sec\theta|=-\ln|\cos\theta|$

78. $\ln|\tan\theta|=\ln|\sin\theta|-\ln|\cos\theta|$

79. $\ln|1+\cos\theta|+\ln|1-\cos\theta|=2\ln|\sin\theta|$

80. $\ln|\sec\theta+\tan\theta|+\ln|\sec\theta-\tan\theta|=0$

81. Write a few paragraphs outlining your strategy for establishing identities.

7.2 SUM AND DIFFERENCE FORMULAS

1. Use Sum and Difference Formulas to Find Exact Values
2. Use Sum and Difference Formulas to Establish Identities

In this section, we continue our derivation of trigonometric identities by obtaining formulas that involve the sum or difference of two angles, such as $\cos(\alpha + \beta)$, $\cos(\alpha - \beta)$, or $\sin(\alpha + \beta)$. These formulas are referred to as the **sum and difference formulas.** We begin with the formulas for $\cos(\alpha + \beta)$ and $\cos(\alpha - \beta)$.

Theorem Sum and Difference Formulas for Cosines

$$\cos(\alpha + \beta) = \cos\alpha\cos\beta - \sin\alpha\sin\beta \qquad (1)$$
$$\cos(\alpha - \beta) = \cos\alpha\cos\beta + \sin\alpha\sin\beta \qquad (2)$$

In words, formula (1) states that the cosine of the sum of two angles equals the cosine of the first times the cosine of the second minus the sine of the first times the sine of the second.

FIGURE 1

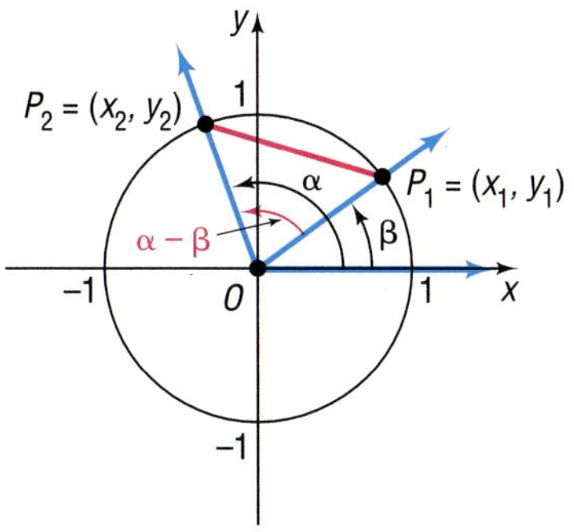

(a)

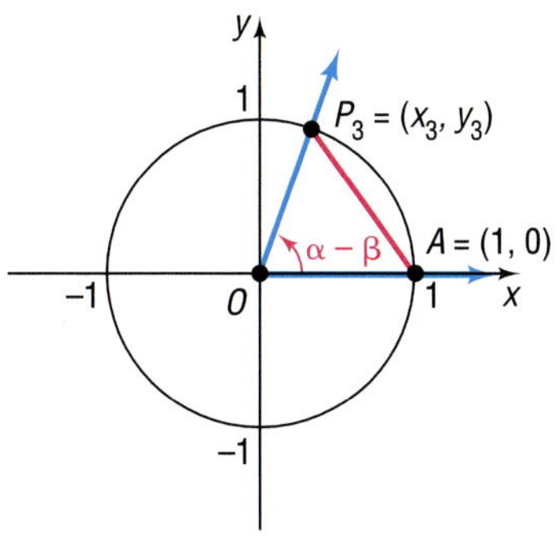

(b)

Proof We will prove formula (2) first. Although this formula is true for all numbers α and β, we shall assume in our proof that $0 < \beta < \alpha < 2\pi$. We begin with a circle with center at the origin (0, 0) and radius of 1 unit (the unit circle), and we place the angles α and β in standard position, as shown in Figure 1(a). The point $P_1 = (x_1, y_1)$ lies on the terminal side of β, and the point $P_2 = (x_2, y_2)$ lies on the terminal side of α.

Now, place the angle $\alpha - \beta$ in standard position, as shown in Figure 1(b), where the point A has coordinates (1, 0) and the point $P_3 = (x_3, y_3)$ is on the terminal side of the angle $\alpha - \beta$.

Looking at triangle OP_1P_2 in Figure 1(a) and triangle OAP_3 in Figure 1(b), we see that these triangles are congruent. (Do you see why? Two sides and the included angle, $\alpha - \beta$, are equal.) Hence, the unknown side of each triangle must be equal; that is,

$$d(A, P_3) = d(P_1, P_2)$$

Using the distance formula, we find that

$$\sqrt{(x_3 - 1)^2 + y_3^2} = \sqrt{(x_2 - x_1)^2 + (y_2 - y_1)^2}$$
$$(x_3 - 1)^2 + y_3^2 = (x_2 - x_1)^2 + (y_2 - y_1)^2 \quad \text{Square each side.}$$
$$x_3^2 - 2x_3 + 1 + y_3^2 = x_2^2 - 2x_1x_2 + x_1^2 + y_2^2 - 2y_1y_2 + y_1^2 \qquad (3)$$

Since $P_1 = (x_1, y_1)$, $P_2 = (x_2, y_2)$, and $P_3 = (x_3, y_3)$ are points on the unit circle $x^2 + y^2 = 1$, it follows that

$$x_1^2 + y_1^2 = 1 \qquad x_2^2 + y_2^2 = 1 \qquad x_3^2 + y_3^2 = 1$$

Consequently, equation (3) simplifies to

$$\begin{aligned} x_3^2 + y_3^2 - 2x_3 + 1 &= (x_2^2 + y_2^2) + (x_1^2 + y_1^2) - 2x_1x_2 - 2y_1y_2 \\ 2 - 2x_3 &= 2 - 2x_1x_2 - 2y_1y_2 \\ x_3 &= x_1x_2 + y_1y_2 \qquad (4) \end{aligned}$$

But $P_1 = (x_1, y_1)$ is on the terminal side of angle β and is a distance of 1 unit from the origin. Thus,

$$\sin\beta = \frac{y_1}{1} = y_1 \qquad \cos\beta = \frac{x_1}{1} = x_1 \qquad (5)$$

Similarly,

$$\sin\alpha = \frac{y_2}{1} = y_2 \qquad \cos\alpha = \frac{x_2}{1} = x_2 \qquad \cos(\alpha - \beta) = \frac{x_3}{1} = x_3 \qquad (6)$$

Using equations (5) and (6) in equation (4), we get

$$\cos(\alpha - \beta) = \cos\alpha\cos\beta + \sin\alpha\sin\beta$$

which is formula (2).

The proof of formula (1) follows from formula (2). We use the fact that $\alpha + \beta = \alpha - (-\beta)$. Then

$$\begin{aligned} \cos(\alpha + \beta) &= \cos[\alpha - (-\beta)] \\ &= \cos\alpha\cos(-\beta) + \sin\alpha\sin(-\beta) \qquad \text{Use formula (2).} \\ &= \cos\alpha\cos\beta - \sin\alpha\sin\beta \qquad \text{Even–odd identities.} \end{aligned}$$

1 One use of formulas (1) and (2) is to obtain the exact value of the cosine of an angle that can be expressed as the sum or difference of angles whose sine and cosine are known exactly.

EXAMPLE 1 Using the Sum Formula to Find Exact Values

Find the exact value of $\cos 75°$.

Solution Since $75° = 45° + 30°$, we use formula (1) to obtain

$$\cos 75° = \cos(45° + 30°) \underset{\text{Formula (1)}}{=} \cos 45°\cos 30° - \sin 45°\sin 30°$$

$$= \frac{\sqrt{2}}{2}\cdot\frac{\sqrt{3}}{2} - \frac{\sqrt{2}}{2}\cdot\frac{1}{2} = \frac{1}{4}(\sqrt{6} - \sqrt{2})$$

EXAMPLE 2 Using the Difference Formula to Find Exact Values

Find the exact value of $\cos(\pi/12)$.

Solution

$$\cos\frac{\pi}{12} = \cos\left(\frac{3\pi}{12} - \frac{2\pi}{12}\right) = \cos\left(\frac{\pi}{4} - \frac{\pi}{6}\right)$$

$$= \cos\frac{\pi}{4}\cos\frac{\pi}{6} + \sin\frac{\pi}{4}\sin\frac{\pi}{6} \qquad \text{Use formula (2).}$$

$$= \frac{\sqrt{2}}{2}\cdot\frac{\sqrt{3}}{2} + \frac{\sqrt{2}}{2}\cdot\frac{1}{2} = \frac{1}{4}(\sqrt{6} + \sqrt{2})$$

Now work Problem 3.

Another use of formulas (1) and (2) is to establish other identities. One important pair of identities is given next:

$$\cos\left(\frac{\pi}{2} - \theta\right) = \sin\theta \qquad (7a)$$

$$\sin\left(\frac{\pi}{2} - \theta\right) = \cos\theta \qquad (7b)$$

Proof To prove formula (7a), we use the formula for $\cos(\alpha - \beta)$ with $\alpha = \pi/2$ and $\beta = \theta$:

$$\begin{aligned}\cos\left(\frac{\pi}{2} - \theta\right) &= \cos\frac{\pi}{2}\cos\theta + \sin\frac{\pi}{2}\sin\theta \\ &= 0 \cdot \cos\theta + 1 \cdot \sin\theta \\ &= \sin\theta\end{aligned}$$

To prove formula (7b), we make use of the identity (7a) just established:

$$\sin\left(\frac{\pi}{2} - \theta\right) \underset{\substack{\uparrow \\ \text{Use (7a).}}}{=} \cos\left[\frac{\pi}{2} - \left(\frac{\pi}{2} - \theta\right)\right] = \cos\theta$$

■

Formulas (7a) and (7b) should look familiar. They are the basis for the theorem stated in Chapter 6: Cofunctions of complementary angles are equal.

Furthermore, since $\cos(\pi/2 - \theta) = \cos(\theta - \pi/2)$, it follows from formula (7a) that $\cos(\theta - \pi/2) = \sin\theta$. Thus, the graphs of $y = \sin x$ and $y = \cos(x - \pi/2)$ are identical, a fact we will use later in Section 8.5.

Now work Problem 29.

Formulas for $\sin(\alpha + \beta)$ and $\sin(\alpha - \beta)$

Having established the identities in formulas (7a) and (7b), we now can derive the sum and difference formulas for $\sin(\alpha + \beta)$ and $\sin(\alpha - \beta)$.

Proof

$$\begin{aligned}\sin(\alpha + \beta) &= \cos\left[\frac{\pi}{2} - (\alpha + \beta)\right] && \text{Formula (7a).} \\ &= \cos\left[\left(\frac{\pi}{2} - \alpha\right) - \beta\right] \\ &= \cos\left(\frac{\pi}{2} - \alpha\right)\cos\beta + \sin\left(\frac{\pi}{2} - \alpha\right)\sin\beta && \text{Formula (2).} \\ &= \sin\alpha\cos\beta + \cos\alpha\sin\beta && \text{Formulas (7a) and (7b).} \\ \sin(\alpha - \beta) &= \sin[\alpha + (-\beta)] \\ &= \sin\alpha\cos(-\beta) + \cos\alpha\sin(-\beta) \\ &= \sin\alpha\cos\beta + \cos\alpha(-\sin\beta) && \text{Even–odd identities.} \\ &= \sin\alpha\cos\beta - \cos\alpha\sin\beta\end{aligned}$$

■

Thus,

Theorem Sum and Difference Formulas for Sines

$$\sin(\alpha + \beta) = \sin\alpha\cos\beta + \cos\alpha\sin\beta \qquad (8)$$
$$\sin(\alpha - \beta) = \sin\alpha\cos\beta - \cos\alpha\sin\beta \qquad (9)$$

In words, formula (8) states that the sine of the sum of two angles equals the sine of the first times the cosine of the second plus the cosine of the first times the sine of the second.

EXAMPLE 3 Using the Sum Formula to Find Exact Values

Find the exact value of $\sin(7\pi/12)$.

Solution

$$\sin\frac{7\pi}{12} = \sin\left(\frac{3\pi}{12} + \frac{4\pi}{12}\right) = \sin\left(\frac{\pi}{4} + \frac{\pi}{3}\right)$$
$$= \sin\frac{\pi}{4}\cos\frac{\pi}{3} + \cos\frac{\pi}{4}\sin\frac{\pi}{3} \quad \text{Formula (8).}$$
$$= \frac{\sqrt{2}}{2}\cdot\frac{1}{2} + \frac{\sqrt{2}}{2}\cdot\frac{\sqrt{3}}{2} = \frac{1}{4}(\sqrt{2} + \sqrt{6})$$

EXAMPLE 4 Using the Difference Formula to Find Exact Values

Find the exact value of $\sin 165°$.

Solution

$$\sin 165° = \sin(225° - 60°)$$
$$= \sin 225°\cos 60° - \cos 225°\sin 60° \quad \text{Formula (9).}$$
$$= \frac{-\sqrt{2}}{2}\cdot\frac{1}{2} - \frac{-\sqrt{2}}{2}\cdot\frac{\sqrt{3}}{2} = \frac{1}{4}(\sqrt{6} - \sqrt{2})$$

Now work Problem 9.

EXAMPLE 5 Using the Difference Formula to Find Exact Values

Find the exact value of $\cos 80°\cos 20° + \sin 80°\sin 20°$.

Solution The form of the expression $\cos 80°\cos 20° + \sin 80°\sin 20°$ is that of the right side of the formula for $\cos(\alpha - \beta)$ with $\alpha = 80°$ and $\beta = 20°$. Thus,

$$\cos 80°\cos 20° + \sin 80°\sin 20° = \cos(80° - 20°) = \cos 60° = \frac{1}{2}$$

EXAMPLE 6 Using the Sum Formula to Find Exact Values

Find the exact value of $\sin\frac{\pi}{9}\cos\frac{\pi}{18} + \cos\frac{\pi}{9}\sin\frac{\pi}{18}$.

Solution We observe that the form of the given expression is that of the right side of the formula for $\sin(\alpha + \beta)$ with $\alpha = \pi/9$ and $\beta = \pi/18$. Thus,

$$\sin\frac{\pi}{9}\cos\frac{\pi}{18} + \cos\frac{\pi}{9}\sin\frac{\pi}{18} = \sin\left(\frac{\pi}{9} + \frac{\pi}{18}\right) = \sin\frac{3\pi}{18} = \sin\frac{\pi}{6} = \frac{1}{2}$$

Now work Problem 19.

EXAMPLE 7 Finding Exact Values

If it is known that $\sin \alpha = \frac{4}{5}$, $\pi/2 < \alpha < \pi$, and that $\sin \beta = -2/\sqrt{5} = -2\sqrt{5}/5$, $\pi < \beta < 3\pi/2$, find the exact value of

(a) $\cos \alpha$ (b) $\cos \beta$ (c) $\cos(\alpha + \beta)$ (d) $\sin(\alpha + \beta)$

Solution (a) See Figure 2. Since $\sin \alpha = \frac{4}{5} = \frac{b}{r}$ and $\frac{\pi}{2} < \alpha < \pi$, we find $b = 4$ and $r = 5$. Since $(a, 4)$ is on the circle $x^2 + y^2 = 25$ and is in Quadrant II, we have

FIGURE 2
Given $\sin \alpha = \frac{4}{5}$, $\pi/2 < \alpha < \pi$

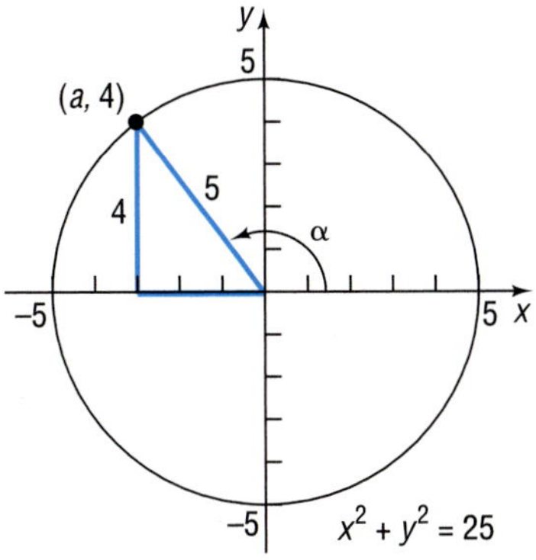

$$\begin{aligned} a^2 + 4^2 &= 25, \quad a < 0 \\ a^2 &= 25 - 16 = 9 \\ a &= -3 \end{aligned}$$

Thus,

$$\cos \alpha = \frac{a}{r} = -\frac{3}{5}$$

(b) See Figure 3. Since $\sin \beta = \frac{-2}{\sqrt{5}} = \frac{b}{r}$ and $\pi < \beta < \frac{3\pi}{2}$, we find $b = -2$ and $r = \sqrt{5}$. Since $(a, -2)$ is on the circle $x^2 + y^2 = 5$ and is in Quadrant III, we have

FIGURE 3
Given $\sin \beta = -2/\sqrt{5}$, $\pi < \beta < 3\pi/2$

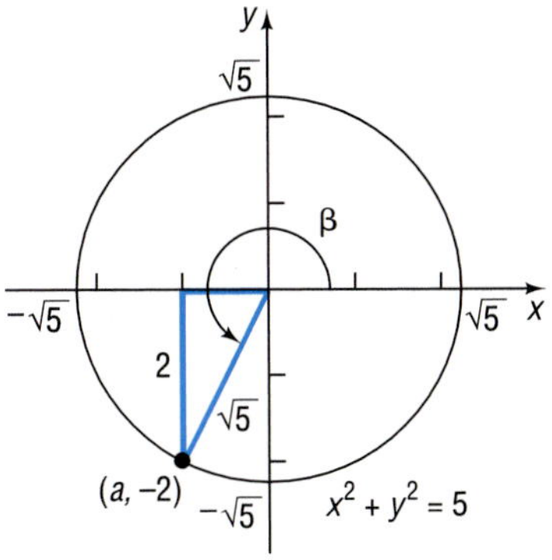

$$\begin{aligned} a^2 + (-2)^2 &= 5, \quad a < 0 \\ a^2 &= 5 - 4 = 1 \\ a &= -1 \end{aligned}$$

Thus,

$$\cos \beta = \frac{a}{r} = -\frac{1}{\sqrt{5}} = -\frac{\sqrt{5}}{5}$$

(c) Using the results found in parts (a) and (b) and formula (1), we have

$$\begin{aligned} \cos(\alpha + \beta) &= \cos \alpha \cos \beta - \sin \alpha \sin \beta \\ &= -\frac{3}{5}\left(-\frac{\sqrt{5}}{5}\right) - \frac{4}{5}\left(-\frac{2\sqrt{5}}{5}\right) = \frac{11\sqrt{5}}{25} \end{aligned}$$

(d)

$$\begin{aligned} \sin(\alpha + \beta) &= \sin \alpha \cos \beta + \cos \alpha \sin \beta \\ &= \frac{4}{5}\left(-\frac{\sqrt{5}}{5}\right) + \left(-\frac{3}{5}\right)\left(-\frac{2\sqrt{5}}{5}\right) = \frac{2\sqrt{5}}{25} \end{aligned}$$

Now work Problem 23, parts (a), (b), and (c).

Formulas for $\tan(\alpha + \beta)$ and $\tan(\alpha - \beta)$

We use the identity $\tan \theta = (\sin \theta)/(\cos \theta)$ and the sum formulas for $\sin(\alpha + \beta)$ and $\cos(\alpha + \beta)$ to derive a formula for $\tan(\alpha + \beta)$.

Proof

$$\tan(\alpha + \beta) = \frac{\sin(\alpha + \beta)}{\cos(\alpha + \beta)} = \frac{\sin \alpha \cos \beta + \cos \alpha \sin \beta}{\cos \alpha \cos \beta - \sin \alpha \sin \beta}$$

Now we divide the numerator and denominator by $\cos \alpha \cos \beta$:

$$\tan(\alpha + \beta) = \frac{\dfrac{\sin \alpha \cos \beta + \cos \alpha \sin \beta}{\cos \alpha \cos \beta}}{\dfrac{\cos \alpha \cos \beta - \sin \alpha \sin \beta}{\cos \alpha \cos \beta}} = \frac{\dfrac{\sin \alpha \cancel{\cos \beta}}{\cos \alpha \cancel{\cos \beta}} + \dfrac{\cancel{\cos \alpha} \sin \beta}{\cancel{\cos \alpha} \cos \beta}}{\dfrac{\cancel{\cos \alpha \cos \beta}}{\cancel{\cos \alpha \cos \beta}} - \dfrac{\sin \alpha \sin \beta}{\cos \alpha \cos \beta}}$$

$$= \frac{\dfrac{\sin \alpha}{\cos \alpha} + \dfrac{\sin \beta}{\cos \beta}}{1 - \dfrac{\sin \alpha \sin \beta}{\cos \alpha \cos \beta}} = \frac{\tan \alpha + \tan \beta}{1 - \tan \alpha \tan \beta}$$

We use the sum formula for $\tan(\alpha + \beta)$ to get the difference formula:

$$\tan(\alpha - \beta) = \tan[\alpha + (-\beta)] = \frac{\tan \alpha + \tan(-\beta)}{1 - \tan \alpha \tan(-\beta)} = \frac{\tan \alpha - \tan \beta}{1 + \tan \alpha \tan \beta}$$

■

Thus, we have proved the following results:

Theorem Sum and Difference Formulas for Tangents

$$\tan(\alpha + \beta) = \frac{\tan \alpha + \tan \beta}{1 - \tan \alpha \tan \beta} \qquad (10)$$

$$\tan(\alpha - \beta) = \frac{\tan \alpha - \tan \beta}{1 + \tan \alpha \tan \beta} \qquad (11)$$

■

In words, formula (10) states that the tangent of the sum of two angles equals the tangent of the first plus the tangent of the second divided by 1 minus their product.

EXAMPLE 8 Establishing an Identity

2

Prove the identity: $\tan(\theta + \pi) = \tan \theta$

Solution

$$\tan(\theta + \pi) = \frac{\tan \theta + \tan \pi}{1 - \tan \theta \tan \pi} = \frac{\tan \theta + 0}{1 - \tan \theta \cdot 0} = \tan \theta$$

■

The result obtained in Example 8 verifies that the tangent function is periodic with period π, a fact we mentioned earlier.

Warning: Be careful when using formulas (10) and (11). These formulas can be used only for angles α and β for which $\tan \alpha$ and $\tan \beta$ are defined, that is, all angles except odd multiples of $\pi/2$.

EXAMPLE 9 Establishing an Identity

Prove the identity: $\tan\left(\theta + \dfrac{\pi}{2}\right) = -\cot\theta$

Solution We cannot use formula (10), since $\tan(\pi/2)$ is not defined. Instead, we proceed as follows:

$$\tan\left(\theta + \frac{\pi}{2}\right) = \frac{\sin\left(\theta + \frac{\pi}{2}\right)}{\cos\left(\theta + \frac{\pi}{2}\right)} = \frac{\sin\theta\cos\frac{\pi}{2} + \cos\theta\sin\frac{\pi}{2}}{\cos\theta\cos\frac{\pi}{2} - \sin\theta\sin\frac{\pi}{2}}$$

$$= \frac{(\sin\theta)(0) + (\cos\theta)(1)}{(\cos\theta)(0) - (\sin\theta)(1)} = \frac{\cos\theta}{-\sin\theta} = -\cot\theta$$

EXAMPLE 10 Finding Exact Values

Find the exact value of $\sin(\cos^{-1}\frac{1}{2} + \sin^{-1}\frac{3}{5})$.

Solution Let $\alpha = \cos^{-1}\frac{1}{2}$ and $\beta = \sin^{-1}\frac{3}{5}$. Then

$$\cos\alpha = \tfrac{1}{2}, \quad 0 \le \alpha \le \pi, \quad \text{and} \quad \sin\beta = \tfrac{3}{5}, \quad -\frac{\pi}{2} \le \beta \le \frac{\pi}{2}$$

Based on Figure 4, we obtain $\sin\alpha = \sqrt{3}/2$ and $\cos\beta = \frac{4}{5}$. Thus,

$$\sin\left(\cos^{-1}\frac{1}{2} + \sin^{-1}\frac{3}{5}\right) = \sin(\alpha + \beta) = \sin\alpha\cos\beta + \cos\alpha\sin\beta$$

$$= \frac{\sqrt{3}}{2}\cdot\frac{4}{5} + \frac{1}{2}\cdot\frac{3}{5} = \frac{4\sqrt{3} + 3}{10}$$

FIGURE 4

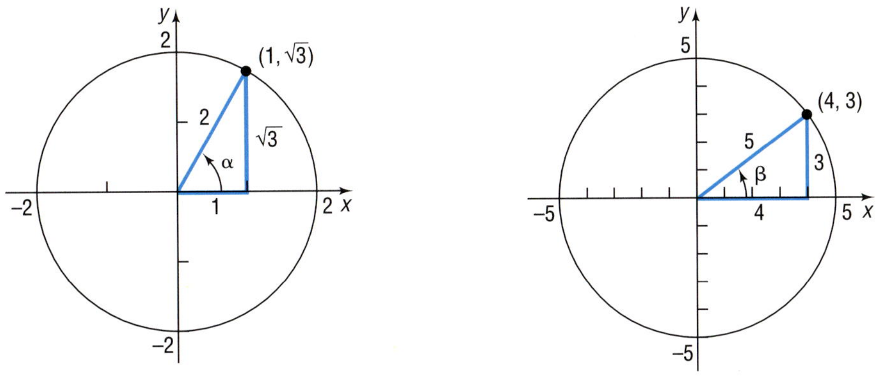

(a) $\cos\alpha = \frac{1}{2}, 0 \le \alpha \le \pi$ **(b)** $\sin\beta = \frac{3}{5}, -\pi/2 \le \beta \le \pi/2$

Now work Problem 57.

EXAMPLE 11 Writing a Trigonometric Expression as an Algebraic Expression

Write $\sin(\sin^{-1} u + \cos^{-1} v)$ as an algebraic expression containing u and v (that is, without any trigonometric functions).

Solution Let $\alpha = \sin^{-1} u$ and $\beta = \cos^{-1} v$. Then

$$\sin \alpha = u, \quad -\frac{\pi}{2} \le \alpha \le \frac{\pi}{2} \text{ and } \quad \cos \beta = v, \quad 0 \le \beta \le \pi$$

Since $-\pi/2 \le \alpha \le \pi/2$, we know that $\cos \alpha \ge 0$. As a result,

$$\cos \alpha = \sqrt{1 - \sin^2 \alpha} = \sqrt{1 - u^2}$$

Similarly, since $0 \le \beta \le \pi$, we know that $\sin \beta \ge 0$. Thus,

$$\sin \beta = \sqrt{1 - \cos^2 \beta} = \sqrt{1 - v^2}$$

Now,

$$\begin{aligned} \sin(\sin^{-1} u + \cos^{-1} v) = \sin(\alpha + \beta) &= \sin \alpha \cos \beta + \cos \alpha \sin \beta \\ &= uv + \sqrt{1 - u^2}\sqrt{1 - v^2} \end{aligned}$$

SUMMARY

The following box summarizes the sum and difference formulas:

Sum and Difference Formulas

$$\cos(\alpha + \beta) = \cos \alpha \cos \beta - \sin \alpha \sin \beta$$
$$\cos(\alpha - \beta) = \cos \alpha \cos \beta + \sin \alpha \sin \beta$$
$$\sin(\alpha + \beta) = \sin \alpha \cos \beta + \cos \alpha \sin \beta$$
$$\sin(\alpha - \beta) = \sin \alpha \cos \beta - \cos \alpha \sin \beta$$
$$\tan(\alpha + \beta) = \frac{\tan \alpha + \tan \beta}{1 - \tan \alpha \tan \beta}$$
$$\tan(\alpha - \beta) = \frac{\tan \alpha - \tan \beta}{1 + \tan \alpha \tan \beta}$$

7.2 EXERCISES

In Problems 1–12, find the exact value of each trigonometric function.

1. $\sin \frac{5\pi}{12}$ **2.** $\sin \frac{\pi}{12}$ **3.** $\cos \frac{7\pi}{12}$ **4.** $\tan \frac{7\pi}{12}$ **5.** $\cos 165°$ **6.** $\sin 105°$

7. $\tan 15°$ **8.** $\tan 195°$ **9.** $\sin \frac{17\pi}{12}$ **10.** $\tan \frac{19\pi}{12}$ **11.** $\sec\left(-\frac{\pi}{12}\right)$ **12.** $\cot\left(-\frac{5\pi}{12}\right)$

In Problems 13–22, find the exact value of each expression.

13. $\sin 20° \cos 10° + \cos 20° \sin 10°$

14. $\sin 20° \cos 80° - \cos 20° \sin 80°$

15. $\cos 70° \cos 20° - \sin 70° \sin 20°$

16. $\cos 40° \cos 10° + \sin 40° \sin 10°$

17. $\dfrac{\tan 20° + \tan 25°}{1 - \tan 20° \tan 25°}$

18. $\dfrac{\tan 40° - \tan 10°}{1 + \tan 40° \tan 10°}$

19. $\sin \frac{\pi}{12} \cos \frac{7\pi}{12} - \cos \frac{\pi}{12} \sin \frac{7\pi}{12}$

20. $\cos \frac{5\pi}{12} \cos \frac{7\pi}{12} - \sin \frac{5\pi}{12} \sin \frac{7\pi}{12}$

21. $\sin \frac{\pi}{12} \cos \frac{5\pi}{12} - \sin \frac{5\pi}{12} \cos \frac{\pi}{12}$

22. $\sin \frac{\pi}{18} \cos \frac{5\pi}{18} + \cos \frac{\pi}{18} \sin \frac{5\pi}{18}$

In Problems 23–28, find the exact value of each of the following under the given conditions:

(a) $\sin(\alpha + \beta)$ *(b)* $\cos(\alpha + \beta)$ *(c)* $\sin(\alpha - \beta)$ *(d)* $\tan(\alpha - \beta)$

23. $\sin \alpha = \frac{3}{5}, 0 < \alpha < \pi/2;$ $\cos \beta = 2/\sqrt{5}, -\pi/2 < \beta < 0$
24. $\cos \alpha = 1/\sqrt{5}, 0 < \alpha < \pi/2;$ $\sin \beta = -\frac{4}{5}, -\pi/2 < \beta < 0$
25. $\tan \alpha = -\frac{4}{3}, \pi/2 < \alpha < \pi;$ $\cos \beta = \frac{1}{2}, 0 < \beta < \pi/2$
26. $\tan \alpha = \frac{5}{12}, \pi < \alpha < 3\pi/2;$ $\sin \beta = -\frac{1}{2}, \pi < \beta < 3\pi/2$
27. $\sin \alpha = \frac{5}{13}, -3\pi/2 < \alpha < -\pi;$ $\tan \beta = -\sqrt{3}, \pi/2 < \beta < \pi$
28. $\cos \alpha = \frac{1}{2}, -\pi/2 < \alpha < 0;$ $\sin \beta = \frac{1}{3}, 0 < \beta < \pi/2$

In Problems 29–54, establish each identity.

29. $\sin\left(\frac{\pi}{2} + \theta\right) = \cos \theta$
30. $\cos\left(\frac{\pi}{2} + \theta\right) = -\sin \theta$
31. $\sin(\pi - \theta) = \sin \theta$
32. $\cos(\pi - \theta) = -\cos \theta$
33. $\sin(\pi + \theta) = -\sin \theta$
34. $\cos(\pi + \theta) = -\cos \theta$
35. $\tan(\pi - \theta) = -\tan \theta$
36. $\tan(2\pi - \theta) = -\tan \theta$
37. $\sin\left(\frac{3\pi}{2} + \theta\right) = -\cos \theta$
38. $\cos\left(\frac{3\pi}{2} + \theta\right) = \sin \theta$
39. $\sin(\alpha + \beta) + \sin(\alpha - \beta) = 2 \sin \alpha \cos \beta$
40. $\cos(\alpha + \beta) + \cos(\alpha - \beta) = 2 \cos \alpha \cos \beta$
41. $\dfrac{\sin(\alpha + \beta)}{\sin \alpha \cos \beta} = 1 + \cot \alpha \tan \beta$
42. $\dfrac{\sin(\alpha + \beta)}{\cos \alpha \cos \beta} = \tan \alpha + \tan \beta$
43. $\dfrac{\cos(\alpha + \beta)}{\cos \alpha \cos \beta} = 1 - \tan \alpha \tan \beta$
44. $\dfrac{\cos(\alpha - \beta)}{\sin \alpha \cos \beta} = \cot \alpha + \tan \beta$
45. $\dfrac{\sin(\alpha + \beta)}{\sin(\alpha - \beta)} = \dfrac{\tan \alpha + \tan \beta}{\tan \alpha - \tan \beta}$
46. $\dfrac{\cos(\alpha + \beta)}{\cos(\alpha - \beta)} = \dfrac{1 - \tan \alpha \tan \beta}{1 + \tan \alpha \tan \beta}$
47. $\cot(\alpha + \beta) = \dfrac{\cot \alpha \cot \beta - 1}{\cot \beta + \cot \alpha}$
48. $\cot(\alpha - \beta) = \dfrac{\cot \alpha \cot \beta + 1}{\cot \beta - \cot \alpha}$
49. $\sec(\alpha + \beta) = \dfrac{\csc \alpha \csc \beta}{\cot \alpha \cot \beta - 1}$
50. $\sec(\alpha - \beta) = \dfrac{\sec \alpha \sec \beta}{1 + \tan \alpha \tan \beta}$
51. $\sin(\alpha - \beta) \sin(\alpha + \beta) = \sin^2 \alpha - \sin^2 \beta$
52. $\cos(\alpha - \beta) \cos(\alpha + \beta) = \cos^2 \alpha - \sin^2 \beta$
53. $\sin(\theta + k\pi) = (-1)^k \sin \theta$, k any integer
54. $\cos(\theta + k\pi) = (-1)^k \cos \theta$, k any integer

In Problems 55–64, find the exact value for each expression.

55. $\sin(\sin^{-1} \frac{1}{2} + \cos^{-1} 0)$
56. $\sin(\sin^{-1} \frac{\sqrt{3}}{2} + \cos^{-1} 1)$
57. $\sin[\sin^{-1} \frac{3}{5} - \cos^{-1}(-\frac{4}{5})]$
58. $\sin[\sin^{-1}(-\frac{4}{5}) - \tan^{-1} \frac{3}{4}]$
59. $\cos(\tan^{-1} \frac{4}{3} + \cos^{-1} \frac{5}{13})$
60. $\sin[\tan^{-1} \frac{5}{12} - \sin^{-1}(-\frac{3}{5})]$
61. $\sec(\sin^{-1} \frac{5}{13} - \tan^{-1} \frac{3}{4})$
62. $\sec(\tan^{-1} \frac{4}{3} + \cot^{-1} \frac{5}{12})$
63. $\cot(\sec^{-1} \frac{5}{3} + \frac{\pi}{6})$
64. $\cos(\frac{\pi}{4} - \csc^{-1} \frac{5}{3})$

In Problems 65–70, write each trigonometric expression as an algebraic expression containing u and v.

65. $\cos(\cos^{-1} u + \sin^{-1} v)$
66. $\sin(\sin^{-1} u - \cos^{-1} v)$
67. $\sin(\tan^{-1} u - \sin^{-1} v)$
68. $\cos(\tan^{-1} u + \tan^{-1} v)$
69. $\tan(\sin^{-1} u - \cos^{-1} v)$
70. $\sec(\tan^{-1} u + \cos^{-1} v)$

71. **Calculus** Show that the difference quotient for $f(x) = \sin x$ is given by

$$\frac{f(x+h) - f(x)}{h} = \frac{\sin(x+h) - \sin x}{h}$$
$$= \cos x \cdot \frac{\sin h}{h} - \sin x \cdot \frac{1 - \cos h}{h}$$

72. **Calculus** Show that the difference quotient for $f(x) = \cos x$ is given by

$$\frac{f(x+h) - f(x)}{h} = \frac{\cos(x+h) - \cos x}{h}$$
$$= -\sin x \cdot \frac{\sin h}{h} - \cos x \cdot \frac{1 - \cos h}{h}$$

73. Show that $\sin(\sin^{-1} u + \cos^{-1} u) = 1$.

74. Show that $\cos(\sin^{-1} u + \cos^{-1} u) = 0$.

75. Explain why formula (11) cannot be used to show that

$$\tan\left(\frac{\pi}{2} - \theta\right) = \cot\theta$$

Establish this identity by using formulas (7a) and (7b).

76. If $\tan\alpha = x + 1$ and $\tan\beta = x - 1$, show that $2\cot(\alpha - \beta) = x^2$.

77. **Geometry: Angle between Two Lines** Let L_1 and L_2 denote two nonvertical intersecting lines, and let θ denote the acute angle between L_1 and L_2 (see the figure). Show that

$$\tan\theta = \frac{m_2 - m_1}{1 + m_1 m_2}$$

where m_1 and m_2 are the slopes of L_1 and L_2, respectively.

[**Hint:** Use the facts that $\tan\theta_1 = m_1$ and $\tan\theta_2 = m_2$.]

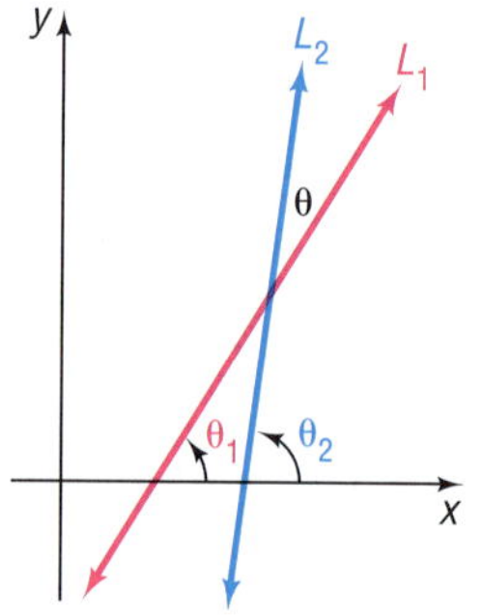

78. If $\alpha + \beta + \gamma = 180°$ and

$\cot\theta = \cot\alpha + \cot\beta + \cot\gamma,\ 0 < \theta < 90°$,

show that

$\sin^3\theta = \sin(\alpha - \theta)\sin(\beta - \theta)\sin(\gamma - \theta)$

79. Discuss the following derivation:

$$\tan\left(\theta + \frac{\pi}{2}\right) = \frac{\tan\theta + \tan(\pi/2)}{1 - \tan\theta\tan(\pi/2)}$$
$$= \frac{\dfrac{\tan\theta}{\tan(\pi/2)} + 1}{\dfrac{1}{\tan(\pi/2)} - \tan\theta} = \frac{0 + 1}{0 - \tan\theta}$$
$$= \frac{1}{-\tan\theta} = -\cot\theta$$

Can you justify each step?

7.3 DOUBLE-ANGLE AND HALF-ANGLE FORMULAS

1 Use Double-Angle Formulas to Find Exact Values
2 Use Double-Angle and Half-Angle Formulas to Establish Identities
3 Use Half-Angle Formulas to Find Exact Values

In this section we derive formulas for $\sin 2\theta$, $\cos 2\theta$, $\sin\left(\frac{1}{2}\theta\right)$, and $\cos\left(\frac{1}{2}\theta\right)$ in terms of $\sin\theta$ and $\cos\theta$. They are easily derived using the sum formulas.

Double-Angle Formulas

In the sum formulas for $\sin(\alpha + \beta)$ and $\cos(\alpha + \beta)$, let $\alpha = \beta = \theta$. Then

$$\sin(\alpha + \beta) = \sin\alpha\cos\beta + \cos\alpha\sin\beta$$
$$\sin(\theta + \theta) = \sin\theta\cos\theta + \cos\theta\sin\theta$$
$$\sin 2\theta = 2\sin\theta\cos\theta \qquad (1)$$

and

$$\cos(\alpha + \beta) = \cos \alpha \cos \beta - \sin \alpha \sin \beta$$
$$\cos(\theta + \theta) = \cos \theta \cos \theta - \sin \theta \sin \theta$$
$$\cos 2\theta = \cos^2 \theta - \sin^2 \theta \qquad (2)$$

An application of the Pythagorean identity $\sin^2 \theta + \cos^2 \theta = 1$ results in two other ways to write formula (2) for $\cos 2\theta$:

$$\cos 2\theta = \cos^2 \theta - \sin^2 \theta = (1 - \sin^2 \theta) - \sin^2 \theta = 1 - 2 \sin^2 \theta$$

and

$$\cos 2\theta = \cos^2 \theta - \sin^2 \theta = \cos^2 \theta - (1 - \cos^2 \theta) = 2 \cos^2 \theta - 1$$

Thus, we have established the following **double-angle formulas:**

Theorem Double-Angle Formulas

$$\sin 2\theta = 2 \sin \theta \cos \theta \qquad (3)$$
$$\cos 2\theta = \cos^2 \theta - \sin^2 \theta \qquad (4a)$$
$$\cos 2\theta = 1 - 2 \sin^2 \theta \qquad (4b)$$
$$\cos 2\theta = 2 \cos^2 \theta - 1 \qquad (4c)$$

EXAMPLE 1 Finding Exact Values Using the Double-Angle Formula

1

If $\sin \theta = \frac{3}{5}$, $\pi/2 < \theta < \pi$, find the exact value of

(a) $\sin 2\theta$ (b) $\cos 2\theta$

Solution

(a) Because $\sin 2\theta = 2 \sin\theta \cos\theta$ and we already know that $\sin \theta = \frac{3}{5}$, we only need to find $\cos \theta$. See Figure 5. Since $\sin \theta = \frac{3}{5} = \frac{b}{r}$, we find $b = 3$ and $r = 5$. Since $\pi/2 < \theta < \pi$, it follows that $(a, 3)$ is on the circle $x^2 + y^2 = 25$ and is in Quadrant II. Thus,

$$a^2 + 3^2 = 25, \quad a < 0$$
$$a^2 = 25 - 9 = 16$$
$$a = -4$$

Thus, $\cos \theta = -\frac{4}{5}$. Now we use formula (3) to obtain

$$\sin 2\theta = 2 \sin \theta \cos \theta = 2\left(\tfrac{3}{5}\right)\left(-\tfrac{4}{5}\right) = -\frac{24}{25}$$

(b) Because we are given $\sin \theta = \frac{3}{5}$, it is easiest to use formula (4b) to get $\cos 2\theta$:

$$\cos 2\theta = 1 - 2 \sin^2 \theta = 1 - 2\left(\tfrac{9}{25}\right) = 1 - \frac{18}{25} = \frac{7}{25}$$

FIGURE 5

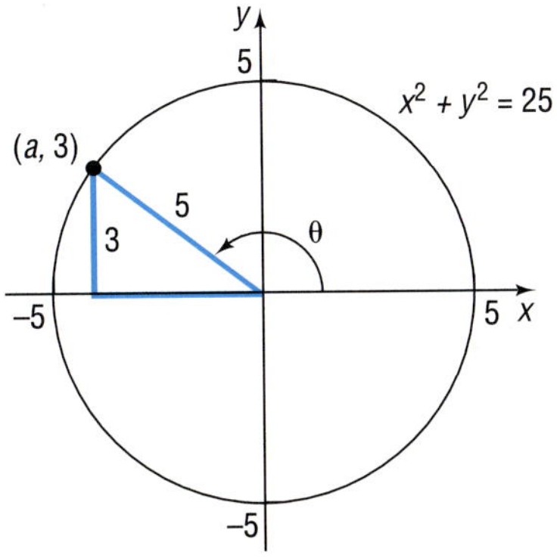

Warning: In finding $\cos 2\theta$ in Example 1(b), we chose to use a version of the double-angle formula, formula (4b). Note that we are unable to use the Pythagorean identity $\cos 2\theta = \pm\sqrt{1 - \sin^2 2\theta}$, with $\sin 2\theta = -\frac{24}{25}$, because we have no way of knowing which sign to choose.

Now work Problems 1(a) and (b).

EXAMPLE 2 Establishing Identities

2

(a) Develop a formula for $\tan 2\theta$ in terms of $\tan \theta$.
(b) Develop a formula for $\sin 3\theta$ in terms of $\sin \theta$ and $\cos \theta$.

Solution (a) In the sum formula for $\tan(\alpha + \beta)$, let $\alpha = \beta = \theta$. Then

$$\tan(\alpha + \beta) = \frac{\tan \alpha + \tan \beta}{1 - \tan \alpha \tan \beta}$$
$$\tan(\theta + \theta) = \frac{\tan \theta + \tan \theta}{1 - \tan \theta \tan \theta}$$

$$\tan 2\theta = \frac{2 \tan \theta}{1 - \tan^2 \theta} \qquad (5)$$

(b) To get a formula for $\sin 3\theta$, we use the sum formula and write 3θ as $2\theta + \theta$.

$$\sin 3\theta = \sin(2\theta + \theta) = \sin 2\theta \cos \theta + \cos 2\theta \sin \theta$$

Now use the Double-Angle Formulas to get

$$\begin{aligned} \sin 3\theta &= (2 \sin \theta \cos \theta)(\cos \theta) + (\cos^2 \theta - \sin^2 \theta)(\sin \theta) \\ &= 2 \sin \theta \cos^2 \theta + \sin \theta \cos^2 \theta - \sin^3 \theta \\ &= 3 \sin \theta \cos^2 \theta - \sin^3 \theta \end{aligned}$$

The formula obtained in Example 2(b) can also be written as

$$\begin{aligned} \sin 3\theta = 3 \sin \theta \cos^2 \theta - \sin^3 \theta &= 3 \sin \theta(1 - \sin^2 \theta) - \sin^3 \theta \\ &= 3 \sin \theta - 4 \sin^3 \theta \end{aligned}$$

That is, $\sin 3\theta$ is a third-degree polynomial in the variable $\sin \theta$. In fact, $\sin n\theta$, n a positive odd integer, can always be written as a polynomial of degree n in the variable $\sin \theta$.*

Now work Problem 47.

Other Variations of the Double-Angle Formulas

By rearranging the Double-Angle Formulas (4b) and (4c), we obtain other formulas that we will use a little later in this section.

If we solve formula (4b) for $\sin^2 \theta$, we get

$$\cos 2\theta = 1 - 2 \sin^2 \theta$$
$$2 \sin^2 \theta = 1 - \cos 2\theta$$

$$\sin^2 \theta = \frac{1 - \cos 2\theta}{2} \qquad (6)$$

*Due to the work done by P. L. Chebyshëv, these polynomials are sometimes called *Chebyshëv polynomials*.

Similarly, we can solve for $\cos^2\theta$ in formula (4c):

$$\cos 2\theta = 2\cos^2\theta - 1$$
$$2\cos^2\theta = 1 + \cos 2\theta$$

$$\cos^2\theta = \frac{1 + \cos 2\theta}{2} \qquad (7)$$

Formulas (6) and (7) can be used to develop a formula for $\tan^2\theta$:

$$\tan^2\theta = \frac{\sin^2\theta}{\cos^2\theta} = \frac{\dfrac{1 - \cos 2\theta}{2}}{\dfrac{1 + \cos 2\theta}{2}}$$

$$\tan^2\theta = \frac{1 - \cos 2\theta}{1 + \cos 2\theta} \qquad (8)$$

Formulas (6) through (8) do not have to be memorized since their derivations are so straightforward.

Formulas (6) and (7) are important in calculus. The next example illustrates a problem that arises in calculus requiring the use of formula (7).

EXAMPLE 3 Establishing an Identity

Write an equivalent expression for $\cos^4\theta$ that does not involve any powers of sine or cosine greater than 1.

Solution The idea here is to apply formula (7) twice:

$$\begin{aligned}
\cos^4\theta = (\cos^2\theta)^2 &= \left(\frac{1 + \cos 2\theta}{2}\right)^2 && \text{Formula (7).}\\
&= \frac{1}{4}(1 + 2\cos 2\theta + \cos^2 2\theta)\\
&= \frac{1}{4} + \frac{1}{2}\cos 2\theta + \frac{1}{4}\cos^2 2\theta\\
&= \frac{1}{4} + \frac{1}{2}\cos 2\theta + \frac{1}{4}\left[\frac{1 + \cos 2(2\theta)}{2}\right] && \text{Formula (7).}\\
&= \frac{1}{4} + \frac{1}{2}\cos 2\theta + \frac{1}{8}(1 + \cos 4\theta)\\
&= \frac{3}{8} + \frac{1}{2}\cos 2\theta + \frac{1}{8}\cos 4\theta
\end{aligned}$$

Now work Problem 23.

Identities, such as the Double-Angle Formulas, can sometimes be used to rewrite expressions in a more suitable form. Let's look at an example.

EXAMPLE 4 Projectile Motion

FIGURE 6

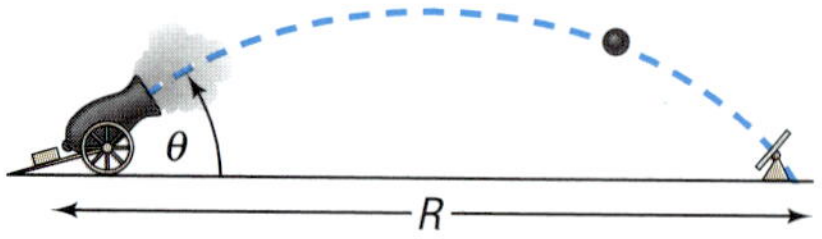

An object is propelled upward at an angle θ to the horizontal with an initial velocity of v_0 feet per second. See Figure 6. If air resistance is ignored, the horizontal distance R it travels, the **range,** is given by

$$R = \frac{1}{16} v_0^2 \sin\theta\cos\theta$$

(a) Show that $R = \dfrac{1}{32} v_0^2 \sin 2\theta$.

(b) Find the angle θ for which R is a maximum.

Solution (a) We rewrite the given expression for the range using the Double-Angle Formulas $\sin 2\theta = 2\sin\theta\cos\theta$. Then

$$R = \frac{1}{16} v_0^2 \sin\theta\cos\theta = \frac{1}{16} v_0^2 \frac{2\sin\theta\cos\theta}{2} = \frac{1}{32} v_0^2 \sin 2\theta$$

(b) In this form, the largest value for the range R can be easily found. For a fixed initial speed v_0, the angle θ of inclination to the horizontal determines the value of R. Since the largest value of a sine function is 1, occurring when the argument is 90°, it follows that for maximum R we must have

$$2\theta = 90^\circ$$
$$\theta = 45^\circ$$

An inclination to the horizontal of 45° results in maximum range.

Half-Angle Formulas

Another important use of formulas (6) through (8) is to prove the **Half-Angle Formulas.** In formulas (6) through (8), let $\theta = \alpha/2$. Then

$$\sin^2\frac{\alpha}{2} = \frac{1-\cos\alpha}{2} \qquad \cos^2\frac{\alpha}{2} = \frac{1+\cos\alpha}{2}$$
$$\tan^2\frac{\alpha}{2} = \frac{1-\cos\alpha}{1+\cos\alpha} \qquad (9)$$

If we solve for the trigonometric functions on the left sides of equations (9), we obtain the Half-Angle Formulas:

Theorem Half-Angle Formulas

$$\sin\frac{\alpha}{2} = \pm\sqrt{\frac{1-\cos\alpha}{2}} \quad (10a)$$

$$\cos\frac{\alpha}{2} = \pm\sqrt{\frac{1+\cos\alpha}{2}} \quad (10b)$$

$$\tan\frac{\alpha}{2} = \pm\sqrt{\frac{1-\cos\alpha}{1+\cos\alpha}} \quad (10c)$$

where the + or − sign is determined by the quadrant of the angle $\alpha/2$.

We use the Half-Angle Formulas in the next example.

EXAMPLE 5 Finding Exact Values Using Half-Angle Formulas

3

Find the exact value of

(a) $\cos 15°$ (b) $\sin(-15°)$

Solution (a) Because $15° = 30°/2$, we can use the Half-Angle Formula for $\cos(\alpha/2)$ with $\alpha = 30°$. Also, because 15° is in quadrant I, $\cos 15° > 0$, and we choose the + sign in using formula (10b):

$$\cos 15° = \cos\frac{30°}{2} = \sqrt{\frac{1+\cos 30°}{2}}$$

$$= \sqrt{\frac{1+\sqrt{3}/2}{2}} = \sqrt{\frac{2+\sqrt{3}}{4}} = \frac{\sqrt{2+\sqrt{3}}}{2}$$

(b) We use the fact that $\sin(-15°) = -\sin 15°$ and then apply formula (10a):

$$\sin(-15°) = -\sin\frac{30°}{2} = -\sqrt{\frac{1-\cos 30°}{2}}$$

$$= -\sqrt{\frac{1-\sqrt{3}/2}{2}} = -\sqrt{\frac{2-\sqrt{3}}{4}} = -\frac{\sqrt{2-\sqrt{3}}}{2}$$

It is interesting to compare the answer found in Example 5(a) with the answer to Example 2 of Section 7.2. There we calculated

$$\cos\frac{\pi}{12} = \cos 15° = \frac{1}{4}(\sqrt{6}+\sqrt{2})$$

Based on these results, we conclude that

$$\frac{1}{4}(\sqrt{6}+\sqrt{2}) \quad \text{and} \quad \frac{\sqrt{2+\sqrt{3}}}{2}$$

are equal. (Since each expression is positive, you can verify this equality by squaring each expression.) Thus, two very different looking, yet correct, answers can be obtained, depending on the approach taken to solve a problem.

Now work Problem 13.

EXAMPLE 6 Finding Exact Values Using Half-Angle Formulas

If $\cos\alpha = -\frac{3}{5}$, $\pi < \alpha < 3\pi/2$, find the exact value of

(a) $\sin\dfrac{\alpha}{2}$ (b) $\cos\dfrac{\alpha}{2}$ (c) $\tan\dfrac{\alpha}{2}$

Solution First, we observe that if $\pi < \alpha < 3\pi/2$, then $\pi/2 < \alpha/2 < 3\pi/4$. As a result, $\alpha/2$ lies in quadrant II.

(a) Because $\alpha/2$ lies in quadrant II, $\sin(\alpha/2) > 0$. Thus, we use the + sign in formula (10a) to get

$$\sin\frac{\alpha}{2} = \sqrt{\frac{1-\cos\alpha}{2}} = \sqrt{\frac{1-\left(-\frac{3}{5}\right)}{2}}$$
$$= \sqrt{\frac{\frac{8}{5}}{2}} = \sqrt{\frac{4}{5}} = \frac{2}{\sqrt{5}} = \frac{2\sqrt{5}}{5}$$

(b) Because $\alpha/2$ lies in quadrant II, $\cos(\alpha/2) < 0$. Thus, we use the $-$ sign in formula (10b) to get

$$\cos\frac{\alpha}{2} = -\sqrt{\frac{1+\cos\alpha}{2}} = -\sqrt{\frac{1+\left(-\frac{3}{5}\right)}{2}}$$
$$= -\sqrt{\frac{\frac{2}{5}}{2}} = -\frac{1}{\sqrt{5}} = -\frac{\sqrt{5}}{5}$$

(c) Because $\alpha/2$ lies in quadrant II, $\tan(\alpha/2) < 0$. Thus, we use the $-$ sign in formula (10c) to get

$$\tan\frac{\alpha}{2} = -\sqrt{\frac{1-\cos\alpha}{1+\cos\alpha}} = -\sqrt{\frac{1-\left(-\frac{3}{5}\right)}{1+\left(-\frac{3}{5}\right)}} = -\sqrt{\frac{\frac{8}{5}}{\frac{2}{5}}} = -2$$

Another way to solve Example 6(c) is to use the solutions found in parts (a) and (b):

$$\tan\frac{\alpha}{2} = \frac{\sin(\alpha/2)}{\cos(\alpha/2)} = \frac{2\sqrt{5}/5}{-\sqrt{5}/5} = -2$$

Now work Problems 1(c) and (d).

There is a formula for $\tan(\alpha/2)$ that does not contain + and $-$ signs, making it more useful than Formula 10(c). Because

$$1-\cos\alpha = 2\sin^2\left(\frac{\alpha}{2}\right) \quad \text{Formula 9.}$$

and

$$\sin\alpha = \sin 2\left(\frac{\alpha}{2}\right) = 2\sin\left(\frac{\alpha}{2}\right)\cos\left(\frac{\alpha}{2}\right) \quad \text{Double-angle formula.}$$

we have

$$\frac{1-\cos\alpha}{\sin\alpha} = \frac{2\sin^2\left(\frac{\alpha}{2}\right)}{2\sin\left(\frac{\alpha}{2}\right)\cos\left(\frac{\alpha}{2}\right)} = \frac{\sin\left(\frac{\alpha}{2}\right)}{\cos\left(\frac{\alpha}{2}\right)} = \tan\left(\frac{\alpha}{2}\right)$$

Since it also can be shown that

$$\frac{1-\cos\alpha}{\sin\alpha} = \frac{\sin\alpha}{1+\cos\alpha}$$

we have the following two Half-Angle Formulas:

Half-Angle Formulas for tan $\alpha/2$

$$\tan\left(\frac{\alpha}{2}\right) = \frac{1-\cos\alpha}{\sin\alpha} = \frac{\sin\alpha}{1+\cos\alpha} \qquad (11)$$

With this formula, the solution to Example 6(c) can be given as

$$\cos\alpha = -\tfrac{3}{5}$$

$$\sin\alpha = -\sqrt{1-\cos^2\alpha} = -\sqrt{1-\tfrac{9}{25}} = -\sqrt{\tfrac{16}{25}} = -\tfrac{4}{5}$$

Thus, by equation (11),

$$\tan\frac{\alpha}{2} = \frac{1-\cos\alpha}{\sin\alpha} = \frac{1-\left(-\frac{3}{5}\right)}{-\frac{4}{5}} = \frac{\frac{8}{5}}{-\frac{4}{5}} = -2$$

The next example illustrates a problem that arises in calculus.

EXAMPLE 7 Using Half-Angle Formulas in Calculus

If $z = \tan(\alpha/2)$, show that

(a) $\sin\alpha = \dfrac{2z}{1+z^2}$ (b) $\cos\alpha = \dfrac{1-z^2}{1+z^2}$

Solution (a) $$\frac{2z}{1+z^2} = \frac{2\tan(\alpha/2)}{1+\tan^2(\alpha/2)} = \frac{2\tan(\alpha/2)}{\sec^2(\alpha/2)} = \frac{\dfrac{2\cdot\sin(\alpha/2)}{\cos(\alpha/2)}}{\dfrac{1}{\cos^2(\alpha/2)}}$$

$$= \frac{2\sin(\alpha/2)}{\cos(\alpha/2)}\cdot\cos^2\frac{\alpha}{2} = 2\sin\frac{\alpha}{2}\cos\frac{\alpha}{2}$$

$$= \sin 2\left(\frac{\alpha}{2}\right) = \sin\alpha$$

↑ Double-Angle Formula (3)

$$\text{(b)}\ \frac{1-z^2}{1+z^2} = \frac{1-\tan^2(\alpha/2)}{1+\tan^2(\alpha/2)} = \frac{1-\dfrac{\sin^2(\alpha/2)}{\cos^2(\alpha/2)}}{1+\dfrac{\sin^2(\alpha/2)}{\cos^2(\alpha/2)}}$$

$$= \frac{\cos^2(\alpha/2)-\sin^2(\alpha/2)}{\cos^2(\alpha/2)+\sin^2(\alpha/2)} = \frac{\cos 2(\alpha/2)}{1} = \cos\alpha$$

↑ Double-Angle Formula (4a) and Pythagorean Identity

7.3 EXERCISES

In Problems 1–10, use the information given about the angle θ to find the exact value of

(a) $\sin 2\theta$ *(b)* $\cos 2\theta$ *(c)* $\sin\frac{\theta}{2}$ *(d)* $\cos\frac{\theta}{2}$

1. $\sin\theta = \frac{3}{5},\ 0<\theta<\pi/2$
2. $\cos\theta = \frac{3}{5},\ 0<\theta<\pi/2$
3. $\tan\theta = \frac{4}{3},\ \pi<\theta<3\pi/2$
4. $\tan\theta = \frac{1}{2},\ \pi<\theta<3\pi/2$
5. $\cos\theta = -\sqrt{2}/\sqrt{3},\ \pi/2<\theta<\pi$
6. $\sin\theta = -1/\sqrt{3},\ 3\pi/2<\theta<2\pi$
7. $\sec\theta = 3,\ \sin\theta>0$
8. $\csc\theta = -\sqrt{5},\ \cos\theta<0$
9. $\cot\theta = -2,\ \sec\theta<0$
10. $\sec\theta = 2,\ \csc\theta<0$
11. $\tan\theta = -3,\ \sin\theta<0$
12. $\cot\theta = 3,\ \cos\theta<0$

In Problems 13–22, use the half-angle formulas to find the exact value of each trigonometric function.

13. $\sin 22.5°$
14. $\cos 22.5°$
15. $\tan\frac{7\pi}{8}$
16. $\tan\frac{9\pi}{8}$
17. $\cos 165°$
18. $\sin 195°$
19. $\sec\frac{15\pi}{8}$
20. $\csc\frac{7\pi}{8}$
21. $\sin\left(-\frac{\pi}{8}\right)$
22. $\cos\left(-\frac{3\pi}{8}\right)$

23. Show that $\sin^4\theta = \frac{3}{8} - \frac{1}{2}\cos 2\theta + \frac{1}{8}\cos 4\theta$.
24. Develop a formula for $\cos 3\theta$ as a third-degree polynomial in the variable $\cos\theta$.
25. Show that $\sin 4\theta = (\cos\theta)(4\sin\theta - 8\sin^3\theta)$.
26. Develop a formula for $\cos 4\theta$ as a fourth-degree polynomial in the variable $\cos\theta$.
27. Find an expression for $\sin 5\theta$ as a fifth-degree polynomial in the variable $\sin\theta$.
28. Find an expression for $\cos 5\theta$ as a fifth-degree polynomial in the variable $\cos\theta$

In Problems 29–48, establish each identity.

29. $\cos^4\theta - \sin^4\theta = \cos 2\theta$
30. $\dfrac{\cot\theta - \tan\theta}{\cot\theta + \tan\theta} = \cos 2\theta$
31. $\cot 2\theta = \dfrac{\cot^2\theta - 1}{2\cot\theta}$
32. $\cot 2\theta = \frac{1}{2}(\cot\theta - \tan\theta)$
33. $\sec 2\theta = \dfrac{\sec^2\theta}{2-\sec^2\theta}$
34. $\csc 2\theta = \frac{1}{2}\sec\theta\csc\theta$
35. $\cos^2 2\theta - \sin^2 2\theta = \cos 4\theta$
36. $(4\sin\theta\cos\theta)(1 - 2\sin^2\theta) = \sin 4\theta$
37. $\dfrac{\cos 2\theta}{1+\sin 2\theta} = \dfrac{\cot\theta - 1}{\cot\theta + 1}$
38. $\sin^2\theta\cos^2\theta = \frac{1}{8}(1 - \cos 4\theta)$
39. $\sec^2\frac{\theta}{2} = \dfrac{2}{1+\cos\theta}$
40. $\csc^2\frac{\theta}{2} = \dfrac{2}{1-\cos\theta}$

41. $\cot^2 \dfrac{\theta}{2} = \dfrac{\sec \theta + 1}{\sec \theta - 1}$

42. $\tan \dfrac{\theta}{2} = \csc \theta - \cot \theta$

43. $\cos \theta = \dfrac{1 - \tan^2(\theta/2)}{1 + \tan^2(\theta/2)}$

44. $1 - \frac{1}{2} \sin 2\theta = \dfrac{\sin^3 \theta + \cos^3 \theta}{\sin \theta + \cos \theta}$

45. $\dfrac{\sin 3\theta}{\sin \theta} - \dfrac{\cos 3\theta}{\cos \theta} = 2$

46. $\dfrac{\cos \theta + \sin \theta}{\cos \theta - \sin \theta} - \dfrac{\cos \theta - \sin \theta}{\cos \theta + \sin \theta} = 2 \tan 2\theta$

47. $\tan 3\theta = \dfrac{3 \tan \theta - \tan^3 \theta}{1 - 3 \tan^2 \theta}$

48. $\tan \theta + \tan(\theta + 120°) + \tan(\theta + 240°) = 3 \tan 3\theta$

In Problems 49–62, find the exact value of each expression.

49. $\sin(2 \sin^{-1} \frac{1}{2})$

50. $\sin[2 \sin^{-1} \frac{\sqrt{3}}{2}]$

51. $\cos(2 \sin^{-1} \frac{3}{5})$

52. $\cos(2 \cos^{-1} \frac{4}{5})$

53. $\tan[2 \cos^{-1}(-\frac{3}{5})]$

54. $\tan(2 \tan^{-1} \frac{3}{4})$

55. $\sin(2 \cos^{-1} \frac{4}{5})$

56. $\cos[2 \tan^{-1}(-\frac{4}{3})]$

57. $\sin^2(\frac{1}{2} \cos^{-1} \frac{3}{5})$

58. $\cos^2(\frac{1}{2} \sin^{-1} \frac{3}{5})$

59. $\sec(2 \tan^{-1} \frac{3}{4})$

60. $\csc[2 \sin^{-1}(-\frac{3}{5})]$

61. $\cot^2(\frac{1}{2} \tan^{-1} \frac{4}{3})$

62. $\cot^2(\frac{1}{2} \cos^{-1} \frac{5}{13})$

63. Find the value of the number C: $\frac{1}{2} \sin^2 x + C = -\frac{1}{4} \cos 2x$

64. Find the value of the number C: $\frac{1}{2} \cos^2 x + C = \frac{1}{4} \cos 2x$

65. Area of an Isosceles Triangle Show that the area A of an isosceles triangle whose equal sides are of length s and the angle between them is θ is

$$A = \frac{1}{2} s^2 \sin \theta$$

[**Hint:** See the illustration. The height h bisects the angle θ and is the perpendicular bisector of the base.]

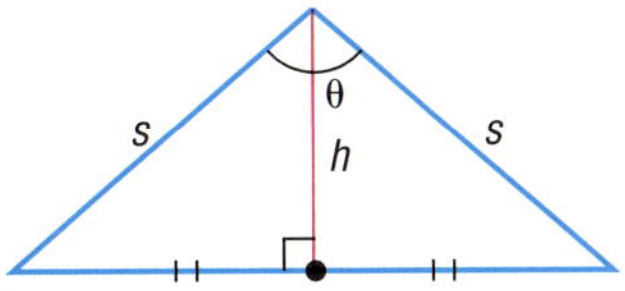

66. Geometry A rectangle is inscribed in a semicircle of radius 1. See the illustration.

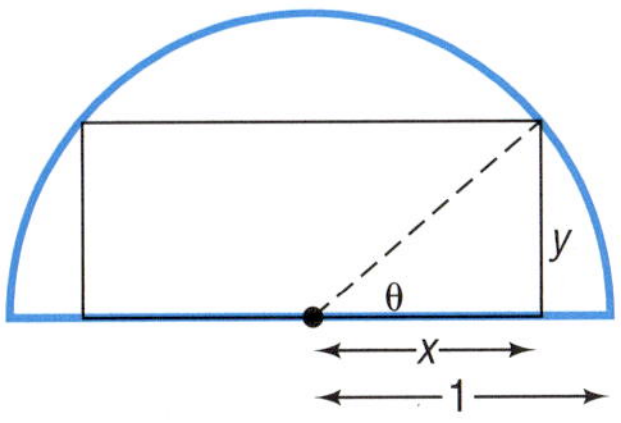

(a) Express the area A of the rectangle as a function of the angle θ shown in the illustration.

(b) Show that $A = \sin 2\theta$.

(c) Find the angle θ that results in the largest area A.

(d) Find the dimensions of this largest rectangle.

67. Graph $f(x) = \sin^2 x = (1 - \cos 2x)/2$ for $0 \le x \le 2\pi$ by using the ideas of shifts, compression, and the like.

68. Repeat Problem 67 for $g(x) = \cos^2 x$.

69. Use the fact that

$$\cos \frac{\pi}{12} = \frac{1}{4}(\sqrt{6} + \sqrt{2})$$

to find $\sin(\pi/24)$ and $\cos(\pi/24)$.

70. Show that

$$\sin \frac{\pi}{8} = \frac{\sqrt{2 - \sqrt{2}}}{2}$$

and use it to find $\sin(\pi/16)$ and $\cos(\pi/16)$.

71. Show that

$$\sin^3 \theta + \sin^3(\theta + 120°) + \sin^3(\theta + 240°) = -\frac{3}{4} \sin 3\theta.$$

72. If $\tan \theta = a \tan(\theta/3)$, express $\tan(\theta/3)$ in terms of a.

In Problems 73 and 74, establish each identity.

73. $\ln|\sin\theta| = \frac{1}{2}(\ln|1 - \cos 2\theta| - \ln 2)$

74. $\ln|\cos\theta| = \frac{1}{2}(\ln|1 + \cos 2\theta| - \ln 2)$

75. Projectile Motion An object is propelled upward at an angle θ, $45° < \theta < 90°$, to the horizontal with an initial velocity of v_0 feet per second from the base of a plane that makes an angle of 45° with the horizontal. See the illustration. If air resistance is ignored, the distance R it travels up the inclined plane is given by

$$R = \frac{v_0^2\sqrt{2}}{16}\cos\theta(\sin\theta - \cos\theta)$$

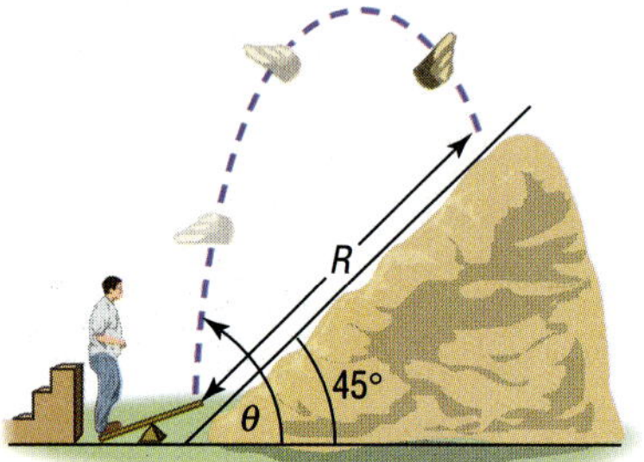

(a) Show that

$$R = \frac{v_0^2\sqrt{2}}{32}(\sin 2\theta - \cos 2\theta - 1)$$

(b) Graph $R = R(\theta)$. (Use $v_0 = 32$ ft/sec.)

(c) What value of θ makes R the largest? (Use $v_0 = 32$ ft/sec.)

76. Sawtooth Curve An oscilloscope often displays a sawtooth curve. This curve can be approximated by sinusoidal curves of varying periods and amplitudes. A first approximation to the sawtooth curve is given by

$$y = \frac{1}{2}\sin 2\pi x + \frac{1}{4}\sin 4\pi x$$

Show that $y = \sin 2\pi x \cos^2 \pi x$.

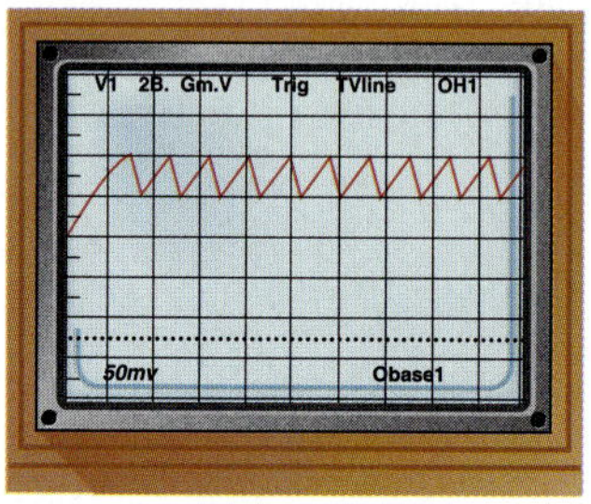

77. Go to the library and research Chebyshëv polynomials. Write a report on your findings.

7.4 PRODUCT-TO-SUM AND SUM-TO-PRODUCT FORMULAS

1 Express Products as Sums

2 Express Sums as Products

1 The sum and difference formulas can be used to derive formulas for writing the products of sines and/or cosines as sums or differences. These identities are usually called the **Product-to-Sum Formulas.**

Theorem Product-to-Sum Formulas

$$\sin\alpha\sin\beta = \tfrac{1}{2}[\cos(\alpha - \beta) - \cos(\alpha + \beta)] \quad (1)$$

$$\cos\alpha\cos\beta = \tfrac{1}{2}[\cos(\alpha - \beta) + \cos(\alpha + \beta)] \quad (2)$$

$$\sin\alpha\cos\beta = \tfrac{1}{2}[\sin(\alpha + \beta) + \sin(\alpha - \beta)] \quad (3)$$

MISSION POSSIBLE

How Far and How High Does a Baseball Need to Go for an Out-of-the-Park Home Run?

The sportscasters for the Cleveland Indians decided they had better be prepared for any eventuality when the new ballpark, Jacob's Field, opened in 1994. One eventuality they worried about was Jim Thome hitting a home run that went so high and so far that it left the ball park. What would they tell their listeners about the height and distance the ball went?

So they called in the Mission Possible team. The sportscasters wanted to know the distance from homeplate to the highest point of the stadium and the distance from the base of the outfield fence to the highest point of the stadium for every 5° from the 3rd base line around to the 1st base line. The problem was complicated by the fact that the distance from homeplate to the outfield fence varied from 325 feet to 410 feet. Furthermore, the height that would have to be cleared also varied, depending on where the ball was hit.

1. First, how many distances and heights do the sportscasters want?
2. To see one solution, use the distance from homeplate across 2nd base to the outfield fence, 410 feet. Using a transit, you find the angle of elevation from homeplate to the highest point of the stadium is 10° and the angle of elevation from the base of the outfield fence to the highest point of the stadium is 32.5°. Use this information to find the minimum distance from homeplate to the highest point of the stadium and the distance from the base of the outfield fence to the highest point of the stadium.
3. You tell the sportscasters that if they provide you the two angles of elevation for each 5° movement around the stadium, you can solve their problem. After receiving the two angles of elevation for each 5° movement around the stadium, you decide that it would be faster to develop a general formula, so that by inputting the two angles and the distance from homeplate to the fence, the distance from homeplate to the top of the roof and the height of the roof would be computed. Let α and β denote the angles of elevation to the top of the roof from homeplate and from the base of the outfield fence and let L be the distance from homeplate to the outfield fence. What are the correct formulas?
4. Compare your formulas with those of other groups. Are they all the same? Are they all equivalent? Which ones are simplest?
5. Now write each correct formula in the form shown below:

$$\text{height} = \frac{L \sin \alpha \sin \beta}{\sin(\beta - \alpha)} \qquad \text{distance} = \frac{L \sin \beta}{\sin(\beta - \alpha)}$$

6. What is the probable trajectory of a hit baseball on a windless day? How might the wind and other factors affect the path of the baseball?
7. If a player actually did hit an out-of-the-park home run, how would the distance the ball traveled compare to the figures you have been developing for the sportscasters? What other factors will add to the distance?
8. Could there be a longest home run? How would you measure it?

The formulas on page 554 do not have to be memorized. Instead, you should remember how they are derived. Then, when you want to use them, either look them up or derive them, as needed.

To derive formulas (1) and (2), write down the sum and difference formulas for the cosine:

$$\cos(\alpha - \beta) = \cos\alpha\cos\beta + \sin\alpha\sin\beta \tag{4}$$

$$\cos(\alpha + \beta) = \cos\alpha\cos\beta - \sin\alpha\sin\beta \tag{5}$$

Subtract equation (5) from equation (4) to get

$$\cos(\alpha - \beta) - \cos(\alpha + \beta) = 2\sin\alpha\sin\beta$$

from which

$$\sin\alpha\sin\beta = \tfrac{1}{2}[\cos(\alpha - \beta) - \cos(\alpha + \beta)]$$

Now, add equations (4) and (5) to get

$$\cos(\alpha - \beta) + \cos(\alpha + \beta) = 2\cos\alpha\cos\beta$$

from which

$$\cos\alpha\cos\beta = \tfrac{1}{2}[\cos(\alpha - \beta) + \cos(\alpha + \beta)]$$

To derive Product-to-Sum Formula (3), use the sum and difference formulas for sine in a similar way. (You are asked to do this in Problem 41 at the end of this section.)

EXAMPLE 1 Expressing Products as Sums

Express each of the following products as a sum containing only sines or cosines:

(a) $\sin 6\theta \sin 4\theta$ (b) $\cos 3\theta \cos\theta$ (c) $\sin 3\theta \cos 5\theta$

Solution (a) We use formula (1) to get

$$\begin{aligned}\sin 6\theta \sin 4\theta &= \tfrac{1}{2}[\cos(6\theta - 4\theta) - \cos(6\theta + 4\theta)]\\ &= \tfrac{1}{2}(\cos 2\theta - \cos 10\theta)\end{aligned}$$

(b) We use formula (2) to get

$$\begin{aligned}\cos 3\theta \cos\theta &= \tfrac{1}{2}[\cos(3\theta - \theta) + \cos(3\theta + \theta)]\\ &= \tfrac{1}{2}(\cos 2\theta + \cos 4\theta)\end{aligned}$$

(c) We use formula (3) to get

$$\begin{aligned}\sin 3\theta \cos 5\theta &= \tfrac{1}{2}[\sin(3\theta + 5\theta) + \sin(3\theta - 5\theta)]\\ &= \tfrac{1}{2}[\sin 8\theta + \sin(-2\theta)] = \tfrac{1}{2}(\sin 8\theta - \sin 2\theta)\end{aligned}$$

Now work Problem 1.

2

The **Sum-to-Product Formulas** are given next.

Theorem Sum-to-Product Formulas

$$\sin \alpha + \sin \beta = 2 \sin \frac{\alpha + \beta}{2} \cos \frac{\alpha - \beta}{2} \qquad (6)$$

$$\sin \alpha - \sin \beta = 2 \sin \frac{\alpha - \beta}{2} \cos \frac{\alpha + \beta}{2} \qquad (7)$$

$$\cos \alpha + \cos \beta = 2 \cos \frac{\alpha + \beta}{2} \cos \frac{\alpha - \beta}{2} \qquad (8)$$

$$\cos \alpha - \cos \beta = -2 \sin \frac{\alpha + \beta}{2} \sin \frac{\alpha - \beta}{2} \qquad (9)$$

We will derive formula (6) and leave the derivations of formulas (7) through (9) as exercises (see Problems 42 through 44).

Proof

$$2 \sin \frac{\alpha + \beta}{2} \cos \frac{\alpha - \beta}{2} \underset{\uparrow}{=} 2 \cdot \frac{1}{2}\left[\sin\left(\frac{\alpha + \beta}{2} + \frac{\alpha - \beta}{2}\right) + \sin\left(\frac{\alpha + \beta}{2} - \frac{\alpha - \beta}{2}\right)\right]$$

Product-to-Sum Formula (3).

$$= \sin \frac{2\alpha}{2} + \sin \frac{2\beta}{2} = \sin \alpha + \sin \beta$$

EXAMPLE 2 Expressing Sums (or Differences) as a Product

Express each sum or difference as a product of sines and/or cosines:

(a) $\sin 5\theta - \sin 3\theta$ (b) $\cos 3\theta + \cos 2\theta$

Solution (a) We use formula (7) to get

$$\sin 5\theta - \sin 3\theta = 2 \sin \frac{5\theta - 3\theta}{2} \cos \frac{5\theta + 3\theta}{2}$$
$$= 2 \sin \theta \cos 4\theta$$

(b) $$\cos 3\theta + \cos 2\theta = 2 \cos \frac{3\theta + 2\theta}{2} \cos \frac{3\theta - 2\theta}{2} \qquad \text{Formula (8).}$$
$$= 2 \cos \frac{5\theta}{2} \cos \frac{\theta}{2}$$

Now work Problem 11.

7.4 EXERCISES

In Problems 1–10, express each product as a sum containing only sines or cosines.

1. $\sin 4\theta \sin 2\theta$
2. $\cos 4\theta \cos 2\theta$
3. $\sin 4\theta \cos 2\theta$
4. $\sin 3\theta \sin 5\theta$
5. $\cos 3\theta \cos 5\theta$
6. $\sin 4\theta \cos 6\theta$
7. $\sin \theta \sin 2\theta$
8. $\cos 3\theta \cos 4\theta$
9. $\sin \dfrac{3\theta}{2} \cos \dfrac{\theta}{2}$
10. $\sin \dfrac{\theta}{2} \cos \dfrac{5\theta}{2}$

In Problems 11–18, express each sum or difference as a product of sines and/or cosines.

11. $\sin 4\theta - \sin 2\theta$
12. $\sin 4\theta + \sin 2\theta$
13. $\cos 2\theta + \cos 4\theta$
14. $\cos 5\theta - \cos 3\theta$
15. $\sin \theta + \sin 3\theta$
16. $\cos \theta + \cos 3\theta$
17. $\cos \dfrac{\theta}{2} - \cos \dfrac{3\theta}{2}$
18. $\sin \dfrac{\theta}{2} - \sin \dfrac{3\theta}{2}$

In Problems 19–36, establish each identity.

19. $\dfrac{\sin \theta + \sin 3\theta}{2 \sin 2\theta} = \cos \theta$
20. $\dfrac{\cos \theta + \cos 3\theta}{2 \cos 2\theta} = \cos \theta$
21. $\dfrac{\sin 4\theta + \sin 2\theta}{\cos 4\theta + \cos 2\theta} = \tan 3\theta$
22. $\dfrac{\cos \theta - \cos 3\theta}{\sin 3\theta - \sin \theta} = \tan 2\theta$
23. $\dfrac{\cos \theta - \cos 3\theta}{\sin \theta + \sin 3\theta} = \tan \theta$
24. $\dfrac{\cos \theta - \cos 5\theta}{\sin \theta + \sin 5\theta} = \tan 2\theta$
25. $\sin \theta(\sin \theta + \sin 3\theta) = \cos \theta(\cos \theta - \cos 3\theta)$
26. $\sin \theta(\sin 3\theta + \sin 5\theta) = \cos \theta(\cos \theta - \cos 5\theta)$
27. $\dfrac{\sin 4\theta + \sin 8\theta}{\cos 4\theta + \cos 8\theta} = \tan 6\theta$
28. $\dfrac{\sin 4\theta - \sin 8\theta}{\cos 4\theta - \cos 8\theta} = -\cot 6\theta$
29. $\dfrac{\sin 4\theta + \sin 8\theta}{\sin 4\theta - \sin 8\theta} = -\dfrac{\tan 6\theta}{\tan 2\theta}$
30. $\dfrac{\cos 4\theta - \cos 8\theta}{\cos 4\theta + \cos 8\theta} = \tan 2\theta \tan 6\theta$
31. $\dfrac{\sin \alpha + \sin \beta}{\sin \alpha - \sin \beta} = \tan \dfrac{\alpha + \beta}{2} \cot \dfrac{\alpha - \beta}{2}$
32. $\dfrac{\cos \alpha + \cos \beta}{\cos \alpha - \cos \beta} = -\cot \dfrac{\alpha + \beta}{2} \cot \dfrac{\alpha - \beta}{2}$
33. $\dfrac{\sin \alpha + \sin \beta}{\cos \alpha + \cos \beta} = \tan \dfrac{\alpha + \beta}{2}$
34. $\dfrac{\sin \alpha - \sin \beta}{\cos \alpha - \cos \beta} = -\cot \dfrac{\alpha + \beta}{2}$
35. $1 + \cos 2\theta + \cos 4\theta + \cos 6\theta = 4 \cos \theta \cos 2\theta \cos 3\theta$
36. $1 - \cos 2\theta + \cos 4\theta - \cos 6\theta = 4 \sin \theta \cos 2\theta \sin 3\theta$

37. **Touch-Tone Phones** On a touch-tone phone, each button produces a unique sound. The sound produced is the sum of two tones, given by

$$y = \sin 2\pi lt \quad \text{and} \quad y = \sin 2\pi ht$$

where l and h are the low and high frequencies (cycles per second) shown on the illustration. For example, if you touch 7, the low frequency is $l = 852$ cycles per second and the high frequency is $h = 1209$ cycles per second. The sound emitted by touching 7 is

$y = \sin 2\pi(852)t + \sin 2\pi(1209)t$

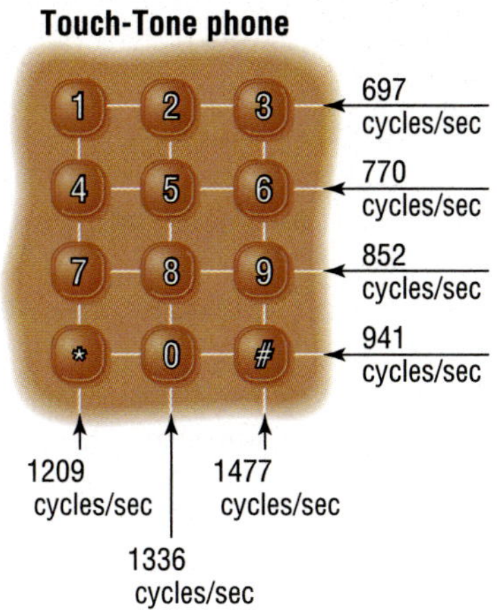

(a) Write this sound as a product of sines and/or cosines.
(b) Graph the sound emitted by touching 7.
(c) Determine the maximum value of y.

38. **Touch-Tone Phones**
(a) Write the sound emitted by touching the # key as a product of sines and/or cosines.
(b) Graph the sound emitted by touching the # key.
(c) Determine the maximum value of y.

39. If $\alpha + \beta + \gamma = \pi$, show that

$\sin 2\alpha + \sin 2\beta + \sin 2\gamma = 4 \sin \alpha \sin \beta \sin \gamma.$

40. If $\alpha + \beta + \gamma = \pi$, show that

$\tan \alpha + \tan \beta + \tan \gamma = \tan \alpha \tan \beta \tan \gamma.$

41. Derive formula (3).

42. Derive formula (7).

43. Derive formula (8).

44. Derive formula (9).

7.5 TRIGONOMETRIC EQUATIONS

1 Solve Trigonometric Equations

1 The previous sections of this chapter were devoted to trigonometric identities—that is, equations involving trigonometric functions that are satisfied by every value in the domain of the variable. In this section, we discuss **trigonometric equations**—that is, equations involving trigonometric functions that are satisfied only by some values of the variable (or, possibly, are not satisfied by any values of the variable). The values that satisfy the equation are called **solutions** of the equation.

EXAMPLE 1 Checking Whether a Given Number Is a Solution of a Trigonometric Equation

Determine whether $\theta = \pi/4$ is a solution of the equation $\sin \theta = \frac{1}{2}$. Is $\theta = \pi/6$ a solution?

Solution Replace θ by $\pi/4$ in the given equation. The result is

$$\sin \frac{\pi}{4} = \frac{\sqrt{2}}{2} \neq \frac{1}{2}$$

We conclude that $\pi/4$ is not a solution.

Next, replace θ by $\pi/6$ in the equation. The result is

$$\sin\frac{\pi}{6} = \frac{1}{2}$$

Thus, $\pi/6$ is a solution of the given equation. ■

The equation given in Example 1 has other solutions besides $\theta = \pi/6$. For example, $\theta = 5\pi/6$ is also a solution, as is $\theta = 13\pi/6$. (You should check this for yourself.) In fact, the equation has an infinite number of solutions due to the periodicity of the sine function, as can be seen in Figure 7.

FIGURE 7

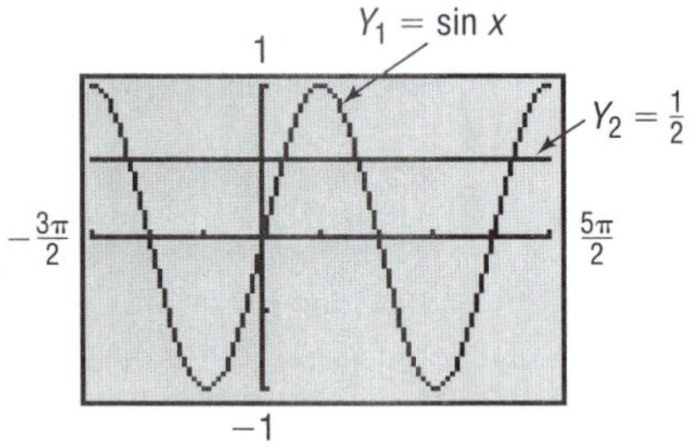

As before, our practice will be to solve equations, whenever possible, by finding exact solutions. In such cases, we will also verify the solution obtained by using a graphing utility. When traditional methods cannot be used, approximate solutions will be obtained using a graphing utility. The reader is encouraged to pay particular attention to the form of equations for which exact solutions are possible.

Unless otherwise indicated, in solving trigonometric equations, we need to find *all* solutions. As the next example illustrates, finding all the solutions can be accomplished by first finding solutions over an interval whose length equals the period of the function and then adding multiples of that period to the solutions found. Let's look at some examples.

EXAMPLE 2 Solving a Trigonometric Equation

Solve the equation: $\cos\theta = \frac{1}{2}$

FIGURE 8 $\cos\theta = \frac{1}{2}$

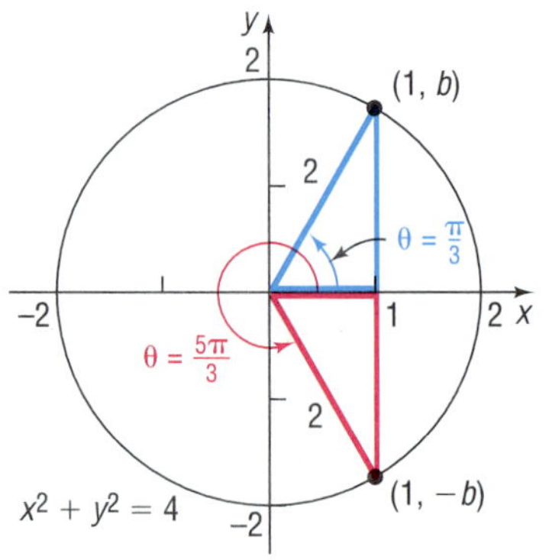

Solution The period of the cosine function is 2π. In the interval $[0, 2\pi)$, there are two angles θ for which $\cos\theta = \frac{1}{2}$: $\theta = \pi/3$ and $\theta = 5\pi/3$. See Figure 8. Because the cosine function has period 2π, all the solutions of $\cos\theta = \frac{1}{2}$ may be given by

$$\theta = \frac{\pi}{3} + 2k\pi \quad \text{or} \quad \theta = \frac{5\pi}{3} + 2k\pi \qquad k \text{ any integer.}$$

Some of the solutions are

$$\underbrace{\frac{\pi}{3}, \frac{5\pi}{3}}_{k=0}, \underbrace{\frac{7\pi}{3}, \frac{11\pi}{3}}_{k=1}, \underbrace{\frac{13\pi}{3}, \frac{17\pi}{3}}_{k=2}, \text{ and so on}$$

■

FIGURE 9

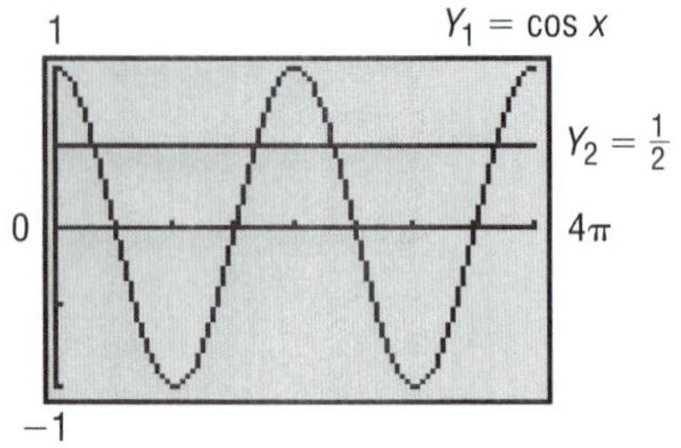

We can verify the solution by graphing $Y_1 = \cos x$ and $Y_2 = \frac{1}{2}$ to determine where the graphs intersect. (Be sure to graph in radian mode.) See Figure 9. The graph of Y_1 intersects the graph of Y_2 at $x = 1.04$ ($\approx\pi/3$), 5.23 ($\approx 5\pi/3$), 7.33 ($\approx 7\pi/3$), and 11.51 ($\approx 11\pi/3$), correct to two decimal places.

Now work Problem 1.

In most of our work, we shall be interested only in finding solutions of trigonometric equations for $0 \le \theta < 2\pi$.

EXAMPLE 3 Solving a Trigonometric Equation

Solve the equation: $\sin 2\theta = 1, 0 \le \theta < 2\pi$

Solution The period of the sine function is 2π. In the interval $[0, 2\pi)$, the sine function has the value 1 only at $\pi/2$. Because the argument is 2θ in the given equation, we have

$$2\theta = \frac{\pi}{2} + 2k\pi \quad \text{k any integer.}$$

$$\theta = \frac{\pi}{4} + k\pi$$

In the interval $[0, 2\pi)$, the solutions of $\sin 2\theta = 1$ are $\pi/4$ $(k = 0)$ and $\pi/4 + \pi = 5\pi/4$ $(k = 1)$. Note that $k = -1$ gives $\theta = -3\pi/4$ and $k = 2$ gives $\theta = 9\pi/4$, both of which are outside $[0, 2\pi)$.

You should verify $\pi/4$ and $5\pi/4$ are solutions by graphing $Y_1 = \sin(2x)$ and $Y_2 = 1$ for $0 \leq x \leq 2\pi$.

Warning: In solving a trigonometric equation for θ, $0 \leq \theta < 2\pi$, in which the argument is not θ (as in Example 3), you must write down all the solutions, first, and then list those that are in the interval $[0, 2\pi)$. Otherwise, solutions may be lost. For example, in solving $\sin 2\theta = 1$, if you merely write the solution $2\theta = \pi/2$, you will find only $\theta = \pi/4$ and miss the solution $\theta = 5\pi/4$.

Now work Problem 7.

EXAMPLE 4 Solving a Trigonometric Equation

Solve the equation: $\sin 2\theta = \frac{1}{2}, 0 \leq \theta < 2\pi$

FIGURE 10
$\sin 2\theta = \frac{1}{2}, 0 \leq \theta < 2\pi$

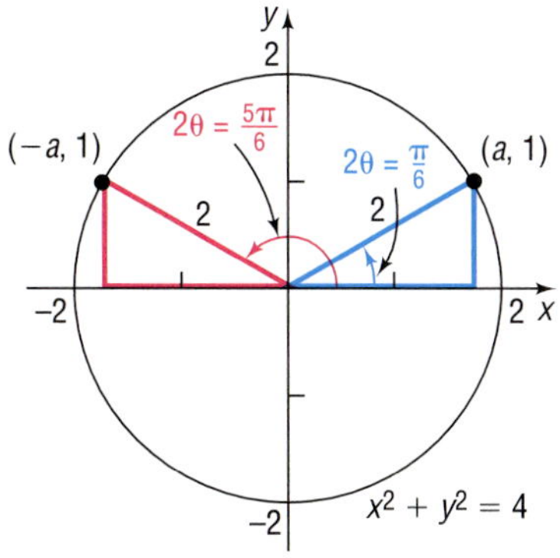

Solution The period of the sine function is 2π. In the interval $[0, 2\pi)$, the sine function has the value $\frac{1}{2}$ at $\pi/6$ and $5\pi/6$. See Figure 10. Consequently, because the argument is 2θ in the equation $\sin 2\theta = \frac{1}{2}$, we have

$$2\theta = \frac{\pi}{6} + 2k\pi \quad \text{or} \quad 2\theta = \frac{5\pi}{6} + 2k\pi \quad \text{k any integer.}$$

$$\theta = \frac{\pi}{12} + k\pi \qquad \theta = \frac{5\pi}{12} + k\pi$$

In the interval $[0, 2\pi)$, the solutions of $\sin 2\theta = \frac{1}{2}$ are $\pi/12$, $\pi/12 + \pi = 13\pi/12$, $5\pi/12$, and $5\pi/12 + \pi = 17\pi/12$.

You should verify the solution using a graphing utility.

EXAMPLE 5 Solving a Trigonometric Equation

Solve the equation: $\tan \frac{\theta}{4} = 1, 0 \leq \theta < 2\pi$

Solution The period of the tangent function is π. In the interval $[0, \pi)$, the tangent function has the value 1 only at $\pi/4$. Because the argument is $\theta/4$ in the given equation, we have

$$\frac{\theta}{4} = \frac{\pi}{4} + k\pi \quad \text{k any integer.}$$

$$\theta = \pi + 4k\pi$$

In the interval $[0, 2\pi)$, $\theta = \pi$ is the only solution.

You should verify the solution using a graphing utility.

EXAMPLE 6 Solving a Trigonometric Equation with a Calculator

Use a calculator to solve the equation: $\sin\theta = 0.3$, $0 \le \theta < 2\pi$

Solution To solve $\sin\theta = 0.3$ on a graphing calculator, we first choose the radian mode. See Figure 11.

FIGURE 11

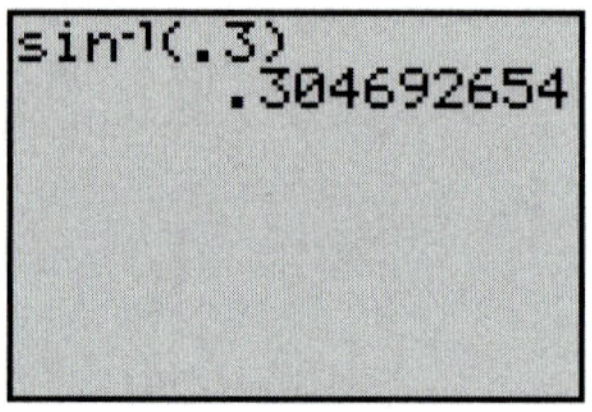

FIGURE 12
$\sin\theta = 0.3$

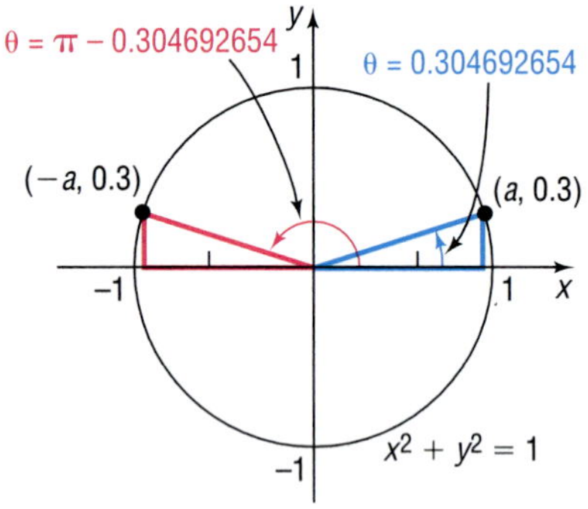

The angle 0.304692654 radian is the angle $-\pi/2 \le \theta \le \pi/2$ for which $\sin\theta = 0.3$. Another angle for which $\sin\theta = 0.3$ is $\pi - 0.304692654$. See Figure 12. The angle $\pi - 0.304692654$ is the angle in quadrant II, where $\sin\theta = 0.3$. Thus, the solutions for $\sin\theta = 0.3$, $0 \le \theta < 2\pi$, are

$$\theta = 0.304692654 \quad \text{or} \quad \theta = \pi - 0.304692654 \approx 2.8369$$

Warning: Example 6 illustrates that caution must be exercised when solving trigonometric equations on a calculator. Remember that the calculator supplies an angle only within the restrictions of the definition of the inverse trigonometric function. To find the remaining solutions, you must identify other quadrants, if any, in which the angle may be located.

Now work Problem 13.

Many trigonometric equations can be solved by applying techniques that we already know, such as applying the quadratic formula (if the equation is a second-degree polynomial) or factoring.

EXAMPLE 7 Solving a Trigonometric Equation Quadratic in Form

Solve the equation: $2\sin^2\theta - 3\sin\theta + 1 = 0$, $0 \le \theta < 2\pi$

Solution The equation we wish to solve is a quadratic equation (in $\sin\theta$) that can be factored:

$$2\sin^2\theta - 3\sin\theta + 1 = 0 \qquad 2x^2 - 3x + 1 = 0, \quad x = \sin\theta$$
$$(2\sin\theta - 1)(\sin\theta - 1) = 0 \qquad (2x-1)(x-1) = 0$$
$$2\sin\theta - 1 = 0 \quad \text{or} \quad \sin\theta - 1 = 0$$
$$\sin\theta = \tfrac{1}{2} \qquad\qquad \sin\theta = 1$$

Thus,

$$\theta = \frac{\pi}{6} \qquad \theta = \frac{5\pi}{6} \qquad \theta = \frac{\pi}{2}$$

Now work Problem 23.

When a trigonometric equation contains more than one trigonometric function, identities sometimes can be used to obtain an equivalent equation that contains only one trigonometric function.

EXAMPLE 8

Solving a Trigonometric Equation Using Identities

Solve the equation: $3 \cos \theta + 3 = 2 \sin^2 \theta, 0 \leq \theta < 2\pi$

Solution The equation in its present form contains sines and cosines. However, a form of the Pythagorean Identity can be used to transform the equation into an equivalent expression containing only cosines:

$$\begin{aligned} 3 \cos \theta + 3 &= 2 \sin^2 \theta \\ 3 \cos \theta + 3 &= 2(1 - \cos^2 \theta) && \sin^2 \theta = 1 - \cos^2 \theta \\ 3 \cos \theta + 3 &= 2 - 2 \cos^2 \theta \\ 2 \cos^2 \theta + 3 \cos \theta + 1 &= 0 && \text{Quadratic in } \cos \theta \\ (2 \cos \theta + 1)(\cos \theta + 1) &= 0 && \text{Factor} \\ 2 \cos \theta + 1 = 0 \quad &\text{or} \quad \cos \theta + 1 = 0 \\ \cos \theta = -\tfrac{1}{2} \quad & \quad \cos \theta = -1 \end{aligned}$$

Thus,

$$\theta = \frac{2\pi}{3} \qquad \theta = \frac{4\pi}{3} \qquad \theta = \pi$$

Check: Graph $Y_1 = 3 \cos x + 3$ and $Y_2 = 2 \sin^2 x, 0 \leq x \leq 2\pi$ and find the points of intersection. How close are your approximate solutions to the exact ones found in this example? ▬

EXAMPLE 9

Solving a Trigonometric Equation Using Identities

Solve the equation: $\cos 2\theta + 3 = 5 \cos \theta, 0 \leq \theta < 2\pi$

Solution First, we observe that the given equation contains two cosine functions, but with different arguments, θ and 2θ. We use the double-angle formula $\cos 2\theta = 2 \cos^2 \theta - 1$ to obtain an equivalent equation containing only $\cos \theta$:

$$\begin{aligned} \cos 2\theta + 3 &= 5 \cos \theta \\ (2 \cos^2 \theta - 1) + 3 &= 5 \cos \theta \\ 2 \cos^2 \theta - 5 \cos \theta + 2 &= 0 \\ (\cos \theta - 2)(2 \cos \theta - 1) &= 0 \\ \cos \theta = 2 \quad \text{or} \quad \cos \theta &= \tfrac{1}{2} \end{aligned}$$

For any angle θ, $-1 \leq \cos \theta \leq 1$; thus, the equation $\cos \theta = 2$ has no solution. The solutions of $\cos \theta = \frac{1}{2}, 0 \leq \theta < 2\pi$, are

$$\theta = \frac{\pi}{3} \qquad \theta = \frac{5\pi}{3}$$

Check: Graph $Y_1 = \cos(2x) + 3$ and $Y_2 = 5 \cos x, 0 \leq x \leq 2\pi$ and find the points of intersection. Compare your results with those of Example 9. ▬

Now work Problem 33.

EXAMPLE 10 Solving a Trigonometric Equation Using Identities

Solve the equation: $\cos^2 \theta + \sin \theta = 2, 0 \le \theta < 2\pi$

Solution We use a form of the Pythagorean identity:

$$\cos^2 \theta + \sin \theta = 2$$
$$(1 - \sin^2 \theta) + \sin \theta = 2 \quad \cos^2 \theta = 1 - \sin^2 \theta$$
$$\sin^2 \theta - \sin \theta + 1 = 0$$

FIGURE 13

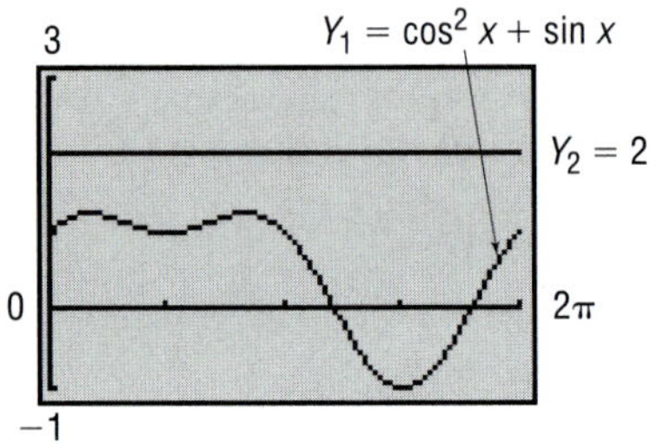

This is a quadratic equation in $\sin \theta$. The discriminant is $b^2 - 4ac = 1 - 4 = -3 < 0$. Therefore, the equation has no real solution.

Figure 13 shows the graphs of $Y_1 = \cos^2 x + \sin x$ and $Y_2 = 2$. The graphs do not intersect anywhere so the equation $Y_1 = Y_2$ has no real solution.

EXAMPLE 11 Solving a Trigonometric Equation Using Identities

Solve the equation: $\sin \theta \cos \theta = -\frac{1}{2}, 0 \le \theta < 2\pi$

Solution The left side of the given equation is in the form of the double-angle formula $2 \sin \theta \cos \theta = \sin 2\theta$, except for a factor of 2. Thus, we multiply each side by 2:

$$\sin \theta \cos \theta = -\frac{1}{2}$$
$$2 \sin \theta \cos \theta = -1$$
$$\sin 2\theta = -1$$

The argument here is 2θ. Thus, we need to write all the solutions of this equation and then list those that are in the interval $[0, 2\pi)$.

$$2\theta = \frac{3\pi}{2} + 2k\pi \quad k \text{ any integer.}$$
$$\theta = \frac{3\pi}{4} + k\pi$$

The solutions in the interval $[0, 2\pi)$ are

$$\theta = \frac{3\pi}{4} \qquad \theta = \frac{7\pi}{4}$$

Sometimes it is necessary to square both sides of an equation in order to obtain expressions that allow the use of identities. Remember, however, that when squaring both sides extraneous solutions may be introduced. As a result, apparent solutions must be checked.

EXAMPLE 12 Other Methods for Solving a Trigonometric Equation

Solve the equation: $\sin \theta + \cos \theta = 1, 0 \le \theta < 2\pi$

Solution A Attempts to use available identities do not lead to equations that are easy to solve. (Try it yourself.) Given the form of this equation, we decide to square each side:

$$\begin{aligned}
\sin\theta + \cos\theta &= 1\\
(\sin\theta + \cos\theta)^2 &= 1\\
\sin^2\theta + 2\sin\theta\cos\theta + \cos^2\theta &= 1\\
2\sin\theta\cos\theta &= 0 \qquad \sin^2\theta + \cos^2\theta = 1\\
\sin\theta\cos\theta &= 0
\end{aligned}$$

Thus,

$$\sin\theta = 0 \quad \text{or} \quad \cos\theta = 0$$

and the apparent solutions are

$$\theta = 0 \qquad \theta = \pi \qquad \theta = \frac{\pi}{2} \qquad \theta = \frac{3\pi}{2}$$

Because we squared both sides of the original equation, we must check these apparent solutions to see if any are extraneous:

$\theta = 0$: $\sin 0 + \cos 0 = 0 + 1 = 1$ — A solution.

$\theta = \pi$: $\sin\pi + \cos\pi = 0 + (-1) = -1$ — Not a solution.

$\theta = \dfrac{\pi}{2}$: $\sin\dfrac{\pi}{2} + \cos\dfrac{\pi}{2} = 1 + 0 = 1$ — A solution.

$\theta = \dfrac{3\pi}{2}$: $\sin\dfrac{3\pi}{2} + \cos\dfrac{3\pi}{2} = -1 + 0 = -1$ — Not a solution.

Thus, $\theta = 3\pi/2$ and $\theta = \pi$ are extraneous. The actual solutions are $\theta = 0$ and $\theta = \pi/2$.

We can solve the equation given in Example 12 another way.

Solution B We start with the equation

$$\sin\theta + \cos\theta = 1$$

and divide each side by $\sqrt{2}$. (The reason for this choice will become apparent shortly.) Then

$$\frac{1}{\sqrt{2}}\sin\theta + \frac{1}{\sqrt{2}}\cos\theta = \frac{1}{\sqrt{2}}$$

The left side now resembles the formula for the sine of the sum of two angles, one of which is θ. The other angle is unknown (call it ϕ.) Then

$$\sin(\theta + \phi) = \sin\theta\cos\phi + \cos\theta\sin\phi = \frac{1}{\sqrt{2}} \qquad (1)$$

where

$$\cos\phi = \frac{1}{\sqrt{2}} \qquad \sin\phi = \frac{1}{\sqrt{2}}, \qquad 0 \le \phi < 2\pi$$

The angle ϕ is therefore $\pi/4$. As a result, equation (1) becomes

$$\sin\left(\theta + \frac{\pi}{4}\right) = \frac{1}{\sqrt{2}}$$

We solve this equation to get

$$\theta + \frac{\pi}{4} = \frac{\pi}{4} \quad \text{or} \quad \theta + \frac{\pi}{4} = \frac{3\pi}{4}$$

$$\theta = 0 \qquad\qquad \theta = \frac{\pi}{2}$$

These solutions agree with the solutions found earlier. ■

This second method of solution can be used to solve any linear equation in the variables $\sin\theta$ and $\cos\theta$.

EXAMPLE 13 Solving a Trigonometric Equation Linear in $\sin\theta$ and $\cos\theta$

Solve:

$$a\sin\theta + b\cos\theta = c, \qquad 0 \le \theta < 2\pi \tag{2}$$

where a, b, and c are constants and either $a \ne 0$ or $b \ne 0$.

Solution We divide each side of equation (2) by $\sqrt{a^2+b^2}$. Then

$$\frac{a}{\sqrt{a^2+b^2}}\sin\theta + \frac{b}{\sqrt{a^2+b^2}}\cos\theta = \frac{c}{\sqrt{a^2+b^2}} \tag{3}$$

There is a unique angle ϕ, $0 \le \phi < 2\pi$, for which

$$\cos\phi = \frac{a}{\sqrt{a^2+b^2}} \quad \text{and} \quad \sin\phi = \frac{b}{\sqrt{a^2+b^2}} \tag{4}$$

FIGURE 14

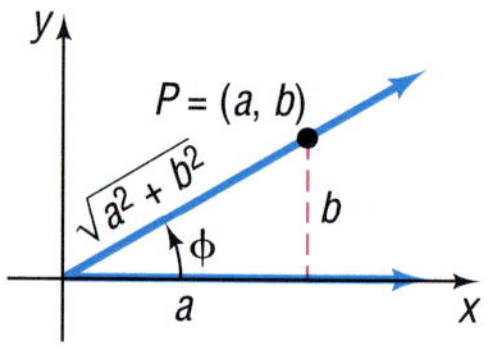

(see Figure 14). Thus, equation (3) may be written as

$$\sin\theta\cos\phi + \cos\theta\sin\phi = \frac{c}{\sqrt{a^2+b^2}}$$

or, equivalently,

$$\sin(\theta+\phi) = \frac{c}{\sqrt{a^2+b^2}} \tag{5}$$

where ϕ satisfies equations (4).

If $|c| > \sqrt{a^2+b^2}$, then $\sin(\theta+\phi) > 1$ or $\sin(\theta+\phi) < -1$, and equation (5) has no solution.

If $|c| \le \sqrt{a^2+b^2}$, then the solutions of equation (5) are

$$\theta + \phi = \sin^{-1}\frac{c}{\sqrt{a^2+b^2}} \quad \text{or} \quad \theta + \phi = \pi - \sin^{-1}\frac{c}{\sqrt{a^2+b^2}}$$

Because the angle ϕ is determined by equations (4), these are the solutions to equation (2). ■

Graphing Utility Solutions

The techniques introduced in this section apply only to certain types of trigonometric equations. Solutions for other types are usually studied in calculus, using numerical methods. In the next example, we show how a graphing utility may be used to obtain solutions.

EXAMPLE 14 Solving Trigonometric Equations Using a Graphing Utility

Solve: $5 \sin x + x = 3$

Express the solution(s) correct to two decimal places.

Solution This type of trigonometric equation cannot be solved by previous methods. A graphing utility, though, can be used here. The solution(s) of this equation are the same as the points of intersection of the graphs of $Y_1 = 5 \sin x + x$ and $Y_2 = 3$. See Figure 15.

FIGURE 15

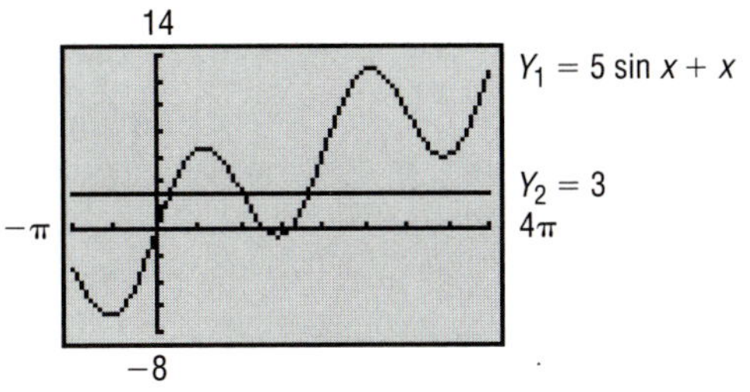

There are three points of intersection; the x-coordinates are the solutions we seek. Using INTERSECT, we find

$$x = 0.51 \qquad x = 3.17 \qquad x = 5.71$$

correct to two decimal places.

7.5 EXERCISES

In Problems 1–12, solve each equation on the interval $0 \le \theta < 2\pi$.

1. $\sin \theta = \frac{1}{2}$
2. $\tan \theta = 1$
3. $\tan \theta = -\frac{1}{\sqrt{3}}$
4. $\cos \theta = -\frac{\sqrt{3}}{2}$
5. $\cos \theta = 0$
6. $\sin \theta = \frac{\sqrt{2}}{2}$
7. $\sin 3\theta = -1$
8. $\tan \frac{\theta}{2} = \sqrt{3}$
9. $\cos\left(2\theta - \frac{\pi}{2}\right) = -1$
10. $\sin\left(3\theta + \frac{\pi}{18}\right) = 1$
11. $\sec \frac{3\theta}{2} = -2$
12. $\cot \frac{2\theta}{3} = -\sqrt{3}$

In Problems 13–20, Use a calculator to solve each equation on the interval $0 \le \theta < 2\pi$.

13. $\sin \theta = 0.4$
14. $\cos \theta = 0.6$
15. $\tan \theta = 5$
16. $\cot \theta = 2$
17. $\cos \theta = -0.9$
18. $\sin \theta = -0.2$
19. $\sec \theta = -4$
20. $\csc \theta = -3$

In Problems 21–50, solve each equation on the interval $0 \le \theta < 2\pi$. Verify your result using a graphing utility.

21. $2 \cos^2 \theta + \cos \theta = 0$
22. $\sin^2 \theta - 1 = 0$
23. $2 \sin^2 \theta - \sin \theta - 1 = 0$
24. $2 \cos^2 \theta + \cos \theta - 1 = 0$
25. $(\tan \theta - 1)(\sec \theta - 1) = 0$
26. $(\cot \theta + 1)(\csc \theta - \frac{1}{2}) = 0$
27. $\cos \theta = \sin \theta$
28. $\cos \theta + \sin \theta = 0$
29. $\tan \theta = 2 \sin \theta$
30. $\sin 2\theta = \cos \theta$
31. $\sin \theta = \csc \theta$
32. $\tan \theta = \cot \theta$
33. $\cos 2\theta = \cos \theta$
34. $\sin 2\theta \sin \theta = \cos \theta$
35. $\sin 2\theta + \sin 4\theta = 0$
36. $\cos 2\theta + \cos 4\theta = 0$
37. $\cos 4\theta - \cos 6\theta = 0$
38. $\sin 4\theta - \sin 6\theta = 0$

39. $1 + \sin\theta = 2\cos^2\theta$

40. $\sin^2\theta = 2\cos\theta + 2$

41. $\tan^2\theta = \frac{3}{2}\sec\theta$

42. $\csc^2\theta = \cot\theta + 1$

43. $3 - \sin\theta = \cos 2\theta$

44. $\cos 2\theta + 5\cos\theta + 3 = 0$

45. $\sec^2\theta + \tan\theta = 0$

46. $\sec\theta = \tan\theta + \cot\theta$

47. $\sin\theta - \sqrt{3}\cos\theta = 1$

48. $\sqrt{3}\sin\theta + \cos\theta = 1$

49. $\tan 2\theta + 2\sin\theta = 0$

50. $\tan 2\theta + 2\cos\theta = 0$

In Problems 51–56, solve each equation for $-\pi \le x \le \pi$. *Express the solution(s) correct to two decimal places.*

51. Solve the equation $\cos x = e^x$ by graphing $Y_1 = \cos x$ and $Y_2 = e^x$ and finding their point(s) of intersection.

52. Solve the equation $\cos x = e^x$ by graphing $Y_1 = \cos x - e^x$ and finding the x-intercept(s).

53. Solve the equation $2\sin x = 0.7x$ by graphing $Y_1 = 2\sin x$ and $Y_2 = 0.7x$ and finding their point(s) of intersection.

54. Solve the equation $2\sin x = 0.7x$ by graphing $Y_1 = 2\sin x - 0.7x$ and finding the x-intercept(s).

55. Solve the equation $\cos x = x^2$ by graphing $Y_1 = \cos x$ and $Y_2 = x^2$ and finding their point(s) of intersection.

56. Solve the equation $\cos x = x^2$ by graphing $Y_1 = \cos x - x^2$ and finding the x-intercept(s).

In Problems 57–68, use a graphing utility to solve each equation. Express the solution(s) correct to two decimal places.

57. $x + 5\cos x = 0$

58. $x - 4\sin x = 0$

59. $22x - 17\sin x = 3$

60. $19x + 8\cos x = 2$

61. $\sin x + \cos x = x$

62. $\sin x - \cos x = x$

63. $x^2 - 2\cos x = 0$

64. $x^2 + 3\sin x = 0$

65. $x^2 - 2\sin 2x = 3x$

66. $x^2 = x + 3\cos 2x$

67. $6\sin x - e^x = 2, x > 0$

68. $4\cos 3x - e^x = 1, x > 0$

69. Constructing a Rain Gutter A rain gutter is to be constructed of aluminum sheets 12 inches wide. After marking off a length of 4 inches from each edge, this length is bent up at an angle θ. See the illustration. The area A of the opening as a function of θ is given by

$$A = 16\sin\theta(\cos\theta + 1),\ 0° < \theta < 90°$$

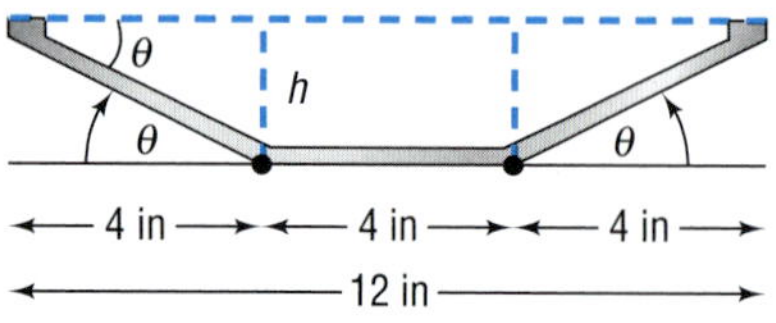

(a) In calculus, you will be asked to find the angle θ that maximizes A by solving the equation

$$\cos 2\theta + \cos\theta = 0,\ 0° < \theta < 90°$$

Solve this equation for θ by using the double-angle formula.

(b) Solve the equation for θ by writing the sum of the two cosines as a product.

(c) What is the maximum area A of the opening?

(d) Graph A, $0° \le \theta \le 90°$, and find the angle θ that maximizes the area A. Also find the maximum area. Compare the results to the answers found earlier.

70. Projectile Motion An object is propelled upward at an angle θ, $45° < \theta < 90°$, to the horizontal with an initial velocity of v_0 feet per second from the base of a plane that makes an angle of 45° with the horizontal. See the illustration. If air resistance is ignored, the distance R it travels up the inclined plane is given by

$$R = \frac{v_0^2\sqrt{2}}{32}(\sin 2\theta - \cos 2\theta - 1)$$

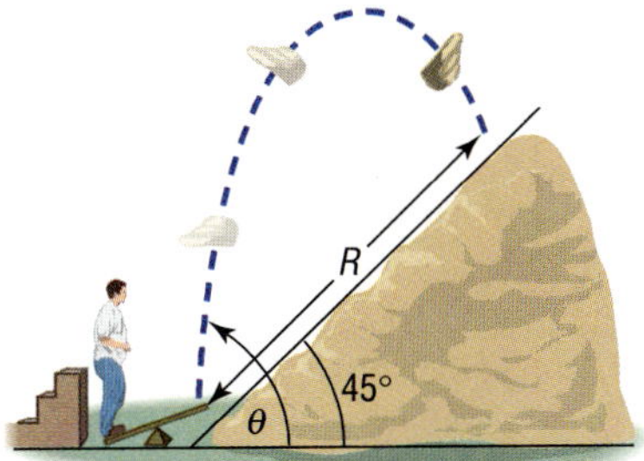

(a) In calculus, you will be asked to find the angle θ that maximizes R by solving the equation

$$\sin 2\theta + \cos 2\theta = 0$$

Solve this equation for θ using the method of Example 13.

(b) Solve this equation for θ by dividing each side by $\cos 2\theta$.

(c) What is the maximum distance R if $v_0 = 32$ feet per second?

(d) Graph R, $45° \le \theta \le 90°$, and find the angle θ that maximizes the distance R. Also find the maximum distance. Use $v_0 = 32$ feet per second. Compare the results with the answers found earlier.

71. Heat Transfer In the study of heat transfer, the equation $x + \tan x = 0$ occurs. Graph $Y_1 = -x$ and $Y_2 = \tan x$ for $x \ge 0$. Conclude that there are an infinite number of points of intersection of these two graphs. Now find the first two positive solutions of $x + \tan x = 0$ correct to two decimal places.

72. Carrying a Ladder around a Corner A ladder of length L is carried horizontally around a corner from a hall 3 feet wide into a hall 4 feet wide. See the illustration.

(a) Express L as a function of θ.

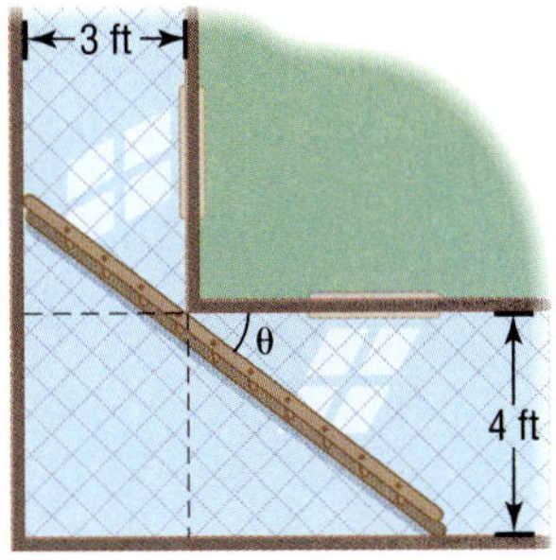

(b) In calculus, you will be asked to find the length of the longest ladder that can turn the corner by solving the equation

$$3 \sec \theta \tan \theta - 4 \csc \theta \cot \theta = 0,$$
$$0° < \theta < 90°$$

Solve this equation for θ.

(c) What is the length of the largest ladder that can be carried around the corner?

(d) Graph L, $0° \le \theta \le 90°$, and find the angle θ that maximizes the length L. Also find the maximum length. Compare the results with the ones found in parts (b) and (c).

73. Projectile Motion The horizontal distance a projectile will travel in the air is given by the equation

$$R = \frac{v_0^2 \sin 2\theta}{g}$$

where v_0 is the initial velocity of the projectile, θ is the angle of elevation, and g is acceleration due to gravity (9.8 m/sec^2).

(a) If you can throw a baseball with an initial speed of 34.8 m/sec, at what angle of elevation θ should you direct the throw so that the ball travels a distance of 107 m before striking the ground.

(b) Determine the maximum distance you can throw the ball.

(c) Graph R, with $v_0 = 34.8$ m/sec.

(d) Verify the results obtained in parts (a) and (b) using ZERO or ROOT.

74. Projectile Motion Refer to Problem 73.

(a) If you can throw a baseball with an initial speed of 40 m/sec, at what angle of elevation θ should you direct the throw so that the ball travels a distance of 110 m before striking the ground.

(b) Determine the maximum distance you can throw the ball.

(c) Graph R, with $v_0 = 40$ m/sec.

(d) Verify the results obtained in parts (a) and (b) using ZERO or ROOT.

*The following discussion of **Snell's Law of Refraction** (named after Willebrod Snell, 1591–1626) is needed for Problems 75–81: Light, sound, and other waves travel at different speeds, depending on the media (air, water, wood, and so on) through which they pass. Suppose that light travels from a point A in one medium, where its speed is v_1, to a point B in another medium, where its speed is v_2. Refer to the figure, where the angle θ_1 is called the **angle of incidence** and the angle θ_2 is the **angle of refraction.** Snell's Law* which can be proved using calculus, states that*

$$\frac{\sin \theta_1}{\sin \theta_2} = \frac{v_1}{v_2}$$

*The ratio v_1/v_2 is called the **index of refraction.** Some values are given in the following table.*

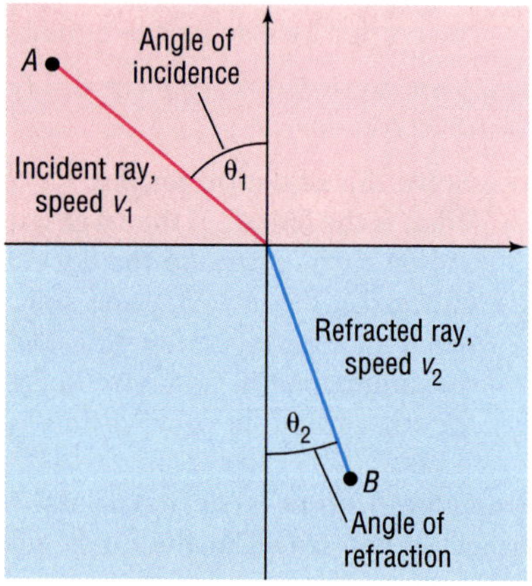

Some Indexes of Refraction

Medium	Index of Refraction
Water	1.33
Ethyl alcohol	1.36
Carbon bisulfide	1.63
Air (1 atm and 20°C)	1.0003
Methylene iodide	1.74
Fused quartz	1.46
Glass, crown	1.52
Glass, dense flint	1.66
Sodium chloride	1.53

For light of wavelength 589 nanometers, measured with respect to a vacuum. The index with respect to air is negligibly different in most cases.

75. The index of refraction of light in passing from a vacuum into water is 1.33. If the angle of incidence is 40°, determine the angle of refraction.

76. The index of refraction of light in passing from a vacuum into dense glass is 1.66. If the angle of incidence is 50°, determine the angle of refraction.

77. Ptolemy, who lived in the city of Alexandria in Egypt during the second century A.D., gave the measured values in the table below for the angle of incidence θ_1 and the angle of refraction θ_2 for a light beam passing from air into water. Do these values agree with Snell's Law? If so, what index of refraction results? (These data are interesting as the oldest recorded physical measurements.)†

θ_1	θ_2
10°	7°45′
20°	15°30′
30°	22°30′
40°	29°0′
50°	35°0′
60°	40°30′
70°	45°30′
80°	50°0′

78. The speed of yellow sodium light (wavelength of 589 nanometers) in a certain liquid is measured to be 1.92×10^8 meters per second. What is the index of refraction of this liquid, with respect to air, for sodium light?‡

79. A beam of light with a wavelength of 589 nanometers traveling in air makes an angle of incidence of 40° upon a slab of transparent material, and the refracted beam makes an angle of refraction of 26°. Find the index of refraction of the material.‡

80. A light ray with a wavelength of 589 nanometers (produced by a sodium lamp) traveling through air makes an angle of incidence of 30° on a smooth, flat slab of crown glass. Find the angle of refraction.‡

81. A light beam passes from one medium to another through a thick slab of material whose index of refraction is n_2. Show that the emerging beam is parallel to the incident beam.‡

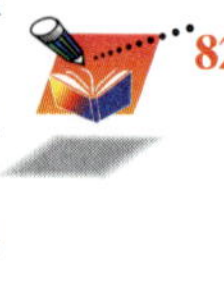

82. Explain in your own words how you would use your calculator to solve the equation $\sin x = 0.3, 0 \le x < 2\pi$. How would you modify your approach in order to solve the equation $\cot x = 5, 0 < x < 2\pi$?

*Because this law was also deduced by René Descartes, in France it is also known as Descartes' Law.

†Adapted from Halliday and Resnick, *Physics, Parts 1 & 2,* 3rd ed. New York: Wiley, 1978, p. 953.

‡Adapted from Serway, *Physics,* 3rd ed. Philadelphia: W.B. Saunders, p. 805.

CHAPTER REVIEW

THINGS TO KNOW

Formulas

Sum and Difference Formulas

$$\cos(\alpha + \beta) = \cos\alpha\cos\beta - \sin\alpha\sin\beta$$
$$\cos(\alpha - \beta) = \cos\alpha\cos\beta + \sin\alpha\sin\beta$$
$$\sin(\alpha + \beta) = \sin\alpha\cos\beta + \cos\alpha\sin\beta$$
$$\sin(\alpha - \beta) = \sin\alpha\cos\beta - \cos\alpha\sin\beta$$
$$\tan(\alpha + \beta) = \frac{\tan\alpha + \tan\beta}{1 - \tan\alpha\tan\beta}$$
$$\tan(\alpha - \beta) = \frac{\tan\alpha - \tan\beta}{1 + \tan\alpha\tan\beta}$$

Double-Angle Formulas

$$\sin 2\theta = 2\sin\theta\cos\theta$$
$$\cos 2\theta = \cos^2\theta - \sin^2\theta$$
$$\cos 2\theta = 1 - 2\sin^2\theta$$
$$\cos 2\theta = 2\cos^2\theta - 1$$
$$\tan 2\theta = \frac{2\tan\theta}{1 - \tan^2\theta}$$

Half-Angle Formulas

$$\sin^2\frac{\alpha}{2} = \frac{1 - \cos\alpha}{2}$$
$$\cos^2\frac{\alpha}{2} = \frac{1 + \cos\alpha}{2}$$
$$\tan^2\frac{\alpha}{2} = \frac{1 - \cos\alpha}{1 + \cos\alpha}$$
$$\sin\frac{\alpha}{2} = \pm\sqrt{\frac{1 - \cos\alpha}{2}}$$
$$\cos\frac{\alpha}{2} = \pm\sqrt{\frac{1 + \cos\alpha}{2}}$$
$$\tan\frac{\alpha}{2} = \pm\sqrt{\frac{1 - \cos\alpha}{1 + \cos\alpha}} = \frac{1 - \cos\alpha}{\sin\alpha} = \frac{\sin\alpha}{1 + \cos\alpha}$$

Where the + or − sign is determined by the quadrant of the angle $\alpha/2$.

Product-to-Sum Formulas

$$\sin\alpha\sin\beta = \tfrac{1}{2}[\cos(\alpha - \beta) - \cos(\alpha + \beta)]$$
$$\cos\alpha\cos\beta = \tfrac{1}{2}[\cos(\alpha - \beta) + \cos(\alpha + \beta)]$$
$$\sin\alpha\cos\beta = \tfrac{1}{2}[\sin(\alpha + \beta) + \sin(\alpha - \beta)]$$

Sum-to-Product Formulas

$$\sin\alpha + \sin\beta = 2\sin\frac{\alpha + \beta}{2}\cos\frac{\alpha - \beta}{2}$$
$$\sin\alpha - \sin\beta = 2\sin\frac{\alpha - \beta}{2}\cos\frac{\alpha + \beta}{2}$$
$$\cos\alpha + \cos\beta = 2\cos\frac{\alpha + \beta}{2}\cos\frac{\alpha - \beta}{2}$$
$$\cos\alpha - \cos\beta = -2\sin\frac{\alpha + \beta}{2}\sin\frac{\alpha - \beta}{2}$$

HOW TO

Establish identities

Solve a trigonometric equation algebraically and graphically

FILL-IN-THE-BLANK ITEMS

1. Suppose that f and g are two functions with the same domain. If $f(x) = g(x)$ for every x in the domain, the equation is called a(n) ______________. Otherwise, it is called a(n) ______________ equation.
2. $\cos(\alpha + \beta) = \cos\alpha\cos\beta$ ______________ $\sin\alpha\sin\beta$
3. $\sin(\alpha + \beta) = \sin\alpha\cos\beta$ ______________ $\cos\alpha\sin\beta$
4. $\cos 2\theta = \cos^2\theta -$ ______________ $=$ ______________ $- 1 = 1 -$ ______________
5. $\sin^2\dfrac{\alpha}{2} = \dfrac{\quad}{2}$

TRUE/FALSE ITEMS

T F 1. $\sin(-\theta) + \sin\theta = 0$ for all θ.

T F 2. $\sin(\alpha + \beta) = \sin\alpha + \sin\beta + 2\sin\alpha\sin\beta$.

T F 3. $\cos 2\theta$ has three equivalent forms: $\cos^2\theta - \sin^2\theta$, $1 - 2\sin^2\theta$, and $2\cos^2\theta - 1$.

T F 4. $\cos\dfrac{\alpha}{2} = \pm\dfrac{\sqrt{1 + \cos\alpha}}{2}$, where the + or − sign depends on the angle $\alpha/2$.

T F 5. Most trigonometric equations have unique solutions.

T F 6. The equation $\tan\theta = \pi/2$ has no solution.

REVIEW EXERCISES

In Problems 1–32, establish each identity.

1. $\tan\theta\cot\theta - \sin^2\theta = \cos^2\theta$
2. $\sin\theta\csc\theta - \sin^2\theta = \cos^2\theta$
3. $\cos^2\theta(1 + \tan^2\theta) = 1$
4. $(1 - \cos^2\theta)(1 + \cot^2\theta) = 1$
5. $4\cos^2\theta + 3\sin^2\theta = 3 + \cos^2\theta$
6. $4\sin^2\theta + 2\cos^2\theta = 4 - 2\cos^2\theta$
7. $\dfrac{1 - \cos\theta}{\sin\theta} + \dfrac{\sin\theta}{1 - \cos\theta} = 2\csc\theta$
8. $\dfrac{\sin\theta}{1 + \cos\theta} + \dfrac{1 + \cos\theta}{\sin\theta} = 2\csc\theta$
9. $\dfrac{\cos\theta}{\cos\theta - \sin\theta} = \dfrac{1}{1 - \tan\theta}$
10. $1 - \dfrac{\cos^2\theta}{1 + \sin\theta} = \sin\theta$
11. $\dfrac{\csc\theta}{1 + \csc\theta} = \dfrac{1 - \sin\theta}{\cos^2\theta}$
12. $\dfrac{1 + \sec\theta}{\sec\theta} = \dfrac{\sin^2\theta}{1 - \cos\theta}$
13. $\csc\theta - \sin\theta = \cos\theta\cot\theta$
14. $\dfrac{\csc\theta}{1 - \cos\theta} = \dfrac{1 + \cos\theta}{\sin^3\theta}$
15. $\dfrac{1 - \sin\theta}{\sec\theta} = \dfrac{\cos^3\theta}{1 + \sin\theta}$
16. $\dfrac{1 - \cos\theta}{1 + \cos\theta} = (\csc\theta - \cot\theta)^2$
17. $\dfrac{1 - 2\sin^2\theta}{\sin\theta\cos\theta} = \cot\theta - \tan\theta$
18. $\dfrac{(2\sin^2\theta - 1)^2}{\sin^4\theta - \cos^4\theta} = 1 - 2\cos^2\theta$
19. $\dfrac{\cos(\alpha + \beta)}{\cos\alpha\sin\beta} = \cot\beta - \tan\alpha$
20. $\dfrac{\sin(\alpha - \beta)}{\sin\alpha\cos\beta} = 1 - \cot\alpha\tan\beta$
21. $\dfrac{\cos(\alpha - \beta)}{\cos\alpha\cos\beta} = 1 + \tan\alpha\tan\beta$
22. $\dfrac{\cos(\alpha + \beta)}{\sin\alpha\cos\beta} = \cot\alpha - \tan\beta$
23. $(1 + \cos\theta)\left(\tan\dfrac{\theta}{2}\right) = \sin\theta$
24. $\sin\theta\tan\dfrac{\theta}{2} = 1 - \cos\theta$
25. $2\cot\theta\cot 2\theta = \cot^2\theta - 1$
26. $2\sin 2\theta(1 - 2\sin^2\theta) = \sin 4\theta$
27. $1 - 8\sin^2\theta\cos^2\theta = \cos 4\theta$
28. $\dfrac{\sin 3\theta\cos\theta - \sin\theta\cos 3\theta}{\sin 2\theta} = 1$

29. $\dfrac{\sin 2\theta + \sin 4\theta}{\cos 2\theta + \cos 4\theta} = \tan 3\theta$

30. $\dfrac{\sin 2\theta + \sin 4\theta}{\sin 2\theta - \sin 4\theta} + \dfrac{\tan 3\theta}{\tan \theta} = 0$

31. $\dfrac{\cos 2\theta - \cos 4\theta}{\cos 2\theta + \cos 4\theta} - \tan\theta \tan 3\theta = 0$

32. $\cos 2\theta - \cos 10\theta = (\tan 4\theta)(\sin 2\theta + \sin 10\theta)$

In Problems 33–40, find the exact value of each expression.

33. $\sin 165°$

34. $\tan 105°$

35. $\cos \dfrac{5\pi}{12}$

36. $\sin\left(-\dfrac{\pi}{12}\right)$

37. $\cos 80° \cos 20° + \sin 80° \sin 20°$

38. $\sin 70° \cos 40° - \cos 70° \sin 40°$

39. $\tan \dfrac{\pi}{8}$

40. $\sin \dfrac{5\pi}{8}$

In Problems 41–50, use the information given about the angles α and β to find the exact value of:

(*a*) $\sin(\alpha + \beta)$ (*b*) $\cos(\alpha + \beta)$ (*c*) $\sin(\alpha - \beta)$ (*d*) $\tan(\alpha + \beta)$

(*e*) $\sin 2\alpha$ (*f*) $\cos 2\beta$ (*g*) $\sin \dfrac{\beta}{2}$ (*h*) $\cos \dfrac{\alpha}{2}$

41. $\sin\alpha = \frac{4}{5}, 0 < \alpha < \pi/2;\quad \sin\beta = \frac{5}{13}, \pi/2 < \beta < \pi$

42. $\cos\alpha = \frac{4}{5}, 0 < \alpha < \pi/2;\quad \cos\beta = \frac{5}{13}, -\pi/2 < \beta < 0$

43. $\sin\alpha = -\frac{3}{5}, \pi < \alpha < 3\pi/2;\quad \cos\beta = \frac{12}{13}, 3\pi/2 < \beta < 2\pi$

44. $\sin\alpha = -\frac{4}{5}, -\pi/2 < \alpha < 0;\quad \cos\beta = -\frac{5}{13}, \pi/2 < \beta < \pi$

45. $\tan\alpha = \frac{3}{4}, \pi < \alpha < 3\pi/2;\quad \tan\beta = \frac{12}{5}, 0 < \beta < \pi/2$

46. $\tan\alpha = -\frac{4}{3}, \pi/2 < \alpha < \pi;\quad \cot\beta = \frac{12}{5}, \pi < \beta < 3\pi/2$

47. $\sec\alpha = 2, -\pi/2 < \alpha < 0;\quad \sec\beta = 3, 3\pi/2 < \beta < 2\pi$

48. $\csc\alpha = 2, \pi/2 < \alpha < \pi;\quad \sec\beta = -3, \pi/2 < \beta < \pi$

49. $\sin\alpha = -\frac{2}{3}, \pi < \alpha < 3\pi/2;\quad \cos\beta = -\frac{2}{3}, \pi < \beta < 3\pi/2$

50. $\tan\alpha = -2, \pi/2 < \alpha < \pi;\quad \cot\beta = -2, \pi/2 < \beta < \pi$

In Problems 51–70, solve each equation on the interval $0 \le \theta < 2\pi$. Use a graphing utility to verify your answer.

51. $\cos\theta = \frac{1}{2}$

52. $\sin\theta = -\sqrt{3}/2$

53. $\cos\theta = -\sqrt{2}/2$

54. $\tan\theta = -\sqrt{3}$

55. $\sin 2\theta = -1$

56. $\cos 2\theta = 0$

57. $\tan 2\theta = 0$

58. $\sin 3\theta = 1$

59. $\sin\theta = 0.9$

60. $\tan\theta = 25$

61. $\sin\theta = \tan\theta$

62. $\cos\theta = \sec\theta$

63. $\sin\theta + \sin 2\theta = 0$

64. $\cos 2\theta = \sin\theta$

65. $\sin 2\theta - \cos\theta - 2\sin\theta + 1 = 0$

66. $\sin 2\theta - \sin\theta - 2\cos\theta + 1 = 0$

67. $2\sin^2\theta - 3\sin\theta + 1 = 0$

68. $2\cos^2\theta + \cos\theta - 1 = 0$

69. $\sin\theta - \cos\theta = 1$

70. $\sin\theta + 2\cos\theta = 1$

In Problems 71–76, use a graphing utility to solve each equation on the interval $0 \le x \le 2\pi$. Approximate any solutions correct to two decimal places.

71. $2x = 5\cos x$

72. $2x = 5\sin x$

73. $2\sin x + 3\cos x = 4x$

74. $3\cos x + x = \sin x$

75. $\sin x = \ln x$

76. $\sin x = e^{-x}$

CHAPTER 8

Applications of Trigonometric Functions

Alternative energy production is being studied and put into practice throughout the world. On the following page you will find your project. You will need to use the Sullivan website at:

www.prenhall.com/sullivan

to help link you to the Internet resources needed to prepare your report.

PREPARING FOR THIS CHAPTER

Before getting started on this chapter, review the following concepts:

Pythagorean Theorem *(Appendix, Section 1, p. 982)*

Right Triangle Trigonometry *(Chapter 6; Section 6.2)*

OUTLINE

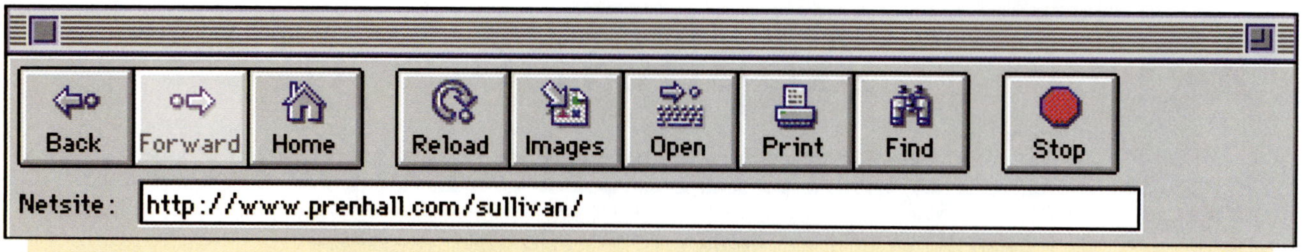

TIDAL ENERGY: MAKING ELECTRICITY FROM OCEAN TIDES

Using tidal flows to generate electrical power seems a viable means of alternative energy production. Suppose you are the project manager for an engineering firm and the city of Hilo, HI. has contacted your firm to do a feasibility study to determine whether or not tidal energy is a practical method for generating electrical power for the city. You begin your study by doing an Internet search on "tidal energy conversion." Your plan is to model Hilo Harbor tides by a mathematical function. You know by Harmonic Analysis of the tidal oscillations that a good model is a function of the form:

$$y = A\sin(bx + c) + D\sin(ex + f)$$

The constants for A, b, c, D, e, f can be calculated using data found at the *School of Ocean and Earth Science and Technology* (SOEST). NOTE: An explanation of Harmonic Analysis can be found at: *The Harmonic Analysis Tutorial;* information on a tidal energy conversion experiment can be found at: *The TOPEX/Poseidon Project*

1. What are the advantages/disadvantages of using tidal energy to produce electrical power? (Clean energy, reduces coal burning . . . only works part time . . . etc.) What are some of the different methods available for tidal energy conversion? (Dams, floating platforms . . .)
2. From the Topex Ocean Tides graph at the SOEST website, does Hawaii seem like a good location for a tidal energy plant?
3. By looking at the SOEST Tide Chart for Hilo Harbor, can you estimate how many hours per day the power plant would produce electricity? What additional information do you need?
4. Using the data from SOEST, determine a function that models the tide at Hilo Harbor. Graph this function. How does your function compare with *the NOAA tide chart*?
5. By using a dam type of power generating plant, your plant will be able to generate 1 Kw per hour if the difference between the high and low tides is 6 meters or more. How many days per month would the plant be able to generate 1 Kw per hour at some time during the day? Use the function found in question 4.
6. If a dam type of power plant could produce electrical power for only 70 percent of the time between high and low tides, how many Kws could be generated in one day if the tides differ by more than 6 meters?
7. What other information would you need before you could determine the monthly electrical output for a Hilo Harbor Tidal Conversion Facility? If 70 percent of the time of the incoming and outgoing tides you would be able to generate 1 Kw of power per hour, how many KWs of power would the plant produce in 1 month? Use the function found in question 4.
8. Before you prepare your final report for the city of Hilo you might want to take a look at the website of the *Office of Scientific and Technical Information (OSTI), an office of the Department of Energy.*

In this chapter we use the trigonometric functions to solve applied problems. The first four sections deal with applications involving right triangles and *oblique triangles,* triangles that do not have a right angle. To solve problems involving oblique triangles, we will develop the Law of Sines and the Law of Cosines. We will also develop formulas for finding the area of a triangle.

The final two sections deal with applications involving the graphs of the sine and cosine functions, simple harmonic motion, and damped motion.

8.1 SOLVING RIGHT TRIANGLES

1 Solve Right Triangles

1 In the discussion that follows, we will always label a right triangle so that side a is opposite angle α, side b is opposite angle β, and side c is the hypotenuse, as shown in Figure 1. **To solve a right triangle** means to find the missing lengths of its sides and the measurements of its angles. We shall follow the practice of expressing the lengths of the sides rounded to two decimal places and of expressing angles in degrees rounded to one decimal place.

FIGURE 1

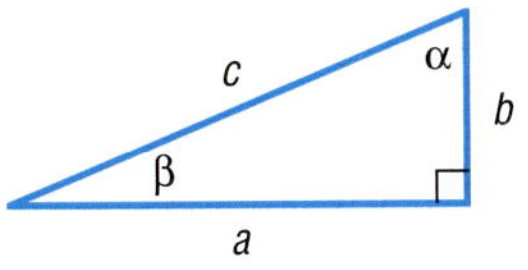

To solve a right triangle, we need to know either an angle (besides the 90° one) and a side or else two sides. Then we make use of the Pythagorean Theorem and the fact that the sum of the angles of a triangle is 180°. The sum of the unknown angles in a right triangle is therefore 90°. Thus, for the right triangle shown in Figure 1, we have

$$c^2 = a^2 + b^2 \qquad \alpha + \beta = 90°$$

EXAMPLE 1 Solving a Right Triangle

FIGURE 2

Use Figure 2. If $b = 2$ and $\alpha = 40°$, find a, c, and β.

Solution Since $\alpha = 40°$ and $\alpha + \beta = 90°$, we find that $\beta = 50°$. To find the sides a and c, we use the facts that

$$\tan 40° = \frac{a}{2} \qquad \text{and} \qquad \cos 40° = \frac{2}{c}$$

Now solve for a and c.

$$a = 2 \tan 40° \approx 1.68 \quad \text{and} \qquad c = \frac{2}{\cos 40°} \approx 2.61$$

Now work Problem 1.

EXAMPLE 2 Solving a Right Triangle

FIGURE 3

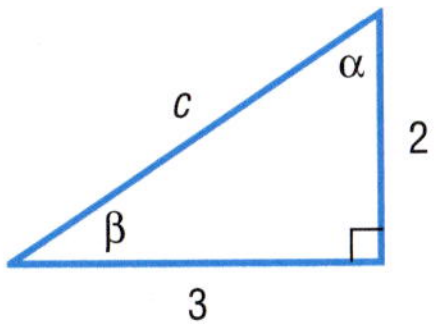

Use Figure 3. If $a = 3$ and $b = 2$, find c, α, and β.

Solution Since $a = 3$ and $b = 2$, then by the Pythagorean Theorem we have

$$c^2 = a^2 + b^2 = 9 + 4 = 13$$
$$c = \sqrt{13} \approx 3.61$$

To find angle α, we use the fact that

$$\tan \alpha = \frac{3}{2}$$

Set the mode on your calculator to degrees. Then, rounded to one decimal place, we find $\alpha = 56.3°$. Since $\alpha + \beta = 90°$, we find that $\beta = 33.7°$.

Note: In subsequent examples and in the exercises that follow, unless otherwise indicated, we will measure angles in degrees and round off to one decimal place; we will round off all sides to two decimal places. To avoid round-off errors when using a calculator, we will store unrounded values in memory for use in subsequent calculations.

Now work Problem 11.

One common use for trigonometry is to measure heights and distances that are either awkward or impossible to measure by ordinary means.

EXAMPLE 3 Finding the Width of a River

A surveyor can measure the width of a river by setting up a transit* at a point C on one side of the river and taking a sighting of a point A on the other side. Refer to Figure 4. After turning through an angle of 90° at C, the surveyor walks a distance of 200 meters to point B. Using the transit at B, the angle β is measured and found to be 20°. What is the width of the river?

FIGURE 4

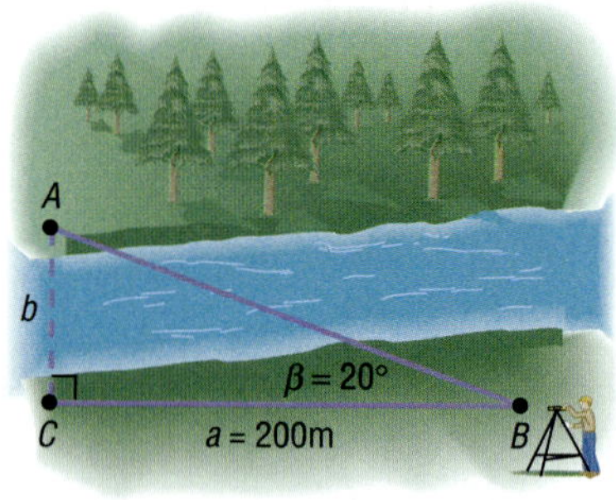

Solution We seek the length of side b. We know a and β, so we use the fact that

$$\tan \beta = \frac{b}{a}$$

to get

$$\tan 20° = \frac{b}{200}$$

$$b = 200 \tan 20° \approx 72.79 \text{ meters}$$

Thus, the width of the river is approximately 73 meters, rounded to the nearest meter. ■

Now work Problem 21.

EXAMPLE 4 Finding the Inclination of a Mountain Trail

A straight trail with a uniform inclination leads from a hotel, elevation 8000 feet, to a scenic overlook, elevation 11,100 feet. The length of the trail is 14,100 feet. What is the inclination (grade) of the trail?

FIGURE 5

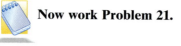

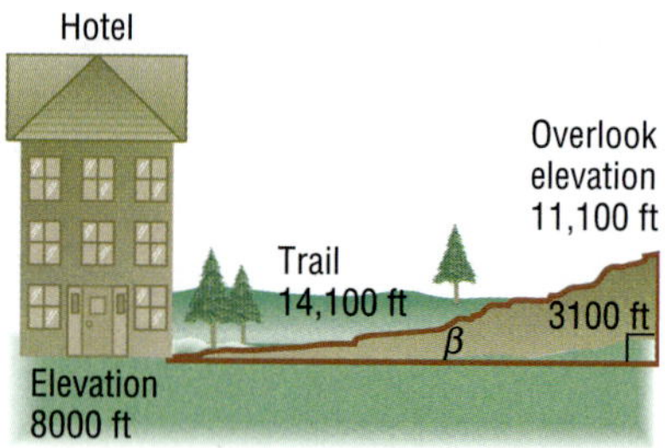

Solution Figure 5 illustrates the situation. We seek the angle β. As the illustration shows,

$$\sin \beta = \frac{3100}{14{,}100}$$

Using a calculator,

$$\beta \approx 12.7°$$

The inclination (grade) of the trail is approximately 12.7°. ■

Vertical heights can sometimes be measured using either the angle of elevation or the angle of depression. If a person is looking up at an object, the

*An instrument used in surveying to measure angles.

acute angle measured from the horizontal to a line-of-sight observation of the object is called the **angle of elevation.** See Figure 6(a).

If a person is standing on a cliff looking down at an object, the acute angle made by the line of sight observation of the object and the horizontal is called the **angle of depression.** See Figure 6(b).

FIGURE 6

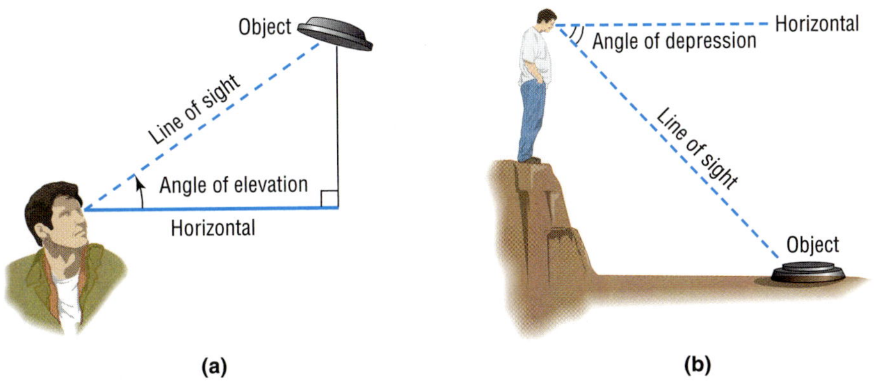

EXAMPLE 5

Finding Heights Using the Angle of Elevation

To determine the height of a radio transmission tower, a surveyor walks off a distance of 300 meters from the base of the tower. The angle of elevation is then measured and found to be 40°. If the transit is 2 meters off the ground when the sighting is taken, how high is the radio tower? See Figure 7(a).

FIGURE 7

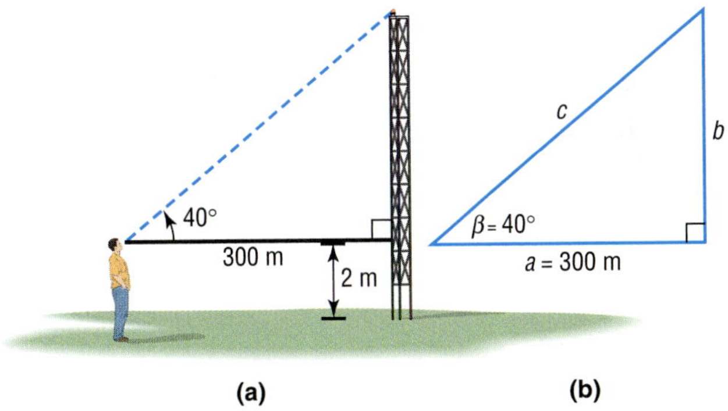

Solution Figure 7(b) shows a triangle that replicates the illustration in Figure 7(a). To find the length b, we use the fact that $\tan \beta = b/a$. Then

$$b = a \tan \beta = 300 \tan 40° = 251.73 \text{ meters}$$

Because the transit is 2 meters high, the actual height of the tower is approximately 254 meters, rounded to the nearest meter. ■

Now work Problem 23.

The idea behind Example 5 can also be used to find the height of an object with a base that is not accessible to the horizontal.

EXAMPLE 6 Finding the Height of a Statue on a Building

Adorning the top of the Board of Trade building in Chicago is a statue of the Greek goddess Ceres, goddess of wheat. From street level, two observations are taken 400 feet from the center of the building. The angle of elevation to the base of the statue is found to be 45°; the angle of elevation to the top of the statue is 47.2°. See Figure 8(a). What is the height of the statue?

FIGURE 8

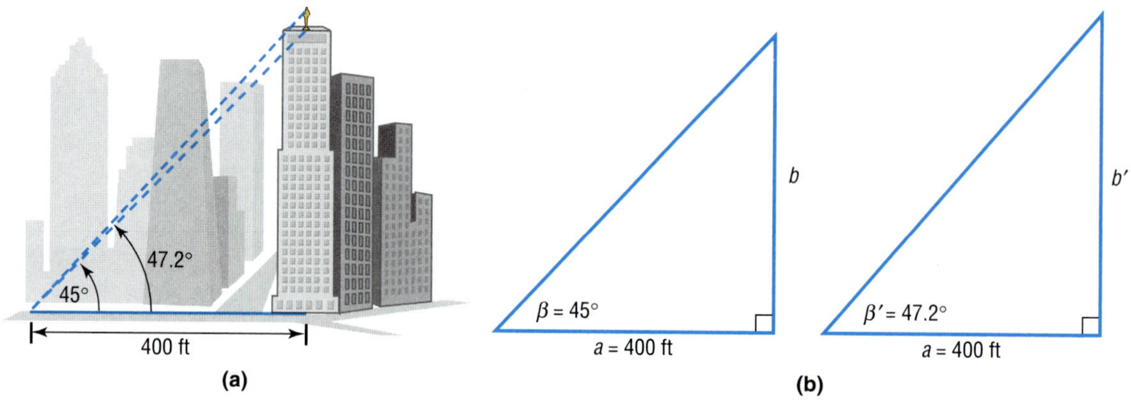

Solution Figure 8(b) shows two triangles that replicate Figure 8(a). The height of the statue of Ceres will be $b' - b$. To find b and b', we refer to Figure 8(b):

$$\tan 45° = \frac{b}{400} \qquad \tan 47.2° = \frac{b'}{400}$$

$$b = 400 \tan 45° = 400 \qquad b' = 400 \tan 47.2° \approx 432$$

The height of the statue is approximately $432 - 400 = 32$ feet. ∎

When it is not possible to walk off a distance from the base of the object whose height we seek, a more imaginative solution is required.

EXAMPLE 7 Finding the Height of a Mountain

To measure the height of a mountain, a surveyor takes two sightings of the peak at a distance 900 meters apart on a direct line to the mountain.* See Figure 9(a). The first observation results in an angle of elevation of 47°, whereas the second results in an angle of elevation of 35°. If the transit is 2 meters high, what is the height h of the mountain?

FIGURE 9

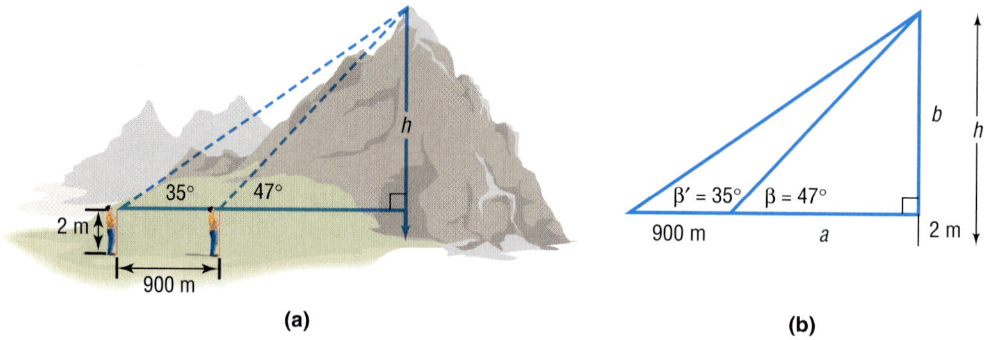

*For simplicity, we assume that these sightings are at the same level.

Solution Figure 9(b) shows two triangles that replicate the illustration in Figure 9(a). From the two right triangles shown, we find

$$\tan \beta' = \frac{b}{a + 900} \qquad \tan \beta = \frac{b}{a}$$

$$\tan 35^\circ = \frac{b}{a + 900} \qquad \tan 47^\circ = \frac{b}{a}$$

This is a system of two equations involving two variables, a and b. Because we seek b, we choose to solve the right-hand equation for a and substitute the result, $a = b/\tan 47^\circ = b \cot 47^\circ$, in the left-hand equation. The result is

$$\tan 35^\circ = \frac{b}{b \cot 47^\circ + 900}$$
$$b = (b \cot 47^\circ + 900) \tan 35^\circ$$
$$b = b \cot 47^\circ \tan 35^\circ + 900 \tan 35^\circ$$
$$b(1 - \cot 47^\circ \tan 35^\circ) = 900 \tan 35^\circ$$
$$b = \frac{900 \tan 35^\circ}{1 - \cot 47^\circ \tan 35^\circ} = \frac{900 \tan 35^\circ}{1 - \dfrac{\tan 35^\circ}{\tan 47^\circ}} \approx 1816$$

The height of the peak from ground level is therefore approximately 1816 + 2 = 1818 meters.

Now work Problem 29.

FIGURE 10
The bearing of P from O is θ

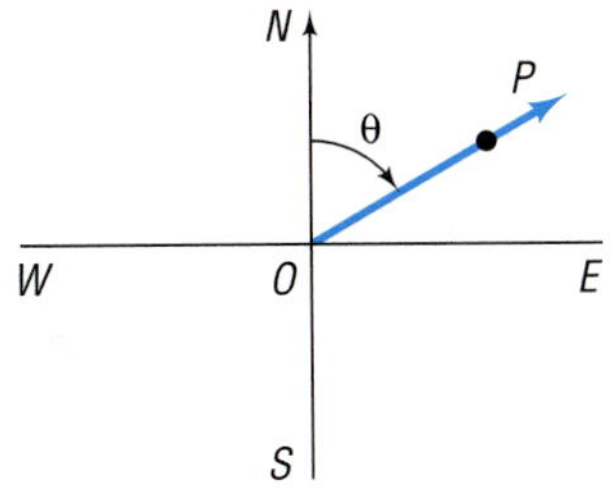

In navigation, the **direction** or **bearing** from O of an object at P means the positive angle measured clockwise* from the north (N) to the ray OP. See Figure 10. Based on Figure 10, we would say that the bearing of P from O is θ degrees.

EXAMPLE 8 Finding the Bearing of an Airplane

A Boeing 777 aircraft takes off from O'Hare Airport on runway 2 LEFT, which has a bearing of 20°. After flying for 1 mile, the pilot of the aircraft requests permission to turn 90° and head toward the northwest. The request is granted.

(a) What is the new bearing?

(b) After the plane goes 2 miles in this direction, what bearing should the control tower use to locate the aircraft?

FIGURE 11

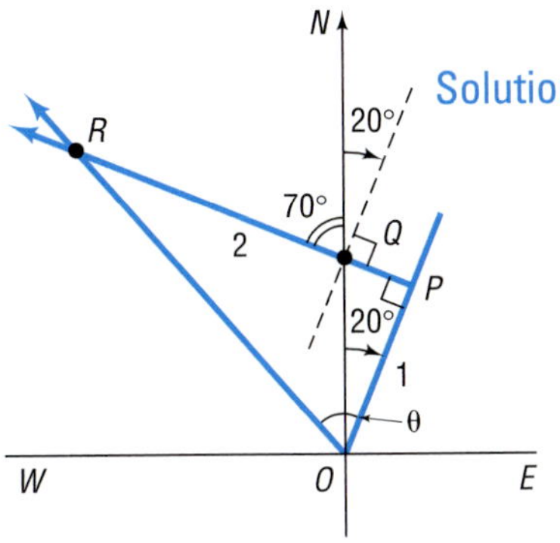

Solution (a) Figure 11 illustrates the situation. After flying 1 mile from the airport O (the control tower), the aircraft is at P. After turning 90° toward the northwest, we see that

$$\text{Angle } NOP = 20^\circ \qquad \text{Angle } RQN = 90^\circ - 20^\circ = 70^\circ$$

The new bearing of the aircraft is therefore

$$360^\circ - \text{Angle } RQN = 360^\circ - 70^\circ = 290^\circ$$

*Notice that this convention is just the opposite of what we are used to.

(b) After flying 2 miles at a bearing of 290°, the aircraft is at R. If $\theta =$ Angle ROP, then

$$\tan\theta = \frac{2}{1} = 2 \quad \text{so} \quad \theta = 63.4°$$

As a result,

$$\text{Angle } RON = \theta - 20° = 43.4°$$

The bearing of the plane from the central tower unit at O is

$$360° - \text{Angle } RON = 360° - 43.4° = 316.6°$$

EXAMPLE 9 The Gibb's Hill Lighthouse, Southampton, Bermuda

In operation since 1846, the Gibb's Hill Lighthouse stands 117 feet high on a hill 245 feet high, so its beam of light is 362 feet above sea level. A brochure states that the light can be seen on the horizon about 26 miles distant. Verify the accuracy of this statement.

Solution Figure 12 illustrates the situation. The central angle θ obeys the equation

FIGURE 12

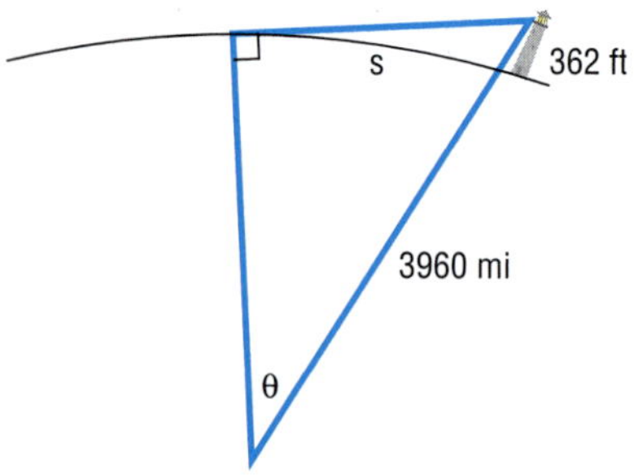

$$\cos\theta = \frac{3960}{3960 + \frac{362}{5280}} = 0.999982687$$

so

$$\theta = 0.00588 \text{ radian} = 0.33715° = 20.23'$$

The brochure does not indicate whether the distance is measured in nautical miles or statute miles. The distance s in nautical miles is 20.23, the measurement of θ in minutes. (Refer to Problem 86, Section 6.1). The distance s in statute miles is

$$s = r\theta = 3960(0.00588) = 23.3 \text{ miles}$$

In either case, it would seem that the brochure overstated the distance somewhat.

8.1 EXERCISES

In Problems 1–14, use the right triangle shown in the margin. Then, using the given information, solve the triangle.

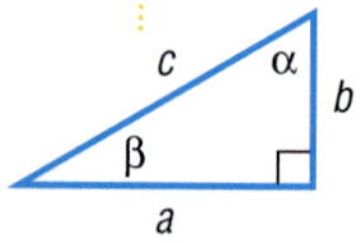

1. $b = 5$, $\beta = 20°$; find a, c, and α
2. $b = 4$, $\beta = 10°$; find a, c, and α
3. $a = 6$, $\beta = 40°$; find b, c, and α
4. $a = 7$, $\beta = 50°$; find b, c, and α
5. $b = 4$, $\alpha = 10°$; find a, c, and β
6. $b = 6$, $\alpha = 20°$; find a, c, and β
7. $a = 5$, $\alpha = 25°$; find b, c, and β
8. $a = 6$, $\alpha = 40°$; find b, c, and β
9. $c = 9$, $\beta = 20°$; find b, a, and α
10. $c = 10$, $\alpha = 40°$; find b, a, and β
11. $a = 5$, $b = 3$; find c, α, and β
12. $a = 2$, $b = 8$; find c, α, and β
13. $a = 2$, $c = 5$; find b, α, and β
14. $b = 4$, $c = 6$; find a, α, and β

15. A right triangle has a hypotenuse of length 3 inches. If one angle is 35°, find the length of each leg.

16. A right triangle has a hypotenuse of length 2 centimeters. If one angle is 40°, find the length of each leg.

17. A right triangle contains a 35° angle. If one leg is of length 5 inches, what is the length of the hypotenuse?
[**Hint:** Two answers are possible.]

18. A right triangle contains an angle of $\pi/10$ radian. If one leg is of length 3 meters, what is the length of the hypotenuse?
[**Hint:** Two answers are possible.]

19. The hypotenuse of a right triangle is 5 inches. If one leg is 2 inches, find the degree measure of each angle.

20. The hypotenuse of a right triangle is 3 feet. If one leg is 1 foot, find the degree measure of each angle.

21. **Finding the Width of a Gorge** Find the distance from A to C across the gorge illustrated in the figure.

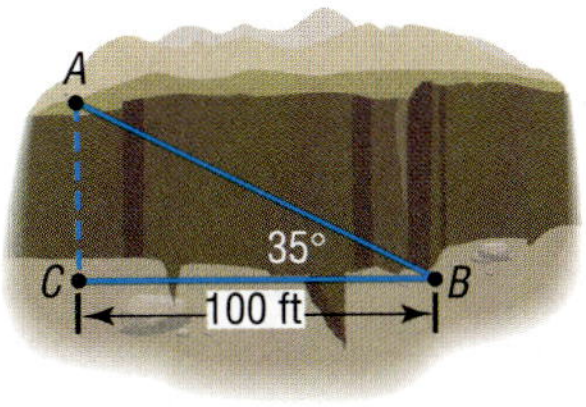

22. **Finding the Distance Across a Pond** Find the distance from A to C across the pond illustrated in the figure.

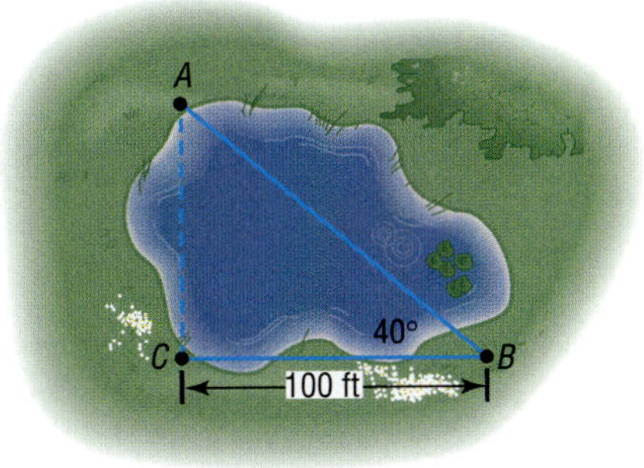

23. **The Eiffel Tower** The tallest tower built before the era of television masts, the Eiffel Tower was completed on March 31, 1889. Find the height of the Eiffel Tower (before a television mast was added to the top) using the information given in the illustration.

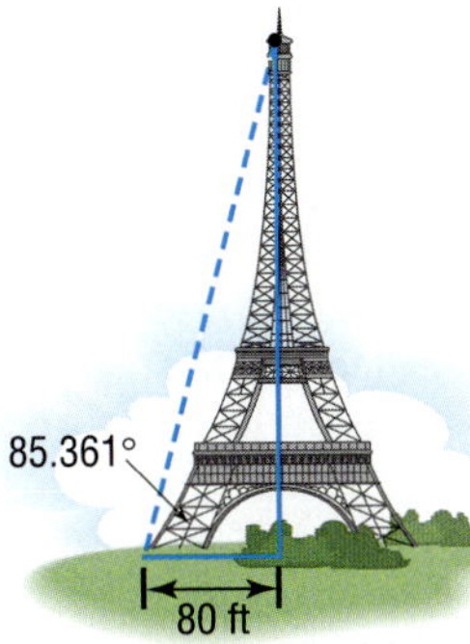

24. **Finding the Distance of a Ship from Shore** A ship, offshore from a vertical cliff known to be 100 feet in height, takes a sighting of the top of the cliff. If the angle of elevation is found to be 25°, how far offshore is the ship?

25. Suppose that you are headed toward a plateau 50 meters high. If the angle of elevation to the top of the plateau is 20°, how far are you from the base of the plateau?

26. **Statue of Liberty** A ship is just offshore of New York City. A sighting is taken of the Statue of Liberty, which is about 305 feet tall. If the angle of elevation to the top of the statue is 20°, how far is the ship from the base of the statue?

27. A 22 foot extension ladder leaning against a building makes a 70° angle with the ground. How far up the building does the ladder touch?

28. **Finding the Height of a Building** To measure the height of a building, two sightings are taken a distance of 50 feet apart. If the first angle of elevation is 40° and the second is 32°, what is the height of the building?

29. **Great Pyramid of Cheops** One of the original Seven Wonders of the World, the Great Pyramid of Cheops was built about 2580 B.C. Its original height was 480 feet 11 inches, but due to the loss of its topmost stones, it is now shorter.* Find the current height of the Great Pyramid, using the information given in the illustration.

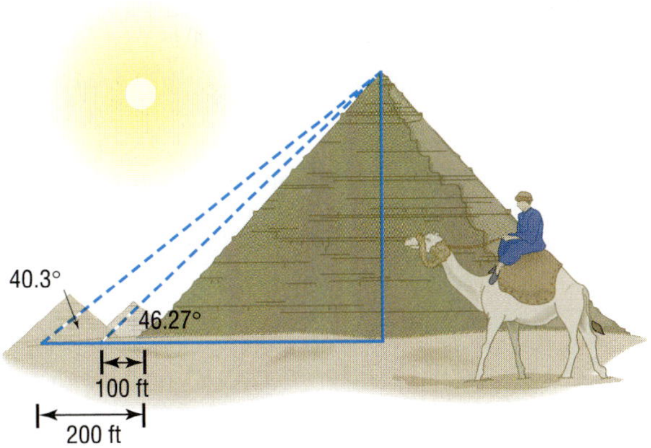

30. A laser beam is to be directed through a small hole in the center of a circle of radius 10 feet. The origin of the beam is 35 feet from the circle (see the figure). At what angle of elevation should the beam be aimed to ensure that it goes through the hole?

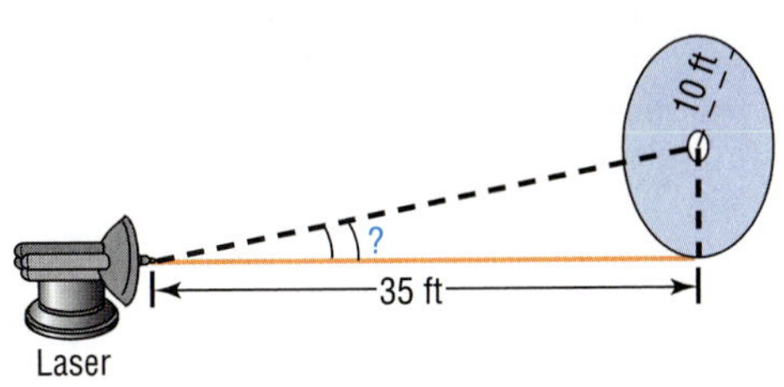

Source: Guinness Book of World Records.

31. Finding the Angle of Elevation of the Sun At 10 A.M. on April 26, 1996, a building 300 feet high casts a shadow 50 feet long. What is the angle of elevation of the Sun?

32. Mt. Rushmore To measure the height of Lincoln's caricature on Mt. Rushmore, two sightings 800 feet from the base of the mountain are taken. If the angle of elevation to the bottom of Lincoln's face is 32° and the angle of elevation to the top is 35°, what is the height of Lincoln's face?

33. Finding the Height of a Helicopter Two observers simultaneously measure the angle of elevation of a helicopter. One angle is measured as 25°, the other as 40° (see the figure). If the observers are 100 feet apart and the helicopter lies over the line joining them, how high is the helicopter?

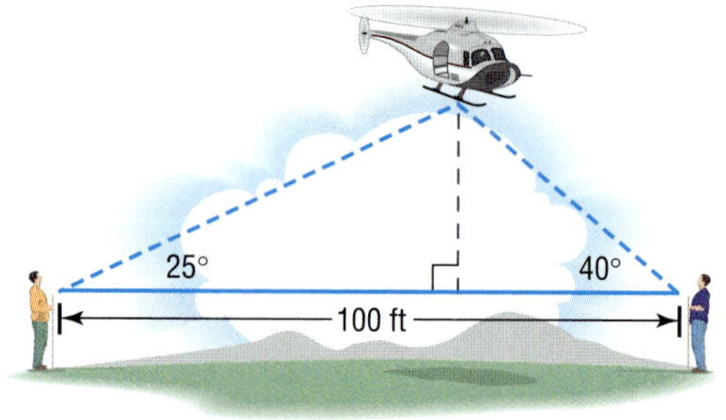

34. Finding the Distance between Two Objects A blimp, suspended in the air at a height of 500 feet, lies directly over a line from Soldier Field to the Adler Planetarium on Lake Michigan (see the figure). If the angle of depression from the blimp to the stadium is 32° and from the blimp to the planetarium is 23°, find the distance between Soldier Field and the Adler Planetarium.

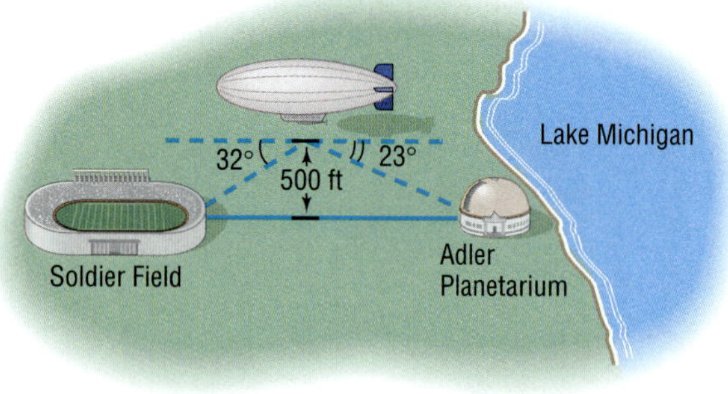

35. Finding the Length of a Guy Wire A radio transmission tower is 200 feet high. How long should a guy wire be if it is to be attached to the tower 10 feet from the top and is to make an angle of 21° with the ground?

36. Finding the Height of a Tower A guy wire 80 feet long is attached to the top of a radio transmission tower, making an angle of 25° with the ground. How high is the tower?

37. Washington Monument The angle of elevation of the Washington Monument is 35.1° at the instant it casts a shadow 789 feet long. Use this information to calculate the height of the monument.

38. Finding the Length of a Mountain Trail A straight trail with a uniform inclination of 17° leads from a hotel at an elevation of 9000 feet to a mountain lake at an elevation of 11,200 feet. What is the length of the trail?

39. Finding the Speed of a Truck A state trooper is hidden 30 feet from a highway. One second after a truck passes, the angle θ between the highway and the line of observation from the patrol car to the truck is measured. See the illustration.

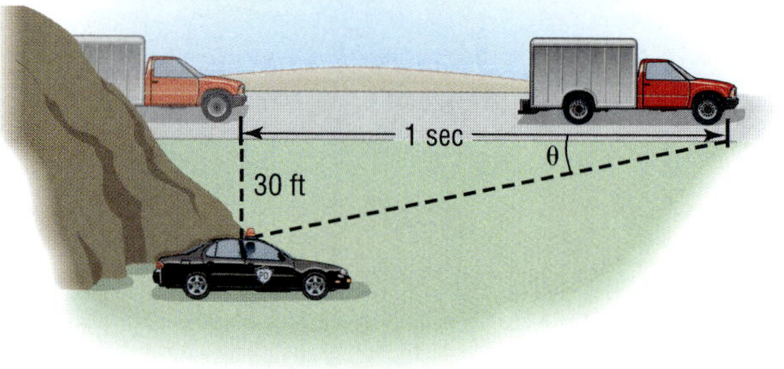

(a) If the angle measures 15°, how fast is the truck traveling? Express the answer in feet per second and in miles per hour.

(b) If the angle measures 20°, how fast is the truck traveling? Express the answer in feet per second and in miles per hour.

(c) If the speed limit is 55 miles per hour and a speeding ticket is issued for speeds of 5 miles per hour or more over the limit, for what angles should the trooper issue a ticket?

40. The Gibb's Hill Lighthouse, Southampton, Bermuda In operation since 1846, the Gibb's Hill Lighthouse stands 117 feet high on a hill 245 feet high, so its beam of light is 362 feet above sea level. A brochure states that ships 40 miles away can see the light and planes flying at 10,000 feet can see it 120 miles away. Verify the accuracy of these statements. What assumption did the brochure make about the height of the ship? (See illustration, Problem 103 in Section 1 of the Appendix.)

41. Finding the Bearing of an Aircraft A DC-9 aircraft leaves Midway Airport from runway 4 RIGHT, whose bearing is 40°. After flying for

1/2 mile, the pilot requests permission to turn 90° and head toward the southeast. The permission is granted.

(a) What is the new bearing?

(b) After the airplane goes 1 mile in this direction, what bearing should the control tower use to locate the aircraft?

42. Finding the Bearing of a Ship A ship leaves the port of Miami with a bearing of 100° and a speed of 15 knots. After 1 hour, the ship turns 90° toward the south.

(a) What is the new bearing?

(b) After 2 hours, maintaining the same speed, what is the bearing of the ship from port?

43. Shooting Free Throws in Basketball The eyes of a basketball player are 6 feet above the floor. The player is at the free-throw line, which is 15 feet from the center of the basket rim (see the figure). What is the angle of elevation from the player's eyes to the center of the rim?

[**Hint:** The rim is 10 feet above the floor.]

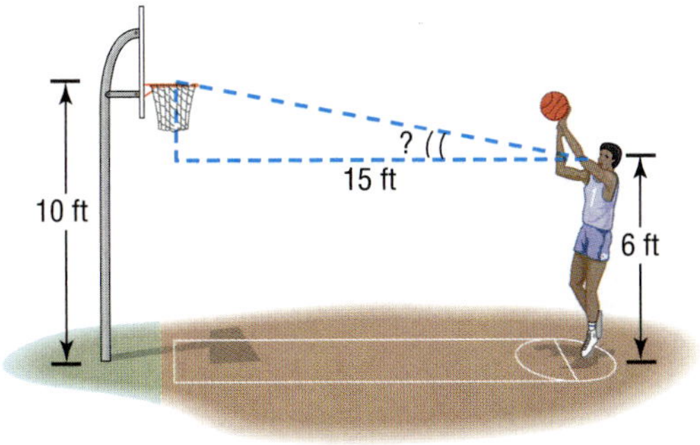

44. Finding the Pitch of a Roof A carpenter is preparing to put a roof on a garage that is 20 feet by 40 feet by 20 feet. A steel support beam 46 feet in length is positioned in the center of the garage. To support the roof, another beam will be attached to the top of the center beam (see the figure). At what angle of elevation is the new beam? In other words, what is the pitch of the roof?

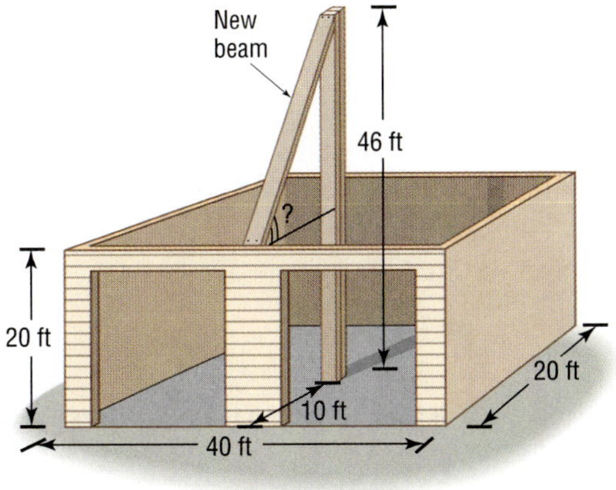

45. Determining Distances at Sea The navigator of a ship at sea spots two lighthouses she recognizes as ones 3 miles apart along a straight seashore. She determines that the angles formed between two line of sight observations of the lighthouses and the line from the ship directly to shore are 15° and 35°. See the illustration.

(a) How far is the ship from shore?

(b) How far is the ship from lighthouse A?

(c) How far is the ship from lighthouse B?

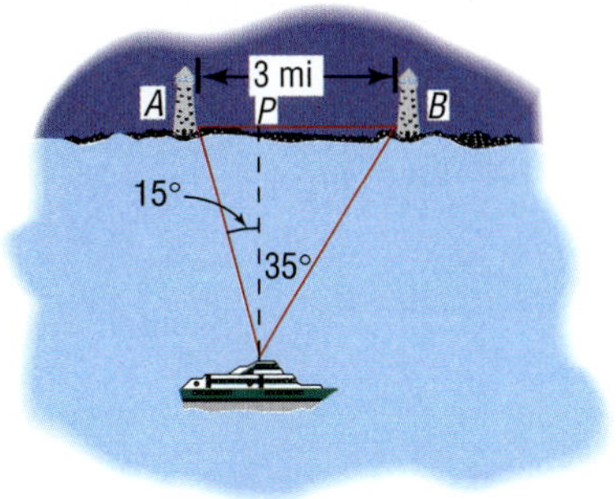

46. Constructing a Highway A highway whose primary directions are North-South is being constructed along the west coast of Florida. Near Naples, a bay obstructs the straight path of the road. Since the cost of a bridge is prohibitive, engineers decide to go around the bay. The illustration shows the path they decide upon and the measurements taken. What is the length of highway needed to go around the bay?

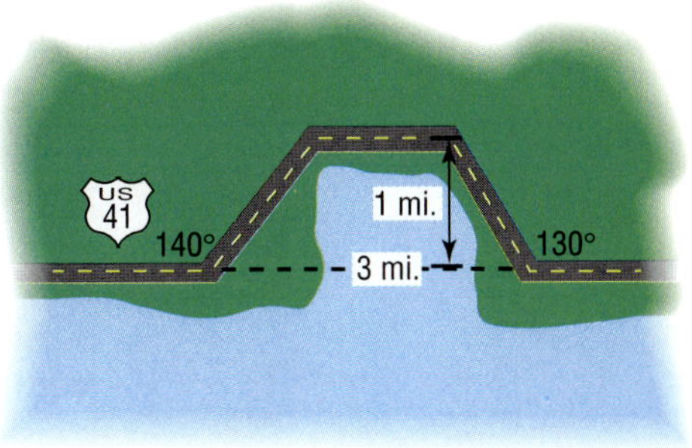

47. Surveillance Satellites A surveillance satellite circles Earth at a height of h miles above the surface. Suppose d is the distance, in miles, on the surface of the Earth that can be observed from the satellite. See the illustration.

(a) Find an equation that relates the central angle θ to the height h.

(b) Find an equation that relates the observable distance d and θ.

(c) Find an equation that relates d and h.

(d) If d is to be 2500 miles, how high must the satellite orbit above Earth?

(e) If the satellite orbits at a height of 300 miles, what distance d on the surface can be observed?

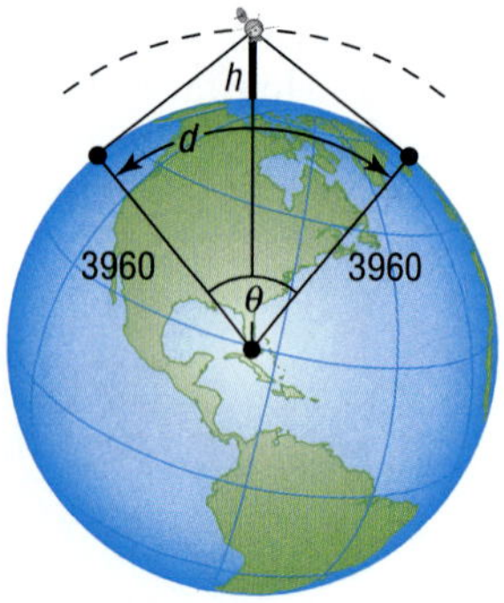

48. Compare Problem 103 in Section 1 of the Appendix with Example 9 and Problem 40 of this section. One uses the Pythagorean Theorem in its solution, while the others use trigonometry. Write a short paper that contrasts the two methods of solution. Be sure to point out the benefits as well as the shortcomings of each method. Then decide which solution you prefer. Be sure to have reasons.

8.2 THE LAW OF SINES

1 Solve SAA or ASA Triangles

2 Solve SSA Triangles

If none of the angles of a triangle is a right angle, the triangle is called **oblique.** Thus, an oblique triangle will have either three acute angles or two acute angles and one obtuse angle (an angle between 90° and 180°). See Figure 13.

FIGURE 13
Oblique triangles

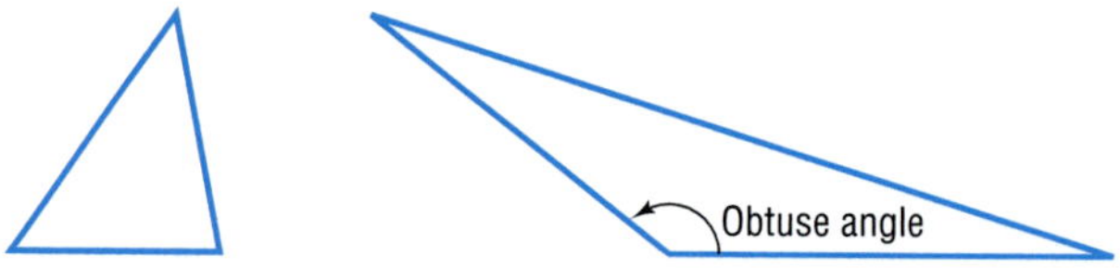

(a) All angles are acute **(b)** Two acute angles and one obtuse angle

In the discussion that follows, we will always label an oblique triangle so that side a is opposite angle α, side b is opposite angle β, and side c is opposite angle γ, as shown in Figure 14.

FIGURE 14

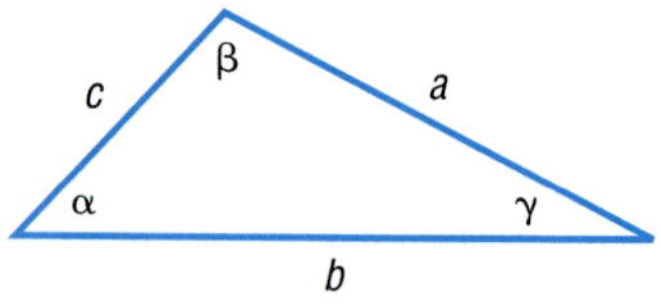

To **solve an oblique triangle** means to find the lengths of its sides and the measurements of its angles. To do this, we shall need to know the length of one side along with two other facts: either two angles, or the other two sides, or one angle and one other side.* Thus, there are four possibilities to consider:

CASE 1: One side and two angles are known (SAA or ASA).
CASE 2: Two sides and the angle opposite one of them are known (SSA).
CASE 3: Two sides and the included angle are known (SAS).
CASE 4: Three sides are known (SSS).

*Recall from plane geometry the fact that knowing three angles of a triangle determines a family of *similar triangles*—that is, triangles that have the same shape but different sizes.

Figure 15 illustrates the four cases.

FIGURE 15

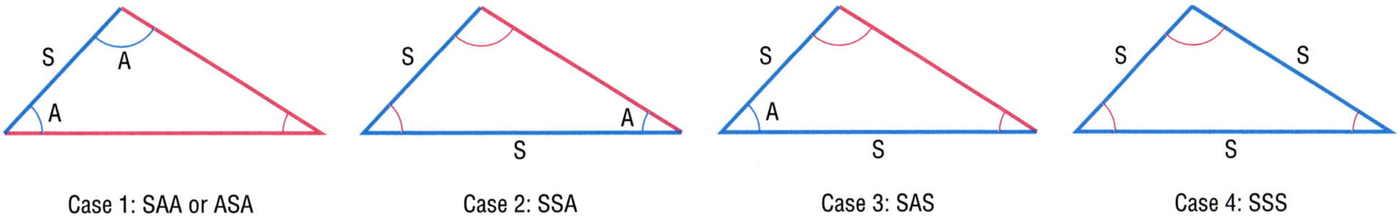

The **Law of Sines** is used to solve triangles for which Case 1 or 2 holds.

Theorem Law of Sines

For a triangle with sides a, b, c and opposite angles α, β, γ, respectively,

$$\frac{\sin \alpha}{a} = \frac{\sin \beta}{b} = \frac{\sin \gamma}{c} \tag{1}$$

FIGURE 16

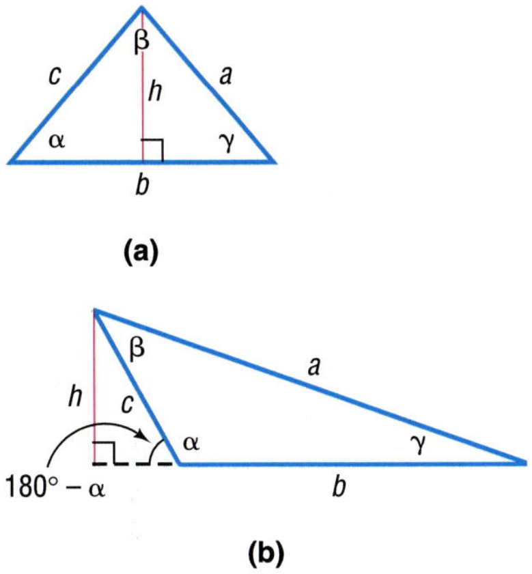

Proof To prove the Law of Sines, we construct an altitude of length h from one of the vertices of such a triangle. Figure 16(a) shows h for a triangle with three acute angles, and Figure 16(b) shows h for a triangle with an obtuse angle. In each case, the altitude is drawn from the vertex at β. Using either illustration, we have

$$\sin \gamma = \frac{h}{a}$$

from which

$$h = a \sin \gamma \tag{2}$$

From Figure 16(a), it also follows that

$$\sin \alpha = \frac{h}{c}$$

from which

$$h = c \sin \alpha \tag{3}$$

From Figure 16(b), it follows that

$$\sin \alpha = \sin(180° - \alpha) = \frac{h}{c}$$

which again gives

$$h = c \sin \alpha$$

Thus, whether the triangle has three acute angles or has two acute angles and one obtuse angle, equations (2) and (3) hold. As a result, we may equate the expressions for h in equations (2) and (3) to get

$$a \sin \gamma = c \sin \alpha$$

from which

$$\frac{\sin \alpha}{a} = \frac{\sin \gamma}{c} \qquad (4)$$

In a similar manner, by constructing the altitude h' from the vertex of angle α as shown in Figure 17, we can show that

FIGURE 17

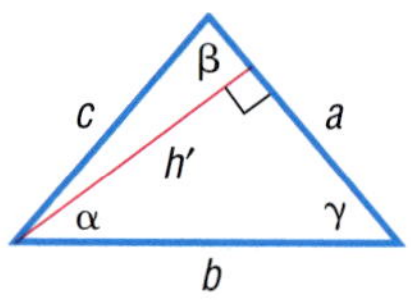

(a)

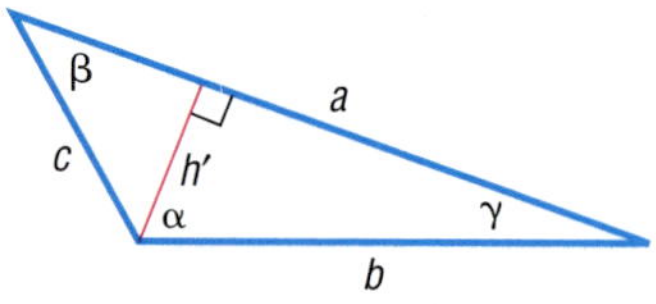

(b)

$$\sin \beta = \frac{h'}{c} \quad \text{and} \quad \sin \gamma = \frac{h'}{b}$$

Thus,

$$h' = c \sin \beta = b \sin \gamma$$

and

$$\frac{\sin \beta}{b} = \frac{\sin \gamma}{c} \qquad (5)$$

When equations (4) and (5) are combined, we have equation (1), the Law of Sines.

In applying the Law of Sines to solve triangles, we use the fact that the sum of the angles of any triangle equals 180°; that is,

$$\alpha + \beta + \gamma = 180° \qquad (6)$$

1 Our first two examples show how to solve a triangle when one side and two angles are known (Case 1: SAA or ASA).

EXAMPLE 1 Using the Law of Sines to Solve a SAA Triangle

Solve the triangle: $\alpha = 40°$, $\beta = 60°$, $a = 4$

FIGURE 18

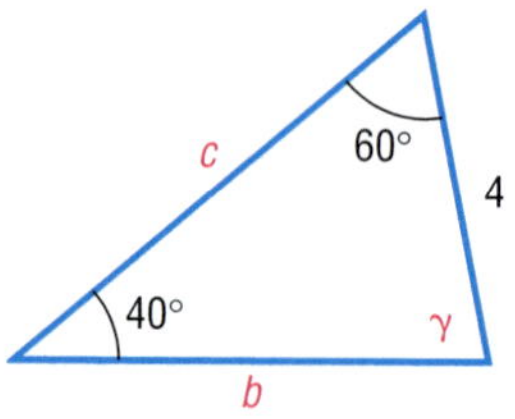

Solution Figure 18 shows the triangle that we want to solve. The third angle γ is easily found using equation (6):

$$\begin{aligned} \alpha + \beta + \gamma &= 180° \\ 40° + 60° + \gamma &= 180° \\ \gamma &= 80° \end{aligned}$$

Now we use the Law of Sines (twice) to find the unknown sides b and c:

$$\frac{\sin \alpha}{a} = \frac{\sin \beta}{b} \qquad \frac{\sin \alpha}{a} = \frac{\sin \gamma}{c}$$

Because $a = 4$, $\alpha = 40°$, $\beta = 60°$, and $\gamma = 80°$, we have

$$\frac{\sin 40°}{4} = \frac{\sin 60°}{b} \qquad \frac{\sin 40°}{4} = \frac{\sin 80°}{c}$$

Thus,

$$b = \frac{4 \sin 60°}{\sin 40°} \approx 5.39 \qquad c = \frac{4 \sin 80°}{\sin 40°} \approx 6.13$$

↑ From a calculator. ↑ From a calculator.

Notice that in Example 1 we found b and c by working with the given side a. This is better than finding b first and working with a rounded value of b to find c.

EXAMPLE 2 Using the Law of Sines to Solve an ASA Triangle

Solve the triangle: $\alpha = 35°$, $\beta = 15°$, $c = 5$

Solution Figure 19 illustrates the triangle that we want to solve. Because we know two angles ($\alpha = 35°$ and $\beta = 15°$), it is easy to find the third angle using equation (6):

FIGURE 19

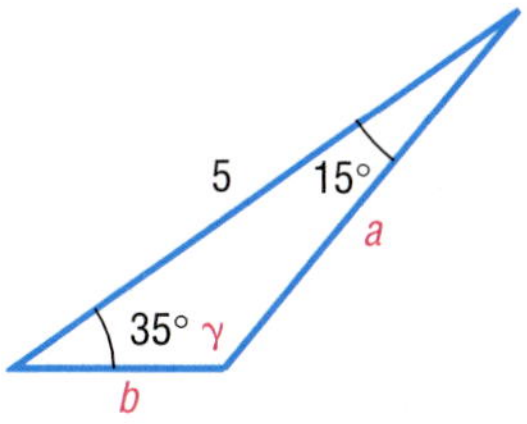

$$\alpha + \beta + \gamma = 180°$$
$$35° + 15° + \gamma = 180°$$
$$\gamma = 130°$$

Now we know the three angles and one side ($c = 5$) of the triangle. To find the remaining two sides a and b, we use the Law of Sines (twice):

$$\frac{\sin \alpha}{a} = \frac{\sin \gamma}{c} \qquad \frac{\sin \beta}{b} = \frac{\sin \gamma}{c}$$

$$\frac{\sin 35°}{a} = \frac{\sin 130°}{5} \qquad \frac{\sin 15°}{b} = \frac{\sin 130°}{5}$$

$$a = \frac{5 \sin 35°}{\sin 130°} \approx 3.74 \qquad b = \frac{5 \sin 15°}{\sin 130°} \approx 1.69$$

Now work Problem 1.

The Ambiguous Case

FIGURE 20

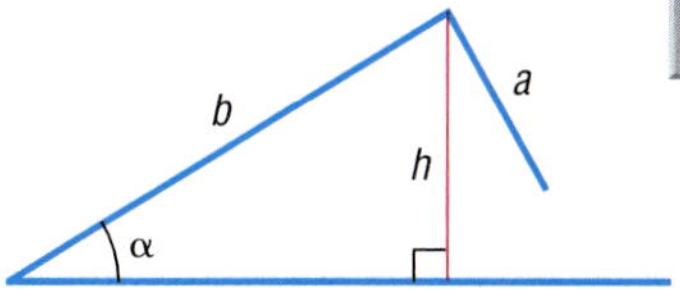

2 Case 2 (SSA), which applies to triangles where two sides and the angle opposite one of them are known, is referred to as the **ambiguous case,** because the known information may result in one triangle, two triangles, or no triangle at all. Suppose that we are given sides a and b and angle α, as illustrated in Figure 20. The key to determining the possible triangles, if any, that

may be formed from the given information, lies primarily with the height h and the fact that $h = b \sin \alpha$.

No Triangle: If $a < h = b \sin \alpha$, then clearly side a is not sufficiently long to form a triangle. See Figure 21.

FIGURE 21
$a < h = b \sin \alpha$

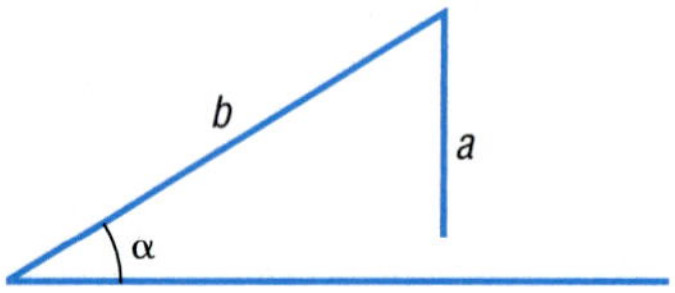

One Right Triangle: If $a = h = b \sin \alpha$, then side a is just long enough to form a right triangle. See Figure 22.

FIGURE 22
$a = b \sin \alpha$

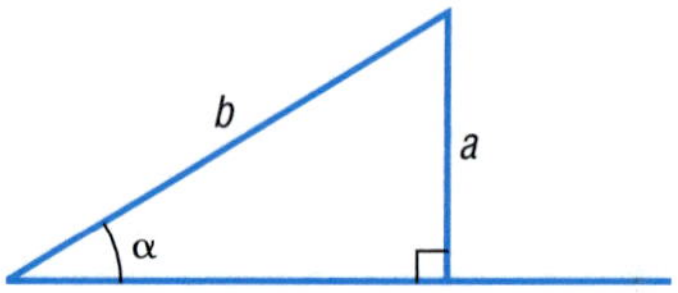

Two Triangles: If $a < b$ and $h = b \sin \alpha < a$, then two distinct triangles can be formed from the given information. See Figure 23.

FIGURE 23
$b \sin \alpha < a$ and $a < b$

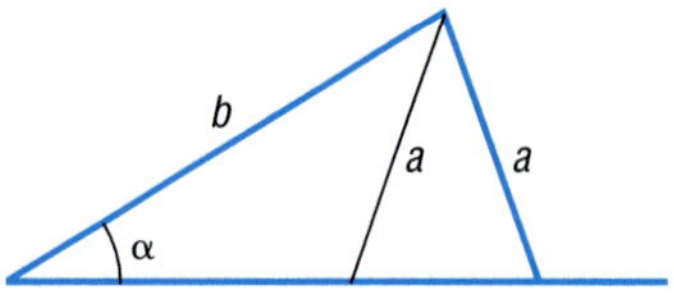

One Triangle: If $a \geq b$, then only one triangle can be formed. See Figure 24.

FIGURE 24
$a \geq b$

Fortunately, we do not have to rely on an illustration to draw the correct conclusion in the ambiguous case. The Law of Sines will lead us to the correct determination. Let's see how.

EXAMPLE 3 Using the Law of Sines to Solve a SSA Triangle (One Solution)

Solve the triangle: $a = 3$, $b = 2$, $\alpha = 40°$.

FIGURE 25(a)

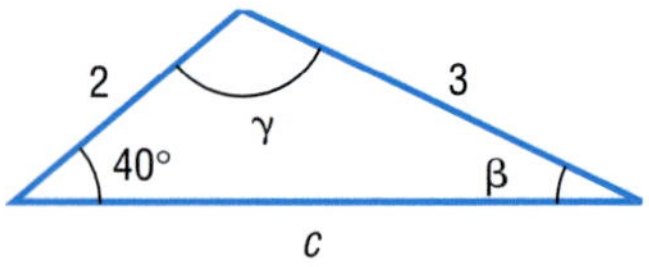

Solution (See Figure 25(a).) Because $a = 3$, $b = 2$, and $\alpha = 40°$ are known, we use the Law of Sines to find the angle β:

$$\frac{\sin \alpha}{a} = \frac{\sin \beta}{b}$$

Then

$$\frac{\sin 40°}{3} = \frac{\sin \beta}{2}$$

$$\sin \beta = \frac{2 \sin 40°}{3} \approx 0.43$$

There are two angles β, $0° < \beta < 180°$, for which $\sin \beta \approx 0.43$:

$$\beta \approx 25.4° \quad \text{and} \quad \beta \approx 154.6°$$

[**Note:** Here we computed β using the stored value of $\sin \beta$. If you use the rounded value, $\sin \beta \approx 0.43$, you will obtain slightly different results.]

The second possibility is ruled out, because $\alpha = 40°$, making $\alpha + \beta \approx 194.6° > 180°$. Now, using $\beta \approx 25.4°$, we find

$$\gamma = 180° - \alpha - \beta \approx 180° - 40° - 25.4° = 114.6°$$

The third side c may now be determined using the Law of Sines:

$$\frac{\sin \gamma}{c} = \frac{\sin \alpha}{a}$$

$$\frac{\sin 114.6°}{c} = \frac{\sin 40°}{3}$$

$$c = \frac{3 \sin 114.6°}{\sin 40°} \approx 4.24$$

FIGURE 25(b)

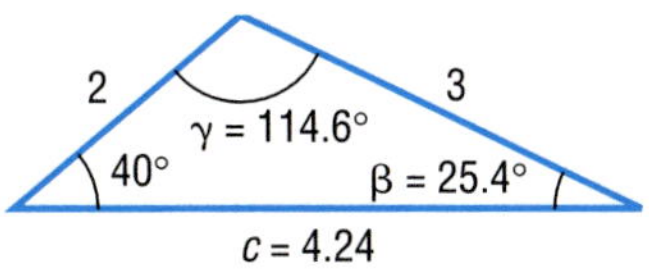

Figure 25(b) illustrates the solved triangle.

EXAMPLE 4 Using the Law of Sines to Solve a SSA Triangle (Two Solutions)

Solve the triangle: $a = 6, b = 8, \alpha = 35°$

Solution Because $a = 6$, $b = 8$, and $\alpha = 35°$ are known, we use the Law of Sines to find the angle β:

$$\frac{\sin \alpha}{a} = \frac{\sin \beta}{b}$$

Then

$$\frac{\sin 35°}{6} = \frac{\sin \beta}{8}$$

$$\sin \beta = \frac{8 \sin 35°}{6} \approx 0.76$$

$$\beta_1 \approx 49.9° \quad \text{or} \quad \beta_2 \approx 130.1°$$

For both possibilities we have $\alpha + \beta < 180°$. Hence, there are two triangles—one containing the angle $\beta = \beta_1 \approx 49.9°$ and the other containing the angle $\beta = \beta_2 \approx 130.1°$. The third angle γ is either

$$\gamma_1 = 180° - \alpha - \beta_1 \approx 95.1° \quad \text{or} \quad \gamma_2 = 180° - \alpha - \beta_2 \approx 14.9°$$

($\alpha = 35°$, $\beta_1 = 49.9°$) ($\alpha = 35°$, $\beta_2 = 130.1°$)

The third side c obeys the Law of Sines, so we have

$$\frac{\sin \gamma}{c} = \frac{\sin \alpha}{a}$$

$$\frac{\sin 95.1°}{c_1} = \frac{\sin 35°}{6} \qquad \text{or} \qquad \frac{\sin 14.9°}{c_2} = \frac{\sin 35°}{6}$$

$$c_1 = \frac{6 \sin 95.1°}{\sin 35°} \approx 10.42 \qquad c_2 = \frac{6 \sin 14.9°}{\sin 35°} \approx 2.69$$

FIGURE 26

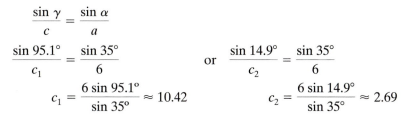

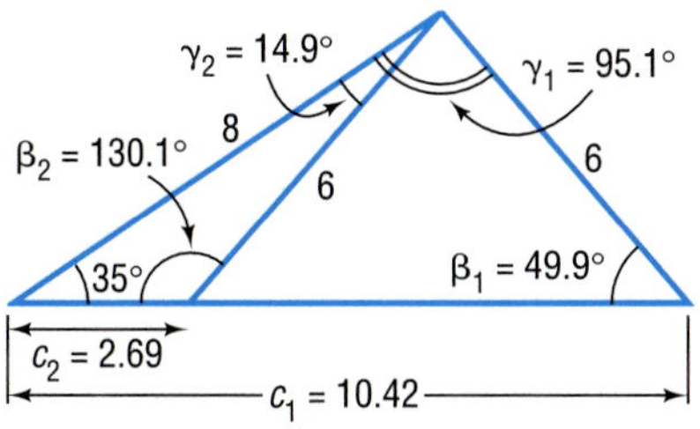

The two solved triangles are illustrated in Figure 26.

EXAMPLE 5 Using the Law of Sines to Solve a SSA Triangle (No Solution)

Solve the triangle: $a = 2, c = 1, \gamma = 50°$

Solution Because $a = 2$, $c = 1$, and $\gamma = 50°$ are known, we use the Law of Sines to find the angle α:

$$\frac{\sin \alpha}{a} = \frac{\sin \gamma}{c}$$

$$\frac{\sin \alpha}{2} = \frac{\sin 50°}{1}$$

$$\sin \alpha = 2 \sin 50° \approx 1.53$$

FIGURE 27

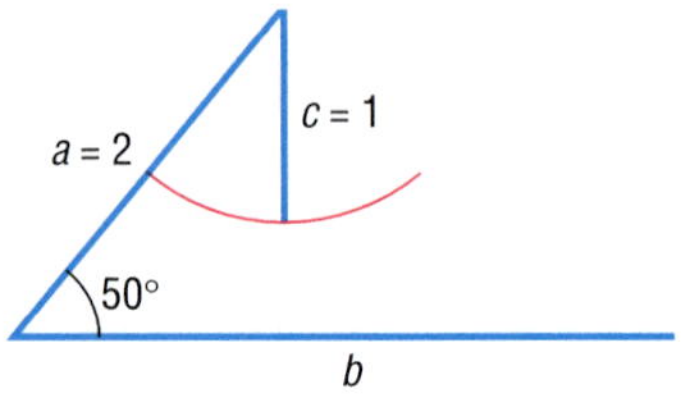

There is no angle α for which $\sin \alpha > 1$. Hence, there can be no triangle with the given measurements. Figure 27 illustrates the measurements given. Notice that, no matter how we attempt to position side c, it will never intersect side b to form a triangle.

Now work Problem 17.

Applied Problems

The Law of Sines is particularly useful for solving certain applied problems.

EXAMPLE 6 Rescue at Sea

Coast Guard Station Zulu is located 120 miles due west of Station X-ray. A ship at sea sends an SOS call that is received by each station. The call to Station Zulu indicates that the location of the ship is 40° east of north; the call to Station X-ray indicates that the location of the ship is 30° west of north.

(a) How far is each station from the ship?

(b) If a helicopter capable of flying 200 miles per hour is dispatched from the nearest station to the ship, how long will it take to reach the ship?

Solution (a) Figure 28 illustrates the situation. The angle γ is found to be

$$\gamma = 180° - 50° - 60° = 70°$$

FIGURE 28

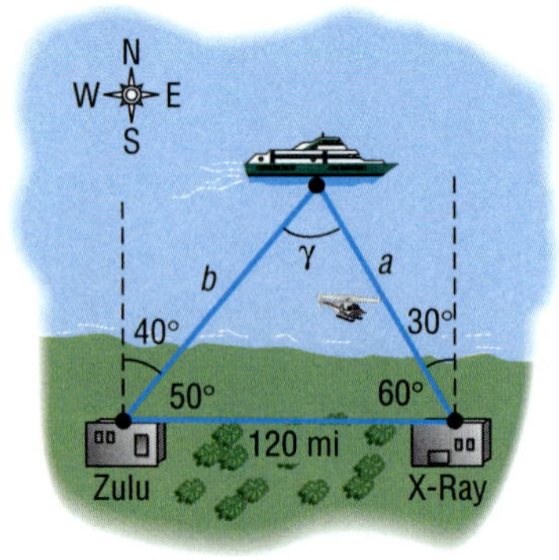

The Law of Sines can now be used to find the two distances a and b that we seek:

$$\frac{\sin 50°}{a} = \frac{\sin 70°}{120}$$

$$a = \frac{120 \sin 50°}{\sin 70°} \approx 97.82 \text{ miles}$$

$$\frac{\sin 60°}{b} = \frac{\sin 70°}{120}$$

$$b = \frac{120 \sin 60°}{\sin 70°} \approx 110.59 \text{ miles}$$

Thus, Station Zulu is about 111 miles from the ship, and Station X-ray is about 98 miles from the ship.

(b) The time t needed for the helicopter to reach the ship from Station X-ray is found by using the formula

(Velocity, v)(Time, t) = Distance, a

Then

$$t = \frac{a}{v} = \frac{97.82}{200} \approx 0.49 \text{ hour} \approx 29 \text{ minutes}$$

It will take about 29 minutes for the helicopter to reach the ship.

Now work Problem 29.

8.2 EXERCISES

In Problems 1–8, solve each triangle.

1.

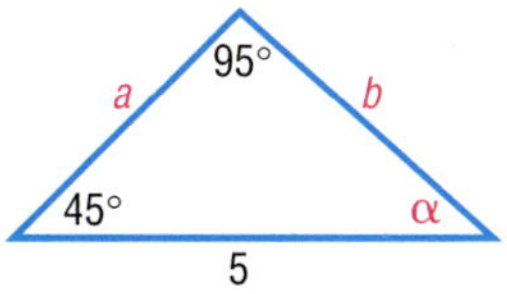

2.

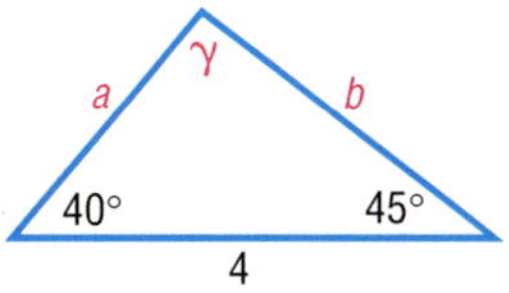

3.

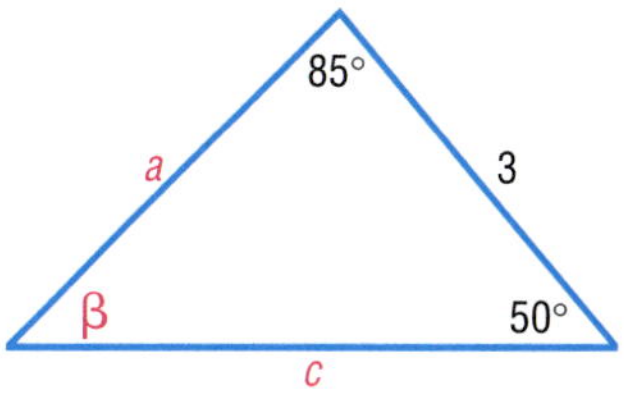

4.

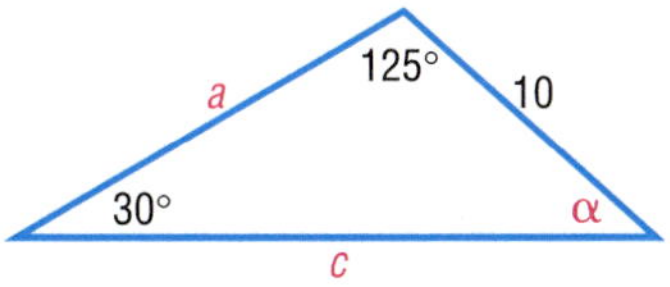

5.

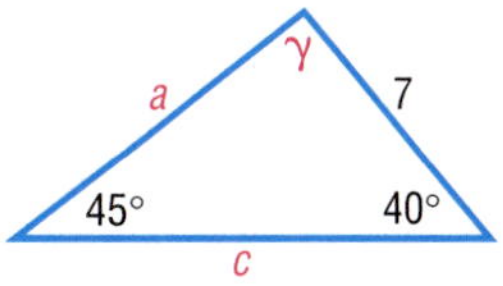

6.

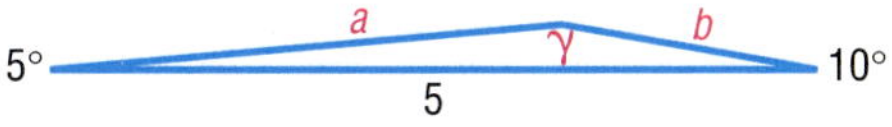

7.

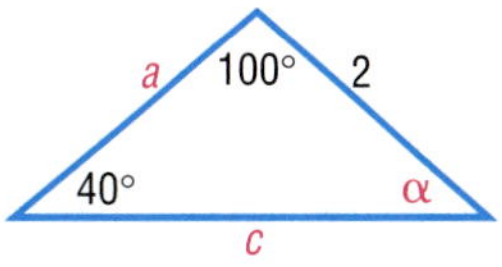

8.

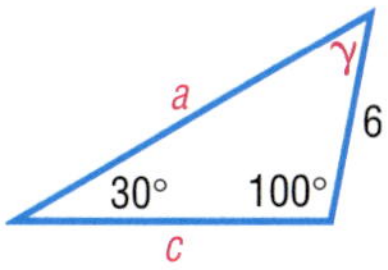

In Problems 9–16, solve each triangle.

9. $\alpha = 40°, \beta = 20°, a = 2$
10. $\alpha = 50°, \gamma = 20°, a = 3$
11. $\beta = 70°, \gamma = 10°, b = 5$
12. $\alpha = 70°, \beta = 60°, c = 4$
13. $\alpha = 110°, \gamma = 30°, c = 3$
14. $\beta = 10°, \gamma = 100°, b = 2$
15. $\alpha = 40°, \beta = 40°, c = 2$
16. $\beta = 20°, \gamma = 70°, a = 1$

In Problems 17–28, two sides and an angle are given. Determine whether the given information results in one triangle, two triangles, or no triangle at all. Solve any triangle(s) that results.

17. $a = 3, b = 2, \alpha = 50°$
18. $b = 4, c = 3, \beta = 40°$
19. $b = 5, \ c = 3, \ \beta = 100°$
20. $a = 2, c = 1, \alpha = 120°$
21. $a = 4, b = 5, \alpha = 60°$
22. $b = 2, c = 3, \beta = 40°$
23. $b = 4, c = 6, \beta = 20°$
24. $a = 3, b = 7, \alpha = 70°$
25. $a = 2, c = 1, \gamma = 100°$
26. $b = 4, c = 5, \beta = 95°$
27. $a = 2, c = 1, \gamma = 25°$
28. $b = 4, c = 5, \beta = 40°$

29. Rescue at Sea Coast Guard Station Able is located 150 miles due south of Station Baker. A ship at sea sends an SOS call that is received by each station. The call to Station Able indicates that the ship is located 35° north of east; the call to Station Baker indicates that the ship is located 30° south of east.

(a) How far is each station from the ship?

(b) If a helicopter capable of flying 200 miles per hour is dispatched from the nearest station to the ship, how long will it take to reach the ship?

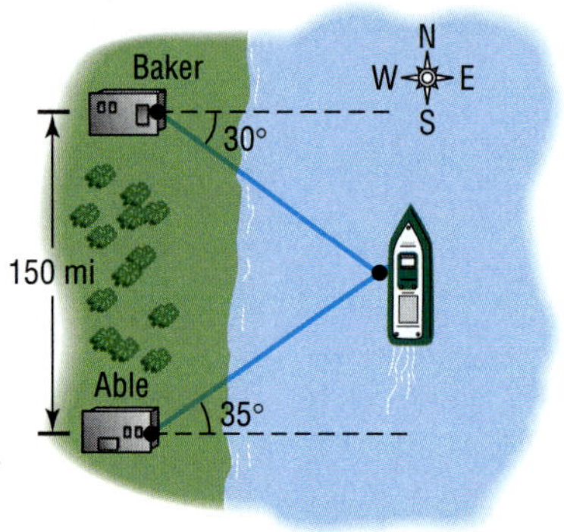

30. Surveying Consult the figure. To find the distance from the house at A to the house at B, a surveyor measures the angle BAC to be 40°, then walks off a distance of 100 feet to C, and measures the angle ACB to be 50°. What is the distance from A to B?

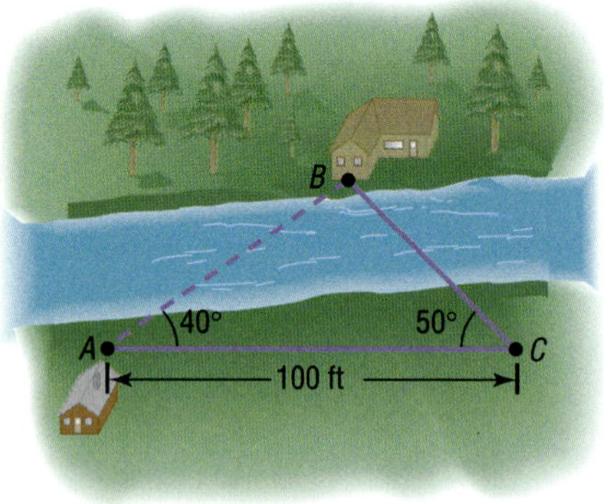

31. Finding the Length of a Ski Lift Consult the figure. To find the length of the span of a proposed ski lift from A to B, a surveyor measures the angle DAB to be 25°, then walks off a distance of 1000 feet to C, and measures the angle ACB to be 15°. What is the distance from A to B?

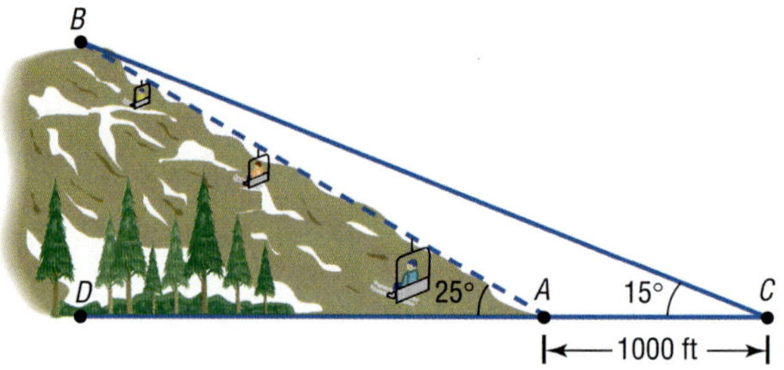

32. Finding the Height of a Mountain Use the illustration in Problem 31 to find the height BD of the mountain at B.

33. Finding the Height of an Airplane An aircraft spotted by two observers who are 1000 feet apart. As the airplane passes over the line joining them, each observer takes a sighting of the angle of elevation to the plane, as indicated in the figure. How high is the airplane?

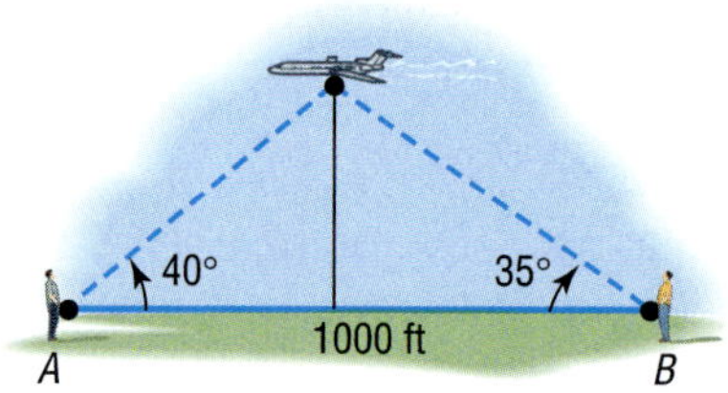

34. Finding the Height of the Bridge over the Royal Gorge The highest bridge in the world is the bridge over the Royal Gorge of the Arkansas River in Colorado.* Sightings to the same point at water level directly under the bridge are taken from each side of the 880-foot-long bridge, as indicated in the figure. How high is the bridge?

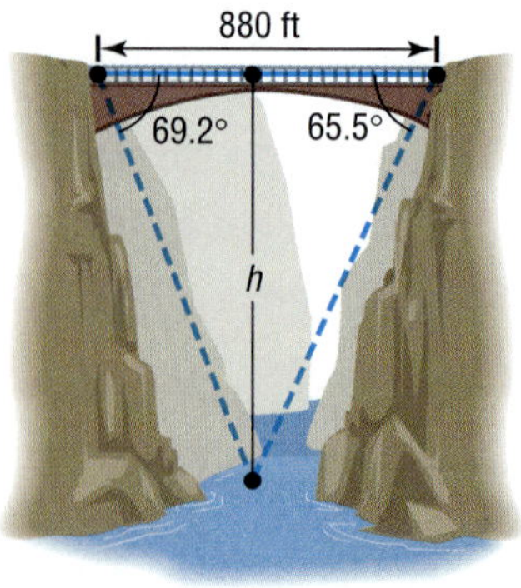

Source: Guinness Book of World Records.

35. Navigation An airplane flies from city A to city B, a distance of 150 miles, then turns through an angle of 40° and heads toward city C, as shown in the figure.

(a) If the distance between cities A and C is 300 miles, how far is it from city B to city C?

(b) Through what angle should the pilot turn at city C to return to City A?

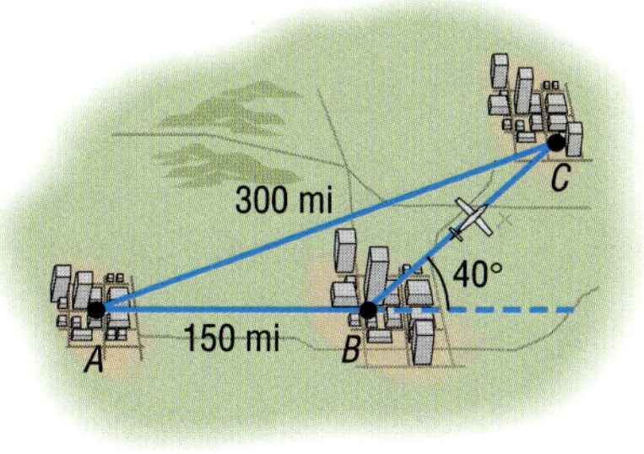

36. Time Lost due to a Navigation Error In attempting to fly from city A to city B, an aircraft followed a course that was 10° in error, as indicated in the figure. After flying a distance of 50 miles, the pilot corrected the course by turning at point C and flying 70 miles farther. If the constant speed of the aircraft was 250 miles per hour, how much time was lost due to the error?

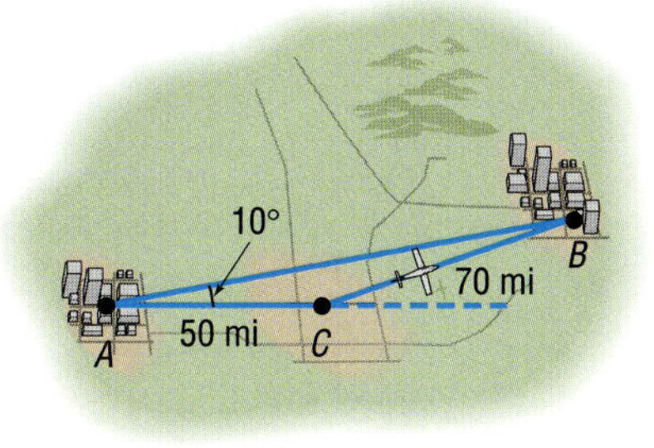

37. Finding the Lean of the Leaning Tower of Pisa The famous Leaning Tower of Pisa was originally 184.5 feet high.* At a distance of 123 feet from the base of the tower, the angle of elevation to the top of the tower is found to be 60°. Find the angle CAB indicated in the figure. Also, find the perpendicular distance from C to AB.

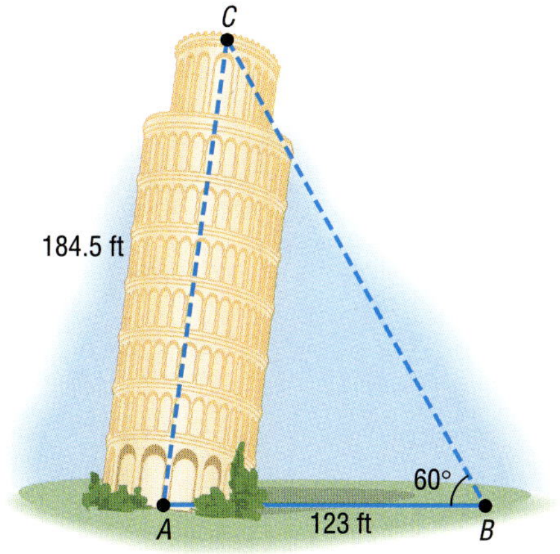

38. Crankshafts on Cars On a certain automobile, the crankshaft is 3 inches long and the connecting rod is 9 inches long (see the figure). At the time when the angle OPA is 15°, how far is the piston (P) from the center (O) of the crankshaft?

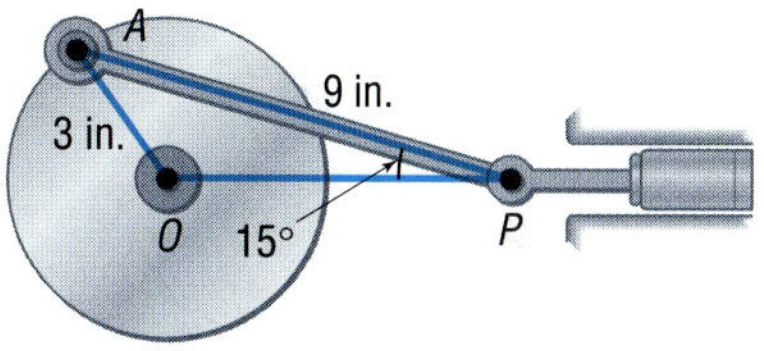

39. Constructing a Highway U.S. 41, a highway whose primary directions are north–south, is being constructed along the west coast of Florida. Near Naples, a bay obstructs the straight path of the road. Since the cost of a bridge is prohibitive, engineers decide to go around the bay. The illustrations show the path that they decide on and the measurements taken. What is the length of highway needed to go around the bay?

*In their 1986 report on the fragile seven-century-old bell tower, scientists in Pisa, Italy, said the Leaning Tower of Pisa had increased its famous lean by 1 millimeter, or 0.04 inch. That is about the annual average, although the tilting had slowed to about half that much in the previous 2 years. (*Source:* United Press International, June 29, 1986.)

Update PISA, ITALY. September, 1995. The Leaning Tower of Pisa has suddenly shifted, jeopardizing years of preservation work to stabilize it, Italian newspapers said Sunday. The tower, built on shifting subsoil between 1174 and 1350 as a belfry for the nearby cathedral, recently moved .07 inches in one night. The tower has been closed to tourists since 1990 but officials hope to partially reopen it next year.

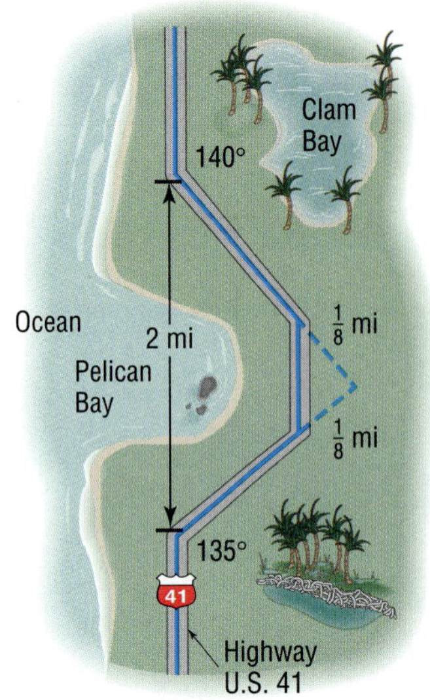

40. Determining Distances at Sea The navigator of a ship at sea spots two lighthouses that she knows to be 3 miles apart along a straight seashore. She determines that the angles formed between two line-of-sight observations of the lighthouses and the line from the ship directly to shore are 15° and 35°. See the illustration.
(a) How far is the ship from lighthouse A?
(b) How far is the ship from lighthouse B?
(c) How far is the ship from shore?

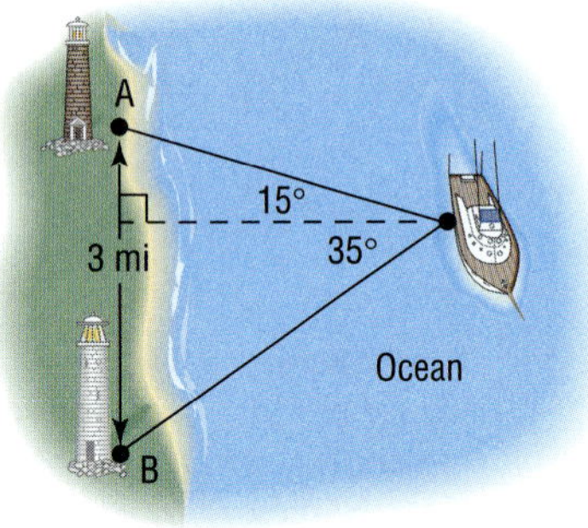

41. Calculating Distances at Sea The navigator of a ship at sea has the harbor in sight at which the ship is to dock. She spots a lighthouse she knows is 1 mile up the coast from the mouth of the harbor, and she measures the angle between the line-of-sight observations of the harbor and lighthouse to be 20°. With the ship heading directly toward the harbor, she repeats this measurement after 5 minutes of traveling at 12 miles per hour. If the new angle is 30°, how far is the ship from the harbor?

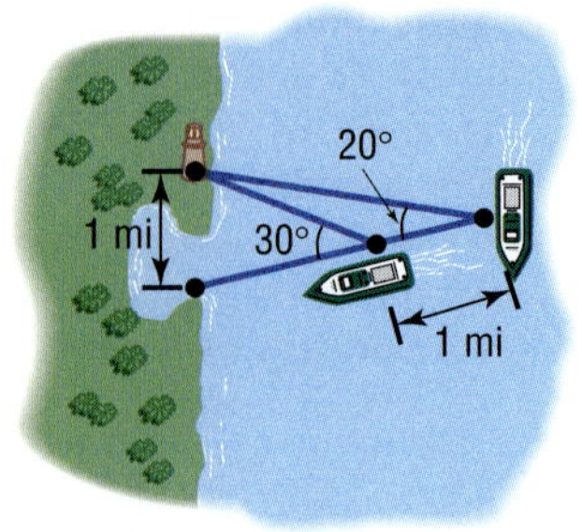

42. Finding Distances A forest ranger is walking on a path inclined at 5° to the horizontal directly toward a 100-foot-tall fire observation tower. The angle of elevation of the top of the tower is 40°. How far is the ranger from the tower at this time?

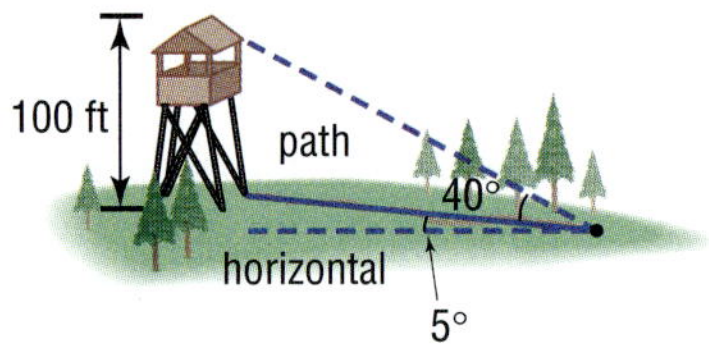

43. Mollweide's Formula For any triangle, **Mollweide's Formula** (named after Karl Mollweide, 1774–1825) states that

$$\frac{a+b}{c} = \frac{\cos \frac{1}{2}(\alpha - \beta)}{\sin \frac{1}{2}\gamma}$$

Derive it.

[**Hint:** Use the Law of Sines and then a sum-to-product formula.] Notice that this formula involves all six parts of a triangle. As a result, it is sometimes used to check the solution of a triangle.

44. Mollweide's Formula Another form of Mollweide's Formula is

$$\frac{a-b}{c} = \frac{\sin \frac{1}{2}(\alpha - \beta)}{\cos \frac{1}{2}\gamma}$$

Derive it.

45. For any triangle, derive the formula

$$a = b \cos \gamma + c \cos \beta$$

[**Hint:** Use the fact that $\sin \alpha = \sin(180° - \beta - \gamma)$.]

46. Law of Tangents For any triangle, derive the Law of Tangents:

$$\frac{a-b}{a+b} = \frac{\tan \frac{1}{2}(\alpha - \beta)}{\tan \frac{1}{2}(\alpha + \beta)}$$

[**Hint:** Use Mollweide's Formula.]

47. Circumscribing a Triangle Show that

$$\frac{\sin \alpha}{a} = \frac{\sin \beta}{b} = \frac{\sin \gamma}{c} = \frac{1}{2r}$$

where r is the radius of the circle circumscribing the triangle ABC whose sides are a, b, c, as shown in the figure in the margin.

[**Hint:** Draw the diameter AB'. Then $\beta =$ angle $ABC =$ angle $AB'C$ and angle $ACB' = 90°$.]

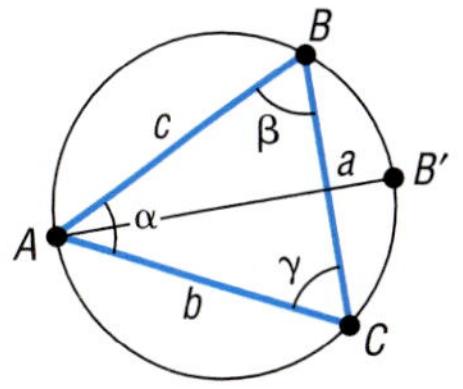

48. Make up three problems involving oblique triangles. One should result in one triangle, the second in two triangles, and the third in no triangle.

8.3 THE LAW OF COSINES

1 Solve SAS Triangles

2 Solve SSS Triangles

In Section 8.2, we used the Law of Sines to solve Case 1 (SAA or ASA) and Case 2 (SSA) of an oblique triangle. In this section, we derive the Law of Cosines and use it to solve the remaining cases, 3 and 4.

CASE 3:	Two sides and the included angle are known (SAS).
CASE 4:	Three sides are known (SSS).

Theorem Law of Cosines

For a triangle with sides a, b, c and opposite angles α, β, γ, respectively.

$$c^2 = a^2 + b^2 - 2ab \cos \gamma \qquad (1)$$

$$b^2 = a^2 + c^2 - 2ac \cos \beta \qquad (2)$$

$$a^2 = b^2 + c^2 - 2bc \cos \alpha \qquad (3)$$

FIGURE 29

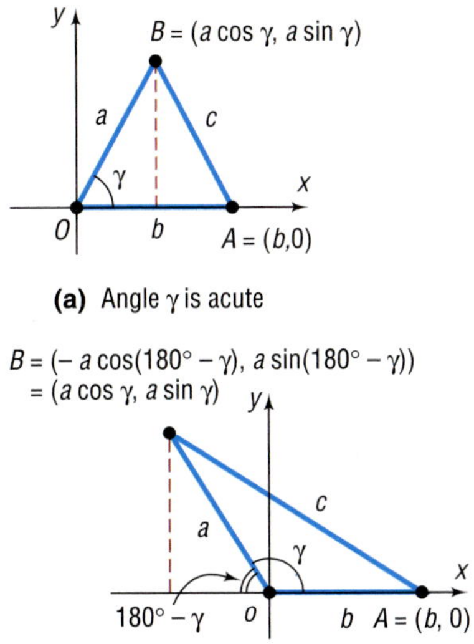

(a) Angle γ is acute

(b) Angle γ is obtuse

Proof We will prove only formula (1) here. Formulas (2) and (3) may be proved using the same argument.

We begin by strategically placing a triangle on a rectangular coordinate system so that the vertex of angle γ is at the origin and side b lies along the positive x-axis. Regardless of whether γ is acute, as in Figure 29(a), or obtuse, as in Figure 29(b), the vertex B has coordinates $(a \cos \gamma, a \sin \gamma)$. Vertex A has coordinates $(b, 0)$.

We can now use the distance formula to compute c^2:

$$\begin{aligned} c^2 &= (b - a \cos \gamma)^2 + (0 - a \sin \gamma)^2 \\ &= b^2 - 2ab \cos \gamma + a^2 \cos^2 \gamma + a^2 \sin^2 \gamma \\ &= a^2(\cos^2 \gamma + \sin^2 \gamma) + b^2 - 2ab \cos \gamma \\ &= a^2 + b^2 - 2ab \cos \gamma \end{aligned}$$

■

Each of formulas (1), (2), and (3) may be stated in words as follows:

Theorem Law of Cosines

The square of one side of a triangle equals the sum of the squares of the other two sides minus twice their product times the cosine of their included angle.

Observe that if the triangle is a right triangle (so that, say, $\gamma = 90°$), then formula (1) becomes the familiar Pythagorean Theorem: $c^2 = a^2 + b^2$. Thus, the Pythagorean Theorem is a special case of the Law of Cosines.

1 Let's see how to use the Law of Cosines to solve Case 3 (SAS), which applies to triangles for which two sides and the included angle are known.

EXAMPLE 1 Using the Law of Cosines to Solve a SAS Triangle

Solve the triangle: $a = 2$, $b = 3$, $\gamma = 60°$

FIGURE 30

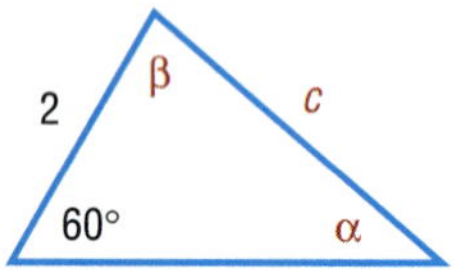

Solution See Figure 30. The Law of Cosines makes it easy to find the third side, c:

$$\begin{aligned} c^2 &= a^2 + b^2 - 2ab\cos\gamma \\ &= 4 + 9 - 2\cdot 2\cdot 3\cdot\cos 60° \\ &= 13 - \left(12\cdot\tfrac{1}{2}\right) = 7 \\ c &= \sqrt{7} \end{aligned}$$

Side c is of length $\sqrt{7}$. To find the angles α and β, we may use either the Law of Sines or the Law of Cosines. It is preferable to use the Law of Cosines, since it will lead to an equation with one solution. Using the Law of Sines would lead to an equation with two solutions that would need to be checked to determine which solution fits the given data. Thus, we choose to use formulas (2) and (3) of the Law of Cosines.

For α:

$$\begin{aligned} a^2 &= b^2 + c^2 - 2bc\cos\alpha \\ 2bc\cos\alpha &= b^2 + c^2 - a^2 \\ \cos\alpha &= \frac{b^2 + c^2 - a^2}{2bc} = \frac{9 + 7 - 4}{2\cdot 3\sqrt{7}} = \frac{12}{6\sqrt{7}} = \frac{2\sqrt{7}}{7} \\ \alpha &\approx 40.9° \end{aligned}$$

For β:

$$\begin{aligned} b^2 &= a^2 + c^2 - 2ac\cos\beta \\ \cos\beta &= \frac{a^2 + c^2 - b^2}{2ac} = \frac{4 + 7 - 9}{4\sqrt{7}} = \frac{1}{2\sqrt{7}} = \frac{\sqrt{7}}{14} \\ \beta &\approx 79.1° \end{aligned}$$

Notice that $\alpha + \beta + \gamma = 40.9° + 79.1° + 60° = 180°$, as required.

Now work Problem 1.

2 The next example illustrates how the Law of Cosines is used when three sides of a triangle are known, Case 4 (SSS).

EXAMPLE 2 Using the Law of Cosines to Solve a SSS Triangle

Solve the triangle: $a = 4, b = 3, c = 6$

FIGURE 31

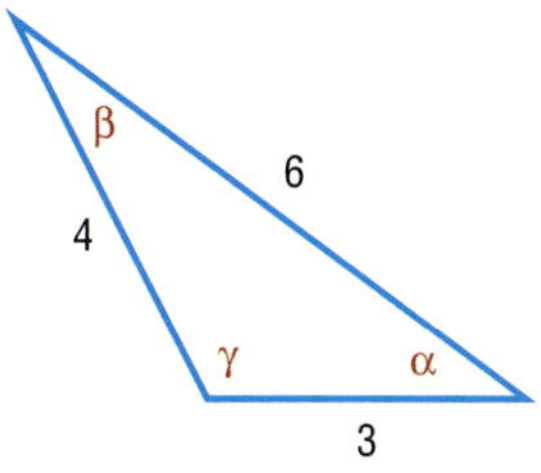

Solution See Figure 31. To find the angles α, β, and γ, we proceed as we did in the latter part of the solution to Example 1.

For α:

$$\cos\alpha = \frac{b^2 + c^2 - a^2}{2bc} = \frac{9 + 36 - 16}{2 \cdot 3 \cdot 6} = \frac{29}{36}$$

$$\alpha \approx 36.3°$$

For β:

$$\cos\beta = \frac{a^2 + c^2 - b^2}{2ac} = \frac{16 + 36 - 9}{2 \cdot 4 \cdot 6} = \frac{43}{48}$$

$$\beta \approx 26.4°$$

Since we know α and β,

$$\gamma = 180° - \alpha - \beta \approx 180° - 36.3° - 26.4° = 117.3°$$

Now work Problem 7.

EXAMPLE 3 Correcting a Navigational Error

A motorized sailboat leaves Naples, Florida, bound for Key West, 150 miles away. Maintaining a constant speed of 15 miles per hour, but encountering heavy cross-winds and strong currents, the crew finds, after 4 hours, that the sailboat is off course by 20°.

(a) How far is the sailboat from Key West at this time?

(b) Through what angle should the sailboat turn to correct its course?

(c) How much time has been added to the trip because of this? (Assume that the speed remains at 15 miles per hour.)

FIGURE 32

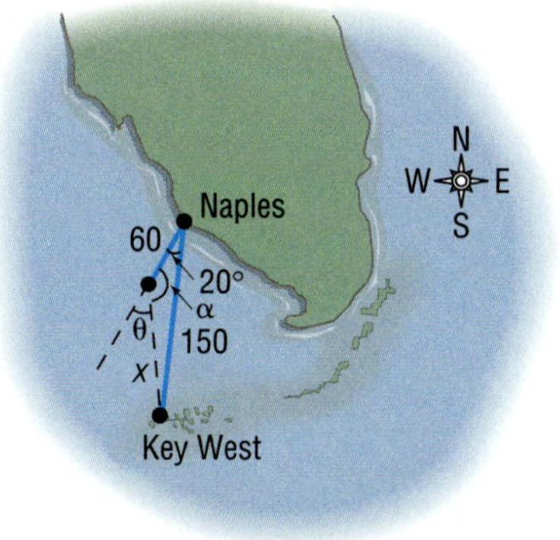

Solution See Figure 32. With a speed of 15 miles per hour, the sailboat has gone 60 miles after 4 hours. We seek the distance x of the sailboat from Key West. We also seek the angle θ that the sailboat should turn through to correct its course.

(a) To find x, we use the Law of Cosines, since we know two sides and the included angle.

$$x^2 = 150^2 + 60^2 - 2(150)(60)\cos 20° = 9186$$

$$x = 95.8$$

The sailboat is about 96 miles from Key West.

(b) We now know three sides of the triangle, so we can use the Law of Cosines again to find the angle α opposite the side of length 150 miles.

$$\begin{aligned} 150^2 &= 96^2 + 60^2 - 2(96)(60)\cos\alpha \\ 9684 &= -11{,}520 \cos\alpha \\ \cos\alpha &\approx -0.8406 \\ \alpha &\approx 147.2^\circ \end{aligned}$$

The sailboat should turn through an angle of

$$\theta = 180^\circ - \alpha \approx 180^\circ - 147.2^\circ = 32.8^\circ$$

The sailboat should turn through an angle of about 33° to correct its course.

(c) The total length of the trip is now $60 + 96 = 156$ miles. The extra 6 miles will only require about 0.4 hours or 24 minutes more if the speed of 15 miles per hour is maintained.

Now work Problem 27.

HISTORICAL FEATURE

The Law of Sines was known vaguely, long before it was explicitly stated by Nasîr ed-dîn (about A.D. 1250). Ptolemy (about A.D. 150) was aware of it in a form using a chord function instead of the sine function. But it was first clearly stated in Europe by Regiomontanus, writing in 1464.

The Law of Cosines appears first in Euclid's *Elements* (Book II), but in a well-disguised form in which squares built on the sides of triangles are added and a rectangle representing the cosine term is subtracted. It was thus known to all mathematicians because of their familiarity with Euclid's work. An early modern form of the Law of Cosines—that for finding the angle when the sides are known—was stated by François Vièta (in 1593).

The Law of Tangents (see Problem 46 of Exercise 8.2) has become obsolete. In the past it was used in place of the Law of Cosines, because the Law of Cosines was very inconvenient for calculation with logarithms or slide rules. Mixing of addition and multiplication is now quite easy on a calculator, however, and the Law of Tangents has been shelved along with the slide rule.

8.3 EXERCISES

In Problems 1–8, solve each triangle.

1. 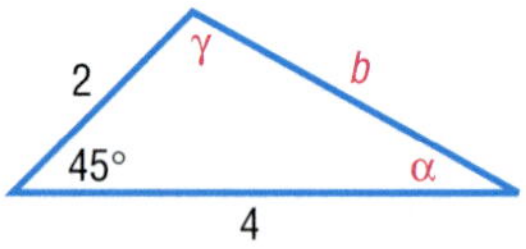

2. 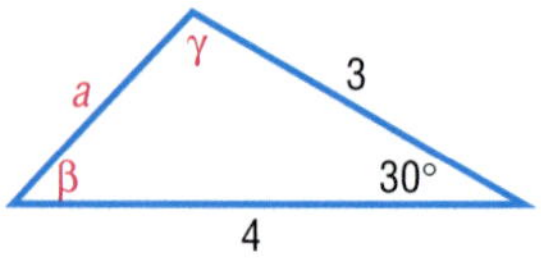

3. 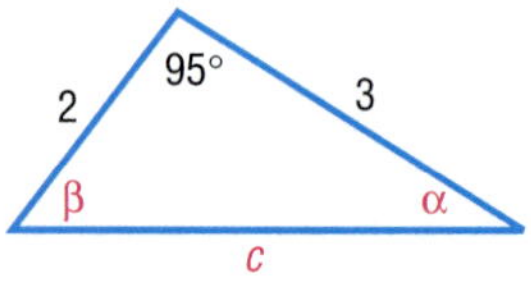

4. 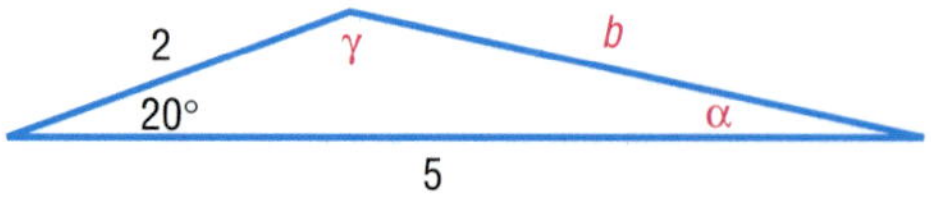

5. 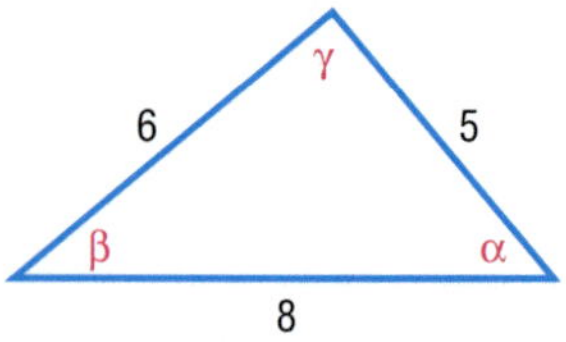

6.

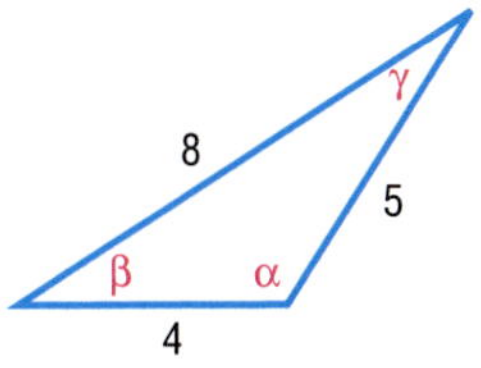

7.

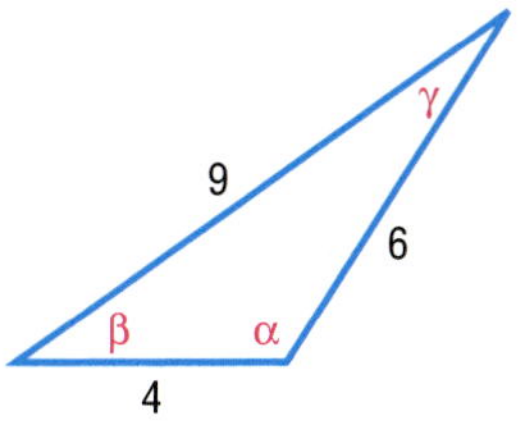

8.

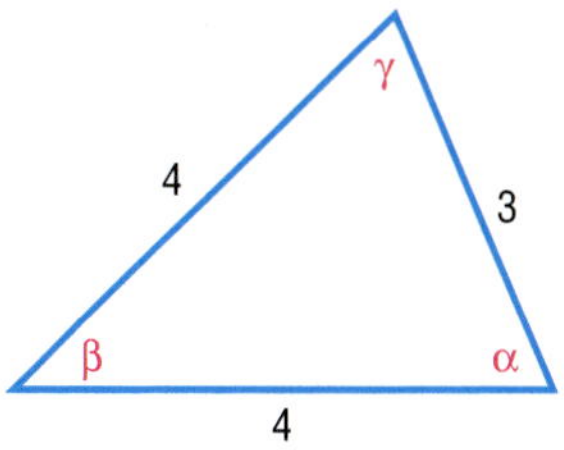

In Problems 9–24, solve each triangle.

9. $a = 3, b = 4, \gamma = 40°$

10. $a = 2, c = 1, \beta = 10°$

11. $b = 1, c = 3, \alpha = 80°$

12. $a = 6, b = 4, \gamma = 60°$

13. $a = 3, c = 2, \beta = 110°$

14. $b = 4, c = 1, \alpha = 120°$

15. $a = 2, b = 2, \gamma = 50°$

16. $a = 3, c = 2, \beta = 90°$

17. $a = 12, b = 13, c = 5$

18. $a = 4, b = 5, c = 3$

19. $a = 2, b = 2, c = 2$

20. $a = 3, b = 3, c = 2$

21. $a = 5, b = 8, c = 9$

22. $a = 4, b = 3, c = 6$

23. $a = 10, b = 8, c = 5$

24. $a = 9, b = 7, c = 10$

25. Surveying Consult the figure. To find the distance from the house at A to the house at B, a surveyor measures the angle ACB, which is found to be 70°, then walks off the distance to each house, 50 feet and 70 feet, respectively. How far apart are the houses?

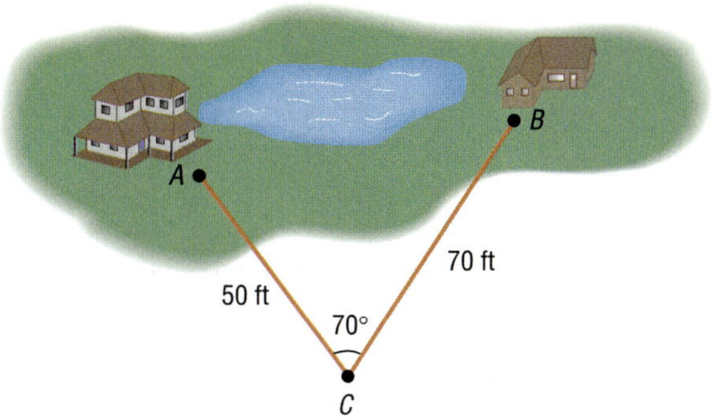

26. Navigation An airplane flies from city A to city B, a distance of 150 miles, then turns through an angle of 50° and flies to city C, a distance of 100 miles (see the figure).

(a) How far is it from city A to city C?

(b) Through what angle should the pilot turn at city C to return to city A?

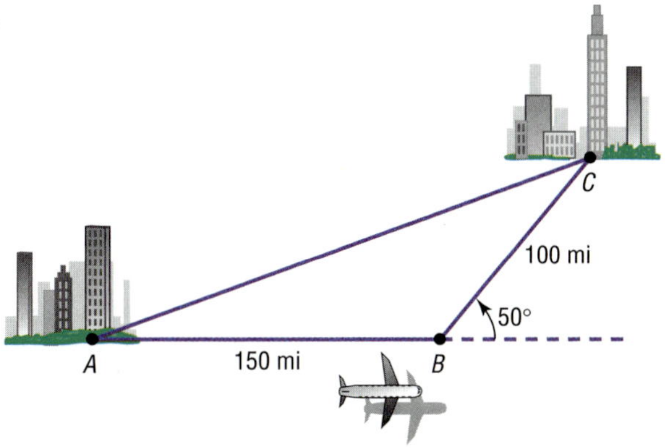

27. Revising a Flight Plan In attempting to fly from city A to city B, a distance of 330 miles, a pilot inadvertently took a course that was 10° in error, as indicated in the figure.

(a) If the aircraft maintains an average speed of 220 miles per hour and if the error in direction is discovered after 15 minutes, through what angle should the pilot turn to head toward city B?

(b) What new average speed should the pilot maintain so that the total time of the trip is 90 minutes?

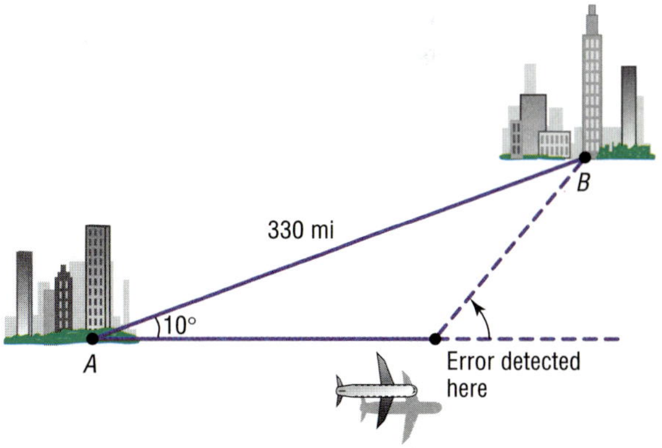

28. Avoiding a Tropical Storm A cruise ship maintains an average speed of 15 knots in going from San Juan, Puerto Rico, to Barbados, West Indies, a distance of 600 nautical miles. To avoid a tropical storm, the captain heads out of San Juan in a direction of 20° off a direct heading to Barbados. The captain maintains the 15 knot speed for 10 hours, after which time the path to Barbados becomes clear of storms.

(a) Through what angle should the captain turn to head directly to Barbados?

(b) How long will it be before the ship reaches Barbados if the same 15 knot speed is maintained?

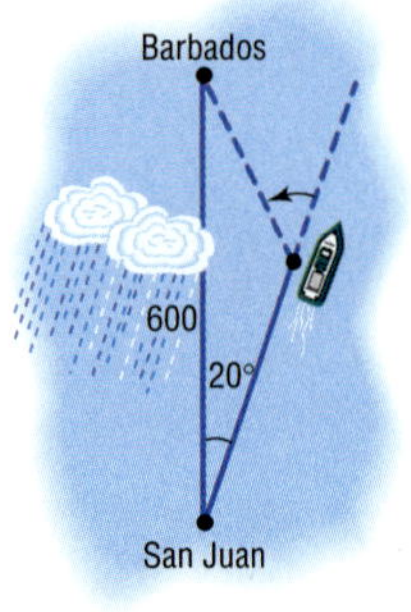

29. Major League Baseball Field A Major League baseball diamond is actually a square 90 feet on a side. The pitching rubber is located 60.5 feet from home plate on a line joining home plate and second base.

(a) How far is it from the pitching rubber to first base?

(b) How far is it from the pitching rubber to second base?

(c) If a pitcher faces home plate, through what angle does he need to turn to face first base?

30. Little League Baseball Field According to Little League baseball official regulations, the diamond is a square 60 feet on a side. The pitching rubber is located 46 feet from home plate on a line joining home plate and second base.

(a) How far is it from the pitching rubber to first base?

(b) How far is it from the pitching rubber to second base?

(c) If a pitcher faces home plate, through what angle does he need to turn to face first base?

31. Finding the Length of a Guy Wire The height of a radio tower is 500 feet, and the ground on one side of the tower slopes upward at an angle of 10° (see the figure).

(a) How long should a guy wire be if it is to connect to the top of the tower and be secured at a point on the sloped side 100 feet from the base of the tower?

(b) How long should a second guy wire be if it is to connect to the middle of the tower and be secured at a point 100 feet from the base on the flat side?

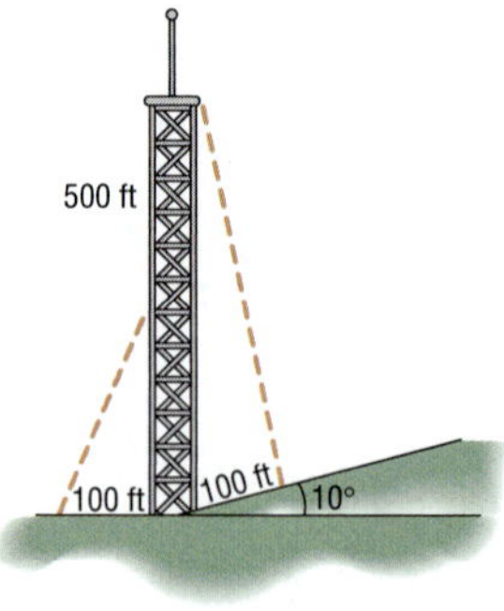

32. Finding the Length of a Guy Wire A radio tower 500 feet high is located on the side of a hill with an inclination to the horizontal of 5° (see the figure). How long should two guy wires be if they are to connect to the top of the tower and be secured at two points 100 feet directly above and directly below the base of the tower?

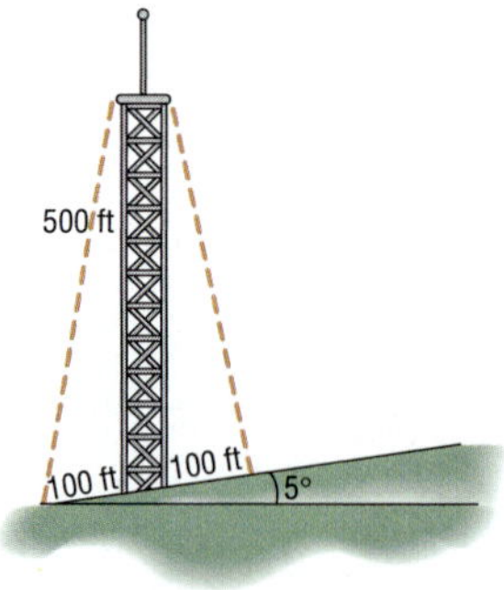

33. Wrigley Field, Home of the Chicago Cubs The distance from home plate to dead center of Wrigley Field is 400 feet (see the figure). How far is it from dead center to third base?

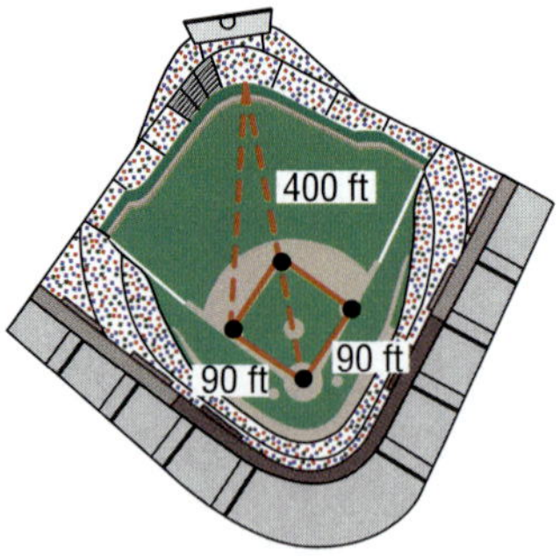

34. Little League Baseball The distance from home plate to dead center at the Oak Lawn Little League field is 280 feet. How far is it from dead center to third base?

[**Hint:** The distance between the bases in Little League is 60 feet.]

35. Rods and Pistons Rod OA (see the figure) rotates about the fixed point O so that point A travels on a circle of radius r. Connected to point A is another rod AB of length $L > r$, and point B is connected to a piston. Show that the distance x between point O and point B is given by

$$x = r\cos\theta + \sqrt{r^2\cos^2\theta + L^2 - r^2}$$

where θ is the angle of rotation of rod OA.

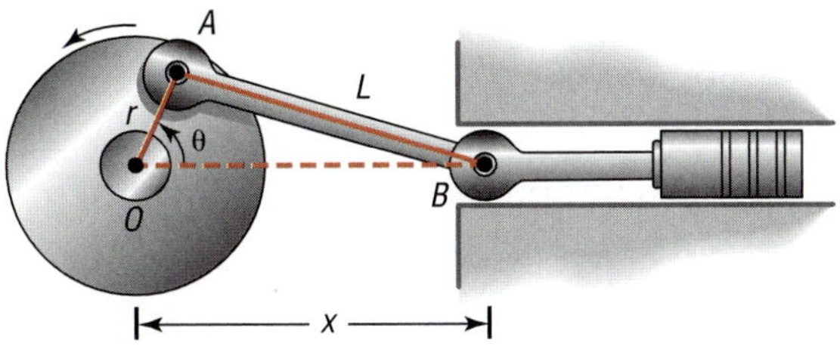

36. **Geometry** Show that the length d of a chord of a circle of radius r is given by the formula

$$d = 2r\sin\frac{\theta}{2}$$

where θ is the central angle formed by the radii to the ends of the chord (see the figure). Use this result to derive the fact that $\sin\theta < \theta$, where $\theta > 0$ is measured in radians.

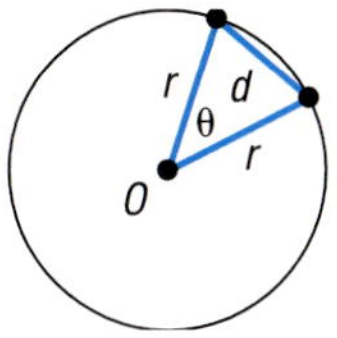

37. For any triangle, show that

$$\cos\frac{\gamma}{2} = \sqrt{\frac{s(s-c)}{ab}}$$

where $s = \frac{1}{2}(a + b + c)$.

[**Hint:** Use a half-angle formula and the Law of Cosines.]

38. For any triangle, show that

$$\sin\frac{\gamma}{2} = \sqrt{\frac{(s-a)(s-b)}{ab}}$$

where $s = \frac{1}{2}(a + b + c)$.

39. Use the Law of Cosines to prove the identity

$$\frac{\cos\alpha}{a} + \frac{\cos\beta}{b} + \frac{\cos\gamma}{c} = \frac{a^2 + b^2 + c^2}{2abc}$$

40. Write down your strategy for solving an oblique triangle.

8.4 THE AREA OF A TRIANGLE

1 Find the Area of SAS Triangles

2 Find the Area of SSS Triangles

In this section, we will derive several formulas for calculating the area A of a triangle. The most familiar of these is the following:

Theorem

The area A of a triangle is

$$A = \tfrac{1}{2}bh \qquad (1)$$

where b is the base and h is an altitude drawn to that base.

FIGURE 33
$A = \frac{1}{2}bh$

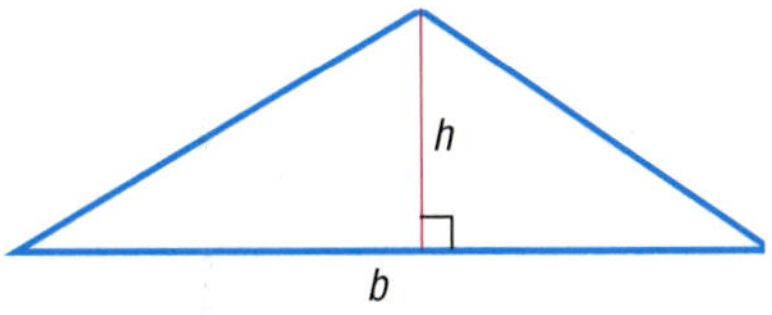

Proof The derivation of this formula is rather easy once a rectangle of base b and height h is constructed around the triangle. See Figures 33 and 34.

Triangles 1 and 2 in Figure 34 are equal in area, as are triangles 3 and 4. Consequently, the area of the triangle with base b and altitude h is exactly half the area of the rectangle, which is bh. ■

FIGURE 34

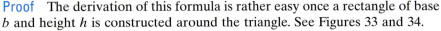

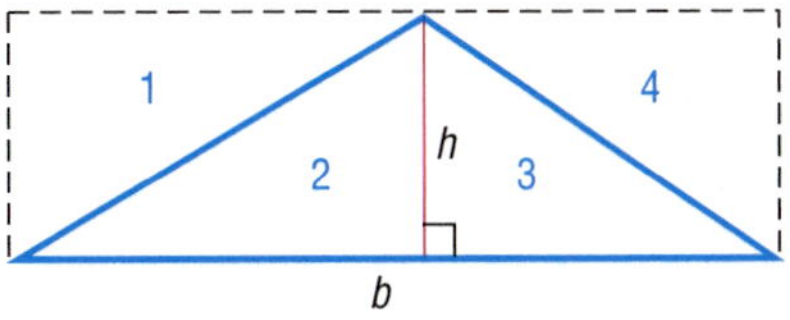

If the base b and altitude h to that base are known, then we can easily find the area of such a triangle using formula (1). Usually, though, the information required to use formula (1) is not given. Suppose, for example,

FIGURE 35
$h = a \sin \gamma$

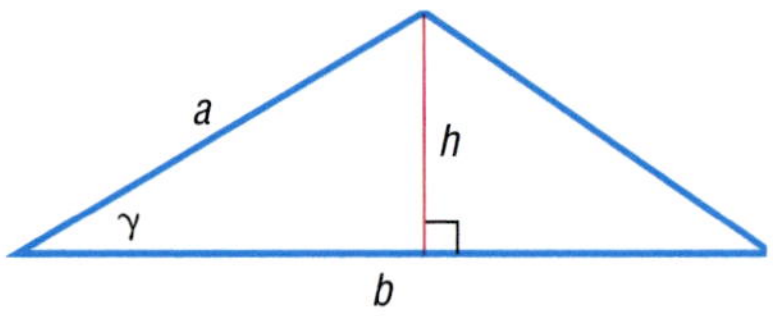

that we know two sides a and b and the included angle γ (see Figure 35). Then the altitude h can be found by noting that

$$\frac{h}{a} = \sin \gamma$$

so that

$$h = a \sin \gamma$$

Using this fact in formula (1) produces

$$A = \tfrac{1}{2}bh = \tfrac{1}{2}b(a \sin \gamma) = \tfrac{1}{2}ab \sin \gamma$$

Thus, we have the formula

$$A = \tfrac{1}{2}ab \sin \gamma \qquad (2)$$

By dropping altitudes from the other two vertices of the triangle, we obtain the following corresponding formulas:

$$A = \tfrac{1}{2}bc \sin \alpha \qquad (3)$$
$$A = \tfrac{1}{2}ac \sin \beta \qquad (4)$$

It is easiest to remember these formulas using the following wording:

Theorem

The area A of a triangle equals one-half the product of two of its sides times the sine of their included angle.

EXAMPLE 1 Finding the Area of a SAS Triangle

1 Find the area A of the triangle for which $a = 8$, $b = 6$, and $\gamma = 30°$.

Solution We use formula (2) to get

$$A = \tfrac{1}{2}ab \sin \gamma = \tfrac{1}{2} \cdot 8 \cdot 6 \sin 30° = 12$$

Now work Problem 1.

2 If the three sides of a triangle are known, another formula, called **Heron's Formula** (named after Heron of Alexandria), can be used to find the area of a triangle.

Theorem Heron's Formula

The area A of a triangle with sides a, b, and c is

$$A = \sqrt{s(s-a)(s-b)(s-c)} \qquad (5)$$

where $s = \tfrac{1}{2}(a + b + c)$.

MISSION POSSIBLE

Locating Lost Treasure

While scuba diving off Wreck Hill in Bermuda, a group of five entrepreneurs discovered a treasure map in a small water-tight cask on a pirate schooner that had sunk in 1747. The map directed them to an area of Bermuda now known as The Flatts, but when they got there, they realized that the most important landmark on the map was gone. They called in the Mission Possible team to help them recreate the map. They promised you 25% of whatever treasure was found.

The directions on the map read as follows:

1. From the tallest palm tree, sight the highest hill. Drop your eyes vertically until you sight the base of the hill.
2. Turn 40° clockwise from that line and walk 70 paces to the big red rock.
3. From the red rock walk 50 paces back to the sight line between the palm tree and the hill. Dig there.

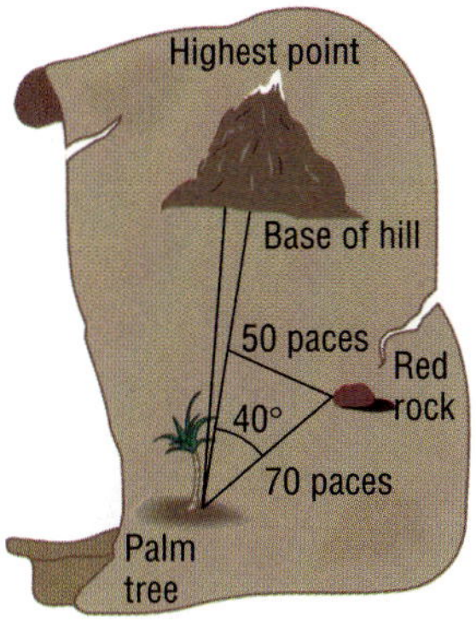

The five entrepreneurs said they believed they had found the red rock and the highest hill in the vicinity, but the "tallest palm tree" had long since fallen and disintegrated. It had occurred to them that the treasure must be located on a circle with radius 50 "paces" centered around the red rock, but they had decided against digging a trench 942 feet in circumference, especially since they had no assurance that the treasure was still there. (They had decided that a "pace" must be about a yard.)

1. Determine a plan to locate the position of the lost palm tree, and write out an explanation of your procedure for the entrepreneurs.
2. Unfortunately, it turns out that the entrepreneurs had more in common with the eighteenth century pirates than you had bargained for. Once you told them the location of the lost palm tree, they tied you all to the red rock, saying they could take it from there. From the location of the palm tree, they sighted 40° counterclockwise from the rock to the hill, then ran about 50 yards to the circle they had traced about the rock and began to dig frantically. Nothing. After about an hour, they drove off shouting back at you, "25% of nothing is nothing!"
3. Fortunately, the entrepreneurs had left the shovels. After you managed to untie yourselves, you went to the correct location and found the treasure. Where was it? How far from the palm tree? Explain.
4. People who scuba dive for sunken treasure have certain legal obligations. What are they? Should you share the treasure with a lawyer, just to make sure you get to keep the rest?

Proof The proof we shall give uses the Law of Cosines and is quite different from the proof given by Heron.

From the Law of Cosines,

$$c^2 = a^2 + b^2 - 2ab\cos\gamma$$

and the two half-angle formulas,

$$\cos^2\frac{\gamma}{2} = \frac{1+\cos\gamma}{2} \qquad \sin^2\frac{\gamma}{2} = \frac{1-\cos\gamma}{2}$$

we find

$$\cos^2\frac{\gamma}{2} = \frac{1+\cos\gamma}{2} = \frac{1+\dfrac{a^2+b^2-c^2}{2ab}}{2}$$

$$= \frac{a^2+2ab+b^2-c^2}{4ab} = \frac{(a+b)^2-c^2}{4ab}$$

$$= \frac{(a+b-c)(a+b+c)}{4ab} = \frac{2(s-c)\cdot 2s}{4ab} = \frac{s(s-c)}{ab} \qquad (6)$$

↑ $a+b-c = a+b+c-2c = 2s-2c$

Similarly,

$$\sin^2\frac{\gamma}{2} = \frac{(s-a)(s-b)}{ab} \qquad (7)$$

Now we use formula (2) for the area:

$$A = \frac{1}{2}ab\sin\gamma$$

$$= \frac{1}{2}ab\cdot 2\sin\frac{\gamma}{2}\cos\frac{\gamma}{2} \qquad \sin\gamma = \sin 2\left(\frac{\gamma}{2}\right) = 2\sin\frac{\gamma}{2}\cos\frac{\gamma}{2}$$

$$= ab\sqrt{\frac{(s-a)(s-b)}{ab}}\sqrt{\frac{s(s-c)}{ab}} \qquad \text{Use equations (6) and (7).}$$

$$= \sqrt{s(s-a)(s-b)(s-c)}$$

■

EXAMPLE 2 Finding the Area of a SSS Triangle

Find the area of a triangle whose sides are 4, 5, and 7.

Solution We let $a = 4$, $b = 5$, and $c = 7$. Then

$$s = \tfrac{1}{2}(a+b+c) = \tfrac{1}{2}(4+5+7) = 8$$

Heron's Formula then gives the area A as

$$A = \sqrt{s(s-a)(s-b)(s-c)} = \sqrt{8\cdot 4\cdot 3\cdot 1} = \sqrt{96} = 4\sqrt{6}$$

■

Now work Problem 7.

Historical Feature

Heron's formula (also known as *Hero's Formula*) is due to Heron of Alexandria (about A.D. 75), who had, besides his mathematical talents, a good deal of engineering skills. In various temples his mechanical devices produced effects that seemed supernatural, and visitors presumably were thus influenced to generosity. Heron's book *Metrica,* on making such devices, has survived and was discovered in 1896 in the city of Constantinople.

Heron's Formula for the area of a triangle caused some mild discomfort in Greek mathematics, because a product with two factors was an area while one with three factors was a volume, but four factors seemed contradictory in Heron's time.

Karl Mollweide (1774–1875), a mathematician and astronomer, discovered the formulas named for him (see Problems 43 and 44 of Exercise 8.2). These formulas are not too important in themselves but often simplify the derivation of other formulas, as demonstrated in Historical Problems 1 and 2, which follow.

Historical Problems

1. This derivation of Heron's formula uses Mollweide's formula.

(a) Show that

$$\frac{s}{c} = \frac{\cos(\alpha/2)\cos(\beta/2)}{\sin(\gamma/2)}$$

where $s = \frac{1}{2}(a + b + c)$.

Hint: Use Mollweide's formula (see Problem 43 in Exercise 8.2) and add 1 to both sides. Then use the fact that

$$\sin\frac{\gamma}{2} = \sin\frac{180° - (\alpha + \beta)}{2} = \cos\frac{\alpha + \beta}{2}$$

(b) Similarly, show that

$$\frac{s}{a} = \frac{\cos(\beta/2)\cos(\gamma/2)}{\sin(\alpha/2)} \quad \text{and} \quad \frac{s}{b} = \frac{\cos(\alpha/2)\cos(\gamma/2)}{\sin(\beta/2)}$$

(c) Use the results of parts (a) and (b) to show that

$$\frac{s-a}{a} = \frac{\sin(\beta/2)\sin(\gamma/2)}{\sin(\alpha/2)} \qquad \frac{s-b}{b} = \frac{\sin(\alpha/2)\sin(\gamma/2)}{\sin(\beta/2)}$$

$$\frac{s-c}{c} = \frac{\sin(\alpha/2)\sin(\beta/2)}{\sin(\gamma/2)}$$

(d) Now form the product

$$\frac{s}{c}\cdot\frac{s-a}{a}\cdot\frac{s-b}{b}\cdot\frac{s-c}{c}$$

After cancellations, multiply each side by *cabc*, use the double-angle formulas, and use the fact that $A = \frac{1}{2}bc\sin\alpha = \frac{1}{2}ac\sin\beta$. Heron's Formula will then follow.

2. We again use Mollweide's Formula to derive some other interesting formulas. (These were derived in another way in Problems 37 and 38 of Exercise 8.3.)

(a) Add 1 to both sides of the second form of Mollweide's Formula (see Problem 44 in Exercise 8.2) and simplify to obtain

$$\frac{s-b}{c} = \frac{\sin(\alpha/2)\cos(\beta/2)}{\cos(\gamma/2)}$$

(b) Similarly, show that

$$\frac{s-c}{b} = \frac{\sin(\alpha/2)\cos(\gamma/2)}{\cos(\beta/2)}$$

(c) Use the results of part (b) and Problem 1(b) and (c) to show that

$$\cos^2\frac{\gamma}{2} = \frac{s(s-c)}{ab} \quad \text{and} \quad \sin^2\frac{\gamma}{2} = \frac{(s-a)(s-b)}{ab}$$

(d) Show that

$$\tan^2\frac{\gamma}{2} = \frac{(s-a)(s-b)}{s(s-c)}$$

3. (a) If h_1, h_2, and h_3 are the altitudes dropped from A, B, and C, respectively, in a triangle (see the figure), show that

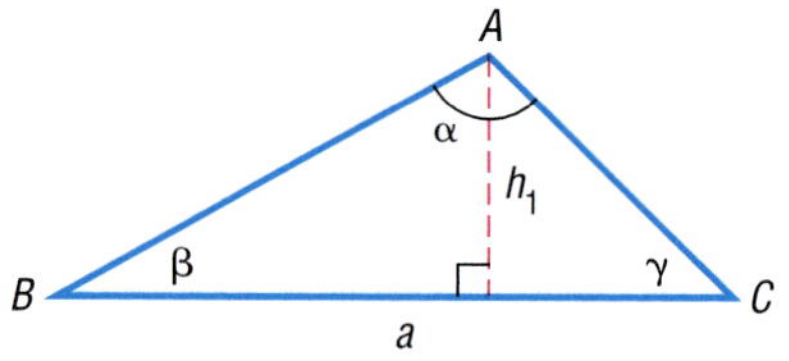

$$\frac{1}{h_1} + \frac{1}{h_2} + \frac{1}{h_3} = \frac{s}{K}$$

where K is the area of the triangle and $s = \frac{1}{2}(a+b+c)$.

[**Hint:** $h_1 = 2K/a$.]

(b) Show that a formula for the altitude h from a vertex to the opposite side a of a triangle is

$$h = \frac{a\sin\beta\sin\gamma}{\sin\alpha}$$

4. **Inscribed Circle.** The lines that bisect each angle of a triangle meet in a single point O, and the perpendicular distance r from O to each side of the triangle is the same. The circle with center at O and radius r is called the *inscribed circle* of the triangle (see the figure).

(a) Apply Problem 3(b) to triangle OAB to show that

$$r = \frac{c\sin(\alpha/2)\sin(\beta/2)}{\cos(\gamma/2)}$$

(b) Use the results of part (a) and Problem 1(c) to show that

$$r = (s-c)\tan\frac{\gamma}{2}$$

(c) Show that

$$\cot\frac{\alpha}{2} + \cot\frac{\beta}{2} + \cot\frac{\gamma}{2} = \frac{s}{r}$$

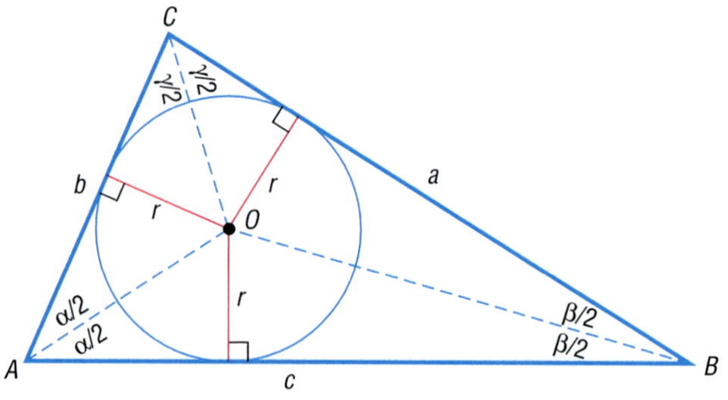

(d) Show that the area K of triangle ABC is $K = rs$. Then show that

$$r = \sqrt{\frac{(s-a)(s-b)(s-c)}{s}}$$

where $s = \frac{1}{2}(a + b + c)$.

5. Find the coordinates of the center O of the inscribed circle in terms of a, b, c, α, β, and γ. Then write the general equation of the inscribed circle.

[**Hint:** Use a system of rectangular coordinates with the vertex A at the origin and side c along the positive x-axis.]

8.4 EXERCISES

In Problems 1–8, find the area of each triangle. Round off answers to two decimal places.

1.

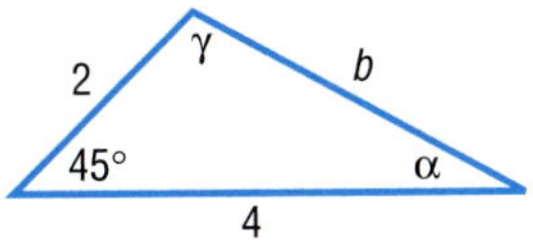

2.

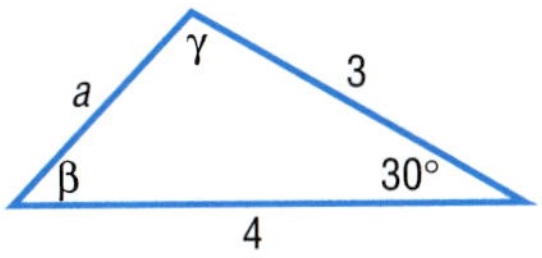

3.

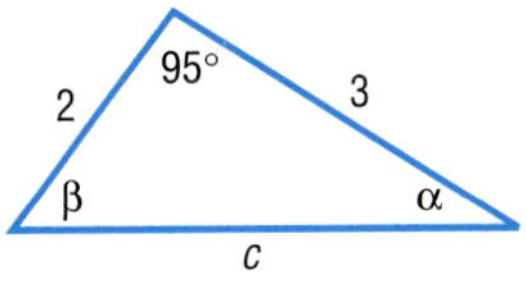

4.

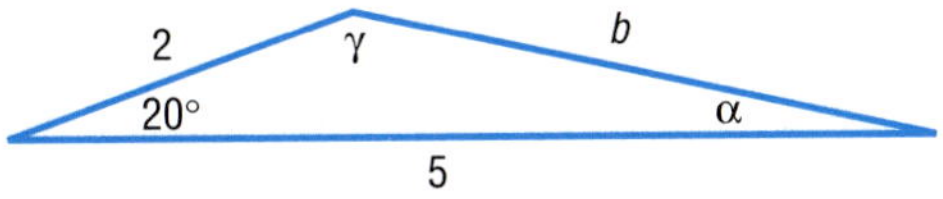

5.

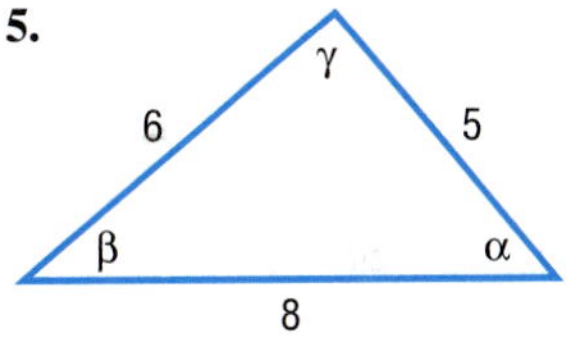

6.

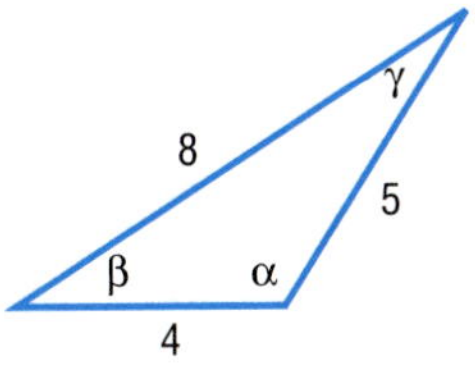

7.

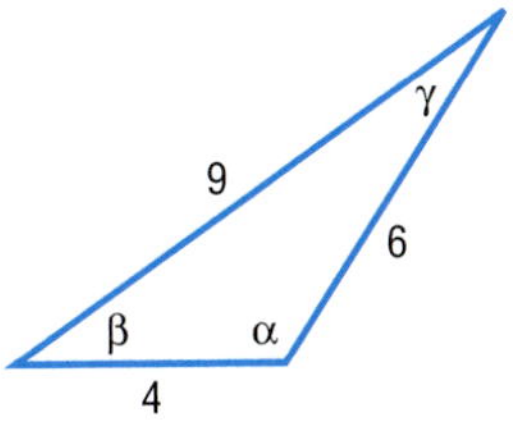

8.

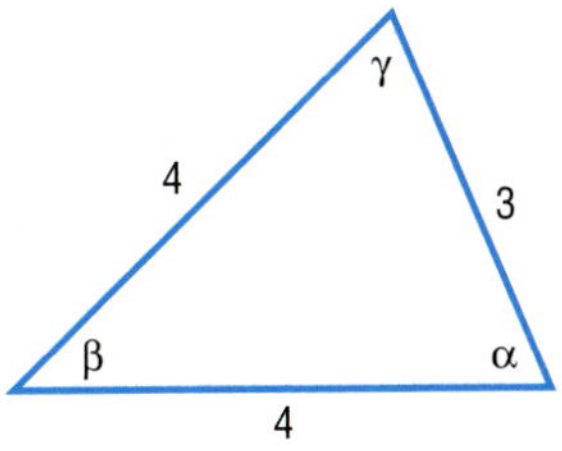

In Problems 9–24, find the area of each triangle. Round off answers to two decimal places.

9. $a = 3, b = 4, \gamma = 40°$

10. $a = 2, c = 1, \beta = 10°$

11. $b = 1, c = 3, \alpha = 80°$

12. $a = 6, b = 4, \gamma = 60°$

13. $a = 3, c = 2, \beta = 110°$

14. $b = 4, c = 1, \alpha = 120°$

15. $a = 2, b = 2, \gamma = 50°$

16. $a = 3, c = 2, \beta = 90°$

17. $a = 12, b = 13, c = 5$

18. $a = 4, b = 5, c = 3$

19. $a = 2, b = 2, c = 2$

20. $a = 3, b = 3, c = 2$

21. $a = 5, b = 8, c = 9$

22. $a = 4, b = 3, c = 6$

23. $a = 10, b = 8, c = 5$

24. $a = 9, b = 7, c = 10$

25. Cost of a Triangular Lot The dimensions of a triangular lot are 100 feet by 50 feet by 75 feet. If the price of such land is \$3 per square foot, how much does the lot cost?

26. Approximating the Area of a Lake To approximate the area of a lake, a surveyor walks around the perimeter of the lake, taking the measurements shown in the illustration. Using this technique, what is the approximate area of the lake?

[**Hint:** Use the Law of Cosines on the three triangles shown and then find the sum of their areas.]

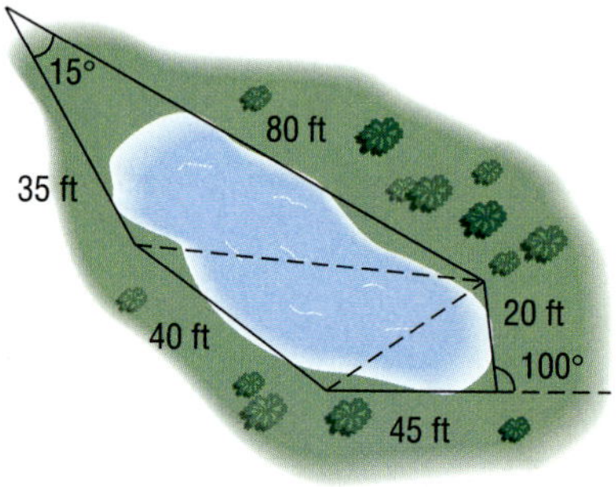

27. The Cow Problem* A cow is tethered to one corner of a square barn, 10 feet by 10 feet, with a rope 100 feet long. What is the maximum grazing area of the cow?

[**Hint:** See the illustration.]

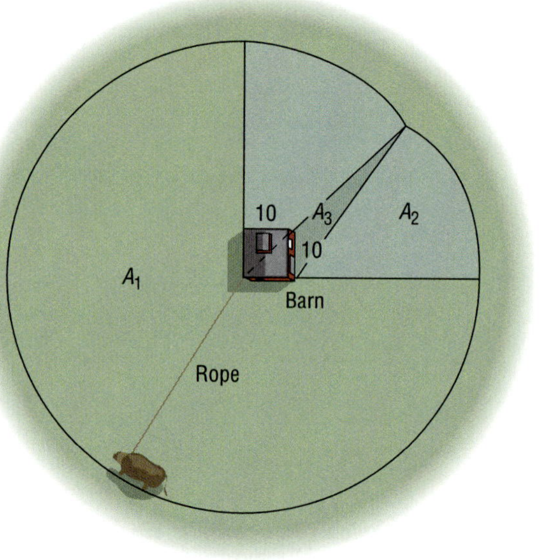

28. Another Cow Problem If the barn in Problem 27 is rectangular, 10 feet by 20 feet, what is the maximum grazing area for the cow?

29. Area of a Triangle Prove that the area A of a triangle is given by the formula

$$A = \frac{a^2 \sin \beta \sin \gamma}{2 \sin \alpha}$$

30. Area of a Triangle Prove the two other forms of the formula given in Problem 29,

$$A = \frac{b^2 \sin \alpha \sin \gamma}{2 \sin \beta} \quad \text{and} \quad A = \frac{c^2 \sin \alpha \sin \beta}{2 \sin \gamma}$$

In Problems 31–38, use the results of Problem 29 or 30 to find the area of each triangle. Round off answers to two decimal places.

31. $\alpha = 40°, \beta = 20°, a = 2$
32. $\alpha = 50°, \gamma = 20°, a = 3$
33. $\beta = 70°, \gamma = 10°, b = 5$
34. $\alpha = 70°, \beta = 60°, c = 4$
35. $\alpha = 110°, \gamma = 30°, c = 3$
36. $\beta = 10°, \gamma = 100°, b = 2$
37. $\alpha = 40°, \beta = 40°, c = 2$
38. $\beta = 20°, \gamma = 70°, a = 1$

39. Geometry Consult the figure on the right, which shows a circle of radius r with center at O. Find the area A of the shaded region as a function of the central angle θ.

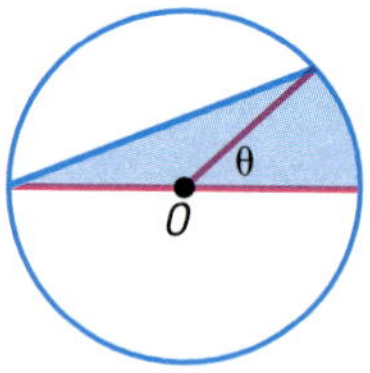

*Suggested by Professor Teddy Koukounas of SUNY at Old Westbury, who learned of it from an old farmer in Virginia. Solution provided by Professor Kathleen Miranda of SUNY at Old Westbury.

8.5 SINUSOIDAL GRAPHS; SINUSOIDAL CURVE FITTING

1. Determine the Amplitude and Period of Sinusoidal Functions
2. Find an Equation for a Sinusoidal Graph
3. Determine the Phase Shift of a Sinusoidal Function
4. Graph Sinusoidal Functions
5. Find a Sinusoidal Function from Data

Because $\sin x = \cos(x - \pi/2)$, it follows that the graph of $y = \sin x$ is the same as the graph of $y = \cos x$ after a horizontal shift of $\pi/2$ units to the right. See Figure 36.

FIGURE 36

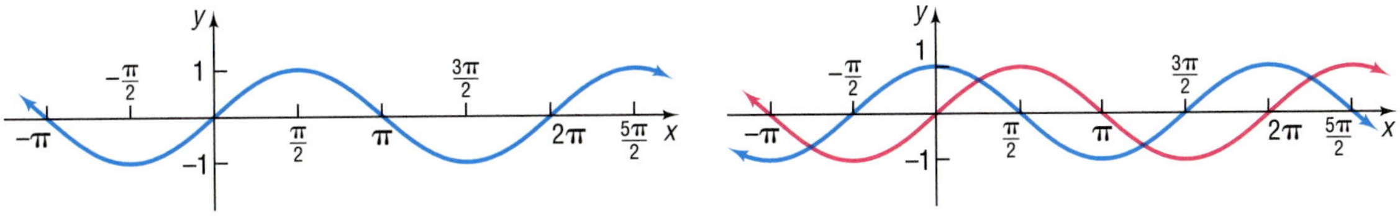

(a) $y = \sin x$ **(b)** $y = \cos x$ $y = \cos\left(x - \frac{\pi}{2}\right)$

Because of this, the graphs of sine functions and cosine functions are referred to as **sinusoidal graphs.**

Let's look at some general characteristics of sinusoidal graphs.

1 In Figure 37, we show the graph of $y = 2\cos x$. Notice that the values of $y = 2\cos x$ lie between -2 and 2, inclusive.

FIGURE 37
$y = 2\cos x$

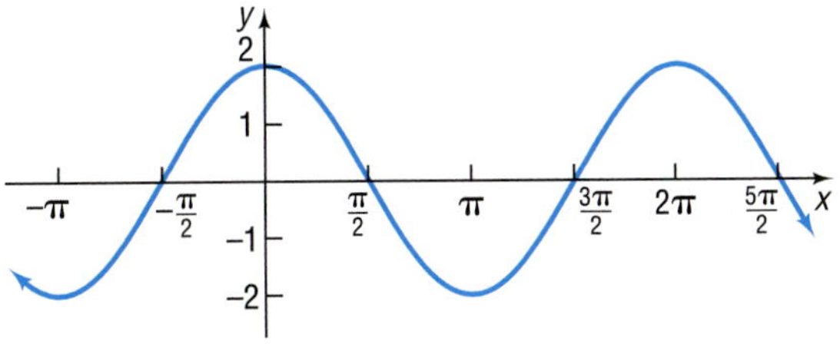

In general, the values of the functions $y = A\sin x$ and $y = A\cos x$, where $A \neq 0$, will always satisfy the inequalities.

$$-|A| \le A\sin x \le |A| \quad \text{and} \quad -|A| \le A\cos x \le |A|$$

respectively. The number $|A|$ is called the **amplitude** of $y = A\sin x$ or $y = A\cos x$. See Figure 38.

FIGURE 38

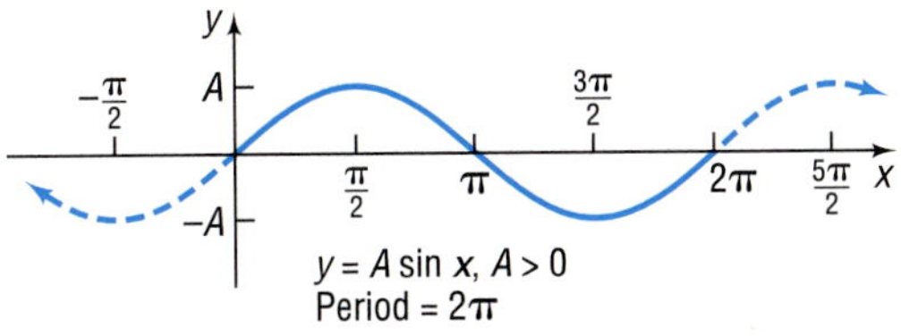

In Figure 39 we show the graph of $y = \cos 3x$. Notice that the period of this function is $2\pi/3$.

FIGURE 39
$y = \cos 3x$

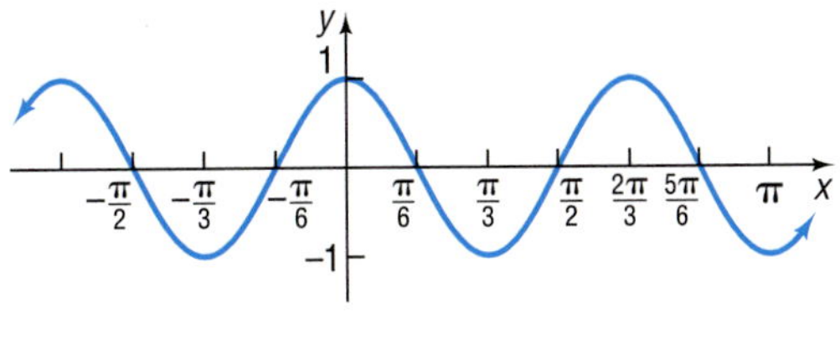

In general, if $\omega > 0$, the functions $y = \sin \omega x$ and $y = \cos \omega x$ will have period $T = 2\pi/\omega$. To see why, recall that the graph of $y = \sin \omega x$ is obtained from the graph of $y = \sin x$ by performing an appropriate horizontal compression or stretch. This horizontal compression replaces the interval $[0, 2\pi]$, which contains one period of the graph of $y = \sin x$, by the interval $[0, 2\pi/\omega]$, which contains one period of the graph of $y = \sin \omega x$. Thus, the period of the functions $y = \sin \omega x$ and $y = \cos \omega x$, $\omega > 0$, is $2\pi/\omega$. See Figure 40.

FIGURE 40

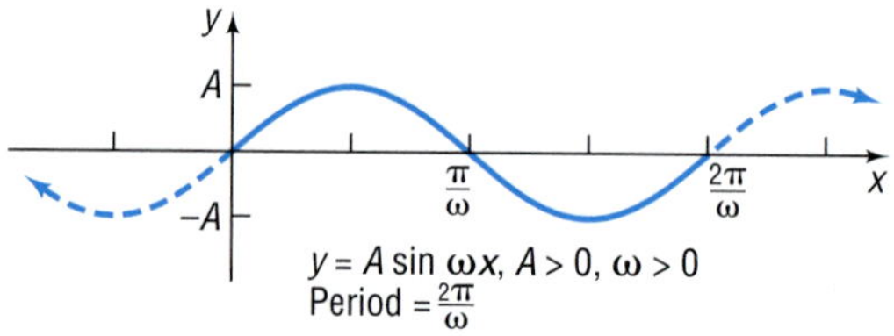

If $\omega < 0$ in $y = \sin \omega x$ or $y = \cos \omega x$, we use the facts that

$$\sin \omega x = -\sin(-\omega x) \quad \text{and} \quad \cos \omega x = \cos(-\omega x)$$

to obtain an equivalent form in which the coefficient of x is positive.

Theorem

If $\omega > 0$, the amplitude and period of $y = A \sin \omega x$ and $y = A \cos \omega x$ are given by

$$\text{Amplitude} = |A| \qquad \text{Period} = T = \frac{2\pi}{\omega} \tag{1}$$

EXAMPLE 1 Finding the Amplitude and Period of a Sinusoidal Function

Determine the amplitude and period of $y = 3 \sin 4x$, and graph the function.

FIGURE 41

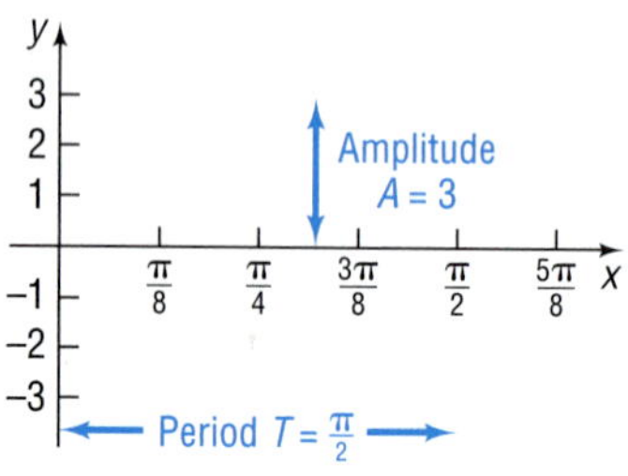

Solution Comparing $y = 3 \sin 4x$ to $y = A \sin \omega x$, we find that $A = 3$ and $\omega = 4$. Thus, from equation (1), the amplitude is $|A| = 3$ and the period is $T = 2\pi/\omega = 2\pi/4 = \pi/2$. We can use this information to graph $y = 3 \sin 4x$ by beginning as shown in Figure 41.

To graph $y = 3 \sin 4x$ using a graphing utility, we use the amplitude to set Ymin and Ymax and use the period to set Xmin and Xmax. See Figure 42(a). To graph $y = 3 \sin 4x$ by hand, we use the amplitude to scale the y-axis and the period to scale the x-axis. Then we fill in the graph of the sine function. See Figure 42(b).

FIGURE 42

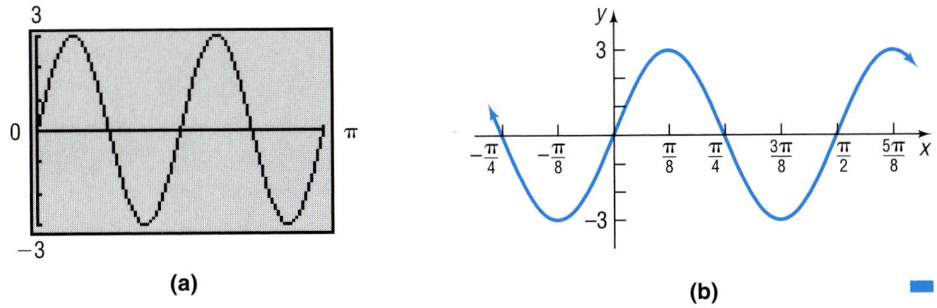

Now work Problem 5.

EXAMPLE 2 Finding the Amplitude and Period of a Sinusoidal Function

Determine the amplitude and period of $y = -4 \cos \pi x$, and graph the function.

Solution Comparing $y = -4 \cos \pi x$ with $y = A \cos \omega x$, we find that $A = -4$ and $\omega = \pi$. Thus, the amplitude is $|A| = |-4| = 4$, and the period is $T = 2\pi/\omega = 2\pi/\pi = 2$.

FIGURE 43

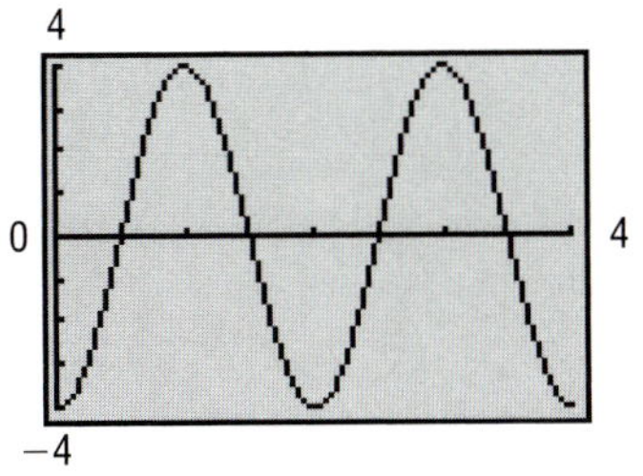

Figure 43 shows the graph of $y = -4 \cos \pi x$ using a graphing utility.

To graph $y = -4 \cos \pi x$ by hand, we use the amplitude to scale the y-axis, the period to scale the x-axis and fill in the graph of the cosine function, thus obtaining the graph of $y = 4 \cos \pi x$ shown in Figure 44(a). Now, since we want the graph of $y = -4 \cos \pi x$, we reflect the graph of $y = 4 \cos \pi x$ about the x-axis, as shown in Figure 44(b).

FIGURE 44

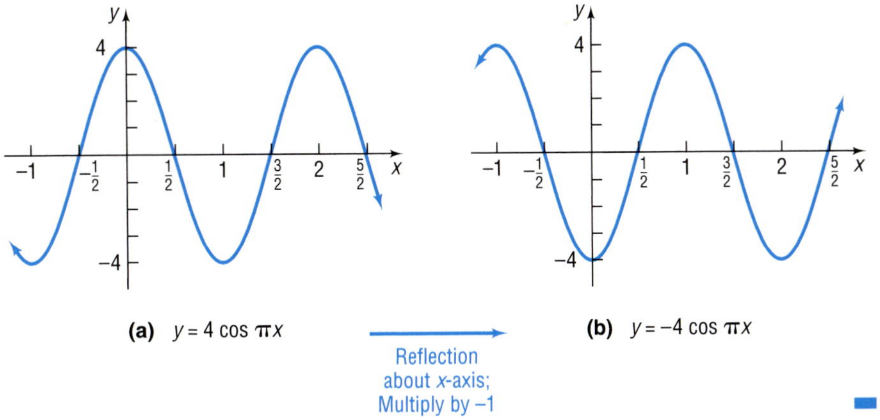

(a) $y = 4 \cos \pi x$ (b) $y = -4 \cos \pi x$

EXAMPLE 3 Finding the Amplitude and Period of a Sinusoidal Function

Determine the amplitude and period of $y = 2 \sin(-\pi x)$, and graph the function.

Solution Since the sine function is odd, we use the equivalent form

$$y = -2 \sin \pi x$$

Comparing $y = -2 \sin \pi x$ to $y = A \sin \omega x$, we find that $A = -2$ and $\omega = \pi$. Thus, the amplitude is $|A| = 2$, and the period is $T = 2\pi/\omega = 2\pi/\pi = 2$.

FIGURE 45

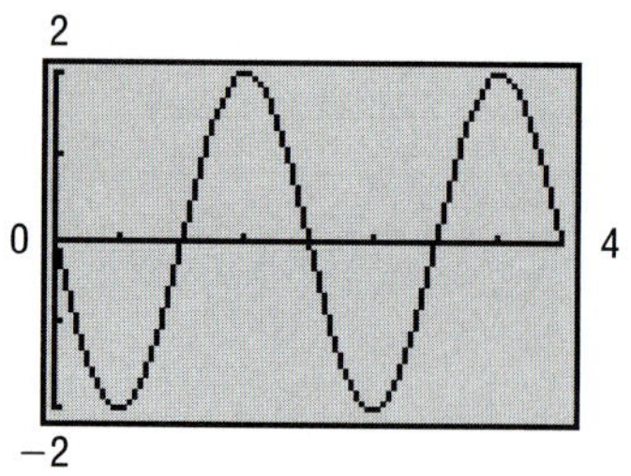

Figure 45 shows the graph of $y = 2\sin(-\pi x)$ using a graphing utility.

To graph $y = -2\sin \pi x$ by hand, we use the amplitude to scale the y-axis, the period to scale the x-axis, and fill in the graph of the sine function. Figure 46(a) shows the resulting graph of $y = 2\sin \pi x$. To obtain the graph of $y = -2\sin \pi x$, we reflect the graph of $y = 2\sin \pi x$ about the x-axis, as shown in Figure 46(b).

FIGURE 46

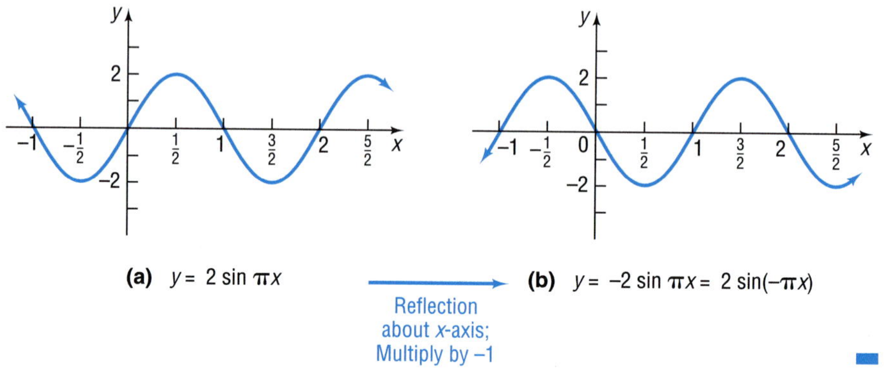

2 We also can use the ideas of amplitude and period to identify a sinusoidal function when its graph is given.

EXAMPLE 4 Finding an Equation for a Sinusoidal Graph

Find an equation for the graph shown in Figure 47.

FIGURE 47

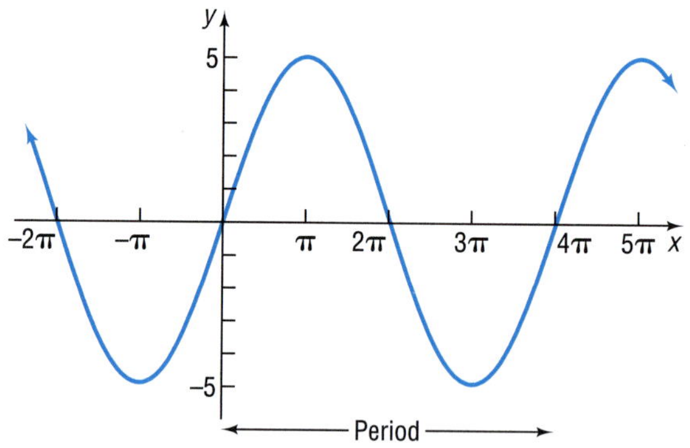

Solution This graph can be viewed as the graph of a sine function* with amplitude $A = 5$. The period T is observed to be 4π. Thus, by equation (1),

$$T = \frac{2\pi}{\omega}$$

$$4\pi = \frac{2\pi}{\omega}$$

$$\omega = \frac{2\pi}{4\pi} = \frac{1}{2}$$

*The equation could also be viewed as a cosine function with a horizontal shift, but viewing it as a sine function is easier.

A sine function whose graph is given in Figure 47 is

$$y = A \sin \omega x = 5 \sin \frac{x}{2}$$

Check: Graph $Y_1 = 5 \sin \frac{x}{2}$ and compare the result with Figure 47.

EXAMPLE 5

Finding an Equation for a Sinusoidal Graph

Find an equation for the graph shown in Figure 48.

FIGURE 48

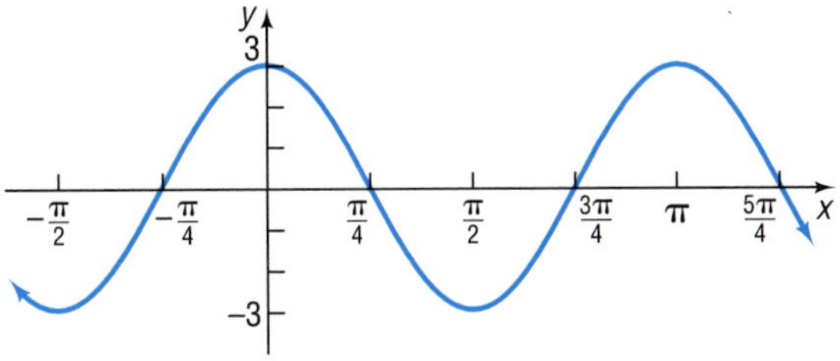

Solution From the graph we conclude that it is easier to view the equation as a cosine function with amplitude $A = 3$ and period $T = \pi$. Thus, $2\pi/\omega = \pi$, so $\omega = 2$. A cosine function whose graph is given in Figure 48 is

$$y = A \cos \omega x = 3 \cos 2x$$

Check: Graph $Y_1 = 3 \cos 2x$ and compare the result with Figure 48.

EXAMPLE 6

Finding an Equation for a Sinusoidal Graph

Find an equation for the graph shown in Figure 49.

FIGURE 49

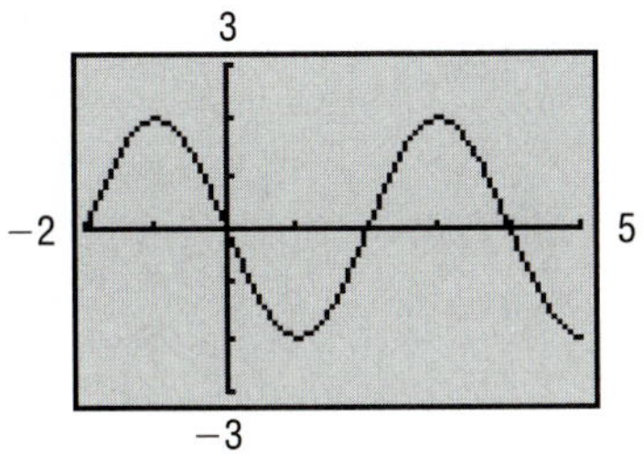

Solution The graph is sinusoidal, with amplitude $A = 2$. The period is 4, so $2\pi/\omega = 4$ or $\omega = \pi/2$. Since the graph passes through the origin, it is easiest to view the equation as a sine function, but notice that the graph is actually the reflection of a sine function about the x-axis (since the graph is decreasing near the origin). Thus, we have

$$y = -A \sin \omega x = -2 \sin \frac{\pi}{2} x$$

Check: Graph $Y_1 = -2 \sin \frac{\pi}{2} x$ and compare the result with Figure 49.

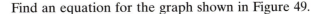

Now work Problem 37.

Phase Shift

3 We have seen that the graph of $y = A \sin \omega x$, $\omega > 0$, has amplitude $|A|$ and period $T = 2\pi/\omega$. Thus, one period can be drawn as x varies from 0 to $2\pi/\omega$ or, equivalently, as ωx varies from 0 to 2π. See Figure 50.

We now want to discuss the graph of

FIGURE 50
One period of $y = A \sin \omega x$, $A > 0$, $\omega > 0$

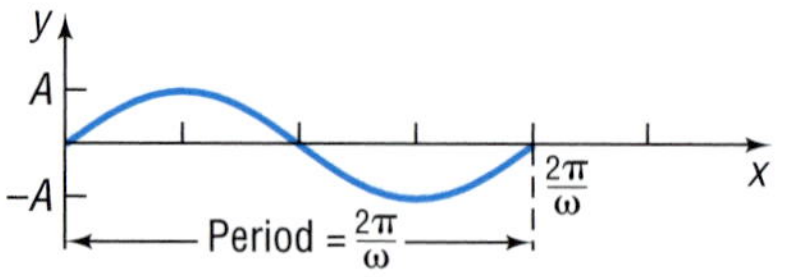

$$y = A \sin(\omega x - \phi) = A \sin \omega\left(x - \frac{\phi}{\omega}\right)$$

where $\omega > 0$ and ϕ (the Greek letter phi) are real numbers. The graph will be a sine curve of amplitude $|A|$. As $\omega x - \phi$ varies from 0 to 2π, one period will be traced out. This period will begin when

$$\omega x - \phi = 0 \quad \text{or} \quad x = \frac{\phi}{\omega}$$

and will end when

$$\omega x - \phi = 2\pi \quad \text{or} \quad x = \frac{2\pi}{\omega} + \frac{\phi}{\omega}$$

FIGURE 51
One period of $y = A \sin(\omega x - \phi)$, $A > 0$, $\omega > 0$, $\phi > 0$

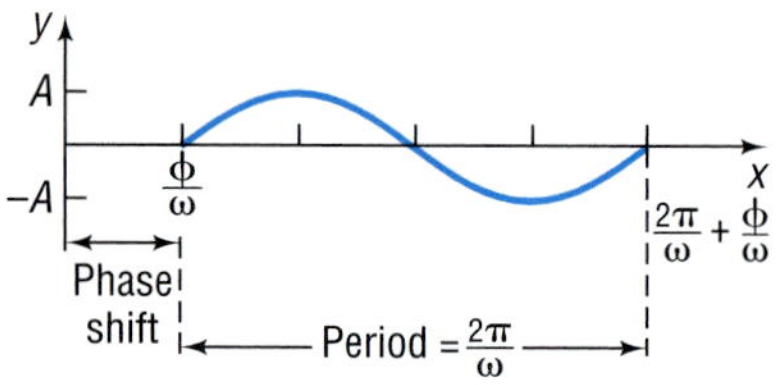

See Figure 51.

Thus, we see the graph of $y = A \sin(\omega x - \phi) = A \sin \omega\left(x - \frac{\phi}{\omega}\right)$ is the same as the graph of $y = A \sin \omega x$, except that it has been shifted ϕ/ω units (to the right if $\phi > 0$ and to the left if $\phi < 0$). This number ϕ/ω is called the **phase shift** of the graph of $y = A \sin(\omega x - \phi)$.

For the graphs of $y = A \sin(\omega x - \phi)$ or $y = A \cos(\omega x - \phi)$, $\omega > 0$,

$$\text{Amplitude} = |A| \qquad \text{Period} = T = \frac{2\pi}{\omega} \qquad \text{Phase shift} = \frac{\phi}{\omega}$$

EXAMPLE 7 Finding the Amplitude, Period, and Phase Shift of a Sinusoidal Function

4 Find the amplitude, period, and phase shift of $y = 3 \sin(2x - \pi)$, and graph the function.

Solution Comparing $y = 3 \sin(2x - \pi)$ to $y = A \sin(\omega x - \phi)$, we find that $A = 3$, $\omega = 2$, and $\phi = \pi$. The graph is a sine curve with amplitude $A = 3$, period $T = 2\pi/\omega = 2\pi/2 = \pi$, and phase shift $= \frac{\phi}{\omega} = \frac{\pi}{2}$.

The graph of $y = 3\sin(2x - \pi) = 3\sin 2\left(x - \dfrac{\pi}{2}\right)$ is obtained by using the graphing techniques discussed in Chapter 2. See Figure 52.

FIGURE 52

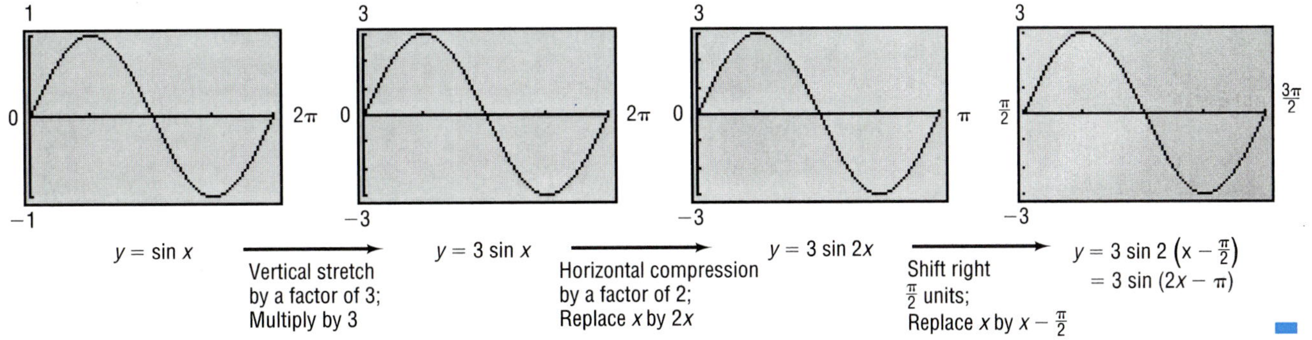

EXAMPLE 8 Finding the Amplitude, Period, and Phase Shift of a Sinusoidal Function

Find the amplitude, period, and phase shift of $y = 2\cos(4x + 3\pi)$, and graph the function.

Solution Comparing $y = 2\cos(4x + 3\pi)$ to $y = A\cos(\omega x - \phi)$, we see that $A = 2$, $\omega = 4$, and $\phi = -3\pi$. The graph is a cosine curve with amplitude $A = 2$, period $T = 2\pi/\omega = 2\pi/4 = \pi/2$, and phase shift $= \dfrac{\phi}{\omega} = \dfrac{-3\pi}{4}$. The stages to graph $y = 2\cos(4x + 3\pi) = 2\cos 4\left(x + \dfrac{3\pi}{4}\right)$ are shown in Figure 53.

FIGURE 53

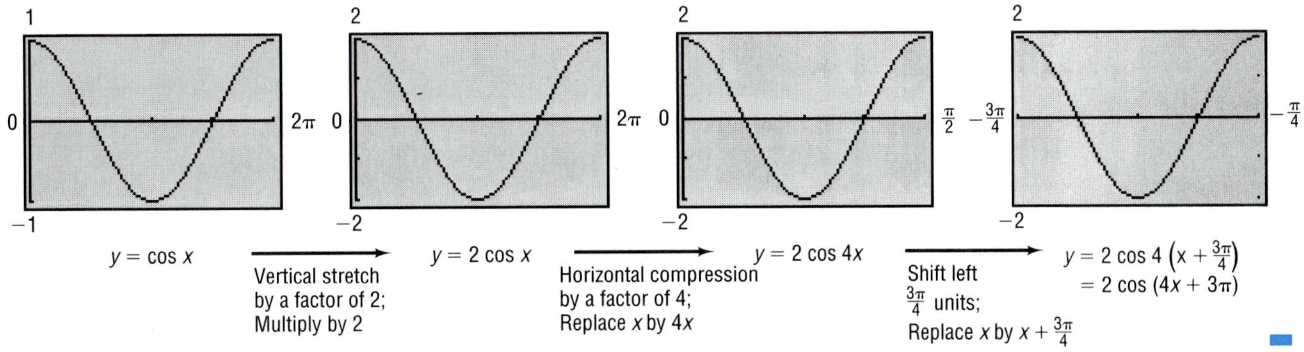

Now work Problem 51.

Finding Sinusoidal Functions from Data by Hand

5 Scatter diagrams of data sometimes take the form of a sinusoidal function. Let's look at an example.

The data given in Table 1 represents the average monthly temperatures in Denver, Colorado. Since the data represents *average* monthly temperatures collected over so many years, the data will not vary much from year to year and so will essentially repeat itself each year. In other words, the data is periodic. Figure 54 shows the scatter diagram of this data repeated over two years, where $x = 1$ represents January, $x = 2$ represents February, and so on.

TABLE 1

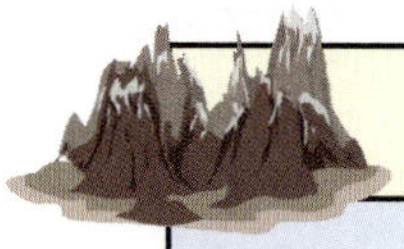

Month, x	Average Monthly Temperature, °F
January, 1	29.7
February, 2	33.4
March, 3	39.0
April, 4	48.2
May, 5	57.2
June, 6	66.9
July, 7	73.5
August, 8	71.4
September, 9	62.3
October, 10	51.4
November, 11	39.0
December, 12	31.0

Source: U.S. National Oceanic and Atmospheric Administration

FIGURE 54

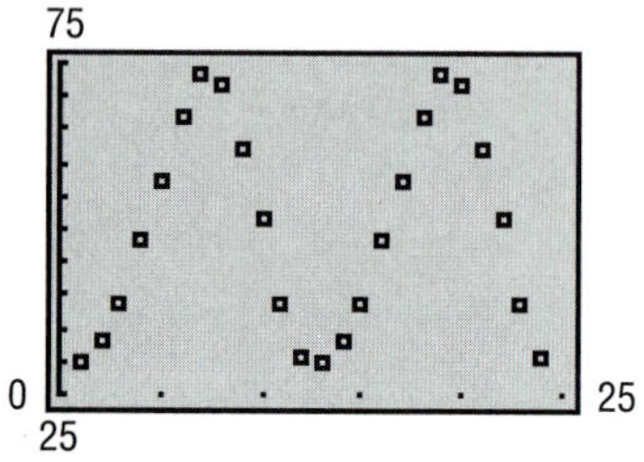

Notice the scatter diagram looks like the graph of a sinusoidal function. We choose to fit the data to a sine function of the form

$$y = A\sin(\omega x - \phi) + B$$

where A, B, ω, and ϕ are constants.

EXAMPLE 9

Finding a Sinusoidal Function from Temperature Data

By hand, fit a sine function to the data in Table 1.

FIGURE 55

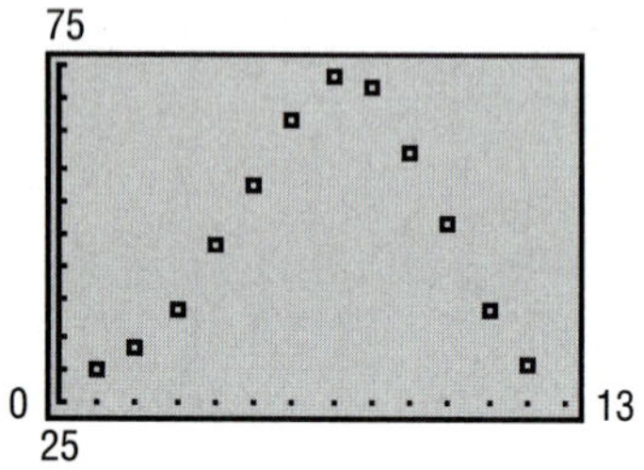

Solution We begin with a scatter diagram of the data for one year. See Figure 55. The data will be fitted to a sine function of the form

$$y = A\sin(\omega x - \phi) + B$$

STEP 1: To find the amplitude A, we compute

$$\begin{aligned}\text{Amplitude} &= (\text{Largest Data Value} - \text{Smallest Data Value})/2\\ &= (73.5 - 29.7)/2 = 21.9\end{aligned}$$

FIGURE 56

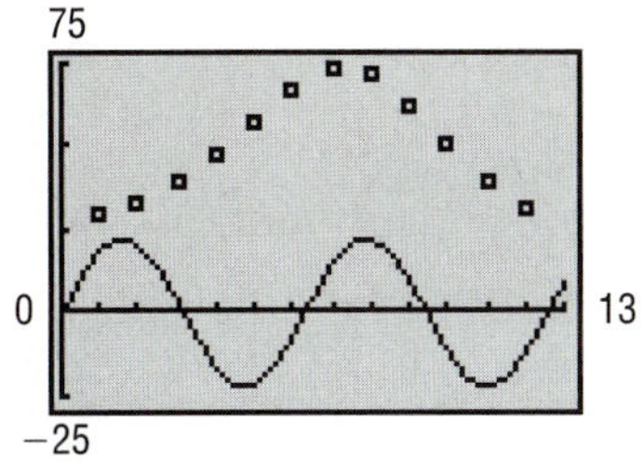

To see the remaining steps in this process, we superimpose the graph of the function $Y_1 = 21.9 \sin x$, where x represents months, on the scatter diagram. To see both, the viewing window has been adjusted. Figure 56 shows the two graphs.

To fit the data, the graph needs to be shifted vertically, shifted horizontally, and stretched horizontally.

STEP 2: We determine the vertical shift by finding the average of the highest and lowest data value.

$$\text{Vertical Shift} = (73.5 + 29.7)/2 = 51.6$$

FIGURE 57

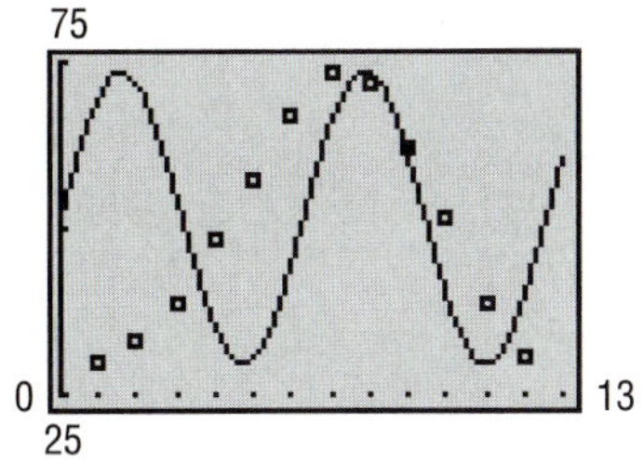

Now we superimpose the graph of $Y_1 = 21.9 \sin x + 51.6$ on the scatter diagram. See Figure 57.

We see the graph needs to be shifted horizontally and stretched horizontally.

STEP 3: It is easier to find the horizontal stretch factor first. Since the temperatures repeat every 12 months, the period of the function is $T = 12$. Since $T = \dfrac{2\pi}{\omega} = 12$,

$$\omega = \frac{2\pi}{12} = \frac{\pi}{6}$$

FIGURE 58

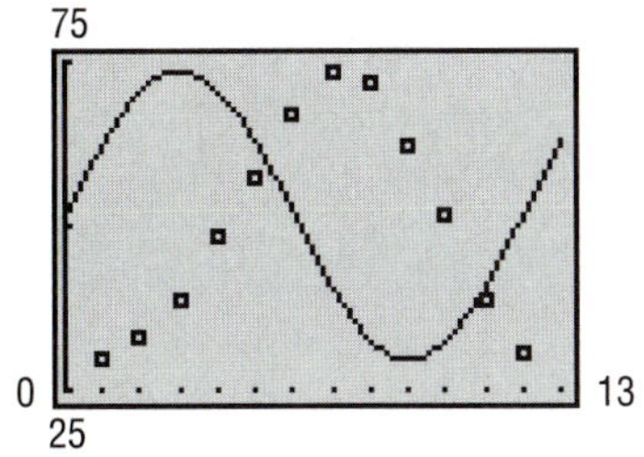

Now we superimpose the graph of $Y_1 = 21.9 \sin \dfrac{\pi x}{6} + 51.6$ on the scatter diagram. See Figure 58.

We see the graph still needs to be shifted horizontally.

STEP 4: To determine the horizontal shift we solve the equation

$$y = 21.9 \sin\left(\frac{\pi}{6}x - \phi\right) + 51.6$$

for ϕ by letting $y = 29.7$ and $x = 1$ (the average temperature in Denver in January).*

$$29.7 = 21.9 \sin\left(\frac{\pi}{6} \cdot 1 - \phi\right) + 51.6$$

$$-21.9 = 21.9 \sin\left(\frac{\pi}{6} - \phi\right)$$ Subtract 51.6 from both sides of the equation.

$$-1 = \sin\left(\frac{\pi}{6} - \phi\right)$$ Divide both sides of the equation by 21.9.

$$-\frac{\pi}{2} = \left(\frac{\pi}{6} - \phi\right)$$ Take the inverse sine of both sides of the equation.

$$\phi = \frac{2\pi}{3}$$ Solve for ϕ.

FIGURE 59

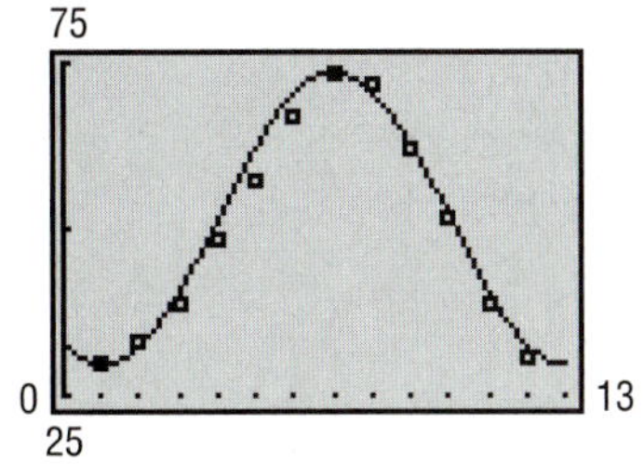

Thus,

$$y = 21.9 \sin\left(\frac{\pi}{6}x - \frac{2\pi}{3}\right) + 51.6$$

The graph of $Y_1 = 21.9 \sin(\pi/6\, x - 2\pi/3) + 51.6$ and the scatter diagram of data are shown in Figure 59.

*The data point selected here to find ϕ is arbitrary. Selecting a different data point will usually result in a different value for ϕ. To maintain consistency, we will always choose the data point for which y is smallest (in this case, January gives the lowest temperature).

The steps to fit a sine function

$$y = A \sin(\omega x - \phi) + B$$

to sinusoidal data by hand follow:

Steps for Fitting Data to a Sine Function by Hand

STEP 1: Determine the amplitude, A, of the function.

$$\text{Amplitude} = \frac{\text{Largest Data Value} - \text{Smallest Data Value}}{2}$$

STEP 2: Determine the vertical shift, B, of the function.

$$\text{Vertical Shift} = \frac{\text{Largest Data Value} + \text{Smallest Data Value}}{2}$$

STEP 3: Determine ω. The period, T, the time it takes for the data to repeat, is $T = \frac{2\pi}{\omega}$, so

$$\omega = \frac{2\pi}{T}$$

STEP 4: Determine the horizontal shift of the function by solving the equation

$$y = A \sin(\omega x - \phi) + B$$

for ϕ by choosing an ordered pair (x, y) from the data. Since answers will vary depending upon the ordered pair selected, we will always choose the ordered pair for which y is smallest in order to maintain consistency.

Certain graphing utilities (such as a TI-83 and TI-86) have the capability of finding the sine function of best fit for sinusoidal data. At least four data points are required for this process.

EXAMPLE 10 Finding the Sine Function of Best Fit

Use a graphing utility to find the sine function of best fit for the data in Table 1. Graph this function with the scatter diagram of the data.

Solution Enter the data from Table 1 and execute the SINe REGression program. The result is shown in Figure 60.

FIGURE 60

The output the utility provides shows us the equation,

$$y = a \sin(bx + c) + d.$$

The sinusoidal function of best fit is

$$y = 21.15 \sin(0.55x - 2.35) + 51.19,$$

where x represents the month and y represents the average temperature.

FIGURE 61

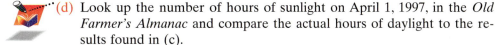

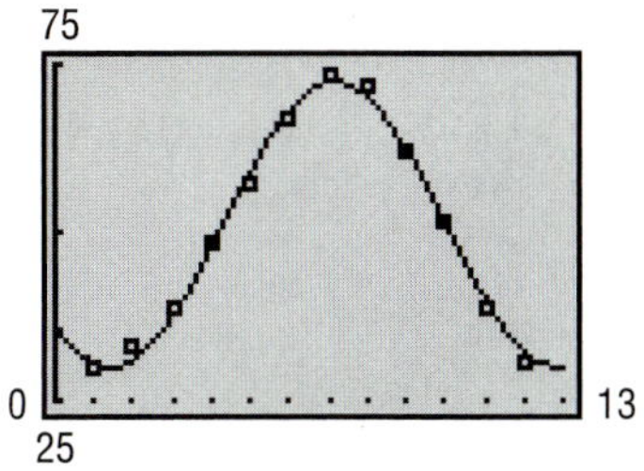

Figure 61 shows the graph of the sinusoidal function of best fit on the scatter diagram. ■

Since the number of hours of sunlight in a day cycles annually, the number of hours of sunlight in a day for a given location can be modeled by a sinusoidal function.

The longest day of the year (in terms of hours of sunlight) occurs during the summer solstice. The summer solstice is the time when the sun is farthest north (for locations in the Northern Hemisphere). In 1997, the summer solstice occurs on June 21 (the 172nd day of the year) at 8:21 A.M. (G.M.T.) The shortest day of the year occurs during the winter solstice. The winter solstice is the time when the sun is farthest south (again, for locations in the Northern Hemisphere). In 1997, the winter solstice occurs on December 21 (the 355th day of the year) at 8:09 P.M. (G.M.T.)

EXAMPLE 11 Finding a Sinusoidal Function for Hours of Daylight

According to the *Old Farmer's Almanac,* the number of hours of sunlight in Boston on the summer solstice is 15.283 and the number of hours of sunlight on the winter solstice is 9.067.

(a) By hand, find a sinusoidal function of the form $y = A\sin(\omega x - \phi) + B$ that fits the data.*

(b) Draw a graph of the function found in (a).

(c) Use the function found in (b) to predict the number of hours of sunlight on April 1, 1997, the 91st day of the year.

(d) Look up the number of hours of sunlight on April 1, 1997, in the *Old Farmer's Almanac* and compare the actual hours of daylight to the results found in (c).

Solution (a) STEP 1: Amplitude = (Largest Data Value − Smallest Data Value)/2
$= (15.283 - 9.067)/2 = 3.108$

STEP 2: Vertical Shift = (Largest Data Value + Smallest Data Value)/2
$= (15.283 + 9.067)/2 = 12.175$

STEP 3: The data repeats every 365 days. Since $T = \dfrac{2\pi}{\omega} = 365$, we find

$$\omega = \frac{2\pi}{365}$$

STEP 4: To determine the horizontal shift we solve the equation

$$y = 3.108\sin\left(\frac{2\pi}{365}x - \phi\right) + 12.175$$

for ϕ by letting $y = 9.067$ and $x = 355$ (the number of hours of daylight in Boston on December 21).

*Notice that only two data points are given so a graphing utility cannot be used to find the sine function of best fit.

$$9.067 = 3.108 \sin\left(\frac{2\pi}{365} \cdot 355 - \phi\right) + 12.175$$

$$-3.108 = 3.108 \sin\left(\frac{2\pi}{365} \cdot 355 - \phi\right) \quad \text{Subtract 12.175 from both sides of the equation.}$$

$$-1 = \sin\left(\frac{2\pi}{365} \cdot 355 - \phi\right) \quad \text{Divide both sides of the equation by 3.108.}$$

$$-\frac{\pi}{2} = \left(\frac{2\pi}{365} \cdot 355 - \phi\right) \quad \text{Take the inverse sine of both sides of the equation.}$$

$$\phi = 2.4\pi \quad \text{Solve for } \phi.$$

Thus, the function that provides the number of hours of daylight in Boston for any day, x, is given by

$$y = 3.108 \sin\left(\frac{2\pi}{365}x - 2.4\pi\right) + 12.175$$

FIGURE 62

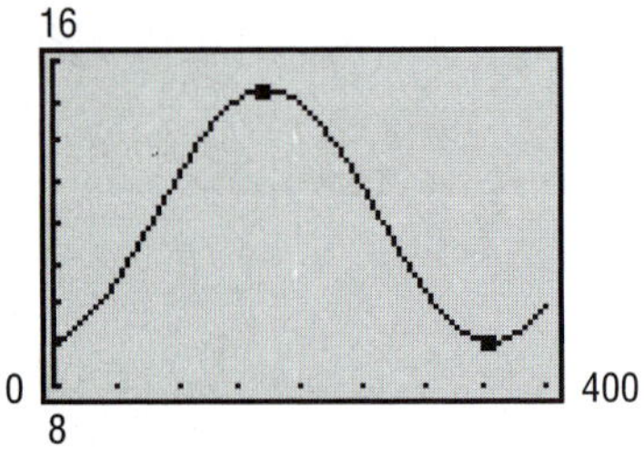

(b) The graph of the function found in (a) is given in Figure 62.

(c) To predict the number of hours of daylight on April 1, we let $x = 91$ in the function found in (a) and obtain:

$$y = 3.108 \sin\left(\frac{2\pi}{365} \cdot 91 - 2.4\pi\right) + 12.175$$
$$= 3.108 \sin(-1.95\pi) + 12.175$$
$$\approx 12.69$$

So, we predict there will be about 12.69 hours of sunlight on April 1, 1997, in Boston.

(d) According to the *Old Farmer's Almanac,* there will be 12 hours 43 minutes of sunlight on April 1, 1997, in Boston. Our prediction of 12.69 hours converts to 12 hours 42 minutes.

8.5 EXERCISES

In Problems 1–10, determine the amplitude and period of each function without graphing.

1. $y = 2 \sin x$ **2.** $y = 3 \cos x$ **3.** $y = -4 \cos 2x$ **4.** $y = -\sin \frac{1}{2}x$

5. $y = 6 \sin \pi x$ **6.** $y = -3 \cos 3x$ **7.** $y = -\frac{1}{2} \cos \frac{3}{2}x$ **8.** $y = \frac{4}{3} \sin \frac{2}{3}x$

9. $y = \frac{5}{3} \sin(-\frac{2\pi}{3}x)$ **10.** $y = \frac{9}{5} \cos(-\frac{3\pi}{2} x)$

In Problems 11–20, match the given function to one of the graphs (A)–(J).

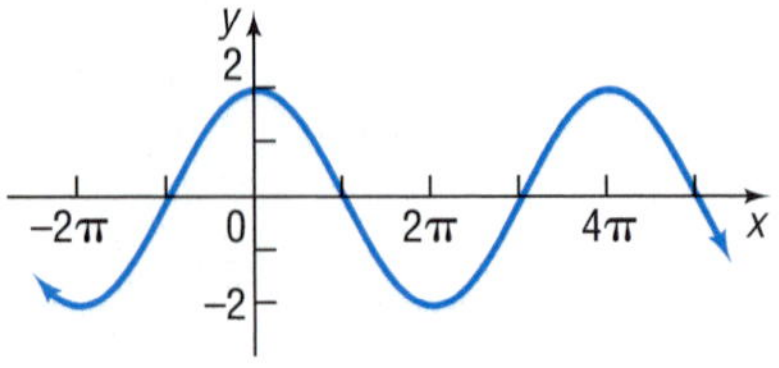

(A)

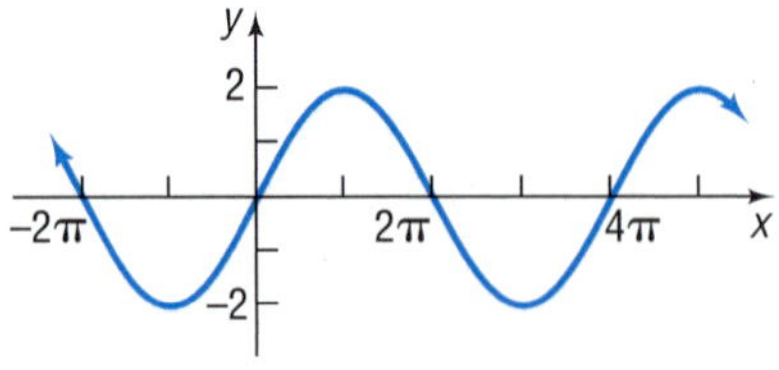

(B)

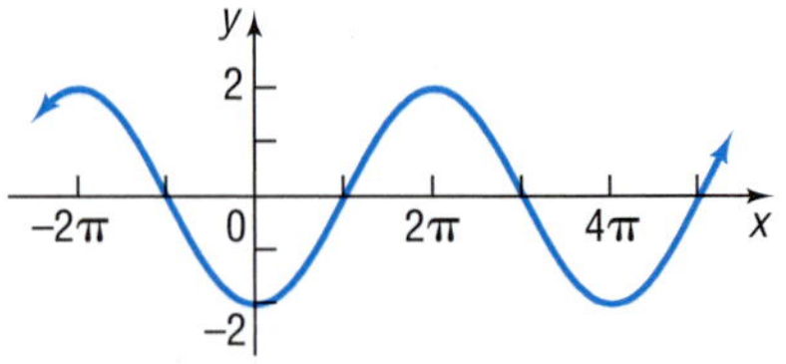

(C)

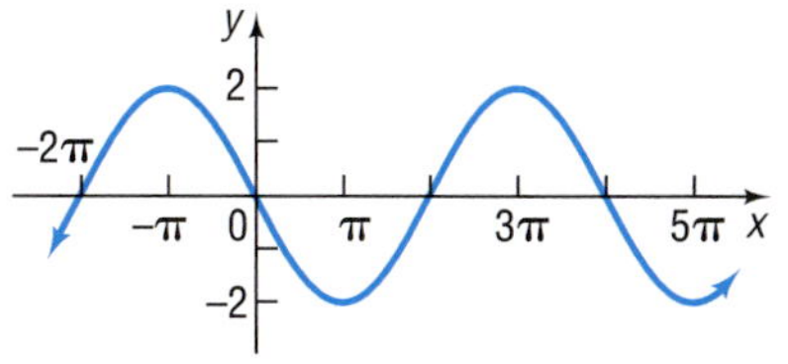

(D)

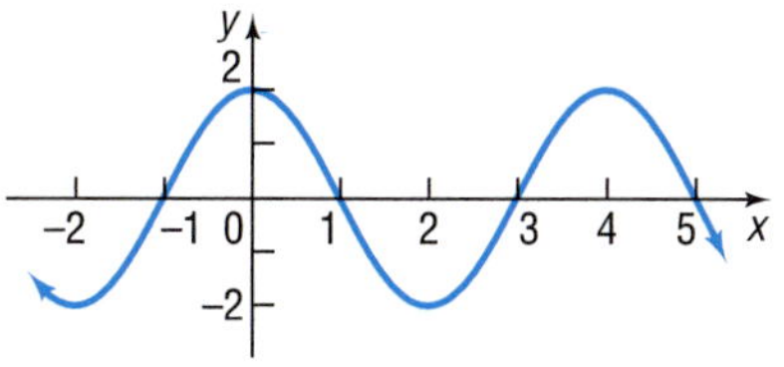

(E)

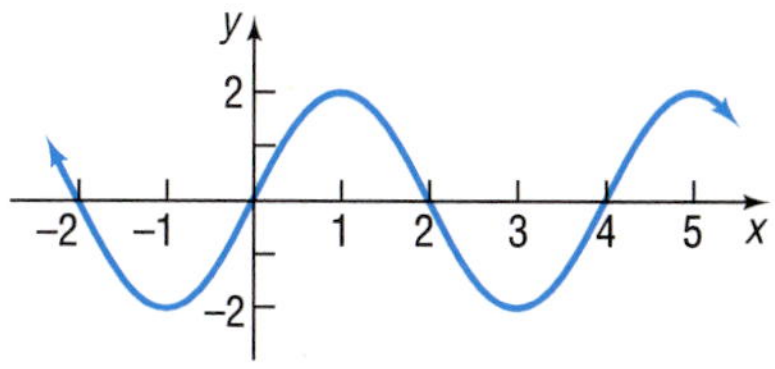

(F)

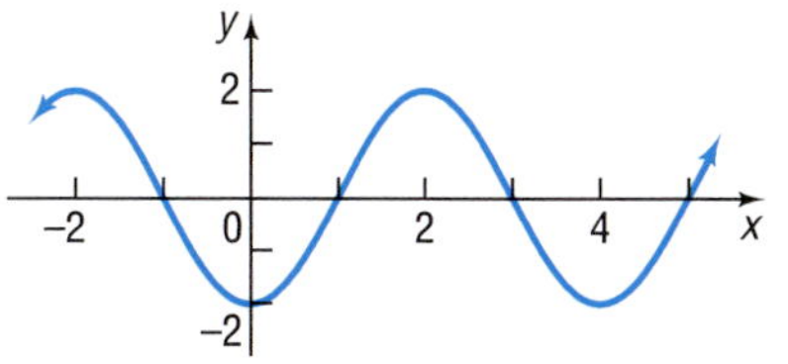

(G)

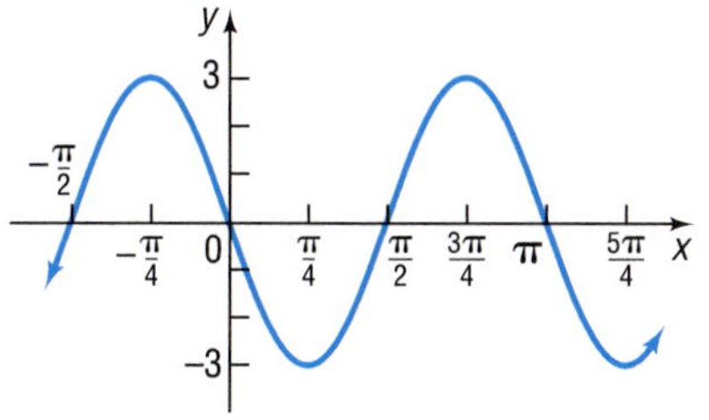

(H)

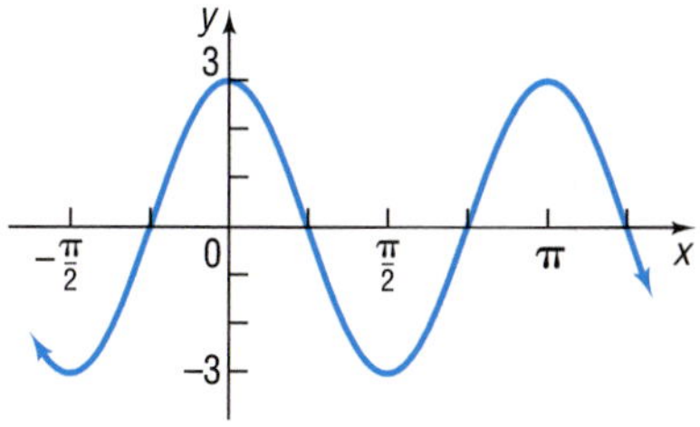

(I)

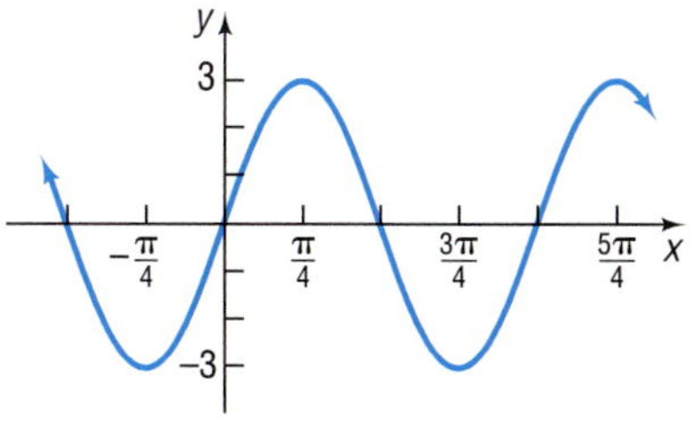

(J)

11. $y = 2\sin\dfrac{\pi}{2}x$ **12.** $y = 2\cos\dfrac{\pi}{2}x$ **13.** $y = 2\cos\frac{1}{2}x$

14. $y = 3\cos 2x$ **15.** $y = -3\sin 2x$ **16.** $y = 2\sin\frac{1}{2}x$

17. $y = -2\cos\frac{1}{2}x$ **18.** $y = -2\cos\dfrac{\pi}{2}x$ **19.** $y = 3\sin 2x$

20. $y = -2\sin\frac{1}{2}x$

In Problems 21–26, match the given function to one of the graphs (A)–(F).

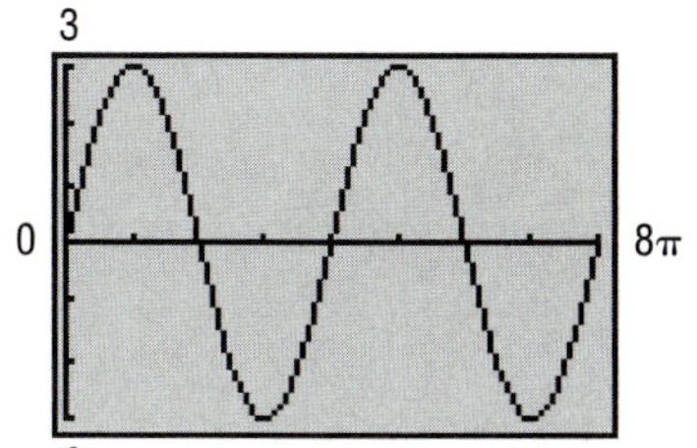

(A)

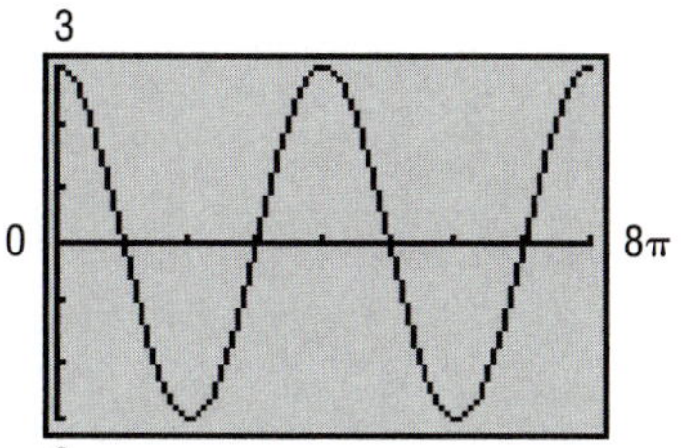

(B)

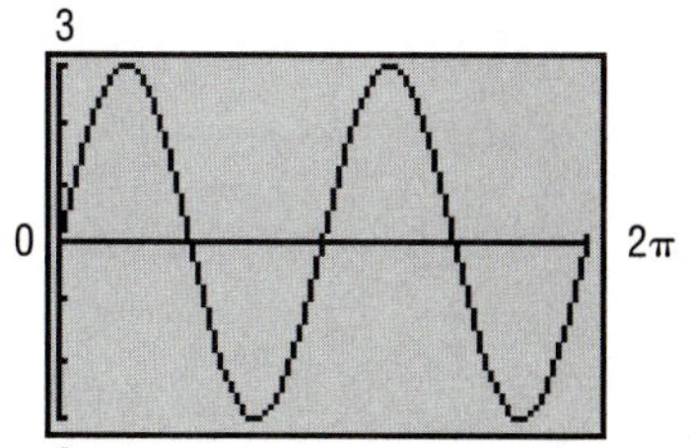

(C)

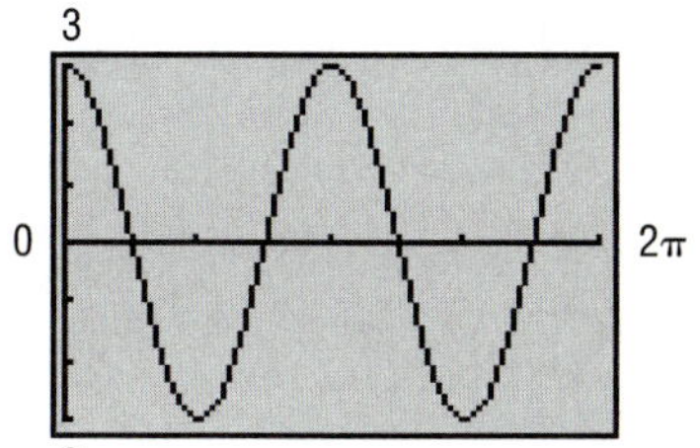

(D)

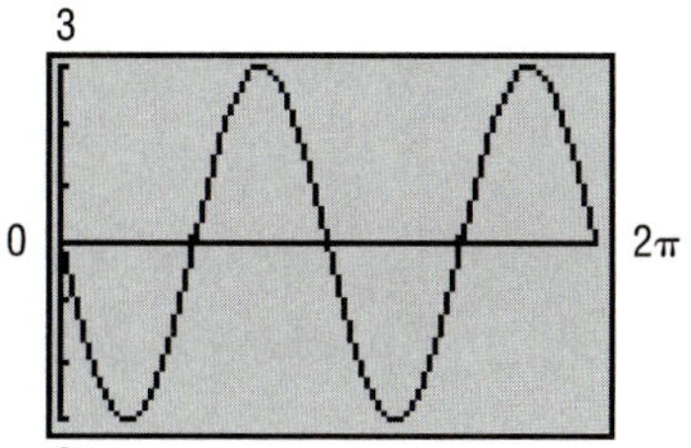

(E)

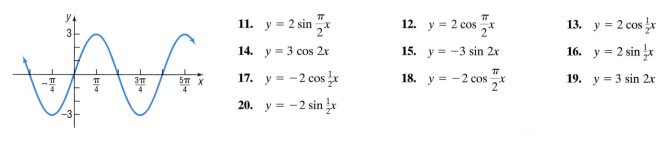

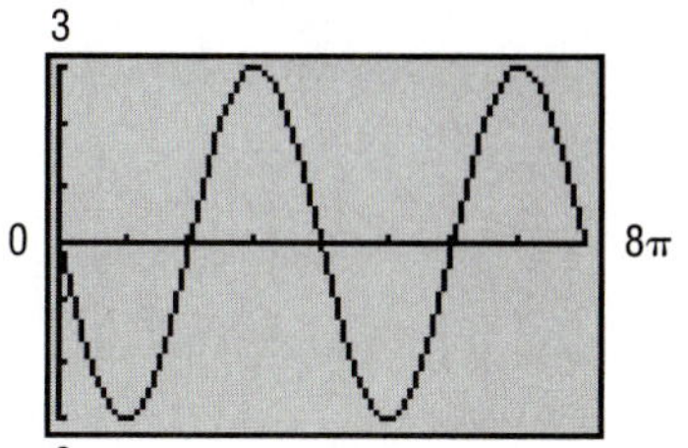

(F)

21. $y = 3\sin\frac{1}{2}x$ **22.** $y = -3\sin 2x$ **23.** $y = 3\cos 2x$

24. $y = 3\sin 2x$ **25.** $y = 3\cos\frac{1}{2}x$ **26.** $y = -3\sin\frac{1}{2}x$

In Problems 27–36, graph each function by hand. Verify the result using a graphing utility.

27. $y = 5\sin 4x$ **28.** $y = 4\cos 6x$ **29.** $y = 5\cos \pi x$ **30.** $y = 2\sin \pi x$

31. $y = -2\cos 2\pi x$ **32.** $y = -5\cos 2\pi x$ **33.** $y = -4\sin\frac{1}{2}x$ **34.** $y = -2\cos\frac{1}{2}x$

35. $y = \frac{3}{2}\sin\left(-\frac{2}{3}x\right)$ **36.** $y = \frac{4}{3}\cos\left(-\frac{1}{3}x\right)$

In Problems 37–50, find an equation for each graph.

37.

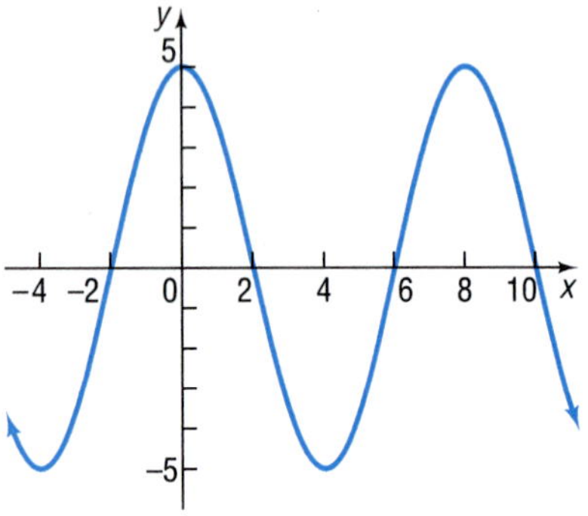

38.

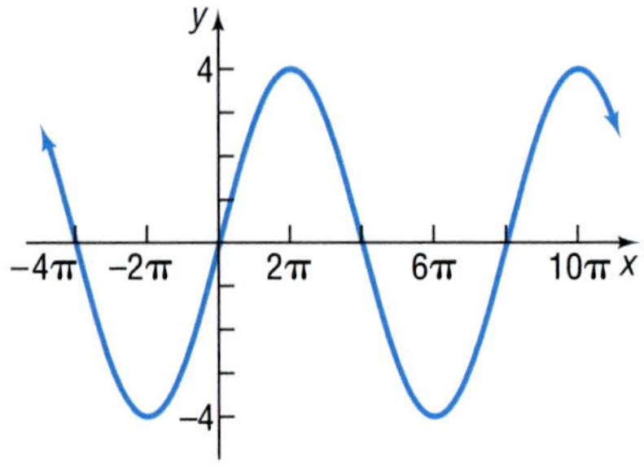

39.

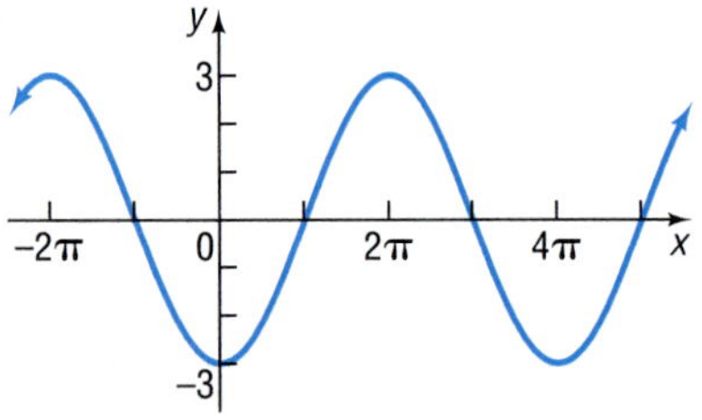

40.

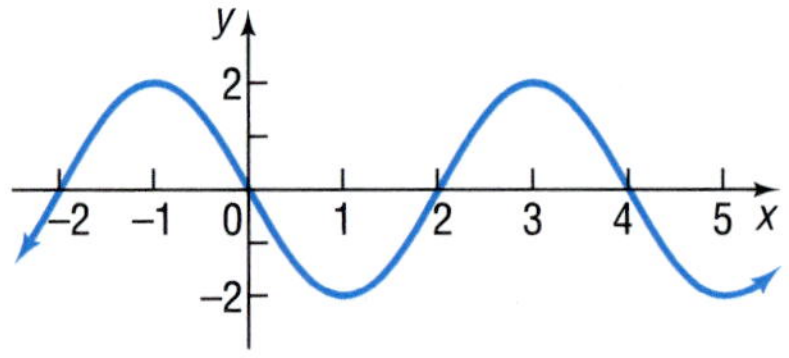

41.

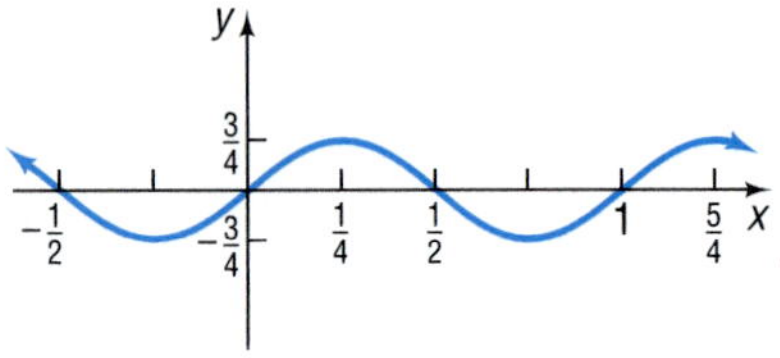

42.

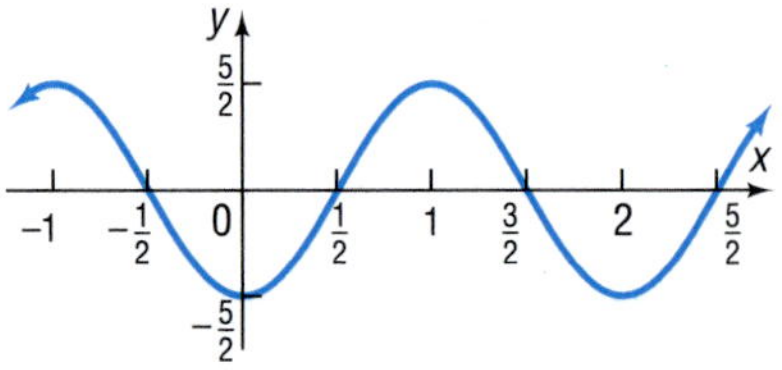

43.

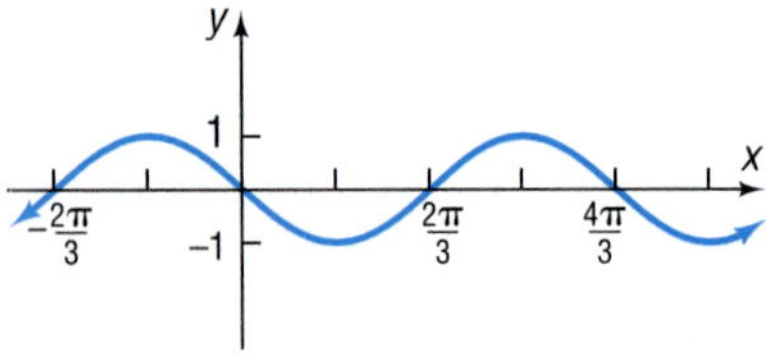

44.

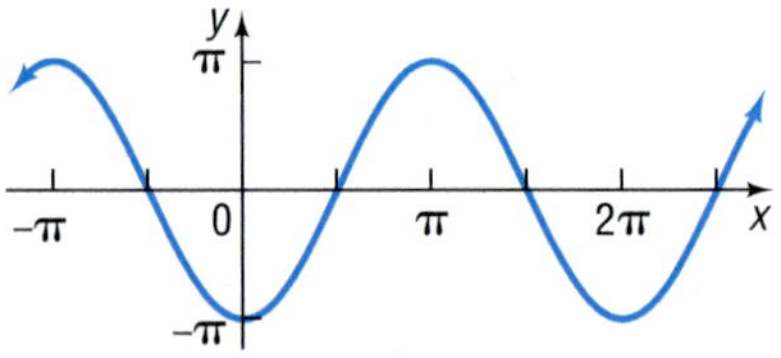

45.

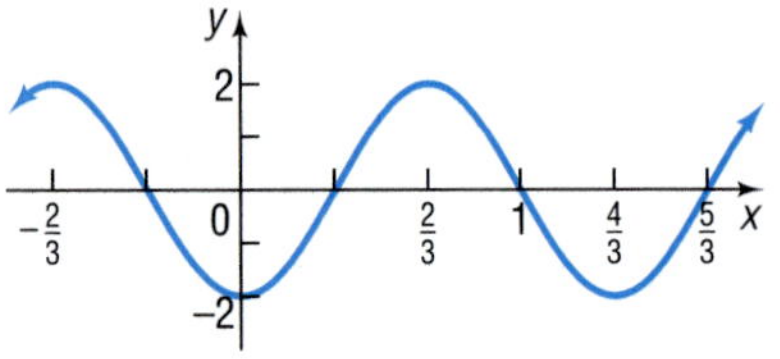

46.

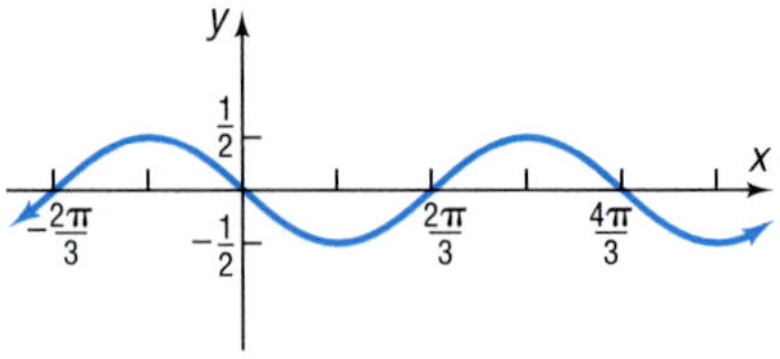

47.

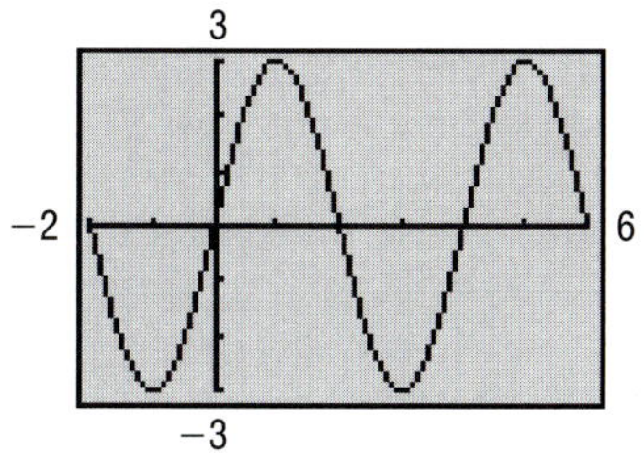

48.

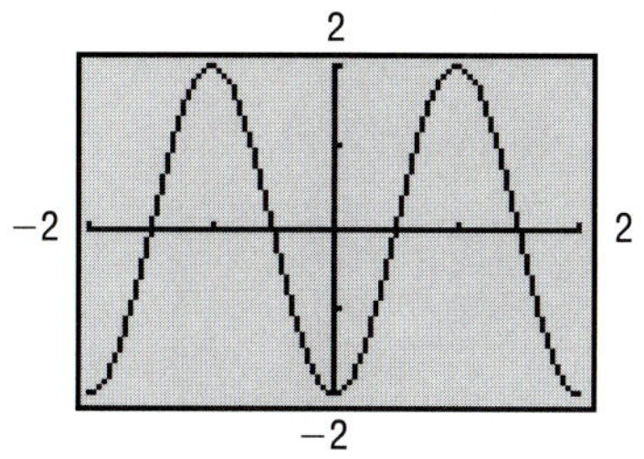

49.

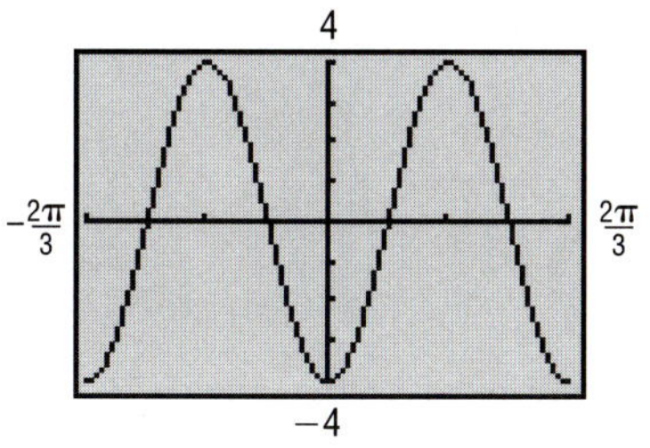

50.

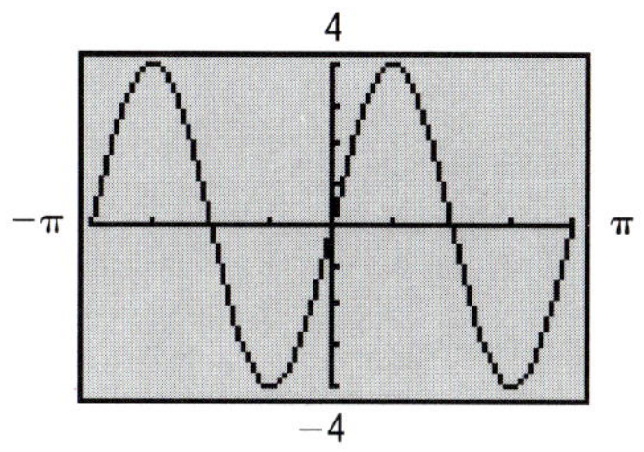

In Problems 51–62, find the amplitude, period, and phase shift of each function. Use a graphing utility to show the stages to graph the function. Show at least one period.

51. $y = 4\sin(2x - \pi)$

52. $y = 3\sin(3x - \pi)$

53. $y = 2\cos\left(3x + \dfrac{\pi}{2}\right)$

54. $y = 3\cos(2x + \pi)$

55. $y = -3\sin\left(2x + \dfrac{\pi}{2}\right)$

56. $y = -2\cos\left(2x - \dfrac{\pi}{2}\right)$

57. $y = 4\sin(\pi x + 2)$

58. $y = 2\cos(2\pi x + 4)$

59. $y = 3\cos(\pi x - 2)$

60. $y = 2\cos(2\pi x - 4)$

61. $y = 3\sin\left(-2x + \dfrac{\pi}{2}\right)$

62. $y = 3\cos\left(-2x + \dfrac{\pi}{2}\right)$

63. Alternating Current (ac) Circuits The current I, in amperes, flowing through an ac (alternating current) circuit at time t is

$$I = 220\sin 60\pi t, \qquad t \ge 0$$

What is the period? What is the amplitude? Graph this function over two periods.

64. Alternating Current (ac) Circuits The current I, in amperes, flowing through an ac (alternating current) circuit at time t is

$$I = 120\sin 30\pi t, \qquad t \ge 0$$

What is the period? What is the amplitude? Graph this function over two periods.

65. Alternating Current (ac) Circuits The current I, in amperes, flowing through an ac (alternating current) circuit at time t is

$$I = 120\sin\left(30\pi t - \frac{\pi}{3}\right), \qquad t \ge 0$$

What is the period? What is the amplitude? What is the phase shift? Graph this function over two periods.

66. Alternating Current (ac) Circuits The current I, in amperes, flowing through an ac (alternating current) circuit at time t is

$$I = 220\sin\left(60\pi t - \frac{\pi}{6}\right), \qquad t \ge 0$$

What is the period? What is the amplitude? What is the phase shift? Graph this function over two periods.

67. Alternating Current (ac) Generators The voltage V produced by an ac generator is

$$V = 220\sin 120\pi t$$

(a) What is the amplitude? What is the period?
(b) Graph V over two periods, beginning at $t = 0$.
(c) If a resistance of $R = 10$ ohms is present, what is the current I?

[**Hint:** Use Ohm's Law, $V = IR$.]

(d) What is the amplitude and period of the current I?
(e) Graph I over two periods, beginning at $t = 0$.

68. Alternating Current (ac) Generators The voltage V produced by an ac generator is

$$V = 120 \sin 120\pi t$$

(a) What is the amplitude? What is the period?
(b) Graph V over two periods, beginning at $t = 0$.
(c) If a resistance of $R = 20$ ohms is present, what is the current I?
[**Hint:** Use Ohm's Law, $V = IR$.]
(d) What is the amplitude and period of the current I?
(e) Graph I over two periods, beginning at $t = 0$.

69. Alternating Current (ac) Generators The voltage V produced by an ac generator is sinusoidal. As a function of time, the voltage V is

$$V = V_0 \sin 2\pi f t$$

where f is the **frequency,** the number of complete oscillations (cycles) per second. [In the United States and Canada, f is 60 hertz (Hz).] The **power** P delivered to a resistance R at any time t is defined as

$$P = \frac{V^2}{R}$$

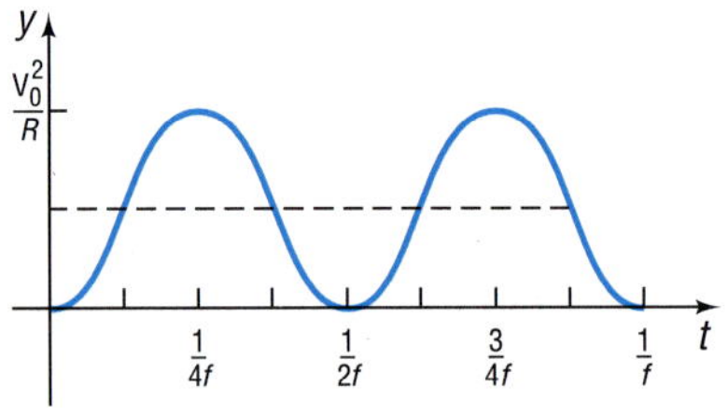

Power in an ac generator

(a) Show that $P = \frac{V_0^2}{R}\sin^2 2\pi f t$.
(b) The graph of P is shown in the figure. Express P as a sinusoidal function.
(c) Deduce that

$$\sin^2 2\pi f t = \frac{1}{2}(1 - \cos 4\pi f t)$$

70. Biorhythms In the theory of biorhythms, a sine function of the form

$$P = 100 \sin \omega t$$

is used to measure the percent P of a person's potential at time t, where t is measured in days and $t = 0$ is the person's birthday. Three characteristics are commonly measured:

Physical potential: period of 23 days
Emotional potential: period of 28 days
Intellectual potential: period of 33 days

(a) Find ω for each characteristic.
(b) Graph all three functions.
(c) Is there a time t when all three characteristics have 100% potential? When is it?
(d) Suppose that you are 20 years old today ($t = 7305$ days). Describe your physical, emotional, and intellectual potential for the next 30 days.

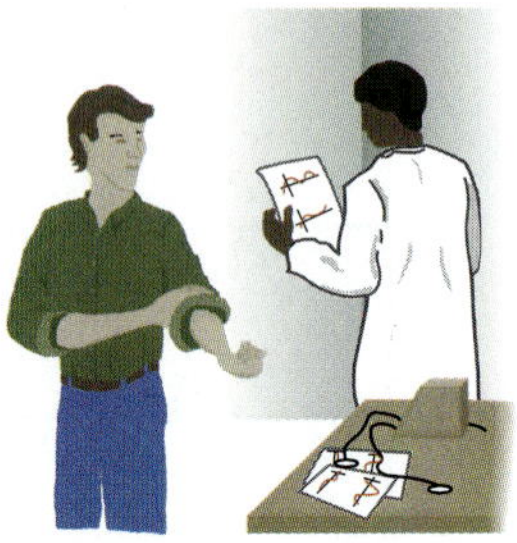

71. Monthly Temperature The data below represents the average monthly temperatures for Juneau, AL.

Month, x	Average Monthly Temperature, °F
January, 1	24.2
February, 2	28.4
March, 3	32.7
April, 4	39.7
May, 5	47.0
June, 6	53.0
July, 7	56.0
August, 8	55.0
September, 9	49.4
October, 10	42.2
November, 11	32.0
December, 12	27.1

Source: U.S. National Oceanic and Atmospheric Administration

(a) Using a graphing utility, draw a scatter diagram of the data for one period.
(b) By hand, find a sinusoidal function of the form $y = A\sin(\omega x - \phi) + B$ that fits the data.
(c) Draw the sinusoidal function found in (b) on the scatter diagram.
(d) Use a graphing utility to find the sinusoidal function of best fit.
(e) Draw the sinusoidal function of best fit on the scatter diagram.

72. Monthly Temperature The data on the top of page 627 represents the average monthly temperatures for Washington, DC.

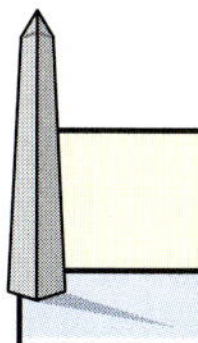

Month, x	Average Monthly Temperature, °F
January, 1	34.6
February, 2	37.5
March, 3	47.2
April, 4	56.5
May, 5	66.4
June, 6	75.6
July, 7	80.0
August, 8	78.5
September, 9	71.3
October, 10	59.7
November, 11	49.8
December, 12	39.4

Source: U.S. National Oceanic and Atmospheric Administration.

(a) Using a graphing utility, draw a scatter diagram of the data for one period.
(b) By hand, find a sinusoidal function of the form $y = A\sin(\omega x - \phi) + B$ that fits the data.
(c) Draw the sinusoidal function found in (b) on the scatter diagram.
(d) Use a graphing utility to find the sinusoidal function of best fit.
(e) Draw the sinusoidal function of best fit on the scatter diagram.

73. Monthly Temperature The data below represents the average monthly temperatures for Indianapolis, IN.

Month, x	Average Monthly Temperature, °F
January, 1	25.5
February, 2	29.6
March, 3	41.4
April, 4	52.4
May, 5	62.8
June, 6	71.9
July, 7	75.4
August, 8	73.2
September, 9	66.6
October, 10	54.7
November, 11	43.0
December, 12	30.9

Source: U.S. National Oceanic and Atmospheric Administration.

(a) Using a graphing utility, draw a scatter diagram of the data for one period.
(b) By hand, find a sinusoidal function of the form $y = A\sin(\omega x - \phi) + B$ that fits the data.
(c) Draw the sinusoidal function found in (b) on the scatter diagram.
(d) Use a graphing utility to find the sinusoidal function of best fit.
(e) Draw the sinusoidal function of best fit on the scatter diagram.

74. Monthly Temperature The data below represents the average monthly temperatures for Baltimore, MD.

Month, x	Average Monthly Temperature, °F
January, 1	31.8
February, 2	34.8
March, 3	44.1
April, 4	53.4
May, 5	63.4
June, 6	72.5
July, 7	77.0
August, 8	75.6
September, 9	68.5
October, 10	56.6
November, 11	46.8
December, 12	36.7

Source: U.S. National Oceanic and Atmospheric Administration.

(a) Using a graphing utility, draw a scatter diagram of the data for one period.
(b) By hand, find a sinusoidal function of the form $y = A\sin(\omega x - \phi) + B$ that fits the data.
(c) Draw the sinusoidal function found in (b) on the scatter diagram.
(d) Use a graphing utility to find the sinusoidal function of best fit.
(e) Draw the sinusoidal function of best fit on the scatter diagram.

75. Tides Suppose the length of time between consecutive high tides is approximately 12.5 hours. According to the National Oceanic and Atmospheric Administration, on Saturday, June 28, 1997, in Savannah, GA, high tide occurred at 3:38 A.M. (03.6333 hours) and low tide occurred at 10:08 A.M. (10.1333 hours). Water heights are measured as the amounts above or below the mean lower low water. The height of the water at high tide was 8.2 feet and the height of the water at low tide was -0.6 feet.

(a) Approximately, when will the next high tide occur?

(b) By hand, find a sinusoidal function of the form $y = A\sin(\omega x - \phi) + B$ that fits the data.

(c) Draw a graph of the function found in (b).

(d) Use the function found in (b) to predict the height of the water at the next high tide.

76. **Tides** Suppose the length of time between consecutive high tides is approximately 12.5 hours. According to the National Oceanic and Atmospheric Administration, on Saturday, June 28, 1997, in Juneau, AL, high tide occurred at 8:11 A.M. (08.1833 hours) and low tide occurred at 2:14 P.M. (14.2333 hours). Water heights are measured as the amounts above or below the mean lower low water. The height of the water at high tide was 13.2 feet and the height of the water at low tide was 2.2 feet.

(a) Approximately when will the next high tide occur?

(b) By hand, find a sinusoidal function of the form $y = A\sin(\omega x - \phi) + B$ that fits the data.

(c) Draw a graph of the function found in (b).

(d) Use the function found in (b) to predict the height of the water at the next high tide.

77. **Hours of Daylight** According to the *Old Farmer's Almanac,* in Miami, FL, the number of hours of sunlight on the summer solstice is 12.75 and the number of hours of sunlight on the winter solstice is 10.583.

(a) By hand, find a sinusoidal function of the form $y = A\sin(\omega x - \phi) + B$ that fits the data.

(b) Draw a graph of the function found in (a).

(c) Use the function found in (a) to predict the number of hours of sunlight on April 1, the 91st day of the year.

(d) Look up the number of hours of sunlight on April 1, 1997, in the *Old Farmer's Almanac,* and compare the actual hours of daylight to the results found in (c).

78. **Hours of Daylight** According to the *Old Farmer's Almanac,* in Detroit, MI, the number of hours of sunlight on the summer solstice is 13.65 and the number of hours of sunlight on the winter solstice is 9.067.

(a) By hand, find a sinusoidal function of the form $y = A\sin(\omega x - \phi) + B$ that fits the data.

(b) Draw a graph of the function found in (a).

(c) Use the function found in (a) to predict the number of hours of sunlight on April 1, the 91st day of the year.

(d) Look up the number of hours of sunlight on April 1, 1997, in the *Old Farmer's Almanac,* and compare the actual hours of daylight to the results found in (c).

79. **Hours of Daylight** According to the *Old Farmer's Almanac,* in Anchorage, AL, the number of hours of sunlight on the summer solstice is 16.233 and the number of hours of sunlight on the winter solstice is 5.45.

(a) By hand, find a sinusoidal function of the form $y = A\sin(\omega x - \phi) + B$ that fits the data.

(b) Draw a graph of the function found in (a).

(c) Use the function found in (a) to predict the number of hours of sunlight on April 1, the 91st day of the year.

(d) Look up the number of hours of sunlight on April 1, 1997, in the *Old Farmer's Almanac,* and compare the actual hours of daylight to the results found in (c).

80. **Hours of Daylight** According to the *Old Farmer's Almanac,* in Honolulu, HI, the numbers of hours of sunlight on the summer solstice is 12.767 and the number of hours of sunlight on the winter solstice is 10.783.

(a) By hand, find a sinusoidal function of the form $y = A\sin(\omega x - \phi) + B$ that fits the data.

(b) Draw a graph of the function found in (a).

(c) Use the function found in (a) to predict the number of hours of sunlight on April 1, the 91st day of the year.

(d) Look up the number of hours of sunlight on April 1, 1997 in the *Old Farmer's Almanac* and compare the actual hours of daylight to the results found in (c).

81. Explain how the amplitude and period of a sinusoidal graph are used to establish the scale on each coordinate axis.

82. Find an application in your major field that leads to a sinusoidal graph. Write a paper about your findings.

83. **CBL Experiment** Part 1: A student puts his/her thumb over a light probe and points it at a light source. The student then begins lifting his/her thumb from the sensor and replacing it repeatedly. Light intensity is plotted over time and the period and frequency are determined. Part 2: A light probe is pointed toward a fluorescent light bulb and its light intensity is recorded. The frequency and period of the light bulb are recorded. (Activity 15, Real-World Math with the CBL System, 1994.)

8.6 SIMPLE HARMONIC MOTION; DAMPED MOTION

1 Find an Equation for an Object in Simple Harmonic Motion
2 Analyze Simple Harmonic Motion
3 Analyze an Object in Damped Motion

Simple Harmonic Motion

There are many physical phenomena that can be described as simple harmonic motion. Radio and television waves, light waves, sound waves, and water waves exhibit motion that is simple harmonic.

The swinging of a pendulum, the vibrations of a tuning fork, and the bobbing of a weight attached to a coiled spring are examples of vibrational motion. In this type of motion, an object swings back and forth over the same path. In each illustration in Figure 63, the point B is the **equilibrium (rest) position** of the vibrating object. The **amplitude** of vibration is the distance from the object's rest position to its point of greatest displacement (either point A or point C in Figure 63). The **period** of a vibrating object is the time required to complete one vibration—that is, the time it takes to go from, say, point A through B to C and back to A.

FIGURE 63

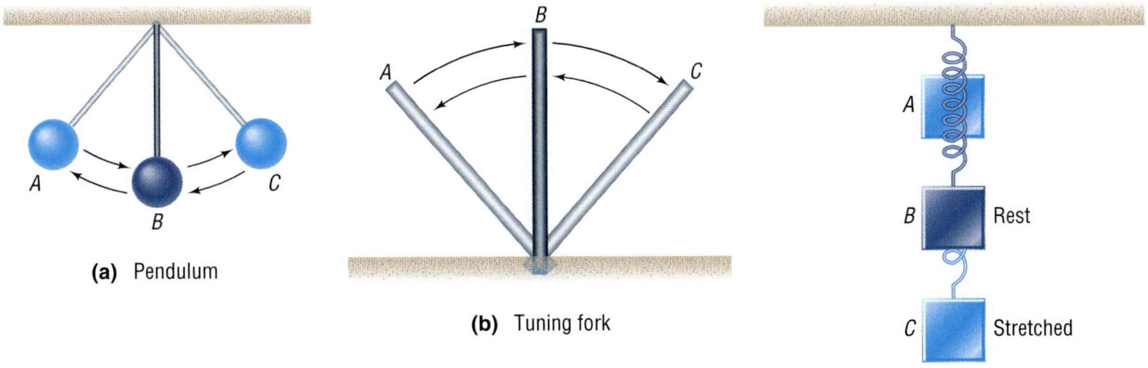

(a) Pendulum **(b)** Tuning fork **(c)** Coiled spring

Simple harmonic motion is a special kind of vibrational motion in which the acceleration a of the object is directly proportional to the negative of its displacement d from its rest position. That is, $a = -kd$, $k > 0$.

For example, when the mass hanging from the spring in Figure 63(c) is pulled down from its rest position B to the point C, the force of the spring tries to restore the mass to its rest position. Assuming that there is no frictional force* to retard the motion, the amplitude will remain constant. The force increases in direct proportion to the distance the mass is pulled from its rest position. Since the force increases directly, the acceleration of the mass of the object must do likewise, because (by Newton's Second Law of Motion) force is directly proportional to acceleration. Thus, the acceleration of the object varies directly with its displacement, and the motion is an example of simple harmonic motion.

*If friction is present, the amplitude will decrease with time to 0. This type of motion is an example of **damped motion,** which is discussed later in this section.

FIGURE 64

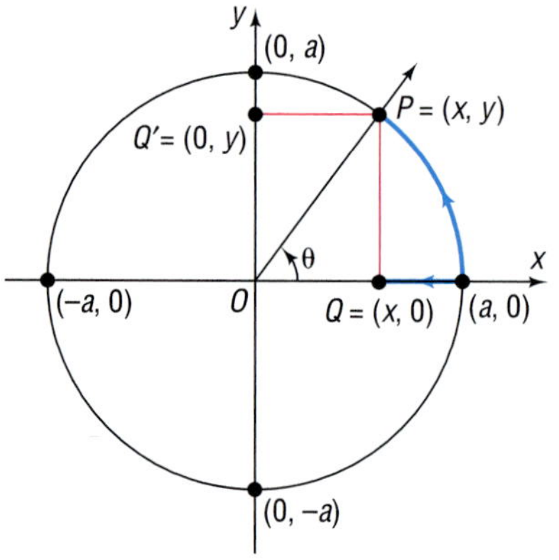

Simple harmonic motion is related to circular motion. To see this relationship, consider a circle of radius a, with center at (0, 0). See Figure 64. Suppose that an object initially placed at $(a, 0)$ moves counterclockwise around the circle at constant angular speed ω. Suppose further that after time t has elapsed, the object is at the point $P = (x, y)$ on the circle. The angle θ, in radians, swept out by the ray $\overrightarrow{OP}$ in this time t is

$$\theta = \omega t$$

The coordinates of the point P at time t are

$$x = a \cos \theta = a \cos \omega t$$
$$y = a \sin \theta = a \sin \omega t$$

Corresponding to each position $P = (x, y)$ of the object moving about the circle, there is the point $Q = (x, 0)$, called the **projection of P on the x-axis.** As P moves around the circle at a constant rate, the point Q moves back and forth between the points $(a, 0)$ and $(-a, 0)$ along the x-axis with a motion that is simple harmonic. Similarly, for each point P there is a point $Q' = (0, y)$, called the **projection of P on the y-axis.** As P moves around the circle, the point Q' moves back and forth between the points $(0, a)$ and $(0, -a)$ on the y-axis with a motion that is simple harmonic. Thus, simple harmonic motion can be described as the projection of constant circular motion on a coordinate axis.

To put it another way, again consider a mass hanging from a spring where the mass is pulled down from its rest position to the point C and then released. See Figure 65(a). The graph shown in Figure 65(b) describes the distance the object is from its rest position, d, as a function of time, t, assuming no frictional force is present.

FIGURE 65

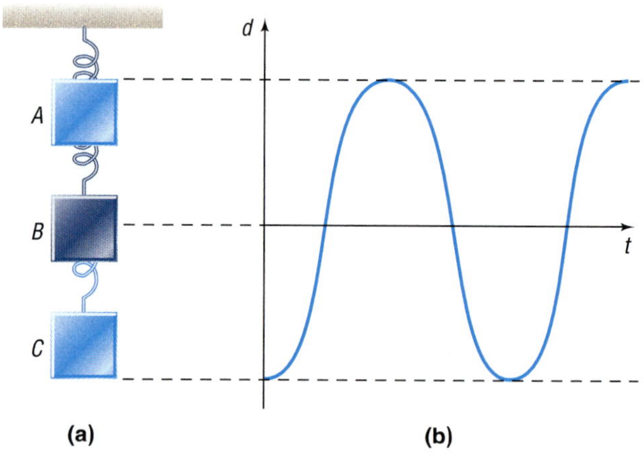

Theorem Simple Harmonic Motion

An object that moves on a coordinate axis so that its distance d from the origin at time t is given by either

$$d = a \cos \omega t \quad \text{or} \quad d = a \sin \omega t$$

where a and $\omega > 0$ are constants, moves with simple harmonic motion. The motion has amplitude $|a|$ and period $2\pi/\omega$.

The **frequency** f of an object in simple harmonic motion is the number of oscillations per unit time. Since the period is the time required for one oscillation, it follows that frequency is the reciprocal of the period; that is,

$$f = \frac{\omega}{2\pi} \qquad \omega > 0$$

EXAMPLE 1

Finding an Equation for an Object in Harmonic Motion

1

Suppose that the object attached to the coiled spring in Figure 63(c) is pulled down a distance of 5 inches from its rest position and then released. If the time for one oscillation is 3 seconds, write an equation that relates the distance d of the object from its rest position after time t (in seconds). Assume no friction.

Solution The motion of the object is simple harmonic. When the object is released ($t = 0$), the distance of the object from the rest position is -5 units (since the object was pulled down). Because $d = -5$ when $t = 0$, it is easier* to use the cosine function

$$d = a \cos \omega t$$

to describe the motion. Now the amplitude is $|-5| = 5$ and the period is 3. Thus,

$$a = -5 \quad \text{and} \quad \frac{2\pi}{\omega} = \text{Period} = 3 \quad \text{or} \quad \omega = \frac{2\pi}{3}$$

An equation of the motion of the object is

$$d = -5 \cos \frac{2\pi}{3} t$$

Note: In the solution to Example 1, we let $a = -5$, since the initial motion is down. If the initial direction was up, we would let $a = 5$.

Now work Problem 1.

EXAMPLE 2

Analyzing the Motion of an Object

2

Suppose that the distance d (in meters) an object travels in time t (in seconds) satisfies the equation

$$d = 10 \sin 5t$$

(a) Describe the motion of the object.
(b) What is the maximum displacement from its resting position?
(c) What is the time required for one oscillation?
(d) What is the frequency?

*No phase shift is required if a cosine function is used.

Solution We observe that the given equation is of the form

$$d = a \sin \omega t \qquad d = 10 \sin 5t$$

where $a = 10$ and $\omega = 5$.

(a) The motion is simple harmonic.

(b) The maximum displacement of the object from its resting position is the amplitude: $a = 10$ meters.

(c) The time required for one oscillation is the period:

$$\text{Period} = \frac{2\pi}{\omega} = \frac{2\pi}{5} \text{ seconds}$$

(d) The frequency is the reciprocal of the period. Thus,

$$\text{Frequency} = f = \frac{5}{2\pi} \text{ oscillation per second}$$

Now work Problem 9.

Damped Motion

3 Most physical phenomena are affected by friction or other resistive forces. These forces remove energy from a moving system and thereby damp its motion. For example, when a mass hanging from a spring is pulled down a distance a and released, the friction in the spring causes the distance the mass moves from its at rest position to decrease over time. Thus, the amplitude of any real oscillating spring or swinging pendulum decreases with time due to air resistance, friction, and so forth. See Figure 66.

FIGURE 66

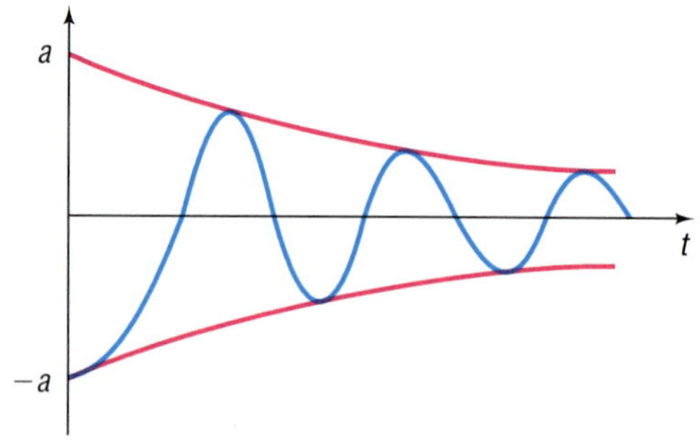

A function that describes this phenomenon maintains a sinusoidal component, but the amplitude of this component will decrease with time in order to account for the damping effect. In addition, the period of the oscillating component will be affected by the damping. The next result, from physics, describes damped motion.

Theorem Damped Motion

The displacement d of an oscillating object from its at rest position at time t is given by

$$d = ae^{-bt/2m}\cos\left(\sqrt{\omega^2 - \frac{b^2}{4m^2}}\, t\right)$$

where b is a **damping factor** (most physics texts call this a **damping coefficient**) and m is the mass of the oscillating object.

Notice for $b = 0$ (zero damping), that we have the formula for simple harmonic motion with amplitude $|a|$ and period $2\pi/\omega$.

EXAMPLE 3 Analyzing the Motion of a Pendulum with Damped Motion

Suppose a simple pendulum with a bob of mass 10 grams and a damping factor of 0.8 grams/second is pulled 20 centimeters from its at rest position and released, see Figure 67. The period of the pendulum without the damping effect is 4 seconds.

FIGURE 67

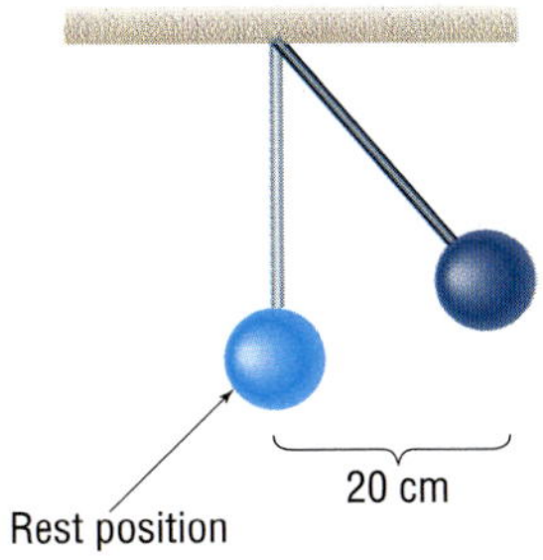

(a) Find an equation that describes the position of the pendulum bob.
(b) Using a graphing utility, graph the function found in (a).
(c) Determine the maximum displacement of the bob after the first oscillation.
(d) What happens to the displacement of the bob as time increases without bound?

Solution (a) We have $m = 10$, $a = 20$, and $b = 0.8$. Since the period of the pendulum under simple harmonic motion is 4 seconds, we have

$$4 = \frac{2\pi}{\omega}$$

$$\omega = \frac{2\pi}{4} = \frac{\pi}{2}$$

Substituting these values into the equation for damped motion we obtain

$$d = 20e^{-0.8t/2(10)}\cos\left(\sqrt{\left(\frac{\pi}{2}\right)^2 - \frac{0.8^2}{4(10)^2}}\, t\right)$$

$$d = 20e^{-0.8t/20}\cos\left(\sqrt{\frac{\pi^2}{4} - \frac{0.64}{400}}\, t\right)$$

FIGURE 68

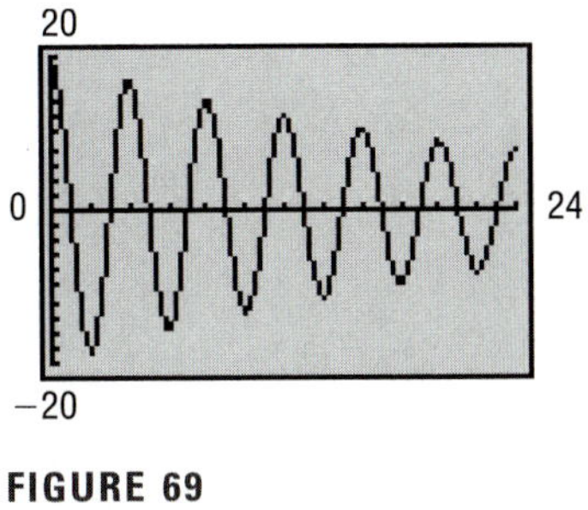

FIGURE 69

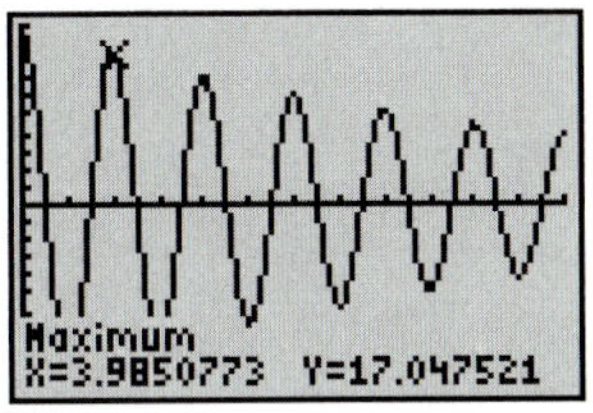

(b) See Figure 68.
(c) See Figure 69. After the first oscillation, the maximum displacement is approximately 17.05 centimeters.
(d) As t increases without bound $e^{-0.8t/20} \to 0$, so the displacement of the bob approaches zero. As a result, the pendulum will eventually come to rest.

8.6 EXERCISES

In Problems 1–4, an object attached to a coiled spring is pulled down a distance a from its rest position and then released. Assuming that the motion is simple harmonic with period T, write an equation that relates the distance d of the object from its rest position after t seconds. Also assume that the positive direction of the motion is up.

1. $a = 5$; $T = 2$ seconds

2. $a = 10$; $T = 3$ seconds

3. $a = 6$; $T = \pi$ seconds

4. $a = 4$; $T = \pi/2$ seconds

5. Rework Problem 1 under the same conditions except that, at time $t = 0$, the object is at its resting position and moving down.

6. Rework Problem 2 under the same conditions except that, at time $t = 0$, the object is at its resting position and moving down.

7. Rework Problem 3 under the same conditions except that, at time $t = 0$, the object is at its resting position and moving down.

8. Rework Problem 4 under the same conditions except that, at time $t = 0$, the object is at its resting position and moving down.

In Problems 9–16, the distance d (in meters) an object travels in time t (in seconds) is given.

(a) Describe the motion of the object.

(b) What is the maximum displacement from its resting position?

(c) What is the time required for one oscillation?

(d) What is the frequency?

9. $d = 5 \sin 3t$

10. $d = 4 \sin 2t$

11. $d = 6 \cos \pi t$

12. $d = 5 \cos \frac{\pi}{2} t$

13. $d = -3 \sin \frac{1}{2} t$

14. $d = -2 \cos 2t$

15. $d = 6 + 2 \cos 2\pi t$

16. $d = 4 + 3 \sin \pi t$

In Problems 17–22, an object of mass m attached to a coiled spring with damping factor b is pulled down a distance a from its rest position and then released. Assume the positive direction of the motion is up and the period of the first oscillation is T.

(a) Write an equation that relates the distance d of the object from its rest position after t seconds.

(b) Graph the equation found in (b) for 5 oscillations using a graphing utility.

17. $m = 25$ grams; $a = 10$ centimeters; $b = 0.7$ grams/second; $T = 5$ seconds

18. $m = 20$ grams; $a = 15$ centimeters; $b = 0.75$ grams/second; $T = 6$ seconds

19. $m = 30$ grams; $a = 18$ centimeters; $b = 0.6$ grams/second; $T = 4$ seconds

20. $m = 15$ grams; $a = 16$ centimeters; $b = 0.65$ grams/second; $T = 5$ seconds

21. $m = 10$ grams; $a = 5$ centimeters; $b = 0.8$ grams/second; $T = 3$ seconds

22. $m = 10$ grams; $a = 5$ centimeters; $b = 0.7$ grams/second; $T = 3$ seconds

In Problems 23–28, the distance d (in meters) of the bob of a pendulum of mass m (in kilograms) from its rest position at time t (in seconds) is given.

(a) Describe the motion of the object. Be sure to give the mass and damping factor.

(b) What is the initial displacement of the bob? That is, what is the displacement at $t = 0$?

(c) Graph the motion.

(d) What is the maximum displacement after the first oscillation?

(e) What happens to the displacement of the bob as time increases without bound?

23. $d = -20e^{-0.7t/40}\cos\left(\sqrt{\left(\frac{2\pi}{5}\right)^2 - \frac{0.49}{1600}}\,t\right)$

24. $d = -20e^{-0.8t/40}\cos\left(\sqrt{\left(\frac{2\pi}{5}\right)^2 - \frac{0.64}{1600}}\,t\right)$

25. $d = -30e^{-0.6t/80}\cos\left(\sqrt{\left(\frac{2\pi}{7}\right)^2 - \frac{0.36}{6400}}\,t\right)$

26. $d = -30e^{-0.5t/70}\cos\left(\sqrt{\left(\frac{\pi}{2}\right)^2 - \frac{0.25}{4900}}\,t\right)$

27. $d = -15e^{-0.9t/30}\cos\left(\sqrt{\left(\frac{\pi}{3}\right)^2 - \frac{0.81}{900}}\,t\right)$

28. $d = -10e^{-0.8t/50}\cos\left(\sqrt{\left(\frac{2\pi}{3}\right)^2 - \frac{0.64}{2500}}\,t\right)$

29. Simple Pendulum The following data represents the distance the "bob" of a simple pendulum is from a motion detector.

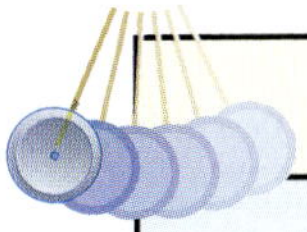

Time (Seconds)	Distance (Centimeters)
0.0	22.59
0.2	20.51
0.4	19.06
0.6	19.52
0.8	21.81
1.0	25.52
1.2	30.01
1.4	34.49
1.6	38.19
1.8	40.47
2.0	40.93
2.2	39.50
2.4	36.41
2.6	32.20
2.8	27.62
3.0	23.45
3.2	20.41

(a) Using a graphing utility, draw a scatter diagram of the data.
(b) By hand, find a sinusoidal function of the form $y = A\sin(\omega x - \phi) + B$ that fits the data.
(c) Graph the sinusoidal function found in (b) on the scatter diagram.
(d) Use a graphing utility to find the sinusoidal function of best fit.
(e) Graph the sinusoidal function of best fit on the scatter diagram.
(f) Based on your answer to (d), what is the period of the pendulum?

30. Simple Pendulum The following data represents the distance the "bob" of a simple pendulum is from a motion detector. See the table.

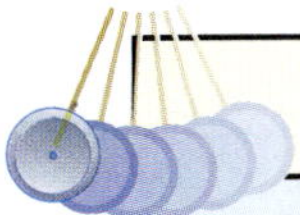

Time (Seconds)	Distance (Centimeters)
0.0	21.79
0.2	26.32
0.4	32.66
0.6	39.48
0.8	45.37
1.0	49.10
1.2	49.91
1.4	47.61
1.6	42.69
1.8	36.17
2.0	29.41
2.2	23.81
2.4	20.54
2.6	20.72

(a) Using a graphing utility, draw a scatter diagram of the data.
(b) By hand, find a sinusoidal function of the form $y = A\sin(\omega x - \phi) + B$ that fits the data.
(c) Graph the sinusoidal function found in (b) on the scatter diagram.
(d) Use a graphing utility to find the sinusoidal function of best fit.
(e) Graph the sinusoidal function of best fit on the scatter diagram.
(f) Based on your answer to (d), what is the period of the pendulum?

31. Charging a Capacitor If a charged capacitor is connected to a coil by closing a switch (see the figure), energy is transferred to the coil and then back to the capacitor in an oscillatory motion. The voltage V (in volts) across the capacitor will gradually diminish to 0 with time t.
(a) Graph the equation relating V and t:

$V(t) = e^{-1.9t}\cos \pi t \qquad 0 \le t \le 3$

(b) At what times t will the graph of V touch the graph of $y = e^{-1.9t}$? When does V touch the graph of $y = -e^{-1.9t}$?

(c) When will the voltage V be between -0.1 and 0.1 volt?

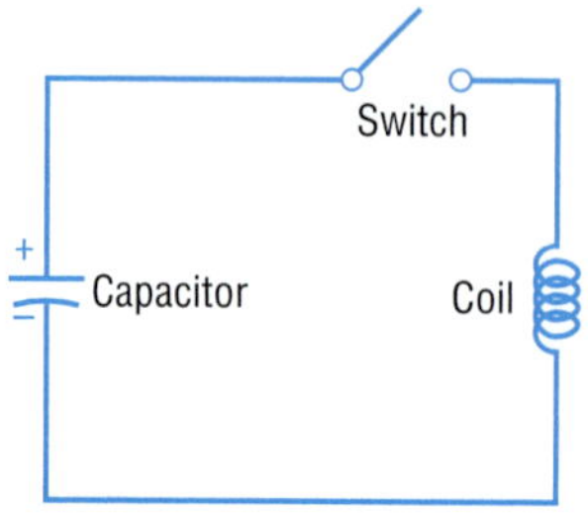

32. Graph the function $f(x) = (\sin x)/x$, $x > 0$. Based on the graph, what do you conjecture about the value of $(\sin x)/x$ for x close to 0?

33. Graph $y = x \sin x$, $y = x^2 \sin x$, and $y = x^3 \sin x$ for $x > 0$. What patterns do you observe?

34. Graph $y = \dfrac{1}{x} \sin x$, $y = \dfrac{1}{x^2} \sin x$, and $y = \dfrac{1}{x^3} \sin x$ for $x > 0$. What patterns do you observe?

35. How would you explain to a friend what simple harmonic motion is? Damped motion?

36. CBL Experiment Pendulum motion is analyzed to estimate simple harmonic motion. A plot is generated with the position of the pendulum over time. The graph is used to find a sinusoidal curve of the form $y = A \cos B(x - C) + D$. Determine the amplitude, period and frequency. (Activity 16, Real-World Math with the CBL System.)

37. CBL Experiment The sound from a tuning fork is collected over time. The amplitude, frequency and period of the graph are determined. A model of the form $y = A \cos B(x - C)$ is fitted to the data. (Activity 23, Real-World Math with the CBL System.)

38. CBL Experiment A weight is attached to a spring and allowed to "bob" up and down. The position of the weight is recorded over time. A model of the form $f(t) = a(t) \sin(bt + c) + d$ is developed from the data. (Experiment M3, CBL System Experiment Workbook, 1994.)

CHAPTER REVIEW

THINGS TO KNOW

Formulas

Law of Sines

$$\frac{\sin \alpha}{a} = \frac{\sin \beta}{b} = \frac{\sin \gamma}{c}$$

Law of Cosines

$$c^2 = a^2 + b^2 - 2ab \cos \gamma$$

$$b^2 = a^2 + c^2 - 2ac \cos \beta$$

$$a^2 = b^2 + c^2 - 2bc \cos \alpha$$

Area of a triangle

$$A = \tfrac{1}{2}bh$$

$$A = \tfrac{1}{2}ab \sin \gamma$$

$$A = \tfrac{1}{2}bc \sin \alpha$$

$$A = \tfrac{1}{2}ac \sin \beta$$

$$A = \sqrt{s(s-a)(s-b)(s-c)}, \text{ where } s = \tfrac{1}{2}(a + b + c)$$

Sinusoidal graphs

$y = A \sin \omega x$, $\omega > 0$ — Period $= 2\pi/\omega$

$y = A \cos \omega x$, $\omega > 0$ — Amplitude $= |A|$

$y = A \sin(\omega x - \phi)$ — Phase shift $= \phi/\omega$

$y = A \cos(\omega x - \phi)$

HOW TO

Solve right triangles

Use the Law of Sines to solve a SAA, ASA, or SSA triangle

Use the Law of Cosines to solve a SAS or SSS triangle

Find the area of a triangle

Find the period and amplitude of a sinusoidal function and use them to graph the function

Find a function whose sinusoidal graph is given

Find a sinusoidal function from data by hand

Find a sinusoidal function of best fit using a graphing utility

Analyze simple harmonic motion

Analyze damped motion

FILL-IN-THE-BLANK ITEMS

1. If two sides and the angle opposite one of them are known, the Law of _______________ is used to determine whether the known information results in no triangle, one triangle, or two triangles.

2. If three sides of a triangle are given, the Law of _______________ is used to solve the triangle.

3. If three sides of a triangle are given, _______________ Formula is used to find the area of the triangle.

4. The following function has amplitude 3 and period 2:

$y =$ _______________ sin _______________ x

5. The function $y = 3 \sin 6x$ has amplitude _______________ and period _______________.

6. The mass and damping factor of $d = 5e^{-0.5t/10} \cos\left(\sqrt{\pi^2 - \frac{(0.5)^2}{100}}\, t\right)$ are _______________ and _______________.

7. The motion of an object obeys the equation $d = 4 \cos 6t$. Such motion is described as _______________ _______________ _______________.

TRUE/FALSE ITEMS

T F **1.** An oblique triangle in which two sides and an angle are given always results in at least one triangle.

T F **2.** Given three sides of a triangle, there is a formula for finding its area.

T F **3.** The graphs of $y = \sin x$ and $y = \cos x$ are identical except for a horizontal shift.

T F **4.** For $y = -2 \sin \pi x$, the amplitude is 2 and the period is $\pi/2$.

REVIEW EXERCISES

In Problems 1–4, solve each triangle.

1.

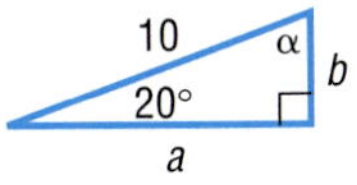

2.

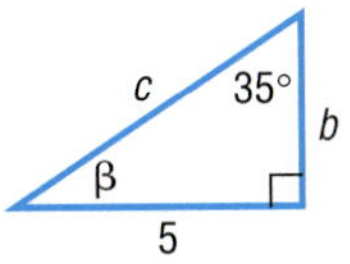

3.

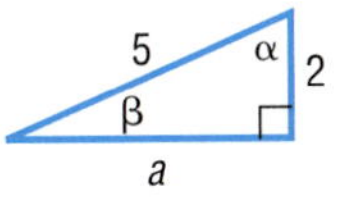

4.

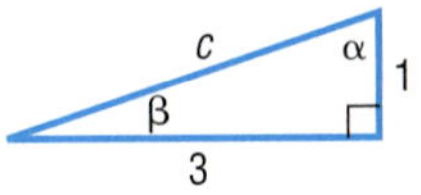

In Problems 5–24, find the remaining angle(s) and side(s) of each triangle, if it exists. If no triangle exists, say "No triangle."

5. $\alpha = 50°, \beta = 30°, a = 1$

6. $\alpha = 10°, \gamma = 40°, c = 2$

7. $\alpha = 100°, a = 5, c = 2$

8. $a = 2, c = 5, \alpha = 60°$

9. $a = 3, c = 1, \gamma = 110°$

10. $a = 3, c = 1, \gamma = 20°$

11. $a = 3, c = 1, \beta = 100°$

12. $a = 3, b = 5, \beta = 80°$

13. $a = 2, b = 3, c = 1$

14. $a = 10, b = 7, c = 8$

15. $a = 1, b = 3, \gamma = 40°$

16. $a = 4, b = 1, \gamma = 100°$

17. $a = 5, b = 3, \alpha = 80°$

18. $a = 2, b = 3, \alpha = 20°$

19. $a = 1, b = \frac{1}{2}, c = \frac{4}{3}$

20. $a = 3, b = 2, c = 2$

21. $a = 3, \alpha = 10°, b = 4$

22. $a = 4, \alpha = 20°, \beta = 100°$

23. $c = 5, b = 4, \alpha = 70°$

24. $a = 1, b = 2, \gamma = 60°$

In Problems 25–34, find the area of each triangle.

25. $a = 2, b = 3, \gamma = 40°$

26. $b = 5, c = 4, \alpha = 20°$

27. $b = 4, c = 10, \alpha = 70°$

28. $a = 2, b = 1, \gamma = 100°$

29. $a = 4, b = 3, c = 5$

30. $a = 10, b = 7, c = 8$

31. $a = 4, b = 2, c = 5$

32. $a = 3, b = 2, c = 2$

33. $\alpha = 50°, \beta = 30°, a = 1$

34. $\alpha = 10°, \gamma = 40°, c = 3$

In Problems 35–38, determine the amplitude and period of each function without graphing.

35. $y = 4 \cos x$

36. $y = \sin 2x$

37. $y = -8 \sin \frac{\pi}{2}x$

38. $y = -2 \cos 3\pi x$

In Problems 39–46, find the amplitude, period, and phase shift of each function. Use a graphing utility to graph each function. Show at least one period.

39. $y = 4 \sin 3x$

40. $y = 2 \cos \frac{1}{3}x$

41. $y = -2 \sin\left(\frac{\pi}{2}x + \frac{1}{2}\right)$

42. $y = -6 \sin(2\pi x - 2)$

43. $y = \frac{1}{2} \sin(\frac{3}{2}x - \pi)$

44. $y = \frac{3}{2} \cos(6x + 3\pi)$

45. $y = -\frac{2}{3} \cos(\pi x - 6)$

46. $y = -7 \sin\left(\frac{\pi}{3}x + \frac{4}{3}\right)$

In Problems 47–50, find a function whose graph is given.

47.

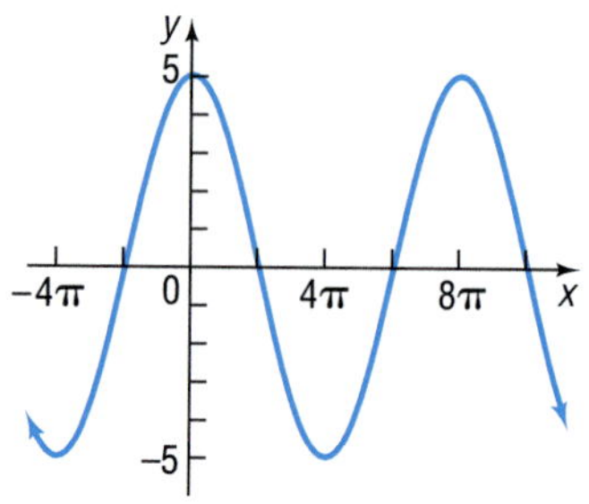

48.

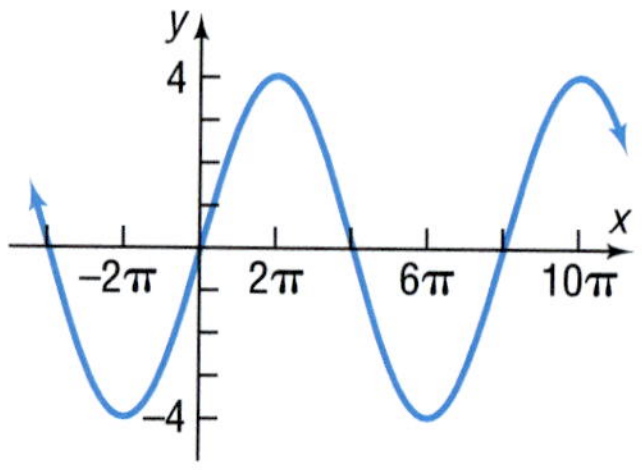

49.

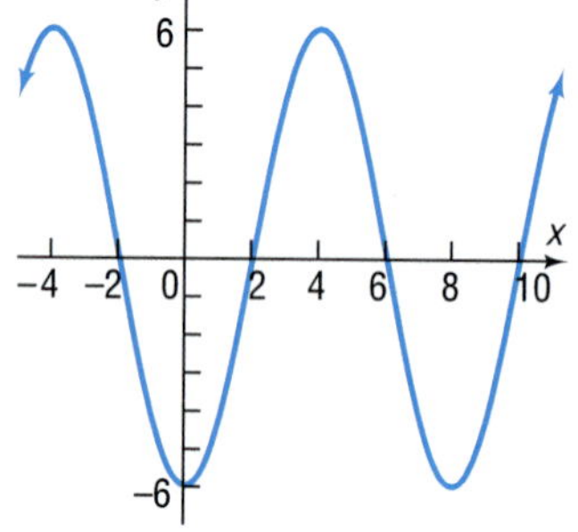

50.

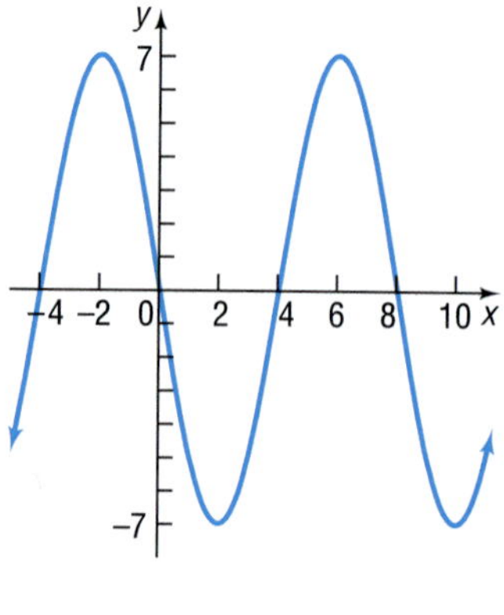

In Problems 51–54, the distance d (in feet) an object travels in time t (in seconds) is given.

(a) Describe the motion of the object.

(b) What is the maximum displacement from its resting position?

(c) What is the time required for one oscillation?

(d) What is the frequency?

51. $d = 6 \sin 2t$ **52.** $d = 2 \cos 4t$ **53.** $d = -2 \cos \pi t$ **54.** $d = -3 \sin \frac{\pi}{2} t$

In Problems 55–56, an object of mass m attached to a coiled spring with damping factor b is pulled down a distance a from its rest position and then released. Assume the positive direction of the motion is up and the period of the first oscillation is T.

(a) Write an equation that relates the distance d of the object from its rest position after t seconds.

(b) Graph the equation found in (a) for 5 oscillations.

55. $m = 40$ grams; $a = 15$ centimeters; $b = 0.75$ grams/second; $T = 5$ seconds

56. $m = 25$ grams; $a = 13$ centimeters; $b = 0.65$ grams/second; $T = 4$ seconds

In Problems 57–58, the distance d (in meters) of the bob of a pendulum of mass m (in kilograms) from its rest position at time t (in seconds) is given.

(a) Describe the motion of the object.

(b) What is the initial displacement of the bob? That is, what is the displacement at t = 0?

(c) Graph the motion.

(d) What is the maximum displacement after the first oscillation?

(e) What happens to the displacement of the bob as time increases without bound?

57. $d = -15e^{-0.6t/40} \cos\left(\sqrt{\left(\frac{2\pi}{5}\right)^2 - \frac{0.36}{1600}}\, t\right)$ **58.** $d = -20e^{-0.5t/60} \cos\left(\sqrt{\left(\frac{2\pi}{3}\right)^2 - \frac{0.25}{3600}}\, t\right)$

59. Alternating Voltage The electromotive force E, in volts in a certain ac circuit obeys the equation

$$E = 120 \sin 120\pi t,\ t \ge 0$$

where t is measured in seconds.

(a) What is the maximum value of E?
(b) What is the period?
(c) Graph this function over two periods.

60. Alternating Current The current I, in amperes, flowing through an ac (alternating current) circuit at time t is

$$I = 220 \sin\left(30\pi t + \frac{\pi}{6}\right), \qquad t \ge 0$$

(a) What is the period?
(b) What is the amplitude?
(c) What is the phase shift?
(d) Graph this function over two periods.

61. Measuring the Length of a Lake From a stationary hot air balloon 500 feet above the ground, two sightings of a lake are made (see the figure). How long is the lake?

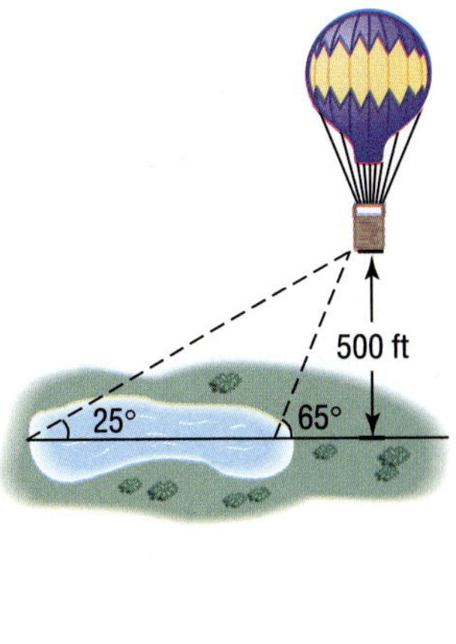

62. Finding the Speed of a Glider From a glider 200 feet above the ground, two sightings of a stationary object directly in front are taken

1 minute apart (see the figure). What is the speed of the glider?

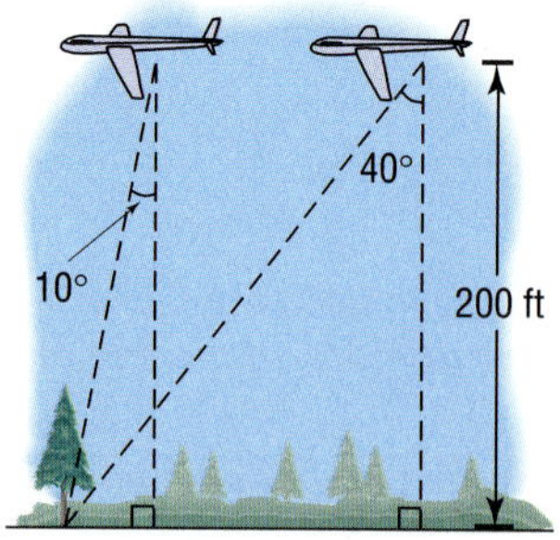

63. Finding the Width of a River Find the distance from A to C across the river illustrated in the figure.

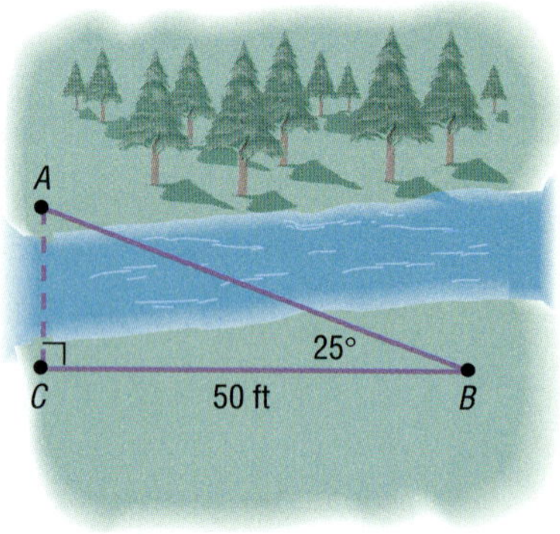

64. Finding the Height of a Building Find the height of the building shown in this figure.

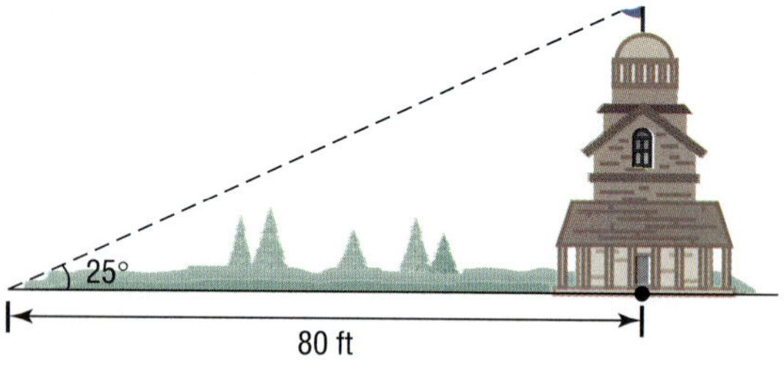

65. Finding the Distance to Shore The Sears Tower in Chicago is 1454 feet tall and is situated about 1 mile inland from the shore of Lake Michigan, as indicated in the figure. An observer in a pleasure boat on the lake directly in front of the Sears Tower looks at the top of the tower and measures the angle of elevation as 5°. How far offshore is the boat?

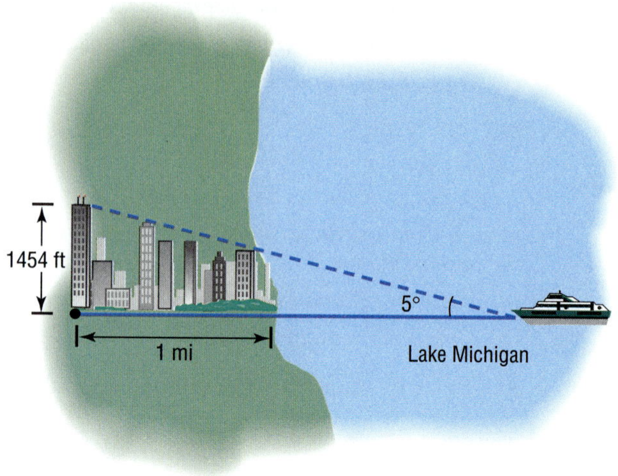

66. Finding the Grade of a Mountain Trail A straight trail with a uniform inclination leads from a hotel, elevation 5000 feet, to a lake in a valley, elevation 4100 feet. The length of the trail is 4100 feet. What is the inclination (grade) of the trail?

67. Navigation An airplane flies from city A to city B, a distance of 100 miles, then turns through an angle of 20° and heads toward city C, as indicated in the figure. If the distance from A to C is 300 miles, how far is it from city B to city C?

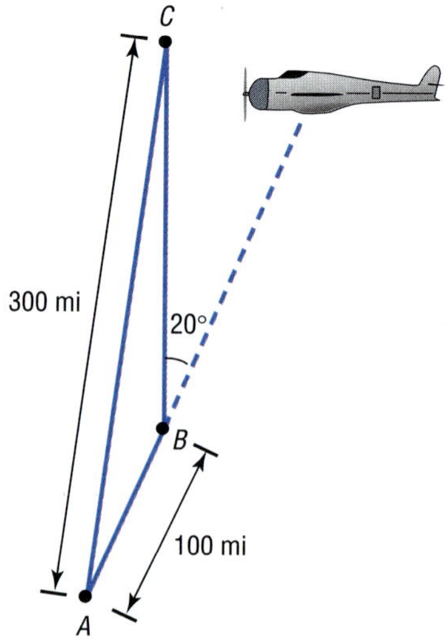

68. Correcting a Navigation Error Two cities A and B are 300 miles apart. In flying from city A to city B, a pilot inadvertently took a course that was 5° in error.

(a) If the error was discovered after flying 10 minutes at a constant speed of 420 miles per hour, through what angle should the pilot turn to correct the course? (Consult the figure below.)

(b) What new constant speed should be maintained so that no time is lost due to the error? (Assume that the speed would have been a constant 420 miles per hour if no error had occurred.)

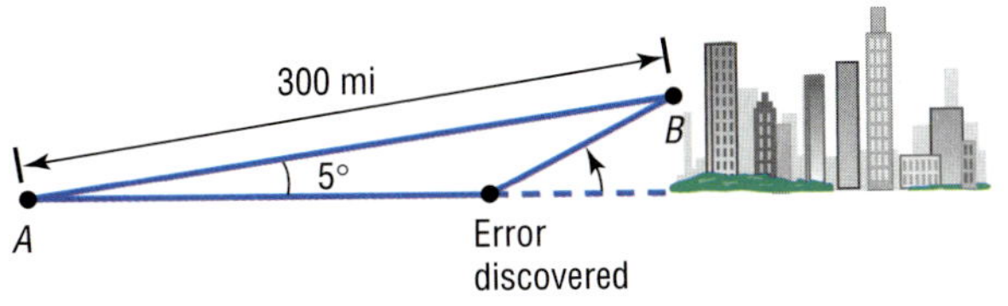

69. **Determining Distances at Sea** Rebecca, the navigator of a ship at sea, spots two lighthouses that she knows to be 2 miles apart along a straight seashore. She determines that the angles formed between two line-of-sight observations of the lighthouses and the line from the ship directly to shore are 12° and 30°. See the illustration.

(a) How far is the ship from lighthouse A?

(b) How far is the ship from lighthouse B?

(c) How far is the ship from shore?

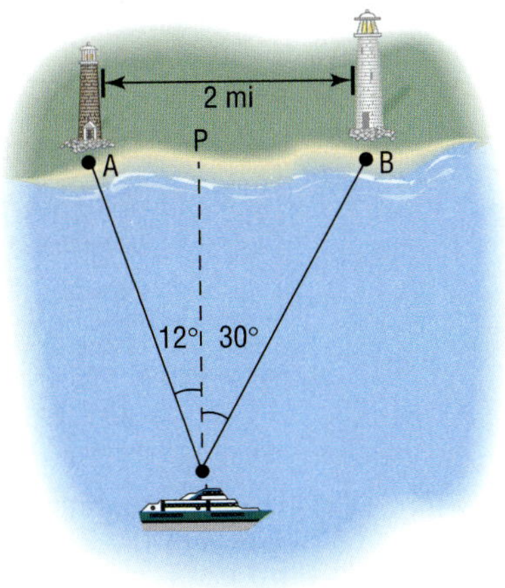

70. **Constructing a Highway** A highway whose primary directions are north–south is being constructed along the west coast of Florida. Near Naples, a bay obstructs the straight path of the road. Since the cost of a bridge is prohibitive, engineers decide to go around the bay. The illustration shows the path they decide on and the measurements taken. What is the length of highway needed to go around the bay?

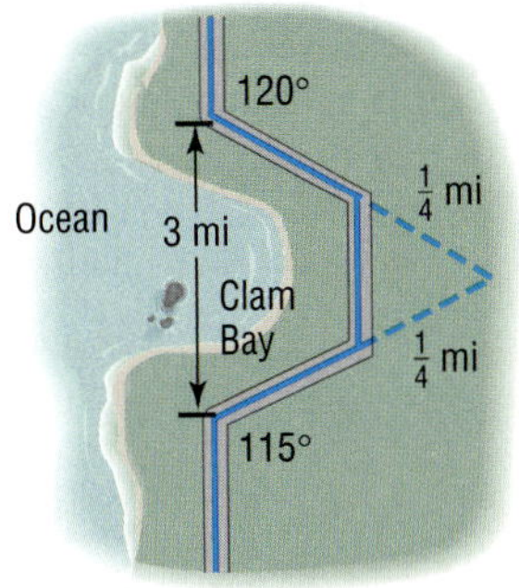

71. **Correcting a Navigational Error** A yacht leaves St. Thomas bound for an island in the British West Indies, 200 miles away. Maintaining a constant speed of 18 miles per hour, but encountering heavy crosswinds and strong currents, the crew finds after 4 hours that the sailboat is off course by 15°.

(a) How far is the sailboat from the island at this time?

(b) Through what angle should the sailboat turn to correct its course?

(c) How much time has been added to the trip because of this? (Assume that the speed remains at 18 miles per hour.)

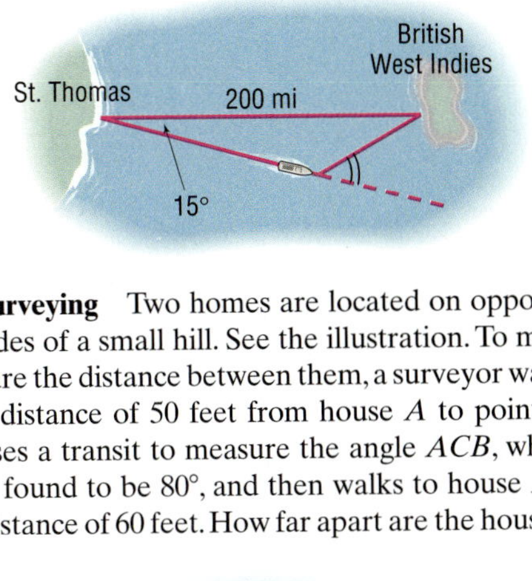

72. **Surveying** Two homes are located on opposite sides of a small hill. See the illustration. To measure the distance between them, a surveyor walks a distance of 50 feet from house A to point C, uses a transit to measure the angle ACB, which is found to be 80°, and then walks to house B, a distance of 60 feet. How far apart are the houses?

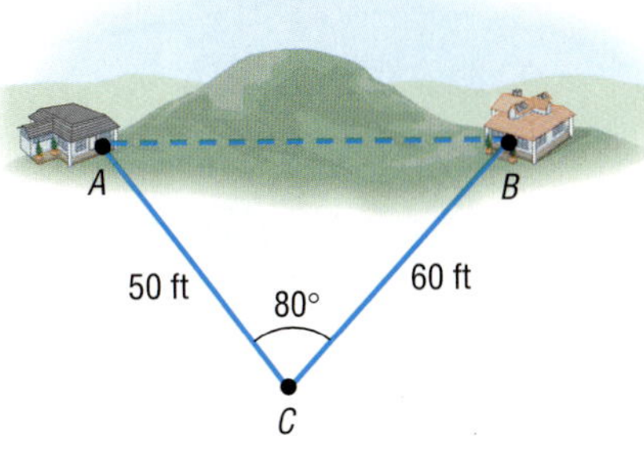

73. Approximating the Area of a Lake To approximate the area of a lake, Cindy walks around the perimeter of the lake, taking the measurements shown in the illustration. Using this technique, what is the approximate area of the lake?

[**Hint:** Use the Law of Cosines on the three triangles shown and then find the sum of their areas.]

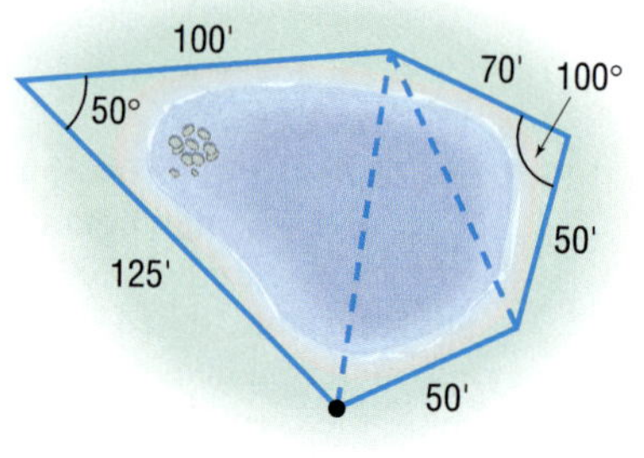

74. Calculating the Cost of Land The irregular parcel of land shown in the figure is being sold for $100 per square foot. What is the cost of this parcel?

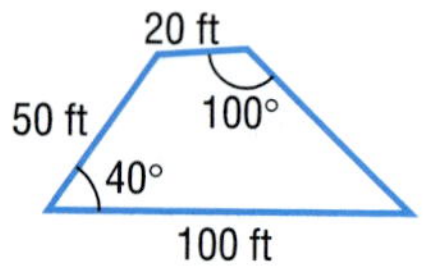

75. The drive wheel of an engine is 13 inches in diameter, and the pulley on the rotary pump is 5 inches in diameter. If the shafts of the drive wheel and the pulley are 2 feet apart, what length of belt is required to join them as shown in the figure below?

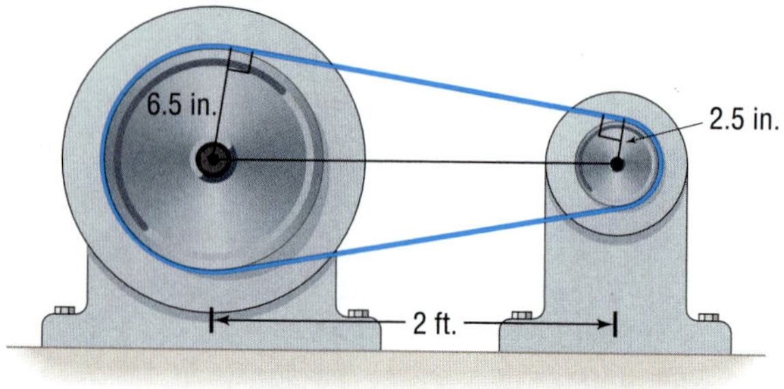

76. Rework Problem 75 if the belt is crossed, as shown in the figure below.

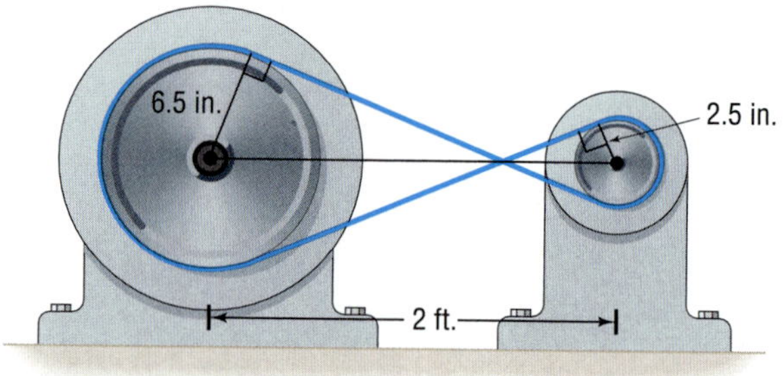

77. Monthly Temperature The data below represents the average monthly temperatures for Phoenix, AZ.

Month, x	Average Monthly Temperature, °F
January, 1	51
February, 2	55
March, 3	63
April, 4	67
May, 5	77
June, 6	86
July, 7	90
August, 8	90
September, 9	84
October, 10	71
November, 11	59
December, 12	52

Source: U.S. National Oceanic and Atmospheric Administration.

(a) Using a graphing utility, draw a scatter diagram of the data for one period.
(b) By hand, find a sinusoidal function of the form $y = A\sin(\omega x - \phi) + B$ that fits the data.
(c) Draw the sinusoidal function found in (b) on the scatter diagram.
(d) Use a graphing utility to find the sinusoidal function of best fit.
(e) Draw the sinusoidal function of best fit on the scatter diagram.

78. Monthly Temperature The data below represents the average monthly temperatures for Chicago, IL.

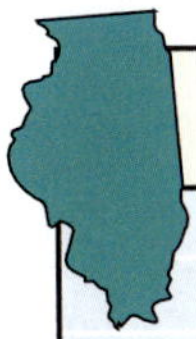

Month, x	Average Monthly Temperature, °F
January, 1	25
February, 2	28
March, 3	36
April, 4	48
May, 5	61
June, 6	72
July, 7	74
August, 8	75
September, 9	66
October, 10	55
November, 11	39
December, 12	28

Source: U.S. National Oceanic and Atmospheric Administration.

(a) Using a graphing utility, draw a scatter diagram of the data for one period.
(b) By hand, find a sinusoidal function of the form $y = A\sin(\omega x - \phi) + B$ that fits the data.
(c) Draw the sinusoidal function found in (b) on the scatter diagram.
(d) Use a graphing utility to find the sinusoidal function of best fit.
(e) Draw the sinusoidal function of best fit on the scatter diagram.

79. Hours of Daylight According to the *Old Farmer's Almanac*, in Las Vegas, NV, the number of hours of sunlight on the summer solstice is 13.367 and the number of hours of sunlight on the winter solstice is 9.667.

(a) By hand, find a sinusoidal function of the form $y = A\sin(\omega x - \phi) + B$ that fits the data.
(b) Draw a graph of the function found in (a).
(c) Use the function found in (a) to predict the number of hours of sunlight on April 1, the 91st day of the year.
(d) Look up the number of hours of sunlight on April 1, 1997, in the *Old Farmer's Almanac* and compare the actual hours of daylight to the results found in (c).

80. Hours of Daylight According to the *Old Farmer's Almanac*, in Seattle, WA, the number of hours of sunlight on the summer solstice is 13.967 and the number of hours of sunlight on the winter solstice is 8.417.

(a) By hand, find a sinusoidal function of the form $y = A\sin(\omega x - \phi) + B$ that fits the data.
(b) Draw a graph of the function found in (a).
(c) Use the function found in (a) to predict the number of hours of sunlight on April 1, the 91st day of the year.
(d) Look up the number of hours of sunlight on April 1, 1997, in the *Old Farmer's Almanac* and compare the actual hours of daylight to the results found in (c).

CHAPTER 9

Polar Coordinates; Vectors

Using computer models is one way structural engineers, architects, and others try out various designs without the costs of building materials and unforeseen construction issues. On the following page you will try your hand at becoming a master bridge builder through computer modeling. The programs and other information you will need are found at:

www.prenhall.com/sullivan

PREPARING FOR THIS CHAPTER

Before getting started on this chapter, review the following concepts:

For Sections 9.1–9.3: Rectangular Coordinates *(pp. 14–15)*

Graphs of Equations *(Section 1.3)*

Complex Numbers *(pp. 317–322)*

For Sections 9.4–9.5: Rectangular Coordinates *(pp. 14–15)*

Law of Cosines *(Section 8.3)*

OUTLINE

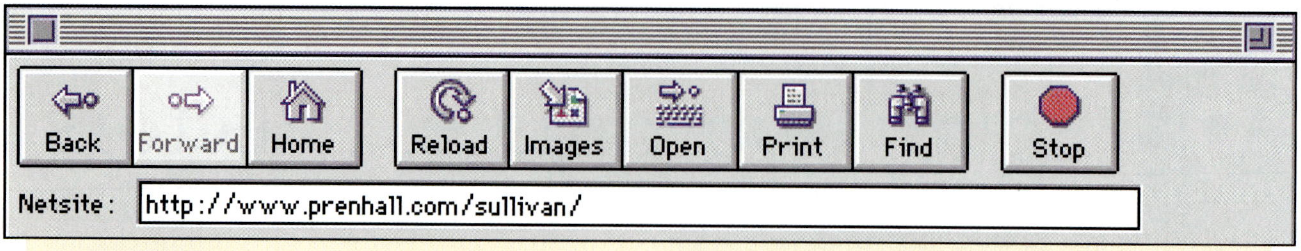

THE GREAT BALSA BRIDGE CONTEST

Your goal is high—you intend to become The Master Balsa Bridge Builder. Over the last two years you have performed countless hours of stress testing on balsa struts to determine to strongest and lightest grains. You have performed wood joint experiments with a host of ultrastrong superglues. You have read and reread the literature on the previous contests, analyzing those bridge designs that have failed, and those that were heads above all the *competition*. You kept careful notes and have entered the data into your *bridge modeling software*. For the last few months you tested your designs on your computer. Only two survived.

You are now ready to enter into the next phase. Yours will be no *bridge disaster*. You are ready to see if your computer generated designs match reality.

You have created a set of microtools so that your joinery will be perfect. Slowly and meticulously you build your first set of two test bridges. You need to determine how accurately your calculations match your craftsmanship. You have laminated the balsa struts, and, at last, your designs come to life! Now you place the two bridges across two chairs and begin to place weights on them . . .

1. If you think of the weights as *force vectors* acting along the struts, does it make sense that we can model the structural design for a bridge?
2. Try the *Design a Bridge* website. How did you do? Did you just overbuild everything?
3. Do you think that engineers, when they are designing a bridge, might rely a little too heavily on the computer? Do you think that might be a mistake?
4. Get out a piece of paper and make a very simple bridge design. Estimate how much weight x and y can hold. Then do a simple vector analysis using your assumptions. Are you surprised at how much your simple bridge can support?
5. Get some pieces of balsa and redo the calculations using your assumptions about the strength of the balsa. How good are you at calculating the actual values?
6. When you look carefully at a complicated bridge like the Golden Gate Bridge, how long do you think that it would take you to do all those calculations?

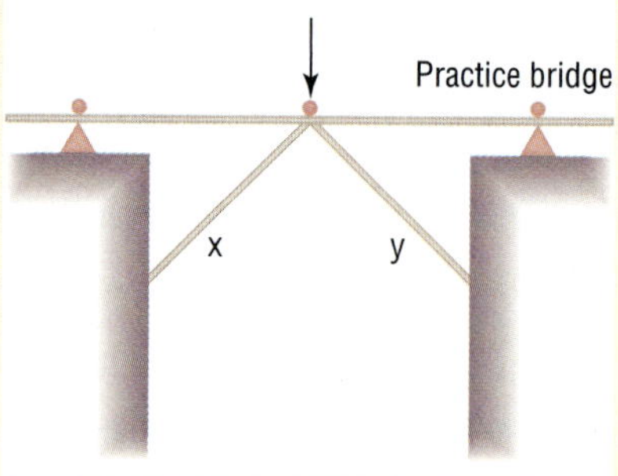

This chapter is in two parts: Polar Coordinates, Sections 9.1–9.3, and Vectors, Sections 9.4–9.5. They are independent of each other and may be covered in any order.

Sections 9.1–9.3 deal with polar coordinates (an alternative to rectangular coordinates for plotting points), graphing in polar coordinates, and finding roots of complex numbers (Demoivre's Theorem).

Sections 9.4–9.5 provide an introduction to the idea of a vector, an extremely important topic in engineering and physics.

9.1 POLAR COORDINATES

1 Plot Points Using Polar Coordinates

2 Convert between Polar Coordinates and Rectangular Coordinates

So far, we have always used a system of rectangular coordinates to plot points in the plane. Now we are ready to describe another system called *polar coordinates.* As we shall soon see, in many instances polar coordinates offer certain advantages over rectangular coordinates.

In a rectangular coordinate system, you will recall, a point in the plane is represented by an ordered pair of numbers (x, y), where x and y equal the signed distance of the point from the y-axis and x-axis, respectively. In a polar coordinate system, we select a point, called the **pole,** and then a ray with vertex at the pole, called the **polar axis.** Comparing the rectangular and polar coordinate systems, we see (in Figure 1) that the origin in rectangular coordinates coincides with the pole in polar coordinates, and the positive x-axis in rectangular coordinates coincides with the polar axis in polar coordinates.

FIGURE 1

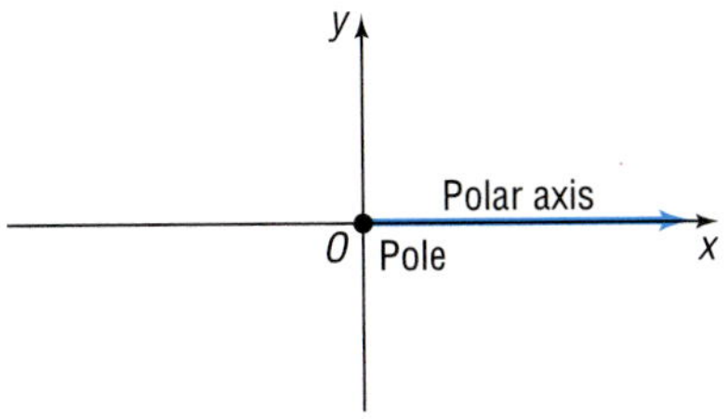

1

A point P in a polar coordinate system is represented by an ordered pair of numbers (r, θ). The number r is the distance of the point from the pole, and θ is an angle (in degrees or radians) formed by the polar axis and a ray from the pole through the point. We call the ordered pair (r, θ) the **polar coordinates** of the point. See Figure 2.

FIGURE 2

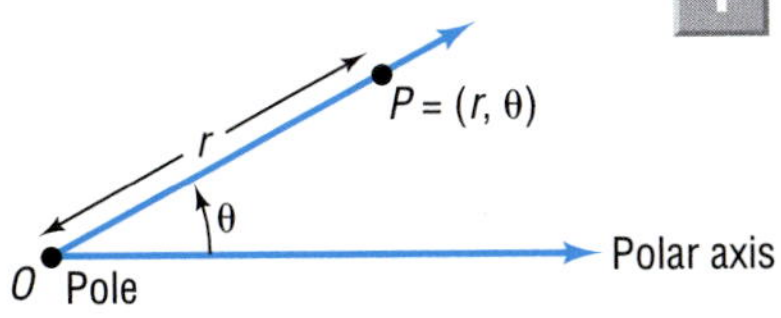

As an example, suppose that the polar coordinates of a point P are $(2, \pi/4)$. We locate P by first drawing an angle of $\pi/4$ radian, placing its vertex at the pole and its initial side along the polar axis. Then we go out a distance of 2 units along the terminal side of the angle to reach the point P. See Figure 3.

FIGURE 3

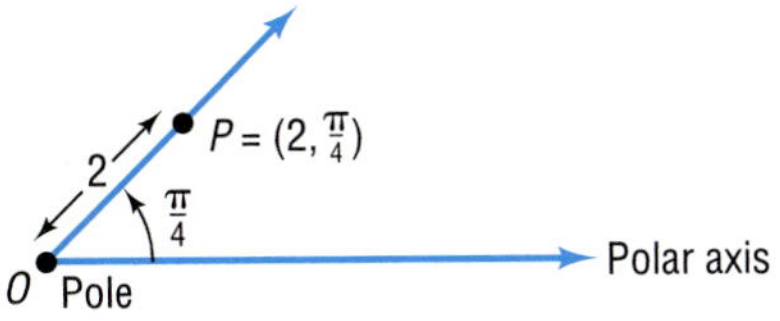

Now work Problem 1.

Recall that an angle measured counterclockwise is positive, whereas one measured clockwise is negative. This convention has some interesting consequences relating to polar coordinates. Let's see what these consequences are.

EXAMPLE 1 Finding Several Polar Coordinates of a Single Point

Consider again a point P with polar coordinates $(2, \pi/4)$, as shown in Figure 4(a). Because $\pi/4$, $9\pi/4$, and $-7\pi/4$ all have the same terminal side, we also could have located this point P by using the polar coordinates $(2, 9\pi/4)$ or $(2, -7\pi/4)$, as shown in Figures 4(b) and (c).

FIGURE 4

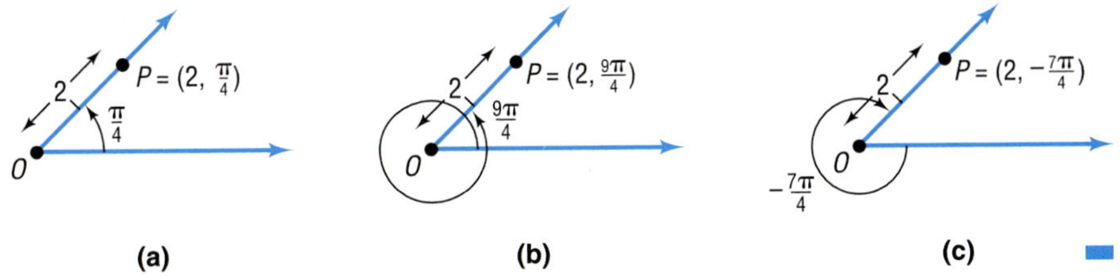

In using polar coordinates (r, θ), it is possible for the first entry r to be negative. When this happens, we follow the convention that the location of the point, instead of being on the terminal side of θ, is on the ray from the pole extending in the direction *opposite* the terminal side of θ at a distance $|r|$ from the pole. See Figure 5 for an illustration.

FIGURE 5

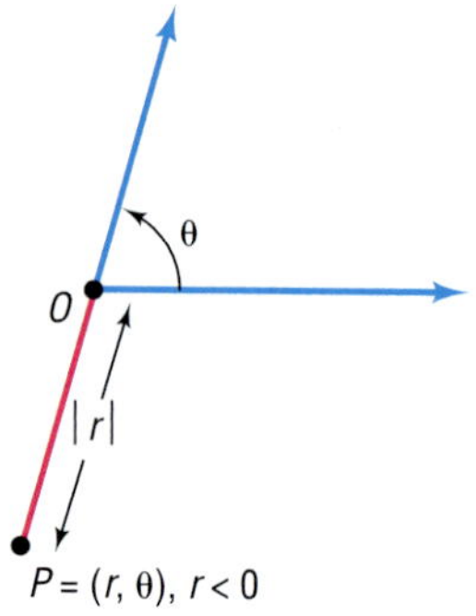

EXAMPLE 2

Polar Coordinates (r, θ), $r < 0$

Consider again the point P with polar coordinates $(2, \pi/4)$, as shown in Figure 6(a). This same point P can be assigned the polar coordinates $(-2, 5\pi/4)$, as indicated in Figure 6(b). To locate the point $(-2, 5\pi/4)$, we use the ray in the opposite direction of $5\pi/4$ and go out 2 units along that ray to find the point P.

FIGURE 6

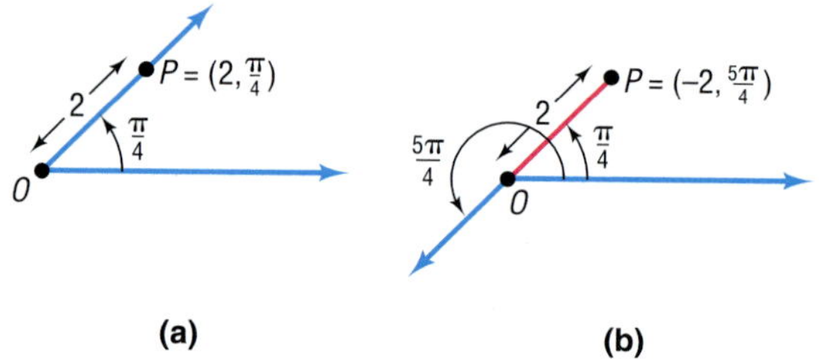

These examples show a major difference between rectangular coordinates and polar coordinates. In the former, each point has exactly one pair of rectangular coordinates; in the latter, a point can have infinitely many pairs of polar coordinates.

SUMMARY

A point with polar coordinates (r, θ) also can be represented by any of the following:

$$(r, \theta + 2k\pi) \quad \text{or} \quad (-r, \theta + \pi + 2k\pi), \qquad k \text{ any integer}$$

The polar coordinates of the pole are $(0, \theta)$, where θ can be any angle.

EXAMPLE 3

Plotting Points Using Polar Coordinates

Plot the points with the following polar coordinates:

(a) $(3, 5\pi/3)$ (b) $(2, -\pi/4)$ (c) $(3, 0)$ (d) $(-2, \pi/4)$

Solution Figure 7 shows the points.

FIGURE 7

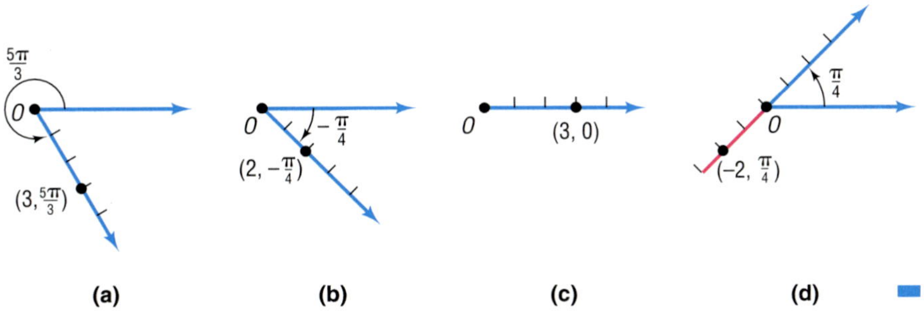

Now work Problem 9.

EXAMPLE 4 Finding Other Polar Coordinates of a Given Point

Plot the point P with polar coordinates $(3, \pi/6)$, and find other polar coordinates (r, θ) of this same point for which

(a) $r > 0, \quad 2\pi \leq \theta < 4\pi$ (b) $r < 0, \quad 0 \leq \theta < 2\pi$

(c) $r > 0, \quad -2\pi \leq \theta < 0$

FIGURE 8

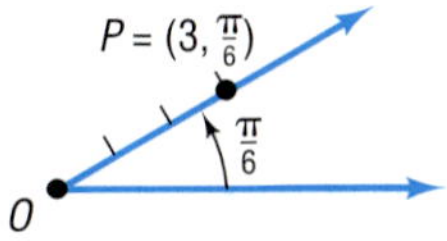

Solution The point $(3, \pi/6)$ is plotted in Figure 8.

(a) We add 1 revolution (2π radians) to the angle $\pi/6$ to get $P = (3, \pi/6 + 2\pi) = (3, 13\pi/6)$. See Figure 9.

(b) We add $\frac{1}{2}$ revolution (π radians) to the angle and replace r by $-r$ to get $P = (-3, \pi/6 + \pi) = (-3, 7\pi/6)$. See Figure 10.

(c) We subtract 2π from the angle $\pi/6$ to get $P = (3, \pi/6 - 2\pi) = (3, -11\pi/6)$. See Figure 11.

FIGURE 9

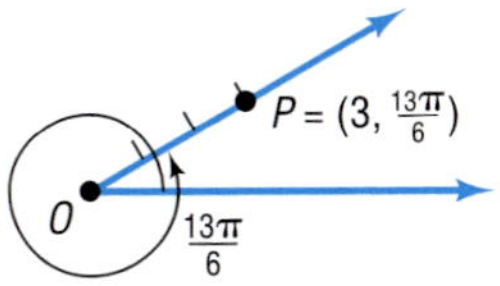

FIGURE 10

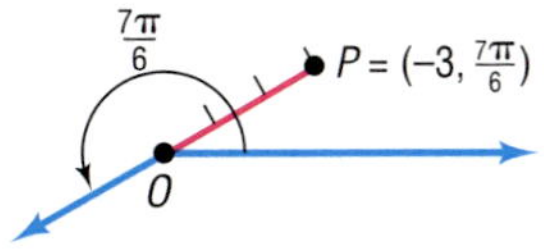

FIGURE 11

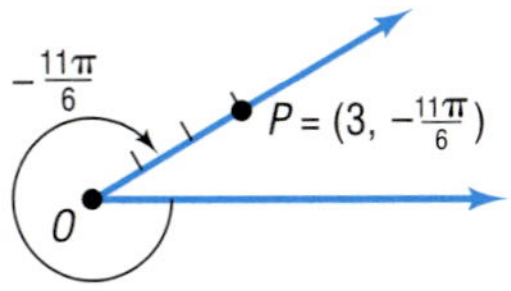

Now work Problem 13.

Conversion between Polar Coordinates and Rectangular Coordinates

2 It is sometimes convenient and, indeed, necessary to be able to convert coordinates or equations in rectangular form to polar form, and vice versa. To do this, we recall that the origin in rectangular coordinates is the pole in polar coordinates and that the positive x-axis in rectangular coordinates is the polar axis in polar coordinates.

Theorem Conversion from Polar Coordinates to Rectangular Coordinates

If P is a point with polar coordinates (r, θ), the rectangular coordinates (x, y) of P are given by

$$x = r\cos\theta \qquad y = r\sin\theta \tag{1}$$

Note: Most graphing calculators have the capability of converting from polar coordinates to rectangular coordinates. Consult your owner's manual to learn the proper key strokes.

Proof Suppose that P has the polar coordinates (r, θ). We seek the rectangular coordinates (x, y) of P. Refer to Figure 12.

FIGURE 12

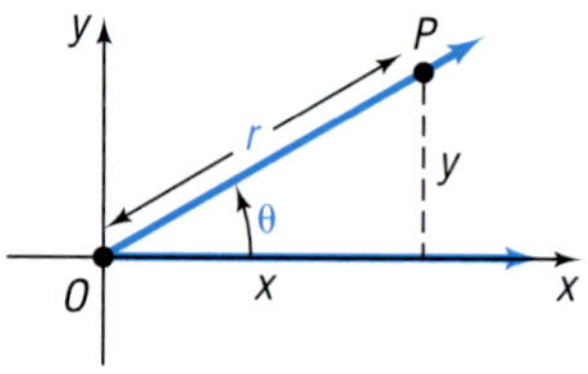

If $r = 0$, then, regardless of θ, the point P is the pole, for which the rectangular coordinates are $(0, 0)$. Thus, formula (1) is valid for $r = 0$.

If $r > 0$, the point P is on the terminal side of θ and $r = d(O, P) = \sqrt{x^2 + y^2}$. Thus,

$$\cos\theta = \frac{x}{r} \qquad \sin\theta = \frac{y}{r}$$

so

$$x = r\cos\theta \qquad y = r\sin\theta$$

If $r < 0$, then the point $P = (r, \theta)$ can be represented as $(-r, \pi + \theta)$, where $-r > 0$. Thus,

$$\cos(\pi + \theta) = -\cos\theta = \frac{x}{-r} \qquad \sin(\pi + \theta) = -\sin\theta = \frac{y}{-r}$$

so

$$x = r\cos\theta \qquad y = r\sin\theta$$

EXAMPLE 5 Converting from Polar Coordinates to Rectangular Coordinates

Find the rectangular coordinates of the points with the following polar coordinates:

(a) $(6, \pi/6)$ (b) $(-2, 5\pi/4)$ (c) $(-4, -\pi/4)$

Solution We use formula (1): $x = r\cos\theta$ and $y = r\sin\theta$.

(a) See Figure 13(a).

$$x = r\cos\theta = 6\cos\frac{\pi}{6} = 6\cdot\frac{\sqrt{3}}{2} = 3\sqrt{3}$$

$$y = r\sin\theta = 6\sin\frac{\pi}{6} = 6\cdot\frac{1}{2} = 3$$

The rectangular coordinates of the point $(6, \pi/6)$ are $(3\sqrt{3}, 3)$.

(b) See Figure 13(b).

$$x = r\cos\theta = -2\cos\frac{5\pi}{4} = -2\left(-\frac{\sqrt{2}}{2}\right) = \sqrt{2}$$

$$y = r\sin\theta = -2\sin\frac{5\pi}{4} = -2\left(-\frac{\sqrt{2}}{2}\right) = \sqrt{2}$$

The rectangular coordinates of the point $(-2, 5\pi/4)$ are $(\sqrt{2}, \sqrt{2})$.

(c) See Figure 13(c).

$$x = r\cos\theta = -4\cos\left(-\frac{\pi}{4}\right) = -4\cdot\frac{\sqrt{2}}{2} = -2\sqrt{2}$$

$$y = r\sin\theta = -4\sin\left(-\frac{\pi}{4}\right) = -4\left(-\frac{\sqrt{2}}{2}\right) = 2\sqrt{2}$$

The rectangular coordinates of the point $(-4, -\pi/4)$ are $(-2\sqrt{2}, 2\sqrt{2})$.

FIGURE 13

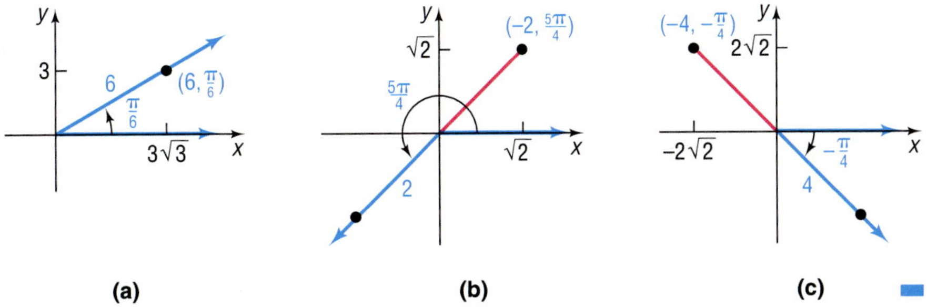

Now work Problems 21 and 33.

To convert from rectangular coordinates (x, y) to polar coordinates (r, θ) is a little more complicated. Let's look at some examples.

EXAMPLE 6 Converting from Rectangular Coordinates to Polar Coordinates

Find polar coordinates of a point whose rectangular coordinates are $(0, 3)$.

FIGURE 14

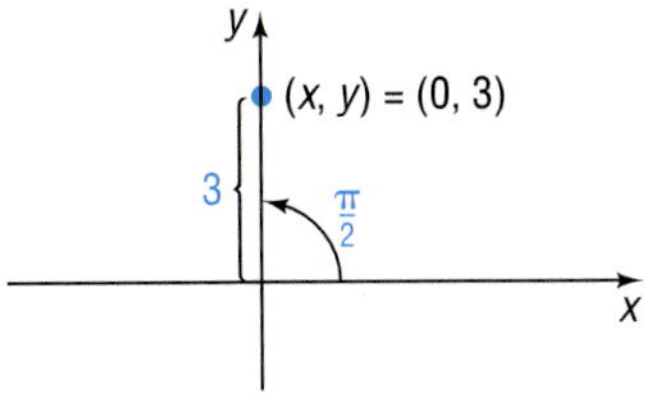

Solution See Figure 14. The point $(0, 3)$ lies on the y-axis a distance of 3 units from the origin (pole) so $r = 3$. A ray with vertex at the pole through $(0, 3)$ forms an angle $\theta = \dfrac{\pi}{2}$ with the polar axis. Polar coordinates for this point can be given by $(3, \pi/2)$.

Figure 15 shows polar coordinates of points that lie on either the x-axis or the y-axis.

FIGURE 15

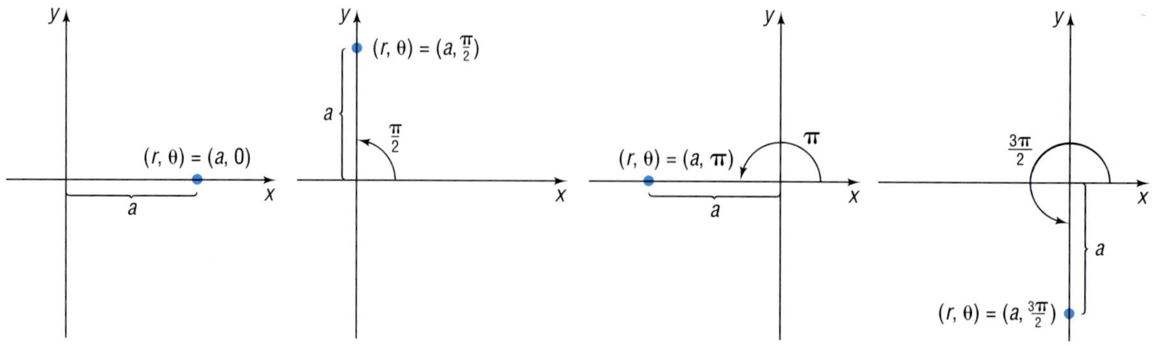

(a) $(x, y) = (a, 0)$, $a > 0$ **(b)** $(x, y) = (0, a)$, $a > 0$ **(c)** $(x, y) = (-a, 0)$, $a > 0$ **(d)** $(x, y) = (0, -a)$, $a > 0$

Now work Problem 37.

EXAMPLE 7 Converting from Rectangular Coordinates to Polar Coordinates

Find polar coordinates of a point whose rectangular coordinates are $(2, -2)$.

FIGURE 16

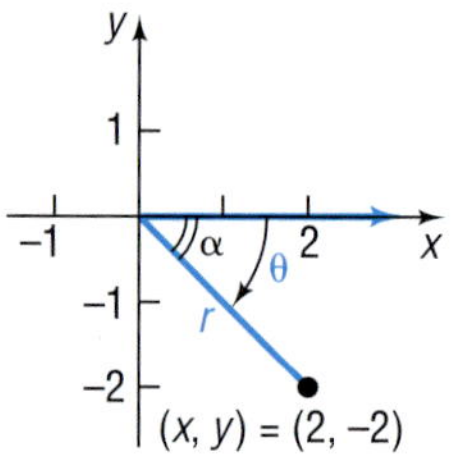

Solution See Figure 16. The distance r from the origin to the point $(2, -2)$ is

$$r = \sqrt{x^2 + y^2} = \sqrt{(2)^2 + (-2)^2} = \sqrt{8} = 2\sqrt{2}$$

To find θ, we use the reference angle α. Then

$$\alpha = \tan^{-1}\left|\frac{y}{x}\right| = \tan^{-1}\left|\frac{-2}{2}\right| = \tan^{-1} 1 = \frac{\pi}{4}$$

Thus, $\theta = -\pi/4$, and a set of polar coordinates for this point is $(2\sqrt{2}, -\pi/4)$. Other possible representations include $(2\sqrt{2}, 7\pi/4)$ and $(-2\sqrt{2}, 3\pi/4)$.

EXAMPLE 8 Converting from Rectangular Coordinates to Polar Coordinates

Find polar coordinates of a point whose rectangular coordinates are $(-1, -\sqrt{3})$.

FIGURE 17

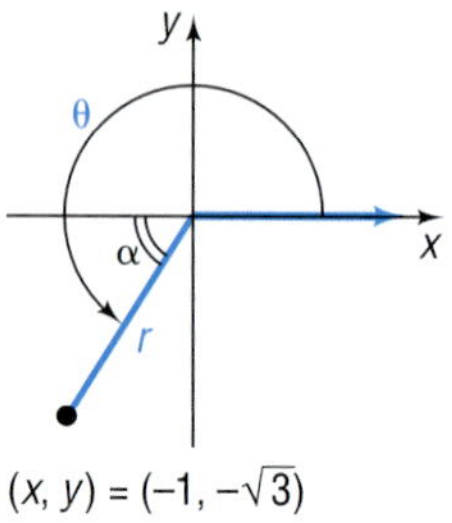

Solution See Figure 17. The distance r from the origin to the point $(-1, -\sqrt{3})$ is

$$r = \sqrt{x^2 + y^2} = \sqrt{(-1)^2 + (-\sqrt{3})^2} = \sqrt{4} = 2$$

To find θ, we use the reference angle α. Then

$$\alpha = \tan^{-1}\left|\frac{y}{x}\right| = \tan^{-1}\left|\frac{-\sqrt{3}}{-1}\right| = \tan^{-1}\sqrt{3} = \frac{\pi}{3}$$

Thus,

$$\theta = \pi + \alpha = \pi + \frac{\pi}{3} = \frac{4\pi}{3}$$

and a set of polar coordinates is $(2, 4\pi/3)$. Other possible representations include $(-2, \pi/3)$ and $(2, -2\pi/3)$.

Figure 18 shows how to find polar coordinates of a point that lies in a quadrant when its rectangular coordinates (x, y) are given.

FIGURE 18

(a) $r = \sqrt{x^2 + y^2}$
$\theta = \alpha = \tan^{-1} \left| \frac{y}{x} \right|$

(b) $r = \sqrt{x^2 + y^2}$
$\theta = \pi - \alpha = \pi - \tan^{-1} \left| \frac{y}{x} \right|$

(c) $r = \sqrt{x^2 + y^2}$
$\theta = \pi + \alpha = \pi + \tan^{-1} \left| \frac{y}{x} \right|$

(d) $r = \sqrt{x^2 + y^2}$
$\theta = -\alpha = -\tan^{-1} \left| \frac{y}{x} \right|$

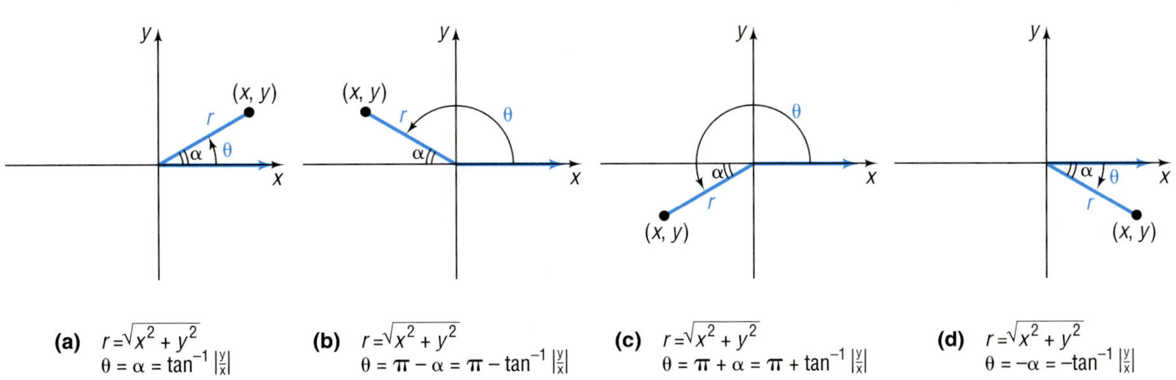

Now work Problem 41.

Based on the preceding discussion, we have the formulas

$$r^2 = x^2 + y^2 \qquad \tan \theta = \frac{y}{x} \qquad \text{if } x \neq 0 \qquad (2)$$

Warning: Be careful when using formula (2). Always plot the point (x, y) first, as we did in Examples 7 and 8, so that r and θ are chosen to reflect the correct quadrant. Look again at Figure 18.

Formulas (1) and (2) may also be used to transform equations.

E X A M P L E   9

Transforming Equations from Polar to Rectangular Form

Transform the equation $r = 4 \sin \theta$ from polar coordinates to rectangular coordinates, and identify the graph.

Solution   If we multiply each side by r, it will be easier to apply formulas (1) and (2):

$$r = 4 \sin \theta$$
$$r^2 = 4r \sin \theta \qquad \text{Multiply each side by } r.$$
$$x^2 + y^2 = 4y \qquad \text{Apply formulas (1) and (2).}$$

This is the equation of a circle:

$$x^2 + (y^2 - 4y) = 0$$
$$x^2 + (y^2 - 4y + 4) = 4 \qquad \text{Complete the squaring.}$$
$$x^2 + (y - 2)^2 = 4$$

Its center is at $(0, 2)$, and its radius is 2.

Now work Problem 57.

EXAMPLE 10 Transforming an Equation from Rectangular to Polar Form

Transform the equation $4xy = 9$ from rectangular coordinates to polar coordinates.

Solution We use formula (1):

$$4xy = 9$$
$$4(r\cos\theta)(r\sin\theta) = 9 \quad \text{Formula (1).}$$
$$4r^2\cos\theta\sin\theta = 9$$
$$2r^2\sin 2\theta = 9 \quad \text{Double-angle formula.}$$

9.1 EXERCISES

In Problems 1–12, plot each point given in polar coordinates.

1. $(3, 90°)$ **2.** $(4, 270°)$ **3.** $(-2, 0)$ **4.** $(-3, \pi)$

5. $(6, \pi/6)$ **6.** $(5, 5\pi/3)$ **7.** $(-2, 135°)$ **8.** $(-3, 120°)$

9. $(-1, -\pi/3)$ **10.** $(-3, -3\pi/4)$ **11.** $(-2, -\pi)$ **12.** $(-3, -\pi/2)$

In Problems 13–20, plot each point given in polar coordinates, and find other polar coordinates (r, θ) of the point for which

(a) $r > 0$, $-2\pi \le \theta < 0$ (b) $r < 0$, $0 \le \theta < 2\pi$ (c) $r > 0$, $2\pi \le \theta < 4\pi$

13. $(5, 2\pi/3)$ **14.** $(4, 3\pi/4)$ **15.** $(-2, 3\pi)$ **16.** $(-3, 4\pi)$

17. $(1, \pi/2)$ **18.** $(2, \pi)$ **19.** $(-3, -\pi/4)$ **20.** $(-2, -2\pi/3)$

In Problems 21–36, polar coordinates of a point are given. Find the rectangular coordinates of each point.

21. $(3, \pi/2)$ **22.** $(4, 3\pi/2)$ **23.** $(-2, 0)$ **24.** $(-3, \pi)$

25. $(6, 150°)$ **26.** $(5, 300°)$ **27.** $(-2, 3\pi/4)$ **28.** $(-3, 2\pi/3)$

29. $(-1, -\pi/3)$ **30.** $(-3, -3\pi/4)$ **31.** $(-2, -180°)$ **32.** $(-3, -90°)$

33. $(7.5, 110°)$ **34.** $(-3.1, 182°)$ **35.** $(6.3, 3.8)$ **36.** $(8.1, 5.2)$

In Problems 37–48, the rectangular coordinates of a point are given. Find polar coordinates for each point.

37. $(3, 0)$ **38.** $(0, 2)$ **39.** $(-1, 0)$ **40.** $(0, -2)$

41. $(1, -1)$ **42.** $(-3, 3)$ **43.** $(\sqrt{3}, 1)$ **44.** $(-2, -2\sqrt{3})$

45. $(1.3, -2.1)$ **46.** $(-0.8, -2.1)$ **47.** $(8.3, 4.2)$ **48.** $(-2.3, 0.2)$

In Problems 49–56, the letters x and y represent rectangular coordinates. Write each equation using polar coordinates (r, θ).

49. $2x^2 + 2y^2 = 3$ **50.** $x^2 + y^2 = x$ **51.** $x^2 = 4y$ **52.** $y^2 = 2x$

53. $2xy = 1$ **54.** $4x^2y = 1$ **55.** $x = 4$ **56.** $y = -3$

In Problems 57–64, the letters r and θ represent polar coordinates. Write each equation using rectangular coordinates (x, y).

57. $r = \cos\theta$ **58.** $r = \sin\theta + 1$ **59.** $r^2 = \cos\theta$ **60.** $r = \sin\theta - \cos\theta$

61. $r = 2$ **62.** $r = 4$ **63.** $r = \dfrac{4}{1 - \cos\theta}$ **64.** $r = \dfrac{3}{3 - \cos\theta}$

65. Show that the formula for the distance d between two points $P_1 = (r_1, \theta_1)$ and $P_2 = (r_2, \theta_2)$ is
$$d = \sqrt{r_1^2 + r_2^2 - 2r_1r_2\cos(\theta_2 - \theta_1)}$$

9.2 POLAR EQUATIONS AND GRAPHS

1 Graph and Identify Polar Equations by Converting to Rectangular Equations
2 Graph Polar Equations Using a Graphing Utility
3 Test a Polar Equation for Symmetry
4 Graph Polar Equations by Hand by Plotting Points

An equation whose variables are polar coordinates is called a **polar equation.** The **graph of a polar equation** consists of all points whose polar coordinates satisfy the equation.

Just as a rectangular grid may be used to plot points given by rectangular coordinates, as in Figure 19(a), we can use a grid consisting of concentric circles (with centers at the pole) and rays (with vertices at the pole) to plot points given by polar coordinates, as shown in Figure 19(b). We shall use such **polar grids** to graph polar equations by hand.

FIGURE 19

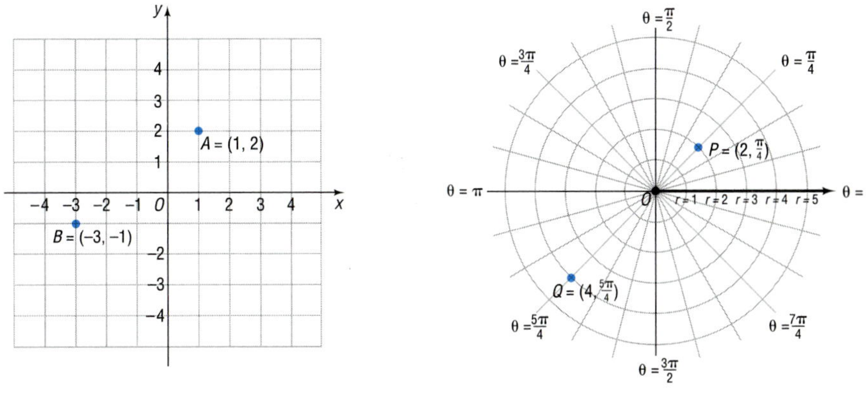

(a) Rectangular Grid **(b)** Polar Grid

1 One method we can use to graph a polar equation by hand is to convert the equation to rectangular coordinates. In the discussion that follows, (x, y) represent the rectangular coordinates of a point P and (r, θ) represent polar coordinates of the point P.

EXAMPLE 1 Identifying and Graphing a Polar Equation by Hand (Circle)

Identify and graph the equation: $r = 3$

Solution We convert the polar equation to a rectangular equation:

$$r = 3$$
$$r^2 = 9$$
$$x^2 + y^2 = 9$$

Thus, the graph of $r = 3$ is a circle, with center at the pole and radius 3. See Figure 20.

FIGURE 20
$r = 3$ or $x^2 + y^2 = 9$

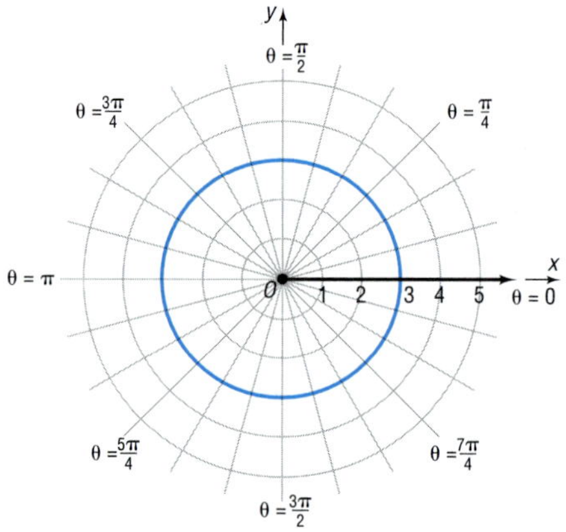

Now work Problem 1.

EXAMPLE 2 Identifying and Graphing a Polar Equation by Hand (Line)

Identify and graph the equation: $\theta = \pi/4$

Solution We convert the polar equation to a rectangular equation:

$$\theta = \frac{\pi}{4}$$

$$\tan \theta = \tan \frac{\pi}{4} = 1$$

$$\frac{y}{x} = 1 \quad \text{Formula (2), Section 9.1.}$$

$$y = x$$

The graph of $\theta = \pi/4$ is a line passing through the pole making an angle of $\pi/4$ with the polar axis. See Figure 21.

FIGURE 21

$\theta = \frac{\pi}{4}$ or $y = x$

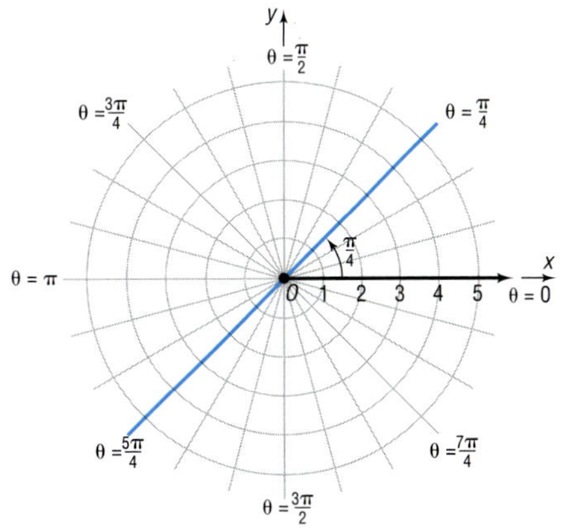

EXAMPLE 3 Identifying and Graphing a Polar Equation by Hand (Line)

Identify and graph the equation: $\theta = -\pi/6$

Solution We convert the polar equation to a rectangular equation:

$$\theta = -\frac{\pi}{6}$$

$$\tan\theta = \tan\left(-\frac{\pi}{6}\right) = -\frac{\sqrt{3}}{3}$$

$$\frac{y}{x} = -\frac{\sqrt{3}}{3}$$

$$y = -\frac{\sqrt{3}}{3}x$$

The graph of $\theta = -\pi/6$ is a line passing through the pole making an angle of $-\pi/6$ with the polar axis. See Figure 22.

FIGURE 22

$\theta = -\frac{\pi}{6}$ or $y = -\frac{\sqrt{3}}{3}x$

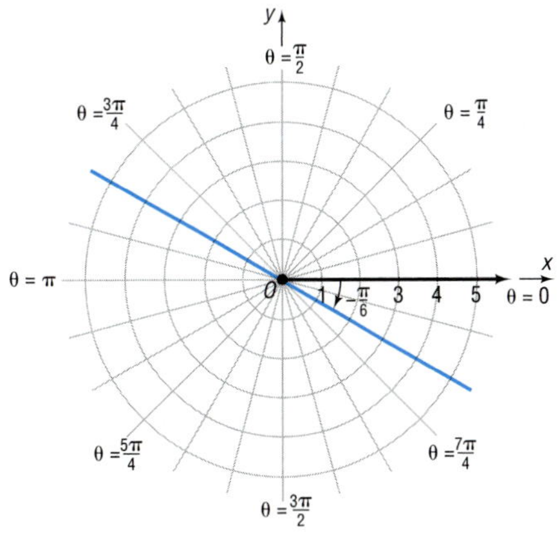

Now work Problem 3.

EXAMPLE 4 Identifying and Graphing a Polar Equation by Hand (Horizontal Line)

Identify and graph the equation: $r \sin \theta = 2$

Solution Since $y = r \sin \theta$, we can simply write the equation as

$$y = 2$$

We conclude that the graph of $r \sin \theta = 2$ is a horizontal line 2 units above the pole. See Figure 23.

FIGURE 23
$r \sin \theta = 2$ or $y = 2$

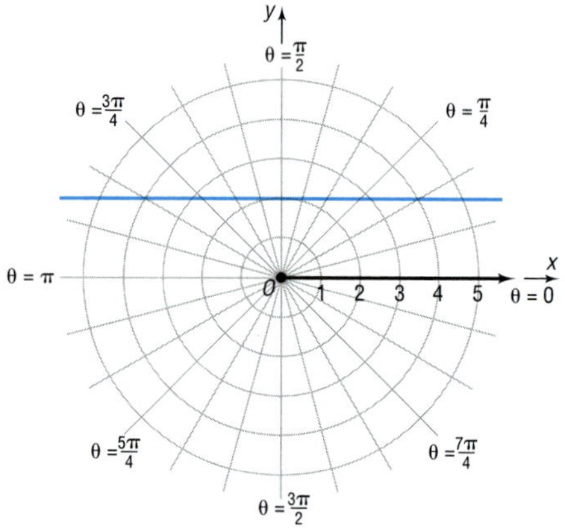

2 A second method we can use to graph a polar equation is to graph the equation using a graphing utility.

Most graphing utilities require the following steps in order to obtain the graph of an equation:

Graphing a Polar Equation Using a Graphing Utility

STEP 1: Solve the equation for r in terms of θ.
STEP 2: Select the viewing rectangle in polar mode. In addition to setting Xmin, Xmax, Xscl, and so forth, the viewing rectangle in polar mode requires setting minimum and maximum values for θ and an increment setting for θ (θ step). In addition, a square screen and radian measure should be used.
STEP 3: Enter the expression involving θ that you found in STEP 1. (Consult your manual for the correct way to enter the expression.)
STEP 4: Execute.

EXAMPLE 5 Graphing a Polar Equation Using a Graphing Utility

Use a graphing utility to graph the polar equation $r \sin \theta = 2$.

Solution STEP 1: We solve the equation for r in terms of θ.

$$r \sin \theta = 2$$

$$r = \frac{2}{\sin \theta}$$

STEP 2: From the polar mode, select the viewing rectangle. We will use the one given next.

$$\begin{aligned}\theta\text{min} &= 0\\ \theta\text{max} &= 2\pi\\ \theta step &= \pi/24\\ X\text{min} &= -9\\ X\text{max} &= 9\\ Xscl &= 1\\ Y\text{min} &= -6\\ Y\text{max} &= 6\\ Yscl &= 1\end{aligned}$$

$\theta step$ determines the number of points the graphing utility will plot. For example, if $\theta step$ is $\pi/24$, then the graphing utility will evaluate r at $\theta = 0(\theta\text{min})$, $\pi/24$, $2\pi/24$, $3\pi/24$, and so forth, up to $2\pi(\theta\text{max})$. The smaller $\theta step$, the more points the graphing utility will plot. The student is encouraged to experiment with different values for θmin, θmax, and θstep to see how the graph is affected.

FIGURE 24

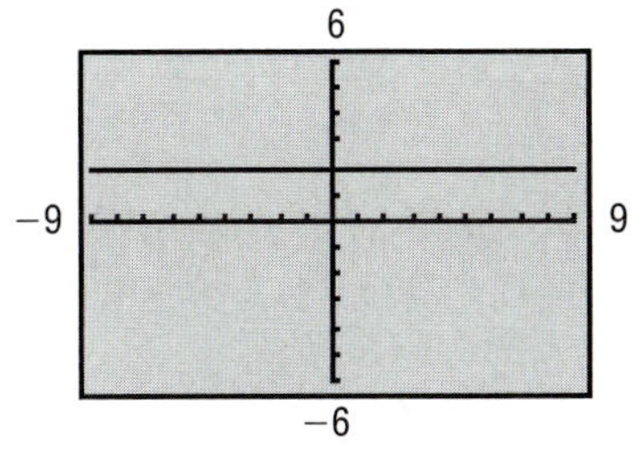

STEP 3: Enter the expression $\dfrac{2}{\sin\theta}$ after the prompt $r =$.

STEP 4: Execute.

The graph is shown in Figure 24.

EXAMPLE 6 Identifying and Graphing a Polar Equation (Vertical Line)

Identify and graph the equation: $r\cos\theta = -3$

Solution Since $x = r\cos\theta$, we can simply write the equation as

$$x = -3$$

We conclude that the graph of $r\cos\theta = -3$ is a vertical line 3 units to the left of the pole. Figure 25(a) shows the graph drawn by hand. Figure 25(b) shows the graph of $r\cos\theta = -3$ using a graphing utility with θmin $= 0$, θmax $= 2\pi$, and θstep $= \pi/24$.

FIGURE 25
$r\cos\theta = -3$ or $x = -3$

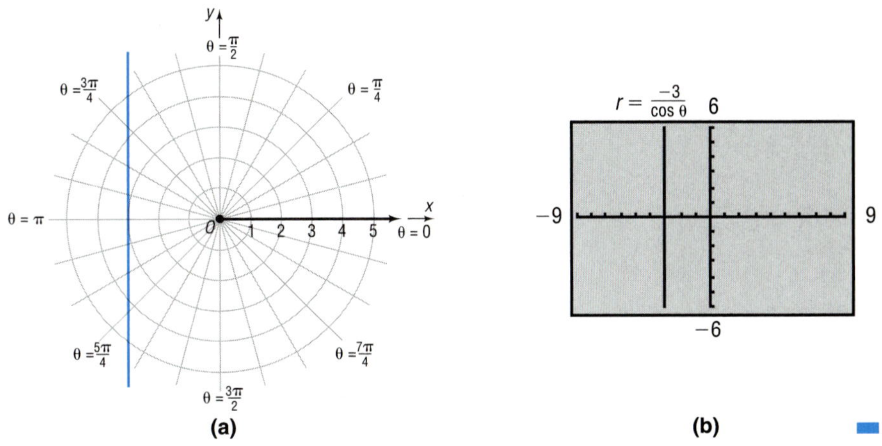

Based on Examples 4, 5, and 6, we are led to the following results. (The proofs are left as exercises.)

Theorem

Let a be a nonzero real number, the graph of the equation

$$r \sin \theta = a$$

is a horizontal line a units above the pole if $a > 0$ and $|a|$ units below the pole if $a < 0$.

The graph of the equation

$$r \cos \theta = a$$

is a vertical line a units to the right of the pole if $a > 0$ and $|a|$ units to the left of the pole if $a < 0$.

Now work Problem 7.

EXAMPLE 7 Identifying and Graphing a Polar Equation (Circle)

Identify and graph the equation: $r = 4 \sin \theta$

Solution To transform the equation to rectangular coordinates, we multiply each side by r:

$$r^2 = 4r \sin \theta$$

Now we use the facts that $r^2 = x^2 + y^2$ and $y = r \sin \theta$. Then

$$x^2 + y^2 = 4y$$
$$x^2 + (y^2 - 4y) = 0$$
$$x^2 + (y - 2)^2 = 4$$

This is the equation of a circle with center at (0, 2) in rectangular coordinates and radius 2. Figure 26(a) shows the graph drawn by hand. Figure 26(b) shows the graph of $r = 4 \sin \theta$ using a graphing utility with $\theta\text{min} = 0$, $\theta\text{max} = 2\pi$, and $\theta\text{step} = \pi/24$.

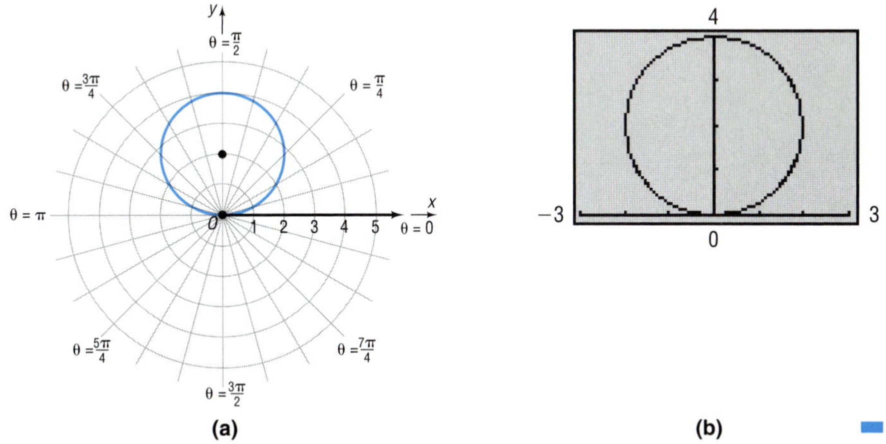

FIGURE 26
$r = 4 \sin \theta$ or
$x^2 + (y - 2)^2 = 4$

EXAMPLE 8 Identifying and Graphing a Polar Equation (Circle)

Identify and graph the equation: $r = -2 \cos \theta$

Solution We proceed as in Example 7:

$$r^2 = -2r \cos \theta$$
$$x^2 + y^2 = -2x$$
$$x^2 + 2x + y^2 = 0$$
$$(x + 1)^2 + y^2 = 1$$

This is the equation of a circle with center at $(-1, 0)$ in rectangular coordinates and radius 1. Figure 27(a) shows the graph drawn by hand. Figure 27(b) shows the graph of $r = -2 \cos \theta$ using a graphing utility with θmin $= 0$, θmax $= 2\pi$, and θstep $= \pi/24$.

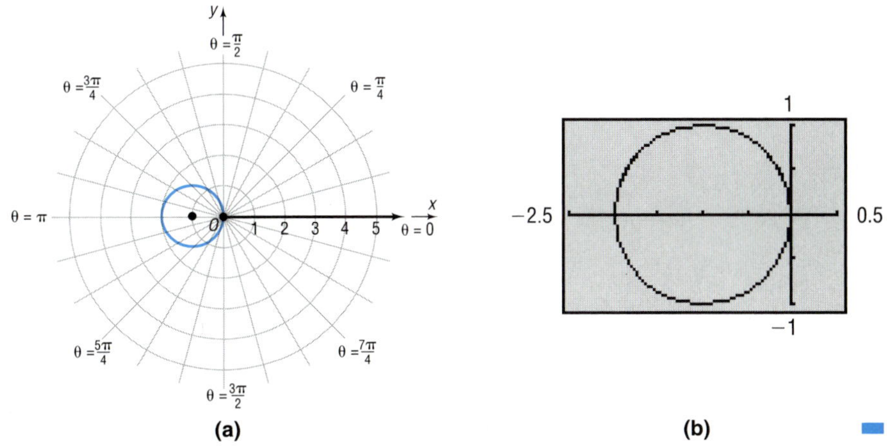

FIGURE 27
$r = -2 \cos \theta$ or
$(x + 1)^2 + y^2 = 1$

Exploration Graph $r_1 = \sin\theta$, $r_2 = 2\sin\theta$, and $r_3 = 3\sin\theta$. Do you see the pattern? Clear the screen and graph $r_1 = -\sin\theta$, $r_2 = -2\sin\theta$, and $r_3 = -3\sin\theta$. Do you see the pattern? Clear the screen and graph $r_1 = \cos\theta$, $r_2 = 2\cos\theta$, and $r_3 = 3\cos\theta$. Do you see the pattern? Clear the screen and graph $r_1 = -\cos\theta$, $r_2 = -2\cos\theta$, and $r_3 = -3\cos\theta$. Do you see the pattern?

Based on Examples 7, 8, and the Exploration, we are led to the following results. (The proofs are left as exercises.)

Theorem

Let a be a positive real number. Then,

Equation	**Description**
(a) $r = 2a\sin\theta$	Circle: radius a; center at $(0, a)$ in rectangular coordinates
(b) $r = -2a\sin\theta$	Circle: radius a; center at $(0, -a)$ in rectangular coordinates
(c) $r = 2a\cos\theta$	Circle: radius a; center at $(a, 0)$ in rectangular coordinates
(d) $r = -2a\cos\theta$	Circle: radius a; center at $(-a, 0)$ in rectangular coordinates

Each circle contains the pole.

The method of converting a polar equation to an identifiable rectangular equation in order to graph it is not always helpful, nor is it always necessary. Usually, we set up a table that lists several points on the graph. By checking for symmetry, it may be possible to reduce the number of points needed to draw the graph by hand.

Symmetry

3 In polar coordinates the points (r, θ) and $(r, -\theta)$ are symmetric with respect to the polar axis (and to the x-axis). See Figure 28(a). The points (r, θ) and $(r, \pi - \theta)$ are symmetric with respect to the line $\theta = \pi/2$ (y-axis). See Figure 28(b). The points (r, θ) and $(-r, \theta)$ are symmetric with respect to the pole (origin). See Figure 28(c).

FIGURE 28

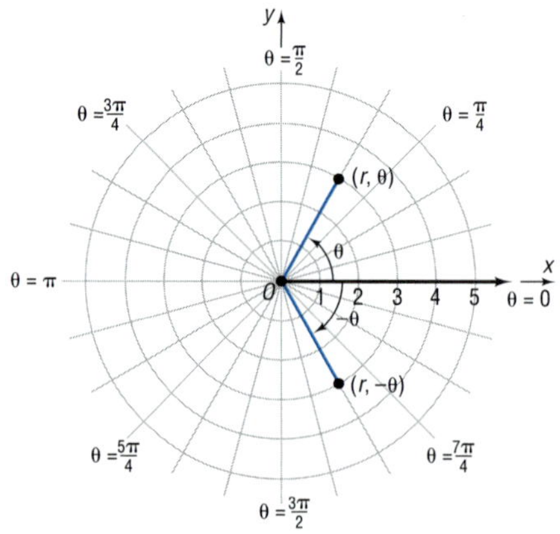

(a) Points symmetric with respect to the polar axis

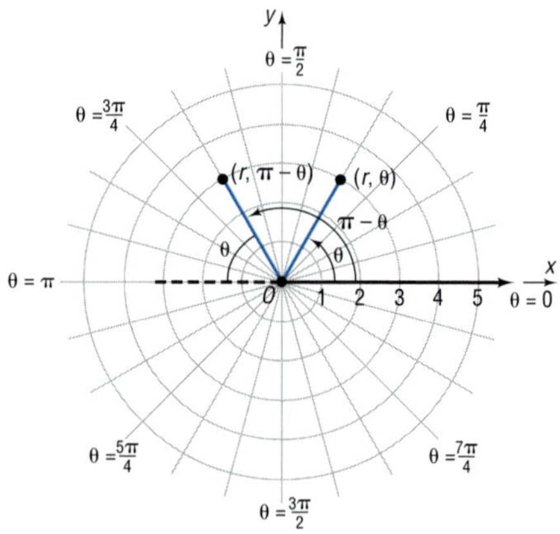

(b) Points symmetric with respect to the line $\theta = \frac{\pi}{2}$

FIGURE 28
(continued)

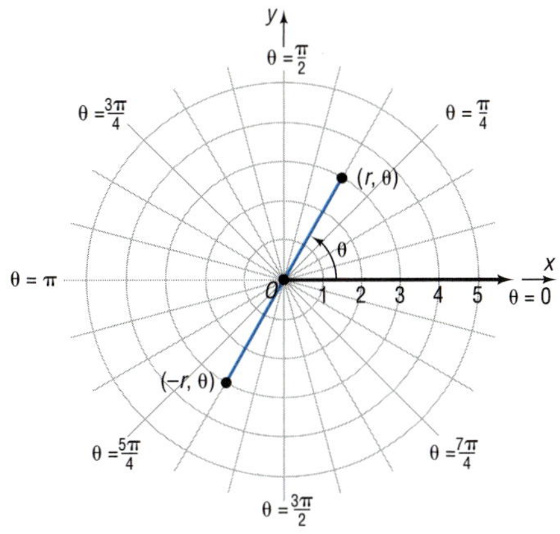

(c) Points symmetric with respect to the pole

The following tests are a consequence of these observations.

Theorem Tests for Symmetry

Symmetry with Respect to the Polar Axis (x-Axis):
In a polar equation, replace θ by $-\theta$. If an equivalent equation results, the graph is symmetric with respect to the polar axis.

Symmetry with Respect to the Line $\theta = \pi/2$ (y-Axis):
In a polar equation, replace θ by $\pi - \theta$. If an equivalent equation results, the graph is symmetric with respect to the line $\theta = \pi/2$.

Symmetry with Respect to the Pole (Origin):
In a polar equation, replace r by $-r$. If an equivalent equation results, the graph is symmetric with respect to the pole.

The three tests for symmetry given here are *sufficient* conditions for symmetry, but they are not *necessary* conditions. That is, an equation may fail these tests and still have a graph that is symmetric with respect to the polar axis, the line $\theta = \pi/2$, or the pole. For example, the graph of $r = \sin 2\theta$ turns out to be symmetric with respect to the polar axis, the line $\theta = \pi/2$, and the pole, but all three tests given here fail. Refer to Problems 71, 72, and 73 in Exercise 9.2.

EXAMPLE 9 Graphing a Polar Equation (Cardioid)

4

Graph the equation: $r = 1 - \sin\theta$

Solution We check for symmetry first:

Polar Axis: Replace θ by $-\theta$. The result is

$$r = 1 - \sin(-\theta) = 1 + \sin\theta$$

The test fails, so the graph may or may not be symmetric with respect to the polar axis.

The Line $\theta = \pi/2$: Replace θ by $\pi - \theta$. The result is

$$\begin{aligned} r = 1 - \sin(\pi - \theta) &= 1 - (\sin\pi\cos\theta - \cos\pi\sin\theta) \\ &= 1 - [0 \cdot \cos\theta - (-1)\sin\theta] = 1 - \sin\theta \end{aligned}$$

Thus, the graph is symmetric with respect to the line $\theta = \pi/2$.

The Pole: Replace r by $-r$. Then the result is $-r = 1 - \sin\theta$, so $r = -1 + \sin\theta$. The test fails, so the graph may or may not be symmetric with respect to the pole.

Next, we identify points on the graph by assigning values to the angle θ and calculating the corresponding values of r. Due to the symmetry with respect to the line $\theta = \pi/2$, we only need to assign values to θ from $-\pi/2$ to $\pi/2$, as given in Table 1.

TABLE 1

θ	$r = 1 - \sin\theta$
$-\pi/2$	$1 + 1 = 2$
$-\pi/3$	$1 + \sqrt{3}/2 \approx 1.87$
$-\pi/6$	$1 + \frac{1}{2} = \frac{3}{2}$
0	1
$\pi/6$	$1 - \frac{1}{2} = \frac{1}{2}$
$\pi/3$	$1 - \sqrt{3}/2 \approx 0.13$
$\pi/2$	0

Now we plot the points (r, θ) from Table 1 and trace out the graph, beginning at the point $(2, -\pi/2)$ and ending at the point $(0, \pi/2)$. Then we reflect this portion of the graph about the line $\theta = \pi/2$ (y-axis) to obtain the complete graph.

Figure 29(a) shows the graph drawn by hand. Figure 29(b) shows the graph of $r = 1 - \sin\theta$ using a graphing utility with θmin $= 0$, θmax $= 2\pi$, and θstep $= \pi/24$.

FIGURE 29
$r = 1 - \sin\theta$

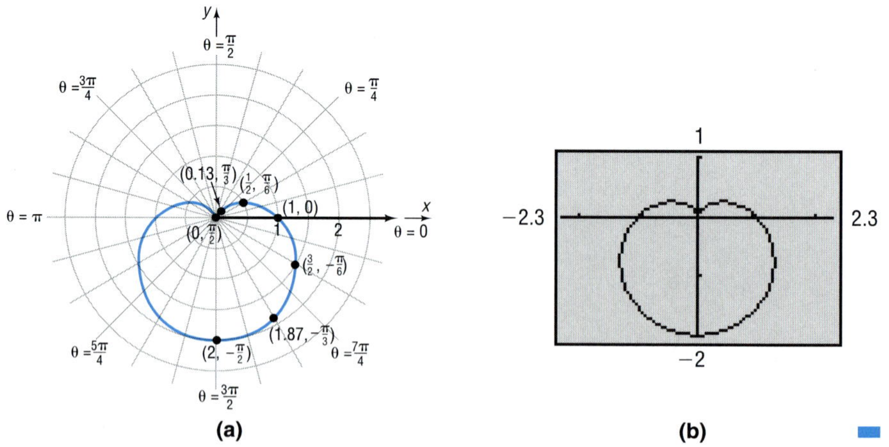

Exploration Graph $r_1 = 1 + \sin\theta$. Clear the screen and graph $r_1 = 1 - \cos\theta$. Clear the screen and graph $r_1 = 1 + \cos\theta$. Do you see a pattern?

The curve in Figure 29 is an example of a *cardioid* (a heart-shaped curve).

Cardioids are characterized by equations of the form

$$r = a(1 + \cos\theta) \qquad r = a(1 + \sin\theta)$$
$$r = a(1 - \cos\theta) \qquad r = a(1 - \sin\theta)$$

where $a > 0$. The graph of a cardioid contains the pole.

Now work Problem 31.

EXAMPLE 10

Graphing a Polar Equation (Limaçon)

Graph the equation: $r = 3 + 2\cos\theta$

Solution We check for symmetry first:

Polar Axis: Replace θ by $-\theta$. The result is

$$r = 3 + 2\cos(-\theta) = 3 + 2\cos\theta$$

Thus, the graph is symmetric with respect to the polar axis.

The Line $\theta = \pi/2$: Replace θ by $\pi - \theta$. The result is

$$\begin{aligned} r = 3 + 2\cos(\pi - \theta) &= 3 + 2(\cos\pi\cos\theta + \sin\pi\sin\theta) \\ &= 3 - 2\cos\theta \end{aligned}$$

The test fails, so the graph may or may not be symmetric with respect to the line $\theta = \pi/2$.

The Pole: Replace r by $-r$. The test fails, so the graph may or may not be symmetric with respect to the pole.

Next, we identify points on the graph by assigning values to the angle θ and calculating the corresponding values of r. Due to the symmetry with respect to the polar axis, we only need to assign values to θ from 0 to π, as given in Table 2.

TABLE 2

θ	$r = 3 + 2\cos\theta$
0	5
$\pi/6$	$3 + \sqrt{3} \approx 4.73$
$\pi/3$	4
$\pi/2$	3
$2\pi/3$	2
$5\pi/6$	$3 - \sqrt{3} \approx 1.27$
π	1

Now we plot the points (r, θ) from Table 2 and trace out the graph, beginning at the point $(5, 0)$ and ending at the point $(1, \pi)$. Then we reflect this portion of the graph about the polar axis (x-axis) to obtain the complete graph.

Figure 30(a) shows the graph drawn by hand. Figure 30(b) shows the graph of $r = 3 + 2\cos\theta$ using a graphing utility with $\theta\text{min} = 0$, $\theta\text{max} = 2\pi$, and $\theta\text{step} = \pi/24$.

FIGURE 30
$r = 3 + 2\cos\theta$

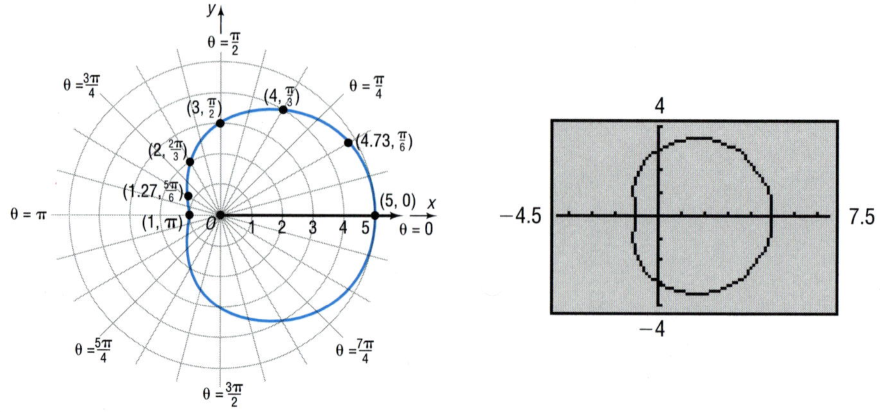

(a) (b)

Exploration Graph $r_1 = 3 - 2\cos\theta$. Clear the screen and graph $r_1 = 3 + 2\sin\theta$. Clear the screen and graph $r_1 = 3 - 2\sin\theta$. Do you see a pattern?

The curve in Figure 30 is an example of a *limaçon* (the French word for *snail*) without an inner loop.

Limaçons without an inner loop are characterized by equations of the form

$$r = a + b\cos\theta \qquad r = a + b\sin\theta$$
$$r = a - b\cos\theta \qquad r = a - b\sin\theta$$

where $a > 0$, $b > 0$, and $a > b$. The graph of a limaçon without an inner loop does not contain the pole.

Now work Problem 37.

EXAMPLE 11 Graphing a Polar Equation (Limaçon with Inner Loop)

Graph the equation: $r = 1 + 2\cos\theta$

Solution First, we check for symmetry:

Polar Axis: Replace θ by $-\theta$. The result is

$$r = 1 + 2\cos(-\theta) = 1 + 2\cos\theta$$

Thus, the graph is symmetric with respect to the polar axis.

The Line $\theta = \pi/2$: Replace θ by $\pi - \theta$. The result is

$$r = 1 + 2\cos(\pi - \theta) = 1 + 2(\cos\pi\cos\theta + \sin\pi\sin\theta)$$
$$= 1 - 2\cos\theta$$

The test fails, so the graph may or may not be symmetric with respect to the line $\theta = \pi/2$.

The Pole: Replace r by $-r$. The test fails, so the graph may or may not be symmetric with respect to the pole.

TABLE 3

θ	$r = 1 + 2\cos\theta$
0	3
$\pi/6$	$1 + \sqrt{3} \approx 2.73$
$\pi/3$	2
$\pi/2$	1
$2\pi/3$	0
$5\pi/6$	$1 - \sqrt{3} \approx -0.73$
π	-1

Next, we identify points on the graph of $r = 1 + 2\cos\theta$ by assigning values to the angle θ and calculating the corresponding values of r. Due to the symmetry with respect to the polar axis, we only need to assign values to θ from 0 to π, as given in Table 3.

Now we plot the points (r, θ) from Table 3, beginning at $(3, 0)$ and ending at $(-1, \pi)$. See Figure 31(a). Finally, we reflect this portion of the graph about the polar axis (x-axis) to obtain the complete graph. Figure 31(b) shows a complete graph drawn by hand. Figure 31(c) shows the graph of $r = 1 + 2\cos\theta$ using a graphing utility with $\theta\text{min} = 0$, $\theta\text{max} = 2\pi$, and $\theta\text{step} = \pi/24$.

FIGURE 31

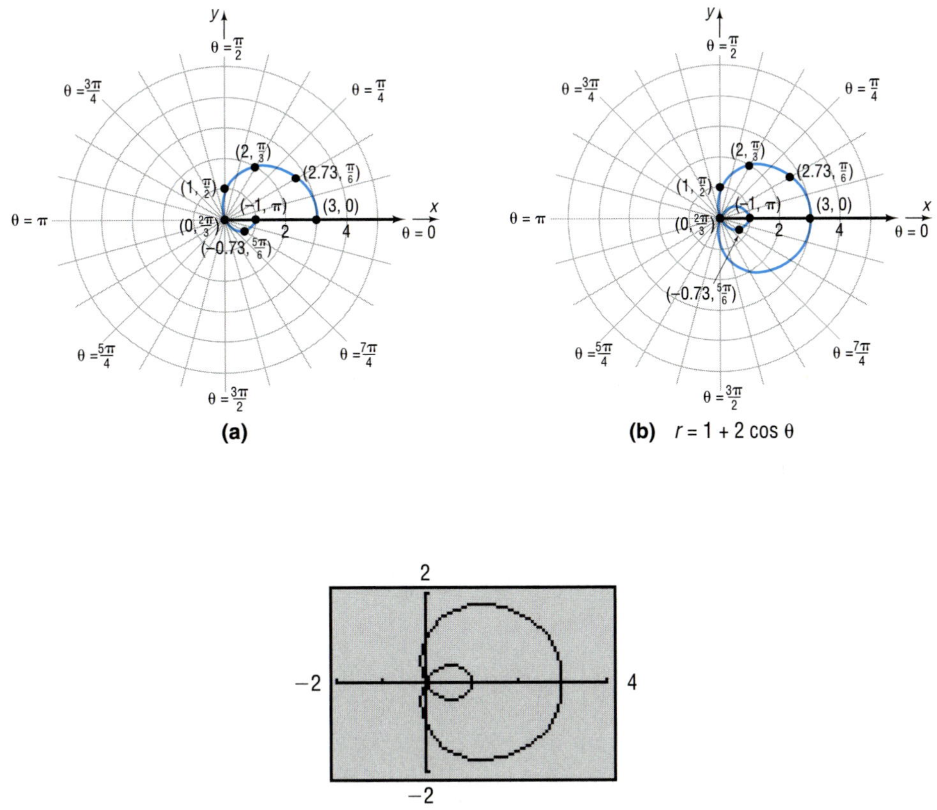

(a) (b) $r = 1 + 2\cos\theta$ (c)

Exploration Graph $r_1 = 1 - 2\cos\theta$. Clear the screen and graph $r_1 = 1 + 2\sin\theta$. Clear the screen and graph $r_1 = 1 - 2\sin\theta$. Do you see a pattern?

The curve in Figure 31(b) is an example of a limaçon with an inner loop.

Limaçons with an inner loop are characterized by equations of the form

$$r = a + b\cos\theta \qquad r = a + b\sin\theta$$
$$r = a - b\cos\theta \qquad r = a - b\sin\theta$$

where $a > 0$, $b > 0$, and $a < b$. The graph of a limaçon with an inner loop will pass through the pole twice.

Now work Problem 39.

EXAMPLE 12 Graphing a Polar Equation (Rose)

Graph the equation: $r = 2\cos 2\theta$

Solution We check for symmetry:

Polar Axis: If we replace θ by $-\theta$, the result is

$$r = 2\cos 2(-\theta) = 2\cos 2\theta$$

Thus, the graph is symmetric with respect to the polar axis.

The Line $\theta = \pi/2$: If we replace θ by $\pi - \theta$, we obtain

$$r = 2\cos 2(\pi - \theta) = 2\cos(2\pi - 2\theta) = 2\cos(-2\theta) = 2\cos 2\theta$$

Thus, the graph is symmetric with respect to the line $\theta = \pi/2$.

The Pole: Since the graph is symmetric with respect to both the polar axis and the line $\theta = \pi/2$, it must be symmetric with respect to the pole.

Next, we construct Table 4. Due to the symmetry with respect to the polar axis, the line $\theta = \pi/2$, and the pole, we consider only values of θ from 0 to $\pi/2$.

TABLE 4

θ	$r = 2\cos 2\theta$
0	2
$\pi/6$	1
$\pi/4$	0
$\pi/3$	-1
$\pi/2$	-2

We plot and connect these points in Figure 32(a). Finally, because of symmetry, we reflect this portion of the graph first about the polar axis (x-axis) and then about the line $\theta = \pi/2$ (y-axis) to obtain the complete graph. Figure 32(b) shows a complete graph drawn by hand. Figure 32(c) shows the graph of $r = 2\cos 2\theta$ using a graphing utility with $\theta\text{min} = 0$, $\theta\text{max} = 2\pi$, and $\theta\text{step} = \pi/24$.

FIGURE 32

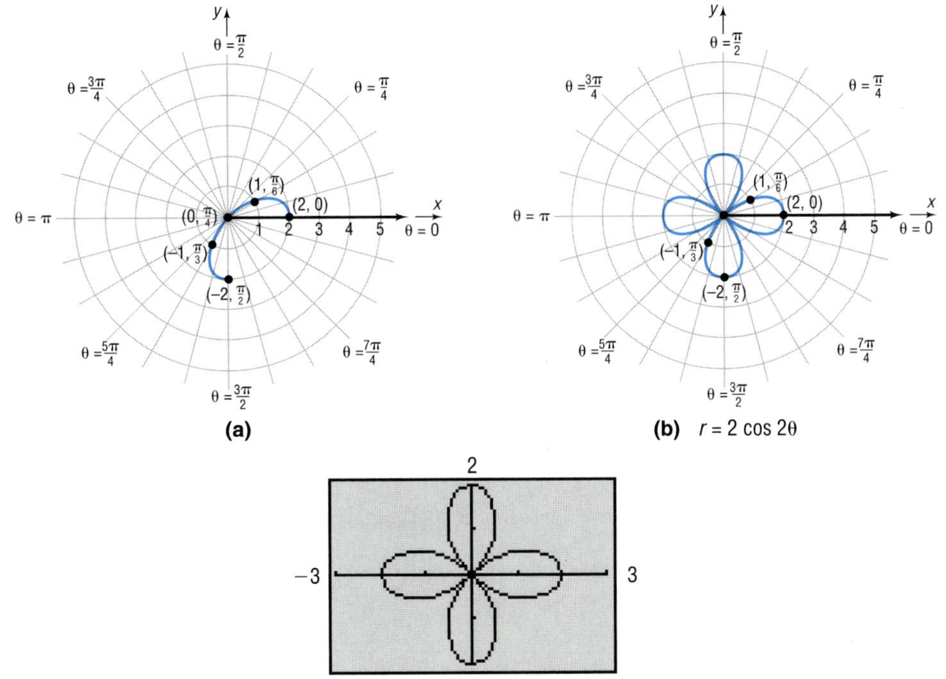

(a) (b) $r = 2\cos 2\theta$ (c)

The curve in Figure 32(b) is called a *rose* with four petals.

Rose curves are characterized by equations of the form

$$r = a\cos n\theta \qquad r = a\sin n\theta, \qquad a > 0$$

and have graphs that are rose-shaped. If n is even, the rose has $2n$ petals; if n is odd, the rose has n petals.

Now work Problem 43.

EXAMPLE 13

Graphing a Polar Equation (Lemniscate)

Graph the equation: $r^2 = 4\sin 2\theta$

Solution

We leave it to you to verify that the graph is symmetric with respect to the pole. Table 5 lists points on the graph for values of $\theta = 0$ through $\theta = \pi/2$. Note that there are no points on the graph for $\pi/2 < \theta < \pi$ (quadrant II), since $\sin 2\theta < 0$ for such values. The points from Table 5 where $r \geq 0$ are plotted in Figure 33(a). The remaining points on the graph may be obtained by using symmetry. Figure 33(b) shows the final graph drawn by hand. Figure 33(c) shows the graph of $r^2 = 4\sin 2\theta$ by entering $r_1 = 2\sqrt{\sin 2\theta}$ and $r_2 = -2\sqrt{\sin 2\theta}$ with $\theta\text{min} = 0$, $\theta\text{max} = \pi/2$, and $\theta\text{step} = \pi/48$.*

TABLE 5

θ	$r^2 = 4\sin 2\theta$	r
0	0	0
$\pi/6$	$2\sqrt{3}$	± 1.9
$\pi/4$	4	± 2
$\pi/3$	$2\sqrt{3}$	± 1.9
$\pi/2$	0	0

*When the argument of a trigonometric function is a multiple of θ, such as 2θ, more points can be plotted and a better graph obtained by choosing a smaller θstep, such as $\pi/48$. In general, if the argument is $n\theta$, choose θstep as $\pi/(24n)$. Experiment with various θsteps and compare the results.

FIGURE 33

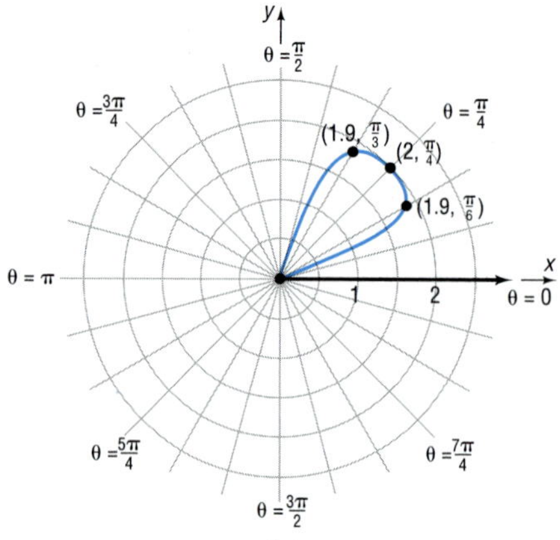

(a)

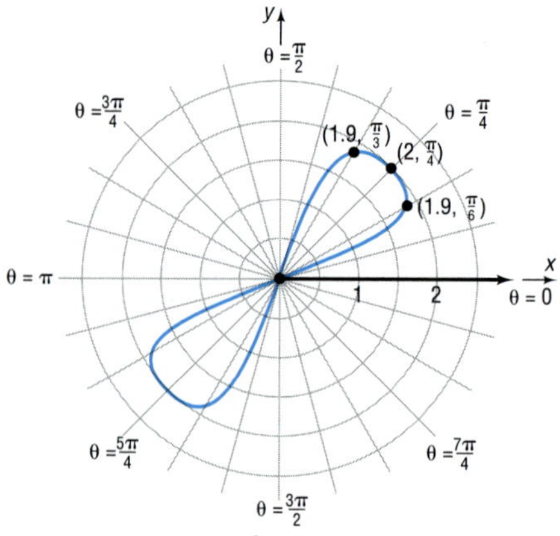

(b) $r^2 = 4 \sin 2\theta$

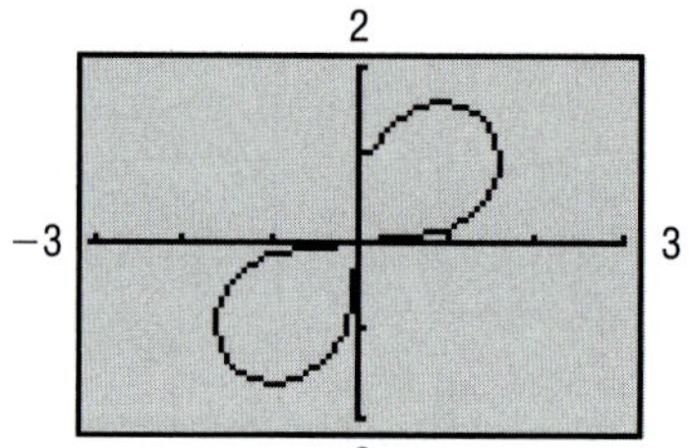

(c)

The curve in Figure 33(b) is an example of a *lemniscate.*

Lemniscates are characterized by equations of the form

$$r^2 = a^2 \sin 2\theta \qquad r^2 = a^2 \cos 2\theta$$

where $a \neq 0$, and have graphs that are propeller shaped.

Now work Problem 47.

EXAMPLE 14 Graphing a Polar Equation (Spiral)

Sketch the graph of: $r = e^{\theta/5}$

Solution The tests for symmetry with respect to the pole, the polar axis, and the line $\theta = \pi/2$ fail. Furthermore, there is no number θ for which $r = 0$. Hence, the graph does not pass through the pole. We observe that r is positive for all θ, r increases as θ increases, $r \to 0$ as $\theta \to -\infty$, and $r \to \infty$ as $\theta \to \infty$. With the help of a calculator, we obtain the values in Table 6. See Figure 34(a) for the graph drawn by hand. Figure 34(b) shows the graph of $r = e^{\theta/5}$ using a graphing utility with $\theta\text{min} = -5\pi$, $\theta\text{max} = 3\pi$, and $\theta\text{step} = \pi/24$.

TABLE 6

θ	$r = e^{\theta/5}$
$-3\pi/2$	0.39
$-\pi$	0.53
$-\pi/2$	0.73
$-\pi/4$	0.85
0	1
$\pi/4$	1.17
$\pi/2$	1.37
π	1.87
$3\pi/2$	2.57
2π	3.51

FIGURE 34 $r = e^{\theta/5}$

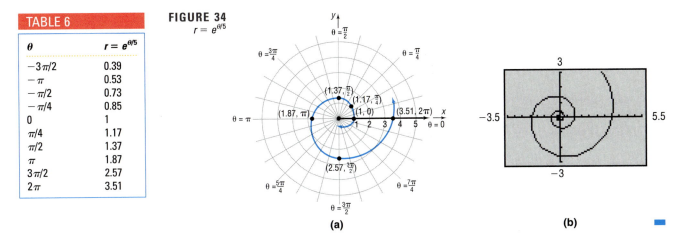

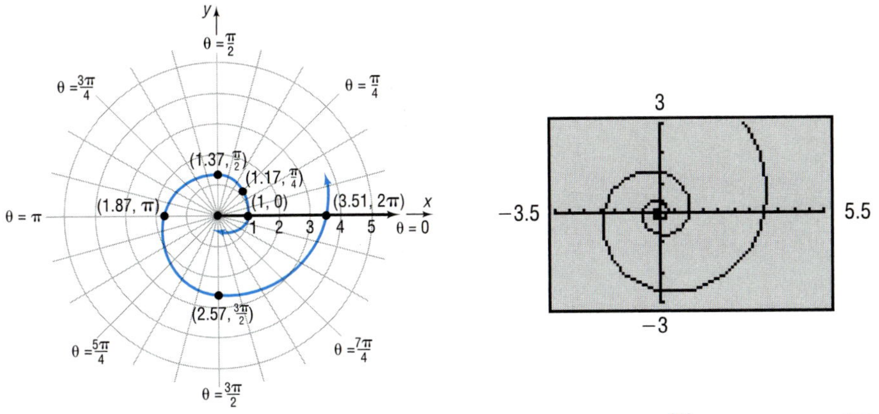

(a) (b)

The curve in Figure 34 is called a **logarithmic spiral,** since its equation may be written as $\theta = 5 \ln r$ and it spirals infinitely both toward the pole and away from it.

Classification of Polar Equations

The equations of some lines and circles in polar coordinates and their corresponding equations in rectangular coordinates are given in Table 7 on page 672. Also included are the names and the graphs of a few of the more frequently encountered polar equations.

Calculus Comment

For those of you who are planning to study calculus, a comment about one important role of polar equations is in order.

In rectangular coordinates, the equation $x^2 + y^2 = 1$, whose graph is the unit circle, does not define a function. In fact, on the interval $[-1, 1]$ it defines two functions,

$$y = \sqrt{1 - x^2} \quad \text{Upper semicircle.} \qquad y = -\sqrt{1 - x^2} \quad \text{Lower semicircle.}$$

In polar coordinates, the equation $r = 1$, whose graph is also the unit circle, does define a function. That is, for each choice of θ there is only one corresponding value of r, that is, $r = 1$. Since many uses of calculus require that functions be used, the opportunity to express nonfunctions in rectangular coordinates as functions in polar coordinates becomes extremely useful.

Note also that the vertical line test for functions is valid only for equations in rectangular coordinates.

Historical Feature

Polar coordinates seem to have been invented by Jacob Bernoulli (1654–1705) about 1691, although, as with most such ideas, earlier traces of the notion exist. Early users of calculus remained committed to rectangular coordinates, and polar coordinates did not become widely used until the early

TABLE 7

Lines

Description	Line passing through the pole making an angle α with the polar axis	Vertical line	Horizontal line
Rectangular equation	$y = (\tan \alpha)x$	$x = a$	$y = b$
Polar equation	$\theta = \alpha$	$r \cos \theta = a$	$r \sin \theta = b$
Typical graph	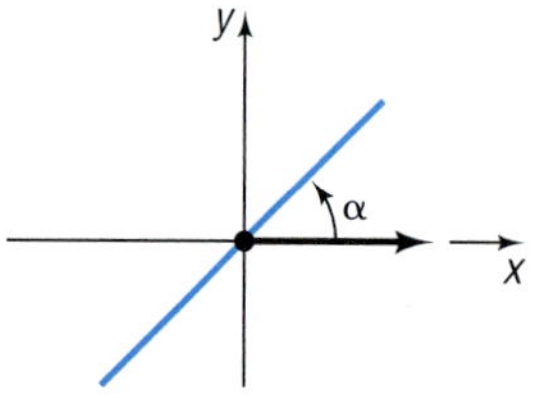	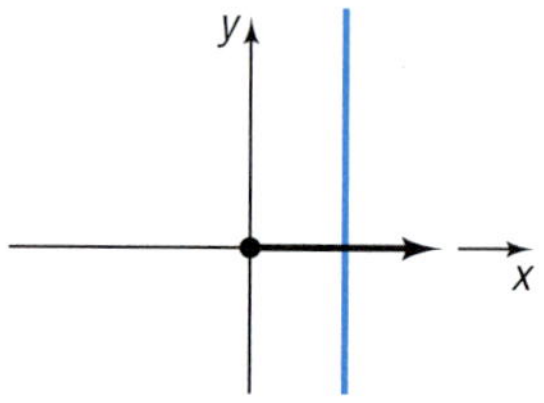	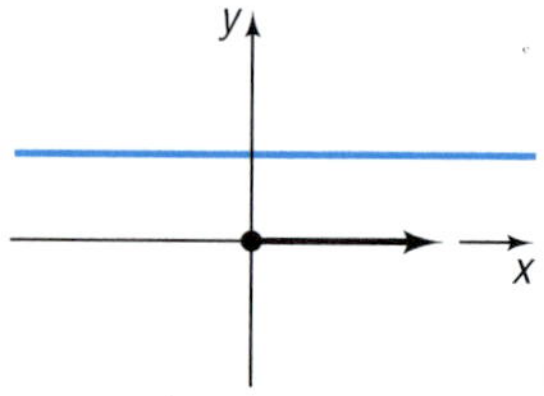

Circles

Description	Center at the pole, radius a	Passing through the pole, tangent to the line $\theta = \pi/2$, center on the polar axis, radius a	Passing through the pole, tangent to the polar axis, center on the line $\theta = \pi/2$, radius a
Rectangular equation	$x^2 + y^2 = a^2, a > 0$	$x^2 + y^2 = \pm 2ax, a > 0$	$x^2 + y^2 = \pm 2ay, a > 0$
Polar equation	$r = a, a > 0$	$r = \pm 2a \cos \theta, a > 0$	$r = \pm 2a \sin \theta, a > 0$
Typical graph	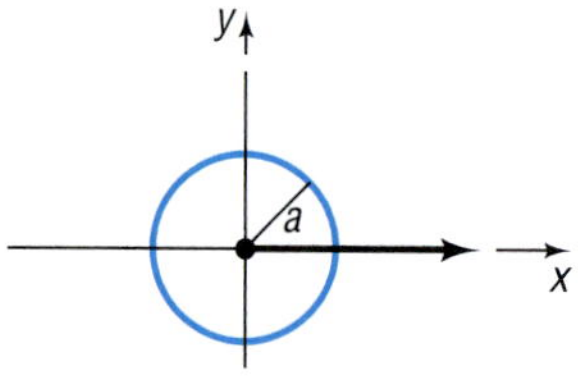	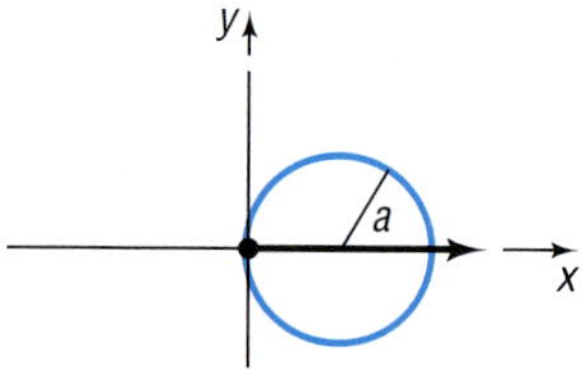	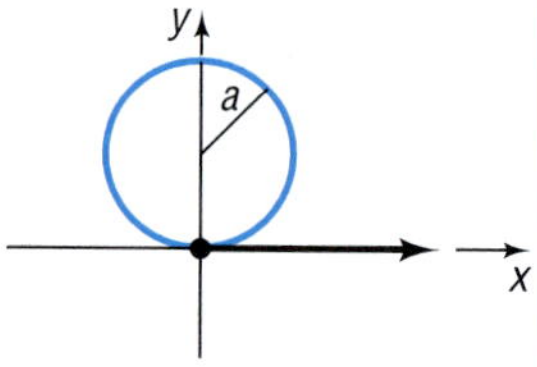

Other Equations

Name	Cardioid	Limaçon without inner loop	Limaçon with inner loop
Polar equations	$r = a \pm a \cos \theta, a > 0$ $r = a \pm a \sin \theta, a > 0$	$r = a \pm b \cos \theta, 0 < b < a$ $r = a \pm b \sin \theta, 0 < b < a$	$r = a \pm b \cos \theta, 0 < a < b$ $r = a \pm b \sin \theta, 0 < a < b$
Typical graph	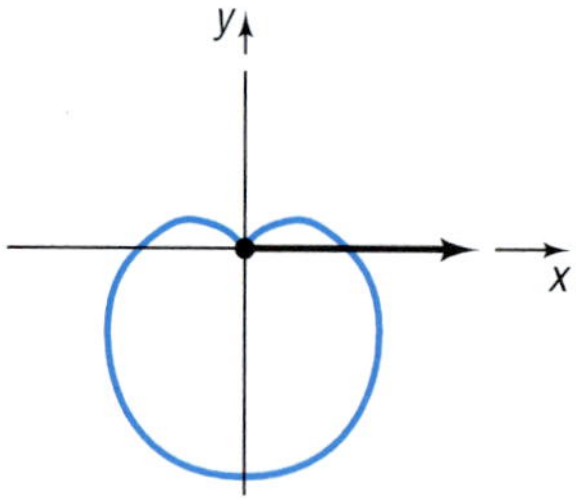	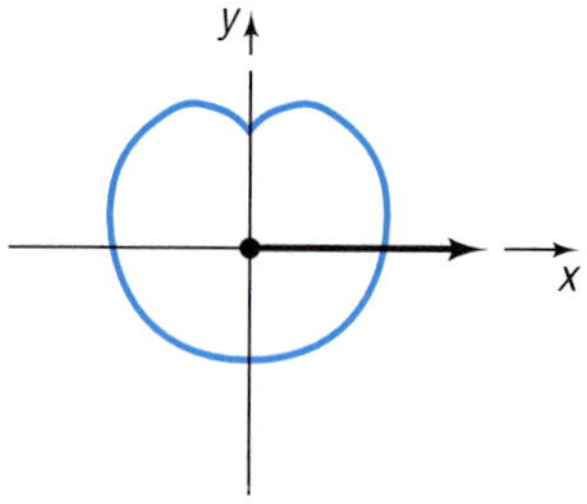	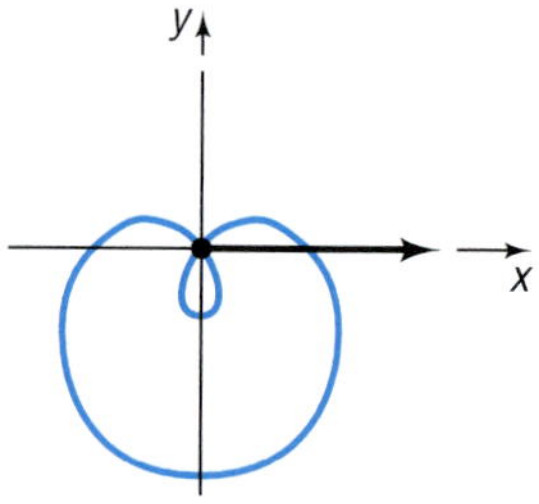
Name	Lemniscate	Rose with three petals	Rose with four petals
Polar equations	$r^2 = a^2 \cos 2\theta, a > 0$ $r^2 = a^2 \sin 2\theta, a > 0$	$r = a \sin 3\theta, a > 0$ $r = a \cos 3\theta, a > 0$	$r = a \sin 2\theta, a > 0$ $r = a \cos 2\theta, a > 0$
Typical graph	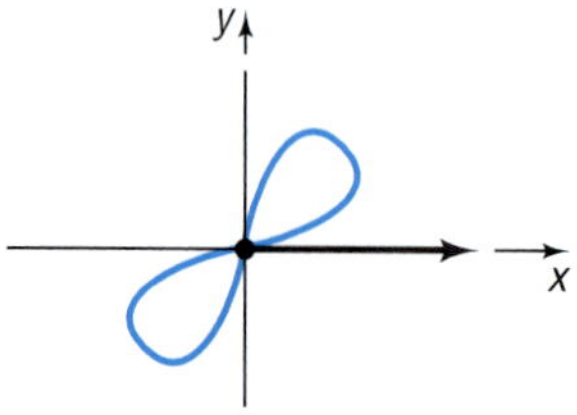	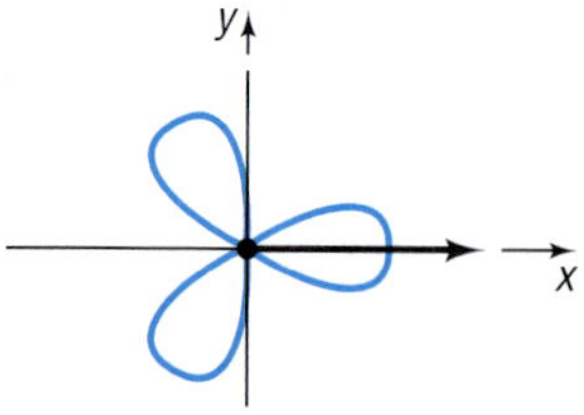	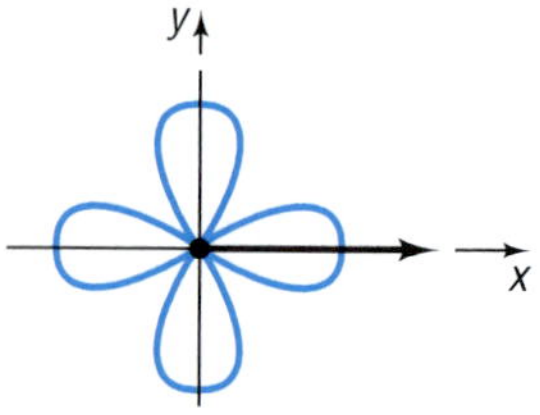

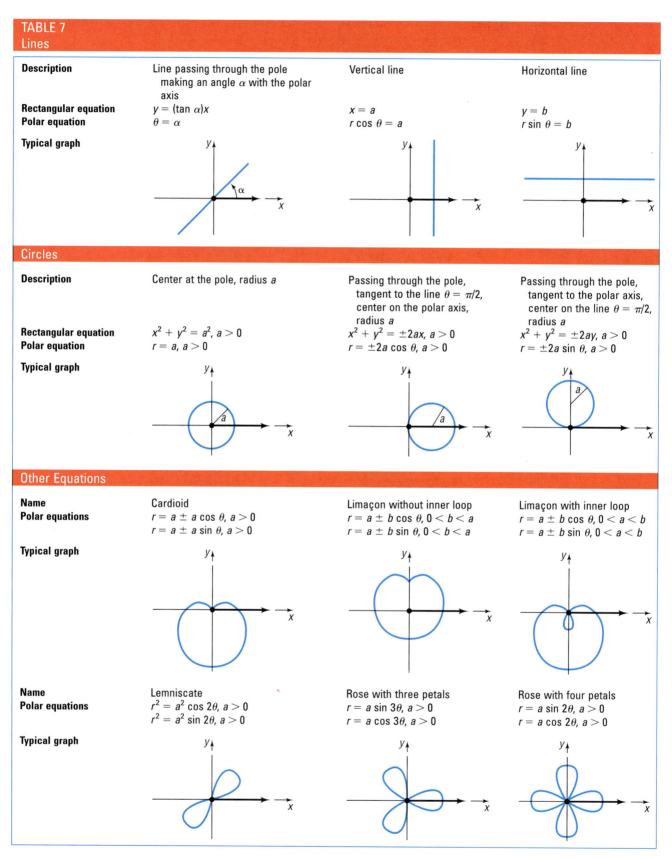

1800s. Even then, it was mostly geometers who used them for describing odd curves. Finally, about the mid-1800s, applied mathematicians realized the tremendous simplification polar coordinates make possible in the description of objects with circular or cylindrical symmetry. From then on their use became widespread.

9.2 EXERCISES

In Problems 1–16, identify each polar equation by transforming the equation to rectangular coordinates. Graph each polar equation by hand. Verify your results using a graphing utility.

1. $r = 4$

2. $r = 2$

3. $\theta = \pi/3$

4. $\theta = -\pi/4$

5. $r \sin \theta = 4$

6. $r \cos \theta = 4$

7. $r \cos \theta = -2$

8. $r \sin \theta = -2$

9. $r = 2 \cos \theta$

10. $r = 2 \sin \theta$

11. $r = -4 \sin \theta$

12. $r = -4 \cos \theta$

13. $r \sec \theta = 4$

14. $r \csc \theta = 8$

15. $r \csc \theta = -2$

16. $r \sec \theta = -4$

In Problems 17–24, match each of the graphs (A) through (H) to one of the following polar equations.

17. $r = 2$

18. $\theta = \pi/4$

19. $r = 2 \cos \theta$

20. $r \cos \theta = 2$

21. $r = 1 + \cos \theta$

22. $r = 2 \sin \theta$

23. $\theta = 3\pi/4$

24. $r \sin \theta = 2$

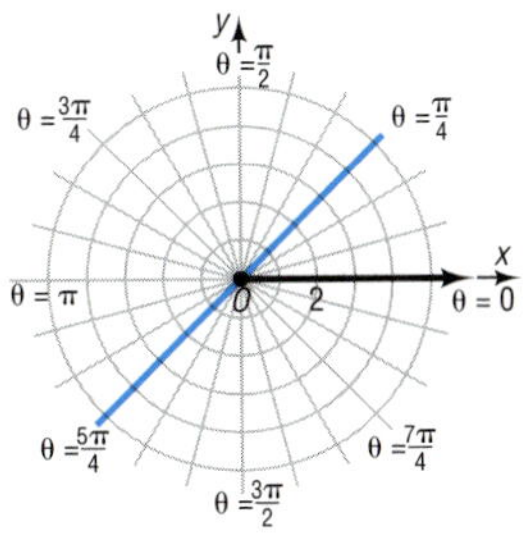

(A)

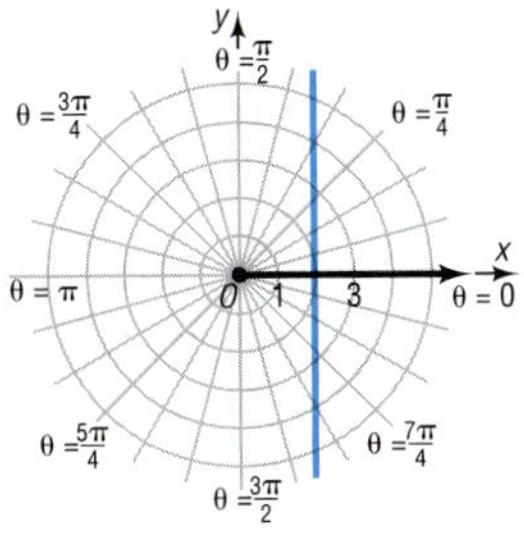

(B)

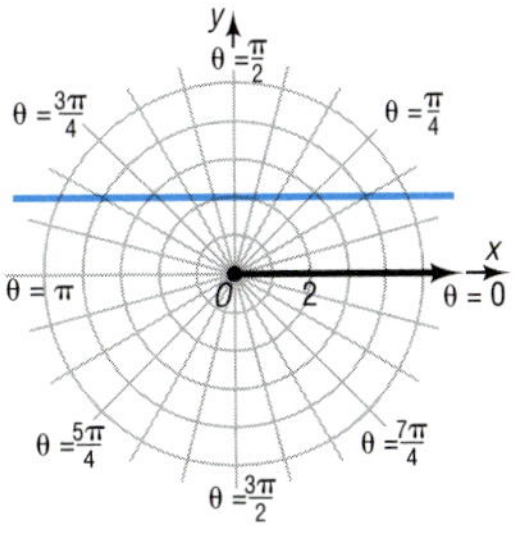

(C)

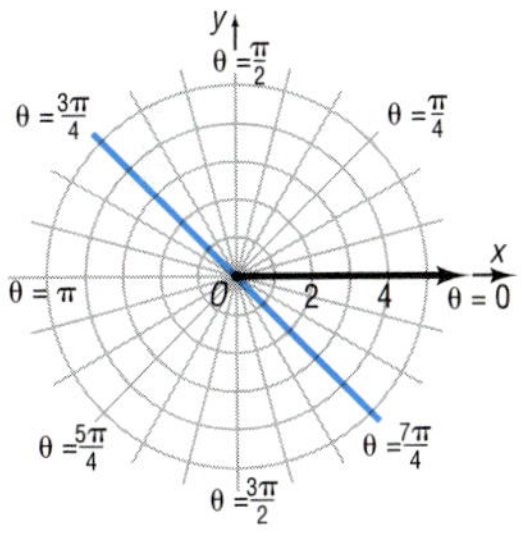

(D)

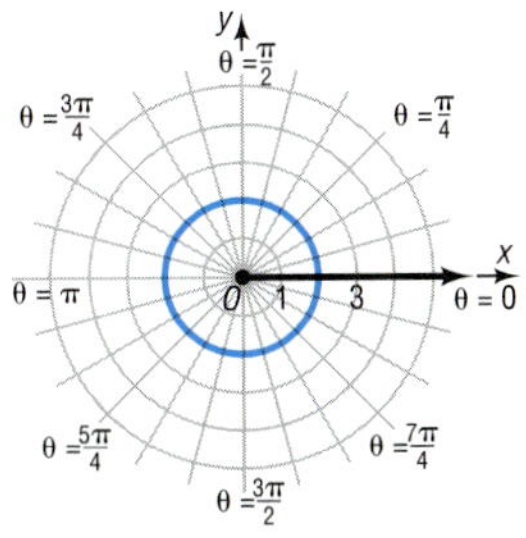

(E)

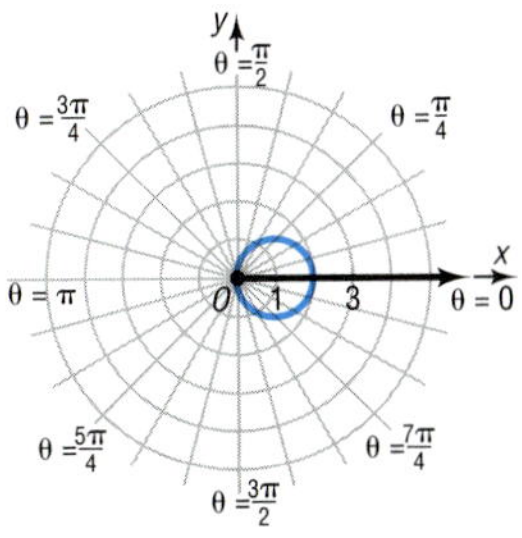

(F)

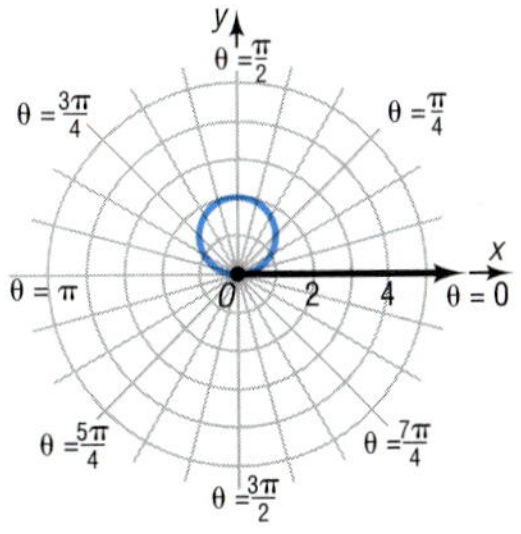

(G)

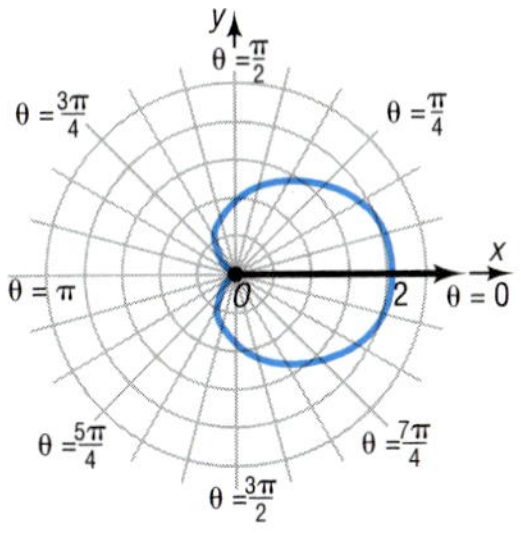

(H)

In Problems 25–30, match each of the graphs (A) through (F) to one of the following polar equations.

25. $r = 4$

26. $r = 3 \cos \theta$

27. $r = 3 \sin \theta$

28. $r \sin \theta = 3$

29. $r \cos \theta = 3$

30. $r = 2 + \sin \theta$

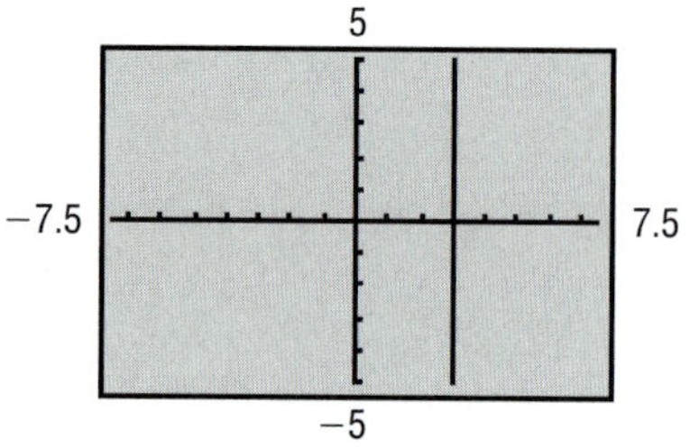

(A)

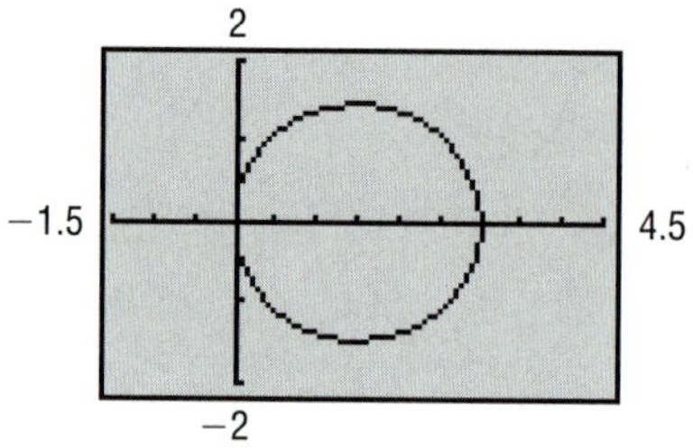

(B)

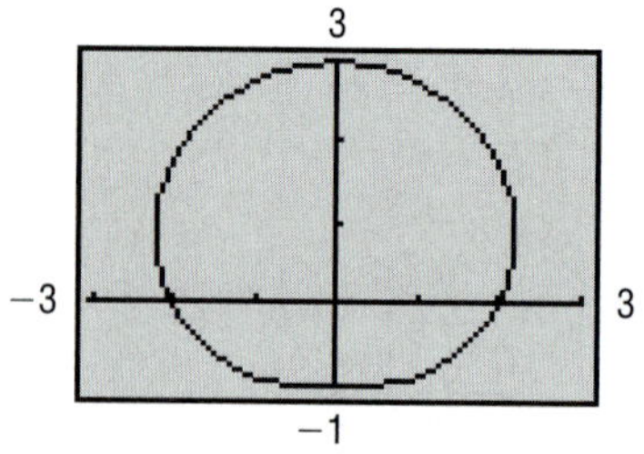

(C)

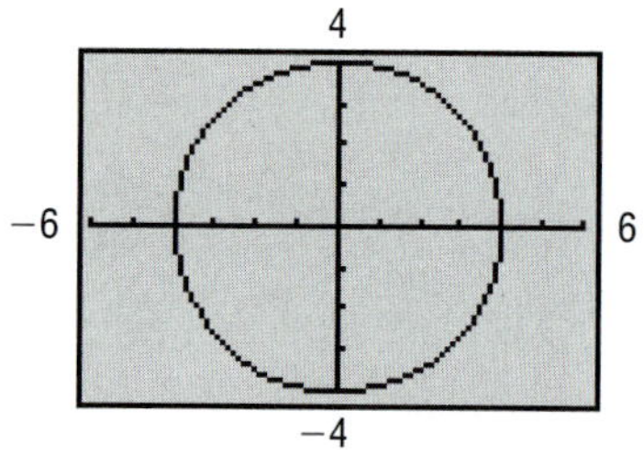

(D)

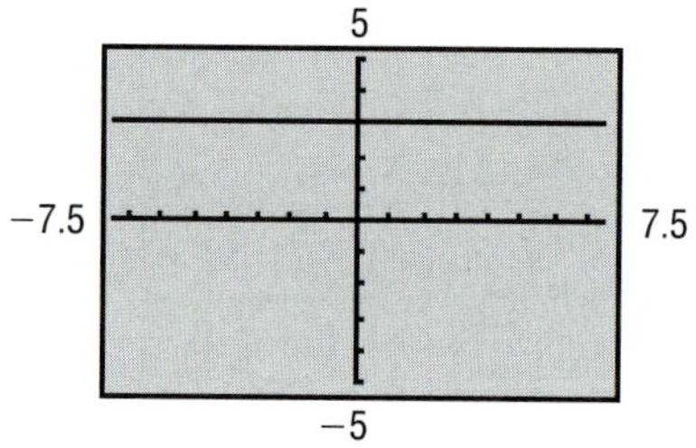

(E)

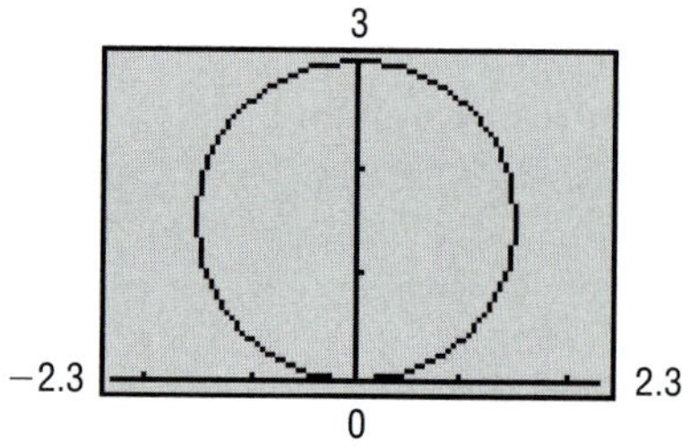

(F)

In Problems 31–54, identify and graph each polar equation by hand. Be sure to test for symmetry. Verify your results using a graphing utility.

31. $r = 2 + 2 \cos \theta$

32. $r = 1 + \sin \theta$

33. $r = 3 - 3 \sin \theta$

34. $r = 2 - 2 \cos \theta$

35. $r = 2 + \sin \theta$

36. $r = 2 - \cos \theta$

37. $r = 4 - 2 \cos \theta$

38. $r = 4 + 2 \sin \theta$

39. $r = 1 + 2 \sin \theta$

40. $r = 1 - 2 \sin \theta$

41. $r = 2 - 3 \cos \theta$

42. $r = 2 + 4 \cos \theta$

43. $r = 3 \cos 2\theta$

44. $r = 2 \sin 2\theta$

45. $r = 4 \sin 3\theta$

46. $r = 3 \cos 4\theta$

47. $r^2 = 9 \cos 2\theta$

48. $r^2 = \sin 2\theta$

49. $r = 2^\theta$

50. $r = 3^\theta$

51. $r = 1 - \cos \theta$

52. $r = 3 + \cos \theta$

53. $r = 1 - 3 \cos \theta$

54. $r = 4 \cos 3\theta$

In Problems 55–64, graph each polar equation by hand. Verify your results using a graphing utility.

55. $r = \dfrac{2}{1 - \cos \theta}$ (parabola)

56. $r = \dfrac{2}{1 - 2 \cos \theta}$ (hyperbola)

57. $r = \dfrac{1}{3 - 2 \cos \theta}$ (ellipse)

58. $r = \dfrac{1}{1 - \cos \theta}$ (parabola)

59. $r = \theta, \quad \theta \geq 0$ (spiral of Archimedes)

60. $r = \dfrac{3}{\theta}$ (reciprocal spiral)

61. $r = \csc \theta - 2, \quad 0 < \theta < \pi$ (conchoid)

62. $r = \sin \theta \tan \theta$ (cissoid)

63. $r = \tan \theta$ (kappa curve)

64. $r = \cos \dfrac{\theta}{2}$

65. Show that the graph of the equation $r \sin \theta = a$ is a horizontal line a units above the pole if $a > 0$ and $|a|$ units below the pole if $a < 0$.

66. Show that the graph of the equation $r \cos \theta = a$ is a vertical line a units to the right of the pole if $a > 0$ and $|a|$ units to the left of the pole if $a < 0$.

67. Show that the graph of the equation $r = 2a \sin \theta$, $a > 0$, is a circle of radius a with center at $(0, a)$ in rectangular coordinates.

68. Show that the graph of the equation $r = -2a \sin \theta$, $a > 0$, is a circle of radius a with center at $(0, -a)$ in rectangular coordinates.

69. Show that the graph of the equation $r = 2a \cos \theta$ $a > 0$, is a circle of radius a with center at $(a, 0)$ in rectangular coordinates.

70. Show that the graph of the equation $r = -2a \cos \theta$, $a > 0$, is a circle of radius a with center at $(-a, 0)$ in rectangular coordinates.

71. Explain why the following test for symmetry is valid: Replace r by $-r$ and θ by $-\theta$ in a polar equation. If an equivalent equation results, the graph is symmetric with respect to the line $\theta = \pi/2$ (y-axis).
(a) Show that the test on page 663 fails for $r^2 = \cos \theta$, but this new test works.
(b) Show that the test on page 663 works for $r^2 = \sin \theta$, yet this new test fails.

72. Develop a new test for symmetry with respect to the pole.
(a) Find a polar equation for which this new test fails, yet the test on page 663 works.
(b) Find a polar equation for which the test on page 663 fails, yet the new test works.

73. Write down two different tests for symmetry with respect to the polar axis. Find examples in which one works and the other fails. Which test do you prefer to use? Justify your position.

9.3 THE COMPLEX PLANE; DEMOIVRE'S THEOREM

1 Convert a Complex Number from Rectangular Form to Polar Form
2 Plot Points in the Complex Plane
3 Find Products and Quotients of Complex Numbers in Polar Form
4 Use DeMoivre's Theorem
5 Find Complex Roots

When we first introduced complex numbers, we were not prepared to give a geometric interpretation of a complex number. Now we are ready. Although there are several such interpretations we could give, the one that follows is the easiest to understand.

A complex number $z = x + yi$ can be interpreted geometrically as the point (x, y) in the xy-plane. Thus, each point in the plane corresponds to a complex number and, conversely, each complex number corresponds to a point in the plane. We shall refer to the collection of such points as the **complex plane.** The x-axis will be referred to as the **real axis,** because any point that lies on the real axis is of the form $z = x + 0i = x$, a real number. The y-axis is called the **imaginary axis,** because any point that lies on it is of the form $z = 0 + yi = yi$, a pure imaginary number. See Figure 35.

FIGURE 35
Complex plane

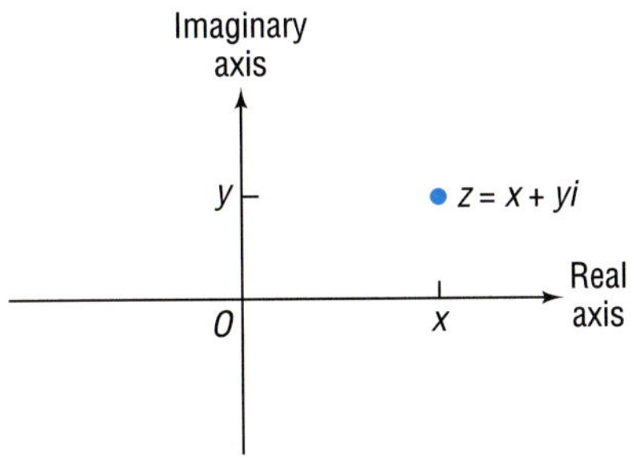

Let $z = x + yi$ be a complex number. The **magnitude** or **modulus** of z, denoted by $|z|$, is defined as the distance from the origin to the point (x, y). Thus,

$$|z| = \sqrt{x^2 + y^2} \quad (1)$$

FIGURE 36

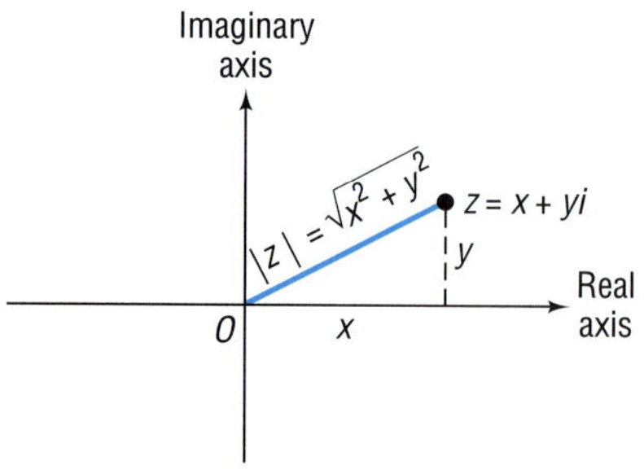

See Figure 36 for an illustration.

This definition for $|z|$ is consistent with the definition for the absolute value of a real number: If $z = x + yi$ is real, then $z = x + 0i$ and

$$|z| = \sqrt{x^2 + 0^2} = \sqrt{x^2} = |x|$$

Recall (Section 4.6) that if $z = x + yi$, then its **conjugate,** denoted by $\bar{z}$, is $\bar{z} = x - yi$. Because $z\bar{z} = x^2 + y^2$, it follows from equation (1) that the magnitude of z can be written as

$$|z| = \sqrt{z\bar{z}} \qquad (2)$$

Polar Form of a Complex Number

1 When a complex number is written in the form $z = x + yi$, we say that it is in **rectangular,** or **Cartesian, form,** because (x, y) are the rectangular coordinates of the corresponding point in the complex plane. Suppose that (r, θ) are the polar coordinates of this point. Then

$$x = r\cos\theta \qquad y = r\sin\theta \qquad (3)$$

If $r \geq 0$ and $0 \leq \theta < 2\pi$, the complex number $z = x + yi$ may be written in **polar form** as

$$z = x + yi = (r\cos\theta) + (r\sin\theta)i = r(\cos\theta + i\sin\theta) \qquad (4)$$

FIGURE 37

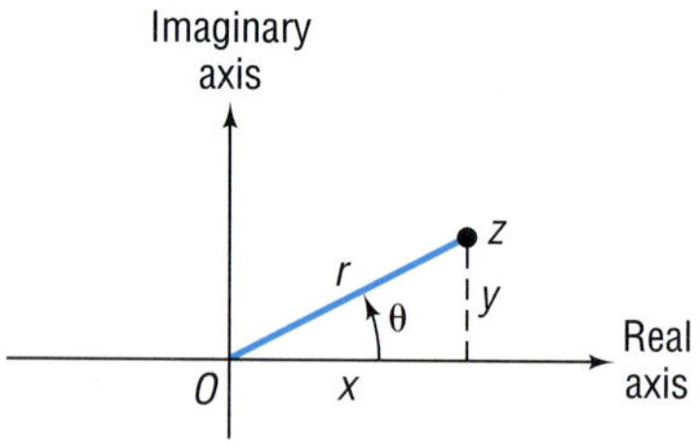

$z = x + yi = r(\cos\theta + i\sin\theta)$, $r \geq 0$, $0 \leq \theta < 2\pi$

See Figure 37.

If $z = r(\cos\theta + i\sin\theta)$ is the polar form of a complex number, the angle θ, $0 \leq \theta < 2\pi$, is called the **argument of *z*.** Also, because $r \geq 0$, from equation (3) we have $r = \sqrt{x^2 + y^2}$. Thus, from equation (1) it follows that the magnitude of $z = r(\cos\theta + i\sin\theta)$ is

$$|z| = r$$

EXAMPLE 1

Plotting a Point in the Complex Plane and Writing a Complex Number in Polar Form

2 Plot the point corresponding to $z = \sqrt{3} - i$ in the complex plane, and write an expression for z in polar form.

MISSION POSSIBLE

Mapping Indianapolis

Supplies needed: Graph paper with squares 1/4 inch or 1 cm. or larger per side; compass, protractor, ruler; graphing calculator; map of Indianapolis (from any atlas)

Unlike many major American cities which grew up next to large bodies of water and hence were limited in their ability to spread out in all directions, Indianapolis is surrounded by land on all sides. Major roads radiate out from the city center and intersect the ring road (Rt. 465). A map of the city looks somewhat like a rectangular grid overlaying a polar coordinate system.

The Mission Possible Team has been called to Indianapolis by a car rental company that wishes to create a computer system of interactive maps to be installed in their vehicles. The customers renting their cars would be able to indicate to the computer where they would like to go and the computer screen would then show them a map giving them the quickest route from their present location to their chosen destination. Today's project represents the beginning of a study to create such a computer system.

As a first step, your team needs to set up a polar coordinate system. Using a simplified model, we designate the intersection of Meridian St. and Tenth St. as the pole. The ring road is (roughly) a circle 7 miles out from the center.

1. On graph paper sketch a large polar model of the streets of Indianapolis, using 1 square of the graph paper to represent 1 mile. The ring road would be represented by a circle with radius 7 miles. Rt. 40 is the road that intersects the ring road due east of the pole, so we will designate Rt. 40 as 0°. The other roads and their approximate angle from Rt. 40 are as follows:

Rt. 36: 30°	Rt. 65: 135°	Mann Rd.: 240°
Rt. 37: 60°	Rt. 74: 165°	Rt. 31 (Meridian St.): 270°
Rt. 31 (Meridian St.): 90°	Tenth St.: 180°	Rt. 65: 300°
Rt. 421: 105°	Rt. 70: 210°	Rt. 421: 330°

2. All but 3 of these angles are multiples of 30°. What is the greatest common factor of *all* the angles?
3. In order to establish a mathematical structure for the computer map, your team decides to create a polar equation of the petalled rose variety that would pass through all these points of intersection with the ring road and connect each to the center. What would the equation be?
4. Sketch the graph of your rose on the polar map you created for question 1. By using your graphing calculator, you can determine the order in which the petals are drawn. Number the petals on your sketch to indicate that order.
5. Find the xy equation determined by taking your rose equation and replacing the r by y and the θ by x. (Example: If your polar equation were $r = 5\cos 7\theta$, your corresponding rectangular equation would be $y = 5\cos 7x$. Warning: These two equations are not equivalent in the sense that they will not give the same graph.) Sketch the resulting rectangular equation in the xy-plane, using the domain $0 \le x \le 2\pi/3$. Label the maximum and minimum points with the numbers that correspond to the rose petals of your polar graph.
6. If your original sketch for question 1 was done on graph paper, you can now see the lines of the graph paper as representing a rectangular grid overlaying the polar graph. If not, you will need to sketch vertical and horizontal lines on your polar graph to represent the major city streets. Again, for purposes of simplifying the problem, the major streets are assumed to be occurring at 1 mile intervals.
7. Suppose that travel on the Interstate (Rt. 465, the ring road) has an average speed of 55 mph and travel on city streets has an average speed of 20 mph. Estimate the times it would take to go from the intersection of Rt. 74 and Rt. 465 to the intersection of Rt. 36 and Rt. 465 by each route. (Recall that traveling on city streets limits you to vertical and horizontal grid lines.)
8. What would be the longest time it would take to go from any intersection to another on Rt. 465, assuming you chose the shortest distance along the circle? How long would it take to go between those same two points using the city streets?

FIGURE 38

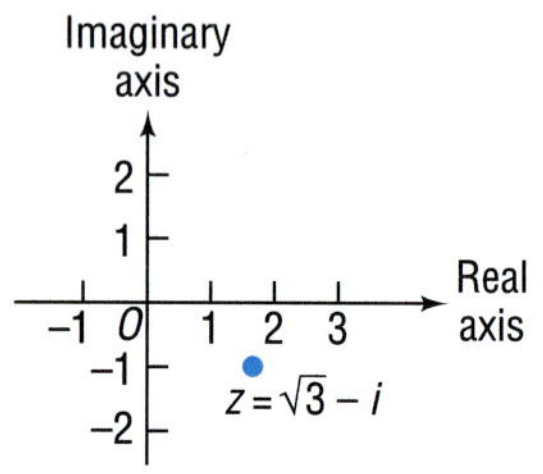

Solution The point corresponding to $z = \sqrt{3} - i$ has the rectangular coordinates $(\sqrt{3}, -1)$. The point, located in quadrant IV, is plotted in Figure 38. Because $x = \sqrt{3}$ and $y = -1$, it follows that

$$r = \sqrt{x^2 + y^2} = \sqrt{(\sqrt{3})^2 + (-1)^2} = \sqrt{4} = 2$$

and

$$\sin\theta = \frac{y}{r} = \frac{-1}{2} \qquad \cos\theta = \frac{x}{r} = \frac{\sqrt{3}}{2}, \qquad 0 \le \theta < 2\pi$$

Thus, $\theta = 11\pi/6$ and $r = 2$, so the polar form of $z = \sqrt{3} - i$ is

$$z = r(\cos\theta + i\sin\theta) = 2\left(\cos\frac{11\pi}{6} + i\sin\frac{11\pi}{6}\right)$$

Now work Problem 1.

EXAMPLE 2 Plotting a Point in the Complex Plane and Converting from Polar to Rectangular Form

Plot the point corresponding to $z = 2(\cos 30° + i\sin 30°)$ in the complex plane, and write an expression for z in rectangular form.

Solution To plot the complex number $z = 2(\cos 30° + i\sin 30°)$, we plot the point whose polar coordinates are $(r, \theta) = (2, 30°)$, as shown in Figure 39. In rectangular form,

$$z = 2(\cos 30° + i\sin 30°) = 2\left(\frac{\sqrt{3}}{2} + \frac{1}{2}i\right) = \sqrt{3} + i$$

FIGURE 39

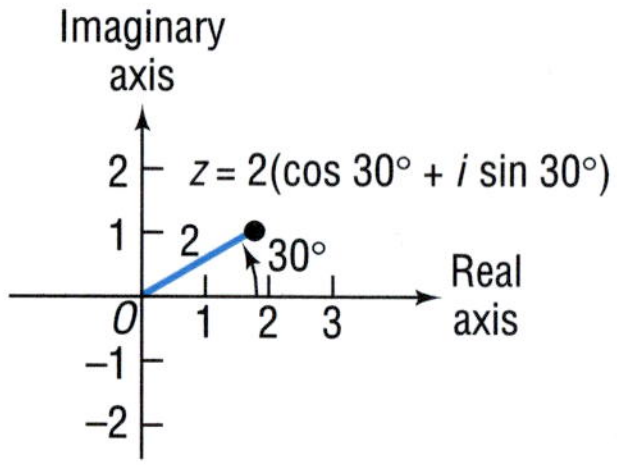

Now work Problem 13.

3 The polar form of a complex number provides an alternative for finding products and quotients of complex numbers.

Theorem

Let $z_1 = r_1(\cos\theta_1 + i\sin\theta_1)$ and $z_2 = r_2(\cos\theta_2 + i\sin\theta_2)$ be two complex numbers. Then

$$z_1 z_2 = r_1 r_2[\cos(\theta_1 + \theta_2) + i\sin(\theta_1 + \theta_2)] \tag{5}$$

If $z_2 \ne 0$, then

$$\frac{z_1}{z_2} = \frac{r_1}{r_2}[\cos(\theta_1 - \theta_2) + i\sin(\theta_1 - \theta_2)] \tag{6}$$

Proof We will prove formula (5). The proof of formula (6) is left as an exercise (see Problem 56).

$$\begin{aligned} z_1 z_2 &= [r_1(\cos\theta_1 + i\sin\theta_1)][r_2(\cos\theta_2 + i\sin\theta_2)] \\ &= r_1 r_2[(\cos\theta_1 + i\sin\theta_1)(\cos\theta_2 + i\sin\theta_2)] \\ &= r_1 r_2[(\cos\theta_1\cos\theta_2 - \sin\theta_1\sin\theta_2) + i(\sin\theta_1\cos\theta_2 + \cos\theta_1\sin\theta_2)] \\ &= r_1 r_2[(\cos(\theta_1 + \theta_2) + i\sin(\theta_1 + \theta_2)] \end{aligned}$$

Because the magnitude of a complex number z is r and its argument is θ, when $z = r(\cos\theta + i\sin\theta)$, we can restate this theorem as follows:

Theorem

The magnitude of the product (quotient) of two complex numbers equals the product (quotient) of their magnitudes; the argument of the product (quotient) of two complex numbers equals the sum (difference) of their arguments.

Let's look at an example of how this theorem can be used.

EXAMPLE 3 Finding Products and Quotients of Complex Numbers in Polar Form

If $z = 3(\cos 20° + i\sin 20°)$ and $w = 5(\cos 100° + \sin 100°)$, find the following (leave your answers in polar form):

(a) zw (b) z/w

Solution (a)
$$\begin{aligned} zw &= [3(\cos 20° + i\sin 20°)][5(\cos 100° + i\sin 100°)] \\ &= (3 \cdot 5)[\cos(20° + 100°) + i\sin(20° + 100°)] \\ &= 15(\cos 120° + i\sin 120°) \end{aligned}$$

(b)
$$\begin{aligned} \frac{z}{w} &= \frac{3(\cos 20° + i\sin 20°)}{5(\cos 100° + i\sin 100°)} \\ &= \tfrac{3}{5}[\cos(20° - 100°) + i\sin(20° - 100°)] \\ &= \tfrac{3}{5}[\cos(-80°) + i\sin(-80°)] \\ &= \tfrac{3}{5}(\cos 280° + i\sin 280°) \end{aligned}$$

Argument must lie between 0° and 360°.

Now work Problem 23.

Demoivre's Theorem

4 Demoivre's Theorem, stated by Abraham Demoivre (1667–1754) in 1730, but already known to many people by 1710, is important for the following reason: The fundamental processes of algebra are the four operations of addition, subtraction, multiplication, and division, together with powers and the extraction of roots. Demoivre's Theorem allows these latter fundamental algebraic operations to be applied to complex numbers.

Demoivre's Theorem, in its most basic form, is a formula for raising a complex number z to the power n, where $n \geq 1$ is a positive integer. Let's see if we can guess the form of the result.

Let $z = r(\cos\theta + i\sin\theta)$ be a complex number. Then, based on equation (5), we have

$$\begin{aligned}
n = 2{:}\quad z^2 &= r^2(\cos 2\theta + i\sin 2\theta)\\
n = 3{:}\quad z^3 &= z^2 \cdot z\\
&= [r^2(\cos 2\theta + i\sin 2\theta)][r(\cos\theta + i\sin\theta)]\\
&= r^3(\cos 3\theta + i\sin 3\theta) \quad \text{Equation (5).}\\
n = 4{:}\quad z^4 &= z^3 \cdot z\\
&= [r^3(\cos 3\theta + i\sin 3\theta)][r(\cos\theta + i\sin\theta)]\\
&= r^4(\cos 4\theta + i\sin 4\theta) \quad \text{Equation (5).}
\end{aligned}$$

The pattern should now be clear.

Theorem Demoivre's Theorem

If $z = r(\cos\theta + i\sin\theta)$ is a complex number, then

$$z^n = r^n(\cos n\theta + i\sin n\theta) \tag{7}$$

where $n \geq 1$ is a positive integer.

We will not prove Demoivre's Theorem because it requires mathematical induction (which is not discussed until Section 12.4).

Let's look at some examples.

EXAMPLE 4 Using Demoivre's Theorem

Write $[2(\cos 20° + i\sin 20°)]^3$ in the standard form $a + bi$.

Solution

$$\begin{aligned}
[2(\cos 20° + i\sin 20°)]^3 &= 2^3[\cos(3 \cdot 20°) + i\sin(3 \cdot 20°)]\\
&= 8(\cos 60° + i\sin 60°)\\
&= 8\left(\frac{1}{2} + \frac{\sqrt{3}}{2}i\right) = 4 + 4\sqrt{3}i
\end{aligned}$$

Now work Problem 31.

EXAMPLE 5 Using Demoivre's Theorem

Write $(1 + i)^5$ in the standard form $a + bi$.

Solution To apply Demoivre's Theorem, we must first write the complex number in polar form. Thus, since the magnitude of $1 + i$ is $\sqrt{1^2 + 1^2} = \sqrt{2}$, we begin by writing

$$1 + i = \sqrt{2}\left(\frac{1}{\sqrt{2}} + \frac{1}{\sqrt{2}}i\right) = \sqrt{2}\left(\cos\frac{\pi}{4} + i\sin\frac{\pi}{4}\right)$$

Now

$$\begin{aligned}(1+i)^5 &= \left[\sqrt{2}\left(\cos\frac{\pi}{4} + i\sin\frac{\pi}{4}\right)\right]^5 \\ &= (\sqrt{2})^5\left[\cos\left(5\cdot\frac{\pi}{4}\right) + i\sin\left(5\cdot\frac{\pi}{4}\right)\right] \\ &= 4\sqrt{2}\left(\cos\frac{5\pi}{4} + i\sin\frac{5\pi}{4}\right) \\ &= 4\sqrt{2}\left[-\frac{1}{\sqrt{2}} + \left(-\frac{1}{\sqrt{2}}\right)i\right] = -4-4i\end{aligned}$$

EXAMPLE 6

Using a Calculator with Demoivre's Theorem

Write $(3+4i)^3$ in the standard form $a+bi$.

Solution Again, we start by writing $3+4i$ in polar form. This time we will use degrees for the argument. The magnitude of $3+4i$ is $\sqrt{3^2+4^2} = \sqrt{25} = 5$, so we write

$$3+4i = 5\left(\frac{3}{5} + \frac{4}{5}i\right) \approx 5(\cos 53.1^\circ + i\sin 53.1^\circ)$$

Although we have written the angle rounded to one decimal place (53.1°), we keep the actual value of the angle in memory. Now

$$\begin{aligned}(3+4i)^3 &\approx [5(\cos 53.1^\circ + i\sin 53.1^\circ)]^3 \\ &= 5^3[\cos(3\cdot 53.1^\circ) + i\sin(3\cdot 53.1^\circ)] \\ &= 125(\cos 159.4^\circ + i\sin 159.4^\circ) \\ &\approx 125[-0.936 + i(0.352)] = -117+44i\end{aligned}$$

In this computation, we used the actual values STOred in memory, not the rounded values shown. The final answer, $-117+44i$, is exact as you can verify by cubing $3+4i$.

Complex Roots

5 Let w be a given complex number, and let $n \geq 2$ denote a positive integer. Any complex number z that satisfies the equation

$$z^n = w$$

is called a **complex nth root** of w. In keeping with previous usage, if $n=2$, the solutions of the equation $z^2 = w$ are called **complex square roots** of w, and if $n=3$, the solutions of the equation $z^3=w$ are called **complex cube roots** of w.

Theorem Finding Complex Roots

Let $w = r(\cos\theta + i\sin\theta)$ be a complex number. If $w \neq 0$, there are n distinct complex nth roots of w, given by the formula

$$z_k = \sqrt[n]{r}\left[\cos\left(\frac{\theta}{n} + \frac{2k\pi}{n}\right) + i\sin\left(\frac{\theta}{n} + \frac{2k\pi}{n}\right)\right] \tag{8}$$

where $k = 0, 1, 2, \ldots, n - 1$.

Proof (Outline) We will not prove this result in its entirety. Instead, we shall show only that each z_k in equation (8) obeys the equation $z_k^n = w$ and, hence, each z_k is a complex nth root of w.

$$\begin{aligned}
z_k^n &= \left\{\sqrt[n]{r}\left[\cos\left(\frac{\theta}{n} + \frac{2k\pi}{n}\right) + i\sin\left(\frac{\theta}{n} + \frac{2k\pi}{n}\right)\right]\right\}^n \\
&= (\sqrt[n]{r})^n\left[\cos n\left(\frac{\theta}{n} + \frac{2k\pi}{n}\right) + i\sin n\left(\frac{\theta}{n} + \frac{2k\pi}{n}\right)\right] \quad \text{Demoivre's Theorem.} \\
&= r[\cos(\theta + 2k\pi) + i\sin(\theta + 2k\pi)] \\
&= r(\cos\theta + i\sin\theta) = w
\end{aligned}$$

Thus, each z_k, $k = 0, 1, \ldots, n - 1$, is a complex nth root of w. To complete the proof, we would need to show that each z_k, $k = 0, 1, 2, \ldots, n - 1$, is, in fact, distinct and that there are no complex nth roots of w other than those given by equation (8). ▬

EXAMPLE 7 Finding Complex Cube Roots

Find the complex cube roots of $-1 + \sqrt{3}\,i$. Leave your answers in polar form, with θ in degrees.

Solution First, we express $-1 + \sqrt{3}\,i$ in polar form using degrees:

$$-1 + \sqrt{3}\,i = 2\left(-\frac{1}{2} + \frac{\sqrt{3}}{2}i\right) = 2(\cos 120° + i\sin 120°)$$

The three complex cube roots of $-1 + \sqrt{3}\,i = 2(\cos 120° + i\sin 120°)$ are

$$z_k = \sqrt[3]{2}\left[\cos\left(\frac{120°}{3} + \frac{360°k}{3}\right) + i\sin\left(\frac{120°}{3} + \frac{360°k}{3}\right)\right], \qquad k = 0, 1, 2$$

Thus,

$$\begin{aligned}
z_0 &= \sqrt[3]{2}(\cos 40° + i\sin 40°) \\
z_1 &= \sqrt[3]{2}(\cos 160° + i\sin 160°) \\
z_2 &= \sqrt[3]{2}(\cos 280° + i\sin 280°)
\end{aligned}$$

▬

Notice that each of the three complex cube roots of $-1 + \sqrt{3}\,i$ has the same magnitude, $\sqrt[3]{2}$. This means that the points corresponding to each cube root lie the same distance from the origin; hence, the three points lie on a circle with center at the origin and radius $\sqrt[3]{2}$. Furthermore, the arguments of these cube roots are 40°, 160°, and 280°, the difference of consecutive pairs being 120°. This means that the three points are equally spaced on the circle, as shown in Figure 40. These results are not coincidental. In fact, you are asked to show that these results hold for complex nth roots in Problems 53 through 55.

FIGURE 40

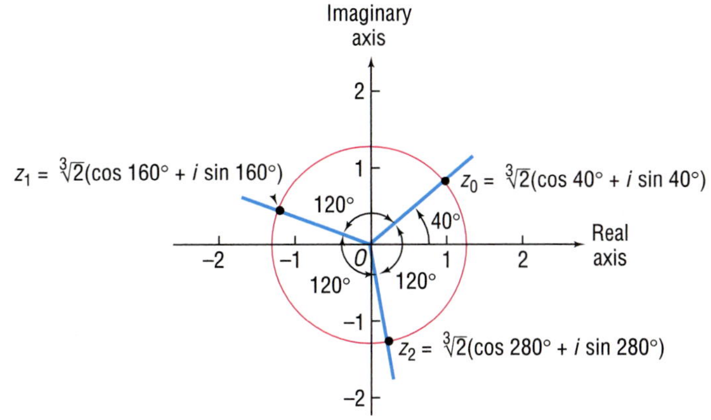

Now work Problem 43.

HISTORICAL FEATURE

The Babylonians, Greeks, and Arabs considered square roots of negative quantities to be impossible and equations with complex solutions to be unsolvable. The first hint that there was some connection between real solutions of equations and complex numbers came when Girolamo Cardano (1501–1576) and Tartaglia (1499–1557) found *real* roots of cubic equations by taking cube roots of *complex* quantities. For centuries thereafter, mathematicians worked with complex numbers without much belief in their actual existence. In 1673, John Wallis appears to have been the first to suggest the graphical representation of complex numbers, a truly significant idea that was not pursued further until about 1800. Several people, including Karl Friedrich Gauss (1777–1855), then rediscovered the idea, and the graphical representation helped to establish complex numbers as equal members of the number family. In practical applications, complex numbers have found their greatest uses in the area of alternating current, where they are a commonplace tool, and subatomic physics.

HISTORICAL PROBLEMS

1. The quadratic formula will work perfectly well if the coefficients are complex numbers. Solve the following, using Demoivre's Theorem where necessary.

   [**Hint:** The answers are "nice."]

   (a) $z^2 - (2 + 5i)z - 3 + 5i = 0$ (b) $z^2 - (1 + i)z - 2 - i = 0$

9.3 EXERCISES

In Problems 1–12, plot each complex number in the complex plane and write it in polar form. Express the argument in degrees.

1. $1 + i$ **2.** $-1 + i$ **3.** $\sqrt{3} - i$ **4.** $1 - \sqrt{3}i$

5. $-3i$ **6.** -2 **7.** $4 - 4i$ **8.** $9\sqrt{3} + 9i$

9. $3 - 4i$ **10.** $2 + \sqrt{3}i$ **11.** $-2 + 3i$ **12.** $\sqrt{5} - i$

In Problems 13–22, write each complex number in rectangular form.

13. $2(\cos 120° + i \sin 120°)$

14. $3(\cos 210° + i \sin 210°)$

15. $4\left(\cos \frac{7\pi}{4} + i \sin \frac{7\pi}{4}\right)$

16. $2\left(\cos \frac{5\pi}{6} + i \sin \frac{5\pi}{6}\right)$

17. $3\left(\cos \frac{3\pi}{2} + i \sin \frac{3\pi}{2}\right)$

18. $4\left(\cos \frac{\pi}{2} + i \sin \frac{\pi}{2}\right)$

19. $0.2(\cos 100° + i \sin 100°)$

20. $0.4(\cos 200° + i \sin 200°)$

21. $2\left(\cos \frac{\pi}{18} + i \sin \frac{\pi}{18}\right)$

22. $3\left(\cos \frac{\pi}{10} + i \sin \frac{\pi}{10}\right)$

In Problems 23–30, find zw and z/w. Leave your answer in polar form.

23. $z = 2(\cos 40° + i \sin 40°)$
$w = 4(\cos 20° + i \sin 20°)$

24. $z = \cos 120° + i \sin 120°$
$w = \cos 100° + i \sin 100°$

25. $z = 3(\cos 130° + i \sin 130°)$
$w = 4(\cos 270° + i \sin 270°)$

26. $z = 2(\cos 80° + i \sin 80°)$
$w = 6(\cos 200° + i \sin 200°)$

27. $z = 2\left(\cos \frac{\pi}{8} + i \sin \frac{\pi}{8}\right)$
$w = 2\left(\cos \frac{\pi}{10} + i \sin \frac{\pi}{10}\right)$

28. $z = 4\left(\cos \frac{3\pi}{8} + i \sin \frac{3\pi}{8}\right)$
$w = 2\left(\cos \frac{9\pi}{16} + i \sin \frac{9\pi}{16}\right)$

29. $z = 2 + 2i$
$w = \sqrt{3} - i$

30. $z = 1 - i$
$w = 1 - \sqrt{3}\,i$

In Problems 31–42, write each expression in the standard from a + bi.

31. $[4(\cos 40° + i \sin 40°)]^3$

32. $[3(\cos 80° + i \sin 80°)]^3$

33. $\left[2\left(\cos \frac{\pi}{10} + i \sin \frac{\pi}{10}\right)\right]^5$

34. $\left[\sqrt{2}\left(\cos \frac{5\pi}{16} + i \sin \frac{5\pi}{16}\right)\right]^4$

35. $[\sqrt{3}(\cos 10° + i \sin 10°)]^6$

36. $[\frac{1}{2}(\cos 72° + i \sin 72°)]^5$

37. $\left[\sqrt{5}\left(\cos \frac{3\pi}{16} + i \sin \frac{3\pi}{16}\right)\right]^4$

38. $\left[\sqrt{3}\left(\cos \frac{5\pi}{18} + i \sin \frac{5\pi}{18}\right)\right]^6$

39. $(1 - i)^5$

40. $(\sqrt{3} - i)^6$

41. $(\sqrt{2} - i)^6$

42. $(1 - \sqrt{5}\,i)^8$

In Problems 43–50, find all the complex roots. Leave your answers in polar form with the argument in degrees.

43. The complex cube roots of $1 + i$

44. The complex fourth roots of $\sqrt{3} - i$

45. The complex fourth roots of $4 - 4\sqrt{3}\,i$

46. The complex cube roots of $-8 - 8i$

47. The complex fourth roots of $-16i$

48. The complex cube roots of -8

49. The complex fifth roots of i

50. The complex fifth roots of $-i$

51. Find the four complex roots of unity (1). Plot each one.

52. Find the six complex roots of unity (1). Plot each one.

53. Show that each complex nth root of a nonzero complex number w has the same magnitude.

54. Use the result of Problem 53 to draw the conclusion that each complex nth root lies on a circle with center at the origin. What is the radius of this circle?

55. Refer to Problem 54. Show that the complex nth roots of a nonzero complex number w are equally spaced on the circle.

56. Prove formula (6).

9.4 VECTORS

1. Graph Vectors
2. Find a Position Vector
3. Add and Subtract Vectors
4. Find a Scalar Product and the Magnitude of a Vector
5. Find a Unit Vector

In simple terms, a **vector** (derived from the Latin *vehere,* meaning "to carry") is a quantity that has both magnitude and direction. For a vector in the plane, which is the only type we will discuss, it is convenient to represent a vector by using an arrow. The length of the arrow represents the **magnitude** of the vector, and the arrowhead indicates the **direction** of the vector.

FIGURE 41

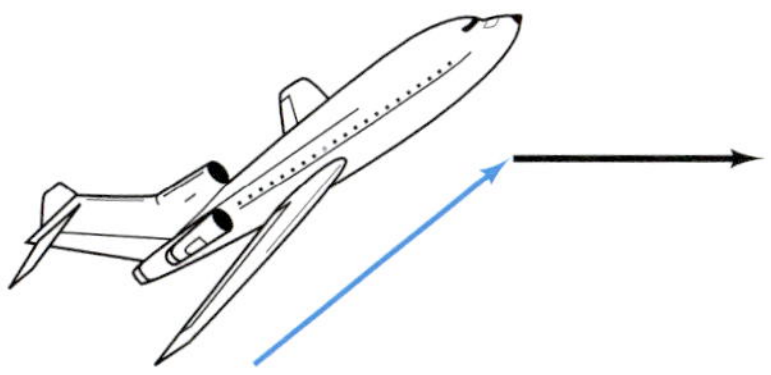

Many quantities in physics can be represented by vectors. For example, the velocity of an aircraft can be represented by an arrow that points in the direction of movement; the length of the arrow represents speed. Thus, if the aircraft speeds up, we lengthen the arrow; if the aircraft changes direction, we introduce an arrow in the new direction. See Figure 41. Based on this representation, it is not surprising that vectors and directed line segments are somehow related.

Directed Line Segments

If P and Q are two distinct points in the xy-plane, there is exactly one line containing both P and Q. The points on that part of the line that joins P to Q, including P and Q, form what is called the **line segment** $\overline{PQ}$. If we order the points so that they proceed from P to Q, we have a **directed line segment** from P to Q, which we denote by $\overrightarrow{PQ}$. In a directed line segment $\overrightarrow{PQ}$, we call P the **initial point** and Q the **terminal point,** as indicated in Figure 42.

FIGURE 42

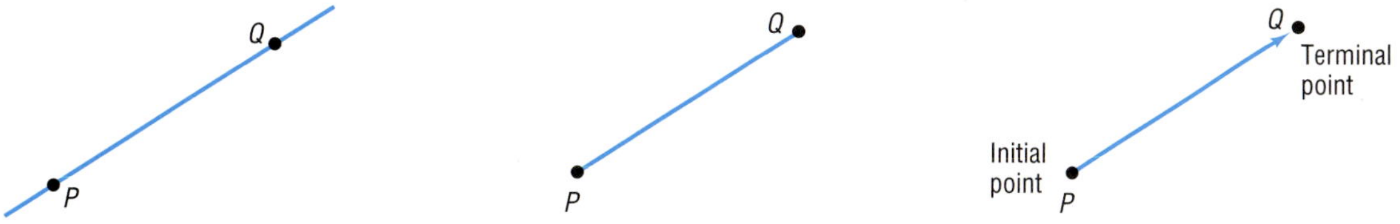

(a) Line containing P and Q **(b)** Line segment $\overline{PQ}$ **(c)** Directed line segment $\overrightarrow{PQ}$

The magnitude of the directed line segment $\overrightarrow{PQ}$ is the distance from the point P to the point Q; that is, it is the length of the line segment. The direction of $\overrightarrow{PQ}$ is from P to Q. If a vector **v*** has the same magnitude and the same direction as the directed line segment PQ, then we write

$$\mathbf{v} = \overrightarrow{PQ}$$

*Boldface letters will be used to denote vectors, in order to distinguish them from numbers. For handwritten work, an arrow is placed over the letter to signify a vector.

The vector **v** whose magnitude is 0 is called the **zero vector, 0.** The zero vector is assigned no direction.

Two vectors **v** and **w** are **equal,** written

$$\mathbf{v} = \mathbf{w}$$

if they have the same magnitude and the same direction.

FIGURE 43

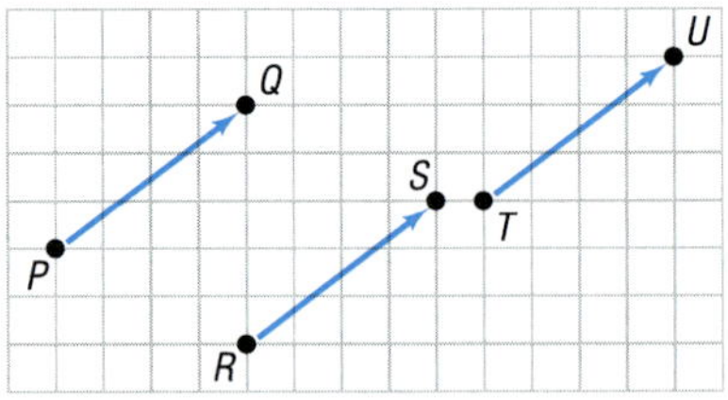

For example, the vectors shown in Figure 43 have the same magnitude and the same direction, so they are equal, even though they have different initial points and different terminal points. As a result, we find it useful to think of a vector simply as an arrow, keeping in mind that two arrows (vectors) are equal if they have the same direction and the same magnitude (length).

Adding Vectors

FIGURE 44

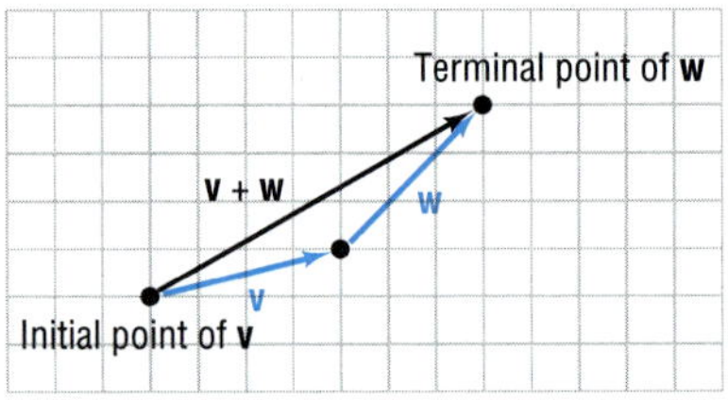

The **sum v + w** of two vectors is defined as follows: We position the vectors **v** and **w** so that the terminal point of **v** coincides with the initial point of **w**, as shown in Figure 44. The vector **v + w** is then the unique vector whose initial point coincides with the initial point of **v** and whose terminal point coincides with the terminal point of **w.**

Vector addition is **commutative.** That is, if **v** and **w** are any two vectors, then

$$\mathbf{v} + \mathbf{w} = \mathbf{w} + \mathbf{v}$$

FIGURE 45

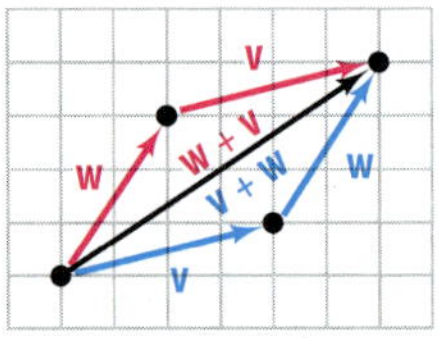

Figure 45 illustrates this fact. (Observe that the commutative property is another way of saying that opposite sides of a parallelogram are equal and parallel.)

Vector addition is also **associative.** That is, if **u, v,** and **w** are vectors, then

$$\mathbf{u} + (\mathbf{v} + \mathbf{w}) = (\mathbf{u} + \mathbf{v}) + \mathbf{w}$$

Figure 46 illustrates the associative property for vectors.

FIGURE 46
$(\mathbf{u} + \mathbf{v}) + \mathbf{w} = \mathbf{u} + (\mathbf{v} + \mathbf{w})$

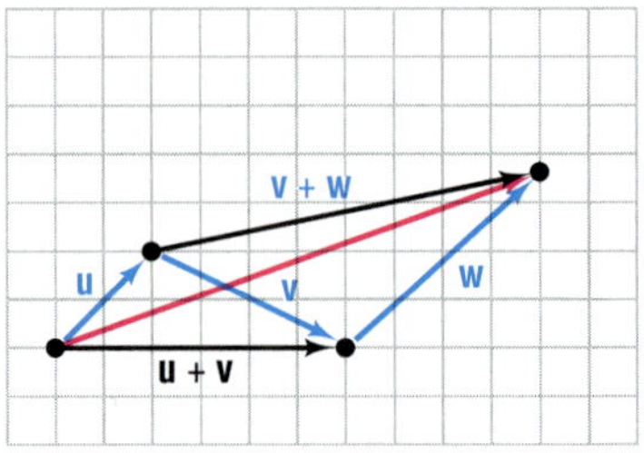

The zero vector has the property that

$$\mathbf{v} + \mathbf{0} = \mathbf{0} + \mathbf{v} = \mathbf{v}$$

FIGURE 47

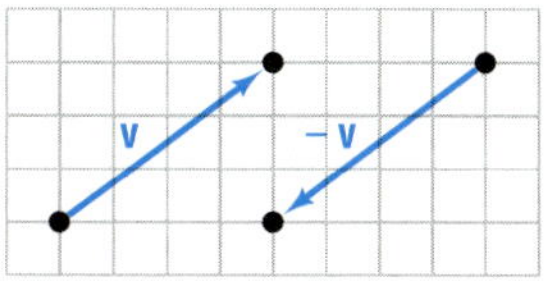

for any vector **v.**

If **v** is a vector, then $-\mathbf{v}$ is the vector having the same magnitude as **v,** but whose direction is opposite to **v,** as shown in Figure 47.

Furthermore,

$$\mathbf{v} + (-\mathbf{v}) = \mathbf{0}$$

FIGURE 48

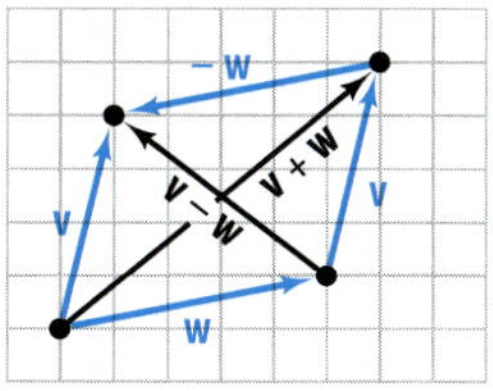

If **v** and **w** are two vectors, we define the **difference** $\mathbf{v} - \mathbf{w}$ as

$$\mathbf{v} - \mathbf{w} = \mathbf{v} + (-\mathbf{w})$$

Figure 48 illustrates the relationships among **v, w,** $\mathbf{v} + \mathbf{w}$, and $\mathbf{v} - \mathbf{w}$.

Multiplying Vectors by Numbers

When dealing with vectors, we refer to real numbers as **scalars.** Scalars are quantities that have only magnitude. Examples from physics of scalar quantities are temperature, speed, and time. We now define how to multiply a vector by a scalar.

If α is a scalar and **v** is a vector, the **scalar product** $\alpha\mathbf{v}$ is defined as

1. If $\alpha > 0$, the product $\alpha\mathbf{v}$ is the vector whose magnitude is α times the magnitude of **v** and whose direction is the same as **v**.
2. If $\alpha < 0$, the product $\alpha\mathbf{v}$ is the vector whose magnitude is $|\alpha|$ times the magnitude of **v** and whose direction is opposite that of **v**.
3. If $\alpha = 0$ or if $\mathbf{v} = \mathbf{0}$, then $\alpha\mathbf{v} = \mathbf{0}$.

FIGURE 49

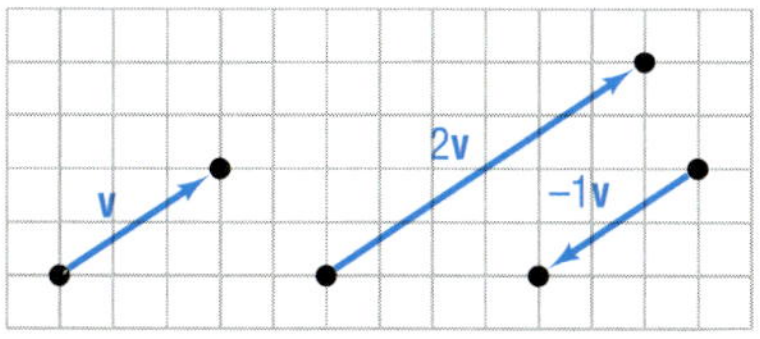

See Figure 49 for some illustrations.

For example, if **a** is the acceleration of an object of mass m due to a force **F** being exerted on it, then, by Newton's second law of motion, $\mathbf{F} = m\mathbf{a}$. Here, $m\mathbf{a}$ is the product of the scalar m and the vector **a.**

Scalar products have the following properties:

$$0\mathbf{v} = \mathbf{0} \qquad 1\mathbf{v} = \mathbf{v} \qquad -1\mathbf{v} = -\mathbf{v}$$
$$(\alpha + \beta)\mathbf{v} = \alpha\mathbf{v} + \beta\mathbf{v} \qquad \alpha(\mathbf{v} + \mathbf{w}) = \alpha\mathbf{v} + \alpha\mathbf{w}$$
$$\alpha(\beta\mathbf{v}) = (\alpha\beta)\mathbf{v}$$

EXAMPLE 1 Graphing Vectors

Use the vectors illustrated in Figure 50 to graph each expression.

FIGURE 50

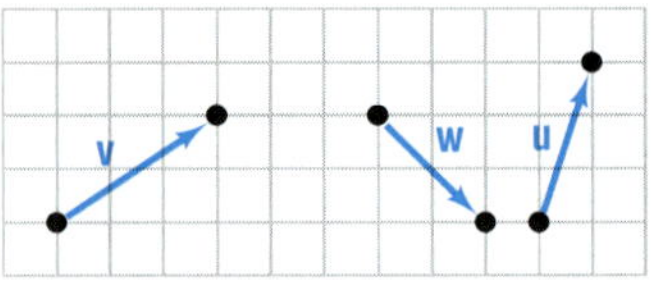

(a) $\mathbf{v} - \mathbf{w}$ (b) $2\mathbf{v} + 3\mathbf{w}$ (c) $2\mathbf{v} - \mathbf{w} + 3\mathbf{u}$

Solution Figure 51 illustrates each graph.

FIGURE 51

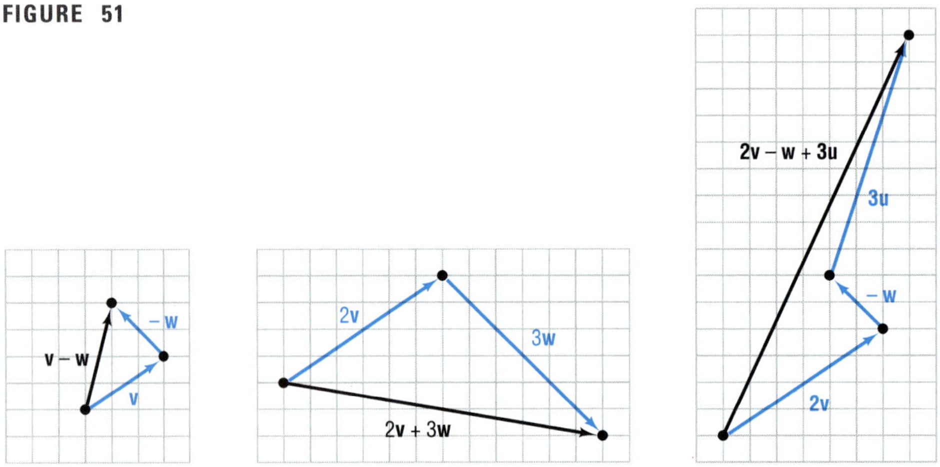

(a) v – w (b) 2v + 3w (c) 2v – w + 3u

Now work Problems 1 and 3.

Magnitudes of Vectors

If $\mathbf{v}$ is a vector, we use the symbol $\|\mathbf{v}\|$ to represent the **magnitude** of $\mathbf{v}$. Since $\|\mathbf{v}\|$ equals the length of a directed line segment, it follows that $\|\mathbf{v}\|$ has the following properties:

Theorem Properties of $\|\mathbf{v}\|$

If $\mathbf{v}$ is a vector and if α is a scalar, then

(a) $\|\mathbf{v}\| \geq 0$
(b) $\|\mathbf{v}\| = 0$ if and only if $\mathbf{v} = \mathbf{0}$
(c) $\|-\mathbf{v}\| = \|\mathbf{v}\|$
(d) $\|\alpha\mathbf{v}\| = |\alpha|\ \|\mathbf{v}\|$

Property (a) is a consequence of the fact that distance is a nonnegative number. Property (b) follows, because the length of the directed line segment $\overrightarrow{PQ}$ is positive unless P and Q are the same point, in which case the length is 0. Property (c) follows, because the length of the line segment $\overline{PQ}$ equals the length of the line segment $\overline{QP}$. Property (d) is a direct consequence of the definition of a scalar product.

A vector $\mathbf{u}$ for which $\|\mathbf{u}\| = 1$ is called a **unit vector.**

To compute the magnitude and direction of a vector, we need an algebraic way of representing vectors.

Representing Vectors in the Plane

We use a rectangular coordinate system to represent vectors in the plane. Let $\mathbf{i}$ denote a unit vector whose direction is along the positive x-axis; let $\mathbf{j}$ denote a unit vector whose direction is along the positive y-axis. If $\mathbf{v}$ is a vector with initial point at the origin O and terminal point at $P = (a, b)$, then we can represent $\mathbf{v}$ in terms of the vectors $\mathbf{i}$ and $\mathbf{j}$ as

$$\mathbf{v} = a\mathbf{i} + b\mathbf{j}$$

FIGURE 52

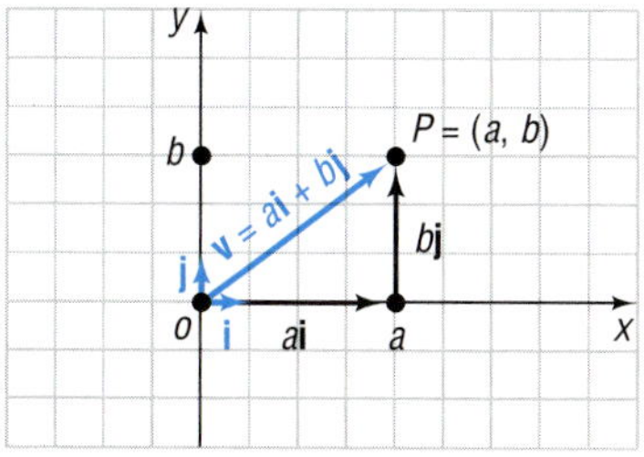

See Figure 52. The scalars a and b are called the **components** of the vector $\mathbf{v} = a\mathbf{i} + b\mathbf{j}$, with a being the component in the direction $\mathbf{i}$ and b being the component in the direction $\mathbf{j}$.

A vector whose initial point is at the origin is called a **position vector.** The next result states that any vector whose initial point is not at the origin is equal to a unique position vector.

Theorem

Suppose that $\mathbf{v}$ is a vector with initial point $P_1 = (x_1, y_1)$, not necessarily the origin, and terminal point $P_2 = (x_2, y_2)$. If $\mathbf{v} = \overrightarrow{P_1P_2}$, then $\mathbf{v}$ is equal to the position vector

$$\mathbf{v} = (x_2 - x_1)\mathbf{i} + (y_2 - y_1)\mathbf{j} \tag{1}$$

To see why this is true, look at Figure 53. Triangle OPA and triangle P_1P_2Q are congruent. (Do you see why?) The line segments have the same magnitude, so $d(O, P) = d(P_1, P_2)$; and they have the same direction, so $\angle POA = \angle P_2P_1Q$. Since the triangles are right triangles, we have

FIGURE 53
$\mathbf{v} = a\mathbf{i} + b\mathbf{j} = (x_2 - x_1)\mathbf{i} + (y_2 - y_1)\mathbf{j}$

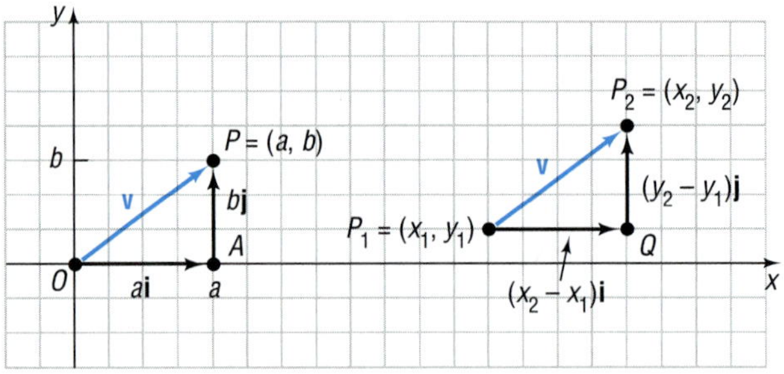

angle–side–angle. Thus, it follows that corresponding sides are equal. As a result, $x_2 - x_1 = a$ and $y_2 - y_1 = b$, so **v** may be written as

$$\mathbf{v} = a\mathbf{i} + b\mathbf{j} = (x_2 - x_1)\mathbf{i} + (y_2 - y_1)\mathbf{j}$$

EXAMPLE 2 Finding a Position Vector

2

Find the position vector of the vector $\mathbf{v} = \overrightarrow{P_1P_2}$ if $P_1 = (-1, 2)$ and $P_2 = (4, 6)$.

Solution By equation (1), the position vector equal to **v** is

$$\mathbf{v} = [4 - (-1)]\mathbf{i} + (6 - 2)\mathbf{j} = 5\mathbf{i} + 4\mathbf{j}$$

See Figure 54.

FIGURE 54

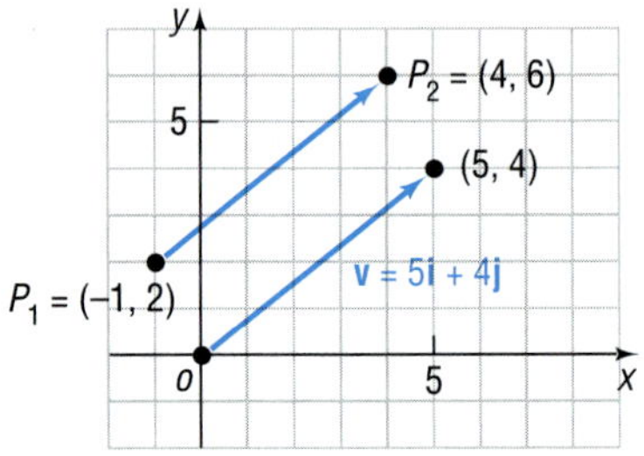

Now work Problem 21.

Two position vectors **v** and **w** are equal if and only if the terminal point of **v** is the same as the terminal point of **w**. This leads to the following result:

Theorem Equality of Vectors

Two vectors **v** and **w** are equal if and only if their corresponding components are equal. That is,

If $\mathbf{v} = a_1\mathbf{i} + b_1\mathbf{j}$ and $\mathbf{w} = a_2\mathbf{i} + b_2\mathbf{j}$,
then $\mathbf{v} = \mathbf{w}$ if and only if $a_1 = a_2$ and $b_1 = b_2$.

Because of the above result, we can replace any vector (directed line segment) by a unique position vector, and vice versa. This flexibility is one of the main reasons for the wide use of vectors. Unless otherwise specified, from now on the term *vector* will mean the unique position vector equal to it.

Next, we define addition, subtraction, scalar product, and magnitude in terms of the components of a vector.

Let $\mathbf{v} = a_1\mathbf{i} + b_1\mathbf{j}$ and $\mathbf{w} = a_2\mathbf{i} + b_2\mathbf{j}$ be two vectors, and let α be a scalar. Then

$$\mathbf{v} + \mathbf{w} = (a_1 + a_2)\mathbf{i} + (b_1 + b_2)\mathbf{j} \quad (2)$$
$$\mathbf{v} - \mathbf{w} = (a_1 - a_2)\mathbf{i} + (b_1 - b_2)\mathbf{j} \quad (3)$$
$$\alpha\mathbf{v} = (\alpha a_1)\mathbf{i} + (\alpha b_1)\mathbf{j} \quad (4)$$
$$\|\mathbf{v}\| = \sqrt{a_1^2 + b_1^2} \quad (5)$$

These definitions are compatible with the geometric ones given earlier in this section. See Figure 55.

FIGURE 55

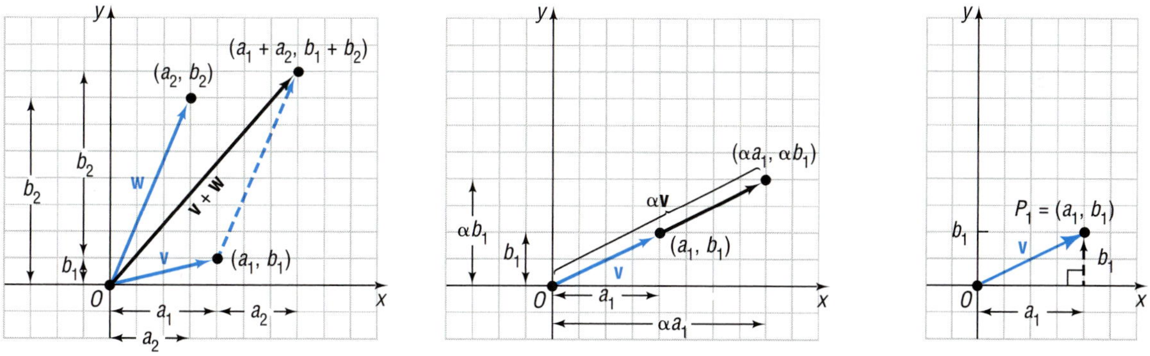

(a) Illustration of property (2) **(b)** Illustration of property (4) **(c)** Illustration of property (5): $\|\mathbf{v}\|$ = Distance of O to P_1; $\|\mathbf{v}\| = \sqrt{a_1^2 + b_1^2}$

Thus, to add two vectors, simply add corresponding components. To subtract two vectors, subtract corresponding components.

EXAMPLE 3 Adding and Subtracting Vectors

3

If $\mathbf{v} = 2\mathbf{i} + 3\mathbf{j}$ and $\mathbf{w} = 3\mathbf{i} - 4\mathbf{j}$, find

(a) $\mathbf{v} + \mathbf{w}$ (b) $\mathbf{v} - \mathbf{w}$

Solution (a) $\mathbf{v} + \mathbf{w} = (2\mathbf{i} + 3\mathbf{j}) + (3\mathbf{i} - 4\mathbf{j}) = (2 + 3)\mathbf{i} + (3 - 4)\mathbf{j} = 5\mathbf{i} - \mathbf{j}$

(b) $\mathbf{v} - \mathbf{w} = (2\mathbf{i} + 3\mathbf{j}) - (3\mathbf{i} - 4\mathbf{j}) = (2 - 3)\mathbf{i} + [3 - (-4)]\mathbf{j} = -\mathbf{i} + 7\mathbf{j}$

EXAMPLE 4 Finding Scalar Products and Magnitudes

4

If $\mathbf{v} = 2\mathbf{i} + 3\mathbf{j}$ and $\mathbf{w} = 3\mathbf{i} - 4\mathbf{j}$, find

(a) $3\mathbf{v}$ (b) $2\mathbf{v} - 3\mathbf{w}$ (c) $\|\mathbf{v}\|$

Solution (a) $3\mathbf{v} = 3(2\mathbf{i} + 3\mathbf{j}) = 6\mathbf{i} + 9\mathbf{j}$

(b) $2\mathbf{v} - 3\mathbf{w} = 2(2\mathbf{i} + 3\mathbf{j}) - 3(3\mathbf{i} - 4\mathbf{j}) = 4\mathbf{i} + 6\mathbf{j} - 9\mathbf{i} + 12\mathbf{j}$
$= -5\mathbf{i} + 18\mathbf{j}$

(c) $\|\mathbf{v}\| = \|2\mathbf{i} + 3\mathbf{j}\| = \sqrt{2^2 + 3^2} = \sqrt{13}$

Now work Problems 27 and 33.

Recall that a unit vector $\mathbf{u}$ is one for which $\|\mathbf{u}\| = 1$. In many applications, it is useful to be able to find a unit vector $\mathbf{u}$ that has the same direction as a given vector $\mathbf{v}$.

Theorem Unit Vector in Direction of v

For any nonzero vector **v**, the vector

$$\mathbf{u} = \frac{\mathbf{v}}{\|\mathbf{v}\|}$$

is a unit vector that has the same direction as **v**.

Proof Let $\mathbf{v} = a\mathbf{i} + b\mathbf{j}$. Then $\|\mathbf{v}\| = \sqrt{a^2 + b^2}$ and

$$\mathbf{u} = \frac{\mathbf{v}}{\|\mathbf{v}\|} = \frac{a\mathbf{i} + b\mathbf{j}}{\sqrt{a^2 + b^2}} = \frac{a}{\sqrt{a^2 + b^2}}\mathbf{i} + \frac{b}{\sqrt{a^2 + b^2}}\mathbf{j}$$

The vector $\mathbf{u}$ is in the same direction as $\mathbf{v}$, since $\|\mathbf{v}\| > 0$. Furthermore,

$$\|\mathbf{u}\| = \sqrt{\frac{a^2}{a^2 + b^2} + \frac{b^2}{a^2 + b^2}} = \sqrt{\frac{a^2 + b^2}{a^2 + b^2}} = 1$$

Thus, $\mathbf{u}$ is a unit vector in the direction of $\mathbf{v}$.

As a consequence of this theorem, if $\mathbf{u}$ is a unit vector in the same direction as a vector $\mathbf{v}$, then $\mathbf{v}$ may be expressed as

$$\mathbf{v} = \|\mathbf{v}\|\,\mathbf{u} \tag{6}$$

This way of expressing a vector is useful in many applications.

EXAMPLE 5 Finding a Unit Vector

5

Find a unit vector in the same direction as $\mathbf{v} = 4\mathbf{i} - 3\mathbf{j}$.

Solution We find $\|\mathbf{v}\|$ first:

$$\|\mathbf{v}\| = \|4\mathbf{i} - 3\mathbf{j}\| = \sqrt{16 + 9} = 5$$

Now we multiply $\mathbf{v}$ by the scalar $1/\|\mathbf{v}\| = \frac{1}{5}$. The result is

$$\frac{\mathbf{v}}{\|\mathbf{v}\|} = \frac{4\mathbf{i} - 3\mathbf{j}}{5} = \frac{4}{5}\mathbf{i} - \frac{3}{5}\mathbf{j}$$

Check: This vector is, in fact, a unit vector because

$$\left(\tfrac{4}{5}\right)^2 + \left(-\tfrac{3}{5}\right)^2 = \tfrac{16}{25} + \tfrac{9}{25} = \tfrac{25}{25} = 1.$$

Now work Problem 43.

Applications

Forces provide an example of physical quantities that may be conveniently represented by vectors; two forces "combine" the way vectors "add." How do we know this? Laboratory experiments bear it out. Thus, if $\mathbf{F}_1$ and $\mathbf{F}_2$ are two forces simultaneously acting on an object, the vector sum $\mathbf{F}_1 + \mathbf{F}_2$ is the force that produces the same effect on the object as that obtained when the forces $\mathbf{F}_1$ and $\mathbf{F}_2$ act on the object. The force $\mathbf{F}_1 + \mathbf{F}_2$ is sometimes called the **resultant** of $\mathbf{F}_1$ and $\mathbf{F}_2$. See Figure 56.

FIGURE 56

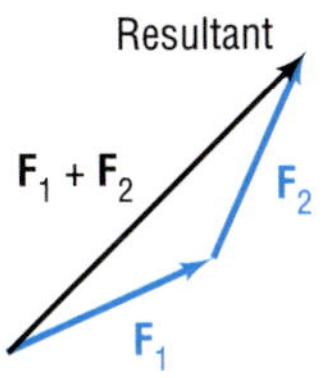

Two important applications of the resultant of two vectors occur with aircraft flying in the presence of a wind and with boats cruising across a river with a current. For example, consider the velocity of wind acting on the velocity of an airplane (see Figure 57). Suppose that $\mathbf{w}$ is a vector describing the velocity of the wind; that is, $\mathbf{w}$ represents the direction and speed of the wind. If $\mathbf{v}$ is the velocity of the airplane in the absence of wind (called its **velocity relative to the air**), then $\mathbf{v} + \mathbf{w}$ is the vector equal to the actual velocity of the airplane (called its **velocity relative to the ground**).

Our next example illustrates this use of vectors in navigation.

FIGURE 57

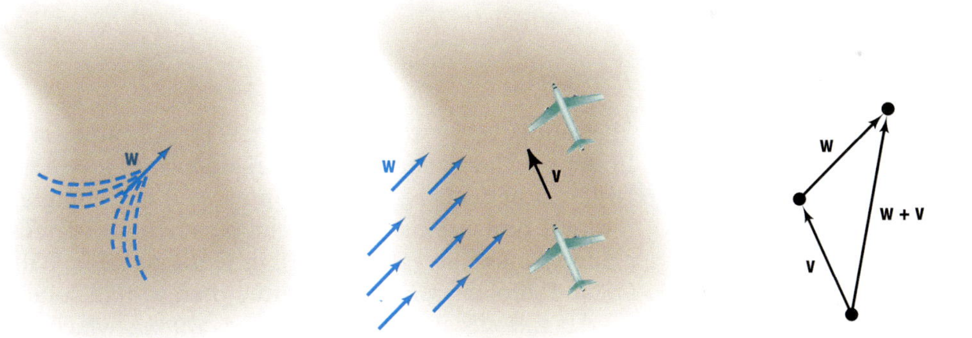

(a) Velocity **w** of wind relative to ground

(b) Velocity **v** of airplane relative to air

(c) Resultant **w + v** equals velocity of airplane relative to the ground

EXAMPLE 6

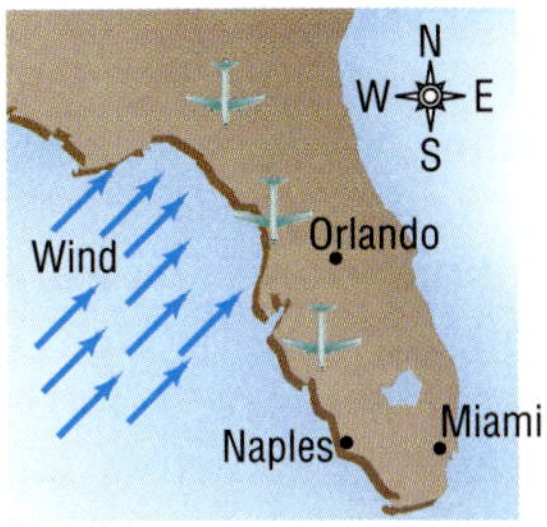

Finding the Actual Speed of an Aircraft

A Boeing 737 aircraft maintains a constant airspeed of 500 miles per hour in the direction due south. The velocity of the jet stream is 80 miles per hour in a north-easterly direction.

(a) Find a unit vector having northeast as direction.

(b) Find a vector 80 units in magnitude having the same direction as the unit vector found in part (a).

(c) Find the actual speed of the aircraft relative to the ground.

FIGURE 58

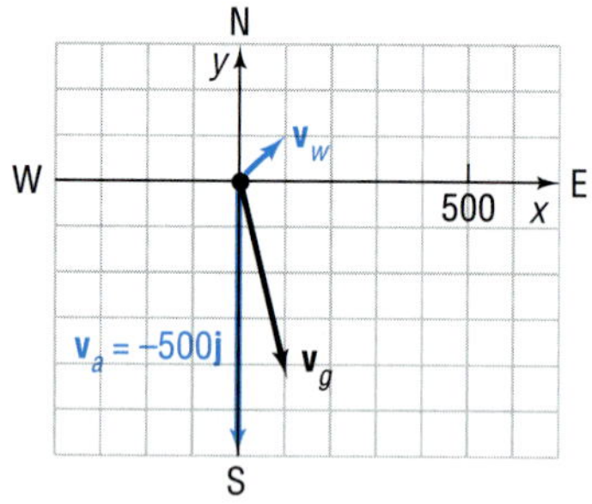

Solution We set up a coordinate system in which north (N) is along the positive y-axis. See Figure 58. Let

$$\mathbf{v}_a = \text{Velocity of aircraft relative to the air} = -500\mathbf{j}$$
$$\mathbf{v}_g = \text{Velocity of aircraft relative to ground}$$
$$\mathbf{v}_w = \text{Velocity of jet stream}$$

(a) A vector having northeast as direction is $\mathbf{i} + \mathbf{j}$. The unit vector in this direction is

$$\frac{\mathbf{i}+\mathbf{j}}{\|\mathbf{i}+\mathbf{j}\|} = \frac{\mathbf{i}+\mathbf{j}}{\sqrt{1+1}} = \frac{1}{\sqrt{2}}(\mathbf{i}+\mathbf{j})$$

(b) The velocity $\mathbf{v}_w$ of the jet stream is a vector with magnitude 80 in the direction of the unit vector $(1/\sqrt{2})(\mathbf{i}+\mathbf{j})$. Thus, from (6),

$$\mathbf{v}_w = 80\left[\frac{1}{\sqrt{2}}(\mathbf{i}+\mathbf{j})\right] = 40\sqrt{2}(\mathbf{i}+\mathbf{j})$$

(c) The velocity $\mathbf{v}_g$ of the aircraft relative to the ground is the resultant of the vectors $\mathbf{v}_a$ and $\mathbf{v}_w$. Thus,

$$\begin{aligned}\mathbf{v}_g = \mathbf{v}_a + \mathbf{v}_w &= -500\mathbf{j} + 40\sqrt{2}(\mathbf{i}+\mathbf{j})\\ &= 40\sqrt{2}\mathbf{i} + (40\sqrt{2} - 500)\mathbf{j}\end{aligned}$$

The actual speed (speed relative to the ground) of the aircraft is

$$\|\mathbf{v}_g\| = \sqrt{(40\sqrt{2})^2 + (40\sqrt{2} - 500)^2} \approx 447 \text{ miles per hour}$$

We will find the actual direction of the aircraft in Example 4 of the next section.

Historical Feature

The history of vectors is surprisingly complicated for such a natural concept. In the xy-plane, complex numbers do a good job of imitating vectors. About 1840, mathematicians became interested in finding a system that would do for three dimensions what the complex numbers do for two dimensions. Hermann Grassmann (1809–1877), in Germany, and William Rowan Hamilton (1805–1865), in Ireland, both attempted to find solutions.

Hamilton's system was the *quaternions,* which are best thought of as a real number plus a vector, and do for four dimensions what complex numbers do for two dimensions. In this system the order of multiplication matters; that is, $\mathbf{ab} \neq \mathbf{ba}$. Hamilton spent the rest of his life working out quaternion theory and trying to get it accepted in applied mathematics, but he encountered fierce resistance due to the complicated nature of quaternion multiplication. In the work with quaternions, two products of vectors emerged, the scalar (or dot) and the vector (or cross) products.

Grassmann fared even worse than Hamilton; if people did not like Hamilton's work, at least they understood it. Grassmann's abstract style, although easily read today, was almost impenetrable during the previous century, and only a few of his ideas were appreciated. Among those few were the same scalar and vector products that Hamilton had found.

About 1880, the American physicist Josiah Willard Gibbs (1839–1903) worked out an algebra involving only the simplest concepts—the vectors and the two products. He then added some calculus, and the resulting system was simple, flexible, and well adapted to expressing a large number of physical laws. This system remains in use essentially unchanged. Hamilton's and Grassmann's more extensive systems each gave birth to much interesting mathematics, but little of this mathematics is seen at elementary levels.

9.4 EXERCISES

In Problems 1–8, use the vectors in the figure to the right to graph each expression.

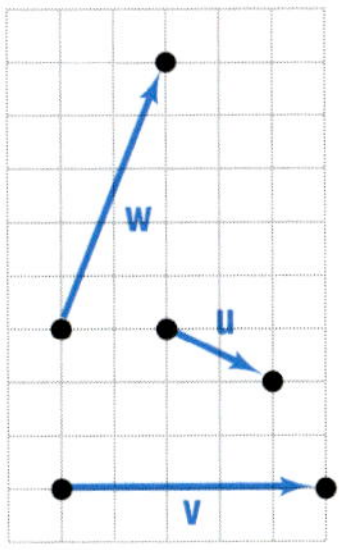

1. **v** + **w** 2. **u** + **v** 3. 3**v** 4. 4**w**
5. **v** − **w** 6. **u** − **v** 7. 3**v** + **u** − 2**w** 8. 2**u** − 3**v** + **w**

In Problems 9–16, use the figure below to find each vector.

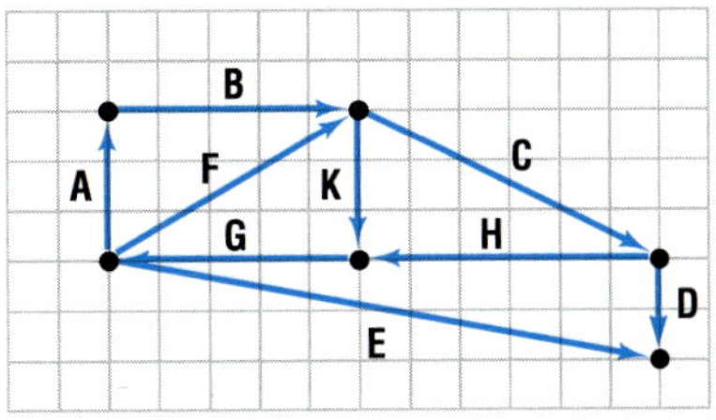

9. **x**, if **x** + **B** = **F**
10. **x**, if **x** + **D** = **E**
11. **C** in terms of **E**, **D**, and **F**
12. **G** in terms of **C**, **D**, **E**, and **K**
13. **E** in terms of **G**, **H**, and **D**
14. **E** in terms of **A**, **B**, **C**, and **D**
15. **x**, if **x** = **A** + **B** + **K** + **G**
16. **x**, if **x** = **A** + **B** + **C** + **H** + **G**
17. If $\|\mathbf{v}\| = 4$, what is $\|3\mathbf{v}\|$?
18. If $\|\mathbf{v}\| = 2$, what is $\|-4\mathbf{v}\|$?

In Problems 19–26, the vector **v** *has initial point P and terminal point Q. Write* **v** *in the form a***i** + *b***j**, *that is, find its position vector.*

19. $P = (0, 0); Q = (3, 4)$
20. $P = (0, 0); Q = (-3, -5)$
21. $P = (3, 2); Q = (5, 6)$
22. $P = (-3, 2); Q = (6, 5)$
23. $P = (-2, -1); Q = (6, -2)$
24. $P = (-1, 4); Q = (6, 2)$
25. $P = (1, 0); Q = (0, 1)$
26. $P = (1, 1); Q = (2, 2)$

In Problems 27–32, find $\|\mathbf{v}\|$.

27. $\mathbf{v} = 3\mathbf{i} - 4\mathbf{j}$ 28. $\mathbf{v} = -5\mathbf{i} + 12\mathbf{j}$ 29. $\mathbf{v} = \mathbf{i} - \mathbf{j}$
30. $\mathbf{v} = -\mathbf{i} - \mathbf{j}$ 31. $\mathbf{v} = -2\mathbf{i} + 3\mathbf{j}$ 32. $\mathbf{v} = 6\mathbf{i} + 2\mathbf{j}$

In Problems 33–38, find each quantity if $\mathbf{v} = 3\mathbf{i} - 5\mathbf{j}$ *and* $\mathbf{w} = -2\mathbf{i} + 3\mathbf{j}$.

33. $2\mathbf{v} + 3\mathbf{w}$ 34. $3\mathbf{v} - 2\mathbf{w}$ 35. $\|\mathbf{v} - \mathbf{w}\|$
36. $\|\mathbf{v} + \mathbf{w}\|$ 37. $\|\mathbf{v}\| - \|\mathbf{w}\|$ 38. $\|\mathbf{v}\| + \|\mathbf{w}\|$

In Problems 39–44, find the unit vector having the same direction as **v**.

39. $\mathbf{v} = 5\mathbf{i}$ 40. $\mathbf{v} = -3\mathbf{j}$ 41. $\mathbf{v} = 3\mathbf{i} - 4\mathbf{j}$
42. $\mathbf{v} = -5\mathbf{i} + 12\mathbf{j}$ 43. $\mathbf{v} = \mathbf{i} - \mathbf{j}$ 44. $\mathbf{v} = 2\mathbf{i} - \mathbf{j}$

45. Find a vector **v** whose magnitude is 4 and whose component in the **i** direction is twice the component in the **j** direction.

46. Find a vector **v** whose magnitude is 3 and whose component in the **i** direction is equal to the component in the **j** direction.

47. If $\mathbf{v} = 2\mathbf{i} - \mathbf{j}$ and $\mathbf{w} = x\mathbf{i} + 3\mathbf{j}$, find all numbers x for which $\|\mathbf{v} + \mathbf{w}\| = 5$.

48. If $P = (-3, 1)$ and $Q = (x, 4)$, find all numbers x such that the vector represented by $\overrightarrow{PQ}$ has length 5.

49. Finding Ground Speed An airplane has an airspeed of 500 kilometers per hour in an easterly direction. If the wind velocity is 60 kilometers per hour in a northwesterly direction, find the speed of the airplane relative to the ground.

50. Finding Airspeed After 1 hour in the air, an airplane arrives at a point 200 miles due south of its departure point. If there was a steady wind of 30 miles per hour from the northwest during the entire flight, what was the average airspeed of the airplane?

51. Finding Speed without Wind An airplane travels in a northwesterly direction at a constant speed of 250 miles per hour, due to an easterly wind of 50 miles per hour. How fast would the plane have gone if there had been no wind?

52. Finding the True Speed of a Motorboat A small motorboat in still water maintains a speed of 10 miles per hour. In heading directly across a river (that is, perpendicular to the current) whose current is 4 miles per hour, what will be the true speed of the motorboat?

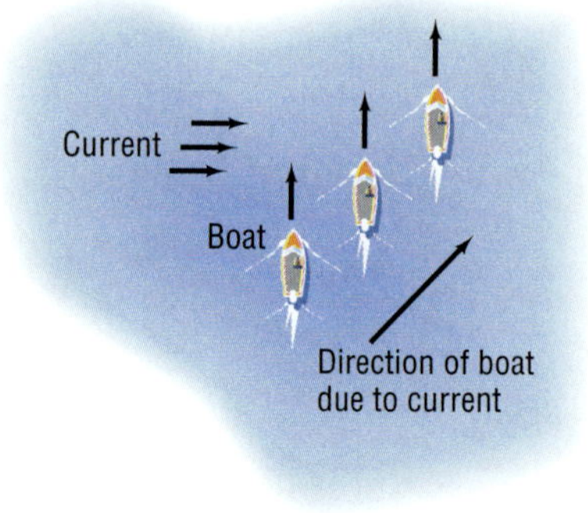

53. Show on the graph below the force needed to prevent an object at P from moving.

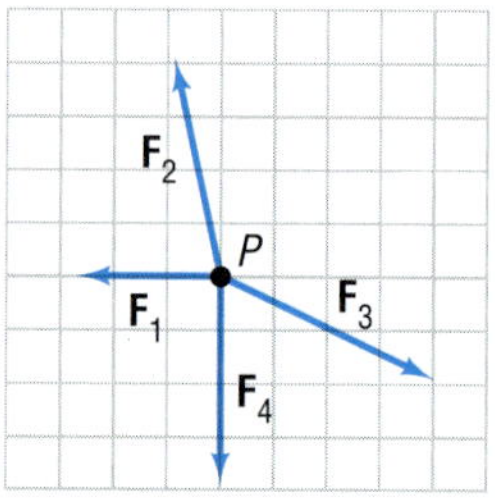

54. Explain in your own words what a vector is. Give an example of a vector.

55. Write a brief paragraph comparing the algebra of complex numbers and the algebra of vectors.

9.5 THE DOT PRODUCT

1. Find the Dot Product of Two Vectors
2. Find the Angle between Two Vectors
3. Find a Vector from Its Magnitude and Direction
4. Determine Whether Two Vectors Are Parallel
5. Determine Whether Two Vectors Are Orthogonal
6. Decompose a Vector into Two Orthogonal Vectors
7. Compute Work

The definition for a product of two vectors is somewhat unexpected. However, such a product has meaning in many geometric and physical applications.

If $\mathbf{v} = a_1\mathbf{i} + b_1\mathbf{j}$ and $\mathbf{w} = a_2\mathbf{i} + b_2\mathbf{j}$ are two vectors, the **dot product $\mathbf{v} \cdot \mathbf{w}$** is defined as

$$\mathbf{v} \cdot \mathbf{w} = a_1a_2 + b_1b_2 \tag{1}$$

EXAMPLE 1

Finding Dot Products

1

If $\mathbf{v} = 2\mathbf{i} - 3\mathbf{j}$ and $\mathbf{w} = 5\mathbf{i} + 3\mathbf{j}$, find

(a) $\mathbf{v} \cdot \mathbf{w}$ (b) $\mathbf{w} \cdot \mathbf{v}$ (c) $\mathbf{v} \cdot \mathbf{v}$
(d) $\mathbf{w} \cdot \mathbf{w}$ (e) $\|\mathbf{v}\|$ (f) $\|\mathbf{w}\|$

Solution

(a) $\mathbf{v} \cdot \mathbf{w} = 2(5) + (-3)3 = 1$ (b) $\mathbf{w} \cdot \mathbf{v} = 5(2) + 3(-3) = 1$
(c) $\mathbf{v} \cdot \mathbf{v} = 2(2) + (-3)(-3) = 13$ (d) $\mathbf{w} \cdot \mathbf{w} = 5(5) + 3(3) = 34$
(e) $\|\mathbf{v}\| = \sqrt{2^2 + (-3)^2} = \sqrt{13}$ (f) $\|\mathbf{w}\| = \sqrt{5^2 + 3^2} = \sqrt{34}$

Since the dot product $\mathbf{v} \cdot \mathbf{w}$ of two vectors $\mathbf{v}$ and $\mathbf{w}$ is a real number (scalar), we sometimes refer to it as the **scalar product.**

Properties

The results obtained in Example 1 suggest some general properties.

Theorem Properties of Dot Product

If $\mathbf{u}$, $\mathbf{v}$, and $\mathbf{w}$ are vectors, then

Commutative Property

$$\mathbf{u} \cdot \mathbf{v} = \mathbf{v} \cdot \mathbf{u} \tag{2}$$

Distributive Property

$$\mathbf{u} \cdot (\mathbf{v} + \mathbf{w}) = \mathbf{u} \cdot \mathbf{v} + \mathbf{u} \cdot \mathbf{w} \tag{3}$$

$$\mathbf{v} \cdot \mathbf{v} = \|\mathbf{v}\|^2 \tag{4}$$

$$\mathbf{0} \cdot \mathbf{v} = 0 \tag{5}$$

Proof We will prove properties (2) and (4) here and leave properties (3) and (5) as exercises (see Problems 29 and 30 at the end of this section).

To prove property (2), we let $\mathbf{u} = a_1\mathbf{i} + b_1\mathbf{j}$ and $\mathbf{v} = a_2\mathbf{i} + b_2\mathbf{j}$. Then

$$\mathbf{u} \cdot \mathbf{v} = a_1a_2 + b_1b_2 = a_2a_1 + b_2b_1 = \mathbf{v} \cdot \mathbf{u}$$

To prove property (4), we let $\mathbf{v} = a\mathbf{i} + b\mathbf{j}$. Then

$$\mathbf{v} \cdot \mathbf{v} = a^2 + b^2 = \|\mathbf{v}\|^2$$

One use of the dot product is to calculate the angle between two vectors.

Angle between Vectors

Let **u** and **v** be two vectors with the same initial point A. Then the vectors **u**, **v**, and $\mathbf{u} - \mathbf{v}$ form a triangle. The angle θ at vertex A of the triangle is the **angle between the vectors u and v.** See Figure 59. We wish to find a formula for calculating the angle θ.

FIGURE 59

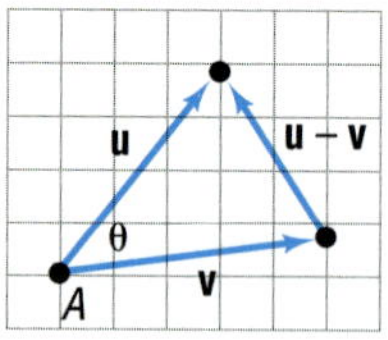

The sides of the triangle are of lengths $\|\mathbf{v}\|$, $\|\mathbf{u}\|$, and $\|\mathbf{u} - \mathbf{v}\|$, and θ is the included angle between the sides of length $\|\mathbf{v}\|$ and $\|\mathbf{u}\|$. The Law of Cosines (Section 8.3) can be used to find the cosine of the included angle:

$$\|\mathbf{u} - \mathbf{v}\|^2 = \|\mathbf{u}\|^2 + \|\mathbf{v}\|^2 - 2\,\|\mathbf{u}\|\,\|\mathbf{v}\|\cos\theta$$

Now we use property (4) to rewrite this equation in terms of dot products:

$$(\mathbf{u} - \mathbf{v}) \cdot (\mathbf{u} - \mathbf{v}) = \mathbf{u} \cdot \mathbf{u} + \mathbf{v} \cdot \mathbf{v} - 2\,\|\mathbf{u}\|\,\|\mathbf{v}\|\cos\theta \tag{6}$$

and then apply the distributive property (3) twice on the left side of (6) to obtain

$$\begin{aligned}(\mathbf{u} - \mathbf{v}) \cdot (\mathbf{u} - \mathbf{v}) &= \mathbf{u} \cdot (\mathbf{u} - \mathbf{v}) - \mathbf{v} \cdot (\mathbf{u} - \mathbf{v})\\ &= \mathbf{u} \cdot \mathbf{u} - \mathbf{u} \cdot \mathbf{v} - \mathbf{v} \cdot \mathbf{u} + \mathbf{v} \cdot \mathbf{v}\\ &= \mathbf{u} \cdot \mathbf{u} + \mathbf{v} \cdot \mathbf{v} - 2\,\mathbf{u} \cdot \mathbf{v}\end{aligned} \tag{7}$$

↑ Property (2).

Combining equations (6) and (7), we have

$$\begin{aligned}\mathbf{u} \cdot \mathbf{u} + \mathbf{v} \cdot \mathbf{v} - 2\,\mathbf{u} \cdot \mathbf{v} &= \mathbf{u} \cdot \mathbf{u} + \mathbf{v} \cdot \mathbf{v} - 2\,\|\mathbf{u}\|\,\|\mathbf{v}\|\cos\theta\\ \mathbf{u} \cdot \mathbf{v} &= \|\mathbf{u}\|\,\|\mathbf{v}\|\cos\theta\end{aligned}$$

Thus, we have proved the following result:

Theorem Angle between Vectors

If **u** and **v** are two nonzero vectors, the angle θ, $0 \le \theta \le \pi$, between **u** and **v** is determined by the formula

$$\cos\theta = \frac{\mathbf{u} \cdot \mathbf{v}}{\|\mathbf{u}\|\,\|\mathbf{v}\|} \tag{8}$$

EXAMPLE 2 Finding the Angle θ between Two Vectors

2

Find the angle θ between $\mathbf{u} = 4\mathbf{i} - 3\mathbf{j}$ and $\mathbf{v} = 2\mathbf{i} + 5\mathbf{j}$.

Solution We compute the quantities $\mathbf{u} \cdot \mathbf{v}$, $\|\mathbf{u}\|$, and $\|\mathbf{v}\|$:

$$\begin{aligned}\mathbf{u} \cdot \mathbf{v} &= 4(2) + (-3)(5) = -7\\ \|\mathbf{u}\| &= \sqrt{4^2 + (-3)^2} = 5\\ \|\mathbf{v}\| &= \sqrt{2^2 + 5^2} = \sqrt{29}\end{aligned}$$

By formula (8), if θ is the angle between **u** and **v**, then

$$\cos\theta = \frac{\mathbf{u} \cdot \mathbf{v}}{\|\mathbf{u}\|\,\|\mathbf{v}\|} = \frac{-7}{5\sqrt{29}} \approx -0.26$$

Using a calculator, we find that $\theta \approx 105°$. See Figure 60.

FIGURE 60

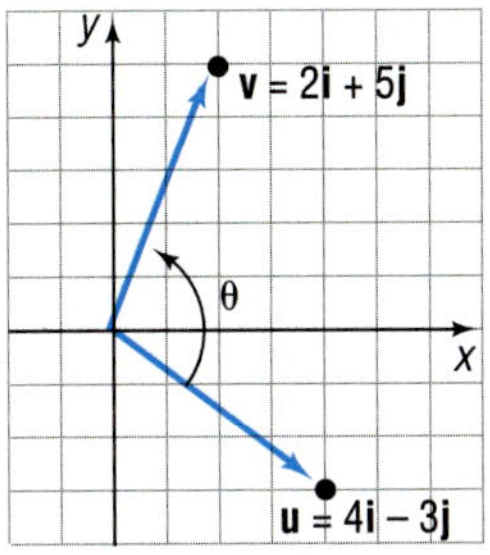

Now work Problem 1.

Writing a Vector in Terms of Its Magnitude and Direction

Many applications describe a vector in terms of its magnitude and direction, rather than in terms of its components. Suppose that we are given the magnitude $\|\mathbf{v}\|$ of a nonzero vector $\mathbf{v}$ and the angle α between $\mathbf{v}$ and $\mathbf{i}$. To express $\mathbf{v}$ in terms of $\|\mathbf{v}\|$ and α, we first find the unit vector $\mathbf{u}$ having the same direction as $\mathbf{v}$:

FIGURE 61

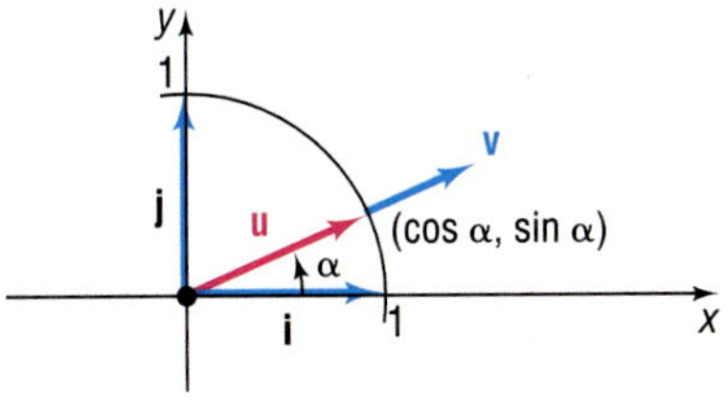

$$\mathbf{u} = \frac{\mathbf{v}}{\|\mathbf{v}\|} \quad \text{or} \quad \mathbf{v} = \|\mathbf{v}\|\mathbf{u} \tag{9}$$

Look at Figure 61. The coordinates of the terminal point of $\mathbf{u}$ are $(\cos\alpha, \sin\alpha)$. Thus, $\mathbf{u} = \cos\alpha\mathbf{i} + \sin\alpha\mathbf{j}$ and, from (9),

$$\mathbf{v} = \|\mathbf{v}\|(\cos\alpha\mathbf{i} + \sin\alpha\mathbf{j}) \tag{10}$$

EXAMPLE 3

Writing a Vector When Its Magnitude and Direction Are Given

3

A force $\mathbf{F}$ of 5 pounds is applied in a direction that makes an angle of 30° with the positive x-axis. Express $\mathbf{F}$ in terms of $\mathbf{i}$ and $\mathbf{j}$.

Solution The magnitude of $\mathbf{F}$ is $\|\mathbf{F}\| = 5$ and the angle between the direction of $\mathbf{F}$ and $\mathbf{i}$, the positive x-axis, is $\alpha = 30°$. Thus, by (10),

$$\mathbf{F} = \|\mathbf{F}\|(\cos\alpha\mathbf{i} + \sin\alpha\mathbf{j}) = 5(\cos 30°\mathbf{i} + \sin 30°\mathbf{j})$$
$$= 5\left(\frac{\sqrt{3}}{2}\mathbf{i} + \frac{1}{2}\mathbf{j}\right) = \frac{5}{2}(\sqrt{3}\,\mathbf{i} + \mathbf{j})$$

EXAMPLE 4

Finding the Actual Direction of an Aircraft

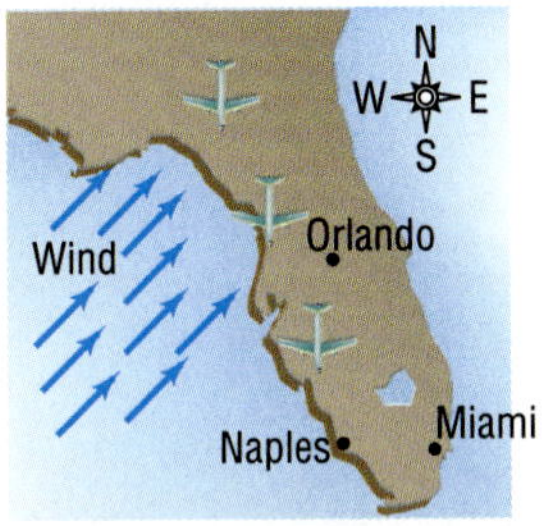

A Boeing 737 aircraft maintains a constant airspeed of 500 miles per hour in the direction due south. The velocity of the jet stream is 80 miles per hour in a north-easterly direction. Find the actual direction of the aircraft relative to the ground.

FIGURE 62

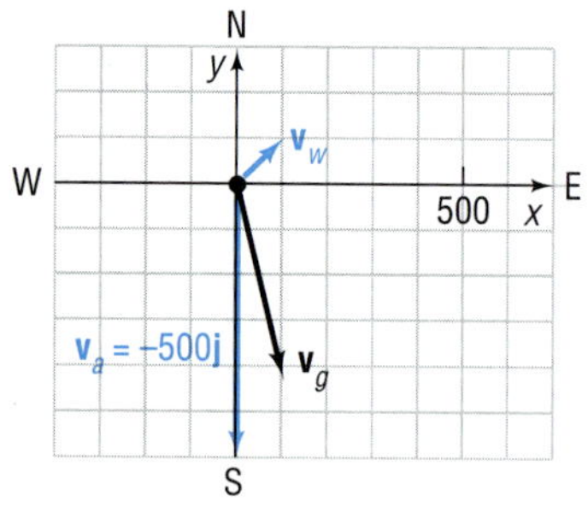

Solution This is the same information given in Example 6 of Section 9.4. We repeat the figure from that example as Figure 62.

The velocity of the aircraft relative to the air is

$$\mathbf{v}_a = -500\mathbf{j}$$

The wind has magnitude 80 and direction $\alpha = 45°$. Thus, the velocity of the wind is

$$\mathbf{v}_w = 80(\cos 45°\mathbf{i} + \sin 45°\mathbf{j}) = 80\left(\frac{\sqrt{2}}{2}\mathbf{i} + \frac{\sqrt{2}}{2}\mathbf{j}\right)$$
$$= 40\sqrt{2}(\mathbf{i} + \mathbf{j})$$

The velocity of the aircraft relative to the ground is

$$\mathbf{v}_g = \mathbf{v}_a + \mathbf{v}_w = -500\mathbf{j} + 40\sqrt{2}(\mathbf{i} + \mathbf{j}) = 40\sqrt{2}\mathbf{i} + (40\sqrt{2} - 500)\mathbf{j}$$

The angle θ between $\mathbf{v}_g$ and the vector $\mathbf{v}_a = -500\mathbf{j}$ (the velocity of the aircraft relative to the air) is determined by the equation

$$\cos\theta = \frac{\mathbf{v}_g \cdot \mathbf{v}_a}{\|\mathbf{v}_g\|\,\|\mathbf{v}_a\|} = \frac{(40\sqrt{2} - 500)(-500)}{(447)(500)} \approx 0.9920$$
$$\theta \approx 7.2°$$

The direction of the aircraft relative to the ground is about 7.2° east of south.

Now work Problem 19.

Parallel and Orthogonal Vectors

Two vectors $\mathbf{v}$ and $\mathbf{w}$ are said to be **parallel** if there is a nonzero scalar α so that $\mathbf{v} = \alpha\mathbf{w}$. In this case, the angle θ between $\mathbf{v}$ and $\mathbf{w}$ is 0 or π.

EXAMPLE 5 Determining Whether Vectors Are Parallel

4

The vectors $\mathbf{v} = 3\mathbf{i} - \mathbf{j}$ and $\mathbf{w} = 6\mathbf{i} - 2\mathbf{j}$ are parallel, since $\mathbf{v} = \frac{1}{2}\mathbf{w}$. Furthermore, since

$$\cos\theta = \frac{\mathbf{v} \cdot \mathbf{w}}{\|\mathbf{v}\|\,\|\mathbf{w}\|} = \frac{18 + 2}{\sqrt{10}\sqrt{40}} = \frac{20}{\sqrt{400}} = 1$$

the angle θ between $\mathbf{v}$ and $\mathbf{w}$ is 0.

If the angle θ between two nonzero vectors $\mathbf{v}$ and $\mathbf{w}$ is $\pi/2$, the vectors $\mathbf{v}$ and $\mathbf{w}$ are called **orthogonal.***

It follows from formula (8) that if $\mathbf{v}$ and $\mathbf{w}$ are orthogonal then $\mathbf{v} \cdot \mathbf{w} = 0$, since $\cos(\pi/2) = 0$.

**Orthogonal, perpendicular,* and *normal* are all terms that mean "meet at a right angle." It is customary to refer to two vectors as being *orthogonal,* two lines as being *perpendicular,* and a line and a plane or a vector and a plane as being *normal.*

FIGURE 63
$\mathbf{v} \cdot \mathbf{w} = 0$; $\mathbf{v}$ is orthogonal to $\mathbf{w}$

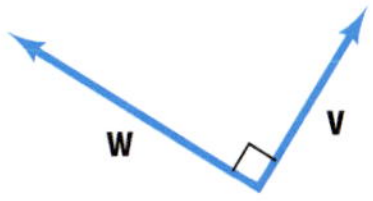

On the other hand, if $\mathbf{v} \cdot \mathbf{w} = 0$, then either $\mathbf{v} = \mathbf{0}$ or $\mathbf{w} = \mathbf{0}$ or $\cos\theta = 0$. In the latter case, $\theta = \pi/2$ and $\mathbf{v}$ and $\mathbf{w}$ are orthogonal. See Figure 63. If $\mathbf{v}$ or $\mathbf{w}$ is the zero vector, then, since the zero vector has no specific direction, we adopt the convention that the zero vector is orthogonal to every vector.

Theorem

Two vectors $\mathbf{v}$ and $\mathbf{w}$ are orthogonal if and only if

$$\mathbf{v} \cdot \mathbf{w} = 0$$

EXAMPLE 6 Determining Whether Two Vectors Are Orthogonal

FIGURE 64

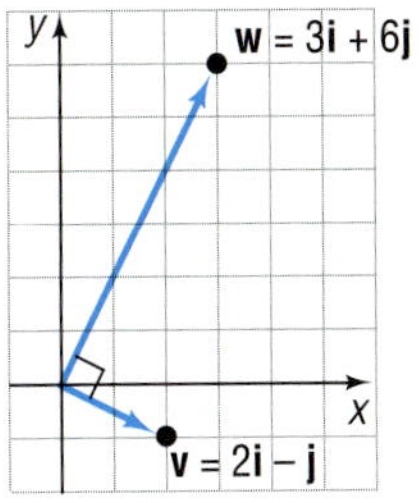

5

The vectors

$$\mathbf{v} = 2\mathbf{i} - \mathbf{j} \quad \text{and} \quad \mathbf{w} = 3\mathbf{i} + 6\mathbf{j}$$

are orthogonal, since

$$\mathbf{v} \cdot \mathbf{w} = 6 - 6 = 0$$

See Figure 64.

Now work Problem 11.

FIGURE 65

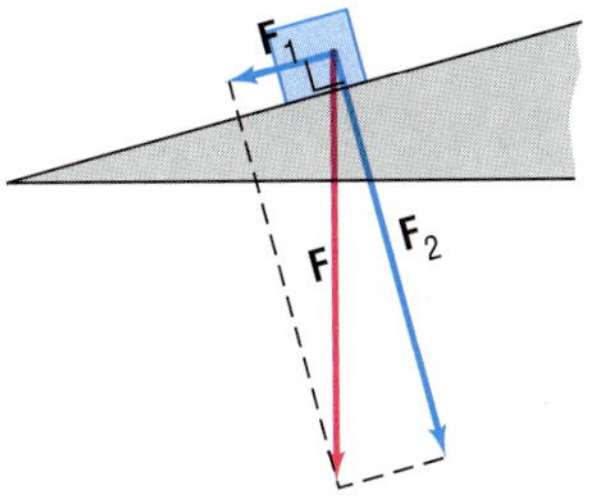

Projection of a Vector onto Another Vector

6

In many physical applications, it is necessary to find "how much" of a vector is applied in a given direction. Look at Figure 65. The force $\mathbf{F}$ due to gravity is pulling straight down (toward the center of Earth) on the block. To study the effect of gravity on the block, it is necessary to determine how much of $\mathbf{F}$ is actually pushing the block down the incline ($\mathbf{F}_1$) and how much is pressing the block against the incline ($\mathbf{F}_2$), at a right angle to the incline. Knowing the **decomposition** of $\mathbf{F}$ often will allow us to determine when friction is overcome and the block will slide down the incline.

FIGURE 66

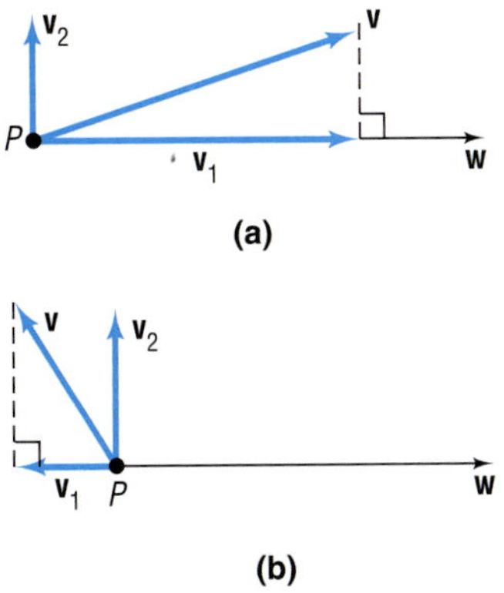

Suppose that $\mathbf{v}$ and $\mathbf{w}$ are two nonzero vectors with the same initial point P. We seek to decompose $\mathbf{v}$ into two vectors: $\mathbf{v}_1$, which is parallel to $\mathbf{w}$, and $\mathbf{v}_2$, which is orthogonal to $\mathbf{w}$. See Figure 66(a) and 66(b). The vector $\mathbf{v}_1$ is called the **vector projection of v onto w** and is denoted by $\text{proj}_{\mathbf{w}}\,\mathbf{v}$.

The vector $\mathbf{v}_1$ is obtained as follows: From the terminal point of $\mathbf{v}$, drop a perpendicular to the line containing $\mathbf{w}$. The vector $\mathbf{v}_1$ is the vector from P to the foot of this perpendicular. The vector $\mathbf{v}_2$ is given by $\mathbf{v}_2 = \mathbf{v} - \mathbf{v}_1$. Note that $\mathbf{v} = \mathbf{v}_1 + \mathbf{v}_2$, $\mathbf{v}_1$ is parallel to $\mathbf{w}$, and $\mathbf{v}_2$ is orthogonal to $\mathbf{w}$. This is the decomposition of $\mathbf{v}$ that we wanted.

Now we seek a formula for $\mathbf{v}_1$ that is based on a knowledge of the vectors $\mathbf{v}$ and $\mathbf{w}$. Since $\mathbf{v} = \mathbf{v}_1 + \mathbf{v}_2$, we have

$$\mathbf{v} \cdot \mathbf{w} = (\mathbf{v}_1 + \mathbf{v}_2) \cdot \mathbf{w} = \mathbf{v}_1 \cdot \mathbf{w} + \mathbf{v}_2 \cdot \mathbf{w} \tag{11}$$

Since $\mathbf{v}_2$ is orthogonal to $\mathbf{w}$, we have $\mathbf{v}_2 \cdot \mathbf{w} = 0$. Since $\mathbf{v}_1$ is parallel to $\mathbf{w}$, we have $\mathbf{v}_1 = \alpha\mathbf{w}$ for some scalar α. Thus, equation (11) can be written as

$$\mathbf{v} \cdot \mathbf{w} = \alpha\mathbf{w} \cdot \mathbf{w} = \alpha\|\mathbf{w}\|^2$$

$$\alpha = \frac{\mathbf{v} \cdot \mathbf{w}}{\|\mathbf{w}\|^2}$$

Thus,

$$\mathbf{v}_1 = \alpha\mathbf{w} = \frac{\mathbf{v} \cdot \mathbf{w}}{\|\mathbf{w}\|^2}\mathbf{w}$$

Theorem

If $\mathbf{v}$ and $\mathbf{w}$ are two nonzero vectors, the vector projection of $\mathbf{v}$ onto $\mathbf{w}$ is

$$\text{proj}_{\mathbf{w}}\mathbf{v} = \frac{\mathbf{v} \cdot \mathbf{w}}{\|\mathbf{w}\|^2}\mathbf{w}$$

The decomposition of $\mathbf{v}$ into $\mathbf{v}_1$ and $\mathbf{v}_2$, where $\mathbf{v}_1$ is parallel to $\mathbf{w}$ and $\mathbf{v}_2$ is perpendicular to $\mathbf{w}$, is

$$\mathbf{v}_1 = \text{proj}_{\mathbf{w}}\mathbf{v} = \frac{\mathbf{v} \cdot \mathbf{w}}{\|\mathbf{w}\|^2}\mathbf{w} \qquad \mathbf{v}_2 = \mathbf{v} - \mathbf{v}_1 \tag{12}$$

EXAMPLE 7 Decomposing a Vector into Two Orthogonal Vectors

Find the vector projection of $\mathbf{v} = \mathbf{i} + 3\mathbf{j}$ onto $\mathbf{w} = \mathbf{i} + \mathbf{j}$. Decompose $\mathbf{v}$ into two vectors $\mathbf{v}_1$ and $\mathbf{v}_2$, where $\mathbf{v}_1$ is parallel to $\mathbf{w}$ and $\mathbf{v}_2$ is orthogonal to $\mathbf{w}$.

FIGURE 67

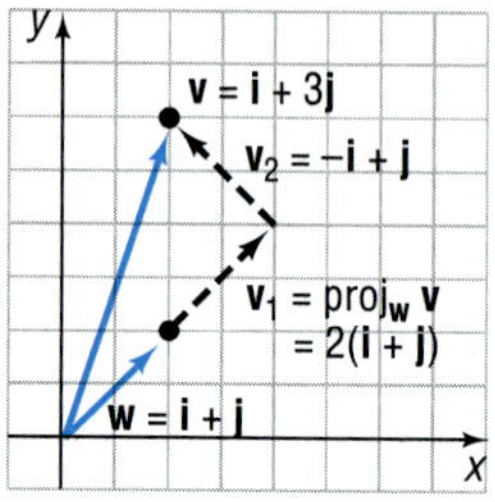

Solution We use formulas (12).

$$\mathbf{v}_1 = \text{proj}_{\mathbf{w}}\mathbf{v} = \frac{\mathbf{v} \cdot \mathbf{w}}{\|\mathbf{w}\|^2}\mathbf{w} = \frac{1+3}{(\sqrt{2})^2}\mathbf{w} = 2\mathbf{w} = 2(\mathbf{i} + \mathbf{j})$$

$$\mathbf{v}_2 = \mathbf{v} - \mathbf{v}_1 = (\mathbf{i} + 3\mathbf{j}) - 2(\mathbf{i} + \mathbf{j}) = -\mathbf{i} + \mathbf{j}$$

See Figure 67.

Now work Problem 13.

Work Done by a Constant Force

7 In elementary physics, the **work** W done by a constant force $\mathbf{F}$ in moving an object from a point A to a point B is defined as

$$W = (\text{Magnitude of force})(\text{Distance}) = \|\mathbf{F}\|\,\|\overrightarrow{AB}\|$$

(Work is commonly measured in foot-pounds or in Newton-meters.)

FIGURE 68

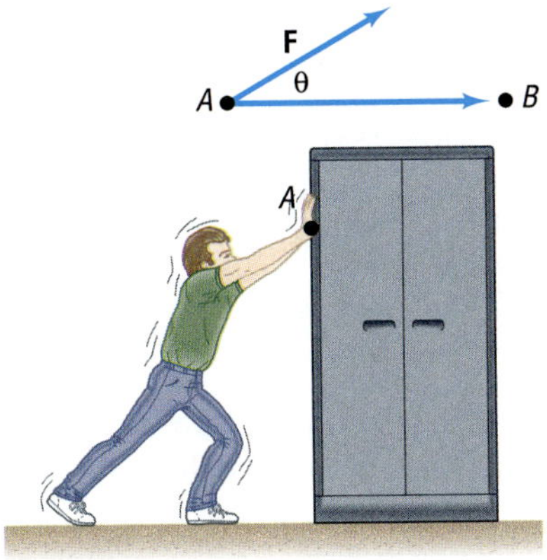

In this definition, it is assumed that the force **F** is applied along the line of motion. If the constant force **F** is not along the line of motion, but, instead, is at an angle θ to the direction of motion, as illustrated in Figure 68, then the **work** W **done by F** in moving an object from A to B is defined as

$$W = \mathbf{F} \cdot \overrightarrow{AB} \tag{13}$$

This definition is compatible with the force times distance definition given above, since

$$\begin{aligned} W &= (\text{Amount of force in direction of } \overrightarrow{AB})(\text{Distance}) \\ &= \|\text{proj}_{\overrightarrow{AB}}\mathbf{F}\|\,\|\overrightarrow{AB}\| = \frac{\mathbf{F} \cdot \overrightarrow{AB}}{\|\overrightarrow{AB}\|^2}\|\overrightarrow{AB}\|\,\|\overrightarrow{AB}\| = \mathbf{F} \cdot \overrightarrow{AB} \end{aligned}$$

EXAMPLE 8

Computing Work

Find the work done by a force of 5 pounds acting in the direction $\mathbf{i} + \mathbf{j}$ in moving an object 1 foot from $(0, 0)$ to $(1, 0)$.

Solution First, we must express the force **F** as a vector. The force has magnitude 5 and direction $\mathbf{i} + \mathbf{j}$. The direction of **F** therefore makes an angle of 45° with **i**. Thus, the force **F** is

$$\mathbf{F} = 5(\cos 45°\mathbf{i} + \sin 45°\mathbf{j}) = 5\left(\frac{\sqrt{2}}{2}\mathbf{i} + \frac{\sqrt{2}}{2}\mathbf{j}\right) = \frac{5\sqrt{2}}{2}(\mathbf{i} + \mathbf{j})$$

The line of motion of the object is from $A = (0, 0)$ to $B = (1, 0)$, so $\overrightarrow{AB} = \mathbf{i}$. The work W is therefore

$$W = \mathbf{F} \cdot \overrightarrow{AB} = \frac{5\sqrt{2}}{2}(\mathbf{i} + \mathbf{j}) \cdot \mathbf{i} = \frac{5\sqrt{2}}{2} \text{ foot-pounds}$$

Now work Problem 25.

EXAMPLE 9

Computing Work

Figure 69(a) shows a girl pulling a wagon with a force of 50 pounds. How much work is done in moving the wagon 100 feet if the handle makes an angle of 30° with the ground?

FIGURE 69

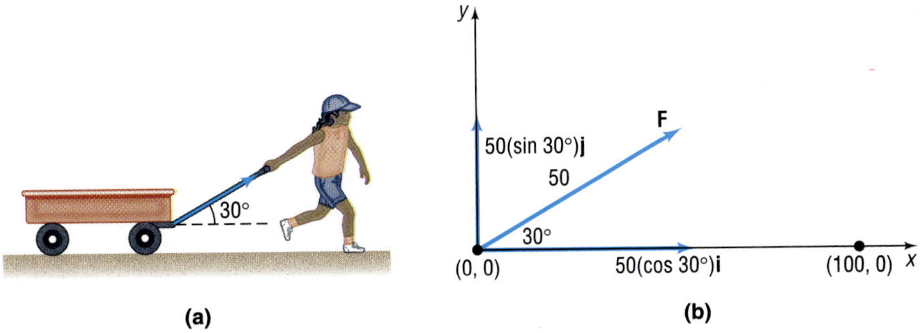

Solution We position the vectors in a coordinate system in such a way that the wagon is moved from (0, 0) to (100, 0). Thus, the motion is from $A = (0, 0)$ to $B = (100, 0)$, so $\overrightarrow{AB} = 100\mathbf{i}$. The force vector **F**, as shown in Figure 69(b), is

$$\mathbf{F} = 50(\cos 30°\mathbf{i} + \sin 30°\mathbf{j}) = 50\left(\frac{\sqrt{3}}{2}\mathbf{i} + \frac{1}{2}\mathbf{j}\right) = 25\sqrt{3}\mathbf{i} + 25\mathbf{j}$$

By formula (13), the work W done is

$$W = \mathbf{F} \cdot \overrightarrow{AB} = (25\sqrt{3}\mathbf{i} + 25\mathbf{j}) \cdot 100\mathbf{i} = 2500\sqrt{3} \text{ foot-pounds}$$

HISTORICAL PROBLEM

1. We stated in an earlier Historical Feature that complex numbers were used as vectors in the plane before the general notion of vector was clarified. Suppose that we make the correspondence

$$\begin{aligned} &\text{Vector} \longleftrightarrow \text{Complex number} \\ &a\mathbf{i} + b\mathbf{j} \longleftrightarrow a + bi \\ &c\mathbf{i} + d\mathbf{j} \longleftrightarrow c + di \end{aligned}$$

Show that

$$(a\mathbf{i} + b\mathbf{j}) \cdot (c\mathbf{i} + d\mathbf{j}) = \text{Real part}[\overline{(a + bi)}(c + di)]$$

This is how the dot product was found originally. The imaginary part is also interesting. It is a determinant (see Section 11.3) and represents the area of the parallelogram whose edges are the vectors. This is close to some of Hermann Grassmann's ideas and is also connected with the scalar triple product of three-dimensional vectors.

9.5 EXERCISES

In Problems 1–10, find the dot product $\mathbf{v} \cdot \mathbf{w}$ *and the angle between* **v** *and* **w**.

1. $\mathbf{v} = \mathbf{i} - \mathbf{j}, \quad \mathbf{w} = \mathbf{i} + \mathbf{j}$
2. $\mathbf{v} = \mathbf{i} + \mathbf{j}, \quad \mathbf{w} = -\mathbf{i} + \mathbf{j}$
3. $\mathbf{v} = 2\mathbf{i} + \mathbf{j}, \quad \mathbf{w} = \mathbf{i} + 2\mathbf{j}$
4. $\mathbf{v} = 2\mathbf{i} + 2\mathbf{j}, \quad \mathbf{w} = \mathbf{i} + 2\mathbf{j}$
5. $\mathbf{v} = \sqrt{3}\mathbf{i} - \mathbf{j}, \quad \mathbf{w} = \mathbf{i} + \mathbf{j}$
6. $\mathbf{v} = \mathbf{i} + \sqrt{3}\mathbf{j}, \quad \mathbf{w} = \mathbf{i} - \mathbf{j}$
7. $\mathbf{v} = 3\mathbf{i} + 4\mathbf{j}, \quad \mathbf{w} = 4\mathbf{i} + 3\mathbf{j}$
8. $\mathbf{v} = 3\mathbf{i} - 4\mathbf{j}, \quad \mathbf{w} = 4\mathbf{i} - 3\mathbf{j}$
9. $\mathbf{v} = 4\mathbf{i}, \quad \mathbf{w} = \mathbf{j}$
10. $\mathbf{v} = \mathbf{i}, \quad \mathbf{w} = -3\mathbf{j}$
11. Find a such that the angle between $\mathbf{v} = a\mathbf{i} - \mathbf{j}$ and $\mathbf{w} = 2\mathbf{i} + 3\mathbf{j}$ is $\pi/2$.
12. Find b such that the angle between $\mathbf{v} = \mathbf{i} + \mathbf{j}$ and $\mathbf{w} = \mathbf{i} + b\mathbf{j}$ is $\pi/2$.

In Problems 13–18, decompose **v** *into two vectors* $\mathbf{v}_1$ *and* $\mathbf{v}_2$*, where* $\mathbf{v}_1$ *is parallel to* **w** *and* $\mathbf{v}_2$ *is orthogonal to* **w**.

13. $\mathbf{v} = 2\mathbf{i} - 3\mathbf{j}, \quad \mathbf{w} = \mathbf{i} - \mathbf{j}$
14. $\mathbf{v} = -3\mathbf{i} + 2\mathbf{j}, \quad \mathbf{w} = 2\mathbf{i} + \mathbf{j}$
15. $\mathbf{v} = \mathbf{i} - \mathbf{j}, \quad \mathbf{w} = \mathbf{i} + 2\mathbf{j}$
16. $\mathbf{v} = 2\mathbf{i} - \mathbf{j}, \quad \mathbf{w} = \mathbf{i} - 2\mathbf{j}$
17. $\mathbf{v} = 3\mathbf{i} + \mathbf{j}, \quad \mathbf{w} = -2\mathbf{i} - \mathbf{j}$
18. $\mathbf{v} = \mathbf{i} - 3\mathbf{j}, \quad \mathbf{w} = 4\mathbf{i} - \mathbf{j}$
19. **Finding the Actual Speed and Direction of an Aircraft** A DC-10 jumbo jet maintains an airspeed of 550 miles per hour in a southwesterly direction. The velocity of the jet stream is a constant 80 miles per hour from the west. Find the actual speed and direction of the aircraft.

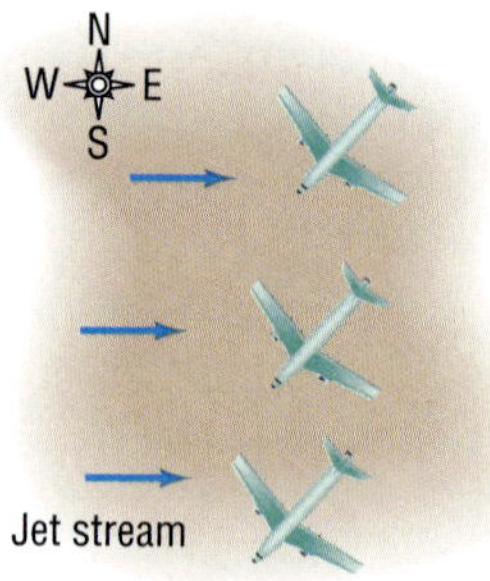

20. **Finding the Correct Compass Heading** The pilot of an aircraft wishes to head directly east but is faced with a wind speed of 40 miles per hour from the northwest. If the pilot maintains an airspeed of 250 miles per hour, what compass heading should be maintained? What is the actual speed of the aircraft?

21. **Correct Direction for Crossing a Stream** A small stream has a constant current of 3 kilometers per hour. At what angle to a boat dock should a motorboat—capable of maintaining a constant speed of 20 kilometers per hour—be headed in order to reach a point directly opposite the dock? If the stream is $\frac{1}{2}$ kilometer wide, how long will it take to cross?

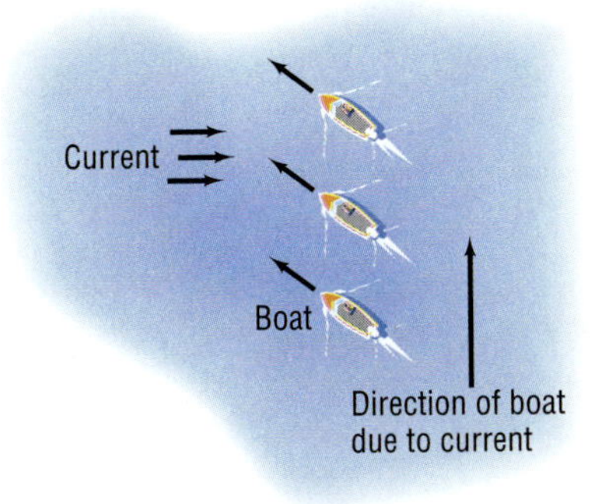

22. **Correct Direction for Crossing a Stream** Repeat Problem 21 if the current is 5 kilometers per hour.

23. **Correct Direction for Crossing a Stream** A river is 500 meters wide and has a current of 1 kilometer per hour. If Sharon can swim at a rate of 2 kilometers per hour, at what angle to the shore should she swim if she wishes to cross the river to a point directly opposite? How long will it take to swim across the river?

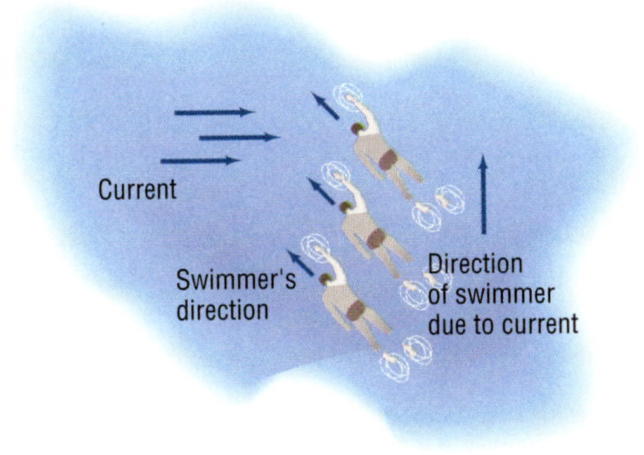

24. An airplane travels 200 miles due west and then 150 miles 60° north of west. Determine the resultant displacement.

25. **Computing Work** Find the work done by a force of 3 pounds acting in the direction $2\mathbf{i} + \mathbf{j}$ in moving an object 2 feet from $(0, 0)$ to $(0, 2)$.

26. **Computing Work** Find the work done by a force of 1 pound acting in the direction $2\mathbf{i} + 2\mathbf{j}$ in moving an object 5 feet from $(0, 0)$ to $(3, 4)$.

27. **Computing Work** A wagon is pulled horizontally by exerting a force of 20 pounds on the handle at an angle of 30° with the horizontal. How much work is done in moving the wagon 100 feet?

28. Find the acute angle that a constant unit force vector makes with the positive x-axis if the work done by the force in moving a particle from $(0, 0)$ to $(4, 0)$ equals 2.

29. Prove the distributive property,

$$\mathbf{u} \cdot (\mathbf{v} + \mathbf{w}) = \mathbf{u} \cdot \mathbf{v} + \mathbf{u} \cdot \mathbf{w}.$$

30. Prove property (5), $\mathbf{0} \cdot \mathbf{v} = 0$.

31. If $\mathbf{v}$ is a unit vector and the angle between $\mathbf{v}$ and $\mathbf{i}$ is α, show that $\mathbf{v} = \cos \alpha \mathbf{i} + \sin \alpha \mathbf{j}$.

32. Suppose that $\mathbf{v}$ and $\mathbf{w}$ are unit vectors. If the angle between $\mathbf{v}$ and $\mathbf{i}$ is α and if the angle between $\mathbf{w}$ and $\mathbf{i}$ is β, use the idea of the dot product $\mathbf{v} \cdot \mathbf{w}$ to prove that

$$\cos(\alpha - \beta) = \cos \alpha \cos \beta + \sin \alpha \sin \beta$$

33. Show that the projection of $\mathbf{v}$ onto $\mathbf{i}$ is $(\mathbf{v} \cdot \mathbf{i})\mathbf{i}$. In fact, show that we can always write a vector $\mathbf{v}$ as

$$\mathbf{v} = (\mathbf{v} \cdot \mathbf{i})\mathbf{i} + (\mathbf{v} \cdot \mathbf{j})\mathbf{j}$$

34. (a) If $\mathbf{u}$ and $\mathbf{v}$ have the same magnitude, then show that $\mathbf{u} + \mathbf{v}$ and $\mathbf{u} - \mathbf{v}$ are orthogonal.
(b) Use this to prove that an angle inscribed in a semicircle is a right angle (see the figure).

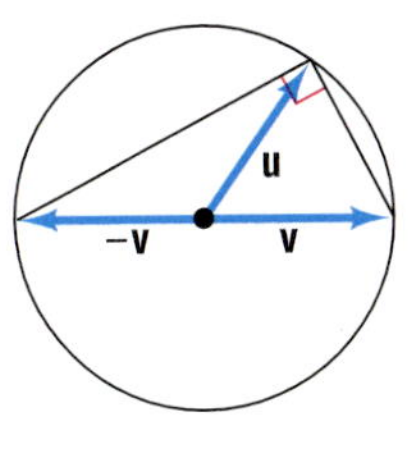

35. Let $\mathbf{v}$ and $\mathbf{w}$ denote two nonzero vectors. Show that the vector $\mathbf{v} - \alpha\mathbf{w}$ is orthogonal to $\mathbf{w}$ if $\alpha = (\mathbf{v} \cdot \mathbf{w})/\|\mathbf{w}\|^2$.

36. Let $\mathbf{v}$ and $\mathbf{w}$ denote two nonzero vectors. Show that the vectors $\|\mathbf{w}\|\mathbf{v} + \|\mathbf{v}\|\mathbf{w}$ and $\|\mathbf{w}\|\mathbf{v} - \|\mathbf{v}\|\mathbf{w}$ are orthogonal.

37. In the definition of work given in this section, what is the work done if $\mathbf{F}$ is orthogonal to $\overrightarrow{AB}$?

38. Prove the **polarization identity**

$$\|\mathbf{u} + \mathbf{v}\|^2 - \|\mathbf{u} - \mathbf{v}\|^2 = 4(\mathbf{u} \cdot \mathbf{v}).$$

39. Make up an application different from any found in the text that requires the dot product.

CHAPTER REVIEW

THINGS TO KNOW

Relationship between polar coordinates (r, θ) and rectangular coordinates (x, y)	$x = r\cos\theta,\ y = r\sin\theta$ $x^2 + y^2 = r^2,\ \tan\theta = \dfrac{y}{x},\ x \neq 0$
Polar form of a complex number	If $z = x + iy$, then $z = r(\cos\theta + i\sin\theta)$, where $r = \lvert z \rvert = \sqrt{x^2 + y^2},\ \sin\theta = \dfrac{y}{r},\ \cos\theta = \dfrac{x}{r},\ 0 \le \theta < 2\pi$
Demoivre's Theorem	If $z = r(\cos\theta + i\sin\theta)$, then $z^n = r^n(\cos n\theta + i\sin n\theta)$, where $n \ge 1$ is a positive integer
nth root of a complex number	$\sqrt[n]{z} = \sqrt[n]{r}\left[\cos\left(\dfrac{\theta}{n} + \dfrac{2k\pi}{n}\right) + i\sin\left(\dfrac{\theta}{n} + \dfrac{2k\pi}{n}\right)\right],\ k = 0, \ldots, n-1$
Vector	Quantity having magnitude and direction; equivalent to a directed line segment $\overrightarrow{PQ}$
Position vector	Vector whose initial point is at the origin
Unit vector	Vector whose magnitude is 1
Dot product	If $\mathbf{v} = a_1\mathbf{i} + b_1\mathbf{j}$ and $\mathbf{w} = a_2\mathbf{i} + b_2\mathbf{j}$, then $\mathbf{v} \cdot \mathbf{w} = a_1a_2 + b_1b_2$.
Angle θ between two nonzero vectors **u** and **v**	$\cos\theta = \dfrac{\mathbf{u} \cdot \mathbf{v}}{\lVert\mathbf{u}\rVert\,\lVert\mathbf{v}\rVert}$

HOW TO

- Plot polar coordinates
- Convert from polar to rectangular coordinates
- Convert from rectangular to polar coordinates
- Graph polar equations by hand and by using a graphing utility (see Table 7)
- Write a complex number in polar form, $z = r(\cos\theta + i\sin\theta),\ 0° \le \theta < 360°$
- Use Demoivre's Theorem to find powers of complex numbers
- Find the nth roots of a complex number
- Add and subtract vectors
- Form scalar multiples of vectors
- Find the magnitude of a vector
- Solve problems involving vectors
- Find the dot product of two vectors
- Find the angle between two vectors
- Determine whether two vectors are parallel
- Determine whether two vectors are orthogonal
- Find the vector projection of **v** onto **w**

FILL-IN-THE-BLANK ITEMS

1. In polar coordinates, the origin is called the ______________, and the positive x-axis is referred to as the ______________ ______________.

2. Another representation in polar coordinates for the point $(2, \pi/3)$ is (______________, $4\pi/3$).

3. Using polar coordinates (r, θ), the circle $x^2 + y^2 = 2x$ takes the form ______________.

4. In a polar equation, replace θ by $-\theta$. If an equivalent equation results, the graph is symmetric with respect to ______________ ______________.

5. When a complex number z is written in the polar form $z = r(\cos\theta + i\sin\theta)$, the nonnegative number r is the ______________ ______________ of z, and the angle θ, $0 \le \theta < 2\pi$, is the ______________ of z.

6. A vector whose magnitude is 1 is called a(n) ______________ vector.

7. If the angle between two vectors **v** and **w** is $\pi/2$, then the dot product $\mathbf{v} \cdot \mathbf{w}$ equals ______________.

TRUE/FALSE ITEMS

T F 1. The polar coordinates of a point are unique.

T F 2. The rectangular coordinates of a point are unique.

T F 3. The tests for symmetry in polar coordinates are conclusive.

T F 4. Demoivre's Theorem is useful for raising a complex number to a positive integer power.

T F 5. Vectors are quantities that have magnitude and direction.

T F 6. Force is a physical example of a vector.

T F 7. If **u** and **v** are orthogonal vectors, then $\mathbf{u} \cdot \mathbf{v} = 0$.

REVIEW EXERCISES

In Problems 1–6, plot each point given in polar coordinates, and find its rectangular coordinates.

1. $(3, \pi/6)$ **2.** $(4, 2\pi/3)$ **3.** $(-2, 4\pi/3)$

4. $(-1, 5\pi/4)$ **5.** $(-3, -\pi/2)$ **6.** $(-4, -\pi/4)$

In Problems 7–12, the rectangular coordinates of a point are given. Find two pairs of polar coordinates (r, θ) for each point, one with $r > 0$ and the other with $r < 0$. Express θ in radians.

7. $(-3, 3)$ **8.** $(1, -1)$ **9.** $(0, -2)$

10. $(2, 0)$ **11.** $(3, 4)$ **12.** $(-5, 12)$

In Problems 13–18, the letters x and y represent rectangular coordinates. Write each equation using polar coordinates (r, θ).

13. $3x^2 + 3y^2 = 6y$ **14.** $2x^2 - 2y^2 = 5y$ **15.** $2x^2 - y^2 = \dfrac{y}{x}$

16. $x^2 + 2y^2 = \dfrac{y}{x}$ **17.** $x(x^2 + y^2) = 4$ **18.** $y(x^2 - y^2) = 3$

In Problems 19–24, write each polar equation as an equation in rectangular coordinates (x, y).

19. $r = 2 \sin \theta$ **20.** $3r = \sin \theta$ **21.** $r = 5$

22. $\theta = \pi/4$ **23.** $r \cos \theta + 3r \sin \theta = 6$ **24.** $r^2 \tan \theta = 1$

In Problems 25–30, sketch the graph of each polar equation by hand. Be sure to test for symmetry. Verify your results using a graphing utility.

25. $r = 4 \cos \theta$ **26.** $r = 3 \sin \theta$ **27.** $r = 3 - 3 \sin \theta$

28. $r = 2 + \cos \theta$ **29.** $r = 4 - \cos \theta$ **30.** $r = 1 - 2 \sin \theta$

In Problems 31–34, write each complex number in polar form. Express each argument in degrees.

31. $-1 - i$ **32.** $-\sqrt{3} + i$ **33.** $4 - 3i$ **34.** $3 - 2i$

In Problems 35–40, write each complex number in the standard form $a + bi$.

35. $2(\cos 150° + i \sin 150°)$ **36.** $3(\cos 60° + i \sin 60°)$ **37.** $3\left(\cos \dfrac{2\pi}{3} + i \sin \dfrac{2\pi}{3}\right)$

38. $4\left(\cos \dfrac{3\pi}{4} + i \sin \dfrac{3\pi}{4}\right)$ **39.** $0.1(\cos 350° + i \sin 350°)$ **40.** $0.5(\cos 160° + i \sin 160°)$

In Problems 41–46, find zw and z/w. Leave your answers in polar form.

41. $z = \cos 80° + i \sin 80°$
$w = \cos 50° + i \sin 50°$

42. $z = \cos 205° + i \sin 205°$
$w = \cos 85° + i \sin 85°$

43. $z = 3\left(\cos \dfrac{9\pi}{5} + i \sin \dfrac{9\pi}{5}\right)$
$w = 2\left(\cos \dfrac{\pi}{5} + i \sin \dfrac{\pi}{5}\right)$

44. $z = 2\left(\cos\frac{5\pi}{3} + i\sin\frac{5\pi}{3}\right)$
$w = 3\left(\cos\frac{\pi}{3} + i\sin\frac{\pi}{3}\right)$

45. $z = 5(\cos 10° + i\sin 10°)$
$w = \cos 355° + i\sin 355°$

46. $z = 4(\cos 50° + i\sin 50°)$
$w = \cos 340° + i\sin 340°$

In Problems 47–54, write each expression in the standard form $a + bi$.

47. $[3(\cos 20° + i\sin 20°)]^3$

48. $[2(\cos 50° + i\sin 50°)]^3$

49. $\left[\sqrt{2}\left(\cos\frac{5\pi}{8} + i\sin\frac{5\pi}{8}\right)\right]^4$

50. $\left[2\left(\cos\frac{5\pi}{16} + i\sin\frac{5\pi}{16}\right)\right]^4$

51. $(1 - \sqrt{3}i)^6$

52. $(2 - 2i)^8$

53. $(3 + 4i)^4$

54. $(1 - 2i)^4$

55. Find all the complex cube roots of 27.

56. Find all the complex fourth roots of -16.

In Problems 57–60, the vector $\mathbf{v}$ *is represented by the directed line segment* $\overrightarrow{PQ}$. *Write* $\mathbf{v}$ *in the form* $a\mathbf{i} + b\mathbf{j}$ *and find* $\|\mathbf{v}\|$.

57. $P = (1, -2); Q = (3, -6)$

58. $P = (-3, 1); Q = (4, -2)$

59. $P = (0, -2); Q = (-1, 1)$

60. $P = (3, -4); Q = (-2, 0)$

In Problems 61–68, use the vectors $\mathbf{v} = -2\mathbf{i} + \mathbf{j}$ *and* $\mathbf{w} = 4\mathbf{i} - 3\mathbf{j}$.

61. Find $4\mathbf{v} - 3\mathbf{w}$.

62. Find $-\mathbf{v} + 2\mathbf{w}$.

63. Find $\|\mathbf{v}\|$.

64. Find $\|\mathbf{v} + \mathbf{w}\|$.

65. Find $\|\mathbf{v}\| + \|\mathbf{w}\|$.

66. Find $\|2\mathbf{v}\| - 3\|\mathbf{w}\|$.

67. Find a unit vector having the same direction as $\mathbf{v}$.

68. Find a unit vector having the opposite direction of $\mathbf{w}$.

In Problems 69–72, find the dot product $\mathbf{v} \cdot \mathbf{w}$ *and the cosine of the angle between* $\mathbf{v}$ *and* $\mathbf{w}$.

69. $\mathbf{v} = -2\mathbf{i} + \mathbf{j}, \mathbf{w} = 4\mathbf{i} - 3\mathbf{j}$

70. $\mathbf{v} = 3\mathbf{i} - \mathbf{j}, \mathbf{w} = \mathbf{i} + \mathbf{j}$

71. $\mathbf{v} = \mathbf{i} - 3\mathbf{j}, \mathbf{w} = -\mathbf{i} + \mathbf{j}$

72. $\mathbf{v} = \mathbf{i} + 4\mathbf{j}, \mathbf{w} = 3\mathbf{i} - 2\mathbf{j}$

73. Find the vector projection of $\mathbf{v} = 2\mathbf{i} + 3\mathbf{j}$ onto $\mathbf{w} = 3\mathbf{i} + \mathbf{j}$.

74. Find the vector projection of $\mathbf{v} = -\mathbf{i} + 2\mathbf{j}$ onto $\mathbf{w} = 3\mathbf{i} - \mathbf{j}$.

75. Find the angle between the vectors $\mathbf{v} = 3\mathbf{i} - 4\mathbf{j}$ and $\mathbf{w} = 12\mathbf{i} - 5\mathbf{j}$.

76. Find the angle between the vectors $\mathbf{v} = \mathbf{i} - \mathbf{j}$ and $\mathbf{w} = 2\mathbf{i} + \mathbf{j}$.

77. Actual Speed and Direction of a Swimmer A swimmer can maintain a constant speed of 5 miles per hour. If the swimmer heads directly across a river that has a current moving at the rate of 2 miles per hour, what is the actual speed of the swimmer? (See the figure.) If the river is 1 mile wide, how far downstream will the swimmer end up from the point directly across the river?

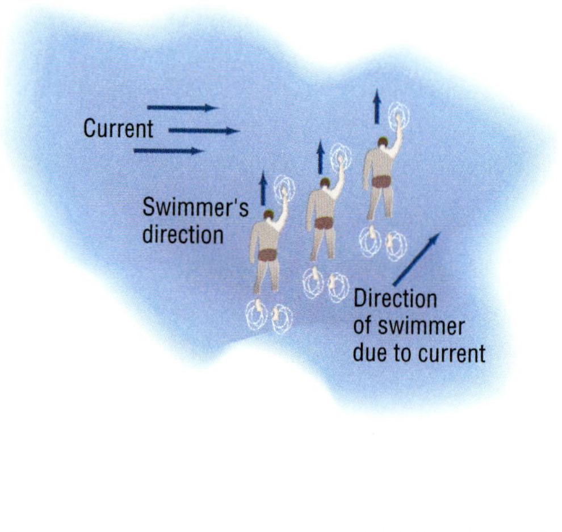

78. Actual Speed and Direction of a Motorboat A small motorboat is moving at a true speed of 11 miles per hour in a southerly direction. The current is known to be from the northeast at 3 miles per hour. What is the speed of the motorboat relative to the water? In what direction does the compass indicate that the boat is headed?

79. Correct Direction for Crossing a Stream A stream 1 kilometer wide has a constant current of 5 kilometers per hour. At what angle to the shore should a person head a boat that is capable of maintaining a constant speed of 15 kilometers per hour in order to reach a point directly opposite?

80. Actual Speed and Direction of an Airplane An airplane has an airspeed of 500 kilometers per hour in a northerly direction. The wind velocity is 60 kilometers per hour in a southeasterly direction. Find the actual speed and direction of the plane relative to the ground.

CHAPTER 10

Analytic Geometry

Many computer users enjoy collecting and putting on their computer the latest "screen savers". As a creative entrepreneur, you have come up with an idea for a new piece of software. You will need to use the Sullivan website at:

www.prenhall.com/sullivan

to find the computer programs and information to build your screen saver.

PREPARING FOR THIS CHAPTER

Before getting started on this chapter, review the following concepts:

Distance Formula *(p. 17)*
Completing the Square *(Appendix, Section 5)*
Intercepts *(pp. 36–40)*
Symmetry *(pp. 40–43)*
Circles *(pp. 66–69)*
Double-Angle and Half-Angle Formulas *(Section 7.3)*
Polar Coordinates *(Section 9.1)*
Amplitude and Period of Sinusoidal Graphs *(p. 612)*

OUTLINE

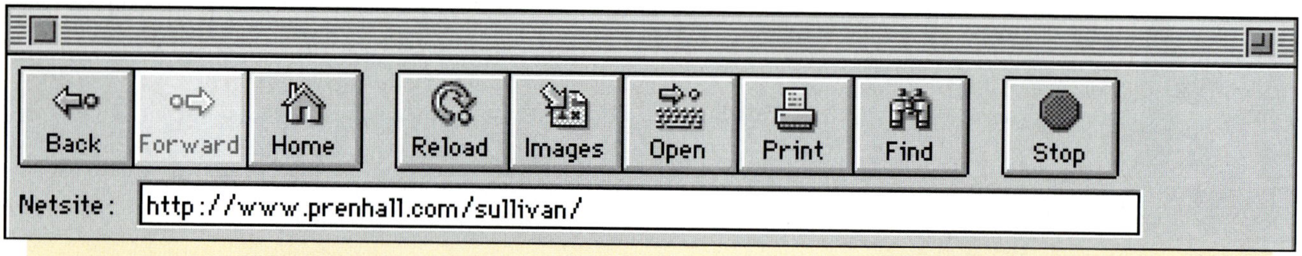

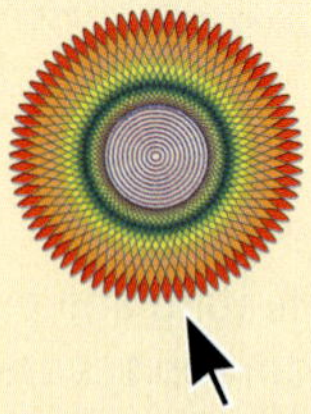

WHEELS OF FIRE

You have a great idea for a computer screen saver—an electronic spirograph! You are certain that you could sell such a hot piece of software. The great thing about making a screen saver spirograph is that you will be working with light, color blending, and high-resolution images. You let your imagination go . . .

1. In order to produce your screen saver you first want to know if there are equations for a *spirograph*. If there are such equations, how could you tweak them to produce special effects? Can you make a spirograph on the surface of a 3-D object like a ball or a donut? Can you make 3-D spirographs like some wild celtic knot? How would you add shading?

   By changing the lengths of the arms and the gear ratios, the spirograph can produce many different patterns. For example you could generate a *trochoid.* In this lesson we will use the principles of analytic geometry to compose a collection of new curves, called **Roulettes.**
2. You need to understand how these equations work in order to adapt them to your screen saver, so you try to take them apart. Experiment by graphing the following equations for various values of R and r.

$$x = (R + r)\cos(t)$$
$$y = (R + r)\sin(t)$$

3. What kinds of designs can be made with a *spirograph*? How do the symbolic equations simulate the patterns made with the toy? Experiment by graphing the following equations for various values of R, r, and O.

$$x = (R + r)\cos(t) - (r + O)\cos(((R + r)/r)\,t)$$
$$y = (R + r)\sin(t) - (r + O)\sin(((R + r)/r)\,t)$$

4. What would happen if you were to change the equations to:

$$x = A\cos(t) - (r + O)\cos(((R + r)/r)\,t)$$
$$y = B\sin(t) - (r + O)\sin(((R + r)/r)\,t)$$

   where A is not equal to B? What does the graph look like?
5. Suppose you wanted to change the Roulette base curve to an asteroid. That is, now you want to roll the circle on a star instead of another circle. Write the equations for your "Astrograph."
6. By now you should start to see that the key idea is simply the composition of parametric equations. Once you have this idea then you are ready to play. You want to create a set of great looking designs for the advertisement of your screen saver. Start with a base curve that you like, from The MacTutor Famous Curves Index, and then try adding two motions. Write the equations and discuss the possible types of patterns that you might expect to see. Does this give you any further designs?
7. Could you extend these curves into surfaces in 3-D? Can you think of a better name than Wheels of Fire?

Historically, Apollonius (200 B.C.) was among the first to study *conics* and discover some of their interesting properties. Today, conics are still studied because of their many uses. *Paraboloids of revolution* (parabolas rotated about their axes of symmetry) are used as signal collectors (the satellite dishes used with radar and cable

TV, for example), as solar energy collectors, and as reflectors (telescopes, light projection, and so on). The planets circle the Sun in approximately *elliptical* orbits. Elliptical surfaces can be used to reflect signals such as light and sound from one place to another. And *hyperbolas* can be used to determine the positions of ships at sea.

The Greeks used the methods of Euclidean geometry to study conics. We shall use the more powerful methods of analytic geometry, bringing to bear both algebra and geometry, for our study of conics. Thus, we shall give a geometric description of each conic, and then, using rectangular coordinates and the distance formula, we shall find equations that represent conics. We used this same development, you may recall, when we first defined a circle in Section 1.5.

The chapter concludes with a section on equations of conics in polar coordinates, followed by a discussion of plane curves and parametric equations.

10.1 CONICS

1 Know the names of the conics

1 The word *conic* derives from the word *cone,* which is a geometric figure that can be constructed in the following way: Let a and g be two distinct lines that intersect at a point V. Keep the line a fixed. Now rotate the line g about a while maintaining the same angle between a and g. The collection of points swept out (generated) by the line g is called a (**right circular**) **cone.** See Figure 1. The fixed line a is called the **axis** of the cone; the point V is called its **vertex;** the lines that pass through V and make the same angle with a as g are called **generators** of the cone. Thus, each generator is a line that lies entirely on the cone. The cone consists of two parts, called **nappes,** that intersect at the vertex.

FIGURE 1

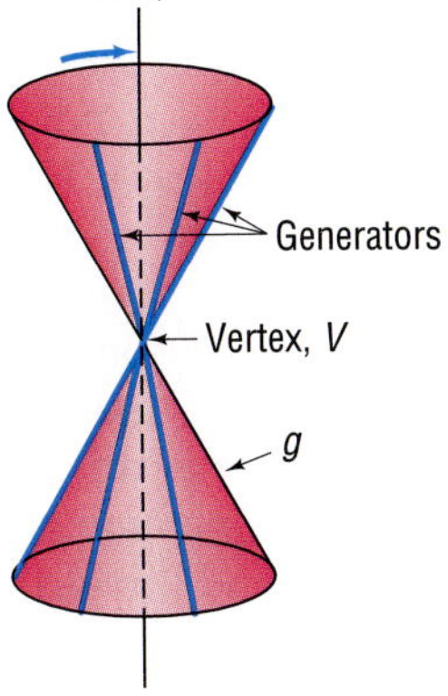

Conics, an abbreviation for **conic sections,** are curves that result from the intersection of a (right circular) cone and a plane. The conics we shall study arise when the plane does not contain the vertex, as shown in Figure 2. These conics are **circles** when the plane is perpendicular to the axis of the cone and intersects each generator; **ellipses** when the plane is tilted slightly so that it intersects each generator, but intersects only one nappe of the cone; **parabolas** when the plane is tilted further so that it is parallel to one (and only one) generator and intersects only one nappe of the cone; and **hyperbolas** when the plane intersects both nappes.

If the plane does contain the vertex, the intersection of the plane and the cone is a point, a line, or a pair of intersecting lines. These are usually called **degenerate conics.**

FIGURE 2

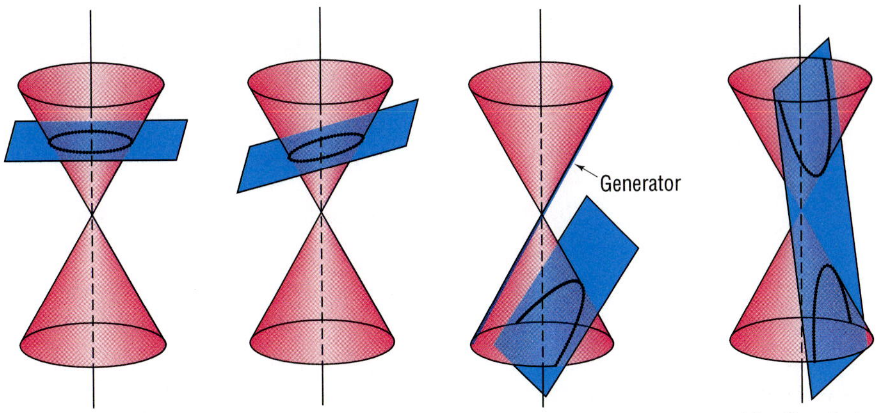

(a) Circle **(b)** Ellipse **(c)** Parabola **(d)** Hyperbola

10.2 THE PARABOLA

1 Find the Equation of a Parabola
2 Graph Parabolas
3 Discuss the Equation of a Parabola
4 Work with Parabolas with Vertex at (h, k)

We stated earlier (Section 4.1) that the graph of a quadratic function is a parabola. In this section, we begin with a geometric definition of a parabola and use it to obtain an equation.

A **parabola** is defined as the collection of all points P in the plane that are the same distance from a fixed point F as they are from a fixed line D. The point F is called the **focus** of the parabola, and the line D is its **directrix**. As a result, a parabola is the set of points P for which

$$d(F, P) = d(P, D) \qquad (1)$$

FIGURE 3

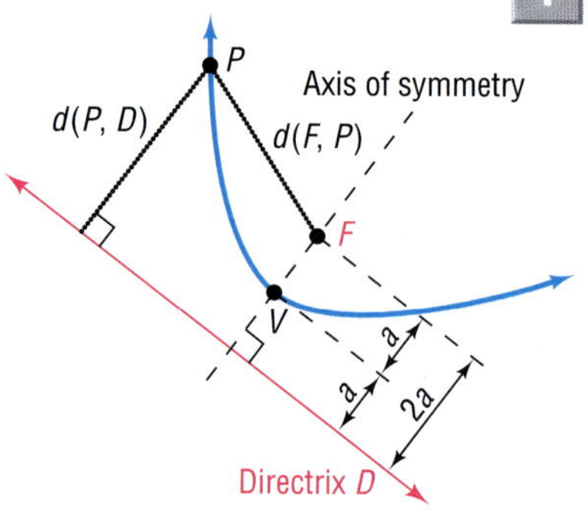

1 Figure 3 shows a parabola. The line through the focus F and perpendicular to the directrix D is called the **axis of symmetry** of the parabola. The point of intersection of the parabola with its axis of symmetry is called the **vertex** V.

Because the vertex V lies on the parabola, it must satisfy equation (1): $d(F, V) = d(V, D)$. Thus, the vertex is midway between the focus and the directrix. We shall let a equal the distance $d(F, V)$ from F to V. Now we are ready to derive an equation for a parabola. To do this, we use a rectangular system of coordinates, positioned so that the vertex V, focus F, and directrix D of the parabola are conveniently located. If we choose to locate the vertex V at the origin $(0, 0)$, then we can conveniently position the focus F on either the x-axis or the y-axis.

FIGURE 4
$y^2 = 4ax$

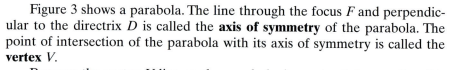

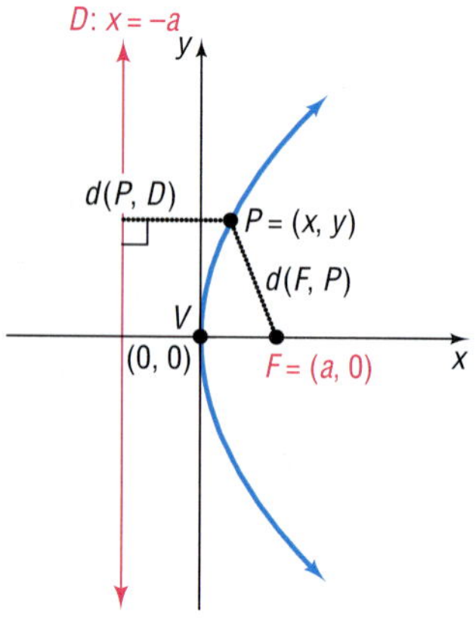

First, we consider the case where the focus F is on the positive x-axis, as shown in Figure 4. Because the distance from F to V is a, the coordinates of F will be $(a, 0)$ with $a > 0$. Similarly, because the distance from V to the directrix D is also a and because D must be perpendicular to the x-axis (since the x-axis is the axis of symmetry), the equation of the directrix D must be $x = -a$. Now, if $P = (x, y)$ is any point on the parabola, then P must obey equation (1):

$$d(F, P) = d(P, D)$$

So, we have

$$\begin{aligned} \sqrt{(x-a)^2 + y^2} &= |x + a| && \text{Use the distance formula.} \\ (x-a)^2 + y^2 &= (x+a)^2 && \text{Square both sides.} \\ x^2 - 2ax + a^2 + y^2 &= x^2 + 2ax + a^2 \\ y^2 &= 4ax \end{aligned}$$

> **Theorem Equation of a Parabola; Vertex at (0, 0), Focus at $(a, 0)$, $a > 0$**
>
> The equation of a parabola with vertex at (0, 0), focus at $(a, 0)$, and directrix $x = -a$, $a > 0$, is
>
> $$y^2 = 4ax \qquad (2)$$

EXAMPLE 1 Finding the Equation of a Parabola

2

Find an equation of the parabola with vertex at (0, 0) and focus at (3, 0). Graph the equation by hand.

Solution The distance from the vertex (0, 0) to the focus (3, 0) is $a = 3$. Based on equation (2), the equation of this parabola is

$$y^2 = 4ax$$
$$y^2 = 12x \qquad a = 3.$$

To graph this parabola by hand, it is helpful to plot the two points on the graph above and below the focus. To locate them, we let $x = 3$. Then,

$$y^2 = 12x = 36$$
$$y = \pm 6$$

The points on the parabola above and below the focus are $(3, -6)$ and $(3, 6)$. See Figure 5.

FIGURE 5
$y^2 = 12x$

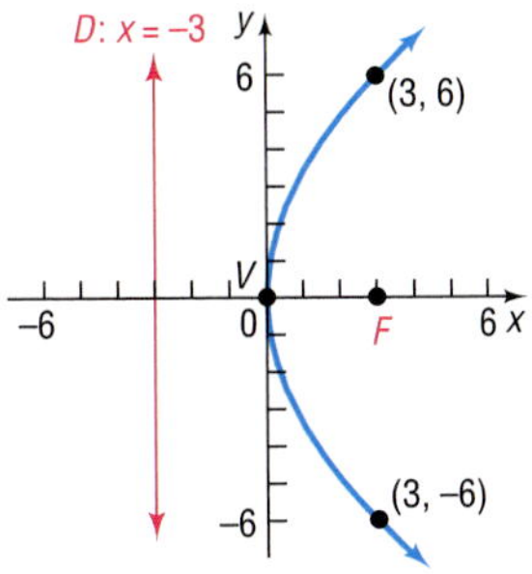

In general, the points on a parabola $y^2 = 4ax$ that lie above and below the focus $(a, 0)$ are each at a distance $2a$ from the focus. This follows from the fact that if $x = a$ then $y^2 = 4ax = 4a^2$, or $y = \pm 2a$. The line segment joining these two points is called the **latus rectum;** its length is $4a$.

Now work Problem 17.

EXAMPLE 2 Graphing a Parabola Using a Graphing Utility

Graph the parabola $y^2 = 12x$.

Solution To graph the parabola $y^2 = 12x$, we need to graph the two functions $Y_1 = \sqrt{12x}$ and $Y_2 = -\sqrt{12x}$ on a square screen. Figure 6 shows the graph of $y^2 = 12x$. Notice that the graph fails the vertical line test, so $y^2 = 12x$ is not a function.

FIGURE 6
$y^2 = 12x$

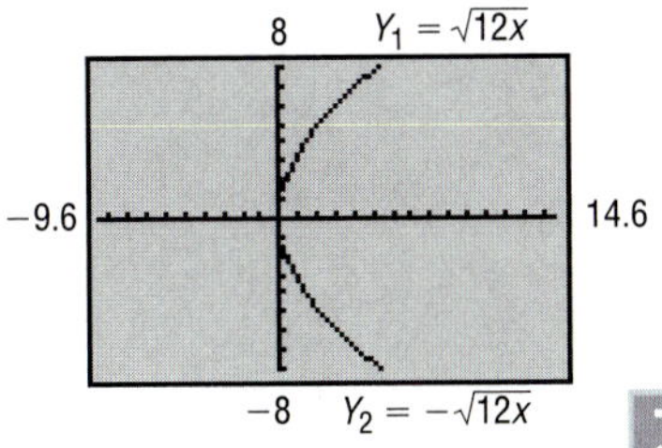

By reversing the steps we used to obtain equation (2), it follows that the graph of an equation of the form of equation (2), $y^2 = 4ax$, is a parabola; its vertex is at (0, 0), its focus is at $(a, 0)$, its directrix is the line $x = -a$, and its axis of symmetry is the x-axis.

For the remainder of this section, the direction "Discuss the equation" will mean to find the vertex, focus, and directrix of the parabola and graph it.

3

EXAMPLE 3 Discussing the Equation of a Parabola

Discuss the equation $y^2 = 8x$.

Solution Figure 7(a) shows the graph of $y^2 = 8x$ using a graphing utility. We now proceed to analyze the equation.

The equation $y^2 = 8x$ is of the form $y^2 = 4ax$, where $4a = 8$. Thus, $a = 2$. Consequently, the graph of the equation is a parabola with vertex at (0, 0) and focus on the positive x-axis at (2, 0). The directrix is the vertical line $x = -2$. The two points defining the latus rectum are obtained by letting $x = 2$. Then $y^2 = 16$, or $y = \pm 4$. These points help in graphing the parabola by hand since they determine the "opening" of the graph. See Figure 7(b).

FIGURE 7
$y^2 = 8x$

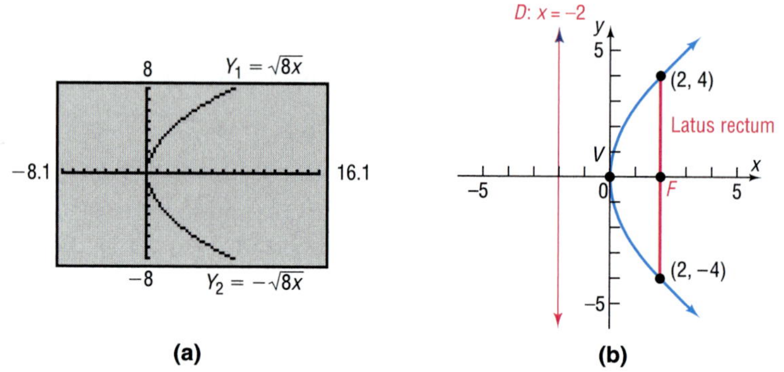

Recall that we arrived at equation (2) after placing the focus on the positive x-axis. If the focus is placed on the negative x-axis, positive y-axis, or negative y-axis, a different form of the equation for the parabola results. The four forms of the equation of a parabola with vertex at (0, 0) and focus on a coordinate axis a distance a from (0, 0) are given in Table 1, and their graphs are given in Figure 8. Notice that each graph is symmetric with respect to its axis of symmetry.

TABLE 1 Equations of a Parabola: Vertex at (0, 0); Focus on Axis; $a > 0$

Vertex	Focus	Directrix	Equation	Description
(0, 0)	$(a, 0)$	$x = -a$	$y^2 = 4ax$	Parabola, axis of symmetry is the x-axis, opens to right
(0, 0)	$(-a, 0)$	$x = a$	$y^2 = -4ax$	Parabola, axis of symmetry is the x-axis, opens to left
(0, 0)	$(0, a)$	$y = -a$	$x^2 = 4ay$	Parabola, axis of symmetry is the y-axis, opens up
(0, 0)	$(0, -a)$	$y = a$	$x^2 = -4ay$	Parabola, axis of symmetry is the y-axis, opens down

FIGURE 8

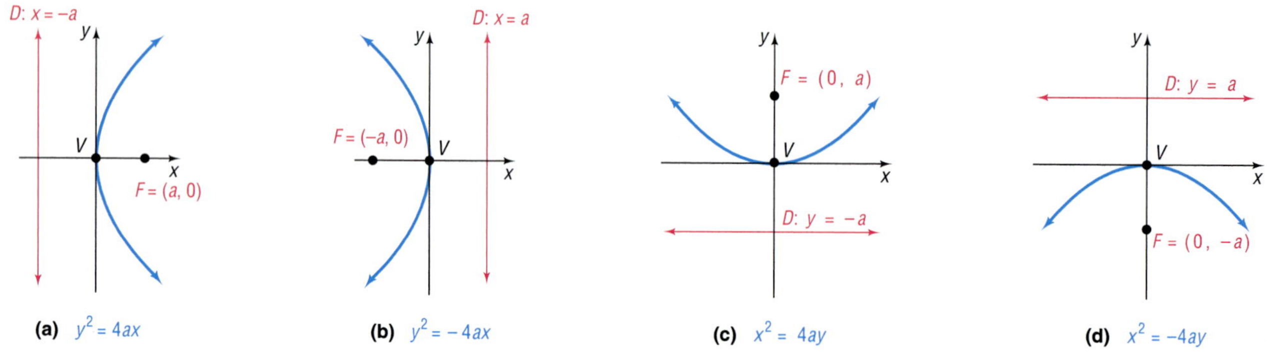

EXAMPLE 4 Discussing the Equation of a Parabola

Discuss the equation $x^2 = -12y$.

Solution Figure 9(a) shows the graph of $x^2 = -12y$ using a graphing utility. We now proceed to analyze the equation.

The equation $x^2 = -12y$ is of the form $x^2 = -4ay$, with $a = 3$. Consequently, the graph of the equation is a parabola with vertex at $(0, 0)$, focus at $(0, -3)$, and directrix the line $y = 3$. The parabola opens down, and its axis of symmetry is the y-axis. To obtain the points defining the latus rectum, let $y = -3$. Then $x^2 = 36$, or $x = \pm 6$. See Figure 9(b).

FIGURE 9 $x^2 = -12y$

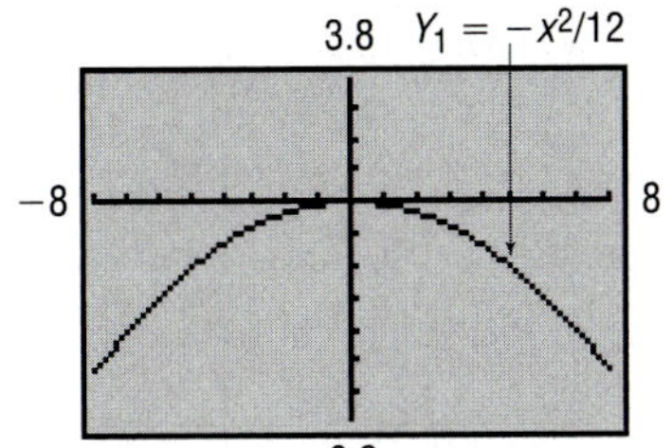

(a)

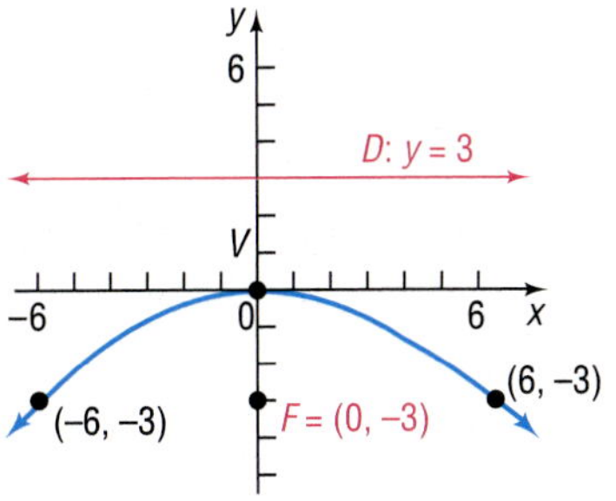

(b)

Now work Problem 35.

EXAMPLE 5 Finding the Equation of a Parabola

Find the equation of the parabola with focus at $(0, 4)$ and directrix the line $y = -4$. Graph the equation by hand.

Solution A parabola whose focus is at $(0, 4)$ and whose directrix is the horizontal line $y = -4$ will have its vertex at $(0, 0)$. (Do you see why? The vertex is midway between the focus and the directrix.) Thus, the equation of this parabola is of the form $x^2 = 4ay$, with $a = 4$; that is,

$$x^2 = 16y$$

Figure 10 shows the graph.

FIGURE 10 $x^2 = 16y$

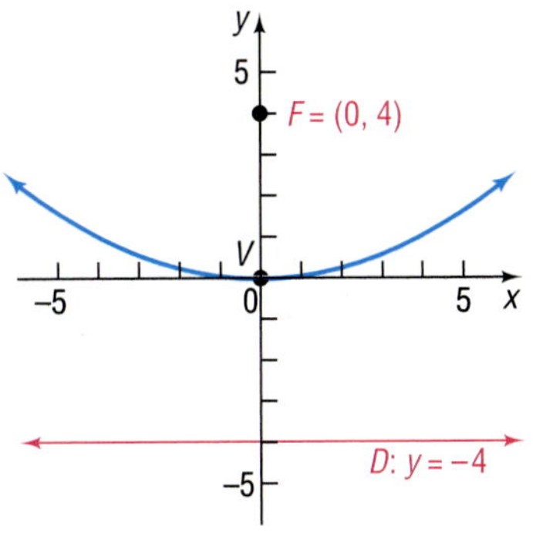

EXAMPLE 6 Finding the Equation of a Parabola

Find the equation of a parabola with vertex at $(0, 0)$ if its axis of symmetry is the x-axis and its graph contains the point $(-\frac{1}{2}, 2)$. Find its focus and directrix, and graph the equation by hand.

Solution Because the vertex is at the origin, the axis of symmetry is the x-axis, and the graph contains a point in the second quadrant, we see from Table 1 that the form of the equation is

$$y^2 = -4ax$$

FIGURE 11
$y^2 = -8x$

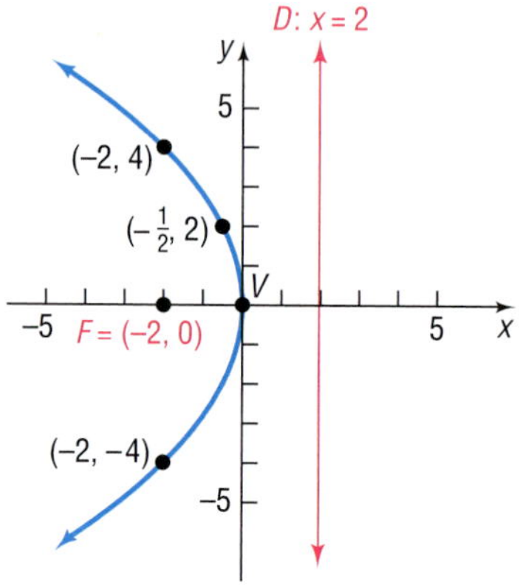

Because the point $(-\frac{1}{2}, 2)$ is on the parabola, the coordinates $x = -\frac{1}{2}$, $y = 2$ must satisfy the equation. Putting $x = -\frac{1}{2}$ and $y = 2$ into the equation, we find

$$4 = -4a(-\tfrac{1}{2})$$
$$a = 2$$

Thus, the equation of the parabola is

$$y^2 = -8x$$

The focus is at $(-2, 0)$ and the directrix is the line $x = 2$. Letting $x = -2$, we find $y^2 = 16$ or $y = \pm 4$. The points $(-2, 4)$ and $(-2, -4)$ define the latus rectum. See Figure 11.

Now work Problem 27.

Vertex at (h, k)

4 If a parabola with vertex at the origin and axis of symmetry along a coordinate axis is shifted horizontally h units and then vertically k units, the result is a parabola with vertex at (h, k) and axis of symmetry parallel to a coordinate axis. The equations of such parabolas have the same forms as those in Table 1, but with x replaced by $x - h$ and y replaced by $y - k$. Table 2 gives the forms of the equations of such parabolas. Figure 12(a)–(d) illustrates the graphs for $h > 0$, $k > 0$.

TABLE 2 Parabolas with Vertex at (h, k), Axis of Symmetry Parallel to a Coordinate Axis, $a > 0$

Vertex	Focus	Directrix	Equation	Description
(h, k)	$(h + a, k)$	$x = -a + h$	$(y - k)^2 = 4a(x - h)$	Parabola, axis of symmetry parallel to x-axis, opens to right
(h, k)	$(h - a, k)$	$x = a + h$	$(y - k)^2 = -4a(x - h)$	Parabola, axis of symmetry parallel to x-axis, opens to left
(h, k)	$(h, k + a)$	$y = -a + k$	$(x - h)^2 = 4a(y - k)$	Parabola, axis of symmetry parallel to y-axis, opens up
(h, k)	$(h, k - a)$	$y = a + k$	$(x - h)^2 = -4a(y - k)$	Parabola, axis of symmetry parallel to y-axis, opens down

FIGURE 12

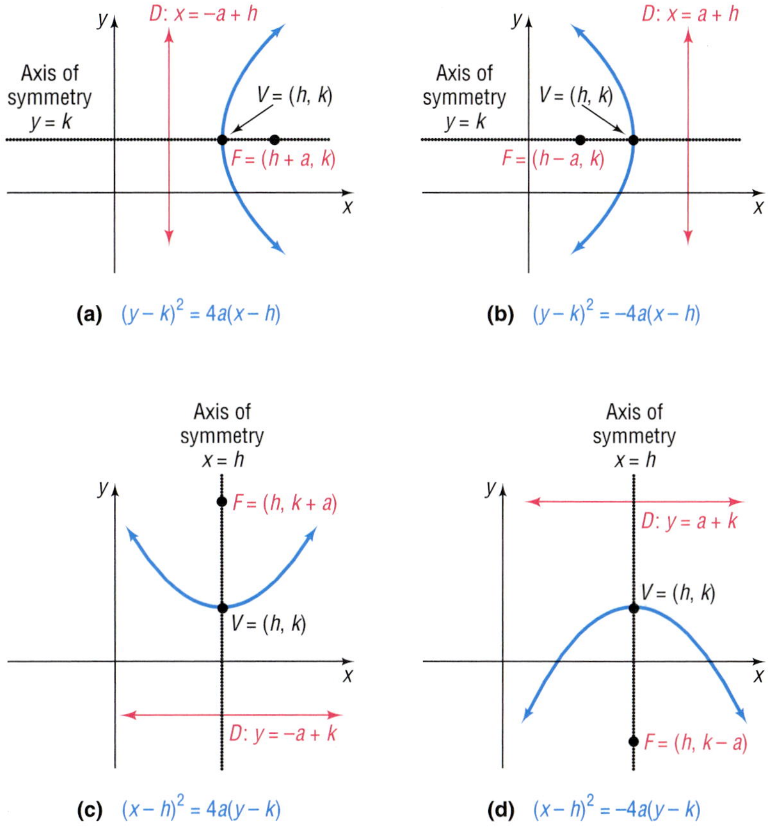

EXAMPLE 7 Finding the Equation of a Parabola, Vertex Not at Origin

Find an equation of the parabola with vertex at $(-2, 3)$ and focus at $(0, 3)$. Graph the equation by hand.

Solution The vertex $(-2, 3)$ and focus $(0, 3)$ both lie on the horizontal line $y = 3$ (the axis of symmetry). The distance a from $(-2, 3)$ to $(0, 3)$, is $a = 2$. Also, because the focus lies to the right of the vertex, we know the parabola opens to the right. Consequently, the form of the equation is

$$(y - k)^2 = 4a(x - h)$$

where $(h, k) = (-2, 3)$ and $a = 2$. Therefore, the equation is

$$(y - 3)^2 = 4 \cdot 2[x - (-2)]$$
$$(y - 3)^2 = 8(x + 2)$$

If $x = 0$, then $(y - 3)^2 = 16$. Thus, $y - 3 = \pm 4$ and $y = -1, y = 7$. The points $(0, -1)$ and $(0, 7)$ define the latus rectum; the line $x = -4$ is the directrix. See Figure 13.

FIGURE 13
$(y - 3)^2 = 8(x + 2)$

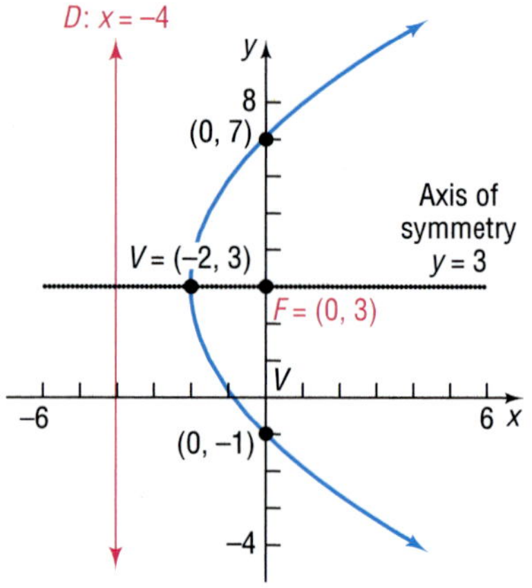

Now work Problem 25.

EXAMPLE 8 Using a Graphing Utility to Graph a Parabola, Vertex Not at Origin

Using a graphing utility, graph the equation $(y - 3)^2 = 8(x + 2)$.

Solution First, we must solve the equation for y.

FIGURE 14

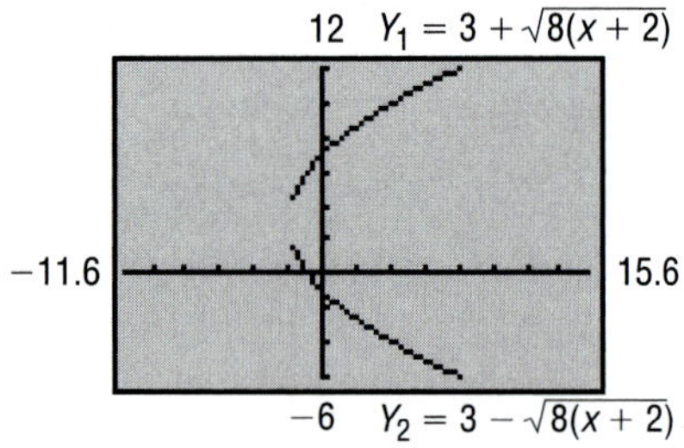

$$(y - 3)^2 = 8(x + 2)$$

$$y - 3 = \pm\sqrt{8(x + 2)} \quad \text{Take the square root of each side.}$$

$$y = 3 \pm \sqrt{8(x + 2)} \quad \text{Solve for } y.$$

Figure 14 shows the graphs of the equations $Y_1 = 3 + \sqrt{8(x + 2)}$ and $Y_2 = 3 - \sqrt{8(x + 2)}$.

Now work Problem 39.

Polynomial equations define parabolas whenever they involve two variables that are quadratic in one variable and linear in the other. To discuss this type of equation, we first complete the square of the quadratic variable.

EXAMPLE 9 Discussing the Equation of a Parabola

Discuss the equation $x^2 + 4x - 4y = 0$.

Solution Figure 15(a) shows the graph of $x^2 + 4x - 4y = 0$ using a graphing utility. We now proceed to analyze the equation.

To discuss the equation $x^2 + 4x - 4y = 0$, we complete the square involving the variable x. Thus,

$$x^2 + 4x - 4y = 0$$

$$x^2 + 4x = 4y \quad \text{Isolate the terms involving } x \text{ on the left side.}$$

$$x^2 + 4x + 4 = 4y + 4 \quad \text{Complete the square on the left side.}$$

$$(x + 2)^2 = 4(y + 1)$$

This equation is of the form $(x - h)^2 = 4a(y - k)$, with $h = -2$, $k = -1$, and $a = 1$. The graph is a parabola with vertex at $(h, k) = (-2, -1)$ that opens up. The focus is at $(-2, 0)$, and the directrix is the line $y = -2$. See Figure 15(b).

FIGURE 15
$x^2 + 4x - 4y = 0$

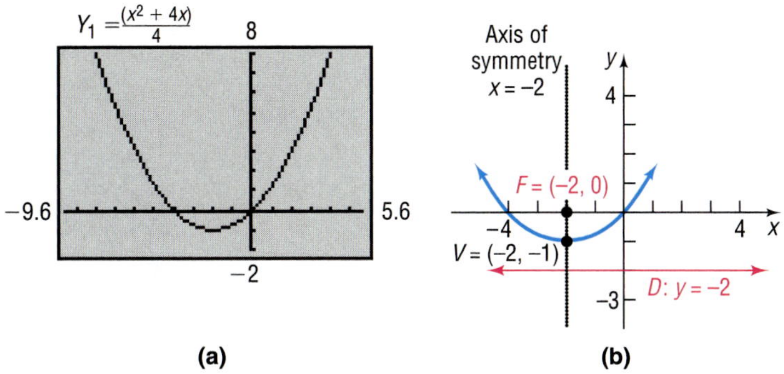

Parabolas find their way into many applications. For example, as we discussed in Section 4.1, suspension bridges have cables in the shape of a parabola. Another property of parabolas that is used in applications is their reflecting property.

Reflecting Property

FIGURE 16
Searchlight

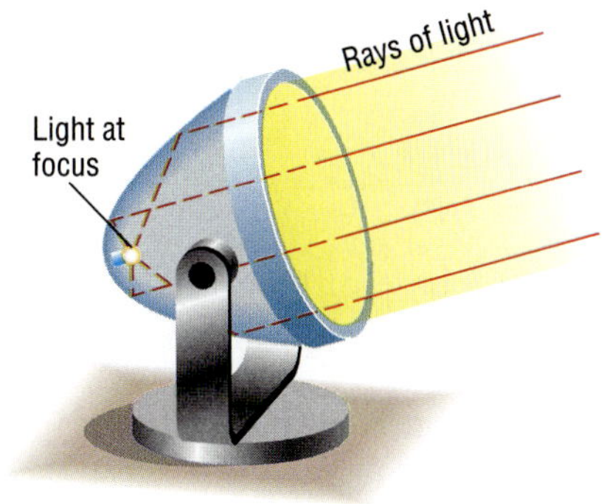

Suppose that a mirror is shaped like a **paraboloid of revolution,** a surface formed by rotating a parabola about its axis of symmetry. If a light (or any other emitting source) is placed at the focus of the parabola, all the rays emanating from the light will reflect off the mirror in lines parallel to the axis of symmetry. This principle is used in the design of searchlights, flashlights, certain automobile headlights, and other such devices. See Figure 16.

Conversely, suppose rays of light (or other signals) emanate from a distant source so that they are essentially parallel. When these rays strike the surface of a parabolic mirror whose axis of symmetry is parallel to these rays, they are reflected to a single point at the focus. This principle is used in the design of some solar energy devices, satellite dishes, and the mirrors used in some types of telescopes. See Figure 17.

FIGURE 17

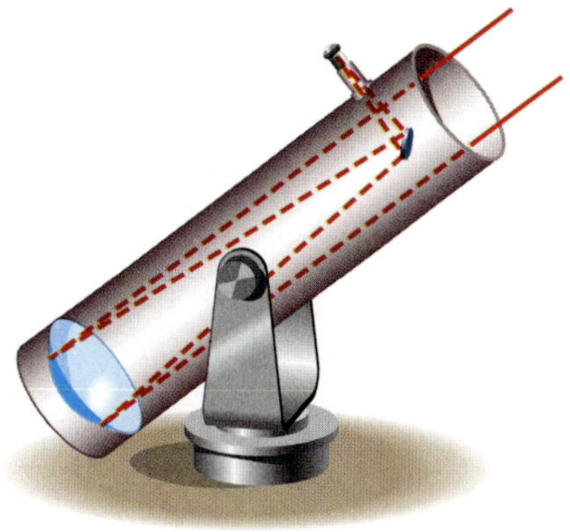

EXAMPLE 10 Satellite Dish

A satellite dish is shaped like a paraboloid of revolution. The signals that emanate from a satellite strike the surface of the dish and are reflected to a single point, where the receiver is located. If the dish is 8 feet across at its opening and is 3 feet deep at its center, at what position should the receiver be placed?

FIGURE 18

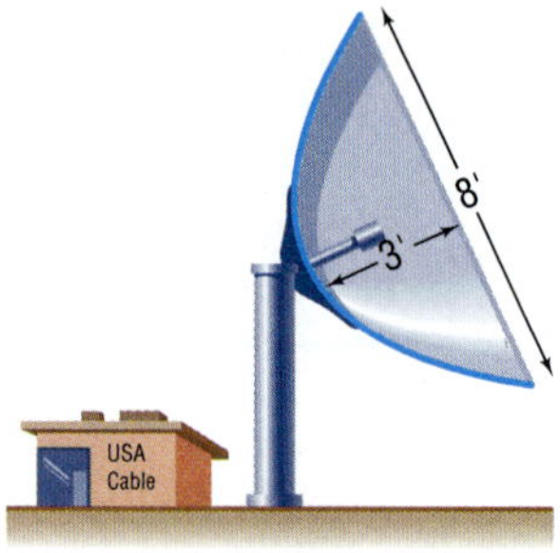

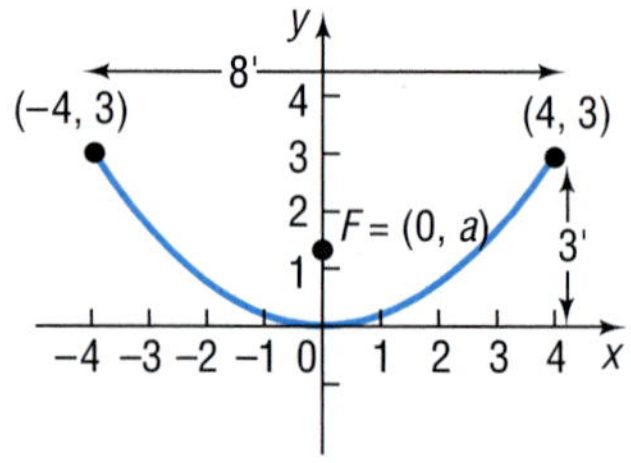

Solution Figure 18(a) shows the satellite dish. We draw the parabola used to form the dish on a rectangular coordinate system so that the vertex of the parabola is at the origin and its focus is on the positive y-axis. See Figure 18(b). The form of the equation of the parabola is

$$x^2 = 4ay$$

and its focus is at $(0, a)$. Since $(4, 3)$ is a point on the graph, we have

$$4^2 = 4a(3)$$

$$a = \frac{4}{3}$$

The receiver should be located $1\frac{1}{3}$ feet from the base of the dish, along its axis of symmetry.

10.2 EXERCISES

In Problems 1–8, the graph of a parabola is given. Match each graph to its equation.

A. $y^2 = 4x$ B. $x^2 = 4y$ C. $y^2 = -4x$

D. $x^2 = -4y$ E. $(y-1)^2 = 4(x-1)$ F. $(x+1)^2 = 4(y+1)$

G. $(y-1)^2 = -4(x-1)$ H. $(x+1)^2 = -4(y+1)$

1.

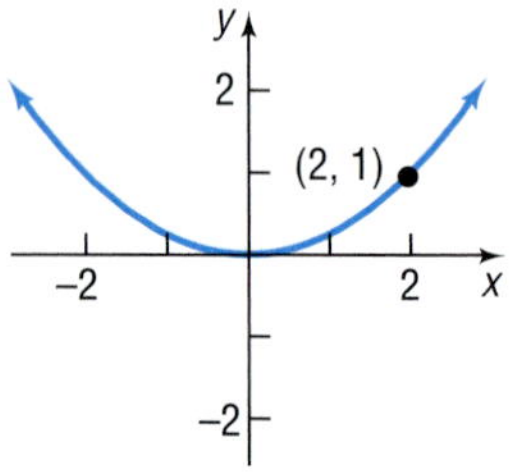

2.

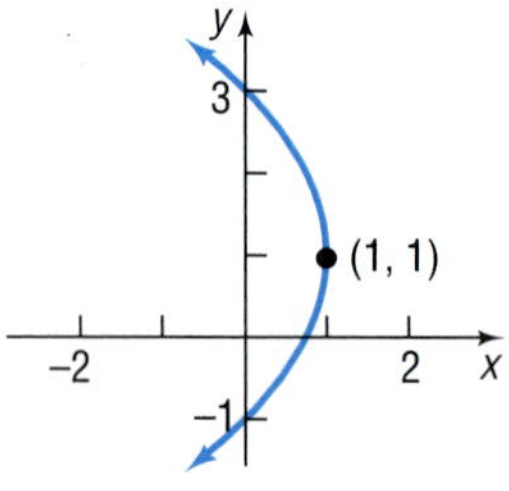

3.

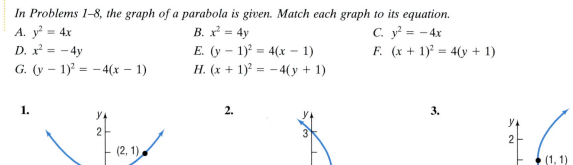

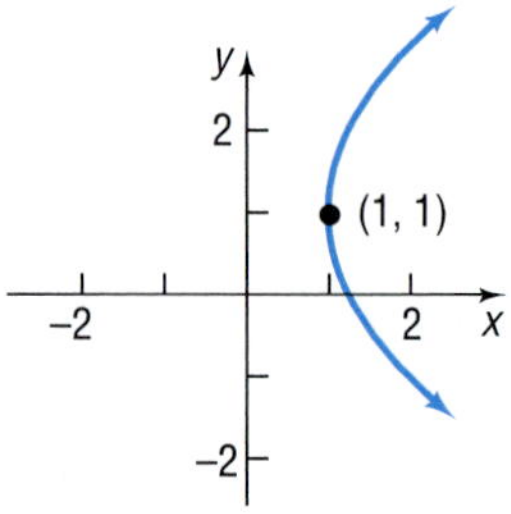

4.

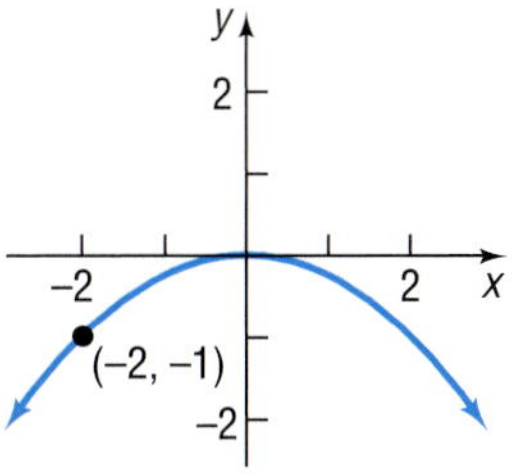

5.

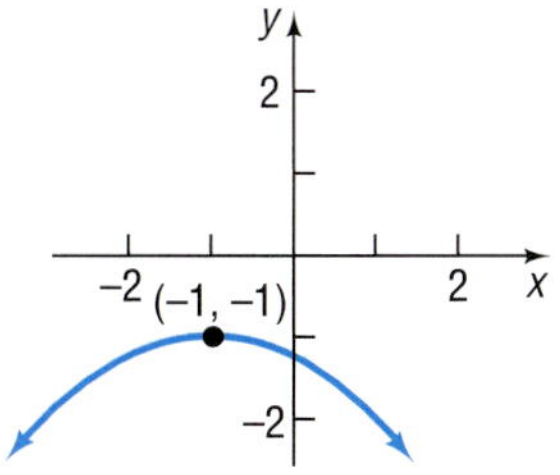

6.

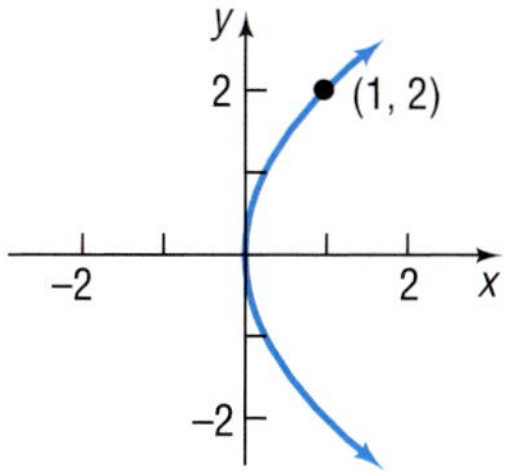

7.

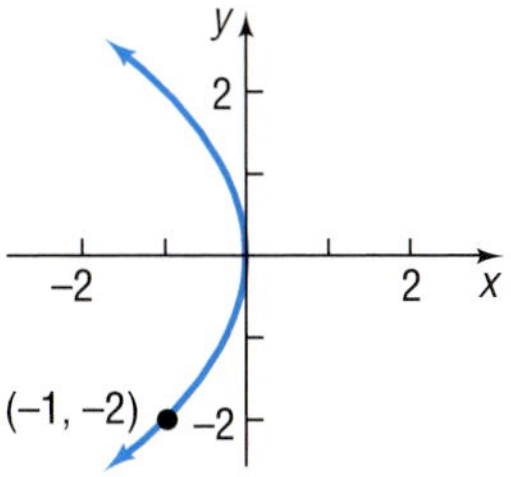

8.

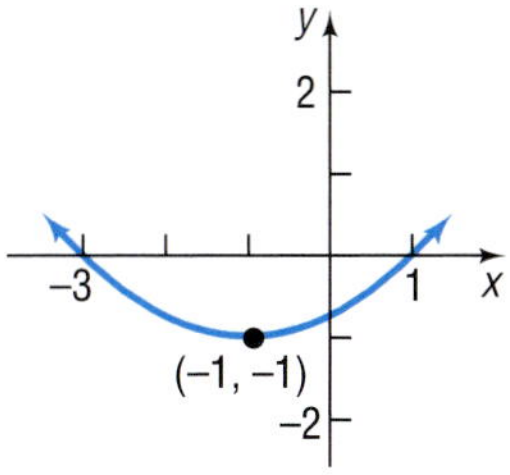

In Problems 9–16, the graph of a parabola is given. Match each graph to its equation.

A. $x^2 = 6y$ *B.* $x^2 = -6y$ *C.* $y^2 = 6x$

D. $y^2 = -6x$ *E.* $(y - 2)^2 = -6(x + 2)$ *F.* $(y - 2)^2 = 6(x + 2)$

G. $(x + 2)^2 = -6(y - 2)$ *H.* $(x + 2)^2 = 6(y - 2)$

9.

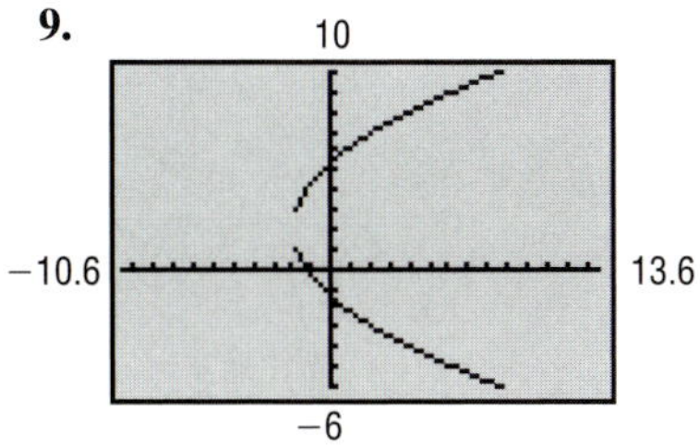

10.

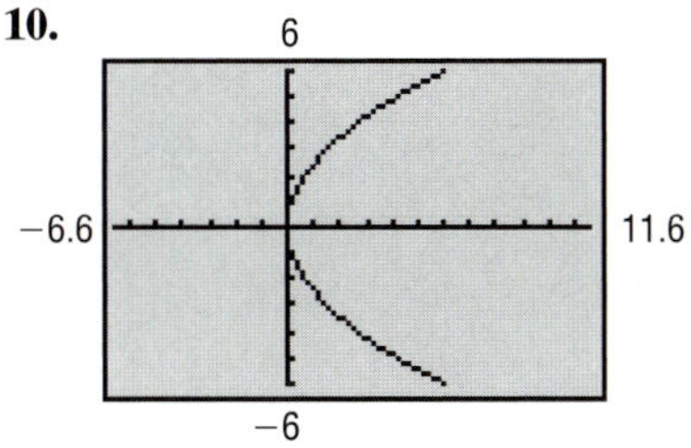

11.

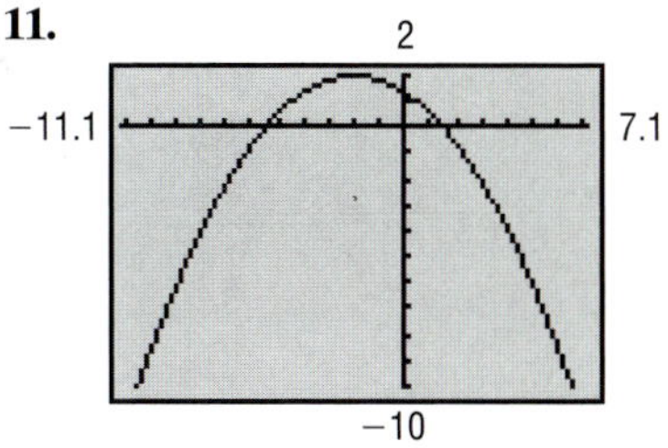

12.

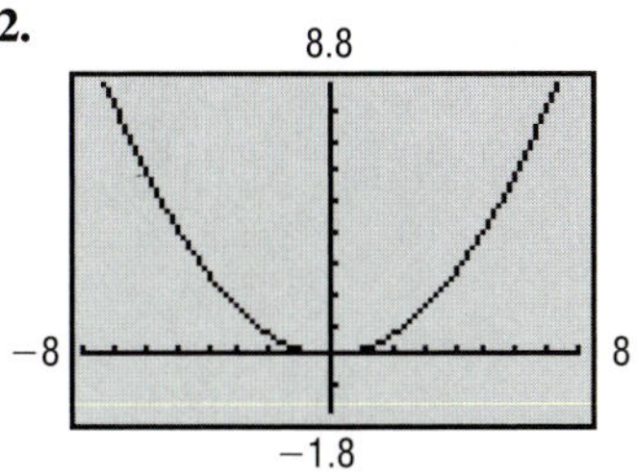

13.

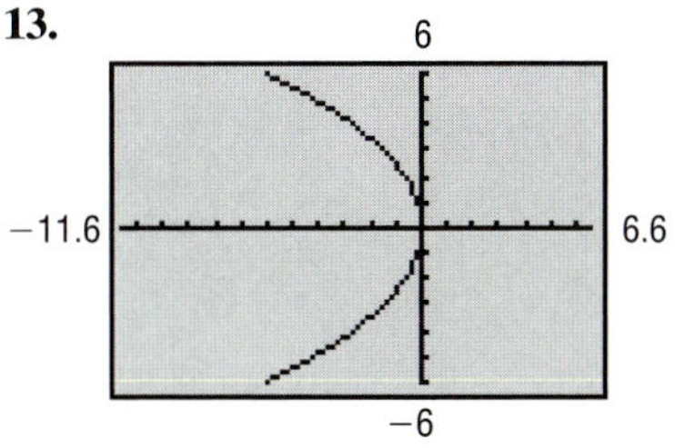

14.

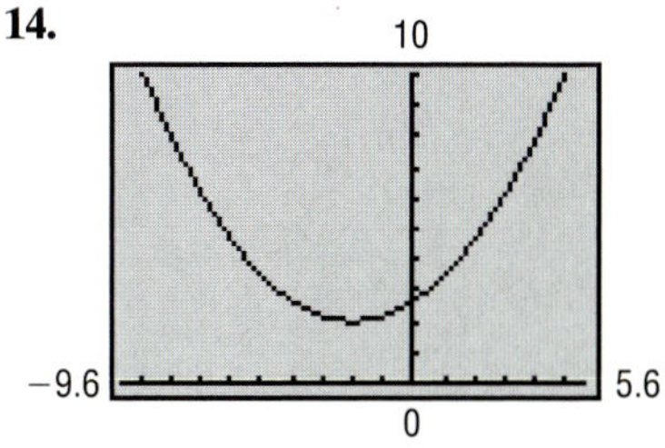

15.

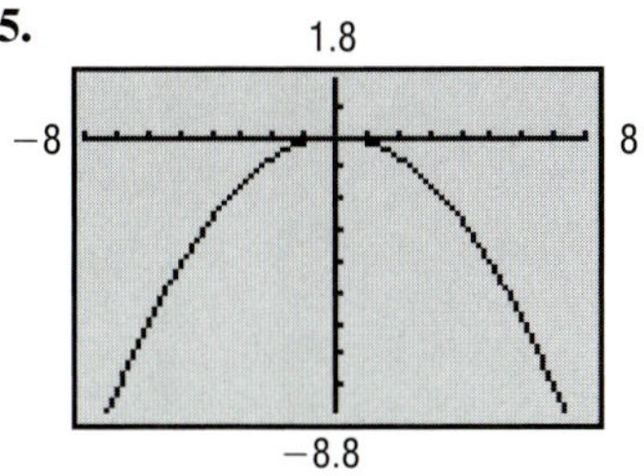

16.

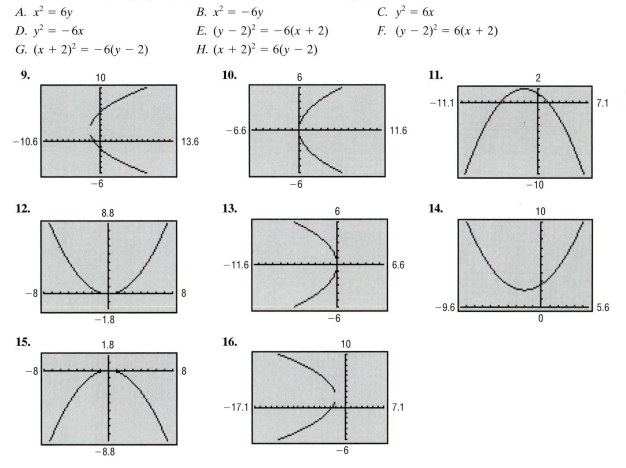

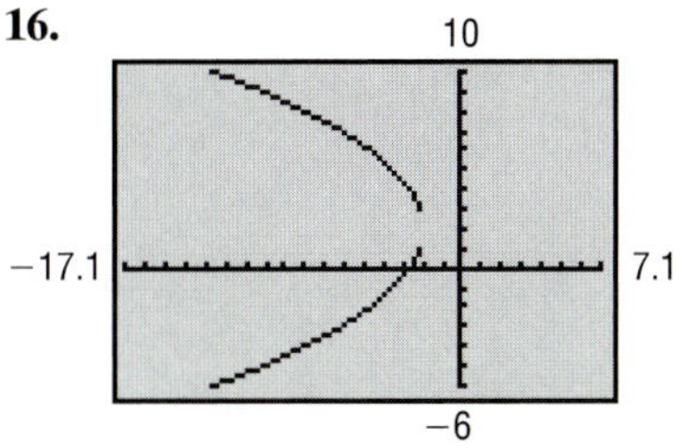

In Problems 17–32, find the equation of the parabola described. Find the two points that define the latus rectum, and graph the equation by hand.

17. Focus at $(4, 0)$; vertex at $(0, 0)$

18. Focus at $(0, 2)$; vertex at $(0, 0)$

19. Focus at $(0, -3)$; vertex at $(0, 0)$

20. Focus at $(-4, 0)$; vertex at $(0, 0)$

21. Focus at $(-2, 0)$; directrix the line $x = 2$

22. Focus at $(0, -1)$; directrix the line $y = 1$

23. Directrix the line $y = -\frac{1}{2}$; vertex at $(0, 0)$

24. Directrix the line $x = -\frac{1}{2}$; vertex at $(0, 0)$

25. Vertex at $(2, -3)$; focus at $(2, -5)$

26. Vertex at $(4, -2)$; focus at $(6, -2)$

27. Vertex at $(0, 0)$; axis of symmetry the y-axis; containing the point $(2, 3)$

28. Vertex at $(0, 0)$; axis of symmetry the x-axis; containing the point $(2, 3)$

29. Focus at $(-3, 4)$; directrix the line $y = 2$

30. Focus at $(2, 4)$; directrix the line $x = -4$

31. Focus at $(-3, -2)$; directrix the line $x = 1$

32. Focus at $(-4, 4)$; directrix the line $y = -2$

In Problems 33–50, find the vertex, focus, and directrix of each parabola. Graph the equation using a graphing utility.

33. $x^2 = 4y$

34. $y^2 = 8x$

35. $y^2 = -16x$

36. $x^2 = -4y$

37. $(y - 2)^2 = 8(x + 1)$

38. $(x + 4)^2 = 16(y + 2)$

39. $(x - 3)^2 = -(y + 1)$

40. $(y + 1)^2 = -4(x - 2)$

41. $(y + 3)^2 = 8(x - 2)$

42. $(x - 2)^2 = 4(y - 3)$

43. $y^2 - 4y + 4x + 4 = 0$

44. $x^2 + 6x - 4y + 1 = 0$

45. $x^2 + 8x = 4y - 8$

46. $y^2 - 2y = 8x - 1$

47. $y^2 + 2y - x = 0$

48. $x^2 - 4x = 2y$

49. $x^2 - 4x = y + 4$

50. $y^2 + 12y = -x + 1$

In Problems 51–58, write an equation for each parabola.

51.

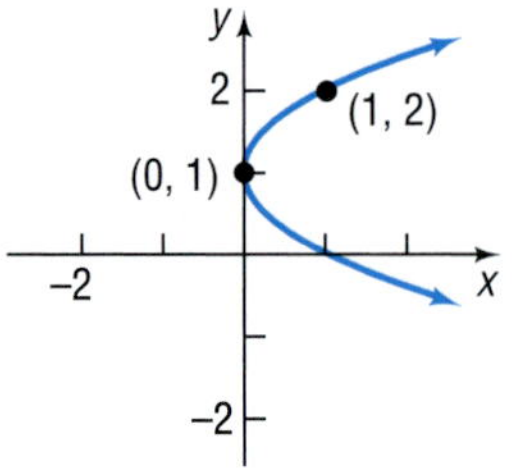

52.

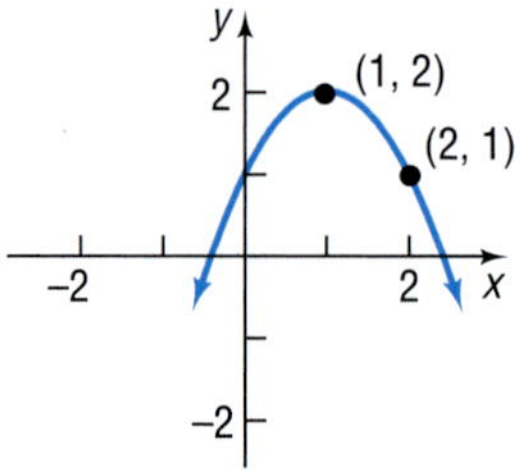

53.

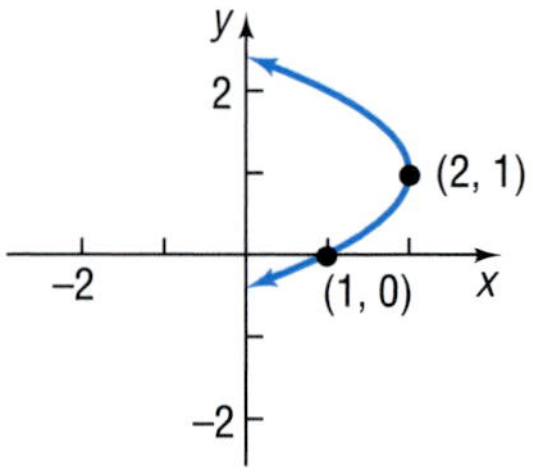

54.

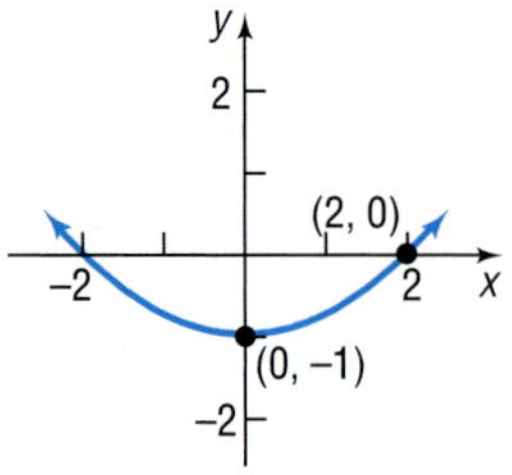

55.

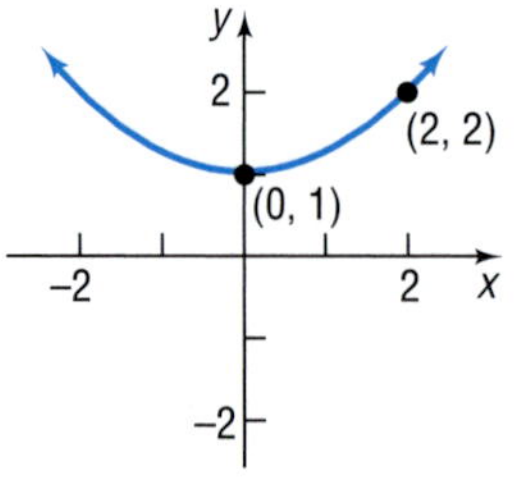

56.

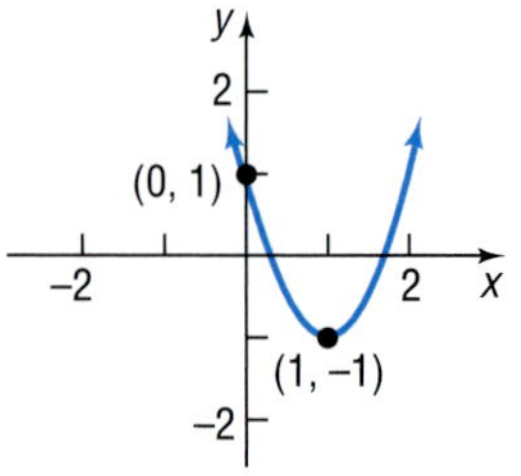

57.

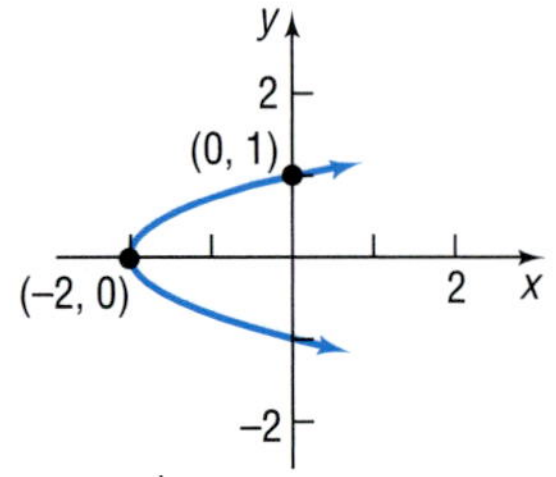

58.

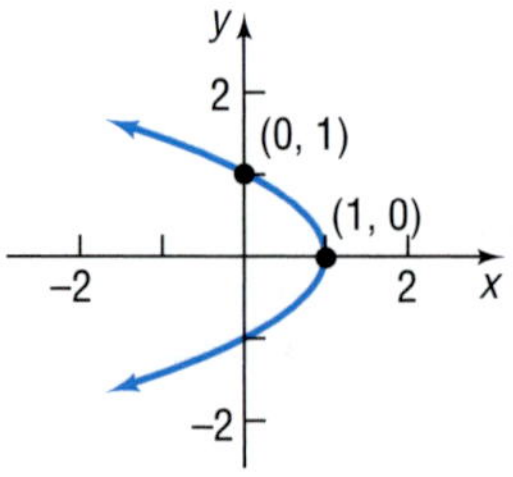

59. Satellite Dish A satellite dish is shaped like a paraboloid of revolution. The signals that emanate from a satellite strike the surface of the dish and are reflected to a single point, where the receiver is located. If the dish is 10 feet across at its opening and is 4 feet deep at is center, at what position should the receiver be placed?

60. Constructing a TV Dish A cable TV receiving dish is in the shape of a paraboloid of revolution. Find the location of the receiver, which is placed at the focus, if the dish is 6 feet across at its opening and 2 feet deep.

61. Constructing a Flashlight The reflector of a flashlight is in the shape of a paraboloid of revolution. Its diameter is 4 inches and its depth is 1 inch. How far from the vertex should the light bulb be placed so that the rays will be reflected parallel to the axis?

62. Constructing a Headlight A sealed-beam headlight is in the shape of a paraboloid of revolution. The bulb, which is placed at the focus, is 1 inch from the vertex. If the depth is to be 2 inches, what is the diameter of the headlight at its opening?

63. Suspension Bridges The cables of a suspension bridge are in the shape of a parabola, as shown in the figure. The towers supporting the cable are 600 feet apart and 80 feet high. If the cables touch the road surface midway between the towers, what is the height of the cable at a point 150 feet from the center of the bridge?

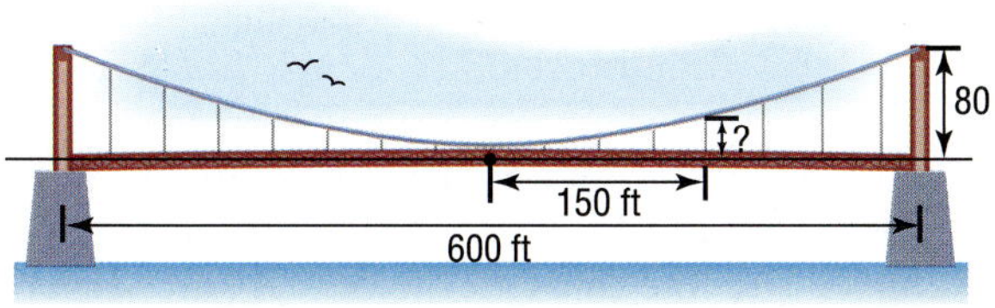

64. Suspension Bridges The cables of a suspension bridge are in the shape of a parabola. The towers supporting the cable are 400 feet apart and 100 feet high. If the cables are at a height of 10 feet midway between the towers, what is the height of the cable at a point 50 feet from the center of the bridge?

65. Searchlights A searchlight is shaped like a paraboloid of revolution. If the light source is located 2 feet from the base along the axis of symmetry and the opening is 5 feet across, how deep should the searchlight be?

66. Searchlights A searchlight is shaped like a paraboloid of revolution. If the light source is located 2 feet from the base along the axis of symmetry and the depth of the searchlight is 4 feet, what should the width of the opening be?

67. Solar Heat A mirror is shaped like a paraboloid of revolution and will be used to concentrate the rays of the sun at its focus, creating a heat source. If the mirror is 20 feet across at its opening and is 6 feet deep, where will the heat source be concentrated?

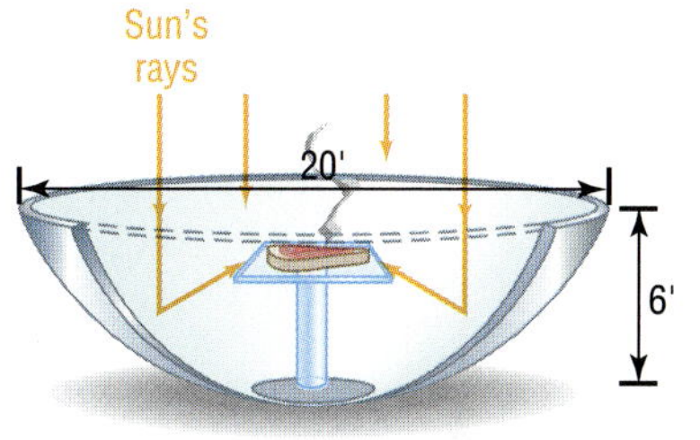

68. Reflecting Telescopes A reflecting telescope contains a mirror shaped like a paraboloid of revolution. If the mirror is 4 inches across at its opening and is 3 feet deep, where will the light collected be concentrated?

69. Parabolic Arch Bridge A bridge is built in the shape of a parabolic arch. The bridge has a span of 120 feet and a maximum height of 25 feet. See the illustration. Choose a suitable rectangular coordinate system and find the height of the arch at distances of 10, 30, and 50 feet from the center.

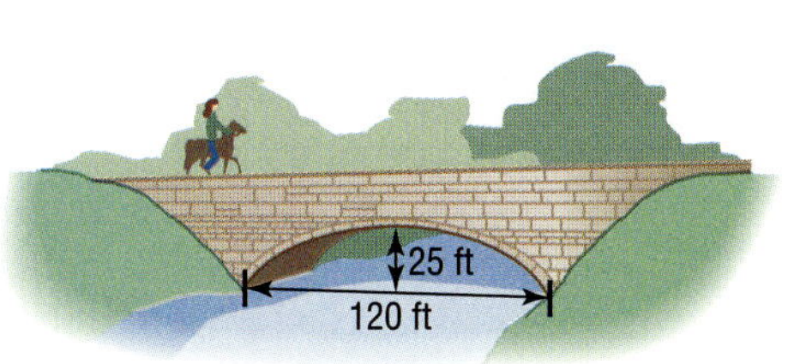

70. Parabolic Arch Bridge A bridge is to be built in the shape of a parabolic arch and is to have a span of 100 feet. The height of the arch a distance of 40 feet from the center is to be 10 feet. Find the height of the arch at its center.

71. Show that an equation of the form

$$Ax^2 + Ey = 0 \qquad A \neq 0, E \neq 0$$

is the equation of a parabola with vertex at $(0, 0)$ and axis of symmetry the y-axis. Find its focus and directrix.

72. Show that an equation of the form

$$Cy^2 + Dx = 0 \qquad C \neq 0, D \neq 0$$

is the equation of a parabola with vertex at $(0,0)$ and axis of symmetry the x-axis. Find its focus and directrix.

73. Show that the graph of an equation of the form

$$Ax^2 + Dx + Ey + F = 0 \qquad A \neq 0$$

(a) Is a parabola if $E \neq 0$.
(b) Is a vertical line if $E = 0$ and $D^2 - 4AF = 0$.
(c) Is two vertical lines if $E = 0$ and $D^2 - 4AF > 0$.
(d) Contains no points if $E = 0$ and $D^2 - 4AF < 0$.

74. Show that the graph of an equation of the form

$$Cy^2 + Dx + Ey + F = 0 \qquad C \neq 0$$

(a) Is a parabola if $D \neq 0$.
(b) Is a horizontal lines if $D = 0$ and $E^2 - 4CF = 0$.
(c) Is two horizontal lines if $D = 0$ and $E^2 - 4CF > 0$.
(d) Contains no points if $D = 0$ and $E^2 - 4CF < 0$.

10.3 THE ELLIPSE

1 Find the Equation of an Ellipse
2 Graph Ellipses
3 Discuss the Equation of an Ellipse
4 Work with Ellipses with Center at (h, k)

An **ellipse** is the collection of all points in the plane the sum of whose distances from two fixed points, called the **foci,** is a constant.

FIGURE 19

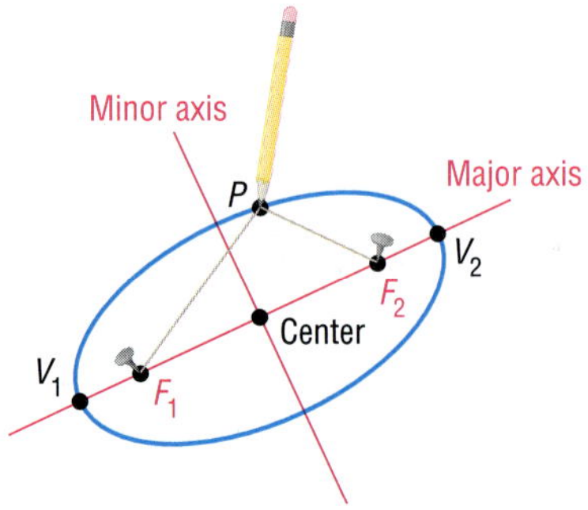

The definition actually contains within it a physical means for drawing an ellipse. Find a piece of string (the length of this string is the constant referred to in the definition). Then take two thumbtacks (the foci) and stick them on a piece of cardboard so that the distance between them is less than the length of the string. Now attach the ends of the string to the thumbtacks and, using the point of a pencil, pull the string taut. Keeping the string taut, rotate the pencil around the two thumbtacks. The pencil traces out an ellipse. as shown in Figure 19.

1 In Figure 19, the foci are labeled F_1 and F_2. The line containing the foci is called the **major axis.** The midpoint of the line segment joining the foci is called the **center** of the ellipse. The line through the center and perpendicular to the major axis is called the **minor axis.**

The two points of intersection of the ellipse and the major axis are the **vertices,** V_1 and V_2, of the ellipse. The distance from one vertex to the other is called the **length of the major axis.** The ellipse is symmetric with respect to its major axis and with respect to its minor axis.

FIGURE 20
$d(F_1, P) + d(F_2, P) = 2a$

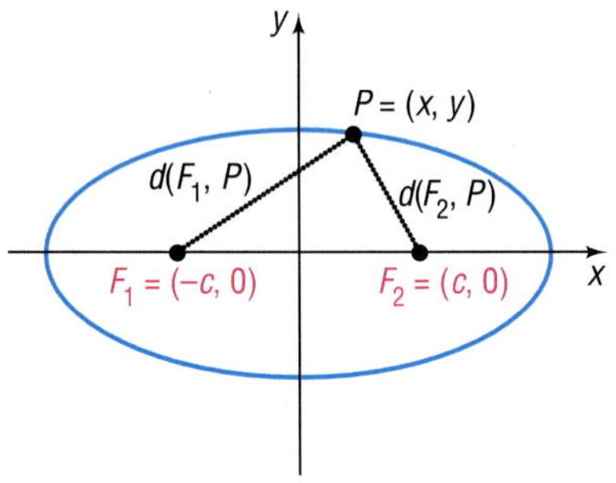

With these ideas in mind, we are now ready to find the equation of an ellipse in a rectangular coordinate system. First, we place the center of the ellipse at the origin. Second, we position the ellipse so that its major axis coincides with a coordinate axis. Suppose that the major axis coincides with the x-axis, as shown in Figure 20. If c is the distance from the center to a focus, then one focus will be at $F_1 = (-c, 0)$ and the other at $F_2 = (c, 0)$. As we shall see, it is convenient to let $2a$ denote the constant distance referred to in the definition. Thus, if $P = (x, y)$ is any point on the ellipse, we have

MISSION POSSIBLE

Building a Bridge over the East River

Your team is working for the transportation authority in New York City. You have been asked to study the construction plans for a new bridge over the East River in New York City. The space between supports needs to be 1050 feet; the height at the center of the arch needs to be 350 feet. One company has suggested the support be in the shape of a parabola; another company suggests a semi-ellipse. The engineering team will determine the relative strengths of the two plans; your job is to find out if there are any differences in the channel widths.

An empty tanker needs a 280-foot clearance to pass beneath the bridge. You need to find the width of the channel for each of the two different plans.

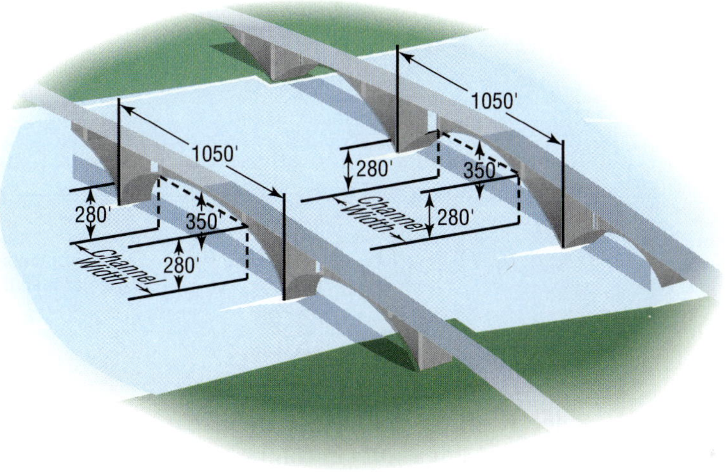

1. To determine the equation of a parabola with these characteristics, first place the parabola on coordinate axes in a convenient location and sketch it.
2. What is the equation of the parabola? (If using a decimal in the equation, you may want to carry 6 decimal places. If using a fraction, your answer will be more exact.)
3. How wide is the channel the tanker can pass through if the shape of the support is parabolic?
4. To determine the equation of a semi-ellipse with these characteristics, place the semi-ellipse on coordinate axes in a convenient location and sketch it.
5. What is the equation of the ellipse? How wide is the channel the tanker can pass through?
6. Now that you know which of the two provides the wider channel, consider some other factors. Your department is also in charge of channel depth, amount of traffic on the river, and various other factors. For example, if the river were to flood and the water level rose by 10 feet, how would the clearances be affected? Make a decision about which plan you think would be the better one as far as your department is concerned, and explain why you think so.

$$\begin{aligned}
d(F_1, P) + d(F_2, P) &= 2a && \text{Sum of the distances from } P \text{ to the foci equals a constant.}\\
\sqrt{(x+c)^2 + y^2} + \sqrt{(x-c)^2 + y^2} &= 2a && \text{Use the distance formula.}\\
\sqrt{(x+c)^2 + y^2} &= 2a - \sqrt{(x-c)^2 + y^2} && \text{Isolate one radical.}\\
(x+c)^2 + y^2 &= 4a^2 - 4a\sqrt{(x-c)^2 + y^2} && \text{Square both sides.}\\
&\quad + (x-c)^2 + y^2 &&\\
x^2 + 2cx + c^2 + y^2 &= 4a^2 - 4a\sqrt{(x-c)^2 + y^2} && \text{Simplify.}\\
&\quad + x^2 - 2cx + c^2 + y^2 &&\\
4cx - 4a^2 &= -4a\sqrt{(x-c)^2 + y^2} && \text{Isolate the radical.}\\
cx - a^2 &= -a\sqrt{(x-c)^2 + y^2} && \text{Divide each side by 4.}\\
c^2x^2 - 2a^2cx + a^4 &= a^2[(x-c)^2 + y^2] && \text{Square both sides again.}\\
c^2x^2 - 2a^2cx + a^4 &= a^2(x^2 - 2cx + c^2 + y^2) &&\\
(c^2 - a^2)x^2 - a^2y^2 &= a^2c^2 - a^4 &&\\
(a^2 - c^2)x^2 + a^2y^2 &= a^2(a^2 - c^2) && \text{Multiply each side by } -1\text{; factor } a^2 \text{ on the right side.} \quad (1)
\end{aligned}$$

To obtain points on the ellipse off the x-axis, it must be that $a > c$. To see why, look again at Figure 20:

$$\begin{aligned}
d(F_1, P) + d(F_2, P) &> d(F_1, F_2) && \text{The sum of the lengths of two sides of a triangle is greater than the length of the third side.}\\
2a &> 2c && d(F_1, P) + d(F_2, P) = 2a;\ d(F_1, F_2) = 2c.\\
a &> c &&
\end{aligned}$$

Since $a > c$, we also have $a^2 > c^2$, so $a^2 - c^2 > 0$. Let $b^2 = a^2 - c^2$, $b > 0$. Then $a > b$ and equation (1) can be written as

$$\begin{aligned}
b^2x^2 + a^2y^2 &= a^2b^2\\
\frac{x^2}{a^2} + \frac{y^2}{b^2} &= 1 && \text{Divide each side by } a^2b^2.
\end{aligned}$$

Theorem Equation of an Ellipse; Center at (0, 0); Foci at ($\pm c$, 0); Major Axis along the x-Axis

An equation of the ellipse with center at (0, 0) and foci at $(-c, 0)$ and $(c, 0)$ is

$$\frac{x^2}{a^2} + \frac{y^2}{b^2} = 1 \quad \text{where } a > b > 0 \text{ and } b^2 = a^2 - c^2 \qquad (2)$$

The major axis is the x-axis.

2 As you can verify, the ellipse defined by equation (2) is symmetric with respect to the x-axis, y-axis, and origin.

FIGURE 21

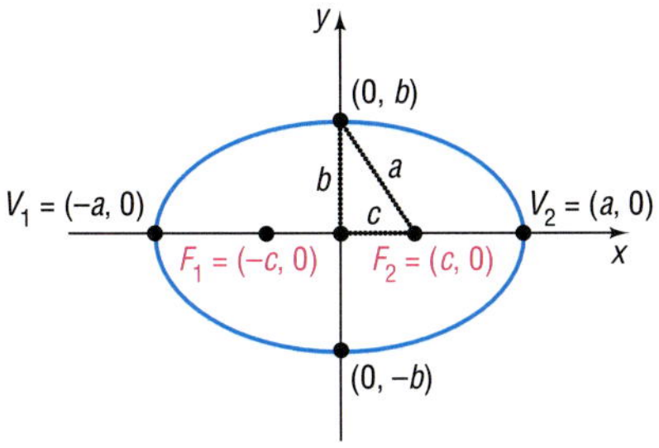

To find the vertices of the ellipse defined by equation (2), let $y = 0$. The vertices satisfy the equation $x^2/a^2 = 1$, the solutions of which are $x = \pm a$. Consequently, the vertices of the ellipse given by equation (2) are $V_1 = (-a, 0)$ and $V_2 = (a, 0)$. The y-intercepts of the ellipse, found by letting $x = 0$, have coordinates $(0, -b)$ and $(0, b)$. These four intercepts, $(a, 0)$, $(-a, 0)$, $(0, b)$, and $(0, -b)$, are used to graph the ellipse by hand. See Figure 21.

Notice in Figure 21 the right triangle formed with the points $(0, 0)$, $(c, 0)$, and $(0, b)$. Because $b^2 = a^2 - c^2$ (or $b^2 + c^2 = a^2$), the distance from the focus at $(c, 0)$ to the point $(0, b)$ is a.

EXAMPLE 1 Finding an Equation of an Ellipse

Find an equation of the ellipse with center at the origin, one focus at $(3, 0)$, and a vertex at $(-4, 0)$. Graph the equation by hand.

Solution The ellipse has its center at the origin, and the major axis coincides with the x-axis. One focus is at $(c, 0) = (3, 0)$, so $c = 3$. One vertex is at $(-a, 0) = (-4, 0)$, so $a = 4$. From equation (2), it follows that

$$b^2 = a^2 - c^2 = 16 - 9 = 7$$

so an equation of the ellipse is

$$\frac{x^2}{16} + \frac{y^2}{7} = 1$$

Figure 22 shows the graph drawn by hand.

FIGURE 22 $\frac{x^2}{16} + \frac{y^2}{7} = 1$

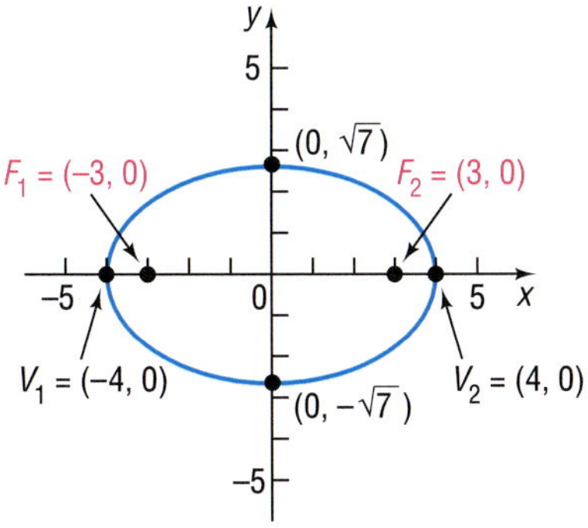

Notice in Figure 22 how we used the intercepts of the equation to graph the ellipse. Following this practice will make it easier for you to obtain an accurate graph of an ellipse when graphing by hand. It also tells you how to set the viewing rectangle when using a graphing utility.

EXAMPLE 2 Graphing an Ellipse Using a Graphing Utility

Use a graphing utility to graph the ellipse $\frac{x^2}{16} + \frac{y^2}{7} = 1$.

Solution First, we must solve $\frac{x^2}{16} + \frac{y^2}{7} = 1$ for y.

$$\frac{y^2}{7} = 1 - \frac{x^2}{16} \quad \text{Subtract } \frac{x^2}{16} \text{ from each side.}$$

$$y^2 = 7\left(1 - \frac{x^2}{16}\right) \quad \text{Multiply both sides by 7.}$$

$$y = \pm\sqrt{7\left(1 - \frac{x^2}{16}\right)} \quad \text{Take the square root of each side.}$$

FIGURE 23

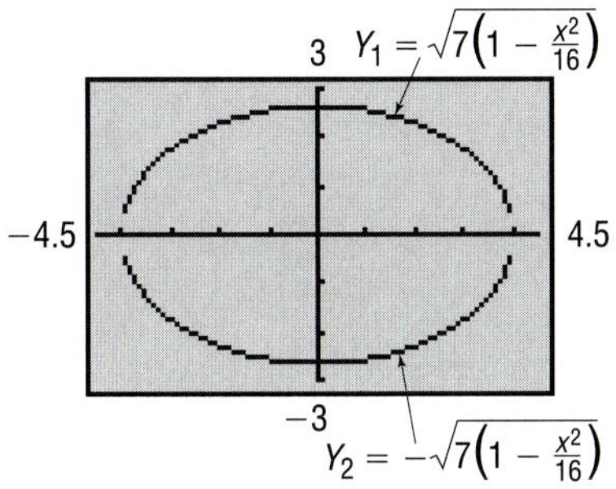

Figure 23 shows the graphs of $Y_1 = \sqrt{7\left(1 - \frac{x^2}{16}\right)}$ and $Y_2 = -\sqrt{7\left(1 - \frac{x^2}{16}\right)}$.

Notice in Figure 23 that we used a square screen. As with circles and parabolas, this is done to avoid a distorted view of the graph.

An equation of the form of equation (2), with $a > b$, is the equation of an ellipse with center at the origin, foci on the x-axis at $(-c, 0)$ and $(c, 0)$, where $c^2 = a^2 - b^2$, and major axis along the x-axis.

3

For the remainder of this section, the direction "Discuss the equation" will mean to find the center, major axis, foci, and vertices of the ellipse and graph it.

EXAMPLE 3 Discussing the Equation of an Ellipse

Discuss the equation $\dfrac{x^2}{25} + \dfrac{y^2}{9} = 1$.

Solution Figure 24(a) shows the graph of $\dfrac{x^2}{25} + \dfrac{y^2}{9} = 1$ using a graphing utility. We now proceed to analyze the equation. The given equation is of the form of equation (2), with $a^2 = 25$ and $b^2 = 9$. The equation is that of an ellipse with center $(0, 0)$ and major axis along the x-axis. The vertices are at $(\pm a, 0) = (\pm 5, 0)$. Because $b^2 = a^2 - c^2$, we find

$$c^2 = a^2 - b^2 = 25 - 9 = 16$$

The foci are at $(\pm c, 0) = (\pm 4, 0)$. Figure 24(b) shows the graph drawn by hand.

FIGURE 24 $\dfrac{x^2}{25} + \dfrac{y^2}{9} = 1$

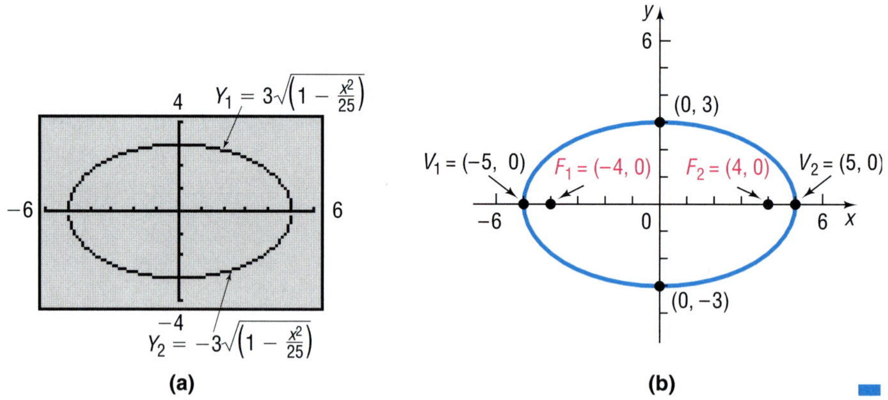

Now work Problems 9 and 19.

If the major axis of an ellipse with center at $(0, 0)$ coincides with the y-axis, then the foci are at $(0, -c)$ and $(0, c)$. Using the same steps as before, the definition of an ellipse leads to the following result:

Theorem Equation of an Ellipse; Center at (0, 0); Foci at (0, $\pm c$); Major Axis along the y-Axis

An equation of the ellipse with center at (0, 0) and foci at (0, $-c$) and (0, c) is

$$\frac{x^2}{b^2} + \frac{y^2}{a^2} = 1 \qquad \text{where } a > b > 0 \text{ and } b^2 = a^2 - c^2 \qquad (3)$$

The major axis is the y-axis; the vertices are at (0, $-a$) and (0, a).

Figure 25 illustrates the graph of such an ellipse. Again, notice the right triangle with the points at (0, 0), (b, 0), and (0, c).

FIGURE 25

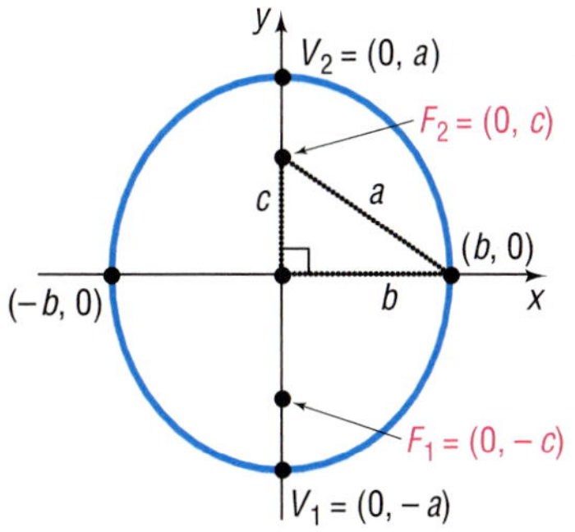

Look closely at equations (2) and (3). Although they may look alike, there is a difference! In equation (2), the larger number, a^2, is in the denominator of the x^2-term, so the major axis of the ellipse is along the x-axis. In equation (3), the larger number, a^2, is in the denominator of the y^2-term, so the major axis is along the y-axis.

EXAMPLE 4 Discussing the Equation of an Ellipse

Discuss the equation $9x^2 + y^2 = 9$.

Solution Figure 26(a) shows the graph of $9x^2 + y^2 = 9$ using a graphing utility. We now proceed to analyze the equation. To put the equation in proper form, we divide each side by 9:

$$x^2 + \frac{y^2}{9} = 1$$

The larger number, 9, is in the denominator of the y^2-term so, based on equation (3), this is the equation of an ellipse with center at the origin and major axis along the y-axis. Also, we conclude that $a^2 = 9$, $b^2 = 1$, and

$c^2 = a^2 - b^2 = 9 - 1 = 8$. The vertices are at $(0, \pm a) = (0, \pm 3)$, and the foci are at $(0, \pm c) = (0, \pm 2\sqrt{2})$. The graph, drawn by hand, is given in Figure 26(b).

FIGURE 26 $x^2 + \frac{y^2}{9} = 1$

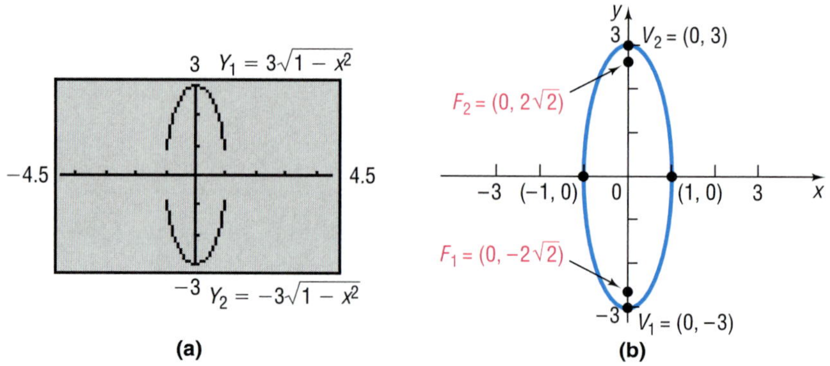

(a) (b)

EXAMPLE 5 Finding an Equation of an Ellipse

Find an equation of the ellipse having one focus at (0, 2) and vertices at (0, −3) and (0, 3). Graph the equation by hand.

Solution Because the vertices are at (0, −3) and (0, 3), the center of this ellipse is at the origin. Also, its major axis coincides with the y-axis. The given information also reveals that $c = 2$ and $a = 3$, so $b^2 = a^2 - c^2 = 9 - 4 = 5$. The form of the equation of this ellipse is given by equation (3):

$$\frac{x^2}{b^2} + \frac{y^2}{a^2} = 1$$

$$\frac{x^2}{5} + \frac{y^2}{9} = 1$$

Figure 27 shows the graph.

FIGURE 27 $\frac{x^2}{5} + \frac{y^2}{9} = 1$

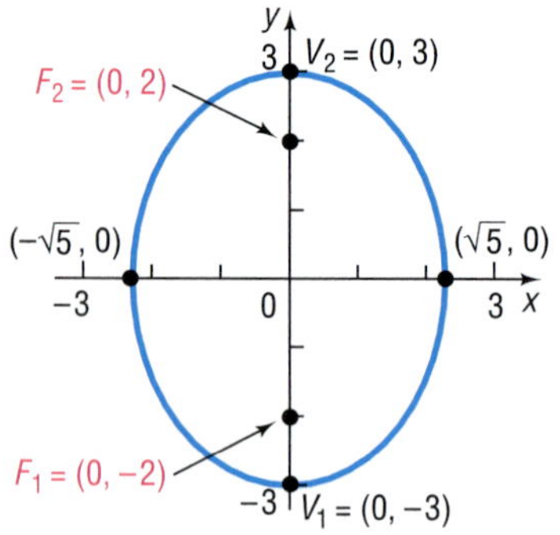

Now work Problems 13 and 21.

The circle may be considered a special kind of ellipse. To see why, let $a = b$ in equation (2) or in equation (3). Then

$$\frac{x^2}{a^2} + \frac{y^2}{a^2} = 1$$

$$x^2 + y^2 = a^2$$

This is the equation of a circle with center at the origin and radius a. The value of c is

$$c^2 = a^2 - b^2 = 0$$

We conclude that the closer the two foci of an ellipse are, the more the ellipse will look like a circle.

Center at (h, k)

If an ellipse with center at the origin and major axis coinciding with a coordinate axis is shifted horizontally h units and then vertically k units, the result is an ellipse with center at (h, k) and major axis parallel to a coordinate axis. Table 3 gives the forms of the equations of such ellipses, and Figure 28 shows their graphs.

TABLE 3	Ellipses with Center at (h, k) and Major Axis Parallel to a Coordinate Axis			
Center	Major Axis	Foci	Vertices	Equation
(h, k)	Parallel to x-axis	$(h \pm c, k)$	$(h \pm a, k)$	$\dfrac{(x-h)^2}{a^2} + \dfrac{(y-k)^2}{b^2} = 1,$ $a > b$ and $b^2 = a^2 - c^2$
(h, k)	Parallel to y-axis	$(h, k \pm c)$	$(h, k \pm a)$	$\dfrac{(x-h)^2}{b^2} + \dfrac{(y-k)^2}{a^2} = 1,$ $a > b$ and $b^2 = a^2 - c^2$

FIGURE 28

(a) $\dfrac{(x-h)^2}{a^2} + \dfrac{(y-k)^2}{b^2} = 1$ (b) $\dfrac{(x-h)^2}{b^2} + \dfrac{(y-k)^2}{a^2} = 1$

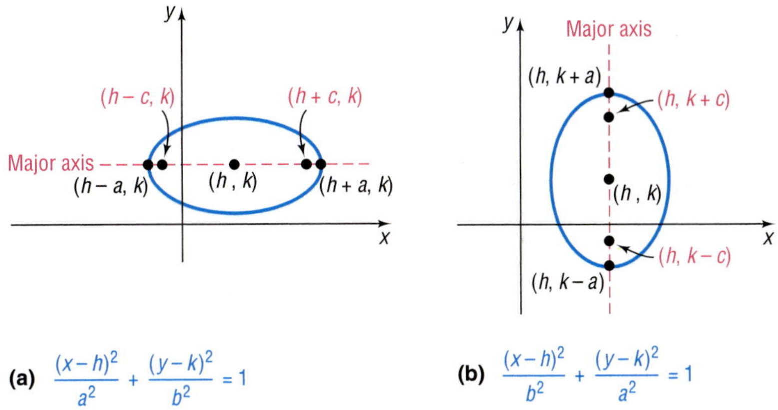

EXAMPLE 6   Finding an Equation of an Ellipse, Center Not at the Origin

Find an equation for the ellipse with center at $(2, -3)$, one focus at $(3, -3)$, and one vertex at $(5, -3)$. Graph the equation by hand.

Solution The center is at $(h, k) = (2, -3)$, so $h = 2$ and $k = -3$. The major axis is parallel to the x-axis. The distance from the center $(2, -3)$ to a focus $(3, -3)$ is $c = 1$; the distance from the center $(2, -3)$ to a vertex $(5, -3)$ is $a = 3$. Thus, $b^2 = a^2 - c^2 = 9 - 1 = 8$. The form of the equation is

$$\frac{(x-h)^2}{a^2} + \frac{(y-k)^2}{b^2} = 1 \text{ where } h = 2, k = -3, a = 3, b = 2\sqrt{2}$$

$$\frac{(x-2)^2}{9} + \frac{(y+3)^2}{8} = 1$$

Figure 29 shows the graph.

FIGURE 29
$\frac{(x-2)^2}{9} + \frac{(y+3)^2}{8} = 1$

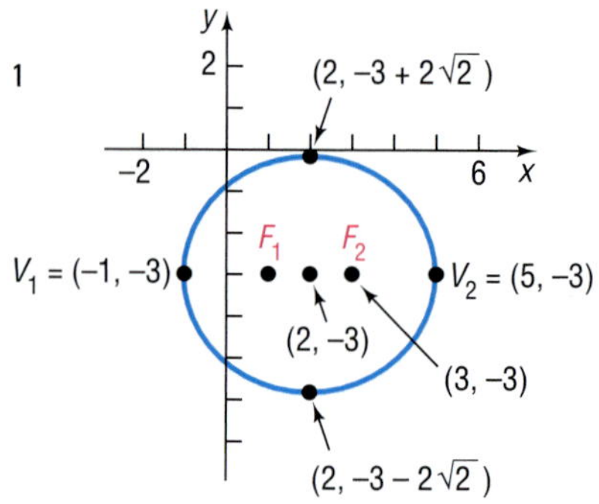

EXAMPLE 7 Using a Graphing Utility to Graph an Ellipse, Center Not at the Origin

Using a graphing utility, graph the ellipse: $\frac{(x-2)^2}{9} + \frac{(y+3)^2}{8} = 1$

Solution First, we must solve the equation $\frac{(x-2)^2}{9} + \frac{(y+3)^2}{8} = 1$ for y.

$$\frac{(y+3)^2}{8} = 1 - \frac{(x-2)^2}{9} \qquad \text{Subtract } \frac{(x-2)^2}{9} \text{ from each side.}$$

$$(y+3)^2 = 8\left[1 - \frac{(x-2)^2}{9}\right] \qquad \text{Multiply each side by 8.}$$

$$y + 3 = \pm\sqrt{8\left[1 - \frac{(x-2)^2}{9}\right]} \qquad \text{Take the square root of each side.}$$

$$y = -3 \pm \sqrt{8\left[1 - \frac{(x-2)^2}{9}\right]} \qquad \text{Subtract 3 from each side.}$$

FIGURE 30

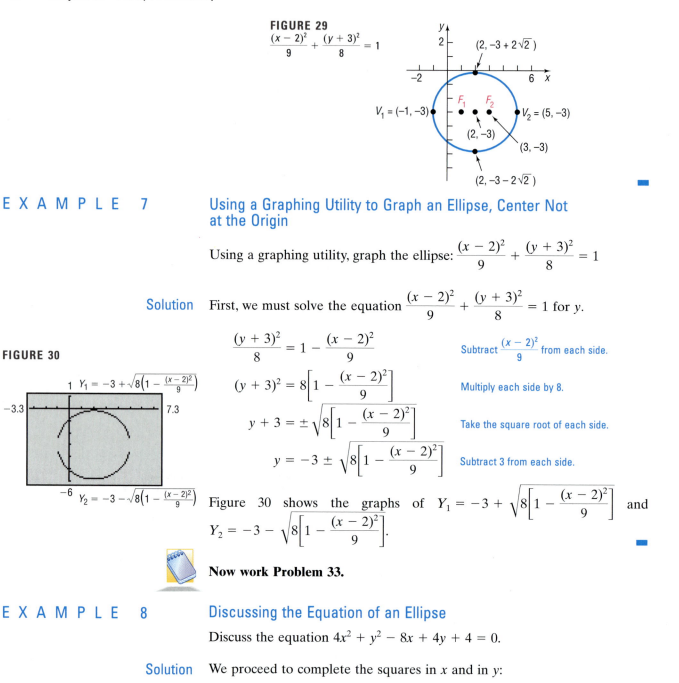

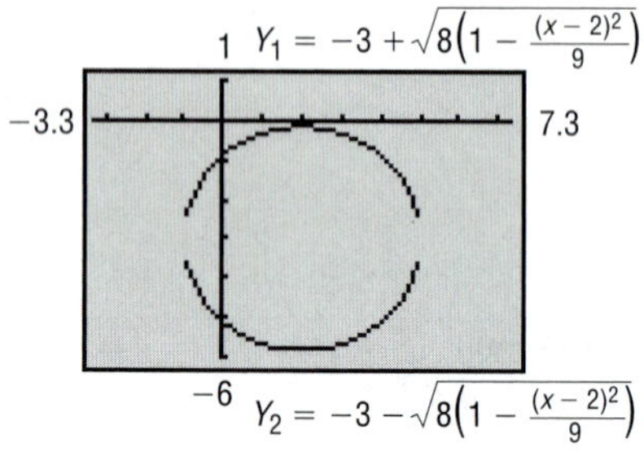

Figure 30 shows the graphs of $Y_1 = -3 + \sqrt{8\left[1 - \frac{(x-2)^2}{9}\right]}$ and $Y_2 = -3 - \sqrt{8\left[1 - \frac{(x-2)^2}{9}\right]}$.

Now work Problem 33.

EXAMPLE 8 Discussing the Equation of an Ellipse

Discuss the equation $4x^2 + y^2 - 8x + 4y + 4 = 0$.

Solution We proceed to complete the squares in x and in y:

$$4x^2 + y^2 - 8x + 4y + 4 = 0$$

$$4x^2 - 8x + y^2 + 4y = -4$$

$$4(x^2 - 2x) + (y^2 + 4y) = -4$$

$$4(x^2 - 2x + 1) + (y^2 + 4y + 4) = -4 + 4 + 4 \qquad \text{Complete each square.}$$

$$4(x-1)^2 + (y+2)^2 = 4$$

$$(x-1)^2 + \frac{(y+2)^2}{4} = 1 \qquad \text{Divide each side by 4.}$$

Figure 31(a) shows the graph of $(x-1)^2 + \frac{(y+2)^2}{4} = 1$ using a graphing utility. We now proceed to analyze the equation.

This is the equation of an ellipse with center at $(1, -2)$ and major axis parallel to the y-axis. Since $a^2 = 4$ and $b^2 = 1$, we have $c^2 = a^2 - b^2 = 4 - 1 = 3$. The vertices are at $(h, k \pm a) = (1, -2 \pm 2)$ or $(1, 0)$ and $(1, -4)$. The foci are at $(h, k \pm c) = (1, -2 \pm \sqrt{3})$ or $(1, -2 - \sqrt{3})$ and $(1, -2 + \sqrt{3})$. Figure 31(b) shows the graph drawn by hand.

FIGURE 31
$(x-1)^2 + \dfrac{(y+2)^2}{4} = 1$

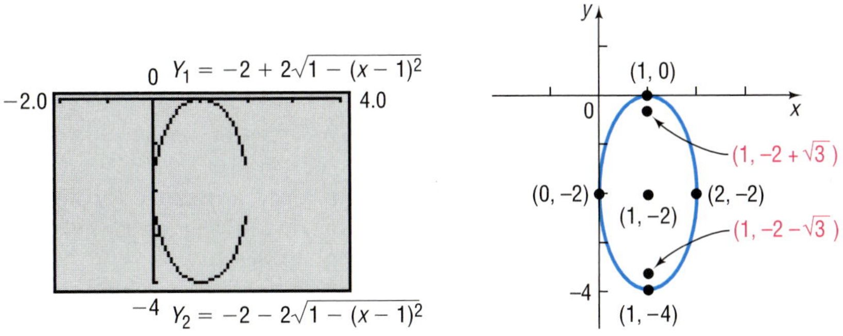

(a) (b)

Applications

Ellipses are found in many applications in science and engineering. For example, the orbits of the planets around the Sun are elliptical, with the Sun's position at a focus. See Figure 32.

FIGURE 32

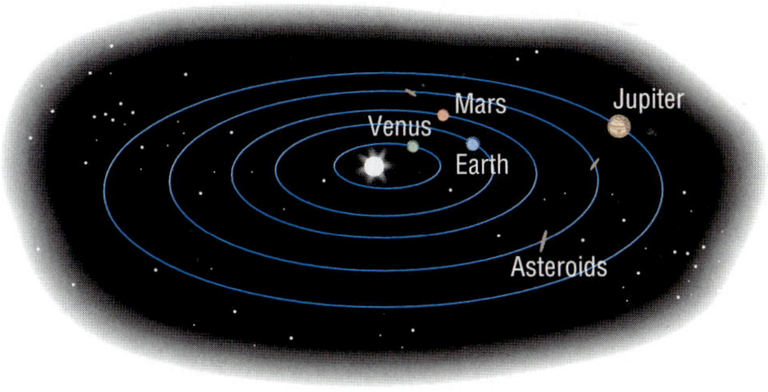

Stone and concrete bridges are often shaped as semielliptical arches. Elliptical gears are used in machinery when a variable rate of motion is required.

Ellipses also have an interesting reflection property. If a source of light (or sound) is placed at one focus, the waves transmitted by the source will reflect off the ellipse and concentrate at the other focus. This is the principal behind "whispering galleries," which are rooms designed with elliptical ceilings. A person standing at one focus of the ellipse can whisper and be heard by a person standing at the other focus, because all the sound waves that reach the ceiling are reflected to the other person.

EXAMPLE 9

Whispering Galleries

Figure 33 shows the specifications for an elliptical ceiling in a hall designed to be a whispering gallery. In a whispering gallery, a person standing at one focus of the ellipse can whisper and be heard by another person standing at the other focus, because all the sound waves that reach the ceiling from one focus are reflected to the other focus. Where in the hall are the foci located?

FIGURE 33

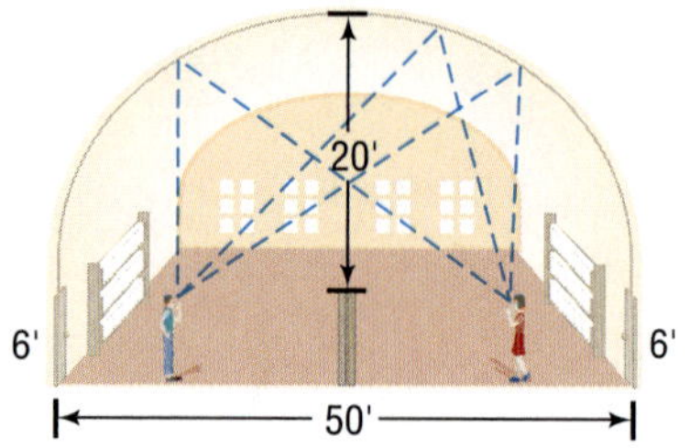

Solution We set up a rectangular coordinate system so that the center of the ellipse is at the origin and the major axis is along the x-axis. See Figure 34. The equation of the ellipse is

$$\frac{x^2}{a^2} + \frac{y^2}{b^2} = 1$$

where $a = 25$ and $b = 20$. Since

$$c^2 = a^2 - b^2 = 25^2 - 20^2 = 625 - 400 = 225$$

we have $c = 15$. Thus, the foci are located 15 feet from the center of the ellipse along the major axis.

FIGURE 34

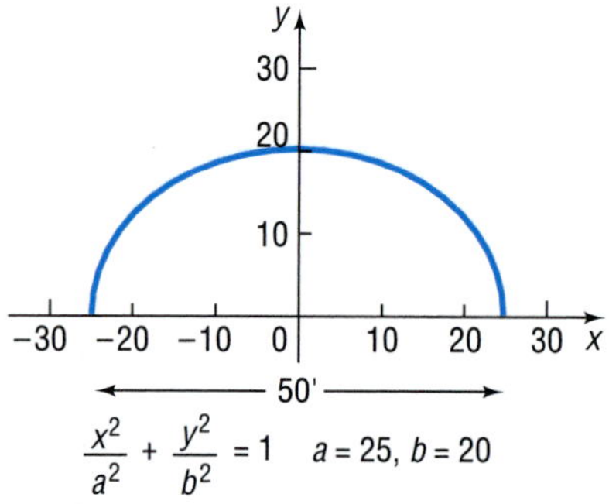

10.3 EXERCISES

In Problems 1–4, the graph of an ellipse is given. Match each graph to its equation.

A. $\frac{x^2}{4} + y^2 = 1$ *B.* $x^2 + \frac{y^2}{4} = 1$ *C.* $\frac{x^2}{16} + \frac{y^2}{4} = 1$ *D.* $\frac{x^2}{4} + \frac{y^2}{16} = 1$

1.

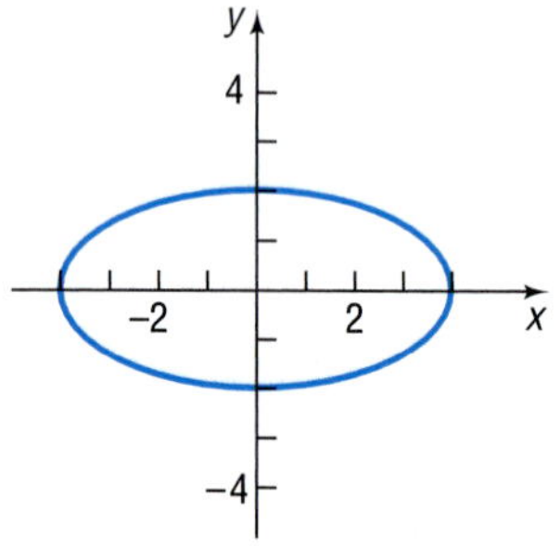

2.

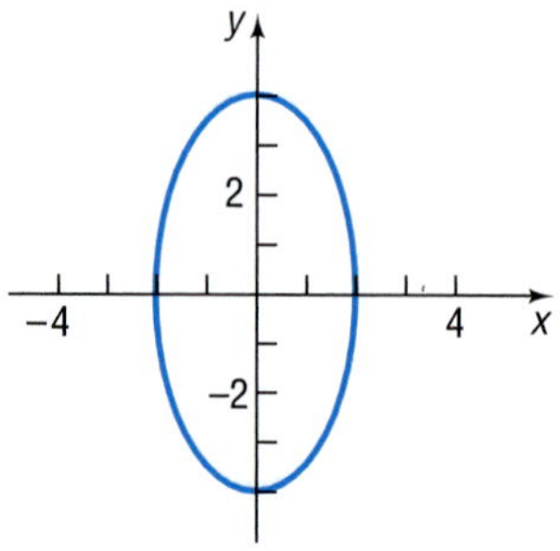

3.

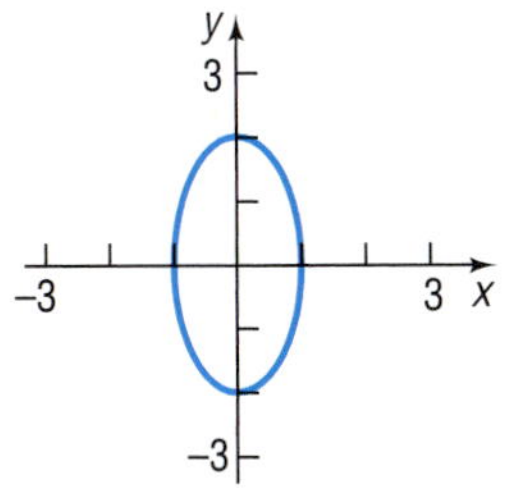

4.

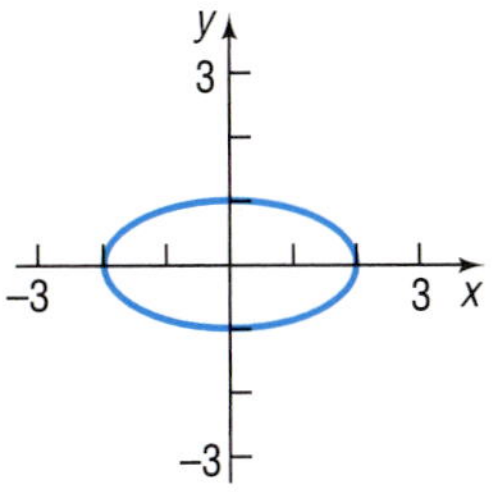

In Problems 5–8, the graph of an ellipse is given. Match each graph to its equation.

A. $\dfrac{(x+1)^2}{4} + \dfrac{(y-1)^2}{9} = 1$ *B.* $\dfrac{(x-1)^2}{4} + \dfrac{(y+1)^2}{9} = 1$

C. $\dfrac{(x-1)^2}{9} + \dfrac{(y+1)^2}{4} = 1$ *D.* $\dfrac{(x+1)^2}{9} + \dfrac{(y-1)^2}{4} = 1$

5.

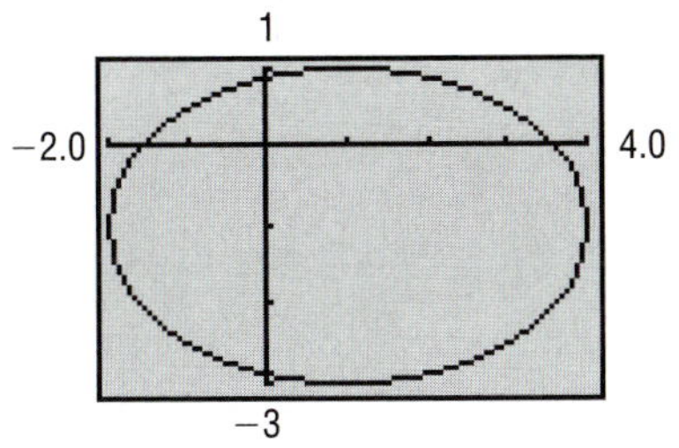

6.

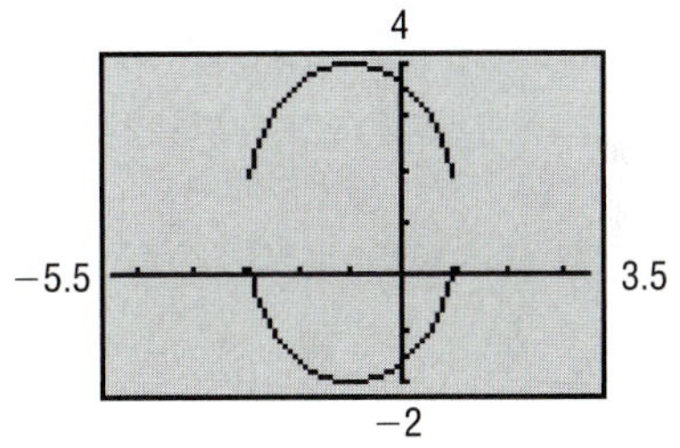

7.

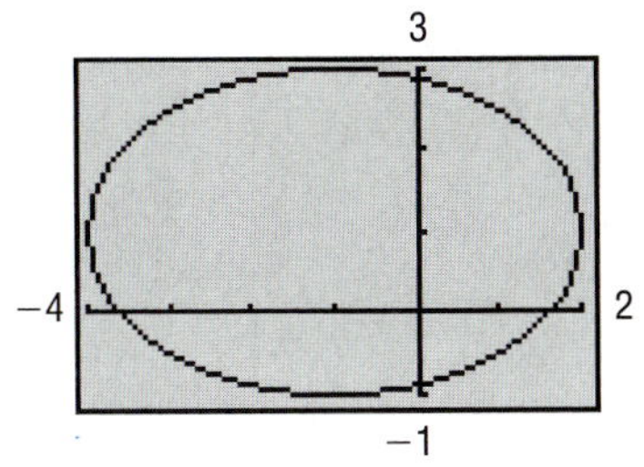

8.

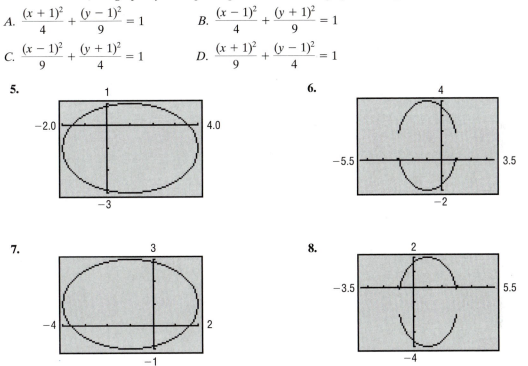

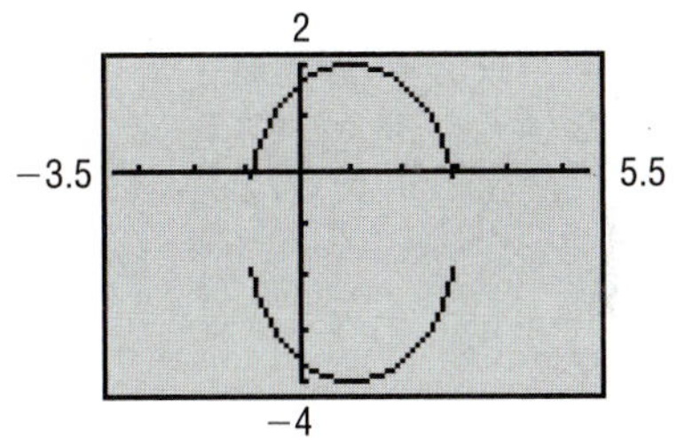

In Problems 9–18, find the vertices and foci of each ellipse. Graph each equation using a graphing utility.

9. $\dfrac{x^2}{25} + \dfrac{y^2}{4} = 1$ **10.** $\dfrac{x^2}{9} + \dfrac{y^2}{4} = 1$ **11.** $\dfrac{x^2}{9} + \dfrac{y^2}{25} = 1$ **12.** $\dfrac{x^2}{4} + \dfrac{y^2}{16} = 1$

13. $4x^2 + y^2 = 16$ **14.** $x^2 + 9y^2 = 18$ **15.** $4y^2 + x^2 = 8$ **16.** $4y^2 + 9x^2 = 36$

17. $x^2 + y^2 = 16$ **18.** $x^2 + y^2 = 4$

In Problems 19–28, find an equation for each ellipse. Graph the equation by hand.

19. Center at $(0, 0)$; focus at $(3, 0)$; vertex at $(5, 0)$

20. Center at $(0, 0)$; focus at $(-1, 0)$; vertex at $(3, 0)$

21. Center at $(0, 0)$; focus at $(0, -4)$; vertex at $(0, 5)$

22. Center at $(0, 0)$; focus at $(0, 1)$; vertex at $(0, -2)$

23. Foci at $(\pm 2, 0)$; length of the major axis is 6

24. Focus at $(0, -4)$; vertices at $(0, \pm 8)$

25. Foci at $(0, \pm 3)$; x-intercepts are ± 2

26. Foci at $(0, \pm 2)$; length of the major axis is 8

27. Center at $(0, 0)$; vertex at $(0, 4)$; $b = 1$

28. Vertices at $(\pm 5, 0)$; $c = 2$

In Problems 29–32, write an equation for each ellipse.

29.

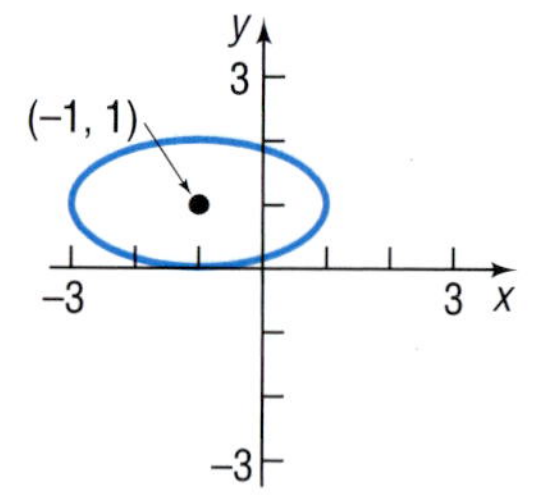

30.

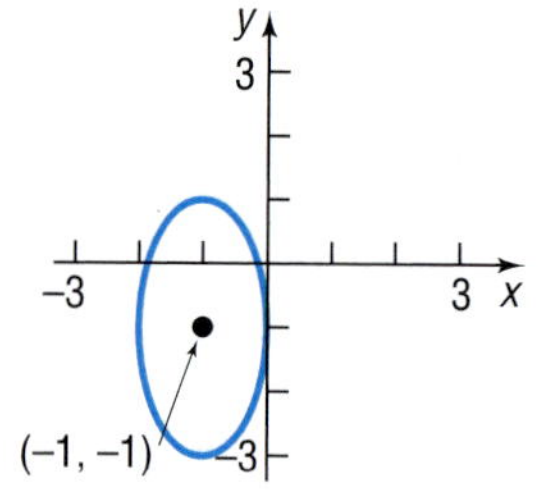

31.

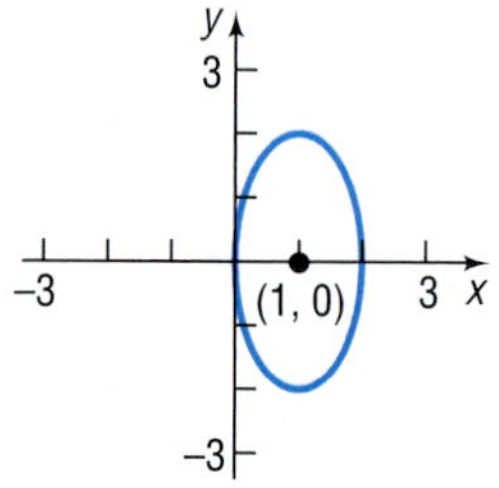

32.

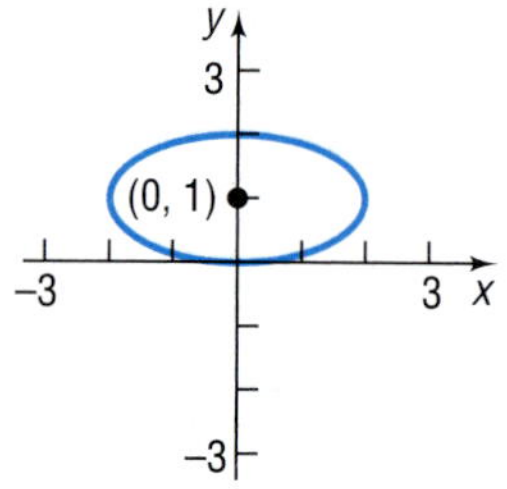

In Problems 33–44, find the center, foci, and vertices of each ellipse. Graph each equation using a graphing utility.

33. $\dfrac{(x-3)^2}{4} + \dfrac{(y+1)^2}{9} = 1$

34. $\dfrac{(x+4)^2}{9} + \dfrac{(y+2)^2}{4} = 1$

35. $(x+5)^2 + 4(y-4)^2 = 16$

36. $9(x-3)^2 + (y+2)^2 = 18$

37. $x^2 + 4x + 4y^2 - 8y + 4 = 0$

38. $x^2 + 3y^2 - 12y + 9 = 0$

39. $2x^2 + 3y^2 - 8x + 6y + 5 = 0$

40. $4x^2 + 3y^2 + 8x - 6y = 5$

41. $9x^2 + 4y^2 - 18x + 16y - 11 = 0$

42. $x^2 + 9y^2 + 6x - 18y + 9 = 0$

43. $4x^2 + y^2 + 4y = 0$

44. $9x^2 + y^2 - 18x = 0$

In Problems 45–54, find an equation for each ellipse. Graph the equation by hand.

45. Center at $(2, -2)$; vertex at $(7, -2)$; focus at $(4, -2)$

46. Center at $(-3, 1)$; vertex at $(-3, 3)$; focus at $(-3, 0)$

47. Vertices at $(4, 3)$ and $(4, 9)$; focus at $(4, 8)$

48. Foci at $(1, 2)$ and $(-3, 2)$; vertex at $(-4, 2)$

49. Foci at $(5, 1)$ and $(-1, 1)$; length of the major axis is 8

50. Vertices at $(2, 5)$ and $(2, -1)$; $c = 2$

51. Center at $(1, 2)$; focus at $(4, 2)$; contains the point $(1, 3)$

52. Center at $(1, 2)$; focus at $(1, 4)$; contains the point $(2, 2)$

53. Center at $(1, 2)$; vertex at $(4, 2)$; contains the point $(1, 3)$

54. Center at $(1, 2)$; vertex at $(1, 4)$; contains the point $(2, 2)$

In Problems 55–58, graph each function by hand. Use a graphing utility to verify your results.

[**Hint:** Notice that each function is half an ellipse.]

55. $f(x) = \sqrt{16 - 4x^2}$ **56.** $f(x) = \sqrt{9 - 9x^2}$ **57.** $f(x) = -\sqrt{64 - 16x^2}$ **58.** $f(x) = -\sqrt{4 - 4x^2}$

59. **Semielliptical Arch Bridge** An arch in the shape of the upper half of an ellipse is used to support a bridge that is to span a river 20 meters wide. The center of the arch is 6 meters above the center of the river (see the figure). Write an equation for the ellipse in which the x-axis coincides with the water level and the y-axis passes through the center of the arch.

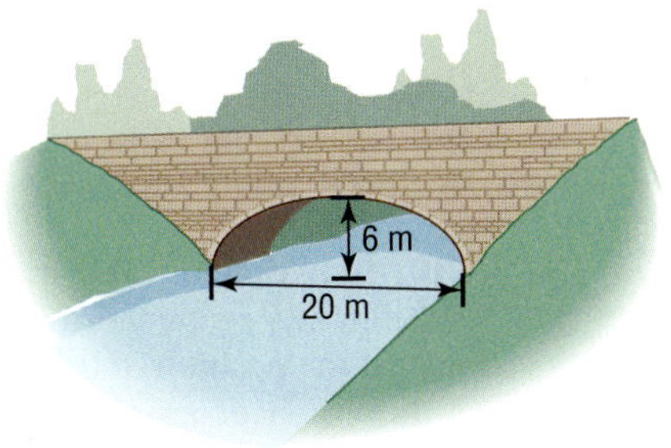

60. **Semielliptical Arch Bridge** The arch of a bridge is a semiellipse with a horizontal major axis. The span is 30 feet, and the top of the arch is 10 feet above the major axis. The roadway is horizontal and is 2 feet above the top of the arch. Find the vertical distance from the roadway to the arch at 5 foot intervals along the roadway.

61. **Whispering Galleries** A hall 100 feet in length is to be designed as a whispering gallery. If the foci are located 25 feet from the center, how high will the ceiling be at the center?

62. **Whispering Galleries** Jim, standing at one focus of a whispering gallery, is 6 feet from the nearest wall. His friend is standing at the other focus, 100 feet away. What is the length of this whispering gallery? How high is its elliptical ceiling at the center?

63. **Semielliptical Arch Bridge** A bridge is built in the shape of a semielliptical arch. The bridge has a span of 120 feet and a maximum height of 25 feet. Choose a suitable rectangular coordinate system and find the height of the arch at distances of 10, 30, and 50 feet from the center.

64. **Semielliptical Arch Bridge** A bridge is built in the shape of a semielliptical arch and is to have a span of 100 feet. The height of the arch, at a distance of 40 feet from the center, is to be 10 feet. Find the height of the arch at its center.

65. **Semielliptical Arch** An arch in the form of half an ellipse is 40 feet wide and 15 feet high at the center. Find the height of the arch at intervals of 10 feet along its width.

66. **Semielliptical Arch Bridge** An arch for a bridge over a highway is in the form of half an ellipse. The top of the arch is 20 feet above the ground level (the major axis). The highway has four lanes, each 12 feet wide; a center safety strip 8 feet wide; and two side strips, each 4 feet wide. What should the span of the bridge be (the length of its major axis) if the height 28 feet from the center is to be 13 feet?

In Problems 67–70, use the fact that the orbit of a planet about the Sun is an ellipse, with the Sun at one focus. The ***aphelion*** *of a planet is its greatest distance from the Sun and the* ***perihelion*** *is its shortest distance. The* ***mean distance*** *of a planet from the Sun is the length of the semimajor axis of the elliptical orbit. See the illustration.*

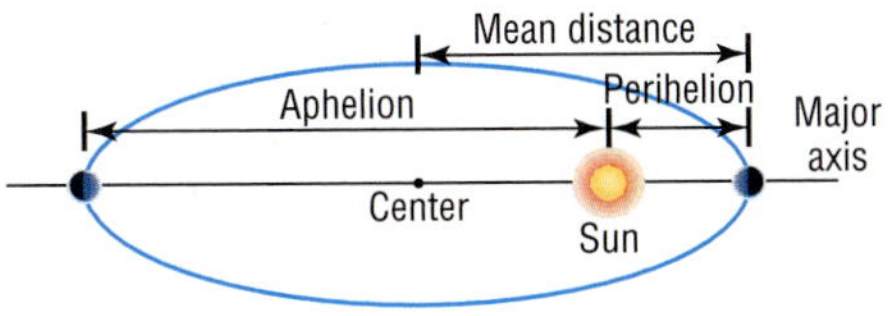

67. **Earth** The mean distance of Earth from the Sun is 93 million miles. If the aphelion of Earth is 94.5 million miles, what is the perihelion? Write an equation for the orbit of Earth around the Sun.

68. **Mars** The mean distance of Mars from the Sun is 142 million miles. If the perihelion of Mars is 128.5 million miles, what is the aphelion? Write an equation for the orbit of Mars about the Sun.

69. **Jupiter** The aphelion of Jupiter is 507 million miles. If the distance from the Sun to the center of its elliptical orbit is 23.2 million miles, what is the perihelion? What is the mean distance? Write an equation for the orbit of Jupiter around the Sun.

70. **Pluto** The perihelion of Pluto is 4551 million miles and the distance of the Sun from the center of its elliptical orbit is 897.5 million miles. Find the aphelion of Pluto. What is the mean distance of Pluto from the Sun? Write an equation for the orbit of Pluto about the Sun.

71. Consult the figure. A racetrack is in the shape of an ellipse, 100 feet long and 50 feet wide. What is the width 10 feet from the side?

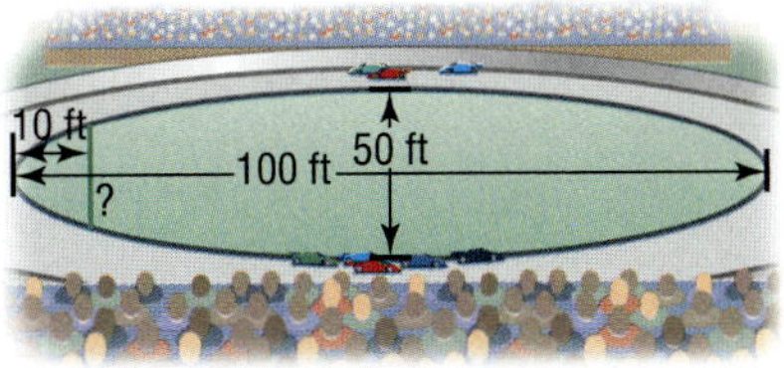

72. A racetrack is in the shape of an ellipse 80 feet long and 40 feet wide. What is the width 10 feet from the side?

73. Show that an equation of the form

$$Ax^2 + Cy^2 + F = 0 \qquad A \neq 0, C \neq 0, F \neq 0$$

where A and C are of the same sign and F is of opposite sign.

(a) Is the equation of an ellipse with center at $(0, 0)$ if $A \neq C$.

(b) Is the equation of a circle with center $(0, 0)$ if $A = C$.

74. Show that the graph of an equation of the form

$$Ax^2 + Cy^2 + Dx + Ey + F = 0 \qquad A \neq 0, C \neq 0$$

where A and C are of the same sign:

(a) Is an ellipse if $(D^2/4A) + (E^2/4C) - F$ is the same sign as A.

(b) Is a point if $(D^2/4A) + (E^2/4C) - F = 0$.

(c) Contains no points if $(D^2/4A) + (E^2/4C) - F$ is of opposite sign to A.

75. The **eccentricity** e of an ellipse is defined as the number c/a, where a and c are the numbers given in equation (2). Because $a > c$, it follows that $e < 1$. Write a brief paragraph about the general shape of each of the following ellipses. Be sure to justify your conclusions.

(a) Eccentricity close to 0

(b) Eccentricity = 0.5

(c) Eccentricity close to 1

76. **CBL Experiment** The motion of a swinging pendulum is examined by plotting the velocity against the position of the pendulum. The result is the graph of an ellipse. (Activity 17, Real-World Math with the CBL System)

10.4 THE HYPERBOLA

1 Find the Equation of a Hyperbola
2 Graph Hyperbolas
3 Discuss the Equation of a Hyperbola
4 Find the Asymptotes of a Hyperbola
5 Work with Hyperbolas with Center at (h, k)

A **hyperbola** is the collection of all points in the plane the difference of whose distances from two fixed points, called the **foci,** is a constant.

1

FIGURE 35

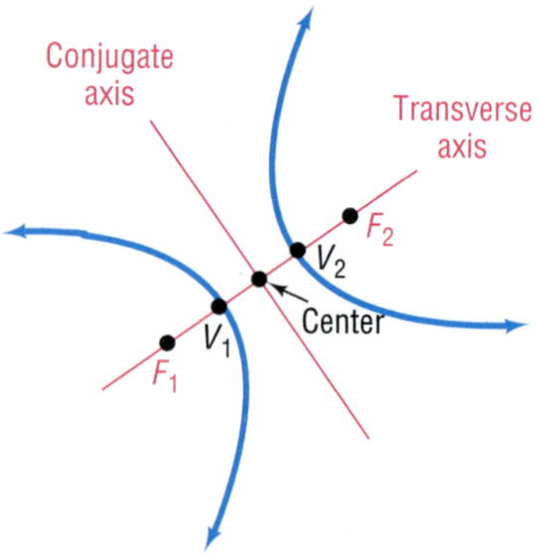

Figure 35 illustrates a hyperbola with foci F_1 and F_2. The line containing the foci is called the **transverse axis.** The midpoint of the line segment joining the foci is called the **center** of the hyperbola. The line through the center and perpendicular to the transverse axis is called the **conjugate axis.** The hyperbola consists of two separate curves, called **branches,** that are symmetric with respect to the transverse axis, conjugate axis, and center. The two points of intersection of the hyperbola and the transverse axis are the **vertices,** V_1 and V_2, of the hyperbola.

With these ideas in mind, we are now ready to find the equation of a hyperbola in a rectangular coordinate system. First, we place the center at the origin. Next, we position the hyperbola so that its transverse axis coincides with a coordinate axis. Suppose that the transverse axis coincides with the x-axis, as shown in Figure 36.

If c is the distance from the center to a focus, then one focus will be at $F_1 = (-c, 0)$ and the other at $F_2 = (c, 0)$. Now we let the constant difference of the distances from any point $P = (x, y)$ on the hyperbola to the foci F_1 and F_2 be denoted by $\pm 2a$. (If P is on the right branch, the + sign is used; if P is on the left branch, the − sign is used.) The coordinates of P must satisfy the equation

FIGURE 36
$d(F_1, P) - d(F_2, P) = \pm 2a$

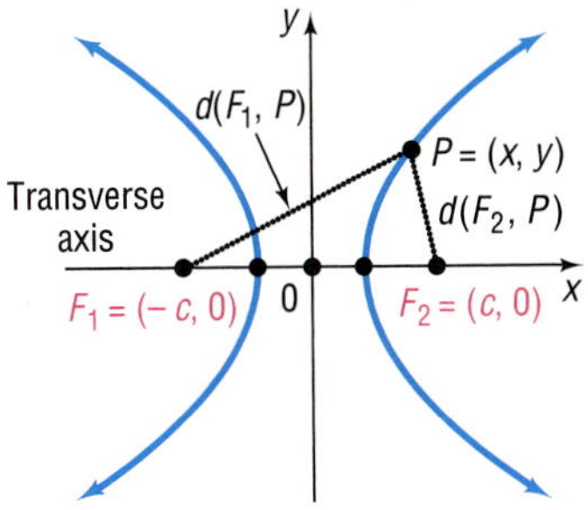

$$d(F_1, P) - d(F_2, P) = \pm 2a$$ Difference of the distances from P to the foci equals $\pm 2a$.

$$\sqrt{(x+c)^2 + y^2} - \sqrt{(x-c)^2 + y^2} = \pm 2a$$ Use the distance formula.

$$\sqrt{(x+c)^2 + y^2} = \pm 2a + \sqrt{(x-c)^2 + y^2}$$ Isolate one radical.

$$(x+c)^2 + y^2 = 4a^2 \pm 4a\sqrt{(x-c)^2 + y^2} + (x-c)^2 + y^2$$ Square both sides.

Next, we remove the parentheses:

$$x^2 + 2cx + c^2 + y^2 = 4a^2 \pm 4a\sqrt{(x-c)^2 + y^2} + x^2 - 2cx + c^2 + y^2$$

$$4cx - 4a^2 = \pm 4a\sqrt{(x-c)^2 + y^2}$$ Isolate the radical.

$$cx - a^2 = \pm a\sqrt{(x-c)^2 + y^2}$$ Divide each side by 4.

$$(cx - a^2)^2 = a^2[(x-c)^2 + y^2]$$ Square both sides.

$$c^2x^2 - 2ca^2x + a^4 = a^2(x^2 - 2cx + c^2 + y^2)$$

$$c^2x^2 + a^4 = a^2x^2 + a^2c^2 + a^2y^2$$

$$(c^2 - a^2)x^2 - a^2y^2 = a^2c^2 - a^4$$

$$(c^2 - a^2)x^2 - a^2y^2 = a^2(c^2 - a^2) \qquad (1)$$

To obtain points on the hyperbola off the x-axis, it must be that $a < c$. To see why, look again at Figure 36.

$$d(F_1, P) < d(F_2, P) + d(F_1, F_2)$$ Use triangle F_1PF_2.

$$d(F_1, P) - d(F_2, P) < d(F_1, F_2)$$ P is on the right branch, so $d(F_1, P) - d(F_2, P) = 2a$.

$$2a < 2c$$

$$a < c$$

Since $a < c$, we also have $a^2 < c^2$, so $c^2 - a^2 > 0$. Let $b^2 = c^2 - a^2$, $b > 0$. Then equation (1) can be written as

$$b^2x^2 - a^2y^2 = a^2b^2$$

$$\frac{x^2}{a^2} - \frac{y^2}{b^2} = 1$$

To find the vertices of the hyperbola defined by this equation, let $y = 0$. The vertices satisfy the equation $x^2/a^2 = 1$, the solutions of which are $x = \pm a$. Consequently, the vertices of the hyperbola are $V_1 = (-a, 0)$ and $V_2 = (a, 0)$.

Theorem Equation of a Hyperbola; Center at (0, 0); Foci at ($\pm c$, 0); Vertices at ($\pm a$, 0); Transverse Axis along x-Axis

An equation of the hyperbola with center at (0, 0), foci at $(-c, 0)$ and $(c, 0)$, and vertices at $(-a, 0)$ and $(a, 0)$ is

$$\frac{x^2}{a^2} - \frac{y^2}{b^2} = 1 \quad \text{where } b^2 = c^2 - a^2 \qquad (2)$$

The transverse axis is the x-axis.

2 As you can verify, the hyperbola defined by equation (2) is symmetric with respect to the x-axis, y-axis, and origin. To find the y-intercepts, if any,

let $x = 0$ in equation (2). This results in the equation $y^2/b^2 = -1$, which has no solution. We conclude that the hyperbola defined by equation (2) has no y-intercepts. In fact, since $x^2/a^2 - 1 = y^2/b^2 \geq 0$, it follows that $x^2/a^2 \geq 1$. Thus, there are no points on the graph for $-a < x < a$. See Figure 37.

FIGURE 37 $\dfrac{x^2}{a^2} - \dfrac{y^2}{b^2} = 1,\ b^2 = c^2 - a^2$

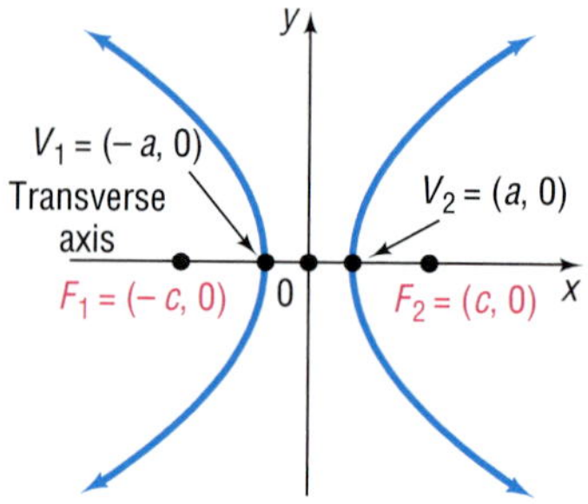

EXAMPLE 1 Finding an Equation of a Hyperbola

Find an equation of the hyperbola with center at the origin, one focus at (3, 0), and one vertex at (−2, 0). Graph the equation by hand.

Solution The hyperbola has its center at the origin, and the transverse axis coincides with the x-axis. One focus is at $(c, 0) = (3, 0)$, so $c = 3$. One vertex is at $(-a, 0) = (-2, 0)$, so $a = 2$. From equation (2), it follows that $b^2 = c^2 - a^2 = 9 - 4 = 5$, so an equation of the hyperbola is

$$\frac{x^2}{4} - \frac{y^2}{5} = 1$$

See Figure 38.

FIGURE 38 $\dfrac{x^2}{4} - \dfrac{y^2}{5} = 1$

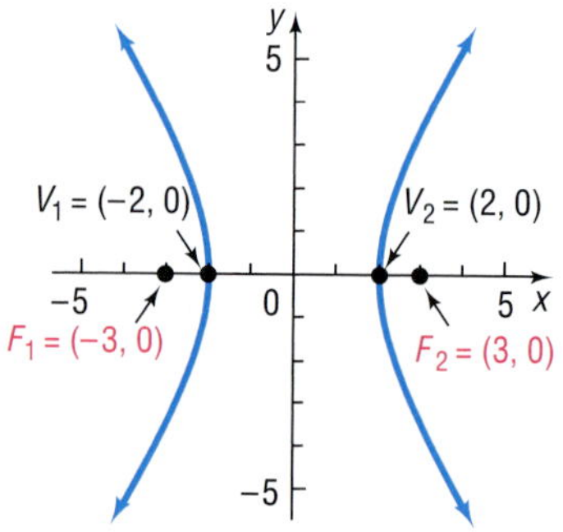

EXAMPLE 2 Using a Graphing Utility to Graph a Hyperbola

Use a graphing utility to graph the hyperbola $\dfrac{x^2}{4} - \dfrac{y^2}{5} = 1$.

Solution To graph the hyperbola $\dfrac{x^2}{4} - \dfrac{y^2}{5} = 1$, we need to graph the two functions $Y_1 = \sqrt{5}\sqrt{\dfrac{x^2}{4} - 1}$ and $Y_2 = -\sqrt{5}\sqrt{\dfrac{x^2}{4} - 1}$. As with graphing circles, parabolas, and ellipses on a graphing utility, we use a square screen setting so that the graph is not distorted. Figure 39 shows the graph of the hyperbola.

FIGURE 39

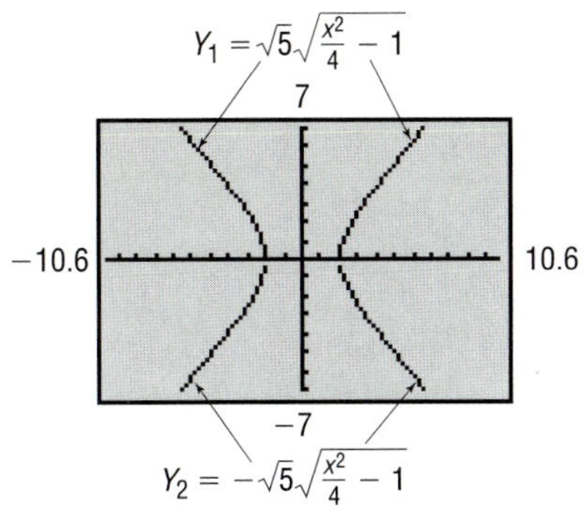

Now work Problem 9.

An equation of the form of equation (2) is the equation of a hyperbola with center at the origin, foci on the x-axis at $(-c, 0)$ and $(c, 0)$, where $c^2 = a^2 + b^2$, and transverse axis along the x-axis.

3 For the remainder of this section, the direction "Discuss the equation" will mean to find the center, transverse axis, vertices, and foci of the hyperbola and graph it.

EXAMPLE 3 Discussing the Equation of a Hyperbola

Discuss the equation $\frac{x^2}{16} - \frac{y^2}{4} = 1$.

Solution Figure 40(a) shows the graph of $\frac{x^2}{16} - \frac{y^2}{4} = 1$ using a graphing utility. We now proceed to analyze the equation.

The given equation is of the form of equation (2), with $a^2 = 16$ and $b^2 = 4$. Thus, the graph of the equation is a hyperbola with center at $(0, 0)$ and transverse axis along the x-axis. Also, we know that $c^2 = a^2 + b^2 = 16 + 4 = 20$. The vertices are at $(\pm a, 0) = (\pm 4, 0)$, and the foci are at $(\pm c, 0) = (\pm 2\sqrt{5}, 0)$. Figure 40(b) shows the graph drawn by hand.

FIGURE 40 $\frac{x^2}{16} - \frac{y^2}{4} = 1$

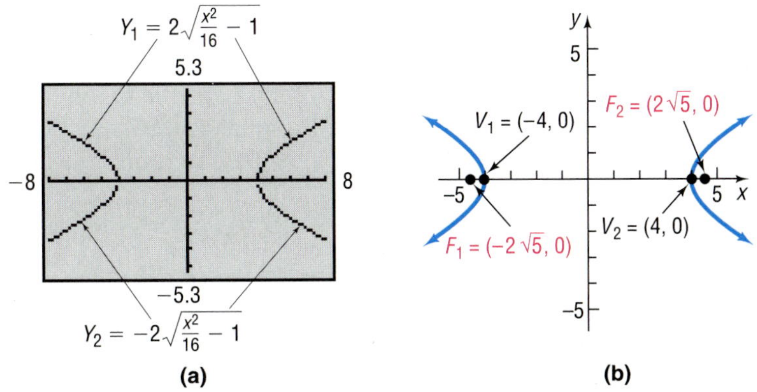

The next result gives the form of the equation of a hyperbola with center at the origin and transverse axis along the y-axis.

Theorem **Equation of a Hyperbola; Center at (0, 0); Foci at $(0, \pm c)$; Vertices at $(0, \pm a)$; Transverse Axis along y-Axis**

An equation of the hyperbola with center at (0, 0), foci at $(0, -c)$ and $(0, c)$, and vertices at $(0, -a)$ and $(0, a)$ is

$$\frac{y^2}{a^2} - \frac{x^2}{b^2} = 1 \qquad \text{where } b^2 = c^2 - a^2 \tag{3}$$

The transverse axis is the y-axis.

Figure 41 shows the graph of a typical hyperbola defined by equation (3).

FIGURE 41
$\frac{y^2}{a^2} - \frac{x^2}{b^2} = 1,\ b^2 = c^2 - a^2$

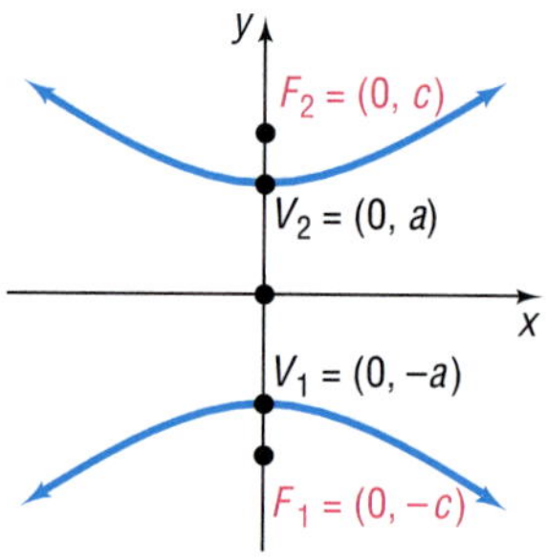

Notice the difference in the form of equations (2) and (3). When the y^2-term is subtracted from the x^2-term, the transverse axis is the x-axis. When the x^2-term is subtracted from the y^2-term, the transverse axis is the y-axis.

EXAMPLE 4 Discussing the Equation of a Hyperbola

Discuss the equation: $y^2 - 4x^2 = 4$

Solution Figure 42(a) shows the graph of $y^2 - 4x^2 = 4$ using a graphing utility. We now proceed to analyze the equation.

To put the equation in proper form, we divide each side by 4:

$$\frac{y^2}{4} - x^2 = 1$$

Since the x^2-term is subtracted from the y^2-term, the equation is that of a hyperbola with center at the origin and transverse axis along the y-axis. Also, comparing the above equation to equation (3), we find $a^2 = 4$, $b^2 = 1$, and $c^2 = a^2 + b^2 = 5$. The vertices are at $(0, \pm a) = (0, \pm 2)$, and the foci are at $(0, \pm c) = (0, \pm\sqrt{5})$. The graph, drawn by hand, is given in Figure 42(b).

FIGURE 42
$\frac{y^2}{4} - x^2 = 1$

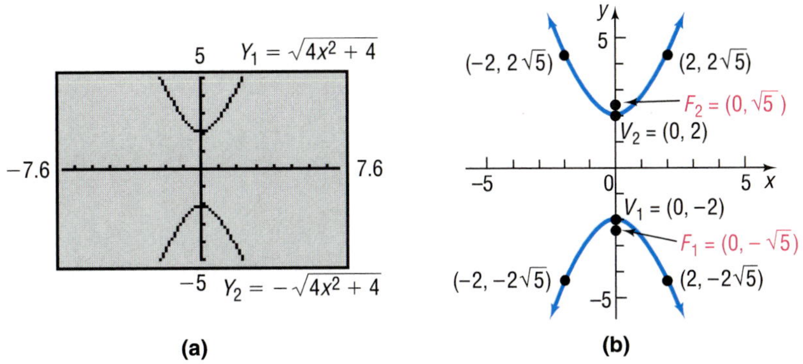

EXAMPLE 5 Finding an Equation of a Hyperbola

Find an equation of the hyperbola having one vertex at (0, 2) and foci at (0, −3) and (0, 3). Graph the equation by hand.

Solution Since the foci are at (0, −3) and (0, 3), the center of the hyperbola is at the origin. Also, the transverse axis is along the y-axis. The given information also reveals that $c = 3$, $a = 2$, and $b^2 = c^2 - a^2 = 9 - 4 = 5$. The form of the equation of the hyperbola is given by equation (3):

$$\frac{y^2}{a^2} - \frac{x^2}{b^2} = 1$$

$$\frac{y^2}{4} - \frac{x^2}{5} = 1$$

See Figure 43.

FIGURE 43
$\frac{y^2}{4} - \frac{x^2}{5} = 1$

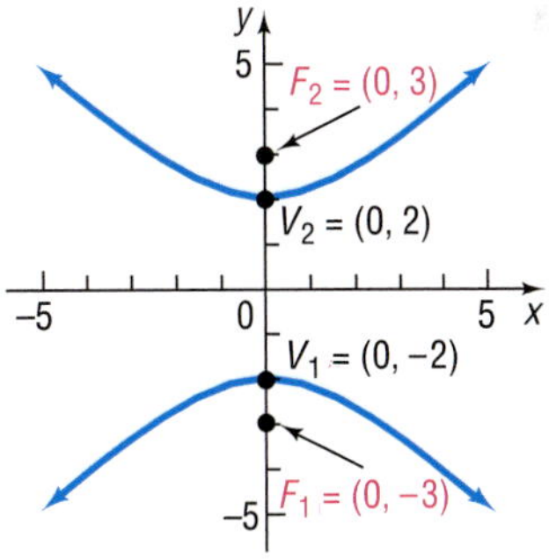

Now work Problem 13.

Look at the equations of the hyperbolas in Examples 4 and 5. For the hyperbola in Example 4, $a^2 = 4$ and $b^2 = 1$, so $a > b$; for the hyperbola in Example 5, $a^2 = 4$ and $b^2 = 5$, so $a < b$. We conclude that, for hyperbolas, there are no requirements involving the relative sizes of a and b. Contrast this situation to the case of an ellipse, in which the relative sizes of a and b dictate which axis is the major axis. Hyperbolas have another feature to distinguish them from ellipses and parabolas: Hyperbolas have asymptotes.

Asymptotes

4 Recall from Section 4.4 that a horizontal or oblique asymptote of a graph is a line with the property that the distance from the line to points on the graph approaches 0 as $x \to -\infty$ or as $x \to \infty$.

Theorem Asymptotes of a Hyperbola

The hyperbola $\dfrac{x^2}{a^2} - \dfrac{y^2}{b^2} = 1$ has the two oblique asymptotes

$$y = \frac{b}{a}x \quad \text{and} \quad y = -\frac{b}{a}x$$

Proof We begin by solving for y in the equation of the hyperbola:

$$\frac{x^2}{a^2} - \frac{y^2}{b^2} = 1$$

$$\frac{y^2}{b^2} = \frac{x^2}{a^2} - 1$$

$$y^2 = b^2\left(\frac{x^2}{a^2} - 1\right)$$

If $x \neq 0$, we can rearrange the right side in the form

$$y^2 = \frac{b^2x^2}{a^2}\left(1 - \frac{a^2}{x^2}\right)$$

$$y = \pm\frac{bx}{a}\sqrt{1 - \frac{a^2}{x^2}}$$

Now, as $x \to -\infty$ or as $x \to \infty$, the term a^2/x^2 approaches 0, so the expression under the radical approaches 1. Thus, as $x \to -\infty$ or as $x \to \infty$, the value of y approaches $\pm bx/a$; that is, the graph of the hyperbola approaches the lines

$$y = -\frac{b}{a}x \quad \text{and} \quad y = \frac{b}{a}x$$

Thus, these lines are oblique asymptotes of the hyperbola. ■

The asymptotes of a hyperbola are not part of the hyperbola, but they do serve as a guide for graphing a hyperbola by hand. For example, suppose that we want to graph the equation

$$\frac{x^2}{a^2} - \frac{y^2}{b^2} = 1$$

We begin by plotting the vertices $(-a, 0)$ and $(a, 0)$. Then we plot the points $(0, -b)$ and $(0, b)$ and use these four points to construct a rectangle, as shown in Figure 44. The diagonals of this rectangle have slopes b/a and $-b/a$, and their extensions are the asymptotes $y = (b/a)x$ and $y = -(b/a)x$ of the hyperbola.

FIGURE 44 $\dfrac{x^2}{a^2} - \dfrac{y^2}{b^2} = 1$

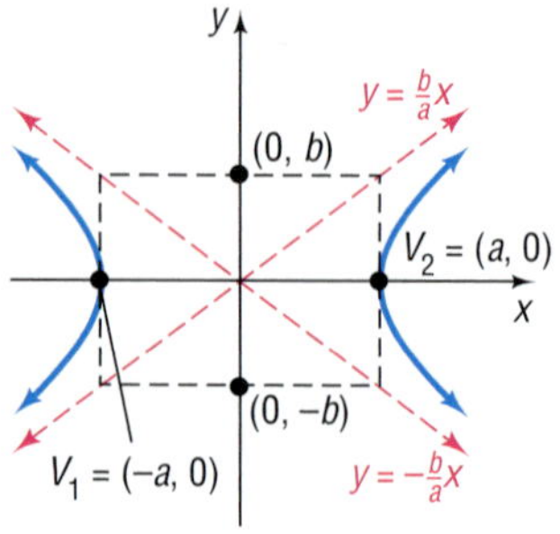

Theorem Asymptotes of a Hyperbola

The hyperbola $\dfrac{y^2}{a^2} - \dfrac{x^2}{b^2} = 1$
has the two oblique asymptotes

$$y = \frac{a}{b}x \quad \text{and} \quad y = -\frac{a}{b}x$$

You are asked to prove this result in Problem 64.

EXAMPLE 6 Discussing the Equation of a Hyperbola

Discuss the equation: $9x^2 - 4y^2 = 36$.

Solution Divide each side by 36 to put the equation in proper form:

$$\frac{x^2}{4} - \frac{y^2}{9} = 1$$

Figure 45(a) shows the graph of $\dfrac{x^2}{4} - \dfrac{y^2}{9} = 1$ and its asymptotes, $y = \dfrac{3}{2}x$ and $y = -\dfrac{3}{2}x$, using a graphing utility. We now proceed to analyze the equation. This is the equation of a hyperbola with center at the origin and transverse axis along the x-axis. Using $a^2 = 4$ and $b^2 = 9$, we find $c^2 = a^2 + b^2 = 13$. The vertices are at $(\pm a, 0) = (\pm 2, 0)$, the foci are at $(\pm c, 0) = (\pm\sqrt{13}, 0)$, and the asymptotes have the equations

$$y = \frac{3}{2}x \quad \text{and} \quad y = -\frac{3}{2}x$$

Now form the rectangle containing the points $(\pm a, 0)$ and $(0, \pm b)$, that is, $(-2, 0)$, $(2, 0)$, $(0, -3)$, and $(0, 3)$. The extensions of the diagonals of this rectangle are the asymptotes. See Figure 45(b) for the graph drawn by hand.

FIGURE 45 $\dfrac{x^2}{4} - \dfrac{y^2}{9} = 1$

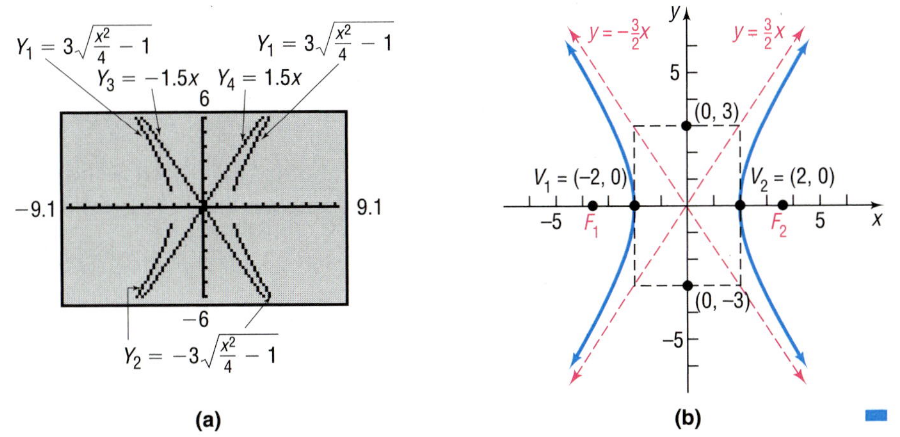

Now work Problem 21.

Exploration Use TRACE to see what happens as x becomes unbounded in the positive direction for both the upper and lower portions of the hyperbola in Example 6.

Center at (h, k)

5 If a hyperbola with center at the origin and transverse axis coinciding with a coordinate axis is shifted horizontally h units and then vertically k units, the result is a hyperbola with center at (h, k) and transverse axis parallel to a coordinate axis. Table 4 gives the forms of the equations of such hyperbolas. See Figure 46 for the graphs.

TABLE 4 Hyperbolas with Center at (h, k) and Transverse Axis Parallel to a Coordinate Axis

Center	Transverse Axis	Foci	Vertices	Equation	Asymptotes
(h, k)	Parallel to x-axis	$(h \pm c, k)$	$(h \pm a, k)$	$\dfrac{(x - h)^2}{a^2} - \dfrac{(y - k)^2}{b^2} = 1,$ $b^2 = c^2 - a^2$	$y - k = \pm\dfrac{b}{a}(x - h)$
(h, k)	Parallel to y-axis	$(h, k \pm c)$	$(h, k \pm a)$	$\dfrac{(y - k)^2}{a^2} - \dfrac{(x - h)^2}{b^2} = 1,$ $b^2 = c^2 - a^2$	$y - k = \pm\dfrac{a}{b}(x - h)$

FIGURE 46

(a) $\dfrac{(x-h)^2}{a^2} - \dfrac{(y-k)^2}{b^2} = 1$

(b) $\dfrac{(y-k)^2}{a^2} - \dfrac{(x-h)^2}{b^2} = 1$

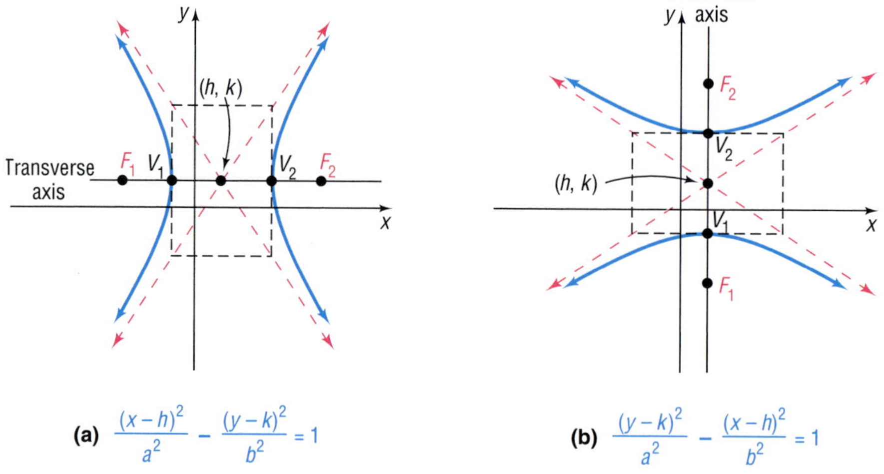

E X A M P L E 7

Finding an Equation of a Hyperbola, Center Not at the Origin

Find an equation for the hyperbola with center at $(1, -2)$, one focus at $(4, -2)$, and one vertex at $(3, -2)$. Graph the equation by hand.

Solution The center is at $(h, k) = (1, -2)$, so $h = 1$ and $k = -2$. The transverse axis is parallel to the x-axis. The distance from the center $(1, -2)$ to the focus $(4, -2)$ is $c = 3$; the distance from the center $(1, -2)$ to the vertex $(3, -2)$ is $a = 2$. Thus, $b^2 = c^2 - a^2 = 9 - 4 = 5$. The equation is

$$\frac{(x - h)^2}{a^2} - \frac{(y - k)^2}{b^2} = 1$$

$$\frac{(x - 1)^2}{4} - \frac{(y + 2)^2}{5} = 1$$

See Figure 47.

FIGURE 47 $\frac{(x-1)^2}{4} - \frac{(y+2)^2}{5} = 1$

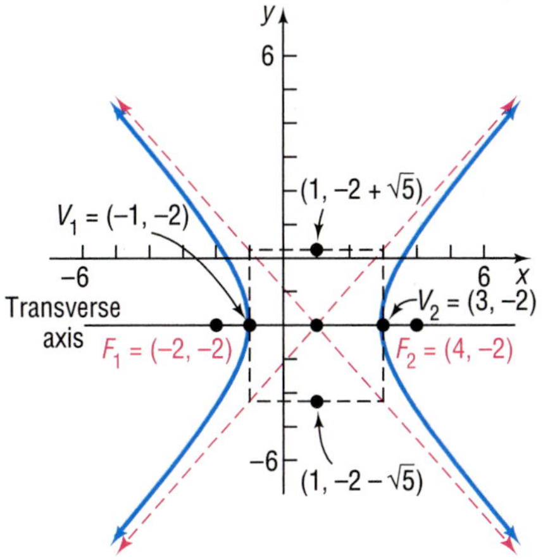

Now work Problem 31.

EXAMPLE 8 Using a Graphing Utility to Graph a Hyperbola, Center Not at the Origin

Use a graphing utility to graph the hyperbola $\frac{(x-1)^2}{4} - \frac{(y+2)^2}{5} = 1$.

Solution First, we must solve the equation for y:

$$\frac{(x-1)^2}{4} - \frac{(y+2)^2}{5} = 1$$

$$\frac{(y+2)^2}{5} = \frac{(x-1)^2}{4} - 1$$

$$(y+2)^2 = 5\left[\frac{(x-1)^2}{4} - 1\right]$$

$$y + 2 = \pm\sqrt{5\left[\frac{(x-1)^2}{4} - 1\right]}$$

$$y = -2 \pm \sqrt{5\left[\frac{(x-1)^2}{4} - 1\right]}$$

FIGURE 48

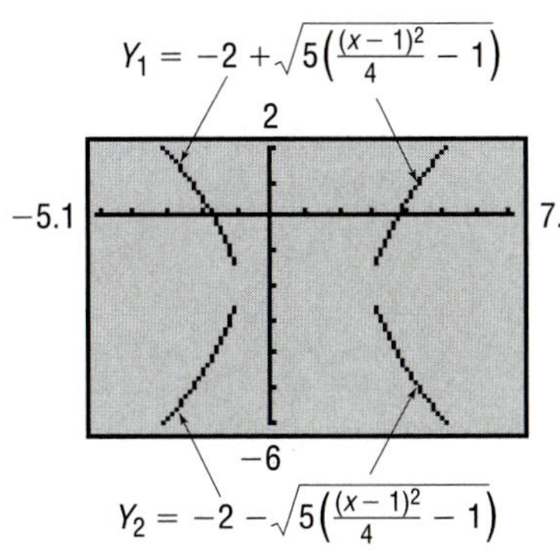

Figure 48 shows the graph of $Y_1 = -2 + \sqrt{5\left[\frac{(x-1)^2}{4} - 1\right]}$ and $Y_2 = -2 - \sqrt{5\left[\frac{(x-1)^2}{4} - 1\right]}$.

EXAMPLE 9 Discussing the Equation of a Hyperbola

Discuss the equation: $-x^2 + 4y^2 - 2x - 16y + 11 = 0$.

Solution We complete the squares in x and in y:

$$\begin{aligned}
-x^2 + 4y^2 - 2x - 16y + 11 &= 0 \\
-(x^2 + 2x) + 4(y^2 - 4y) &= -11 && \text{Group terms.} \\
-(x^2 + 2x + 1) + 4(y^2 - 4y + 4) &= -1 + 16 - 11 && \text{Complete each square.} \\
-(x + 1)^2 + 4(y - 2)^2 &= 4 \\
(y - 2)^2 - \frac{(x + 1)^2}{4} &= 1 && \text{Divide by 4.}
\end{aligned}$$

Figure 49(a) shows the graph of $(y - 2)^2 - \frac{(x + 1)^2}{4} = 1$ using a graphing utility. We now proceed to analyze the equation.

It is the equation of a hyperbola with center at $(-1, 2)$ and transverse axis parallel to the y-axis. Also, $a^2 = 1$ and $b^2 = 4$, so $c^2 = a^2 + b^2 = 5$. The vertices are at $(h, k \pm a) = (-1, 2 \pm 1)$, or $(-1, 1)$ and $(-1, 3)$. The foci are at $(h, k \pm c) = (-1, 2 \pm \sqrt{5})$. The asymptotes are $y - 2 = \frac{1}{2}(x + 1)$ and $y - 2 = -\frac{1}{2}(x + 1)$. Figure 49(b) shows the graph drawn by hand.

FIGURE 49 $(y - 2)^2 - \frac{(x + 1)^2}{4} = 1$

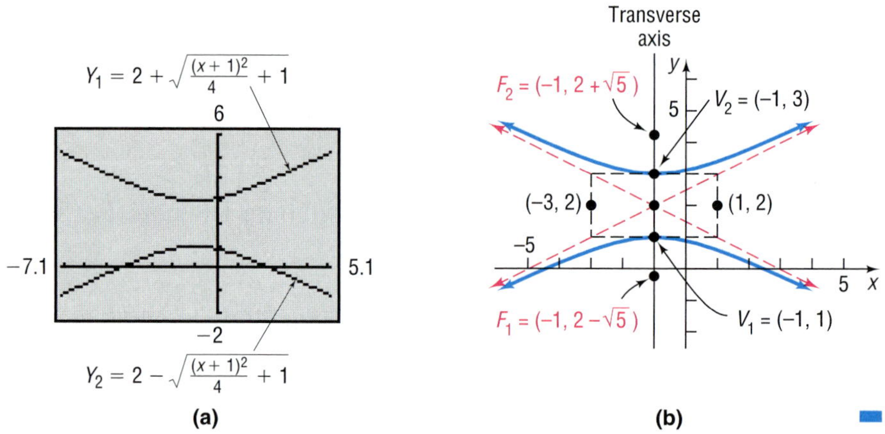

FIGURE 50

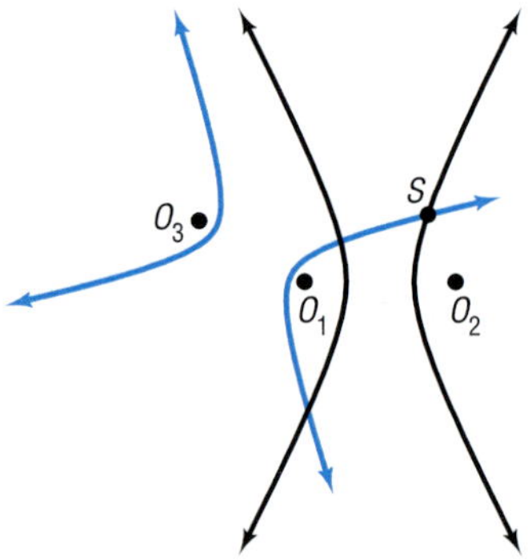

Applications

Suppose that a gun is fired from an unknown source S. An observer at O_1 hears the report (sound of gun shot) 1 second after another observer at O_2. Because sound travels at about 1100 feet per second, it follows that the point S must be 1100 feet closer to O_2 than to O_1. Thus, S lies on one branch of a hyperbola with foci at O_1 and O_2. (Do you see why? The difference of the distances from S to O_1 and from S to O_2 is the constant 1100.) If a third observer at O_3 hears the same report 2 seconds after O_1 hears it, then S will lie on a branch of a second hyperbola with foci at O_1 and O_3. The intersection of the two hyperbolas will pinpoint the location of S. See Figure 50 for an illustration.

LORAN

In the LOng RAnge Navigation system (LORAN), a master radio sending station and a secondary sending station emit signals that can be received by a ship at sea. (See Figure 51.) Because a ship monitoring the two signals will usually be nearer to one of the two stations, there will be a difference in the distance the two signals travel, which will register as a slight time difference between the signals. As long as the time difference remains constant, the dif-

FIGURE 51

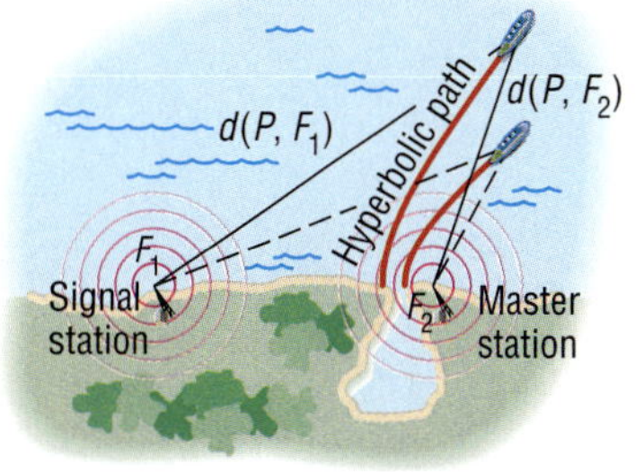

$d(P, F_1) - d(P, F_2) = \text{constant}$

ference of the two distances will also be constant. If the ship follows a path corresponding to the fixed time difference, it will follow the path of a hyperbola whose foci are located at the positions of the two sending stations. So for each time difference a different hyperbolic path results, each bringing the ship to a different shore location. Navigation charts show the various hyperbolic paths corresponding to different time differences.

EXAMPLE 10 LORAN

Two LORAN stations are positioned 250 miles apart along a straight shore.

(a) A ship records a time difference of 0.00086 second between the LORAN signals. Set up an appropriate rectangular coordinate system to determine where the ship would reach shore if it were to follow the hyperbola corresponding to this time difference.

(b) If the ship wants to enter a harbor located between the two stations 25 miles from the master station, what time difference should it be looking for?

(c) If the ship is 80 miles off shore when the desired time difference is obtained, what is the exact location of the ship?

[**Note:** The speed of each radio signal is 186,000 miles per second.]

Solution

FIGURE 52

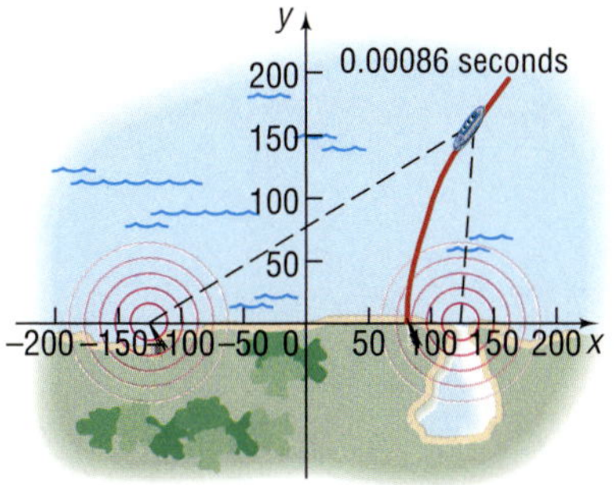

(a) We set up a rectangular coordinate system so that the two stations lie on the x-axis and the origin is midway between them. See Figure 52. The ship lies on a hyperbola whose foci are the locations of the two stations. The reason for this is that the constant time difference of the signals from each station results in a constant difference in the distance of the ship from each station. Since the time difference is 0.00086 second and the speed of the signal is 186,000 miles per second, the difference of the distances from the ship to each station (foci) is

$$\text{Distance} = \text{Speed} \times \text{Time} = 186{,}000 \times 0.00086 = 160 \text{ miles}$$

The difference of the distances from the ship to each station, 160, equals $2a$, so $a = 80$ and the vertex of the corresponding hyperbola is at $(80, 0)$. Since the focus is at $(125, 0)$, following this hyperbola the ship would reach shore 45 miles from the master station.

(b) To reach shore 25 miles from the master station, the ship should follow a hyperbola with vertex at $(100, 0)$. For this hyperbola, $a = 100$, so the constant difference of the distances from the ship to each station is 200. The time difference the ship should look for is

$$\text{Time} = \frac{\text{Distance}}{\text{Speed}} = \frac{200}{186{,}000} = 0.001075 \text{ second}$$

(c) To find the exact location of the ship, we need to find the equation of the hyperbola with vertex at (100, 0) and a focus at (125, 0). The form of the equation of this hyperbola is

$$\frac{x^2}{a^2} - \frac{y^2}{b^2} = 1$$

where $a = 100$. Since $c = 125$, we have

$$b^2 = c^2 - a^2 = 125^2 - 100^2 = 5625$$

The equation of the hyperbola is

$$\frac{x^2}{100^2} - \frac{y^2}{5625} = 1$$

Since the ship is 80 miles from shore, we use $y = 80$ in the equation and solve for x.

$$\frac{x^2}{100^2} - \frac{80^2}{5625} = 1$$

$$\frac{x^2}{100^2} = 1 + \frac{80^2}{5625} = 2.14$$

$$x^2 = 100^2(2.14)$$

$$x = 146$$

The ship is at the position (146, 80). See Figure 53.

FIGURE 53

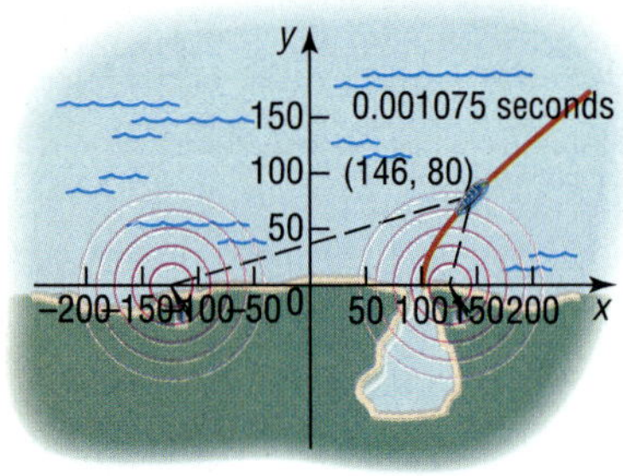

10.4 EXERCISES

In Problems 1–4, the graph of a hyperbola is given. Match each graph to its equation.

A. $\frac{x^2}{4} - y^2 = 1$ B. $x^2 - \frac{y^2}{4} = 1$ C. $\frac{y^2}{4} - x^2 = 1$ D. $y^2 - \frac{x^2}{4} = 1$

1.

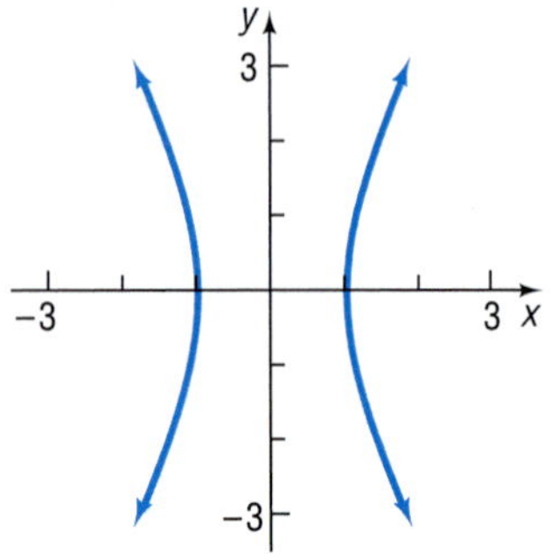

2.

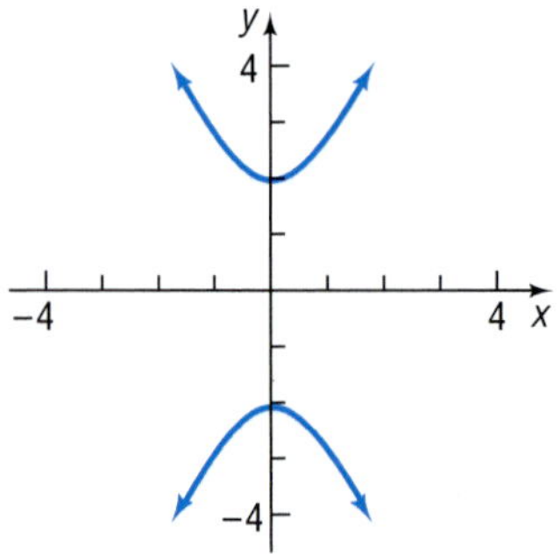

3.

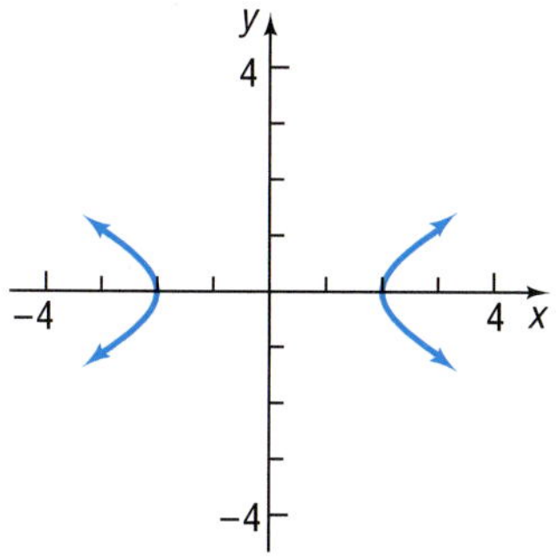

4.

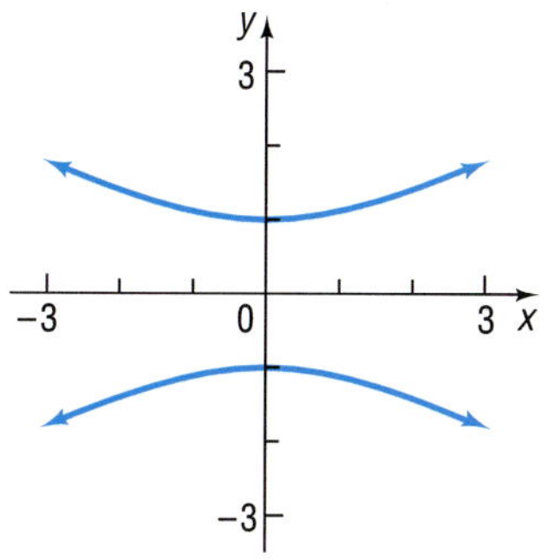

In Problems 5–8, the graph of a hyperbola is given. Match each graph to its equation.

A. $\frac{x^2}{16} - \frac{y^2}{9} = 1$ *B.* $\frac{x^2}{9} - \frac{y^2}{16} = 1$ *C.* $\frac{y^2}{16} - \frac{x^2}{9} = 1$ *D.* $\frac{y^2}{9} - \frac{x^2}{16} = 1$

5.

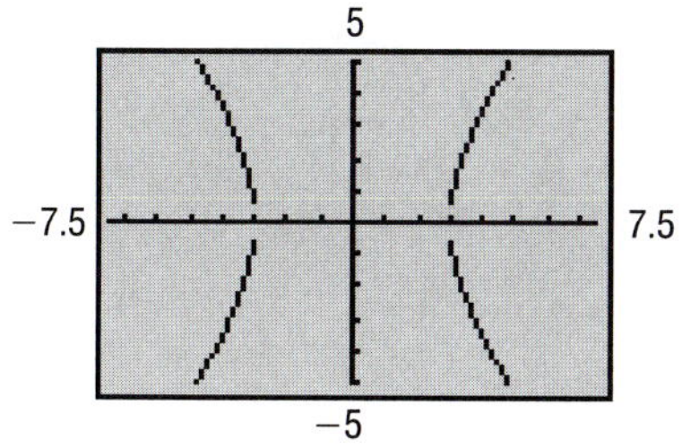

6.

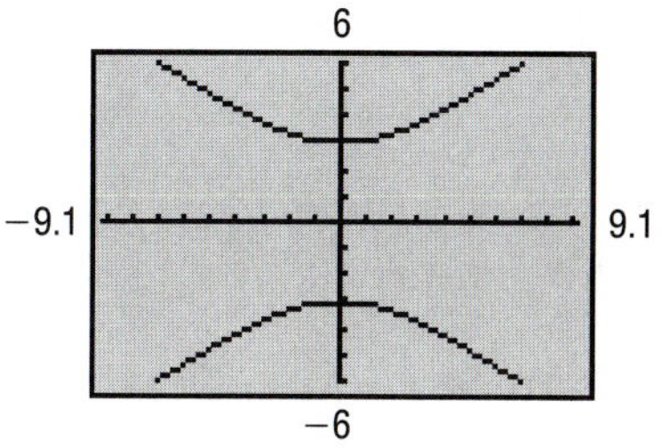

7.

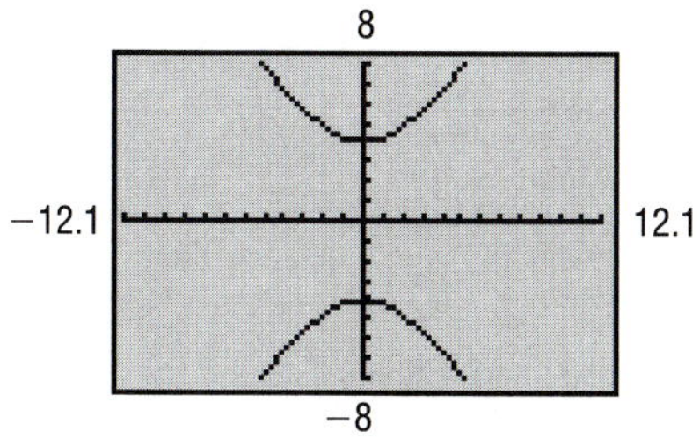

8.

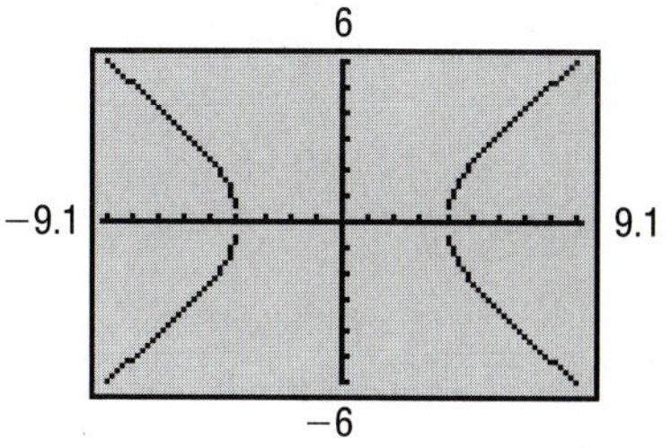

In Problems 9–18, find an equation for the hyperbola described. Graph the equation by hand.

9. Center at (0, 0); focus at (3, 0); vertex at (1, 0)

10. Center at (0, 0); focus at (0, 5); vertex at (0, 3)

11. Center at (0, 0); focus at (0, −6); vertex at (0, 4)

12. Center at (0, 0); focus at (−3, 0); vertex at (2, 0)

13. Foci at (−5, 0) and (5, 0); vertex at (3, 0)

14. Focus at (0, 6); vertices at (0, −2) and (0, 2)

15. Vertices at (0, −6) and (0, 6); asymptote the line $y = 2x$

16. Vertices at (−4, 0) and (4, 0); asymptote the line $y = 2x$

17. Foci at (−4, 0) and (4, 0); asymptote the line $y = -x$

18. Foci at (0, −2) and (0, 2); asymptote the line $y = -x$

In Problems 19–26, find the center, transverse axis, vertices, foci, and asymptotes. Graph each equation using a graphing utility.

19. $\dfrac{x^2}{25} - \dfrac{y^2}{9} = 1$ **20.** $\dfrac{y^2}{16} - \dfrac{x^2}{4} = 1$ **21.** $4x^2 - y^2 = 16$ **22.** $y^2 - 4x^2 = 16$

23. $y^2 - 9x^2 = 9$ **24.** $x^2 - y^2 = 4$ **25.** $y^2 - x^2 = 25$ **26.** $2x^2 - y^2 = 4$

In Problems 27–30, write an equation for each hyperbola.

27.

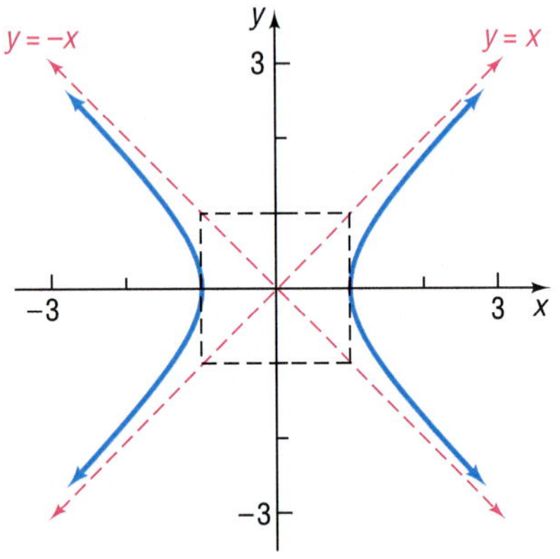

28.

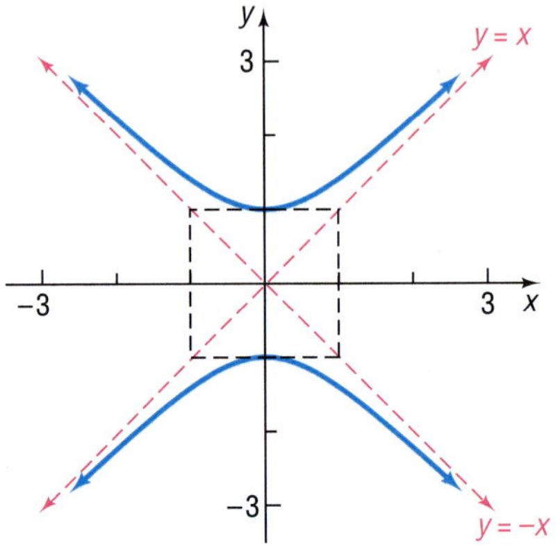

29.

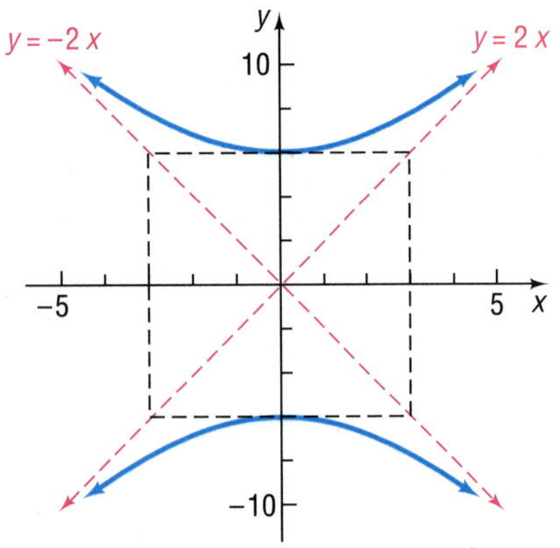

30.

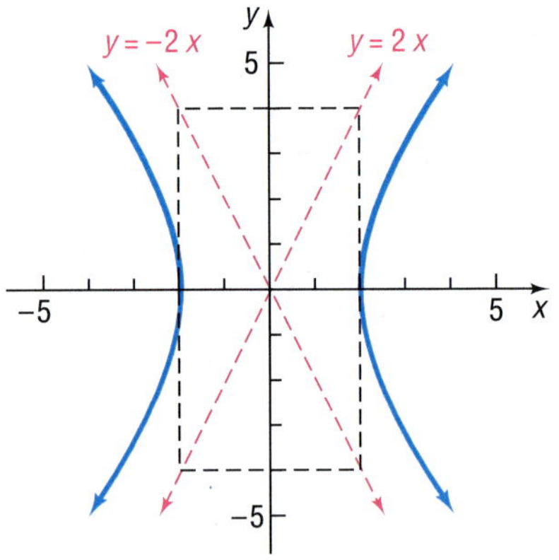

In Problems 31–38, find an equation for the hyperbola described. Graph the equation by hand.

31. Center at $(4, -1)$; focus at $(7, -1)$; vertex at $(6, -1)$

32. Center at $(-3, 1)$; focus at $(-3, 6)$; vertex at $(-3, 4)$

33. Center at $(-3, -4)$; focus at $(-3, -8)$; vertex at $(-3, -2)$

34. Center at $(1, 4)$; focus at $(-2, 4)$; vertex at $(0, 4)$

35. Foci at $(3, 7)$ and $(7, 7)$; vertex at $(6, 7)$

36. Focus at $(-4, 0)$; vertices at $(-4, 4)$ and $(-4, 2)$

37. Vertices at $(-1, -1)$ and $(3, -1)$; asymptote the line $(x - 1)/2 = (y + 1)/3$

38. Vertices at $(1, -3)$ and $(1, 1)$; asymptote the line $(x - 1)/2 = (y + 1)/3$

In Problems 39–52, find the center, transverse axis, vertices, foci, and asymptotes. Graph each equation using a graphing utility.

39. $\dfrac{(x-2)^2}{4} - \dfrac{(y+3)^2}{9} = 1$

40. $\dfrac{(y+3)^2}{4} - \dfrac{(x-2)^2}{9} = 1$

41. $(y-2)^2 - 4(x+2)^2 = 4$

42. $(x+4)^2 - 9(y-3)^2 = 9$

43. $(x+1)^2 - (y+2)^2 = 4$

44. $(y-3)^2 - (x+2)^2 = 4$

45. $x^2 - y^2 - 2x - 2y - 1 = 0$

46. $y^2 - x^2 - 4y + 4x - 1 = 0$

47. $y^2 - 4x^2 - 4y - 8x - 4 = 0$

48. $2x^2 - y^2 + 4x + 4y - 4 = 0$

49. $4x^2 - y^2 - 24x - 4y + 16 = 0$

50. $2y^2 - x^2 + 2x + 8y + 3 = 0$

51. $y^2 - 4x^2 - 16x - 2y - 19 = 0$

52. $x^2 - 3y^2 + 8x - 6y + 4 = 0$

In Problems 53–56, graph each function by hand. Verify your answer using a graphing utility.
[**Hint:** Notice that each function is "half" a hyperbola.]

53. $f(x) = \sqrt{16 + 4x^2}$

54. $f(x) = -\sqrt{9 + 9x^2}$

55. $f(x) = -\sqrt{-25 + x^2}$

56. $f(x) = \sqrt{-1 + x^2}$

57. **LORAN** Two LORAN stations are positioned 200 miles apart along a straight shore.

(a) A ship records a time difference of 0.00038 second between the LORAN signals. Set up an appropriate rectangular coordinate system to determine where the ship would reach shore if it were to follow the hyperbola corresponding to this time difference.

(b) If the ship wants to enter a harbor located between the two stations 20 miles from the master station, what time difference should it be looking for?

(c) If the ship is 50 miles off shore when the desired time difference is obtained, what is the exact location of the ship?

[**Note:** The speed of each radio signal is 186,000 miles per second.]

58. **LORAN** Two LORAN stations are positioned 100 miles apart along a straight shore.

(a) A ship records a time difference of 0.00032 second between the LORAN signals. Set up an appropriate rectangular coordinate system to determine where the ship would reach shore if it were to follow the hyperbola corresponding to this time difference.

(b) If the ship wants to enter a harbor located between the two stations 10 miles from the master station, what time difference should it be looking for?

(c) If the ship is 20 miles off shore when the desired time difference is obtained, what is the exact location of the ship?

[**Note:** The speed of each radio signal is 186,000 miles per second.]

59. **Calibrating Instruments** In a test of their recording devices a team of seismologists positioned two of the devices 2000 feet apart, with the device at point A to the west of the device at point B. At a point between the devices and 200 feet from point B, a small amount of explosive was detonated and a note made of the time at which the sound reached each device. A second explosion is to be carried out at a point directly north of point B.

(a) How far north should the site of the second explosion be chosen so that the measured time difference recorded by the devices for the second detonation is the same as that recorded for the first detonation?

(b) Explain why this experiment can be used to calibrate the instruments.

60. Explain in your own words the LORAN system of navigation.

61. The **eccentricity** e of a hyperbola is defined as the number c/a, where a and c are the numbers given in equation (2). Because $c > a$, it follows that $e > 1$. Describe the general shape of a hyperbola whose eccentricity is close to 1. What is the shape if e is very large?

62. A hyperbola for which $a = b$ is called an **equilateral hyperbola.** Find the eccentricity e of an equilateral hyperbola.

[**Note:** The eccentricity of a hyperbola is defined in Problem 61.]

63. Two hyperbolas that have the same set of asymptotes are called **conjugate.** Show that the hyperbolas

$$\frac{x^2}{4} - y^2 = 1 \quad \text{and} \quad y^2 - \frac{x^2}{4} = 1$$

are conjugate. Graph each hyperbola on the same set of coordinate axes.

64. Prove that the hyperbola

$$\frac{y^2}{a^2} - \frac{x^2}{b^2} = 1$$

has the two oblique asymptotes

$$y = \frac{a}{b}x \quad \text{and} \quad y = -\frac{a}{b}x$$

65. Show that the graph of an equation of the form

$$Ax^2 + Cy^2 + F = 0 \qquad A \neq 0, C \neq 0, F \neq 0$$

where A and C are of opposite sign, is a hyperbola with center at $(0, 0)$.

66. Show that the graph of an equation of the form

$$Ax^2 + Cy^2 + Dx + Ey + F = 0 \quad A \neq 0, C \neq 0$$

where A and C are of opposite sign:

(a) Is a hyperbola if $(D^2/4A) + (E^2/4C) - F \neq 0$.

(b) Is two intersecting lines if $(D^2/4A) + (E^2/4C) - F = 0$.

10.5 ROTATION OF AXES; GENERAL FORM OF A CONIC

1 Identify a Conic

2 Use a Rotation of Axes to Transform Equations

3 Discuss an Equation Using a Rotation of Axes

4 Identify Conics Without a Rotation of Axes

In this section, we show that the graph of a general second-degree polynomial containing two variables x and y, that is, an equation of the form

$$Ax^2 + Bxy + Cy^2 + Dx + Ey + F = 0 \qquad (1)$$

where A, B, and C are not simultaneously 0, is a conic. We shall not concern ourselves here with the degenerate cases of equation (1), such as $x^2 + y^2 = 0$, whose graph is a single point $(0, 0)$; or $x^2 + 3y^2 + 3 = 0$, whose graph contains no points; or $x^2 - 4y^2 = 0$, whose graph is two lines, $x - 2y = 0$ and $x + 2y = 0$.

We begin with the case where $B = 0$. In this case, the term containing xy is not present, so equation (1) has the form

$$Ax^2 + Cy^2 + Dx + Ey + F = 0$$

where either $A \neq 0$ or $C \neq 0$.

1 We have already discussed the procedure for identifying the graph of this kind of equation; we complete the squares of the quadratic expressions in x or y, or both. Once this has been done, the conic can be identified by comparing it to one of the forms studied in Sections 10.2 through 10.4.

In fact, though, we can identify the conic directly from the equation without completing the squares.

Theorem Identifying Conics without Completing the Squares

Excluding degenerate cases, the equation

$$Ax^2 + Cy^2 + Dx + Ey + F = 0 \qquad (2)$$

where either $A \neq 0$ or $C \neq 0$:

(a) Defines a parabola if $AC = 0$.

(b) Defines an ellipse (or a circle) if $AC > 0$.

(c) Defines a hyperbola if $AC < 0$.

Proof (a) If $AC = 0$, then either $A = 0$ or $C = 0$, but not both, so the form of equation (2) is either

$$Ax^2 + Dx + Ey + F = 0, \qquad A \neq 0$$

or

$$Cy^2 + Dx + Ey + F = 0, \qquad C \neq 0$$

Using the results of Problems 73 and 74 in Exercise 10.2, it follows that, except for the degenerate cases, the equation is a parabola.

(b) If $AC > 0$, then A and C are of the same sign. Using the results of Problems 73 and 74 in Exercise 10.3, except for the degenerate cases, the equation is an ellipse if $A \neq C$ or a circle if $A = C$.

(c) If $AC < 0$, then A and C are of opposite sign. Using the results of Problems 65 and 66 in Exercise 10.4, except for the degenerate cases, the equation is a hyperbola. ■

We will not be concerned with the degenerate cases of equation (2). However, in practice, you should be alert to the possibility of degeneracy.

EXAMPLE 1

Identifying a Conic without Completing the Squares

Identify each equation without completing the squares.

(a) $3x^2 + 6y^2 + 6x - 12y = 0$ (b) $2x^2 - 3y^2 + 6y + 4 = 0$
(c) $y^2 - 2x + 4 = 0$

Solution (a) We compare the given equation to equation (2) and conclude that $A = 3$ and $C = 6$. Since $AC = 18 > 0$, the equation is an ellipse.

(b) Here, $A = 2$ and $C = -3$, so $AC = -6 < 0$. The equation is a hyperbola.

(c) Here, $A = 0$ and $C = 1$, so $AC = 0$. The equation is a parabola. ■

Now work Problem 1.

Although we can now identify the type of conic represented by any equation of the form of equation (2) without completing the squares, we will still need to complete the squares if we desire additional information about a conic.

Now we turn our attention to equations of the form of equation (1), where $B \neq 0$. To discuss this case, we first need to investigate a new procedure: *rotation of axes*.

FIGURE 54

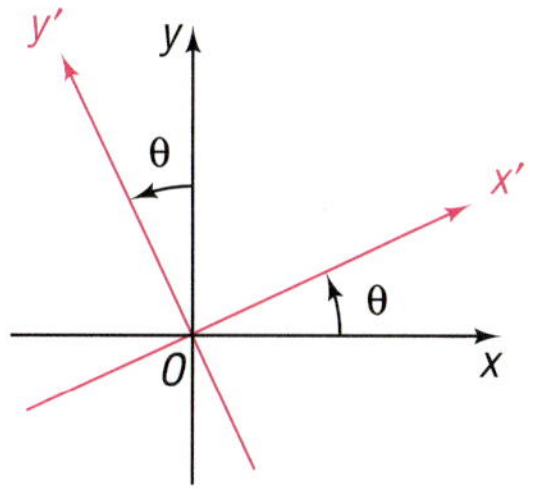

(a)

Rotation of Axes

2

In a **rotation of axes,** the origin remains fixed while the x-axis and y-axis are rotated through an angle θ to a new position; the new positions of the x- and y-axes are denoted by x' and y', respectively, as shown in Figure 54(a).

Now look at Figure 54(b) on page 756. There the point P has the coordinates (x, y) relative to the xy-plane, while the same point P has coordinates (x', y') relative to the $x'y'$-plane. We seek relationships that will enable us to express x and y in terms of x', y', and θ.

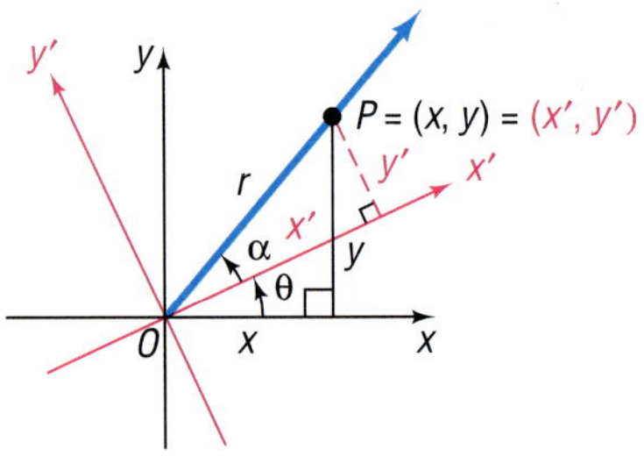

(b)

As Figure 54(b) shows, r denotes the distance from the origin O to the point P, and α denotes the angle between the positive x'-axis and the ray from O through P. Then, using the definitions of sine and cosine, we have

$$x' = r\cos\alpha \qquad y' = r\sin\alpha \tag{3}$$

$$x = r\cos(\theta + \alpha) \qquad y = r\sin(\theta + \alpha) \tag{4}$$

Now

$$\begin{aligned} x &= r\cos(\theta+\alpha) \\ &= r(\cos\theta\cos\alpha - \sin\theta\sin\alpha) && \text{Sum formula.} \\ &= (r\cos\alpha)(\cos\theta) - (r\sin\alpha)(\sin\theta) \\ &= x'\cos\theta - y'\sin\theta && \text{By equation (3).} \end{aligned}$$

Similarly,

$$\begin{aligned} y &= r\sin(\theta+\alpha) \\ &= r(\sin\theta\cos\alpha + \cos\theta\sin\alpha) \\ &= x'\sin\theta + y'\cos\theta \end{aligned}$$

Theorem Rotation Formulas

If the x- and y-axes are rotated through an angle θ, the coordinates (x, y) of a point P relative to the xy-plane and the coordinates (x', y') of the same point relative to the new x'- and y'-axes are related by the formulas

$$x = x'\cos\theta - y'\sin\theta \qquad y = x'\sin\theta + y'\cos\theta \tag{5}$$

EXAMPLE 2 Rotating Axes

Express the equation $xy = 1$ in terms of new $x'y'$-coordinates by rotating the axes through a 45° angle. Discuss the new equation.

Solution Figure 55(a) shows the graph of $xy = 1$ using a graphing utility. We now proceed to express the equation in terms of new $x'y'$-coordinates and discuss the new equation.

Let $\theta = 45°$ in equation (5). Then

$$x = x'\cos 45° - y'\sin 45° = x'\frac{\sqrt{2}}{2} - y'\frac{\sqrt{2}}{2} = \frac{\sqrt{2}}{2}(x' - y')$$

$$y = x'\sin 45° + y'\cos 45° = x'\frac{\sqrt{2}}{2} + y'\frac{\sqrt{2}}{2} = \frac{\sqrt{2}}{2}(x' + y')$$

Substituting these expressions for x and y in $xy = 1$ gives

$$\left[\frac{\sqrt{2}}{2}(x' - y')\right]\left[\frac{\sqrt{2}}{2}(x' + y')\right] = 1$$

$$\frac{1}{2}(x'^2 - y'^2) = 1$$

$$\frac{x'^2}{2} - \frac{y'^2}{2} = 1$$

FIGURE 55 $xy = 1$ or $\frac{x'^2}{2} - \frac{y'^2}{2} = 1$

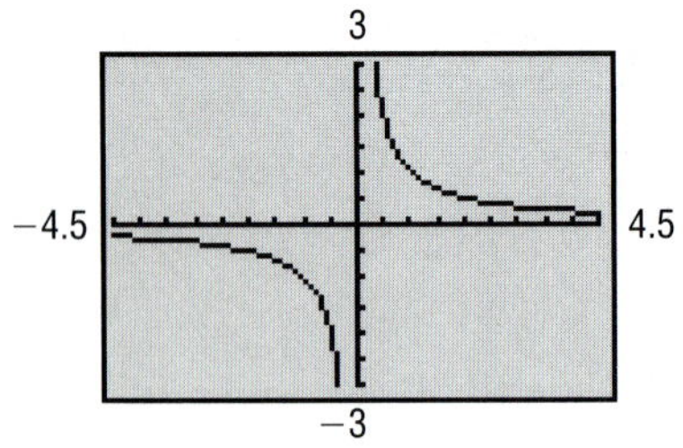

(a)

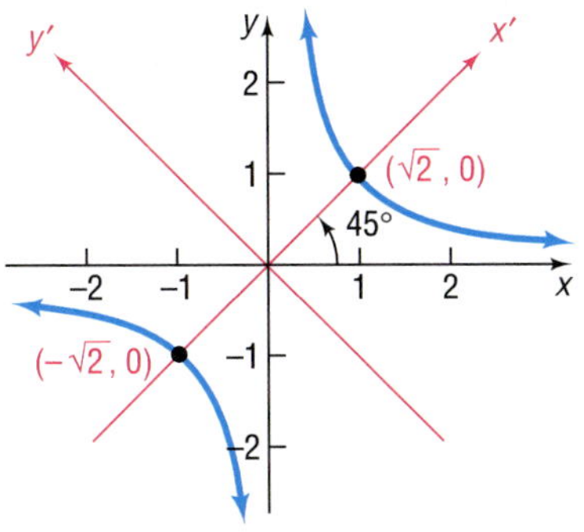

(b)

This is the equation of a hyperbola with center at $(0, 0)$ and transverse axis along the x'-axis. The vertices are at $(\pm\sqrt{2}, 0)$ on the x'-axis; the asymptotes are $y' = x'$ and $y' = -x'$ (which correspond to the original x- and y-axes). See Figure 55(b) for the graph drawn by hand.

As Example 2 illustrates, a rotation of axes through an appropriate angle can transform a second-degree equation in x and y containing an xy-term into one in x' and y' in which no $x'y'$-term appears. In fact, we will show that a rotation of axes through an appropriate angle will transform any equation of the form of equation (1) into an equation in x' and y' without an $x'y'$-term.

To find the formula for choosing an appropriate angle θ through which to rotate the axes, we begin with equation (1),

$$Ax^2 + Bxy + Cy^2 + Dx + Ey + F = 0, \qquad B \neq 0$$

Next we rotate through an angle θ using rotation formulas (5):

$$\begin{aligned} A(x' \cos\theta - y' \sin\theta)^2 &+ B(x' \cos\theta - y' \sin\theta)(x' \sin\theta + y' \cos\theta) \\ &+ C(x' \sin\theta + y' \cos\theta)^2 + D(x' \cos\theta - y' \sin\theta) \\ &+ E(x' \sin\theta + y' \cos\theta) + F = 0 \end{aligned}$$

By expanding and collecting like terms, we obtain

$$\begin{aligned} (A\cos^2\theta + B\sin\theta\cos\theta + C\sin^2\theta)x'^2 &+ [B(\cos^2\theta - \sin^2\theta) + 2(C - A)(\sin\theta\cos\theta)]x'y' \\ &+ (A\sin^2\theta - B\sin\theta\cos\theta + C\cos^2\theta)y'^2 \\ &+ (D\cos\theta + E\sin\theta)x' \\ &+ (-D\sin\theta + E\cos\theta)y' + F = 0 \end{aligned} \qquad (6)$$

In equation (6), the coefficient of $x'y'$ is

$$B' = 2(C - A)(\sin\theta\cos\theta) + B(\cos^2\theta - \sin^2\theta)$$

Since we want to eliminate the $x'y'$-term, we select an angle θ so that $B' = 0$. Thus,

$$\begin{aligned} 2(C - A)(\sin\theta\cos\theta) + B(\cos^2\theta - \sin^2\theta) &= 0 \\ (C - A)(\sin 2\theta) + B\cos 2\theta &= 0 \quad \text{Double-angle formulas.} \\ B\cos 2\theta &= (A - C)(\sin 2\theta) \\ \cot 2\theta &= \frac{A - C}{B}, \qquad B \neq 0 \end{aligned}$$

Theorem

To transform the equation

$$Ax^2 + Bxy + Cy^2 + Dx + Ey + F = 0, \qquad B \neq 0$$

into an equation in x' and y' without an $x'y'$-term, rotate the axes through an angle θ that satisfies the equation

$$\cot 2\theta = \frac{A - C}{B} \qquad (7)$$

Equation (7) has an infinite number of solutions for θ. We shall adopt the convention of choosing the acute angle θ that satisfies (7). Then we have the following two possibilities:

If $\cot 2\theta > 0$, then $0 < 2\theta \le \pi/2$ so that $0 < \theta \le \pi/4$.

If $\cot 2\theta < 0$, then $\pi/2 < 2\theta < \pi$ so that $\pi/4 < \theta < \pi/2$.

Each of these results in a counterclockwise rotation of the axes through an acute angle θ.*

Warning: Be careful if you use a calculator to solve equation (7).

1. If $\cot 2\theta = 0$, then $2\theta = \pi/2$ and $\theta = \pi/4$.
2. If $\cot 2\theta \neq 0$, first find $\cos 2\theta$. Then use the inverse cosine function key(s) to obtain 2θ, $0 < 2\theta < \pi$. Finally, divide by 2 to obtain the correct acute angle θ.

EXAMPLE 3 Discussing an Equation Using a Rotation of Axes

3

Discuss the equation $x^2 + \sqrt{3}xy + 2y^2 - 10 = 0$.

Solution We need to solve the equation for y in order to graph the equation using a graphing utility. Rearranging the terms we observe the equation is quadratic in the variable y: $2y^2 + \sqrt{3}xy + (x^2 - 10) = 0$. We can solve the equation for y using the quadratic formula with $a = 2$, $b = \sqrt{3}x$, and $c = x^2 - 10$.

$$Y_1 = \frac{-\sqrt{3}x + \sqrt{(\sqrt{3}x)^2 - 4(2)(x^2 - 10)}}{2(2)} = \frac{-\sqrt{3}x + \sqrt{-5x^2 + 80}}{4}$$

and

$$Y_2 = \frac{-\sqrt{3}x - \sqrt{(\sqrt{3}x)^2 - 4(2)(x^2 - 10)}}{2(2)} = \frac{-\sqrt{3}x - \sqrt{-5x^2 + 80}}{4}$$

Figure 56(a) shows the graph of Y_1 and Y_2.

We now proceed to analyze the equation. Since an xy-term is present, we must rotate the axes. Using $A = 1$, $B = \sqrt{3}$, and $C = 2$ in equation (7), the appropriate acute angle θ through which to rotate the axes satisfies the equation

$$\cot 2\theta = \frac{A - C}{B} = \frac{-1}{\sqrt{3}} = \frac{-\sqrt{3}}{3}, \qquad 0° < 2\theta < 180°$$

Since $\cot 2\theta = -\sqrt{3}/3$, we find $2\theta = 120°$, so $\theta = 60°$. Using $\theta = 60°$ in rotation formulas (5), we find

$$x = \frac{1}{2}x' - \frac{\sqrt{3}}{2}y' = \frac{1}{2}(x' - \sqrt{3}y')$$
$$y = \frac{\sqrt{3}}{2}x' + \frac{1}{2}y' = \frac{1}{2}(\sqrt{3}x' + y')$$

*Any rotation (clockwise or counterclockwise) through an angle θ that satisfies $\cot 2\theta = (A - C)/B$ will eliminate the $x'y'$-term. However, the final form of the transformed equation may be different (but be equivalent), depending on the angle chosen.

Substituting these values into the original equation and simplifying, we have

$$x^2 + \sqrt{3}xy + 2y^2 - 10 = 0$$

$$\frac{1}{4}(x' - \sqrt{3}y')^2 + \sqrt{3}\left[\frac{1}{2}(x' - \sqrt{3}y')\right]\left[\frac{1}{2}(\sqrt{3}x' + y')\right] + 2\left[\frac{1}{4}(\sqrt{3}x' + y')^2\right] = 10$$

Multiply both sides by 4 and expand to obtain

$$x'^2 - 2\sqrt{3}x'y' + 3y'^2 + \sqrt{3}(\sqrt{3}x'^2 - 2x'y' - \sqrt{3}y'^2) + 2(3x'^2 + 2\sqrt{3}x'y' + y'^2) = 40$$

$$10x'^2 + 2y'^2 = 40$$

$$\frac{x'^2}{4} + \frac{y'^2}{20} = 1$$

This is the equation of an ellipse with center at (0, 0) and major axis along the y'-axis. The vertices are at $(0, \pm 2\sqrt{5})$ on the y'-axis. See Figure 56(b) for the graph drawn by hand.

FIGURE 56

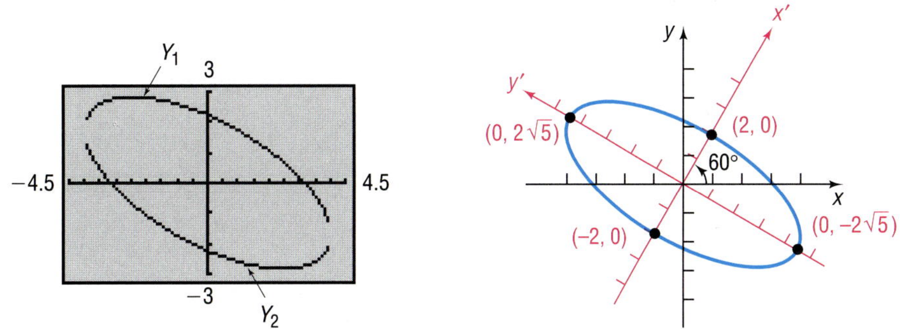

(a) (b)

Now work Problem 21.

In Example 3, the acute angle θ through which to rotate the axes was easy to find because of the numbers we used in the given equation. In general, the equation $\cot 2\theta = (A - C)/B$ will not have such a "nice" solution. As the next example shows, we can still find the appropriate rotation formulas without using a calculator approximation by applying half-angle formulas.

EXAMPLE 4 Discussing an Equation Using a Rotation of Axes

Discuss the equation: $4x^2 - 4xy + y^2 + 5\sqrt{5}x + 5 = 0$.

Solution We need to solve the equation for y in order to graph the equation using a graphing utility. Rearranging the terms we observe the equation is quadratic in the variable y: $y^2 - 4xy + (4x^2 + 5\sqrt{5}x + 5) = 0$. We can solve the equation for y using the quadratic formula with $a = 1$, $b = -4x$, and $c = 4x^2 + 5\sqrt{5}x + 5$.

$$Y_1 = \frac{-(-4x) + \sqrt{(-4x)^2 - 4(1)(4x^2 + 5\sqrt{5}x + 5)}}{2(1)} = 2x + \sqrt{-5(\sqrt{5}x + 1)}$$

$$Y_2 = \frac{-(-4x) - \sqrt{(-4x)^2 - 4(1)(4x^2 + 5\sqrt{5}x + 5)}}{2(1)} = 2x - \sqrt{-5(\sqrt{5}x + 1)}$$

FIGURE 57

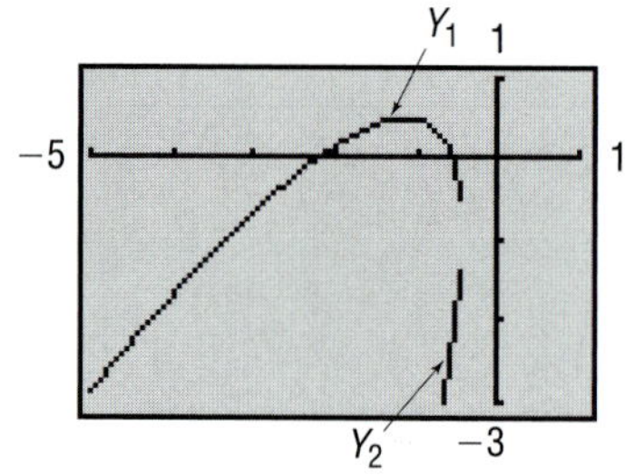

(a)

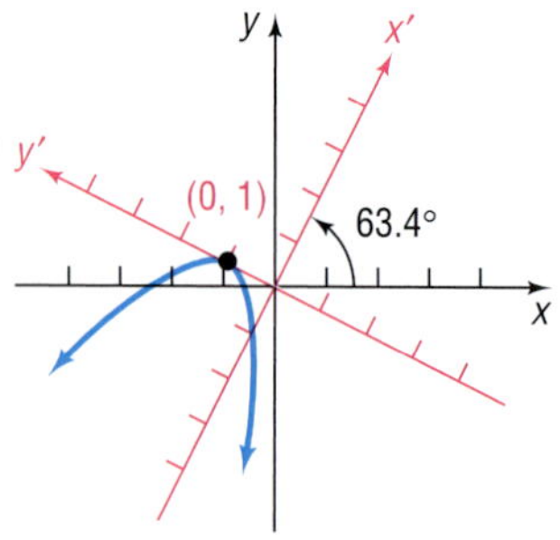

(b)

Figure 57(a) shows the graph of Y_1 and Y_2.

We now proceed to analyze the equation. Letting $A = 4$, $B = -4$, and $C = 1$ in equation (7), the appropriate angle θ through which to rotate the axes satisfies

$$\cot 2\theta = \frac{A - C}{B} = \frac{3}{-4}$$

In order to use rotation formulas (5), we need to know the values of $\sin\theta$ and $\cos\theta$. Since we seek an acute angle θ, we know that $\sin\theta > 0$ and $\cos\theta > 0$. Thus, we use the half-angle formulas in the form

$$\sin\theta = \sqrt{\frac{1 - \cos 2\theta}{2}} \qquad \cos\theta = \sqrt{\frac{1 + \cos 2\theta}{2}}$$

Now we need to find the value of $\cos 2\theta$. Since $\cot 2\theta = -\frac{3}{4}$ and $\pi/2 < 2\theta < \pi$, it follows that $\cos 2\theta = -\frac{3}{5}$. Thus,

$$\sin\theta = \sqrt{\frac{1 - \cos 2\theta}{2}} = \sqrt{\frac{1 - \left(-\frac{3}{5}\right)}{2}} = \sqrt{\frac{4}{5}} = \frac{2}{\sqrt{5}} = \frac{2\sqrt{5}}{5}$$

$$\cos\theta = \sqrt{\frac{1 + \cos 2\theta}{2}} = \sqrt{\frac{1 + \left(-\frac{3}{5}\right)}{2}} = \sqrt{\frac{1}{5}} = \frac{1}{\sqrt{5}} = \frac{\sqrt{5}}{5}$$

With these values, rotation formulas (5) give us

$$x = \frac{\sqrt{5}}{5}x' - \frac{2\sqrt{5}}{5}y' = \frac{\sqrt{5}}{5}(x' - 2y')$$

$$y = \frac{2\sqrt{5}}{5}x' + \frac{\sqrt{5}}{5}y' = \frac{\sqrt{5}}{5}(2x' + y')$$

Substituting these values in the original equation and simplifying, we obtain

$$4x^2 - 4xy + y^2 + 5\sqrt{5}x + 5 = 0$$

$$4\left[\frac{\sqrt{5}}{5}(x' - 2y')\right]^2 - 4\left[\frac{\sqrt{5}}{5}(x' - 2y')\right]\left[\frac{\sqrt{5}}{5}(2x' + y')\right] + \left[\frac{\sqrt{5}}{5}(2x' + y')\right]^2 + 5\sqrt{5}\left[\frac{\sqrt{5}}{5}(x' - 2y')\right] = -5$$

Multiply both sides by 5 and expand to obtain

$$\begin{aligned}
4(x'^2 - 4x'y' + 4y'^2) - 4(2x'^2 - 3x'y' - 2y'^2) \\
+ 4x'^2 + 4x'y' + y'^2 + 25(x' - 2y') &= -25 \\
25y'^2 - 50y' + 25x' &= -25 \\
y'^2 - 2y' + x' &= -1 \\
y'^2 - 2y' + 1 &= -x' \\
(y' - 1)^2 &= -x'
\end{aligned}$$

Complete the square in y'.

This is the equation of a parabola with vertex at $(0, 1)$ in the $x'y'$-plane. The axis of symmetry is parallel to the x'-axis. Using a calculator to solve $\sin\theta = 2\sqrt{5}/5$, we find that $\theta \approx 63.4°$. See Figure 57(b) for the graph drawn by hand.

Now work Problem 27.

Identifying Conics without a Rotation of Axes

Suppose that we are required only to identify (rather than discuss) an equation of the form

$$Ax^2 + Bxy + Cy^2 + Dx + Ey + F = 0, \qquad B \neq 0 \tag{8}$$

If we apply rotation formulas (5) to this equation, we obtain an equation of the form

$$A'x'^2 + B'x'y' + C'y'^2 + D'x' + E'y' + F' = 0 \tag{9}$$

where A', B', C', D', E', and F' can be expressed in terms of A, B, C, D, E, F, and the angle θ of rotation (see Problem 43 at the end of this section). It can be shown that the value of $B^2 - 4AC$ in equation (8) and the value of $B'^2 - 4A'C'$ in equation (9) are equal no matter what angle θ of rotation is chosen (see Problem 45). In particular, if the angle θ of rotation satisfies equation (7), then $B' = 0$ in equation (9), and $B^2 - 4AC = -4A'C'$. Since equation (9) then has the form of equation (2),

$$A'x'^2 + C'y'^2 + D'x' + E'y' + F' = 0$$

we can identify it without completing the squares, as we did in the beginning of this section. In fact, now we can identify the conic described by any equation of the form of equation (8) without a rotation of axes.

Theorem Identifying Conics without a Rotation of Axes

Except for degenerate cases, the equation

$$Ax^2 + Bxy + Cy^2 + Dx + Ey + F = 0$$

(a) Defines a parabola if $B^2 - 4AC = 0$.
(b) Defines an ellipse (or a circle) if $B^2 - 4AC < 0$.
(c) Defines a hyperbola if $B^2 - 4AC > 0$.

You are asked to prove this theorem in Problem 46.

EXAMPLE 5 Identifying a Conic without a Rotation of Axes

Identify the equation $8x^2 - 12xy + 17y^2 - 4\sqrt{5}x - 2\sqrt{5}y - 15 = 0$.

Solution Here, $A = 8$, $B = -12$, and $C = 17$, so that $B^2 - 4AC = -400$. Since $B^2 - 4AC < 0$, the equation defines an ellipse.

Now work Problem 33.

10.5 EXERCISES

In Problems 1–10, identify each equation without completing the squares.

1. $x^2 + 4x + y + 3 = 0$

2. $2y^2 - 3y + 3x = 0$

3. $6x^2 + 3y^2 - 12x + 6y = 0$

4. $2x^2 + y^2 - 8x + 4y + 2 = 0$

5. $3x^2 - 2y^2 + 6x + 4 = 0$

6. $4x^2 - 3y^2 - 8x + 6y + 1 = 0$

7. $2y^2 - x^2 - y + x = 0$

8. $y^2 - 8x^2 - 2x - y = 0$

9. $x^2 + y^2 - 8x + 4y = 0$

10. $2x^2 + 2y^2 - 8x + 8y = 0$

In Problems 11–20, determine the appropriate rotation formulas to use so that the new equation contains no xy-term.

11. $x^2 + 4xy + y^2 - 3 = 0$

12. $x^2 - 4xy + y^2 - 3 = 0$

13. $5x^2 + 6xy + 5y^2 - 8 = 0$

14. $3x^2 - 10xy + 3y^2 - 32 = 0$

15. $13x^2 - 6\sqrt{3}xy + 7y^2 - 16 = 0$

16. $11x^2 + 10\sqrt{3}xy + y^2 - 4 = 0$

17. $4x^2 - 4xy + y^2 - 8\sqrt{5}x - 16\sqrt{5}y = 0$

18. $x^2 + 4xy + 4y^2 + 5\sqrt{5}y + 5 = 0$

19. $25x^2 - 36xy + 40y^2 - 12\sqrt{13}x - 8\sqrt{13}y = 0$

20. $34x^2 - 24xy + 41y^2 - 25 = 0$

In Problems 21–32, graph the equation using a graphing utility. Rotate the axes so that the new equation contains no xy-term. Discuss and, by hand, graph the new equation. (Refer to Problems 11–20 for Problems 21–30).

21. $x^2 + 4xy + y^2 - 3 = 0$

22. $x^2 - 4xy + y^2 - 3 = 0$

23. $5x^2 + 6xy + 5y^2 - 8 = 0$

24. $3x^2 - 10xy + 3y^2 - 32 = 0$

25. $13x^2 - 6\sqrt{3}xy + 7y^2 - 16 = 0$

26. $11x^2 + 10\sqrt{3}xy + y^2 - 4 = 0$

27. $4x^2 - 4xy + y^2 - 8\sqrt{5}x - 16\sqrt{5}y = 0$

28. $x^2 + 4xy + 4y^2 + 5\sqrt{5}y + 5 = 0$

29. $25x^2 - 36xy + 40y^2 - 12\sqrt{13}x - 8\sqrt{13}y = 0$

30. $34x^2 - 24xy + 41y^2 - 25 = 0$

31. $16x^2 + 24xy + 9y^2 - 130x + 90y = 0$

32. $16x^2 + 24xy + 9y^2 - 60x + 80y = 0$

In Problems 33–42, identify each equation without applying a rotation of axes.

33. $x^2 + 3xy - 2y^2 + 3x + 2y + 5 = 0$

34. $2x^2 - 3xy + 4y^2 + 2x + 3y - 5 = 0$

35. $x^2 - 7xy + 3y^2 - y - 10 = 0$

36. $2x^2 - 3xy + 2y^2 - 4x - 2 = 0$

37. $9x^2 + 12xy + 4y^2 - x - y = 0$

38. $10x^2 + 12xy + 4y^2 - x - y + 10 = 0$

39. $10x^2 - 12xy + 4y^2 - x - y - 10 = 0$

40. $4x^2 + 12xy + 9y^2 - x - y = 0$

41. $3x^2 - 2xy + y^2 + 4x + 2y - 1 = 0$

42. $3x^2 + 2xy + y^2 + 4x - 2y + 10 = 0$

In Problems 43–46, apply rotation formulas (5) to

$$Ax^2 + Bxy + Cy^2 + Dx + Ey + F = 0$$

to obtain the equation

$$A'x'^2 + B'x'y' + C'y'^2 + D'x' + E'y' + F' = 0$$

43. Express A', B', C', D', E', and F' in terms of A, B, C, D, E, F, and the angle θ of rotation.

44. Show that $A + C = A' + C'$, and thus show that $A + C$ is **invariant;** that is, its value does not change under a rotation of axes.

45. Refer to Problem 44. Show that $B^2 - 4AC$ is invariant.

46. Prove that, except for degenerate cases, the equation

$$Ax^2 + Bxy + Cy^2 + Dx + Ey + F = 0$$

(a) Defines a parabola if $B^2 - 4AC = 0$.

(b) Defines an ellipse (or a circle) if $B^2 - 4AC < 0$.

(c) Defines a hyperbola if $B^2 - 4AC > 0$.

47. Use rotation formulas (5) to show that distance is invariant under a rotation of axes. That is, show that the distance from $P_1 = (x_1, y_1)$ to $P_2 = (x_2, y_2)$ in the xy-plane equals the distance from $P_1 = (x_1', y_1')$ to $P_2 = (x_2', y_2')$ in the $x'y'$-plane.

48. Show that the graph of the equation $x^{1/2} + y^{1/2} = a^{1/2}$ is part of the graph of a parabola.

49. Formulate a strategy for discussing and graphing an equation of the form

$$Ax^2 + Cy^2 + Dx + Ey + F = 0.$$

How does your strategy change if the equation is of the form

$$Ax^2 + Bxy + Cy^2 + Dx + Ey + F = 0?$$

10.6 POLAR EQUATIONS OF CONICS

1 Discuss and Graph Polar Equations of Conics
2 Convert a Polar Equation of a Conic to a Rectangular Equation

In Sections 10.2, 10.3, and 10.4, we gave separate definitions for the parabola, ellipse, and hyperbola based on geometric properties and the distance formula. In this section, we present an alternative definition that simultaneously defines all these conics. As we shall see, this approach is well suited to polar coordinate representation. (Refer to Section 9.1.)

Let D denote a fixed line called the **directrix;** let F denote a fixed point called the **focus,** which is not on D; and let e be a fixed positive number called the **eccentricity.** A **conic** is the set of points P in the plane such that the ratio of the distance from F to P to the distance from D to P equals e. Thus, a conic is the collection of points P for which

$$\frac{d(F, P)}{d(D, P)} = e \qquad (1)$$

If $e = 1$, the conic is a **parabola.**
If $e < 1$, the conic is an **ellipse.**
If $e > 1$, the conic is a **hyperbola.**

Observe that if $e = 1$ the definition of a parabola in equation (1) is exactly the same as the definition used earlier in Section 10.2.

In the case of an ellipse, the **major axis** is a line through the focus perpendicular to the directrix. In the case of a hyperbola, the **transverse axis** is a line through the focus perpendicular to the directrix. For both an ellipse and a hyperbola, the eccentricity e satisfies

$$e = \frac{c}{a} \qquad (2)$$

FIGURE 58

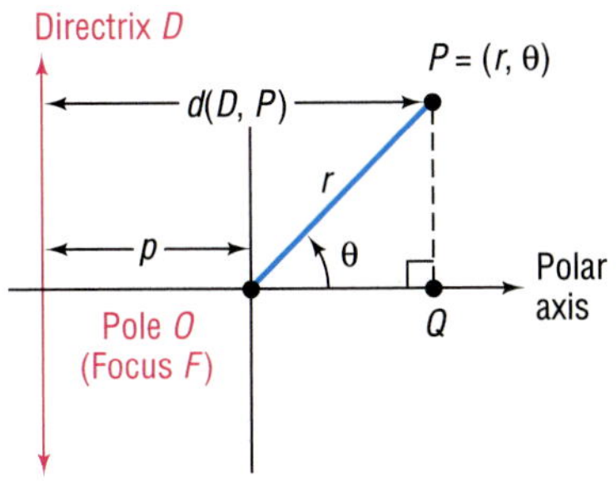

where c is the distance from the center to the focus and a is the distance from the center to a vertex.

Just as we did earlier using rectangular coordinates, we derive equations for the conics in polar coordinates by choosing a convenient position for the focus F and the directrix D. The focus F is positioned at the pole, and the directrix D is either parallel to the polar axis or perpendicular to it.

Suppose that we start with the directrix D perpendicular to the polar axis at a distance p units to the left of the pole (the focus F). See Figure 58.

If $P = (r, \theta)$ is any point on the conic, then, by equation (1),

$$\frac{d(F, P)}{d(D, P)} = e \quad \text{or} \quad d(F, P) = e \cdot d(D, P) \tag{3}$$

Now we use the point Q obtained by dropping the perpendicular from P to the polar axis to calculate $d(D, P)$:

$$d(D, P) = p + d(O, Q) = p + r \cos \theta$$

Using this expression and the fact that $d(F, P) = d(O, P) = r$ in equation (3), we get

$$\begin{aligned} d(F, P) &= e \cdot d(D, P) \\ r &= e(p + r \cos \theta) \\ r &= ep + er \cos \theta \\ r - er \cos \theta &= ep \\ r(1 - e \cos \theta) &= ep \\ r &= \frac{ep}{1 - e \cos \theta} \end{aligned}$$

Theorem **Polar Equation of a Conic; Focus at Pole; Directrix Perpendicular to Polar Axis a Distance p to the Left of the Pole**

The polar equation of a conic with focus at the pole and directrix perpendicular to the polar axis at a distance p to the left of the pole is

$$r = \frac{ep}{1 - e \cos \theta} \tag{4}$$

where e is the eccentricity of the conic.

EXAMPLE 1 Discussing and Graphing the Polar Equation of a Conic

1 Discuss and graph the equation: $r = \dfrac{4}{2 - \cos \theta}$

Solution Figure 59(a) shows the graph of the equation using a graphing utility in POLar mode with θmin $= 0$, θmax $= 2\pi$, and θstep $= \pi/24$. We now proceed to discuss the equation.

The given equation is not quite in the form of equation (4), since the first term in the denominator is 2 instead of 1. Thus, we divide the numerator and denominator by 2 to obtain

$$r = \frac{2}{1 - \frac{1}{2} \cos \theta}$$

This equation is in the form of equation (4), with

$$e = \frac{1}{2} \quad \text{and} \quad ep = \frac{1}{2}p = 2$$

Thus, $e = \frac{1}{2}$ and $p = 4$. We conclude that the conic is an ellipse, since $e = \frac{1}{2} < 1$. One focus is at the pole, and the directrix is perpendicular to the polar axis, a distance of 4 units to the left of the pole. It follows that the major axis is along the polar axis. To find the vertices, we let $\theta = 0$ and $\theta = \pi$. Thus, the vertices of the ellipse are $(4, 0)$ and $\left(\frac{4}{3}, \pi\right)$. At the midpoint of the vertices, we locate the center of the ellipse at $\left(\frac{4}{3}, 0\right)$. [Do you see why? The vertices $(4, 0)$ and $\left(\frac{4}{3}, \pi\right)$ in polar coordinates are $(4, 0)$ and $\left(-\frac{4}{3}, 0\right)$ in rectangular coordinates. The midpoint in rectangular coordinates is $\left(\frac{4}{3}, 0\right)$, which is also $\left(\frac{4}{3}, 0\right)$ in polar coordinates.] Thus, a = distance from the center to a vertex $= \frac{8}{3}$. Using $a = \frac{8}{3}$ and $e = \frac{1}{2}$ in equation (2), $e = c/a$, we find $c = \frac{4}{3}$. Finally, using $a = \frac{8}{3}$ and $c = \frac{4}{3}$ in $b^2 = a^2 - c^2$, we have

$$b^2 = a^2 - c^2 = \frac{64}{9} - \frac{16}{9} = \frac{48}{9}$$

$$b = \frac{4\sqrt{3}}{3}$$

Figure 59(b) shows the graph drawn by hand.

FIGURE 59

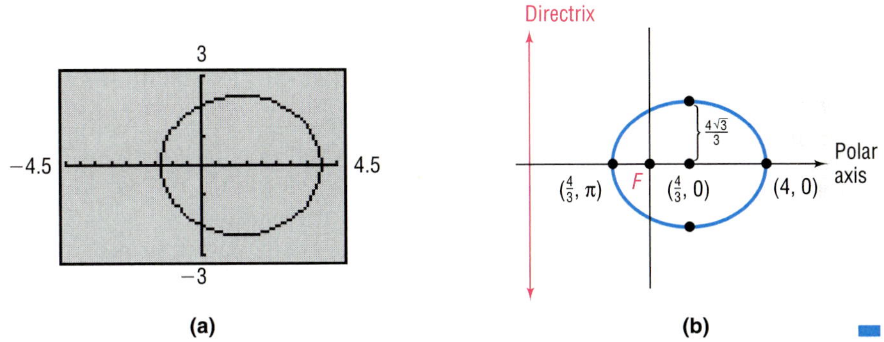

(a) (b)

Exploration: Graph $r_1 = 4/(2 + \cos\theta)$ and compare the result with Figure 59(a). What do you conclude? Clear the screen and graph $r_1 = 4/(2 - \sin\theta)$ and then $r_1 = 4/(2 + \sin\theta)$. Compare each of these graphs with Figure 59(a). What do you conclude?

Now work Problem 5.

Equation (4) was obtained under the assumption that the directrix was perpendicular to the polar axis at a distance p units to the left of the pole. A similar derivation (see Problem 37), in which the directrix is perpendicular to the polar axis at a distance p units to the right of the pole, results in the equation

$$r = \frac{ep}{1 + e\cos\theta}$$

In Problems 38 and 39 you are asked to derive the polar equations of conics with focus at the pole and directrix parallel to the polar axis. Table 5 summarizes the polar equations of conics.

TABLE 5 Polar Equations of Conics (Focus at the Pole, Eccentricity e)

Equation	Description
(a) $r = \dfrac{ep}{1 - e\cos\theta}$	Directrix is perpendicular to the polar axis at a distance p units to the left of the pole.
(b) $r = \dfrac{ep}{1 + e\cos\theta}$	Directrix is perpendicular to the polar axis at a distance p units to the right of the pole.
(c) $r = \dfrac{ep}{1 + e\sin\theta}$	Directrix is parallel to the polar axis at a distance p units above the pole.
(d) $r = \dfrac{ep}{1 - e\sin\theta}$	Directrix is parallel to the polar axis at a distance p units below the pole.

Eccentricity

If $e = 1$, the conic is a parabola; the axis of symmetry is perpendicular to the directrix.
If $e < 1$, the conic is an ellipse; the major axis is perpendicular to the directrix.
If $e > 1$, the conic is a hyperbola; the transverse axis is perpendicular to the directrix.

EXAMPLE 2 Discussing and Graphing the Polar Equation of a Conic

Discuss and graph the equation $r = \dfrac{6}{3 + 3\sin\theta}$.

Solution Figure 60(a) shows the graph of the equation using a graphing utility in POLar mode with θmin $= 0$, θmax $= 2\pi$, and θstep $= \pi/24$. We now proceed to discuss the equation.

To place the equation in proper form, we divide the numerator and denominator by 3 to get

$$r = \frac{2}{1 + \sin\theta}$$

Referring to Table 5, we conclude that this equation is in the form of equation (c) with

$$e = 1 \quad \text{and} \quad ep = 2$$

Thus, $e = 1$ and $p = 2$. The conic is a parabola with focus at the pole. The directrix is parallel to the polar axis at a distance 2 units above the pole; the axis of symmetry is perpendicular to the polar axis. The vertex of the parabola is at $(1, \pi/2)$. (Do you see why?) See Figure 60(b) for the graph drawn by hand. Notice that we plotted two additional points, $(2, 0)$ and $(2, \pi)$, to assist in graphing.

FIGURE 60

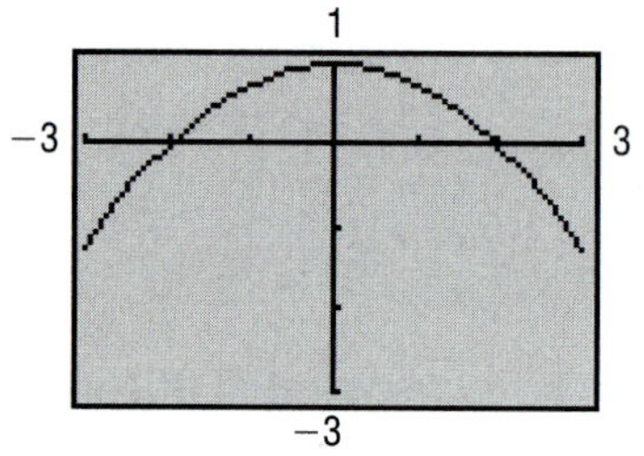

(a)

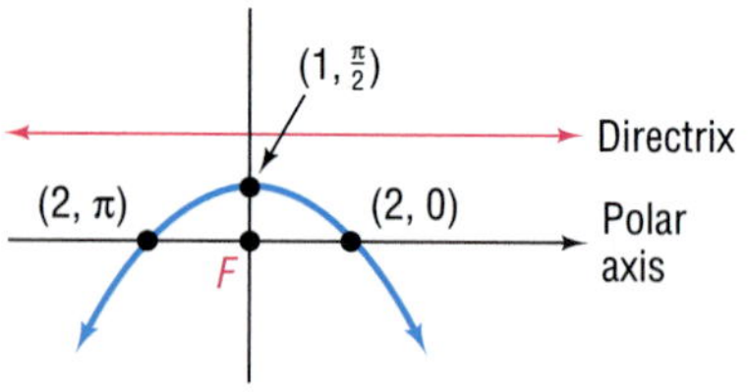

(b)

Now work Problem 7.

EXAMPLE 3 Discussing and Graphing the Polar Equation of a Conic

Discuss and graph the equation: $r = \dfrac{3}{1 + 3\cos\theta}$.

Solution Figure 61(a) and (b) show the graph of the equation using a graphing utility in POLar mode with $\theta\text{min} = 0$, $\theta\text{max} = 2\pi$, and $\theta\text{step} = \pi/24$, using both dot mode and connected mode. Notice the extraneous asymptotes in the connected mode. We now proceed to discuss the equation.

FIGURE 61

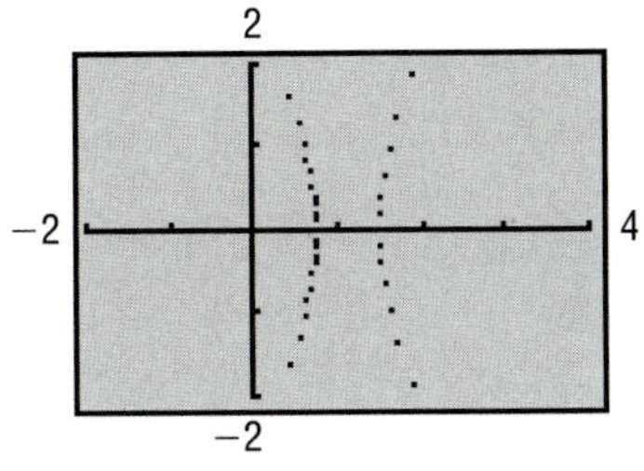

(a) Dot Mode

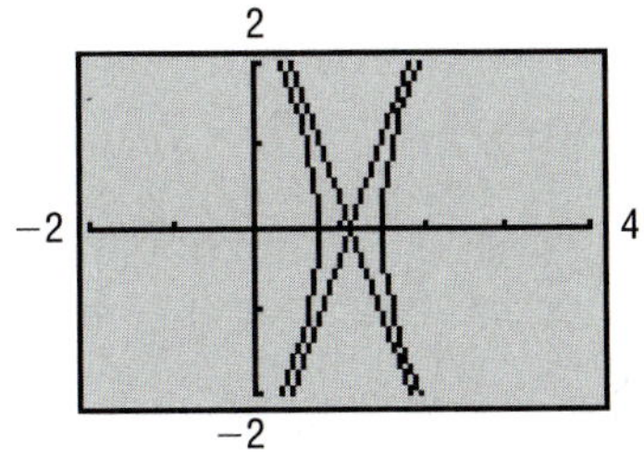

(b) Connected Mode

This equation is in the form of equation (b) in Table 5. We conclude that

$$e = 3 \qquad \text{and} \qquad ep = 3p = 3$$

Thus, $e = 3$ and $p = 1$. This is the equation of a hyperbola with a focus at the pole. The directrix is perpendicular to the polar axis, 1 unit to the right of the pole. The transverse axis is along the polar axis. To find the vertices, we let $\theta = 0$ and $\theta = \pi$. Thus, the vertices are $(\frac{3}{4}, 0)$ and $(-\frac{3}{2}, \pi)$. The center, which is at the midpoint of $(\frac{3}{4}, 0)$ and $(-\frac{3}{2}, \pi)$, is $(\frac{9}{8}, 0)$. Thus, c = distance from the center to a focus $= \frac{9}{8}$. Since $e = 3$, it follows from equation (2), $e = c/a$, that $a = \frac{3}{8}$. Finally, using $a = \frac{3}{8}$ and $c = \frac{9}{8}$ in $b^2 = c^2 - a^2$, we find

$$b^2 = c^2 - a^2 = \frac{81}{64} - \frac{9}{64} = \frac{72}{64} = \frac{9}{8}$$

$$b = \frac{3}{2\sqrt{2}} = \frac{3\sqrt{2}}{4}$$

Figure 61(c) shows the graph drawn by hand. Notice that we plotted two additional points, $(3, \pi/2)$ and $(3, 3\pi/2)$, on the left branch and used symmetry to obtain the right branch. The asymptotes of this hyperbola were found in the usual way by constructing the rectangle shown.

FIGURE 61

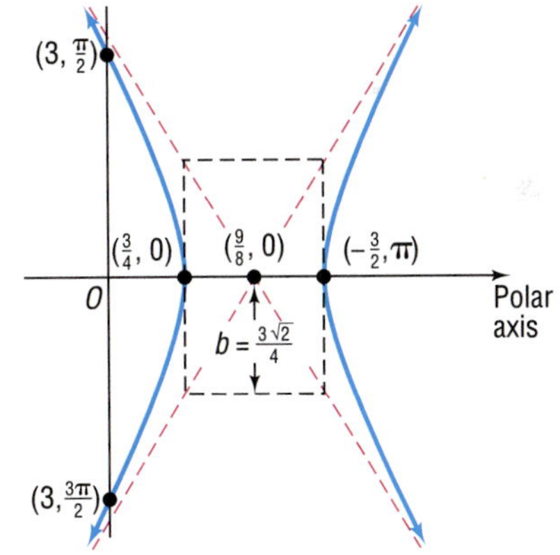

(c)

Now work Problem 11.

EXAMPLE 4 Converting a Polar Equation to a Rectangular Equation

2

Convert the polar equation

$$r = \frac{1}{3 - 3\cos\theta}$$

to a rectangular equation.

Solution The strategy here is first to rearrange the equation and square each side, before using the transformation equations:

$$r = \frac{1}{3 - 3\cos\theta}$$

$$3r - 3r\cos\theta = 1$$

$$3r = 1 + 3r\cos\theta \quad \text{Rearrange the equation.}$$

$$9r^2 = (1 + 3r\cos\theta)^2 \quad \text{Square each side.}$$

$$9(x^2 + y^2) = (1 + 3x)^2 \quad \text{Use the transformation equations.}$$

$$9x^2 + 9y^2 = 9x^2 + 6x + 1$$

$$9y^2 = 6x + 1$$

This is the equation of a parabola in rectangular coordinates.

Now work Problem 19.

10.6 EXERCISES

In Problems 1–6, identify the conic that each polar equation represents. Also, give the position of the directrix.

1. $r = \dfrac{1}{1 + \cos\theta}$ **2.** $r = \dfrac{3}{1 - \sin\theta}$ **3.** $r = \dfrac{4}{2 - 3\sin\theta}$

4. $r = \dfrac{2}{1 + 2\cos\theta}$ **5.** $r = \dfrac{3}{4 - 2\cos\theta}$ **6.** $r = \dfrac{6}{8 + 2\sin\theta}$

In Problems 7–18, graph each equation using a graphing utility. Discuss each equation and graph it by hand.

7. $r = \dfrac{1}{1 + \cos\theta}$ **8.** $r = \dfrac{3}{1 - \sin\theta}$ **9.** $r = \dfrac{8}{4 + 3\sin\theta}$

10. $r = \dfrac{10}{5 + 4\cos\theta}$ **11.** $r = \dfrac{9}{3 - 6\cos\theta}$ **12.** $r = \dfrac{12}{4 + 8\sin\theta}$

13. $r = \dfrac{8}{2 - \sin\theta}$ **14.** $r = \dfrac{8}{2 + 4\cos\theta}$ **15.** $r(3 - 2\sin\theta) = 6$

16. $r(2 - \cos\theta) = 2$ **17.** $r = \dfrac{6\sec\theta}{2\sec\theta - 1}$ **18.** $r = \dfrac{3\csc\theta}{\csc\theta - 1}$

In Problems 19–30, convert each polar equation to a rectangular equation.

19. $r = \dfrac{1}{1 + \cos\theta}$ **20.** $r = \dfrac{3}{1 - \sin\theta}$ **21.** $r = \dfrac{8}{4 + 3\sin\theta}$

22. $r = \dfrac{10}{5 + 4\cos\theta}$ **23.** $r = \dfrac{9}{3 - 6\cos\theta}$ **24.** $r = \dfrac{12}{4 + 8\sin\theta}$

25. $r = \dfrac{8}{2 - \sin\theta}$ **26.** $r = \dfrac{8}{2 + 4\cos\theta}$ **27.** $r(3 - 2\sin\theta) = 6$

28. $r(2 - \cos\theta) = 2$ **29.** $r = \dfrac{6\sec\theta}{2\sec\theta - 1}$ **30.** $r = \dfrac{3\csc\theta}{\csc\theta - 1}$

In Problems 31–36, find a polar equation for each conic. For each, a focus is at the pole.

31. $e = 1$; directrix is parallel to the polar axis 1 unit above the pole

32. $e = 1$; directrix is parallel to the polar axis 2 units below the pole

33. $e = \frac{4}{5}$; directrix is perpendicular to the polar axis 3 units to the left of the pole

34. $e = \frac{2}{3}$; directrix is parallel to the polar axis 3 units above the pole

35. $e = 6$; directrix is parallel to the polar axis 2 units below the pole

36. $e = 5$; directrix is perpendicular to the polar axis 5 units to the right of the pole

37. Derive equation (b) in Table 5:

$$r = \frac{ep}{1 + e \cos \theta}$$

38. Derive equation (c) in Table 5:

$$r = \frac{ep}{1 + e \sin \theta}$$

39. Derive equation (d) in Table 5:

$$r = \frac{ep}{1 - e \sin \theta}$$

40. Orbit of Mercury The planet Mercury travels around the Sun in an elliptical orbit given approximately by

$$r = \frac{(3.442)10^7}{1 - 0.206 \cos \theta}$$

where r is measured in miles and the Sun is at the pole. Find the distance from Mercury to the Sun at *aphelion* (greatest distance from the Sun) and at *perihelion* (shortest distance from the Sun). See the figure below. Use the aphelion and perihelion to graph the orbit of Mercury using a graphing utility.

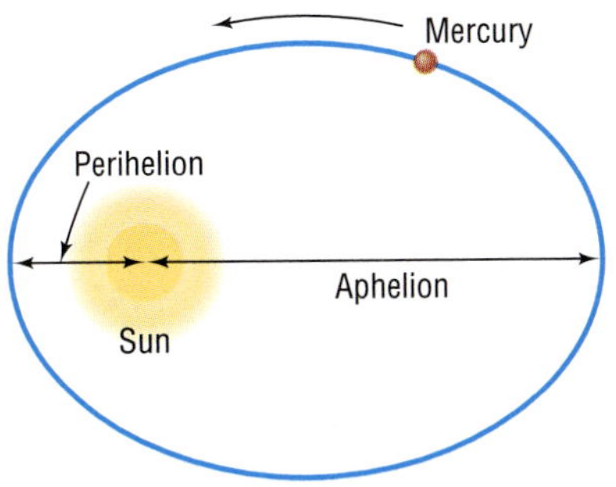

10.7 PLANE CURVES AND PARAMETRIC EQUATIONS

1 Graph Parametric Equations by Hand
2 Graph Parametric Equations Using a Graphing Utility
3 Find a Rectangular Equation for a Curve Defined Parametrically
4 Use Time as a Parameter in Parametric Equations
5 Find Parametric Equations for Curves Defined by Rectangular Equations

Equations of the form $y = f(x)$, where f is a function, have graphs that are intersected no more than once by any vertical line. The graphs of many of the conics and certain other, more complicated graphs do not have this characteristic. Yet each graph, like the graph of a function, is a collection of points (x, y) in the xy-plane; that is, each is a *plane curve.* In this section, we discuss another way of representing such graphs.

Let $x = f(t)$ and $y = g(t)$, where f and g are two functions whose common domain is some interval I. The collection of points defined by

$$(x, y) = (f(t), g(t))$$

is called a **plane curve.** The equations

$$x = f(t) \qquad y = g(t)$$

where t is in I, are called **parametric equations** of the curve. The variable t is called a **parameter.**

1

Parametric equations are particularly useful in describing movement along a curve. Suppose that a curve is defined by the parametric equations

$$x = f(t) \qquad y = g(t)$$

where f and g are each defined over some interval I. For a given value of t in I, we can find the value of $x = f(t)$ and $y = g(t)$, thus obtaining a point (x, y) on the curve. In fact, as t varies over the interval I in some order, say from left to right, successive values of t give rise to a directed movement along the curve. That is, as t varies over I in some order, the curve is traced out in a certain direction by the corresponding succession of points (x, y). Let's look at an example.

EXAMPLE 1 Discussing a Curve Defined by Parametric Equations

Discuss the curve defined by the parametric equations

$$x = 3t^2 \qquad y = 2t, \qquad -2 \le t \le 2 \tag{1}$$

Solution For each number t, $-2 \le t \le 2$, there corresponds a number x and a number y. For example, when $t = -2$, then $x = 12$ and $y = -4$. When $t = 0$, then $x = 0$ and $y = 0$. Indeed, we can set up a table listing various choices of the parameter t and the corresponding values for x and y, as shown in Table 6. Plotting these points and connecting them with a smooth curve leads to Figure 62.

TABLE 6

t	x	y	(x, y)
−2	12	−4	(12, −4)
−1	3	−2	(3, −2)
0	0	0	(0, 0)
1	3	2	(3, 2)
2	12	4	(12, 4)

FIGURE 62

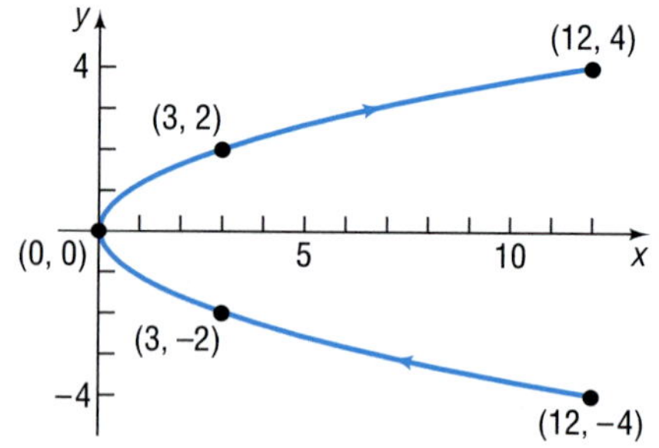

Notice the arrows on the curve in Figure 62. They indicate the direction, or **orientation,** of the curve for increasing values of the parameter t.

2

Most graphing utilities have the capability of graphing parametric equations. The following steps are usually required in order to obtain the graph of parametric equations. Check your owner's manual to see how yours works.

Graphing Parametric Equations Using a Graphing Utility

STEP 1: Set the mode to PARametric. Enter $x(t)$ and $y(t)$.

STEP 2: Select the viewing rectangle. In addition to setting Xmin, Xmax, Xscl, and so on, the viewing rectangle in parametric mode requires setting minimum and maximum values for the parameter t and an increment setting for t (Tstep).

STEP 3: Execute.

EXAMPLE 2 Graphing a Curve Defined by Parametric Equations Using a Graphing Utility

Graph the curve defined by the parametric equations

$$x = 3t^2 \qquad y = 2t \qquad -2 \le t \le 2$$

Solution STEP 1: Enter the equations $x(t) = 3t^2, y(t) = 2t$ with the graphing utility in PARametric mode.

STEP 2: Select the viewing rectangle. The interval I is $-2 \le t \le 2$, so we select the following square viewing rectangle:

$T\text{min} = -2$
$T\text{max} = 2$
$T\text{step} = 0.1$
$X\text{min} = 0$
$X\text{max} = 15$
$X\text{scl} = 1$
$Y\text{min} = -5$
$Y\text{max} = 5$
$Y\text{scl} = 1$

We choose $T\text{min} = -2$ and $T\text{max} = 2$ because $-2 \le t \le 2$. Finally, the choice for Tstep will determine the number of points the graphing utility will plot. For example, with Tstep at 0.1, the graphing utility will evaluate x and y at $t = -2, -1.9, -1.8$, and so on. The smaller the Tstep, the more points the graphing utility will plot. The reader is encouraged to experiment with different values of Tstep to see how the graph is affected.

STEP 3: Execute. Notice the direction the graph is drawn. This direction shows the orientation of the curve.

The graph shown in Figure 63 is complete.

FIGURE 63

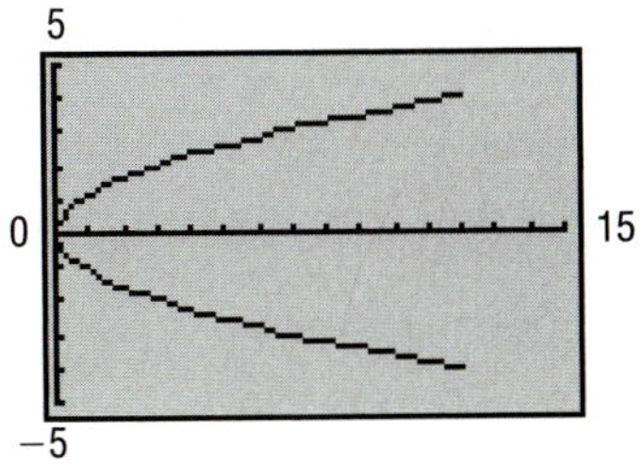

Now work Problem 1.

Exploration Graph the following parametric equations using a graphing utility with $X\text{min} = 0$, $X\text{max} = 15$, $Y\text{min} = -5$, $Y\text{max} = 5$, and $T\text{step} = 0.1$:

1. $x = \dfrac{3t^2}{4}, y = t, -4 \le t \le 4$
2. $x = 3t^2 + 12t + 12, y = 2t + 4, -4 \le t \le 0$
3. $x = 3t^{2/3}, y = 2\sqrt[3]{t}, -8 \le t \le 8$

Compare these graphs to the graph in Figure 63. Conclude that parametric equations defining a curve are not unique, that is, different parametric equations can represent the same graph.

3 The curve given in Example 1 and 2 should be familiar. To identify it accurately, we find the corresponding rectangular equation by eliminating the parameter t from the parametric equations (1) given in Example 1,

$$x = 3t^2 \qquad y = 2t, \qquad -2 \le t \le 2$$

Noting that we can readily solve for t in $y = 2t$, obtaining $t = y/2$, we substitute this expression in the other equation:

$$x = 3t^2 = 3\left(\frac{y}{2}\right)^2 = \frac{3y^2}{4}$$

$$\uparrow \quad t = \frac{y}{2}$$

This equation, $x = 3y^2/4$, is the equation of a parabola with vertex at (0, 0) and axis of symmetry along the x-axis.

Exploration In FUNCtion mode graph $x = \frac{3y^2}{4}\left(Y_1 = \sqrt{\frac{4x}{3}}\right.$ and $\left.Y_2 = -\sqrt{\frac{4x}{3}}\right)$ with Xmin = 0, Xmax = 15, Ymin = −5, Ymax = 5. Compare this graph with Figure 63. Why do the graphs differ?

Note that the parameterized curve defined by equation (1) and shown in Figure 62 (or 63) is only a part of the parabola $x = 3y^2/4$. Thus, the graph of the rectangular equation obtained by eliminating the parameter will, in general, contain more points than the original parameterized curve. Care must therefore be taken when a parameterized curve is sketched by hand after eliminating the parameter. Even so, the process of eliminating the parameter t of a parameterized curve in order to identify it accurately is sometimes a better approach than merely plotting points. However, the elimination process sometimes requires a little ingenuity.

EXAMPLE 3 Finding the Rectangular Equation of a Curve Defined Parametrically

Find the rectangular equation of the curve whose parametric equations are

$$x = a\cos t \qquad y = a\sin t$$

where $a > 0$ is a constant. Graph this curve, indicating its orientation.

Solution The presence of sines and cosines in the parametric equations suggests that we use a Pythagorean identity. In fact, since

$$\cos t = \frac{x}{a} \qquad \sin t = \frac{y}{a}$$

we find that

$$\cos^2 t + \sin^2 t = 1$$

$$\left(\frac{x}{a}\right)^2 + \left(\frac{y}{a}\right)^2 = 1$$

$$x^2 + y^2 = a^2$$

FIGURE 64

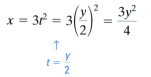

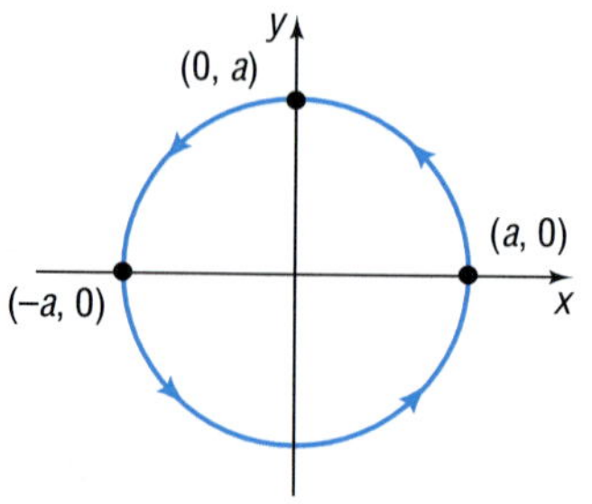

Thus, the curve is a circle with center at (0, 0) and radius a. As the parameter t increases, say from $t = 0$ [the point $(a, 0)$] to $t = \pi/2$ [the point $(0, a)$] to $t = \pi$ [the point $(-a, 0)$], we see that the corresponding points are traced in a counterclockwise direction around the circle. Hence, the orientation is as indicated in Figure 64.

Now work Problem 13.

Let's discuss the curve in Example 3 further. The domain of each parameteric equation is $-\infty < t < \infty$. Thus, the graph in Figure 64 is actually being repeated each time t increases by 2π. If we wanted the curve to consist of exactly 1 revolution in the counterclockwise direction, we could write

$$x = a\cos t \qquad y = a\sin t, \qquad 0 \le t \le 2\pi$$

This curve starts at $t = 0$ [the point $(a, 0)$] and, proceeding counterclockwise around the circle, ends at $t = 2\pi$ [also the point $(a, 0)$].

If we wanted the curve to consist of exactly three revolutions in the counterclockwise direction, we could write

$$x = a\cos t \qquad y = a\sin t, \qquad -2\pi \le t \le 4\pi$$

or

$$x = a\cos t \qquad y = a\sin t, \qquad 0 \le t \le 6\pi$$

or

$$x = a\cos t \qquad y = a\sin t, \qquad 2\pi \le t \le 8\pi$$

If we wanted the curve to consist of the upper semicircle of radius a with a counterclockwise orientation, we could write

$$x = a\cos t \qquad y = a\sin t, \qquad 0 \le t \le \pi$$

See Figure 65.

FIGURE 65

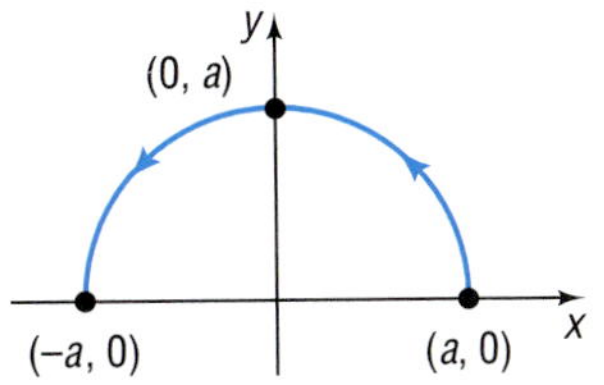

If we wanted the curve to consist of the left semicircle of radius a with a clockwise orientation, we could write

$$x = -a\sin t \qquad y = -a\cos t, \qquad 0 \le t \le \pi$$

See Figure 66.

FIGURE 66

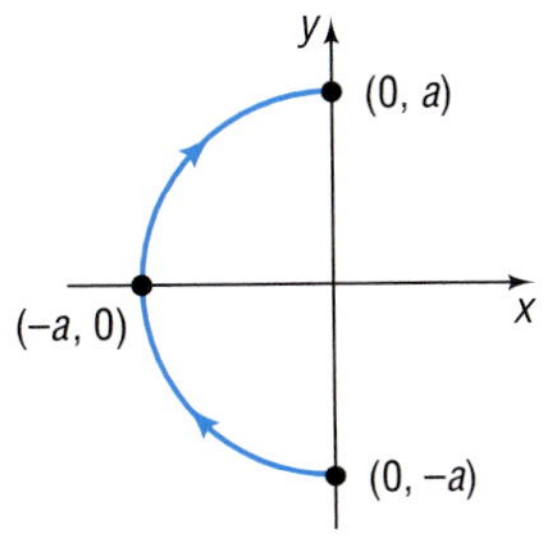

Seeing the Concept (a) Graph $x = \cos t, y = \sin t$ for $0 \le t \le 2\pi$. Compare to Figure 64. Graph $x = \cos t, y = \sin t$ for $0 \le t \le \pi$. Compare to Figure 65. Graph $x = -\sin t, y = -\cos t$ for $0 \le t \le \pi$. Compare to Figure 66.

Time as a Parameter: Projectile Motion; Simulated Motion

4 If we think of the parameter t as time, then the parametric equations $x = f(t)$ and $y = g(t)$ of a curve C specify how the x- and y-coordinates of a moving point vary with time.

For example, we can use parametric equations to describe the motion of an object, sometimes referred to as **curvilinear motion.** Using parametric equations, we can specify not only where the object travels, that is, its location (x, y), but also when it gets there, that is, the time t.

When an object is propelled upward at an inclination θ to the horizontal with initial speed v_0, the resulting motion is called **projectile motion.** See Figure 67(a).

In calculus it is shown that the parametric equations of the path of a projectile fired at an inclination θ to the horizontal, with an initial speed v_0, from a height h above the horizontal are

$$x = (v_0\cos\theta)t \qquad y = -\frac{1}{2}gt^2 + (v_0\sin\theta)t + h \qquad (2)$$

where t is the time and g is the constant acceleration due to gravity (approximately 32 ft/sec/sec or 9.8 m/sec/sec). See Figure 67(b).

FIGURE 67

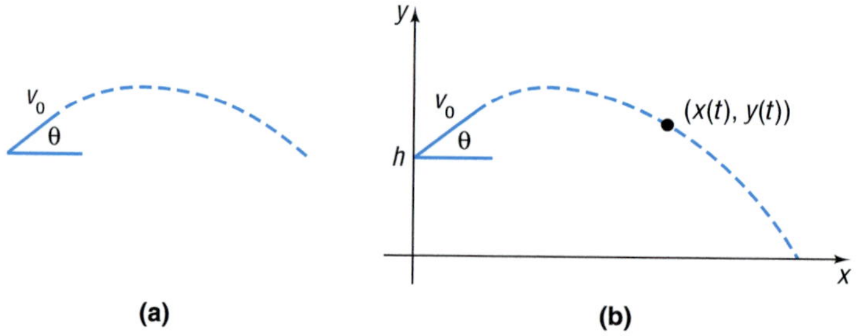

EXAMPLE 4 Projectile Motion

FIGURE 68

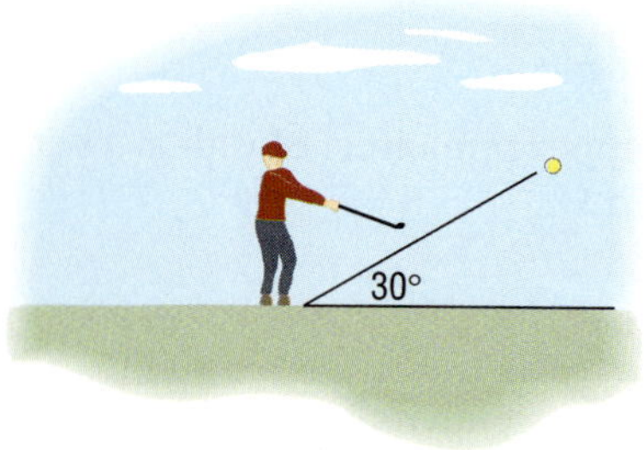

Suppose Jim hit a golf ball with an initial velocity of 150 feet per second at an angle of 30° to the horizontal. See Figure 68.

(a) Find parametric equations that describe the position of the ball as a function of time.

(b) How long is the golf ball in the air?

(c) When is the ball at its maximum height? Determine the maximum height of the ball.

(d) Determine the distance the ball traveled.

(e) Using a graphing utility, simulate the motion of the golfball by simultaneously graphing the equations found in (a).

Solution (a) We have $v_0 = 150$, $\theta = 30°$, $h = 0$ (the ball is on the ground) and $g = 32$ (since units are in feet and seconds). Substituting these values into equations, (2) we find

$$x = (150\cos 30°)t = 75\sqrt{3}\,t$$

$$y = \frac{-1}{2}(32)t^2 + (150\sin 30°)t + 0 = -16t^2 + (150\sin 30°)t$$

$$= -16t^2 + 75t$$

(b) To determine the length of time the ball is in the air, we solve the equation $y = 0$:

$$-16t^2 + 75t = 0$$

$$t(-16t + 75) = 0$$

$$t = 0 \text{ seconds or } t = \frac{75}{16} = 4.6875 \text{ seconds}$$

The ball will strike the ground after 4.6875 seconds.

(c) Notice the height y of the ball is a quadratic function of t. Thus, the maximum height of the ball can be found by determining the vertex of $y = -16t^2 + 75t$. The value of t at the vertex is

$$t = \frac{-b}{2a} = \frac{-75}{-32} = 2.34375 \text{ seconds}$$

The ball is at its maximum height after 2.34375 seconds. The maximum height of the ball is found by evaluating the function y at $t = 2.34375$ seconds.

$$\text{Maximum height} = -16(2.34375)^2 + (75)2.34375 \approx 87.89 \text{ feet}$$

FIGURE 69

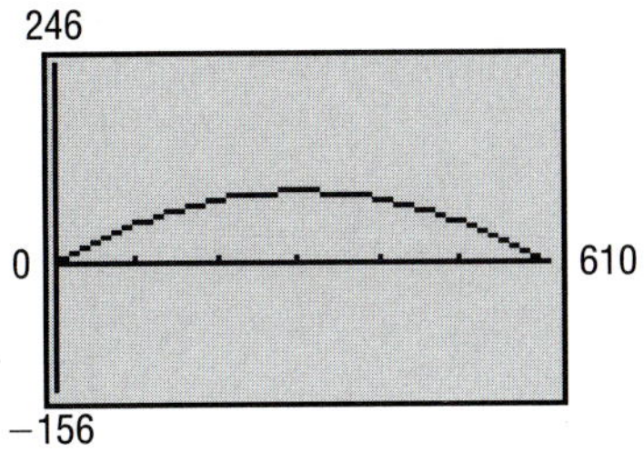

(d) Since the ball is in the air for 4.6875 seconds, the horizontal distance the ball travels is

$$x = (75\sqrt{3})4.6875 \approx 608.92 \text{ feet}$$

(e) We enter the equations from (a) into a graphing utility with Tmin = 0, Tmax = 4.7 and Tstep = 0.1. See Figure 69. ■

Exploration Simulate the motion of a ball thrown straight up with an initial speed of 100 feet per second from a height of 5 feet above the ground. Use PARametric mode with Tmin = 0, Tmax = 6.5, Tstep = 0.1, Xmin = 0, Xmax = 5, Ymin = 0, and Ymax = 180. What happens to the speed at which the graph is drawn as the ball goes up and then comes back down? How do you interpret this physically? Repeat the experiment using other values for Tstep. How does this affect the experiment?

Hint: In the projectile motion equations, let $\theta = 90°$, $v_0 = 100$, $h = 5$, and $g = 32$.) We use $x = 3$ instead of $x = 0$ to see the vertical motion better.

Result See Figure 70. In Figure 70(a) the ball is going up. In Figure 70(b), the ball is near its highest point. Finally, in Figure 70(c), the ball is coming back down.

FIGURE 70

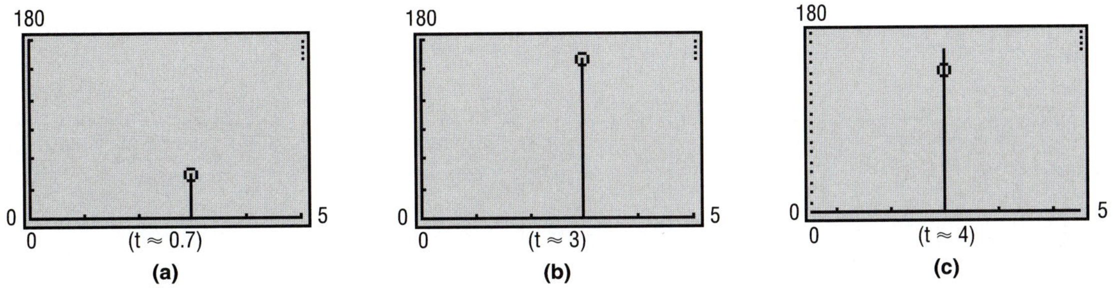

Notice that, as the ball goes up, its speed decreases, until at the highest point it is zero. Then the speed increases as the ball comes back down. ■

A graphing utility can be used to simulate other kinds of motion as well. Let's work again Example 5 from Section 3.3.

EXAMPLE 5 Simulating Motion

Tanya, who is a long distance runner, runs at an average velocity of 8 miles per hour. Two hours after Tanya leaves your house, you leave in your Honda and follow the same route. If your average velocity is 40 miles per hour, how long will it be before you catch up to Tanya? See Figure 71. Use a simulation of the two motions to answer the question.

FIGURE 71

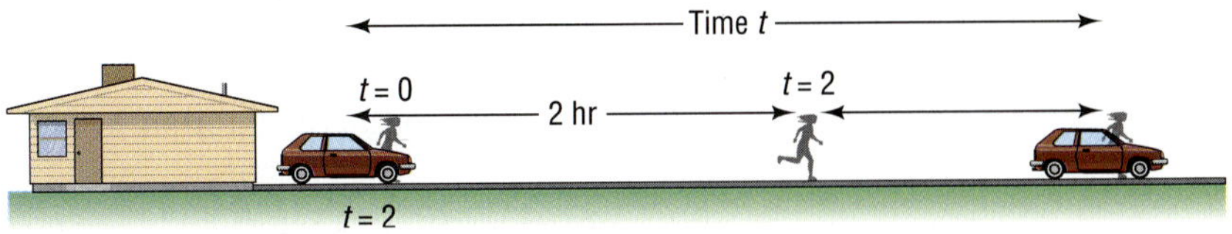

Solution We begin by setting up two sets of parametric equations: one to describe Tanya's motion, the other to describe the motion of the Honda. We choose time $t = 0$ to be when Tanya leaves the house. If we choose $y_1 = 2$ as Tanya's path, then we can use $y_2 = 4$ as the parallel path of the Honda. The horizontal distances traversed in time t are:

$$\text{Tanya: } x_1 = 8t \qquad \text{Honda: } x_2 = 40(t - 2)$$

The Honda catches up to Tanya when $x_1 = x_2$.

$$\begin{aligned} 8t &= 40(t - 2) \\ 8t &= 40t - 80 \\ -32t &= -80 \\ t &= \frac{-80}{-32} = 2.5 \end{aligned}$$

The Honda catches up to Tanya after 2.5 hours.

In PARametric mode with *T*step = 0.01, we simultaneously graph

$$\begin{aligned} &\text{Tanya: } x_1 = 8t \qquad &&\text{Honda: } x_2 = 40(t - 2) \\ &\qquad\quad y_1 = 2 &&\qquad\quad y_2 = 4 \end{aligned}$$

for $0 \le t \le 3$.

Figure 72 shows the relative position of Tanya and the Honda for $t = 0$, $t = 2$, $t = 2.25$, $t = 2.5$, and $t = 2.75$.

FIGURE 72

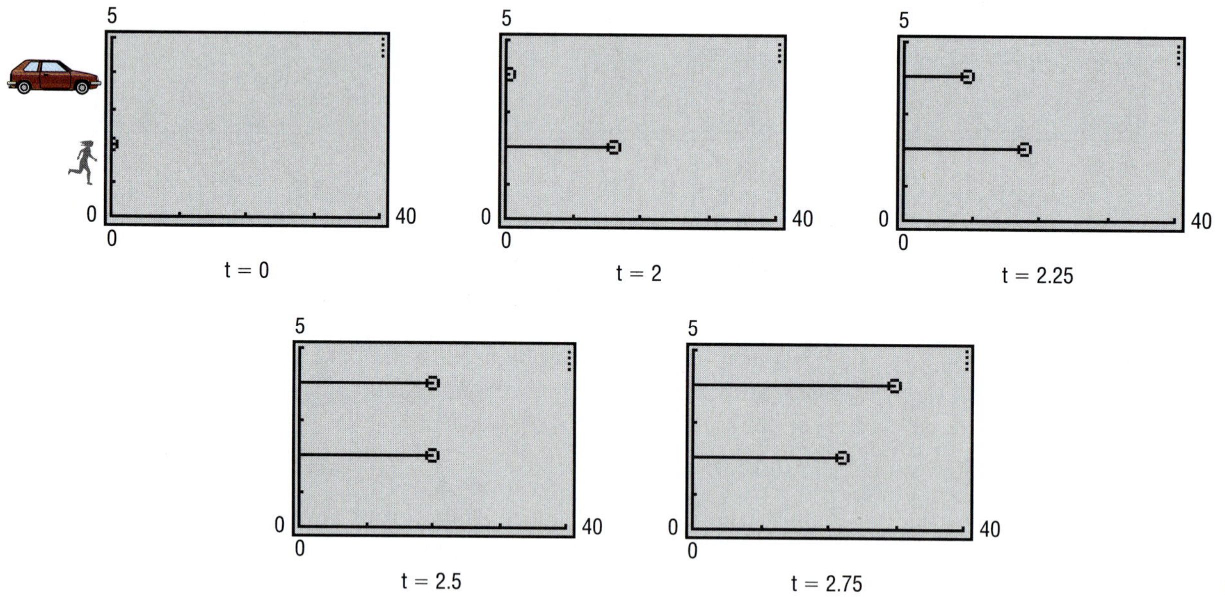

Now work Problem 27.

Finding Parametric Equations

We now take up the question of how to find parametric equations of a given curve.

5 If a curve is defined by the equation $y = f(x)$, where f is a function, one way of finding parametric equations is simply to let $x = t$. Then $y = f(t)$. Thus,

$$x = t \qquad y = f(t) \qquad t \text{ in the domain of } f$$

are parametric equations of the curve.

EXAMPLE 6 Finding Parametric Equations for a Curve Defined by a Rectangular Equation

Find parametric equations for the equation $y = x^2 - 4$.

Solution Let $x = t$. Then the parametric equations are

$$x = t \qquad y = t^2 - 4, \qquad -\infty < t < \infty$$

Another less obvious approach to Example 6 is to let $x = t^3$. Then the parametric equations become

$$x = t^3 \qquad y = t^6 - 4, \qquad -\infty < t < \infty$$

Care must be taken when using this approach, since the substitution for x must be a function that allows x to take on all the values stipulated by the domain of f. Thus, for example, letting $x = t^2$ so that $y = t^4 - 4$ does not result in equivalent parametric equations for $y = x^2 - 4$, since only points for which $x \geq 0$ are obtained.

EXAMPLE 7 Finding Parametric Equations for an Object in Motion

Find parametric equations for the ellipse

$$x^2 + \frac{y^2}{9} = 1$$

where the parameter t is time (in seconds) and

(a) The motion around the ellipse is clockwise, begins at the point (0, 3), and requires 1 second for a complete revolution.

(b) The motion around the ellipse is counterclockwise, begins at the point (1, 0), and requires 2 seconds for a complete revolution.

Solution (a) Since the motion begins at the point (0, 3), we want $x = 0$ and $y = 3$ when $t = 0$. Furthermore, since the given equation is an ellipse, we begin by letting

$$x = \sin \omega t \qquad \frac{y}{3} = \cos \omega t$$

for some constant ω. These parametric equations clearly satisfy the equation. Furthermore, with this choice, when $t = 0$, we have $x = 0$ and $y = 3$. For the motion to be clockwise, the motion will have to begin with the value of x increasing and y decreasing as t increases. Thus, $\omega > 0$. Finally, since 1 revolution requires 1 second, the period $2\pi/\omega = 1$, so $\omega = 2\pi$. Thus, parametric equations that satisfy the conditions stipulated are

$$x = \sin 2\pi t \qquad y = 3 \cos 2\pi t, \qquad 0 \leq t \leq 1 \qquad (3)$$

(b) Since the motion begins at the point (1, 0), we want $x = 1$ and $y = 0$ when $t = 0$. Furthermore, since the given equation is an ellipse, we begin by letting

$$x = \cos \omega t \qquad \frac{y}{3} = \sin \omega t$$

for some constant ω. These parametric equations clearly satisfy the equation. Furthermore, with this choice, when $t = 0$, we have $x = 1$ and $y = 0$. For the motion to be counterclockwise, the motion will have to begin with the value of x decreasing and y increasing as t increases. Thus, $\omega > 0$. Finally, since 1 revolution requires 2 seconds, the period is $2\pi/\omega = 2$, so $\omega = \pi$. Thus, the parametric equations that satisfy the conditions stipulated are

$$x = \cos \pi t \qquad y = 3 \sin \pi t, \qquad 0 \leq t \leq 2 \tag{4}$$

Either of equations (3) or (4) can serve as parametric equations for the ellipse $x^2 + y^2/9 = 1$ given in Example 6. The direction of the motion, the beginning point, and the time for 1 revolution merely serve to help arrive at a particular parametric representation.

Now work Problem 35.

The Cycloid

Suppose that a circle radius a rolls along a horizontal line without slipping. As the circle rolls along the line, a point P on the circle will trace out a curve called a **cycloid** (see Figure 73). We now seek parametric equations* for a cycloid.

We begin with a circle of radius a and take the fixed line on which the circle rolls as the x-axis. Let the origin be one of the points at which the point P comes in contact with the x-axis. Figure 73 illustrates the position of this

FIGURE 73

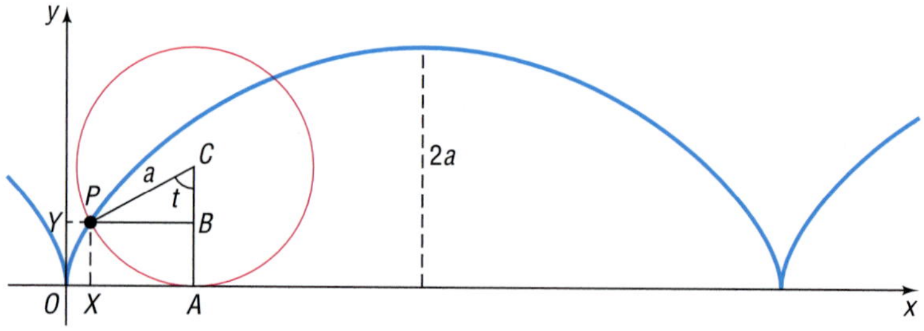

*Any attempt to derive the rectangular equation of a cycloid would soon demonstrate how complicated the task is.

point P after the circle has rolled somewhat. The angle t (in radians) measures the angle through which the circle has rolled.

Since we require no slippage, it follows that

$$\text{Arc } AP = d(O, A)$$

Therefore,

$$at = d(O, A)$$

The x-coordinate of the point P is

$$d(O, X) = d(O, A) - d(X, A) = at - a \sin t = a(t - \sin t)$$

The y-coordinate of the point P is equal to

$$d(O, Y) = d(A, C) - d(B, C) = a - a \cos t = a(1 - \cos t)$$

Thus, the parametric equations of the cycloid are

$$x = a(t - \sin t) \qquad y = a(1 - \cos t) \tag{5}$$

Exploration Graph $x = t - \sin t, y = 1 - \cos t, 0 \leq t \leq 3\pi$ using your graphing utility with Tstep $= \pi/36$ and a square screen. Compare your results with Figure 73.

Applications to Mechanics

If a is negative in equation (5), we obtain an inverted cycloid, as shown in Figure 74(a). The inverted cycloid occurs as a result of some remarkable applications in the field of mechanics. We shall mention two of them: the *brachistochrone* and the *tautochrone.**

The **brachistochrone** is the curve of quickest descent. If a particle is constrained to follow some path from one point A to a lower point B (not on the same vertical line) and is acted on only by gravity, the time needed to make the descent is least if the path is an inverted cycloid. See Figure 74(b). This remarkable discovery, which is attributed to many famous mathemati-

*In Greek, *brachistochrone* means "the shortest time," and *tautochrone* means "equal time."

FIGURE 74

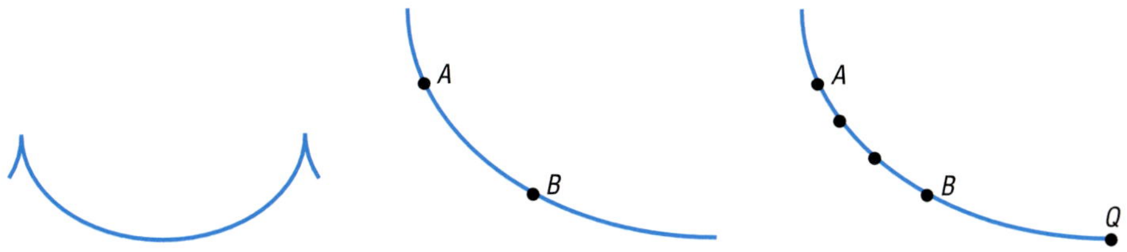

(a) Inverted cycloid **(b)** Curve of quickest descent **(c)** All reach Q at the same time

FIGURE 75
A flexible pendulum constrained by cycloids swings in a cycloid.

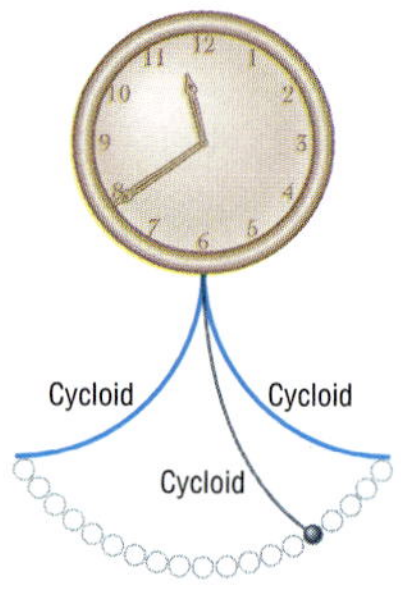

cians (including Johann Bernoulli and Blaise Pascal), was a significant step in creating the branch of mathematics known as the *calculus of variations.*

To define the **tautochrone,** let Q be the lowest point on an inverted cycloid. If several particles placed at various positions on an inverted cycloid simultaneously begin to slide down the cycloid, they will reach the point Q at the same time, as indicated in Figure 74(c). The tautochrone property of the cycloid was used by Christiaan Huygens (1629–1695), the Dutch mathematician, physicist, and astronomer, to construct a pendulum clock with a bob that swings along a cycloid (see Figure 75). In Huygen's clock, the bob was made to swing along a cycloid by suspending the bob on a thin wire constrained by two plates shaped like cycloids. In a clock of this design, the period of the pendulum is independent of its amplitude.

10.7 EXERCISES

In Problems 1–20, graph the curve whose parametric equations are given using a graphing utility. Find the rectangular equation of each curve. Graph the curve by hand and show its orientation.

1. $x = 3t + 2, \quad y = t + 1; \quad 0 \le t \le 4$
2. $x = t - 3, \quad y = 2t + 4; \quad 0 \le t \le 2$
3. $x = t + 2, \quad y = \sqrt{t}; \quad t \ge 0$
4. $x = \sqrt{2t}, \quad y = 4t; \quad t \ge 0$
5. $x = t^2 + 4, \quad y = t^2 - 4; \quad -\infty < t < \infty$
6. $x = \sqrt{t} + 4, \quad y = \sqrt{t} - 4; \quad t \ge 0$
7. $x = 3t^2, \quad y = t + 1; \quad -\infty < t < \infty$
8. $x = 2t - 4, \quad y = 4t^2; \quad -\infty < t < \infty$
9. $x = 2e^t, \quad y = 1 + e^t; \quad t \ge 0$
10. $x = e^t, \quad y = e^{-t}; \quad t \ge 0$
11. $x = \sqrt{t}, \quad y = t^{3/2}; \quad t \ge 0$
12. $x = t^{3/2} + 1, \quad y = \sqrt{t}; \quad t \ge 0$
13. $x = 2\cos t, \quad y = 3\sin t; \quad 0 \le t \le 2\pi$
14. $x = 2\cos t, \quad y = 3\sin t; \quad 0 \le t \le \pi$
15. $x = 2\cos t, \quad y = 3\sin t; \quad -\pi \le t \le 0$
16. $x = 2\cos t, \quad y = \sin t; \quad 0 \le t \le \pi/2$
17. $x = \sec t, \quad y = \tan t; \quad 0 \le t \le \pi/4$
18. $x = \csc t, \quad y = \cot t; \quad \pi/4 \le t \le \pi/2$
19. $x = \sin^2 t, \quad y = \cos^2 t; \quad 0 \le t \le 2\pi$
20. $x = t^2, \quad y = \ln t; \quad t > 0$

21. **Projectile Motion** Bob throws a ball straight up with an initial speed of 50 feet per second from a height of 6 feet.
    (a) Find parametric equations that describe the motion of the ball as a function of time.
    (b) How long is the ball in the air?
    (c) When is the ball at its maximum height? Determine the maximum height of the ball.
    (d) Simulate the motion of the ball by graphing the equations found in (a).

22. **Projectile Motion** Alice throws a ball straight up with an initial speed of 40 feet per second from a height of 5 feet.
    (a) Find parametric equations that describe the motion of the ball as a function of time.
    (b) How long is the ball in the air?
    (c) When is the ball at its maximum height? Determine the maximum height of the ball.
    (d) Simulate the motion of the ball by graphing the equations found in (a).

23. **Catching a Train** Bill's train leaves at 8:06 A.M. and accelerates at the rate of 2 meters/second/second. Bill arrives at the train station 5 seconds after the train had left. Bill can run 5 meters/second. Will Bill catch the train?
    (a) Find parametric equations that describe the motion of the train and Bill as a function of time.
    (b) Determine algebraically whether Bill will catch the train. If so, when?
    (c) Simulate the motion of the train and Bill by simultaneously graphing the equations found in (a).

24. **Catching a Bus** Jodi's bus leaves at 5:03 P.M. and accelerates at the rate of 3 meters/sec-

ond/second. Jodi arrives at the bus station 2 seconds after the bus had left. Jodi can run 5 meters/second. Will Jodi catch the bus?

(a) Find parametric equations that describe the motion of the bus and Jodi as a function of time.
(b) Determine algebraically whether Jodi will catch the bus. If so, when?
(c) Simulate the motion of the bus and Jodi by simultaneously graphing the equations found in (a).

25. **Projectile Motion** Nolan Ryan throws a baseball with an initial speed of 145 feet per second at an angle of 20° to the horizontal. The ball leaves Nolan Ryan's hand at a height of 5 feet.
(a) Find parametric equations that describe the position of the ball as a function of time.
(b) How long is the ball in the air?
(c) When is the ball at its maximum height? Determine the maximum height of the ball.
(d) Determine the distance the ball traveled.
(e) Using a graphing utility, simultaneously graph the equations found in (a).

26. **Projectile Motion** Mark McGuire hit a baseball with an initial speed of 180 feet per second at an angle of 40° to the horizontal. The ball was hit at a height of 3 feet off the ground.
(a) Find parametric equations that describe the position of the ball as a function of time.
(b) How long is the ball in the air?
(c) When is the ball at its maximum height? Determine the maximum height of the ball.
(d) Determine the distance the ball traveled.
(e) Using a graphing utility, simultaneously graph the equations found in (a).

27. **Projectile Motion** Suppose Adam throws a tennis ball off a cliff 300 meters high with an initial speed of 40 meters per second at an angle of 45° to the horizontal.
(a) Find parametric equations that describe the position of the ball as a function of time.
(b) How long is the ball in the air?
(c) When is the ball at its maximum height? Determine the maximum height of the ball.
(d) Determine the distance the ball traveled.
(e) Using a graphing utility, simultaneously graph the equations found in (a).

28. **Projectile Motion** Suppose Adam throws a tennis ball off a cliff 300 meters high with an initial speed of 40 meters per second at an angle of 45° to the horizontal on the moon (gravity on the Moon is 1/6 of that on earth).
(a) Find parametric equations that describe the position of the ball as a function of time.
(b) How long is the ball in the air?
(c) When is the ball at its maximum height? Determine the maximum height of the ball.
(d) Determine the distance the ball traveled.
(e) Using a graphing utility, simultaneously graph the equations found in (a).

29. **Uniform Motion** A Toyota Paseo (traveling East at 40 mph) and Pontiac Bonneville (traveling North at 30 mph) are heading toward the same intersection. The Paseo is 5 miles from the intersection when the Bonneville is 4 miles from the intersection. See the figure.

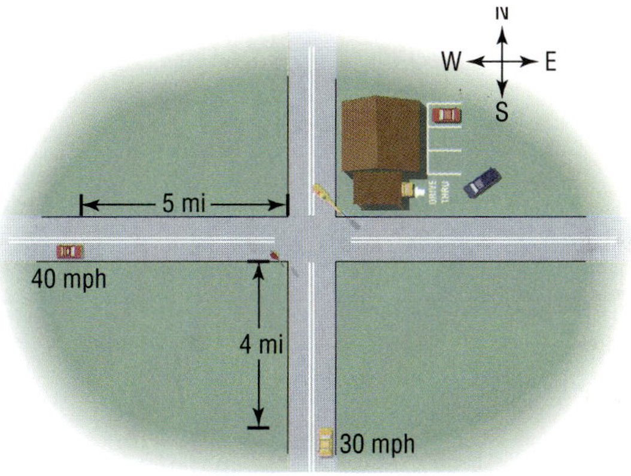

(a) Find a formula for the distance between the cars as a function of time.
(b) Graph the function in (a) using a graphing utility.
(c) What is the minimum distance between the cars? When are the cars closest?
(d) Find parametric equations that describe the motion of the Paseo and Bonneville.
(e) Simulate the motion of the cars by simultaneously graphing the equations found in (d).

30. **Uniform Motion** A Cesna (heading South at 120 mph) and a Boeing 747 (heading West at 600 mph) are flying toward each other at the same altitude. The Cesna is 100 miles from the point the flight patterns intersect and the 747 is 550 miles from this intersection point. See the figure.

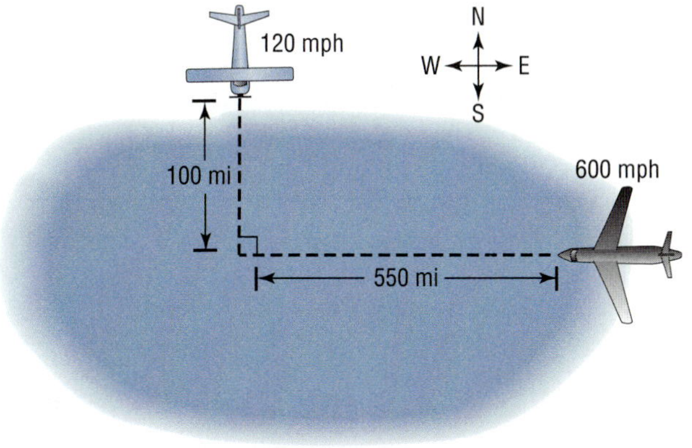

(a) Find a formula for the distance between the planes as a function of time.
(b) Graph the function in (a) using a graphing utility.
(c) What is the minimum distance between the planes? When are the planes closest?
(d) Find parametric equations that describe the motion of the Cesna and 747.
(e) Simulate the motion of the planes by simultaneously graphing the equations found in (d).

In Problems 31–34, find two different parametric equations for each rectangular equation.

31. $y = x^3$ **32.** $y = x^4 + 1$ **33.** $x = y^{3/2}$ **34.** $x = \sqrt{y}$

In Problems 35–38, find parametric equations for an object that moves along the ellipse $x^2/4 + y^2/9 = 1$ with the motion described.

35. The motion begins at $(2, 0)$, is clockwise, and requires 2 seconds for a complete revolution.

36. The motion begins at $(0, 3)$, is clockwise, and requires 1 second for a complete revolution.

37. The motion begins at $(0, 3)$, is counterclockwise, and requires 1 second for a complete revolution.

38. The motion begins at $(2, 0)$, is counterclockwise, and requires 3 seconds for a complete revolution.

In Problems 39 and 40, the parametric equations of four curves are given. Graph each of them by hand, indicating the orientation.

39. C_1: $x = t,\ \ y = t^2;\ \ -4 \le t \le 4$
C_2: $x = \cos t,\ \ y = 1 - \sin^2 t;\ \ 0 \le t \le \pi$
C_3: $x = e^t,\ \ y = e^{2t};\ \ 0 \le t \le \ln 4$
C_4: $x = \sqrt{t},\ y = t;\ \ 0 \le t \le 16$

40. C_1: $x = t,\ \ y = \sqrt{1 - t^2};\ \ -1 \le t \le 1$
C_2: $x = \sin t,\ \ y = \cos t;\ \ 0 \le t \le 2\pi$
C_3: $x = \cos t,\ \ y = \sin t;\ \ 0 \le t \le 2\pi$
C_4: $x = \sqrt{1 - t^2},\ \ y = t;\ \ -1 \le t \le 1$

41. Show that the parametric equations for a line passing through the points (x_1, y_1) and (x_2, y_2) are

$$x = (x_2 - x_1)t + x_1 \qquad y = (y_2 - y_1)t + y_1, \quad -\infty < t < \infty$$

What is the orientation of this line?

42. Projectile Motion The position of a projectile fired with an initial velocity v_0 feet per second and at an angle θ to the horizontal at the end of t seconds is given by the parametric equations

$$x = (v_0 \cos\theta)t \qquad y = (v_0 \sin\theta)t - 16t^2$$

See the following illustration.

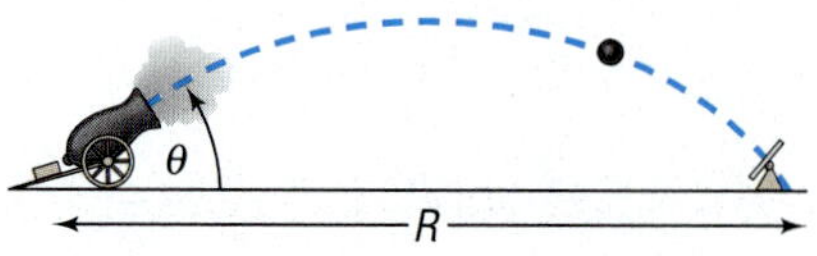

(a) Obtain the rectangular equation of the trajectory and identify the curve.
(b) Show that the projectile hits the ground $(y = 0)$ when $t = \frac{1}{16} v_0 \sin\theta$.
(c) How far has the projectile traveled (horizontally) when it strikes the ground? In other words, find the **range R.**
(d) Find the time t when $x = y$. Then find the horizontal distance x and the vertical distance y traveled by the projectile in this time. Then compute $\sqrt{x^2 + y^2}$. This is the distance R, the range, that the projectile travels up a plane inclined at 45° to the horizontal $(x = y)$. See the following illustration. (See also Problem 75 in Exercise 7.3.)

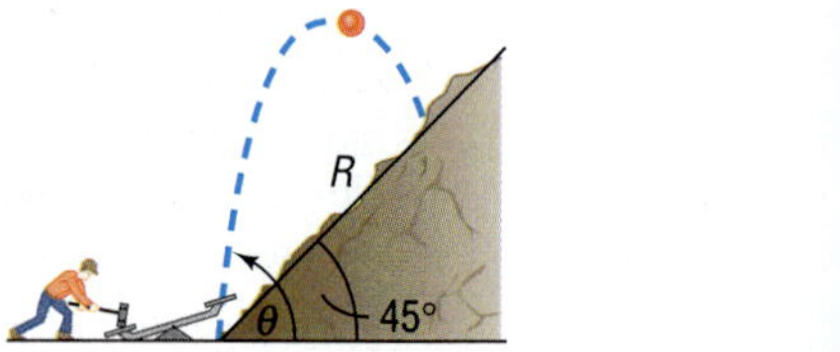

In Problems 43–46, use a graphing utility to graph the curve defined by the given parametric equations.

43. $x = t \sin t, \quad y = t \cos t$

44. $x = \sin t + \cos t, \quad y = \sin t - \cos t$

45. $x = 4 \sin t - 2 \sin 2t$
$y = 4 \cos t - 2 \cos 2t$

46. $x = 4 \sin t + 2 \sin 2t$
$y = 4 \cos t + 2 \cos 2t$

47. The Hypocycloid The hypocycloid is a curve defined by the parametric equations

$$x(t) = \cos^3 t \qquad y(t) = \sin^3 t, \qquad 0 \le t \le 2\pi$$

(a) Graph the hypocycloid using a graphing utility.
(b) Find rectangular equations of the hypocycloid.

48. In Problem 47, we graphed the hypocycloid. Now graph the rectangular equations of the hypocycloid. Did you obtain a complete graph? If not, experiment until you do.

49. Look up the curves called *hypocycloid* and *epicycloid*. Write a report on what you find. Be sure to draw comparisons with the cycloid.

CHAPTER REVIEW

THINGS TO KNOW

Equations

Parabola	See Tables 1 and 2.	
Ellipse	See Table 3.	
Hyperbola	See Table 4.	
General equation of a conic	$Ax^2 + Bxy + Cy^2 + Dx + Ey + F = 0$	Parabola if $B^2 - 4AC = 0$ Ellipse (or circle) if $B^2 - 4AC < 0$ Hyperbola if $B^2 - 4AC > 0$
Conic in polar coordinates	$\dfrac{d(F, P)}{d(P, D)} = e$	Parabola if $e = 1$ Ellipse if $e < 1$ Hyperbola if $e > 1$
Polar equations of a conic	See Table 5.	
Plane curve	$(x, y) = (f(t), g(t))$, t a parameter	

Definitions

Parabola	Set of points P in the plane for which $d(F, P) = d(P, D)$, where F is the focus and D is the directrix
Ellipse	Set of points P in the plane, the sum of whose distances from two fixed points (the foci) is a constant
Hyperbola	Set of points P in the plane, the difference of whose distances from two fixed points (the foci) is a constant

Formulas

Rotation formulas	$x = x' \cos\theta - y' \sin\theta$ $y = x' \sin\theta + y' \cos\theta$	
Angle θ of rotation that eliminates the $x'y'$-term	$\cot 2\theta = \dfrac{A - C}{B}$	$0 < \theta < \pi/2$

HOW TO

Find the vertex, focus, and directrix of a parabola given its equation

Graph a parabola given its equation (by hand and by using a graphing utility)

Find an equation of a parabola given certain information about the parabola

Find the center, foci, and vertices of an ellipse given its equation

Graph an ellipse given its equation (by hand and by using a graphing utility)

Find an equation of an ellipse given certain information about the ellipse

Find the center, foci, vertices, and asymptotes of a hyperbola given its equation

Graph a hyperbola given its equation (by hand and by using a graphing utility)

Find an equation of a hyperbola given certain information about the hyperbola

Identify conics without completing the square

Identify conics without a rotation of axes

Use rotation formulas to transform second-degree equations so that no xy-term is present

Identify and graph (by hand and by using a graphing utility) conics given by a polar equation

Graph parametric equations by hand and by using a graphing utility

Find the rectangular equation given the parametric equations

Simulate motion problems

FILL-IN-THE-BLANK ITEMS

1. A(n) ______________ is the collection of all points in the plane such that the distance from each point to a fixed point equals its distance to a fixed line.

2. A(n) ______________ is the collection of all points in the plane the sum of whose distances from two fixed points is a constant.

3. A(n) ______________ is the collection of all points in the plane the difference of whose distances from two fixed points is a constant.

4. For an ellipse, the foci lie on the ______________ axis; for a hyperbola, the foci lie on the ______________ axis.

5. For the ellipse $(x^2/9) + (y^2/16) = 1$, the major axis is along the ______________.

6. The equations of the asymptotes of the hyperbola $(y^2/9) - (x^2/4) = 1$ are ______________ and ______________.

7. To transform the equation

$$Ax^2 + Bxy + Cy^2 + Dx + Ey + F = 0, \qquad B \neq 0$$

into one in x' and y' without an $x'y'$-term, rotate the axes through an acute angle θ that satisfies the equation ______________.

8. The polar equation

$$r = \frac{8}{4 - 2\sin\theta}$$

is a conic whose eccentricity is ______________. It is a(n) ______________ whose directrix is ______________ to the polar axis at a distance ______________ units ______________ the pole.

9. The parametric equations $x = 2\sin t$ and $y = 3\cos t$ represent a(n) ______________.

TRUE/FALSE ITEMS

T F **1.** On a parabola, the distance from any point to the focus equals the distance from that point to the directrix.

T F **2.** The foci of an ellipse lie on its minor axis.

T F **3.** The foci of a hyperbola lie on its transverse axis.

T F **4.** Hyperbolas always have asymptotes, and ellipses never have asymptotes.

T F **5.** A hyperbola never intersects its conjugate axis.

T F **6.** A hyperbola always intersects its transverse axis.

T F **7.** The equation $ax^2 + 6y^2 - 12y = 0$ defines an ellipse if $a > 0$.

T F **8.** The equation $3x^2 + bxy + 12y^2 = 10$ defines a parabola if $b = -12$.

T F **9.** If (r, θ) are polar coordinates, the equation $r = 2/(2 + 3 \sin \theta)$ defines a hyperbola.

T F **10.** Parametric equations defining a curve are unique.

REVIEW EXERCISES

In Problems 1–20, identify each equation. If it is a parabola, give its vertex, focus, and directrix; if it is an ellipse, give its center, vertices, and foci; if it is a hyperbola, give its center, vertices, foci, and asymptotes.

1. $y^2 = -16x$

2. $16x^2 = y$

3. $\dfrac{x^2}{25} - y^2 = 1$

4. $\dfrac{y^2}{25} - x^2 = 1$

5. $\dfrac{y^2}{25} + \dfrac{x^2}{16} = 1$

6. $\dfrac{x^2}{9} + \dfrac{y^2}{16} = 1$

7. $x^2 + 4y = 4$

8. $3y^2 - x^2 = 9$

9. $4x^2 - y^2 = 8$

10. $9x^2 + 4y^2 = 36$

11. $x^2 - 4x = 2y$

12. $2y^2 - 4y = x - 2$

13. $y^2 - 4y - 4x^2 + 8x = 4$

14. $4x^2 + y^2 + 8x - 4y + 4 = 0$

15. $4x^2 + 9y^2 - 16x - 18y = 11$

16. $4x^2 + 9y^2 - 16x + 18y = 11$

17. $4x^2 - 16x + 16y + 32 = 0$

18. $4y^2 + 3x - 16y + 19 = 0$

19. $9x^2 + 4y^2 - 18x + 8y = 23$

20. $x^2 - y^2 - 2x - 2y = 1$

In Problems 21–36, obtain an equation of the conic described. Graph the equation by hand.

21. Parabola; focus at $(-2, 0)$; directrix the line $x = 2$

22. Ellipse; center at $(0, 0)$; focus at $(0, 3)$; vertex at $(0, 5)$

23. Hyperbola; center at $(0, 0)$; focus at $(0, 4)$; vertex at $(0, -2)$

24. Parabola; vertex at $(0, 0)$; directrix the line $y = -3$

25. Ellipse; foci at $(-3, 0)$ and $(3, 0)$; vertex at $(4, 0)$

26. Hyperbola; vertices at $(-2, 0)$ and $(2, 0)$; focus at $(4, 0)$

27. Parabola; vertex at $(2, -3)$; focus at $(2, -4)$

28. Ellipse; center at $(-1, 2)$; focus at $(0, 2)$; vertex at $(2, 2)$

29. Hyperbola; center at $(-2, -3)$; focus at $(-4, -3)$; vertex at $(-3, -3)$

30. Parabola; focus at $(3, 6)$; directrix the line $y = 8$

31. Ellipse; foci at $(-4, 2)$ and $(-4, 8)$; vertex at $(-4, 10)$

32. Hyperbola; vertices at $(-3, 3)$ and $(5, 3)$; focus at $(7, 3)$

33. Center at $(-1, 2)$; $a = 3$; $c = 4$; transverse axis parallel to the x-axis

34. Center at $(4, -2)$; $a = 1$; $c = 4$; transverse axis parallel to the y-axis

35. Vertices at $(0, 1)$ and $(6, 1)$; asymptote the line $3y + 2x - 9 = 0$

36. Vertices at $(4, 0)$ and $(4, 4)$; asymptote the line $y + 2x - 10 = 0$

In Problems 37–46, identify each conic without completing the squares and without applying a rotation of axes.

37. $y^2 + 4x + 3y - 8 = 0$

38. $2x^2 - y + 8x = 0$

39. $x^2 + 2y^2 + 4x - 8y + 2 = 0$

40. $x^2 - 8y^2 - x - 2y = 0$

41. $9x^2 - 12xy + 4y^2 + 8x + 12y = 0$

42. $4x^2 + 4xy + y^2 - 8\sqrt{5}x + 16\sqrt{5}y = 0$

43. $4x^2 + 10xy + 4y^2 - 9 = 0$

44. $4x^2 - 10xy + 4y^2 - 9 = 0$

45. $x^2 - 2xy + 3y^2 + 2x + 4y - 1 = 0$

46. $4x^2 + 12xy - 10y^2 + x + y - 10 = 0$

In Problems 47–52, rotate the axes so that the new equation contains no xy-term. Discuss and graph the new equation by hand. Verify your result using a graphing utility.

47. $2x^2 + 5xy + 2y^2 - \frac{9}{2} = 0$

48. $2x^2 - 5xy + 2y^2 - \frac{9}{2} = 0$

49. $6x^2 + 4xy + 9y^2 - 20 = 0$

50. $x^2 + 4xy + 4y^2 + 16\sqrt{5}x - 8\sqrt{5}y = 0$

51. $4x^2 - 12xy + 9y^2 + 12x + 8y = 0$

52. $9x^2 - 24xy + 16y^2 + 80x + 60y = 0$

In Problems 53–58, identify the conic that each polar equation represents, and graph it by hand. Verify your result using a graphing utility.

53. $r = \dfrac{4}{1 - \cos\theta}$

54. $r = \dfrac{6}{1 + \sin\theta}$

55. $r = \dfrac{6}{2 - \sin\theta}$

56. $r = \dfrac{2}{3 + 2\cos\theta}$

57. $r = \dfrac{8}{4 + 8\cos\theta}$

58. $r = \dfrac{10}{5 + 20\sin\theta}$

In Problems 59–62, convert each polar equation to a rectangular equation.

59. $r = \dfrac{4}{1 - \cos\theta}$

60. $r = \dfrac{6}{2 - \sin\theta}$

61. $r = \dfrac{8}{4 + 8\cos\theta}$

62. $r = \dfrac{2}{3 + 2\cos\theta}$

In Problems 63–68, graph the curve whose parametric equations are given using a graphing utility. Find the rectangular equation of each curve. Graph the curve by hand and show its orientation.

63. $x = 4t - 2, \quad y = 1 - t; \quad -\infty < t < \infty$

64. $x = 2t^2 + 6, \quad y = 5 - t; \quad -\infty < t < \infty$

65. $x = 3\sin t, \quad y = 4\cos t + 2; \quad 0 \le t \le 2\pi$

66. $x = \ln t, \quad y = t^3; \quad t > 0$

67. $x = \sec^2 t, \quad y = \tan^2 t; \quad 0 \le t \le \pi/4$

68. $x = t^{3/2}, \quad y = 2t + 4; \quad t \ge 0$

69. Find an equation of the hyperbola whose foci are the vertices of the ellipse $4x^2 + 9y^2 = 36$ and whose vertices are the foci of this ellipse.

70. Find an equation of the ellipse whose foci are the vertices of the hyperbola $x^2 - 4y^2 = 16$ and whose vertices are the foci of this hyperbola.

71. Describe the collection of points in a plane so that the distance from each point to the point $(3, 0)$ is three-fourths of its distance from the line $x = \frac{16}{3}$.

72. Describe the collection of points in a plane so that the distance from each point to the point $(5, 0)$ is five-fourths of its distance from the line $x = \frac{16}{5}$.

73. Mirrors A mirror is shaped like a paraboloid of revolution. If a light source is located 1 foot from the base along the axis of symmetry and the opening is 2 feet across, how deep should the mirror be?

74. Parabolic Arch Bridge A bridge is built in the shape of a parabolic arch. The bridge has a span of 60 feet and a maximum height of 20 feet. Find the height of the arch at distances of 5, 10, and 20 feet from the center.

75. Semielliptical Arch Bridge A bridge is built in the shape of a semielliptical arch. The bridge has a span of 60 feet and a maximum height of 20 feet. Find the height of the arch at distances of 5, 10, and 20 feet from the center.

76. Whispering Galleries The figure shows the specifications for an elliptical ceiling in a hall designed to be a whispering gallery. Where in the hall are the foci located?

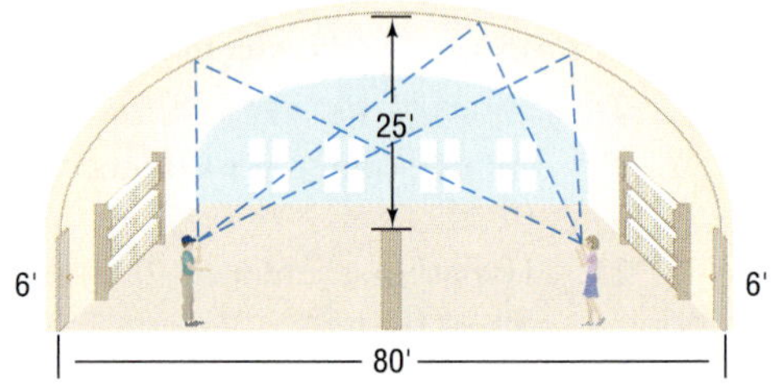

77. LORAN Two LORAN stations are positioned 150 miles apart along a straight shore.

(a) A ship records a time difference of 0.00032 second between the LORAN signals. Set up an appropriate rectangular coordinate system to determine where the ship would reach shore if it were to follow the hyperbola corresponding to this time difference.

(b) If the ship wants to enter a harbor located between the two stations 15 miles from the master station, what time difference should it be looking for?

(c) If the ship is 20 miles offshore when the desired time difference is obtained, what is the exact location of the ship?

[**Note:** The speed of each radio signal is 186,000 miles per second.]

78. **Uniform Motion** Mary's train leaves at 7:15 A.M. and accelerates at the rate of 3 meters/second/second. Mary arrives at the train station 2 seconds after the train had left. Mary can run 6 meters/second. Will Mary catch the train?
(a) Find parametric equations that describe the motion of the train and Mary as a function of time.
(b) Determine algebraically whether Mary will catch the train. If so, when?
(c) Simulate the motion of the train and Mary by simultaneously graphing the equations found in (a).

79. **Projectile Motion** Drew Bledsoe throws a football with an initial speed of 100 feet per second at an angle of 35° to the horizontal. The ball leaves Drew Bledsoe's hand at a height of 6 feet.
(a) Find parametric equations that describe the position of the ball as a function of time.
(b) How long is the ball in the air?
(c) When is the ball at its maximum height? Determine the maximum height of the ball.
(d) Determine the distance the ball traveled.
(e) Using a graphing utility, simultaneously graph the equations found in (a).

80. Formulate a strategy for discussing and graphing an equation of the form
$Ax^2 + Bxy + Cy^2 + Dx + Ey + F = 0$.

CHAPTER 11

Systems of Equations and Inequalities

Internet usage and the security of the information passing through the Internet is a major concern for corporations and other organizations. On the following page, you and your friends decide to start an encryption software company. You will need to use the Sullivan website at:

www.prenhall.com/sullivan

in order to find the resources available to build your encryption program.

PREPARING FOR THIS CHAPTER

Before getting started on this chapter, review the following concepts:

For Sections 11.1, 11.2, 11.7, 11.8: Lines *(Section 1.4)*

For Section 11.5: Identity *(p. 529)*

Factoring Polynomials *(Appendix, Section 2, pp. 992–994)*

Proper and Improper Rational Functions *(p. 285)*

Fundamental Theorem of Algebra *(p. 326)*

For Section 11.6: Circles *(pp. 66–69)*

Parabolas *(pp. 712–719)*

Ellipses *(pp. 724–733)*

Hyperbolas *(pp. 738–748)*

For Section 11.7: Inequalities *(Section 3.5)*

OUTLINE

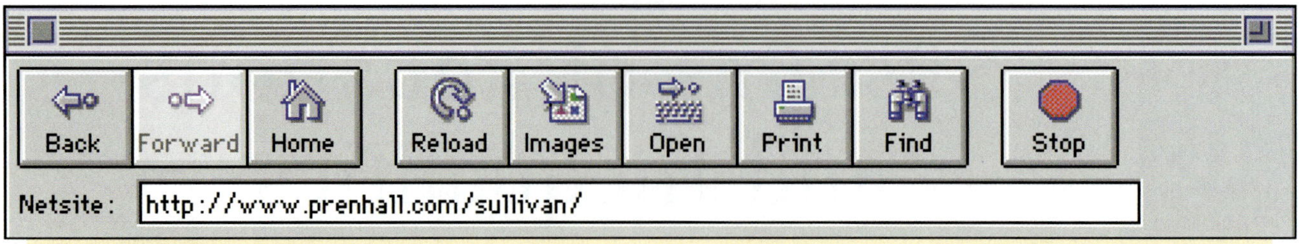

CRYPTOGRAPHY

Cryptography has a long and fascinating *history*. For example, some believe that Hitler would have won World War II if the German secret codes produced by the *Enigma machine* were not broken!

Today, in the age of the Internet, electronic commerce, and banking, cryptography has once again become indispensable. Companies are investing large sums of money to insure their security.

You and some of your friends have decided to start your own software company, Phinx-s Security. You want to market your system to small businesses with PC's, as this market is wide open at the present moment. Your part of the project is to design the encryption algorithm for the programmer. You have a very clever idea about how to use a dynamic matrix key that will be hardwired into a circuit board.

Now for some of the *basics.*

1. What are some of the current *encryption schemes*?
2. Who are the "cyberpunks"?
3. Are you surprised to see the *$10,000 RSA and DES contests*?
4. The kernel of your scheme uses *matrix encryption.* What size of a matrix might you want to use? 64 by 64? 128 by 128? What are the advantages of using a larger matrix? The disadvantages?
5. How many 64 by 64 size matrices are possible mod(26)? Is this equal to the number of possible keys?
6. Would this method work if you were to use mod(71)? Why might you want to use mod(71) over mod(26)? Would you put up a $10,000 challenge?

In this chapter we take up the problem of solving equations and inequalities containing two or more variables. As the section titles suggest, there are various ways to solve such problems.

The *method of substitution* for solving equations in several unknowns goes back to ancient times.

The *method of elimination,* although it had existed for centuries, was put into systematic order by Karl Friedrich Gauss (1777–1855) and by Camille Jordan (1838–1922). This method is now used for solving large systems by computer.

The theory of *matrices* was developed in 1857 by Arthur Cayley (1821–1895), although only later were matrices used as we use them in this chapter. Matrices have become a very flexible instrument, useful in almost all areas of mathematics.

The method of *determinants* was invented by Seki Kōwa (1642–1708) in 1683 in Japan and by Gottfried Wilhelm von Leibniz (1646–1716) in 1693 in Germany. Both used them only in relation to linear equations. *Cramer's Rule* is named after Gabriel Cramer (1704–1752) of Switzerland, who popularized the use of determinants for solving linear systems.

Section 11.5, *partial fraction decomposition,* provides an application of systems of equations. This particular application is one that is used in integral calculus.

Section 11.8 introduces *linear programming,* a modern application of linear inequalities to certain types of problems. This topic is particularly useful for students interested in operations research.

11.1 SYSTEMS OF LINEAR EQUATIONS: SUBSTITUTION; ELIMINATION

1 Solve Systems of Equations by Substitution
2 Solve Systems of Equations by Elimination
3 Inconsistent Systems and Dependent Equations

We begin with an example.

EXAMPLE 1 Movie Theater Ticket Sales

A movie theater sells tickets for \$8.00 each, with Seniors receiving a discount of \$2.00. One evening the theater took in \$3580 in revenue. If x represents the number of tickets sold at \$8.00 and y the number of tickets sold at the discounted price of \$6.00, write an equation that relates these variables.

Solution Each nondiscounted ticket brings in \$8.00, so x tickets will bring in $8x$ dollars. Similarly, y discounted tickets bring in $6y$ dollars. If the total brought in is \$3580, we must have

$$8x + 6y = 3580$$

In Example 1, suppose we also know that 525 tickets were sold that evening. Then we have another equation relating the variables x and y:

$$x + y = 525$$

The two equations

$$\begin{aligned} 8x + 6y &= 3580 \\ x + y &= 525 \end{aligned}$$

form a *system* of equations.

In general, a **system of equations** is a collection of two or more equations, each containing one or more variables. Example 2 gives some samples of systems of equations.

EXAMPLE 2 Examples of Systems of Equations

(a) $\begin{cases} 2x + y = 5 & (1) \\ -4x + 6y = -2 & (2) \end{cases}$ Two equations containing two variables, x and y.

(b) $\begin{cases} x + y^2 = 5 & (1) \\ 2x + y = 4 & (2) \end{cases}$ Two equations containing two variables, x and y.

(c) $\begin{cases} x + y + z = 6 & (1) \\ 3x - 2y + 4z = 9 & (2) \\ x - y - z = 0 & (3) \end{cases}$ Three equations containing three variables, x, y, and z.

(d) $\begin{cases} x + y + z = 5 & (1) \\ x - y = 2 & (2) \end{cases}$ Two equations containing three variables, x, y, and z.

(e) $\begin{cases} x + y + z = 6 & (1) \\ 2x + 2z = 4 & (2) \\ y + z = 2 & (3) \\ x = 4 & (4) \end{cases}$ Four equations containing three variables, x, y, and z.

We use a brace, as shown, to remind us that we are dealing with a system of equations. We also will find it convenient to number each equation in the system.

A **solution** of a system of equations consists of values for the variables that reduce each equation of the system to a true statement. To **solve** a system of equations means to find all solutions of the system.

For example, $x = 2$, $y = 1$ is a solution of the system in Example 2(a), because

$$2(2) + 1 = 5 \quad \text{and} \quad -4(2) + 6(1) = -2$$

A solution of the system in Example 2(b) is $x = 1$, $y = 2$, because

$$1 + 2^2 = 5 \quad \text{and} \quad 2(1) + 2 = 4$$

Another solution of the system in Example 2(b) is $x = \frac{11}{4}$, $y = -\frac{3}{2}$ which you can check for yourself. A solution of the system in Example 2(c) is $x = 3$, $y = 2$, $z = 1$, because

$$\begin{cases} 3 + 2 + 1 = 6 & \text{(1) } x = 3,\ y = 2,\ z = 1. \\ 3(3) - 2(2) + 4(1) = 9 & \text{(2)} \\ 3 - 2 - 1 = 0 & \text{(3)} \end{cases}$$

Note that $x = 3$, $y = 3$, $z = 0$ is not a solution of the system in Example 2(c):

$$\begin{cases} 3 + 3 + 0 = 6 & \text{(1) } x = 3,\ y = 3,\ z = 0. \\ 3(3) - 2(3) + 4(0) = 3 \neq 9 & \text{(2)} \\ 3 - 3 - 0 = 0 & \text{(3)} \end{cases}$$

Although these values satisfy equations (1) and (3), they do not satisfy equation (2). Any solution of the system must satisfy *each* equation of the system.

Now work Problem 3.

When a system of equations has at least one solution, it is said to be **consistent;** otherwise, it is called **inconsistent.**

An equation in n variables is said to be **linear** if it is equivalent to an equation of the form

$$a_1x_1 + a_2x_2 + \cdots + a_nx_n = b$$

where $x_1, x_2, \ldots, x_n$ are n distinct variables, $a_1, a_2, \ldots, a_n$, b are constants, and at least one of the a's is not 0.

Some examples of linear equations are

$$2x + 3y = 2 \qquad 5x - 2y + 3z = 10 \qquad 8x + 8y - 2z + 5w = 0$$

If each equation in a system of equations is linear, then we have a **system of linear equations.** Thus, the systems in Examples 2(a), (c), (d), and (e) are linear, whereas the system in Example 2(b) is nonlinear. We concentrate on solving linear systems in Sections 11.1–11.3, and we will take up nonlinear systems in Section 11.6.

Two Linear Equations Containing Two Variables

We can view the problem of solving a system of two linear equations containing two variables as a geometry problem. The graph of each equation in such a system is a straight line. Thus, a system of two equations containing two variables represents a pair of lines. The lines either (1) intersect or (2) are parallel or (3) are **coincident** (that is, identical).

1. If the lines intersect, then the system of equations has one solution, given by the point of intersection. The system is **consistent** and the equations are **independent.**
2. If the lines are parallel, then the system of equations has no solution, because the lines never intersect. The system is **inconsistent.**
3. If the lines are coincident, then the system of equations has infinitely many solutions, represented by the totality of points on the line. The system is **consistent** and the equations are **dependent.**

Figure 1 illustrates these conclusions.

FIGURE 1

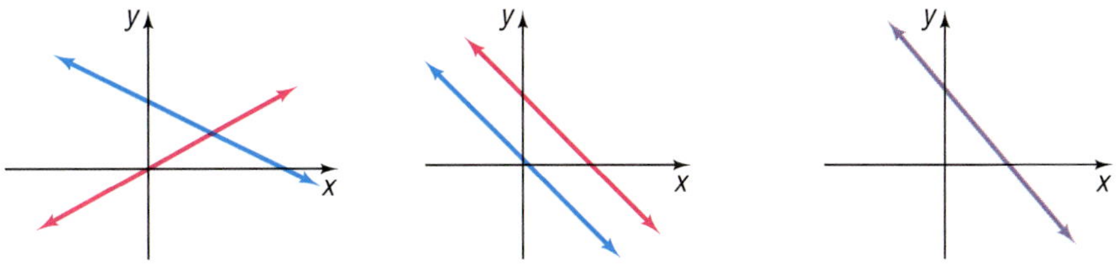

(a) Intersecting lines; system has one solution **(b)** Parallel lines; system has no solution **(c)** Coincident lines; system has infinitely many solutions

EXAMPLE 3 Solving a System of Linear Equations Using a Graphing Utility

Solve: $\begin{cases} 2x + y = 5 & (1) \\ -4x + 6y = 12 & (2) \end{cases}$

Solution First, we solve each equation for y. This is equivalent to writing each equation in slope-intercept form. Equation (1) in slope-intercept form is $Y_1 = -2x + 5$. Equation (2) in slope-intercept form is $Y_2 = \frac{2}{3}x + 2$. Figure 2 shows the graphs using a graphing utility. From the graph in Figure 2, we see that the lines intersect, so the system is consistent and the equations are independent. Using INTERSECT, we obtain the solution (1.125, 2.75).

FIGURE 2

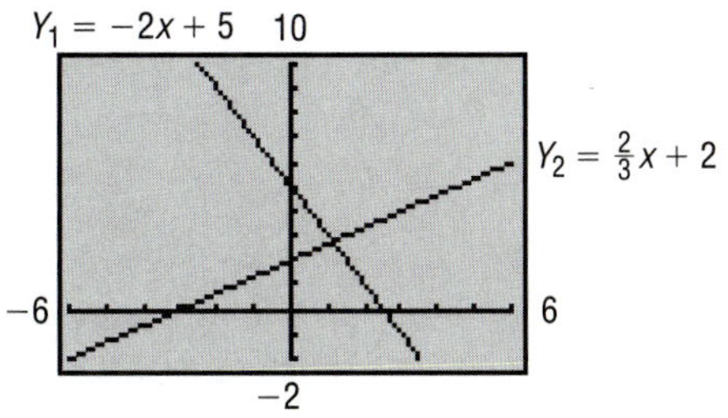

Sometimes, in order to obtain exact solutions, we must use algebraic methods. The first algebraic method we discuss is the *method of substitution.*

Method of Substitution

1 We illustrate the **method of substitution** by solving the system given in Example 3.

EXAMPLE 4 Solving a System of Linear Equations by Substitution

Solve: $\begin{cases} 2x + y = 5 & (1) \\ -4x + 6y = 12 & (2) \end{cases}$

Solution We solve the first equation for y, obtaining

$$y = 5 - 2x$$

We substitute this result for y in the second equation. The result is an equation containing just the variable x, which we can then solve for:

$$\begin{aligned} -4x + 6y &= 12 \\ -4x + 6(5 - 2x) &= 12 \\ -4x + 30 - 12x &= 12 \\ -16x &= -18 \\ x &= \frac{-18}{-16} = \frac{9}{8} \end{aligned}$$

Once we know that $x = \frac{9}{8}$, we can easily find the value of y by **back-substitution,** that is, by substituting $\frac{9}{8}$ for x in one of the original equations. We use the first one:

$$\begin{aligned} 2x + y &= 5 \\ 2\left(\frac{9}{8}\right) + y &= 5 \\ \frac{9}{4} + y &= 5 \\ y &= 5 - \frac{9}{4} = \frac{20}{4} - \frac{9}{4} = \frac{11}{4} \end{aligned}$$

The solution of the system is $x = \frac{9}{8} = 1.125$, $y = \frac{11}{4} = 2.75$. ■

The method used to solve the system in Example 4 is called **substitution.** The steps to be used are outlined next.

Steps for Solving by Substitution

STEP 1: Pick one of the equations and solve for one of the variables in terms of the remaining variables.
STEP 2: Substitute the result in the remaining equations.
STEP 3: If one equation in one variable results, solve this equation. Otherwise, repeat step 1 until a single equation with one variable remains.
STEP 4: Find the values of the remaining variables by back-substitution.
STEP 5: Check the solution found.

EXAMPLE 5 Solving a System of Linear Equations by Substitution

Solve: $\begin{cases} 3x - 2y = 5 & (1) \\ 5x - y = 6 & (2) \end{cases}$

Solution STEP 1: After looking at the two equations, we conclude that it is easiest to solve for the variable y in equation (2):

$$\begin{aligned} 5x - y &= 6 \\ y &= 5x - 6 \end{aligned}$$

STEP 2: We substitute this result into equation (1) and simplify:

$$\begin{aligned} 3x - 2y &= 5 \\ 3x - 2(5x - 6) &= 5 \\ -7x + 12 &= 5 \\ -7x &= -7 \\ x &= 1 \end{aligned}$$

STEP 3: Because we now have one solution, $x = 1$, we proceed to step 4.
STEP 4: Knowing $x = 1$, we can find y from the equation

$$y = 5x - 6 = 5(1) - 6 = -1$$

STEP 5: *Check:* $\begin{cases} 3(1) - 2(-1) = 3 + 2 = 5 \\ 5(1) - (-1) \;\;= 5 + 1 = 6 \end{cases}$

The solution of the system is $x = 1, y = -1$.

EXAMPLE 6 Solving a System of Linear Equations by Substitution

Solve: $\begin{cases} 2x - 3y = 7 & (1) \\ 4x + 5y = 3 & (2) \end{cases}$

Solution STEP 1: In looking over the system, we conclude that there is no way to solve for one of the variables without introducing fractions. We solve for the variable x in equation (1):

$$\begin{aligned} 2x - 3y &= 7 \\ 2x &= 3y + 7 \\ x &= \frac{3}{2}y + \frac{7}{2} \end{aligned}$$

STEP 2: We substitute this result for x in equation (2) and simplify:

$$\begin{aligned} 4x + 5y &= 3 \\ 4\left(\frac{3}{2}y + \frac{7}{2}\right) + 5y &= 3 \\ 6y + 14 + 5y &= 3 \\ 11y + 14 &= 3 \\ 11y &= -11 \end{aligned}$$

STEP 3: $y = -1$

STEP 4: $x = \frac{3}{2}y + \frac{7}{2} = \frac{3}{2}(-1) + \frac{7}{2} = \frac{4}{2} = 2$

STEP 5: *Check:* $\begin{cases} 2(2) - 3(-1) = 4 + 3 = 7 \\ 4(2) + 5(-1) = 8 - 5 = 3 \end{cases}$

The solution is $x = 2, y = -1$.

Now use substitution to work Problem 13.

Method of Elimination

2 A second method for solving a system of linear equations is the *method of elimination.* This method is usually preferred over substitution if substitution

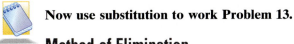

leads to fractions or if the system contains more than two variables. Elimination also provides the necessary motivation for solving systems using matrices (the subject of the next section).

The idea behind the method of elimination is to keep replacing the original equations in the system with equivalent equations until a system of equations with an obvious solution is reached. When we proceed in this way, we obtain **equivalent systems of equations.** The rules for obtaining equivalent equations are the same as those studied earlier. However, we may also interchange any two equations of the system and/or replace any equation in the system by the sum (or difference) of that equation and any other equation in the system.

Rules for Obtaining an Equivalent System of Equations

1. Interchange any two equations of the system.
2. Multiply (or divide) each side of an equation by the same nonzero constant.
3. Replace any equation in the system by the sum (or difference) of that equation and any other equation in the system.

An example will give you the idea. As you work through the example, pay particular attention to the pattern being followed.

EXAMPLE 7 Solving a System of Linear Equations by Elimination

Solve: $\begin{cases} 2x + 3y = 1 & (1) \\ -x + y = -3 & (2) \end{cases}$

Solution We multiply each side of equation (2) by 2 so that the coefficients of x in the two equations are negatives of one another. The result is the equivalent system

$$\begin{cases} 2x + 3y = 1 & (1) \\ -2x + 2y = -6 & (2) \end{cases}$$

If we now replace equation (2) of this system by the sum of the two equations, we obtain an equation containing just the variable y, which we can solve for:

$$\begin{cases} 2x + 3y = 1 & (1) \\ -2x + 2y = -6 & (2) \end{cases}$$
$$5y = -5$$
$$y = -1$$

We back-substitute by using this value for y in equation (1) and simplify to get

$$2x + 3(-1) = 1$$
$$2x = 4$$
$$x = 2$$

Thus, the solution of the original system is $x = 2$, $y = -1$. We leave it to you to check the solution.

The procedure used in Example 7 is called the **method of elimination.** Notice the pattern of the solution. First, we eliminated the variable x from the second equation. Then we back-substituted; that is, we substituted the value found for y back into the first equation to find x.

Let's return to the movie theater example (Example 1).

EXAMPLE 8 Movie Theater Ticket Sales

A movie theater sells tickets for \$8.00 each, with Seniors receiving a discount of \$2.00. One evening the theater sold 525 tickets and took in \$3580 in revenue. How many of each type of ticket was sold?

Solution If x represents the number of tickets sold at \$8.00 and y the number of tickets sold at the discounted price of \$6.00, then the given information results in the system of equations

$$\begin{cases} 8x + 6y = 3580 \\ x + y = 525 \end{cases}$$

We use elimination and multiply the second equation by -6 and then add the equations.

$$\begin{cases} 8x + 6y = 3580 \\ -6x - 6y = -3150 \end{cases}$$

$$2x = 430$$

$$x = 215$$

Since $x + y = 525$, then $y = 525 - x = 525 - 215 = 310$. Thus, 215 nondiscounted tickets and 310 Senior discount tickets were sold.

Now use elimination to work Problem 13.

3 The previous examples dealt with consistent systems of equations that had a unique solution. The next two examples deal with two other possibilities that may occur, the first being a system that has no solution.

EXAMPLE 9 An Inconsistent System of Linear Equations

Solve: $\begin{cases} 2x + y = 5 & (1) \\ 4x + 2y = 8 & (2) \end{cases}$

Solution We choose to use the method of substitution and solve equation (1) for y:

$$2x + y = 5$$

$$y = 5 - 2x$$

Substituting in equation (2), we get

$$4x + 2y = 8$$

$$4x + 2(5 - 2x) = 8$$

$$4x + 10 - 4x = 8$$

$$0 \cdot x = -2$$

This equation has no solution. Thus, we conclude that the system itself has no solution and is therefore inconsistent.

Figure 3 illustrates the pair of lines whose equations form the system in Example 9. Notice that the graphs of the two equations are lines, each with slope -2; one has a y-intercept of 5, the other a y-intercept of 4. Thus, the lines are parallel and have no point of intersection. This geometric statement is equivalent to the algebraic statement that the system has no solution.

FIGURE 3

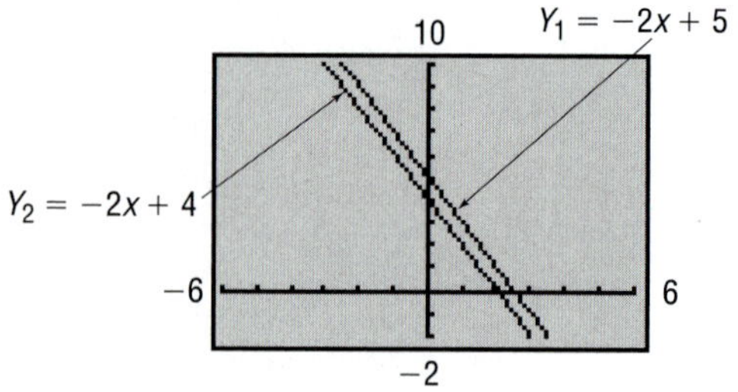

The next example is an illustration of a system with infinitely many solutions.

EXAMPLE 10 Solving a System of Linear Equations with Infinitely Many Solutions

Solve: $\begin{cases} 2x + y = 4 & (1) \\ -6x - 3y = -12 & (2) \end{cases}$

Solution We choose to use the method of elimination:

$$\begin{cases} 2x + y = 4 & (1) \\ -6x - 3y = -12 & (2) \end{cases}$$

$$\begin{cases} 6x + 3y = 12 & (1) \\ -6x - 3y = -12 & (2) \end{cases} \quad \text{Multiply each side of equation (1) by 3.}$$

$$\begin{cases} 6x + 3y = 12 & (1) \\ 0 = 0 & (2) \end{cases} \quad \text{Replace equation (2) by the sum of equations (1) and (2).}$$

The original system is thus equivalent to a system containing one equation, so the equations are dependent. This means that any values of x and y for which $6x + 3y = 12$ or, equivalently, $2x + y = 4$ are solutions. For example, $x = 2, y = 0; x = 0, y = 4; x = -2, y = 8; x = 4, y = -4$; and so on, are solutions. There are, in fact, infinitely many values of x and y for which $2x + y = 4$, so the original system has infinitely many solutions. We will write the solutions of the original systems either as

$$y = 4 - 2x$$

where x can be any real number, or as

$$x = 2 - \frac{1}{2}y$$

where y can be any real number. ■

FIGURE 4
$Y_1 = Y_2 = -2x + 4$

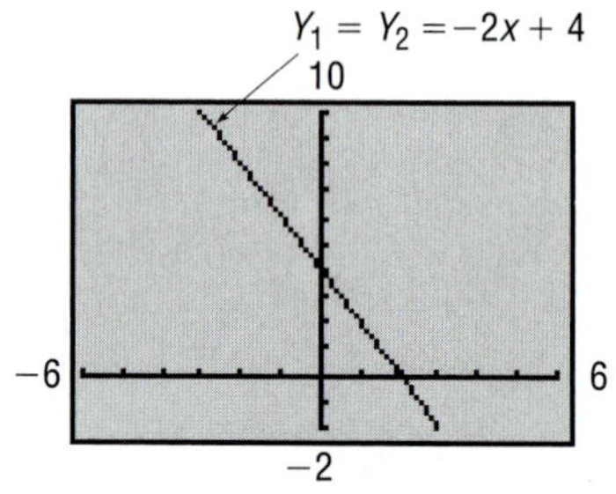

Figure 4 illustrates the situation presented in Example 10. Notice that the graphs of the two equations are lines, each with slope -2 and each with y-intercept 4. Thus, the lines are coincident. Notice also that equation (2) in the original system is just -3 times equation (1), indicating that the two equations are dependent.

For the system in Example 10, we can write down some of the infinite number of solutions by assigning values to x and then finding $y = 4 - 2x$. Thus,

If $x = -2$, then $y = 8$.
If $x = 0$, then $y = 4$.
If $x = 2$, then $y = 0$.

The pairs (x, y) are points on the line in Figure 4.

Now work Problems 19 and 23.

Three Linear Equations Containing Three Variables

Just as with a system of two linear equations containing two variables, a system of three linear equations containing three variables also has either exactly one solution, no solution, or infinitely many solutions.

Let's see how elimination works on a system of three equations containing three variables.

EXAMPLE 11 Solving a System of Three Linear Equations with Three Variables

Use the method of elimination to solve the system of equations:

$$\begin{cases} x + y - z = -1 & (1) \\ 4x - 3y + 2z = 16 & (2) \\ 2x - 2y - 3z = 5 & (3) \end{cases}$$

Solution For a system of three equations, we attempt to eliminate one variable at a time, using pairs of equations. Our plan of attack on this system will be to first eliminate the variable x from equations (2) and (3). Next, we will eliminate the variable y from equation (3), leaving only the variable z. Back-substitution can then be used to obtain the values of y and then x.

We begin by multiplying each side of equation (1) by -2, in anticipation of eliminating the variable x from equation (3) by adding equations (1) and (3):

$$\begin{cases} -2x - 2y + 2z = 2 & (1) \\ 4x - 3y + 2z = 16 & (2) \\ 2x - 2y - 3z = 5 & (3) \end{cases}$$

Multiply each side of equation (1) by -2.

$$\begin{cases} -2x - 2y + 2z = 2 & (1) \\ 4x - 3y + 2z = 16 & (2) \\ -4y - z = 7 & (3) \end{cases}$$

Replace equation (3) by the sum of equations (1) and (3).

We now eliminate the variable x from equation (2):

$$\begin{cases} -4x - 4y + 4z = 4 & (1) \\ 4x - 3y + 2z = 16 & (2) \\ -4y - z = 7 & (3) \end{cases}$$

Multiply each side of equation (1) by 2.

$$\begin{cases} -4x - 4y + 4z = 4 & (1) \\ -7y + 6z = 20 & (2) \\ -4y - z = 7 & (3) \end{cases}$$

Replace equation (2) by the sum of equations (1) and (2).

We now eliminate y from equation (3):

$$\begin{cases} -4x - 4y + 4z = 4 & (1) \\ -28y + 24z = 80 & (2) \\ 28y + 7z = -49 & (3) \end{cases}$$

(2) Multiply each side of equation (2) by 4.
(3) Multiply each side of equation (3) by −7.

$$\begin{cases} -4x - 4y + 4z = 4 & (1) \\ -28y + 24z = 80 & (2) \\ 31z = 31 & (3) \end{cases}$$

(3) Replace equation (3) by the sum of equations (2) and (3).

$$\begin{cases} -4x - 4y + 4z = 4 & (1) \\ -28y + 24z = 80 & (2) \\ z = 1 & (3) \end{cases}$$

(3) Multiply each side of equation (3) by $\frac{1}{31}$.

$$\begin{cases} -4x - 4y + 4 = 4 & (1) \\ -28y + 24 = 80 & (2) \\ z = 1 & (3) \end{cases}$$

Back-substitute; replace z by 1 in equations (1) and (2).

$$\begin{cases} -4x - 4y = 0 & (1) \\ y = -2 & (2) \\ z = 1 & (3) \end{cases}$$

(2) Solve equation (2) for y.

$$\begin{cases} -4x + 8 = 0 & (1) \\ y = -2 & (2) \\ z = 1 & (3) \end{cases}$$

(1) Back-substitute; replace y by −2.

$$\begin{cases} x = 2 & (1) \\ y = -2 & (2) \\ z = 1 & (3) \end{cases}$$

The solution of the original system is $x = 2, y = -2, z = 1$. (You should check this.)

Look back over the solution given in Example 11. Note the pattern of making equation (3) contain only the variable z, followed by making equation (2) contain only the variable y and equation (1) contain only the variable x. Although which variables to isolate is your choice, the methodology remains the same for all systems.

11.1 EXERCISES

In Problems 1–10, verify that the values of the variables listed are solutions of the system of equations.

1. $\begin{cases} 2x - y = 5 \\ 5x + 2y = 8 \end{cases}$ $x = 2, y = -1$

2. $\begin{cases} 3x + 2y = 2 \\ x - 7y = -30 \end{cases}$ $x = -2, y = 4$

3. $\begin{cases} 3x - 4y = 4 \\ \frac{1}{2}x - 3y = -\frac{1}{2} \end{cases}$ $x = 2, y = \frac{1}{2}$

4. $\begin{cases} 2x + \frac{1}{2}y = 0 \\ 3x - 4y = -\frac{19}{2} \end{cases}$ $x = -\frac{1}{2}, y = 2$

5. $\begin{cases} x^2 - y^2 = 3 \\ xy = 2 \end{cases}$ $x = 2, y = 1$

6. $\begin{cases} x^2 - y^2 = 3 \\ xy = 2 \end{cases}$ $x = -2, y = -1$

7. $\begin{cases} \dfrac{x}{1 + x} + 3y = 6 \\ x + 9y^2 = 36 \end{cases}$ $x = 0, y = 2$

8. $\begin{cases} \dfrac{x}{x - 1} + y = 5 \\ 3x - y = 3 \end{cases}$ $x = 2, y = 3$

9. $\begin{cases} 3x + 3y + 2z = 4 \\ x - y - z = 0 \\ 2y - 3z = -8 \end{cases}$

$x = 1, y = -1, z = 2$

10. $\begin{cases} 4x - z = 7 \\ 8x + 5y - z = 0 \\ -x - y + 5z = 6 \end{cases}$

$x = 2, y = -3, z = 1$

In Problems 11–46, solve each system of equations. If the system has no solution, say it is inconsistent. Use either substitution or elimination. Verify your solution using a graphing utility (when possible).

11. $\begin{cases} x + y = 8 \\ x - y = 4 \end{cases}$

12. $\begin{cases} x + 2y = 5 \\ x + y = 3 \end{cases}$

13. $\begin{cases} 5x - y = 13 \\ 2x + 3y = 12 \end{cases}$

14. $\begin{cases} x + 3y = 5 \\ 2x - 3y = -8 \end{cases}$

15. $\begin{cases} 3x = 24 \\ x + 2y = 0 \end{cases}$

16. $\begin{cases} 4x + 5y = -3 \\ -2y = -4 \end{cases}$

17. $\begin{cases} 3x - 6y = 2 \\ 5x + 4y = 1 \end{cases}$

18. $\begin{cases} 2x + 4y = \frac{2}{3} \\ 3x - 5y = -10 \end{cases}$

19. $\begin{cases} 2x + y = 1 \\ 4x + 2y = 3 \end{cases}$

20. $\begin{cases} x - y = 5 \\ -3x + 3y = 2 \end{cases}$

21. $\begin{cases} 2x - y = 0 \\ 3x + 2y = 7 \end{cases}$

22. $\begin{cases} 3x + 3y = -1 \\ 4x + y = \frac{8}{3} \end{cases}$

23. $\begin{cases} x + 2y = 4 \\ 2x + 4y = 8 \end{cases}$

24. $\begin{cases} 3x - y = 7 \\ 9x - 3y = 21 \end{cases}$

25. $\begin{cases} 2x - 3y = -1 \\ 10x + y = 11 \end{cases}$

26. $\begin{cases} 3x - 2y = 0 \\ 5x + 10y = 4 \end{cases}$

27. $\begin{cases} 2x + 3y = 6 \\ x - y = \frac{1}{2} \end{cases}$

28. $\begin{cases} \frac{1}{2}x + y = -2 \\ x - 2y = 8 \end{cases}$

29. $\begin{cases} \frac{1}{2}x + \frac{1}{3}y = 3 \\ \frac{1}{4}x - \frac{2}{3}y = -1 \end{cases}$

30. $\begin{cases} \frac{1}{3}x - \frac{3}{2}y = -5 \\ \frac{3}{4}x + \frac{1}{3}y = 11 \end{cases}$

31. $\begin{cases} 3x - 5y = 3 \\ 15x + 5y = 21 \end{cases}$

32. $\begin{cases} 2x - y = -1 \\ x + \frac{1}{2}y = \frac{3}{2} \end{cases}$

33. $\begin{cases} x - y = 6 \\ 2x - 3z = 16 \\ 2y + z = 4 \end{cases}$

34. $\begin{cases} 2x + y = -4 \\ -2y + 4z = 0 \\ 3x - 2z = -11 \end{cases}$

35. $\begin{cases} x - 2y + 3z = 7 \\ 2x + y + z = 4 \\ -3x + 2y - 2z = -10 \end{cases}$

36. $\begin{cases} 2x + y - 3z = 0 \\ -2x + 2y + z = -7 \\ 3x - 4y - 3z = 7 \end{cases}$

37. $\begin{cases} x - y - z = 1 \\ 2x + 3y + z = 2 \\ 3x + 2y = 0 \end{cases}$

38. $\begin{cases} 2x - 3y - z = 0 \\ -x + 2y + z = 5 \\ 3x - 4y - z = 1 \end{cases}$

39. $\begin{cases} x - y - z = 1 \\ -x + 2y - 3z = -4 \\ 3x - 2y - 7z = 0 \end{cases}$

40. $\begin{cases} 2x - 3y - z = 0 \\ 3x + 2y + 2z = 2 \\ x + 5y + 3z = 2 \end{cases}$

41. $\begin{cases} 2x - 2y + 3z = 6 \\ 4x - 3y + 2z = 0 \\ -2x + 3y - 7z = 1 \end{cases}$

42. $\begin{cases} 3x - 2y + 2z = 6 \\ 7x - 3y + 2z = -1 \\ 2x - 3y + 4z = 0 \end{cases}$

43. $\begin{cases} x + y - z = 6 \\ 3x - 2y + z = -5 \\ x + 3y - 2z = 14 \end{cases}$

44. $\begin{cases} x - y + z = -4 \\ 2x - 3y + 4z = -15 \\ 5x + y - 2z = 12 \end{cases}$

45. $\begin{cases} x + 2y - z = -3 \\ 2x - 4y + z = -7 \\ -2x + 2y - 3z = 4 \end{cases}$

46. $\begin{cases} x + 4y - 3z = -8 \\ 3x - y + 3z = 12 \\ x + y + 6z = 1 \end{cases}$

47. Solve $\begin{cases} \dfrac{1}{x} + \dfrac{1}{y} = 8 \\ \dfrac{3}{x} - \dfrac{5}{y} = 0 \end{cases}$

48. Solve $\begin{cases} \dfrac{4}{x} - \dfrac{3}{y} = 0 \\ \dfrac{6}{x} + \dfrac{3}{2y} = 2 \end{cases}$

[**Hint:** Let $u = 1/x$ and $v = 1/y$, and solve for u and v. Then $x = 1/u$ and $y = 1/v$.]

In Problems 49–54, use a graphing utility to solve each system of equations. Approximate the solution correct to two decimal places.

49. $\begin{cases} y = \sqrt{2}x - 20\sqrt{7} \\ y = -0.1x + 20 \end{cases}$

50. $\begin{cases} y = -\sqrt{3}x + 100 \\ y = 0.2x + \sqrt{19} \end{cases}$

51. $\begin{cases} \sqrt{2}x + \sqrt{3}y + \sqrt{6} = 0 \\ \sqrt{3}x - \sqrt{2}y + 60 = 0 \end{cases}$

52. $\begin{cases} \sqrt{5}x - \sqrt{6}y + 60 = 0 \\ 0.2x + 0.3y + \sqrt{5} = 0 \end{cases}$

53. $\begin{cases} \sqrt{3}x + \sqrt{2}y = \sqrt{0.3} \\ 100x - 95y = 20 \end{cases}$

54. $\begin{cases} \sqrt{6}x - \sqrt{5}y + \sqrt{1.1} = 0 \\ y = -0.2x + 0.1 \end{cases}$

55. The sum of two numbers is 81. The difference of twice one and three times the other is 62. Find the two numbers.

56. The difference of two numbers is 40. Six times the smaller one less the larger one is 5. Find the two numbers.

57. The perimeter of a rectangular floor is 90 feet. Find the dimensions of the floor if the length is twice the width.

58. The length of fence required to enclose a rectangular field is 3000 meters. What are the dimensions of the field if it is known that the difference between its length and width is 50 meters?

59. Cost of Fast Food Four large cheeseburgers and two chocolate shakes cost a total of $7.90. Two shakes cost 15¢ more than one cheeseburger. What is the cost of a cheeseburger? A shake?

60. Movie Theater Tickets A movie theater charges $9.00 for adults and $7.00 for senior citizens. On a day when 325 people paid an admission, the total receipts were $2495. How many who paid were adults? How many were seniors?

61. Mixing Nuts A store sells cashews for $5.00 per pound and peanuts for $1.50 per pound. The manager decides to mix 30 pounds of peanuts with some cashews and sell the mixture for $3.00 per pound. How many pounds of cashews should be mixed with the peanuts so that the mixture will produce the same revenue as would selling the nuts separately?

62. Financial Planning A recently retired couple need $12,000 per year to supplement their Social Security. They have $150,000 to invest to obtain this income. They have decided on two investment options: AA bonds yielding 10% per annum and a Bank Certificate yielding 5%.

(a) How much should be invested in each to realize exactly $12,000?

(b) If, after two years, the couple requires $14,000 per year in income, how should they reallocate their investment to achieve the new amount?

63. Computing Wind Speed With a tail wind, a small Piper aircraft can fly 600 miles in 3 hours. Against this same wind, the Piper can fly the same distance in 4 hours. Find the average wind speed and the average airspeed of the Piper.

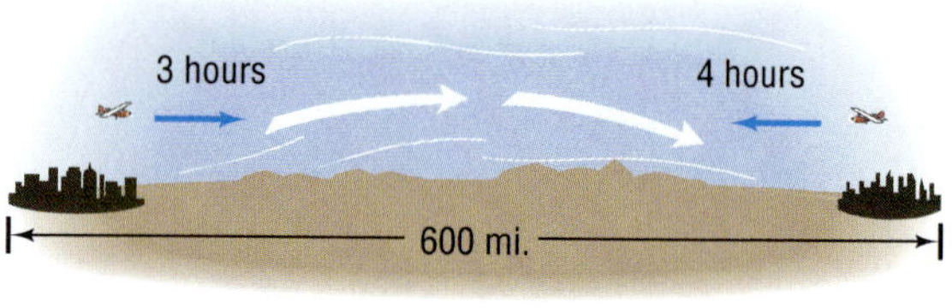
3 hours
4 hours
600 mi.

64. Computing Wind Speed The average airspeed of a single-engine aircraft is 150 miles per hour. If the aircraft flew the same distance in 2 hours with the wind as it flew in 3 hours against the wind, what was the wind speed?

65. Restaurant Management A restaurant manager wants to purchase 200 sets of dishes. One design costs $25 per set, while another costs $45 per set. If she only has $7400 to spend, how many of each design should be ordered?

66. Cost of Fast Food One group of people purchased 10 hot dogs and 5 soft drinks at a cost of $12.50. A second bought 7 hot dogs and 4 soft drinks at a cost of $9.00. What is the cost of a single hot dog? A single soft drink?

We paid $12.50.
How much is one hot dog?
How much is one cola?
We paid $9.00.
How much is one hot dog?
How much is one cola?
HOT DOGS
SODA
HOT DOGS
SODA

67. Computing a Refund The grocery store we use does not mark prices on its goods. My wife went to this store, bought three 1 pound packages of bacon and two cartons of eggs, and paid a total of $7.45. Not knowing she went to the store, I also went to the same store, purchased two 1 pound packages of bacon and three cartons of eggs, and paid a total of \$6.45. Now we want to return two 1 pound packages of bacon and two cartons of eggs. How much will be refunded?

68. Finding the Current of a Stream Pamela requires 3 hours to swim 15 miles downstream on the Illinois River. The return trip upstream takes 5 hours. Find Pamela's average speed in still water. How fast is the current? (Assume that Pamela's speed is the same in each direction.)

69. The sum of three numbers is 48. The sum of the two larger numbers is three times the smallest. The sum of the two smaller numbers is 6 more than the largest. Find the numbers.

70. A coin collection consists of 37 coins (nickels, dimes, and quarters). If the collection has a face value of \$3.25 and there are 5 more dimes than there are nickels, how many of each coin are in the collection?

71. Electricity: Kirchoff's Rules An application of *Kirchoff's Rules* to the circuit shown results in the following system of equations:

$$\begin{cases} I_2 = I_1 + I_3 \\ 5 - 3I_1 - 5I_2 = 0 \\ 10 - 5I_2 - 7I_3 = 0 \end{cases}$$

Find the currents I_1, I_2, and I_3.*

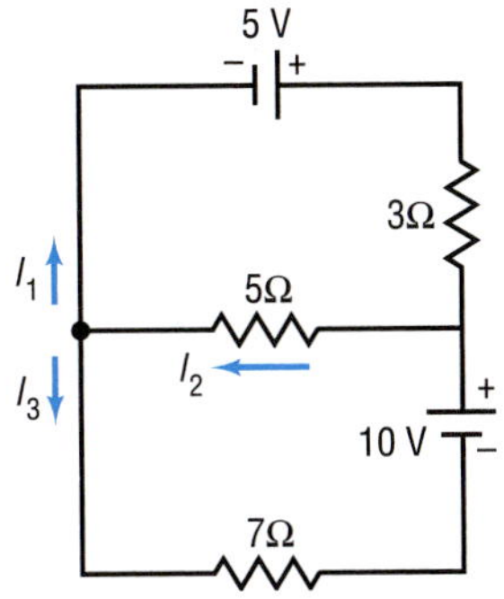

72. Electricity: Kirchoff's Rules An application of Kirchoff's Rules to the circuit shown results in the following system of equations:

$$\begin{cases} I_3 = I_1 + I_2 \\ 8 = 4I_3 + 6I_2 \\ 8I_1 = 4 + 6I_2 \end{cases}$$

Find the currents I_1, I_2, and I_3.†

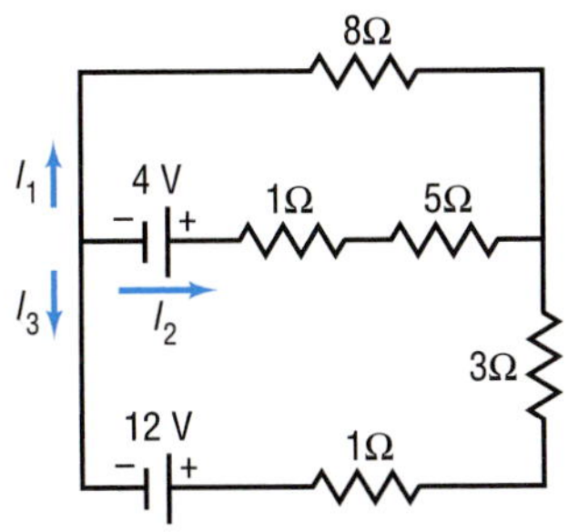

73. Theater Revenues A Broadway theater has 500 seats, divided into orchestra, main, and balcony seating. Orchestra seats sell for \$50, main seats for \$35, and balcony seats for \$25. If all the seats are sold, the gross revenue to the theater is \$17,100. If all the main and balcony seats are sold, but only half the orchestra seats are sold, the gross revenue is \$14,600. How many are there of each kind of seat?

74. Laboratory Workstations A chemistry laboratory can be used by 38 students at one time. The laboratory has 16 workstations, some set up for 2 students each and the others set up for 3 students each. How many are there of each kind of workstation?

75. Make up a system of two linear equations containing two variables that has
(a) No solution
(b) Exactly one solution
(c) Infinitely many solutions

Give the three systems to a friend to solve and critique.

76. Write a brief paragraph outlining your strategy for solving a system of two linear equations containing two variables.

77. Do you prefer the method of substitution or the method of elimination for solving a system of two linear equations containing two variables? What about for solving a system of three linear equations containing three variables? Give reasons.

78. Curve Fitting Find real numbers b and c such that the parabola $y = x^2 + bx + c$ passes through the points (1, 3) and (3, 5).

**Source:* Based on Raymond Serway, *Physics,* 3rd ed. (Philadelphia: Saunders, 1990), Prob. 26, p. 790.

†*Source:* Ibid., Prob. 27, p. 790.

79. **Curve Fitting** Find real numbers b and c such that the parabola $y = x^2 + bx + c$ passes through the points $(1, 2)$ and $(-1, 3)$.

80. **Curve Fitting** Find real numbers b and c such that the parabola $y = x^2 + bx + c$ passes through the points (x_1, y_1) and (x_2, y_2).

81. **Curve Fitting** Find real numbers a, b, and c such that the parabola $y = ax^2 + bx + c$ passes through the points $(-1, 4)$, $(2, 3)$, and $(0, 1)$.

82. **Curve Fitting** Find real numbers a, b, and c such that the parabola $y = ax^2 + bx + c$ passes through the points $(-1, -2)$, $(1, -4)$, and $(2, 4)$.

83. Solve $\begin{cases} y = m_1x + b_1 \\ y = m_2x + b_2 \end{cases}$

where $m_1 \neq m_2$.

84. Solve $\begin{cases} y = m_1x + b_1 \\ y = m_2x + b_2 \end{cases}$

where $m_1 = m_2 = m$ and $b_1 \neq b_2$.

85. Solve $\begin{cases} y = m_1x + b_1 \\ y = m_2x + b_2 \end{cases}$

where $m_1 = m_2 = m$ and $b_1 = b_2 = b$.

86. **CBL Experiment** Two students walk toward each other at a *constant* rate. Plotting distance against time on the same viewing window for each student results in a system of linear equations. By finding the intersection point of the two graphs, the time and location where the students pass each other is determined. (Activity 4, Real-World Math with the CBL System, 1994.)

11.2 Systems of Linear Equations: Matrices

1 Write the Augmented Matrix of a System of Linear Equations
2 Write the System from the Augmented Matrix
3 Perform Row Operations on a Matrix
4 Solve Systems of Linear Equations Using Matrices
5 Use a Graphing Utility to Perform Row Operations

The systematic approach of the method of elimination for solving a system of linear equations provides another method of solution that involves a simplified notation.

Consider the following system of linear equations:

$$\begin{cases} x + 4y = 14 \\ 3x - 2y = 0 \end{cases}$$

If we choose not to write the symbols used for the variables, we can represent this system as

$$\left[\begin{array}{cc|c} 1 & 4 & 14 \\ 3 & -2 & 0 \end{array}\right]$$

where it is understood that the first column represents the coefficients of the variable x, the second column the coefficients of y, and the third column the constants on the right side of the equal signs. The vertical line serves as a reminder of the equal signs. The large square brackets are the traditional symbols used to denote a *matrix* in algebra.

A **matrix** is defined as a rectangular array of numbers,

$$\begin{array}{l} \\ \text{Row } 1 \\ \text{Row } 2 \\ \vdots \\ \text{Row } i \\ \vdots \\ \text{Row } m \end{array} \begin{array}{c} \begin{array}{cccccc} \text{Column 1} & \text{Column 2} & & \text{Column } j & & \text{Column } n \end{array} \\ \begin{bmatrix} a_{11} & a_{12} & \cdots & a_{1j} & \cdots & a_{1n} \\ a_{21} & a_{22} & \cdots & a_{2j} & \cdots & a_{2n} \\ \vdots & \vdots & & \vdots & & \vdots \\ a_{i1} & a_{i2} & \cdots & a_{ij} & \cdots & a_{in} \\ \vdots & \vdots & & \vdots & & \vdots \\ a_{m1} & a_{m2} & \cdots & a_{mj} & \cdots & a_{mn} \end{bmatrix} \end{array} \quad (1)$$

Each number a_{ij} of the matrix has two indices: the **row index** i and the **column index** j. The matrix shown in display (1) has m rows and n columns. The numbers a_{ij} are usually referred to as the **entries** of the matrix.

1 Now we will use matrix notation to represent a system of linear equations. The matrices used to represent systems of linear equations are called **augmented matrices.**

EXAMPLE 1 Writing the Augmented Matrix of a System of Linear Equations

Write the augmented matrix of each system of equations.

(a) $\begin{cases} 3x - 4y = -6 & (1) \\ 2x - 3y = -5 & (2) \end{cases}$ (b) $\begin{cases} 2x - y + z = 0 & (1) \\ x + z - 1 = 0 & (2) \\ x + 2y - 8 = 0 & (3) \end{cases}$

Solution (a) The augmented matrix is

$$\left[\begin{array}{rr|r} 3 & -4 & -6 \\ 2 & -3 & -5 \end{array}\right]$$

(b) Care must be taken that the system is written with the coefficients of all variables present (if any variable is missing, its coefficient is 0) and with all constants to the right of the equal sign. Thus, we need to rearrange the given system as follows:

$$\begin{cases} 2x - y + z = 0 & (1) \\ x + z - 1 = 0 & (2) \\ x + 2y - 8 = 0 & (3) \end{cases}$$

$$\begin{cases} 2x - y + z = 0 & (1) \\ x + 0 \cdot y + z = 1 & (2) \\ x + 2y + 0 \cdot z = 8 & (3) \end{cases}$$

The augmented matrix is

$$\left[\begin{array}{rrr|r} 2 & -1 & 1 & 0 \\ 1 & 0 & 1 & 1 \\ 1 & 2 & 0 & 8 \end{array}\right]$$

If we do not include the constants to the right of the equal sign, that is, to the right of the vertical bar in the augmented matrix of a system of equations, the resulting matrix is called the **coefficient matrix** of the system. For the systems discussed in Example 1, the coefficient matrices are

$$\begin{bmatrix} 3 & -4 \\ 2 & -3 \end{bmatrix} \quad \text{and} \quad \begin{bmatrix} 2 & -1 & 1 \\ 1 & 0 & 1 \\ 1 & 2 & 0 \end{bmatrix}$$

Now work Problem 3.

EXAMPLE 2 Writing the System of Linear Equations from the Augmented Matrix

2 Write the system of linear equations corresponding to each augmented matrix.

$$\text{(a)} \left[\begin{array}{cc|c} 5 & 2 & 13 \\ -3 & 1 & -10 \end{array}\right] \qquad \text{(b)} \left[\begin{array}{ccc|c} 3 & -1 & -1 & 7 \\ 2 & 0 & 2 & 8 \\ 0 & 1 & 1 & 0 \end{array}\right]$$

Solution (a) The matrix has two rows and so represents a system of two equations. The two columns to the left of the vertical bar indicate that the system has two variables. If x and y are used to denote these variables, the system of equations is

$$\begin{cases} 5x + 2y = 13 & (1) \\ -3x + y = -10 & (2) \end{cases}$$

(b) This matrix represents a system of three equations containing three variables. If x, y, and z are the three variables, this system is

$$\begin{cases} 3x - y - z = 7 & (1) \\ 2x + 2z = 8 & (2) \\ y + z = 0 & (3) \end{cases}$$

Row Operations on a Matrix

EXAMPLE 3 Solving a System of Equations

3 Solve the system of equations

$$\begin{cases} 4x - 3y = 11 & (1) \\ 3x + 2y = 4 & (2) \end{cases}$$

Solution We will use a variation of the method of elimination to solve the system. First, multiply each side of equation (2) by -1 and add it to equation (1). Replace equation (1) with the result:

$$\begin{cases} x - 5y = 7 & (1) \\ 3x + 2y = 4 & (2) \end{cases}$$

Multiply each side of equation (1) by -3 and add it to equation (2). Replace equation (2) with the result:

$$\begin{cases} x - 5y = 7 & (1) \\ 0 \cdot x + 17y = -17 & (2) \end{cases}$$

Multiply each side of equation (2) by $\frac{1}{17}$:

$$\begin{cases} x - 5y = 7 & (1) \\ y = -1 & (2) \end{cases}$$

Now we back-substitute $y = -1$ into equation (1) to get

$$\begin{aligned} x - 5y &= 7 \\ x - 5(-1) &= 7 \\ x &= 2 \end{aligned}$$

The solution of the system is $x = 2$, $y = -1$.

The pattern of solution shown above provides a systematic way to solve any system of equations. The idea is to start with the augmented matrix of the system,

$$\left[\begin{array}{cc|c} 4 & -3 & 11 \\ 3 & 2 & 4 \end{array}\right] \qquad \begin{cases} 4x - 3y = 11 & (1) \\ 3x + 2y = 4 & (2) \end{cases}$$

and eventually arrive at the matrix,

$$\left[\begin{array}{cc|c} 1 & -5 & 7 \\ 0 & 1 & -1 \end{array}\right] \qquad \begin{cases} x - 5y = 7 & (1) \\ y = -1 & (2) \end{cases}$$

Let's go through the procedure again, this time starting with the augmented matrix and keeping the final augmented matrix given above in mind:

$$\left[\begin{array}{cc|c} 4 & -3 & 11 \\ 3 & 2 & 4 \end{array}\right] \qquad \begin{cases} 4x - 3y = 11 & (1) \\ 3x + 2y = 4 & (2) \end{cases}$$

As before, we start by multiplying each side of equation (2) by -1 and adding it to equation (1). This is equivalent to multiplying each entry in the second row of the matrix by -1, adding the result to the corresponding entries in row 1, and replacing row 1 by these entries. The result of this step is that the number 1 appears in row 1, column 1:

$$\left[\begin{array}{cc|c} 1 & -5 & 7 \\ 3 & 2 & 4 \end{array}\right] \qquad \begin{cases} x - 5y = 7 & (1) \\ 3x + 2y = 4 & (2) \end{cases}$$

Multiply each entry in the first row by -3, add the result to the entries in the second row, and replace the second row by these entries. The result of this step is that the number 0 appears in row 2, column 1:

$$\left[\begin{array}{cc|c} 1 & -5 & 7 \\ 0 & 17 & -17 \end{array}\right] \qquad \begin{cases} x - 5y = 7 & (1) \\ 0 \cdot x + 17y = -17 & (2) \end{cases}$$

Multiply each entry in the second row by $\frac{1}{17}$. The result of this step is that the number 1 appears in row 2, column 2:

$$\left[\begin{array}{cc|c} 1 & -5 & 7 \\ 0 & 1 & -1 \end{array}\right] \qquad \begin{cases} x - 5y = 7 & (1) \\ y = -1 & (2) \end{cases}$$

Now that we know that $y = -1$, we can back-substitute to find that $x = 2$.

The manipulations just performed on the augmented matrix are called **row operations.** There are three basic row operations:

Row Operations

1. Interchange any two rows.
2. Replace a row by a nonzero multiple of that row.
3. Replace a row by the sum of that row and a constant multiple of some other row.

These three row operations correspond to the three rules given earlier for obtaining an equivalent system of equations. Thus, when a row operation is performed on a matrix, the resulting matrix represents a system of equations equivalent to the system represented by the original matrix.

For example, consider the augmented matrix

$$\left[\begin{array}{cc|c} 1 & 2 & 3 \\ 4 & -1 & 2 \end{array}\right]$$

Suppose we want to apply a row operation to this matrix that results in a matrix whose entry in row 2, column 1 is a 0. The row operation to use is

Multiply each entry in row 1 by -4 and add the result to the corresponding entries in row 2. (2)

If we use R_2 to represent the new entries in row 2 and we use r_1 and r_2 to represent the original entries in rows 1 and 2, respectively, then we can represent the row operation in statement (2) by

$$R_2 = -4r_1 + r_2$$

Then

$$\left[\begin{array}{cc|c} 1 & 2 & 3 \\ 4 & -1 & 2 \end{array}\right] \underset{R_2 = -4r_1 + r_2}{\overset{\rightarrow}{\uparrow}} \left[\begin{array}{cc|c} 1 & 2 & 3 \\ -4(1)+4 & -4(2)+(-1) & -4(3)+2 \end{array}\right] = \left[\begin{array}{cc|c} 1 & 2 & 3 \\ 0 & -9 & -10 \end{array}\right]$$

As desired, we now have the entry 0 in row 2, column 1.

EXAMPLE 4 Applying a Row Operation to an Augmented Matrix

Apply the row operation $R_2 = -3r_1 + r_2$ to the augmented matrix

$$\left[\begin{array}{cc|c} 1 & -2 & 2 \\ 3 & -5 & 9 \end{array}\right]$$

Solution The row operation $R_2 = -3r_1 + r_2$ tells us that the entries in row 2 are to be replaced by the entries obtained after multiplying each entry in row 1 by -3 and adding the result to the corresponding entries in row 2. Thus,

$$\left[\begin{array}{cc|c} 1 & -2 & 2 \\ 3 & -5 & 9 \end{array}\right] \underset{R_2 = -3r_1 + r_2}{\overset{\rightarrow}{\uparrow}} \left[\begin{array}{cc|c} 1 & -2 & 2 \\ -3(1)+3 & (-3)(-2)+(-5) & -3(2)+9 \end{array}\right] = \left[\begin{array}{cc|c} 1 & -2 & 2 \\ 0 & 1 & 3 \end{array}\right]$$

■

EXAMPLE 5 Finding a Particular Row Operation

Using the matrix

$$\left[\begin{array}{cc|c} 1 & -2 & 2 \\ 0 & 1 & 3 \end{array}\right]$$

find a row operation that will result in a matrix with a 0 in row 1, column 2.

Solution We want a 0 in row 1, column 2. This result can be accomplished by multiplying row 2 by 2 and adding the result to row 1. That is, we apply the row operation $R_1 = 2r_2 + r_1$:

$$\left[\begin{array}{cc|c} 1 & -2 & 2 \\ 0 & 1 & 3 \end{array}\right] \underset{\substack{\uparrow \\ R_1 = 2r_2 + r_1}}{\rightarrow} \left[\begin{array}{cc|c} 2(0)+1 & 2(1)+(-2) & 2(3)+2 \\ 0 & 1 & 3 \end{array}\right] = \left[\begin{array}{cc|c} 1 & 0 & 8 \\ 0 & 1 & 3 \end{array}\right].$$

A word about the notation we have introduced. A row operation such as $R_1 = 2r_2 + r_1$ changes the entries in row 1. Note also that to change the entries in a given row we multiply the entries in some other row by an appropriate number and add the results to the original entries of the row to be changed.

4 Now let's see how we use row operations to solve a system of linear equations.

EXAMPLE 6 Solving a System of Linear Equations Using Matrices

Solve: $\begin{cases} 4x + 3y = 11 & (1) \\ x - 3y = -1 & (2) \end{cases}$

Solution First, we write the augmented matrix that represents this system:

$$\left[\begin{array}{cc|c} 4 & 3 & 11 \\ 1 & -3 & -1 \end{array}\right]$$

The first step requires getting the entry 1 in row 1, column 1. An interchange of rows 1 and 2 is the easiest way to do this:

$$\left[\begin{array}{cc|c} 1 & -3 & -1 \\ 4 & 3 & 11 \end{array}\right]$$

Next, we want a 0 under the entry 1 in column 1. We use the row operation $R_2 = -4r_1 + r_2$:

$$\left[\begin{array}{cc|c} 1 & -3 & -1 \\ 4 & 3 & 11 \end{array}\right] \underset{\substack{\uparrow \\ R_2 = -4r_1 + r_2}}{\rightarrow} \left[\begin{array}{cc|c} 1 & -3 & -1 \\ 0 & 15 & 15 \end{array}\right]$$

Now we want the entry 1 in row 2, column 2. We use $R_2 = \frac{1}{15}r_2$:

$$\left[\begin{array}{cc|c} 1 & -3 & -1 \\ 0 & 15 & 15 \end{array}\right] \underset{\substack{\uparrow \\ R_2 = \frac{1}{15}r_2}}{\rightarrow} \left[\begin{array}{cc|c} 1 & -3 & -1 \\ 0 & 1 & 1 \end{array}\right]$$

The second row of the matrix on the right represents the equation $y = 1$. Thus, using $y = 1$, we back-substitute into the equation $x - 3y = -1$ (from the first row) to get

$$x - 3(1) = -1 \quad y = 1$$
$$x = 2$$

The solution of the system is $x = 2$, $y = 1$.

Now work Problem 31.

The steps we used to solve the system of linear equations in Example 6 can be summarized as follows:

Matrix Method for Solving a System of Linear Equations

STEP 1: Write the augmented matrix that represents the system.

STEP 2: Perform row operations that place the entry 1 in row 1, column 1.

STEP 3: Perform row operations that leave the entry 1 in row 1, column 1 unchanged, while causing 0's to appear below it in column 1.

STEP 4: Perform row operations that place the entry 1 in row 2, column 2 and leave the entries in columns to the left unchanged. If it is impossible to place a 1 in row 2, column 2, then proceed to place a 1 in row 2, column 3. Once a 1 is in place, perform row operations to place 0's under it.

STEP 5: Now repeat Step 4, placing a 1 in the next row, but one column to the right. Continue until the bottom row or the vertical bar is reached.

STEP 6: If any rows are obtained that contain only 0's on the left side of the vertical bar, then place such rows at the bottom of the matrix.

After Steps 1 to 6 have been completed, the matrix is said to be in **echelon form.** A little thought should convince you that a matrix is in echelon form when

1. The entry in row 1, column 1 is a 1, and 0's appear below it.
2. The first nonzero entry in each row after the first row is a 1, 0's appear below it, and it appears to the right of the first nonzero entry in any row above.
3. Any rows that contain all 0's to the left of the vertical bar appear at the bottom.

Two advantages of solving a system of equations by writing the augmented matrix in echelon form are the following:

1. The process is algorithmic; that is, it consists of repetitive steps that can be programmed on a computer.
2. The process works on any system of linear equations, no matter how many equations or variables are present.

The next example shows how to write a matrix in echelon form.

EXAMPLE 7 Solving a System of Linear Equations Using Matrices

Solve: $\begin{cases} x - y + z = 8 & (1) \\ 2x + 3y - z = -2 & (2) \\ 3x - 2y - 9z = 9 & (3) \end{cases}$

Solution STEP 1: The augmented matrix of the system is

$$\left[\begin{array}{rrr|r} 1 & -1 & 1 & 8 \\ 2 & 3 & -1 & -2 \\ 3 & -2 & -9 & 9 \end{array}\right]$$

STEP 2: Because the entry 1 is already present in row 1, column 1, we can go to Step 3.

STEP 3: Perform the row operations $R_2 = -2r_1 + r_2$ and $R_3 = -3r_1 + r_3$. Each of these leaves the entry 1 in row 1, column 1 unchanged, while causing 0's to appear under it:

$$\left[\begin{array}{rrr|r} 1 & -1 & 1 & 8 \\ 2 & 3 & -1 & -2 \\ 3 & -2 & -9 & 9 \end{array}\right] \underset{\substack{R_2 = -2r_1 + r_2 \\ R_3 = -3r_1 + r_3}}{\rightarrow} \left[\begin{array}{rrr|r} 1 & -1 & 1 & 8 \\ 0 & 5 & -3 & -18 \\ 0 & 1 & -12 & -15 \end{array}\right]$$

STEP 4: The easiest way to obtain the entry 1 in row 2, column 2 without altering column 1 is to interchange rows 2 and 3 (another way would be to multiply row 2 by $\frac{1}{5}$, but this introduces fractions):

$$\left[\begin{array}{rrr|r} 1 & -1 & 1 & 8 \\ 0 & 1 & -12 & -15 \\ 0 & 5 & -3 & -18 \end{array}\right]$$

To get 0's under the 1 in row 2, column 2, perform the row operation $R_3 = -5r_2 + r_3$:

$$\left[\begin{array}{rrr|r} 1 & -1 & 1 & 8 \\ 0 & 1 & -12 & -15 \\ 0 & 5 & -3 & -18 \end{array}\right] \underset{R_3 = -5r_2 + r_3}{\rightarrow} \left[\begin{array}{rrr|r} 1 & -1 & 1 & 8 \\ 0 & 1 & -12 & -15 \\ 0 & 0 & 57 & 57 \end{array}\right]$$

STEP 5: Continuing, we place a 1 in row 3, column 3 by using $R_3 = \frac{1}{57}r_3$:

$$\left[\begin{array}{rrr|r} 1 & -1 & 1 & 8 \\ 0 & 1 & -12 & -15 \\ 0 & 0 & 57 & 57 \end{array}\right] \underset{R_3 = \frac{1}{57}r_3}{\rightarrow} \left[\begin{array}{rrr|r} 1 & -1 & 1 & 8 \\ 0 & 1 & -12 & -15 \\ 0 & 0 & 1 & 1 \end{array}\right]$$

Because we have reached the bottom row, the matrix is in echelon form and we can stop.

The system of equations represented by the matrix in echelon form is

$$\begin{cases} x - y + z = 8 \\ y - 12z = -15 \\ z = 1 \end{cases}$$

Using $z = 1$, we back-substitute to get

$$\begin{cases} x - y + 1 = 8 & \text{From row 1 of the matrix.} \\ y - 12(1) = -15 & \text{From row 2 of the matrix.} \end{cases}$$

Thus, we get $y = -3$, and back-substituting into $x - y = 7$, we find $x = 4$. The solution of the system is $x = 4$, $y = -3$, $z = 1$.

Sometimes, it is advantageous to write a matrix in **reduced echelon form.** In this form, row operations are used to obtain entries that are 0 above (as well as below) the leading 1 in a row. For example, the echelon form obtained in the solution to Example 7 is

$$\left[\begin{array}{ccc|c} 1 & -1 & 1 & 8 \\ 0 & 1 & -12 & -15 \\ 0 & 0 & 1 & 1 \end{array}\right]$$

To write this matrix in reduced echelon form, we proceed as follows:

$$\left[\begin{array}{ccc|c} 1 & -1 & 1 & 8 \\ 0 & 1 & -12 & -15 \\ 0 & 0 & 1 & 1 \end{array}\right] \underset{\uparrow \; R_1 = r_2 + r_1}{\rightarrow} \left[\begin{array}{ccc|c} 1 & 0 & -11 & -7 \\ 0 & 1 & -12 & -15 \\ 0 & 0 & 1 & 1 \end{array}\right] \underset{\substack{\uparrow \; R_1 = 11r_3 + r_1 \\ R_2 = 12r_3 + r_2}}{\rightarrow} \left[\begin{array}{ccc|c} 1 & 0 & 0 & 4 \\ 0 & 1 & 0 & -3 \\ 0 & 0 & 1 & 1 \end{array}\right]$$

The matrix is now written in reduced echelon form. The advantage of writing the matrix in this form is that the solution to the system, $x = 4$, $y = -3$, $z = 1$, is readily found, without the need to back-substitute. Another advantage will be seen in Section 11.4, where the inverse of a matrix is discussed.

Now work Problem 49.

The matrix method for solving a system of linear equations also identifies systems that have infinitely many solutions and systems that are inconsistent. Let's see how.

EXAMPLE 8 Solving a System of Linear Equations Using Matrices

$$\text{Solve:} \begin{cases} 6x - y - z = 4 & (1) \\ -12x + 2y + 2z = -8 & (2) \\ 5x + y - z = 3 & (3) \end{cases}$$

Solution We start with the augmented matrix of the system:

$$\left[\begin{array}{ccc|c} 6 & -1 & -1 & 4 \\ -12 & 2 & 2 & -8 \\ 5 & 1 & -1 & 3 \end{array}\right] \underset{\uparrow \; R_1 = 1r_3 + r_1}{\rightarrow} \left[\begin{array}{ccc|c} 1 & -2 & 0 & 1 \\ -12 & 2 & 2 & -8 \\ 5 & 1 & -1 & 3 \end{array}\right] \underset{\substack{\uparrow \; R_2 = 12r_1 + r_2 \\ R_3 = -5r_1 + r_3}}{\rightarrow} \left[\begin{array}{ccc|c} 1 & -2 & 0 & 1 \\ 0 & -22 & 2 & 4 \\ 0 & 11 & -1 & -2 \end{array}\right]$$

Obtaining a 1 in row 2, column 2 without altering column 1 can be accomplished only by $R_2 = -\frac{1}{22}r_2$ or by $R_3 = \frac{1}{11}r_3$ and interchanging rows. (Do you see why?) We shall use the first of these:

$$\left[\begin{array}{rrr|r} 1 & -2 & 0 & 1 \\ 0 & -22 & 2 & 4 \\ 0 & 11 & -1 & -2 \end{array}\right] \underset{R_2 = -\frac{1}{22}r_2}{\rightarrow} \left[\begin{array}{rrr|r} 1 & -2 & 0 & 1 \\ 0 & 1 & -\frac{1}{11} & -\frac{2}{11} \\ 0 & 11 & -1 & -2 \end{array}\right] \underset{R_3 = -11r_2 + r_3}{\rightarrow} \left[\begin{array}{rrr|r} 1 & -2 & 0 & 1 \\ 0 & 1 & -\frac{1}{11} & -\frac{2}{11} \\ 0 & 0 & 0 & 0 \end{array}\right]$$

This matrix is in echelon form. Because the bottom row consists entirely of 0's, the system actually consists of only two equations:

$$\begin{cases} x - 2y = 1 & (1) \\ y - \frac{1}{11}z = -\frac{2}{11} & (2) \end{cases}$$

We shall back-substitute the solution for y from the second equation, $y = \frac{1}{11}z - \frac{2}{11}$, into the first equation to get

$$x = 2y + 1 = 2\left(\tfrac{1}{11}z - \tfrac{2}{11}\right) + 1 = \tfrac{2}{11}z + \tfrac{7}{11}$$

Thus, the original system is equivalent to the system

$$\begin{cases} x = \frac{2}{11}z + \frac{7}{11} & (1) \\ y = \frac{1}{11}z - \frac{2}{11} & (2) \end{cases}$$

where z can be any real number.

Let's look at the situation. The original system of three equations is equivalent to a system containing two equations. This means any values of x, y, z that satisfy both

$$x = \tfrac{2}{11}z + \tfrac{7}{11} \quad \text{and} \quad y = \tfrac{1}{11}z - \tfrac{2}{11}$$

will be solutions. For example, $z = 0, x = \frac{7}{11}, y = -\frac{2}{11}$; $z = 1, x = \frac{9}{11}, y = -\frac{1}{11}$; and $z = -1, x = \frac{5}{11}, y = -\frac{3}{11}$ are some of the solutions of the original system. There are, in fact, infinitely many values of x, y, and z for which the two equations are satisfied. That is, the original system has infinitely many solutions. We will write the solution of the original system as

$$\begin{cases} x = \frac{2}{11}z + \frac{7}{11} \\ y = \frac{1}{11}z - \frac{2}{11} \end{cases}$$

where z can be any real number.

We can also find the solution by writing the augmented matrix in reduced echelon form. Starting with the echelon form, we have

$$\left[\begin{array}{rrr|r} 1 & -2 & 0 & 1 \\ 0 & 1 & -\frac{1}{11} & -\frac{2}{11} \\ 0 & 0 & 0 & 0 \end{array}\right] \underset{R_1 = 2r_2 + r_1}{\rightarrow} \left[\begin{array}{rrr|r} 1 & 0 & -\frac{2}{11} & \frac{7}{11} \\ 0 & 1 & -\frac{1}{11} & -\frac{2}{11} \\ 0 & 0 & 0 & 0 \end{array}\right]$$

The matrix on the right is in reduced echelon form. The corresponding system of equations is

$$\begin{cases} x - \frac{2}{11}z = \frac{7}{11} & (1) \\ y - \frac{1}{11}z = -\frac{2}{11} & (2) \end{cases}$$

or, equivalently,

$$\begin{cases} x = \frac{2}{11}z + \frac{7}{11} & (1) \\ y = \frac{1}{11}z - \frac{2}{11} & (2) \end{cases}$$

where z can be any real number.

Now work Problem 53.

EXAMPLE 9 Solving a System of Linear Equations Using Matrices

$$\text{Solve: } \begin{cases} x + y + z = 6 \\ 2x - y - z = 3 \\ x + 2y + 2z = 0 \end{cases}$$

Solution The augmented matrix is

$$\left[\begin{array}{rrr|r} 1 & 1 & 1 & 6 \\ 2 & -1 & -1 & 3 \\ 1 & 2 & 2 & 0 \end{array}\right] \rightarrow \left[\begin{array}{rrr|r} 1 & 1 & 1 & 6 \\ 0 & -3 & -3 & -9 \\ 0 & 1 & 1 & -6 \end{array}\right] \rightarrow \left[\begin{array}{rrr|r} 1 & 1 & 1 & 6 \\ 0 & 1 & 1 & -6 \\ 0 & -3 & -3 & -9 \end{array}\right] \rightarrow \left[\begin{array}{rrr|r} 1 & 1 & 1 & 6 \\ 0 & 1 & 1 & -6 \\ 0 & 0 & 0 & -27 \end{array}\right]$$

↑ $R_2 = -2r_1 + r_2$, $R_3 = -1r_1 + r_3$ ↑ Interchange rows 2 and 3. ↑ $R_3 = 3r_2 + r_3$.

This matrix is in echelon form. The bottom row is equivalent to the equation

$$0x + 0y + 0z = -27$$

which has no solution. Hence, the original system is inconsistent. ▬

Now work Problems 23 and 29.

The matrix method is especially effective for systems of equations for which the number of equations and the number of variables are unequal. Here, too, such a system is either inconsistent or consistent. If it is consistent, it will have either exactly one solution or infinitely many solutions.

Let's look at a system of four equations containing three variables.

EXAMPLE 10 Solving a System of Linear Equations Using Matrices

$$\text{Solve: } \begin{cases} x - 2y + z = 0 & (1) \\ 2x + 2y - 3z = -3 & (2) \\ y - z = -1 & (3) \\ -x + 4y + 2z = 13 & (4) \end{cases}$$

Solution The augmented matrix is

$$\left[\begin{array}{rrr|r} 1 & -2 & 1 & 0 \\ 2 & 2 & -3 & -3 \\ 0 & 1 & -1 & -1 \\ -1 & 4 & 2 & 13 \end{array}\right] \rightarrow \left[\begin{array}{rrr|r} 1 & -2 & 1 & 0 \\ 0 & 6 & -5 & -3 \\ 0 & 1 & -1 & -1 \\ 0 & 2 & 3 & 13 \end{array}\right] \rightarrow \left[\begin{array}{rrr|r} 1 & -2 & 1 & 0 \\ 0 & 1 & -1 & -1 \\ 0 & 6 & -5 & -3 \\ 0 & 2 & 3 & 13 \end{array}\right]$$

↑ $R_2 = -2r_1 + r_2$, $R_4 = r_1 + r_4$ ↑ Interchange rows 2 and 3.

$$\rightarrow \left[\begin{array}{rrr|r} 1 & -2 & 1 & 0 \\ 0 & 1 & -1 & -1 \\ 0 & 0 & 1 & 3 \\ 0 & 0 & 5 & 15 \end{array}\right] \rightarrow \left[\begin{array}{rrr|r} 1 & -2 & 1 & 0 \\ 0 & 1 & -1 & -1 \\ 0 & 0 & 1 & 3 \\ 0 & 0 & 0 & 0 \end{array}\right]$$

$\uparrow$ $R_3 = -6r_2 + r_3$, $R_4 = -2r_2 + r_4$ $\qquad$ $\uparrow$ $R_4 = -5r_3 + r_4$

We could stop here, since the matrix is in echelon form, and back-substitute $z = 3$ to find x and y. Or we can continue to obtain the reduced echelon form:

$$\rightarrow \left[\begin{array}{rrr|r} 1 & 0 & -1 & -2 \\ 0 & 1 & -1 & -1 \\ 0 & 0 & 1 & 3 \\ 0 & 0 & 0 & 0 \end{array}\right] \rightarrow \left[\begin{array}{rrr|r} 1 & 0 & 0 & 1 \\ 0 & 1 & 0 & 2 \\ 0 & 0 & 1 & 3 \\ 0 & 0 & 0 & 0 \end{array}\right]$$

$\uparrow$ $R_1 = 2r_2 + r_1$ $\qquad$ $\uparrow$ $R_1 = r_3 + r_1$, $R_2 = r_3 + r_2$

The matrix is now in reduced echelon form, and we can see that the solution is $x = 1, y = 2, z = 3$.

EXAMPLE 11 Mixing Acids

A chemistry laboratory has three containers of nitric acid, HNO_3. One container holds a solution with a concentration of 10% HNO_3, the second holds 20% HNO_3, and the third holds 40% HNO_3. How many liters of each solution should be mixed to obtain 100 liters of a solution whose concentration is 25% HNO_3?

Solution Let x, y, and z represent the number of liters of 10%, 20%, and 40% concentrations of HNO_3, respectively. We want 100 liters in all, and the concentration of HNO_3 from each solution must sum to 25% of 100 liters. Thus, we find that

$$\begin{cases} x + y + z = 100 \\ 0.10x + 0.20y + 0.40z = 0.25(100) \end{cases}$$

Now, the augmented matrix is

$$\left[\begin{array}{ccc|c} 1 & 1 & 1 & 100 \\ 0.10 & 0.20 & 0.40 & 25 \end{array}\right] \rightarrow \left[\begin{array}{ccc|c} 1 & 1 & 1 & 100 \\ 0 & 0.10 & 0.30 & 15 \end{array}\right]$$

$\uparrow$ $R_2 = -0.10r_1 + r_2$

$$\rightarrow \left[\begin{array}{ccc|c} 1 & 1 & 1 & 100 \\ 0 & 1 & 3 & 150 \end{array}\right] \rightarrow \left[\begin{array}{ccc|c} 1 & 0 & -2 & -50 \\ 0 & 1 & 3 & 150 \end{array}\right]$$

$\uparrow$ $R_2 = 10r_2$ $\qquad$ $\uparrow$ $R_1 = -1r_2 + r_1$

The matrix is now in reduced echelon form. The final matrix represents the system

$$\begin{cases} x - 2z = -50 & (1) \\ y + 3z = 150 & (2) \end{cases}$$

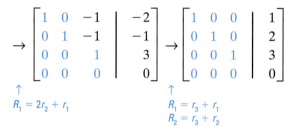

which has infinitely many solutions given by

$$\begin{cases} x = \quad 2z - 50 & (1) \\ y = -3z + 150 & (2) \end{cases}$$

where z is any real number. However, the practical considerations of this problem require us to restrict the solutions to $x \geq 0$, $y \geq 0$, $z \geq 0$. Furthermore, we require $25 \leq z \leq 50$, because otherwise $x < 0$ or $y < 0$. Some of the possible solutions are given in Table 1. The final determination of what solution the laboratory will pick very likely depends on availability, cost differences, and other considerations.

TABLE 1

Liters of 10% Solution	Liters of 20% Solution	Liters of 40% Solution	Liters of 25% Solution
0	75	25	100
10	60	30	100
12	57	31	100
16	51	33	100
26	36	38	100
38	18	44	100
46	6	48	100
50	0	50	100

Row Operations on a Graphing Utility

5 Graphing utilities can be used to perform row operations on an augmented matrix in order to obtain a matrix in reduced echelon form.

EXAMPLE 12

Using a Graphing Utility and Matrices to Solve a System of Equations

Solve the system of Example 7 using a graphing utility:

$$\begin{cases} x - y + z = 8 & (1) \\ 2x + 3y - z = -2 & (2) \\ 3x - 2y - 9z = 9 & (3) \end{cases}$$

Solution First, we enter the augmented matrix into our graphing utility. See Figure 5. Now perform the row operations from Example 7 until the augmented matrix is in reduced echelon form. They were:

FIGURE 5

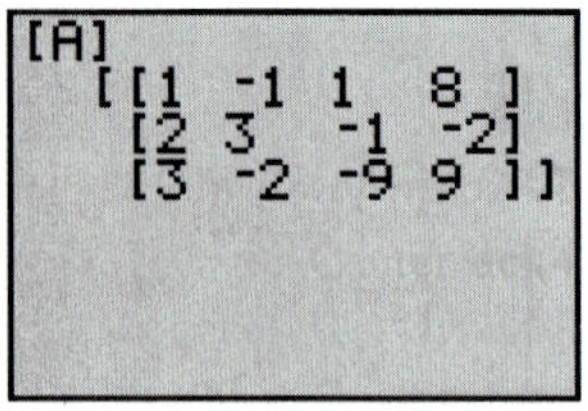

1. $R_2 = -2r_1 + r_2$
2. $R_3 = -3r_1 + r_3$
3. Interchange rows 2 and 3
4. $R_3 = -5r_2 + r_3$
5. $R_3 = \frac{1}{57}r_3$

For example, to perform the row operation $R_2 = -2r_1 + r_2$, we enter the following command into a T1-83 graphing calculator.*

*Check your manual to see how it is done on your utility.

FIGURE 6

From the command listed in Figure 6, we obtain the augmented matrix given in Figure 7.

FIGURE 7

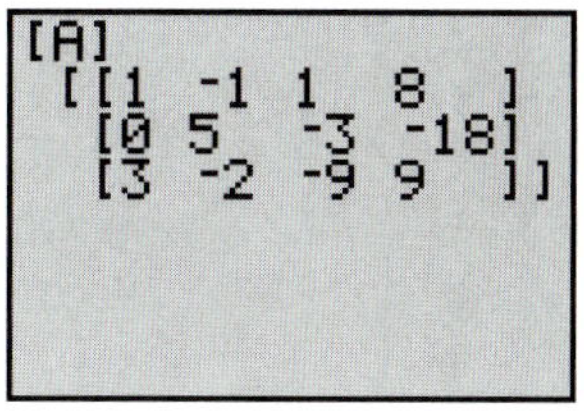

Notice row 2 is the same as row 2 in Step 3 from Example 7. We continue to enter row operations until we obtain a matrix in reduced echelon form. See Figure 8.

FIGURE 8

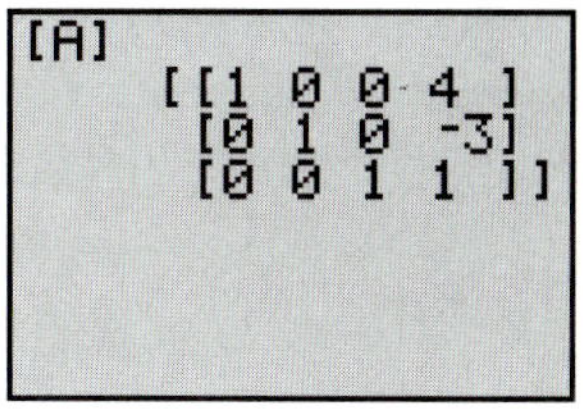

From the matrix shown in Figure 8, we see the solution to the system is $x = 4$, $y = -3$, $z = 1$. ■

11.2 EXERCISES

In Problems 1–10, write the augmented matrix of the given system of equations.

1. $\begin{cases} x - 5y = 5 \\ 4x + 3y = 6 \end{cases}$

2. $\begin{cases} 3x + 4y = 7 \\ 4x - 2y = 5 \end{cases}$

3. $\begin{cases} 2x + 3y - 6 = 0 \\ 4x - 6y + 2 = 0 \end{cases}$

4. $\begin{cases} 9x - y = 0 \\ 3x - y - 4 = 0 \end{cases}$

5. $\begin{cases} 0.01x - 0.03y = 0.06 \\ 0.13x + 0.10y = 0.20 \end{cases}$

6. $\begin{cases} \frac{4}{3}x - \frac{3}{2}y = \frac{3}{4} \\ -\frac{1}{4}x + \frac{1}{3}y = \frac{2}{3} \end{cases}$

7. $\begin{cases} x - y + z = 10 \\ 3x + 2y = 5 \\ x + y + 2z = 2 \end{cases}$

8. $\begin{cases} 5x - y - z = 0 \\ x + y = 5 \\ 2x - 3z = 2 \end{cases}$

9. $\begin{cases} x + y - z = 2 \\ 3x - 2y = 2 \end{cases}$

10. $\begin{cases} 2x + 3y - 4z = 0 \\ x - 5z + 2 = 0 \end{cases}$

In Problems 11–20, use a graphing utility to perform, in order (a), followed by (b), followed by (c) on the given augmented matrix.

11. $\left[\begin{array}{rrr|r} 1 & -3 & -5 & -2 \\ 2 & -5 & -4 & 5 \\ -3 & 5 & 4 & 6 \end{array}\right]$ (a) $R_2 = -2r_1 + r_2$ (b) $R_3 = 3r_1 + r_3$ (c) $R_3 = 4r_2 + r_3$

12. $\left[\begin{array}{rrr|r} 1 & -3 & -3 & -3 \\ 2 & -5 & 2 & -4 \\ -3 & 2 & 4 & 6 \end{array}\right]$ (a) $R_2 = -2r_1 + r_2$ (b) $R_3 = 3r_1 + r_3$ (c) $R_3 = 7r_2 + r_3$

13. $\left[\begin{array}{rrr|r} 1 & -3 & 4 & 3 \\ 2 & -5 & 6 & 6 \\ -3 & 3 & 4 & 6 \end{array}\right]$ (a) $R_2 = -2r_1 + r_2$ (b) $R_3 = 3r_1 + r_3$ (c) $R_3 = 6r_2 + r_3$

14. $\left[\begin{array}{rrr|r} 1 & -3 & 3 & -5 \\ 2 & -5 & -3 & -5 \\ -3 & -2 & 4 & 6 \end{array}\right]$ (a) $R_2 = -2r_1 + r_2$ (b) $R_3 = 3r_1 + r_3$ (c) $R_3 = 11r_2 + r_3$

15. $\left[\begin{array}{rrr|r} 1 & -3 & 2 & -6 \\ 2 & -5 & 3 & -4 \\ -3 & -6 & 4 & 6 \end{array}\right]$ (a) $R_2 = -2r_1 + r_2$ (b) $R_3 = 3r_1 + r_3$ (c) $R_3 = 15r_2 + r_3$

16. $\left[\begin{array}{rrr|r} 1 & -3 & -4 & -6 \\ 2 & -5 & 6 & -6 \\ -3 & 1 & 4 & 6 \end{array}\right]$ (a) $R_2 = -2r_1 + r_2$ (b) $R_3 = 3r_1 + r_3$ (c) $R_3 = 8r_2 + r_3$

17. $\left[\begin{array}{rrr|r} 1 & -3 & 1 & -2 \\ 2 & -5 & 6 & -2 \\ -3 & 1 & 4 & 6 \end{array}\right]$ (a) $R_2 = -2r_1 + r_2$ (b) $R_3 = 3r_1 + r_3$ (c) $R_3 = 8r_2 + r_3$

18. $\left[\begin{array}{rrr|r} 1 & -3 & -1 & 2 \\ 2 & -5 & 2 & 6 \\ -3 & -6 & 4 & 6 \end{array}\right]$ (a) $R_2 = -2r_1 + r_2$ (b) $R_3 = 3r_1 + r_3$ (c) $R_3 = 15r_2 + r_3$

19. $\left[\begin{array}{rrr|r} 1 & -3 & -2 & 3 \\ 2 & -5 & 2 & -1 \\ -3 & -2 & 4 & 6 \end{array}\right]$ (a) $R_2 = -2r_1 + r_2$ (b) $R_3 = 3r_1 + r_3$ (c) $R_3 = 11r_2 + r_3$

20. $\left[\begin{array}{rrr|r} 1 & -3 & 5 & -3 \\ 2 & -5 & 1 & -4 \\ -3 & 3 & 4 & 6 \end{array}\right]$ (a) $R_2 = -2r_1 + r_2$ (b) $R_3 = 3r_1 + r_3$ (c) $R_3 = 6r_2 + r_3$

In Problems 21–30, the reduced echelon form of a system of linear equations is given. Write the system of equations corresponding to the given matrix. Use x, y, or x, y, z or x_1, x_2, x_3, x_4 as variables. Determine whether the system is consistent or inconsistent. If it is consistent, give the solution.

21. $\left[\begin{array}{rr|r} 1 & 0 & 5 \\ 0 & 1 & -1 \end{array}\right]$

22. $\left[\begin{array}{rr|r} 1 & 0 & -4 \\ 0 & 1 & 0 \end{array}\right]$

23. $\left[\begin{array}{rrr|r} 1 & 0 & 0 & 1 \\ 0 & 1 & 0 & 2 \\ 0 & 0 & 0 & 3 \end{array}\right]$

24. $\left[\begin{array}{rrr|r} 1 & 0 & 0 & 0 \\ 0 & 1 & 0 & 0 \\ 0 & 0 & 0 & 2 \end{array}\right]$

25. $\left[\begin{array}{rrr|r} 1 & 0 & 2 & -1 \\ 0 & 1 & -4 & -2 \\ 0 & 0 & 0 & 0 \end{array}\right]$

26. $\left[\begin{array}{rrr|r} 1 & 0 & 4 & 4 \\ 0 & 1 & 3 & 2 \\ 0 & 0 & 0 & 0 \end{array}\right]$

27. $\left[\begin{array}{rrrr|r} 1 & 0 & 0 & 0 & 1 \\ 0 & 1 & 0 & 1 & 2 \\ 0 & 0 & 1 & 2 & 3 \end{array}\right]$

28. $\left[\begin{array}{rrrr|r} 1 & 0 & 0 & 0 & 1 \\ 0 & 1 & 0 & 2 & 2 \\ 0 & 0 & 1 & 3 & 0 \end{array}\right]$

29. $\left[\begin{array}{rrrr|r} 1 & 0 & 0 & 4 & 2 \\ 0 & 1 & 1 & 3 & 3 \\ 0 & 0 & 0 & 0 & 0 \end{array}\right]$

30. $\left[\begin{array}{rrrr|r} 1 & 0 & 0 & 0 & 1 \\ 0 & 1 & 0 & 0 & 2 \\ 0 & 0 & 1 & 2 & 3 \end{array}\right]$

In Problems 31–72, solve each system of equations using matrices (row operations). If the system has no solution, say it is inconsistent.

31. $\begin{cases} x + y = 8 \\ x - y = 4 \end{cases}$

32. $\begin{cases} x + 2y = 5 \\ x + y = 3 \end{cases}$

33. $\begin{cases} x - 5y = -13 \\ 3x + 2y = 12 \end{cases}$

34. $\begin{cases} x + 3y = 5 \\ 2x - 3y = -8 \end{cases}$

35. $\begin{cases} 3x - 6y = 24 \\ 5x + 4y = 12 \end{cases}$

36. $\begin{cases} 2x + 4y = 16 \\ 3x - 5y = -9 \end{cases}$

37. $\begin{cases} 2x + y = 1 \\ 4x + 2y = 6 \end{cases}$

38. $\begin{cases} x - y = 5 \\ -3x + 3y = 2 \end{cases}$

39. $\begin{cases} 2x - 4y = -2 \\ 3x + 2y = 3 \end{cases}$

40. $\begin{cases} 3x + 3y = 3 \\ 4x + 2y = \frac{8}{3} \end{cases}$

41. $\begin{cases} x + 2y = 4 \\ 2x + 4y = 8 \end{cases}$

42. $\begin{cases} 3x - y = 7 \\ 9x - 3y = 21 \end{cases}$

43. $\begin{cases} 2x + 3y = 6 \\ x - y = \frac{1}{2} \end{cases}$

44. $\begin{cases} \frac{1}{2}x + y = -2 \\ x - 2y = 8 \end{cases}$

45. $\begin{cases} 3x - 5y = 3 \\ 15x + 5y = 21 \end{cases}$

46. $\begin{cases} 2x - y = -1 \\ x + \frac{1}{2}y = \frac{3}{2} \end{cases}$

47. $\begin{cases} x - y = 6 \\ 2x - 3z = 16 \\ 2y + z = 4 \end{cases}$

48. $\begin{cases} 2x + y = -4 \\ -2y + 4z = 0 \\ 3x - 2z = -11 \end{cases}$

49. $\begin{cases} x - 2y + 3z = 7 \\ 2x + y + z = 4 \\ -3x + 2y - 2z = -10 \end{cases}$

50. $\begin{cases} 2x + y - 3z = 0 \\ -2x + 2y + z = -7 \\ 3x - 4y - 3z = 7 \end{cases}$

51. $\begin{cases} 2x - 2y - 2z = 2 \\ 2x + 3y + z = 2 \\ 3x + 2y = 0 \end{cases}$

52. $\begin{cases} 2x - 3y - z = 0 \\ -x + 2y + z = 5 \\ 3x - 4y - z = 1 \end{cases}$

53. $\begin{cases} -x + y + z = -1 \\ -x + 2y - 3z = -4 \\ 3x - 2y - 7z = 0 \end{cases}$

54. $\begin{cases} 2x - 3y - z = 0 \\ 3x + 2y + 2z = 2 \\ x + 5y + 3z = 2 \end{cases}$

55. $\begin{cases} 2x - 2y + 3z = 6 \\ 4x - 3y + 2z = 0 \\ -2x + 3y - 7z = 1 \end{cases}$

56. $\begin{cases} 3x - 2y + 2z = 6 \\ 7x - 3y + 2z = -1 \\ 2x - 3y + 4z = 0 \end{cases}$

57. $\begin{cases} x + y - z = 6 \\ 3x - 2y + z = -5 \\ x + 3y - 2z = 14 \end{cases}$

58. $\begin{cases} x - y + z = -4 \\ 2x - 3y + 4z = -15 \\ 5x + y - 2z = 12 \end{cases}$

59. $\begin{cases} x + 2y - z = -3 \\ 2x - 4y + z = -7 \\ -2x + 2y - 3z = 4 \end{cases}$

60. $\begin{cases} x + 4y - 3z = -8 \\ 3x - y + 3z = 12 \\ x + y + 6z = 1 \end{cases}$

61. $\begin{cases} 3x + y - z = \frac{2}{3} \\ 2x - y + z = 1 \\ 4x + 2y = \frac{8}{3} \end{cases}$

62. $\begin{cases} x + y = 1 \\ 2x - y + z = 1 \\ x + 2y + z = \frac{8}{3} \end{cases}$

63. $\begin{cases} x + y + z + w = 4 \\ 2x - y + z = 0 \\ 3x + 2y + z - w = 6 \\ x - 2y - 2z + 2w = -1 \end{cases}$

64. $\begin{cases} x + y + z + w = 4 \\ -x + 2y + z = 0 \\ 2x + 3y + z - w = 6 \\ -2x + y - 2z + 2w = -1 \end{cases}$

65. $\begin{cases} x + 2y + z = 1 \\ 2x - y + 2z = 2 \\ 3x + y + 3z = 3 \end{cases}$

66. $\begin{cases} x + 2y - z = 3 \\ 2x - y + 2z = 6 \\ x - 3y + 3z = 4 \end{cases}$

67. $\begin{cases} x - y + z = 5 \\ 3x + 2y - 2z = 0 \end{cases}$

68. $\begin{cases} 2x + y - z = 4 \\ -x + y + 3z = 1 \end{cases}$

69. $\begin{cases} 2x + 3y - z = 3 \\ x - y - z = 0 \\ -x + y + z = 0 \\ x + y + 3z = 5 \end{cases}$

70. $\begin{cases} x - 3y + z = 1 \\ 2x - y - 4z = 0 \\ x - 3y + 2z = 1 \\ x - 2y = 5 \end{cases}$

71. $\begin{cases} 4x + y + z - w = 4 \\ x - y + 2z + 3w = 3 \end{cases}$

72. $\begin{cases} -4x + y = 5 \\ 2x - y + z - w = 5 \\ z + w = 4 \end{cases}$

73. Curve Fitting Find the parabola $y = ax^2 + bx + c$ that passes through the points $(1, 2)$, $(-2, -7)$, and $(2, -3)$.

74. Curve Fitting Find the parabola $y = ax^2 + bx + c$ that passes through the points $(1, -1)$, $(3, -1)$, and $(-2, 14)$.

75. Curve Fitting Find the function $f(x) = ax^3 + bx^2 + cx + d$ for which $f(-3) = -112$, $f(-1) = -2$, $f(1) = 4$, and $f(2) = 13$.

76. Curve Fitting Find the function $f(x) = ax^3 + bx^2 + cx + d$ for which $f(-2) = -10$, $f(-1) = 3$, $f(1) = 5$, and $f(3) = 15$.

77. Mixing Acids A chemistry laboratory has three containers of sulfuric acid, H_2SO_4. One container holds a solution with a concentration of 15% H_2SO_4, the second holds 25% H_2SO_4, and the third holds 50% H_2SO_4. How many liters of each solution should be mixed to obtain 100 liters of a solution with a concentration of 40% H_2SO_4? Construct a table similar to Table 1 illustrating some of the possible combinations.

78. Painting a House Three painters, Mike, Dan, and Katy, working together can paint the exterior of a home in 10 hours. Dan and Katy together have painted a similar house in 15 hours. One day, all three worked on this same kind of house for 4 hours, after which Katy left. Mike and Dan required 8 more hours to finish. Assuming no gain or loss in efficiency, how long should it take each person to complete such a job alone?

79. Prices of Fast Food One group of customers bought 8 deluxe hamburgers, 6 orders of large fries, and 6 large colas for \$26.10. A second group ordered 10 deluxe hamburgers, 6 large fries, and 8 large colas and paid \$31.60. Is there sufficient information to determine the price of each food item? If not, construct a table showing the various possibilities. Assume that the hamburgers cost between \$1.75 and \$2.25, the fries between \$0.75 and \$1.00, and the colas between \$0.60 and \$0.90.

80. Prices of Fast Food Use the information given in Problem 79, and suppose that a third group purchased 3 deluxe hamburgers, 2 large fries, and 4 large colas for \$10.95. Now is there sufficient information to determine the price of each food item?

81. Financial Planning Three retired couples each require an additional annual income of \$2000 per year. As their financial consultant, you recommend that they invest some money in Treasury bills that yield 7%, some money in corporate bonds that yield 9%, and some money in junk bonds that yield 11%. Prepare a table for each couple showing the various ways their goals can be achieved:

(a) If the first couple has \$20,000 to invest.
(b) If the second couple has \$25,000 to invest.
(c) If the third couple has \$30,000 to invest.
(d) What advice would you give each couple regarding the amount to invest and the choices available?

[Higher yields generally carry more risk].

82. Financial Planning A retired couple has \$25,000 to invest. As their financial consultant, you recommend that they invest some money in Treasury bills that yield 7%, some money in corporate bonds that yield 9%, and some money in junk bonds that yield 11%. Prepare a table showing the various ways this couple can achieve the following goals:

(a) The couple wants \$1500 per year in income.
(b) The couple wants \$2000 per year in income.
(c) The couple wants \$2500 per year in income.
(d) What advice would you give this couple regarding the income they require and the choices available?

[Higher yields generally carry more risk].

83. Electricity: Kirchoff's Rules An application of Kirchoff's Rules to the circuit shown results in the following system of equations:

$$\begin{cases} I_1 + I_2 = I_3 \\ 16 - 8 - 9I_3 - 3I_1 = 0 \\ 16 - 4 - 9I_3 - 9I_2 = 0 \\ 8 - 4 - 9I_2 + 3I_1 = 0 \end{cases}$$

Find the currents I_1, I_2, and I_3.*

*Source: Based on Raymond Serway, *Physics,* 3rd ed. (Philadelphia: Saunders, 1990), Prob. 31, p. 790.

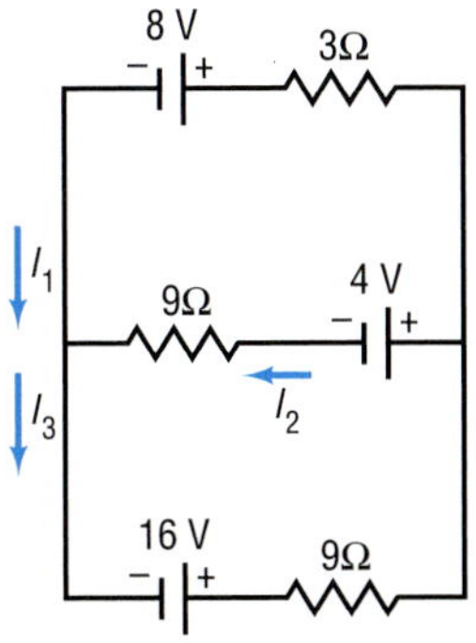

84. Electricity: Kirchoff's Rules An application of Kirchoff's Rules to the circuit shown results in the following system of equations:

$$\begin{cases} -4 + 8 - 2I_2 = 0 \\ 8 = 5I_4 + I_1 \\ 4 = 3I_3 + I_1 \\ I_3 + I_4 = I_1 \end{cases}$$

Find the currents I_1, I_2, I_3, and I_4.*

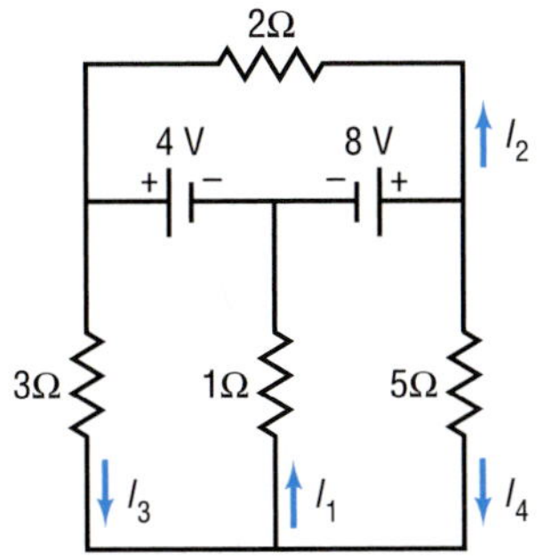

85. Electricity: Kirchoff's Rules An application of Kirchoff's Rules to the circuit shown results in the following system of equations:

$$\begin{cases} I_1 = I_3 + I_2 \\ 24 - 6I_1 - 3I_3 = 0 \\ 12 + 24 - 6I_1 - 6I_2 = 0 \end{cases}$$

Find the currents I_1, I_2, and I_3.†

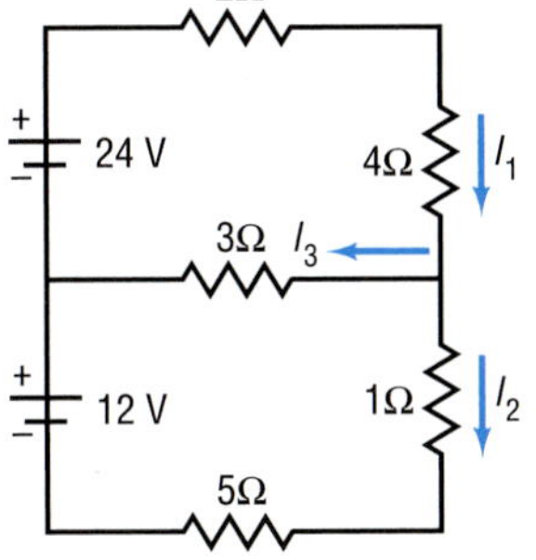

*Source: Ibid., Prob. 34, p. 791.

†Source: Ibid., Prob. 38, p. 791.

86. Write a brief paragraph or two that outlines your strategy for solving a system of linear equations using matrices.

87. When solving a system of linear equations using matrices, do you prefer to place the augmented matrix in echelon form or in reduced echelon form? Give reasons for your choice.

88. Make up a system of three linear equations containing three variables that has:

(a) No solution
(b) Exactly one solution
(c) Infinitely many solutions

Give the three systems to a friend to solve and critique.

89. Consider the system of equations

$$\begin{cases} a_1x + b_1y = c_1 \\ a_2x + b_2y = c_2 \end{cases}$$

If $D = a_1b_2 - a_2b_1 \neq 0$, use matrices to show that the solution is

$$x = \frac{1}{D}(c_1b_2 - c_2b_1), \qquad y = \frac{1}{D}(a_1c_2 - a_2c_1)$$

90. For the system in Problem 89, suppose that $D = a_1b_2 - a_2b_1 = 0$. Use matrices to show that the system is inconsistent if either $a_1c_2 \neq a_2c_1$ or $b_1c_2 \neq b_2c_1$ and has infinitely many solutions if both $a_1c_2 = a_2c_1$ and $b_1c_2 = b_2c_1$.

91. The graph of a linear equation containing three variables is a plane. Give a geometrical argument for what can result when solving a system of two linear equations containing three variables.

[**Hint:** Two planes in a three-dimensional space are either coincident (the same), parallel, or intersect in a line.]

92. Refer to Problem 91. Give a geometrical argument for what can result when solving a system of three linear equations containing three variables.

93. Refer to Problem 91. Give a geometrical argument for what can result when solving a system of four linear equations containing three variables.

11.3 SYSTEMS OF LINEAR EQUATIONS: DETERMINANTS

1 Evaluate 2 by 2 Determinants
2 Use Cramer's Rule to Solve a System of Two Equations, Two Variables
3 Evaluate 3 by 3 Determinants
4 Use Cramer's Rule to Solve a System of Three Equations, Three Variables
5 Know Some Properties of Determinants

1 In the preceding section, we described a method of using matrices to solve any system of linear equations. This section deals with yet another method for solving systems of linear equations; however, it can be used only when the number of equations equals the number of variables. Although the method will work for any system (provided the number of equations equals the number of variables), it is most often used for systems of two equations containing two variables or three equations containing three variables. This method, called *Cramer's Rule,* is based on the concept of a *determinant.*

2 by 2 Determinants

If a, b, c, and d are four real numbers, the symbol

$$D = \begin{vmatrix} a & b \\ c & d \end{vmatrix}$$

is called a **2 by 2 determinant.** Its value is the number $ad - bc$; that is,

$$D = \begin{vmatrix} a & b \\ c & d \end{vmatrix} = ad - bc \qquad (1)$$

A device that may be helpful for remembering the value of a 2 by 2 determinant is the following:

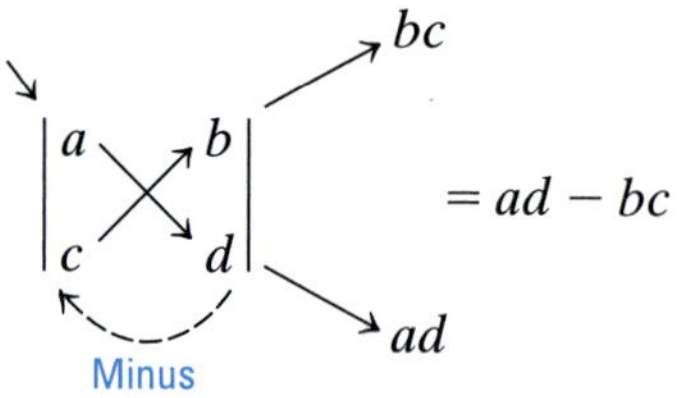

EXAMPLE 1

Evaluating a 2 × 2 Determinant

$$\begin{vmatrix} 3 & -2 \\ 6 & 1 \end{vmatrix} = (3)(1) - (6)(-2) = 3 - (-12) = 15$$

Graphing utilities can also be used to evaluate determinants.

EXAMPLE 2

Using a Graphing Utility to Evaluate a 2 × 2 Determinant

Use a graphing utility to evaluate the determinant from Example 1: $\begin{vmatrix} 3 & -2 \\ 6 & 1 \end{vmatrix}$.

Solution First, we enter the matrix whose entries are those of the determinant into the graphing utility and name it A. Using the determinant command, we obtain the result shown in Figure 9. ■

FIGURE 9

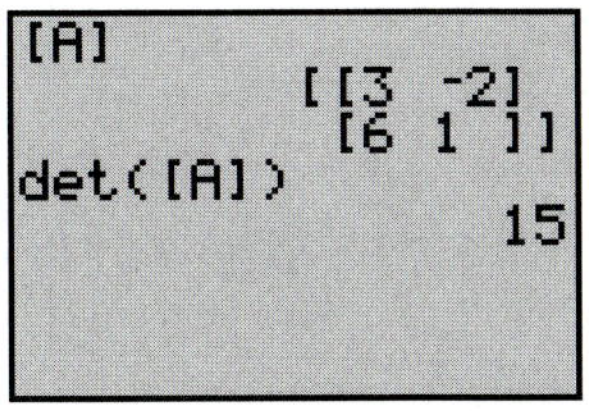

Now work Problem 3.

2 Let's now see the role that a 2 by 2 determinant plays in the solution of a system of two equations containing two variables. Consider the system

$$\begin{cases} ax + by = s & (1) \\ cx + dy = t & (2) \end{cases} \qquad (2)$$

We shall use the method of elimination to solve this system.

Provided $d \neq 0$ and $b \neq 0$, this system is equivalent to the system

$$\begin{cases} adx + bdy = sd & (1) \text{ Multiply by } d. \\ bcx + bdy = tb & (2) \text{ Multiply by } b. \end{cases}$$

On subtracting the second equation from the first equation, we get

$$\begin{cases} (ad - bc)x + 0 \cdot y = sd - tb & (1) \\ \quad bcx \quad + bdy = tb & (2) \end{cases}$$

Now, the first equation can be rewritten using determinant notation:

$$\begin{vmatrix} a & b \\ c & d \end{vmatrix} x = \begin{vmatrix} s & b \\ t & d \end{vmatrix}$$

If $D = \begin{vmatrix} a & b \\ c & d \end{vmatrix} = ad - bc \neq 0$, we can solve for x to get

$$x = \frac{\begin{vmatrix} s & b \\ t & d \end{vmatrix}}{\begin{vmatrix} a & b \\ c & d \end{vmatrix}} = \frac{\begin{vmatrix} s & b \\ t & d \end{vmatrix}}{D} \qquad (3)$$

Return now to the original system (2). Provided $a \neq 0$ and $c \neq 0$, the system is equivalent to

$$\begin{cases} acx + bcy = cs & (1) \text{ Multiply by } c. \\ acx + ady = at & (2) \text{ Multiply by } a. \end{cases}$$

On subtracting the first equation from the second equation, we get

$$\begin{cases} acx + \quad bcy \quad = cs & (1) \\ 0 \cdot x + (ad - bc)y = at - cs & (2) \end{cases}$$

The second equation now can be rewritten using determinant notation:

$$\begin{vmatrix} a & b \\ c & d \end{vmatrix} y = \begin{vmatrix} a & s \\ c & t \end{vmatrix}$$

If $D = \begin{vmatrix} a & b \\ c & d \end{vmatrix} = ad - bc \neq 0$, we can solve for y to get

$$y = \frac{\begin{vmatrix} a & s \\ c & t \end{vmatrix}}{\begin{vmatrix} a & b \\ c & d \end{vmatrix}} = \frac{\begin{vmatrix} a & s \\ c & t \end{vmatrix}}{D} \qquad (4)$$

Equations (3) and (4) lead us to the following result, called **Cramer's Rule:**

Theorem Cramer's Rule for Two Equations Containing Two Variables

The solution to the system of equations

$$\begin{cases} ax + by = s & (1) \\ cx + dy = t & (2) \end{cases} \qquad (5)$$

is given by

$$x = \frac{\begin{vmatrix} s & b \\ t & d \end{vmatrix}}{\begin{vmatrix} a & b \\ c & d \end{vmatrix}}, \qquad y = \frac{\begin{vmatrix} a & s \\ c & t \end{vmatrix}}{\begin{vmatrix} a & b \\ c & d \end{vmatrix}} \qquad (6)$$

provided that

$$D = \begin{vmatrix} a & b \\ c & d \end{vmatrix} = ad - bc \neq 0$$

In the derivation given for Cramer's Rule above, we assumed that none of the numbers a, b, c, and d were 0. In Problem 58 at the end of this section you will be asked to complete the proof under the less stringent conditions that $D = ad - bc \neq 0$.

Now look carefully at the pattern in Cramer's Rule. The denominator in the solution (6) is the determinant of the coefficients of the variables:

$$\begin{cases} ax + by = s \\ cx + dy = t \end{cases} \qquad D = \begin{vmatrix} a & b \\ c & d \end{vmatrix}$$

In the solution for x, the numerator is the determinant, denoted by D_x, formed by replacing the entries in the first column (the coefficients of x) in D by the constants on the right side of the equal sign:

$$D_x = \begin{vmatrix} s & b \\ t & d \end{vmatrix}$$

In the solution for y, the numerator is the determinant, denoted by D_y, formed by replacing the entries in the second column (the coefficients of y) in D by the constants on the right side of the equal sign:

$$D_y = \begin{vmatrix} a & s \\ c & t \end{vmatrix}$$

Cramer's Rule then states that, if $D \neq 0$,

$$x = \frac{D_x}{D}, \qquad y = \frac{D_y}{D} \qquad (7)$$

EXAMPLE 3

Solving a System of Linear Equations Using Determinants

Use Cramer's Rule, if applicable, to solve the system

$$\begin{cases} 3x - 2y = 4 & (1) \\ 6x + y = 13 & (2) \end{cases}$$

Algebraic Solution

The determinant D of the coefficients of the variables is

$$D = \begin{vmatrix} 3 & -2 \\ 6 & 1 \end{vmatrix} = (3)(1) - (6)(-2) = 15$$

Because $D \neq 0$, Cramer's Rule (7) can be used:

$$x = \frac{D_x}{D} = \frac{\begin{vmatrix} 4 & -2 \\ 13 & 1 \end{vmatrix}}{15} = \frac{30}{15} = 2, \qquad y = \frac{D_y}{D} = \frac{\begin{vmatrix} 3 & 4 \\ 6 & 13 \end{vmatrix}}{15} = \frac{15}{15} = 1$$

The solution is $x = 2$, $y = 1$.

Graphing Solution

FIGURE 10

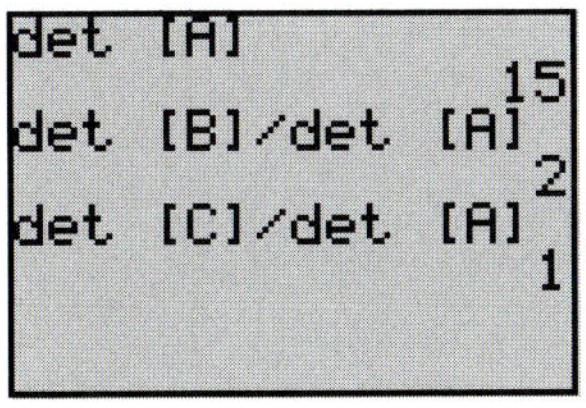

We enter the coefficient matrix into our graphing utility and call it A. We enter the matrices D_x and D_y into our graphing utility and call them B and C, respectively. Evaluate det $[A]$. Since det $[A] \neq 0$, we can use Cramer's Rule. Finally, we find x by calculating det$[B]$/det$[A]$ and y by calculating det$[C]$/det$[A]$. The results are shown in Figure 10.

The solution is $x = 2$, $y = 1$. ■

If, in attempting to use Cramer's Rule, the determinant D of the coefficients of the variables is found to equal 0 (so that Cramer's Rule is not applicable), then the system either is inconsistent or has infinitely many solutions. (Refer to Problem 90 in Exercise 11.2.)

Now work Problem 11.

3 by 3 Determinants

3 In order to use Cramer's Rule to solve a system of three equations containing three variables, we need to define a 3 by 3 determinant.

A **3 by 3 determinant** is symbolized by

$$\begin{vmatrix} a_{11} & a_{12} & a_{13} \\ a_{21} & a_{22} & a_{23} \\ a_{31} & a_{32} & a_{33} \end{vmatrix} \tag{8}$$

in which $a_{11}, a_{12}, \ldots$ are real numbers.

As with matrices, we use a double subscript to identify an entry by indicating its row and column numbers. For example, the entry a_{23} is in row 2, column 3.

The value of a 3 by 3 determinant may be defined in terms of 2 by 2 determinants by the following formula:

$$\begin{vmatrix} a_{11} & a_{12} & a_{13} \\ a_{21} & a_{22} & a_{23} \\ a_{31} & a_{32} & a_{33} \end{vmatrix} = a_{11}\begin{vmatrix} a_{22} & a_{23} \\ a_{32} & a_{33} \end{vmatrix} - a_{12}\begin{vmatrix} a_{21} & a_{23} \\ a_{31} & a_{33} \end{vmatrix} + a_{13}\begin{vmatrix} a_{21} & a_{22} \\ a_{31} & a_{32} \end{vmatrix} \tag{9}$$

Minus (pointing to the − sign before a_{12})

2 by 2 determinant left after removing row and column containing a_{11}.

2 by 2 determinant left after removing row and column containing a_{12}.

2 by 2 determinant left after removing row and column containing a_{13}.

Be sure to take note of the minus sign that appears with the second term—it's easy to forget it! Formula (9) is best remembered by noting that each entry in row 1 is multiplied by the 2 by 2 determinant that remains after the row and column containing the entry have been removed, as follows:

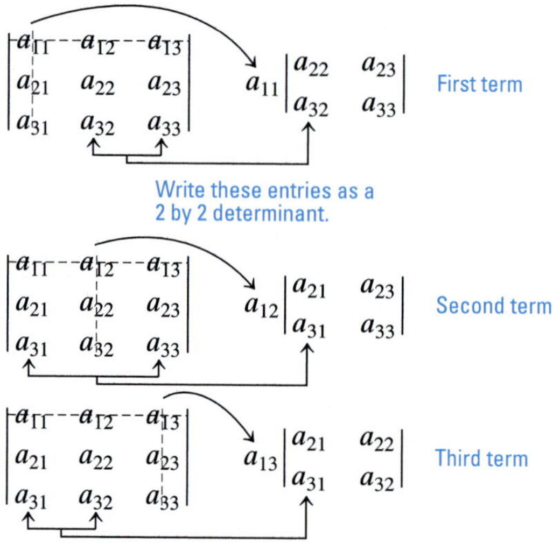

Now insert the minus sign before the middle expression and add

$$\begin{vmatrix} a_{11} & a_{12} & a_{13} \\ a_{21} & a_{22} & a_{23} \\ a_{31} & a_{32} & a_{33} \end{vmatrix} = a_{11}\begin{vmatrix} a_{22} & a_{23} \\ a_{32} & a_{33} \end{vmatrix} \overset{\text{Minus}\downarrow}{-} a_{12}\begin{vmatrix} a_{21} & a_{23} \\ a_{31} & a_{33} \end{vmatrix} + a_{13}\begin{vmatrix} a_{21} & a_{22} \\ a_{31} & a_{32} \end{vmatrix}$$

Formula (9) exhibits one way to find the value of a 3 by 3 determinant, *by expanding across row 1.* In fact, the expansion can take place across any row or down any column. The terms to be added or subtracted consist of the row (or column) entry times the value of the 2 by 2 determinant that remains after removing the row and column entry. The value of the determinant is found by adding or subtracting the terms according to the following scheme:

$$\begin{matrix} + & - & + \\ - & + & - \\ + & - & + \end{matrix}$$

For example, if we choose to expand down column 2, we obtain

$$\begin{vmatrix} a_{11} & a_{12} & a_{13} \\ a_{21} & a_{22} & a_{23} \\ a_{31} & a_{32} & a_{33} \end{vmatrix} = -a_{12}\begin{vmatrix} a_{21} & a_{23} \\ a_{31} & a_{33} \end{vmatrix} + a_{22}\begin{vmatrix} a_{11} & a_{13} \\ a_{31} & a_{33} \end{vmatrix} - a_{32}\begin{vmatrix} a_{11} & a_{13} \\ a_{21} & a_{23} \end{vmatrix}$$

↑ Expand down column 2 $(-, +, -)$.

If we choose to expand across row 3, we obtain

$$\begin{vmatrix} a_{11} & a_{12} & a_{13} \\ a_{21} & a_{22} & a_{23} \\ a_{31} & a_{32} & a_{33} \end{vmatrix} = a_{31}\begin{vmatrix} a_{12} & a_{13} \\ a_{22} & a_{23} \end{vmatrix} - a_{32}\begin{vmatrix} a_{11} & a_{13} \\ a_{21} & a_{23} \end{vmatrix} + a_{33}\begin{vmatrix} a_{11} & a_{12} \\ a_{21} & a_{22} \end{vmatrix}$$

↑ Expand across row 3 $(+, -, +)$.

MISSION POSSIBLE

Keeping Secrets

Your group has discovered that a rival consulting firm has been intercepting and reading the reports you are sending one another. You need a way to encode your final decisions so that this rival firm does not take business away from you by stealing your ideas. You have decided to find a way to encode your most important messages to each other so that prying eyes are not able to read them.

Your method of coding begins with the simple one used by children, that is, assigning to each letter of the alphabet the number which represents its position in the order. For example, A = 1, B = 2, . . . , M = 13, N = 14, . . . , Z = 26. A space would be represented by 27; a period by 0. Then the message is translated into 2×1 matrices as in the example which follows.

Suppose you wish to send the message: *Choose Dealer C.* You would write this out first as letters in groups of two, then as matrices.

Ch	*oo*	*se*	*D*	*ea*	*le*	*r*	*C*
$\begin{bmatrix} 3 \\ 8 \end{bmatrix}$	$\begin{bmatrix} 15 \\ 15 \end{bmatrix}$	$\begin{bmatrix} 19 \\ 5 \end{bmatrix}$	$\begin{bmatrix} 27 \\ 4 \end{bmatrix}$	$\begin{bmatrix} 5 \\ 1 \end{bmatrix}$	$\begin{bmatrix} 12 \\ 5 \end{bmatrix}$	$\begin{bmatrix} 18 \\ 27 \end{bmatrix}$	$\begin{bmatrix} 3 \\ 0 \end{bmatrix}$

Next you would use a coding matrix, a 2×2 square matrix with numbers chosen so that the inverse exists. We'll use $\begin{bmatrix} 6 & 4 \\ -2 & 7 \end{bmatrix}$ as our coding matrix. To encode a message, you multiply the coding matrix times each matrix of the message, as follows:.

$$\begin{bmatrix} 6 & 4 \\ -2 & 7 \end{bmatrix} \cdot \begin{bmatrix} 3 \\ 8 \end{bmatrix} = \begin{bmatrix} 50 \\ 50 \end{bmatrix}, \begin{bmatrix} 6 & 4 \\ -2 & 7 \end{bmatrix} \cdot \begin{bmatrix} 15 \\ 15 \end{bmatrix} = \begin{bmatrix} 150 \\ 75 \end{bmatrix}, \text{ and so on}$$

The resulting set of matrices would be sent along to your partner as a sequence of numbers: 50, 50, 150, 75, and so on. Your partner would use the inverse of the encoding matrix to decipher your message. Obviously you would have to guard the encoding matrix and its inverse very carefully!

1. Find the inverse of the coding matrix above. Make sure it is correct by multiplying the two encoded matrices times your matrix to see if you can recover the original two matrices.
2. Next, create your own message (not too long!) and encrypt it using the coding matrix above. Copy the resulting code onto a separate piece of paper as a sequence of numbers.
3. Exchange coded messages with another group.
4. Use the inverse matrix to decode the message sent to you by the other group.
5. Cryptography is a science practiced by governments at war, or businesses which are competitive, in situations where secrecy is crucial. It is especially important in transferring valuable information, such as credit card numbers, over the Internet. If your consulting firm is serious about keeping secrets, you may need to upgrade your coding procedures. For example, the message *Choose Dealer C,* given above, could be broken up into four 4×1 matrices. Then the coding matrix would have to be 4×4 with an inverse to match. A government might use a 20×20 coding matrix to decrease the probability that it could be deciphered. Create your own 4×4 coding matrix, find its inverse, and encode a message of your choosing in 4×1 matrices. Then pass the code and the key to a neighboring group for decoding.

It can be shown that the value of a determinant does not depend on the choice of the row or column used in the expansion.

EXAMPLE 4 Evaluating a 3 × 3 Determinant

Find the value of the 3 by 3 determinant $\begin{vmatrix} 3 & 4 & -1 \\ 4 & 6 & 2 \\ 8 & -2 & 3 \end{vmatrix}$

Solution We choose to expand across row 1.

Remember the minus sign.

$$\begin{vmatrix} 3 & 4 & -1 \\ 4 & 6 & 2 \\ 8 & -2 & 3 \end{vmatrix} = 3\begin{vmatrix} 6 & 2 \\ -2 & 3 \end{vmatrix} - 4\begin{vmatrix} 4 & 2 \\ 8 & 3 \end{vmatrix} + (-1)\begin{vmatrix} 4 & 6 \\ 8 & -2 \end{vmatrix}$$
$$= 3(18 + 4) - 4(12 - 16) + (-1)(-8 - 48)$$
$$= 3(22) - 4(-4) + (-1)(-56)$$
$$= 66 + 16 + 56 = 138$$

We could also find the value of the 3 by 3 determinant in Example 5 by expanding down column 3 (the signs are $+$, $-$, $+$):

$$\begin{vmatrix} 3 & 4 & -1 \\ 4 & 6 & 2 \\ 8 & -2 & 3 \end{vmatrix} = (-1)\begin{vmatrix} 4 & 6 \\ 8 & -2 \end{vmatrix} - 2\begin{vmatrix} 3 & 4 \\ 8 & -2 \end{vmatrix} + 3\begin{vmatrix} 3 & 4 \\ 4 & 6 \end{vmatrix}$$
$$= -1(-8 - 48) - 2(-6 - 32) + 3(18 - 16)$$
$$= 56 + 76 + 6 = 138$$

Evaluating 3×3 determinants on a graphing utility follows the same procedure as evaluating 2×2 determinants.

Now work Problem 7.

Systems of Three Equations Containing Three Variables

4 Consider the following system of three equations containing three variables:

$$\begin{cases} a_{11}x + a_{12}y + a_{13}z = c_1 \\ a_{21}x + a_{22}y + a_{23}z = c_2 \\ a_{31}x + a_{32}y + a_{33}z = c_3 \end{cases} \tag{10}$$

It can be shown that if the determinant D of the coefficients of the variables is not 0, that is, if

$$D = \begin{vmatrix} a_{11} & a_{12} & a_{13} \\ a_{21} & a_{22} & a_{23} \\ a_{31} & a_{32} & a_{33} \end{vmatrix} \neq 0$$

then the unique solution of system (10) is given by

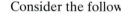

Cramer's Rule for Three Equations Containing Three Variables

$$x = \frac{D_x}{D}, \qquad y = \frac{D_y}{D}, \qquad z = \frac{D_z}{D}$$

where

$$D_x = \begin{vmatrix} c_1 & a_{12} & a_{13} \\ c_2 & a_{22} & a_{23} \\ c_3 & a_{32} & a_{33} \end{vmatrix} \qquad D_y = \begin{vmatrix} a_{11} & c_1 & a_{13} \\ a_{21} & c_2 & a_{23} \\ a_{31} & c_3 & a_{33} \end{vmatrix} \qquad D_z = \begin{vmatrix} a_{11} & a_{12} & c_1 \\ a_{21} & a_{22} & c_2 \\ a_{31} & a_{32} & c_3 \end{vmatrix}$$

The similarity of this pattern and the pattern observed earlier for a system of two equations containing two variables should be apparent.

EXAMPLE 5 Using Cramer's Rule

Use Cramer's Rule, if applicable, to solve the following system:

$$\begin{cases} 2x + y - z = 3 & (1) \\ -x + 2y + 4z = -3 & (2) \\ x - 2y - 3z = 4 & (3) \end{cases}$$

Solution The value of the determinant D of the coefficients of the variables is

$$D = \begin{vmatrix} 2 & 1 & -1 \\ -1 & 2 & 4 \\ 1 & -2 & -3 \end{vmatrix} = 2\begin{vmatrix} 2 & 4 \\ -2 & -3 \end{vmatrix} - 1\begin{vmatrix} -1 & 4 \\ 1 & -3 \end{vmatrix} + (-1)\begin{vmatrix} -1 & 2 \\ 1 & -2 \end{vmatrix}$$
$$= 2(2) - 1(-1) + (-1)(0)$$
$$= 4 + 1 = 5$$

Because $D \neq 0$, we proceed to find the values of D_x, D_y, and D_z:

$$D_x = \begin{vmatrix} 3 & 1 & -1 \\ -3 & 2 & 4 \\ 4 & -2 & -3 \end{vmatrix} = 3\begin{vmatrix} 2 & 4 \\ -2 & -3 \end{vmatrix} - 1\begin{vmatrix} -3 & 4 \\ 4 & -3 \end{vmatrix} + (-1)\begin{vmatrix} -3 & 2 \\ 4 & -2 \end{vmatrix}$$
$$= 3(2) - 1(-7) + (-1)(-2) = 15$$

$$D_y = \begin{vmatrix} 2 & 3 & -1 \\ -1 & -3 & 4 \\ 1 & 4 & -3 \end{vmatrix} = 2\begin{vmatrix} -3 & 4 \\ 4 & -3 \end{vmatrix} - 3\begin{vmatrix} -1 & 4 \\ 1 & -3 \end{vmatrix} + (-1)\begin{vmatrix} -1 & -3 \\ 1 & 4 \end{vmatrix}$$
$$= 2(-7) - 3(-1) + (-1)(-1)$$
$$= -14 + 3 + 1 = -10$$

$$D_z = \begin{vmatrix} 2 & 1 & 3 \\ -1 & 2 & -3 \\ 1 & -2 & 4 \end{vmatrix} = 2\begin{vmatrix} 2 & -3 \\ -2 & 4 \end{vmatrix} - 1\begin{vmatrix} -1 & -3 \\ 1 & 4 \end{vmatrix} + 3\begin{vmatrix} -1 & 2 \\ 1 & -2 \end{vmatrix}$$
$$= 2(2) - 1(-1) + 3(0) = 5$$

As a result,

$$x = \frac{D_x}{D} = \frac{15}{5} = 3, \qquad y = \frac{D_y}{D} = \frac{-10}{5} = -2, \qquad z = \frac{D_z}{D} = \frac{5}{5} = 1$$

The solution is $x = 3$, $y = -2$, $z = 1$.

If the determinant of the coefficients of the variables of a system of three linear equations containing three variables is 0, then Cramer's Rule is not applicable. In such a case, the system either is inconsistent or has infinitely many solutions.

Solving systems of three equations containing three variables using Cramer's Rule on a graphing utility follows the same procedure as that for solving systems of two equations containing two variables.

Now work Problem 29.

More about Determinants

5 Determinants have several properties that are sometimes helpful for obtaining their value. We list some of them here.

Theorem

The value of a determinant changes sign if any two rows (or any two columns) are interchanged. (11)

Proof for 2 by 2 Determinants

$$\begin{vmatrix} a & b \\ c & d \end{vmatrix} = ad - bc \quad \text{and} \quad \begin{vmatrix} c & d \\ a & b \end{vmatrix} = bc - ad = -(ad - bc)$$

EXAMPLE 6

Demonstrating a Theorem (11)

$$\begin{vmatrix} 3 & 4 \\ 1 & 2 \end{vmatrix} = 6 - 4 = 2 \qquad \begin{vmatrix} 1 & 2 \\ 3 & 4 \end{vmatrix} = 4 - 6 = -2$$

Theorem

If all the entries in any row (or any column) equal 0, the value of the determinant is 0. (12)

Proof Merely expand across the row (or down the column) containing the 0's.

Theorem

If any two rows (or any two columns) of a determinant have corresponding entries that are equal, the value of the determinant is 0. (13)

You are asked to prove this result for a 3 by 3 determinant in which the entries in column 1 equal the entries in column 3 in Problem 61 at the end of this section.

EXAMPLE 7 Demonstrating a Theorem (13)

$$\begin{vmatrix} 1 & 2 & 3 \\ 1 & 2 & 3 \\ 4 & 5 & 6 \end{vmatrix} = 1\begin{vmatrix} 2 & 3 \\ 5 & 6 \end{vmatrix} - 2\begin{vmatrix} 1 & 3 \\ 4 & 6 \end{vmatrix} + 3\begin{vmatrix} 1 & 2 \\ 4 & 5 \end{vmatrix}$$
$$= 1(-3) - 2(-6) + 3(-3)$$
$$= -3 + 12 - 9 = 0$$

Theorem

If any row (or any column) of a determinant is multiplied by a nonzero number k, the value of the determinant is also changed by a factor of k. (14)

You are asked to prove this result for a 3 by 3 determinant using row 2 in Problem 60 at the end of this section.

EXAMPLE 8 Demonstrating a Theorem (14)

$$\begin{vmatrix} 1 & 2 \\ 4 & 6 \end{vmatrix} = 6 - 8 = -2$$

$$\begin{vmatrix} k & 2k \\ 4 & 6 \end{vmatrix} = 6k - 8k = -2k = k(-2) = k\begin{vmatrix} 1 & 2 \\ 4 & 6 \end{vmatrix}$$

Theorem

If the entries of any row (or any column) of a determinant are multiplied by a nonzero number k and the result is added to the corresponding entries of another row (or column), the value of the determinant remains unchanged. (15)

In Problem 62 at the end of this section, you are asked to prove this result for a 3 by 3 determinant using rows 1 and 2.

EXAMPLE 9 Demonstrating a Theorem (15)

$$\begin{vmatrix} 3 & 4 \\ 5 & 2 \end{vmatrix} \underset{\uparrow}{=} \begin{vmatrix} -7 & 0 \\ 5 & 2 \end{vmatrix} = -14$$

Multiply row 2 by -2 and add to row 1.

11.3 EXERCISES

In Problems 1–10, find the value of each determinant (a) by hand; (b) using a graphing utility.

1. $\begin{vmatrix} 3 & 1 \\ 4 & 2 \end{vmatrix}$ **2.** $\begin{vmatrix} 6 & 1 \\ 5 & 2 \end{vmatrix}$ **3.** $\begin{vmatrix} 6 & 4 \\ -1 & 3 \end{vmatrix}$ **4.** $\begin{vmatrix} 8 & -3 \\ 4 & 2 \end{vmatrix}$ **5.** $\begin{vmatrix} -3 & -1 \\ 4 & 2 \end{vmatrix}$

6. $\begin{vmatrix} -4 & 2 \\ -5 & 3 \end{vmatrix}$ **7.** $\begin{vmatrix} 3 & 4 & 2 \\ 1 & -1 & 5 \\ 1 & 2 & -2 \end{vmatrix}$ **8.** $\begin{vmatrix} 1 & 3 & -2 \\ 6 & 1 & -5 \\ 8 & 2 & 3 \end{vmatrix}$ **9.** $\begin{vmatrix} 4 & -1 & 2 \\ 6 & -1 & 0 \\ 1 & -3 & 4 \end{vmatrix}$ **10.** $\begin{vmatrix} 3 & -9 & 4 \\ 1 & 4 & 0 \\ 8 & -3 & 1 \end{vmatrix}$

In Problems 11–38, solve each system of equations using Cramer's Rule if it is applicable (a) by hand; (b) using a graphing utility. If Cramer's Rule is not applicable, say so.

11. $\begin{cases} x + y = 8 \\ x - y = 4 \end{cases}$ **12.** $\begin{cases} x + 2y = 5 \\ x - y = 3 \end{cases}$ **13.** $\begin{cases} 5x - y = 13 \\ 2x + 3y = 12 \end{cases}$ **14.** $\begin{cases} x + 3y = 5 \\ 2x - 3y = -8 \end{cases}$

15. $\begin{cases} 3x = 24 \\ x + 2y = 0 \end{cases}$ **16.** $\begin{cases} 4x + 5y = -3 \\ -2y = -4 \end{cases}$ **17.** $\begin{cases} 3x - 6y = 24 \\ 5x + 4y = 12 \end{cases}$ **18.** $\begin{cases} 2x + 4y = 16 \\ 3x - 5y = -9 \end{cases}$

19. $\begin{cases} 3x - 2y = 4 \\ 6x - 4y = 0 \end{cases}$ **20.** $\begin{cases} -x + 2y = 5 \\ 4x - 8y = 6 \end{cases}$ **21.** $\begin{cases} 2x - 4y = -2 \\ 3x + 2y = 3 \end{cases}$ **22.** $\begin{cases} 3x + 3y = 3 \\ 4x + 2y = \frac{8}{3} \end{cases}$

23. $\begin{cases} 2x - 3y = -1 \\ 10x + 10y = 5 \end{cases}$ **24.** $\begin{cases} 3x - 2y = 0 \\ 5x + 10y = 4 \end{cases}$ **25.** $\begin{cases} 2x + 3y = 6 \\ x - y = \frac{1}{2} \end{cases}$ **26.** $\begin{cases} \frac{1}{2}x + y = -2 \\ x - 2y = 8 \end{cases}$

27. $\begin{cases} 3x - 5y = 3 \\ 15x + 5y = 21 \end{cases}$ **28.** $\begin{cases} 2x - y = -1 \\ x + \frac{1}{2}y = \frac{3}{2} \end{cases}$ **29.** $\begin{cases} x + y - z = 6 \\ 3x - 2y + z = -5 \\ x + 3y - 2z = 14 \end{cases}$

30. $\begin{cases} x - y + z = -4 \\ 2x - 3y + 4z = -15 \\ 5x + y - 2z = 12 \end{cases}$ **31.** $\begin{cases} x + 2y - z = -3 \\ 2x - 4y + z = -7 \\ -2x + 2y - 3z = 4 \end{cases}$ **32.** $\begin{cases} x + 4y - 3z = -8 \\ 3x - y + 3z = 12 \\ x + y + 6z = 1 \end{cases}$

33. $\begin{cases} x - 2y + 3z = 1 \\ 3x + y - 2z = 0 \\ 2x - 4y + 6z = 2 \end{cases}$ **34.** $\begin{cases} x - y + 2z = 5 \\ 3x + 2y = 4 \\ -2x + 2y - 4z = -10 \end{cases}$ **35.** $\begin{cases} x + 2y - z = 0 \\ 2x - 4y + z = 0 \\ -2x + 2y - 3z = 0 \end{cases}$

36. $\begin{cases} x + 4y - 3z = 0 \\ 3x - y + 3z = 0 \\ x + y + 6z = 0 \end{cases}$ **37.** $\begin{cases} x - 2y + 3z = 0 \\ 3x + y - 2z = 0 \\ 2x - 4y + 6z = 0 \end{cases}$ **38.** $\begin{cases} x - y + 2z = 0 \\ 3x + 2y = 0 \\ -2x + 2y - 4z = 0 \end{cases}$

39. Solve $\begin{cases} \frac{1}{x} + \frac{1}{y} = 8 \\ \frac{3}{x} - \frac{5}{y} = 0 \end{cases}$ **40.** Solve $\begin{cases} \frac{4}{x} - \frac{3}{y} = 0 \\ \frac{6}{x} + \frac{3}{2y} = 2 \end{cases}$

[**Hint:** Let $u = 1/x$ and $v = 1/y$ and solve for u and v.]

In Problems 41–46, solve for x.

41. $\begin{vmatrix} x & x \\ 4 & 3 \end{vmatrix} = 5$ **42.** $\begin{vmatrix} x & 1 \\ 3 & x \end{vmatrix} = -2$ **43.** $\begin{vmatrix} x & 1 & 1 \\ 4 & 3 & 2 \\ -1 & 2 & 5 \end{vmatrix} = 2$

44. $\begin{vmatrix} 3 & 2 & 4 \\ 1 & x & 5 \\ 0 & 1 & -2 \end{vmatrix} = 0$ **45.** $\begin{vmatrix} x & 2 & 3 \\ 1 & x & 0 \\ 6 & 1 & -2 \end{vmatrix} = 7$ **46.** $\begin{vmatrix} x & 1 & 2 \\ 1 & x & 3 \\ 0 & 1 & 2 \end{vmatrix} = -4x$

In Problems 47–54, use properties of determinants to find the value of each determinant if it is known that

$$\begin{vmatrix} x & y & z \\ u & v & w \\ 1 & 2 & 3 \end{vmatrix} = 4$$

47. $\begin{vmatrix} 1 & 2 & 3 \\ u & v & w \\ x & y & z \end{vmatrix}$

48. $\begin{vmatrix} x & y & z \\ u & v & w \\ 2 & 4 & 6 \end{vmatrix}$

49. $\begin{vmatrix} x & y & z \\ -3 & -6 & -9 \\ u & v & w \end{vmatrix}$

50. $\begin{vmatrix} 1 & 2 & 3 \\ x-u & y-v & z-w \\ u & v & w \end{vmatrix}$

51. $\begin{vmatrix} 1 & 2 & 3 \\ x-3 & y-6 & z-9 \\ 2u & 2v & 2w \end{vmatrix}$

52. $\begin{vmatrix} x & y & z-x \\ u & v & w-u \\ 1 & 2 & 2 \end{vmatrix}$

53. $\begin{vmatrix} 1 & 2 & 3 \\ 2x & 2y & 2z \\ u-1 & v-2 & w-3 \end{vmatrix}$

54. $\begin{vmatrix} x+3 & y+6 & z+9 \\ 3u-1 & 3v-2 & 3w-3 \\ 1 & 2 & 3 \end{vmatrix}$

55. Geometry: Equation of a Line An equation of the line containing the two points (x_1, y_1) and (x_2, y_2) may be expressed as the determinant

$$\begin{vmatrix} x & y & 1 \\ x_1 & y_1 & 1 \\ x_2 & y_2 & 1 \end{vmatrix} = 0$$

Prove this result by expanding the determinant and comparing the result to the two-point form of the equation of a line.

56. Geometry: Collinear Points Using the result obtained in Problem 55, show that three distinct points (x_1, y_1), (x_2, y_2), and (x_3, y_3) are collinear (lie on the same line) if and only if

$$\begin{vmatrix} x_1 & y_1 & 1 \\ x_2 & y_2 & 1 \\ x_3 & y_3 & 1 \end{vmatrix} = 0$$

57. Show that $\begin{vmatrix} x^2 & x & 1 \\ y^2 & y & 1 \\ z^2 & z & 1 \end{vmatrix} = (y-z)(x-y)(x-z)$.

58. Complete the proof of Cramer's Rule for two equations containing two variables.

[**Hint:** In system (5), page 824, if $a = 0$, then $b \neq 0$ and $c \neq 0$, since $D = -bc \neq 0$. Now show that equation (6) provide a solution of the system when $a = 0$. There are then three remaining cases: $b = 0$, $c = 0$, and $d = 0$.]

59. Interchange columns 1 and 3 of a 3 by 3 determinant. Show that the value of the new determinant is -1 times the value of the original determinant.

60. Multiply each entry in row 2 of a 3 by 3 determinant by the number k, $k \neq 0$. Show that the value of the new determinant is k times the value of the original determinant.

61. Prove that a 3 by 3 determinant in which the entries in column 1 equal those in column 3 has the value 0.

62. Prove that, if row 2 of a 3 by 3 determinant is multiplied by k, $k \neq 0$, and the result is added to the entries in row 1, then there is no change in the value of the determinant.

11.4 MATRIX ALGEBRA

1 Work with Equality and Addition of Matrices
2 Know Properties of Matrices
3 Know How to Multiply Matrices
4 Find the Inverse of a Matrix
5 Solve Systems of Equations Using Inverse Matrices

In Section 11.2, we defined a matrix as an array of real numbers and used an augmented matrix to represent a system of linear equations. There is, however, a branch of mathematics, called **linear algebra,** that deals with matrices in such a way that an algebra of matrices is permitted. In this section, we provide a survey of how this **matrix algebra** is developed.

Before getting started, we restate the definition of a matrix.

A **matrix** is defined as a rectangular array of numbers:

$$
\begin{array}{l} \\ \text{Row } 1 \\ \text{Row } 2 \\ \vdots \\ \text{Row } i \\ \vdots \\ \text{Row } m \end{array}
\begin{array}{c} \begin{array}{cccccc} \text{Column } 1 & \text{Column } 2 & & \text{Column } j & & \text{Column } n \end{array} \\
\begin{bmatrix} a_{11} & a_{12} & \cdots & a_{1j} & \cdots & a_{1n} \\ a_{21} & a_{22} & \cdots & a_{2j} & \cdots & a_{2n} \\ \vdots & \vdots & & \vdots & & \vdots \\ a_{i1} & a_{i2} & \cdots & a_{ij} & \cdots & a_{in} \\ \vdots & \vdots & & \vdots & & \vdots \\ a_{m1} & a_{m2} & \cdots & a_{mj} & \cdots & a_{mn} \end{bmatrix} \end{array}
$$

Each number a_{ij} of the matrix has two indices: the **row index** i and the **column index** j. The matrix shown above has m rows and n columns. The $m \cdot n$ numbers a_{ij} are usually referred to as the **entries** of the matrix.

Let's begin with an example that illustrates how matrices can be used to conveniently represent an array of information.

EXAMPLE 1

Arranging Data in a Matrix

In a survey of 900 people, the following information was obtained:

200 males	Thought federal defense spending was too high
150 males	Thought federal defense spending was too low
45 males	Had no opinion
315 females	Thought federal defense spending was too high
125 females	Thought federal defense spending was too low
65 females	Had no opinion

We can arrange the above data in a rectangular array as follows:

	Too High	Too Low	No Opinion
Male	200	150	45
Female	315	125	65

or as

$$\begin{bmatrix} 200 & 150 & 45 \\ 315 & 125 & 65 \end{bmatrix}$$

This matrix has two rows (representing males and females) and three columns (representing "too high," "too low," and "no opinion").

The matrix we developed in Example 1 has 2 rows and 3 columns. In general, a matrix with m rows and n columns is called an **m by n matrix.** Thus, the matrix we developed in Example 1 is a 2 by 3 matrix. Notice that an m by n matrix will contain $m \cdot n$ entries.

If an m by n matrix has the same number of rows as columns, that is, if $m = n$, then the matrix is referred to as a **square matrix.**

EXAMPLE 2

Examples of Matrices

(a) $\begin{bmatrix} 5 & 0 \\ -6 & 1 \end{bmatrix}$ A 2 by 2 square matrix (b) $[1 \quad 0 \quad 3]$ A 1 by 3 matrix

(c) $\begin{bmatrix} 6 & -2 & 4 \\ 4 & 3 & 5 \\ 8 & 0 & 1 \end{bmatrix}$ A 3 by 3 square matrix

Equality and Addition of Matrices

1 We begin our discussion of matrix algebra by first defining what is meant by two matrices being equal and then defining the operations of addition and subtraction. It is important to note that these definitions require each matrix to have the same number of rows *and* the same number of columns as a prerequisite for equality and for addition and subtraction.

We usually represent matrices by capital letters, such as *A*, *B*, *C*, and so on.

Two *m* by *n* matrices *A* and *B* are said to be **equal,** written as

$$A = B$$

provided each entry a_{ij} in *A* is equal to the corresponding entry b_{ij} in *B*.

For example,

$$\begin{bmatrix} 2 & 1 \\ 0.5 & -1 \end{bmatrix} = \begin{bmatrix} \sqrt{4} & 1 \\ \frac{1}{2} & -1 \end{bmatrix} \quad \text{and} \quad \begin{bmatrix} 3 & 2 & 1 \\ 0 & 1 & -2 \end{bmatrix} = \begin{bmatrix} \sqrt{9} & \sqrt{4} & 1 \\ 0 & 1 & \sqrt[3]{-8} \end{bmatrix}$$

$$\begin{bmatrix} 4 & 1 \\ 6 & 1 \end{bmatrix} \neq \begin{bmatrix} 4 & 0 \\ 6 & 1 \end{bmatrix}$$ Because the entries in row 1, column 2 are not equal.

$$\begin{bmatrix} 4 & 1 & 2 \\ 6 & 1 & 2 \end{bmatrix} \neq \begin{bmatrix} 4 & 1 & 2 & 3 \\ 6 & 1 & 2 & 4 \end{bmatrix}$$ Because the matrix on the left is 2 by 3 and the matrix on the right is 2 by 4.

If each of *A* and *B* is an *m* by *n* matrix (and each therefore contains $m \cdot n$ entries), the statement $A = B$ actually represents a system of $m \cdot n$ ordinary equations. We will make use of this fact a little later.

Suppose *A* and *B* represent two *m* by *n* matrices. We define their **sum** $\boldsymbol{A + B}$ to be the *m* by *n* matrix formed by adding the corresponding entries a_{ij} of *A* and b_{ij} of *B*. The **difference** $\boldsymbol{A - B}$ is defined as the *m* by *n* matrix formed by subtracting the entries b_{ij} in *B* from the corresponding entries a_{ij} in *A*. Addition and subtraction of matrices are allowed only for matrices having the same number *m* of rows and the same number *n* of columns. Thus, for example, a 2 by 3 matrix and a 2 by 4 matrix cannot be added or subtracted.

The advent of graphing utilities has made the sometimes tedious process of matrix algebra easy. Let's compare how a graphing utility adds and subtracts matrices with doing it by hand.

EXAMPLE 3 Adding and Subtracting Matrices

Suppose that

$$A = \begin{bmatrix} 2 & 4 & 8 & -3 \\ 0 & 1 & 2 & 3 \end{bmatrix} \quad \text{and} \quad B = \begin{bmatrix} -3 & 4 & 0 & 1 \\ 6 & 8 & 2 & 0 \end{bmatrix}.$$

Find (a) $A + B$ (b) $A - B$

Graphing Solution Enter the matrices into a graphing utility. Name them $[A]$ and $[B]$. Figure 11 shows the results of adding and subtracting $[A]$ and $[B]$.

FIGURE 11

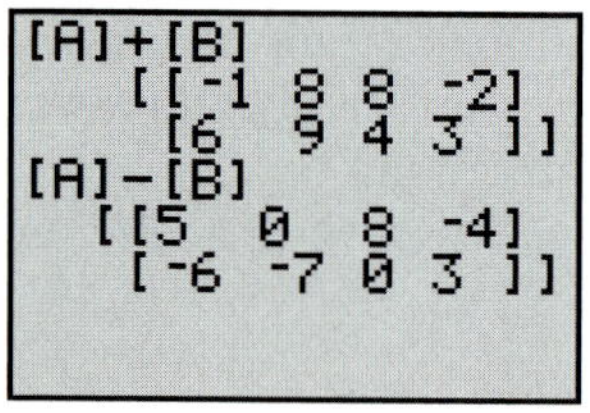

Algebraic Solution

(a) $$A + B = \begin{bmatrix} 2 & 4 & 8 & -3 \\ 0 & 1 & 2 & 3 \end{bmatrix} + \begin{bmatrix} -3 & 4 & 0 & 1 \\ 6 & 8 & 2 & 0 \end{bmatrix}$$

$$= \begin{bmatrix} 2 + (-3) & 4 + 4 & 8 + 0 & -3 + 1 \\ 0 + 6 & 1 + 8 & 2 + 2 & 3 + 0 \end{bmatrix}$$ Add corresponding entries.

$$= \begin{bmatrix} -1 & 8 & 8 & -2 \\ 6 & 9 & 4 & 3 \end{bmatrix}$$

(b) $$A - B = \begin{bmatrix} 2 & 4 & 8 & -3 \\ 0 & 1 & 2 & 3 \end{bmatrix} - \begin{bmatrix} -3 & 4 & 0 & 1 \\ 6 & 8 & 2 & 0 \end{bmatrix}$$

$$= \begin{bmatrix} 2 - (-3) & 4 - 4 & 8 - 0 & -3 - 1 \\ 0 - 6 & 1 - 8 & 2 - 2 & 3 - 0 \end{bmatrix}$$ Subtract corresponding entries.

$$= \begin{bmatrix} 5 & 0 & 8 & -4 \\ -6 & -7 & 0 & 3 \end{bmatrix}$$

Now work Problem 1.

2 Many of the algebraic properties of sums of real numbers are also true for sums of matrices. Suppose that A, B, and C are m by n matrices. Then matrix addition is **commutative.** That is,

Commutative Property

$$A + B = B + A$$

Matrix addition is also **associative.** That is,

Associative Property

$$(A + B) + C = A + (B + C)$$

Although we shall not prove these results, the proofs, as the following example illustrates, are based on the commutative and associative properties for real numbers.

EXAMPLE 4 Demonstrating the Commutative Property

$$\begin{bmatrix} 2 & 3 & -1 \\ 4 & 0 & 7 \end{bmatrix} + \begin{bmatrix} -1 & 2 & 1 \\ 5 & -3 & 4 \end{bmatrix} = \begin{bmatrix} 2 + (-1) & 3 + 2 & -1 + 1 \\ 4 + 5 & 0 + (-3) & 7 + 4 \end{bmatrix}$$

$$= \begin{bmatrix} -1+2 & 2+3 & 1+(-1) \\ 5+4 & -3+0 & 4+7 \end{bmatrix}$$

$$= \begin{bmatrix} -1 & 2 & 1 \\ 5 & -3 & 4 \end{bmatrix} + \begin{bmatrix} 2 & 3 & -1 \\ 4 & 0 & 7 \end{bmatrix}$$ ▪

A matrix whose entries are all equal to 0 is called a **zero matrix.** Each of the following matrices is a zero matrix:

$\begin{bmatrix} 0 & 0 \\ 0 & 0 \end{bmatrix}$ 2 by 2 square zero matrix. $\begin{bmatrix} 0 & 0 & 0 \\ 0 & 0 & 0 \end{bmatrix}$ 2 by 3 zero matrix. $[0 \quad 0 \quad 0]$ 1 by 3 zero matrix.

Zero matrices have properties similar to the real number 0. Thus, if A is an m by n matrix and 0 is an m by n zero matrix, then

$$A + 0 = A$$

In other words, the zero matrix is the additive identity in matrix algebra.

We also can multiply a matrix by a real number. If k is a real number and A is an m by n matrix, the matrix kA is the m by n matrix formed by multiplying each entry a_{ij} in A by k. The number k is sometimes referred to as a **scalar,** and the matrix kA is called a **scalar multiple** of A.

EXAMPLE 5 Operations Using Matrices

Suppose that

$$A = \begin{bmatrix} 3 & 1 & 5 \\ -2 & 0 & 6 \end{bmatrix} \quad B = \begin{bmatrix} 4 & 1 & 0 \\ 8 & 1 & -3 \end{bmatrix} \quad C = \begin{bmatrix} 9 & 0 \\ -3 & 6 \end{bmatrix}$$

Find (a) $4A$ (b) $\frac{1}{3}C$ (c) $3A - 2B$

Graphing Solution Enter the matrices [A], [B], and [C] into a graphing utility. Figure 12 shows the required computations.

FIGURE 12

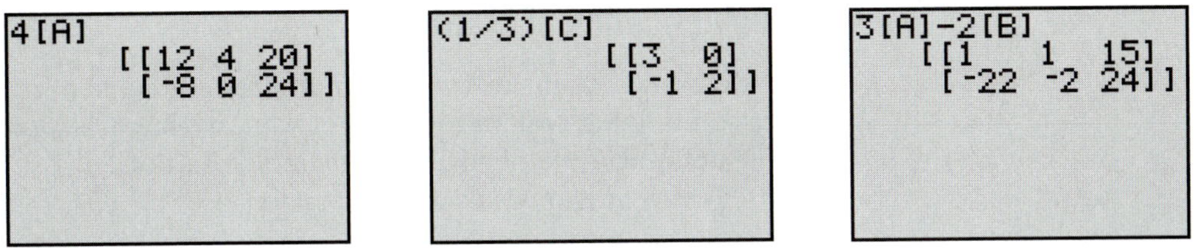

(a) (b) (c)

Algebraic Solution (a) $4A = 4\begin{bmatrix} 3 & 1 & 5 \\ -2 & 0 & 6 \end{bmatrix} = \begin{bmatrix} 4\cdot 3 & 4\cdot 1 & 4\cdot 5 \\ 4(-2) & 4\cdot 0 & 4\cdot 6 \end{bmatrix} = \begin{bmatrix} 12 & 4 & 20 \\ -8 & 0 & 24 \end{bmatrix}$

(b) $\frac{1}{3}C = \frac{1}{3}\begin{bmatrix} 9 & 0 \\ -3 & 6 \end{bmatrix} = \begin{bmatrix} \frac{1}{3}\cdot 9 & \frac{1}{3}\cdot 0 \\ \frac{1}{3}(-3) & \frac{1}{3}\cdot 6 \end{bmatrix} = \begin{bmatrix} 3 & 0 \\ -1 & 2 \end{bmatrix}$

$$(c)\ 3A - 2B = 3\begin{bmatrix} 3 & 1 & 5 \\ -2 & 0 & 6 \end{bmatrix} - 2\begin{bmatrix} 4 & 1 & 0 \\ 8 & 1 & -3 \end{bmatrix}$$

$$= \begin{bmatrix} 3 \cdot 3 & 3 \cdot 1 & 3 \cdot 5 \\ 3(-2) & 3 \cdot 0 & 3 \cdot 6 \end{bmatrix} - \begin{bmatrix} 2 \cdot 4 & 2 \cdot 1 & 2 \cdot 0 \\ 2 \cdot 8 & 2 \cdot 1 & 2(-3) \end{bmatrix}$$

$$= \begin{bmatrix} 9 & 3 & 15 \\ -6 & 0 & 18 \end{bmatrix} - \begin{bmatrix} 8 & 2 & 0 \\ 16 & 2 & -6 \end{bmatrix}$$

$$= \begin{bmatrix} 9-8 & 3-2 & 15-0 \\ -6-16 & 0-2 & 18-(-6) \end{bmatrix}$$

$$= \begin{bmatrix} 1 & 1 & 15 \\ -22 & -2 & 24 \end{bmatrix}$$

Now work Problem 5.

We list next some of the algebraic properties of scalar multiplication. Let h and k be real numbers, and let A and B be m by n matrices. Then

Properties of Scalar Multiplication

$$k(hA) = (kh)A$$
$$(k + h)A = kA + hA$$
$$k(A + B) = kA + kB$$

The proofs of these properties are based on properties of real numbers. For example, if A and B are 2 by 2 matrices, then

$$k(A + B) = k\left(\begin{bmatrix} a_{11} & a_{12} \\ a_{21} & a_{22} \end{bmatrix} + \begin{bmatrix} b_{11} & b_{12} \\ b_{21} & b_{22} \end{bmatrix}\right) = k\begin{bmatrix} a_{11} + b_{11} & a_{12} + b_{12} \\ a_{21} + b_{21} & a_{22} + b_{22} \end{bmatrix}$$

$$= \begin{bmatrix} k(a_{11} + b_{11}) & k(a_{12} + b_{12}) \\ k(a_{21} + b_{21}) & k(a_{22} + b_{22}) \end{bmatrix} = \begin{bmatrix} ka_{11} + kb_{11} & ka_{12} + kb_{12} \\ ka_{21} + kb_{21} & ka_{22} + kb_{22} \end{bmatrix}$$

$$= \begin{bmatrix} ka_{11} & ka_{12} \\ ka_{21} & ka_{22} \end{bmatrix} + \begin{bmatrix} kb_{11} & kb_{12} \\ kb_{21} & kb_{22} \end{bmatrix} = k\begin{bmatrix} a_{11} & a_{12} \\ a_{21} & a_{22} \end{bmatrix} + k\begin{bmatrix} b_{11} & b_{12} \\ b_{21} & b_{22} \end{bmatrix} = kA + kB$$

Multiplication of Matrices

3 Unlike the straightforward definition for adding two matrices, the definition for multiplying two matrices is not what we might expect. In preparation for this definition, we need the following definitions:

A **row vector** R is a 1 by n matrix

$$R = [r_1 \quad r_2 \quad \cdots \quad r_n]$$

A **column vector** C is an n by 1 matrix

$$C = \begin{bmatrix} c_1 \\ c_2 \\ \vdots \\ c_n \end{bmatrix}$$

The **product** RC of R times C is defined as the number

$$RC = [r_1 \; r_2 \; \cdots \; r_n]\begin{bmatrix} c_1 \\ c_2 \\ \vdots \\ c_n \end{bmatrix} = r_1c_1 + r_2c_2 + \cdots + r_nc_n$$

Notice that a row vector and a column vector can be multiplied only if they contain the same number of entries.

EXAMPLE 6 The Product of a Row Vector by a Column Vector

If $R = [3 \quad -5 \quad 2]$ and $C = \begin{bmatrix} 3 \\ 4 \\ -5 \end{bmatrix}$, then

$$RC = [3 \quad -5 \quad 2]\begin{bmatrix} 3 \\ 4 \\ -5 \end{bmatrix} = 3 \cdot 3 + (-5)4 + 2(-5)$$
$$= 9 - 20 - 10 = -21$$

Let's look at an application of the product of a row vector by a column vector.

EXAMPLE 7 Using Matrices to Compute Revenue

A clothing store sells men's shirts for \$25, silk ties for \$8, and wool suits for \$300. Last month, the store had sales consisting of 100 shirts, 200 ties, and 50 suits. What was the total revenue due to these sales?

Solution We set up a row vector R to represent the prices of each item and a column vector C to represent the corresponding number of items sold.

Then

$$\underset{\text{Shirts Ties Suits}}{\overset{\text{Prices}}{R = [25 \quad 8 \quad 300]}} \qquad \overset{\text{Number sold}}{C = \begin{bmatrix} 100 \\ 200 \\ 50 \end{bmatrix}} \begin{matrix} \text{Shirts} \\ \text{Ties} \\ \text{Suits} \end{matrix}$$

The total revenue obtained is the product RC. That is,

$$RC = [25 \quad 8 \quad 300]\begin{bmatrix} 100 \\ 200 \\ 50 \end{bmatrix}$$
$$= \underbrace{25 \cdot 100}_{\text{Shirt revenue}} + \underbrace{8 \cdot 200}_{\text{Tie revenue}} + \underbrace{300 \cdot 50}_{\text{Suit revenue}} = \underbrace{\$19{,}100}_{\text{Total revenue}}$$

The definition for multiplying two matrices is based on the definition of a row vector times a column vector.

> Let A denote an m by r matrix and let B denote an r by n matrix. The **product** AB is defined as the m by n matrix whose entry in row i, column j is the product of the ith row of A and the jth column of B.

The definition of the product AB of two matrices A and B, in this order, requires that the number of columns of A equal the number of rows of B; otherwise, no product is defined:

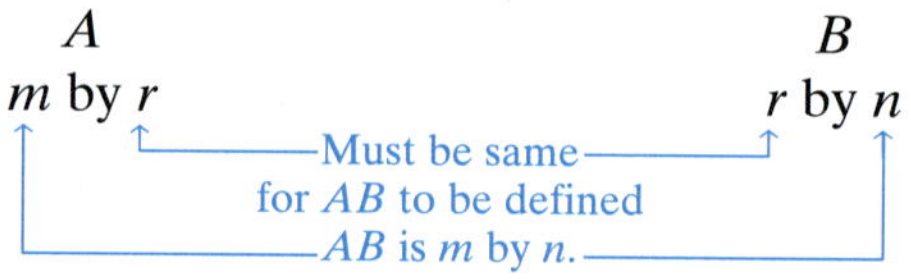

An example will help clarify the definition.

EXAMPLE 8 Multiplying Two Matrices

Find the product AB if

$$A = \begin{bmatrix} 2 & 4 & -1 \\ 5 & 8 & 0 \end{bmatrix} \quad \text{and} \quad B = \begin{bmatrix} 2 & 5 & 1 & 4 \\ 4 & 8 & 0 & 6 \\ -3 & 1 & -2 & -1 \end{bmatrix}$$

Solution First, we note that A is 2 by 3 and B is 3 by 4, so the product AB is defined and will be a 2 by 4 matrix.

Graphing Solution Enter the matrices A and B into a graphing utility. Figure 13 shows the product AB.

FIGURE 13

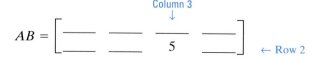

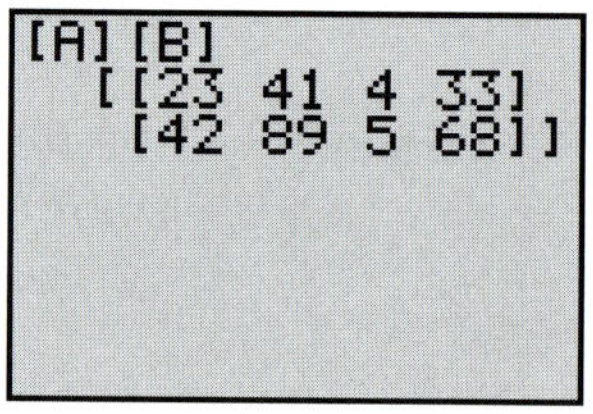

Algebraic Solution Suppose we want the entry in row 2, column 3 of AB. To find it, we find the product of the row vector from row 2 of A and the column vector from column 3 of B:

$$\underset{\text{Row 2 of } A}{[5 \quad 8 \quad 0]} \overset{\text{Column 3 of } B}{\begin{bmatrix} 1 \\ 0 \\ -2 \end{bmatrix}} = 5 \cdot 1 + 8 \cdot 0 + 0(-2) = 5$$

So far, we have

$$AB = \begin{bmatrix} \_\_ & \_\_ & \_\_ & \_\_ \\ \_\_ & \_\_ & 5 & \_\_ \end{bmatrix} \begin{matrix} \\ \leftarrow \text{Row 2} \end{matrix}$$

(Column 3 ↓ points to the third column.)

Now, to find the entry in row 1, column 4 of AB, we find the product of row 1 of A and column 4 of B:

$$\underset{\text{Row 1 of } A}{\begin{bmatrix} 2 & 4 & -1 \end{bmatrix}} \overset{\text{Column 4 of } B}{\begin{bmatrix} 4 \\ 6 \\ -1 \end{bmatrix}} = 2 \cdot 4 + 4 \cdot 6 + (-1)(-1) = 33$$

Continuing in this fashion, we find AB:

$$AB = \begin{bmatrix} 2 & 4 & -1 \\ 5 & 8 & 0 \end{bmatrix} \begin{bmatrix} 2 & 5 & 1 & 4 \\ 4 & 8 & 0 & 6 \\ -3 & 1 & -2 & -1 \end{bmatrix}$$

$$= \begin{bmatrix} \text{Row 1 of } A \text{ times column 1 of } B & \text{Row 1 of } A \text{ times column 2 of } B & \text{Row 1 of } A \text{ times column 3 of } B & \text{Row 1 of } A \text{ times column 4 of } B \\ \text{Row 2 of } A \text{ times column 1 of } B & \text{Row 2 of } A \text{ times column 2 of } B & \text{Row 2 of } A \text{ times column 3 of } B & \text{Row 2 of } A \text{ times column 4 of } B \end{bmatrix}$$

$$= \begin{bmatrix} 2 \cdot 2 + 4 \cdot 4 + (-1)(-3) & 2 \cdot 5 + 4 \cdot 8 + (-1)1 & 2 \cdot 1 + 4 \cdot 0 + (-1)(-2) & 33 \text{ (from earlier)} \\ 5 \cdot 2 + 8 \cdot 4 + 0(-3) & 5 \cdot 5 + 8 \cdot 8 + 0 \cdot 1 & 5 \text{ (from earlier)} & 5 \cdot 4 + 8 \cdot 6 + 0(-1) \end{bmatrix}$$

$$= \begin{bmatrix} 23 & 41 & 4 & 33 \\ 42 & 89 & 5 & 68 \end{bmatrix}$$

Now work Problem 17.

Notice that for the matrices given in Example 8 the product BA is not defined, because B is 3 by 4 and A is 2 by 3. Try calculating BA on a graphing utility. What do you notice?

Another result that can occur when multiplying two matrices is illustrated in the next example.*

EXAMPLE 9 Multiplying Two Matrices

If

$$A = \begin{bmatrix} 2 & 1 & 3 \\ 1 & -1 & 0 \end{bmatrix} \quad \text{and} \quad B = \begin{bmatrix} 1 & 0 \\ 2 & 1 \\ 3 & 2 \end{bmatrix}$$

find (a) AB (b) BA

Solution (a) $$AB = \underset{\text{2 by 3}}{\begin{bmatrix} 2 & 1 & 3 \\ 1 & -1 & 0 \end{bmatrix}} \underset{\text{3 by 2}}{\begin{bmatrix} 1 & 0 \\ 2 & 1 \\ 3 & 2 \end{bmatrix}} = \underset{\text{2 by 2}}{\begin{bmatrix} 13 & 7 \\ -1 & -1 \end{bmatrix}}$$

*For most of the examples that follow, we will multiply matrices by hand. You should verify each result using a graphing utility.

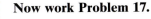

$$\text{(b) } BA = \underset{3 \text{ by } 2}{\begin{bmatrix} 1 & 0 \\ 2 & 1 \\ 3 & 2 \end{bmatrix}} \underset{2 \text{ by } 3}{\begin{bmatrix} 2 & 1 & 3 \\ 1 & -1 & 0 \end{bmatrix}} = \underset{3 \text{ by } 3}{\begin{bmatrix} 2 & 1 & 3 \\ 5 & 1 & 6 \\ 8 & 1 & 9 \end{bmatrix}}$$

Notice in Example 9 that AB is 2 by 2 and BA is 3 by 3. Thus, it is possible for both AB and BA to be defined, yet be unequal. In fact, even if A and B are both n by n matrices, so that AB and BA are each defined and n by n, AB and BA will nearly always be unequal.

EXAMPLE 10 Multiplying Two Square Matrices

If

$$A = \begin{bmatrix} 2 & 1 \\ 0 & 4 \end{bmatrix} \quad \text{and} \quad B = \begin{bmatrix} -3 & 1 \\ 1 & 2 \end{bmatrix}$$

find (a) AB (b) BA

Solution (a)

$$AB = \begin{bmatrix} 2 & 1 \\ 0 & 4 \end{bmatrix}\begin{bmatrix} -3 & 1 \\ 1 & 2 \end{bmatrix}$$

$$= \begin{bmatrix} 2(-3) + 1 \cdot 1 & 2 \cdot 1 + 1 \cdot 2 \\ 0(-3) + 4 \cdot 1 & 0 \cdot 1 + 4 \cdot 2 \end{bmatrix} = \begin{bmatrix} -5 & 4 \\ 4 & 8 \end{bmatrix}$$

(b)

$$BA = \begin{bmatrix} -3 & 1 \\ 1 & 2 \end{bmatrix}\begin{bmatrix} 2 & 1 \\ 0 & 4 \end{bmatrix}$$

$$= \begin{bmatrix} (-3)2 + 1 \cdot 0 & (-3)1 + 1 \cdot 4 \\ 1 \cdot 2 + 2 \cdot 0 & 1 \cdot 1 + 2 \cdot 4 \end{bmatrix} = \begin{bmatrix} -6 & 1 \\ 2 & 9 \end{bmatrix}$$

The preceding examples demonstrate that an important property of real numbers, the commutative property of multiplication, is not shared by matrices. Thus, in general:

Theorem

Matrix multiplication is not commutative.

Now work Problems 7 and 9.

Next we give two of the properties of real numbers that are shared by matrices. Assuming each product and sum is defined, we have:

Associative Property

$$A(BC) = (AB)C$$

Distributive Property

$$A(B + C) = AB + AC$$

The Identity Matrix

For an n by n square matrix, the entries located in row i, column i, $1 \le i \le n$, are called the **diagonal entries.** An n by n square matrix whose diagonal entries are 1's, while all other entries are 0's, is called the **identity matrix I_n.** For example,

$$I_2 = \begin{bmatrix} 1 & 0 \\ 0 & 1 \end{bmatrix} \quad I_3 = \begin{bmatrix} 1 & 0 & 0 \\ 0 & 1 & 0 \\ 0 & 0 & 1 \end{bmatrix}$$

and so on.

EXAMPLE 11

Multiplication with an Identity Matrix

Let

$$A = \begin{bmatrix} -1 & 2 & 0 \\ 0 & 1 & 3 \end{bmatrix} \quad \text{and} \quad B = \begin{bmatrix} 3 & 2 \\ 4 & 6 \\ 5 & 2 \end{bmatrix}$$

Find (a) AI_3 (b) I_2A (c) BI_2

Solution (a) $AI_3 = \begin{bmatrix} -1 & 2 & 0 \\ 0 & 1 & 3 \end{bmatrix}\begin{bmatrix} 1 & 0 & 0 \\ 0 & 1 & 0 \\ 0 & 0 & 1 \end{bmatrix} = \begin{bmatrix} -1 & 2 & 0 \\ 0 & 1 & 3 \end{bmatrix} = A$

(b) $I_2A = \begin{bmatrix} 1 & 0 \\ 0 & 1 \end{bmatrix}\begin{bmatrix} -1 & 2 & 0 \\ 0 & 1 & 3 \end{bmatrix} = \begin{bmatrix} -1 & 2 & 0 \\ 0 & 1 & 3 \end{bmatrix} = A$

(c) $BI_2 = \begin{bmatrix} 3 & 2 \\ 4 & 6 \\ 5 & 2 \end{bmatrix}\begin{bmatrix} 1 & 0 \\ 0 & 1 \end{bmatrix} = \begin{bmatrix} 3 & 2 \\ 4 & 6 \\ 5 & 2 \end{bmatrix} = B$ ▪

Example 11 demonstrates the following property:

Identity Property

If A is an m by n matrix, then

$$I_m A = A \quad \text{and} \quad AI_n = A$$

If A is an n by n square matrix, then $AI_n = I_n A = A$.

Thus, an identity matrix has properties analogous to those of the real number 1. In other words, the identity matrix is a multiplicative identity in matrix algebra.

4 The Inverse of a Matrix

Let A be a square n by n matrix. If there exists an n by n matrix A^{-1}, read "A inverse," for which

$$AA^{-1} = A^{-1}A = I_n$$

then A^{-1} is called the **inverse** of the matrix A.

As we shall soon see, not every square matrix has an inverse. When a matrix A does have an inverse A^{-1}, then A is said to be **nonsingular.** If a matrix A has no inverse, it is called **singular.**

EXAMPLE 12 Multiplying a Matrix by Its Inverse

Show that the inverse of

$$A = \begin{bmatrix} 3 & 1 \\ 2 & 1 \end{bmatrix} \quad \text{is} \quad A^{-1} = \begin{bmatrix} 1 & -1 \\ -2 & 3 \end{bmatrix}$$

Solution We need to show that $AA^{-1} = A^{-1}A = I_2$.

$$AA^{-1} = \begin{bmatrix} 3 & 1 \\ 2 & 1 \end{bmatrix}\begin{bmatrix} 1 & -1 \\ -2 & 3 \end{bmatrix} = \begin{bmatrix} 3 \cdot 1 + 1(-2) & 3(-1) + 1 \cdot 3 \\ 2 \cdot 1 + 1(-2) & 2(-1) + 1 \cdot 3 \end{bmatrix}$$

$$= \begin{bmatrix} 1 & 0 \\ 0 & 1 \end{bmatrix} = I_2$$

$$A^{-1}A = \begin{bmatrix} 1 & -1 \\ -2 & 3 \end{bmatrix}\begin{bmatrix} 3 & 1 \\ 2 & 1 \end{bmatrix} = \begin{bmatrix} 3-2 & 1-1 \\ -6+6 & -2+3 \end{bmatrix} = \begin{bmatrix} 1 & 0 \\ 0 & 1 \end{bmatrix} = I_2$$

We now show one way to find the inverse of

$$A = \begin{bmatrix} 3 & 1 \\ 2 & 1 \end{bmatrix}$$

Suppose that A^{-1} is given by

$$A^{-1} = \begin{bmatrix} x & y \\ z & w \end{bmatrix} \tag{1}$$

where x, y, z, and w are four variables. Based on the definition of an inverse, if, indeed, A has an inverse, we have

$$AA^{-1} = I_2$$
$$\begin{bmatrix} 3 & 1 \\ 2 & 1 \end{bmatrix}\begin{bmatrix} x & y \\ z & w \end{bmatrix} = \begin{bmatrix} 1 & 0 \\ 0 & 1 \end{bmatrix}$$
$$\begin{bmatrix} 3x + z & 3y + w \\ 2x + z & 2y + w \end{bmatrix} = \begin{bmatrix} 1 & 0 \\ 0 & 1 \end{bmatrix}$$

Because corresponding entries must be equal, it follows that this matrix equation is equivalent to four ordinary equations:

$$\begin{cases} 3x + z = 1 \\ 2x + z = 0 \end{cases} \qquad \begin{cases} 3y + w = 0 \\ 2y + w = 1 \end{cases}$$

The augmented matrix of each system is

$$\left[\begin{array}{cc|c} 3 & 1 & 1 \\ 2 & 1 & 0 \end{array}\right] \qquad \left[\begin{array}{cc|c} 3 & 1 & 0 \\ 2 & 1 & 1 \end{array}\right] \tag{2}$$

The usual procedure would be to transform each augmented matrix into reduced echelon form. Notice, though, that the left sides of the augmented matrices are equal, so the same row operations (see Section 11.2) can be used to reduce each one. Thus, we find it more efficient to combine the two augmented matrices (2) into a single matrix, as shown next, and then transform it into reduced echelon form:

$$\left[\begin{array}{cc|cc} 3 & 1 & 1 & 0 \\ 2 & 1 & 0 & 1 \end{array}\right]$$

Now we attempt to transform the left side into an identity matrix:

$$\left[\begin{array}{cc|cc} 3 & 1 & 1 & 0 \\ 2 & 1 & 0 & 1 \end{array}\right] \underset{\substack{\uparrow \\ R_1 = -1r_2 + r_1}}{\rightarrow} \left[\begin{array}{cc|cc} 1 & 0 & 1 & -1 \\ 2 & 1 & 0 & 1 \end{array}\right]$$

$$\underset{\substack{\uparrow \\ R_2 = -2r_1 + r_2}}{\rightarrow} \left[\begin{array}{cc|cc} 1 & 0 & 1 & -1 \\ 0 & 1 & -2 & 3 \end{array}\right] \tag{3}$$

Matrix (3) is in reduced echelon form. Now we reverse the earlier step of combining the two augmented matrices in (2) and write the single matrix (3) as two augmented matrices:

$$\left[\begin{array}{cc|c} 1 & 0 & 1 \\ 0 & 1 & -2 \end{array}\right] \text{ and } \left[\begin{array}{cc|c} 1 & 0 & -1 \\ 0 & 1 & 3 \end{array}\right]$$

We conclude from these matrices that $x = 1$, $z = -2$, and $y = -1$, $w = 3$. Substituting these values into matrix (1), we find that

$$A^{-1} = \begin{bmatrix} 1 & -1 \\ -2 & 3 \end{bmatrix}$$

Notice in display (3) that the 2 by 2 matrix to the right of the vertical bar is, in fact, the inverse of A. Also notice that the identity matrix I_2 is the matrix that appears to the left of the vertical bar. These observations and the procedures followed above will work in general.

Procedure for Finding the Inverse of a Nonsingular Matrix

To find the inverse of an n by n nonsingular matrix A, proceed as follows:

STEP 1: Form the matrix $[A|I_n]$.

STEP 2: Transform the matrix $[A|I_n]$ into reduced echelon form.

STEP 3: The reduced echelon form of $[A|I_n]$ will contain the identity matrix I_n on the left of the vertical bar; the n by n matrix on the right of the vertical bar is the inverse of A.

In other words, if A is nonsingular, we begin with the matrix $[A|I_n]$ and, after transforming it into reduced echelon form, we end up with the matrix $[I_n|A^{-1}]$.

Let's look at another example.

EXAMPLE 13 Finding the Inverse of a Matrix

The matrix

$$A = \begin{bmatrix} 1 & 1 & 0 \\ -1 & 3 & 4 \\ 0 & 4 & 3 \end{bmatrix}$$

is nonsingular. Find its inverse.

Graphing Solution Enter the matrix A into a graphing utility. Figure 14 shows A^{-1}.

FIGURE 14

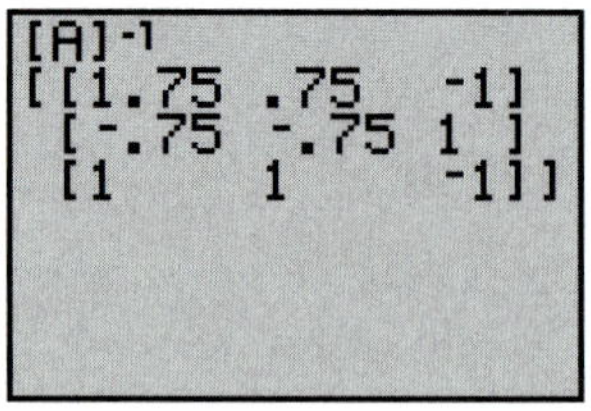

Algebraic Solution First, we form the matrix

$$[A|I_3] = \left[\begin{array}{rrr|rrr} 1 & 1 & 0 & 1 & 0 & 0 \\ -1 & 3 & 4 & 0 & 1 & 0 \\ 0 & 4 & 3 & 0 & 0 & 1 \end{array}\right]$$

Next, we use row operations to transform $[A|I_3]$ into reduced echelon form:

$$\left[\begin{array}{rrr|rrr} 1&1&0&1&0&0\\ -1&3&4&0&1&0\\ 0&4&3&0&0&1\end{array}\right] \to \left[\begin{array}{rrr|rrr} 1&1&0&1&0&0\\ 0&4&4&1&1&0\\ 0&4&3&0&0&1\end{array}\right] \to \left[\begin{array}{rrr|rrr} 1&1&0&1&0&0\\ 0&1&1&\frac{1}{4}&\frac{1}{4}&0\\ 0&4&3&0&0&1\end{array}\right]$$

$\uparrow$ $R_2 = r_1 + r_2$ $\qquad$ $\uparrow$ $R_2 = \frac{1}{4}r_2$

$$\to \left[\begin{array}{rrr|rrr} 1&0&-1&\frac{3}{4}&-\frac{1}{4}&0\\ 0&1&1&\frac{1}{4}&\frac{1}{4}&0\\ 0&0&-1&-1&-1&1\end{array}\right] \to \left[\begin{array}{rrr|rrr} 1&0&-1&\frac{3}{4}&-\frac{1}{4}&0\\ 0&1&1&\frac{1}{4}&\frac{1}{4}&0\\ 0&0&1&1&1&-1\end{array}\right]$$

$\uparrow$ $R_1 = -1r_2 + r_1$, $R_3 = -4r_2 + r_3$ $\qquad$ $\uparrow$ $R_3 = -1r_3$

$$\to \left[\begin{array}{rrr|rrr} 1&0&0&\frac{7}{4}&\frac{3}{4}&-1\\ 0&1&0&-\frac{3}{4}&-\frac{3}{4}&1\\ 0&0&1&1&1&-1\end{array}\right]$$

$\uparrow$ $R_1 = r_3 + r_1$, $R_2 = -1r_3 + r_2$

The matrix $[A|I_3]$ is now in reduced echelon form, and the identity matrix I_3 is on the left of the vertical bar. Hence, the inverse of A is

$$A^{-1} = \begin{bmatrix} \frac{7}{4} & \frac{3}{4} & -1 \\ -\frac{3}{4} & -\frac{3}{4} & 1 \\ 1 & 1 & -1 \end{bmatrix}$$

You can (and should) verify that this is the correct inverse by showing that $AA^{-1} = A^{-1}A = I_3$.

Now work Problem 23.

If transforming the matrix $[A|I_n]$ into reduced echelon form does not result in the identity matrix I_n to the left of the vertical bar, then A has no inverse. The next example demonstrates such a matrix.

EXAMPLE 14 Showing a Matrix Has No Inverse

Show that the following matrix has no inverse.

$$A = \begin{bmatrix} 4 & 6 \\ 2 & 3 \end{bmatrix}$$

Graphing Solution Enter the matrix A. Figure 15 shows the result when we try to find its inverse. The ERRor comes about because A is singular.

FIGURE 15

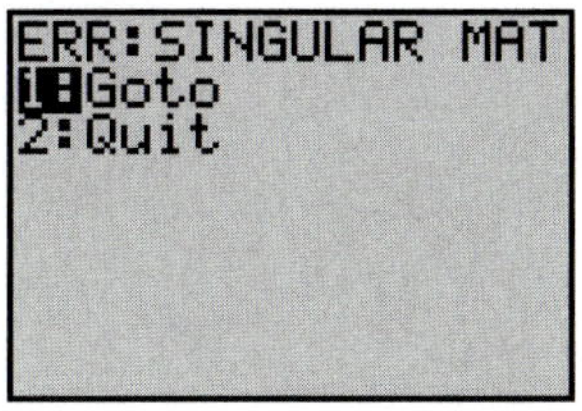

Algebraic Solution Proceeding as in Example 13, we form the matrix

$$[A|I_2] = \left[\begin{array}{cc|cc} 4 & 6 & 1 & 0 \\ 2 & 3 & 0 & 1 \end{array}\right]$$

Then we use row operations to transform $[A \mid I_2]$ into reduced echelon form:

$$[A|I_2] = \left[\begin{array}{cc|cc} 4 & 6 & 1 & 0 \\ 2 & 3 & 0 & 1 \end{array}\right] \underset{R_1 = \frac{1}{4}r_1}{\rightarrow} \left[\begin{array}{cc|cc} 1 & \frac{3}{2} & \frac{1}{4} & 0 \\ 2 & 3 & 0 & 1 \end{array}\right] \underset{R_2 = -2r_1 + r_2}{\rightarrow} \left[\begin{array}{cc|cc} 1 & \frac{3}{2} & \frac{1}{4} & 0 \\ 0 & 0 & -\frac{1}{2} & 1 \end{array}\right]$$

The matrix $[A|I_2]$ is sufficiently reduced for us to see the identity matrix cannot appear to the left of the vertical bar. We conclude that A has no inverse. ■

Now work Problem 51.

Solving Systems of Linear Equations

5 Inverse matrices can be used to solve systems of equations in which the number of equations is the same as the number of variables.

EXAMPLE 15 Using the Inverse Matrix to Solve a System of Linear Equations

Solve the system of equations: $\begin{cases} x + y = 3 \\ -x + 3y + 4z = -3 \\ 4y + 3z = 2 \end{cases}$

Solution If we let

$$A = \begin{bmatrix} 1 & 1 & 0 \\ -1 & 3 & 4 \\ 0 & 4 & 3 \end{bmatrix}, \quad X = \begin{bmatrix} x \\ y \\ z \end{bmatrix}, \quad \text{and} \quad B = \begin{bmatrix} 3 \\ -3 \\ 2 \end{bmatrix}$$

then the original system of equations can be written compactly as the matrix equation

$$AX = B \tag{4}$$

We know from Example 13 that the matrix A has the inverse A^{-1}, so we multiply each side of equation (4) by A^{-1}:

$$\begin{aligned} AX &= B \\ A^{-1}(AX) &= A^{-1}B \\ (A^{-1}A)X &= A^{-1}B \quad \text{Associative property of multiplication.} \\ I_3X &= A^{-1}B \quad \text{Definition of inverse matrix.} \\ X &= A^{-1}B \quad \text{Property of identity matrix.} \end{aligned} \tag{5}$$

Now we use (5) to find $X = \begin{bmatrix} x \\ y \\ z \end{bmatrix}$.

Graphing Solution Enter the matrices A and B into a graphing utility. Figure 16 shows the solution to the system of equations.

FIGURE 16

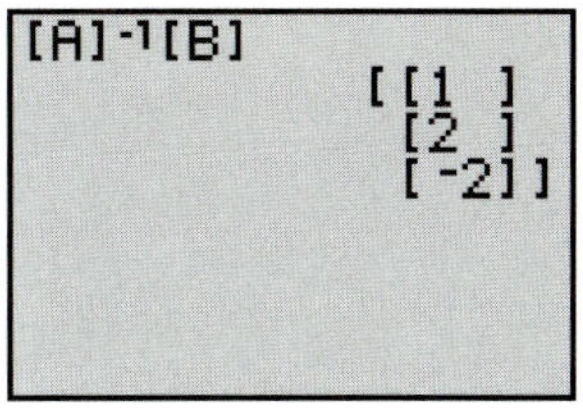

Algebraic Solution

$$X = \begin{bmatrix} x \\ y \\ z \end{bmatrix} = A^{-1}B = \begin{bmatrix} \frac{7}{4} & \frac{3}{4} & -1 \\ -\frac{3}{4} & -\frac{3}{4} & 1 \\ 1 & 1 & -1 \end{bmatrix} \begin{bmatrix} 3 \\ -3 \\ 2 \end{bmatrix} = \begin{bmatrix} 1 \\ 2 \\ -2 \end{bmatrix}$$

Example 13 (points to A^{-1})

Thus, $x = 1, y = 2, z = -2$.

The method used in Example 15 to solve a system of equations is particularly useful when it is necessary to solve several systems of equations in which the constants appearing to the right of the equal signs change, while the coefficients of the variables on the left side remain the same. See Problems 31–50 for some illustrations.

Historical Feature

Matrices were invented in 1857 by Arthur Cayley (1821–1895) as a way of efficiently computing the result of substituting one linear system into another (see Historical Problem 2). The resulting system had incredible richness, in the sense that a very wide variety of mathematical systems could be mimicked by the matrices. Cayley and his friend J.J. Sylvester (1814–1897) spent much of the rest of their lives elaborating the theory. The torch was then passed to G. Frobenius (1849–1917), whose deep investigations established a central place for matrices in modern mathematics. In 1924, rather to the surprise of physicists, it was found that matrices (with complex numbers in them) were exactly the right tool for describing the behavior of atomic systems. Today, matrices are used in a wide variety of applications.

Historical Problems

1. *Matrices and Complex Numbers* Frobenius emphasized in his research how matrices could be used to mimic other mathematical systems. Here, we mimic the behavior of complex numbers using matrices. Mathematicians call such a relationship an *isomorphism.*

Complex number ⟷ Matrix

$$a + bi \longleftrightarrow \begin{bmatrix} a & b \\ -b & a \end{bmatrix}$$

Note that the complex number can be read off the top line of the matrix. Thus,

$$2 + 3i \longleftrightarrow \begin{bmatrix} 2 & 3 \\ -3 & 2 \end{bmatrix} \quad \text{and} \quad \begin{bmatrix} 4 & -2 \\ 2 & 4 \end{bmatrix} \longleftrightarrow 4 - 2i$$

(a) Find the matrices corresponding to $2 - 5i$ and $1 + 3i$.
(b) Multiply the two matrices.
(c) Find the corresponding complex number for the matrix found in part (b).
(d) Multiply $2 - 5i$ by $1 + 3i$. The result should be the same as that found in part (c).

The process also works for addition and subtraction. Try it for yourself.

2. *Cayley's Definition of Matrix Multiplication* Cayley invented matrix multiplication to simplify the following problem:

$$\begin{cases} u = ar + bs \\ v = cr + ds \end{cases} \qquad \begin{cases} x = ku + lv \\ y = mu + nv \end{cases}$$

(a) Find x and y in terms of r and s by substituting u and v from the first system of equations into the second system of equations.
(b) Use the result of part (a) to find the 2 by 2 matrix A in

$$\begin{bmatrix} x \\ y \end{bmatrix} = A\begin{bmatrix} r \\ s \end{bmatrix}$$

(c) Now look at the following way to do it. Write the equations in matrix form

$$\begin{bmatrix} u \\ v \end{bmatrix} = \begin{bmatrix} a & b \\ c & d \end{bmatrix}\begin{bmatrix} r \\ s \end{bmatrix} \qquad \begin{bmatrix} x \\ y \end{bmatrix} = \begin{bmatrix} k & l \\ m & n \end{bmatrix}\begin{bmatrix} u \\ v \end{bmatrix}$$

So

$$\begin{bmatrix} x \\ y \end{bmatrix} = \begin{bmatrix} k & l \\ m & n \end{bmatrix}\begin{bmatrix} a & b \\ c & d \end{bmatrix}\begin{bmatrix} r \\ s \end{bmatrix}$$

Do you see how Cayley defined matrix multiplication?

11.4 EXERCISES

In Problems 1–16, use the following matrices to compute the given expression: (a) by hand; (b) using a graphing utility.

$$A = \begin{bmatrix} 0 & 3 & -5 \\ 1 & 2 & 6 \end{bmatrix} \qquad B = \begin{bmatrix} 4 & 1 & 0 \\ -2 & 3 & -2 \end{bmatrix} \qquad C = \begin{bmatrix} 4 & 1 \\ 6 & 2 \\ -2 & 3 \end{bmatrix}$$

1. $A + B$ **2.** $A - B$ **3.** $4A$ **4.** $-3B$

5. $3A - 2B$ **6.** $2A + 4B$ **7.** AC **8.** BC

9. CA **10.** CB **11.** $C(A + B)$ **12.** $(A + B)C$

13. $AC - 3I_2$ **14.** $CA + 5I_3$ **15.** $CA - CB$ **16.** $AC + BC$

In Problems 17–20, compute each product: (a) by hand; (b) using a graphing utility.

17. $\begin{bmatrix} 2 & -2 \\ 1 & 0 \end{bmatrix}\begin{bmatrix} 2 & 1 & 4 & 6 \\ 3 & -1 & 3 & 2 \end{bmatrix}$

18. $\begin{bmatrix} 4 & 1 \\ 2 & 1 \end{bmatrix}\begin{bmatrix} -6 & 6 & 1 & 0 \\ 2 & 5 & 4 & -1 \end{bmatrix}$

19. $\begin{bmatrix} 1 & 0 & 1 \\ 2 & 4 & 1 \\ 3 & 6 & 1 \end{bmatrix}\begin{bmatrix} 1 & 3 \\ 6 & 2 \\ 8 & -1 \end{bmatrix}$

20. $\begin{bmatrix} 4 & -2 & 3 \\ 0 & 1 & 2 \\ -1 & 0 & 1 \end{bmatrix}\begin{bmatrix} 2 & 6 \\ 1 & -1 \\ 0 & 2 \end{bmatrix}$

In Problems 21–30, each matrix is nonsingular. Find the inverse of each matrix. Be sure to check your answer using a graphing utility (when possible).

21. $\begin{bmatrix} 2 & 1 \\ 1 & 1 \end{bmatrix}$ **22.** $\begin{bmatrix} 3 & -1 \\ -2 & 1 \end{bmatrix}$ **23.** $\begin{bmatrix} 6 & 5 \\ 2 & 2 \end{bmatrix}$ **24.** $\begin{bmatrix} -4 & 1 \\ 6 & -2 \end{bmatrix}$

25. $\begin{bmatrix} 2 & 1 \\ a & a \end{bmatrix}, \quad a \neq 0$ **26.** $\begin{bmatrix} b & 3 \\ b & 2 \end{bmatrix}, \quad b \neq 0$ **27.** $\begin{bmatrix} 1 & -1 & 1 \\ 0 & -2 & 1 \\ -2 & -3 & 0 \end{bmatrix}$ **28.** $\begin{bmatrix} 1 & 0 & 2 \\ -1 & 2 & 3 \\ 1 & -1 & 0 \end{bmatrix}$

29. $\begin{bmatrix} 1 & 1 & 1 \\ 3 & 2 & -1 \\ 3 & 1 & 2 \end{bmatrix}$ **30.** $\begin{bmatrix} 3 & 3 & 1 \\ 1 & 2 & 1 \\ 2 & -1 & 1 \end{bmatrix}$

In Problems 31–50, use the inverses found in Problems 21–30 to solve each system of equations by hand.

31. $\begin{cases} 2x + y = 8 \\ x + y = 5 \end{cases}$ **32.** $\begin{cases} 3x - y = 8 \\ -2x + y = 4 \end{cases}$ **33.** $\begin{cases} 2x + y = 0 \\ x + y = 5 \end{cases}$

34. $\begin{cases} 3x - y = 4 \\ -2x + y = 5 \end{cases}$ **35.** $\begin{cases} 6x + 5y = 7 \\ 2x + 2y = 2 \end{cases}$ **36.** $\begin{cases} -4x + y = 0 \\ 6x - 2y = 14 \end{cases}$

37. $\begin{cases} 6x + 5y = 13 \\ 2x + 2y = 5 \end{cases}$ **38.** $\begin{cases} -4x + y = 5 \\ 6x - 2y = -9 \end{cases}$ **39.** $\begin{cases} 2x + y = -3 \\ ax + ay = -a \end{cases} \quad a \neq 0$

40. $\begin{cases} bx + 3y = 2b + 3 \\ bx + 2y = 2b + 2 \end{cases} \quad b \neq 0$ **41.** $\begin{cases} 2x + y = 7/a \\ ax + ay = 5 \end{cases} \quad a \neq 0$ **42.** $\begin{cases} bx + 3y = 14 \\ bx + 2y = 10 \end{cases} \quad b \neq 0$

43. $\begin{cases} x - y + z = 0 \\ -2y + z = -1 \\ -2x - 3y = -5 \end{cases}$ **44.** $\begin{cases} x + 2z = 6 \\ -x + 2y + 3z = -5 \\ x - y = 6 \end{cases}$ **45.** $\begin{cases} x - y + z = 2 \\ -2y + z = 2 \\ -2x - 3y = \frac{1}{2} \end{cases}$

46. $\begin{cases} x + 2z = 2 \\ -x + 2y + 3z = -\frac{3}{2} \\ x - y = 2 \end{cases}$ **47.** $\begin{cases} x + y + z = 9 \\ 3x + 2y - z = 8 \\ 3x + y + 2z = 1 \end{cases}$ **48.** $\begin{cases} 3x + 3y + z = 8 \\ x + 2y + z = 5 \\ 2x - y + z = 4 \end{cases}$

49. $\begin{cases} x + y + z = 2 \\ 3x + 2y - z = \frac{7}{3} \\ 3x + y + 2z = \frac{10}{3} \end{cases}$ **50.** $\begin{cases} 3x + 3y + z = 1 \\ x + 2y + z = 0 \\ 2x - y + z = 4 \end{cases}$

In Problems 51–56, by hand, show that each matrix has no inverse.

51. $\begin{bmatrix} 4 & 2 \\ 2 & 1 \end{bmatrix}$ **52.** $\begin{bmatrix} -3 & \frac{1}{2} \\ 6 & -1 \end{bmatrix}$ **53.** $\begin{bmatrix} 15 & 3 \\ 10 & 2 \end{bmatrix}$

54. $\begin{bmatrix} -3 & 0 \\ 4 & 0 \end{bmatrix}$ **55.** $\begin{bmatrix} -3 & 1 & -1 \\ 1 & -4 & -7 \\ 1 & 2 & 5 \end{bmatrix}$ **56.** $\begin{bmatrix} 1 & 1 & -3 \\ 2 & -4 & 1 \\ -5 & 7 & 1 \end{bmatrix}$

In Problems 57–60, use a graphing utility to find the inverse, if it exists, of each matrix. Round answers to two decimal places.

57. $\begin{bmatrix} 25 & 61 & -12 \\ 18 & -2 & 4 \\ 8 & 35 & 21 \end{bmatrix}$ **58.** $\begin{bmatrix} 18 & -3 & 4 \\ 6 & -20 & 14 \\ 10 & 25 & -15 \end{bmatrix}$

59. $\begin{bmatrix} 44 & 21 & 18 & 6 \\ -2 & 10 & 15 & 5 \\ 21 & 12 & -12 & 4 \\ -8 & -16 & 4 & 9 \end{bmatrix}$ **60.** $\begin{bmatrix} 16 & 22 & -3 & 5 \\ 21 & -17 & 4 & 8 \\ 2 & 8 & 27 & 20 \\ 5 & 15 & -3 & -10 \end{bmatrix}$

In Problems 61–64, use the idea behind Example 15 with a graphing utility to solve the following systems of equations. Round answers to two decimal places.

61. $\begin{cases} 25x + 61y - 12z = 10 \\ 18x - 12y + 7z = -9 \\ 3x + 4y - z = 12 \end{cases}$

62. $\begin{cases} 25x + 61y - 12z = 15 \\ 18x - 12y + 7z = -3 \\ 3x + 4y - z = 12 \end{cases}$

63. $\begin{cases} 25x + 61y - 12z = 21 \\ 18x - 12y + 7z = 7 \\ 3x + 4y - z = -2 \end{cases}$

64. $\begin{cases} 25x + 61y - 12z = 25 \\ 18x - 12y + 7z = 10 \\ 3x + 4y - z = -4 \end{cases}$

65. Computing the Cost of Production The Acme Steel Company is a producer of stainless steel and aluminum containers. On a certain day, the following stainless steel containers were manufactured: 500 with 10-gallon capacity, 350 with 5-gallon capacity, and 400 with 1-gallon capacity. On the same day, the following aluminum containers were manufactured: 700 with 10-gallon capacity, 500 with 5-gallon capacity, and 850 with 1-gallon capacity.

(a) Find a 2 by 3 matrix representing the above data. Find a 3 by 2 matrix to represent the same data.

(b) If the amount of material used in the 10-gallon containers is 15 pounds, the amount used in the 5-gallon containers is 8 pounds, and the amount used in the 1-gallon containers is 3 pounds, find a 3 by 1 matrix representing the amount of material.

(c) Multiply the 2 by 3 matrix found in part (a) and the 3 by 1 matrix found in part (b) to get a 2 by 1 matrix showing the day's usage of material.

(d) If stainless steel costs Acme \$0.10 per pound and aluminum costs \$0.05 per pound, find a 1 by 2 matrix representing cost.

(e) Multiply the matrices found in parts (c) and (d) to determine what the total cost of the day's production was.

66. Computing Profit Rizza Ford has two locations, one in the city and the other in the suburbs. In January, the city location sold 400 subcompacts, 250 intermediate-size cars, and 50 station wagons; in February, it sold 350 subcompacts, 100 intermediates, and 30 station wagons. At the suburban location in January, 450 subcompacts, 200 intermediates, and 140 station wagons were sold. In February, the suburban location sold 350 subcompacts, 300 intermediates, and 100 station wagons.

(a) Find 2 by 3 matrices that summarize the sales data for each location for January and February (one matrix for each month).

(b) Use matrix addition to obtain total sales for the two-month period.

(c) The profit on each kind of car is \$100 per subcompact, \$150 per intermediate, and \$200 per station wagon. Find a 3 by 1 matrix representing this profit.

(d) Multiply the matrices found in parts (b) and (c) to get a 2 by 1 matrix showing the profit at each location.

67. Consider the 2 by 2 square matrix

$$A = \begin{bmatrix} a & b \\ c & d \end{bmatrix}$$

If $D = ad - bc \neq 0$, show that A is nonsingular and that

$$A^{-1} = \frac{1}{D}\begin{bmatrix} d & -b \\ -c & a \end{bmatrix}$$

68. Make up a situation different from any found in the text that can be represented by a matrix.

11.5 PARTIAL FRACTION DECOMPOSITION

1 Decompose *P/Q*, Where *Q* Has Only Nonrepeated Linear Factors

2 Decompose *P/Q*, Where *Q* Has Repeated Linear Factors

3 Decompose *P/Q*, Where *Q* Has Only Nonrepeated Irreducible Quadratic Factors

4 Decompose *P/Q*, Where *Q* Has Repeated Irreducible Quadratic Factors

Consider the problem of adding two fractions

$$\frac{3}{x+4} \quad \text{and} \quad \frac{2}{x-3}$$

The result is

$$\frac{3}{x+4} + \frac{2}{x-3} = \frac{3(x-3) + 2(x+4)}{(x+4)(x-3)} = \frac{5x-1}{x^2+x-12}$$

The reverse procedure, of starting with the rational expression $(5x - 1)/(x^2 + x - 12)$ and writing it as the sum (or difference) of the two simpler fractions $3/(x + 4)$ and $2/(x - 3)$ is referred to as **partial fraction decomposition,** and the two simpler fractions are called **partial fractions.** Decomposing a rational expression into a sum of partial fractions is important in solving certain types of calculus problems. This section presents a systematic way to decompose rational expressions.

We begin by recalling that a rational expression is the ratio of two polynomials, say, P and $Q \neq 0$, that have no common factors. Recall also that a rational expression P/Q is called **proper** if the degree of the polynomial in the numerator is less than the degree of the polynomial in the denominator. Otherwise, the rational expression is termed **improper.**

Because any improper rational expression can be reduced by long division to a mixed form consisting of the sum of a polynomial and a proper rational expression, we shall restrict the discussion that follows to proper rational expressions.

The partial fraction decomposition of the rational expression P/Q depends on the factors of the denominator Q. Recall (from Section 4.5) that any polynomial whose coefficients are real numbers can be factored (over the real numbers) into products of linear and/or irreducible quadratic factors. Thus, the denominator Q of the rational expression P/Q will contain only factors of one or both of the following types:

1. *Linear factors* of the form $x - a$, where a is a real number.
2. *Irreducible quadratic factors* of the form $ax^2 + bx + c$, where a, b, and c are real numbers, $a \neq 0$, and $b^2 - 4ac < 0$ (which guarantees that $ax^2 + bx + c$ cannot be written as the product of two linear factors with real coefficients).

1 As it turns out, there are four cases to be examined. We begin with the case for which Q has only nonrepeated linear factors.

CASE 1: Q **has only nonrepeated linear factors.**

Under the assumption that Q has only nonrepeated linear factors, the polynomial Q has the form

$$Q(x) = (x - a_1)(x - a_2) \cdot \ \ldots \ \cdot (x - a_n)$$

where none of the numbers $a_1, a_2, \dots, a_n$ are equal. In this case, the partial fraction decomposition of P/Q is of the form

$$\frac{P(x)}{Q(x)} = \frac{A_1}{x - a_1} + \frac{A_2}{x - a_2} + \cdots + \frac{A_n}{x - a_n} \tag{1}$$

where the numbers $A_1, A_2, \dots, A_n$ are to be determined.

We show how to find these numbers in the example that follows.

EXAMPLE 1 Nonrepeated Linear Factors

Write the partial fraction decomposition of $\dfrac{x}{x^2 - 5x + 6}$.

Solution First, we factor the denominator,

$$x^2 - 5x + 6 = (x - 2)(x - 3)$$

and conclude that the denominator contains only nonrepeated linear factors. Then we decompose the rational expression according to equation (1):

$$\frac{x}{x^2 - 5x + 6} = \frac{A}{x - 2} + \frac{B}{x - 3} \tag{2}$$

where A and B are to be determined. To find A and B, we clear the fractions by multiplying each side by $(x - 2)(x - 3) = x^2 - 5x + 6$. The result is

$$x = A(x - 3) + B(x - 2) \tag{3}$$

or

$$x = (A + B)x + (-3A - 2B)$$

This equation is an identity in x. Thus, we may equate the coefficients of like powers of x to get

$$\begin{cases} 1 = A + B \\ 0 = -3A - 2B \end{cases}$$

Equate coefficients of x: $1x = (A + B)x$.
Equate coefficients of x^0, the constants: $0x^0 = (-3A - 2B)x^0$.

This system of two equations containing two variables, A and B, can be solved using whatever method you wish. Solving it, we get

$$A = -2 \qquad B = 3$$

Thus, from equation (2), the partial fraction decomposition is

$$\frac{x}{x^2 - 5x + 6} = \frac{-2}{x - 2} + \frac{3}{x - 3}$$

Check: The decomposition can be checked by adding the fractions:

$$\frac{-2}{x - 2} + \frac{3}{x - 3} = \frac{-2(x - 3) + 3(x - 2)}{(x - 2)(x - 3)} = \frac{x}{(x - 2)(x - 3)}$$
$$= \frac{x}{x^2 - 5x + 6}$$

Now work Problem 9.

The numbers to be found in the partial fraction decomposition can sometimes be found more readily by using suitable choices for x (which may include complex numbers) in the identity obtained after fractions have been cleared. In Example 1, the identity after clearing fractions, equation (3), is

$$x = A(x - 3) + B(x - 2)$$

If we let $x = 2$ in this expression, the term containing B drops out, leaving $2 = A(-1)$, or $A = -2$. Similarly, if we let $x = 3$, the term containing A drops out, leaving $3 = B$. Thus, as before $A = -2$ and $B = 3$.

We use this method in the next example.

2

CASE 2: Q has repeated linear factors.

If the polynomial Q has a repeated factor, say, $(x - a)^n$, $n \geq 2$ an integer, then, in the partial fraction decomposition of P/Q, we allow for the terms

$$\frac{A_1}{x - a} + \frac{A_2}{(x - a)^2} + \cdots + \frac{A_n}{(x - a)^n}$$

where the numbers $A_1, A_2, \ldots, A_n$ are to be determined.

EXAMPLE 2 Repeated Linear Factors

Write the partial fraction decomposition of $\dfrac{x + 2}{x^3 - 2x^2 + x}$.

Solution First, we factor the denominator,

$$x^3 - 2x^2 + x = x(x^2 - 2x + 1) = x(x - 1)^2$$

and find that the denominator has the nonrepeated linear factor x and the repeated linear factor $(x - 1)^2$. By Case 1, we must allow for the term A/x in the decomposition; and, by Case 2, we must allow for the terms $B/(x - 1) + C/(x - 1)^2$ in the decomposition.

Thus, we write

$$\frac{x + 2}{x^3 - 2x^2 + x} = \frac{A}{x} + \frac{B}{x - 1} + \frac{C}{(x - 1)^2} \tag{4}$$

Again, we clear fractions by multiplying each side by $x^3 - 2x^2 + x = x(x - 1)^2$. The result is the identity

$$x + 2 = A(x - 1)^2 + Bx(x - 1) + Cx \tag{5}$$

If we let $x = 0$ in this expression, the terms containing B and C drop out leaving $2 = A(-1)^2$, or $A = 2$. Similarly, if we let $x = 1$, the terms containing A and B drop out, leaving $3 = C$. Thus, equation (5) becomes

$$x + 2 = 2(x - 1)^2 + Bx(x - 1) + 3x$$

Now, let $x = 2$ (any choice other than 0 or 1 will work as well). The result is

$$\begin{aligned} 4 &= 2(1)^2 + B(2)(1) + 3(2) \\ 2B &= 4 - 2 - 6 = -4 \\ B &= -2 \end{aligned}$$

Thus, we have $A = 2$, $B = -2$, and $C = 3$.

From equation (4), the partial fraction decomposition is

$$\frac{x+2}{x^3 - 2x^2 + x} = \frac{2}{x} + \frac{-2}{x-1} + \frac{3}{(x-1)^2}$$

EXAMPLE 3 Repeated Linear Factors

Write the partial fraction decomposition of $\dfrac{x^3 - 8}{x^2(x-1)^3}$.

Solution The denominator contains the repeated linear factor x^2 and the repeated linear factor $(x-1)^3$. Thus, the partial fraction decomposition takes the form

$$\frac{x^3 - 8}{x^2(x-1)^3} = \frac{A}{x} + \frac{B}{x^2} + \frac{C}{x-1} + \frac{D}{(x-1)^2} + \frac{E}{(x-1)^3} \tag{6}$$

As before, we clear fractions and obtain the identity

$$x^3 - 8 = Ax(x-1)^3 + B(x-1)^3 + Cx^2(x-1)^2 + Dx^2(x-1) + Ex^2 \tag{7}$$

Let $x = 0$. (Do you see why this choice was made?) Then,

$$\begin{aligned} -8 &= B(-1) \\ B &= 8 \end{aligned}$$

Now let $x = 1$ in equation (7). Then

$$-7 = E$$

Use $B = 8$ and $E = -7$ in equation (7) and collect like terms:

$$\begin{aligned} x^3 - 8 &= Ax(x-1)^3 + 8(x-1)^3 \\ &\quad + Cx^2(x-1)^2 + Dx^2(x-1) - 7x^2 \\ x^3 - 8 - 8(x^3 - 3x^2 + 3x - 1) + 7x^2 &= Ax(x-1)^3 + Cx^2(x-1)^2 + Dx^2(x-1) \\ -7x^3 + 31x^2 - 24x &= x(x-1)[A(x-1)^2 + Cx(x-1) + Dx] \\ x(x-1)(-7x+24) &= x(x-1)[A(x-1)^2 + Cx(x-1) + Dx] \\ -7x + 24 &= A(x-1)^2 + Cx(x-1) + Dx \end{aligned} \tag{8}$$

We now work with equation (8). Let $x = 0$. Then

$$24 = A$$

Now, let $x = 1$ in equation (8). Then,

$$17 = D$$

Use $A = 24$ and $D = 17$ in equation (8) and collect like terms:

$$\begin{aligned} -7x + 24 &= 24(x-1)^2 + Cx(x-1) + 17x \\ -24x^2 + 48x - 24 - 17x - 7x + 24 &= Cx(x-1) \\ -24x^2 + 24x &= Cx(x-1) \\ -24x(x-1) &= Cx(x-1) \\ -24 &= C \end{aligned}$$

We now know all the numbers A, B, C, D, and E, so, from equation (6), we have the decomposition

$$\frac{x^3 - 8}{x^2(x-1)^3} = \frac{24}{x} + \frac{8}{x^2} + \frac{-24}{x-1} + \frac{17}{(x-1)^2} + \frac{-7}{(x-1)^3}$$

The method employed in Example 3, although somewhat tedious, is still preferable to solving the system of five equations containing five variables that the expansion of equation (6) leads to.

Now work Problem 15.

3

The final two cases involve irreducible quadratic factors. As mentioned in Section 4.5, a quadratic factor is irreducible if it cannot be factored into linear factors with real coefficients. A quadratic expression $ax^2 + bx + c$ is irreducible whenever $b^2 - 4ac < 0$. For example, $x^2 + x + 1$ and $x^2 + 4$ are irreducible.

CASE 3: Q contains a nonrepeated irreducible quadratic factor.

If Q contains a nonrepeated irreducible quadratic factor of the form $ax^2 + bx + c$, then, in the partial fraction decomposition of P/Q, allow for the term

$$\frac{Ax + B}{ax^2 + bx + c}$$

where the numbers A and B are to be determined.

EXAMPLE 4 Nonrepeated Irreducible Quadratic Factor

Write the partial fraction decomposition of: $\dfrac{3x - 5}{x^3 - 1}$.

Solution We factor the denominator,

$$x^3 - 1 = (x - 1)(x^2 + x + 1)$$

and find that is has a nonrepeated linear factor $x - 1$ and a nonrepeated irreducible quadratic factor $x^2 + x + 1$. Thus, we allow for the term $A/(x - 1)$ by Case 1, and we allow for the term $(Bx + C)/(x^2 + x + 1)$ by Case 3. Hence, we write

$$\frac{3x - 5}{x^3 - 1} = \frac{A}{x - 1} + \frac{Bx + C}{x^2 + x + 1} \tag{9}$$

We clear fractions by multiplying each side of equation (9) by $x^3 - 1 = (x - 1)(x^2 + x + 1)$ to obtain the identity

$$3x - 5 = A(x^2 + x + 1) + (Bx + C)(x - 1) \tag{10}$$

Now let $x = 1$. Then equation (10) gives $-2 = A(3)$, or $A = -\frac{2}{3}$. We use this value of A in equation (10) and simplify:

$$\begin{aligned}
3x - 5 &= -\frac{2}{3}(x^2 + x + 1) + (Bx + C)(x - 1) \\
3(3x - 5) &= -2(x^2 + x + 1) + 3(Bx + C)(x - 1) && \text{Multiply each side by 3.} \\
9x - 15 &= -2x^2 - 2x - 2 + 3(Bx + C)(x - 1) \\
2x^2 + 11x - 13 &= 3(Bx + C)(x - 1) && \text{Collect terms.} \\
(2x + 13)(x - 1) &= 3(Bx + C)(x - 1) && \text{Factor the left side.} \\
2x + 13 &= 3Bx + 3C \\
2 = 3B \quad &\text{and} \quad 13 = 3C && \text{Equate coefficients.} \\
B = \frac{2}{3} \quad & \quad C = \frac{13}{3}
\end{aligned}$$

Thus, from equation (9), we see that

$$\frac{3x - 5}{x^3 - 1} = \frac{-\frac{2}{3}}{x - 1} + \frac{\frac{2}{3}x + \frac{13}{3}}{x^2 + x + 1}$$

Now work Problem 17.

4

CASE 4: *Q* contains repeated irreducible quadratic factors.

If the polynomial Q contains a repeated irreducible quadratic factor $(ax^2 + bx + c)^n$, $n \geq 2$, n an integer, then, in the partial fraction decomposition of P/Q, allow for the terms

$$\frac{A_1x + B_1}{ax^2 + bx + c} + \frac{A_2x + B_2}{(ax^2 + bx + c)^2} + \cdots + \frac{A_nx + B_n}{(ax^2 + bx + c)^n}$$

where the numbers $A_1, B_1, A_2, B_2, \ldots, A_n, B_n$ are to be determined.

EXAMPLE 5 Repeated Irreducible Quadratic Factor

Write the partial fraction decomposition of: $\dfrac{x^3 + x^2}{(x^2 + 4)^2}$.

Solution The denominator contains the repeated irreducible quadratic factor $(x^2 + 4)^2$, so we write

$$\frac{x^3 + x^2}{(x^2 + 4)^2} = \frac{Ax + B}{x^2 + 4} + \frac{Cx + D}{(x^2 + 4)^2} \tag{11}$$

We clear fractions to obtain

$$x^3 + x^2 = (Ax + B)(x^2 + 4) + Cx + D$$

Collecting like terms yields

$$x^3 + x^2 = Ax^3 + Bx^2 + (4A + C)x + D + 4B$$

Equating coefficients, we arrive at the system

$$\begin{cases} A = 1 \\ B = 1 \\ 4A + C = 0 \\ D + 4B = 0 \end{cases}$$

The solution is $A = 1, B = 1, C = -4, D = -4$. Hence, from equation (11),

$$\frac{x^3 + x^2}{(x^2+4)^2} = \frac{x+1}{x^2+4} + \frac{-4x-4}{(x^2+4)^2}$$

11.5 EXERCISES

In Problems 1–8, tell whether the given rational expression is proper or improper. If improper, rewrite it as the sum of a polynomial and a proper rational expression.

1. $\frac{x}{x^2-1}$ **2.** $\frac{5x+2}{x^3-1}$ **3.** $\frac{x^2+5}{x^2-4}$ **4.** $\frac{3x^2-2}{x^2-1}$

5. $\frac{5x^3+2x-1}{x^2-4}$ **6.** $\frac{3x^4+x^2-2}{x^3+8}$ **7.** $\frac{x(x-1)}{(x+4)(x-3)}$ **8.** $\frac{2x(x^2+4)}{x^2+1}$

In Problems 9–42, write the partial fraction decomposition of each rational expression.

9. $\frac{4}{x(x-1)}$ **10.** $\frac{3x}{(x+2)(x-1)}$ **11.** $\frac{1}{x(x^2+1)}$

12. $\frac{1}{(x+1)(x^2+4)}$ **13.** $\frac{x}{(x-1)(x-2)}$ **14.** $\frac{3x}{(x+2)(x-4)}$

15. $\frac{x^2}{(x-1)^2(x+1)}$ **16.** $\frac{x+1}{x^2(x-2)}$ **17.** $\frac{1}{x^3-8}$

18. $\frac{2x+4}{x^3-1}$ **19.** $\frac{x^2}{(x-1)^2(x+1)^2}$ **20.** $\frac{x+1}{x^2(x-2)^2}$

21. $\frac{x-3}{(x+2)(x+1)^2}$ **22.** $\frac{x^2+x}{(x+2)(x-1)^2}$ **23.** $\frac{x+4}{x^2(x^2+4)}$

24. $\frac{10x^2+2x}{(x-1)^2(x^2+2)}$ **25.** $\frac{x^2+2x+3}{(x+1)(x^2+2x+4)}$ **26.** $\frac{x^2-11x-18}{x(x^2+3x+3)}$

27. $\frac{x}{(3x-2)(2x+1)}$ **28.** $\frac{1}{(2x+3)(4x-1)}$ **29.** $\frac{x}{x^2+2x-3}$

30. $\frac{x^2-x-8}{(x+1)(x^2+5x+6)}$ **31.** $\frac{x^2+2x+3}{(x^2+4)^2}$ **32.** $\frac{x^3+1}{(x^2+16)^2}$

33. $\frac{7x+3}{x^3-2x^2-3x}$ **34.** $\frac{x^5+1}{x^6-x^4}$ **35.** $\frac{x^2}{x^3-4x^2+5x-2}$

36. $\frac{x^2+1}{x^3+x^2-5x+3}$ **37.** $\frac{x^3}{(x^2+16)^3}$ **38.** $\frac{x^2}{(x^2+4)^3}$

39. $\frac{4}{2x^2-5x-3}$ **40.** $\frac{4x}{2x^2+3x-2}$ **41.** $\frac{2x+3}{x^4-9x^2}$

42. $\frac{x^2+9}{x^4-2x^2-8}$

11.6 SYSTEMS OF NONLINEAR EQUATIONS

1 Solve a System of Nonlinear Equations Using Substitution

2 Solve a System of Nonlinear Equations Using Elimination

1 There is no general methodology for solving a system of nonlinear equations by hand. There are times when substitution is best; other times, elimination is best; and there are times when neither of these methods works. Experience and a certain degree of imagination are your allies here.

Before we begin, two comments are in order:

1. If the system contains two variables, then graph them. By graphing each equation in the system, we can get an idea of how many solutions a system has and approximately where they are located.
2. Extraneous solutions can creep in when solving nonlinear systems algebraically, so it is imperative that all apparent solutions be checked.

EXAMPLE 1 Solving a System of Nonlinear Equations

Solve the following system of equations:

$$\begin{cases} 3x - y = -2 & \text{(1) A line.} \\ 2x^2 - y = 0 & \text{(2) A parabola.} \end{cases}$$

Graphing Solution We use a graphing utility to graph $Y_1 = 3x + 2$ and $Y_2 = 2x^2$. From Figure 17, we see that the system apparently has two solutions. Using INTERSECT; the solutions to the system of equations are $(-0.5, 0.5)$ and $(2, 8)$.

FIGURE 17

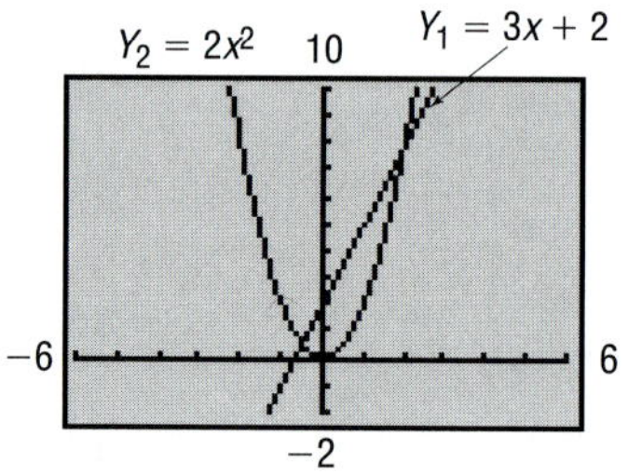

Algebraic Solution Using Substitution First, we notice that the system contains two variables and that we know how to graph each equation by hand. In Figure 18, we see that the system apparently has two solutions.

We will use substitution to solve the system. Equation (1) is easily solved for y:

$$3x - y = -2$$
$$y = 3x + 2$$

We substitute this expression for y in equation (2). The result is an equation containing just the variable x, which we can then solve:

$$2x^2 - y = 0$$
$$2x^2 - (3x + 2) = 0$$
$$2x^2 - 3x - 2 = 0$$
$$(2x + 1)(x - 2) = 0$$
$$2x + 1 = 0 \quad \text{or} \quad x - 2 = 0$$
$$x = -\frac{1}{2} \qquad x = 2$$

FIGURE 18

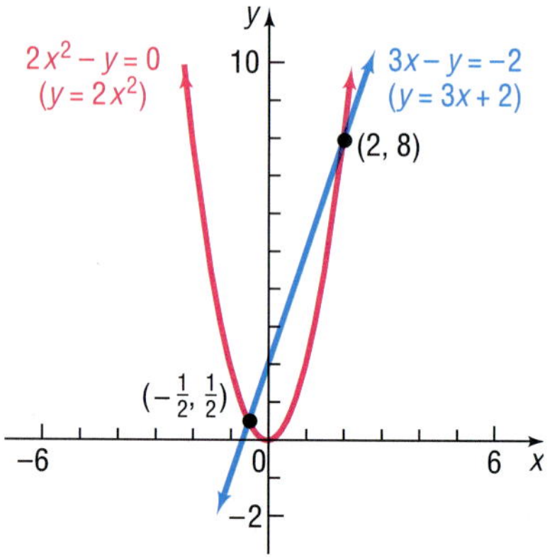

Using these values for x in $y = 3x + 2$, we find

$$y = 3\left(-\frac{1}{2}\right) + 2 = \frac{1}{2} \quad \text{or} \quad y = 3(2) + 2 = 8$$

The apparent solutions are $x = -\frac{1}{2}, y = \frac{1}{2}$ and $x = 2, y = 8$.

Check: For $x = -\frac{1}{2}, y = \frac{1}{2}$:

$$\begin{cases} 3\left(-\frac{1}{2}\right) - \frac{1}{2} = -\frac{3}{2} - \frac{1}{2} = -2 & (1) \\ 2\left(-\frac{1}{2}\right)^2 - \frac{1}{2} = 2\left(\frac{1}{4}\right) - \frac{1}{2} = 0 & (2) \end{cases}$$

For $x = 2, y = 8$:

$$\begin{cases} 3(2) - 8 = 6 - 8 = -2 & (1) \\ 2(2)^2 - 8 = 2(4) - 8 = 0 & (2) \end{cases}$$

Each solution checks. Now we know the graphs in Figure 18 intersect at $(-\frac{1}{2}, \frac{1}{2})$ and at $(2, 8)$.

Now work Problem 11.

2 Our next example illustrates how the method of elimination works for nonlinear systems.

EXAMPLE 2 Solving a System of Nonlinear Equations

Solve: $\begin{cases} x^2 + y^2 = 13 & \text{(1) A circle.} \\ x^2 - y = 7 & \text{(2) A parabola.} \end{cases}$

Graphing Solution We use a graphing utility to graph $x^2 + y^2 = 13$ and $x^2 - y = 7$. (Remember that to graph $x^2 + y^2 = 13$ requires two functions, $Y_1 = \sqrt{13 - x^2}$ and $Y_2 = -\sqrt{13 - x^2}$, and a square screen.) From Figure 19 we see that the system apparently has four solutions. Using INTERSECT, the solutions to the system of equations are $(-3, 2)$, $(3, 2)$, $(-2, -3)$, and $(2, -3)$.

Algebraic Solution Using Elimination First, we graph each equation, as shown in Figure 20. Based on the graph, we expect four solutions. By subtracting equation (2) from equation (1), the variable x is eliminated, leaving

$$y^2 + y = 6$$

FIGURE 19

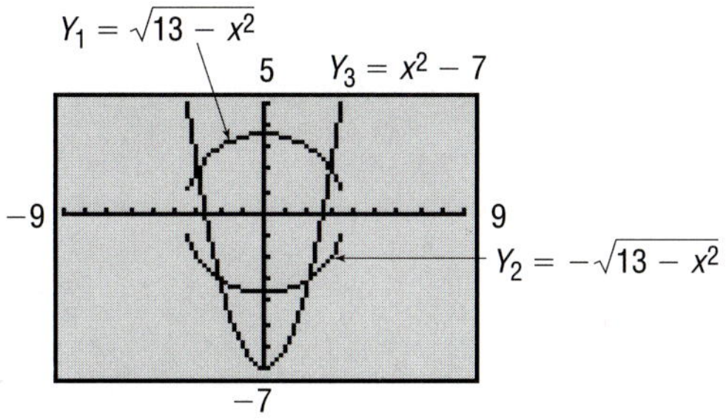

FIGURE 20

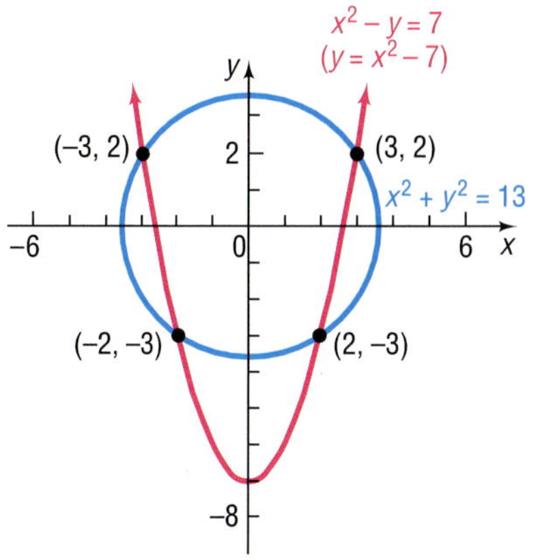

This quadratic equation in y is easily solved by factoring:

$$y^2 + y - 6 = 0$$
$$(y + 3)(y - 2) = 0$$
$$y = -3 \quad \text{or} \quad y = 2$$

We use these values for y in equation (2) to find x. If $y = 2$, then $x^2 = y + 7 = 9$ and $x = 3$ or -3. If $y = -3$, then $x^2 = y + 7 = 4$ and $x = 2$ or -2. Thus, we have four solutions: $x = 3$, $y = 2$; $x = -3$, $y = 2$; $x = 2$, $y = -3$; and $x = -2$, $y = -3$. You should verify that, in fact, these four solutions also satisfy equation (1), so that all four are solutions of the system. The four points, (3, 2), (−3, 2), (2, −3), and (−2, −3), are the points of intersection of the graphs. Look again at Figure 20.

Now work Problem 9.

EXAMPLE 3 Solving a System of Nonlinear Equations

Solve: $\begin{cases} x^2 - y^2 = 1 & (1) \\ x^3 - y^2 = x & (2) \end{cases}$

Graphing Solution We use a graphing utility to graph $x^2 - y^2 = 1$ and $x^3 - y^2 = x$. (You will need to graph four functions: $Y_1 = \sqrt{x^2 - 1}$, $Y_2 = -\sqrt{x^2 - 1}$, $Y_3 = \sqrt{x^3 - x}$, and $Y_4 = -\sqrt{x^3 - x}$). From Figure 21 we see that the system apparently has two solutions. Using INTERSECT, the solutions to the system of equations are (−1, 0) and (1, 0).

FIGURE 21

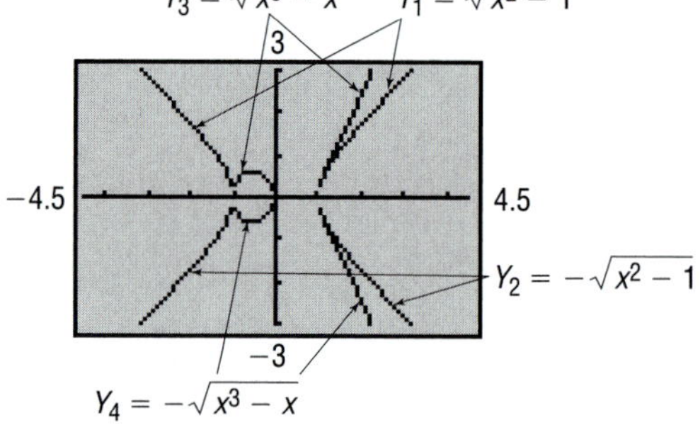

Algebraic Solution Using Elimination Because the second equation is not so easy to graph by hand, we omit the graphing step. We use elimination, subtracting equation (2) from equation (1), to obtain

$$x^2 - x^3 = 1 - x$$
$$x^2(1 - x) = 1 - x$$
$$x^2(1 - x) - (1 - x) = 0$$
$$(x^2 - 1)(1 - x) = 0$$
$$x^2 - 1 = 0 \quad \text{or} \quad 1 - x = 0$$
$$x = \pm 1 \qquad\qquad x = 1$$

We now use equation (1) to get y. If $x = 1$, then $1 - y^2 = 1$ and $y = 0$. If $x = -1$, then $1 - y^2 = 1$ and $y = 0$. There are two apparent solutions: $x = 1$, $y = 0$ and $x = -1$, $y = 0$. Because each of these solutions also satisfies equation (2), the system has two solutions: $x = 1$, $y = 0$ and $x = -1$, $y = 0$.

EXAMPLE 4 Solving a System of Nonlinear Equations

Solve: $\begin{cases} x^2 + x + y^2 - 3y + 2 = 0 & (1) \\ x + 1 + \dfrac{y^2 - y}{x} = 0 & (2) \end{cases}$

Graphing Solution First, we multiply equation (2) by x to eliminate the fraction. The result is an equivalent system because x cannot be 0 [look at equation (2) to see why]:

$$\begin{cases} x^2 + x + y^2 - 3y + 2 = 0 & (1) \\ x^2 + x + y^2 - y = 0 & (2) \end{cases}$$

We need to solve each equation for y. First, we solve equation (1) for y:

$$x^2 + x + y^2 - 3y + 2 = 0$$

$$y^2 - 3y = -x^2 - x - 2 \qquad \text{Rearrange so terms involving } y \text{ are on left side.}$$

$$y^2 - 3y + \frac{9}{4} = -x^2 - x - 2 + \frac{9}{4} \qquad \text{Complete the square involving } y.$$

$$\left(y - \frac{3}{2}\right)^2 = -x^2 - x + \frac{1}{4}$$

$$y - \frac{3}{2} = \pm\sqrt{-x^2 - x + \frac{1}{4}} \qquad \text{Solve for the squared term.}$$

$$y = \frac{3}{2} \pm \sqrt{-x^2 - x + \frac{1}{4}} \qquad \text{Solve for } y.$$

Now we solve equation (2) for y:

$$x^2 + x + y^2 - y = 0$$

$$y^2 - y = -x^2 - x \qquad \text{Rearrange so terms involving } y \text{ are on left side.}$$

$$y^2 - y + \frac{1}{4} = -x^2 - x + \frac{1}{4} \qquad \text{Complete the square involving } y.$$

$$\left(y - \frac{1}{2}\right)^2 = -x^2 - x + \frac{1}{4}$$

$$y - \frac{1}{2} = \pm\sqrt{-x^2 - x + \frac{1}{4}} \qquad \text{Solve for the squared term.}$$

$$y = \frac{1}{2} \pm \sqrt{-x^2 - x + \frac{1}{4}} \qquad \text{Solve for } y.$$

Now graph each equation using a graphing utility. See Figure 22.

FIGURE 22

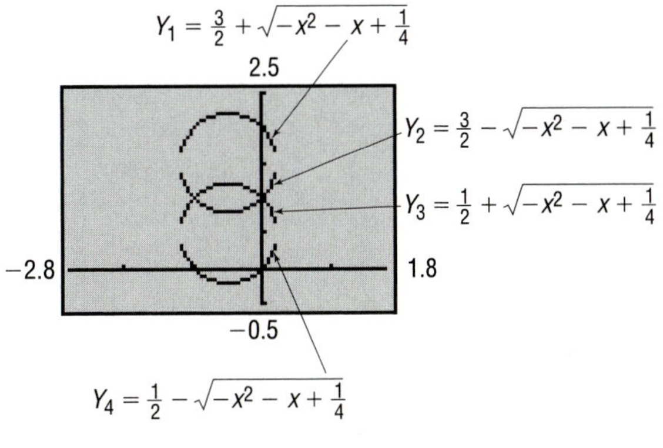

Using INTERSECT, the points of intersection are $(-1, 1)$ and $(0, 1)$. Because x cannot be 0, the value $x = 0$ is extraneous, and we discard it. Thus, the only solution is $x = -1$ and $y = 1$.

Algebraic Solution Using Elimination First, we multiply equation (2) by x to eliminate the fraction. The result is an equivalent system because x cannot be 0 [look at equation (2) to see why]:

$$\begin{cases} x^2 + x + y^2 - 3y + 2 = 0 & (1) \\ x^2 + x + y^2 - y = 0 & (2) \end{cases}$$

Now subtract equation (2) from equation (1) to eliminate x. The result is

$$-2y + 2 = 0$$
$$y = 1$$

To find x, we back-substitute $y = 1$ in equation (1):

$$x^2 + x + 1 - 3 + 2 = 0$$
$$x^2 + x = 0$$
$$x(x + 1) = 0$$
$$x = 0 \quad \text{or} \quad x = -1$$

Because x cannot be 0, the value $x = 0$ is extraneous, and we discard it. Thus, the solution is $x = -1, y = 1$.

Check: We now check $x = -1, y = 1$:

$$\begin{cases} (-1)^2 + (-1) + 1^2 - 3(1) + 2 = 1 - 1 + 1 - 3 + 2 = 0 & (1) \\ -1 + 1 + \dfrac{1^2 - 1}{-1} = 0 + \dfrac{0}{-1} = 0 & (2) \end{cases}$$

Thus, the only solution to the system is $x = -1, y = 1$. ▪

Now work Problems 25 and 49.

EXAMPLE 5

Solving a System of Nonlinear Equations

Solve: $\begin{cases} x^2 - y^2 = 4 & \text{(1) A hyperbola.} \\ y = x^2 & \text{(2) A parabola.} \end{cases}$

Graphing Solution We graph $Y_1 = x^2$ and $x^2 - y^2 = 4$ in Figure 23(a). You will need to graph $x^2 - y^2 = 4$ as two functions:

$$Y_2 = \sqrt{x^2 - 4} \qquad \text{and} \qquad Y_3 = -\sqrt{x^2 - 4}$$

From Figure 23(a) we see that the graphs of these two equations do not intersect. Thus, the system is inconsistent.

FIGURE 23(a)

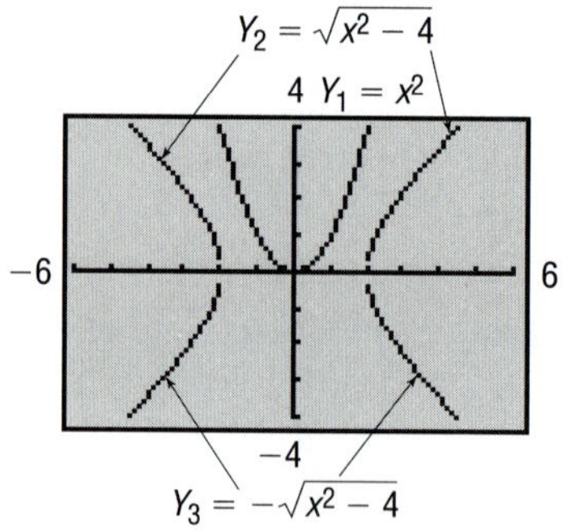

Algebraic Solution Either substitution or elimination can be used here. We use substitution and replace x^2 by y in equation (1). The result is

$$y - y^2 = 4$$
$$y^2 - y + 4 = 0$$

This is a quadratic equation whose discriminant is $1 - 4 \cdot 4 = -15 < 0$. Thus, the equation has no real solutions, and hence, the system is inconsistent. The graphs of these two equations will not intersect. See Figure 23(b).

FIGURE 23(b)

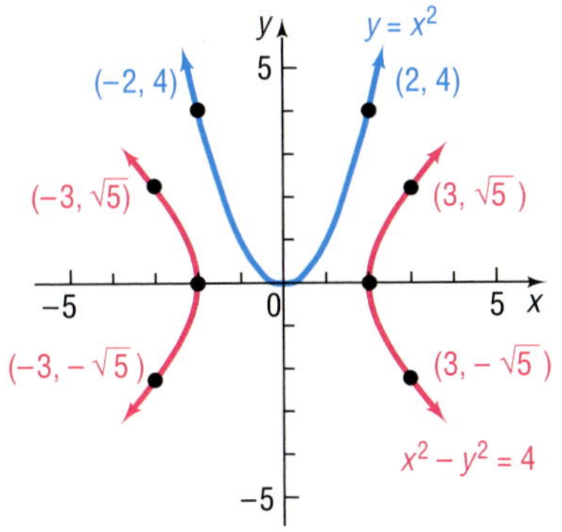

The following examples illustrate two of the more imaginative ways to solve systems of nonlinear equations algebraically.

EXAMPLE 6 Solving a System of Nonlinear Equations

Solve: $\begin{cases} 4x^2 - 9xy - 28y^2 = 0 & (1) \\ 16x^2 - 4xy = 16 & (2) \end{cases}$

Graphing Solution Instead of solving for y by completing the square, we notice that equation (1) can be factored:

$$4x^2 - 9xy - 28y^2 = 0$$
$$(4x + 7y)(x - 4y) = 0$$

This results in the two equations:

$$4x + 7y = 0 \qquad \text{or} \qquad x - 4y = 0$$
$$Y_1 = -\frac{4}{7}x \qquad\qquad Y_2 = \frac{1}{4}x$$

We graph these two equations and equation (2): $Y_3 = \dfrac{4(x^2 - 1)}{x}$. See Figure 24. Using INTERSECT, we find the intersection points to be $(-0.93, 0.53)$, $(-1.03, -0.25)$, $(0.93, -0.53)$, and $(1.03, 0.25)$, each correct to two decimal places.

FIGURE 24

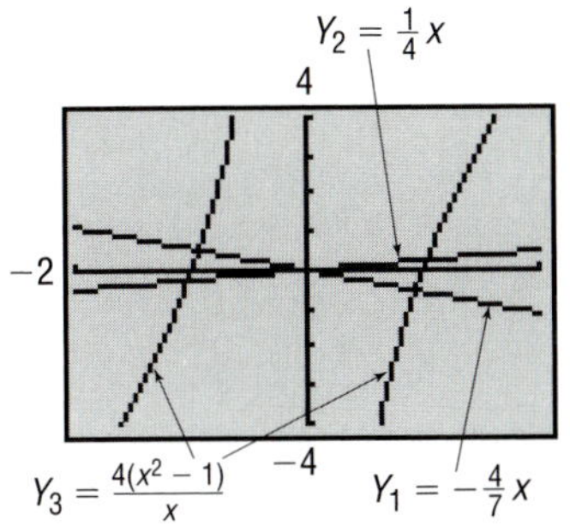

Algebraic Solution We take note of the fact that equation (1) can be factored:

$$4x^2 - 9xy - 28y^2 = 0$$
$$(4x + 7y)(x - 4y) = 0$$

This results in the two equations

$$\begin{aligned} 4x + 7y &= 0 \\ x &= -\tfrac{7}{4}y \end{aligned} \qquad \text{or} \qquad \begin{aligned} x - 4y &= 0 \\ x &= 4y \end{aligned}$$

We substitute each of these values for x in equation (2):

$$\begin{aligned} 16x^2 - 4xy &= 16 \\ 16(-\tfrac{7}{4}y)^2 - 4(-\tfrac{7}{4}y)y &= 16 \\ 49y^2 + 7y^2 &= 16 \\ 56y^2 &= 16 \\ 7y^2 &= 2 \\ y^2 &= \tfrac{2}{7} \end{aligned} \qquad \begin{aligned} 16x^2 - 4xy &= 16 \\ 16(4y)^2 - 4(4y)y &= 16 \\ 16(16y^2) - 16y^2 &= 16 \\ 15y^2 &= 1 \\ y^2 &= \tfrac{1}{15} \end{aligned}$$

Thus, we have

$$y = \pm\sqrt{\frac{2}{7}} = \pm\frac{\sqrt{14}}{7} \qquad y = \pm\frac{\sqrt{15}}{15}$$

$$x = -\frac{7}{4}y = \pm\frac{\sqrt{14}}{4} \qquad x = 4y = \pm\frac{4\sqrt{15}}{15}$$

You should verify for yourself that, in fact, the four solutions $x = -\sqrt{14}/4$, $y = \sqrt{14}/7$; $x = \sqrt{14}/4$, $y = -\sqrt{14}/7$; $x = 4\sqrt{15}/15$, $y = \sqrt{15}/15$; and $x = -4\sqrt{15}/15$, $y = -\sqrt{15}/15$ are actually solutions of the system. ■

EXAMPLE 7 Solving a System of Nonlinear Equations

Solve: $\begin{cases} 3xy - 2y^2 = -2 & (1) \\ 9x^2 + 4y^2 = 10 & (2) \end{cases}$

Graphing Solution To graph $3xy - 2y^2 = -2$, we need to solve for y. In this instance, it is easier to view the equation as a quadratic equation in the variable y.

$$\begin{aligned} 3xy - 2y^2 &= -2 \\ 2y^2 - 3xy - 2 &= 0 && \text{Place in standard form.} \\ y &= \frac{-(-3x) \pm \sqrt{(-3x)^2 - 4(2)(-2)}}{2(2)} && \text{Use the quadratic formula } a = 2,\ b = -3x,\ c = -2. \\ y &= \frac{3x \pm \sqrt{9x^2 + 16}}{4} && \text{Simplify.} \end{aligned}$$

FIGURE 25

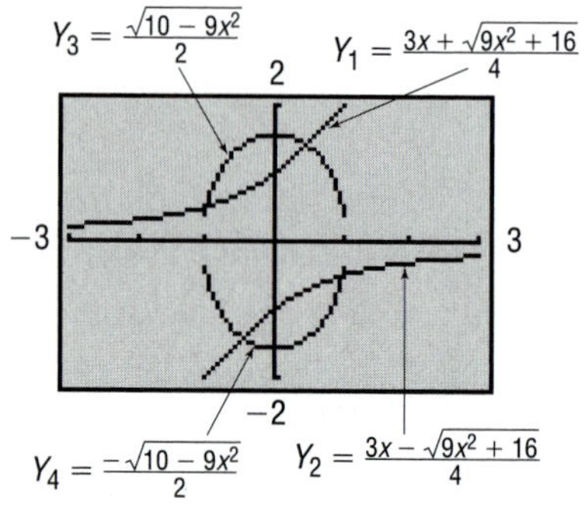

Using a graphing utility, we graph $Y_1 = \dfrac{3x + \sqrt{9x^2 + 16}}{4}$, and $Y_2 = \dfrac{3x - \sqrt{9x^2 + 16}}{4}$. From (2), We graph $Y_3 = \dfrac{\sqrt{10 - 9x^2}}{2}$ and $Y_4 = \dfrac{-\sqrt{10 - 9x^2}}{2}$. See Figure 25.

Using INTERSECT, the solutions to the system of equations are $(-1, 0.5)$, $(0.47, 1.41)$, $(1, -0.5)$, and $(-0.47, -1.41)$, each correct to two decimal places.

Algebraic Solution We multiply equation (1) by 2 and add the result to equation (2) to eliminate the y^2-terms:

$$\begin{cases} 6xy - 4y^2 = -4 & \text{(1)} \\ 9x^2 + 4y^2 = 10 & \text{(2)} \end{cases}$$
$$9x^2 + 6xy = 6$$
$$3x^2 + 2xy = 2 \qquad \text{Divide each side by 3.}$$

Since $x \neq 0$ (do you see why?), we can solve for y in this equation to get

$$y = \frac{2 - 3x^2}{2x}, \qquad x \neq 0 \qquad (1)$$

Now substitute for y in equation (2) of the system:

$$9x^2 + 4y^2 = 10$$
$$9x^2 + 4\left(\frac{2 - 3x^2}{2x}\right)^2 = 10$$
$$9x^2 + \frac{4 - 12x^2 + 9x^4}{x^2} = 10$$
$$9x^4 + 4 - 12x^2 + 9x^4 = 10x^2$$
$$18x^4 - 22x^2 + 4 = 0$$
$$9x^4 - 11x^2 + 2 = 0$$

This quadratic equation (in x^2) can be factored:

$$(9x^2 - 2)(x^2 - 1) = 0$$
$$9x^2 - 2 = 0 \qquad \text{or} \qquad x^2 - 1 = 0$$
$$x^2 = \tfrac{2}{9} \qquad\qquad x^2 = 1$$
$$x = \pm\frac{\sqrt{2}}{3} \qquad\qquad x = \pm 1$$

To find y, we use equation (1):

$$\text{If } x = \frac{\sqrt{2}}{3}: \quad y = \frac{2 - 3x^2}{2x} = \frac{2 - \frac{2}{3}}{2(\sqrt{2}/3)} = \frac{4}{2\sqrt{2}} = \sqrt{2}$$
$$\text{If } x = -\frac{\sqrt{2}}{3}: \quad y = \frac{2 - 3x^2}{2x} = \frac{2 - \frac{2}{3}}{-2(\sqrt{2}/3)} = \frac{4}{-2\sqrt{2}} = -\sqrt{2}$$
$$\text{If } x = 1: \quad y = \frac{2 - 3x^2}{2x} = \frac{2 - 3}{2} = -\frac{1}{2}$$
$$\text{If } x = -1: \quad y = \frac{2 - 3x^2}{2x} = \frac{2 - 3}{-2} = \frac{1}{2}$$

The system has four solutions. Check them for yourself.

Now work Problem 45.

EXAMPLE 8 Running a Race

In a 50-mile race, the winner crosses the finish line 1 mile ahead of the second place runner and 4 miles ahead of the third place runner. Assuming that each runner maintains a constant speed throughout the race, by how many miles does the second place runner beat the third place runner?

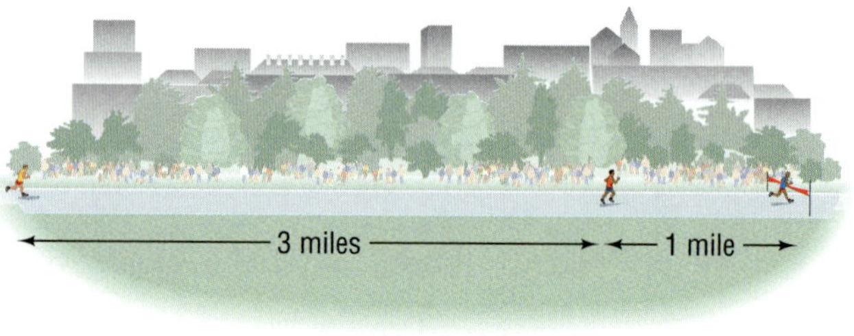

Solution Let v_1, v_2, v_3 denote the speeds of the first, second, and third place runners, respectively. Let t_1 and t_2 denote the times (in hours) required for the first place runner and second place runner to finish the race. Then we have the system of equations

$$\begin{cases} 50 = v_1 t_1 & \text{(1) First place runner goes 50 miles in } t_1 \text{ hours.} \\ 49 = v_2 t_1 & \text{(2) Second place runner goes 49 miles in } t_1 \text{ hours.} \\ 46 = v_3 t_1 & \text{(3) Third place runner goes 46 miles in } t_1 \text{ hours.} \\ 50 = v_2 t_2 & \text{(4) Second place runner goes 50 miles in } t_2 \text{ hours.} \end{cases}$$

We seek the distance of the third place runner from the finish at time t_2. That is, we seek

$$\begin{aligned} 50 - v_3 t_2 &= 50 - v_3\left(t_1 \cdot \frac{t_2}{t_1}\right) \\ &= 50 - (v_3 t_1) \cdot \frac{t_2}{t_1} \\ &= 50 - 46 \cdot \frac{50/v_2}{50/v_1} && \begin{cases} \text{From (3), } v_3 t_1 = 46; \\ \text{from (4), } t_2 = 50/v_2; \\ \text{from (1), } t_1 = 50/v_1. \end{cases} \\ &= 50 - 46 \cdot \frac{v_1}{v_2} \\ &= 50 - 46 \cdot \frac{50}{49} && \text{Form the quotient of (1) and (2).} \\ &\approx 3.06 \text{ miles} \end{aligned}$$

■

HISTORICAL FEATURE

Recall that, in the beginning of this section, we said imagination and experience are important in solving simultaneous nonlinear equations. Indeed, these kinds of problems lead into some of the deepest and most difficult parts of modern mathematics. Look again at the graphs in Examples 1 and 2 of this section (Figures 17 or 18 and 19 or 20). We see that Example 1 has two solutions, and Example 2 has four solutions. We might conjecture that the number of solutions is equal to the product of the degrees of the equations involved. This conjecture was indeed made by Etienne Bezout (1730–1783), but working out the details took about 150 years. It turns out that, to arrive at the correct number of intersections, we must count not only the complex intersections, but also the intersections that, in a certain sense, lie at infinity. For example, a parabola and a line lying on the axis of the parabola intersect at the vertex and at infinity. This topic is part of the study of algebraic geometry.

HISTORICAL PROBLEM

1. A papyrus dating back to 1950 B.C. contains the following problem: A given surface area of 100 units of area shall be represented as the sum of two squares whose sides are to each other as $1:\frac{3}{4}$. Solve for the sides by solving the system of equations

$$\begin{cases} x^2 + y^2 = 100 \\ x = \frac{3}{4}y \end{cases}$$

11.6 EXERCISES

In Problems 1–20, use a graphing utility to graph each equation of the system. Then solve the system by finding the intersection points. Express your answer correct to two decimal places. Also solve each system algebraically.

1. $\begin{cases} y = x^2 + 1 \\ y = x + 1 \end{cases}$

2. $\begin{cases} y = x^2 + 1 \\ y = 4x + 1 \end{cases}$

3. $\begin{cases} y = \sqrt{36 - x^2} \\ y = 8 - x \end{cases}$

4. $\begin{cases} y = \sqrt{4 - x^2} \\ y = 2x + 4 \end{cases}$

5. $\begin{cases} y = \sqrt{x} \\ y = 2 - x \end{cases}$

6. $\begin{cases} y = \sqrt{x} \\ y = 6 - x \end{cases}$

7. $\begin{cases} x = 2y \\ x = y^2 - 2y \end{cases}$

8. $\begin{cases} y = x - 1 \\ y = x^2 - 6x + 9 \end{cases}$

9. $\begin{cases} x^2 + y^2 = 4 \\ x^2 + 2x + y^2 = 0 \end{cases}$

10. $\begin{cases} x^2 + y^2 = 8 \\ x^2 + y^2 + 4y = 0 \end{cases}$

11. $\begin{cases} y = 3x - 5 \\ x^2 + y^2 = 5 \end{cases}$

12. $\begin{cases} x^2 + y^2 = 10 \\ y = x + 2 \end{cases}$

13. $\begin{cases} x^2 + y^2 = 4 \\ y^2 - x = 4 \end{cases}$

14. $\begin{cases} x^2 + y^2 = 16 \\ x^2 - 2y = 8 \end{cases}$

15. $\begin{cases} xy = 4 \\ x^2 + y^2 = 8 \end{cases}$

16. $\begin{cases} x^2 = y \\ xy = 1 \end{cases}$

17. $\begin{cases} x^2 + y^2 = 4 \\ y = x^2 - 9 \end{cases}$

18. $\begin{cases} xy = 1 \\ y = 2x + 1 \end{cases}$

19. $\begin{cases} y = x^2 - 4 \\ y = 6x - 13 \end{cases}$

20. $\begin{cases} x^2 + y^2 = 10 \\ xy = 3 \end{cases}$

In Problems 21–52, solve each system. Use any method you wish.

21. $\begin{cases} 2x^2 + y^2 = 18 \\ xy = 4 \end{cases}$

22. $\begin{cases} x^2 - y^2 = 21 \\ x + y = 7 \end{cases}$

23. $\begin{cases} y = 2x + 1 \\ 2x^2 + y^2 = 1 \end{cases}$

24. $\begin{cases} x^2 - 4y^2 = 16 \\ 2y - x = 2 \end{cases}$

25. $\begin{cases} x + y + 1 = 0 \\ x^2 + y^2 + 6y - x = -5 \end{cases}$

26. $\begin{cases} 2x^2 - xy + y^2 = 8 \\ xy = 4 \end{cases}$

27. $\begin{cases} 4x^2 - 3xy + 9y^2 = 15 \\ 2x + 3y = 5 \end{cases}$

28. $\begin{cases} 2y^2 - 3xy + 6y + 2x + 4 = 0 \\ 2x - 3y + 4 = 0 \end{cases}$

29. $\begin{cases} x^2 - 4y^2 + 7 = 0 \\ 3x^2 + y^2 = 31 \end{cases}$

30. $\begin{cases} 3x^2 - 2y^2 + 5 = 0 \\ 2x^2 - y^2 + 2 = 0 \end{cases}$

31. $\begin{cases} 7x^2 - 3y^2 + 5 = 0 \\ 3x^2 + 5y^2 = 12 \end{cases}$

32. $\begin{cases} x^2 - 3y^2 + 1 = 0 \\ 2x^2 - 7y^2 + 5 = 0 \end{cases}$

33. $\begin{cases} x^2 + 2xy = 10 \\ 3x^2 - xy = 2 \end{cases}$

34. $\begin{cases} 5xy + 13y^2 + 36 = 0 \\ xy + 7y^2 = 6 \end{cases}$

35. $\begin{cases} 2x^2 + y^2 = 2 \\ x^2 - 2y^2 + 8 = 0 \end{cases}$

36. $\begin{cases} y^2 - x^2 + 4 = 0 \\ 2x^2 + 3y^2 = 6 \end{cases}$

37. $\begin{cases} x^2 + 2y^2 = 16 \\ 4x^2 - y^2 = 24 \end{cases}$

38. $\begin{cases} 4x^2 + 3y^2 = 4 \\ 2x^2 - 6y^2 = -3 \end{cases}$

39. $\begin{cases} \dfrac{5}{x^2} - \dfrac{2}{y^2} + 3 = 0 \\ \dfrac{3}{x^2} + \dfrac{1}{y^2} = 7 \end{cases}$

40. $\begin{cases} \dfrac{2}{x^2} - \dfrac{3}{y^2} + 1 = 0 \\ \dfrac{6}{x^2} - \dfrac{7}{y^2} + 2 = 0 \end{cases}$

41. $\begin{cases} \dfrac{1}{x^4} + \dfrac{6}{y^4} = 6 \\ \dfrac{2}{x^4} - \dfrac{2}{y^4} = 19 \end{cases}$

42. $\begin{cases} \dfrac{1}{x^4} - \dfrac{1}{y^4} = 1 \\ \dfrac{1}{x^4} + \dfrac{1}{y^4} = 4 \end{cases}$

43. $\begin{cases} x^2 - 3xy + 2y^2 = 0 \\ x^2 + xy = 6 \end{cases}$

44. $\begin{cases} x^2 - xy - 2y^2 = 0 \\ xy + x + 6 = 0 \end{cases}$

45. $\begin{cases} xy - x^2 + 3 = 0 \\ 3xy - 4y^2 = 2 \end{cases}$

46. $\begin{cases} 5x^2 + 4xy + 3y^2 = 36 \\ x^2 + xy + y^2 = 9 \end{cases}$

47. $\begin{cases} x^3 - y^3 = 26 \\ x - y = 2 \end{cases}$

48. $\begin{cases} x^3 + y^3 = 26 \\ x + y = 2 \end{cases}$

49. $\begin{cases} y^2 + y + x^2 - x - 2 = 0 \\ y + 1 + \dfrac{x - 2}{y} = 0 \end{cases}$

50. $\begin{cases} x^3 - 2x^2 + y^2 + 3y - 4 = 0 \\ x - 2 + \dfrac{y^2 - y}{x^2} = 0 \end{cases}$

51. $\begin{cases} \log_x y = 3 \\ \log_x(4y) = 5 \end{cases}$

52. $\begin{cases} \log_x(2y) = 3 \\ \log_x(4y) = 2 \end{cases}$

In Problems 53–60, use a graphing utility to solve each system of equations. Express the solution(s) correct to two decimal places.

53. $\begin{cases} y = x^{2/3} \\ y = e^{-x} \end{cases}$

54. $\begin{cases} y = x^{3/2} \\ y = e^{-x} \end{cases}$

55. $\begin{cases} x^2 + y^3 = 2 \\ x^3 y = 4 \end{cases}$

56. $\begin{cases} x^3 + y^2 = 2 \\ x^2 y = 4 \end{cases}$

57. $\begin{cases} x^4 + y^4 = 12 \\ xy^2 = 2 \end{cases}$

58. $\begin{cases} x^4 + y^4 = 6 \\ xy = 1 \end{cases}$

59. $\begin{cases} xy = 2 \\ y = \ln x \end{cases}$

60. $\begin{cases} x^2 + y^2 = 4 \\ y = \ln x \end{cases}$

61. The difference of two numbers is 2 and the sum of their squares is 10. Find the numbers.

62. The sum of two numbers is 7 and the difference of their squares is 21. Find the numbers.

63. The product of two numbers is 4 and the sum of their squares is 8. Find the numbers.

64. The product of two numbers is 10 and the difference of their squares is 21. Find the numbers.

65. The difference of two numbers is the same as their product, and the sum of their reciprocals is 5. Find the numbers.

66. The sum of two numbers is the same as their product, and the difference of their reciprocals is 3. Find the numbers.

67. The ratio of a to b is $\frac{2}{3}$. The sum of a and b is 10. What is the ratio of $a + b$ to $b - a$?

68. The ratio of a to b is $\frac{4}{3}$. The sum of a and b is 14. What is the ratio of $a - b$ to $a + b$?

In Problems 69–74, graph each equation by hand and find the point(s) of intersection, if any.

69. The line $x + 2y = 0$ and the circle $(x - 1)^2 + (y - 1)^2 = 5$

70. The line $x + 2y + 6 = 0$ and the circle $(x + 1)^2 + (y + 1)^2 = 5$

71. The circle $(x - 1)^2 + (y + 2)^2 = 4$ and the parabola $y^2 + 4y - x + 1 = 0$

72. The circle $(x + 2)^2 + (y - 1)^2 = 4$ and the parabola $y^2 - 2y - x - 5 = 0$

73. The graph of $y = \dfrac{4}{x - 3}$ and the circle $x^2 - 6x + y^2 + 1 = 0$

74. The graph of $y = \dfrac{4}{x + 2}$ and the circle $x^2 + 4x + y^2 - 4 = 0$

75. Geometry The perimeter of a rectangle is 16 inches and its area is 15 square inches. What are its dimensions?

76. Geometry An area of 52 square feet is to be enclosed by two squares whose sides are in the ratio of 2:3. Find the sides of the squares.

77. Geometry Two circles have perimeters that add up to 12π centimeters and areas that add up to 20π square centimeters. Find the radius of each circle.

78. Geometry The altitude of an isosceles triangle drawn to its base is 3 centimeters, and its perimeter is 18 centimeters. Find the length of its base.

79. The Tortoise and the Hare In a 21-meter race between a tortoise and a hare, the tortoise leaves 9 minutes before the hare. The hare, by running at an average speed of 0.5 meter per hour faster than the tortoise, crosses the finish line 3 minutes before the tortoise. What are the average speeds of the tortoise and the hare?

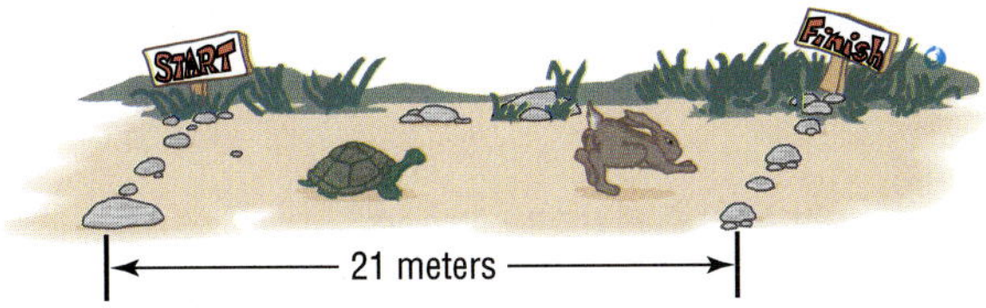

80. Running a Race In a 1-mile race, the winner crosses the finish line 10 feet ahead of the second place runner and 20 feet ahead of the third place runner. Assuming that each runner maintains a constant speed throughout the race, by how many feet does the second place runner beat the third place runner?

81. Constructing a Box A rectangular piece of cardboard, whose area is 216 square centimeters, is made into an open box by cutting a 2-centimeter square from each corner and turning up the sides. See the figure. If the box is to have a volume of 224 cubic centimeters, what size cardboard should you start with?

82. Constructing a Cylindrical Tube A rectangular piece of cardboard, whose area is 216 square centimeters, is made into a cylindrical tube by joining together two sides of the rectangle. (See the figure.) If the tube is to have a volume of 224 cubic centimeters, what size cardboard should you start with?

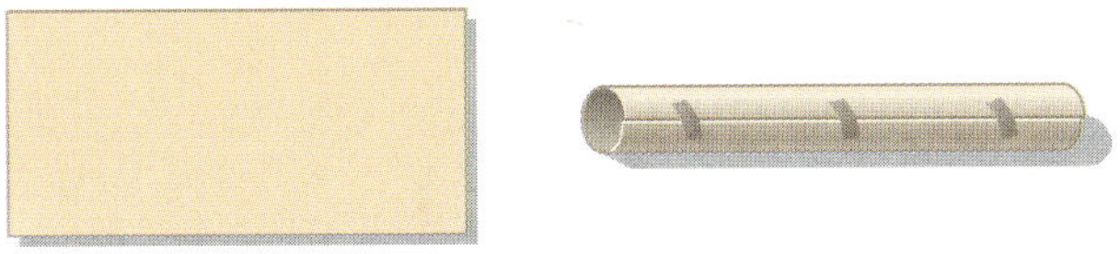

83. Fencing A farmer has 300 feet of fence available to enclose 4500 square feet in the shape of adjoining squares, with sides of length x and y. See the figure. Find x and y.

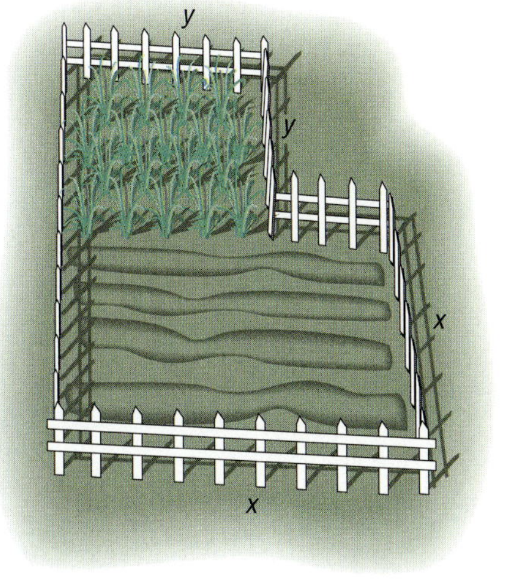

84. Bending Wire A wire 60 feet long is cut into two pieces. Is it possible to bend one piece into the shape of a square and the other into the shape of a circle so that the total area enclosed by the two pieces is 100 square feet? If this is possible, find the length of the side of the square and the radius of the circle.

85. Geometry Find formulas for the length l and width w of a rectangle in terms of its area A and perimeter P.

86. Geometry Find formulas for the base b and one of the equal sides l of an isosceles triangle in terms of its altitude h and perimeter P.

87. Descartes' Method of Equal Roots Descartes' method for finding tangents depends on the idea that, for many graphs, the tangent line at a given point is the *unique* line that intersects the graph at that point only. We will apply his method to find an equation of the tangent line to the parabola $y = x^2$ at the point $(2, 4)$; see the figure. First, we know the equation of the tangent line must be in the form $y = mx + b$. Using the fact that the point $(2, 4)$ is on the line, we can solve for b in terms of m and get the equation $y = mx + (4 - 2m)$. Now we want $(2, 4)$ to be the *unique* solution to the system

$$\begin{cases} y = x^2 \\ y = mx + 4 - 2m \end{cases}$$

From this system, we get $x^2 = mx + 4 - 2m$ or $x^2 - mx + (2m - 4) = 0$. By using the quadratic formula, we get

$$x = \frac{m \pm \sqrt{m^2 - 4(2m - 4)}}{2}$$

To obtain a unique solution for x, the two roots must be equal; in other words, the discriminant $m^2 - 4(2m - 4)$ must be 0. Complete the work to get m, and write an equation of the tangent line.

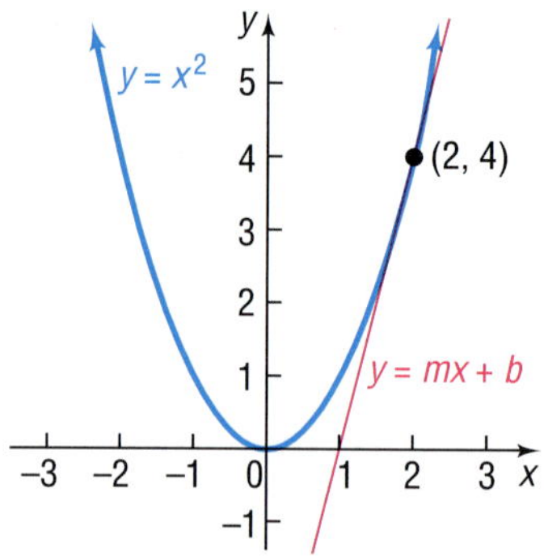

In Problems 88–94, use Descartes' method from Problem 87 to find the equation of the line tangent to each graph at the given point.

88. $x^2 + y^2 = 10$; at $(1, 3)$

89. $y = x^2 + 2$; at $(1, 3)$

90. $x^2 + y = 5$; at $(-2, 1)$

91. $2x^2 + 3y^2 = 14$; at $(1, 2)$

92. $3x^2 + y^2 = 7$; at $(-1, 2)$

93. $x^2 - y^2 = 3$; at $(2, 1)$

94. $2y^2 - x^2 = 14$; at $(2, 3)$

95. If r_1 and r_2 are two solutions of a quadratic equation $ax^2 + bx + c = 0$, then it can be shown that

$$r_1 + r_2 = -\frac{b}{a} \quad \text{and} \quad r_1 r_2 = \frac{c}{a}$$

Solve this system of equations for r_1 and r_2.

96. A circle and a line intersect at most twice. A circle and a parabola intersect at most four times. Deduce that a circle and the graph of a polynomial of degree 3 intersect at most six times. What do you conjecture about a polynomial of degree 4? What about a polynomial of degree n? Can you explain your conclusions using an algebraic argument?

97. Suppose that you are the manager of a sheet metal shop. A customer asks you to manufacture 10,000 boxes, each box being open on top. The boxes are required to have a square base and a 9-cubic foot capacity. You construct the boxes by cutting a square out from each corner of a square piece of sheet metal and folding along the edges.

(a) What are the dimensions of the square to be cut if the area of the square piece of sheet metal is 100 square feet?

(b) Could you make the box using a smaller piece of sheet metal? Make a list of the dimensions of the box for various pieces of sheet metal.

11.7 SYSTEMS OF INEQUALITIES

1 Graph an Inequality by Hand

2 Graph an Inequality Using a Graphing Utility

3 Graph a System of Linear Inequalities

In Chapter 3, we discussed inequalities in one variable. In this section, we discuss inequalities in two variables. Samples are given in Example 1.

EXAMPLE 1 Samples of Inequalities in Two Variables

(a) $3x + y - 6 < 0$ (b) $x^2 + y^2 < 4$ (c) $y^2 \le x$

1

An inequality in two variables x and y is **satisfied** by an ordered pair (a, b) if, when x is replaced by a and y by b, a true statement results. A **graph of an inequality in two variables** x and y consists of all points (x, y) whose coordinates satisfy the inequality.

Let's look at an example.

EXAMPLE 2 Graphing a Linear Inequality

Graph the linear inequality $3x + y - 6 \leq 0$.

Solution We begin with the associated problem of the graph of the linear equation

$$3x + y - 6 = 0$$

formed by replacing (for now) the $\leq$ symbol with an $=$ sign. The graph of the linear equation is a line. See Figure 26(a). This line is part of the graph of the inequality we seek because the inequality is nonstrict. (Do you see why? We are seeking points for which $3x + y - 6$ is less than *or equal to* 0.)

Now, let's test a few randomly selected points to see whether they belong to the graph of the inequality

	$3x + y - 6$	**Conclusion**
$(4, -1)$	$3(4) + (-1) - 6 = 5 > 0$	Does not belong to graph
$(5, 5)$	$3(5) + 5 - 6 = 14 > 0$	Does not belong to graph
$(-1, 2)$	$3(-1) + 2 - 6 = -7 < 0$	Belongs to graph
$(-2, -2)$	$3(-2) + (-2) - 6 = -14 < 0$	Belongs to graph

Look again at Figure 26(a). Notice that the two points that belong to the graph both lie on the same side of the line, and the two points that do not belong to the graph lie on the opposite side. As it turns out, this is always the case. Thus, the graph we seek consists of all points that lie on the same side of the line as do $(-1, 2)$ and $(-2, -2)$. The graph we seek is the shaded region in Figure 26(b).

FIGURE 26

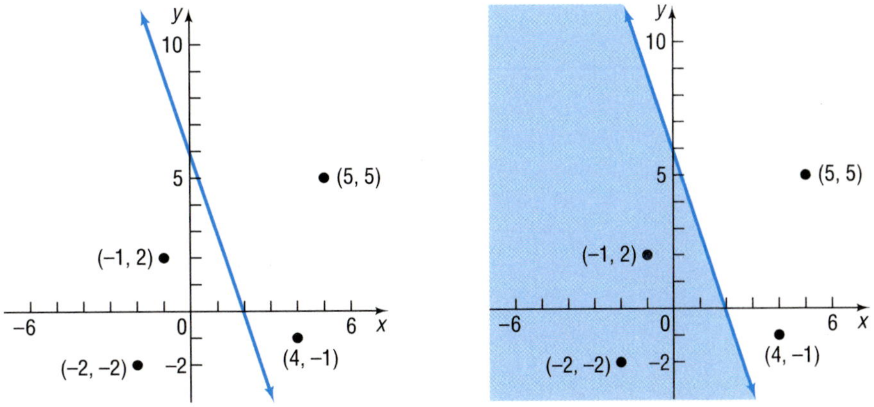

(a) $3x + y - 6 = 0$

(b) Graph of $3x + y - 6 \leq 0$

The graph of any inequality in two variables may be obtained in a like way. First, the equation corresponding to the inequality is graphed, using dashes if the inequality is strict and solid marks if it is nonstrict. This graph, in almost every case, will separate the xy-plane into two or more regions. In each region either all points satisfy the inequality or no points satisfy the inequality. Thus, the use of a single test point in each region is all that is required to determine whether the points of that region are part of the graph or not. The steps to follow are given next.

Steps for Graphing an Inequality By Hand

STEP 1: Replace the inequality symbol by an equal sign and graph the resulting equation. If the inequality is strict, use dashes; if it is nonstrict, use a solid mark. This graph separates the xy-plane into two or more regions.

STEP 2: In each of the regions, select a test point P.
(a) If the coordinates of P satisfy the inequality, then so do all the points in that region. Indicate this by shading the region.
(b) If the coordinates of P do not satisfy the inequality, then none of the points in that region do.

2 Graphing utilities can also be used to graph inequalities.

EXAMPLE 3 Graphing an Inequality Using a Graphing Utility

Use a graphing utility to graph $3x + y - 6 \leq 0$.

Solution We begin by graphing the equation $3x + y - 6 = 0$ ($Y_1 = -3x + 6$). See Figure 27.

FIGURE 27

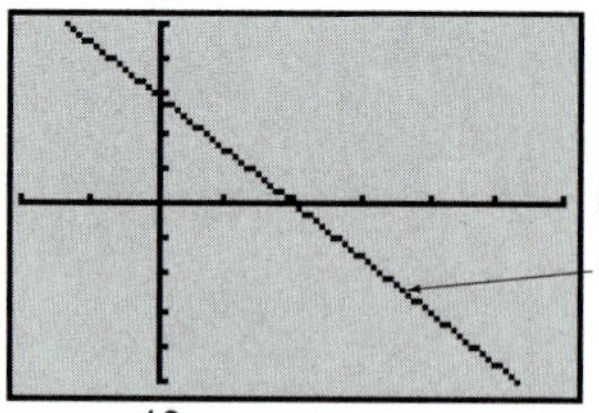

As with graphing by hand, we need to test points selected from each region and determine whether or not they satisfy the inequality. To test the point $(-1, 2)$, for example, enter 3∗−1+2−6≤0. See Figure 28(a). The 1 that appears indicates the statement entered (the inequality) is true. When the point $(5, 5)$ is tested, a 0 appears, indicating the statement entered is false. Thus, $(-1, 2)$ is a part of the graph of the inequality and $(5, 5)$ is not. Figure 28(b) shows the graph of the inequality on a TI-83.*

*Consult your owner's manual for shading techniques.

FIGURE 28

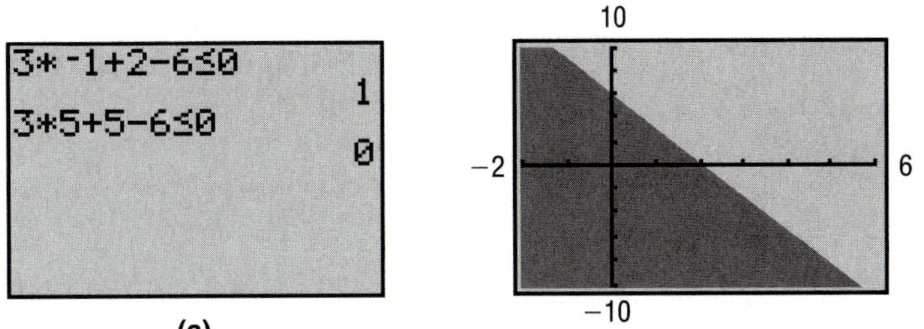

(a) (b)

The steps to follow to graph an inequality using a graphing utility are given next.

Steps for Graphing an Inequality Using a Graphing Utility

> STEP 1: Replace the inequality symbol by an equal sign and graph the resulting equation.
>
> STEP 2: In each of the regions, select a test point P.
>
> (a) Use a graphing utility to determine if the test point P satisfies the inequality. If the test point satisfies the inequality, then so do all the points in the region. Indicate this by using the graphing utility to shade the region.
>
> (b) If the coordinates of P do not satisfy the inequality, then none of the points in that region do.

EXAMPLE 4 Graphing an Inequality

Graph $x^2 + y^2 \le 4$.

FIGURE 29

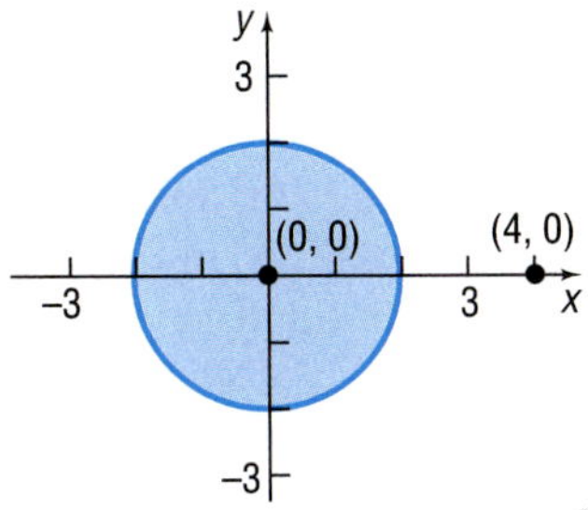

Solution First we graph the equation $x^2 + y^2 = 4$, a circle of radius 2, center at the origin. A solid circle will be used because the inequality is not strict. We use two test points, one inside the circle, the other outside;

Inside $(0, 0)$: $x^2 + y^2 = 0^2 + 0^2 = 0 \le 4$ Belongs to the graph.

Outside $(4, 0)$: $x^2 + y^2 = 4^2 + 0^2 = 16 > 4$ Does not belong to graph.

All the points inside and on the circle satisfy the inequality. See Figure 29.

Now work Problem 7.

Linear inequalities are inequalities in one of the forms

$$Ax + By < C, \qquad Ax + By > C, \qquad Ax + By \le C, \qquad Ax + By \ge C$$

The graph of the corresponding equation of a linear inequality is a line, which separates the xy-plane into two regions, called **half-planes.** See Figure 30.

FIGURE 30

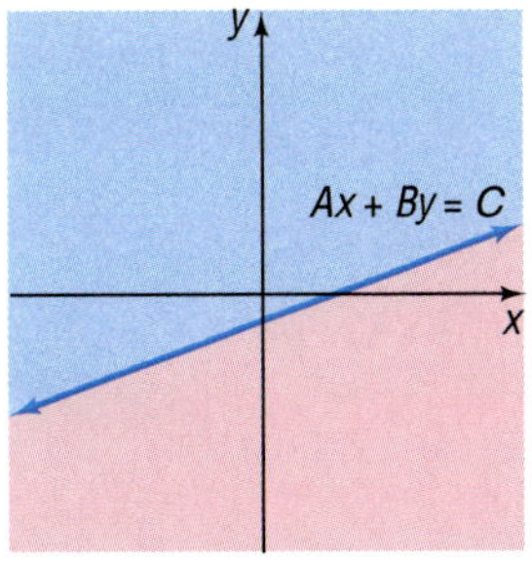

As shown, if $Ax + By = C$ is the equation of the boundary line, then it divides the plane into two half-planes; one for which $Ax + By < C$ and the other for which $Ax + By > C$. Because of this, for linear inequalities, only one test point is required.

EXAMPLE 5 Graphing Linear Inequalities

Graph (a) $y < 2$ (b) $y \geq 2x$

Solution (a) The graph of the equation $y = 2$ is a horizontal line and is not part of the graph of the inequality. Since $(0, 0)$ satisfies the inequality, the graph consists of the half-plane below the line $y = 2$. See Figure 31.

(b) The graph of the equation $y = 2x$ is a line and is part of the graph of the inequality. Using $(3, 0)$ as a test point, we find it does not satisfy the inequality $[0 < 2 \cdot 3]$. Thus, points in the half-plane on the opposite side of $y = 2x$ satisfy the inequality. See Figure 32.

FIGURE 31

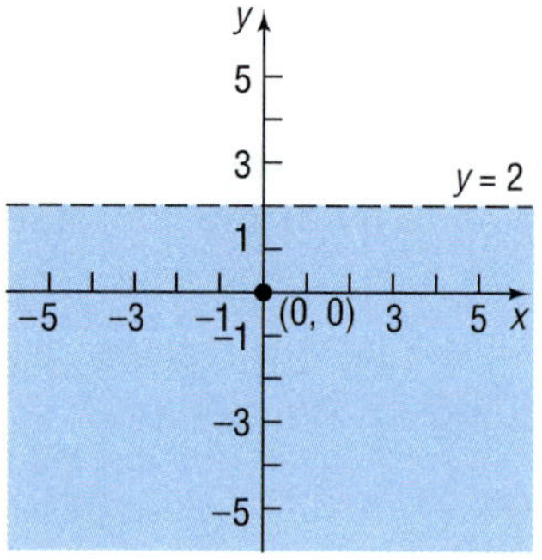

FIGURE 32

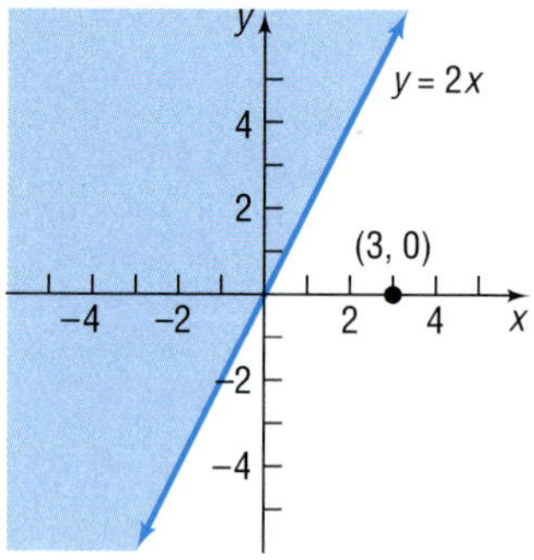

Now work Problem 3.

Systems of Inequalities in Two Variables

3 The **graph of a system of inequalities** in two variables x and y is the set of all points (x, y) that simultaneously satisfy *each* of the inequalities in the system. Thus, the graph of a system of inequalities can be obtained by graphing each inequality individually and then determining where, if at all, they intersect.

EXAMPLE 6 Graphing a System of Linear Inequalities Using a Graphing Utility

Graph the system: $\begin{cases} x + y \geq 2 \\ 2x - y \leq 4 \end{cases}$

Solution First, we graph the lines $x + y = 2$ $(Y_1 = -x + 2)$ and $2x - y = 4$ $(Y_2 = 2x - 4)$. See Figure 33.

FIGURE 33

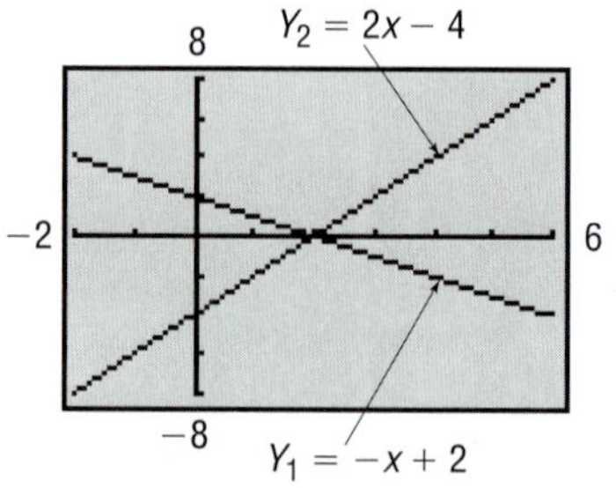

Notice the graphs divide the viewing window into four regions. We select a test point for each region and determine whether the point makes *both* inequalities true. We choose to test $(0, 0)$, $(2, 3)$, $(4, 0)$, and $(2, -2)$. Figure 34(a) shows that $(2, 3)$ is the only point for which both inequalities are true. Thus, we obtain the graph shown in Figure 34(b).

FIGURE 34

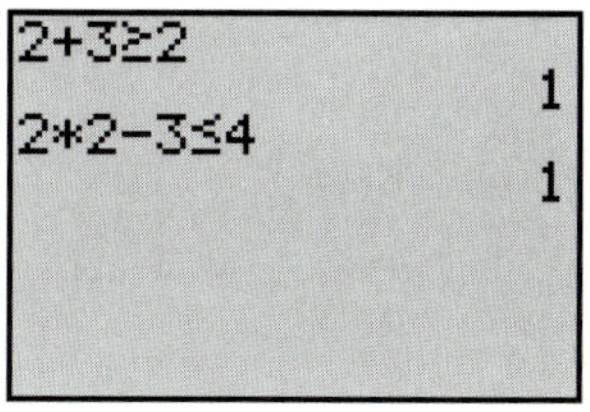

(a)

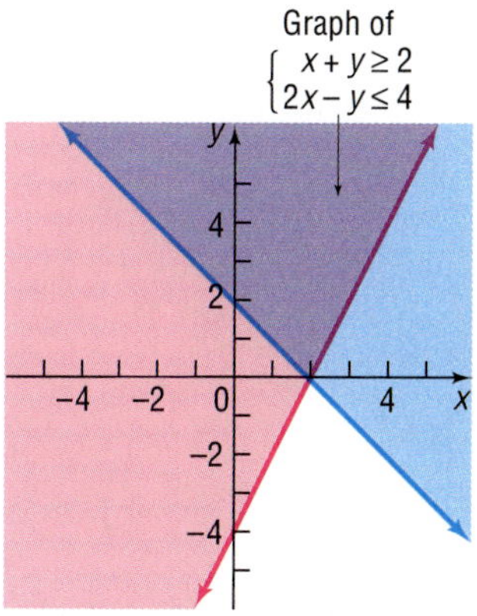

(b)

Notice in Figure 34(b) that the darker region, the graph of the system of inequalities, is the intersection of the graphs of the single inequalities $x + y \geq 2$ and $2x - y \leq 4$. Obtaining Figure 34(b) by this method is sometimes faster than using test points.

EXAMPLE 7 Graphing a System of Linear Inequalities

Graph the system: $\begin{cases} x + y \leq 2 \\ x + y \geq 0 \end{cases}$

Solution See Figure 35. The overlapping, darker shaded region between the two boundary lines is the graph of the system.

FIGURE 35

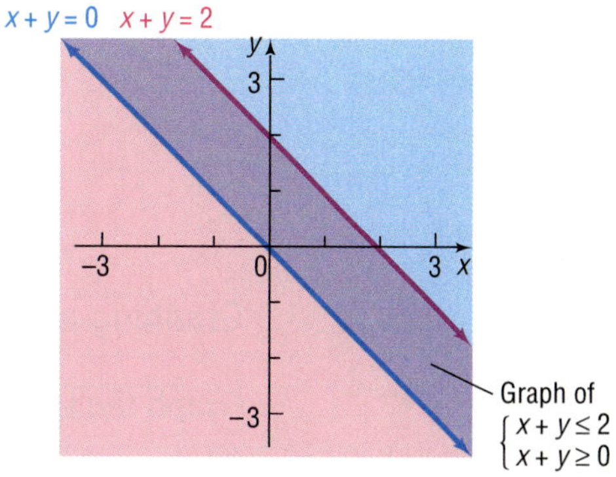

Now work Problem 25.

EXAMPLE 8 Graphing a System of Linear Inequalities

Graph the system: $\begin{cases} 2x - y \geq 2 \\ 2x - y \geq 0 \end{cases}$

Solution See Figure 36. The overlapping, darker shaded region is the graph of the system. Note that the graph of the system is identical to the graph of the single inequality $2x - y \geq 2$.

FIGURE 36

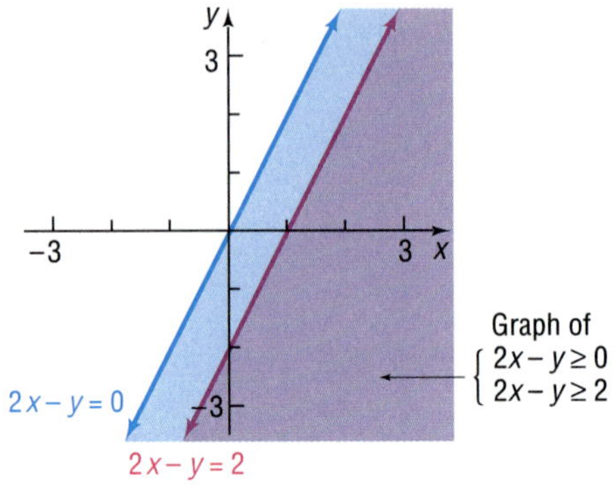

EXAMPLE 9 Graphing a System of Linear Inequalities

Graph the system: $\begin{cases} x + 2y \leq 2 \\ x + 2y \geq 6 \end{cases}$

Solution See Figure 37. Because no overlapping region results, there are no points in the xy-plane that simultaneously satisfy each inequality. Hence, the system has no solution.

FIGURE 37

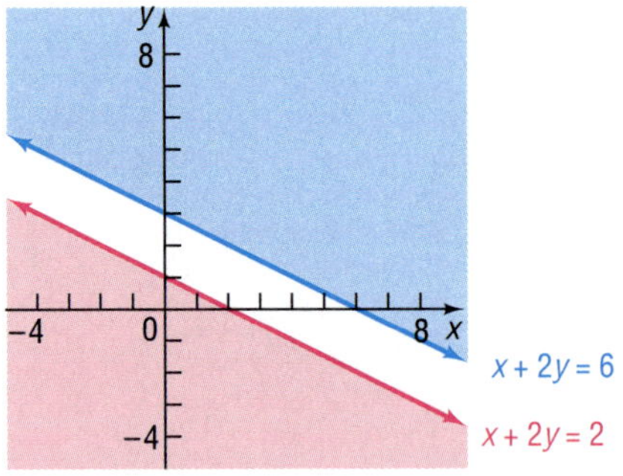

EXAMPLE 10 Graphing a System of Inequalities

Graph the system: $\begin{cases} y \geq x^2 - 4 \\ x + y \leq 2 \end{cases}$

Solution Figure 38 shows the graph of the equation $Y_1 = x^2 - 4$ and $x + y = 2$ $(y_2 = 2 - x)$. The graphs divide the viewing window into five regions. Instead of selecting test points in each region, we determine the graph of the system consists of the intersection of the graphs of the single inequalities $y \geq x^2 - 4$ and $x + y \leq 2$. See Figure 39.

FIGURE 38

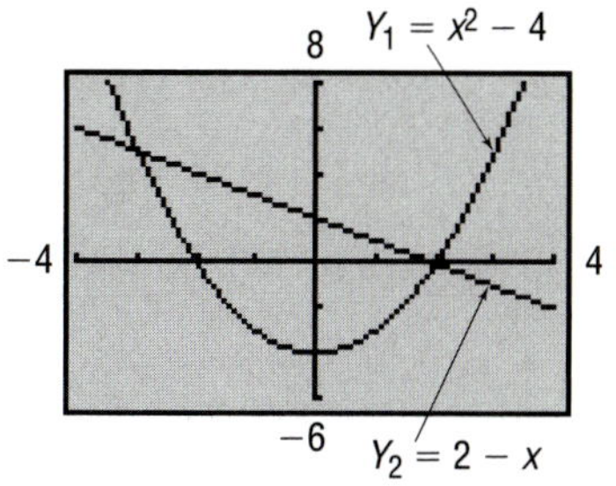

FIGURE 39

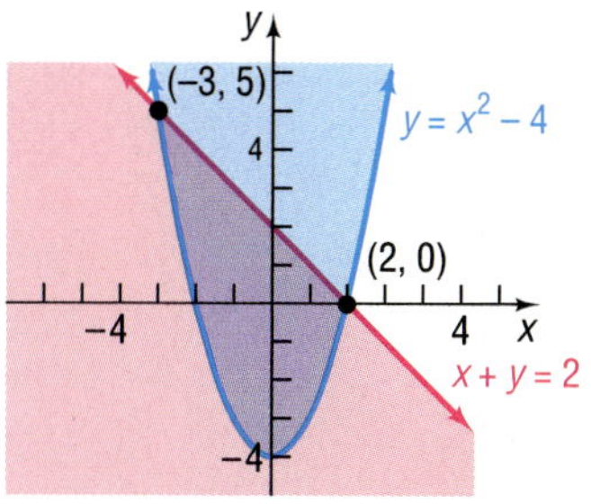

The points of intersection of the two equations are found by solving the system of equations

$$\begin{cases} y = x^2 - 4 \\ x + y = 2 \end{cases}$$

Using substitution, we find

$$\begin{aligned} x + (x^2 - 4) &= 2 \\ x^2 + x - 6 &= 0 \\ (x + 3)(x - 2) &= 0 \\ x = -3, \qquad x &= 2 \end{aligned}$$

The two points of intersection are $(-3, 5)$ and $(2, 0)$.

Now work Problem 19.

EXAMPLE 11 Graphing a System of Four Linear Inequalities

Graph the system: $\begin{cases} x + y \geq 3 \\ 2x + y \geq 4 \\ x \geq 0 \\ y \geq 0 \end{cases}$

Solution The two inequalities $x \geq 0$ and $y \geq 0$ require that the graph be in quadrant I. Thus, we set our viewing window accordingly. Figure 40 shows the graph of $x + y = 3$ ($Y_1 = -x + 3$) and $2x + y = 4$ ($Y_2 = -2x + 4$). The graphs divide the viewing window into four regions.

FIGURE 40

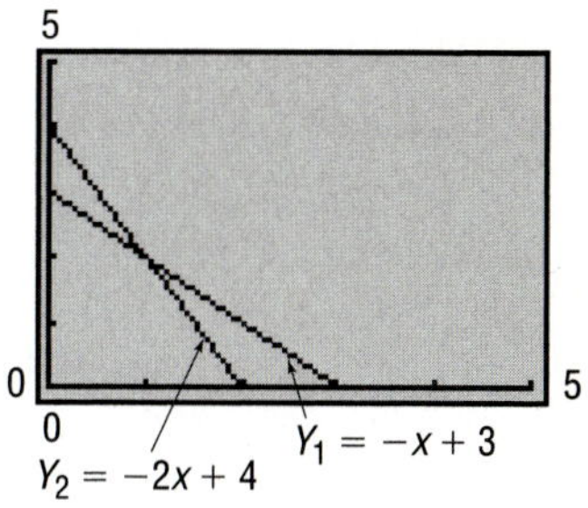

Instead of selecting test points in each region, we determine that the graph of the system is the intersection of the graphs of the single inequalities $x + y \geq 3$ and $2x + y \geq 4$ in quadrant I. See Figure 41.

FIGURE 41

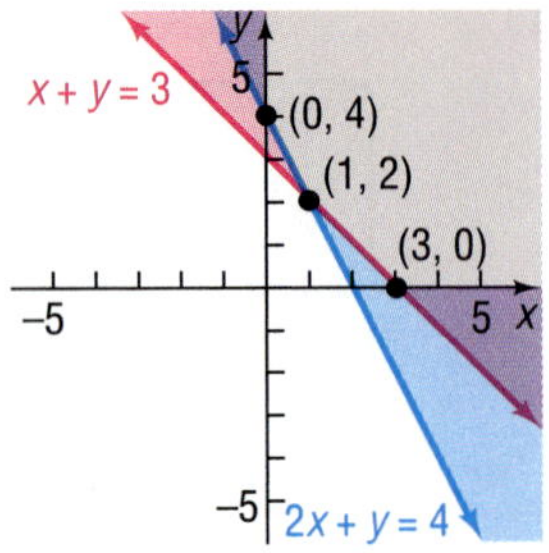

EXAMPLE 12 Financial Planning

A retired couple has up to \$25,000 to invest. As their financial adviser, you recommend that they place at least \$15,000 in Treasury bills yielding 6% and at most \$5000 in corporate bonds yielding 9%.

(a) Using x to denote the amount of money invested in Treasury bills and y the amount invested in corporate bonds, write a system of linear inequalities that describes the possible amounts of each investment.

(b) Graph the system.

Solution (a) The system of linear inequalities is

$$\begin{cases} x \geq 0 & \text{x and y are nonnegative variables since they represent money invested.} \\ y \geq 0 & \\ x + y \leq 25{,}000 & \text{The total of the two investments, } x + y\text{, cannot exceed \$25,000.} \\ x \geq 15{,}000 & \text{At least \$15,000 in Treasury bills.} \\ y \leq 5000 & \text{At most \$5000 in corporate bonds.} \end{cases}$$

(b) See the shaded region in Figure 42. Note that the inequalities $x \geq 0$ and $y \geq 0$ again require that the graph of the system be in quadrant I.

FIGURE 42

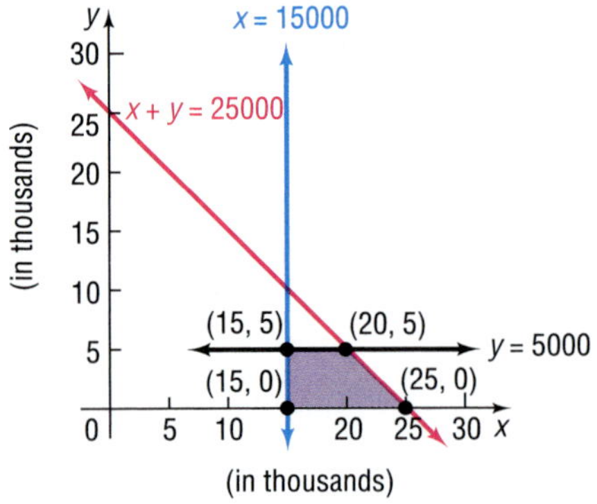

The graph of the system of linear inequalities in Figure 42 is said to be **bounded,** because it can be contained within some circle of sufficiently large radius. A graph that cannot be contained in any circle is said to be **unbounded.** For example, the graph of the system of linear inequalities in Figure 41 is unbounded, since it extends indefinitely in a particular direction.

Notice in Figures 41 and 42 that those points belonging to the graph that are also points of intersection of boundary lines have been plotted. Such points are referred to as **vertices** or **corner points** of the graph. Thus, the system graphed in Figure 41 has three corner points: (0, 4), (1, 2), and (3, 0). The system graphed in Figure 42 has four corner points: (15, 0), (25, 0), (20, 5), and (15, 5).

These ideas will be used in the next section in developing a method for solving linear programming problems, an important application of linear inequalities.

Now work Problem 33.

11.7 EXERCISES

In Problems 1–12, graph each inequality.

1. $x \geq 0$ **2.** $y \geq 0$ **3.** $x \geq 4$ **4.** $y \leq 2$

5. $2x + y \geq 6$ **6.** $3x + 2y \leq 6$ **7.** $x^2 + y^2 > 1$ **8.** $x^2 + y^2 \leq 9$

9. $y \leq x^2 - 1$ **10.** $y > x^2 + 2$ **11.** $xy \geq 4$ **12.** $xy \leq 1$

In Problems 13–30, graph each system of inequalities.

13. $\begin{cases} x + y \leq 2 \\ 2x + y \geq 4 \end{cases}$ **14.** $\begin{cases} 3x - y \geq 6 \\ x + 2y \leq 2 \end{cases}$ **15.** $\begin{cases} 2x - y \leq 4 \\ 3x + 2y \geq -6 \end{cases}$ **16.** $\begin{cases} 4x - 5y \leq 0 \\ 2x - y \geq 2 \end{cases}$

17. $\begin{cases} 2x - 3y \leq 0 \\ 3x + 2y \leq 6 \end{cases}$ **18.** $\begin{cases} 4x - y \geq 2 \\ x + 2y \geq 2 \end{cases}$ **19.** $\begin{cases} x^2 + y^2 \leq 9 \\ x + y \geq 3 \end{cases}$ **20.** $\begin{cases} x^2 + y^2 \geq 9 \\ x + y \leq 3 \end{cases}$

21. $\begin{cases} y \geq x^2 - 4 \\ y \leq x - 2 \end{cases}$ **22.** $\begin{cases} y^2 \leq x \\ y \geq x \end{cases}$ **23.** $\begin{cases} xy \geq 4 \\ y \geq x^2 + 1 \end{cases}$ **24.** $\begin{cases} y + x^2 \leq 1 \\ y \geq x^2 - 1 \end{cases}$

25. $\begin{cases} x - 2y \leq 6 \\ 2x - 4y \geq 0 \end{cases}$ **26.** $\begin{cases} x + 4y \leq 8 \\ x + 4y \geq 4 \end{cases}$ **27.** $\begin{cases} 2x + y \geq -2 \\ 2x + y \geq 2 \end{cases}$ **28.** $\begin{cases} x - 4y \leq 4 \\ x - 4y \geq 0 \end{cases}$

29. $\begin{cases} 2x + 3y \geq 6 \\ 2x + 3y \leq 0 \end{cases}$ **30.** $\begin{cases} 2x + y \geq 0 \\ 2x + y \geq 2 \end{cases}$

In Problems 31–40, graph each system of linear inequalities. Tell whether the graph is bounded or unbounded, and label the corner points.

31. $\begin{cases} x \geq 0 \\ y \geq 0 \\ 2x + y \leq 6 \\ x + 2y \leq 6 \end{cases}$ **32.** $\begin{cases} x \geq 0 \\ y \geq 0 \\ x + y \geq 4 \\ 2x + 3y \geq 6 \end{cases}$ **33.** $\begin{cases} x \geq 0 \\ y \geq 0 \\ x + y \geq 2 \\ 2x + y \geq 4 \end{cases}$ **34.** $\begin{cases} x \geq 0 \\ y \geq 0 \\ 3x + y \leq 6 \\ 2x + y \leq 2 \end{cases}$ **35.** $\begin{cases} x \geq 0 \\ y \geq 0 \\ x + y \geq 2 \\ 2x + 3y \leq 12 \\ 3x + y \leq 12 \end{cases}$

36. $\begin{cases} x \geq 0 \\ y \geq 0 \\ x + y \geq 2 \\ x + y \leq 10 \\ 2x + y \leq 3 \end{cases}$ **37.** $\begin{cases} x \geq 0 \\ y \geq 0 \\ x + y \geq 2 \\ x + y \leq 8 \\ 2x + y \leq 10 \end{cases}$ **38.** $\begin{cases} x \geq 0 \\ y \geq 0 \\ x + y \geq 2 \\ x + y \leq 8 \\ x + 2y \geq 1 \end{cases}$ **39.** $\begin{cases} x \geq 0 \\ y \geq 0 \\ x + 2y \geq 1 \\ x + 2y \leq 10 \end{cases}$ **40.** $\begin{cases} x \geq 0 \\ y \geq 0 \\ x + 2y \geq 1 \\ x + 2y \leq 10 \\ x + y \geq 2 \\ x + y \leq 8 \end{cases}$

In Problems 41–44, write a system of linear inequalities that has the given graph.

41.

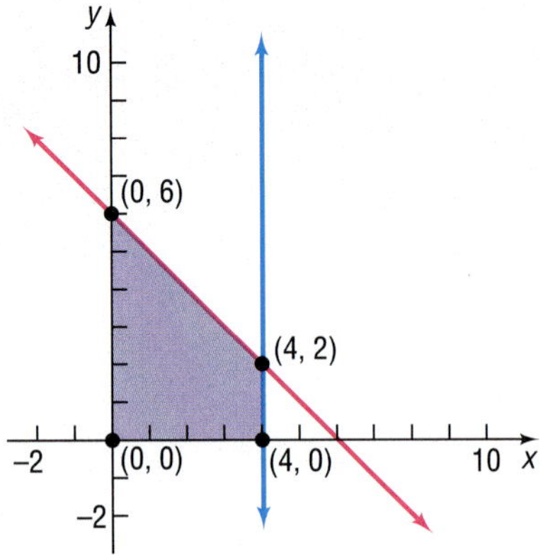

42.

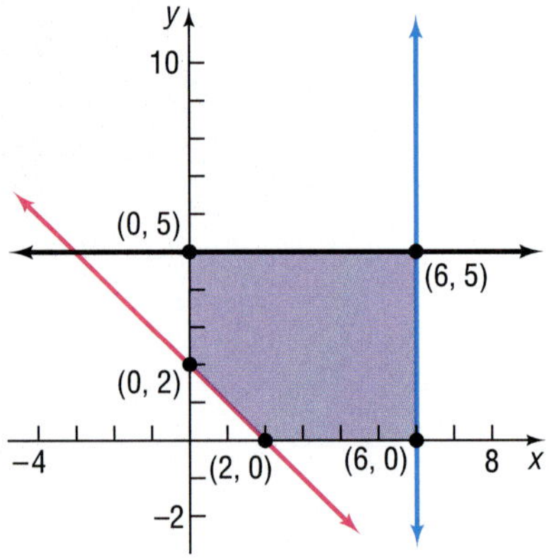

43.

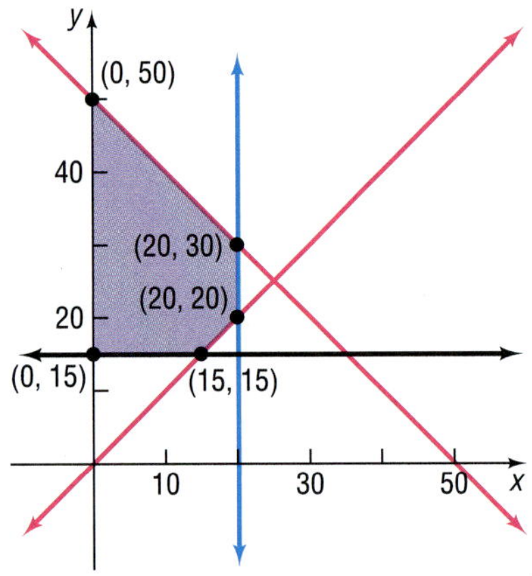

44.

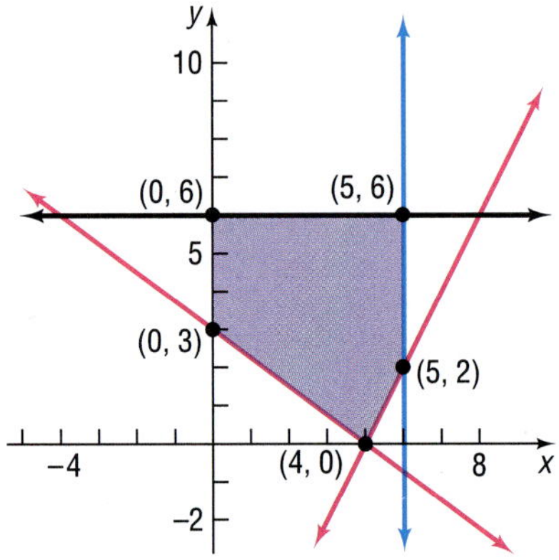

45. Financial Planning A retired couple has up to \$50,000 to invest. As their financial adviser, you recommend they place at least \$35,000 in Treasury bills yielding 7% and at most \$10,000 in corporate bonds yielding 10%.

(a) Using x to denote the amount of money invested in Treasury bills and y the amount invested in corporate bonds, write a system of linear inequalities that describes the possible amounts of each investment.

(b) Graph the system and label the corner points.

46. Manufacturing Trucks Mike's Toy Truck Company manufactures two models of toy trucks, a standard model and a deluxe model. Each standard model requires 2 hours for painting and 3 hours for detail work; each deluxe model requires 3 hours for painting and 4 hours for detail work. Two painters and three detail workers are employed by the company, and each works 40 hours per week.

(a) Using x to denote the number of standard model trucks and y to denote the number of deluxe model trucks, write a system of linear inequalities that describes the possible number of each model of truck that can be manufactured in a week.

(b) Graph the system and label the corner points.

47. Blending Coffee Bill's Coffee House, a store that specializes in coffee, has available 75 pounds of A grade coffee and 120 pounds of B grade coffee. These will be blended into 1 pound packages as follows: An economy blend that contains 4 ounces of A grade coffee and 12 ounces of B grade coffee and a superior blend that contains 8 ounces of A grade coffee and 8 ounces of B grade coffee.

(a) Using x to denote the number of packages of the economy blend and y to denote the number of packages of the superior blend, write a system of linear inequalities that describes the possible number of packages of each kind of blend.
(b) Graph the system and label the corner points.

48. Mixed Nuts Nola's Nuts, a store that specializes in selling nuts, has 90 pounds of cashews and 120 pounds of peanuts available. These are to be mixed in 12 ounce packages as follows: a lower priced package containing 8 ounces of peanuts and 4 ounces of cashews and a quality package containing 6 ounces of peanuts and 6 ounces of cashews.
(a) Use x to denote the number of lower priced packages and use y to denote the number of quality packages. Write a system of linear inequalities that describes the possible number of each kind of package.
(b) Graph the system and label the corner points.

49. Transporting Goods A small truck can carry no more than 1600 pounds of cargo or 150 cubic feet of cargo. A printer weighs 20 pounds and occupies 3 cubic feet of space. A microwave oven weighs 30 pounds and occupies 2 cubic feet of space.
(a) Using x to represent the number of microwave ovens and y to represent the number of printers, write a system of linear inequalities that describes the number of ovens and printers that can be hauled by the truck.
(b) Graph the system and label the corner points.

11.8 LINEAR PROGRAMMING

1 Set Up a Linear Programming Problem
2 Solve a Linear Programming Problem

1 Historically, linear programming evolved as a technique for solving problems involving resource allocation of goods and materials for the U.S. Air Force during World War II. Today, linear programming techniques are used to solve a wide variety of problems, such as optimizing airline scheduling and establishing telephone lines. Although most practical linear programming problems involve systems of several hundred linear inequalities containing several hundred variables, we will limit our discussion to problems containing only two variables, because we can solve such problems using graphing techniques.*

We begin by returning to Example 12 of the previous section.

EXAMPLE 1 Financial Planning

A retired couple has up to \$25,000 to invest. As their financial adviser, you recommend that they place at least \$15,000 in Treasury bills yielding 6% and at most \$5000 in corporate bonds yielding 9%. How much money should be placed in each investment so that income is maximized?

The problem given here is typical of a *linear programming problem.* The problem requires that a certain linear expression, the income, be maximized. If I represents income, x the amount invested in Treasury bills at 6%, and y the amount invested in corporate bonds at 9%, then

$$I = 0.06x + 0.09y$$

*The **simplex method** is a way to solve linear programming problems involving many inequalities and variables. This method was developed by George Dantzig in 1946 and is particularly well suited for computerization. In 1984, Narendra Karmarkar of Bell Laboratories discovered a way of solving large linear programming problems that improves on the simplex method.

This linear expression is called the **objective function.** Furthermore, the problem requires that the maximum income be achieved under certain conditions or **constraints,** each of which is a linear inequality involving the variables. (See Example 12 in Section 11.7.) The linear programming problem given in Example 1 may be restated as

$$\text{Maximize} \qquad I = 0.06x + 0.09y$$

subject to the conditions that

$$\begin{aligned} x \geq 0, \qquad y &\geq 0 \\ x + y &\leq 25{,}000 \\ x &\geq 15{,}000 \\ y &\leq 5000 \end{aligned}$$

In general, every linear programming problem has two components:

1. A linear objective function that is to be maximized or minimized.
2. A collection of linear inequalities that must be satisfied simultaneously.

A **linear programming problem** in two variables x and y consists of maximizing (or minimizing) a linear objective function

$$z = Ax + By, \quad A \text{ and } B \text{ are real numbers, not both } 0$$

subject to certain conditions, or constraints, expressible as linear inequalities in x and y.

To maximize (or minimize) the quantity $z = Ax + By$, we need to identify points (x, y) that make the expression for z the largest (or smallest) possible. But not all points (x, y) are eligible; only those that also satisfy each linear inequality (constraint) can be used. We refer to each point (x, y) that satisfies the system of linear inequalities (the constraints) as a **feasible point.** Thus, in a linear programming problem, we seek the feasible point(s) that maximizes (or minimizes) the objective function.

Let's look again at the linear programming problem in Example 1.

EXAMPLE 2 Analyzing a Linear Programming Problem

Consider the linear programming problem:

$$\text{Maximize} \qquad I = 0.06x + 0.09y$$

subject to the conditions that

$$\begin{aligned} x \geq 0, \qquad y &\geq 0 \\ x + y &\leq 25{,}000 \\ x &\geq 15{,}000 \\ y &\leq 5000 \end{aligned}$$

Graph the constraints. Then graph the objective function for $I = 0, 0.9, 1.35, 1.65$, and 1.8.

Solution Figure 43 shows the graph of the constraints. We superimpose on this graph the graph of the objective function for the given values of I.

For $I = 0$, the objective function is the line $0 = 0.06x + 0.09y$.

For $I = 0.9$, the objective function is the line $0.9 = 0.06x + 0.09y$.

For $I = 1.35$, the objective function is the line $1.35 = 0.06x + 0.09y$.

For $I = 1.65$, the objective function is the line $1.65 = 0.06x + 0.09y$.

For $I = 1.8$, the objective function is the line $1.8 = 0.06x + 0.09y$.

FIGURE 43

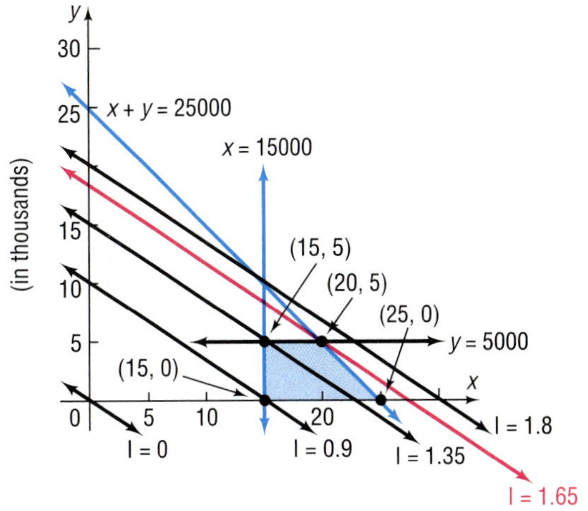

2 A **solution** to a linear programming problem consists of the feasible point(s) that maximizes (or minimizes) the objective function, together with the corresponding value of the objective function. One condition for a linear programming problem in two variables to have a solution is that the graph of the feasible points be bounded. (Refer to page 880).

If none of the feasible points maximizes (or minimizes) the objective function or if there are no feasible points, then the linear programming problem has no solution.

Consider the linear programming problem stated in Example 2, and look again at Figure 43. The feasible points are the points that lie in the shaded region. For example, (20, 3) is a feasible point, as is (15, 5), (20, 5), (18, 4), etc. To find the solution of the problem requires that we find a feasible point (x, y) that makes $I = 0.06x + 0.09y$ as large as possible. Notice that as I increases in value from $I = 0$ to $I = 0.9$ to $I = 1.35$ to $I = 1.65$ to $I = 1.8$, we obtain a collection of parallel lines. Furthermore, notice that the largest value of I that can be obtained while feasible points are present is $I = 1.65$, which corresponds to the line $1.65 = 0.06x + 0.09y$. Any larger value of I results in a line that does not pass through any feasible points. Finally, notice that the feasible point that yields $I = 1.65$ is the point (20, 5), a corner point. These observations form the basis of the following result, which we state without proof.

Theorem Location of the Solution of a Linear Programming Problem

If a linear programming problem has a solution, it is located at a corner point of the graph of the feasible points.

If a linear programming problem has multiple solutions, at least one of them is located at a corner point of the graph of the feasible points.

In either case, the corresponding value of the objective function is unique.

We shall not consider here linear programming problems that have no solution. As a result, we can outline the procedure for solving a linear programming problem as follows:

Procedure for Solving a Linear Programming Problem

STEP 1: Write an expression for the quantity to be maximized (or minimized). This expression is the objective function.
STEP 2: Write all the constraints as a system of linear inequalities and graph the system.
STEP 3: List the corner points of the graph of the feasible points.
STEP 4: List the corresponding values of the objective function at each corner point. The largest (or smallest) of these is the solution.

EXAMPLE 3 Solving a Minimum Linear Programming Problem

Minimize the expression

$$z = 2x + 3y$$

subject to the constraints

$$y \le 5, \quad x \le 6, \quad x + y \ge 2, \quad x \ge 0, \quad y \ge 0$$

Solution The objective function is $z = 2x + 3y$. We seek the smallest value of z that can occur if x and y are solutions of the system of linear inequalities

$$\begin{cases} y \le 5 \\ x \le 6 \\ x + y \ge 2 \\ x \ge 0 \\ y \ge 0 \end{cases}$$

The graph of this system (the feasible points) is shown as the shaded region in Figure 44. We have also plotted the corner points. Table 2 lists the corner points and the corresponding values of the objective function. From the table, we can see that the minimum value of z is 4, and it occurs at the point $(2, 0)$.

FIGURE 44

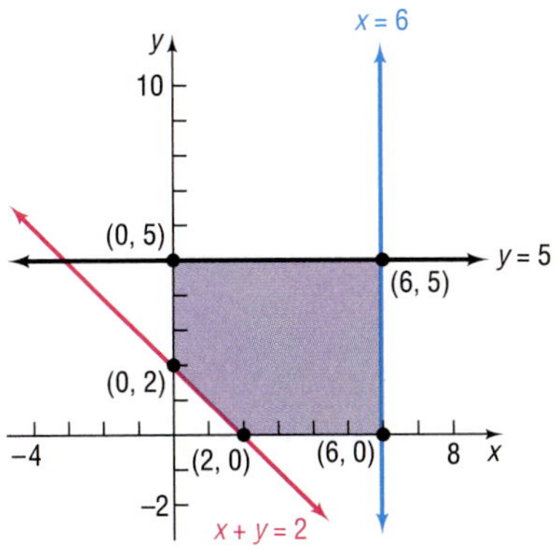

TABLE 2

Corner Point	Value of the Objective Function
(x, y)	$z = 2x + 3y$
(0, 2)	$z = 2(0) + 3(2) = 6$
(0, 5)	$z = 2(0) + 3(5) = 15$
(6, 5)	$z = 2(6) + 3(5) = 27$
(6, 0)	$z = 2(6) + 3(0) = 12$
(2, 0)	$z = 2(2) + 3(0) = 4$

Now work Problems 3 and 9.

EXAMPLE 4

Maximizing Profit

At the end of every month, after filling orders for its regular customers, a coffee company has some pure Colombian coffee and some special-blend coffee remaining. The practice of the company has been to package a mixture of the two coffees into 1-pound packages as follows: a low-grade mixture containing 4 ounces of Colombian coffee and 12 ounces of special-blend coffee and a high-grade mixture containing 8 ounces of Colombian and 8 ounces of special-blend coffee. A profit of \$0.30 per package is made on the low-grade mixture, whereas a profit of \$0.40 per package is made on the high-grade mixture. This month, 120 pounds of special-blend coffee and 100 pounds of pure Colombian coffee remain. How many packages of each mixture should be prepared to achieve a maximum profit? Assume that all packages prepared can be sold.

Solution We begin by assigning symbols for the two variables:

x = Number of packages of the low-grade mixture
y = Number of packages of the high-grade mixture

If P denotes the profit, then

$$P = \$0.30x + \$0.40y$$

This expression is the objective function. We seek to maximize P subject to certain constraints on x and y. Because x and y represent numbers of packages, the only meaningful values for x and y are nonnegative integers. Thus, we have the two constraints

$$x \geq 0, \qquad y \geq 0$$ Nonnegative constraints.

We also have only so much of each type of coffee available. For example, the total amount of Colombian coffee used in the two mixtures cannot exceed 100 pounds, or 1600 ounces. Because we use 4 ounces in each low-grade package and 8 ounces in each high-grade package, we are led to the constraint

$$4x + 8y \leq 1600$$ Colombian coffee constraint.

Similarly, the supply of 120 pounds, or 1920 ounces, of special-blend coffee leads to the constraint

$$12x + 8y \le 1920 \quad \text{Special-blend coffee constraint.}$$

The linear programming problem may be stated as

$$\text{Maximize} \quad P = 0.3x + 0.4y$$

subject to the constraints

$$x \ge 0, \quad y \ge 0, \quad 4x + 8y \le 1600, \quad 12x + 8y \le 1920$$

The graph of the constraints (the feasible points) is illustrated in Figure 45. We list the corner points and evaluate the objective function at each one. In Table 3, we can see that the maximum profit, \$84, is achieved with 40 packages of the low-grade mixture and 180 packages of the high-grade mixture.

FIGURE 45

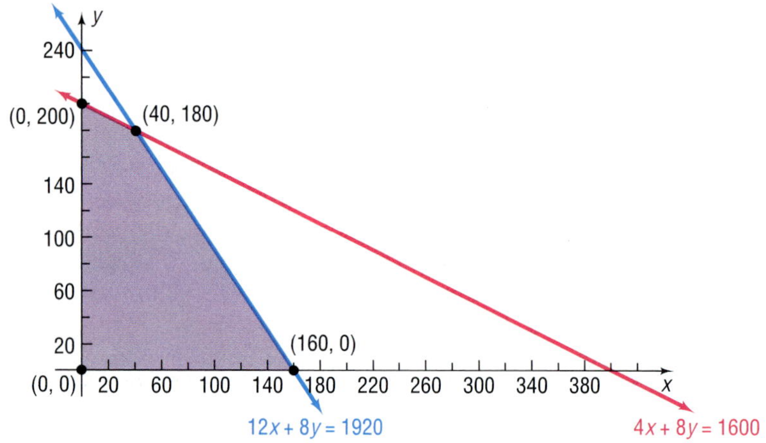

TABLE 3

Corner Point	Value of Profit
(x, y)	$P = 0.3x + 0.4y$
(0, 0)	$P = 0$
(0, 200)	$P = 0.3(0) + 0.4(200) = \$80$
(40, 180)	$P = 0.3(40) + 0.4(180) = \$84$
(160, 0)	$P = 0.3(160) + 0.4(0) = \$48$

Now work Problem 17.

11.8 EXERCISES

In Problems 1–6, find the maximum and minimum value of the given objective function.

The figure illustrates the graph of the feasible points of a linear programming problem.

1. $z = x + y$ **2.** $z = 2x + 3y$ **3.** $z = x + 10y$

4. $z = 10x + y$ **5.** $z = 5x + 7y$ **6.** $z = 7x + 5y$

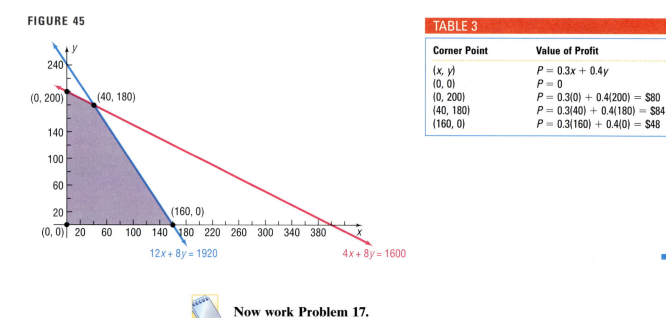

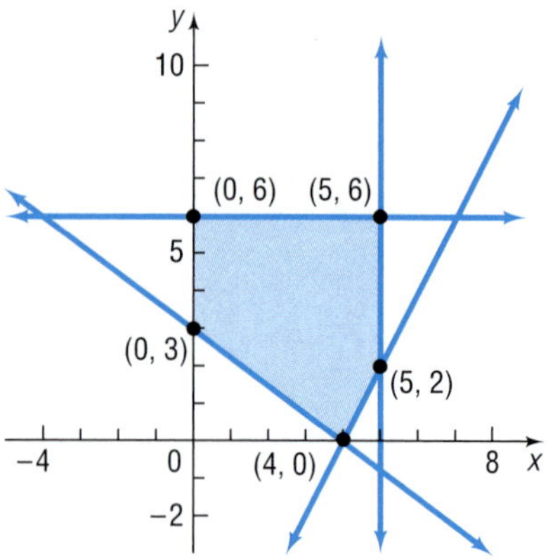

In Problems 7–16, solve each linear programming problem.

7. Maximize $z = 2x + y$ subject to $x \geq 0,\quad y \geq 0,\quad x + y \leq 6,\quad x + y \geq 1$

8. Maximize $z = x + 3y$ subject to $x \geq 0,\quad y \geq 0,\quad x + y \geq 3,\quad x \leq 5, y \leq 7$

9. Minimize $z = 2x + 5y$ subject to $x \geq 0,\quad y \geq 0,\quad x + y \geq 2,\quad x \leq 5, y \leq 3$

10. Minimize $z = 3x + 4y$ subject to $x \geq 0,\quad y \geq 0,\quad 2x + 3y \geq 6,\quad x + y \leq 8$

11. Maximize $z = 3x + 5y$ subject to $x \geq 0,\quad y \geq 0,\quad x + y \geq 2,\quad 2x + 3y \leq 12,\quad 3x + 2y \leq 12$

12. Maximize $z = 5x + 3y$ subject to $x \geq 0,\quad y \geq 0,\quad x + y \geq 2,\quad x + y \leq 8,\quad 2x + y \leq 10$

13. Minimize $z = 5x + 4y$ subject to $x \geq 0,\quad y \geq 0,\quad x + y \geq 2,\quad 2x + 3y \leq 12,\quad 3x + y \leq 12$

14. Minimize $z = 2x + 3y$ subject to $x \geq 0,\quad y \geq 0,\quad x + y \geq 3,\quad x + y \leq 9,\quad x + 3y \geq 6$

15. Maximize $z = 5x + 2y$ subject to $x \geq 0,\quad y \geq 0,\quad x + y \leq 10,\quad 2x + y \geq 10,\quad x + 2y \geq 10$

16. Maximize $z = 2x + 4y$ subject to $x \geq 0,\quad y \geq 0,\quad 2x + y \geq 4,\quad x + y \leq 9$

17. **Maximizing Profit** A manufacturer of skis produces two types: downhill and cross-country. Use the following table to determine how many of each kind of ski should be produced to achieve a maximum profit. What is the maximum profit? What would the maximum profit be if the maximum time available for manufacturing is increased to 48 hours?

	Downhill	Cross-Country	Maximum Time Available
Manufacturing time per ski	2 hours	1 hour	40 hours
Finishing time per ski	1 hour	1 hour	32 hours
Profit per ski	$70	$50	

18. **Farm Management** A farmer has 70 acres of land available for planting either soybeans or wheat. The cost of preparing the soil, the workdays required, and the expected profit per acre planted for each type of crop are given in the following table:

	Soybeans	Wheat
Preparation cost per acre	$60	$30
Workdays required per acre	3	4
Profit per acre	$180	$100

The farmer cannot spend more than $1800 in preparation costs nor more than a total of 120 workdays. How many acres of each crop should be planted in order to maximize the profit? What is the maximum profit? What is the maximum profit if the farmer is willing to spend no more than $2400 on preparation?

19. **Farm Management** A small farm in Illinois has 100 acres of land available on which to grow corn and soybeans. The following table shows the cultivation cost per acre, the labor cost per acre, and the expected profit per acre. The column on the right shows the amount of money available for each of these expenses. Find the number of acres of each crop that should be planted in order to maximize profit.

	Soybeans	Corn	Money Available
Cultivation cost per acre	$40	$60	$1800
Labor cost per acre	$60	$60	$2400
Profit per acre	$200	$250	

20. **Dietary Requirements** A certain diet requires at least 60 units of carbohydrates, 45 units of protein, and 30 units of fat each day. Each ounce of Supplement A provides 5 units of carbohydrates, 3 units of protein, and 4 units of fat. Each ounce of Supplement B provides 2 units of carbohydrates, 2 units of protein, and 1 unit of fat. If Supplement A costs $1.50 per ounce and Supplement B costs $1.00 per ounce, how many ounces of each supplement should be taken daily to minimize the cost of the diet?

21. **Production Scheduling** In a factory, machine 1 produces 8″ plyers at the rate of 60 units per hour, and 6″ plyers at the rate of 70 units per hour. Machine 2 produces 8″ plyers at the rate of 40 units per hour and 6″ plyers at the rate of 20 units per hour. It costs $50 per hour to operate machine 1, while machine 2 costs $30 per hour to operate. The production schedule requires that at least 240 units of 8″ plyers and at least 140 units of 6″ plyers must be produced during each 10 hour day. Which combination of machines will cost the least money to operate?

22. **Farm Management** An owner of a fruit orchard hires a crew of workers to prune at least 25 of his 50 fruit trees. Each newer tree requires one hour to prune, while each older tree needs one-and-a-half hours. The crew contracts to work for at least 30 hours and charge \$15 for each newer tree and \$20 for each older tree. To minimize his cost, how many of each kind of tree will the orchard owner have pruned? What will be the cost?

23. **Managing a Meat Market** A meat market combines ground beef and ground pork in a single package for meat loaf. The ground beef is 75% lean (75% beef, 25% fat) and costs the market \$0.75 per pound. The ground pork is 60% lean and costs the market \$0.45 per pound. The meat loaf must be at least 70% lean. If the market wants to use at least 50 lbs of its available pork, but no more than 200 lbs of its available ground beef, how much ground beef should be mixed with ground pork so that the cost is minimized?

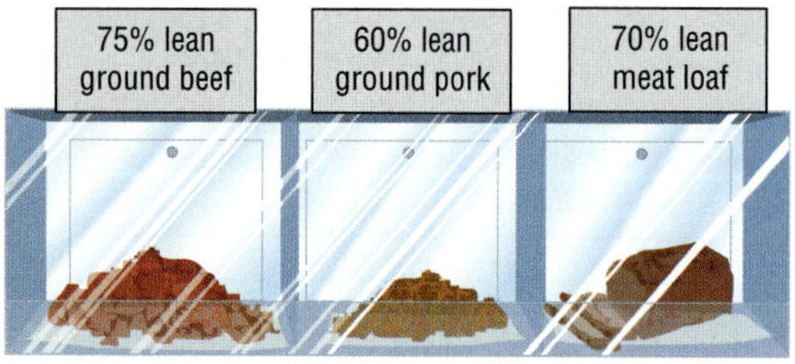

24. **Return on Investment** An investment broker is instructed by her client to invest up to \$20,000, some in a junk bond yielding 9% per annum and some in Treasury bills yielding 7% per annum. The client wants to invest at least \$8000 in T-bills and no more than \$12,000 in the junk bond.
    (a) How much should the broker recommend that the client place in each investment to maximize income if the client insists that the amount invested in T-bills must equal or exceed the amount placed in junk bonds?
    (b) How much should the broker recommend the client place in each investment to maximize income if the client insists that the amount invested in T-bills must not exceed the amount placed in junk bonds?

25. **Maximizing Profit on Ice Skates** A factory manufactures two kinds of ice skates: racing skates and figure skates. The racing skates require 6 work-hours in the fabrication department, whereas the figure skates require 4 work-hours there. The racing skates require 1 work-hour in the finishing department, whereas the figure skates require 2 work-hours there. The fabricating department has available at most 120 work-hours per day, and the finishing department has no more than 40 work-hours per day available. If the profit on each racing skate is \$10 and the profit on each figure skate is \$12, how many of each should be manufactured each day to maximize profit? (Assume that all skates made are sold.)

26. **Financial Planning** A retired couple has up to \$50,000 to place in fixed-income securities. Their financial adviser suggests two securities to them: one is an AAA bond that yields 8% per annum; the other is a Certificate of Deposit (CD) that yields 4%. After careful consideration of the alternatives, the couple decides to place at most \$20,000 in the AAA bond and at least \$15,000 in the CD. They also instruct the financial adviser to place at least as much in the CD as in the AAA bond. How should the financial adviser proceed to maximize the return on their investment?

27. **Product Design** An entrepreneur is having a design group produce at least six samples of a new kind of fastener that he wants to market. It costs \$9.00 to produce each metal fastener and \$4.00 to produce each plastic fastener. He wants to have at least two of each version of the fastener, and needs to have all the samples 24 hours from now. It takes 4 hours to produce each metal sample and 2 hours to produce each plastic sample. To minimize the cost of the samples, how many of each kind should the entrepreneur order? What will be the cost of the samples?

28. **Animal Nutrition** Kevin's dog Amadeus likes two kinds of canned dog food. "Gourmet Dog" costs 40 cents a can and has 20 units of a vitamin complex; the calorie content is 75 calories. "Chow Hound" costs 32 cents a can and has 35 units of vitamins and 50 calories. Kevin likes Amadeus to have at least 1175 units of vitamins a month and at least 2375 calories during the same time period. Kevin has space to store only 60 cans of dog food at a time. How much of each kind of dog food should Kevin buy each month in order to minimize his cost?

29. **Airline Revenue** An airline has two classes of service: first class and coach. Management's experience has been that each aircraft should have at least 8 but not more than 16 first-class seats and at least 80 but not more than 120 coach seats.
    (a) If management decides that the ratio of first-class to coach seats should never exceed 1:12, with how many of each type seat should an aircraft be configured to maximize revenue?
    (b) If management decides that the ratio of first-class to coach seats should never exceed 1:8, with how many of each type seat should an aircraft be configured to maximize revenue?

(c) If you were management, what would you do?

[**Hint:** Assume that the airline charges $\$C$ for a coach seat and $\$F$ for a first-class seat; $C > 0, F > 0$.]

30. Minimizing Cost A farm that specializes in raising frying chickens supplements the regular chicken feed with four vitamins. The owner wants the supplemental food to contain at least 50 units of vitamin I, 90 units of vitamin II, 60 units of vitamin III, and 100 units of vitamin IV per 100 ounces of feed. Two supplements are available: supplement A, which contains 5 units of vitamin I, 25 units of vitamin II, 10 units of vitamin III, and 35 units of vitamin IV per ounce; and supplement B, which contains 25 units of vitamin I, 10 units of vitamin II, 10 units of vitamin III, and 20 units of vitamin IV per ounce. If supplement A costs \$0.06 per ounce and supplement B costs \$0.08 per ounce, how much of each supplement should the manager of the farm buy to add to each 100 ounces of feed in order to keep the total cost at a minimum, while still meeting the owner's vitamin specifications?

31. Explain in your own words what a linear programming problem is and how it can be solved.

CHAPTER REVIEW

THINGS TO KNOW

Systems of equations

Systems with no solutions are inconsistent. Systems with a solution are consistent.

Consistent systems have either a unique solution or an infinite number of solutions.

Matrix	Rectangular array of numbers, called entries
m by n matrix	Matrix with m rows and n columns
Identity matrix I	Square matrix whose diagonal entries are 1's, while all other entries are 0's
Inverse of a matrix	A^{-1} is the inverse of A if $AA^{-1} = A^{-1}A = I$
Nonsingular matrix	A matrix that has an inverse

Linear programming problem

Maximize (or minimize) a linear objective function, $z = Ax + By$, subject to certain conditions, or constraints, expressible as linear inequalities in x and y.

Feasible point

A point (x, y) that satisfies the constraints of a linear programming problem

Location of solution

If a linear programming problem has a solution, it is located at a corner point of the graph of the feasible points.

If a linear programming problem has multiple solutions, at least one of them is located at a corner point of the graph of the feasible points.

In either case, the corresponding value of the objective function is unique.

HOW TO

Solve a system of linear equations using the method of substitution

Solve a system of linear equations using the method of elimination

Solve a system of linear equations using matrices

Solve a system of linear equations using determinants

Recognize equal matrices

Add and subtract matrices

Multiply matrices

Find the inverse of a nonsingular matrix

Solve a system of linear equations using the inverse of a matrix

Write the partial fraction decomposition of a rational expression

Solve a system of nonlinear equations

Graph a system of inequalities

Find the corner points of the graph of a system of linear inequalities

Solve linear programming problems

FILL-IN-THE-BLANK ITEMS

1. If a system of equations has no solution, it is said to be ______________.

2. An m by n rectangular array of numbers is called a(n) ______________.

3. Cramer's Rules uses ______________ to solve a system of linear equations.

4. The matrix used to represent a system of linear equations is called a(n) ______________ matrix.

5. A matrix B, for which $AB = I_n$, the identity matrix, is called the ______________ of A.

6. A matrix that has the same number of rows as columns is called a(n) ______________ matrix.

7. In the algebra of matrices, the matrix that has the properties similar to the number 1 is called the ______________ matrix.

8. A rational function is called ______________ if the degree of its numerator is less than the degree of its denominator.

9. The graph of a linear inequality is called a(n) ______________.

10. A linear programming problem requires that a linear expression, called the ______________ ______________, be maximized or minimized.

11. Each point that satisfies the constraints of a linear programming problem is called a(n) ______________ ______________.

TRUE/FALSE ITEMS

T F **1.** A system of two linear equations containing two unknowns always has at least one solution.

T F **2.** The augmented matrix of a system of two equations containing three variables has two rows and four columns.

T F **3.** A 3 by 3 determinant can never equal 0.

T F **4.** A consistent system of equations will have exactly one solution.

T F **5.** Every square matrix has an inverse.

T F **6.** Matrix multiplication is commutative.

T F **7.** Any pair of matrices can be multiplied.

T F **8.** The factors of the denominator of a rational expression are used to arrive at the partial fraction decomposition.

T F **9.** The graph of a linear inequality is a half-plane.

T F **10.** The graph of a system of linear inequalities is sometimes unbounded.

T F **11.** If a linear programming problem has a solution, it is located at a corner point of the graph of the feasible points.

REVIEW EXERCISES

In Problems 1–20, solve each system of equations algebraically using the method of substitution or the method of elimination. If the system has no solution, say it is inconsistent. Verify your result using a graphing utility.

1. $\begin{cases} 2x - y = 5 \\ 5x + 2y = 8 \end{cases}$

2. $\begin{cases} 2x + 3y = 2 \\ 7x - y = 3 \end{cases}$

3. $\begin{cases} 3x - 4y = 4 \\ x - 3y = \frac{1}{2} \end{cases}$

4. $\begin{cases} 2x + y = 0 \\ 5x - 4y = -\frac{13}{2} \end{cases}$

5. $\begin{cases} x - 2y - 4 = 0 \\ 3x + 2y - 4 = 0 \end{cases}$

6. $\begin{cases} x - 3y + 5 = 0 \\ 2x + 3y - 5 = 0 \end{cases}$

7. $\begin{cases} y = 2x - 5 \\ x = 3y + 4 \end{cases}$

8. $\begin{cases} x = 5y + 2 \\ y = 5x + 2 \end{cases}$

9. $\begin{cases} x - y + 4 = 0 \\ \frac{1}{2}x + \frac{1}{6}y + \frac{2}{5} = 0 \end{cases}$

10. $\begin{cases} x + \frac{1}{4}y = 2 \\ y + 4x + 2 = 0 \end{cases}$

11. $\begin{cases} x - 2y - 8 = 0 \\ 2x + 2y - 10 = 0 \end{cases}$

12. $\begin{cases} x - 3y + \frac{7}{2} = 0 \\ \frac{1}{2}x + 3y - 5 = 0 \end{cases}$

13. $\begin{cases} y - 2x = 11 \\ 2y - 3x = 18 \end{cases}$

14. $\begin{cases} 3x - 4y - 12 = 0 \\ 5x + 2y + 6 = 0 \end{cases}$

15. $\begin{cases} 2x + 3y - 13 = 0 \\ 3x - 2y = 0 \end{cases}$

16. $\begin{cases} 4x + 5y = 21 \\ 5x + 6y = 42 \end{cases}$

17. $\begin{cases} 3x - 2y = 8 \\ x - \frac{2}{3}y = 12 \end{cases}$

18. $\begin{cases} 2x + 5y = 10 \\ 4x + 10y = 15 \end{cases}$

19. $\begin{cases} x + 2y - z = 6 \\ 2x - y + 3z = -13 \\ 3x - 2y + 3z = -16 \end{cases}$

20. $\begin{cases} x + 5y - z = 2 \\ 2x + y + z = 7 \\ x - y + 2z = 11 \end{cases}$

In Problems 21–28, use the following matrices to compute each expression. Verify your result using a graphing utility.

$$A = \begin{bmatrix} 1 & 0 \\ 2 & 4 \\ -1 & 2 \end{bmatrix} \quad B = \begin{bmatrix} 4 & -3 & 0 \\ 1 & 1 & -2 \end{bmatrix} \quad C = \begin{bmatrix} 3 & -4 \\ 1 & 5 \\ 5 & -2 \end{bmatrix}$$

21. $A + C$ **22.** $A - C$ **23.** $6A$ **24.** $-4B$

25. AB **26.** BA **27.** CB **28.** BC

In Problems 29–34, find the inverse of each matrix algebraically, if there is one. If there is not an inverse, say that the matrix is singular. Verify your result using a graphing utility.

29. $\begin{bmatrix} 4 & 6 \\ 1 & 3 \end{bmatrix}$ **30.** $\begin{bmatrix} -3 & 2 \\ 1 & -2 \end{bmatrix}$ **31.** $\begin{bmatrix} 1 & 3 & 3 \\ 1 & 2 & 1 \\ 1 & -1 & 2 \end{bmatrix}$

32. $\begin{bmatrix} 3 & 1 & 2 \\ 3 & 2 & -1 \\ 1 & 1 & 1 \end{bmatrix}$ **33.** $\begin{bmatrix} 4 & -8 \\ -1 & 2 \end{bmatrix}$ **34.** $\begin{bmatrix} -3 & 1 \\ -6 & 2 \end{bmatrix}$

In Problems 35–44, solve each system of equations algebraically using matrices. If the system has no solution, say it is inconsistent. Verify your result using a graphing utility.

35. $\begin{cases} 3x - 2y = 1 \\ 10x + 10y = 5 \end{cases}$ **36.** $\begin{cases} 3x + 2y = 6 \\ x - y = -\frac{1}{2} \end{cases}$ **37.** $\begin{cases} 5x + 6y - 3z = 6 \\ 4x - 7y - 2z = -3 \\ 3x + y - 7z = 1 \end{cases}$

38. $\begin{cases} 2x + y + z = 5 \\ 4x - y - 3z = 1 \\ 8x + y - z = 5 \end{cases}$ **39.** $\begin{cases} x - 2z = 1 \\ 2x + 3y = -3 \\ 4x - 3y - 4z = 3 \end{cases}$ **40.** $\begin{cases} x + 2y - z = 2 \\ 2x - 2y + z = -1 \\ 6x + 4y + 3z = 5 \end{cases}$

41. $\begin{cases} x - y + z = 0 \\ x - y - 5z - 6 = 0 \\ 2x - 2y + z - 1 = 0 \end{cases}$ **42.** $\begin{cases} 4x - 3y + 5z = 0 \\ 2x + 4y - 3z = 0 \\ 6x + 2y + z = 0 \end{cases}$ **43.** $\begin{cases} x - y - z - t = 1 \\ 2x + y + z + 2t = 3 \\ x - 2y - 2z - 3t = 0 \\ 3x - 4y + z + 5t = -3 \end{cases}$

44. $\begin{cases} x - 3y + 3z - t = 4 \\ x + 2y - z = -3 \\ x + 3z + 2t = 3 \\ x + y + 5z = 6 \end{cases}$

In Problems 45–50, find the value of each determinant algebraically. Verify your result using a graphing utility.

45. $\begin{vmatrix} 3 & 4 \\ 1 & 3 \end{vmatrix}$ **46.** $\begin{vmatrix} -4 & 0 \\ 1 & 3 \end{vmatrix}$ **47.** $\begin{vmatrix} 1 & 4 & 0 \\ -1 & 2 & 6 \\ 4 & 1 & 3 \end{vmatrix}$ **48.** $\begin{vmatrix} 2 & 3 & 10 \\ 0 & 1 & 5 \\ -1 & 2 & 3 \end{vmatrix}$ **49.** $\begin{vmatrix} 2 & 1 & -3 \\ 5 & 0 & 1 \\ 2 & 6 & 0 \end{vmatrix}$ **50.** $\begin{vmatrix} -2 & 1 & 0 \\ 1 & 2 & 3 \\ -1 & 4 & 2 \end{vmatrix}$

In Problems 51–56, use Cramer's Rule, if applicable, to solve each system.

51. $\begin{cases} x - 2y = 4 \\ 3x + 2y = 4 \end{cases}$ **52.** $\begin{cases} x - 3y = -5 \\ 2x + 3y = 5 \end{cases}$ **53.** $\begin{cases} 2x + 3y - 13 = 0 \\ 3x - 2y = 0 \end{cases}$

54. $\begin{cases} 3x - 4y - 12 = 0 \\ 5x + 2y + 6 = 0 \end{cases}$ **55.** $\begin{cases} x + 2y - z = 6 \\ 2x - y + 3z = -13 \\ 3x - 2y + 3z = -16 \end{cases}$ **56.** $\begin{cases} x - y + z = 8 \\ 2x + 3y - z = -2 \\ 3x - y - 9z = 9 \end{cases}$

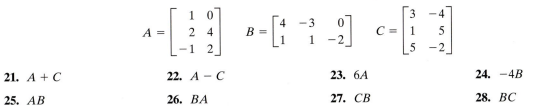

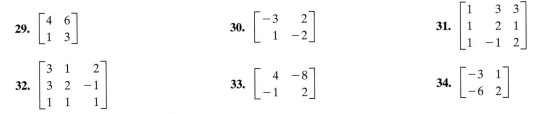

In Problems 57–66, write the partial fraction decomposition of each rational expression.

57. $\dfrac{6}{x(x-4)}$ **58.** $\dfrac{x}{(x+2)(x-3)}$ **59.** $\dfrac{x-4}{x^2(x-1)}$ **60.** $\dfrac{2x-6}{(x-2)^2(x-1)}$

61. $\dfrac{x}{(x^2+9)(x+1)}$ **62.** $\dfrac{3x}{(x-2)(x^2+1)}$ **63.** $\dfrac{x^3}{(x^2+4)^2}$ **64.** $\dfrac{x^3+1}{(x^2+16)^2}$

65. $\dfrac{x^2}{(x^2+1)(x^2-1)}$ **66.** $\dfrac{4}{(x^2+4)(x^2-1)}$

In Problems 67–76, solve each system of equations algebraically. Verify your result using a graphing utility.

67. $\begin{cases} 2x+y+3=0 \\ x^2+y^2=5 \end{cases}$ **68.** $\begin{cases} x^2+y^2=16 \\ 2x-y^2=-8 \end{cases}$ **69.** $\begin{cases} 2xy+y^2=10 \\ 3y^2-xy=2 \end{cases}$ **70.** $\begin{cases} 3x^2-y^2=1 \\ 7x^2-2y^2-5=0 \end{cases}$

71. $\begin{cases} x^2+y^2=6y \\ x^2=3y \end{cases}$ **72.** $\begin{cases} 2x^2+y^2=9 \\ x^2+y^2=9 \end{cases}$ **73.** $\begin{cases} 3x^2+4xy+5y^2=8 \\ x^2+3xy+2y^2=0 \end{cases}$ **74.** $\begin{cases} 3x^2+2xy-2y^2=6 \\ xy-2y^2+4=0 \end{cases}$

75. $\begin{cases} x^2-3x+y^2+y=-2 \\ \dfrac{x^2-x}{y}+y+1=0 \end{cases}$ **76.** $\begin{cases} x^2+x+y^2=y+2 \\ x+1=\dfrac{2-y}{x} \end{cases}$

In Problems 77–82, graph each system of inequalities. Tell whether the graph is bounded or unbounded, and label the corner points.

77. $\begin{cases} -2x+y\le 2 \\ x+y\ge 2 \end{cases}$ **78.** $\begin{cases} x-2y\le 6 \\ 2x+y\ge 2 \end{cases}$ **79.** $\begin{cases} x\ge 0 \\ y\ge 0 \\ x+y\le 4 \\ 2x+3y\le 6 \end{cases}$ **80.** $\begin{cases} x\ge 0 \\ y\ge 0 \\ 3x+y\ge 6 \\ 2x+y\ge 2 \end{cases}$

81. $\begin{cases} x\ge 0 \\ y\ge 0 \\ 2x+y\le 8 \\ x+2y\ge 2 \end{cases}$ **82.** $\begin{cases} x\ge 0 \\ y\ge 0 \\ 3x+y\le 9 \\ 2x+3y\ge 6 \end{cases}$

In Problems 83–86, graph each system of inequalities.

83. $\begin{cases} x^2+y^2\le 16 \\ x+y\ge 2 \end{cases}$ **84.** $\begin{cases} y^2\le x-1 \\ x-y\le 3 \end{cases}$ **85.** $\begin{cases} y\le x^2 \\ xy\le 4 \end{cases}$ **86.** $\begin{cases} x^2+y^2\ge 1 \\ x^2+y^2\le 4 \end{cases}$

In Problems 87–92, solve each linear programming problem.

87. Maximize $z=3x+4y$ subject to $x\ge 0,\ y\ge 0,\ 3x+2y\ge 6,\ x+y\le 8$

88. Maximize $z=2x+4y$ subject to $x\ge 0,\ y\ge 0,\ x+y\le 6,\ x\ge 2$

89. Minimize $z=3x+5y$ subject to $x\ge 0,\ y\ge 0,\ x+y\ge 1,\ 3x+2y\le 12,\ x+3y\le 12$

90. Minimize $z=3x+y$ subject to $x\ge 0,\ y\ge 0,\ x\le 8,\ y\le 6,\ 2x+y\ge 4$

91. Maximize $z=5x+4y$ subject to $x\ge 0,\ y\ge 0,\ x+2y\ge 2,\ 3x+4y\le 12,\ y\ge x$

92. Maximize $z=4x+5y$ subject to $x\ge 0,\ y\ge 0,\ 2x+3y\ge 6,\ x\ge y,\ 2x+y\le 12$

93. Find A such that the system of equations has infinitely many solutions.

$$\begin{cases} 2x+5y=5 \\ 4x+10y=A \end{cases}$$

94. Find A such that the system in Problem 93 is inconsistent.

95. Curve Fitting Find the quadratic function $y=ax^2+bx+c$ that passes through the three points $(0,1)$, $(1,0)$, and $(-2,1)$.

96. Curve Fitting Find the general equation of the circle that passes through the three points $(0,1)$, $(1,0)$, and $(-2,1)$.

[**Hint:** The general equation of a circle is $x^2+y^2+Dx+Ey+F=0$.]

97. Blending Coffee A coffee distributor is blending a new coffee that will cost \$3.90 per pound. It will consist of a blend of \$3.00 per pound coffee and \$6.00 per pound coffee. What amounts of each type of coffee should be mixed to achieve the desired blend?

[**Hint:** Assume that the weight of the blended coffee is 100 pounds.]

98. Farming A 1000 acre farm in Illinois is used to raise corn and soy beans. The cost per acre for raising corn is \$65 and the cost per acre for soy beans is \$45. If \$54,325 has been budgeted for costs and all the acreage is to be used, how many acres should be allocated for each crop?

99. Cookie Orders A cookie company makes three kinds of cookies, oatmeal raisin, chocolate chip, and shortbread, packaged in small, medium, and large boxes. The small box contains 1 dozen oatmeal raisin and 1 dozen chocolate chip; the medium box has 2 dozen oatmeal raisin, 1 dozen chocolate chip, and 1 dozen shortbread; the large box contains 2 dozen oatmeal raisin, 2 dozen chocolate chip, and 3 dozen shortbread. If you require exactly 15 dozen oatmeal raisin, 10 dozen chocolate chip, and 11 dozen shortbread, how many of each size box should you buy?

100. Mixed Nuts A store that specializes in selling nuts has 72 pounds of cashews and 120 pounds of peanuts available. These are to be mixed in 12-ounce packages as follows: a lower-priced package containing 8 ounces of peanuts and 4 ounces of cashews and a quality package containing 6 ounces of peanuts and 6 ounces of cashews.

(a) Use x to denote the number of lower-priced packages and use y to denote the number of quality packages. Write a system of linear inequalities that describes the possible number of each kind of package.

(b) Graph the system and label the corner points.

101. A small rectangular lot has a perimeter of 68 feet. If its diagonal is 26 feet, what are the dimensions of the lot?

102. The area of a rectangular window is 4 square feet. If the diagonal measures $2\sqrt{2}$ feet, what are the dimensions of the window?

103. Geometry A certain right triangle has a perimeter of 14 inches. If the hypotenuse is 6 inches long, what are the lengths of the legs?

104. Geometry A certain isosceles triangle has a perimeter of 18 inches. If the altitude is 6 inches, what is the length of the base?

105. Building a Fence How much fence is required to enclose 5000 square feet by two squares whose sides are in the ratio of 1:2?

106. Mixing Acids A chemistry laboratory has three containers of hydrochloric acid, HCl. One container holds a solution with a concentration of 10% HCl, the second holds 25% HCl, and the third holds 40% HCl. How many liters of each should be mixed to obtain 100 liters of a solution with a concentration of 30% HCl? Construct a table showing some of the possible combinations.

107. Calculating Allowances Katy, Mike, Danny, and Colleen agreed to do yard work at home for \$45 to be split among them. After they finished, their father determined that Mike deserves twice what Katy gets, Katy and Colleen deserve the same amount, and Danny deserves half of what Katy gets. How much does each one receive?

108. Finding the Speed of the Jet Stream On a flight between Midway Airport in Chicago and Ft. Lauderdale, Florida, a Boeing 737 jet maintains an airspeed of 475 miles per hour. If the trip from Chicago to Ft. Lauderdale takes 2 hours, 30 minutes and the return flight takes 2 hours, 50 minutes, what is the speed of the jet stream? (Assume the speed of the jet stream remains constant at the various altitudes of the plane.)

109. Constant Rate Jobs If Bruce and Bryce work together for 1 hour and 20 minutes, they will finish a certain job. If Bryce and Marty work together for 1 hour and 36 minutes, the same job can be finished. If Marty and Bruce work together, they can complete this job in 2 hours and 40 minutes. How long will it take each of them working alone to finish the job?

110. Maximizing Profit on Figurines A factory manufactures two kinds of ceramic figurines: a dancing girl and a mermaid, each requiring three processes—molding, painting, and glazing. The daily labor available for molding is no more than 90 work-hours, labor available for painting does not exceed 120 work-hours, and labor available for glazing is no more than 60 work-hours. The dancing girl requires 3 work-hours for molding, 6 work-hours for painting, and 2 work-hours for glazing. The mermaid requires 3 work-hours for molding, 4 work-hours for painting, and 3 work-hours for glazing. If the profit on each figurine is \$25 for dancing girls and \$30 for mermaids, how many of each should be produced each day to maximize profit? If management decides to produce the number of each figurine that maximizes profit, determine which of these processes has excess work-hours assigned to it.

111. Minimizing Production Cost A factory produces gasoline engines and diesel engines. Each week the factory is obligated to deliver at least 20 gasoline engines and at least 15 diesel engines. Due to physical limitations, however, the factory cannot make more than 60 gasoline engines nor more than 40 diesel engines. Finally, to prevent layoffs, a total of at least 50 engines must be produced. If gasoline engines cost \$450 each to produce and diesel engines cost \$550 each to produce, how many of each should be produced per week to minimize the cost? What is the excess capacity of the factory; that is, how many of each kind of engine are being produced in excess of the number the factory is obligated to deliver?

112. Describe four ways of solving a system of three linear equations containing three variables. Which method do you prefer? Why?

CHAPTER 12

Sequences; Induction; Counting; Probability

Bicycle riding has grown in popularity in the past few years. With the increase of riders also comes an increase in bicycle accidents. On the following page you are placed in the position of being a law student and having to argue a case regarding helmet safety. You will need to use the Sullivan website at:

www.prenhall.com/sullivan

to find the relevant data upon which to base your case.

PREPARING FOR THIS CHAPTER

Before getting started on this chapter, review the following concepts:

For Section 12.3: Compound Interest *(Section 5.6)*

OUTLINE

Back Forward Home Reload Images Open Print Find Stop

Netsite: http://www.prenhall.com/sullivan/

BICYCLE HELMET SAFETY

Should wearing a bicycle helmet be required by law or would that violate an individual's civil liberty? As a law student you are required to analyze this point of law and defend your position in the people's court. In order to base your case on relevant data you decide to do a statistical analysis of the data on helmet safety.

1. On the radio you hear "If every child wore a helmet, one brain injury could be prevented every five minutes and one death every day." Is this statement true? Should a bicycler be required to wear a helmet? Is this an issue of a civil liberty, as some say? What do you think justifies a mandatory helmet law?
2. Before you analyze the data, do you think that your bias will cause you to be selective in the data that you accept as valid? How can you keep bias from entering into your evaluation?
3. By contrast, do you think that the arguments for bicycle helmets based on the data also apply for motorcycles? Can you find any examples of bias affecting judgment as you look at the apparently contradictory data offered by *the AAOS* and *Opposing Facts*?
4. Review the *statistical data* from various sources. Determine the data sources that you believe are most reliable.
5. Compare the data for bicycles to that for motorcycles. Is there any correlation in the data?
6. What is the probability that you will be involved in a bicycle accident if you ride your bike about two hours per day?
7. Using your most reliable sources of data on bicycle fatalities, could one death every day be prevented?

This chapter introduces topics that are covered in more detail in courses titled *Finite Mathematics* or *Discrete Mathematics.* Applications of these topics can be found in the fields of computer science, engineering, business and economics, the social sciences, and the physical and biological sciences.

The chapter may be divided into four independent parts:

Sections 12.1–12.3, Sequences, which are functions whose domain is the set of natural numbers. Sequences form the basis for the *recursively defined functions* and *recursive procedures* used in computer programming.

Section 12.4, Mathematical Induction, a technique for proving theorems involving the natural numbers.

Section 12.5, the Binomial Theorem, a formula for the expansion of $(x + a)^n$, where n is any natural number.

Sections 12.6–12.8, The first two sections deal with techniques and formulas for counting the number of objects in a set, a part of the branch of mathematics called *combinatorics.* These formulas are used in computer science to analyze algorithms and recursive functions and to study stacks and queues. They are also used to determine *probabilities,* the likelihood that a certain outcome of a random experiment will occur.

12.1 SEQUENCES

1 Write the First Several Terms of a Sequence
2 Write the Terms of a Sequence Defined by a Recursion Formula
3 Find the Sum of a Sequence by Hand and by Using a Graphing Utility
4 Write a Sequence in Summation Notation

A **sequence** is a function whose domain is the set of positive integers.

Because a sequence is a function, it will have a graph. In Figure 1(a), you will recognize the graph of the function $f(x) = 1/x, x > 0$. If all the points on this graph were removed except those whose x-coordinates are positive integers—that is, if all points were removed except $(1, 1)$, $(2, \frac{1}{2})$, $(3, \frac{1}{3})$, and so on—the remaining points would be the graph of the sequence $f(n) = 1/n$, as shown in Figure 1(b).

FIGURE 1

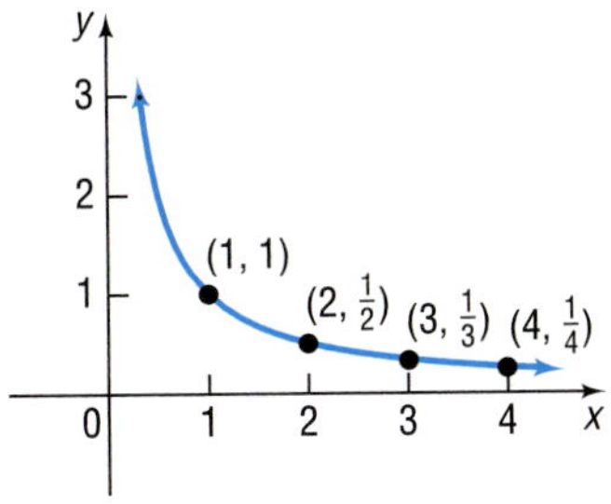

(a) $f(x) = \frac{1}{x}, x > 0$

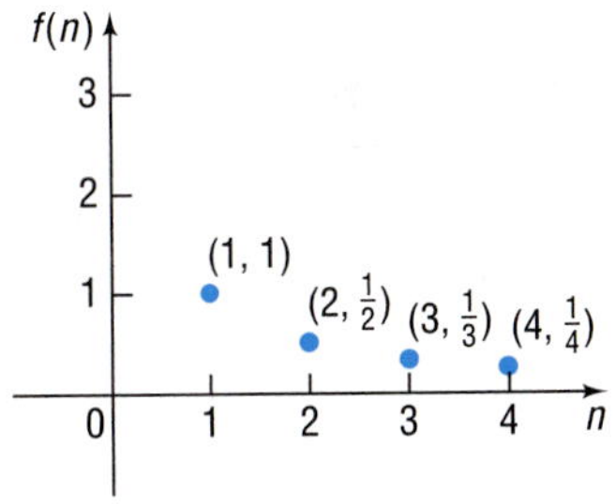

(b) $f(n) = \frac{1}{n}$

A sequence is usually represented by listing its values in order. For example, the sequence whose graph is given in Figure 1(b) might be represented as

$$f(1),\ f(2),\ f(3),\ f(4), \ldots \quad \text{or} \quad 1, \frac{1}{2}, \frac{1}{3}, \frac{1}{4}, \ldots$$

The list never ends, as the ellipsis dots indicate. The numbers in this ordered list are called the **terms** of the sequence.

1 In dealing with sequences, we usually use subscripted letters, for example, a_1 to represent the first term, a_2 for the second term, a_3 for the third term, and so on. Thus, for the sequence $f(n) = 1/n$, we write

$$a_1 = f(1) = 1 \quad a_2 = f(2) = \frac{1}{2} \quad a_3 = f(3) = \frac{1}{3} \quad a_4 = f(4) = \frac{1}{4} \quad \ldots \quad a_n = f(n) = \frac{1}{n} \quad \ldots$$

In other words, we usually do not use the traditional function notation $f(n)$ for sequences. For this particular sequence, we have a rule for the nth term, namely, $a_n = 1/n$, so it is easy to find any term of the sequence.

When a formula for the nth term of a sequence is known, rather than write out the terms of the sequence, we usually represent the entire sequence by placing braces around the formula for the nth term. For example, the sequence whose nth term is $b_n = (\frac{1}{2})^n$ may be represented as

$$\{b_n\} = \left\{\left(\frac{1}{2}\right)^n\right\}$$

or by

$$b_1 = \frac{1}{2}$$
$$b_2 = \frac{1}{4}$$
$$b_3 = \frac{1}{8}$$
$$\vdots$$
$$b_n = \left(\frac{1}{2}\right)^n$$
$$\vdots$$

EXAMPLE 1 Writing the First Several Terms of a Sequence

Write down the first six terms of the following sequence and graph it.

$$\{a_n\} = \left\{\frac{n-1}{n}\right\}$$

Solution

$$a_1 = 0$$
$$a_2 = \frac{1}{2}$$
$$a_3 = \frac{2}{3}$$
$$a_4 = \frac{3}{4}$$
$$a_5 = \frac{4}{5}$$
$$a_6 = \frac{5}{6}$$

See Figure 2. ■

FIGURE 2

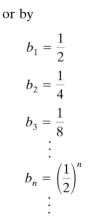

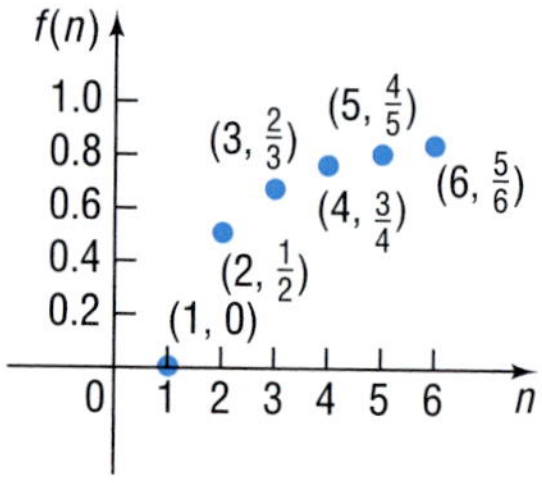

Graphing utilities can be used to write the terms of a sequence and graph them as the following example illustrates.

EXAMPLE 2 Using a Graphing Utility to Write the First Several Terms of a Sequence

Use a graphing utility to write the first six terms of the following sequence and graph it.

$$\{a_n\} = \left\{\frac{n-1}{n}\right\}$$

Solution Figure 3 shows the sequence generated on a TI-83 graphing calculator. We can see the first few terms of the sequence on the viewing window. You need to press the right arrow key to scroll right in order to see the remaining terms of the sequence.

FIGURE 3

Figure 4 shows a graph of the sequence. Notice the first term of the sequence is not visible since it lies on the x-axis.

FIGURE 4

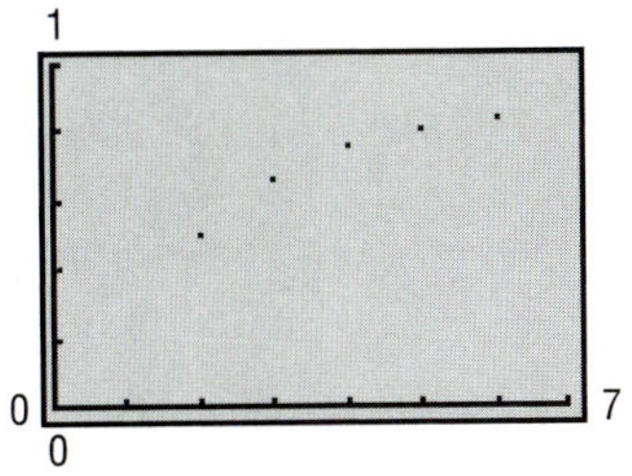

We will usually provide solutions done by hand. The reader is encouraged to verify solutions using a graphing utility.

EXAMPLE 3

Writing the First Several Terms of a Sequence

Write down the first six terms of the following sequence and graph it by hand.

$$\{b_n\} = \left\{(-1)^{n-1}\left(\frac{2}{n}\right)\right\}$$

Solution

$$b_1 = 2$$
$$b_2 = -1$$
$$b_3 = \frac{2}{3}$$
$$b_4 = -\frac{1}{2}$$
$$b_5 = \frac{2}{5}$$
$$b_6 = -\frac{1}{3}$$

See Figure 5.

FIGURE 5

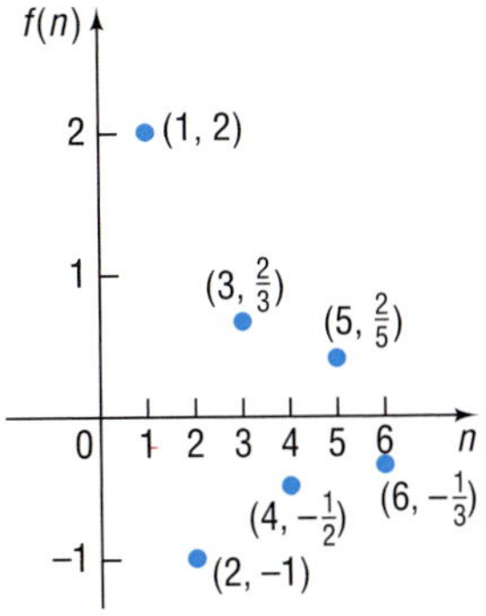

EXAMPLE 4

Writing the First Several Terms of a Sequence

Write down the first six terms of the following sequence and graph it by hand.

$$\{c_n\} = \begin{Bmatrix} n & \text{if } n \text{ is even} \\ 1/n & \text{if } n \text{ is odd} \end{Bmatrix}$$

FIGURE 6

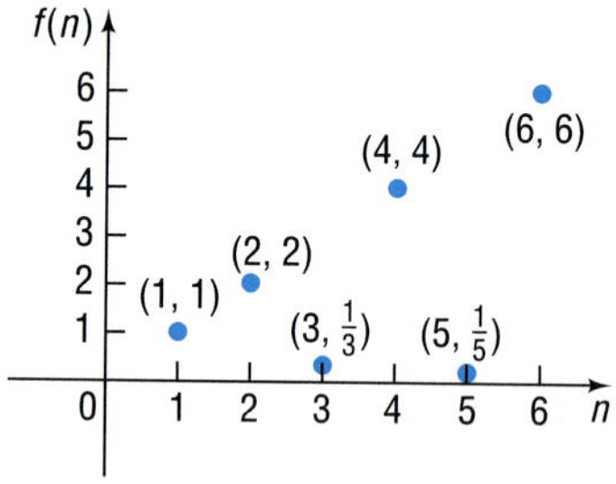

Solution

$$c_1 = 1$$
$$c_2 = 2$$
$$c_3 = \frac{1}{3}$$
$$c_4 = 4$$
$$c_5 = \frac{1}{5}$$
$$c_6 = 6$$

See Figure 6.

Now work Problems 3 and 5.

Sometimes a sequence is indicated by an observed pattern in the first few terms that makes it possible to infer the makeup of the nth term. In the example that follows, a sufficient number of terms of the sequence is given so that a natural choice for the nth term is suggested.

EXAMPLE 5

Determining a Sequence from a Pattern

(a) $e, \frac{e^2}{2}, \frac{e^3}{3}, \frac{e^4}{4}, \ldots$ $\quad a_n = \frac{e^n}{n}$

(b) $1, \frac{1}{3}, \frac{1}{9}, \frac{1}{27}, \ldots$ $\quad b_n = \frac{1}{3^{n-1}}$

(c) $1, 3, 5, 7, \ldots$ $\quad c_n = 2n - 1$

(d) $1, 4, 9, 16, 25, \ldots$ $\quad d_n = n^2$

(e) $1, -\frac{1}{2}, \frac{1}{3}, -\frac{1}{4}, \frac{1}{5}, \ldots$ $\quad e_n = (-1)^{n+1}\left(\frac{1}{n}\right)$

Notice in the sequence $\{e_n\}$ in Example 5(e) that the signs of the terms **alternate.** When this occurs, we use factors such as $(-1)^{n+1}$, which equals 1 if n is odd and -1 if n is even, or $(-1)^n$, which equals -1 if n is odd and 1 if n is even.

Now work Problem 13.

The Factorial Symbol

If $n \geq 0$ is an integer, the **factorial symbol** $n!$ is defined as follows:

$$0! = 1 \qquad 1! = 1$$
$$n! = n(n-1) \cdot \ldots \cdot 3 \cdot 2 \cdot 1 \qquad \text{if } n \geq 2$$

For example, $2! = 2 \cdot 1 = 2$, $3! = 3 \cdot 2 \cdot 1 = 6$, $4! = 4 \cdot 3 \cdot 2 \cdot 1 = 24$, and so on. Table 1 lists the values of $n!$ for $0 \le n \le 6$.

TABLE 1

n	0	1	2	3	4	5	6
$n!$	1	1	2	6	24	120	720

Because

$$n! = n\underbrace{(n-1)(n-2) \cdot \ldots \cdot 3 \cdot 2 \cdot 1}_{(n-1)!}$$

we can use the formula

$$n! = n(n-1)!$$

to find successive factorials. For example, because $6! = 720$, we have

$$7! = 7 \cdot 6! = 7(720) = 5040$$

and

$$8! = 8 \cdot 7! = 8(5040) = 40{,}320$$

Comment: Your calculator may have a factorial key. Use it to see how fast factorials increase in value. Find the value of 69!. What happens when you try to find 70!? In fact, 70! is larger than 10^{100} (a *googol*), the largest number most calculators can display.

Recursion Formulas

2 A second way of defining a sequence is to assign a value to the first (or the first few) terms and specify the nth term by a formula or equation that involves one or more of the terms preceding it. Sequences defined this way are said to be defined **recursively,** and the rule or formula is called a **recursive formula.**

EXAMPLE 6

Writing the Terms of a Recursively Defined Sequence

Write down the first five terms of the following recursively defined sequence.

$$s_1 = 1, \qquad s_n = 4s_{n-1}$$

Solution The first term is given as $s_1 = 1$. To get the second term, we use $n = 2$ in the formula to get $s_2 = 4s_1 = 4 \cdot 1 = 4$. To get the third term, we use $n = 3$ in the formula to get $s_3 = 4s_2 = 4 \cdot 4 = 16$. To get a new term requires that we know the value of the preceding term. The first five terms are

$$s_1 = 1$$
$$s_2 = 4 \cdot 1 = 4$$
$$s_3 = 4 \cdot 4 = 16$$
$$s_4 = 4 \cdot 16 = 64$$
$$s_5 = 4 \cdot 64 = 256$$

Graphing utilities can be used to generate recursively defined sequences when the nth term depends upon the term preceding it.

EXAMPLE 7 Using a Graphing Utility to Write the Terms of a Recursively Defined Sequence

Use a graphing utility to write down the first five terms of the following recursively defined sequence

$$s_1 = 1, \qquad s_n = 4s_{n-1}$$

Solution First, put the graphing utility into sequence mode and dot mode. Using $Y =$, enter the recursive formula into the graphing utility. Next, set up the viewing window to generate the desired sequence. Finally, graph the recursion relation and use TRACE to determine the terms in the sequence. For example, in Figure 7 we see the fourth term of the sequence is 64.

Note: If your graphing utility is capable of generating tables, the TABLE feature displays the terms in the sequence. See Table 2.

FIGURE 7

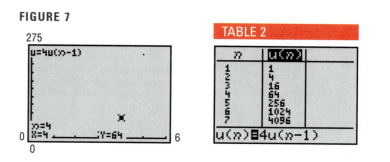

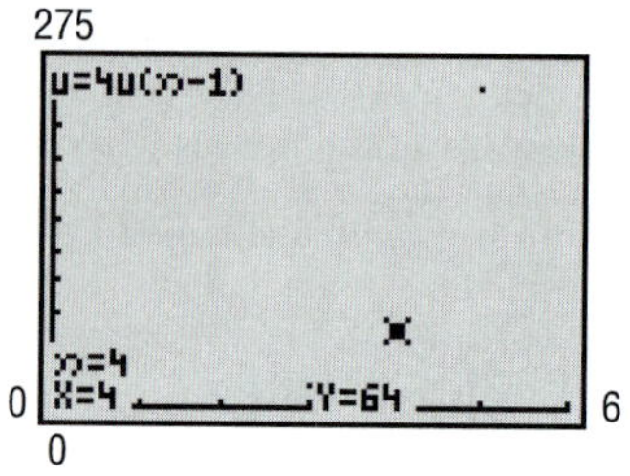

TABLE 2

n	u(n)
1	1
2	4
3	16
4	64
5	256
6	1024
7	4096

u(n)=4u(n−1)

EXAMPLE 8 Writing the Terms of a Recursively Defined Sequence

Write down the first five terms of the following recursively defined sequence.

$$u_1 = 1, \qquad u_2 = 1, \qquad u_{n+2} = u_n + u_{n+1}$$

Solution We are given the first two terms. To get the third term requires that we know each of the previous two terms. Thus,

$$u_1 = 1$$
$$u_2 = 1$$
$$u_3 = u_1 + u_2 = 2$$
$$u_4 = u_2 + u_3 = 1 + 2 = 3$$
$$u_5 = u_3 + u_4 = 2 + 3 = 5$$

The sequence defined in Example 8 is called a **Fibonacci sequence,** and the terms of this sequence are called **Fibonacci numbers.** These numbers appear in a wide variety of applications (see Problems 67 and 68).

EXAMPLE 9

Writing the Terms of a Recursively Defined Sequence

Write down the first five terms of the following recursively defined sequence.

$$f_1 = 1, \qquad f_{n+1} = (n + 1)f_n$$

Solution Here

$$\begin{aligned} f_1 &= 1 \\ f_2 &= 2f_1 = 2 \cdot 1 = 2 \\ f_3 &= 3f_2 = 3 \cdot 2 = 6 \\ f_4 &= 4f_3 = 4 \cdot 6 = 24 \\ f_5 &= 5f_4 = 5 \cdot 24 = 120 \end{aligned}$$

You should recognize the nth term of the sequence in Example 9 as $n!$.

Now work Problems 21 and 29.

Adding the First n Terms of a Sequence; Summation Notation

3 It is often important to be able to find the sum of the first n terms of a sequence $\{a_n\}$, namely,

$$a_1 + a_2 + a_3 + \cdots + a_n \tag{1}$$

Rather than write down all these terms, we introduce a more concise way to express the sum, called **summation notation.** Using summation notation, we would write the sum (1) as

$$a_1 + a_2 + a_3 + \cdots + a_n = \sum_{k=1}^{n} a_k$$

The symbol Σ (a stylized version of the Greek letter sigma, which is an S in our alphabet) is simply an instruction to sum, or add up, the terms. The integer k is called the **index** of the sum; it tells you where to start the sum and where to end it. Therefore, the expression

$$\sum_{k=1}^{n} a_k \tag{2}$$

is an instruction to add the terms a_k of the sequence $\{a_n\}$ from $k = 1$ through $k = n$. We read expression (2) as "the sum of a_k from $k = 1$ to $k = n$."

Next, we list some properties of sequences using summation notation.

Theorem Properties of Sequences

If $\{a_n\}$ and $\{b_n\}$ are two sequences and c is a real number, then

1. $\displaystyle\sum_{k=1}^{n} c = c \cdot n$

2. $\displaystyle\sum_{k=1}^{n} ca_k = c\sum_{k=1}^{n} a_k$

3. $\sum_{k=1}^{n} (a_k + b_k) = \sum_{k=1}^{n} a_k + \sum_{k=1}^{n} b_k$

4. $\sum_{k=1}^{n} (a_k - b_k) = \sum_{k=1}^{n} a_k - \sum_{k=1}^{n} b_k$

5. $\sum_{k=1}^{n} a_k = \sum_{k=1}^{j} a_k + \sum_{k=j+1}^{n} a_k$, when $1 < j < n$

Although we shall not prove these properties, the proofs are based on properties of real numbers.

EXAMPLE 10 Finding the Sum of a Sequence

Find the sum of each sequence.

(a) $\sum_{k=1}^{5} 3k$ (b) $\sum_{k=1}^{4} (k^2 - 7k + 2)$

Solution (a) $\sum_{k=1}^{5} 3k = 3\sum_{k=1}^{5} k = 3(1 + 2 + 3 + 4 + 5) = 3(15) = 45$

↑ Property 2

(b)
$$\sum_{k=1}^{4} (k^2 - 7k + 2) = \sum_{k=1}^{4} k^2 - \sum_{k=1}^{4} 7k + \sum_{k=1}^{4} 2 \qquad \text{Properties 3, 4.}$$
$$= \sum_{k=1}^{4} k^2 - 7\sum_{k=1}^{4} k + \sum_{k=1}^{4} 2 \qquad \text{Property 2.}$$
$$= (1^2 + 2^2 + 3^2 + 4^2) - 7(1 + 2 + 3 + 4) + 2(4) \qquad \text{Property 1.}$$
$$= (1 + 4 + 9 + 16) - 7(10) + 8$$
$$= 30 - 70 + 8$$
$$= -32$$

EXAMPLE 11 Using a Graphing Utility to Find the Sum of a Sequence

Using a graphing utility, find the sum of each sequence.

(a) $\sum_{k=1}^{5} 3k$ (b) $\sum_{k=1}^{4} (k^2 - 7k + 2)$

Solution (a) Figure 8 shows the solution using a TI-83 graphing calculator.

FIGURE 8

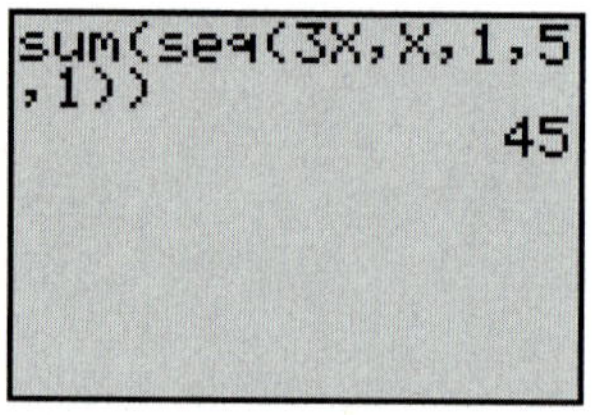

Thus, $\sum_{k=1}^{5} 3k = 45$.

(b) Figure 9 shows the solution using a TI-83 graphing calculator.

FIGURE 9

Thus, $\sum_{k=1}^{4} (k^2 - 7k + 2) = -32$. ■

Now work Problem 39.

EXAMPLE 12 Expanding Summation Notation

Write out each sum.

(a) $\sum_{k=1}^{n} \frac{1}{k}$ (b) $\sum_{k=1}^{n} k!$

Solution (a) $\sum_{k=1}^{n} \frac{1}{k} = \frac{1}{1} + \frac{1}{2} + \frac{1}{3} + \cdots + \frac{1}{n}$ (b) $\sum_{k=1}^{n} k! = 1! + 2! + \cdots + n!$ ■

EXAMPLE 13 Writing a Sum in Summation Notation

Express each sum using summation notation.

(a) $1^2 + 2^2 + 3^2 + \cdots + 9^2$ (b) $1 + \frac{1}{2} + \frac{1}{4} + \frac{1}{8} + \cdots + \frac{1}{2^{n-1}}$

Solution (a) The sum $1^2 + 2^2 + 3^2 + \cdots + 9^2$ has 9 terms, each of the form k^2 and starts at $k = 1$ and ends at $k = 9$. Thus,

$$1^2 + 2^2 + 3^2 + \cdots + 9^2 = \sum_{k=1}^{9} k^2$$

(b) The sum

$$1 + \frac{1}{2} + \frac{1}{4} + \frac{1}{8} + \cdots + \frac{1}{2^{n-1}}$$

has n terms, each of the form $1/2^{k-1}$ and starts at $k = 1$ and ends at $k = n$. Thus,

$$1 + \frac{1}{2} + \frac{1}{4} + \frac{1}{8} + \cdots + \frac{1}{2^{n-1}} = \sum_{k=1}^{n} \frac{1}{2^{k-1}}$$ ■

The index of summation need not always begin at 1 or end at n; for example,

$$\sum_{k=0}^{n-1} \frac{1}{2^k} = 1 + \frac{1}{2} + \frac{1}{4} + \cdots + \frac{1}{2^{n-1}}$$

Letters other than k may be used as the index. For example,

$$\sum_{j=1}^{n} j! \quad \text{and} \quad \sum_{i=1}^{n} i!$$

each represent the same sum as the one given in Example 12(b).

Now work Problems 49 and 59.

12.1 EXERCISES

In Problems 1–12, write down the first five terms of each sequence by hand. Verify your results using a graphing utility.

1. $\{n\}$

2. $\{n^2 + 1\}$

3. $\left\{\dfrac{n}{n+2}\right\}$

4. $\left\{\dfrac{2n+1}{2n}\right\}$

5. $\{(-1)^{n+1}n^2\}$

6. $\left\{(-1)^{n-1}\left(\dfrac{n}{2n-1}\right)\right\}$

7. $\left\{\dfrac{2^n}{3^n+1}\right\}$

8. $\left\{\left(\dfrac{4}{3}\right)^n\right\}$

9. $\left\{\dfrac{(-1)^n}{(n+1)(n+2)}\right\}$

10. $\left\{\dfrac{3^n}{n}\right\}$

11. $\left\{\dfrac{n}{e^n}\right\}$

12. $\left\{\dfrac{n^2}{2^n}\right\}$

In Problems 13–20, the given pattern continues. Write down the nth term of each sequence suggested by the pattern.

13. $\dfrac{1}{2}, \dfrac{2}{3}, \dfrac{3}{4}, \dfrac{4}{5}, \ldots$

14. $\dfrac{1}{1\cdot 2}, \dfrac{1}{2\cdot 3}, \dfrac{1}{3\cdot 4}, \dfrac{1}{4\cdot 5}, \ldots$

15. $1, \dfrac{1}{2}, \dfrac{1}{4}, \dfrac{1}{8}, \ldots$

16. $\dfrac{2}{3}, \dfrac{4}{9}, \dfrac{8}{27}, \dfrac{16}{81}, \ldots$

17. $1, -1, 1, -1, 1, -1, \ldots$

18. $1, \dfrac{1}{2}, 3, \dfrac{1}{4}, 5, \dfrac{1}{6}, 7, \dfrac{1}{8}, \ldots$

19. $1, -2, 3, -4, 5, -6, \ldots$

20. $2, -4, 6, -8, 10, \ldots$

In Problems 21–34, a sequence is defined recursively. Write the first five terms by hand. When possible, use a graphing utility to verify your results.

21. $a_1 = 2;\ a_{n+1} = 3 + a_n$

22. $a_1 = 3;\ a_{n+1} = 4 - a_n$

23. $a_1 = -2;\ a_{n+1} = n + a_n$

24. $a_1 = 1;\ a_{n+1} = n - a_n$

25. $a_1 = 5;\ a_{n+1} = 2a_n$

26. $a_1 = 2;\ a_{n+1} = -a_n$

27. $a_1 = 3;\ a_{n+1} = \dfrac{a_n}{n}$

28. $a_1 = -2;\ a_{n+1} = n + 3a_n$

29. $a_1 = 1;\ a_2 = 2;\ a_{n+2} = a_n a_{n+1}$

30. $a_1 = -1;\ a_2 = 1;\ a_{n+2} = a_{n+1} + na_n$

31. $a_1 = A;\ a_{n+1} = a_n + d$

32. $a_1 = A;\ a_{n+1} = ra_n,\ r \neq 0$

33. $a_1 = \sqrt{2};\ a_{n+1} = \sqrt{2 + a_n}$

34. $a_1 = \sqrt{2};\ a_{n+1} = \sqrt{a_n/2}$

In Problems 35–46, find the sum of each sequence. Verify your results using a graphing utility.

35. $\sum_{k=1}^{10} 5$

36. $\sum_{k=1}^{20} 8$

37. $\sum_{k=1}^{6} k$

38. $\sum_{k=1}^{4} (-k)$

39. $\sum_{k=1}^{5} (5k + 3)$

40. $\sum_{k=1}^{6} (3k - 7)$

41. $\sum_{k=1}^{3} (k^2 + 4)$

42. $\sum_{k=0}^{4} (k^2 - 4)$

43. $\sum_{k=1}^{6} (-1)^k 2^k$

44. $\sum_{k=1}^{4} (-1)^k 3^k$

45. $\sum_{k=1}^{4} (k^3 - 1)$

46. $\sum_{k=0}^{3} (k^3 + 2)$

In Problems 47–56, write out each sum.

47. $\sum_{k=1}^{n}(k+2)$

48. $\sum_{k=1}^{n}(2k+1)$

49. $\sum_{k=1}^{n}\frac{k^2}{2}$

50. $\sum_{k=1}^{n}(k+1)^2$

51. $\sum_{k=0}^{n}\frac{1}{3^k}$

52. $\sum_{k=0}^{n}\left(\frac{3}{2}\right)^k$

53. $\sum_{k=0}^{n-1}\frac{1}{3^{k+1}}$

54. $\sum_{k=0}^{n-1}(2k+1)$

55. $\sum_{k=2}^{n}(-1)^k \ln k$

56. $\sum_{k=3}^{n}(-1)^{k+1}2^k$

In Problems 57–66, express each sum using summation notation.

57. $1+2+3+\cdots+20$

58. $1^3+2^3+3^3+\cdots+8^3$

59. $\frac{1}{2}+\frac{2}{3}+\frac{3}{4}+\cdots+\frac{13}{13+1}$

60. $1+3+5+7+\cdots+[2(12)-1]$

61. $1-\frac{1}{3}+\frac{1}{9}-\frac{1}{27}+\cdots+(-1)^6\left(\frac{1}{3^6}\right)$

62. $\frac{2}{3}-\frac{4}{9}+\frac{8}{27}-\cdots+(-1)^{11+1}\left(\frac{2}{3}\right)^{11}$

63. $3+\frac{3^2}{2}+\frac{3^3}{3}+\cdots+\frac{3^n}{n}$

64. $\frac{1}{e}+\frac{2}{e^2}+\frac{3}{e^3}+\cdots+\frac{n}{e^n}$

65. $a+(a+d)+(a+2d)+\cdots+(a+nd)$

66. $a+ar+ar^2+\cdots+ar^{n-1}$

67. Growth of a Rabbit Colony A colony of rabbits begins with one pair of mature rabbits, which will produce a pair of offspring (one male, one female) each month. Assume that all rabbits mature in 1 month and produce a pair of offspring (one male, one female) after 2 months. If no rabbits ever die, how many pairs of mature rabbits are there after 7 months?

[**Hint:** A Fibonacci sequence models this colony. Do you see why?]

68. Fibonacci Sequence Let

$$u_n = \frac{(1+\sqrt{5})^n-(1-\sqrt{5})^n}{2^n\sqrt{5}}$$

define the nth term of a sequence.

(a) Show that $u_1 = 1$ and $u_2 = 1$.

(b) Show that $u_{n+2} = u_{n+1} + u_n$.

(c) Draw the conclusion that $\{u_n\}$ is a Fibonacci sequence.

In Problems 69 and 70, we use the fact that in some programming languages it is possible to have a function subroutine include a call to itself.

69. Programming Exercise Write a program that accepts an integer as input and prints the number and its factorial. Use a recursively defined function, that is, use a function subroutine that calls itself.

70. Programming Exercise Write a program that accepts a positive integer N as input and outputs the Nth Fibonacci number. Use a recursively defined subroutine.

71. Pascal's Triangle Divide the triangular array shown (called Pascal's triangle) using diagonal lines as indicated. Find the sum of the numbers in each of these diagonal rows. Do you recognize this sequence?

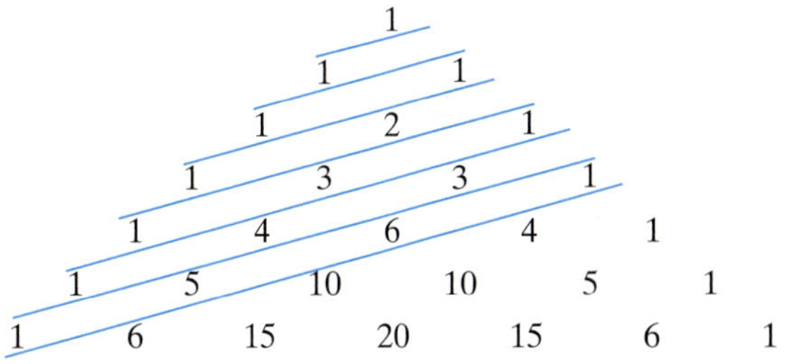

72. Fibonacci Sequence Use the result of Problem 68 to do the following problems:

(a) Write the first 10 terms of the Fibonacci sequence.

(b) Compute the ratio $\frac{u_{n+1}}{u_n}$ for the first 10 terms.

(c) As n gets large, what number does the ratio approach? This number is referred to as the **golden ratio.** Rectangles whose sides are in this ratio were considered pleasing to the eye by the Greeks. For example, the facade of the Parthenon was constructed using the "golden ratio."

(d) Compute the ratio $\frac{u_n}{u_{n+1}}$ for the first 10 terms.

(e) As n gets large, what number does the ratio approach? This number is also referred to as the **golden ratio.** This ratio is believed to have been used in the construction of the Great Pyramid in Egypt. The ratio equals the sum of the areas of the four face triangles divided by the total surface area of the Great Pyramid.

73. Investigate various applications that lead to a Fibonacci sequence such as art, architecture, or financial markets. Write an essay on these applications.

12.2 ARITHMETIC SEQUENCES

1 Determine if a Sequence Is Arithmetic

2 Find a Formula for an Arithmetic Sequence

3 Find the Sum of an Arithmetic Sequence

1 When the difference between successive terms of a sequence is always the same number, the sequence is called **arithmetic.** Thus, an **arithmetic sequence*** may be defined recursively as $a_1 = a$, $a_{n+1} - a_n = d$, or as

$$a_1 = a, \qquad a_{n+1} = a_n + d \tag{1}$$

where $a = a_1$ and d are real numbers. The number a is the first term, and the number d is called the **common difference.**

Thus, the terms of an arithmetic sequence with first term a and common difference d follow the pattern

$$a, \quad a + d, \quad a + 2d, \quad a + 3d, \quad \cdots$$

EXAMPLE 1 Determining if a Sequence is Arithmetic

The sequence

$$4, 7, 10, 13, \ldots$$

is arithmetic since the difference of successive terms is 3. The first term is 4, and the common difference is 3.

EXAMPLE 2 Determining if a Sequence is Arithmetic

Show that the following sequence is arithmetic. Find the first term and the common difference.

$$\{s_n\} = \{3n + 5\}$$

Solution The first term is $s_1 = 3 \cdot 1 + 5 = 8$. The $(n + 1)$st and nth terms of the sequence $\{s_n\}$ are

$$s_{n+1} = 3(n + 1) + 5 = 3n + 8 \quad \text{and} \quad s_n = 3n + 5$$

*Sometimes called an **arithmetic progression.**

Their difference is

$$s_{n+1} - s_n = (3n + 8) - (3n + 5) = 8 - 5 = 3$$

Since the difference of two successive terms does not depend on n, the common difference is 3 and the sequence is arithmetic. ■

EXAMPLE 3 Determining if a Sequence Is Arithmetic

Show that the sequence $\{t_n\} = \{4 - n\}$ is arithmetic. Find the first term and the common difference.

Solution The first term is $t_1 = 4 - 1 = 3$. The $(n + 1)$st and nth terms are

$$t_{n+1} = 4 - (n + 1) = 3 - n \quad \text{and} \quad t_n = 4 - n$$

Their difference is

$$t_{n+1} - t_n = (3 - n) - (4 - n) = 3 - 4 = -1$$

The difference of two successive terms does not depend on n; it always equals the same number, -1. Hence, $\{t_n\}$ is an arithmetic sequence whose common difference is -1. ■

Now work Problem 3.

2 Suppose that a is the first term of an arithmetic sequence whose common difference is d. We seek a formula for the nth term, a_n. To see the pattern, we write down the first few terms:

$$a_1 = a$$
$$a_2 = a + d = a + 1 \cdot d$$
$$a_3 = a_2 + d = (a + d) + d = a + 2 \cdot d$$
$$a_4 = a_3 + d = (a + 2 \cdot d) + d = a + 3 \cdot d$$
$$a_5 = a_4 + d = (a + 3 \cdot d) + d = a + 4 \cdot d$$
$$\vdots$$
$$a_n = a_{n-1} + d = [a + (n - 2)d] + d = a + (n - 1)d$$

We are led to the following result:

Theorem *n*th Term of an Arithmetic Sequence

For an arithmetic sequence $\{a_n\}$ whose first term is a and whose common difference is d, the nth term is determined by the formula

$$a_n = a + (n - 1)d \qquad (2)$$

EXAMPLE 4 Finding a Particular Term of an Arithmetic Sequence

Find the 13th term of the arithmetic sequence: 2, 6, 10, 14, 18, ...

Solution The first term of this arithmetic sequence is $a = 2$, and the common difference is 4. By formula (2), the nth term is

$$a_n = 2 + (n - 1)4$$

Hence, the 13th term is

$$a_{13} = 2 + 12 \cdot 4 = 50$$

Exploration Use a graphing utility to find the 13th term of the sequence given in Example 4. Use it to find the 20th term and the 50th term.

EXAMPLE 5

Finding a Recursive Formula for an Arithmetic Sequence

The 8th term of an arithmetic sequence is 75, and the 20th term is 39. Find the first term and the common difference. Give a recursive formula for the sequence.

Solution By equation (2), we know that $a_n = a + (n-1)d$. As a result,

$$\begin{cases} a_8 = a + 7d = 75 \\ a_{20} = a + 19d = 39 \end{cases}$$

This is a system of two linear equations containing two variables, a and d, which we can solve by elimination. Thus, subtracting the second equation from the first equation, we get

$$\begin{aligned} -12d &= 36 \\ d &= -3 \end{aligned}$$

With $d = -3$, we find $a = 75 - 7d = 75 - 7(-3) = 96$. The first term is $a = 96$ and the common difference is $d = -3$. A recursive formula for this sequence is found using (1).

$$a_1 = 96, \quad a_{n+1} = a_n - 3$$

Based on formula (2), a formula for the nth term of the sequence $\{a_n\}$ in Example 5 is

$$a_n = a + (n-1)d = 96 + (n-1)(-3) = 99 - 3n$$

Now work Problems 19 and 25.

Adding the First *n* Terms of an Arithmetic Sequence

3 The next result gives a formula for finding the sum of the first n terms of an arithmetic sequence.

Theorem Sum of *n* Terms of an Arithmetic Sequence

Let $\{a_n\}$ be an arithmetic sequence with first term a and common difference d. The sum S_n of the first n terms of $\{a_n\}$ is

$$S_n = \frac{n}{2}[2a + (n-1)d] = \frac{n}{2}(a + a_n) \qquad (3)$$

Proof

$$
\begin{aligned}
S_n &= a_1 + a_2 + a_3 + \cdots + a_n \\
&= a + (a + d) + (a + 2d) + \cdots + [a + (n-1)d] \\
&= \underbrace{(a + a + \cdots + a)}_{n \text{ terms}} + [d + 2d + \cdots + (n-1)d] \\
&= na + d[1 + 2 + \cdots + (n-1)] \\
&= na + d\left[\frac{(n-1)n}{2}\right] \\
&= na + \frac{n}{2}(n-1)d \\
&= \frac{n}{2}[2a + (n-1)d] && \text{Factor out } n/2. \qquad (4) \\
&= \frac{n}{2}[a + a + (n-1)d] \\
&= \frac{n}{2}(a + a_n) && \text{Formula (2).} \qquad (5)
\end{aligned}
$$

Formula (3) provides two ways to find the sum of the first n terms of an arithmetic sequence. Notice that (4) involves the first term and common difference, while (5) involves the first term and the nth term. Use whichever form is easier.

EXAMPLE 6 Finding the Sum of n Terms of an Arithmetic Sequence

Find the sum S_n of the first n terms of the sequence $\{3n + 5\}$; that is, find

$$8 + 11 + 14 + \cdots + (3n + 5)$$

Solution The sequence $\{3n + 5\}$ is an arithmetic sequence with first term $a = 8$ and the nth term $(3n + 5)$. To find the sum S_n we use formula (3), as given in (5):

$$S_n = \frac{n}{2}(a + a_n) = \frac{n}{2}[8 + (3n + 5)] = \frac{n}{2}(3n + 13)$$

Now work Problem 33.

EXAMPLE 7 Using a Graphing Utility to Find the Sum of 20 Terms of an Arithmetic Sequence

Use a graphing utility to find the sum S_n of the first 20 terms of the sequence $\{9.5n + 2.6\}$.

Solution Figure 10 shows the results obtained using a TI-83 graphing calculator.

FIGURE 10

Thus, the sum of the first 20 terms of the sequence $\{9.5n + 2.6\}$ is 2047.

Now work Problem 41.

EXAMPLE 8

Creating a Floor Design

A ceramic tile floor is designed in the shape of a trapezoid 20 feet wide at the base and 10 feet wide at the top. See Figure 11. The tiles, 12 inches by 12 inches, are to be placed so that each successive row contains one less tile than the row below. How many tiles will be required?

FIGURE 11

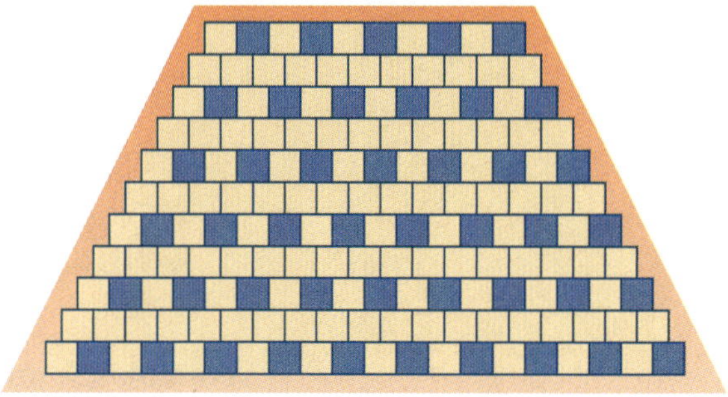

Solution The bottom row requires 20 tiles and the top row 10 tiles. Since each successive row requires one less tile, the total number of tiles required is

$$S = 20 + 19 + 18 + \cdots + 11 + 10$$

This is the sum of an arithmetic sequence; the common difference is -1. The number of terms to be added is $n = 11$, with the first term $a = 20$ and the last term $a_{11} = 10$. The sum S is

$$S = \frac{n}{2}(a + a_{11}) = \frac{11}{2}(20 + 10) = 165$$

Thus, 165 tiles will be required.

12.2 EXERCISES

In Problems 1–10, an arithmetic sequence is given. Find the common difference and write out the first four terms.

1. $\{n + 4\}$ **2.** $\{n - 5\}$ **3.** $\{2n - 5\}$ **4.** $\{3n + 1\}$ **5.** $\{6 - 2n\}$

6. $\{4 - 2n\}$ **7.** $\left\{\frac{1}{2} - \frac{1}{3}n\right\}$ **8.** $\left\{\frac{2}{3} + \frac{n}{4}\right\}$ **9.** $\{\ln 3^n\}$ **10.** $\{e^{\ln n}\}$

In Problems 11–18, find the nth term of the arithmetic sequence whose initial term a and common difference d are given. What is the fifth term?

11. $a = 2; d = 3$ **12.** $a = -2; d = 4$ **13.** $a = 5; d = -3$

14. $a = 6; d = -2$ **15.** $a = 0; d = \frac{1}{2}$ **16.** $a = 1; d = -\frac{1}{3}$

17. $a = \sqrt{2}; d = \sqrt{2}$ **18.** $a = 0; d = \pi$

In Problems 19–24, find the indicated term in each arithmetic sequence.

19. 12th term of $2, 4, 6, \ldots$

20. 8th term of $-1, 1, 3, \ldots$

21. 10th term of $1, -2, -5, \ldots$

22. 9th term of $5, 0, -5, \ldots$

23. 8th term of $a, a + b, a + 2b, \ldots$

24. 7th term of $2\sqrt{5}, 4\sqrt{5}, 6\sqrt{5}, \ldots$

In Problems 25–32, find the first term and the common difference of the arithmetic sequence described. Give a recursive formula for the sequence.

25. 8th term is 8; 20th term is 44

26. 4th term is 3; 20th term is 35

27. 9th term is -5; 15th term is 31

28. 8th term is 4; 18th term is -96

29. 15th term is 0; 40th term is -50

30. 5th term is -2; 13th term is 30

31. 14th term is -1; 18th term is -9

32. 12th term is 4; 18th term is 28

In Problems 33–40, find the sum.

33. $1 + 3 + 5 + \cdots + (2n - 1)$

34. $2 + 4 + 6 + \cdots + 2n$

35. $7 + 12 + 17 + \cdots + (2 + 5n)$

36. $-1 + 3 + 7 + \cdots + (4n - 5)$

37. $2 + 4 + 6 + \cdots + 70$

38. $1 + 3 + 5 + \cdots + 59$

39. $5 + 9 + 13 + \cdots + 49$

40. $2 + 5 + 8 + \cdots + 41$

For Problems 41–46, use a graphing utility to find the sum of each sequence.

41. $\{3.45n + 4.12\}$ $n = 20$

42. $\{2.67n - 1.23\}$ $n = 25$

43. $2.8 + 5.2 + 7.6 + \cdots + 36.4$

44. $5.4 + 7.3 + 9.2 + \cdots + 32$

45. $4.9 + 7.48 + 10.06 + \cdots + 66.82$

46. $3.71 + 6.9 + 10.09 + \cdots + 80.27$

47. Find x so that $x + 3$, $2x + 1$, and $5x + 2$ are terms of an arithmetic sequence.

48. Find x so that $2x$, $3x + 2$, and $5x + 3$ are terms of an arithmetic sequence.

49. **Drury Lane Theater** The Drury Lane Theater has 25 seats in the first row and 30 rows in all. Each successive row contains one additional seat. How many seats are in the theater?

50. **Football Stadium** The corner section of a football stadium has 15 seats in the first row and 40 rows in all. Each successive row contains two additional seats. How many seats are in this section?

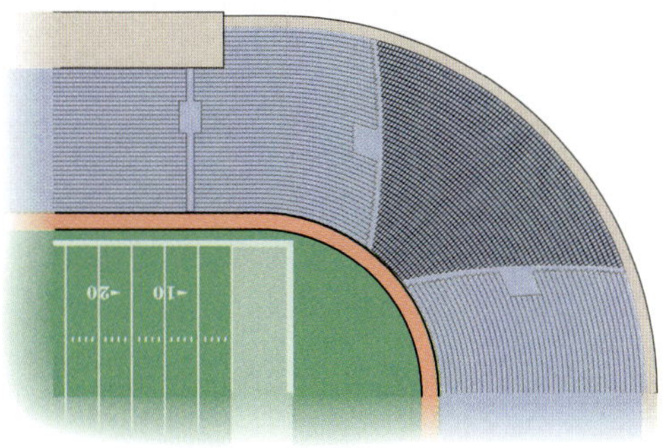

51. **Creating a Mosaic** A mosaic is designed in the shape of an equilateral triangle, 20 feet on each side. Each tile in the mosaic is in the shape of an equilateral triangle, 12 inches to a side. The tiles are to alternate in color as shown in the illustration. How many tiles of each color will be required?

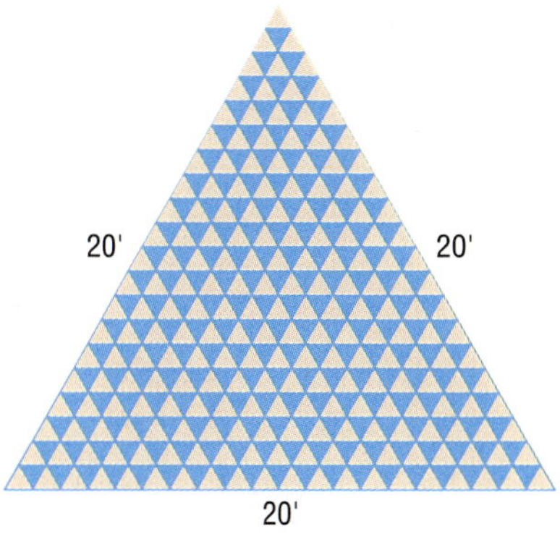

52. **Constructing a Brick Staircase** A brick staircase has a total of 30 steps. The bottom step requires 100 bricks. Each successive step requires two less bricks than the prior step.

(a) How many bricks are required for the top step?

(b) How many bricks are required to build the staircase?

53. Make up an arithmetic sequence. Give it to a friend and ask for its 20th term.

12.3 GEOMETRIC SEQUENCES; GEOMETRIC SERIES

1 Determine if a Sequence Is Geometric
2 Find a Formula for a Geometric Sequence
3 Find the Sum of a Geometric Sequence
4 Solve Annuity Problems
5 Find the Sum of a Geometric Series

1 When the ratio of successive terms of a sequence is always the same nonzero number, the sequence is called **geometric.** Thus, a **geometric sequence*** may be defined recursively as $a_1 = a$, $a_{n+1}/a_n = r$, or as

$$a_1 = a, \quad a_{n+1} = ra_n \tag{1}$$

where $a_1 = a$ and $r \neq 0$ are real numbers. The number a is the first term, and the nonzero number r is called the **common ratio.**

Thus, the terms of a geometric sequence with first term a and common ratio r follow the pattern

$$a, ar, ar^2, ar^3, \ldots$$

EXAMPLE 1 Determining if a Sequence Is Geometric

The sequence

$$2, 6, 18, 54, 162, \ldots$$

is geometric since the ratio of successive terms is 3. The first term is 2, and the common ratio is 3.

EXAMPLE 2 Determining if a Sequence Is Geometric

Show that the following sequence is geometric. Find the first term and the common ratio.

$$\{s_n\} = 2^{-n}$$

Solution The first term is $s_1 = 2^{-1} = \frac{1}{2}$. The $(n + 1)$st and nth terms of the sequence $\{s_n\}$ are

$$s_{n+1} = 2^{-(n+1)} \quad \text{and} \quad s_n = 2^{-n}$$

Their ratio is

$$\frac{s_{n+1}}{s_n} = \frac{2^{-(n+1)}}{2^{-n}} = 2^{-n-1+n} = 2^{-1} = \frac{1}{2}$$

Because the ratio of successive terms is a nonzero number independent of n, the sequence $\{s_n\}$ is geometric with common ratio $\frac{1}{2}$.

*Sometimes called a **geometric progression.**

EXAMPLE 3 Determining if a Sequence Is Geometric

Show that the following sequence is geometric. Find the first term and the common ratio.

$$\{t_n\} = \{4^n\}$$

Solution The first term is $t_1 = 4^1 = 4$. The $(n + 1)$st and nth terms are

$$t_{n+1} = 4^{n+1} \quad \text{and} \quad t_n = 4^n$$

Their ratio is

$$\frac{t_{n+1}}{t_n} = \frac{4^{n+1}}{4^n} = 4$$

Thus, $\{t_n\}$ is a geometric sequence with common ratio 4. ▬

Now work Problem 3.

2 Suppose a is the first term of a geometric sequence with common ratio $r \neq 0$. We seek a formula for the nth term a_n. To see the pattern, we write down the first few terms:

$$\begin{aligned}
a_1 &= 1 \cdot a = ar^0 \\
a_2 &= ra = ar^1 \\
a_3 &= ra_2 = r(ar) = ar^2 \\
a_4 &= ra_3 = r(ar^2) = ar^3 \\
a_5 &= ra_4 = r(ar^3) = ar^4 \\
&\vdots \\
a_n &= ra_{n-1} = r(ar^{n-2}) = ar^{n-1}
\end{aligned}$$

We are led to the following result:

Theorem ***n*th Term of a Geometric Sequence**

For a geometric sequence $\{a_n\}$ whose first term is a and whose common ratio is r, the nth term is determined by the formula

$$a_n = ar^{n-1}, \quad r \neq 0 \qquad (2)$$

▬

EXAMPLE 4 Finding a Particular Term of a Geometric Sequence

Find the 9th term of the geometric sequence $2, \frac{2}{3}, \frac{2}{9}, \frac{2}{27}, \ldots$.

Solution The first term of this geometric sequence is $a = 2$, and the common ratio is $\frac{1}{3}$. (Use $\frac{2}{3}/2 = \frac{1}{3}$, or $\frac{2}{9}/\frac{2}{3} = \frac{1}{3}$, or any two successive terms.) By formula (2), the nth term is

$$a_n = 2\left(\frac{1}{3}\right)^{n-1}$$

Hence, the 9th term is

$$a_9 = 2\left(\frac{1}{3}\right)^8 = \frac{2}{3^8} = \frac{2}{6561} \approx 0.0003$$

Exploration Use a graphing utility to find the 9th term of the sequence given in Example 4. Use it to find the 20th term and the 50th term.

Now work Problems 25 and 33.

Adding the First *n* Terms of a Geometric Sequence

3 The next result gives us a formula for finding the sum of the first n terms of a geometric sequence.

Theorem Sum of *n* Terms of a Geometric Sequence

Let $\{a_n\}$ be a geometric sequence with first term a and common ratio r. The sum S_n of the first n terms of $\{a_n\}$ is

$$S_n = a\frac{1 - r^n}{1 - r}, \quad r \neq 0, 1 \qquad (3)$$

Proof

$$S_n = a + ar + \cdots + ar^{n-1} \qquad (4)$$

Multiply each side by r to obtain

$$rS_n = ar + ar^2 + \cdots + ar^n \qquad (5)$$

Now, subtract (5) from (4). The result is

$$S_n - rS_n = a - ar^n$$
$$(1 - r)S_n = a(1 - r^n)$$

Since $r \neq 1$, we can solve for S_n:

$$S_n = a\frac{1 - r^n}{1 - r}$$

EXAMPLE 5 **Finding the Sum of *n* Terms of a Geometric Sequence**

Find the sum S_n of the first n terms of the sequence $\{(\frac{1}{2})^n\}$; that is, find

$$\frac{1}{2} + \frac{1}{4} + \frac{1}{8} + \cdots + \left(\frac{1}{2}\right)^n$$

Solution The sequence $\{(\frac{1}{2})^n\}$ is a geometric sequence with $a = \frac{1}{2}$ and $r = \frac{1}{2}$. The sum S_n that we seek is the sum of the first n terms of the sequence, so we use formula (3) to get

$$S_n = \sum_{k=1}^{n} \left(\frac{1}{2}\right)^k = \frac{1}{2} + \frac{1}{4} + \frac{1}{8} + \cdots + \left(\frac{1}{2}\right)^n$$

$$= \frac{1}{2}\left[\frac{1 - (\frac{1}{2})^n}{1 - \frac{1}{2}}\right]$$

$$= \frac{1}{2}\left[\frac{1 - (\frac{1}{2})^n}{\frac{1}{2}}\right]$$

$$= 1 - \left(\frac{1}{2}\right)^n$$

Now work Problem 39.

EXAMPLE 6 Using a Graphing Utility to Find the Sum of a Geometric Sequence

Use a graphing utility to find the sum S_n of the first 15 terms of the sequence $\left\{\left(\frac{1}{3}\right)^n\right\}$; that is, find

$$\frac{1}{3} + \frac{1}{9} + \frac{1}{27} + \cdots + \left(\frac{1}{3}\right)^{15}$$

Solution Figure 12 shows the result obtained using a TI-83 graphing calculator.

FIGURE 12

Thus, the sum of the first 15 terms of the sequence $\left\{\left(\frac{1}{3}\right)^n\right\}$ is 0.4999999652.

Now work Problem 45.

Annuities

4 In Section 5.6 we developed the compound interest formula which gives the future value when a fixed amount of money is deposited in an account that pays interest compounded periodically. Often, though, money is invested in small amounts at periodic intervals. An **annuity** is a sequence of equal periodic deposits. The periodic deposits may be made annually, quarterly, monthly, or daily.

When deposits are made at the same time the interest is credited, the annuity is called **ordinary.** We will only deal with ordinary annuities here.

The **amount of an annuity** is the sum of all deposits made plus all interest paid.

Suppose the interest an account earns is i percent per payment period (expressed as a decimal). For example, if an account pays 12% compounded monthly (12 times a year) then $i = 0.12/12 = 0.01$. If an account pays 8% compounded quarterly (4 times a year) then $i = 0.08/4 = 0.02$. To develop a formula for the amount of an annuity, suppose \$$P$ is deposited each payment period for n payment periods in an account that earns i percent per payment period. When the last deposit is made at the nth payment period, the first deposit of \$$P$ has earned interest compounded for $n - 1$ payment periods, the second deposit of \$$P$ has earned interest compounded for $n - 2$ payment periods, and so on. Table 3 shows the value of each deposit after n deposits have been made.

TABLE 3

Deposit	1	2	3	$n-1$	n
Amount	$P(1+i)^{n-1}$	$P(1+i)^{n-2}$	$P(1+i)^{n-3}$	$P(1+i)$	P

The amount A of the annuity is the sum of the amounts shown in Table 3, namely,

$$\begin{aligned} A &= P(1+i)^{n-1} + P(1+i)^{n-2} + \cdots + P(1+i) + P \\ &= P[1 + (1+i) + \cdots + (1+i)^{n-1}] \end{aligned}$$

The expression in brackets is the sum of a geometric sequence with n terms and a common ratio of $(1 + i)$. As a result,

$$\begin{aligned} A &= P[1 + (1+i) + \cdots + (1+i)^{n-2} + (1+i)^{n-1}] \\ &= P\frac{1-(1+i)^n}{1-(1+i)} = P\frac{1-(1+i)^n}{-i} = P\frac{(1+i)^n - 1}{i} \end{aligned}$$

We have established the following result.

Theorem Amount of an Annuity

If P represents the deposit in dollars made at each payment period for an annuity at i percent interest per payment period, the amount A of the annuity after n payment periods is

$$A = P\frac{(1+i)^n - 1}{i} \qquad (6)$$

EXAMPLE 7

Determining the Amount of an Annuity

To save for retirement, Brett decides to place \$2,000 into an Individual Retirement Account (IRA) each year for the next 30 years. What will the value of the IRA be when Bill retires in 30 years if the rate of return of the IRA is assumed to be 10% per annum compounded annually?

Solution This is an ordinary annuity with 30 annual deposits of $P = \$2000$. The rate of interest per payment period is $i = 0.10/1 = 0.10$. The number of payment periods is $n = 30$. The amount A of the annuity in 30 years is

$$A = 2000\left[\frac{(1 + 0.10)^{30} - 1}{0.10}\right] = \$2000(164.49402) = \$328{,}988.05$$

EXAMPLE 8

To save for his daughter's college education, Mr. McGowen decides to put \$50 aside every month in a credit union account paying 10% interest compounded monthly. If he begins this savings program when his daughter is 3 years old, how much will he have saved by the time his daughter is 18 years old?

Solution When his daughter is 18 years old, Mr. McGowen will have made his 180th payment (15 years × 12 payments per year). This is an annuity with $P = \$50$, $n = 180$, and $i = \frac{0.10}{12}$. The amount A saved is

$$A = 50\left[\frac{\left(1 + \frac{0.10}{12}\right)^{180} - 1}{\frac{0.10}{12}}\right] = \$50(414.4703) = \$20{,}723.52$$

Geometric Series

An infinite sum of the form

$$a + ar + ar^2 + \cdots + ar^{n-1} + \cdots$$

with first term a and common ratio r, is called an **infinite geometric series** and is denoted by

$$\sum_{k=1}^{\infty} ar^{k-1}$$

5

Based on formula (3), the sum S_n of the first n terms of a geometric series is

$$S_n = a\frac{1 - r^n}{1 - r} = \frac{a}{1 - r} - \frac{ar^n}{1 - r} \tag{7}$$

If this finite sum S_n approaches a number L as $n \to \infty$, then we call L the **sum of the infinite geometric series,** and we write

$$L = \sum_{k=1}^{\infty} ar^{k-1}$$

Theorem Sum of an Infinite Geometric Series

If $|r| < 1$, the sum of the infinite geometric series $\sum_{k=1}^{\infty} ar^{k-1}$ is

$$\sum_{k=1}^{\infty} ar^{k-1} = \frac{a}{1 - r} \tag{8}$$

Intuitive Proof Since $|r| < 1$, it follows that $|r^n|$ approaches 0 as $n \to \infty$. Then, based on formula (7), the sum S_n approaches $a/(1 - r)$ as $n \to \infty$. ■

EXAMPLE 9 Finding the Sum of a Geometric Series

Find the sum of the geometric series $2 + \frac{4}{3} + \frac{8}{9} + \cdots$.

Solution The first term is $a = 2$ and the common ratio is

$$r = \frac{\frac{4}{3}}{2} = \frac{4}{6} = \frac{2}{3}$$

Since $|r| < 1$, we use formula (8) to find that

$$2 + \frac{4}{3} + \frac{8}{9} + \cdots = \frac{2}{1 - \frac{2}{3}} = 6$$

■

Now work Problem 51.

Exploration Use a graphing utility to graph $U_n = 2\left(\frac{2}{3}\right)^{n-1} + U_{n-1}$ in sequence mode. TRACE the graph for large values of n. What happens to the value of U_n as n increases without bound? What can you conclude about $\sum_{n=1}^{\infty} 2\left(\frac{2}{3}\right)^{n-1}$? ■

EXAMPLE 10 Repeating Decimals

Show that the repeating decimal 0.999 . . . equals 1.

Solution

$$0.999\ldots = \frac{9}{10} + \frac{9}{100} + \frac{9}{1000} + \cdots$$

Thus, 0.999 . . . is a geometric series with first term $\frac{9}{10}$ and common ratio $\frac{1}{10}$. Hence,

$$0.999\cdots = \frac{\frac{9}{10}}{1 - \frac{1}{10}} = \frac{\frac{9}{10}}{\frac{9}{10}} = 1$$

■

EXAMPLE 11 Pendulum Swings

FIGURE 13

Initially, a pendulum swings through an arc of 18 inches. See Figure 13. On each successive swing, the length of the arc is 0.98 of the previous length.

(a) What is the length of arc after 10 swings?
(b) On which swing is the length of arc first less than 12 inches?
(c) After 15 swings, what total length will the pendulum have swung?
(d) When it stops, what total length will the pendulum have swung?

Solution (a) The length of the first swing is 18 inches. The length of the second swing is 0.98(18) inches; the length of the third swing is $0.98(0.98)(18) = 0.98^2(18)$ inches. The length of arc of the 10th swing is

$$(0.98)^9(18) = 15.007 \text{ inches}$$

(b) The length of arc of the nth swing is $(0.98)^{n-1}(18)$. For this to be exactly 12 inches requires

$$
\begin{aligned}
(0.98)^{n-1}(18) &= 12 \\
(0.98)^{n-1} &= \frac{12}{18} = \frac{2}{3} \\
n - 1 &= \log_{0.98}\left(\frac{2}{3}\right) \\
n &= 1 + \frac{\ln\left(\frac{2}{3}\right)}{\ln 0.98} = 1 + 20.07 = 21.07
\end{aligned}
$$

The length of arc of the pendulum exceeds 12 inches on the 21st swing and is first less than 12 inches on the 22nd swing.

(c) After 15 swings, the pendulum will have swung the following total length L:

$$
L = \underset{\text{1st}}{18} + \underset{\text{2nd}}{0.98(18)} + \underset{\text{3rd}}{(0.98)^2(18)} + \underset{\text{4th}}{(0.98)^3(18)} + \cdots + \underset{\text{15th}}{(0.98)^{14}(18)}
$$

This is the sum of a geometric sequence. The common ratio is 0.98; the first term is 18. The sum has 15 terms, so

$$
L = 18\,\frac{1 - 0.98^{15}}{1 - 0.98} = 18(13.07) = 235.29 \text{ inches}
$$

The pendulum will have swung through 235.29 inches after 15 swings.

(d) When the pendulum stops, it will have swung the following total length T:

$$
T = 18 + 0.98(18) + (0.98)^2(18) + (0.98)^3(18) + \cdots
$$

This is the sum of a geometric series. The common ratio is $r = 0.98$; the first term is $a = 18$. The sum is

$$
T = \frac{a}{1 - r} = \frac{18}{1 - 0.98} = 900
$$

The pendulum will have swung a total of 900 inches when it finally stops.

Historical Feature

Sequences are among the oldest objects of mathematical investigation, having been studied for over 3500 years. After the initial steps, however, little progress was made until about 1600.

Arithmetic and geometric sequences appear in the Rhind papyrus, a mathematical text containing 85 problems copied around 1650 B.C. by the Egyptian scribe Ahmes from an earlier work (see Historical Problems 1). Fibonacci (A.D. 1220) wrote about problems similar to those found in the Rhind papyrus, leading one to suspect that Fibonacci may have had material available that is now lost. This material would have been in the non-Euclidean Greek tradition of Heron (about A.D. 75) and Diophantus (about A.D. 250). One problem, again modified slightly, is still with us in the familiar puzzle rhyme "As I was going to St. Ives . . ." (see Historical Problem 2).

The Rhind papyrus indicates that the Egyptians knew how to add up the terms of an arithmetic or geometric sequence, as did the Babylonians. The rule for summing up a geometric sequence is found in Euclid's *Elements* (book IX, 35, 36), where, like all of Euclid's algebra, it is presented in a geometric form.

Investigations of other kinds of sequences began in the 1500s, when algebra became sufficiently developed to handle the more complicated problems. The development of calculus in the 1600s added a powerful new tool, especially for finding the sum of infinite series, and the subject continues to flourish today.

Historical Problems

1. *Arithmetic sequence problem from the Rhind papyrus (statement modified slightly for clarity)* One hundred loaves of bread are to be divided among five people so that the amounts they receive form an arithmetic sequence. The first two together receive one-seventh of what the last three receive. How many does each receive? [*Partial answer:* First person receives $1\frac{2}{3}$ loaves.]
2. The following old English children's rhyme resembles one of the Rhind papyrus problems:

   As I was going to St. Ives
   I met a man with seven wives
   Each wife had seven sacks
   Each sack had seven cats
   Each cat had seven kits [kittens]
   Kits, cats, sacks, wives
   How many were going to St. Ives?

   (a) Assuming that the speaker and the cat fanciers met by traveling in opposite directions, what is the answer?
   (b) How many kittens are being transported?
   (c) Kits, cats, sacks, wives; how many?

   [**Hint:** It is easier to include the man, find the sum with the formula, and then subtract 1 for the man.]

12.3 EXERCISES

In Problems 1–10, a geometric sequence is given. Find the common ratio and write out the first four terms.

1. $\{3^n\}$ **2.** $\{(-5)^n\}$ **3.** $\left\{-3\left(\frac{1}{2}\right)^n\right\}$ **4.** $\left\{\left(\frac{5}{2}\right)^n\right\}$ **5.** $\left\{\frac{2^{n-1}}{4}\right\}$

6. $\left\{\frac{3^n}{9}\right\}$ **7.** $\{2^{n/3}\}$ **8.** $\{3^{2n}\}$ **9.** $\left\{\frac{3^{n-1}}{2^n}\right\}$ **10.** $\left\{\frac{2^n}{3^{n-1}}\right\}$

In Problems 11–24, determine whether the given sequence is arithmetic, geometric, or neither. If the sequence is arithmetic, find the common difference; if it is geometric, find the common ratio.

11. $\{n+2\}$ **12.** $\{2n-5\}$ **13.** $\{4n^2\}$ **14.** $\{5n^2+1\}$ **15.** $\{3-\frac{2}{3}n\}$

16. $\{8-\frac{3}{4}n\}$ **17.** $1, 3, 6, 10, \ldots$ **18.** $2, 4, 6, 8, \ldots$ **19.** $\{(\frac{2}{3})^n\}$ **20.** $\{(\frac{5}{4})^n\}$

21. $-1, -2, -4, -8, \ldots$ **22.** $1, 1, 2, 3, 5, 8, \ldots$ **23.** $\{3^{n/2}\}$ **24.** $\{(-1)^n\}$

In Problems 25–32, find the fifth term and the nth term of the geometric sequence whose initial term a and common ratio r are given.

25. $a=2; r=3$ **26.** $a=-2; r=4$ **27.** $a=5; r=-1$ **28.** $a=6; r=-2$

29. $a=0; r=\frac{1}{2}$ **30.** $a=1; r=-\frac{1}{3}$ **31.** $a=\sqrt{2}; r=\sqrt{2}$ **32.** $a=0; r=1/\pi$

In Problems 33–38, find the indicated term of each geometric sequence.

33. 7th term of $1, \frac{1}{2}, \frac{1}{4}, \ldots$

34. 8th term of $1, 3, 9, \ldots$

35. 9th term of $1, -1, 1, \ldots$

36. 10th term of $-1, 2, -4, \ldots$

37. 8th term of $0.4, 0.04, 0.004, \ldots$

38. 7th term of $0.1, 1.0, 10.0, \ldots$

In Problems 39–44, find the sum.

39. $\frac{1}{4} + \frac{2}{4} + \frac{2^2}{4} + \frac{2^3}{4} + \cdots + \frac{2^{n-1}}{4}$

40. $\frac{3}{9} + \frac{3^2}{9} + \frac{3^3}{9} + \cdots + \frac{3^n}{9}$

41. $\sum_{k=1}^{n} \left(\frac{2}{3}\right)^k$

42. $\sum_{k=1}^{n} 4 \cdot 3^{k-1}$

43. $-1 - 2 - 4 - 8 - \cdots - (2^{n-1})$

44. $2 + \frac{6}{5} + \frac{18}{25} + \cdots + 2\left(\frac{3}{5}\right)^n$

For Problems 45–50, use a graphing utility to find the sum of each geometric sequence.

45. $\frac{1}{4} + \frac{2}{4} + \frac{2^2}{4} + \frac{2^3}{4} + \cdots + \frac{2^{14}}{4}$

46. $\frac{3}{9} + \frac{3^2}{9} + \frac{3^3}{9} + \cdots + \frac{3^{15}}{9}$

47. $\sum_{n=1}^{15} \left(\frac{2}{3}\right)^n$

48. $\sum_{n=1}^{15} 4 \cdot 3^{n-1}$

49. $-1 - 2 - 4 - 8 - \cdots - 2^{14}$

50. $2 + \frac{6}{5} + \frac{18}{25} + \cdots + 2\left(\frac{3}{5}\right)^{15}$

In Problems 51–60, find the sum of each infinite geometric series.

51. $1 + \frac{1}{3} + \frac{1}{9} + \cdots$

52. $2 + \frac{4}{3} + \frac{8}{9} + \cdots$

53. $8 + 4 + 2 + \cdots$

54. $6 + 2 + \frac{2}{3} + \cdots$

55. $2 - \frac{1}{2} + \frac{1}{8} - \frac{1}{32} + \cdots$

56. $1 - \frac{3}{4} + \frac{9}{16} - \frac{27}{64} + \cdots$

57. $\sum_{k=1}^{\infty} 5\left(\frac{1}{4}\right)^{k-1}$

58. $\sum_{k=1}^{\infty} 8\left(\frac{1}{3}\right)^{k-1}$

59. $\sum_{k=1}^{\infty} 6\left(-\frac{2}{3}\right)^{k-1}$

60. $\sum_{k=1}^{\infty} 4\left(-\frac{1}{2}\right)^{k-1}$

61. Find x so that x, $x + 2$, and $x + 3$ are terms of a geometric sequence.

62. Find x so that $x - 1$, x, and $x + 2$ are terms of a geometric sequence.

63. Retirement Christine contributes \$100 each month to her 401(k). What will be the value of Christine's 401(k) in 30 years if the per annum rate of return is assumed to be 12% compounded monthly?

64. Saving for a Home Jolene wants to purchase a new home. Suppose she invests \$400 per month into a mutual fund. If the per annum rate of return of the mutual fund is assumed to be 10% compounded monthly, how much will Jolene have for a down payment after 3 years?

65. Tax Sheltered Annuity Don contributes \$500 at the end of each quarter to a Tax Sheltered Annuity (TSA). What will the value of the TSA be in 20 years if the per annum rate of return is assumed to be 8% compounded quarterly?

66. Retirement Ray, planning on retiring in 15 years, contributes \$1000 to an Individual Retirement Account (IRA) semiannually. What will the value of the IRA be when Ray retires if the per annum rate of return is assumed to be 10% compounded semiannually?

67. Sinking Fund Scott and Alice want to purchase a vacation home in 10 years and have \$50,000 for a down payment. How much should they place in a savings account each month if the per annum rate of return is assumed to be 6% compounded monthly?

68. Sinking Fund For a child born in 1996, a 4-year college education at a public university is projected to be \$150,000. Assuming an 8% per annum rate of return compounded monthly how much must be contributed to a college fund every month in order to have \$150,000 in 18 years when the child begins college.

69. Pendulum Swings Initially, a pendulum swings through an arc of 2 feet. On each successive swing, the length of arc is 0.9 of the previous length.

(a) What is the length of arc after 10 swings?

(b) On which swing is the length of arc first less than 1 foot?

(c) After 15 swings, what total length will the pendulum have swung?

(d) When it stops, what total length will the pendulum have swung?

70. **Bouncing Balls** A ball is dropped from a height of 30 feet. Each time it strikes the ground, it bounces up to 0.8 of the previous height.

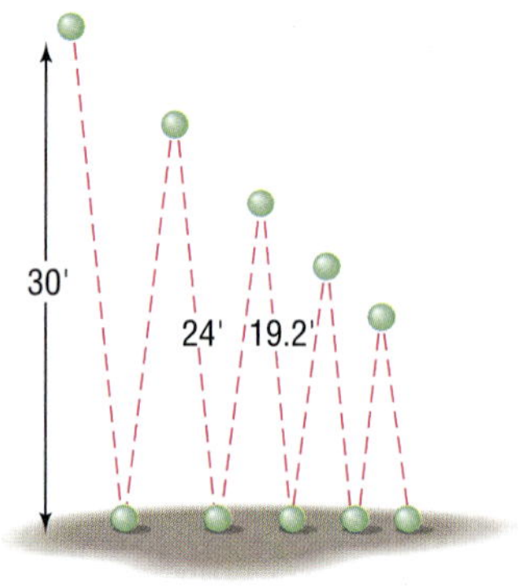

(a) What height will the ball bounce up to after it strikes the ground for the third time?
(b) What is its height after it strikes the ground for the nth time?
(c) How many times does the ball need to strike the ground before its height is less than 6 inches?
(d) What total distance does the ball travel before it stops bouncing?

71. **Salary Increases** Suppose you have just been hired at an annual salary of \$18,000 and expect to receive annual increases of 5%. What will your salary be when you begin your fifth year?

72. **Equipment Depreciation** A new piece of equipment cost a company \$15,000. Each year, for tax purposes, the company depreciates the value by 15%. What value should the company give the equipment after 5 years?

73. **Critical Thinking** You have just signed a 7-year professional football league contract with a beginning salary of \$2,000,000 per year. Management gives you the following options with regard to your salary over the 7 years.
(1) A bonus of \$100,000 each year
(2) An annual increase of 4.5% per year beginning after 1 year
(3) An annual increase of \$95,000 per year beginning after 1 year
Which option provides the most money over the 7 year period? Which the least? Which would you choose? Why?

74. **A Rich Man's Promise** A rich man promises to give you \$1000 on September 1, 1998. Each day thereafter he will give you $\frac{9}{10}$ of what he gave you the previous day. What is the first date on which the amount you receive is less than 1¢? How much have you received when this happens?

75. **Grains of Wheat on a Chess Board** In an old fable, a commoner who had just saved the king's life was told he could ask the king for any just reward. Being a shrewd man, the commoner said, "A simple wish, sire. Place one grain of wheat on the first square of a chessboard, two grains on the second square, four grains on the third square, continuing until you have filled the board. This is all I seek." Compute the total number of grains needed to do this to see why the request, seemingly simple, could not be granted. (A chessboard consists of $8 \times 8 = 64$ squares.)

76. Can a sequence be both arithmetic and geometric? Give reasons for your answer.

77. Make up a geometric sequence. Give it to a friend and ask for its 20th term.

78. Make up two infinite geometric series, one that has a sum and one that does not. Give them to a friend and ask for the sum of each series.

79. If $x < 1$, then $1 + x + x^2 + x^3 + \cdots + x^n + \cdots = 1/(1 - x)$. Make up a table of values using $x = 0.1$, $x = 0.25$, $x = 0.5$, $x = 0.75$, and $x = 0.9$ to compute $1/(1 - x)$. Now determine how many terms are needed in the expansion $1 + x + x^2 + x^3 + \cdots + x^n + \cdots$ before it approximates $1/(1 - x)$ correct to two decimal places. For example, if $x = 0.1$, then $1/(1 - x) = 10/9 = 1.111\ldots$. The expansion requires three terms.

80. Which of the following choices, A or B, results in more money?

A: To receive \$1000 on day 1, \$999 on day 2, \$998 on day 3, with the process to end after 1000 days

B: To receive \$1 on day 1, \$2 on day 2, \$4 on day 3, for 19 days

81. You are interviewing for a job and receive two offers:

A: \$20,000 to start with guaranteed annual increases of 6% for the first 5 years

B: \$22,000 to start with guaranteed increases of 3% for the first 5 years

Which offer is best if your goal is to be making as much as possible after 5 years? Which is best if your goal is to make as much money as possible over the contract (5 years)?

82. Look at the figure below. What fraction of the square is eventually shaded if the indicated shading process continues indefinitely?

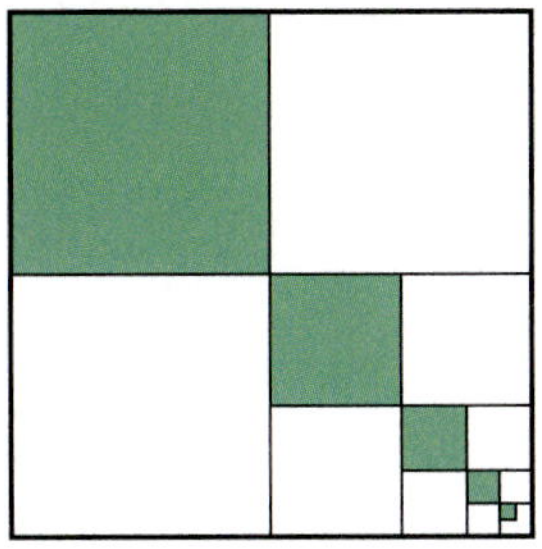

12.4 MATHEMATICAL INDUCTION

1 Prove Statements Using Mathematical Induction

1 *Mathematical induction* is a method for proving that statements involving natural numbers are true for all natural numbers.* For example, the statement, "$2n$ is always an even integer" can be proved true for all natural numbers by using mathematical induction. Also, the statement "the sum of the first n positive odd integers equals n^2," that is,

$$1 + 3 + 5 + \cdots + (2n - 1) = n^2 \tag{1}$$

can be proved for all natural numbers n by using mathematical induction.

Before stating the method of mathematical induction, let's try to gain a sense of the power of the method. We shall use the statement in equation (1) for this purpose by restating it for various values of $n = 1, 2, 3, \ldots$:

$n = 1$ The sum of the first positive odd integer is 1^2; $1 = 1^2$.

$n = 2$ The sum of the first 2 positive odd integers is 2^2; $1 + 3 = 4 = 2^2$.

$n = 3$ The sum of the first 3 positive odd integers is 3^2; $1 + 3 + 5 = 9 = 3^2$.

$n = 4$ The sum of the first 4 positive odd integers is 4^2; $1 + 3 + 5 + 7 = 16 = 4^2$.

Although from this pattern we might conjecture that statement (1) is true for any choice of n, can we really be sure that it does not fail for some choice of n? The method of proof by mathematical induction will, in fact, prove that the statement is true for all n.

*Recall from the Appendix, Section 1, that the natural numbers are the numbers $1, 2, 3, 4, \ldots$. In other words, the terms *natural numbers* and *positive integers* are synonymous.

Theorem The Principle of Mathematical Induction

Suppose the following two conditions are satisfied with regard to a statement about natural numbers:

CONDITION I: The statement is true for the natural number 1.

CONDITION II: If the statement is true for some natural number k, it is also true for the next natural number $k + 1$.

Then the statement is true for all natural numbers.

We shall not prove this principle. However, we can provide a physical interpretation that will help us see why the principle works. Think of a collection of natural numbers obeying a statement as a collection of infinitely many dominoes (see Figure 14).

FIGURE 14

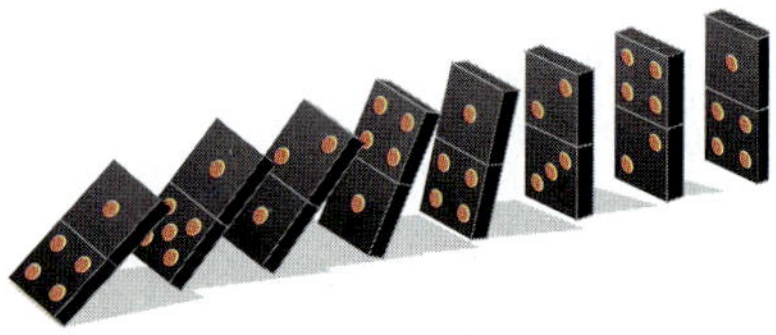

Now, suppose we are told two facts:

1. The first domino is pushed over.
2. If one of the dominoes falls over, say the kth domino, then so will the next one, the $(k + 1)$st domino.

Is it safe to conclude that *all* the dominoes fall over? The answer is yes, because, if the first one falls (Condition I), then the second one does also (by Condition II); and if the second one falls, then so does the third (by Condition II); and so on.

Now let's prove some statements about natural numbers using mathematical induction.

EXAMPLE 1 Using Mathematical Induction

Show that the following statement is true for all natural numbers n:

$$1 + 3 + 5 + \cdots + (2n - 1) = n^2 \qquad (2)$$

Solution We need to show first that statement (2) holds for $n = 1$. Because $1 = 1^2$, statement (2) is true for $n = 1$. Thus, Condition I holds.

Next, we need to show that Condition II holds. Suppose we know for some k that

$$1 + 3 + \cdots + (2k - 1) = k^2 \qquad (3)$$

We wish to show that, based on equation (3), statement (2) holds for $k + 1$. Thus, we look at the sum of the first $k + 1$ positive odd integers to determine whether this sum equals $(k + 1)^2$:

$$\begin{aligned} 1 + 3 + \cdots + (2k - 1) + (2k + 1) &= \underbrace{[1 + 3 + \cdots + (2k - 1)]}_{= k^2 \text{ by equation (3)}} + (2k + 1) \\ &= k^2 + (2k + 1) \\ &= k^2 + 2k + 1 = (k + 1)^2 \end{aligned}$$

Conditions I and II are satisfied; thus, by the Principle of Mathematical Induction, statement (2) is true for all natural numbers.

EXAMPLE 2 Using Mathematical Induction

Show that the following statement is true for all natural numbers n:

$$2^n > n$$

Solution First, we show that the statement $2^n > n$ holds when $n = 1$. Because $2^1 = 2 > 1$, the inequality is true for $n = 1$. Thus, Condition I holds.

Next, we assume, for some natural number k, that $2^k > k$. We wish to show that the formula holds for $k + 1$; that is, we wish to show that $2^{k+1} > k + 1$. Now,

$$2^{k+1} = 2 \cdot 2^k > 2 \cdot k = k + k \geq k + 1$$

(We know that $2^k > k$.) ($k \geq 1$.)

Thus, if $2^k > k$, then $2^{k+1} > k + 1$, so Condition II of the Principle of Mathematical Induction is satisfied. Hence, the statement $2^n > n$ is true for all natural numbers n. ■

EXAMPLE 3 Using Mathematical Induction

Show that the following formula is true for all natural numbers n:

$$1 + 2 + 3 + \cdots + n = \frac{n(n+1)}{2} \tag{4}$$

Solution First, we show that formula (4) is true when $n = 1$. Because

$$\frac{1(1+1)}{2} = \frac{1(2)}{2} = 1$$

Condition I of the Principle of Mathematical Induction holds.

Next, we assume that formula (4) holds for some k, and we determine whether the formula then holds for $k + 1$. Thus, we assume that

$$1 + 2 + 3 + \cdots + k = \frac{k(k+1)}{2} \quad \text{for some } k \tag{5}$$

Now, we need to show that

$$1 + 2 + 3 + \cdots + k + (k+1) = \frac{(k+1)(k+1+1)}{2} = \frac{(k+1)(k+2)}{2}$$

We do this as follows:

$$\begin{aligned} 1 + 2 + 3 + \cdots + k + (k+1) &= \underbrace{[1 + 2 + 3 + \cdots + k]}_{= \frac{k(k+1)}{2} \text{ by equation (5)}} + (k+1) \\ &= \frac{k(k+1)}{2} + (k+1) \\ &= \frac{k^2 + k + 2k + 2}{2} \\ &= \frac{k^2 + 3k + 2}{2} = \frac{(k+1)(k+2)}{2} \end{aligned}$$

Thus, Condition II also holds. As a result, formula (4) is true for all natural numbers. ■

Now work Problem 1.

EXAMPLE 4 Using Mathematical Induction

Show that $3^n - 1$ is divisible by 2 for all natural numbers n.

Solution First, we show that the statement is true when $n = 1$. Because $3^1 - 1 = 3 - 1 = 2$ is divisible by 2, the statement is true when $n = 1$. Thus, Condition I is satisfied.

Next, we assume that the statement holds for some k, and we determine whether the statement then holds for $k + 1$. Thus, we assume that $3^k - 1$ is divisible by 2 for some k. We need to show that $3^{k+1} - 1$ is divisible by 2. Now,

$$3^{k+1} - 1 = 3^{k+1} - 3^k + 3^k - 1$$
$$= 3^k(3 - 1) + (3^k - 1) = 3^k \cdot 2 + (3^k - 1)$$

Because $3^k \cdot 2$ is divisible by 2 and $3^k - 1$ is divisible by 2, it follows that $3^k \cdot 2 + (3^k - 1) = 3^{k+1} - 1$ is divisible by 2. Thus, Condition II is also satisfied. As a result, the statement, "$3^n - 1$ is divisible by 2" is true for all natural numbers n.

Warning: The conclusion that a statement involving natural numbers is true for all natural numbers is made only after *both* Conditions I and II of the Principle of Mathematical Induction have been satisfied. Problem 27 (on page 931) demonstrates a statement for which only Condition I holds, but the statement is not true for all natural numbers. Problem 28 demonstrates a statement for which only Condition II holds, but the statement is *not* true for any natural number.

12.4 EXERCISES

In Problems 1–26, use the principle of mathematical induction to show that the given statement is true for all natural numbers.

1. $2 + 4 + 6 + \cdots + 2n = n(n + 1)$

2. $1 + 5 + 9 + \cdots + (4n - 3) = n(2n - 1)$

3. $3 + 4 + 5 + \cdots + (n + 2) = \frac{1}{2}n(n + 5)$

4. $3 + 5 + 7 + \cdots + (2n + 1) = n(n + 2)$

5. $2 + 5 + 8 + \cdots + (3n - 1) = \frac{1}{2}n(3n + 1)$

6. $1 + 4 + 7 + \cdots + (3n - 2) = \frac{1}{2}n(3n - 1)$

7. $1 + 2 + 2^2 + \cdots + 2^{n-1} = 2^n - 1$

8. $1 + 3 + 3^2 + \cdots + 3^{n-1} = \frac{1}{2}(3^n - 1)$

9. $1 + 4 + 4^2 + \cdots + 4^{n-1} = \frac{1}{3}(4^n - 1)$

10. $1 + 5 + 5^2 + \cdots + 5^{n-1} = \frac{1}{4}(5^n - 1)$

11. $\dfrac{1}{1 \cdot 2} + \dfrac{1}{2 \cdot 3} + \dfrac{1}{3 \cdot 4} + \cdots + \dfrac{1}{n(n + 1)} = \dfrac{n}{n + 1}$

12. $\dfrac{1}{1 \cdot 3} + \dfrac{1}{3 \cdot 5} + \dfrac{1}{5 \cdot 7} + \cdots + \dfrac{1}{(2n - 1)(2n + 1)} = \dfrac{n}{2n + 1}$

13. $1^2 + 2^2 + 3^2 + \cdots + n^2 = \frac{1}{6}n(n + 1)(2n + 1)$

14. $1^3 + 2^3 + 3^3 + \cdots + n^3 = \frac{1}{4}n^2(n + 1)^2$

15. $4 + 3 + 2 + \cdots + (5 - n) = \frac{1}{2}n(9 - n)$

16. $-2 - 3 - 4 - \cdots - (n + 1) = -\frac{1}{2}n(n + 3)$

17. $1 \cdot 2 + 2 \cdot 3 + 3 \cdot 4 + \cdots + n(n + 1) = \frac{1}{3}n(n + 1)(n + 2)$

18. $1 \cdot 2 + 3 \cdot 4 + 5 \cdot 6 + \cdots + (2n - 1)(2n) = \frac{1}{3}n(n + 1)(4n - 1)$

19. $n^2 + n$ is divisible by 2.

20. $n^3 + 2n$ is divisible by 3.

21. $n^2 - n + 2$ is divisible by 2.

22. $n(n + 1)(n + 2)$ is divisible by 6.

23. If $x > 1$, then $x^n > 1$.

24. If $0 < x < 1$, then $0 < x^n < 1$.

25. $a - b$ is a factor of $a^n - b^n$.
[**Hint:** $a^{k+1} - b^{k+1} = a(a^k - b^k) + b^k(a - b)$]

26. $a + b$ is a factor of $a^{2n+1} + b^{2n+1}$.

27. Show that the statement "$n^2 - n + 41$ is a prime number," is true for $n = 1$, but is not true for $n = 41$.

28. Show that the formula

$$2 + 4 + 6 + \cdots + 2n = n^2 + n + 2$$

obeys Condition II of the Principle of Mathematical Induction. That is, show that if the formula is true for some k it is also true for $k + 1$. Then show that the formula is false for $n = 1$ (or for any other choice of n).

29. Use mathematical induction to prove that if $r \neq 1$ then

$$a + ar + ar^2 + \cdots + ar^{n-1} = a\frac{1 - r^n}{1 - r}$$

30. Use mathematical induction to prove that

$$a + (a + d) + (a + 2d) + \cdots + [a + (n - 1)d] = na + d\frac{n(n - 1)}{2}$$

31. Geometry Use mathematical induction to show that the sum of the interior angles of a convex polygon of n sides equals $(n - 2) \cdot 180°$.

32. The Extended Principle of Mathematical Induction The Extended Principle of Mathematical Induction states that if conditions I and II hold, that is,

(I) A statement is true for a natural number j.

(II) If the statement is true for some natural number $k > j$, then it is also true for the next natural number $k + 1$.

Then the statement is true for *all* natural numbers $\geq j$.

Use the Extended Principle of Mathematical Induction to show that the number of diagonals in a convex polygon of n sides is $\frac{1}{2}n(n - 3)$.

[**Hint:** Begin by showing that the result is true when $n = 4$ (Condition I).]

33. How would you explain to a friend the principle of mathematical induction?

12.5 THE BINOMIAL THEOREM

1 Evaluate a Binomial Coefficient

2 Expand a Binomial

In the Appendix, Section 2, we listed some special products. Among these were formulas for expanding $(x + a)^n$ for $n = 2$ and $n = 3$. The *binomial theorem** is a formula for the expansion of $(x + a)^n$ for n any positive integer. If $n = 1, 2, 3$, and 4, the expansion of $(x + a)^n$ is straightforward:

$$(x + a)^1 = x + a$$ 2 terms, beginning with x^1 and ending with a^1.

$$(x + a)^2 = x^2 + 2ax + a^2$$ 3 terms, beginning with x^2 and ending with a^2.

$$(x + a)^3 = x^3 + 3ax^2 + 3a^2x + a^3$$ 4 terms, beginning with x^3 and ending with a^3.

$$(x + a)^4 = x^4 + 4ax^3 + 6a^2x^2 + 4a^3x + a^4$$ 5 terms, beginning with x^4 and ending with a^4.

Notice that each expansion of $(x + a)^n$ begins with x^n and ends with a^n. As you read from left to right, the powers of x are decreasing, while the powers of a are increasing. Also, the number of terms that appear equals $n + 1$. Notice, too, that the degree of each monomial in the expansion equals n. For example, in the expansion of $(x + a)^3$, each monomial $(x^3, 3ax^2, 3a^2x, a^3)$ is of degree 3. As a result, we might conjecture that the expansion of $(x + a)^n$ would look like this:

$$(x + a)^n = x^n + \_ax^{n-1} + \_a^2x^{n-2} + \cdots + \_a^{n-1}x + a^n$$

*The name *binomial* derives from the fact that $x + a$ is a binomial, that is, contains two terms.

where the blanks are numbers to be found. This is, in fact, the case, as we shall see shortly.

First, we need to introduce a symbol.

The Symbol $\binom{n}{j}$

We define the symbol $\binom{n}{j}$, read "n taken j at a time," as follows:

If j and n are integers with $0 \le j \le n$, the **symbol** $\binom{n}{j}$ is defined as

$$\binom{n}{j} = \frac{n!}{j!(n-j)!} \tag{1}$$

Comment: On a graphing calculator, the symbol $\binom{n}{j}$ may be denoted by the key nCr.

EXAMPLE 1 Evaluating $\binom{n}{j}$

Find:

(a) $\binom{3}{1}$ (b) $\binom{4}{2}$ (c) $\binom{8}{7}$ (d) $\binom{65}{15}$

Solution (a) $\binom{3}{1} = \frac{3!}{1!(3-1)!} = \frac{3!}{1!2!} = \frac{3 \cdot 2 \cdot 1}{1(2 \cdot 1)} = \frac{6}{2} = 3$

(b) $\binom{4}{2} = \frac{4!}{2!(4-2)!} = \frac{4!}{2!2!} = \frac{4 \cdot 3 \cdot 2 \cdot 1}{(2 \cdot 1)(2 \cdot 1)} = \frac{24}{4} = 6$

(c) $\binom{8}{7} = \frac{8!}{7!(8-7)!} = \frac{8!}{7!1!} = \frac{8 \cdot \cancel{7!}}{\cancel{7!} \cdot 1!} = \frac{8}{1} = 8$

$8! = 8 \cdot 7!$

(d) Figure 15 shows the solution using a TI-83 graphing calculator.

Thus, $\binom{65}{15} = 2.073746998 \times 10^{14}$.

FIGURE 15

Now work Problem 1.

Two useful formulas involving the symbol $\binom{n}{j}$ are

$$\binom{n}{0} = 1 \quad \text{and} \quad \binom{n}{n} = 1$$

MISSION POSSIBLE

Getting the Most Out of Your Contract

You and your band have just signed a recording contract with the nationally acclaimed recording studio, NASHBURG TENS. They've promised $200,000 a year for six years, plus a choice of one of the following four options:

(a) a bonus of $10,000 per year
(b) an annual increase of 4.5% per year (starting after the first year)
(c) an annual increase of 6% per year (starting after the second year)
(d) an annual increase of $9,500 per year (starting after the first year)

You have to tell them today which option you want to take. Your agent is out of town and out of cellular phone range, so you'll have to do the math yourselves.

1. For each option, find out what your payment would be every year for the six years. Round to the nearest dollar.
2. Identify which options are examples of arithmetic or geometric sequences.
3. For each option, find out how much the recording studio will have paid in total over the whole six years. Do you have a formula that will shorten your work?
4. Which option pays your band the most over all? Are there any considerations or circumstances that would make other options better choices even though they pay less money? Are there advantages to being paid more at the beginning of the contract? What are they?

Proof

$$\binom{n}{0} = \frac{n!}{0!(n-0)!} = \frac{n!}{0!n!} = \frac{1}{1} = 1$$

You are asked to show that $\binom{n}{n} = 1$ in Problem 41 at the end of this section. ∎

Suppose we arrange the various values of the symbol $\binom{n}{j}$ in a triangular display, as shown next and in Figure 16.

$$\begin{array}{c} \binom{0}{0} \\ \binom{1}{0} \quad \binom{1}{1} \\ \binom{2}{0} \quad \binom{2}{1} \quad \binom{2}{2} \\ \binom{3}{0} \quad \binom{3}{1} \quad \binom{3}{2} \quad \binom{3}{3} \\ \binom{4}{0} \quad \binom{4}{1} \quad \binom{4}{2} \quad \binom{4}{3} \quad \binom{4}{4} \\ \binom{5}{0} \quad \binom{5}{1} \quad \binom{5}{2} \quad \binom{5}{3} \quad \binom{5}{4} \quad \binom{5}{5} \end{array}$$

This display is called the **Pascal triangle,** named after Blaise Pascal (1623–1662), a French mathematician.

FIGURE 16
Pascal triangle

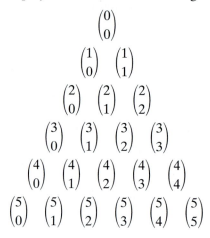

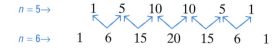

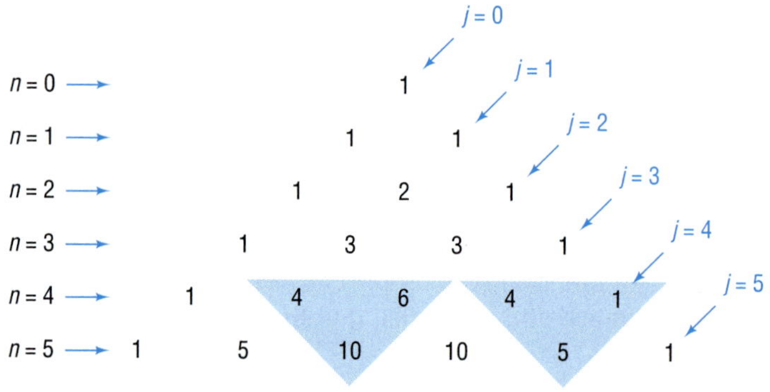

The Pascal triangle has 1's down the sides. To get any other entry, merely add the two nearest entries in the row above it. The shaded triangles in Figure 16 serve to illustrate this feature of the Pascal triangle. Based on this feature, the row corresponding to $n = 6$ is found as follows:

$n = 5 \rightarrow$ 1 5 10 10 5 1

$n = 6 \rightarrow$ 1 6 15 20 15 6 1

Later, we shall prove that this addition always works (see the theorem on page 937).

Although the Pascal triangle provides an interesting and organized display of the symbol $\binom{n}{j}$, in practice it is not all that helpful. For example, if you wanted to know the value of $\binom{12}{5}$, you would need to produce twelve rows of the triangle before seeing the answer. It is much faster instead to use the definition (1).

The Binomial Theorem

2 Now we are ready to state the **Binomial Theorem.** A proof is given at the end of this section.

Theorem Binomial Theorem

Let x and a be real numbers. For any positive integer n, we have

$$(x+a)^n = \binom{n}{0}x^n + \binom{n}{1}ax^{n-1} + \cdots + \binom{n}{j}a^jx^{n-j} + \cdots + \binom{n}{n}a^n$$
$$= \sum_{j=0}^{n}\binom{n}{j}x^{n-j}a^j \qquad (2)$$

Now you know why we needed to introduce the symbol $\binom{n}{j}$; these symbols are the numerical coefficients that appear in the expansion of $(x+a)^n$. Because of this, the symbol $\binom{n}{j}$ is called the **binomial coefficient.**

EXAMPLE 2 Expanding a Binomial

Use the Binomial Theorem to expand $(x+2)^5$.

Solution In the Binomial Theorem, let $a = 2$ and $n = 5$. Then

$$(x+2)^5 = \binom{5}{0}x^5 + \binom{5}{1}2x^4 + \binom{5}{2}2^2x^3 + \binom{5}{3}2^3x^2 + \binom{5}{4}2^4x + \binom{5}{5}2^5$$

↑ Use equation (2).

$$= 1\cdot x^5 + 5\cdot 2x^4 + 10\cdot 4x^3 + 10\cdot 8x^2 + 5\cdot 16x + 1\cdot 32$$

↑ Use row $n = 5$ of the Pascal triangle or formula (1) for $\binom{n}{j}$.

$$= x^5 + 10x^4 + 40x^3 + 80x^2 + 80x + 32$$

EXAMPLE 3 Expanding a Binomial

Expand $(2y-3)^4$ using the Binomial Theorem.

Solution First, we rewrite the expression $(2y-3)^4$ as $[2y+(-3)]^4$. Now we use the Binomial Theorem with $n = 4$, $x = 2y$, and $a = -3$:

$$[2y+(-3)]^4 = \binom{4}{0}(2y)^4 + \binom{4}{1}(-3)(2y)^3 + \binom{4}{2}(-3)^2(2y)^2$$
$$+ \binom{4}{3}(-3)^3(2y) + \binom{4}{4}(-3)^4$$
$$= 1\cdot 16y^4 + 4(-3)8y^3 + 6\cdot 9\cdot 4y^2 + 4(-27)2y + 1\cdot 81$$

↑ Use row $n = 4$ of the Pascal triangle or formula (1) for $\binom{n}{j}$.

$$= 16y^4 - 96y^3 + 216y^2 - 216y + 81$$

In this expansion, note that the signs alternate due to the fact that $a = -3 < 0$.

Now work Problem 17.

EXAMPLE 4 Finding a Particular Coefficient in a Binomial Expansion

Find the coefficient of y^8 in the expansion of $(2y + 3)^{10}$.

Solution We write out the expansion using the Binomial Theorem:

$$(2y+3)^{10} = \binom{10}{0}(2y)^{10} + \binom{10}{1}(2y)^9(3)^1 + \binom{10}{2}(2y)^8(3)^2 + \binom{10}{3}(2y)^7(3)^3$$
$$+ \binom{10}{4}(2y)^6(3)^4 + \cdots + \binom{10}{9}(2y)(3)^9 + \binom{10}{10}(3)^{10}$$

From the third term in the expansion, the coefficient of y^8 is

$$\binom{10}{2}(2)^8(3)^2 = \frac{10!}{2!8!}\cdot 2^8 \cdot 9 = \frac{10 \cdot 9 \cdot 8!}{2 \cdot 8!}\cdot 2^8 \cdot 9 = 103{,}680$$

As this solution demonstrates, we can use the Binomial Theorem to write a particular term in an expansion without writing the entire expansion. Based on the expansion of $(x + a)^n$, the term containing x^j is

$$\binom{n}{n-j}a^{n-j}x^j \qquad (3)$$

For example, we can solve Example 4 by using formula (3) with $n = 10$, $a = 3$, $x = 2y$, and $j = 8$. Then the term containing y^8 is

$$\binom{10}{10-8}3^{10-8}(2y)^8 = \binom{10}{2}\cdot 3^2 \cdot 2^8 \cdot y^8 = \frac{10!}{2!8!}\cdot 9 \cdot 2^8 y^8$$
$$= \frac{10 \cdot 9 \cdot 8!}{2!8!}\cdot 9 \cdot 2^8 y^8 = 103{,}680y^8$$

EXAMPLE 5 Finding a Particular Term in a Binomial Expansion

Find the sixth term in the expansion of $(x + 2)^9$.

Solution A We expand using the Binomial Theorem until the sixth term is reached:

$$(x+2)^9 = \binom{9}{0}x^9 + \binom{9}{1}x^8 \cdot 2 + \binom{9}{2}x^7 \cdot 2^2 + \binom{9}{3}x^6 \cdot 2^3 + \binom{9}{4}x^5 \cdot 2^4$$
$$+ \binom{9}{5}x^4 \cdot 2^5 + \cdots$$

The sixth term is

$$\binom{9}{5}x^4 \cdot 2^5 = \frac{9!}{5!4!}\cdot x^4 \cdot 32 = 4032x^4$$

Solution B The sixth term in the expansion of $(x + 2)^9$, which has ten terms total, contains x^4. (Do you see why?) Thus, by formula (3), the sixth term is

$$\binom{9}{9-4}2^{9-4}x^4 = \binom{9}{5}2^5x^4 = \frac{9!}{5!4!}\cdot 32x^4 = 4032x^4$$

Now work Problems 25 and 31.

Next we show that the "triangular addition" feature of the Pascal triangle illustrated in Figure 16 always works.

Theorem

If n and j are integers with $1 \le j \le n$, then

$$\binom{n}{j-1} + \binom{n}{j} = \binom{n+1}{j} \tag{4}$$

Proof

$$\binom{n}{j-1} + \binom{n}{j} = \frac{n!}{(j-1)![n-(j-1)]!} + \frac{n!}{j!(n-j)!}$$

$$= \frac{n!}{(j-1)!(n-j+1)!} + \frac{n!}{j!(n-j)!} \qquad \begin{array}{l}\text{Multiply the first term}\\ \text{by } j/j \text{ and the second}\\ \text{term by } (n-j+1)/\\ (n-j+1).\end{array}$$

$$= \frac{jn!}{j(j-1)!(n-j+1)!} + \frac{(n-j+1)n!}{j!(n-j+1)(n-j)!}$$

$$= \frac{jn!}{j!(n-j+1)!} + \frac{(n-j+1)n!}{j!(n-j+1)!} \qquad \begin{array}{l}\text{Now the denominators}\\ \text{are equal.}\end{array}$$

$$= \frac{jn! + (n-j+1)n!}{j!(n-j+1)!}$$

$$= \frac{n!(j+n-j+1)}{j!(n-j+1)!}$$

$$= \frac{n!(n+1)}{j!(n-j+1)!} = \frac{(n+1)!}{j![(n+1)-j]!} = \binom{n+1}{j} \qquad \blacksquare$$

Proof of the Binomial Theorem We use mathematical induction to prove the Binomial Theorem. First, we show that formula (2) is true for $n = 1$:

$$(x+a)^1 = x + a = \binom{1}{0}x^1 + \binom{1}{1}a^1$$

Next we suppose that formula (2) is true for some k. That is, we assume that

$$(x+a)^k = \binom{k}{0}x^k + \binom{k}{1}ax^{k-1} + \cdots + \binom{k}{j-1}a^{j-1}x^{k-j+1} + \binom{k}{j}a^j x^{k-j} + \cdots + \binom{k}{k}a^k \tag{5}$$

Now we calculate $(x+a)^{k+1}$:

$$(x+a)^{k+1} = (x+a)(x+a)^k = x(x+a)^k + a(x+a)^k$$

Use equation (5).

$$= x\left[\binom{k}{0}x^k + \binom{k}{1}ax^{k-1} + \cdots + \binom{k}{j-1}a^{j-1}x^{k-j+1} + \binom{k}{j}a^j x^{k-j} + \cdots + \binom{k}{k}a^k\right]$$

$$+ a\left[\binom{k}{0}x^k + \binom{k}{1}ax^{k-1} + \cdots + \binom{k}{j-1}a^{j-1}x^{k-j+1} + \binom{k}{j}a^j x^{k-j} + \cdots + \binom{k}{k-1}a^{k-1}x + \binom{k}{k}a^k\right]$$

$$= \binom{k}{0}x^{k+1} + \binom{k}{1}ax^k + \cdots + \binom{k}{j-1}a^{j-1}x^{k-j+2} + \binom{k}{j}a^j x^{k-j+1} + \cdots + \binom{k}{k}a^k x$$

$$+ \binom{k}{0}ax^k + \binom{k}{1}a^2 x^{k-1} + \cdots + \binom{k}{j-1}a^j x^{k-j+1} + \binom{k}{j}a^{j+1}x^{k-j} + \cdots + \binom{k}{k-1}a^k x + \binom{k}{k}a^{k+1}$$

$$= \binom{k}{0}x^{k+1} + \left[\binom{k}{1} + \binom{k}{0}\right]ax^k + \cdots + \left[\binom{k}{j} + \binom{k}{j-1}\right]a^j x^{k-j+1} + \cdots + \left[\binom{k}{k} + \binom{k}{k-1}\right]a^k x + \binom{k}{k}a^{k+1}$$

Because

$$\binom{k}{0} = 1 = \binom{k+1}{0}, \quad \binom{k}{1} + \binom{k}{0} \underset{\uparrow\,(4)}{=} \binom{k+1}{1}, \quad \ldots,$$

$$\binom{k}{j} + \binom{k}{j-1} \underset{\uparrow\,(4)}{=} \binom{k+1}{j}, \quad \ldots, \quad \binom{k}{k} = 1 = \binom{k+1}{k+1}$$

we have

$$(x+a)^{k+1} = \binom{k+1}{0}x^{k+1} + \binom{k+1}{1}ax^{k} + \cdots + \binom{k+1}{j}a^{j}x^{k-j+1} + \cdots + \binom{k+1}{k+1}a^{k+1}$$

Thus, Conditions I and II of the Principle of Mathematical Induction are satisfied, and formula (2) is therefore true for all n. ■

Historical Feature

The case $n = 2$ of the binomial theorem, $(a + b)^2$, was known to Euclid in 300 B.C., but the general law seems to have been discovered by the Persian mathematician and astronomer Omar Khayyám (1044?–1123?), who is also well known as the author of the *Rubaiyat,* a collection of four-line poems making observations on the human condition. Omar Khayyám did not state the binomial theorem explicitly, but he claimed to have a method for extracting third, fourth, fifth roots, and so on. A little study shows that one must know the binomial theorem to create such a method.

The heart of the binomial theorem is the formula for the numerical coefficients, and, as we saw, they can be written out in a symmetric triangular form. The Pascal triangle appears first in the books of Yang Hui (about 1270) and Chu Shihchie (1303). Pascal's name is attached to the triangle because of the many applications he made of it, especially to counting and probability. In establishing these results, he was one of the earliest users of mathematical induction.

Many people worked on the proof of the binomial theorem, which was finally completed for all n (including complex numbers) by Niels Abel (1802–1829). ■

12.5 EXERCISES

In Problems 1–12, evaluate each expression by hand. Use a graphing utility to verify your answer.

1. $\binom{5}{3}$ **2.** $\binom{7}{3}$ **3.** $\binom{7}{5}$ **4.** $\binom{9}{7}$ **5.** $\binom{50}{49}$ **6.** $\binom{100}{98}$

7. $\binom{1000}{1000}$ **8.** $\binom{1000}{0}$ **9.** $\binom{55}{23}$ **10.** $\binom{60}{20}$ **11.** $\binom{47}{25}$ **12.** $\binom{37}{19}$

In Problems 13–24, expand each expression using the binomial theorem.

13. $(x + 1)^5$ **14.** $(x - 1)^5$ **15.** $(x - 2)^6$ **16.** $(x + 3)^4$

17. $(3x + 1)^4$ **18.** $(2x + 3)^5$ **19.** $(x^2 + y^2)^5$ **20.** $(x^2 - y^2)^6$

21. $(\sqrt{x} + \sqrt{2})^6$ **22.** $(\sqrt{x} - \sqrt{3})^4$ **23.** $(ax + by)^5$ **24.** $(ax - by)^4$

In Problems 25–38, use the binomial theorem to find the indicated coefficient or term.

25. The coefficient of x^6 in the expansion of $(x + 3)^{10}$

26. The coefficient of x^3 in the expansion of $(x - 3)^{10}$

27. The coefficient of x^7 in the expansion of $(2x - 1)^{12}$

28. The coefficient of x^3 in the expansion of $(2x + 1)^{12}$

29. The coefficient of x^7 in the expansion of $(2x + 3)^9$

30. The coefficient of x^2 in the expansion of $(2x - 3)^9$

31. The fifth term in the expansion of $(x + 3)^7$

32. The third term in the expansion of $(x - 3)^7$

33. The third term in the expansion of $(3x - 2)^9$

34. The sixth term in the expansion of $(3x + 2)^8$

35. The coefficient of x^0 in the expansion of $\left(x^2 + \frac{1}{x}\right)^{12}$

36. The coefficient of x^0 in the expansion of $\left(x - \frac{1}{x^2}\right)^9$

37. The coefficient of x^4 in the expansion of $\left(x - \frac{2}{\sqrt{x}}\right)^{10}$

38. The coefficient of x^2 in the expansion of $\left(\sqrt{x} + \frac{3}{\sqrt{x}}\right)^8$

39. Use the Binomial Theorem to find the numerical value of $(1.001)^5$ correct to five decimal places. [**Hint:** $(1.001)^5 = (1 + 10^{-3})^5$]

40. Use the Binomial Theorem to find the numerical value of $(0.998)^6$ correct to five decimal places.

41. Show that $\binom{n}{n} = 1$.

42. Show that, if n and j are integers with $0 \le j \le n$, then

$$\binom{n}{j} = \binom{n}{n - j}$$

Thus, conclude that the Pascal triangle is symmetric with respect to a vertical line drawn from the topmost entry.

43. If n is a positive integer, show that

$$\binom{n}{0} + \binom{n}{1} + \cdots + \binom{n}{n} = 2^n$$

[**Hint:** $2^n = (1 + 1)^n$; now use the binomial theorem.]

44. If n is a positive integer, show that

$$\binom{n}{0} - \binom{n}{1} + \binom{n}{2} - \cdots + (-1)^n\binom{n}{n} = 0$$

45. $$\binom{5}{0}\left(\frac{1}{4}\right)^5 + \binom{5}{1}\left(\frac{1}{4}\right)^4\left(\frac{3}{4}\right) + \binom{5}{2}\left(\frac{1}{4}\right)^3\left(\frac{3}{4}\right)^2 + \binom{5}{3}\left(\frac{1}{4}\right)^2\left(\frac{3}{4}\right)^3 + \binom{5}{4}\left(\frac{1}{4}\right)\left(\frac{3}{4}\right)^4 + \binom{5}{5}\left(\frac{3}{4}\right)^5 = ?$$

46. *Stirling's Formula* for approximating $n!$ when n is large is given by

$$n! \approx \sqrt{2n\pi}\left(\frac{n}{e}\right)^n\left(1 + \frac{1}{12n - 1}\right)$$

Calculate 12!, 20!, and 25!. Then use Stirling's formula to approximate 12!, 20!, and 25!.

12.6 SETS AND COUNTING

1. Find All the Subsets of a Set
2. Find the Intersection and Union of Sets
3. Find the Complement of a Set
4. Count the Number of Elements in a Set

Sets

A **set** is a well-defined collection of distinct objects. The objects of a set are called its **elements.** By **well-defined,** we mean that there is a rule that enables us to determine whether a given object is an element of the set. If a set has no elements, it is called the **empty set,** or **null set,** and is denoted by the symbol $\varnothing$.

Because the elements of a set are distinct, we never repeat elements. Thus, we would never write {1, 2, 3, 2}; the correct listing is {1, 2, 3}. Furthermore, because a set is a collection, the order in which the elements are listed is immaterial. Thus, {1, 2, 3}, {1, 3, 2}, {2, 1, 3}, and so on, all represent the same set.

EXAMPLE 1 Writing the Elements of a Set

Write the set consisting of the possible outcomes from tossing a coin twice. Use H for "heads" and T for "tails."

Solution In tossing a coin twice, we can get heads each time, HH; or heads the first time and tails the second, HT; or tails the first time and heads the second, TH; or tails each time, TT. Because no other possibilities exist, the set of outcomes is

$$\{HH, HT, TH, TT\}$$

1

If two sets A and B have precisely the same elements, then we say that A and B are **equal** and write $A = B$.

If each element of a set A is also an element of a set B, then we say that A is a **subset** of B and write $A \subseteq B$.

If $A \subseteq B$ and $A \neq B$, then we say that A is a **proper subset** of B and write $A \subset B$.

Thus, if $A \subseteq B$, every element in set A is also in set B, but B may or may not have additional elements. If $A \subset B$, every element in A is also in B, and B has at least one element not found in A.

Finally, we agree that the empty set is a subset of every set; that is,

$$\varnothing \subseteq A \qquad \text{for any set } A$$

EXAMPLE 2 Finding All the Subsets of a Set

Write down all the subsets of the set $\{a, b, c\}$.

Solution To organize our work, we write down all the subsets with no elements, then those with one element, then those with two elements, and finally those with three elements. These will give us all the subsets. Do you see why?

0 Elements	1 Element	2 Elements	3 Elements
$\varnothing$	$\{a\}, \{b\}, \{c\}$	$\{a, b\}, \{b, c\}, \{a, c\}$	$\{a, b, c\}$

Now work Problem 21.

If A and B are sets, the **intersection** of A with B, denoted $A \cap B$, is the set consisting of elements that belong to both A and B. The **union** of A with B, denoted $A \cup B$, is the set consisting of elements that belong to *either* A or B, or both.

EXAMPLE 3 Finding the Intersection and Union of Sets

Let $A = \{1, 3, 5, 8]$, $B = \{3, 5, 7\}$, and $C = \{2, 4, 6, 8\}$. Find:

(a) $A \cap B$ (b) $A \cup B$ (c) $B \cap (A \cup C)$

Solution

(a) $A \cap B = \{1, 3, 5, 8\} \cap \{3, 5, 7\} = \{3, 5\}$

(b) $A \cup B = \{1, 3, 5, 8\} \cup \{3, 5, 7\} = \{1, 3, 5, 7, 8\}$

(c) $B \cap (A \cup C) = \{3, 5, 7\} \cap [\{1, 3, 5, 8\} \cup \{2, 4, 6, 8\}]$
$= \{3, 5, 7\} \cap \{1, 2, 3, 4, 5, 6, 8\} = \{3, 5\}$

Now work Problem 5.

Usually, in working with sets, we designate a **universal set,** the set consisting of all the elements we wish to consider. Once a universal set has been designated, we can consider elements of the universal set not found in a given set.

If A is a set, the **complement** of A, denoted A', is the set consisting of all the elements not in A.

EXAMPLE 4 Finding the Complement of a Set

If the universal set is $U = \{1, 2, 3, 4, 5, 6, 7, 8, 9\}$, and if $A = \{1, 3, 5, 7, 9\}$, then $A' = \{2, 4, 6, 8\}$.

Notice that: $A \cup A' = U$ and $A \cap A' = \varnothing$.

Now work Problem 13.

FIGURE 17

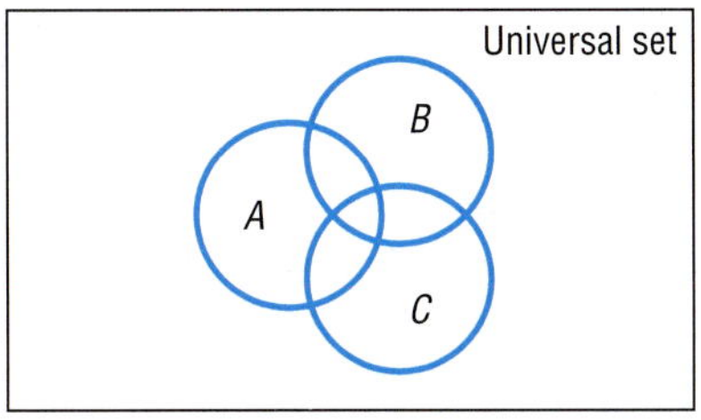

It is often helpful to draw pictures of sets. Such pictures, called **Venn diagrams,** represent sets as circles enclosed in a rectangle, which represents the universal set. Such diagrams often help us to visualize various relationships among sets. See Figure 17.

If we know that $A \subseteq B$, we might use the Venn diagram in Figure 18(a). If we know that A and B have no elements in common, that is, if $A \cap B = \varnothing$, we might use the Venn diagram in Figure 18(b).

FIGURE 18

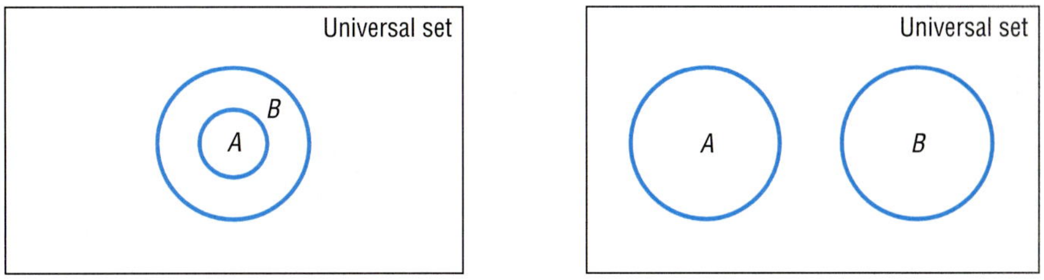

(a) $A \subseteq B$ **(b)** $A \cap B = \varnothing$

Figures 19(a), 19(b), and 19(c) use Venn diagrams to illustrate the definitions of intersection, union, and complement, respectively.

FIGURE 19

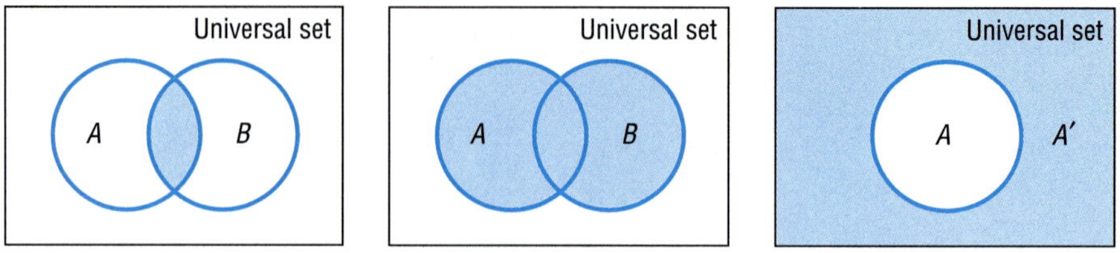

(a) $A \cap B$ (b) $A \cup B$ (c) A'

Counting

4 As you count the number of students in a classroom or the number of pennies in your pocket, what you are really doing is matching, on a one-to-one basis, each object to be counted with the counting numbers $1, 2, 3, \ldots, n$, for some number n. If a set A matched up in this fashion with the set $\{1, 2, \ldots, 25\}$, you would conclude that there are 25 elements in the set A. We use the notation $n(A) = 25$ to indicate that there are 25 elements in the set A.

Because the empty set has no elements, we write

$$n(\varnothing) = 0$$

If the number of elements in a set is a nonnegative integer, we say the set is **finite.** Otherwise, it is **infinite.** We shall concern ourselves only with finite sets.

From Example 2, we can see that a set with 3 elements has $2^3 = 8$ subsets. In fact, it can be shown that a set with n elements has exactly 2^n subsets. This fact has an important application to computers, which we take up at the end of this section.

EXAMPLE 5 Analyzing Survey Data

In a survey of 100 college students, 35 were registered in College Algebra, 52 were registered in Introduction to Computer Science, and 18 were in both courses. How many were registered in neither course?

FIGURE 20

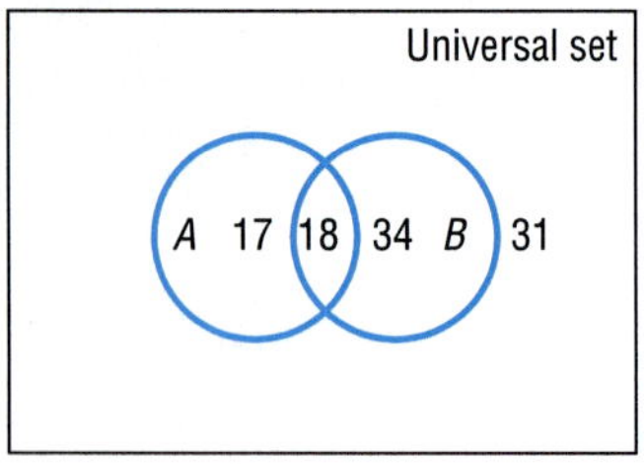

Solution First, we let A = Set of students in College Algebra
B = Set of students in Introduction to Computer Science

Then the given information tells us that

$$n(A) = 35 \qquad n(B) = 52 \qquad n(A \cap B) = 18$$

Refer to Figure 20. Do you see how the numerical entries were determined? Based on the diagram, we conclude that $17 + 18 + 34 = 69$ students were registered in at least one of the two courses. Since 100 students were surveyed, it follows that $100 - 69 = 31$ were registered in neither course.

Now work Problem 35.

The conclusions drawn in Example 5 lead us to formulate a general counting formula. If we count the elements in each of two sets A and B, we necessarily count twice any elements that are in both A and B, that is, those elements in $A \cap B$. Thus, to count correctly the elements that are in A or B, that is, to find $n(A \cup B)$, we need to subtract those in $A \cap B$ from $n(A) + n(B)$.

Theorem Counting Formula

If A and B are finite sets, then

$$n(A \cup B) = n(A) + n(B) - n(A \cap B) \tag{1}$$

A special case of the counting formula (1) occurs if A and B have no elements in common. In this case, $A \cap B = \varnothing$ so that $n(A \cap B) = 0$.

Theorem Addition Principle of Counting

If two sets A and B have no elements in common, then

$$n(A \cup B) = n(A) + n(B) \tag{2}$$

EXAMPLE 6 Counting the Number of Possible Codes

A certain code is to consist of either a letter of the alphabet or a digit, but not both. How many codes are possible?

Solution Let the sets A and B be defined as

A = Set of letters in the alphabet
B = Set of digits $\{0, 1, 2, \ldots, 9\}$

Then

$$n(A) = 26 \qquad n(B) = 10$$

Because letters and digits are different, $A \cap B = \varnothing$. The number of ways either a letter or a digit can be chosen is, therefore,

$$n(A \cup B) = n(A) + n(B) = 26 + 10 = 36$$

Application to Computers

Information stored in a computer may be thought of as a series of switches, which are either on or off and are denoted by either the number 0 (off) or the number 1 (on). These numbers are the binary digits, or **bits.** A **register** holds a certain fixed number of bits. For example, the Z-80 microprocessor has 8-bit registers; the PDP-11 minicomputer has 16-bit registers, and the IBM-370 computer has 32-bit registers. Thus, a Z-80 register may hold an entry that looks like this: 01111001 (8 bits). We wish to find out how many different representations are possible in a given register.

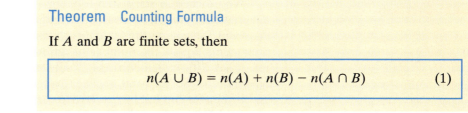

TABLE 4

a	b	c	Subset
0	0	0	∅
1	0	0	{a}
0	1	0	{b}
0	0	1	{c}
1	1	0	{a, b}
0	1	1	[b, c}
1	0	1	{a, c}
1	1	1	{a, b, c}

We proceed in steps, first looking at a hypothetical 3-bit register. Look again at the solution to Example 2 and arrange all the subsets of $\{a, b, c\}$ as shown in Table 4. As the table illustrates, the number of subsets of a set with 3 elements equals the number of different representations in a 3-bit register. A set with n elements has 2^n subsets; thus, an n-bit register has 2^n representations. So an 8-bit register can hold $2^8 = 256$ different symbols, a 16-bit register can hold $2^{16} = 65{,}536$ different symbols, and a 32-bit register can hold $2^{32} \approx 4.3 \times 10^9$ different symbols.

12.6 EXERCISES

In Problems 1–10, use $A = \{1, 3, 5, 7, 9\}$, $B = \{1, 5, 6, 7\}$, *and* $C = \{1, 2, 4, 6, 8, 9\}$ *to find each set.*

1. $A \cup B$ **2.** $A \cup C$ **3.** $A \cap B$ **4.** $A \cap C$

5. $(A \cup B) \cap C$ **6.** $(A \cap C) \cup (B \cap C)$ **7.** $(A \cap B) \cup C$ **8.** $(A \cup B) \cup C$

9. $(A \cup C) \cap (B \cup C)$ **10.** $(A \cap B) \cap C$

In Problems 11–20, use U = *Universal set* = $\{0, 1, 2, 3, 4, 5, 6, 7, 8, 9\}$, $A = \{1, 3, 4, 5, 9\}$, $B = \{2, 4, 6, 7, 8\}$, *and* $C = \{1, 3, 4, 6\}$ *to find each set.*

11. A' **12.** C' **13.** $(A \cap B)'$ **14.** $(B \cup C)'$ **15.** $A' \cup B'$

16. $B' \cap C'$ **17.** $(A \cap C')'$ **18.** $(B' \cup C)'$ **19.** $(A \cup B \cup C)'$ **20.** $(A \cap B \cap C)'$

21. Write down all the subsets of $\{a, b, c, d\}$.

22. Write down all the subsets of $\{a, b, c, d, e\}$.

23. If $n(A) = 15$, $n(B) = 20$, and $n(A \cap B) = 10$, find $n(A \cup B)$.

24. If $n(A) = 20$, $n(B) = 40$, and $n(A \cup B) = 35$, find $n(A \cap B)$.

25. If $n(A \cup B) = 50$, $n(A \cap B) = 10$, and $n(B) = 20$, find $n(A)$.

26. If $n(A \cup B) = 60$, $n(A \cap B) = 40$, and $n(A) = n(B)$, find $n(A)$.

In Problems 27–34, use the information given in the figure.

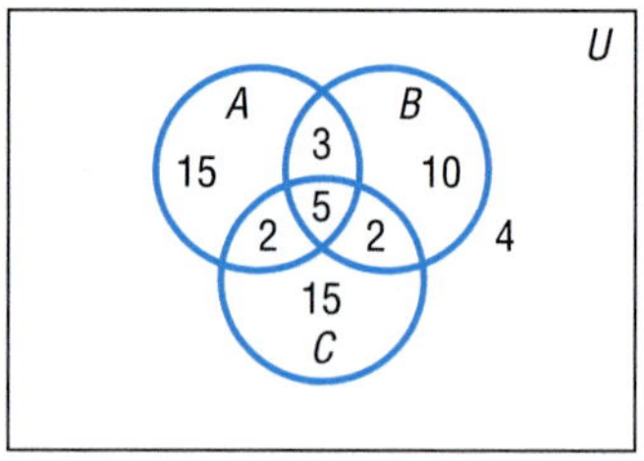

27. How many are in set A?

28. How many are in set B?

29. How many are in A or B?

30. How many are in A and B?

31. How many are in A but not C?

32. How many are not in A?

33. How many are in A and B and C?

34. How many are in A or B or C?

35. Analyzing Survey Data In a consumer survey of 500 people, 200 indicated that they would be buying a major appliance within the next month; 150 indicated they would buy a car, and 25 said they would purchase both a major appliance and a car. How many will purchase neither? How many will purchase only a car?

36. Analyzing Survey Data In a student survey, 200 indicated that they would attend Summer Session I and 150 indicated Summer Session II. If 75 students plan to attend both summer sessions and 275 indicated that they would attend neither session, how many students participated in the survey?

37. Analyzing Survey Data In a survey of 100 investors in the stock market,

50 owned shares in IBM

40 owned shares in AT&T

45 owned shares in GE

20 owned shares in both IBM and GE

15 owned shares in both AT&T and GE

20 owned shares in both IBM and AT&T

5 owned shares in all three

(a) How many of the investors surveyed did not have shares in any of the three companies?
(b) How many owned just IBM shares?
(c) How many owned just GE shares?
(d) How many owned neither IBM nor GE?
(e) How many owned either IBM or AT&T but no GE?

38. Classifying Blood Types Human blood is classified as either Rh+ or Rh−. Blood is also classified by type: A, if it contains an A antigen; B, if it contains a B antigen; AB, if it contains both A and B antigens; and O, if it contains neither antigen. Draw a Venn diagram illustrating the various blood types. Based on this classification, how many different kinds of blood are there?

39. Make up a problem different from any found in the text that requires the addition principle of counting to solve. Give it to a friend to solve and critique.

40. Investigate the notion of counting as it relates to infinite sets. Write an essay on your findings.

12.7 PERMUTATIONS AND COMBINATIONS

1. Solve Counting Problems Using the Multiplication Principle
2. Solve Counting Problems Using Permutations
3. Solve Counting Problems Using Combinations

1 Counting plays a major role in many diverse areas, such as probability, statistics, and computer science. In this section we shall look at special types of counting problems and develop general formulas for solving them.

We begin with an example that will demonstrate a general counting principle.

EXAMPLE 1 Counting the Number of Possible Meals

The fixed-price dinner at a restaurant provides the following choices:

Appetizer: soup or salad
Entree: baked chicken, broiled beef patty, baby beef liver, or roast beef au jus
Dessert: ice cream or cheese cake

How many different meals can be ordered?

Solution Ordering such a meal requires three separate decisions:

Choose an Appetizer	Choose an Entree	Choose a Dessert
2 choices	4 choices	2 choices

Look at the **tree diagram** in Figure 21. We see that, for each choice of appetizer, there are 4 choices of entrees. And for each of these $2 \cdot 4 = 8$ choices, there are 2 choices for dessert. Thus, there are a total of

$$2 \cdot 4 \cdot 2 = 16$$

different meals that can be ordered.

FIGURE 21

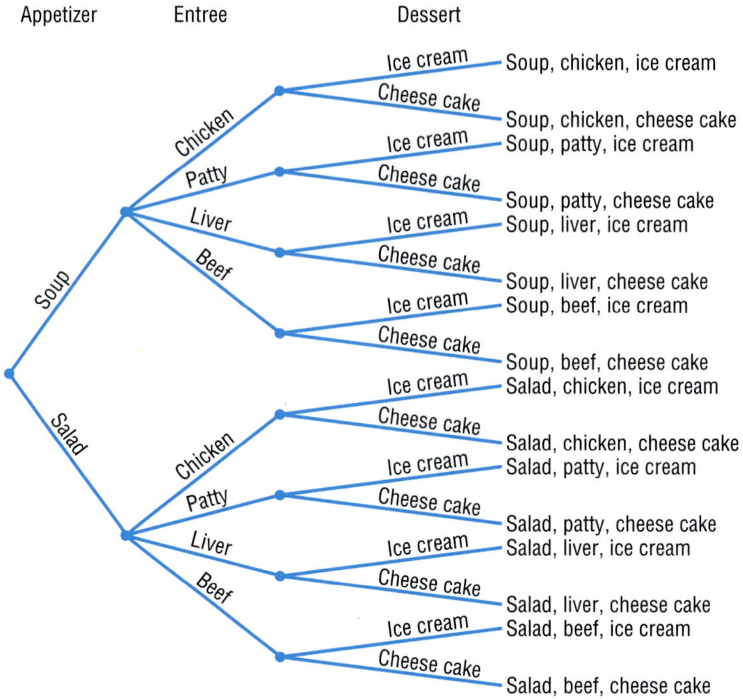

Example 1 illustrates a general counting principle.

Theorem Multiplication Principle of Counting

If a task consists of a sequence of choices in which there are p selections for the first choice, q selections for the second choice, r selections for the third choice, and so on, then the task of making these selections can be done in

$$p \cdot q \cdot r \cdot \ldots$$

different ways.

EXAMPLE 2 Counting Airport Codes

The International Airline Transportation Association (IATA) assigns three-letter codes to represent airport locations. For example, JFK represents Kennedy International in New York. How many different airport codes are possible?

Solution The task consists of making three selections. Each selection requires choosing a letter of the alphabet (26 choices). Thus, by the multiplication principle, there are

$$26 \cdot 26 \cdot 26 = 17{,}576$$

different airport codes.

2 In Example 2, we were allowed to repeat a letter. For example, a valid airport code is FLL (Ft. Lauderdale International Airport), in which the letter L appears twice. In the next example, such repetition is not allowed.

EXAMPLE 3 Counting without Repetition

Suppose that we wish to establish a three-letter code using any of the 26 letters of the alphabet, but we require that no letter be used more than once. How many different three-letter codes are there?

Solution The task consists of making three selections. The first selection requires choosing from 26 letters. Because no letter can be used more than once, the second selection requires choosing from 25 letters. The third selection requires choosing from 24 letters. (Do you see why?) By the multiplication principle, there are

$$26 \cdot 25 \cdot 24 = 15{,}600$$

different three-letter codes with no letter repeated.

Now work Problems 25 and 29.

Example 3 illustrates a type of counting problem referred to as a *permutation.*

> A **permutation** is an ordered arrangement of n distinct objects without repetitions. The symbol $P(n, r)$ represents the number of permutations of n distinct objects, taken r at a time, where $r \leq n$.

For example, the question posed in Example 3 asks for the number of ways the 26 letters of the alphabet can be arranged using three nonrepeated letters. The answer is

$$P(26, 3) = 26 \cdot 25 \cdot 24 = 15{,}600$$

To arrive at a formula for $P(n, r)$, we note that the task of obtaining an ordered arrangement of n objects in which only $r \leq n$ of them are used, without repeating any of them, requires making r selections. For the first selection, there are n choices; for the second selection, there are $n - 1$ choices; for the third selection, there are $n - 2$ choices; . . . ; for the rth selection, there are $n - (r - 1)$ choices. By the multiplication principle, we have

$$\begin{aligned} P(n, r) &= \underset{\text{1st}}{n} \cdot \underset{\text{2nd}}{(n-1)} \cdot \underset{\text{3rd}}{(n-2)} \cdot \ldots \cdot \underset{r\text{th}}{[n-(r-1)]} \\ &= n \cdot (n-1) \cdot (n-2) \cdot \ldots \cdot (n-r+1) \end{aligned}$$

This formula for $P(n, r)$ can be compactly written using factorial notation.*

$$\begin{aligned} P(n, r) &= n \cdot (n-1) \cdot (n-2) \cdot \ldots \cdot (n-r+1) \\ &= n \cdot (n-1) \cdot (n-2) \cdot \ldots \cdot (n-r+1) \cdot \frac{(n-r) \cdot \ldots \cdot 3 \cdot 2 \cdot 1}{(n-r) \cdot \ldots \cdot 3 \cdot 2 \cdot 1} = \frac{n!}{(n-r)!} \end{aligned}$$

*Recall that $0! = 1, 1! = 1, 2! = 2 \cdot 1, \cdots, n! = n(n-1) \cdot \ldots \cdot 3 \cdot 2 \cdot 1$.

Theorem Number of Permutations of n Distinct Objects Taken r at a Time

The number of different arrangements of n objects using $r \leq n$ of them, in which

1. the n objects are distinct,
2. once an object is used it cannot be repeated, and
3. order is important,

is given by the formula

$$P(n, r) = \frac{n!}{(n-r)!} \tag{1}$$

EXAMPLE 4

Evaluate: (a) $P(7, 3)$ (b) $P(6, 1)$ (c) $P(52, 5)$

Solution We shall work parts (a) and (b) in two ways.

(a) $P(7, 3) = \underbrace{7 \cdot 6 \cdot 5}_{\text{3 factors}} = 210$

or

$$P(7, 3) = \frac{7!}{(7-3)!} = \frac{7!}{4!} = \frac{7 \cdot 6 \cdot 5 \cdot \cancel{4!}}{\cancel{4!}} = 210$$

(b) $P(6, 1) = \underbrace{6}_{\text{1 factor}} = 6$

or

$$P(6, 1) = \frac{6!}{(6-1)!} = \frac{6!}{5!} = \frac{6 \cdot \cancel{5!}}{\cancel{5!}} = 6$$

(c) Figure 22 shows the solution using a TI-83 graphing calculator. Thus, $P(52, 5) = 311{,}875{,}200$.

FIGURE 22

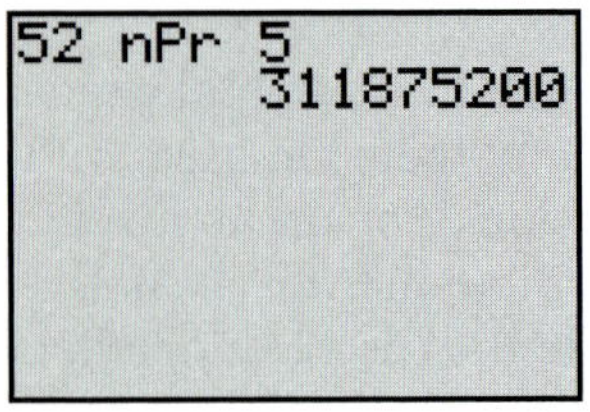

Now work Problem 1.

EXAMPLE 5

In how many ways can 5 people be lined up?

Solution The 5 people are obviously distinct. Once a person is in line, that person will not be repeated elsewhere in the line; and, in lining up people, order is important. Thus, we have a permutation of 5 objects taken 5 at a time. We can line up 5 people in

$$P(5, 5) = \underbrace{5 \cdot 4 \cdot 3 \cdot 2 \cdot 1}_{\text{5 factors}} = 5! = 120 \text{ ways}$$

Now work Problem 31.

Combinations

3 In a permutation, order is important; for example, the arrangements ABC, CAB, BAC, . . . are considered different arrangements of the letters A, B, and C. In many situations, though, order is unimportant. For example, in the card game of poker, the order in which the cards are received does not matter; it is the *combination* of the cards that matters.

A **combination** is an arrangement, without regard to order, of n distinct objects without repetitions. The symbol $C(n, r)$ represents the number of combinations of n distinct objects taken r at a time, where $r \leq n$.

EXAMPLE 6

Listing Combinations

List all the combinations of the 4 objects a, b, c, d taken 2 at a time. What is $C(4, 2)$?

Solution One combination of a, b, c, d taken 2 at a time is

$$ab$$

The object ba is excluded, because order is not important in a combination. The list of all such combinations (convince yourself of this) is

$$ab, \quad ac, \quad ad, \quad bc, \quad bd, \quad cd$$

Thus,

$$C(4, 2) = 6$$

We can find a formula for $C(n, r)$ by noting that the only difference between a permutation and a combination is that we disregard order in combinations. Thus, to determine $C(n, r)$, we need only eliminate from the formula for $P(n, r)$ the number of permutations that were simply rearrangements of a given set of r objects. But that is easily determined from the formula for $P(n, r)$ by calculating $P(r, r) = r!$. So, if we divide $P(n, r)$ by $r!$, we will have the desired formula for $C(n, r)$:

$$C(n,r) = \frac{P(n,r)}{r!} \underset{\uparrow}{=} \frac{n!/(n-r)!}{r!} = \frac{n!}{(n-r)!r!}$$

Use formula (1).

We have proved the following result.

Theorem Number of Combinations of n Distinct Objects Taken r at a Time

The number of different arrangements of n objects using $r \leq n$ of them, in which

1. the n objects are distinct,
2. once an object is used, it cannot be repeated, and
3. order is not important

is given by the formula

$$C(n, r) = \frac{n!}{(n-r)!r!} \qquad (2)$$

Based on formula (2), we discover that the symbol $C(n, r)$ and the symbol $\binom{n}{r}$ for the binomial coefficients are, in fact, the same. Thus, the Pascal triangle (see Section 12.5) can be used to find the value of $C(n, r)$. However, because it is more practical and convenient, we will use formula (2) instead.

EXAMPLE 7 Using Formula (2)

Use formula (2) to find the value of each expression.

(a) $C(3, 1)$ (b) $C(6, 3)$ (c) $C(n, n)$ (d) $C(n, 0)$ (e) $C(52, 5)$

Solution

(a) $C(3, 1) = \dfrac{3!}{(3-1)!1!} = \dfrac{3!}{2!1!} = \dfrac{3 \cdot 2 \cdot 1}{2 \cdot 1 \cdot 1} = 3$

(b) $C(6, 3) = \dfrac{6!}{(6-3)!3!} = \dfrac{6 \cdot 5 \cdot 4 \cdot 3!}{3! \cdot 3!} = \dfrac{6 \cdot 5 \cdot 4}{6} = 20$

(c) $C(n, n) = \dfrac{n!}{(n-n)!n!} = \dfrac{n!}{0!n!} = \dfrac{1}{1} = 1$

(d) $C(n, 0) = \dfrac{n!}{(n-0)!0!} = \dfrac{n!}{n!0!} = \dfrac{1}{1} = 1$

(e) Figure 23 shows the solution using a TI-83 graphing calculator.

Thus, $C(52, 5) = 2{,}598{,}960$.

FIGURE 23

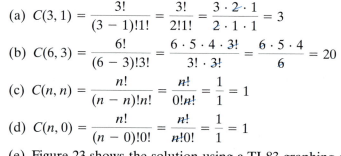

Now work Problem 9.

EXAMPLE 8 Forming Committees

How many different committees of 3 people can be formed from a pool of 7 people?

Solution The 7 people are, of course, distinct. More important, though, is the observation that the order of being selected for a committee is not significant. Thus, the problem asks for the number of combinations of 7 objects taken 3 at a time:

$$C(7, 3) = \frac{7!}{4!3!} = \frac{7 \cdot 6 \cdot 5 \cdot 4!}{4!3!} = \frac{7 \cdot 6 \cdot 5}{6} = 35$$

EXAMPLE 9 Forming Committees

In how many ways can a committee consisting of 2 faculty members and 3 students be formed if there are 6 faculty members and 10 students eligible to serve on the committee?

Solution The problem can be separated into two parts: the number of ways the faculty members can be chosen, $C(6, 2)$, and the number of ways the student members can be chosen, $C(10, 3)$. By the Multiplication Principle, the committee can be formed in

$$C(6, 2) \cdot C(10, 3) = \frac{6!}{4!2!} \cdot \frac{10!}{7!3!} = \frac{6 \cdot 5 \cdot \cancel{4!}}{\cancel{4!}2!} \cdot \frac{10 \cdot 9 \cdot 8 \cdot \cancel{7!}}{\cancel{7!}3!}$$
$$= \frac{30}{2} \cdot \frac{720}{6} = 1800 \text{ ways}$$

Now work Problem 47.

Permutations with Repetition

Recall that a permutation involves counting *distinct* objects. A permutation in which some of the objects are repeated is called a **permutation with repetition.** Some books refer to this as a **nondistinguishable permutation.**

Let's look at an example.

EXAMPLE 10 Forming Different Words

How many different words can be formed using all the letters in the word REARRANGE?

Solution Each word formed will have 9 letters: 3 R's, 2 A's, 2 E's, 1 N, and 1 G. To construct each word, we need to fill in 9 positions with the 9 letters:

$$\overline{1} \quad \overline{2} \quad \overline{3} \quad \overline{4} \quad \overline{5} \quad \overline{6} \quad \overline{7} \quad \overline{8} \quad \overline{9}$$

The process of forming a word consists of five tasks:

Task 1: Choose the positions for the 3 R's.
Task 2: Choose the positions for the 2 A's.
Task 3: Choose the positions for the 2 E's.
Task 4: Choose the position for the 1 N.
Task 5: Choose the position for the 1 G.

Task 1 can be done in $C(9, 3)$ ways. There then remain 6 positions to be filled, so Task 2 can be done in $C(6, 2)$ ways. There remain 4 positions to be filled, so Task 3 can be done in $C(4, 2)$ ways. There remain 2 positions to be filled, so Task 4 can be done in $C(2, 1)$ ways. The last position can be filled in $C(1, 1)$ way. Using the Multiplication Principle, the number of possible words that can be formed is

$$C(9, 3) \cdot C(6, 2) \cdot C(4, 2) \cdot C(2, 1) \cdot C(1, 1) = \frac{9!}{3! \cdot \cancel{6!}} \cdot \frac{\cancel{6!}}{2! \cdot \cancel{4!}} \cdot \frac{\cancel{4!}}{2! \cdot \cancel{2!}} \cdot \frac{\cancel{2!}}{1! \cdot \cancel{1!}} \cdot \frac{\cancel{1!}}{0! \cdot 1!}$$
$$= \frac{9!}{3! \cdot 2! \cdot 2! \cdot 1! \cdot 1!}$$

The form of the answer to Example 10 is suggestive of a general result. Had the letters in REARRANGE each been different, there would have been $P(9, 9) = 9!$ possible words formed. This is the numerator of the answer. The presence of 3 R's, 2 A's, and 2 E's reduces the number of different words, as the entries in the denominator illustrate. We are led to the following result:

Theorem Permutations with Repetition

The number of permutations of n objects of which n_1 are of one kind, n_2 are of a second kind, . . . , and n_k are of a kth kind is given by

$$\frac{n!}{n_1! \cdot n_2! \cdot \ldots \cdot n_k!} \tag{3}$$

where $n = n_1 + n_2 + \cdots + n_k$.

EXAMPLE 11 Arranging Flags

How many different vertical arrangements are there of 8 flags if 4 are white, 3 are blue, and 1 is red?

Solution We seek the number of permutations of 8 objects, of which 4 are of one kind, 3 of a second kind, and 1 of a third kind. Using formula (3), we find that there are

$$\frac{8!}{4! \cdot 3! \cdot 1!} = \frac{8 \cdot 7 \cdot 6 \cdot 5 \cdot \cancel{4!}}{\cancel{4!} \cdot 3! \cdot 1!} = 280 \text{ different arrangements:}$$

Now work Problem 53.

12.7 EXERCISES

In Problems 1–8, find the value of each permutation. Verify your results using a graphing calculator.

1. $P(6, 2)$
2. $P(7, 2)$
3. $P(5, 5)$
4. $P(4, 4)$
5. $P(8, 0)$
6. $P(9, 0)$
7. $P(8, 3)$
8. $P(8, 5)$

In Problems 9–16, use formula (2) to find the value of each combination. Verify your results using a graphing calculator.

9. $C(8, 2)$
10. $C(8, 6)$
11. $C(6, 4)$
12. $C(6, 2)$
13. $C(15, 15)$
14. $C(18, 1)$
15. $C(26, 13)$
16. $C(18, 9)$

17. List all the permutations of 5 objects a, b, c, d, and e taken 3 at a time. What is $P(5, 3)$?
18. List all the permutations of 5 objects a, b, c, d, and e taken 2 at a time. What is $P(5, 2)$?
19. List all the permutations of 4 objects 1, 2, 3, and 4 taken 3 at a time. What is $P(4, 3)$?
20. List all the permutations of 6 objects, 1, 2, 3, 4, 5, and 6 taken 3 at a time. What is $P(6, 3)$?
21. List all the combinations of the 5 objects a, b, c, d, and e taken 3 at a time. What is $C(5, 3)$?
22. List all the combinations of the 5 objects a, b, c, d, and e taken 2 at a time. What is $C(5, 2)$?

23. List all the combinations of the 4 objects 1, 2, 3, and 4 taken 3 at a time. What is $C(4, 3)$?

24. List all the combinations of the 6 objects 1, 2, 3, 4, 5, and 6 taken 3 at a time. What is $C(6, 3)$?

25. A man has 5 shirts and 3 ties. How many different shirt and tie combinations can he wear?

26. A woman has 3 blouses and 5 skirts. How many different outfits can she wear?

27. **Forming Codes** How many two-letter codes can be formed using the letters *A, B, C,* and *D*? Repeated letters are allowed.

28. **Forming Codes** How many two-letter codes can be formed using the letters *A, B, C, D,* and *E*? Repeated letters are allowed.

29. **Forming Numbers** How many three-digit numbers can be formed using the digits 0 and 1? Repeated digits are allowed.

30. **Forming Numbers** How many three-digit numbers can be formed using the digits 0, 1, 2, 3, 4, 5, 6, 7, 8, and 9? Repeated digits are allowed.

31. In how many ways can 4 people be lined up?

32. In how many ways can 5 different boxes be stacked?

33. **Forming Codes** How many different three-letter codes are there if only the letters *A, B, C, D,* and *E* can be used and no letter can be used more than once?

34. **Forming Codes** How many different four-letter codes are there if only the letters *A, B, C, D, E,* and *F* can be used and no letter can be used more than once?

35. **Arranging Letters** How many arrangements are there of the letters in the word MONEY?

36. **Arranging Digits** How many arrangements are there of the digits in the number 51,342?

37. **Establishing Committees** In how many ways can a committee of 4 students be formed from a pool of 7 students?

38. **Establishing Committees** In how many ways can a committee of 3 professors be formed from a department having 8 professors?

39. **Possible Answers on a True/False Test** How many arrangements of answers are possible for a true/false test with 10 questions?

40. **Possible Answers on a Multiple-choice Test** How many arrangements of answers are possible in a multiple-choice test with 5 questions, each of which has 4 possible answers?

41. How many four-digit numbers can be formed using the digits 0, 1, 2, 3, 4, 5, 6, 7, 8, and 9 if the first digit cannot be 0? Repeated digits are allowed.

42. How many five-digit numbers can be formed using the digits 0, 1, 2, 3, 4, 5, 6, 7, 8, and 9 if the first digit cannot be 0 or 1? Repeated digits are allowed.

43. **Arranging Books** Five different mathematics books are to be arranged on a student's desk. How many arrangements are possible?

44. **Forming License Plate Numbers** How many different license plate numbers can be made using 2 letters followed by 4 digits selected from the digits 0 through 9, if
(a) Letters and digits may be repeated?
(b) Letters may be repeated, but digits are not repeated?
(c) Neither letters nor digits may be repeated?

45. **Stock Portfolios** As a financial planner, you are asked to select one stock each from the following groups: 8 DOW stocks, 15 NASDAQ stocks, and 4 global stocks. How many different portfolios are possible?

46. **Combination Locks** A combination lock has 50 numbers on it. To open it, you turn to a number, then rotate clockwise to a second number, and then counterclockwise to the third number. How many different lock combinations are there?

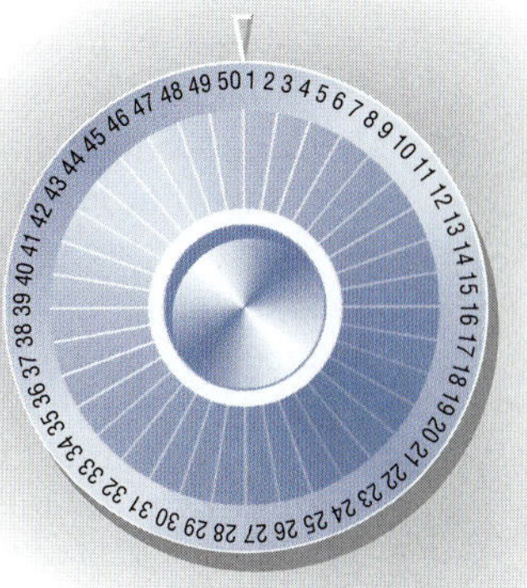

47. A student dance committee is to be formed consisting of 2 boys and 3 girls. If the membership is to be chosen from 4 boys and 8 girls, how many different committees are possible?

48. Baseball Teams A baseball team has 15 members. Four of the players are pitchers, and the remaining 11 members can play any position. How many different teams of 9 players can be formed?

49. The student relations committee of a college consists of 2 administrators, 3 faculty members, and 5 students. There are 4 administrators, 8 faculty members, and 20 students eligible to serve. How many different committees are possible?

50. Football Teams A defensive football squad consists of 25 players. Of these, 10 are linemen, 10 are linebackers, and 5 are safeties. How many different teams of 5 linemen, 3 linebackers, and 3 safeties can be formed?

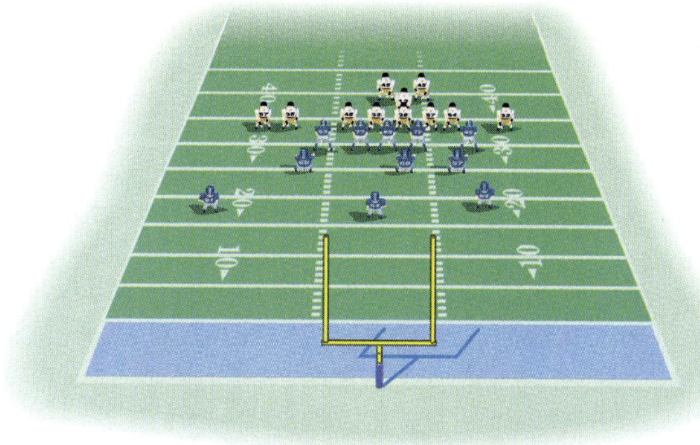

51. Baseball In the American Baseball League, a designated hitter may be used. How many batting orders is it possible for a manager to use? (There are 9 regular players on a team.)

52. Baseball In the National Baseball League, the pitcher usually bats ninth. If this is the case, how many batting orders are possible for a manager to use?

53. Forming Words How many different 9-letter words (real or imaginary) can be formed from the letters in the word ECONOMICS?

54. Forming Words How many different 11-letter words (real or imaginary) can be formed from the letters in the word MATHEMATICS?

55. Senate Committees The U.S. Senate has 100 members. Suppose it is desired to place each senator on exactly 1 of 7 possible committees. The first committee has 22 members, the second has 13, the third has 10, the fourth has 5, the fifth has 16, and the sixth and seventh have 17 apiece. In how many ways can these committees be formed?

56. World Series In the World Series the American League team (A) and the National League team (N) play until one team wins four games. If the sequence of winners is designated by letters (for example, $NAAAA$ means the National League team won the first game and the American League won the next four), how many different sequences are possible?

57. Basketball Teams A basketball team has 6 players who play guard (2 of 5 starting positions). How many different teams are possible, assuming that the remaining 3 positions are filled and it is not possible to distinguish a left guard from a right guard?

58. Basketball Teams On a basketball team of 12 players, 2 only play center, 3 only play guard, and the rest play forward (5 players on a team: 2 forwards, 2 guards, and 1 center). How many different teams are possible, assuming that it is not possible to distinguish left and right guards and left and right forwards?

59. Selecting Objects An urn contains 7 white balls and 3 red balls. Three balls are selected. In how many ways can the 3 balls be drawn from the total of 10 balls:

(a) If 2 balls are white and 1 is red?
(b) If all 3 balls are white?
(c) If all 3 balls are red?

60. Selecting Objects An urn contains 15 red balls and 10 white balls. Five balls are selected. In how many ways can the 5 balls be drawn from the total of 25 balls:

(a) If all balls are red?
(b) If 3 balls are red and 2 are white?
(c) If at least 4 are red balls?

61. Programming Exercise When both n and r are large, finding $C(n, r)$ on a computer may lead to integers too large to compute. To avoid this, we can approximate the values of $C(n, r)$. One way to do this is the following:

$$\begin{aligned} C(40, 20) &= \frac{40!}{20!20!} = \frac{40 \cdot 39 \cdot 38 \cdot \ldots \cdot 21}{20 \cdot 19 \cdot 18 \cdot \ldots \cdot 1} \\ &= \frac{40}{20} \cdot \frac{39}{19} \cdot \frac{38}{18} \cdot \ldots \cdot \frac{21}{1} \\ &\approx 2.000 \cdot 2.053 \cdot 2.111 \cdot \ldots \cdot 21.000 \\ &= 1.3784652 \times 10^{11} \end{aligned}$$

(a) Write a program that inputs two integers N and R and computes $C(N, R)$ using formula (1).
(b) Use the program to determine where overflow occurs on your computer.
(c) Write a program that inputs two integers N and R and computes $C(N, R)$ by the approximation technique shown above.

(d) Compare the answers found in parts (a) and (c).

62. Make up a problem different from any found in the text that requires the Multiplication Principle of counting to solve. Give it to a friend to solve and critique.

63. Make up a problem different from any found in the text that requires a permutation to solve. Give it to a friend to solve and critique.

64. Make up a problem different from any found in the text that requires a combination to solve. Give it to a friend to solve and critique.

65. Explain the difference between a permutation and a combination. Give an example to illustrate your explanation.

12.8 PROBABILITY

1 Construct Probability Models

2 Utilize the Additive Rule to Find Probabilities

3 Compute Probabilities of Equally Likely Outcomes

Probability is an area of mathematics that deals with experiments that yield random results yet admit a certain regularity. Such experiments do not always produce the same result or outcome, so the result of any one observation is not predictable. However, the results of the experiment over a long period do produce regular patterns that enable us to predict with remarkable accuracy.

EXAMPLE 1

Tossing a Fair Coin

In tossing a fair coin, we know that the outcome is either a head or a tail. On any particular throw, we cannot predict what will happen, but, if we toss the coin many times, we observe that the number of times a head comes up is approximately equal to the number of times we get a tail. It seems reasonable, therefore, to assign a probability of $\frac{1}{2}$ that a head comes up and a probability of $\frac{1}{2}$ that a tail comes up.

Probability Models

1 The discussion in Example 1 constitutes the construction of a **probability model** for the experiment of tossing a fair coin once. A probability model has two components: a sample space and an assignment of probabilities. A **sample space** S is a set whose elements represent all the possibilities that can occur as a result of the experiment. Each element of S is called an **outcome.** To each outcome, we assign a number, called the **probability** of that outcome, which has two properties:

1. Each probability is nonnegative.
2. The sum of all the probabilities equals 1.

Thus, if a probability model has the sample space

$$S = \{e_1, e_2, \ldots, e_n\}$$

where $e_1, e_2, \ldots, e_n$ are the possible outcomes, and if $P(e_1), P(e_2), \ldots, P(e_n)$ denote the respective probabilities of these outcomes, then

$$P(e_1) \geq 0, \quad P(e_2) \geq 0, \quad \ldots, \quad P(e_n) \geq 0 \qquad (1)$$
$$P(e_1) + P(e_2) + \cdots + P(e_n) = 1 \qquad (2)$$

Let's look at an example.

EXAMPLE 2 Constructing a Probability Model

An experiment consists of rolling a fair die once.* Construct a probability model for this experiment.

FIGURE 24

Solution A sample space S consists of all the possibilities that can occur. Because rolling the die will result in one of six faces showing, the sample space S consists of

$$S = \{1, 2, 3, 4, 5, 6\}$$

Because the die is fair, one face is no more likely to occur than another. As a result, our assignment of probabilities is

$$P(1) = \tfrac{1}{6} \qquad P(2) = \tfrac{1}{6}$$
$$P(3) = \tfrac{1}{6} \qquad P(4) = \tfrac{1}{6}$$
$$P(5) = \tfrac{1}{6} \qquad P(6) = \tfrac{1}{6}$$

Suppose that a die is loaded so that the probability assignments are

$$P(1) = 0, \quad P(2) = 0, \quad P(3) = \frac{1}{3}, \quad P(4) = \frac{2}{3}, \quad P(5) = 0, \quad P(6) = 0$$

This assignment would be made if the die was loaded so that only a 3 or 4 could occur and the 4 is twice as likely as the 3 to occur. This assignment is consistent with the definition since each assignment is between 0 and 1, and the sum of all the probability assignments equals 1.

Now work Problem 13.

EXAMPLE 3 Constructing a Probability Model

An experiment consists of tossing a coin. The coin is weighted so that heads (H) is three times as likely to occur as tails (T). Construct a probability model for this experiment.

Solution The sample space S is $S = \{\text{H}, \text{T}\}$. If x denotes the probability that a tail occurs, then

$$P(\text{T}) = x \quad \text{and} \quad P(\text{H}) = 3x$$

Since the sum of the probabilities of the possible outcomes must equal 1, we have

$$\begin{aligned} P(\text{T}) + P(\text{H}) = x + 3x &= 1 \\ 4x &= 1 \\ x &= \frac{1}{4} \end{aligned}$$

*A die is a cube with each face having either 1, 2, 3, 4, 5, or 6 dots on it. See Figure 24.

Thus, we assign the probabilities

$$P(\mathrm{T}) = \frac{1}{4} \qquad P(\mathrm{H}) = \frac{3}{4}$$

Now work Problem 17.

EXAMPLE 4

An experiment consists of tossing a fair die and then a fair coin. Construct a probability model for this experiment.

Solution A tree diagram is helpful in listing all the possible outcomes. See Figure 25. The sample space consists of the outcomes

FIGURE 25

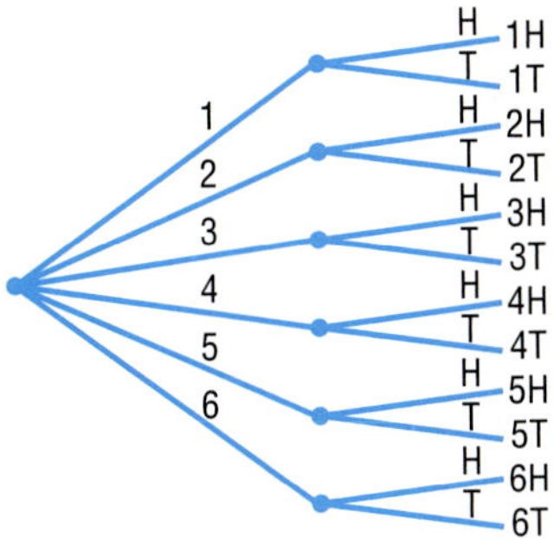

$$S = \{1\mathrm{H}, 1\mathrm{T}, 2\mathrm{H}, 2\mathrm{T}, 3\mathrm{H}, 3\mathrm{T}, 4\mathrm{H}, 4\mathrm{T}, 5\mathrm{H}, 5\mathrm{T}, 6\mathrm{H}, 6\mathrm{T}\}$$

The die and the coin are fair; thus, no one outcome is more likely to occur than another. As a result, we assign the probability $\frac{1}{12}$ to each of the 12 outcomes.

In working with probability models, the term **event** is used to describe a set of possible outcomes of the experiment. Thus, an event E is some subset of the sample space S. The **probability of an event** E, $E \neq 0$, denoted by $P(E)$, is defined as the sum of the probabilities of the outcomes in E. If $E = \varnothing$, then $P(E) = 0$; if $E = S$, then $P(E) = P(S) = 1$.

EXAMPLE 5

Finding the Probability of an Event

For the experiment described in Example 4, what is the probability that an even number followed by a head occurs?

Solution The event E, an even number followed by a head, consists of

$$E = \{2\mathrm{H}, 4\mathrm{H}, 6\mathrm{H}\}$$

The probability of E is

$$P(E) = P(2\mathrm{H}) + P(4\mathrm{H}) + P(6\mathrm{H}) = \tfrac{1}{12} + \tfrac{1}{12} + \tfrac{1}{12} = \tfrac{1}{4}$$

2 The next result, called the **additive rule,** may be used to find the probability of the union of two events.

Theorem Additive Rule

For any two events E and F,

$$P(E \cup F) = P(E) + P(F) - P(E \cap F) \tag{3}$$

If E and F are disjoint, so that $E \cap F = \varnothing$, then formula (3) takes the form

Mutually Exclusive Events

$$P(E \cup F) = P(E) + P(F) \qquad (4)$$

When formula (4) applies, we say that E and F are **mutually exclusive events.**

EXAMPLE 6

Using Formulas (3) and (4)

(a) If $P(E) = 0.2$, $P(F) = 0.3$, and $P(E \cap F) = 0.1$, find $P(E \cup F)$.
(b) If $P(E) = 0.2$, $P(F) = 0.3$, and E, F are mutually exclusive, find $P(E \cup F)$.

Solution (a) We use the additive rule, formula (3).

$$P(E \cup F) = P(E) + P(F) - P(\mathrm{E} \cap \mathrm{F}) = 0.2 + 0.3 - 0.1 = 0.4$$

(b) Since E, F are mutually exclusive, we use formula (4).

$$P(E \cup F) = P(E) + P(F) = 0.2 + 0.3 = 0.5$$

A Venn diagram can sometimes be used to obtain probabilities. To construct a Venn diagram representing the information in Example 6(a), we draw two sets E and F. We begin with the fact that $P(E \cap F) = 0.1$. See Figure 26(a). Then, since $P(E) = 0.2$ and $P(F) = 0.3$, we fill in E with $0.2 - 0.1 = 0.1$ and F with $0.3 - 0.1 = 0.2$. See Figure 26(b). Since $P(S) = 1$, we complete the diagram by inserting $1 - [0.1 + 0.1 + 0.2] = 0.6$. See Figure 26(c). Now it is easy to see, for example, that the probability of F, but not E, is 0.2. Also, the probability of neither E nor F is 0.6.

FIGURE 26

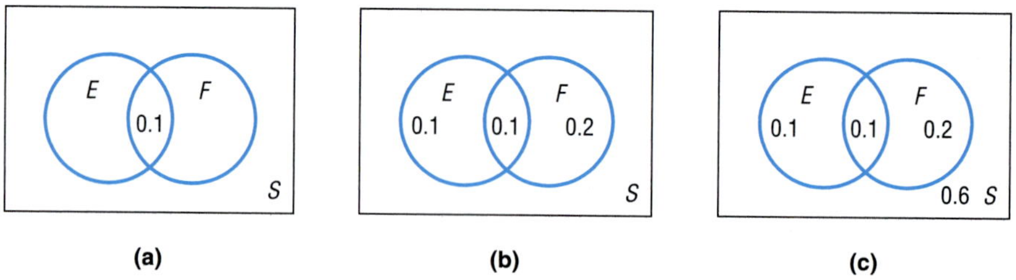

Now work Problem 23.

Equally Likely Outcomes

3 When the same probability is assigned to each outcome of the sample space, the experiment is said to have **equally likely outcomes.**

Theorem Probability for Equally Likely Outcomes

If an experiment has n equally likely outcomes, and if the number of ways an event E can occur is m, then the probability of E is

$$P(E) = \frac{\text{Number of ways that } E \text{ can occur}}{\text{Number of all logical possibilities}} = \frac{m}{n} \qquad (5)$$

Thus, if S is the sample space of this experiment, then

$$P(E) = \frac{n(E)}{n(S)} \tag{6}$$

Based on (6), an alternative method of solution of Example 5 is

$$P(E) = \frac{n(E)}{n(S)} = \frac{3}{12} = \frac{1}{4}$$

EXAMPLE 7

Computing Probabilities for Equally Likely Outcomes

A jar contains 10 marbles; 5 are solid color, 4 are speckled, and 1 is clear.

(a) If one marble is picked at random, what is the probability it is speckled?
(b) If one marble is picked at random, what is the probability it is clear or a solid color?

Solution The experiment is an example of one in which the outcomes are equally likely; that is, no one marble is more likely to be picked than another. If S is the sample space, then there are 10 possible outcomes in S, so $n(S) = 10$.

(a) Define the event E: Speckled marble is picked. There are 4 ways E can occur. Thus,

$$P(E) = \frac{n(E)}{n(S)} = \frac{4}{10} = 0.4$$

(b) Define the events F: Clear marble is picked and G: Solid color marble is picked. Then there is 1 way for F to occur and 5 ways for G to occur. Thus,

$$P(F) = \frac{n(F)}{n(S)} = \frac{1}{10}, \qquad P(G) = \frac{n(G)}{n(S)} = \frac{5}{10}$$

We seek the probability of the event F or G, that is, $P(F \cup G)$. Since F, G are mutually exclusive, we use (4).

$$P(F \cup G) = P(F) + P(G) = \frac{1}{10} + \frac{5}{10} = \frac{6}{10} = 0.6$$

Now work Problem 27.

EXAMPLE 8

The Game of Craps

In the game of "craps," two fair dice are rolled. If the total of the faces equals 7 or 11, you win. If the totals are 2, 3, or 12, you have thrown craps and you lose. In all other cases, you throw again.

(a) What is the probability that you will win?
(b) What is the probability that you will lose?
(c) What is the probability that you will need to throw again?

FIGURE 27

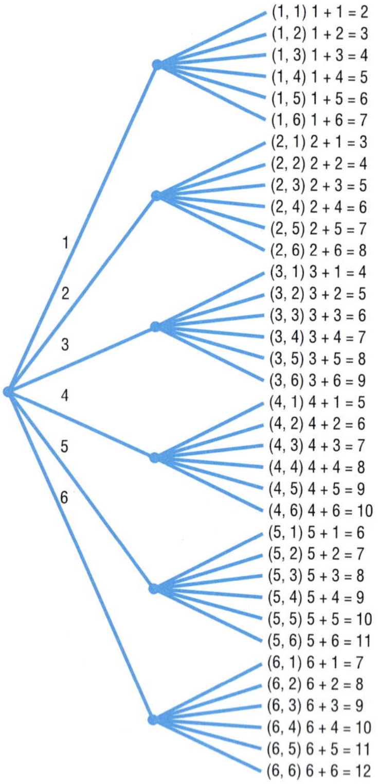

Solution We begin by constructing a probability model for the experiment. The tree diagram in Figure 27 will help us to see all the possibilities.

Because the dice are fair, no one of the 36 possible outcomes in the sample space S is more likely to occur than any other. Thus, we have equally likely outcomes with $n(S) = 36$.

(a) The event E, "the dice total 7 or 11," consists of the outcomes

$$E = \{(1, 6), (2, 5), (3, 4), (4, 3), (5, 2), (6, 1), (5, 6), (6, 5)\}$$

Because $n(E) = 8$, we have

$$P(E) = \frac{n(E)}{n(S)} = \frac{8}{36} = \frac{2}{9} \approx 0.222$$

(b) The event F, "the dice total 2, 3, or 12," consists of the outcomes

$$F = \{(1, 1), (1, 2), (2, 1), (6, 6)\}$$

Because $n(F) = 4$,

$$P(F) = \frac{n(F)}{n(S)} = \frac{4}{36} = \frac{1}{9} \approx 0.111$$

(c) The number of possibilities that require that you throw again is $36 - n(E) - n(F) = 36 - 8 - 4 = 24$. Thus, the probability that another throw is required is

$$\frac{24}{36} = \frac{2}{3} \approx 0.667$$

Now work Problem 29.

Applications Involving Permutations and Combinations

EXAMPLE 9 Computing Probabilities

Because of a mistake in packaging, 5 defective phones were packaged with 15 good ones. All phones look alike and have equal probability of being chosen. Three phones are selected.

(a) What is the probability that all 3 are defective?

(b) What is the probability that exactly 2 are defective?

(c) What is the probability that at least 2 are defective?

Solution The sample space S consists of the number of ways 3 objects can be selected from 20 objects, that is, the number of combinations of 20 things taken 3 at a time.

$$n(S) = C(20, 3) = \frac{20!}{17! \cdot 3!} = \frac{20 \cdot 19 \cdot 18}{6} = 1140$$

Each of these outcomes is equally likely to occur.

(a) If E is the event "3 are defective," then the number of elements in E is the number of ways the 3 defective phones can be chosen from the 5 defective phones: $C(5, 3) = 10$. Thus, the probability of E is

$$P(E) = \frac{n(E)}{n(S)} = \frac{10}{1140} \approx 0.0088$$

(b) If F is the event "exactly 2 are defective" and 3 phones are selected, then the number of elements in F is the number of ways to select 2 defective phones from the 5 defective phones and 1 good phone from the 15 good ones. The first of these can be done in $C(5, 2)$ ways and the second in $C(15, 1)$ ways. By the multiplication principle, the event F can occur in

$$C(5, 2) \cdot C(15, 1) = \frac{5!}{3! \cdot 2!} \cdot \frac{15!}{14! \cdot 1!} = 10 \cdot 15 = 150 \text{ ways}$$

The probability of F is therefore

$$P(F) = \frac{n(F)}{n(S)} = \frac{150}{1140} \approx 0.1316$$

(c) The event G, "at least two are defective," when 3 are chosen, is equivalent to requiring that either exactly 2 defective are chosen or exactly 3 defective are chosen. That is, $G = E \cup F$. Since E and F are mutually exclusive (it is not possible to select 2 defective phones and, at the same time, select 3 defective phones), we find

$$P(G) = P(E) + P(F) \approx 0.0088 + 0.1316 = 0.1404$$

Now work Problem 43.

EXAMPLE 10 Tossing a Coin

A fair coin is tossed 6 times.

(a) What is the probability of obtaining exactly 5 heads and one tail?

(b) What is the probability of obtaining between 4 and 6 heads, inclusive?

Solution The number of elements in the sample space S is found using the Multiplication Principle. Each toss results in a head (H) or a tail (T). Since the coin is tossed 6 times, we have

$$n(S) = \underbrace{2 \cdot 2 \cdot \ldots \cdot 2}_{\text{6 tosses}} = 2^6 = 64$$

The outcomes are equally likely since the coin is fair.

(a) Any sequence that contains 5 heads and 1 tail is determined once the position of the 5 heads (or 1 tail) is known. The number of ways we can position 5 heads in a sequence of 6 slots is $C(6, 5) = 6$. The probability of the event E—exactly 5 heads and one tail—is

$$P(E) = \frac{n(E)}{n(S)} = \frac{C(6, 5)}{2^6} = \frac{6}{64} \approx 0.0938$$

(b) Let F be the event: between 4 and 6 heads, inclusive. To obtain between 4 and 6 heads is equivalent to the event: either 4 heads or 5 heads or 6 heads. Since each of these is mutually exclusive (it is impossible to obtain both 4 heads and 5 heads when tossing a coin 6 times), we have

$$\begin{aligned} P(F) &= P(\text{4 heads or 5 heads or 6 heads}) \\ &= P(\text{4 heads}) + P(\text{5 heads}) + P(\text{6 heads}) \end{aligned}$$

The probabilities on the right are obtained as in part (a). Thus

$$P(F) = \frac{C(6, 4)}{2^6} + \frac{C(6, 5)}{2^6} + \frac{C(6, 6)}{2^6} = \frac{15}{64} + \frac{6}{64} + \frac{1}{64} = \frac{22}{64} \approx 0.3438$$

HISTORICAL FEATURE

Set theory, counting, and probability first took form as a systematic theory in the exchange of letters (1654) between Pierre de Fermat (1601–1665) and Blaise Pascal (1623–1662). They discussed the problem of how to divide the stakes in a game that is interrupted before completion, knowing how many points each player needs to win. Fermat solved the problem by listing all possibilities and counting the favorable ones, whereas Pascal made use of the triangle that now bears his name. As mentioned in the text, the entries in Pascal's triangle are equivalent to $C(n, r)$. This recognition of the role of $C(n, r)$ in counting is the foundation of all further developments.

The first book on probability, the work of Christiaan Huygens (1629–1695), appeared in 1657. In it, the notion of mathematical expectations is explored. This allows the calculation of the profit or loss a gambler may expect, knowing the probabilities involved in the game (see the Historical Problems that follow).

It is interesting to note that Girolamo Cardano (1501–1576) wrote a treatise on probability, but it was not published until 1663 in Cardano's collected works, and this was too late to have any effect on the development of the theory.

In 1713, the posthumously published *Ars Conjectandi* of Jacob Bernoulli gave the theory the form it would have until 1900. In the current century, both combinatorics (counting) and probability have undergone rapid development due to the use of computers.

A final comment about notation. The notations $C(n, r)$ and $P(n, r)$ are variants of a form of notation developed in England after 1830. The notation $\binom{n}{r}$ for $C(n, r)$ goes back to Leonhard Euler (1707–1783) but is now losing ground because it has no clearly related symbolism of the same type for permutations. The set symbols $\cup$ and $\cap$ were introduced by Giuseppe Peano (1858–1932) in 1888 in a slightly different context. The inclusion symbol $\subset$ was introduced by E. Schroeder (1841–1902) about 1890. The treatment of set theory in the text is due to George Boole (1815–1864), who wrote $A + B$ for $A \cup B$ and AB for $A \cap B$ (statisticians still use AB for $A \cap B$).

HISTORICAL PROBLEMS

1. *The Problem Discussed by Fermat and Pascal* A game between two equally skilled players, A and B, is interrupted when A needs 2 points to win and B needs 3 points. In what proportion should the stakes be divided?

   [**Note:** If each play results in 1 point for either player, at most four more plays will decide the game.]

   (a) *Fermat's solution* List all possible outcomes that will end the game to form the sample space (for example, *ABAA*, *ABBB*, etc.). The probabilities for A to win and B to win then determine how the stakes should be divided.

   (b) *Pascal's solution* Use combinations to determine the number of ways the 2 points needed for A to win could occur in four plays. Then use combinations to determine the number of ways the 3 points needed for B to win could occur. This is trickier than it looks, since A can win with 2 points in either two plays, three plays, or four plays. Compute the probabilities and compare with the results in part (a).

2. *Huygen's mathematical expectation* In a game with n possible outcomes with probabilities $p_1, p_2, \dots, p_n$, suppose that the *net* winnings are $w_1, w_2, \dots, w_n$, respectively. Then the mathematical expectation is

$$E = p_1w_1 + p_2w_2 + \cdots + p_nw_n$$

The number E represents the profit or loss per game in the long run. The following problems are a modification of those of Huygens:

(a) A fair die is tossed. A gambler wins \$3 if he throws a 6 and \$6 if he throws a 5. What is his expectation?

[**Note:** $w_1 = w_2 = w_3 = w_4 = 0$]

(b) A gambler plays the same game as in part (a), but now the gambler must pay \$1 to play. This means $w_5 = \$5$, $w_6 = \$2$, and $w_1 = w_2 = w_3 = w_4 = -\$1$. What is the expectation?

12.8 EXERCISES

In Problems 1–6, construct a probability model for each experiment.

1. Tossing a fair coin twice

2. Tossing two fair coins once

3. Tossing two fair coins, then a fair die

4. Tossing a fair coin, a fair die, and then a fair coin

5. Tossing three fair coins once

6. Tossing one fair coin three times

In Problems 7–12, use the spinners shown below, and construct a probability model for each experiment.

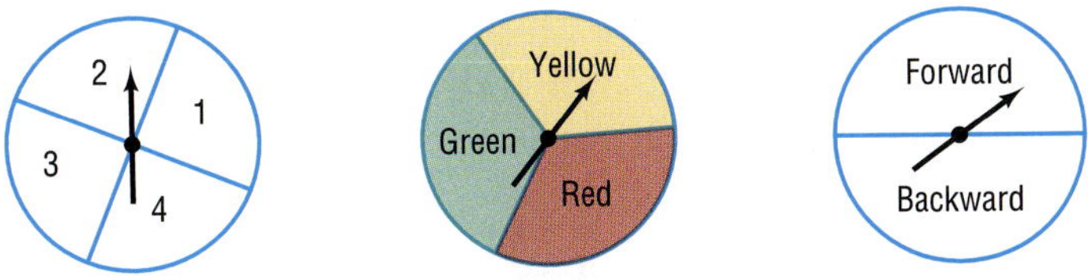

Spinner I   Spinner II   Spinner III

7. Spin spinner I, then spinner II. What is the probability of getting a 2 or a 4, followed by Red?

8. Spin spinner III, then spinner II. What is the probability of getting Forward, followed by Yellow or Green?

9. Spin spinner I, then II, then III. What is the probability of getting a 1, followed by Red or Green, followed by Backward?

10. Spin spinner II, then I, then III. What is the probability of getting Yellow, followed by a 2 or a 4, followed by Forward?

11. Spin spinner I twice, then spinner II. What is the probability of getting a 2, followed by a 2 or a 4, followed by Red or Green?

12. Spin spinner III, then spinner I twice. What is the probability of getting Forward, followed by a 1 or a 3, followed by a 2 or a 4?

In Problems 13–16, consider the experiment of tossing a coin twice. The table lists six possible assignments of probabilities for this experiment. Using this table, answer the following questions.

Assignments	**Sample Space**			
	HH	**HT**	**TH**	**TT**
A	$\frac{1}{4}$	$\frac{1}{4}$	$\frac{1}{4}$	$\frac{1}{4}$
B	0	0	0	1
C	$\frac{3}{16}$	$\frac{5}{16}$	$\frac{5}{16}$	$\frac{3}{16}$
D	$\frac{1}{2}$	$\frac{1}{2}$	$-\frac{1}{2}$	$\frac{1}{2}$
E	$\frac{1}{8}$	$\frac{1}{4}$	$\frac{1}{4}$	$\frac{1}{8}$
F	$\frac{1}{9}$	$\frac{2}{9}$	$\frac{2}{9}$	$\frac{4}{9}$

13. Which of the assignments of probabilities are consistent with the definition of the probability of an outcome?

14. Which of the assignments of probabilities should be used if the coin is known to be fair?

15. Which of the assignments of probabilities should be used if the coin is known to always come up tails?

16. Which of the assignments of probabilities should be used if tails is twice as likely as heads to occur?

17. **Assigning Probabilities** A coin is weighted so that heads is four times as likely as tails to occur. What probability should we assign to heads? to tails?

18. **Assigning Probabilities** A coin is weighted so that tails is twice as likely as heads to occur. What probability should we assign to heads? to tails?

19. **Assigning Probabilities** A die is weighted so that an odd-numbered face is twice as likely as an even-numbered face. What probability should we assign to each face?

20. **Assigning Probabilities** A die is weighted so that a six cannot appear. The other faces occur with the same probability. What probability should we assign to each face?

In Problems 21–24, find the probability of the indicated event if $P(A) = 0.30$ and $P(B) = 0.40$.

21. $P(A \cup B)$ if A, B are mutually exclusive

22. $P(A \cap B)$ if A, B are mutually exclusive

23. $P(A \cup B)$ if $P(A \cap B) = 0.15$

24. $P(A \cap B)$ if $P(A \cup B) = 0.6$

Problems 25–28, a golf ball is selected at random from a container. If the container has 9 white balls, 8 green balls, and 3 orange ones, find the probability of each event.

25. The golf ball is white.

26. The golf ball is green.

27. The golf ball is white or green.

28. The golf ball is not white.

29. What is the probability of throwing a 6 or an 8 in a game of craps? (Consult Example 8.)

30. What is the probability of throwing a 5 or a 9 in a game of craps? (Consult Example 8.)

In Problems 31–34 are based on a consumer survey of annual incomes in 100 households. The following table gives the data:

Income	\$0–9999	\$10,000–19,999	\$20,000–29,999	\$30,000—39,999	\$40,000 or more
Number of households	5	35	30	20	10

31. What is the probability that a household has an annual income of \$30,000 or more?

32. What is the probability that a household has an annual income between \$10,000 and \$29,999, inclusive?

33. What is the probability that a household has an annual income less than \$20,000?

34. What is the probability that a household has an annual income of \$20,000 or more?

35. **Surveys** In a survey about the number of TV sets in a house, the following probability table was constructed:

Number of TV sets	0	1	2	3	4 or more
Probability	0.05	0.24	0.33	0.21	0.17

Find the probability of a house having:
(a) 1 or 2 TV sets
(b) 1 or more TV sets
(c) 3 or fewer TV sets
(d) 3 or more TV sets
(e) Less than 2 TV sets
(f) Less than 1 TV set
(g) 1, 2, or 3 TV sets
(h) 2 or more TV sets

36. **Checkout Lines** Through observation it has been determined that the probability for a given number of people waiting in line at the "5 items or less" checkout register of a supermarket is:

Number waiting in line	0	1	2	3	4 or more
Probability	0.10	0.15	0.20	0.24	0.31

Find the probability of
(a) At most 2 people in line
(b) At least 2 people in line
(c) At least 1 person in line

37. **Winning a Lottery** In a certain lottery, there are ten balls, numbered 1, 2, 3, 4, 5, 6, 7, 8, 9, 10. Of these, five are drawn in order. If you pick five numbers that match those drawn in the

correct order, you win \$1,000,000. What is the probability of winning such a lottery?

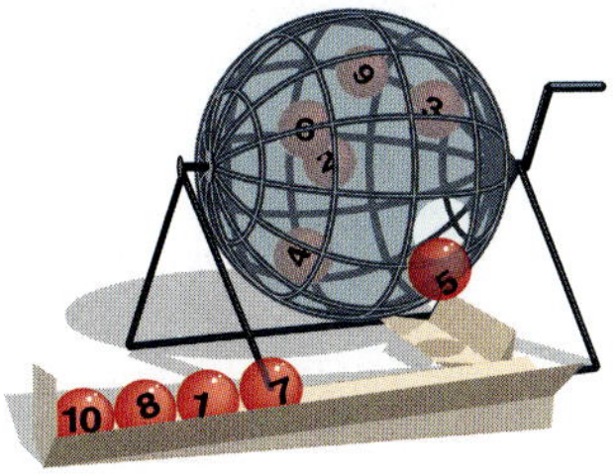

38. A committee of 6 people is to be chosen at random from a group of 14 people consisting of 2 supervisors, 5 skilled laborers, and 7 unskilled laborers. What is the probability that the committee chosen consists of 2 skilled and 4 unskilled laborers?

39. A fair coin is tossed 5 times.
(a) Find the probability that exactly 3 heads appear.
(b) Find the probability that no heads appear.

40. A fair coin is tossed 4 times.
(a) Find the probability that exactly 1 tail appears.
(b) Find the probability that no more than 1 tail appears.

41. A pair of fair dice are tossed 3 times.
(a) Find the probability that the sum of seven appears 3 times.
(b) Find the probability that a sum of 7 or 11 appears at least twice.

42. A pair of fair dice are tossed 5 times.
(a) Find the probability that the sum is never 2.
(b) Find the probability that the sum is never 7.

43. Through a mix-up on the production line, 5 defective TV's were shipped out with 25 good ones. If 5 are selected at random, what is the probability that all 5 are defective? What is the probability that at least 2 of them are defective?

44. In a shipment of 50 transformers, 10 are known to be defective. If 30 transformers are picked at random, what is the probability that all 30 are nondefective? Assume that all transformers look alike and have an equal probability of being chosen.

45. In a promotion, 50 silver dollars are placed in a bag, one of which is valued at more than \$10,000. The winner of the promotion is given the opportunity to reach into the bag, while blindfolded, and pull out 5 coins. What is the probability that one of the 5 coins is the one valued at more than \$10,000?

46. Go to the library and look up the "birthday problem" in a book on probability. Write a brief essay about this problem and its solution.

CHAPTER REVIEW

THINGS TO KNOW

Sequence	A function whose domain is the set of positive integers.
Factorials	$0! = 1, 1! = 1, n! = n(n-1) \cdot \ldots \cdot 3 \cdot 2 \cdot 1$ if $n \geq 2$
Arithmetic sequence	$a_1 = a$, $a_{n+1} = a_n + d$, where a = first term, d = common difference, $a_n = a + (n-1)d$
Sum of the first n terms of an arithmetic sequence	$S_n = \frac{n}{2}[2a + (n-1)d] = \frac{n}{2}(a + a_n)$
Geometric sequence	$a_1 = a$, $a_{n+1} = ra_n$; where a = first term, r = common ratio, $a_n = ar^{n-1}$, $r \neq 0$
Sum of the first n terms of a geometric sequence	$S_n = a\frac{1-r^n}{1-r}$, $r \neq 0, 1$
Amount of an annuity	$A = P\frac{(1+i)^n - 1}{i}$
Infinite geometric series	$a + ar + \cdots + ar^{n-1} + \cdots = \sum_{k=1}^{\infty} ar^{k-1}$
Sum of an infinite geometric series	$\sum_{k=1}^{\infty} ar^{k-1} = \frac{a}{1-r}$, $\lvert r \rvert < 1$

Principle of Mathematical Induction	Condition I: The statement is true for the natural number 1. Condition II: If the statement is true for some natural number k, it is also true for $k + 1$ Then the statement is true for all natural numbers.	
Binomial coefficient	$\binom{n}{j} = \dfrac{n!}{j!(n-j)!}$	
Pascal triangle	See Figure 16.	
Binomial Theorem	$(x+a)^n = \binom{n}{0}x^n + \binom{n}{1}ax^{n-1} + \cdots + \binom{n}{j}a^j x^{n-j} + \cdots + \binom{n}{n}a^n$	
Set		Well-defined collection of distinct objects, called elements
Null set	$\varnothing$	Set that has no elements
Equality	$A = B$	A and B have the same elements
Subset	$A \subseteq B$	Each element of A is also an element of B.
Intersection	$A \cap B$	Set consisting of elements that belong to both A and B
Union	$A \cup B$	Set consisting of elements that belong to either A or B, or both
Universal set	U	Set consisting of all the elements we wish to consider
Complement	A'	Set consisting of elements of the universal set that are not in A
Finite set		The number of elements in the set is a nonnegative integer
Infinite set		A set that is not finite
Counting formula	$n(A \cup B) = n(A) + n(B) - n(A \cap B)$	
Addition Principle		If $A \cap B = \varnothing$, then $n(A \cup B) = n(A) + n(B)$.
Multiplication Principle		If a task consists of a sequence of choices in which there are p selections for the first choice, q selections for the second choice, and so on, then the task of making these selections can be done in $p \cdot q \cdot \ldots$ different ways.
Permutation	$P(n, r) = n(n-1) \cdot \ldots \cdot [n - (r-1)]$ $= \dfrac{n!}{(n-r)!}$	An ordered arrangement of n distinct objects without repetition
Combination	$C(n, r) = \dfrac{P(n, r)}{r!}$ $= \dfrac{n!}{(n-r)!r!}$	An arrangement, without regard to order, of n distinct objects without repetition
Permutations with repetition	$\dfrac{n!}{n_1!n_2! \cdots n_k!}$	The number of permutations of n objects of which n_1 are of one kind, n_2 are of a second kind, . . . , and n_k are of a kth kind, where $n = n_1 + n_2 + \cdots + n_k$
Sample space		Set whose elements represent all the logical possibilities that can occur as a result of an experiment

Probability		A number assigned to each outcome of a sample space; the sum of all the probabilities of the outcomes equals 1
Additive rule	$P(E \cup F) = P(E) + P(F) - P(E \cap F)$	
Equally likely outcomes	$P(E) = \dfrac{n(E)}{n(S)}$	The same probability is assigned to each outcome.

HOW TO

Write down the terms of a sequence
Use summation notation
Identify an arithmetic sequence
Find the sum of the first n terms of an arithmetic sequence
Identify a geometric sequence
Solve annuity problems
Apply the binomial theorem
Find unions, intersections, and complements of sets
Use Venn diagrams to illustrate sets
Recognize a permutation problem
Find the sum of arithmetic and geometric sequences using a graphing utility
Find the sum of the first n terms of a geometric sequence
Find the sum of an infinite geometric series
Prove statements about natural numbers using mathematical induction
Recognize a combination problem
Solve certain probability problems
Count the elements in a sample space
Draw a tree diagram

FILL-IN-THE-BLANK ITEMS

1. A(n) ______________ is a function whose domain is the set of positive integers.

2. In a(n) ______________ sequence, the difference between successive terms is always the same number.

3. In a(n) ______________ sequence, the ratio of successive terms is always the same number.

4. The ______________ ______________ is a triangular display of the binomial coefficients.

5. $\binom{6}{2} =$ ______________

6. The ______________ of A with B consists of all elements in either A or B or both; the ______________ of A with B consists of all elements in both A and B.

7. $P(5, 2) =$ ______________; $C(5, 2) =$ ______________.

8. A(n) ______________ is an ordered arrangement of n distinct objects.

9. A(n) ______________ is an arrangement of n distinct objects without regard to order.

10. When the same probability is assigned to each outcome of a sample space, the experiment is said to have ______________ ______________ outcomes.

TRUE/FALSE ITEMS

T F **1.** A sequence is a function.

T F **2.** For arithmetic sequences, the difference of successive terms is always the same number.

T F **3.** For geometric sequences, the ratio of successive terms is always the same number.

T F **4.** Mathematical induction can sometimes be used to prove theorems that involve natural numbers.

T F **5.** $\binom{n}{j} = \dfrac{j!}{n!(n-j)!}$

T F **6.** The expansion of $(x + a)^n$ contains n terms.

T F **7.** $\sum_{i=1}^{n+1} i = 1 + 2 + 3 + \cdots + n$

T F **8.** The intersection of two sets is always a subset of their union.

T F **9.** $P(n, r) = \dfrac{n!}{r!}$

T F **10.** In a combination problem, order is not important.

T F **11.** In a permutation problem, once an object is used, it cannot be repeated.

T F **12.** The probability of an event can never equal 0.

REVIEW EXERCISES

In Problems 1–8, evaluate each expression by hand. Verify your results using a graphing utility.

1. $5!$ **2.** $6!$ **3.** $\binom{5}{2}$ **4.** $\binom{8}{6}$ **5.** $P(8, 3)$ **6.** $P(7, 3)$ **7.** $C(8, 3)$ **8.** $C(7, 3)$

In Problems 9–16, write down the first five terms of each sequence.

9. $\left\{(-1)^n\left(\dfrac{n+3}{n+2}\right)\right\}$ **10.** $\{(-1)^{n+1}(2n+3)\}$ **11.** $\left\{\dfrac{2^n}{n^2}\right\}$

12. $\left\{\dfrac{e^n}{n}\right\}$ **13.** $a_1 = 3;\ a_{n+1} = \frac{2}{3}a_n$ **14.** $a_1 = 4;\ a_{n+1} = -\frac{1}{4}a_n$

15. $a_1 = 2;\ a_{n+1} = 2 - a_n$ **16.** $a_1 = -3;\ a_{n+1} = 4 + a_n$

In Problems 17–28, determine whether the given sequence is arithmetic, geometric, or neither. If the sequence is arithmetic, find the common difference and the sum of the first n terms. If the sequence is geometric, find the common ratio and the sum of the first n terms.

17. $\{n + 5\}$ **18.** $\{4n + 3\}$ **19.** $\{2n^3\}$ **20.** $\{2n^2 - 1\}$

21. $\{2^{3n}\}$ **22.** $\{3^{2n}\}$ **23.** $0, 4, 8, 12, \ldots$ **24.** $1, -3, -7, -11, \ldots$

25. $3, \frac{3}{2}, \frac{3}{4}, \frac{3}{8}, \frac{3}{16}, \ldots$ **26.** $5, -\frac{5}{3}, \frac{5}{9}, -\frac{5}{27}, \frac{5}{81}, \ldots$ **27.** $\frac{2}{3}, \frac{3}{4}, \frac{4}{5}, \frac{5}{6}, \ldots$ **28.** $\frac{3}{2}, \frac{5}{4}, \frac{7}{6}, \frac{9}{8}, \frac{11}{10}, \ldots$

In Problems 29–34, evaluate each sum. Verify your results using a graphing utility.

29. $\sum_{k=1}^{5} (k^2 + 12)$ **30.** $\sum_{k=1}^{3} (k + 2)^2$ **31.** $\sum_{k=1}^{10} (3k - 9)$ **32.** $\sum_{k=1}^{9} (-2k + 8)$

33. $\sum_{k=1}^{7} \left(\frac{1}{3}\right)^k$ **34.** $\sum_{k=1}^{10} (-2)^k$

In Problems 35–40, find the indicated term in each sequence: (a) by hand; (b) using a graphing utility.

35. 9th term of $3, 7, 11, 15, \ldots$ **36.** 8th term of $1, -1, -3, -5, \ldots$

37. 11th term of $1, \frac{1}{10}, \frac{1}{100}, \ldots$ **38.** 11th term of $1, 2, 4, 8, \ldots$

39. 9th term of $\sqrt{2}, 2\sqrt{2}, 3\sqrt{2}, \ldots$ **40.** 9th term of $\sqrt{2}, 2, 2^{3/2}, \ldots$

In Problems 41–44, find a general formula for each arithmetic sequence.

41. 7th term is 31; 20th term is 96 **42.** 8th term is -20; 17th term is -47

43. 10th term is 0; 18th term is 8 **44.** 12th term is 30; 22nd term is 50

In Problems 45–50, find the sum of each infinite geometric series.

45. $3 + 1 + \frac{1}{3} + \frac{1}{9} + \cdots$ **46.** $2 + 1 + \frac{1}{2} + \frac{1}{4} + \cdots$ **47.** $2 - 1 + \frac{1}{2} - \frac{1}{4} + \cdots$

48. $6 - 4 + \frac{8}{3} - \frac{16}{9} + \cdots$ **49.** $\sum_{k=1}^{\infty} 4\left(\frac{1}{2}\right)^{k-1}$ **50.** $\sum_{k=1}^{\infty} 3\left(-\frac{3}{4}\right)^{k-1}$

In Problems 51–56, use the Principle of Mathematical Induction to show that the given statement is true for all natural numbers.

51. $3 + 6 + 9 + \cdots + 3n = \frac{3n}{2}(n + 1)$

52. $2 + 6 + 10 + \cdots + (4n - 2) = 2n^2$

53. $2 + 6 + 18 + \cdots + 2 \cdot 3^{n-1} = 3^n - 1$

54. $3 + 6 + 12 + \cdots + 3 \cdot 2^{n-1} = 3(2^n - 1)$

55. $1^2 + 4^2 + 7^2 + \cdots + (3n - 2)^2 = \frac{1}{2}n(6n^2 - 3n - 1)$

56. $1 \cdot 3 + 2 \cdot 4 + 3 \cdot 5 + \cdots + n(n + 2) = \frac{n}{6}(n + 1)(2n + 7)$

In Problems 57–60, expand each expression using the Binomial Theorem.

57. $(x + 2)^5$

58. $(x - 3)^4$

59. $(2x + 3)^5$

60. $(3x - 4)^4$

61. Find the coefficient of x^7 in the expansion of $(x + 2)^9$.

62. Find the coefficient of x^3 in the expansion of $(x - 3)^8$.

63. Find the coefficient of x^2 in the expansion of $(2x + 1)^7$.

64. Find the coefficient of x^6 in the expansion of $(2x + 1)^8$.

In Problems 65–72, use U = Universal set = $\{1, 2, 3, 4, 5, 6, 7, 8, 9\}$, $A = \{1, 3, 5, 7\}$, $B = \{3, 5, 6, 7, 8\}$, and $C = \{2, 3, 7, 8, 9\}$ to find each set.

65. $A \cup B$

66. $B \cup C$

67. $A \cap C$

68. $A \cap B$

69. $A' \cup B'$

70. $B' \cap C'$

71. $(B \cap C)'$

72. $(A \cup B)'$

73. If $n(A) = 8$, $n(B) = 12$ and $n(A \cap B) = 3$, find $n(A \cup B)$.

74. If $n(A) = 12$, $n(A \cup B) = 30$, and $n(A \cap B) = 6$, find $n(B)$.

In Problems 75–80, use the information supplied in the figure:

75. How many are in A?

76. How many are in A or B?

77. How many are in A and C?

78. How many are not in B?

79. How many are in neither A nor C?

80. How many are in B but not in C?

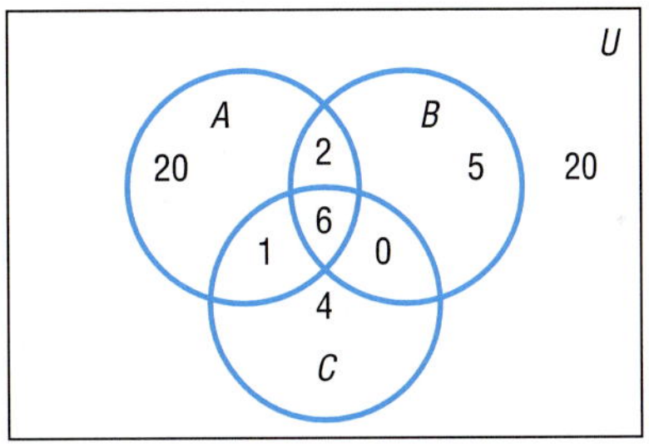

81. A clothing store sells pure wool and polyester/wool suits. Each suit comes in 3 colors and 10 sizes. How many suits are required for a complete assortment?

82. In connecting a certain electrical device, 5 wires are to be connected to 5 different terminals. How many different wirings are possible if 1 wire is connected to each terminal?

83. Baseball On a given day, the American Baseball League schedules 7 games. How many different outcomes are possible, assuming that each game is played to completion?

84. Baseball On a given day, the National Baseball League schedules 6 games. How many different outcomes are possible, assuming that each game is played to completion?

85. If 4 people enter a bus having 9 vacant seats, in how many ways can they be seated?

86. How many different arrangements are there of the letters in the word ROSE?

87. In how many ways can a squad of 4 relay runners be chosen from a track team of 8 runners?

88. A professor has 10 similar problems to put on a test with 3 problems. How many different tests can she design?

89. Baseball In how many different ways can the 14 baseball teams in the American League be paired without regard to which team is at home?

90. Arranging Books on a Shelf There are 5 different French books and 5 different Spanish books. How many ways are there to arrange them on a shelf if

(a) Books of the same language must be grouped together, French on the left, Spanish on the right?

(b) French and Spanish books must alternate in the grouping, beginning with a French book?

91. Telephone Numbers Using the digits 0, 1, 2, ..., 9, how many 7-digit numbers can be formed if the first digit cannot be 0 or 9 and if the last digit is greater than or equal to 2 and less than or equal to 3? Repeated digits are allowed.

92. Home Choices A contractor constructs homes with 5 different choices of exterior finish, 3 different roof arrangements, and 4 different window designs. How many different types of homes can be built?

93. License Plate Possibilities A license plate consists of 1 letter, excluding O and I, followed by a 4-digit number that cannot have a 0 in the lead position. How many different plates are possible?

94. Using the digits 0 and 1, how many different numbers consisting of 8 digits can be formed?

95. Forming Different Words How many different words can be formed using all the letters in the word MISSING?

96. Arranging Flags How many different vertical arrangements are there of 10 flags, if 4 are white, 3 are blue, 2 are green, and 1 is red?

97. Forming Committees A group of 9 people is going to be formed into committees of 4, 3, and 2 people. How many committees can be formed if

(a) A person can serve on any number of committees?

(b) No person can serve on more than one committee?

98. Forming Committees A group consists of 5 men and 8 women. A committee of 4 is to be formed from this group, and policy dictates that at least 1 woman be on this committee.

(a) How many committees can be formed that contain exactly 1 man?

(b) How many committees can be formed that contain exactly 2 women?

(c) How many committees can be formed that contain at least 1 man?

99. From a box containing three 40-watt bulbs, six 60-watt bulbs, and eleven 75-watt bulbs, a bulb is drawn at random. What is the probability that the bulb is 40 watts? What is the probability that it is not a 75 watt bulb?

100. You have four \$1 bills, three \$5 bills, and two \$10 bills in your wallet. If you pick a bill at random, what is the probability it will be a \$1 bill?

101. Each of the letters in the word ROSE is written on an index card and the cards are then shuffled. What is the probability that, when the cards are dealt out, they spell the word ROSE?

102. Each of the numbers, $1, 2, \ldots, 100$ is written on an index card and the cards are then shuffled. If a card is selected at random, what is the probability that the number on the card is divisible by 5? What is the probability that the card selected either is a 1 or names a prime number?

103. Computing Probabilities Because of a mistake in packaging, a case of 12 bottles of red wine contained 5 Merlot and 7 Cabernet, each without labels. All the bottles look alike and have equal probability of being chosen. Three bottles are selected.

(a) What is the probability all 3 are Merlot?

(b) What is the probability exactly 2 are Merlot?

(c) What is the probability none is a Merlot?

104. Tossing a Coin A fair coin is tossed 10 times.

(a) What is the probability of obtaining exactly 5 heads?

(b) What is the probability of obtaining all heads?

105. Constructing a Brick Staircase A brick staircase has a total of 25 steps. The bottom step requires 80 bricks. Each successive step requires three less bricks than the prior step.

(a) How many bricks are required for the top step?

(b) How many bricks are required to build the staircase?

106. Creating a Floor Design A mosaic tile floor is designed in the shape of a trapezoid 30 feet wide at the base and 15 feet wide at the top. See Figure 11, p. 914. The tiles, 12 inches by 12 inches, are to be placed so that each successive row contains one less tile than the row below. How many tiles will be required?

107. Retirement Planning Chris gets paid once a month and contributes \$200 each pay period into his 401(k). If Chris plans on retiring in 20 years, what will the value of his 401(k) be if the per annum rate of return of the 401(k) is 10% compounded monthly?

108. Retirement Planning Jacky contributes \$500 every quarter into an IRA. If Jacky plans on retiring in 30 years, what will the value of the IRA be if the per annum rate of return of the IRA is 8% compounded quarterly?

109. Bouncing Balls A ball is dropped from a height of 20 feet. Each time it strikes the ground, it bounces up to $\frac{3}{4}$ of the previous height.

(a) What height will the ball bounce up to after it strikes the ground for the third time?

(b) What is its height after it strikes the ground for the nth time?

(c) How many times does the ball need to strike the ground before its height is less than 6 inches?

(d) What total distance does the ball travel before it stops bouncing?

110. Salary Increases Your friend has just been hired at an annual salary of \$20,000. If she expects to receive annual increases of 4%, what will her salary be as she begins her fifth year?

111. At the Milex tune-up and brake repair shop, the manager has found that a car will require a tune-up with a probability of 0.6, a brake job with a probability of 0.1, and both with a probability of 0.02.

(a) What is the probability that a car requires either a tune-up or a brake job?

(b) What is the probability that a car requires a tune-up but not a brake job?

(c) What is the probability that a car requires neither type of repair?

APPENDIX

Review

1 TOPICS FROM ALGEBRA AND GEOMETRY

1 Interval Notation
2 Absolute Value
3 Exponents
4 Polynomials
5 Pythagorean Theorem

Sets

When we want to treat a collection of similar but distinct objects as a whole, we use the idea of a **set.** For example, the set of *digits* consists of the collection of numbers 0, 1, 2, 3, 4, 5, 6, 7, 8, and 9. If we use the symbol D to denote the set of digits, then we can write

$$D = \{0, 1, 2, 3, 4, 5, 6, 7, 8, 9\}$$

In this notation, the braces $\{\ \}$ are used to enclose the objects, or **elements,** in the set. This method of denoting a set is called the **roster method.** A second way to denote a set is to use **set-builder notation,** where the set D of digits is written as

$$D = \{\ x \mid x \text{ is a digit}\}$$

Read as "D is the set of all x such that x is a digit."

EXAMPLE 1

Using Set-builder Notation and the Roster Method

(a) $E = \{x|x \text{ is an even digit}\} = \{0, 2, 4, 6, 8\}$
(b) $O = \{x|x \text{ is an odd digit}\} = \{1, 3, 5, 7, 9\}$

In listing the elements of a set, we do not list an element more than once because the elements of a set are distinct. Also, the order in which the elements are listed is not relevant. Thus, for example, $\{2, 3\}$ and $\{3, 2\}$ both represent the same set.

If every element of a set A is also an element of a set B, then we say that A is a **subset** of B. If two sets A and B have the same elements, then we say that A **equals** B. For example, $\{1, 2, 3\}$ is a subset of $\{1, 2, 3, 4, 5\}$; and $\{1, 2, 3\}$ equals $\{2, 3, 1\}$.

Real Numbers

Real numbers are represented by symbols such as

$$25, \quad 0, \quad -3, \quad \tfrac{1}{2}, \quad -\tfrac{5}{4}, \quad 0.125, \quad \sqrt{2}, \quad \pi, \quad \sqrt[3]{-2}, \quad 0.666\ldots$$

The set of **counting numbers,** or **natural numbers,** is the set $\{1, 2, 3, 4, \ldots\}$. (The three dots, called an **ellipsis,** indicate that the pattern continues indefinitely.) The set of **integers** is the set $\{\ldots, -3, -2, -1, 0, 1, 2, 3, \ldots\}$. A **rational number** is a number that can be expressed as a *quotient* a/b of two integers, where the integer b cannot be 0. Examples of rational numbers are $\frac{3}{4}, \frac{5}{2}, \frac{0}{4}$, and $-\frac{2}{3}$. Since $a/1 = a$ for any integer a, every integer is also a rational number. Real numbers that are not rational are called **irrational.** Examples of irrational numbers are $\sqrt{2}$ and π (the Greek letter pi), which equals the constant ratio of the circumference to the diameter of a circle. See Figure 1.

FIGURE 1

$\pi = \dfrac{C}{d}$

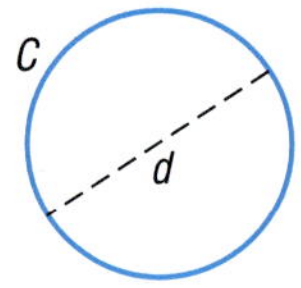

Real numbers can be represented as **decimals.** Rational real numbers have decimal representations that either **terminate** or are nonterminating with **repeating** blocks of digits. For example, $\frac{3}{4} = 0.75$, which terminates; and $\frac{2}{3} = 0.666\ldots$, in which the digit 6 repeats indefinitely. Irrational real numbers have decimal representations that neither repeat nor terminate. For example, $\sqrt{2} = 1.414213\ldots$ and $\pi = 3.14159\ldots$. In practice, irrational numbers are generally represented by approximations. We use the symbol $\approx$ (read as "approximately equal to") to write $\sqrt{2} \approx 1.4142$ and $\pi \approx 3.1416$.

Often, letters are used to represent numbers. If the letter used is to represent *any* number from a given set of numbers, it is referred to as a **variable.** A **constant** is either a fixed number, such as 5, $\sqrt{2}$, and so on, or a letter that represents a fixed (possibly unspecified) number. In general, we will follow the practice of using letters near the beginning of the alphabet, such as a, b, and c, for constants and using those near the end, such as x, y, and z, as variables.

In working with expressions or formulas involving variables, the variables may only be allowed to take on values from a certain set of numbers, called the **domain of the variable.** For example, in the expression $1/x$, the variable x cannot take on the value 0, since division by 0 is not allowed.

It can be shown that there is a one-to-one correspondence between real numbers and points on a line. That is, every real number corresponds to a point on the line and, conversely, each point on the line has a unique real number associated with it. We establish this correspondence of real numbers with points on a line in the following manner.

We start with a line that is, for convenience, drawn horizontally. Pick a point on the line and label it O, for **origin.** Then pick another point some fixed distance to the right of O and label it U, for **unit.** The fixed distance, which may be 1 inch, 1 centimeter, 1 light-year, or any unit distance, determines the **scale.** We associate the real number 0 with the origin O and the

FIGURE 2
Real number line

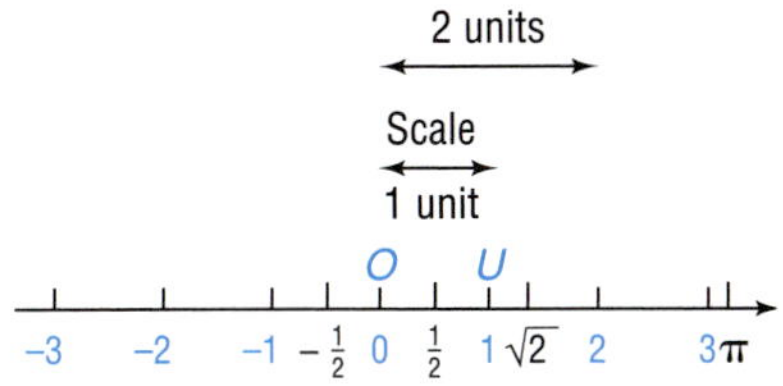

number 1 with the point U. Refer to Figure 2. The point to the right of U that is twice as far from O as U is associated with the number 2. The point to the right of U that is three times as far from O as U is associated with the number 3. The point midway between O and U is assigned the number 0.5, or $\frac{1}{2}$. Corresponding points to the left of the origin O are assigned the numbers $-\frac{1}{2}$, -1, -2, -3, and so on. The real number x associated with a point P is called the **coordinate** of P, and the line whose points have been assigned coordinates is called the **real number line.** Notice in Figure 2 that we placed an arrowhead on the right end of the line to indicate the direction in which the assigned numbers increase. Figure 2 also shows the points associated with the irrational numbers $\sqrt{2}$ and π.

The real number line divides the real numbers into three classes: the **negative real numbers** are the coordinates of points to the left of the origin O; the real number **zero** is the coordinate of the origin O; the **positive real numbers** are the coordinates of points to the right of the origin O.

Let a and b be two real numbers. If the difference $a - b$ is positive, then we say that a is **greater than** b and write $a > b$. Alternatively, if $a - b$ is positive, we can also say that b is **less than** a and write $b < a$. Thus, $a > b$ and $b < a$ are equivalent statements.

On the real number line, if $a > b$, the point with coordinate a is to the right of the point with coordinate b. For example, $0 > -1$, $\pi > 3$, and $\sqrt{2} < 2$. Furthermore,

> $a > 0$ is equivalent to a is positive
> $a < 0$ is equivalent to a is negative

If the difference $a - b$ of two real numbers is positive or 0, that is, if $a > b$ or $a = b$, then we say that a is **greater than or equal to** b and write $a \geq b$. Alternatively, if $a \geq b$, we can also say that b is **less than or equal to** a and write $b \leq a$.

Statements of the form $a < b$ or $b > a$ are called **strict inequalities;** statements of the form $a \leq b$ or $b \geq a$ are called **nonstrict inequalities.** The symbols $>$, $<$, $\geq$, and $\leq$ are called **inequality signs.**

If x is a real number and $x \geq 0$, then x is either positive or 0. As a result, we describe the inequality $x \geq 0$ by saying that x is nonnegative.

Inequalities are useful in representing certain subsets of real numbers. In so doing, though, other variations of the inequality notation may be used.

EXAMPLE 2 Graphing Inequalities

(a) In the inequality $x > 4$, x is any number greater than 4. In Figure 3, we use a left parenthesis to indicate that the number 4 is not part of the graph.

FIGURE 3
$x > 4$

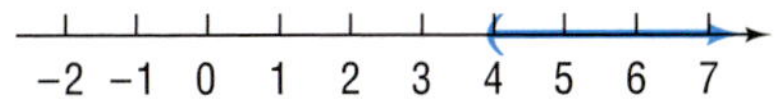

(b) In the inequality $4 < x \leq 6$, x is any number between 4 and 6, including 6 but not excluding 4. In Figure 4, we use a right bracket to indicate that 6 is part of the graph.

FIGURE 4
$x > 4$ and $x \leq 6$

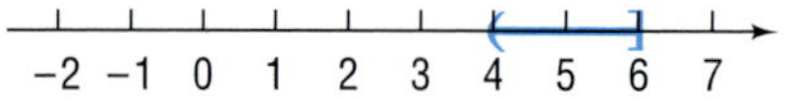

Now work Problem 15 (in the exercise set at the end of this section).

1 Let a and b represent two real numbers with $a < b$: A **closed interval,** denoted by **[*a*, *b*],** consists of all real numbers x for which $a \leq x \leq b$. An **open interval,** denoted by **(*a*, *b*),** consists of all real numbers x for which $a < x < b$. The **half-open,** or **half-closed intervals** are **(*a*, *b*],** consisting of all real numbers x for which $a < x \leq b$, and **[*a*, *b*),** consisting of all real numbers x for which $a \leq x < b$. In each of these definitions, a is called the **left end point** and b the **right end point** of the interval. Figure 5 illustrates each type of interval.

FIGURE 5

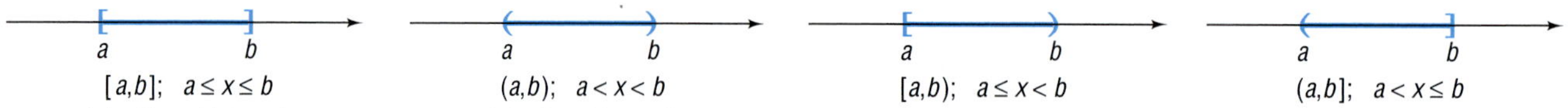

(a) Closed interval **(b)** Open interval **(c)** Half-open (half-closed) intervals

The symbol ∞ (read as "infinity") is not a real number, but a notational device used to indicate unboundedness in the positive direction. The symbol $-\infty$ (read as "minus infinity") also is not a real number, but a notational device used to indicate unboundedness in the negative direction. Using the symbols ∞ and $-\infty$, we can define five other kinds of intervals:

$[a, \infty)$	consists of all real numbers x for which $a \leq x < \infty$ $(x \geq a)$
(a, ∞)	consists of all real numbers x for which $a < x < \infty$ $(x > a)$
$(-\infty, a]$	consists of all real numbers x for which $-\infty < x \leq a$ $(x \leq a)$
$(-\infty, a)$	consists of all real numbers x for which $-\infty < x < a$ $(x < a)$
$(-\infty, \infty)$	consists of all real numbers x for which $-\infty < x < \infty$ (all real numbers)

Figure 6 illustrates these types of intervals.

FIGURE 6

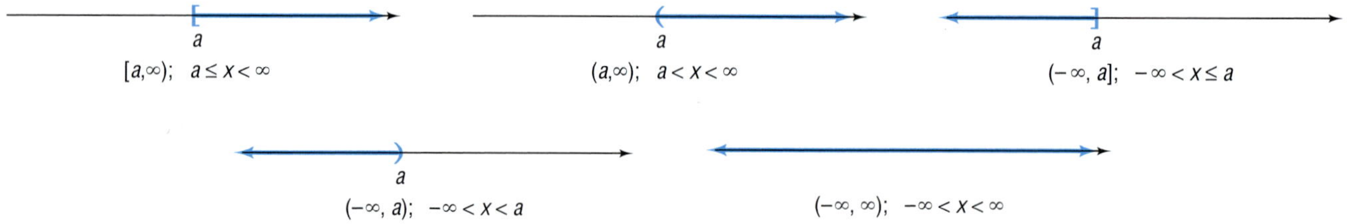

Now work Problems 21 and 25.

2 The *absolute value* of a number a is the distance from the point whose coordinate is a to the origin. For example, the point whose coordinate is -4 is 4 units from the origin. The point whose coordinate is 3 is 3 units from the origin. See Figure 7. Thus, the absolute value of -4 is 4, and the absolute value of 3 is 3.

FIGURE 7

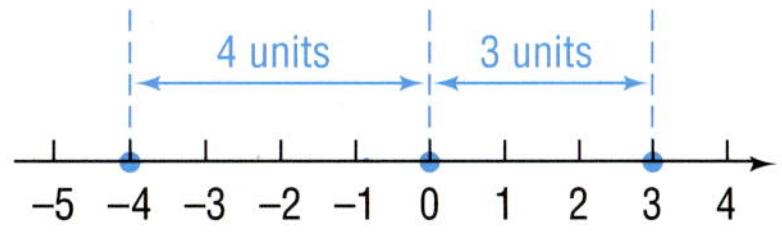

A more formal definition of absolute value is given next.

The **absolute value** of a real number a, denoted by the symbol $|a|$, is defined by the rules

$$|a| = a \quad \text{if } a \geq 0 \quad \text{and} \quad |a| = -a \quad \text{if } a < 0$$

For example, since $-4 < 0$, then the second rule must be used to get $|-4| = -(-4) = 4$.

EXAMPLE 3 Computing Absolute Value

(a) $|8| = 8$ (b) $|0| = 0$ (c) $|-15| = 15$

Now work Problem 29.

Look again at Figure 7. The distance from the point whose coordinate is -4 to the point whose coordinate is 3 is 7 units. This distance is the difference $3 - (-4)$, obtained by subtracting the smaller coordinate from the larger. However, since $|3 - (-4)| = |7| = 7$ and $|-4 - 3| = |-7| = 7$, we can use the absolute value to calculate the distance between two points without being concerned about which coordinate is smaller.

If P and Q are two points on a real number line with coordinates a and b, respectively, the **distance between P and Q**, denoted by $d(P, Q)$ is

$$d(P, Q) = |b - a|$$

Since $|b - a| = |a - b|$, it follows that $d(P, Q) = d(Q, P)$.

EXAMPLE 4 Finding Distance on a Number Line

Let P, Q, and R be points on the real number line with coordinates -5, 7, and -3, respectively. Find the distance:

(a) Between P and Q (b) Between Q and R

Solution (a) $d(P, Q) = |7 - (-5)| = |12| = 12$ (See Figure 8.)

(b) $d(Q, R) = |-3 - 7| = |-10| = 10$

FIGURE 8

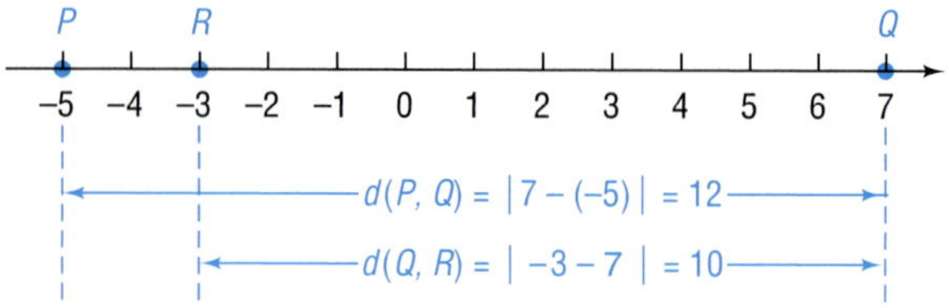

Exponents

3 Integer exponents provide a shorthand device for representing repeated multiplications of a real number.

If a is a real number and n is a positive integer, then the symbol $\boldsymbol{a^n}$ represents the product of n factors of a. That is,

$$a^n = \underbrace{a \cdot a \cdot \ldots \cdot a}_{n \text{ factors}}$$

where it is understood that $a^1 = a$. Thus, $a^2 = a \cdot a$, $a^3 = a \cdot a \cdot a$, and so on. In the expression a^n, a is called the **base** and n is called the **exponent,** or **power.** We read a^n as "a raised to the power n" or as "a to the nth power." We usually read a^2 as "a squared" and a^3 as "a cubed."

Care must be taken when parentheses are used in conjunction with exponents. For example, $-2^4 = -(2 \cdot 2 \cdot 2 \cdot 2) = -16$, whereas $(-2)^4 = (-2) \cdot (-2) \cdot (-2) \cdot (-2) = 16$. Notice the difference: The exponent applies only to the number or parenthetical expression immediately preceding it.

If $a \neq 0$, we define

$$a^0 = 1 \qquad \text{if } a \neq 0$$

If $a \neq 0$ and if n is a positive integer, then we define

$$a^{-n} = \frac{1}{a^n} \qquad \text{if } a \neq 0$$

With these definitions, the symbol a^n is defined for any integer n.

The following properties, called the **laws of exponents,** can be proved using the preceding definitions. In the list, a and b are real numbers, and m and n are integers.

Laws of Exponents

$$a^m a^n = a^{m+n} \qquad (a^m)^n = a^{mn} \qquad (ab)^n = a^n b^n$$

$$\frac{a^m}{a^n} = a^{m-n} = \frac{1}{a^{n-m}}, \quad \text{if } a \neq 0 \qquad \left(\frac{a}{b}\right)^n = \frac{a^n}{b^n}, \quad \text{if } b \neq 0$$

EXAMPLE 5 Using the Laws of Exponents

Write each expression so that all exponents are positive.

(a) $\dfrac{x^5y^{-2}}{x^3y}, \quad x \neq 0, y \neq 0$ (b) $\dfrac{xy}{x^{-1} - y^{-1}}, \quad x \neq 0, y \neq 0$

Solution (a) $\dfrac{x^5y^{-2}}{x^3y} = \dfrac{x^5}{x^3} \cdot \dfrac{y^{-2}}{y} = x^{5-3} \cdot y^{-2-1} = x^2y^{-3} = x^2 \cdot \dfrac{1}{y^3} = \dfrac{x^2}{y^3}$

(b) $\dfrac{xy}{x^{-1} - y^{-1}} = \dfrac{xy}{\dfrac{1}{x} - \dfrac{1}{y}} = \dfrac{xy}{\dfrac{y - x}{xy}} = \dfrac{(xy)(xy)}{y - x} = \dfrac{x^2y^2}{y - x}$

Now work Problem 57.

The **principal *n*th root of a number *a*,** symbolized by $\sqrt[n]{a}$, is defined as follows:

> $\sqrt[n]{a} = b$ means $a = b^n$ where $a \geq 0$ and $b \geq 0$ if n is even and a, b are any real numbers if n is odd

Notice that if a is negative and n is even, then $\sqrt[n]{a}$ is not defined. When it is defined, the principal nth root of a number is unique.

The symbol $\sqrt[n]{a}$ for the principal nth root of a is sometimes called a **radical;** the integer n is called the **index,** and a is called the **radicand.** If the index of a radical is 2, we call $\sqrt[2]{a}$ the **square root** of a and omit the index 2 by simply writing $\sqrt{a}$. If the index is 3, we call $\sqrt[3]{a}$ the **cube root** of a.

EXAMPLE 6 Simplifying Principal *n*th Roots

(a) $\sqrt[3]{8} = 2$ because $8 = 2^3$

(b) $\sqrt{64} = 8$ because $64 = 8^2$

(c) $\sqrt[3]{-64} = -4$ because $-64 = (-4)^3$

(d) $\sqrt[4]{\frac{1}{16}} = \frac{1}{2}$ because $\frac{1}{16} = (\frac{1}{2})^4$

(e) $\sqrt{0} = 0$ because $0 = 0^2$

These are examples of **perfect roots.** Thus, 8 and -64 are perfect cubes, since $8 = 2^3$ and $-64 = (-4)^3$; 64 and 0 are perfect squares, since $64 = 8^2$ and $0 = 0^2$; and $\frac{1}{2}$ is a perfect 4th root of $\frac{1}{16}$, since $\frac{1}{16} = (\frac{1}{2})^4$.

In general, if $n \geq 2$ is a positive integer and a is a real number,

$$\sqrt[n]{a^n} = a \quad \text{if } n \text{ is odd} \qquad \text{(1a)}$$

$$\sqrt[n]{a^n} = |a| \quad \text{if } n \text{ is even} \qquad \text{(1b)}$$

Notice the need for the absolute value in equation (1b). If n is even, then a^n is positive whether $a > 0$ or $a < 0$. But if n is even, the principal nth root must be nonnegative. Hence, the reason for using the absolute value—it gives a nonnegative result.

EXAMPLE 7

Simplifying Radicals

(a) $\sqrt{8} = \sqrt{4 \cdot 2} = \sqrt{4} \cdot \sqrt{2} = 2\sqrt{2}$

(b) $\sqrt[3]{-16} = \sqrt[3]{-8 \cdot 2} = \sqrt[3]{-8} \cdot \sqrt[3]{2} = -2\sqrt[3]{2}$

(c) $\sqrt{x^2} = |x|$

Now work Problem 49.

Radicals are used to define **rational exponents.** If a is a real number and $n \geq 2$ is an integer, then

$$a^{1/n} = \sqrt[n]{a}$$

provided $\sqrt[n]{a}$ exists.

If a is a real number and m and n are integers containing no common factors with $n \geq 2$, then

$$a^{m/n} = \sqrt[n]{a^m} = (\sqrt[n]{a})^m \qquad (2)$$

provided $\sqrt[n]{a}$ exists.

In simplifying $a^{m/n}$, either $\sqrt[n]{a^m}$ or $(\sqrt[n]{a})^m$ may be used. Generally, taking the root first, as in $(\sqrt[n]{a})^m$, is preferred.

EXAMPLE 8

Using Equation (2)

(a) $8^{2/3} = (\sqrt[3]{8})^2 = 2^2 = 4$ (b) $16^{3/2} = (\sqrt{16})^3 = 4^3 = 64$

(c) $(-8x^5)^{1/3} = \sqrt[3]{-8x^3 \cdot x^2} = \sqrt[3]{(-2x)^3 \cdot x^2} = \sqrt[3]{(-2x)^3}\,\sqrt[3]{x^2}$

$= -2x\sqrt[3]{x^2}$

A more detailed discussion of radicals and rational exponents is given in Section 3 of this Appendix.

Now work Problem 47.

4 Polynomials

A **monomial** in one variable is the product of a constant times a variable raised to a nonnegative integer power. Thus, a monomial is of the form

$$ax^k$$

where a is a constant, x is a variable, and $k \geq 0$ is an integer.

Two monomials ax^k and bx^k, when added or subtracted, can be combined into a single monomial by using the distributive property. For example,

$$2x^2 + 5x^2 = (2 + 5)x^2 = 7x^2 \quad \text{and} \quad 8x^3 - 5x^3 = (8 - 5)x^3 = 3x^3$$

A **polynomial** in one variable is an algebraic expression of the form

$$a_n x^n + a_{n-1}x^{n-1} + \cdots + a_1 x + a_0$$

where $a_n, a_{n-1}, \ldots, a_1, a_0$ are constants,* called the **coefficients** of the polynomial, $n \geq 0$ is an integer, and x is a variable. If $a_n \neq 0$, it is called the **leading coefficient,** and n is called the **degree** of the polynomial.

The monomials that make up a polynomial are called its **terms.** If all the coefficients are 0, the polynomial is called the **zero polynomial,** which has no degree.

Polynomials are usually written in **standard form,** beginning with the nonzero term of highest degree and continuing with terms in descending order according to degree. Examples of polynomials are

Polynomial	Coefficients	Degree
$3x^2 - 5 = 3x^2 + 0 \cdot x + (-5)$	3, 0, −5	2
$8 - 2x + x^2 = 1 \cdot x^2 - 2x + 8$	1, −2, 8	2
$5x + \sqrt{2} = 5x^1 + \sqrt{2}$	5, $\sqrt{2}$	1
$3 = 3 \cdot 1 = 3 \cdot x^0$	3	0
0	0	No degree

Although we have been using x to represent the variable, letters such as y or z are also commonly used. Thus,

$3x^4 - x^2 + 2$ is a polynomial (in x) of degree 4.

$9y^3 - 2y^2 + y - 3$ is a polynomial (in y) of degree 3.

$z^5 + \pi$ is a polynomial (in z) of degree 5.

A more detailed discussion of polynomials is given in Section 2 of this Appendix.

Pythagorean Theorem

5 The *Pythagorean Theorem* is a statement about *right triangles.* A **right triangle** is one that contains a **right angle,** that is, an angle of 90°. The side of the triangle opposite the 90° angle is called the **hypotenuse;** the remaining two sides are called **legs.** In Figure 9 we have used c to represent the length of the hypotenuse and a and b to represent the lengths of the legs. Notice the use of the symbol ⌐ to show the 90° angle. We now state the Pythagorean Theorem.

FIGURE 9

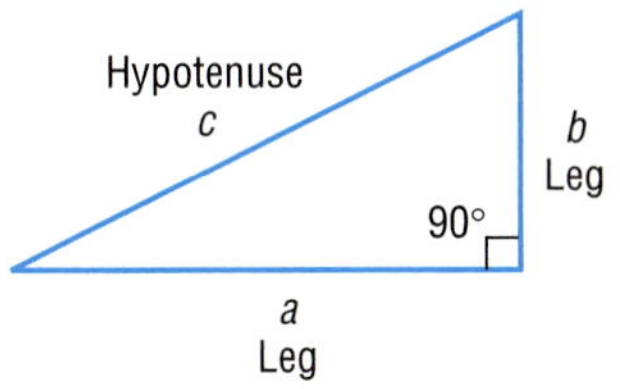

*The notation a_n is read as "a sub n." The number n is called a **subscript** and should not be confused with an exponent. We use subscripts to distinguish one constant from another when a large or undetermined number of constants is required.

Pythagorean Theorem

In a right triangle, the square of the length of the hypotenuse is equal to the sum of the squares of the lengths of the legs. That is, in the right triangle shown in Figure 9,

$$c^2 = a^2 + b^2 \qquad (3)$$

EXAMPLE 9 Finding the Hypotenuse of a Right Triangle

In a right triangle, one leg is of length 4 and the other is of length 3. What is the length of the hypotenuse?

Solution Since the triangle is a right triangle, we use the Pythagorean Theorem with $a = 4$ and $b = 3$ to find the length c of the hypotenuse. Thus, from equation (3), we have

$$c^2 = a^2 + b^2$$
$$c^2 = 4^2 + 3^2 = 16 + 9 = 25$$
$$c = 5$$

Now work Problem 69.

The converse of the Pythagorean Theorem is also true.

Converse of the Pythagorean Theorem

In a triangle, if the square of the length of one side equals the sum of the squares of the lengths of the other two sides, then the triangle is a right triangle. The 90° angle is opposite the longest side.

EXAMPLE 10 Verifying That a Triangle Is a Right Triangle

Show that a triangle whose sides are of lengths 5, 12, and 13 is a right triangle. Identify the hypotenuse.

Solution We square the lengths of the sides:

$$5^2 = 25, \quad 12^2 = 144, \quad 13^2 = 169$$

Notice that the sum of the first two squares (25 and 144) equals the third square (169). Hence, the triangle is a right triangle. The longest side, 13, is the hypotenuse. See Figure 10.

FIGURE 10

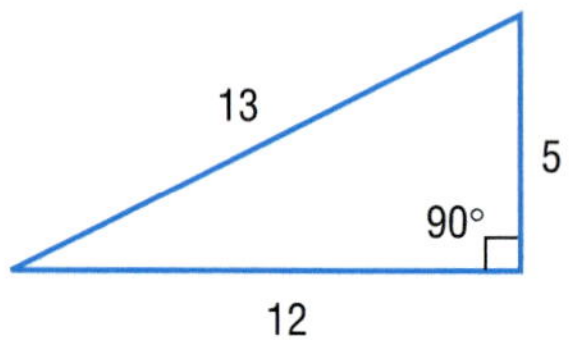

Now work Problem 79.

Geometry Formulas

Certain formulas from geometry are useful in solving algebra problems. We list some of these formulas next.

For a rectangle of length l and width w,

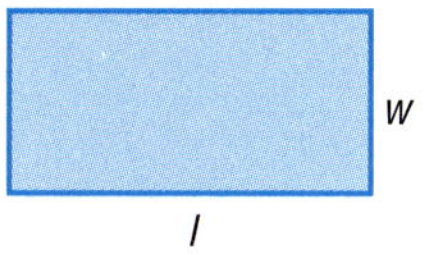

$$\text{Area} = lw \qquad \text{Perimeter} = 2l + 2w$$

For a triangle with base b and altitude h,

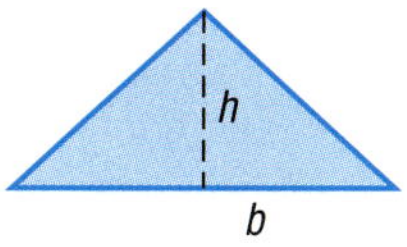

$$\text{Area} = \frac{1}{2}bh$$

For a circle of radius r (diameter $d = 2r$),

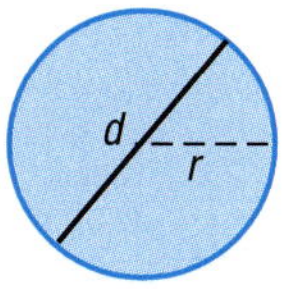

$$\text{Area} = \pi r^2 \qquad \text{Circumference} = 2\pi r = \pi d$$

For a rectangular box of length l, width w, and height h,

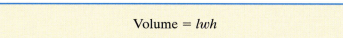

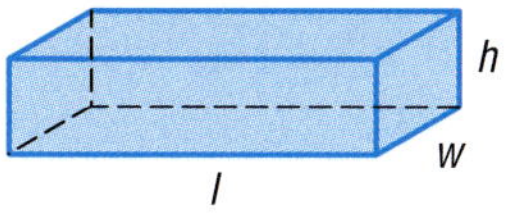

$$\text{Volume} = lwh$$

1 EXERCISES

In Problems 1–10, replace the question mark by $<$, $>$, *or* $=$, *whichever is correct.*

1. $\frac{1}{2}\ ?\ 0$ **2.** $5\ ?\ 6$ **3.** $-1\ ?\ -2$ **4.** $-3\ ?\ -\frac{5}{2}$ **5.** $\pi\ ?\ 3.14$

6. $\sqrt{2}\ ?\ 1.41$ **7.** $\frac{1}{2}\ ?\ 0.5$ **8.** $\frac{1}{3}\ ?\ 0.33$ **9.** $\frac{2}{3}\ ?\ 0.67$ **10.** $\frac{1}{4}\ ?\ 0.25$

11. On the real number line, label the points with coordinates $0, 1, -1, \frac{5}{2}, -2.5, \frac{3}{4}$, and 0.25.

12. Repeat Problem 11 for the coordinates $0, -2, 2, -1.5, \frac{3}{2}, \frac{1}{3}$, and $\frac{2}{3}$.

In Problems 13–20, write each statement as an inequality.

13. x is positive

14. z is negative

15. x is less than 2

16. y is greater than -5

17. x is less than or equal to 1

18. x is greater than or equal to 2

19. x is less than 5 and x is greater than 2

20. y is less than or equal to 2 and y is greater than 0

In Problems 21–24, write each inequality using interval notation, and illustrate each inequality using the real number line.

21. $0 \le x \le 4$ **22.** $-1 < x < 5$ **23.** $4 \le x < 6$ **24.** $-2 < x \le 0$

In Problems 25–28, write each interval as an inequality involving x, and illustrate each inequality using the real number line.

25. $[2, 5]$ **26.** $(1, 2)$ **27.** $[4, \infty)$ **28.** $(-\infty, 2]$

In Problems 29–32, find the value of each expression if $x = 2$ and $y = -3$.

29. $|x + y|$ **30.** $|x - y|$ **31.** $|x| + |y|$ **32.** $|x| - |y|$

In Problems 33–52, simplify each expression.

33. 3^0 **34.** 3^2 **35.** 4^{-2} **36.** $(-3)^2$

37. $(\frac{2}{3})^2$ **38.** $(\frac{-4}{5})^3$ **39.** $3^{-6} \cdot 3^4$ **40.** $4^{-2} \cdot 4^3$

41. $(\frac{2}{3})^{-2}$ **42.** $(\frac{3}{2})^{-3}$ **43.** $\dfrac{2^3 \cdot 3^2}{2 \cdot 3^{-2}}$ **44.** $\dfrac{3^{-2} \cdot 5^3}{3 \cdot 5}$

45. $9^{3/2}$ **46.** $16^{3/4}$ **47.** $(-8)^{4/3}$ **48.** $(-27)^{2/3}$

49. $\sqrt{32}$ **50.** $\sqrt[3]{24}$ **51.** $\sqrt[3]{-\frac{8}{27}}$ **52.** $\sqrt{\frac{4}{9}}$

In Problems 53–68, simplify each expression so that all exponents are positive. Whenever an exponent is negative or 0, we assume that the base does not equal 0.

53. x^0y^2 **54.** $x^{-1}y$ **55.** $x^{-2}y$ **56.** x^4y^0

57. $\dfrac{x^{-2}y^3}{xy^4}$ **58.** $\dfrac{x^{-2}y}{xy^2}$ **59.** $\left(\dfrac{4x}{5y}\right)^{-2}$ **60.** $(xy)^{-2}$

61. $\dfrac{x^{-1}y^{-2}z}{x^2yz^3}$ **62.** $\dfrac{3x^{-2}yz^2}{x^4y^{-3}z}$ **63.** $\dfrac{(-2)^3x^4(yz)^2}{3^2xy^3z^4}$ **64.** $\dfrac{4x^{-2}(yz)^{-1}}{(-5)^2x^4y^2z^{-2}}$

65. $\dfrac{x^{-2}}{\dfrac{1}{x^2} + \dfrac{1}{y^2}}$ **66.** $\dfrac{x^{-1} + y^{-1}}{x^{-1} - y^{-1}}$ **67.** $\left(\dfrac{3x^{-1}}{4y^{-1}}\right)^{-2}$ **68.** $\left(\dfrac{5x^{-2}}{6y^{-2}}\right)^{-3}$

In Problems 69–78, a and b are the lengths of the legs of a right triangle and c is the length of the hypotenuse. Find the missing length.

69. $a = 5, b = 12, c = ?$ **70.** $a = 6, b = 8, c = ?$ **71.** $a = 10, b = 24, c = ?$

72. $a = 4, b = 3, c = ?$ **73.** $a = 7, b = 24, c = ?$ **74.** $a = 14, b = 48, c = ?$

75. $a = 3, c = 5, b = ?$ **76.** $b = 6, c = 10, a = ?$ **77.** $b = 7,\ c = 25,\ a = ?$

78. $a = 10, c = 13, b = ?$

In Problems 79–84, the lengths of the sides of a triangle are given. Determine which are right triangles. For those that are right triangles, identify the hypotenuse.

79. 3, 4, 5 **80.** 6, 8, 10 **81.** 4, 5, 6

82. 2, 2, 3 **83.** 7, 24, 25 **84.** 10, 24, 26

85. Geometry Find the diagonal of a rectangle whose length is 8 inches and whose width is 5 inches.

86. Geometry Find the length of a rectangle of width 3 inches if its diagonal is 20 inches long.

87. Finding the Length of a Guy Wire A radio transmission tower is 100 feet high. How long does a guy wire need to be if it is to connect a point halfway up the tower to a point 30 feet from the base?

88. Answer Problem 87 if the guy wire is attached to the top of the tower.

89. How Far Can You See? The tallest inhabited building in North America is the Sears Tower in Chicago.* If the observation tower is 1454 feet above ground level, use the figure to determine how far a person standing in the observation tower can see (with the aid of a telescope). Use 3960 miles for the radius of Earth. [**Note:** 1 mile = 5280 feet]

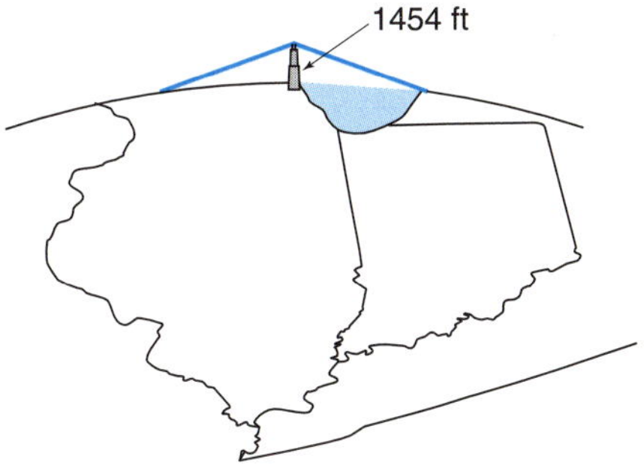

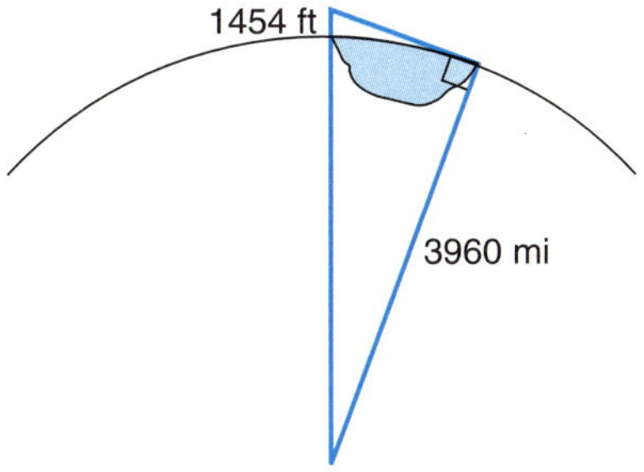

In Problems 90–92, use the fact that the radius of Earth is 3960 miles.

90. How Far Can You See? The conning tower of the USS *Silversides,* a World War II submarine now permanently stationed in Muskegon, Michigan, is approximately 20 feet above sea level. How far can one see from the conning tower?

91. How Far Can You See? A person who is 6 feet tall is standing on the beach in Fort Lauderdale, Florida and looks out into the Atlantic Ocean. Suddenly, a ship appears on the horizon. How far is the ship from shore?

92. How Far Can You See? The deck of a destroyer is 100 feet above sea level. How far can a person see from the deck? How far can a person see from the bridge, which is 150 feet above sea level?

93. If $a \le b$ and $c > 0$, show that $ac \le bc$.

[**Hint:** Since $a \le b$, it follows that $a - b \le 0$. Now multiply each side by c.]

94. If $a \le b$ and $c < 0$, show that $ac \ge bc$.

**Source: Guinness Book of World Records.*

95. If $a < b$, show that $a < (a + b)/2 < b$. The number $(a + b)/2$ is called the **arithmetic mean** of a and b.

96. Refer to Problem 95. Show that the arithmetic mean of a and b is equidistant from a and b.

97. Are there any real numbers that are both rational and irrational? Are there any real numbers that are neither? Explain your reasoning.

98. Explain why the sum of a rational number and an irrational number must be irrational.

99. What rational number does the repeating decimal 0.9999 . . . equal?

100. Is there a positive real number "closest" to 0?

101. I'm thinking of a number! It lies between 1 and 10; its square is rational and lies between 1 and 10. The number is larger than π. Correct to two decimal places, name the number. Now think of your own number, describe it, and challenge a fellow student to name it.

102. Write a brief paragraph that illustrates the similarities between "less than" ($<$) and "less than or equal" ($\le$).

103. The Gibb's Hill Lighthouse, Southampton, Bermuda, in operation since 1846, stands 117 feet high on a hill 245 feet high, so its beam of light is 362 feet above sea level. A brochure states that the light itself can be seen on the horizon about 26 miles distant. Verify the correctness of this information. The brochure further states that ships 40 miles away can see the light and planes flying at 10,000 feet can see it 120 miles away. Verify the accuracy of these statements. What assumption did the brochure make about the height of the ship?

104. You have 1000 feet of flexible pool siding and wish to construct a swimming pool. Experiment with rectangular-shaped pools with perimeters of 1000 feet. How do their areas vary? What is the shape of the rectangle with the largest area? Now compute the area enclosed by a circular pool with a perimeter (circumference) of 1000 feet. What would be your choice of shape for the pool? If rectangular, what is your preference for dimensions? Justify your choice. If your only consideration is to have a pool that encloses the most area, what shape should you use?

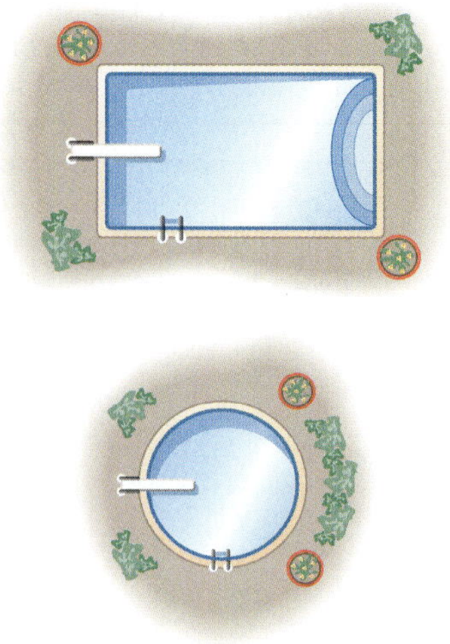

2 POLYNOMIALS AND RATIONAL EXPRESSIONS

1 Operations with Polynomials
2 Factoring Polynomials
3 Simplify Rational Expressions
4 Operations with Rational Expressions

Algebra may be described as a generalization of arithmetic in which letters are used to represent real numbers. We shall use the letters at the end of the alphabet, such as x, y, and z, to represent variables and the letters at the beginning of the alphabet, such as a, b, and c, to represent constants. Thus, in the expressions $3x + 5$ and $ax + b$, it is understood that x is a variable and that a and b are constants, even though the constants a and b are unspecified. As you will find out, the context usually makes the intended meaning clear.

Now we introduce some basic vocabulary.

A **monomial** in one variable is the product of a constant times a variable raised to a nonnegative integer power. Thus, a monomial is of the form

$$ax^k$$

where a is a constant, x is a variable, and $k \geq 0$ is an integer. The constant a is called the **coefficient** of the monomial. If $a \neq 0$, then k is called the **degree** of the monomial.

Examples of monomials are as follows:

Monomial	Coefficient	Degree	
$6x^2$	6	2	
$-\sqrt{2}x^3$	$-\sqrt{2}$	3	
3	3	0	Since $3 = 3 \cdot 1 = 3x^0$.
$-5x$	-5	1	Since $-5x = -5x^1$.
x^4	1	4	Since $x^4 = 1 \cdot x^4$.

Two monomials ax^k and bx^k with the same degree and the same variable are called **like terms.** Such monomials when added or subtracted can be combined into a single monomial by using the distributive property. For example,

$$2x^2 + 5x^2 = (2 + 5)x^2 = 7x^2 \quad \text{and} \quad 8x^3 - 5x^3 = (8 - 5)x^3 = 3x^3$$

The sum or difference of two monomials having different degrees is called a **binomial.** The sum or difference of three monomials with three different degrees is called a **trinomial.** For example,

$x^2 - 2$ is a binomial.
$x^3 - 3x + 5$ is a trinomial.
$2x^2 + 5x^2 + 2 = 7x^2 + 2$ is a binomial.

A **polynomial** in one variable is an algebraic expression of the form

$$a_nx^n + a_{n-1}x^{n-1} + \cdots + a_1x + a_0 \tag{1}$$

where $a_n, a_{n-1}, \ldots, a_1, a_0$ are constants* called the **coefficients** of the polynomial, $n \geq 0$ is an integer, and x is a variable. If $a_n \neq 0$, it is called the **leading coefficient** and n is called the **degree** of the polynomial.

The monomials that make up a polynomial are called its **terms.** If all the coefficients are 0, the polynomial is called the **zero polynomial,** which has no degree.

Polynomials are usually written in **standard form,** beginning with the nonzero term of highest degree and continuing with terms in descending order according to degree. Examples of polynomials are the following:

Polynomial	Coefficients	Degree
$3x^2 - 5 = 3x^2 + 0 \cdot x + (-5)$	$3, 0, -5$	2
$8 - 2x + x^2 = 1 \cdot x^2 - 2x + 8$	$1, -2, 8$	2
$5x + \sqrt{2} = 5x^1 + \sqrt{2}$	$5, \sqrt{2}$	1
$3 = 3 \cdot 1 = 3 \cdot x^0$	3	0
0	0	No degree

*The notation a_n is read as "a sub n." The number n is called a **subscript** and should not be confused with an exponent. We use subscripts in order to distinguish one constant from another when a large or undetermined number of constants is required.

Although we have been using x to represent the variable, letters such as y or z are also commonly used. Thus,

$3x^4 - x^2 + 2$ is a polynomial (in x) of degree 4.
$9y^3 - 2y^2 + y - 3$ is a polynomial (in y) of degree 3.
$z^5 + \pi$ is a polynomial (in z) of degree 5.

Algebraic expressions such as

$$\frac{1}{x} \quad \text{and} \quad \frac{x^2 + 1}{x + 5}$$

are not polynomials. The first is not a polynomial because $1/x = x^{-1}$ has an exponent that is not a nonnegative integer. The second expression is the quotient of two polynomials, since the polynomial in the denominator has degree greater than 0. The expression cannot be a polynomial.

Adding and Subtracting Polynomials

1 Polynomials are added and subtracted by combining like terms.

EXAMPLE 1

Finding Sums and Differences of Polynomials

Find

(a) $(8x^3 - 2x^2 + 6x - 2) + (3x^4 - 2x^3 + x^2 + x)$
(b) $(3x^4 - 4x^3 + 6x^2 - 1) - (2x^4 - 8x^2 - 6x + 5)$

Solution (a) The idea here is to group the like terms and then combine them.

$$\begin{aligned}(8x^3 - 2x^2 + 6x - 2) &+ (3x^4 - 2x^3 + x^2 + x) \\ &= 3x^4 + (8x^3 - 2x^3) + (-2x^2 + x^2) + (6x + x) - 2 \\ &= 3x^4 + 6x^3 - x^2 + 7x - 2\end{aligned}$$

(b) $(3x^4 - 4x^3 + 6x^2 - 1) - (2x^4 - 8x^2 - 6x + 5)$

$$= 3x^4 - 4x^3 + 6x^2 - 1 \underbrace{- 2x^4 + 8x^2 + 6x - 5}$$

Be sure to change the sign of each term in the second polynomial.

$$= (3x^4 - 2x^4) + (-4x^3) + (6x^2 + 8x^2) + 6x + (-1 - 5)$$

↑ Group like terms.

$$= x^4 - 4x^3 + 14x^2 + 6x - 6$$

Now work Problem 1.

Multiplying Polynomials

Products of polynomials are found by repeated use of the distributive property and the laws of exponents.

EXAMPLE 2

Finding the Product of Two Polynomials

Find the product: $(2x + 5)(x^2 - x + 2)$.

Solution *Horizontal Multiplication*

$$\begin{aligned}(2x + 5)(x^2 - x + 2) &= 2x(x^2 - x + 2) + 5(x^2 - x + 2) \\ &\uparrow \text{ Distributive property} \\ &= 2x \cdot x^2 - 2x \cdot x + 2x \cdot 2 + 5 \cdot x^2 - 5 \cdot x + 5 \cdot 2 \\ &\uparrow \text{ Distributive property} \\ &= 2x^3 - 2x^2 + 4x + 5x^2 - 5x + 10 \\ &\uparrow \text{ Law of exponents} \\ &= 2x^3 + 3x^2 - x + 10 \\ &\uparrow \text{ Combine like terms}\end{aligned}$$

Vertical Multiplication: The idea here is very much like multiplying a two-digit number by a three-digit number.

$$\begin{array}{rrrrl} & x^2 & -\ x & +\ 2 & \\ & & 2x & +\ 5 & \\ \hline & 2x^3 - 2x^2 & +\ 4x & & \text{This line is } 2x(x^2 - x + 2). \\ (+) & 5x^2 & -\ 5x & +\ 10 & \text{This line is } 5(x^2 - x + 2). \\ \hline & 2x^3 + 3x^2 & -\ x & +\ 10 & \text{The sum of the preceding two lines.} \end{array}$$

■

Now work Problem 5.

Certain products, which we call **special products,** occur frequently in algebra. In the list that follows, x, a, b, c, and d are real numbers:

Difference of Two Squares

$$(x - a)(x + a) = x^2 - a^2 \tag{2}$$

Squares of Binomials, or Perfect Squares

$$(x + a)^2 = x^2 + 2ax + a^2 \tag{3a}$$
$$(x - a)^2 = x^2 - 2ax + a^2 \tag{3b}$$

Miscellaneous Trinomials

$$(x + a)(x + b) = x^2 + (a + b)x + ab \tag{4a}$$
$$(ax + b)(cx + d) = acx^2 + (ad + bc)x + bd \tag{4b}$$

Cubes of Binomials, or Perfect Cubes

$$(x + a)^3 = x^3 + 3ax^2 + 3a^2x + a^3 \qquad (5a)$$
$$(x - a)^3 = x^3 - 3ax^2 + 3a^2x - a^3 \qquad (5b)$$

Difference of Two Cubes

$$(x - a)(x^2 + ax + a^2) = x^3 - a^3 \qquad (6)$$

Sum of Two Cubes

$$(x + a)(x^2 - ax + a^2) = x^3 + a^3 \qquad (7)$$

The formulas in equations (2) through (7) are used often and their patterns should be committed to memory. But if you forget one or are unsure of its form, you should be able to derive it as needed.

A **polynomial in two variables** x and y is the sum of one or more monomials of the form ax^ny^m, where a is a constant called the **coefficient,** x and y are variables, and n and m are nonnegative integers. The **degree** of the monomial ax^ny^m is $n + m$. The **degree** of a polynomial in two variables x and y is the highest degree of all the monomials with nonzero coefficients that appear.

Polynomials in three variables x, y, and z and polynomials in more than three variables are defined in a similar way. Here are some examples:

$3x^2 + 2x^3y + 5$ Two variables, degree is 4.	$\pi x^3 - y^2$ Two variables, degree is 3.	$x^4 + 4x^3y - xy^3 + y^4$ Two variables, degree is 4.
$x^2 + y^2 - z^2 + 4$ Three variables, degree is 2.	x^3y^2z Three variables, degree is 6.	$5x^2 - 4y^2 + z^3y + 2w^2x$ Four variables, degree is 4.

Adding and multiplying polynomials in two or more variables is handled in the same way as for polynomials in one variable.

Now work Problem 9.

Dividing Polynomials

The procedure for dividing two polynomials is similar to the procedure for dividing two integers. This process should be familiar to you, but we review it briefly next.

EXAMPLE 3

Dividing Two Integers

Divide 842 by 15.

Solution

$$
\begin{array}{rrl}
 & 56 & \leftarrow \text{Quotient} \\
\text{Divisor} \rightarrow & 15\overline{)842} & \leftarrow \text{Dividend} \\
 & \underline{75} & \leftarrow 5 \cdot 15 \quad \text{(Subtract)} \\
 & 92 & \\
 & \underline{90} & \leftarrow 6 \cdot 15 \quad \text{(Subtract)} \\
 & 2 & \leftarrow \text{Remainder}
\end{array}
$$

Thus, $\frac{842}{15} = 56 + \frac{2}{15}$.

In the long division process detailed in Example 3, the number 15 is called the **divisor,** the number 842 is called the **dividend,** the number 56 is called the **quotient,** and the number 2 is called the **remainder.**

To check the answer obtained in a division problem, multiply the quotient by the divisor and add the remainder. The answer should be the dividend.

> (Quotient)(Divisor) + Remainder = Dividend

For example, we can check the results obtained in Example 3 as follows:

$$(56)(15) + 2 = 840 + 2 = 842$$

To divide two polynomials, we first must write each polynomial in standard form. The process then follows a pattern similar to that of Example 3. The next example illustrates the procedure.

EXAMPLE 4

Dividing Two Polynomials

Find the quotient and the remainder when

$$3x^3 + 4x^2 + x + 7 \quad \text{is divided by} \quad x^2 + 1$$

Solution Each polynomial is in standard form. The dividend is $3x^3 + 4x^2 + x + 7$, and the divisor is $x^2 + 1$.

STEP 1: Divide the leading term of the dividend, $3x^3$, by the leading term of the divisor, x^2. Enter the result, $3x$, over the term $3x^3$, as follows:

$$
\begin{array}{rl}
 & 3x \\
x^2 + 1 & \overline{)3x^3 + 4x^2 + x + 7}
\end{array}
$$

STEP 2: Multiply $3x$ by $x^2 + 1$ and enter the result below the dividend.

$$
\begin{array}{rll}
 & 3x & \\
x^2 + 1 & \overline{)3x^3 + 4x^2 + x + 7} & \\
 & \underline{3x^3 + 3x } & \leftarrow 3x \cdot (x^2 + 1) = 3x^3 + 3x
\end{array}
$$

↑ Notice that we align the $3x$ term under the x to make the next step easier.

STEP 3: Subtract and bring down the remaining terms.

$$
\begin{array}{rll}
 & 3x & \\
x^2 + 1 & \overline{)3x^3 + 4x^2 + x + 7} & \\
 & \underline{3x^3 + 3x } & \leftarrow \text{Subtract.} \\
 & 4x^2 - 2x + 7 & \leftarrow \text{Bring down the } 4x^2 \text{ and the } 7.
\end{array}
$$

STEP 4: Repeat Steps 1 through 3 using $4x^2 - 2x + 7$ as the dividend.

$$\begin{array}{r}
3x + 4 \qquad\qquad \\
x^2 + 1\overline{)3x^3 + 4x^2 + x + 7} \\
\underline{3x^3 \qquad\quad + 3x \qquad} \\
4x^2 - 2x + 7 \\
\underline{4x^2 \qquad\quad + 4} \\
-2x + 3
\end{array}$$

← Divide $4x^2$ by x^2 to get 4.

← Multiply $x^2 + 1$ by 4; subtract.

Since x^2 does not divide $-2x$ evenly (that is, the result is not a monomial), the process ends. The quotient is $3x + 4$, and the remainder is $-2x + 3$.

Check: (Quotient)(Divisor) + Remainder

$$\begin{aligned}
&= (3x + 4)(x^2 + 1) + (-2x + 3) \\
&= 3x^3 + 4x^2 + 3x + 4 + (-2x + 3) \\
&= 3x^3 + 4x^2 + x + 7 = \text{Dividend}
\end{aligned}$$

Thus,

$$\frac{3x^3 + 4x^2 + x + 7}{x^2 + 1} = 3x + 4 + \frac{-2x + 3}{x^2 + 1}$$

The process for dividing two polynomials leads to the following result.

Theorem

The remainder after dividing two polynomials is either the zero polynomial or a polynomial of degree less than the degree of the divisor.

Now work Problem 13.

Factoring

2 Consider the following product:

$$(2x + 3)(x - 4) = 2x^2 - 5x - 12$$

The two polynomials on the left are called **factors** of the polynomial on the right. Expressing a given polynomial as a product of other polynomials, that is, finding the factors of a polynomial, is called **factoring.**

We shall restrict our discussion here to factoring polynomials in one variable into products of polynomials in one variable, where all coefficients are integers. We call this **factoring over the integers.** There will be times, though, when we will want to **factor the rational numbers** and even **factor over the real numbers.** Factoring over the rational numbers means to write a given polynomial whose coefficients are rational numbers as a product of polynomials whose coefficients are also rational numbers. Factoring over the real numbers means to write a given polynomial whose coefficients are real numbers as a product of polynomials whose coefficients are also real numbers. Unless specified otherwise, we will be factoring over the integers.

Any polynomial can be written as the product of 1 times itself or as -1 times its additive inverse. If a polynomial cannot be written as the product of two other polynomials (excluding 1 and -1), then the polynomial is said to be **prime.** When a polynomial has been written as a product consisting only of prime factors, then it is said to be **factored completely.** Examples of prime polynomials are

$$2, \quad 3, \quad 5, \quad x, \quad x+1, \quad x-1, \quad 3x+4$$

The first factor to look for in a factoring problem is a common monomial factor present in each term of the polynomial. If one is present, use the distributive property to factor it out. For example,

Polynomial	Common Monomial Factor	Remaining Factor	Factored Form
$2x+4$	2	$x+2$	$2x+4=2(x+2)$
$3x-6$	3	$x-2$	$3x-6=3(x-2)$
$2x^2-4x+8$	2	x^2-2x+4	$2x^2-4x+8=2(x^2-2x+4)$
$8x-12$	4	$2x-3$	$8x-12=4(2x-3)$
x^2+x	x	$x+1$	$x^2+x=x(x+1)$
x^3-3x^2	x^2	$x-3$	$x^3-3x^2=x^2(x-3)$
$6x^2+9x$	$3x$	$2x+3$	$6x^2+9x=3x(2x+3)$

The list of special products (2) through (7) given on pages 989–990 provides a list of factoring formulas when the equations are read from right to left. For example, equation (2) states that if the polynomial is the difference of two squares, x^2-a^2, it can be factored into $(x-a)(x+a)$. The following example illustrates several factoring techniques.

EXAMPLE 5 Factoring Polynomials

Factor completely each polynomial.

(a) x^4-16 (b) x^3-1 (c) $9x^2-6x+1$ (d) $x^2+4x-12$
(e) $3x^2+10x-8$ (f) x^3-4x^2+2x-8

Solution (a) $x^4-16=(x^2-4)(x^2+4)=(x-2)(x+2)(x^2+4)$

↑ Difference of squares ↑ Difference of squares

(b) $x^3-1=(x-1)(x^2+x+1)$

↑ Difference of cubes

(c) $9x^2-6x+1=(3x-1)^2$

↑ Perfect square

(d) $x^2+4x-12=(x+6)(x-2)$

↑ 6 and -2 are factors of -12, and the sum of 6 and -2 is 4.

$12x-2x=10x$

(e) $3x^2+10x-8=(3x-2)(x+4)$

$3x^2$ -8

(f) $x^3 - 4x^2 + 2x - 8 = (x^3 - 4x^2) + (2x - 8)$

↑ Regroup

$= x^2(x - 4) + 2(x - 4) = (x^2 + 2)(x - 4)$

↑ Distributive property

The technique used in Example 5(f) is called **factoring by grouping.**

Now work Problem 21.

Rational Expressions

3 If we form the quotient of two polynomials, the result is called a **rational expression.** Some examples of rational expressions are

(a) $\dfrac{x^3 + 1}{x}$ (b) $\dfrac{3x^2 + x - 2}{x^2 + 5}$ (c) $\dfrac{x}{x^2 - 1}$ (d) $\dfrac{xy^2}{(x - y)^2}$

Expressions (a), (b), and (c) are rational expressions in one variable, x, whereas (d) is a rational expression in two variables, x and y.

Rational expressions are described in the same manner as rational numbers. Thus, in expression (a), the polynomial $x^3 + 1$ is called the **numerator,** and x is called the **denominator.** When the numerator and denominator of a rational expression contain no common factors (except 1 and -1), we say that the rational expression is **reduced to lowest terms,** or **simplified.**

A rational expression is reduced to lowest terms by completely factoring the numerator and the denominator and canceling any common factors by using the cancellation property,

$$\frac{ac}{bc} = \frac{a}{b}, \qquad b \neq 0, c \neq 0$$

We shall follow the common practice of using a slash mark to indicate cancellation. For example,

$$\frac{x^2 - 1}{x^2 - 2x - 3} = \frac{(x - 1)\cancel{(x + 1)}}{(x - 3)\cancel{(x + 1)}} = \frac{x - 1}{x - 3}$$

EXAMPLE 6

Simplifying Rational Expressions

Reduce each rational expression to lowest terms.

(a) $\dfrac{x^2 + 4x + 4}{x^2 + 3x + 2}$ (b) $\dfrac{x^3 - 8}{x^3 - 2x^2}$ (c) $\dfrac{8 - 2x}{x^2 - x - 12}$

Solution (a) $\dfrac{x^2 + 4x + 4}{x^2 + 3x + 2} = \dfrac{\cancel{(x + 2)}(x + 2)}{\cancel{(x + 2)}(x + 1)} = \dfrac{x + 2}{x + 1}, \quad x \neq -2, -1$

(b) $\dfrac{x^3 - 8}{x^3 - 2x^2} = \dfrac{\cancel{(x - 2)}(x^2 + 2x + 4)}{x^2\cancel{(x - 2)}} = \dfrac{x^2 + 2x + 4}{x^2}, \quad x \neq 0, 2$

(c) $\dfrac{8 - 2x}{x^2 - x - 12} = \dfrac{2(4 - x)}{(x - 4)(x + 3)} = \dfrac{2(-1)\cancel{(x - 4)}}{\cancel{(x - 4)}(x + 3)} = \dfrac{-2}{x + 3}, \quad x \neq -3, 4$

4 The rules for multiplying and dividing rational expressions are the same as the rules for multiplying and dividing rational numbers:

$$\frac{a}{b} \cdot \frac{c}{d} = \frac{ac}{bd}, \quad \text{if } b \neq 0,\ d \neq 0 \tag{8}$$

$$\frac{\frac{a}{b}}{\frac{c}{d}} = \frac{a}{b} \cdot \frac{d}{c} = \frac{ad}{bc}, \quad \text{if } b \neq 0,\ c \neq 0,\ d \neq 0 \tag{9}$$

In using equations (8) and (9) with rational expressions, be sure first to factor each polynomial completely so that common factors can be canceled. We shall follow the practice of leaving our answers in factored form.

EXAMPLE 7 Finding Products and Quotients of Rational Expressions

Perform the indicated operation and simplify the result. Leave your answer in factored form.

(a) $\dfrac{x^2 - 2x + 1}{x^3 + x} \cdot \dfrac{4x^2 + 4}{x^2 + x - 2}$ (b) $\dfrac{\frac{x + 3}{x^2 - 4}}{\frac{x^2 - x - 12}{x^3 - 8}}$

Solution (a)

$$\frac{x^2 - 2x + 1}{x^3 + x} \cdot \frac{4x^2 + 4}{x^2 + x - 2} = \frac{(x - 1)^2}{x(x^2 + 1)} \cdot \frac{4(x^2 + 1)}{(x + 2)(x - 1)}$$

$$= \frac{(x - 1)^{\cancel{2}}(4)\cancel{(x^2 + 1)}}{x\cancel{(x^2 + 1)}(x + 2)\cancel{(x - 1)}} = \frac{4(x - 1)}{x(x + 2)}, \quad x \neq -2, 0, 1$$

(b)

$$\frac{\frac{x + 3}{x^2 - 4}}{\frac{x^2 - x - 12}{x^3 - 8}} = \frac{x + 3}{x^2 - 4} \cdot \frac{x^3 - 8}{x^2 - x - 12}$$

$$= \frac{x + 3}{(x - 2)(x + 2)} \cdot \frac{(x - 2)(x^2 + 2x + 4)}{(x - 4)(x + 3)}$$

$$= \frac{\cancel{(x + 3)}\cancel{(x - 2)}(x^2 + 2x + 4)}{\cancel{(x - 2)}(x + 2)(x - 4)\cancel{(x + 3)}} = \frac{x^2 + 2x + 4}{(x + 2)(x - 4)}, \quad x \neq -3, -2, 2, 4$$

Note: Slanting the cancellation marks in different directions for different factors, as in Example 7, is a good practice to follow, since it will help in checking for errors.

Now work Problem 31.

If the denominators of two rational expressions to be added (or subtracted) are equal, we add (or subtract) the numerators and keep the common denominator. That is, if a/b and c/b are two rational expressions, then

$$\frac{a}{b} + \frac{c}{b} = \frac{a + c}{b} \qquad \frac{a}{b} - \frac{c}{b} = \frac{a - c}{b}, \qquad \text{if } b \neq 0 \qquad (10)$$

EXAMPLE 8

Finding the Sum of Two Rational Expressions

Perform the indicated operation and simplify the result. Leave your answer in factored form.

$$\frac{2x^2 - 4}{2x + 5} + \frac{x + 3}{2x + 5}, \qquad x \neq -\tfrac{5}{2}$$

Solution

$$\frac{2x^2 - 4}{2x + 5} + \frac{x + 3}{2x - 5} = \frac{(2x^2 - 4) + (x + 3)}{2x + 5}$$
$$= \frac{2x^2 + x - 1}{2x + 5} = \frac{(2x - 1)(x + 1)}{2x + 5}$$

If the denominators of two rational expressions to be added or subtracted are not equal, we can use the general formulas for adding and subtracting quotients:

$$\frac{a}{b} + \frac{c}{d} = \frac{a \cdot d}{b \cdot d} + \frac{b \cdot c}{b \cdot d} = \frac{ad + bc}{bd}, \qquad \text{if } b \neq 0,\ d \neq 0$$
$$\frac{a}{b} - \frac{c}{d} = \frac{a \cdot d}{b \cdot d} - \frac{b \cdot c}{b \cdot d} = \frac{ad - bc}{bd}, \qquad \text{if } b \neq 0,\ d \neq 0 \qquad (11)$$

EXAMPLE 9

Finding the Difference of Two Rational Expressions

Perform the indicated operation and simplify the result. Leave your answer in factored form.

$$\frac{x^2}{x^2 - 4} - \frac{1}{x}, \qquad x \neq -2, 0, 2$$

Solution

$$\frac{x^2}{x^2 - 4} - \frac{1}{x} = \frac{x^2(x) - (x^2 - 4)(1)}{(x^2 - 4)(x)} = \frac{x^3 - x^2 + 4}{(x - 2)(x + 2)(x)}$$

Least Common Multiple (LCM)

If the denominators of two rational expressions to be added (or subtracted) have common factors, we usually do not use the general rules given by equation (11), since, in doing so, we make the problem more complicated than it needs to be. Instead, just as with fractions, we apply the **least common multiple (LCM) method** by using the polynomial of least degree that contains each denominator polynomial as a factor. Then we rewrite each rational expression using the LCM as the common denominator and use equation (10) to do the addition (or subtraction).

To find the least common multiple of two or more polynomials, first factor completely each polynomial. The LCM is the product of the different prime factors of each polynomial, each factor appearing the greatest number of times it occurs in each polynomial. The next example will give you the idea.

EXAMPLE 10 Finding the Least Common Multiple

Find the least common multiple of the following pair of polynomials:

$$x(x-1)^2(x+1) \quad \text{and} \quad 4(x-1)(x+1)^3$$

Solution The polynomials are already factored completely as

$$x(x-1)^2(x+1) \quad \text{and} \quad 4(x-1)(x+1)^3$$

Start by writing the factors of the left-hand polynomial. (Alternatively, you could start with the one on the right.)

$$x(x-1)^2(x+1)$$

Now look at the right-hand polynomial. Its first factor, 4, does not appear in our list, so we insert it:

$$4x(x-1)^2(x+1)$$

The next factor, $x-1$, is already in our list, so no change is necessary. The final factor is $(x+1)^3$. Since our list has $x+1$ to the first power only, we replace $x+1$ in the list by $(x+1)^3$. The LCM is

$$4x(x-1)^2(x+1)^3$$

Notice that the LCM is, in fact, the polynomial of least degree that contains $x(x-1)^2(x+1)$ and $4(x-1)(x+1)^3$ as factors. ▬

The next example illustrates how the LCM is used for adding and subtracting rational expressions.

EXAMPLE 11 Using the LCM to Add Rational Expressions

Perform the indicated operation and simplify the result. Leave your answer in factored form.

$$\frac{x}{x^2+3x+2} + \frac{2x-3}{x^2-1}, \qquad x \neq -2, -1, 1$$

Solution First, we find the LCM of the denominators:

$$x^2+3x+2 = (x+2)(x+1)$$
$$x^2-1 = (x-1)(x+1)$$

The LCM is $(x + 2)(x + 1)(x - 1)$. Next, we rewrite each rational expression using the LCM as the denominator:

$$\frac{x}{x^2 + 3x + 2} = \frac{x}{(x + 2)(x + 1)} \underset{\uparrow}{=} \frac{x(x - 1)}{(x + 2)(x + 1)(x - 1)}$$

Multiply numerator and denominator by $x - 1$ to get the LCM in the denominator.

$$\frac{2x - 3}{x^2 - 1} = \frac{2x - 3}{(x - 1)(x + 1)} \underset{\uparrow}{=} \frac{(2x - 3)(x + 2)}{(x - 1)(x + 1)(x + 2)}$$

Multiply numerator and denominator by $x + 2$ to get the LCM in the denominator.

Now we can add by using equation (10).

$$\begin{aligned}\frac{x}{x^2 + 3x + 2} + \frac{2x - 3}{x^2 - 1} &= \frac{x(x - 1)}{(x + 2)(x + 1)(x - 1)} + \frac{(2x - 3)(x + 2)}{(x + 2)(x + 1)(x - 1)} \\ &= \frac{(x^2 - x) + (2x^2 + x - 6)}{(x + 2)(x + 1)(x - 1)} \\ &= \frac{3x^2 - 6}{(x + 2)(x + 1)(x - 1)} = \frac{3(x^2 - 2)}{(x + 2)(x + 1)(x - 1)}\end{aligned}$$

If we had not used the LCM technique to add the quotients in Example 11, but decided instead to use the general rule of equation (11), we would have obtained a more complicated expression, as follows:

$$\begin{aligned}\frac{x}{x^2 + 3x + 2} + \frac{2x - 3}{x^2 - 1} &= \frac{x(x^2 - 1) + (x^2 + 3x + 2)(2x - 3)}{(x^2 + 3x + 2)(x^2 - 1)} \\ &= \frac{3x^3 + 3x^2 - 6x - 6}{(x^2 + 3x + 2)(x^2 - 1)} = \frac{3(x^3 + x^2 - 2x - 2)}{(x^2 + 3x + 2)(x^2 - 1)}\end{aligned}$$

Now we are faced with a more complicated problem of expressing this quotient in lowest terms. It is always best to first look for common factors in the denominators of expressions to be added or subtracted and to use the LCM if any common factors are found.

Now work Problem 35.

Mixed Quotients

When sums and/or differences of rational expressions appear as the numerator and/or denominator of a quotient, the quotient is called a **mixed quotient.** For example,

$$\frac{1 + \dfrac{1}{x}}{1 - \dfrac{1}{x}} \quad \text{and} \quad \frac{\dfrac{x^2}{x^2 - 4} - 3}{\dfrac{x - 3}{x + 2} - 1}$$

are mixed quotients. To **simplify** a mixed quotient means to write it as a rational expression reduced to lowest terms. This can be accomplished by treating the numerator and denominator of the mixed quotient separately, performing whatever operations are indicated and simplifying the results. Follow this by simplifying the resulting rational expression.

EXAMPLE 12 Simplifying Mixed Quotients

Simplify the mixed quotient: $\dfrac{1 + \dfrac{1}{x}}{1 - \dfrac{1}{x}}$.

Solution

$$\frac{1 + \dfrac{1}{x}}{1 - \dfrac{1}{x}} = \frac{\dfrac{x}{x} + \dfrac{1}{x}}{\dfrac{x}{x} - \dfrac{1}{x}} = \frac{\dfrac{x+1}{x}}{\dfrac{x-1}{x}} = \frac{x+1}{x} \cdot \frac{x}{x-1}$$

$$= \frac{(x+1)x}{x(x-1)} = \frac{x+1}{x-1}$$

Now work Problem 39.

2 EXERCISES

In Problems 1–10, perform the indicated operations. Express each answer as a polynomial.

1. $(10x^5 - 8x^2) + (3x^3 - 2x^2 + 6)$
2. $3(x^2 - 3x + 1) + 2(3x^2 + x - 4)$
3. $(x + a)^2 - x^2$
4. $(x - a)^2 - x^2$
5. $(x + 8)(2x + 1)$
6. $(2x - 1)(x + 2)$
7. $(x^2 + x - 1)(x^2 - x + 1)$
8. $(x^2 + 2x + 1)(x^2 - 3x + 4)$
9. $(x + 1)^3 - (x - 1)^3$
10. $(x + 1)^3 - (x + 2)^3$

In Problems 11–20, find the quotient and the remainder. Check your work by verifying that

(Quotient)(Divisor) + Remainder = Dividend

11. $4x^3 - 3x^2 + x + 1$ divided by x
12. $3x^3 - x^2 + x - 2$ divided by x
13. $4x^3 - 3x^2 + x + 1$ divided by $x + 2$
14. $3x^3 - x^2 + x - 2$ divided by $x + 2$
15. $4x^3 - 3x^2 + x + 1$ divided by $x - 4$
16. $3x^3 - x^2 + x - 2$ divided by $x - 4$
17. $4x^3 - 3x^2 + x + 1$ divided by x^2
18. $3x^3 - x^2 + x - 2$ divided by x^2
19. $4x^3 - 3x^2 + x + 1$ divided by $x^2 + 2$
20. $3x^3 - x^2 + x - 2$ divided by $x^2 + 2$

In Problems 21–30, factor completely each polynomial. If the polynomial cannot be factored, say it is prime.

21. $x^2 - 2x - 15$
22. $x^2 - 6x - 14$
23. $ax^2 - 4a^2x - 45a^3$
24. $bx^2 + 14b^2x + 45b^3$
25. $x^3 - 27$
26. $x^3 + 27$
27. $3x^2 + 4x + 1$
28. $4x^2 + 3x - 1$
29. $x^7 - x^5$
30. $x^8 - x^5$

In Problems 31–34, perform the indicated operation and simplify the result. Leave your answer in factored form.

31. $\dfrac{3x - 6}{5x} \cdot \dfrac{x^2 - x - 6}{x^2 - 4}$
32. $\dfrac{9x - 25}{2x - 2} \cdot \dfrac{1 - x^2}{6x - 10}$
33. $\dfrac{4x^2 - 1}{x^2 - 16} \cdot \dfrac{x^2 - 4x}{2x + 1}$
34. $\dfrac{12}{x^2 - x} \cdot \dfrac{x^2 - 1}{4x - 2}$

In Problems 35–42, perform the indicated operations and simplify the result. Leave your answer in factored form.

35. $\dfrac{x}{x^2 - 7x + 6} - \dfrac{x}{x^2 - 2x - 24}$

36. $\dfrac{x}{x - 3} - \dfrac{x + 1}{x^2 + 5x - 24}$

37. $\dfrac{4}{x^2 - 4} - \dfrac{2}{x^2 + x - 6}$

38. $\dfrac{3}{x - 1} - \dfrac{x - 4}{x^2 - 2x + 1}$

39. $\dfrac{x - \dfrac{1}{x}}{x + \dfrac{1}{x}}$

40. $\dfrac{1 - \dfrac{x}{x + 1}}{2 - \dfrac{x - 1}{x}}$

41. $\dfrac{3 - \dfrac{x^2}{x + 1}}{1 + \dfrac{x}{x^2 - 1}}$

42. $\dfrac{3x - \dfrac{3}{x^2}}{\dfrac{1}{(x - 1)^2} - 1}$

3 RADICALS; RATIONAL EXPONENTS

1. Evaluate Square Roots
2. Simplify Radicals
3. Simplify Rational Exponents

Square Roots

1 A real number is squared when it is raised to the power 2. The inverse of squaring is finding a **square root.** For example, since $6^2 = 36$ and $(-6)^2 = 36$, the numbers 6 and -6 are square roots of 36.

The symbol $\sqrt{\ }$, called a **radical sign,** is used to denote the *principal,* or nonnegative, square root. Thus, $\sqrt{36} = 6$.

> In general, if a is a nonnegative real number, the nonnegative number b such that $b^2 = a$ is the **principal square root** of a and is denoted by $b = \sqrt{a}$.

The following comments are noteworthy:

1. Negative numbers do not have square roots (in the real number system), because the square of any real number is *nonnegative.* For example, $\sqrt{-4}$ is not a real number, because there is no real number whose square is -4.
2. The principal square root of 0 is 0, since $0^2 = 0$. That is, $\sqrt{0} = 0$.
3. The principal square root of a positive number is positive.
4. If $c \geq 0$, then $(\sqrt{c})^2 = c$. For example, $(\sqrt{2})^2 = 2$ and $(\sqrt{3})^2 = 3$.

EXAMPLE 1 **Evaluating Square Roots**

(a) $\sqrt{64} = 8$ (b) $\sqrt{\frac{1}{16}} = \frac{1}{4}$ (c) $(\sqrt{1.4})^2 = 1.4$

Examples 1(a) and (b) are examples of **perfect square roots.** Thus, 64 is a **perfect square,** since $64 = 8^2$; and $\frac{1}{16}$ is a perfect square, since $\frac{1}{16} = \left(\frac{1}{4}\right)^2$. In general, we have

$$\sqrt{a^2} = |a| \qquad (1)$$

Notice the need for the absolute value in equation (1). Since $a^2 \geq 0$, the principal square root of a^2 is defined whether $a > 0$ or $a < 0$. However, since the principal square root is nonnegative, we need the absolute value to ensure the nonnegative result.

EXAMPLE 2

Using Equation (1)

(a) $\sqrt{(2.3)^2} = |2.3| = 2.3$ (b) $\sqrt{(-2.3)^2} = |-2.3| = 2.3$

(c) $\sqrt{x^2} = |x|$

*n*th Roots

The **principal *n*th root of a real number *a*,** symbolized by $\sqrt[n]{a}$, is defined as follows:

$\sqrt[n]{a} = b$ means $a = b^n$, where $a \geq 0$ and $b \geq 0$ if n is even and a, b are any real numbers if n is odd

Notice that if a is negative and n is even then $\sqrt[n]{a}$ is not defined. When it is defined, the principal nth root of a number is unique.

The symbol $\sqrt[n]{a}$ for the principal nth root of a is sometimes called a **radical;** the integer n is called the **index,** and a is called the **radicand.** If the index of a radical is 2, we call $\sqrt[2]{a}$ the **square root** of a and omit the index 2 by simply writing $\sqrt{a}$. If the index is 3, we call $\sqrt[3]{a}$ the **cube root** of a.

EXAMPLE 3

Evaluating Principal *n*th Roots

$$\sqrt[3]{8} = 2 \qquad \sqrt[6]{64} = 2 \qquad \sqrt[3]{-64} = -4 \qquad \sqrt[4]{\frac{1}{16}} = \frac{1}{2}$$

because

$$8 = 2^3 \qquad 64 = 2^6 \qquad -64 = (-4)^3 \qquad \tfrac{1}{16} = (\tfrac{1}{2})^4$$

These are examples of **perfect roots.** Thus, 8 and -64 are perfect cubes, since $8 = 2^3$ and $-64 = (-4)^3$; 2 is a perfect sixth root of 64, since $64 = 2^6$; and $\frac{1}{2}$ is a perfect fourth root of $\frac{1}{16}$, since $\frac{1}{16} = (\frac{1}{2})^4$.

In general, if $n \geq 2$ is a positive integer and a is a real number, we have

$$\sqrt[n]{a^n} = a, \quad \text{if } n \text{ is odd} \tag{2a}$$
$$\sqrt[n]{a^n} = |a|, \quad \text{if } n \text{ is even} \tag{2b}$$

Notice the need for the absolute value in equation (2b). If n is even, then a^n is positive whether $a > 0$ or $a < 0$. But if n is even, the principal nth root must be nonnegative. Hence, the reason for using the absolute value—it gives a nonnegative result.

EXAMPLE 4

Using Equations (2a) and (2b)

(a) $\sqrt[3]{4^3} = 4$ (b) $\sqrt[5]{(-3)^5} = -3$ (c) $\sqrt[4]{2^4} = 2$

(d) $\sqrt[4]{(-3)^4} = |-3| = 3$ (e) $\sqrt[4]{x^4} = |x|$

Properties of Radicals

2 Let $n \geq 2$ and $m \geq 2$ denote positive integers, and let a and b represent real numbers. Assuming that all radicals are defined, we have the following properties:

$$\sqrt[n]{ab} = \sqrt[n]{a}\sqrt[n]{b} \qquad (3a)$$

$$\sqrt[n]{\frac{a}{b}} = \frac{\sqrt[n]{a}}{\sqrt[n]{b}} \qquad (3b)$$

$$\sqrt[n]{a^m} = (\sqrt[n]{a})^m \qquad (3c)$$

$$\sqrt[m]{\sqrt[n]{a}} = \sqrt[mn]{a} \qquad (3d)$$

When used in reference to radicals, the direction to "simplify" will mean to remove from the radicals any perfect roots that occur as factors. Let's look at some examples of how the rules listed in the box are applied to simplify radicals.

EXAMPLE 5

Simplifying Radicals

Simplify each expression. Assume that all radicals containing variables are defined.

(a) $\sqrt{32}$ (b) $\sqrt[3]{8x^4}$ (c) $\sqrt{\sqrt[3]{x^7}}$ (d) $\sqrt[3]{\frac{8x^5}{27y^2}}$ (e) $\frac{\sqrt{x^5y}}{\sqrt{x^3y^3}}$

Solution (a) $\sqrt{32} = \sqrt{16 \cdot 2} = \sqrt{16}\sqrt{2} = 4\sqrt{2}$

↑ 16 is a perfect square. ↑ (3a)

(b) $\sqrt[3]{8x^4} = \sqrt[3]{8x^3 \cdot x} = \sqrt[3]{(2x)^3 \cdot x} = \sqrt[3]{(2x)^3}\,\sqrt[3]{x} = 2x\sqrt[3]{x}$

↑ Factor out perfect cube. ↑ (3a) ↑ (2a)

(c) $\sqrt{\sqrt[3]{x^7}} = \sqrt[6]{x^7} = \sqrt[6]{x^6 \cdot x} = \sqrt[6]{x^6} \cdot \sqrt[6]{x} = |x|\sqrt[6]{x}$

↑ (3d) ↑ (2b)

(d) $\sqrt[3]{\frac{8x^5}{27y^2}} = \sqrt[3]{\frac{2^3x^3x^2}{3^3y^2}} = \sqrt[3]{\left(\frac{2x}{3}\right)^3 \cdot \frac{x^2}{y^2}} = \sqrt[3]{\left(\frac{2x}{3}\right)^3} \cdot \sqrt[3]{\frac{x^2}{y^2}} = \frac{2x}{3}\sqrt[3]{\frac{x^2}{y^2}}$

(e) $\frac{\sqrt{x^5y}}{\sqrt{x^3y^3}} = \sqrt{\frac{x^5y}{x^3y^3}} = \sqrt{\frac{x^2}{y^2}} = \sqrt{\left(\frac{x}{y}\right)^2} = \left|\frac{x}{y}\right|$

Now work Problems 1 and 15.

Rationalizing

When radicals occur in quotients, it has become common practice to rewrite the quotient so that the denominator contains no radicals. This process is referred to as **rationalizing the denominator.**

The idea is to find an appropriate expression so that, when it is multiplied by the radical in the denominator, the new denominator that results contains no radicals. For example,

If Radical Is	Multiply By	To Get Product Free of Radicals
$\sqrt{3}$	$\sqrt{3}$	$\sqrt{9} = 3$
$\sqrt[3]{4}$	$\sqrt[3]{2}$	$\sqrt[3]{8} = 2$
$\sqrt{3} + 1$	$\sqrt{3} - 1$	$(\sqrt{3})^2 - 1^2 = 3 - 1 = 2$
$\sqrt{2} - 3$	$\sqrt{2} + 3$	$(\sqrt{2})^2 - 3^2 = 2 - 9 = -7$
$\sqrt{5} - \sqrt{3}$	$\sqrt{5} + \sqrt{3}$	$(\sqrt{5})^2 - (\sqrt{3})^2 = 5 - 3 = 2$

You are correct if you observed in this list that, after the second type of radical, the special product for differences of squares is the basis for determining by what to multiply.

EXAMPLE 6

Rationalizing Denominators

Rationalize the denominator of each expression.

(a) $\dfrac{4}{\sqrt{2}}$ (b) $\dfrac{\sqrt{3}}{\sqrt[3]{2}}$ (c) $\dfrac{\sqrt{x} - 2}{\sqrt{x} + 2}$, $x \ge 0$

Solution (a) $$\frac{4}{\sqrt{2}} = \frac{4}{\sqrt{2}} \cdot \frac{\sqrt{2}}{\sqrt{2}} = \frac{4\sqrt{2}}{(\sqrt{2})^2} = \frac{4\sqrt{2}}{2} = 2\sqrt{2}$$

Multiply by $\dfrac{\sqrt{2}}{\sqrt{2}}$.

(b) $$\frac{\sqrt{3}}{\sqrt[3]{2}} = \frac{\sqrt{3}}{\sqrt[3]{2}} \cdot \frac{\sqrt[3]{4}}{\sqrt[3]{4}} = \frac{\sqrt{3}\sqrt[3]{4}}{\sqrt[3]{8}} = \frac{\sqrt{3}\sqrt[3]{4}}{2}$$

Multiply by $\dfrac{\sqrt[3]{4}}{\sqrt[3]{4}}$.

(c) $$\frac{\sqrt{x} - 2}{\sqrt{x} + 2} = \frac{\sqrt{x} - 2}{\sqrt{x} + 2} \cdot \frac{\sqrt{x} - 2}{\sqrt{x} - 2} = \frac{(\sqrt{x} - 2)^2}{(\sqrt{x})^2 - 2^2}$$
$$= \frac{(\sqrt{x})^2 - 4\sqrt{x} + 4}{x - 4} = \frac{x - 4\sqrt{x} + 4}{x - 4}$$

In calculus, sometimes the numerator must be rationalized.

EXAMPLE 7

Rationalizing Numerators

Rationalize the numerator: $\dfrac{\sqrt{x} - 2}{\sqrt{x} + 1}$, $x \ge 0$.

Solution We multiply by $\dfrac{\sqrt{x} + 2}{\sqrt{x} + 2}$:

$$\frac{\sqrt{x} - 2}{\sqrt{x} + 1} = \frac{\sqrt{x} - 2}{\sqrt{x} + 1} \cdot \frac{\sqrt{x} + 2}{\sqrt{x} + 2} = \frac{(\sqrt{x})^2 - 2^2}{(\sqrt{x} + 1)(\sqrt{x} + 2)} = \frac{x - 4}{x + 3\sqrt{x} + 2}$$

Now work Problem 31.

3 Rational Exponents

Radicals are used to define rational exponents.

If a is a real number and $n \geq 2$ is an integer, then

$$a^{1/n} = \sqrt[n]{a} \tag{4}$$

provided $\sqrt[n]{a}$ exists.

EXAMPLE 8

Using Equation (4)

(a) $4^{1/2} = \sqrt{4} = 2$ (b) $(-27)^{1/3} = \sqrt[3]{-27} = -3$
(c) $8^{1/2} = \sqrt{8} = 2\sqrt{2}$ (d) $16^{1/3} = \sqrt[3]{16} = 2\sqrt[3]{2}$

If a is a real number and m and n are integers containing no common factors with $n \geq 2$, then

$$a^{m/n} = \sqrt[n]{a^m} = (\sqrt[n]{a})^m \tag{5}$$

provided $\sqrt[n]{a}$ exists.

We have two comments about equation (5):

1. The exponent m/n must be in lowest terms and n must be positive.
2. In simplifying $a^{m/n}$, either $\sqrt[n]{a^m}$ or $(\sqrt[n]{a})^m$ may be used. Generally, taking the root first is preferred.

It can be shown that the laws of exponents hold for rational exponents.

EXAMPLE 9

Simplifying Expressions with Rational Exponents

Simplify each expression. Express your answer so that only positive exponents occur. Assume that the variables are positive.

(a) $\left(\dfrac{2x^{1/3}}{y^{2/3}}\right)^{-3}$ (b) $(x^{2/3}y^{-3/4})(x^{-2}y)^{1/2}$

(c) $\left(\dfrac{9x^2y^{1/3}}{x^{1/3}y}\right)^{1/2}$ (d) $\dfrac{(2x+5)^{1/3}(2x+5)^{-1/2}}{(2x+5)^{-3/4}}$

Solution (a) $\left(\dfrac{2x^{1/3}}{y^{2/3}}\right)^{-3} = \left(\dfrac{y^{2/3}}{2x^{1/3}}\right)^{3} = \dfrac{(y^{2/3})^3}{(2x^{1/3})^3} = \dfrac{y^2}{2^3(x^{1/3})^3} = \dfrac{y^2}{8x}$

(b) $(x^{2/3}y^{-3/4})(x^{-2}y)^{1/2} = (x^{2/3}y^{-3/4})[(x^{-2})^{1/2}y^{1/2}]$

$= x^{2/3}y^{-3/4}x^{-1}y^{1/2} = (x^{2/3}x^{-1})(y^{-3/4}y^{1/2})$

$= x^{-1/3}y^{-1/4} = \dfrac{1}{x^{1/3}y^{1/4}}$

(c) $\left(\dfrac{9x^2y^{1/3}}{x^{1/3}y}\right)^{1/2} = \left(\dfrac{9x^{2-(1/3)}}{y^{1-(1/3)}}\right)^{1/2} = \left(\dfrac{9x^{5/3}}{y^{2/3}}\right)^{1/2} = \dfrac{9^{1/2}(x^{5/3})^{1/2}}{(y^{2/3})^{1/2}} = \dfrac{3x^{5/6}}{y^{1/3}}$

(d) $\dfrac{(2x+5)^{1/3}(2x+5)^{-1/2}}{(2x+5)^{-3/4}} = (2x+5)^{(1/3)-(1/2)-(-3/4)}$

$= (2x+5)^{(4-6+9)/12} = (2x+5)^{7/12}$ ▪

Now work Problem 53.

The next two examples illustrate some algebra that you will need to know for certain calculus problems.

EXAMPLE 10 Writing an Expression as a Single Quotient

Write the expression as a single quotient in which only positive exponents appear.

$$(x^2+1)^{1/2} + x \cdot \frac{1}{2}(x^2+1)^{-1/2} \cdot 2x$$

Solution

$$\begin{aligned}(x^2+1)^{1/2} + x \cdot \frac{1}{2}(x^2+1)^{-1/2} \cdot 2x &= (x^2+1)^{1/2} + \frac{x^2}{(x^2+1)^{1/2}}\\ &= \frac{(x^2+1)^{1/2}(x^2+1)^{1/2} + x^2}{(x^2+1)^{1/2}}\\ &= \frac{(x^2+1) + x^2}{(x^2+1)^{1/2}}\\ &= \frac{2x^2+1}{(x^2+1)^{1/2}}\end{aligned}$$ ▪

EXAMPLE 11 Factoring an Expression Containing Rational Exponents

Factor: $4x^{1/3}(2x+1) + 2x^{4/3}$.

Solution We begin by looking for factors that are common to the two terms. Notice that 2 and $x^{1/3}$ are common factors. Thus,

$$\begin{aligned}4x^{1/3}(2x+1) + 2x^{4/3} &= 2x^{1/3}[2(2x+1) + x]\\ &= 2x^{1/3}(5x+2)\end{aligned}$$ ▪

3 EXERCISES

In Problems 1–20, simplify each expression. Assume that all radicals containing variables are defined.

1. $\sqrt{8}$ **2.** $\sqrt[4]{32}$ **3.** $\sqrt[3]{16x^4}$ **4.** $\sqrt{27x^3}$

5. $\sqrt[3]{\sqrt{x^6}}$ **6.** $\sqrt{\sqrt{x^6}}$ **7.** $\sqrt{\dfrac{32x^3}{9x}}$ **8.** $\sqrt[3]{\dfrac{x}{8x^4}}$

9. $\sqrt[4]{x^{12}y^8}$ **10.** $\sqrt[5]{x^{10}y^5}$ **11.** $\sqrt[4]{\dfrac{x^9y^7}{xy^3}}$ **12.** $\sqrt[3]{\dfrac{3xy^2}{81x^4y^2}}$

13. $\sqrt{36x}$ **14.** $\sqrt{9x^5}$ **15.** $\sqrt{3x^2}\sqrt{12x}$ **16.** $\sqrt{5x}\sqrt{20x^3}$

17. $(\sqrt{5}\sqrt[3]{9})^2$ **18.** $(\sqrt[3]{3}\sqrt{10})^4$

19. $\sqrt{\dfrac{2x-3}{2x^4+3x^3}}\sqrt{\dfrac{x}{4x^2-9}}$ **20.** $\sqrt[3]{\dfrac{x-1}{x^2+2x+1}}\sqrt[3]{\dfrac{(x-1)^2}{x+1}}$

In Problems 21–26, perform the indicated operation and simplify the result. Assume that all variables are positive when they appear.

21. $(3\sqrt{6})(2\sqrt{2})$ **22.** $(5\sqrt{8})(-3\sqrt{3})$ **23.** $(\sqrt{3}+3)(\sqrt{3}-1)$

24. $(\sqrt{5}-2)(\sqrt{5}+3)$ **25.** $(\sqrt{x}-1)^2$ **26.** $(\sqrt{x}+\sqrt{5})^2$

In Problems 27–36, rationalize the denominator of each expression. Assume that all variables are positive when they appear.

27. $\dfrac{1}{\sqrt{2}}$ **28.** $\dfrac{6}{\sqrt[3]{4}}$ **29.** $\dfrac{-\sqrt{3}}{\sqrt{5}}$ **30.** $\dfrac{-\sqrt[3]{3}}{\sqrt{8}}$

31. $\dfrac{\sqrt{3}}{5-\sqrt{2}}$ **32.** $\dfrac{\sqrt{2}}{\sqrt{7}+2}$ **33.** $\dfrac{2-\sqrt{5}}{2+3\sqrt{5}}$ **34.** $\dfrac{\sqrt{3}-1}{2\sqrt{3}+3}$

35. $\dfrac{\sqrt{x+h}-\sqrt{x}}{\sqrt{x+h}+\sqrt{x}}$ **36.** $\dfrac{\sqrt{x+h}+\sqrt{x-h}}{\sqrt{x+h}-\sqrt{x-h}}$

In Problems 37–48, simplify each expression.

37. $8^{2/3}$ **38.** $4^{3/2}$ **39.** $(-27)^{1/3}$ **40.** $16^{3/4}$ **41.** $16^{3/2}$

42. $64^{3/2}$ **43.** $9^{-3/2}$ **44.** $25^{-5/2}$ **45.** $\left(\dfrac{9}{8}\right)^{3/2}$ **46.** $\left(\dfrac{27}{8}\right)^{2/3}$

47. $\left(\dfrac{8}{9}\right)^{-3/2}$ **48.** $\left(\dfrac{8}{27}\right)^{-2/3}$

In Problems 49–56, simplify each expression. Express your answer so that only positive exponents occur. Assume that the variables are positive.

49. $x^{5/4}x^{2/3}x^{-1/2}$ **50.** $x^{4/3}x^{1/2}x^{-1/4}$ **51.** $(x^3y^6)^{2/3}$ **52.** $(x^4y^8)^{5/4}$

53. $(x^2y)^{1/3}(xy^2)^{2/3}$ **54.** $(xy)^{1/4}(x^2y^2)^{1/2}$ **55.** $(16x^2y^{-1/3})^{3/4}$ **56.** $(4x^{-1}y^{1/3})^{3/2}$

In Problems 57–62, write each expression as a single quotient in which only positive exponents and/or radicals appear.

57. $\dfrac{x}{(1+x)^{1/2}}+2(1+x)^{1/2}$ **58.** $\dfrac{1+x}{2x^{1/2}}+x^{1/2}$

59. $\dfrac{\sqrt{1+x}-x\cdot\dfrac{1}{2\sqrt{1+x}}}{1+x}$ **60.** $\dfrac{\sqrt{x^2+1}-x\cdot\dfrac{2x}{2\sqrt{x^2+1}}}{x^2+1}$

61. $\dfrac{(x+4)^{1/2}-2x(x+4)^{-1/2}}{x+4}$ **62.** $\dfrac{(9-x^2)^{1/2}+x^2(9-x^2)^{-1/2}}{9-x^2}$

In Problems 63–66, factor each expression.

63. $(x+1)^{3/2}+x\cdot\dfrac{3}{2}(x+1)^{1/2}$ **64.** $(x^2+4)^{4/3}+x\cdot\dfrac{4}{3}(x^2+4)^{1/3}\cdot 2x$

65. $6x^{1/2}(x^2+x)-8x^{3/2}-8x^{1/2}$ **66.** $6x^{1/2}(2x+3)+x^{3/2}\cdot 8$

4 SOLVING EQUATIONS

1 Solve an Equation in One Variable

An **equation in one variable** is a statement in which two expressions, at least one containing the variable, are equal. The expressions are called the **sides** of the equation. Since an equation is a statement, it may not be true, depending on the value of the variable. Unless otherwise restricted, the admissible values of the variable are those in the domain of the variable. Those admissible values of the variable, if any, that result in a true statement are called **solutions,** or **roots,** of the equation. To **solve an equation** means to find all the solutions of the equation.

For example, the following are all equations in one variable, x:

$$x + 5 = 9 \qquad x^2 + 5x = 2x - 2 \qquad \frac{x^2 - 4}{x + 1} = 0 \qquad x^2 + 9 = 5$$

The first of these statements, $x + 5 = 9$, is true when $x = 4$ and false for any other choice of x. Thus, 4 is a solution of the equation $x + 5 = 9$. We also say that 4 **satisfies** the equation $x + 5 = 9$, because, when x is replaced by 4, a true statement results.

Sometimes an equation will have more than one solution. For example, the equation

$$\frac{x^2 - 4}{x + 1} = 0$$

has either $x = -2$ or $x = 2$ as a solution.

Sometimes we will write the solution of an equation in set notation. This set is called the **solution set** of the equation. For example, the solution set of the equation $x^2 - 9 = 0$ is $\{-3, 3\}$.

Unless indicated otherwise, we will limit ourselves to real solutions. Some equations have no real solution. For example, $x^2 + 9 = 5$ has no real solution, because there is no real number whose square when added to 9 equals 5.

An equation that is satisfied for every choice of the variable for which both sides are defined is called an **identity.** For example, the equation

$$3x + 5 = x + 3 + 2x + 2$$

is an identity, because this statement is true for any real number x.

1 Two or more equations that have precisely the same solutions are called **equivalent equations.** For example, all the following equations are equivalent, because each has only the solution $x = 5$:

$$\begin{aligned} 2x + 3 &= 13 \\ 2x &= 10 \\ x &= 5 \end{aligned}$$

These three equations illustrate one method for solving many types of equations: Replace the original equation by an equivalent equation, and continue until an equation with an obvious solution, such as $x = 5$, is reached. The question though, is: "How do I obtain an equivalent equation?" In general, there are five ways to do so.

Procedures That Result in Equivalent Equations

1. Interchange the two sides of the equation:

   Replace: $3 = x$ by $x = 3$

2. Simplify the sides of the equation by combining like terms, eliminating parentheses, and so on:

   Replace $(x + 2) + 6 = 2x + (x + 1)$
   by $x + 8 = 3x + 1$

3. Add or subtract the same expression on both sides of the equation:

   Replace $3x - 5 = 4$
   by $(3x - 5) + 5 = 4 + 5$

4. Multiply or divide both sides of the equation by the same nonzero expression:

   Replace $\frac{3x}{x - 1} = \frac{6}{x - 1} \qquad x \neq 1$
   by $\frac{3x}{x - 1} \cdot (x - 1) = \frac{6}{x - 1} \cdot (x - 1)$

5. If one side of the equation is 0 and the other side can be factored, then we may use the zero-product law* and set each factor equal to 0:

   Replace $x(x - 3) = 0$
   by $x = 0$ or $x - 3 = 0$

Whenever it is possible to solve an equation in your head, do so. For example:

The solution of $2x = 8$ is $x = 4$.
The solution of $3x - 15 = 0$ is $x = 5$.

Often, though, some rearrangement is necessary.

EXAMPLE 1 Solving an Equation

Solve the equation: $(x + 1)(2x) = (x + 1)(2)$.

Solution We begin by collecting all terms on the left side:

$$\begin{aligned} (x + 1)(2x) &= (x + 1)(2) \\ (x + 1)(2x) - (x + 1)(2) &= 0 \\ (x + 1)(2x - 2) &= 0 \quad \text{Factor.} \\ x + 1 = 0 \quad \text{or} \quad 2x - 2 &= 0 \quad \text{Apply the product law.} \\ x = -1 \qquad\qquad 2x &= 2 \\ x &= 1 \end{aligned}$$

The solution set is $\{-1, 1\}$.

*The zero-product law states that if $ab = 0$ then $a = 0$ or $b = 0$ or both equal 0.

EXAMPLE 2 Solving an Equation

Solve the equation: $\dfrac{3x}{x-1} + 2 = \dfrac{3}{x-1}$.

Solution First, we note that the domain of the variable is $\{x|x \neq 1\}$. Since the two quotients in the equation have the same denominator, $x - 1$, we can simplify by multiplying both sides by $x - 1$. The resulting equation is equivalent to the original equation, since we are multiplying by $x - 1$, which is not 0 (remember, $x \neq 1$).

$$\frac{3x}{x-1} + 2 = \frac{3}{x-1}$$

$$\left(\frac{3x}{x-1} + 2\right) \cdot (x-1) = \frac{3}{\cancel{x-1}} \cdot \cancel{(x-1)}$$ Multiply both sides by $x - 1$; cancel on the right.

$$\frac{3x}{\cancel{x-1}} \cdot \cancel{(x-1)} + 2 \cdot (x-1) = 3$$ Use the distributive property on the left side; cancel on the left.

$$3x + (2x - 2) = 3$$ Simplify.

$$5x - 2 = 3$$

$$5x = 5$$ Add 2 to each side.

$$x = 1$$ Divide both sides by 5.

The solution appears to be 1. But recall that $x = 1$ is not in the domain of the variable. Thus, the equation has no solution. ■

Now work Problems 5 and 19.

Steps for Solving Equations

STEP 1: List any restrictions on the domain of the variable.
STEP 2: Simplify the equation by replacing the original equation by a succession of equivalent equations following the procedures listed earlier.
STEP 3: If the result of STEP 2 is a product of factors equal to 0, use the product law and set each factor equal to 0 (procedure 5).
STEP 4: Check your solution(s).

EXAMPLE 3 Solving an Equation

Solve the equation: $x^3 = 25x$.

Solution We first rearrange the equation to get 0 on the right side:

$$x^3 = 25x$$

$$x^3 - 25x = 0$$

We notice that x is a factor of each term on the left:

$$x(x^2 - 25) = 0$$

$$x(x+5)(x-5) = 0$$ Difference of two squares.

$$x = 0 \quad \text{or} \quad x + 5 = 0 \quad \text{or} \quad x - 5 = 0$$ Set each factor equal to 0.

$$x = -5 \quad \text{or} \quad x = 5$$ Solve.

The solution set is $\{-5, 0, 5\}$. ■

Now work Problem 21.

4 ## EXERCISES

In Problems 1–36, solve each equation.

1. $6 - x = 2x + 9$

2. $3 - 2x = 2 - x$

3. $2(3 + 2x) = 3(x - 4)$

4. $3(2 - x) = 2x - 1$

5. $8x - (2x + 1) = 3x - 10$

6. $5 - (2x - 1) = 10$

7. $\frac{1}{2}x - 4 = \frac{3}{4}x$

8. $1 - \frac{1}{2}x = 5$

9. $0.9t = 0.4 + 0.1t$

10. $0.9t = 1 + t$

11. $\dfrac{2}{y} + \dfrac{4}{y} = 3$

12. $\dfrac{4}{y} - 5 = \dfrac{5}{2y}$

13. $(x + 7)(x - 1) = (x + 1)^2$

14. $(x + 2)(x - 3) = (x - 3)^2$

15. $x(2x - 3) = (2x + 1)(x - 4)$

16. $x(1 + 2x) = (2x - 1)(x - 2)$

17. $z(z^2 + 1) = 3 + z^3$

18. $w(4 - w^2) = 8 - w^3$

19. $\dfrac{x}{x - 3} + 3 = \dfrac{3}{x - 3}$

20. $\dfrac{3x}{x + 2} = \dfrac{-6}{x + 2} - 2$

21. $x^2 = 9x$

22. $x^3 = x^2$

23. $t^3 - 9t^2 = 0$

24. $4z^3 - 8z^2 = 0$

25. $\dfrac{2x}{x^2 - 4} = \dfrac{4}{x^2 - 4} - \dfrac{1}{x + 2}$

26. $\dfrac{x}{x^2 - 9} + \dfrac{1}{x + 3} = \dfrac{3}{x^2 - 9}$

27. $\dfrac{x}{x + 2} = \dfrac{1}{2}$

28. $\dfrac{3x}{x - 1} = 2$

29. $\dfrac{3}{2x - 3} = \dfrac{2}{x + 5}$

30. $\dfrac{-2}{x + 4} = \dfrac{-3}{x + 1}$

31. $(x + 2)(3x) = (x + 2)(6)$

32. $(x - 5)(2x) = (x - 5)(4)$

33. $\dfrac{6t + 7}{4t - 1} = \dfrac{3t + 8}{2t - 4}$

34. $\dfrac{8w + 5}{10w - 7} = \dfrac{4w - 3}{5w + 7}$

35. $\dfrac{2}{x - 2} = \dfrac{3}{x + 5} + \dfrac{10}{(x + 5)(x - 2)}$

36. $\dfrac{1}{2x + 3} + \dfrac{1}{x - 1} = \dfrac{1}{(2x + 3)(x - 1)}$

5 ## COMPLETING THE SQUARE

1 Complete the Square of an Equation

The idea behind the method of **completing the square** is to "adjust" the left side of a second degree polynomial, $ax^2 + bx + c$, so it becomes a perfect square—the square of a first-degree polynomial. For example, $x^2 + 6x + 9$ and $x^2 - 4x + 4$ are perfect squares* because

$$x^2 + 6x + 9 = (x + 3)^2 \quad \text{and} \quad x^2 - 4x + 4 = (x - 2)^2$$

How do we adjust the second degree polynomial? We do it by adding the appropriate number to create a perfect square. For example, to make $x^2 + 6x$ a perfect square, we add 9.

*Perfect squares are discussed in Section 2 of this Appendix.

Let's look at several examples of completing the square when the coefficient of x^2 is 1:

Start	**Add**	**Result**
$x^2 + 4x$	4	$x^2 + 4x + 4 = (x + 2)^2$
$x^2 + 12x$	36	$x^2 + 12x + 36 = (x + 6)^2$
$x^2 - 6x$	9	$x^2 - 6x + 9 = (x - 3)^2$
$x^2 + x$	$\frac{1}{4}$	$x^2 + x + \frac{1}{4} = (x + \frac{1}{2})^2$

Do you see the pattern? Provided the coefficient of x^2 is 1, we complete the square by adding the square of one-half the coefficient of x:

Start	**Add**	**Result**
$x^2 + mx$	$\left(\frac{m}{2}\right)^2$	$x^2 + mx + \left(\frac{m}{2}\right)^2 = \left(x + \frac{m}{2}\right)^2$

Now work Problem 1.

EXAMPLE 1 Completing the Square of an Equation Containing Two Variables

Complete the squares of x and y in the equation

$$x^2 + y^2 - 2x + 4y - 4 = 0$$

Solution We rearrange the equation, grouping the terms involving the variable x and the variable y.

$$(x^2 - 2x) + (y^2 + 4y) = 4$$

Next, we complete the square of each parenthetical expression. Of course, since we want an equivalent equation, whatever we add to the left side, we also add to the right side.

$$(x^2 - 2x + 1) + (y^2 + 4y + 4) = 4 + 1 + 4$$
$$(x - 1)^2 + (y + 2)^2 = 9$$

The terms involving the variables x and y now appear as perfect squares.

Now work Problem 7.

The next example illustrates how the procedure of completing the square can be used to solve a quadratic equation.

EXAMPLE 2 Solving a Quadratic Equation by Completing the Square

Solve by completing the square: $x^2 + 5x + 4 = 0$.

Solution We always begin this procedure by rearranging the equation so that the constant is on the right side:

$$x^2 + 5x + 4 = 0$$
$$x^2 + 5x = -4$$

Since the coefficient of x^2 is 1, we can complete the square on the left side by adding $(\frac{1}{2} \cdot 5)^2 = \frac{25}{4}$. Of course, in an equation, whatever we add to the left side must also be added to the right side. Thus, we add $\frac{25}{4}$ to *both* sides:

$$x^2 + 5x + \frac{25}{4} = -4 + \frac{25}{4}$$
$$\left(x + \frac{5}{2}\right)^2 = \frac{9}{4}$$
$$x + \frac{5}{2} = \pm\sqrt{\frac{9}{4}}$$
$$x + \frac{5}{2} = \pm\frac{3}{2}$$
$$x = -\frac{5}{2} \pm \frac{3}{2}$$
$$x = -\frac{5}{2} + \frac{3}{2} = -1 \quad \text{or} \quad x = -\frac{5}{2} - \frac{3}{2} = -4$$

The solution set is $\{-4, -1\}$.

Now work Problem 13.

EXAMPLE 3

Solving a Quadratic Equation by Completing the Square

Solve by completing the square: $2x^2 - 8x - 5 = 0$.

Solution First, we rewrite the equation:

$$2x^2 - 8x - 5 = 0$$
$$2x^2 - 8x = 5$$

Next, we divide by 2 so that the coefficient of x^2 is 1. (This enables us to complete the square at the next step.)

$$x^2 - 4x = \frac{5}{2}$$

Finally, we complete the square by adding 4 to each side:

$$x^2 - 4x + 4 = \frac{5}{2} + 4$$
$$(x - 2)^2 = \frac{13}{2}$$
$$x - 2 = \pm\frac{\sqrt{13}}{\sqrt{2}} = \pm\frac{\sqrt{26}}{2}$$
$$x = 2 \pm \frac{\sqrt{26}}{2}$$

We choose to leave our answer in this compact form. Thus, the solution set is $\{2 - \sqrt{26}/2, 2 + \sqrt{26}/2\}$.

Note: If we wanted an approximation, say, to two decimal places, of these solutions, we would use a calculator to get $\{-0.55, 4.55\}$.

Now work Problem 17.

5 EXERCISES

In Problems 1–6, tell what number should be added to complete the square of each expression.

1. $x^2 - 4x$ **2.** $x^2 - 2x$ **3.** $x^2 + \frac{1}{2}x$

4. $x^2 - \frac{1}{3}x$ **5.** $x^2 - \frac{2}{3}x$ **6.** $x^2 - \frac{2}{5}x$

In Problems 7–12, complete the squares of x and y in each equation.

7. $x^2 + y^2 - 4x + 4y - 1 = 0$
8. $x^2 + y^2 + 4x + 4y - 8 = 0$
9. $x^2 + y^2 + 6x - 2y + 1 = 0$
10. $x^2 + y^2 - 8x + 2y + 1 = 0$
11. $x^2 + y^2 + x - y - \frac{1}{2} = 0$
12. $x^2 + y^2 - x + y - \frac{3}{2} = 0$

In Problems 13–18, solve each equation by completing the square.

13. $x^2 + 4x - 21 = 0$
14. $x^2 - 6x = 13$
15. $x^2 - \frac{1}{2}x = \frac{3}{16}$
16. $x^2 + \frac{2}{3}x = \frac{1}{3}$
17. $3x^2 + x - \frac{1}{2} = 0$
18. $2x^2 - 3x = 1$

6 SYNTHETIC DIVISION

1 Use Synthetic Division

To find the quotient as well as the remainder when a polynomial function f of degree 1 or higher is divided by $g(x) = x - c$, a shortened version of long division, called **synthetic division,** makes the task simpler.

1 To see how synthetic division works, we will use long division* to divide the polynomial $f(x) = 2x^3 - x^2 + 3$ by $g(x) = x - 3$.

$$\begin{array}{r}
2x^2 + 5x + 15 \\
x - 3 \overline{) 2x^3 - x^2 + 3} \\
\underline{2x^3 - 6x^2} \\
5x^2 \\
\underline{5x^2 - 15x} \\
15x + 3 \\
\underline{15x - 45} \\
48
\end{array}$$

The process of synthetic division arises from rewriting the long division in a more compact form, using simpler notation. For example, in the long division above, the terms in color are not really necessary because they are identical to the terms directly above them. With these terms removed, we have

$$\begin{array}{r}
2x^2 + 5x + 15 \\
x - 3 \overline{) 2x^3 - x^2 + 3} \\
\underline{- 6x^2} \\
5x^2 \\
\underline{- 15x} \\
15x \\
\underline{- 45} \\
48
\end{array}$$

Most of the x's that appear in this process can also be removed, provided we are careful about positioning each coefficient. In this regard, we will need to

*Long division is discussed in Section 2 of this Appendix.

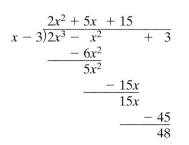

use 0 as the coefficient of x in the dividend, because that power of x is missing. Now we have

$$
\begin{array}{rrrrr}
 & 2x^2 & +\,5x & +\,15 & \\ \cline{2-5}
x-3) & 2 & -1 & 0 & 3 \\
 & & -6 & & \\ \cline{2-3}
 & & 5 & & \\
 & & & -15 & \\ \cline{3-5}
 & & & 15 & \\
 & & & & -45 \\ \cline{4-5}
 & & & & 48
\end{array}
$$

We can make this display more compact by moving the lines up until the numbers in color align horizontally:

$$
\begin{array}{rrrrrl}
 & 2x^2 & +\,5x & +\,15 & & \text{Row 1.} \\ \cline{2-5}
x-3) & 2 & -1 & 0 & 3 & \text{Row 2.} \\
 & & -6 & -15 & -45 & \text{Row 3.} \\ \cline{2-5}
 & \bigcirc & 5 & 15 & 48 & \text{Row 4.}
\end{array}
$$

Now, if we place the leading coefficient of the quotient (2) in the circled position, the first three numbers in row 4 are precisely the coefficients of the quotient, and the last number in row 4 is the remainder. Thus, row 1 is not really needed, so we can compress the process to three rows, where the bottom row contains the coefficients of both the quotient and the remainder:

$$
\begin{array}{rrrrrl}
x-3) & 2 & -1 & 0 & 3 & \text{Row 1.} \\
 & & -6 & -15 & -45 & \text{Row 2 (subtract).} \\ \cline{2-5}
 & 2 & 5 & 15 & 48 & \text{Row 3.}
\end{array}
$$

Recall that the entries in row 3 are obtained by subtracting the entries in row 2 from those in row 1. Rather than subtracting the entries in row 2, we can change the sign of each entry and add. With this modification, our display will look like this:

$$
\begin{array}{rrrrrl}
x-3) & 2 & -1 & 0 & 3 & \text{Row 1.} \\
 & & 6 & 15 & 45 & \text{Row 2 (add).} \\ \cline{2-5}
 & 2 & 5 & 15 & 48 & \text{Row 3.}
\end{array}
$$

Notice that the entries in row 2 are three times the prior entries in row 3. Our last modification to the display replaces the $x - 3$ by 3. The entries in row 3 give the quotient and the remainder as shown next.

$$
\begin{array}{rrrrrl}
3) & 2 & -1 & 0 & 3 & \text{Row 1.} \\
 & & 6 & 15 & 45 & \text{Row 2 (add).} \\ \cline{2-5}
 & 2 \xrightarrow{\times 3} & 5 \xrightarrow{\times 3} & 15 \xrightarrow{\times 3} & 48 & \text{Row 3.}
\end{array}
$$

Quotient: $2x^2 + 5x + 15$ Remainder: $R = 48$

Let's go through another example step by step.

EXAMPLE 1 Using Synthetic Division to Find the Quotient and Remainder

Use synthetic division to find the quotient and remainder when

$$f(x) = 3x^4 + 8x^2 - 7x + 4 \quad \text{is divided by} \quad g(x) = x - 1$$

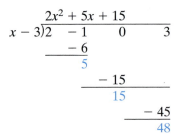

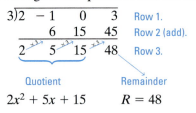

Solution STEP 1: Write the dividend in descending powers of x. Then copy the coefficients, remembering to insert a 0 for any missing powers of x:

$$3 \quad 0 \quad 8 \quad -7 \quad 4 \qquad \text{Row 1.}$$

STEP 2: Insert the usual division symbol. Since the divisor is $x - 1$, we insert 1 to the left of the division symbol

$$1\overline{)3 \quad 0 \quad 8 \quad -7 \quad 4} \qquad \text{Row 1.}$$

STEP 3: Bring the 3 down two rows, and enter it in row 3:

1)	3	0	8	−7	4	Row 1.
	↓					Row 2.
	3					Row 3.

STEP 4: Multiply the latest entry in row 3 by 1 and place the result in row 2, but one column over to the right:

1)	3	0	8	−7	4	Row 1.
		3				Row 2.
	3					Row 3.

STEP 5: Add the entry in row 2 to the entry above it in row 1, and enter the sum to row 3:

1)	3	0	8	−7	4	Row 1.
		3				Row 2.
	3	3				Row 3.

STEP 6: Repeat steps 4 and 5 until no more entries are available in row 1:

1)	3	0	8	−7	4	Row 1.
		3	3	11	4	Row 2 (add).
	3	3	11	4	8	Row 3.

STEP 7: The final entry in row 3, an 8, is the remainder; the other entries in row 3 (3, 3, 11, and 4) are the coefficients (in descending order) of a polynomial whose degree is 1 less than that of the dividend; this is the quotient. Thus,

$$\text{Quotient} = 3x^3 + 3x^2 + 11x + 4 \qquad \text{Remainder} = 8$$

Check: (Divisor)(Quotient) + Remainder

$$= (x - 1)(3x^3 + 3x^2 + 11x + 4) + 8$$
$$= 3x^4 + 3x^3 + 11x^2 + 4x - 3x^3 - 3x^2 - 11x - 4 + 8$$
$$= 3x^4 + 8x^2 - 7x + 4 = \text{Dividend}$$

Let's do an example in which all seven steps are combined.

EXAMPLE 2 Using Synthetic Division to Verify a Factor

Use synthetic division to show that $g(x) = x + 3$ is a factor of

$$f(x) = 2x^5 + 5x^4 - 2x^3 + 2x^2 - 2x + 3$$

Solution The divisor is $x + 3 = x - (-3)$, so the row 3 entries will be multiplied by -3 entered in row 2, and added to row 1:

$$\begin{array}{r|rrrrrrl} -3 & 2 & 5 & -2 & 2 & -2 & 3 & \text{Row 1.} \\ & & -6 & 3 & -3 & 3 & -3 & \text{Row 2.} \\ \hline & 2 & -1 & 1 & -1 & 1 & 0 & \text{Row 3.} \end{array}$$

Because the remainder is 0, it follows that $f(-3) = 0$. Hence, by the Factor Theorem (Section 4.5), $x - (-3) = x + 3$ is a factor of $f(x)$.

Now work Problem 3.

One important use of synthetic division is to find the value of a polynomial.

EXAMPLE 3 Using Synthetic Division to Find the Value of a Polynomial

Use synthetic division to find the value of $f(x) = -3x^4 + 2x^3 - x + 1$ at $x = -2$; that is, find $f(-2)$.

Solution The Remainder Theorem (Section 4.5) tells us that the value of a polynomial function at c equals the remainder when the polynomial is divided by $x - c$. This remainder is the final entry of the third row in the process of synthetic division. We want $f(-2)$, so we divide by $x - (-2)$:

$$\begin{array}{r|rrrrr} -2 & -3 & 2 & 0 & -1 & 1 \\ & & 6 & -16 & 32 & -62 \\ \hline & -3 & 8 & -16 & 31 & -61 \end{array}$$

The quotient is $q(x) = -3x^3 + 8x^2 - 16x + 31$; the reminder is $R = -61$. Because the remainder was found to be -61, it follows from the Remainder Theorem that $f(-2) = -61$.

Now work Problem 23.

As Example 3 illustrates, we can use the process of synthetic division to find the value of a polynomial function at a number c as an alternative to merely substituting c for x. Compare the work required in Example 3 with the arithmetic involved in substituting:

$$\begin{aligned} f(-2) &= -3(-2)^4 + 2(-2)^3 - (-2) + 1 \\ &= -3(16) + 2(-8) + 2 + 1 \\ &= -48 - 16 + 2 + 1 = -61 \end{aligned}$$

As you can see, finding $f(-2)$ may be easier using synthetic division.

6 EXERCISES

In Problems 1–12, use synthetic division to find the quotient $q(x)$ and remainder R when $f(x)$ is divided by $g(x)$.

1. $f(x) = x^3 - x^2 + 2x + 4$; $g(x) = x - 2$

2. $f(x) = x^3 + 2x^2 - 3x + 1$; $g(x) = x + 1$

3. $f(x) = 3x^3 + 2x^2 - x + 3$; $g(x) = x - 3$

4. $f(x) = -4x^3 + 2x^2 - x + 1$; $g(x) = x + 2$

5. $f(x) = x^5 - 4x^3 + x$; $g(x) = x + 3$

6. $f(x) = x^4 + x^2 + 2$; $g(x) = x - 2$

7. $f(x) = 4x^6 - 3x^4 + x^2 + 5; \quad g(x) = x - 1$

8. $f(x) = x^5 + 5x^3 - 10; \quad g(x) = x + 1$

9. $f(x) = 0.1x^3 + 0.2x; \quad g(x) = x + 1.1$

10. $f(x) = 0.1x^2 - 0.2; \quad g(x) = x + 2.1$

11. $f(x) = x^5 - 1; \quad g(x) = x - 1$

12. $f(x) = x^5 + 1; \quad g(x) = x + 1$

In Problems 13–22, use synthetic division to determine whether $x - c$ is a factor of $f(x)$.

13. $f(x) = 4x^3 - 3x^2 - 8x + 4; \quad c = 2$

14. $f(x) = -4x^3 + 5x^2 + 8; \quad c = -3$

15. $f(x) = 3x^4 - 6x^3 - 5x + 10; \quad c = 2$

16. $f(x) = 4x^4 - 15x^2 - 4; \quad c = 2$

17. $f(x) = 3x^6 + 82x^3 + 27; \quad c = -3$

18. $f(x) = 2x^6 - 18x^4 + x^2 - 9; \quad c = -3$

19. $f(x) = 4x^6 - 64x^4 + x^2 - 15; \quad c = -4$

20. $f(x) = x^6 - 16x^4 + x^2 - 16; \quad c = -4$

21. $f(x) = 2x^4 - x^3 + 2x - 1; \quad c = \frac{1}{2}$

22. $f(x) = 3x^4 + x^3 - 3x + 1; \quad c = -\frac{1}{3}$

In Problems 23–28, use synthetic division to find $f(c)$.

23. $f(x) = 5x^4 - 3x^2 + 1; \quad c = 2$

24. $f(x) = -2x^3 + 3x^2 + 5; \quad c = -2$

25. $f(x) = 4x^5 - 3x^3 + 2x - 1; \quad c = -1$

26. $f(x) = -3x^4 + 3x^3 - 2x^2 + 5; \quad c = -1$

27. $f(x) = 9x^{17} - 8x^{10} + 9x^8 + 5; \quad c = 1$

28. $f(x) = 10x^{15} + 4x^{12} - 2x^5 + x^2; \quad c = -1$

ANSWERS

CHAPTER 1 *1.1 Exercises*

1. qualitative **3.** quantitative **5.** quantitative **7.** qualitative **9. (a)** Northwest **(b)** TWA **(c)** About 78%
11. (a) Housing, fuel, and utilities **(b)** Misc. goods and services **13. (a)** 13 **(b)** 20; 24 **(c)** 5 **(d)** About 145,000 **(e)** 30–34
(f) 80–84 **(g)**

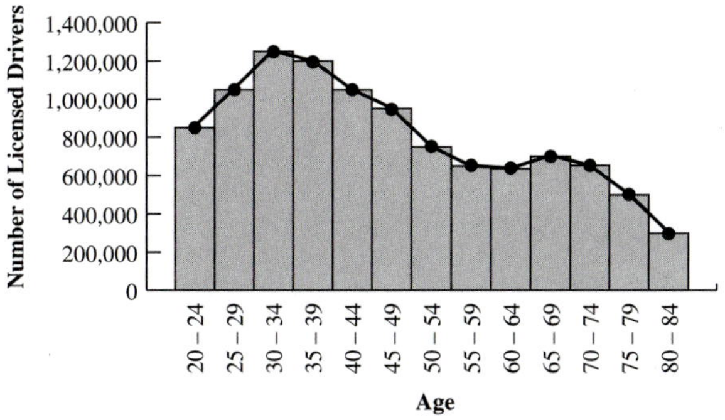

15. (a)

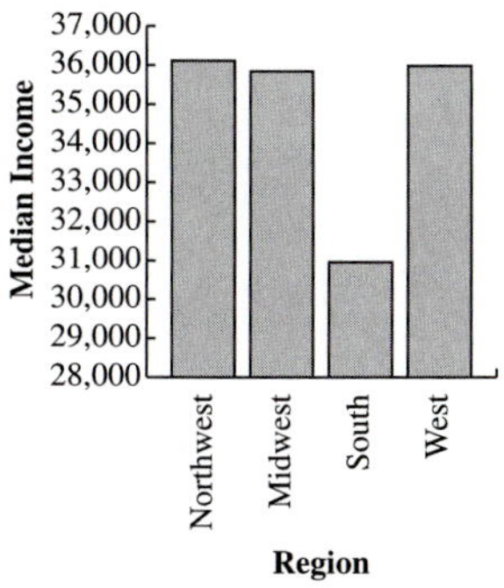

(b)

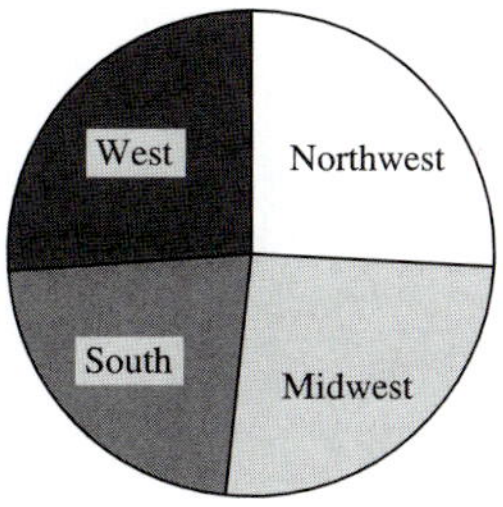

(d) Northwest **(e)** South

17. (a)

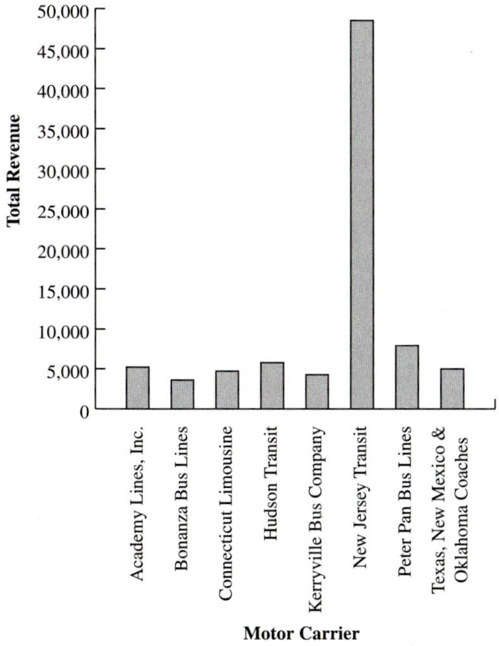

(b)

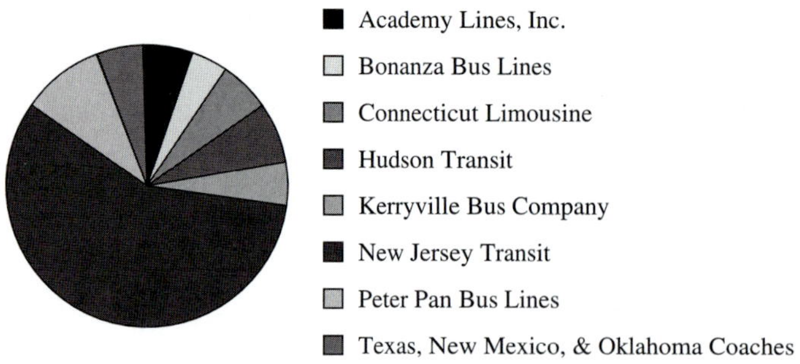
Academy Lines, Inc.
Bonanza Bus Lines
Connecticut Limousine
Hudson Transit
Kerryville Bus Company
New Jersey Transit
Peter Pan Bus Lines
Texas, New Mexico, & Oklahoma Coaches

(d) New Jersey Transit **(e)** Bonanza Bus Lines

19. (a)

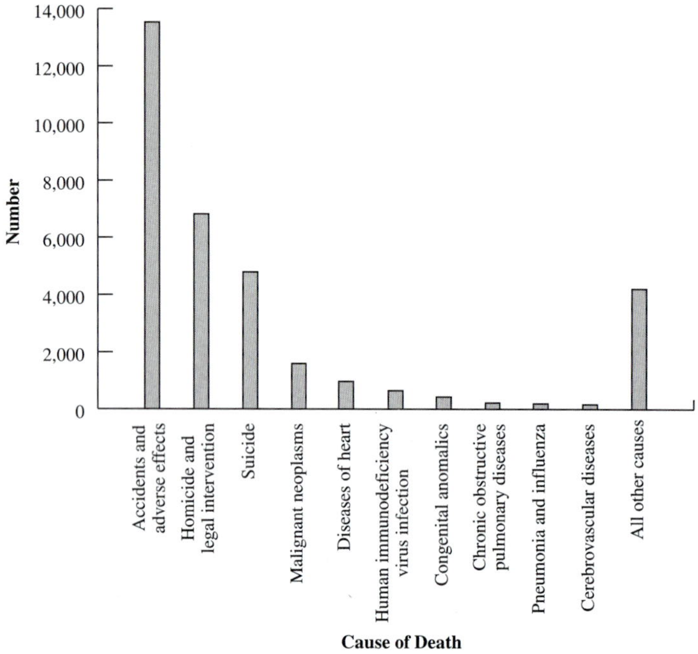
14,000
12,000
10,000
8,000
6,000
4,000
2,000
0
Number
Accidents and adverse effects
Homicide and legal intervention
Suicide
Malignant neoplasms
Diseases of heart
Human immunodeficiency virus infection
Congenital anomalics
Chronic obstructive pulmonary diseases
Pneumonia and influenza
Cerebrovascular diseases
All other causes
Cause of Death

(b)

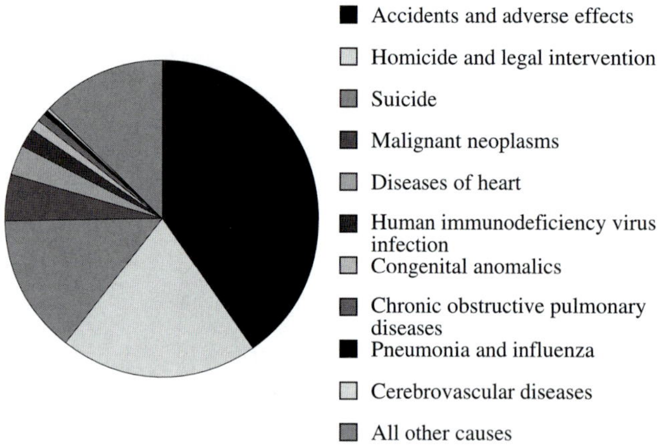
Accidents and adverse effects
Homicide and legal intervention
Suicide
Malignant neoplasms
Diseases of heart
Human immunodeficiency virus infection
Congenital anomalics
Chronic obstructive pulmonary diseases
Pneumonia and influenza
Cerebrovascular diseases
All other causes

(d) Accidents and adverse effects.

21. (a) 13 **(b)** 20; 24 **(c)** 5 **(d)**

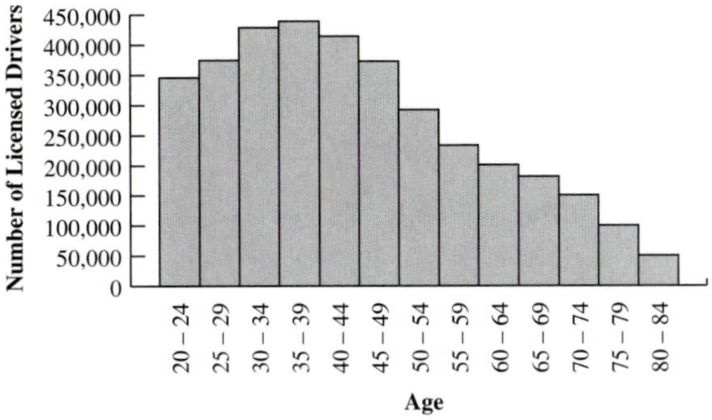

(e)

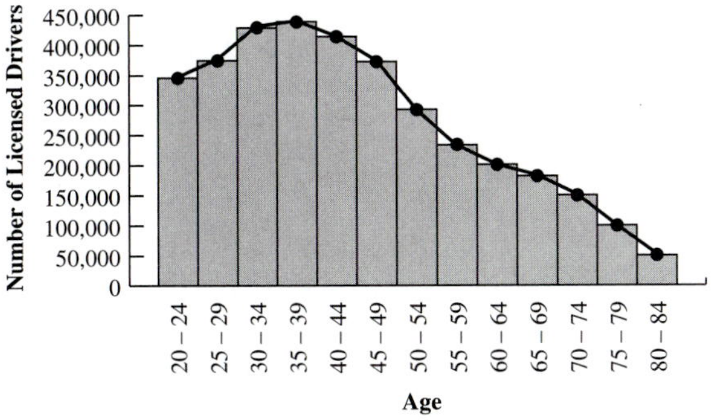

(f) 35–39 **(g)** 80–84

23. (a) 15 **(b)** 0; 999 **(c)** 1000 **(d)**

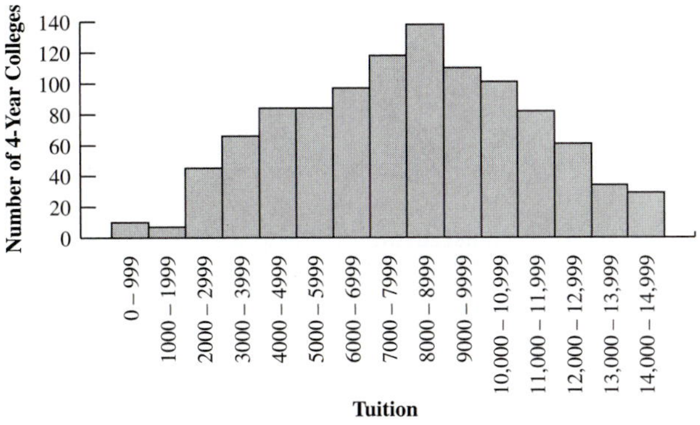

(e)

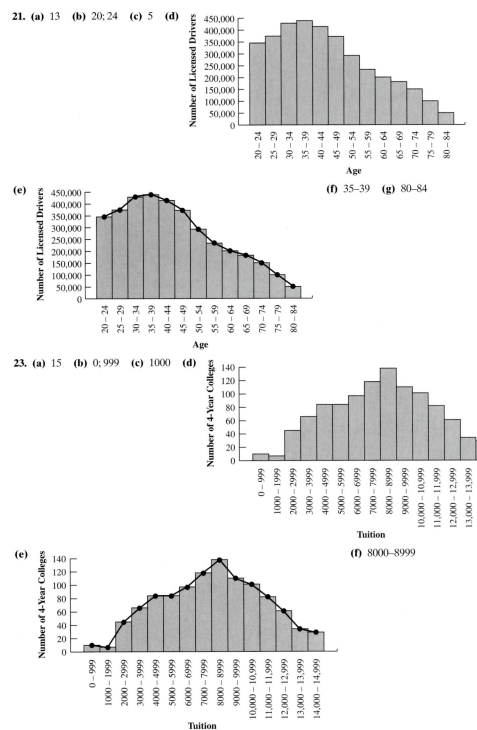

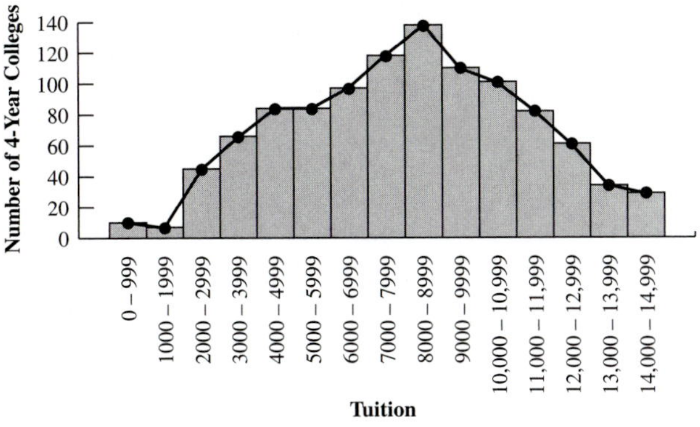

(f) 8000–8999

1.2 Exercises

1. **(a)** Quadrant II **(b)** Positive x-axis **(c)** Quadrant III **(d)** Quadrant I **(e)** Negative y-axis **(f)** Quadrant IV

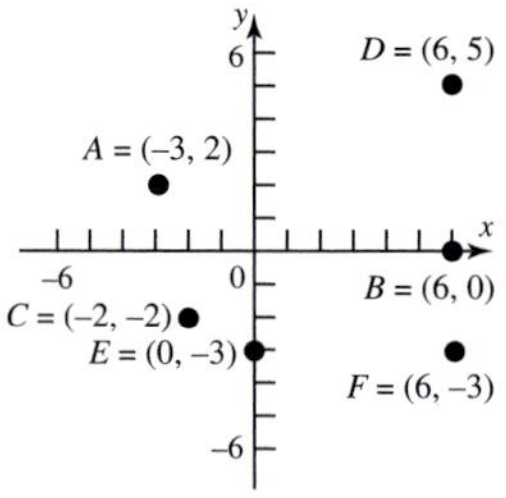

3. The points will be on a vertical line that is 2 units to the right of the y-axis.

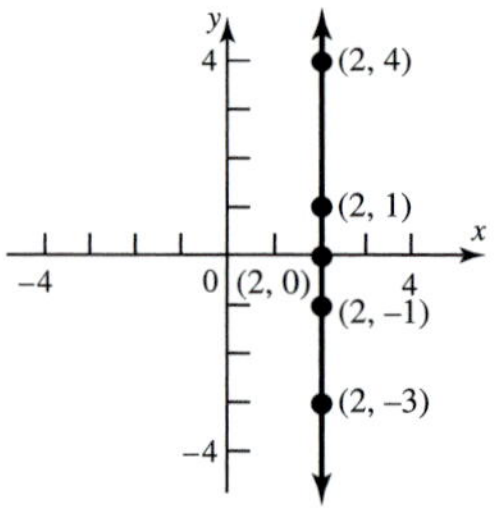

5. $(-1, 4)$ **7.** $(3, 1)$ **9.** $Xmin = -11, Xmax = 5, Xscl = 1, Ymin = -3, Ymax = 6, Yscl = 1$ **11.** $Xmin = -30, Xmax = 50, Xscl = 10, Ymin = -90, Ymax = 50, Yscl = 10$ **13.** $Xmin = -10, Xmax = 110, Xscl = 10, Ymin = -10, Ymax = 160, Yscl = 10$ **15.** $Xmin = -6, Xmax = 6, Xscl = 2, Ymin = -4, Ymax = 4, Yscl = 2$ **17.** $Xmin = -9, Xmax = 9, Xscl = 3, Ymin = -4, Ymax = 4, Yscl = 2$ **19.** $Xmin = -6, Xmax = 6, Xscl = 1, Ymin = -8, Ymax = 8, Yscl = 2$ **21.** $Xmin = -6, Xmax = 6, Xscl = 2, Ymin = -1, Ymax = 3, Yscl = 1$ **23.** $Xmin = 3, Xmax = 9, Xscl = 1, Ymin = 2, Ymax = 10, Yscl = 2$ **25.** $\sqrt{5}$ **27.** $2\sqrt{2}$ **29.** $2\sqrt{17}$ **31.** $\sqrt{85}$ **33.** $\sqrt{53}$ **35.** 2.625 **37.** $\sqrt{a^2 + b^2}$ **39.** $4\sqrt{10}$ **41.** $2\sqrt{65}$

43. $d(A, B) = \sqrt{13}$
$d(B, C) = \sqrt{13}$
$d(A, C) = \sqrt{26}$
$(\sqrt{13})^2 + (\sqrt{13})^2 = (\sqrt{26})^2$
Area $= \dfrac{13}{2}$ square units

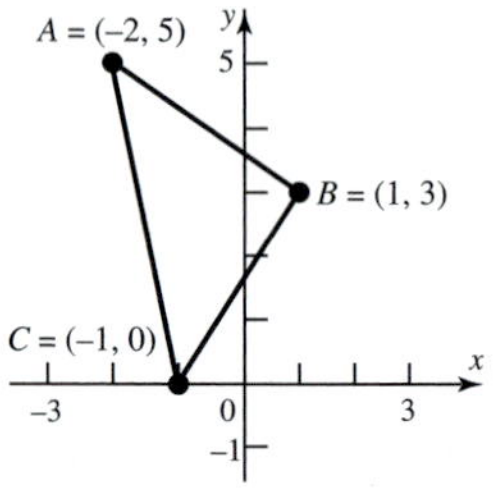

45. $d(A, B) = \sqrt{130}$
$d(B, C) = \sqrt{26}$
$d(A, C) = \sqrt{104}$
$(\sqrt{26})^2 + (\sqrt{104})^2 = (\sqrt{130})^2$
Area = 26 square units

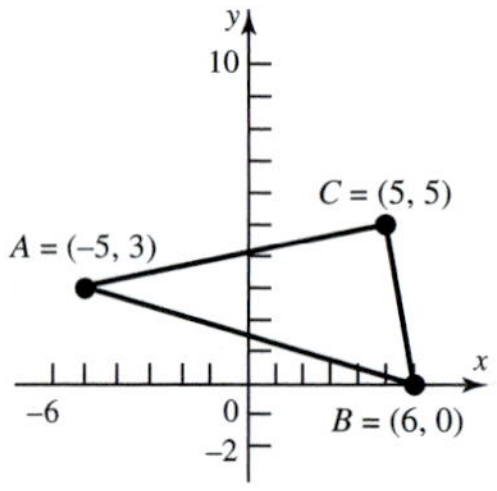

47. $d(A, B) = 4$
$d(A, C) = 5$
$d(B, C) = \sqrt{41}$
$4^2 + 5^2 = 16 + 25 = (\sqrt{41})^2$
Area = 10 square units

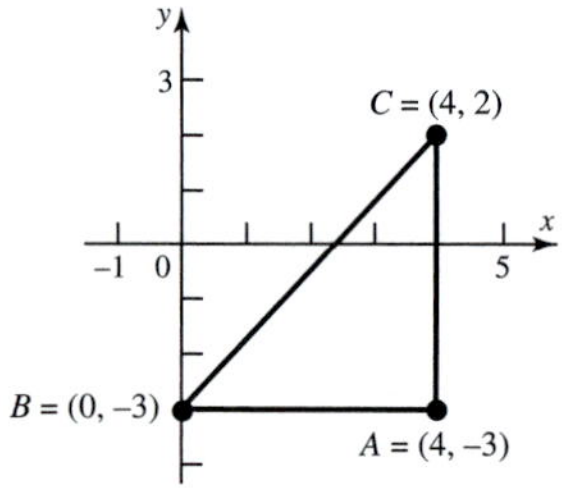

49. $(2,2); (2, -4)$ **51.** $(0, 0); (8, 0)$ **53.** $(4, -1)$ **55.** $\left(\dfrac{3}{2}, 1\right)$ **57.** $(5, -1)$ **59.** $(1.05, 0.7)$ **61.** $\left(\dfrac{a}{2}, \dfrac{b}{2}\right)$ **63.** **(a)** December 1994 **(b)** January 1996 **(c)** 7.5%

65. **(a)**

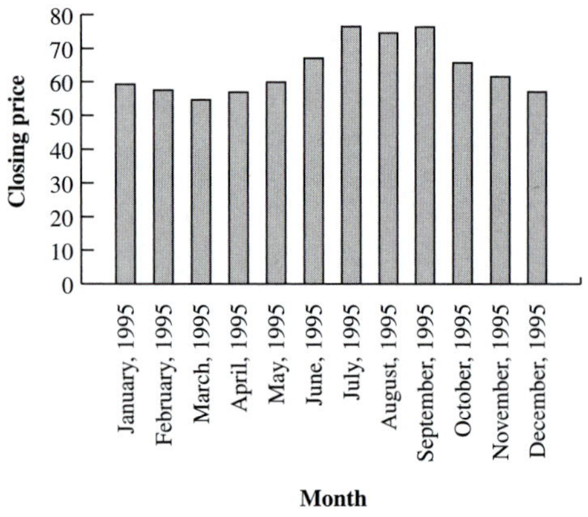

(b)

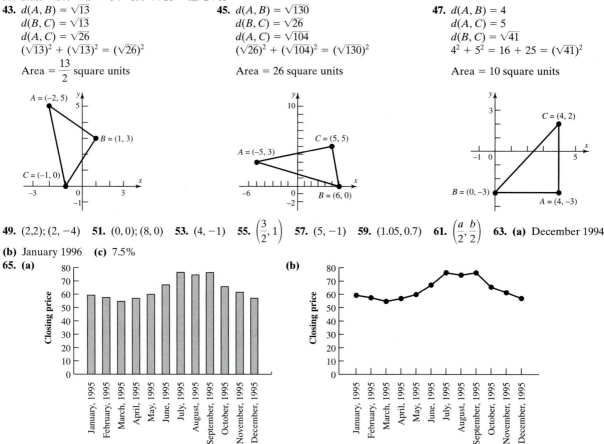

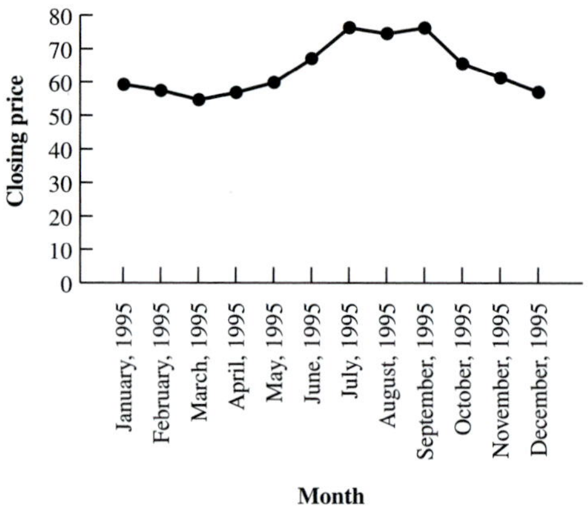

(c)

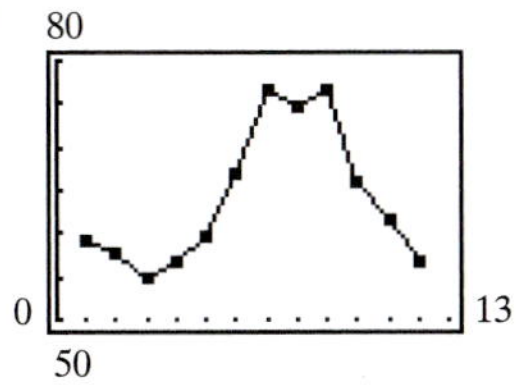

(d) July, 1995 **(e)** March, 1995

67. (a)

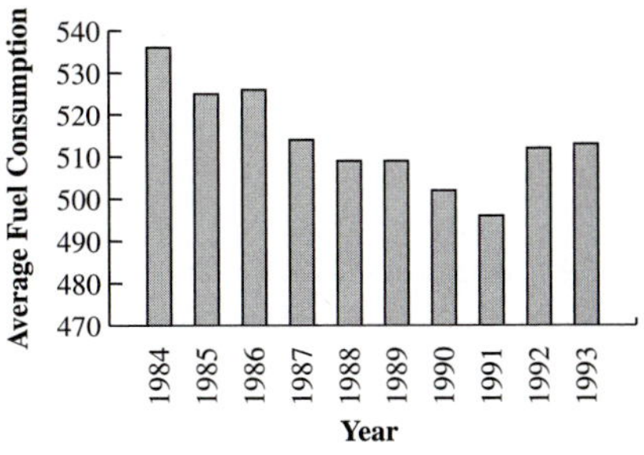

(b)

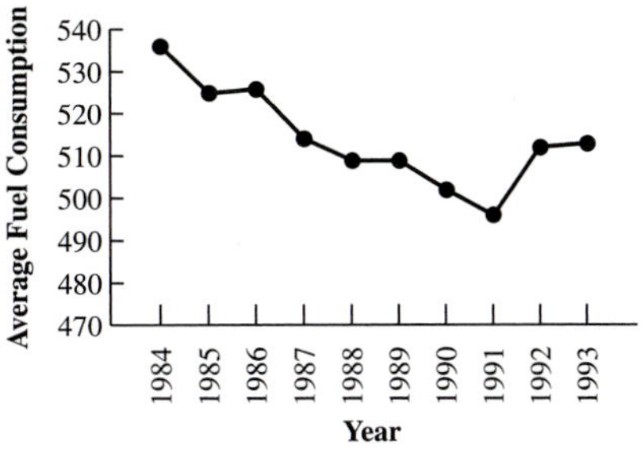

(c)

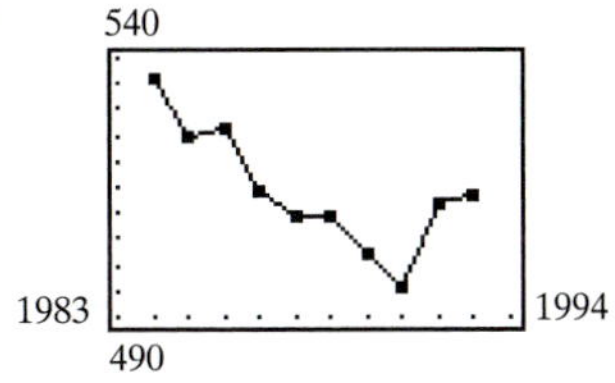

(d) 1984 **(e)** 1991

69. (a)

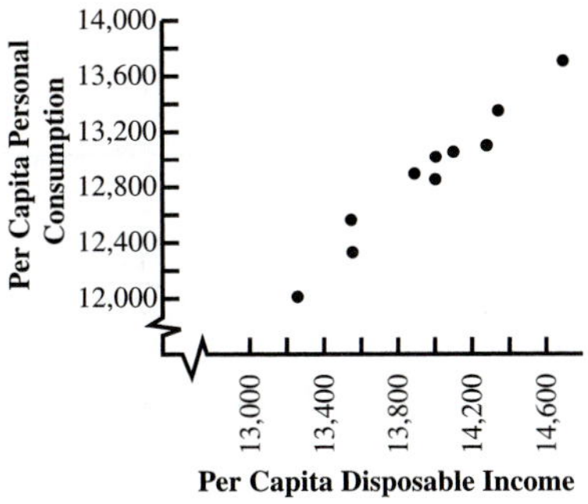

(b)

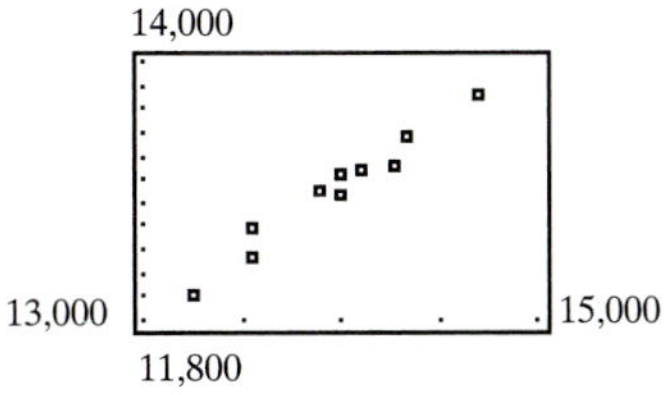

71. (a)

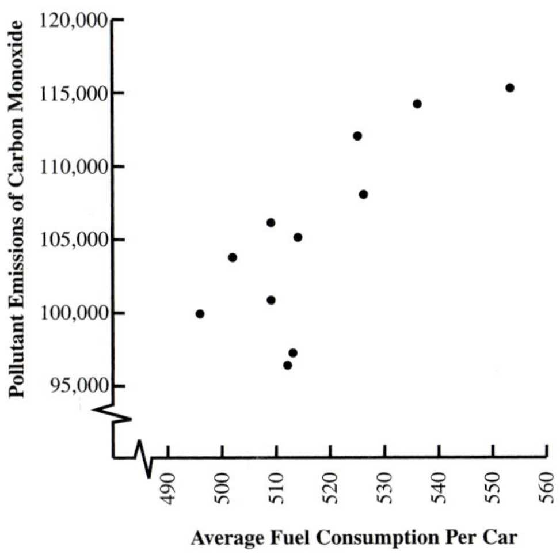

(b)

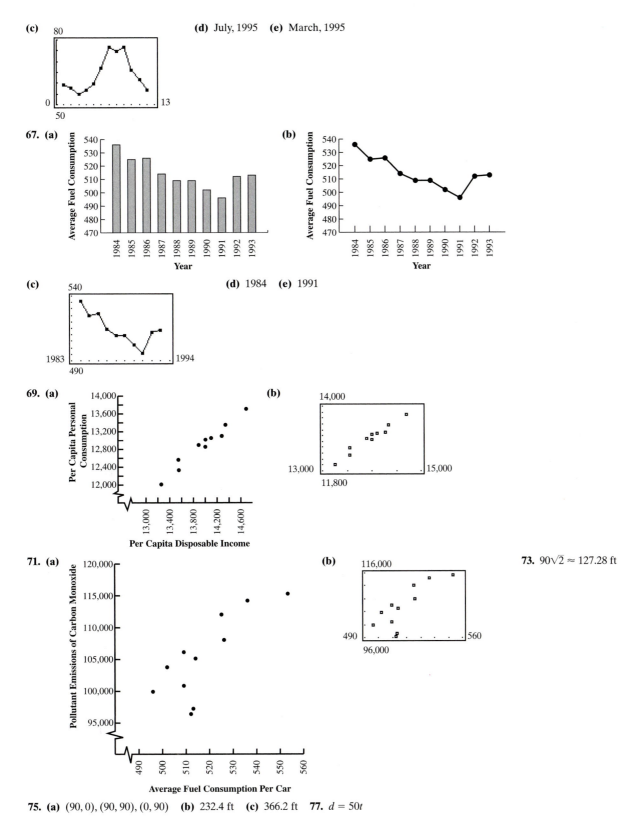

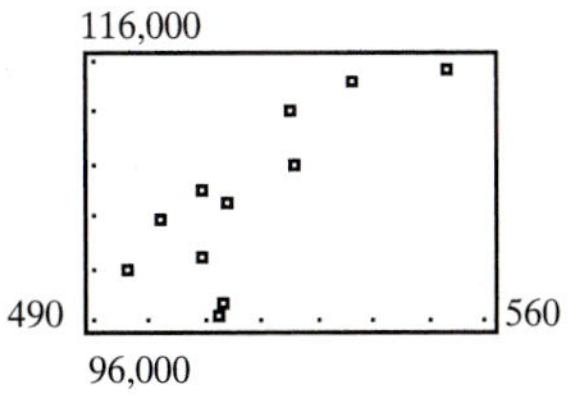

73. $90\sqrt{2} \approx 127.28$ ft

75. (a) (90, 0), (90, 90), (0, 90) **(b)** 232.4 ft **(c)** 366.2 ft **77.** $d = 50t$

1.3 Exercises

1.
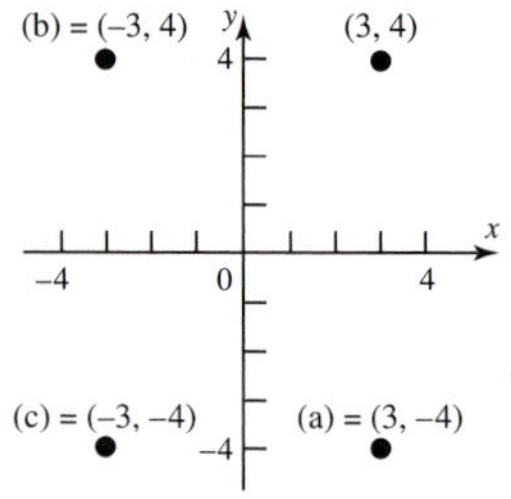

3.
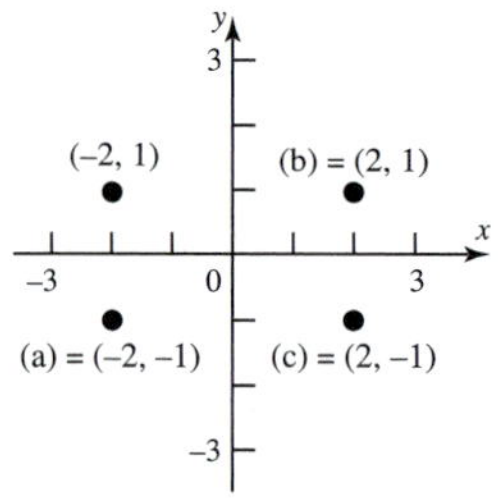

5.
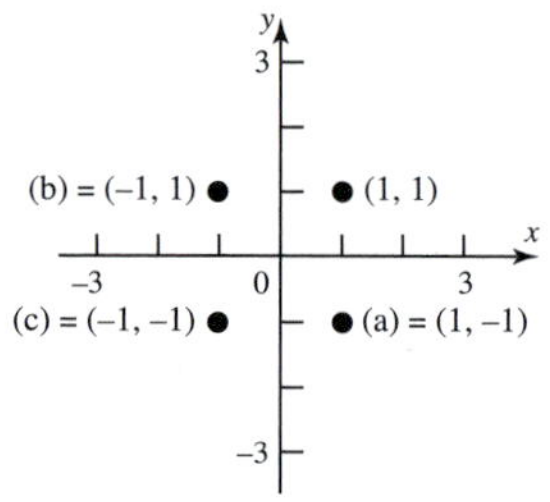

7.
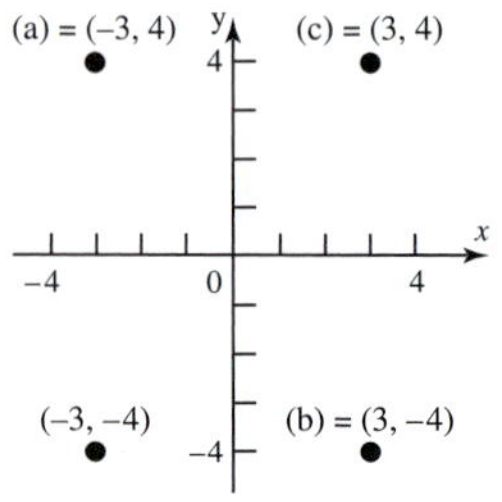

9.
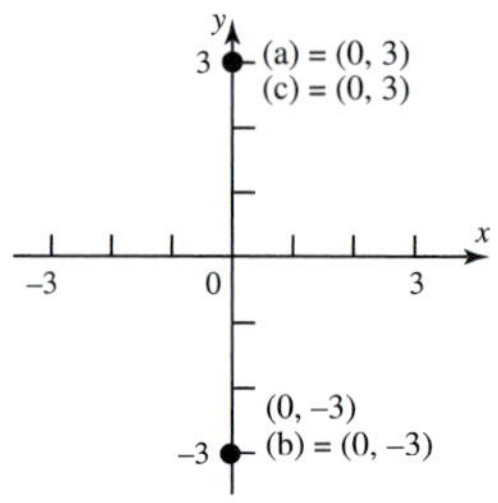

11. (a)
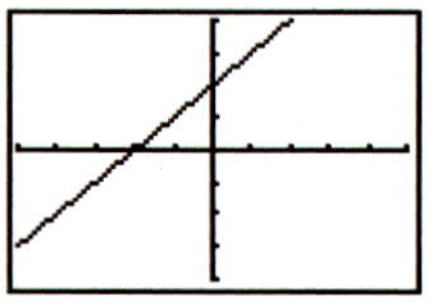
(b)
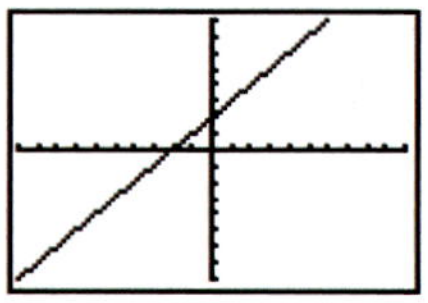
(c)
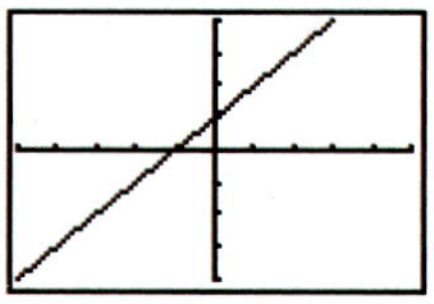
(d)
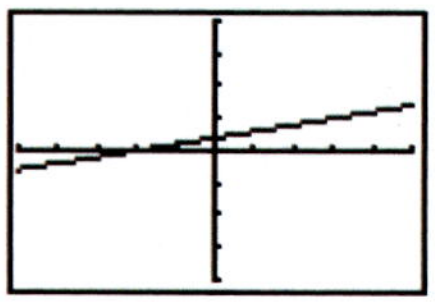

13. (a)
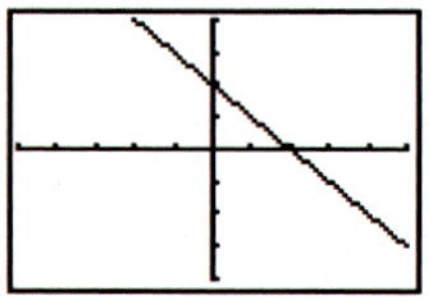
(b)
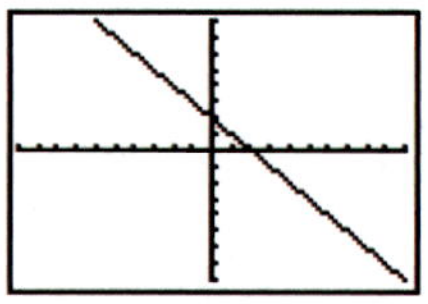
(c)
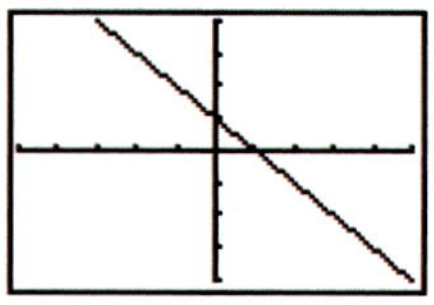
(d)
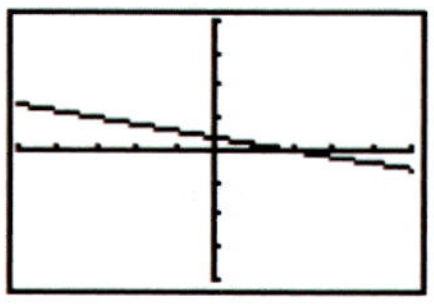

15. (a)
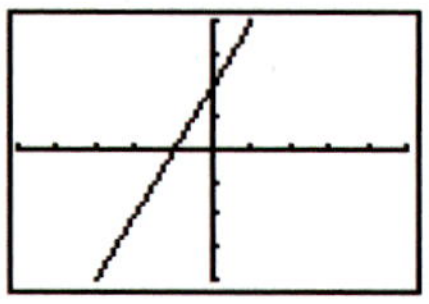
(b)
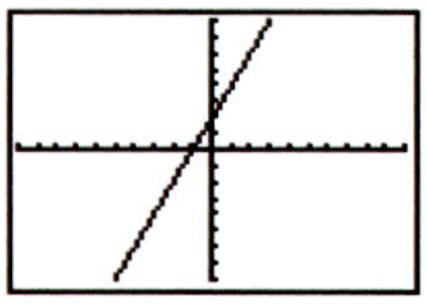
(c)
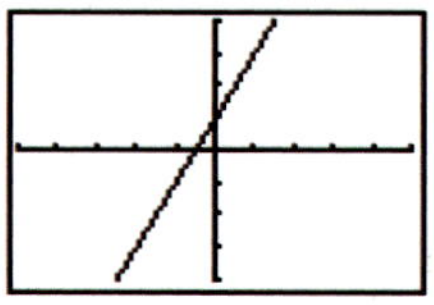
(d)
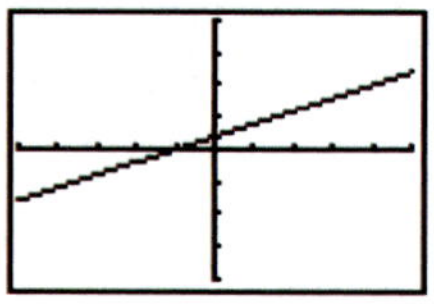

17. (a)
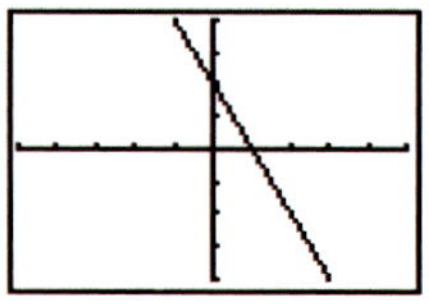
(b)
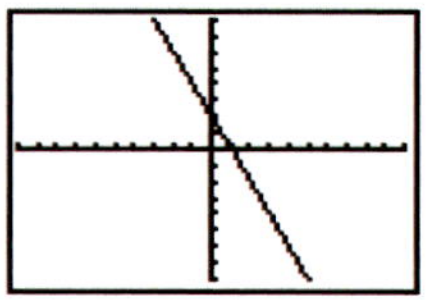
(c)
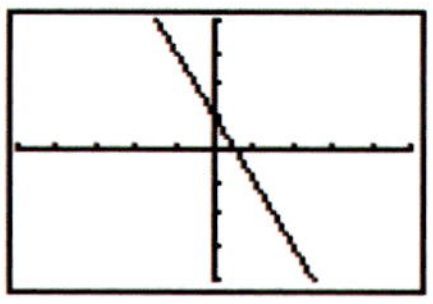
(d)
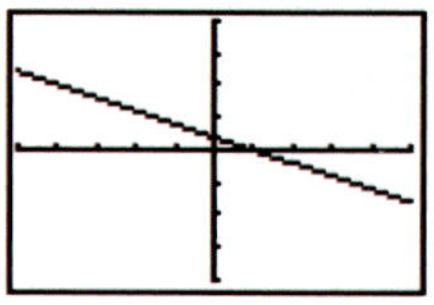

19. (a)
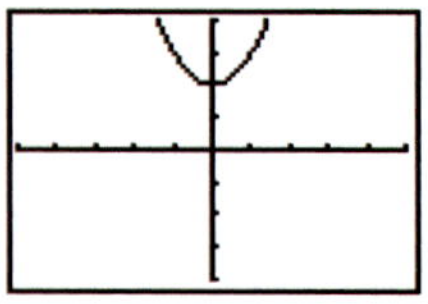
(b)
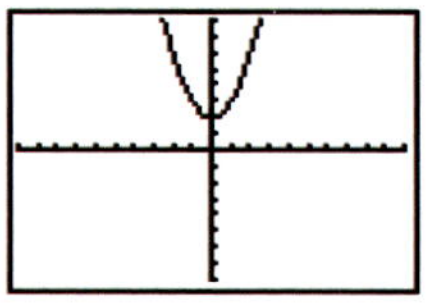
(c)
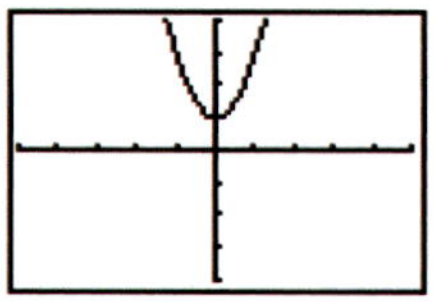
(d)
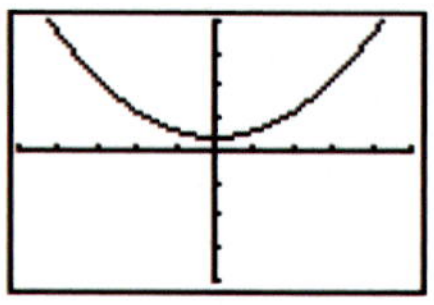

21. (a)
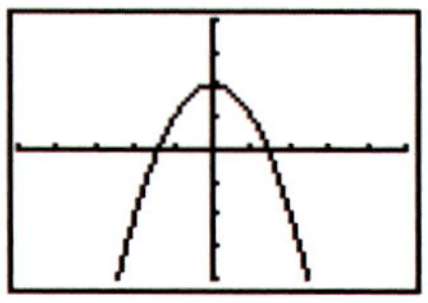
(b)
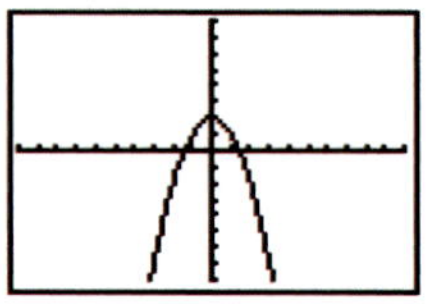
(c)
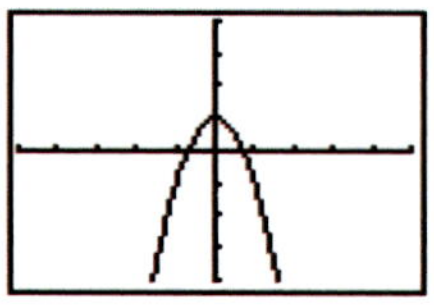
(d)
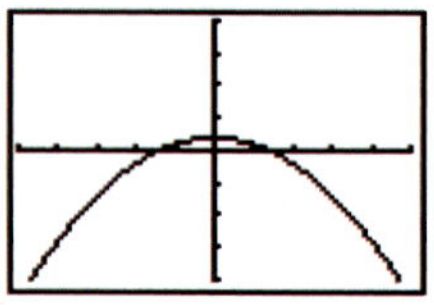

23. **(a)** 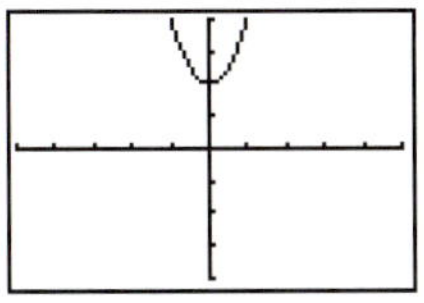 **(b)** 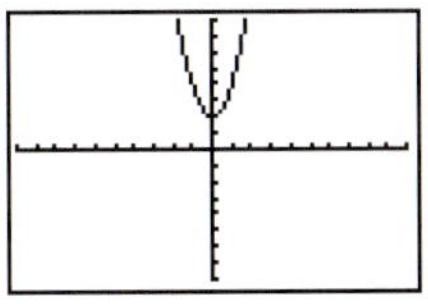 **(c)** 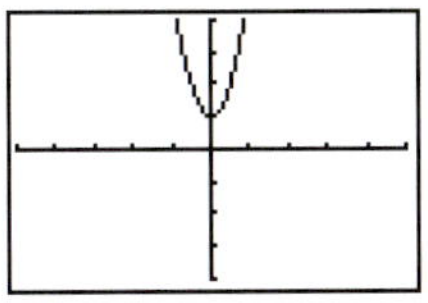 **(d)** 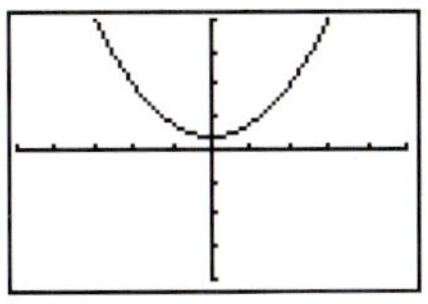

25. **(a)** 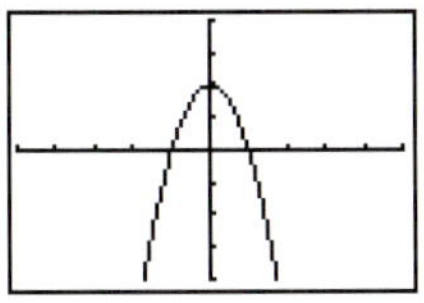 **(b)** 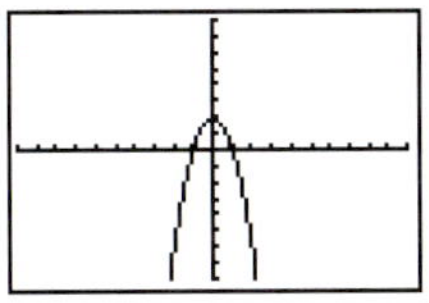 **(c)** 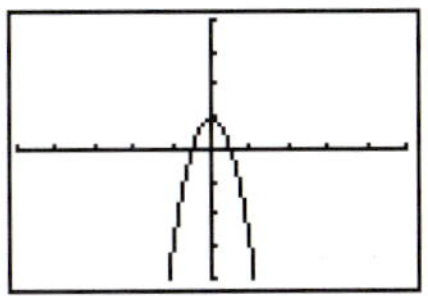 **(d)** 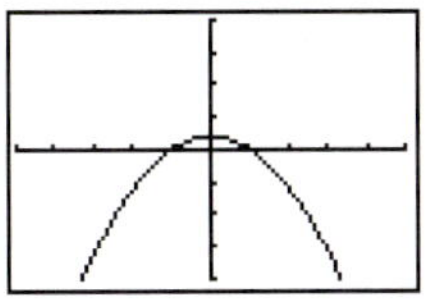

27. **(a)** 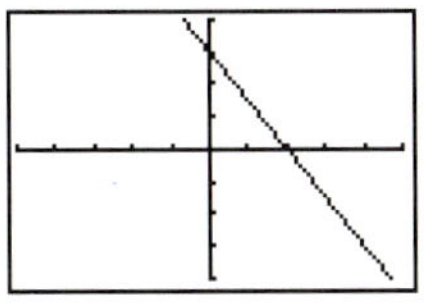 **(b)** 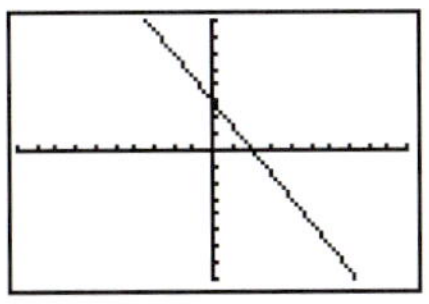 **(c)** 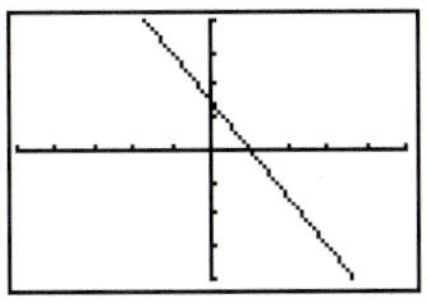 **(d)** 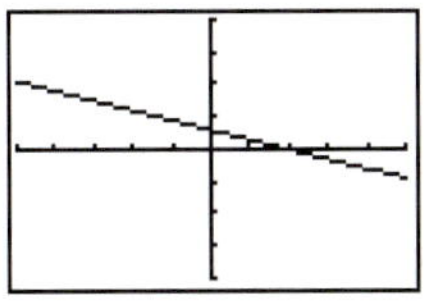

29. **(a)** 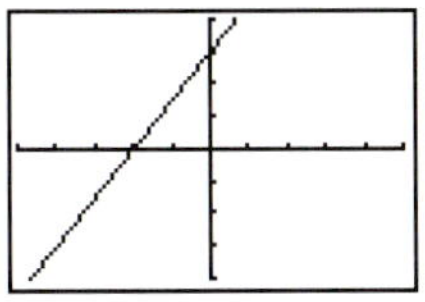 **(b)** 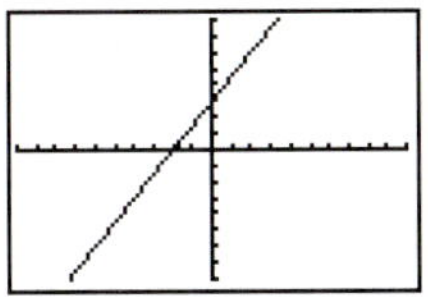 **(c)** 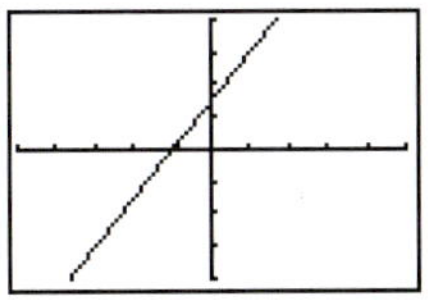 **(d)** 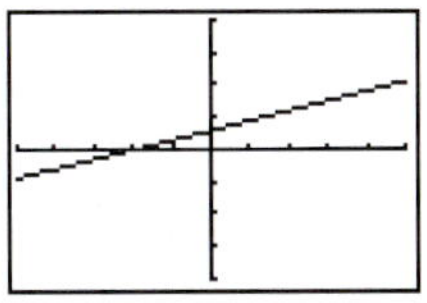

31. **(a)** $(-1, 0), (1, 0)$ **(b)** x-axis, y-axis, origin **33.** **(a)** $\left(-\frac{\pi}{2}, 0\right), (0, 1), \left(\frac{\pi}{2}, 0\right)$ **(b)** y-axis **35.** **(a)** $(0, 0)$ **(b)** x-axis
37. **(a)** $(1, 0)$ **(b)** none **39.** **(a)** $(-3, 0), (0, 2), (3, 0)$ **(b)** y-axis **41.** **(a)** $(x, 0), 0 \le x < 2$ **(b)** none
43. **(a)** $(-1.5, 0), (0, -2), (1.5, 0)$ **(b)** y-axis **45.** **(a)** none **(b)** origin **47.** $(0, 0)$ is on the graph **49.** $(0, 2)$ is on the graph
51. $(0, 2)$ and $(\sqrt{2}, \sqrt{2})$ are on the graph **53.** $-\frac{2}{5}$ **55.** $2a + 3b = 6$

57.
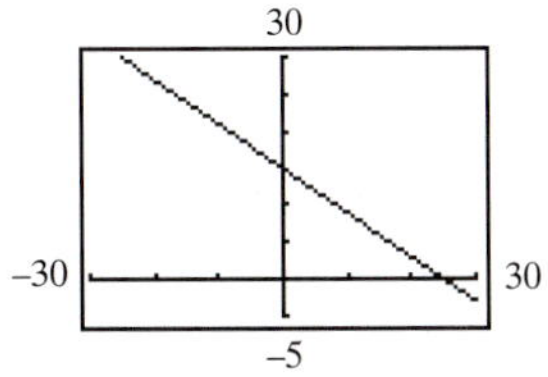

59.
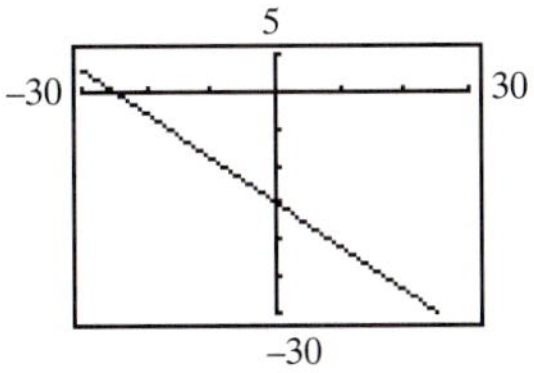

61.
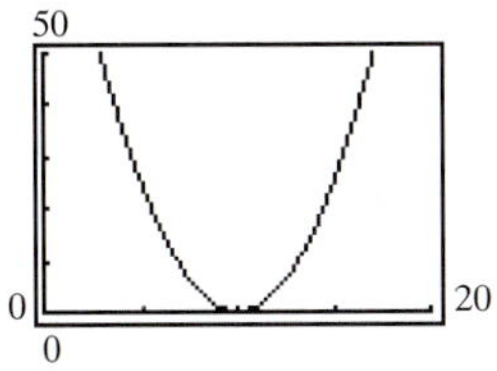

63.
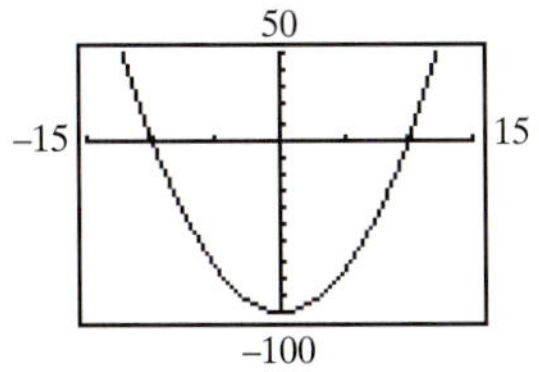

65.
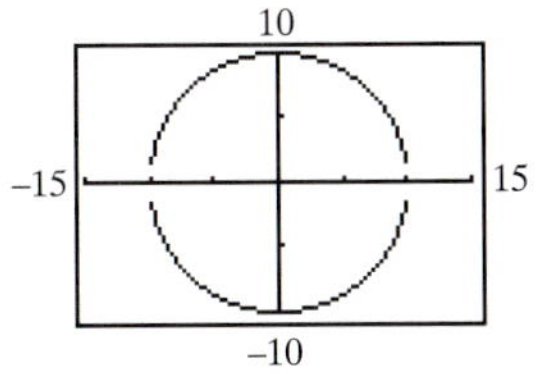

67.
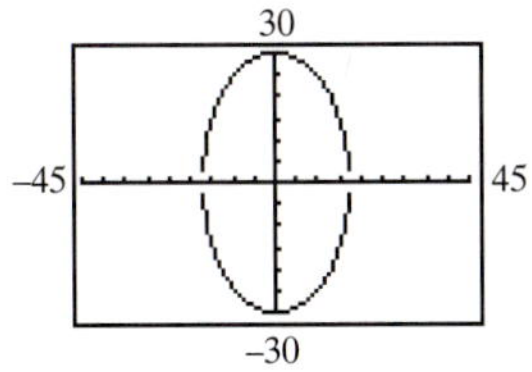

69.
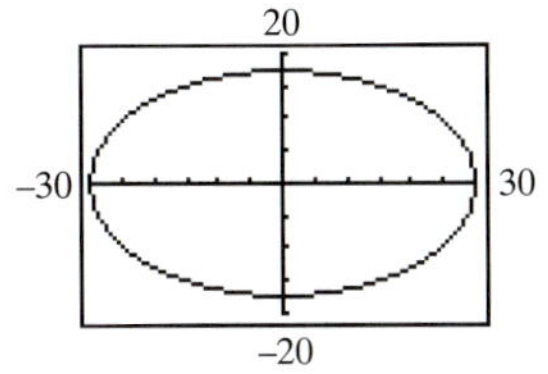

71.
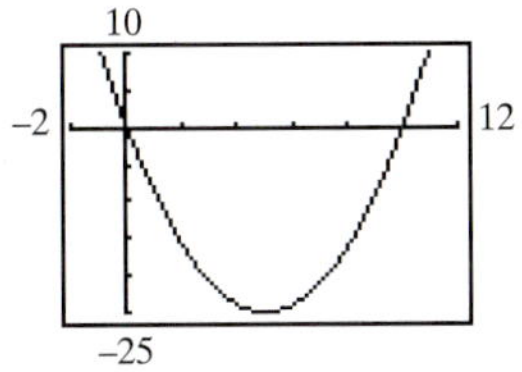

73.
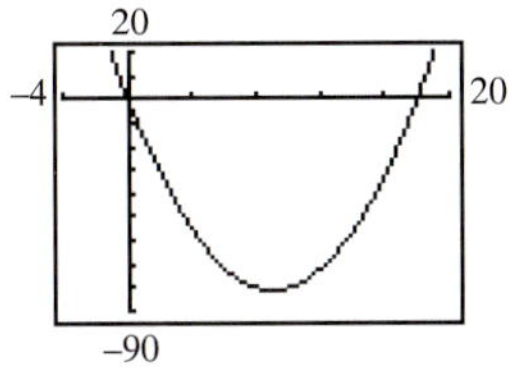

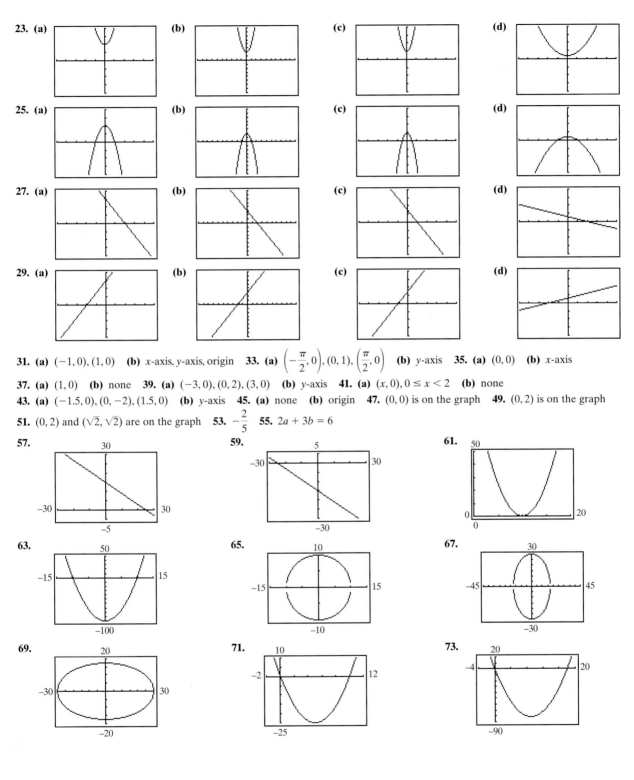

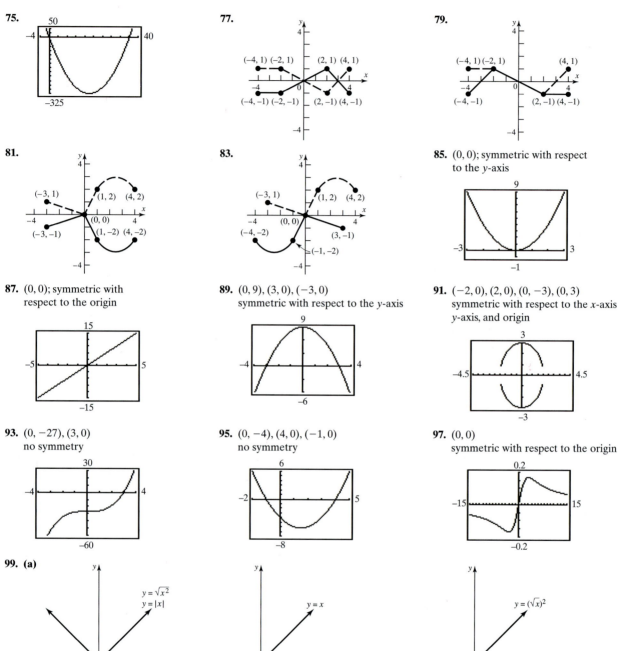

75.

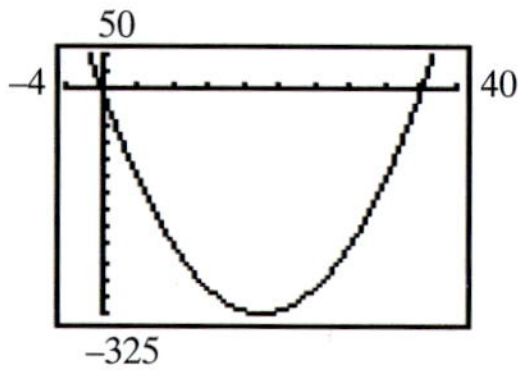

77.

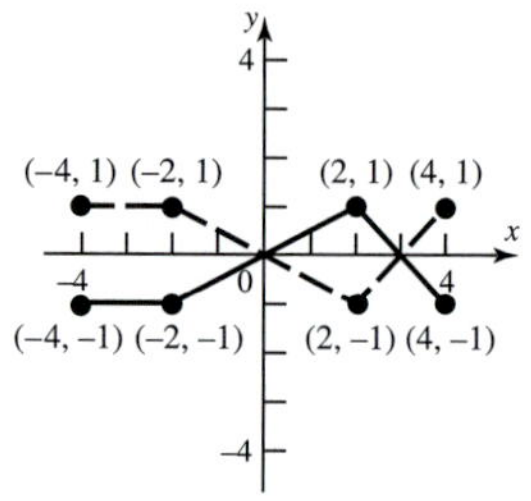

79.

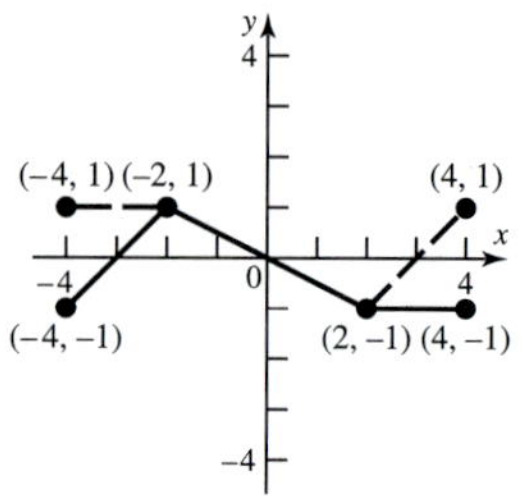

81.

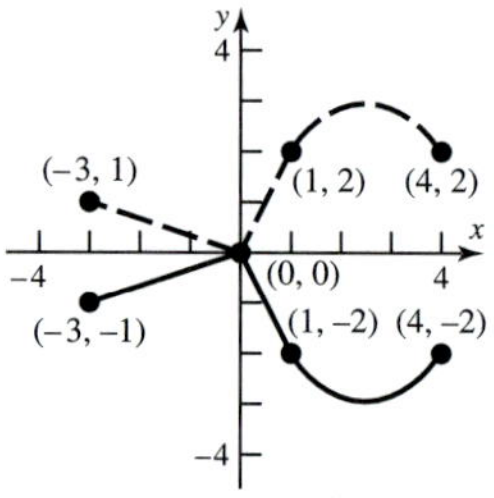

83.

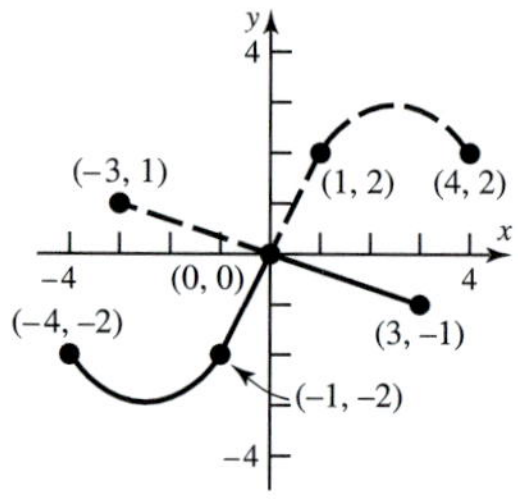

85. (0, 0); symmetric with respect to the y-axis

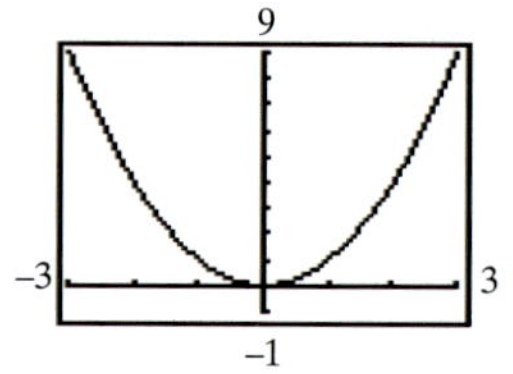

87. (0, 0); symmetric with respect to the origin

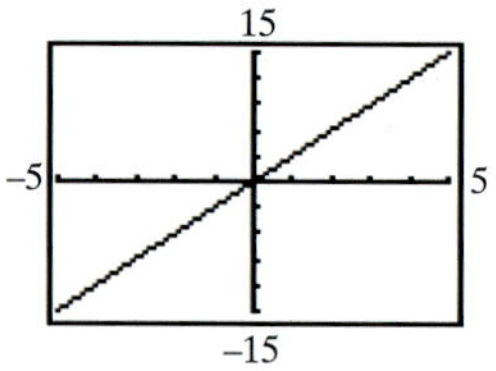

89. (0, 9), (3, 0), (−3, 0) symmetric with respect to the y-axis

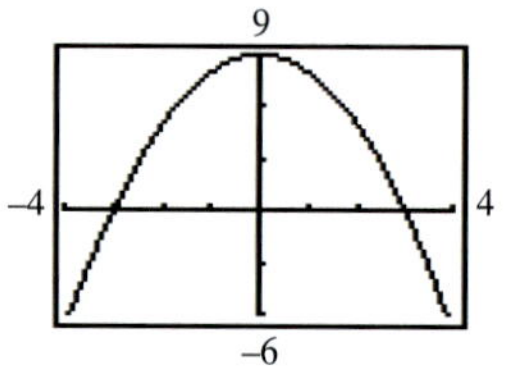

91. (−2, 0), (2, 0), (0, −3), (0, 3) symmetric with respect to the x-axis y-axis, and origin

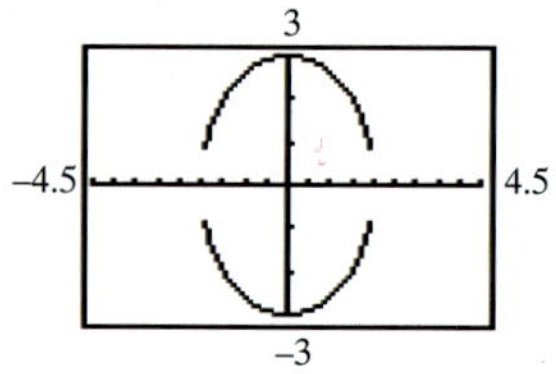

93. (0, −27), (3, 0) no symmetry

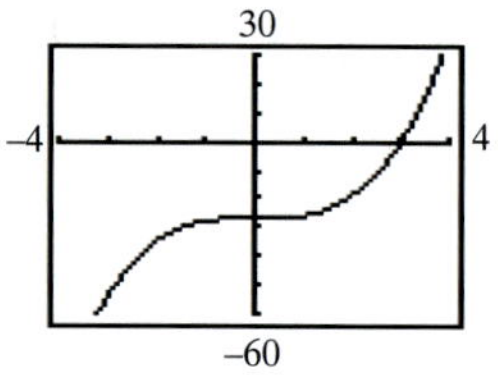

95. (0, −4), (4, 0), (−1, 0) no symmetry

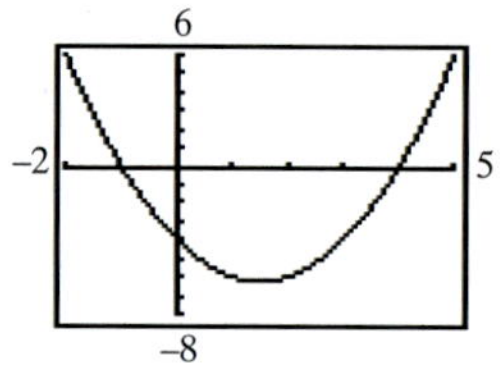

97. (0, 0) symmetric with respect to the origin

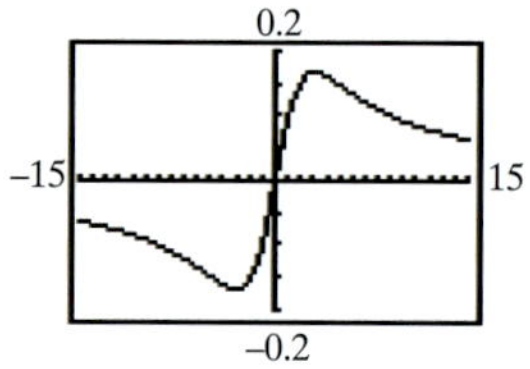

99. **(a)**

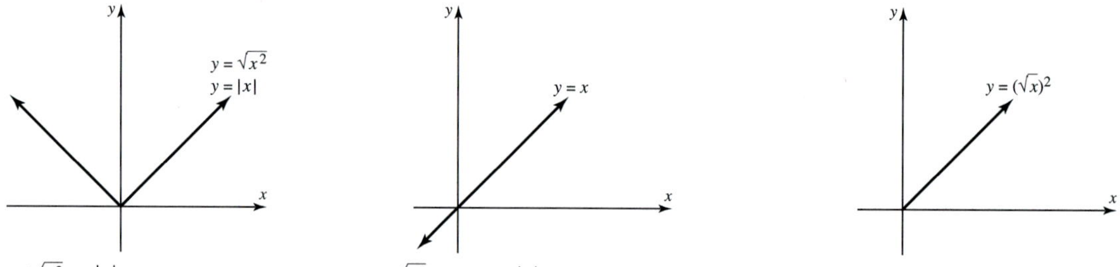

(b) Since $\sqrt{x^2} = |x|$, for all x, the graphs of $y = \sqrt{x^2}$ and $y = |x|$ are the same. **(c)** For $y = (\sqrt{x})^2$, the domain of the variable x is $x \geq 0$; for $y = x$, the domain of the variable x is all real numbers. Thus, $(\sqrt{x})^2 = x$ only for $x \geq 0$. **(d)** For $y = \sqrt{x^2}$, the range of the variable y is $y \geq 0$; for $y = x$, the range of the variable y is all real numbers. Also, $\sqrt{x^2} = |x|$, which equals x only if $x \geq 0$.

1.4 Exercises

1. **(a)** $\frac{1}{2}$ **(b)** For every 2 unit change in x, y will change 1 unit; if x increases by 2 units, y will increase by 1 unit.

3. **(a)** -1 **(b)** If x increases by 1 unit, y will decrease by 1 unit.

5. Slope $= -\frac{3}{2}$

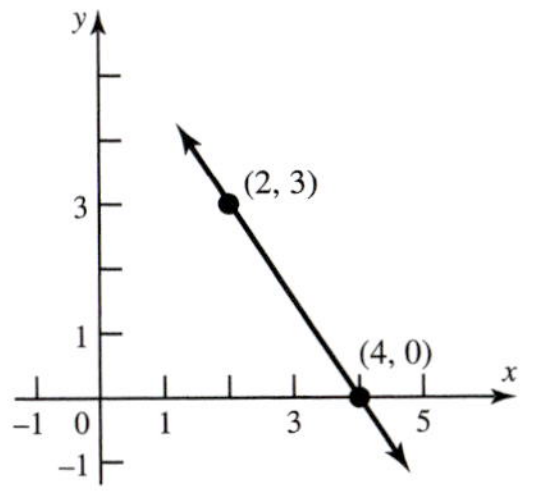

7. Slope $= -\frac{1}{2}$

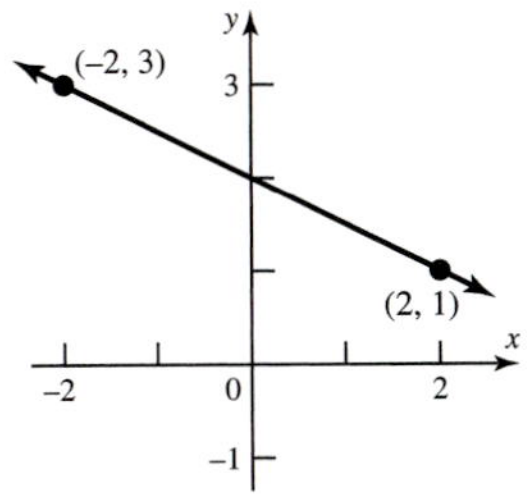

9. Slope $= 0$

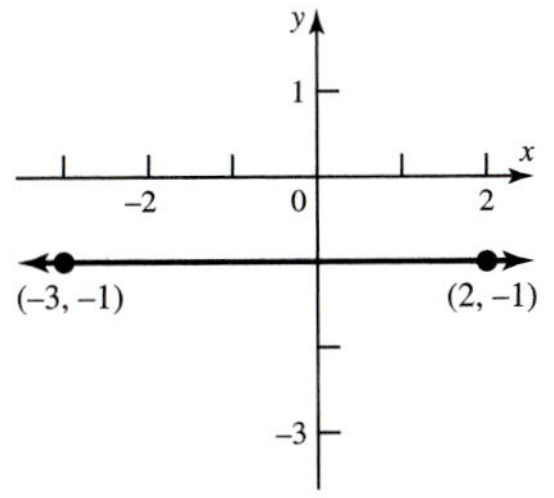

11. Slope undefined

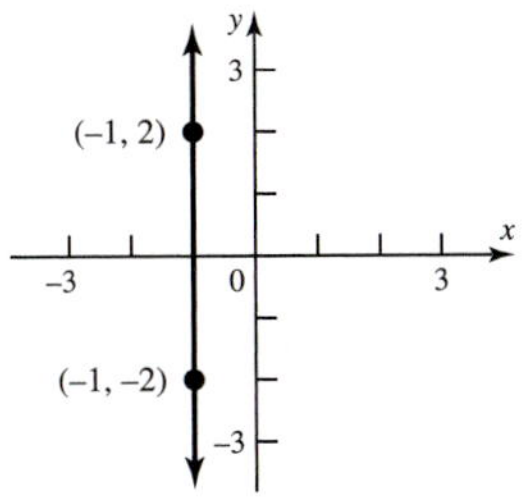

13. Slope $= \frac{\sqrt{3} - 3}{1 - \sqrt{2}} \approx 3.06$

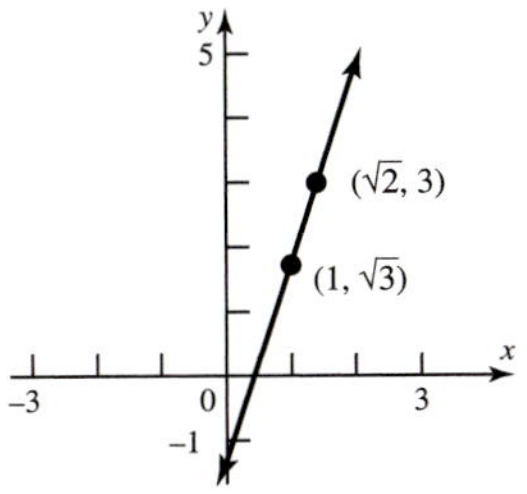

15.

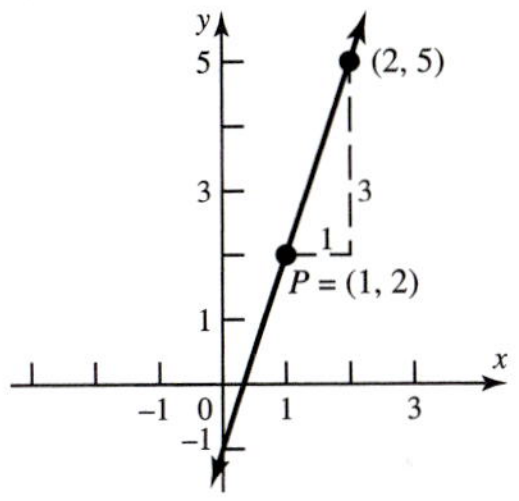

17.

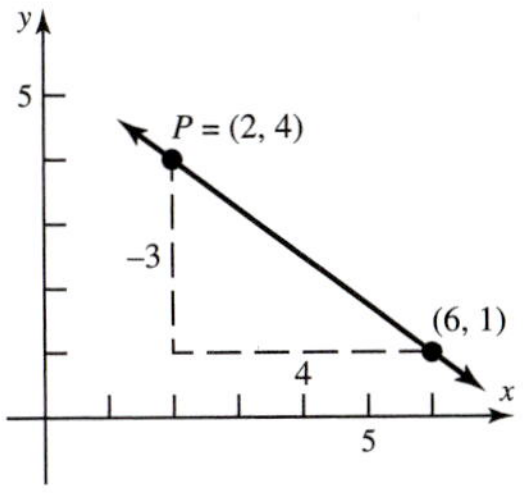

19.

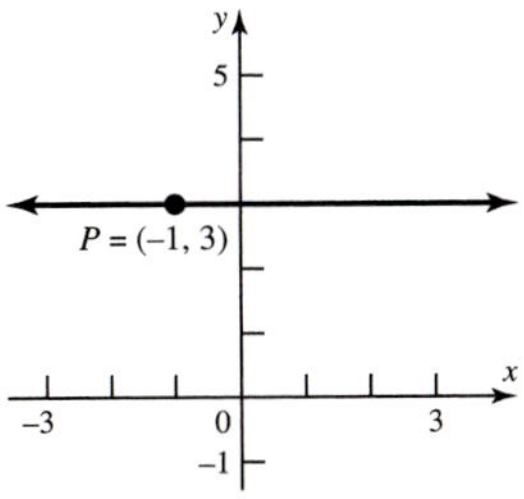

21.

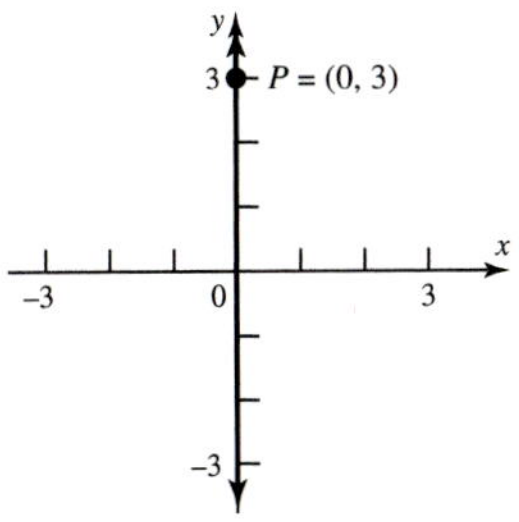

23. $x - 2y = 0$ or $y = \frac{1}{2}x$ **25.** $x + y - 2 = 0$ or $y = -x + 2$ **27.** $3x - y + 9 = 0$ or $y = 3x + 9$

29. $2x + 3y + 1 = 0$ or $y = -\frac{2}{3}x - \frac{1}{3}$ **31.** $x - 2y + 5 = 0$ or $y = \frac{1}{2}x + \frac{5}{2}$

33. $3x + y - 3 = 0$ or $y = -3x + 3$ **35.** $x - 2y - 2 = 0$ or $y = \frac{1}{2}x - 1$ **37.** $x - 2 = 0$; no slope-intercept form

39. Slope $= 2$; y-intercept $= 3$

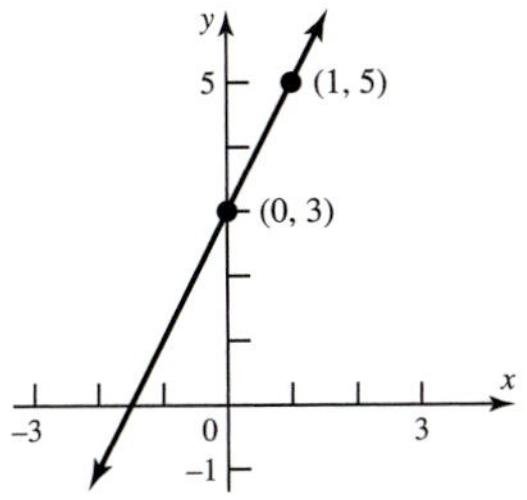

41. $y = 2x - 2$; Slope $= 2$; y-intercept $= -2$

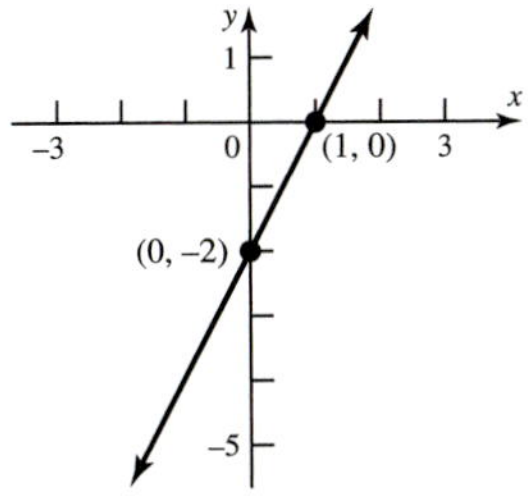

43. Slope $= \frac{1}{2}$; y-intercept $= 2$

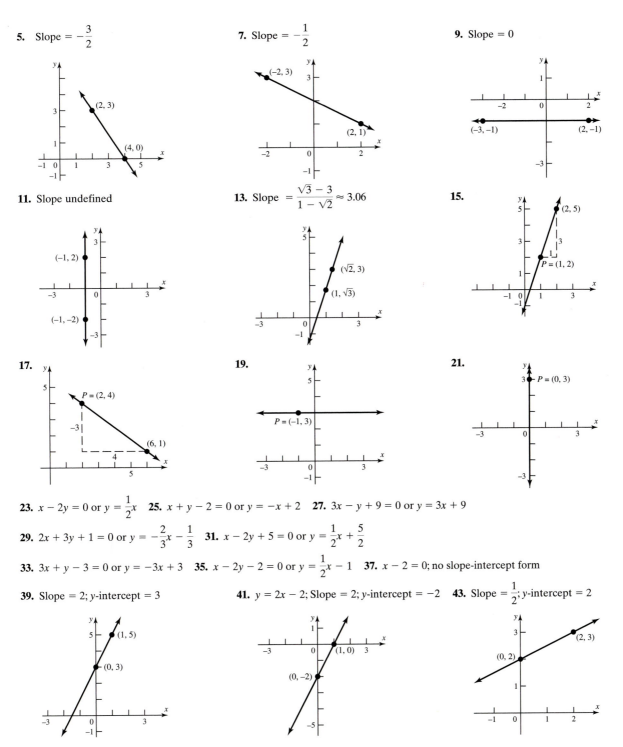

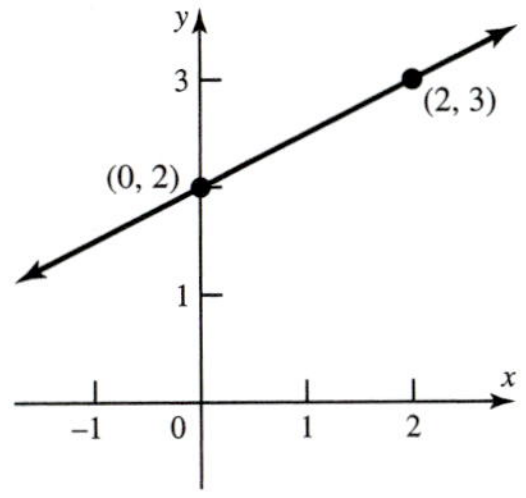

45. $y = -\frac{1}{2}x + 2$; Slope $= -\frac{1}{2}$
y-intercept $= 2$

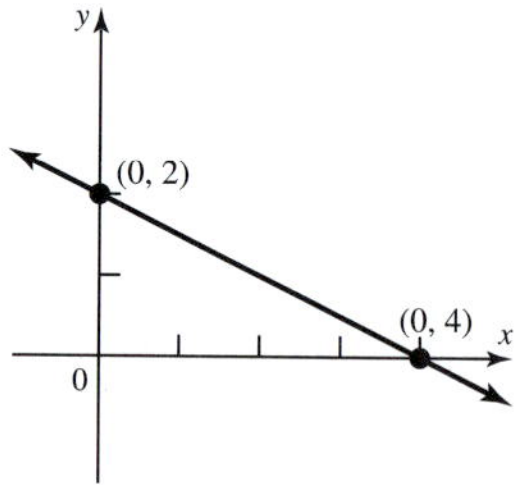

47. $y = \frac{2}{3}x - 2$; Slope $= \frac{2}{3}$
y-intercept $= -2$

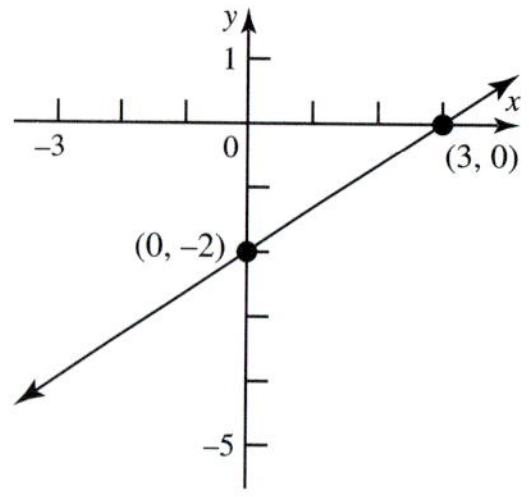

49. $y = -x + 1$; Slope $= -1$
y-intercept $= 1$

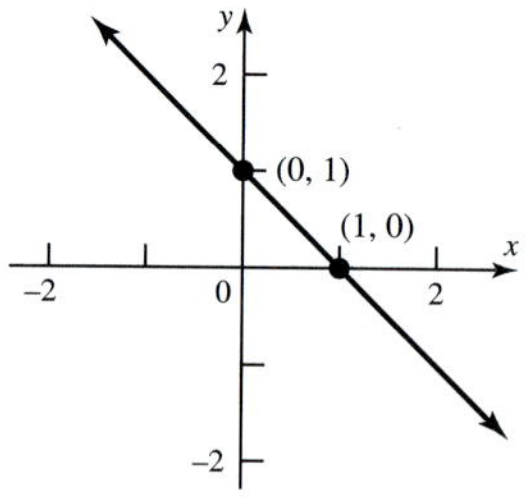

51. Slope undefined; no y-intercept

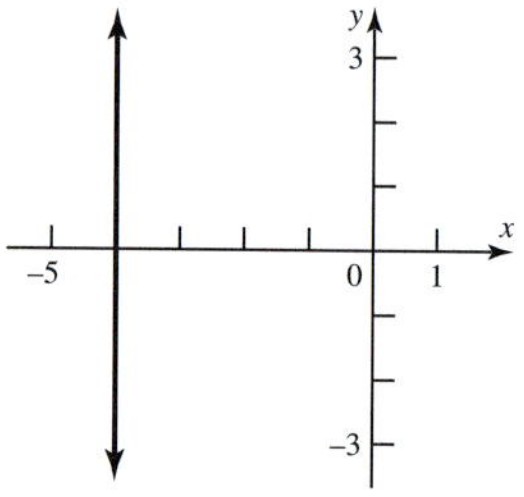

53. Slope $= 0$; y-intercept $= 5$

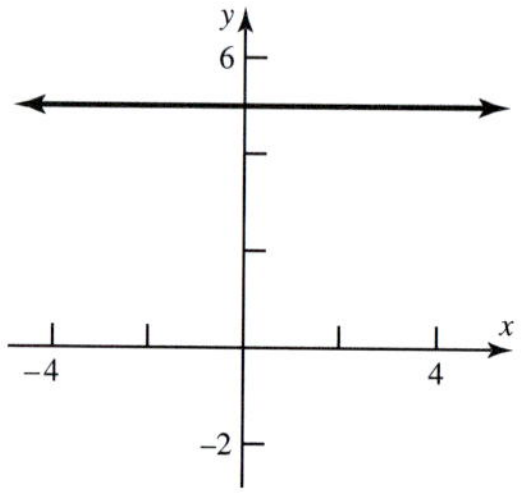

55. $y = x$; Slope $= 1$; y-intercept $= 0$

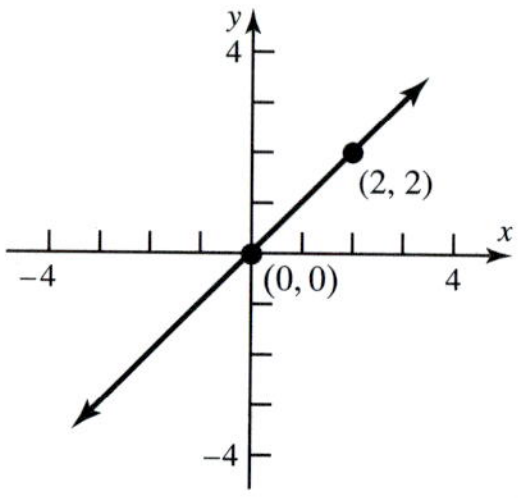

57. $y = \frac{3}{2}x$; Slope $= \frac{3}{2}$
y-intercept $= 0$

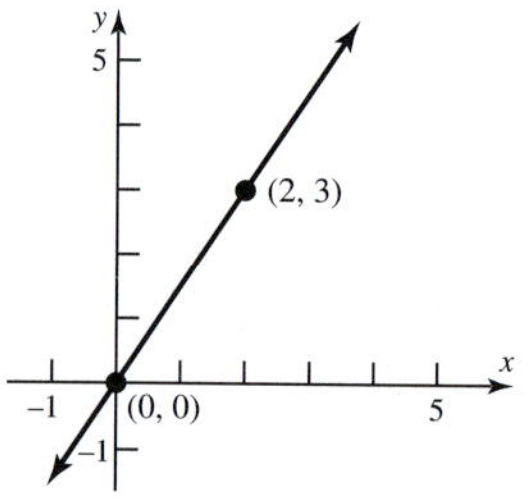

59. $y = 0$

61. $°C = \frac{5}{9}(°F - 32)$; approx. 21°C

63. **(a)** $P = 0.5x - 100$ **(b)** $400 **(c)** $2400

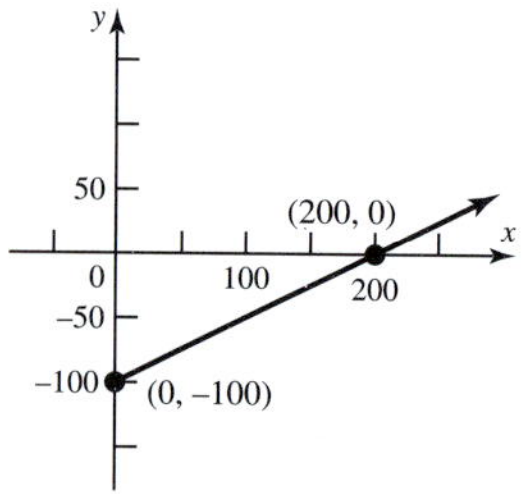

65. $C = 0.10819x + 9.06$; $C = \$41.52$; $C = \$106.43$

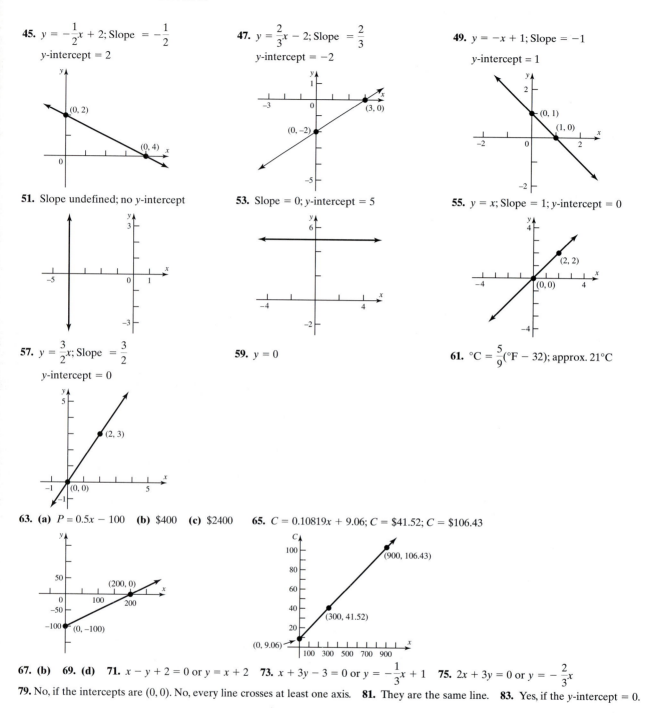

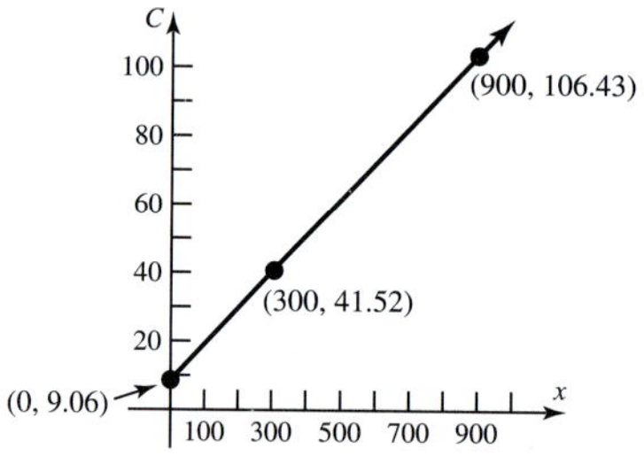

67. **(b)** **69.** **(d)** **71.** $x - y + 2 = 0$ or $y = x + 2$ **73.** $x + 3y - 3 = 0$ or $y = -\frac{1}{3}x + 1$ **75.** $2x + 3y = 0$ or $y = -\frac{2}{3}x$

79. No, if the intercepts are (0, 0). No, every line crosses at least one axis. **81.** They are the same line. **83.** Yes, if the y-intercept $= 0$.

1.5 Exercises

1. **(a)** 4 **(b)** $-\frac{1}{4}$ **3.** **(a)** $-\frac{1}{2}$ **(b)** 2 **5.** **(a)** $\frac{1}{2}$ **(b)** -2 **7.** **(a)** $-\frac{3}{5}$ **(b)** $\frac{5}{3}$ **9.** **(a)** Undefined **(b)** 0

11. $2x - y - 3 = 0$ or $y = 2x - 3$ **13.** $x + 2y - 5 = 0$ or $y = -\frac{1}{2}x + \frac{5}{2}$ **15.** $2x - y + 4 = 0$ or $y = 2x + 4$

17. $2x - y = 0$ or $y = 2x$ **19.** $x - 4 = 0$; no slope intercept form **21.** $2x + y = 0$ or $y = -2x$ **23.** $x - 2y + 3 = 0$ or $y = \frac{1}{2}x + \frac{3}{2}$

25. $y - 4 = 0$ or $y = 4$ **27.** Center $(2, 1)$; Radius 2; $(x - 2)^2 + (y - 1)^2 = 4$ **29.** Center $\left(\frac{5}{2}, 2\right)$; Radius $\frac{3}{2}$; $\left(x - \frac{5}{2}\right)^2 + (y - 2)^2 = \frac{9}{4}$

31. $(x - 1)^2 + (y + 1)^2 = 1$; $x^2 + y^2 - 2x + 2y + 1 = 0$

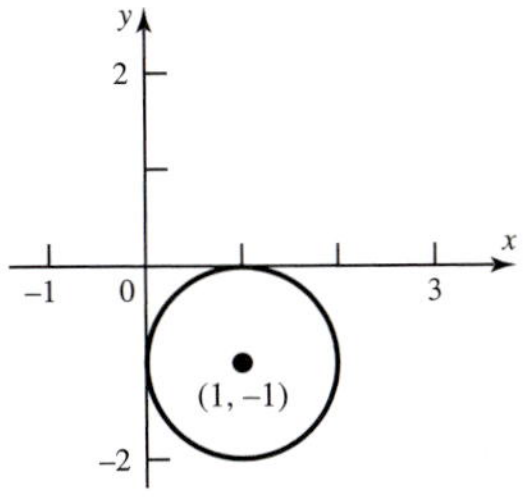

33. $x^2 + (y - 2)^2 = 4$; $x^2 + y^2 - 4y = 0$

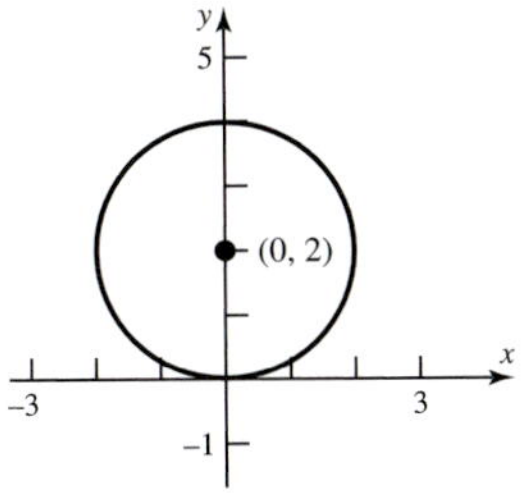

35. $(x - 4)^2 + (y + 3)^2 = 25$; $x^2 + y^2 - 8x + 6y = 0$

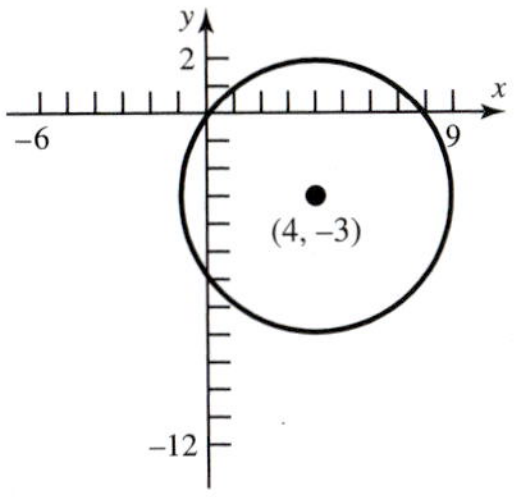

37. $x^2 + y^2 = 4$; $x^2 + y^2 - 4 = 0$

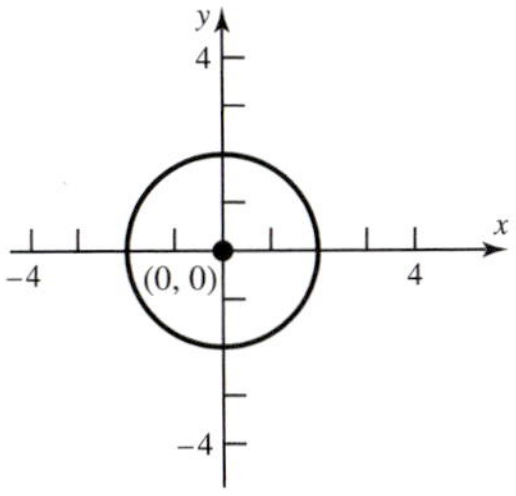

39. $\left(x - \frac{1}{2}\right)^2 + y^2 = \frac{1}{4}$
$x^2 + y^2 - x = 0$

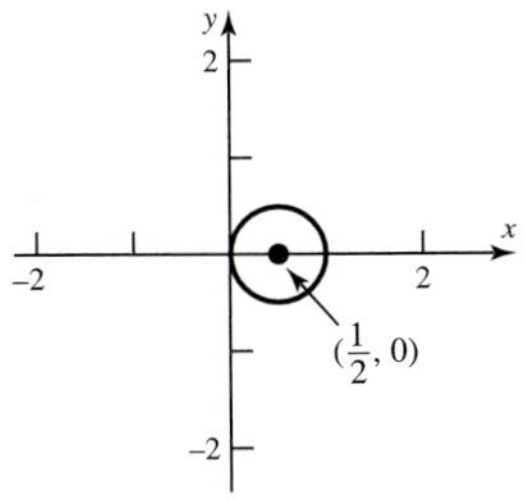

41. $r = 2$

$(h, k) = (0, 0)$

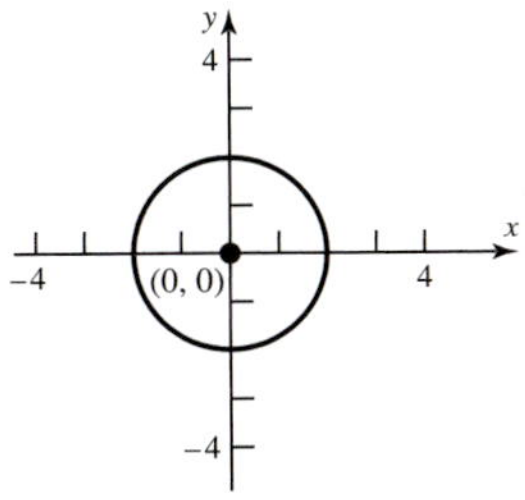

43. $r = 2$

$(h, k) = (3, 0)$

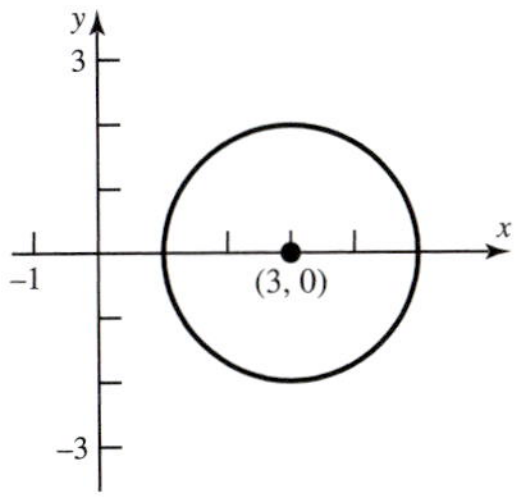

45. $r = 3$

$(h, k) = (-2, 2)$

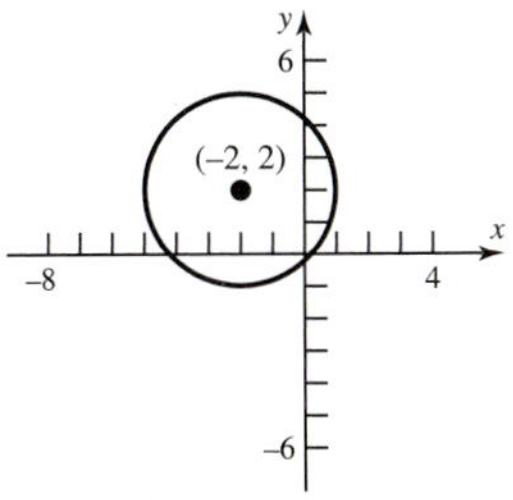

47. $r = \frac{1}{2}$

$(h, k) = \left(\frac{1}{2}, -1\right)$

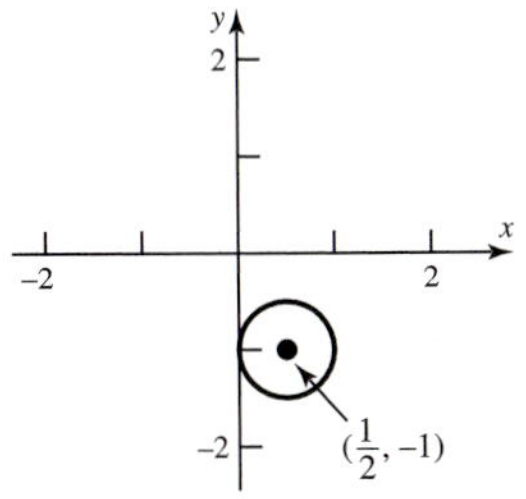

49. $r = 5$

$(h, k) = (3, -2)$

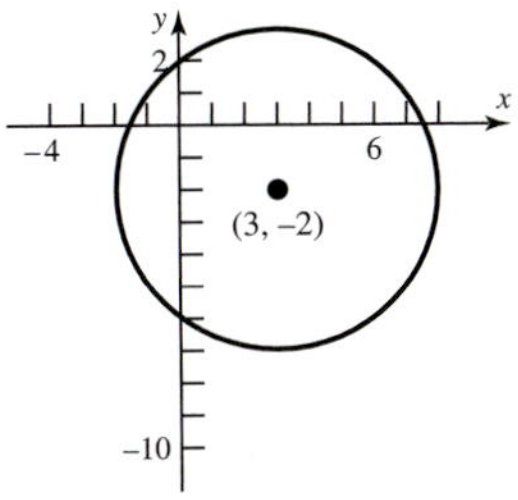

51. $x^2 + y^2 - 13 = 0$ **53.** $x^2 + y^2 - 4x - 6y + 4 = 0$ **55.** $x^2 + y^2 + 2x - 6y + 5 = 0$ **57.** **(c)** **59.** **(b)**
61. $(x + 3)^2 + (y - 1)^2 = 16$ **63.** $(x - 2)^2 + (y - 2)^2 = 9$

1.6 Exercises

1. Linear relation **3.** Linear relation **5.** Nonlinear relation **7.** **(a)**

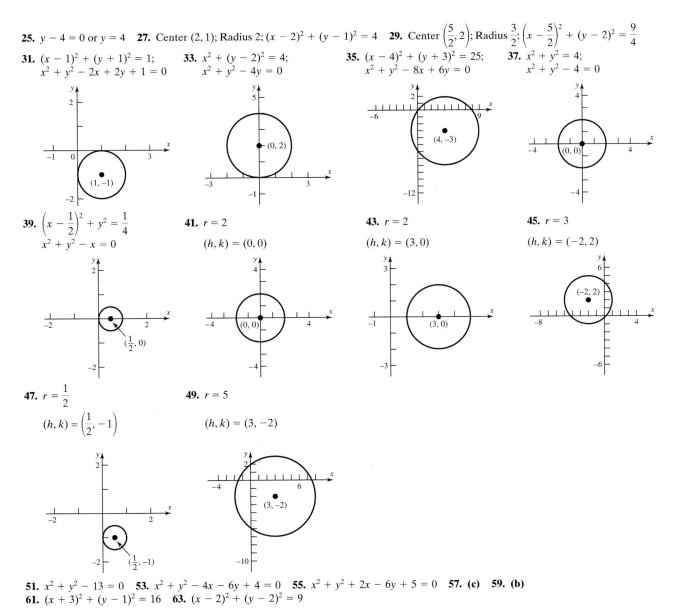

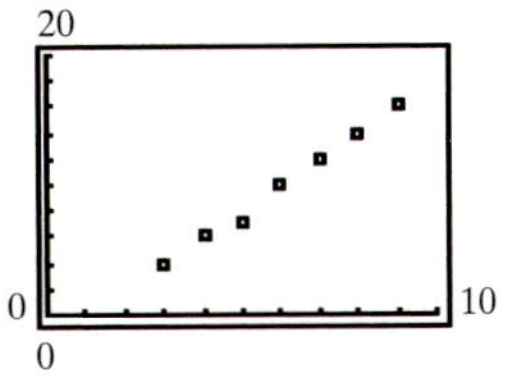

(b) Answers will vary. Using (4, 6) and (8, 14) : $y = 2x - 2$ **(c)**

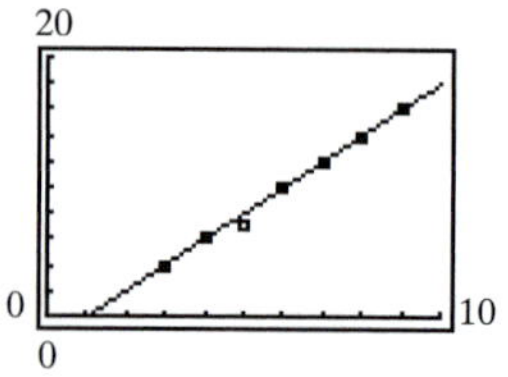

(d) $y = 2.0357x - 2.3571$ **(e)**

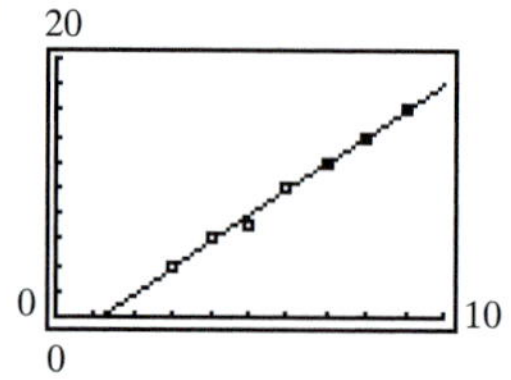

9. (a)

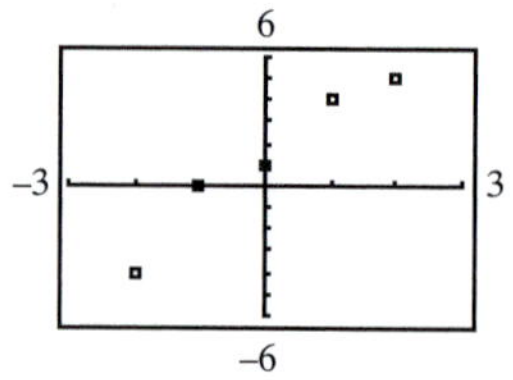

(b) Answers will vary. Using (−2, −4) and (1, 4): $y = \frac{8}{3}x + \frac{4}{3}$ **(c)**

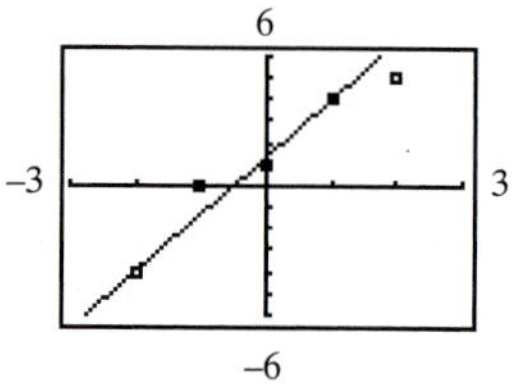

(d) $y = 2.2x + 1.2$

(e)

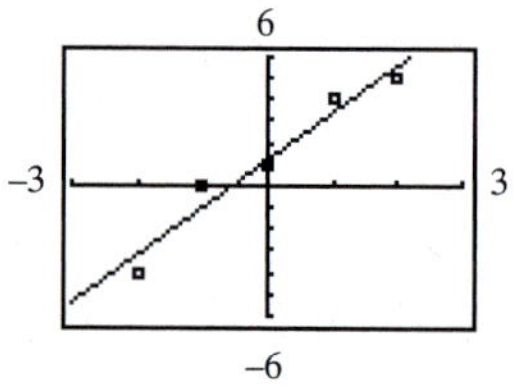

11. (a)

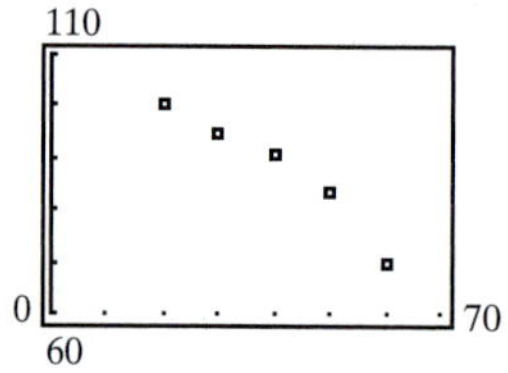

(b) Answers will vary. Using (30, 95) and (60, 70): $y = -\frac{5}{6}x + 120$ **(c)**

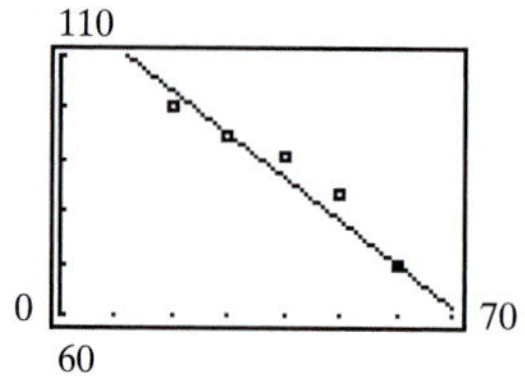

(d) $y = -0.72x + 116.6$

(e)

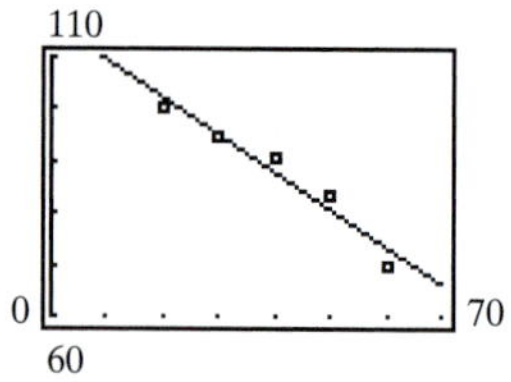

13. (a)

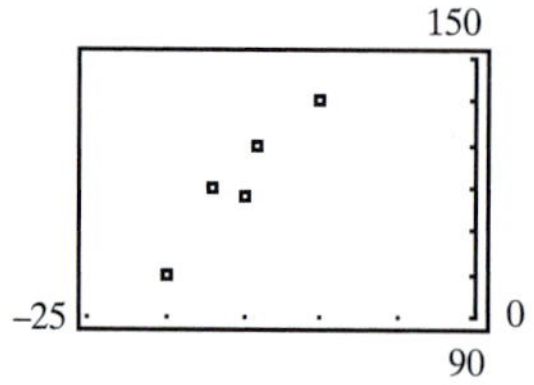

(b) Answers will vary. Using (−20, 100) and (−15, 118): $y = \frac{18}{5}x + 172$

(c)

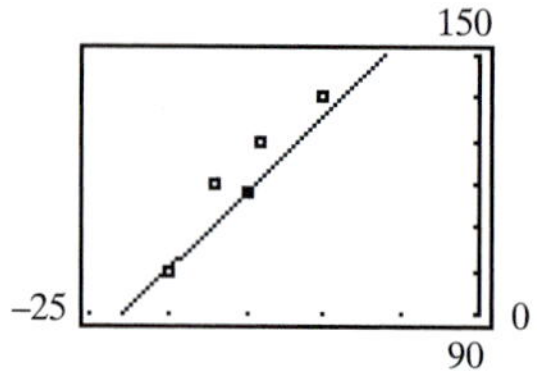

(d) $y = 3.8613x + 180.292$ **(e)**

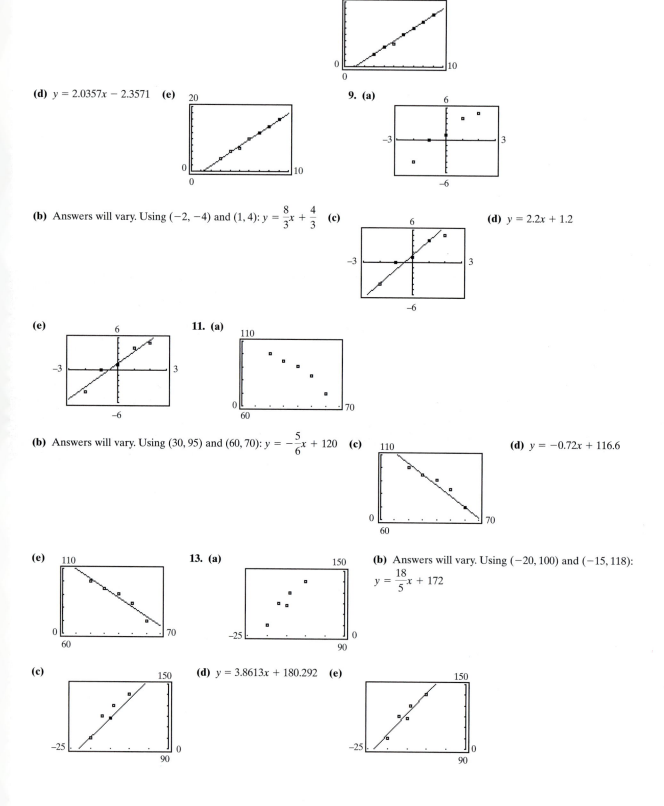

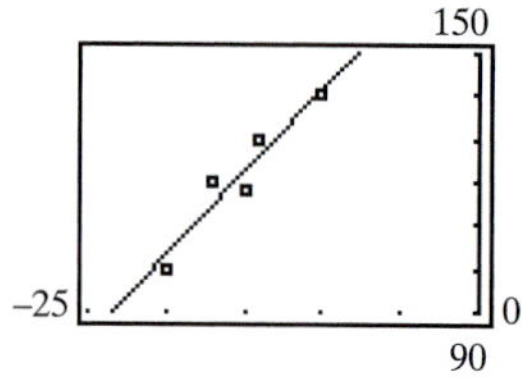

15. (a)

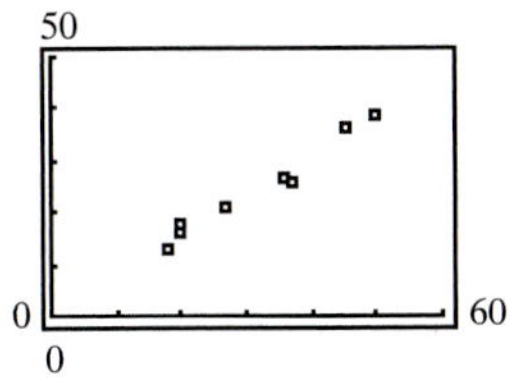

(b) $C = 0.755I + 0.6266$
(c) As disposable income increases by \$1, consumption increases by about \$0.755
(d) \$32,337

17. (a)

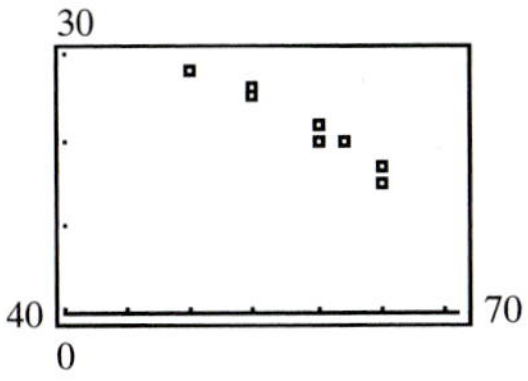

(b) $y = -0.8265x + 70.3903$
(c) As speed increases by 1 mph, mpg decreases by 0.8265.
(d) 20 mpg

19. (a)

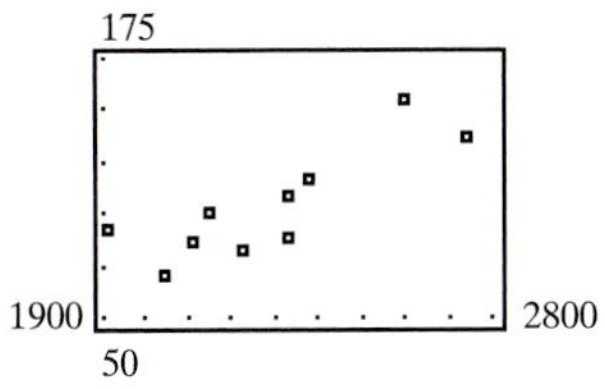

(b) NIBT = 0.0842 sales − 88.5776
(c) As sales increase by \$1, NIBT increases by \$0.0842
(d) \$118.3 billion

21. (a) $y = 1.13x - 2841.69$ **(b)** If per capita disposable income increases \$1, per capita personal consumption increases \$1.13. **(c)** \$14,095.88 **23. (a)** $y = 323.42x - 62{,}080.03$ **(b)** If average fuel consumption per car increases by 1, pollutant emissions of carbon monoxide increase 323.42. **(c)** 101,247.

1.7 Exercises

1. $y = \frac{1}{5}x$ **3.** $A = \pi x^2$ **5.** $F = \frac{250}{d^2}$ **7.** $z = \frac{1}{5}(x^2 + y^2)$ **9.** $M = \frac{9d^2}{2\sqrt{x}}$ **11.** $T^2 = 8\frac{a^3}{d^2}$ **13.** $V = \frac{4\pi}{3}r^3$ **15.** $A = \frac{1}{2}bh$ **17.** $V = \pi r^2 h$

19. $F = 6.67 \times 10^{-11}\left(\frac{mM}{d^2}\right)$ **21.** 144 ft; 2 sec **23.** 2.25 **25.** 45.45 lb **27.** $\sqrt[3]{6} \approx 1.82$ in. **29.** 900 ft-lb **31.** 384 psi

33. $\frac{720}{49} \approx 14.69$ ohms **35.** $v = \sqrt{gr}$ **37.** 17,913 mph **39.** Approx. 545 mi **41.** $F = \frac{mv^2}{r}$ **43.** by 21% **45.** 9 times

Fill-in-the-Blank Items

1. x-coordinate; y-coordinate **2.** midpoint **3.** y-axis **4.** circle; radius; center **5.** undefined; 0 **6.** $m_1 = m_2$; $m_1m_2 = -1$ **7.** $\frac{kx^2y^3}{\sqrt{t}}$

True/False Items

1. F **2.** T **3.** T **4.** F **5.** F **6.** T **7.** T

Review Exercises

1. (a)

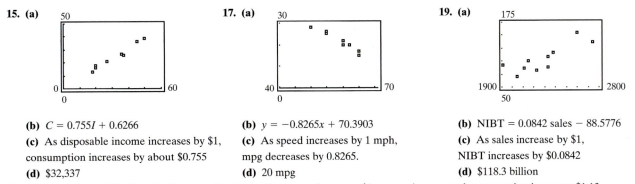

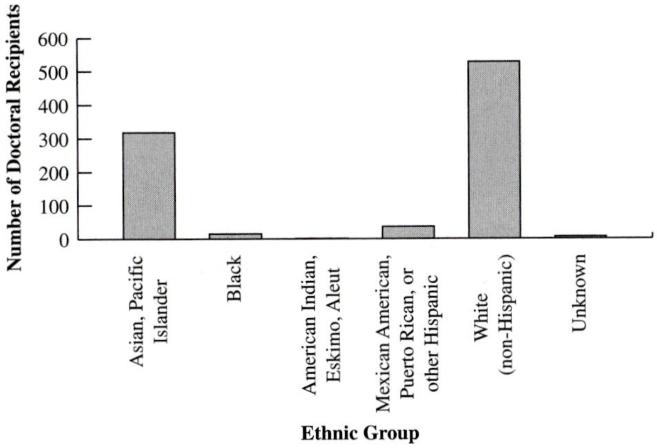

(b)

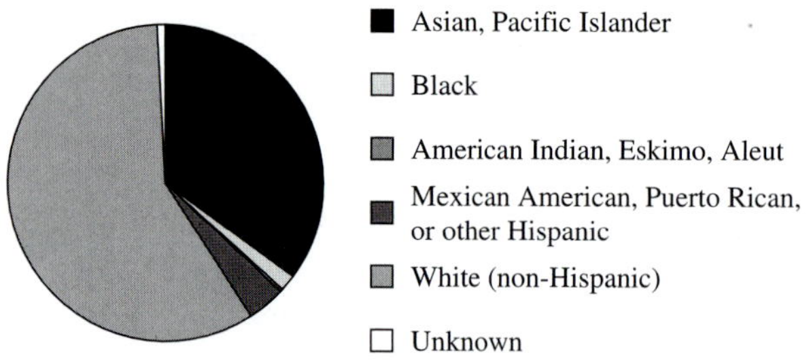

(d) White (non-Hispanic)

3. (a) 5 **(b)**

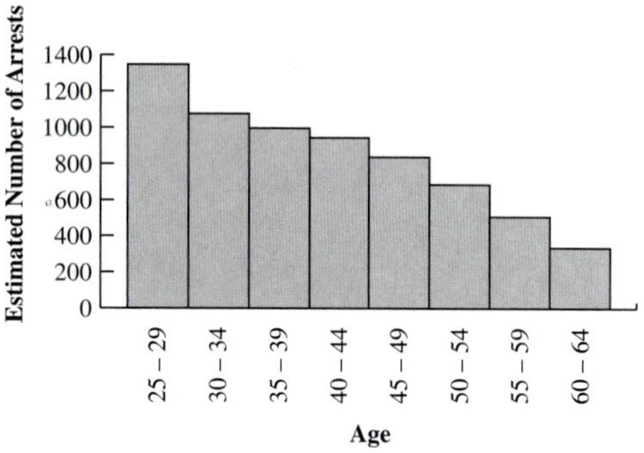

(c)

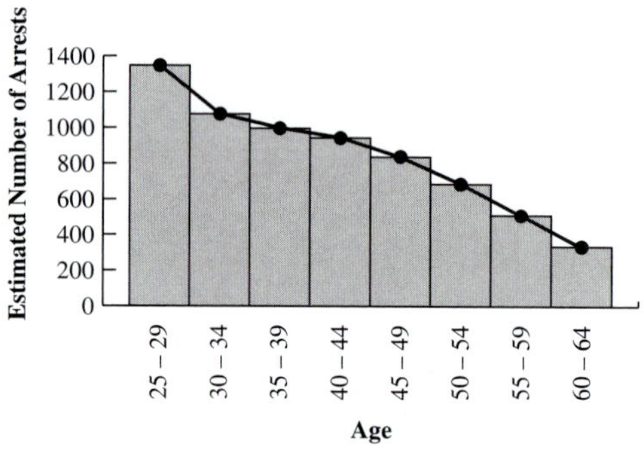

(d) 25 – 29 year olds **5. (a)**

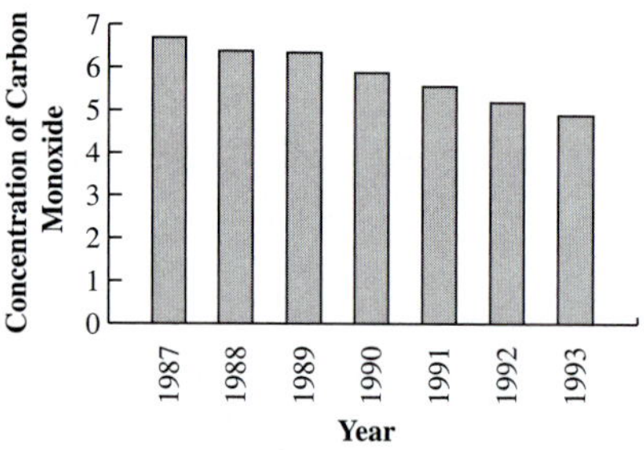

(b)

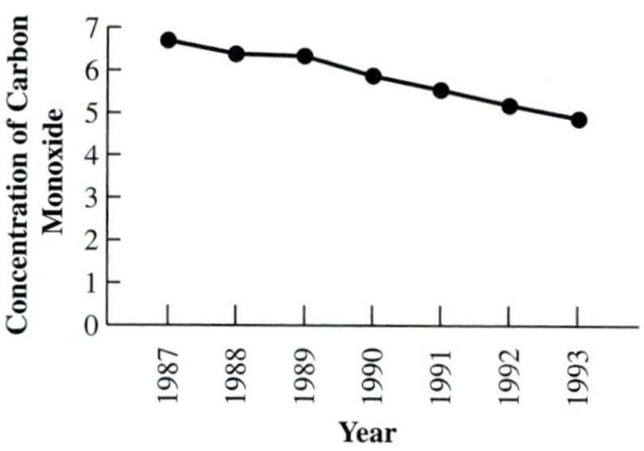

(c) 1987

(d) 1993 **(f)**

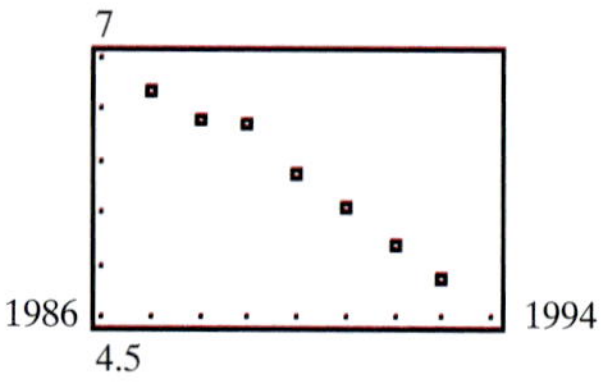

(g) $\dfrac{5.87 - 6.69}{1990 - 1987} = -\dfrac{0.82}{3} \approx -0.27$

(h) Between the years 1987 and 1990, as x increases by 1 year, the level of carbon monoxide decreases by about 0.27 ppm **(i)** $-\dfrac{0.99}{3} = -0.33$

(j) As x increases by 1 year, the level of carbon monoxide decreases by 0.33 ppm **(k)** ≈ -0.308 **(l)** Between the years 1987 and 1993, as x increases by 1 year, the level of carbon monoxide decreases by about 0.308 ppm.

7. $2x + y - 5 = 0$ or $y = -2x + 5$ **9.** $x + 3 = 0$; no slope-intercept form **11.** $x + 5y + 10 = 0$ or $y = -\dfrac{1}{5}x - 2$

13. $2x - 3y + 19 = 0$ or $y = \dfrac{2}{3}x + \dfrac{19}{3}$ **15.** $-x + y + 7 = 0$ or $y = x - 7$

17.

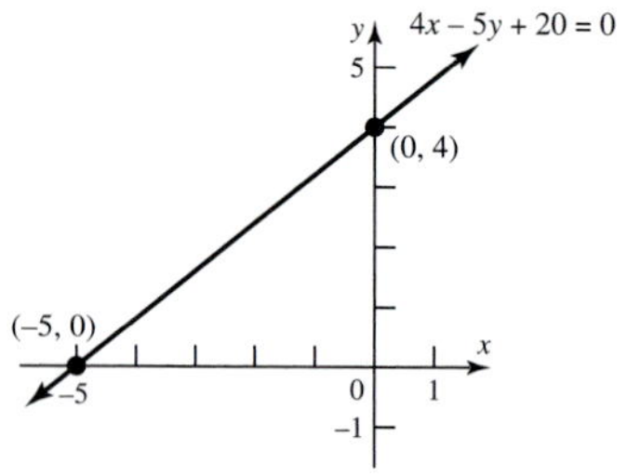

19.

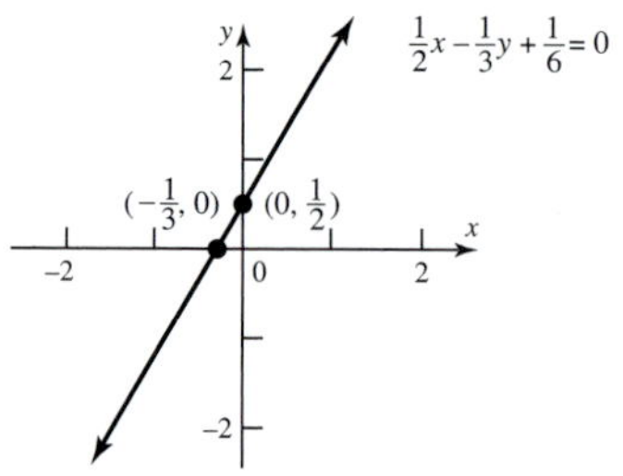

21.

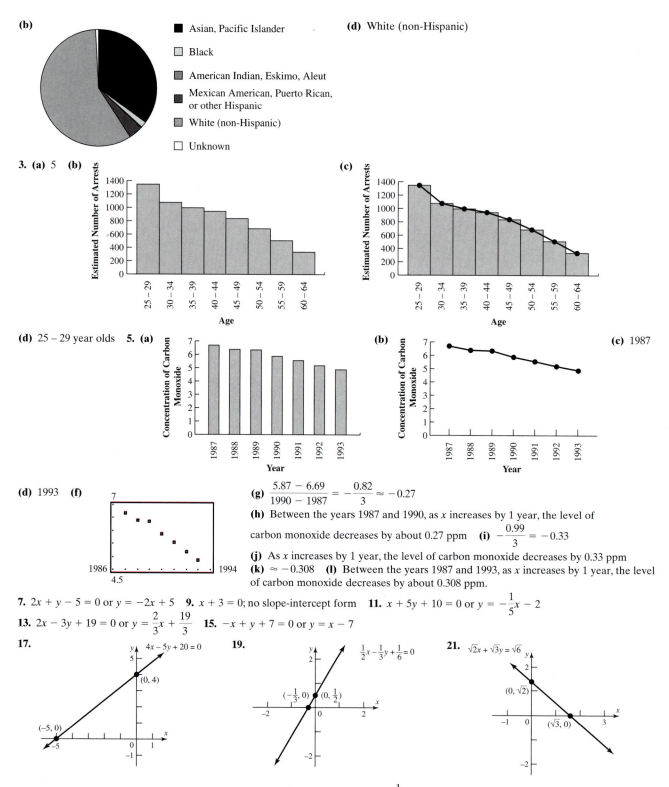

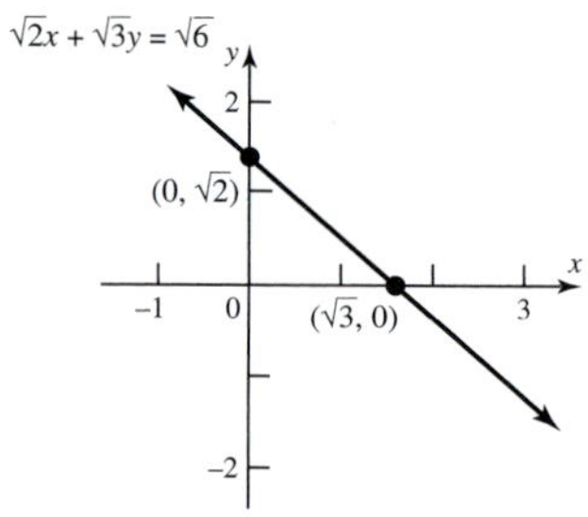

23. Center $(1, -2)$; Radius $= 3$ **25.** Center $(1, -2)$; Radius $= \sqrt{5}$ **27.** Slope $= \dfrac{1}{5}$; distance $= 2\sqrt{26}$; midpoint $= (2, 3)$

29. Equations in **(c)** **31.** Intercept: $(0, 0)$; symmetric with respect to the x-axis **33.** Intercepts: $(0, -2)$, $(0, 2)$, $(4, 0)$, $(-4, 0)$; symmetric with respect to the x-axis, y-axis, and origin **35.** Intercept: $(0, 1)$; symmetric with respect to the y-axis

37. Intercepts: $(0, 0)$, $(0, -2)$, $(-1, 0)$; no symmetry

39. (a)

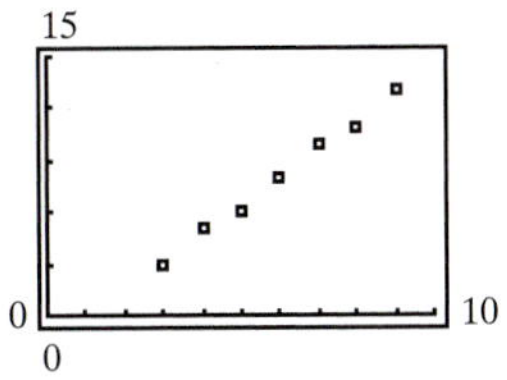

(b) $y = 1.643x - 1.857$
(c) As x increases by 1, y increases by 1.643.

41. (a)

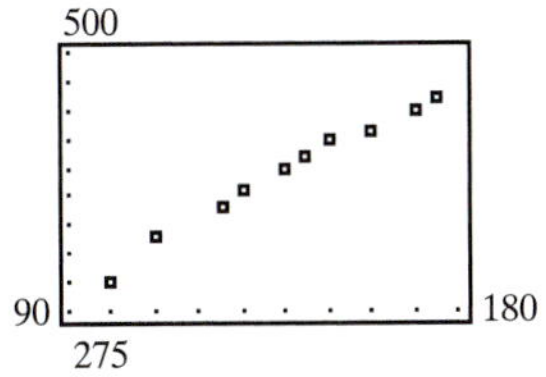

(b) $y = 2.018x + 112.433$
(c) As x increases by 1, y increases by 2.018.

43. (a)

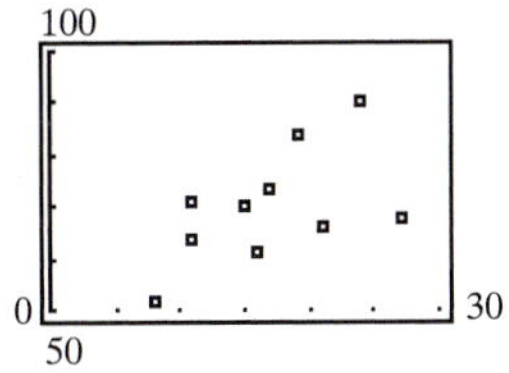

(b) Calculus $= 1.018$algebra $+ 52.69$
(c) As the algebra score increases by 1 point, the calculus score increases by about 1.018 points.
(d) 73

45. $\frac{8}{\sqrt[4]{3}}$ **47.** $a \approx 36$ million miles **49.** $d(A, B) = \sqrt{13}$; $d(B, C) = \sqrt{13}$ **51.** Slope using A *and* B is -1; slope using B and C is -1

53. Center $(1, -2)$; radius $= 4\sqrt{2}$; $x^2 + y^2 - 2x + 4y - 27 = 0$

CHAPTER 2 *2.1 Exercises*

1. Function **3.** Not a function **5.** Function **7.** Function **9.** Not a function **11.** Function **13. (a)** -4 **(b)** -5 **(c)** -9 **(d)** $-3x^2 - 2x - 4$ **(e)** $3x^2 - 2x + 4$ **(f)** $-3x^2 - 4x - 5$ **15. (a)** 0 **(b)** $\frac{1}{2}$ **(c)** $-\frac{1}{2}$ **(d)** $\frac{-x}{x^2+1}$ **(e)** $\frac{-x}{x^2+1}$ **(f)** $\frac{x+1}{x^2+2x+2}$ **17. (a)** 4 **(b)** 5 **(c)** 5 **(d)** $|x| + 4$ **(e)** $-|x| - 4$ **(f)** $|x+1| + 4$ **19. (a)** $-\frac{1}{5}$ **(b)** $-\frac{3}{2}$ **(c)** $\frac{1}{8}$ **(d)** $\frac{-2x+1}{-3x-5}$ **(e)** $\frac{-2x-1}{3x-5}$ **(f)** $\frac{2x+3}{3x-2}$ **21.** $f(0) = 3$; $f(-6) = -3$ **23.** Positive **25.** -3, 6 and 10 **27.** $\{x \mid -6 \le x \le 11\}$ **29.** $(-3, 0)$, $(6, 0)$, $(10, 0)$ **31.** 3 times **33. (a)** No **(b)** -3; $(4, -3)$ **(c)** 14; $(14, 2)$ **(d)** $\{x \mid x \ne 6\}$ **35. (a)** Yes **(b)** $\frac{8}{17}$; $\left(2, \frac{8}{17}\right)$ **(c)** $-1, 1$; $(-1, 1)$, $(1, 1)$ **(d)** All real numbers **37.** Not a function **39.** Function **(a)** Domain: $\{x \mid -\pi \le x \le \pi\}$; Range: $\{y \mid -1 \le y \le 1\}$ **(b)** Intercepts: $\left(-\frac{\pi}{2}, 0\right)$, $\left(\frac{\pi}{2}, 0\right)$, $(0, 1)$ **(c)** y-axis **41.** Not a function **43.** Function **(a)** Domain: $\{x \mid x > 0\}$; Range: All real numbers **(b)** Intercept: $(1, 0)$ **(c)** none **45.** Function **(a)** Domain: all real numbers; Range: $\{y \mid y \le 2\}$ **(b)** Intercepts: $(-3, 0)$, $(3, 0)$, $(0, 2)$ **(c)** y-axis **47.** Function **(a)** Domain: $\{x \mid x \ne 2\}$; Range: $\{y \mid y \ne 1\}$ **(b)** Intercept: $(0, 0)$ **(c)** None **49.** All real numbers **51.** All real numbers **53.** $\{x \mid x \ne -1, x \ne 1\}$ **55.** $\{x \mid x \ne 0\}$ **57.** $\{x \mid x \ge 4\}$ **59.** $\{x \mid x > 9\}$ **61.** $\{x \mid x \ne 1\}$ **63. (a)** III **(b)** IV **(c)** I **(d)** V **(e)** II **65.**

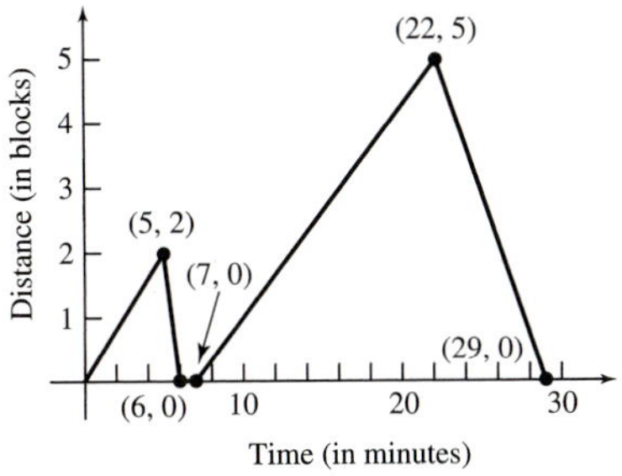

67. $A = -\frac{7}{2}$ **69.** $A = -4$ **71.** $A = 8$; undefined at $x = 3$

73. (a) No **(b)**

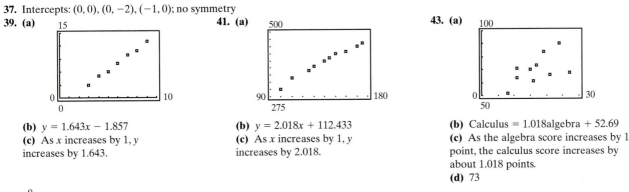

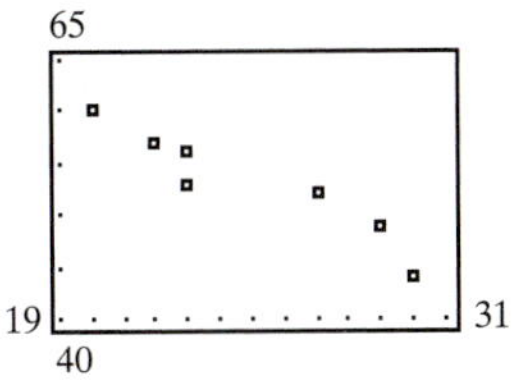

(c) $D = -1.34p + 86.20$
(d) If price increases \$1, quantity demanded will decrease 1.34.
(e) $D(p) = -1.34p + 86.20$
(f) $\{p \mid p > 0\}$
(g) About 49 pairs

75. (a) Yes **(b)**

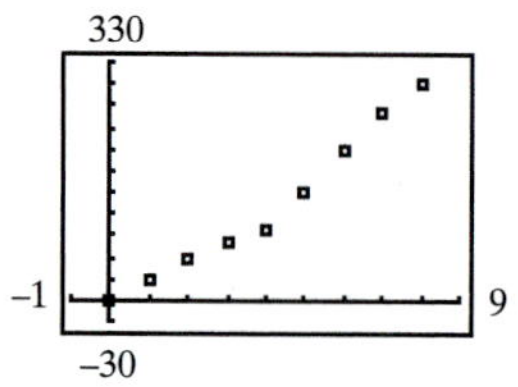

(c) $s = 37.78t - 19.13$
(d) If t increases by 1 hour, distance increases by 37.78 miles.
(e) $s(t) = 37.78t - 19.13$
(f) Since time must be nonnegative, the domain is $\{t \mid t \geq 0\}$.
(g) About 396 miles.

77. (a)

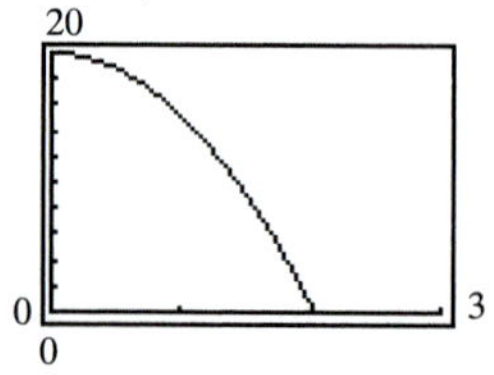

(b) 15.1 m, 14.07 m, 12.94 m, 11.72 m
(c) 1.01 seconds, 1.42 seconds, 1.74 seconds
(d) 2.02 sec

79. $A(x) = \frac{1}{2}x^2$ **81.** $G(x) = 10x$ **83. (a)** $C(x) = 10x + 14\sqrt{x^2 - 10x + 29}, 0 < x < 5$

(b) $C(1) = \$72.61$ **(c)** $C(3) = \$69.60$

(d)

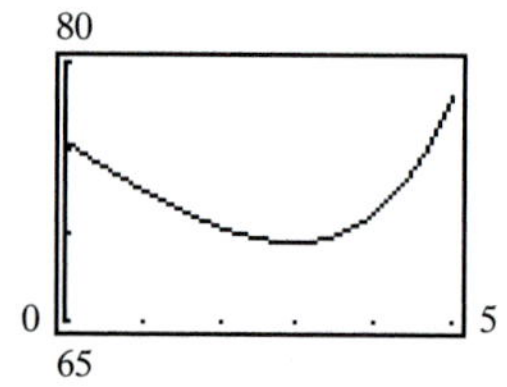

(e) least cost: $x \approx 2.95$ miles

85. (a) $A(x) = (8.5 - 2x)(11 - 2x)$
(b) $0 \leq x \leq 4.25, 0 \leq A \leq 93.5$
(c) $A(1) = 58.5$ in.^{2}, $A(1.2) = 52.46$ in.^{2}, $A(1.5) = 44$ in.^{2}
(d)

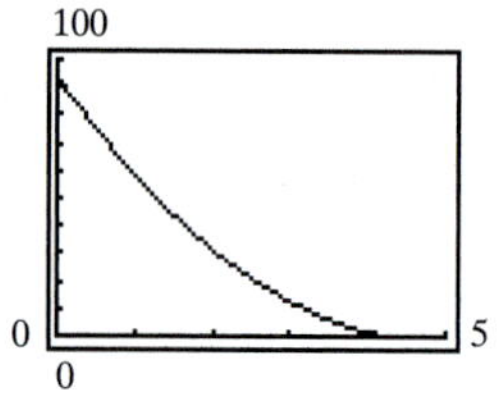

87. (a)

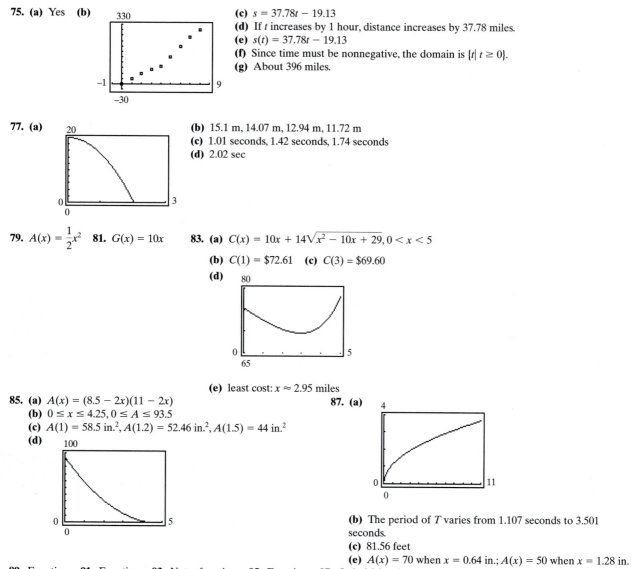

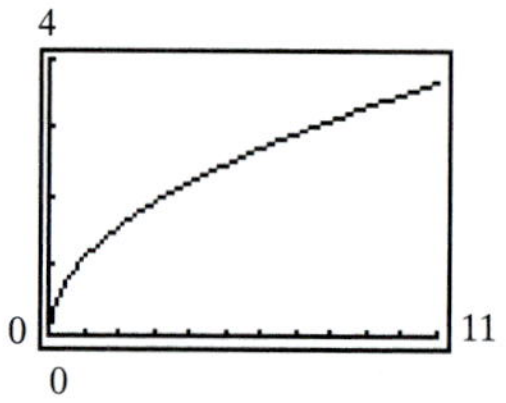

(b) The period of T varies from 1.107 seconds to 3.501 seconds.
(c) 81.56 feet
(e) $A(x) = 70$ when $x = 0.64$ in.; $A(x) = 50$ when $x = 1.28$ in.

89. Function **91.** Function **93.** Not a function **95.** Function **97.** Only $h(x) = 2x$

2.2 Exercises

1. *C* **3.** *E* **5.** *B* **7.** *F* **9. (a)** Domain: $\{x \mid -3 \leq x \leq 4\}$; Range: $\{y \mid 0 \leq y \leq 3\}$ **(b)** Increasing on $(-3, 0)$ and on $(2, 4)$; decreasing on $(0, 2)$ **(c)** Neither **(d)** $(-3, 0), (0, 3), (2, 0)$
11. (a) Domain: all real numbers; Range: $\{y \mid 0 < y < \infty\}$ **(b)** Increasing on $(-\infty, \infty)$ **(c)** Neither **(d)** $(0, 1)$
13. (a) Domain: $\{x \mid -\pi \leq x \leq \pi\}$; Range: $\{y \mid -1 \leq y \leq 1\}$ **(b)** Increasing on $\left(-\frac{\pi}{2}, \frac{\pi}{2}\right)$; decreasing on $\left(-\pi, -\frac{\pi}{2}\right)$ and on $\left(\frac{\pi}{2}, \pi\right)$ **(c)** Odd (symmetric with respect to the origin) **(d)** $(-\pi, 0), (0, 0), (\pi, 0)$ **15. (a)** Domain: $\{x \mid x \neq 2\}$; Range: $\{y \mid y \neq 1\}$ **(b)** Decreasing on $(-\infty, 2)$ and on $(2, \infty)$ **(c)** Neither **(d)** $(0, 0)$ **17. (a)** Domain: $\{x \mid x \neq 0\}$; Range: all real numbers **(b)** Increasing on $(-\infty, 0)$ and on $(0, \infty)$ **(c)** Odd **(d)** $(-1, 0), (1, 0)$
19. (a) Domain: $\{x \mid x \neq -2, x \neq 2\}$; Range: $\{y \mid -\infty < y \leq 0 \text{ and } 1 < y < \infty\}$ **(b)** Increasing on $(-\infty, -2)$ and on $(-2, 0)$; decreasing on $(0, 2)$ and on $(2, \infty)$ **(c)** Even **(d)** $(0, 0)$
21. (a) Domain: $\{x \mid -4 \leq x \leq 4\}$; Range: $\{y \mid 0 \leq y \leq 2\}$ **(b)** Increasing on $(-2, 0)$ and $(2, 4)$; Decreasing on $(-4, -2)$ and $(0, 2)$ **(c)** Even **(d)** $(-2, 0), (0, 2), (2, 0)$ **23. (a)** Domain: $\{x \mid -4 \leq x \leq 4\}$; Range: $\{y \mid 0 < y \leq 4\}$ **(b)** Increasing on $(-4, 0)$; Decreasing on $(0, 4)$ **(c)** Even **(d)** None
25. (a) 4 **(b)** 2 **(c)** 5 **27. (a)** 2 **(b)** 3 **(c)** −4 **29.** 3 **31.** −3 **33.** $3x + 1$ **35.** $x(x + 1)$ **37.** $\frac{-1}{x + 1}$ **39.** $\frac{1}{\sqrt{x} + 1}$ **41.** Odd

43. Even **45.** Odd **47.** Neither **49.** Even **51.** Odd **53.** At most one

55. **(a)** All real numbers
(b) $(0, -3), (1, 0)$
(c)

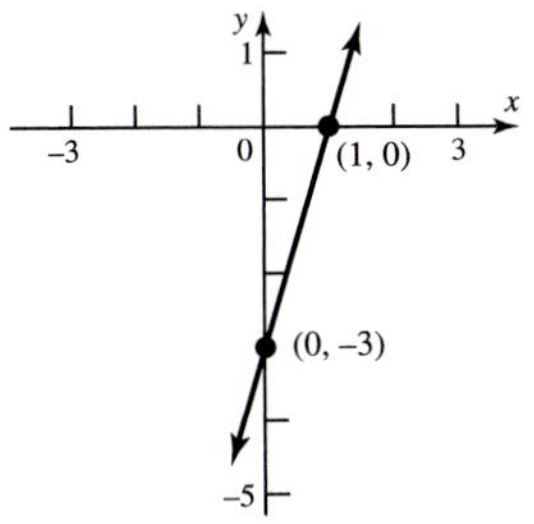

(d) All real numbers

57. **(a)** All real numbers
(b) $(-2, 0), (2, 0), (0, -4)$
(c)

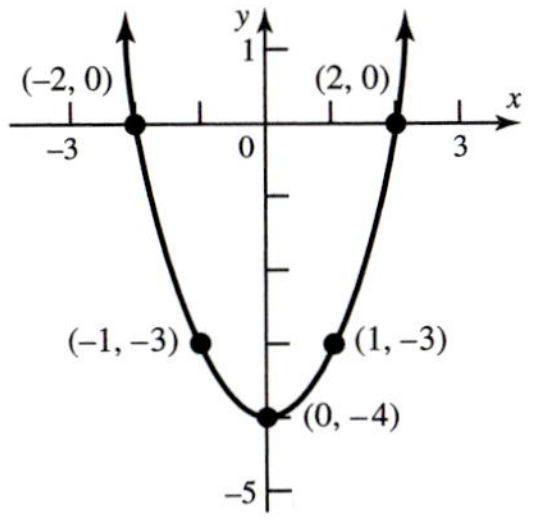

(d) $\{y \mid -4 \le y < \infty\}$

59. **(a)** All real numbers
(b) $(0, 0)$
(c)

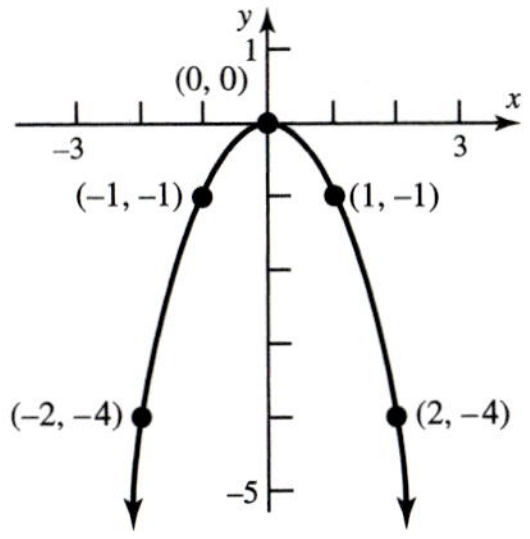

(d) $\{y \mid -\infty < y \le 0\}$

61. **(a)** $\{x \mid 2 \le x < \infty\}$
(b) $(2, 0)$
(c)

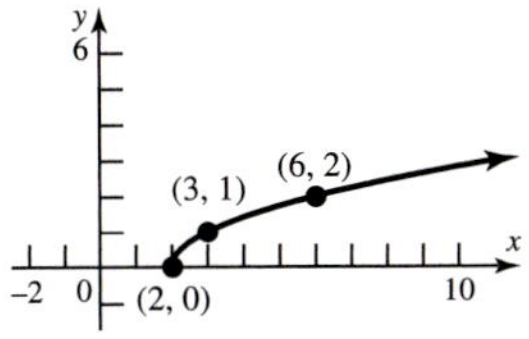

(d) $\{y \mid 0 \le y < \infty\}$

63. **(a)** $\{x \mid -\infty < x \le 2\}$
(b) $(2, 0), (0, \sqrt{2})$
(c)

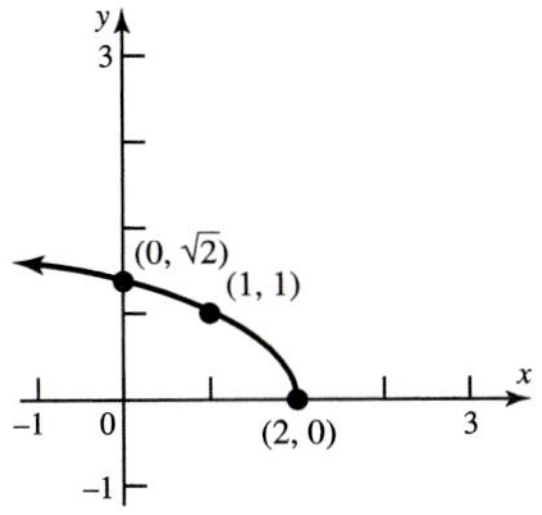

(d) $\{y \mid 0 \le y < \infty\}$

65. **(a)** All real numbers
(b) $(0, 3)$
(c)

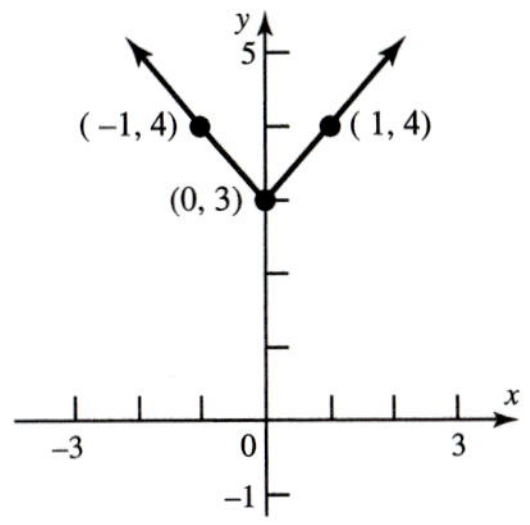

(d) $\{y \mid 3 \le y < \infty\}$

67. **(a)** All real numbers
(b) $(0, 0)$
(c)

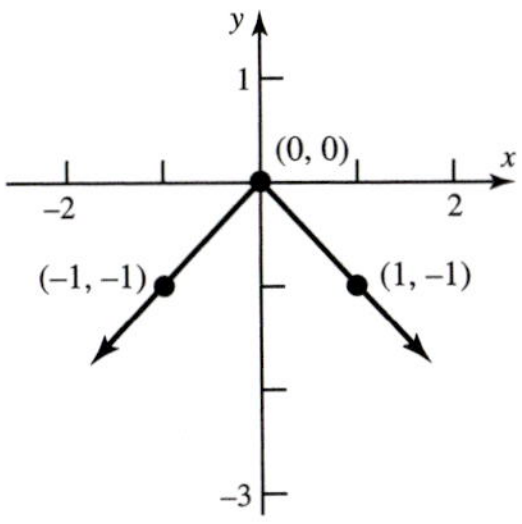

(d) $\{y \mid -\infty < y \le 0\}$

69. **(a)** All real numbers
(b) $(0, 0)$
(c)

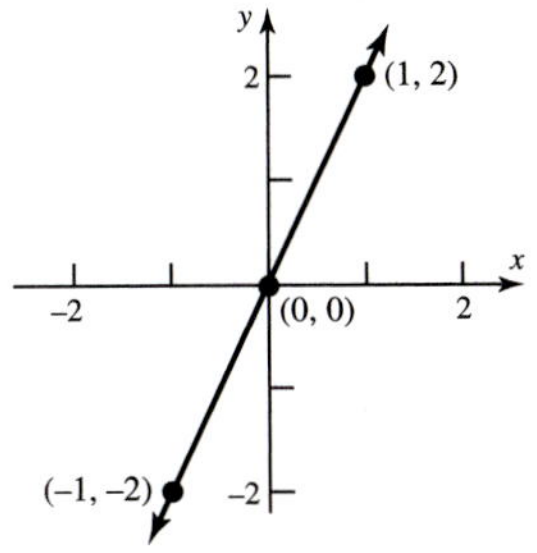

(d) All real numbers

71. **(a)** All real numbers
(b) $(0, 0), (-1, 0)$
(c)

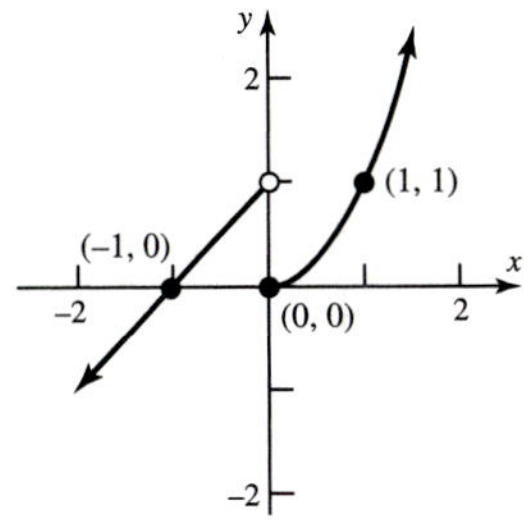

(d) All real numbers

73. **(a)** $\{x \mid -2 \le x < \infty\}$
(b) $(0, 1)$
(c)

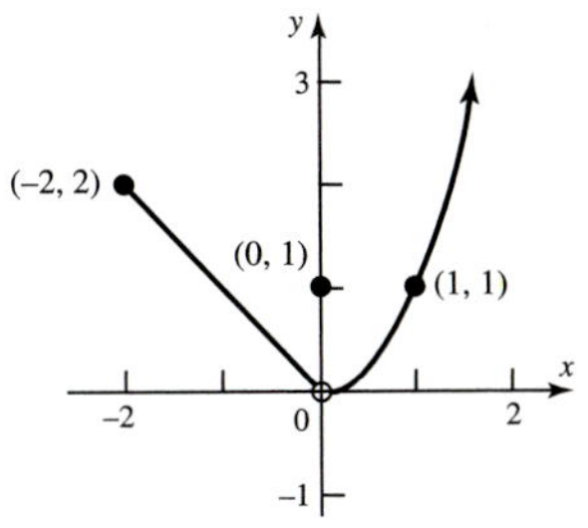

(d) $\{y \mid 0 < y < \infty\}$

75. **(a)** All real numbers
(b) $(0, 1)$
(c)

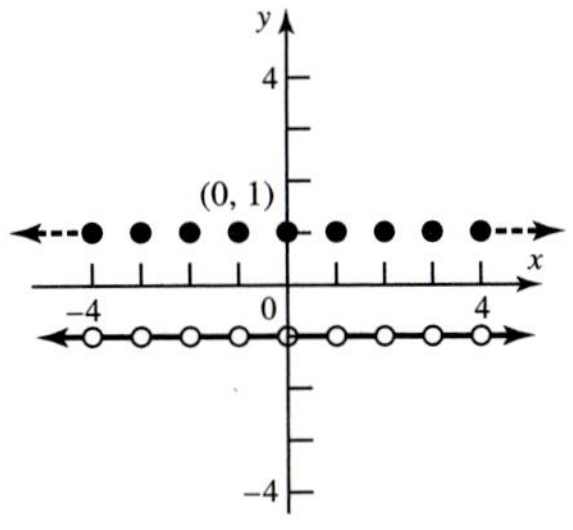

(d) $\{-1, 1\}$

77. **(a)** All real numbers
(b) $(x, 0)$ for $0 \le x < 1$
(c)

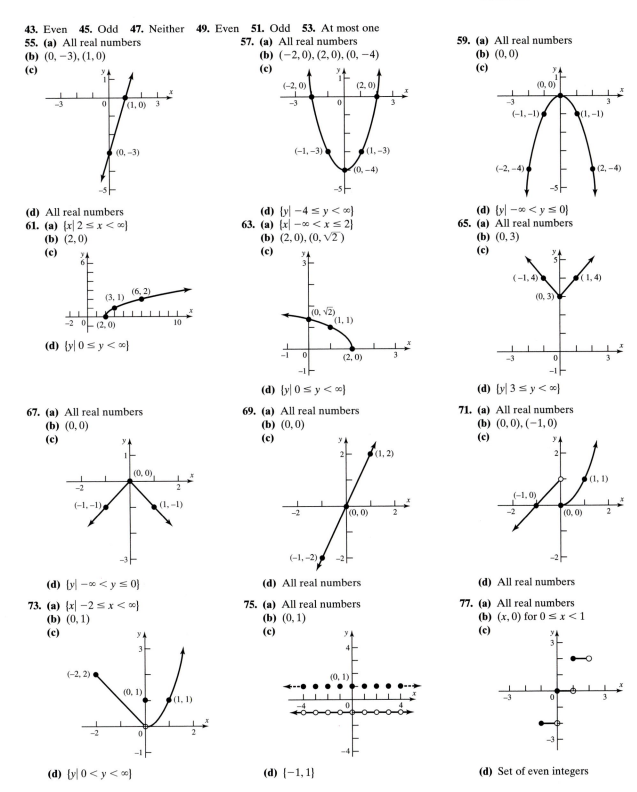

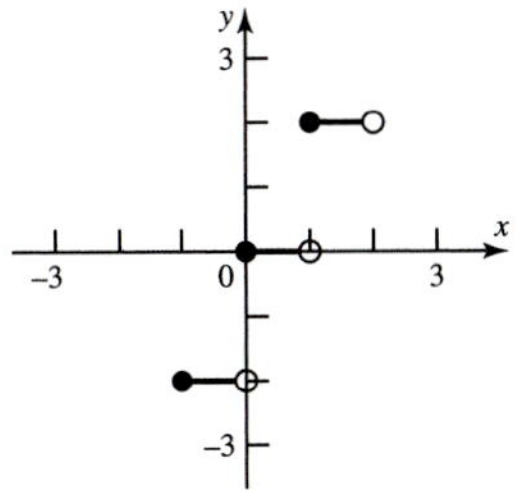

(d) Set of even integers

79. **(a)** All real numbers
(b) $(-2, 0), (0, 4), (2, 0)$
(c)

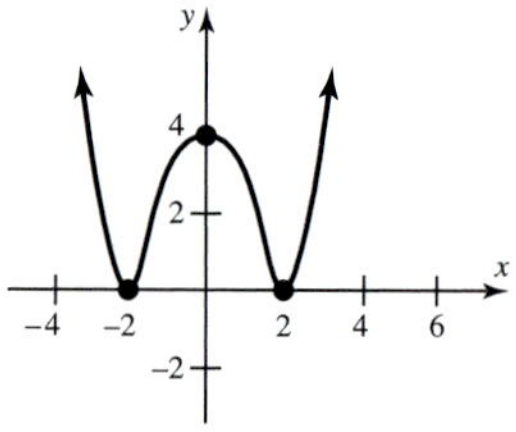

(d) $\{y \mid y \geq 0\}$

81. 2

83. $2x + h + 2$

85. $f(x) = \begin{cases} -x & \text{if } -1 \leq x \leq 0 \\ \frac{1}{2}x & \text{if } 0 < x \leq 2 \end{cases}$ (Other answers are possible.)

87. $f(x) = \begin{cases} -x & \text{if } x \leq 0 \\ -x + 2 & \text{if } 0 < x \leq 2 \end{cases}$ (Other answers are possible.)

89. **(a), (b), (e)**

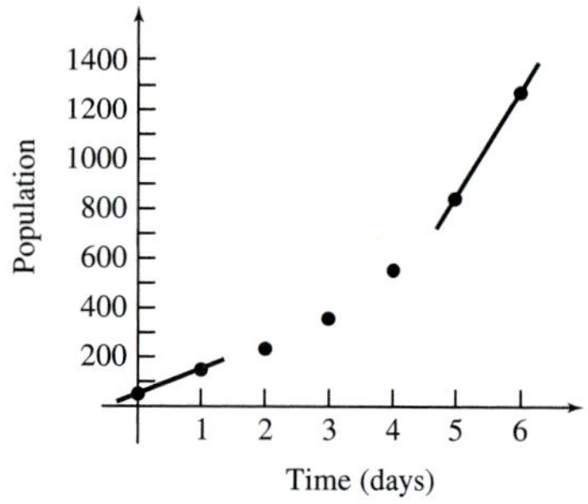

(c) 103
(d) The population is increasing at a rate of 103 per day between day 0 and day 1.
(f) 441
(g) The population is increasing at a rate of 441 per day between day 5 and day 6.
(h) The population is increasing at an increasing rate.

91. **(a), (b), (e)**

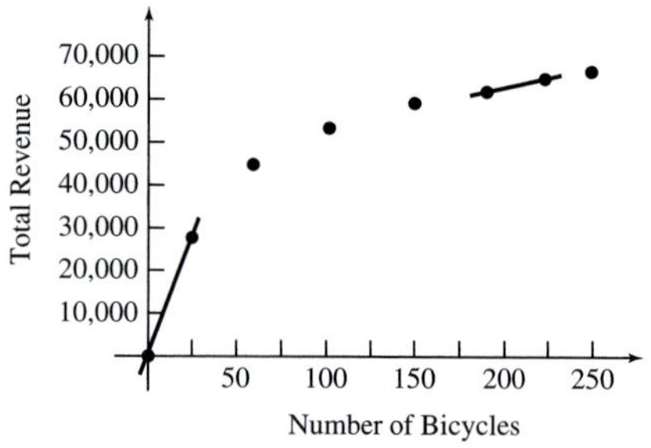

(c) 1120
(d) For each additional bicycle sold between 0 and 25, total revenue increases by \$1120.
(f) 75
(g) For each additional bicycle sold between 190 and 223, total revenue increases by \$75.
(h) As the number of bicycles sold increases, marginal revenue decreases.
(i) 150 bicycles

93. No; $f(-2) = 8$ and $f(2) = 6$

95.

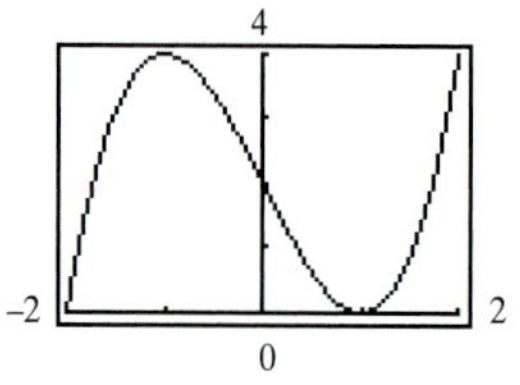

Increasing: $(-2, -1), (1, 2)$
Decreasing: $(-1, 1)$
Local maximum: $(-1, 4)$
Local minimum: $(1, 0)$

97.

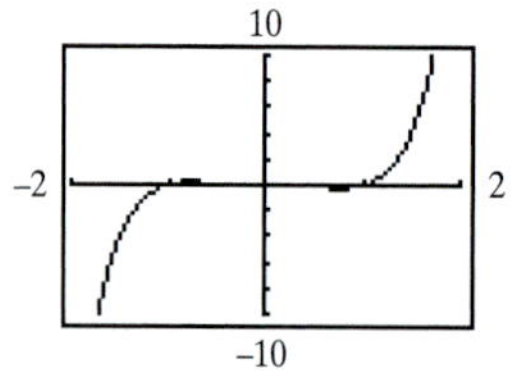

Increasing: $(-2, -0.77), (0.77, 2)$
Decreasing: $(-0.77, 0.77)$
Local maximum: $(-0.77, 0.18)$
Local minimum: $(0.77, -0.18)$

99.

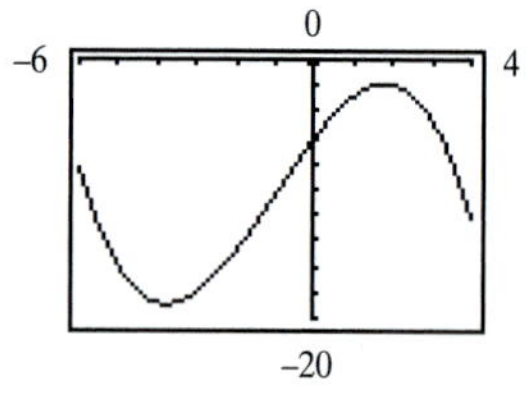

Increasing: $(-3.76, 1.76)$
Decreasing: $(-6, -3.76), (1.76, 4)$
Local maximum: $(1.76, -1.90)$
Local minimum: $(-3.76, -18.89)$

101.

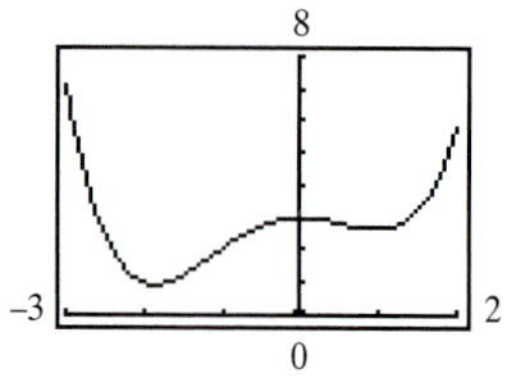

Increasing: $(-1.86, 0), (0.96, 2)$
Decreasing: $(-3, -1.86), (0, 0.96)$
Local maximum: $(0, 3)$
Local minima: $(-1.86, 0.94), (0.96, 2.64)$

103.

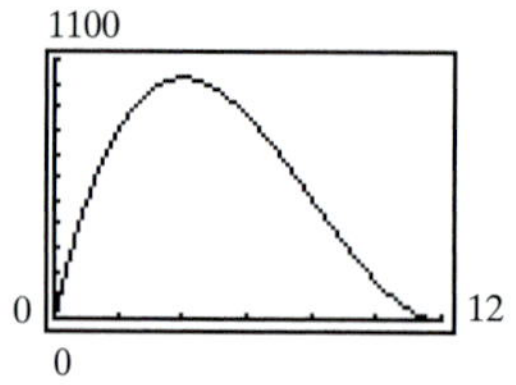

The volume is largest at $x = 4$ inches

105. **(a)**

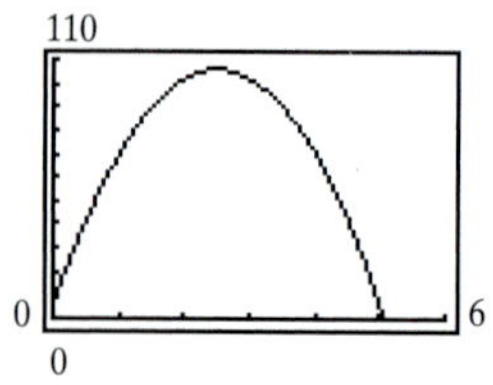

(b) 2.5 seconds
(c) 106 feet

107. Each graph is that of $y = x^2$, but shifted vertically. If $y = x^2 + k$, $k > 0$, the shift is up k units; if $y = x^2 + k$, $k < 0$, the shift is down $|k|$ units. **109.** Each graph is that of $y = |x|$, but either compressed or stretched. If $y = k|x|$ and $k > 1$, the graph is stretched; if $y = k|x|$, $0 < k < 1$, the graph is compressed. **111.** The graph of $y = f(-x)$ is the reflection about the y-axis of the graph of $y = f(x)$.

113. They are all ∪-shaped and open up. All three go through the point (0, 0). As the exponent increases, the steepness of the curve increases. **115. (a)** \$43.47 **(b)** \$241.49 **(c)** $C = \begin{cases} 9 + 0.68935x & \text{if } 0 \le x \le 50 \\ 21.465 + 0.44005x & \text{if } x > 50 \end{cases}$ **(d)**

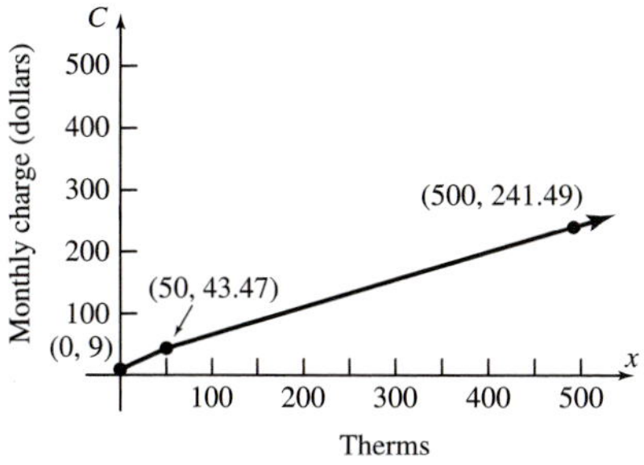

2.3 Exercises

1. B **3.** H **5.** I **7.** L **9.** F **11.** G **13.** C **15.** B **17.** $y = (x - 4)^3$ **19.** $y = x^3 + 4$ **21.** $y = -x^3$ **23.** $y = 4x^3$

25.

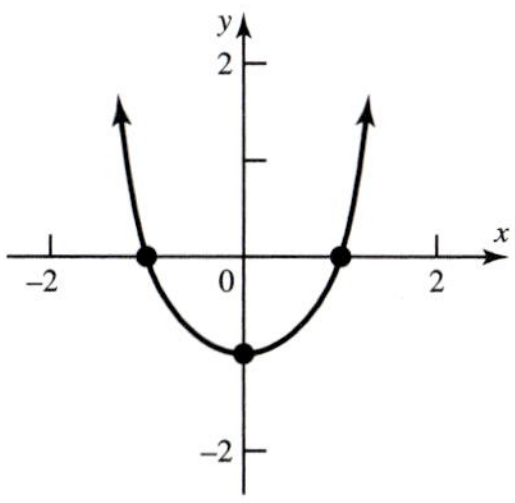

27.

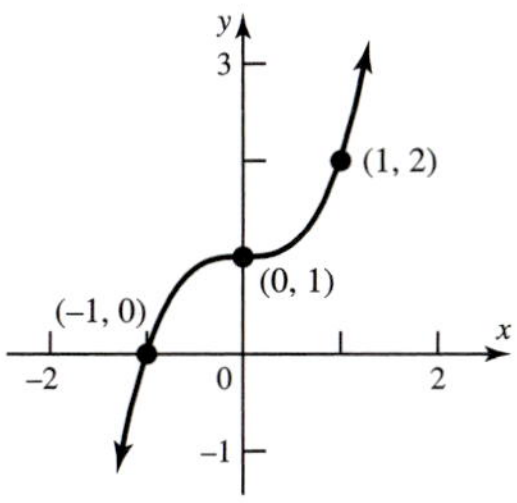

29.

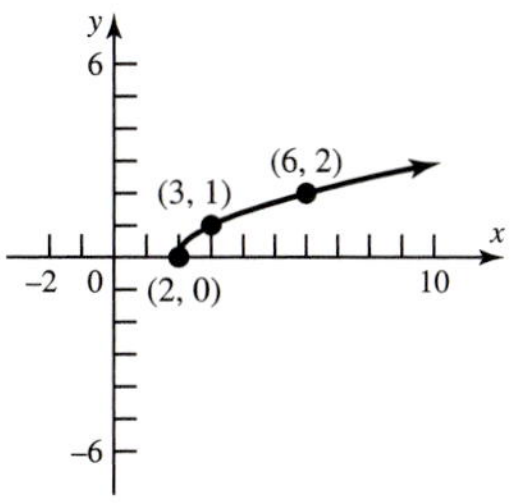

31.

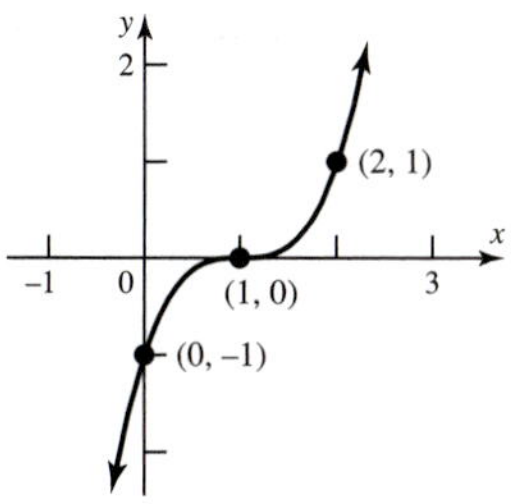

33.

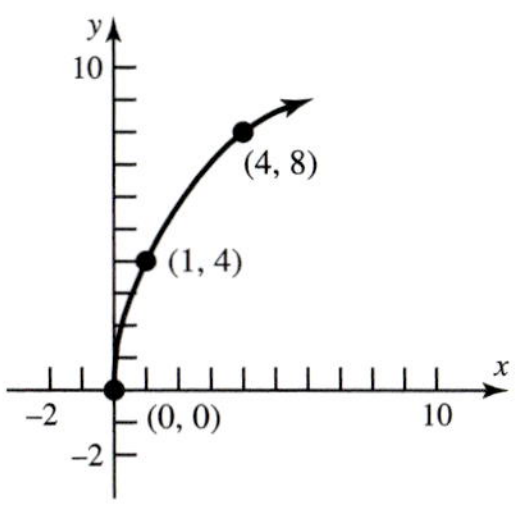

35.

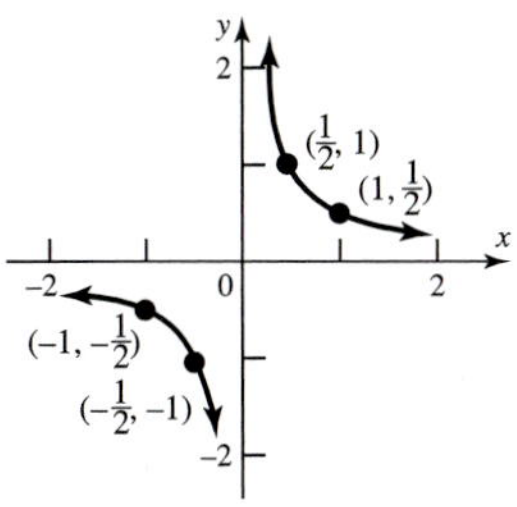

37.

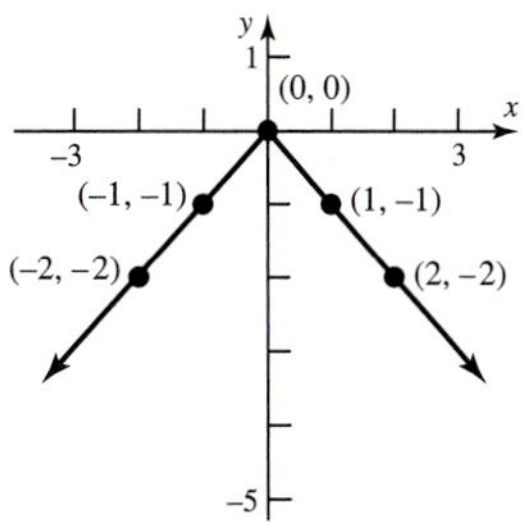

39.

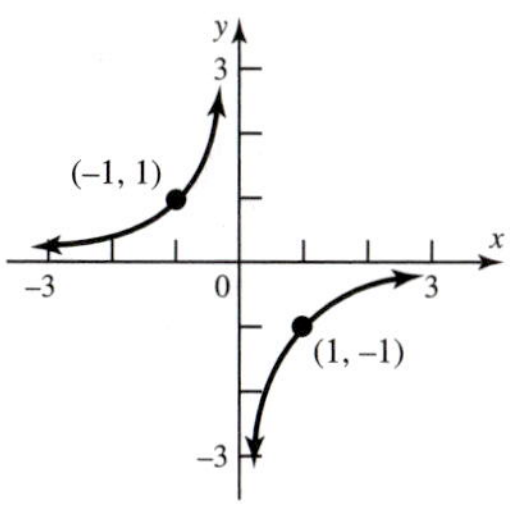

41.

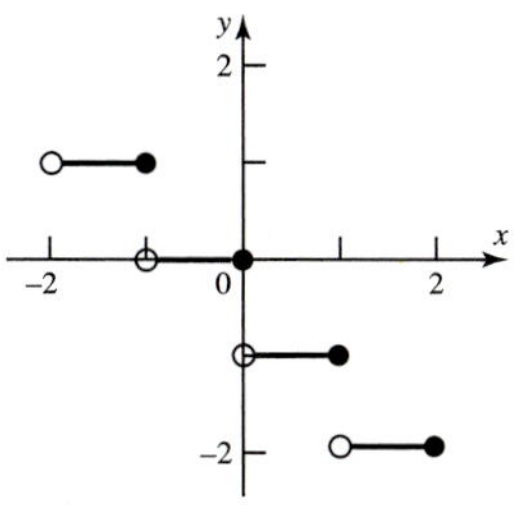

43.

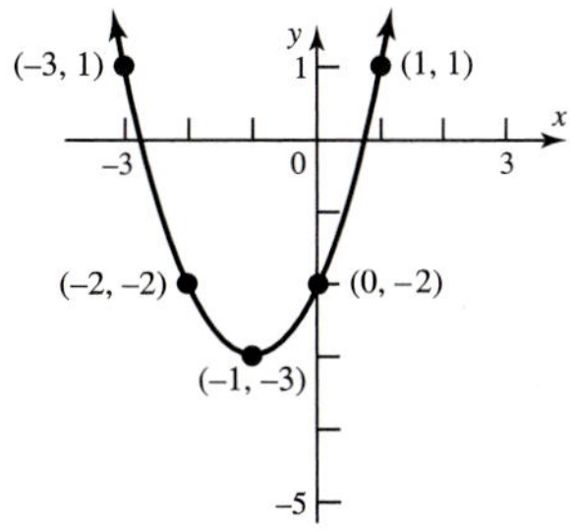

45.

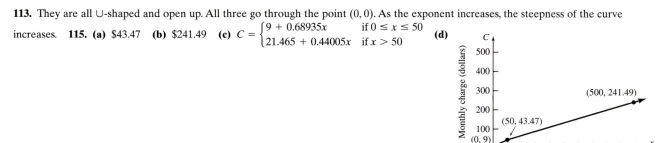

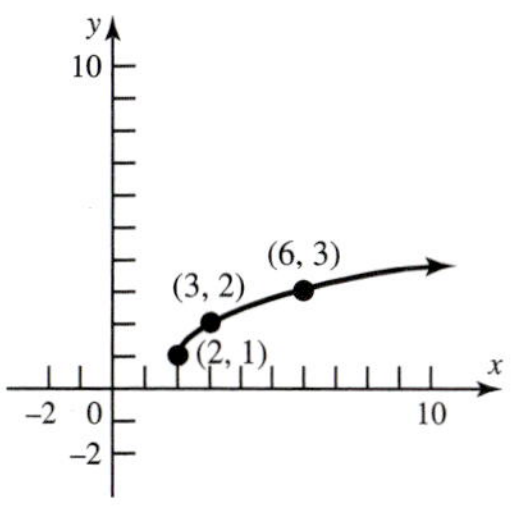

47.

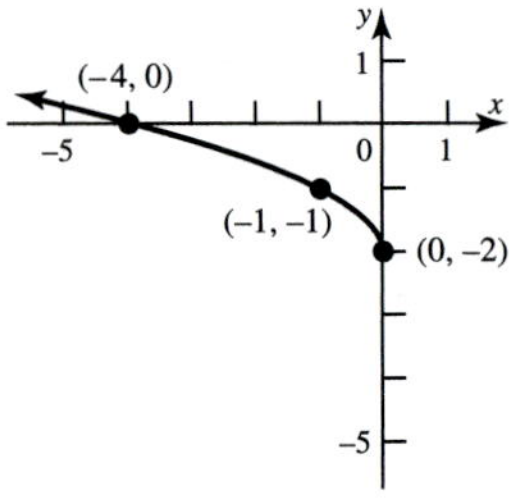

49.

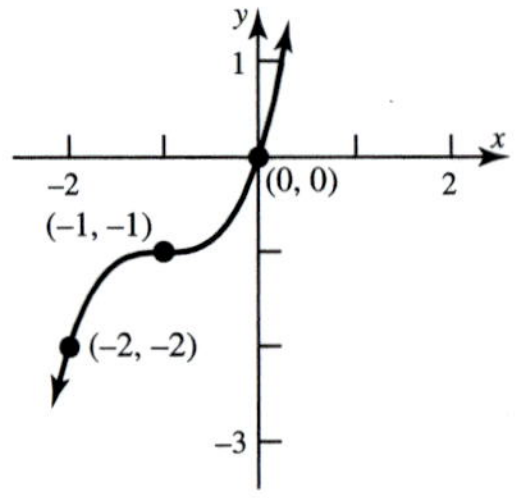

51.

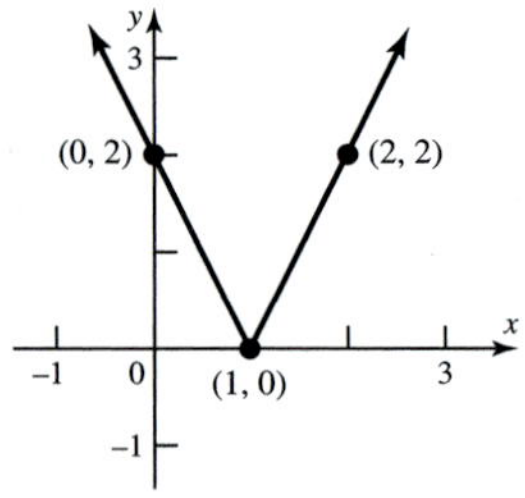

53.

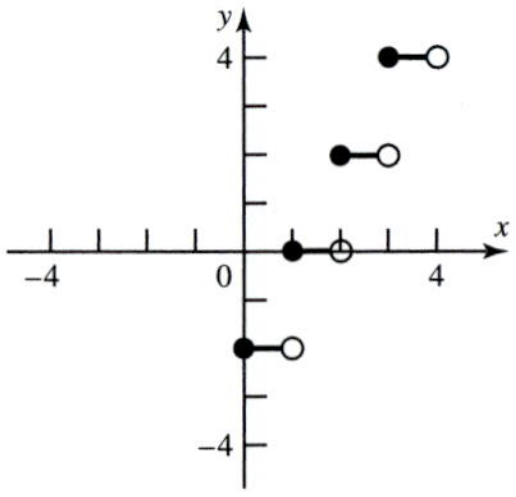

55. **(a)** $F(x) = f(x) + 3$

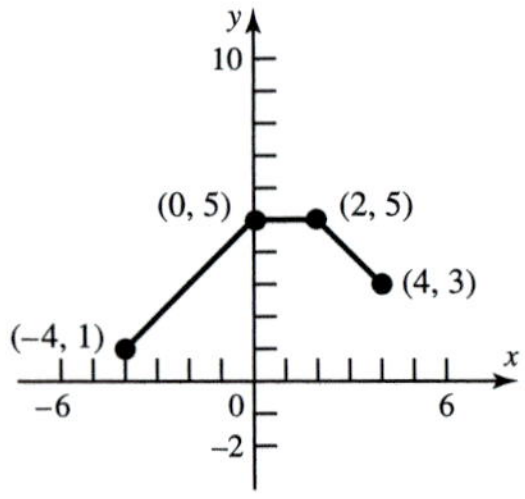

(b) $G(x) = f(x + 2)$

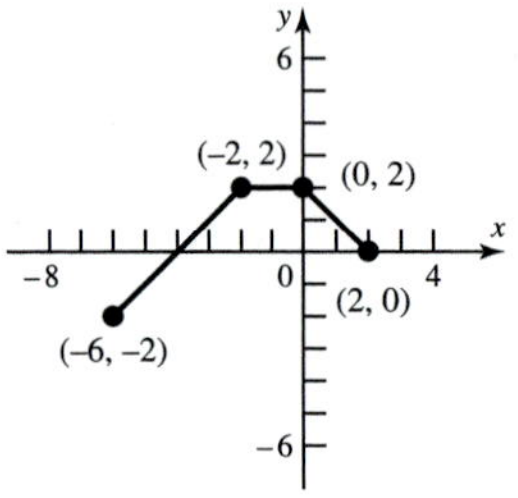

(c) $P(x) = -f(x)$

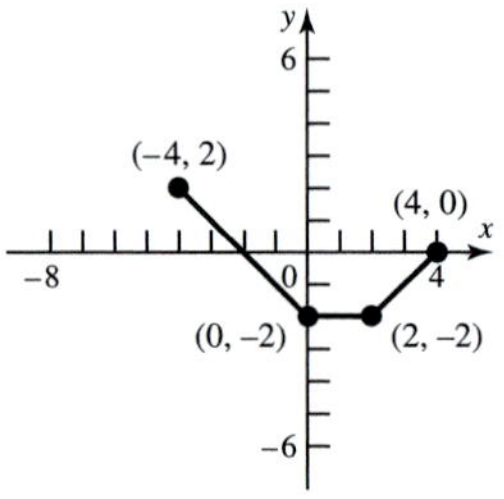

(d) $Q(x) = \frac{1}{2}f(x)$

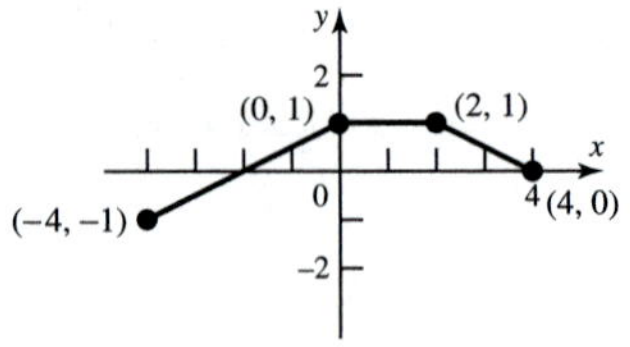

(e) $g(x) = f(-x)$

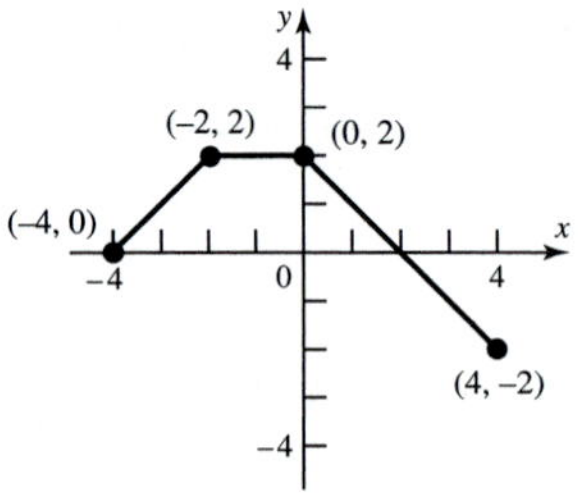

(f) $h(x) = 3f(x)$

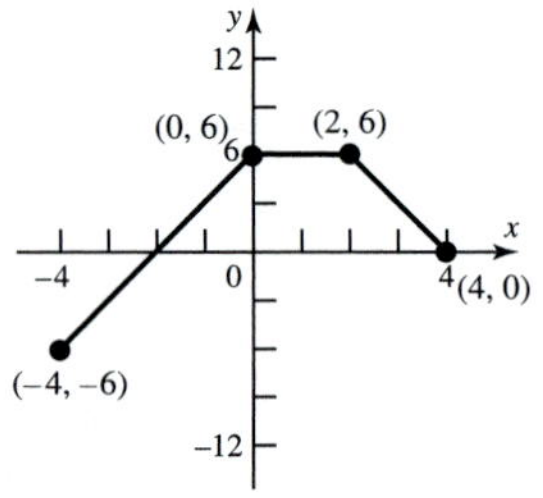

57. **(a)** $F(x) = f(x) + 3$

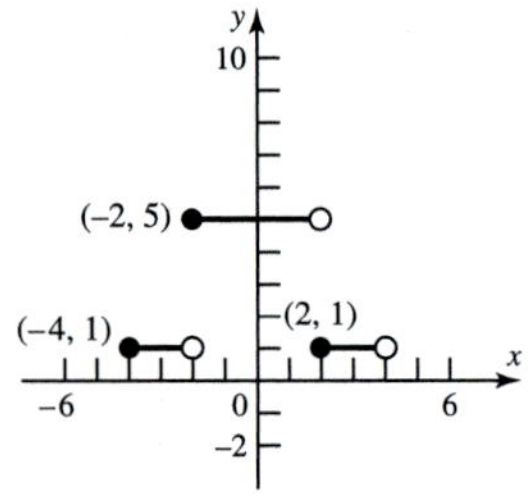

(b) $G(x) = f(x + 2)$

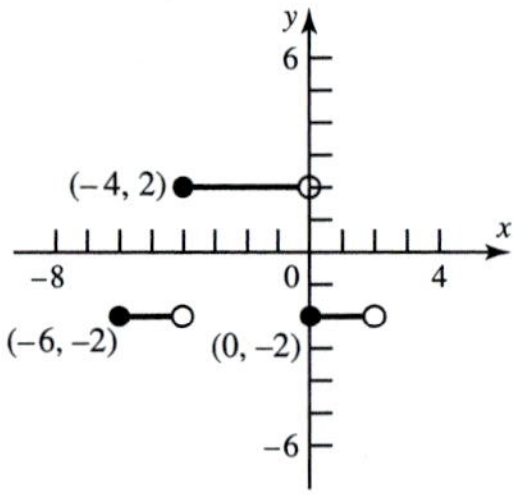

(c) $P(x) = -f(x)$

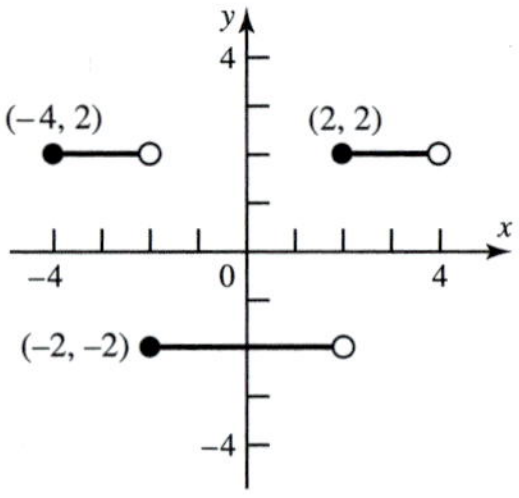

(d) $Q(x) = \frac{1}{2}f(x)$

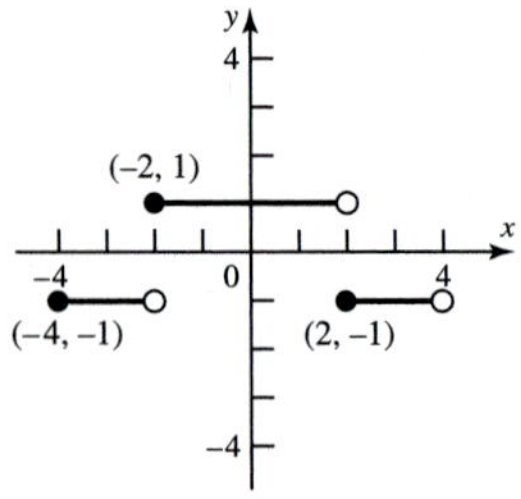

(e) $g(x) = f(-x)$

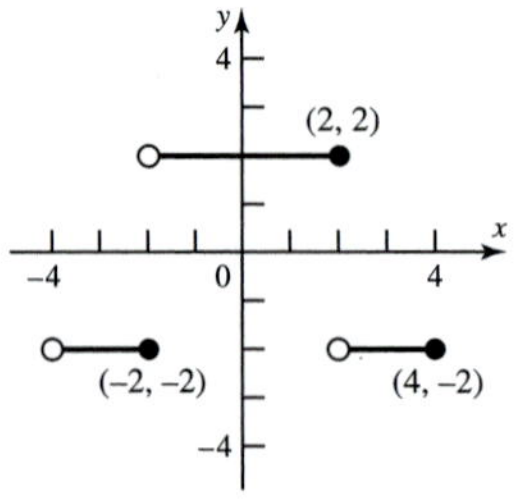

(f) $h(x) = 3f(x)$

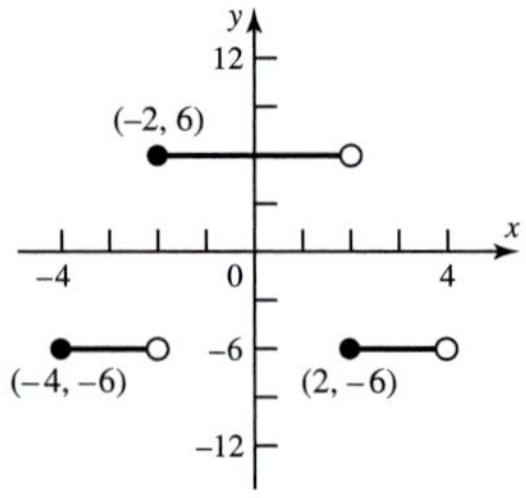

59. (a) $F(x) = f(x) + 3$

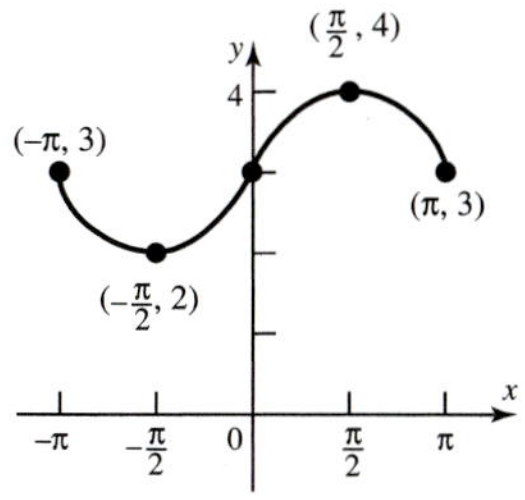

(b) $G(x) = f(x + 2)$

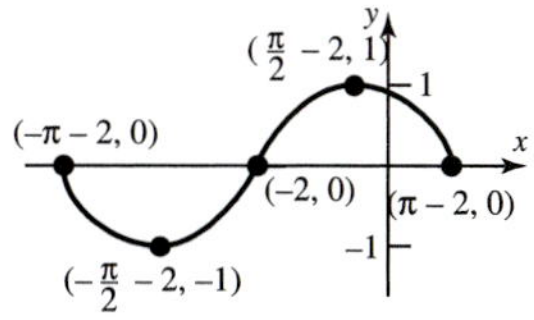

(c) $P(x) = -f(x)$

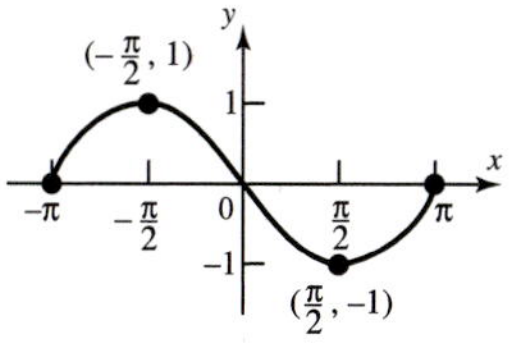

(d) $Q(x) = \frac{1}{2}f(x)$

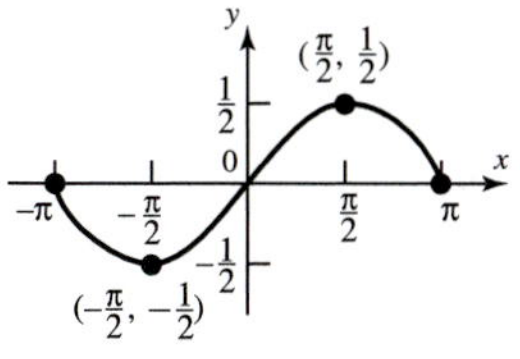

(e) $g(x) = f(-x)$

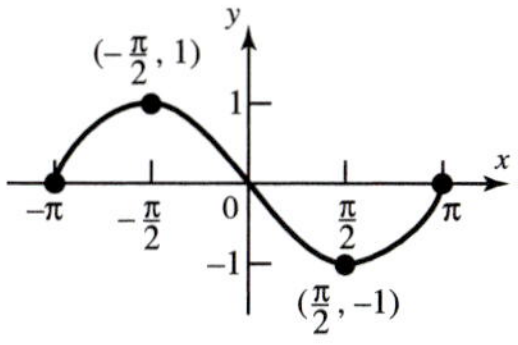

(f) $h(x) = 3f(x)$

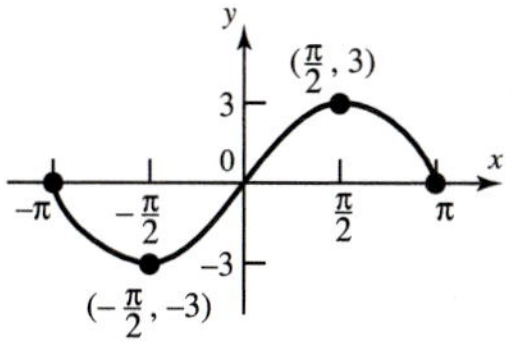

61. (a)

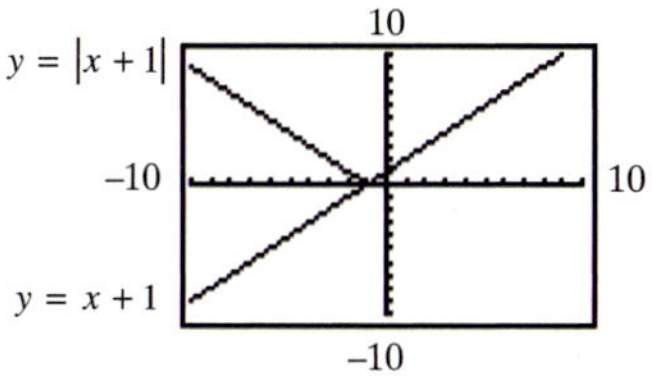

(b)

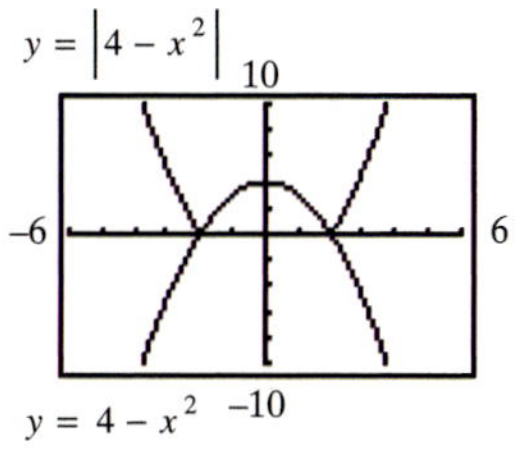

(c)

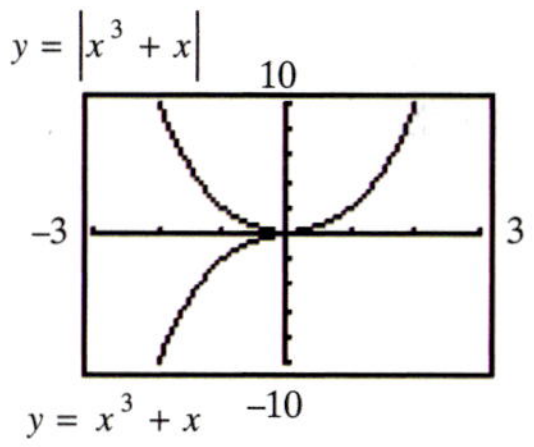

(d) Any part of the graph of $y = f(x)$ that lies below the x-axis is reflected about the x-axis to obtain the graph of $y = |f(x)|$

63. (a)

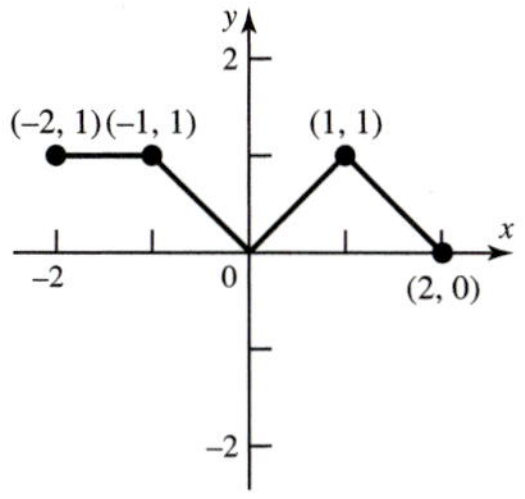

(b)

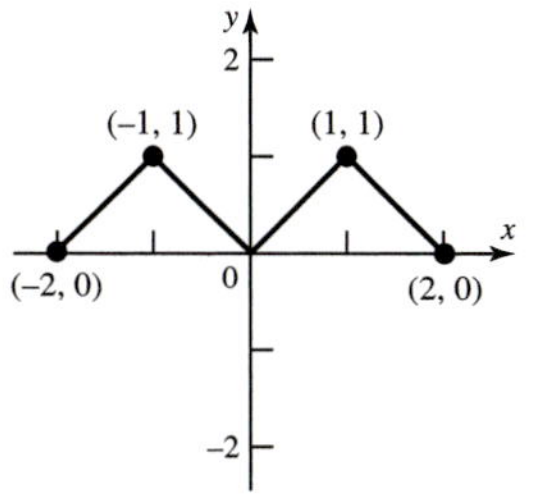

65. $f(x) = (x + 1)^2 - 1$

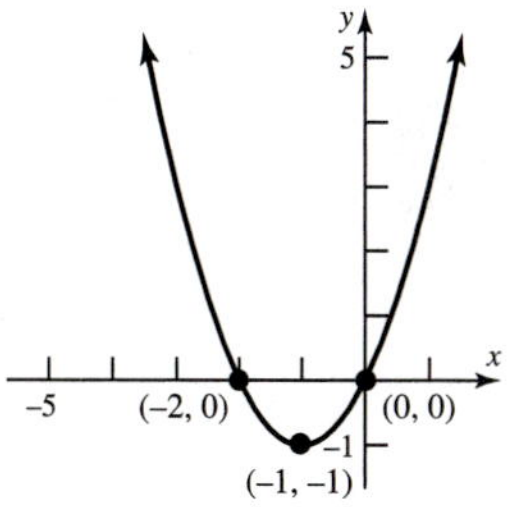

67. $f(x) = (x - 4)^2 - 15$

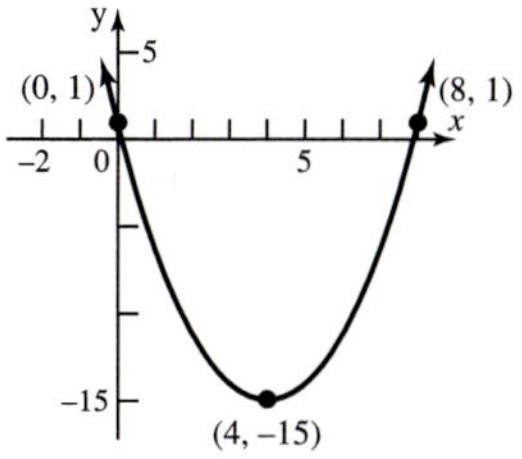

69. $f(x) = \left(x + \frac{1}{2}\right)^2 + \frac{3}{4}$

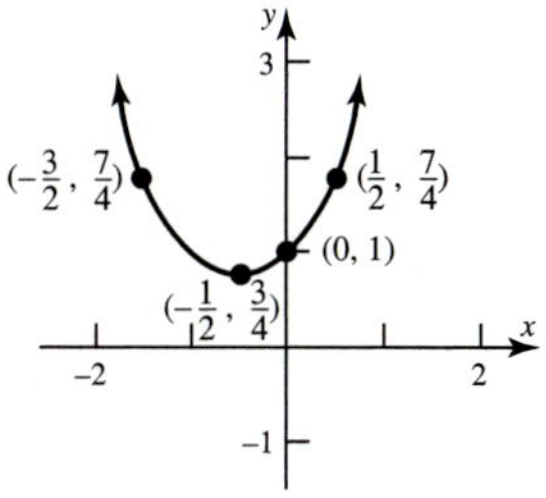

71.

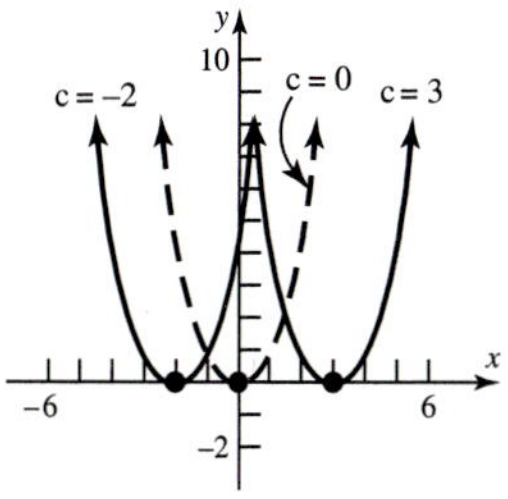

73.

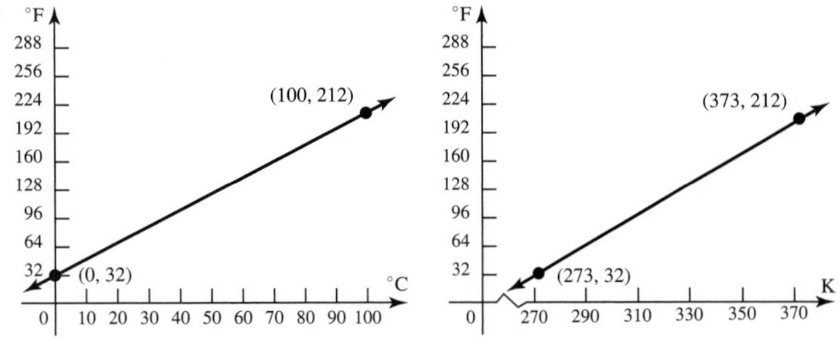

75. (a)

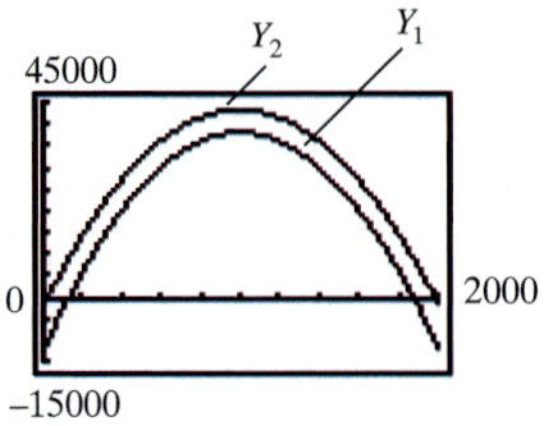

(b) 10% tax **(c)** Y_1 is the graph of $p(x)$ shifted down vertically 10,000 units. Y_2 is the graph of $p(x)$ vertically compressed by a factor of 0.9. **(d)** 10% tax

2.4 Exercises

1. (a) $(f + g)(x) = 5x + 1$; all real numbers **(b)** $(f - g)(x) = x + 7$; all real numbers **(c)** $(f \cdot g)(x) = 6x^2 - x - 12$; all real numbers **(d)** $\left(\dfrac{f}{g}\right)(x) = \dfrac{3x + 4}{2x - 3}$; $\left\{x \middle| x \neq \dfrac{3}{2}\right\}$ **3. (a)** $(f + g)(x) = 2x^2 + x - 1$; all real numbers **(b)** $(f - g)(x) = -2x^2 + x - 1$; all real numbers **(c)** $(f \cdot g)(x) = 2x^3 - 2x^2$; all real numbers **(d)** $\left(\dfrac{f}{g}\right)(x) = \dfrac{x - 1}{2x^2}$; $\{x| x \neq 0\}$

5. (a) $(f + g)(x) = \sqrt{x} + 3x - 5$; $\{x| x \geq 0\}$ **(b)** $(f - g)(x) = \sqrt{x} - 3x + 5$; $\{x| x \geq 0\}$ **(c)** $(f \cdot g)(x) = 3x\sqrt{x} - 5\sqrt{x}$; $\{x| x \geq 0\}$ **(d)** $\left(\dfrac{f}{g}\right)(x) = \dfrac{\sqrt{x}}{3x - 5}$; $\left\{x \middle| x \geq 0, x \neq \dfrac{5}{3}\right\}$ **7. (a)** $(f + g)(x) = 1 + \dfrac{2}{x}$; $\{x| x \neq 0\}$ **(b)** $(f - g)(x) = 1$; $\{x| x \neq 0\}$ **(c)** $(f \cdot g)(x) = \dfrac{1}{x} + \dfrac{1}{x^2}$; $\{x| x \neq 0\}$ **(d)** $\left(\dfrac{f}{g}\right)(x) = x + 1$; $\{x| x \neq 0\}$ **9. (a)** $(f + g)(x) = \dfrac{6x + 3}{3x - 2}$; $\left\{x \middle| x \neq \dfrac{2}{3}\right\}$ **(b)** $(f - g)(x) = \dfrac{-2x + 3}{3x - 2}$; $\left\{x \middle| x \neq \dfrac{2}{3}\right\}$ **(c)** $(f \cdot g)(x) = \dfrac{8x^2 + 12x}{(3x - 2)^2}$; $\left\{x \middle| x \neq \dfrac{2}{3}\right\}$ **(d)** $\left(\dfrac{f}{g}\right)(x) = \dfrac{2x + 3}{4x}$; $\{x| x \neq 0, x \neq \dfrac{2}{3}\}$ **11.** $g(x) = 5 - \dfrac{7}{2}x$

13. $f + g = x + \dfrac{1}{x}$

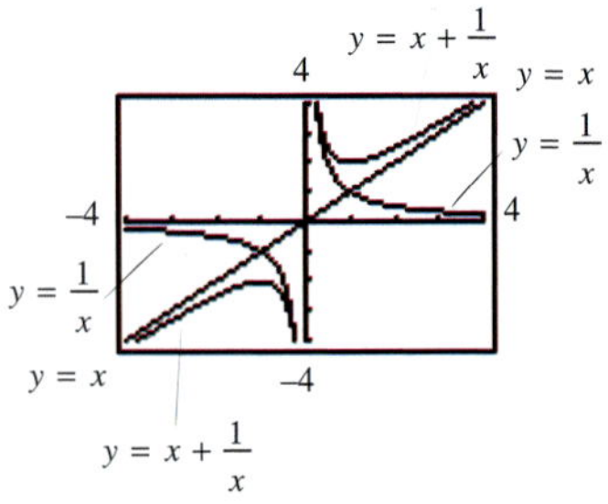

15. $f + g = x^2 + \dfrac{1}{x}$

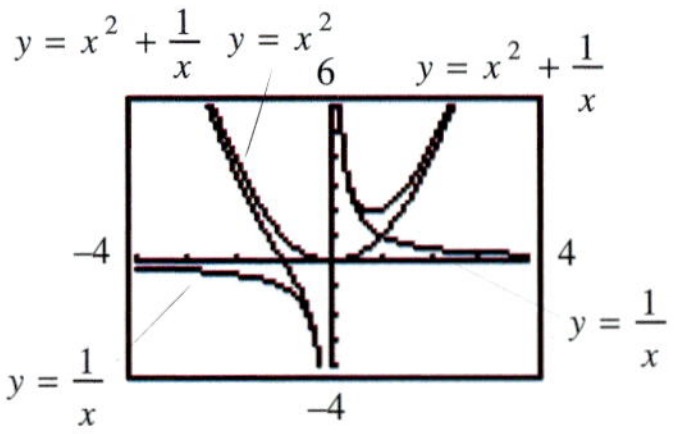

17. $f \cdot g = \dfrac{x}{x^2 + 1}$

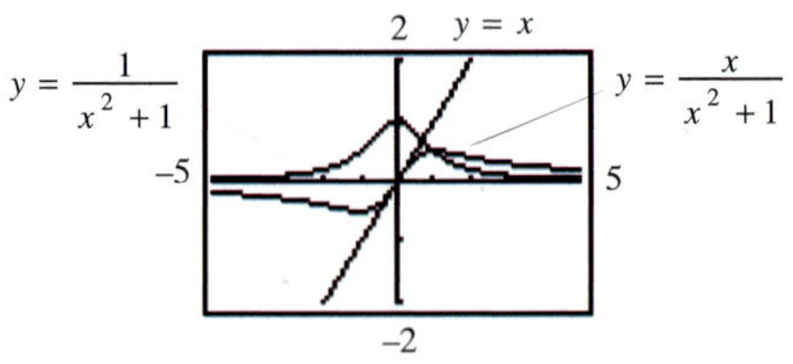

19. $f \cdot g = \dfrac{x^2}{x^2 + 1}$

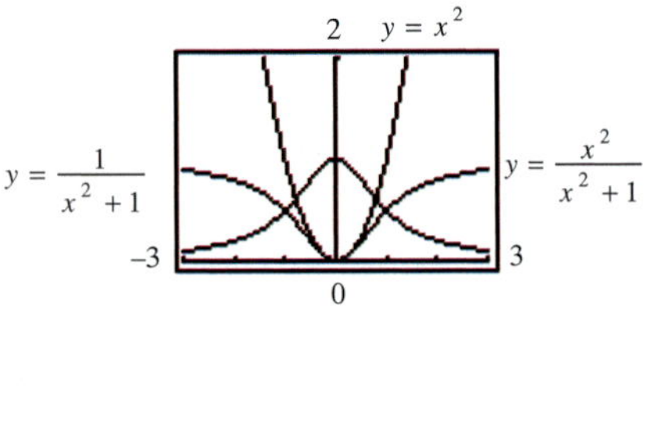

21. **(a)** 98 **(b)** 49 **(c)** 4 **(d)** 4 **23.** **(a)** 97 **(b)** $-\frac{163}{2}$ **(c)** 1 **(d)** $-\frac{3}{2}$ **25.** **(a)** $2\sqrt{2}$ **(b)** $2\sqrt{2}$ **(c)** 1 **(d)** 0 **27.** **(a)** $\frac{1}{17}$ **(b)** $\frac{1}{5}$ **(c)** 1 **(d)** $\frac{1}{2}$ **29.** **(a)** $\frac{3}{5}$ **(b)** $\frac{\sqrt{15}}{5}$ **(c)** $\frac{12}{13}$ **(d)** 0 **31.** **(a)** $(f \circ g)(x) = 6x + 3$; All real numbers **(b)** $(g \circ f)(x) = 6x + 9$; All real numbers **(c)** $(f \circ f)(x) = 4x + 9$; All real numbers **(d)** $(g \circ g)(x) = 9x$; All real numbers

33. **(a)** $(f \circ g)(x) = 3x^2 + 1$; All real numbers **(b)** $(g \circ f)(x) = 9x^2 + 6x + 1$; All real numbers **(c)** $(f \circ f)(x) = 9x + 4$; All real numbers **(d)** $(g \circ g)(x) = x^4$; All real numbers **35.** **(a)** $(f \circ g)(x) = \sqrt{x^2 - 1}$; $\{x \mid x \le -1 \text{ or } x \ge 1\}$ **(b)** $(g \circ f)(x) = x - 1$; $\{x \mid x \ge 0\}$ **(c)** $(f \circ f)(x) = \sqrt[4]{x}$; $\{x \mid x \ge 0\}$ **(d)** $(g \circ g)(x) = x^4 - 2x^2$; All real numbers **37.** **(a)** $(f \circ g)(x) = \frac{1 - x}{1 + x}$; $\{x \mid x \ne -1, x \ne 0\}$ **(b)** $(g \circ f)(x) = \frac{x + 1}{x - 1}$; $\{x \mid x \ne -1, x \ne 1\}$ **(c)** $(f \circ f)(x) = -\frac{1}{x}$; $\{x \mid x \ne -1, x \ne 0\}$ **(d)** $(g \circ g)(x) = x$; $\{x \mid x \ne 0\}$

39. **(a)** $(f \circ g)(x) = x$; $\{x \mid x \ge 0\}$ **(b)** $(g \circ f)(x) = |x|$; All real numbers **(c)** $(f \circ f)(x) = x^4$; All real numbers **(d)** $(g \circ g)(x) = \sqrt[4]{x}$; $\{x \mid x \ge 0\}$ **41.** **(a)** $(f \circ g)(x) = \frac{1}{4x + 9}$; $\{x \mid x \ne -\frac{9}{4}\}$ **(b)** $(g \circ f)(x) = \frac{2}{2x + 3} + 3 = \frac{6x + 11}{2x + 3}$; $\{x \mid x \ne -\frac{3}{2}\}$ **(c)** $(f \circ f)(x) = \frac{2x + 3}{6x + 11}$; $\{x \mid x \ne -\frac{3}{2}, x \ne -\frac{11}{6}\}$ **(d)** $(g \circ g)(x) = 4x + 9$; All real numbers

43. **(a)** $(f \circ g)(x) = acx + ad + b$; All real numbers **(b)** $(g \circ f)(x) = acx + bc + d$; All real numbers **(c)** $(f \circ f)(x) = a^2x + ab + b$; All real numbers **(d)** $(g \circ g)(x) = c^2x + cd + d$; All real numbers

45. $(f \circ g)(x) = f(g(x)) = f\left(\frac{1}{2}x\right) = 2\left(\frac{1}{2}x\right) = x$; $(g \circ f)(x) = g(f(x)) = g(2x) = \frac{1}{2}(2x) = x$

47. $(f \circ g)(x) = f(\sqrt[3]{x}) = (\sqrt[3]{x})^3 = x$; $(g \circ f)(x) = g(x^3) = \sqrt[3]{x^3} = x$

49. $(f \circ g)(x) = f\left(\frac{1}{2}(x + 6)\right) = 2\left[\frac{1}{2}(x + 6)\right] - 6 = x + 6 - 6 = x$; $(g \circ f)(x) = g(2x - 6) = \frac{1}{2}(2x - 6 + 6) = x$

51. $(f \circ g)(x) = f\left(\frac{1}{a}(x - b)\right) = a\left[\frac{1}{a}(x - b)\right] + b = x$; $(g \circ f)(x) = g(ax + b) = \frac{1}{a}(ax + b - b) = x$

53. $(f \circ g)(x) = 11$; $(g \circ f)(x) = 2$ **55.** $-3, 3$ **57.** $S(r(t)) = \frac{16}{9}\pi t^6$ **59.** $C(N(t)) = 15{,}000 + 800{,}000t - 40{,}000t^2$

61. $C = \frac{2\sqrt{100 - p}}{25} + 600$

63. Since f and g are odd, $f(-x) = -f(x)$ and $g(-x) = -g(x)$; $(f \circ g)(-x) = f(g(-x)) = f(-g(x)) = -f(g(x)) = -(f \circ g)(x)$; thus, $f \circ g$ is odd.

2.5 Exercises

1. $V(r) = 2\pi r^3$ **3.** **(a)** $R(x) = -\frac{1}{6}x^2 + 100x$ **(b)** \$13,333 **(c)**

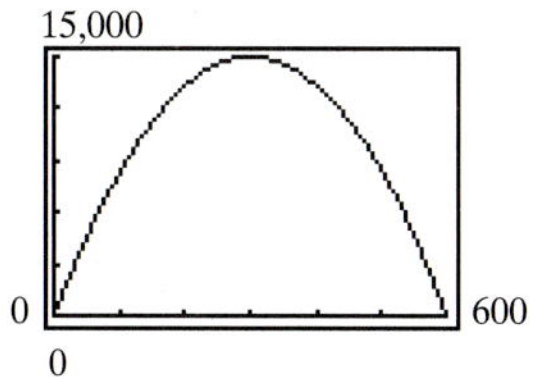

(d) 300; \$15,000 **(e)** \$50

5. **(a)** $R(x) = -\frac{1}{5}x^2 + 20x$ **(b)** \$255 **(c)**

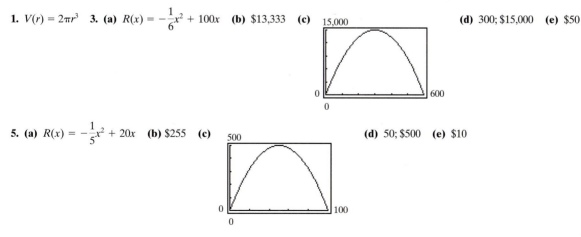

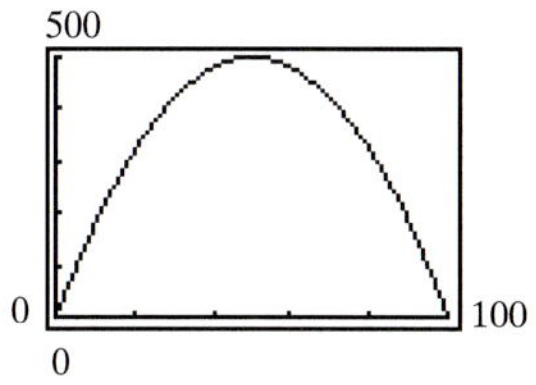

(d) 50; \$500 **(e)** \$10

7. **(a)** $A(x) = -x^2 + 200x$ **(b)** $0 < x < 200$ **(c)** A is largest when $x = 100$ yards **9.** **(a)** $C(x) = x$ **(b)** $A(x) = \frac{x^2}{4\pi}$

11. $A(x) = \frac{1}{2}x^4$ **13.** **(a)** $d(x) = \sqrt{x^4 - 15x^2 + 64}$ **(b)** $d(0) = 8$ **(c)** $d(1) = \sqrt{50} \approx 7.07$

(d)

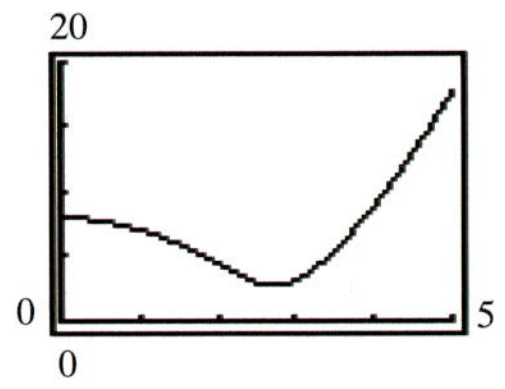

(e) 2.73 **15. (a)** $d(x) = \sqrt{x^2 - x + 1}$ **(b)**

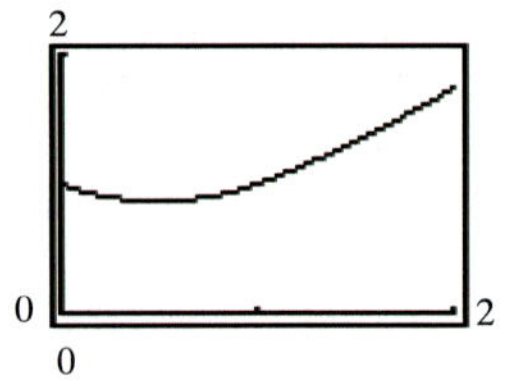

(c) 0.50

17. $d(t) = 50t$

19. (a) $V(x) = x(24 - 2x)^2$

(b) The volume is largest when x is 4

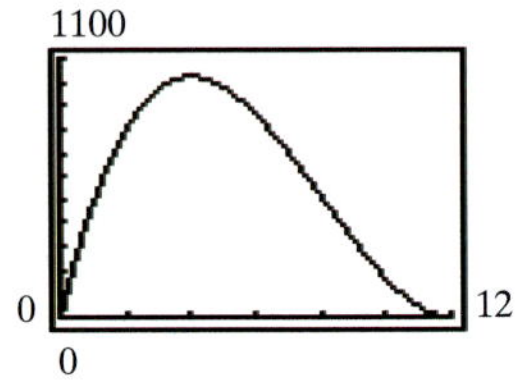

21. (a) $A(x) = 2x^2 + \dfrac{40}{x}$

(b) The area is smallest when x is about 2.15

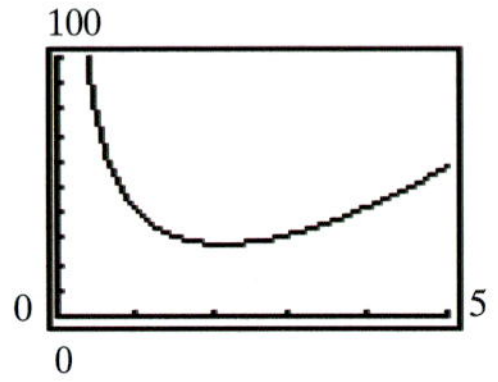

23. (a) $A(x) = x(16 - x^2)$
(b) Domain: $\{x \mid 0 < x < 4\}$
(c) The area is largest for x about 2.31

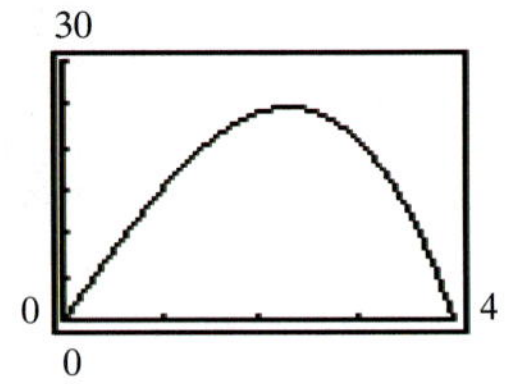

25. (a) $A(x) = 4x(4 - x^2)^{1/2}$
(b) $p(x) = 4x + 4(4 - x^2)^{1/2}$
(c) The area is largest for x about 1.41

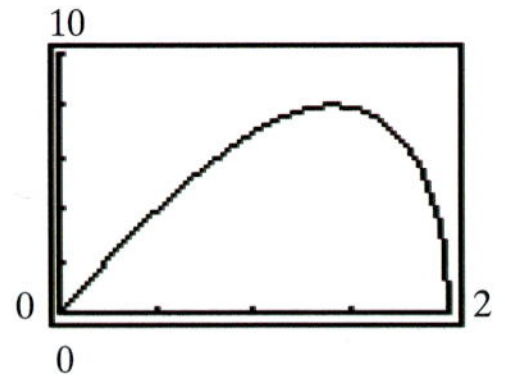

(d) The perimeter is largest for x about 1.41

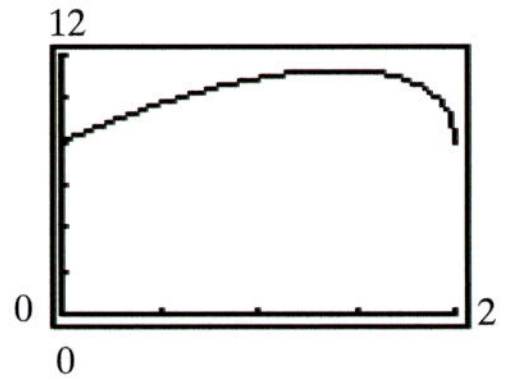

27. (a) $A(x) = x^2 + \dfrac{25 - 20x + 4x^2}{\pi}$ **(b)** Domain $\{x \mid 0 < x < 2.5\}$ **(c)** The area is smallest for x about 1.40 meters

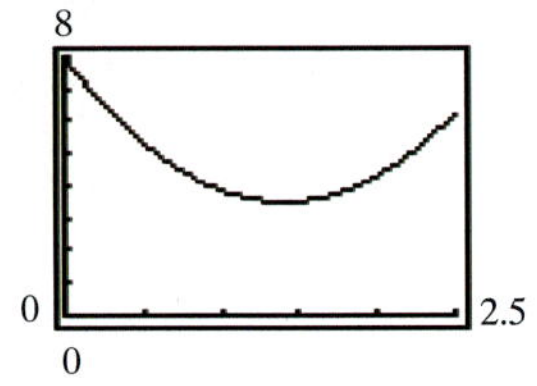

29. (a) $A(r) = 2r^2$ **(b)** $p(r) = 6r$ **31.** $A(x) = \left(\dfrac{\pi}{3} - \dfrac{\sqrt{3}}{4}\right)x^2$ **33.** $C = \begin{cases} 95 & \text{if } x = 7 \\ 119 & \text{if } 7 < x \le 8 \\ 143 & \text{if } 8 < x \le 9 \\ 167 & \text{if } 9 < x \le 10 \\ 190 & \text{if } 10 < x \le 14 \end{cases}$

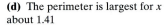

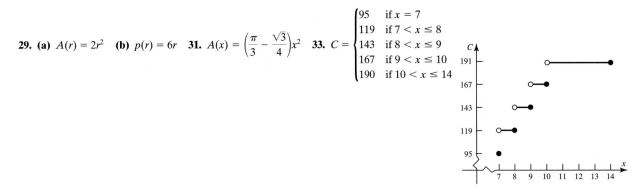

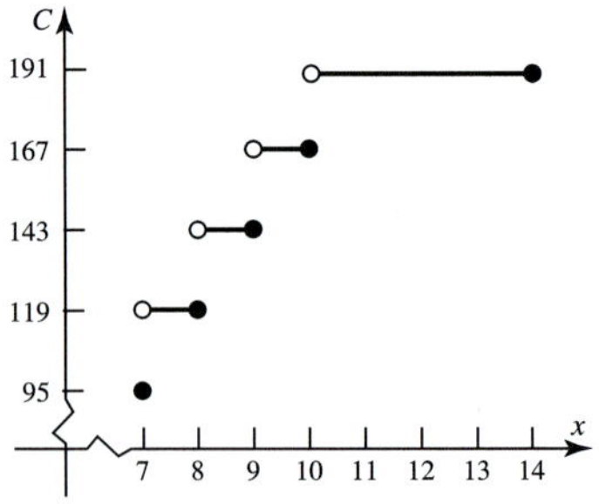

35. $V(h) = \dfrac{\pi}{48}h^3$

37. (a) $Q = -37.6355p + 1571.7196$ **(b)** $R = -37.6355p^2 + 1571.7196p$ **(c)**

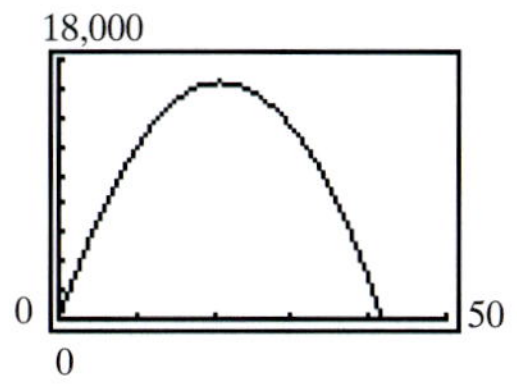

(d) \$20.88
(e) 786; \$16,409.39

Fill-in-the-Blank Items

1. independent; dependent **2.** vertical **3.** even; odd **4.** horizontal; right **5.** $g(f(x)) = (g \circ f)(x)$

True/False Items

1. T **2.** T **3.** F **4.** T **5.** F

Review Exercises

1. $f(x) = -2x + 3$ **3.** $A = 11$ **5.** b, c, d **7. (a)** $f(-x) = \dfrac{-3x}{x^2 - 4}$ **(b)** $-f(x) = \dfrac{-3x}{x^2 - 4}$ **(c)** $f(x + 2) = \dfrac{3x + 6}{x^2 + 4x}$ **(d)** $f(x - 2) = \dfrac{3x - 6}{x^2 - 4x}$ **9. (a)** $f(-x) = \sqrt{x^2 - 4}$ **(b)** $-f(x) = -\sqrt{x^2 - 4}$ **(c)** $f(x + 2) = \sqrt{x^2 + 4x}$ **(d)** $f(x - 2) = \sqrt{x^2 - 4x}$ **11. (a)** $f(x) = \dfrac{x^2 - 4}{x^2}$ **(b)** $-f(x) = -\dfrac{x^2 - 4}{x^2}$ **(c)** $f(x + 2) = \dfrac{x^2 + 4x}{x^2 + 4x + 4}$ **(d)** $f(x - 2) = \dfrac{x^2 - 4x}{x^2 - 4x + 4}$ **13.** Odd **15.** Even

17. Neither **19.** $\{x \mid x \neq -3, x \neq 3\}$ **21.** $(-\infty, 2]$ **23.** $(0, \infty)$ **25.** $\{x \mid x \neq -3, x \neq 1\}$ **27.** $[-1, \infty)$ **29.** $[0, \infty)$

31. (a) All real numbers
(b) $(0, -4), (-4, 0), (4, 0)$
(c)

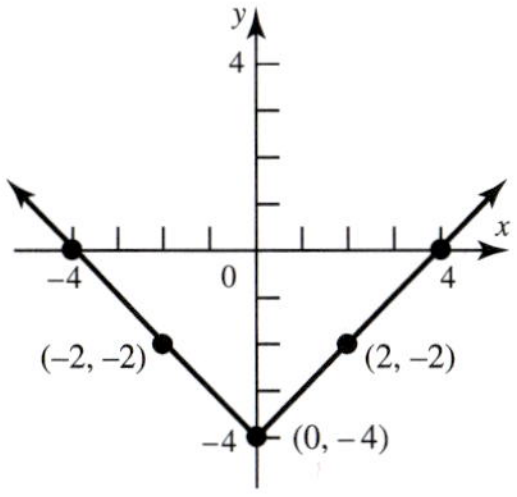

(d) $\{y \mid -4 \leq y < \infty\}$

33. (a) All real numbers
(b) $(0, 0)$
(c)

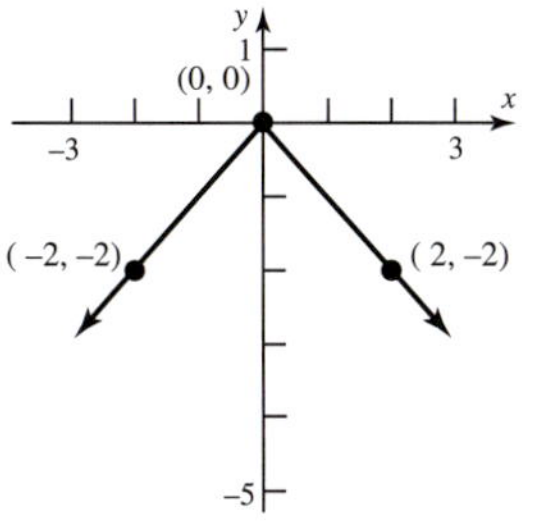

(d) $\{y \mid -\infty < y \leq 0\}$

35. (a) $\{x \mid 1 \leq x < \infty\}$
(b) $(1, 0)$
(c)

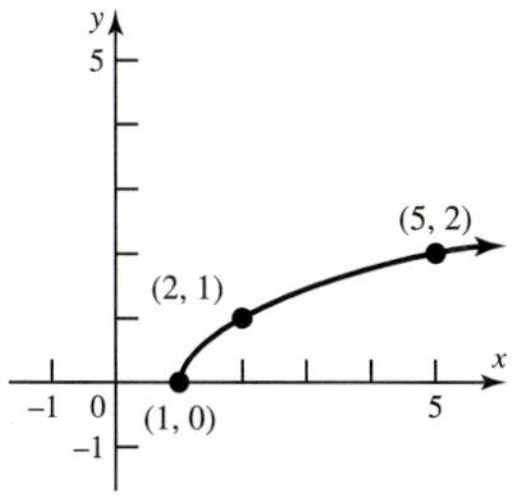

(d) $\{y \mid 0 \leq y < \infty\}$

37. (a) $\{x \mid -\infty < x \leq 1\}$
(b) $(1, 0), (0, 1)$
(c)

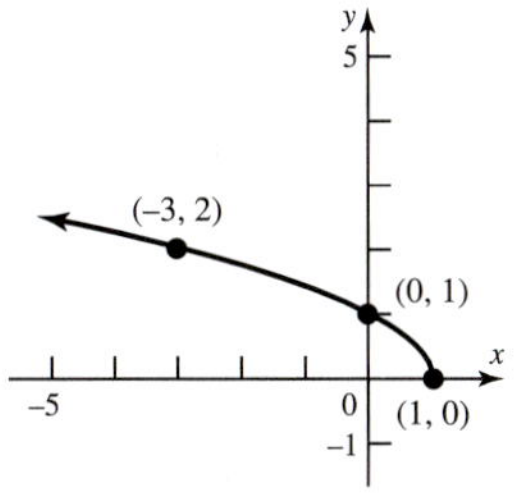

(d) $\{y \mid 0 \leq y < \infty\}$

39. (a) All real numbers
(b) $(0, 4), (2, 0)$
(c)

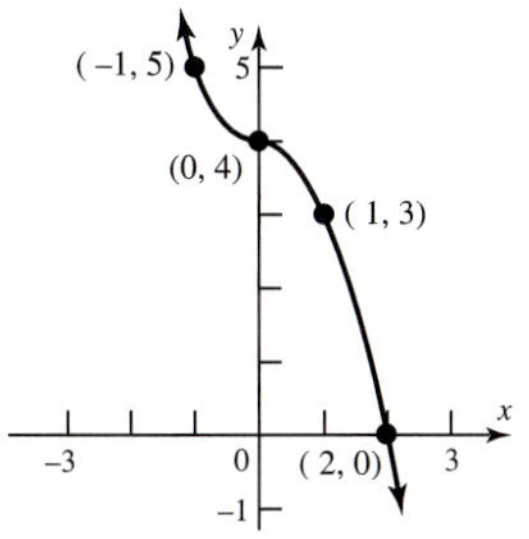

(d) All real numbers

41. (a) All real numbers
(b) $(0, 3)$
(c)

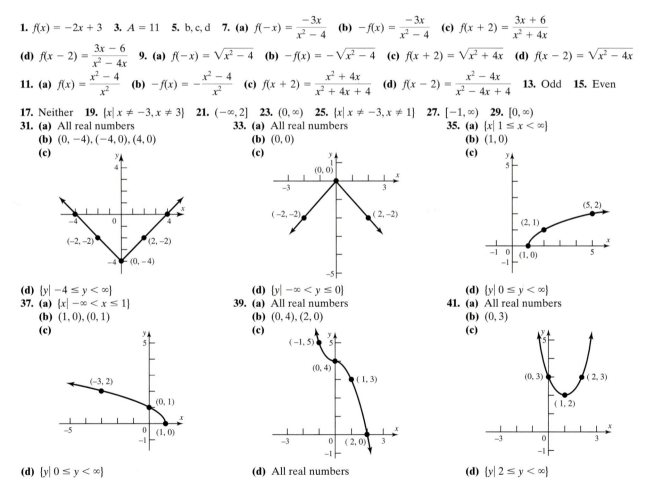

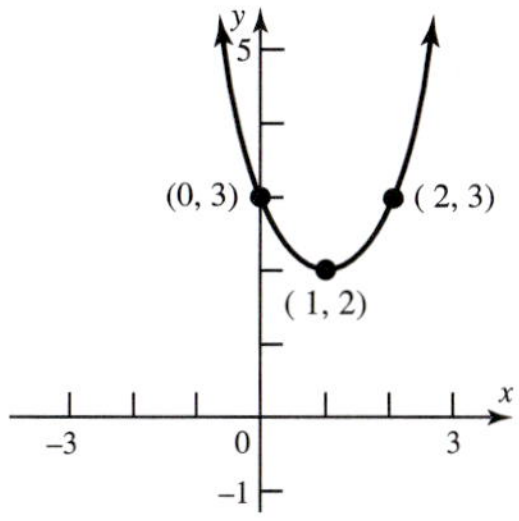

(d) $\{y \mid 2 \leq y < \infty\}$

43. (a) All real numbers
(b) $(0, 0)$
(c)

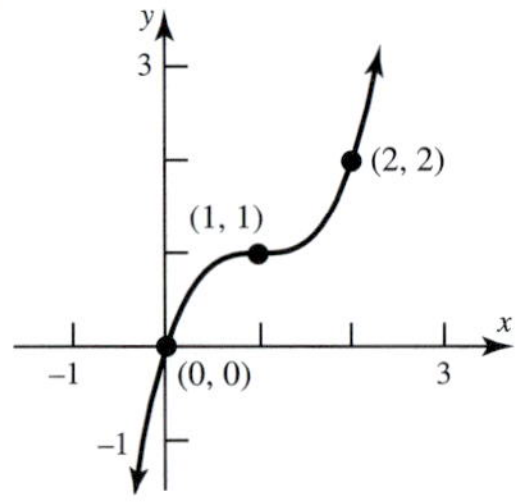

(d) All real numbers

45. (a) $\{x \mid 0 < x < \infty\}$
(b) None
(c)

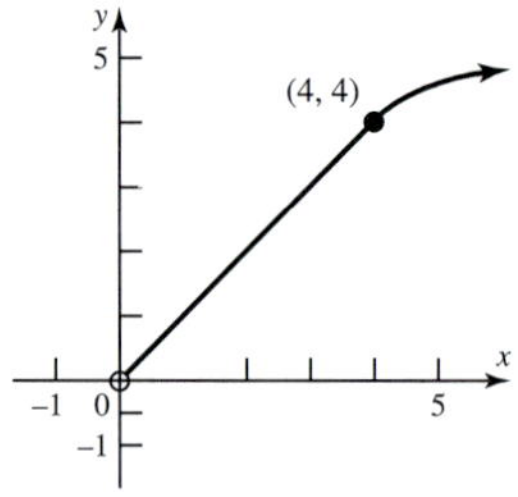

(d) $\{y \mid 0 < y < \infty\}$

47. (a) $\{x \mid x \neq 1\}$
(b) $(0, 0)$
(c)

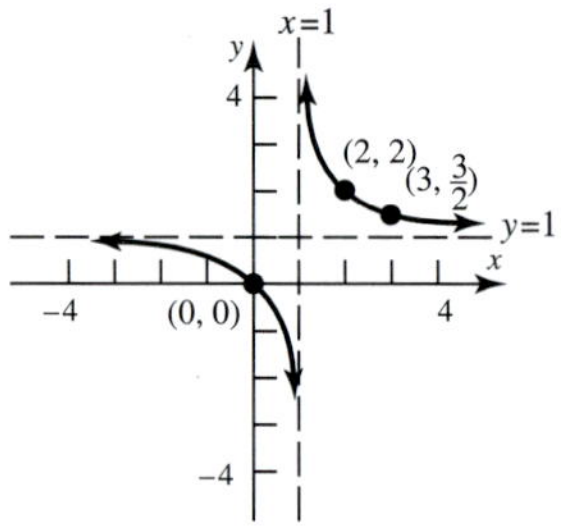

(d) $\{y \mid y \neq 1\}$

49. (a) All real numbers
(b) $-1 < x \leq 0$ are x-intercepts 0 is the y-intercept
(c)

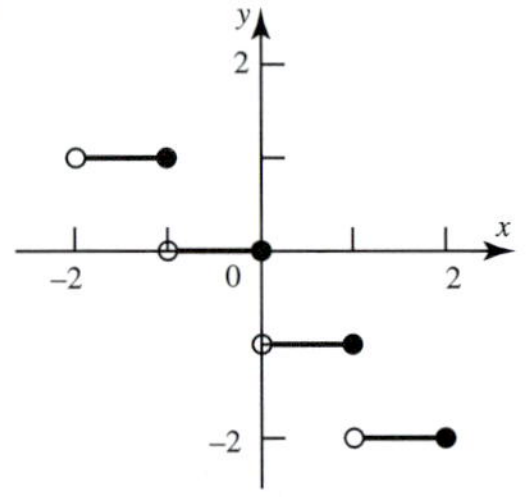

(d) Set of integers

51. (a) -26
(b) -241
(c) 16
(d) -1

53. (a) $\sqrt{11}$ **(b)** 1 **(c)** $\sqrt{\sqrt{6} + 2}$ **(d)** 19 **55. (a)** $\frac{1}{20}$ **(b)** $-\frac{13}{8}$ **(c)** $\frac{400}{1601}$ **(d)** -17

57. $(f \circ g)(x) = \frac{-3x + 1}{3x + 1}$; $\{x \mid x \neq -\frac{1}{3}\}$; $(g \circ f)(x) = \frac{6 - 2x}{x}$; $\{x \mid x \neq 0\}$; $(f \circ f)(x) = \frac{3x - 2}{2 - x}$; $\{x \mid x \neq 0, x \neq 2\}$; $(g \circ g)(x) = 9x + 4$; All real numbers **59.** $(f \circ g)(x) = 27x^2 + 3|x| + 1$; All real numbers; $(g \circ f)(x) = 3|3x^2 + x + 1|$; All real numbers; $(f \circ f)(x) = 3(3x^2 + x + 1)^2 + 3x^2 + x + 2$; All real numbers; $(g \circ g)(x) = 9|x|$; All real numbers

61. $(f \circ g)(x) = \frac{1 + x}{1 - x}$; $\{x \mid x \neq 0, \neq 1\}$; $(g \circ f)(x) = \frac{x - 1}{x + 1}$; $\{x \mid x \neq -1, x \neq 1\}$; $(f \circ f)(x) = x$; $\{x \mid x \neq 1\}$; $(g \circ g)(x) = x$; $\{x \mid x \neq 0\}$

63. (a)

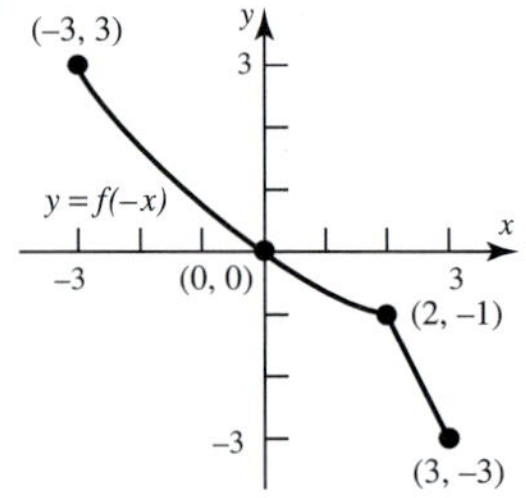

(b)

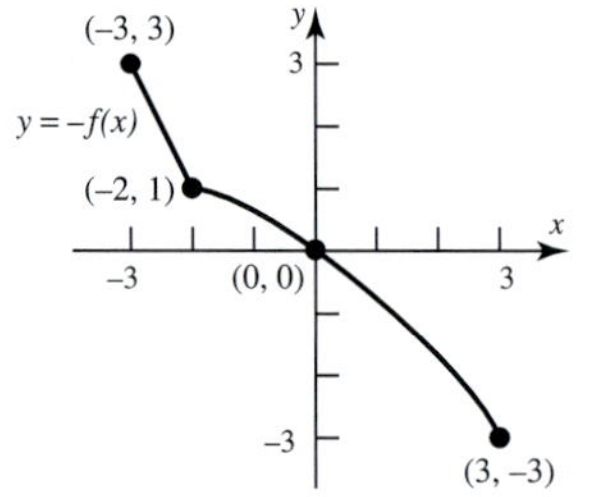

(c)

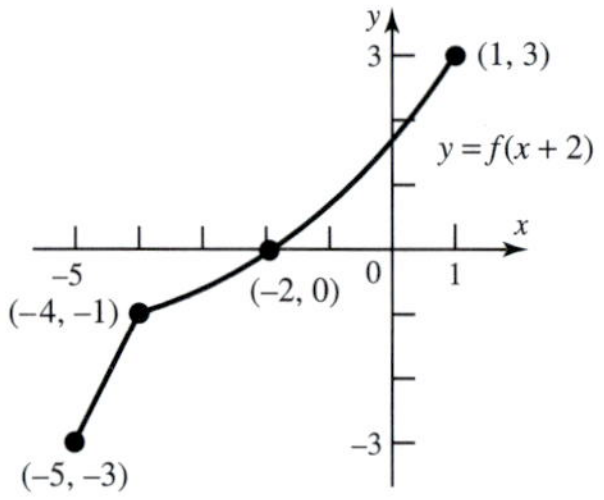

(d)

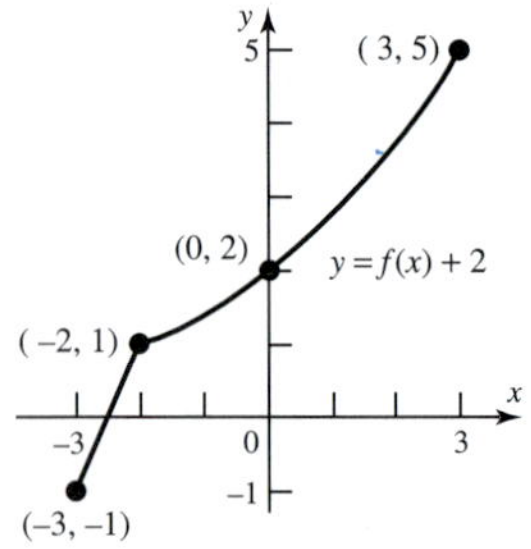

(e)

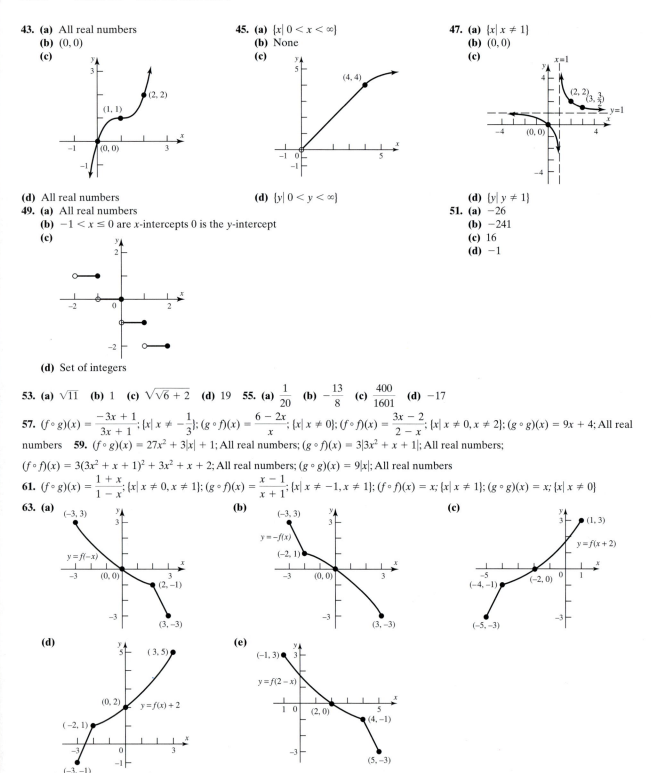

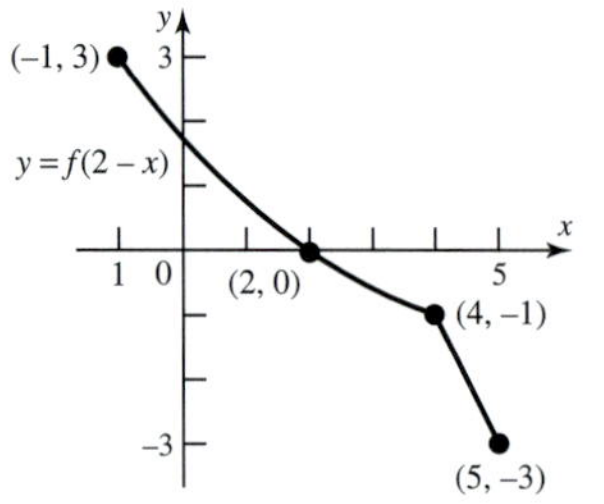

65. $T(h) = -0.0025h + 30, 0 \leq x \leq 10{,}000$ **67.** $S(x) = kx(36 - x^2)^{3/2}$; Domain: $\{x \mid 0 < x < 6\}$

69. (a), (b), (e)

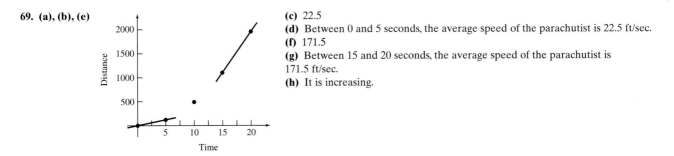

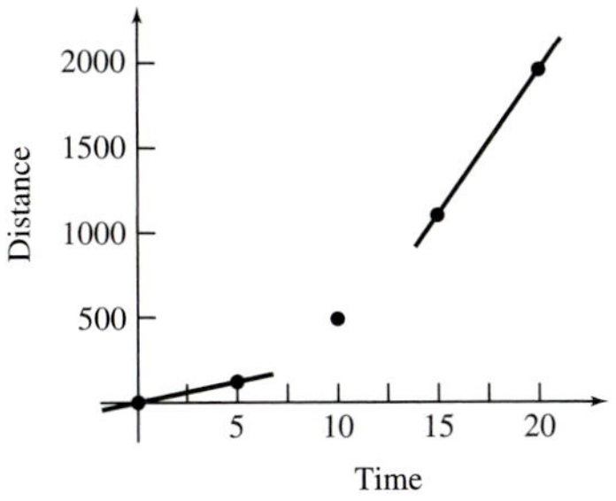

(c) 22.5
(d) Between 0 and 5 seconds, the average speed of the parachutist is 22.5 ft/sec.
(f) 171.5
(g) Between 15 and 20 seconds, the average speed of the parachutist is 171.5 ft/sec.
(h) It is increasing.

CHAPTER 3 *3.1 Exercises*

1. 0.42 **3.** 2.23 **5.** 1.25 **7.** $f(0) = -1, f(1) = 10$; 0.21 **9.** $f(-5) = -58, f(-4) = 2$; -4.04
11. $f(1.4) = -0.17536, f(1.5) = 1.40625$; 1.41 **13.** -3.41 **15.** -1.70 **17.** -0.28 **19.** 3 **21.** 4.5 **23.** 0.31, 12.30 **25.** 1.00, 23.00

3.2 Exercises

1. 7 **3.** -3 **5.** 4 **7.** $\frac{5}{4}$ **9.** 2 **11.** 3 **13.** -1 **15.** $-\frac{4}{3}$ **17.** -18 **19.** -3 **21.** $-\frac{3}{4}$ **23.** -20 **25.** 2 **27.** 0.5 **29.** $\frac{46}{5}$ **31.** 2
33. 20 **35.** $\{0, 9\}$ **37.** $\{-5, 5\}$ **39.** $\{-4, 3\}$ **41.** $\left\{-\frac{1}{2}, 3\right\}$ **43.** $\{-4, 4\}$ **45.** $\{3, 4\}$ **47.** $\frac{3}{2}$ **49.** $\left\{-\frac{2}{3}, \frac{3}{2}\right\}$ **51.** $\left\{-\frac{2}{3}, \frac{3}{2}\right\}$ **53.** $\left\{-\frac{3}{4}, 2\right\}$
55. 2 **57.** -1 **59.** 3 **61.** No solution **63.** $\{0, 4\}$ **65.** $\{0, 9\}$ **67.** 5.90 **69.** 0.40 **71.** $x = \frac{b + c}{a}$ **73.** $x = \frac{abc}{a + b}$ **75.** $x = a^2$
77. $\{2 - \sqrt{2}, 2 + \sqrt{2}\}$ **79.** $\{2 - \sqrt{5}, 2 + \sqrt{5}\}$ **81.** $\left\{1, \frac{3}{2}\right\}$ **83.** No real solution **85.** $\left\{\frac{-1 - \sqrt{5}}{4}, \frac{-1 + \sqrt{5}}{4}\right\}$ **87.** $\left\{0, \frac{9}{4}\right\}$ **89.** $\frac{1}{3}$
91. $\left\{\frac{1 - \sqrt{7}}{3}, \frac{1 + \sqrt{7}}{3}\right\}$ **93.** $\left\{\frac{1 - \sqrt{33}}{8}, \frac{1 + \sqrt{33}}{8}\right\}$ **95.** No real solution **97.** $\{0.58, 3.41\}$; $\{2 - \sqrt{2}, 2 + \sqrt{2}\}$
99. $\{-2.80, 1.07\}$; $\left\{\frac{-\sqrt{3} - \sqrt{15}}{2}, \frac{-\sqrt{3} + \sqrt{15}}{2}\right\}$ **101.** $\{-0.85, 1.17\}$; $\left\{\frac{1 \pm \sqrt{1 + 4\pi^2}}{2\pi}\right\}$
103. $\{-8.15, -0.22\}$; $\left\{\frac{-4\pi \pm \sqrt{16\pi^2 - 3\sqrt{29}}}{3}\right\}$ **105.** $\{-2.64, 2.64\}$; $\{-\sqrt{7}, \sqrt{7}\}$ **107.** 0.25, $\frac{1}{4}$ **109.** $\{-0.60, 2.50\}$; $\left\{-\frac{3}{5}, \frac{5}{2}\right\}$
111. $\{-0.50, 0.66\}$; $\left\{-\frac{1}{2}, \frac{2}{3}\right\}$ **113.** $\{-1.70, 0.29\}$; $\left\{\frac{-\sqrt{2} + 2}{2}, \frac{-\sqrt{2} - 2}{2}\right\}$ **115.** $\{-2.56, 1.56\}$; $\left\{\frac{-1 - \sqrt{17}}{2}, \frac{-1 + \sqrt{17}}{2}\right\}$
117. No real solution **119.** Repeated real solution **121.** Two unequal real solutions **123.** $R = \frac{R_1R_2}{R_1 + R_2}$ **125.** $R = \frac{mv^2}{F}$
127. $r = \frac{S - a}{S}$

3.3 Exercises

1. $A = \pi r^2$; r = Radius, A = Area **3.** $A = s^2$, A = Area, s = Length of a side **5.** $F = ma$; F = Force, m = Mass, a = Acceleration
7. $W = Fd$; W = Work, F = Force, d = Distance **9.** $C = 150x$; C = Total cost, x = number of dishwashers
11. $11,000 will be invested in bonds and $9000 in CD's. **13.** Scott will receive $400,000, Alice $300,000, and Tricia $200,000.
15. The regular hourly rate is $8.50. **17.** The Bears got 5 touchdowns. **19.** The length is 19ft; the width is 11 ft.
21. **(a)** The dimensions are 10 ft by 5 ft. **(b)** The area is 50 sq ft. **(c)** The dimensions are 7.5 ft by 7.5 ft. **(d)** The area is 56.25 sq ft.
23. Invest $31,250 in bonds and $18,750 in CD's. **25.** $11,600 was loaned out at 8%. **27.** The dimensions are 11 ft by 13 ft.
29. The dimensions are 5 m by 8 m. **31.** The dimensions of the sheet metal should be 4 ft by 4 ft.
33. **(a)** The ball strikes the ground after 6 seconds. **(b)** The ball passes the top of the building on its way down after 5 seconds.
35. The original price was $147,058.82; purchasing the model saves $22,058.82. **37.** The bookstore paid $44.80.
39. Working together, it takes 12 min. **41.** Bridgette needs a score of 85.
43. The defensive back catches up to the tight end at the tight end's 45 yard line.
45. The border will be 2.56 ft wide. **47.** The dimensions of the reduced candy bar are 11.55 cm by 6.55 cm by 3 cm.

49. The border will be 2.71 ft wide. **51.** The current is 2.286 mi/hr. **53.** Mix 40 pounds of cashews with the peanuts. **55.** Mike passes Dan $\frac{1}{3}$ mile from the start, 2 minutes from the time Mike started to race. **57.** The rescue craft reaches the ship in 2 hrs. **59.** Start the auxiliary pump at 9:45 AM. **61.** The tub will fill in 1 hr. **63.** The most you can invest in the CD is \$66,667. **65.** The average speed is 49.5 mi/hr. **67.** Set the original price at \$40. At 50% off, there will be no profit at all.

3.4 Exercises

1. 1 **3.** No real solution **5.** -13 **7.** 3 **9.** 2 **11.** $-\frac{8}{5}$ **13.** 8 **15.** $\{-1, 3\}$ **17.** $\{1, 5\}$ **19.** 1 **21.** 5 **23.** 9 **25.** $\{-4, 4\}$ **27.** $\{-5, -4\}$ **29.** $-\frac{1}{3}$ **31.** $\left\{-\frac{3}{2}, 2\right\}$ **33.** $\{0, 16\}$ **35.** 16 **37.** 1 **39.** $\left\{\left(\frac{9-\sqrt{17}}{8}\right)^4, \left(\frac{9+\sqrt{17}}{8}\right)^4\right\}$ **41.** $\{\sqrt{2}, \sqrt{3}\}$ **43.** $\{-4, 1\}$ **45.** $\left\{-2, -\frac{1}{2}\right\}$ **47.** $\left\{-\frac{3}{2}, \frac{1}{3}\right\}$ **49.** $\left\{-\frac{1}{8}, 27\right\}$ **51.** $\left\{-2, -\frac{4}{5}\right\}$ **53.** $\{-1, 0, 1\}$ **55.** $\{-2, -1, 1, 2\}$ **57.** $\{-1, 1\}$ **59.** $\{-2, 1\}$ **61.** $\{0, 36\}$ **63.** $\{0, 2\}$ **65.** $\{-5, 0, 4\}$ **67.** -1 **69.** $\{-2, 2, 3\}$ **71.** $\{-1, 1\}$ **73.** $\{0.34, 11.65\}$ **75.** $\{-1.03, 1.03\}$ **77.** $\{-1.85, 0.17\}$ **79.** $\left\{\frac{3}{2}, 5\right\}$ **81. (a)** The depth of the well is about 230 ft. **(b)**

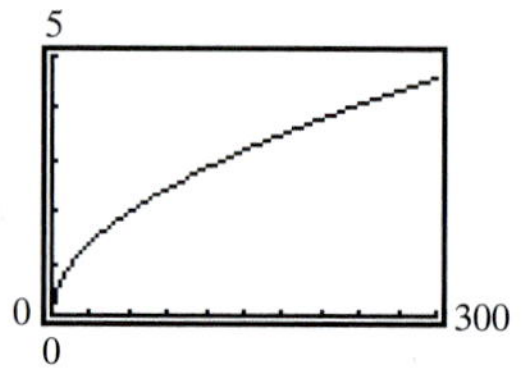

(c)

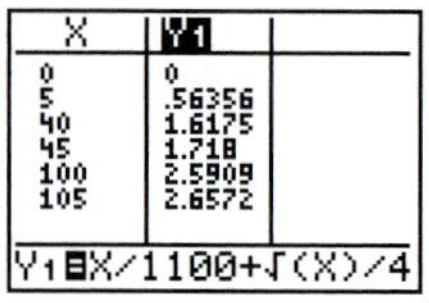

X	Y1
0	0
5	.56356
40	1.6175
45	1.718
100	2.5909
105	2.6572

Y1■X/1100+√(X)/4

The slope is decreasing as x increases

(d) As the depth of the well increases, the time that elapses before a sound is heard increases at a decreasing rate.

3.5 Exercises

1. $<$ **3.** $>$ **5.** $>$ **7.** $>$ **9. (a)** $0 < 2$ **(b)** $-2 < 0$ **(c)** $9 < 15$ **(d)** $-6 > -10$ **11. (a)** $2x - 2 < -1$ **(b)** $2x - 4 < -3$ **(c)** $6x + 3 < 6$ **(d)** $-4x - 2 > -4$

13. $\{x \mid x < 4\}$ or $(-\infty, 4)$

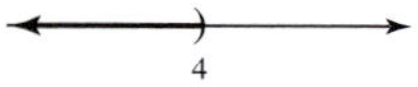

15. $\{x \mid x \ge -1\}$ or $[-1, \infty)$

17. $\{x \mid x > 3\}$ or $(3, \infty)$

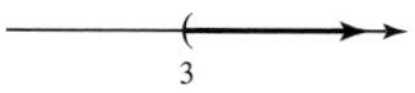

19. $\{x \mid x \ge 2\}$ or $[2, \infty)$

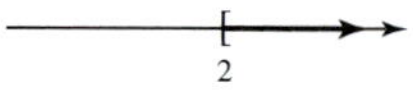

21. $\{x \mid x > -7\}$ or $(-7, \infty)$

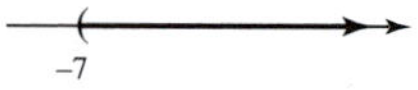

23. $\left\{x \middle| x \le \frac{2}{3}\right\}$ or $\left(-\infty, \frac{2}{3}\right]$

25. $\{x \mid x < -20\}$ or $(-\infty, -20)$

27. $\left\{x \middle| x \ge \frac{4}{3}\right\}$ or $\left[\frac{4}{3}, \infty\right)$

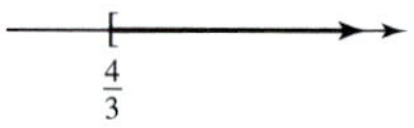

29. $\{x \mid 3 \le x \le 5\}$ or $[3, 5]$

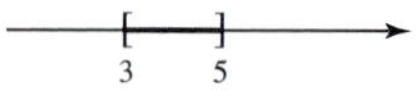

31. $\left\{x \middle| \frac{2}{3} \le x \le 3\right\}$ or $\left[\frac{2}{3}, 3\right]$

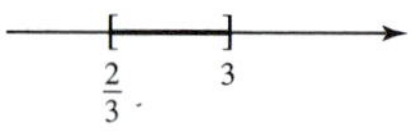

33. $\left\{x \middle| -\frac{11}{2} < x < \frac{1}{2}\right\}$ or $\left(-\frac{11}{2}, \frac{1}{2}\right)$

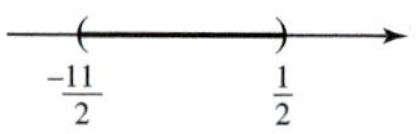

35. $\{x \mid -6 < x < 0\}$ or $(-6, 0)$

37. $\{x \mid x < -5\}$ or $(-\infty, -5)$

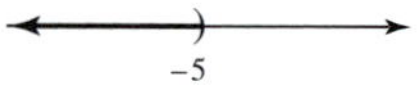

39. $\{x \mid x \ge -1\}$ or $[-1, \infty)$

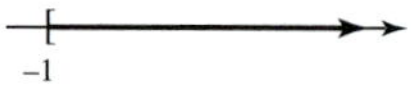

41. $\left\{x \middle| \frac{1}{2} \le x < \frac{5}{4}\right\}$ or $\left[\frac{1}{2}, \frac{5}{4}\right)$

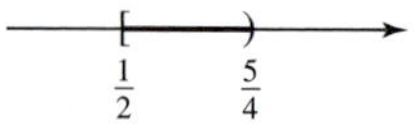

43. $\left\{x \middle| x < \frac{1}{2}\right\}$ or $\left(-\infty, -\frac{1}{2}\right)$

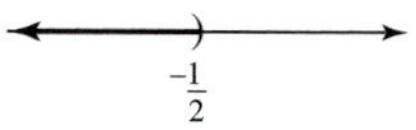

45. $\{x \mid x > 5\}$ or $(5, \infty)$

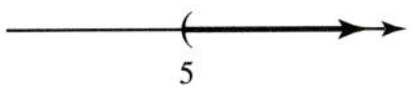

47. $\{x \mid x > 3\}$ or $(3, \infty)$

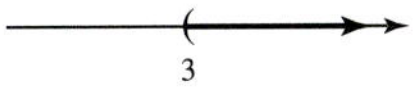

49. $a = 3, b = 5$ **51.** $a = -12, b = -8$

53. $a = 3, b = 11$ **55.** $a = \frac{1}{4}, b = 1$ **57.** $a = 4, b = 16$ **59.** $\{x \mid -10 < x < 0\}$ **61.** $\{x \mid x \geq -2\}$ or $[-2, \infty)$

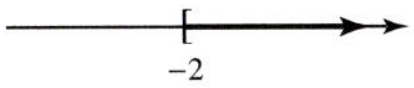

63. $a \leq b, c > 0;$ $\begin{aligned} a - b &\leq 0 \\ (a - b)c &\leq 0(c) \\ ac - bc &\leq 0 \\ ac &\leq bc \end{aligned}$

65. $\frac{a+b}{2} - a = \frac{a + b - 2a}{2} = \frac{b-a}{2} > 0$; therefore, $a < \frac{a+b}{2}$

$b - \frac{a+b}{2} = \frac{2b - a - b}{2} = \frac{b-a}{2} > 0$; therefore, $b > \frac{a+b}{2}$

67. $(\sqrt{ab})^2 - a^2 = ab - a^2 = a(b - a) > 0$; thus, $(\sqrt{ab})^2 > a^2$ and $\sqrt{ab} > a$
$b^2 - (\sqrt{ab})^2 = b^2 - ab = b(b - a) > 0$; thus, $b^2 > (\sqrt{ab})^2$ and $b > \sqrt{ab}$

69. $h - a = \frac{2ab}{a+b} - a = \frac{ab - a^2}{a+b} = \frac{a(b-a)}{a+b} > 0$; thus, $h > a$

$b - h = b - \frac{2ab}{a+b} = \frac{b^2 - ab}{a+b} = \frac{b(b-a)}{a+b} > 0$; thus, $h < b$

71. $21 < \text{Age} < 30$ **73.** **(a)** Male ≥ 73.4 **(b)** Female ≥ 79.7 **(c)** A female can expect to live at least 6.3 years longer.
75. The agent's commission ranges from \$45,000 to \$95,000, inclusive. As a percent of selling price, the commission ranges from 5% to 8.6%, inclusive. **77.** The amount withheld varies from \$70.62 to \$84.62, inclusive.
79. The usage varies from 675.48 kWhr to 2500.88 kWhr, inclusive. **81.** The dealer's cost varies from \$7457.63 to \$7857.14, inclusive.
83. Fifth test score ≥ 74 **85.** The amount of gasoline ranged from 12 to 20 gal, inclusive.

3.6 Exercises

1. $\{-4, 4\}$ **3.** $\{-4, 1\}$ **5.** $\left\{-1, \frac{3}{2}\right\}$ **7.** $\{-4, 4\}$ **9.** 2 **11.** $\{-12, 12\}$ **13.** $\left\{-\frac{36}{5}, \frac{24}{5}\right\}$ **15.** No real solution **17.** $\{-4, 4\}$ **19.** $\{-1, 3\}$

21. $\{-2, -1, 0, 1\}$

23. $(-4, 4)$

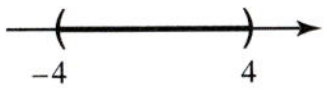

25. $(-\infty, -4)$ or $(4, \infty)$

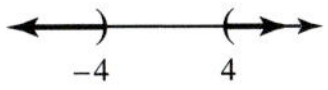

27. $(1, 3)$

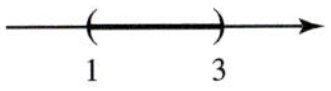

29. $\left[-\frac{2}{3}, 2\right]$

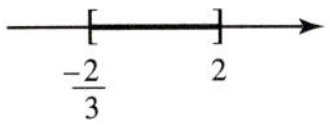

31. $(-\infty, 1]$ or $[5, \infty)$

33. $\left(-1, \frac{3}{2}\right)$

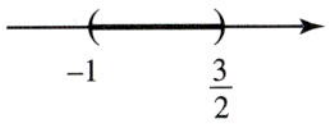

35. $(-\infty, -1)$ or $(2, \infty)$

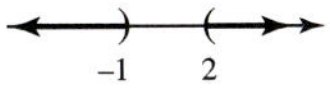

37. $(-\infty, -3)$ or $(-3, \infty)$

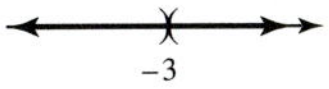

39. $(-\infty, \infty)$

41. $(0.49, 0.51)$

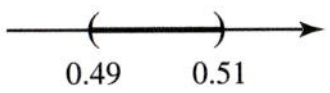

43. $a = 2, b = 8$ **45.** $a = -15, b = -7$ **47.** $a = -1, b = -\frac{1}{15}$

49. $\left|\frac{a}{b}\right| = \sqrt{\left(\frac{a}{b}\right)^2} = \sqrt{\frac{a^2}{b^2}} = \frac{\sqrt{a^2}}{\sqrt{b^2}} = \frac{|a|}{|b|}$

51. $(a + b)^2 = a^2 + 2ab + b^2 \leq |a|^2 + 2|a||b| + |b|^2 = [|a| + |b|]^2$; therefore, $\sqrt{(a+b)^2} \leq \sqrt{(|a| + |b|)^2}$ or $|a + b| \leq |a| + |b|$
53. $|x - 3| < \frac{1}{2}; \frac{5}{2} < x < \frac{7}{2}$ **55.** $|x + 3| > 2; x < -5$ or $x > -1$ **57.** $|x - 98.6| \geq 1.5; x \leq 97.1°\text{F}$ or $x \geq 100.1°\text{F}$
59. $x^2 - a < 0; (x - \sqrt{a})(x + \sqrt{a}) < 0$; therefore, $-\sqrt{a} < x < \sqrt{a}$ **61.** $-1 < x < 1$ **63.** $x \geq 3$ or $x \leq -3$ **65.** $-4 \leq x \leq 4$
67. $x > 2$ or $x < -2$ **69.** $\{-1, 5\}$

Fill-in-the-Blank Items

1. equivalent **2.** add; $\frac{25}{4}$ **3.** discriminant; negative **4.** $-a$ **5.** double; multiplicity 2 **6.** extraneous **7.** negative **8.** $-2; 2$

True/False Items

1. T **2.** F **3.** T **4.** T **5.** T **6.** **(a)** T **(b)** F **(c)** T **7.** **(a)** F **(b)** T **(c)** T

Review Exercises

1. -12 **3.** 6 **5.** $\frac{1}{5}$ **7.** -5 **9.** No real solution **11.** $\frac{11}{8}$ **13.** $\left\{-2, \frac{3}{2}\right\}$ **15.** $\left\{\frac{1-\sqrt{13}}{4}, \frac{1+\sqrt{13}}{4}\right\}$ **17.** $\{-3, 3\}$

19. No real solution **21.** $\{-2, -1, 1, 2\}$ **23.** 2 **25.** 0 **27.** $\frac{\sqrt{5}}{2}$ **29.** $\left\{-\frac{1}{8}, 1\right\}$ **31.** $\left\{-1, \frac{1}{2}\right\}$ **33.** $-\frac{9}{5}$ **35.** $\{-5, 2\}$ **37.** $\left\{-\frac{5}{3}, 3\right\}$

39. $\left\{\frac{m}{1-n}, \frac{m}{1+n}\right\}$ **41.** $\left\{-\frac{9b}{5a}, \frac{2b}{a}\right\}$

43. $\{x \mid 14 \le x < \infty\}$

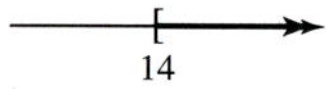

45. $\left\{x \,\middle|\, -\frac{31}{2} \le x \le \frac{33}{2}\right\}$

47. $\{x \mid -23 < x < -7\}$

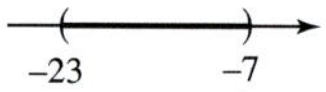

49. $\left\{x \,\middle|\, -\frac{3}{2} < x < -\frac{7}{6}\right\}$

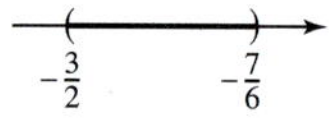

51. $\{x \mid -\infty < x \le -2 \text{ or } 7 \le x < \infty\}$ **53.** 3.13 **55.** $\{0.31, 12.29\}$ **57.** The storm is 3300 ft away.

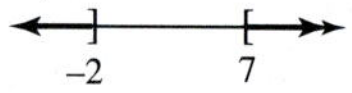

59. The search plane can go as far as 616 mi. **61.** The helicopter will reach the life raft in a little less than 1 hr, 35 min.
63. It takes Clarissa 10 days by herself. **65.** Mix 30cc of 15% HCl with 70cc of 5% HCl. **67.** There were 3260 adults.
69. The freight train is 190.67 ft long. **71.** 36 seniors went on the trip, each one paid \$13.40. **73.** **(a)** No **(b)** Todd wins again.
(c) Todd wins by $\frac{1}{4}$ meter. **(d)** Todd should line up 5.26316 meters behind the start line. **(e)** Yes

CHAPTER 4 *4.1 Exercises*

1. D **3.** A **5.** B **7.** E **9.** D **11.** B

13. $f(x) = \frac{1}{4}(x-0)^2 + 0$

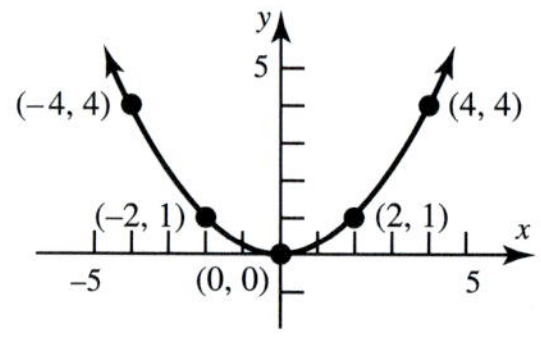

15. $f(x) = \frac{1}{4}(x-0)^2 - 2$

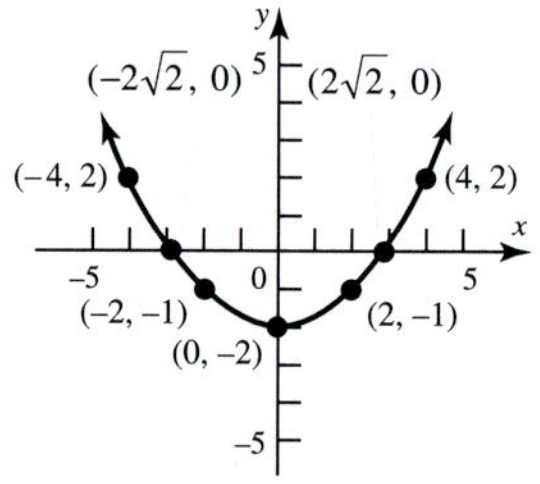

17. $f(x) = \frac{1}{4}(x-0)^2 + 2$

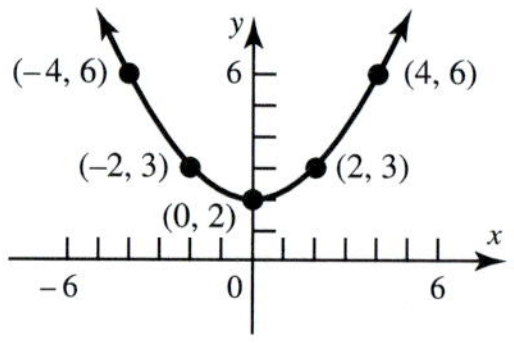

19. $f(x) = -\frac{1}{4}(x-0)^2 + 1$

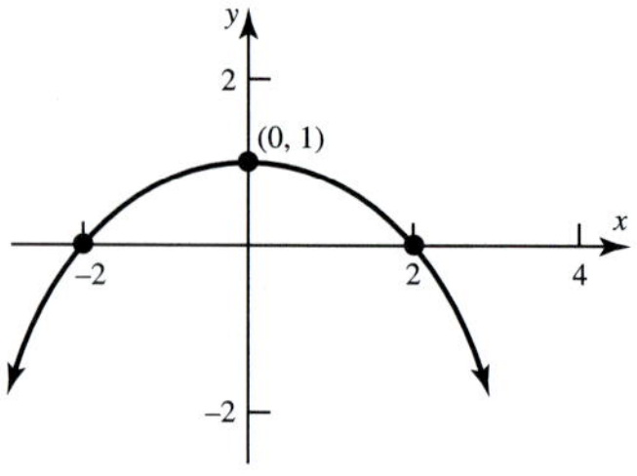

21. $f(x) = (x+2)^2 - 2$

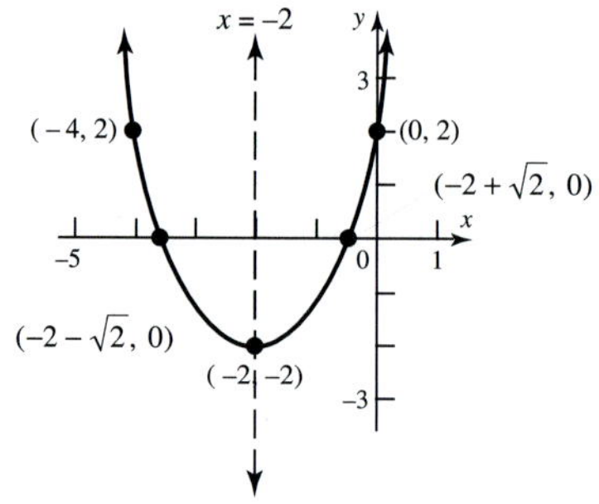

23. $f(x) = 2(x-1)^2 - 1$

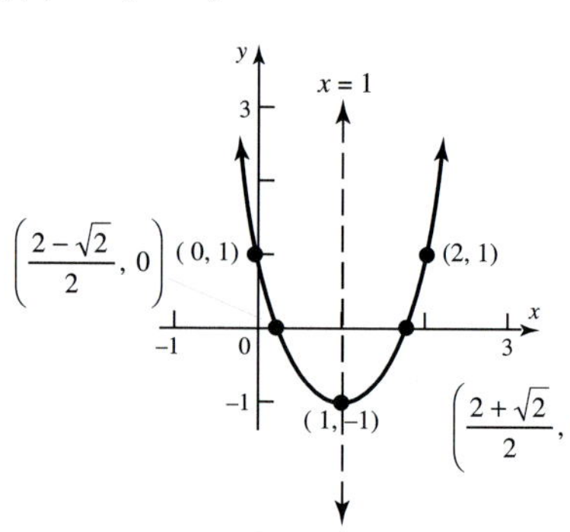

25. $f(x) = -(x+1)^2 + 1$

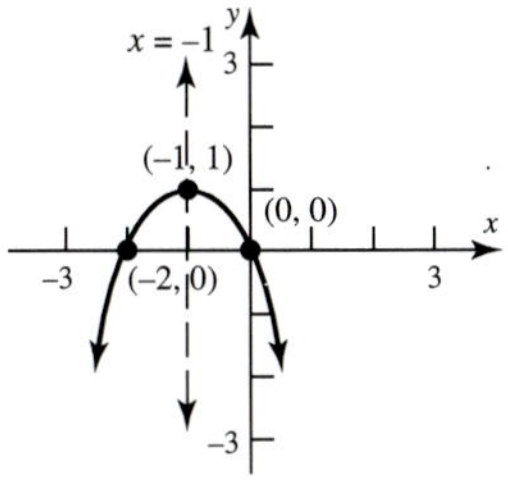

27. $f(x) = \frac{1}{2}(x+1)^2 - \frac{3}{2}$

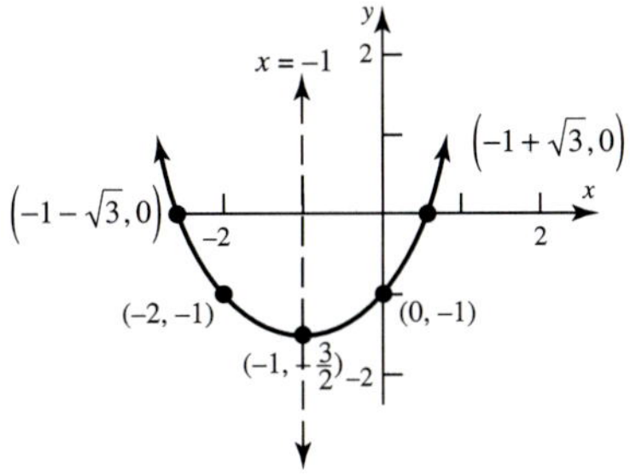

29. $f(x) = (x+1)^2 - 9$

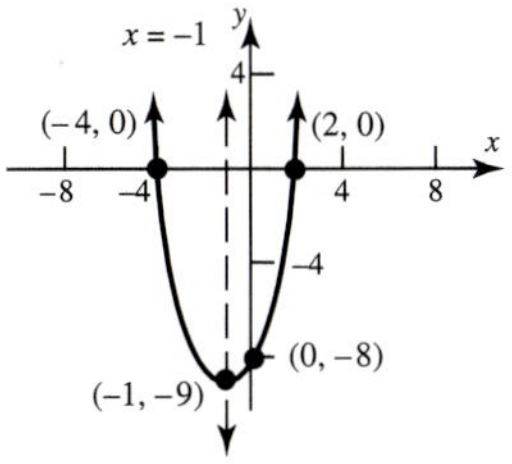

31. $f(x) = -\left(x + \frac{3}{2}\right)^2 + \frac{25}{4}$

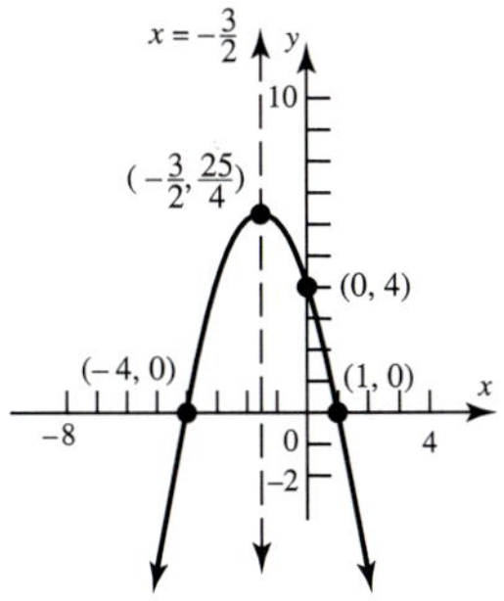

33. $f(x) = (x+1)^2$

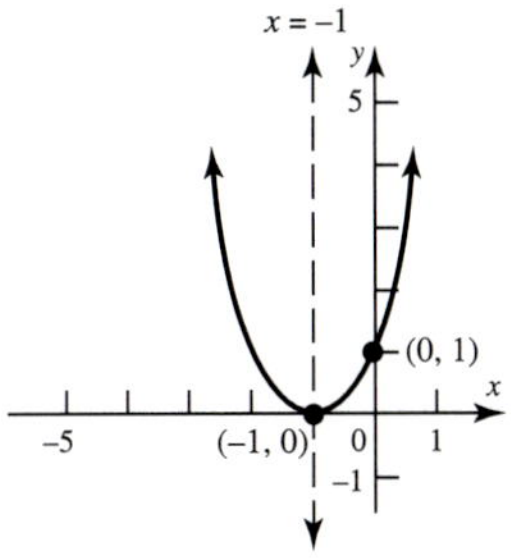

35. $f(x) = 2\left(x - \frac{1}{4}\right)^2 + \frac{15}{8}$

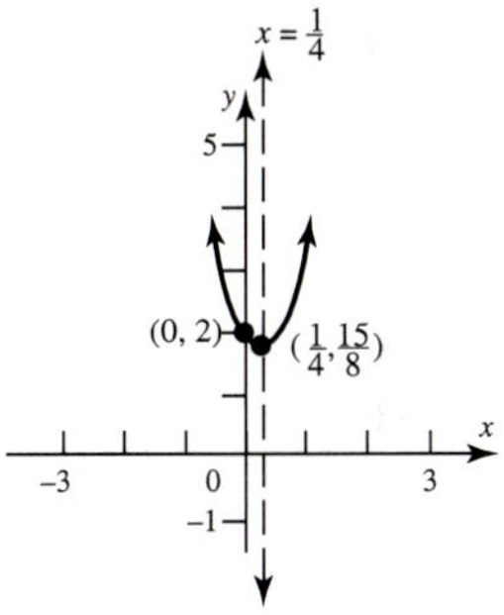

37. $f(x) = -2\left(x - \frac{1}{2}\right)^2 - \frac{5}{2}$

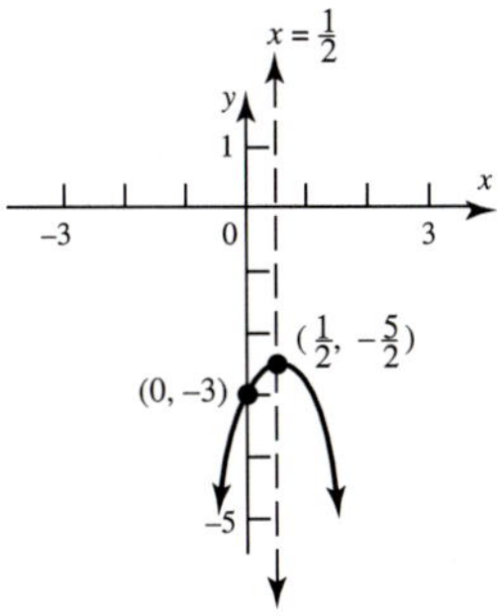

39. $f(x) = 3(x+1)^2 - 1$

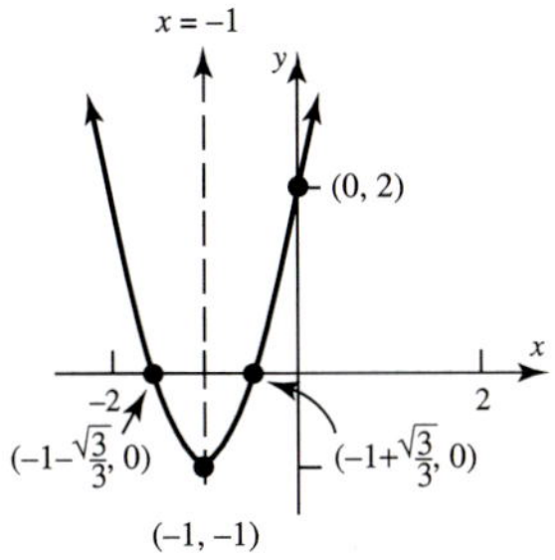

41. $f(x) = -4\left(x + \frac{3}{4}\right)^2 + \frac{17}{4}$

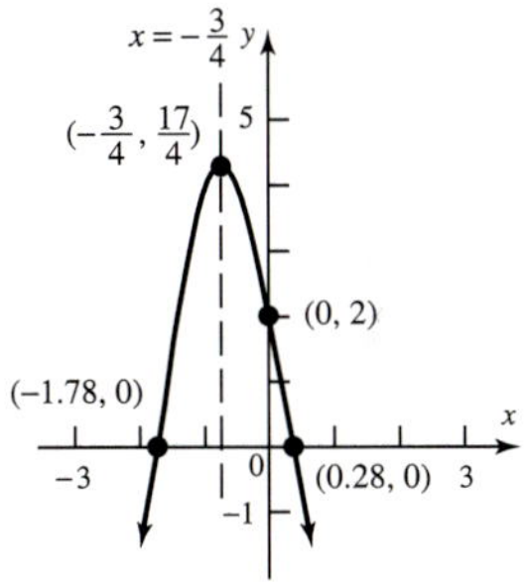

43. Minimum value; -21 **45.** Maximum value; 21 **47.** Maximum value; 13

49. Opens up; vertex at $(-1, f(-1))$; axis of symmetry $x = -1$

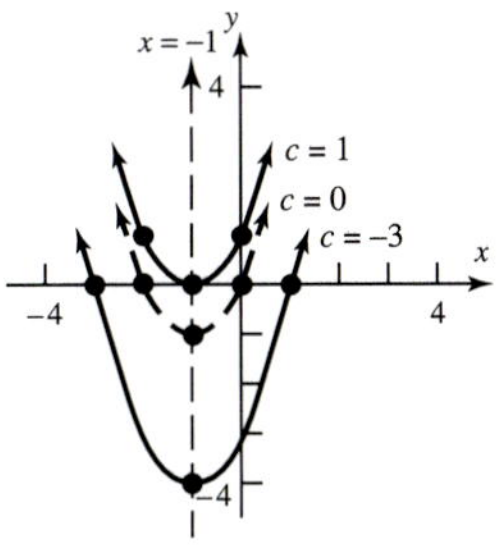

51. Each parabola opens up and passes through $(0, 1)$. Each one has the same shape. **53.** Price: \$500; maximum revenue: \$1,000,000

55. 10,000 ft^2; 100 ft by 100 ft **57.** 2,000,000 m^2 **59.** 4,166,666.7 m^2 **61. (a)**

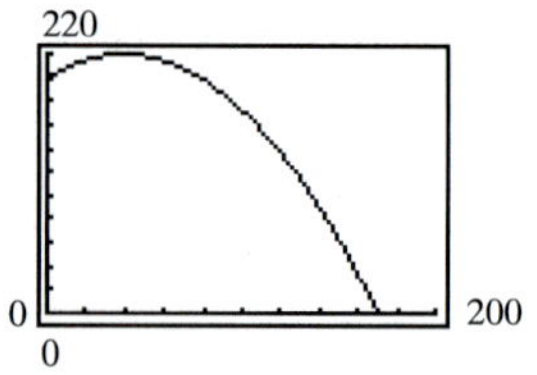

(b) 219.53 ft

(c) 170.02 ft **(e)** When the height is 100 ft, the projectile is 135.69 ft from the cliff. **63.** 8 P.M. **65.** 18.75 m **67.** 3 in.

69. Width $= \dfrac{40}{\pi + 4} \approx 5.6$ ft; length ≈ 2.8 ft. **71.** $x = \dfrac{16}{6 - \sqrt{3}} \approx 3.75$ ft.; other side ≈ 2.38 ft **73.** 70 members

75. (a) Quadratic, $a < 0$.

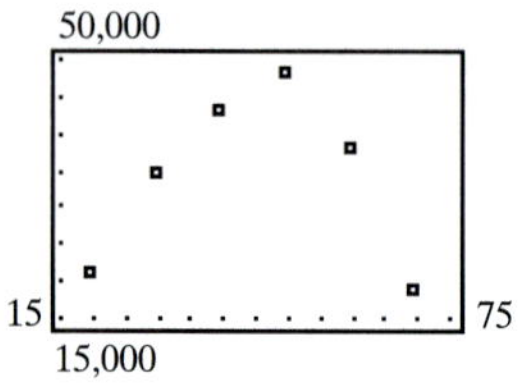

(b) $I(x) = -42.5884x^2 + 3805.5270x - 38{,}526.0037$

(c)

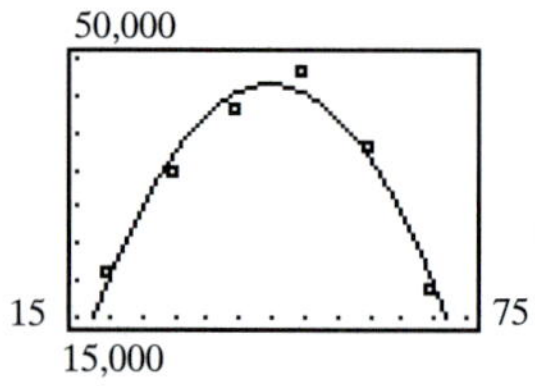

(d) 44.7 years old **(e)** \$46,486 **77. (a)** Quadratic, $a > 0$.

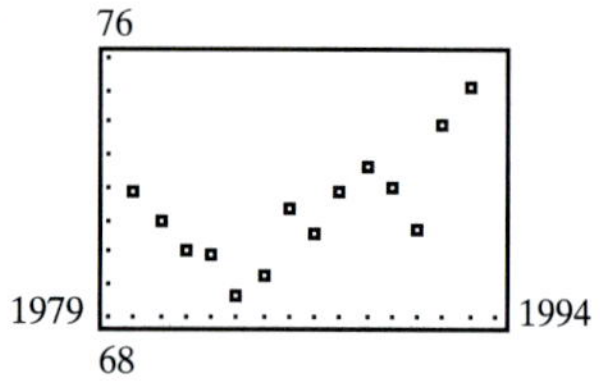

(b) $C(t) = 0.0629807692t^2 - 249.9575412t + 248{,}077.41$ **(c)**

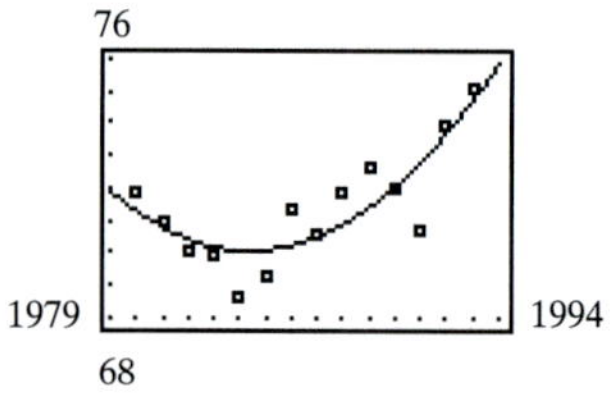

(d) 1984 **(f)** 75.9 billion gallons

79. (a) Quadratic, $a < 0$.

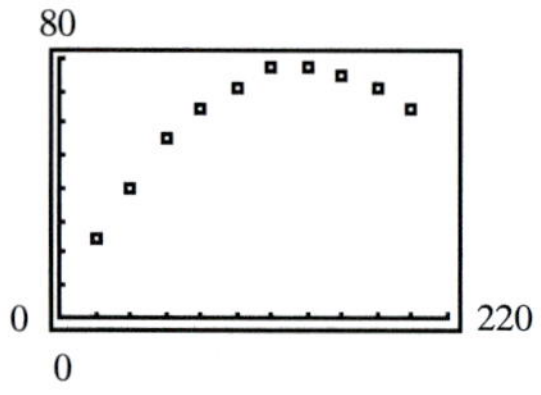

(b) $h(x) = -0.0037x^2 + 1.0318x + 5.667$ **(c)**

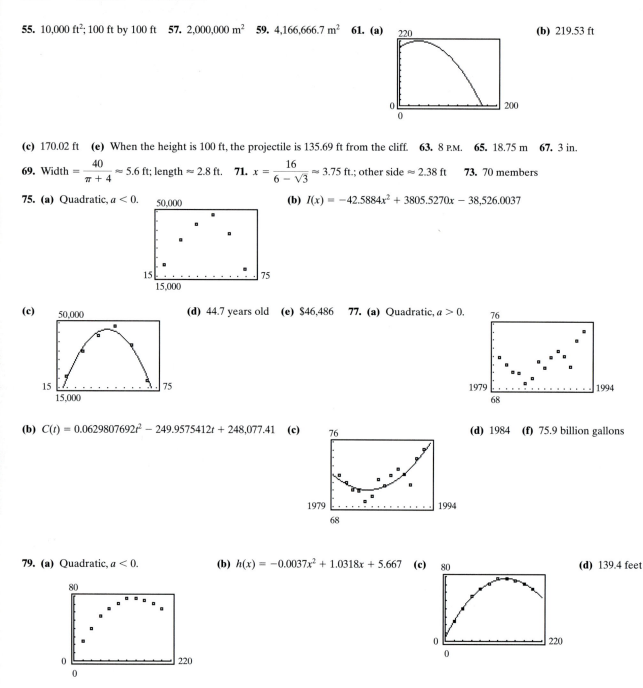

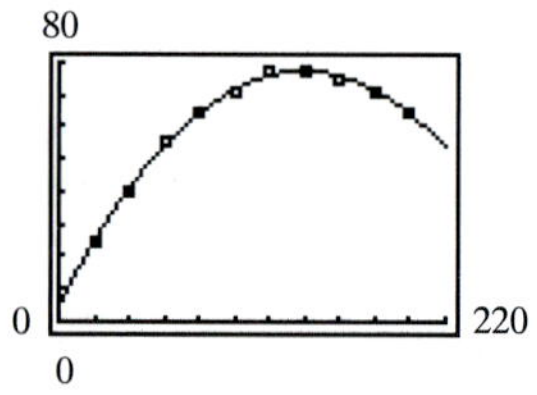

(d) 139.4 feet

(e) 77.6 feet **(g)** 284.3 feet **81.** $x = \dfrac{a}{2}$

4.2 Exercises

1. 8; 5 **3.** 4; 0 **5.** −7; 1 **7.**

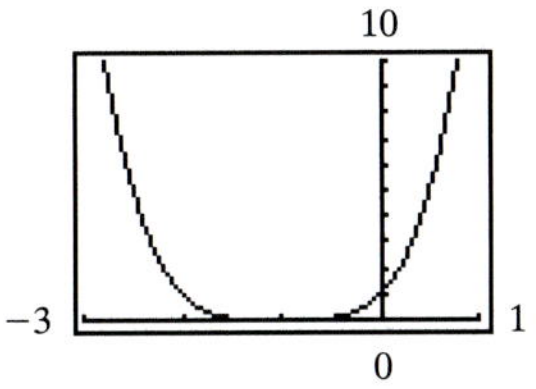

9.

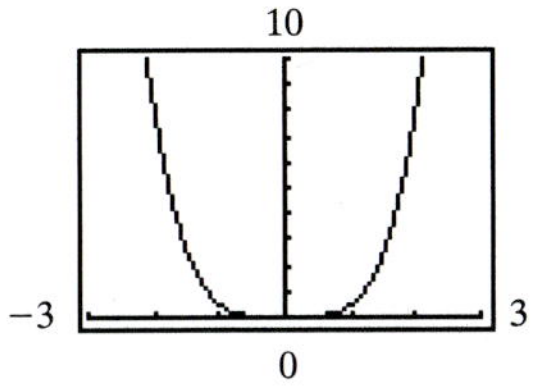

11.

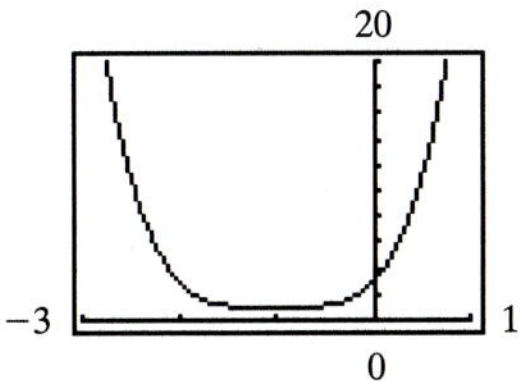

13.

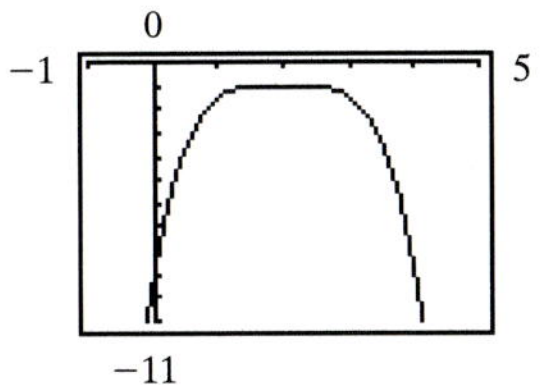

15. (a)

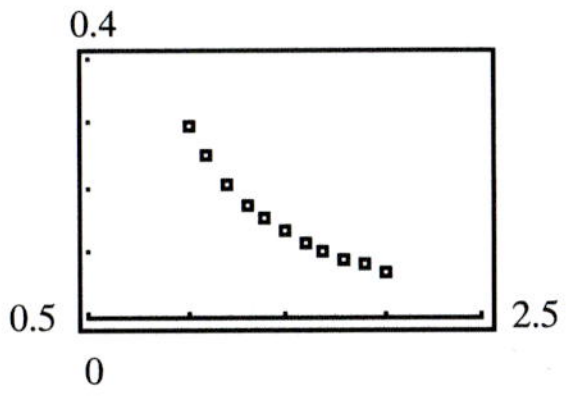

(b) $I = 0.2995x^{-2.0120}$ **(c)**

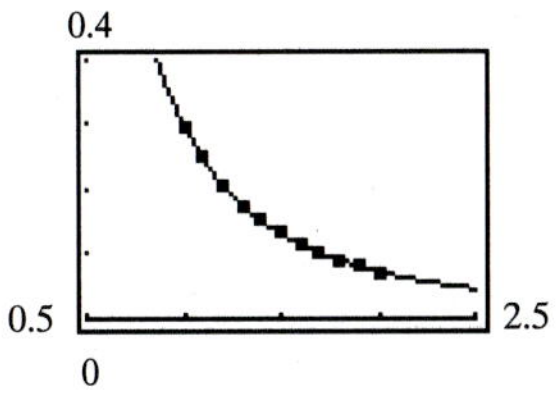

(d) 0.05605 **(e)** Close; −2.01 versus −2 **17. (a)**

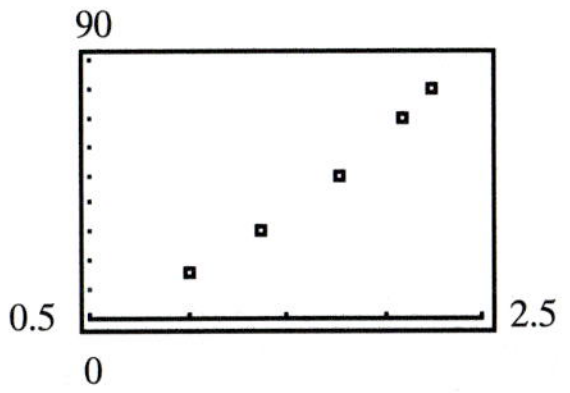

(b) $s = 15.9727t^{2.0018}$

(c)

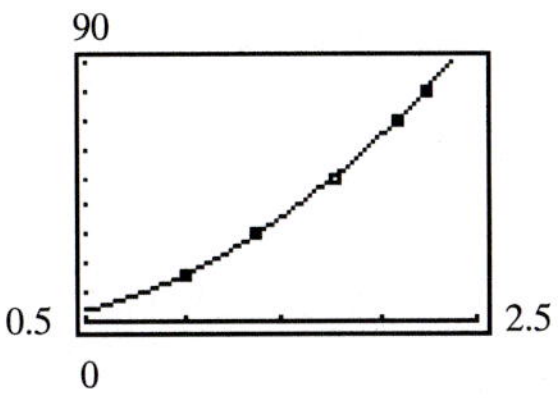

(d) $t \approx 2.5$ sec **(e)** $s = \frac{1}{2} \cdot 31.9454t^{2.0018}$, $g \approx 31.9454$ feet/sec^2

4.3 Exercises

1. Yes; degree 3 **3.** Yes; degree 2 **5.** No; x is raised to the −1 power. **7.** No; x is raised to the $\frac{3}{2}$ power. **9.** Yes; degree 4 **11.** 7, multiplicity 1; −3, multiplicity 2; graph touches the x-axis at −3 and crosses it at 7 **13.** 2, multiplicity 3; graph crosses the x-axis at 2 **15.** $-\frac{1}{2}$, multiplicity 2; graph touches the x-axis at $-\frac{1}{2}$ **17.** 5, multiplicity 3; −4, multiplicity 2; graph touches the x-axis at −4 and crosses it at 5 **19.** No real zeros; graph neither crosses nor touches the x-axis

21. (a)

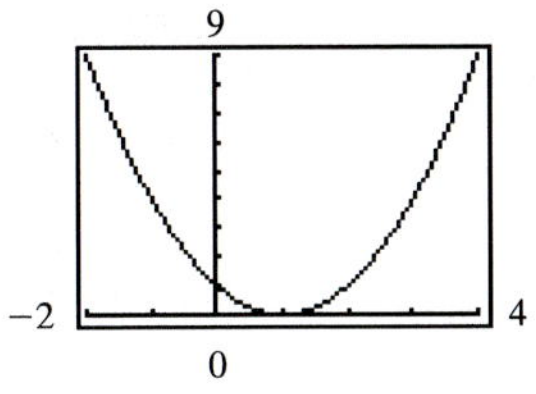

(b) x-intercept: 1; y-intercept: 1
(c) 1: Even
(d) $y = x^2$
(e) 1
(f) Local minimum: (1, 0)

23. (a)

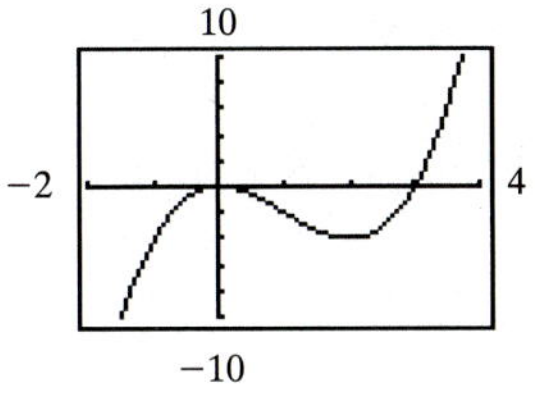

(b) x-intercepts: 0, 3; y-intercept; 0
(c) 0: Even; 3: Odd
(d) $y = x^3$
(e) 2
(f) Local maximum: (0, 0);
Local minimum: (2, −4)

25. (a)

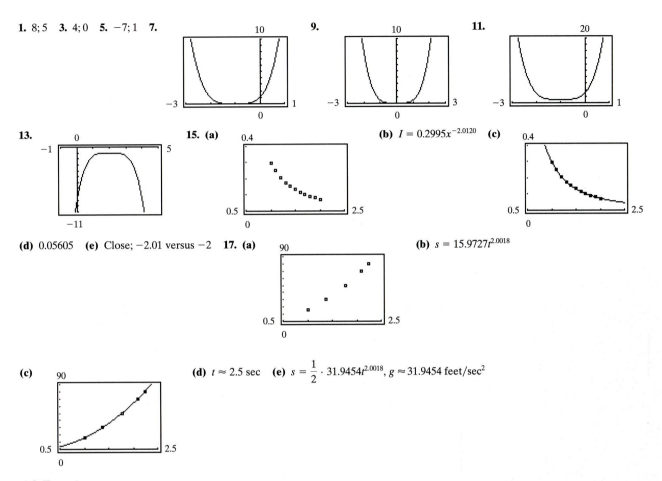

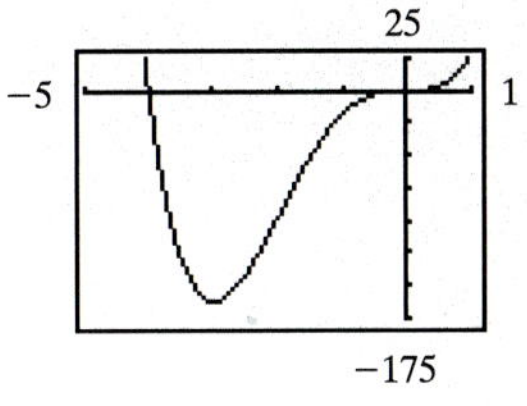

(b) x-intercepts: −4, 0; y-intercept: 0
(c) −4, 0: Odd
(d) $y = 6x^4$
(e) 1
(f) Local minimum: (−3, −162)

27. (a)

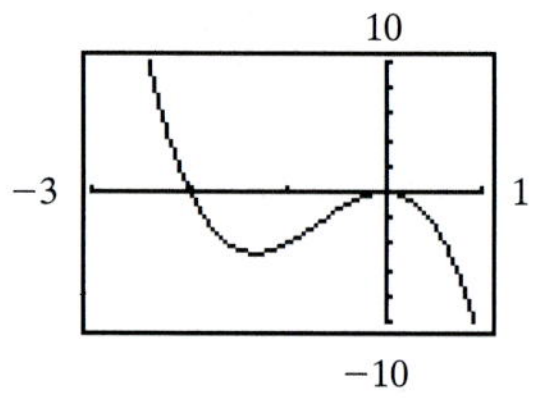

(b) x-intercepts: $-2, 0$; y-intercept: 0
(c) -2: Odd; 0: Even
(d) $y = -4x^3$
(e) 2
(f) Local minimum: $(-1.33, -4.74)$;
Local maximum: $(0, 0)$

29. (a)

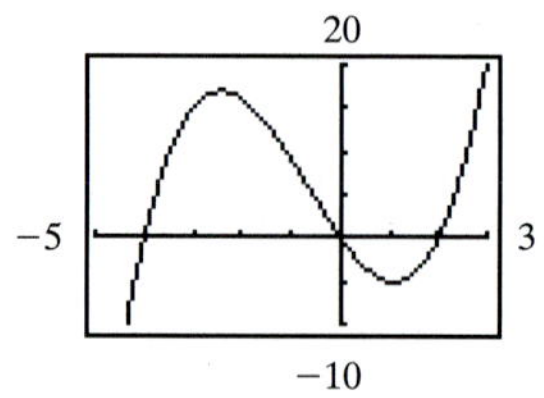

(b) x-intercepts: $-4, 0, 2$; y-intercept: 0
(c) $-4, 0, 2$: Odd
(d) $y = x^3$
(e) 2
(f) Local maximum: $(-2.43, 16.90)$;
Local minimum: $(1.09, -5.04)$

31. (a)

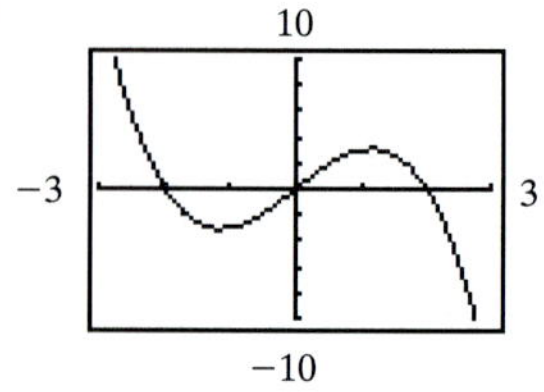

(b) x-intercepts: $-2, 0, 2$; y-intercept: 0
(c) $-2, 0, 2$: Odd
(d) $y = -x^3$
(e) 2
(f) Local minimum: $(-1.15, -3.07)$;
Local maximum: $(1.15, 3.07)$

33. (a)

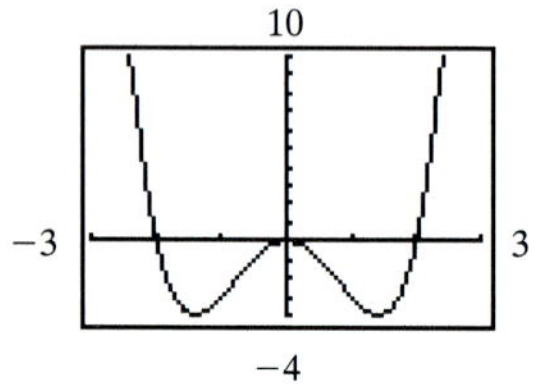

(b) x-intercepts: $-2, 0, 2$; y-intercept: 0
(c) $-2, 2$: Odd; 0: Even
(d) $y = x^4$
(e) 3
(f) Local minima: $(-1.41, -4), (1.41, -4)$;
Local maximum: $(0, 0)$

35. (a)

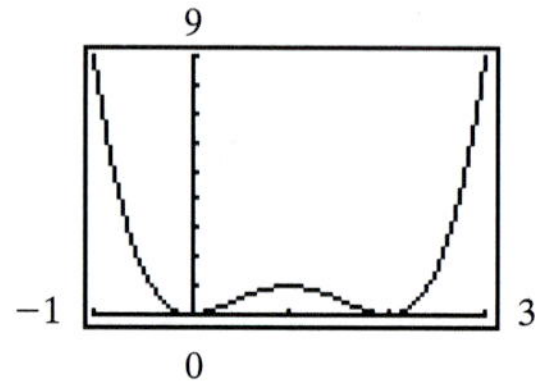

(b) x-intercepts: $0, 2$; y-intercept: 0
(c) $0, 2$: Even
(d) $y = x^4$
(e) 3
(f) Local minima: $(0, 0), (2, 0)$;
Local maximum: $(1, 1)$

37. (a)

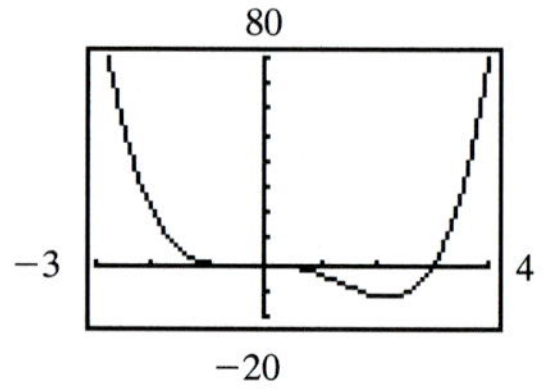

(b) x-intercepts: $-1, 0, 3$; y-intercept: 0
(c) $-1, 3$: Odd; 0: Even
(d) $y = x^4$
(e) 3
(f) Local minima: $(2.18, -12.39)$,
$(-0.68, -0.54)$;
Local maximum: $(0, 0)$

39. (a)

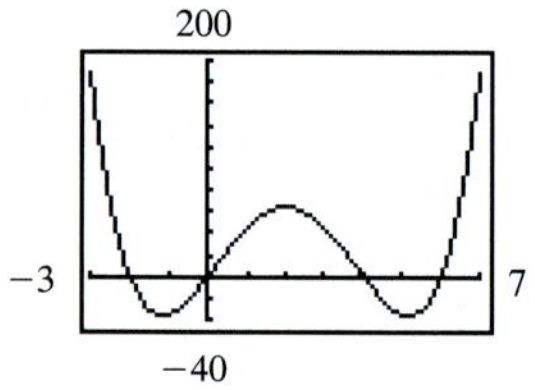

(b) x-intercepts: $-2, 0, 4, 6$; y-intercept: 0
(c) $-2, 0, 4, 6$: Odd
(d) $y = x^4$
(e) 3
(f) Local minima: $(-1.16, -36), (5.16, -36)$;
Local maximum: $(2, 64)$

41. (a)

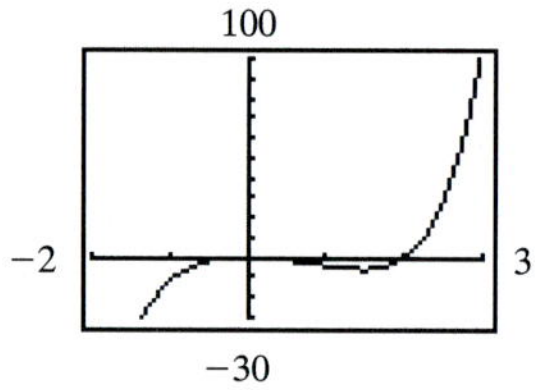

(b) x-intercepts: $0, 2$; y-intercept: 0
(c) 0: Even; 2: Odd
(d) $y = x^5$
(e) 2
(f) Local minimum: $(1.47, -5.91)$;
Local maximum: $(0, 0)$

43. (a)

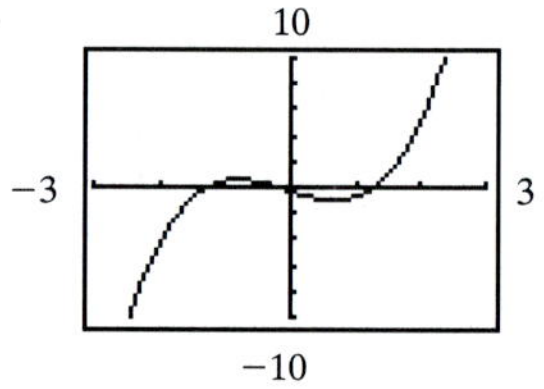

(b) x-intercepts: $-1.26, -0.20, 1.26$;
y-intercept: -0.31752
(c) $-1.26, -0.20, 1.26$: Odd
(d) $y = x^3$
(e) 2
(f) Local minimum: $(0.66, -0.99)$;
Local maximum: $(-0.79, 0.56)$

45. (a)

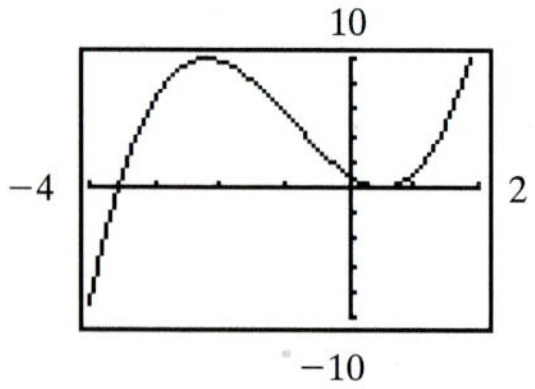

(b) x-intercepts: $-3.56, 0.50$;
y-intercept: 0.89
(c) -3.56: Odd; 0.50: Even
(d) $y = x^3$
(e) 2
(f) Local minimum: $(0.50, 0)$;
Local maximum: $(-2.20, 9.91)$

47. (a)

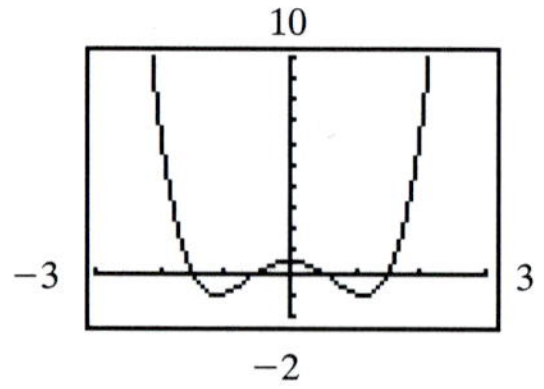

(b) x-intercepts: $-1.50, -0.50, 0.50, 1.50$;
y-intercept: 0.5625
(c) $-1.50, -0.50, 0.50, 1.50$: Odd
(d) $y = x^4$
(e) 3
(f) Local minima: $(-1.11, -1), (1.11, -1)$
Local maximum: $(0, 0.5625)$

49. (a)

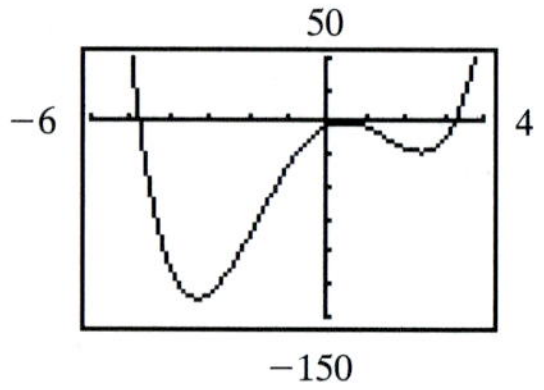

(b) x-intercepts: $-4.78, 0.45, 3.23$;
y-intercept: -3.1264785
(c) $-4.78, 3.23$: Odd; 0.45: Even
(d) $y = x^4$
(e) 3
(f) Local minima: $(-3.31, -135.91)$
$(2.37, -22.66)$;
Local maximum: $(0.45, 0)$

51. (a)

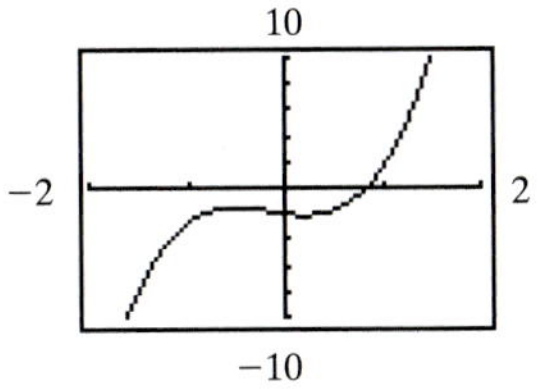

(b) x-intercept: 0.83; y-intercept: -2
(c) 0.83: Odd
(d) $y = \pi x^3$
(e) 2
(f) Local maximum: $(-0.50, -1.53)$;
Local maximum: $(0.20, -2.11)$

53. (a)

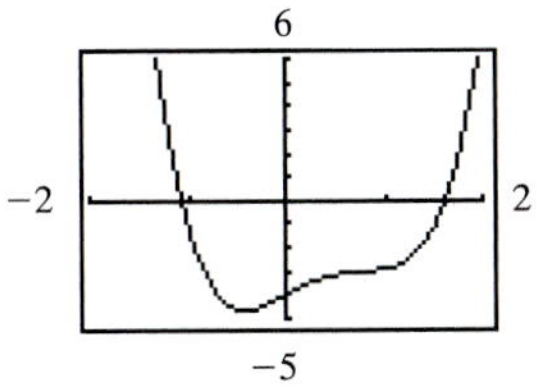

(b) x-intercepts: $-1.06, 1.61$; y-intercept: -4
(c) $-1.06, 1.61$: Odd
(d) $y = 2x^4$
(e) 1
(f) Local minimum: $(-0.41, -4.64)$

55. (a)

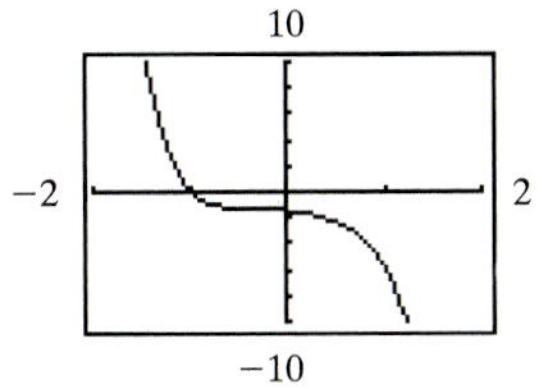

(b) x-intercept: -0.97; y-intercept: $-\sqrt{2}$
(c) -0.97: Odd
(d) $y = -2x^5$
(e) None
(f) None

57. c, e, f **59.** c, e **61. (a)** Cubic, $a < 0$

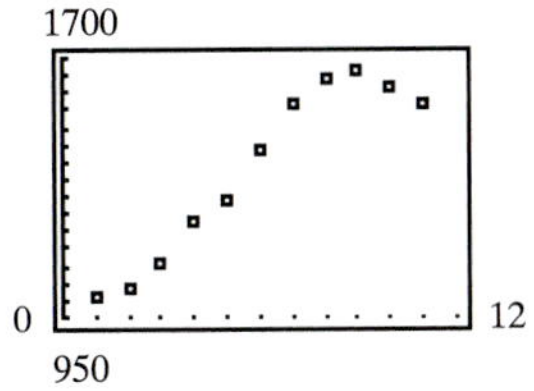

(b) $M(x) = -2.4324x^3 + 37.3170x^2 - 70.3718x + 1043.7879$

(c)

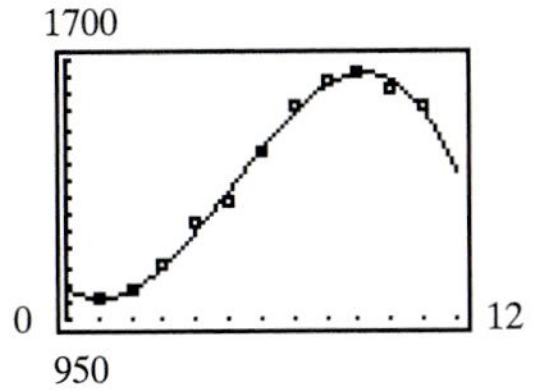

(d) $\approx$ 1370 motor vehicle thefts **63. (a)** Cubic, $a > 0$

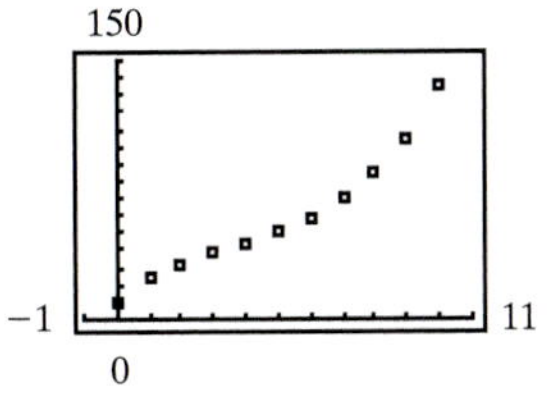

(b) 7 **(c)** 20

(d) $C(x) = 0.2156x^3 - 2.3473x^2 + 14.3275x + 10.2238$ **(e)**

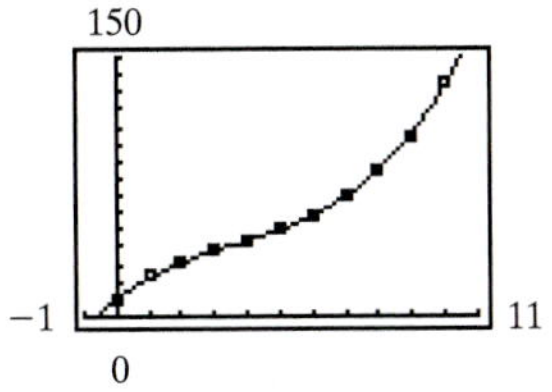

(f) \$170,767 **(g)** Fixed costs of \$10,224.

4.4 Exercises

1. All real numbers except 3. **3.** All real numbers except 2 and -4. **5.** All real numbers except $-\frac{1}{2}$ and 3.
7. All real numbers except 2. **9.** All real numbers. **11. (a)** Domain: $\{x \mid x \neq 2\}$; Range: $\{y \mid y \neq 1\}$ **(b)** $(0, 0)$ **(c)** $y = 1$ **(d)** $x = 2$ **(e)** None **13. (a)** Domain: $\{x \mid x \neq 0\}$; Range: all real numbers **(b)** $(-1, 0), (1, 0)$ **(c)** None **(d)** None **(e)** $y = 2x$
15. (a) Domain: $\{x \mid x \neq -2, x \neq 2\}$; Range: $\{y \mid -\infty < y \leq 0, 1 < y < \infty\}$ **(b)** $(0, 0)$ **(c)** $y = 1$ **(d)** $x = -2, x = 2$ **(e)** None
17. (a) Domain: $\{x \mid x \neq -1\}$; Range: $\{y \mid y \neq 2\}$ **(b)** $(-1.5, 0), (0, 3)$ **(c)** $y = 2$ **(d)** $x = -1$ **(e)** None
19. (a) Domain: $\{x \mid x \neq -4, x \neq 3\}$; Range: All real numbers **(b)** $(0, 0)$ **(c)** $y = 0$ **(d)** $x = -4; x = 3$ **(e)** None

21.

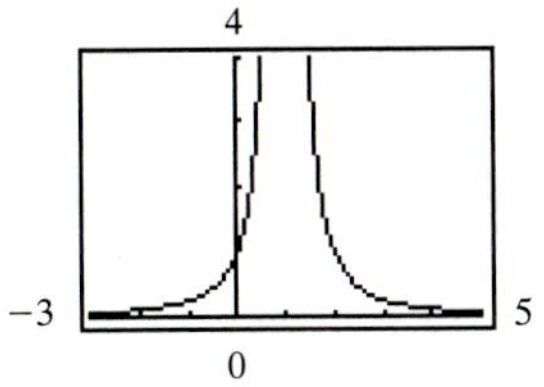

23.

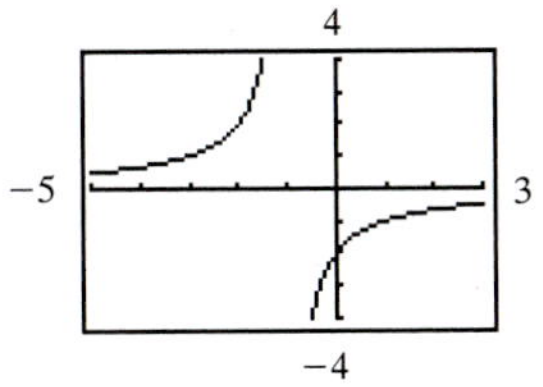

25.

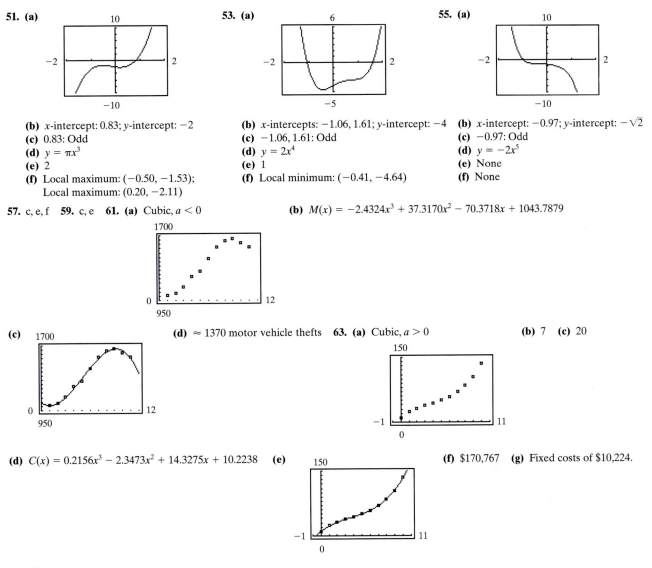

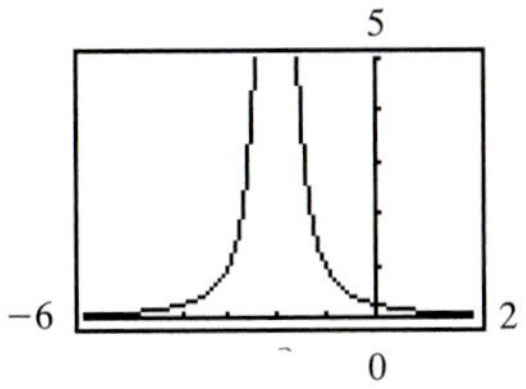

27.

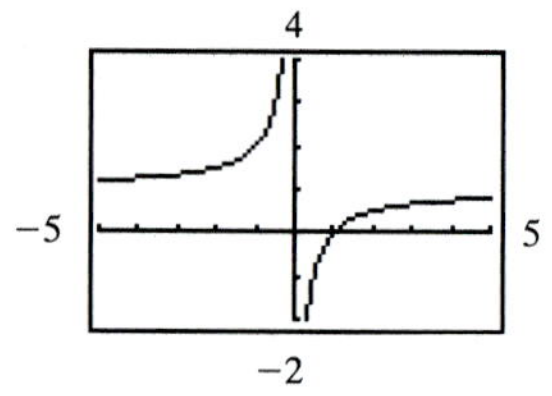

29.

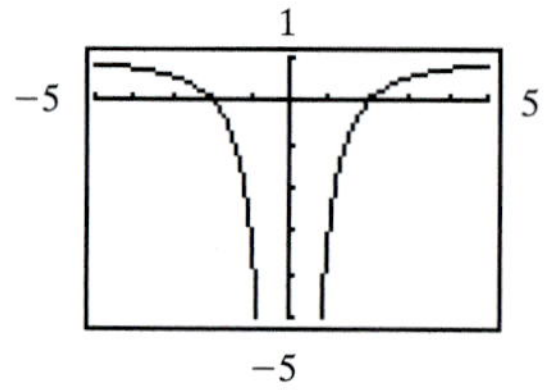

31. Horizontal asymptote: $y = 3$; vertical asymptote: $x = -4$ **33.** No asymptotes
35. Horizontal asymptote: $y = 0$; vertical asymptotes: $x = 1, x = -1$ **37.** Horizontal asymptote: $y = 0$, vertical asymptote: $x = 0$
39. Oblique asymptote: $y = 3x$; vertical asymptote: $x = 0$

41. 1. x-intercept: -1; no y-intercept
2. No symmetry
3. Vertical asymptotes: $x = 0, x = -4$
4. Horizontal asymptote: $y = 0$
intersected at $(-1, 0)$
5.

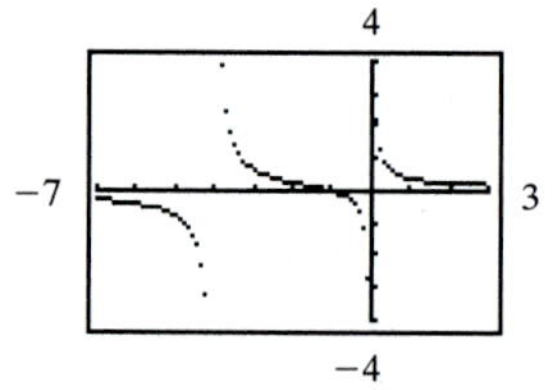

43. 1. x-intercept: -1; y-intercept: $\frac{3}{4}$
2. No symmetry
3. Vertical asymptote: $x = -2$
4. Horizontal asymptote: $y = \frac{3}{2}$
not intersected
5.

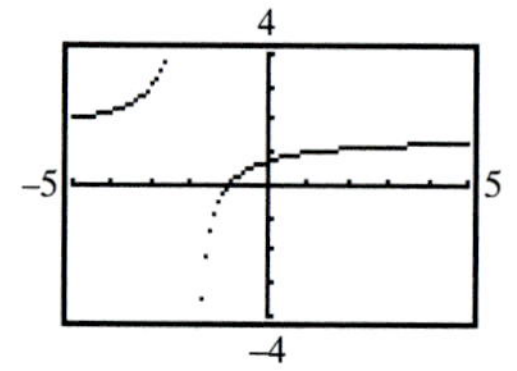

45. 1. No x-intercept; y-intercept: $-\frac{3}{4}$
2. Symmetric with respect to y-axis
3. Vertical asymptotes: $x = 2, x = -2$
4. Horizontal asymptote: $y = 0$
not intersected
5.

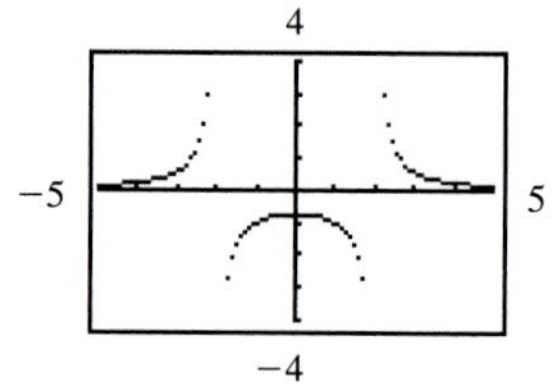

47. 1. No x-intercept; y-intercept: -1
2. Symmetric with respect to y-axis
3. Vertical asymptotes: $x = -1, x = 1$
4. No horizontal or oblique asymptotes
5.

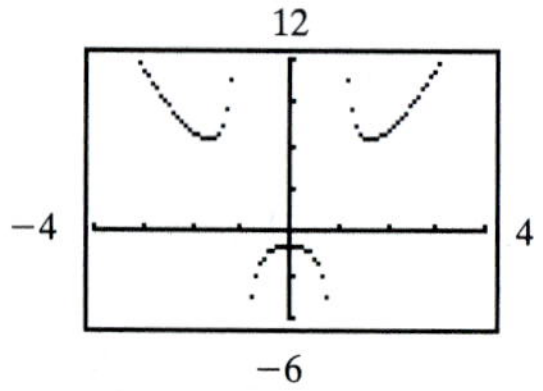

49. 1. x-intercept: 1; y-intercept: $\frac{1}{9}$
2. No symmetry
3. Vertical asymptotes: $x = 3, x = -3$
4. Oblique asymptote: $y = x$
intersected at $\left(\frac{1}{9}, \frac{1}{9}\right)$
5.

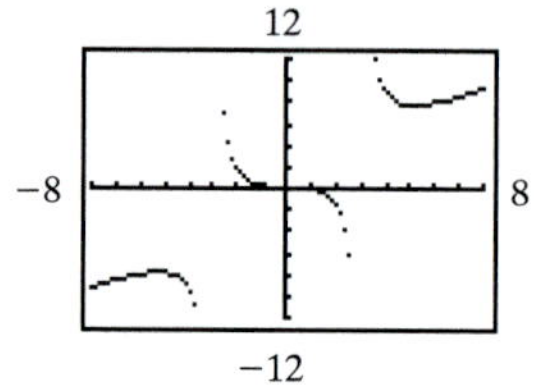

51. 1. Intercept $(0, 0)$
2. No symmetry
3. Vertical asymptotes: $x = 2, x = -3$
4. Horizontal asymptote: $y = 1$
intersected at $(6, 1)$
5.

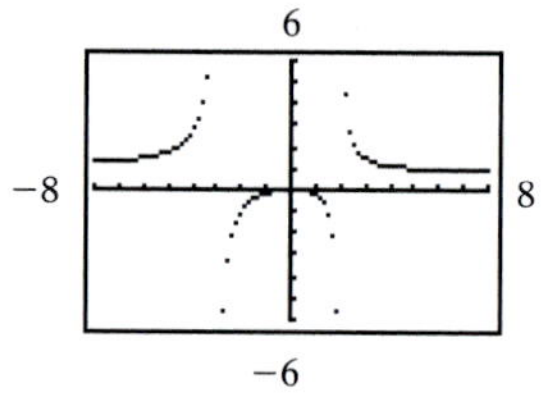

53. 1. Intercept $(0, 0)$
2. Symmetry with respect to origin
3. Vertical asymptotes: $x = -2, x = 2$
4. Horizontal asymptote: $y = 0$
intersected at $(0, 0)$
5.

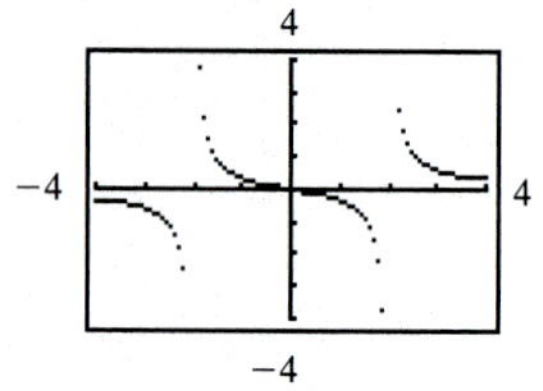

55. 1. No x-intercept; y-intercept: $\frac{3}{4}$
2. No symmetry
3. Vertical asymptotes: $x = -2, x = 1, x = 2$
4. Horizontal asymptote: $y = 0$
not intersected
5.

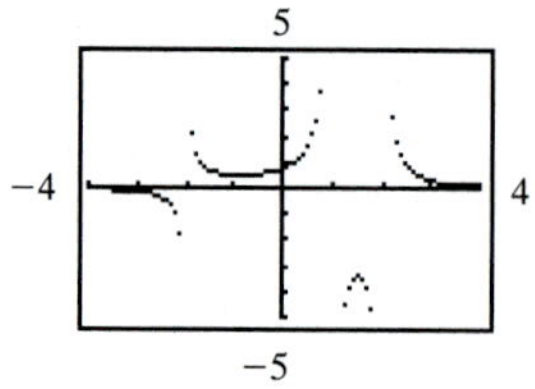

57. 1. x-intercepts: $-1, 1$; y-intercept: $\frac{1}{4}$
2. Symmetric with respect to y-axis
3. Vertical asymptotes: $x = -2, x = 2$
4. Horizontal asymptote: $y = 0$
intersected at $(-1, 0)$ and $(1, 0)$
5.

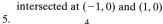

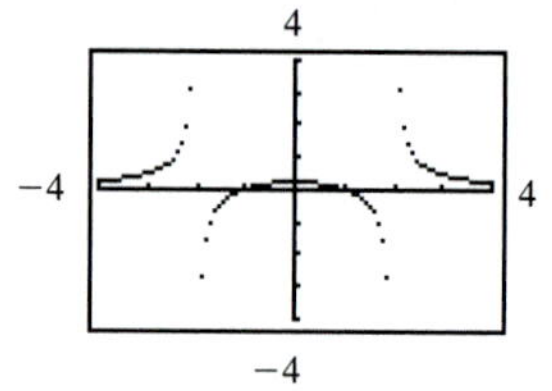

59. 1. x-intercepts: $-1, 4$, y-intercept: -2
2. No symmetry
3. Vertical asymptote: $x = -2$
4. Oblique asymptote: $y = x - 5$ not intersected
5. 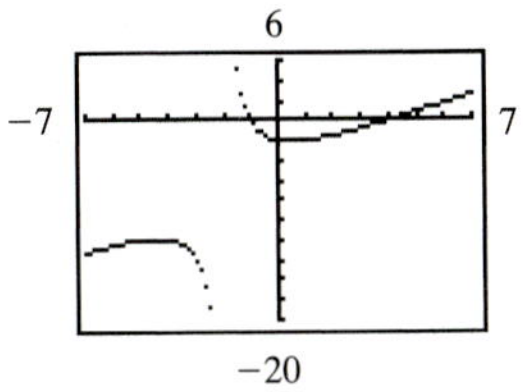

61. 1. x-intercepts: $-4, 3$; y-intercept: 3
2. No symmetry
3. Vertical asymptote: $x = 4$
4. Oblique asymptote: $y = x + 5$ not intersected
5. 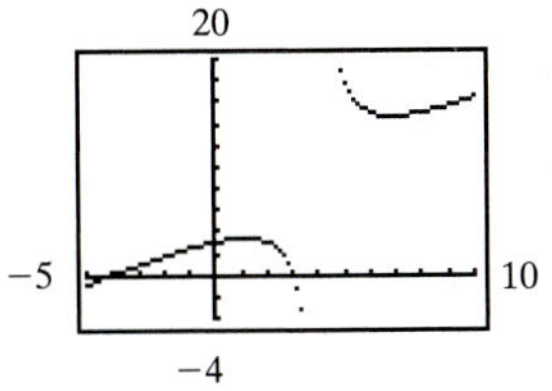

63. 1. x-intercepts: $-4, 3$; y-intercept: -6
2. No symmetry
3. Vertical asymptote: $x = -2$
4. Oblique asymptote: $y = x - 1$ not intersected
5. 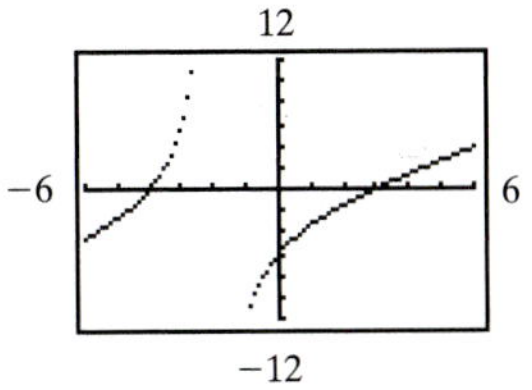

65. 1. x-intercept: 0, 1; y-intercept: 0
2. No symmetry
3. Vertical asymptote: $x = -3$
4. Horizontal asymptote: $y = 1$ not intersected
5. 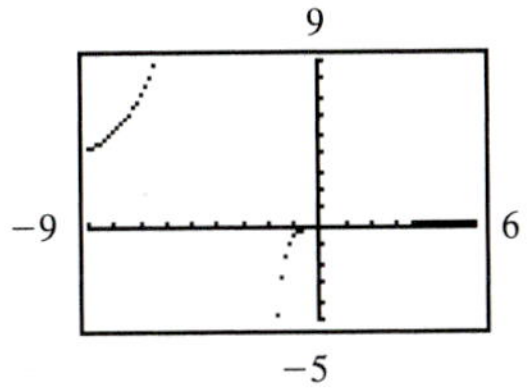

67. 1. x-intercepts: -0.84; y-intercept: -2
2. No symmetry
3. Vertical asymptote: $x = 0.74$
4. Horizontal asymptote: $y = 2$ intersected at $(2.19, 2)$ and $(-3.03, 2)$
5. 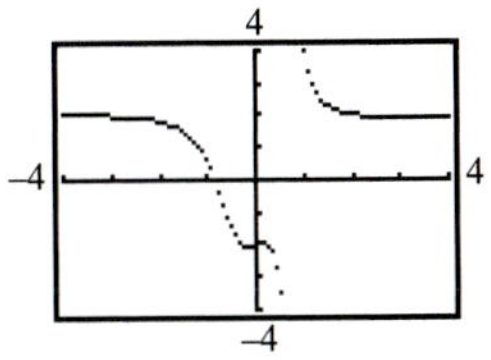

69. 1. x-intercepts: 0.65; y-intercept: -3
2. No symmetry
3. Vertical asymptotes: $x = -4.41$; $x = -0.71$; $x = 0.71$; $x = 4.41$
4. Horizontal asymptote: $y = 0$ intersected at $(0.65, 0)$
5. 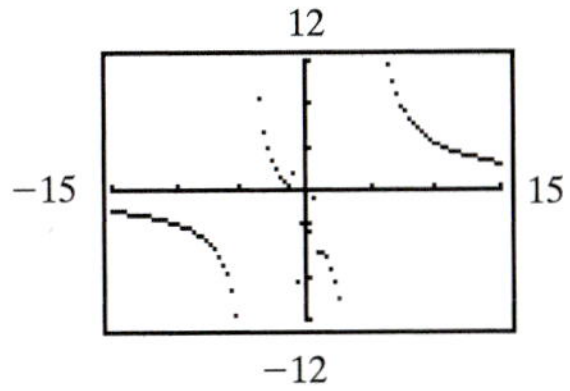

71. 1. No x-intercept; y-intercept: 1
2. No symmetry
3. Vertical asymptotes: $x = -5.75$; $x = 0.31$; $x = 5.43$
4. Oblique asymptotes: $y = 5x - 10$ intersected at $(0.35, -8.21)$ and $(1.96, -0.19)$
5. 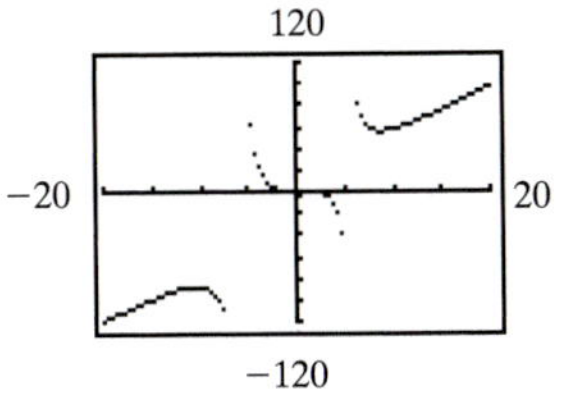

73. 1. x-intercept: -4; y-intercept: 2
2. No symmetry
3. Vertical asymptote: $x = -2$
4. Horizontal asymptote: $y = 1$ not intersected
5. 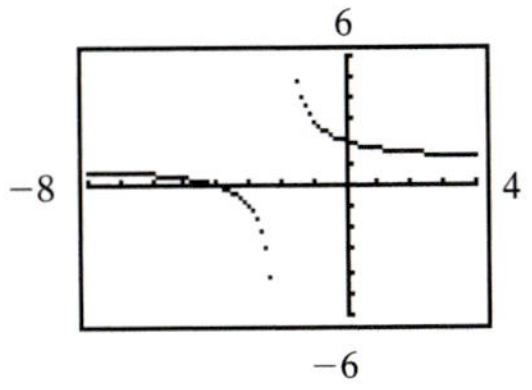

75. 1. x-intercept: $-\frac{1}{3}$; y-intercept: $-\frac{1}{2}$
2. No symmetry
3. Vertical asymptote: $x = 2$
4. Horizontal asymptote: $y = 3$ not intersected
5. 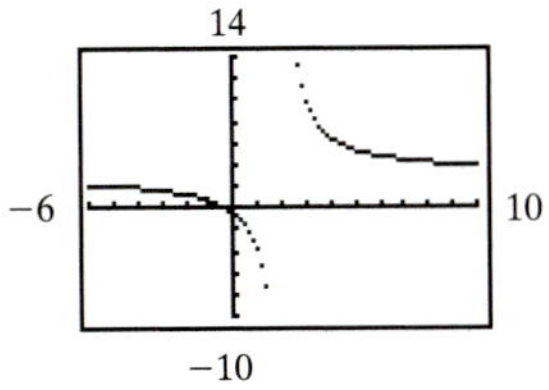

77. 1. x-intercept: $-1, 2$; y-intercept: $\frac{2}{5}$
2. No symmetry
3. Vertical asymptote: $x = -5$; $x = 1$
4. Horizontal asymptote: $y = 1$ intersected at $\left(\frac{3}{5}, 1\right)$
5. 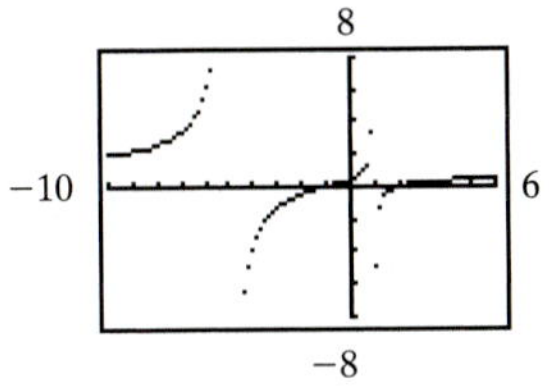

79. 1. No x-intercept; no y-intercept
2. Symmetric about the origin
3. Vertical asymptote: $x = 0$
4. Oblique asymptote: $y = x$ not intersected
5. 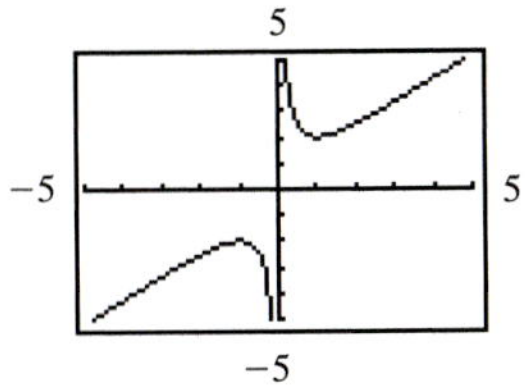

81. x-intercept: -1; no y-intercept
2. No symmetry
3. Vertical asymptote: $x = 0$
4. No horizontal or oblique asymptotes
5. 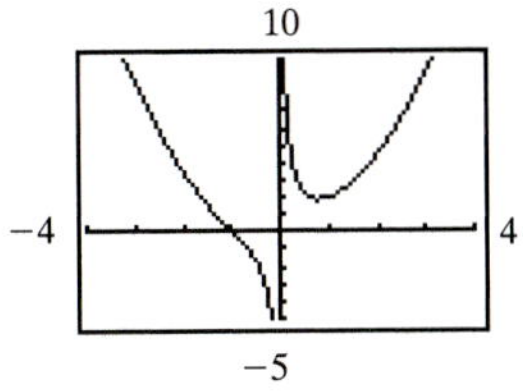

83. 1. No x-intercept; no y-intercept
2. Symmetric about the origin
3. Vertical asymptote: $x = 0$
4. Oblique asymptote: $y = x$ not intersected
5.

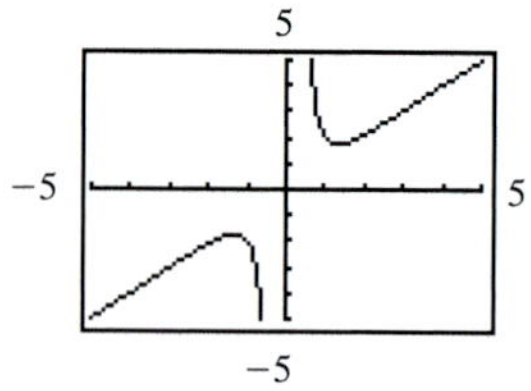

85. 4 must be a zero of the denominator, hence, $x - 4$ must be a factor. **87.** c, d

89. **(a)** 9.82 m/sec^2 **(b)** 9.8195 m/sec^2 **(c)** 9.7936 m/sec^2 **(d)** h-axis **(e)**

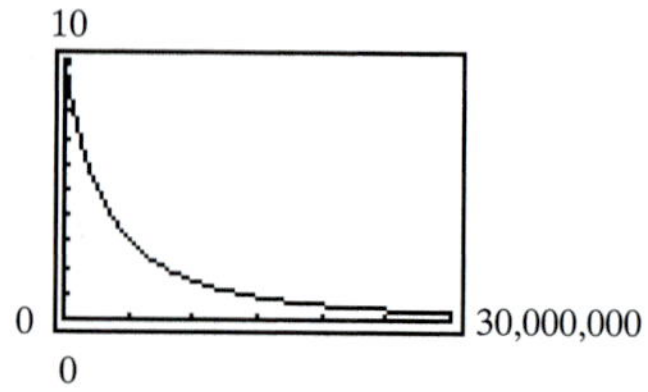

91. **(a)**

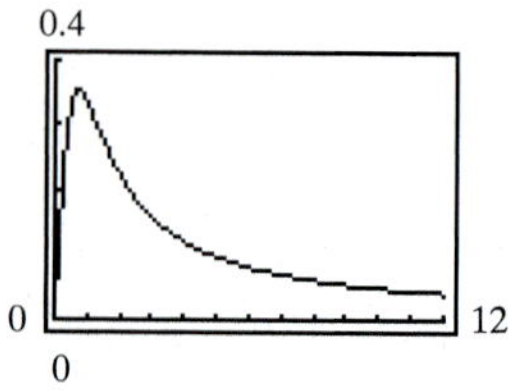

(b) 0.70 hours **(c)** t-axis; $C(t) \to 0$

93. **(a)** $\overline{C}(x) = \dfrac{0.2156x^3 - 2.3473x^2 + 14.3275x + 10.2238}{x}$ **(b)** $\overline{C}(6) = \$9709$ **(c)** $\overline{C}(9) = \$11{,}801$ **(d)**

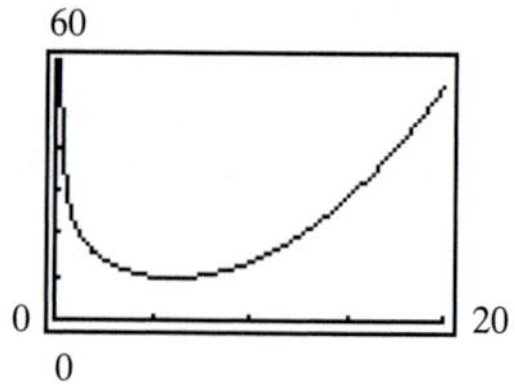

(e) 6 **(f)** \$9709 **95.** **(a)** $S(x) = 2x^2 + \dfrac{40{,}000}{x}$ **(b)**

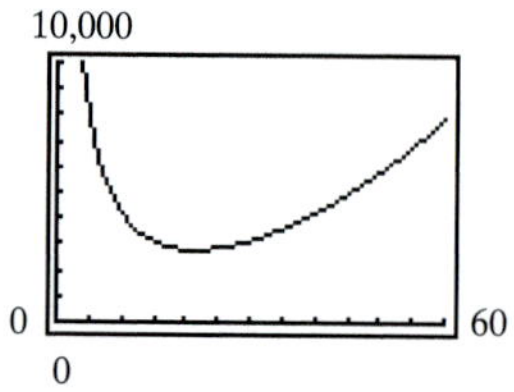

(c) 2784.95 **(d)** 21.54″ × 21.54″ × 21.54″

97. **(a)** $C(r) = 12\pi r^2 + \dfrac{4000}{r}$ **(b)** The cost is least for r about 3.75 cm.

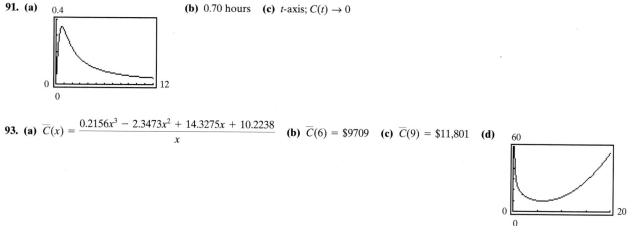

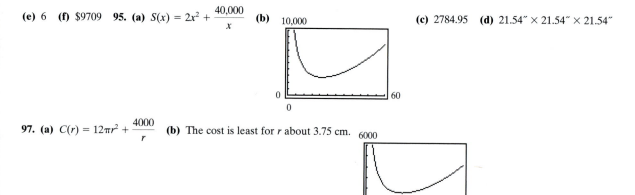

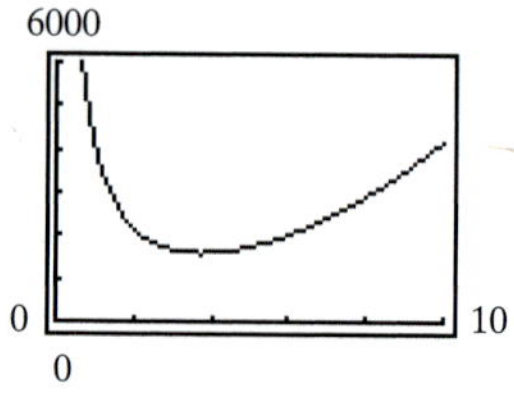

4.5 Exercises

1. No; $f(2) = 8$ **3.** Yes; $f(2) = 0$ **5.** Yes; $f(-3) = 0$ **7.** No; $f(-4) = 1$ **9.** Yes; $f\left(\frac{1}{2}\right) = 0$ **11.** 7; 3 or 1 positive; 2 or 0 negative **13.** 6; 2 or 0 positive; 2 or 0 negative **15.** 3; 2 or 0 positive; 1 negative **17.** 4; 2 or 0 positive; 2 or 0 negative

19. 5; 0 positive; 3 or 1 negative **21.** 6; 1 positive; 1 negative **23.** $\pm 1, \pm\frac{1}{3}$ **25.** $\pm 1, \pm 3$ **27.** $\pm 1, \pm 2, \pm\frac{1}{4}, \pm\frac{1}{2}$ **29.** $\pm\frac{1}{3}, \pm\frac{2}{3}, \pm 1, \pm 2$

31. $\pm\frac{1}{2}, \pm 1, \pm 2, \pm 4$ **33.** $\pm\frac{1}{6}, \pm\frac{1}{3}, \pm\frac{1}{2}, \pm\frac{2}{3}, \pm 1, \pm 2$ **35.** -1 and 1 **37.** -3 and 7 **39.** -5 and 2

41. $-3, -1, 2; f(x) = (x+3)(x+1)(x-2)$ **43.** $\frac{1}{2}; f(x) = 2\left(x - \frac{1}{2}\right)(x^2+1)$ **45.** $-1, 1; f(x) = (x+1)(x-1)(x^2+2)$

47. $-\frac{1}{2}, \frac{1}{2}; f(x) = 4\left(x + \frac{1}{2}\right)\left(x - \frac{1}{2}\right)(x^2+2)$ **49.** $-2, -1, 1, 1; f(x) = (x+2)(x+1)(x-1)^2$

51. $-0.70, 0.70, 2; f(x) = 4\left(x + \frac{\sqrt{2}}{2}\right)\left(x - \frac{\sqrt{2}}{2}\right)(x-2)\left(x^2 + \frac{1}{2}\right)$ **53.** $-5.90, -0.30, 3.00$ **55.** $-3.80, 4.50$ **57.** $-3.42, 0.31, 12.30$

59. $-43.50, 1.00, 23.00$ **61.** $\{-1, 2\}$ **63.** $\left\{\frac{2}{3}, -1+\sqrt{2}, -1-\sqrt{2}\right\}$ **65.** $\left\{\frac{1}{3}, \sqrt{5}, -\sqrt{5}\right\}$ **67.** $\{-3, -2\}$ **69.** $-\frac{1}{3}$

71. Approximately 4 or 8 Cavaliers can be produced at an average cost of \$10,500. **73.** $k = 5$ **75.** -7
77. No (use the Rational Zeros Theorem) **79.** No (use the Rational Zeros Theorem) **81.** 7 in.

4.6 Exercises

1. $8 + 5i$ **3.** $-7 + 6i$ **5.** $-6 - 11i$ **7.** $6 - 18i$ **9.** $6 + 4i$ **11.** $10 - 5i$ **13.** 37 **15.** $\frac{6}{5} + \frac{8}{5}i$ **17.** $1 - 2i$ **19.** $\frac{5}{2} - \frac{7}{2}i$

21. $-\frac{1}{2} + \frac{\sqrt{3}}{2}i$ **23.** $2i$ **25.** $-i$ **27.** i **29.** -6 **31.** $-10i$ **33.** $-2 + 2i$ **35.** 0 **37.** 0 **39.** $2i$ **41.** $5i$ **43.** $5i$ **45.** $\{-2i, 2i\}$

47. $\{-4, 4\}$ **49.** $\{3 - 2i, 3 + 2i\}$ **51.** $\{3 - i, 3 + i\}$ **53.** $\left\{\frac{1}{4} - \frac{1}{4}i, \frac{1}{4} + \frac{1}{4}i\right\}$ **55.** $\left\{-\frac{1}{5} - \frac{2}{5}i, -\frac{1}{5} + \frac{2}{5}i\right\}$ **57.** $\left\{-\frac{1}{2} - \frac{\sqrt{3}}{2}i, -\frac{1}{2} + \frac{\sqrt{3}}{2}i\right\}$
59. $\{2, -1 - \sqrt{3}i, -1 + \sqrt{3}i\}$ **61.** $\{-2, 2, -2i, 2i\}$ **63.** $\{-3i, -2i, 2i, 3i\}$ **65.** Two complex solutions. **67.** Two unequal real solutions.
69. A repeated real solution. **71.** $2 - 3i$ **73.** 6 **75.** 25 **77.** $z + \bar{z} = (a + bi) + (a - bi) = 2a; z - \bar{z} = (a + bi) - (a - bi) = 2bi$
79. $\overline{z + w} = \overline{(a + bi) + (c + di)} = \overline{(a + c) + (b + d)i} = (a + c) - (b + d)i = (a - bi) + (c - di) = \bar{z} + \bar{w}$

4.7 Exercises

1. $4 + i$ **3.** $-i, 1 - i$ **5.** $-i, -2i$ **7.** $-i$ **9.** $2 - i, -3 + i$ **11.** $f(x) = x^4 - 14x^3 + 77x^2 - 200x + 208; a = 1$
13. $f(x) = x^5 - 4x^4 + 7x^3 - 8x^2 + 6x - 4; a = 1$ **15.** $f(x) = x^4 - 6x^3 + 10x^2 - 6x + 9; a = 1$ **17.** $-2i, 4$ **19.** $2i, -3, \frac{1}{2}$

21. $3 + 2i, -2, 5$ **23.** $4i, -\sqrt{11}, \sqrt{11}, -\frac{2}{3}$ **25.** $1, -\frac{1}{2} - \frac{\sqrt{3}}{2}i, -\frac{1}{2} + \frac{\sqrt{3}}{2}i$ **27.** $2, 3 - 2i, 3 + 2i$ **29.** $-i, i, -2i, 2i$ **31.** $-5i, 5i, -3, 1$

33. $-4, \frac{1}{3}, 2 - 3i, 2 + 3i$ **35.** Zeros that are complex numbers must occur in conjugate pairs; or a polynomial with real coefficients of odd degree must have at least one real zero. **37.** If the remaining zero were a complex number, then its conjugate would also be a zero.

4.8 Exercises

1. $\{x \mid -2 < x < 5\}$ **3.** $\{x \mid -\infty < x < 0 \text{ or } 4 < x < \infty\}$ **5.** $\{x \mid -3 < x < 3\}$ **7.** $\{x \mid -\infty < x < -4 \text{ or } 3 < x < \infty\}$ **9.** $\{x \mid -\frac{1}{2} < x < 3\}$
11. $\{x \mid -\infty < x < -1 \text{ or } 8 < x < \infty\}$ **13.** No real solution **15.** $\{x \mid -\infty < x < -\frac{2}{3} \text{ or } \frac{3}{2} < x < \infty\}$ **17.** $\{x \mid 1 < x < \infty\}$
19. $\{x \mid -\infty < x < 1 \text{ or } 2 < x < 3\}$ **21.** $\{x \mid -1 < x < 0 \text{ or } 3 < x < \infty\}$ **23.** $\{x \mid -\infty < x < -1 \text{ or } 1 < x < \infty\}$ **25.** $\{x \mid 1 < x < \infty\}$
27. $\{x \mid -\infty < x < -1 \text{ or } 1 < x < \infty\}$ **29.** $\{x \mid -1 < x < 8\}$ **31.** $\{x \mid 2.14 \le x < \infty\}$ **33.** $\{x \mid -\infty < x < -2 \text{ or } 2 < x < \infty\}$
35. $\{x \mid -1 \le x \le -0.23 \text{ or } 4.23 \le x < \infty\}$ **37.** $\{x \mid -\infty < x < -1 \text{ or } 1 < x < \infty\}$ **39.** $\{x \mid -\infty < x < -1 \text{ or } 0 < x < 1\}$
41. $\{x \mid -\infty < x < -1 \text{ or } 1 < x < \infty\}$ **43.** $\{x \mid -\infty < x < -\frac{2}{3} \text{ or } 0 < x < \frac{3}{2}\}$ **45.** $\{x \mid -\infty < x < 2\}$ **47.** $\{x \mid -2 < x \le 9\}$
49. $\{x \mid -\infty < x < 2 \text{ or } 3 < x < 5\}$ **51.** $\{x \mid -\infty < x < -3 \text{ or } -1 < x < 1 \text{ or } 2 < x < \infty\}$
53. $\{x \mid -\infty < x < -5 \text{ or } -4 < x < -3 \text{ or } 1 < x < \infty\}$ **55.** $\{x \mid -\infty < x \le -0.5 \text{ or } 1 \le x < 4\}$ **57.** $\{x \mid -3.30 < x < -3 \text{ or } 0.30 < x < \infty\}$
59. $\{x \mid 4 < x < \infty\}$ **61.** $\{x \mid -\infty < x \le -4 \text{ or } 4 \le x < \infty\}$ **63.** $\{x \mid -\infty < x < -4 \text{ or } 2 \le x < \infty\}$

65. (a) The ball is more than 96 feet above the ground for time t between 2 and 3 seconds, $2 < t < 3$. **(b)**

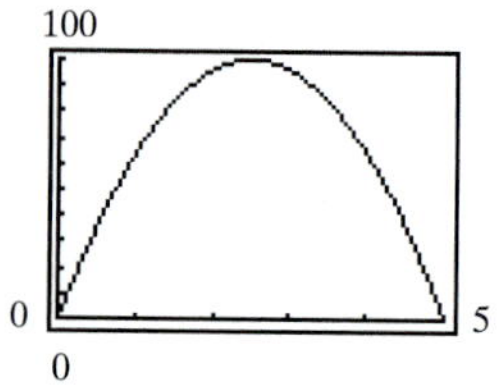

(c) 100 feet **(d)** 2.5 seconds **67. (a)** For a profit of at least \$50, between 8 and 32 watches must be sold, $8 \le x \le 32$.

(b)

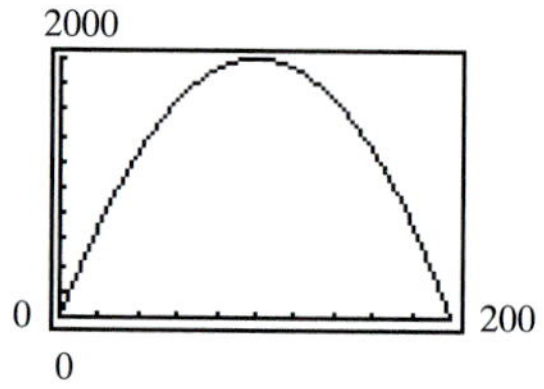

(c) \$2000 **(d)** 100 **(e)**

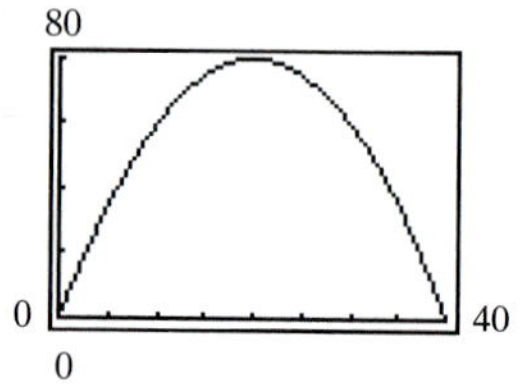

(f) \$80 **(g)** 20

69. Chevy can produce at most 8 Cavaliers in a day, assuming that cars cannot be partially completed in a day.

71. (a)

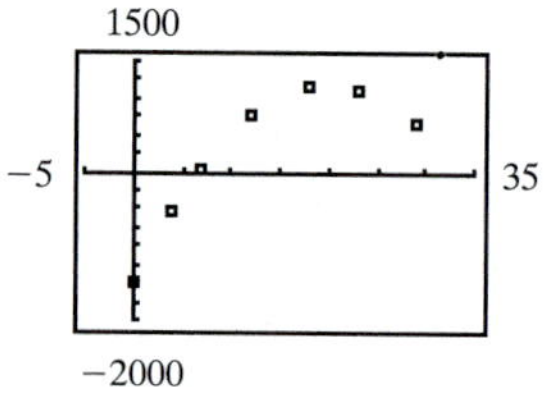

(b) $p(x) = -6.7760x^2 + 270.6678x - 1500.2015$ **(c)** Between 15 and 25 computers. **(d)** 20 computers **(e)** \$1202.75

73. $b - a = (\sqrt{b} - \sqrt{a})(\sqrt{b} + \sqrt{a})$; since $a \ge 0$ and $b \ge 0$, then $\sqrt{a} \ge 0$ and $\sqrt{b} \ge 0$ so that $\sqrt{b} + \sqrt{a} \ge 0$; thus, $b - a \ge 0$ is equivalent to $\sqrt{b} - \sqrt{a} \ge 0$ and $a \le b$ is equivalent to $\sqrt{a} \le \sqrt{b}$

Fill-in-the-Blank Items

1. parabola; vertex **2.** Remainder; Dividend **3.** $f(c)$ **4.** $f(c) = 0$ **5.** zero **6.** three; one; two; no **7.** $\pm 1, \pm\frac{1}{2}$ **8.** $y = 1$ **9.** $x = -1$ **10.** $3 - 4i$

True/False Items

1. F **2.** F **3.** T **4.** T **5.** T

Review Exercises

1.

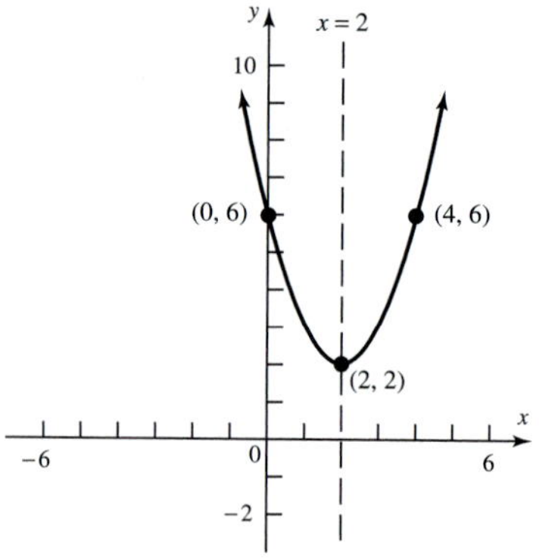

3.

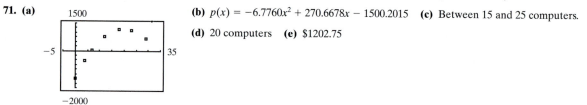

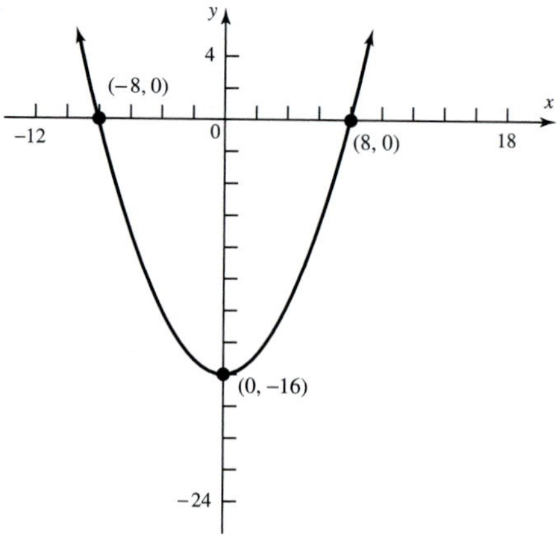

5.

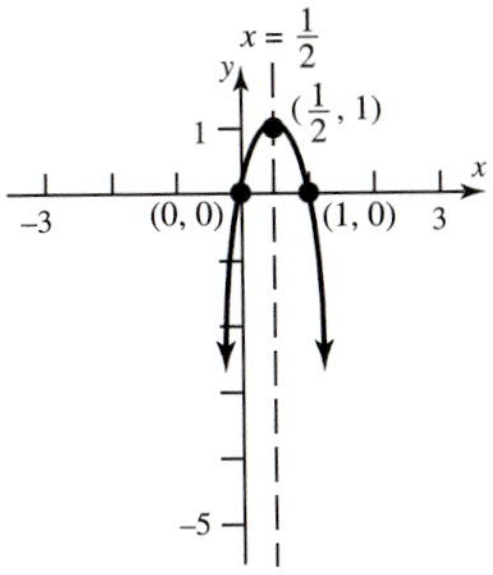

7.

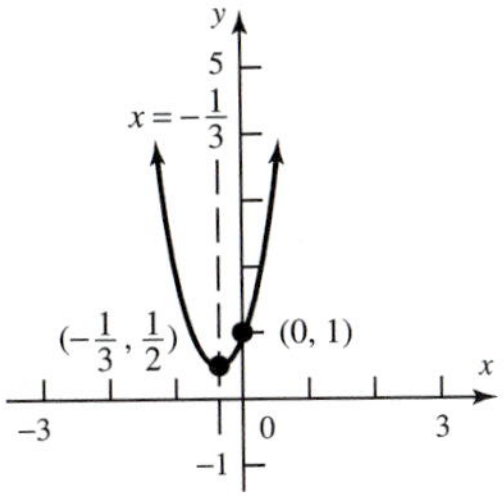

9.

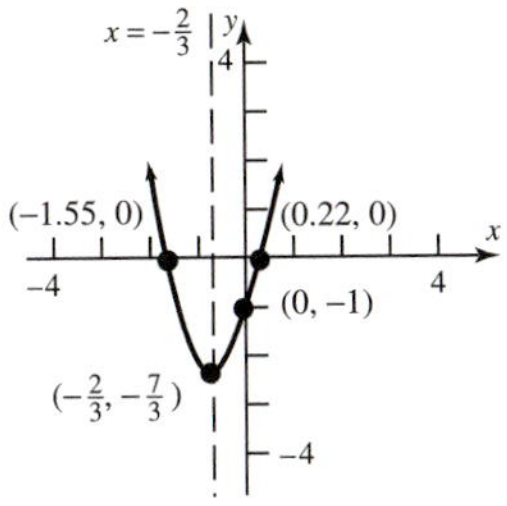

11.

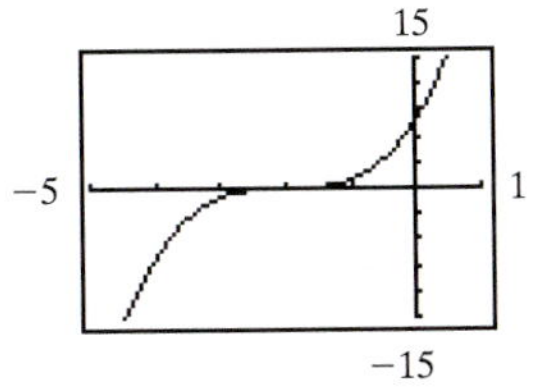

13.

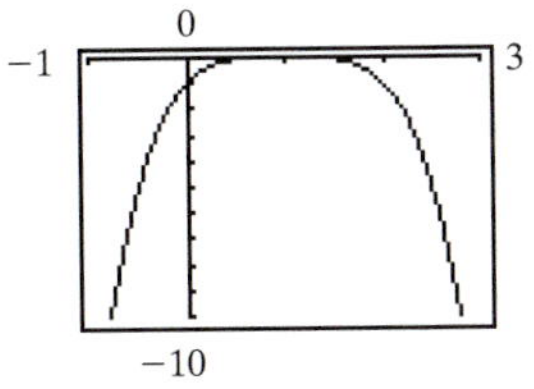

15.

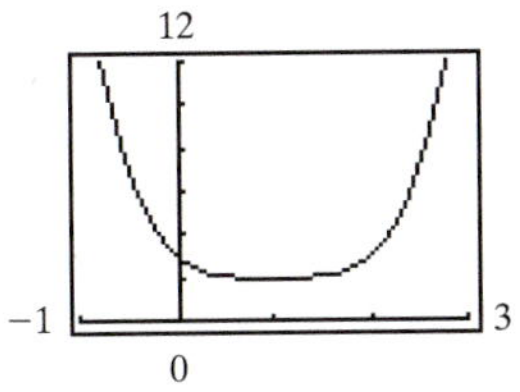

17. Minimum value; 1 **19.** Maximum value; 12 **21.** Maximum value; 16

23. **(a)**

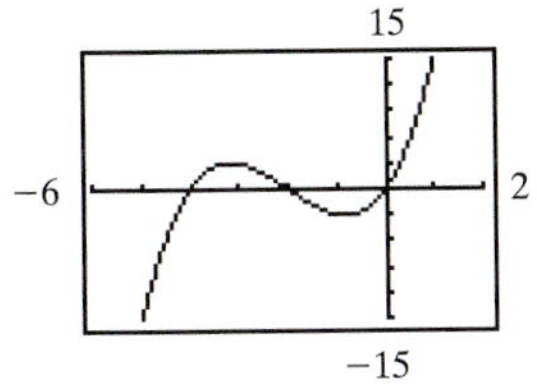

(b) x-intercepts: $-4, -2, 0$; y-intercept: 0
(c) $-4, -2, 0$: odd
(d) $y = x^3$
(e) 2
(f) Local minimum: $(-0.84, -3.07)$
Local maximum: $(-3.15, 3.07)$

25. **(a)**

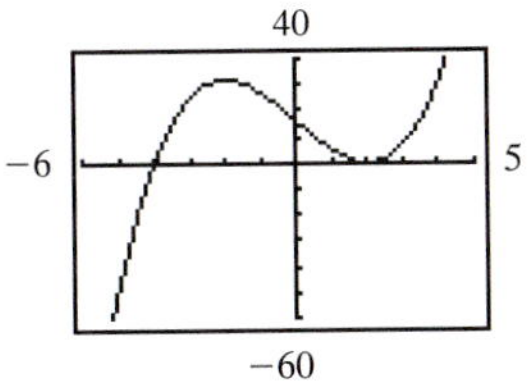

(b) x-intercepts: $-4, 2$; y-intercept: 16
(c) -4: odd; 2: even
(d) $y = x^3$
(e) 2
(f) Local minimum: $(2, 0)$
Local maximum: $(-2, 32)$

27. $f(x) = x^3 - 4x^2 = x^2(x - 4)$

(a)

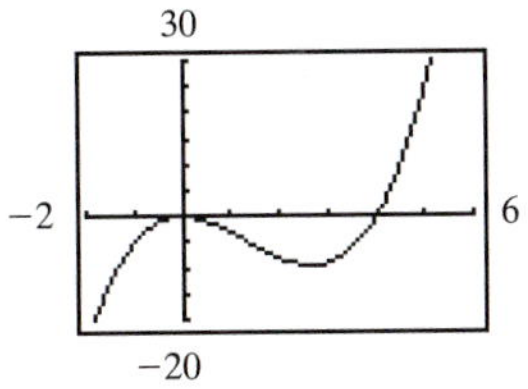

(b) x-intercepts: 0, 4; y-intercept: 0
(c) 0: even; 4: odd
(d) $y = x^3$
(e) 2
(f) Local minimum: $(2.66, -9.48)$
Local maximum: $(0.00, 0.00)$

29. **(a)**

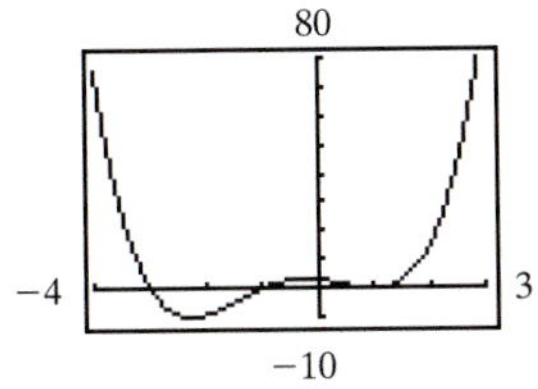

(b) x-intercepts: $-3, -1, 1$; y-intercept: 3
(c) $-3, -1$: odd; 1: even
(d) $y = x^4$
(e) 3
(f) Local minima: $(-2.28, -9.91)$, $(1.00, 0.00)$
Local maximum: $(-0.21, 3.22)$

31.
1. x-intercept: 3; no y-intercept
2. No symmetry
3. Vertical asymptote: $x = 0$
4. Horizontal asymptote: $y = 2$ not intersected
5.

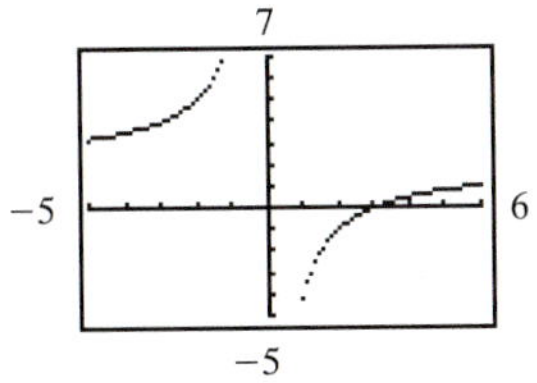

33.
1. x-intercept: -2; no y-intercept
2. No symmetry
3. Vertical asymptotes: $x = 0, x = 2$
4. Horizontal asymptote: $y = 0$ intersected at $(-2, 0)$
5.

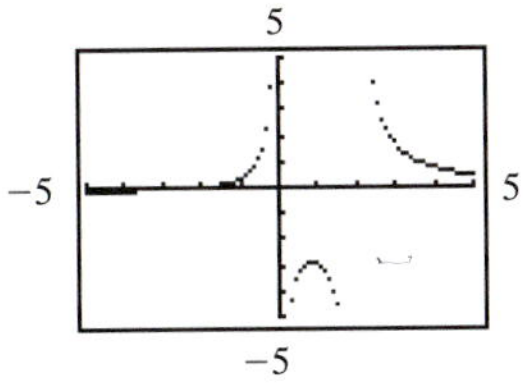

35. 1. Intercepts: $(-3, 0), (2, 0), (0, 1)$
2. No symmetry
3. Vertical asymptote: $x = -2, x = 3$
4. Horizontal asymptote: $y = 1$ intersected at $(0, 1)$
5.

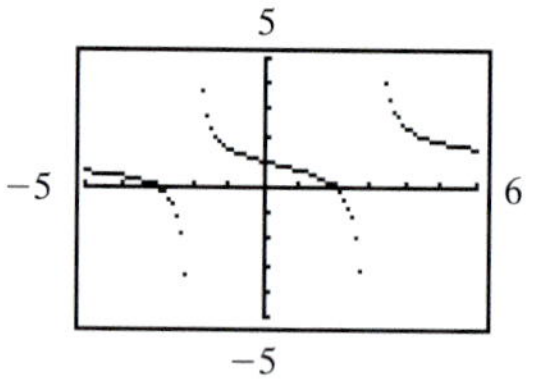

37. 1. Intercept $(0, 0)$
2. Symmetric with respect to the origin
3. Vertical asymptotes: $x = -2, x = 2$
4. Oblique asymptote: $y = x$ intersected at $(0, 0)$
5.

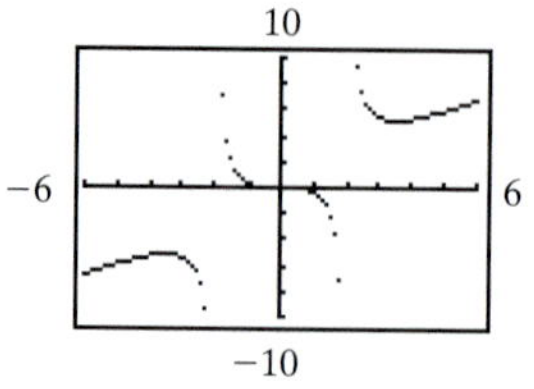

39. 1. Intercept $(0, 0)$
2. No symmetry
3. Vertical asymptote: $x = 1$
4. No oblique or horizontal asymptote
5.

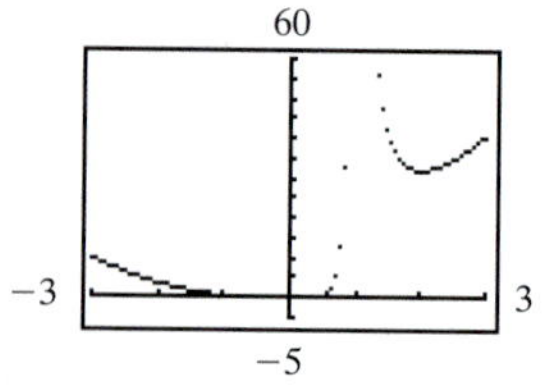

41. 4, 2, or 0 positive; 2 or 0 negative **43.** $\pm\frac{1}{12}, \pm\frac{1}{6}, \pm\frac{1}{4}, \pm\frac{1}{3}, \pm\frac{1}{2}, \pm\frac{3}{4}, \pm 1, \pm\frac{3}{2}, \pm 3$ **45.** $-2, 1, 4; f(x) = (x + 2)(x - 1)(x - 4)$ **47.** $\frac{1}{2}$, multiplicity 2; -2; $f(x) = 4\left(x - \frac{1}{2}\right)^2(x + 2)$ **49.** 2, multiplicity 2; $f(x) = (x - 2)^2(x^2 + 5)$ **51.** $-2.50, 3.10, 5.32$ **53.** $-11.30, -0.60, 4.00, 9.33$ **55.** $-3.67, 1.33$ **57.** $\{-3, 2\}$ **59.** $\left\{-3, -1, -\frac{1}{2}, 1\right\}$ **61.** -2 and 2 **63.** -3 and 5 **65.** 1.52 **67.** 0.93 **69.** $4 - i$ **71.** $-i, 1 - i$ **73.** $q(x) = x^2 + 5x + 6; R = 0$ **75.** $\left\{-\frac{1}{2} - \frac{\sqrt{3}}{2}i, -\frac{1}{2} + \frac{\sqrt{3}}{2}i\right\}$ **77.** $\left\{\frac{-1 - \sqrt{17}}{4}, \frac{-1 + \sqrt{17}}{4}\right\}$ **79.** $\left\{\frac{1}{2} - \frac{\sqrt{11}}{2}i, \frac{1}{2} + \frac{\sqrt{11}}{2}i\right\}$ **81.** $\left\{\frac{1}{2} - \frac{\sqrt{23}}{2}i, \frac{1}{2} + \frac{\sqrt{23}}{2}i\right\}$ **83.** $\{-\sqrt{2}, \sqrt{2}, -2i, 2i\}$ **85.** $\{-3, 2\}$ **87.** $\left\{\frac{1}{3}, 1, -i, i\right\}$ **89.** $-5, -6i, 6i$ **91.** $1, -8, 5 - i, 5 + i$ **93.** 1 is an upper bound; -2 is a lower bound. **95.** $\left\{x \middle| -4 < x < \frac{3}{2}\right\}$ **97.** $\{x \mid -3 < x \le 3\}$ **99.** $\{x \mid -\infty < x < 1 \text{ or } 2 < x < \infty\}$ **101.** $\{x \mid 1 < x < 2 \text{ or } 3 < x < \infty\}$ **103.** $\{x \mid -\infty < x < -4 \text{ or } 2 < x < 4 \text{ or } 6 < x < \infty\}$

105. (a)

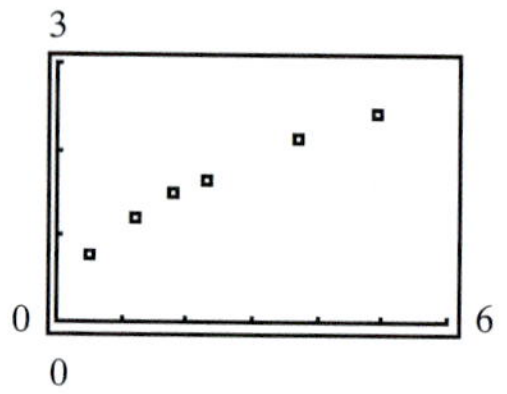

(b) $T(l) = 1.1094l^{0.4952}$ **(c)**

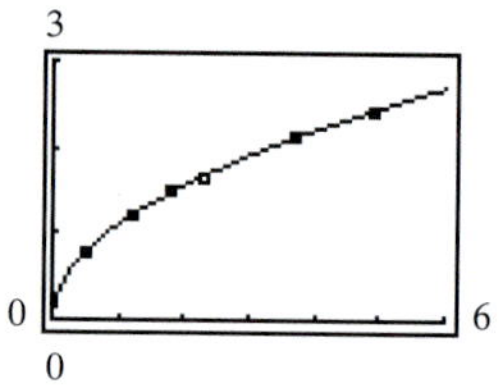

(d) Approximately 1.5 ft

107. (a)

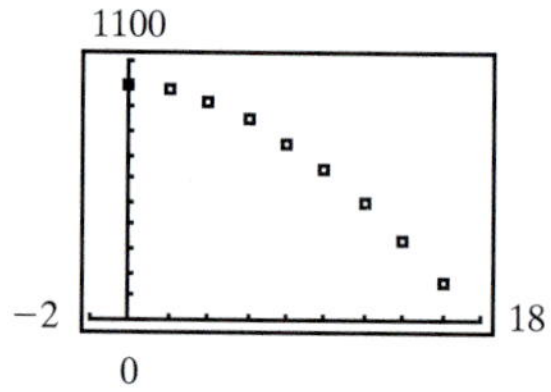

(b) $s(t) = -2.6664t^2 - 10.0127t + 999.8545$ **(c)**

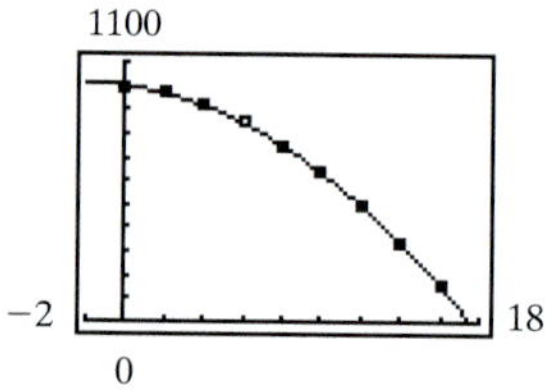

(d) -5.3328 ft/sec^2 **(e)** -10.0127 ft/sec **(f)** 17.577 seconds

109. (a)

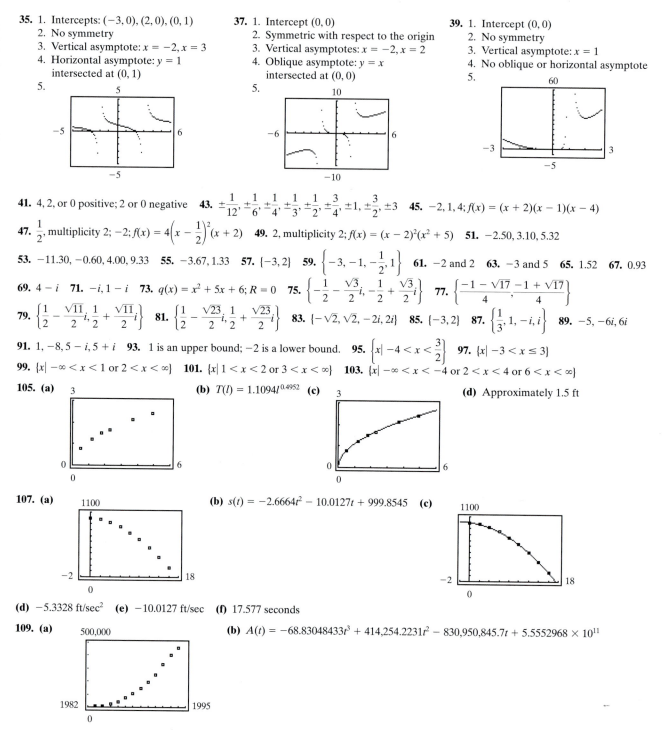

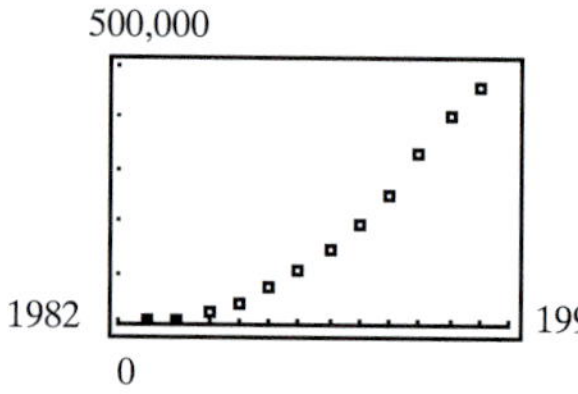

(b) $A(t) = -68.83048433t^3 + 414{,}254.2231t^2 - 830{,}950{,}845.7t + 5.5552968 \times 10^{11}$

(c)

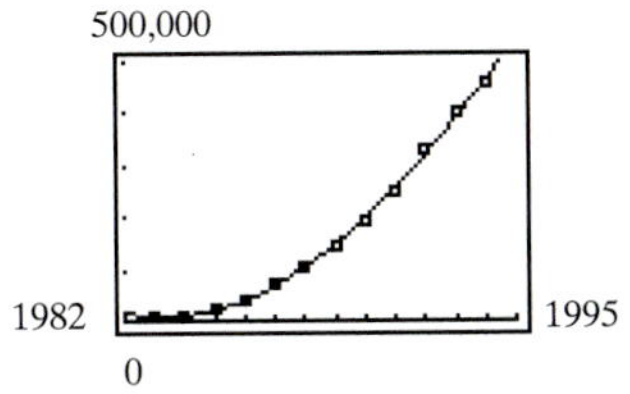

(d) 545,573 **111.** (2, 2) **113.** 3.6 ft

117. (a) even **(b)** positive **(c)** even **(d)** 0 is a zero of even multiplicity. **(e)** 8

CHAPTER 5 *5.1 Exercises*

1. (a)

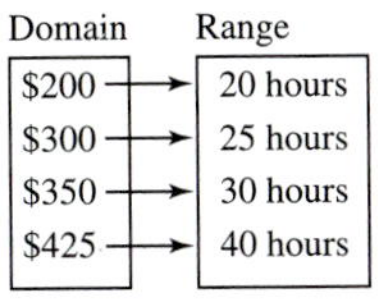

(b) Inverse is a function **3. (a)**

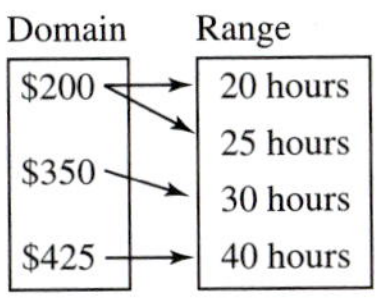

(b) Not a function

5. (a) $\{(6, 2), (6, -3), (9, 4), (10, 1)\}$ **(b)** Not a function **7. (a)** $\{(0, 0), (1, 1), (16, 2), (81, 3)\}$ **(c)** Inverse is a function

9. One-to-one **11.** Not one-to-one **13.** One-to-one

15.

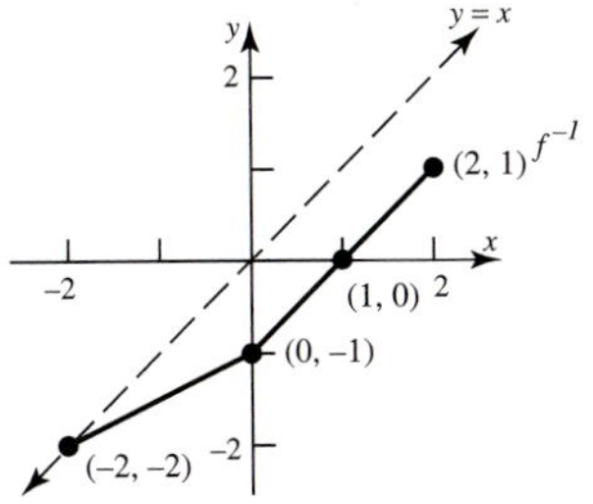

17.

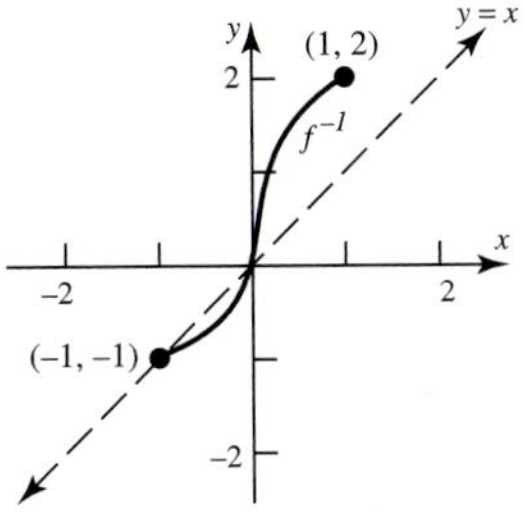

19.

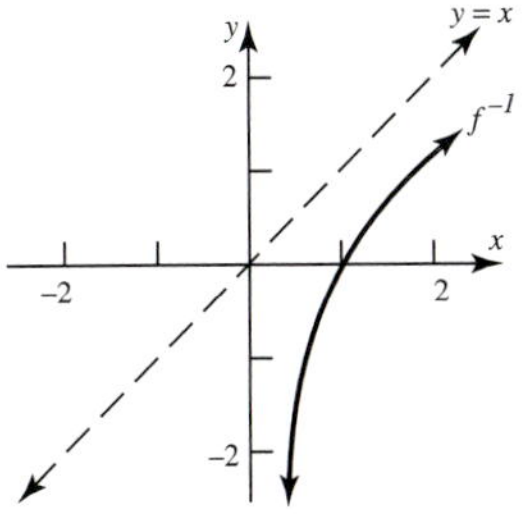

21. $f(g(x)) = f\left(\frac{1}{3}(x-4)\right) = 3\left[\frac{1}{3}(x-4)\right] + 4 = x$; $g(f(x)) = g(3x+4) = \frac{1}{3}[(3x+4)-4] = x$

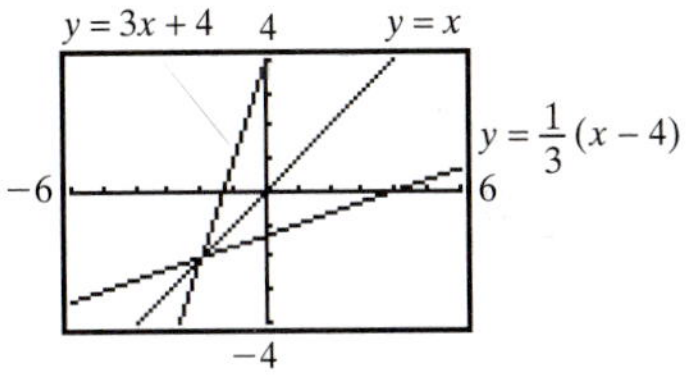

23. $f(g(x)) = 4\left[\frac{x}{4} + 2\right] - 8 = x$; $g(f(x)) = \frac{4x-8}{4} + 2 = x$

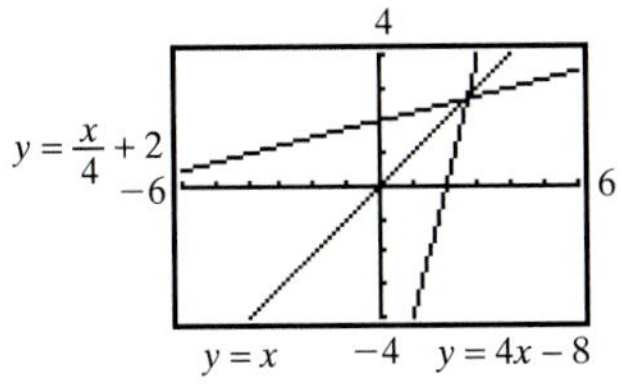

25. $f(g(x)) = (\sqrt[3]{x+8})^3 - 8 = x$; $g(f(x)) = \sqrt[3]{(x^3-8)+8} = x$

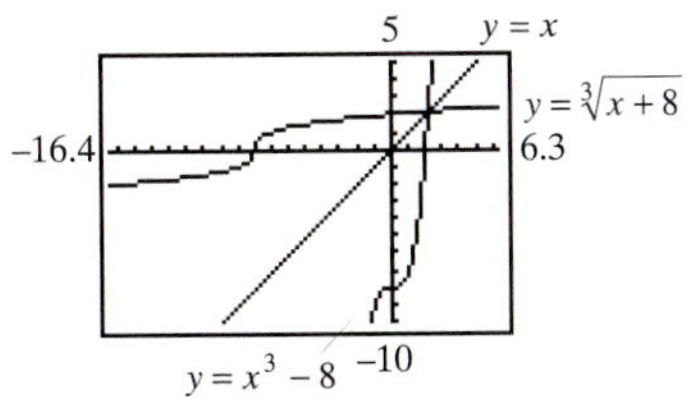

27. $f(g(x)) = \dfrac{1}{\left(\dfrac{1}{x}\right)} = x$; $g(f(x)) = \dfrac{1}{\left(\dfrac{1}{x}\right)} = x$

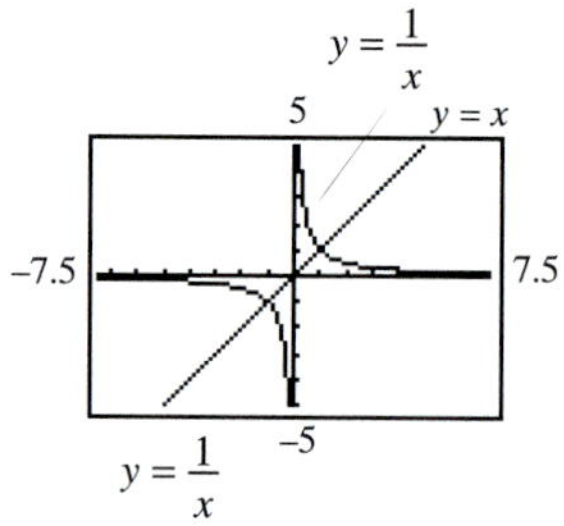

29. $f(g(x)) = \dfrac{2\left(\dfrac{4x-3}{2-x}\right)+3}{\dfrac{4x-3}{2-x}+4} = x$; $g(f(x)) = \dfrac{4\left(\dfrac{2x+3}{x+4}\right)-3}{2-\dfrac{2x+3}{x+4}} = x$

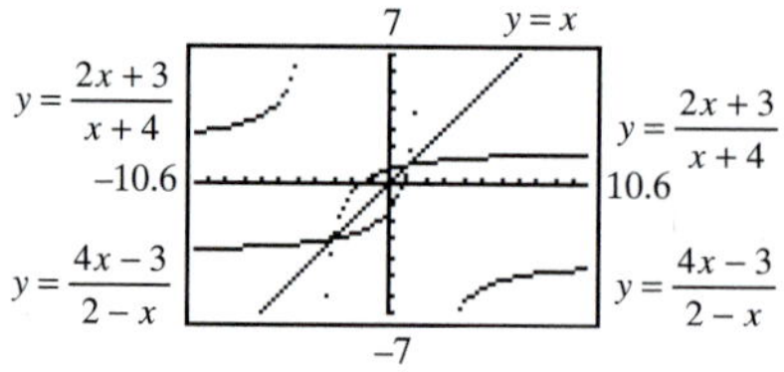

31. $f^{-1}(x) = \frac{1}{3}x$; $f(f^{-1}(x)) = 3\left(\frac{1}{3}x\right) = x$

$f^{-1}(f(x)) = \frac{1}{3}(3x) = x$

Domain f = Range $f^{-1} = (-\infty, \infty)$

Range f = Domain $f^{-1} = (-\infty, \infty)$

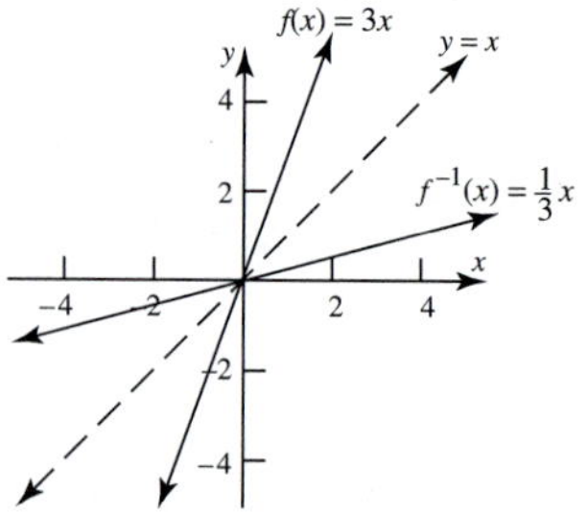

33. $f^{-1}(x) = \frac{x}{4} - \frac{1}{2}$; $f(f^{-1}(x)) = 4\left(\frac{x}{4} - \frac{1}{2}\right) + 2 = x$

$f^{-1}(f(x)) = \dfrac{4x+2}{4} - \dfrac{1}{2} = x$

Domain f = Range $f^{-1} = (-\infty, \infty)$

Range f = Domain $f^{-1} = (-\infty, \infty)$

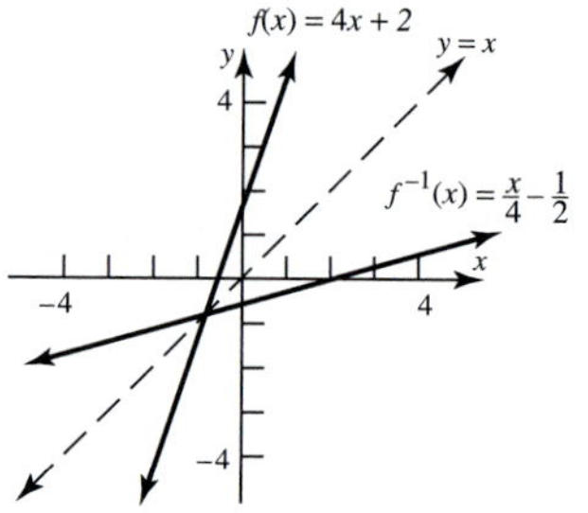

35. $f^{-1}(x) = \sqrt[3]{x+1}$; $f(f^{-1}(x)) = (\sqrt[3]{x+1})^3 - 1 = x$

$f^{-1}(f(x)) = \sqrt[3]{x^3 - 1 + 1} = x$;

Domain f = Range $f^{-1} = (-\infty, \infty)$

Range f = Domain $f^{-1} = (-\infty, \infty)$

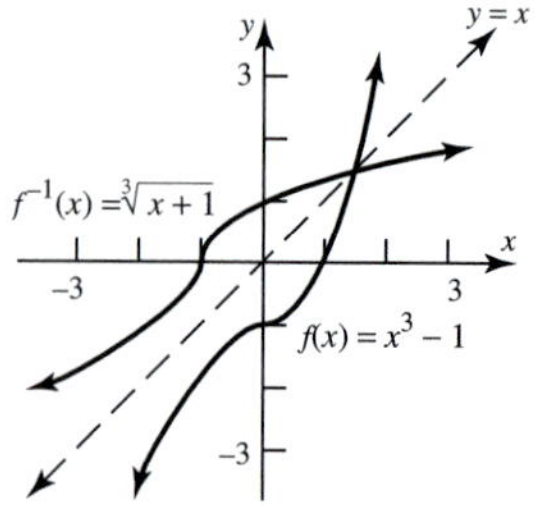

37. $f^{-1}(x) = \sqrt{x-4}, x \geq 4$; $f(f^{-1}(x)) = (\sqrt{x-4})^2 + 4 = x$

$f^{-1}(f(x)) = \sqrt{(x^2+4) - 4} = \sqrt{x^2} = |x| = x$

Domain f = Range $f^{-1} = [0, \infty)$

Range f = Domain $f^{-1} = [4, \infty)$

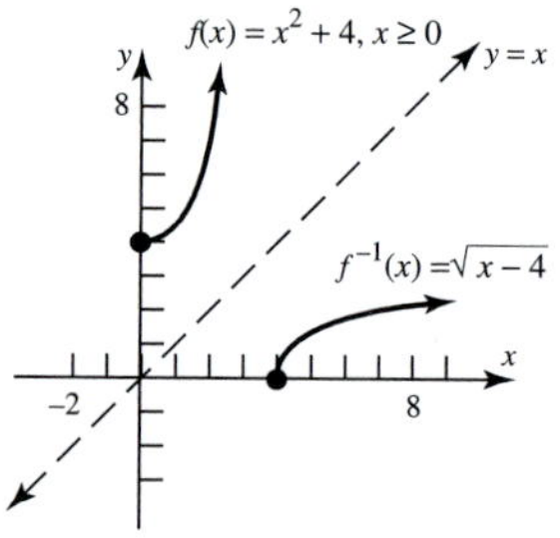

39. $f^{-1}(x) = \dfrac{4}{x}$

$f(f^{-1}(x)) = \dfrac{4}{4/x} = x$

$f^{-1}(f(x)) = \dfrac{4}{4/x} = x$

41. $f^{-1}(x) = \dfrac{2x+1}{x}$

$f(f^{-1}(x)) = \dfrac{1}{\dfrac{2x+1}{x} - 2} = x$

$f^{-1}(f(x)) = \dfrac{2\left(\dfrac{1}{x-2}\right)+1}{\dfrac{1}{x-2}} = x$

Domain f = Range f^{-1} = All real numbers except 0
Range f = Domain f^{-1} = All real numbers except 0

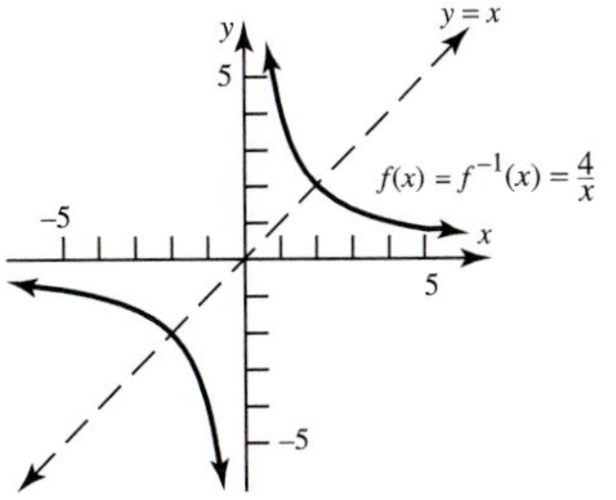

43. $f^{-1}(x) = \dfrac{2 - 3x}{x}$

$$f(f^{-1}(x)) = \frac{2}{3 + \dfrac{2 - 3x}{x}} = x$$

$$f^{-1}(f(x)) = \frac{2 - 3\left(\dfrac{2}{3 + x}\right)}{\dfrac{2}{3 + x}} = x$$

Domain f = Range f^{-1} = All real numbers except -3
Range f = Domain f^{-1} = All real numbers except 0

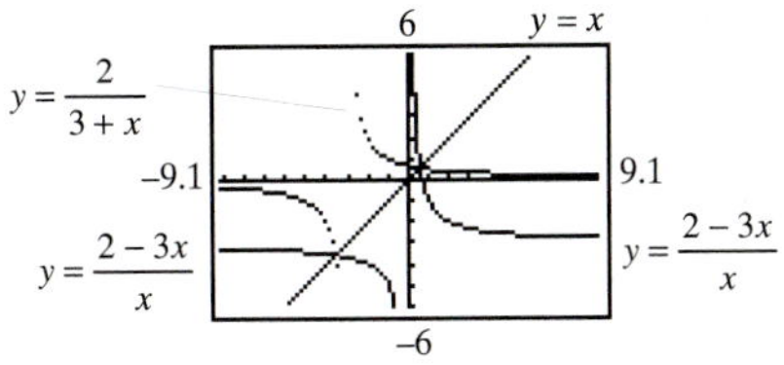

47. $f^{-1}(x) = \dfrac{x}{x - 2}$

$$f(f^{-1}(x)) = \frac{2\left(\dfrac{x}{x - 2}\right)}{\dfrac{x}{x - 2} - 1} = x$$

$$f^{-1}(f(x)) = \frac{\dfrac{2x}{x - 1}}{\dfrac{2x}{x - 1} - 2} = x$$

Domain f = Range f^{-1} = All real numbers except 1

Range f = Domain f^{-1} = All real numbers except 2

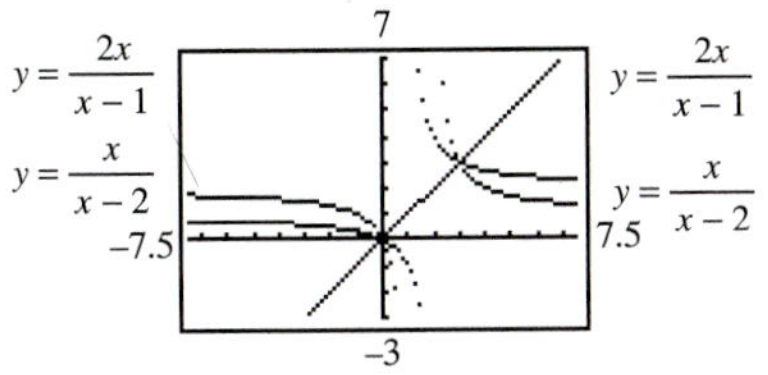

Domain f = Range f^{-1} = All real numbers except 2
Range f = Domain f^{-1} = All real numbers except 0

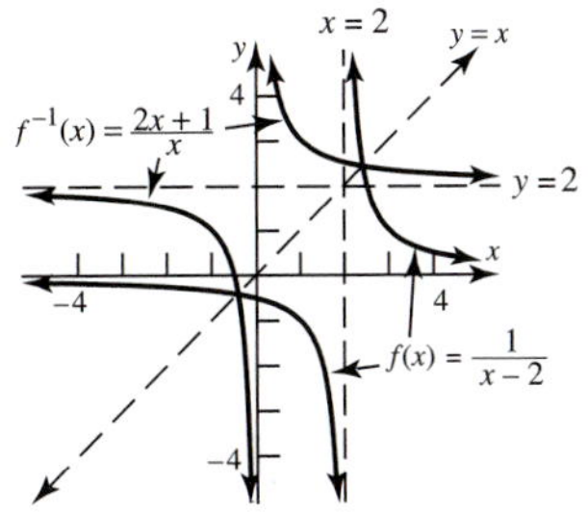

45. $f^{-1}(x) = \sqrt{x} - 2, x \geq 0$

$$f(f^{-1}(x)) = (\sqrt{x} - 2 + 2)^2 = x$$

$$f^{-1}(f(x)) = \sqrt{(x + 2)^2} - 2 = |x + 2| - 2 = x$$

Domain f = Range f^{-1} = $[-2, \infty)$
Range f = Domain f^{-1} = $[0, \infty)$

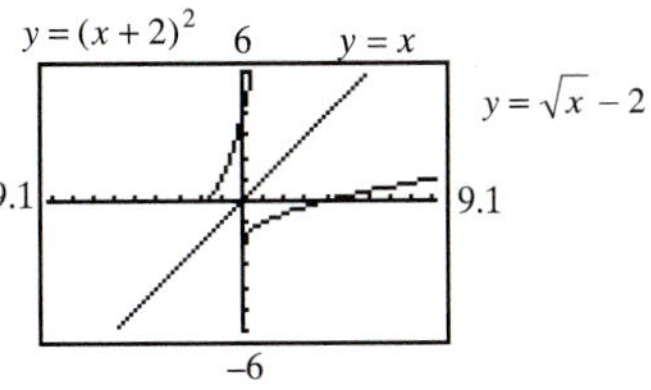

49. $f^{-1}(x) = \dfrac{3x + 4}{2x - 3}$

$$f(f^{-1}(x)) = \frac{3\left(\dfrac{3x + 4}{2x - 3}\right) + 4}{2\left(\dfrac{3x + 4}{2x - 3}\right) - 3} = x$$

$$f^{-1}(f(x)) = \frac{3\left(\dfrac{3x + 4}{2x - 3}\right) + 4}{2\left(\dfrac{3x + 4}{2x - 3}\right) - 3} = x$$

Domain f = Range f^{-1} = All real numbers except $\dfrac{3}{2}$

Range f = Domain f^{-1} = All real numbers except $\dfrac{3}{2}$

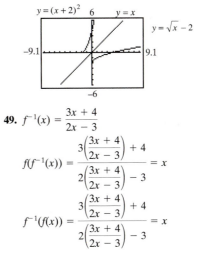

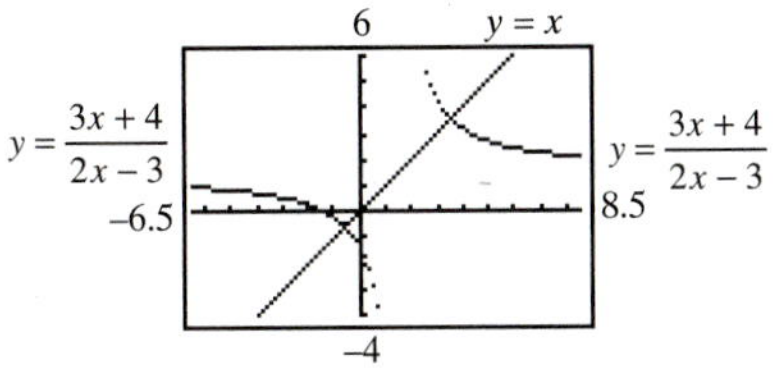

51. $f^{-1}(x) = \dfrac{-2x+3}{x-2}$

$$f(f^{-1}(x)) = \frac{2\left(\dfrac{-2x+3}{x-2}\right)+3}{\dfrac{-2x+3}{x-2}+2} = x$$

$$f^{-1}(f(x)) = \frac{-2\left(\dfrac{2x+3}{x+2}\right)+3}{\dfrac{2x+3}{x+2}-2} = x$$

Domain f = Range f^{-1} = All real numbers except -2
Range f = Domain f^{-1} = All real numbers except 2

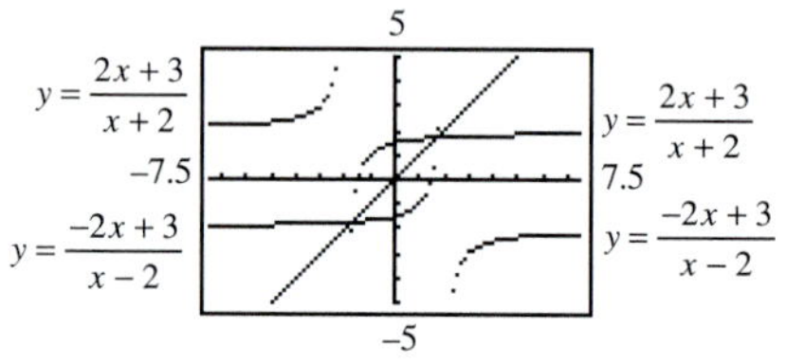

53. $f^{-1}(x) = \dfrac{x^3}{8}$

$$f(f^{-1}(x)) = 2\sqrt[3]{\frac{x^3}{8}} = x$$

$$f^{-1}(f(x)) = \frac{(2\sqrt[3]{x})^3}{8} = x$$

Domain f = Range f^{-1} = $(-\infty, \infty)$
Range f = Domain f^{-1} = $(-\infty, \infty)$

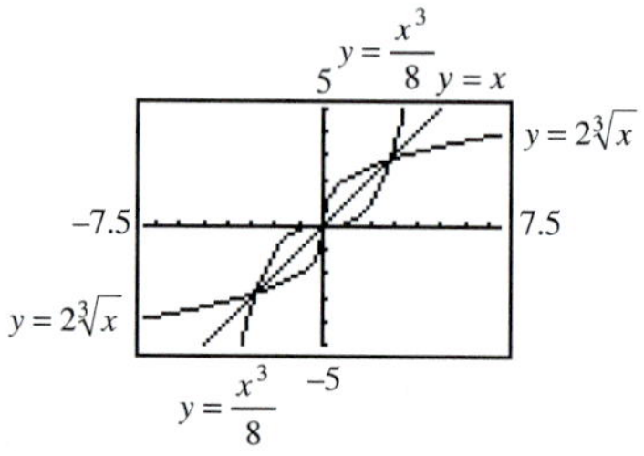

55. $f^{-1}(x) = \dfrac{1}{m}(x-b), m \neq 0$ **57.** No; whenever x and $-x$ are in the domain of f, two equal y values, $f(x)$ and $f(-x)$, are present.
59. Quadrant I **61.** $f(x) = |x|, x \geq 0$ is one-to-one; this may be written as $f(x) = x; f^{-1}(x) = x$
63. $f(g(x)) = \dfrac{9}{5}\left[\dfrac{5}{9}(x-32)\right] + 32 = x; g(f(x)) = \dfrac{5}{9}\left[\left(\dfrac{9}{5}x + 32\right) - 32\right] = x$ **65.** $l(T) = gT^2/4\pi^2$ **67.** $f^{-1}(x) = \dfrac{-dx+b}{cx-a}; f = f^{-1}$ if $a = -d$

5.2 Exercises

1. (a) 11.212 **(b)** 11.587 **(c)** 11.664 **(d)** 11.665 **3. (a)** 8.815 **(b)** 8.821 **(c)** 8.824 **(d)** 8.825 **5. (a)** 21.217 **(b)** 22.217 **(c)** 22.440 **(d)** 22.459 **7.** 3.320 **9.** 0.427 **11.** B **13.** D **15.** A **17.** E **19.** A **21.** E **23.** B

25.

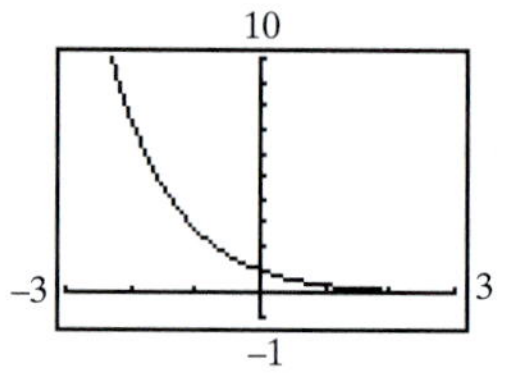

Domain: $(-\infty, \infty)$
Range: $(0, \infty)$
Horizontal Asymptote: $y = 0$

27.

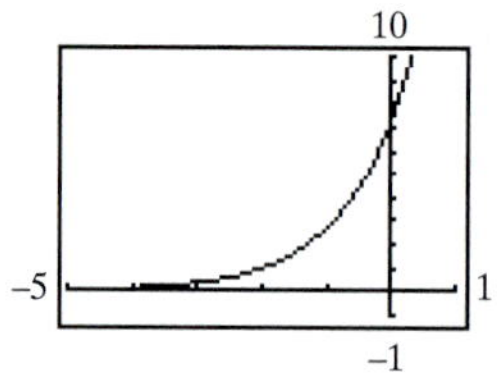

Domain: $(-\infty, \infty)$
Range: $(0, \infty)$
Horizontal Asymptote: $y = 0$

29.

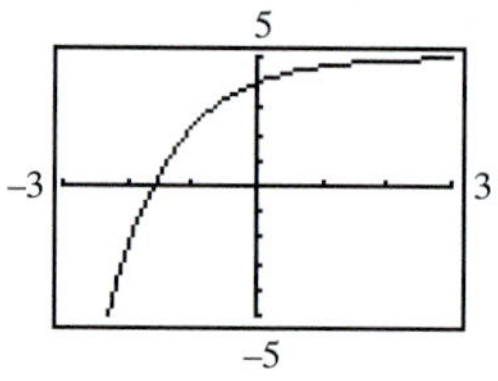

Domain: $(-\infty, \infty)$
Range: $(-\infty, 5)$
Horizontal Asymptote: $y = 5$

31.

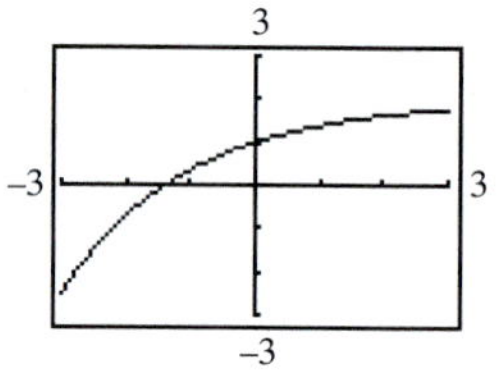

Domain: $(-\infty, \infty)$
Range: $(-\infty, 2)$
Horizontal Asymptote: $y = 2$

33. $\dfrac{1}{49}$ **35.** $\dfrac{1}{4}$ **37. (a)** 74% **(b)** 47% **39. (a)** 44 watts **(b)** 11.6 watts **41.** 3.35 milligrams; 0.45 milligrams

43. (a) 0.63 **(b)** 0.98 **(c)**

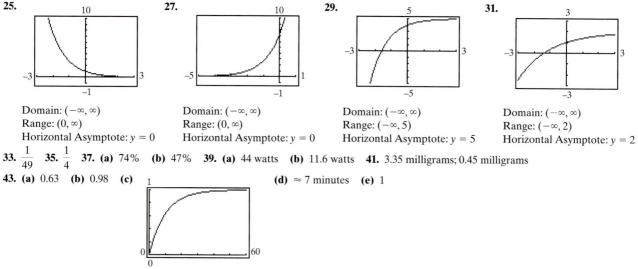

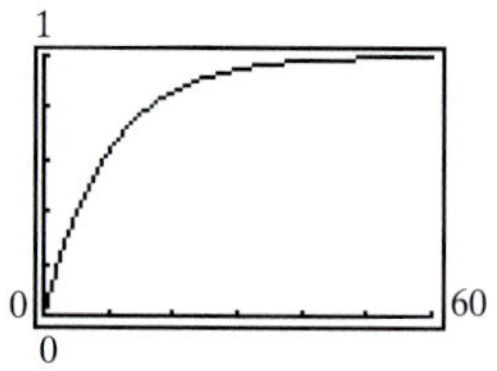

(d) $\approx$ 7 minutes **(e)** 1

45. (a) 56% **(b)** 68% **(c)** 70%

(d) $R = 40\%$ just after 6 days

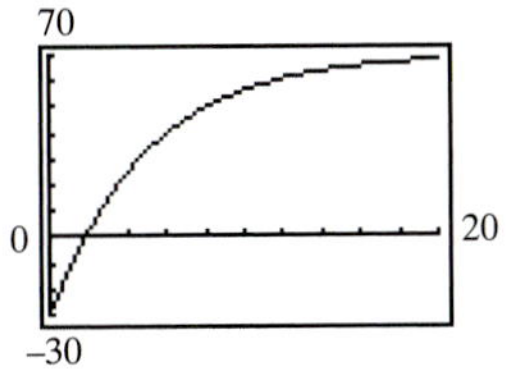

47. (a) 5.414 amperes, 7.5854 amperes, 10.38 amperes **(b)** 12 amperes **(c)** $I_1(t) = 12(1 - e^{-2t})$
(d) 3.343 amperes, 5.309 amperes, 9.443 amperes **(e)** 24 amperes **(f)** $I_2(t) = 24(1 - e^{-1/2t})$

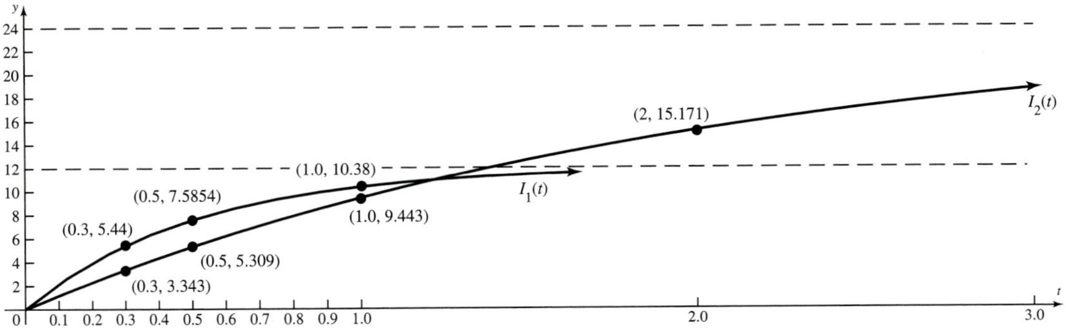

49. (a) 9.23×10^{-3} **(b)** 0.81 **(c)** 5 **(d)** 57.91°, 43.98°, 30.06°

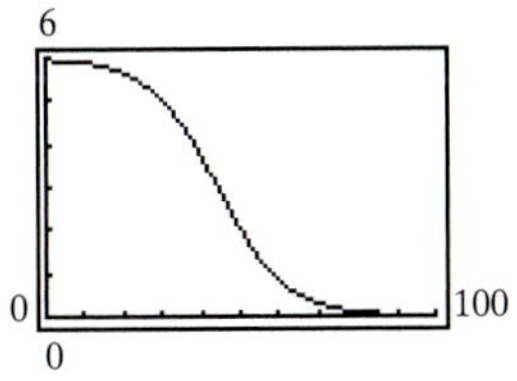

51. $f(1) = 5, f(2) = 17, f(3) = 257, f(4) = 65{,}537,\ f(5) = 4{,}294{,}967{,}297 = 641 \times 6{,}700{,}417$

5.3 Exercises

1. $2 = \log_3 9$ **3.** $2 = \log_a 1.6$ **5.** $2 = \log_{1.1} M$ **7.** $x = \log_2 7.2$ **9.** $\sqrt{2} = \log_x \pi$ **11.** $x = \ln 8$ **13.** $2^3 = 8$ **15.** $a^6 = 3$ **17.** $3^x = 2$ **19.** $2^{1.3} = M$ **21.** $(\sqrt{2})^x = \pi$ **23.** $e^x = 4$ **25.** 0 **27.** 2 **29.** −4 **31.** $\frac{1}{2}$ **33.** 4 **35.** $\frac{1}{2}$ **37.** $\{x \mid x < 3\}$ **39.** All real numbers except 0 **41.** $\{x \mid x < -2 \text{ or } x > 3\}$ **43.** $\{x \mid x > 0, x \neq 1\}$ **45.** $\{x \mid x < -1 \text{ or } x > 0\}$ **47.** 0.511 **49.** 30.099 **51.** $\sqrt{2}$ **53.** B **55.** D **57.** A **59.** E **61.** C **63.** A **65.** D

67.

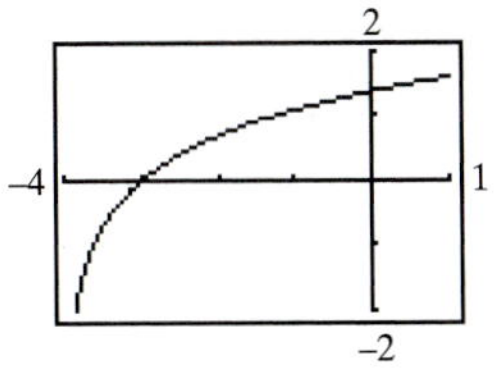

Domain: $(-4, \infty)$
Range: $(-\infty, \infty)$
Vertical Asymptote: $x = -4$

69.

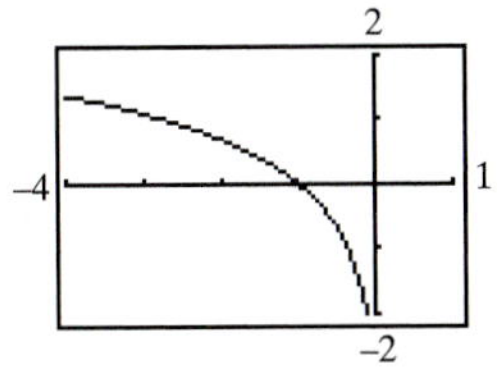

Domain: $(-\infty, 0)$
Range: $(-\infty, \infty)$
Vertical Asymptote: $x = 0$

71.

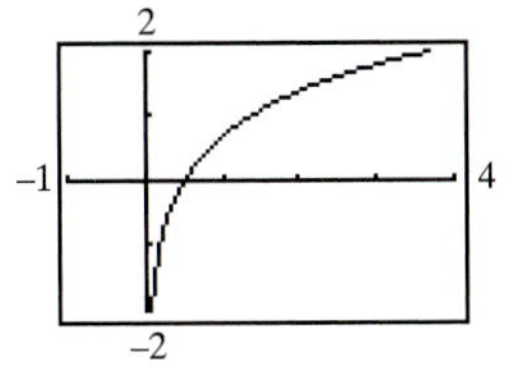

Domain: $(0, \infty)$
Range: $(-\infty, \infty)$
Vertical Asymptote: $x = 0$

73.

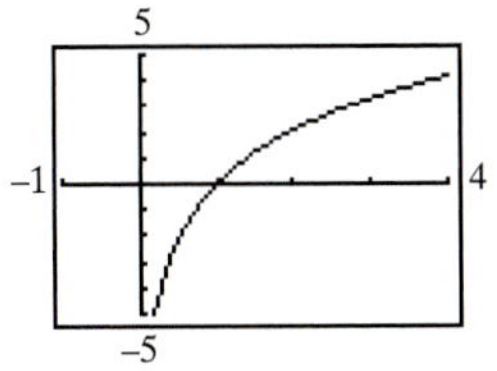

Domain: $(0, \infty)$
Range: $(-\infty, \infty)$
Vertical Asymptote: $x = 0$

75.

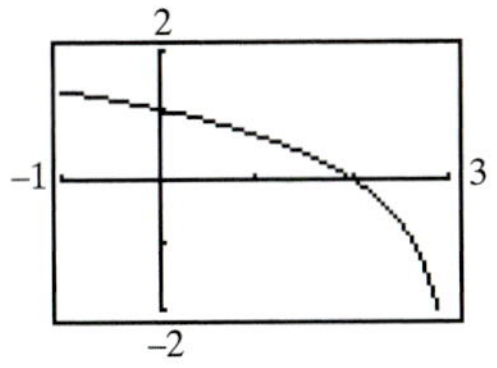

Domain: $(-\infty, 3)$
Range: $(-\infty, \infty)$
Vertical Asymptote: $x = 3$

77. (a) $n = 6.93$ so 7 panes are necessary **(b)** $n = 13.86$ so 14 panes are necessary **79. (a)** $d = 127.7$ so it takes about 128 days **(b)** $d = 575.6$ so it takes about 576 days **81.** $h = 2.29$ so the time between injections is about 2 hours, 20 minutes

83. 0.2695 sec
0.8959 sec

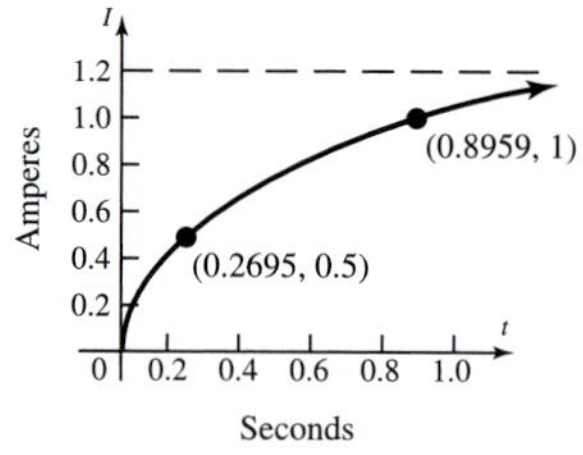

85. (a) $k = 20.07$
(b) 91%
(c) 0.175
(d) 0.08

87. $y = 20e^{0.023t}$
$y = 89.2$ is predicted

5.4 Exercises

1. $a + b$ **3.** $b - a$ **5.** $a + 1$ **7.** $2a + b$ **9.** $\frac{1}{5}(a + 2b)$ **11.** $\frac{b}{a}$ **13.** $2 \ln x + \frac{1}{2}\ln(1 - x)$ **15.** $3 \log_2 x - \log_2(x - 3)$
17. $\log x + \log(x + 2) - 2 \log(x + 3)$ **19.** $\frac{1}{3}\ln(x - 2) + \frac{1}{3}\ln(x + 1) - \frac{2}{3}\ln(x + 4)$ **21.** $\ln 5 + \ln x + \frac{1}{2}\ln(1 - 3x) - 3 \ln(x - 4)$
23. $\log_5 u^3v^4$ **25.** $-\frac{5}{2}\log_{1/2}x$ **27.** $-2 \ln(x - 1)$ **29.** $\log_2[x(3x - 2)^4]$ **31.** $\log_a\left[\frac{25x^6}{(2x + 3)^{1/2}}\right]$ **33.** $\ln y = \ln a + (\ln b)x$
35. 2.771 **37.** −3.880 **39.** 5.615 **41.** 0.874

43.

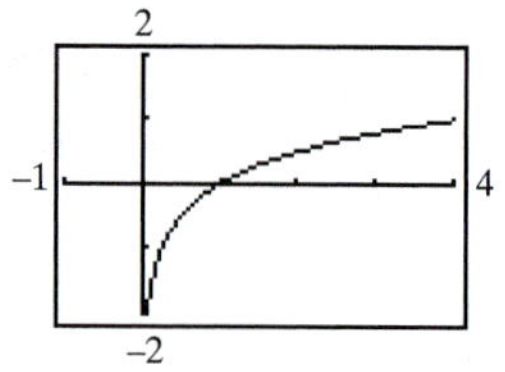

45.

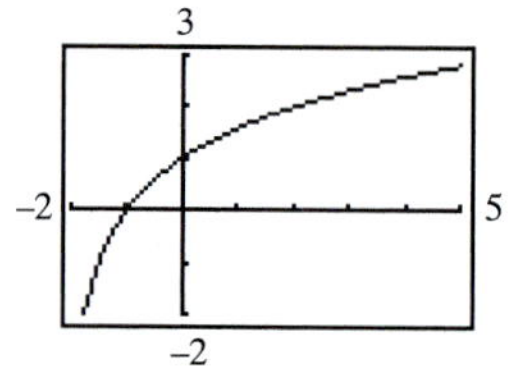

47.

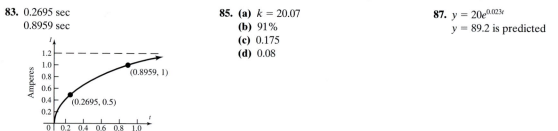

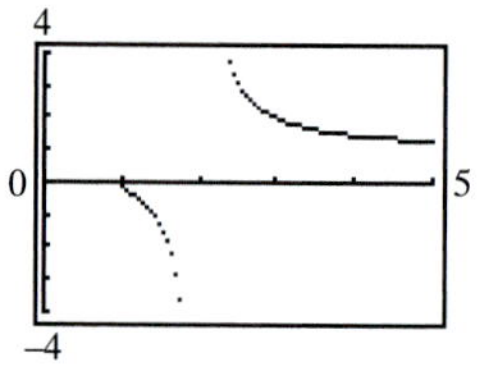

49. $y = Cx$ **51.** $y = Cx(x + 1)$ **53.** $y = Ce^{3x}$ **55.** $y = Ce^{-4x} + 3$ **57.** $y = \frac{\sqrt[3]{C}(2x + 1)^{1/6}}{(x + 4)^{1/9}}$ **59.** 3 **61.** 1
63. If $A = \log_a M$ and $B = \log_a N$, then $a^A = M$ and $a^B = N$. Then $\log_a (M/N) = \log_a (a^A/a^B) = \log_a a^{A - B} = A - B = \log_a M - \log_a N$.

5.5 Exercises

1. $\frac{7}{2}$ **3.** $\{-2\sqrt{2}, 2\sqrt{2}\}$ **5.** 16 **7.** 8 **9.** 3 **11.** 5 **13.** 2 **15.** $\{-2, 4\}$ **17.** 21 **19.** $\frac{1}{2}$ **21.** $\{-\sqrt{2}, 0, \sqrt{2}\}$ **23.** $\left\{1 - \frac{\sqrt{6}}{3}, 1 + \frac{\sqrt{6}}{3}\right\}$
25. 0 **27.** ≈ 1.585 **29.** 0 **31.** $\frac{3}{2}$ **33.** 3.322 **35.** −0.088 **37.** 0.307 **39.** 1.356 **41.** 0 **43.** 0.534 **45.** 0.226 **47.** 2.027 **49.** $\frac{9}{2}$ **51.** 2
53. −1 **55.** 1 **57.** 16 **59.** 1.92 **61.** 2.78 **63.** −0.56 **65.** −0.70 **67.** 0.56 **69.** $\{0.39, 1.00\}$ **71.** 1.31 **73.** 1.30

5.6 Exercises

1. \$108.29 **3.** \$609.50 **5.** \$697.09 **7.** \$12.46 **9.** \$125.23 **11.** \$88.72 **13.** \$860.72 **15.** \$554.09 **17.** \$59.71 **19.** \$361.93
21. 5.35% **23.** 26% **25.** $6\frac{1}{4}$% compounded annually **27.** 9% compounded monthly **29.** 104.32 mo; 103.97 mo
31. 61.02 mo; 60.82 mo **33.** 15.27 yrs or 15 yrs, 4 months **35.** \$104,335 **37.** \$12,910.62 **39.** About \$30.17 per share or \$3017
41. 9.35% **43.** Not quite. Jim will have \$1057.60. The second bank gives a better deal, since Jim will have \$1060.62 after 1 year
45. Will has \$11,632.73; Henry has \$10,947.89
47. (a) Interest is \$30,000 (b) Interest is \$38,613.59 (c) Interest is \$37,752.73. Simple interest at 12% is best

49. **(a)** \$1364.62 **(b)** \$1353.35 **51.** \$4631.93 **59.** **(a)** 6.1 years **(b)** 18.45 years **(c)**

$$mP = P\left(1 + \frac{r}{n}\right)^{nt}$$

$$m = \left(1 + \frac{r}{n}\right)^{nt}$$

$$\ln m = \ln\left(1 + \frac{r}{n}\right)^{nt} = nt\ln\left(1 + \frac{r}{n}\right)$$

$$t = \frac{\ln m}{n\ln\left(1 + \frac{r}{n}\right)}$$

5.7 Exercises

1. **(a)** 34.7 days; 69.3 days **(b)**

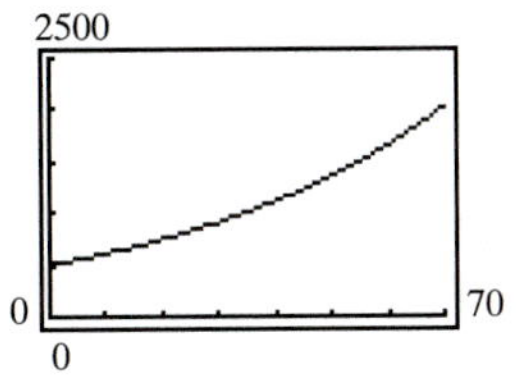

3. 28.4 years **5.** **(a)** 94.4 years **(b)**

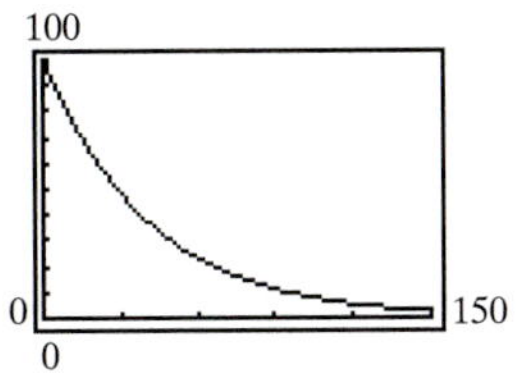

7. 5832; 3.9 days **9.** 25,198 **11.** 9.797 g **13.** **(a)** 9727 years ago **(b)**

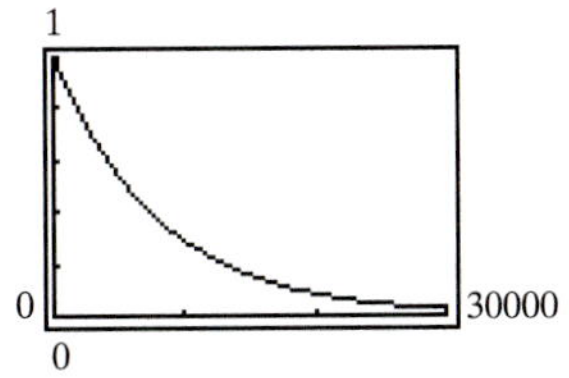

(c) 5600 years

15. **(a)** 5:18 PM **(b)**

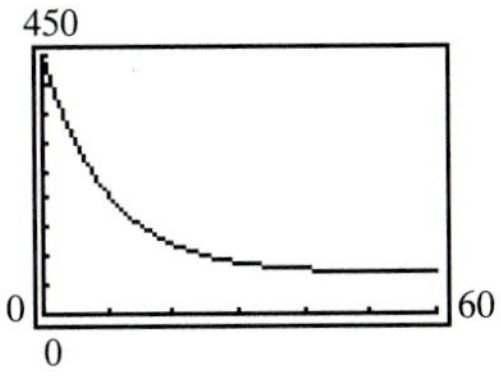

(c) After 14.3 minutes, the pizza will be 160°F.
(d) As time passes, the temperature of the pizza gets closer to 70°F.

17. **(a)** 18.63°C; 25.1°C **(b)**

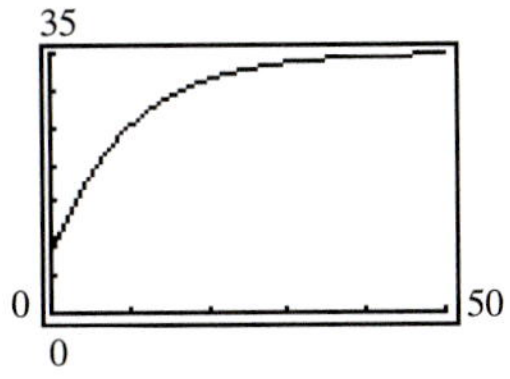

19. 7.34 kg; 76.6 hr **21.** 26.5 days.

23. **(a)** 0.1286 **(b)** 0.9 **(c)**

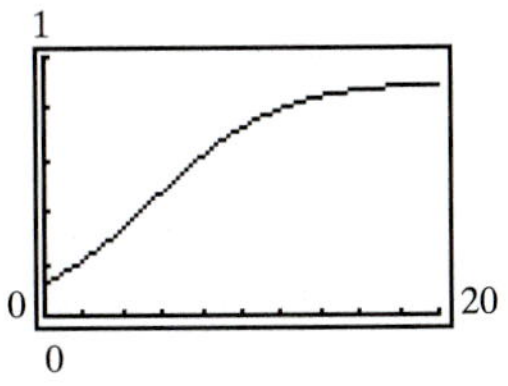

(d) 1996 **25.** **(a)**

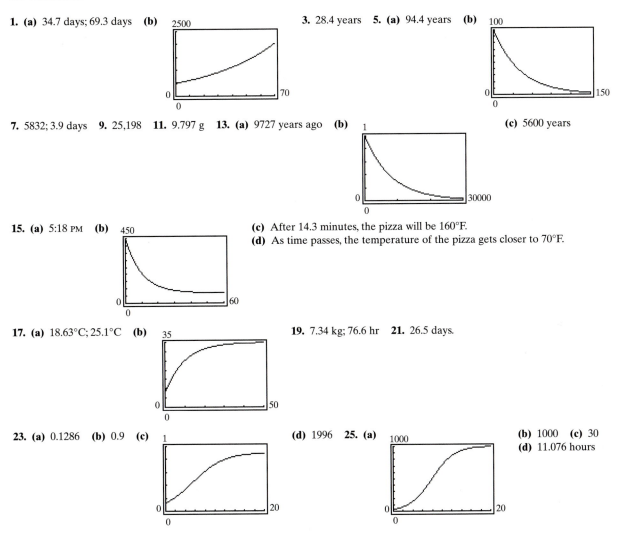

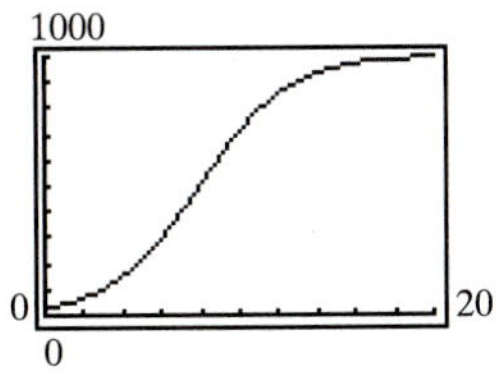

(b) 1000 **(c)** 30
(d) 11.076 hours

5.8 Exercises

1. (a)

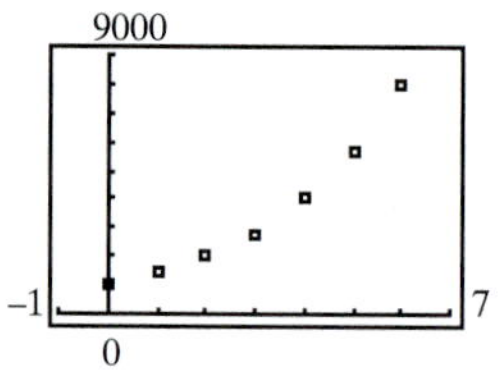

(b) $y = 1000.188\,(1.414)^x$ **(c)** $N = 1000.188e^{0.3465t}$ **(d)**

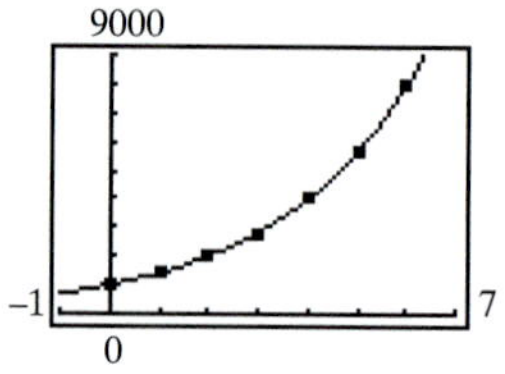

(e) 11,304

3. (a)

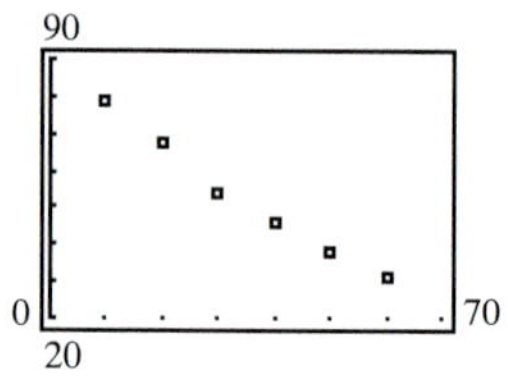

(b) $y = 96.01\,(0.9815)^x$ **(c)**

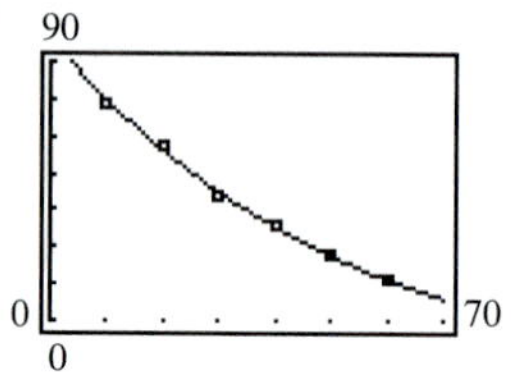

(d) $Q = 25$

5. (a)

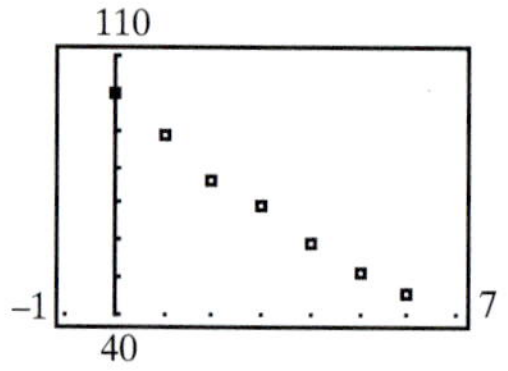

(b) $y = 100.326\,(0.877)^x$ **(c)** $A = 100.326e^{-0.1314t}$ **(d)**

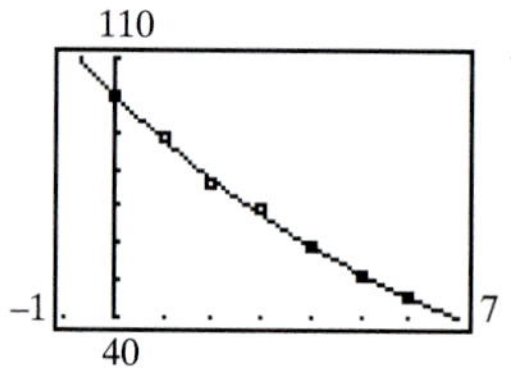

(e) 5.3 weeks **(f)** 0.14 grams

7. (a) x is the number of years after 1985

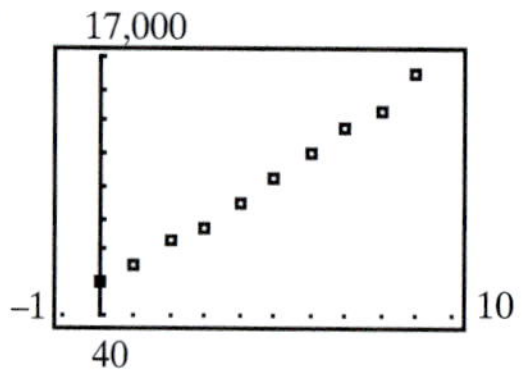

(b) $y = 10{,}014.4963\,(1.05655)^x$ **(c)** 5.655% **(d)** $68,672.21

9. (a)

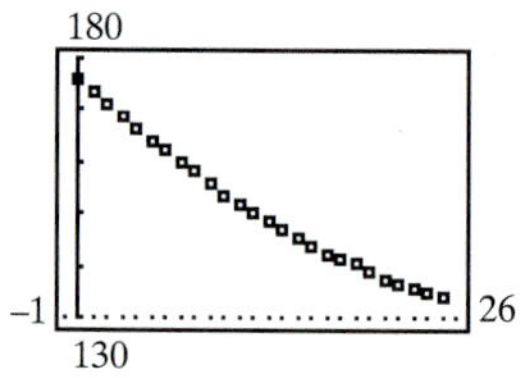

(b) $y = 173.2867\,(0.98899)^x$ **(c)**

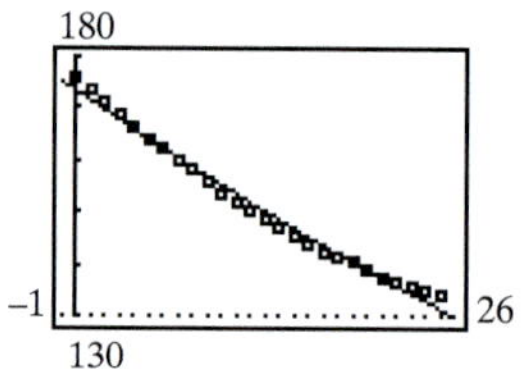

(d) 41.05

11. (a)

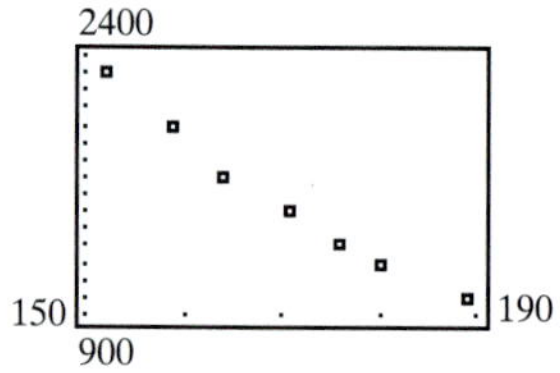

(b) $y = 32{,}741.02 - 6070.96 \ln x$ **(c)**

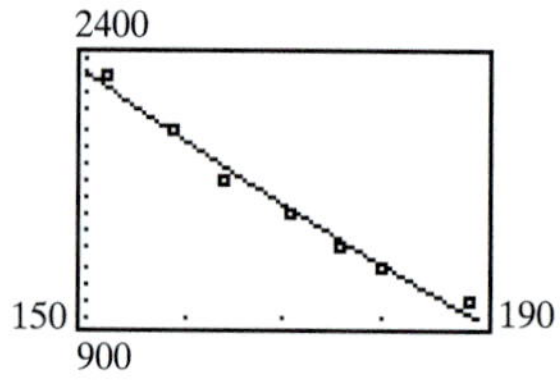

(d) Approximately 168 computers

13. (a)

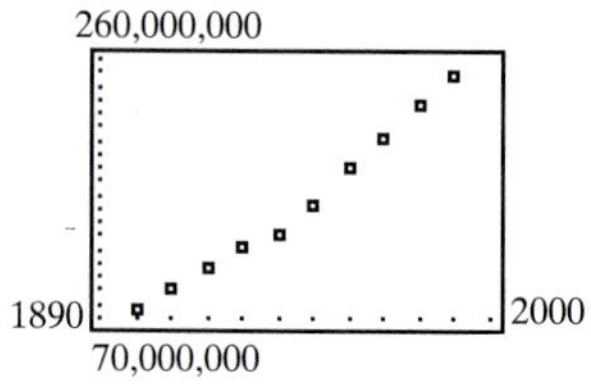

(b) $y = \dfrac{695{,}129{,}657.2}{1 + 4.1136 \times 10^{14} e^{-0.0166x}}$ **(c)**

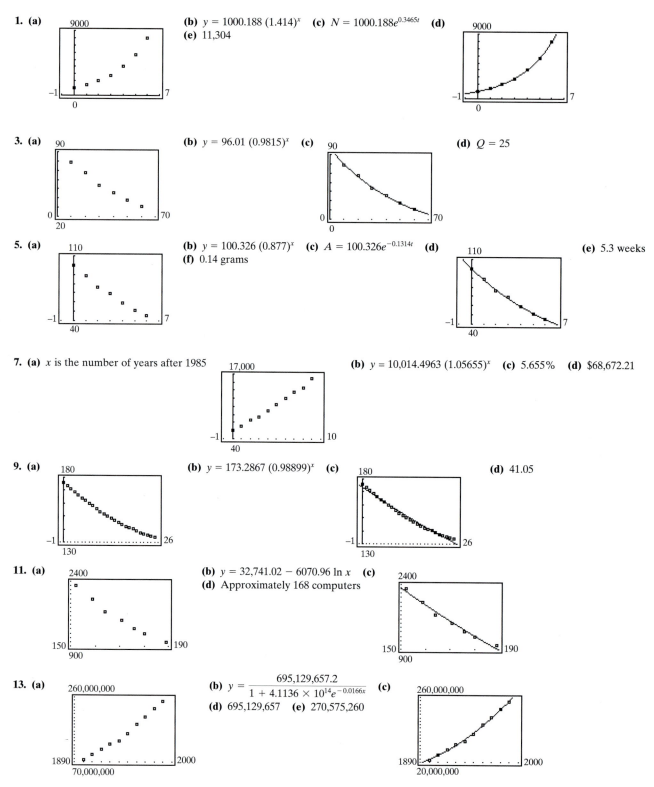

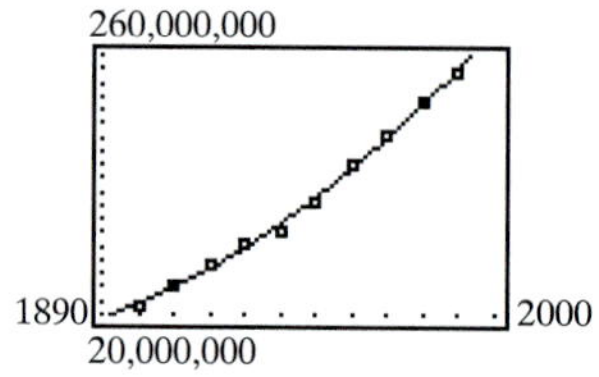

(d) 695,129,657 **(e)** 270,575,260

15. (a)

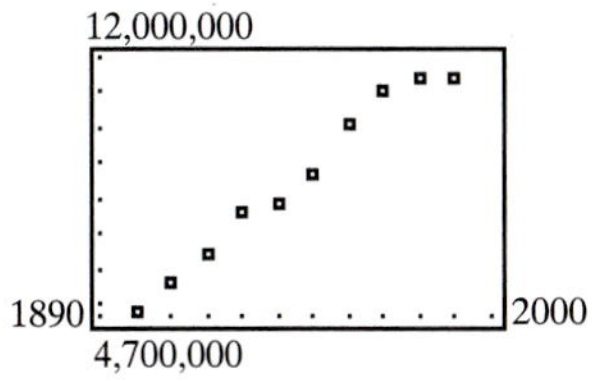

(b) $y = \dfrac{14{,}168{,}687.34}{1 + 1.5075745 \times 10^{21} e^{-0.02531x}}$ **(c)**

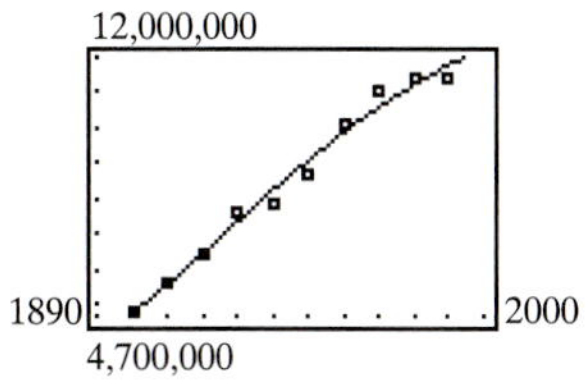

(d) 14,168,687 **(e)** Approximately 12,252,199

17. (a)

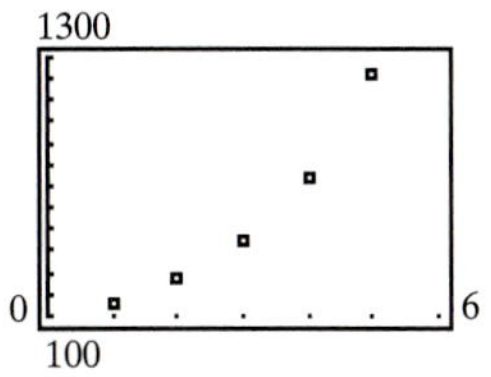

(b) Exponential model: $y = 99.06645375\,(1.652917358)^x$

19. (a)

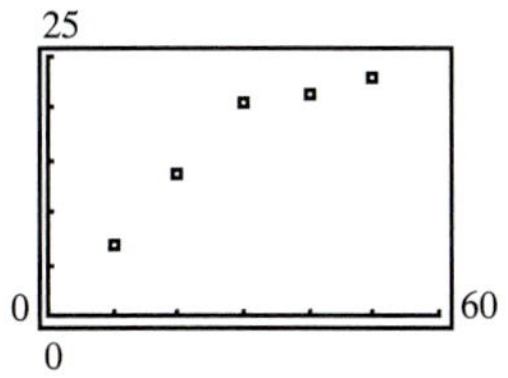

(b) Logarithmic model: $y = -16.55688464 + 10.31473128 \ln x$

21. (a)

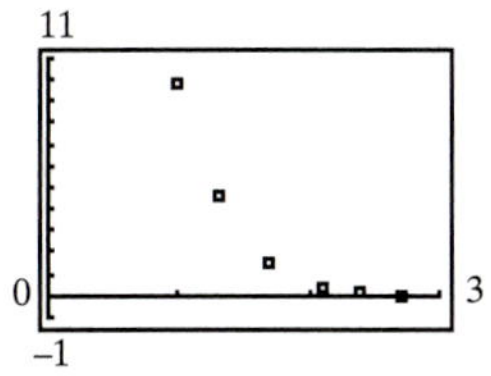

(b) Exponential model: $y = 346.113717\,(0.0352268374)^x$

23. (a)

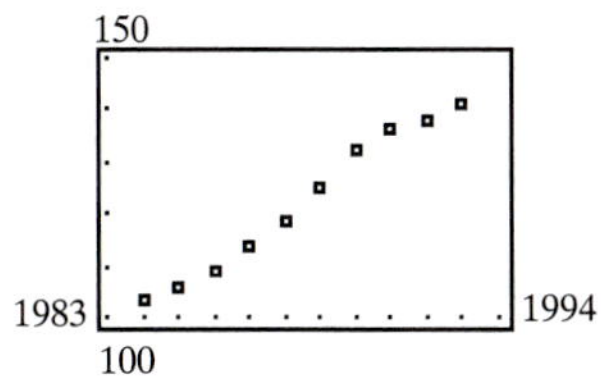

(b) Logarithmic model: $\text{CPI} = -69{,}946.44499 + 9225.465596 \ln t$ **(c)** 147.7

25. (a)

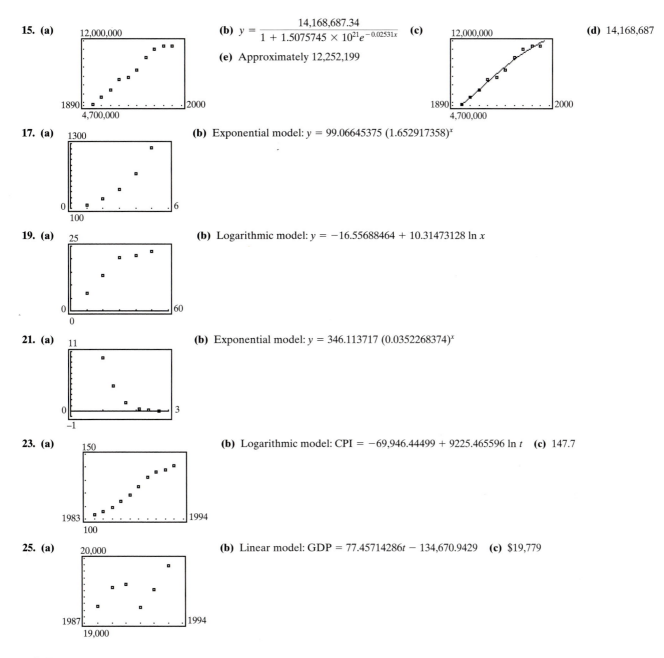

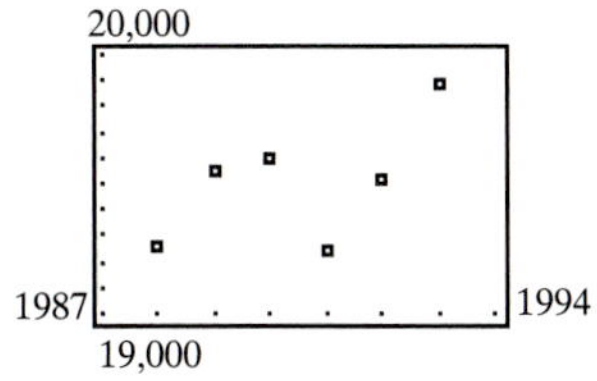

(b) Linear model: $\text{GDP} = 77.45714286t - 134{,}670.9429$ **(c)** $19,779

5.9 Exercises

1. 70 decibels **3.** 111.76 decibels **5.** 10 watts/m^2 **7.** 4.0 on the Richter scale
9. 125,892.54 mm; the Mexico City earthquake was 15.85 times as intense as the one in San Francisco.

Fill-in-the-Blank Items

1. one-to-one **2.** $y = x$ **3.** $(0, 1)$ and $(1, a)$ **4.** 1 **5.** 4 **6.** sum **7.** 1 **8.** 7 **9.** $x > 0$ **10.** $(1, 0)$ and $(a, 1)$ **11.** 1 **12.** 7

True/False Items

1. F **2.** T **3.** T **4.** T **5.** F **6.** F **7.** T **8.** F **9.** F **10.** T

Review Exercises

1. $f^{-1}(x) = \dfrac{2x+3}{5x-2}$; $f(f^{-1}(x)) = \dfrac{2\left(\dfrac{2x+3}{5x-2}\right)+3}{5\left(\dfrac{2x+3}{5x-2}\right)-2} = x$; $f^{-1}(f(x)) = \dfrac{2\left(\dfrac{2x+3}{5x-2}\right)+3}{5\left(\dfrac{2x+3}{5x-2}\right)-2} = x$;

Domain f = Range f^{-1} = All real numbers except $\frac{2}{5}$; Range f = Domain f^{-1} = All real numbers except $\frac{2}{5}$

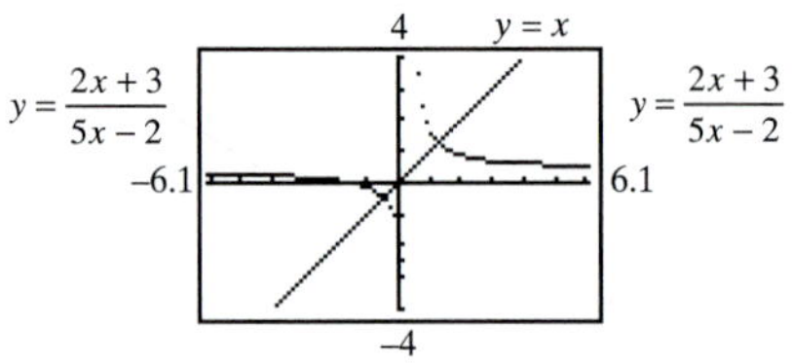

3. $f^{-1}(x) = \dfrac{x+1}{x}$; $f(f^{-1}(x)) = \dfrac{1}{\dfrac{x+1}{x}-1} = x$; $f^{-1}(f(x)) = \dfrac{\dfrac{1}{x-1}+1}{\dfrac{1}{x-1}} = x$;

Domain f = Range f^{-1} = All real numbers except 1; Range f = Domain f^{-1} = All real numbers except 0

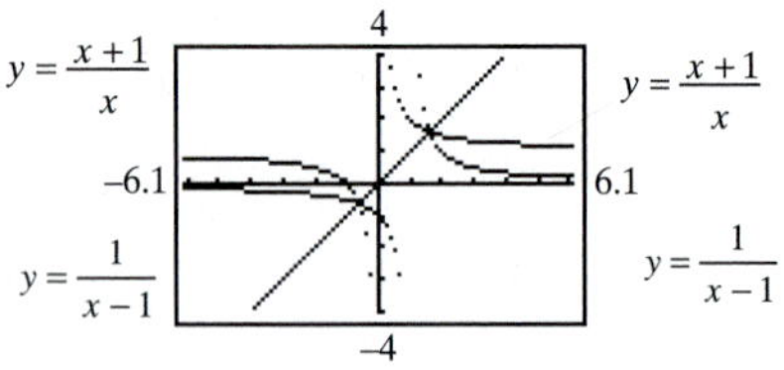

5. $f^{-1}(x) = \dfrac{27}{x^3}$; $f(f^{-1}(x)) = \dfrac{3}{(27/x^3)^{1/3}} = x$; $f^{-1}(f(x)) = \dfrac{27}{(3/x^{1/3})^3} = x$;

Domain f = Range f^{-1} = All real numbers except 0; Range f = Domain f^{-1} = All real numbers except 0

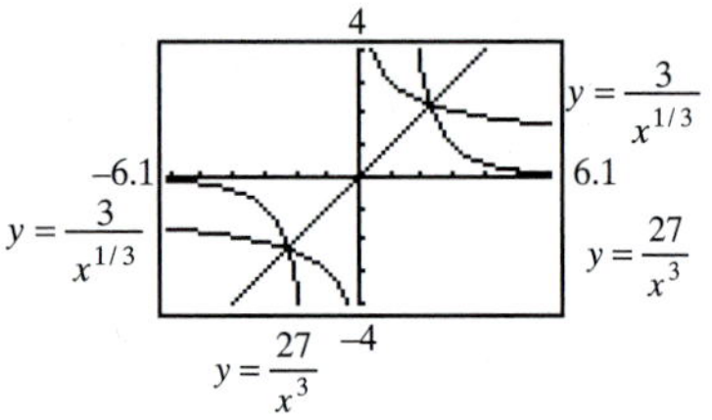

7. -3 **9.** $\sqrt{2}$ **11.** 0.4 **13.** $\frac{25}{4}\log_4 x$ **15.** $\ln\left[\dfrac{1}{(x+1)^2}\right] = -2\ln(x+1)$ **17.** $\log\left(\dfrac{4x^3}{[(x+3)(x-2)]^{1/2}}\right)$ **19.** $y = Ce^{2x^2}$

21. $y = (Ce^{3x^2})^2$ **23.** $y = \sqrt{e^{x+C}+9}$ **25.** $y = \ln(x^2+4) - C$

27.

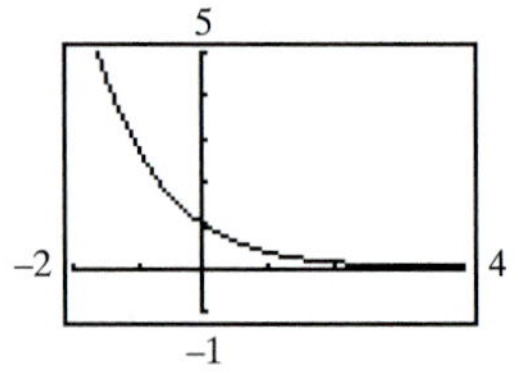

Domain: $(-\infty, \infty)$
Range: $(0, \infty)$
Asymptote: $y = 0$

29.

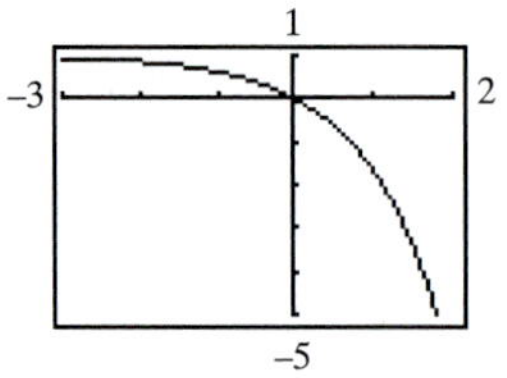

Domain: $(-\infty, \infty)$
Range: $(-\infty, 1)$
Asymptote: $y = 1$

31.

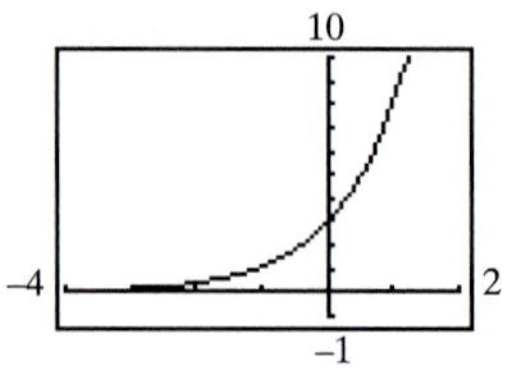

Domain: $(-\infty, \infty)$
Range: $(0, \infty)$
Asymptote: $y = 0$

33.

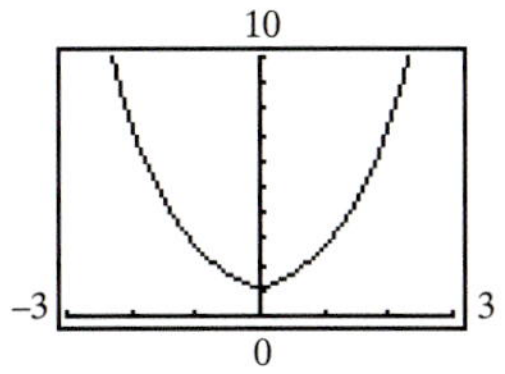

Domain: $(-\infty, \infty)$
Range: $[1, \infty)$
Asymptote: none

35.

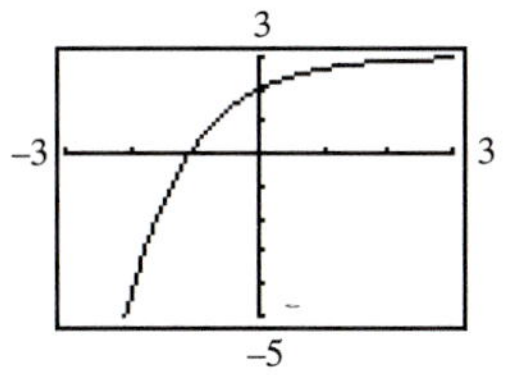

Domain: $(-\infty, \infty)$
Range: $(-\infty, 3)$
Asymptote: $y = 3$

37. $\frac{1}{4}$ **39.** $\left\{\frac{-1-\sqrt{3}}{2}, \frac{-1+\sqrt{3}}{2}\right\}$ **41.** $\frac{1}{4}$ **43.** 4.301 **45.** $\frac{12}{5}$ **47.** 83 **49.** $\left\{-3, \frac{1}{2}\right\}$ **51.** -1 **53.** -0.609 **55.** -9.327
57. 3229.5 m **59.** 7.6 mm of mercury **61.** **(a)** 37.3 watts **(b)** 6.9 decibels
63. **(a)** 71% **(b)** 85.5% **(c)** 90% **(d)** About 1.6 months **(e)** About 4.8 months
65. **(a)** 9.85 years **(b)** 4.27 **67.** \$41,669 **69.** 80 decibels **71.** 24,203 years ago
73. **(a)**

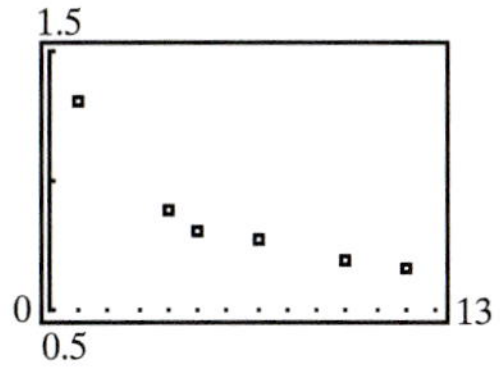

(b) Power model: $y = 1.298985429x^{-0.2721702954}$

75. **(a)**

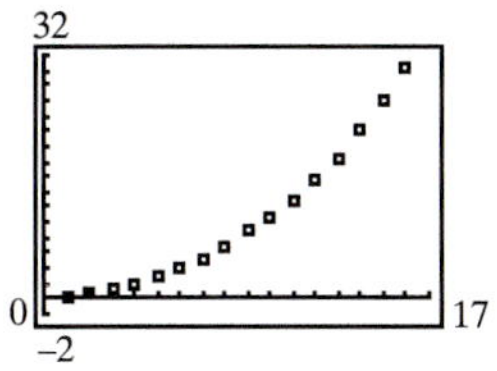

(b) $y = \dfrac{55.24286446}{1 + 69.63173349e^{-0.2760889967x}}$ **(c)**

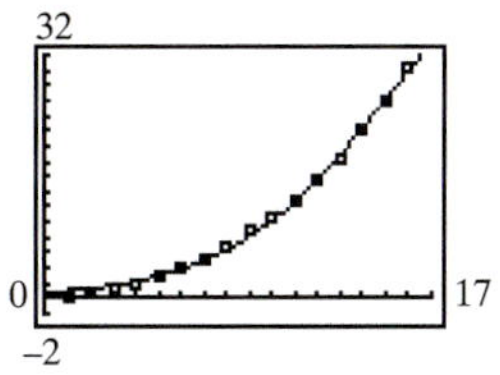

(d) Approximately 55.24 million **(e)** $y = 0.4510626632\,(1.339709)^x$ **(f)**

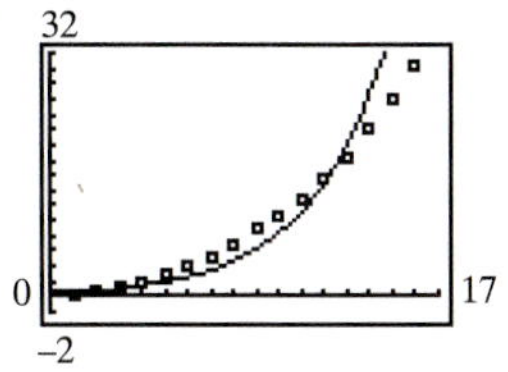

(g) Logistic model

CHAPTER 6 *6.1 Exercises*

1.

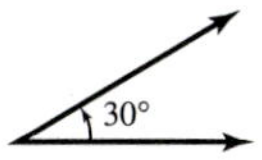

3.

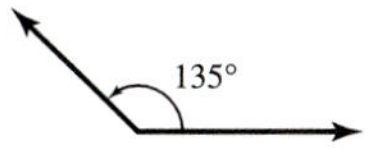

5.

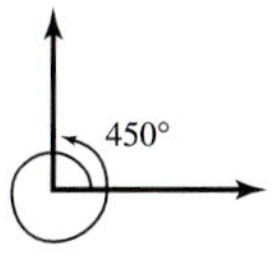

7.

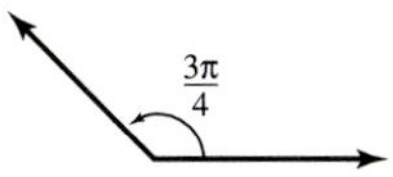

9.

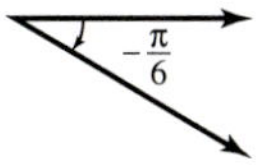

11.

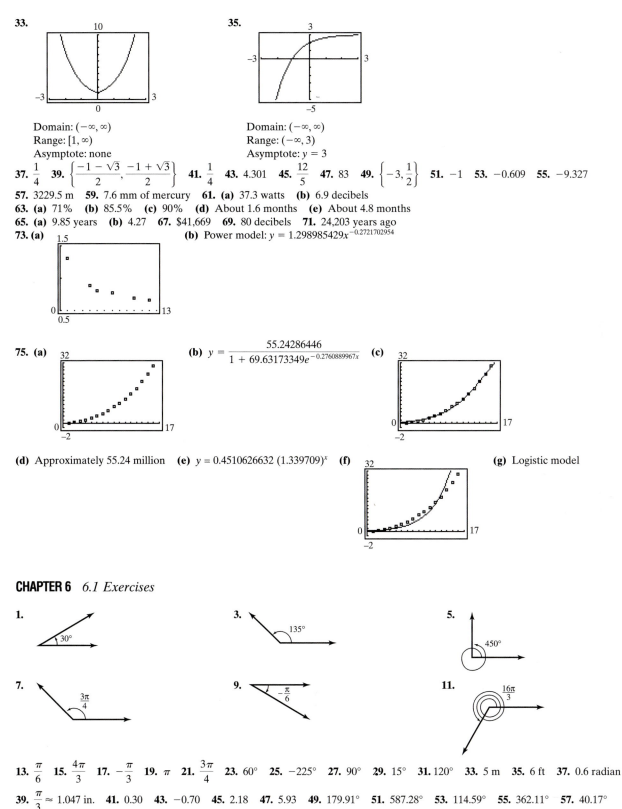

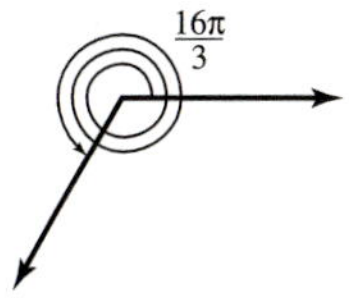

13. $\frac{\pi}{6}$ **15.** $\frac{4\pi}{3}$ **17.** $-\frac{\pi}{3}$ **19.** π **21.** $\frac{3\pi}{4}$ **23.** 60° **25.** $-225°$ **27.** 90° **29.** 15° **31.** 120° **33.** 5 m **35.** 6 ft **37.** 0.6 radian
39. $\frac{\pi}{3} \approx 1.047$ in. **41.** 0.30 **43.** -0.70 **45.** 2.18 **47.** 5.93 **49.** 179.91° **51.** 587.28° **53.** 114.59° **55.** 362.11° **57.** 40.17°
59. 1.03° **61.** 9.15° **63.** 40°19′12″ **65.** 18°15′18″ **67.** 19°59′24″ **69.** $3\pi \approx 9.4248$ in.; $5\pi \approx 15.7080$ in.

71. $\omega = \frac{1}{60}$ radian/sec; $v = \frac{1}{12}$ cm/sec **73.** 452.5 rpm **75.** 37.7 in. **77.** 2292 mph **79.** $\frac{3}{4}$ rpm **81.** 2.86 mph
83. 31.47 revolutions/minute **85.** 1037 mi/hr

6.2 Exercises

1. $\sin\theta = \frac{5}{13}$; $\cos\theta = \frac{12}{13}$; $\tan\theta = \frac{5}{12}$; $\csc\theta = \frac{13}{5}$; $\sec\theta = \frac{13}{12}$; $\cot\theta = \frac{12}{5}$
3. $\sin\theta = \frac{2\sqrt{13}}{13}$; $\cos\theta = \frac{3\sqrt{13}}{13}$; $\tan\theta = \frac{2}{3}$; $\csc\theta = \frac{\sqrt{13}}{2}$; $\sec\theta = \frac{\sqrt{13}}{3}$; $\cot\theta = \frac{3}{2}$
5. $\sin\theta = \frac{\sqrt{3}}{2}$; $\cos\theta = \frac{1}{2}$; $\tan\theta = \sqrt{3}$; $\csc\theta = \frac{2\sqrt{3}}{3}$; $\sec\theta = 2$; $\cot\theta = \frac{\sqrt{3}}{3}$
7. $\sin\theta = \frac{\sqrt{6}}{3}$; $\cos\theta = \frac{\sqrt{3}}{3}$; $\tan\theta = \sqrt{2}$; $\csc\theta = \frac{\sqrt{6}}{2}$; $\sec\theta = \sqrt{3}$; $\cot\theta = \frac{\sqrt{2}}{2}$
9. $\sin\theta = \frac{\sqrt{5}}{5}$; $\cos\theta = \frac{2\sqrt{5}}{5}$; $\tan\theta = \frac{1}{2}$; $\csc\theta = \sqrt{5}$; $\sec\theta = \frac{\sqrt{5}}{2}$; $\cot\theta = 2$ **11.** $\tan\theta = \frac{\sqrt{3}}{3}$; $\csc\theta = 2$; $\sec\theta = \frac{2\sqrt{3}}{3}$; $\cot\theta = \sqrt{3}$
13. $\tan\theta = \frac{2\sqrt{5}}{5}$; $\csc\theta = \frac{3}{2}$; $\sec\theta = \frac{3\sqrt{5}}{5}$; $\cot\theta = \frac{\sqrt{5}}{2}$ **15.** $\cos\theta = \frac{\sqrt{2}}{2}$; $\tan\theta = 1$; $\csc\theta = \sqrt{2}$; $\sec\theta = \sqrt{2}$; $\cot\theta = 1$
17. $\sin\theta = \frac{2\sqrt{2}}{3}$; $\tan\theta = 2\sqrt{2}$; $\csc\theta = \frac{3\sqrt{2}}{4}$; $\sec\theta = 3$; $\cot\theta = \frac{\sqrt{2}}{4}$ **19.** $\sin\theta = \frac{\sqrt{5}}{5}$; $\cos\theta = \frac{2\sqrt{5}}{5}$; $\csc\theta = \sqrt{5}$; $\sec\theta = \frac{\sqrt{5}}{2}$; $\cot\theta = 2$
21. $\sin\theta = \frac{2\sqrt{2}}{3}$; $\cos\theta = \frac{1}{3}$; $\tan\theta = 2\sqrt{2}$; $\csc\theta = \frac{3\sqrt{2}}{4}$; $\cot\theta = \frac{\sqrt{2}}{4}$ **23.** $\sin\theta = \frac{\sqrt{6}}{3}$; $\cos\theta = \frac{\sqrt{3}}{3}$; $\csc\theta = \frac{\sqrt{6}}{2}$; $\sec\theta = \sqrt{3}$; $\cot\theta = \frac{\sqrt{2}}{2}$
25. 1 **27.** 1 **29.** 0 **31.** 0 **33.** 1 **35.** 0 **37.** 0 **39.** 1 **41.** **(a)** $\frac{1}{2}$ **(b)** $\frac{3}{4}$ **(c)** 2 **(d)** 2 **43.** **(a)** 17 **(b)** $\frac{1}{4}$ **(c)** 4 **(d)** $\frac{17}{16}$
45. **(a)** $\frac{1}{4}$ **(b)** 15 **(c)** 4 **(d)** $\frac{16}{15}$ **47.** **(a)** 0.78 **(b)** 0.79 **(c)** 1.27 **(d)** 1.27 **(e)** 1.61 **(f)** 0.78 **(g)** 0.62 **(h)** 1.27
49. 0.6 **51.** 20° **53.** **(a)** 10 min **(b)** 20 min **(c)** $T(\theta) = 5\left(1 - \frac{1}{3\tan\theta} + \frac{1}{\sin\theta}\right)$ **(d)** 15.8 min **(e)** 10.4 min
(f) 70.5°; 176.8 ft, 9.7 min

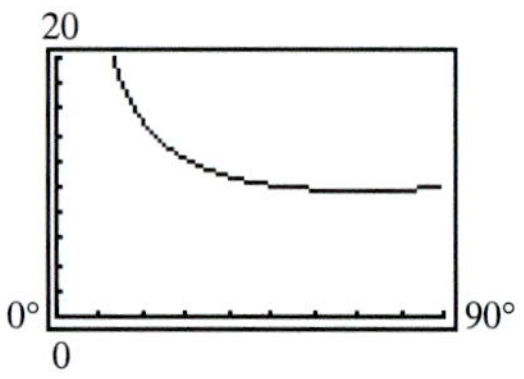

55. **(a)** $|OA| = |OC| = 1$; angle OAC + angle $OAC + 180° - \theta = 180°$; angle $OAC = \frac{\theta}{2}$
(b) $\sin\theta = \frac{|CD|}{|OC|} = |CD|$; $\cos\theta = \frac{|OD|}{|OC|} = |OD|$ **(c)** $\tan\frac{\theta}{2} = \frac{|CD|}{|AD|} = \frac{\sin\theta}{1 + |OD|} = \frac{\sin\theta}{1 + \cos\theta}$

57. $h = x\tan\theta$ and $h = (1 - x)\tan n\theta$; thus, $x\tan\theta = (1 - x)\tan n\theta$

$$x = \frac{\tan n\theta}{\tan\theta + \tan n\theta}$$

59. **(a)** Area $\triangle OAC = \frac{1}{2}|OC||AC| = \frac{1}{2}\cdot\frac{|OC|}{1}\cdot\frac{|AC|}{1} = \frac{1}{2}\sin\alpha\cos\alpha$
(b) Area $\triangle OCB = \frac{1}{2}|BC|\,|OC| = \frac{1}{2}|OB|^2\frac{|BC|}{|OB|}\cdot\frac{|OC|}{|OB|} = \frac{1}{2}|OB|^2\sin\beta\cos\beta$
(c) Area $\triangle OAB = \frac{1}{2}|BD|\,|OA| = \frac{1}{2}|OB|\frac{|BD|}{|OB|} = \frac{1}{2}|OB|\sin(\alpha + \beta)$
(d) $\frac{\cos\alpha}{\cos\beta} = \frac{\frac{|OC|}{1}}{\frac{|OC|}{|OB|}} = |OB|$ **(e)** Use the hint and above results.

61. $\sin\alpha = \tan\alpha\cos\alpha = \cos\beta\cos\alpha = \cos\beta\tan\beta = \sin\beta$;

$$\begin{aligned} \sin^2\alpha + \cos^2\alpha &= 1 \\ \sin^2\alpha + \tan^2\beta &= 1 \\ \sin^2\alpha + \frac{\sin^2\beta}{\cos^2\beta} &= 1 \\ \sin^2\alpha + \frac{\sin^2\alpha}{1-\sin^2\alpha} &= 1 \\ \sin^2\alpha - \sin^4\alpha + \sin^2\alpha &= 1 - \sin^2\alpha \\ \sin^4\alpha - 3\sin^2\alpha + 1 &= 0 \\ \sin^2\alpha &= \frac{3 \pm \sqrt{5}}{2} \\ \sin^2\alpha &= \frac{3 - \sqrt{5}}{2} \\ \sin\alpha &= \sqrt{\frac{3-\sqrt{5}}{2}} \end{aligned}$$

63. Since $a^2 + b^2 = c^2$, $a > 0$, $b > 0$, then $0 < a^2 < c^2$ or $0 < a < c$. Thus, $0 < \frac{a}{c} < 1$ and $0 < \sin\theta < 1$.

6.3 Exercises

1. $\sin 45° = \frac{\sqrt{2}}{2}$; $\cos 45° = \frac{\sqrt{2}}{2}$; $\tan 45° = 1$; $\cot 45° = 1$; $\sec 45° = \sqrt{2}$; $\csc 45° = \sqrt{2}$

3. $\sin 60° = \frac{\sqrt{3}}{2}$; $\cos 60° = \frac{1}{2}$; $\tan 60° = \sqrt{3}$; $\cot 60° = \frac{\sqrt{3}}{3}$; $\sec 60° = 2$; $\csc 60° = \frac{2\sqrt{3}}{3}$ **5.** 2 **7.** $\sqrt{2} + \frac{4\sqrt{3}}{3}$ **9.** $-\frac{8}{3}$ **11.** 1 **13.** 0

15. $\frac{\sqrt{3}}{2}$ **17.** $\frac{1}{2}$ **19.** $\frac{3}{4}$ **21.** $\sqrt{3}$ **23.** $\frac{\sqrt{3}}{4}$ **25.** 0.47 **27.** 0.38 **29.** 1.33 **31.** 0.36 **33.** 0.31 **35.** 3.73 **37.** 1.04 **39.** 5.67 **41.** 0.84

43. 0.02 **45.** 0.93 **47.** 0.31 **49.** $R \approx 310.56$ ft; $H \approx 77.64$ ft **51.** $R \approx 19{,}542$ m; $H \approx 2278$ m

53. (a) 1.2 sec **(b)** 1.12 sec **(c)** 1.2 sec **55. (a)** $T(\theta) = 1 + \frac{2}{3\sin\theta} - \frac{1}{4\tan\theta}$ **(b)** 1.9 hrs; 0.57 hrs **(c)** 1.69 hrs; 0.75 hrs

(d) 1.63 hrs; 0.86 hrs **(e)** 1.67 hrs **(f)** 2.75 hrs **(g)**

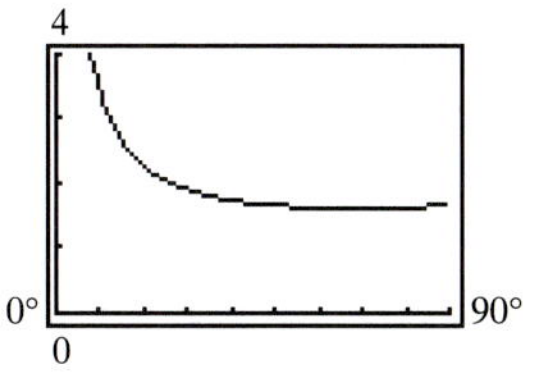

$\theta = 67.97°$ for least time; the least time is $T = 1.61$ hrs Sally is 0.9 hr on the road

57.

θ	0.5	0.4	0.2	0.1	0.01	0.001	0.0001	0.00001
$\sin\theta$	0.4794	0.3894	0.1987	0.0998	0.0100	0.0010	0.0001	0.00001
$\frac{\sin\theta}{\theta}$	0.9589	0.9735	0.9933	0.9983	1.0000	1.0000	1.0000	1.0000

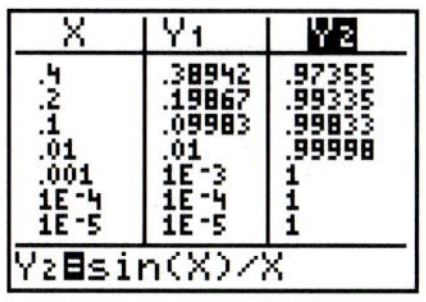

$\frac{(\sin\theta)}{\theta}$ approaches 1 as $\theta \to 0$.

59. 1 **61.** $\frac{\sqrt{2}}{2}$

6.4 Exercises

1. $\sin\theta = \frac{4}{5}$; $\cos\theta = -\frac{3}{5}$; $\tan\theta = -\frac{4}{3}$; $\csc\theta = \frac{5}{4}$; $\sec\theta = -\frac{5}{3}$; $\cot\theta = -\frac{3}{4}$

3. $\sin\theta = -\frac{3\sqrt{13}}{13}$; $\cos\theta = \frac{2\sqrt{13}}{13}$; $\tan\theta = -\frac{3}{2}$; $\csc\theta = -\frac{\sqrt{13}}{3}$; $\sec\theta = \frac{\sqrt{13}}{2}$; $\cot\theta = -\frac{2}{3}$

5. $\sin\theta = -\frac{\sqrt{2}}{2}$; $\cos\theta = -\frac{\sqrt{2}}{2}$; $\tan\theta = 1$; $\csc\theta = -\sqrt{2}$; $\sec\theta = -\sqrt{2}$; $\cot\theta = 1$

7. $\sin\theta = -\frac{2\sqrt{13}}{13}$; $\cos\theta = -\frac{3\sqrt{13}}{13}$; $\tan\theta = \frac{2}{3}$; $\csc\theta = -\frac{\sqrt{13}}{2}$; $\sec\theta = -\frac{\sqrt{13}}{3}$; $\cot\theta = \frac{3}{2}$

9. II **11.** IV **13.** IV **15.** III **17.** 30° **19.** 60° **21.** 30° **23.** $\frac{\pi}{4}$ **25.** $\frac{\pi}{3}$ **27.** 45° **29.** $\frac{\pi}{3}$ **31.** 80° **33.** $\frac{\sqrt{2}}{2}$ **35.** 1 **37.** 1

39. $\sqrt{3}$ **41.** $\frac{\sqrt{2}}{2}$ **43.** 0 **45.** $\sqrt{2}$ **47.** $\frac{\sqrt{3}}{3}$ **49.** $\frac{1}{2}$ **51.** $\frac{\sqrt{2}}{2}$ **53.** -2 **55.** $-\sqrt{3}$ **57.** $\frac{\sqrt{2}}{2}$ **59.** $\sqrt{3}$ **61.** $\frac{1}{2}$ **63.** $-\frac{\sqrt{3}}{2}$ **65.** $-\sqrt{3}$

67. $\sqrt{2}$ **69.** $\cos\theta = -\frac{5}{13}$; $\tan\theta = -\frac{12}{5}$; $\csc\theta = \frac{13}{12}$; $\sec\theta = -\frac{13}{5}$; $\cot\theta = -\frac{5}{12}$

71. $\sin\theta = -\frac{3}{5}$; $\tan\theta = \frac{3}{4}$; $\csc\theta = -\frac{5}{3}$; $\sec\theta = -\frac{5}{4}$; $\cot\theta = \frac{4}{3}$ **73.** $\cos\theta = -\frac{12}{13}$; $\tan\theta = -\frac{5}{12}$; $\csc\theta = \frac{13}{5}$; $\sec\theta = -\frac{13}{12}$; $\cot\theta = -\frac{12}{5}$

75. $\sin\theta = \frac{2\sqrt{2}}{3}$; $\tan\theta = -2\sqrt{2}$; $\csc\theta = \frac{3\sqrt{2}}{4}$; $\sec\theta = -3$; $\cot\theta = -\frac{\sqrt{2}}{4}$

77. $\cos\theta = -\frac{\sqrt{5}}{3}$; $\tan\theta = -\frac{2\sqrt{5}}{5}$; $\csc\theta = \frac{3}{2}$; $\sec\theta = -\frac{3\sqrt{5}}{5}$; $\cot\theta = -\frac{\sqrt{5}}{2}$

79. $\sin\theta = -\frac{\sqrt{3}}{2}$; $\cos\theta = \frac{1}{2}$; $\tan\theta = -\sqrt{3}$; $\csc\theta = -\frac{2\sqrt{3}}{3}$; $\cot\theta = -\frac{\sqrt{3}}{3}$

81. $\sin\theta = -\frac{3}{5}$; $\cos\theta = -\frac{4}{5}$; $\csc\theta = -\frac{5}{3}$; $\sec\theta = -\frac{5}{4}$; $\cot\theta = \frac{4}{3}$

83. $\sin\theta = \frac{\sqrt{10}}{10}$; $\cos\theta = -\frac{3\sqrt{10}}{10}$; $\csc\theta = \sqrt{10}$; $\sec\theta = -\frac{\sqrt{10}}{3}$; $\cot\theta = -3$ **85.** 0 **87.** -0.2 **89.** 3 **91.** 5 **93.** 0

95. **(a)** 16.6 ft **(b)**

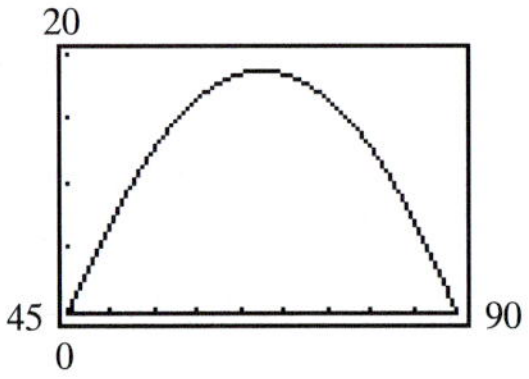

(c) 67.5°

6.5 Exercises

1. $\sin t = \frac{1}{2}$; $\cos t = \frac{\sqrt{3}}{2}$; $\tan t = \frac{\sqrt{3}}{3}$; $\csc t = 2$; $\sec t = \frac{2\sqrt{3}}{3}$; $\cot t = \sqrt{3}$

3. $\sin t = -\frac{\sqrt{3}}{2}$; $\cos t = -\frac{1}{2}$; $\tan t = \sqrt{3}$; $\csc t = -\frac{2\sqrt{3}}{3}$; $\sec t = -2$; $\cot t = \frac{\sqrt{3}}{3}$

5. $\sin t = \frac{4}{5}$; $\cos t = -\frac{3}{5}$; $\tan t = -\frac{4}{3}$; $\csc t = \frac{5}{4}$; $\sec t = -\frac{5}{3}$; $\cot t = -\frac{3}{4}$

7. $\sin t = -\frac{2\sqrt{5}}{5}$; $\cos t = \frac{\sqrt{5}}{5}$; $\tan t = -2$; $\csc t = -\frac{\sqrt{5}}{2}$; $\sec t = \sqrt{5}$; $\cot t = -\frac{1}{2}$

9. $\cos\theta = -\frac{\sqrt{5}}{3}$; $\tan\theta = -\frac{2\sqrt{5}}{5}$; $\csc\theta = \frac{3}{2}$; $\sec\theta = -\frac{3\sqrt{5}}{5}$; $\cot\theta = -\frac{\sqrt{5}}{2}$

11. $\sin\theta = -\frac{\sqrt{5}}{5}$; $\cos\theta = -\frac{2\sqrt{5}}{5}$; $\csc\theta = -\sqrt{5}$; $\sec\theta = -\frac{\sqrt{5}}{2}$; $\cot\theta = 2$

13. $\cos\theta = \frac{\sqrt{15}}{4}$; $\tan\theta = -\frac{\sqrt{15}}{15}$; $\csc\theta = -4$; $\sec\theta = \frac{4\sqrt{15}}{15}$; $\cot\theta = -\sqrt{15}$

15. $\sin\theta = -\frac{2\sqrt{2}}{3}$; $\cos\theta = \frac{1}{3}$; $\tan\theta = -2\sqrt{2}$; $\csc\theta = -\frac{3\sqrt{2}}{4}$; $\cot\theta = -\frac{\sqrt{2}}{4}$ **17.** All real numbers **19.** At odd multiples of $\frac{\pi}{2}$

21. At odd multiples of $\frac{\pi}{2}$ **23.** $[-1, 1]$ **25.** $(-\infty, \infty)$ **27.** $(-\infty, -1]$ or $[1, \infty)$ **29.** Odd; yes; origin **31.** Odd; yes; origin

33. Even; yes; y-axis

35. Let $P = (x, y)$ be the point on the unit circle that corresponds to t. Consider the equation $\tan t = \frac{y}{x} = a$. Then $y = ax$. But $x^2 + y^2 = 1$ so that $x^2 + a^2x^2 = 1$. Thus, $x = \pm\frac{1}{\sqrt{1 + a^2}}$ and $y = \pm\frac{a}{\sqrt{1 + a^2}}$; that is, for any real number a, there is a point $P = (x, y)$ on the unit circle for which $\tan t = a$. In other words, $-\infty < \tan t < \infty$, and the range of the tangent function is the set of all real numbers.

37. Suppose there is a number p, $0 < p < 2\pi$, for which $\sin(\theta + p) = \sin\theta$ for all θ. If $\theta = 0$, then $\sin(0 + p) = \sin p = \sin 0 = 0$; so that $p = \pi$. If $\theta = \frac{\pi}{2}$, the $\sin\left(\frac{\pi}{2} + p\right) = \sin\left(\frac{\pi}{2}\right)$. But $p = \pi$. Thus, $\sin\left(\frac{3\pi}{2}\right) = -1 = \sin\left(\frac{\pi}{2}\right) = 1$. This is impossible. The smallest positive number p for which $\sin(\theta + p) = \sin\theta$ for all θ is therefore $p = 2\pi$.

39. $\sec\theta = \frac{1}{\cos\theta}$; since $\cos\theta$ has period 2π, so does $\sec\theta$.

41. If $P = (a, b)$ is the point on the unit circle corresponding to θ, then $Q = (-a, -b)$ is the point on the unit circle corresponding to $\theta + \pi$. Thus, $\tan(\theta + \pi) = \frac{(-b)}{(-a)} = \frac{b}{a} = \tan\theta$; that is, the period of the tangent function is π.

43. $m = \frac{\sin\theta - 0}{\cos\theta - 0} = \frac{\sin\theta}{\cos\theta} = \tan\theta$

6.6 Exercises

1. 0 **3.** $-\frac{\pi}{2} \le x \le \frac{\pi}{2}$ **5.** 1 **7.** $0, \pi, 2\pi$ **9.** $\sin x = 1$ for $x = -\frac{3\pi}{2}, \frac{\pi}{2}$; $\sin x = -1$ for $x = -\frac{\pi}{2}, \frac{3\pi}{2}$ **11.** 0 **13.** 1

15. $\sec x = 1$ for $x = -2\pi, 0, 2\pi$; $\sec x = -1$ for $x = -\pi, \pi$ **17.** $-\frac{3\pi}{2}, -\frac{\pi}{2}, \frac{\pi}{2}, \frac{3\pi}{2}$ **19.** $-\frac{3\pi}{2}, -\frac{\pi}{2}, \frac{\pi}{2}, \frac{3\pi}{2}$ **21.** B, C, F **23.** C **25.** D

27. B **29.** A

31.

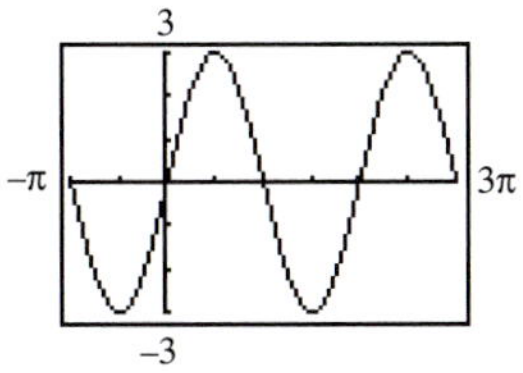

33.

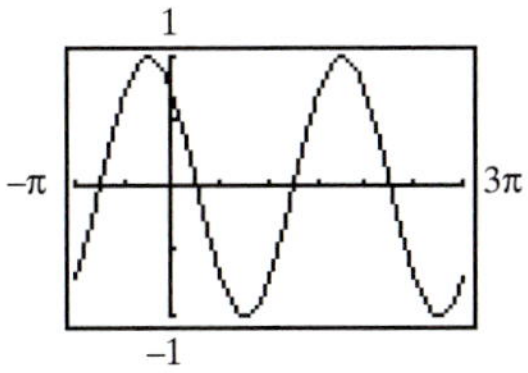

35.

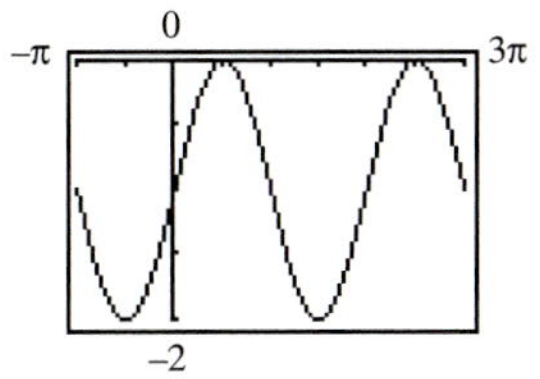

37.

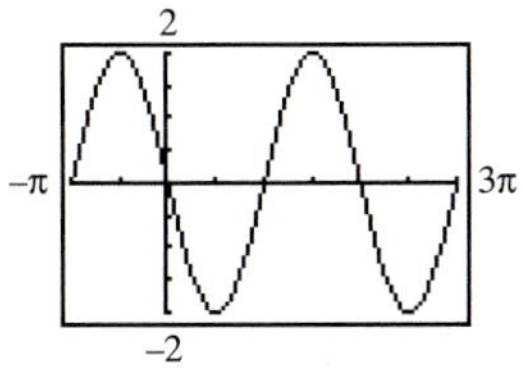

39.

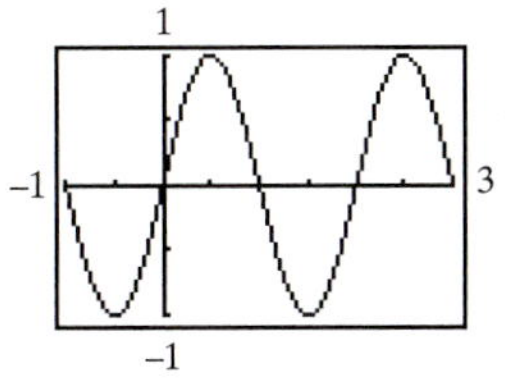

41.

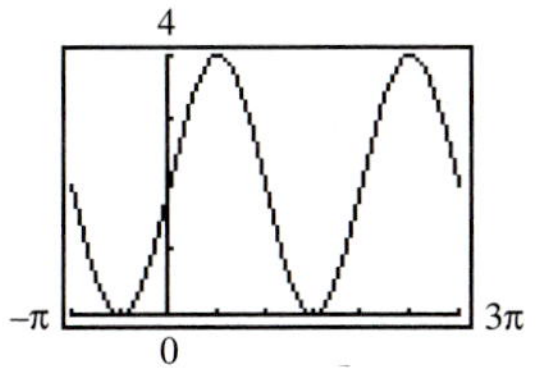

43.

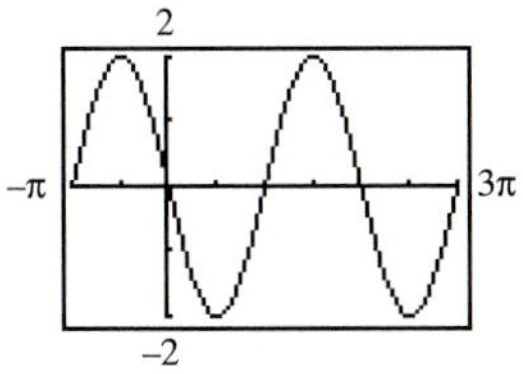

45.

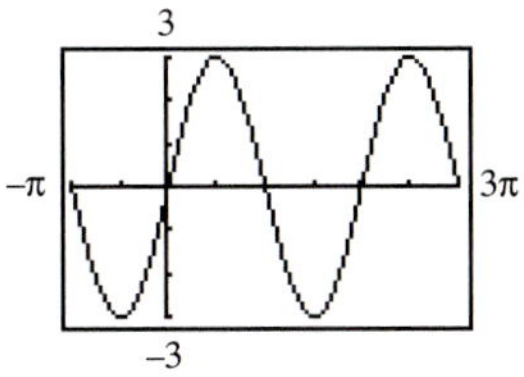

47.

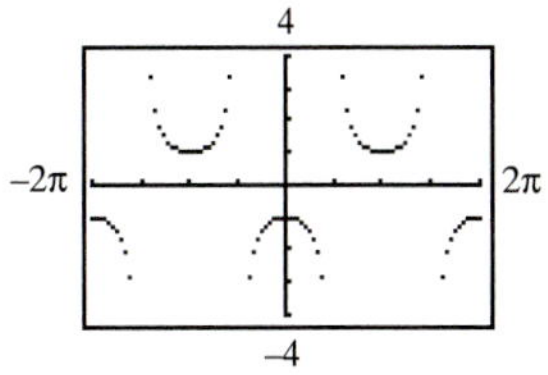

49.

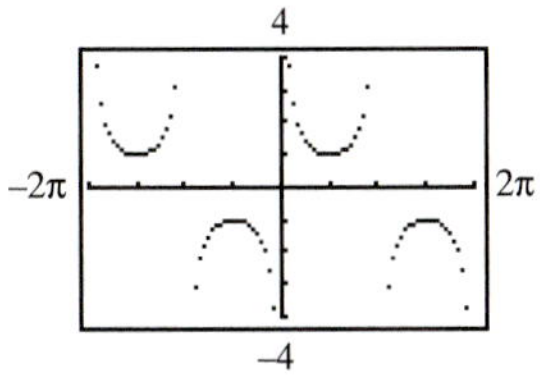

51.

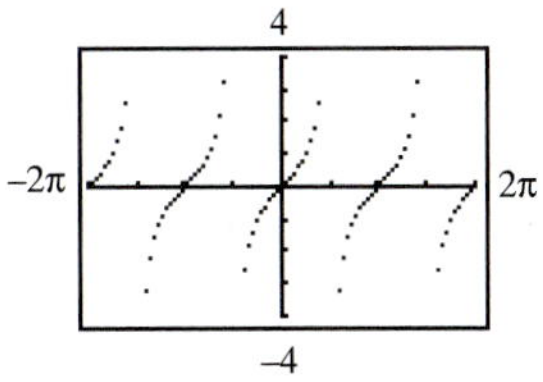

53.

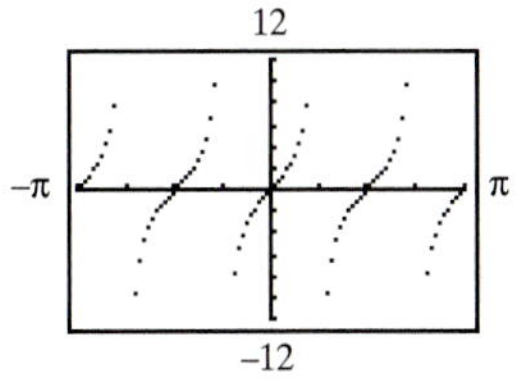

55.

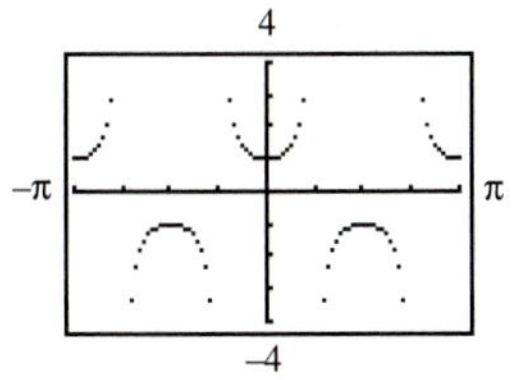

57.

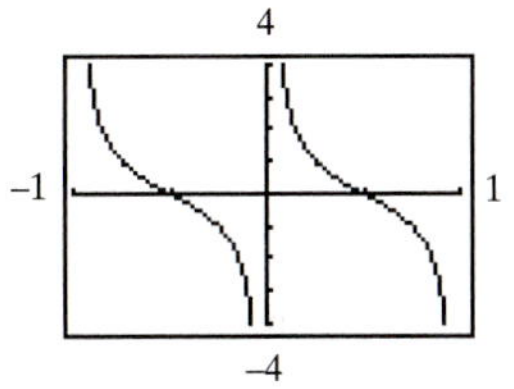

59.

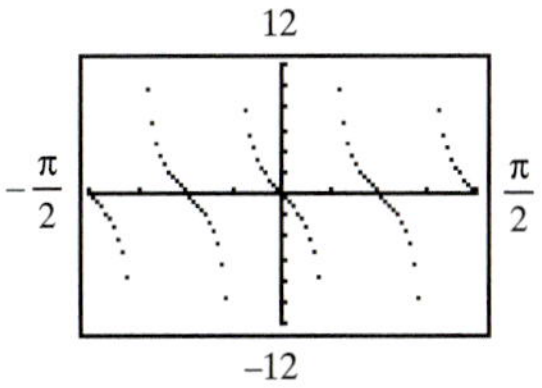

61.

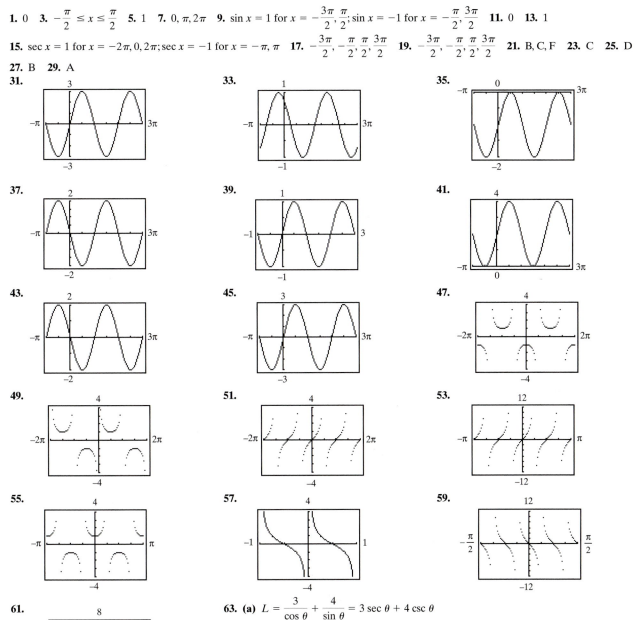

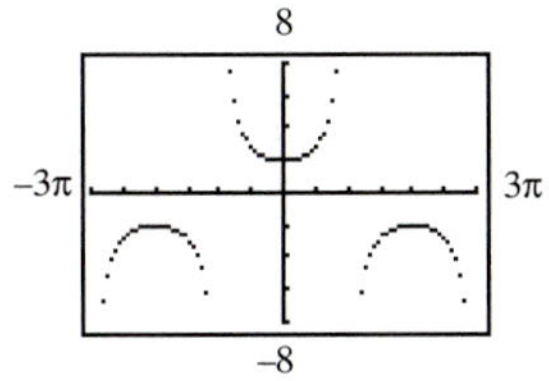

63. (a) $L = \frac{3}{\cos\theta} + \frac{4}{\sin\theta} = 3\sec\theta + 4\csc\theta$

(b)

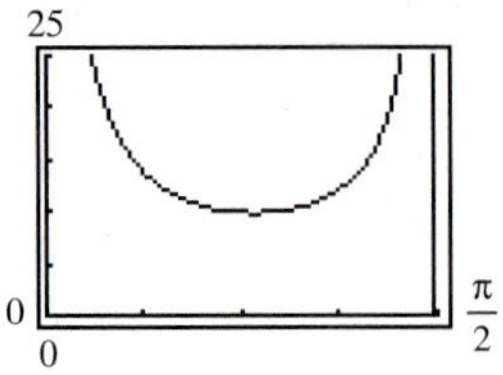

(c) L is smallest when $\theta = 0.83$ **(d)** $L(0.83) = 9.86$ ft.

65. $y = \sin \omega x$ has period $\frac{2\pi}{\omega}$ **67.** The graphs are the same; yes

6.7 Exercises

1. 0 **3.** $-\frac{\pi}{2}$ **5.** 0 **7.** $\frac{\pi}{4}$ **9.** $\frac{\pi}{3}$ **11.** $\frac{5\pi}{6}$ **13.** 0.10 **15.** 1.37 **17.** 0.51 **19.** −0.38 **21.** −0.12 **23.** 1.08 **25.** $\frac{\sqrt{2}}{2}$ **27.** $-\frac{\sqrt{3}}{3}$ **29.** 2 **31.** $\sqrt{2}$ **33.** $-\frac{\sqrt{2}}{2}$ **35.** $\frac{2\sqrt{3}}{3}$ **37.** $\frac{\sqrt{2}}{4}$ **39.** $\frac{\sqrt{5}}{2}$ **41.** $-\frac{\sqrt{14}}{2}$ **43.** $-\frac{3\sqrt{10}}{10}$ **45.** $\sqrt{5}$ **47.** 0.58 **49.** 0.10 **51.** 0.57 **53.** 0.43 **55.** 0.37

57. Let $\theta = \tan^{-1} v$. Then $\tan\theta = v$, $-\frac{\pi}{2} < \theta < \frac{\pi}{2}$. Now, $\sec\theta > 0$ and $\tan^2\theta + 1 = \sec^2\theta$. Thus $\sec\theta = \sec(\tan^{-1} v) = \sqrt{1 + v^2}$.

59. Let $\theta = \cos^{-1} v$. Then $\cos\theta = v$, $0 \le \theta \le \pi$, and $\tan(\cos^{-1} v) = \tan\theta = \frac{\sin\theta}{\cos\theta} = \frac{\sqrt{1-\cos^2\theta}}{\cos\theta} = \frac{\sqrt{1-v^2}}{v}$.

61. Let $\theta = \sin^{-1} v$. Then $\sin\theta = v$, $-\frac{\pi}{2} \le \theta \le \frac{\pi}{2}$, and $\cos(\sin^{-1} v) = \cos\theta = \sqrt{1-\sin^2\theta} = \sqrt{1-v^2}$.

63. Let $\alpha = \sin^{-1} v$ and $\beta = \cos^{-1} v$. Then $\sin\alpha = v = \cos\beta$, so α and β are complementary angles. Thus, $\alpha + \beta = \frac{\pi}{2}$.

65. Let $\alpha = \tan^{-1}\frac{1}{v}$. Then $\frac{1}{v} = \tan\alpha$, $\frac{-\pi}{2} < \alpha < \frac{\pi}{2}$, $\alpha \neq 0$. Let $\beta = \tan^{-1} v$. Then $v = \tan\beta$, $\frac{-\pi}{2} < \beta < \frac{\pi}{2}$, $\beta \neq 0$. Thus, $\tan\alpha\tan\beta = 1$ so that $\tan\alpha = \cot\beta$. Thus $\alpha + \beta = \frac{\pi}{2}$ and $\tan^{-1}\left(\frac{1}{v}\right) = \frac{\pi}{2} - \tan^{-1} v$.

67. 1.32 **69.** 0.46 **71.** −0.34 **73.** 2.72 **75.** −0.73 **77.** 2.55 **79.** 2.77 minutes **81.** $-1 \le x \le 1$ **83.** $0 \le x \le \pi$

85. $y = \sec^{-1}(x) = \cos^{-1}\left(\frac{1}{x}\right)$

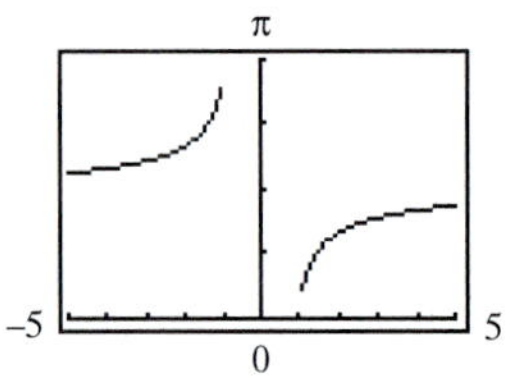

Fill-in-the-Blanks

1. angle; initial side; terminal side **2.** radians **3.** π **4.** complementary **5.** cosine **6.** standard position **7.** 45° **8.** 2π; π **9.** $y = \cos x$, $y = \sec x$ **10.** $y = \sin x$, $y = \tan x$, $y = \csc x$, $y = \cot x$ **11.** $-1 \le x \le 1$; $-\frac{\pi}{2} \le y \le \frac{\pi}{2}$ **12.** 0

True/False Items

1. F **2.** T **3.** F **4.** T **5.** F **6.** F **7.** T **8.** F **9.** T

Review Exercises

1. $\frac{3\pi}{4}$ **3.** $\frac{\pi}{10}$ **5.** 135° **7.** −450° **9.** $\frac{1}{2}$ **11.** $\frac{3\sqrt{2}}{2} - \frac{4\sqrt{3}}{3}$ **13.** $-3\sqrt{2} - 2\sqrt{3}$ **15.** 3 **17.** 0 **19.** 0 **21.** 1 **23.** 1 **25.** 1 **27.** −1 **29.** 1 **31.** $\cos\theta = \frac{3}{5}$, $\tan\theta = -\frac{4}{3}$, $\csc\theta = -\frac{5}{4}$, $\sec\theta = \frac{5}{3}$, $\cot\theta = -\frac{3}{4}$

33. $\sin\theta = -\frac{12}{13}$, $\cos\theta = -\frac{5}{13}$, $\csc\theta = -\frac{13}{12}$, $\sec\theta = -\frac{13}{5}$, $\cot\theta = \frac{5}{12}$ **35.** $\sin\theta = \frac{3}{5}$, $\cos\theta = -\frac{4}{5}$, $\tan\theta = -\frac{3}{4}$, $\csc\theta = \frac{5}{3}$, $\cot\theta = -\frac{4}{3}$

37. $\cos\theta = -\frac{5}{13}$, $\tan\theta = -\frac{12}{5}$, $\csc\theta = \frac{13}{12}$, $\sec\theta = -\frac{13}{5}$, $\cot\theta = -\frac{5}{12}$

39. $\cos\theta = \frac{12}{13}, \tan\theta = -\frac{5}{12}, \csc\theta = -\frac{13}{5}, \sec\theta = \frac{13}{12}, \cot\theta = -\frac{12}{5}$

41. $\sin\theta = -\frac{\sqrt{10}}{10}, \cos\theta = -\frac{3\sqrt{10}}{10}, \csc\theta = -\sqrt{10}, \sec\theta = -\frac{\sqrt{10}}{3}, \cot\theta = 3$

43. $\sin\theta = -\frac{2\sqrt{2}}{3}, \cos\theta = \frac{1}{3}, \tan\theta = -2\sqrt{2}, \csc = -\frac{3\sqrt{2}}{4}, \cot\theta = -\frac{\sqrt{2}}{4}$

45. $\sin\theta = \frac{\sqrt{5}}{5}, \cos\theta = -\frac{2\sqrt{5}}{5}, \tan\theta = -\frac{1}{2}, \csc\theta = \sqrt{5}, \sec\theta = -\frac{\sqrt{5}}{2}$

47.

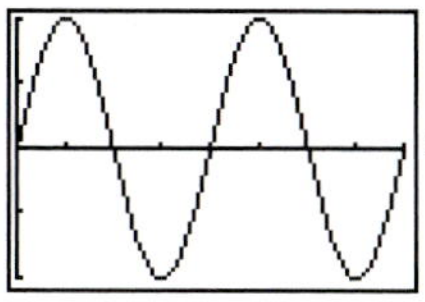

49.

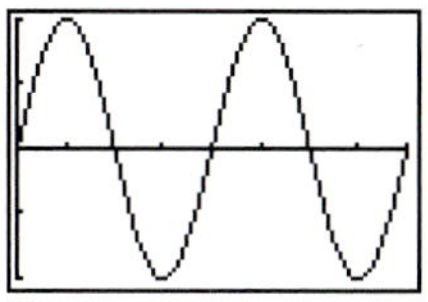

51.

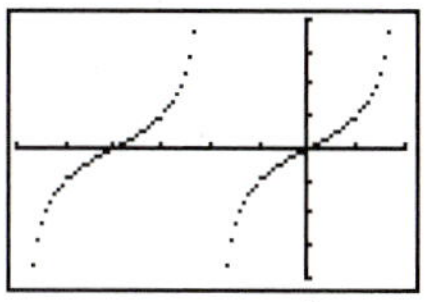

53.

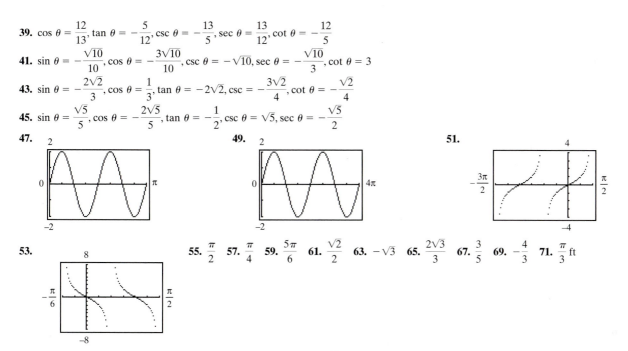

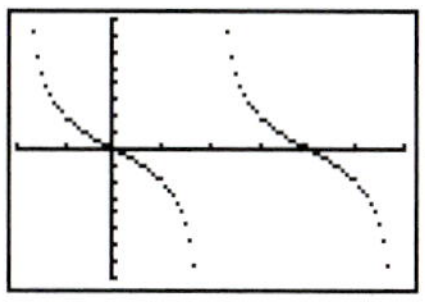

55. $\frac{\pi}{2}$ **57.** $\frac{\pi}{4}$ **59.** $\frac{5\pi}{6}$ **61.** $\frac{\sqrt{2}}{2}$ **63.** $-\sqrt{3}$ **65.** $\frac{2\sqrt{3}}{3}$ **67.** $\frac{3}{5}$ **69.** $-\frac{4}{3}$ **71.** $\frac{\pi}{3}$ ft

73. 114.59 revolutions per hour **75.** 0.1 revolutions per second $= \frac{\pi}{5}$ radians/second

CHAPTER 7 *Exercises 7.1*

1. $\csc\theta\cdot\cos\theta = \frac{1}{\sin\theta}\cdot\cos\theta = \frac{\cos\theta}{\sin\theta} = \cot\theta$ **3.** $1+\tan^2(-\theta) = 1 + (-\tan\theta)^2 = 1+\tan^2\theta = \sec^2\theta$

5. $\cos\theta(\tan\theta+\cot\theta) = \cos\theta\left(\frac{\sin\theta}{\cos\theta}+\frac{\cos\theta}{\sin\theta}\right) = \cos\theta\left(\frac{\sin^2\theta+\cos^2\theta}{\cos\theta\sin\theta}\right) = \frac{1}{\sin\theta} = \csc\theta$

7. $\tan\theta\cot\theta - \cos^2\theta = \frac{\sin\theta}{\cos\theta}\cdot\frac{\cos\theta}{\sin\theta} - \cos^2\theta = 1-\cos^2\theta = \sin^2\theta$ **9.** $(\sec\theta-1)(\sec\theta+1) = \sec^2\theta - 1 = \tan^2\theta$

11. $(\sec\theta+\tan\theta)(\sec\theta-\tan\theta) = \sec^2\theta-\tan^2\theta = 1$ **13.** $\sin^2\theta(1+\cot^2\theta) = \sin^2\theta+\sin^2\theta\cdot\frac{\cos^2\theta}{\sin^2\theta} = \sin^2\theta+\cos^2\theta = 1$

15. $(\sin\theta+\cos\theta)^2+(\sin\theta-\cos\theta)^2 = \sin^2\theta+2\sin\theta\cos\theta+\cos^2\theta+\sin^2\theta-2\sin\theta\cos\theta+\cos^2\theta$
$= \sin^2\theta+\cos^2\theta+\sin^2\theta+\cos^2\theta = 1+1 = 2$

17. $\sec^4\theta-\sec^2\theta = \sec^2\theta(\sec^2\theta-1) = (1+\tan^2\theta)\tan^2\theta = \tan^4\theta+\tan^2\theta$

19. $\sec\theta-\tan\theta = \frac{1}{\cos\theta}-\frac{\sin\theta}{\cos\theta} = \frac{1-\sin\theta}{\cos\theta}\cdot\frac{1+\sin\theta}{1+\sin\theta} = \frac{1-\sin^2\theta}{\cos\theta(1+\sin\theta)} = \frac{\cos^2\theta}{\cos\theta(1+\sin\theta)} = \frac{\cos\theta}{1+\sin\theta}$

21. $3\sin^2\theta+4\cos^2\theta = 3\sin^2\theta+3\cos^2\theta+\cos^2\theta = 3(\sin^2\theta+\cos^2\theta)+\cos^2\theta = 3+\cos^2\theta$

23. $1-\frac{\cos^2\theta}{1+\sin\theta} = 1-\frac{1-\sin^2\theta}{1+\sin\theta} = 1-(1-\sin\theta) = \sin\theta$ **25.** $\frac{1+\tan\theta}{1-\tan\theta} = \frac{1+\frac{1}{\cot\theta}}{1-\frac{1}{\cot\theta}} = \frac{\frac{\cot\theta+1}{\cot\theta}}{\frac{\cot\theta-1}{\cot\theta}} = \frac{\cot\theta+1}{\cot\theta-1}$

27. $\frac{\sec\theta}{\csc\theta}+\frac{\sin\theta}{\cos\theta} = \frac{\frac{1}{\cos\theta}}{\frac{1}{\sin\theta}}+\tan\theta = \frac{\sin\theta}{\cos\theta}+\tan\theta = \tan\theta+\tan\theta = 2\tan\theta$ **29.** $\frac{1+\sin\theta}{1-\sin\theta} = \frac{1+\frac{1}{\csc\theta}}{1-\frac{1}{\csc\theta}} = \frac{\frac{\csc\theta+1}{\csc\theta}}{\frac{\csc\theta-1}{\csc\theta}} = \frac{\csc\theta+1}{\csc\theta-1}$

31. $\frac{1-\sin\theta}{\cos\theta}+\frac{\cos\theta}{1-\sin\theta} = \frac{(1-\sin\theta)^2+\cos^2\theta}{\cos\theta(1-\sin\theta)} = \frac{1-2\sin\theta+\sin^2\theta+\cos^2\theta}{\cos\theta(1-\sin\theta)} = \frac{2-2\sin\theta}{\cos\theta(1-\sin\theta)} = \frac{2(1-\sin\theta)}{\cos\theta(1-\sin\theta)} = \frac{2}{\cos\theta}$
$= 2\sec\theta$

33. $\frac{\sin\theta}{\sin\theta-\cos\theta} = \frac{1}{\frac{\sin\theta-\cos\theta}{\sin\theta}} = \frac{1}{1-\frac{\cos\theta}{\sin\theta}} = \frac{1}{1-\cot\theta}$

35. $(\sec\theta-\tan\theta)^2=\sec^2\theta-2\sec\theta\tan\theta+\tan^2\theta=\dfrac{1}{\cos^2\theta}-\dfrac{2\sin\theta}{\cos^2\theta}+\dfrac{\sin^2\theta}{\cos^2\theta}=\dfrac{1-2\sin\theta+\sin^2\theta}{\cos^2\theta}=\dfrac{(1-\sin\theta)^2}{1-\sin^2\theta}$
$=\dfrac{(1-\sin\theta)^2}{(1-\sin\theta)(1+\sin\theta)}=\dfrac{1-\sin\theta}{1+\sin\theta}$

37. $\dfrac{\cos\theta}{1-\tan\theta}+\dfrac{\sin\theta}{1-\cot\theta}=\dfrac{\cos\theta}{1-\dfrac{\sin\theta}{\cos\theta}}+\dfrac{\sin\theta}{1-\dfrac{\cos\theta}{\sin\theta}}=\dfrac{\cos\theta}{\dfrac{\cos\theta-\sin\theta}{\cos\theta}}+\dfrac{\sin\theta}{\dfrac{\sin\theta-\cos\theta}{\sin\theta}}=\dfrac{\cos^2\theta}{\cos\theta-\sin\theta}+\dfrac{\sin^2\theta}{\sin\theta-\cos\theta}=\dfrac{\cos^2\theta-\sin^2\theta}{\cos\theta-\sin\theta}$
$=\dfrac{(\cos\theta-\sin\theta)(\cos\theta+\sin\theta)}{\cos\theta-\sin\theta}=\sin\theta+\cos\theta$

39. $\tan\theta+\dfrac{\cos\theta}{1+\sin\theta}=\dfrac{\sin\theta}{\cos\theta}+\dfrac{\cos\theta}{1+\sin\theta}=\dfrac{\sin\theta(1+\sin\theta)+\cos^2\theta}{\cos\theta(1+\sin\theta)}=\dfrac{\sin\theta+\sin^2\theta+\cos^2}{\cos\theta(1+\sin\theta)}=\dfrac{\sin\theta+1}{\cos\theta(1+\sin\theta)}=\dfrac{1}{\cos\theta}=\sec\theta$

41. $\dfrac{\tan\theta+\sec\theta-1}{\tan\theta-\sec\theta+1}=\dfrac{\tan\theta+(\sec\theta-1)}{\tan\theta-(\sec\theta-1)}\cdot\dfrac{\tan\theta+(\sec\theta-1)}{\tan\theta+(\sec\theta-1)}=\dfrac{\tan^2\theta+2\tan\theta(\sec\theta-1)+\sec^2\theta-2\sec\theta+1}{\tan^2\theta-(\sec^2\theta-2\sec\theta+1)}$
$=\dfrac{\sec^2\theta-1+2\tan\theta(\sec\theta-1)+\sec^2\theta-2\sec\theta+1}{\sec^2\theta-1-\sec^2\theta+2\sec\theta-1}=\dfrac{2\sec^2\theta-2\sec\theta+2\tan\theta(\sec\theta-1)}{-2+2\sec\theta}$
$=\dfrac{2\sec\theta(\sec\theta-1)+2\tan\theta(\sec\theta-1)}{2(\sec\theta-1)}=\dfrac{2(\sec\theta-1)(\sec\theta+\tan\theta)}{2(\sec\theta-1)}=\tan\theta+\sec\theta$

43. $\dfrac{\tan\theta-\cot\theta}{\tan\theta+\cot\theta}=\dfrac{\dfrac{\sin\theta}{\cos\theta}-\dfrac{\cos\theta}{\sin\theta}}{\dfrac{\sin\theta}{\cos\theta}+\dfrac{\cos\theta}{\sin\theta}}=\dfrac{\dfrac{\sin^2\theta-\cos^2\theta}{\cos\theta\sin\theta}}{\dfrac{\sin^2\theta+\cos^2\theta}{\cos\theta\sin\theta}}=\dfrac{\sin^2\theta-\cos^2\theta}{1}=\sin^2\theta-\cos^2\theta$

45. $\dfrac{\tan\theta-\cot\theta}{\tan\theta+\cot\theta}=\dfrac{\dfrac{\sin\theta}{\cos\theta}-\dfrac{\cos\theta}{\sin\theta}}{\dfrac{\sin\theta}{\cos\theta}+\dfrac{\cos\theta}{\sin\theta}}=\dfrac{\dfrac{\sin^2\theta-\cos^2\theta}{\cos\theta\sin\theta}}{\dfrac{\sin^2\theta+\cos^2\theta}{\cos\theta\sin\theta}}=\sin^2\theta-\cos^2\theta=\sin^2\theta-(1-\sin^2\theta)=2\sin^2\theta-1$

47. $\dfrac{\sec\theta+\tan\theta}{\cot\theta+\cos\theta}=\dfrac{\dfrac{1}{\cos\theta}+\dfrac{\sin\theta}{\cos\theta}}{\dfrac{\cos\theta}{\sin\theta}+\dfrac{\cos\theta\sin\theta}{\sin\theta}}=\dfrac{\dfrac{1+\sin\theta}{\cos\theta}}{\dfrac{\cos\theta+\cos\theta\sin\theta}{\sin\theta}}=\dfrac{1+\sin\theta}{\cos\theta}\cdot\dfrac{\sin\theta}{\cos\theta(1+\sin\theta)}=\dfrac{\sin\theta}{\cos\theta}\cdot\dfrac{1}{\cos\theta}=\tan\theta\sec\theta$

49. $\dfrac{1-\tan^2\theta}{1+\tan^2\theta}=\dfrac{1-\tan^2\theta}{\sec^2\theta}=\dfrac{1}{\sec^2\theta}-\dfrac{\tan^2\theta}{\sec^2\theta}=\cos^2\theta-\dfrac{\dfrac{\sin^2\theta}{\cos^2\theta}}{\dfrac{1}{\cos^2\theta}}=\cos^2\theta-\sin^2\theta=\cos^2\theta-(1-\cos^2\theta)=2\cos^2\theta-1$

51. $\dfrac{\sec\theta-\csc\theta}{\sec\theta\csc\theta}=\dfrac{\dfrac{1}{\cos\theta}-\dfrac{1}{\sin\theta}}{\dfrac{1}{\cos\theta}\cdot\dfrac{1}{\sin\theta}}=\dfrac{\dfrac{\sin\theta-\cos\theta}{\cos\theta\sin\theta}}{\dfrac{1}{\cos\theta\sin\theta}}=\sin\theta-\cos\theta$

53. $\sec\theta-\cos\theta=\dfrac{1}{\cos\theta}-\dfrac{\cos^2\theta}{\cos\theta}=\dfrac{1-\cos^2\theta}{\cos\theta}=\dfrac{\sin^2\theta}{\cos\theta}=\sin\theta\cdot\dfrac{\sin\theta}{\cos\theta}=\sin\theta\tan\theta$

55. $\dfrac{1}{1-\sin\theta}+\dfrac{1}{1+\sin\theta}=\dfrac{1+\sin\theta+1-\sin\theta}{(1+\sin\theta)(1-\sin\theta)}=\dfrac{2}{1-\sin^2\theta}=\dfrac{2}{\cos^2\theta}=2\sec^2\theta$

57. $\dfrac{\sec\theta}{1-\sin\theta}=\dfrac{\sec\theta}{1-\sin\theta}\cdot\dfrac{1+\sin\theta}{1+\sin\theta}=\dfrac{\sec\theta(1+\sin\theta)}{1-\sin^2\theta}=\dfrac{\sec\theta(1+\sin\theta)}{\cos^2\theta}=\dfrac{1+\sin\theta}{\cos^3\theta}$

59. $\dfrac{(\sec\theta-\tan\theta)^2+1}{\csc\theta(\sec\theta-\tan\theta)}=\dfrac{\sec^2\theta-2\sec\theta\tan\theta+\tan^2\theta+1}{\dfrac{1}{\sin\theta}\left(\dfrac{1}{\cos\theta}-\dfrac{\sin\theta}{\cos\theta}\right)}=\dfrac{2\sec^2\theta-2\sec\theta\tan\theta}{\dfrac{1}{\sin\theta}\left(\dfrac{1-\sin\theta}{\cos\theta}\right)}=\dfrac{\dfrac{2}{\cos^2\theta}-\dfrac{2\sin\theta}{\cos^2\theta}}{\dfrac{1-\sin\theta}{\sin\theta\cos\theta}}=\dfrac{2-2\sin\theta}{\cos^2\theta}\cdot\dfrac{\sin\theta\cos\theta}{1-\sin\theta}$
$=\dfrac{2(1-\sin\theta)}{\cos\theta}\cdot\dfrac{\sin\theta}{1-\sin\theta}=\dfrac{2\sin\theta}{\cos\theta}=2\tan\theta$

61. $\dfrac{\sin\theta+\cos\theta}{\cos\theta}-\dfrac{\sin\theta-\cos\theta}{\sin\theta}=\dfrac{\sin\theta(\sin\theta+\cos\theta)-\cos\theta(\sin\theta-\cos\theta)}{\cos\theta\sin\theta}=\dfrac{\sin^2\theta+\sin\theta\cos\theta-\sin\theta\cos\theta+\cos^2\theta}{\cos\theta\sin\theta}$
$=\dfrac{1}{\cos\theta\sin\theta}=\sec\theta\csc\theta$

63. $\dfrac{\sin^3\theta+\cos^3\theta}{\sin\theta+\cos\theta}=\dfrac{(\sin\theta+\cos\theta)(\sin^2\theta-\sin\theta\cos\theta+\cos^2\theta)}{\sin\theta+\cos\theta}=\sin^2\theta+\cos^2\theta-\sin\theta\cos\theta=1-\sin\theta\cos\theta$

65. $\dfrac{\cos^2\theta - \sin^2\theta}{1 - \tan^2\theta} = \dfrac{\cos^2\theta - \sin^2\theta}{1 - \dfrac{\sin^2\theta}{\cos^2\theta}} = \dfrac{\cos^2\theta - \sin^2\theta}{\dfrac{\cos^2\theta - \sin^2\theta}{\cos^2\theta}} = \cos^2\theta$

67. $\dfrac{(2\cos^2\theta - 1)^2}{\cos^4\theta - \sin^4\theta} = \dfrac{[2\cos^2\theta - (\sin^2\theta + \cos^2\theta)]^2}{(\cos^2\theta - \sin^2\theta)(\cos^2\theta + \sin^2\theta)} = \cos^2\theta - \sin^2\theta = (1 - \sin^2\theta) - \sin^2\theta = 1 - 2\sin^2\theta$

69. $\dfrac{1 + \sin\theta + \cos\theta}{1 + \sin\theta - \cos\theta} = \dfrac{(1 + \sin\theta) + \cos\theta}{(1 + \sin\theta) - \cos\theta} \cdot \dfrac{(1 + \sin\theta) + \cos\theta}{(1 + \sin\theta) + \cos\theta} = \dfrac{1 + 2\sin\theta + \sin^2\theta + 2(1 + \sin\theta)(\cos\theta) + \cos^2\theta}{1 + 2\sin\theta + \sin^2\theta - \cos^2\theta}$
$= \dfrac{1 + 2\sin\theta + \sin^2\theta + 2(1 + \sin\theta)(\cos\theta) + (1 - \sin^2\theta)}{1 + 2\sin\theta + \sin^2\theta - (1 - \sin^2\theta)} = \dfrac{2 + 2\sin\theta + 2(1 + \sin\theta)(\cos\theta)}{2\sin\theta + 2\sin^2\theta}$
$= \dfrac{2(1 + \sin\theta) + 2(1 + \sin\theta)(\cos\theta)}{2\sin\theta(1 + \sin\theta)} = \dfrac{2(1 + \sin\theta)(1 + \cos\theta)}{2\sin\theta(1 + \sin\theta)} = \dfrac{1 + \cos\theta}{\sin\theta}$

71. $(a\sin\theta + b\cos\theta)^2 + (a\cos\theta - b\sin\theta)^2 = a^2\sin^2\theta + 2ab\sin\theta\cos\theta + b^2\cos^2\theta + a^2\cos^2\theta - 2ab\sin\theta\cos\theta + b^2\sin^2\theta$
$= a^2(\sin^2\theta + \cos^2\theta) + b^2(\cos^2\theta + \sin^2\theta) = a^2 + b^2$

73. $\dfrac{\tan\alpha + \tan\beta}{\cot\alpha + \cot\beta} = \dfrac{\tan\alpha + \tan\beta}{\dfrac{1}{\tan\alpha} + \dfrac{1}{\tan\beta}} = \dfrac{\tan\alpha + \tan\beta}{\dfrac{\tan\beta + \tan\alpha}{\tan\alpha\tan\beta}} = (\tan\alpha + \tan\beta) \cdot \dfrac{\tan\alpha\tan\beta}{\tan\alpha + \tan\beta} = \tan\alpha\tan\beta$

75. $(\sin\alpha + \cos\beta)^2 + (\cos\beta + \sin\alpha)(\cos\beta - \sin\alpha) = (\sin^2\alpha + 2\sin\alpha\cos\beta + \cos^2\beta) + (\cos^2\beta - \sin^2\alpha)$
$= 2\cos^2\beta + 2\sin\alpha\cos\beta = 2\cos\beta(\cos\beta + \sin\alpha)$

77. $\ln|\sec\theta| = \ln|\cos\theta|^{-1} = -\ln|\cos\theta|$

79. $\ln|1 + \cos\theta| + \ln|1 - \cos\theta| = \ln(|1 + \cos\theta|\,|1 - \cos\theta|) = \ln|1 - \cos^2\theta| = \ln|\sin^2\theta| = 2\ln|\sin\theta|$

7.2 Exercises

1. $\frac{1}{4}(\sqrt{6} + \sqrt{2})$ **3.** $\frac{1}{4}(\sqrt{2} - \sqrt{6})$ **5.** $-\frac{1}{4}(\sqrt{2} + \sqrt{6})$ **7.** $\dfrac{\sqrt{3} - 1}{1 + \sqrt{3}} = 2 - \sqrt{3}$ **9.** $-\frac{1}{4}(\sqrt{6} + \sqrt{2})$ **11.** $\dfrac{4}{\sqrt{6} + \sqrt{2}} = \sqrt{6} - \sqrt{2}$
13. $\frac{1}{2}$ **15.** 0 **17.** 1 **19.** -1 **21.** $-\frac{\sqrt{3}}{2}$ **23.** **(a)** $\frac{2\sqrt{5}}{25}$ **(b)** $\frac{11\sqrt{5}}{25}$ **(c)** $\frac{2\sqrt{5}}{5}$ **(d)** 2 **25.** **(a)** $\frac{4 - 3\sqrt{3}}{10}$
(b) $\frac{-3 - 4\sqrt{3}}{10}$ **(c)** $\frac{4 + 3\sqrt{3}}{10}$ **(d)** $\frac{4 + 3\sqrt{3}}{4\sqrt{3} - 3} = \frac{25\sqrt{3} + 48}{39}$ **27.** **(a)** $-\frac{1}{26}(5 + 12\sqrt{3})$ **(b)** $\frac{1}{26}(12 - 5\sqrt{3})$ **(c)** $-\frac{1}{26}(5 - 12\sqrt{3})$
(d) $\frac{-5 + 12\sqrt{3}}{12 + 5\sqrt{3}} = \frac{-240 + 169\sqrt{3}}{69}$ **29.** $\sin\left(\frac{\pi}{2} + \theta\right) = \sin\frac{\pi}{2}\cos\theta + \cos\frac{\pi}{2}\sin\theta = 1 \cdot \cos\theta + 0 \cdot \sin\theta = \cos\theta$

31. $\sin(\pi - \theta) = \sin\pi\cos\theta - \cos\pi\sin\theta = 0 \cdot \cos\theta - (-1)\sin\theta = \sin\theta$

33. $\sin(\pi + \theta) = \sin\pi\cos\theta + \cos\pi\sin\theta = 0 \cdot \cos\theta + (-1)\sin\theta = -\sin\theta$

35. $\tan(\pi - \theta) = \dfrac{\tan\pi - \tan\theta}{1 + \tan\pi\tan\theta} = \dfrac{0 - \tan\theta}{1 + 0} = -\tan\theta$

37. $\sin\left(\dfrac{3\pi}{2} + \theta\right) = \sin\dfrac{3\pi}{2}\cos\theta + \cos\dfrac{3\pi}{2}\sin\theta = (-1)\cos\theta + 0 \cdot \sin\theta = -\cos\theta$

39. $\sin(\alpha + \beta) + \sin(\alpha - \beta) = \sin\alpha\cos\beta + \cos\alpha\sin\beta + \sin\alpha\cos\beta - \cos\alpha\sin\beta = 2\sin\alpha\cos\beta$

41. $\dfrac{\sin(\alpha + \beta)}{\sin\alpha\cos\beta} = \dfrac{\sin\alpha\cos\beta + \cos\alpha\sin\beta}{\sin\alpha\cos\beta} = \dfrac{\sin\alpha\cos\beta}{\sin\alpha\cos\beta} + \dfrac{\cos\alpha\sin\beta}{\sin\alpha\cos\beta} = 1 + \cot\alpha\tan\beta$

43. $\dfrac{\cos(\alpha + \beta)}{\cos\alpha\cos\beta} = \dfrac{\cos\alpha\cos\beta - \sin\alpha\sin\beta}{\cos\alpha\cos\beta} = \dfrac{\cos\alpha\cos\beta}{\cos\alpha\cos\beta} - \dfrac{\sin\alpha\sin\beta}{\cos\alpha\cos\beta} = 1 - \tan\alpha\tan\beta$

45. $\dfrac{\sin(\alpha + \beta)}{\sin(\alpha - \beta)} = \dfrac{\sin\alpha\cos\beta + \cos\alpha\sin\beta}{\sin\alpha\cos\beta - \cos\alpha\sin\beta} = \dfrac{\dfrac{\sin\alpha\cos\beta + \cos\alpha\sin\beta}{\cos\alpha\cos\beta}}{\dfrac{\sin\alpha\cos\beta - \cos\alpha\sin\beta}{\cos\alpha\cos\beta}} = \dfrac{\dfrac{\sin\alpha\cos\beta}{\cos\alpha\cos\beta} + \dfrac{\cos\alpha\sin\beta}{\cos\alpha\cos\beta}}{\dfrac{\sin\alpha\cos\beta}{\cos\alpha\cos\beta} - \dfrac{\cos\alpha\sin\beta}{\cos\alpha\cos\beta}} = \dfrac{\tan\alpha + \tan\beta}{\tan\alpha - \tan\beta}$

47. $\cot(\alpha + \beta) = \dfrac{\cos(\alpha + \beta)}{\sin(\alpha + \beta)} = \dfrac{\cos\alpha\cos\beta - \sin\alpha\sin\beta}{\sin\alpha\cos\beta + \cos\alpha\sin\beta} = \dfrac{\dfrac{\cos\alpha\cos\beta - \sin\alpha\sin\beta}{\sin\alpha\sin\beta}}{\dfrac{\sin\alpha\sin\beta - \cos\alpha\sin\beta}{\sin\alpha\sin\beta}} = \dfrac{\dfrac{\cos\alpha\cos\beta}{\sin\alpha\sin\beta} - \dfrac{\sin\alpha\sin\beta}{\sin\alpha\sin\beta}}{\dfrac{\sin\alpha\cos\beta}{\sin\alpha\sin\beta} + \dfrac{\cos\alpha\sin\beta}{\sin\alpha\sin\beta}} = \dfrac{\cot\alpha\cot\beta - 1}{\cot\beta + \cot\alpha}$

49. $\sec(\alpha + \beta) = \dfrac{1}{\cos(\alpha + \beta)} = \dfrac{1}{\cos\alpha\cos\beta - \sin\alpha\sin\beta} = \dfrac{\dfrac{1}{\sin\alpha\sin\beta}}{\dfrac{\cos\alpha\cos\beta - \sin\alpha\sin\beta}{\sin\alpha\sin\beta}} = \dfrac{\dfrac{1}{\sin\alpha} \cdot \dfrac{1}{\sin\beta}}{\dfrac{\cos\alpha\cos\beta}{\sin\alpha\sin\beta} - \dfrac{\sin\alpha\sin\beta}{\sin\alpha\sin\beta}} = \dfrac{\csc\alpha\csc\beta}{\cot\alpha\cot\beta - 1}$

51. $\sin(\alpha - \beta)\sin(\alpha + \beta) = (\sin\alpha\cos\beta - \cos\alpha\sin\beta)(\sin\alpha\cos\beta + \cos\alpha\sin\beta) = \sin^2\alpha\cos^2\beta - \cos^2\alpha\sin^2\beta$
$= (\sin^2\alpha)(1 - \sin^2\beta) - (1 - \sin^2\alpha)(\sin^2\beta) = \sin^2\alpha - \sin^2\beta$

53. $\sin(\theta + k\pi) = \sin\theta\cos k\pi + \cos\theta\sin k\pi = (\sin\theta)(-1)^k + (\cos\theta)(0) = (-1)^k \cdot \sin\theta$, k any integer

37. **(a)** $y = 2 \sin 2061\, \pi t \cos 357\, \pi t$ **(b)**

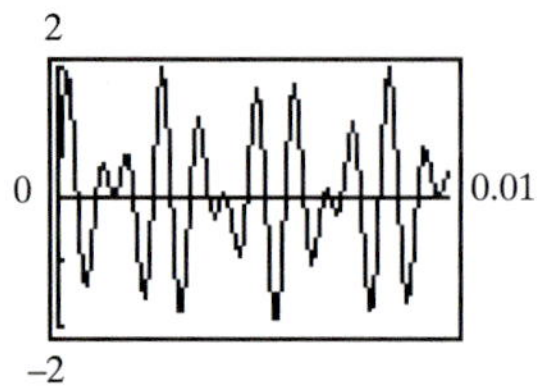

(c) $y_{max} = 2$

39. $\sin 2\alpha + \sin 2\beta + \sin 2\gamma = 2\sin(\alpha+\beta)\cos(\alpha-\beta) + \sin 2\gamma = 2\sin(\alpha+\beta)\cos(\alpha-\beta) + 2\sin\gamma\cos\gamma$
$= 2\sin(\pi-\gamma)\cos(\alpha-\beta) + 2\sin\gamma\cos\gamma = 2\sin\gamma\cos(\alpha-\beta) + 2\sin\gamma\cos\gamma = 2\sin\gamma[\cos(\alpha-\beta)+\cos\gamma]$
$= 2\sin\gamma\left(2\cos\frac{\alpha-\beta+\gamma}{2}\cos\frac{\alpha-\beta-\gamma}{2}\right) = 4\sin\gamma\cos\frac{\pi-2\beta}{2}\cos\frac{2\alpha-\pi}{2} = 4\sin\gamma\cos\left(\frac{\pi}{2}-\beta\right)\cos\left(\alpha-\frac{\pi}{2}\right)$
$= 4\sin\gamma\sin\beta\sin\alpha$

41.
$$\begin{aligned}\sin(\alpha-\beta) &= \sin\alpha\cos\beta - \cos\alpha\sin\beta\\ \sin(\alpha+\beta) &= \sin\alpha\cos\beta + \cos\alpha\sin\beta\\ \sin(\alpha-\beta)+\sin(\alpha+\beta) &= 2\sin\alpha\cos\beta\\ \sin\alpha\cos\beta &= \frac{1}{2}[\sin(\alpha+\beta)+\sin(\alpha-\beta)]\end{aligned}$$

43. $2\cos\frac{\alpha+\beta}{2}\cos\frac{\alpha-\beta}{2} = 2\cdot\frac{1}{2}\left[\cos\left(\frac{\alpha+\beta}{2}+\frac{\alpha-\beta}{2}\right)+\cos\left(\frac{\alpha+\beta}{2}-\frac{\alpha-\beta}{2}\right)\right] = \cos\frac{2\alpha}{2}+\cos\frac{2\beta}{2} = \cos\alpha+\cos\beta$

7.5 Exercises

1. $\frac{\pi}{6}, \frac{5\pi}{6}$ **3.** $\frac{5\pi}{6}, \frac{11\pi}{6}$ **5.** $\frac{\pi}{2}, \frac{3\pi}{2}$ **7.** $\frac{\pi}{2}, \frac{7\pi}{6}, \frac{11\pi}{6}$ **9.** $\frac{3\pi}{4}, \frac{7\pi}{4}$ **11.** $\frac{4\pi}{9}, \frac{8\pi}{9}, \frac{16\pi}{9}$ **13.** 0.4115168, 2.73007581
15. 1.3734008, 4.5149934 **17.** 2.6905658, 3.5926195 **19.** 1.8234766, 4.4597087 **21.** $\frac{\pi}{2}, \frac{2\pi}{3}, \frac{4\pi}{3}, \frac{3\pi}{2}$
23. $\frac{\pi}{2}, \frac{7\pi}{6}, \frac{11\pi}{6}$ **25.** $0, \frac{\pi}{4}, \frac{5\pi}{4}$ **27.** $\frac{\pi}{4}, \frac{5\pi}{4}$ **29.** $0, \frac{\pi}{3}, \pi, \frac{5\pi}{3}$ **31.** $\frac{\pi}{2}, \frac{3\pi}{2}$ **33.** $0, \frac{2\pi}{3}, \frac{4\pi}{3}$ **35.** $0, \frac{\pi}{3}, \frac{2\pi}{3}, \frac{\pi}{2}, \frac{4\pi}{3}, \pi, \frac{5\pi}{3}, \frac{3\pi}{2}$
37. $0, \frac{\pi}{5}, \frac{2\pi}{5}, \frac{3\pi}{5}, \frac{4\pi}{5}, \pi, \frac{6\pi}{5}, \frac{7\pi}{5}, \frac{8\pi}{5}, \frac{9\pi}{5}$ **39.** $\frac{\pi}{6}, \frac{5\pi}{6}, \frac{3\pi}{2}$ **41.** $\frac{\pi}{3}, \frac{5\pi}{3}$ **43.** No real solutions **45.** No real solutions **47.** $\frac{\pi}{2}, \frac{7\pi}{6}$
49. $0, \frac{\pi}{3}, \pi, \frac{5\pi}{3}$

51.

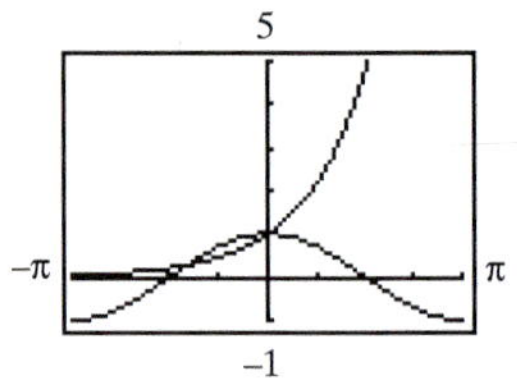

0, −1.29

53.

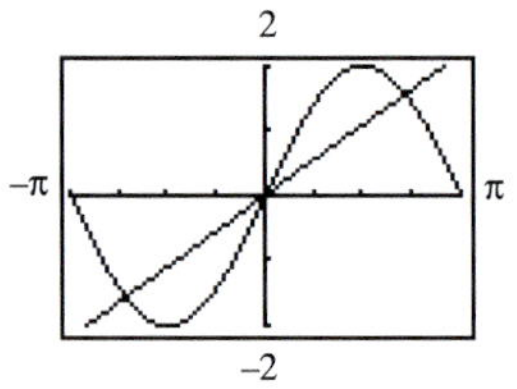

−2.24, 0, 2.24

55.

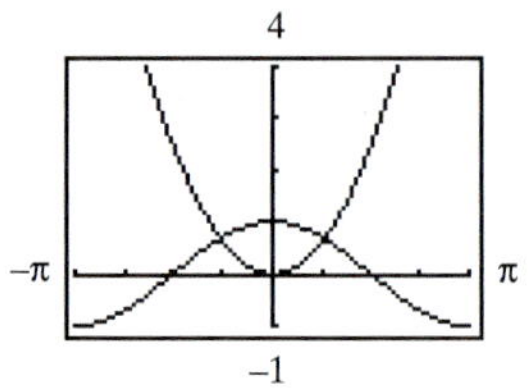

−0.82, 0.82

57. −1.30, 1.97, 3.83 **59.** 0.52 **61.** 1.25 **63.** −1.02, 1.02 **65.** 0, 2.14 **67.** 0.76, 1.34 **69.** **(a)** 60° **(b)** 60° **(c)** $A(60°) = 12\sqrt{3}$ sq in.
(d)

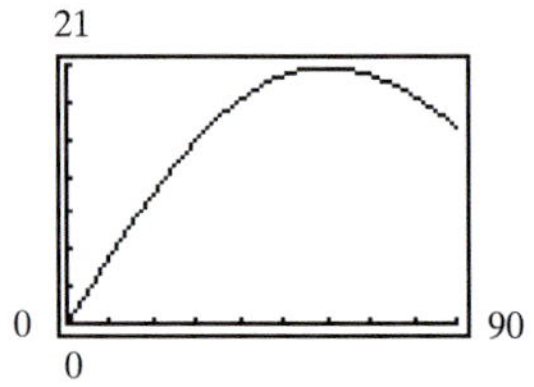

θmax = 60°
Maximum Area = 20.78 sq in.

71. 2.02, 4.91 **73.** **(a)** 29.99° **(b)**

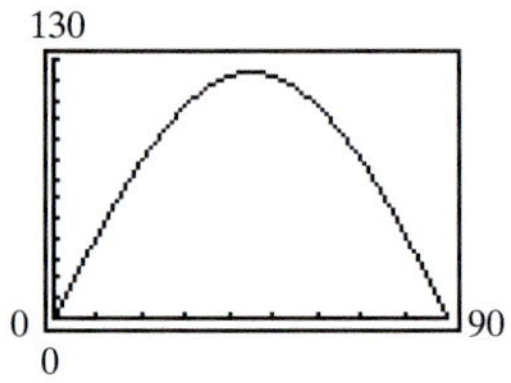

(d) 123.6 meters **75.** 28.9° **77.** Yes; it varies from 1.27 to 1.34 **79.** 1.47
81. If θ is the original angle of incidence and ϕ is the angle of refraction, then $\frac{\sin\theta}{\sin\phi} = n_2$. The angle of incidence of the emerging beam is also ϕ, and the index of refraction is $\frac{1}{n_2}$. Thus, θ is the angle of refraction of the emerging beam.

Fill-in-the-Blank Items

1. identity; conditional **2.** − **3.** + **4.** $\sin^2\theta$; $2\cos^2\theta$; $2\sin^2\theta$ **5.** $1 - \cos\alpha$

True/False Items

1. T **2.** F **3.** T **4.** F **5.** F **6.** F

Review Exercises

1. $\tan\theta\cot\theta - \sin^2\theta = 1 - \sin^2\theta = \cos^2\theta$ **3.** $\cos^2\theta(1 + \tan^2\theta) = \cos^2\theta\sec^2\theta = 1$

5. $4\cos^2\theta + 3\sin^2\theta = \cos^2\theta + 3(\cos^2\theta + \sin^2\theta) = 3 + \cos^2\theta$

7. $\dfrac{1-\cos\theta}{\sin\theta} + \dfrac{\sin\theta}{1-\cos\theta} = \dfrac{(1-\cos\theta)^2 + \sin^2\theta}{\sin\theta(1-\cos\theta)} = \dfrac{1 - 2\cos\theta + \cos^2\theta + \sin^2\theta}{\sin\theta(1-\cos\theta)} = \dfrac{2(1-\cos\theta)}{\sin\theta(1-\cos\theta)} = 2\csc\theta$

9. $\dfrac{\cos\theta}{\cos\theta - \sin\theta} = \dfrac{\dfrac{\cos\theta}{\cos\theta}}{\dfrac{\cos\theta - \sin\theta}{\cos\theta}} = \dfrac{1}{1 - \dfrac{\sin\theta}{\cos\theta}} = \dfrac{1}{1-\tan\theta}$

11. $\dfrac{\csc\theta}{1+\csc\theta} = \dfrac{\dfrac{1}{\sin\theta}}{1 + \dfrac{1}{\sin\theta}} = \dfrac{1}{1+\sin\theta} = \dfrac{1}{1+\sin\theta}\cdot\dfrac{1-\sin\theta}{1-\sin\theta} = \dfrac{1-\sin\theta}{1-\sin^2\theta} = \dfrac{1-\sin\theta}{\cos^2\theta}$

13. $\csc\theta - \sin\theta = \dfrac{1}{\sin\theta} - \sin\theta = \dfrac{1-\sin^2\theta}{\sin\theta} = \dfrac{\cos^2\theta}{\sin\theta} = \cos\theta\cdot\dfrac{\cos\theta}{\sin\theta} = \cos\theta\cot\theta$

15. $\dfrac{1-\sin\theta}{\sec\theta} = \cos\theta(1-\sin\theta)\cdot\dfrac{1+\sin\theta}{1+\sin\theta} = \dfrac{\cos\theta(1-\sin^2\theta)}{1+\sin\theta} = \dfrac{\cos^3\theta}{1+\sin\theta}$

17. $\cot\theta - \tan\theta = \dfrac{\cos\theta}{\sin\theta} - \dfrac{\sin\theta}{\cos\theta} = \dfrac{\cos^2\theta - \sin^2\theta}{\sin\theta\cos\theta} = \dfrac{1 - 2\sin^2\theta}{\sin\theta\cos\theta}$

19. $\dfrac{\cos(\alpha+\beta)}{\cos\alpha\sin\beta} = \dfrac{\cos\alpha\cos\beta - \sin\alpha\sin\beta}{\cos\alpha\sin\beta} = \dfrac{\cos\alpha\cos\beta}{\cos\alpha\sin\beta} - \dfrac{\sin\alpha\sin\beta}{\cos\alpha\sin\beta} = \cot\beta - \tan\alpha$

21. $\dfrac{\cos(\alpha-\beta)}{\cos\alpha\cos\beta} = \dfrac{\cos\alpha\cos\beta + \sin\alpha\sin\beta}{\cos\alpha\cos\beta} = \dfrac{\cos\alpha\cos\beta}{\cos\alpha\cos\beta} + \dfrac{\sin\alpha\sin\beta}{\cos\alpha\cos\beta} = 1 + \tan\alpha\tan\beta$

23. $(1+\cos\theta)\left(\tan\dfrac{\theta}{2}\right) = \left(2\cos^2\dfrac{\theta}{2}\right)\dfrac{\sin\left(\dfrac{\theta}{2}\right)}{\cos\left(\dfrac{\theta}{2}\right)} = 2\sin\dfrac{\theta}{2}\cos\dfrac{\theta}{2} = \sin\theta$

25. $2\cot\theta\cot 2\theta = 2\left(\dfrac{\cos\theta}{\sin\theta}\right)\left(\dfrac{\cos 2\theta}{\sin 2\theta}\right) = \dfrac{2\cos\theta(\cos^2\theta - \sin^2\theta)}{2\sin^2\theta\cos\theta} = \dfrac{\cos^2\theta - \sin^2\theta}{\sin^2\theta} = \cot^2\theta - 1$

27. $1 - 8\sin^2\theta\cos^2\theta = 1 - 2(2\sin\theta\cos\theta)^2 = 1 - 2\sin^2 2\theta = \cos 4\theta$ **29.** $\dfrac{\sin 2\theta + \sin 4\theta}{\cos 2\theta + \cos 4\theta} = \dfrac{2\sin 3\theta\cos(-\theta)}{2\cos 3\theta\cos(-\theta)} = \tan 3\theta$

31. $\dfrac{\cos 2\theta - \cos 4\theta}{\cos 2\theta + \cos 4\theta} - \tan\theta\tan 3\theta = \dfrac{-2\sin 3\theta\sin(-\theta)}{2\cos 3\theta\cos(-\theta)} - \tan\theta\tan 3\theta = \tan 3\theta\tan\theta - \tan\theta\tan 3\theta = 0$ **33.** $\frac{1}{4}(\sqrt{6} - \sqrt{2})$

35. $\frac{1}{4}(\sqrt{6} - \sqrt{2})$ **37.** $\frac{1}{2}$ **39.** $\sqrt{\dfrac{2-\sqrt{2}}{2+\sqrt{2}}} = \sqrt{2} - 1$ **41.** **(a)** $-\frac{33}{65}$ **(b)** $-\frac{56}{65}$ **(c)** $-\frac{63}{65}$ **(d)** $\frac{33}{56}$ **(e)** $\frac{24}{25}$ **(f)** $\frac{119}{169}$ **(g)** $\frac{5\sqrt{26}}{26}$ **(h)** $\frac{2\sqrt{5}}{5}$ **43.** **(a)** $-\frac{16}{65}$ **(b)** $-\frac{63}{65}$ **(c)** $-\frac{56}{65}$ **(d)** $\frac{16}{63}$ **(e)** $\frac{24}{25}$ **(f)** $\frac{119}{169}$ **(g)** $\frac{\sqrt{26}}{26}$ **(h)** $-\frac{\sqrt{10}}{10}$ **45.** **(a)** $-\frac{63}{65}$ **(b)** $\frac{16}{65}$ **(c)** $\frac{33}{65}$ **(d)** $-\frac{63}{16}$ **(e)** $\frac{24}{25}$ **(f)** $-\frac{119}{169}$ **(g)** $\frac{2\sqrt{13}}{13}$ **(h)** $-\frac{\sqrt{10}}{10}$ **47.** **(a)** $\frac{(-\sqrt{3} - 2\sqrt{2})}{6}$ **(b)** $\frac{(1 - 2\sqrt{6})}{6}$ **(c)** $\frac{(-\sqrt{3} + 2\sqrt{2})}{6}$ **(d)** $\frac{(-\sqrt{3} - 2\sqrt{2})}{(1 - 2\sqrt{6})} = \frac{(8\sqrt{2} + 9\sqrt{3})}{23}$ **(e)** $-\frac{\sqrt{3}}{2}$ **(f)** $-\frac{7}{9}$ **(g)** $\frac{\sqrt{3}}{3}$ **(h)** $\frac{\sqrt{3}}{2}$ **49.** **(a)** 1 **(b)** 0 **(c)** $-\frac{1}{9}$ **(d)** Not defined **(e)** $\frac{4\sqrt{5}}{9}$ **(f)** $-\frac{1}{9}$ **(g)** $\frac{\sqrt{30}}{6}$ **(h)** $-\frac{\sqrt{6}\sqrt{3 - \sqrt{5}}}{6}$ **51.** $\frac{\pi}{3}, \frac{5\pi}{3}$ **53.** $\frac{3\pi}{4}, \frac{5\pi}{4}$ **55.** $\frac{3\pi}{4}, \frac{7\pi}{4}$ **57.** $0, \frac{\pi}{2}, \pi, \frac{3\pi}{2}$

59. 1.1197695, $\pi - 1.1197695$ **61.** $0, \pi$ **63.** $0, \frac{2\pi}{3}, \pi, \frac{4\pi}{3}$ **65.** $0, \frac{\pi}{6}, \frac{5\pi}{6}$ **67.** $\frac{\pi}{6}, \frac{\pi}{2}, \frac{5\pi}{6}$ **69.** $\frac{\pi}{2}, \pi$ **71.** 1.11 **73.** 0.86 **75.** 2.21

CHAPTER 8 *8.1 Exercises*

1. $a \approx 13.74, c \approx 14.62, \alpha = 70°$ **3.** $b \approx 5.03, c \approx 7.83, \alpha = 50°$ **5.** $a \approx 0.71, c \approx 4.06, \beta = 80°$ **7.** $b \approx 10.72, c \approx 11.83, \beta = 65°$ **9.** $b \approx 3.08, a \approx 8.46, \alpha = 70°$ **11.** $c \approx 5.83, \alpha \approx 59.0°, \beta \approx 31.0°$ **13.** $b \approx 4.58, \alpha \approx 23.6°, \beta \approx 66.4°$ **15.** 1.72 in., 2.46 in. **17.** 6.10 in. or 8.72 in. **19.** 23.6° and 66.4° **21.** 70.02 ft **23.** 985.91 ft **25.** 137 m **27.** 20.67 ft **29.** 449.36 ft **31.** 80.5° **33.** 30 ft

35. 530 ft **37.** 555 ft **39. (a)** 112 ft/sec or 76.3 mph **(b)** 82.41 ft/sec or 56.2 mph **(c)** under 18.8° **41. (a)** 130° **(b)** 103.4° **43.** 14.9° **45. (a)** 3.1 mi **(b)** 3.2 mi **(c)** 3.8 mi **47. (a)** $\cos\frac{\theta}{2} = \frac{3960}{3960 + h}$ **(b)** $d = 3960\,\theta$ **(c)** $\cos\frac{d}{7920} = \frac{3960}{3960 + h}$ **(d)** 206 mi **(e)** 2990 miles

8.2 Exercises

1. $a = 3.23, b = 3.55, \alpha = 40°$ **3.** $a = 3.25, c = 4.23, \beta = 45°$ **5.** $\gamma = 95°, c = 9.86, a = 6.36$ **7.** $\alpha = 40°, a = 2, c = 3.06$ **9.** $\gamma = 120°, b = 1.06, c = 2.69$ **11.** $\alpha = 100°, a = 5.24, c = 0.92$ **13.** $\beta = 40°, a = 5.64, b = 3.86$ **15.** $\gamma = 100°, a = 1.31, b = 1.31$ **17.** One triangle; $\beta = 30.7°, \gamma = 99.3°, c = 3.86$ **19.** One triangle; $\gamma = 36.2°, \alpha = 43.8°, a = 3.51$ **21.** No triangle **23.** Two triangles; $\gamma_1 = 30.9°, \alpha_1 = 129.1°, a_1 = 9.08$ or $\gamma_2 = 149.1°, \alpha_2 = 10.9°, a_2 = 2.21$ **25.** No triangle **27.** Two triangles; $\alpha_1 = 57.7°, \beta_1 = 97.3°, b_1 = 2.35$ or $\alpha_2 = 122.3°, \beta_2 = 32.7°, b_2 = 1.28$ **29. (a)** Station Able is 143.3 mi from the ship; Station Baker is 135.6 mi from the ship. **(b)** Approx. 41 min **31.** 1490.5 ft **33.** 381.7 ft **35. (a)** 169 mi **(b)** 161.3° **37.** 84.7°; 183.7 ft **39.** 2.64 mi **41.** 1.88 mi

43. $\frac{a+b}{c} = \frac{a}{c} + \frac{b}{c} = \frac{\sin\alpha}{\sin\gamma} + \frac{\sin\beta}{\sin\gamma} = \frac{\sin\alpha + \sin\beta}{\sin\gamma} = \frac{2\sin\frac{\alpha+\beta}{2}\cos\frac{\alpha-\beta}{2}}{2\sin\frac{\gamma}{2}\cos\frac{\gamma}{2}} = \frac{\sin\left(\frac{\pi}{2} - \frac{\gamma}{2}\right)\cos\frac{\alpha-\beta}{2}}{\sin\frac{\gamma}{2}\cos\frac{\gamma}{2}} = \frac{\cos\frac{1}{2}(\alpha - \beta)}{\sin\frac{1}{2}\gamma}$

45. $a = \frac{b\sin\alpha}{\sin\beta} = \frac{b\sin[180° - (\beta + \gamma)]}{\sin\beta} = \frac{b}{\sin\beta}(\sin\beta\cos\gamma + \cos\beta\sin\gamma) = b\cos\gamma + \frac{b\sin\gamma}{\sin\beta}\cos\beta = b\cos\gamma + c\cos\beta$

47. $\sin\beta = \sin(\text{Angle } AB'C) = \frac{b}{2r}; \frac{\sin\beta}{b} = \frac{1}{2r}$; the result follows using the Law of Sines.

8.3 Exercises

1. $b = 2.95, \alpha = 28.7°, \gamma = 106.3°$ **3.** $c = 3.75, \alpha = 32.1°, \beta = 52.9°$ **5.** $\alpha = 48.5°, \beta = 38.6°, \gamma = 92.9°$ **7.** $\alpha = 127.2°, \beta = 32.1°, \gamma = 20.7°$ **9.** $c = 2.57, \alpha = 48.6°, \beta = 91.4°$ **11.** $a = 2.99, \beta = 19.2°, \gamma = 80.8°$ **13.** $b = 4.14, \alpha = 43.0°, \gamma = 27.0°$ **15.** $c = 1.69, \alpha = 65.0°, \beta = 65.0°$ **17.** $\alpha = 67.4°, \beta = 90°, \gamma = 22.6°$ **19.** $\alpha = 60°, \beta = 60°, \gamma = 60°$ **21.** $\alpha = 33.6°, \beta = 62.2°, \gamma = 84.3°$ **23.** $\alpha = 97.9°, \beta = 52.4°, \gamma = 29.7°$ **25.** 70.75 ft **27. (a)** 12° **(b)** 220.8 mph **29. (a)** 63.7 ft **(b)** 66.8 ft **(c)** 92.5° **31. (a)** 492.6 ft **(b)** 269.3 ft **33.** 342.3 ft

35. Using the Law of Cosines:

$L^2 = x^2 + r^2 - 2rx\cos\theta$

$x^2 - 2rx\cos\theta + r^2 - L^2 = 0$

$x = r\cos\theta + \sqrt{r^2\cos^2\theta + L^2 - r^2}$

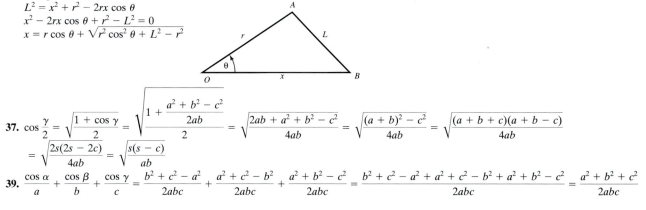

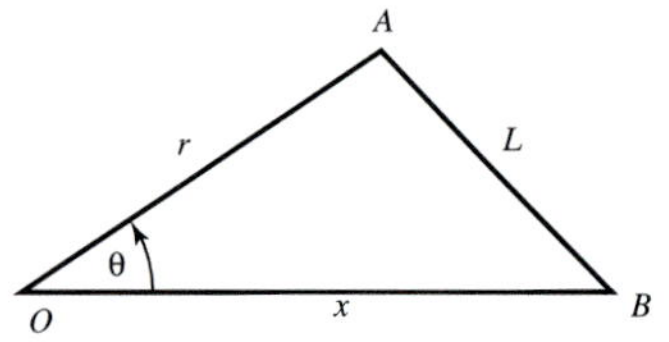

37. $\cos\frac{\gamma}{2} = \sqrt{\frac{1 + \cos\gamma}{2}} = \sqrt{\frac{1 + \frac{a^2 + b^2 - c^2}{2ab}}{2}} = \sqrt{\frac{2ab + a^2 + b^2 - c^2}{4ab}} = \sqrt{\frac{(a + b)^2 - c^2}{4ab}} = \sqrt{\frac{(a + b + c)(a + b - c)}{4ab}}$

$= \sqrt{\frac{2s(2s - 2c)}{4ab}} = \sqrt{\frac{s(s - c)}{ab}}$

39. $\frac{\cos\alpha}{a} + \frac{\cos\beta}{b} + \frac{\cos\gamma}{c} = \frac{b^2 + c^2 - a^2}{2abc} + \frac{a^2 + c^2 - b^2}{2abc} + \frac{a^2 + b^2 - c^2}{2abc} = \frac{b^2 + c^2 - a^2 + a^2 + c^2 - b^2 + a^2 + b^2 - c^2}{2abc} = \frac{a^2 + b^2 + c^2}{2abc}$

8.4 Exercises

1. 2.83 **3.** 2.99 **5.** 14.98 **7.** 9.56 **9.** 3.86 **11.** 1.48 **13.** 2.82 **15.** 1.53 **17.** 30 **19.** 1.73 **21.** 19.90 **23.** 19.81 **25.** $5446.38 **27.** 31,144 sq. ft. **29.** $A = \frac{1}{2}ab\sin\gamma = \frac{1}{2}a\sin\gamma\left(\frac{a\sin\beta}{\sin\alpha}\right) = \frac{a^2\sin\beta\sin\gamma}{2\sin\alpha}$ **31.** 0.92 **33.** 2.27 **35.** 5.44 **37.** 0.84

39. $A = \frac{1}{2}r^2(\theta + \sin\theta)$

8.5 Exercises

1. Amplitude $= 2$; Period $= 2\pi$ **3.** Amplitude $= 4$; Period $= \pi$ **5.** Amplitude $= 6$; Period $= 2$ **7.** Amplitude $= \frac{1}{2}$; Period $= \frac{4\pi}{3}$

9. Amplitude $= \frac{5}{3}$; Period $= 3$ **11.** F **13.** A **15.** H **17.** C **19.** J **21.** A **23.** D **25.** B

27.

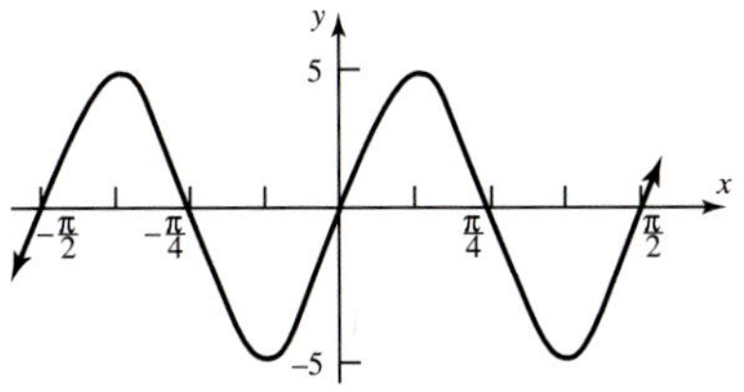

29.

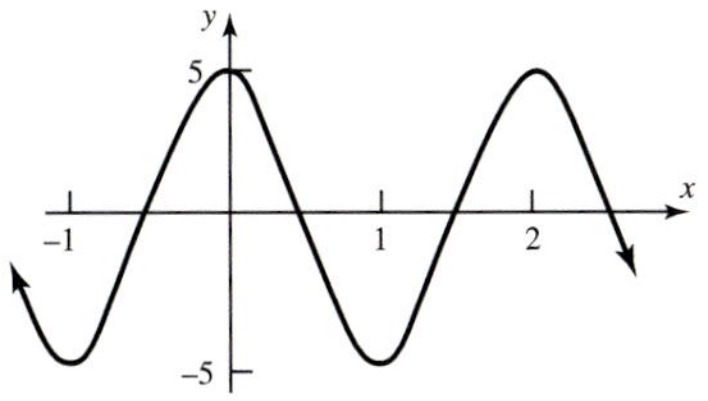

31.

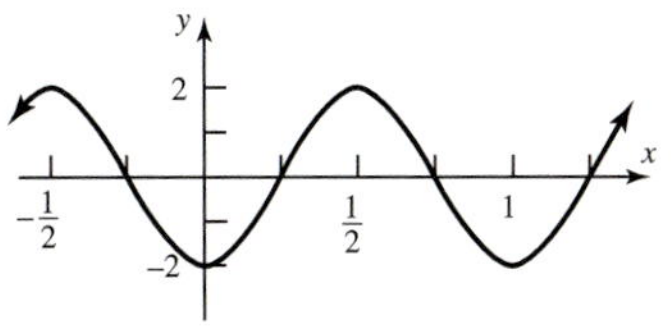

33.

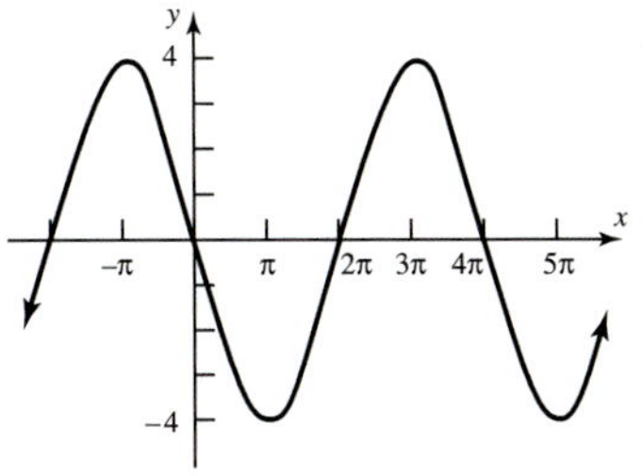

35.

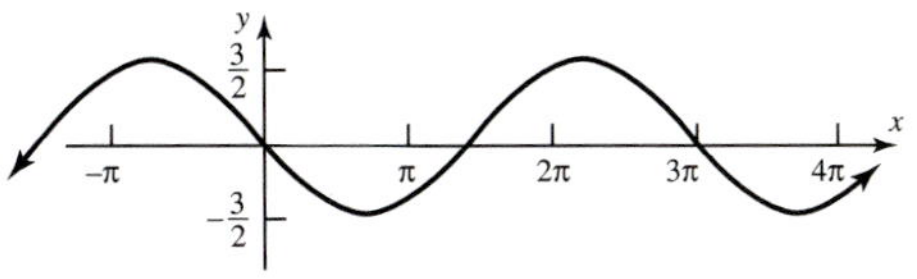

37. $y = 5\cos\frac{\pi}{4}x$ **39.** $y = -3\cos\frac{1}{2}x$ **41.** $y = \frac{3}{4}\sin 2\pi x$ **43.** $y = -\sin\frac{3}{2}x$ **45.** $y = -2\cos\frac{3\pi}{2}x$ **47.** $y = 3\sin\frac{\pi}{2}x$

49. $y = -4\cos 3x$

51. Amplitude = 4

Period = π

Phase shift = $\frac{\pi}{2}$

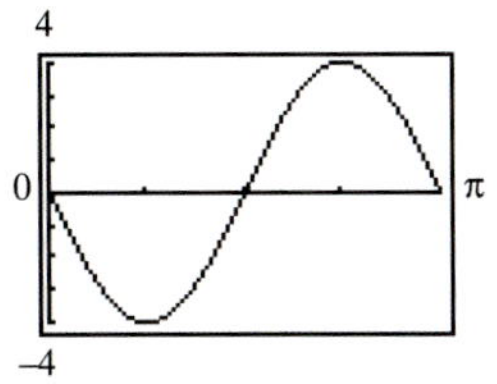

53. Amplitude = 2

Period = $\frac{2\pi}{3}$

Phase shift = $-\frac{\pi}{6}$

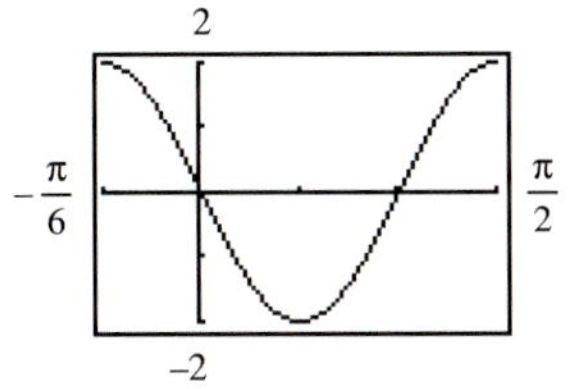

55. Amplitude = 3

Period = π

Phase shift = $-\frac{\pi}{4}$

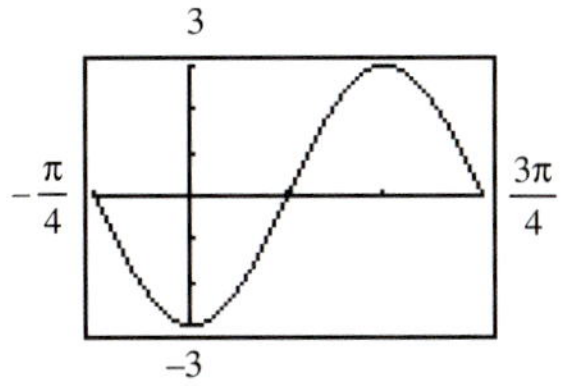

57. Amplitude = 4
Period = 2

Phase shift = $-\frac{2}{\pi}$

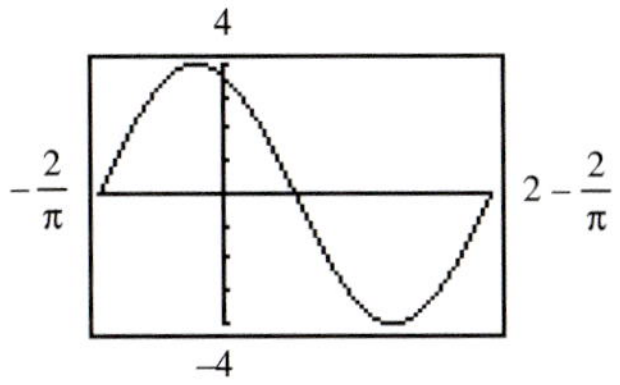

59. Amplitude = 3
Period = 2

Phase shift = $\frac{2}{\pi}$

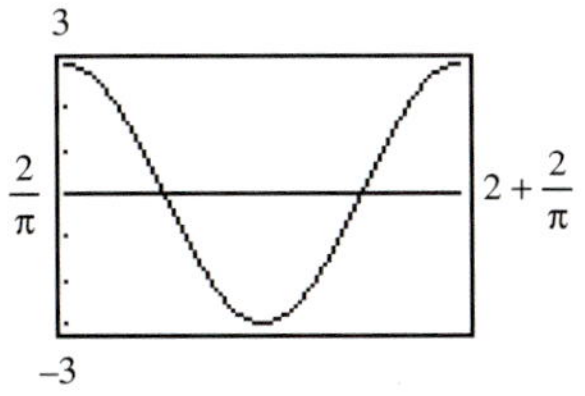

61. Amplitude = 3
Period = π

Phase shift = $\frac{\pi}{4}$

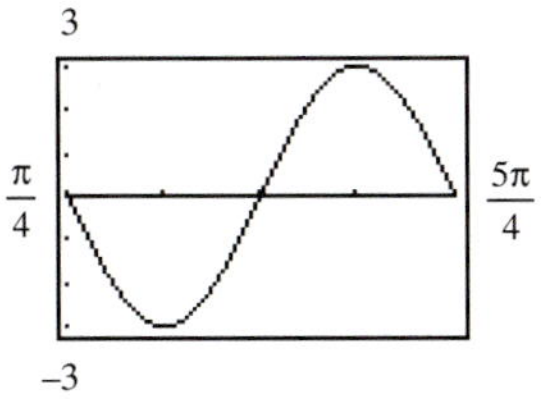

63. Period = $\frac{1}{30}$, Amplitude = 220

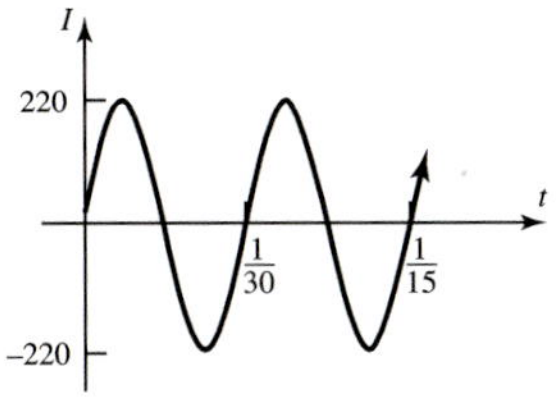

65. Period = $\frac{1}{15}$, Amplitude = 120, Phase shift = $\frac{1}{90}$

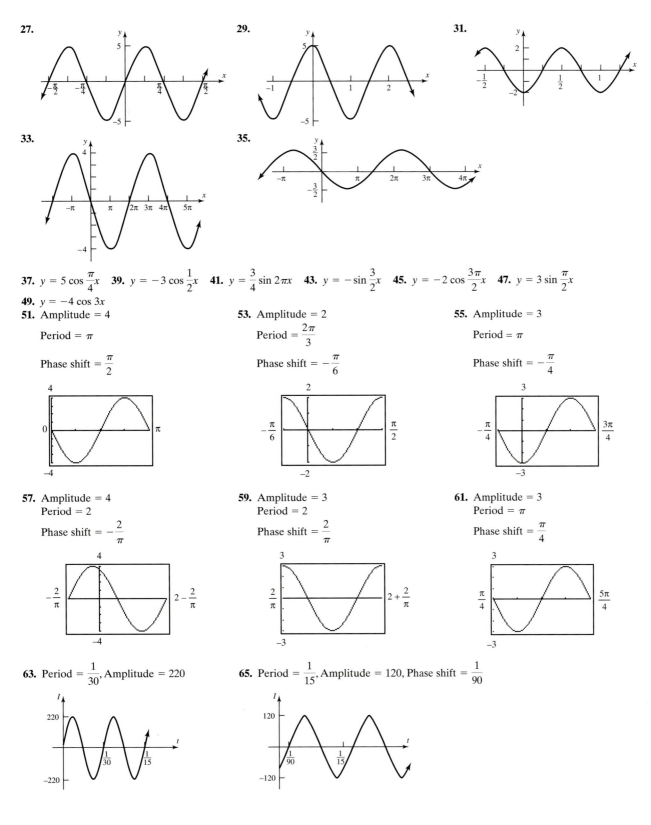

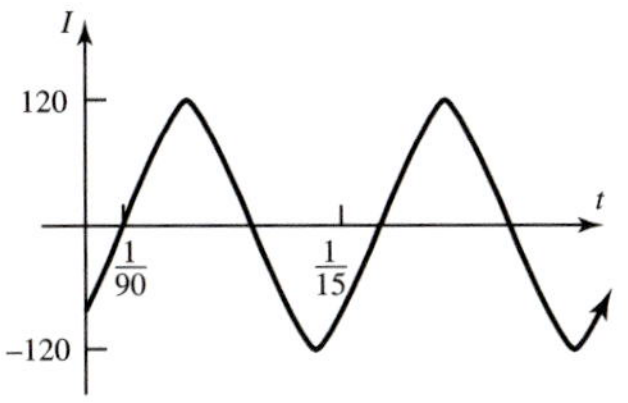

67. (a) Amplitude = 220, period = $\frac{1}{60}$ **(c)** $I = 22 \sin 120\pi t$

(b) & **(e)** **(d)** Amplitude = 22, Period = $\frac{1}{60}$

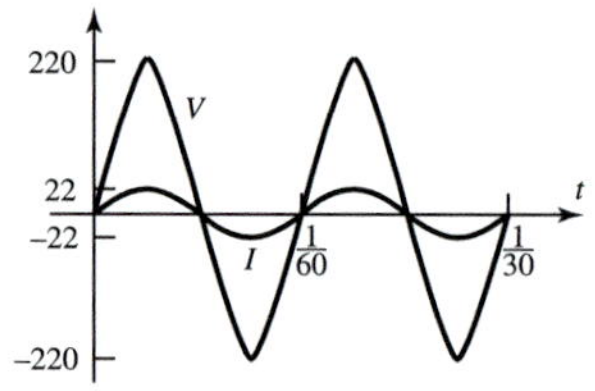

69. (a) $P = \frac{(V_0 \sin 2\pi ft)^2}{R} = \frac{V_0^2}{R} \sin^2 2\pi ft$ **(b)** $P = \frac{V_0^2}{R} \frac{1}{2}(1 - \cos 4\pi ft)$

71. (a)

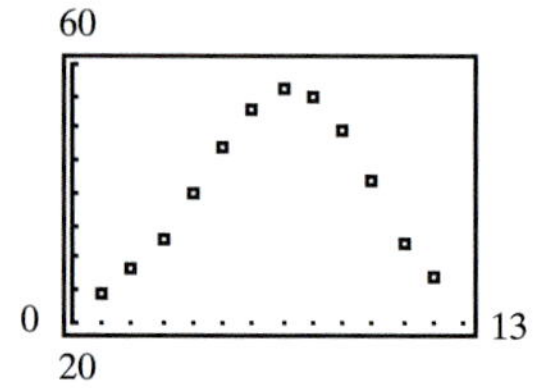

(b) $y = 15.9 \sin\left(\frac{\pi}{6}x - \frac{2\pi}{3}\right) + 40.1$ **(c)**

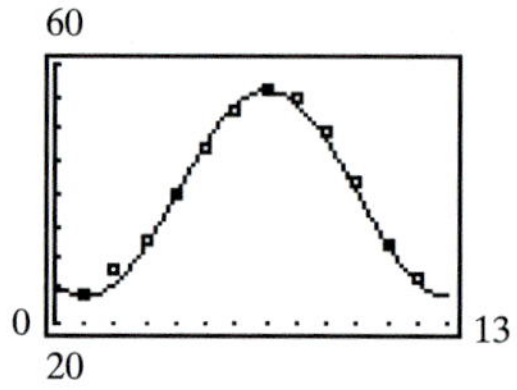

(d) $y = 15.62 \sin(0.517x - 2.096) + 40.377$ **(e)**

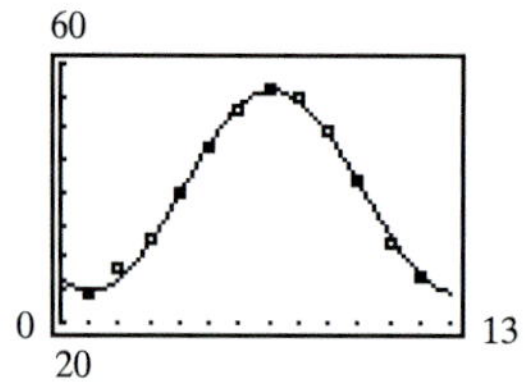

73. (a)

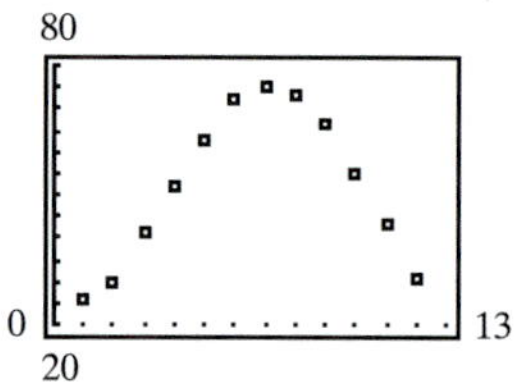

(b) $y = 24.95 \sin\left(\frac{\pi}{6}x - \frac{2\pi}{3}\right) + 50.45$ **(c)**

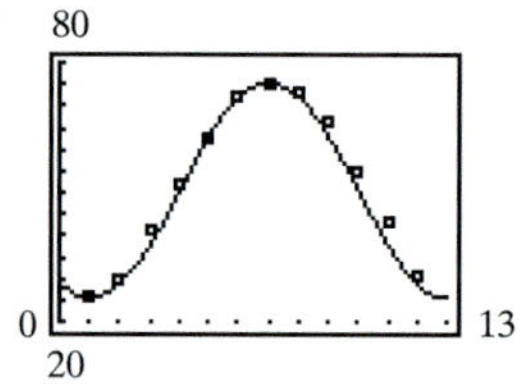

(d) $y = 25.693 \sin(0.476x - 1.814) + 49.854$

(e)

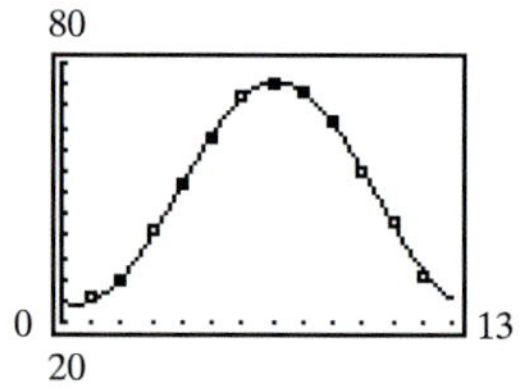

75. (a) 4:08 P.M. **(b)** $y = 4.4 \sin\left(\frac{\pi}{6.25}x - 6.6643\right) + 3.8$ **(c)**

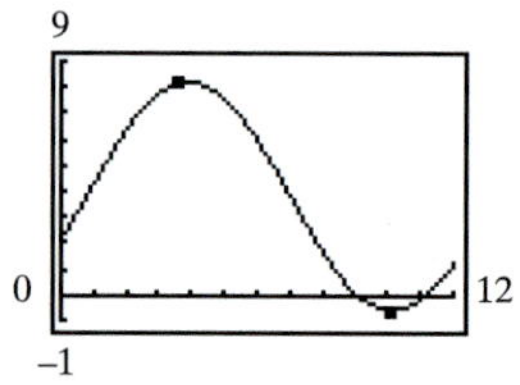

(d) 8.2 feet **77. (a)** $y = 1.0835 \sin\left(\frac{2\pi}{365}x - 7.6818\right) + 11.6665$ **(b)**

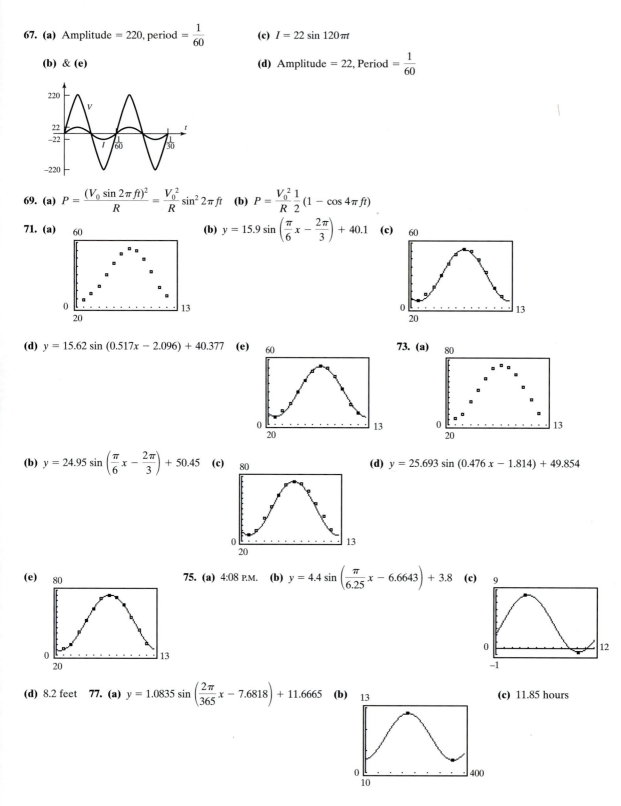

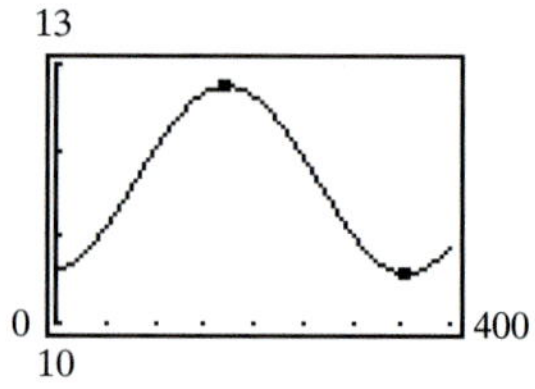

(c) 11.85 hours

79. **(a)** $y = 5.3915 \sin\left(\frac{2\pi}{365}x - 7.6818\right) + 10.8415$ **(b)**

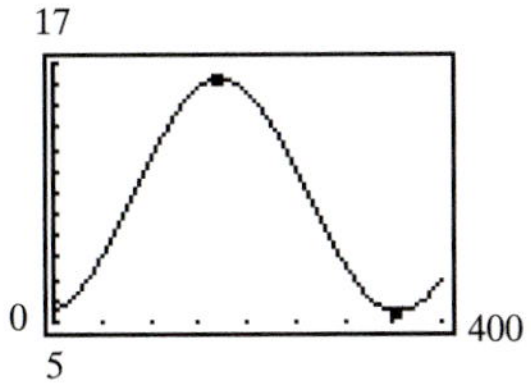

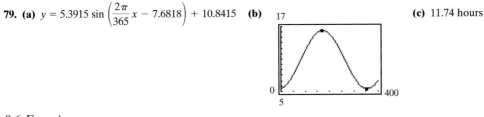

(c) 11.74 hours

8.6 Exercises

1. $d = -5 \cos \pi t$ **3.** $d = -6 \cos 2t$ **5.** $d = -5 \sin \pi t$ **7.** $d = -6 \sin 2t$ **9.** **(a)** Simple harmonic **(b)** 5 m **(c)** $\frac{2\pi}{3}$ sec **(d)** $\frac{3}{2\pi}$ oscillation/sec **11.** **(a)** Simple harmonic **(b)** 6 m **(c)** 2 sec **(d)** $\frac{1}{2}$ oscillation/sec **13.** **(a)** Simple harmonic **(b)** 3 m **(c)** 4π sec **(d)** $\frac{1}{4\pi}$ oscillation/sec **15.** **(a)** Simple harmonic **(b)** 2 m **(c)** 1 sec **(d)** 1 oscillation/sec

17. **(a)** $d = -10\, e^{-0.7t/50} \cos\left[\sqrt{\left(\frac{2\pi}{5}\right)^2 - \frac{(0.7)^2}{4(625)}}\, t\right]$

(b)

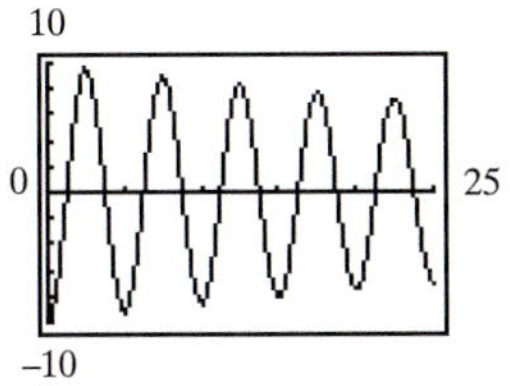

19. **(a)** $d = -18\, e^{-0.6t/60} \cos\left[\sqrt{\left(\frac{\pi}{2}\right)^2 - \frac{(0.6)^2}{4(900)}}\, t\right]$

(b)

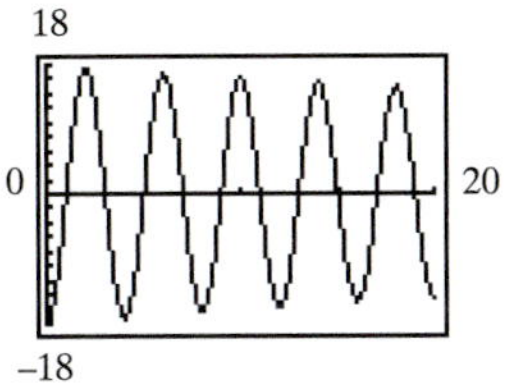

21. **(a)** $d = -5\, e^{-0.8t/20} \cos\left[\sqrt{\left(\frac{2\pi}{3}\right)^2 - \frac{(0.8)^2}{4(100)}}\, t\right]$

(b)

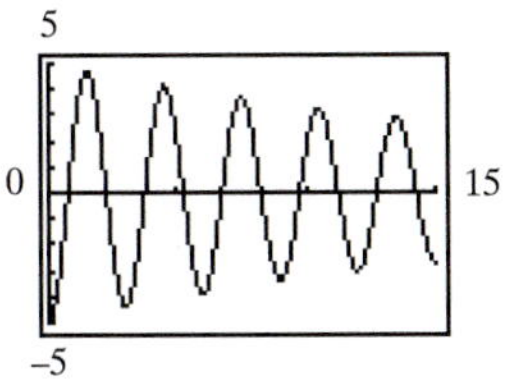

23. **(a)** The motion is damped. The bob has mass $m = 20$ kg with a damping factor of 0.7. **(b)** 20 m downward

(c)

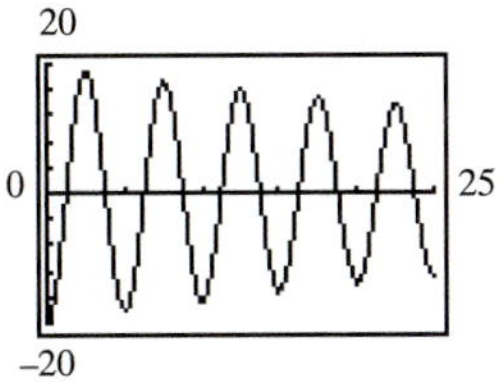

(d) 18.32 meters **(e)** $d \to 0$

25. **(a)** The motion is damped. The bob has mass $m = 40$ kg with a damping factor of 0.6. **(b)** 30 m downward

(c)

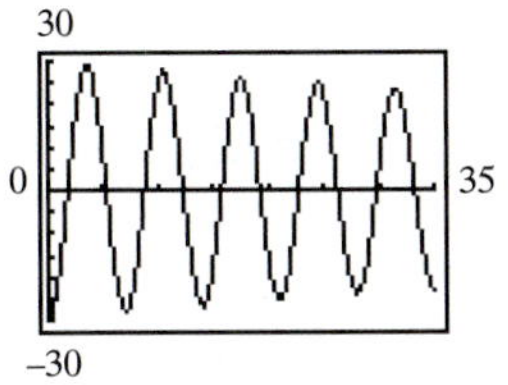

(d) 28.46 meters **(e)** $d \to 0$

27. **(a)** The motion is damped. The bob has mass $m = 15$ kg with a damping factor of 0.9 **(b)** 15 m downward

(c)

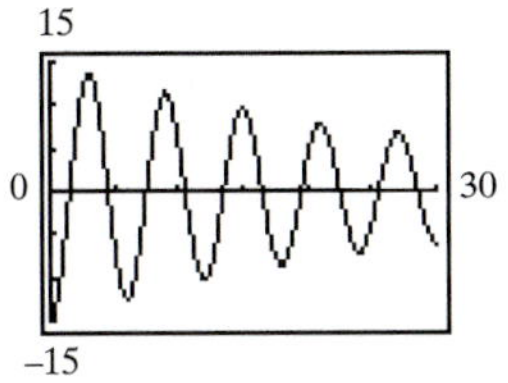

(d) 12.53 meters **(e)** $d \to 0$

29. (a)

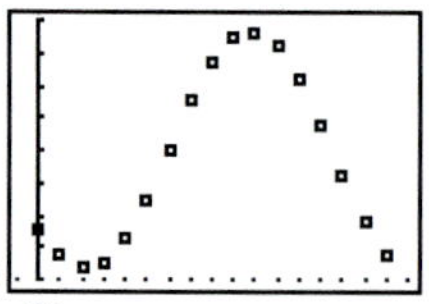

(b) $y = 10.935 \sin\left(\frac{2\pi}{3}x - 2.4086\right) + 29.995$, assuming $T = 3$ seconds

(c)

(d) $y = 11.043 \sin(2.076x - 2.472) + 29.877$ **(e)**

(f) 3.03 seconds **31. (a)**

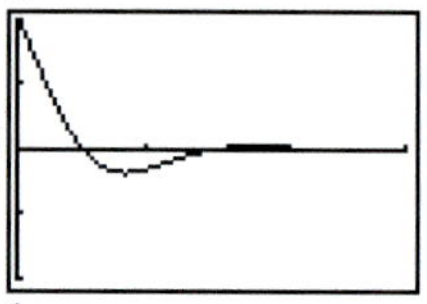

(b) The graph of V touches the graph of $y = e^{-1.9t}$ when $t = 0, 2, 4, \cdots$. The graph of V touches the graph of $y = -e^{-1.9t}$ when $t = 1, 3, 5, \cdots$.
(c) $-0.1 < V < 0.1$ for $t > 1.15$

Fill-In-The Blank Items

1. Sines **2.** Cosines **3.** Heron's **4.** $y = 3 \sin \pi x$ **5.** $3; \frac{\pi}{3}$ **6.** 5; 0.5 **7.** simple harmonic motion

True/False Items

1. F **2.** T **3.** T **4.** F

Review Exercises

1. $\alpha = 70°, b \approx 3.42, a \approx 9.4$ **3.** $a \approx 4.58, \alpha \approx 66.4°, \beta \approx 23.6°$ **5.** $\gamma = 100°, b = 0.65, c = 1.29$ **7.** $\beta = 56.8°, \gamma = 23.2°, b = 4.25$
9. No triangle **11.** $b = 3.32, \alpha = 62.8°, \gamma = 17.2°$ **13.** No triangle **15.** $c = 2.32, \alpha = 16.1°, \beta = 123.9°$
17. $\beta = 36.2°, \gamma = 63.8°, c = 4.56$ **19.** $\alpha = 39.6°, \beta = 18.6°, \gamma = 121.8°$
21. Two triangles: $\beta_1 = 13.4°, \gamma_1 = 156.6°, c_1 = 6.86; \beta_2 = 166.6°, \gamma_2 = 3.4°, c_2 = 1.02$ **23.** $a = 5.23, \beta = 46°, \gamma = 64°$ **25.** 1.93
27. 18.79 **29.** 6 **31.** 3.80 **33.** 0.32 **35.** Amplitude $= 4$; Period $= 2\pi$ **37.** Amplitude $= 8$; Period $= 4$

39. Amplitude $= 4$

Period $= \frac{2\pi}{3}$

Phase shift $= 0$

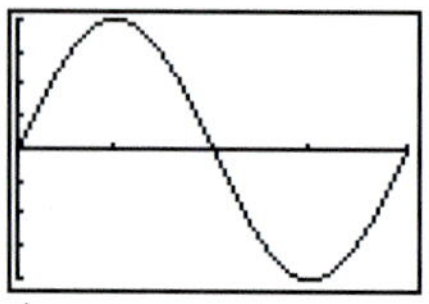

41. Amplitude $= 2$

Period $= 4$

Phase shift $= -\frac{1}{\pi}$

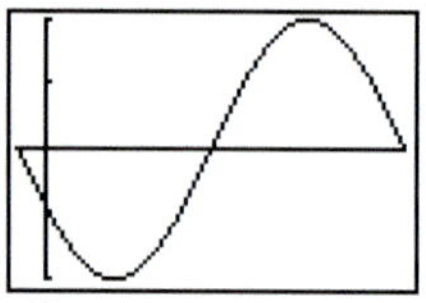

43. Amplitude $= \frac{1}{2}$

Period $= \frac{4\pi}{3}$

Phase shift $= \frac{2\pi}{3}$

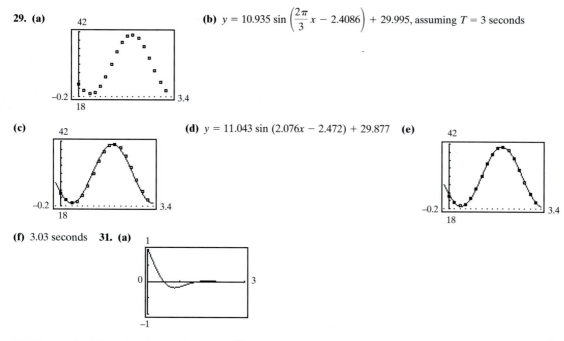

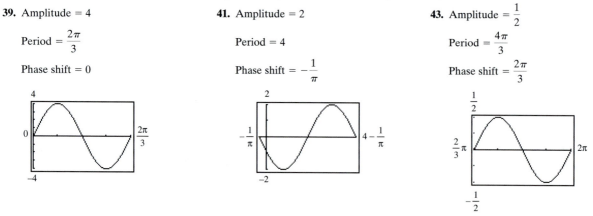

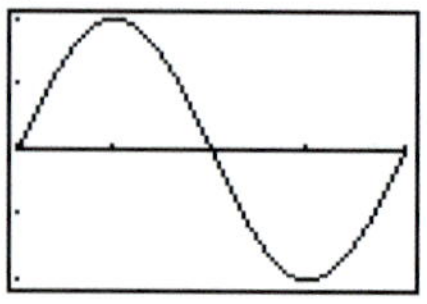

45. Amplitude $= \frac{2}{3}$
Period = 2
Phase shift $= \frac{6}{\pi}$

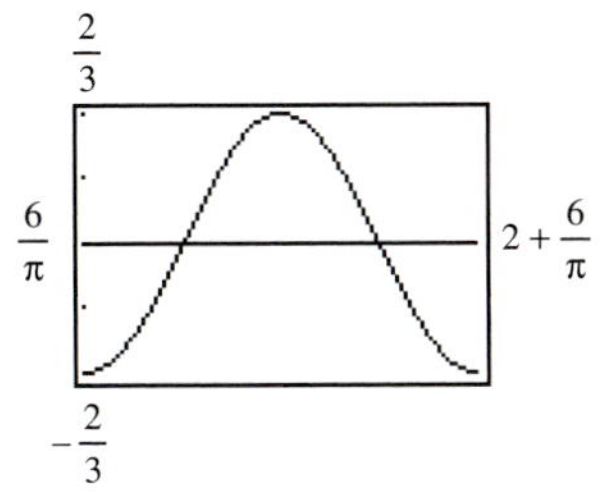

47. $y = 5 \cos \frac{x}{4}$

49. $y = -6 \cos \frac{\pi}{4} x$

51. **(a)** Simple harmonic **(b)** 6 ft **(c)** π sec **(d)** $\frac{1}{\pi}$ oscillation/sec **53.** **(a)** Simple harmonic **(b)** 2 ft **(c)** 2 sec **(d)** $\frac{1}{2}$ oscillation/sec **55.** **(a)** $d = -15 e^{-0.75t/80} \cos\left[\sqrt{\left(\frac{2\pi}{5}\right)^2 - \frac{(0.75)^2}{4(40)^2}}\, t\right]$ **(b)**

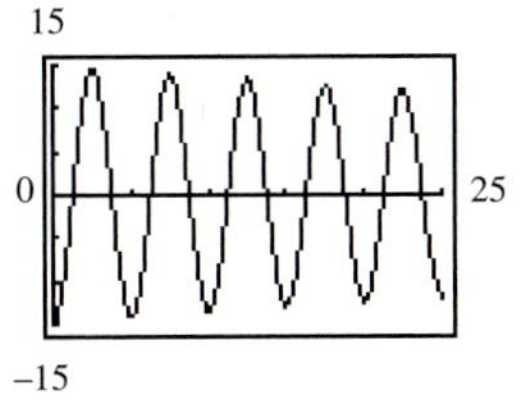

57. **(a)** The motion is damped. The bob has mass $m = 20$ kg with a damping factor of 0.6 **(b)** 15 m downward **(c)**

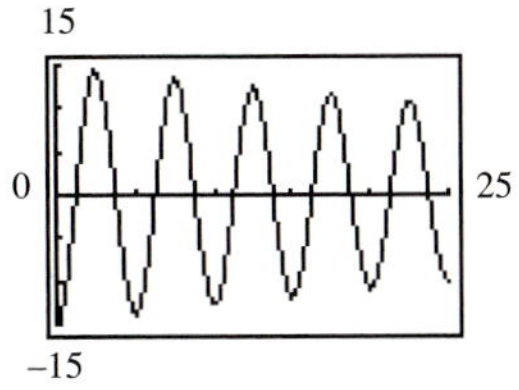

(d) 13.91 meters **(e)** $d \to 0$ **59.** **(a)** 120 V **(b)** $\frac{1}{60}$ s **(c)**

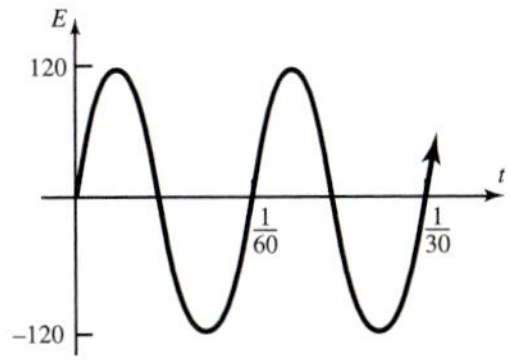

61. 839 ft **63.** 23.32 ft **65.** 2.15 mi **67.** 204.1 mi **69.** **(a)** 2.59 mi **(b)** 2.92 mi **(c)** 2.53 mi **71.** **(a)** 131.8 mi **(b)** 23.1° **(c)** 0.2 hr **73.** 8799 sq ft. **75.** 76.9 in. **77.** **(a)**

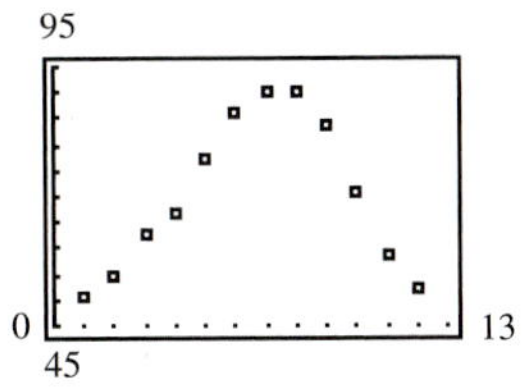

(b) $y = 19.5 \sin\left(\frac{\pi}{6}x - \frac{2\pi}{3}\right) + 70.5$

(c)

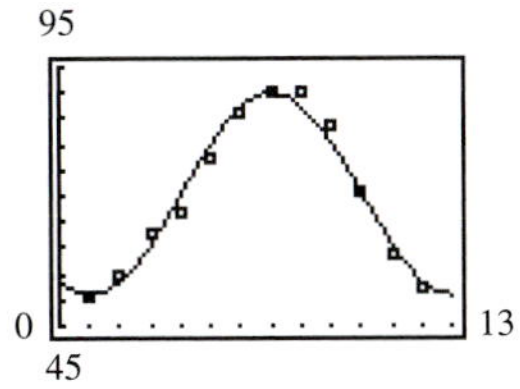

(d) $y = 19.52 \sin(0.54x - 2.28) + 71.01$ **(e)**

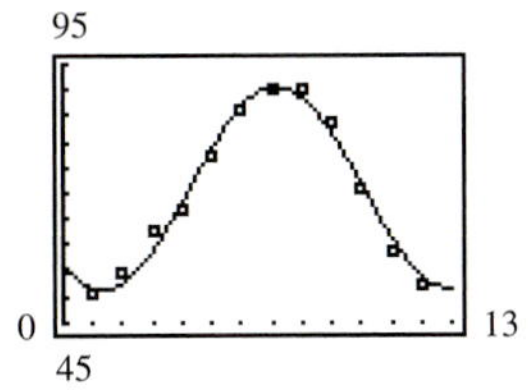

79. **(a)** $y = 1.85 \sin\left(\frac{2\pi}{365}x - 7.6818\right) + 11.517$ **(b)**

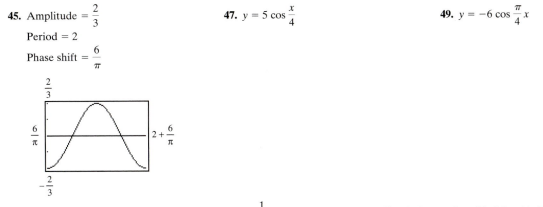

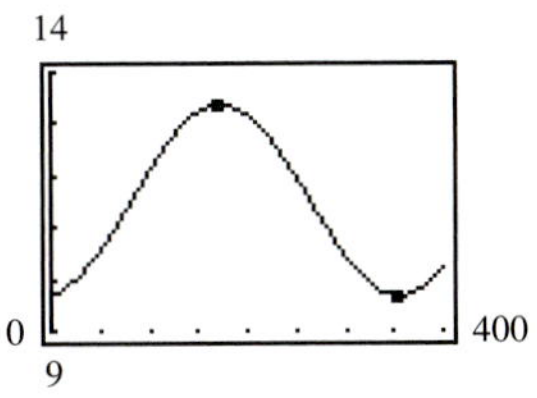

(c) 11.826 hours

CHAPTER 9 *9.1 Exercises*

1.
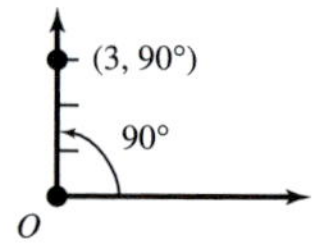

3.
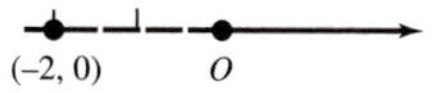

5.
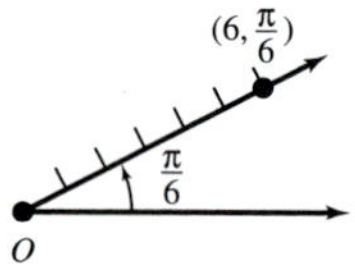

7.
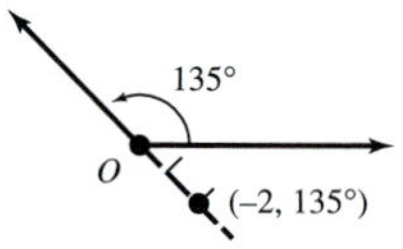

9.
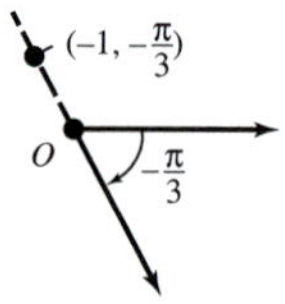

11.
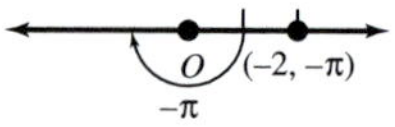

13.
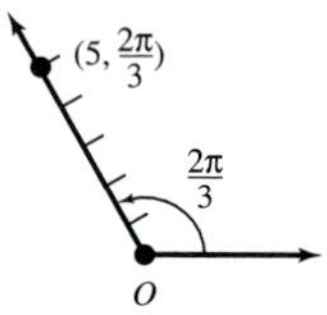

(a) $\left(5, -\dfrac{4\pi}{3}\right)$
(b) $\left(-5, \dfrac{5\pi}{3}\right)$
(c) $\left(5, \dfrac{8\pi}{3}\right)$

15.
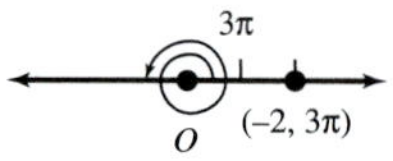

(a) $(2, -2\pi)$
(b) $(-2, \pi)$
(c) $(2, 2\pi)$

17.
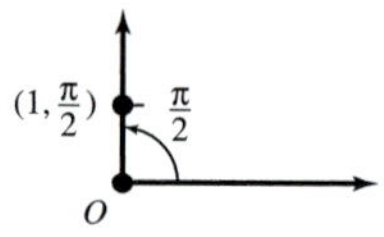

(a) $\left(1, -\dfrac{3\pi}{2}\right)$
(b) $\left(-1, \dfrac{3\pi}{2}\right)$
(c) $\left(1, \dfrac{5\pi}{2}\right)$

19.
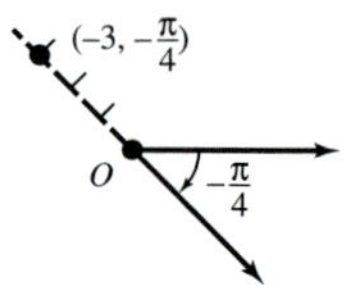

(a) $\left(3, -\dfrac{5\pi}{4}\right)$
(b) $\left(-3, \dfrac{7\pi}{4}\right)$
(c) $\left(3, \dfrac{11\pi}{4}\right)$

21. $(0, 3)$ **23.** $(-2, 0)$ **25.** $(-3\sqrt{3}, 3)$ **27.** $(\sqrt{2}, -\sqrt{2})$ **29.** $\left(-\dfrac{1}{2}, \dfrac{\sqrt{3}}{2}\right)$ **31.** $(2, 0)$ **33.** $(-2.57, 7.05)$ **35.** $(-4.98, -3.86)$ **37.** $(3, 0)$
39. $(1, \pi)$ **41.** $\left(\sqrt{2}, -\dfrac{\pi}{4}\right)$ **43.** $\left(2, \dfrac{\pi}{6}\right)$ **45.** $(2.47, -1.02)$ **47.** $(9.3, 0.47)$ **49.** $r^2 = \dfrac{3}{2}$ **51.** $r^2\cos^2\theta - 4r\sin\theta = 0$ **53.** $r^2\sin 2\theta = 1$
55. $r\cos\theta = 4$ **57.** $x^2 + y^2 - x = 0$ or $\left(x - \dfrac{1}{2}\right)^2 + y^2 = \dfrac{1}{4}$ **59.** $(x^2 + y^2)^{3/2} - x = 0$ **61.** $x^2 + y^2 = 4$ **63.** $y^2 = 8(x + 2)$
65. $d = \sqrt{(r_2\cos\theta_2 - r_1\cos\theta_1)^2 + (r_2\sin\theta_2 - r_1\sin\theta_1)^2} = \sqrt{r_1^2 + r_2^2 - 2r_1r_2\cos(\theta_2 - \theta_1)}$

9.2 Exercises

1. Circle, radius 4, center at pole

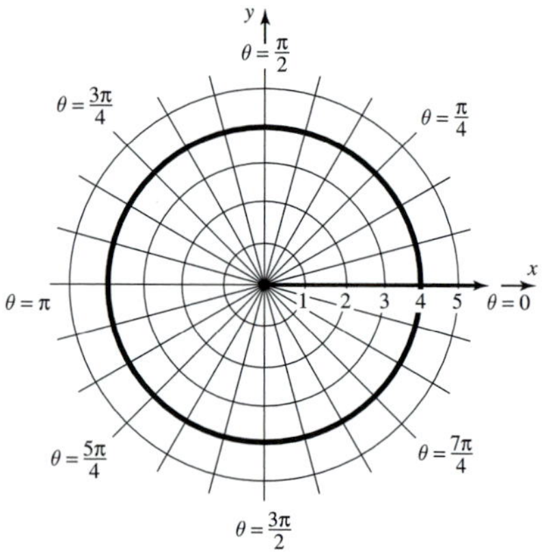

3. Line through pole, making an angle of $\dfrac{\pi}{3}$ with polar axis

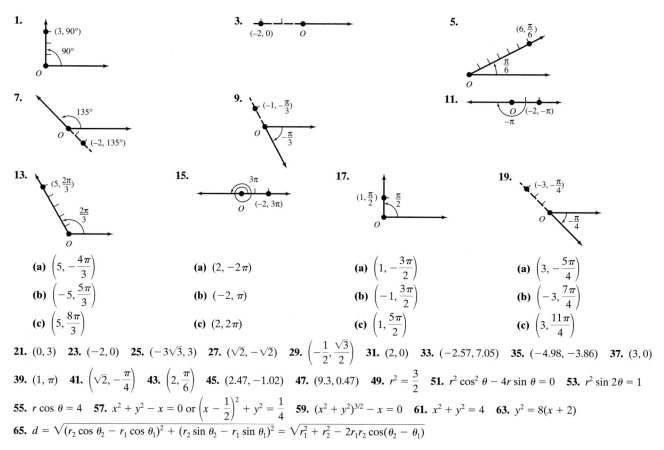

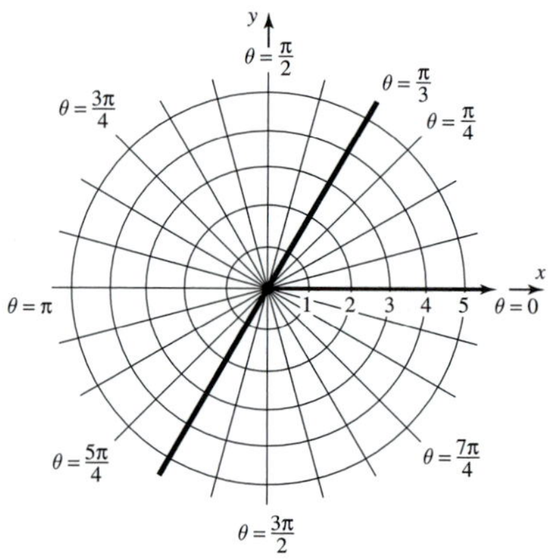

5. Horizontal line 4 units above the pole

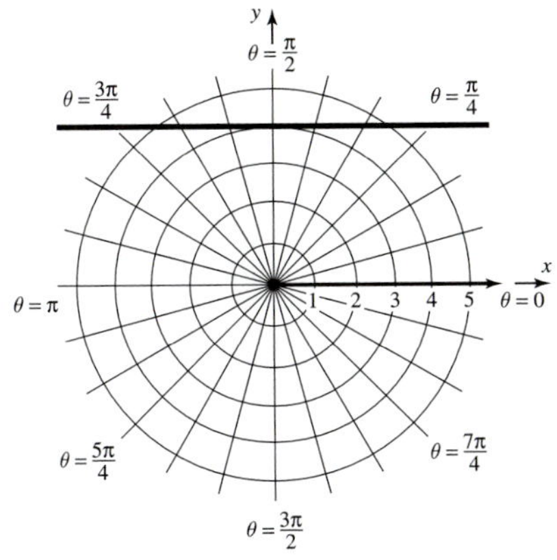

7. Vertical line 2 units to the left of the pole

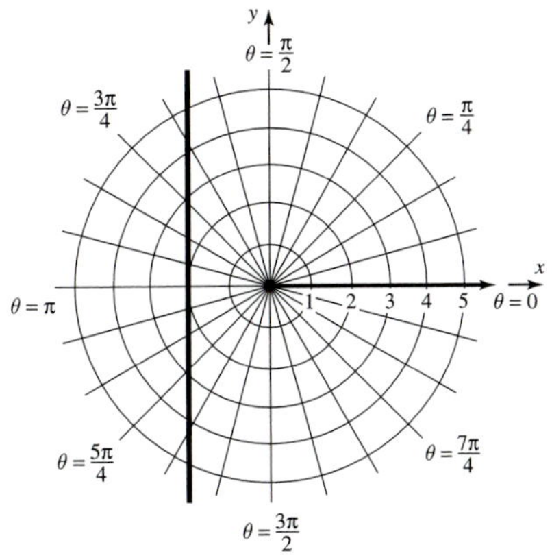

9. Circle, radius 1, center $(1, 0)$ in rectangular coordinates

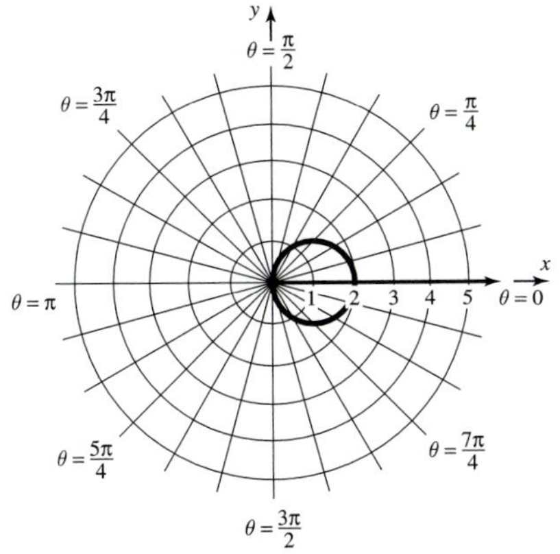

11. Circle, radius 2, center at $(0, -2)$ in rectangular coordinates

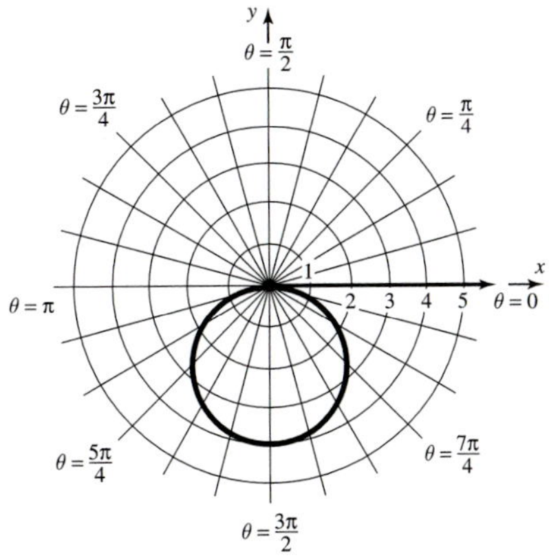

13. Circle, radius 2, center at $(2, 0)$ in rectangular coordinates

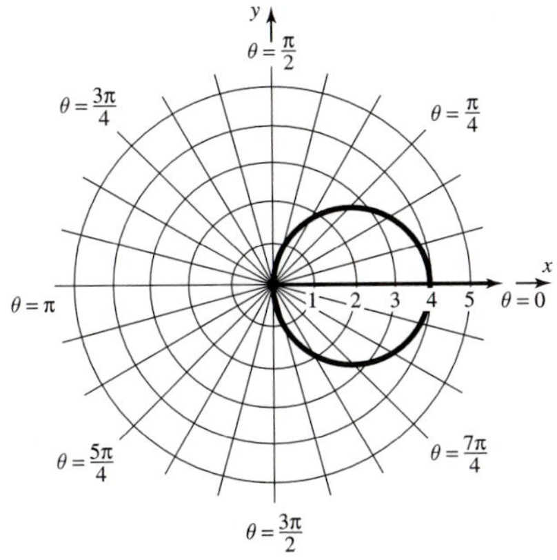

15. Circle, radius 1, center $(0, -1)$ in rectangular coordinates

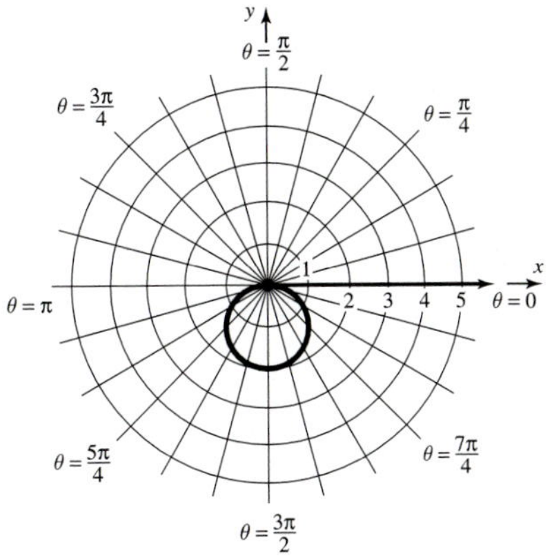

17. E **19.** F **21.** H **23.** D **25.** D **27.** F **29.** A

31. Cardioid

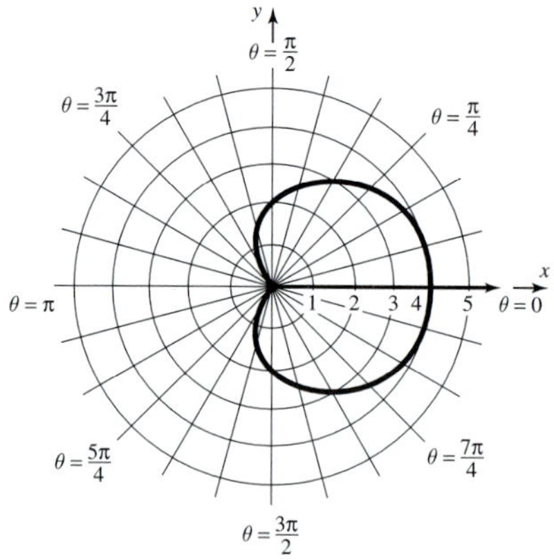

33. Cardioid

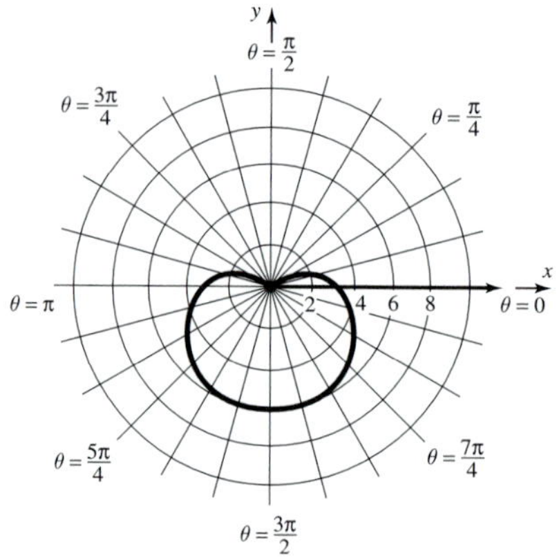

35. Limaçon without inner loop

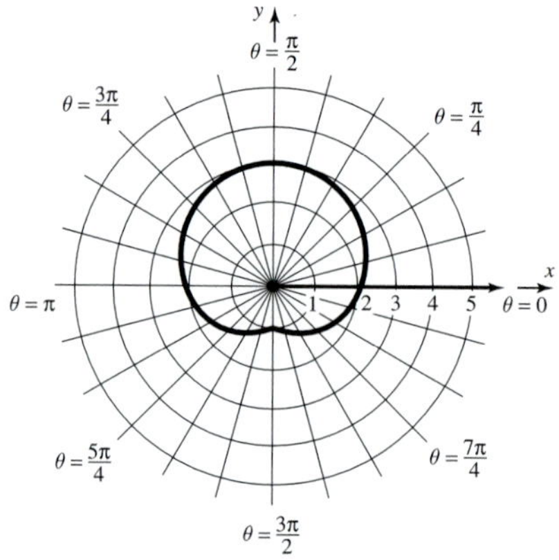

37. Limaçon without inner loop

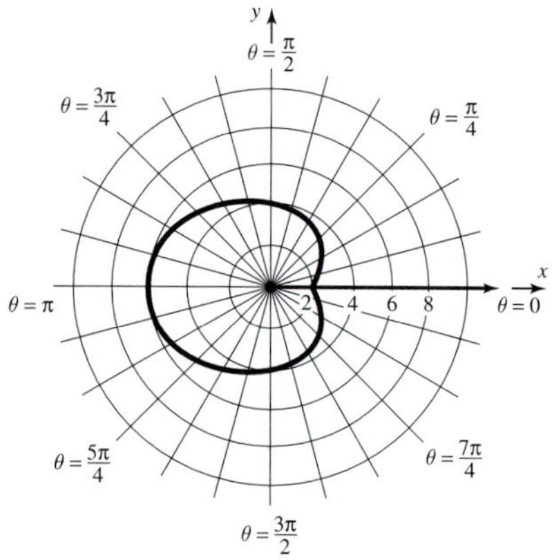

39. Limaçon with inner loop

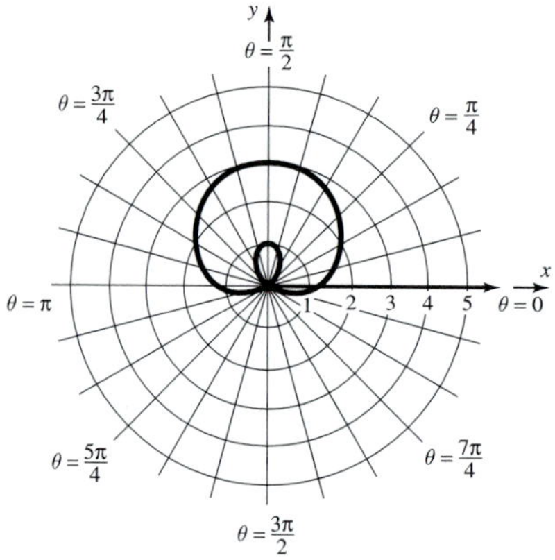

41. Limaçon with inner loop

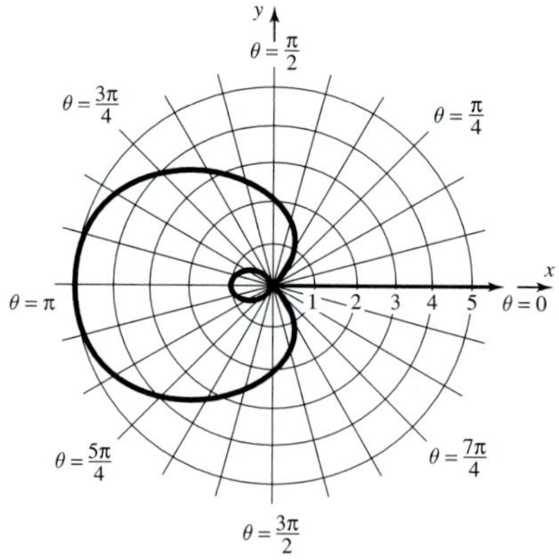

43. Rose

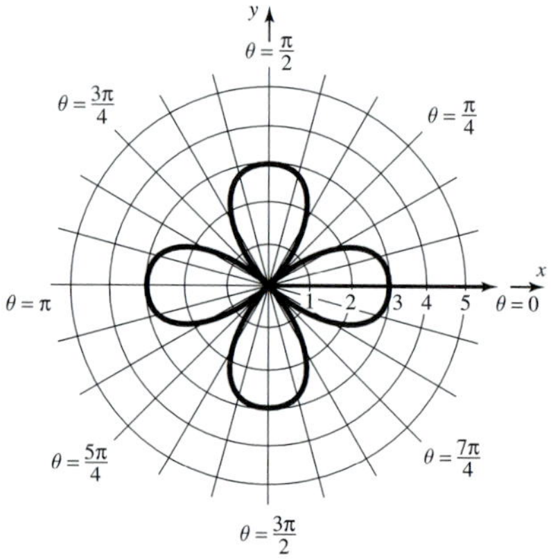

45. Rose

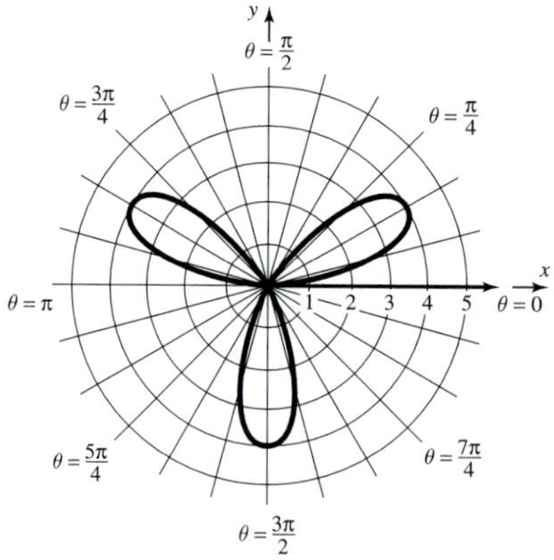

47. Lemniscate

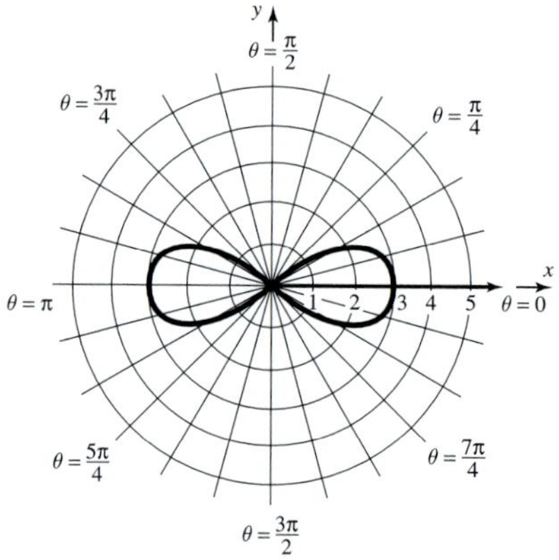

49. Spiral

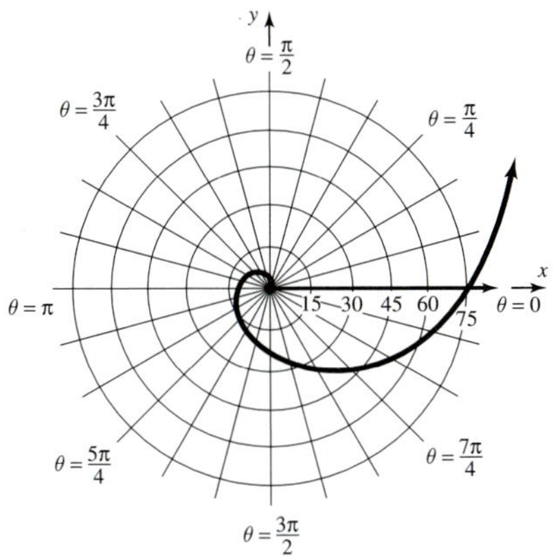

51. Cardioid

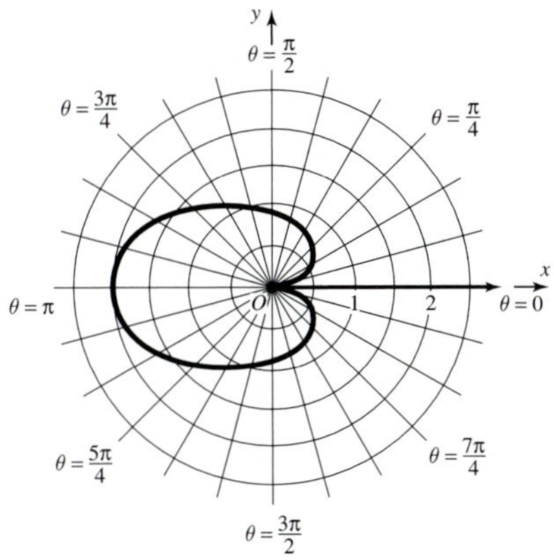

53. Limaçon with inner loop

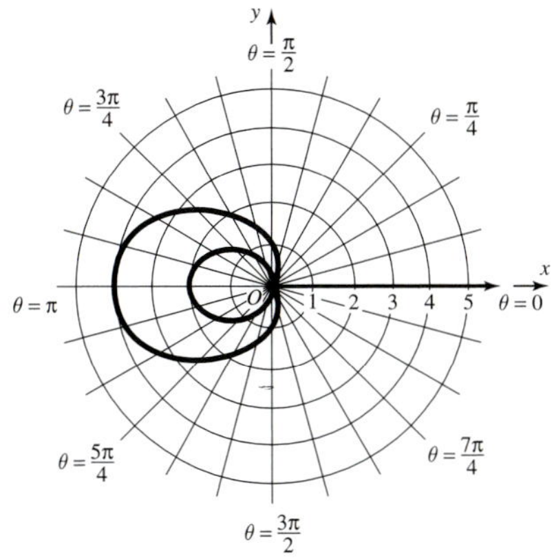

55.

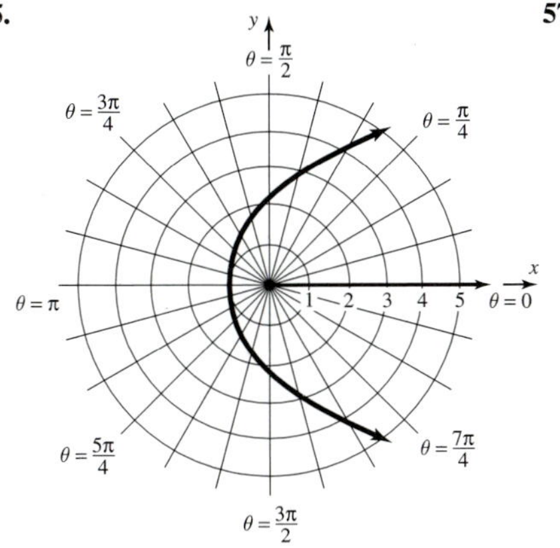

57.

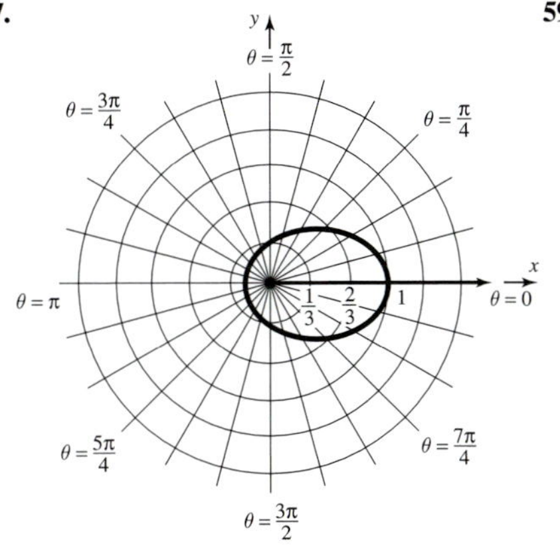

59.

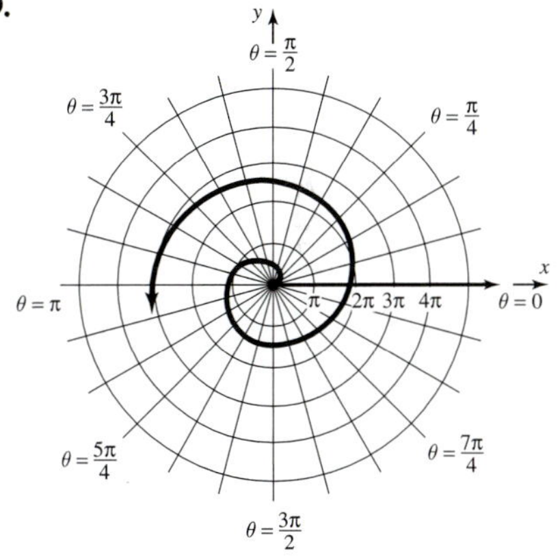

61.

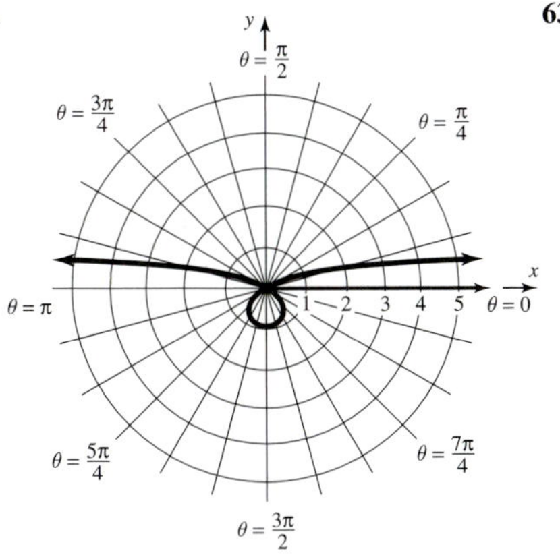

63.

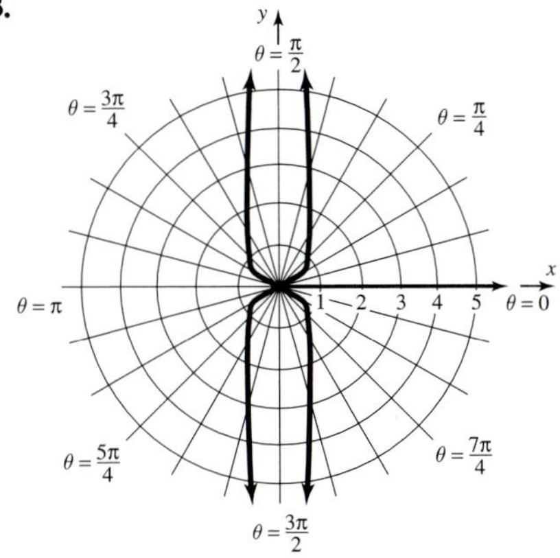

65.

$$\begin{aligned} r \sin \theta &= a \\ y &= a \end{aligned}$$

67.

$$\begin{aligned} r &= 2a \sin \theta \\ r^2 &= 2ar \sin \theta \\ x^2 + y^2 &= 2ay \\ x^2 + y^2 - 2ay &= 0 \\ x^2 + (y - a)^2 &= a^2 \end{aligned}$$

Circle, radius a, center at $(0, a)$ in rectanglar coordinates

69.

$$\begin{aligned} r &= 2a \cos \theta \\ r^2 &= 2ar \cos \theta \\ x^2 + y^2 &= 2ax \\ x^2 - 2ax + y^2 &= 0 \\ (x - a)^2 + y^2 &= a^2 \end{aligned}$$

Circle, radius a, center at $(a, 0)$ in rectangular coordinates

71. **(a)** $r^2 = \cos\theta$: $r^2 = \cos(\pi - \theta)$, $r^2 = -\cos\theta$. Not equivalent; test fails. $(-r)^2 = \cos(-\theta)$, $r^2 = \cos\theta$. New test works

(b) $r^2 = \sin\theta$: $r^2 = \sin(\pi - \theta)$, $r^2 = \sin\theta$. Test Works. $(-r)^2 = \sin(-\theta)$, $r^2 = -\sin\theta$. Not equivalent; new test fails

9.3 Exercises

1.

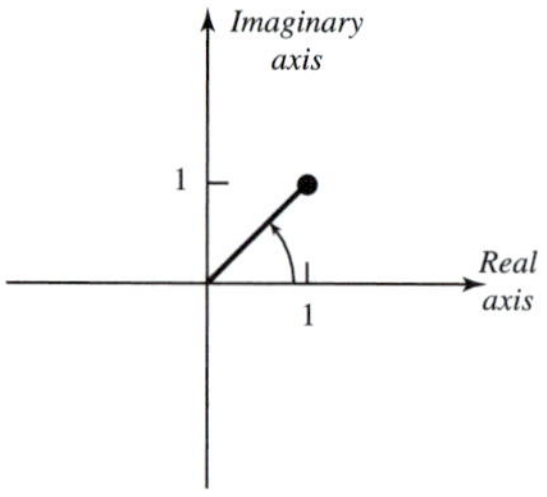

$\sqrt{2}(\cos 45^\circ + i\sin 45^\circ)$

3.

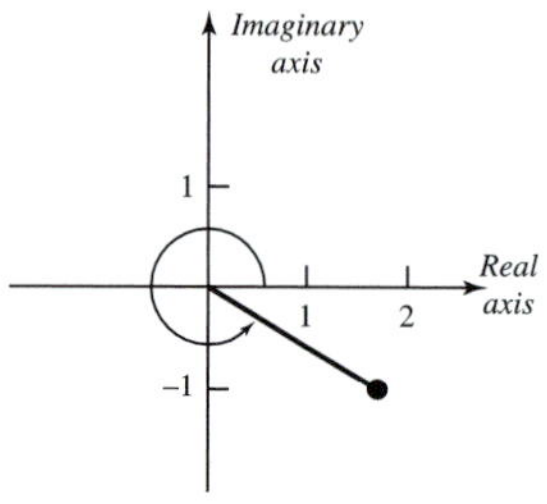

$2(\cos 330^\circ + i\sin 330^\circ)$

5.

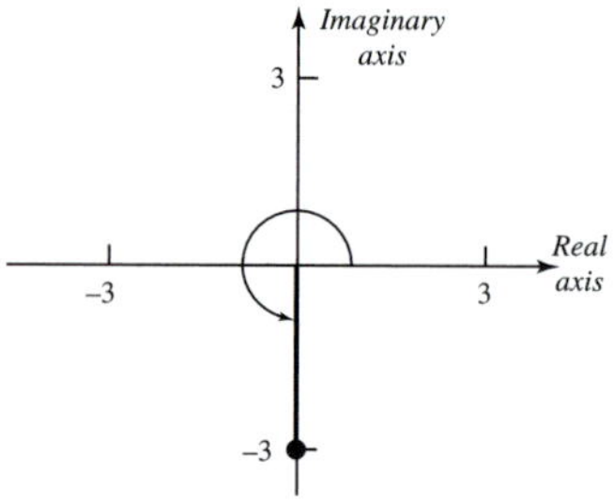

$3(\cos 270^\circ + i\sin 270^\circ)$

7.

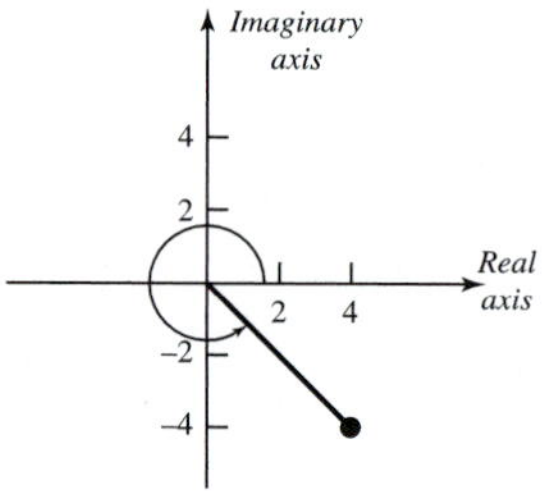

$4\sqrt{2}(\cos 315^\circ + i\sin 315^\circ)$

9.

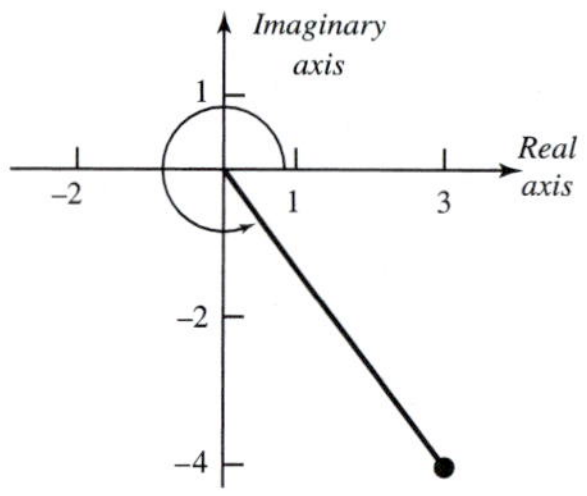

$5(\cos 306.9^\circ + i\sin 306.9^\circ)$

11.

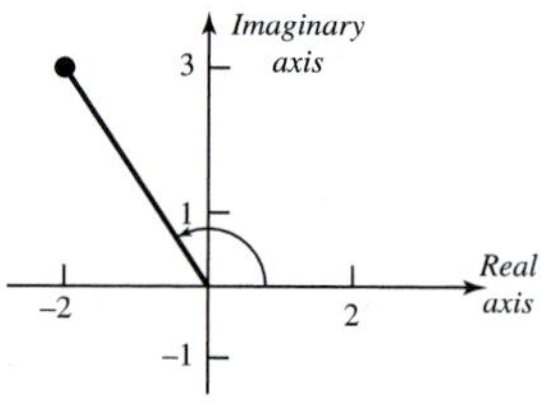

$\sqrt{13}(\cos 123.7^\circ + i\sin 123.7^\circ)$

13. $-1 + \sqrt{3}i$ **15.** $2\sqrt{2} - 2\sqrt{2}i$ **17.** $-3i$ **19.** $-0.035 + 0.197i$ **21.** $1.97 + 0.347i$

23. $zw = 8(\cos 60^\circ + i\sin 60^\circ)$; $\dfrac{z}{w} = \dfrac{1}{2}(\cos 20^\circ + i\sin 20^\circ)$ **25.** $zw = 12(\cos 40^\circ + i\sin 40^\circ)$; $\dfrac{z}{w} = \dfrac{3}{4}(\cos 220^\circ + i\sin 220^\circ)$

27. $zw = 4\left(\cos\dfrac{9\pi}{40} + i\sin\dfrac{9\pi}{40}\right)$; $\dfrac{z}{w} = \cos\dfrac{\pi}{40} + i\sin\dfrac{\pi}{40}$ **29.** $zw = 4\sqrt{2}(\cos 15^\circ + i\sin 15^\circ)$; $\dfrac{z}{w} = \sqrt{2}(\cos 75^\circ + i\sin 75^\circ)$

31. $-32 + 32\sqrt{3}i$ **33.** $32i$ **35.** $\dfrac{27}{2} + \dfrac{27\sqrt{3}}{2}i$ **37.** $-\dfrac{25\sqrt{2}}{2} + \dfrac{25\sqrt{2}}{2}i$ **39.** $-4 + 4i$ **41.** $-23 + 14.15i$

43. $\sqrt[6]{2}(\cos 15^\circ + i\sin 15^\circ)$, $\sqrt[6]{2}(\cos 135^\circ + i\sin 135^\circ)$, $\sqrt[6]{2}(\cos 255^\circ + i\sin 255^\circ)$

45. $\sqrt[4]{8}(\cos 75^\circ + i\sin 75^\circ)$, $\sqrt[4]{8}(\cos 165^\circ + i\sin 165^\circ)$, $\sqrt[4]{8}(\cos 255^\circ + i\sin 255^\circ)$, $\sqrt[4]{8}(\cos 345^\circ + i\sin 345^\circ)$

47. $2(\cos 67.5^\circ + i\sin 67.5^\circ)$, $2(\cos 157.5^\circ + i\sin 157.5^\circ)$, $2(\cos 247.5^\circ + i\sin 247.5^\circ)$, $2(\cos 337.5^\circ + i\sin 337.5^\circ)$

49. $\cos 18^\circ + i\sin 18^\circ$, $\cos 90^\circ + i\sin 90^\circ$, $\cos 162^\circ + i\sin 162^\circ$, $\cos 234^\circ + i\sin 234^\circ$, $\cos 306^\circ + i\sin 306^\circ$

51. $1, i, -1, -i$

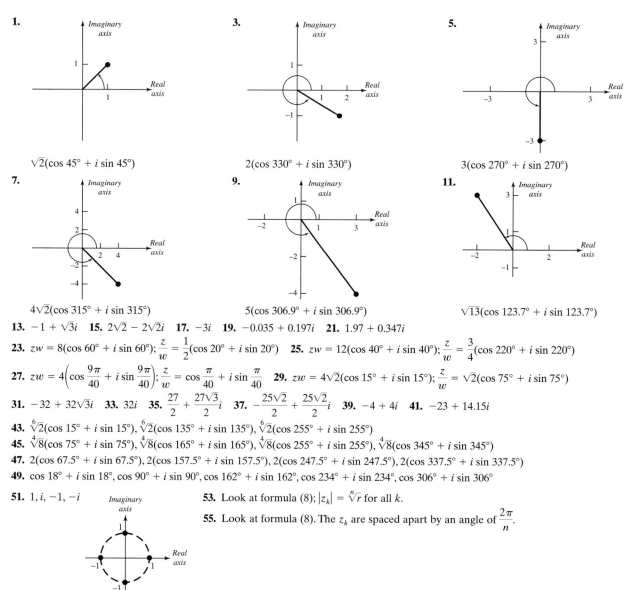

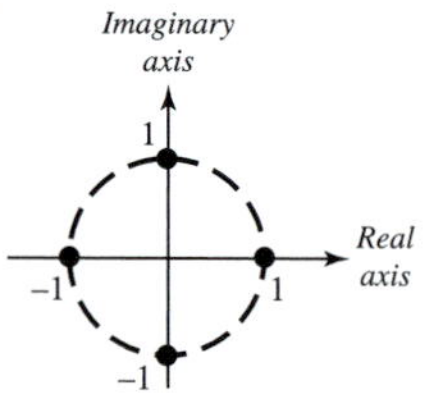

53. Look at formula (8); $|z_k| = \sqrt[n]{r}$ for all k.

55. Look at formula (8). The z_k are spaced apart by an angle of $\dfrac{2\pi}{n}$.

9.4 Exercises

1.

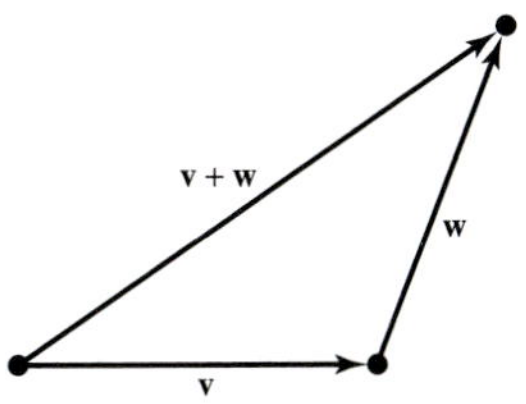

3.

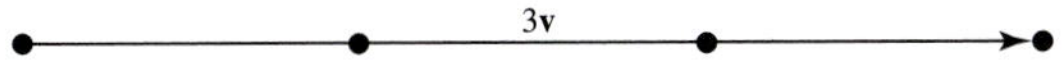

5.

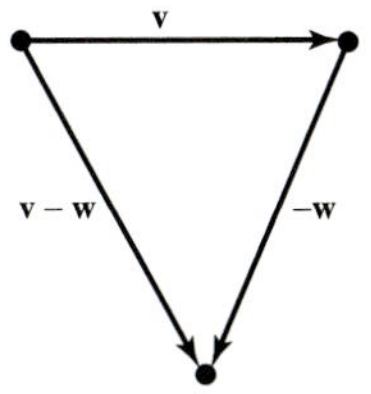

7.

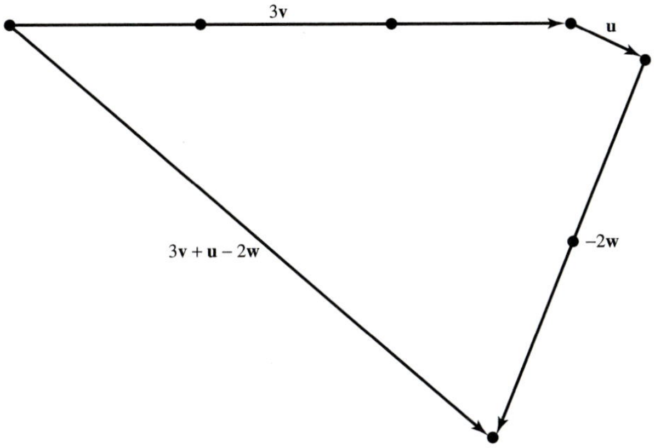

9. $\mathbf{x} = \mathbf{A}$ **11.** $\mathbf{C} = -\mathbf{F} + \mathbf{E} - \mathbf{D}$ **13.** $\mathbf{E} = -\mathbf{G} - \mathbf{H} + \mathbf{D}$ **15.** $x = 0$ **17.** 12 **19.** $\mathbf{v} = 3\mathbf{i} + 4\mathbf{j}$ **21.** $\mathbf{v} = 2\mathbf{i} + 4\mathbf{j}$ **23.** $\mathbf{v} = 8\mathbf{i} - \mathbf{j}$ **25.** $\mathbf{v} = -\mathbf{i} + \mathbf{j}$ **27.** 5 **29.** $\sqrt{2}$ **31.** $\sqrt{13}$ **33.** $-\mathbf{j}$ **35.** $\sqrt{89}$ **37.** $\sqrt{34} - \sqrt{13}$ **39.** $\mathbf{i}$ **41.** $\frac{3}{5}\mathbf{i} - \frac{4}{5}\mathbf{j}$ **43.** $\frac{\sqrt{2}}{2}\mathbf{i} - \frac{\sqrt{2}}{2}\mathbf{j}$ **45.** $\mathbf{v} = \frac{8\sqrt{5}}{5}\mathbf{i} + \frac{4\sqrt{5}}{5}\mathbf{j}$ or $\mathbf{v} = -\frac{8\sqrt{5}}{5}\mathbf{i} - \frac{4\sqrt{5}}{5}\mathbf{j}$ **47.** $\{-2 + \sqrt{21}, -2 - \sqrt{21}\}$ **49.** 460 kph **51.** 218 mph

53.

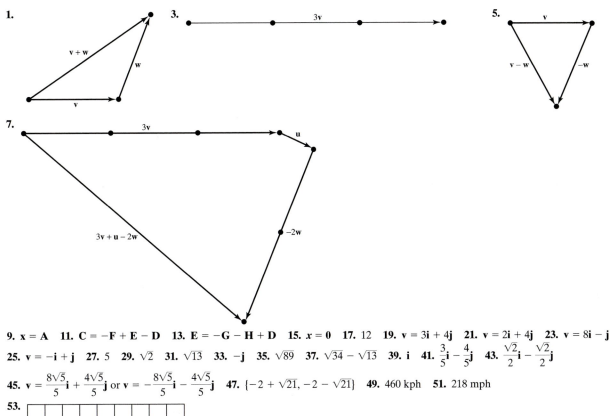

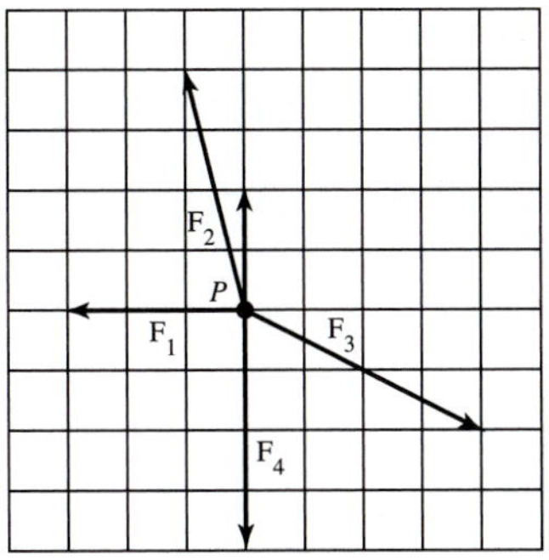

9.5 Exercises

1. 0; 90° **3.** 4; 36.87° **5.** $\sqrt{3} - 1$; 75° **7.** 24; 16.26° **9.** 0; 90° **11.** $\frac{3}{2}$ **13.** $\mathbf{v}_1 = \text{proj}_\mathbf{w}\, \mathbf{v} = \frac{5}{2}(\mathbf{i} - \mathbf{j}),\ \mathbf{v}_2 = -\frac{1}{2}\mathbf{i} - \frac{1}{2}\mathbf{j}$ **15.** $\mathbf{v}_1 = \text{proj}_\mathbf{w}\, \mathbf{v} = -\frac{1}{5}(\mathbf{i} + 2\mathbf{j}),\ \mathbf{v}_2 = \frac{6}{5}\mathbf{i} - \frac{3}{5}\mathbf{j}$ **17.** $\mathbf{v}_1 = \text{proj}_\mathbf{w}\, \mathbf{v} = \frac{7}{5}(2\mathbf{i} + \mathbf{j}),\ \mathbf{v}_2 = \frac{1}{5}\mathbf{i} - \frac{2}{5}\mathbf{j}$ **19.** 496.7 mph; 51.5° south of west **21.** 8.6° off direct heading into the current; 1.5 min **23.** 60°; 17.32 min **25.** $\frac{6\sqrt{5}}{5}$ ft-lb $\approx$ 2.68 ft-lb **27.** $1000\sqrt{3}$ ft-lb $\approx$ 1732 ft-lb

29. Let $\mathbf{u} = a_1\mathbf{i} + b_1\mathbf{j}$, $\mathbf{v} = a_2\mathbf{i} + b_2\mathbf{j}$, $\mathbf{w} = a_3\mathbf{i} + b_3\mathbf{j}$. Compute $\mathbf{u} \cdot (\mathbf{v} + \mathbf{w})$ and $\mathbf{u} \cdot \mathbf{v} + \mathbf{u} \cdot \mathbf{w}$.

31. $\cos\alpha = \frac{\mathbf{v} \cdot \mathbf{i}}{\|\mathbf{v}\|\,\|\mathbf{i}\|} = \mathbf{v} \cdot \mathbf{i}$; if $\mathbf{v} = x\mathbf{i} + y\mathbf{j}$, then $\mathbf{v} \cdot \mathbf{i} = x = \cos\alpha$ and $\mathbf{v} \cdot \mathbf{j} = y = \cos\left(\frac{\pi}{2} - \alpha\right) = \sin\alpha$.

33. $\mathbf{v} = a\mathbf{i} + b\mathbf{j}$; $\text{proj}_\mathbf{i}\, v = \frac{\mathbf{v} \cdot \mathbf{i}}{\|\mathbf{i}\|^2}\mathbf{i} = (\mathbf{v} \cdot \mathbf{i})\mathbf{i}$; $\mathbf{v} \cdot \mathbf{i} = a$, $\mathbf{v} \cdot \mathbf{j} = b$, so $\mathbf{v} = (\mathbf{v} \cdot \mathbf{i})\mathbf{i} + (\mathbf{v} \cdot \mathbf{j})\mathbf{j}$

35. $(\mathbf{v} - \alpha\mathbf{w}) \cdot \mathbf{w} = \mathbf{v} \cdot \mathbf{w} - \alpha\mathbf{w} \cdot \mathbf{w} = \mathbf{v} \cdot \mathbf{w} - \alpha\|\mathbf{w}\|^2 = \mathbf{v} \cdot \mathbf{w} - \frac{\mathbf{v} \cdot \mathbf{w}}{\|\mathbf{w}\|^2}\|\mathbf{w}\|^2 = 0$ **37.** 0

Fill-in-the-Blank Items

1. pole; polar axis **2.** -2 **3.** $r = 2\cos\theta$ **4.** the polar axis (x-axis) **5.** magnitude or modulus; argument **6.** unit **7.** 0

True/False Items

1. F **2.** T **3.** F **4.** T **5.** T **6.** T **7.** T

Review Exercises

1. $\left(\frac{3\sqrt{3}}{2}, \frac{3}{2}\right)$

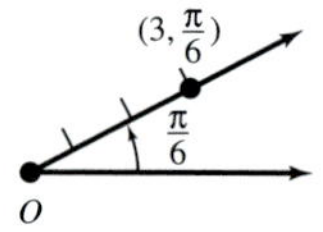

3. $(1, \sqrt{3})$

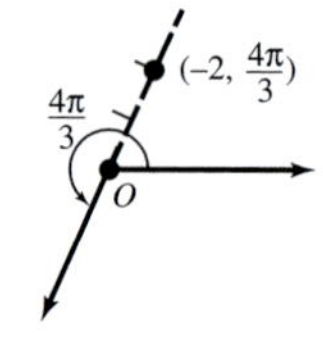

5. $(0, 3)$

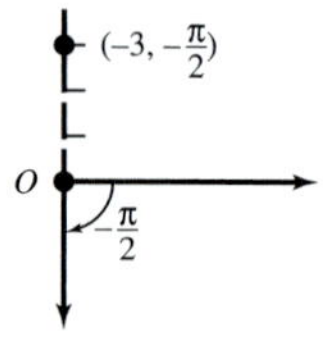

7. $\left(3\sqrt{2}, \frac{3\pi}{4}\right), \left(-3\sqrt{2}, -\frac{\pi}{4}\right)$ **9.** $\left(2, -\frac{\pi}{2}\right), \left(-2, \frac{\pi}{2}\right)$ **11.** $(5, 0.93), (-5, 4.07)$ **13.** $3r^2 - 6r\sin\theta = 0$ **15.** $r^2(2 - 3\sin^2\theta) - \tan\theta = 0$
17. $r^3\cos\theta = 4$ **19.** $x^2 + y^2 - 2y = 0$ **21.** $x^2 + y^2 = 25$ **23.** $x + 3y = 6$

25. Circle; radius 2, center at $(2, 0)$ in rectangular coordinates

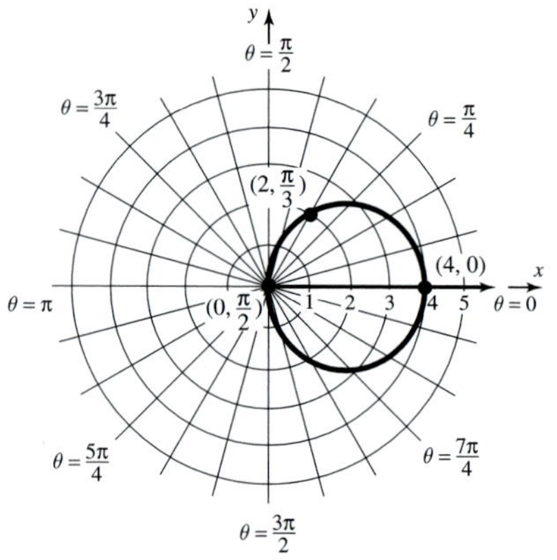

27. Cardioid

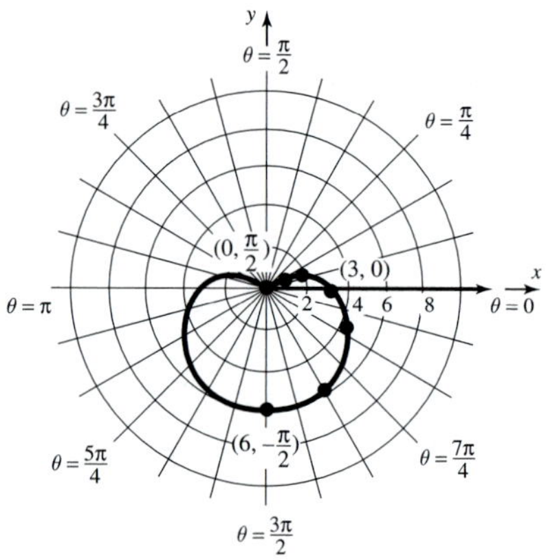

29. Limaçon without inner loop

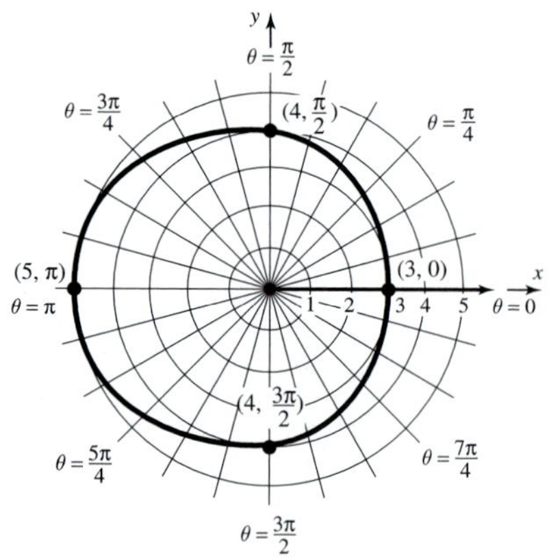

31. $\sqrt{2}(\cos 225° + i\sin 225°)$ **33.** $5(\cos 323.1° + i\sin 323.1°)$ **35.** $-\sqrt{3} + i$ **37.** $-\frac{3}{2} + \left(\frac{3\sqrt{3}}{2}\right)i$ **39.** $0.098 - 0.017i$
41. $zw = \cos 130° + i\sin 130°; \frac{z}{w} = \cos 30° + i\sin 30°$ **43.** $zw = 6(\cos 0 + i\sin 0); \frac{z}{w} = \frac{3}{2}\left(\cos\frac{8\pi}{5} + i\sin\frac{8\pi}{5}\right)$
45. $zw = 5(\cos 5° + i\sin 5°); \frac{z}{w} = 5(\cos 15° + i\sin 15°)$ **47.** $\frac{27}{2} + \frac{27\sqrt{3}}{2}i$ **49.** $4i$ **51.** 64 **53.** $-527 - 336i$
55. $3, 3(\cos 120° + i\sin 120°), 3(\cos 240° + i\sin 240°)$ **57.** $\mathbf{v} = 2\mathbf{i} - 4\mathbf{j}; \|\mathbf{v}\| = 2\sqrt{5}$ **59.** $\mathbf{v} = -\mathbf{i} + 3\mathbf{j}; \|\mathbf{v}\| = \sqrt{10}$ **61.** $-20\mathbf{i} + 13\mathbf{j}$
63. $\sqrt{5}$ **65.** $\sqrt{5} + 5 \approx 7.24$ **67.** $\frac{-2\sqrt{5}}{5}\mathbf{i} + \frac{\sqrt{5}}{5}\mathbf{j}$ **69.** $\mathbf{v}\cdot\mathbf{w} = -11; \cos\theta = -\frac{11\sqrt{5}}{25}$ **71.** $\mathbf{v}\cdot\mathbf{w} = -4; \cos\theta = -\frac{2\sqrt{5}}{5}$
73. $\text{proj}_{\mathbf{w}}\,\mathbf{v} = \frac{9}{10}(3\mathbf{i} + \mathbf{j})$ **75.** 30.5° **77.** $\sqrt{29} \approx 5.39$ mph; 0.4 mi **79.** At an angle of 70.5° to the shore

CHAPTER 10 *10.2 Exercises*

1. B **3.** E **5.** H **7.** C **9.** F **11.** G **13.** D **15.** B

17. $y^2 = 16x$

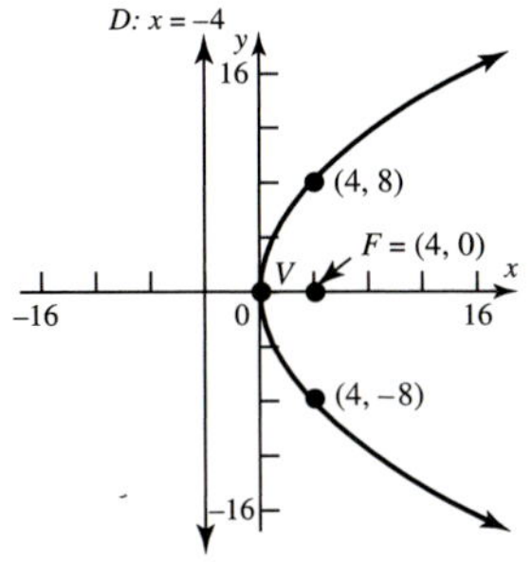

19. $x^2 = -12y$

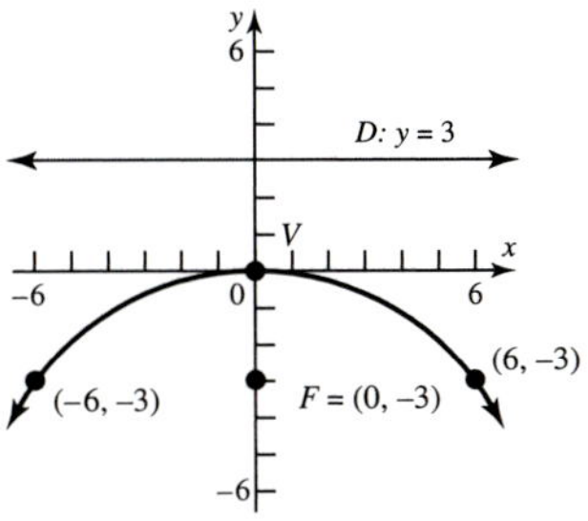

21. $y^2 = -8x$

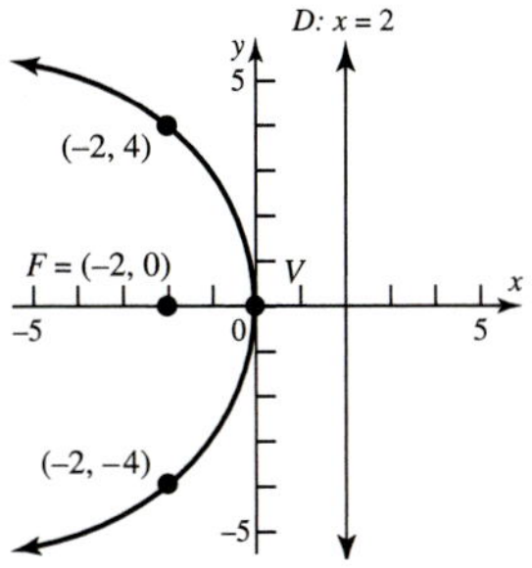

23. $x^2 = 2y$

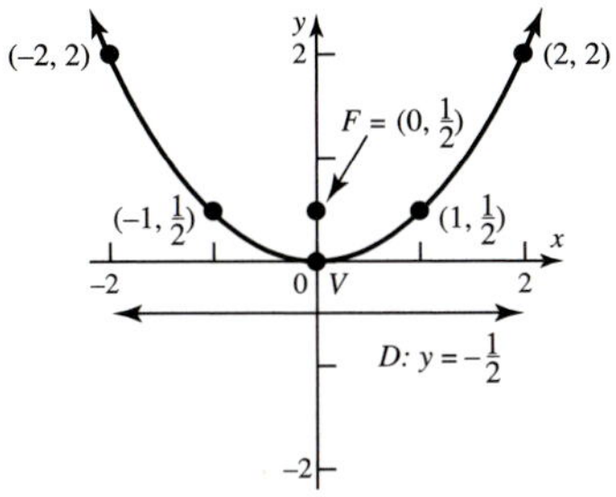

25. $(x - 2)^2 = -8(y + 3)$

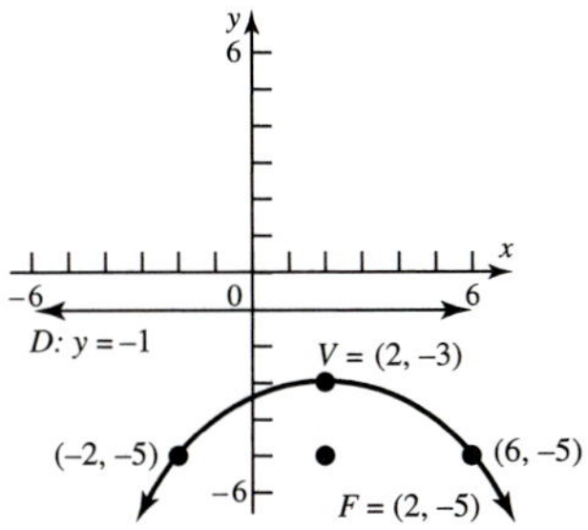

27. $x^2 = \frac{4}{3}y$

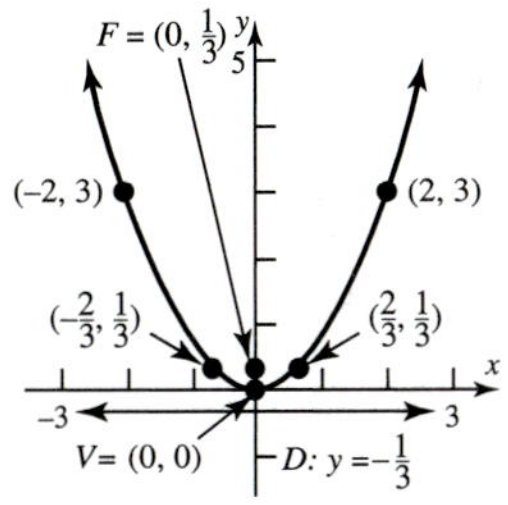

29. $(x + 3)^2 = 4(y - 3)$

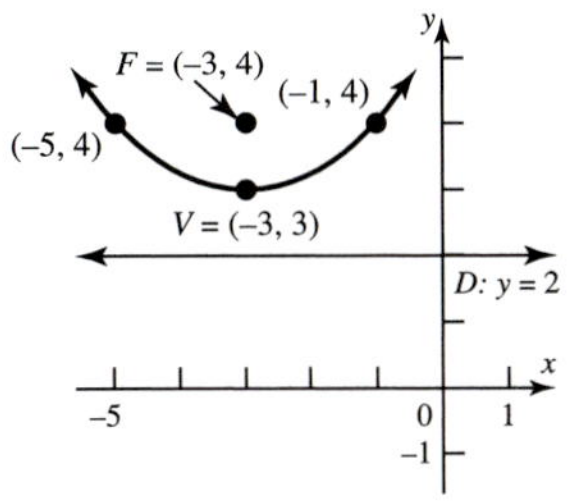

31. $(y + 2)^2 = -8(x + 1)$

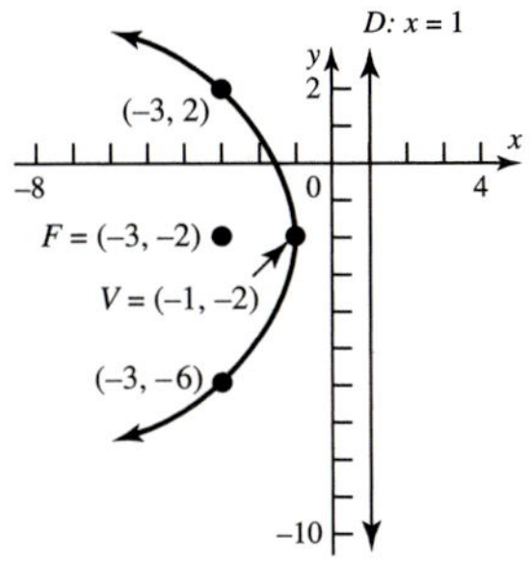

33. Vertex: $(0, 0)$; Focus: $(0, 1)$; Directrix: $y = -1$

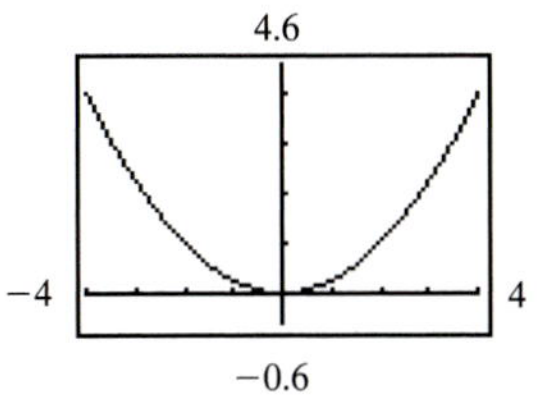

35. Vertex: $(0, 0)$; Focus $(-4, 0)$; Directrix: $x = 4$

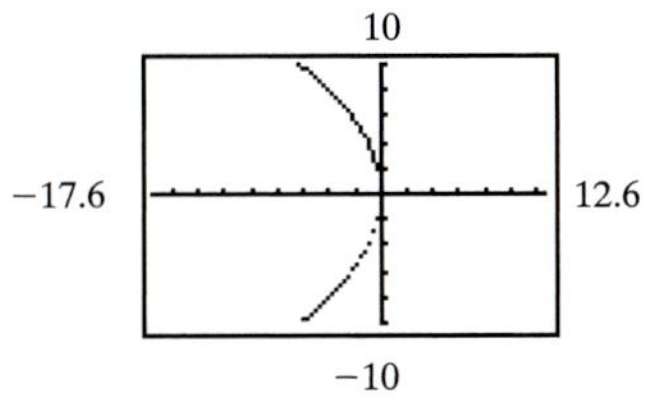

37. Vertex: $(-1, 2)$; Focus: $(1, 2)$; Directrix: $x = -3$

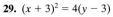

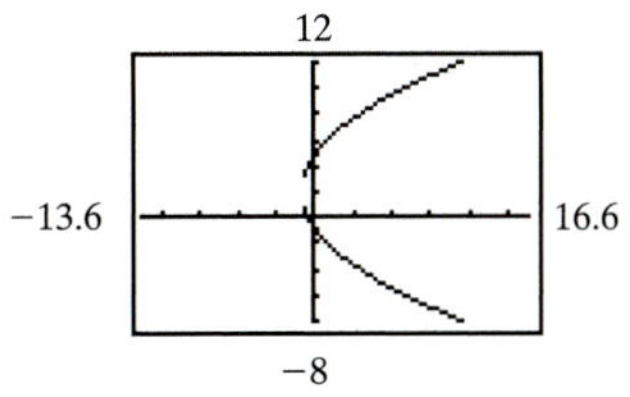

39. Vertex: $(3, -1)$; Focus: $\left(3, -\frac{5}{4}\right)$; Directrix: $y = -\frac{3}{4}$

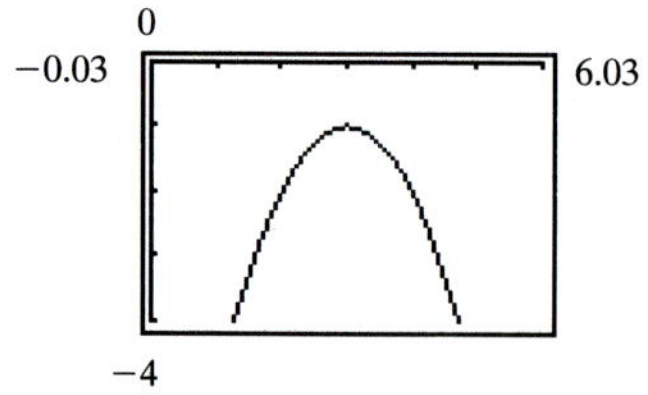

41. Vertex: $(2, -3)$; Focus: $(4, -3)$; Directrix: $x = 0$

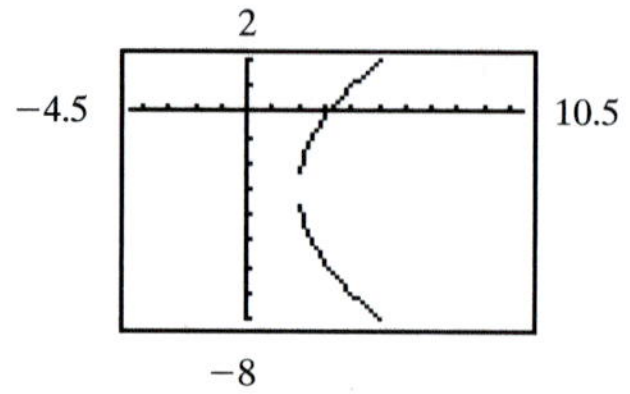

43. Vertex: $(0, 2)$; Focus: $(-1, 2)$; Directrix: $x = 1$

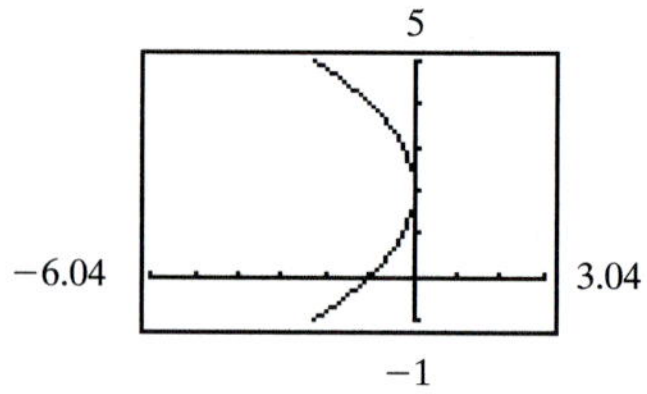

45. Vertex: $(-4, -2)$; Focus: $(-4, -1)$; Directrix: $y = -3$

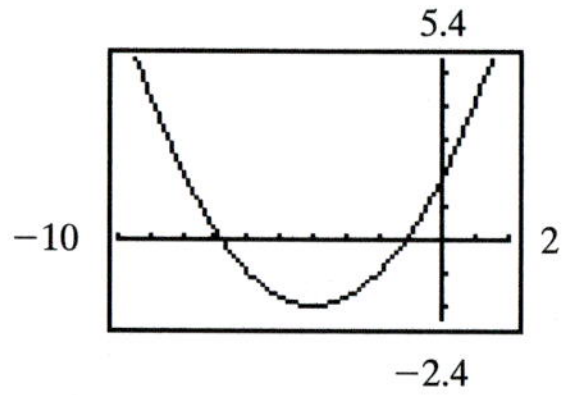

47. Vertex: $(-1, -1)$; Focus: $\left(-\frac{3}{4}, -1\right)$; Directrix: $x = -\frac{5}{4}$

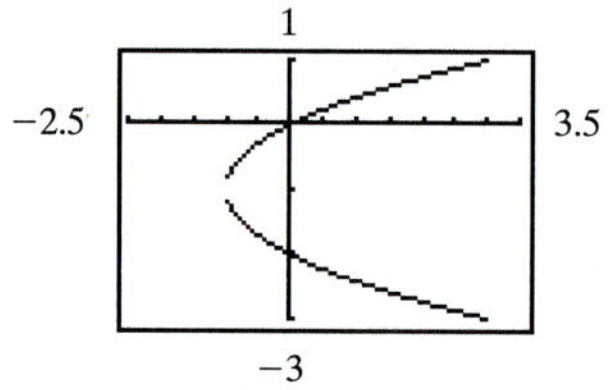

49. Vertex: $(2, -8)$; Focus: $\left(2, -\frac{31}{4}\right)$; Directrix: $y = -\frac{33}{4}$

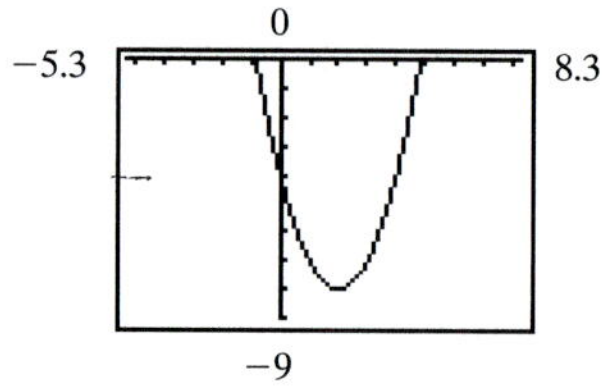

51. $(y - 1)^2 = x$ **53.** $(y - 1)^2 = -(x - 2)$ **55.** $x^2 = 4(y - 1)$ **57.** $y^2 = \frac{1}{2}(x + 2)$

59. 1.5625 ft from the base of the dish, along the axis of symmetry **61.** 1 in from the vertex **63.** 20 ft **65.** 0.78125 ft

67. 4.17 ft from the base along the axis of symmetry **69.** 24.31 ft, 18.75 ft, 7.64 ft

71. $Ax^2 + Ey = 0$

$$x^2 = -\frac{E}{A}y$$

This is the equation of a parabola with vertex at $(0, 0)$ and axis of symmetry the y-axis. The focus is $\left(0, -\frac{E}{4A}\right)$; the directrix is the line $y = \frac{E}{4A}$. The parabola opens up if $-\frac{E}{A} > 0$ and down if $-\frac{E}{A} < 0$.

73. $A^2 + Dx + Ey + F = 0, A \neq 0$

$$Ax^2 + Dx = -Ey - F$$

$$x^2 + \frac{D}{A}x = -\frac{E}{A}y - \frac{F}{A}$$

$$\left(x + \frac{D}{2A}\right)^2 = -\frac{E}{A}y - \frac{F}{A} + \frac{D^2}{4A^2}$$

$$\left(x + \frac{D}{2A}\right)^2 = -\frac{E}{A}y + \frac{D^2 - 4AF}{4A^2}$$

(a) If $E \neq 0$, then the equation may be written as

$$\left(x + \frac{D}{2A}\right)^2 = -\frac{E}{A}\left(y - \frac{D^2 - 4AF}{4AE}\right)$$

This is the equation of a parabola with vertex at $\left(-\frac{D}{2A}, \frac{D^2 - 4AF}{4AE}\right)$ and axis of symmetry parallel to the y-axis.

(b)–(d) If $E = 0$, the graph of the equation contains no points if $D^2 - 4AF < 0$, is a single vertical line if $D^2 - 4AF = 0$, and is two vertical lines if $D^2 - 4AF > 0$.

10.3 Exercises

1. C **3.** B **5.** C **7.** D

9. Vertices: $(-5, 0), (5, 0)$
Foci: $(-\sqrt{21}, 0), (\sqrt{21}, 0)$

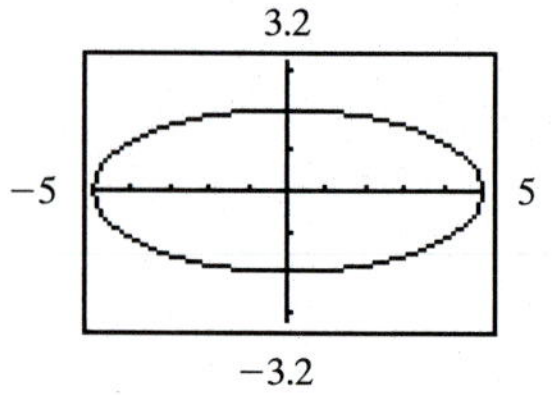

11. Vertices: $(0, -5), (0, 5)$
Foci: $(0, -4), (0, 4)$

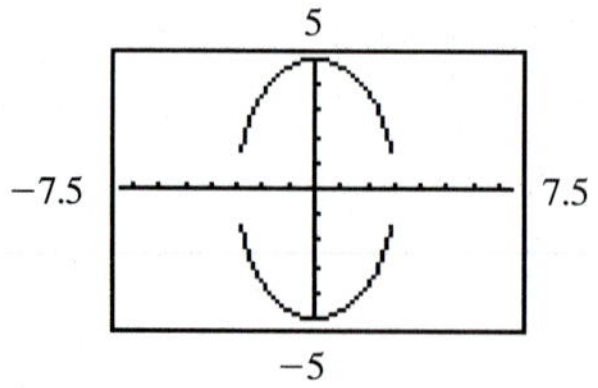

13. Vertices: $(0, -4), (0, 4)$
Foci: $(0, -2\sqrt{3}), (0, 2\sqrt{3})$

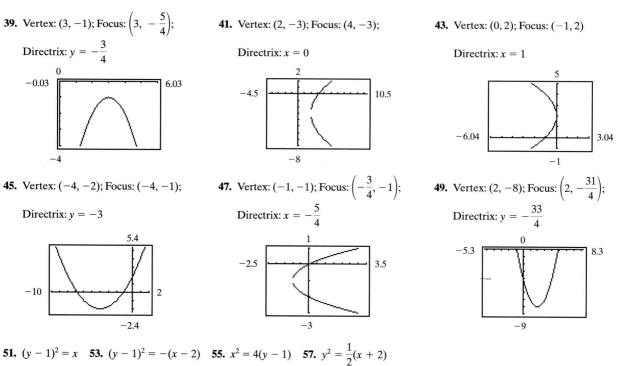

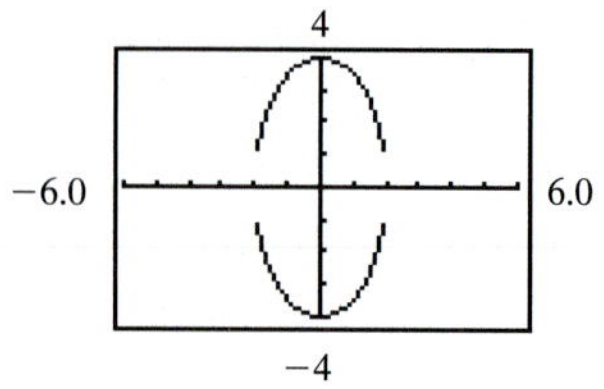

15. Vertices: $(-2\sqrt{2}, 0)$, $(2\sqrt{2}, 0)$
Foci: $(-\sqrt{6}, 0)$, $(\sqrt{6}, 0)$

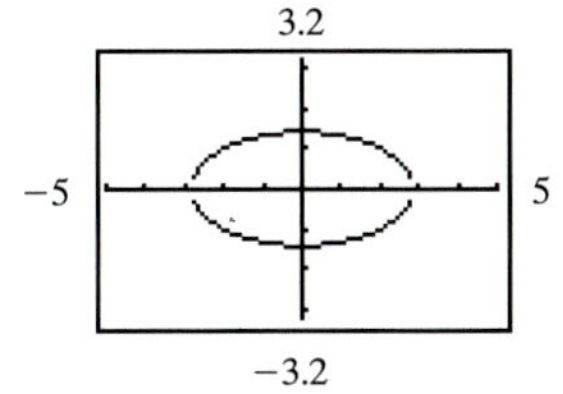

17. Vertices: $(-4, 0)$, $(4, 0)$, $(0, -4)$, $(0, 4)$
Focus: $(0, 0)$

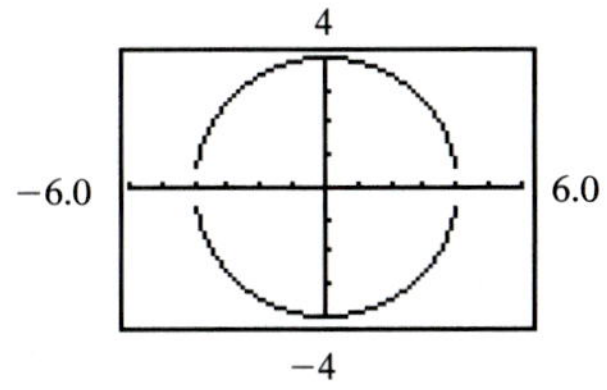

19. $\frac{x^2}{25} + \frac{y^2}{16} = 1$

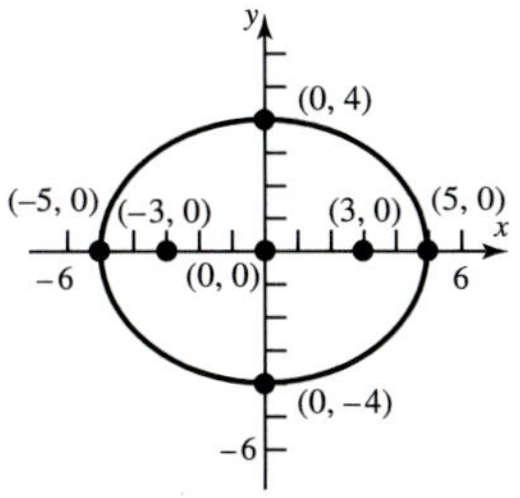

21. $\frac{x^2}{9} + \frac{y^2}{25} = 1$

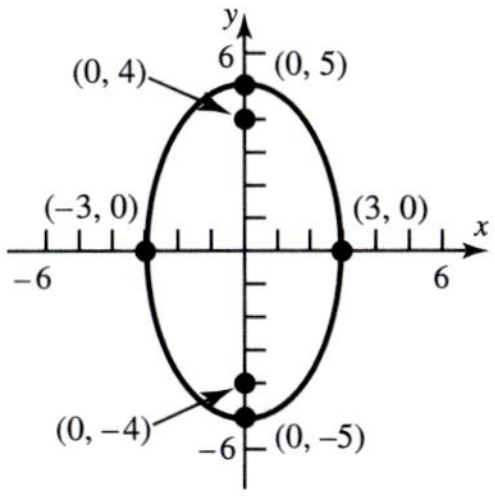

23. $\frac{x^2}{9} + \frac{y^2}{5} = 1$

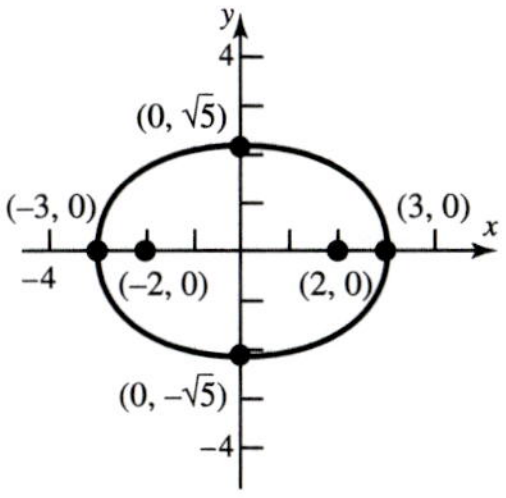

25. $\frac{x^2}{4} + \frac{y^2}{13} = 1$

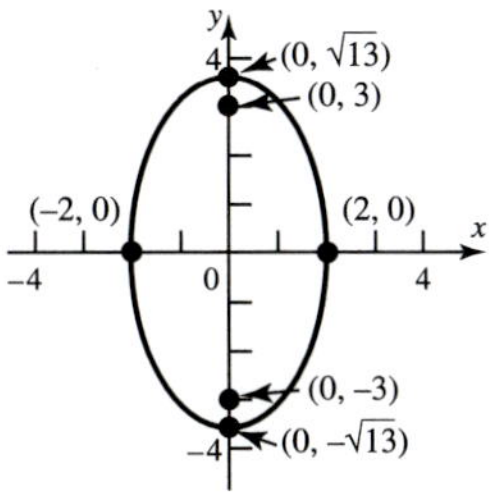

27. $x^2 + \frac{y^2}{16} = 1$

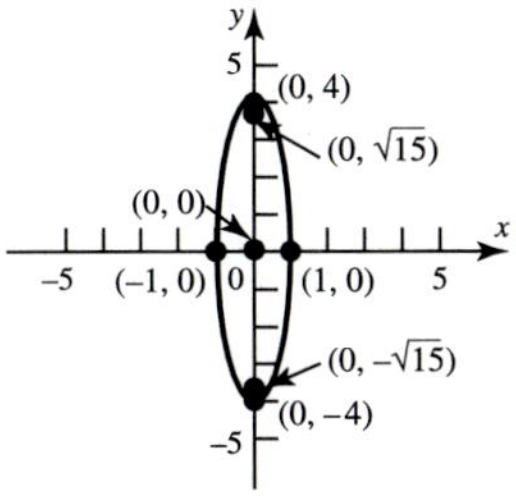

29. $\frac{(x+1)^2}{4} + (y-1)^2 = 1$

31. $(x-1)^2 + \frac{y^2}{4} = 1$

33. Center: $(3, -1)$; Vertices: $(3, -4)$, $(3, 2)$
Foci: $(3, -1-\sqrt{5})$, $(3, -1+\sqrt{5})$

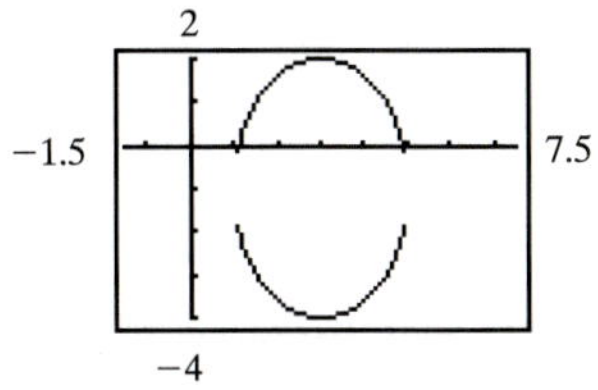

35. Center: $(-5, 4)$; Vertices: $(-9, 4)$, $(-1, 4)$
Foci: $(-5-2\sqrt{3}, 4)$, $(-5+2\sqrt{3}, 4)$

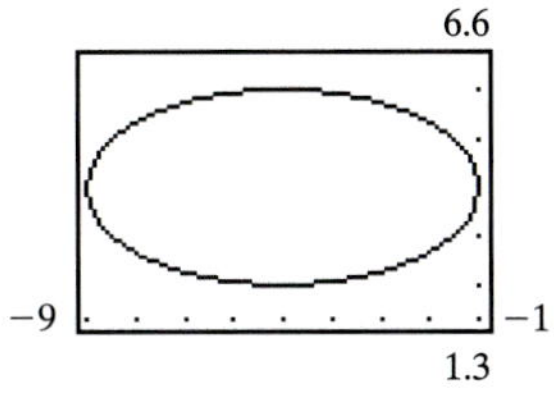

37. Center: $(-2, 1)$; Vertices: $(-4, 1)$, $(0, 1)$
Foci: $(-2-\sqrt{3}, 1)$, $(-2+\sqrt{3}, 1)$

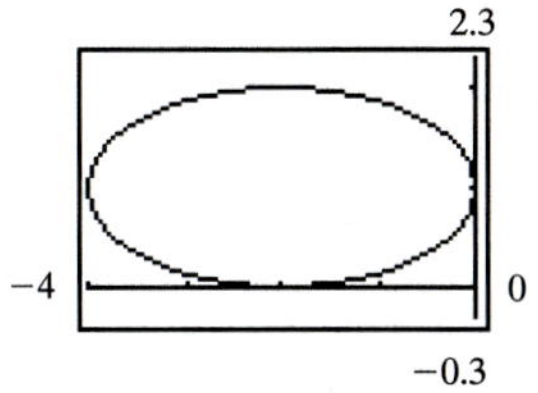

39. Center: $(2, -1)$; Vertices: $(2-\sqrt{3}, -1)$, $(2+\sqrt{3}, -1)$; Foci: $(1, -1)$, $(3, -1)$

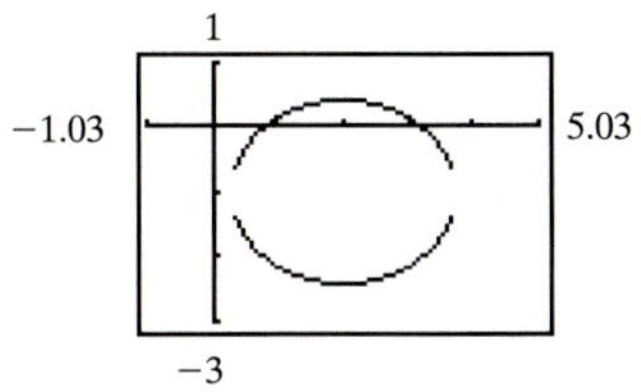

41. Center: $(1, -2)$; Vertices: $(1, -5)$, $(1, 1)$
Foci: $(1, -2-\sqrt{5})$, $(1, -2+\sqrt{5})$

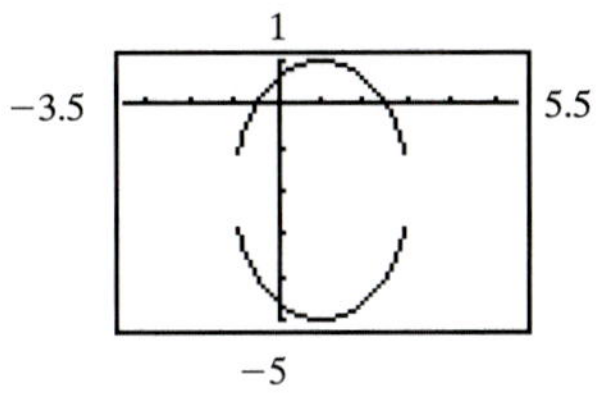

43. Center: $(0, -2)$; Vertices: $(0, -4)$, $(0, 0)$
Foci: $(0, -2-\sqrt{3})$, $(0, -2+\sqrt{3})$

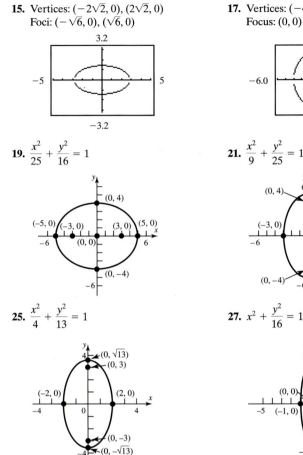

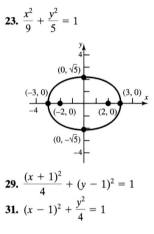

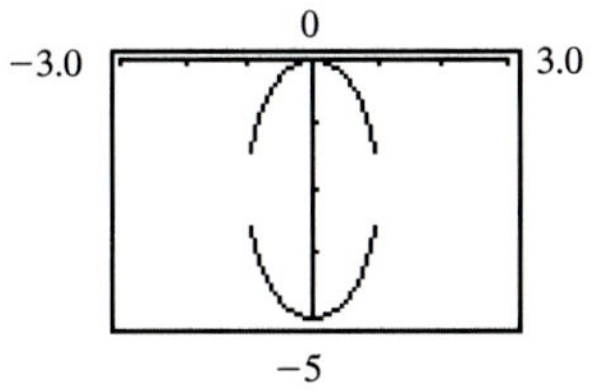

45. $\dfrac{(x-2)^2}{25}+\dfrac{(y+2)^2}{21}=1$

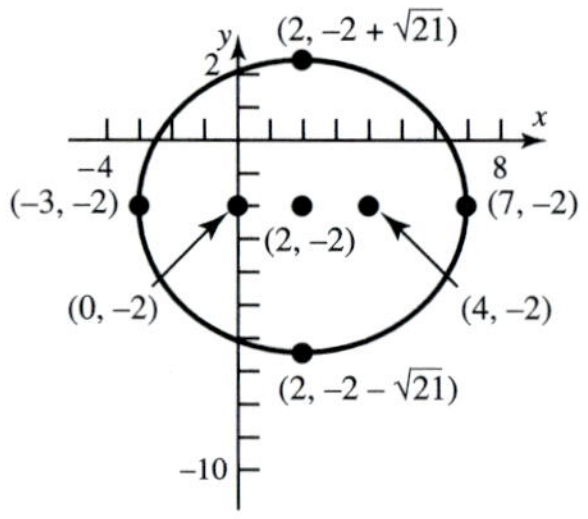

47. $\dfrac{(x-4)^2}{5}+\dfrac{(y-6)^2}{9}=1$

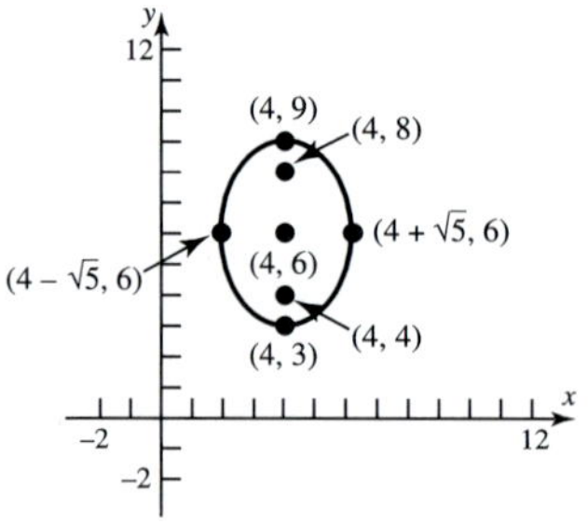

49. $\dfrac{(x-2)^2}{16}+\dfrac{(y-1)^2}{7}=1$

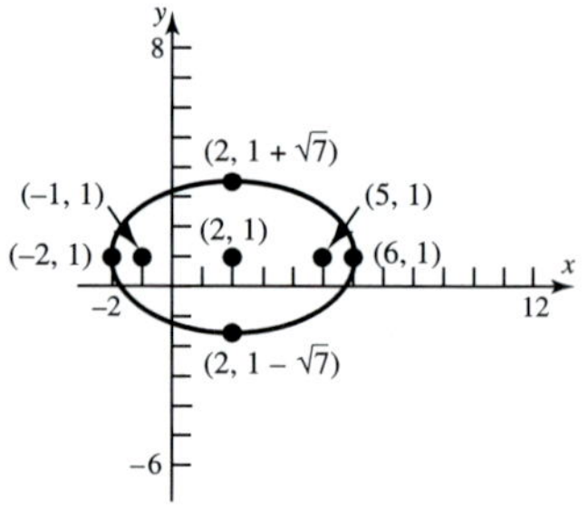

51. $\dfrac{(x-1)^2}{10}+(y-2)^2=1$

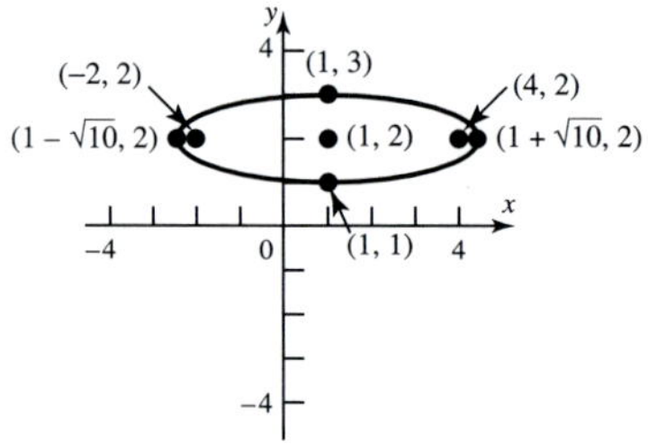

53. $\dfrac{(x-1)^2}{9}+(y-2)^2=1$

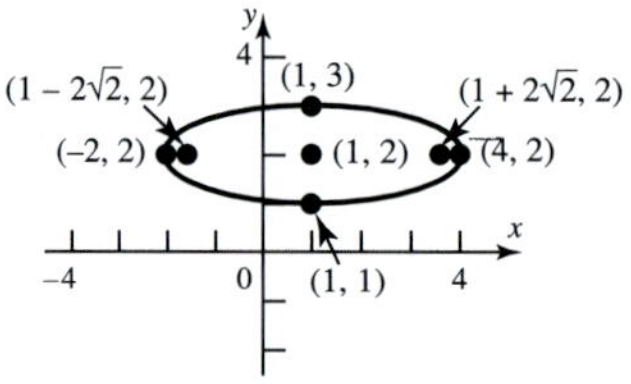

55.

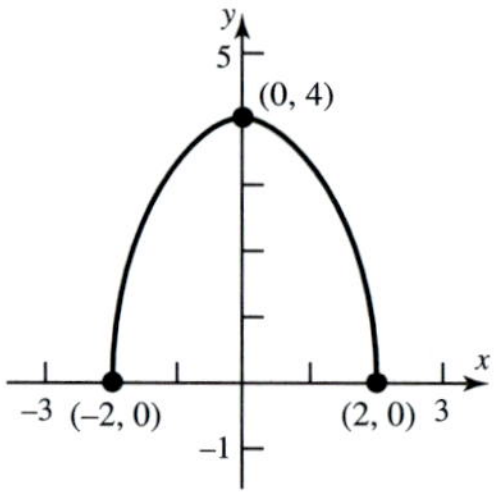

57.

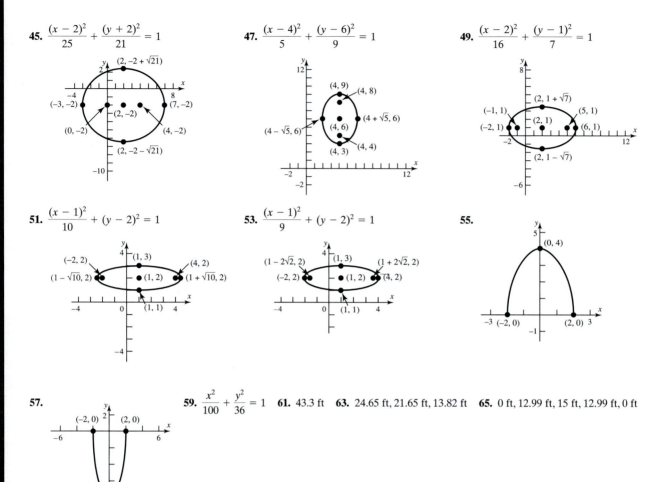

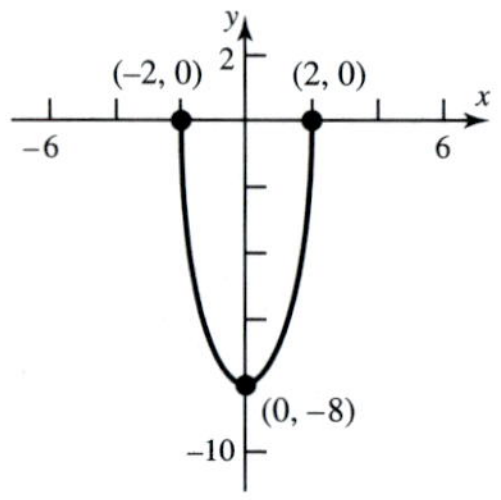

59. $\dfrac{x^2}{100}+\dfrac{y^2}{36}=1$ **61.** 43.3 ft **63.** 24.65 ft, 21.65 ft, 13.82 ft **65.** 0 ft, 12.99 ft, 15 ft, 12.99 ft, 0 ft

67. 91.5 million miles; $\dfrac{x^2}{(93)^2}+\dfrac{y^2}{8646.75}=1$

69. perihelion: 460.6 million miles; mean distance: 483.8 million miles; $\dfrac{x^2}{(483.8)^2}+\dfrac{y^2}{233{,}524.2}=1$

71. 30 ft

73. **(a)** $Ax^2+Cy^2+F=0$

$Ax^2+Cy^2=-F$

If A and C are of the same sign and F is of opposite sign, then the equation takes the form $\dfrac{x^2}{\left(-\dfrac{F}{A}\right)}+\dfrac{y^2}{\left(-\dfrac{F}{C}\right)}=1$, where $-\dfrac{F}{A}$ and $-\dfrac{F}{C}$ are positive. This is the equation of an ellipse with center at (0, 0).

(b) If $A=C$, the equation may be written as $x^2+y^2=-\dfrac{F}{A}$. This is the equation of a circle with center at (0, 0) and radius equal to $\sqrt{-\dfrac{F}{A}}$

10.4 Exercises

1. B **3.** A **5.** B **7.** C

9. $x^2 - \frac{y^2}{8} = 1$

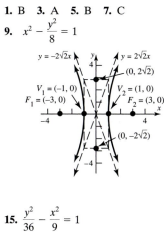

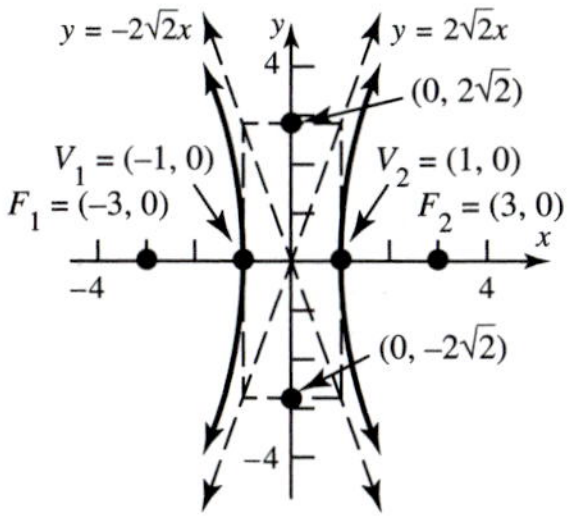

11. $\frac{y^2}{16} - \frac{x^2}{20} = 1$

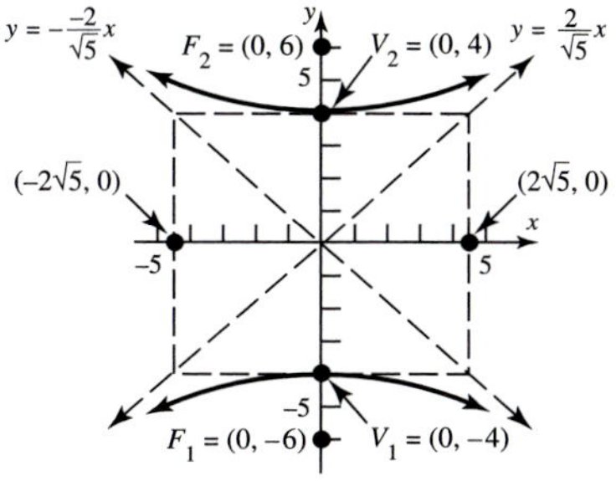

13. $\frac{x^2}{9} - \frac{y^2}{16} = 1$

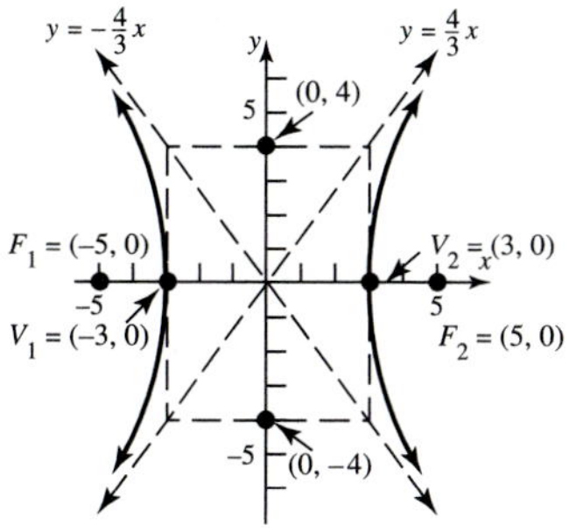

15. $\frac{y^2}{36} - \frac{x^2}{9} = 1$

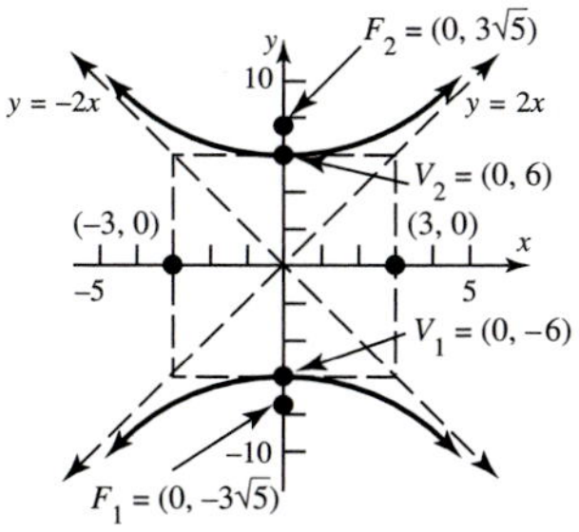

17. $\frac{x^2}{8} - \frac{y^2}{8} = 1$

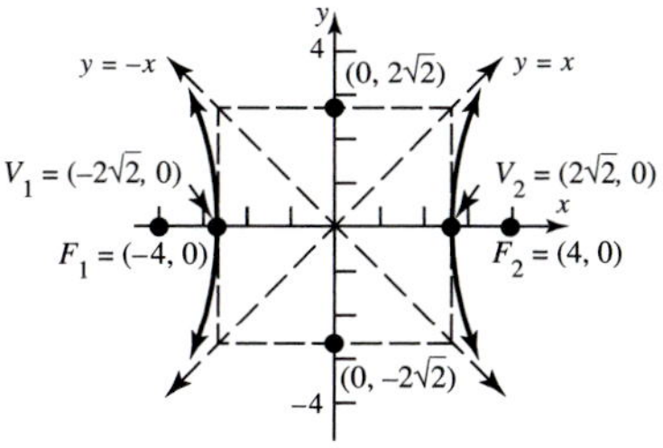

19. Center: (0, 0)
Transverse axis: x-axis
Vertices: $(-5, 0)$, $(5, 0)$
Foci: $(-\sqrt{34}, 0)$, $(\sqrt{34}, 0)$
Asymptotes: $y = \pm\frac{3}{5}x$

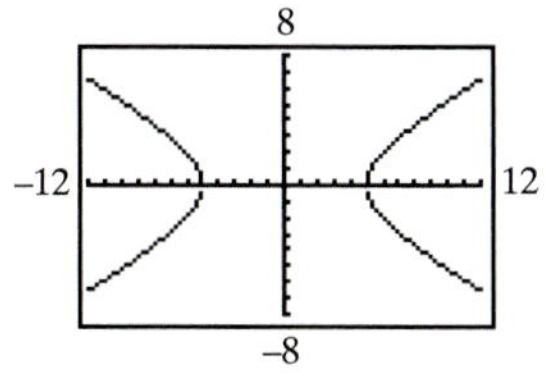

21. Center: (0, 0)
Transverse axis: x-axis
Vertices: $(-2, 0)$, $(2, 0)$
Foci: $(-2\sqrt{5}, 0)$, $(2\sqrt{5}, 0)$
Asymptotes: $y = \pm 2x$

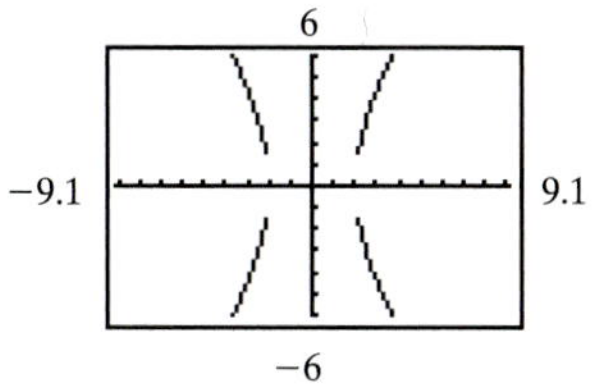

23. Center: (0, 0)
Transverse axis: y-axis
Vertices: $(0, -3)$, $(0, 3)$
Foci: $(0, -\sqrt{10})$, $(0, \sqrt{10})$
Asymptotes: $y = \pm 3x$

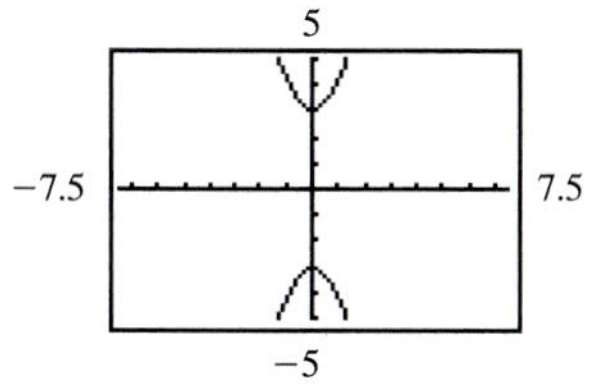

25. Center: (0, 0)
Transverse axis: y-axis
Vertices: $(0, -5)$, $(0, 5)$
Foci: $(0, -5\sqrt{2})$, $(0, 5\sqrt{2})$
Asymptotes: $y = \pm x$

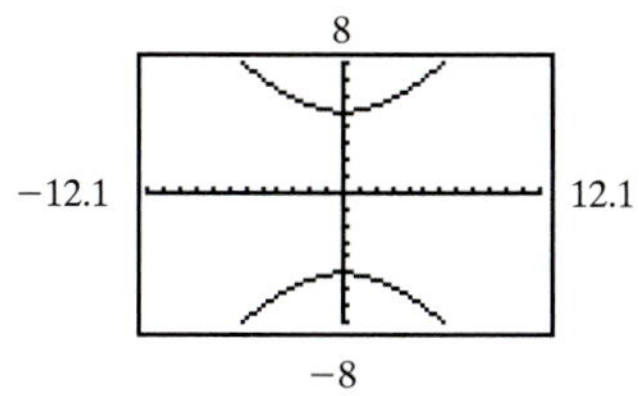

27. $x^2 - y^2 = 1$

29. $\frac{y^2}{36} - \frac{x^2}{9} = 1$

31. $\frac{(x-4)^2}{4} - \frac{(y+1)^2}{5} = 1$

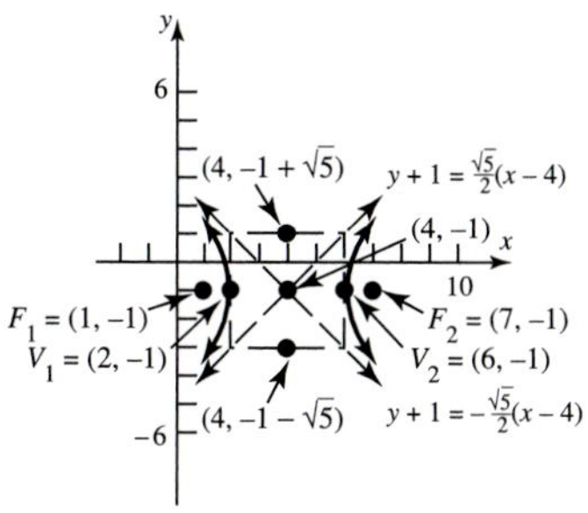

33. $\dfrac{(y+4)^2}{4} - \dfrac{(x+3)^2}{12} = 1$

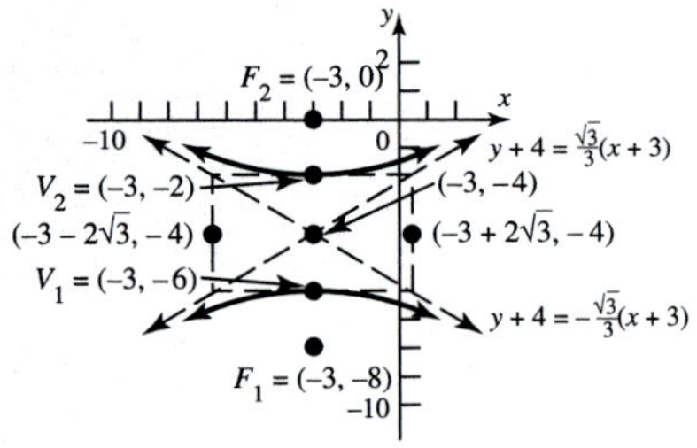

35. $(x-5)^2 - \dfrac{(y-7)^2}{3} = 1$

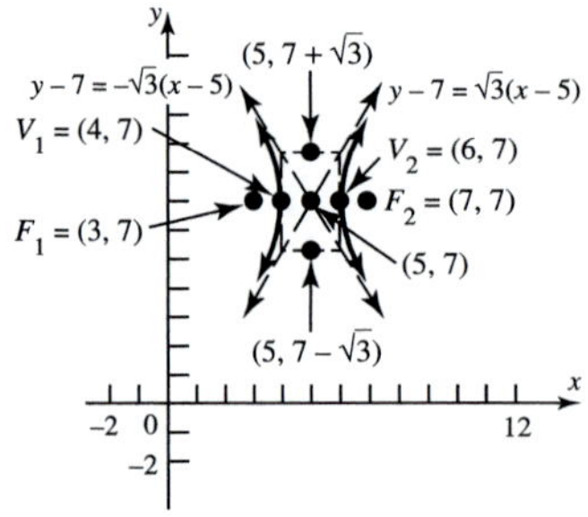

37. $\dfrac{(x-1)^2}{4} - \dfrac{(y+1)^2}{9} = 1$

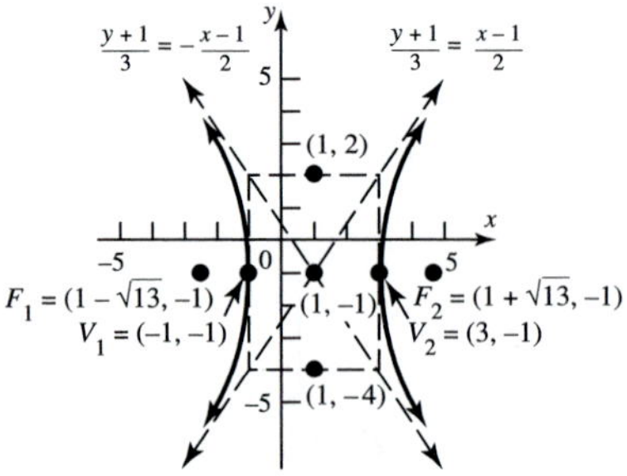

39. Center: $(2, -3)$
Transverse axis: Parallel to x-axis
Vertices: $(0, -3), (4, -3)$
Foci: $(2 - \sqrt{13}, -3), (2 + \sqrt{13}, -3)$
Asymptotes: $y + 3 = \pm\frac{3}{2}(x - 2)$

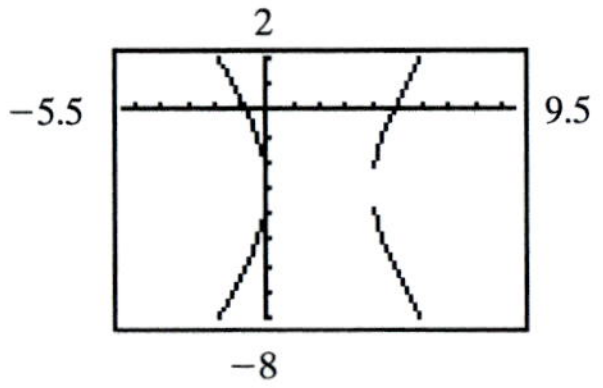

41. Center: $(-2, 2)$
Transverse axis: Parallel to y-axis
Vertices: $(-2, 0), (-2, 4)$
Foci: $(-2, 2 - \sqrt{5}), (-2, 2 + \sqrt{5})$
Asymptotes: $y - 2 = \pm 2(x + 2)$

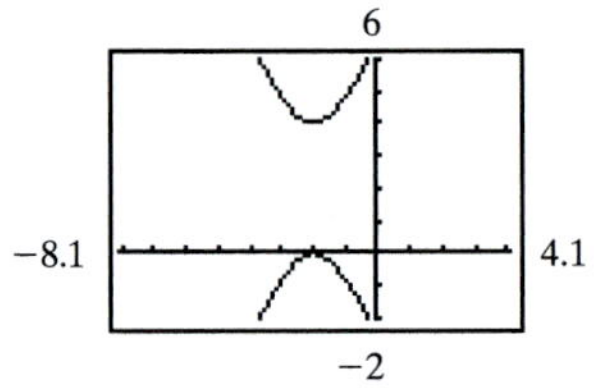

43. Center: $(-1, -2)$
Transverse axis: Parallel to x-axis
Vertices: $(-3, -2), (1, -2)$
Foci: $(-1 - 2\sqrt{2}, -2), (-1 + 2\sqrt{2}, -2)$
Asymptotes: $y + 2 = \pm(x + 1)$

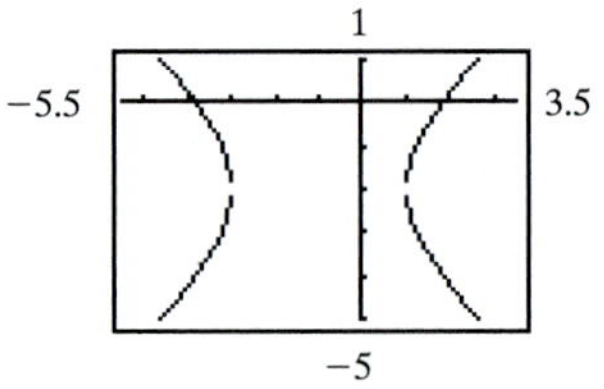

45. Center: $(1, -1)$
Transverse axis: Parallel to x-axis
Vertices: $(0, -1), (2, -1)$
Foci: $(1 - \sqrt{2}, -1), (1 + \sqrt{2}, -1)$
Asymptotes: $y + 1 = \pm(x - 1)$

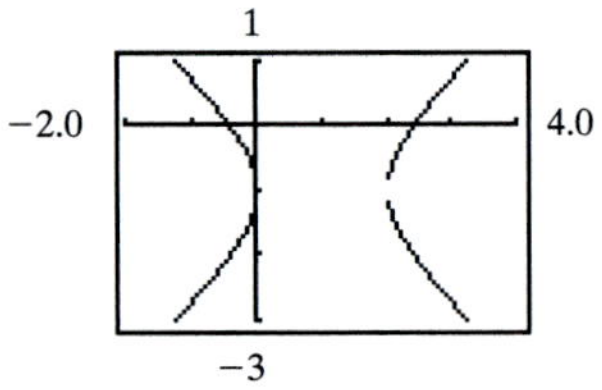

47. Center: $(-1, 2)$
Transverse axis: Parallel to y-axis
Vertices: $(-1, 0), (-1, 4)$
Foci: $(-1, 2 - \sqrt{5}), (-1, 2 + \sqrt{5})$
Asymptotes: $y - 2 = \pm 2(x + 1)$

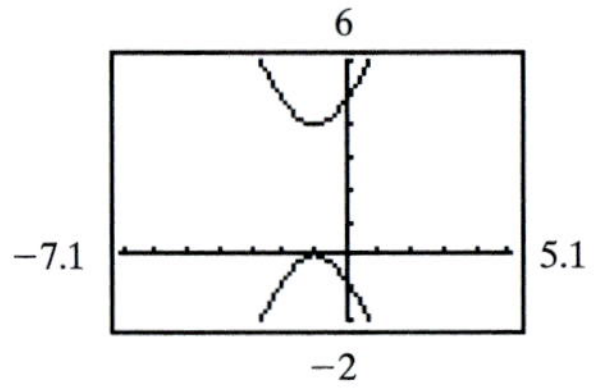

49. Center: $(3, -2)$
Transverse axis: Parallel to x-axis
Vertices: $(1, -2), (5, -2)$
Foci: $(3 - 2\sqrt{5}, -2), (3 + 2\sqrt{5}, -2)$
Asymptotes: $y + 2 = \pm 2(x - 3)$

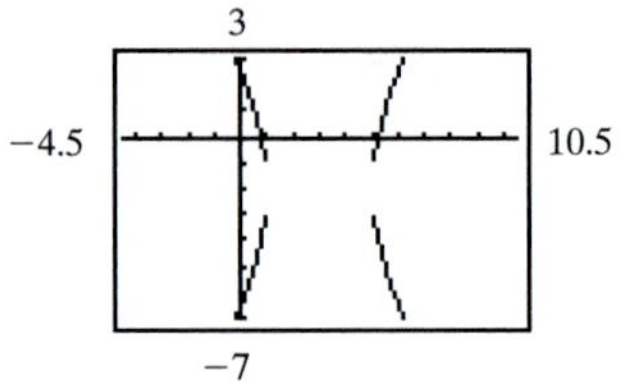

51. Center: $(-2, 1)$
Transverse axis: Parallel to y-axis
Vertices: $(-2, -1), (-2, 3)$
Foci: $(-2, 1 - \sqrt{5}), (-2, 1 + \sqrt{5})$
Asymptotes: $y - 1 = \pm 2(x + 2)$

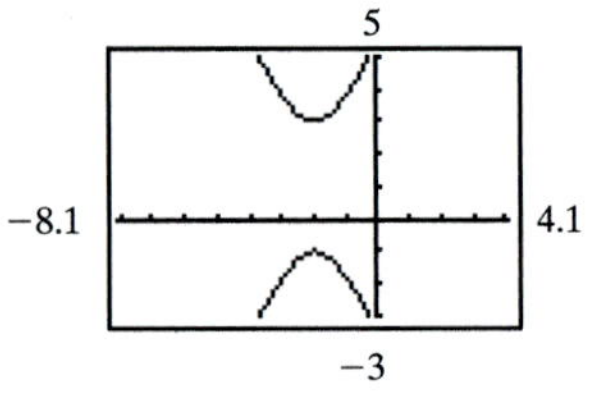

53.

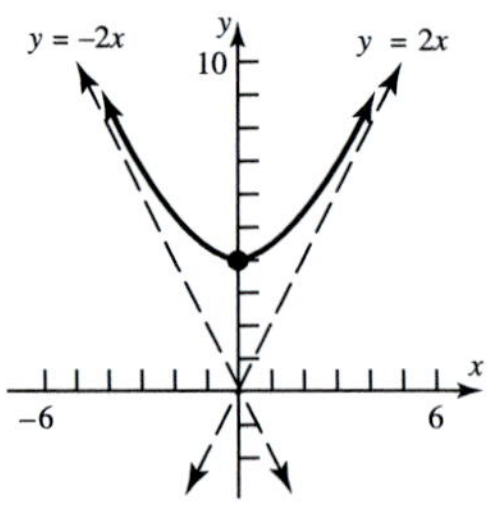

55.

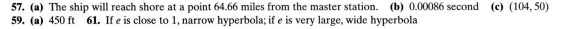

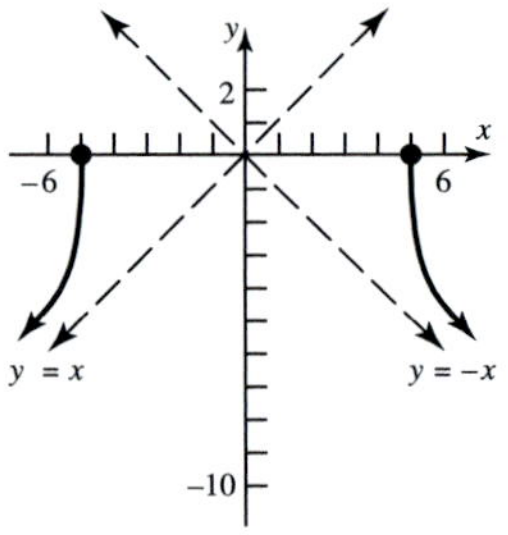

57. **(a)** The ship will reach shore at a point 64.66 miles from the master station. **(b)** 0.00086 second **(c)** (104, 50)
59. **(a)** 450 ft **61.** If e is close to 1, narrow hyperbola; if e is very large, wide hyperbola

63. $\dfrac{x^2}{4} - y^2 = 1$; asymptotes $y = \pm\dfrac{1}{2}x$, $y^2 - \dfrac{x^2}{4} = 1$; asymptotes $y = \pm\dfrac{1}{2}x$

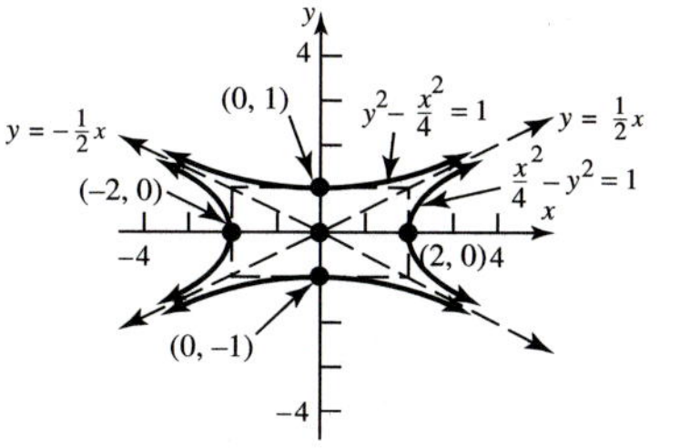

65. $Ax^2 + Cy^2 + F = 0$

$Ax^2 + Cy^2 = -F$

If A and C are of opposite sign and $F \neq 0$, this equation may be written as $\dfrac{x^2}{\left(-\dfrac{F}{A}\right)} + \dfrac{y^2}{\left(-\dfrac{F}{C}\right)} = 1$, where $-\dfrac{F}{A}$ and $-\dfrac{F}{C}$ are opposite in sign. This is the equation of a hyperbola with center $(0, 0)$. The transverse axis is the x-axis if $-\dfrac{F}{A} > 0$; the transverse axis is the y-axis if $-\dfrac{F}{A} < 0$.

Exercises 10.5

1. Parabola **3.** Ellipse **5.** Hyperbola **7.** Hyperbola **9.** Circle **11.** $x = \dfrac{\sqrt{2}}{2}(x' - y')$, $y = \dfrac{\sqrt{2}}{2}(x' + y')$

13. $x = \dfrac{\sqrt{2}}{2}(x' - y')$, $y = \dfrac{\sqrt{2}}{2}(x' + y')$ **15.** $x = \dfrac{1}{2}(x' - \sqrt{3}y')$, $y = \dfrac{1}{2}(\sqrt{3}x' + y')$

17. $x = \dfrac{\sqrt{5}}{5}(x' - 2y')$, $y = \dfrac{\sqrt{5}}{5}(2x' + y')$ **19.** $x = \dfrac{\sqrt{13}}{13}(3x' - 2y')$, $y = \dfrac{\sqrt{13}}{13}(2x' + 3y')$

21.

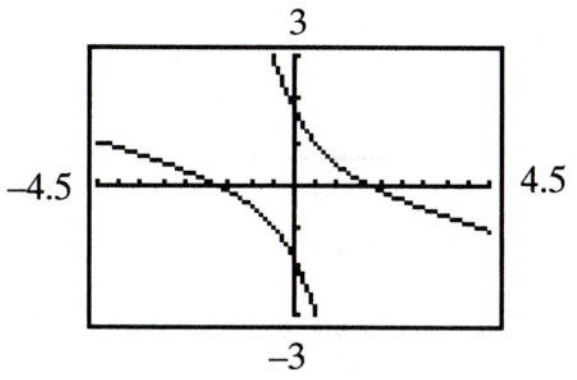

$\theta = 45°$ (see Problem 11)

$x'^2 - \dfrac{y^2}{3} = 1$

Hyperpola
Center at origin
Transverse axis is the x'-axis.
Vertices at $(\pm 1, 0)$

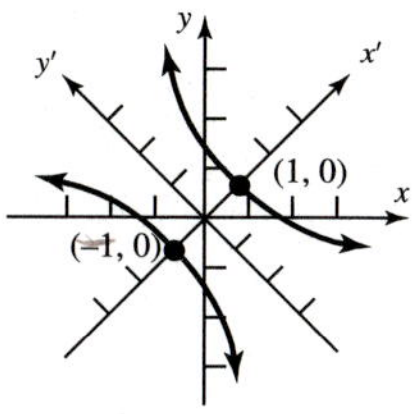

23.

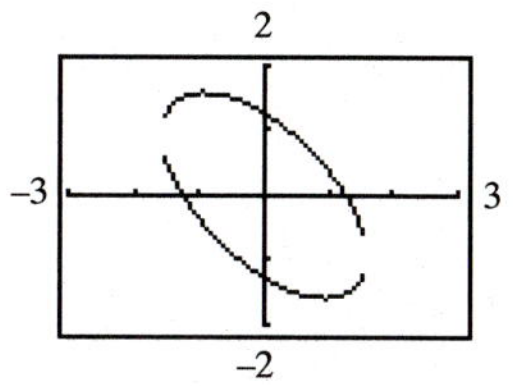

$\theta = 45°$ (see Problem 13)

$x'^2 + \dfrac{y'^2}{4} = 1$

Ellipse
Center at $(0, 0)$
Major axis is the y'-axis.
Vertices at $(0, \pm 2)$

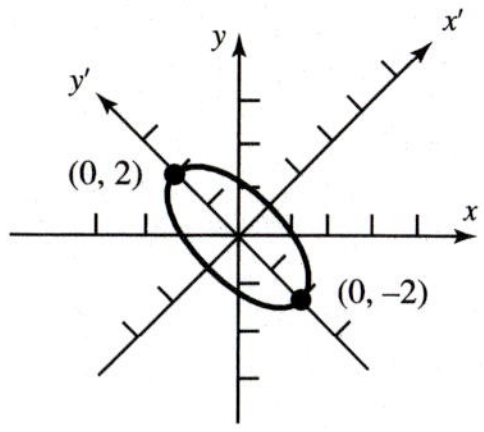

25.

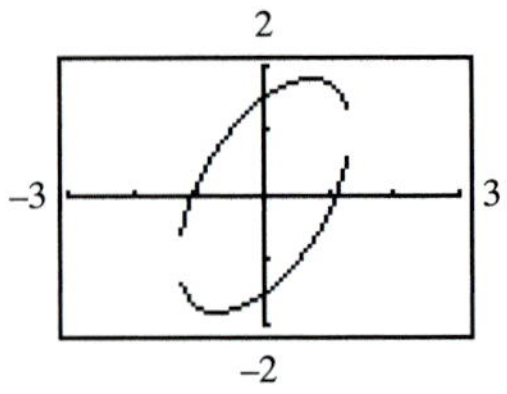

$\theta = 60°$ (see Problem 15)

$\dfrac{x'^2}{4} + y'^2 = 1$

Ellipse
Center at $(0, 0)$
Major axis is the x'-axis.
Vertices at $(\pm 2, 0)$

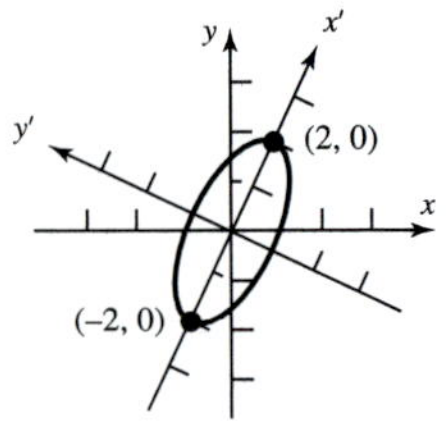

27.

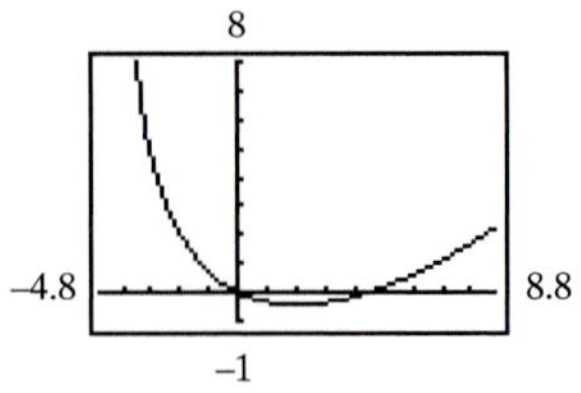

$\theta \approx 63°$ (see Problem 17)

$y'^2 = 8x'$

Parabola
Vertex at $(0, 0)$
Focus at $(2, 0)$

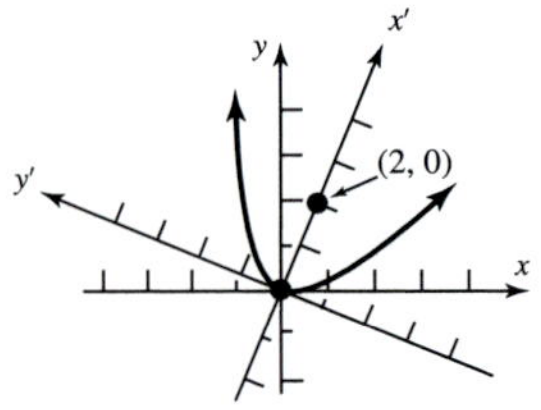

29.

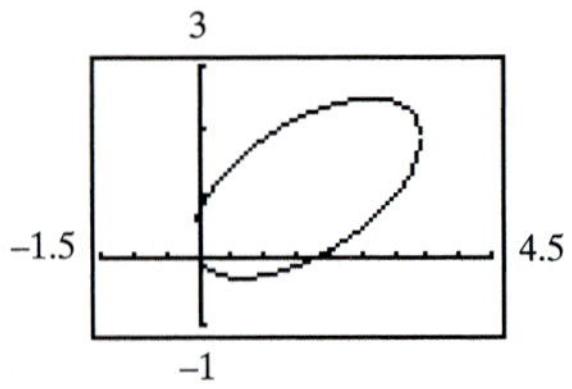

$\theta \approx 34°$ (see Problem 19)

$\dfrac{(x' - 2)^2}{4} + y'^2 = 1$

Ellipse
Center at $(2, 0)$
Major axis is the x'-axis
Vertices at $(4, 0)$ and $(0, 0)$

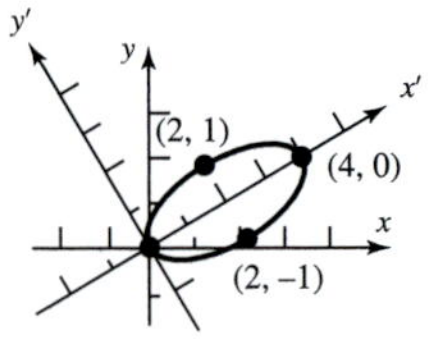

31.

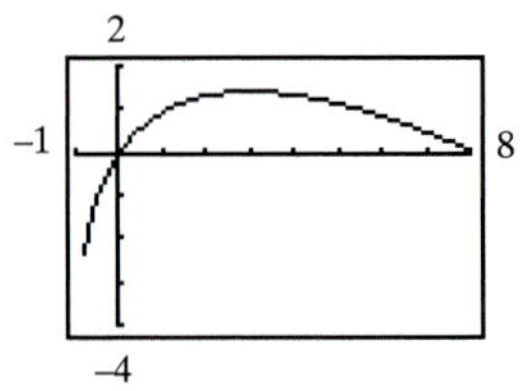

$\cot 2\theta = \dfrac{7}{24}$; $\theta \approx 37°$

$(x' - 1)^2 = -6\left(y' - \dfrac{1}{6}\right)$

Parabola

Vertex at $\left(1, \dfrac{1}{6}\right)$

Focus at $\left(1, -\dfrac{4}{3}\right)$

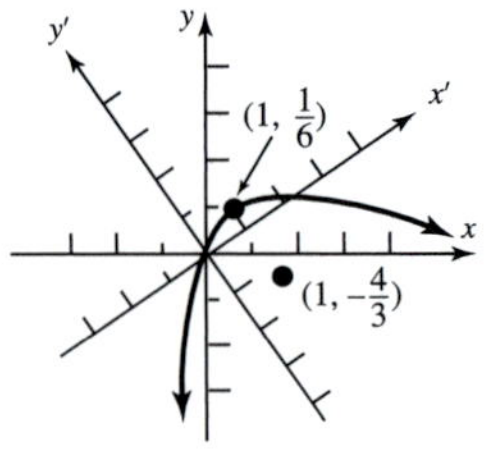

33. Hyperbola **35.** Hyperbola **37.** Parabola **39.** Ellipse **41.** Ellipse

43. Refer to equation (6):

$$A' = A\cos^2\theta + B\sin\theta\cos\theta + C\sin^2\theta$$
$$B' = B(\cos^2\theta - \sin^2\theta) + 2(C - A)(\sin\theta\cos\theta)$$
$$C' = A\sin^2\theta - B\sin\theta\cos\theta + C\cos^2\theta$$
$$D' = D\cos\theta + E\sin\theta$$
$$E' = -D\sin\theta + E\cos\theta$$
$$F' = F$$

45. Use Problem 43 to find $B'^2 - 4A'C'$. After much cancellation, $B'^2 - 4A'C' = B^2 - 4AC$.

47. Use formulas (5) and find $d^2 = (x_2 - x_1)^2 + (y_2 - y_1)^2$. After simplifying, $(x_2 - x_1)^2 + (y_2 - y_1)^2 = (x'_2 - x'_1)^2 + (y'_2 - y'_1)^2$.

Exercises 10.6

1. Parabola; directrix is perpendicular to the polar axis 1 unit to the right of the pole.

3. Hyperbola; directrix is parallel to the polar axis $\frac{4}{3}$ units below the pole.

5. Ellipse; directrix is perpendicular to the polar axis $\frac{3}{2}$ units to the left of the pole.

7.

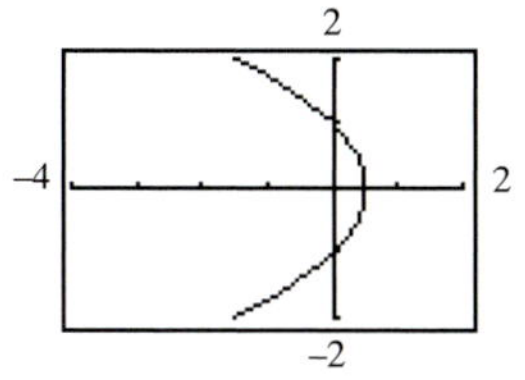

Parabola; directrix is perpendicular to the polar axis 1 unit to the right of the pole.

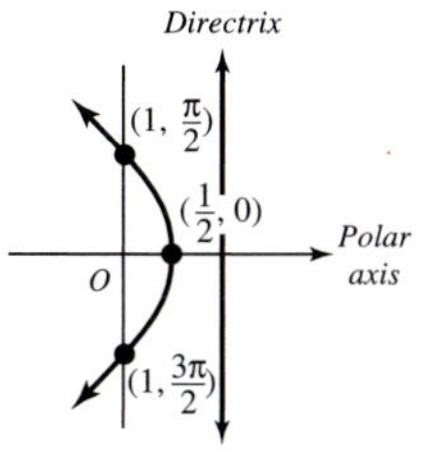

9.

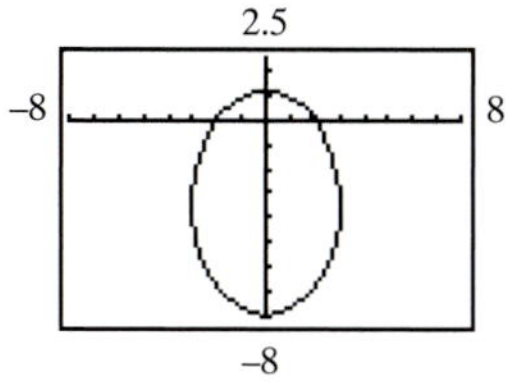

Ellipse; directrix is parallel to the polar axis $\frac{8}{3}$ units above the pole.

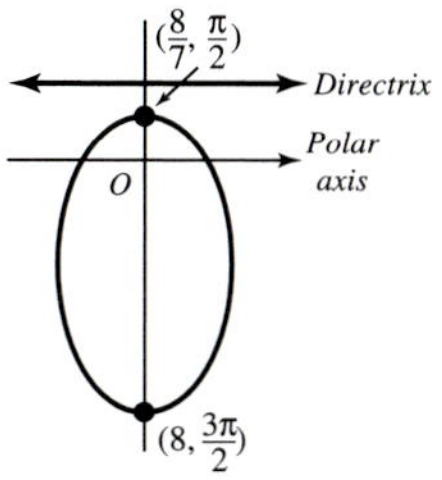

11.

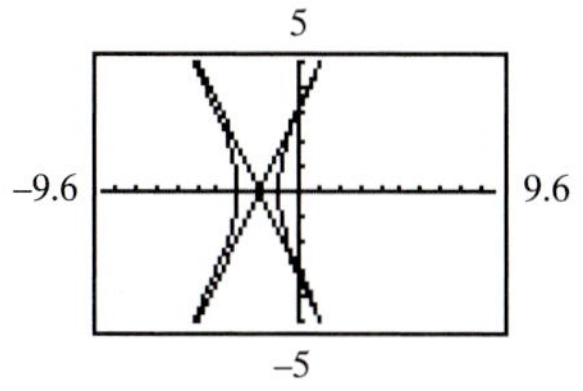

Hyperbola; directrix is perpendicular to the polar axis $\frac{3}{2}$ units to the left of the pole.

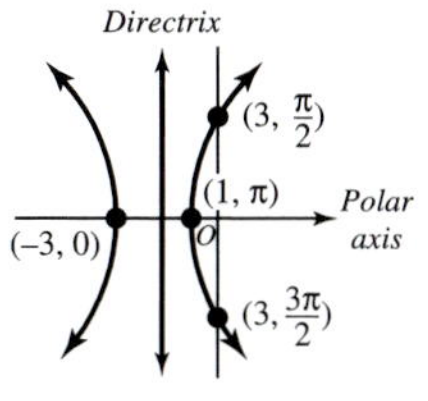

13.

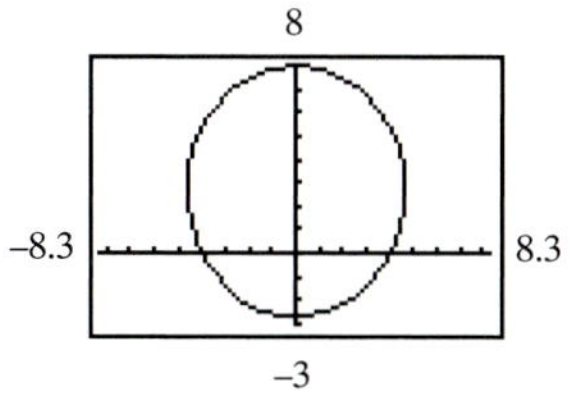

Ellipse; directrix is parallel to the polar axis 8 units below the pole; vertices are at $\left(8, \frac{\pi}{2}\right)$ and $\left(\frac{8}{3}, \frac{3\pi}{2}\right)$.

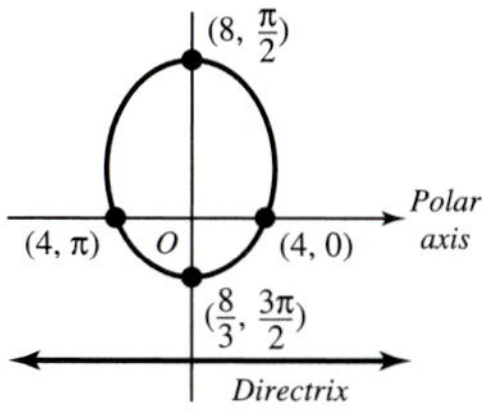

15.

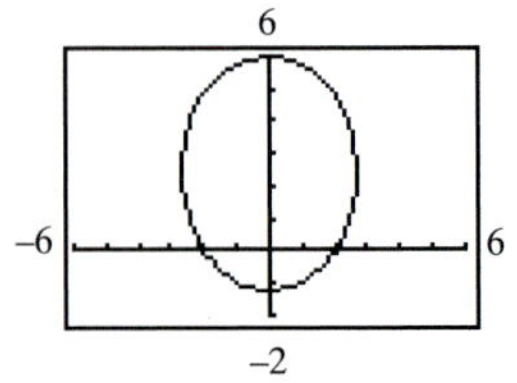

Ellipse; directrix is parallel to the polar axis 3 units below the pole; vertices are at $\left(6, \frac{\pi}{2}\right)$ and $\left(\frac{6}{5}, \frac{3\pi}{2}\right)$.

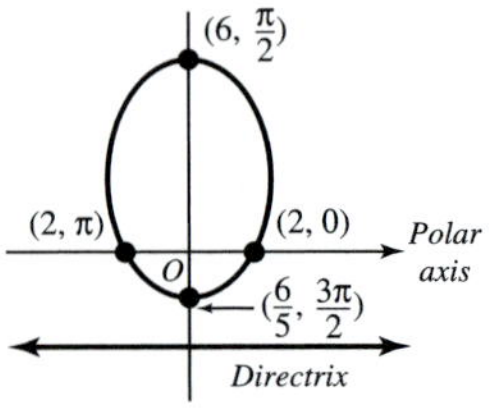

17.

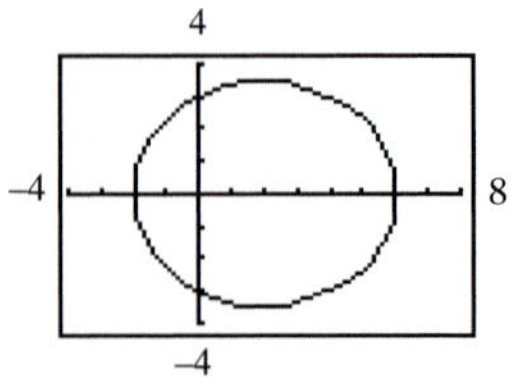

Ellipse; directrix is perpendicular to the polar axis 6 units to the left of the pole; vertices are at $(6, 0)$ and $(2, \pi)$.

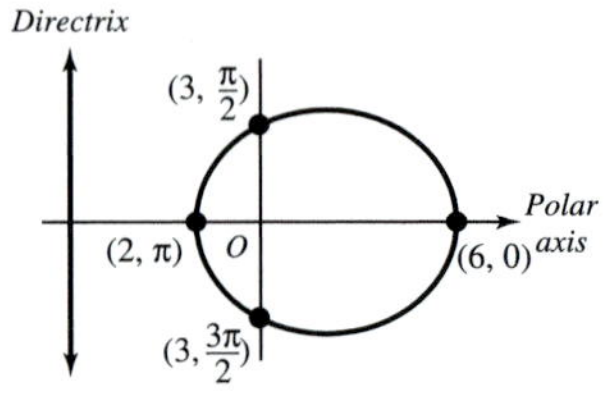

19. $y^2 + 2x - 1 = 0$ **21.** $16x^2 + 7y^2 + 48y - 64 = 0$ **23.** $3x^2 - y^2 + 12x + 9 = 0$ **25.** $4x^2 + 3y^2 - 16y - 64 = 0$
27. $9x^2 + 5y^2 - 24y - 36 = 0$ **29.** $3x^2 + 4y^2 - 12x - 36 = 0$ **31.** $r = \dfrac{1}{1 + \sin\theta}$ **33.** $r = \dfrac{12}{5 - 4\cos\theta}$ **35.** $r = \dfrac{12}{1 - 6\sin\theta}$
37. Use $d(D, P) = p - r\cos\theta$ in the derivation of equation (a) in Table 5.
39. Use $d(D, P) = p + r\sin\theta$ in the derivation of equation (a) in Table 5.

Exercises 10.7

1.

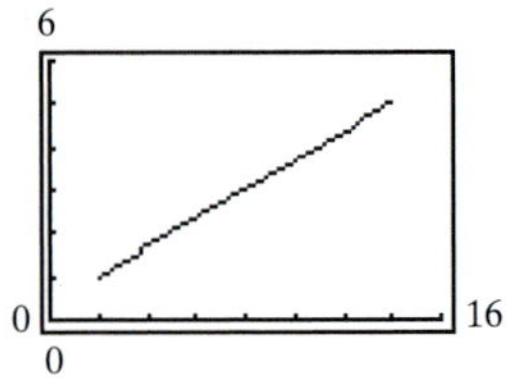

$x - 3y + 1 = 0$

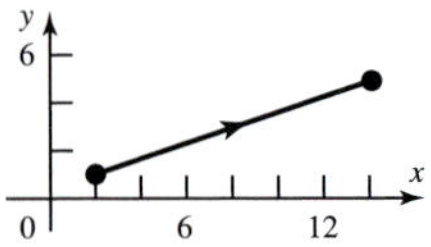

3.

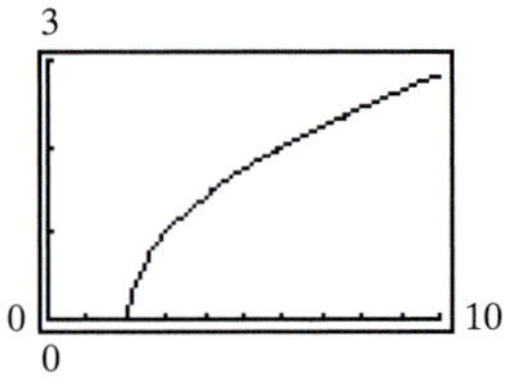

$y = \sqrt{x - 2}$

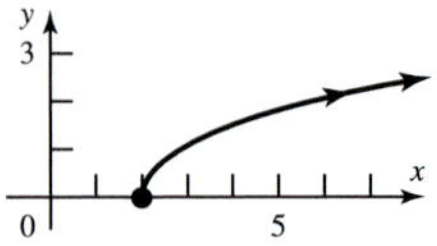

5.

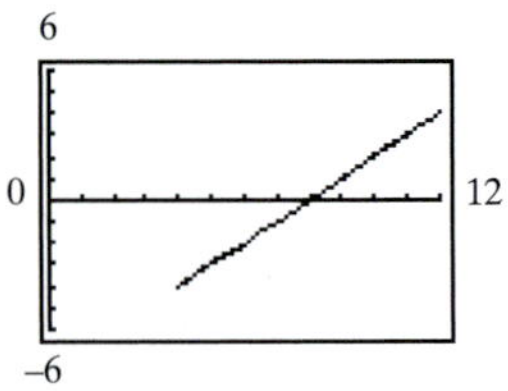

$x = y + 8$

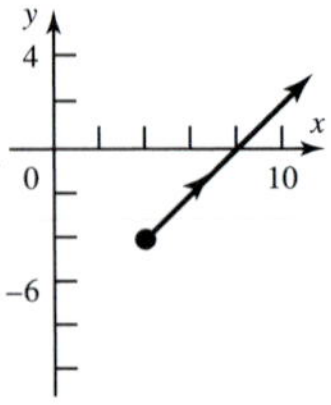

7.

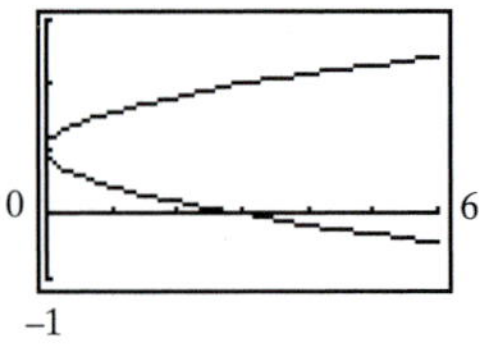

$x = 3(y - 1)^2$

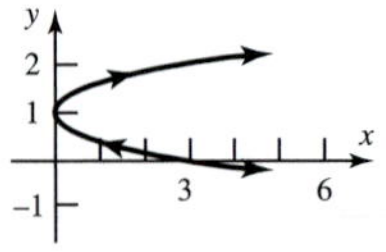

9.

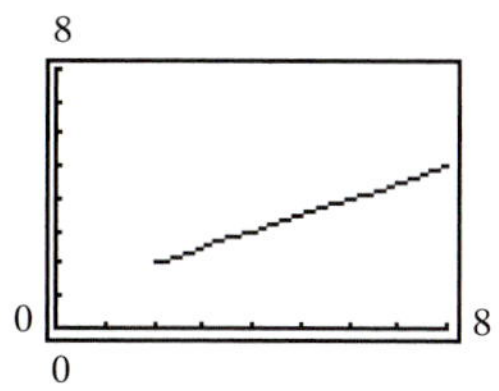

$2y = 2 + x$

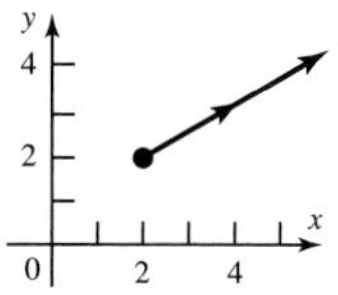

11.

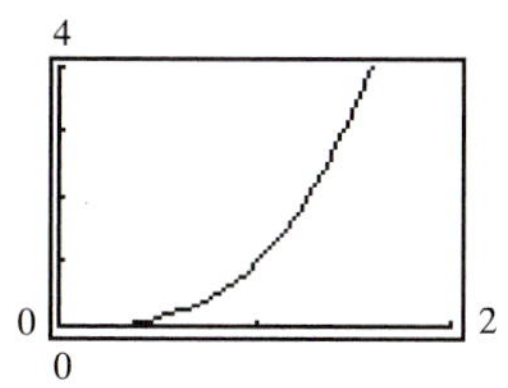

$y = x^3$

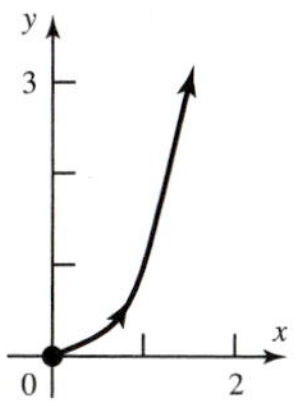

13.

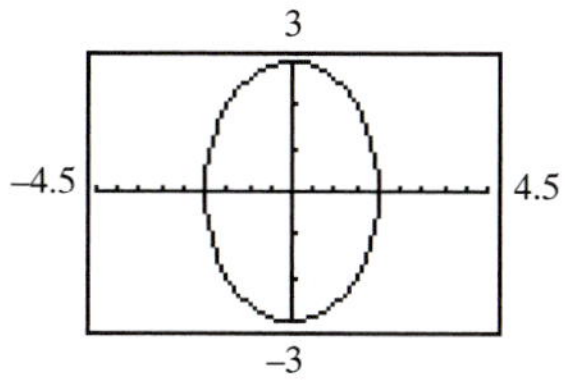

$\frac{x^2}{4} + \frac{y^2}{9} = 1$

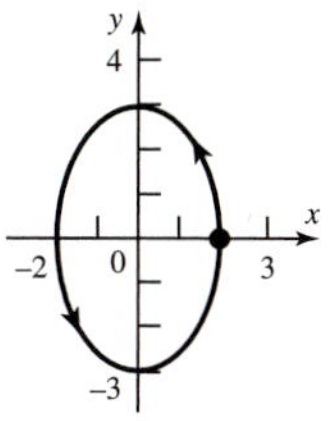

15.

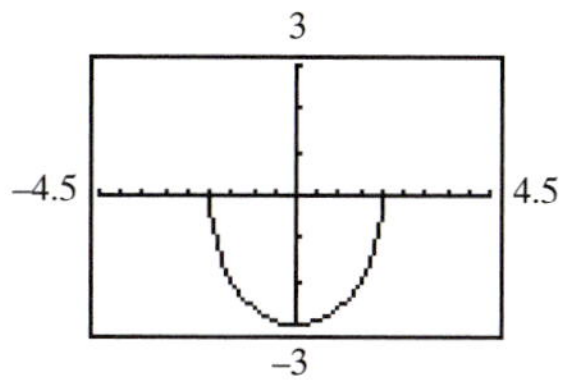

$\frac{x^2}{4} + \frac{y^2}{9} = 1$

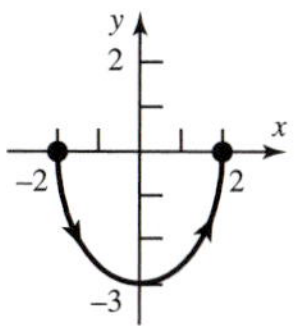

17.

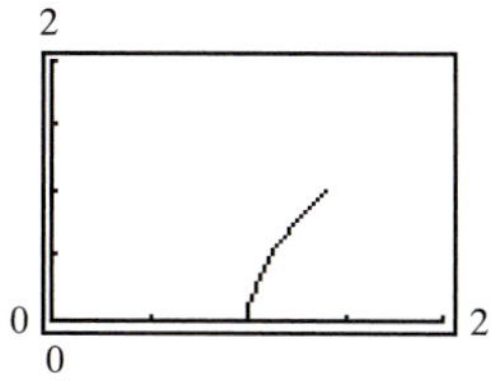

$x^2 - y^2 = 1$

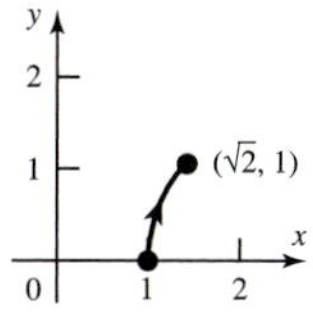

19.

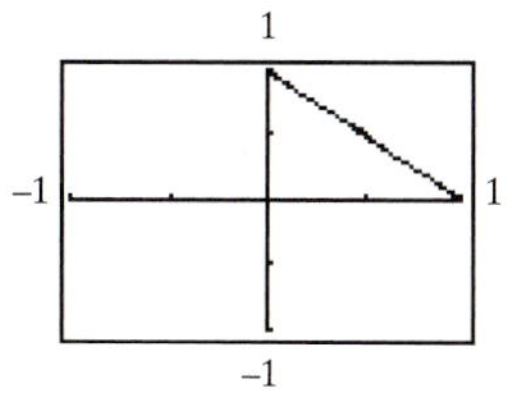

$x + y = 1$

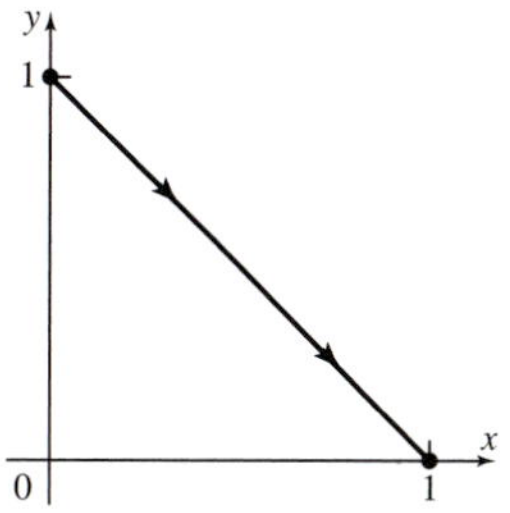

21. **(a)** $x = 3$
$y = -16t^2 + 50t + 6$

(b) 3.24 seconds

(c) 1.5625 seconds; 45.0625 feet

(d)

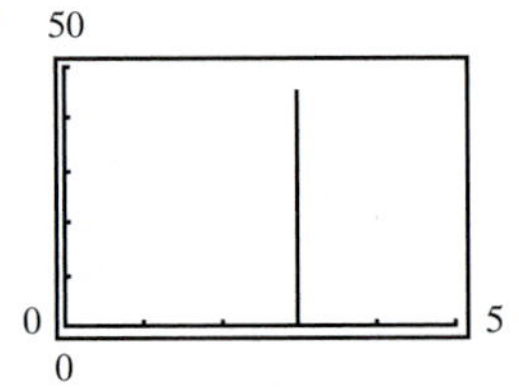

23. **(a)** Train: $x_1 = t^2$, $y_1 = 1$; Bill: $x_2 = 5(t - 5)$, $y_2 = 3$

(b) Bill won't catch the train.

(c)

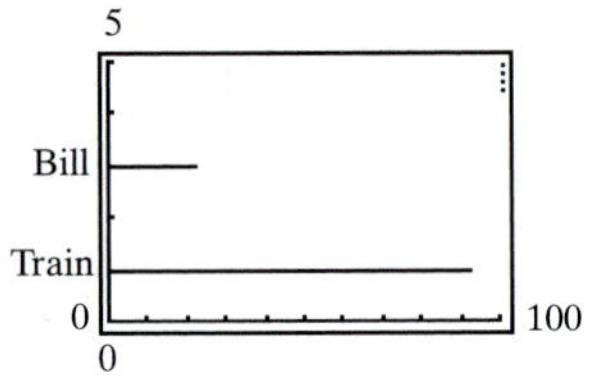

25. **(a)** $x = (145 \cos 20°)t$
$y = -16t^2 + (145 \sin 20°)t + 5$

(b) 3.197 seconds

(c) 1.55 seconds; 43.43 feet

(d) 435.61 feet

(e)

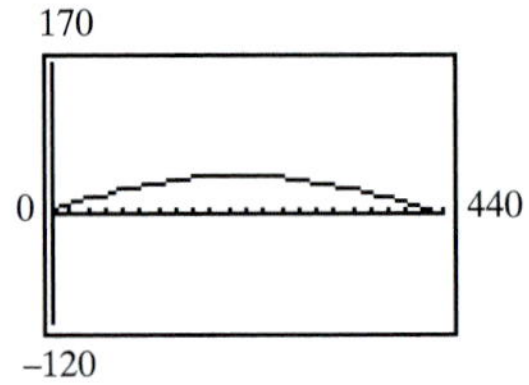

27. **(a)** $x = (40 \cos 45°)t$
$y = -4.9t^2 + (40 \sin 45°)t + 300$

(b) 11.226 seconds

(c) 2.886 seconds; 340.8 meters

(d) 317.5 meters

(e)

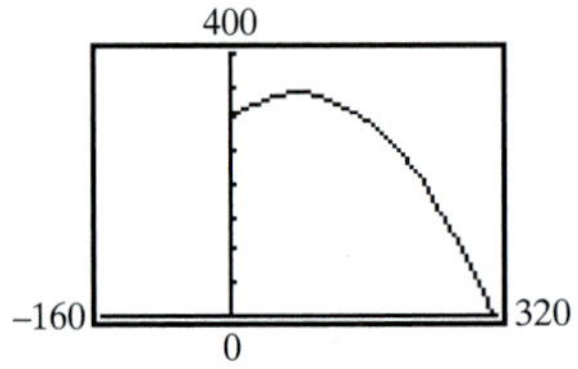

29. **(a)** $d = \sqrt{(40t - 5)^2 + (30t - 4)^2}$

(b)

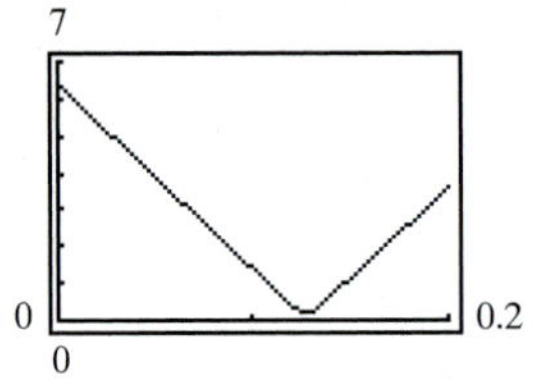

(c) 0.2 miles; 0.128 seconds

(d) Paseo: $x = 40t - 5$, $y = 0$; Bonneville: $x = 0$, $y = 30t - 4$

(e) Turn axes off to see the graph:

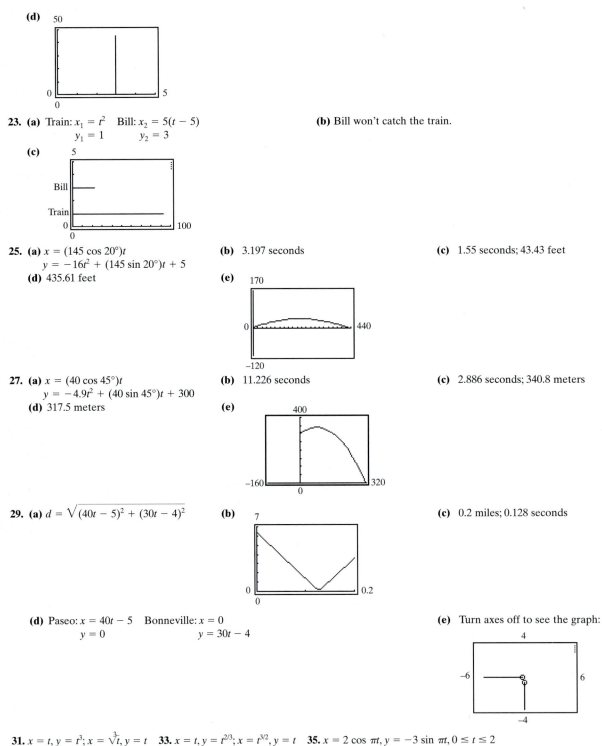

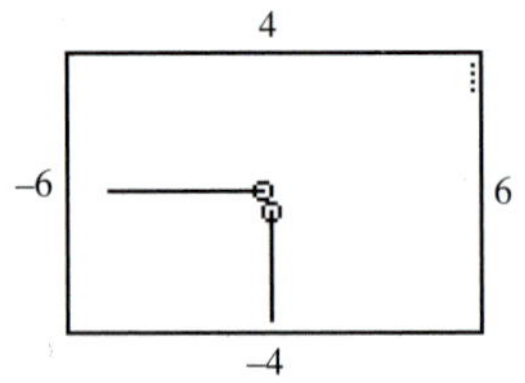

31. $x = t, y = t^3; x = \sqrt[3]{t}, y = t$ **33.** $x = t, y = t^{2/3}; x = t^{3/2}, y = t$ **35.** $x = 2 \cos \pi t, y = -3 \sin \pi t, 0 \le t \le 2$
37. $x = -2 \sin 2\pi t, y = 3 \cos 2\pi t, 0 \le t \le 1$

39.

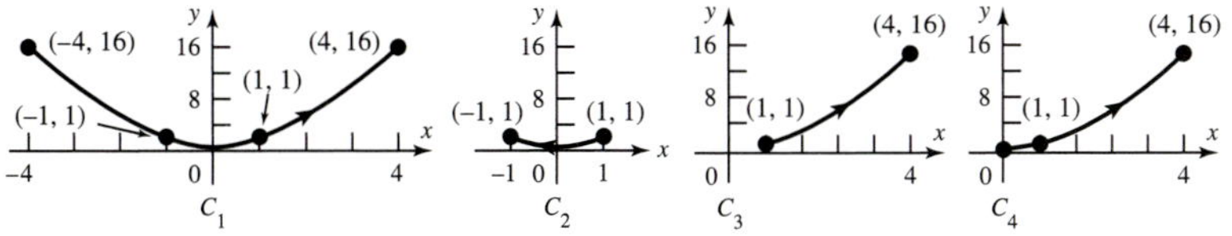

41. The orientation is from (x_1, y_1) to (x_2, y_2).

43.

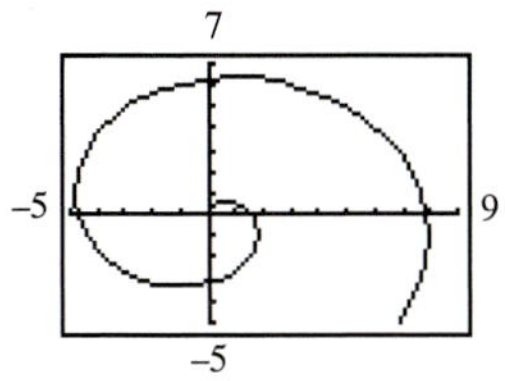

45.

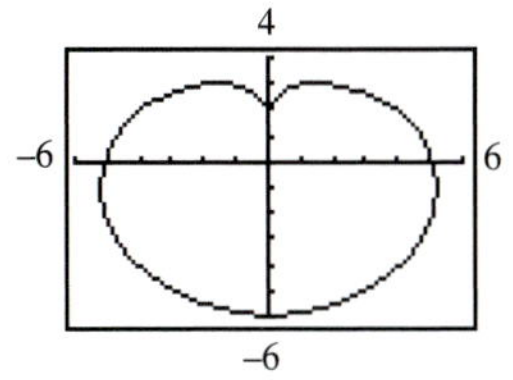

47. (a)

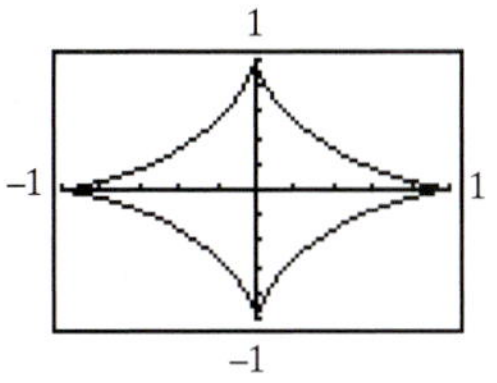

(b) $x^{2/3} + y^{2/3} = 1$

Fill-in-the-Blank Items

1. parabola **2.** ellipse **3.** hyperbola **4.** major; transverse **5.** y-axis **6.** $\frac{y}{3} = \frac{x}{2}; \frac{y}{3} = -\frac{x}{2}$ **7.** $\cot 2\theta = \frac{A - C}{B}$

8. $\frac{1}{2}$; ellipse; parallel; 4; below **9.** ellipse

True/False Items

1. T **2.** F **3.** T **4.** T **5.** T **6.** T **7.** T **8.** T **9.** T **10.** F

Review Exercises

1. Parabola; vertex $(0, 0)$, focus $(-4, 0)$, directrix $x = 4$

3. Hyperbola; center $(0, 0)$, vertices $(5, 0)$ and $(-5, 0)$, foci $(\sqrt{26}, 0)$ and $(-\sqrt{26}, 0)$, asymptotes $y = \frac{1}{5}x$ and $y = -\frac{1}{5}x$

5. Ellipse; center $(0, 0)$, vertices $(0, 5)$ and $(0, -5)$, foci $(0, 3)$ and $(0, -3)$

7. $x^2 = -4(y - 1)$: Parabola; vertex $(0, 1)$, focus $(0, 0)$, directrix $y = 2$

9. $\frac{x^2}{2} - \frac{y^2}{8} = 1$: Hyperbola; center $(0, 0)$, vertices $(\sqrt{2}, 0)$ and $(-\sqrt{2}, 0)$, foci $(\sqrt{10}, 0)$ and $(-\sqrt{10}, 0)$, asymptotes $y = 2x$ and $y = -2x$

11. $(x - 2)^2 = 2(y + 2)$: Parabola; vertex $(2, -2)$, focus $\left(2, -\frac{3}{2}\right)$, directrix $y = -\frac{5}{2}$

13. $\frac{(y - 2)^2}{4} - (x - 1)^2 = 1$: Hyperbola; center $(1, 2)$, vertices $(1, 4)$ and $(1, 0)$, foci $(1, 2 + \sqrt{5})$ and $(1, 2 - \sqrt{5})$,

asymptotes $y - 2 = \pm 2(x - 1)$

15. $\frac{(x - 2)^2}{9} + \frac{(y - 1)^2}{4} = 1$: Ellipse; center $(2, 1)$, vertices $(5, 1)$ and $(-1, 1)$, foci $(2 + \sqrt{5}, 1)$ and $(2 - \sqrt{5}, 1)$

17. $(x - 2)^2 = -4(y + 1)$: Parabola; vertex $(2, -1)$, focus $(2, -2)$, directrix $y = 0$

19. $\frac{(x - 1)^2}{4} + \frac{(y + 1)^2}{9} = 1$: Ellipse; center $(1, -1)$, vertices $(1, 2)$ and $(1, -4)$, foci $(1, -1 + \sqrt{5})$ and $(1, -1 - \sqrt{5})$

21. $y^2 = -8x$

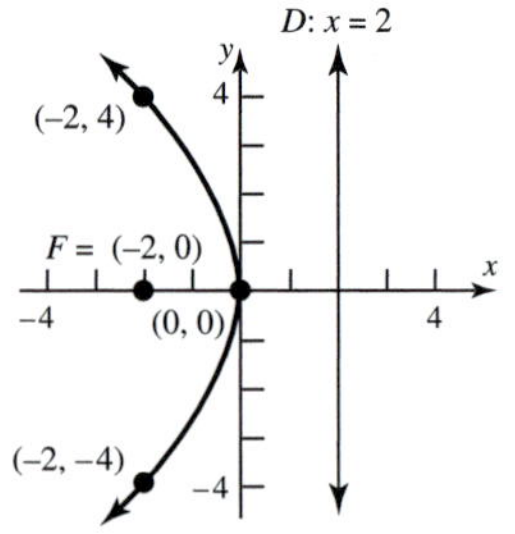

23. $\dfrac{y^2}{4} - \dfrac{x^2}{12} = 1$

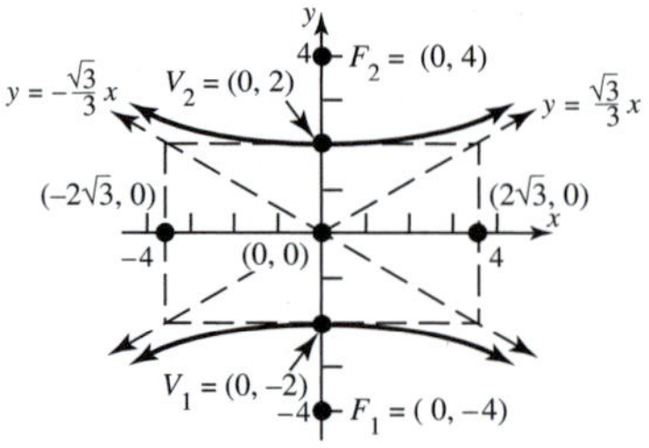

25. $\dfrac{x^2}{16} + \dfrac{y^2}{7} = 1$

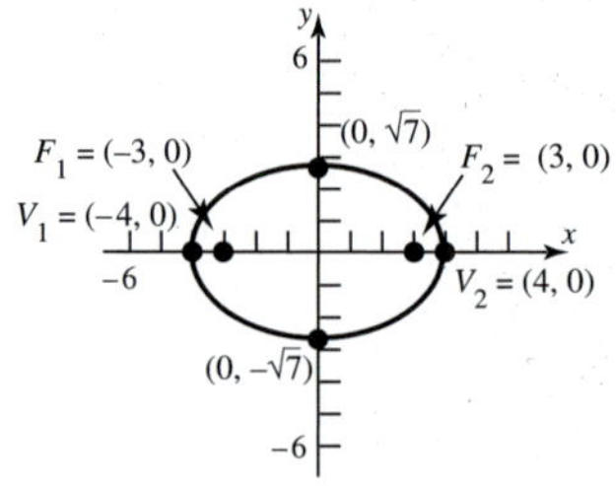

27. $(x - 2)^2 = -4(y + 3)$

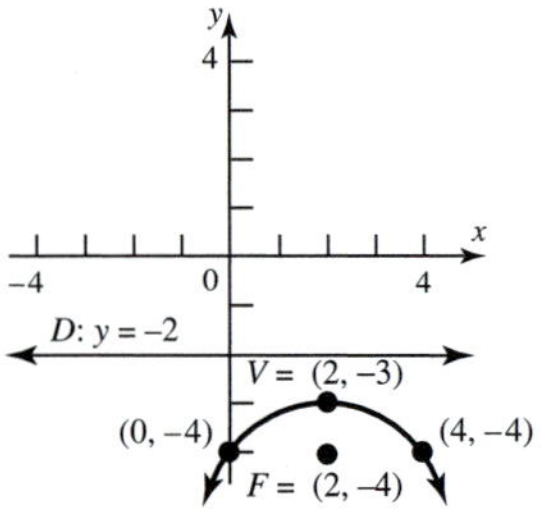

29. $(x + 2)^2 - \dfrac{(y + 3)^2}{3} = 1$

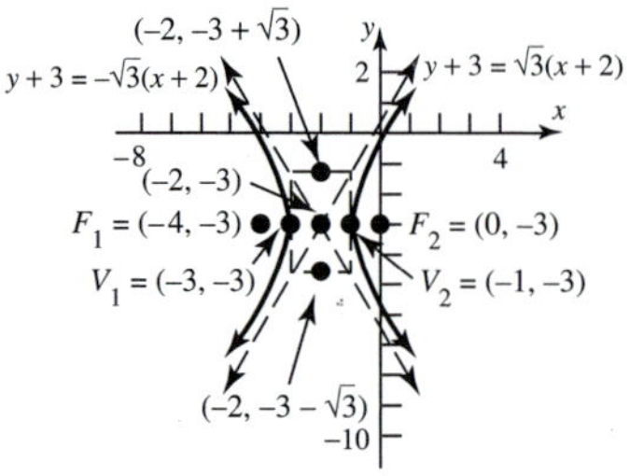

31. $\dfrac{(x + 4)^2}{16} + \dfrac{(y - 5)^2}{25} = 1$

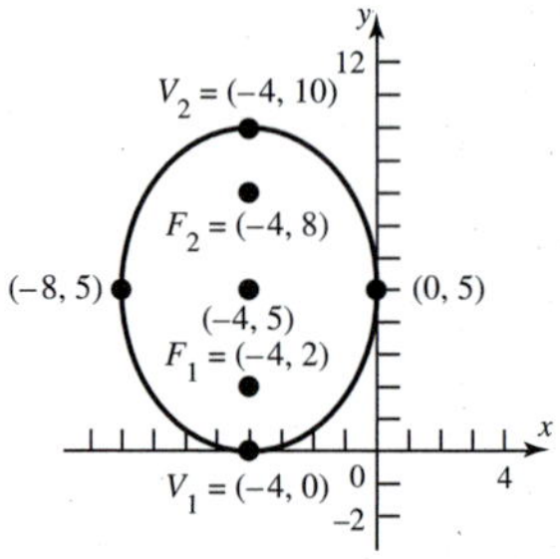

33. $\dfrac{(x + 1)^2}{9} - \dfrac{(y - 2)^2}{7} = 1$

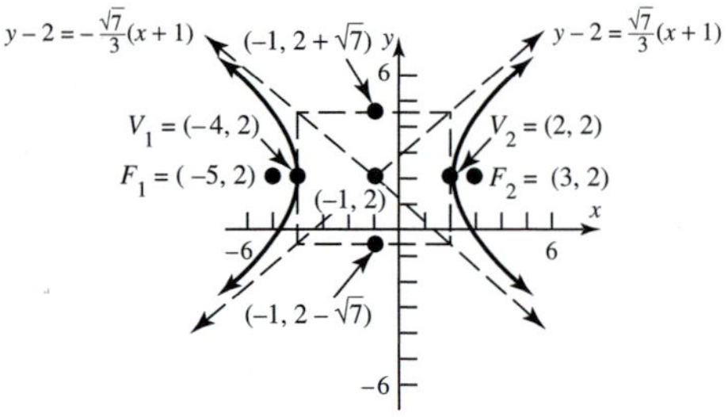

35. $\dfrac{(x - 3)^2}{9} - \dfrac{(y - 1)^2}{4} = 1$

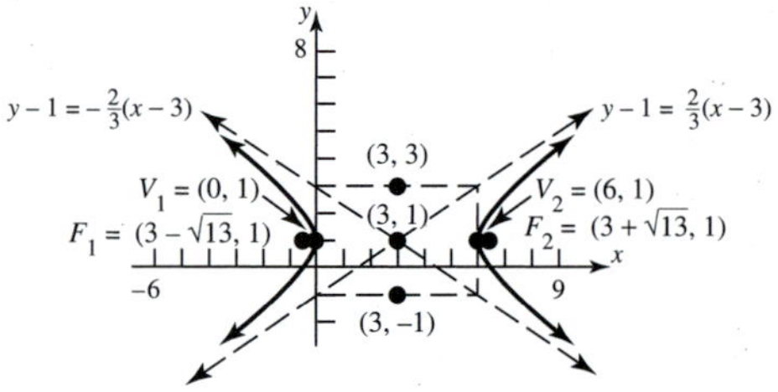

37. Parabola **39.** Ellipse **41.** Parabola **43.** Hyperbola **45.** Ellipse

47. $x'^2 - \dfrac{y'^2}{9} = 1$

Hyperbola
Center at the origin

Transverse axis the x'-axis

Vertices at $(\pm 1, 0)$

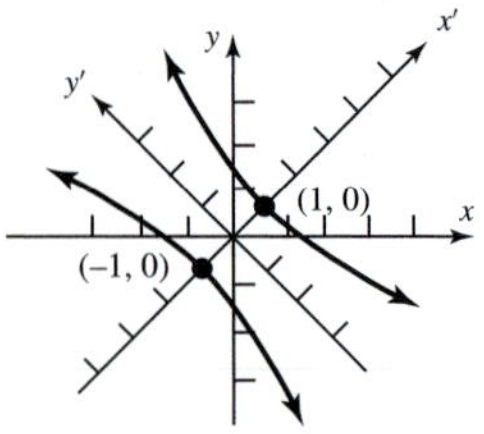

49. $\dfrac{x'^2}{2} + \dfrac{y'^2}{4} = 1$

Ellipse
Center at origin

Major axis the y'-axis

Vertices at $(0, \pm 2)$

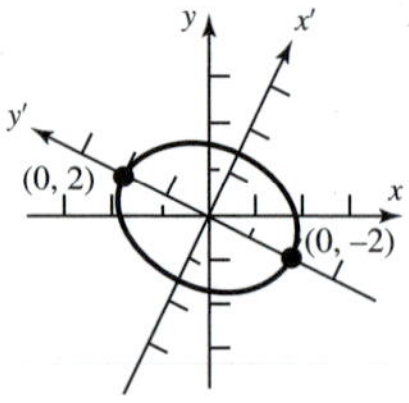

51. $y'^2 = -\dfrac{4\sqrt{13}}{13}x'$

Parabola
Vertex at the origin

Focus on the x'-axis at $\left(-\dfrac{\sqrt{13}}{13}, 0\right)$

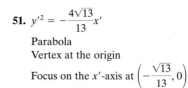

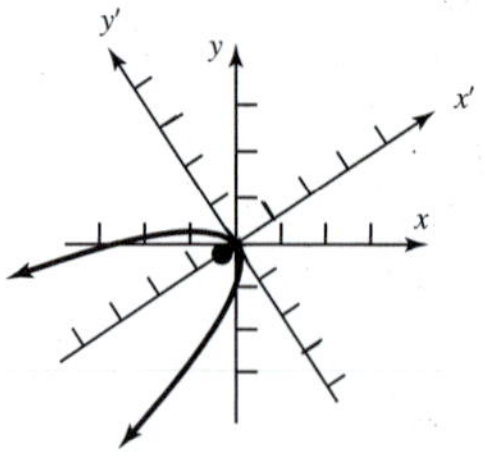

53. Parabola; directrix is perpendicular to the polar axis 4 units to the left of the pole.

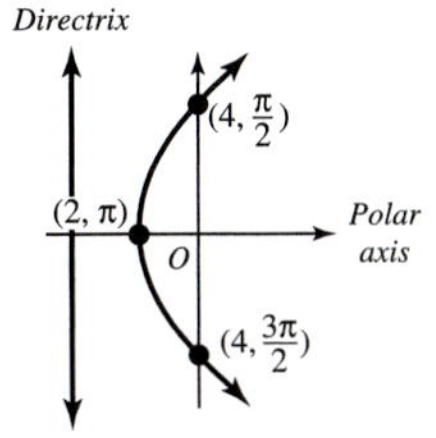

55. Ellipse; directrix is parallel to the polar axis 6 units below the pole; vertices are $\left(6, \frac{\pi}{2}\right)$ and $\left(2, \frac{3\pi}{2}\right)$.

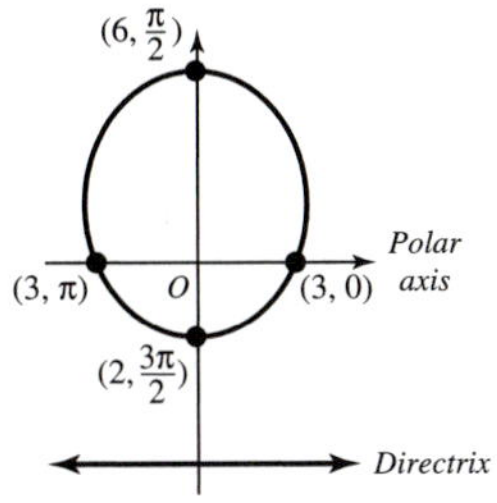

57. Hyperbola; directrix is perpendicular to the polar axis 1 unit to the right of the pole; vertices are at $\left(\frac{2}{3}, 0\right)$ and $(-2, \pi)$.

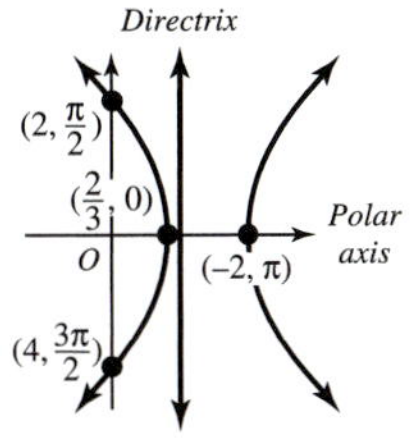

59. $y^2 - 8x - 16 = 0$ **61.** $3x^2 - y^2 - 8x + 4 = 0$

63.

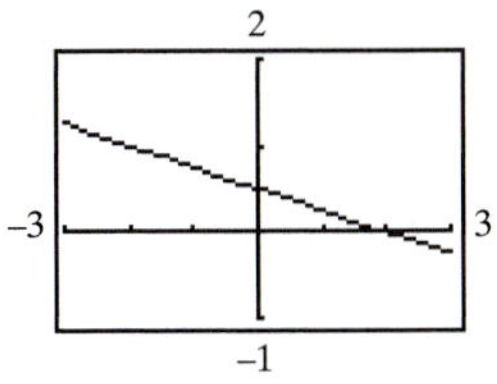

$x + 4y = 2$

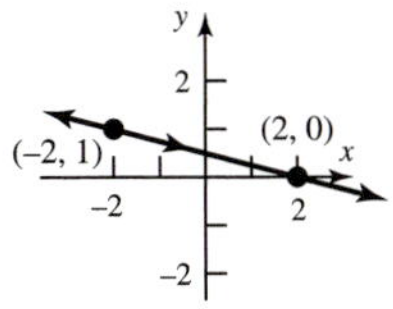

65.

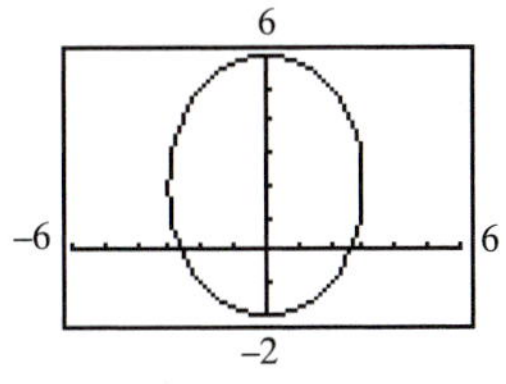

$\frac{x^2}{9} + \frac{(y-2)^2}{16} = 1$

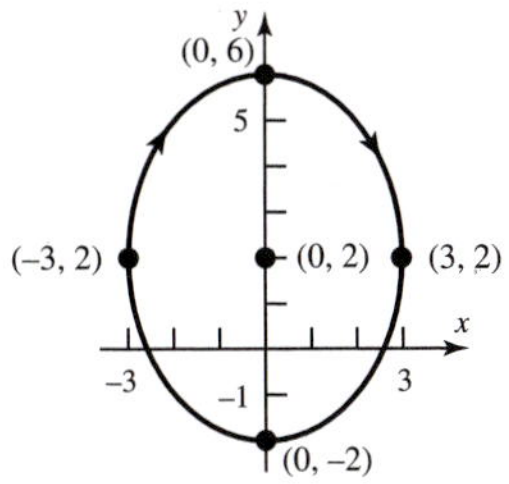

67.

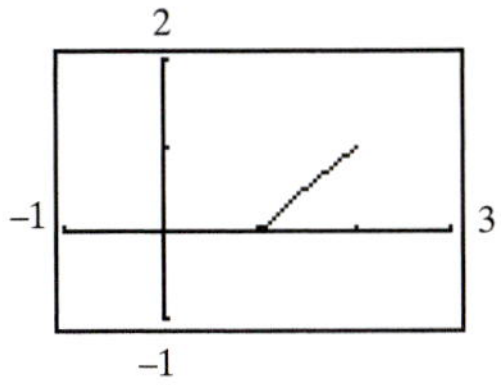

$1 + y = x$

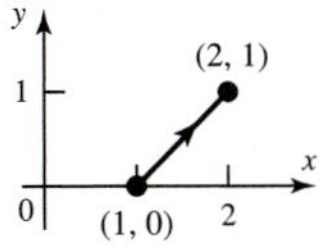

69. $\frac{x^2}{5} - \frac{y^2}{4} = 1$ **71.** The ellipse $\frac{x^2}{16} + \frac{y^2}{7} = 1$ **73.** $\frac{1}{4}$ ft or 3 in.

75. 19.72 ft, 18.86 ft, 14.91 ft

77. (a) 45.24 miles from the Master Station **(b)** 0.000645 second **(c)** (66, 20)

79. (a) $x = (100 \cos 35°)t$
$y = -16t^2 + (100 \sin 35°)t + 6$
(b) 3.6866 seconds **(c)** 1.7924 seconds; 57.4 feet
(d) 302 feet
(e)

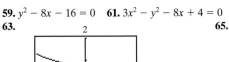

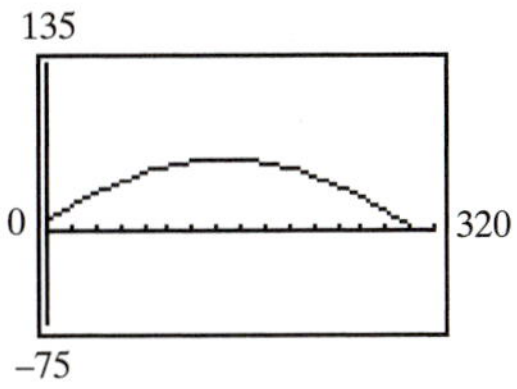

CHAPTER 11 *11.1 Exercises*

1. $2(2) - (-1) = 5$ and $5(2) + 2(-1) = 8$ **3.** $3(2) - 4\left(\frac{1}{2}\right) = 4$ and $\frac{1}{2}(2) - 3\left(\frac{1}{2}\right) = -\frac{1}{2}$ **5.** $2^2 - 1^2 = 3$ and $(2)(1) = 2$
7. $\frac{0}{1+0} + 3(2) = 6$ and $0 + 9(2)^2 = 36$ **9.** $3(1) + 3(-1) + 2(2) = 4$, $1 - (-1) - 2 = 0$, and $2(-1) - 3(2) = -8$ **11.** $x = 6, y = 2$
13. $x = 3, y = 2$ **15.** $x = 8, y = -4$ **17.** $x = \frac{1}{3}, y = -\frac{1}{6}$ **19.** Inconsistent **21.** $x = 1, y = 2$ **23.** $x = 4 - 2y$, y is any real number
25. $x = 1, y = 1$ **27.** $x = \frac{3}{2}, y = 1$ **29.** $x = 4, y = 3$ **31.** $x = \frac{4}{3}, y = \frac{1}{5}$ **33.** $x = 8, y = 2, z = 0$ **35.** $x = 2, y = -1, z = 1$
37. Inconsistent **39.** $x = 5z - 2, y = 4z - 3$ where z is any real number, or $x = \frac{5}{4}y + \frac{7}{4}, z = \frac{1}{4}y + \frac{3}{4}$ where y is any real number, or $y = \frac{4}{5}x - \frac{7}{5}, z = \frac{1}{5}x + \frac{2}{5}$ where x is any real number **41.** Inconsistent **43.** $x = 1, y = 3, z = -2$ **45.** $x = -3, y = \frac{1}{2}, z = 1$
47. $x = \frac{1}{5}, y = \frac{1}{3}$ **49.** $x = 48.15, y = 15.18$ **51.** $x = -21.47, y = 16.12$ **53.** $x = 0.26, y = 0.06$ **55.** 20 and 61
57. Length 30 ft; width 15 ft **59.** Cheeseburger \$1.55; shake \$0.85 **61.** 22.5 lb
63. Average wind speed 25 mph; average airspeed 175 mph **65.** 80 \$25 sets and 120 \$45 sets **67.** \$5.56 **69.** 12, 15, 21
71. $I_1 = \frac{10}{71}, I_2 = \frac{65}{71}, I_3 = \frac{55}{71}$ **73.** 100 orchestra, 210 main, and 190 balcony seats **79.** $b = -\frac{1}{2}; c = \frac{3}{2}$ **81.** $a = \frac{4}{3}, b = -\frac{5}{3}, c = 1$
83. $x = \frac{b_1 - b_2}{m_2 - m_1}, y = \frac{m_2 b_1 - m_1 b_2}{m_2 - m_1}$ **85.** $y = mx + b$, x is any real number

11.2 Exercises

1. $\left[\begin{array}{rr|r} 1 & -5 & 5 \\ 4 & 3 & 6 \end{array}\right]$ **3.** $\left[\begin{array}{rr|r} 2 & 3 & 6 \\ 4 & -6 & -2 \end{array}\right]$ **5.** $\left[\begin{array}{rr|r} 0.01 & -0.03 & 0.06 \\ 0.13 & 0.10 & 0.20 \end{array}\right]$ **7.** $\left[\begin{array}{rrr|r} 1 & -1 & 1 & 10 \\ 3 & 2 & 0 & 5 \\ 1 & 1 & 2 & 2 \end{array}\right]$ **9.** $\left[\begin{array}{rrr|r} 1 & 1 & -1 & 2 \\ 3 & -2 & 0 & 2 \end{array}\right]$

11.
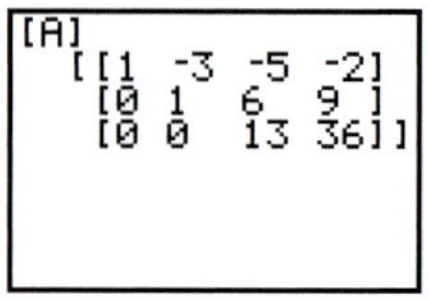

13.
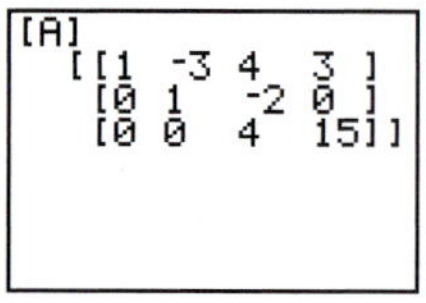

15.
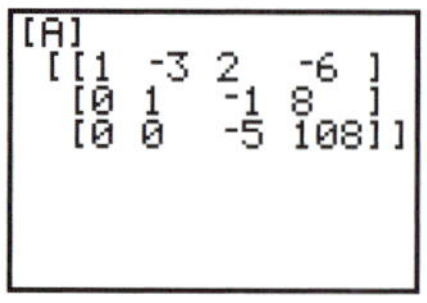

17.
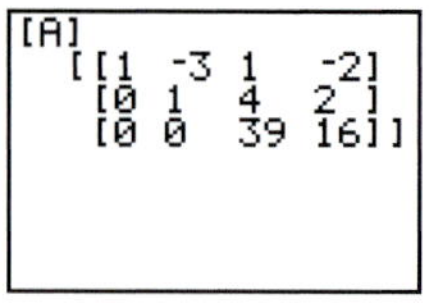

19.
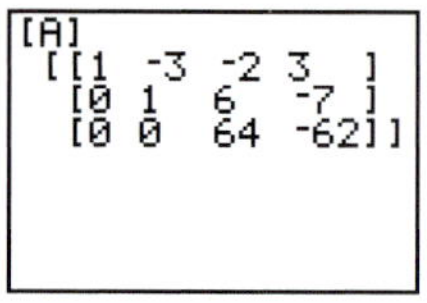

21. $\begin{cases} x = 5 \\ y = -1 \end{cases}$ consistent; $x = 5, y = -1$

23. $\begin{cases} x = 1 \\ y = 2 \\ 0 = 3 \end{cases}$ inconsistent

25. $\begin{cases} x + 2z = -1 \\ y - 4z = -2 \\ 0 = 0 \end{cases}$ consistent; $x = -1 - 2z$, $y = -2 + 4z$, z any real number

27. $\begin{cases} x_1 = 1 \\ x_2 + x_4 = 2 \\ x_3 + 2x_4 = 3 \end{cases}$ consistent; $x_1 = 1, x_2 = 2 - x_4$, $x_3 = 3 - 2x_4$, x_4 any real number

29. $\begin{cases} x_1 + 4x_4 = 2 \\ x_2 + x_3 + 3x_4 = 3 \\ 0 = 0 \end{cases}$ consistent; $x_1 = 2 - 4x_4$, $x_2 = 3 - x_3 - 3x_4$, x_3, x_4 any real numbers

31. $x = 6, y = 2$ **33.** $x = 2, y = 3$ **35.** $x = 4, y = -2$ **37.** Inconsistent **39.** $x = \frac{1}{2}, y = \frac{3}{4}$ **41.** $x = 4 - 2y$, y is any real number
43. $x = \frac{3}{2}, y = 1$ **45.** $x = \frac{4}{3}, y = \frac{1}{5}$ **47.** $x = 8, y = 2, z = 0$ **49.** $x = 2, y = -1, z = 1$ **51.** Inconsistent
53. $x = 5z - 2, y = 4z - 3$, where z is any real number, or $x = \frac{5}{4}y + \frac{7}{4}, z = \frac{1}{4}y + \frac{3}{4}$, where y is any real number, or $y = \frac{4}{5}x - \frac{7}{5}, z = \frac{1}{5}x + \frac{2}{5}$, where x is any real number **55.** Inconsistent **57.** $x = 1, y = 3, z = -2$ **59.** $x = -3, y = \frac{1}{2}, z = 1$
61. $x = \frac{1}{3}, y = \frac{2}{3}, z = 1$ **63.** $x = 1, y = 2, z = 0, w = 1$ **65.** $y = 0, z = 1 - x$, x is any real number
67. $x = 2, y = z - 3$, z is any real number **69.** $x = \frac{13}{9}, y = \frac{7}{18}, z = \frac{19}{18}$

71. $x = \frac{7}{5} - \frac{3}{5}z - \frac{2}{5}w,\ y = -\frac{8}{5} + \frac{7}{5}z + \frac{13}{5}w$, where z and w are any real numbers **73.** $y = -2x^2 + x + 3$ **75.** $f(x) = 3x^3 - 4x^2 + 5$

77. x = liters of 15% H_2SO_4, y = liters of 25% H_2SO_4, z = liters of 50% H_2SO_4: $\begin{cases} x = \frac{5}{2}z - 150 \\ y = 250 - \frac{7}{2}z \end{cases}$

15%	25%	50%	40%
0	40	60	100
10	26	64	100
20	12	68	100

79. If x = Price of hamburgers, y = Price of fries, z = Price of colas, then $x = 2.75 - z,\ y = 0.68 + \frac{1}{3}z$, z any real number.

There is not sufficient information:

x	\$2.15	\$2.00	\$1.85
y	\$0.88	\$0.93	\$0.98
z	\$0.60	\$0.75	\$0.90

81. (a) Amount Invested At

7%	9%	11%
0	10,000	10,000
1000	8000	11,000
2000	6000	12,000
3000	4000	13,000
4000	2000	14,000
5000	0	15,000

(b) Amount Invested At

7%	9%	11%
12,500	12,500	0
14,500	8500	2000
16,500	4500	4000
18,750	0	6250

(c) All the money invested 7% provides \$2100, more than what is required.

83. $I_1 = \frac{4}{15},\ I_2 = \frac{8}{15},\ I_3 = \frac{4}{5}$ **85.** $I_1 = 3.5,\ I_2 = 2.5,\ I_3 = 1$

89. If $a_1 \neq 0$,

$$\left[\begin{array}{cc|c} a_1 & b_1 & c_1 \\ a_2 & b_2 & c_2 \end{array}\right] \rightarrow \left[\begin{array}{cc|c} 1 & \frac{b_1}{a_1} & \frac{c_1}{a_1} \\ a_2 & b_2 & c_2 \end{array}\right] \rightarrow \left[\begin{array}{cc|c} 1 & \frac{b_1}{a_1} & \frac{c_1}{a_1} \\ 0 & \frac{-a_2b_1}{a_1} + b_2 & \frac{-a_2c_1}{a_1} + c_2 \end{array}\right] \rightarrow \left[\begin{array}{cc|c} 1 & \frac{b_1}{a_1} & \frac{c_1}{a_1} \\ 0 & \frac{-a_2b_1 + b_2a_1}{a_1} & \frac{-a_2c_1 + c_2a_1}{a_1} \end{array}\right]$$

$$\rightarrow \left[\begin{array}{cc|c} 1 & \frac{b_1}{a_1} & \frac{c_1}{a_1} \\ 0 & 1 & \frac{-a_2c_1 + c_2a_1}{a_1} \cdot \frac{a_1}{-a_2b_1 + b_2a_1} \end{array}\right] \rightarrow \left[\begin{array}{cc|c} 1 & \frac{b_1}{a_1} & \frac{c_1}{a_1} \\ 0 & 1 & \frac{-a_2c_1 + c_2a_1}{-a_2b_1 + b_2a_1} \end{array}\right] \rightarrow \left[\begin{array}{cc|c} 1 & 0 & \frac{-b_1c_2 + b_2c_1}{-a_2b_1 + b_2a_1} \\ 0 & 1 & \frac{-a_2c_1 + c_2a_1}{-a_2b_1 + b_2a_1} \end{array}\right]$$

$$x = \frac{1}{a_1b_2 - a_2b_1}(c_1b_2 - c_2b_1) = \frac{1}{D}(c_1b_2 - c_2b_1),\ y = \frac{1}{a_1b_2 - a_2b_1}(a_1c_2 - a_2c_1) = \frac{1}{D}(a_1c_2 - a_2c_1)$$

If $a_1 = 0$, then $a_2 \neq 0$, $b_1 \neq 0$, and

$$\left[\begin{array}{cc|c} 0 & b_1 & c_1 \\ a_2 & b_2 & c_2 \end{array}\right] \rightarrow \left[\begin{array}{cc|c} a_2 & b_2 & c_2 \\ 0 & b_1 & c_1 \end{array}\right] \rightarrow \left[\begin{array}{cc|c} 1 & \frac{b_2}{a_2} & \frac{c_2}{a_2} \\ 0 & b_1 & c_1 \end{array}\right] \rightarrow \left[\begin{array}{cc|c} 1 & \frac{b_2}{a_2} & \frac{c_2}{a_2} \\ 0 & 1 & \frac{c_1}{b_1} \end{array}\right] \rightarrow \left[\begin{array}{cc|c} 1 & 0 & \frac{c_2}{a_2} - \frac{b_2c_1}{a_2b_1} = \frac{c_1b_2 - c_2b_1}{-a_2b_1} \\ 0 & 1 & \frac{c_1}{b_1} = \frac{-a_2c_1}{-a_2b_1} \end{array}\right]$$

11.3 Exercises

1. 2 **3.** 22 **5.** -2 **7.** 10 **9.** -26 **11.** $x = 6, y = 2$ **13.** $x = 3, y = 2$ **15.** $x = 8, y = -4$ **17.** $x = 4, y = -2$ **19.** Not applicable **21.** $x = \frac{1}{2}, y = \frac{3}{4}$ **23.** $x = \frac{1}{10}, y = \frac{2}{5}$ **25.** $x = \frac{3}{2}, y = 1$ **27.** $x = \frac{4}{3}, y = \frac{1}{5}$ **29.** $x = 1, y = 3, z = -2$ **31.** $x = -3, y = \frac{1}{2}, z = 1$ **33.** Not applicable **35.** $x = 0, y = 0, z = 0$ **37.** Not applicable **39.** $x = \frac{1}{5}, y = \frac{1}{3}$ **41.** -5 **43.** $\frac{13}{11}$ **45.** 0 or -9 **47.** -4 **49.** 12 **51.** 8 **53.** 8

55. $(y_1 - y_2)x - (x_1 - x_2)y + (x_1y_2 - x_2y_1) = 0$

$(y_1 - y_2)x + (x_2 - x_1)y = x_2y_1 - x_1y_2$

$(x_2 - x_1)y - (x_2 - x_1)y_1 = (y_2 - y_1)x + x_2y_1 - x_1y_2 - (x_2 - x_1)y_1$

$(x_2 - x_1)(y - y_1) = (y_2 - y_1)x - (y_2 - y_1)x_1$

$$y - y_1 = \frac{y_2 - y_1}{x_2 - x_1}(x - x_1)$$

57. $\begin{vmatrix} x^2 & x & 1 \\ y^2 & y & 1 \\ z^2 & z & 1 \end{vmatrix} = x^2\begin{vmatrix} y & 1 \\ z & 1 \end{vmatrix} - x\begin{vmatrix} y^2 & 1 \\ z^2 & 1 \end{vmatrix} + \begin{vmatrix} y^2 & y \\ z^2 & z \end{vmatrix} = x^2(y - z) - x(y^2 - z^2) + yz(y - z)$

$= (y - z)[x^2 - x(y + z) + yz] = (y - z)[(x^2 - xy) - (xz - yz)] = (y - z)[x(x - y) - z(x - y)] = (y - z)(x - y)(x - z)$

59. $\begin{vmatrix} a_{13} & a_{12} & a_{11} \\ a_{23} & a_{22} & a_{21} \\ a_{33} & a_{32} & a_{31} \end{vmatrix} = a_{13}(a_{22}a_{31} - a_{32}a_{21}) - a_{12}(a_{23}a_{31} - a_{33}a_{21}) + a_{11}(a_{23}a_{32} - a_{33}a_{22})$

$= -[a_{11}(a_{22}a_{33} - a_{32}a_{23}) - a_{12}(a_{21}a_{33} - a_{31}a_{23}) + a_{13}(a_{21}a_{32} - a_{31}a_{22})] = -\begin{vmatrix} a_{11} & a_{21} & a_{13} \\ a_{21} & a_{22} & a_{23} \\ a_{31} & a_{32} & a_{33} \end{vmatrix}$

61. $\begin{vmatrix} a_{11} & a_{12} & a_{11} \\ a_{21} & a_{22} & a_{21} \\ a_{31} & a_{32} & a_{31} \end{vmatrix} = a_{11}(a_{22}a_{31} - a_{32}a_{21}) - a_{12}(a_{21}a_{31} - a_{31}a_{21}) + a_{11}(a_{21}a_{32} - a_{31}a_{22})$

$= a_{11}a_{22}a_{31} - a_{11}a_{32}a_{21} - a_{12}(0) + a_{11}a_{21}a_{32} - a_{11}a_{31}a_{22} = 0$

11.4 Exercises

1. $\begin{bmatrix} 4 & 4 & -5 \\ -1 & 5 & 4 \end{bmatrix}$ **3.** $\begin{bmatrix} 0 & 12 & -20 \\ 4 & 8 & 24 \end{bmatrix}$ **5.** $\begin{bmatrix} -8 & 7 & -15 \\ 7 & 0 & 22 \end{bmatrix}$ **7.** $\begin{bmatrix} 28 & -9 \\ 4 & 23 \end{bmatrix}$ **9.** $\begin{bmatrix} 1 & 14 & -14 \\ 2 & 22 & -18 \\ 3 & 0 & 28 \end{bmatrix}$ **11.** $\begin{bmatrix} 15 & 21 & -16 \\ 22 & 34 & -22 \\ -11 & 7 & 22 \end{bmatrix}$ **13.** $\begin{bmatrix} 25 & -9 \\ 4 & 20 \end{bmatrix}$

15. $\begin{bmatrix} -13 & 7 & -12 \\ -18 & 10 & -14 \\ 17 & -7 & 34 \end{bmatrix}$ **17.** $\begin{bmatrix} -2 & 4 & 2 & 8 \\ 2 & 1 & 4 & 6 \end{bmatrix}$ **19.** $\begin{bmatrix} 9 & 2 \\ 34 & 13 \\ 47 & 20 \end{bmatrix}$ **21.** $\begin{bmatrix} 1 & -1 \\ -1 & 2 \end{bmatrix}$ **23.** $\begin{bmatrix} 1 & -\frac{5}{2} \\ -1 & 3 \end{bmatrix}$ **25.** $\begin{bmatrix} 1 & \frac{-1}{a} \\ -1 & \frac{2}{a} \end{bmatrix}$

27. $\begin{bmatrix} 3 & -3 & 1 \\ -2 & 2 & -1 \\ -4 & 5 & -2 \end{bmatrix}$ **29.** $\begin{bmatrix} -\frac{5}{7} & \frac{1}{7} & \frac{3}{7} \\ \frac{9}{7} & \frac{1}{7} & -\frac{4}{7} \\ \frac{3}{7} & -\frac{2}{7} & \frac{1}{7} \end{bmatrix}$ **31.** $x = 3, y = 2$ **33.** $x = -5, y = 10$ **35.** $x = 2, y = -1$ **37.** $x = \frac{1}{2}, y = 2$

39. $x = -2, y = 1$ **41.** $x = \frac{2}{a}, y = \frac{3}{a}$ **43.** $x = -2, y = 3, z = 5$ **45.** $x = \frac{1}{2}, y = -\frac{1}{2}, z = 1$ **47.** $x = -\frac{34}{7}, y = \frac{85}{7}, z = \frac{12}{7}$

49. $x = \frac{1}{3}, y = 1, z = \frac{2}{3}$ **51.** $\left[\begin{array}{cc|cc} 4 & 2 & 1 & 0 \\ 2 & 1 & 0 & 1 \end{array}\right] \rightarrow \left[\begin{array}{cc|cc} 1 & \frac{1}{2} & \frac{1}{4} & 0 \\ 2 & 1 & 0 & 1 \end{array}\right] \rightarrow \left[\begin{array}{cc|cc} 1 & \frac{1}{2} & \frac{1}{4} & 0 \\ 0 & 0 & -\frac{1}{2} & 1 \end{array}\right]$

53. $\left[\begin{array}{cc|cc} 15 & 3 & 1 & 0 \\ 10 & 2 & 0 & 1 \end{array}\right] \rightarrow \left[\begin{array}{cc|cc} 1 & \frac{1}{5} & \frac{1}{15} & 0 \\ 10 & 2 & 0 & 1 \end{array}\right] \rightarrow \left[\begin{array}{cc|cc} 1 & \frac{1}{5} & \frac{1}{15} & 0 \\ 0 & 0 & -\frac{2}{3} & 1 \end{array}\right]$

55. $\left[\begin{array}{ccc|ccc} -3 & 1 & -1 & 1 & 0 & 0 \\ 1 & -4 & -7 & 0 & 1 & 0 \\ 1 & 2 & 5 & 0 & 0 & 1 \end{array}\right] \rightarrow \left[\begin{array}{ccc|ccc} 1 & 2 & 5 & 0 & 0 & 1 \\ 1 & -4 & -7 & 0 & 1 & 0 \\ -3 & 1 & -1 & 1 & 0 & 0 \end{array}\right] \rightarrow \left[\begin{array}{ccc|ccc} 1 & 2 & 5 & 0 & 0 & 1 \\ 0 & -6 & -12 & 0 & 1 & -1 \\ 0 & 7 & 14 & 1 & 0 & 3 \end{array}\right]$

$\rightarrow \left[\begin{array}{ccc|ccc} 1 & 2 & 5 & 0 & 0 & 1 \\ 0 & 1 & 2 & 0 & -\frac{1}{6} & \frac{1}{6} \\ 0 & 1 & 2 & \frac{1}{7} & 0 & \frac{3}{7} \end{array}\right] \rightarrow \left[\begin{array}{ccc|ccc} 1 & 2 & 5 & 0 & 0 & 1 \\ 0 & 1 & 2 & 0 & -\frac{1}{6} & \frac{1}{6} \\ 0 & 0 & 0 & \frac{1}{7} & \frac{1}{6} & \frac{11}{42} \end{array}\right]$

57. $\begin{bmatrix} 0.01 & 0.05 & -0.01 \\ 0.01 & -0.02 & 0.01 \\ -0.02 & 0.01 & 0.03 \end{bmatrix}$ **59.** $\begin{bmatrix} 0.02 & -0.04 & -0.01 & 0.01 \\ -0.02 & 0.05 & 0.03 & -0.03 \\ 0.02 & 0.01 & -0.04 & 0.00 \\ -0.02 & 0.06 & 0.07 & 0.06 \end{bmatrix}$ **61.** $x = 4.57, y = -6.44, z = -24.07$

63. $x = -1.19, y = 2.46, z = 8.27$ **65. (a)** $\begin{bmatrix} 500 & 350 & 400 \\ 700 & 500 & 850 \end{bmatrix}$; $\begin{bmatrix} 500 & 700 \\ 350 & 500 \\ 400 & 850 \end{bmatrix}$ **(b)** $\begin{bmatrix} 15 \\ 8 \\ 3 \end{bmatrix}$ **(c)** $\begin{bmatrix} 11{,}500 \\ 17{,}050 \end{bmatrix}$ **(d)** $[0.10 \;\; 0.05]$ **(e)** \$2002.50

67. If $a \neq 0$, $\left[\begin{array}{cc|cc} a & b & 1 & 0 \\ c & d & 0 & 1 \end{array}\right] \rightarrow \left[\begin{array}{cc|cc} 1 & \frac{b}{a} & \frac{1}{a} & 0 \\ c & d & 0 & 1 \end{array}\right] \rightarrow \left[\begin{array}{cc|cc} 1 & \frac{b}{a} & \frac{1}{a} & 0 \\ 0 & \frac{-cb+da}{a} & -\frac{c}{a} & 1 \end{array}\right] \rightarrow \left[\begin{array}{cc|cc} 1 & \frac{b}{a} & \frac{1}{a} & 0 \\ 0 & 1 & \frac{-c}{-cb+da} & \frac{a}{-cb+da} \end{array}\right] \rightarrow \left[\begin{array}{cc|cc} 1 & 0 & \frac{d}{ad-bc} & \frac{-b}{ad-bc} \\ 0 & 1 & \frac{-c}{ad-bc} & \frac{a}{ad-bc} \end{array}\right].$

Therefore, $A^{-1} = \frac{1}{D}\begin{bmatrix} d & -b \\ -c & a \end{bmatrix}$. If $a = 0$, $\left[\begin{array}{cc|cc} 0 & b & 1 & 0 \\ c & d & 0 & 1 \end{array}\right] \rightarrow \left[\begin{array}{cc|cc} c & d & 0 & 1 \\ 0 & b & 1 & 0 \end{array}\right] \rightarrow \left[\begin{array}{cc|cc} 1 & \frac{d}{c} & 0 & \frac{1}{c} \\ 0 & b & 1 & 0 \end{array}\right] \rightarrow \left[\begin{array}{cc|cc} 1 & 0 & \frac{-d}{cb} & \frac{1}{c} \\ 0 & 1 & \frac{1}{b} & 0 \end{array}\right] \rightarrow \left[\begin{array}{cc|cc} 1 & 0 & \frac{d}{-bc} & \frac{-b}{-bc} \\ 0 & 1 & \frac{-c}{-bc} & 0 \end{array}\right].$

Since $a = 0$, $D = ad - bc = -bc$, so $A^{-1} = \frac{1}{D}\begin{bmatrix} d & -b \\ -c & a \end{bmatrix}$.

11.5 Exercises

1. Proper **3.** Improper: $1 + \frac{9}{x^2 - 4}$ **5.** Improper; $5x + \frac{22x - 1}{x^2 - 4}$ **7.** Improper; $1 + \frac{-2(x - 6)}{(x + 4)(x - 3)}$ **9.** $\frac{-4}{x} + \frac{4}{x - 1}$ **11.** $\frac{1}{x} + \frac{-x}{x^2 + 1}$

13. $\frac{-1}{x - 1} + \frac{2}{x - 2}$ **15.** $\frac{\frac{1}{4}}{x + 1} + \frac{\frac{3}{4}}{x - 1} + \frac{\frac{1}{2}}{(x - 1)^2}$ **17.** $\frac{\frac{1}{12}}{x - 2} + \frac{-\frac{1}{12}(x + 4)}{x^2 + 2x + 4}$ **19.** $\frac{\frac{1}{4}}{x - 1} + \frac{\frac{1}{4}}{(x - 1)^2} + \frac{-\frac{1}{4}}{x + 1} + \frac{\frac{1}{4}}{(x + 1)^2}$

21. $\frac{-5}{x + 2} + \frac{5}{x + 1} + \frac{-4}{(x + 1)^2}$ **23.** $\frac{\frac{1}{4}}{x} + \frac{1}{x^2} + \frac{-\frac{1}{4}(x + 4)}{x^2 + 4}$ **25.** $\frac{\frac{2}{3}}{x + 1} + \frac{\frac{1}{3}(x + 1)}{x^2 + 2x + 4}$ **27.** $\frac{\frac{2}{7}}{3x - 2} + \frac{\frac{1}{7}}{2x + 1}$ **29.** $\frac{\frac{3}{4}}{x + 3} + \frac{\frac{1}{4}}{x - 1}$

31. $\frac{1}{x^2 + 4} + \frac{2x - 1}{(x^2 + 4)^2}$ **33.** $\frac{-1}{x} + \frac{2}{x - 3} + \frac{-1}{x + 1}$ **35.** $\frac{4}{x - 2} + \frac{-3}{x - 1} + \frac{-1}{(x - 1)^2}$ **37.** $\frac{x}{(x^2 + 16)^2} + \frac{-16x}{(x^2 + 16)^3}$ **39.** $\frac{-\frac{8}{7}}{2x + 1} + \frac{\frac{4}{7}}{x - 3}$

41. $\frac{-\frac{2}{9}}{x} + \frac{-\frac{1}{3}}{x^2} + \frac{\frac{1}{6}}{x - 3} + \frac{\frac{1}{18}}{x + 3}$

11.6 Exercises

1

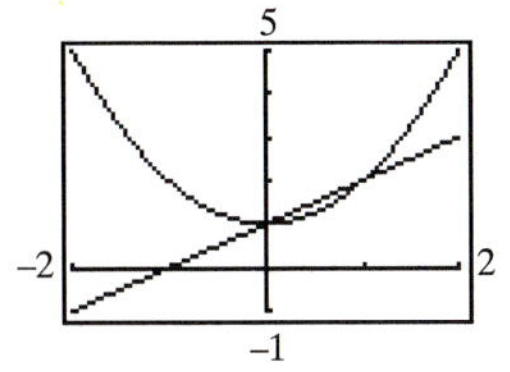

(0.00, 1.00); (1.00, 2.00)
(0, 1); (1, 2)

3.

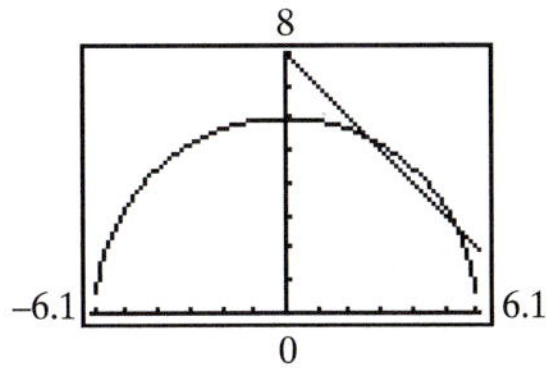

(2.58, 5.41); (5.41, 2.58)
$(4 - \sqrt{2}, 4 + \sqrt{2})$; $(4 + \sqrt{2}, 4 - \sqrt{2})$

5.

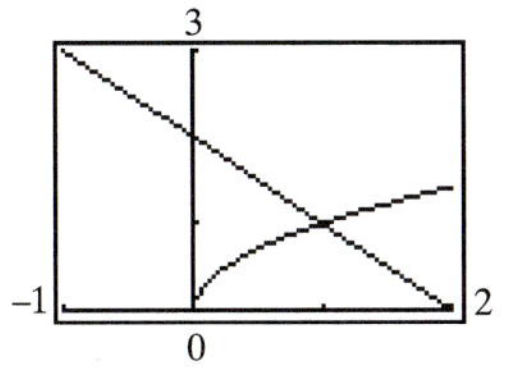

(1.00, 1.00)
(1, 1)

7.

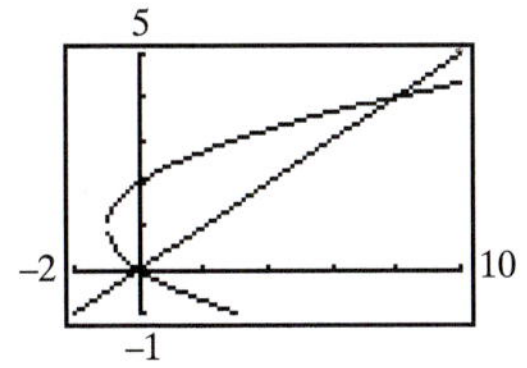

(0.00, 0.00); (8.00, 4.00)
(0, 0); (8, 4)

9.

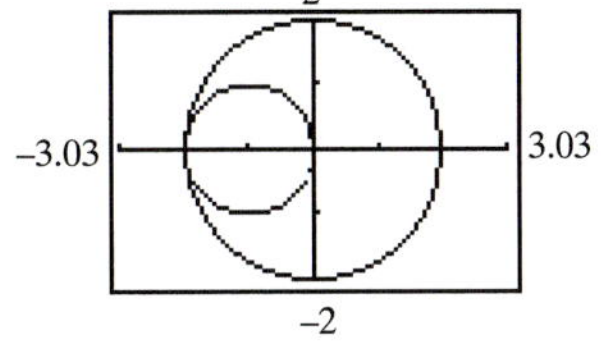

(−2.00, 0.00)
(−2, 0)

11.

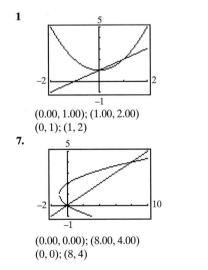

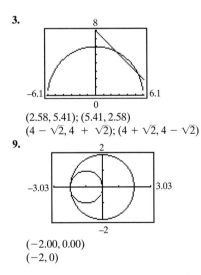

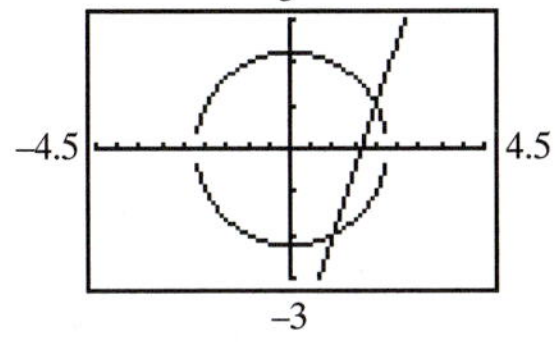

(1.00, −2.00); (2.00, 1.00)
(1, −2); (2, 1)

13.

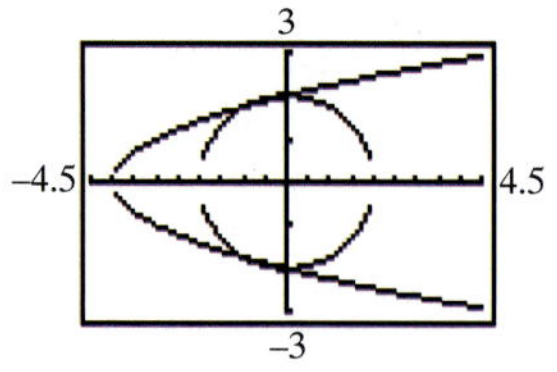

(0.00, 2.00); (0.00, −2.00); (−1, −1.73); (−1, 1.73)
$(0, 2); (0, -2); (-1, -\sqrt{3}); (-1, \sqrt{3})$

15.

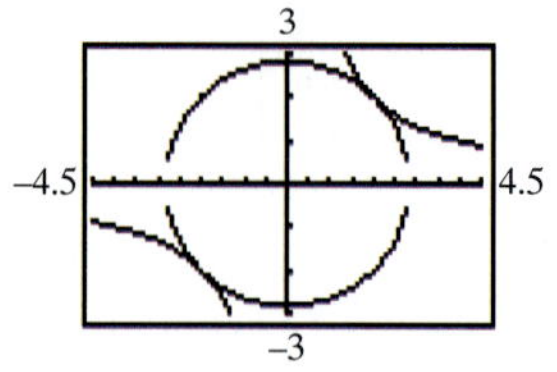

(2.00, 2.00); (−2.00, −2.00)
(2, 2); (−2, −2)

17.

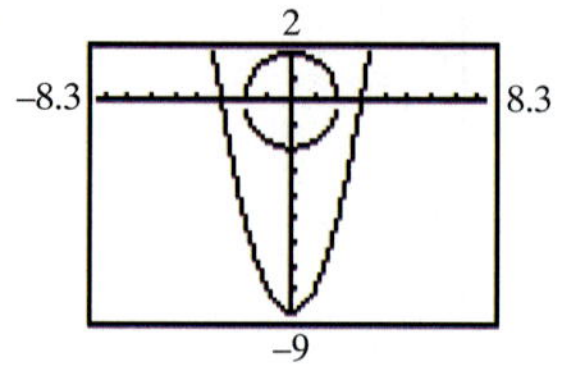

No solution; Inconsistent

19.

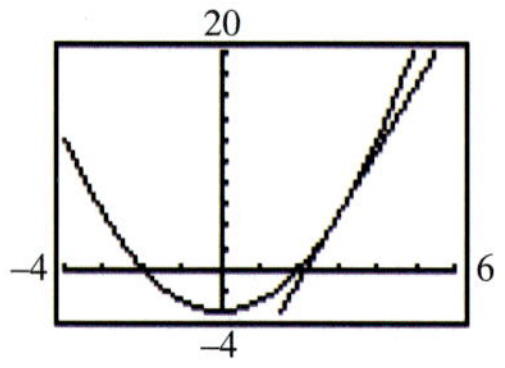

(3.00, 5.00)
(3, 5)

21. $x = 1, y = 4; x = -1, y = -4; x = 2\sqrt{2}, y = \sqrt{2}; x = -2\sqrt{2}, y = -\sqrt{2}$ **23.** $x = 0, y = 1; x = -\frac{2}{3}, y = -\frac{1}{3}$ **25.** $x = 0, y = -1; x = \frac{5}{2}, y = -\frac{7}{2}$ **27.** $x = 2, y = \frac{1}{3}; x = \frac{1}{2}, y = \frac{4}{3}$ **29.** $x = 3, y = 2; x = 3, y = -2; x = -3, y = 2; x = -3, y = -2$ **31.** $x = \frac{1}{2}, y = \frac{3}{2}; x = \frac{1}{2}, y = -\frac{3}{2}; x = -\frac{1}{2}, y = \frac{3}{2}; x = -\frac{1}{2}, y = -\frac{3}{2}$ **33.** $x = \sqrt{2}, y = 2\sqrt{2}; x = -\sqrt{2}, y = -2\sqrt{2}$ **35.** No solution; system is inconsistent

37. $x = \frac{8}{3}, y = \frac{2\sqrt{10}}{3}; x = -\frac{8}{3}, y = \frac{2\sqrt{10}}{3}; x = \frac{8}{3}, y = -\frac{2\sqrt{10}}{3}; x = -\frac{8}{3}, y = -\frac{2\sqrt{10}}{3}$

39. $x = 1, y = \frac{1}{2}; x = -1, y = \frac{1}{2}; x = 1, y = -\frac{1}{2}; x = -1, y = -\frac{1}{2}$ **41.** System is inconsistent

43. $x = \sqrt{3}, y = \sqrt{3}; x = -\sqrt{3}, y = -\sqrt{3}; x = 2, y = 1; x = -2, y = -1$ **45.** $x = 3, y = 2; x = -3, y = -2; x = 2, y = \frac{1}{2}; x = -2, y = -\frac{1}{2}$

47. $x = 3, y = 1; x = -1, y = -3$ **49.** $x = 0, y = -2; x = 0, y = 1; x = 2, y = -1$ **51.** $x = 2, y = 8$ **53.** $x = 0.48, y = 0.61$ **55.** $x = -1.64, y = -0.89$ **57.** $x = 0.58, y = 1.85; x = 1.81, y = 1.05; x = 0.58, y = -1.85; x = 1.81, y = -1.05$ **59.** $x = 2.34, y = 0.85$

61. 3 and 1; −3 and −1 **63.** 2 and 2; −2 and −2 **65.** $\frac{1}{2}$ and $\frac{1}{3}$ **67.** 5

69.

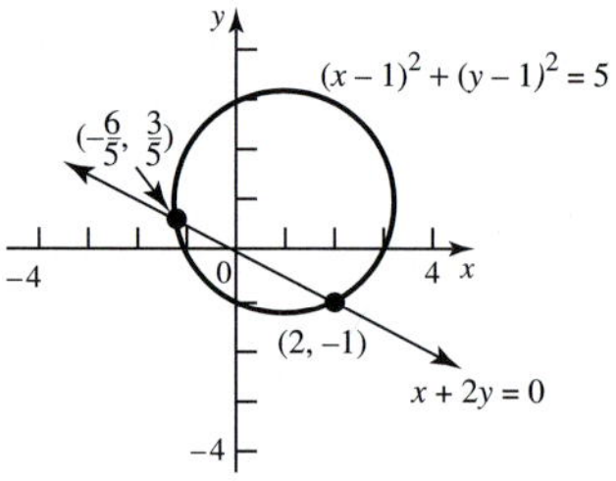

71. Solutions: $(0, -\sqrt{3} - 2), (0, \sqrt{3} - 2), (1, 0), (1, -4)$

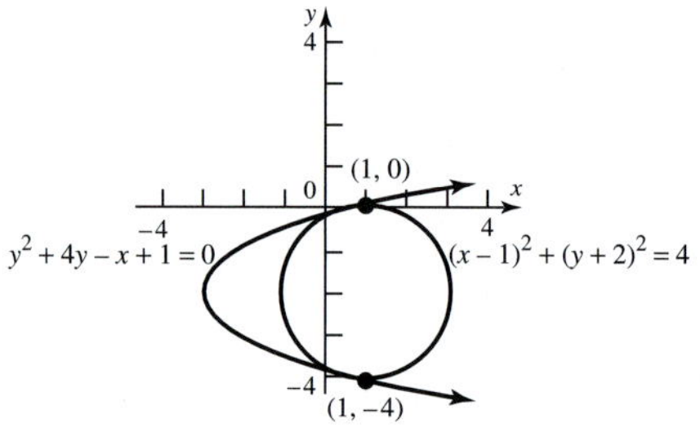

73.

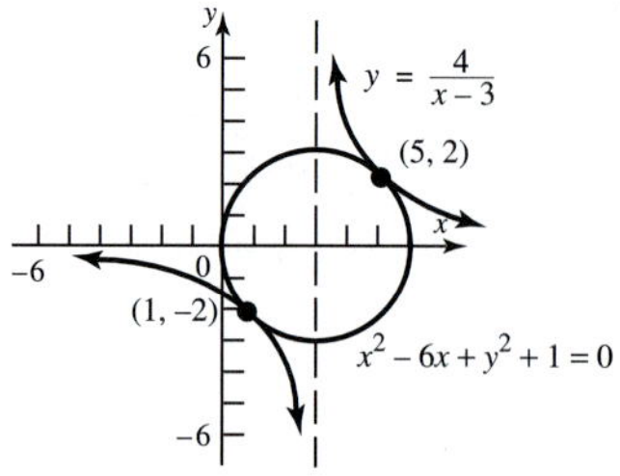

75. 5 in. by 3 in. **77.** 2 cm and 4 cm **79.** tortoise: 7 m/hr, hare: $7\frac{1}{2}$ m/hr **81.** 12 cm by 18 cm **83.** $x = 60$ ft; $y = 30$ ft

85. $l = \frac{P + \sqrt{P^2 - 16A}}{4}; w = \frac{P - \sqrt{P^2 - 16A}}{4}$ **87.** $y = 4x - 4$ **89.** $y = 2x + 1$ **91.** $y = -\frac{1}{3}x + \frac{7}{3}$ **93.** $y = 2x - 3$

95. $r_1 = \frac{-b + \sqrt{b^2 - 4ac}}{2a}; r_2 = \frac{-b - \sqrt{b^2 - 4ac}}{2a}$

11.7 Exercises

1.

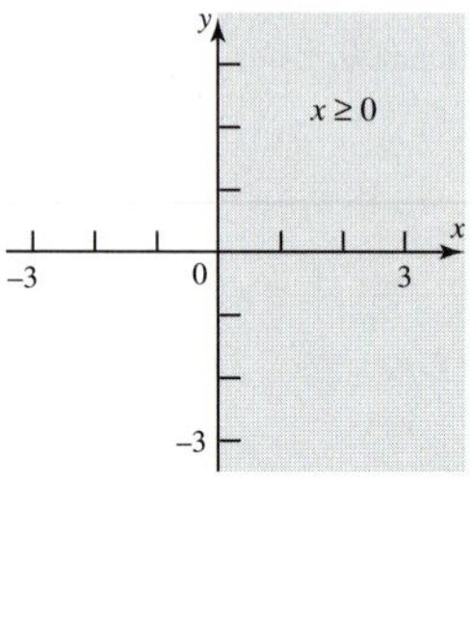

3.

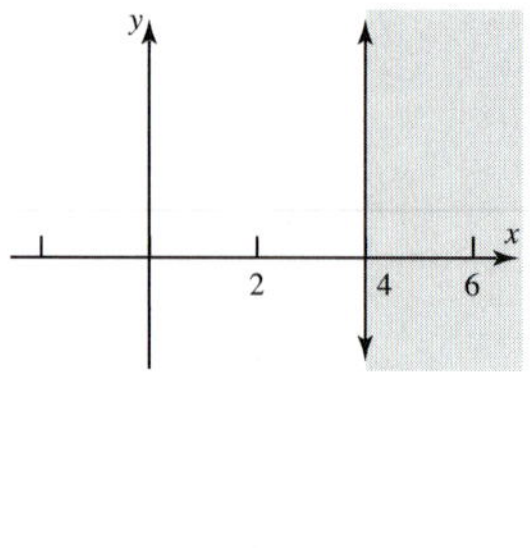

5.

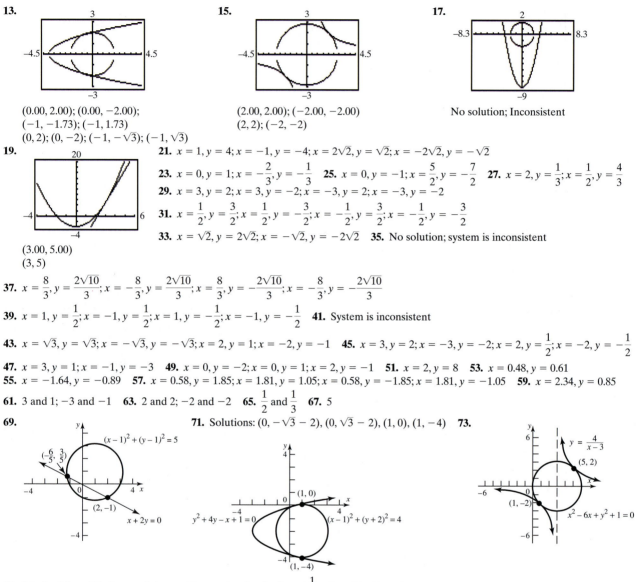

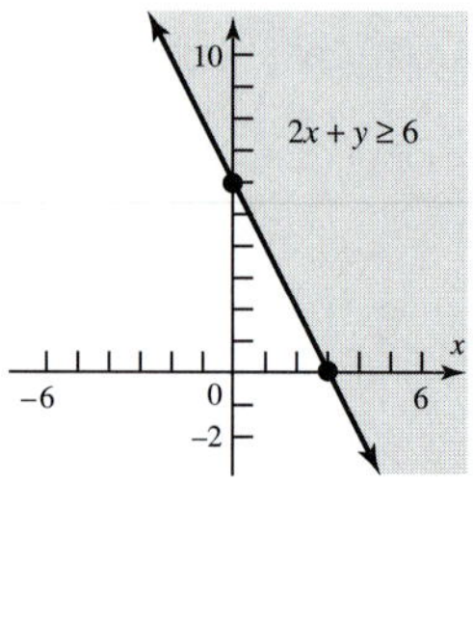

7.

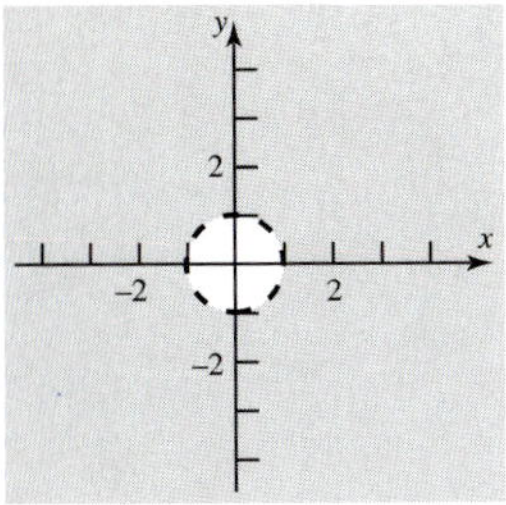

9.

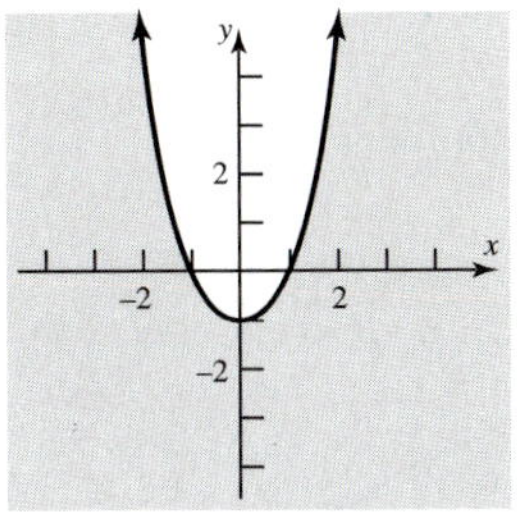

11.

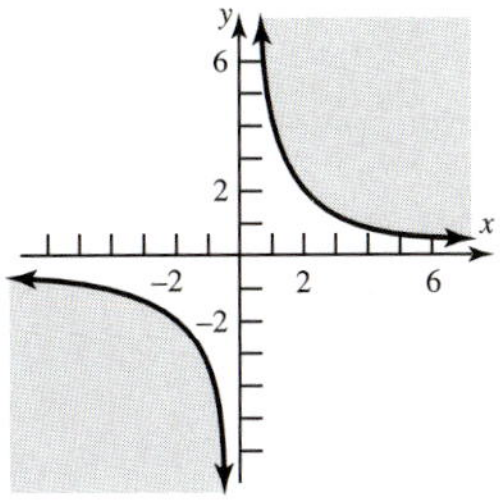

13.

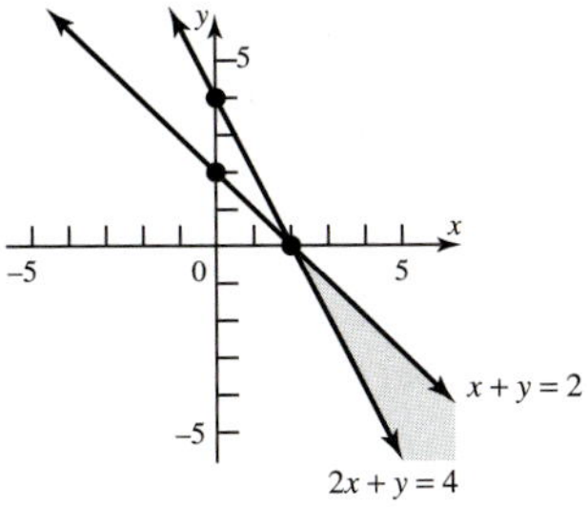

15.

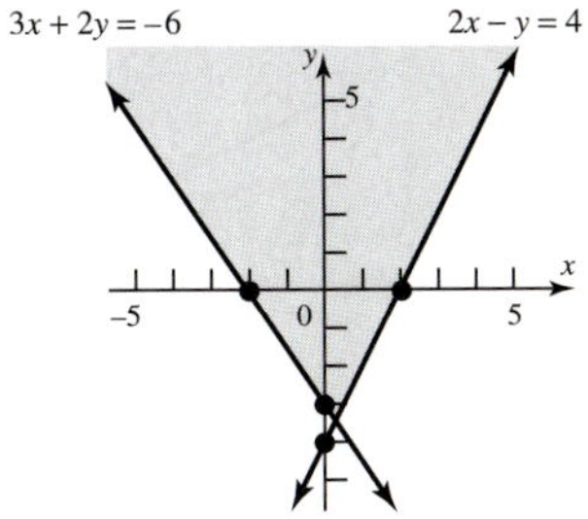

17.

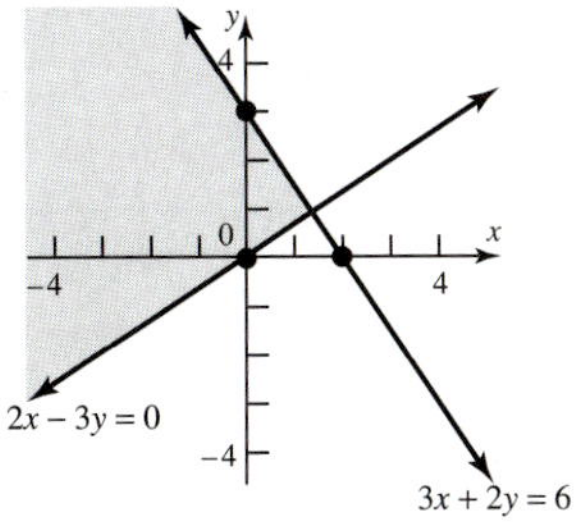

19.

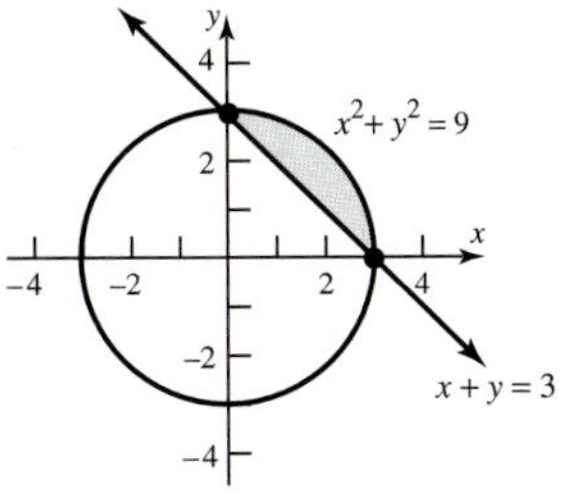

21.

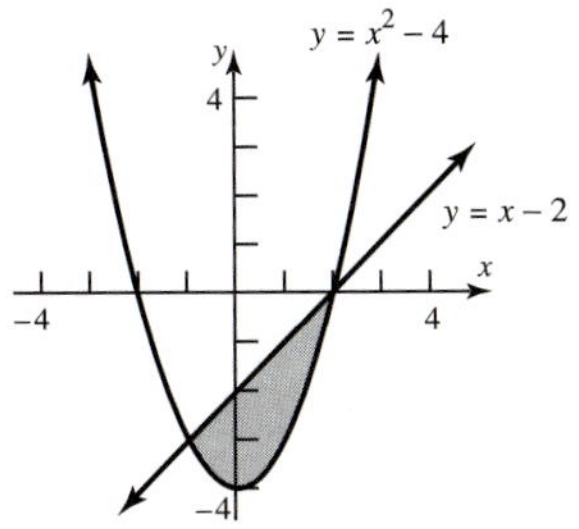

23.

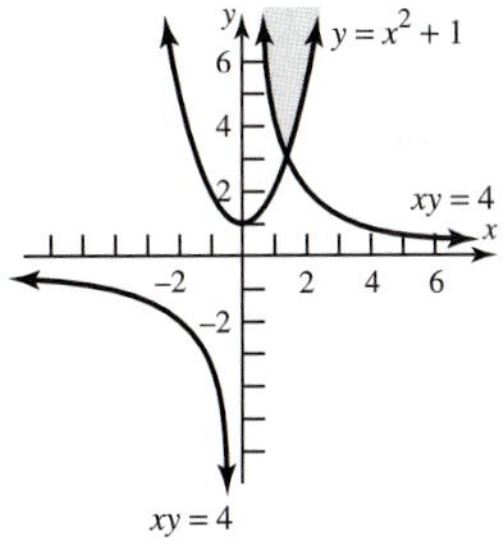

25.

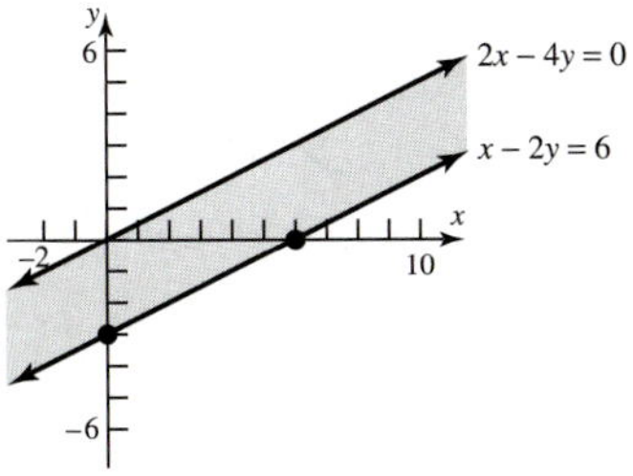

27.

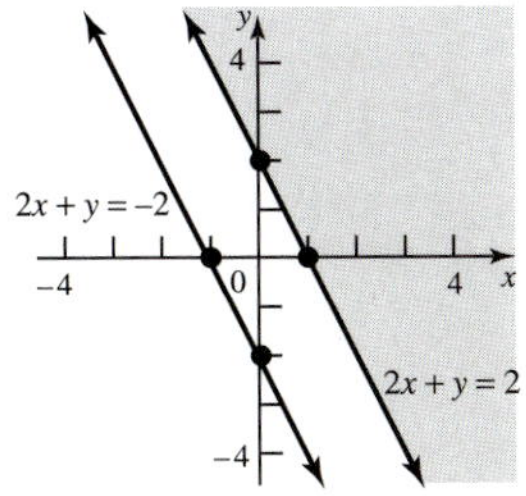

29. No solution

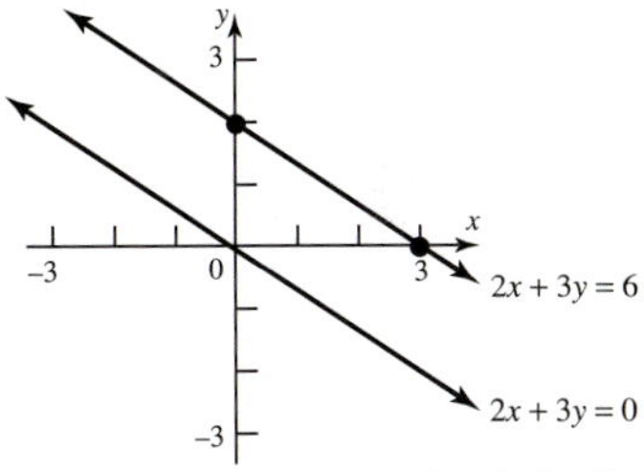

31. Bounded; corner points $(0, 0), (3, 0), (2, 2), (0, 3)$

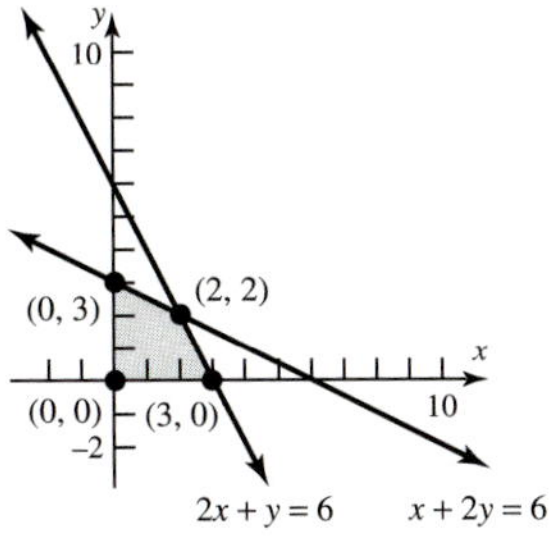

33. Unbounded; corner points $(2, 0), (0, 4)$

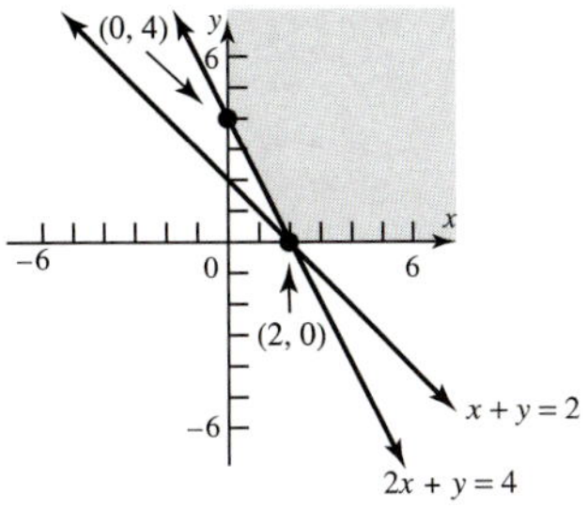

35. Bounded; corner points $(2, 0), (4, 0), \left(\frac{24}{7}, \frac{12}{7}\right), (0, 4), (0, 2)$

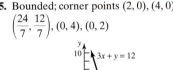

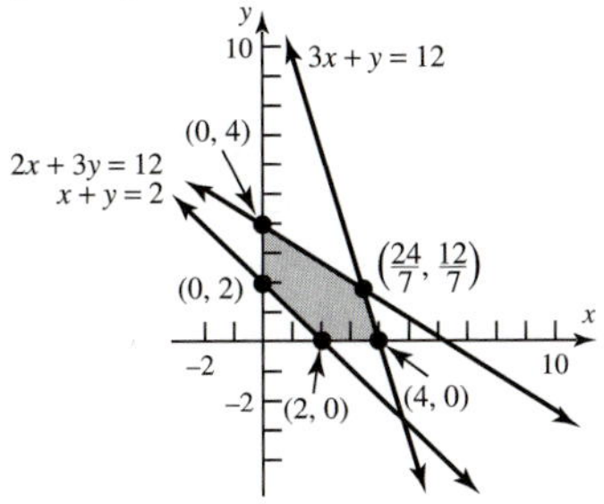

37. Bounded; corner points (2, 0), (5, 0), (2, 6), (0, 8), (0, 2)

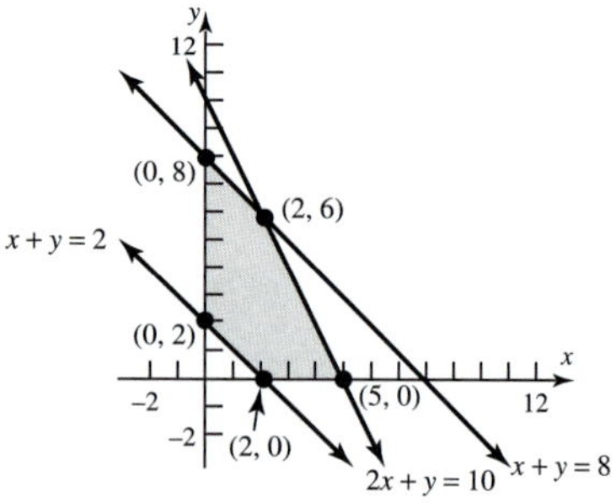

39. Bounded; corner points (1, 0), (10, 0), (0, 5), $\left(0, \frac{1}{2}\right)$

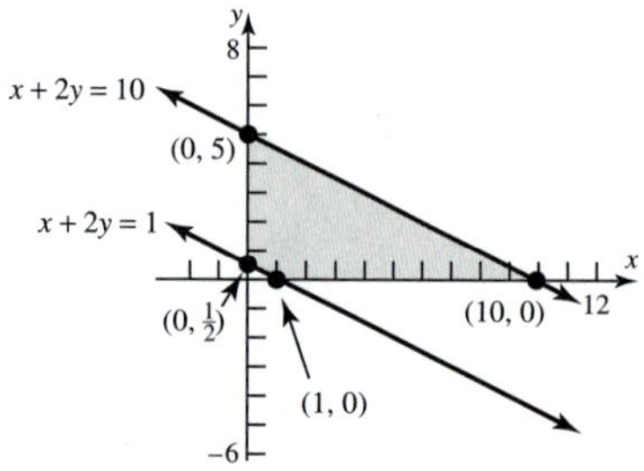

41. $\begin{cases} x \le 4 \\ x + y \le 6 \\ x \ge 0 \\ y \ge 0 \end{cases}$

43. $\begin{cases} x \le 20 \\ y \ge 15 \\ x + y \le 50 \\ x - y \le 0 \\ x \ge 0 \end{cases}$

45. **(a)** $\begin{cases} x + y \le 50{,}000 \\ x \ge 35{,}000 \\ y \le 10{,}000 \\ x \ge 0 \\ y \ge 0 \end{cases}$ **(b)**

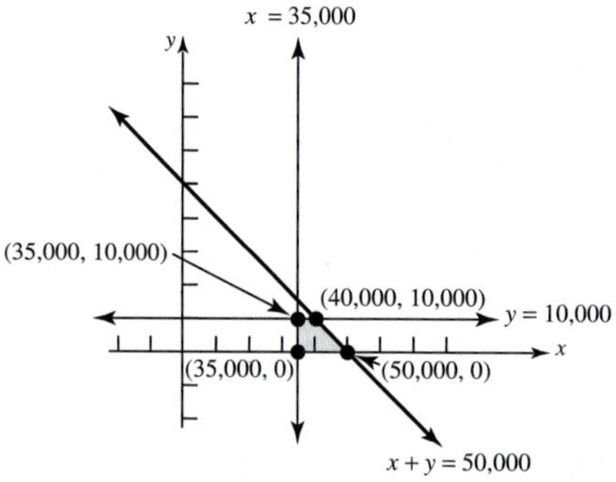

47. **(a)** $\begin{cases} x \ge 0 \\ y \ge 0 \\ x + 2y \le 300 \\ 3x + 2y \le 480 \end{cases}$ **(b)**

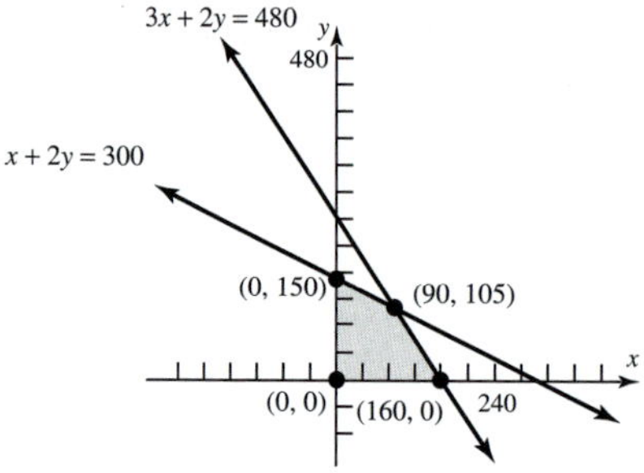

49. **(a)** $\begin{cases} 30x + 20y \le 1600 \\ 2x + 3y \le 150 \\ x \ge 0 \\ y \ge 0 \end{cases}$ **(b)**

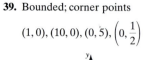

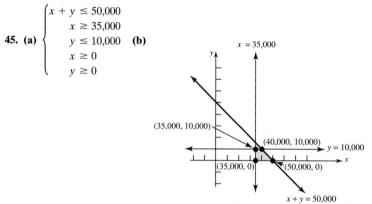

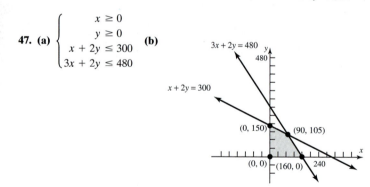

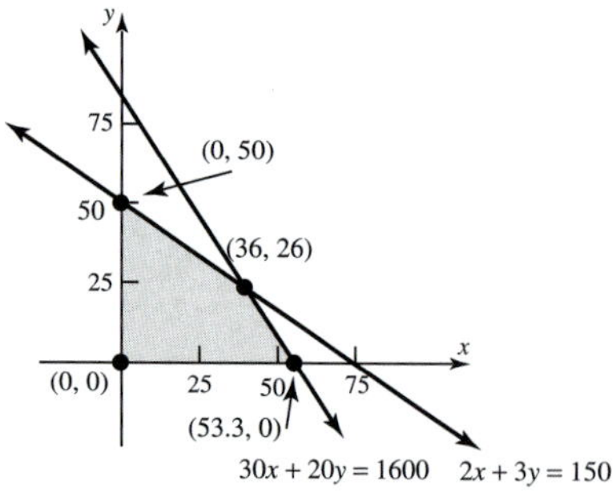

11.8 Exercises

1. Maximum value is 11; minimum value is 3 **3.** Maximum value is 65; minimum value is 4
5. Maximum value is 67; minimum value is 20 **7.** The maximum value of z is 12, and it occurs at the point $(6, 0)$
9. The minimum value of z is 4, and it occurs at the point $(2, 0)$ **11.** The maximum value of z is 20, and it occurs at the point $(0, 4)$
13. The minimum value of z is 8, and it occurs at the point $(0, 2)$ **15.** The maximum value of z is 50, and it occurs at the point $(10, 0)$
17. 8 downhill, 24 cross-country; \$1760; \$1920 **19.** 30 acres of soybeans and 10 acres of corn
21. $\frac{1}{2}$ hour on machine I; $5\frac{1}{4}$ hours on machine II **23.** 100 lbs. of ground beef and 50 lbs of pork **25.** 10 racing skates, 15 figure skates
27. 2 metal samples, 4 plastic samples; \$34 **29. (a)** 10 first class, 120 coach **(b)** 15 first class, 120 coach

Fill-in-the-Blank Items

1. inconsistent **2.** matrix **3.** determinants **4.** augmented **5.** inverse **6.** square **7.** identity **8.** proper **9.** half-plane
10. objective function **11.** feasible point

True/False Items

1. F **2.** T **3.** F **4.** F **5.** F **6.** F **7.** F **8.** T **9.** T **10.** T **11.** T

Review Exercises

1. $x = 2, y = -1$ **3.** $x = 2, y = \frac{1}{2}$ **5.** $x = 2, y = -1$ **7.** $x = \frac{11}{5}, y = -\frac{3}{5}$ **9.** $x = -\frac{8}{5}, y = \frac{12}{5}$ **11.** $x = 6, y = -1$ **13.** $x = -4, y = 3$

15. $x = 2, y = 3$ **17.** Inconsistent **19.** $x = -1, y = 2, z = -3$ **21.** $\begin{bmatrix} 4 & -4 \\ 3 & 9 \\ 4 & 0 \end{bmatrix}$ **23.** $\begin{bmatrix} 6 & 0 \\ 12 & 24 \\ -6 & 12 \end{bmatrix}$ **25.** $\begin{bmatrix} 4 & -3 & 0 \\ 12 & -2 & -8 \\ -2 & 5 & -4 \end{bmatrix}$

27. $\begin{bmatrix} 8 & -13 & 8 \\ 9 & 2 & -10 \\ 18 & -17 & 4 \end{bmatrix}$ **29.** $\begin{bmatrix} \frac{1}{2} & -1 \\ -\frac{1}{6} & \frac{2}{3} \end{bmatrix}$ **31.** $\begin{bmatrix} -\frac{5}{7} & \frac{9}{7} & \frac{3}{7} \\ \frac{1}{7} & \frac{1}{7} & -\frac{2}{7} \\ \frac{3}{7} & -\frac{4}{7} & \frac{1}{7} \end{bmatrix}$ **33.** Singular **35.** $x = \frac{2}{5}, y = \frac{1}{10}$ **37.** $x = \frac{1}{2}, y = \frac{2}{3}, z = \frac{1}{6}$

39. $x = -\frac{1}{2}, y = -\frac{2}{3}, z = -\frac{3}{4}$ **41.** $z = -1, x = y + 1$, y any real number **43.** $x = 1, y = 2, z = -3, t = 1$ **45.** 5

47. 108 **49.** -100 **51.** $x = 2, y = -1$ **53.** $x = 2, y = 3$ **55.** $x = -1, y = 2, z = -3$ **57.** $\frac{-\frac{3}{2}}{x} + \frac{\frac{3}{2}}{x - 4}$

59. $\frac{-3}{x - 1} + \frac{3}{x} + \frac{4}{x^2}$ **61.** $\frac{-\frac{1}{10}}{x + 1} + \frac{\frac{1}{10}x + \frac{9}{10}}{x^2 + 9}$ **63.** $\frac{x}{x^2 + 4} + \frac{-4x}{(x^2 + 4)^2}$ **65.** $\frac{\frac{1}{2}}{x^2 + 1} + \frac{\frac{1}{4}}{x - 1} + \frac{-\frac{1}{4}}{x + 1}$

67. $x = -\frac{2}{5}, y = -\frac{11}{5}; x = -2, y = 1$ **69.** $x = 2\sqrt{2}, y = \sqrt{2}; x = -2\sqrt{2}, y = -\sqrt{2}$ **71.** $x = 0, y = 0; x = -3, y = 3; x = 3, y = 3$

73. $x = \sqrt{2}, y = -\sqrt{2}; x = -\sqrt{2}, y = \sqrt{2}; x = \frac{4}{3}\sqrt{2}, y = -\frac{2}{3}\sqrt{2}; x = -\frac{4}{3}\sqrt{2}, y = \frac{2}{3}\sqrt{2}$ **75.** $x = 1, y = -1$

77. Unbounded; corner point $(0, 2)$

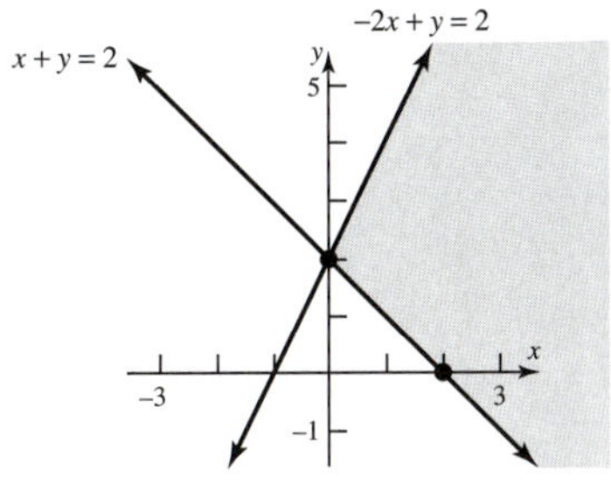

79. Bounded; corner points $(0, 0)$, $(0, 2)$, $(3, 0)$

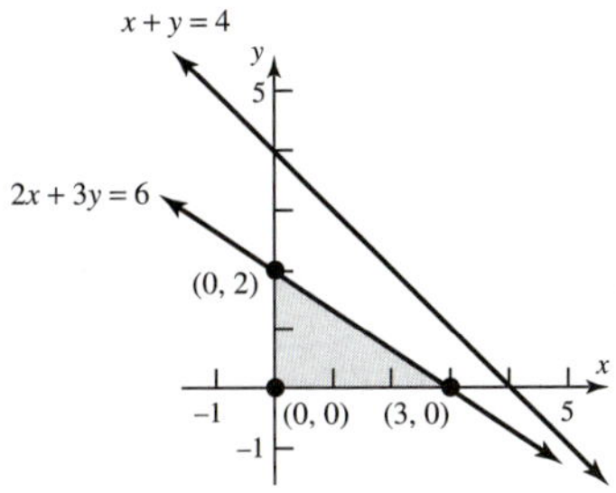

81. Bounded; corner points $(0, 1)$, $(0, 8)$, $(4, 0)$, $(2, 0)$

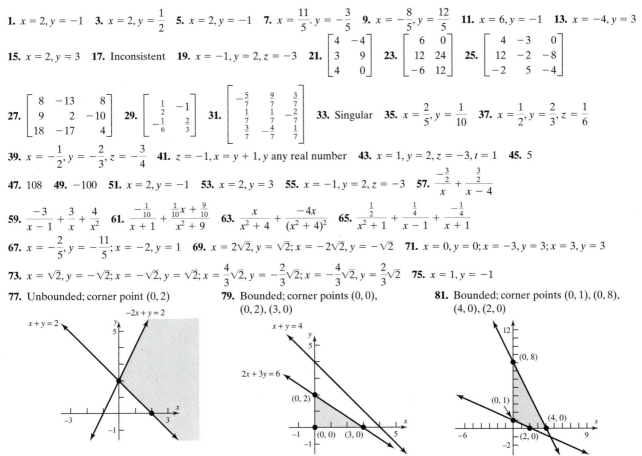

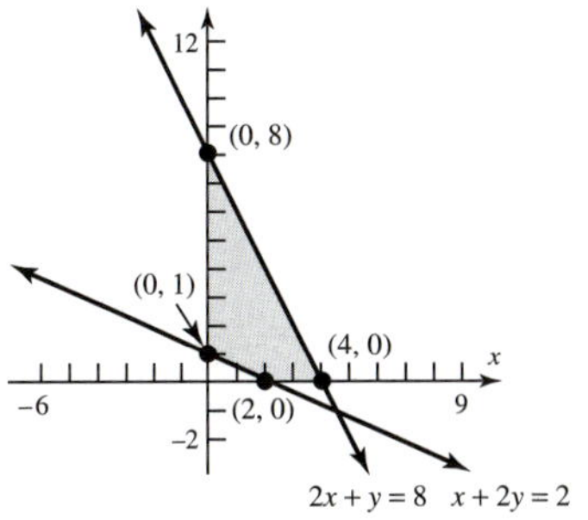

83.

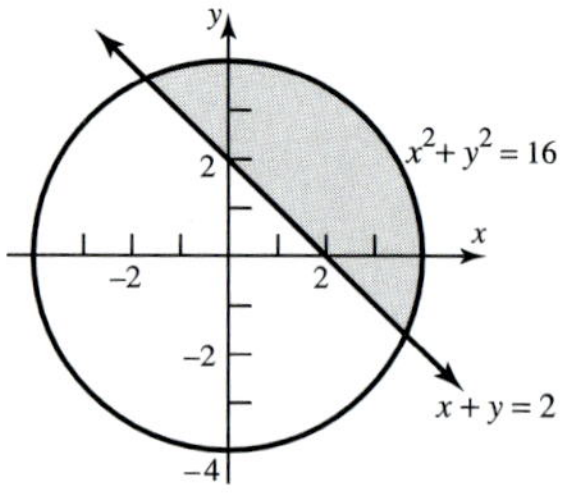

85.

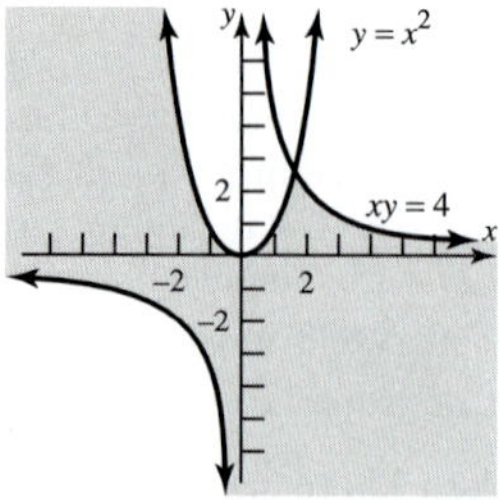

87. The maximum value is 32 when $x = 0$ and $y = 8$ **89.** The minimum value is 3 when $x = 1$ and $y = 0$
91. The maximum value is $\frac{108}{7}$ when $x = \frac{12}{7}$ and $y = \frac{12}{7}$ **93.** 10 **95.** $y = -\frac{1}{3}x^2 - \frac{2}{3}x + 1$
97. 70 lbs of \$3 coffee and 30 lbs of \$6 coffee **99.** 1 small, 5 medium, 2 large **101.** 24 ft by 10 ft **103.** $4 + \sqrt{2}$ in. and $4 - \sqrt{2}$ in.
105. $120\sqrt{10}$ ft **107.** Katy gets \$10, Mike gets \$20, Danny gets \$5, Colleen gets \$10 **109.** Katy: 4 hr; Mike: 2hr; Danny: 8 hr
111. 35 gasoline engines, 15 diesel engines; 15 gasoline engines, 0 diesel engines

CHAPTER 12 *12.1 Exercises*

1. 1, 2, 3, 4, 5 **3.** $\frac{1}{3}, \frac{2}{4} = \frac{1}{2}, \frac{3}{5}, \frac{4}{6} = \frac{2}{3}, \frac{5}{7}$ **5.** 1, −4, 9, −16, 25 **7.** $\frac{1}{2}, \frac{2}{5}, \frac{2}{7}, \frac{8}{41}, \frac{8}{61}$ **9.** $-\frac{1}{6}, \frac{1}{12}, -\frac{1}{20}, \frac{1}{30}, -\frac{1}{42}$
11. $\frac{1}{e}, \frac{2}{e^2}, \frac{3}{e^3}, \frac{4}{e^4}, \frac{5}{e^5}$ **13.** $\frac{n}{(n+1)}$ **15.** $\frac{1}{2^{n-1}}$ **17.** $(-1)^{n+1}$ **19.** $(-1)^{n+1}n$ **21.** $a_1 = 2, a_2 = 5, a_3 = 8, a_4 = 11, a_5 = 14$
23. $a_1 = -2, a_2 = -1, a_3 = 1, a_4 = 4, a_5 = 8$ **25.** $a_1 = 5, a_2 = 10, a_3 = 20, a_4 = 40, a_5 = 80$ **27.** $a_1 = 3, a_2 = 3, a_3 = \frac{3}{2}, a_4 = \frac{1}{2}, a_5 = \frac{1}{8}$
29. $a_1 = 1, a_2 = 2, a_3 = 2, a_4 = 4, a_5 = 8$ **31.** $a_1 = A, a_2 = A + d, a_3 = A + 2d, a_4 = A + 3d, a_5 = A + 4d$
33. $a_1 = \sqrt{2}, a_2 = \sqrt{2 + \sqrt{2}}, a_3 = \sqrt{2 + \sqrt{2 + \sqrt{2}}}, a_4 = \sqrt{2 + \sqrt{2 + \sqrt{2 + \sqrt{2}}}}, a_5 = \sqrt{2 + \sqrt{2 + \sqrt{2 + \sqrt{2 + \sqrt{2}}}}}$
35. 50 **37.** 21 **39.** 90 **41.** 26 **43.** 10 **45.** 96 **47.** $3 + 4 + \cdots + (n + 2)$ **49.** $\frac{1}{2} + 2 + \frac{9}{2} + \cdots + \frac{n^2}{2}$ **51.** $1 + \frac{1}{3} + \frac{1}{9} + \cdots + \frac{1}{3^n}$
53. $\frac{1}{3} + \frac{1}{9} + \cdots + \frac{1}{3^n}$ **55.** $\ln 2 - \ln 3 + \ln 4 - \cdots + (-1)^n \ln n$ **57.** $\sum_{k=1}^{20} k$ **59.** $\sum_{k=1}^{13} \frac{k}{k+1}$ **61.** $\sum_{k=0}^{6} (-1)^k\left(\frac{1}{3^k}\right)$ **63.** $\sum_{k=1}^{n} \frac{3^k}{k}$
65. $\sum_{k=0}^{n} (a + kd)$ **67.** 21 **71.** A Fibonacci sequence

12.2 Exercises

1. $d = 1$; 5, 6, 7, 8 **3.** $d = 2$; −3, −1, 1, 3 **5.** $d = -2$; 4, 2, 0, −2 **7.** $d = -\frac{1}{3}; \frac{1}{6}, -\frac{1}{6}, -\frac{1}{2}, -\frac{5}{6}$ **9.** $d = \ln 3$; $\ln 3, 2\ln 3, 3\ln 3, 4\ln 3$
11. $a_5 = 14; a_n = 3n - 1$ **13.** $a_5 = -7; a_n = 8 - 3n$ **15.** $a_5 = 2; a_n = \frac{1}{2}(n - 1)$ **17.** $a_5 = 5\sqrt{2}; a_n = \sqrt{2}n$ **19.** $a_{12} = 24$ **21.** $a_{10} = -26$
23. $a_8 = a + 7b$ **25.** $a_1 = -13; d = 3; a_{n+1} = a_n + 3$ **27.** $a_1 = -53, d = 6; a_{n+1} = a_n + 6$ **29.** $a_1 = 28; d = -2; a_{n+1} = a_n - 2$
31. $a_1 = 25; d = -2; a_{n+1} = a_n - 2$ **33.** n^2 **35.** $\frac{n}{2}(9 + 5n)$ **37.** 1260 **39.** 324

41.

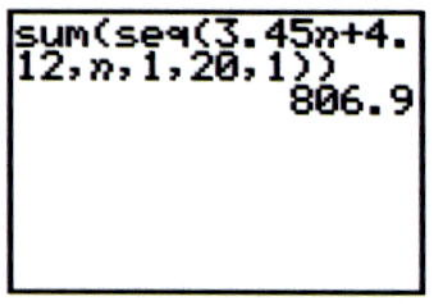

43.

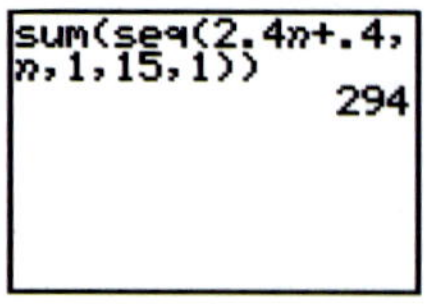

45.

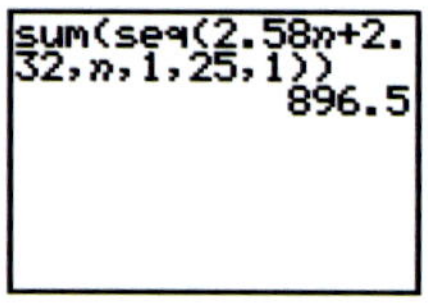

47. $-\frac{3}{2}$ **49.** 1185 seats **51.** 210 of (1st colors name) and 190 (2nd colors name)

12.3 Exercises

1. $r = 3; 3, 9, 27, 81$ **3.** $r = \frac{1}{2}; -\frac{3}{2}, -\frac{3}{4}, -\frac{3}{8}, -\frac{3}{16}$ **5.** $r = 2; \frac{1}{4}, \frac{1}{2}, 1, 2$ **7.** $r = 2^{1/3}; 2^{1/3}, 2^{2/3}, 2, 2^{4/3}$ **9.** $r = \frac{3}{2}; \frac{1}{2}, \frac{3}{4}, \frac{9}{8}, \frac{27}{16}$

11. Arithmetic; $d = 1$ **13.** Neither **15.** Arithmetic; $d = -\frac{2}{3}$ **17.** Neither **19.** Geometric; $r = \frac{2}{3}$ **21.** Geometric; $r = 2$

23. Geometric; $r = 3^{1/2}$ **25.** $a_5 = 162; a_n = 2 \cdot 3^{n-1}$ **27.** $a_5 = 5; a_n = 5 \cdot (-1)^{n-1}$ **29.** $a_5 = 0; a_n = 0$ **31.** $a_5 = 4\sqrt{2}; a_n = (\sqrt{2})^n$

33. $a_7 = \frac{1}{64}$ **35.** $a_9 = 1$ **37.** $a_8 = 0.00000004$ **39.** $-\frac{1}{4}(1 - 2^n)$ **41.** $2\left[1 - \left(\frac{2}{3}\right)^n\right]$ **43.** $1 - 2^n$

45.

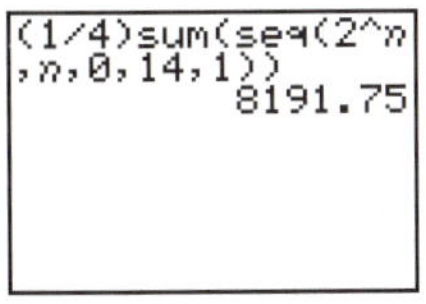
(1/4)sum(seq(2^n
,n,0,14,1))
8191.75

47.

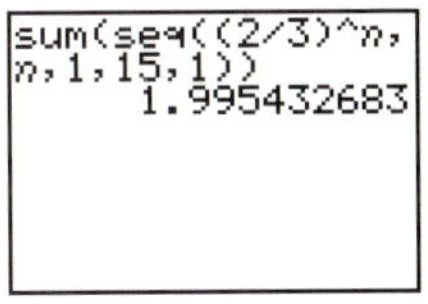
sum(seq((2/3)^n,
n,1,15,1))
1.995432683

49.

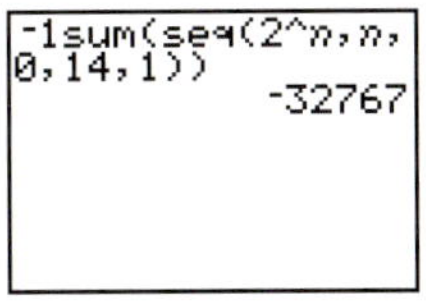
-1sum(seq(2^n,n,
0,14,1))
-32767

51. $\frac{3}{2}$ **53.** 16 **55.** $\frac{8}{5}$ **57.** $\frac{20}{3}$ **59.** $\frac{18}{5}$ **61.** -4 **63.** \$349,496.41 **65.** \$96,885.98 **67.** \$305.10

69. **(a)** 0.775 ft **(b)** 8th **(c)** 15.88 ft **(d)** 20 ft **71.** \$21,879.11

73. Option 2 results in the most: \$16,038,304; Option 1 results in the least: \$14,700,000 **75.** 1.845×10^{19} **79.** 3, 5, 9, 24, 73

81. *A*: \$25,250 per year in 5th year, \$112,742 total; *B*: \$24,761 per year in 5th year, \$116,801 total

12.4 Exercises

1. (I) $n = 1: 2 \cdot 1 = 2$ and $1(1 + 1) = 2$
(II) If $2 + 4 + 6 + \cdots + 2k = k(k + 1)$, then $2 + 4 + 6 + \cdots + 2k + 2(k + 1) = (2 + 4 + 6 + \cdots + 2k) + 2(k + 1)$
$= k(k + 1) + 2(k + 1) = k^2 + 3k + 2 = (k + 1)(k + 2)$.

3. (I) $n = 1: 1 + 2 = 3$ and $\frac{1}{2}(1)(1 + 5) = \frac{1}{2}(6) = 3$
(II) If $3 + 4 + 5 + \cdots + (k + 2) = \frac{1}{2}k(k + 5)$, then $3 + 4 + 5 + \cdots + (k + 2) + [(k + 1) + 2]$
$= [3 + 4 + 5 + \cdots + (k + 2)] + (k + 3) = \frac{1}{2}k(k + 5) + k + 3 = \frac{1}{2}(k^2 + 7k + 6) = \frac{1}{2}(k + 1)(k + 6)$

5. (I) $n = 1: 3 \cdot 1 - 1 = 2$ and $\frac{1}{2}(1)[3(1) + 1] = \frac{1}{2}(4) = 2$
(II) If $2 + 5 + 8 + \cdots + (3k - 1) = \frac{1}{2}k(3k + 1)$, then $2 + 5 + 8 + \cdots + (3k - 1) + [3(k + 1) - 1]$
$= [2 + 5 + 8 + \cdots + (3k - 1)] + 3k + 2 = \frac{1}{2}k(3k + 1) + (3k + 2) = \frac{1}{2}(3k^2 + 7k + 4) = \frac{1}{2}(k + 1)(3k + 4)$

7. (I) $n = 1: 2^{1-1} = 1$ and $2^1 - 1 = 1$
(II) If $1 + 2 + 2^2 + \cdots + 2^{k-1} = 2^k - 1$, then $1 + 2 + 2^2 + \cdots + 2^{k-1} + 2^{(k+1)-1} = (1 + 2 + 2^2 + \cdots + 2^{k-1}) + 2^k$
$= 2^k - 1 + 2^k = 2(2^k) - 1 = 2^{k+1} - 1$.

9. (I) $n = 1: 4^{1-1} = 1$ and $\frac{1}{3}(4^1 - 1) = \frac{1}{3}(3) = 1$
(II) If $1 + 4 + 4^2 + \cdots + 4^{k-1} = \frac{1}{3}(4^k - 1)$, then $1 + 4 + 4^2 + \cdots + 4^{k-1} + 4^{(k+1)-1} = (1 + 4 + 4^2 + \cdots + 4^{k-1}) + 4^k$
$= \frac{1}{3}(4^k - 1) + 4^k = \frac{1}{3}[4^k - 1 + 3(4^k)] = \frac{1}{3}[4(4^k) - 1] = \frac{1}{3}(4^{k+1} - 1)$.

11. (I) $n = 1: \frac{1}{1 \cdot 2} = \frac{1}{2}$ and $\frac{1}{1 + 1} = \frac{1}{2}$
(II) If $\frac{1}{1 \cdot 2} + \frac{1}{2 \cdot 3} + \frac{1}{3 \cdot 4} + \cdots + \frac{1}{k(k + 1)} = \frac{k}{k + 1}$, then $\frac{1}{1 \cdot 2} + \frac{1}{2 \cdot 3} + \frac{1}{3 \cdot 4} + \cdots + \frac{1}{k(k + 1)} + \frac{1}{(k + 1)[(k + 1) + 1]}$
$= \left[\frac{1}{1 \cdot 2} + \frac{1}{2 \cdot 3} + \frac{1}{3 \cdot 4} + \cdots + \frac{1}{k(k + 1)}\right] + \frac{1}{(k + 1)(k + 2)} = \frac{k}{k + 1} + \frac{1}{(k + 1)(k + 2)} = \frac{k + 1}{k + 2}$.

13. (I) $n = 1: 1^2 = 1$ and $\frac{1}{6} \cdot 1 \cdot 2 \cdot 3 = 1$
(II) If $1^2 + 2^2 + 3^2 + \cdots + k^2 = \frac{1}{6}k(k + 1)(2k + 1)$, then $1^2 + 2^2 + 3^2 + \cdots + k^2 + (k + 1)^2$
$= (1^2 + 2^2 + 3^2 + \cdots + k^2) + (k + 1)^2 = \frac{1}{6}k(k + 1)(2k + 1) + (k + 1)^2 = \frac{1}{6}(2k^3 + 9k^2 + 13k + 6) = \frac{1}{6}(k + 1)(k + 2)(2k + 3)$.

15. (I) $n = 1$: $5 - 1 = 4$ and $\frac{1}{2}(1)(9 - 1) = \frac{1}{2} \cdot 8 = 4$

(II) If $4 + 3 + 2 + \cdots + (5 - k) = \frac{1}{2}k(9 - k)$, then $4 + 3 + 2 + \cdots + (5 - k) + 5 - (k + 1)$

$= [4 + 3 + 2 + \cdots + (5 - k)] + 5 - (k + 1) = \frac{1}{2}k(9 - k) + 4 - k = \frac{1}{2}(-k^2 + 7k + 8) = \frac{1}{2}(8 - k)(k + 1) = \frac{1}{2}(k + 1)[9 - (k + 1)]$.

17. (I) $n = 1$: $1 \cdot (1 + 1) = 2$ and $\frac{1}{3} \cdot 1 \cdot 2 \cdot 3 = 2$

(II) If $1 \cdot 2 + 2 \cdot 3 + 3 \cdot 4 + \cdots + k(k + 1) = \frac{1}{3}k(k + 1)(k + 2)$, then

$1 \cdot 2 + 2 \cdot 3 + 3 \cdot 4 + \cdots + k(k + 1) + (k + 1)(k + 2) = [1 \cdot 2 + 2 \cdot 3 + 3 \cdot 4 + \cdots + k(k + 1)] + (k + 1)(k + 2)$

$= \frac{1}{3}k(k + 1)(k + 2) + (k + 1)(k + 2) = \frac{1}{3}(k + 1)(k + 2)(k + 3)$

19. (I) $n = 1$: $1^2 + 1 = 2$ is divisible by 2.
(II) If $k^2 + k$ is divisible by 2, then $(k + 1)^2 + (k + 1) = k^2 + 2k + 1 + k + 1 = (k^2 + k) + 2k + 2$. Since $k^2 + k$ is divisible by 2 and $2k + 2$ is divisible by 2, therefore, $(k + 1)^2 + k + 1$ is divisible by 2.

21. (I) $n = 1$: $1^2 - 1 + 2 = 2$ is divisible by 2.
(II) If $k^2 - k + 2$ is divisible by 2, then $(k + 1)^2 - (k + 1) + 2 = k^2 + 2k + 1 - k - 1 + 2 = (k^2 - k + 2) + 2k$. Since $k^2 - k + 2$ is divisible by 2 and $2k$ is divisible by 2, therefore, $(k + 1)^2 - (k + 1) + 2$ is divisible by 2.

23. (I) $n = 1$: If $x > 1$, then $x^1 = x > 1$.
(II) Assume, for any natural number k, that if $x > 1$, then $x^k > 1$. Show that if $x > 1$, then $x^{k+1} > 1$:

$x^{k+1} = x^k \cdot x^1 > 1 \cdot x = x > 1$
↑
$x^k > 1$

25. (I) $n = 1$: $a - b$ is a factor of $a^1 - b^1 = a - b$.
(II) If $a - b$ is a factor of $a^k - b^k$, show that $a - b$ is a factor of $a^{k+1} - b^{k+1}$: $a^{k+1} - b^{k+1} = a(a^k - b^k) + b^k(a - b)$. Since $a - b$ is a factor of $a^k - b^k$ and $a - b$ is a factor of $a - b$, therefore, $a - b$ is a factor of $a^{k+1} - b^{k+1}$.

27. $n = 1$: $1^2 - 1 + 41 = 41$ is a prime number.
$n = 41$: $41^2 - 41 + 41 = 1681 = 41^2$ is not prime.

29. (I) $n = 1$: $ar^{1-1} = a \cdot 1 = a$ and $a \cdot \frac{1 - r^1}{1 - r} = a$, because $r \neq 1$.

(II) If $a + ar + ar^2 + \cdots + ar^{k-1} = a\left(\frac{1 - r^k}{1 - r}\right)$, then $a + ar + ar^2 + \cdots + ar^{k-1} + ar^{(k+1)-1} = (a + ar + ar^2 + \cdots + ar^{k-1}) + ar^k$

$= a\left(\frac{1 - r^k}{1 - r}\right) + ar^k = \frac{a(1 - r^k) + ar^k(1 - r)}{1 - r} = \frac{a - ar^k + ar^k - ar^{k+1}}{1 - r} = a\left(\frac{1 - r^{k+1}}{1 - r}\right)$

31. (I) $n = 3$: The sum of the angles of a triangle is $(3 - 2) \cdot 180° = 180°$.
(II) Assume for any k that the sum of the angles of a convex polygon of k sides is $(k - 2) \cdot 180°$. A convex poloygon of $k + 1$ sides consists of a convex polygon of k sides plus a triangle (see the illustration). The sum of the angles is $(k - 2) \cdot 180° + 180° = (k - 1) \cdot 180°$. Since Conditions I and II have been met, the result follows.

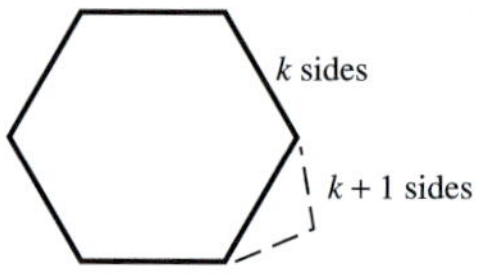

12.5 Exercises

1. 10 **3.** 21 **5.** 50 **7.** 1 **9.** 1.866×10^{15} **11.** 1.483×10^{13} **13.** $x^5 + 5x^4 + 10x^3 + 10x^2 + 5x + 1$
15. $x^6 - 12x^5 + 60x^4 - 160x^3 + 240x^2 - 192x + 64$ **17.** $81x^4 + 108x^3 + 54x^2 + 12x + 1$
19. $x^{10} + 5y^2x^8 + 10y^4x^6 + 10y^6x^4 + 5y^8x^2 + y^{10}$ **21.** $x^3 + 6\sqrt{2}x^{5/2} + 30x^2 + 40\sqrt{2}x^{3/2} + 60x + 24\sqrt{2}x^{1/2} + 8$
23. $(ax)^5 + 5by(ax)^4 + 10(by)^2(ax)^3 + 10(by)^3(ax)^2 + 5(by)^4(ax) + (by)^5$ **25.** 17,010 **27.** −101,376 **29.** 41,472 **31.** $2835x^3$
33. $314{,}928x^7$ **35.** 495 **37.** 3360 **39.** 1.00501 **41.** $\binom{n}{n} = \frac{n!}{n!(n - n)!} = \frac{n!}{n!0!} = \frac{n!}{n!} = 1$
43. $2^n = (1 + 1)^n = \binom{n}{0}1^n + \binom{n}{1}(1)(1)^{n-1} + \cdots + \binom{n}{n}1^n = \binom{n}{0} + \binom{n}{1} + \cdots + \binom{n}{n}$ **45.** 1

12.6 Exercises

1. {1, 3, 5, 6, 7, 9} **3.** {1, 5, 7} **5.** {1, 6, 9} **7.** {1, 2, 4, 5, 6, 7, 8, 9} **9.** {1, 2, 4, 5, 6, 7, 8, 9} **11.** {0, 2, 6, 7, 8} **13.** {0, 1, 2, 3, 5, 6, 7, 8, 9}
15. {0, 1, 2, 3, 5, 6, 7, 8, 9} **17.** {0, 1, 2, 3, 4, 6, 7, 8} **19.** {0}

21. $\varnothing, \{a\}, \{b\}, \{c\}, \{d\}, \{a,b\}, \{a,c\}, \{a,d\}, \{b,c\}, \{b,d\}, \{c,d\}, \{a,b,c\}, \{b,c,d\}, \{a,c,d\}, \{a,b,d\}, \{a,b,c,d\}$ **23.** 25 **25.** 40 **27.** 25 **29.** 37 **31.** 18 **33.** 5 **35.** 175; 125 **37.** **(a)** 15 **(b)** 15 **(c)** 15 **(d)** 25 **(e)** 40

12.7 Exercises

1. 30 **3.** 120 **5.** 1 **7.** 336 **9.** 28 **11.** 15 **13.** 1 **15.** 10,400,600
17. {*abc, abd, abe, acb, acd, ace, adb, adc, ade, aeb, aec, aed*
bac, bad, bae, bca, bcd, bce, bda, bdc, bde, bea, bec, bed
cab, cad, cae, cba, cbd, cbe, cda, cdb, cde, cea, ceb, ced
dab, dac, dae, dba, dbc, dbe, dca, dcb, dce, dea, deb, dec
eab, eac, ead, eba, ebc, ebd, eca, ecb, ecd, eda, edb, edc}; 60
19. {123, 124, 132, 134, 142, 143, 213, 214, 231, 234, 241, 243, 312, 314, 321, 324, 341, 342, 412, 413, 421, 423, 431, 432}; 24
21. {*abc, abd, abe, acd, ace, ade, bcd, bce, bde, cde*}; 10 **23.** {123, 234, 124, 134}; 4 **25.** 15 **27.** 16 **29.** 8 **31.** 24 **33.** 60 **35.** 120 **37.** 35 **39.** 1024 **41.** 9000 **43.** $P(5,5) = 5! = 120$ **45.** $C(8,1)\cdot C(15,1)\cdot C(4,1) = 8\cdot 15\cdot 4 = 480$ **47.** 336 **49.** 5,209,344 **51.** 362,880 **53.** 90,720 **55.** 1.156×10^{76} **57.** 15 **59.** **(a)** 63 **(b)** 35 **(c)** 1

12.8 Exercises

1. $S = \{\text{HH, HT, TH, TT}\}$; $P(\text{HH}) = \frac{1}{4}, P(\text{HT}) = \frac{1}{4}, P(\text{TH}) = \frac{1}{4}, P(\text{TT}) = \frac{1}{4}$

3. S = {HH1, HH2, HH3, HH4, HH5, HH6, HT1, HT2, HT3, HT4, HT5, HT6, TH1, TH2, TH3, TH4, TH5, TH6, TT1, TT2, TT3, TT4, TT5, TT6}; each outcome has the probability of $\frac{1}{24}$.

5. S = {HHH, HHT, HTH, HTT, THH, THT, TTH, TTT}; each outcome has the probability of $\frac{1}{8}$.

7. S = {1 Yellow, 1 Red, 1 Green, 2 Yellow, 2 Red, 2 Green, 3 Yellow, 3 Red, 3 Green, 4 Yellow, 4 Red, 4 Green}; each outcome has the probability of $\frac{1}{12}$; thus, $P(\text{2 Red}) + P(\text{4 Red}) = \frac{1}{12} + \frac{1}{12} = \frac{1}{6}$.

9. S = {1 Yellow Forward, 1 Yellow Backward, 1 Red Forward, 1 Red Backward, 1 Green Forward, 1 Green Backward, 2 Yellow Forward, 2 Yellow Backward, 2 Red Forward, 2 Red Backward, 2 Green Forward, 2 Green Backward, 3 Yellow Forward, 3 Yellow Backward, 3 Red Forward, 3 Red Backward, 3 Green Forward, 3 Green Backward, 4 Yellow Forward, 4 Yellow Backward, 4 Red Forward, 4 Red Backward, 4 Green Forward, 4 Green Backward}; each outcome has the probability of $\frac{1}{24}$; thus, $P(\text{1 Red Backward}) + P(\text{1 Green Backward}) = \frac{1}{24} + \frac{1}{24} = \frac{1}{12}$.

11. S = {11 Red, 11 Yellow, 11 Green, 12 Red, 12 Yellow, 12 Green, 13 Red, 13 Yellow, 13 Green, 14 Red, 14 Yellow, 14 Green, 21 Red, 21 Yellow, 21 Green, 22 Red, 22 Yellow, 22 Green, 23 Red, 23 Yellow, 23 Green, 24 Red, 24 Yellow, 24 Green, 31 Red, 31 Yellow, 31 Green, 32 Red, 32 Yellow, 32 Green, 33 Red, 33 Yellow, 33 Green, 34 Red, 34 Yellow, 34 Green, 41 Red, 41 Yellow, 41 Green, 42 Red, 42 Yellow, 42 Green, 43 Red, 43 Yellow, 43 Green, 44 Red, 44 Yellow, 44 Green}; each outcome has the probability of $\frac{1}{48}$; thus, E = {22 Red, 22 Green, 24 Red, 24 Green}; $P(E) = \frac{n(E)}{n(S)} = \frac{4}{48} = \frac{1}{12}$.

13. A, B, C, F **15.** B **17.** $\frac{4}{5}, \frac{1}{5}$ **19.** $P(1) = P(3) = P(5) = \frac{2}{9}; P(2) = P(4) = P(6) = \frac{1}{9}$ **21.** 0.7 **23.** 0.55 **25.** $\frac{9}{20}$ **27.** $\frac{17}{20}$ **29.** $\frac{5}{18}$

31. $\frac{3}{10}$ **33.** $\frac{2}{5}$ **35.** **(a)** 0.57 **(b)** 0.95 **(c)** 0.83 **(d)** 0.38 **(e)** 0.29 **(f)** 0.05 **(g)** 0.78 **(h)** 0.71 **37.** 0.000033069

39. **(a)** $\frac{10}{32}$ **(b)** $\frac{1}{32}$ **41.** **(a)** 0.00463 **(b)** 0.049 **43.** $\frac{1}{C(30,5)} = 7.02\times 10^{-6}$; 0.183 **45.** 0.1

Fill-in-the-Blank Items

1. sequence **2.** arithmetic **3.** geometric **4.** Pascal triangle **5.** 15 **6.** union; intersection **7.** 20; 10 **8.** permutation **9.** combination **10.** equally likely

True/False Items

1. T **2.** T **3.** T **4.** T **5.** F **6.** F **7.** F **8.** T **9.** F **10.** T **11.** T **12.** F

Review Exercises

1. 120 **3.** 10 **5.** 336 **7.** 56 **9.** $-\frac{4}{3}, \frac{5}{4}, -\frac{6}{5}, \frac{7}{6}, -\frac{8}{7}$ **11.** $2, 1, \frac{8}{9}, 1, \frac{32}{25}$ **13.** $3, 2, \frac{4}{3}, \frac{8}{9}, \frac{16}{27}$ **15.** 2, 0, 2, 0, 2

17. Arithmetic; $d = 1$; $\frac{n}{2}(n + 11)$ **19.** Neither **21.** Geometric; $r = 8$; $\frac{8}{7}(8^n - 1)$ **23.** Arithmetic ; $d = 4$; $2n(n - 1)$

25. Geometric; $r = \frac{1}{2}$; $6\left[1 - \left(\frac{1}{2}\right)^n\right]$ **27.** Neither **29.** 115 **31.** 75 **33.** 0.49977 **35.** 35 **37.** $\frac{1}{10^{10}}$ **39.** $9\sqrt{2}$ **41.** $5n - 4$

43. $n - 10$ **45.** $\frac{9}{2}$ **47.** $\frac{4}{3}$ **49.** 8

51. (I) $n = 1: 3 \cdot 1 = 3$ and $\frac{3 \cdot 1}{2}(2) = 3$

(II) If $3 + 6 + 9 + \cdots + 3k = \frac{3k}{2}(k + 1)$, then $3 + 6 + 9 + \cdots + 3k + 3(k + 1) = (3 + 6 + 9 + \cdots + 3k) + (3k + 3)$

$= \frac{3k}{2}(k + 1) + (3k + 3) = \frac{3k^2}{2} + \frac{9k}{2} + \frac{6}{2} = \frac{3}{2}(k^2 + 3k + 2) = \frac{3}{2}(k + 1)(k + 2)$.

53. (I) $n = 1: 2 \cdot 3^{1-1} = 2$ and $3^1 - 1 = 2$

(II) If $2 + 6 + 18 + \cdots + 2 \cdot 3^{k-1} = 3^k - 1$, then $2 + 6 + 18 + \cdots + 2 \cdot 3^{k-1} + 2 \cdot 3^{(k+1)-1}$

$= (2 + 6 + 18 + \cdots + 2 \cdot 3^{k-1}) + 2 \cdot 3^k = 3^k - 1 + 2 \cdot 3^k = 3 \cdot 3^k - 1 = 3^{k+1} - 1$.

55. (I) $n = 1: 1^2 = 1$ and $\frac{1}{2}(6 - 3 - 1) = \frac{1}{2}(2) = 1$

(II) If $1^2 + 4^2 + 7^2 + \cdots + (3k - 2)^2 = \frac{1}{2}k(6k^2 - 3k - 1)$, then

$1^2 + 4^2 + 7^2 + \cdots + (3k - 2)^2 + [3(k + 1) - 2]^2 = [1^2 + 4^2 + 7^2 + \cdots + (3k - 2)^2] + (3k + 1)^2 = \frac{1}{2}k(6k^2 - 3k - 1) + (3k + 1)^2$

$= \frac{1}{2}(6k^3 + 15k^2 + 11k + 2) = \frac{1}{2}(k + 1)(6k^2 + 9k + 2) = \frac{1}{2}(k + 1)[6(k + 1)^2 - 3(k + 1) - 1]$.

57. $x^5 + 10x^4 + 40x^3 + 80x^2 + 80x + 32$ **59.** $32x^5 + 240x^4 + 720x^3 + 1080x^2 + 810x + 243$ **61.** 144 **63.** 84 **65.** {1, 3, 5, 6, 7, 8}
67. {3, 7} **69.** {1, 2, 4, 6, 8, 9} **71.** {1, 2, 4, 5, 6, 9} **73.** 17 **75.** 29 **77.** 7 **79.** 25 **81.** 60 **83.** 128 **85.** 3024 **87.** 70 **89.** 91

91. 1,600,000 **93.** 216,000 **95.** 1260 **97.** **(a)** 381,024 **(b)** 1260 **99.** $\frac{3}{20}, \frac{9}{20}$ **101.** $\frac{1}{24}$ **103.** **(a)** 0.045 **(b)** 0.318 **(c)** 0.159

105. **(a)** 8 **(b)** 1100 **107.** \$151,873.77 **109.** **(a)** $\left(\frac{3}{4}\right)^3 \cdot 20 = \frac{135}{16}$ ft **(b)** $20\left(\frac{3}{4}\right)^n$ ft **(c)** after the 13th time **(d)** 140 ft

111. **(a)** 0.68 **(b)** 0.58 **(c)** 0.32

APPENDIX *A1 Exercises*

1. > **3.** > **5.** > **7.** = **9.** < **11.**

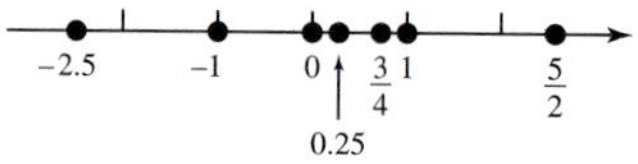

13. $x > 0$ **15.** $x < 2$ **17.** $x \leq 1$ **19.** $2 < x < 5$

21. [0, 4]

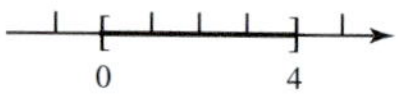

23. [4, 6)

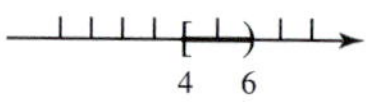

25. $2 \leq x \leq 5$

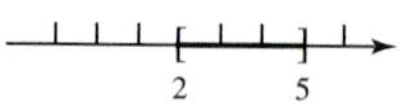

27. $x \geq 4$ or $4 \leq x < \infty$

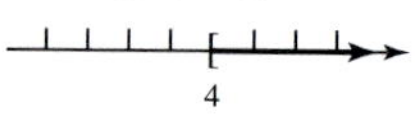

29. 1 **31.** 5 **33.** 1 **35.** $\frac{1}{16}$ **37.** $\frac{4}{9}$ **39.** $\frac{1}{9}$ **41.** $\frac{9}{4}$ **43.** 324 **45.** 27 **47.** 16 **49.** $4\sqrt{2}$ **51.** $-\frac{2}{3}$ **53.** y^2 **55.** $\frac{y}{x^2}$

57. $\frac{1}{x^3y}$ **59.** $\frac{25y^2}{16x^2}$ **61.** $\frac{1}{x^3y^3z^2}$ **63.** $\frac{-8x^3}{9yz^2}$ **65.** $\frac{y^2}{x^2 + y^2}$ **67.** $\frac{16x^2}{9y^2}$ **69.** 13 **71.** 26 **73.** 25 **75.** 4 **77.** 24 **79.** Yes; 5

81. Not a right triangle **83.** Yes; 25 **85.** 9.4 in. **87.** 58.3 ft **89.** 46.7 mi **91.** 3 mi **93.** $a \leq b, c > 0$; $a - b \leq 0$

$(a - b)c \leq 0(c)$

$ac - bc \leq 0$

$ac \leq bc$

95. $\frac{a + b}{2} - a = \frac{a + b - 2a}{2} = \frac{b - a}{2} > 0$; therefore, $a < \frac{a + b}{2}$

$b - \frac{a + b}{2} = \frac{2b - a - b}{2} = \frac{b - a}{2} > 0$; therefore, $b > \frac{a + b}{2}$

97. No; No **99.** 1 **101.** 3.15 or 3.16

103. The light can be seen on the horizon 23.3 miles distant. Planes flying at 10,000 feet can see it 146 miles away. The ship would need to be 185.9 ft tall for the data to be correct.

A2 Exercises

1. $10x^5 + 3x^3 - 10x^2 + 6$ **3.** $2ax + a^2$ **5.** $2x^2 + 17x + 8$ **7.** $x^4 - x^2 + 2x - 1$ **9.** $6x^2 + 2$ **11.** $4x^2 - 3x + 1$; remainder 1 **13.** $4x^2 - 11x + 23$; remainder -45 **15.** $4x^2 + 13x + 53$; remainder 213 **17.** $4x - 3$; remainder $x + 1$ **19.** $4x - 3$; remainder $-7x + 7$ **21.** $(x - 5)(x + 3)$ **23.** $a(x - 9a)(x + 5a)$ **25.** $(x - 3)(x^2 + 3x + 9)$ **27.** $(3x + 1)(x + 1)$ **29.** $x^5(x - 1)(x + 1)$ **31.** $\frac{3(x - 3)}{5x}$ **33.** $\frac{x(2x - 1)}{(x + 4)}$ **35.** $\frac{5x}{(x - 6)(x - 1)(x + 4)}$ **37.** $\frac{2(x + 4)}{(x - 2)(x + 2)(x + 3)}$ **39.** $\frac{(x - 1)(x + 1)}{(x^2 + 1)}$ **41.** $\frac{(x - 1)(-x^2 + 3x + 3)}{(x^2 + x - 1)}$

A3 Exercises

1. $2\sqrt{2}$ **3.** $2x\sqrt[3]{2x}$ **5.** $|x|$ **7.** $\frac{4}{3}x\sqrt{2}$ **9.** x^3y^2 **11.** $x^2|y|$ **13.** $6\sqrt{x}$ **15.** $6x\sqrt{x}$ **17.** $15\sqrt[3]{3}$ **19.** $\frac{1}{x(2x + 3)}$ **21.** $12\sqrt{3}$ **23.** $2\sqrt{3}$ **25.** $x - 2\sqrt{x} + 1$ **27.** $\frac{\sqrt{2}}{2}$ **29.** $\frac{-\sqrt{15}}{5}$ **31.** $\frac{\sqrt{3}(5 + \sqrt{2})}{23}$ **33.** $\frac{-19 + 8\sqrt{5}}{41}$ **35.** $\frac{2x + h - 2\sqrt{x(x + h)}}{h}$ **37.** 4 **39.** -3 **41.** 64 **43.** $\frac{1}{27}$ **45.** $\frac{27\sqrt{2}}{32}$ **47.** $\frac{27\sqrt{2}}{32}$ **49.** $x^{17/12}$ **51.** x^2y^4 **53.** $x^{4/3}y^{5/3}$ **55.** $\frac{8x^{3/2}}{y^{1/4}}$ **57.** $\frac{3x + 2}{(1 + x)^{1/2}}$ **59.** $\frac{2 + x}{2(1 + x)^{3/2}}$ **61.** $\frac{4 - x}{(x + 4)^{3/2}}$ **63.** $\frac{1}{2}(5x + 2)(x + 1)^{1/2}$ **65.** $2x^{1/2}(3x - 4)(x + 1)$

A4 Exercises

1. -1 **3.** -18 **5.** -3 **7.** -16 **9.** 0.5 **11.** 2 **13.** 2 **15.** -1 **17.** 3 **19.** No real solution **21.** $\{0, 9\}$ **23.** $\{0, 9\}$ **25.** No real solution **27.** 2 **29.** 21 **31.** $\{-2, 2\}$ **33.** $-\frac{20}{39}$ **35.** 6

A5 Exercises

1. 4 **3.** $\frac{1}{16}$ **5.** $\frac{1}{9}$ **7.** $(x - 2)^2 + (y + 2)^2 = 9$ **9.** $(x + 3)^2 + (y - 1)^2 = 9$ **11.** $\left(x + \frac{1}{2}\right)^2 + \left(y - \frac{1}{2}\right)^2 = 1$ **13.** $\{-7, 3\}$ **15.** $\left\{-\frac{1}{4}, \frac{3}{4}\right\}$ **17.** $\left\{\frac{-1 - \sqrt{7}}{6}, \frac{-1 + \sqrt{7}}{6}\right\}$

A6 Exercises

1. $q(x) = x^2 + x + 4; R = 12$ **3.** $q(x) = 3x^2 + 11x + 32; R = 99$ **5.** $q(x) = x^4 - 3x^3 + 5x^2 - 15x + 46; R = -138$ **7.** $q(x) = 4x^5 + 4x^4 + x^3 + x^2 + 2x + 2; R = 7$ **9.** $q(x) = 0.1x^2 - 0.11x + 0.321; R = -0.3531$ **11.** $q(x) = x^4 + x^3 + x^2 + x + 1; R = 0$ **13.** No; $f(2) = 8$ **15.** Yes; $f(2) = 0$ **17.** Yes; $f(-3) = 0$ **19.** No; $f(-4) = 1$ **21.** Yes; $f\left(\frac{1}{2}\right) = 0$ **23.** 69 **25.** -4 **27.** 15

INDEX

Conics

Parabola

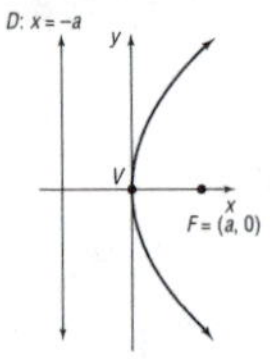

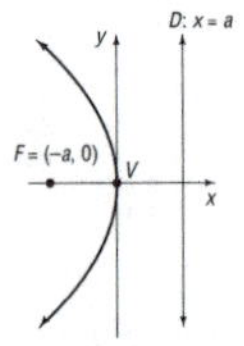

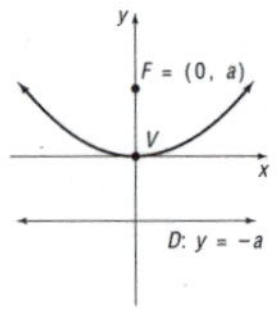

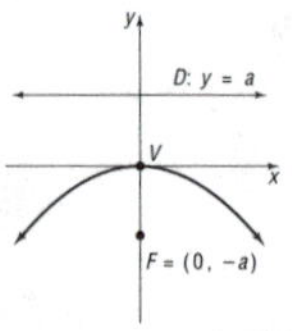

$y^2 = 4ax$ $\qquad$ $y^2 = -4ax$ $\qquad$ $x^2 = 4ay$ $\qquad$ $x^2 = -4ay$

Ellipse

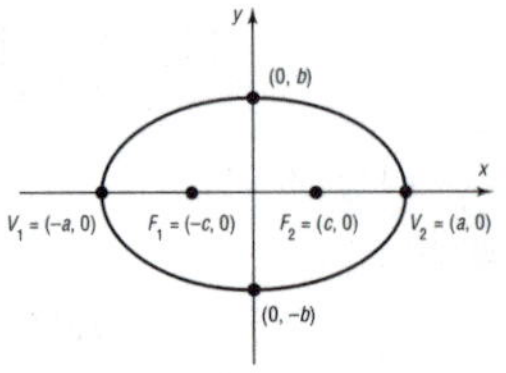

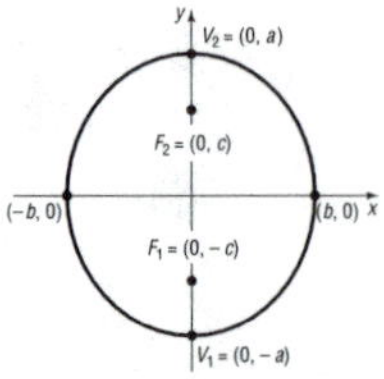

$$\frac{x^2}{a^2} + \frac{y^2}{b^2} = 1, \quad c^2 = a^2 - b^2 \qquad\qquad \frac{x^2}{b^2} + \frac{y^2}{a^2} = 1, \quad c^2 = a^2 - b^2$$

Hyperbola

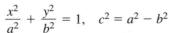

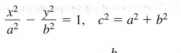

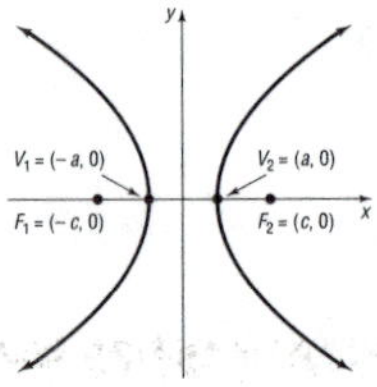

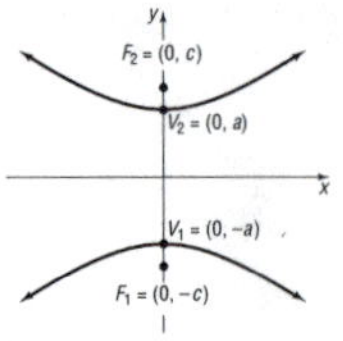

$$\frac{x^2}{a^2} - \frac{y^2}{b^2} = 1, \quad c^2 = a^2 + b^2 \qquad\qquad \frac{y^2}{a^2} - \frac{x^2}{b^2} = 1, \quad c^2 = a^2 + b^2$$

Asymptotes: $y = \frac{b}{a}x, \quad y = -\frac{b}{a}x$ $\qquad$ Asymptotes: $y = \frac{a}{b}x, \quad y = -\frac{a}{b}x$

Properties of Logarithms

$$\log_a MN = \log_a M + \log_a N$$

$$\log_a\left(\frac{M}{N}\right) = \log_a M - \log_a N$$

$$\log_a M^r = r \log^a M$$

$$\log_a M = \frac{\log M}{\log a} = \frac{\ln M}{\ln a}$$

Permutations/Combinations

$0! = 1 \qquad 1! = 1$

$n! = n(n-1) \cdot \ldots \cdot (3)(2)(1)$

$$P(n, r) = \frac{n!}{(n-r)!}$$

$$C(n, r) = \binom{n}{r} = \frac{n!}{(n-r)!r!}$$

Binomial Theorem

$$(a+b)^n = a^n + \binom{n}{1}ba^{n-1} + \binom{n}{2}b^2a^{n-2} + \cdot\cdot\cdot + \binom{n}{1}b^{n-1}a + b^n$$

Arithmetic Sequence

$$a + (a+d) + (a+2d) + \cdot\cdot\cdot + [a + (n-1)d] = na + \frac{n(n-1)}{2}d$$

Geometric Sequence

$$a + ar + ar^2 + \cdot\cdot\cdot + ar^{n-1} = a\frac{1-r^n}{1-r}$$

Geometric Series

$$\text{If } |r| < 1,\ a + ar + ar^2 + \cdot\cdot\cdot = \sum_{k=1}^{\infty} ar^{k-1} = \frac{a}{1-r}$$